Library of Congress Card Catalog Number 2001056870
International Standard Book Number 1-56637-910-5

1 2 3 4 5 6 7 8 9 0 03 06 05 04 03 02

Library of Congress Cataloging-in-Publication Data

Stockel, Martin W.
Auto diagnosis, service & repair / by Martin W. Stockel, Martin T. Stockel, Chris Johanson.
p. cm.
Rev. ed. of: Auto service & repair. c1996.
Includes index.
ISBN 1-56637-910-5
1. Automobiles—Maintenance and repair. I. Stockel, Martin T. II. Johanson, Chris. III. Stockel, Martin W. Auto service & repair. IV. Title.
TL152.S7743 2002
629.28'72—dc21
2001056870

Auto Diagnos
Service, and
Repair

by

Martin W. Stockel
Automotive Writer

Martin T. Stockel
Automotive Writer

Chris Johanson
ASE Certified Master Technician

Publisher
The Goodheart-Willcox Company, Inc.
Tinley Park, Illinois

Introduction

Auto Diagnosis, Service, and Repair tells and shows how to troubleshoot, service, and repair late-model cars, sport-utility vehicles, and light trucks. It includes information on the latest developments in the automotive field, including OBD II diagnostics, enhanced emissions testing, anti-lock brakes, air bags, and R-134a refrigerant. The material in this text is easy to understand and applicable to all makes and models.

The 2003 edition of **Auto Diagnosis, Service, and Repair** is organized around the eight ASE automobile test areas and correlated to the NATEF Task List. This textbook teaches essential automotive repair skills, encourages the development of good work habits, and emphasizes safety. It is comprehensive, detailed, and profusely illustrated. Many of the illustrations were prepared especially for use in this text. All illustrations and photographs are in full color.

The complexity of today's vehicles has changed the nature of automotive repair. In late-model vehicles, most systems are interrelated. A malfunction in one system can cause trouble in a seemingly unrelated system. To prevent unnecessary repairs, a logical approach must be taken to pinpoint the source of the problem. Therefore, effective diagnosis and troubleshooting procedures are emphasized throughout the new edition of **Auto Diagnosis, Service, and Repair.**

This textbook is intended for beginners who need a sound foundation in the fundamentals of automotive repair, as well as those now engaged in automotive service who want to improve their skills and increase their earning potential. The textbook also serves as a training aid for those taking an ASE test in any automotive service area.

Each chapter of **Auto Diagnosis, Service, and Repair** opens with learning objectives that provide focus for the chapter and a list of the important technical terms that will be encountered in the chapter. These terms are printed in ***bold italic type*** and are defined when introduced in the chapter. Each chapter also includes a "Tech Talk" section, a Summary, Review Questions, ASE-Type Questions, and several Suggested Activities. The Suggested Activities are designed to emphasize reading, math, and communications skills.

A **Workbook for Auto Diagnosis, Service, and Repair** is also available. It is a convenient study guide and shop activity manual correlated directly to this textbook.

Martin W. Stockel
Martin T. Stockel
Chris Johanson

Contents

Chapter 5 Jacks, Lifts, and Holding Fixtures 79

Chapter 6 Cleaning Equipment and Techniques 89

Chapter 7 Welding Equipment and Techniques 99

Chapter 8 Fasteners, Gaskets, and Sealants 113

Chapter 9 Tubing and Hose 143

Section 2
Electrical/Electronic Systems

Chapter 14 Charging and Starting System Service 251

Chapter 15 Computer System Diagnosis and Repair 279

Section 3

Engine Performance and Driveability

Chapter 16 Ignition System Service 313

Chapter 21 Emission System Testing and Service 417

Section 4

Engine Repair

Chapter 22 Engine Mechanical Troubleshooting 439

Chapter 23 Cooling System Service 453

Chapter 24 Engine Removal, Disassembly, and Inspection 479

Chapter 25 Cylinder Head and Valve Service 497

Chapter 26 Engine Block, Crankshaft, and Lubrication System Service 517

Chapter 27 Valve Train and Intake Service 551

Chapter 28 Engine Assembly, Installation, and Break-In 575

Section 5
Manual Drive Train and Axles

Section 6
Automatic Transmission and Transaxle

Section 7
Brakes

Chapter 35 Anti-Lock Brake and Traction Control System Service 767

Section 8

Suspension and Steering

Chapter 36 Suspension System Service 785

Chapter 37 Steering System Service 811

Chapter 38 Wheel and Tire Service 841

1

Introduction to the Automotive Service and Repair Industry

After studying this chapter, you will be able to:

- Identify the major sources of employment in the automotive industry.
- List and describe classifications of automotive technicians.
- Identify different types of automotive service facilities.
- Identify advancement possibilities for automotive technicians.
- Identify automotive training opportunities.
- Explain the benefits of ASE technician certification.

Technical Terms

Helper
Installer
Technicians
Shop foreman
Service manager
Sales person
Service advisor
Parts persons
Entrepreneur
New-vehicle dealerships
Flat rate
Department store service centers
Specialty shops
Independent auto repair shops
Government agencies
Trade magazines
Internet
Aftermarket seminars
Automotive trade associations
Certification
National Institute for Automotive Service Excellence (ASE)

Auto Diagnosis, Service, and Repair is designed to show you how to diagnose and repair late-model vehicles. This chapter covers the career opportunities you will most likely find in the automotive service industry. It discusses the various types of automotive technicians and the type of work each performs. This chapter also includes information on the types of repair outlets, work, and working conditions a beginning technician can expect.

Automotive Service

The business of servicing and repairing cars and trucks has been a source of employment for millions of people over the last 100 years. It will continue to provide ample employment opportunities for years to come. Like any career, automotive service and repair has its drawbacks, but it also has its rewards.

Most people in the auto service business work long hours. Diagnosis and repair procedures can be mentally taxing and physically difficult. Working conditions are often hot and dirty. Automotive service has never been a prestigious career, although this is changing as vehicles become more complex and technicians become better trained. The technician must often deal with vehicle owners who can be difficult and condescending.

The advantages of the auto service business include the opportunity to work with your hands and the satisfaction of fixing something that is broken. The auto repair industry offers salaries that are competitive with those for similar jobs. Auto service is a secure profession in which the good technician can always find work. To ensure that you stay employable, always try to learn new things and become ASE certified in as many areas as possible.

Automotive Service Positions

Although the public tends to classify all automotive technicians as "mechanics," there are many types of auto service professionals. The classifications of auto service professionals include the helper, who changes oil or performs other simple tasks; the installer, who removes and installs components; and the technician, who is capable of diagnosing difficult problems and repairing complex automotive systems. Although these classifications are unofficial, they tend to hold true throughout the automotive repair industry.

Helpers

The ***helper*** performs basic service and maintenance tasks, such as installing tires, cleaning parts, changing engine oil and filters, and installing batteries. See **Figure 1-1.** The skills required of the helper are low, and the pay will be less than that of the installer or the technician. However, working as a helper is a good way to break into the automotive service area. In fact, many technicians started out doing this type of automotive service when they were in their teens.

Installers

The ***installer*** is an automotive service person who removes and installs parts. See **Figure 1-2.** Installers seldom do complicated repair work, and they generally do not diagnose vehicle problems. Installers are paid more than helpers

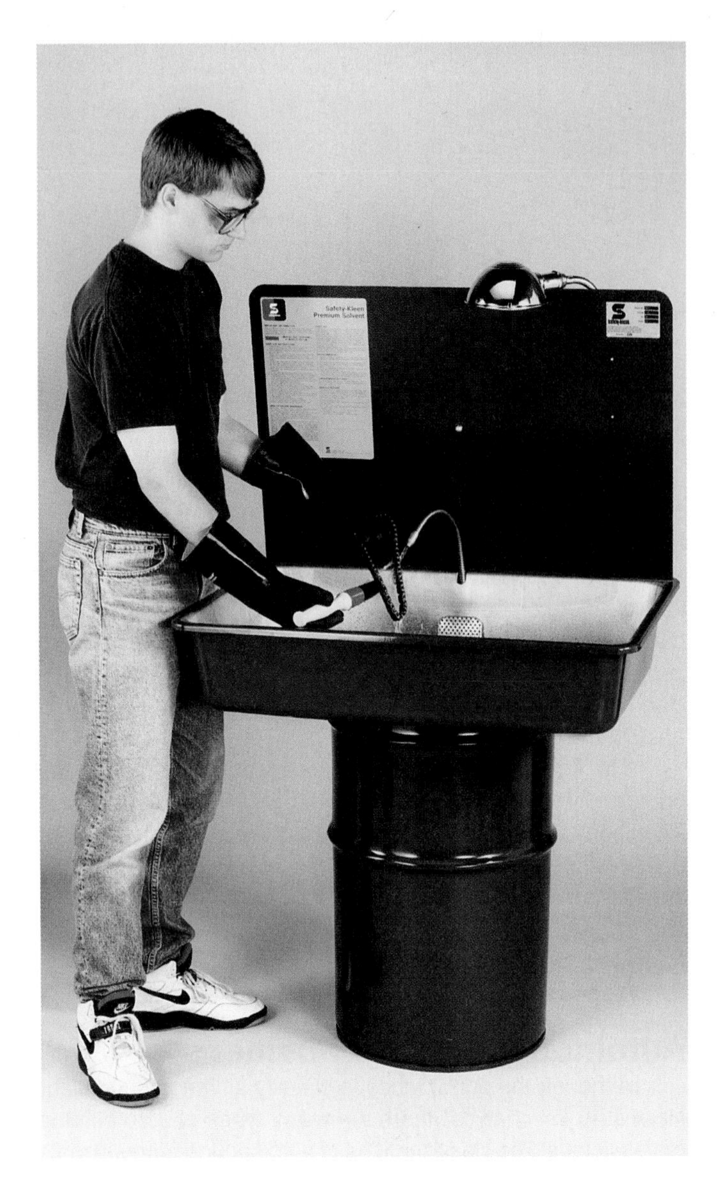

Figure 1-1. The helper is generally responsible for minor tasks, such as cleaning parts. Note the use of safety glasses and rubber gloves. (Safety Kleen)

Figure 1-2. This installer is replacing suspension components. (Ford)

but less than technicians. Many installers take advantage of opportunities to improve their knowledge and skills, and they eventually become technicians.

Technicians

Technicians are at the top of the automotive service profession and have the skills to prove it, **Figure 1-3.** The technician is generally paid more than the helper or the installer. Most technicians have achieved ASE certification in at least one automotive area, and many are certified in all automotive or truck areas. The ASE-certified technician can successfully diagnose and repair every area that he or she is certified in, and can perform many other service jobs. The ASE-certification process is discussed in more detail later in this chapter. Chapter 42, ASE Certification, contains information on applying for and taking the ASE certification tests.

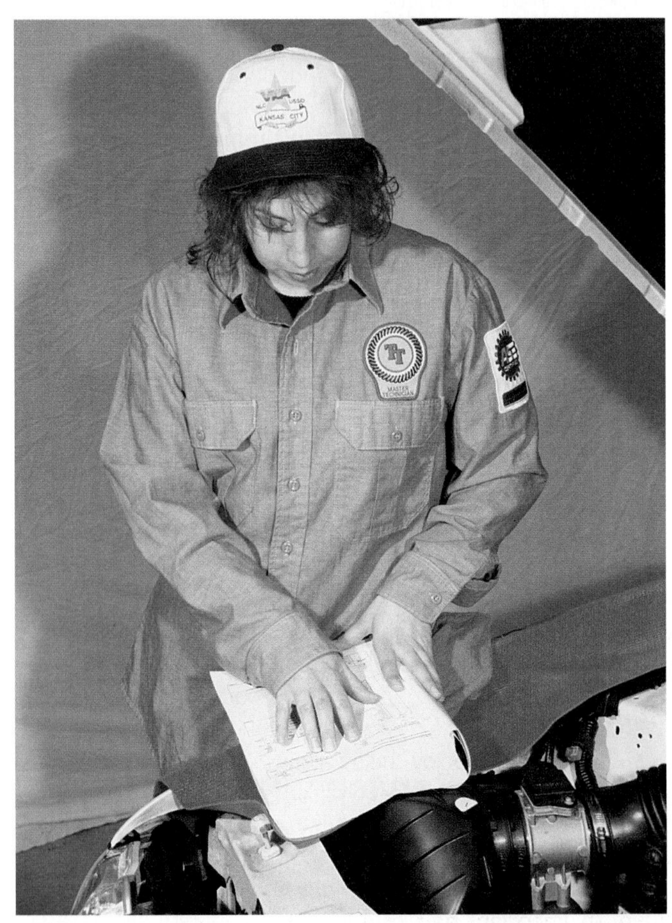

Figure 1-3. The technician has the knowledge and experience to perform most repair jobs. (Hunter Engineering)

Other Opportunities in the Auto Service Industry

Many other job opportunities are available to the automotive technician. These opportunities still involve the servicing of vehicles, but without some of the physical work. If you like cars and trucks but are unsure you want to make a career of repairing them, one of these jobs may be right for you.

Service Manager

As a technician, the automotive promotion you will most likely be offered is to the position of ***service manager,*** or ***shop foreman.*** Many repair facilities are large enough to require one or more service managers. Service managers perform duties similar to those performed by managers in other professions. They carry out administrative duties, such as setting work schedules, ordering supplies, preparing bills, determining pay scales, and dealing with employee problems. Some managers also perform financial record keeping, prepare employee paychecks, and pay suppliers. A few place advertising and perform other marketing and public relations duties.

Service managers closely supervise shop operations and may be called in by technicians when a problem arises. Service managers usually do the hiring. In addition to their administrative duties, most service managers are directly involved with customer relations, such as tending to billing or warranty disputes.

If you move into management from the shop floor, your salary will increase and you will be in a cleaner, less physically demanding position. Many technicians enjoy the management position because it lets them in on the fun part of service—troubleshooting—without the drudgery of making the actual repairs.

The disadvantage of a move to management is that you will no longer deal with the logical principles of troubleshooting and repair. Instead, you will deal with illogical and arbitrary personalities of people. Both customers and technicians will have problems and attitudes you have to contend with. Unlike solving a vehicle problem, resolving these problems may require considerable tact.

The paperwork load is large for any manager, and it may not be something a former technician can get used to. Record keeping requires a great deal of desk and computer time. Automotive record keeping is like balancing a checkbook and writing a term paper every few days. If you do not care to deal with people or keep records, a career in management may not be for you.

Sales Person

Many people enjoy the challenge of selling. The ***sales person,*** or ***service advisor,*** performs a vital service in the automotive shop, since repairs will not be performed unless the owner is sold on the necessity of having them done. Sales persons are not necessary in many small shops, but they are often an important part of dealership service departments, department store service centers, and specialty repair shops. See **Figure 1-4.**

Figure 1-4. Sales persons must be able to work well with customers. (Goodyear)

The sales person may enjoy a large income and may be directly responsible for a large amount of business in the shop. However, selling is a people-oriented job, and it takes a lot of persuasive ability and diplomacy. If you are not interested in dealing with the public, you would probably not be happy in a sales job.

Parts Person

One often-overlooked area of the automotive service business is the process of supplying parts to those performing repairs. There are many types of parts outlets, including dealership parts departments, independent parts stores, parts departments in retail stores, and combination parts and service outlets. All these parts outlets must meet the needs of technicians and shops, as well as those of the do-it-yourselfer. Due to the variety of vehicles available, as well as the complexity of the late-model vehicle, a large number of parts must be kept in stock and located quickly when needed.

Parts persons are trained in the methods of keeping parts flowing through the system until they reach the ultimate endpoint, the vehicle. Parts must be carefully checked into the parts department and stored so they can be found when needed. When a specific part is needed, it must be located and brought to the person requesting it. If the part is not in stock, it must be special ordered. Rebuildable used parts, called cores, must be accounted for and returned to the rebuilder. This can be a challenging job. See **Figure 1-5.**

The job of the parts person appeals to many people. There is an ASE test for parts specialists, and you may want to consider taking it to enhance your employability in this area.

Self-Employment

Many persons dream of going into business for themselves. This can be a good and profitable option for the accomplished technician. However, in addition to mechanical and diagnostic ability, a person who owns a business must have a certain type of personality to be successful. This person must be able to shoulder responsibilities, handle problems, and look for practical ways to increase business and make a profit. This person must also maintain a clear idea of what plans, both long and short term, need to be made. A person who owns a business and possesses these traits is often referred to as an ***entrepreneur.***

When you own a business, the responsibility for repairs, parts ordering, bookkeeping, debt collection, and many other tasks is yours. Starting your own shop requires a large investment in tools, equipment, and workspace. If the money needed to start the business must be borrowed, you will be responsible for paying it back. Nevertheless, many people enjoy the independence that owning a business offers. If you have the personality to deal with the responsibilities involved, you may enjoy owning your own business.

Figure 1-5. The parts person is responsible for making sure technicians are supplied with the correct replacement parts and assemblies. (Triad)

Another possible method of self-employment is to obtain a franchise from a national chain. Owning a franchise operation eliminates some of the headaches of being in business for yourself. Many muffler, tire, transmission, tune-up, and other nationally recognized businesses have local owners. These owners enjoy the advantages of the franchise affiliation, including national advertising, reliable parts supplies, and employee benefit programs. Disadvantages of owning a franchise operation include high franchise fees and startup costs, lack of local advertising, and some loss of control of shop operations to the national headquarters.

Types of Auto Service Facilities

There are many types of auto service facilities. The traditional place to get started in automotive repair—the corner service station—is virtually gone, replaced by self-serve gas stations. However, many opportunities to repair vehicles still exist. Even the smallest community has several automotive repair facilities. Some of these are discussed in the following section.

New-Vehicle Dealerships

All ***new-vehicle dealerships*** must have large service departments to meet the warranty service requirements of the vehicle manufacturer. These service departments are usually well equipped, with all the special testers, tools, and service literature needed to service a specific make of vehicle. Dealership service departments are also equipped with lifts, parts cleaners, hydraulic presses, and other equipment for efficiently servicing vehicles. However, the technician must generally provide his or her own hand and air tools. Dealers stock most common parts and are usually tied into a factory parts network, which allows them to quickly obtain any part. See **Figure 1-6.**

Pay scales are generally competitive between dealerships in a given area and are usually higher than local industry in general. At most dealerships, technicians are paid on a flat rate basis. ***Flat rate*** pay means that the technician is paid by

Figure 1-6. New vehicle dealerships offer excellent working conditions and benefits. However, you must deal with all types of vehicle problems.

the job instead of by the hour. The amount of money the technician makes depends on the amount of work that comes in and how fast he or she can complete it. If you can work fast and enough work comes in, the pay can be excellent. Most dealers offer some sort of benefit package, which often includes health insurance and vacation pay.

Dealership working conditions are relatively good, and most of the vehicles requiring service are new or well-cared-for older models. Since the dealership must be able to fix any part of the vehicle, the technician can perform a large variety of work. Although many dealer service departments have technicians who work only in specific areas, the trend is toward training all technicians to handle any type of work. Most repairs will be on the same make of vehicle, although many large dealerships now handle more than one make. Most dealerships are tied into manufacturer's hotlines. These hotlines are used to access factory diagnosis and repair information, which makes it easier to troubleshoot and correct problems.

The disadvantages of dealership employment are the lack of salary guarantees, low pay rates for warranty repairs, and fast-paced working conditions. If you welcome the challenge of being paid by the job and do not mind working under deadlines, a dealership may be the ideal employer. Also check out the local large-truck dealerships. Although the work is much heavier, the pay is usually somewhat higher and working conditions are not as fast paced.

Chain and Department Store Auto Service Centers

Many national chain and department stores have auto service centers that perform various automotive repairs. See **Figure 1-7.** ***Department store service centers*** often hire technicians for entry-level positions and may offer opportunities for advancement. Technicians in most of these positions receive a base salary (or hourly wage), plus a commission on the work performed. Pay scales for the various work classifications are competitive, and most national companies offer generous benefit packages. One advantage of working for a large company is the chance for advancement into other areas, such as sales or management.

One disadvantage of working at the average auto service center is the lack of variety. Most auto service centers concentrate on a few types of repairs, such as brake work and alignment, and turn down most other repairs. The work can become monotonous due to the lack of variety. Although the job pressures are usually less than those at dealerships, customers still expect their vehicles to be repaired in a reasonable period of time. However, if you enjoy working on only one or two areas of automotive repair, this type of job may be ideal for you.

Specialty Shops

Specialty shops can offer good working conditions and good pay. Most specialty shops concentrate on one area of service, such as automatic transmission and transaxle repair, and are fully equipped to handle all aspects of their particular specialty. See **Figure 1-8.** These shops may occasionally take in other repair work when business in their specialty is slow. Specialty shops can be ideal places to work if you want to concentrate on one area of repair. Specialty shops usually offer a base salary, plus commission for the work performed.

A disadvantage of specialty shops is the lack of variety. Since specialty shops concentrate on only a few types of repairs, working at these facilities can become monotonous. Many specialty shops are franchise operations, and the demands of the franchise can create problems. If, for example, the prime purpose of the shop is to sell tires, the technician hired to perform steering and suspension repairs may spend most of the workday installing tires. This can be annoying to technicians who want to be doing the job they were hired to perform. However, if this does not bother you, a franchise operation can be a good work situation.

Independent Shops

There are millions of ***independent auto repair shops.*** These shops have no affiliation with vehicle manufacturers, chain stores, or franchise operations, **Figure 1-9.** As places to work, independent repair shops range from excellent to terrible. Many are run by competent, fair-minded managers

Figure 1-7. A department store service center can be a good source of employment.

Figure 1-8. Specialty shops often perform only one type of service or repair operation. This particular shop specializes in tire and battery replacement.

Figure 1-9. Independent repair shops have no affiliation with vehicle manufacturers, chain stores, or franchise operations. This particular shop will perform a variety of repairs on all types of vehicles.

and have first-rate equipment and good working conditions. Others, however, have almost no equipment, low pay rates, and extremely poor—even dangerous—working conditions. Technicians at independent shops are usually paid on a salary-plus-commission basis.

There are two major classifications of independent repair shops: the general repair shop and the specialty shop. The general repair shop takes in most types of work and performs a variety of jobs. General repair shops may avoid some jobs that require special equipment, such as automatic transmission repair, alignment, or air conditioning service. However, they will usually perform a variety of other repair work on different makes and types of vehicles. A general repair shop can be a good place to work if you like to be involved in many different types of diagnosis and repair.

As stated earlier, specialty shops confine their work to one area of repair, such as brake repair or tune-ups. These shops can be ideal places to work if you want the concentrate on one area of repair.

Government Agencies

Many local, state, and federal ***government agencies*** maintain their own vehicles. Government-operated repair shops can be good places to work. Employees are generally paid on an hourly or weekly basis. Pay is usually set by law, and there is no commission paid. Although pay scales at government agencies are lower than those in private industry, the benefits are usually excellent. Pay raises, while relatively small, are regular. Most technicians employed in government shops work 35–40 hours a week and have the same holidays as other government employees.

The working conditions in most government-operated shops are good, without the stress of meeting deadlines or dealing with customers. Hiring procedures at government-operated shops are more involved than those at other auto repair facilities. Applicants must take civil service examinations, which often have little to do with automotive subjects. Government agencies may require a certain level of education and a thorough background check. In some cases, the prospective employee must be a registered voter to be considered for the job.

If you are interested in working for a government-owned repair shop, contact your state employment agency for the addresses of local, state, and federal employment offices in your area.

Technician Training Opportunities

Until about 20 years ago, anyone could buy a set of hand tools and call himself or herself an automotive technician. This is no longer possible. Today, technicians must have extensive training in automotive theory, as well as in diagnosis and repair techniques. In addition, every model year brings major changes in vehicle technology. Technicians must study these new systems and learn how to service them. There are many training opportunities available to the automotive technician. The following section briefly describes some of these opportunities.

Training to become a Technician

There are many training programs available to those who want to improve their automotive repair skills. If you are reading this book, you may have already enrolled in an automotive training program. Some school districts offer automotive programs as part of their high-school curriculum. State and local governments often operate junior or community colleges that have various automotive, collision, and diesel courses. These colleges usually give college credits for the courses taken. Many county and city governments support vocational schools. These schools were formerly called trade schools or vo-tech schools.

Some private universities offer automotive courses or conduct training programs for various vehicle manufacturers. In some cases, vehicle manufacturers open their schools to beginning students, but these openings are usually limited to students who are already enrolled in another type of course. Some manufacturers have on-campus schools at local colleges. Examples are the GM ASEP, Ford ASSET, and Toyota T-TEN programs.

Many correspondence (home study) schools offer automotive courses. Taking a correspondence course allows you to work at your own pace. These courses are ideal if you cannot devote yourself to classes full time.

Training for Experienced Technicians

There are many ways for experienced technicians to keep up with changes in the automotive repair industry. Some of the most common and useful are detailed in the following sections.

Trade Magazines

Many technicians subscribe to automotive ***trade magazines.*** These magazines are specifically marketed to automotive technicians and managers. They contain information on the latest vehicle systems and new service equipment. Trade magazines also have articles about business operation and legislative affairs that affect the automotive service industry. These magazines occasionally run articles covering the theory of automotive systems. Trade magazines are published monthly or quarterly. Typical trade magazines are shown in **Figure 1-10.**

Some manufacturers publish their own magazines to address the special service aspects of their vehicles. An example of manufacturer-issued magazines is shown in **Figure 1-11.**

Internet Resources

The ***Internet*** is a source of many automotive websites, forums, and bulletin boards. The International Automotive Technicians Network, usually called iATN, operates one popular website. The iATN website, **Figure 1-12,** contains many resources. It is divided into major areas, such as drivability, transmissions, air conditioning, and engine repair.

Figure 1-10. These trade magazines are designed to help technicians keep up with changes in the automotive field.

Figure 1-11. Some trade magazines are published by vehicle manufacturers. These magazines deal with various aspects of the manufacturers' products.

Manufacturers' Schools

Vehicle manufacturers often have schools that offer classes on technical changes to their latest models. See **Figure 1-13.** These schools are open to the technicians employed at that manufacturer's dealerships. To serve a particular region, manufacturers' schools are usually located in a large city that is centrally located in that region.

Aftermarket Seminars

Aftermarket parts distributors often hold classes, which are often called ***aftermarket seminars.*** These seminars are usually given at night and held at a local repair shop that has agreed to act as host. The seminars usually cover service procedures on a single automotive part or system. Seminars are often advertised through local automotive associations or through fliers displayed at parts stores. Do not turn down the

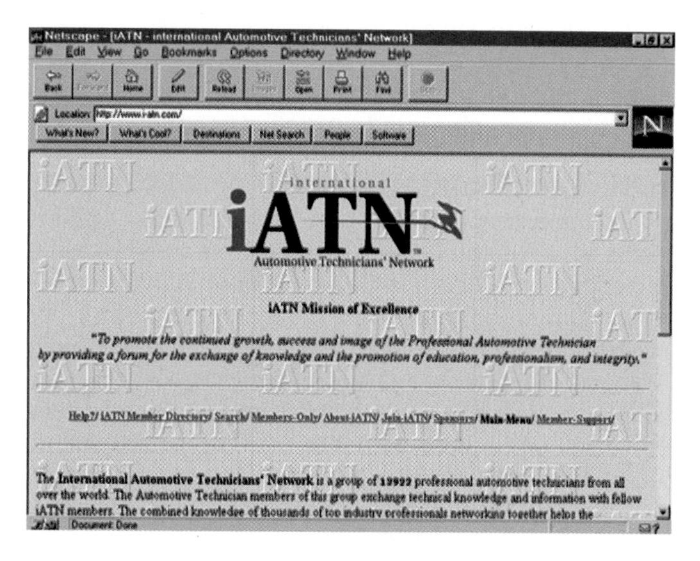

Figure 1-12. The Internet can be a good source of technical information. Some Internet sites allow technicians to ask questions and share information about difficult problems.

Figure 1-13. Manufacturers' training programs help technicians keep up with changes in the field. (Ford)

chance to attend one of these seminars. You always learn at least one thing that will save time or prevent damaged parts.

Automotive Trade Associations

There are many ***automotive trade associations.*** Most publish magazines and have Internet websites. One group that covers the entire range of automotive service is the Service Technicians Society (STS), which is part of the Society of Automotive Engineers (SAE). STS has many local and regional associations. STS also publishes a magazine called *Service TECH*, which contains many useful technical articles.

Other associations are dedicated to special segments of the automotive industry. For example, the Mobile Air Conditioning Society (MACS) is a trade association that addresses the concerns of the automobile air conditioning industry. MACS publishes a magazine called *ACtion.* The American Transmission Rebuilders Association (ATRA) is an association dedicated to technicians and shop owners who rebuild automatic and manual transmissions. ATRA publishes a magazine called *GEARS,* as well as many technical bulletins dealing with specific transmission problems. Many other trade associations are active in the automotive service field.

ASE Certification Tests

Certification is the process by which individuals are tested to determine whether they meet the standards of a specialized profession. Certification is used in many professions. Many skilled trades had a system of certification by the beginning of the 20th century. The automotive service industry, however, was resistant to the idea of certification, and standards were difficult to establish. Many unnecessary or improperly done repairs were performed, and the public came to regard auto mechanics as unintelligent, dishonest, or both.

In 1972, the ***National Institute for Automotive Service Excellence (ASE)*** was established to improve the standards of automotive service and repair by providing a certification process for automobile technicians. ASE offers a series of written tests in various automotive subject areas. These tests are standardized, so everyone who takes the tests on a given date is given the same tests. Any technician who passes one of the tests and meets certain experience requirements is certified in the area covered by that test. If a technician passes all the tests in a certain area, he or she is certified as a master technician in that area. The purposes of the ASE-certification program are to identify and reward skilled and knowledgeable technicians, help technicians advance their careers, and allow potential employers and the public to identify good technicians.

Many repair shops now hire only ASE-certified technicians. Most shops with certified technicians prominently display the ASE logo on their shops, **Figure 1-14.** Close to 500,000 technicians are now ASE certified in one or more areas. The ASE certification program has now been extended to Canada, Brazil, and Mexico. Other countries may become involved in the future.

ASE is also involved in the development of effective training programs, conducting research on the best methods of performing instruction, and publicizing the advantages of technician certification. The advantages that the ASE certification program has brought to the automotive industry include increased respect and trust of automotive technicians. This has resulted in better pay and working conditions for technicians, and increased standing in the community. Thanks to ASE, the automotive technicians are taking their place next to other skilled professionals. Chapter 42, ASE Certification, contains detailed information on how to apply for and take the ASE certification tests.

Figure 1-14. Shops that hire ASE-certified technicians often display the ASE logo.

Tech Talk

As you browse through this book, you may worry that you will never be able to learn everything it contains. However, if you look closely, you will notice that the book is grouped into sections. You can study these sections one at a time, building on the information presented in earlier sections. By the time you finish all the sections in this book, you will wonder why you were ever worried.

Tackling big jobs one "section" at a time is a good practice to follow throughout your career. Any job, no matter how large or complicated, can be broken down into smaller tasks. Then these tasks can be performed one at a time until the job has been completed successfully.

Summary

The automotive service industry provides employment for many people and will continue to do so. Automotive service has some disadvantages, such as long hours, hard work, and lack of status. Advantages include interesting work, the security of a guaranteed career, and the enjoyment of diagnosing and correcting problems. Auto service is a secure profession, and the good technician can always find work.

The three general classes of technicians are the helper, the installer, and the certified technician. The helper does the simplest tasks, such as changing tires and oil. Many helpers move up into other positions after a short time. The installer installs new parts, such as shock absorbers and strut assemblies, and sometimes moves into brake repair and wheel alignment. The certified technician performs the most complex diagnosis and repair jobs, and makes the most money.

Other opportunities in the automotive service field include management and sales positions. Another overlooked employment possibility is the automotive parts business.

There are many places to work as an automotive technician. Among the most popular are new-vehicle dealerships, auto centers affiliated with department or chain stores, specialty shops, independent repair shops, and government agencies. Some people prefer to have their own businesses, either as independent owners or as part of a franchise system.

Today's technicians must have extensive training in theory and in service techniques. Technicians must keep up with new developments. Many training methods are available to the beginning automotive technician. These include high school programs, junior and community colleges, and private schools.

To remain current, the experienced technician can subscribe to trade magazines, attend manufacturer's schools and aftermarket seminars, and join automotive trade associations.

ASE certification tests are a way for technicians to improve themselves financially and add to the professionalism of the entire automotive industry.

Review Questions—Chapter 1

Do not write in this book. Write your answers on a separate sheet of paper.

1. To ensure that you stay employable, always seek to learn _____ things, and become _____ certified in as many areas as possible.
2. Many technicians begin their careers as _____.

Matching

Match the grade of technician to the job they would most likely perform.

3. Changing shock absorbers _____
4. Installing tires _____
5. Aligning a suspension _____
6. Replacing an alternator _____
7. Changing oil _____
8. Checking fuel pressure _____
9. Diagnosing ignition problems _____
10. Adding fluid to a transmission _____

(A) Helper.
(B) Installer.
(C) Certified technician.

11. Shop managers often deal with _____ and _____ instead of making repairs.
12. One advantage of a franchise operation is _____ advertising.
13. How many areas of the vehicle do specialty shops usually confine their work?
14. Employees at government garages are generally paid on a _____ or _____ basis.
15. How many years ago was it possible to buy a set of tools and say that you were an automotive technician?
    (A) 5.
    (B) 10.
    (C) 15.
    (D) 20.
16. A magazine that is specifically marketed to people in the automotive service business is called a _____ magazine.
17. Classes given by aftermarket parts distributors are called _____.
18. List three examples of automotive trade associations.
19. Define a standardized test.
20. What are the three main purposes of the ASE certification test program?

ASE-Type Questions

1. Technician A says that one advantage of working in the automotive service field is the opportunity to work with your hands. Technician B says that one advantage of working in the automotive service field is that it is a prestigious career. Who is right?
    (A) A only.
    (B) B only.
    (C) Both A and B.
    (D) Neither A nor B.

2. All of the following statements about classes of automotive service jobs are true, *except:*
   (A) certified technicians are usually paid the most money.
   (B) helpers may advance to become installers.
   (C) installers are expected to diagnose problems.
   (D) certified technicians are expected to know more than installers.
3. Which of the following management positions is a technician most likely to move into?
   (A) Sales manager.
   (B) Shop manager.
   (C) Parts department manager.
   (D) Entrepreneur.
4. The sales person must enjoy working with_____.
   (A) paper
   (B) figures
   (C) engines
   (D) people
5. All of the following are job duties of the parts specialist, *except:*
   (A) obtaining parts for technicians.
   (B) telling the technicians how to install parts.
   (C) selling parts to the general public.
   (D) special ordering parts.
6. In which of the following business is an automotive technician most likely to get a job doing only a few kinds of repairs?
   (A) Chain stores.
   (B) Corner service stations.
   (C) Government garages.
   (D) New vehicle dealers.
7. All of the following statements about working for specialty shops are true, *except:*
   (A) the variety of repairs is limited.
   (B) the opportunity exists to become very skilled in one area of service.
   (C) the technician is generally paid a base salary, plus a commission on the work performed.
   (D) new vehicle warranty repairs are commonly performed.
8. One drawback to working for government-operated repair shops is the _____.
   (A) lack of benefits
   (B) lack of security
   (C) relatively low pay
   (D) tight deadlines
9. Technician A says that the Internet is a good source of automotive information. Technician B says that many vehicle manufacturers offer classes to familiarize technicians with the technical changes to their latest models. Who is right?
   (A) A only.
   (B) B only.
   (C) Both A and B.
   (D) Neither A nor B.
10. Which of the following is *not* a reason to become ASE certified?
    (A) To increase your knowledge.
    (B) To be paid by the hour.
    (C) To raise your standing in the community.
    (D) To improve the image of automotive technicians.

2

Safety and Environmental Protection

After studying this chapter, you will be able to:

- Identify the major causes of accidents.
- Explain why accidents must be avoided.
- Recognize unsafe conditions in the shop.
- Give examples of unsafe work procedures.
- Use personal protective equipment.
- Describe types of environmental damage caused by improper auto shop practices.
- Identify ways to prevent environmental damage.

Technical Terms

Accidents
Shopkeeping
Work procedures
Material safety data sheets (MSDS)
Personal protective equipment
Eye protection
Safety shoes
Respiratory protection
Protective gloves
Ear protection
Pollution
Environmental Protection Agency (EPA)

This chapter emphasizes the importance of working safely in the automotive shop. Covered are the safety procedures to follow when performing service and repair work. Also covered are proper procedures for handling and disposal of waste products to safeguard both the technician and the environment.

Accidents

Accidents are unplanned events that often cause damage or injury. The following sections examine the causes of accidents and explain how knowing what causes accidents can help you avoid them.

There are many causes of accidents. Accidents may occur when technicians take shortcuts instead of following proper repair procedures. Other accidents may occur when technicians fail to identify and correct dangerous conditions in the work area.

In many cases, accidents happen when more than one unsafe condition or act occurs simultaneously. For example, a technician may remove the ground prong from the electrical plug of a portable drill because it is difficult to find a three-prong extension cord or outlet. See **Figure 2-1.** Also, a technician may use the drill when standing on a wet concrete floor without wearing insulated rubber-soled shoes. They may feel that it is too much trouble to move to a dry area or change shoes. Neither of these unsafe acts causes an accident. However, when the technician uses the ungrounded drill while standing on a wet floor without the insulated rubber-soled shoes, an accident is possible. If the drill motor develops a short to the drill body, current will flow through the user to ground. This leads to severe electrical shock, which can cause electrical burns, damage to internal organs, or even death.

In this example, the technician did not set out to cause an accident. However, the technician thought it to be too much trouble to do things correctly. The result of thinking this way and the combination of unsafe acts is an accident.

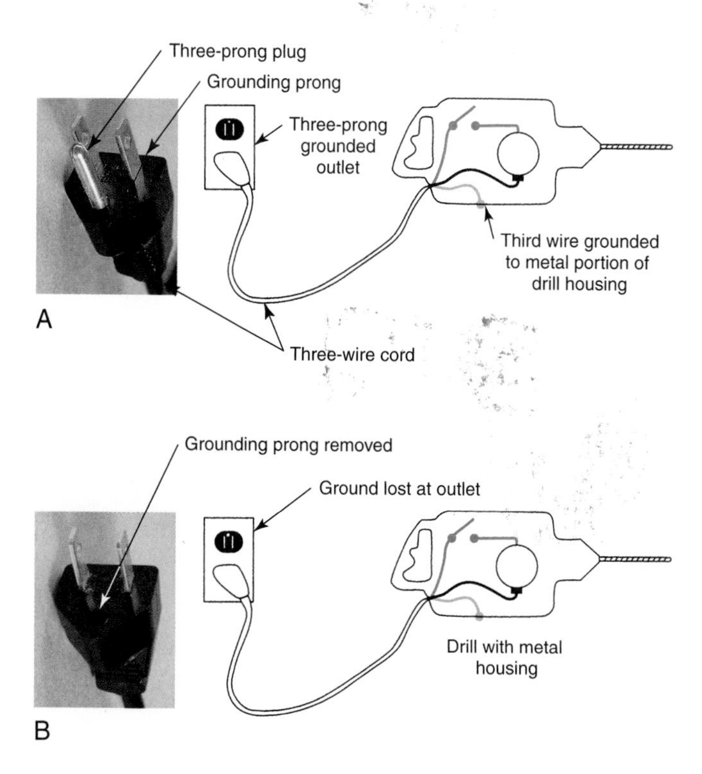

Figure 2-1. A—Properly grounded drill. B—The grounding prong was removed from the plug, making the drill unsafe.

An accident may cause injuries that keep the technician from working or enjoying free time. Some accidents can be fatal. Even slight injuries are painful and annoying. An accident may not cause personal injury, but property damage may result. Repairing damage caused by an accident can be expensive. In some cases, a technician can lose their job as a result of a careless accident.

No automotive technician wants to be injured or cause property damage. However, sometimes even the most experienced technicians become rushed and careless. As a result, falls, injuries to hands and feet, fires, explosions, electric shocks, and even poisonings may occur in auto repair shops. See **Figure 2-2.** Additionally, carelessness in the shop can lead to long-term bodily harm from prolonged exposure to harmful liquids, vapors, and dust. Lung damage, skin disorders, and even cancer can result from contact from these substances. For this reason, technicians must keep safety in mind at all times, especially when conditions tend to make it the last thing on their minds.

Preventing Accidents

Most accidents occur because of improper shopkeeping, incorrect work procedures, or a combination of both. ***Shopkeeping*** means maintaining the shop premises in order by keeping all tools and equipment in good working order and storing them in their proper places. Shopkeeping also involves properly disposing of dirt, oil, and other refuse. Examples of improper shopkeeping include failing to maintain tools and equipment; allowing old parts, containers, or other trash to accumulate in the shop; and ignoring fluid spills.

Work procedures are the actual diagnosis and repair operations performed in the shop. Improper work procedures include using the wrong tools or methods to perform repairs, using defective tools, not wearing protective equipment when necessary, and not paying close attention while performing the job.

Figure 2-2. This is what happens when a broken lightbulb meets spilled gasoline. This fire was caused by a large gasoline spill meeting an open source of ignition.

The best way to prevent accidents is to maintain a neat and clean work area, use safe methods and common sense when making repairs, and wear protective equipment when needed. The following sections cover suggestions for reducing the possibility of accidents.

Proper Shopkeeping

Shopkeeping does not mean just keeping the shop neat. It also refers to the process of identifying and correcting unsafe conditions. Many of the statements covered in this section will seem to be matters of common sense, but they are often disregarded during service operations. The following list indicates proper shopkeeping techniques:

- *Maintain tools and equipment*—Return all tools and equipment to their proper storage places. This saves time in the long run and reduces the chance of accidents and theft. Do not leave equipment out where others can trip on it. Closely monitor the condition of tools and equipment, and make repairs when necessary. This includes such varied tasks as replacing damaged leads on test equipment, checking and adding oil to hydraulic jacks, and regrinding the tips on screwdrivers and chisels.
- *Organize work area*—Keep workbenches clean. This reduces the chances of tools or parts falling from the bench to the floor, where they could be lost or damaged. Of course, this reduces the chances a tool or part will fall on your foot. A clean workbench also reduces the possibility critical parts will be lost in the clutter and reduces the chance of a fire from oily debris.
- *Handle liquids properly*—Clean up spills immediately to avoid tracking them through the shop. Many people are injured when they slip on floors coated with oil, antifreeze, or water. Gasoline spills can be extremely dangerous because a flame or spark can ignite the gasoline vapors, causing a major explosion and fire.
- *Handle chemicals properly*—Become familiar with the chemicals stored in the shop. Chemicals commonly used in the automotive shop include carburetor cleaners, hot tank solutions, parts cleaners, motor oil, and antifreeze. Chemical manufacturers provide ***material safety data sheets (MSDS)*** for every chemical they produce. See **Figure 2-3.** These sheets list all the known dangers of a specific chemical, as well as first aid procedures for skin or respiratory system contact. There should be an MSDS for every type of chemical used in the shop. Always read the appropriate MSDS before working with an unfamiliar chemical. Store chemicals in properly labeled, approved containers.
- *Use proper lighting*—Make sure the shop is well lit. Poor lighting makes it hard to see what you are doing. Not only does poor lighting make the job take longer, it can lead to accidental contact with moving parts or hot surfaces. Overhead lights should be bright and centrally located. Portable droplights should be kept in

ACME Chemical Company

Material Safety Data Sheet

Product Name: Acetylene

24-hour Emergency Phone: Chemtrec 1-800-424-9300 Outside United States 1-905-501-0802

Trade Name/Syn: Acetylene
Chemical Name/Syn: Acetylene, Ethyne, Acetylen, Ethine
CAS Number: 74-86-2
Formula: C_2H_2

NFPA Ratings
Health: 0
Flammability: 4
Reactivity: 0

Hazards Identification

Simple Asphyxiant. This product does not contain oxygen and may cause asphyxia if released in a confined area. Maintain oxygen levels above 19.5%. May cause anesthetic effect. Highly flammable under pressure. Spontaneous combustion in air at pressures above 15 psig. Acetylene liquid is shock sensitive.

Effects of Exposure-Toxicity-Route of Entry

Toxic by inhalation. May cause irritation of the eyes and skin. May cause an anesthetic effect. At high concentrations, excludes an adequate oxygen supply to the lungs. Inhalation of high vapor concentrations causes rapid breathing, diminished mental alertness, impaired muscle coordination, faulty judgment, depression of sensations, emotional instability, and fatigue. Continued exposure may cause nausea, vomiting, prostration, loss of consciousness, eventually leading to convulsions, coma, and death.

Hazardous Decomposition Product

Carbon, hydrogen, carbon monoxide may be produced from burning.

Hazardous Polymerization

Can occur if acetylene is exposed to 250°F (121°C) at high pressures or at low pressures in the presence of a catalyst. Polymerization can lead to heat release, possibly causing ignition and decomposition.

Stability

Unstable--shock sensitive in its liquid form. Do not expose cylinder to shock or heat, do not allow free gas to exceed 15 psig.

Fire and Explosion Hazard

Pure acetylene can explode by decomposition above 15 psig; therefore the UEL is 100% if the ignition source is of sufficient intensity. Spontaneously combustible in air at pressures above 15 psi (207 kPa). Requires very low ignition source. Does not readily dissipate, has density similar to air. Gas may travel to source of ignition and flash back, possibly with explosive force.

Conditions to Avoid

Contact with open flame and hot surfaces, physical shock. Contact with copper, mercury, silver, brasses containing >66% copper and brazing materials containing silver or copper.

Accidental Release Measures

Evacuate all personnel from affected areas. Use appropriate protective equipment. Shut off all ignition sources. Stop leak by closing valve. Keep cylinders cool.

Ventilation, Respiratory, and Protective Equipment

General room ventilation and local exhaust to prevent accumulation and to maintain oxygen levels above 19.5%. Mechanical ventilation should be designed in accordance with electrical codes. Positive pressure air line with full face mask or SCBA. Safety goggles or glasses, PVC or rubber gloves in laboratory; and as required for cutting or welding, safety shoes.

Figure 2-3. The MSDS shown here contains all information needed to safely use the listed chemical.

proper operating condition. Always use a "rough service" bulb in incandescent droplights. See **Figure 2-4.** These bulbs are more rugged than normal lightbulbs and will not shatter if they break. Do not use a high-wattage bulb in a droplight. High-wattage bulbs get very hot and can melt the light socket or burn anyone who touches the droplight's safety cover.

- *Use care when using electrical devices*—Do not overload electrical outlets or extension cords by operating several electrical devices from one outlet. See **Figure 2-5.** Never pair up high-current electrical devices or operate them through extension cords. Examples of high-current electrical devices include drills, grinders, and electric heaters.
- *Inspect electrical devices*—Periodically inspect electrical cords and compressed air lines to ensure they are in good condition. Do not close vehicle doors on electrical cords or air lines. Do not run electrical cords through water puddles or use them outside when it is raining.
- *Use guards on shop equipment*—Make sure all shop equipment, such as grinders and drill presses, is equipped with appropriate safety guards. These guards should only be removed for service operations, such as changing the grinding wheels.
- *Work under safe conditions*—When servicing any piece of equipment, be sure it is unplugged. Read the equipment's service literature before beginning repairs.
- *Keep chemicals secure*—Never leave open containers of antifreeze inside or outside the shop. Ethylene glycol antifreeze will poison any animal (or person) that drinks it, and antifreeze spills will create an extremely slippery floor.

Following Proper Work Procedures

Whenever working, safety should be the primary consideration. The following safe work procedures may seem simple. However, they are often disregarded, with tragic results. The following list indicates proper work procedures:

- *Use protective equipment*—Wear proper clothing and use protective equipment when necessary. Various types of protective equipment are explained in the next section.
- *Understand the procedure*—Study proper work procedures before beginning any job that is unfamiliar. Never assume the procedure you have used in the past will work with a different type of vehicle. Always work carefully. Speed is not as important as doing the job properly and avoiding injury. Avoid people who will not work carefully.
- *Use the proper tool*—Use the right tool for a given job. Using a screwdriver as a chisel or a wrench as a hammer often leads to an accident. Never use a hand socket (12-point) with an impact wrench. Hand sockets are not designed to withstand the torque produced by an impact wrench and many shatter. If necessary, impact sockets can be used with a hand ratchet.
- *Understand how to operate equipment*—Learn how to operate new equipment before using it. This is especially true of air-powered tools, such as impact wrenches and chisels, and large electrical devices, such as drill presses, boring bars, and brake lathes. These tools are very powerful and can cause severe injury if used improperly. A good way to learn about new equipment is to read the manufacturer's instructions.
- *Avoid short circuiting equipment*—When working on electrical systems, avoid creating a short circuit with a jumper wire or a metal tool. Not only will this damage the vehicle components or wiring, but it will generate enough heat to cause a severe burn or to start a fire.

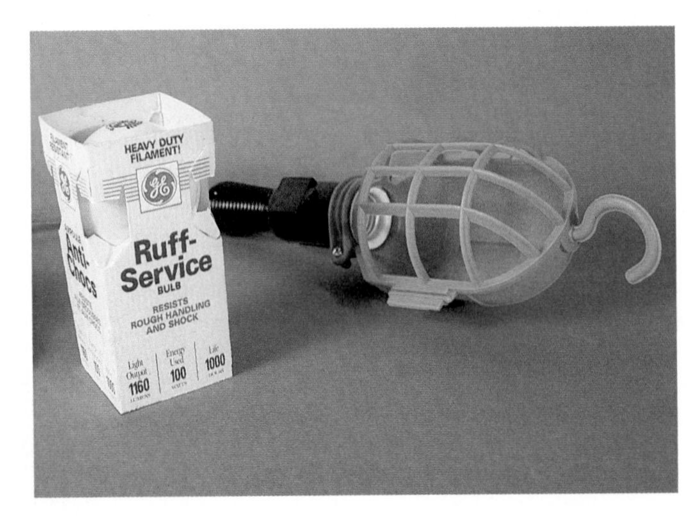

Figure 2-4. Always use a rough service bulb in droplights, or any portable light. A rough service bulb is much cooler and reduces the chances of burns. (Jack Klasey)

Figure 2-5. This overloaded outlet is a common cause of fires. A two socket electrical fixture cannot safely handle the electrical load caused by attaching this many electrical devices.

- *Lift objects properly*—Take proper precautions when lifting objects. Make sure you are strong enough to lift the object to be moved. Always lift with your legs, not your back. If an object is too heavy to lift alone, get help.
- *Do not smoke in the shop*—You may accidentally ignite an undetected gasoline leak. A burning cigarette can also ignite oily rags or paper debris.
- *Use proper techniques to lift and support a vehicle*—Never attempt to raise a vehicle with a jack that is unsafe or too small. Always support a raised vehicle with quality jack stands. Never use boards or cement blocks to support a vehicle. Boards can break and cement blocks can crumble under the weight of the vehicle.
- *Always have proper ventilation*—Never run an engine in a closed area without proper ventilation, even for a short time. Carbon monoxide builds up quickly, is odorless, and is deadly.
- *Take care around moving parts*—When working on or near a running engine, keep away from moving parts. Never reach between moving engine parts. A rotating engine part can cause serious injury.
- *Never leave a running vehicle unattended*—The vehicle may slip into gear or overheat while you are away. When working on a running vehicle, set the parking brake.
- *Obey traffic laws*—When road testing a vehicle, be alert and obey all traffic laws. Do not become so absorbed in diagnosing a problem that you forget to watch the road. Always watch the actions of other drivers carefully. If necessary, take another technician along to assist you during the road test.
- *Know where safety equipment is located*—Know where the shop fire extinguishers are located and how to operate them. Make sure you know what type of fire extinguisher is used on each type of fire. See **Figure 2-6.** Periodically check each fire extinguisher to ensure they are in working order and have them inspected periodically by qualified personnel.

Personal Protective Equipment

The proper use of ***personal protective equipment*** is vital. Some of the equipment described in the following sections are not only for protection from immediate injury but also for preventing damage and disease caused from long-term exposure to harmful substances. For example, long-term exposure to brake dust, used oil, or exhaust fumes can cause skin cancer, emphysema, and lung cancer. Various types of protective equipment are shown in **Figure 2-7.**

Do not wear open jackets, scarves, or shirts with long, loose sleeves. Loose clothing can become caught in moving parts of a machine or engine. Long hair should be tied back or secured under a hat to keep it away from moving parts. Jewelry can also be caught in moving parts, so rings and other jewelry should be removed. Metal jewelry can cause a short circuit if it completes a path between a positive source and ground. The short circuit can generate enough heat to cause serious burns.

Eye protection is very important when working in any situation that could result in airborne dirt, metal, or liquids. This includes working around running engines; using drills, saws, grinders or tire changers; and working around batteries and hot cooling system parts. See Figure 2-7A.

Safety shoes, preferably with steel toe inserts, should be worn when there is any danger of falling parts or tools. See Figure 2-7B. Since such an incident can occur at any time, it is a good idea to wear safety shoes whenever you are in the shop.

Respiratory protection, such as an approved respirator, should be worn whenever you are working on brake systems or clutches. See Figure 2-7C. The dust from the friction lining materials used in these devices can cause lung damage or cancer. Respiratory protection is also recommended when working around any equipment that gives off fumes, such as a hot tank or steam cleaner.

Protective gloves should be worn when working with solvents, such as parts cleaner. If oil, gasoline, cleaning solvents, or any other substance comes in contact with skin, it should be cleaned off immediately. Prolonged exposure to even mild solvents or petroleum products can cause severe skin rashes or chemical burns. If there is any question about the toxicity of a particular substance, refer to the proper MSDS.

Ear Protection should be worn to protect ears when working with air wrenches, engines under load, or engines running in an enclosed area. *Earphone-type protectors* and *earplugs* are two effective forms of ear protection. See Figure 2-7D.

Remember after all other factors have been figured in, it is still up to you to correct safety hazards, to work safely, and to prevent accidents. Always use common sense when working on vehicles.

Preventing Environmental Damage

Controversy continues over the extent of the environmental damage caused by ***pollution.*** However, it is clear that some damage is definitely occurring as a result of careless disposal of wastes.

Unfortunately, the automotive service industry is a major source of waste materials. These wastes can be liquids (antifreeze and oil), solids (scrap parts, tires, and paper containers), or gases (refrigerant or carbon monoxide). Preventing the accidental spilling or careless disposal of these wastes is important if we are to protect the environment.

It is important to remember when we protect the earth's environment we are protecting ourselves. Pollutants can kill—sometimes quickly, sometimes slowly. They will also contaminate the air we breathe, the water we drink, and the soil in which our food grows. The economic burden of dealing with waste will grow ever larger. If we do not take responsibility for the wastes we generate, our descendants will suffer the effects of our irresponsible acts.

Fire Extinguishers and Fire Classifications

Fires
Class A Fires Ordinary Combustibles (Materials such as wood, paper, textiles.) *Requires... cooling-quenching.* A
Class B Fires Flammable Liquids (Liquids such as grease, gasoline, oils, and paints.) *Requires...blanketing or smothering.* B
Class C Fires Electrical Equipment (Motors, switches, etc.). *Requires... a nonconducting agent.* C
Class D Fires Combustible Metals (Flammable metals such as magnesium and lithium.) *Requires...blanketing or smothering.* D

Type	Use	Operation
Soda-acid Bicarbonate of soda solution and sulfuric acid 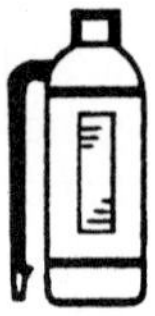	Okay for use on: A Not for use on: B C D	Direct stream at base of flame.
Pressurized Water Water under pressure 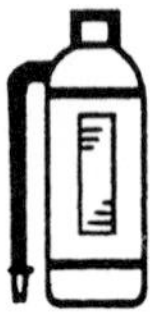	Okay for use on: A Not for use on: B C D	Direct stream at base of flame.
Carbon Dioxide (CO_2) Carbon dioxide (CO_2) gas under pressure 	Okay for use on: B C Not for use on: A D	Direct discharge as close to fire as possible, first at edge of flames and gradually forward and upward.
Foam Solution of aluminum sulfate and bicarbonate of soda 	Okay for use on: A B Not for use on: C D	Direct stream into the burning material or liquid. Allow foam to fall lightly on fire.
Dry Chemical 	Multi-purpose type — Okay for: A B C; Not okay for: D Ordinary BC type — Okay for: B C; Not okay for: A D	Direct stream at base of flames. Use rapid left-to-right motion toward flames.
Dry Chemical *Granular-type material*	Okay for use on: D Not for use on: A B C	Smother flames by scooping granular material from bucket onto burning metal.

Figure 2-6. Fire and fire extinguisher classification. Be sure to use the right extinguisher for each fire.

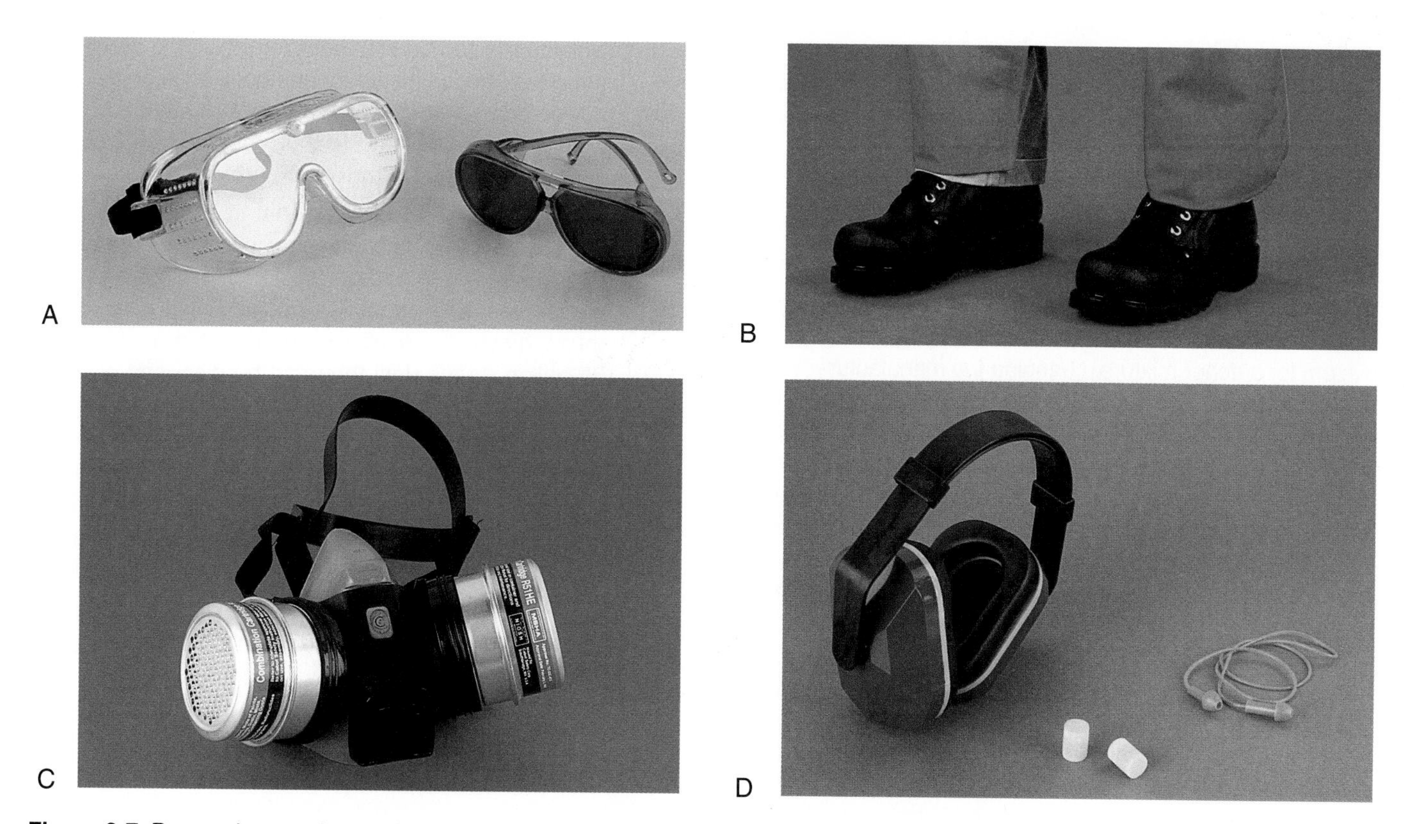

Figure 2-7. Personal protective equipment. A—Safety goggles and glasses. B—Steel-toed safety boots. C—Respirator. D—Earphone-type protector and two variations of earplugs.

Ways of Damaging the Environment

It is common to identify pollution with large companies. However, anyone can be a polluter. The ways in which the automotive technician can cause environmental damage can be divided into main areas:

- Carelessly disposing of wastes generated in the shop.
- Repairing vehicles in such a way they cause increased air pollution.

Careless Waste Disposal

One of the most common ways that automotive technicians cause environmental damage is by improperly disposing of liquid wastes. Pouring oil, transmission fluid, brake fluid, antifreeze, and used cleaning solutions on the ground contaminates the soil. In addition, these liquids sink farther into the ground every time it rains, eventually contaminating the water table. Your local drinking water may come from this water table. Another way in which liquids can contaminate the soil and the water table is through leaking storage tanks. Although this problem is confined primarily to underground gasoline tanks, any type of tank can leak.

Liquid wastes should never be disposed of by pouring them into the local drainage system. Municipal waste treatment plants cannot handle petroleum products or the heavy metals they absorb from the vehicle during use. In most areas, such dumping is illegal. In addition, the ***Environmental Protection Agency (EPA)*** has established strict guidelines for disposing of toxic waste. In some cases, you may be liable for cleaning up a contaminated area years after the actual violation occurred.

In many areas, there are companies that specialize in recycling liquid wastes. These companies will accept used oil and used antifreeze for recycling. The oil and antifreeze are refined again and reused. Some used oil is burned by power plants to produce electricity. This eliminates the waste oil and reduces the dependence on imported crude oil.

Another way that auto technicians improperly dispose of wastes is by discharging air conditioner refrigerant into the atmosphere. Studies have shown refrigerants cause extensive damage to the earth's environment in the form of ozone layer reduction and global warming. Even so-called "safe" refrigerants, such as R-134a, contribute to environmental damage. Therefore, federal and, in some cases, state laws require all refrigerants to be recycled. In addition to damaging the environment, replacing refrigerants is expensive. Consequently, it makes good economic sense to recover and reuse refrigerants.

Automotive technicians also increase waste problems by failing to recycle. Parts that can be rebuilt, paper products, old tires, and salable scrap metals are often carelessly discarded, increasing the amount of solid waste that must be disposed of in landfills. Landfill space is becoming scarce. In many parts of the country, local trash departments will not take certain materials for disposal. Burning is illegal in most areas of the country and simply turns solid wastes into airborne wastes. It makes good economic sense to recycle because almost every rebuildable part has a return, or "core," value. The value of paper, old tires, and scrap metals will depend on the current market conditions.

Improper Vehicle Repair Procedures

Another major way in which automotive technicians may cause damage to the environment is by making adjustments or modifications that defeat the purpose of the vehicle emission control systems. This can be done by adjusting the fuel system for a richer mixture, changing the manufacturer's initial timing settings or advance curve, retrofitting older model cylinder heads to a late-model vehicle, disconnecting the engine control computer, or removing the air injection (smog) pump or the catalytic converter. In addition, some seemingly harmless actions, such as installing a lower temperature cooling system thermostat or a nonstock air cleaner, can also cause a rise in emissions. Not only are these actions illegal, but they almost never increase power or mileage as much as hoped.

Any increase in power and economy is almost always offset by increased engine wear, decreased driveability, and other problems. Automobile manufacturers have carefully designed the function of the emission controls as part of the overall engine and vehicle design, and the technician can rarely outguess them. This is especially true of vehicles with electronic engine controls.

Modifying, disabling, removing, or otherwise tampering with engine emission controls is a crime, which carries harsh penalties. The Environmental Protection Agency (EPA), as well as state agencies, enforces vehicle emission laws. The EPA and many states investigate suspected violations and often conducts "sting" operations to catch violators. Some states, such as California, have additional laws protecting the environment.

Following Proper Environmental Procedures

The following basic rules will help make the shop and the vehicles serviced more environmentally responsible. Some of these rules are enforced by federal and state laws.

- Do not make adjustments or modifications to any vehicle system without determining the effect on emissions. It may surprise you to find out how service operations affect emissions.
- Do not discharge any air conditioner refrigerant into the atmosphere. Refrigerant recovery equipment must be used at every shop servicing air conditioning systems.
- Recycle parts and scrap materials whenever possible. Check with your local parts supplier to determine which parts can be sent back for rebuilding. Recyclers are often listed in the telephone book and can give you advice on what to do with recyclable materials. If solid wastes cannot be recycled, dispose of them responsibly, not by illegal dumping or burning.
- Do not pour used motor oil, transmission fluid, antifreeze, brake fluid, or gear oil on the ground or into municipal drainage systems. Many recycling companies will pick up used oil and antifreeze. Store these materials in above-ground storage tanks or 55-gallon drums until they can be disposed of properly, **Figure 2-8.**

Additional information about waste disposal and vehicle emissions can be obtained from the Environmental Protection Agency. The EPA has ten regional offices throughout the United States. For the address of the nearest EPA office, write to:

Office of Transportation and Air Quality
United States Environmental Protection Agency
1200 Pennsylvania Ave, N.W.
Washington, DC 20460
Or go online to:
http://www.epa.gov/otaq

Figure 2-8. Above-ground tanks should be used to store liquid waste. (Justrite Manufacturing Co.)

Tech Talk

These days, there is a lot of talk about job security and how you must stay on your toes to keep your job. While many people feel their jobs are in danger, automotive technicians are generally not among them. There is a strong demand for qualified automotive technicians. Whether you want to fix cars, light trucks, tractor-trailers, construction equipment, or boats, you can find a job if you have the necessary skills.

Although the demand for qualified technicians is high, you must remember no job is 100% guaranteed. Even if your skills are in demand, you must do the work right, show up on time, and meet the minimum standards of attitude and appearance.

Summary

Many accidents occur when technicians try to take shortcuts instead of following proper repair procedures. Accidents also occur when dangerous conditions in the work area are not corrected.

An accident may result in personal injuries, long-term bodily harm, or damage to equipment or property. No automotive technician wants to be injured or cause property damage.

Improper shopkeeping and improper working procedures are the two major areas of unsafe acts. The best way to prevent accidents is to maintain a neat work area, use proper repair methods, and use protective equipment when needed. It is up to the technician to study the job beforehand, work safely, and prevent accidents. Always use common sense when working on vehicles.

Various types of protective equipment are needed to protect the eyes, feet, lungs, and skin. Protective equipment should guard against immediate injury and the long-term effects of exposure to toxic substances.

Much environmental damage is caused by careless production and disposal of wastes. Wastes can take the form of liquids, solids, or gases. Anyone can be a cause of pollution.

Automotive shops are a major source of environmental problems. The two main ways that an automotive shop can cause environmental damage are carelessly disposing of wastes and repairing vehicles in such a way they pollute the atmosphere.

Environmental rules should be followed in all instances to prevent damage to the air, water, and soil. In many cases, proper disposal of wastes and proper vehicle repairs are required by federal and state law.

Review Questions—Chapter 2

Do not write in this book. Write your answers on a separate sheet of paper.

1. The two major areas of unsafe acts are improper _____ and improper work _____.
2. You should always wear respiratory protection when working on _____ and _____.
3. Creating a short circuit with a jumper wire or a metal tool can cause _____.
   (A) damage to vehicle electrical components
   (B) fires
   (C) personal injury
   (D) All of the above.
4. Technicians should lift heavy objects with their _____, not their backs.
5. If jack stands are not available, is it ever okay to support a vehicle with good quality cement blocks? Explain your answer.
6. Each shop electrical outlet should be used to operate _____ high-current tools.
   (A) 1
   (B) 2
   (C) 3
   (D) 4 or more
7. Material Safety Data Sheets are provided for all dangerous _____.
   (A) procedures
   (B) tools
   (C) chemicals
   (D) working conditions
8. The most important part of working safely is using _____ _____.
9. Waste can take three forms. What are they?
10. State the two main ways that an automotive technician can cause environmental damage.

ASE-Type Questions

1. Technician A says accidents occur in auto repair shops when technicians try to take shortcuts. Technician B says accidents occur in auto repair shops when technicians fail to correct dangerous conditions in the work area. Who is right?
   (A) A only.
   (B) B only.
   (C) Both A and B.
   (D) Neither A nor B.
2. Some of the common accidents that occur in automotive shops are _____.
   (A) falls
   (B) fires
   (C) electric shocks
   (D) All of the above.

3. Technician A says the best way to prevent accidents is to maintain a neat shop. Technician B says the best way to prevent accidents is to keep an orderly workplace and to use proper methods of repair. Who is right?
   (A) A only.
   (B) B only.
   (C) Both A and B.
   (D) Neither A nor B.
4. If all equipment and tools are returned to their proper storage places, all of the following will occur, *except:*
   (A) they will be hard to find the next time.
   (B) the chance of tool theft will be reduced.
   (C) the chance of tripping will be reduced.
   (D) All of the above.
5. If the safety guards have been removed from a grinder, what should you do?
   (A) Be very careful when using the grinder.
   (B) Let someone else do the grinding.
   (C) Do not use the grinder until the guards are replaced.
   (D) Wear eye protection.
6. Always use a "rough service" bulb in _____.
   (A) fluorescent droplights
   (B) incandescent droplights
   (C) every light in the shop
   (D) the ceiling lights
7. Technician A says the technician should never use a hand socket with an impact wrench. Technician B says the technician should never use an impact socket with a ratchet or pull handle. Who is right?
   (A) A only.
   (B) B only.
   (C) Both A and B.
   (D) Neither A or B.
8. A high-wattage bulb in a droplight can cause _____.
   (A) blindness
   (B) burns
   (C) electric shock
   (D) All of the above.
9. Spilled ethylene glycol antifreeze can cause all of the following, *except:*
   (A) slipping.
   (B) poisoning.
   (C) environmental damage.
   (D) fires.
10. The technician should wear eye protection when _____.
    (A) working around running engines
    (B) using drills, saws, or grinders
    (C) working around batteries
    (D) All of the above.
11. Carbon monoxide can cause all of the following, *except:*
    (A) death from asphyxiation.
    (B) damage to paint and rubber materials.
    (C) long-term respiratory system damage.
    (D) environmental damage.
12. Used antifreeze should be _____.
    (A) poured into municipal drains
    (B) poured into a pit dug in the ground
    (C) left in the sun to evaporate
    (D) stored in drums until it can be recycled
13. Technician A says the Environmental Protection Agency (EPA) enforces rules concerning waste disposal. Technician B says the Environmental Protection Agency enforces laws concerning vehicle emissions. Who is right?
    (A) A only.
    (B) B only.
    (C) Both A and B.
    (D) Neither A nor B.
14. All of the following actions could cause vehicle emissions to increase, *except:*
    (A) adjusting the carburetor.
    (B) changing the cooling system thermostat.
    (C) changing the alternator.
    (D) adjusting the ignition timing.
15. Technician A says recycling parts makes good economic sense. Technician B says recycling parts increases waste. Who is right?
    (A) A only.
    (B) B only.
    (C) Both A and B.
    (D) Neither A nor B.

Suggested Activities

1. Think back to any minor accidents you have had in your life, especially accidents that happened while working on a vehicle or doing another kind of repair work. Based on these accidents, answer the following questions:
   A. What were the results of the accidents in the way of pain and inconvenience?
   B. Was any property damaged?
   C. Could any of the accidents have been prevented if I had recognized an unsafe condition?
   D. Could any of the accidents have been prevented if I had used safe work procedures?

   Discuss your answers to these questions with the other class members.
2. Using an Internet search engine, look for news stories describing accidents that occurred in the workplace. Answer the following questions for each accident.
   A. How were people hurt?
   B. How was property damaged?
   C. What caused the accident?
   D. What accident prevention steps covered in this chapter might have prevented the accident?
3. Draw a floor plan of your automotive shop. Decide where equipment should be placed to avoid accidents and provide escape routes in case of fire. Discuss your layout with other class members.
4. Locate every shop fire extinguisher in the shop and list the types of fires each is designed to put out. Draw a "map" showing the location of each fire extinguisher.

Cutaway of a late-model pickup truck. An extensive set of quality tools is needed to properly service today's complex vehicles. (DaimlerChrysler)

3

Basic Tools and Service Information

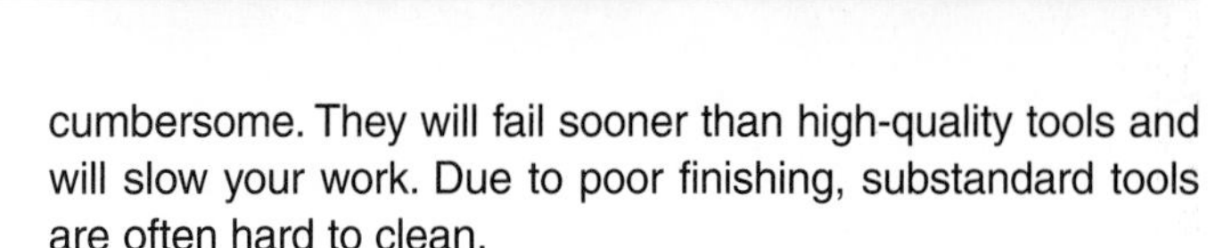

After studying this chapter, you will be able to:

- Identify and describe the proper use of the hand tools commonly used by the automotive technician.
- List the safety rules for hand tools.
- Identify and describe the proper use of the power tools commonly used by the automotive technician.
- List the safety rules for using hand and power tools.
- Identify types of service information and training materials.
- Select the correct tools and service information for a given job.

Technical Terms

Hand tools	Screw extractor
Screwdrivers	Probing tools
Wrench	Pullers
Specialty wrenches	Vises
Sockets	Drill bits
Drive handles	Power hand drills
Extensions	Drill press
Universal joint	Grinder
Adapter	Impact wrench
Pliers	Air hammers
Hammer	Hydraulic presses
Chisels	Factory service manuals
Punches	General service manuals
Files	Specialized manuals
Hacksaw	Schematic
Taps	Troubleshooting charts
Dies	Training materials
Reamers	

Access to a large selection of high-quality tools will help you work more efficiently. Proper tools enable you to quickly perform the jobs encountered by the automotive technician. This chapter covers the hand tools, shop equipment, and service information commonly used when servicing late-model vehicles.

Buy High-Quality Tools

Rule out inferior tools. Second-rate tools are usually made of substandard material, and they are often thick and cumbersome. They will fail sooner than high-quality tools and will slow your work. Due to poor finishing, substandard tools are often hard to clean.

High-quality tools, on the other hand, are made of alloy steel and are carefully heat-treated for great strength and long life. They are less bulky than second-rate tools and generally have a smooth finish. This makes them easy on the hands and quick to clean. The working surfaces of quality tools are made to close tolerances. Repair parts and service facilities will be available. Additionally, quality tools are often guaranteed.

Many tool manufacturers produce excellent products. The selection of a specific brand must be left to the individual technician. The initial cost of good tools may be high. However, considering pride of ownership, dependability, life span, and ease of use, high-quality tools tend to be more economical than tools of lesser quality.

Proper Care Is Essential

Efficiency and confusion cannot exist together. Therefore, it is important to keep your tools clean, orderly, and accessible. A roll cabinet, a tool chest, and a "tote" tray (small tray containing a few selected tools) will provide proper storage and accessibility. See **Figure 3-1.**

Place delicate measuring tools in protective cases. Separate cutting tools (files, chisels, drills, etc.) to prevent damage to cutting edges. Lightly oil tools that are likely to rust. Store heavy tools by themselves. In general, keep frequently used tools handy. Store sockets, open-end wrenches, and box-end wrenches together. The time it takes to keep your tools clean and orderly will be greatly offset by the time saved on the job.

Types of Tools

The following sections detail the tools and equipment needed by the typical automotive technician. Keep in mind that other tools and equipment will be needed for specialized tasks. Also, some of the tools described here will not be needed if the technician works in a specialized shop and does only one type of work.

Hand Tools

The simplest tools that a technician uses are called ***hand tools.*** As the name implies, hand tools are operated

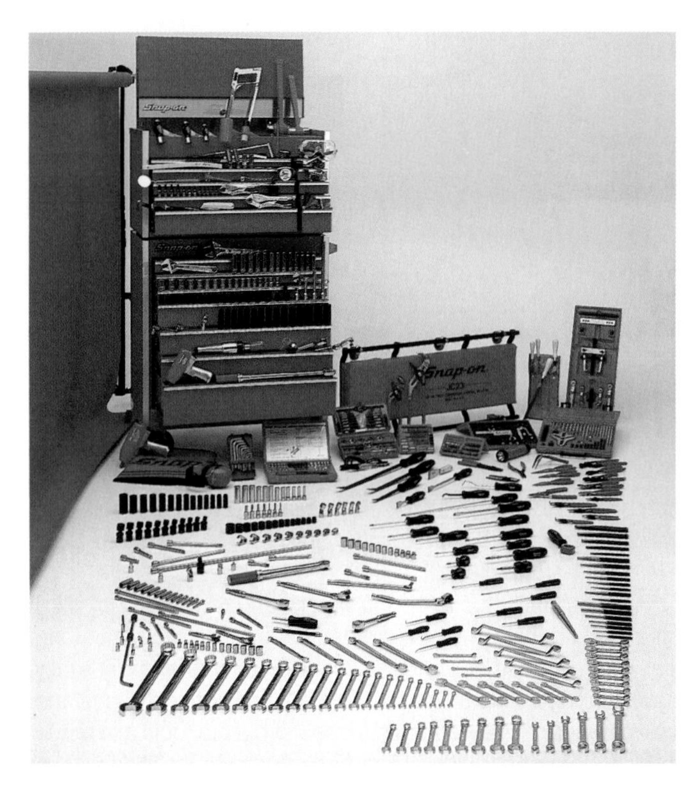

Figure 3-1. Keep your tools clean and store them in a roll cabinet tool chest, such as the one shown above. These tools are arranged for display. They will fit neatly into drawers, carry boxes, etc. (Snap-on Tools)

manually, without any type of power assist. Most hand tools are used to remove or install fasteners, or to cut or otherwise modify metal parts. Some hand tools are used to perform specialized functions. These tools will be discussed in detail later in this chapter.

Technicians are generally expected to have their own hand tools. Many of the hand tools discussed in this chapter can be purchased in sets. Screwdrivers, sockets, and wrenches are usually available in sets containing all the commonly needed sizes. Buying sets is often the most economical way to obtain the hand tools you need.

Screwdrivers

The technician should own several different sizes and types of ***screwdrivers,*** including the *standard, Reed & Prince, Phillips, Torx,* and *clutch* types. Different lengths are also desirable because screws are often located in areas only accessible to long- or short-shank screwdrivers. See **Figure 3-2.** The *offset screwdriver,* shown in **Figure 3-3,** is useful in tight quarters where even a short-shank screwdriver cannot be used.

Most screwdrivers are designed to remove and install screws only. Avoid prying with or hammering on a screwdriver.

Note: Some large screwdrivers are made so minor prying and hammering will not harm them.

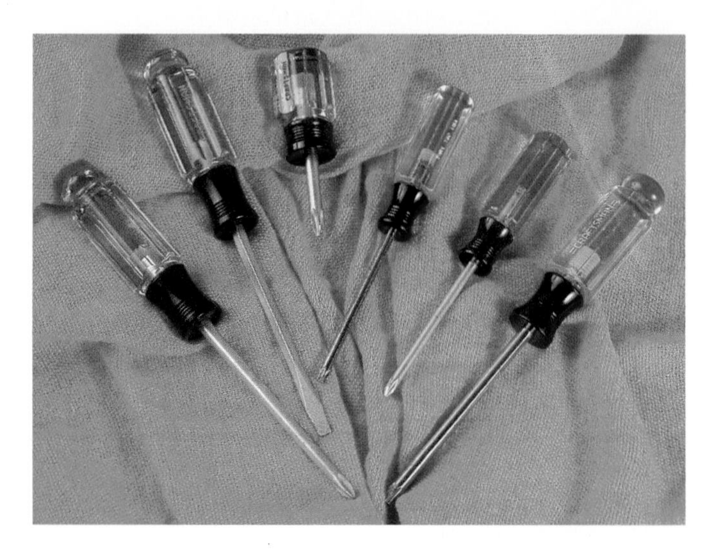

Figure 3-2. Screwdriver types. When using screwdrivers, select right type and size. A good assortment is essential.

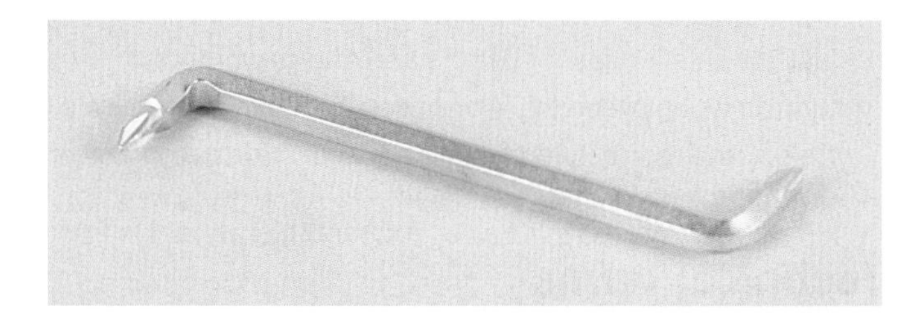

Figure 3-3. An offset screwdriver can be used in tight spaces.

Wrenches

The term ***wrench*** refers to a group of tools used to install and remove bolts, nuts, and capscrews. The *open-end wrench* grasps a fastener on only two flats. See **Figure 3-4.** Unless the wrench fits well, it can slip, rounding off the corners of the bolt or nut. There are many places where you can use open-end wrenches. Whenever possible, however, use a box-end wrench or socket instead.

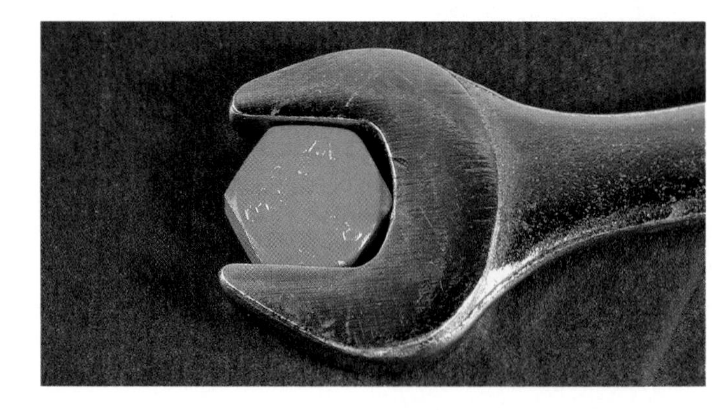

Figure 3-4. This open-end wrench is being used to remove a bolt. This type of wrench contacts the fastener on two sides.

The heads of most open-end wrenches are set at an angle to facilitate fastener removal. In tight quarters where handle swing is limited, pull the handle as far as it will go. Then, flip the wrench over and replace it on the fastener. Using this method, a wrench with an angled head will operate in a wider arc than a wrench with a straight head.

Box-end wrenches completely surround the fastener, **Figure 3-5.** They are available with 6-point or 12-point openings. Although the 12-point opening allows a shorter swing of the tool, the 6-point opening provides superior holding power. One design uses a double offset to give more handle clearance. Another has a 15° offset.

The *combination wrench* is closed on one end and open on the other. Both ends are the same size. A variety of lengths and a complete range of opening sizes are needed for automotive work, **Figure 3-6.**

Specialty Wrenches

Specialty wrenches are used on nonstandard fasteners or when clearances will not allow the use of the wrenches discussed above.

The *flare-nut wrench,* or *tubing wrench,* is similar to the box-end wrench. However, it has a small section cut out so it can be slipped around tubing and dropped over the tubing nut. The tubing wrench has either 6-point or 12-point openings and should be used on fuel, vacuum, brake, and other fittings.

The *flex-head wrench* is a valuable addition to the toolbox. It can be used through various angles and in cramped quarters. See **Figure 3-7.** Several types of *stud wrenches* are available to remove studs from large parts, such as blocks, heads, or exhaust manifolds. When using a stud wrench, be careful not to damage the threads on the stud.

The *adjustable wrench* is a useful tool. Its size can be readily changed to match that of the fastener. However, the adjustable wrench will tend to loosen and slip. Therefore, it should be used only when other wrenches are not available. Refer to **Figure 3-8.**

The *pipe wrench* is used to grasp round or irregular surfaces. It provides great gripping power. Both inside and outside pipe wrenches are available. Look at **Figure 3-9.**

Allen wrenches and *fluted wrenches* are used to turn setscrews and capscrews with recessed, or internal, heads. See **Figure 3-10.**

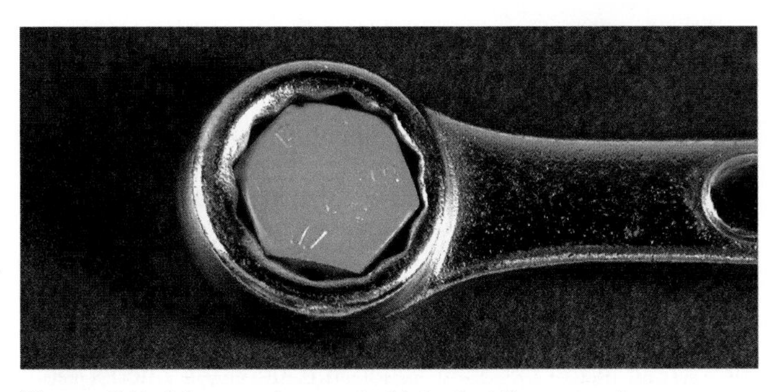

Figure 3-5. A box-end wrench. Note that the wrench completely surrounds the fastener.

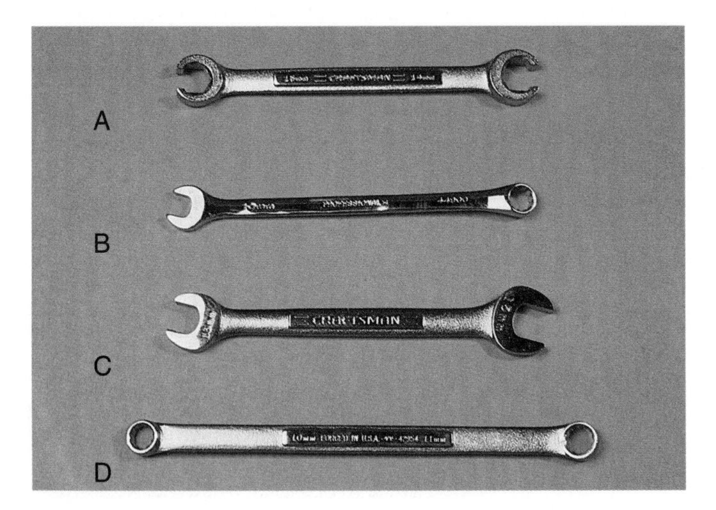

Figure 3-6. Various hand wrenches. Many other types of wrenches are available for special uses. A—Flare-nut. B—Combination. C—Open-end. D—Box-end.

Figure 3-7. Flex wrenches are a type of combination wrench. The socket end is used to loosen the fastener, after which the open end is used to remove it.

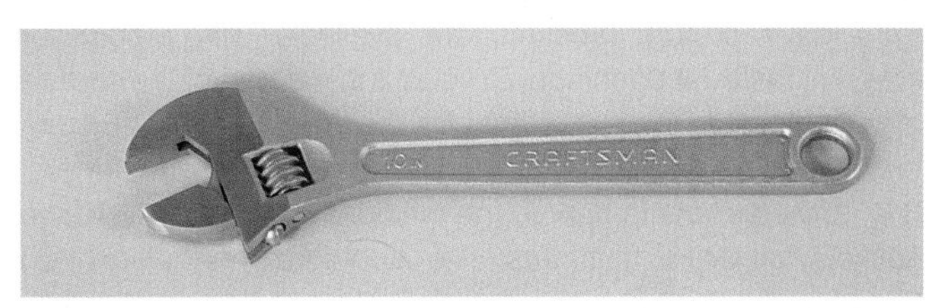

Figure 3-8. Adjustable wrenches often come in handy when the right size wrench is not available. Adjustable wrenches should be used carefully. Improper use, such as not adjusting the jaws to exactly the right size, can damage the fastener head.

Figure 3-9. A pipe wrench is useful for grasping large, irregularly shaped objects, such as exhaust system parts. Pipe wrenches can also be used to loosen fasteners when the flats are rounded off and cannot be turned by a six-point wrench.

Figure 3-10. Since various sizes of Allen wrenches are used on modern vehicles, the technician should have a set of all common sizes.

Figure 3-11. This illustration shows various types of sockets. Deep sockets are used to remove nuts from long threaded shafts, and for removing various sensors. The standard depth sockets are used for other fasteners. Deep and standard depth sockets are available with 12- or 6-point openings. 12-point openings allow the socket to be placed in more positions, while 6-point sockets hold the fastener more securely.

STOP **Warning: When using any wrench, make sure that it is the correct size and that it is securely engaged. Always pull on the wrench handle. If pushing is absolutely necessary, open your hand and push with your palm. Be careful. If a wrench slips, you can be seriously cut.**

Figure 3-12. Swivel socket sets are useful when a fastener is blocked and a standard socket cannot be used. Sometimes the swivel can be used when an extension or a driver cannot be installed on a standard socket once it is in place on the fastener. Most swivel sockets are 6-point types that hold the fastener securely.

Sockets

Sockets are used with drive handles, such as a ratchet handle, to loosen and tighten fasteners. One opening of the socket fits over the fastener and the other opening is attached to a drive handle. Sockets are available with 6-point or 12-point fastener openings. *Drive size* is the size of the square drive opening in the socket that fits the drive handle. The technician should have 1/4″-drive sockets for small fasteners, 3/8″-drive sockets to handle the medium sizes, and 1/2″-drive sockets for larger fasteners. The technician may also need 3/4″-drive sockets for very large fasteners.

Sockets are manufactured in two depths: standard and deep. *Standard sockets* will handle the bulk of the work, while the extra reach of the *deep socket* is occasionally needed for such things as oxygen sensors and oil pressure switches. See **Figure 3-11.**

The *swivel socket* allows the user to turn a fastener that is at an angle other than 90° to the drive handle. As a result, it is handy for many jobs where clearances are tight, **Figure 3-12.**

Drive Handles

Several different ***drive handles*** are used with sockets. See **Figure 3-13.** The *ratchet handle,* or *ratchet,* allows both heavy turning force and speed. The ratchet is also useful in areas where a limited swing is necessary, Figure 3-13A.

The ratchet mechanism allows a socket to be turned through many revolutions without being removed from the fastener. When the handle is moved in one direction, the ratchet teeth engage and turn the socket and fastener. Moving the handle in the opposite direction causes the ratchet teeth to overrun. The handle can be returned to its original position without turning the socket or fastener in the wrong direction. The ratchet teeth are strong enough to allow substantial force to be placed on them when a fastener is loosened or tightened. The direction of fastener rotation can be changed by simply moving a lever on the ratchet handle.

Breaker bars of different lengths provide heavy turning leverage and may be used at many angles, Figure 3-13B. The *speed handle* is used whenever possible because it can be turned more rapidly than other drive handles. See Figure 3-13C.

The *sliding T-handle*, Figure 3-13D, can be used on a fastener in a confined space. T-handles can also be useful when the fastener is extremely tight and clearances prevent other types of drivers from being used. The bar is placed in the centered position and the technician places a hand on each side to break the fastener loose.

Torque wrenches are used to accurately tighten fasteners, Figure 3-13E. *Spinner handles* are used in the same manner as screwdrivers and will accept all sockets or attachments. See Figure 3-13F.

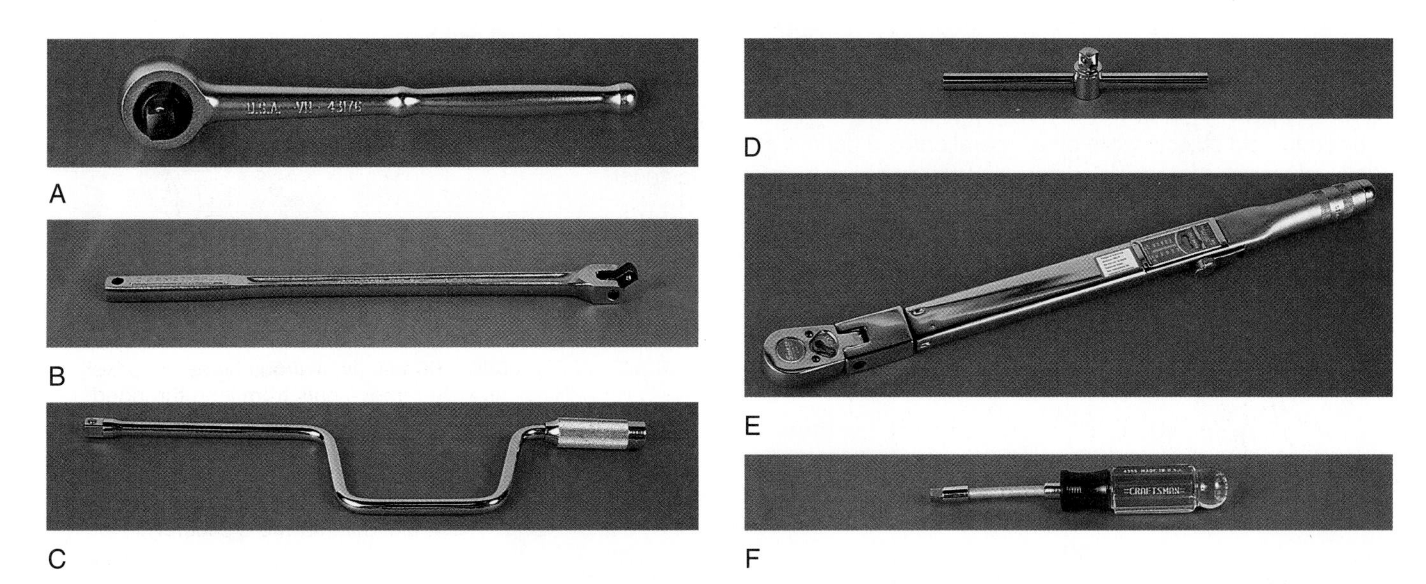

Figure 3-13. This illustration shows various drivers used to turn sockets. A—Standard ratchet handle. B—Breaker bar. C—Speed handle. D—T-handle. E—Torque wrench. F—Spinner handle. All these drivers are useful in specific situations.

Socket Attachments

A variety of socket attachments can be used to make the job easier. Long, medium, and short ***extensions*** allow the user to extend the reach of the tool. They may be used singly or snapped together to provide greater reach, **Figure 3-14.**

The ***universal joint*** allows the technician to drive fasteners at various angles in relation to the socket handle. See **Figure 3-15.** An ***adapter*** allows sockets of one particular drive size to be turned with a drive handle of another size,

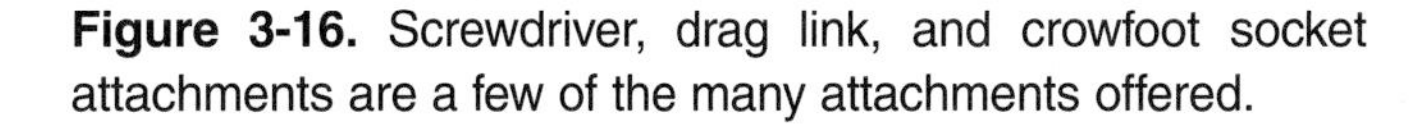

Figure 3-16. Screwdriver, drag link, and crowfoot socket attachments are a few of the many attachments offered.

Pliers

Standard ***pliers*** are used for common gripping, crimping, and bending jobs. *Slip joint pliers* are adjustable pliers with large jaws. They are used for grasping large objects. The size of the jaw opening can be adjusted to accommodate different

Figure 3-14. Socket extensions in various lengths make fastener removal easier by raising the driver above any obstructions. Extension lengths range from less than one inch (25.4 mm) to over 3 feet (1 m). Extensions are often sold in sets.

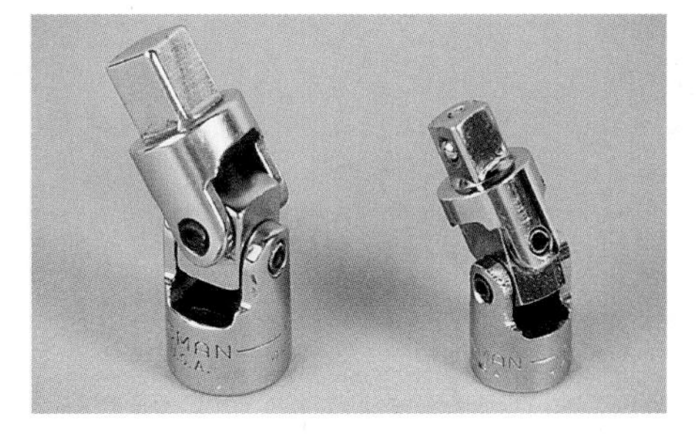

Figure 3-15. Universal joints allow a socket and extension to be used on fasteners that would otherwise be unreachable. The universal joint may be installed between the socket and extension, or between two extensions.

Figure 3-16. A socket adapter allows a socket to be used with a driver of a different size.

size parts. To tightly hold a small part where clearances are tight, *needle nose pliers* are handy. *Diagonal pliers* are useful for cutting cotter pins or wires. *Vise grip pliers* can be locked in place to hold objects while other operations are performed. *Snap ring pliers* are special pliers used to remove a type of fastener called a snap ring. All these pliers are shown in **Figure 3-17.** Other pliers for special purposes, such as removing or installing brake springs, will be covered in later chapters.

Avoid using pliers to cut hardened objects. Never use pliers to turn nuts, bolts, or tubing fittings. Pliers cannot hold the sides of these fasteners properly and will damage them. In most cases, the user cannot hold the pliers tightly enough against the fastener while turning, and the pliers will slip. The jaws of the pliers will round off the sides of the fastener.

Hammers

A ***hammer*** is a basic striking tool. Ball peen, plastic, brass, lead, rawhide, dead blow, and rubber hammers should be included in every technician's selection. See **Figure 3-18.** Various sizes of each type are desirable.

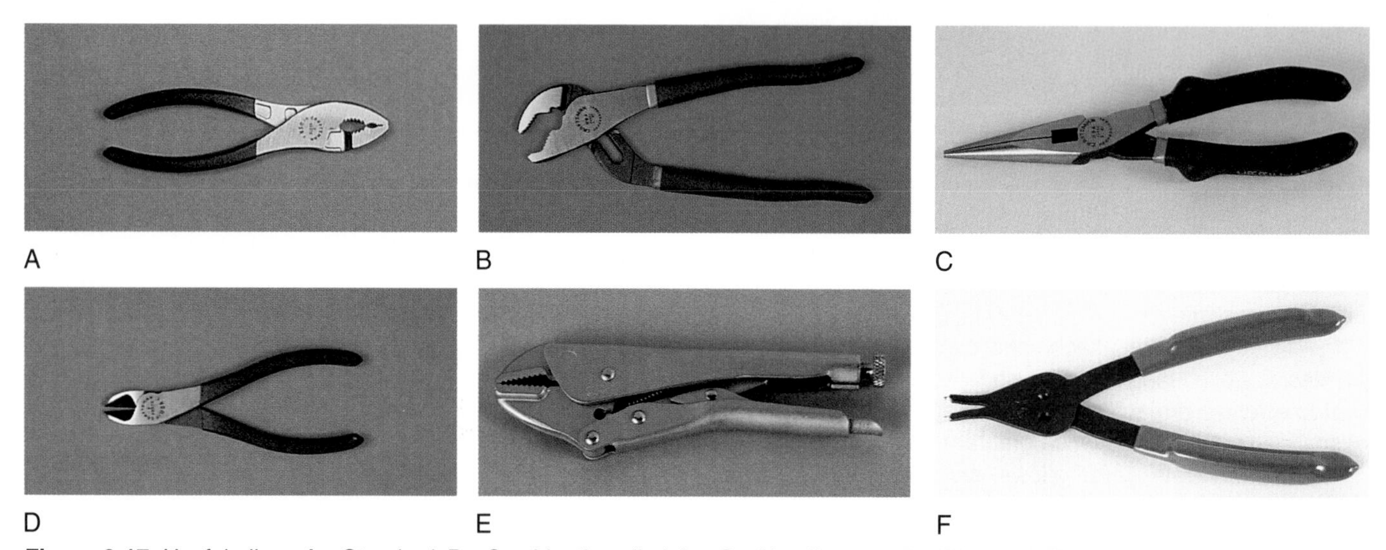

Figure 3-17. Useful pliers. A—Standard. B—Combination slip joint. C—Needle-nose. D—Diagonal. E— Vise-grips. F—Snap ring.

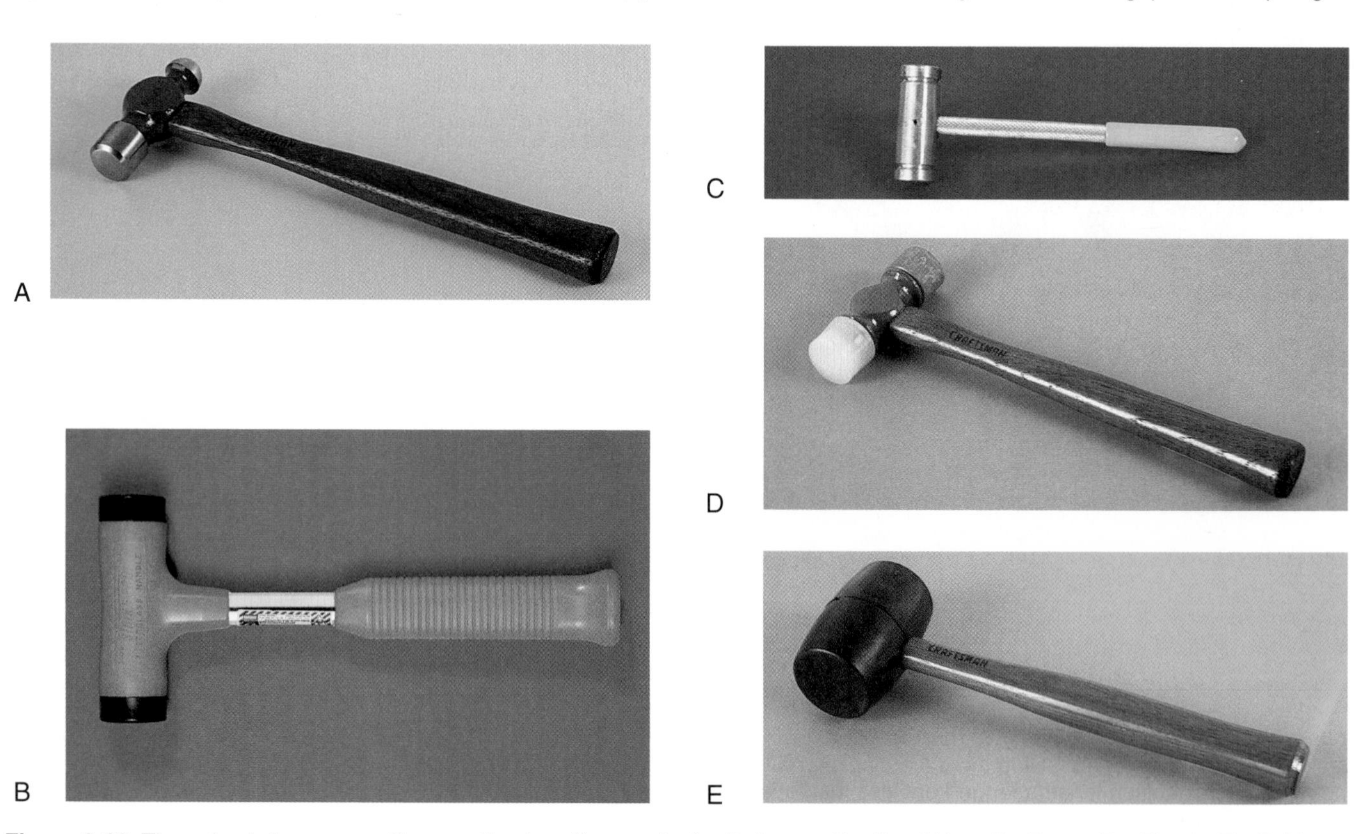

Figure 3-18. These basic hammers will cover all automotive needs. A—Ball-peen. B—Dead-blow. C—Brass. D—Plastic. E—Rubber.

The *ball peen hammer* is used for general striking, installing bushings, and cutting gaskets. *Plastic, lead,* and *brass hammers* are used to prevent marring of part surfaces. A *dead blow hammer* has a hollow head that is filled with shot (small pieces of metal). The advantage of a dead blow hammer is that it will not bounce back. Instead, the shot absorbs the force of the blow. When using a hammer, grasp the handle firmly. Place your hand near the handle end. Strike so the face of the hammer engages the work squarely, **Figure 3-19.**

STOP Warning: Use a hammer with care. Do not swing it in a direction that would allow it to strike someone if it slips from your grasp. Make sure the handle is tight in the head and that it is clean and dry.

Chisels

Chisels are used to cut off rivet heads, bolts, and rusted nuts. Flat, cape, diamond, half-round, and "rivet buster" chisels should be available, **Figure 3-20.**

When using a chisel, hold it securely but not tightly. Grasp it as far from the top as practical. This will help protect your fingers if the hammer slips from the chisel head. For heavy hammering, use a chisel holder.

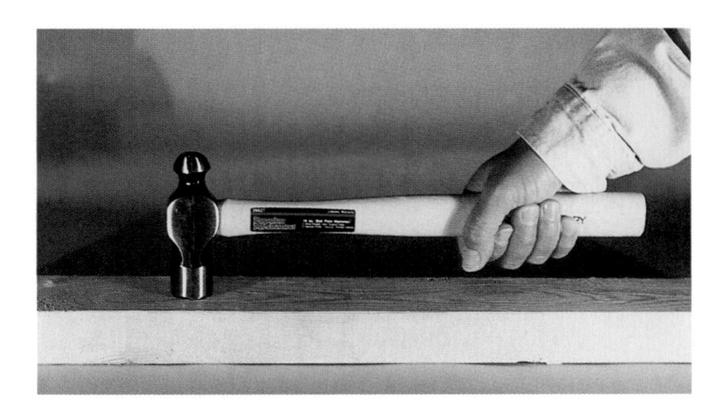

Figure 3-19. The hammer should be held close to the end of the handle. Note the head contacts the workpiece squarely.

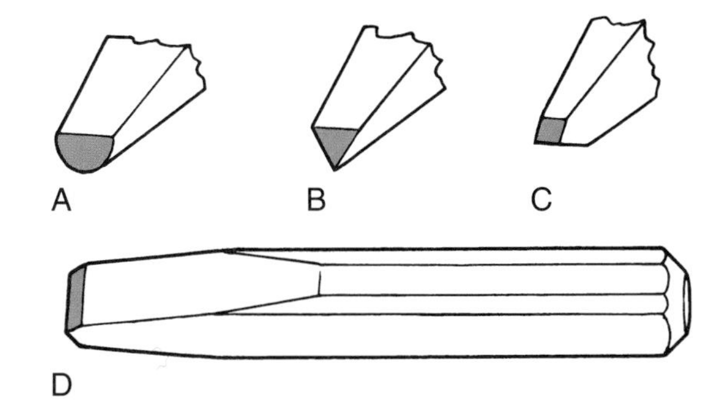

Figure 3-20. The following selection of chisels will handle most automotive needs. A—Half round. B—Diamond. C—Cape. D—Flat. Other chisels are available for special purposes.

Keep the cutting edge of the chisel sharp. Also, make sure the top of the chisel is chamfered (tapered along the edges). Refer to **Figure 3-21.** This reduces the possibility of small chisel segments breaking off and flying outward when struck with the hammer. Always wear goggles when using a chisel.

Punches

Punches are designed to be used with hammers when driving small parts into or out of holes, marking spots before drilling, **Figure 3-22,** and aligning holes in adjacent parts.

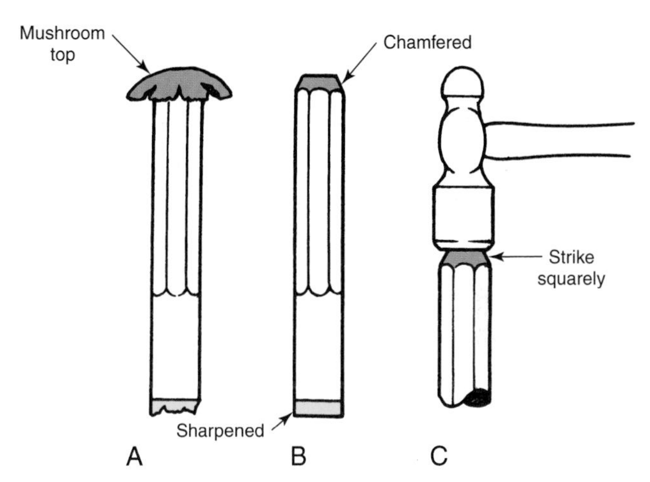

Figure 3-21. A—This chisel is dangerous to use. B—The same chisel after chamfering and sharpening. C—Be sure to strike the chisel squarely.

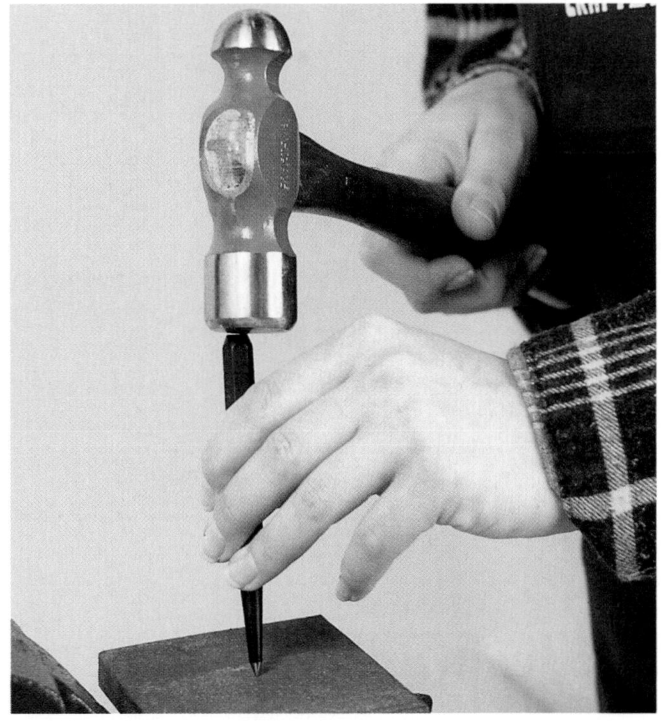

Figure 3-22. When using a punch, hold it in the middle and be sure to strike it squarely.

Starting, drift, and pin punches are essential. A few sections of round brass stock in varying diameters are useful for driving parts that may be damaged with steel punches.

A *starting punch* is used to begin driving rivets, bolts, and pins from a hole. Due to its taper, the starting punch may fill the hole before the part has been completely removed. When this occurs, the job can be completed with a *drift punch*. See **Figure 3-23.** A *pin punch* is similar to a drift punch, but its driving shank is smaller in diameter. Pin punches are useful for removing pins and other fasteners installed in small holes.

A *center punch* is used to mark work before drilling, **Figure 3-24.** The small V-shaped indentation created by the punch will help align the drill bit. The center punch is also useful for marking parts so they can be assembled in their original positions. The *aligning punch* is very helpful when shifting parts so the holes line up, **Figure 3-25.**

Sharpening Chisels and Punches

Use care when sharpening chisels and punches. Grind the tools slowly, keeping correct angles, and quench them (dip them in cold water) to prevent drawing the temper (overheating, turning the metal blue and rendering it soft). Wear goggles whenever using grinding equipment.

Files

Files are used for removing burrs from metal, final fitting after hacksawing, and smoothing out surfaces. The most frequently used files are the flat mill, round, square, triangular, and point files. Many other special shapes are available, **Figure 3-26.**

A file's relative size and the number of cutting edges per inch determine the file's *cut*. Use a cut suitable for the work at hand. Coarse cuts are best for soft metals, such as aluminum, brass, and lead. Finer cuts work well on steel. Your choice of cut will also depend on the finish desired.

A file can be either single cut (single row of diagonal cutting edges at same angle) or double cut (two rows of diagonal cutting edges that cross each other at an angle). Files may also be rasp cut or curve cut, **Figure 3-27.**

Every file should have a handle. Be sure the file handle is firmly affixed to the file's tang before using the tool. This will provide a firm grip and eliminate the possibility of the tang piercing your hand.

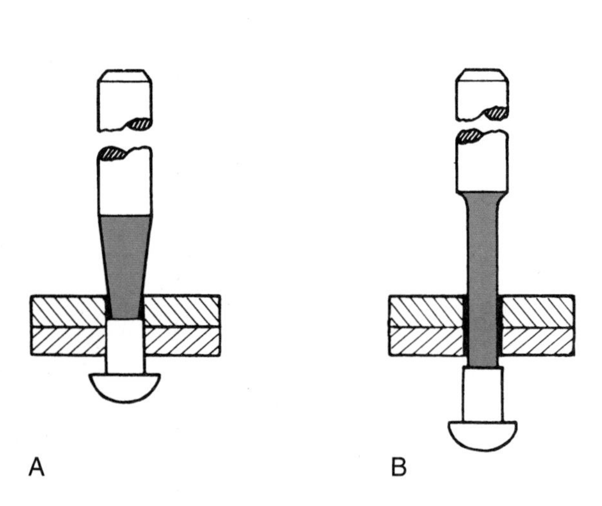

Figure 3-23. A—A starting punch is used to begin driving a rivet from a hole. B— A drift punch is used after the starting punch to drive the rivet out of the hole.

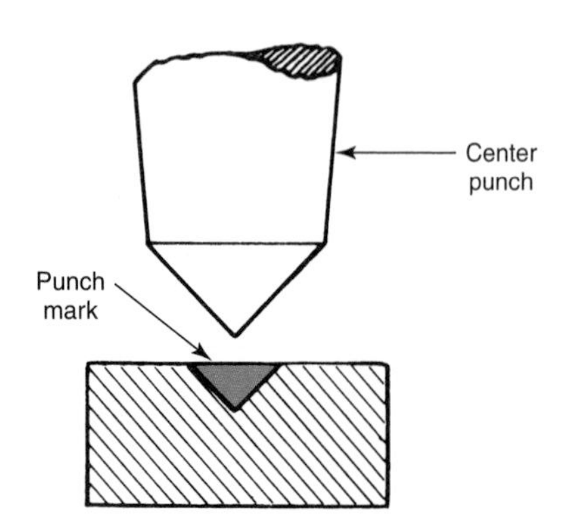

Figure 3-24. A center punch is used to mark work for drilling.

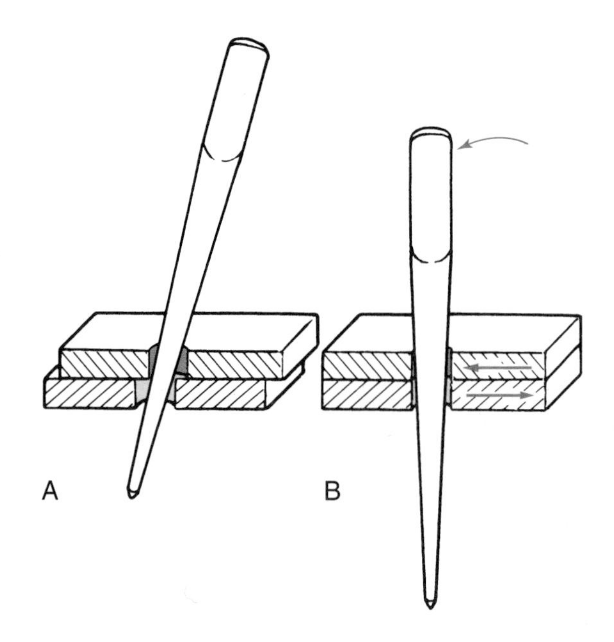

Figure 3-25. An aligning punch can be used to align parts. A—Run the punch through the holes as far as possible. B—Pull the punch upright and force it into the holes. This will cause the parts to shift into alignment.

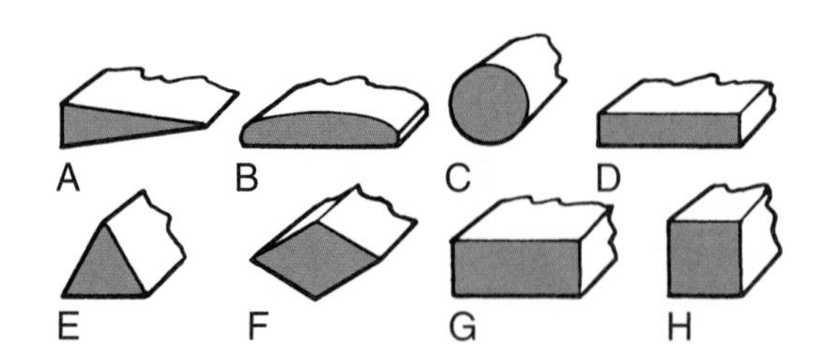

Figure 3-26. The file shapes shown will handle most automotive filing jobs. A—Knife. B—Half round. C—Round. D—Flat. E—Triangle. F—Slitting. G—Pillar. H—Square.

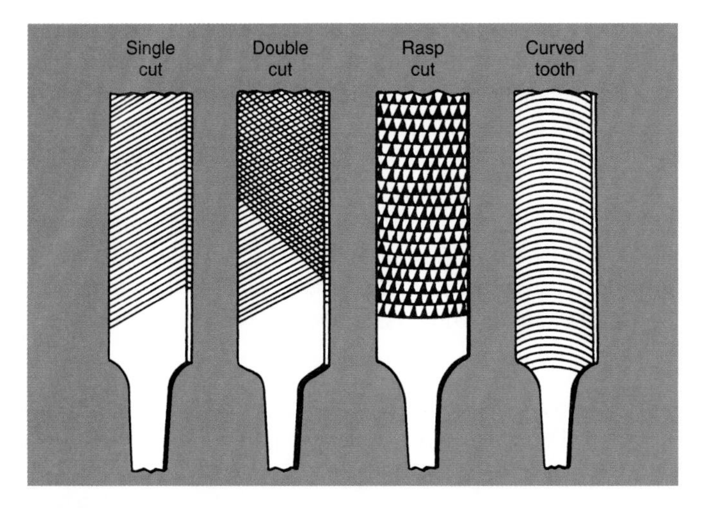

Figure 3-27. Different file cuts are intended for different jobs. Rougher cuts remove more metal, while finer cuts produce a smoother finish. Each type is designed for certain jobs. (Deere & Co.)

The *rotary file* is a circular file that is chucked in a drill or a power grinder. This type of file is very handy for blind holes or recesses where a regular file will not work. Several useful shapes are shown in **Figure 3-28.**

Using Files

Grasp the file handle with your right hand (for right-handed persons) and hold the tip of the file with the fingers of your left hand. On the forward stroke, bear down with enough pressure to produce a good cutting action. On the return stroke, raise the file to avoid damaging the cutting edges.

Control the file to prevent rocking (unless you are filing round stock). It takes practice to become skilled at filing. In the hands of a professional, a file can be used to do amazingly accurate work.

Keep the file clean and free of oil. A special wire brush, called a ***card file,*** should be used to clean the file's teeth. Chalk may be rubbed into the file to help prevent clogging.

Hacksaws

A ***hacksaw*** is used to cut tubing, bolts, and other parts. The technician should have blades with 18, 24, and 32 teeth per inch. The 18-tooth blade is used for cutting thick metal, the 24-tooth blade is used for cutting steel of medium thickness, and the 32-tooth blade is used for cutting thin sheet metal and tubing. For very thick work, a 14-tooth blade can be used. The blades should be made of high-quality steel. High-quality blades will cut faster and last longer than low-quality blades. **Figure 3-29** illustrates a typical hacksaw.

Taps and Dies

Taps are used to cut internal threads. ***Dies,*** on the other hand, are used to cut external threads, **Figure 3-30.** The technician should have a set of taps and dies covering both Unified National Fine threads and Unified National Coarse threads (sometimes called SAE or English sizes), as well as a full set

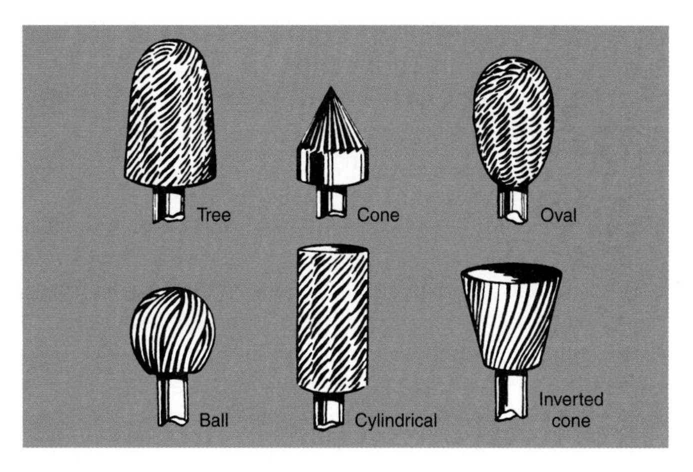

Figure 3-28. Rotary files.

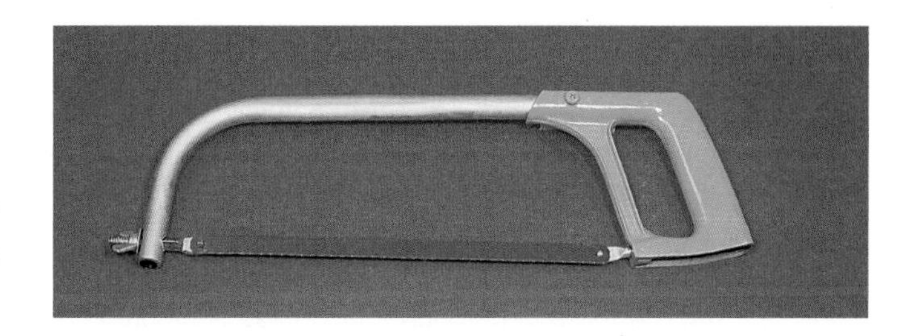

Figure 3-29. Typical hacksaw. Hacksaw blade tooth size is important to the speed and quality of the finished cut. (Snap-on Tools)

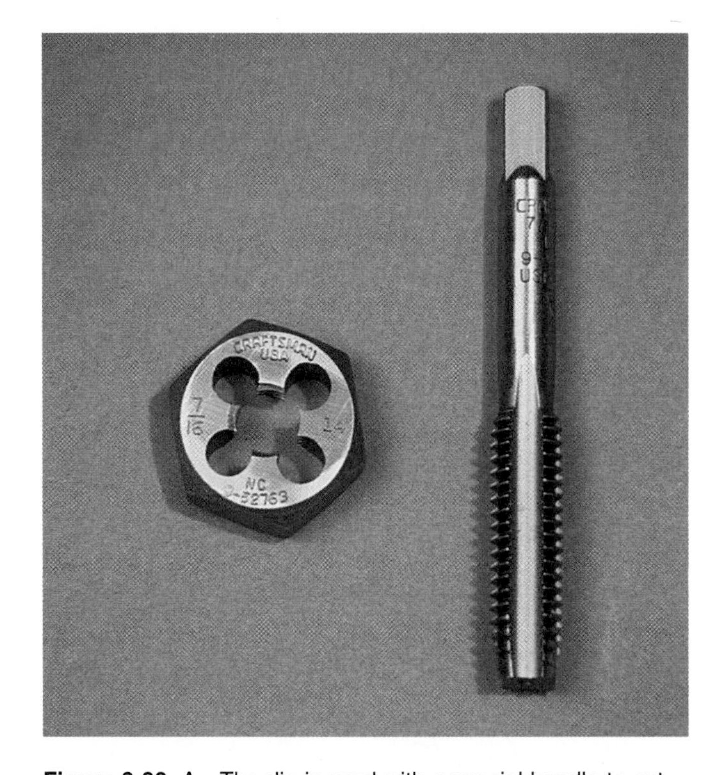

Figure 3-30. A—The die is used with a special handle to cut external threads. B— A tap is used to cut threads in drilled holes. Both taps and dies are sometimes used to clean up existing threads.

covering both fine and coarse metric threads. Thread sizes are covered in the chapter on fasteners.

There are many specialized taps. Taper, plug, bottoming, and pipe taps are commonly needed for automotive service. The *taper tap* has a long chamfer (about 10 threads) that allows it to start easily. It cannot, however, be used when threads must run almost to the bottom of a blind hole. The *plug tap* has a shorter chamfer (about 5 threads). With care, it can be started successfully. It is useful for open holes and blind holes. The *bottoming tap* has a short chamfer (about 1 thread) and is used in blind holes to finish the thread to the bottom of the hole. The *plug tap* should be used first. When it strikes bottom, the bottoming tap should be used. The *pipe tap* is tapered over its entire length (about 3/4″ per ft.). It is used to tap holes for pipe fittings. Taps should always be turned with a tap handle to reduce the chance of tap breakage. See **Figure 3-31.**

Using Taps

To produce a tapped hole, determine the exact number of threads per inch and the diameter of the screw that will enter the hole. Referring to a tap drill size chart, select the proper tap size drill.

If, for example, you desire a threaded hole for a 7/16″ capscrew with 20 threads per inch, you will find that a 7/16″ capscrew with 20 threads per inch is a Unified National Fine size. Going directly across the chart from the 7/16″ UNF, notice the column marked "Tap Drill Size." In this case, the tap drill size for a 7/16″ x 20 capscrew is a 25/64″ drill.

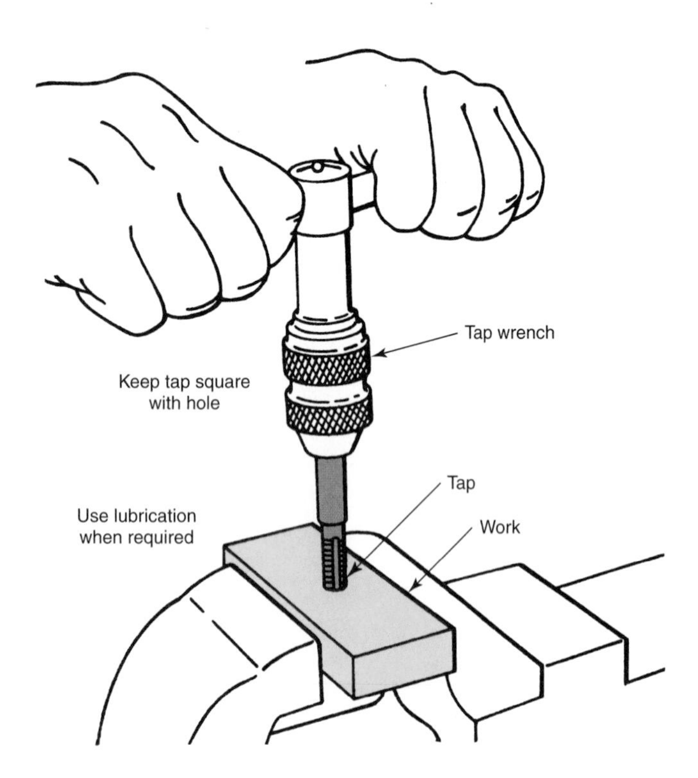

Figure 3-31. The T-handle tap wrench is the best way to turn a tap. It allows the tap to enter easily. The T-handle permits turning in tight quarters. (Deere & Co.)

Drill the hole with the proper size tap drill. Holes larger than 1/4″ should be drilled in at least two operations. Start with a small pilot drill of about 1/8″ in diameter. Then work up to the appropriate tap drill size.

Using a suitable tap wrench, carefully start the tap in the hole. Cutting oil will help when tapping steel. After running the tap in one or two turns, back it up about one-half turn to break the chip. Repeat this process until the hole is fully tapped. Remember that taps are very brittle. Do not strain them and be sure to keep the hole from clogging with chips.

Using Dies

The die is used much like the tap. After selecting a die of the correct size, place it in a die stock (handle). Apply cutting oil to the part to be threaded and start the die. Turn the die in the same manner as the tap, backing it up occasionally to break the chip.

Dies are often adjustable so the thread fit can be changed. A small screw is often used to change the diameter of the die threads. It is often necessary to adjust the die to a larger diameter to make the initial cut. After the threads have been formed, the thread diameter can be reduced to cut the threads more deeply. Variations in metal hardness may also require a different die diameter. When using a die, frequently check the threads with a thread pitch gauge or the proper size nut to ensure that the threads are not being cut too deeply. A cut that is too shallow can be recut more deeply, but a cut that is too deep cannot be fixed. Keep taps and dies clean, oiled, and in a box.

Special-Purpose Taps and Dies

There are many special-purpose taps and dies, most of which are used to restore existing threads that have been damaged. The *axle rethreader* is opened up and placed around the good threads. It is then backed off to restore damaged threads. In addition to repairing damaged threads, *spark plug hole thread restorers* are handy for removing rust and carbon. See **Figure 3-32.**

Reamers

Reamers are used to enlarge, shape, and smooth holes. They produce a finish much smoother and more accurate than that produced by drilling. Some reamers can be adjusted, and others are of a fixed size. Both straight and tapered reamers are needed. They can use either straight or spiral flutes, **Figure 3-33.**

Use cutting oil when reaming. Turn the reamers in a *clockwise* direction only on entering and leaving the hole. Take small cuts, removing approximately 0.001″–0.002″ (0.025 mm –0.050 mm) during each pass. Reamers are very hard, and the cutting edges chip readily. Wipe reamers down with oil and keep them in a protective container for storage.

Screw Extractors

A ***screw extractor*** is often needed to remove the threaded portion of a bolt that has broken off inside a hole. One type of screw extractor is shown in **Figure 3-34.** To use this type of extractor, drill a hole into the broken bolt. Then hammer the extractor into the drilled hole. Next, turn the extractor with a wrench to remove the broken bolt.

Another type of thread extractor is threaded into a hole drilled in the bolt. The extractor threads are the reverse of the bolt threads. When a wrench is used to turn the extractor in the normal loosening direction, it is forced deeper into the drilled hole. At the same time, the turning effort unthreads the broken bolt.

Warning: Screw extractors are made of extremely hard metal. Sharp shocks can break them. Always place steady pressure on an extractor when turning it.

Probing Tools

Probing tools, such as mechanical fingers, extension magnets, and mirror devices, help the technician to retrieve parts in hard-to-reach places and to see into blind areas, **Figure 3-35.**

A

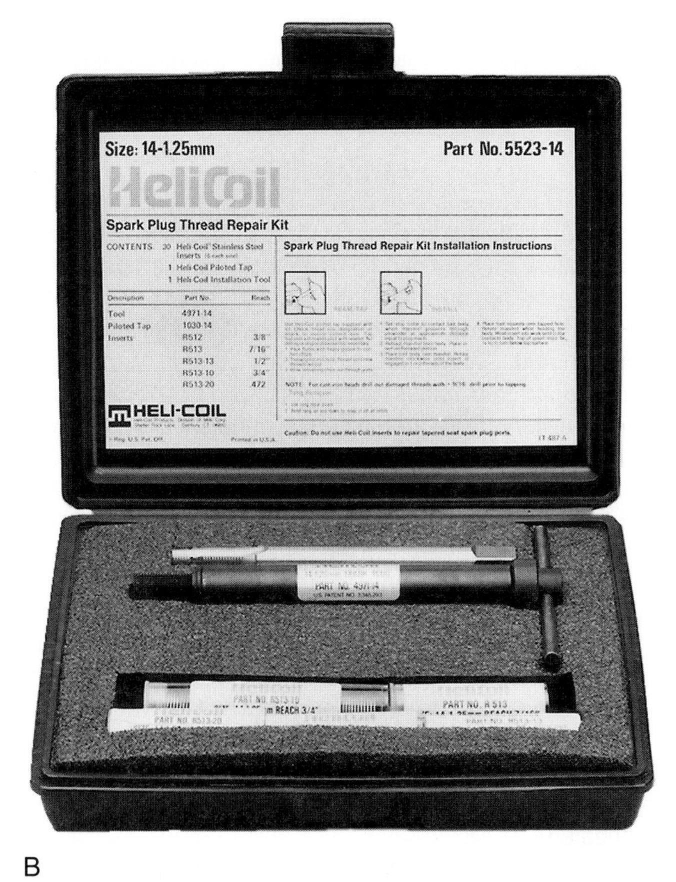

B

Figure 3-32. Special-purpose taps and dies. A—External rethreading set. B—Internal thread restorer.

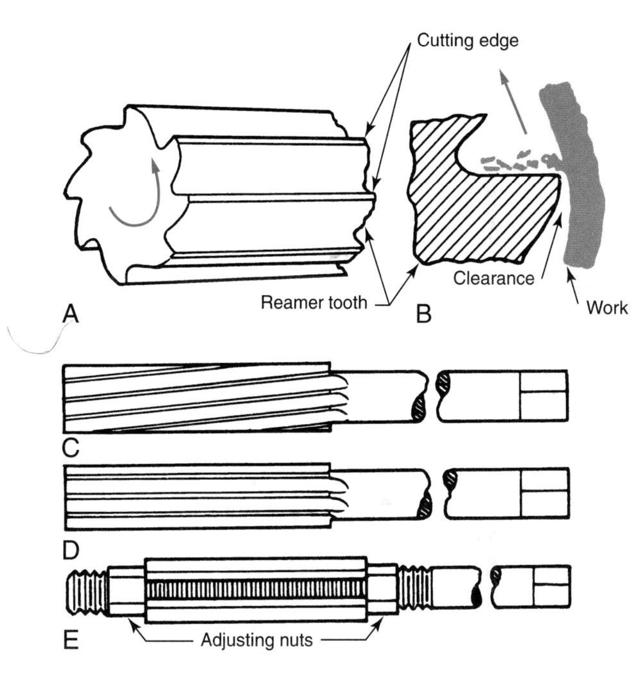

Figure 3-33. A—Enlarged section showing reamer tooth construction. B—Reamer tooth removing stock. C—Nonadjustable, spiral flute reamer. D—Nonadjustable, straight flute reamer. E—Adjustable straight reamer. It is opened and closed by removing adjusting nuts.

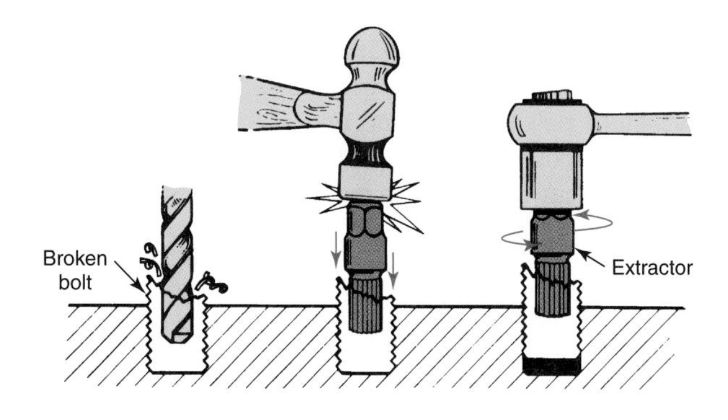

Figure 3-34. Screw extractor being used to remove a broken bolt. (Snap-on Tools)

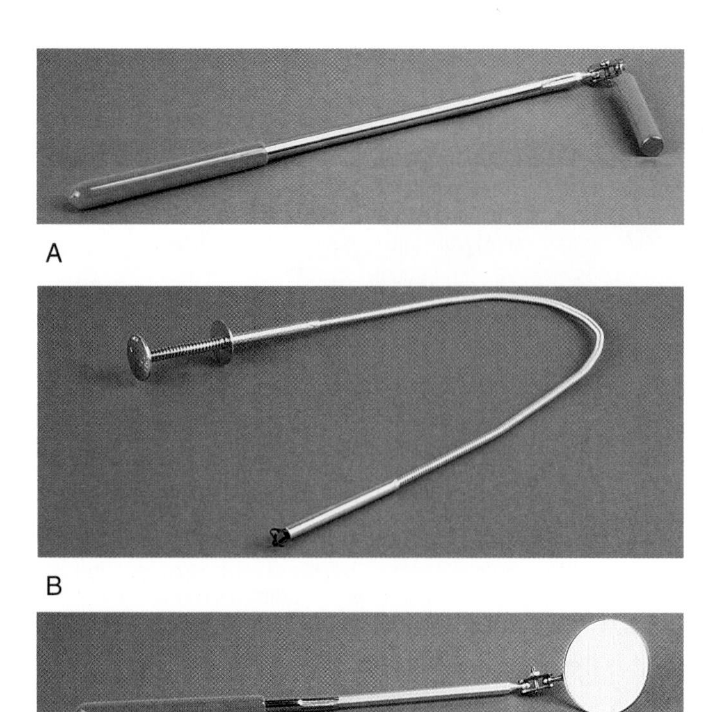

Figure 3-35. The tools shown here are used to locate and retrieve small parts or to observe components in inaccessible places. A—Telescoping magnet. B—Mechanical finger pickup. C—Telescoping mirror.

Pullers

Pullers are used to remove interference-fit (pressed-on) parts. Pullers can be operated manually or hydraulically. A good assortment of pullers is important. An attempt to "get by" with a few pullers will result in wasted time and damaged parts. Many jobs are almost impossible without proper pullers.

Three Types of Pulling Jobs

All pulling jobs can be classified as one of three types:

- Pulling an object (such as a gear, pulley, bearing, or retainer) from a shaft.
- Pulling a shaft (such as an axle, transmission, or pinion shaft) from an object.
- Pulling an object (such as bearing outer rings, cylinder sleeves, or camshaft bearings) from a housing bore.

Figures 3-36 through **3-38** illustrate these three basic pulling jobs. A typical hydraulic puller is shown in **Figure 3-39.** Always store pullers, adapters, and related parts together. Some shops mount individual puller sets on "tote" boards so that all parts may be carried to the job.

Vises

Vises are used to hold parts securely so other work, such as drilling, cutting, or filing, can be done. A vise suitable for automotive work is shown in **Figure 3-40.** Keep the vise clean. Use copper, lead, or nylon *jaw covers* for work that is easily marred. Oil the working parts of the vise and avoid hammering on the handle or other surfaces.

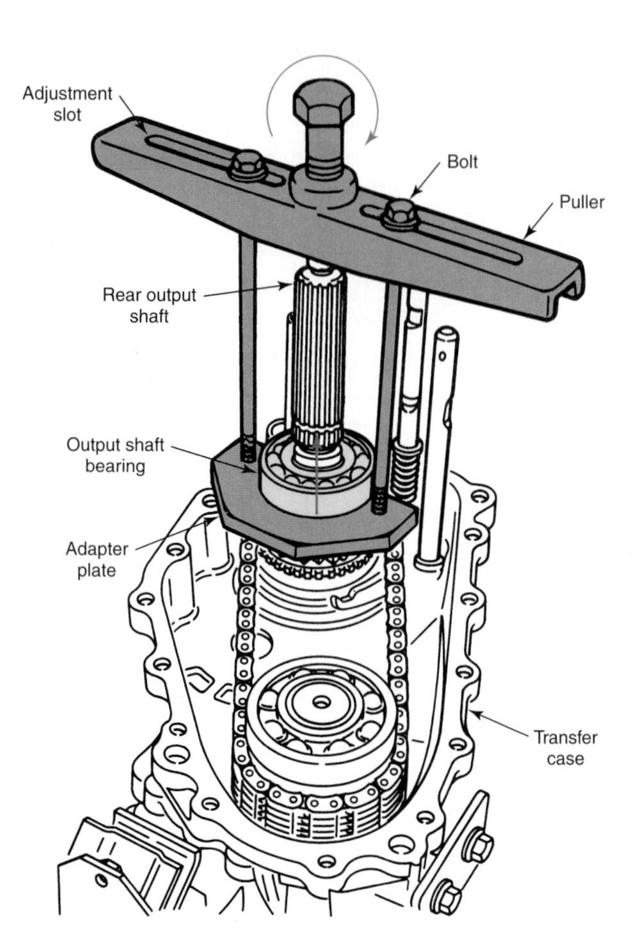

Figure 3-36. Removing a transfer case output shaft bearing with a puller. Note the adjustment slots, which enable the puller to be used on a variety of bearings, gears, etc. (Chevrolet)

Power Tools

Power tools are those tools that are operated by a power source rather than by hand. Power tools can be operated by electricity or compressed air. Electric power tools often operate from standard 120-volt current, sometimes called house current. A few require 220-volt current. Many newer electric tools are cordless and are powered by a rechargeable battery pack. Air tools are operated by the shop air supply.

Using power tools reduces the amount of physical labor involved in automotive repair operations. Power tools also reduce the time needed to perform a job. Using power tools makes the job faster and easier.

Drills and Drill Bits

The technician will often need to drill a hole in a vehicle part. Quality ***drill bits,*** which are made of high-speed steel, will do a good job of drilling on most parts of the vehicle. They can be readily ground without drawing their temper. Carbon-steel drill bits are cheaper than high-speed bits, but they require frequent sharpening and lose their temper if slightly overheated.

The technician should have a set of fractional-size drill bits (1/16–1/2″) and a set of metric drill bits (1–13 mm). See **Figure 3-41.**

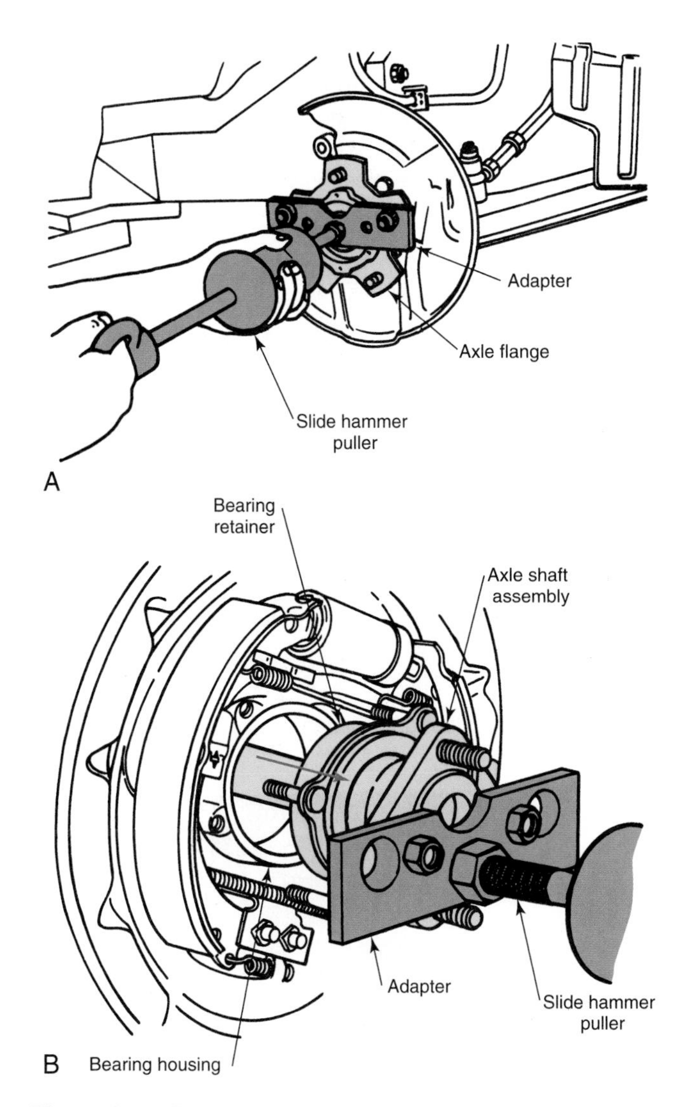

Figure 3-37. Removing an axle shaft with a slide hammer puller. A—The technician has fastened the puller to the axle shaft flange with an adapter. B—The axle is being removed from its housing with the puller. (Chevrolet)

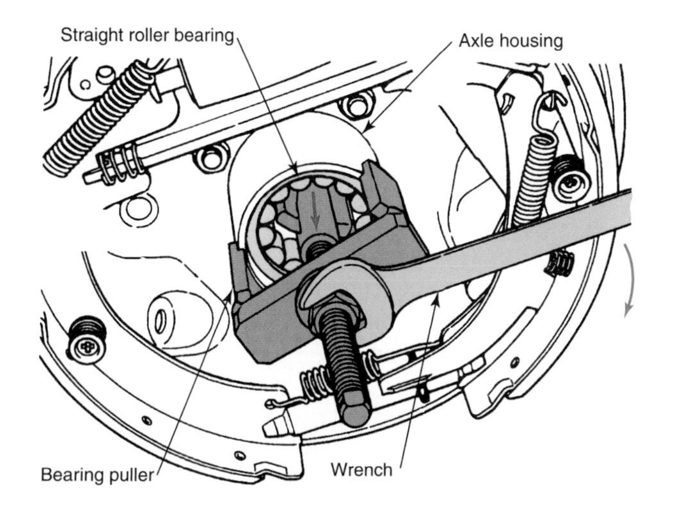

Figure 3-38. Pulling a bearing from the hub bore. The internal jaws of the puller are inserted into the bearing and tightened to hold the bearing. Then, the nut is tightened to draw the bearing from the hub.

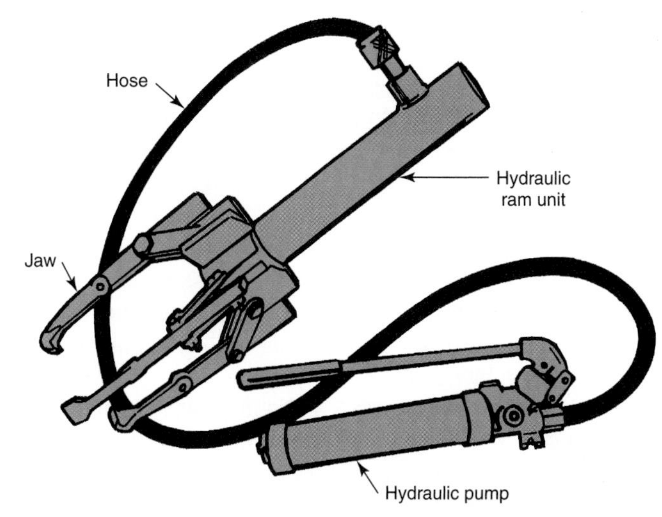

Figure 3-39. The puller shown here is operated by hydraulic pressure. The hand pump shown at the bottom of the illustration produces the needed hydraulic pressure.

Figure 3-40. Vises are used to hold parts firmly while repair operations are performed.

Figure 3-41. A typical twist drill set.

All technicians should have ***power hand drills.*** These drills can be powered by compressed air or electricity, **Figure 3-42.** The drill contains a chuck to securely hold the drill bit in position. See **Figure 3-43.** The 3/8″ hand drill is handy for most shop drilling jobs. A 1/2″ drill will handle heavy drilling, honing, and other shop tasks requiring high drill torque.

Newer drills have a *keyless chuck.* This type of chuck can be tightened and loosened by hand. Keyless chucks can be identified by the lack of a key-type tightening mechanism. Most keyless chucks are smooth, with slight indentations that can be gripped by hand.

The ***drill press,*** **Figure 3-44,** is another type of electric drill. The drill motor and chuck are mounted on a stand and can be moved up and down. The workpiece is mounted on a table directly under the drill motor. The drill bit, chuck, and motor are lowered by a lever until the bit contacts the workpiece. The drill press eliminates the need to hold the drill and ensures the drill bit will produce a hole that is properly aligned and the exact size needed. The disadvantage of the drill press is it can only accept parts up to a certain size, and it cannot be taken to the vehicle for drilling parts that cannot be removed.

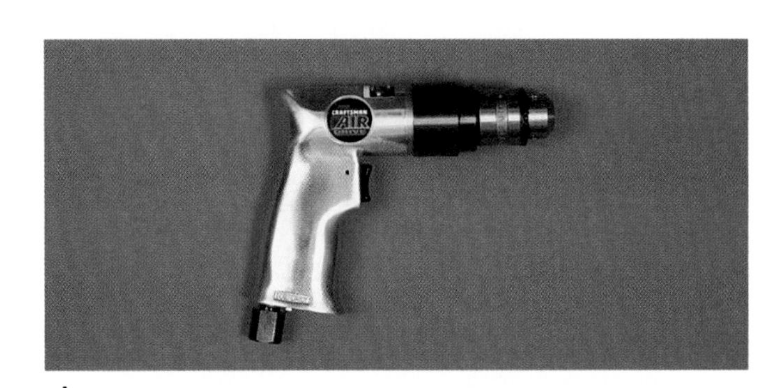

A

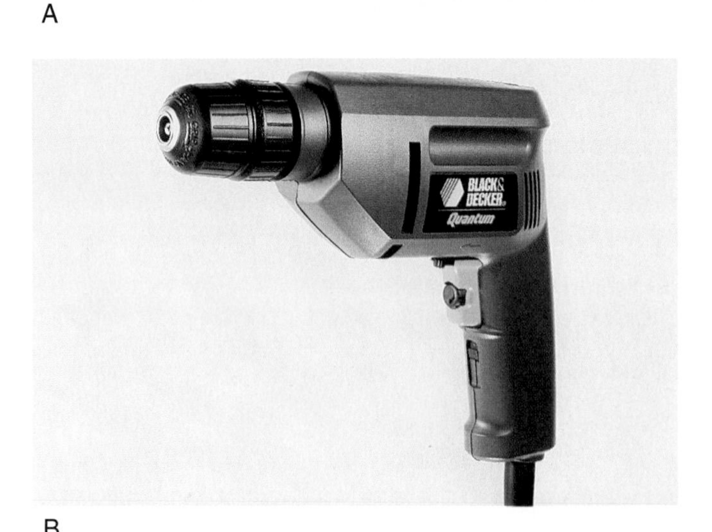

B

Figure 3-42. A—Air-powered hand drill. This type of drill is useful for lengthy drilling jobs that would overheat an electric drill. B—Electric drill. This electric drill is powered from a standard 120-volt outlet.

Using Drills

Drills are relatively easy to use, but care is required to prevent injury or damage to the bit or the workpiece. Securely fasten the workpiece to the workbench or to the drill press table. Do not try to hold the piece in your hand. When using a

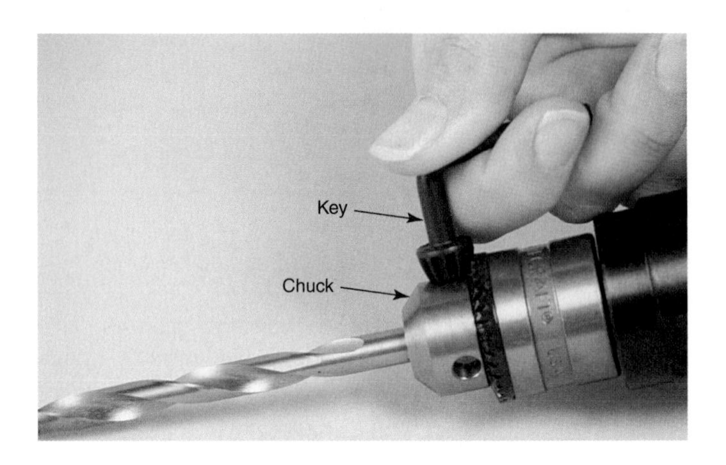

Figure 3-43. A key is used to tighten the bit in this chuck.

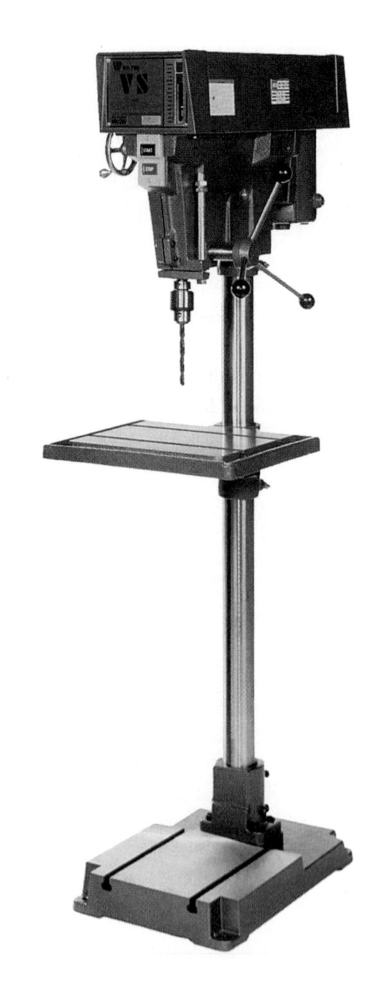

Figure 3-44. Drill presses are useful for making precisely positioned holes in metal parts. Almost all drill presses are operated by electric motors. The part is clamped to the table and positioned under the drill bit. The drill and motor assembly is then lowered until the bit contacts the part. (Wilton)

hand drill, center punch the spot to be drilled. Chuck the drill bit tightly in the hand drill or drill press. When drilling cast iron, pot metal, aluminum, and thin body metal, cutting oil is not necessary. When drilling steel, however, a small quantity of cutting oil is helpful.

When using a hand drill, keep the drill at the proper angle and apply only enough pressure to produce good cutting action. When drilling thin stock, be careful to hold the workpiece down, as it has a tendency to climb up the flutes of the drill bit. Just before the bit breaks through, reduce pressure on the drill to prevent grabbing and possible bit breakage.

Safety Rules for the Use of Drills

Any power tool can cause damage or injury. This is especially true of drills, which spin rapidly and can throw metal that will cut the skin or cause eye injury. Electric drills can also cause shocks. Before using any drill, carefully read over the following safety rules.

- Unplug the cord before inserting or removing a drill bit from the chuck. (If the drill starts while you are holding the chuck wrench in the chuck, it might cut your hand badly.)
- Keep loose clothing, ties, jewelry, and hair away from the drill.
- When using an electric drill, make sure the drill is properly grounded. All electric drills have the ability to shock you.
- Never use power tools of any kind while standing on a wet surface.
- Properly secure the workpiece to be drilled. (If the drill grabs and the workpiece is loose, it can spin with a vicious cutting force.)
- Always wear safety goggles when drilling.
- Do not use any power tool in the presence of explosive vapors.

Grinders

The automotive technician will often use a ***grinder*** to sharpen tools and rework parts. Grinding, like all shop operations, requires skill and patience. Several types of grinders are found in automotive service facilities. You should be familiar with them all.

The *bench grinder,* **Figure 3-45,** is mounted on the workbench and is commonly used to sharpen tools and remove stock from various parts. It is often fitted with a grinding wheel on one side and a wire wheel on the other side. A grinder mounted on a stand instead of on a bench is called a *pedestal grinder.*

Hand-held grinders and sanders can be fitted with grinding stones, wire wheels, or abrasive discs. They are used in bodywork, carbon and rust removal, weld smoothing and cleaning, etc. **Figure 3-46** shows several of these power tools.

Grinding stones are often used to smooth down welds and other parts. Sometimes a hand-held grinder is used with the proper stones to refinish valve seats. Wire wheels and abrasive discs are used to remove rust and paint from metal surfaces. Abrasive discs are used to sand metal or painted surfaces.

Figure 3-45. A bench grinder is a useful shop tool. Most bench grinders have coarse and fine abrasive wheels. On some grinders, one of the abrasive wheels is replaced with a wire wheel. To prevent injury or property damage, always follow grinder safety rules.

Figure 3-46. Hand-held wheel grinders and disk grinders. These tools can perform cutting, grinding, sanding, buffing, and other tasks. (Ingersoll-Rand)

Specialized grinders are also found in automotive shops. The use of these tools will be discussed in the chapters relating to the work they are designed to perform.

Safety Rules for Grinders, Sanders, and Wire Wheels

If improperly used, grinders can be extremely dangerous. They can cause many serious injuries to the eyes, hands, and face. Realizing this, the technician must *always* observe the following safety rules when using grinding equipment:

- Protect your eyes by wearing an approved face shield or goggles.
- Keep abrasive stones tight, clean, and true.

- Allow the grinder to reach full speed before using it.
- Stand to one side of the grinder until full wheel speed is reached.
- When used, keep the tool rest as close to the wheel as possible.
- Stand to one side of the stone as much as possible.
- Keep persons without goggles away from the tool you are using.
- Hold small objects with vise-grip pliers. This will help you avoid grinding your fingers or having the object seized by the wheel and thrown violently.
- Wear leather gloves for heavy grinding.
- Never strike a grinding wheel while it is rotating. It may shatter and explode.
- Avoid grinding in the presence of explosive vapors, such as gasoline, paint thinner, or the gases from batteries.
- When installing a new stone, make certain it is designed for the speed of the grinder.
- Keep the grinding wheel guard in place to minimize the danger of flying parts.

Power Cutting Tools

Some repair jobs require that the technician cut a metal part. Hand-held power cutting tools can quickly cut metal and, when used properly, will make a smooth, accurate cut. Many shops use small power cutting tools for precise cutting jobs.

Impact Wrenches

An ***impact wrench,*** which is used in conjunction with impact sockets, speeds up most jobs. Using an impact wrench also makes fastener removal less tiring. Impact wrenches can be the familiar air gun type or the ratchet handle type, **Figure 3-47.** Impact wrenches used in most shops are pneumatic (powered by air pressure).

Impact wrenches are usually equipped with 3/8″ or 1/2″ drives to accept 3/8″ and 1/2″ drive sockets respectively. Heavy-duty impact wrenches often have 3/4″ or 1″ drives.

Always use impact sockets when using an impact wrench. These sockets are available in either six- or twelve-point configurations and are made with heavy sidewalls, **Figure 3-48.** The heavy sidewalls help protect the socket against the extreme strains caused by using them with an impact wrench.

A

B

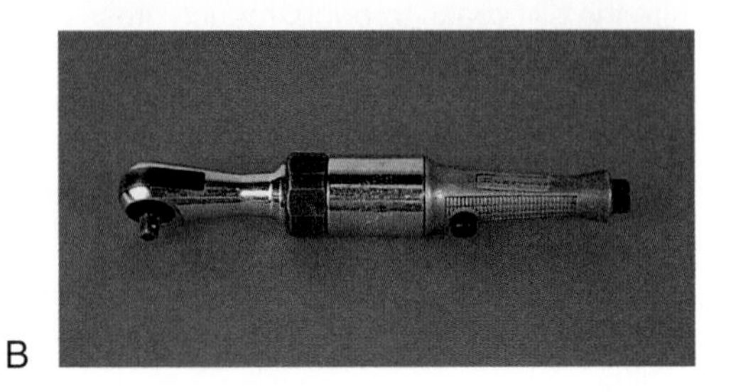

Figure 3-47. A—This gun-type 1/2″ drive impact wrench is used for heavy work. B—Ratchet-type 3/8″ drive impact wrench is used for lighter work and in close quarters.

Figure 3-48. Impact sockets are designed to withstand the tremendous torque produced by the impact wrench. Never use a standard socket with an impact wrench. It can shatter and cause injury. (Snap-on Tools)

STOP Warning: Never use standard sockets with an impact wrench. The strains put on the socket can cause it to break apart violently. Impact sockets can, however, be used with non-powered socket handles.

Air Hammers

Air hammers, sometimes called air chisels, are pneumatic tools that use air pressure to produce a series of fast, heavy blows. When equipped with a chisel attachment, the air hammer is handy for removing exhaust parts and spot welds, and for other metal cutting jobs. When used with a flat attachment, air hammers will loosen stuck parts or remove pressed-in parts, such as front suspension bushings.

Compressed Air Systems

To operate air tools, a *compressed air system* is needed. A shop compressed air system consists of:

- *Air compressor*—Stationary compressors are driven by an electric motor. The motor usually drives the compressor through one or more belts. The compressor intake has a filter to keep out dust.
- *Air pressure regulator*—The regulator usually limits shop air pressure to between 150 and 175 psi (1034 to 1227 kPa). Most compressed air systems have separate regulators at each workstation to further control pressure.

- *Compressed air storage tank*—The minimum for a shop with several bays is a 150-gallon (568 liter) tank.
- *Water separators*—One or more water separators are used to remove moisture from the system. Moisture is always present in the air and can damage air tools if it is not removed. Sometimes the water separator is part of the workstation pressure regulator. Condensed water often builds up in the storage tank. To keep this water out of the other parts of the system, the tank should be drained every few days.
- *Lines and hoses*—These components deliver the compressed air to the work area. Metal pipe should always be used for stationary air lines. Plastic pipe cannot stand up to the shock loads in a compressed air system. Air hoses are reinforced with fiber to withstand hard use. Hoses are also oil and heat resistant.

Air tools have couplings that allow the tool to be quickly attached to the air hose. There are several types of couplings, and the right coupling must be used. Mismatched couplings will not connect.

Hydraulically Operated Tools

Some pulling or pressing tools are operated by hydraulic pressure. Hydraulic pullers were already discussed earlier in this chapter. Other hydraulically operated devices include the hydraulic press and the hydraulic power unit. The technician creates the needed hydraulic power with a hand- or foot-operated pump. Hydraulic presses and power units are discussed below.

Hydraulic Presses

Hydraulic presses can be used to remove bearings, straighten shafts, and press bushings. The hydraulic press is superior to striking tools because the pressure exerted by the press is smooth and controlled. The workpiece is not "upset" by the shock of hammer blows, and enormous pressures can be generated. **Figure 3-49** illustrates a typical hydraulic press with a movable table. When using a hydraulic press, make sure the table pins are in place and the table winch is slacked off. Failure to do this can break the winch gear or cable.

Press Safety Precautions

All hydraulic pressing and pulling tools are potentially dangerous if improperly used. General safety rules applicable to all types of hydraulic tools include:

- Stand free while pressure is applied.
- Apply only enough pressure to do the job.
- Shield brittle parts, such as bearings, to protect against flying debris.
- Engage the ram securely and make sure it is in line with the work.
- Wear goggles to protect your eyes from flying debris.
- If work must be performed while maintaining pressure, be careful to keep out of line with the tool.
- Be careful of part snap-back if the tool slips.

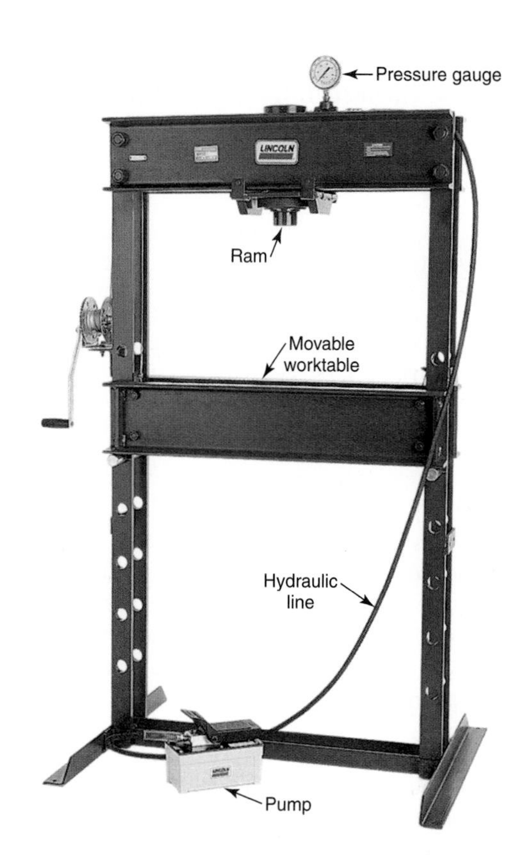

Figure 3-49. This hydraulic press is used with various adapters to remove bearings, collars, and other interference fit parts. The worktable can be moved up or down to accommodate various parts. The worktable is also open in the center to accommodate axle shafts and other long assemblies. (Ammco Tools)

Other Tools Will Follow

As mentioned, many specialty tools will be discussed in this text. When you come across one, pay particular attention to its name and its operating instructions. Many potentially difficult jobs can be made easier if the proper tools are used. Proper tool selection and use is very important; learn all you can about the tools available to the technician.

Mark Your Tools

As you obtain tools, scribe your name on them. An electric engraving pen will work well for this purpose. Be sure to mark the tools in an area that will be difficult to grind off.

Service Information

Modern cars, trucks, and SUVs are increasingly complex, with electronic controls operating or receiving information from almost every vehicle system. Engines range from inline three-, four-, and six-cylinder designs to V-6, V-8, V-10, and V-12s. Drive train variations include front-wheel drive, rear-wheel drive, and four-wheel drive. In addition to the complexity of modern vehicles, there are more vehicle manufacturers than ever before. Each one has its own diagnosis and service requirements. In addition, there are many different models, engines, drive trains, and body types within each make. To make sense of the complexity and diversity of late-model vehicles, the modern technician must have access to many kinds of service information. Major types of service information are described below.

Note: Much of today's service information is available on CD-ROM, Figure 3-50. These discs can be inserted in a computer equipped with a CD drive. The information on the disc is then accessed by the computer and displayed on the computer screen. If the computer has an attached printer, the information contained on the CD-ROM can be printed out. The printout can then be carried to the vehicle.

Factory Service Manuals

Factory service manuals contain the information needed to repair one manufacturer's vehicles. They contain information specific to a particular vehicle and model year, **Figure 3-51.** In some cases, factory service manuals are divided into diagnosis and overhaul manuals.

Diagnostic Manuals

Many vehicle manufacturers publish separate *diagnostic manuals*. These manuals contain only the information needed to diagnose problems. The diagnostic manual consists of troubleshooting procedures and discussion of common problems. Diagnostic manuals also contain scan tool data and other specifications. Actual repair procedures are covered in the overhaul manual.

Overhaul Manuals

Overhaul manuals contain comprehensive information needed to disassemble, inspect, and reassemble various vehicle components. This information is confined to disassembly and reassembly only, and does not cover diagnosis.

Figure 3-50. Service information is often available on CD-ROM. A single disc can hold the equivalent of thousands of printed pages. (Expertec)

General Service Manual

General service manuals are less detailed than factory service manuals, but they contain service information for many types of vehicles. Motor, Chilton, and Mitchell are among those publishing general service manuals. Like manufacturers' manuals, these manuals may be available in diagnostic, overhaul, and other variations.

Specialized Manual

Specialized manuals generally cover one common system for one vehicle or many types of vehicles, **Figure 3-52.** These manuals often cover such topics as emission controls, automatic transmissions, computerized engine controls, electrical systems, brakes, or suspension systems. They

Figure 3-51. All vehicle manufacturers print manuals containing all the information needed to service their vehicles. These manuals may cover one model or several models that are mechanically similar. Manufacturer's service manuals are sometimes called factory manuals. Many of these manuals are divided into two volumes: diagnosis and overhaul. (DaimerChrysler, MVCC, Jack Klasey)

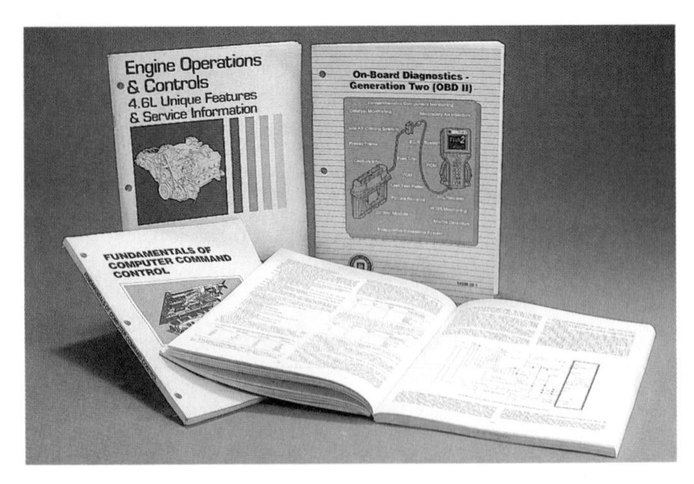

Figure 3-52. Specialized manuals cover a system or component used on many vehicles.

combine some of the best features of both factory manuals and general manuals.

Schematics

It is almost impossible to repair modern vehicles without occasionally referring to a ***schematic.*** A schematic is a graphic representation of a hydraulic, pneumatic, or electrical system. Tracing the flow of power along the path shown in the schematic allows the technician to determine which components should be energized under specific conditions. A typical electrical schematic is shown in **Figure 3-53.**

Troubleshooting Charts

Troubleshooting charts contain the logical steps used to determine the cause of a problem. A typical troubleshooting chart is shown in **Figure 3-54.** To use a troubleshooting chart, begin at the top and follow the instructions. In most cases, the chart will call for a yes or no answer at certain points. Depending on your answer, you will be directed to further diagnostic steps. If you follow the chart closely, you will reach the proper diagnostic conclusion.

Telephone Hotlines

When all other sources of information have been exhausted, many technicians call a manufacturer's or part supplier's *telephone hotline.* Some general service manual manufacturers provide technical support services over a technical hotline. Calling these hotlines will connect you with a technical support person. Hotline personnel often have information gathered from actual repair and diagnosis situations. This allows the technician to obtain real-life information that would not otherwise be available. Hotline personnel will also have access to the latest update information from manufacturers' engineering departments.

Some vehicle manufacturers sponsor hotlines that are available only to the technicians who work for their dealerships. Other hotlines are available by subscription. These hotlines can be accessed only after a fee is paid. Some parts manufacturers' hotlines are available to anyone. These are primarily aimed at the technician who has questions about the manufacturer's parts.

Another use for technical hotlines is computer reprogramming. If a vehicle's computer needs updated software,

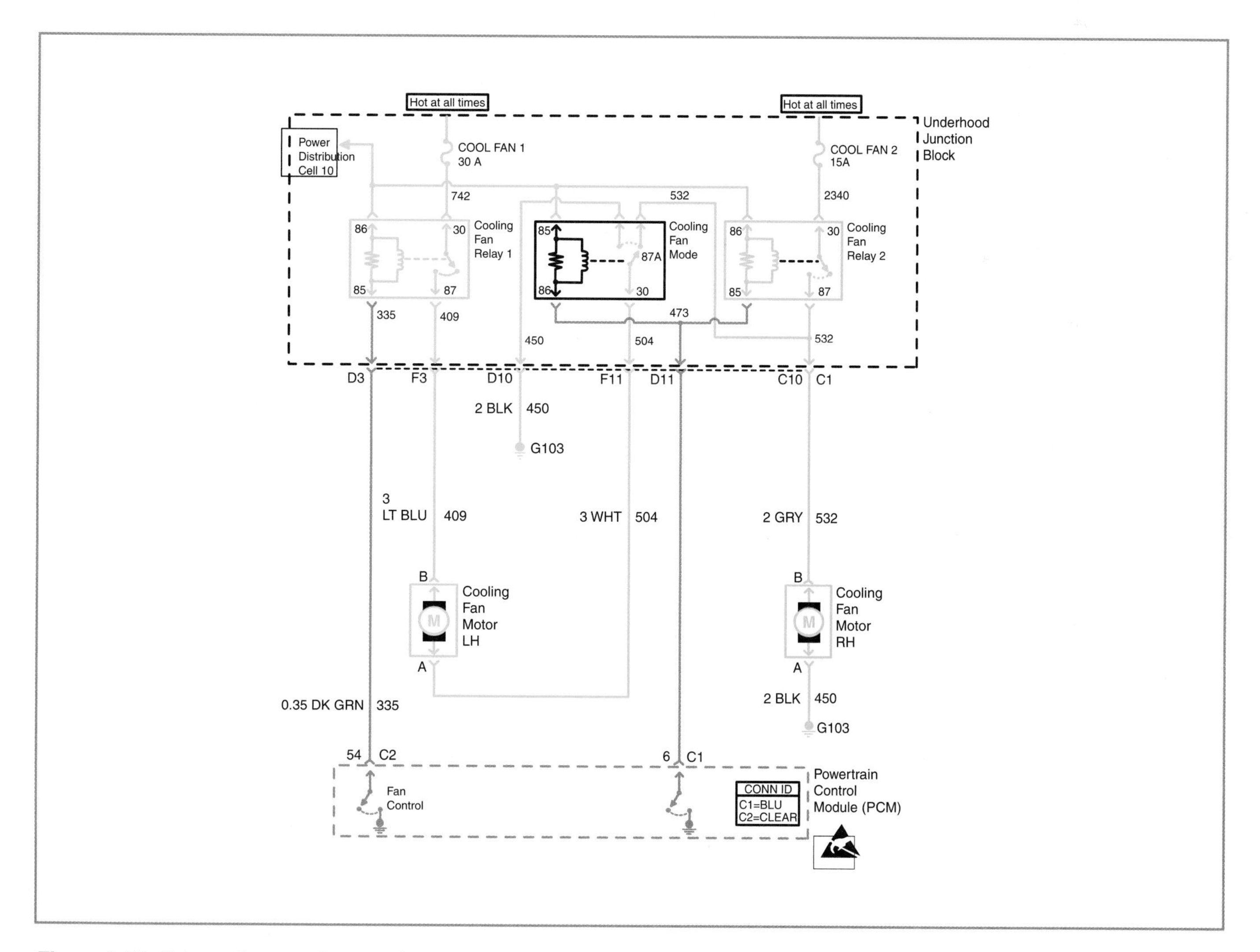

Figure 3-53. Schematics are often used to trace electrical or vacuum problems. (General Motors)

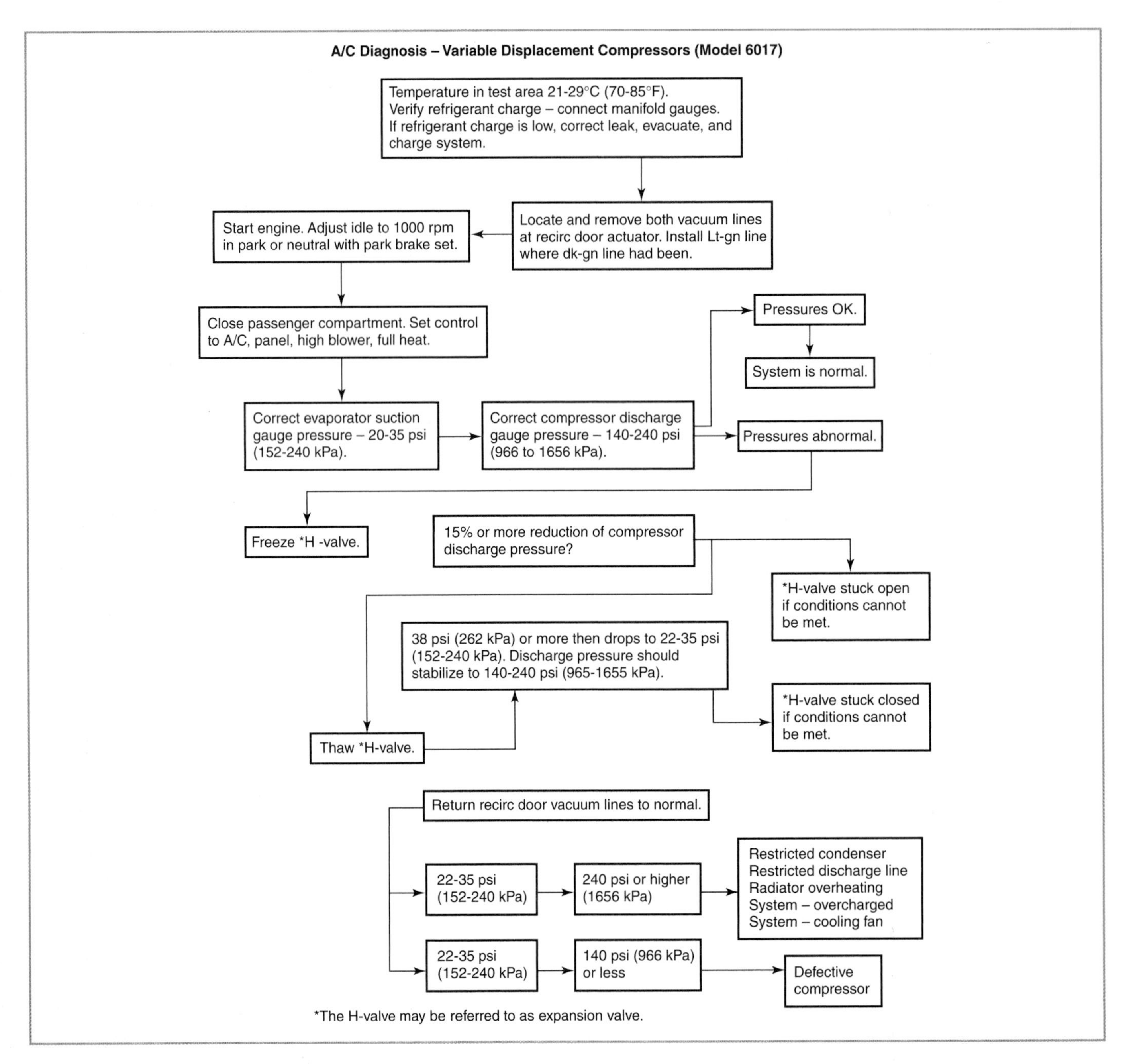

Figure 3-54. Most manuals have diagnostic flowcharts that guide the technician through the diagnostic process, step by step.

the software can sometimes be obtained from the manufacturer over the phone line.

Internet Resources

The Internet has become a valuable source of automotive repair information. Many manufacturers have websites (locations on the Internet) containing a great deal of technical information. Organizations such as such as The International Automotive Technicians Network (iATN) provide a way for technicians from around the world to help each other by way of e-mail, **Figure 3-55.**

Training Materials

Training materials include basic sources of automotive information and update information. Basic information sources include this textbook and related materials, such as workbooks and videotapes, which allow the student to learn about automotive systems and how they operate. Update information is supplied to working technicians to keep them informed of changes that occur as vehicles or their parts are redesigned and improved. Much of the best update information is supplied by vehicle manufacturers and aftermarket (non-factory) parts suppliers.

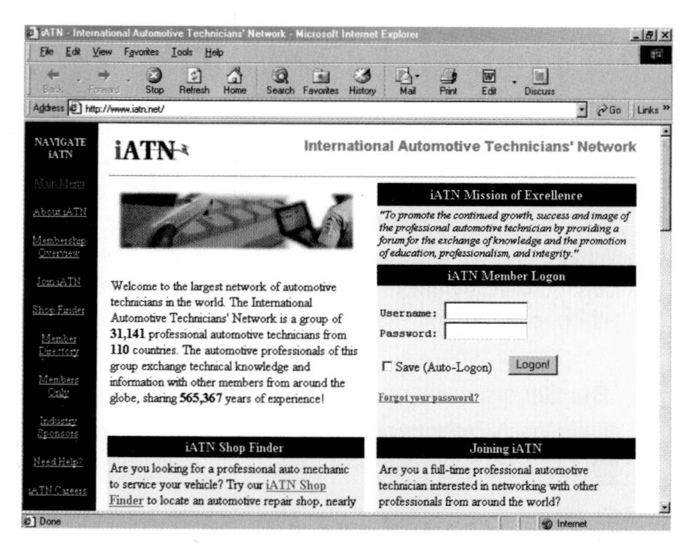

Figure 3-55. A wealth of automotive service information is available through the Internet. (iATN)

Care of Service Information

Unlike other tools, most service information is in the form of printed pages and, therefore, can be damaged by grease, dirt, or rough handling. Always store service manuals and other service information in a clean, dry, and oil-free environment, not on top of the workbench where they can be ruined quickly. To prevent scratches, handle CDs carefully and return them to their sleeves or cases immediately after use.

Tech Talk

To keep your tools from rattling and possibly being damaged when you open and close the drawers of your tool cabinet, line the bottom of each drawer with cushioning material. The roughness of the material will keep the tools in place. A piece of cardboard will work, but it is better to use oil-resistant plastic or rubber matting. Cut the matting to fit the drawer and simply lay the tools on the matting.

Many tools are sold in flat plastic display packs. These are excellent for tool storage. Instead of throwing the display pack away after you remove the tools, trim the display pack to fit in the bottom of one of your tool drawers. The original tools will fit perfectly, and they will stay in place when the drawer is opened and closed.

Summary

The automotive technician must have a large selection of high-quality tools to service late-model vehicles properly and efficiently. Quality tools are stronger and less bulky than inferior tools. They are also easier to clean. All tools must be cared for properly. Tools should be kept clean, and they should be stored in an orderly fashion.

Common tools used by the automotive technician include hammers, chisels, punches, files, grinders, drills, saws, vises, screwdrivers, pliers, wrenches, presses, and pullers.

The technician must have access to the latest service information when repairing today's complex vehicles. Service manuals describe service and repair operations. There are three types of service manuals: factory manuals, general manuals, and specialized manuals.

Review Questions—Chapter 3

Do not write in this book. Write your answers on a separate sheet of paper.

1. All the following statements about high-quality tools are true, *except:*
   (A) they are hard to clean.
   (B) they are less bulky than inferior tools.
   (C) they are made of alloy steel.
   (D) they are often guaranteed.
2. Explain how to store tools properly.
3. Standard size screwdrivers can be used for which of the following purposes?
   (A) Removing and installing screws.
   (B) Prying of oil pans.
   (C) Hammering on tight parts.
   (D) All of the above.
4. What advantage does a 6-point box-end wrench have over a 12-point box-end wrench?
5. Flare nut wrenches can be used to remove and install _____ fittings.
6. A hammer handle should be _____.
   (A) clean and dry
   (B) tight in the hammer head
   (C) held firmly
   (D) All of the above.
7. A chisel can be used to _____.
   (A) mark work before drilling
   (B) cut off rivet heads
   (C) drive pins from holes
   (D) All of the above.
8. Never use a file if it does not have a _____.
   (A) single-cut surface
   (B) double-cut surface
   (C) handle
   (D) None of the above.
9. Files should be _____.
   (A) lightly oiled
   (B) heavily oiled
   (C) clean and dry
   (D) sprayed with penetrating oil

10. A _____ is used to clean files.
    (A) file card
    (B) wire wheel
    (C) wire brush
    (D) soap solution
11. Which of the following hacksaw blades would be best for cutting sheet metal?
    (A) 14 tooth.
    (B) 18 tooth.
    (C) 24 tooth.
    (D) 32 tooth.
12. What kind of tap would be used in a blind hole when the threads must run almost to the bottom?
    (A) Plug.
    (B) Taper.
    (C) Bottoming.
    (D) Pipe.
13. Cutting oil is helpful when you are trying to cut threads in _____.
    (A) steel
    (B) brass
    (C) cast iron
    (D) aluminum
14. A reamer should remove about _____ of stock with each cut.
    (A) 0.0001″–0.0002″ (0.0025 mm–0.005 mm)
    (B) 0.001″–0.002″ (0.025 mm–0.05 mm)
    (C) 0.01″–0.02″ (0.25 mm–0.5 mm)
    (D) 0.1″–0.2″ (2.5 mm–5.0 mm)
15. Extractors are used to remove _____.
    (A) stripped tubing fittings
    (B) broken cap screws
    (C) pressed-on bearings
    (D) All of the above.
16. Briefly describe the three types of pulling jobs. (any order)
17 Carbon steel drill bits _____.
    (A) are cheaper than bits made of high-speed steel
    (B) stay sharp longer than high-speed steel bits
    (C) will not lose their temper if overheated
    (D) All of the above.
18. Compressed air and electricity are two power sources for _____.
    (A) drills
    (B) presses
    (C) pullers
    (D) All of the above.
19. Which of the following information sources requires that the technician have access to a computer?
    (A) Hotline
    (B) CD-ROM
    (C) Specialized manual
    (D) None of the above.
20. Which of the following types of manuals would you consult if you want to know how to repair heaters and air conditioners on many different makes of vehicles?
    (A) Factory manual.
    (B) General manual.
    (C) Specialized manual.
    (D) Update information manual.

ASE-Type Questions

1. Technician A says chisels, files, drills, and similar tools are very hard and can all be piled together for storage. Technician B says every type of tool should be carefully stored to prevent damage. Who is right?
    (A) A only.
    (B) B only.
    (C) Both A and B.
    (D) Neither A nor B.
2. Technician A says that technicians are generally expected to have their own hand tools. Technician B says that it is often best to purchase hand tools in sets. Who is right?
    (A) A only.
    (B) B only.
    (C) Both A and B.
    (D) Neither A nor B.
3. Which of the following wrench types should be used to remove fuel fittings?
    (A) Pipe wrench.
    (B) Flare-nut wrench.
    (C) Adjustable wrench.
    (D) Fluted wrench.
4. Pliers can be used for all of the following, *except:*
    (A) tightening tubing fittings.
    (B) bending cotter pins.
    (C) cutting wire.
    (D) crimping connections.
5. All the following hammers can be used when it is important not to damage the struck surfaces, *except:*
    (A) brass.
    (B) plastic.
    (C) ball peen.
    (D) rubber.

6. Which of the following would be the best cut to rough file aluminum?
   (A) Coarse cut.
   (B) Fine cut.
   (C) Smooth cut.
   (D) You must try each one to see which cuts best.
7. All the following statements about dies are true, *except:*
   (A) special dies are used to rethread axle shafts.
   (B) dies are used to cut internal threads.
   (C) when cutting, the die should be backed up to break the chip.
   (D) some dies are adjustable.
8. All the following are safety rules for grinding, *except:*
   (A) keep the tool as close to the wheel as possible.
   (B) wear safety glasses.
   (C) hold small objects with pliers.
   (D) stand directly in front of the grinding wheel when grinding.
9. Technician A says using an impact wrench will reduce the amount of time a job takes. Technician B says impact sockets should always be used with an impact wrench. Who is right?
   (A) A only.
   (B) B only.
   (C) Both A and B.
   (D) Neither A nor B.
10. Technician A says that service information for late-model vehicles is available on CD-ROM. Technician B says that the Internet is a valuable source of automotive repair information. Who is right?
    (A) A only.
    (B) B only.
    (C) Both A and B.
    (D) Neither A nor B.

Suggested Activities

1. Obtain a complete set of box-end or socket wrenches in both customary and metric sizes. Try both sets of wrenches on various bolts or nuts. Which sizes will interchange?

   Sample: Do not write in this book.
   a. ______ inch = ______ mm
   b. ______ inch = ______ mm
   c. ______ inch = ______ mm
   d. ______ inch = ______ mm
   e. ______ inch = ______ mm

   Discuss with your instructor or the other members of your class whether it is a good idea to use metric wrenches on customary bolts and vice versa.
2. Obtain catalogs from various tool manufacturers and make a list of the tools you will need to get started as an automotive technician. List additional tools you would need to become a(n):
   a. drivability technician.
   b. engine overhaul technician.
   c. transmission/transaxle repair technician.
   d. air conditioning technician.
   e. brake technician.
   f. steering and suspension technician.
3. If possible, figure out the costs of the tools listed in Activity 2. Determine the total investment needed to get started as an automotive technician.
4. Show your classmates how to look up a particular service operation in one or more factory service manuals.
5. Use a computer to enter a CD-ROM based information system and look up service and repair information. After looking up the information, demonstrate the process to your classmates.
6. Access the Internet; then access the website of an online automotive service, such as iATN. Study the site carefully, noting what services are available and whether some parts of the website require you to be a registered member. Write a short report summarizing your findings.

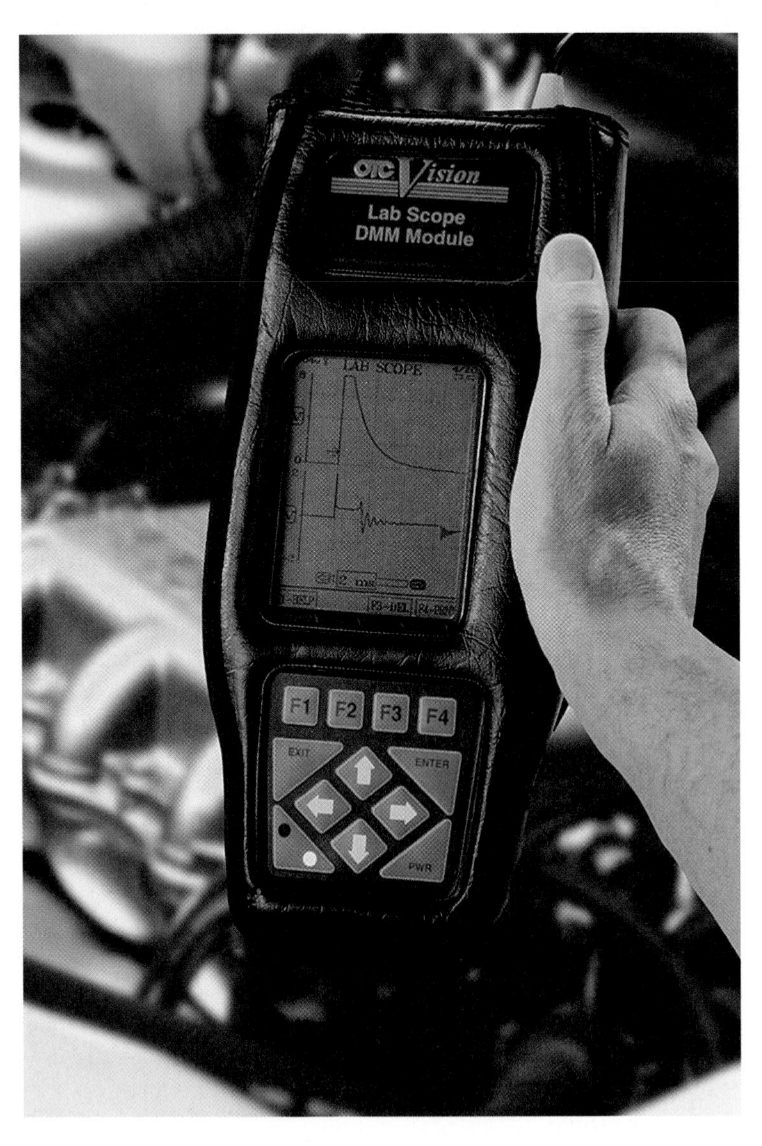

A variety of test equipment is used to troubleshoot today's vehicles. This lab scope can display the waveforms produced by electrical and electronic components. (OTC)

4

Precision Tools and Test Equipment

After studying this chapter, you will be able to:
- Identify common measuring tools.
- Select the appropriate measuring tool for a given job.
- Use precision measuring tools.
- Properly maintain precision measuring tools.

Technical Terms

Outside micrometer
Range
Thimble
Spindle
Sleeve
Vernier micrometer
Metric micrometer
Feel
Digital micrometer
Inside micrometer
Micrometer depth gauge
Dial indicator
Valve seat runout gauge
Cylinder gauge
Calipers
Dial caliper
Feeler gauges
Wire gauge
Screw pitch gauge
Telescoping gauge
Steel rules
Spring scales
Steel straightedge
Compression gauge
Vacuum gauge
Oil pressure tester
Fuel pressure tester
Test light
Jumper wires
Multimeters
Ohmmeters
Voltmeters
Ammeter
Tachometers
Scan tool
Waveform meters
Digital storage oscilloscope (DSO)

The automotive technician should be familiar with the precision measuring tools used in the shop. Many jobs involve checking sizes, clearances, and alignments. Other jobs require the technician to check such things as engine vacuum and compression, battery voltage or other electrical system values, or various pressures in vehicle components. A careless or inaccurate measurement can be costly in terms of money and customer relations, to say nothing of damaging a technician's reputation. Remember, the top-notch technician must be competent in the use of measuring tools. You can be proud of your ability to make precise measurements—it is the mark of a fine technician.

Tool Storage

Keep your measuring tools and test equipment in protective cases. See **Figure 4-1.** Also, store them in an area that will not be subjected to excessive moisture or heavy traffic. Never dip a precision measuring tool in solvent (unless it is being completely dismantled). Do not use compressed air for cleaning precision measuring tools.

Handling Tools

When using a measuring tool, place it in a clean area where it will not fall or be struck by other tools. Never pry, hammer, or force any measuring tool. Remember, measuring tools are precision tools—keep them that way.

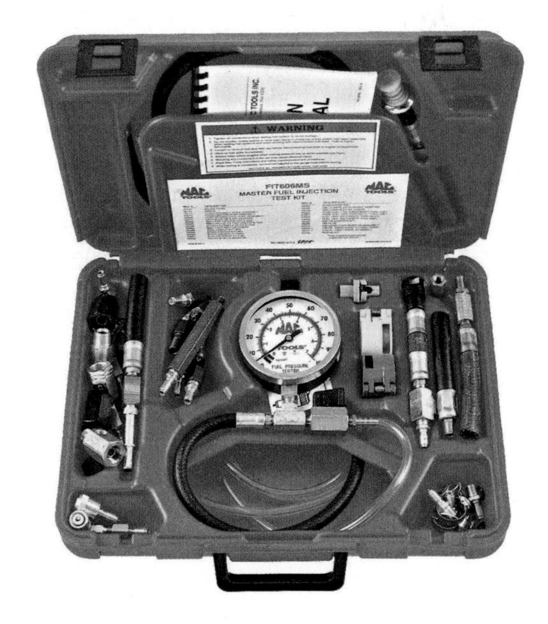

Figure 4-1. Precision measuring equipment, such as this fuel pressure test set, should be stored in a protective case. (Mac Tools)

Checking for Accuracy

Precision tools should be checked for accuracy on a regular basis. They can be checked against another tool of known accuracy, or they can be checked using special gauges. If a tool is accidentally dropped or struck by some object, immediately check it for accuracy. Adjustments for wear or very minor damage can be made to many measuring tools. Follow the manufacturer's instructions for making these adjustments.

Precision Measuring Tools

The most common precision measuring tools are described in the following sections. These tools can be used to check the condition of engine parts and other vehicle components.

Outside Micrometer

The ***outside micrometer,*** often referred to as a "mike," is used to check the diameter of pistons, pins, crankshafts, and other machined parts. The most commonly used micrometer reads in thousandths of an inch. Some micrometers are accurate to one ten-thousandth of an inch. Learn the names of the parts and their relationships to the operation of the tool.

Micrometer Range

Many micrometers are designed to produce readings over a 1″ ***range.*** Ideally, the automotive technician should obtain a set of six micrometers covering 0–1″, 1–2″, 2–3″, 3–4″, 4–5″, and 5–6″ range. Micrometers with interchangeable anvils are available, however, these multirange micrometers are bulky and inconvenient.

Reading the Micrometer

Micrometers are made so every turn of the ***thimble*** will move the ***spindle*** 0.025″ (twenty-five thousandths of an inch). You will notice the sleeve is marked with a series of lines. Each of these lines represents 0.025″. Every fourth one of these 0.025 markings is marked 1, 2, 3, 4, 5, 6, 7, 8, or 9. These sleeve numbers indicate 0.100″ (one hundred thousandths), 0.200″, 0.300″, and so on. The micrometer ***sleeve*** then is marked out for one inch in 0.025″ markings. They will read from 0.000″ to 1.000″.

The tapered end of the thimble has 25 lines around it. Each line represents 1/25 of a revolution. The lines are numbered 1 through 25. One complete turn of the thimble moves the thimble edge exactly 0.025, or one mark on the sleeve. The distance between marks is determined by reading the thimble line that is even with the long line drawn the length of the sleeve markings. Each line on the thimble edge represents 0.001″ (one thousandth of an inch), **Figure 4-2.**

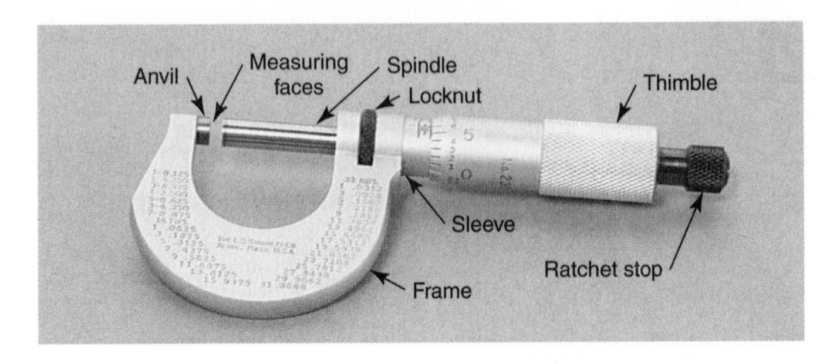

Figure 4-2. A 0–1″ outside micrometer. Study the markings and the part names.

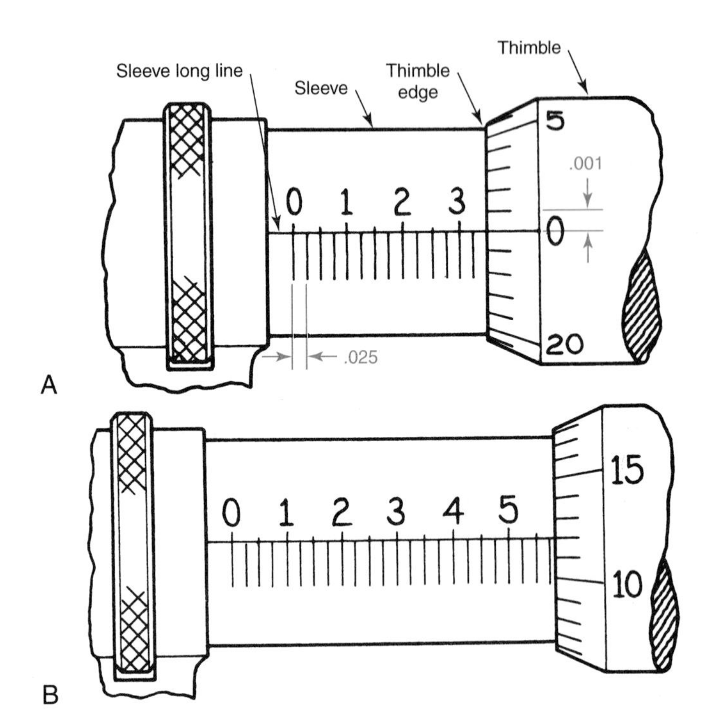

Figure 4-3. Study the numbering system used with an outside micrometer. A—The thimble edge has moved across the sleeve up to the 3 (0.300″), plus two more sleeve marks (0.050″). The thimble 0 mark is in line with the sleeve long line so the reading is 0.300″ + 0.050″ + 0″ = 0.350″. B—The thimble edge has moved up to the 5 (0.500″) plus three more sleeve marks (0.075″) plus 12 thimble marks (0.012″), resulting in a total reading of 0.587″.

Close-ups of a micrometer scale are shown in **Figure 4-3.** Look at the markings on the micrometer section in Figure 4-3A. Three numbers are visible on the sleeve. This indicates the mike is open at least 0.300″ (three hundred thousandths of an inch). You can see the thimble edge is actually past the 0.300″ mark, but not to the 0.400″ mark. By careful study, you will see the thimble edge has moved exactly two additional marks past 0.300″. Since each sleeve mark represents 0.025″ and two marks are showing, the edge is actually stopped at 0.300″ plus 0.050″, or 0.350″ (three hundred and fifty thousandths of an inch). Since the thimble edge 0 marking is aligned with the long sleeve line, the mike is set exactly on 0.350″. If this is a 0–1″ mike, the reading is 0.350″. If this is a 2–3″ mike, the reading is 2″ plus 0.350″, or 2.350″.

In Figure 4-3B, the micrometer has been opened to a wider measurement. You will see the thimble edge is no longer on a sleeve marking, but is somewhere in between.

How many numbers are visible on the sleeve? There are five, or 0.500″ (five hundred thousandths). The thimble edge has moved three marks, or 0.075″, past the 0.500″ mark. This makes a total of 0.575″. The thimble edge has moved past the third mark. Since the fourth mark is not visible, we know the measurement is somewhere between the third and fourth marks.

By examining the thimble edge marks shown in Figure 4-3B, you will see the twelfth mark is aligned with the long line on the sleeve. This means the thimble edge has moved twelve thimble marks past the third sleeve mark. Since each thimble mark equals 0.001″, the thimble has actually moved 0.012″ past the third sleeve mark.

Your reading would be 0.500″ (largest sleeve number visible) plus 0.075″ (three sleeve marks past sleeve number) plus 0.012″ (twelve thimble marks past the third sleeve mark), making a total reading of 0.587″. If this were a 3–4″ micrometer, the actual measurement would be 3.587″.

Study the readings shown in **Figure 4-4.** Compare your answers with those shown.

Make your readings in four steps. See **Figure 4-5.**

1. Read the largest sleeve number visible—each one indicates 0.100″.
2. Count the number of full sleeve marks past this number—each one indicates 0.025″.
3. Count the number of thimble marks past this last sleeve number. Each one indicates 0.001″. If the thimble marks are not quite aligned with the sleeve long line, estimate the fraction of a mark.
4. Add the readings in steps 1, 2, and 3. The total is the correct micrometer reading. Add this reading to the starting size of the micrometer being used. If the mike range was 1–2″, add the total reading to 1.000″.

Reading a Vernier Micrometer

A ***vernier micrometer*** is similar to a standard micrometer, but it can be used to take accurate readings to one ten-thousandth of an inch, **Figure 4-6.** Instead of estimating fractions of a thousandth between thimble marks, a vernier scale on the sleeve is used to accurately divide each thousandth into ten parts, each part equaling one ten-thousandth of an inch (0.0001″).

The vernier scale consists of eleven thin lines scribed parallel to the sleeve long line. They are marked 0–10. When the thimble marks do not fall in line with the long sleeve line, indicating a fraction of one thousandth of an inch, carefully examine the vernier lines. One of the vernier lines will be aligned with one of the thimble marks. When you locate the vernier line that is aligned, the number of that line indicates the number of ten-thousandths to be added to your initial thimble reading. Look at **Figure 4-7.**

Examine the readings shown in **Figure 4-8.** In both instances, a fraction of a thousandth is obvious by examining the thimble marks. By checking the vernier, you can see one of the vernier lines is in alignment with a thimble mark, thus indicating the number of ten-thousandths over the thimble thousandth reading. Compare your readings with those shown.

Reading a Metric Micrometer

The ***metric micrometer*** closely resembles the standard type. See **Figure 4-9.** The basic differences between them are the thimble and sleeve markings. The metric micrometer is marked to measure hundredths of a millimeter instead of thousandths of an inch.

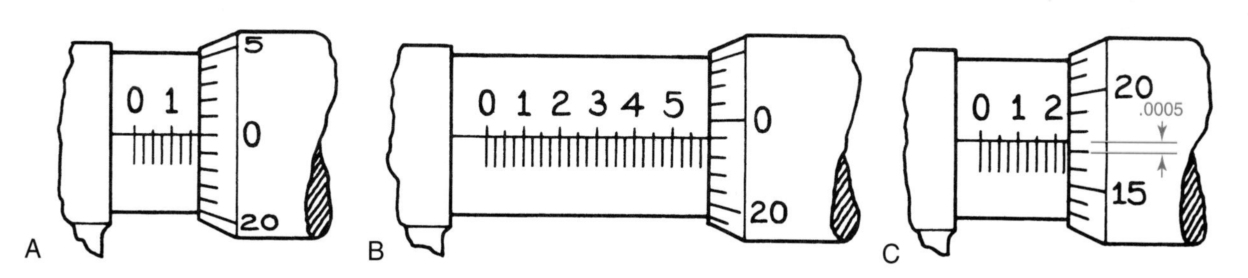

Figure 4-4. A = 0.175″. B = 0.599″. C = 0.242″ + 1/2 thimble mark, or 0.2425″. Note in C, the fraction ten-thousandths is estimated as indicated by the thimble mark.

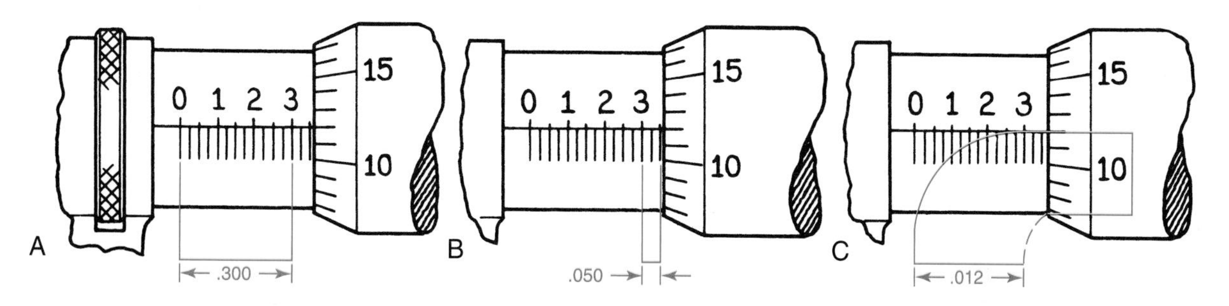

Figure 4-5. Four steps in reading the micrometer. The first reading, in A = 0.300″. The second reading, in B = 0.050″. The third reading, in C = 0.012″. The total reading is 0.362″ (three hundred sixty-two thousandths).

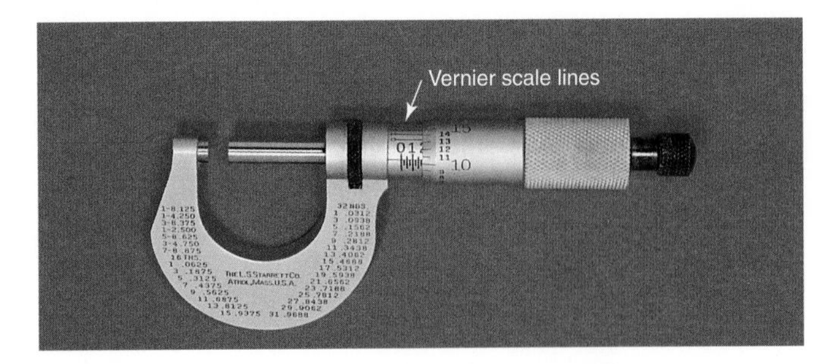

Figure 4-6. Vernier micrometer. The scale lines are at the top of the tool (L. S. Starrett)

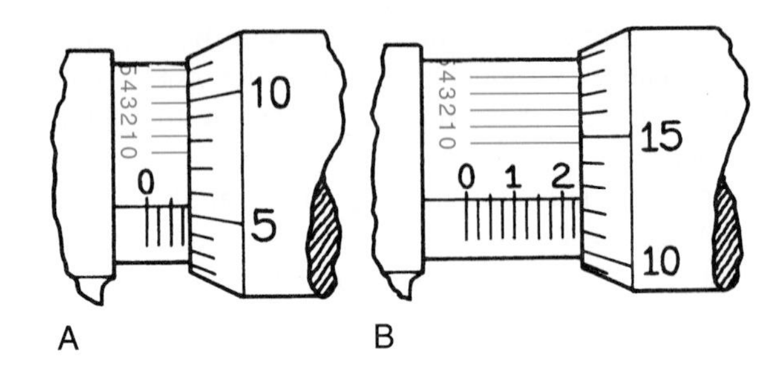

Figure 4-8. A—Vernier line No. 5 is aligned with a thimble mark. Reading is 0.075″ + 0.005″ + 0.0005″ = 0.0805″. B—Vernier line No. 4 is aligned. Reading is 0.200″ + 0.025″ + 0.012″ + 0.0004″ = 0.2374″.

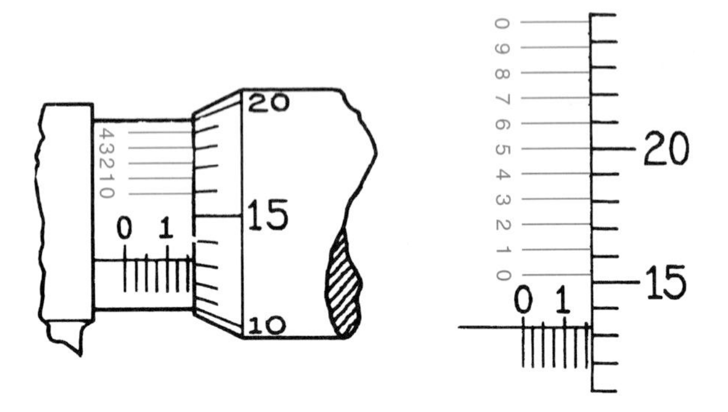

Figure 4-7. Vernier lines are shown in color. Note that vernier line No. 4 is the only one exactly in line with a thimble mark. Your reading would then be 0.100″ + 0.050″ + 0.013″ + 0.0004″ (four ten-thousandths) = 0.1634″.

Each line on the sleeve equals 0.5 millimeters (abbreviated mm). Every thimble line equals 0.01 mm. Each two full revolutions of the thimble advance the spindle 1.00 mm.

To obtain a reading, follow these four steps. Use the micrometer in Figure 4-9B as an example.

1. The highest line showing on the sleeve ....10.0 mm
2. The number of lines showing on the sleeve past the highest figure.....................................0.5 mm
3. The line on the thimble aligning with the sleeve long line ....0.0 mm
4. Add the numbers to obtain the measurement ....10.5 mm

What reading do you get for Figure 4-9C?

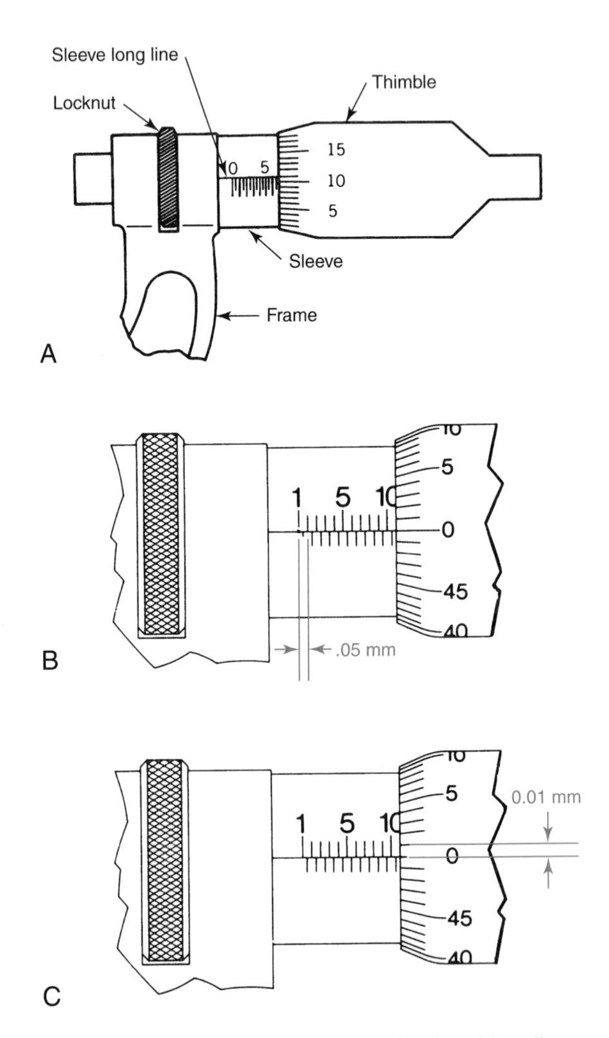

Figure 4-9. Outside metric micrometer. Notice this mike requires two full revolutions of the thimble to equal one millimeter. (TRW, Deere & Co.)

Using an Outside Micrometer

When measuring small objects, grasp the micrometer in your right hand. At the same time, insert the object to be measured between the anvil and spindle.

Note: Before using any measuring tool, always thoroughly clean the work to be measured. This ensures accurate readings and reduces wear on the working tips of the tool.

While holding the work against the anvil, turn the thimble with your thumb and forefinger until the spindle touches the object. Do not clamp the micrometer tightly. Use only enough pressure on the thimble to cause the work to *just fit* between the anvil and spindle. Slip the object in and out of the micrometer while giving the thimble a final adjustment. The work must slip through the micrometer with a very light force. When satisfied your adjustment is correct, read the micrometer setting. Be careful not to move the adjustment, **Figure 4-10.**

Placing the proper force on the micrometer is often called ***feel.*** The technician should develop the proper feel by making practice measurements with a micrometer whenever possible. Some micrometers have a ratchet clutch knob on the end of the thimble. It allows the user to bring the spindle down against the work with the same amount of tension each time.

To measure larger objects, grasp the frame of the micrometer and slip the micrometer over the work. Slide the micrometer back and forth over the workpiece while adjusting the thimble until very light resistance is felt, **Figure 4-11.** As the micrometer is slid back and forth over the work, it should also be rocked from side to side slightly. Continue adjusting the thimble until the micrometer cannot be rocked as it is slid over the work. This will help ensure that the spindle is completely closed against the work, **Figure 4-12.**

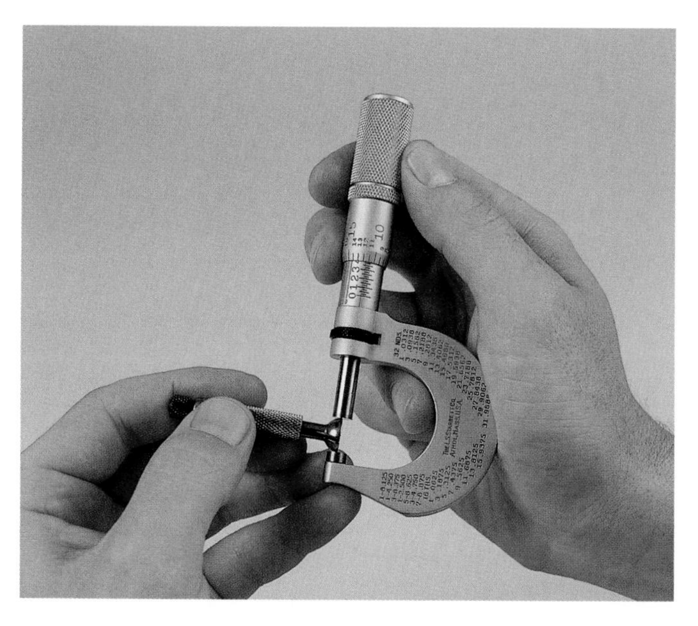

Figure 4-10. Miking a small hole gauge. The heel of the hand supports the micrometer frame while the thumb and forefinger turn the thimble. (L. S. Starrett)

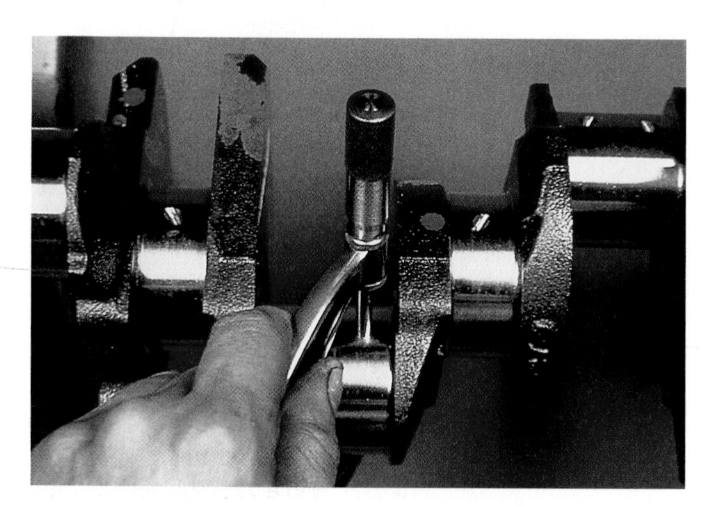

Figure 4-11. This technician is miking a crankshaft. Notice how the micrometer is held. (Federal-Mogul)

Digital Micrometers

The ***digital micrometer*** provides precise measurement readings at a glance. It is similar to a standard micrometer, but the readings appear in a window instead of on the hub and thimble, **Figure 4-13.** Some micrometers provide both standard and digital readings.

Caution: Handle all micrometers with care. Never store a micrometer with the tips of the anvil and the spindle touching, as this may cause rusting between the tips and warping due to temperature changes.

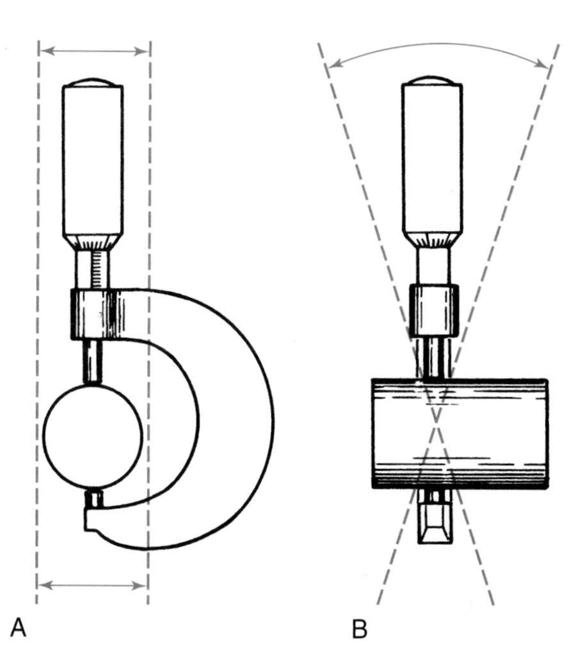

Figure 4-12. A—The micrometer is slid back and forth over the object. B—The micrometer is rocked from side to side to make certain the smallest diameter is found. Rocking is actually very slight.

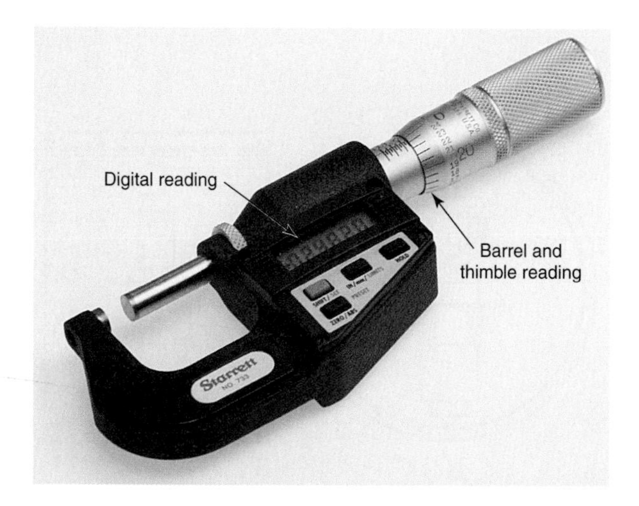

Figure 4-13. Digital micrometer. Four numbers are shown in the window. It can be used to take readings that are accurate to one thousandth of an inch (0.001″). (L. S. Starrett)

Inside Micrometer

The ***inside micrometer*** is used to take measurements in cylinder bores, brake drums, large bushings, etc., **Figure 4-14.** An extension handle permits the use of an inside micrometer in a bore too small to allow you to hold the tool by hand.

An inside micrometer is read in the same manner as the outside micrometer. The same "feel" is required. When measuring, keep the anvil firmly against one side of the bore and rock the free end from side to side. While the free end is being rocked, it should also be tipped in and out. The rocking allows you to locate the widest part of the bore. The tipping ensures that the micrometer is at right angles to the bore, **Figure 4-15.** Continue adjusting the inside micrometer as you rock and tip it until it cannot be moved with light pressure.

Micrometer Depth Gauge

The ***micrometer depth gauge*** is a handy tool for reading the depth of slots, splines, counterbores, and holes. To use this tool, press the base against the work (after cleaning) and run the spindle into the hole to be measured. The depth gauge is read like an outside micrometer, however, the sleeve marks run in the reverse direction, **Figure 4-16.**

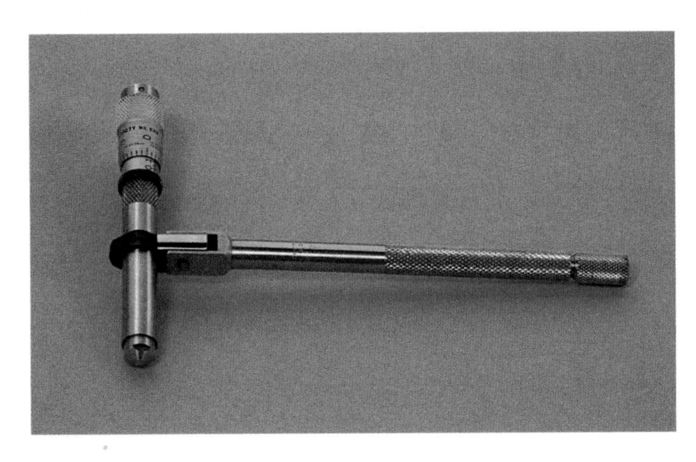

Figure 4-14. Inside micrometer. This tool is used to measure internal cylinder size. (L. S. Starrett)

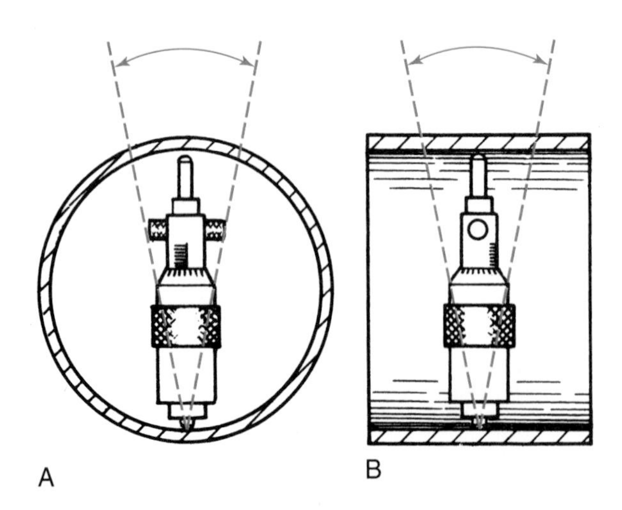

Figure 4-15. A—The inside micrometer must be rocked from side to side. B—At the same time, it must be tipped. Both movements are relatively slight.

Dial Indicator

The ***dial indicator,*** sometimes called a dial gauge, is a precision tool designed to measure movements in thousandths of an inch. Some common uses of the dial indicator include checking shaft endplay, gear backlash, valve lift, shaft run-out, and cylinder taper.

Dial indicator faces are calibrated either in thousandths of an inch or in millimeters. Various dial face markings are available. Ranges (distances over which the indicators can be used) vary, depending upon the instrument, **Figure 4-17.**

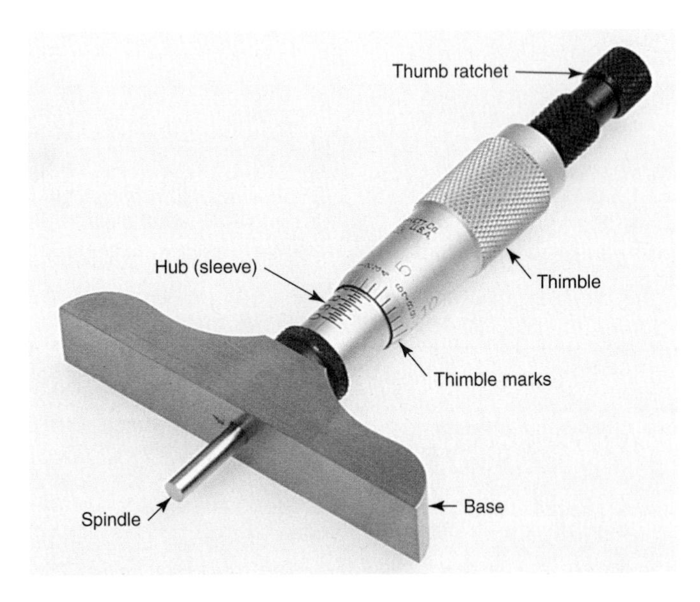

Figure 4-16. Micrometer depth gauge. Study the part names. (L. S. Starrett)

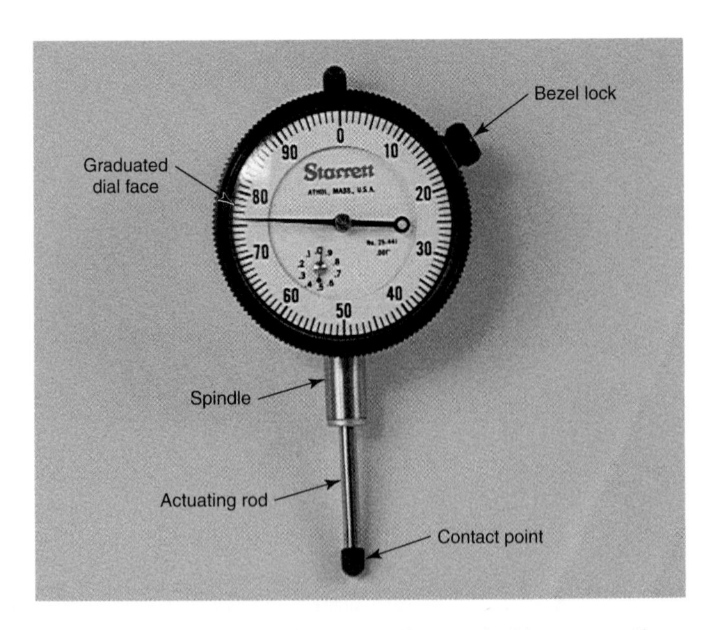

Figure 4-17. This dial indicator can be used with a magnetic base and other holding attachments. (L. S. Starrett)

Various mounting arms, swivels, and adapters are provided so the indicator can be used on various setups. Use care when handling this tool. It is sensitive and easily damaged. Keep it in a protective case when it is not in use.

When using a dial indicator, make sure it is firmly mounted and the standard (actuating rod) is parallel to the plane (direction) of movement to be measured, **Figure 4-18.**

Place the contact point against the work to be measured. Force the indicator toward the work so the indicator needle travels far enough around the dial that movement in either direction can be measured. Securely mount the indicator in this position. The dial face can then be turned to align the 0 mark with the indicator needle. Be sure the indicator range (limit of travel) will cover the movement anticipated. Ranges usually run from around 0.200″ to 1.000″ (one inch), depending on the instrument. **Figure 4-19** illustrates typical dial indicator setups.

Other Dial Indicator Tools

Two other valuable measuring tools utilize a dial indicator as part of their construction. They are the valve seat runout gauge and the cylinder gauge. The ***valve seat runout gauge*** is used to check valve seat concentricity (runout), **Figure 4-20.** The ***cylinder gauge*** provides a quick and accurate way to check cylinder bore size, taper, and out-of-round, **Figure 4-21.**

Other Useful Measuring Tools

In addition to the precision tools already discussed, there are a number of other tools a technician should own. In your work as an automotive technician, a number of measurements, varying from a few thousandths of an inch to several feet, will be required.

Inside and Outside Calipers

Calipers are useful tools for taking quick measurements when accuracy is not critical. See **Figure 4-22.** Figure 4-22A illustrates an *outside caliper.*

Figure 4-22B shows an *inside caliper.* The inside caliper is used to measure the diameter of holes. To determine the reading, hold the calipers on an accurate steel rule and note

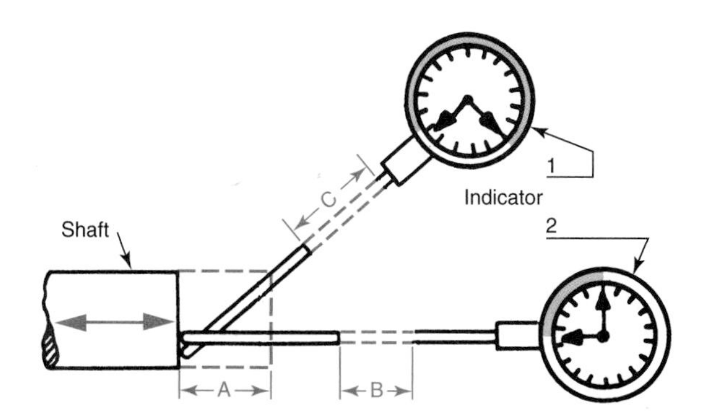

Figure 4-18. Indicator 1 setup is *not* parallel to the movement of the shaft. When the shaft moves distance A, the indicator rod moves distance C, giving a false reading for shaft endplay. Indicator 2 is parallel, and shaft movement A causes the indicator rod to move distance B, producing an accurate reading.

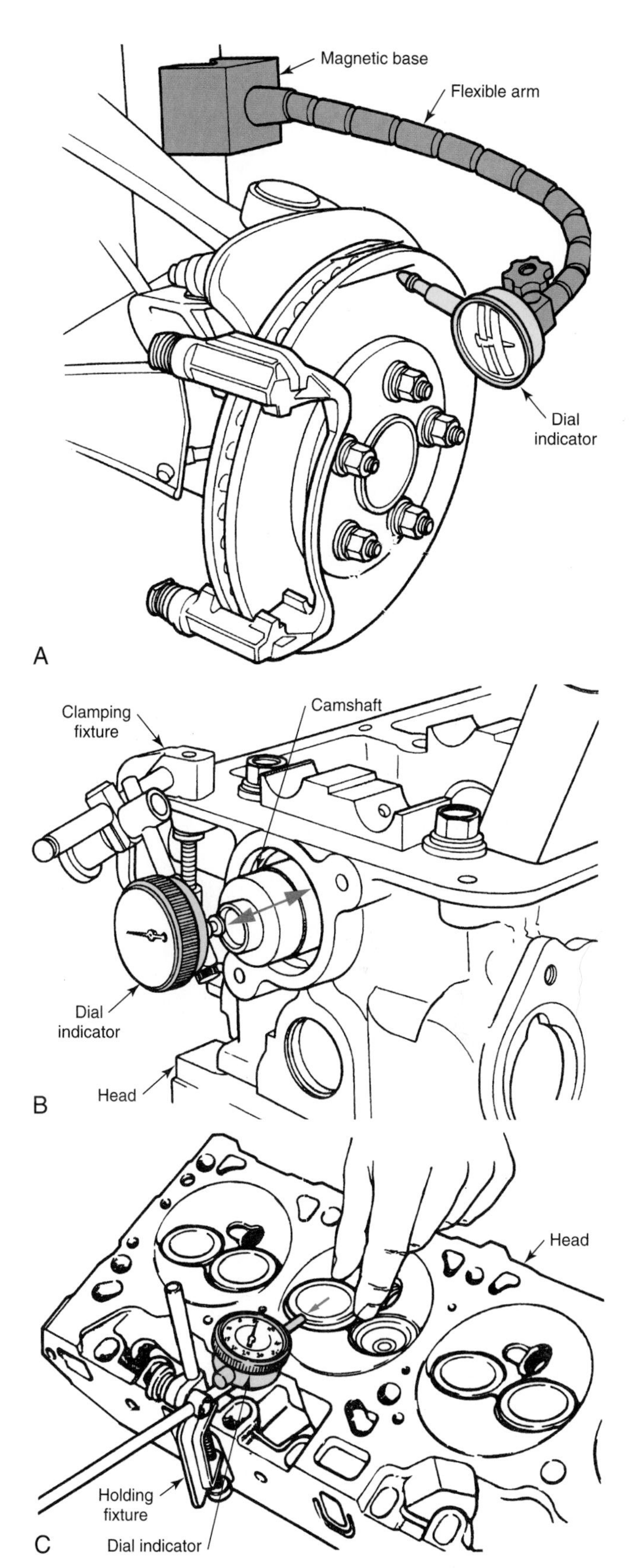

Figure 4-19. A—Checking brake rotor runout with a dial indicator. A magnetic base and a flexible arm are used to hold the indicator in place. B—In this setup, a dial indicator is used with a clamping fixture to determine camshaft endplay. C—This technician is using a dial indicator to measure valve guide wear. (Honda, DaimlerChrysler)

Figure 4-20. Valve seat runout gauge. (Central Tools)

Figure 4-21. This technician is using a cylinder gauge to measure the diameter of a cylinder bore.

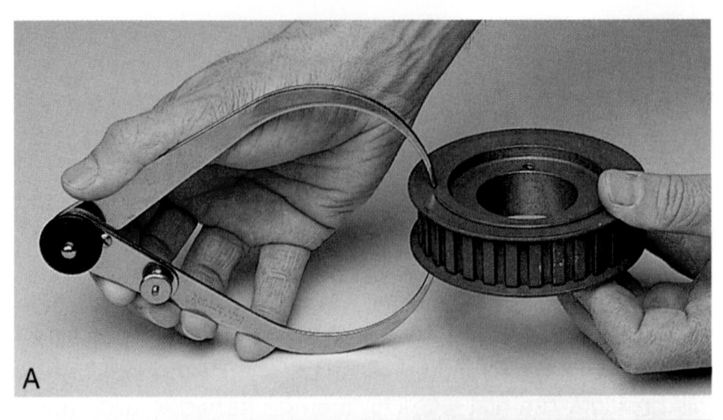

Figure 4-22. A—Outside caliper. B—Inside caliper. (L. S. Starrett)

the distance between the points. For a more accurate reading, carefully measure across the points (using a very light touch) with an outside micrometer.

Dial Calipers

The ***dial caliper*** is a very useful precision measuring instrument capable of obtaining inside, outside, and depth readings. See **Figure 4-23.** Because these calipers are highly accurate, they should be handled with great care. Always store them in protective cases when not in use.

The dial caliper shown in Figure 4-23A will measure objects up to 6″ (152.4 mm). The graduation lines on the bar scale (body) are each equal to 0.100″ (2.54 mm). The dial is calibrated in 0.001″ (0.025 mm) increments. Every full revolution of the dial needle equals 0.100″ on the bar scale. The caliper is equipped with a thumb-operated roll knob that aids in obtaining fine adjustments. Once the measurement is taken, the lockscrew can be tightened. This prevents the caliper from opening or closing and altering the reading.

A newer type of caliper is the digital caliper shown in Figure 4-23B. Instead of a dial, this caliper uses an electronic digital readout. The digital readout eliminates the chance of making an error and can be easily recalibrated before each use.

Feeler Gauges

Feeler gauges, or ***thickness gauges,*** consist of specially hardened and ground steel strips. Each strip is marked with its thickness in thousandths of an inch and/or millimeters. Feeler gauges are used to check clearances between two parts, **Figure 4-24.** They come in sets. Some feeler gauges are made of copper or brass for checking clearances in places where a magnetic field would cause a steel feeler gauge to give an inaccurate reading.

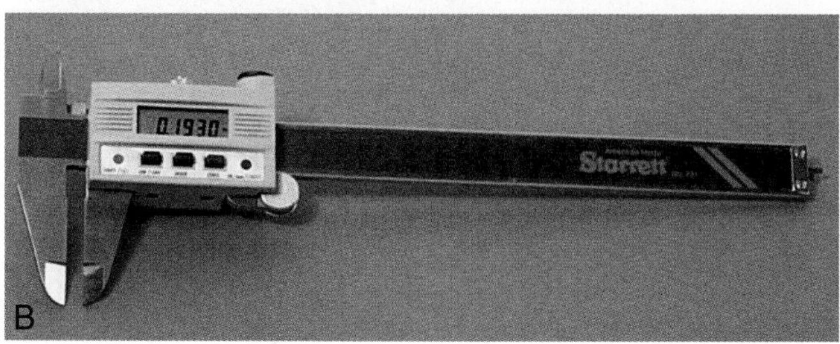

Figure 4-23. A—Dial caliper. This particular instrument will measure objects, holes, etc., up to 6″ (152.4mm) in diameter and depth. B—Digital caliper. This caliper can provide measurements in either standard or metric units. (L. S. Starrett)

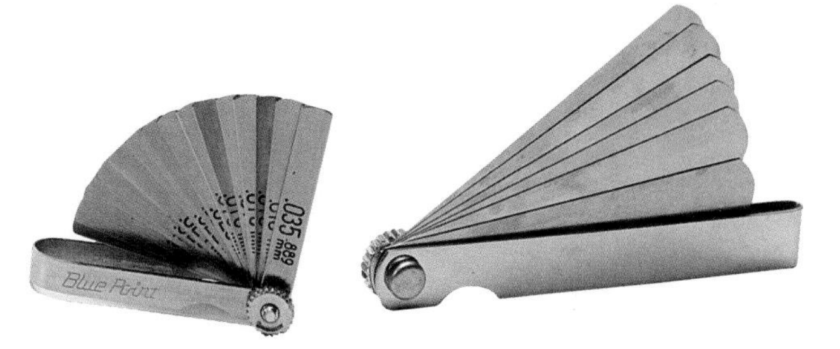

Figure 4-24. Feeler gauges. You will need both steel and brass feeler gauges for automotive work. (Snap-on Tools)

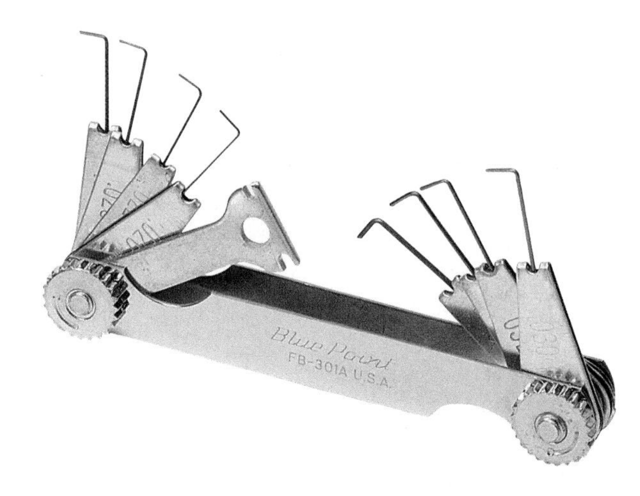

Figure 4-25. This wire gauge set is used to check spark plug gap. (Snap-on Tools)

Figure 4-26. Screw pitch gauge. (L. S. Starrett)

Wire Gauge

The ***wire gauge*** is a thickness gauge using wires of varying diameters instead of thin strips of steel. It is primarily used to check distributor or voltage regulator point gaps on older vehicles or to check spark plug gaps. See **Figure 4-25.**

Screw Pitch Gauge

The ***screw pitch gauge*** is a handy tool for determining the number of threads per inch on bolts, screws, and studs, **Figure 4-26.** It is often used when new threads must be cut. Using a screw pitch gauge will allow the technician to determine which tap or die to use. Taps and dies were discussed in Chapter 3, Basic Tools, Equipment, and Service Information.

Telescoping Gauge

The ***telescoping gauge*** is an accurate tool for measuring inside bores of connecting rods and main bearings. To use this tool, compress the plungers and lock them in place by turning the knurled screw on the gauge's handle. Place the gauge in the bore and release the plungers. When the plungers contact the bore walls, lock the plungers in place and remove the tool. Measure across the plungers with an outside micrometer to accurately determine bore size. Telescoping gauges are available in several ranges and may be purchased in sets. The proper feel for using this tool will be the same as for the inside micrometer, **Figure 4-27.**

Steel Rules

Steel rules are handy when measurements must be made quickly and exact precision is not necessary. There are two major types of steel rules:

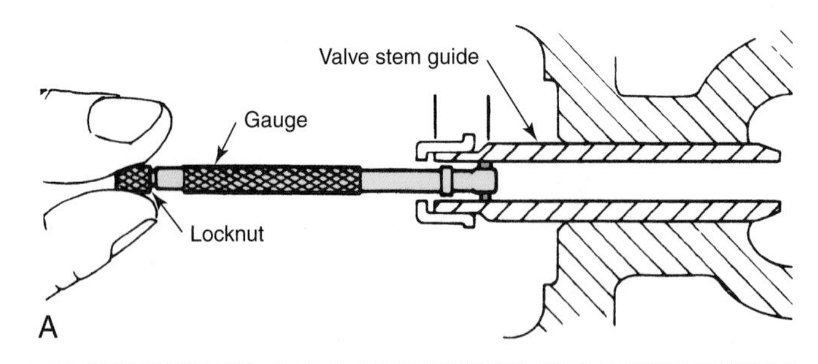

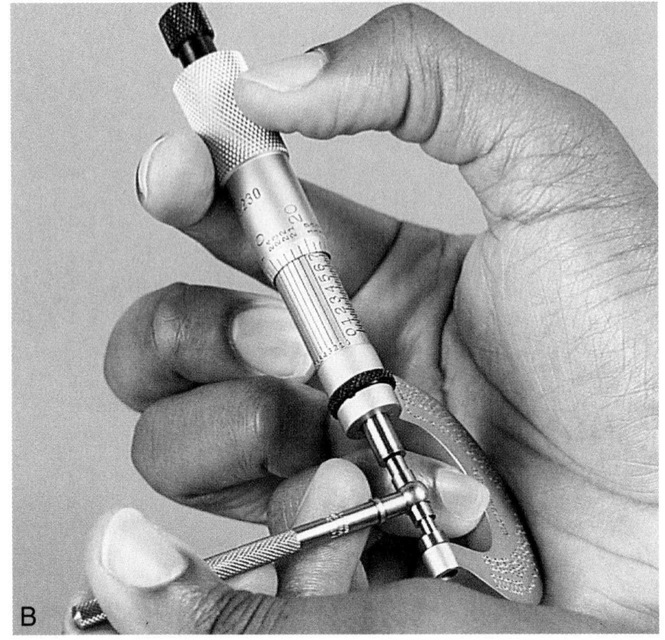

Figure 4-27. A—This technician is using a telescoping gauge to measure the diameter of a valve guide. B—Measuring the telescoping gauge with a micrometer. (DaimlerChrysler, L. S. Starrett)

- A hook rule with a sliding steel head. These are usually marked in thirty-seconds and sixty-fourths of an inch or in millimeters. Some rules are manufactured with inch markings on one side and metric markings on the other. This rule is handy for making quick measurements, such as those taken when comparing old and new parts.
- A steel tape. Most steel tapes can be carried in a pocket or clipped to a belt buckle. Like the hook rule, the steel tape is also marked in thirty-seconds and sixty-fourths of an inch or in millimeters. The tape is handy for making large measurements.

Spring Scale

Spring scales are often used to measure the "pull" on feeler strips when fitting pistons and in any situation where the precise weight or amount of drag must be determined. A typical spring scale is shown in **Figure 4-28.**

Steel Straightedge

An accurate ***steel straightedge*** is frequently used when checking parts for warpage. It should be long enough to span the length of an engine block, head, or flywheel. Be careful when handling and storing a straightedge. It must not be damaged. There are two ways to properly store a straightedge. The best method is to lay it flat in a storage cabinet or other secure place where it cannot be knocked off or be damaged by heavy objects. A straightedge can also be hung vertically on a wall. Be sure to hang it in a location where it cannot be knocked off. Never lean a straightedge against a wall. This places excess weight on one end of the straightedge and can cause it to bend.

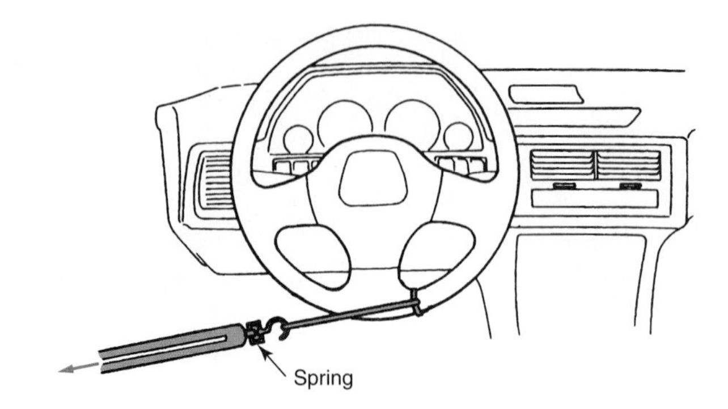

Figure 4-28. A spring scale is a must in every tool kit. (Honda)

Temperature Is Important

Many specifications for measurements will state they apply at room temperature, at an exact temperature, or at engine operating temperature. Remember all metals contract and expand in direct proportion to their temperatures. This makes it imperative that temperature specifications be followed when making precision measurements and settings. Your measuring tools themselves can be affected by extremes of heat and cold. If your tools must be used when very cold or very hot, check them for accuracy before use.

Other Automotive Measuring Equipment

The following section deals with measuring equipment used to perform many diagnostic operations. Additional test and measurement equipment will be covered in later chapters.

Pressure Gauges

When diagnosing many automotive systems, the automotive technician must accurately measure pressures. Therefore, the technician should have several pressure gauges. These gauges measure air pressures or the pressure of automotive liquids, such as engine oil, transmission fluid, power steering fluid, and fuel. The most common pressure gauges are discussed in the following paragraphs.

Compression Gauge

The ***compression gauge,*** **Figure 4-29,** measures the pressure developed in an engine cylinder when the piston moves up on the compression stroke. The compression gauge measures this compression in pounds per square inch, or psi. By using a compression gauge to check the condition of each

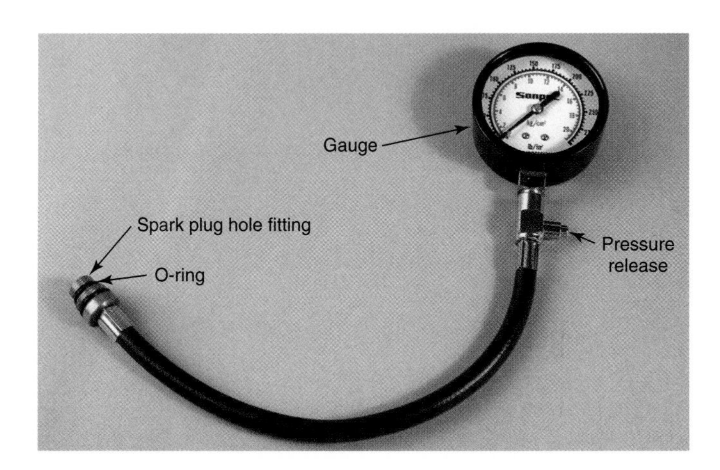

Figure 4-29. The compression gauge is used to measure the pressure developed in the engine cylinders. (Ken Tool)

cylinder, the technician can identify many engine problems. The gauge is connected to the cylinder through a high-pressure hose threaded into the spark plug hole of a gasoline engine (or the fuel injector hole of a diesel engine). When the engine is cranked, the upward movement of the piston in the cylinder compresses the air, and the pressure created by this action is read on the gauge. Use of the compression tester is covered in detail in Chapter 20, Drivability Diagnosis.

Vacuum Gauge

The ***vacuum gauge,*** **Figure 4-30,** measures vacuum, or negative air pressure. Vacuum is the difference between the air pressure in the engine intake manifold and atmospheric pressure. Vacuum is a reliable indicator of engine load and condition. Variations in vacuum readings indicate various engine problems. Vacuum gauges can also be used to test for proper vacuum to and from vacuum-operated components, such as vacuum-operated heater doors.

The vacuum gauge may be used with a vacuum pump to test various vacuum-operated devices. The pump creates a vacuum on the device to be tested, and the gauge measures the vacuum and indicates if the unit can hold the vacuum. The technician may also be able to observe the action of the vacuum device as the vacuum is applied.

Oil Pressure Tester

An ***oil pressure tester*** measures the pressure developed by the engine oil pump or by the pumps in automatic transmissions and power-steering systems. The oil pressure tester is used to determine whether the pump is providing the specified pressure. The pressure tester is attached to the unit to be tested through a high-pressure hose threaded into a pressure port, **Figure 4-31.** The pressure gauge scale is calibrated according to the output of the system being tested. Engine oil pressure seldom exceeds about 80 psi (551 kPa), while automatic transmission pressures can reach 300 psi (2068 kPa) and power steering pressures can reach 2000 psi (13 789 kPa).

Fuel Pressure Tester

The ***fuel pressure tester*** measures the pressure developed by the fuel pump. This tester is used to determine whether the fuel pump is providing enough pressure and flow to keep the engine supplied with fuel. It can also be used to detect clogged fuel filters or other fuel line problems. A typical fuel pressure tester is shown in **Figure 4-32.**

Electrical Testers

The automotive technician must be familiar with several electrical testers in order to troubleshoot electrical systems.

Test Light

A ***test light*** is a simple electrical tester composed of a 12-volt lightbulb, two terminals, and connecting wiring, **Figure 4-33**. Test lights cannot be used to measure exact electrical units; they only indicate whether electricity is present. A test light should not be used to check computers and other solid-state devices. The current flowing through the light can ruin such devices.

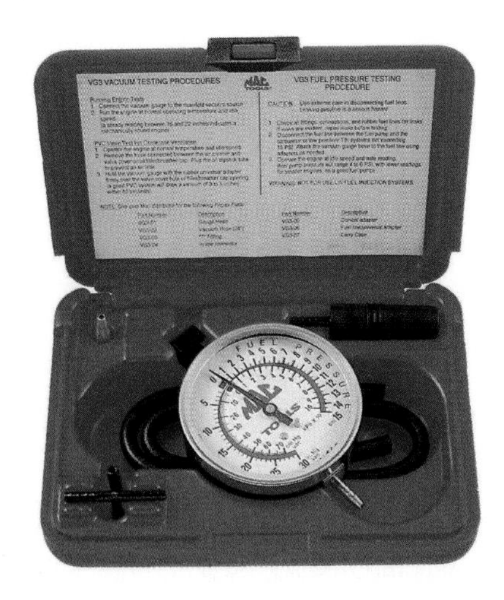

Figure 4-30. A vacuum gauge is used to measure negative air pressure, or vacuum. (Mac Tools)

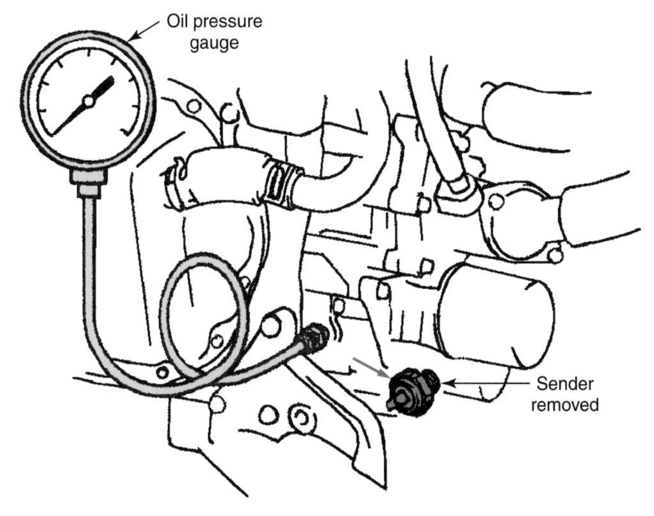

Figure 4-31. Typical setup for using an engine oil pressure gauge. (Toyota)

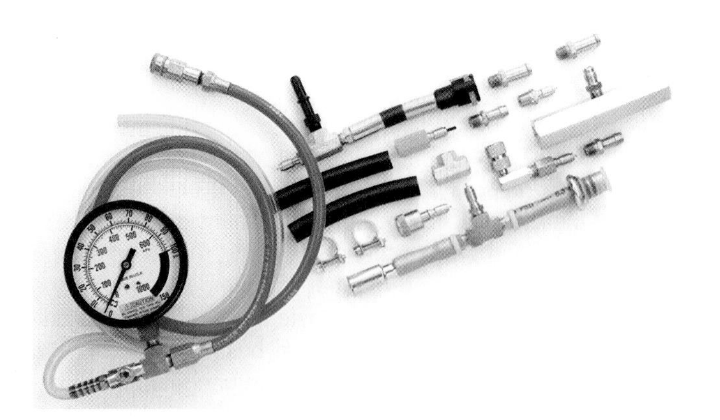

Figure 4-32. Fuel pressure tester is used when diagnosing fuel system problems. (OTC)

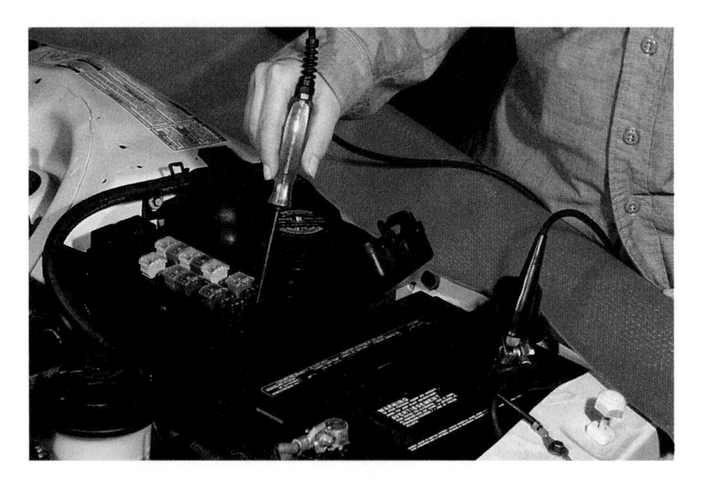

Figure 4-33. Nonpowered test lights are useful for confirming the presence of voltage as well as performing other tests that will be discussed later. (Jack Klasey)

There are two types of test lights. The powered test light contains a battery and is used to check for a complete circuit when no current is flowing in the circuit. The non-powered test light is connected between a powered circuit and ground. It will light up if power is present in the circuit.

Jumper Wire

Jumper wires are short lengths of wire with alligator clips on each end, such as the one shown in **Figure 4-34.** They are used to make temporary electrical connections during vehicle testing procedures. Although they can be purchased, jumper wires can be made easily from a length of wire and alligator clips. Use copper wire that has a large enough gage (wire size) to carry several amps of current, but small enough to be easily handled. 12 or 14 gage wire is a good size for most jumper wires. Cut about 2–3′ (60.96–91.44 cm) of wire from the roll, then strip about .5″ (12.7 mm) of insulation from each end of the wire. Crimp the alligator clips to the wire, making a secure joint at each end. For additional holding power and conductivity, the clips and wires can be lightly soldered together. It is also possible to use special connectors on the end of the wires for special uses.

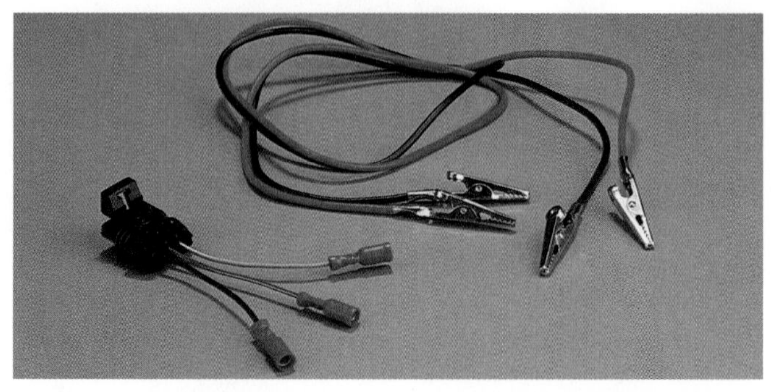

Figure 4-34. Jumper wires are used for bypassing normal circuit pathways to apply voltage or ground.

Note: It is a good idea to install an inline fuse or other circuit protection device in the jumper wire. This helps to protect the circuit wiring and any electronic components should a short circuit be created accidentally.

Another good source for jumper wires is the wiring harness in a wrecked vehicle. These can be found easily and inexpensively in most auto salvage yards. The advantage to having these is they come with the original terminals, which ensures a good connection to the device to be tested. They are handy for testing sensors, fuel injectors, and other parts hard to reach with a meter's test leads. Remove the terminal from the sensor, injector, or part and then cut the terminal from the harness. Strip and crimp a terminal on the wire(s) as described earlier. Additional wire can be added if the connector is used in a hard-to-reach location.

Jumper wires are often used to bypass electrical devices, such as relays and solenoids. Bypassing an electrical device will enable the technician to determine whether it is defective. Jumper wires can also be used to access the computer memory on many vehicles by grounding the diagnostic connector. They are useful for making an extended electrical connection, such as connecting an engine test lead to a test device inside the passenger compartment for a road test.

Caution: Electronic components can be destroyed by careless use of jumper wires or test lights. Do not use a jumper wire or test light to check any electronic devices unless the test is specifically allowed by the manufacturer's service instructions. When testing any nonelectronic device with a jumper wire or test light, be sure it is disconnected from all electronic components.

Multimeter

Modern ***multimeters,*** such as the one in **Figure 4-35,** can read all major electrical values (voltage, resistance, and amperage) and may be able to read voltage waveforms and provide other information. All modern multimeters are digital types that display the reading as a number. Older multimeters used a needle, which moved against a calibrated background. Multimeters usually have at least one positive lead (red) and

Figure 4-35. Typical multimeter used for automotive applications. (OTC)

one negative lead (black). Some multimeters have additional leads for special functions. Polarity should be carefully noted when making some tests with multimeters. The modern multimeter will contain the individual meters discussed in the following paragraphs.

Ohmmeter

Ohmmeters are used to measure electrical resistance. Resistance, which is the opposition to current, exists in every electrical circuit or device. It is measured in ohms. Ohmmeters can only check an electrical circuit when no current is flowing in the circuit. An ohmmeter has two leads, which are connected to each side of the unit or circuit to be tested. Most ohmmeters have selector knobs for checking various ranges of resistance values. Analog ohmmeters have a special knob to adjust the needle to zero before checking resistance.

To make an ohmmeter check, turn on the multimeter and set it to ohms. Most modern digital ohmmeters will select the correct range automatically. Next, attach the leads to the correct wires or terminals as explained in the service manual. Polarity (direction of current flow) is not important when checking resistance, except in the case of diodes. When checking wires or relay contacts for continuity, the resistance should be at or near zero. Other parts, such as solenoid windings, temperature sensors, or throttle position sensors, should have a specific amount of resistance. If the reading is zero or infinity, the part is defective. Resistance of temperature sensors should change with changes in temperature. Resistance of throttle position sensors should change when the throttle is opened.

Voltmeter

Voltmeters are used to check voltage, or electrical potential, between two points in an energized circuit. The circuit to be checked must have a source of electricity available. On some voltmeters, different scales can be selected, depending on the voltage level being measured. Always observe proper polarity when attaching voltmeter leads. The negative lead should always be connected to ground, and the positive lead should be attached to the positive part of the circuit.

The voltmeter can also be connected to read the voltage across a connection as current flows through it. If the connection has high resistance, current will try to flow through the meter, creating a voltage reading. Voltage higher than the specified figure means the connection must be cleaned or replaced.

Ammeter

An ***ammeter*** is used to check the amperage (current) in a circuit. The ammeter is used to check the amperage draw of starters or other motors and to check battery condition. Ammeters can also be used to check the amperage draw of ignition coils, solenoids, and other electrical devices.

The multimeter may also have an amp setting capable of handling amperage up to ten amps. For measuring greater amperage, many modern ammeters are equipped with an inductive pickup. This pickup is clamped over the current carrying wire. The pickup reads the magnetic field created by current in the wire and converts it into an amperage reading.

Note: Many electronic components can be severely damaged by careless use of multimeters. Always check the manufacturer's literature before testing any electronic part.

Tachometer

Tachometers are used to measure engine speed. Modern tachometers are available in either analog or digital versions. Tachometers have at least two leads. One lead is connected to the distributor side of the ignition coil, and the other lead is connected to ground. Some tachometers have a clamp-on pickup that is placed around a plug wire to obtain the engine-speed reading.

The tachometer may have a low range to set idle speeds and a high range for making various tests at high engine speeds. The ranges may be selected with a control knob, or they may switch automatically as engine speed varies. Some tachometers have a provision for checking distributor dwell, or the amount of time primary current is flowing in the ignition coil. This is useful when working on older vehicles with point-type ignition systems.

Scan Tools

The ***scan tool*** is used to obtain trouble codes and other information from the vehicle's on-board computer. Scan tools are small computers that can talk to the vehicle computer. See **Figure 4-36.** The housing has a small screen that displays information. Menus are used to access most of the information. Menus are lists of available test procedures and diagnostic operations. Below the screen is a keypad or a series of pushbuttons that allows the technician to select various menu screens and make tests. The screen displays trouble codes and diagnostic test results as the technician requests them.

The scan tool is attached to the vehicle diagnostic connector and communicates with the vehicle computer or

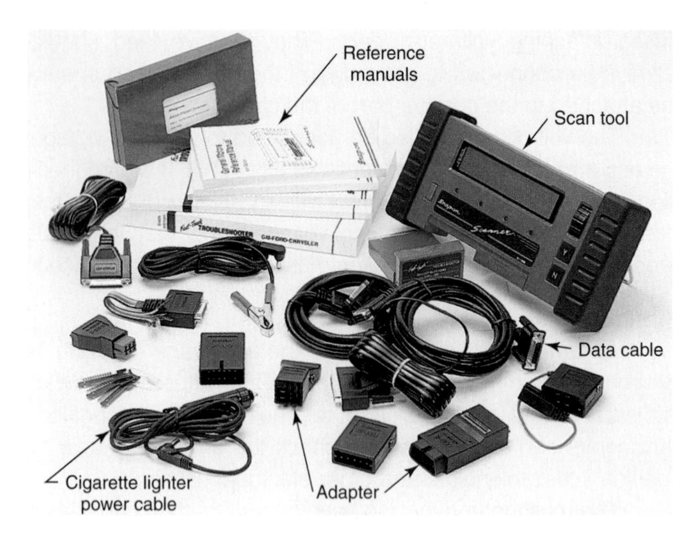

Figure 4-36. Scan tool. This is a generic scan tool, which can be used on more than one manufacturer's line of vehicles. (Snap-on Tools)

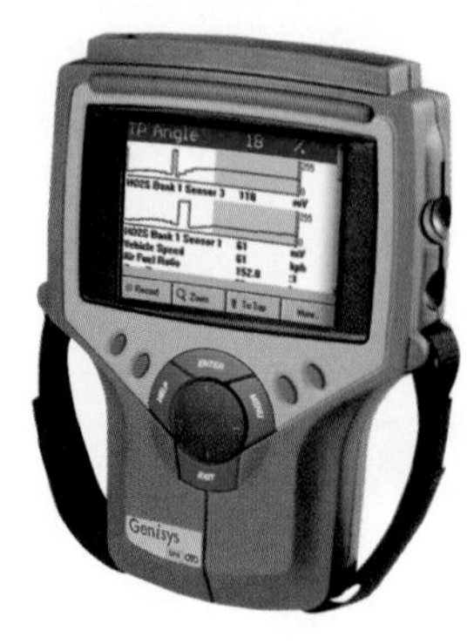

Figure 4-37. Waveform meter. This one has a dual trace ability, which allows more than one waveform to be shown. (OTC)

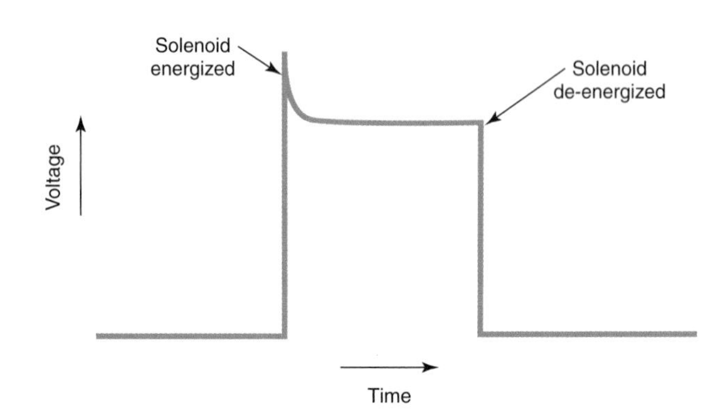

Figure 4-38. This waveform was produced during the normal operation of a transmission solenoid. The voltage is at zero until the solenoid is energized. Voltage rises sharply when the solenoid is energized. It then drops slightly once the solenoid plunger is pulled in. Voltage stays at this level until the solenoid is de-energized.

computers. The simplest scan tools will display trouble codes. A more sophisticated scan tool will display trouble codes and check the operation of the input sensors and output devices. Some scan tools are able to reprogram the vehicle computer with updated information.

There are two main types of scan tools:

- The ***dedicated scan tool,*** which will interface (communicate) with the computer systems of one manufacturer. These tools check all of the computer-operated systems produced by one manufacturer, but are useless for checking the systems produced by other manufacturers.
- The ***generic scan tool,*** which will interface with the computer systems of many manufacturers. This is sometimes called a *multi-system scan tool.* It may be able to diagnose all parts of every system, or only selected portions.

To use any scan tool, make sure the ignition switch is turned off, and then plug the tool connector into the diagnostic connector. Connect the tool power source if necessary. Then follow the instructions on the scan tool screen to retrieve trouble codes or perform other diagnostic routines.

Waveform Meters

The waveform created by the operation of some electrical and electronic devices can be observed to obtain various types of diagnostic information. ***Waveform meters*** can be used to observe electronic voltage and signal waveforms. Waveform meters may be part of a multimeter, or may be separate units. Another device for displaying waveforms is a ***digital storage oscilloscope (DSO),*** which can also store waveform patterns to be compared to known good patterns. A typical waveform meter is shown in **Figure 4-37.**

A typical waveform is shown in **Figure 4-38.** This waveform is produced by a solenoid as it is being energized (turned on) and de-energized (turned off). The advantage of waveforms is they can indicate a problem more accurately than just voltage readings. Note that the solenoid waveform in **Figure 4-39** is not shaped exactly like the standard waveform in Figure 4-38. This could indicate a problem in the solenoid windings or connections.

What Is Your Opinion?

A person has just applied for a job as a technician. The garage has a reputation for excellent work. The owner is interested, there is an opening, and the pay is good. The owner introduces the applicant to you, who, as shop supervisor, will be expected to evaluate this person's worth as a technician.

You walk to a nearby service bench, open your tool chest, and lay out a selection of measuring tools. You indicate a specific cylinder bore you would like miked, and inform the applicant to choose the tools and make the measurement.

The applicant picks up an inside caliper and a six-inch steel rule, adjusts the caliper in the bore, and then places the caliper on the face of the six-inch rule. After some squinting, the applicant informs you the bore diameter is "just a whisker over four inches." The actual bore diameter is 4.030″. What do you think of the applicant's ability? Will you recommend hiring this person? If not, why?

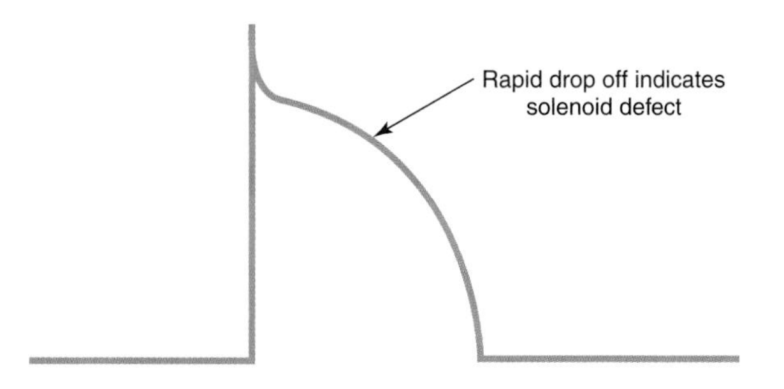

Figure 4-39. Compare this waveform with the waveform shown in Figure 4-42. The voltage rises when the solenoid is energized, but drops off before the solenoid is de-energized. The drop may indicate that the solenoid winding is shorted or that it is drawing excessive current. It may also indicate a problem with the computer solenoid driver. This type of problem would not be caught using a traditional multimeter.

Tech Talk

An office copier is a handy tool. If you are constantly referring to a certain page of a service manual for engine or electrical specifications, you can copy the page and place it where you can see it, such as on the inside cover of your toolbox. This will save time in looking up the specification and will also save wear and tear on the service manual.

Other good subjects for copying are electrical schematics. Instead of tracing out a circuit on the schematic, make a copy and use that to trace the circuit. This saves the original schematic for later use.

Also, when you take a set of measurements, such as compression readings, copy them and give the original to the customer. This allows the customer to see exactly what must be done to the vehicle.

Summary

It is important for you to select and correctly use the measuring tools. This will help you take accurate measurements, which are extremely important when servicing late-model vehicles. Precision measuring tools require careful handling and proper storage.

The technician should be able to use outside and inside micrometers, micrometer depth gauges, dial indicators, inside and outside calipers, dial calipers, dividers, feeler gauges, wire gauges, screw pitch gauges, telescoping gauges, steel rules, straightedges, spring scales, and various pressure and electrical testers. Other specialized measuring tools may be acquired as the need dictates.

Review Questions—Chapter 4

Do not write in this book. Write your answers on a separate sheet of paper.

1. To measure an object 3.500″ (88.90 mm) in diameter, you would use a micrometer with a range of _____.
   (A) 1–2″
   (B) 2–3″
   (C) 3–4″
   (D) 4–5″

Select the correct (some are wrong) readings for the following 0–1″ micrometer settings.

2. _____
3. _____
4. _____
5. _____
6. _____

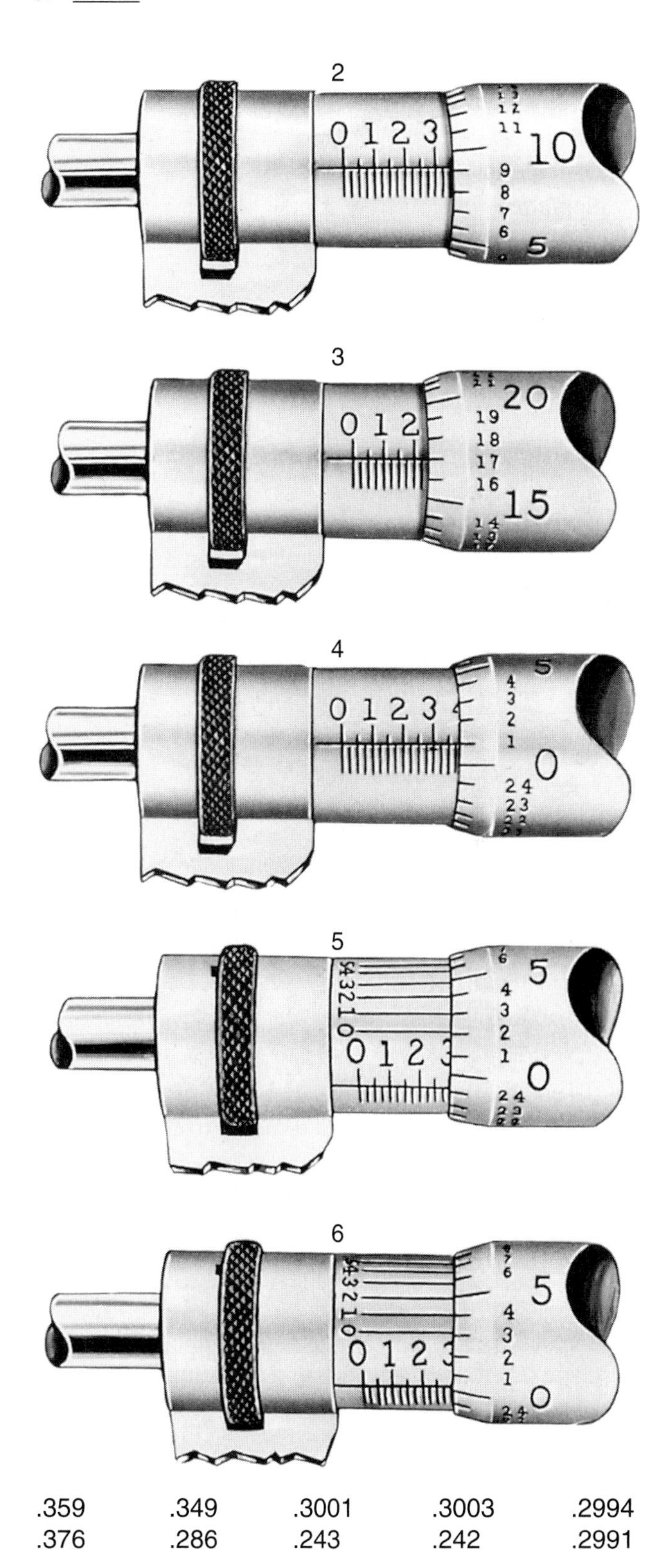

.359	.349	.3001	.3003	.2994
.376	.286	.243	.242	.2991

Name the best tool to handle each of the following measurements:

7. Diameter of a wrist pin.
8. Diameter of a cylinder bore.
9. Distance from face of cylinder head to valve guide top.
10. Endplay in crankshaft.
11. Diameter of wrist pin bore in a piston.
12. Connecting rod big end bore diameter.
13. Lash (free movement or play) between two gears.
14. Teeth per inch on a bolt.
15. Clearance between the valve stem and rocker.
16. Diameter of an exhaust pipe.
17. Spark plug gap.
18. Disc brake rotor runout.
19. Length of a muffler.
20. Distance between the fan blades and radiator.
21. Engine block surface for warpage.

Write the correct decimal readings for the following:

22. _______ Two and three hundred twenty-five thousandths inches.
23. _______ Eight hundred seventy-eight and one-half thousandths inches.
24. _______ Four and six hundred thirteen and one-quarter thousandths inches.
25. _______ Three and one-half inches.
26. _______ One ten-thousandth of an inch.
27. _______ One thousandth of an inch.
28. _______ One hundredth of an inch.
29. _______ One tenth of an inch.
30. _______ One inch.

ASE-Type Questions

1. Each line on the sleeve of a standard micrometer represents _____ inch.
   (A) 0.0025
   (B) 0.025
   (C) 0.25
   (D) 1
2. Each line on the sleeve of a metric micrometer represents _____ mm.
   (A) 0.005
   (B) 0.05
   (C) 0.5
   (D) 5
3. Technician A says a micrometer should be tightly clamped around the work being measured. Technician B says a micrometer should be occasionally checked for accuracy. Who is right?
   (A) A only.
   (B) B only.
   (C) Both A and B.
   (D) Neither A nor B.
4. Technician A says dial calipers are very precise and are capable of obtaining inside, outside, and depth readings. Technician B says digital calipers do not have to be calibrated before each use. Who is right?
   (A) A only.
   (B) B only.
   (C) Both A and B.
   (D) Neither A nor B.
5. Technician A says a compression tester is an air pressure gauge. Technician B says engine compression is negative air pressure. Who is right?
   (A) A only.
   (B) B only.
   (C) Both A and B.
   (D) Neither A nor B.
6. A vacuum gauge will not detect which of the following engine conditions?
   (A) Burned valve.
   (B) Late ignition timing.
   (C) Excessive oil consumption.
   (D) Clogged exhaust.
7. Which of the following engine problems can be detected with an oil pressure gauge?
   (A) Worn rings.
   (B) Worn valves.
   (C) Worn bearings.
   (D) Sticking hydraulic lifter.
8. A fuel pressure tester can measure all of the following fuel pump problems *except:*
   (A) pressure.
   (B) noise.
   (C) vacuum.
   (D) volume.
9. A non-powered test light can be used to check all of the following *except:*
   (A) a circuit for presence of voltage.
   (B) fuse condition.
   (C) an ECU or other solid state component.
   (D) resistor continuity.
10. Technician A says a voltmeter measures resistance to the flow of electricity. Technician B says a multimeter can measure many kinds of electrical properties. Who is right?
    (A) A only.
    (B) B only.
    (C) Both A and B.
    (D) Neither A nor B.

Suggested Activities

1. Obtain several lengths of steel rod, flat stock, and tubing. Measure them with different measuring devices, such as a ruler, steel tape, calipers, vernier caliper, and micrometer. Create a chart showing your readings. Which was the most accurate way of measuring? Why? Display your chart in the classroom or shop.
2. List at least three areas of automotive repair where accurate measurements are absolutely necessary. List some areas of automotive repair where accurate measurements are less critical. Discuss your list with other members of your class.
3. Explain how to read a micrometer to a friend who does not know how to use one. Have your friend try a reading and continue to help until he or she does it correctly. By doing this, you will reinforce your knowledge.
4. Place a wrist pin in the freezer compartment of a refrigerator. When thoroughly cold, remove the pin, wipe it dry, and quickly measure its diameter and length using an outside micrometer. (Hold the wrist pin with a cloth.) Write down your readings. Now place the wrist pin in boiling water. When the pin is hot, remove it (taking care not to get burned), wipe it dry, and quickly recheck its diameter and length. Is there a difference? If so, how much? What does this indicate? Write a short report summarizing your findings.

Lifting equipment, such as this double-post frame lift, makes automotive service and repair operations easier. (Eagle Equipment)

5

Jacks, Lifts, and Holding Fixtures

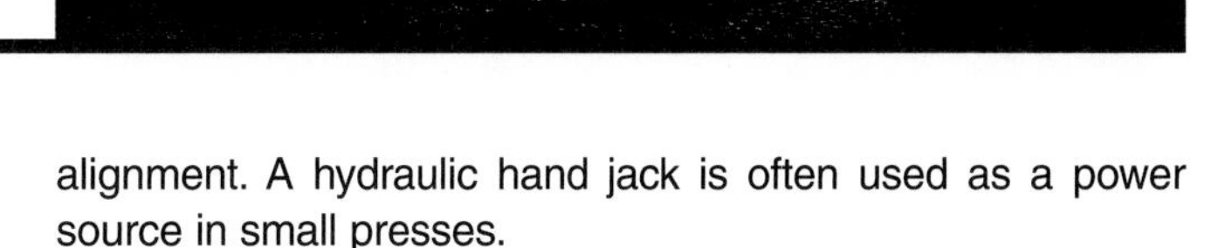

After studying this chapter, you will be able to:

- Identify the most commonly used lifting and holding equipment.
- Select the correct type of lifting or holding equipment for a given job.
- Describe safety precautions for jacks, lifts, and holding fixtures.

Technical Terms

Lifting equipment
Hand jack
Bottle jack
Floor jack
Jack stands
End lift
Bumper lift
Saddles
Single-post frame lift
Double-post frame lift
Double-post suspension lift
Parallelogram lifts
Scissors lifts
Drive-on lift
Power train lift
Transmission jack
Portable crane
Chain hoist
Extension jack
Repair stand

A wide assortment of lifting equipment and holding fixtures is available in most automotive shops. Proper use of this equipment will facilitate repair work. Extreme care must be taken when using lifting equipment and all tools capable of developing high pressures, stresses, and tensions. Never use equipment without familiarizing yourself with its use. There are many safety rules in this chapter. Study them carefully.

Lifting Equipment

Lifting equipment includes many devices operated mechanically, hydraulically (oil pressure), or pneumatically (air pressure). In some cases, lifting devices are powered by a combination of hydraulic and pneumatic pressures. The following section details the common types of lifting equipment used in the automotive shop.

Hand Jack

Jacks are often used to raise vehicles when other lifting equipment is not available. The hydraulic ***hand jack,*** or ***bottle jack,*** has many useful applications. It is short, compact, and capable of producing great pressure. It can be used to raise heavy weights, to bend parts, and to pull or push parts into alignment. A hydraulic hand jack is often used as a power source in small presses.

When using a hand jack, make sure it is positioned so it will not slip as pressure is developed. Be careful not to drop the jack, as it is quite heavy, **Figure 5-1.**

Hydraulic Floor Jack

A ***floor jack*** is used to raise a portion of a vehicle from the ground. It can be used to raise the entire front, back, or side of a vehicle. It is also handy for maneuvering vehicles into tight quarters. The jack is placed under the front or back of the vehicle, and the vehicle is lifted. By pulling the jack in the direction desired, the vehicle can be moved forward, backward, or sideways.

Floor jacks are available in many sizes, with lifting capacities varying from 1–20 tons (900–18,000 kg). **Figure 5-2** illustrates a typical floor jack.

Proper Placement Is Important

When positioning the jack saddle for lifting, make certain it is securely engaged. Select a spot under the vehicle that is strong enough to support the load, such as the frame or the differential housing. Never try raising a vehicle by jacking on

Figure 5-1. Hydraulic hand jacks. (Blackhawk Automotive)

Figure 5-2. Hydraulic floor jack. This particular jack is capable of lifting 4000 lb. (1814 kg). (Lincoln Automotive)

the oil pan, clutch housing, transmission, tie rods, gas tank, or other weak components.

Proper placement requires care. Get down and take a good look at the jack saddle location before raising a vehicle. If the vehicle is part way up and the jack saddle slips, serious damage can occur. On some vehicles, jacking at a corner or near the center of the chassis on one side can cause damage. Vehicle manufacturers illustrate correct lifting points in their manuals. You must follow the manufacturer's specifications carefully. **Figure 5-3** shows a saddle positioned under a vehicle.

Once the vehicle is raised to the desired height, place jack stands in the desired locations and lower the weight onto the stands. The jack may then be removed if desired. If not needed in another area, the jack may be left in position with a very light lifting pressure to keep it in place.

Warning: Never work under a vehicle supported only by a floor jack. If the jack were to fail, the vehicle would crash to the ground with deadly force.

Jack Stands

Jack stands (often called safety stands) are made in numerous heights and are usually adjustable. The jack stand shown in **Figure 5-4** is typical. When inserting jack stands, place them in contact with the vehicle's frame or some other component capable of supporting the load. Do not place the jack stand saddles in contact with tapered edges, which may allow them to slip. Make sure the saddles have a secure "bite." **Figure 5-5** shows a pair of jack stands in place.

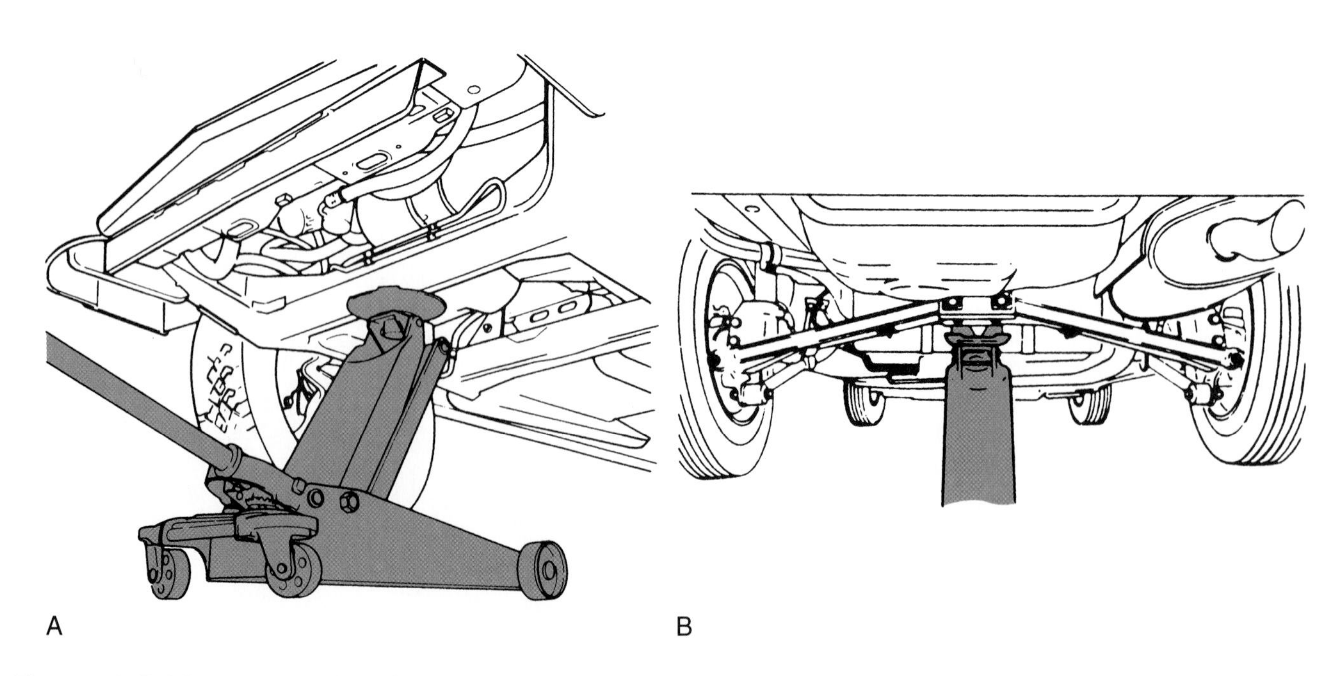

Figure 5-3. Raising a vehicle with a floor jack. A—Front cradle. B—Rear axle. Make certain the saddle is properly positioned. (General Motors)

End Lifts

The ***end lift*** can be operated by pneumatic pressure or hydraulic pressure. Two basic designs are used. One type will reach far enough under the vehicle to contact the rear axle housing. The other design engages the bumper only. End lifts are capable of raising the vehicle much higher than a floor jack.

The ***bumper lift*** shown in **Figure 5-6** is air operated. Note the twin ***saddles.*** The distance between the saddles can be adjusted to engage the bumper where desired. Bumper lifts should be used on larger trucks and vans only, as the bumpers on most modern cars will not support the vehicle's weight. Remember that bumpers are not very strong near their outer ends. If the bumper can be used, place the saddles at the main bumper-to-frame attachment points.

End lifts are generally provided with strong safety locks, so the technician can safely work beneath the vehicle without jack stands. Make sure the safety lock is fully engaged and the lift contact points are solid. If there is even the slightest doubt about the stability of an end lift, use jack stands for additional protection.

Figure 5-4. Typical adjustable jack stands. (Lincoln Automotive)

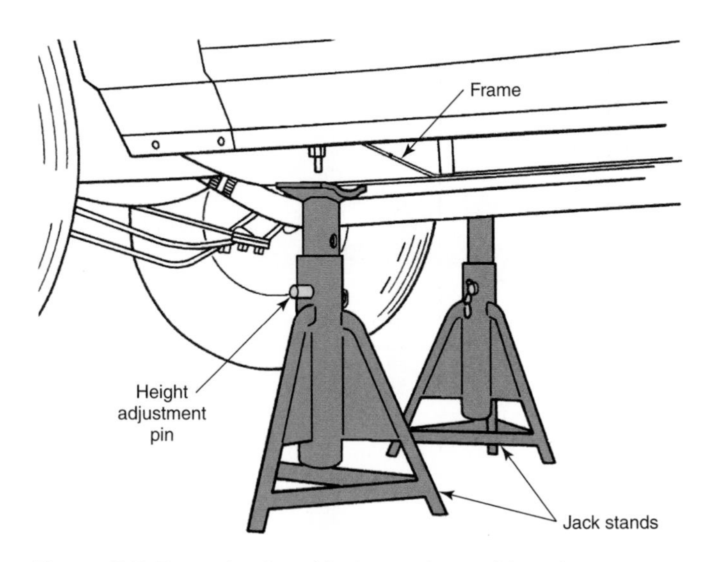

Figure 5-5. Properly placed jack stands provide safe support. (General Motors)

Single-Post Frame Lift

A ***single-post frame lift*** leaves both the front and rear of the vehicle completely exposed. It does, however, create some obstruction in the central portion of the vehicle. **Figure 5-7** shows a vehicle in the raised position on a single-post frame lift. Note the lift contact points on the frame.

Note: Proper lift contact points vary from one type of vehicle to another. Follow the manufacturer's instructions.

Double-Post Frame Lift

The ***double-post frame lift*** eliminates the need for a single, central post. This leaves the center portion of the vehicle more accessible. As with the single-post lift, the vehicle must be carefully centered on a double-post frame lift. After the technician has centered the vehicle, the swivel

Figure 5-6. A typical air-operated bumper lift. Do not use on vehicles with energy-absorbing bumpers. (Lincoln Automotive)

lift arms must be positioned carefully under the vehicle. **Figure 5-8** shows a double-post frame lift.

Double-Post Suspension Lift

The ***double-post suspension lift*** contacts the front suspension arms and either the rear axle housing or the rear wheels. The front lift column can be moved forward or backward to adjust for various wheelbase lengths. This type of lift presents a minimal amount of under-car obstruction. On some models, a single column can be raised. This allows the lift to be used as an end lift, if desired.

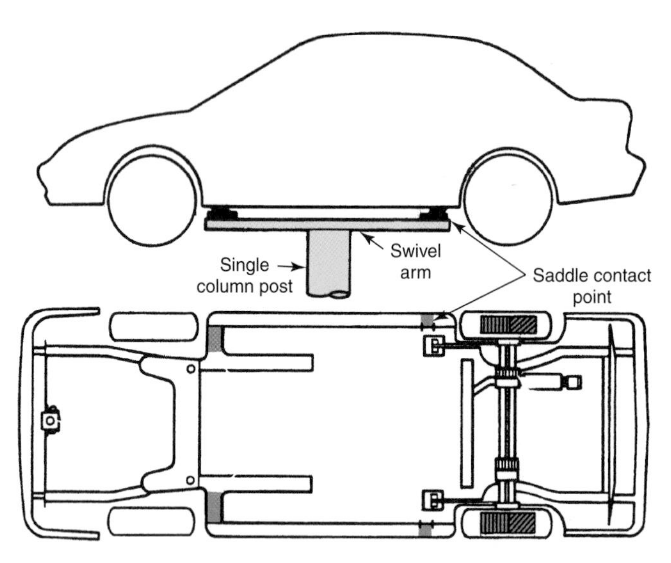

Figure 5-7. A single-post frame lift. The vehicle must be properly centered and the swivel arm pads must contact the vehicle chassis properly. Always double-check the lift contact points before lifting the vehicle. A contact pad not properly located can allow the vehicle to fall. Be careful!

Figure 5-8. A double-post frame lift. This SUV was carefully centered before lifting. Because the lifting posts are on the outside of the vehicle, the undercarriage is completely exposed for service. (Eagle Equipment)

Parallelogram Lifts

Parallelogram lifts, or ***scissors lifts,*** offer the same benefits as double post lifts, but they optimize floor space by eliminating the need for posts and cross beams. This type of lift provides full undercarriage accessibility. See **Figure 5-9.**

Drive-On Lift

The ***drive-on lift,* Figure 5-10,** offers placement speed, but it does have a relatively large obstruction area. Additionally, the wheels cannot be removed from the vehicle without further jacking of the suspension members—a practice that is not recommended.

Choice of Lifts

As you have noticed, each lift offers certain advantages and disadvantages. The type of lift needed will depend on the

Figure 5-9. Parallelogram lifts eliminate the need for posts and cross beams, providing excellent undercarriage accessibility. (Eagle Equipment)

Figure 5-10. Drive-on lifts do not require the technician to position lift saddles. (Eagle Equipment)

work to be performed. Many shops provide several types of lifts, allowing the technician to choose the most appropriate lift for the job at hand.

Safety Considerations

Floor jacks, end lifts, and frame lifts must be used with extreme care. Remember many vehicles can weigh 3000 lb (1360 kg) or more. Each year, a number of technicians are killed or injured by the careless use of lift equipment. When using any type of shop equipment, remember to learn and respect the dangers involved. Consistently follow all safety rules and learn to think before you act. Apply these rules to each and every task. Apply them over and over until they become habits that may someday save you from serious injury or death. In addition to following safe operating procedures, it is imperative that lift equipment be kept in sound operating condition. Cracked or bent parts, faulty safety locks, leaking cylinders, and other problems must be corrected.

The following safety precautions apply to all types of lifting equipment. Study the precautions carefully.

- Position the lift saddles so they securely contact the vehicle's chassis.
- When using a floor jack, always use jack stands.
- Be sure you account for weight distribution when lifting any vehicle. Front-wheel drive vehicles are heavier toward the front axle than rear-wheel drive vehicles.
- Once saddles are located, raise the vehicle until the wheels are just off the ground. Then try to rock the vehicle by applying some pressure to one end. Stop and examine the saddles again before lifting the vehicle. If the vehicle rocks excessively or the saddles are not positioned properly, immediately lower the vehicle and reposition the saddles before proceeding.
- When raising an entire vehicle, watch for side or overhead obstructions.
- Be sure the lift safety lock is securely engaged before going under a vehicle that has been raised on a lift.
- Never remove a lift or jack from another technician's setup without asking permission first.
- If it is necessary to change the raised height of a vehicle, do not raise or lower the vehicle until all persons are out from under it.
- Always check for equipment, parts, and personnel beneath a vehicle before lowering it.
- Lower a vehicle slowly and watch it closely during the entire descent.

Specialty Lifting Equipment

The following section covers some special lifting and holding equipment with which the technician should be familiar.

Power Train Lift

A hydraulic ***power train lift*** is shown in **Figure 5-11.** The lift can be used to remove or install a complete engine/transaxle assembly from under the vehicle. Note the use of safety straps.

Figure 5-11. Power train lift. (Weaver)

Transmission Jack

A ***transmission jack*** is used to remove and install transmissions, transaxles, and transfer cases. The jack's saddle is securely attached to the transmission or transaxle with a series of adapters and a binder chain or strap. The saddle can be raised and lowered hydraulically, and it can be tipped in any direction through the use of adjusting screws. **Figure 5-12** shows a typical transmission jack. **Figure 5-13** illustrates a high-reach jack with the transmission in place. When using a transmission jack, be sure to attach the transmission or transaxle securely. It is heavy, and if it slips, it could cause serious injury.

Portable Crane and Chain Hoist

The ***portable crane*** (often referred to as a *cherry picker*) and the ***chain hoist*** are excellent tools for engine removal. They can also be used to lift heavy parts to bench tops and truck beds. **Figure 5-14** shows a heavy-duty portable crane.

Always observe the following safety rules when using a crane or chain hoist:

- Stand clear of the crane or hoist at all times.
- Lower the engine as soon as it is clear of the vehicle.
- Never roll the crane with the load high in the air. Keep it just above the floor.

- Never leave an engine suspended while working on it. Lower it to the floor or place it on a suitable engine stand.
- Never leave the crane or hoist with the load suspended. If you must leave, even temporarily, lower the load.

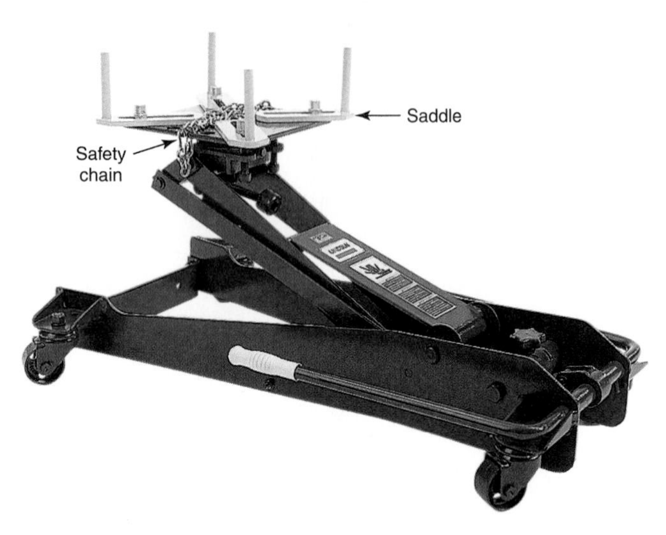

Figure 5-12. Typical transmission jack, which is constructed like a floor jack. (Weaver)

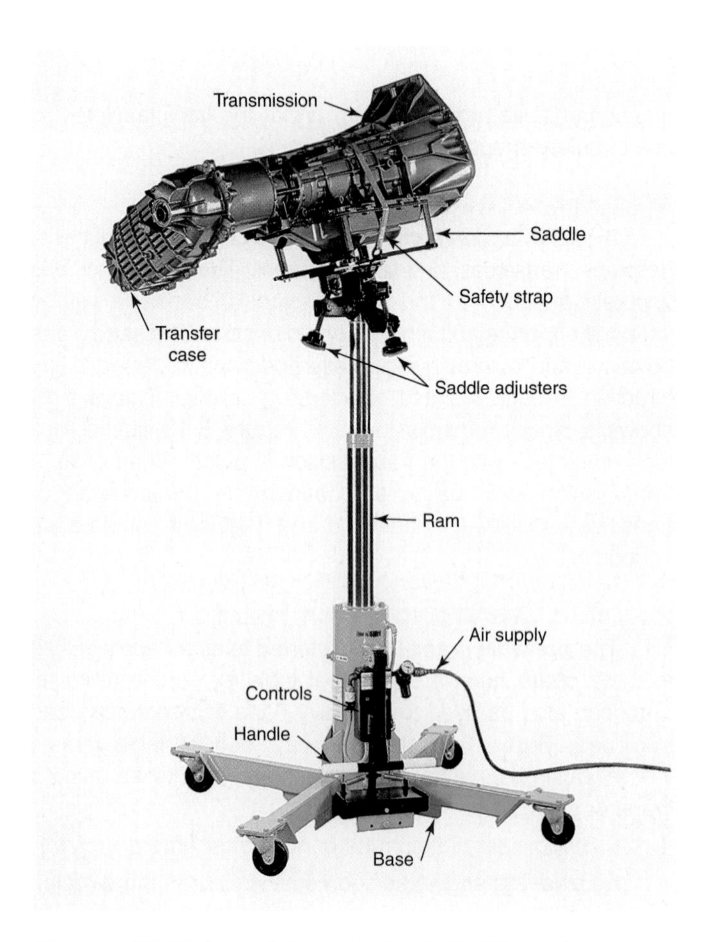

Figure 5-13. High-reach transmission jack. (Weaver)

- When moving heavy loads, alert your fellow technicians.
- When using a chain hoist attached to an overhead track, never give the load a hard shove and let it coast along the track. Move the load slowly and stay with it.
- Cables, chains, bolts, and other lifting devices must have ample strength.
- When using nuts to attach lift cables, each nut must be threaded fully on its fastener. When using capscrews, they must have a thread-engagement depth one and one-half times the diameter of the capscrew.
- If the crane or chain hoist fails, allow the engine, transmission, or other part to fall.

More information on the use of this equipment for engine work will be given in the section on engine removal and installation.

Extension Jack

An ***extension jack*** is a valuable tool for exerting mild pressure and for holding parts in place. Some extension jacks are adjusted up and down manually. Others are hydraulically operated. Two types of extension jacks are shown in **Figure 5-15.**

Repair Stands

Engine block, cylinder head, transmission, and differential repairs can be greatly facilitated by using a ***repair stand.*** Many types are available. When using repair stands, attach the unit being serviced securely to the stand. Carelessness here can be costly.

Figure 5-16 shows an engine mounted in a stand. Note that a crank is used to move the block to various positions.

Figure 5-14. Portable crane can be used to pull an engine or other heavy lifting. (Lincoln Automotive)

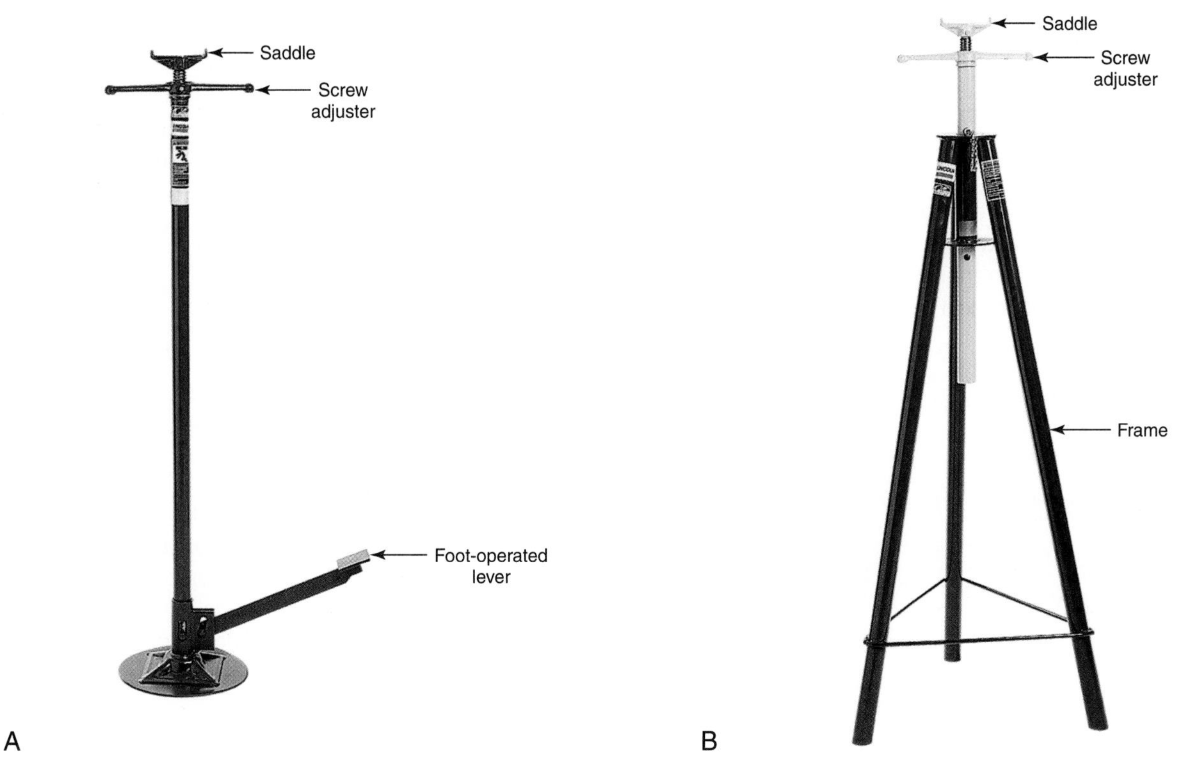

Figure 5-15. Typical extension jacks. A—This jack is raised hydraulically using a foot-operated pump. B—This extension jack is similar to a jack stand. It is adjusted manually and held in place with a locking pin. Note the screw adjusters on both jacks. These are used to make final adjustments. (Lincoln Automotive)

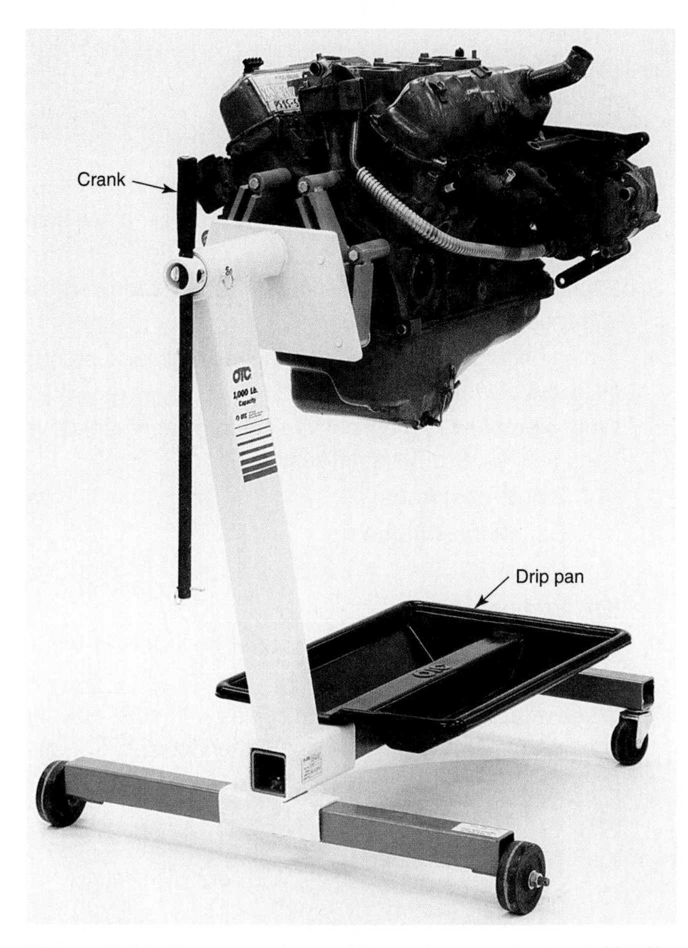

Figure 5-16. One type of movable engine repair stand. (OTC)

This stand has a drip pan, which can catch oil, coolant, and dropped fasteners. Not all engine stands have drip pans. A transmission repair stand is shown in **Figure 5-17.** As with all stands, tighten the holding screws securely.

Hand Lifting

Occasionally, a technician will want to lift an object by hand. There are several important points to remember in order to avoid injury.

- Do not "show off" by attempting to lift heavy objects. If necessary, ask for help or use a lift.
- Keep your back straight and lift with your legs. Keep your legs as close together as possible, **Figure 5-18.**
- Unless you know you can handle the weight, never hold a part with one hand while removing the last fastener with your other hand.
- Get a firm grip to prevent dropping objects.

Tech Talk

It is often necessary to operate a vehicle on a lift when checking for oil leaks or problems with the exhaust system or drivetrain. When operating a vehicle in the shop, it is very important to provide proper ventilation to prevent carbon monoxide poisoning. However, many people close the shop doors on cold days and do not realize they are being poisoned. Symptoms of carbon monoxide poisoning include:

- Headaches or throbbing head.
- Roaring in the ears.

- Nausea.
- Rapid heartbeat.
- Impaired vision.
- Drowsiness.
- Mental confusion.

If you develop any of these symptoms, get out into the fresh air *immediately.* Do not waste time lowering the lift to turn off the engine. Once outside, arrange to open all doors and windows and turn on ventilators until the exhaust gases are removed.

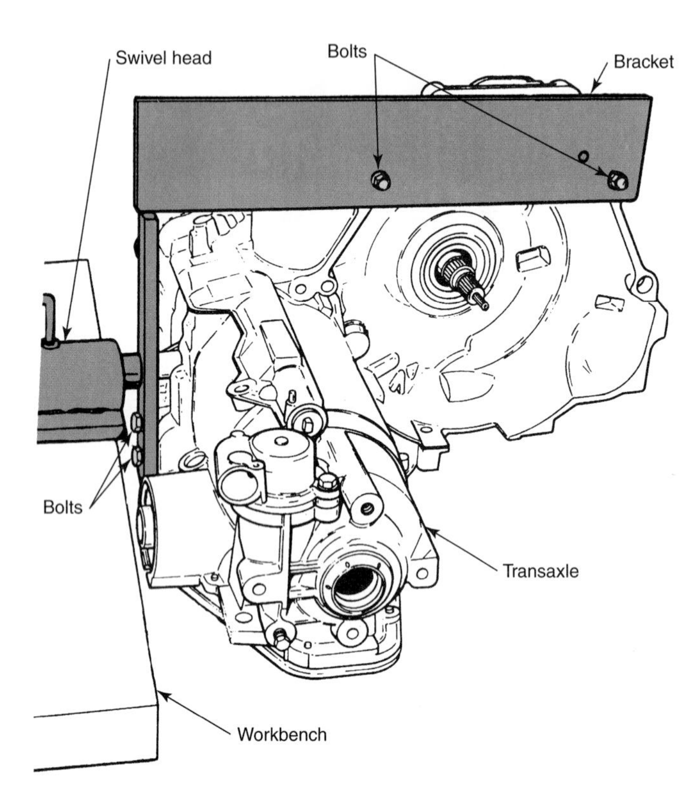

Figure 5-17. This transaxle is mounted in a repair stand.

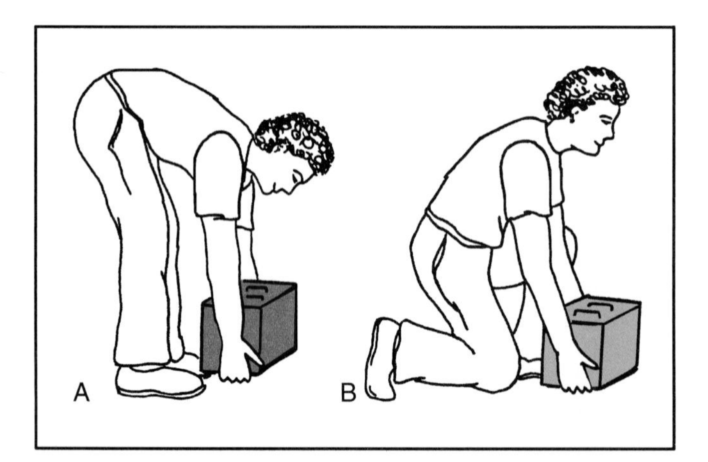

Figure 5-18. When lifting heavy objects, keep your back straight and use your legs to lift the weight. A—Incorrect lifting procedure. B—Correct lifting procedure.

Summary

Technicians should be familiar with various lifting and holding tools available to make their work easier and more efficient. The tools covered in this chapter must be used with extreme caution. Observe all recommended safety precautions.

Hand jacks have many applications. Floor jacks are very handy for raising and positioning vehicles. Never get under a vehicle supported only by a floor jack. Use jack stands to support the vehicle. Be careful not to damage parts when lifting. End lifts have a fairly high reach and support the vehicle safely. Make sure the safety lock is in position. Single- and double-post lifts can be designed to engage either the frame, the suspension system, or the tires. All have advantages and disadvantages.

Vehicles must be centered on the lift, and the lifting brackets should be properly and securely placed. Use care when determining lift points to prevent distortion or part damage.

Transmission jacks, and portable cranes facilitate the removal and installation of heavy parts. Repair stands for engines, transmissions, and other assemblies make repairs faster, safer, and easier. Always fasten the unit securely to the stand.

Review Questions—Chapter 5

Do not write in this book. Write your answers on a separate sheet of paper.

1. Which of these locations is acceptable for placing a jack saddle?
   (A) Oil pan.
   (B) Transmission housing.
   (C) Differential housing.
   (D) Fuel tank.
2. A technician should never work under a vehicle supported only by a _____ _____.
3. Explain why bumper lifts should not be used on most late-model cars.
4. An end lift does not require the use of jack stands because of what feature?
5. What type of lift is most convenient for rear-wheel drive transmission or drive shaft removal?
   (A) Single-post frame lift.
   (B) Double-post frame lift.
   (C) End lift.
   (D) Bumper lift.
6. List seven safety rules that should be followed when using jacks and lifts.
7. When attaching lift chains and cables with nuts, how far should the nuts be threaded onto the studs?
   (A) One turn.
   (B) Two turns.
   (C) Three turns.
   (D) All the way.
8. If a crane or chain hoist fails, allow the engine, transmission, or other part to _____.

9. Working on engines, transmissions, and cylinder heads is made much easier by the use of _____ _____.
10. List three safety precautions that should be followed when hand lifting.

ASE-Type Questions

1. When placing jack stands under a vehicle, they should contact _____.
   (A) the vehicle's frame
   (B) a suspension part capable of supporting the load
   (C) tapered or slanted edges
   (D) Either A or B.
2. Technician A says it is safe to work under a vehicle supported on a good, well-placed floor jack. Technician B says end lifts, if properly designed, provide sufficient holding power to allow the technician to work beneath the vehicle without jack stands. Who is right?
   (A) A only.
   (B) B only.
   (C) Both A and B.
   (D) Neither A nor B.
3. When using a single- or double-post frame lift or a drive-on lift, all the following safety precautions should be followed, *except:*
   (A) place jack stands under the frame.
   (B) watch for any side or overhead obstructions when raising the vehicle.
   (C) make certain the lift safety lock is engaged before getting under the vehicle.
   (D) raise and lower the lift slowly.
4. Each of the following lifts makes it easy to remove wheels, *except:*
   (A) double-post frame lift.
   (B) end lift.
   (C) drive-on lift.
   (D) bumper lift.
5. Which type of lift does not have to be checked for proper positioning before lifting a vehicle?
   (A) Single-post frame lift.
   (B) Double-post frame lift.
   (C) Drive-on lift.
   (D) Bumper lift.
6. A vehicle that has just been raised is observed to be rocking on the lift. Technician A says the lift saddles can be hammered in their proper positions without lowering the vehicle. Technician B says the vehicle should be lowered immediately and the lift saddles repositioned before proceeding. Who is right?
   (A) A only.
   (B) B only.
   (C) Both A and B.
   (D) Neither A nor B.
7. Technician A says lift height should not be varied if anyone is under the vehicle. Technician B says jack contact points are not important as long as the jack gets a good grip. Who is right?
   (A) A only.
   (B) B only.
   (C) Both A and B.
   (D) Neither A nor B.
8. Technician A says when moving an object with a portable crane or chain hoist, the load should be kept as low as possible. Technician B says it is a good safety practice to work on an engine suspended from a crane or a chain hoist. Who is right?
   (A) A only.
   (B) B only.
   (C) Both A and B.
   (D) Neither A nor B.
9. A portable crane is used to remove an engine from a vehicle. Technician A says the engine should immediately be lowered to just above the floor. Technician B says the engine should immediately be wheeled to a repair stand. Who is right?
   (A) A only.
   (B) B only.
   (C) Both A and B.
   (D) Neither A nor B.
10. Technician A says that most extension jacks are pneumatically operated. Technician B says that when removing a transmission or transaxle, the unit should be secured to the transmission jack with a chain or strap. Who is right?
   (A) A only.
   (B) B only.
   (C) Both A and B.
   (D) Neither A nor B.

Suggested Activities

1. Send letters requesting up-to-date catalogs to several lifting equipment manufacturers. Study the catalogs carefully. Although your shop may not have all the equipment shown in the catalogs, you should be familiar with the various types of equipment available.
2. Draw a schematic of the hydraulic lift used in your shop, showing how hydraulic pressure is developed and controlled.
3. Make a list of lift safety rules. Make copies of the list and post one copy at each of the shop lifts.
4. List the series of steps that should be performed to correctly raise a vehicle with a floor jack and secure it with safety stands. Discuss the steps with other members of your class.

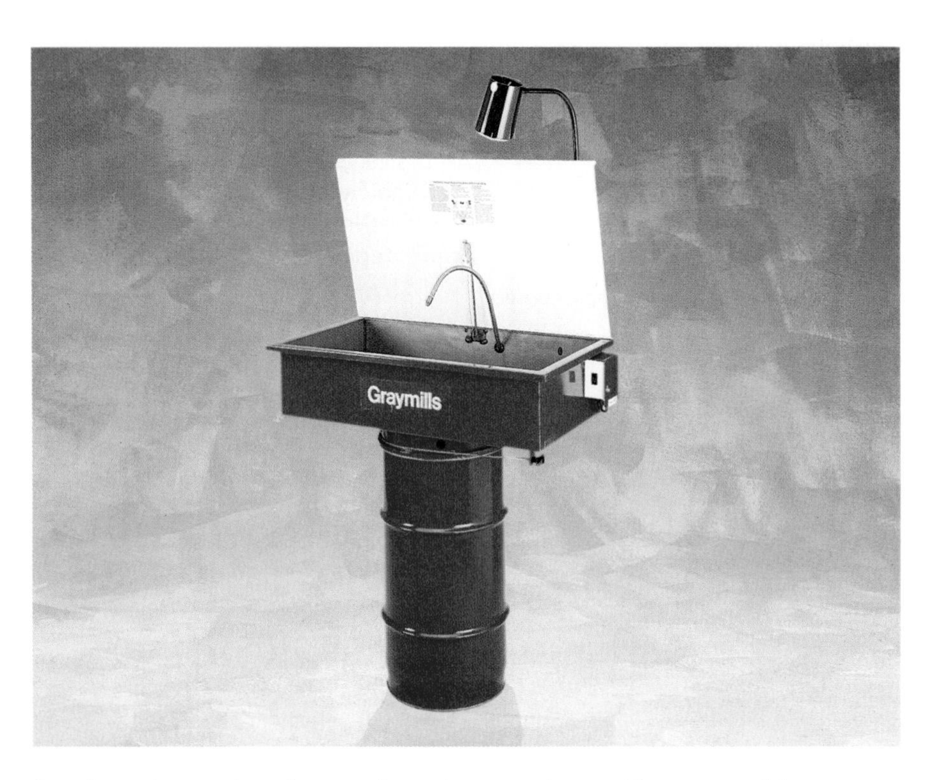

A parts washer, such as the one shown here, can be used to remove, dirt, grease, and oil from automotive components. (Graymills)

6

Cleaning Equipment and Techniques

After studying this chapter, you will be able to:
- List the most common automotive cleaning techniques.
- Compare the advantages and disadvantages of different cleaning methods.
- Select the correct cleaning method for a given job.
- Describe the safety rules that apply to various cleaning techniques.

Technical Terms

Abrasive discs
Wire brushes
Bead blast cleaner
Sandblasted
Toxic
Caustic
Flash point
Steam cleaner
High-pressure spray cleaning equipment
Low-pressure spray cleaning
Parts washer
Hot tank
Alkaline compound
Cold soak cleaning
Vapor cleaner
Brake washer

Cleaning parts is a slow, tedious task that can often account for almost half the time spent on a job. Using improper equipment and techniques will make cleaning even more time-consuming. In order to minimize repair charges, the technician must be familiar with the cleaning equipment available, and how it is used.

On an in-vehicle engine cleaning or an undercarriage cleaning job, missing a few spots may displease the customer, but it will not damage the vehicle. On the other hand, careless cleaning of parts during engine, transmission, or transaxle teardowns may ruin the job, causing expensive comebacks and angry customers. The only safe course is to be meticulous in your cleaning. Remove *all* foreign materials from parts and protect the parts against contamination during storage and handling.

Types of Cleaning Equipment

Cleaning equipment and techniques vary with the size and type of job, **Figure 6-1.** You are obviously not going to use a steam cleaner to clean one universal joint when you could handle the task quickly and efficiently with solvent, a brush, and an air hose. On the other hand, attempting to clean the outside of an engine before disassembly with a brush and solvent would be a foolish waste of time. You must tailor the equipment and technique to the job at hand. This chapter will deal with the most widely used cleaning methods. Study them carefully, so that you will be able to choose wisely.

Get Advice

There are various types of cleaning equipment and solutions designed to perform specific tasks, such as vehicle body washing, in-vehicle engine cleaning, carburetor cleaning, block cleaning, hard carbon removal, etc. There are hot solutions, cold solutions, high- and low-pressure sprays, and

Figure 6-1. A small parts washer such as this one is ideal for cleaning single parts or small components. (Snap-on Tools)

agitators. So many products are available that it can be very confusing. When choosing a cleaning solution or a piece of equipment, it is wise to ask other technicians for their opinions. Also, discuss the problem with sales representatives from reliable companies offering products in this field.

Cleaning Precautions

In today's vehicles, a wider variety of materials is being used to achieve greater operating and fuel efficiency through weight savings. Some of these materials, especially aluminum and different plastic formulations, can be damaged if improper cleaning methods are used.

Aluminum Parts

Many engine, transmission, and other vehicle parts are made from aluminum. Aluminum parts should not be cleaned in hot tanks. Special emulsion cleaners must be used to clean aluminum parts. Also, wire brushes should not be used to clean aluminum engine parts. The wire brush will distort the part's surface, possibly resulting in leaks.

Magnesium Parts

Never clean magnesium parts with any solvent that could remove the plastic coating from the material. Also, do not steam clean magnesium parts, as boiling water can dissolve magnesium.

Plastic Engine Parts

Plastic engine parts are used in most modern vehicles. Timing belt covers, dust and appearance covers, and brackets have been used for many years to save weight. In addition, many modern engines use plastic intake manifolds, water pump housings, and valve covers.

Cleaning plastic engine parts is similar to cleaning metal parts. Plastic parts, however should not be steam-cleaned or soaked in hot tanks unless the manufacturer specifically allows it. Very harsh acid or caustic cleaners should not be used on plastic parts. Never clean plastic parts using a wire brush or abrasive disc.

Types of Cleaning Systems

While there are a number of types of cleaning systems, they can generally be divided into two groups: mechanical and liquid. The mechanical systems include the use of abrasives such as wire brushes, discs, glass beads, or sand. Liquid systems use special solvents, water (hot or cold), or steam. The liquid systems may be further divided into those that are used to clean large assemblies (such as an engine) in place in the vehicle and those that employ a tank for individual parts and disassembled components.

Mechanical Cleaning Systems

Mechanical cleaning systems include hand-held cleaning tools and blast medium cleaning equipment. Various mechanical cleaning systems are discussed in the following sections.

Hand-Held Cleaning

Engine parts often have gasket or gasket residue that remains after part removal. Combustion chambers, piston heads, and piston grooves are subject to accumulations of hard carbon. Even after these components are soaked in powerful cleaning solutions, they may need to be cleaned with scrapers or brushes, **Figure 6-2.**

In many cases, an abrasive disc or wire brush in a hand-held electric- or air-powered drill can be used for final cleaning. Disposable ***abrasive discs,*** **Figure 6-3,** are attached to a support pad that is held by the drill chuck. Different degrees of abrasive coarseness (*grits*) are available.

Wire brushes can quickly remove gasket residue, carbon, and other debris from a part surface. However, both abrasive discs and wire brushes must be used carefully, since

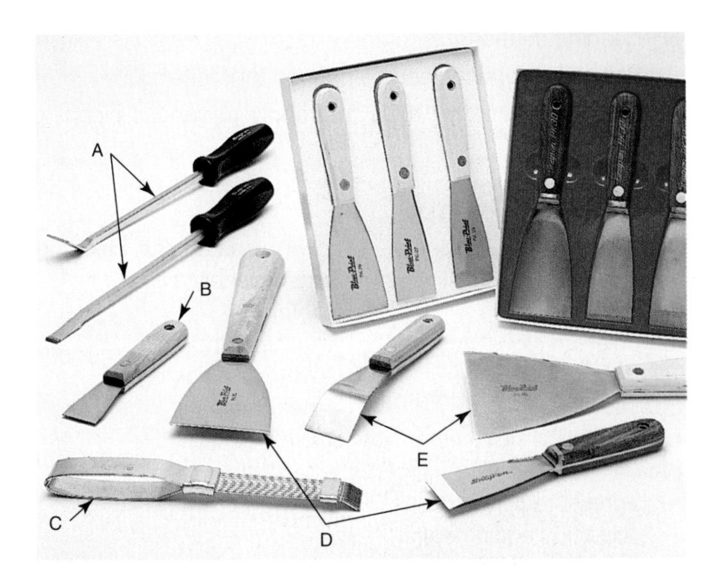

Figure 6-2. Scrapers are used to remove carbon deposits, gasket residue, and other debris. A—Rigid carbon scrapers. B—Gasket scraper. C—Flexible carbon scraper. D—General-purpose scrapers. E—Putty knives. (Snap-on Tools)

Figure 6-3. Abrasive discs are attached to a flexible backing that is chucked into an electric drill or air-operated tool.

they can also remove metal from the part. **Figure 6-4** shows carbon deposits in a cylinder head combustion chamber being removed with a rotary wire brush that is chucked into an air-powered grinder. Wire brushes are often used in bench grinders, as well.

Caution: Never use a power-driven wire brush on soft articles, such as pistons, carburetors, or bearing inserts.

Blast Medium Cleaning

Because fewer shops are reconditioning engine and other parts, blast medium cleaning systems using glass beads, walnut shells, or sand are no longer common. However, they are used in some shops.

The ***bead blast cleaner,*** or *bead blaster,* consists of a cabinet with an internal blaster, **Figure 6-5.**The blaster uses fine glass beads, which clean the part with minimal removal of the original material. Newer bead blast cleaners use crushed walnut shells or another biodegradable medium. Some units have a pair of rubber gloves that extends into the cabinet. The gloves allow the technician to handle the parts during the blasting procedure. Bead blast cleaners are often used to remove rust from small parts and to remove the glaze from automatic transmission clutch plates and drums.

Automotive parts are rarely ***sandblasted.*** The body shop and welding shop occasionally will use a sandblaster to quickly remove paint, rust, and welding scale. A special blast gun, operating under air pressure of around 50 psi–200 psi (345–379 kPa) propels a metered amount of abrasive material with great force against the object to be cleaned.

Always wear a face shield when sandblasting. In addition, a respirator should be worn in situations that are prolonged or produce excessive dust. Never sandblast near an area where repairs are made. The abrasive will contaminate parts, with disastrous results.

Liquid Cleaning Systems

Several types of liquid cleaning systems are used in the automotive shop. Several of these—steam cleaning, high-pressure spray cleaning, and low-pressure spray cleaning—are used for in-vehicle cleaning of engines and similar applications. Other systems, using tanks with either hot or cold solutions, are suitable for cleaning parts individually or in batches.

Figure 6-4. A rotary wire brush in an electric drill is being used to clean deposits from this cylinder head. (Fel-Pro, Inc.)

Figure 6-5. This bead blaster was used to clean the case of an automatic transmission. (Glassinger)

Cleaning Solutions Can Be Dangerous

Because of environmental concerns, liquid cleaning systems have changed in recent years. However, many cleaning solutions are still ***toxic*** (poisonous) and ***caustic*** (will burn skin, eyes). Make sure you know what you are using and follow the manufacturer's recommended handling procedures. General safety rules concerning cleaning solutions include:

- Use cleaning solutions in a well-ventilated area.
- Never use gasoline for cleaning.
- Wear goggles or a face shield when working with cleaning solutions.
- Keep cleaning solutions away from sparks and open flames.
- Do not smoke around solutions.
- Keep solutions covered when not in use. Keep all cleaning solutions in labeled containers.
- Use solutions with a relatively high ***flash point*** (the temperature at which vapors will ignite when brought into contact with an open flame).
- Never heat solutions unless specifically recommended by the manufacturer.
- Do not allow clothing to become soaked with solvent.
- Always read and follow the cleaning solution manufacturer's instructions.
- When brushing parts in solvent, use a nylon or brass brush to avoid sparks.
- A large tank of solvent should have a lid that is held open by a fusible link (a holding device that will melt in the event of fire, allowing the lid to drop).
- Wash hands and arms thoroughly when a cleaning job is complete.
- Avoid prolonged skin exposure to all types of solvents.

STOP **Warning: Whenever doing under-hood cleaning, be sure to remove the battery ground cable. This prevents a possible short circuit caused by grounding a hot wire or terminal with the cleaning gun.**

Steam Cleaning

The ***steam cleaner*** is an excellent tool for many cleaning tasks. Under-car, engine, and transmission cleaning jobs are handled quickly and thoroughly with a steam cleaner. In operation, a water pump forces a solution of water and a metered amount of cleaning solution through a pipe that is formed into a number of coils. A heat source (oil, gas, or electricity) passes heat quickly through the coils, generating steam. From the coils, the steam passes through a flexible hose to a steam gun. The gun has an insulated handle and an adjustable nozzle. **Figure 6-6** shows a typical portable steam cleaner. Some units feed the cleaning solution into the gun instead of into the water supply.

There are a number of steam cleaners on the market. As always, the manufacturer's instructions should be followed regarding specific operating and maintenance procedures. The following operational steps are common to almost all steam cleaners.

Figure 6-6. Portable steam cleaner. Units such as this one provide both steam and hot water under high pressure for efficient cleaning. (Aaladin Industries)

Starting the Cleaner

If used indoors, an oil- or gas-operated steam cleaner must have adequate ventilation. Electric units should be properly grounded. Turn on the water source and then activate the water pump. In a short time, you will notice a stream of water flowing from the gun. This indicates that the heating coils are filled with water and that the burner can be ignited without burning the coils.

Next, ignite the burner. When the gun begins to emit steam, adjust the fuel valve to bring the pressure to the desired limit. If the machine utilizes an integral solution tank, check to see if enough solution is present. Mix the solution by opening the stirring valve for about 30 seconds. If no stirring provision is present, place the gun nozzle into the solution and agitate it with steam pressure. Once the solution is ready, open the solution valve.

Using the Steam Cleaner

The cleaning solution can damage the paint. Cover fenders and windshield area to protect them when doing an engine or other under-hood cleaning job. When finished, flush all painted surfaces with clean water. Also cover all electronic devices, the carburetor or fuel injectors, alternator, distributor, master cylinder, power steering pump, and air conditioning compressor. Avoid prolonged steaming of wiring. Keep spray away from air conditioning lines. **Figure 6-7** shows an operator steam cleaning an engine compartment.

Depending on the nozzle design, the type of dirt to be removed, and the shape of the object being cleaned, hold the gun nozzle from 1″–4″ (25 mm–102 mm) from the surface. If the nozzle is too far from the work, cleaning will be slowed considerably. The steam should be "wet" (include ample hot water along with steam). "Dry" steam will not clean or flush surfaces well.

Do not apply too much steam to the tie rod ends, suspension knuckles, and other under-vehicle bearing areas. Excessive steaming will melt the lubricant in these

Figure 6-7. A steam cleaner can be used for larger tasks, such as engine-compartment or under-body cleaning. (Aaladin Industries)

components and damage the seals. Do not drive dirt and grease from the brake backing plates into the brake drums. Be careful when using steam cleaner on brake lines and flexible hose.

Remember that steam causes condensation. Part and tool rusting will occur if the steam cleaner is operated in a poorly ventilated area.

Shutting Down the Steam Cleaner

When cleaning is finished, shut off the solution control valve and allow the cleaner to operate for a short time. Then, close the fuel valve to shut down the burner, but keep the water pump running. When there is no sign of steam coming from the gun, shut down the water pump. By following this procedure, all solution is removed from the coils and the coils are allowed to cool before the water flow stops. This prevents possible burning and scaling.

Arrange the steam hose so that it is out of the way and will not be kinked or run over. If there is any danger of the temperature dropping below freezing, the machine should be drained after use.

Safety Rules for Steam Cleaning

- Do not operate a steam cleaner without proper burner ventilation.
- Be sure an electric steam cleaner is properly grounded.
- Keep pressure within specified limits.
- Wear a face shield to protect your eyes from splatters.
- Keep other personnel away from the immediate vicinity of the cleaner. When swinging the gun around, watch for bystanders.
- If the burner of a gas- or oil-powered unit does not ignite readily, shut off the fuel valve and have a qualified repair person check the burner fuel and ignition system.
- If the machine must be lighted by hand, keep your face and body away from the burner opening.
- Read the machine instruction book carefully; get "checked out" by an experienced operator.

High-Pressure Spray Cleaning

Effective removal of dirt and grease can be accomplished using ***high-pressure spray cleaning equipment.*** A high-pressure cleaning machine is shown in **Figure 6-8.** This type of equipment forces cold water through a spray gun at high pressures. A cleaning solution is injected into the water as it passes through the gun. Pressure at the gun's nozzle can reach approximately 500 psi (3448 kPa).

By adjusting the gun, a soft mist containing a detergent solution can be sprayed over the object to be cleaned until it is thoroughly saturated. Following a short waiting period to allow the deposits to soften, a fine, fan-shaped high-velocity stream of plain water is used to lift off the dirt. For hard-to-clean corners, the spray can be adjusted to a high-velocity, narrow stream.

When doing an under-hood cleaning job, cover fenders, windshield areas, and sensitive engine and electrical components to protect them.

Low-Pressure Spray Cleaning

Low-pressure spray cleaning is another technique involving the use of an air-operated mixing gun. As air passes through the gun, it draws in a metered amount of cleaning solution and sprays it on the object being cleaned. If desired, the cleaning gun suction hose can be placed in a container of water or cleaning solvent. After deposits soften, the object can be washed down with a water hose.

Special cleaning solutions are generally added to the water or cleaning solvent for the initial cleaning spray. Never use gasoline or any solvent with a low flash point: spraying will atomize the solvent, making it highly explosive.

Figure 6-8. A portable high-pressure spray cleaner. (L & A Products)

Parts Washers

Although small parts can be cleaned in cans or buckets, a faster and more efficient job can be done with a regular cold solution ***parts washer,*** **Figure 6-9.** Parts washers hold a large amount of solvent. They have soaking trays, solvent agitators, and a filter to remove impurities from the solvent for rinsing. Parts washers are available in many different sizes. Some shops use portable parts washers that can be wheeled to the job or placed on a workbench, **Figure 6-10.**

Before using the washer, remove the heaviest deposits from the part with a scraper. Large units, such as engines, should be cleaned before disassembly. The parts are placed in the basket and submerged in the solution. Parts with hollow areas should have the hollows facing up, so that an air trap will not prevent solution entry.

The solution is then agitated (shaken) by air pressure or by passing through nozzles under pressure. Some parts washers have a separate compartment that is air-agitated. The main tank is used for soaking, brushing, and rinsing. Some washers have a separate basket that is used during the agitation cycle. It will hold a few parts for brushing or rinsing while the remainder are still washing.

After a thorough cleaning, the parts should be given a final rinse. After rinsing, let the parts drain and then blow them dry. If there is a possibility of rusting, coat the cleaned parts with oil or grease. Keep parts covered until they are ready to be used.

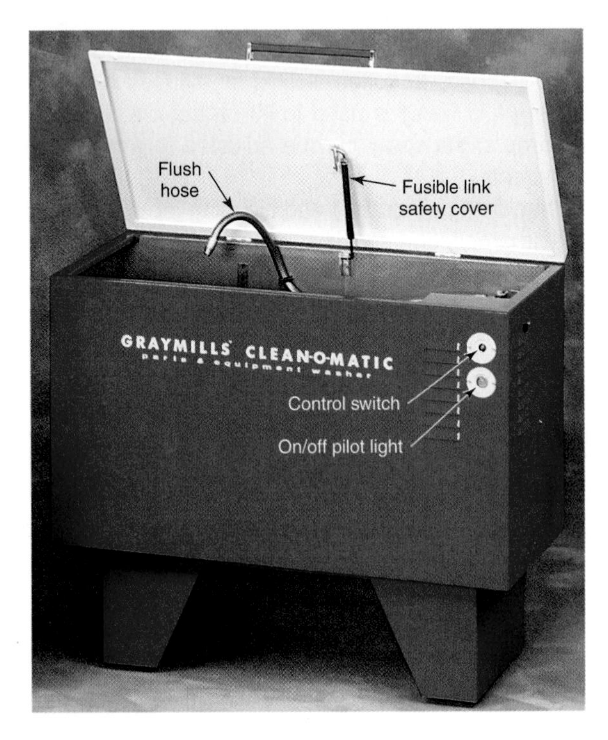

Figure 6-9. A typical cold solution parts washer. (Graymills Co.)

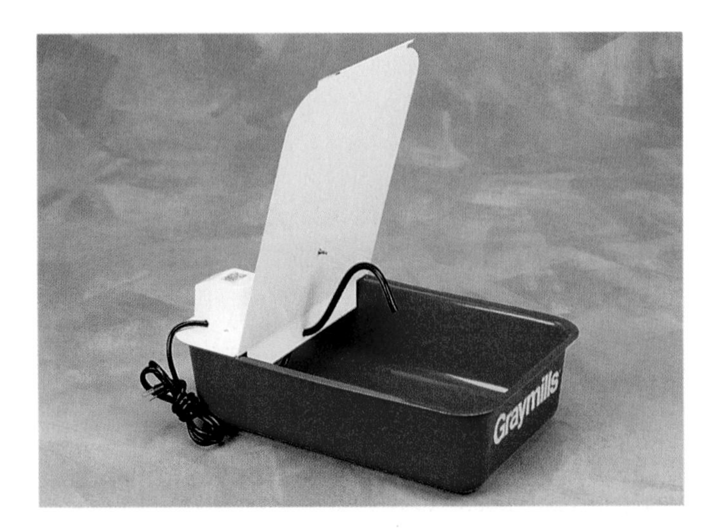

Figure 6-10. This portable parts washer can be placed on the workbench. (Graymills Co.)

High-Pressure Parts Washer

The high-pressure parts washer functions somewhat like a high-performance version of the home dishwasher. Parts are placed in the washer's cabinet and the door latched closed. High-pressure jets of hot water and detergent then scrub the parts clean. Heavy-duty neoprene gloves are built into the cabinet so parts can be manipulated during cleaning, **Figure 6-11.**

Hot Tank Cleaning

Large shops specializing in rebuilding usually have a ***hot tank*** for heavy cleaning tasks. A hot tank is a large container

Figure 6-11. High-pressure jets of heated water and detergent clean and degrease parts quickly in the hot-water parts cleaner. Built-in gloves allow the operator to manipulate parts without coming into contact with the cleaning solution. (Graymills Co.)

filled with hot cleaning solution. Large units such as engine blocks, transmission cases, and radiators can be quickly and thoroughly cleaned in the hot tank, **Figure 6-12.**

The hot tank normally uses a strong ***alkaline compound*** mixed with water to form a solution. Temperatures run from 180°–210°F (82°–99°C). The tank may have an agitator to speed the cleaning process. Most parts are cleaned in thirty minutes or less, depending on tank design, solution strength, solution temperature, and part load.

The alkaline solution is caustic and will attack aluminum. When cleaning aluminum parts, such as modern transmission cases, the solution must be diluted (weakened) to prevent surface erosion of the metal.

When the parts are removed from the tank, they should be thoroughly washed, preferably with hot water. Be careful to flush out oil galleries, water jackets, and other internal passages. Parts or surfaces subject to rusting should be oiled soon after they are dry.

Warning: Be extremely careful when using the hot tank. Observe all safety precautions. Ask someone skilled in the use of the hot tank to give you instructions before you use the tank.

Cold Soak Cleaning

For ***cold soak cleaning,*** the dirty part or parts are placed in a basket and lowered into the cleaning solution. Following a soaking period of 10–30 minutes, the parts are removed and rinsed in solvent or water. They are then blown dry with an air gun. Solutions of various kinds are available for specific tasks.

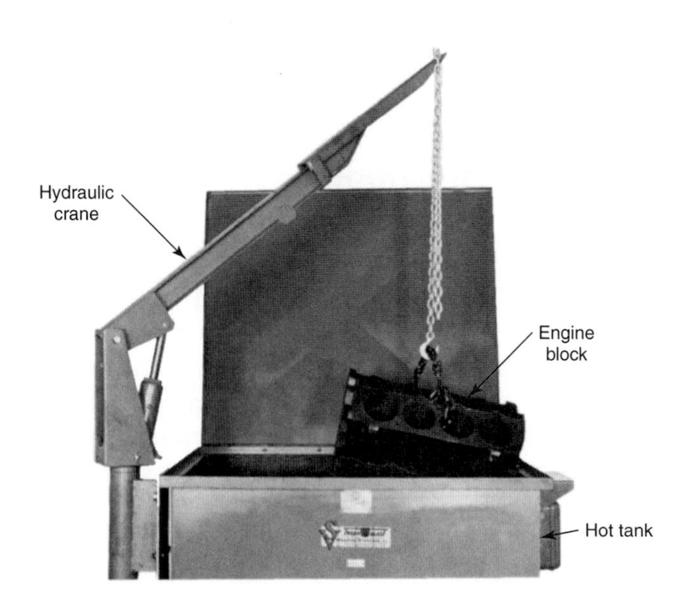

Figure 6-12. An engine block being lowered into a hot tank for cleaning. A hydraulic crane is usually built into such tanks to handle heavy loads. (Storm Vulcan)

Warning: Most of the cold soak solutions are extremely caustic. Keep these solutions away from skin and eyes! Wash the soaked parts thoroughly before handling them.

Cold soak solutions generally come in a special pail or drum that includes a parts basket. The solution is far enough below the top of the pail that a normal load of parts will not cause spillage. A special sealing material floats on top of the cold soak solution to prevent evaporation and excessive odor. When placing parts in the container, make certain that they are completely submerged and are below the sealing material.

Vapor Cleaning

The ***vapor cleaner*** cleans parts by heating a perchloroethylene solution. The resulting vapors remove deposits from the parts suspended in the metal basket. The solution is nonflammable.

Brake Washer

The ***brake washer,*** **Figure 6-13,** is used to clean drum and (in some cases) disc brake systems before they are serviced. It uses water mixed with a solvent or detergent. Air pressure is used to spray the solution onto the wheel brake parts. After the assembly has been thoroughly wet down, the spray force can be increased to wash dust from the brake assembly. The solution will collect in the pan, ready for disposal. This type of wet cleaner collects more dust than spray can cleaners.

Warning: The dust collected by brake washers contains asbestos and should be treated as hazardous waste. Used shop rags, filters, and contaminated cleaning materials should be placed in a closed container and disposed of along with other hazardous waste.

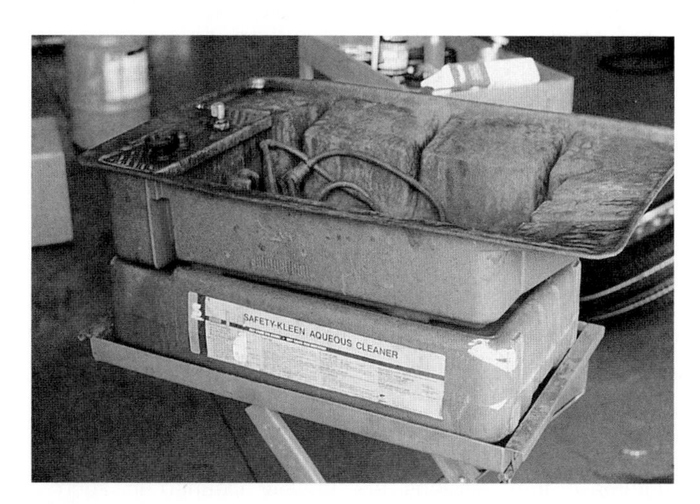

Figure 6-13. A brake washer. The cleaning solution is a mixture of water and organic solvent.

Tech Talk

Have you ever gone into a restaurant and found the tables sticky and littered with dirty plates? Did you want to eat there? Have you ever walked into an electronics repair shop and tripped over disassembled equipment lying around? Did the people who worked there look like they were sleeping on a park bench until five minutes before you got there? Did you trust them to fix your CD player? If you are like most people, you expect a certain level of cleanliness and order, and you probably think that work that comes out of a sloppy place will be sloppy, too.

On the other hand, did you ever walk into a repair shop and find everything clean and neat? Your first impression was probably favorable.

This should give you an idea about how you can create a good first impression. If your work area is clean and neat, it will make a good first impression. If you look like you have given some thought to your appearance, it will make an even better impression. If you make a good first impression, you will not have to spend time overcoming a bad impression.

Elsewhere in this book, we have discussed the importance of good housekeeping and neat clothing to prevent accidents and keep track of tools and parts. Now is a good time to begin thinking about the way that good housekeeping and neat clothing will impress your customers.

Summary

Automotive repair and maintenance work procedures require extensive use of cleaning techniques, equipment, and solutions. You will do better work if you select the best cleaning procedure for the job at hand. As with all work, cleaning must be thorough.

Hand brushes and scrapers are occasionally useful. Abrasive discs and wire brushes are fine for removing hard carbon deposits from some parts.

Blast medium cleaning is useful for removing paint, rust, and weld scale. Do not operate a sandblaster near a repair area. A bead blaster is a type of blast cleaner that uses glass beads. It consists of a cabinet with an internal blaster.

The steam cleaner is a fast, efficient cleaning tool and is especially good for removing heavy dirt and grease deposits. High-pressure spray cleaning also handles dirt and grease very well. Large areas may be cleaned quickly. Low-pressure spray cleaning is effective on many jobs. However, it is generally slower than either steaming or using the high-pressure washer.

A parts washer is excellent for many parts not coated with hard carbon. Parts are soaked in an agitated solution, brushed, rinsed, and blown dry. For larger objects or parts that are hard to clean, a hot tank containing a strong alkaline solution is desirable. Hot tank solutions must be diluted when cleaning aluminum parts.

Cold soak cleaning solutions are widely used for gum, varnish, and hard carbon removal. Pistons, carburetors, and automatic transmissions are usually cleaned in such a cleaner. A parts basket can be furnished with the pail or drum of solution. Vapor cleaning has some advantages and works particularly well on certain parts. Remember that many cleaning solutions are both toxic and caustic and must be handled with care. Observe all safety rules.

Review Questions—Chapter 6

Do not write in this book. Write your answers on a separate sheet of paper.

1. Cleaning often accounts for one _____ of the total repair time.
   (A) tenth
   (B) fifth
   (C) quarter
   (D) half
2. Give some examples of parts that should not be cleaned with a power-driven wire brush.
3. A bead blast cleaner is located inside of a _____.
4. A toxic cleaning solution is _____.
5. If gasoline or any flammable solvent with a low flash point is used for cleaning, what could happen?
   (A) Fire.
   (B) Explosion.
   (C) Both A & B.
   (D) Neither A nor B.
6. When should the technician heat a cleaning solution?
   (A) Never.
   (B) Only when specifically recommended.
   (C) At any time.
   (D) When the parts are very dirty.
7. List six safety rules that should be observed when using the steam cleaner.
8. When submerging a part with a hollow compartment, always place the compartment _____, so that the solution will enter.
9. The hot tank solution will attack _____.
10. Most cold-soak cleaning solutions are extremely _____.

ASE-Type Questions

1. Cleaning means removing _____.
   (A) most soft deposits
   (B) every bit of foreign material
   (C) most hard deposits
   (D) Both A and C.
2. Technician A says that a shop with a steam cleaner really does not need any other cleaning equipment. Technician B says that piston ring grooves are best cleaned with the power wire wheel. Who is right?
   (A) A only.
   (B) B only.
   (C) Both A & B.
   (D) Neither A nor B.

3. Technician A says that carburetors are best cleaned in a bead blast cabinet. Technician B says that machined engine parts may be cleaned satisfactorily with the sandblaster. Who is right?
   (A) A only.
   (B) B only.
   (C) Both A & B.
   (D) Neither A nor B.
4. It is important to remove the _____ before cleaning under the hood.
   (A) battery cable
   (B) air cleaner
   (C) radiator cap
   (D) drive belts
5. Technician A says the water pump should be started before lighting the burner on a steam cleaner. Technician B says that a steam cleaner should be stopped by first shutting off the water pump and, when no water comes from the gun, shutting off the burner. Who is right?
   (A) A only.
   (B) B only.
   (C) Both A & B.
   (D) Neither A nor B.
6. Steam cleaning can be used to clean all of the following *except*:
   (A) engine blocks.
   (B) alternators.
   (C) transmission cases.
   (D) radiators.
7. High-pressure spray cleaning will do a good job of removing _____.
   (A) dirt
   (B) grease
   (C) carbon
   (D) Both A & B.
8. Technician A says that a cold solution parts washer eliminates the need to blow parts dry after cleaning. Technician B says that a hot tank is excellent for cleaning engine blocks. Who is right?
   (A) A only.
   (B) B only.
   (C) Both A & B.
   (D) Neither A nor B.
9. The solution used for hot tank cleaning is _____.
   (A) strongly alkaline
   (B) caustic
   (C) hot
   (D) All of the above.
10. Cold-soak-cleaning solutions are usually supplied with a special _____.
   (A) color code
   (B) rinsing solvent
   (C) parts basket
   (D) Both B & C.

Suggested Activities

1. Make a chart showing various cleaning solutions and the systems they are used on (engine, brakes, body, interior, etc.). Display the chart in the shop or classroom.
2. Using a manufacturer's catalog, determine which cleaning tools are needed in the automotive shop. List the prices of these tools and determine how much it would cost to buy all the necessary items.
3. Visit as many repair shops as possible. Using a video camera, record the cleaning techniques used. Narrate the video as you are taping. Before leaving, thank the technicians concerned, as well as the service manager. Present the video to your classmates.

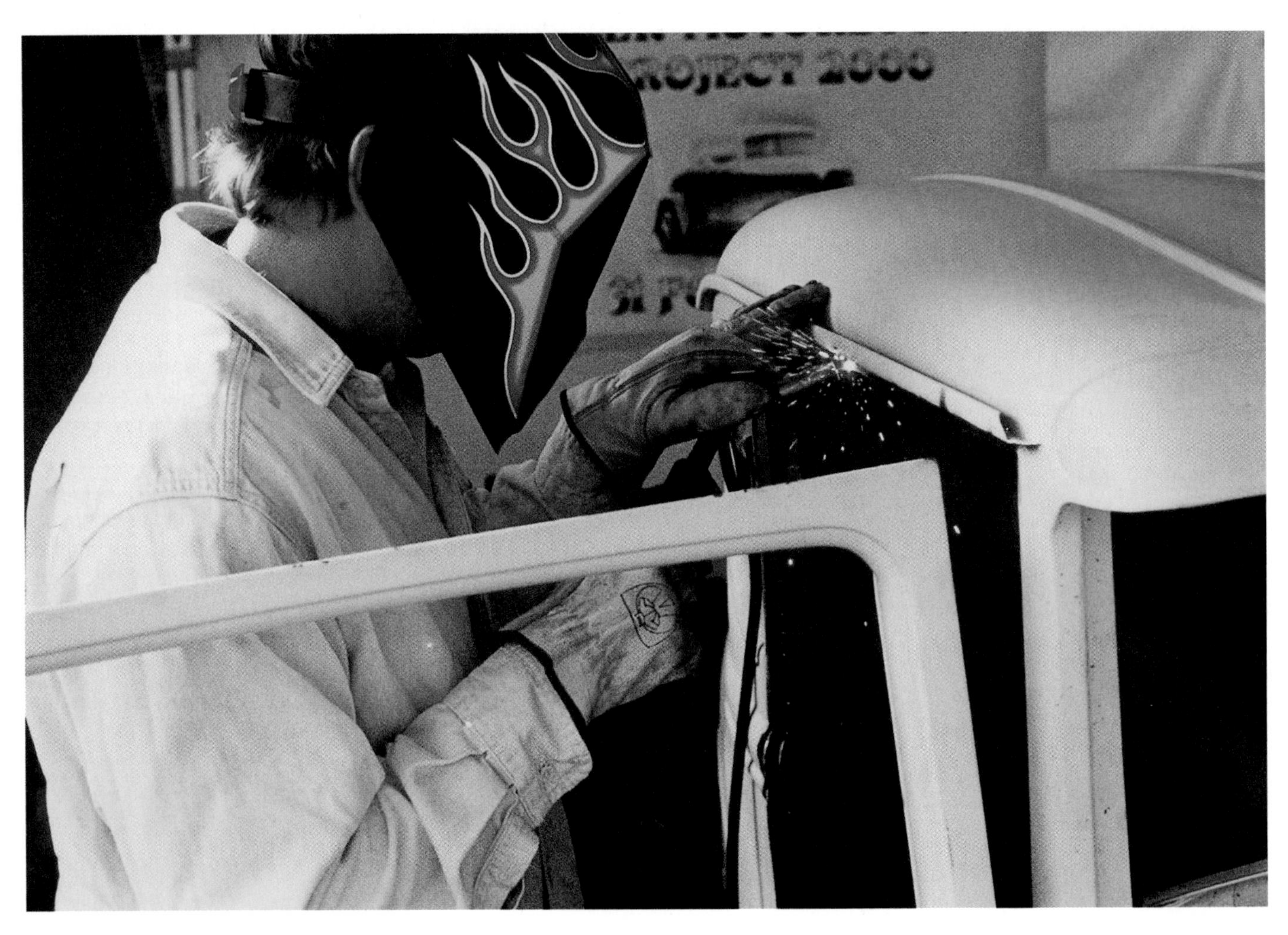

All technicians should have some knowledge of welding and brazing techniques. This technician is using a MIG welder to repair the door opening on a restoration project. (Miller Electric Mfg. Co.)

7

Welding Equipment and Techniques

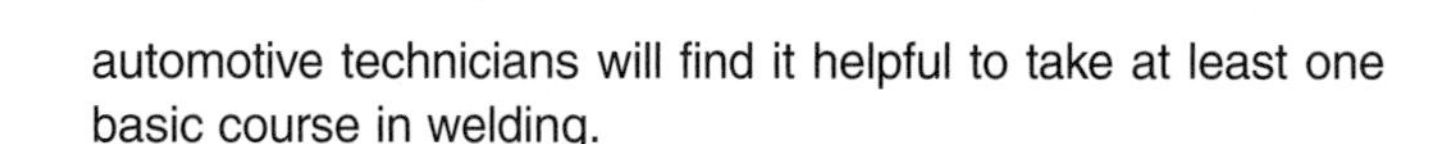

After studying this chapter, you will be able to:
- Describe the equipment needed for welding metal.
- Describe the equipment needed for brazing metal.
- Describe the procedures for brazing metal.

Technical Terms

Welding
Fusion
Forehand method
Backhand method
Uphill welding
Downhill welding
Gas metal arc welding (GMAW)
MIG welding
Inert gases
Gas tungsten arc welding (GTAW)
TIG welding
Plasma cutting
Arc welding
Weld pool
Reverse polarity
Straight polarity
Welding rods
Slag
Inclusions
Blowholes
Oxyfuel gas welding (OFW)
Oxyacetylene welding
Regulators
Hoses
Torch mixing handle
Torch tip
Purge
Spark lighter
Neutral flame
Carburizing flame
Oxidizing flame
Oxyacetylene cutting torch
Kerf
Brazing
Capillary action
Braze welding
Brazing rods
Brazing flux

Becoming a successful automotive technician involves much more than disassembly, inspection, replacement, and reassembly. Numerous basic skills are required, not all of which are commonly associated with automotive work. Quite often, parts must be rebuilt, altered, adapted, or welded. To cope successfully with all these demands, the technician must have some knowledge of welding and brazing techniques, as well as machine shop operations, sheet metal work, electrical work, and many other tasks.

This chapter is designed to help you learn basic techniques, machine operations, and safety rules for welding and brazing. Most often, welding and brazing operations are used in the body shop. However, you will find welding skills can be useful on many different jobs. Students who plan to become automotive technicians will find it helpful to take at least one basic course in welding.

Warning: Use care when welding, heating, or cutting. Fire or explosion can occur. Keep flames, sparks, and heat away from fuel tanks, batteries, and other flammable items.

Preparing the Joint for Welding

Welding is a joining process that involves ***fusion***. This means a portion of the metal of each part being joined is melted. The melted areas flow together and, upon cooling, form one solid part.

Before welding, the joint to be welded must be prepared properly, **Figure 7-1.** When welding metal 1/32″ (0.8 mm) thick or less, the joint is often flanged to protect against heat warpage, as shown in Figure 7-1A. Metals from 1/32″–1/8″ (3.2 mm) in thickness may be welded using a square-edge butt joint, Figure 7-1B. When metal thickness ranges from 1/8″–3/8″ (3.2–9.5 mm), a V-joint is used, Figure 7-1C. Parts thicker than 3/8″ are usually prepared with a double V-joint, Figure 7-1D. In all cases, both the joint and the immediate area must be cleaned of rust, scale, and paint.

Welding Techniques

Several techniques are used in welding metals. Depending on the welding equipment and position of the required weld, one technique may be better than another.

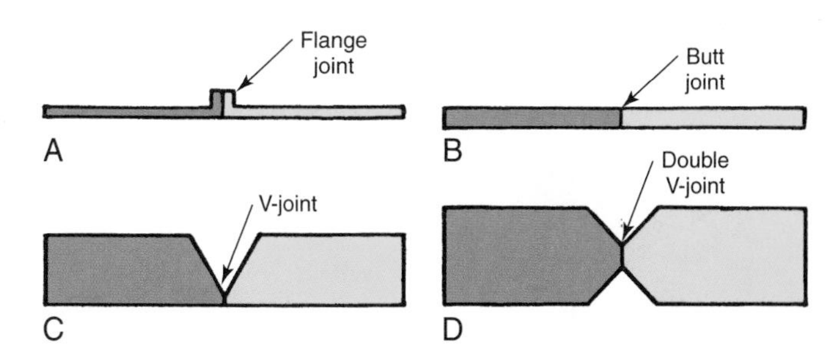

Figure 7-1. Weld joint preparation used for various thicknesses of metal.

Backhand and Forehand Methods

The backhand and forehand welding methods are primarily used when welding in a horizontal position. In the ***forehand method,*** the torch is held so the tip is pointed in the direction of travel, away from the molten weld pool. When using the ***backhand method,*** the torch is held so the tip is pointed opposite the direction of travel and into the pool. The tip of the welding rod is positioned between the flame and the weld. When the *base metal* (metal of the parts being joined) melts and forms a pool, the filler rod is added as the weld progresses. Melt the rod by inserting its end into the pool. Do not hold the rod above the pool, allowing it to melt and drip.

Move the flame along the joint in a steady fashion, causing the base metal to melt continuously. The welding rod can be moved from side to side, in small circles, or in half-circles. The weld should penetrate through the joint. **Figure 7-2** shows both forehand and backhand welding techniques.

Uphill and Downhill Welding

When welding in a vertical position (such as a body panel), two techniques can be used, depending on the job. In ***uphill welding,*** the torch is moved in an upward direction. Uphill welding is generally preferred as it prevents contamination of the weld pool with flux. In ***downhill welding,*** the torch is moved in a downward direction. Downhill welding is used when joining metal that is relatively thin.

Welding Processes

There are more than 100 recognized welding and cutting processes, but almost all of them can be classified into one of two categories: *oxyfuel gas welding* and *arc welding.* Oxyfuel gas welding processes heat the metal with a flame from burning gases. Arc welding processes use the heat of an electric arc (a continuous "spark" between the metal and a welding gun or electrode). There are a number of arc welding processes used in automotive work, but the most common are GMAW (MIG) welding and GTAW (TIG) welding. Arc welding using an electrode (sometimes referred to as "stick welding") also is used for some applications. Oxyfuel gas welding is less widely used than it once was; the most common form of gas welding is oxyacetylene welding, which uses acetylene gas for fuel.

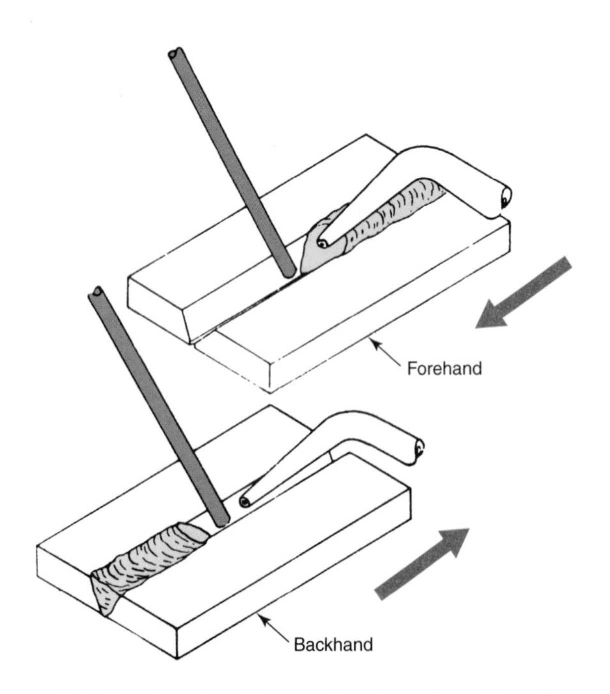

Figure 7-2. Forehand welding and backhand welding. The arrows indicate direction of torch movement. (AIRCO)

MIG Welding

Most shops use a type of arc welding known as ***gas metal arc welding (GMAW),*** commonly called ***MIG welding.*** MIG stands for *metal gas.* MIG welding may also be called *wire feed welding.* This type of welding is used in auto body and exhaust work. The major difference between MIG welding and conventional arc welding is that the MIG process uses a shielding gas to keep the weld from reacting with oxygen and water vapor in the surrounding air. See **Figure 7-3.**

In MIG welding, a thin wire is fed through the center of the welding torch. An electric circuit is created between the wire and the metal to be welded (called the *workpiece*). When the wire contacts the workpiece, current flows and creates an arc. The arc provides the heat that melts the wire and the workpiece. The wire metal takes the place of the filler rod in conventional arc welding. To keep the weld from being contaminated, an inert gas is delivered through the torch to surround the weld. ***Inert gases*** are gases that do not *react* (chemically combine) with other substances. Commonly used inert gases are argon, helium, neon, krypton, and nitrogen. The inert gas is supplied in a pressurized tank. When the tank valve is opened, tank pressure forces the gas through hoses to the torch.

MIG welding is similar to conventional arc welding and is relatively easy to learn. The wire feed and gas supply are semiautomatic processes, allowing the technician to concentrate on welding technique and weld quality. MIG is a relatively rapid process, and is often used to weld large metal sections.

TIG Welding

Another common arc welding process is ***gas tungsten arc welding (GTAW),*** usually referred to as ***TIG welding.*** TIG stands for *tungsten inert gas* welding. TIG welding is sometimes called by the trade name Heliarc. Like MIG welding, this process uses an inert gas to shield the weld from the atmosphere. A TIG welding system uses a nonconsumable tungsten electrode to make the initial arc that creates the weld pool. Filler rod is then fed into the pool. Typical inert gases for TIG welding are argon and nitrogen. **Figure 7-4** illustrates the TIG welding process. TIG welding requires less amperage and heat and therefore is slower than MIG welding. It is used for precise work where a smooth weld is desired. TIG welding is also used for welding on thin metal and welding unusual metal alloys.

TIG welding is more precise than MIG, and the technician may need time to become proficient. The TIG welding

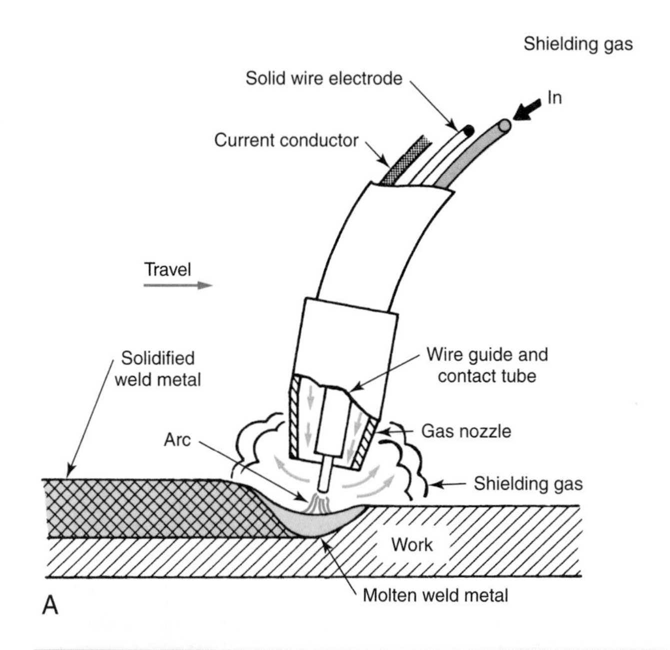

Figure 7-3. Gas metal arc welding, or MIG welding. A—Note protective envelope of shielding gas and the solid wire electrode being fed through nozzle. B—MIG welding is used in many automotive applications. (Miller Electric Mfg. Co.)

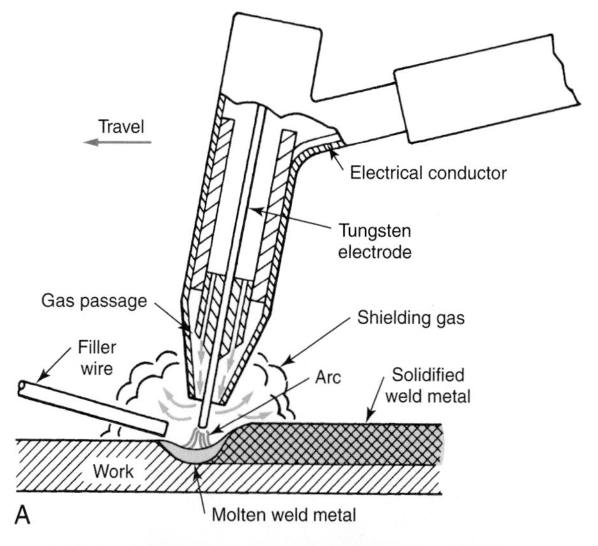

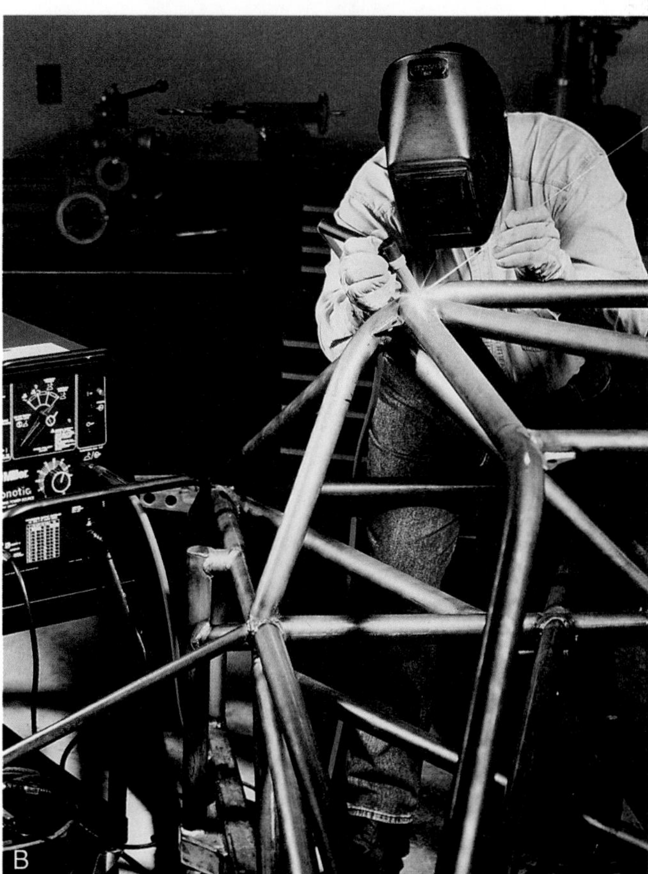

Figure 7-4. Gas tungsten arc welding, or TIG welding. A—Note gas exiting torch nozzle and creating a protective shield. B—TIG welding being used to fabricate a tube frame assembly. (Miller Electric Mfg. Co.)

process is similar to oxyacetylene welding, however, and someone familiar with gas welding may find TIG welding easier to learn than MIG welding.

Plasma Cutting

Plasma cutting is a method of using heat from an electric arc to melt and cut metal. It is similar to cutting done with an oxyacetylene torch, but causes less heat to enter the workpiece, reducing warping and making a clean narrow cut. It is also cheaper since compressed air is used instead of oxygen and acetylene.

The plasma cutter resembles a MIG or TIG welder. Striking an arc begins the plasma cutting process. Once the arc is established, compressed air is forced out of the nozzle in a swirling motion. The fast moving, turbulent air contacts the arc. The electricity in the arc causes the gas to become electrically charged, or *ionized*. Ionization raises the temperature of the gas. This hot ionized gas is called *plasma*. Directing the plasma against the workpiece causes a very small, very hot spot on the metal. The spot melts quickly without transferring heat to the surrounding workpiece metal. Plasma cutting is faster than oxyacetylene cutting. Plasma cutting is being performed in **Figure 7-5.**

Figure 7-5. Plasma cutter. A—Plasma cutting equipment. Note the similarity to an arc welding machine. (ESAB Welding and Cutting Products) B—The plasma cutter permits fast, smooth cutting of metal. (Miller Electric Mfg. Co.)

Arc Welding

Arc welding uses the intense heat (6000–10,000°F or 3318–5542°C) generated by an electric arc between the end of the welding rod and the workpiece. See **Figure 7-6.** Both the base metal and the welding rod quickly reach the fusion state. The molten ***weld pool*** forms as the base metal melts. The arc force actually causes molten globules of metal from the rod to travel through the arc to the pool. This allows the arc welder to be used for overhead welding.

An arc welding machine can be an AC (alternating current) or DC (direct current) machine. Combination AC/DC machines are also available.

An AC or AC/DC machine is generally a power transformer that alters the incoming 220–440 volt current (utility-line voltage) to a low-voltage, high-amperage current for welding. A typical AC/DC machine is shown in **Figure 7-7.**

A DC machine is often driven by a small gas engine. The different types of welding machines have certain advantages and disadvantages. Arc welding machines are rated by maximum output in amperes. The higher the output, the heavier-duty welding work the machine will perform.

Arc Welding Setup

Figure 7-8 shows a typical arc welding setup. Note the different parts — welding machine, rod holder, ground clamp, and connecting cables — and how they fit together.

Two common terms used in DC arc welding are reverse polarity and straight polarity. ***Reverse polarity*** means the current is traveling from the work, through the arc, through the rod, and into the rod holder. ***Straight polarity*** means the current is traveling from the rod holder through the rod, across the arc, and to the work. See **Figure 7-9.** Some machines use a switch to change polarity; others require different cable connections. For a reverse polarity hookup, plug the rod holder cable into the hole with the positive symbol (+). For a straight polarity hookup, connect the rod holder cable to the hole marked with the negative symbol (–).

In AC welding, polarity is not a factor since the direction of alternating current is reversed (or alternated) 60 times per second.

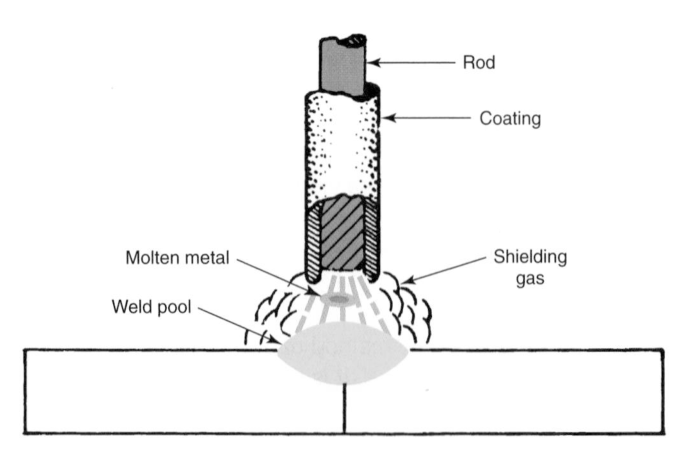

Figure 7-6. Using the electric arc for welding. Note globule of molten metal traveling from rod to pool.

Figure 7-7. A combination AC/DC arc welding machine in use. (Miller Electric Mfg. Co.)

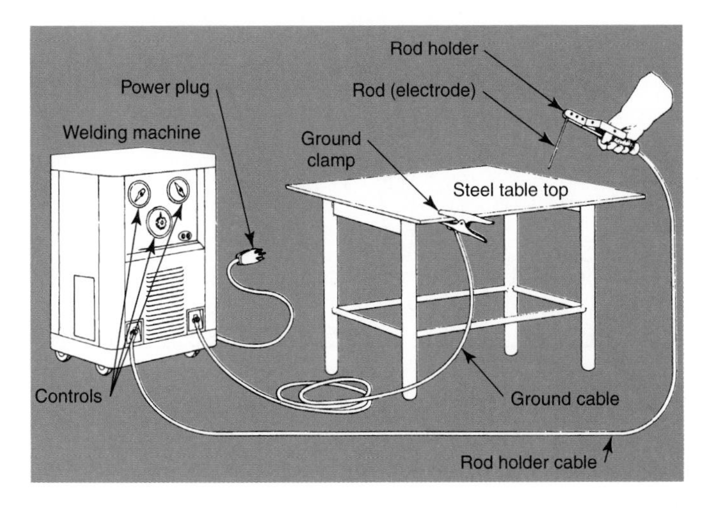

Figure 7-8. Typical setup for arc welding. (Lincoln Electric Co.)

Arc Welding Rods

Welding rods (electrodes) used for arc welding usually range from 12–14″ (305–356 mm) in length and are available in many diameters. They start at 1/16″ (1.5 mm) diameter. For general auto shop use, an assortment of rods in diameters of 1/16″, 3/32″, 1/8″, 5/32″, and 3/16″ (1.5, 2.3, 3.2, 4.0, and 4.8 mm) will be adequate.

Welding rods used in arc welding are usually coated with a material that provides a gaseous shield around the arc. This shield removes impurities and prevents oxidization. A self-starting, self-spacing rod is offered. The coating is kept in contact with the work and maintains the correct distance between the rod and the work.

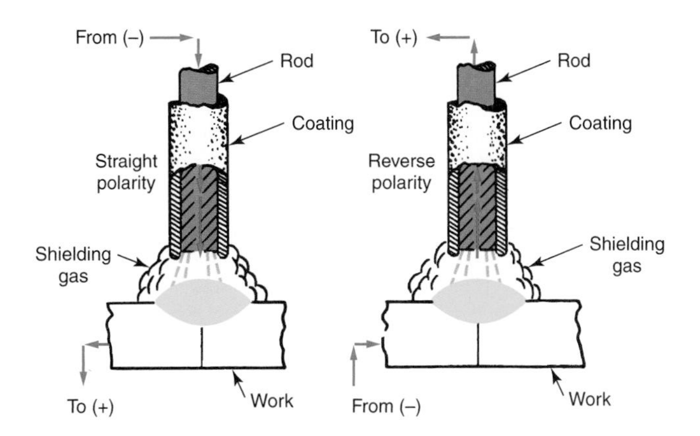

Figure 7-9. Current travel with both straight and reverse polarity.

Rods are available for welding mild steel, carbon steel, cast iron, aluminum, and other metals. Select a rod that is suited to the welding job, both in diameter and material.

Protective Equipment

Always wear a welding helmet to protect your face and eyes while arc welding. A helmet has a dark glass window that will allow you to watch the blinding arc without eyestrain or damage. **Figure 7-10** shows a welder wearing a helmet and other protective equipment.

Leather or Kevlar™ gloves should be used to protect your hands from burns caused by radiant heat or by spatter (flying bits of molten metal). Clothing must be heavy and made from hard-finished cotton (no wool or synthetics) to shed sparks and spatter without igniting. Overhead and horizontal welding can cause a rain of hot spatter to fall on your arms

Figure 7-10. Always wear proper protective equipment when cutting or welding. (Miller Electric Mfg. Co.)

and shoulders. In these cases, a leather cape or jacket should be worn. Pockets must not be open to receive red-hot drops. Shoes must have leather tops and should be high enough to prevent the entry of sparks. When arc welding, do not wear a ring, since it is possible to ground a metal ring between the workpiece and the welding rod. With heavy welding currents, this can heat the ring to a high temperature very quickly.

STOP **Warning: Your eyes can be severely burned by the rays produced during arc welding. Never watch the arc (even for a second) without using a helmet or goggles equipped with the proper dark lenses. Never strike an arc when another person is standing nearby unless that person is wearing protective goggles. Eye burns are "sneaky," because the pain does not immediately follow the exposure.**

Arc Welding Techniques

The procedure for producing an arc weld is as follows:

1. Attach the ground clamp securely to a spot on the work that is dry and free of paint or rust.
2. Select the correct size and type of rod.
3. Set the machine as recommended by the manufacturer.
4. Insert a rod in the rod holder. The holder jaws must grip the uncoated end of the rod to provide a good electrical connection.
5. Turn on the machine and strike an arc by brushing the end of the rod against the work with a short, scratching motion. When the arc forms, pull the rod away the recommended distance. See **Figure 7-11.**
6. When the base metal pools (melts), move the rod forward slowly. Some rods may be held steady, while others require use of a whipping motion.

When whipping, move the rod out of the molten pool until the pool starts to *freeze* (solidify and turn from a shiny, wet look to a dull sheen). Then, immediately move it partially back into the pool. When the pool is fluid again, hold the rod in place for a split second. Then, whip it out again. Repeat this process. Viewed from the top, the whipping process can form either a straight line or a "C" shape, depending on the need.

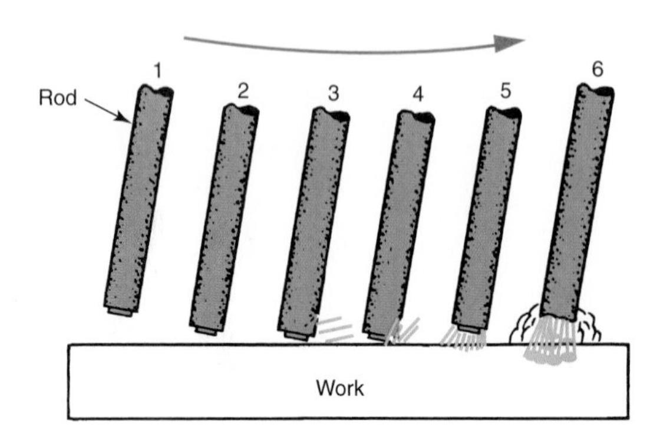

Figure 7-11. Striking an arc.

Whipping is handy in controlling burn-through in thin metal or when working with wide gaps. The rod holder should be gripped so the top of the rod is tilted 5–15° toward the direction of travel. See **Figure 7-12.**

Whipping should be done by flexing the wrist. The whipping motion produces a series of circular ridges along the top of the weld. At first, it may be difficult to maintain correct arc length. Continued practice will enable you to develop skill. Always use the recommended machine settings.

Occasionally, a weaving motion is required. As shown in **Figure 7-13,** this motion helps to bridge wider gaps and will deposit weld metal over a larger surface area.

The sound of the arc is helpful in determining when it is the correct length and of the proper heat. A good arc sounds similar to bacon frying. A short arc will make popping noises and may cause the rod to stick to the work. Excessive arc length will cause a high, humming noise with a lot of spatter. The arc also tends to go out when arc length is too great.

Work Should Be Clean

Despite the fact a good welder can run a bead through rust, paint, and moisture, all weld areas should be dry and

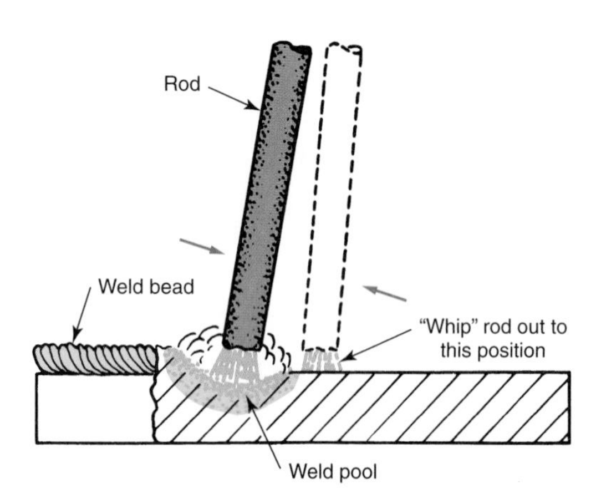

Figure 7-12. Welding with a whipping motion of the electrode (rod).

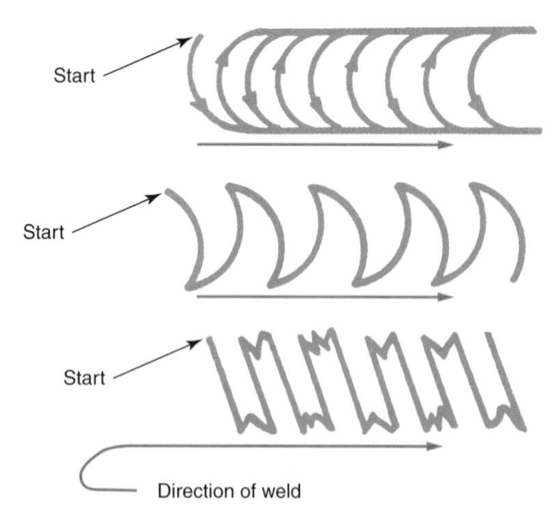

Figure 7-13. Some weaving patterns used in arc welding. (Marquette)

clean. The weld will go faster, look better, and be stronger if the surfaces to be welded are clean.

Some thick parts require a number of welding *passes* (layers of weld metal). After making the first welding pass, chip the ***slag*** (brittle coating left on weld from rod coating material) from the bead. Use a wire brush to complete the cleanup job before making the next pass. If slag is not removed, the joint may be full of slag ***inclusions*** (particles) and ***blowholes*** (air pockets).

STOP **Warning: When chipping or wire brushing, always wear protective goggles or a helmet. Getting a piece of slag in your eye can be serious. Your helmet may be designed so the dark glass can be tilted up, permitting you to look through a piece of clear glass.**

Safety Rules for Arc Welding

The following safety rules should be observed when using any type of arc welding equipment:

- Never look at the arc unless you are wearing a suitable helmet or goggles.
- Do not permit bystanders in the work area unless they are wearing protective gear.
- Wear goggles when chipping or wire brushing.
- Wear protective clothing and gloves.
- Make certain the welding machine is properly grounded.
- Never weld while standing in water or on damp ground.
- Never strike an arc on a fuel tank or on compressed gas cylinders.
- Do not strike an arc on automobile brake lines, fuel lines, or other potentially dangerous parts.
- Weld only in areas with adequate ventilation.
- Be careful when welding metal with coatings such as zinc, cadmium, or beryllium. The fumes from these coatings can be deadly.
- Disconnect the welding machine before attempting any repairs.

Preventing Damage to the Arc Welder, Vehicle, and the Surrounding Area

To prevent damage to arc welding equipment, vehicle, or to the surrounding area:

- Do not adjust machine settings or attempt to change polarity when the machine is under load (welding). This will damage the switch contacts.
- Keep the ground clamp and the tool holder well separated.
- Never start the machine until you are certain the rod holder is not touching the work.
- Make sure the cables are tight in the machine sockets, the clamp, and the rod holder. This will prevent excessive resistance and overheating.
- Protect paint, glass, and upholstery from hot spatter.
- Keep cables coiled when not in use.
- Do not attach the ground clamp to chrome parts. Any looseness will cause arcing that will pit the chrome.

Oxyfuel Gas Welding

Oxyfuel gas welding (OFW), or ***oxyacetylene welding,*** is seeing less use as a means of joining metal in the automotive shop. Many shops only use oxyacetylene to loosen seized exhaust parts or engine bolts. When oxyfuel gas welding is done, a welding rod is often used to supply a filler metal, which helps to reinforce the joint.

Setting up Oxyacetylene Equipment

Support both the acetylene and oxygen cylinders securely in an upright position. Keep cylinders away from heat and flames. Protective tank valve caps must be in place when cylinders are stored. Mark empty cylinders with the letters *MT.* **Figure 7-14** illustrates how cylinders are attached to the welding outfit.

The ***regulators*** reduce cylinder pressures to a controlled and useable amount. **Figure 7-15** illustrates a typical regulator. Note the cylinder and hose fittings. The left-hand gauge reads cylinder pressures. The right-hand gauge indicates tip

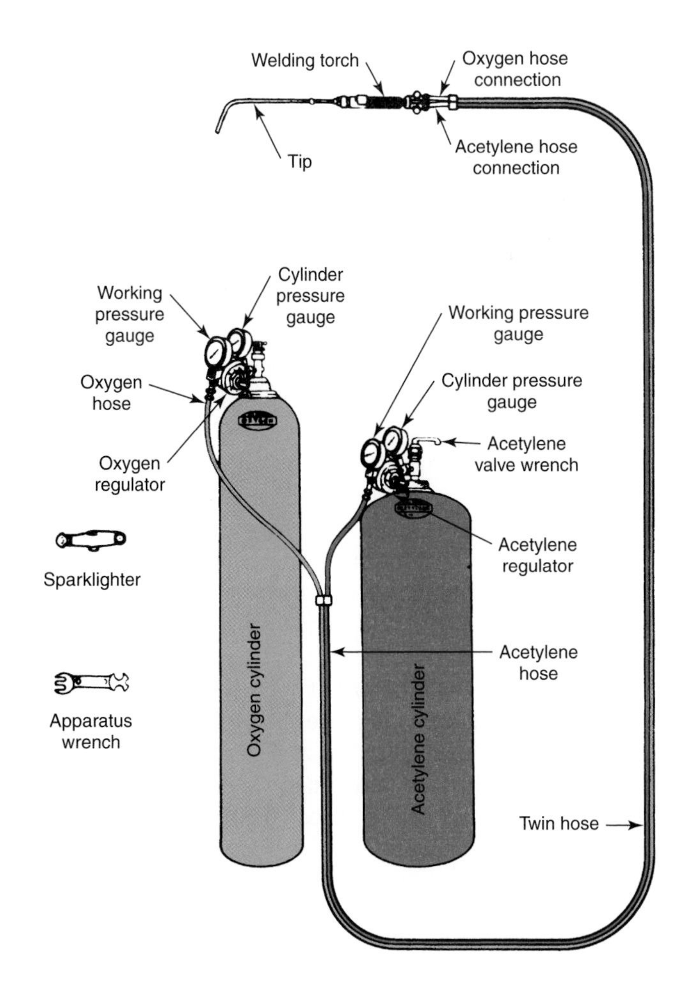

Figure 7-14. An oxyacetylene welding setup. (AIRCO)

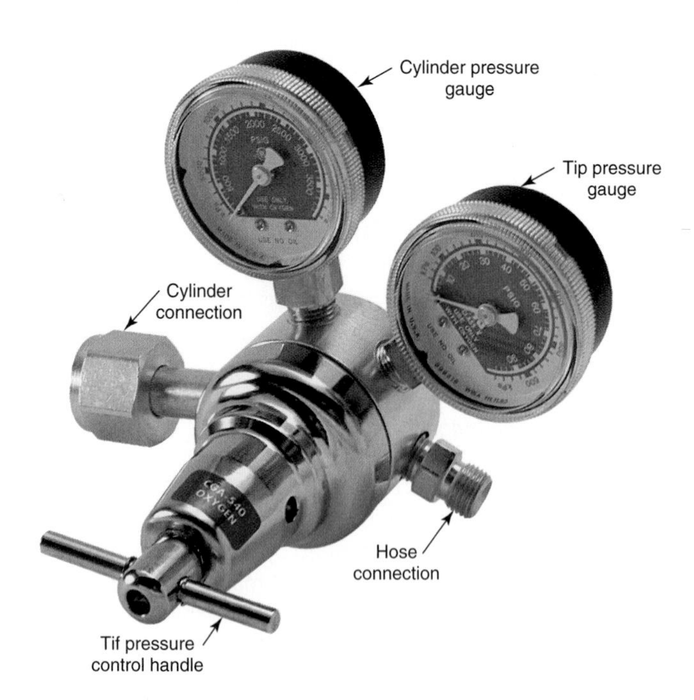

Figure 7-15. Typical oxygen regulator. Note that the cylinder gauge reads much higher pressures compared to the tip gauge. (ESAB Welding and Cutting Products)

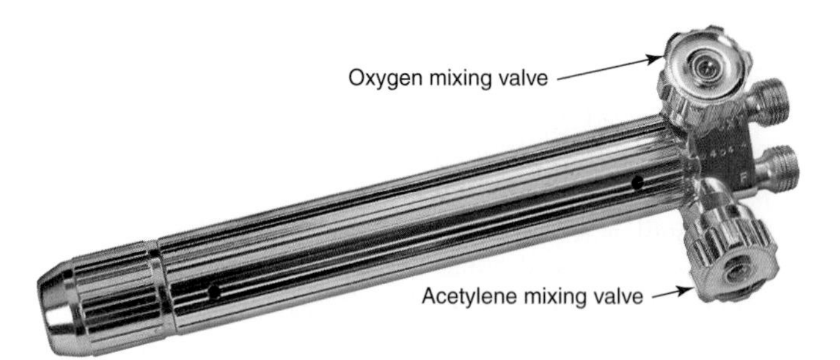

Figure 7-16. Torch mixing handle. (ESAB Welding and Cutting Products)

Approximate Gas Pressures for Operating Welding Torches									
Tip No.	00	0	1	2	3	4	5	6	7
Mixer	00-1	00-1	1-7	1-7	1-7	1-7	1-7	1-7	6-10
Thickness of metal (in.)	1/64	1/32	1/16	3/32	1/8	3/16	1/4	5/16	3/8
Oxygen pressure (psi)	1	1	1	2	3	4	5	6	7
Acetylene pressure (psi)	1	1	1	2	3	4	5	6	7

Figure 7-17. Typical tip sizes and gas pressures for different metal thicknesses. (AIRCO)

operating pressures. Tip pressure is varied by adjusting the control handle of the regulator.

An oxygen regulator has right-hand threads, while an acetylene regulator has left-hand threads. This prevents the technician from installing the regulators on the wrong cylinders.

Before attaching regulators to the cylinders, *crack* (open slightly) the valve on each cylinder for a second to blow out dust or other foreign material. *Do not* crack the cylinder valve near a welding operation or an open flame. Attach the regulators to their respective cylinders and tighten them gently. Finally, back out the pressure control handle on each regulator (turn it counterclockwise) until it moves freely.

Attach the ***hoses*** to the regulators. The acetylene hose is normally red, and the oxygen hose is green. Acetylene fittings have left-hand threads, and oxygen fittings have right-hand threads. When using the equipment, keep hoses away from hot sparks, flames, oil, and grease. Avoid kinking the hoses, and coil them when you finish working.

Attach the ***torch mixing handle*** to the hose end. Do not overtighten either the mixing handle or the regulator hose connections. When rubber O-ring seals are used, hand tightening is sufficient. Note the oxygen and acetylene valves on the mixing handle shown in **Figure 7-16.**

Select the proper ***torch tip*** and install it on the torch mixing handle. Torch tip size must be suited to the job. **Figure 7-17** lists typical tip sizes and gas pressures for different metal thicknesses.

Adjusting Gas Pressure

After installing the desired tip on the mixing handle, adjust the gas pressure as follows:

1. Make sure the regulator pressure control handles are backed off completely.
2. Slowly open the acetylene valve about 1/4 to 1/2 turn.
3. Slowly open the oxygen valve all the way to prevent leakage around the valve stem. Leave the acetylene wrench in place on the valve to facilitate an emergency shutoff.
4. Shut the acetylene mixing valve and open the oxygen mixing valve.
5. Turn the oxygen regulator handle in (clockwise) until the desired working pressure is shown on the tip pressure gauge.
6. ***Purge*** (clear of air or other gases) the oxygen hose line by allowing oxygen to flow from the hose momentarily.
7. Shut off the oxygen mixing valve.
8. Open the acetylene mixing valve and adjust the acetylene regulator to the desired pressure.
9. Purge the acetylene hose, then close the acetylene mixer valve.

Caution: Purging lines is very important. Failure to do so can allow acetylene to enter the oxygen hose and vice versa. This would create a combustible mixture inside the hose and could cause a flashback (fire burning inside hose).

Lighting the Torch

Before attempting to light the torch, put on all needed safety equipment, such as goggles, gloves, protective vest, and helmet. This step is extremely important to prevent possible injury.

To light the torch, open the acetylene mixer valve a small amount while operating a ***spark lighter,*** or *scratcher*, in front of the tip opening. See **Figure 7-18.**

Note: Keep the tip facing in a safe direction. Have your welding goggles in position.

Adjusting the Flame

There are three general types of flames that can be produced by the oxyacetylene torch: a neutral flame, a carburizing flame, and an oxidizing flame. Refer to **Figure 7-19.**

A ***neutral flame*** should generally be used for gas welding. The neutral flame will permit smooth, dense, strong welds. There will be no foaming, sparking, or other problems. A ***carburizing flame*** (one with excess acetylene) will cause molten metal to pick up carbon from the flame. This causes the metal to boil and, upon cooling, to become brittle. An ***oxidizing flame*** (one with an excess of oxygen) will cause the metal to foam and send off a shower of sparks. Also, the excess oxygen combines with the molten metal, causing it to burn. The weld will be porous, weak, and brittle.

When the acetylene ignites, adjust the flame until it is hovering about 1/8″ (3 mm) from the tip, Figure 7-19A. Immediately open the oxygen valve and adjust the flame. By starting with a carburizing flame, Figure 7-19B, and slowly closing the acetylene valve, a neutral flame can be acquired, Figure 7-19C. Watch the yellowish *acetylene feather* to tell when the neutral flame is reached. Figure 7-19D shows an oxidizing flame.

Note: The *inner flame cone* must not touch either the rod or the pool.

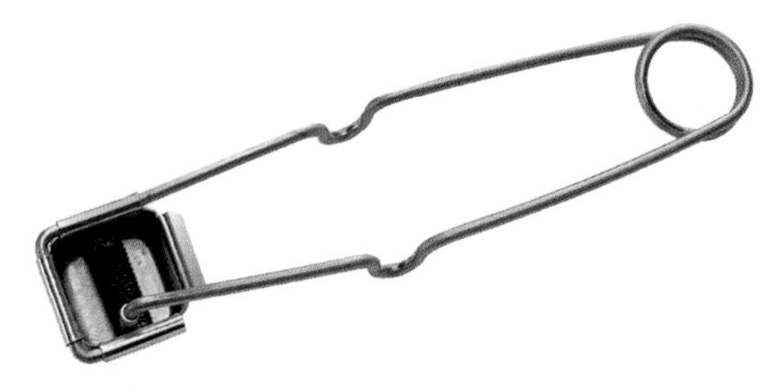

Figure 7-18. Spark lighter. Squeezing handle moves a flint across a rough metal surface, producing a shower of sparks. (ESAB Welding and Cutting Products)

Shutting Down the Torch

To shut down the torch, close the acetylene mixer valve. The oxygen will blow out the flame immediately. Then, shut off the oxygen mixer valve. When using this technique to shut off the flame, make certain the acetylene valve is not leaking. If you will be welding again within a few minutes, hang the torch up out of the way. If it will be some time before the torch is needed, empty the lines. To do so, first shut off both the acetylene and oxygen cylinder valves. Open one of the mixer valves until the tip pressure gauge indicates there is no pressure left in the line. Back off the regulator adjuster handle, close the mixer valve, and repeat the process on the other line.

Using the Oxyacetylene Cutting Torch

There are many applications for the ***oxyacetylene cutting torch*** in the auto shop. In some shops, a separate cutting torch replaces the welding torch. More often, a ***cutting attachment*** is mounted on the welding torch mixing handle in place of the welding tip, **Figure 7-20.** The cutting attachment uses a *preheating flame*, which is maintained at the tip through small *orifices* (openings), located around a larger center orifice. The attachment has two controls: a valve to regulate the oxygen for the preheating flame, and a lever that is used to supply a jet of oxygen for cutting.

To light the cutting torch, set the regulators for the required pressure and then:

1. Close the cutting attachment oxygen valve.
2. Open the mixer oxygen valve all the way.
3. Open the acetylene mixer valve and light the torch.
4. Open the attachment oxygen valve and adjust the preheating flames to neutral.
5. Depress the oxygen jet lever. If preheating flames are altered, readjust them.

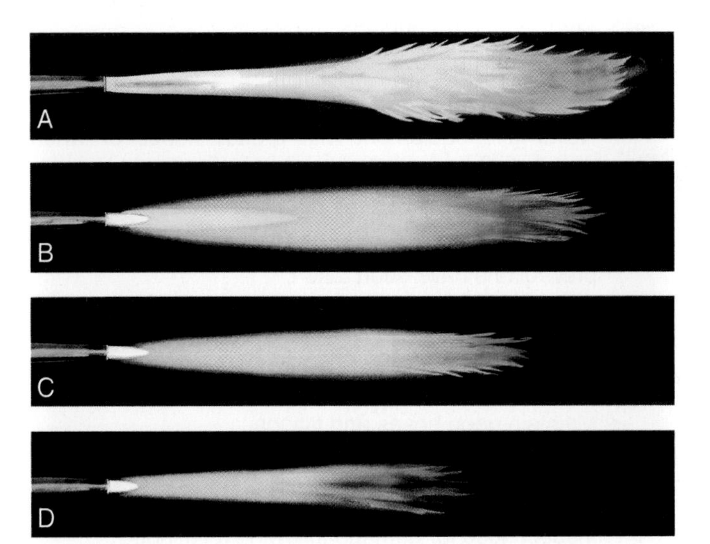

Figure 7-19. Oxyacetylene flames. A—Acetylene-only flame (no added oxygen). B—Carburizing flame (excess acetylene with oxygen). C—Neutral flame (acetylene and oxygen balanced). D—Oxidizing flame (acetylene with excess oxygen) (Smith Equipment, Division of Tescom Corp.)

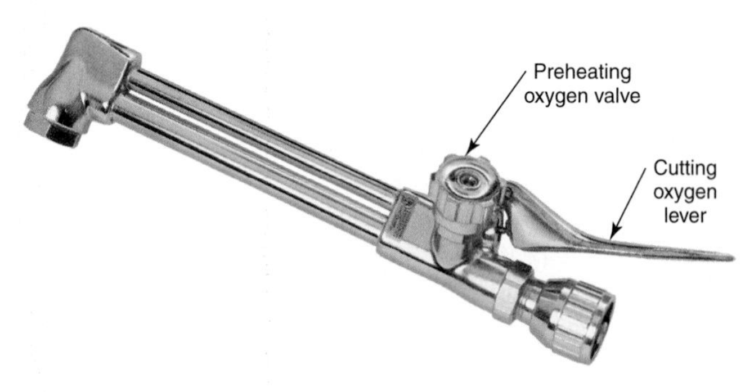

Figure 7-20. A cutting attachment that can be mounted on a welding torch mixing handle. The oxygen valve is used to adjust the preheating flame. (ESAB Welding and Cutting Products)

After lighting the torch, the preheating flames are held close to the workpiece at the point where the cut is to start. When the spot has been heated to a bright cherry red, the oxygen jet lever is pressed. A stream of pure oxygen strikes the heated area, cutting (burning) through the metal.

As soon as the cut starts, move the torch along the work, with the oxygen lever fully depressed. Keep moving the torch as rapidly as the cutting will allow. If the cutting action stops, release the oxygen lever. With the preheat flames (they burn continuously), preheat the workpiece again. Hold the torch so the tip is at a right angle to the work, with the preheat flames just above the surface, **Figure 7-21.** The cutting torch removes a narrow ***kerf*** (cut) and the molten metal (slag) is blown out from beneath the work. See **Figure 7-22.**

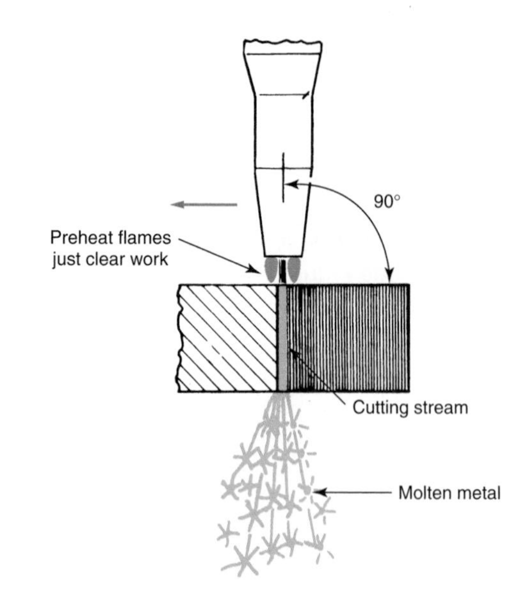

Figure 7-21. Hold the cutting tip at a right angle (90°) to the workpiece. The preheat flames should just clear the work.

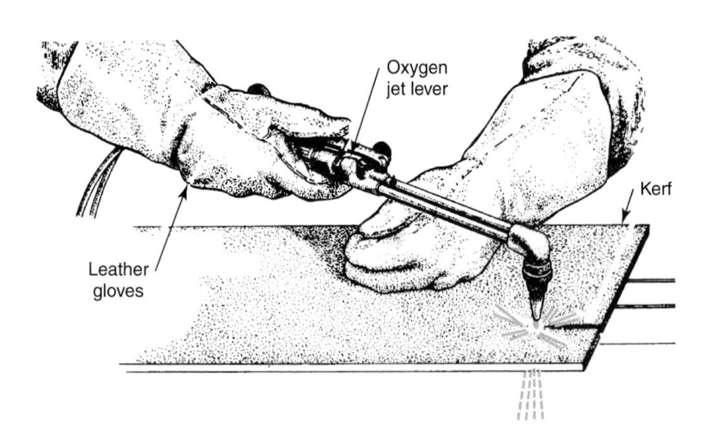

Figure 7-22. Torch cutting action. Note use of gloves. (Lincoln Electric Co.)

Basic Safety Rules for Oxyacetylene Equipment

The following rules should be observed when using oxyacetylene equipment:

- Wear protective goggles.
- Wear welding gloves and other protective clothing.
- Keep all oil and grease away from equipment.
- Never use equipment with greasy hands or when wearing greasy garments.
- Weld only in well-ventilated areas.
- Do not cut, weld, or braze fuel tanks until special precautions have been taken.
- Do not work in an explosive atmosphere.
- Always have a fire extinguisher on the job.
- Open cylinder valves slowly.
- Maintain good hoses and fittings.
- Purge lines before lighting the torch.
- Never use a defective regulator.
- Inspect hose for damage following a flashback.
- Never try to repair hose with tape. If a hose leaks, discard it.
- Stand to one side of the regulators when opening cylinder valves.
- Never open the acetylene cylinder valve more than one turn.
- Never use acetylene at pressures exceeding 15 psi (103 kPa).
- When adjusting either oxygen or acetylene pressures at the mixing handle, make certain the other mixer valve is closed. This will prevent flashbacks.
- Point the torch in a safe direction when lighting.
- Know what materials you are cutting or welding. Some coatings produce deadly gases when heated.

Note: There are many more specific safety rules. Study a booklet on safe practices from one of the companies handling gas welding equipment. Have an experienced operator assist you until you have mastered setting up, lighting, and using welding equipment safely.

Brazing

Brazing consists of heating the work to a temperature high enough to melt the brazing rod metal without melting the workpiece itself. A suitable brazing rod is brought into contact with the heated joint and melted. ***Capillary action*** (attraction between a solid and a liquid) draws the brazing alloy into the joint. This is different from welding, where the work is heated enough to melt it together into a single mass. Brazing is similar to soldering (covered in Chapter 11), but the temperatures involved are higher. Brazing temperatures are above 840°F (450°C).

For successful brazing, the work must be clean, properly fluxed, and brought to the correct temperature. Parts must fit together closely for capillary action to work properly. The joint should be held securely during the brazing operation and while cooling to avoid internal fractures.

Braze Welding

When the joint between parts is loosely fitted, the ***braze welding*** process is used. In braze welding, the metal from the brazing rod is used to fill the joint and build it up until the joint has sufficient strength. See **Figure 7-23** for a comparison of brazed and braze-welded joints.

Brazing Rod

Brazing rods come in a wide variety of *alloys* for different applications. (Alloys are compounds of different metals; bronze, for example, contains copper and tin.). The alloys used for brazing rods are ***nonferrous*** (do not contain iron). A regular bronze or manganese bronze brazing rod is generally acceptable for use on steel, cast iron, and malleable iron. Melting temperature is around 1625°F (886°C), and the rod has a tensile strength (bonded to steel) of around 40,000 psi (275 800 kPa).

Brazing Flux

A ***brazing flux*** is a material used to prevent the formation of oxides or other impurities, or to dissolve and remove impurities. A number of fluxes are available; choose a flux compatible with the brazing rod being used. Flux is available in either powder or liquid form. The uncoated brazing rod tip is heated and dipped into the flux. Enough flux will adhere to the rod to provide proper fluxing for a short time. Brazing rods also are available with flux coatings already applied.

Sources of Heat for Brazing

A blowtorch, propane torch, or oxyacetylene torch all produce sufficient heat for brazing and braze welding. Propane and oxyacetylene torches are well suited for the job and are generally available in the shop. The acetylene torch is similar to a propane torch. A regulator is attached to a tank of gas, the tank valve is opened, and the regulator is set for the desired flow. Because the brazing torch utilizes oxygen from the air, only one tank (acetylene) is required. Several tip sizes are available. An oxyacetylene torch brazing setup is shown in **Figure 7-24.**

Brazing Technique

The technique for brazing is as follows:

1. Select a tip size appropriate to the work. The tip size chart earlier in this chapter will give you an indication of size in relation to metal thickness. Note the recommended gas pressures.
2. Adjust the torch to produce a neutral or slightly carburizing (excess acetylene) flame.
3. With the parts clean, closely fitted (ideal joint gap for brazing is 0.0015–0.003″ or 0.04–0.08 mm), fluxed, and firmly held, apply heat to the joint. Use a brushing motion of the torch tip, as shown in Figure 7-24.
4. Watch the flux. When it starts to turn watery and clear, a little more heat will be sufficient. Touch the brazing rod to the work. When the heat is correct, the brazing metal will melt and be drawn into the joint.
5. Make sure the brazing metal enters the full length of the joint.

Tip Distance and Angle Are Important

The distance the torch tip is held from the work affects the rate and extent of heating. Parts with a low melting point will require you to hold the tip farther from the area to be brazed, **Figure 7-25.**

By holding the tip at an angle, **Figure 7-26,** the work is kept at brazing temperature with minimum danger of

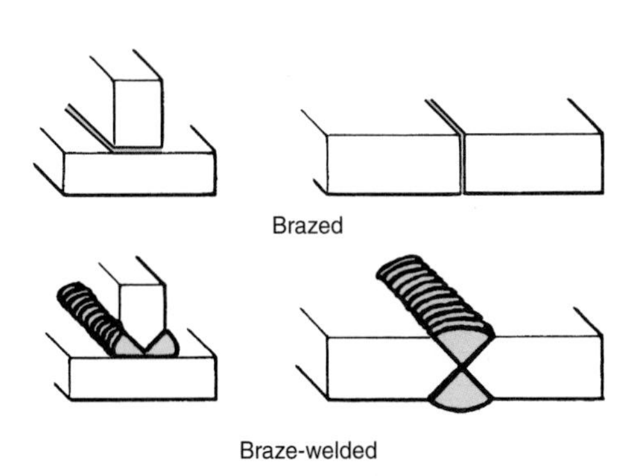

Figure 7-23. Brazed and braze-welded joints.

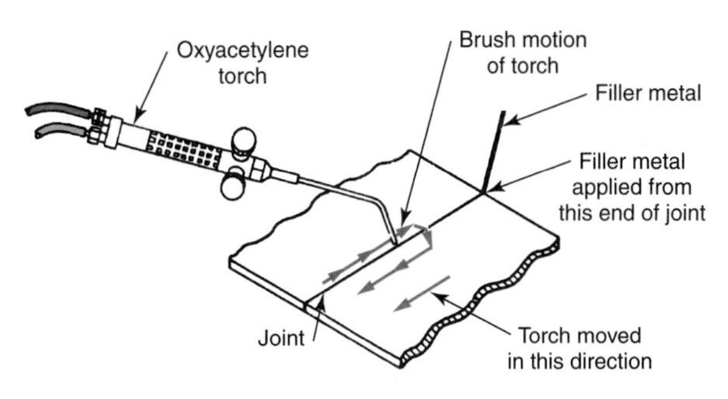

Figure 7-24. Heat joint prior to applying brazing material. When hot, start applying filler metal from one edge. Use the brushing motion of the flame to draw material along and into the joint. (AIRCO)

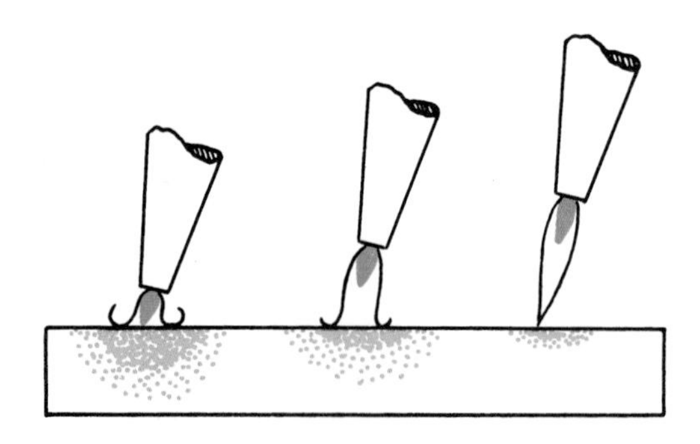

Figure 7-25. Distance from the welding tip to the workpiece affects heat transfer.

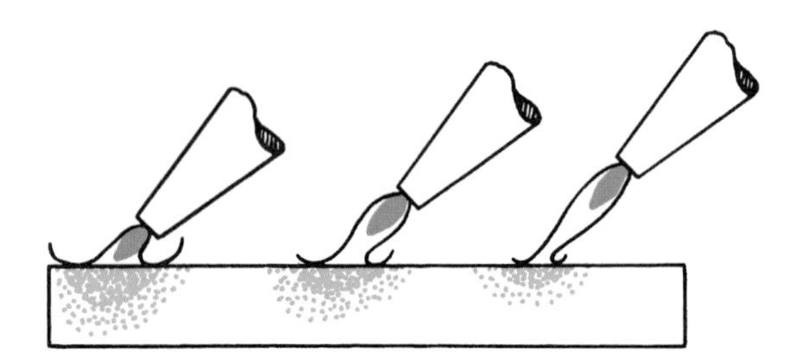

Figure 7-26. Hold torch tip at a consistent angle to the workpiece, regardless of distance.

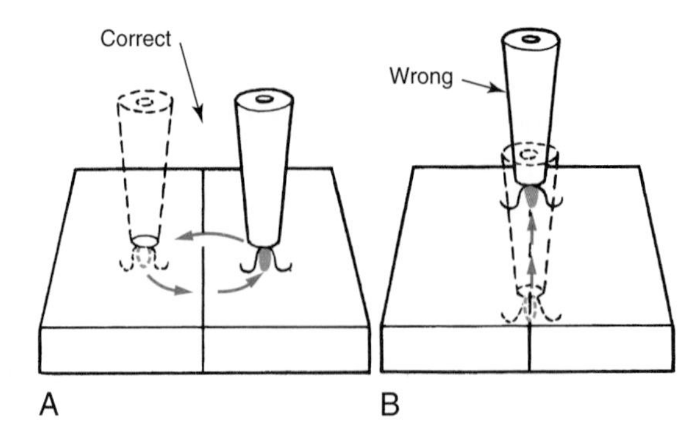

Figure 7-27. Keep torch tip moving; a circular motion is preferred.

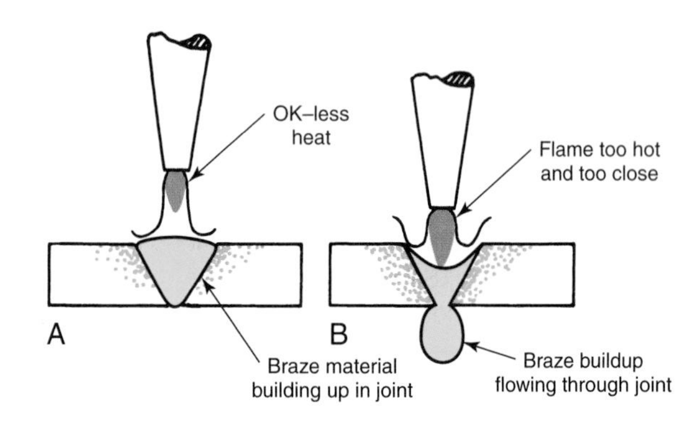

Figure 7-28. Braze welding. A—Correct. B—Too hot. Note how the braze metal base sags.

overheating. Note the distance is varied to suit the work, but the same torch angle is maintained for all brazing jobs.

Keep the tip in motion to spread the heat. If the flame is kept in one spot too long, overheating may result. A circular motion, **Figure 7-27,** is desirable. The size of the circle should be decreased as the joint becomes heated. When brazing temperature is reached, the circles should be quite small. Using a zigzag motion during the application of the welding rod is also satisfactory.

Braze Welding Technique

In braze welding, a groove, fillet, or slot is filled with nonferrous filler metal. The filler metal is not distributed by capillary attraction because of the more-loosely-fitted joint.

The technique used for braze welding is similar to brazing. Once the brazing rod starts melting, the heat should be carefully controlled. This allows the braze metal to build up to the desired thickness. As the filler rod is fed, it must mix with the filler added previously, but it must not cause the buildup to flow. See **Figure 7-28.** The procedures to follow when braze welding include:

- Workpieces must be clean and well fitted.
- Use gas pressures and a tip appropriate for the job at hand.
- Use a neutral or slightly carburizing flame.
- Keep the tip in motion.
- Hold the tip at a constant angle to the work.
- Heat may be controlled by changing distance between the tip and the workpiece.
- Brazing rod alloy should be suited to the job.
- Use a flux compatible with the brazing alloy.
- Braze metal must penetrate the joint.
- Parts must be held in position and must not be disturbed until the braze metal sets.
- Materials must not be overheated.

Tech Talk

The automotive technician is often called on to do welding when fixing a broken part, sealing a tank, or installing a trailer hitch. Some technicians may even get into specialized areas, such as aluminum welding to repair transmission cases. Therefore, it is a good idea to learn at least the basics of welding. You should always try to develop extra skills, since you never know when they will come in handy.

One way to become good at welding is to take a course in this field. Welding may already be part of the automotive curriculum at your school. If not, you might want to consider enrolling in such a course for your own benefit. One course will not make you into a welder, but it will enable you to become better at the welding you will do in the automotive repair shop.

Summary

Metal inert gas (MIG) and tungsten inert gas (TIG) are two types of shielded gas welding. The material being joined is protected from contamination by a shield of inert gas. These welding techniques are used for nonferrous and ferrous metals.

Arc welding is fast and applies a minimum amount of heat to the work. Although the arc temperature is high, the welding process is so rapid the work remains relatively cool. This helps control warpage. Select the appropriate rod size and type.

Adjust the welding machine to the correct polarity and current settings. Tip the top of the rod in the direction of travel (5-15°). A whipping motion will help control the heat, direction, and penetration of the weld.

The weld bead should be smooth and even, have good penetration, and be free of slag and blowholes. Remove slag from a bead before welding another pass over the original bead.

Gas welding involves fusion (melting and mixing) of the metals to be joined. The work should be clean and dry. Thick metal joints should be beveled. Select a torch tip of the size recommended by the manufacturer. Set gas pressures for the selected tip. Adjust the torch to a neutral flame.

When welding, keep the inner flame from touching either the filler rod or the pool. Bring the work to the molten state and, if required, add filler rod. The weld must penetrate the work and should be solid and free of slag and blowholes.

Cutting is fast and easy with an oxyacetylene cutting torch. Follow all safety precautions in setting up the equipment and lighting the torch.

Brazing takes place above a temperature above 840°F (450°C). The work must be clean. Flux the workpieces and heat them until brazing rod melts when in contact with the parts. Capillary action will draw the brazing material into the joints. Do not overheat.

Braze welding is used with less tight-fitting joints than brazing, and involves building up the brazing material to fill joint irregularities and provide strength. Bronze brazing rod may be used on cast iron, malleable iron, and steel. Either a propane torch or an oxyacetylene torch may be used.

Choose a torch tip that is appropriate for the work. Set the gas pressures as recommended by the torch manufacturer. Use a neutral flame (an approximately equal mixture of acetylene and oxygen) to a slightly carburizing flame (a mixture with a slight excess of acetylene). Hold the tip at an angle to the work. Vary the distance between the tip and the work as needed. Keep tip in motion to avoid localized overheating.

Review Questions—Chapter 7

Do not write in this book. Write your answers on a separate sheet of paper.

1. The technician should add filler metal to the weld by _____.
   (A) touching the rod to the pool.
   (B) holding rod above pool and allowing it to drip in.
   (C) laying a length of rod flat on the joint.
   (D) melting and depositing drops of rod all along the joint before melting the base metal.
2. Gas shielded (TIG and MIG) welding processes prevent _____ in the surrounding air from contaminating the weld.
   (A) nitrogen
   (B) argon
   (C) oxygen
   (D) Both A and C.
3. The electrical current output by an arc welding machine is _____.
   (A) AC
   (B) DC
   (C) a choice of AC or DC
   (D) All of the above, depending on the equipment.
4. Always wear _____ when welding, brazing, or cutting.
   (A) eye protection
   (B) gloves
   (C) vest
   (D) All of the above.
5. Acetylene tanks should be used in the _____ position.
6. Before opening tank valves, regulator handles should be _____.
   (A) removed
   (B) backed out until free
   (C) tightened securely
   (D) backed out halfway.
7. When adjusting gas pressure, _______open the valve on the oxygen tank all the way.
8. Before lighting the oxyacetylene torch, the technician should _____ both lines.
   (A) purge
   (B) tighten
   (C) disconnect
   (D) None of the above.
9. Nonferrous metals contain no _____.
   (A) copper
   (B) aluminum
   (C) iron
   (D) zinc
10. To light the oxyacetylene torch, the acetylene mixing valve should be opened _____.
    (A) all the way
    (B) 1/2 turn
    (C) 1/4 turn
    (D) a small amount

ASE-Type Questions

1. Technician A says plasma arc cutting heats the entire workpiece more than oxyacetylene cutting. Technician B says watching the arc without protective equipment can cause serious eye damage. Who is right?
   (A) A only.
   (B) B only.
   (C) Both A and B.
   (D) Neither A nor B.
2. Technician A says the arc temperature can reach 10,000°F. Technician B says one should never weld or braze fuel tanks until special precautions have been taken. Who is right?
   (A) A only.
   (B) B only.
   (C) Both A and B.
   (D) Neither A nor B.
3. Technician A says welding rods are usually coated. Technician B says some metal coatings will give off poisonous fumes when heated. Who is right?
   (A) A only.
   (B) B only.
   (C) Both A and B.
   (D) Neither A nor B.
4. Technician A says using a whipping motion will help control the heat, direction, and penetration of the weld. Technician B says one should remove slag from a bead before welding another pass over the original bead. Who is right?
   (A) A only.
   (B) B only.
   (C) Both A and B.
   (D) Neither A nor B.
5. Technician A says the flame for braze welding should be an oxidizing flame. Technician B says the flame for normal welding should be a neutral flame. Who is right?
   (A) A only.
   (B) B only.
   (C) Both A and B.
   (D) Neither A nor B.
6. Technician A says the inner flame cone must not touch the weld pool. Technician B says the inner flame cone must not touch the rod tip. Who is right?
   (A) A only.
   (B) B only.
   (C) Both A and B.
   (D) Neither A nor B.
7. Technician A says the cutting torch uses a jet of oxygen to produce the cutting action. Technician B says the cutting torch uses a jet of acetylene to produce the cutting action. Who is right?
   (A) A only.
   (B) B only.
   (C) Both A and B.
   (D) Neither A nor B.
8. Technician A says the cutting torch should be held at a right angle to the work. Technician B says oil and grease should be kept away from gas welding equipment. Who is right?
   (A) A only.
   (B) B only.
   (C) Both A and B.
   (D) Neither A nor B.
9. All of the following statements about brazing are true, *except:*
   (A) flux is required for brazing.
   (B) the parent metal does not reach its fusion point.
   (C) brazing and braze welding are one and the same.
   (D) tip size and gas pressures are important.
10. Technician A says heating metal for brazing should be done by keeping the tip moving in a circular pattern. Technician B says the tip should be kept in one spot for long periods for maximum heat penetration. Who is right?
   (A) A only.
   (B) B only.
   (C) Both A and B.
   (D) Neither A nor B.

Suggested Activities

1. Check several vehicles and note places where welds are found. Determine what types of welds are used. Make a sketch of a vehicle showing the areas where welds are used. Be sure to identify the type of welds found.
2. List some situations that might require the automotive technician to weld or braze. Discuss these situations with your instructor or with the other members of your class.
3. Make a poster showing the types of protective clothing that should be used when welding. Display the poster near the welding equipment in your shop.
4. Practice welding on some scrap metal. Write a short description of how you used your mistakes to learn how to weld properly.

8

Fasteners, Gaskets, and Sealants

After studying this chapter, you will be able to:

- Identify automotive fasteners.
- Properly select fasteners.
- Torque fasteners to specifications as necessary.
- Remove damaged or broken fasteners.
- Describe gasket construction, materials, and application.
- Describe the construction and installation of seals.
- Describe the types and selection of sealants and adhesives.

Technical Terms

Fasteners
Machine screws
Sheet metal screws
Bolt
Stud
Screw extractor
Nuts
Penetrating oil
Heli-Coil®
Tensile strength
Major diameter
Minor diameter
Thread pitch
Thread class
Locking devices
Anaerobic sealers
Keys
Splines
Pins
Locking plate
Safety wire
Snap rings
Setscrews
Rivets
Torque
Tension
Elastic limit
Distortion
Tensile strength
Residual tension
Elasticity
Compression
Cold flow
Hooke's law of springs
High-pressure lubricant
Torque specifications
Torque-to-yield bolts
Torque wrench
Blind hole
Tightening sequence
Run-down torque
Gasket
Compressibility
Unit loading
Mating surfaces
Sealant
Oil seal
Seal driver
O-ring seals
Adhesives
Sealants
Form-in-place gaskets

In the modern vehicle, components often are subjected to heavy loads, high frequency vibration, excessive heat, or severe stress. As a result, fastener design, material, and torque settings are extremely important. Gaskets, seals, sealants, and adhesives are used throughout the vehicle. They confine fuel, oil, water, air, and vacuum to specific units or areas. They keep dust, dirt, water, and other foreign materials out of various parts. They affect torque and tension, part alignment and clearance, temperature, compression ratios, and lubrication. Fasteners, gaskets, and seals play an important part in the proper functioning and service life of all components.

Unfortunately, the importance of the proper selection and installation of these items is not always clearly understood. Extensive damage and expense can result from the failure of fasteners, gaskets, sealants, seals, or adhesives. Study the material in this chapter carefully and apply the information to your work.

Types of Fasteners

Fasteners hold vehicle parts together. They can be nuts, bolts, screws, or specialized devices. The technician must become familiar with the types, uses, and installation of various fasteners. Study the fasteners, their markings, and their uses until you can recognize them immediately.

Machine Screws

Machine screws are used without nuts. They are passed through one part and threaded into another, as shown in **Figure 8-1.** When the machine screw is tightened, the two parts are held in firm contact. The fastener shown in Figure 8-1 is a *capscrew*, a machine screw with a hexagonal head. There are many different types of machine screws and screw heads. See **Figure 8-2.**

Sheet Metal Screws

Sheet metal screws are screws with tapering threads, **Figure 8-3.** They are used to fasten thin metal pans together and to attach various items to sheet metal. Sheet metal screws are less expensive than bolts and can be installed more rapidly. To use a sheet metal screw, simply punch a hole in the sheet metal. The hole should be slightly smaller than the screw's minor diameter (diameter of the screw if threads were ground off). A punched hole is better than a drilled hole because the punched hole provides more metal for the screw to grip as it is tightened. See **Figure 8-4.**

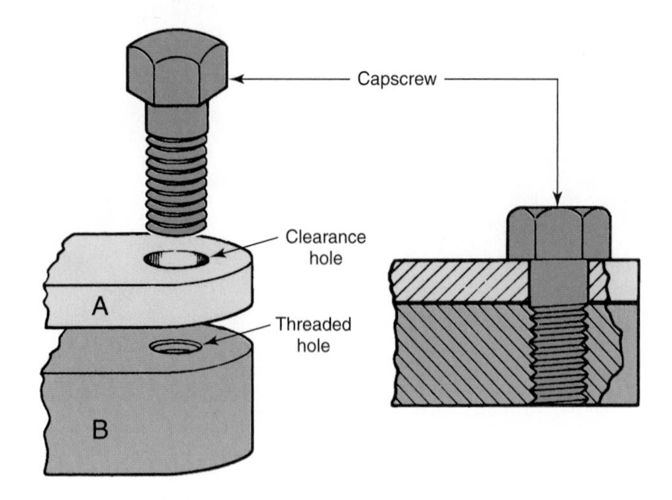

Figure 8-1. A machine screw is passed through a clearance hole in part A and threaded into part B. When the screw is tightened, the parts are pulled tightly together. The hex-headed capscrew is one of a number of types of machine screws used in vehicles.

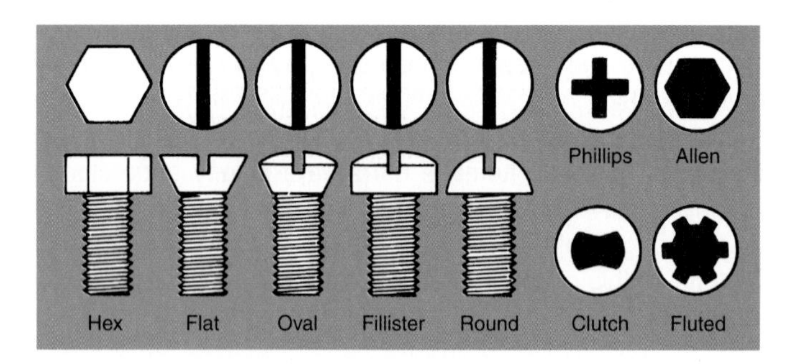

Figure 8-2. Typical machine screws. The four heads on the right illustrate various openings for turning tools in addition to the straight slot.

Bolts

A ***bolt*** is a metal rod with a head at one end and screw threads for a nut at the other. The bolt is passed through holes in the parts to be joined. Then, the nut is installed and tightened. This holds the parts together, **Figure 8-5.**

Studs

A ***stud*** is a metal rod that is threaded on both ends. The stud is turned into a threaded hole in one part, and another part is slipped over the stud. A nut is then turned down on the stud to secure the parts, **Figure 8-6.** Studs are available in many lengths and diameters. Some have a coarse thread on one end and a fine thread on the other. Others have the same thread on both ends. In some cases, this thread may run the full length of the stud.

A *stud wrench* should be used to install or remove studs. Be careful not to damage the threads during removal or installation. If a stud wrench is not available, place two nuts on the stud and "jam" them together (turn the top one clockwise and the bottom one counterclockwise until they are tight). Then, use a wrench on the lower nut to remove the stud. See **Figure 8-7.**

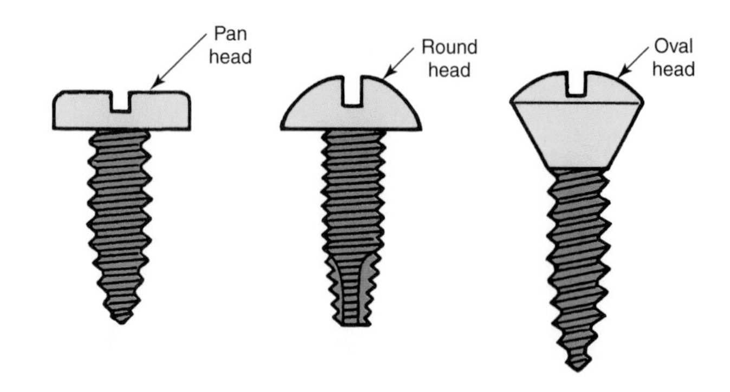

Figure 8-3. Typical sheet metal screws. Note the tapered threads.

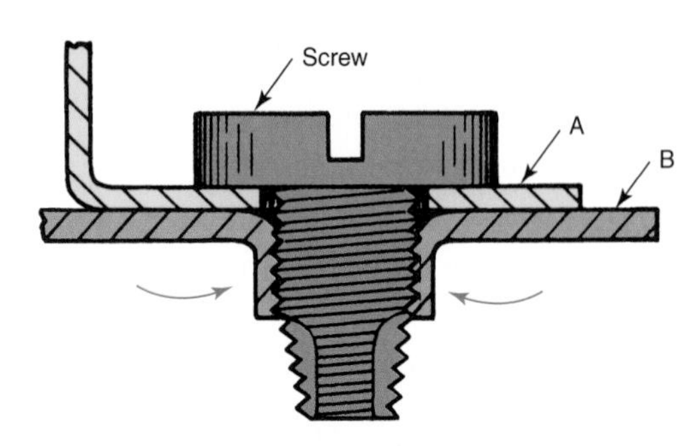

Figure 8-4. The screw passes freely through A and cuts threads in the punched hole in B. When the screw tightens, the punched metal is drawn up and in, providing a secure grip.

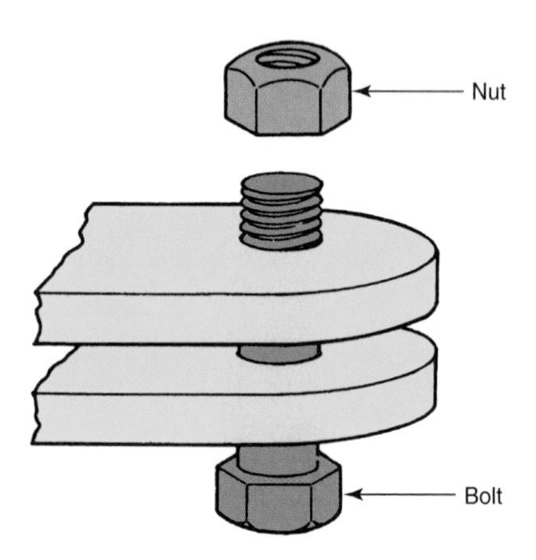

Figure 8-5. Using a bolt and nut to hold two parts together.

To install a stud using the two-nut method, place the wrench on the upper nut.

Removing Broken Studs or Screws

There are several methods for removing broken fasteners. See **Figure 8-8.** If a portion of a broken stud or screw projects above the work, it may be gripped with locking pliers or a small pipe wrench and backed out.

If the portion protruding is not sufficient to grasp with pliers or a wrench, flat surfaces may be filed to accept a wrench. Also, a slot may be cut to allow the use of a screwdriver, Figure 8-8A. A nut large enough to fit over the stud can

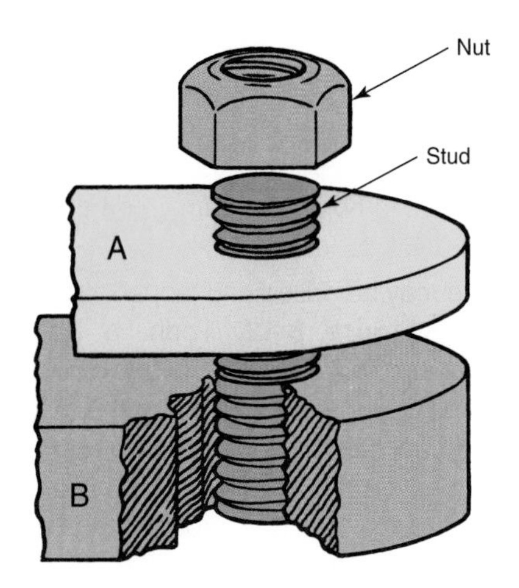

Figure 8-6. The stud is threaded into Part B. Then, Part A can be slipped over the stud. The nut is placed on the stud and tightened.

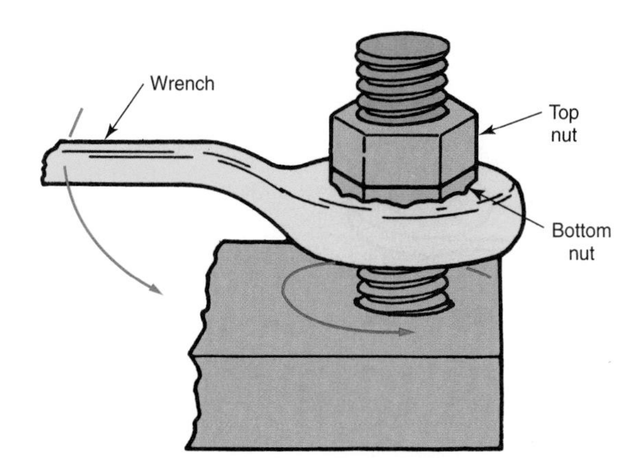

Figure 8-7. Using the jam nut method to remove a stud.

be welded on, Figure 8-8B. The arc welder does this job quickly and with a minimal amount of heating. Another method is to drill a hole in one end of a flat steel strip, slip it over the broken stud, and weld it in place. The steel strip will then provide the leverage to back out the broken fastener.

Warning: Be careful of fire and damage to parts when welding.

When the stud is broken off flush or slightly below the surface, you may use a thin, sharp, pointed punch. Try driving the broken section in a counterclockwise direction, Figure 8-8C. Sometimes the stub will turn out easily. If you are not getting results with a punch, stop and try another method.

A ***screw extractor*** can often be used with good results. Center punch the stub and drill through the stub with a small-diameter drill bit. Then, run a bit through that is slightly smaller than the stud's minor diameter. Lightly tap the extractor into the shell that remains. The sharp edges on the extractor flutes

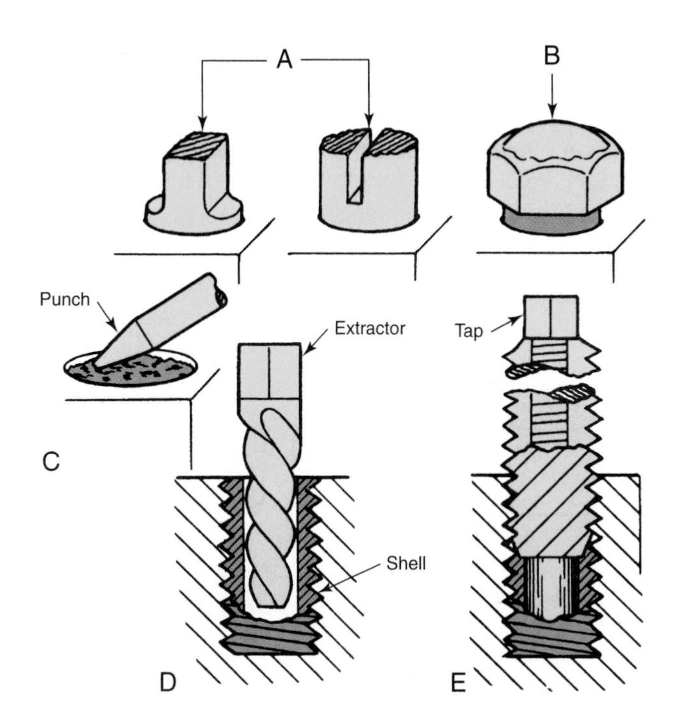

Figure 8-8. Methods used to remove a broken stud. A—The stud is filed flat or slotted. B—A nut is welded on the stud. C—A punch is used to unscrew the broken stud. D—A hole is drilled for a screw extractor. E—A tap is used to remove the shell threads.

will grip the shell, allowing you to back it out with a wrench, Figure 8-8D. Do not exert enough turning force to break the extractor. If the extractor breaks off in the fastener, it could present a problem since it is made of hardened steel.

If all the methods previously described fail, select the proper tap-size drill and drill through the stub shell. Next, carefully tap out the hole. If done properly, the tap will remove the shell threads, leaving the original hole threads undamaged. See Figure 8-8E.

When drilling to use a screw extractor or a tap, be careful to drill through the stub only. Do not drill beyond the stub because you may damage the part. If working on a setup where metal chips may fall into a housing, coat the drill and tap with a heavy coat of sticky grease. The chips will adhere to the grease and tools.

Nuts

Nuts are manufactured in a variety of sizes and styles. Nuts for automotive use are generally hexagonal (six-sided) in shape. They are used on bolts and studs. Nuts must be of the correct diameter and thread pitch (threads per inch), **Figure 8-9.**

Removing Damaged Nuts

Sometimes, a nut will be difficult to remove due to rust, dirt, and corrosion. When this happens, there are several methods you can use to remove the nut. As shown in **Figure 8-10,** the methods include the nut-splitter (cracker), hacksaw, chisel, and heat.

Regardless of the removal method, it is a good idea to apply ***penetrating oil*** (special light oil used to free rusty and dirty parts) to the damaged fastener and give it a few minutes to work in. If heat will not harm the part, applying heat will also help. Be careful not to overheat. Never use a torch near a gas tank, battery, or other flammable materials.

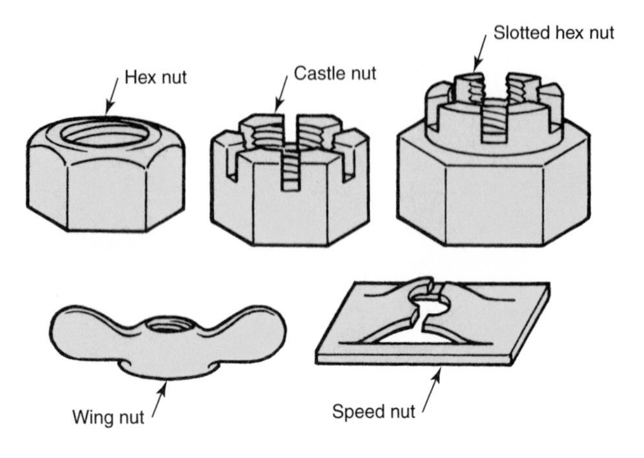

Figure 8-9. Common nuts. The slotted hex nut and castle nut are designed for use with a cotter pin or other locking device. A wing nut is installed and removed with the fingers. The speed nut is used to quickly fasten sheet metal or other parts not requiring the strength of a regular nut.

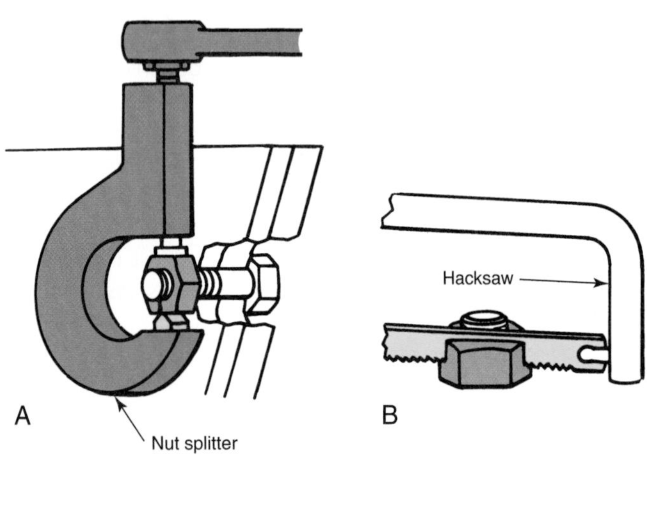

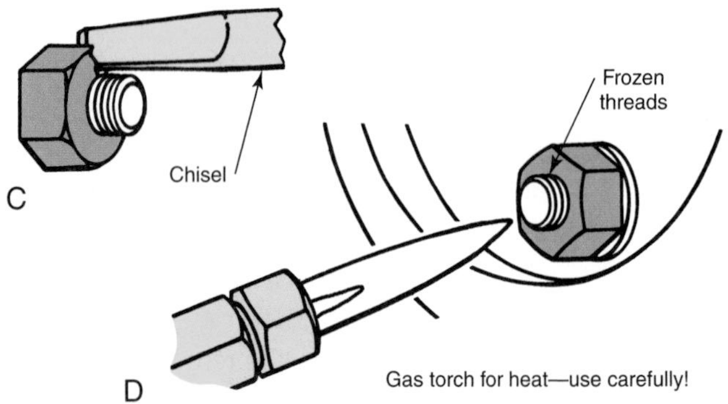

Figure 8-10. Four common methods of removing stubborn nuts. A—Nut splitter. B—Hacksaw. C—Chisel. D—Torch. (Deere & Co.)

Repairing Threads

If threads are only partially stripped, they often can be cleaned up using a thread die, thread chaser, or tap. See **Figure 8-11.**

When threads in holes are damaged beyond repair, one of four things can be done:

- The hole may be drilled and tapped to the next suitable oversize, **Figure 8-12.** Then, a larger diameter capscrew or stud can be installed. Use a chart to determine the proper size (tap size) to use. To allow an oversize capscrew to be used, a clearance or body drill (a drill the size of the bolt's major diameter) must be used to enlarge the hole through the attaching part.
- The hole may be drilled and tapped to accept a threaded plug, **Figure 8-13.** The plug should also be drilled and tapped to the original screw size. A special

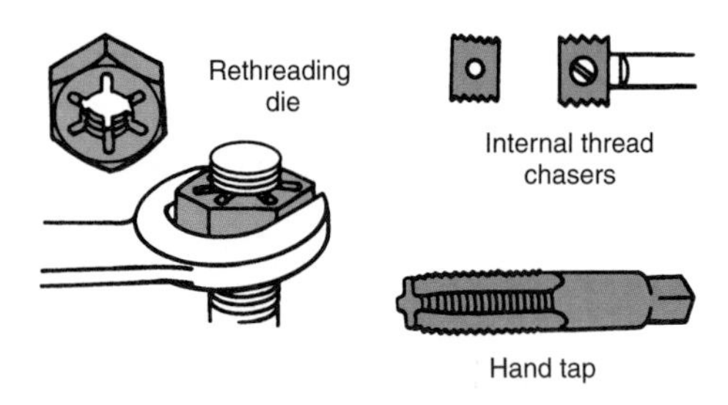

Figure 8-11. Thread restoring tools. (Deere & Co.)

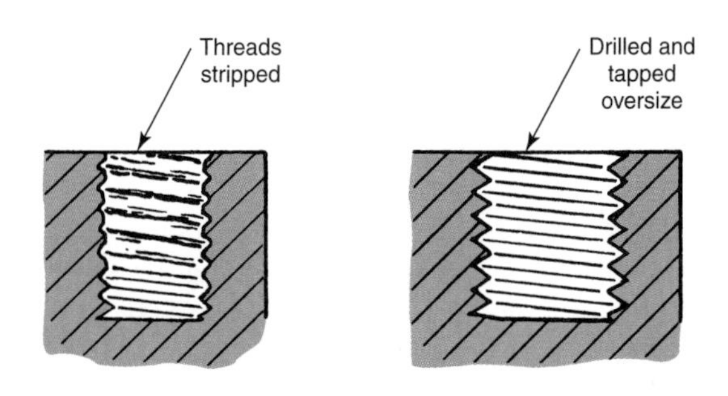

Figure 8-12. Repairing stripped threads by drilling and tapping to next oversize.

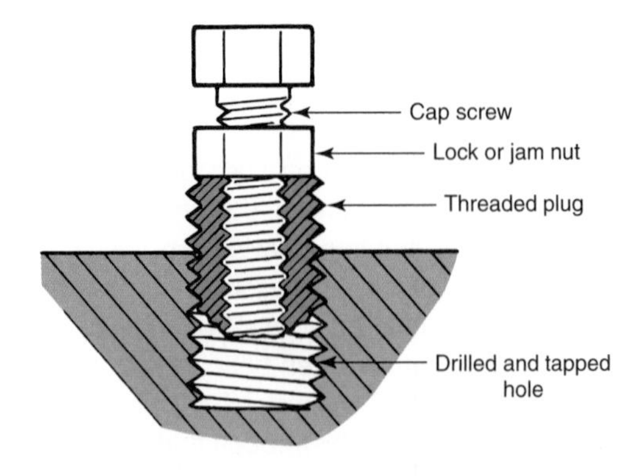

Figure 8-13. Inserting a threaded plug to repair stripped threads.

self-tapping plug that is already threaded to the original size may be used. You merely drill a hole to the specified size and run the threaded plug into a hole using a capscrew and jam nut. When fully seated, the jam nut is loosened and the capscrew removed,

- Another method makes use of a patented coil wire insert called a ***Heli-Coil®.*** A hole is drilled and tapped with a special tap. A Heli-Coil is then inserted. This brings the hole back to its original diameter and thread, **Figure 8-14.**
- Another type of thread repair, similar to the Heli-Coil, is called ***Keensert®.*** Instead of being an open coil insert, the Keensert is a hollow plug with internal and external threads. The damaged threaded hole is drilled and tapped with a special tap, **Figure 8-15.** The Keensert is then installed by hand until it is slightly below the top of the hole. A special punch is then used to force the four locking fingers down the slots that are cut into the insert. The top half of each finger is wider than the bottom, causing it to cut into the threaded hole. This locks the insert firmly into position and brings the hole to the original diameter and thread. If it becomes damaged, the insert may be removed using special tools.

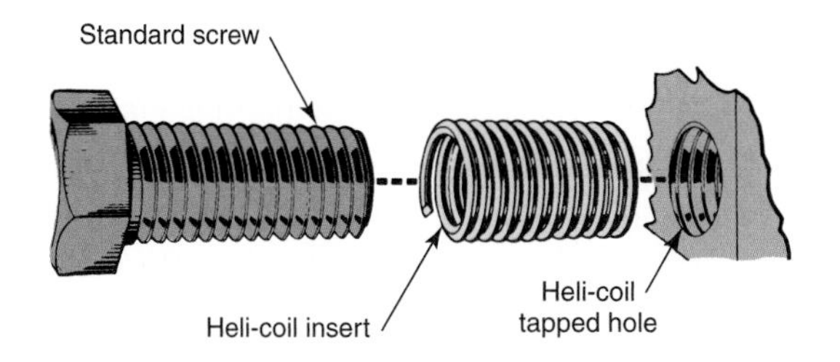

Figure 8-14. Repairing stripped threads using a Heli-Coil®. (DaimlerChrysler)

Caution: When removing a broken screw or repairing stripped threads, proceed carefully. A rushed or careless attempt at repair can often cause serious and costly trouble.

Bolt and Screw Terminology

Bolts and screws can be identified by type, length, major diameter, pitch (threads per inch), length of thread, class or fit, material, or tensile strength. In some cases, these fasteners are identified by the wrench size needed to install or remove them. The modern technician must learn both the unified (inch) thread designations and the metric thread classifications, **Figure 8-16.**

Head Markings

Not all steel bolts and capscrews are made of the same quality material or with the same temper. Current practice utilizes markings on the bolt and screw heads to indicate the ***tensile strength*** of the fastener, **Figure 8-17.**

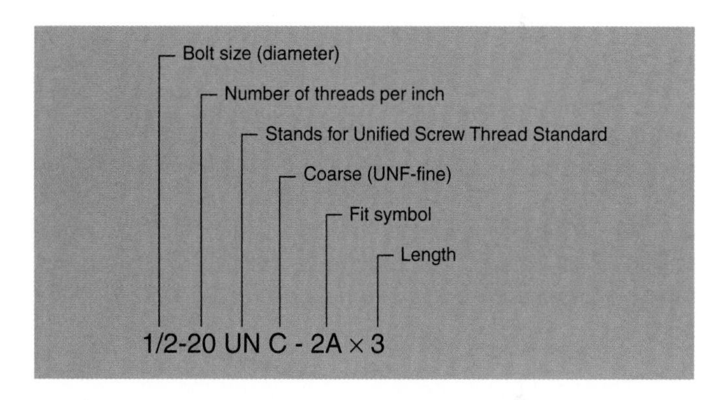

Figure 8-16. Bolt descriptive symbols chart. Learning these symbols and markings will make selecting correct replacement bolts easier and faster. (Deere & Co.)

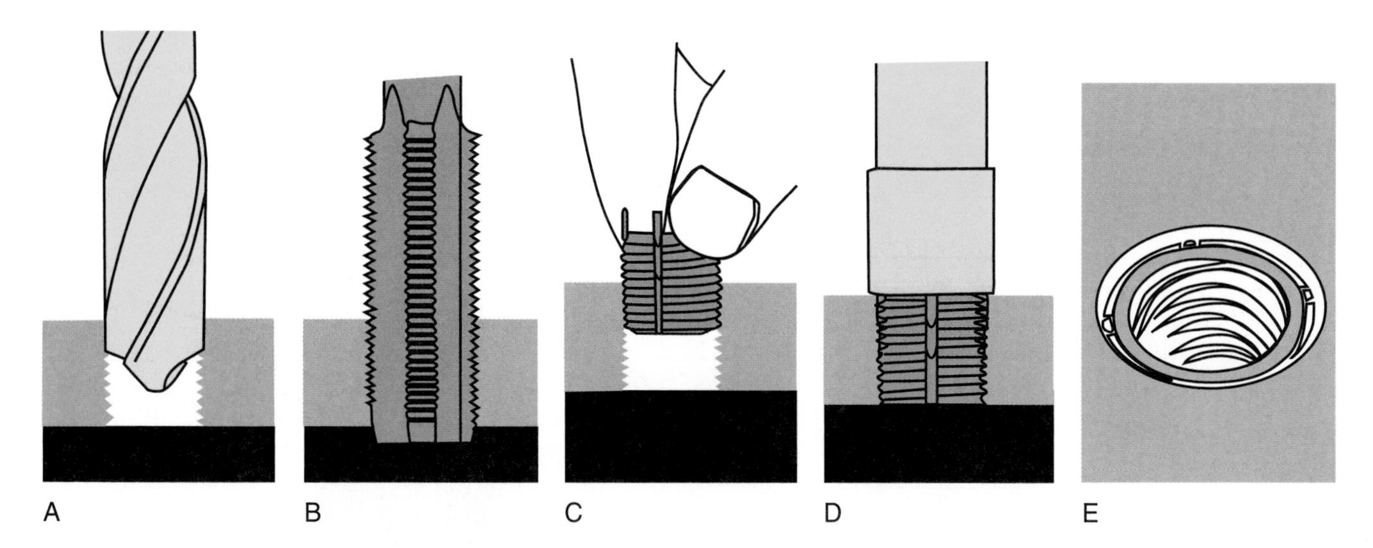

Figure 8-15. Keensert® internal thread repair procedure. A—Drill out damaged threads. B—Cut new threads with the tap. C—Screw insert in until its top is slightly below the hole surface. D—Drive the fingers into place with several light taps on the installation tool. E—Installed insert. The hole is now back to the original inside diameter and is ready to be used. (Rexnord)

Bolt Torquing Chart

Metric Standard						
Grade of bolt		5D	.8G	10K	12K	
Min. tensile strength		71,160 P.S.I.	113,800 P.S.I.	142,200 P.S.I.	170,679 P.S.I.	
Grade markings on head		5D	8G	10K	12K	Size of socket or wrench opening
Metric		Foot pounds				Metric
Bolt dia.	U.S. dec. equiv.					Bolt head
6mm	.2362	5	6	8	10	10mm
8mm	.3150	10	16	22	27	14mm
10mm	.3937	19	31	40	49	17mm
12mm	.4720	34	54	70	86	19mm
14mm	.5512	55	89	117	137	22mm
15mm	.6299	83	132	175	208	24mm
18mm	.709	111	182	236	283	27mm
22mm	.8661	182	284	394	464	32mm

SAE Standard Foot/Pounds						
Grade of bolt	SAE 1 & 2	SAE 5	SAE 6	SAE 8		
Min. tensile strength	64,000 P.S.I.	105,000 P.S.I.	133,000 P.S.I.	150,000 P.S.I.		
Markings on head					Size of socket or wrench opening	
U.S. Standard	Foot pounds				U.S. Standard	
Bolt dia.					Bolt head	Nut
1/4	5	7	10	10.5	3/8	7/16
5/16	9	14	19	22	1/2	9/16
3/8	15	25	34	37	9/16	5/8
7/16	24	40	55	60	5/8	3/4
1/2	37	60	85	92	3/4	13/16
9/16	53	88	120	132	7/8	7/8
5/8	74	120	167	180	15/16	1.
3/4	120	200	280	296	1-1/8	1-1/8

Figure 8-17. Grade markings on bolt heads indicate the tensile strength of the fastener. Different markings are used for Metric Standard and U.S. Standard (Unified) fasteners.

Major and Minor Diameters

The ***major diameter*** is the widest diameter, as measured from the crest, or top, of the threads on one side to the crest of the threads on the other side. The ***minor diameter*** is determined by measuring from the bottom of the threads on one side to the bottom of the threads on the other. If you were to remove all traces of the threads, the diameter of the portion left would be the minor diameter, **Figure 8-18.**

Thread Pitch

Thread pitch is the distance between the crest of one thread to the same spot on the crest of the next thread. The smaller the pitch, the greater number of threads per inch. The pitch, or number of threads per inch, can best be determined by using a thread-pitch gauge, **Figure 8-19.**

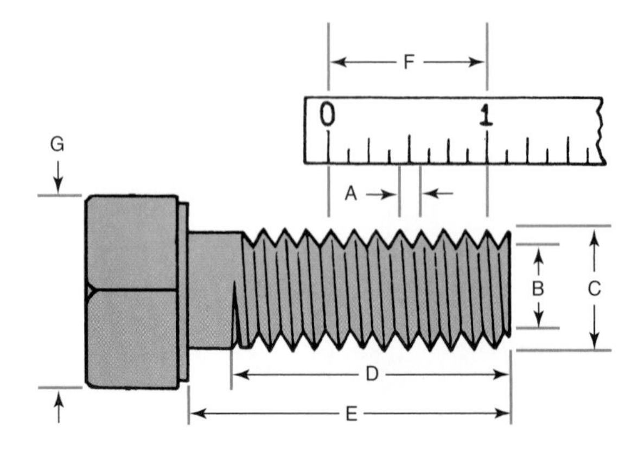

Figure 8-18. Bolt and screw terminology. A—Pitch. B—Minor diameter. C—Major diameter. D—Thread length. E—Screw length. F—Threads per inch. G—Head size measured across flats.

Thread Series

Three kinds of threads are commonly used on modern fasteners:

- Coarse threads (UNC or Unified National Coarse)
- Fine threads (UNF or Unified National Fine)
- Metric threads (SI or Systeme International)

When compared to fine threads, *coarse threads* have a larger and less critical shoulder bearing area, screw in and out faster, and are less subject to stripping and ***galling*** (galling occurs when threads rip particles of metal from each other, thereby damaging both threads).

Fine threads have more holding power than coarse threads. Even though each thread is smaller, there are more threads per inch, so the total gripping area is much more than that of the coarse thread.

The word *unified,* as used in Unified National Coarse and Unified National Fine, indicates that a thread conforms to thread standards used in the United States, Canada, and England. These threads are sometimes called customary or SAE threads. The UNC and UNF threads have been replaced

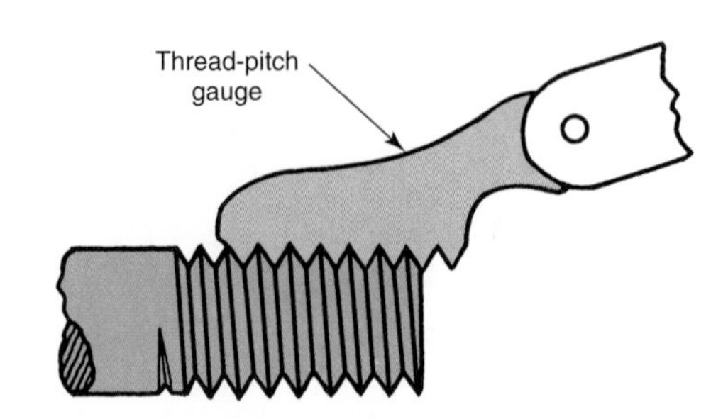

Figure 8-19. Using a thread-pitch gauge to determine the number of threads per inch.

on modern vehicles by *metric* fasteners, which are designated SI. Metric fasteners do not have separate coarse and fine designations.

Right-Hand and Left-Hand Threads

Both right-hand and left-hand threads are used. Almost all fasteners use a right-hand thread. The left-hand thread is reserved for special applications, such as wheel hub nuts or other places where part rotation would tend to loosen a fastener with right-hand threads.

Nut Terminology

Nuts are internally threaded fasteners used with bolts and studs. They are usually hexagonal in shape. Nuts have the same major thread diameter and number of threads per inch as their corresponding bolt, **Figure 8-20.** Wrench size (measured across flats) is standardized, but it does vary for special applications.

Class and Fit

Thread class indicates the operating clearance between a nut's internal threads and a bolt's external threads. Classes are divided into six categories: 1A, 2A, and 3A, for external threads (bolts, studs, screws) and 1B, 2B, and 3B for internal threads (nuts, threaded holes). This, in effect, gives three classes of fit. A Class 1 fit is a relatively loose fit and would be used for ease of assembly and disassembly under adverse conditions. Class 2 provides a fairly accurate fit, with only a small amount of clearance. Class 2 fasteners are commonly used for automotive applications. Class 3 is an extremely close fit and is used where utmost accuracy is essential.

Locking Devices

As screws, bolts, and nuts are subjected to vibration, expansion, and contraction, they tend to work loose. To prevent this, numerous ***locking devices*** have been developed. These may be an integral part of the screw or nut, or they may be a part placed under, through, or around the screw or nut. Epoxy cement or special locking compounds are sometimes used to prevent fasteners from loosening.

Self-Locking Nuts

Some nuts are designed to be self-locking, **Figure 8-21.** This is accomplished in various ways, but all *self-locking nuts* and *prevailing torque nuts* share the same principle. They create friction between the threads of the bolt or stud and the nut. The term "prevailing torque" refers to the fact that these nuts have a resistance to turning even before they are tightened. This torque figure must be added to the tightening torque specification for a particular nut.

The nut shown in Figure 8-21A has a collar of soft metal, fiber, or plastic. As the bolt threads pass up through the nut, they must force their way through the collar. This jams the collar material tightly into the threads, locking the nut in place.

In Figure 8-21B, the upper section of the nut is slotted, and the segments are forced together. When the bolt passes through the nut, it spreads the segments apart, producing a locking action.

Figure 8-21C shows a single slot in the side of a nut. The slot may be forced open or closed during manufacture, distorting the upper thread. This will create a jamming effect when the bolt threads pull the nut threads back into alignment.

Self-Locking Screws

Some capscrews have heads that are designed to spring when tightened, producing a self-locking effect. Occasionally, the threaded end of a capscrew will be split and the halves will be bent outward slightly. When threaded into a hole, the halves are forced together. This creates friction between the threads.

Lock Washers

A *lock washer* is used under a nut and grips both the nut and the part surface. The three basic lock washer designs are the internal lock washer, the external lock washer, and the plain lock washer. When using lock washers with die cast or aluminum parts, a flat steel (nonlocking) washer is often used under the lock washer. This practice prevents damage to the part. See **Figure 8-22.**

Palnuts

The *palnut* locking device is constructed of thin, stamped steel. It is designed to bind against the threads of the bolt

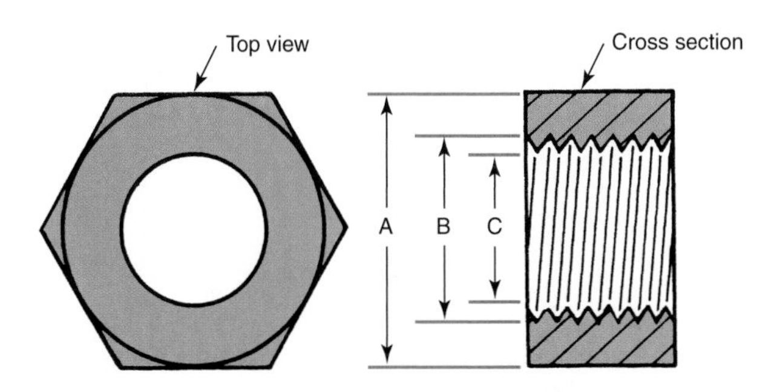

Figure 8-20. Typical nut. A—Size across flats. B—Thread major diameter. C—Thread minor diameter.

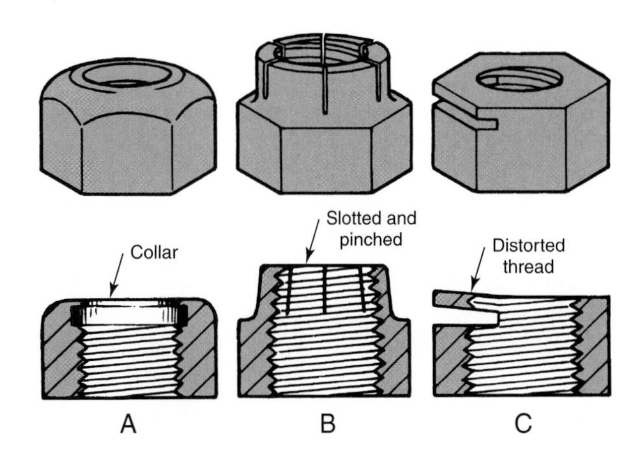

Figure 8-21. Self-locking nuts. A—Soft collar type. B—Top section slotted and pinched together. C—Slotted to distort upper thread area.

when installed. The palnut is spun down into contact with the regular nut (open side of palnut away from regular nut), as shown in **Figure 8-23.** Once firmly in contact with the nut, the palnut is given another half turn. The half turn draws the steel fingers toward the nut, causing them to jam into the threads. Do not tighten the palnut more than a half turn, or its effectiveness will be destroyed.

Cotter Pin

Cotter pins (sometimes called "cotter keys") are used both with slotted nuts and castle nuts. They are also used in clevis pins and linkage ends. Use as thick a cotter pin as possible. Cut off the surplus length and bend the ends as shown in **Figure 8-24.** If necessary, the ends may be bent around the sides of the nut. Make certain the bent ends will not interfere with other parts.

Special Locking Compounds

Special *locking compounds* are sometimes used to hold fasteners in place. These compounds are applied as liquids and harden to hold the fastener in place. They are often called ***anaerobic sealers,*** which means that they remain liquid when exposed to oxygen, but harden after the fastener is tightened and squeezes out all of the air from the threads. A common brand of anaerobic sealer is Locktite®, which is available in many varieties. The type of sealer used will depend on the type of fastener, heat range, or other special circumstances.

Other Types of Fasteners

In addition to the common fasteners described previously, the average vehicle uses many special fasteners. Special fasteners are used in situations where there is no room for a normal fastener or where the use of threaded fasteners would be inconvenient.

Keys, Splines, and Pins

Keys, splines, and ***pins*** are used to attach gears, sprockets, or pulleys to shafts so that they rotate as a unit. When a key or pin is used, the unit being attached to the shaft is generally fixed to eliminate end-to-end, or longitudinal, movement. If desired, splines will allow longitudinal movement while still causing the parts to rotate together. In some cases, pins are used to fix shafts in housings and to prevent end movement and rotation, **Figure 8-25.**

A ***locking plate*** is made of thin sheet metal. The plate is generally arranged so that two or more bolts pass through it. The metal edge or tab is then bent up snugly against the bolt, preventing rotation, **Figure 8-26.** Always dispose of locking plates on which the tabs are fatigued (ready to crack).

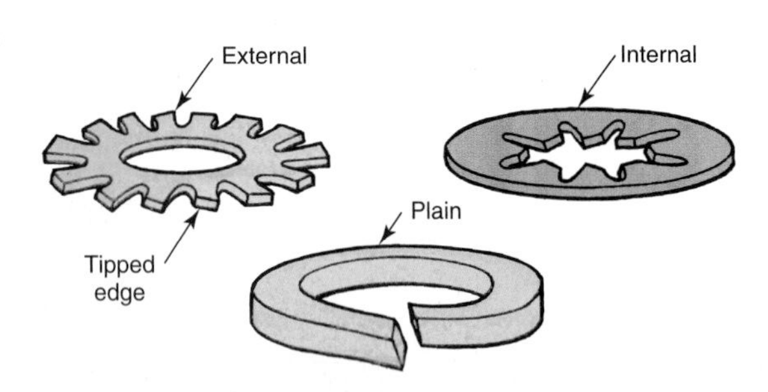

Figure 8-22. Lock washers. Tipped edges provide gripping power in the "off" direction.

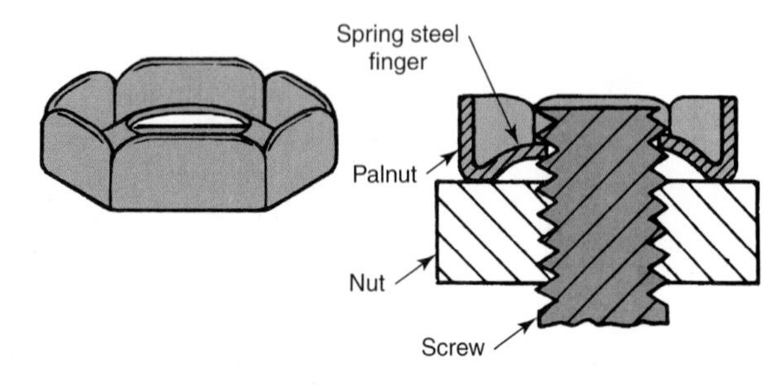

Figure 8-23. Palnut. A half turn jams the steel fingers against the threads.

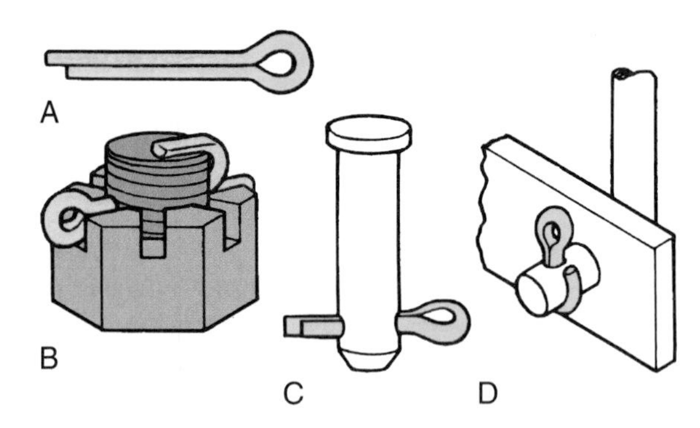

Figure 8-24. Uses of a cotter pin. A— Typical cotter pin. B— Slotted hex nut. C— Clevis pin. D—Linkage.

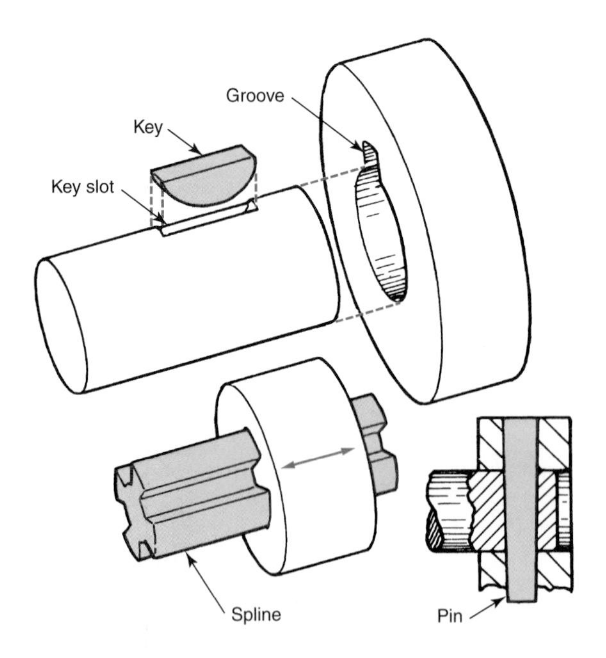

Figure 8-25. Key, spline, and pin. The key is commonly referred to as a Woodruff key or a half-moon key. Note that the spline allows end movement. The pin fixes the shaft to the housing, allowing no end movement.

Occasionally, screws will be locked with ***safety wire*** (soft or ductile wire). The wire is passed from screw to screw so that it exerts a clockwise pull on the fasteners. This prevents counterclockwise (loosening) movement. Never reuse safety wire.

Snap Rings

Snap rings are used to position shafts, bearings, gears, and other similar parts. There are both internal and external snap rings of numerous sizes and shapes, **Figure 8-27.**

The snap ring is made of spring steel. Depending on the type, it must be expanded or contracted to be removed or installed. Special snap ring pliers are used to remove and install the snap rings.

Be careful when installing or removing snap rings: overexpansion or contraction will distort and ruin them. If a snap ring is sprung out of shape, throw it away. Never attempt to pound one back into shape. Never compress or expand snap rings any more than necessary. Above all, do not pry one end free of the groove and slide it along the shaft. This may ruin the ring.

Setscrews

Setscrews are used to both lock and position pulleys and other parts to shafts. The setscrew is hardened and is available with different tips and drive heads.

Keep in mind that setscrews are poor driving devices because they often slip on the shaft. When used in conjunction with a Woodruff key (refer to Figure 8-25), the setscrew merely positions the unit. As a general rule, do not install any driving unit without a Woodruff key.

When a setscrew is used, the shaft will usually have a flat spot to take the screw tip. Make certain this spot is aligned before tightening the screw, **Figure 8-28.**

Rivets

Rivets are made of various metals, including brass, aluminum, and soft steel. They have many applications on an automobile. When using rivets, there are several important considerations. The two parts to be joined must be held tightly together during riveting. The rivet should fit the hole snugly. The rivet material must be in keeping with the job to be done. The rivet must be of the correct type. *Pop rivets*, **Figure 8-29,** are generally used for automotive applications. An important advantage of the pop rivet method is that it can be used where only one side of the part or assembly is accessible.

Figure 8-30 shows pop rivets being used to attach bumper trim. The pop rivet is inserted through the parts to be joined, and a hand-operated setting tool is placed over the rivet anvil pin. When the tool's handles are closed, the anvil pin is pulled outward. As the anvil is drawn outward, the rivet head is forced against the work and the hollow stem is set. The setting process draws the two parts tightly together. Further

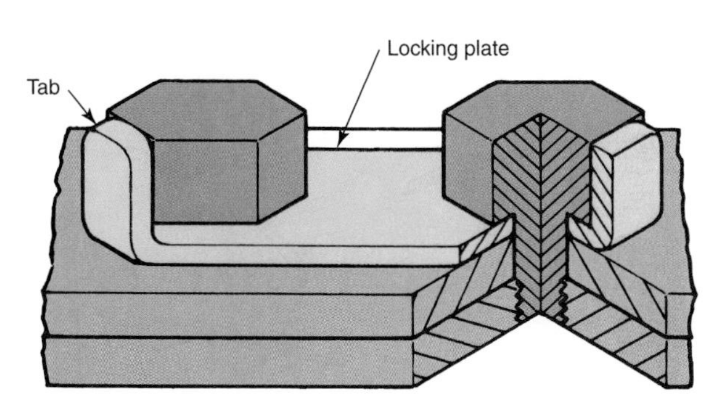

Figure 8-26. The tabs of the locking plate must be bent firmly against the flats of the capscrews to prevent rotation.

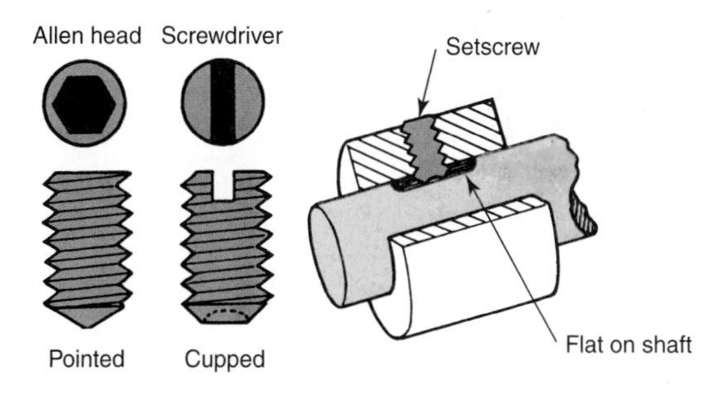

Figure 8-28. Setscrews may have pointed or cupped ends, and commonly have either a screwdriver slot or a recess for an Allen wrench. Setscrews are hardened and should be run up very tightly.

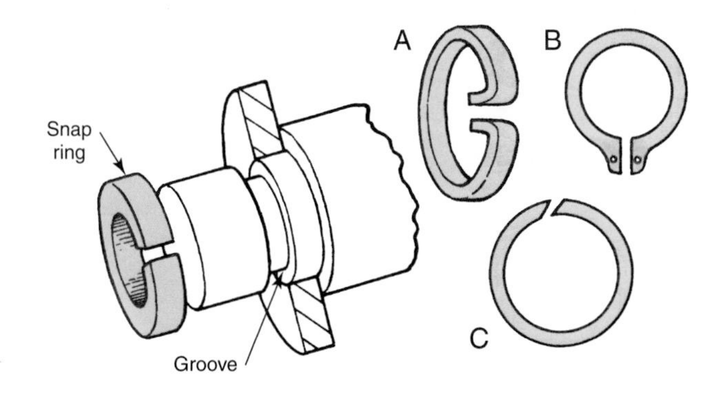

Figure 8-27. A snap ring fits in a groove to position a part, such as a gear or bearing. A—Flat internal snap ring. B—External snap ring. C—Round external snap ring. There are many different shapes and sizes of rings.

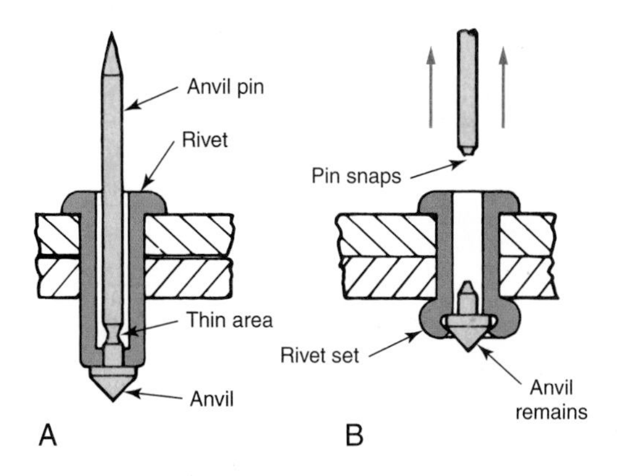

Figure 8-29. Installing a pop rivet. A—Pop rivet in place. B—Rivet tool has pulled the anvil pin outward, pulling the parts together, setting the rivet, and snapping off the pin.

pressure on the tool's handles causes the anvil pin to snap off just ahead of the anvil. The anvil remains in the set area.

Fasteners and Torque

To better understand the reason for and proper application of controlled torque (tightening), the technician should be familiar with several important terms. Read the definitions that follow carefully. These terms will be used a great deal in this section.

- ***Torque***—a turning or twisting force exerted on an object, such as a fastener. Torque is measured in inch-grams, inch-ounces, inch-pounds, foot-pounds, or Newton-meters. See **Figure 8-31.**
- ***Tension***—a pulling force, **Figure 8-32.** When a capscrew is tightened, it actually stretches about 0.001″ (0.025 mm) for every 30,000 lb (13,500 kg) of tension applied.
- ***Elastic limit***—the amount an object can be distorted (compressed, bent, or stretched) and still return to the original dimension when the force is removed, **Figure 8-33.**
- ***Distortion***—change that occurs when the normal shape or configuration of an object is altered due to the application of some force or forces, **Figure 8-34.**
- ***Tensile strength***—the amount of pull an object will withstand before breaking, **Figure 8-35.**
- ***Residual tension***—the stress remaining in an elastic object that has been distorted and not allowed to return to its original dimension, **Figure 8-36.**
- ***Elasticity***—the ability of an object to return, after distortion, to its original shape and dimensions once the distorting force has been removed, **Figure 8-37.**

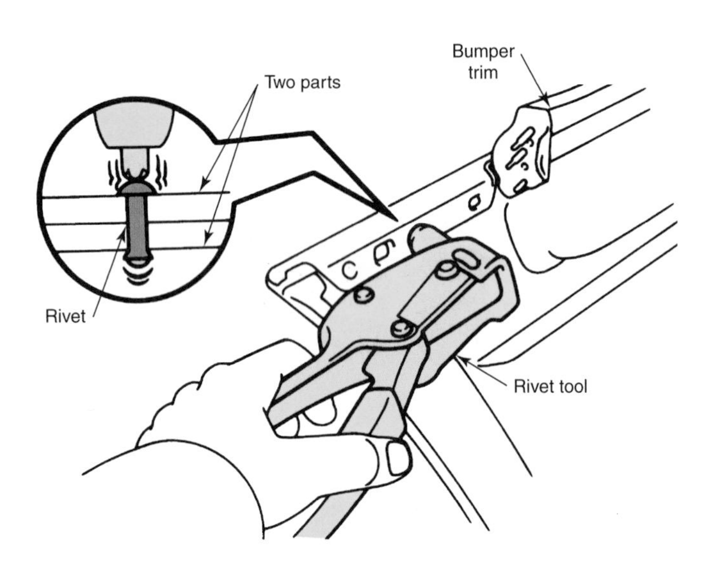

Figure 8-30. Pop rivet tool being used to attach bumper trim parts. (Lexus)

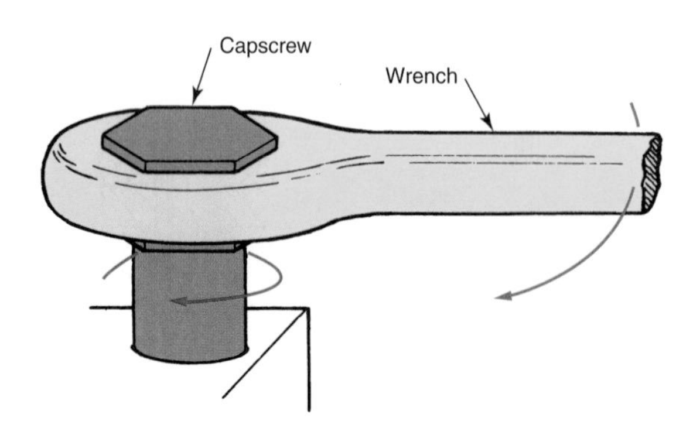

Figure 8-31. Torque, or a twisting force, being applied to a capscrew with a box-end wrench.

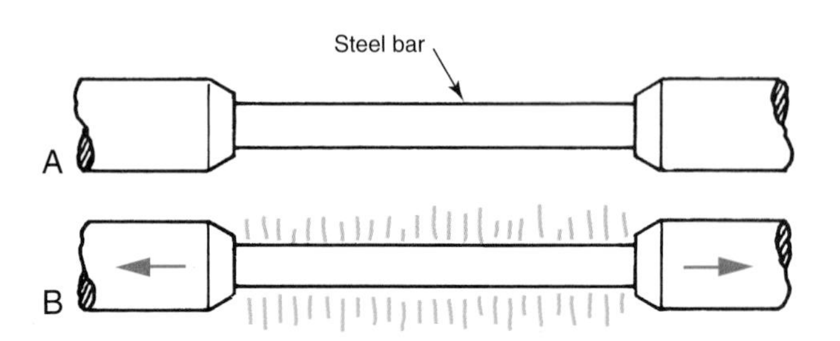

Figure 8-32. Tension. A—A steel bar placed in the jaws of a test machine. B—The jaws are moved apart, creating a pull, or tension, on the bar.

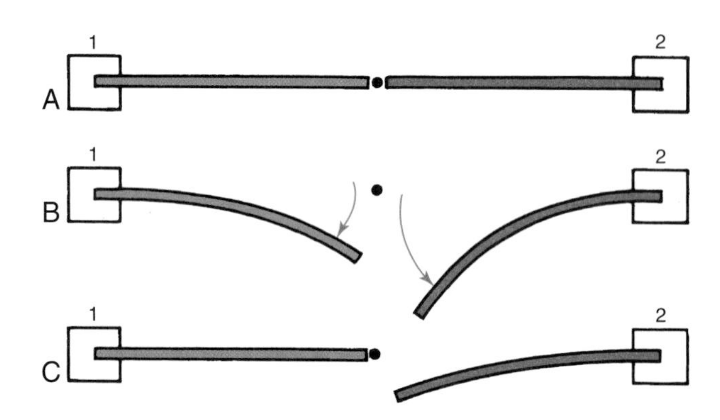

Figure 8-33. Elastic limit. A—Bars 1 and 2 are at rest. Note that they are aligned with the black dot. B—Bar 1 is bent within its elastic limit. Bar 2 is bent beyond its elastic limit. C— When pressure is removed, Bar 1 springs back to its normal position. Bar 2, however, springs only partway back.

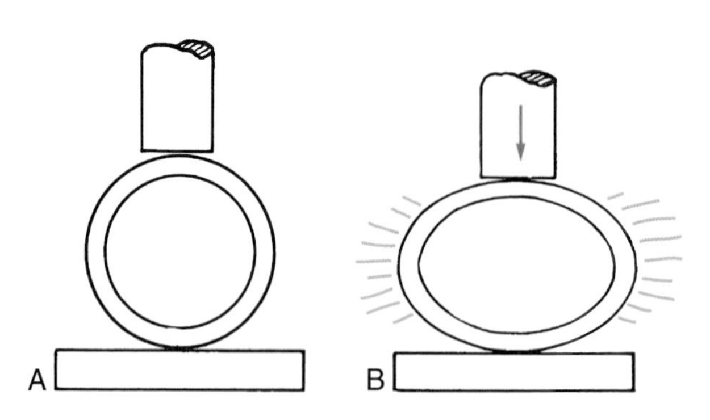

Figure 8-34. Distortion. A—Hydraulic ram about to apply pressure to steel ring. B—Pressure from ram distorts the ring.

- ***Compression***—a force tending to squeeze an object, **Figure 8-38.**
- ***Cold flow***—the tendency of an object under compression to expand outward, thus reducing its thickness in the direction of compression, **Figure 8-39.**
- ***Hooke's law of springs***—this law states that as long as distortion is kept within the elastic limits of a material, the amount of distortion (lengthening, shortening, bending, twisting) will be directly proportional to the applied force. This forms the basis for spring scales and torque wrenches, **Figure 8-40.**
- ***High-pressure lubricant***—a lubricant that continues to reduce friction between two objects even when they are forced together under heavy pressure.

Torquing Fasteners

To understand the need for proper torquing, we should first establish what we want to accomplish by tightening fasteners. Technicians tighten fasteners to hold parts together. Once together, the parts should remain that way. The fasteners should not be tightened to the point at which they will break or distort other parts. However, they must be tightened

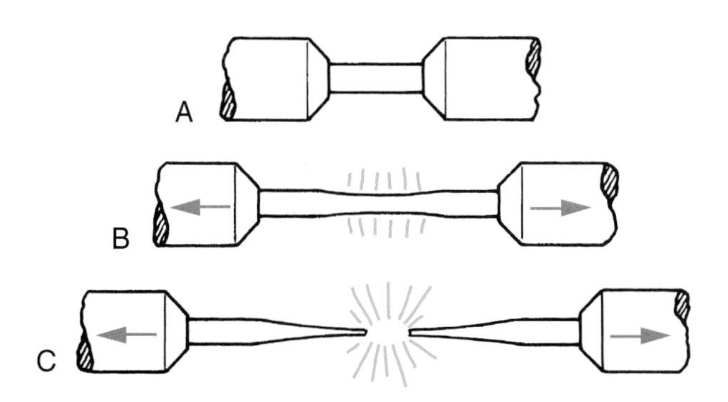

Figure 8-35. Tensile strength. A—Steel bar clamped in a test machine. B—Tension is applied, exceeding the elastic limit and causing the bar to stretch. C—Increased pull snaps the bar as tension exceeds tensile strength.

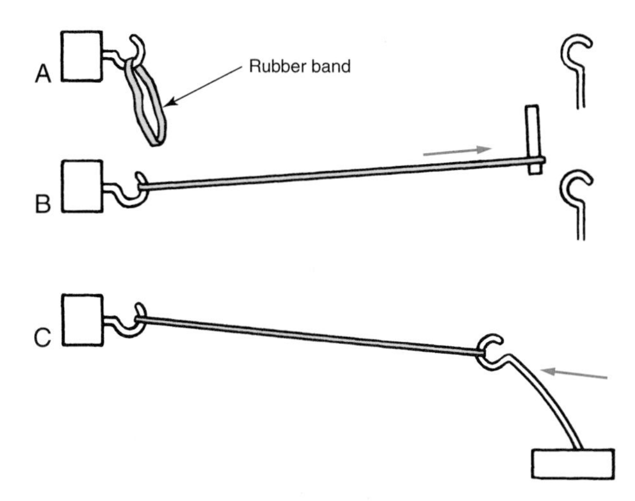

Figure 8-36. Residual tension. A—Rubber band at rest; no residual tension. B—Band being pulled (distorted) out to engage spring steel hook. C—Band attempts to return to original dimensions, creating a pull (residual tension) and bending the hook.

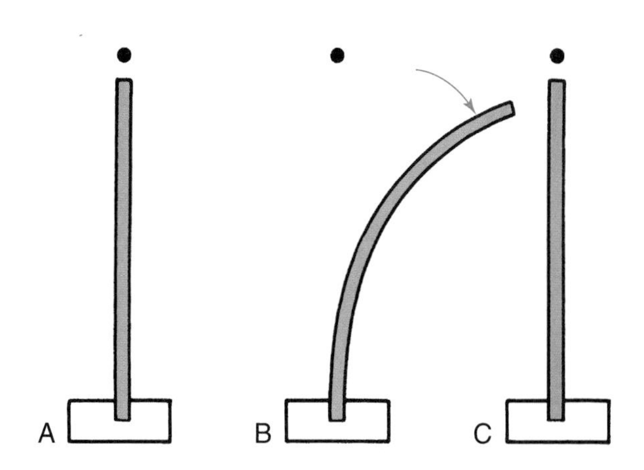

Figure 8-37. Elasticity. A—Original position of bar. B—Bar deflected by pressure. C—Bar returns to its original position when pressure is released.

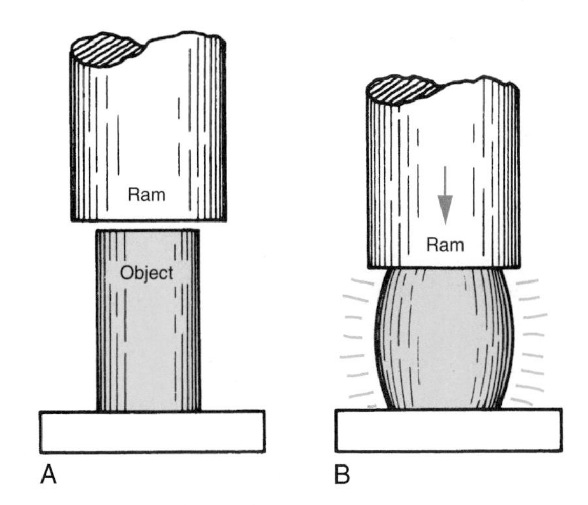

Figure 8-38. Compression. A—Ram about to apply pressure to object. B—Object is distorted by compression as ram builds up pressure.

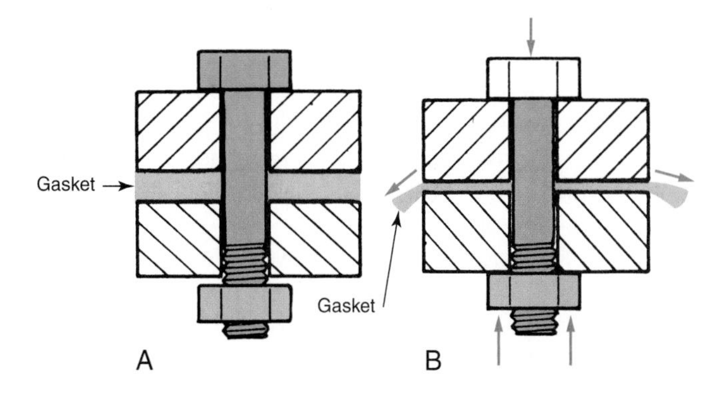

Figure 8-39. Cold flow. A—The nut is loose; there is no compressive force on the gasket. B—The nut is tightened, compressing the gasket and causing it to flow outward as thickness decreases.

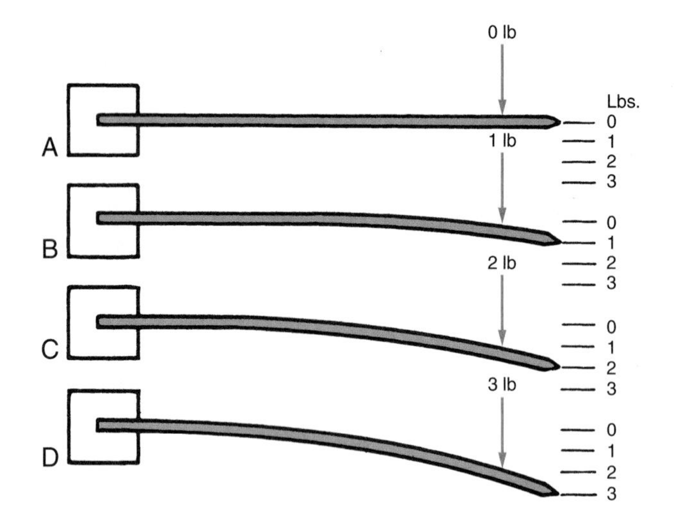

Figure 8-40. Hooke's Law of Springs. As weight on the spring bar is increased, there is a proportional movement across the scale. The proportional increase will continue until the bar is deflected past its elastic limit.

enough to prevent them from working loose or from being sheared or pounded apart. They must also be tightened enough to prevent oil, gas, and water leaks.

Tightening fasteners in a haphazard manner can cause the following problems:

- Out-of-round cylinders.
- Egg-shaped connecting rod and main bearings.
- Warped cylinder head.
- Misaligned valve guides.
- Misaligned camshaft bearings.
- Misaligned crankshaft bearings.

In addition, the engine can suffer blown head gaskets, oil, water, and air leaks, or broken connecting rods if fasteners are improperly tightened. Ring, piston, valve, and bearing wear will be accelerated. The engine will fail in service long before it should.

Proper Fastener Tension

All vehicle manufacturers publish ***torque specifications,*** which should be closely followed. Each company has spent a great deal of time and money determining the fastener torque that will give the best results for their products. When using torque charts, make sure they pertain to the job at hand.

For the vast majority of applications, a fastener should be tightened until it has built up a tension within itself that is around 50%-60% of its elastic limit.

When the fastener has been drawn up to this point, it will not be twisted off. It will retain enough residual tension to continue to exert pressure on the parts and will resist loosening. As mentioned earlier, steel bolts and capscrews will stretch about 0.001″ for each 30,000 pounds of tension. The tendency of the material to return to its normal length (like a rubber band) provides a continuous clamping effect.

Fastener Material

As previously mentioned, most bolts and screws have markings on their heads that indicate tensile strength. When replacing a fastener, always use a fastener that is at least as strong as the original. You will find the more critical the application (main bearing, connecting rod), the higher the tensile strength of the fastener.

Torque-to-Yield Bolts

Some engines and (other vehicle parts) have fasteners called ***torque-to-yield bolts.*** These bolts are made of an alloy that causes them to stretch (or yield) at a specific tension. Torque-to-yield bolts resemble other bolts, but are tightened by a two-step procedure. To correctly torque a torque-to-yield bolt, first tighten it to a relatively low value (usually less than 100 ft lb). The bolt is then tightened an additional number of degrees. This causes the material to slightly stretch, placing the correct amount of torque on the bolt. This method is superior to older torquing methods, especially when used on aluminum or magnesium parts. *Torque-to-angle indicators,* sometimes called "degree wheels" are used to obtain the proper amount of rotation after the starting torque is reached. When torque-to-yield bolts are used on cylinder heads, a specific torquing sequence must be used, just as with other head bolts. The torque-to-angle indicator is a two-piece tool made of clear plastic. One part of the tool moves with the wrench as the bolt is tightened. The other half is held stationary. The relative movement between the two parts allows the technician to accurately tighten the bolt.

Note: Torque-to-yield bolts cannot be reused. When torque-to-yield bolts have been removed, discard them and obtain new fasteners.

Special Precautions for Magnesium Part Fasteners

Magnesium alloys are used to make some drive train parts, since magnesium is lighter than other metals. Magnesium will react chemically with all other metals, except aluminum, however. For this reason, fasteners used with magnesium parts must be made of aluminum, or be separated from the part with aluminum washers. Many fasteners for magnesium parts are plastic-coated to reduce metal-to-metal contact. Other fasteners have seals to keep air out of the fastener area. Such fasteners should always be replaced with the same type of fastener.

Using a Torque Wrench

To tighten a fastener to a recommended torque, a tool called a ***torque wrench*** is used. The torque wrench will measure the twisting force (torque) being applied to the fastener. **Figure 8-41** shows some typical torque wrenches.

Torque recommendations are usually given in inch-grams, inch-ounces, inch-pounds, and foot-pounds. However, some vehicle makers give their torque specifications in

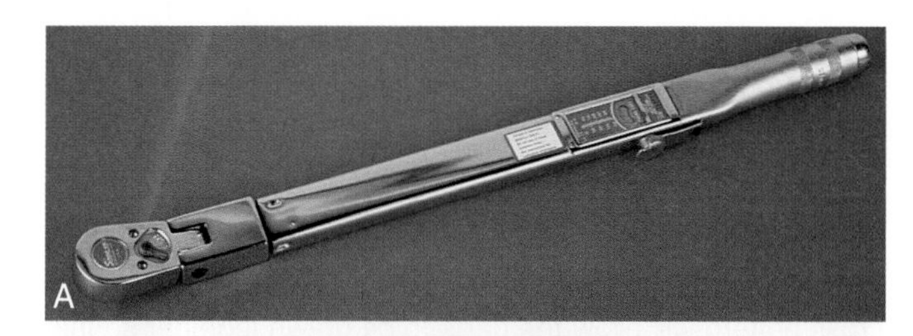

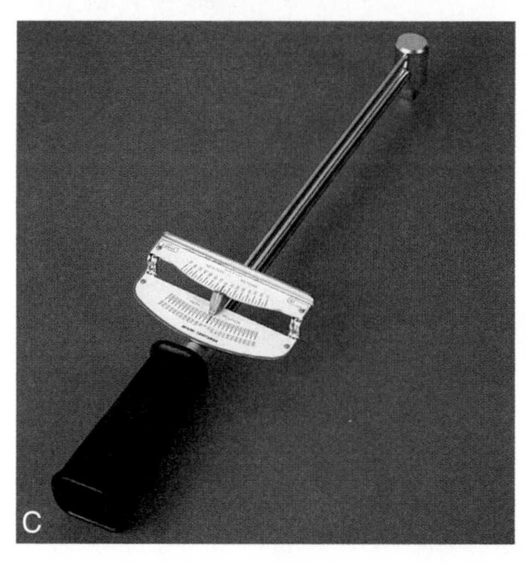

Figure 8-41. Typical torque wrenches. A—Click-type torque wrench. B—Dial-type wrench. C—Beam-type torque wrench. (Fel-Pro)

Newton-meters (N•m). To convert foot-pounds to Newton-meters, multiply by 1.356.

If possible, inch-pound and foot-pound torque wrenches should be used to make direct inch-pound and foot-pound readings. However, if the correct wrench is not available, the readings can be converted. To convert foot-pounds to inch-pounds, multiply the foot-pound reading by 12. To convert inch-pounds to foot-pounds, divide the inch-pound reading by 12.

Torque wrenches are made in different sizes, or *ranges,* as well as in different calibrations. Ideally, the technician should have a 0–200 in lb (0–22.60 N•m) wrench, a 0–50 ft lb (0–67.79 N•m) wrench, a 0–100 ft lb (0–135.58 N•m) wrench, and a 150 ft lb (203.37 N•m) wrench.

A torque wrench will produce the best results if used for readings that fall near the middle half of its range. For example, a 0–100 ft lb wrench would give the most accurate readings from 25 ft lb to 75 ft lb. By having several ranges of wrenches, the technician will also find this will offer several lengths. The shorter wrenches can be useful in restricted areas. After determining the proper torque and selecting a suitable torque wrench, you are ready to proceed.

Threads Must Be Clean

The threads on the bolt or screw and those in the nut or hole must be absolutely clean. Rust, carbon, and dirt will cause galling and improper tension. An accurate torque reading is impossible with dirty threads.

Use High-Pressure Lubricant

Unless the use of a lubricant is specifically forbidden (due to the possibility of area contamination or the need of a special sealant), always apply a lubricant to the threads and to the area where the nut or capscrew head contacts the part. Refer to manufacturer recommendations to find out which lubricant is suitable.

The use of lubricant will prevent or reduce the possibility of galling, seizing (sticking), or stripping. It will ensure that the fastener torque has created the proper tension. It should be mentioned the lubricant, while making the fasteners easier to remove at some future date, will not (if torqued properly) cause them to loosen in service. To the contrary, the increased tensioning for the same torque reading will actually cause the fastener to remain more secure.

Use a Proper Locking Device

Unless a self-locking nut or capscrew is being used, make certain the recommended lock washer is in place.

When tightening a fastener up against the softer metals, the use of a plain flat washer between the lock washer and the part is often specified. This prevents the part from being "chewed up" and allows proper torquing without crushing the part.

Check Fasteners

Always check fasteners for correct diameter, thread type, and length. When installing capscrews, make certain that they will not bottom (strike the bottom of a threaded hole) in a ***blind hole*** (a hole that is not drilled clear through part). Also, make sure that they do not protrude into a housing. This may damage a part of the unit. See **Figure 8-42.**

Caution: Stripped threads, broken screws, loose parts, and damaged units can result from failure to use the correct fastener. Be careful!

In Figure 8-42A, the screw has bottomed, leaving the part loose. Continued torquing could twist off the screw head. In Figure 8-42B, the screw protrudes into the housing, causing damage to a gear. In Figure 8-42C, a coarse thread screw was jammed into a hole with fine threads, cracking the housing.

Some fasteners serve an additional purpose. For instance, a head bolt or capscrew may be drilled for passage of oil, or a capscrew may have a threaded hole in its head to which another assembly is attached. Be careful to insert these fasteners in the correct locations.

Follow the Recommended Tightening Sequence

Where a number of fasteners are used to secure a part (such as a cylinder head), the proper ***tightening sequence*** (order) should be followed. **Figure 8-43** illustrates the head bolt tightening sequence for one engine. Always follow the manufacturer's specifications.

Observe the tightening sequence shown in **Figure 8-44.** Would a good fit be acquired if you followed this sequence? If this sequence is followed, the two ends would be clamped down first. When the center bolts were tightened, the part could not flatten out. In order to flatten, it must spread outward. Therefore, the ends must be free.

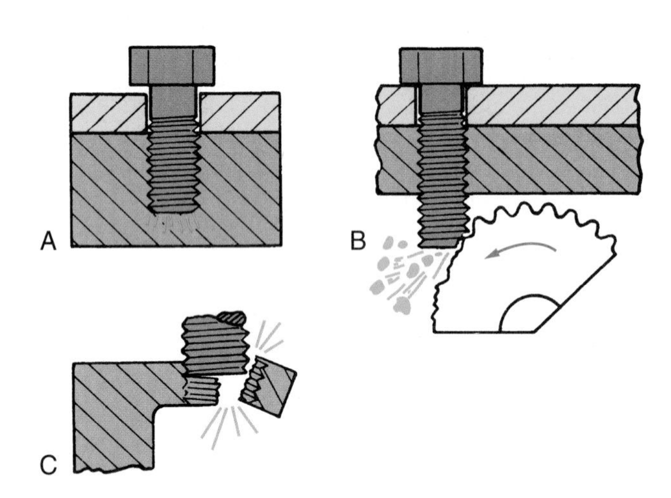

Figure 8-42. Always check fasteners! Make certain the fasteners are of the correct length and diameter and that they have the proper number of threads per inch.

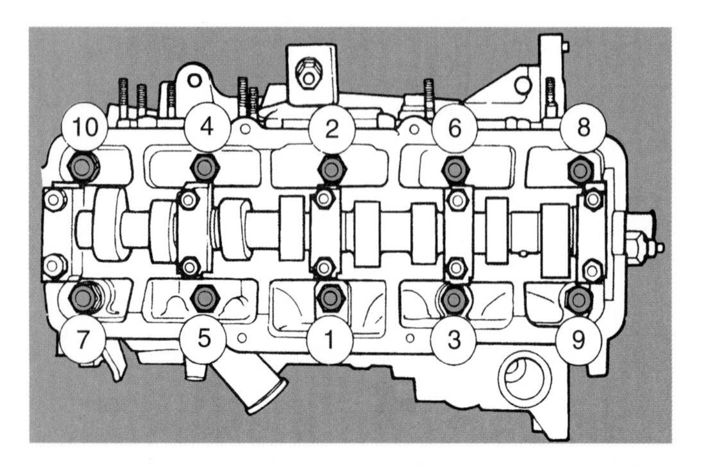

Figure 8-43. Cylinder head bolt tightening sequence for one specific engine. (DaimlerChrysler)

If a sequence chart is not available, it is usually advisable to start in the center of the part and work out to the ends. The chart in **Figure 8-45** illustrates this technique. On some assemblies, it is advisable to use a crisscross sequence, **Figure 8-46.** Always avoid starting in one spot and tightening one fastener after another in a row. Remember the object is to tighten the parts so that an even stress is achieved. A crisscross pattern will allow the parts to be drawn together so that their mating surfaces will contact evenly.

Torque in Four Steps

Begin by tightening fasteners snug (do not overtighten) with a regular wrench. Then, change to a torque wrench for the following four steps:

1. Tighten each fastener, working in the proper sequence, to one-third the recommended torque setting.

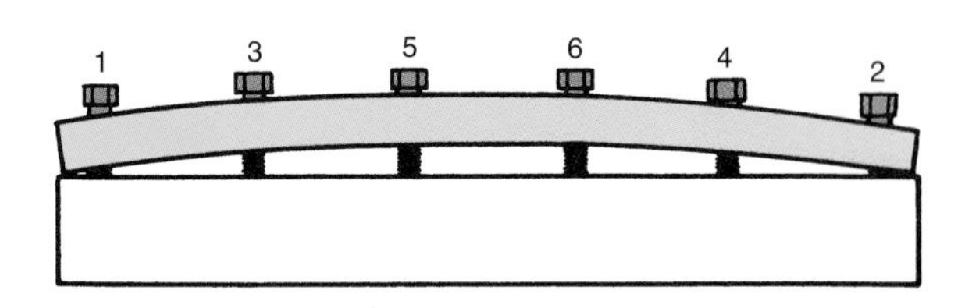

Figure 8-44. Incorrect sequence in tightening fasteners, proceeding from the end toward the middle. This sequence would produce a very poor fit!

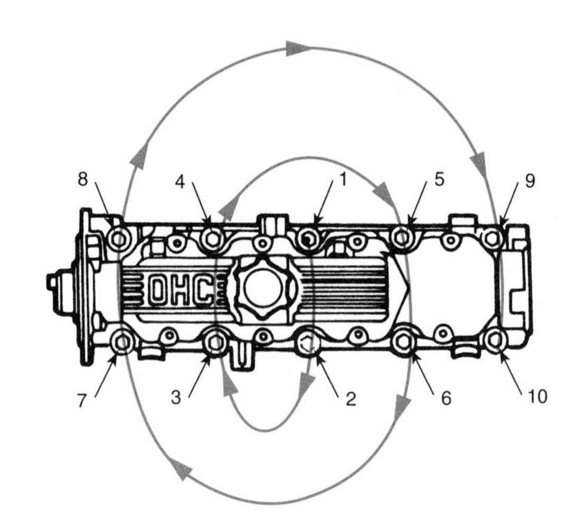

Figure 8-45. This head-bolt tightening sequence can be used when no special recommendation is available. (General Motors)

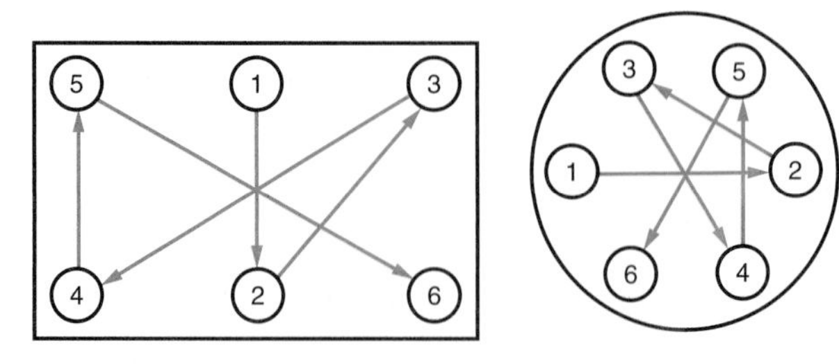

Figure 8-46. Tightening bolts in crisscross sequence.

2. Repeat the process, tightening all fasteners to two-thirds of the full setting.
3. Repeat the process, tightening each fastener to the full torque setting.
4. Repeat step three to be positive you have not missed a fastener.

   Step four is very important and frequently overlooked.

Holding the Torque Wrench

Whenever possible, *pull* on the wrench to prevent skinned knuckles. When using a beam-type torque wrench, be careful to pull so the beam is bent only in the direction of travel. If the wrench is bent up or down while pulling, the indicator point can drag on the scale and affect the reading. Place the palm of your left hand on the head of the wrench to counterbalance the pull on the handle. Allow your palm to turn with the wrench.

Sticking

When nearing full torque value, you will often hear a popping sound. The fastener will seem to stick and stop turning. If you increase pressure on the wrench, it may run up to full torque without moving the fastener.

You will find that when a fastener has stuck, the torque required to start it moving (break-away torque) is much higher than that required to keep it moving. This indicates that break-away torque is not a true fastener torque.

When sticking occurs, loosen the fastener (about one-half turn) until it breaks free. Then, with a smooth and steady pull, sweep the wrench handle around in a tightening direction. Stop when the required torque is reached.

Run-Down Torque

Self-locking nuts, slightly damaged threads, or foreign material on the threads will cause the fastener to turn with some degree of resistance before it begins drawing parts together. This is called ***run-down torque.***

If the run-down torque is noticeable, add it to the recommended torque. Determine run-down torque only during the last one or two turns of the fastener. When a fastener is first started, it may show considerable resistance. However, by the time it reaches bottom, this may have lessened or disappeared.

Note: Whenever a fastener shows undue resistance, remove it and make sure it is the right length and diameter. Also make sure it has the proper number of threads per inch.

When Torque Recommendations Are Not Available

The technician should try to secure the car manufacturer's recommended torque for the specific job. If, however, it is not available, consult a chart such as the one in **Figure 8-47.** You will note that by using the head markings and diameter,

Standard Torque Specifications and Capscrew Markings Chart

Capscrew Head Markings	Capscrew Body Size Inches–thread	SAE Grade 1 or 2 (Used infrequently)		SAE Grade 5 (Used frequently)		SAE Grade 6 or 7 (Used at times)		SAE Grade 8 (Used frequently)	
		Torque		Torque		Torque		Torque	
		Ft Lb	N•m	Ft Lb	N•m	Ft Lb	N•m	Ft Lb	N•m
Manufacturer's marks may vary. Three-line markings on heads shown below, for example, indicate SAE Grade 5.	1/4–20	5	6.7791	8	10.8465	10	13.5582	12	16.2698
	–28	6	8.1349	10	13.5582			14	18.9815
	5/16–18	11	14.9140	17	23.0489	19	25.7605	24	32.5396
	–24	13	17.6256	19	25.7605			27	36.6071
	3/8–16	18	24.4047	31	42.0304	34	46.0978	44	59.6560
	–24	20	27.1164	35	47.4536			49	66.4351
	7/16–14	28	37.9629	49	66.4351	55	74.5700	70	94.9073
	–20	30	40.6745	55	74.5700			78	105.7538
	1/2–13	39	52.8769	75	101.6863	85	115.2445	105	142.3609
	–20	41	55.5885	85	115.2445			120	162.6960
	9/16–12	51	69.1467	110	149.1380	120	162.6960	155	210.1490
	–18	55	74.5700	120	162.6960			170	230.4860
SAE 1 or 2 SAE 5	5/8–11	83	112.5329	150	203.3700	167	226.4186	210	284.7180
	–18	95	128.8027	170	230.4860			240	325.3920
	3/4–10	105	142.3609	270	366.0660	280	379.6240	375	508.4250
	–16	115	155.9170	295	399.9610			420	569.4360
	7/8– 9	160	216.9280	395	535.5410	440	596.5520	605	820.2590
	–14	175	237.2650	435	589.7730			675	915.1650
SAE 6 or 7 SAE 8	1– 8	235	318.6130	590	799.9220	660	894.8280	910	1233.7780
	–14	250	338.9500	660	894.8280			990	1342.2420

Figure 8-47. The chart shows typical torque values for capscrews with clean, dry threads. Reduce torque by 10% if the threads are oiled; reduce torque by 20% if new, plated fasteners are used for various fastener grades. Always follow the manufacturer's torque specifications for the exact job at hand. Capscrews, bolts, and nuts are marked with lines or numbers to indicate their relative strength. (DaimlerChrysler)

an approximate torque setting can be determined. Keep in mind that if the fastener is threaded into aluminum, brass, or thin metal, the torque figures may have to be reduced to prevent stripping.

Retorquing

On some assemblies, such as cylinder heads and manifolds, all fasteners may have to be retorqued after a certain period of operation. Cases such as these will be discussed in the textbook sections covering units to which they apply.

Torque Extensions

A new type of torque-measuring device can be used with ratchets and flex handles. This device, called a *torque extension*, is an electronic torque-measuring device. It is installed between the wrench and the proper size socket. An electrical connection is then attached between the torque extension and a standard multimeter. The multimeter is then turned on and set to the appropriate scale. When the fastener is tightened, the torque extension measures the torque applied and displays it on the multimeter.

Gaskets and Sealants

Gaskets, seals, and sealants are used throughout the vehicle. They confine fuel, oil, water, air, and vacuum to specific units or areas. They also keep dust, dirt, water, and other foreign materials out of various parts. In addition to these duties, they affect torque and tension, part alignment and clearance, temperature, compression ratios, and lubrication. Gaskets, seals, and sealants play an important part in the proper functioning and service life of most components. The importance of the proper selection, preparation, and installation of gaskets and seals is not always clearly understood, however.

Caution: The failure of gaskets, sealants, seals, or adhesives can cause extensive damage and expense. Study the material in this section carefully and apply the information to your work!

Gaskets

A ***gasket*** is a flexible piece of material or (in some cases) a liquid sealant placed between two or more parts. When the parts are drawn together, any irregularities (warped spots, scratches, dents) will be filled by the gasket material to produce a leakproof joint. See **Figure 8-48.**

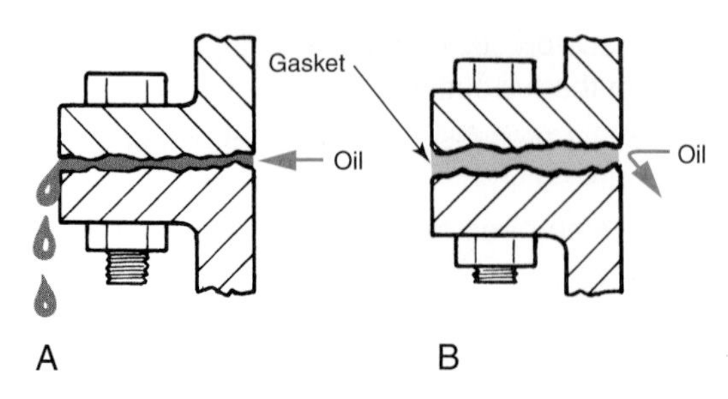

Figure 8-48. Gaskets stop leaks. A—Assembly without a gasket. Irregularities on mating surfaces allow oil to leak through. B—Same assembly, with a gasket installed. Irregularities are filled, and oil cannot leak through.

Gasket Materials

Many materials are used in gasket construction. These include steel, aluminum, copper, cork, synthetic rubber, paper, felt, and liquid silicone. The materials can be used alone or in combination.

Gasket material ***compressibility*** (how easily it flattens under pressure) varies widely. The gasket must compress to some extent to affect a seal. However, excessive compressibility will cause the gasket to extrude (flow outward and become thinner in direction of compression), reducing its thickness beyond a specified point.

The gasket material selected will depend on several variables, including the specific application, temperature, type of fluid to be confined, smoothness of mating parts, fastener tension, pressure of confined fluid, the material used in the construction of mating parts, and the part clearance relationship. All of these affect the choice of gasket material and design.

When constructing or selecting gaskets, give careful thought to these factors and choose wisely. **Figure 8-49** illustrates some of the destructive forces gaskets must resist to function properly.

Gasket Construction

Some gaskets are of very simple construction. The engine top water outlet, for example, can use a medium thickness, chemically treated, fibrous-paper gasket, **Figure 8-50.** Unit loading (pressure between mating parts) is light, temperature is moderate, and coolant pressure is low.

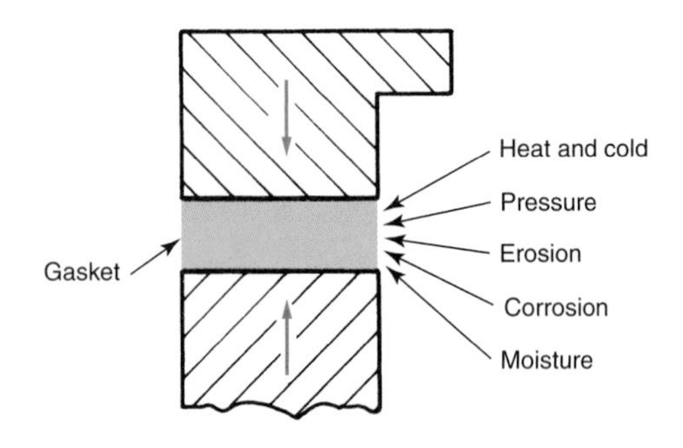

Figure 8-49. A gasket must withstand many forces. The destructive forces shown, in addition to others not illustrated, are constantly attempting to destroy the gasket.

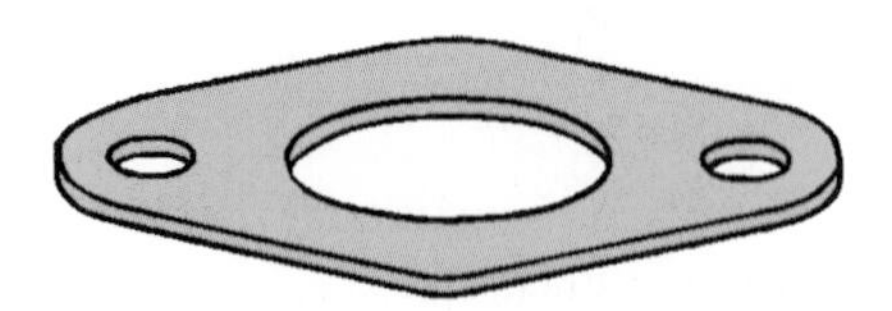

Figure 8-50. A simple paper gasket. The paper is soft, tough, and water resistant.

As the sealing task becomes more difficult, gasket construction becomes more involved. The exhaust manifold-to-exhaust pipe gasket, for example, is more complex than the water outlet gasket. Unit loading pressure is higher. Corrosive flames, gases, and high temperatures attempt to destroy the gasket. This gasket uses a combination of heat-resistant materials and steel in its construction, **Figure 8-51.**

Perhaps the most complicated gasket in terms of materials used and construction techniques is the cylinder head gasket. Unit pressure is tremendous, and combustion temperatures and pressures are very high. The head gasket must also seal against coolant, oil, corrosive gases, and thermal growth. There are several basic cylinder head gasket designs in common use. Steel, copper, and rubber may be used in their construction. See **Figures 8-52** and **8-53.**

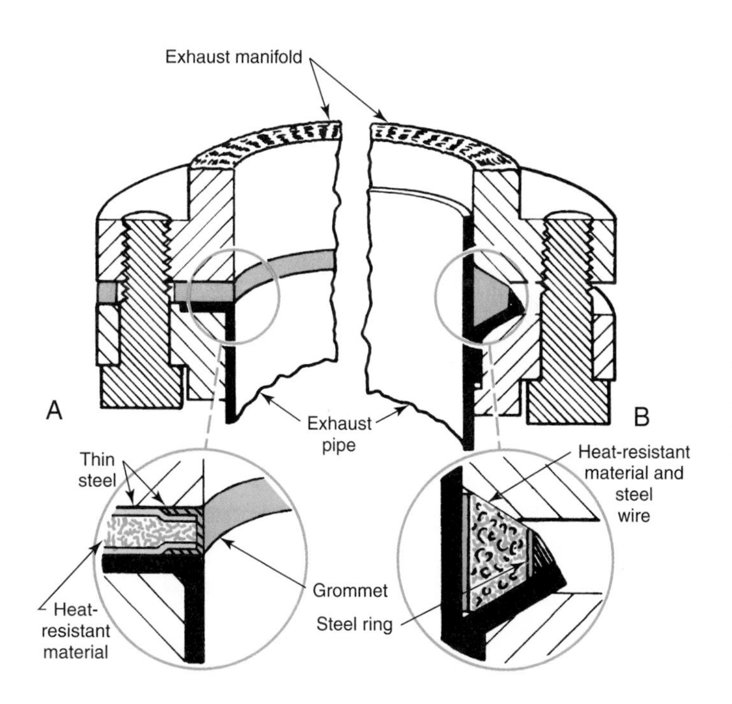

Figure 8-51. Exhaust manifold gaskets. A—A heat-resistant center with a thin steel outer layer is used on this gasket. Note how the inner edge is protected with a steel grommet. B—This gasket is made up of flexible heat-resistant material and steel wire. A thin steel outer ring is used for additional strength.

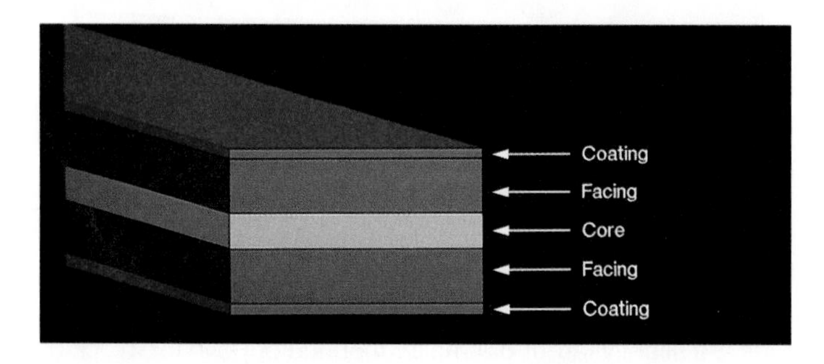

Figure 8-52. Basic construction of one particular multilayered head gasket. The core is generally made of steel, while the facing may be rubber or a dense composite material. The coating may be silicone or another nonstick material. (Fel-Pro)

The embossed steel gasket, also called a "shim" gasket, Figure 8-53A, was once popular with engine manufacturers. This type of gasket, which is produced from sheet steel approximately 0.020″ (0.51 mm) thick, provides good strength in the bead area and fine sealing capabilities when properly installed. However, it does not have the ability to seal irregular surfaces. Therefore, this type of gasket is rarely used in the aftermarket. Because the steel used to make these gaskets is quite resilient, the cylinder head bolts do not require retorquing when embossed steel gaskets are used.

Older cylinder head gaskets used a perforated steel core that was sandwiched between soft facing materials. See Figure 8-53B. The soft facing materials compensated well for the irregularities between mating surfaces, but because the soft material would eventually relax, the cylinder head bolts would have to be retorqued after the first 500 miles of driving.

A modern multilayered head gasket is shown in Figure 8-53C. This gasket has a solid steel core, a rubber facing, and a nonstick coating. The solid steel core eliminates the need for retorquing. The nonstick coating fills irregularities between mating surfaces and allows an aluminum cylinder head to move in relation to a cast iron block, as these components expand and contract at different rates. The head gasket shown in Figure 8-53D has an elastomeric bead to improve the seal in critical areas.

To produce higher ***unit loading*** around the combustion chambers or any other opening, a copper wire or other expansion device is inserted between the top and bottom layers of the gasket near the edges of the openings. The remainder of the gasket tends to compress more readily. This creates the desired pressure around the opening. See Figure 8-53E.

Gasket Sets

Gaskets are often ordered in sets. For engine work, gaskets are available in a head set (includes all gaskets necessary to remove and replace head or heads), a valve grind set (includes all gaskets necessary in doing a valve job), and an overhaul set (includes all gaskets necessary in doing a complete engine overhaul). Gasket sets also include necessary oil seal replacements. Sets for transmissions, and differentials are available separately. Single gaskets for some specific parts are also available.

Gasket Installation Techniques

After selecting a gasket material and construction, there are a few important installation considerations. Regardless of the suitability of the gasket, it will fail if not properly installed.

Never Reuse a Gasket

Once a gasket has been in service, it will lose a great deal of its resiliency. When removed, it will not return to its original thickness. If reused, it will fail to compress and seal properly. Gasket cost, as related to part and labor costs, is small. The professional technician should not even consider using old gaskets. **Figure 8-54** shows how the reuse of an old gasket will result in leaks.

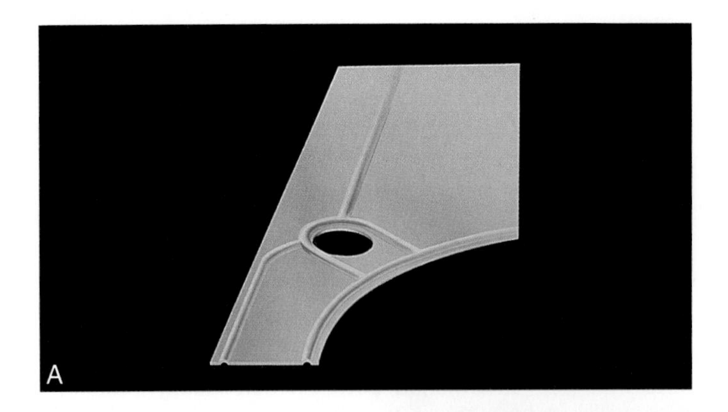

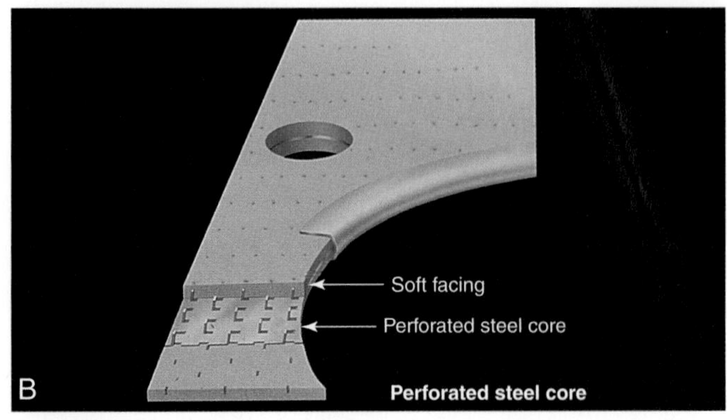

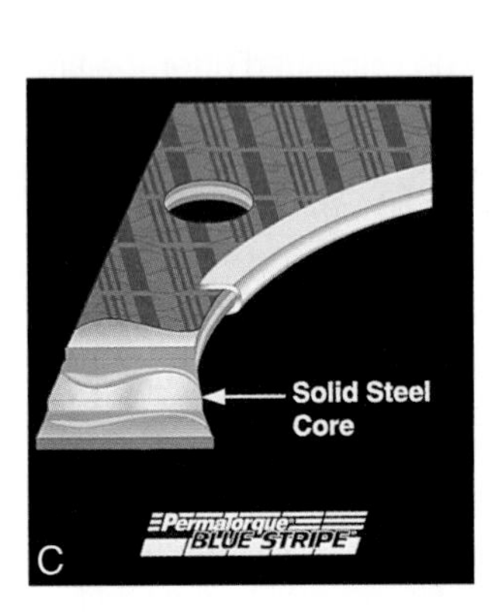

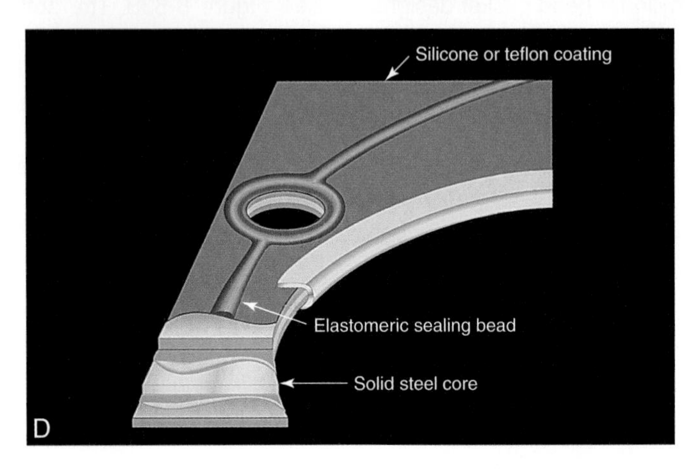

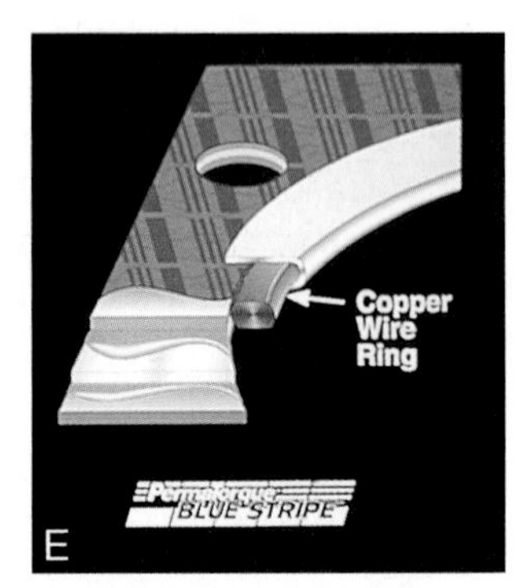

Figure 8-53. Cylinder head gaskets. A—Embossed steel gasket. B—Multilayered gasket with perforated steel core. C—Multilayered gasket with solid steel core. D—Multilayered gasket elastomeric sealing bead. E—Multilayered gasket with copper ring and stainless steel armor around the combustion chamber.

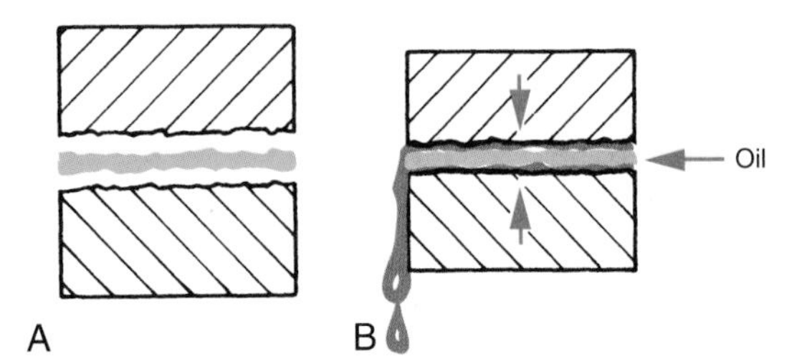

Figure 8-54. Reusing old gaskets will not work! A—Used gasket is positioned. B—When the parts are tightened, old gasket cannot compress and fill irregularities. This results in leaks.

Check Mating Surfaces

After cleaning, inspect the ***mating surfaces*** for damage. If machined, a 90-110 microinch finish is needed for proper sealing. Refer to **Figure 8-55.**

Check Gasket for Proper Fit

Position the gasket on the part to determine if it fits properly. On the more complicated setups, such as cylinder head gaskets, make certain the gasket is right side up, the proper end is forward, and that bolt, coolant, and other openings are clear and in proper alignment.

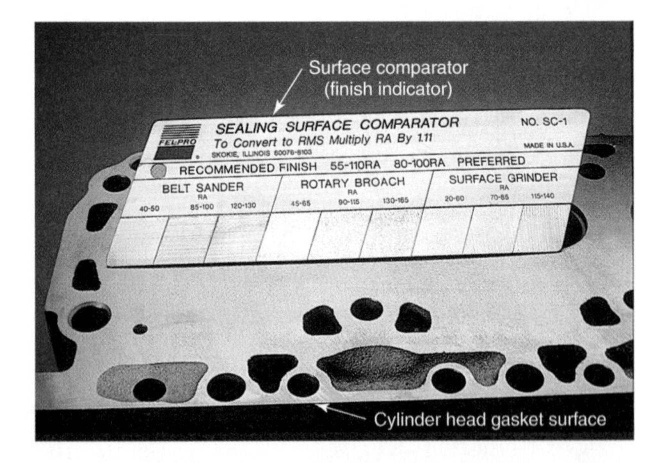

Figure 8-55. Check all mating surfaces. Notice the head-to-block surface is clean and smooth. All openings must also be clean. (General Motors)

Occasionally you may notice the gasket coolant openings are slightly larger or smaller than the ports in the block or head. This gasket may be designed to fit several models or to restrict or improve coolant circulation. Check out these situations carefully.

Head gaskets for the left and right banks on some V-8 engines are interchangeable; others are not. Many head gaskets have the word *top, front,* or *up* stamped on them to facilitate proper installation, **Figure 8-56.**

Paper and Cork Gaskets May Shrink or Expand

Paper and cork gaskets that have been stored for some time tend to either lose or pick up moisture, depending on storage conditions. Loss of moisture can cause gaskets to shrink. Excess moisture can cause them to expand. In either case, they will show signs of misalignment.

This condition can be corrected by soaking shrunken gaskets in water for a few minutes or by placing expanded gaskets in a warm (150°F–200°F or 66°C–93°C) spot. Check them occasionally to prevent overdoing the treatment.

Chamfering Screw Holes

When installing head gaskets, examine the screw holes in the block. If the threads run right up to the very top, it is a good idea to chamfer them lightly, then run the proper size tap in and out of the holes. The chamfer prevents the top thread from being pulled above the block surface. Blow out the holes with compressed air. When using an air hose for cleaning, always wear goggles. Small particles can be thrown into your eyes with great force.

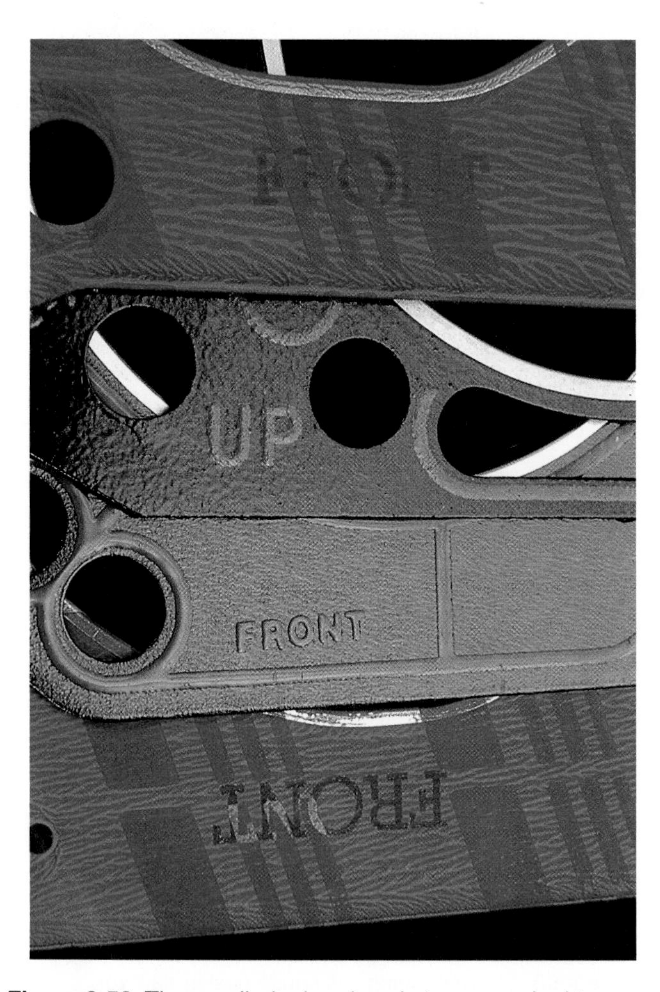

Figure 8-56. These cylinder head gaskets are marked to facilitate proper installation. All openings must align for a proper fit. (Fel-Pro)

Each Gasket Should Be Checked

Carefully inspect the gasket itself for dents, dirt, cracks, or folds. A minor crease in a cork or paper gasket usually does not render it useless. However, a head gasket should have no creases or other damage. A gentle bend will not ruin the gasket. If the gasket has been bent sharply, however, do not attempt to straighten it. The inner layer may be separated and could cause failure. **Figure 8-57** illustrates what happens when a multiple-layer head gasket is creased and then straightened.

Making a Gasket

If necessary, a simple paper or combination cork-and-rubber gasket can be made, **Figure 8-58.** First, trace the pattern. Then, trim the material with scissors or lay the material on the part. Gently tap along the edges with a brass hammer to cut the gasket material. Screw holes can also be tapped lightly with the peen end of the ball peen hammer. Do not tap hard enough to damage the threads. Gasket punches can also be used to make neat screw holes. To help hold the material in place, tap out the corner holes and start these screws before tapping around the edges.

Handle Gaskets with Care

Gaskets should be stored flat, in their containers, and in an area where they will not be bent or crushed. Storage space should not be subjected to extremes of temperature or humidity. Handle gaskets carefully, and do not attempt to force them to fit. If a gasket is accidentally cracked or torn, throw it away.

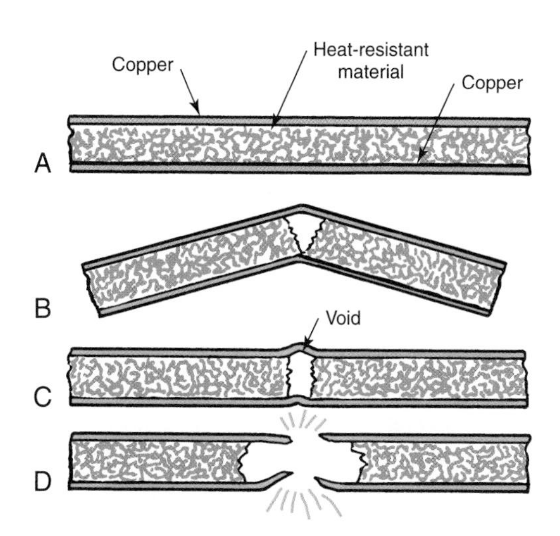

Figure 8-57. Failure of a creased gasket. A—A multiple-layer head gasket. B—Gasket has been creased and the center packing pulled apart. C—Gasket straightened, producing void. D—Gasket has "blown" in service.

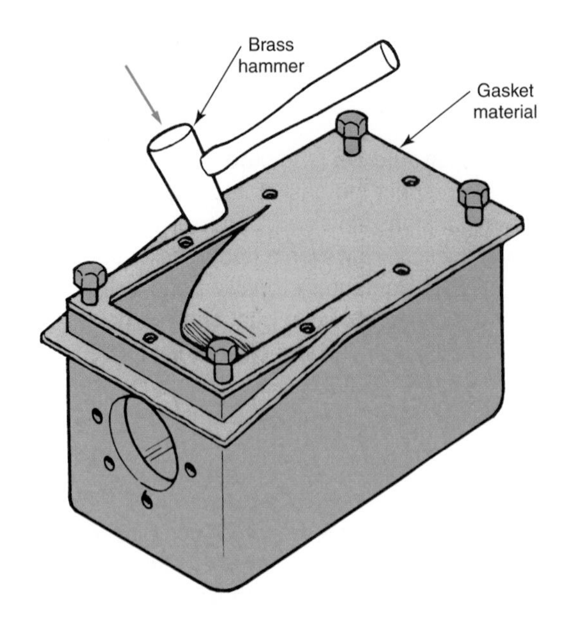

Figure 8-58. Making a gasket. Four corner screws hold the gasket material in place while tapping. A ball peen hammer is used for the holes.

Use of Sealants

A new, properly installed gasket will usually produce a leakproof joint. However, mating surfaces are not always true. Corners can present problems. Torque loss can reduce pressure on the gasket surface. Gaskets may shrink slightly, and minute part shifting can break the seal. Due to engine heat, a small amount of oil seepage will spread over a large area. This produces a messy looking job and will deposit oil on the customer's garage floor. For these reasons, ***sealant*** should be used on some gaskets.

The addition of a sealant helps hold the gaskets in place during assembly. Also, small cracks, indentations, and corner voids are sealed. In short, the use of a good sealant provides additional assurance the joint will be leakproof. Sealants will be covered in more detail later in this chapter.

Using Rubber Gaskets

Rubber gaskets are highly resilient and will usually do a good job of sealing without the addition of a sealer. In fact, rubber gaskets tend to extrude (squeeze out) under pressure when a sealer is used. Unless a sealant is specifically recommended, a rubber gasket should be installed without sealer.

Holding the Gasket during Assembly

When a sealant is used, the gasket will usually stay in place during assembly. If sealant is not being used and the gasket tends to slip, use a thin coat of grease or quick-drying contact adhesive to hold the gasket in place. The use of grease or sealant is not recommended on rubber gaskets, however.

Some parts, such as oil pans, can be difficult to assemble without disturbing gasket position. In some cases, it is advisable to tie the gasket with thin soft string in addition to using a sealant. The parts may be tightened with the string in place. Patented gasket holders are also available and work well.

In other instances, such as cylinder head installation, guide pins are used to hold the gasket in alignment. Make certain the gasket is correctly installed and that it remains in alignment during assembly. See **Figure 8-59.**

Use Proper Sequence and Torque

After running all fasteners up snug, tighten them in the proper sequence as recommended in the section on fasteners. In addition to breaking fasteners and producing distortion, improper sequence and torque will very likely cause the gasket to leak. Excessive torque can place the gasket under too much pressure. This can cause it to extrude badly. **Figure 8-60** shows the results of improper tightening procedures.

Stamped Parts Require Extra Care

If bent along the engaging edge, relatively thin stamped parts must be straightened before installation. Parts such as rocker arm covers, oil pans, and some timing covers often need straightening. Place the part edge on a smooth, solid

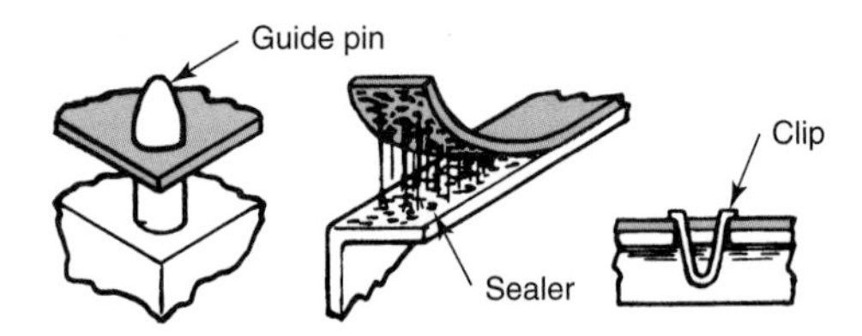

Figure 8-59. It is important that gaskets be held in alignment during assembly. Some methods for holding a gasket in place are shown.

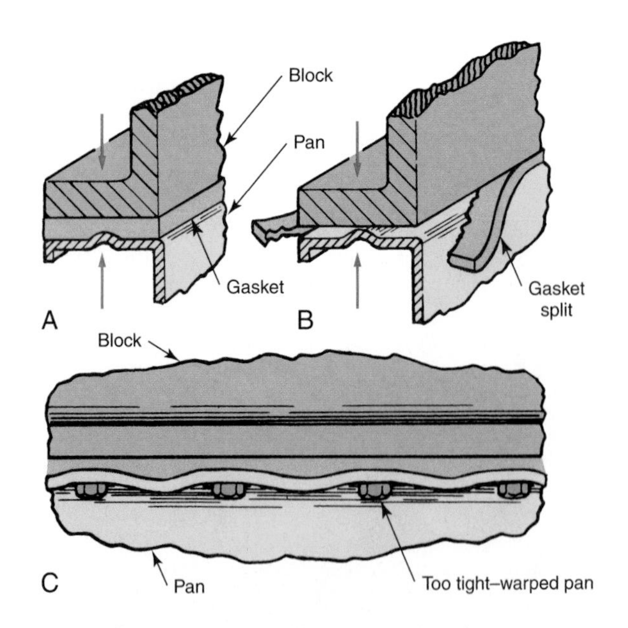

Figure 8-60. Overtightening will cause gasket damage. A—Proper fastener tension. B—Excessive tightening has split this pan gasket. C—Excessive tension has warped this oil pan flange.

metal surface and gently tap the bent sections to straighten them, **Figure 8-61.** When installing thin parts, do not over-tighten the fasteners or the parts will bend again.

Steps in Proper Gasket Installation

The following steps should be followed when installing gaskets:

1. Clean parts, fasteners, and threaded holes.
2. Remove any burrs, bent edges, or excessive warpage and check for dents and scratches.
3. Select a new gasket of the correct size and type.
4. Check the gasket for fit.
5. Where sealant is used, apply a thin coat of the correct sealant on one side of the gasket. Place the gasket with the coated side against the part. Spread a thin coat on the uncoated side. Do not slop sealant into parts. Wipe off excess sealant.
6. If you anticipate difficulty in aligning the gasket during assembly, secure the gasket by additional means.
7. Carefully place the mating part in place.
8. Coat the threads of fasteners with anti-seize lubricant (unless prohibited). Install the fasteners in their proper location and tighten them until they are snug.
9. Torque the fasteners in proper sequence.
10. If necessary, retorque the fasteners after a specified length of time. (These instances will be covered in later chapters.)

Analyze Gasket Failure

When a gasket fails in service, there has to be a reason for the failure. If you do not detect the reason, your own installation might fail also. The following simple steps will help you find the underlying cause of gasket failure:

- Ask the owner about any unusual conditions. Try to determine if the gasket failed suddenly or over a period of time.

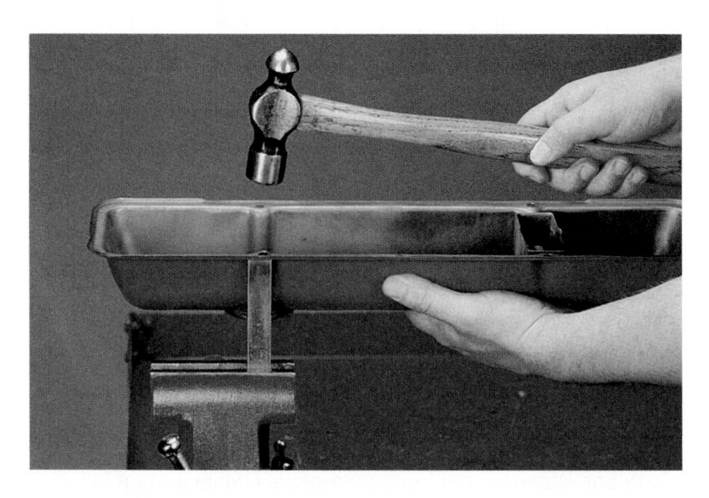

Figure 8-61. Warped flanges cause leaks. Straighten them before installation. (Fel-Pro)

- Before disassembly, check fastener torque with a torque wrench. You can loosen each fastener and notice the reading at break-away (point at which the fastener just begins to unscrew). This reading will be somewhat less than true torque. Another method is to carefully mark the position of the head of the screw or nut in relationship to the part (use a sharp scribe). Back the nut off about one-quarter turn. Carefully retighten the nut until the scribed lines are exactly in alignment. If done properly, this will give you a fair indication of torque at the time of failure. If the torque is significantly below or above specifications, or if torque varies widely between fasteners, this could be the cause of failure.
- Following disassembly, carefully blot off any grease, oil, dirt, and carbon from the gasket. Do not rub or wash the gasket immediately, as this may remove telltale signs. Inspect the gasket for signs of uneven pressure, burning, corrosion, cracks, or voids. Check to determine if the gasket is of the correct material and type for the job.
- Inspect the mating parts for warpage and burrs. Always try to find the cause of gasket failure so you may correct the problem when installing a new gasket.

Retorquing

Constant fastener tension and the expansion and contraction of parts will tend to further compress a gasket. This will leave the fasteners below proper torque. In a critical application, such as a head gasket installation, it can cause gasket failure unless the fasteners are retorqued after a period of time. Situations requiring retorquing will be discussed in later chapters.

Oil Seals

An ***oil seal*** can be used to confine fluids, prevent the entry of foreign materials, and keep two different fluids separate. An oil seal is secured to one part, while the sealing lip allows the other part to rotate or reciprocate (move).

Oil seals are used throughout the mechanical parts of the car. The engine, transmission, driveline, differential, wheels, steering, brakes, and accessories all use seals in their construction.

Oil Seal Construction and Materials

Seals are made up of three basic parts: a metal container or case; a sealing element; and (on most seals) a small spiral spring called a garter spring.

Sealing elements are usually made of synthetic rubber or leather. Synthetic rubber seals are replacing leather in most applications. The rubber seal can be made to close tolerances. It can also be given special shapes and heat-resistant properties.

In the rubber oil seal, the sealing element is bonded to the case. The element rubs against the shaft. The case holds the element in place and in alignment. The garter spring forces the seal lip to conform to minor shaft runout (wobble) and maintains constant and controlled pressure on the lip. **Figure 8-62** illustrates typical oil seal construction.

Oil Seal Designs

Many different element and lip shapes are available. Each is designed to provide the best seal for a specific task. **Figure 8-63** shows several designs. Notice that more than one lip can be used. The outside diameter (OD) may be coated with rubber to provide better OD sealing.

Other Types of Oil and Grease Seals

Engine rear main bearing oil seals are available in both one- and two-piece styles, **Figure 8-64.** They may be made of graphite-impregnated fiber wicking or synthetic rubber. Some grease (not oil) seals use a felt sealing element. Occasionally, a combination will use an inner rubber seal and a felt outer seal.

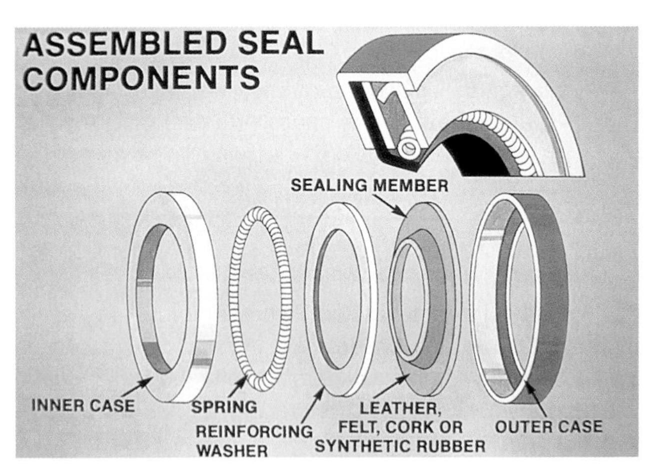

Figure 8-62. Typical oil seal construction.

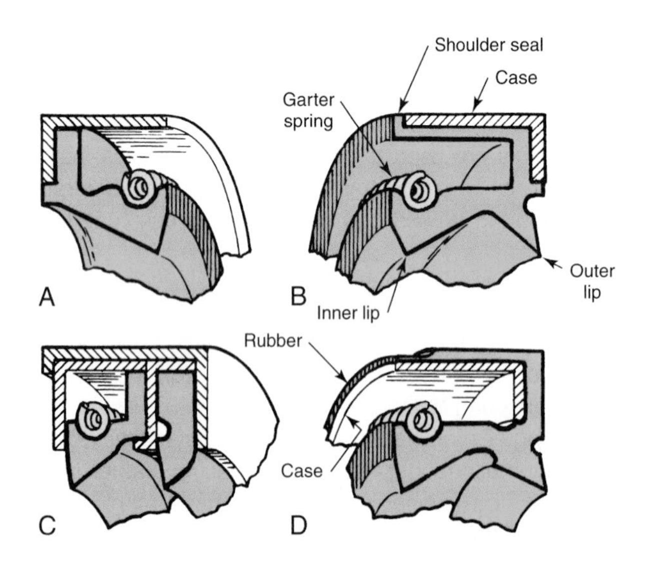

Figure 8-63. Oil seal designs. A—Single lip. B—Double lip with rubber shoulder seal. Inner lip controls oil, and outer lip keeps out dust and water. C—Double lip. Both lips control oil. D—Double lip with rubber outer coat to assist outside diameter sealing.

Oil Seal Removal

Seals may be removed by prying, driving, or pulling, depending on their location. Use care to avoid damage to the seal housing during seal removal. Such damage can cause leaks and make installation difficult. See **Figure 8-65.** Before removal, notice the depth to which the seal was installed. As with a gasket, inspect the seal after removal for signs of unusual wear or hardening.

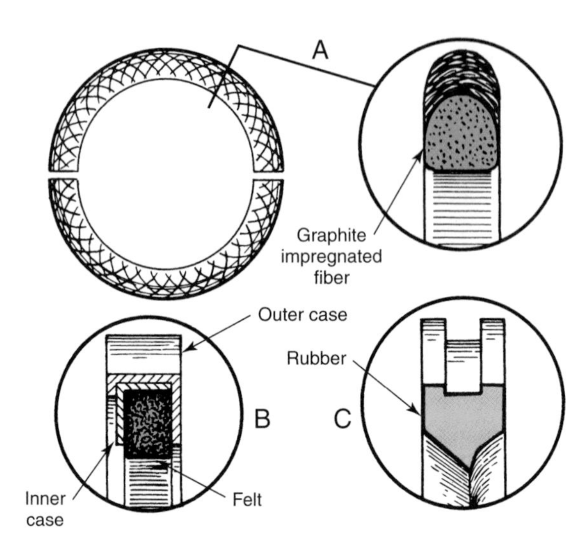

Figure 8-64. Other seal types. A—Main bearing (rear) seal made of fiber wicking. Both upper and lower halves fit into grooves in block and cap. B—Typical grease seal using a felt sealing ring. C—Synthetic rubber main bearing oil seal. Rubber O-rings (not shown) are used in several areas. They are simple round rubber rings.

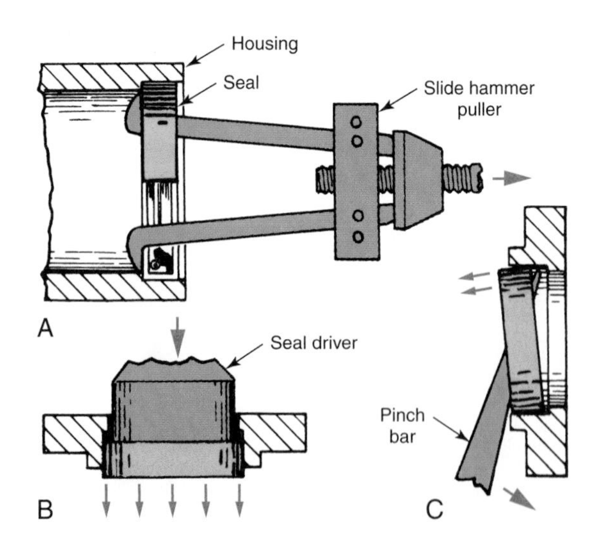

Figure 8-65. Seal removal. A—Slide hammer puller jaws are pushed through the seal and then expanded. Operating the slide hammer will pull the seal out. B—A seal driver can often be used. C—Many seals can be "popped out" with a small pinch bar. When a seal must be removed while a shaft is present, a hollow, threaded cone is threaded into the seal. The cone, which is attached to a slide hammer, will withdraw the seal.

Note: Do not reuse seals. When units are down for service, replace the seals.

Seal Installation

After removing the old seal, carefully clean the seal recess (counterbore). Inspect the seal for nicks or burrs. Compare the old seal with the new one to make certain you have the proper replacement. The outside diameter must be the same. The inside diameter may be slightly smaller in the new seal, as it has not been spread and worn. The width can vary somewhat.

If necessary, coat the inside of the seal counterbore with a thin coat of nonhardening sealer. If there is too much sealer, the seal may scrape it off as it enters, causing the surplus to drip down on the shaft and sealing lip. This can cause seal failure, **Figure 8-66.**

After preparing the seal counterbore, place the seal squarely against the opening with the seal lip facing inward, or toward the area in which the fluid is being confined. If the lip faces the other way, it will probably leak, **Figure 8-67.**

Seals are often damaged through improper installation. The technician should be careful to use the correct ***seal driver*** or, if a driver is not available, treat the seal with care. The seal driver should be just a little smaller (about 0.020″ or 0.51mm) than the seal outside diameter when the seal will be driven below the surface. If the seal is to be driven flush (even with surface), the driver can be somewhat wider. In any case, the driver should contact the seal near the outer edge only. Never strike the inner portion of a seal. This might bend the flange inward and distort the sealing element, **Figure 8-68.**

If a seal driver is not available, a short length of pipe of the correct diameter can be used. Make sure the ends of the pipe are cut square. If a hammer is used to start a seal, follow it up with a drift punch. Be careful to strike the seal at different spots (near the outer edge) each time. If the seal begins to tip, strike the high side.

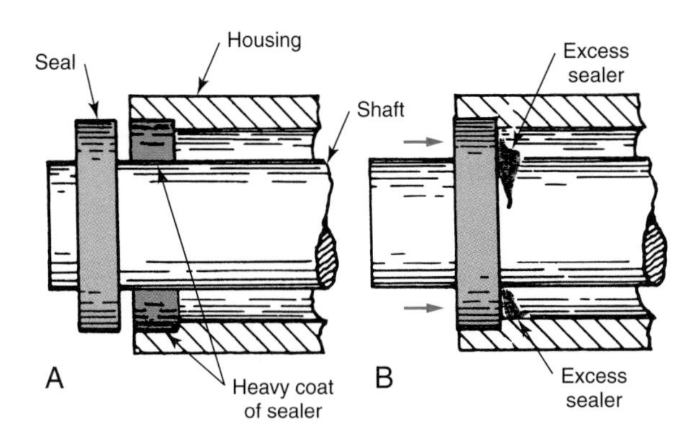

Figure 8-66. Apply sealer sparingly. A—Seal counterbore has been given a heavy coat of sealer. B—When the seal is driven into the counterbore, excess sealer will be forced out onto shaft and seal lips. In addition to ruining the seal, this could clog openings in the mechanism.

If a locating shoulder is used, drive the seal snugly against it. This is especially important if the seal inner edge has a rubber sealing compound designed to flatten against the shoulder.

When no shoulder is used, keep the seal square and stop at the specified depth. If you drive it in too far, you may ruin it while attempting to pull it back.

When driving a seal that must slip over a shaft, use care to see the sealing lip is not nicked or abraded. If a plain shaft (no keyway, splines, or holes) is involved, check the shaft carefully for burrs and nicks.

If any are found, remove them by polishing (using a shoe shine motion) with the very abrasive called *crocus cloth.* Examine the shaft surface where the sealing lips will operate. It must be smooth at this point.

If the end of the shaft is chamfered (beveled), polish the chamfered area. If the chamfer is too steep (30° maximum), either reduce it or use a mounting sleeve or bullet. Once the shaft is chamfered and free of scratches, wipe it clean and

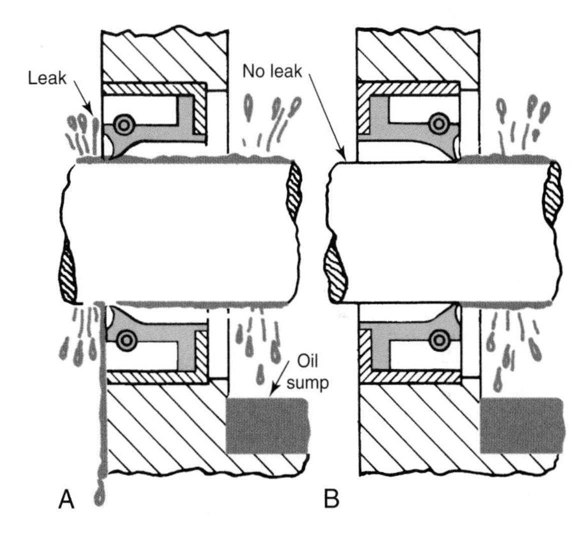

Figure 8-67. To function properly, the seal lip must face the fluid. A—Seal has been installed backward: lip faces away from the fluid. This permits fluid to force the seal lip away from the shaft, causing leakage. B—Seal is correctly installed with the lip facing the fluid. Fluid pressure forces the seal against the shaft, preventing leakage.

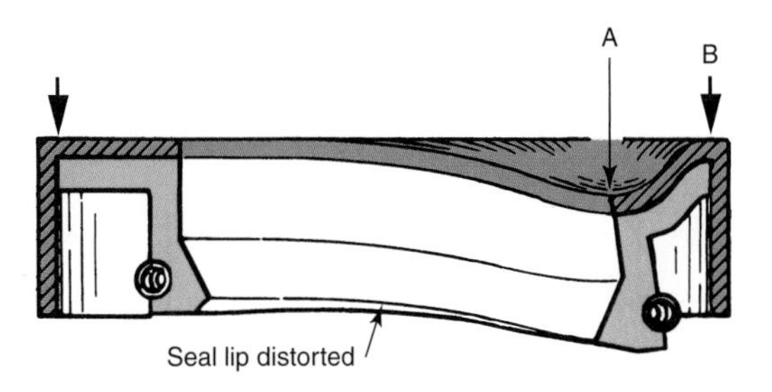

Figure 8-68. Damaged seal. This seal case was badly distorted by careless installation. A punch struck the case at *A.* All driving force should be applied at *B.* This seal would leak badly.

apply a film of oil to the full length of the chamfer. Place a small amount of oil or soft grease on the seal lip and inner face. With the seal lip facing toward the fluid to be confined (counterbore with a thin coat of sealer), carefully slip the sealing lips over the chamfer and onto the shaft. Slide the seal along the shaft until it engages the counterbore. Using a suitable driver, seat the seal, **Figure 8-69.**

Mounting Sleeves

When driving a seal that must first slide over a keyway, drilled hole, splines, or square shaft end, a mounting sleeve, or bullet, should be used. This will prevent damage to the seal lip. **Figure 8-70** illustrates the proper setup. The outside diameter of the mounting sleeve should not be more than 1/32″ (0.8 mm) larger than the shaft, or the seal lips will be spread excessively.

Wrap the stock tightly around the shaft (one wrap with a small lap) and trim the stock off. Tin and solder the lap with a soldering iron. File the lapped edge after soldering. Then, smooth the lapped edge with abrasive cloth. Finally, crimp the leading edge inward to form a cone shape.

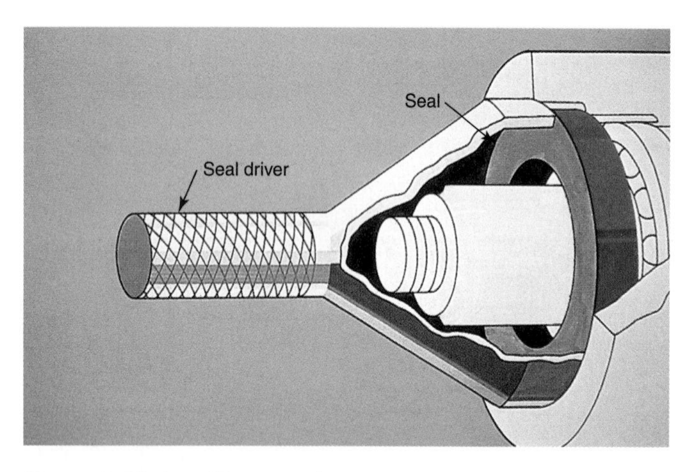

Figure 8-69. Installing a seal over a plain shaft. The seal will start over the chamfered shaft end without damage. The shaft must be smooth, clean, and oiled. (Federal Mogul)

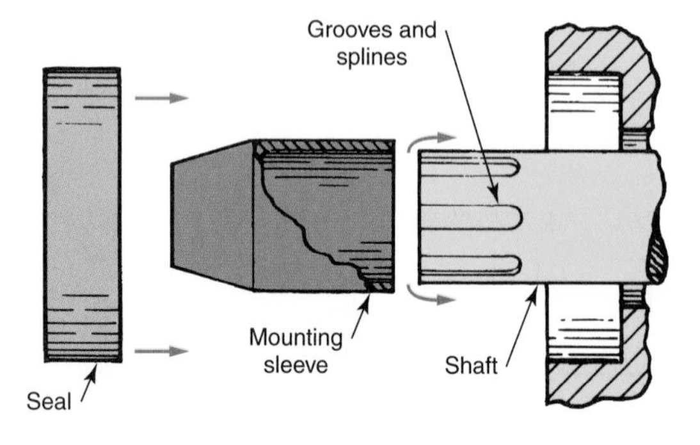

Figure 8-70. Installing a seal using a mounting sleeve. The sleeve, or bullet, is placed over the shaft. The seal can then be installed without lip damage by the spline edges.

Steps in Seal Installation

1. Clean seal counterbore, remove nicks and burrs, and if necessary, coat with a very thin layer of nonhardening sealer.
2. Inspect the shaft and polish out burrs and scratches with crocus cloth. Pay particular attention to the area where the seal lip will be located.
3. Check the new seal for correct size and type.
4. Lubricate the sealing element and shaft.
5. If needed, install the mounting tool on the shaft.
6. Push the seal, lip edge toward fluid, up to the counterbore.
7. Using a suitable driver, seat the seal. Make certain it is inserted to the proper depth and is square with the bore.

Other Information on Seal Installation

The seal must be a drive or press fit in the counterbore. In other words, it must fit tightly. A seal that slides in easily will leak. When the housing has air vents to relieve pressure buildup, make sure they are open. If vents are clogged, pressure within the housing will force the lubricant past the best of seals.

If the shaft is installed after the seal, observe the same precautions against seal damage. Cleanliness here, as in all automotive service operations, is important. If a new seal is improperly installed and must be removed, throw it away. Use another new seal.

Further specific instructions regarding gaskets, sealants, and seals will be given in chapters to which they apply.

O-Ring Seal Construction

O-ring seals are generally solid, doughnut-shaped, and made from an elastic substance (synthetic rubber or plastic). They are used to create a seal between two parts, close off passageways, prevent the loss or transfer of fluids, and help retard the entry of dust and water. See **Figure 8-71.**

O-Ring Operation

Because the O-ring is composed of a soft, pliable material, it seals when slightly squeezed between two surfaces. If the O-ring is also sealing under pressure, the pressure itself will aid in deforming the ring, further making a final seal. O-rings can be used to seal both static (nonmoving) and dynamic (moving) parts, **Figure 8-72.**

O-Ring Installation Steps

1. Make sure the new O-ring is the correct size and that it is compatible with the fluid being sealed.
2. Thoroughly clean the area where the O-ring is to be installed.
3. Inspect the O-ring grooves or notches for burrs or nicks that could damage the new ring. Dress any sharp areas with a fine abrasive stone. Thoroughly clean the area to remove any metal and stone particles.

4. Check the shaft or spool (if used) for sharp edges or nicks. Remove any damaged spots with a fine abrasive stone or cloth. Clean the area thoroughly.
5. Before installation, lubricate the O-ring with the same type of fluid used in the part or system.
6. Install the O-ring. Protect it from sharp edges and other parts. Do not stretch it more than necessary.

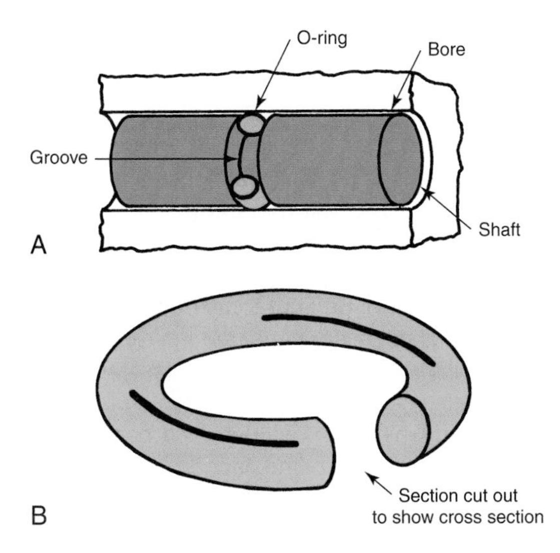

Figure 8-71. O-rings. A—Note how the O-ring is fitted into the groove in the shaft. This allows shaft movement while maintaining a seal. B—Typical O-ring construction. (Parker Seal)

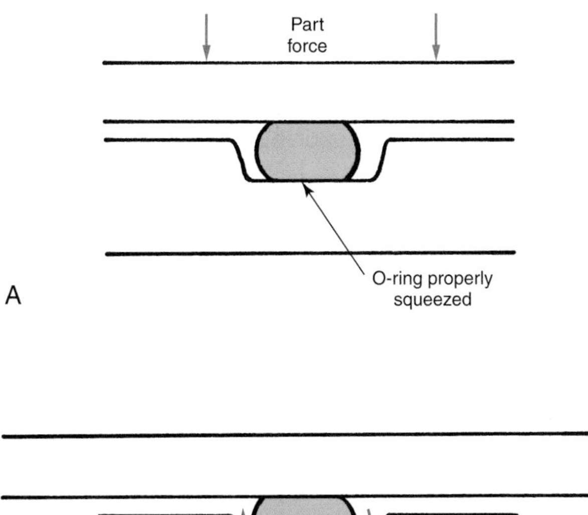

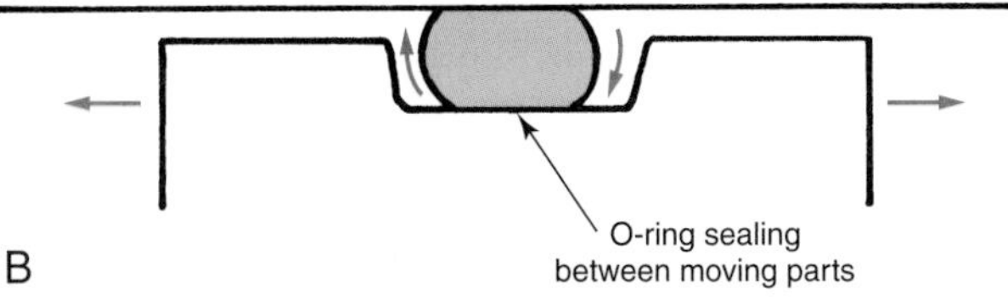

Figure 8-72. When an O-ring is installed, it is slightly squeezed (note its oblong shape). The O-ring attempts to return to its original round cross section and thus maintains constant pressure to form a leakproof seal. A—Static seal. B—Dynamic seal. (Deere & Co.)

7. Be sure the parts are correctly aligned before mating to avoid damage to the O-ring.
8. Make a final check after the O-ring is installed to be sure there are no leaks and that the parts move correctly.

O-Ring Failure Diagnosis

Improper handling, installation, and application will reduce O-ring service life. **Figure 8-73** illustrates some common O-ring failures and their causes. Be sure to follow manufacturer's recommendations when replacing or working with O-rings. When replacing O-rings that have failed in service, try to determine the reason behind the failure.

Adhesives and Sealants

The category of adhesives and sealants is a large one. Modern vehicles use many types of both. The section below explains their types and uses.

Adhesives

The modern automobile uses many types of ***adhesives*** to secure an array of parts. Adhesive uses include securing gaskets, weatherstripping, underhood fiberglass pads, body

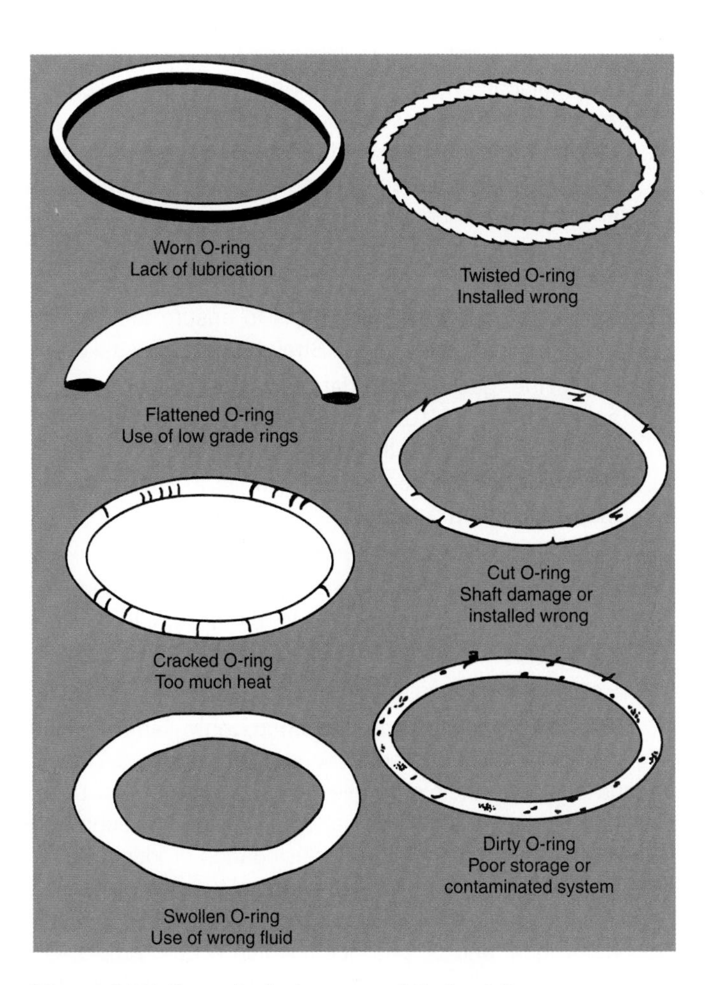

Figure 8-73. Some typical causes of O-ring failure. (Deere & Co.)

side molding, and inside rear view mirror bases (glued to the windshield). The adhesive material is generally a liquid or semi-liquid substance. It can be spread on (with its own dispenser or other suitable tool), brushed on, or sprayed on. When dry, most adhesives form a hard bond. Others remain somewhat pliable (flexible or rubbery). They can be removed with special removers or thinners. Follow the manufacturer's recommendations when using a specific adhesive.

Sealants

Sealants, or gasket sealers, are liquid or semi-liquid materials that may be sprayed, brushed, or spread on the gasket surface. Various types, having different properties, are available. Some set up hard, and others are nonhardening (remain pliable). Some can be used in place of a gasket, since they are able to form the gasket when placed between two mating surfaces and allowed to dry. Most sealants are highly resistant to oil, water, gas, grease, antifreeze, mild acid, and salt solutions. Resistance to heat and cold vary, but in general, most sealers are adequate for all applications other than on exhaust system components, where special high-temperature sealers must be used.

Form-in-Place Gaskets

Form-in-place gaskets are sealers that can be used in place of conventional gaskets. They can be very useful when an exact replacement gasket is not available. Some general rules for using form-in-place gaskets are:

1. The gasket surface should be clean. Use a wire brush on all gasket surfaces to remove loose material. All oil and dirt should be removed, and blind holes should be inspected to ensure that they are free of old gasket material.
2. Inspect all stamped metal parts to ensure the gasket-mounting surfaces are flat. Straighten if necessary.
3. Apply the gasket material in a continuous bead of approximately 1/10″ (3mm). The size of the bead should be even over the entire sealing surface. Too thin a bead will fail to seal properly, while too much sealer may clog a fluid passage.
4. Circle all bolt holes with a bead of sealant.
5. Do not allow the material to dry before installation. For best results, the parts should be assembled within 10 minutes and torqued in place within 15 minutes.
6. Remove all excess material with a rag before it dries.

The technician should be thoroughly familiar with sealants and their properties. The chart in **Figure 8-74** lists various sealants, their properties, and recommended uses. Sealant manufacturers will be happy to provide the technician with specific recommendations for using their products.

The use of too much sealer is generally worse than using none at all. Excess sealer is squeezed out of the joint and can clog water, gas, and oil passages. A thin coat is ample. On some oil pan gaskets with corners that are difficult to seal, a small dab where the gaskets meet is permissible. In general, a nonhardening, flexible sealer will produce the desired results.

Some parts with extremely small holes or ports, such as automatic transmission valve bodies, can be rendered useless if any sealant is squeezed into the openings. In cases such as this, do not use a sealant. In any specific application, be sure to follow the manufacturer's recommendations.

Tech Talk

When you begin a job that involves removing rusty fasteners, spray penetrating oil on the fasteners. Allow the penetrating oil to work as long as possible before trying to loosen the fasteners. Spray the oil when you start the repair, and remove other parts while you are waiting for the oil to work. If the vehicle is being kept overnight, spray the fasteners before going home. This will give the penetrating oil time to enter the threads and overcome much of the corrosion. If penetrating oil is given time to work, rusty bolts will often come off as if they were new. Be sure to use penetrating oil, since lubricating oils will not easily penetrate the rusted threads.

When doing a large job, put all the fasteners and small parts in old coffee cans that are labeled with the name of the component from which the fasteners and parts were taken. This will help you to find all of the needed parts when it is time to reassemble or install a particular component. If the plastic lid is still around, snap it over the can to keep out dirt and moisture, especially if the parts will not be reassembled for a while. Do not use cardboard boxes to hold parts. They will lose their strength if soaked with oil and will fall apart when you try to move them.

Summary

Technicians must be concerned with fastener design, application, and torque. They must realize the success or failure of the work depends upon the proper use of fasteners.

There are many types of fasteners: screws that thread into parts, bolts that pass through the parts and require nuts, studs that thread into parts, and sheet metal screws that cut their own threads.

Unified National Coarse series, Unified National Fine series, and Metric (SI) thread series are commonly used. Threaded fasteners are identified by material, thread pitch, diameter, length of thread, type, etc. Steel bolts and screws have markings on their heads to indicate material and tensile strength.

The removal of broken fasteners can cause difficulty unless done properly. Various methods are used. When threads in a hole are damaged beyond repair, the hole can be drilled and tapped to the next suitable oversize and a larger capscrew can be installed. The stripped threads can also be drilled to accept a patented coil wire insert. This requires special tools.

Snap rings, rivets, clevis pins, keys, and splines are nonthreaded fasteners. Self-locking nuts, various lock washers, safety wire, locking plates, and cotter pins are some of the methods of keeping fasteners tight.

Product	Type of Application	Temp. Range (degrees F.) and Pressure Range	Uses	Resists	Drys Sets Solvent
Form-A-Gasket® No. 1	Spreader cap. spatula or mechanical spreader.	–65 to 400 5000 psi	Permanent assemblies, repair gaskets, fittings, uneven surfaces, thread connections, cracked batteries.	Water, steam, kerosene, gasoline, oil, grease, mild acid, alkali and salt solutions, aliphatic hydrocarbons, antifreeze mixtures.	Fast Hard Alcohol
Form-A-Gasket® No. 2	Spreader cap. spatula or mechanical spreader.	–65 to 400 5000 psi	Semi-permanent reassembly work. Cover plates, threaded and hose connections.	Water, steam, kerosene, gasoline, oil, grease, mild acid, alkali and salt solutions, aliphatic hydrocarbons, antifreeze mixtures.	Slow Flexible Alcohol
Aviation Form-A-Gasket® No. 3	Brush or gun	–65 to 400 5000 psi	Sealing of close fitting parts. Easy to apply on irregular surfaces.	Water, steam, kerosene, gasoline, oil, grease, mild acid, alkali or salt solutions, aliphatic hydrocarbons, antifreeze mixtures.	Slow Flexible Alcohol
Indian Head Gasket Shellac	Brush	–6 to 350 Variable	General assembly work and on gaskets of paper, felt, cardboard, rubber, and metal.	Gasoline, kerosene, greases, oils, water, antifreeze mixtures.	Slow Hard Alcohol
Pipe Joint Compound No. 51	Brushable, viscous liquid	–65 to 400 5000 psi	Threaded fittings, flanges. Can be applied over oil and grease film.	Hot and cold water, steam, illuminating gas, fuel oils, kerosene, lubricating oils, petroleum base hydraulic fluids, antifreeze mixtures.	Slow Flexible Alcohol
Super '300' Form-A-Gasket®	Brush or gun	–65 to 425 5000 psi	Assembly work on hi-compression engines, diesel heads, cover plates, hi-speed turbine, superchargers, automatic transmissions, gaskets.	Hi-detergent oils and lubricants, jet fuels, heat transfer oils, glycols 100%, mild salt solutions, water, steam, aliphatic hydro-carbons, diester, lubricants, antifreeze mixtures, petroleum base hydraulic fluids, aviation fuels.	Slow Flexible Alcohol
Stick-N-Seal®	Brush or gun	–40 to 200 as an adhesive to 400° as a sealant Variable	Seal rubber to rubber, rubber to metal, sealing hydraulic and transmission oils, cork to metal.	Gasoline, grease, oils, aliphatic hydrocarbons, antifreeze mixtures. Glycols, alcohols.	Fast Flexible Methyl Ethyl Ketone and Toulene
Anti-Seize Compound	Stiff brush or spatula	–60 to 1000 –	Threaded connections, cable lubrication, manifolds, nuts and bolts, sliding metal surfaces especially where dissimilar metals meet. Prevents galling and seizure. Excellent on stainless steel.	Water, steam. Primarily designed as anti-binding and anti-corrosion compound.	Flexible Kerosene and light lubricating oil
Silicone Form-A-Gasket® Blue	Tube Cartridge Syringe	–80 to 600 5000 psi	Oil pan, valve covers, timing covers, oil pumps, transmission pans.	Regular and synthetic oil. antifreeze, grease, and transmission fluid.	Fast Flexible Gasket remover
Silicone Form-A-Gasket® Red	Tube Syringe	–80 to 650 5000 psi	Crossover manifolds, exhaust manifolds, hi-temp cam covers, hi-temp timing covers.	Regular and synthetic oil, antifreeze, grease, and transmission fluid.	Fast Flexible Gasket remover
Silicone Form-A-Gasket® Black	Syringe	–80 to 600 5000 psi	Water pumps, thermostat housings.	Regular and synthetic oil, antifreeze, grease, and transmission fluid.	Fast Flexible Gasket remover
Loctite® Master Gasket Flange Sealant	Syringe	–65 to 300 1900 psi	Overhead cam housings, cast metal timing and differential covers, compressors, and transmission assemblies.	Gasoline, gasohol, diesel fuel, regular and synthetic oils, grease, and transmission fluid.	Fast Flexible Gasket remover
Loctite® Quick Metal	Tube	Up to 300 3000 psi shear strength	Restores flywheel to pilot bearing fit. Restores wheel spindle to wheel bearing fit, etc.	Resistant to oils, cutting fluids, chlorinated solvents.	Fast Hard None (when set)
Loctite® Lock N' Seal	Tube Bottle	40 to 300	Starter and alternator mounts, flywheel and gear spine keys. Body and frame bolts. Vibration-prone assemblies. Protects threads from corrosion.	Oils, greases, fuels.	Fast Medium Solvent
Loctite® Super Bonder	Tube Bottle	Up to 175 5000 psi	General purpose. Bonds metals, alloys, plastics, vinyls, rubber, glass, etc.	Cold, mild vibration.	Fast Hard Super glue-type remover
Weatherstrip Adhesive	Tube	Up to 175	Door and windshield weather-stripping.	Freezing and high temperatures.	Fast Tough Solvent

Figure 8-74. Sealant and adhesive chart. (Permatex)

To provide proper tension, fasteners should be torqued sufficiently and correctly. Several types of torque wrenches are available for this purpose. When using a torque wrench, place high-pressure lubricant on the threads and under the head or under the nut area on fasteners. Be certain the fastener is correct for the application. Always follow the manufacturer's recommended torque and tightening sequence.

The selection, preparation, and installation of gaskets is important. Gaskets provide leakproof joints. They are made of paper, cork, rubber, steel, and copper. Different materials or combinations of materials are needed for specific applications.

Gaskets are of single- and multiple-layer construction. Many use steel or copper outer layers with a flexible center. Gaskets may have additional material around the sealing edges to increase unit loading at these points.

Used gaskets should be discarded. Beware of kinked multiple-layer gaskets. Where sealant use is recommended, use sparingly. When a gasket has failed, try to determine why, so you can correct the condition.

Oil seals are used to confine fluids, prevent the entry of foreign material, or to separate two fluids. Seals are generally constructed with a steel case, sealing element, and garter spring. Some specialized seals use fiber wicking or sections of synthetic rubber. Seals use both leather and synthetic rubber sealing elements. Many different seal lip designs are used.

When installing seals, the shaft must be smooth, the counterbore lightly coated with nonhardening sealer, and the seal driven to the proper depth. The seal lip should face toward the fluid to be confined. Protect the seal lip when installing the seal, and always use a suitable driver. Lubricate both the seal and the shaft before installing the seal. Cleanliness must be observed at all times.

O-rings are solid, pliable rings that have a round cross section. They are used to seal between parts and stop the entry of dust and water. They seal best when slightly compressed. Be sure to carefully follow installation instructions so the O-ring will not be damaged. Use care when handling. Never substitute or reuse O-rings.

The modern vehicle makes use of many different adhesives to secure parts. The adhesive material is usually a liquid or semi-liquid. When dry, adhesives can form a solid or flexible bond, depending upon the chemicals used, Follow the manufacturer's recommendations when using or removing adhesives. Sealants of many kinds are available in both hardening and nonhardening types. Select the proper type for the job at hand.

Review Questions—Chapter 8

Do not write in this book. Write your answers on a separate sheet of paper.

1. A stud has threads on _____ end.
   (A) neither
   (B) one
   (C) each
2. Which of the following statements best describes a thread class 3 fit?
   (A) A tight fit.
   (B) An intermediate fit.
   (C) Generally used in automotive work.
   (D) Both B & C.
3. Which of the following statements best describes a thread class 2 fit?
   (A) A tight fit.
   (B) An intermediate fit.
   (C) Generally used in automotive work fasteners.
   (D) Both B & C.
4. All of the following can cause screws, bolts, and nuts to loosen, *except:*
   (A) corrosion.
   (B) vibration.
   (C) expansion.
   (D) contraction.
5. Internal and external are two types of _____.
   (A) palnuts
   (B) lock washers
   (C) cotter pins
   (D) Both A & C.
6. Snap rings are made from _____.
   (A) hardened copper
   (B) ductile iron
   (C) synthetic rubber
   (D) spring steel.
7. What is the size relationship of the head of a setscrew to the major diameter of the threads?
   (A) Smaller.
   (B) Larger.
   (C) The same size.
   (D) There is no head.
8. When one side of a part to be riveted is _____, pop rivets can be used.
   (A) inaccessible
   (B) thick
   (C) plastic
   (D) damaged
9. Torquing should be completed in _____ steps.
   (A) 2
   (B) 3
   (C) 4
   (D) None of the above.

10. Which of the following torque wrenches would you use to tighten a bolt to 50 ft lb?
    (A) 0–200 in lb.
    (B) 0–50 ft lb.
    (C) 0–100 ft lb.
    (D) 0–200 ft lb.
11. All of the following statements about torquing fasteners are true, *except:*
    (A) a popping sound indicates a sticking fastener.
    (B) break-away torque is the true fastener torque.
    (C) run-down torque should be added to the recommended torque.
    (D) increasing numbers on metric bolts indicate increasing tensile strength.
12. Unit loading is accomplished by the use of _____ at certain places in a gasket.
    (A) copper wires
    (B) grommets
    (C) rubber washers
    (D) All of the above.
13. Some parts, such as _____, should be checked for dents where they meet the gasket and straightened if necessary.
    (A) valve covers
    (B) cylinder heads
    (C) engine blocks
    (D) intake manifolds
14. Why should the technician try to determine the reason for gasket failure?
15. Which of the following is not part of a typical oil seal?
    (A) Outer casing.
    (B) Sealing element.
    (C) Locking fingers.
    (D) Garter spring.
16. When removing an oil seal, it is important not to damage the _____.
    (A) seal lip
    (B) seal housing
    (C) seal case
    (D) garter spring
17. What should be placed on the lip of a seal before installation?
    (A) Nonhardening sealer.
    (B) Hardening sealer.
    (C) Lubricant.
    (D) None of the above.
18. O-rings are usually made of _____.
    (A) copper
    (B) rubber
    (C) treated paper
    (D) cork
19. Nonhardening sealers remain _____.
    (A) flexible
    (B) liquid
    (C) brittle
    (D) None of the above.
20. Some sealants are used to make _____.
    (A) seals
    (B) gaskets
    (C) O-rings
    (D) All of the above.

ASE-Type Questions

1. The following are ways to remove broken screws or studs, *except:*
    (A) grasp the broken section with pliers or a wrench.
    (B) cut a slot to allow the use of a screwdriver.
    (C) turn it in further to clear the threads.
    (D) weld a nut to the broken section.
2. Technician A says a stripped hole can be repaired by installing a Heli-coil. Technician B says a stripped hole can be repaired by installing the next larger size capscrew. Who is right?
    (A) A only.
    (B) B only.
    (C) Both A and B.
    (D) Neither A nor B.
3. Technician A says three radial lines on the head of a bolt indicate that it has greater tensile strength than a bolt with six radial lines. Technician B says Unified (customary) and Metric fasteners both have coarse and fine thread series. Who is right?
    (A) A only.
    (B) B only.
    (C) Both A and B.
    (D) Neither A nor B.
4. Technician A says all fasteners have threads. Technician B says splines are a means of driving two parts. Who is right?
    (A) A only.
    (B) B only.
    (C) Both A and B.
    (D) Neither A nor B.

5. Technician A says snap rings should never be reused. Technician B says snap rings should not be reused if they are distorted or sprung out of shape. Who is right?
   (A) A only.
   (B) B only.
   (C) Both A and B.
   (D) Neither A nor B.
6. Technician A says even though fasteners appear to be very strong, they will stretch when tightened. Technician B says stretching a fastener too much will cause it to break. Who is right?
   (A) A only.
   (B) B only.
   (C) Both A and B.
   (D) Neither A nor B.
7. Technician A says to always push a torque wrench. Technician B says once fasteners have been properly torqued, they will never need to be torqued again. Who is right?
   (A) A only.
   (B) B only.
   (C) Both A and B.
   (D) Neither A nor B.
8. Technician A says a used gasket can be reinstalled if it is first soaked in water. Technician B says a gasket that has shrunk can be used if it is first soaked in water. Who is right?
   (A) A only.
   (B) B only.
   (C) Both A and B.
   (D) Neither A nor B.
9. Technician A says the lip of an oil seal should face the fluid to be confined. Technician B says an oil seal should be installed with a special seal driver whenever possible. Who is right?
   (A) A only.
   (B) B only.
   (C) Both A and B.
   (D) Neither A nor B.
10. Technician A says O-rings can be used on parts that move. Technician B says O-rings can be used on parts that do not move. Who is right?
   (A) A only.
   (B) B only.
   (C) Both A and B.
   (D) Neither A nor B.

Suggested Activities

1. Obtain about 20 different nuts, bolts, and other threaded fasteners. List them according to the following specifications:
   a. Part Name
   b. Number
   c. Diameter
   d. Length
   e. Head Shape
   f. Wrench Size
2. Use a thread pitch gauge to measure the threaded fasteners used in Activity 1. Using the thread pitch readings and the information in Activity 1, determine whether the fasteners are:
   a. Coarse SAE thread sizes
   b. Fine SAE thread sizes
   c. Metric thread sizes
   d. Machine screw thread sizes
3. Observe the fasteners in various places on a vehicle and try to develop some general rules that determine what fasteners are used for what purpose. Note any special fasteners and try to determine why they are used. Share your findings with the class.
4. Sketch a simple bolt. Show what dimensions and other information are commonly used to describe a bolt.
5. Make a chart comparing the strength markings of SAE and metric bolts.
6. Take a sheet of paper and wad it into a ball. Pull it back out and lay it on the table. If you were to try to flatten it out, where would you place your hands (fastener) first? In what direction (sequence) would you move them? Try it. Demonstrate this procedure to your classmates and explain how it compares to the recommended tightening sequence for fasteners.
7. Select two 1/4″ bolts of equal length, one with six radial lines on its head and the other with none. Place the two bolts in a vise so that approximately 1/2″ of each bolt is secured by the jaws. Keep the bolts about 2″ apart. Run the vise up tight. With a suitable torque wrench, turn each bolt until it snaps. Watch the scale carefully to determine torque at the moment of failure. Were the readings the same? If not, why? Does it take much effort to snap a 1/4″ bolt? Write a short report summarizing your findings.

9

Tubing and Hose

After studying this chapter, you will be able to:

- Identify different types of tubing, hose, and fittings.
- Select the correct type of tubing, hose, or fitting for the job.
- Properly install new tubing and hose.

Technical Terms

Work hardening
Steel tubing
Double wall construction
Plastic tubing
Flare fitting
SAE (Society of Automotive Engineers)
Inverted
Single-lap flare
Double-lap flare
ISO flare
Flaring tool
Compression fittings
Separate-sleeve compression fitting
Double compression fitting
T-fitting
Swivel fitting
O-ring fitting
Connectors
Union
Elbow
Distribution block
Shutoff cock
Drain cock
Push-on fittings
Tubing cutter
Reamer blade
Bending spring
Flexible hose
Single-ply
Double-ply construction
Molded radiator hose
Flexible radiator hose
Hose clamp
Barrier hose
Impermeable
Skive

Tubing and hose are used in many parts of the car, **Figure 9-1.** Brake systems, fuel delivery, vacuum-powered accessories, air conditioning, heating, transmission fluid cooling, engine cooling, power steering, suspension, lubrication, and instrumentation all use either tubing or hose. In some instances, both are used. Selecting and working with tubing is a part of most repair jobs. A well-trained technician should be thoroughly familiar with the different types, their application, and proper installation.

Tubing Material

Some of the materials used in the manufacturing of tubing include annealed (soft) copper, half-hard copper, steel, aluminum, plastic, and stainless steel. Although all of these are found in the automotive field, the most common types are steel and plastic. While technicians once made up many lines from tubing and fittings, prefabricated replacements are often available today for various applications. This chapter will provide basic information needed for cutting, bending, and flaring tubing when lines must be made to order.

Figure 9-2 shows the pressures that various types of tubing with different wall thickness will withstand. These are safe working pressures when a safety factor of five-to-one (material five times stronger than anticipated working pressure) is desired.

The technician must know what material is used in the tubing. The technician must also have accurate knowledge of the pressures and temperatures produced by the system in which the tubing will be used. Keep in mind that both power steering and braking systems can develop pressures in excess of 1000 psi (6,895 kPa).

Copper Tubing

Copper tubing resists rust, bends easily, and forms good joints. It can be used for vacuum lines, coolant and heater lines, lubrication lines, and for other low-pressure applications. Copper is more easily bent than steel, but is not as strong. In addition, it is subject to ***work hardening*** (becoming hard and brittle from bending). For this reason, copper tubing should be protected from excessive vibration.

STOP **Warning: Never use copper tubing on brake or power steering systems, which generate very high pressures. Never use copper tubing in fuel systems. Vibration and movement can weaken the tubing, resulting in leaks and a potential fire hazard. Use only steel tubing in brake, power steering, and fuel systems.**

Steel Tubing

Steel tubing is suitable for almost all automotive applications. When used in brake and fuel systems, the tubing should be of the double-wrapped, brazed, and tin-plated type. The ***double wall construction*** gives the tubing good strength and makes it easier to bend. The tin-plating protects it from corrosion.

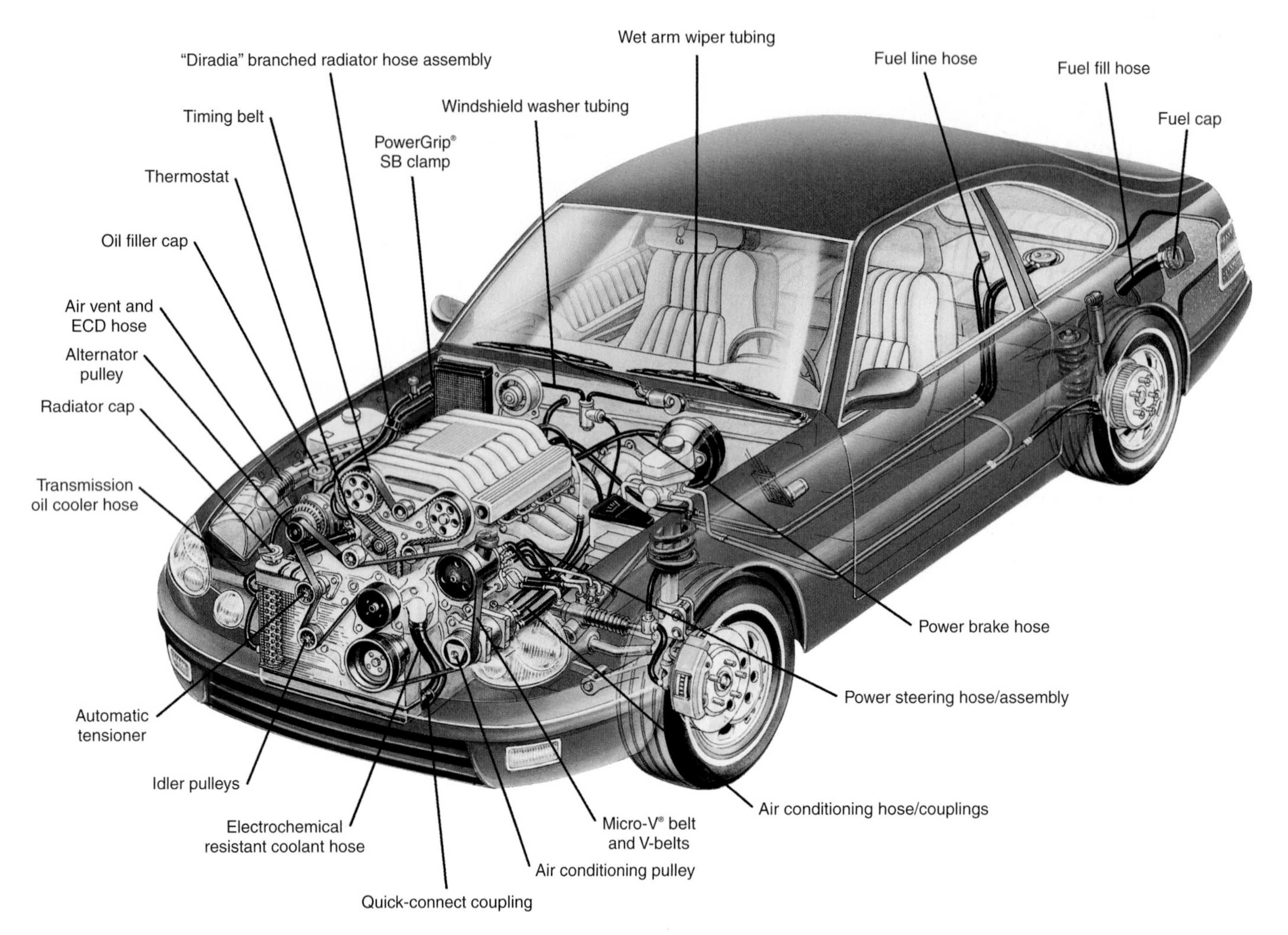

Figure 9-1. Hoses and tubing are used extensively in today's vehicles. Note the many different hoses identified in this illustration. (Gates Rubber Co.)

Material	O.D.	Wall Thickness	Pressure PSI
Polyethylene*	1/4 in.	.062	200
Nylon*	3/16 in.	.023	300
35 Aluminum		.018	500
5250 Aluminum		.018	1,000
Annealed copper		.020	1,000
Half-hard copper		.020	2,000
Double wrap, brazed steel		.020	2,000
1010 Steel		.020	2,000
Annealed stainless steel		.020	3,000
4130 Steel		.018	5,000

* = at 70 deg. F.

Figure 9-2. Tubing pressure comparison chart. Note the variation in safe working pressure for each material.

Plastic Tubing

Polyethylene and nylon are two of the materials used in the construction of plastic tubing. ***Plastic tubing*** has the advantage of flexibility, resistance to corrosion, and work hardening. It will not withstand high pressures and excessive heat, however. It can be used for fuel, vacuum, and some lubrication lines. Special inserts are needed to attach plastic tubing to conventional tube fittings.

Tubing Fittings and Connectors

Proper selection of fittings and connectors is important. The correct choice will speed up the job at hand. Flare and compression fittings are available. Push-on fittings found on some vehicles are designed to be pushed into place. There are fittings in a number of shapes, designed for use in various types of installations. The technician should be familiar with the following basic fittings and their uses.

Flare Fitting

In a ***flare fitting,*** the end of the tubing is spread (flared) outward at an angle. The tube nut securely grasps both sides of the flare to produce a leakproof joint, **Figure 9-3.** The flare fitting can be of the ***SAE*** (***Society of Automotive Engineers***) type, or the ***inverted*** type. Tube nuts for flare (and compression) fittings are available in both standard and long lengths.

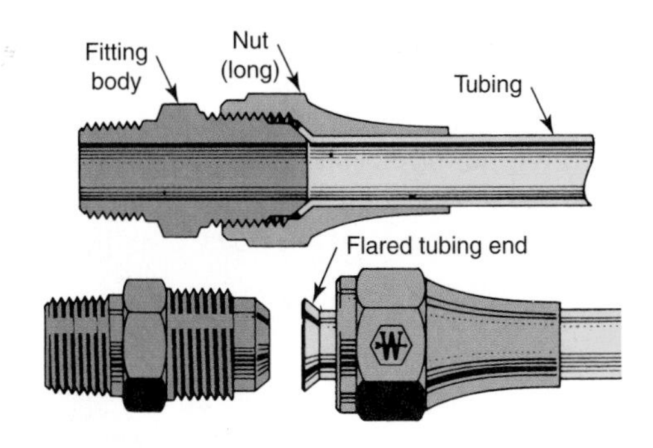

Figure 9-3. A typical flare fitting of the SAE 45° type. The nut threads over fitting body. (Weatherhead Co.)

Where the installation is subject to heavy vibration, use the long nut. This will help support the tubing a greater distance from the actual connection.

There are two flare angles, 37° and 45°. Be certain to determine the one needed before starting. Look at **Figure 9-4.** The flare may be of the ***single-lap, double-lap,*** or ***ISO*** (International Standards Organization) type. See **Figures 9-5** and **9-6.**

Caution: When flaring double-wrapped, brazed steel tubing, always use a double-lap flare or an ISO flare. If a single lap is used with this type of tubing, it will split.

Flare fittings can be used on any type of tubing (copper, aluminum, steel) that will lend itself to flaring. Flared connections *must* be used in high-pressure automotive applications, such as brake and power steering systems.

Forming the Flare

Slide the nut onto the tubing with the threaded end facing the open end of the tubing. Flare the tubing, making certain the flare is of the correct angle and width. The flare must be smooth and square with the centerline of the tubing. Careless cutting or improper use of the ***flaring tool*** will produce weak and uneven flares, which could leak. If you make a flare incorrectly, cut it off and form a new one. See **Figure 9-7.**

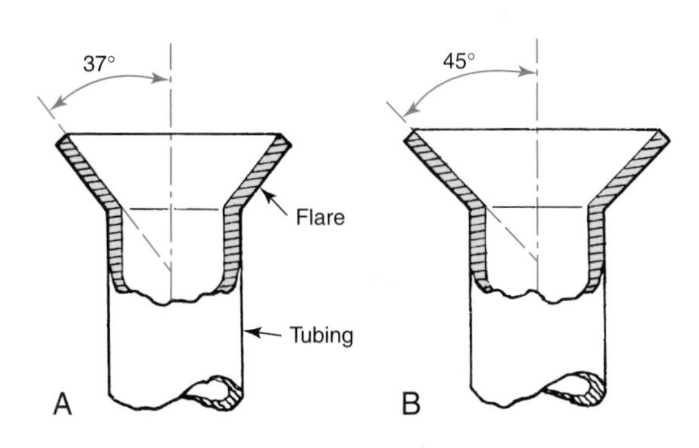

Figure 9-4. These are two typical flare angles. A—37°. B—45°. Know the angle needed before starting your flare.

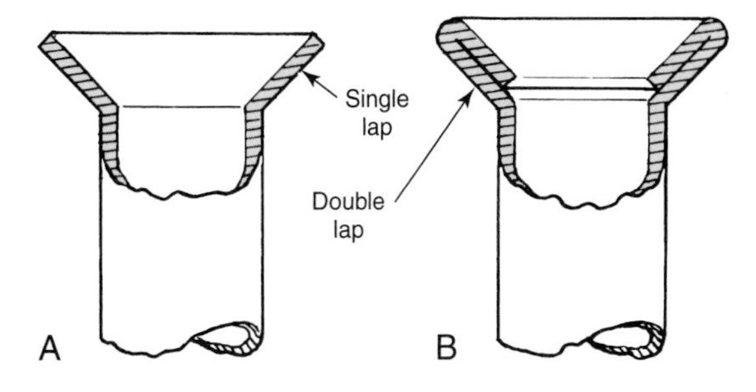

Figure 9-5. Automobiles may use tubing with single-lap or double-lap flares. A—Single lap. B—Double lap. A good technician knows the difference and how to flare either one.

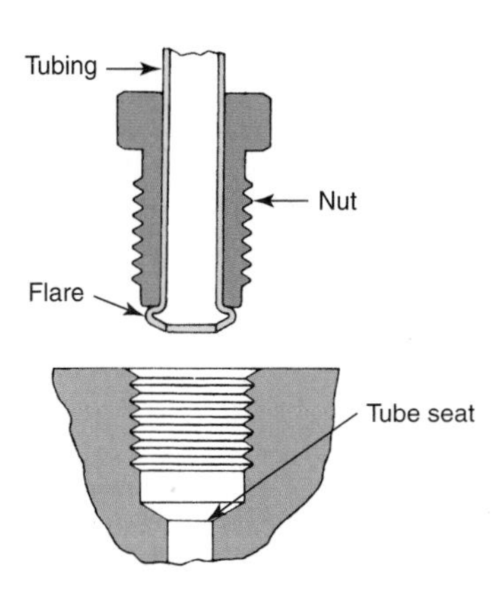

Figure 9-6. The ISO flare is becoming popular with many automotive manufacturers. (General Motors)

Forming a Double-Lap Flare

A double-lap flare should be used on brazed steel tubing, thin-wall tubing, and all tubing that will be subjected to high pressure. After cutting, reaming, and determining the proper flare angle, slide a fitting nut onto the tubing and insert the tubing in a flaring tool. Always follow the instructions provided by the tool's manufacturer.

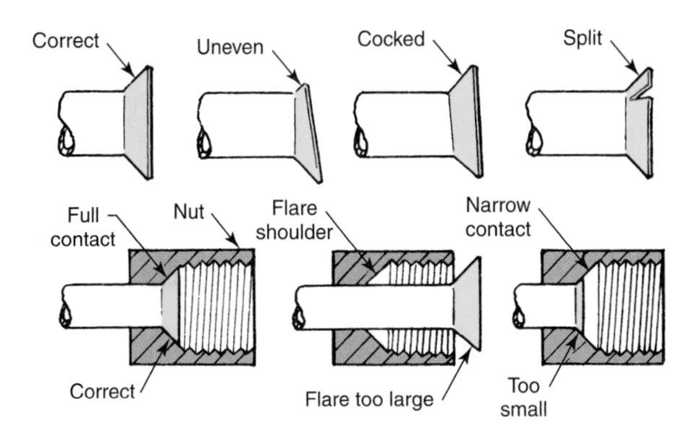

Figure 9-7. All flares must be square with the tubing centerline and of the correct size to provide full contact with the flare shoulder.

The flaring tool shown in **Figure 9-8** will produce either a single- or double-lap flare. To use the tool, arrange the gripping blocks so the correct-size tubing hole is directly beneath the flaring cone. Rotate the adapter plate until the correct size adapter is beneath the cone. Push the tubing through the gripper blocks until it strikes the adapter. Tighten the block securely so the tubing cannot be forced downward under flaring pressure.

Run the flaring cone down until it forces the adapter against the gripping block and bells the end of the tubing, **Figure 9-9.** This is the first step in doing a double-lap flare. Then, turn the flaring cone back. Swing the adapter out of the way, and run the cone tightly down into the belled tubing, **Figure 9-10.** This will form the finished flare.

Forming an ISO Flare

The ISO flare, also called a bubble flare, is produced with a tool similar to that used for making the double flare. The ISO flare is made in just one step, however. When you are making any kind of tubing flare, be sure to follow the toolmaker's directions.

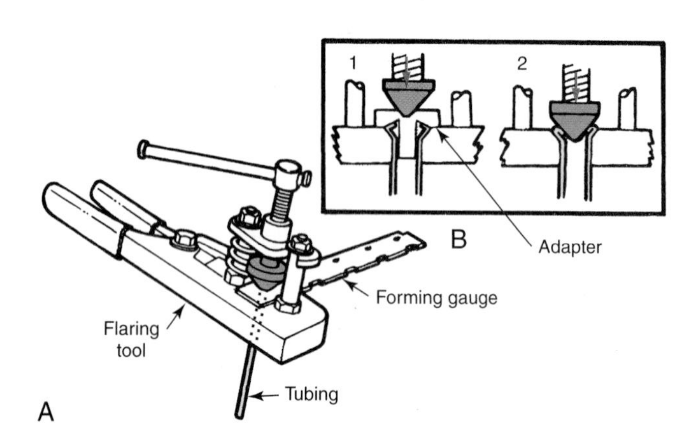

Figure 9-8. A—One type of flaring tool. B—An adapter can be used with this tool to produce double-lap flares. (DaimlerChrysler)

Assembling Flare Fittings

Align the tubing with the fitting. Push the flare against the fitting seat and run the nut up finger tight. Using a flare-nut wrench, bring the nut up solidly. You will feel a firm metal-to-metal contact, indicating that the flare is securely pinched between nut and fitting body. At this point, give the nut an additional 1/6 turn.

Compression Fittings

Compression fittings are used on low-pressure lines for vacuum, fuel, and lubrication. Since no flaring is required, connections are quick and easy to make. In a compression fitting, a sleeve is used. The sleeve may be either a separate unit or part of the nut, **Figure 9-11.** When the fitting and nut are drawn together, the sleeve is compressed against the tubing, fitting, and nut. A ***separate-sleeve compression fitting*** is pictured in Figure 9-11A. A ***double compression fitting,*** using the nose of the nut as the sleeve, is shown in Figure 9-11B.

Caution: Do not use compression fittings on brake and power steering systems. The high pressures may cause them to leak or come apart.

Assembling Compression Fittings

Slide the nut and the sleeve on the tubing. When the tubing is aligned with the fitting, insert the tubing as far as it will go. While holding the tubing in place, tighten the nut until it is finger-tight. Use a flare-nut wrench to tighten the nut until the sleeve just grasps the tubing. For 1/8″, 3/16″, and 1/4″ tubing, give the nut an additional one and one-quarter turns. For 5/16″ tubing, use one and three-quarter turns; for all sizes 3/8″ to 1″, use two and one-quarter turns. While tightening, hold the tubing in the fitting. This tightening procedure applies only to *new* compression fittings. When assembling used fittings, bring the nut up firmly with no additional turns.

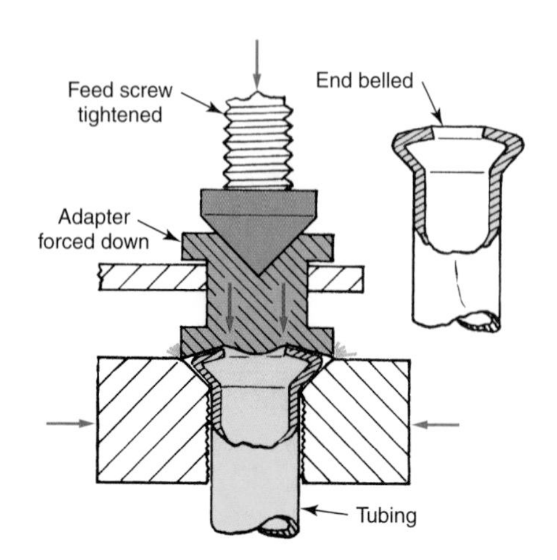

Figure 9-9. Bell the end of the tubing by tightening the cone feed screw until adapter strikes the gripper block.

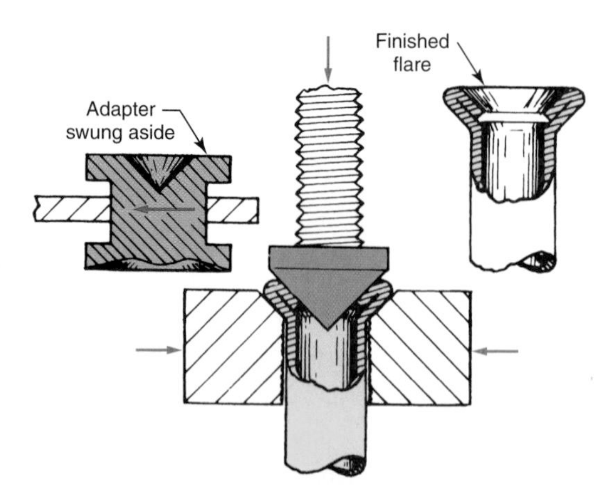

Figure 9-10. The adapter is swung aside and the cone forced into belled end to produce the finished double-lap flare.

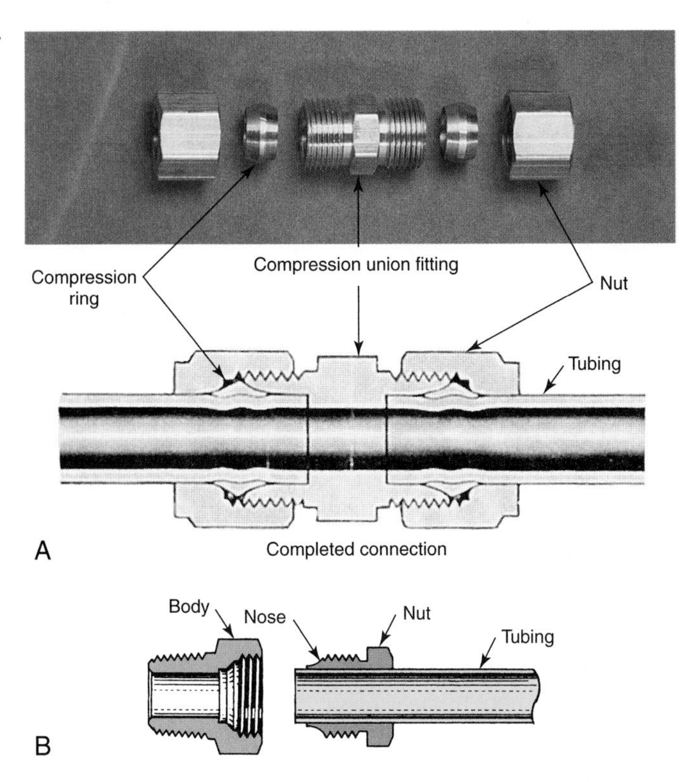

Figure 9-11. Compression fittings. A—The separate sleeve compression fitting uses a compression ring (sleeve). When the nut is tightened on the fitting, the sleeve pinches the tubing. B—When this double compression fitting is tightened, the nose of the nut is forced against the tubing.

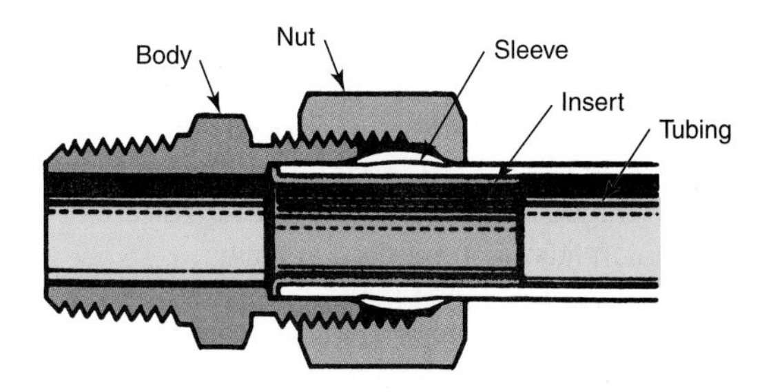

Figure 9-12. In a fitting for soft plastic tubing, an insert is used to prevent the sleeve from crushing the tubing.

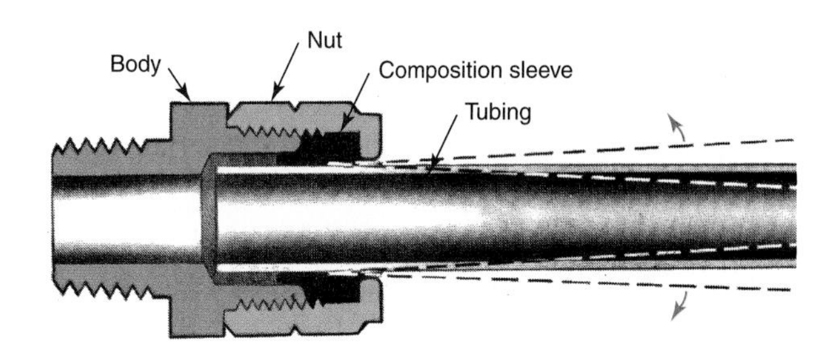

Figure 9-13. A composition sleeve in a flexible compression fitting permits heavy vibration without imposing an undue strain on the tubing.

Plastic Tubing Compression Fittings

When rigid plastic tubing is used, a regular separate sleeve compression fitting will suffice. However, if flexible (soft) tubing is used, a special insert is placed in the tubing so the sleeve will not crush it. See **Figure 9-12.**

Other Specialized Compression Fittings

A compression fitting designed for resistance to extreme vibration is shown in **Figure 9-13.** Instead of the conventional metal sleeve, a composition sleeve material is used. This fitting is for low-pressure use. For some applications involving higher pressure, an Ermeto compression fitting is used. This fitting is designed to withstand high-pressure and can also be used with heavy, difficult-to-flare tubing.

Specialty Fittings

There are a number of tubing fittings that are used for specialized connections or other tasks. These include T-, swivel, and O-ring fittings, connectors, unions, elbows, shutoff cocks, and drain cocks.

T-, Swivel, and O-Ring Fittings

A ***T-fitting*** is used where branch lines are necessary. The two common types are the branch-T and the run-T. Male and female types are available. One end of a ***swivel fitting*** utilizes a swivel nut. This allows the fitting to move. These are available in straight connectors, elbows, and tees.

The ***O-ring fitting*** uses straight threads and depends on an O-ring to prevent leaks. In **Figure 9-14,** the elbow fitting can be positioned at any angle and held as the locknut is tightened. This compresses the O-ring and seals the fitting.

Connectors, Unions, and Elbows

Connectors are used to attach the tubing to a major vehicle unit, such as a carburetor, fuel injector, or fuel pump. Whenever a connector and tubing are assembled, a liquid sealer is used to coat the threads to ensure a good seal. Connectors can also be used to mate the threaded end of a pipe to a flare or compression fitting.

A ***union*** is designed to connect two or more sections of tubing. It can be disassembled without turning the tubing.

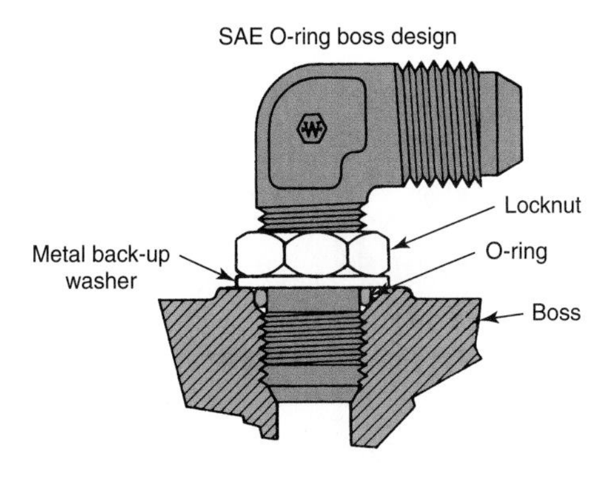

Figure 9-14. *A* 90° O-ring adjustable elbow. The O-ring is crushed against the surface of the boss by the fitting.

The union often uses a compression fitting, which was discussed earlier in this chapter. Compression union fittings are used with older and aftermarket mechanical fuel and oil pressure gauges. Due to advances in electric gauges, mechanical gauges are rarely used. This has all but eliminated the use of union fittings in modern vehicles.

When a line must leave the unit at an angle, a 90° or 45° male or female ***elbow*** is used. Female refers to a fitting with internal threads. The male fitting has external threads. Elbows are normally used with flexible neoprene or braided steel hose. It is important to remember that anytime a fluid must pass through an elbow, the fluid pressure in the system changes. Also, elbows provide a natural point for corrosion to accumulate if they are used with engine coolant. For this reason, elbows are used as little as possible. Whenever an elbow must be installed, be sure to use one with as minimal a bend as possible. It also must be made of material that is correct for the application.

Distribution Blocks

When several branch lines are served by a single feeder line, a ***distribution block*** can be used. One or more distribution blocks are often used with brake lines. They are used to minimize the amount of metal tubing in these systems. Like tubing, they must be secured to prevent damage and leaks. A distribution block is usually fitted with a mounting bracket, **Figure 9-15.**

Shutoff and Drain Cocks

A ***shutoff cock*** is used to stop flow through a line. A ***drain cock*** is used to drain the contents of a system. Always install these fittings so that, when it is in the *off* position, the fluid flow is against the seat and not the threads, **Figure 9-16.** This prevents the threads, especially in radiator drain cocks, from becoming corroded and difficult to turn.

Push-On Fittings

Push-on fittings are pushed into place on the tubing and held by a retaining or locking mechanism. Push-on fittings can be divided into two major classes: *clip* and *spring.*

Many push-on fittings are held in place by retainer clip ears, **Figure 9-17.** To disassemble the fitting, simply push the clip and pull the fitting apart. The spring lock clip, used on many fuel system and air conditioner lines, requires a special tool for removal. The removal processes for many push-on fittings are shown in **Figure 9-18.**

Caution: Release all pressure from a system before removing a push-on fitting. This is especially important when dealing with fuel and air conditioner lines. Do not excessively bend or kink plastic fuel lines. This can cause a permanent restriction in the line.

Working with Tubing

When removing metal tubing from a roll, place the roll in an upright position on a clean bench. Hold the free end of the tube with one hand while rotating the roll over the bench with the other, as shown in **Figure 9-19.** Never lay the roll flat and pull the tubing upward. This will cause the tubing to become twisted.

Avoid working (bending) the tubing more than necessary. Store tubing where no heavy tools or parts are liable to cause dents. Keep the open end of the tubing covered with tape to prevent the entry of foreign material.

Cutting Tubing

Tubing must be cut squarely, and all burrs must be removed, **Figure 9-20.** This is especially important when the ends are to be flared. Although a fine-toothed hacksaw can be

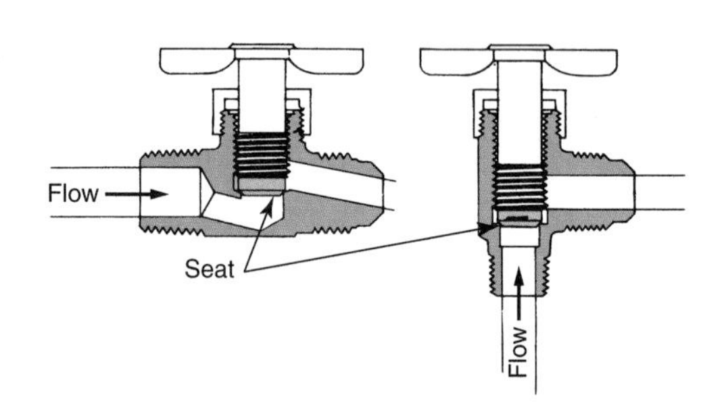

Figure 9-16. In these shutoff cocks, flow is against seat. Both shutoff cocks are in the closed position. (Weatherhead Co.)

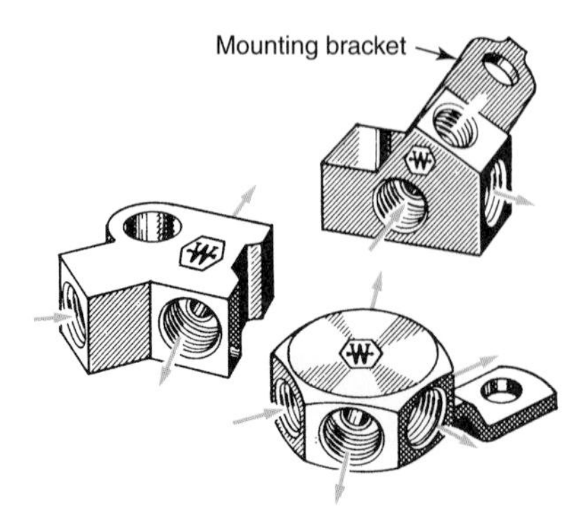

Figure 9-15. Typical distribution blocks. Always use a mounting bracket with a distribution block.

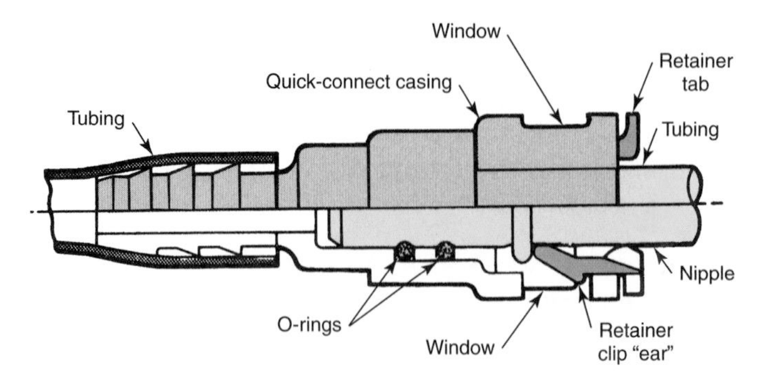

Figure 9-17. A cutaway of a plastic push-on fitting. Note how the retainer mechanism holds the tubing in place.

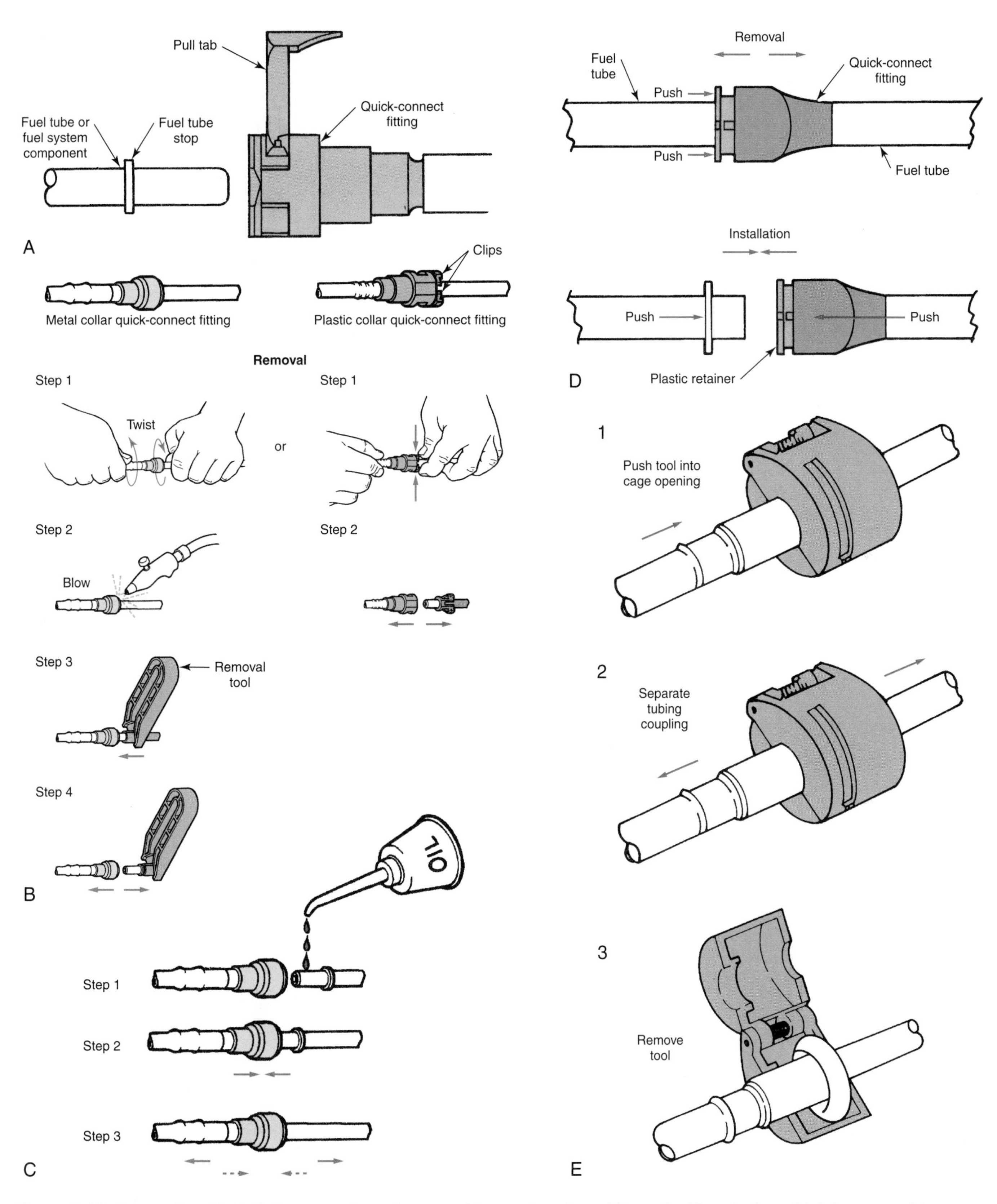

Figure 9-18. Removal and installation procedures for several types of push-on fittings. A—The plastic pull tab is removed to separate the line. B—The metal collar type requires a tool for removal, while the plastic collar unit may be disconnected by pressing the clips down with your fingers and pulling joint apart at the same time. C—Installation of the metal collar type requires several drops of engine oil to ease installation. D—The plastic retainer ring fitting can be removed with a simple push of the retainer ring. E—Spring lock coupling used on air conditioning lines can be separated with the use of a special tool that will release the garter spring when closed and pushed into the joint. 1. Push tool into spring cage opening to release garter spring. 2. Pull tubing coupling apart. 3. Remove tool. Always discharge pressure from any system before disconnecting the coupling. (General Motors, DaimlerChrysler, Kia Motors)

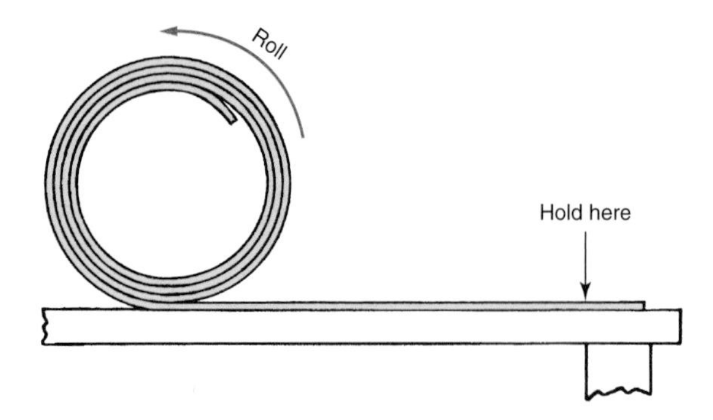

Figure 9-19. Always use the proper method of removing metal tubing from roll. Twisting can result if the tube is pulled from the center of the roll.

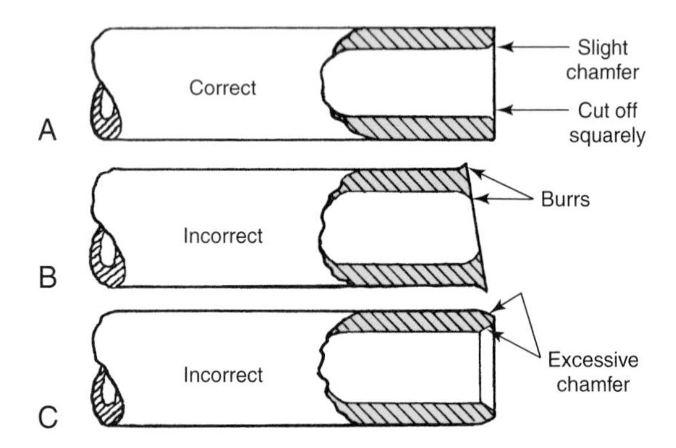

Figure 9-20. Cutting tubing properly. A—Correct cut for metal tubing is square and properly reamed. B—This tubing was cut at an angle and heavily burred. C—This tubing was cut squarely but reamed too much, resulting in an excessive chamfer.

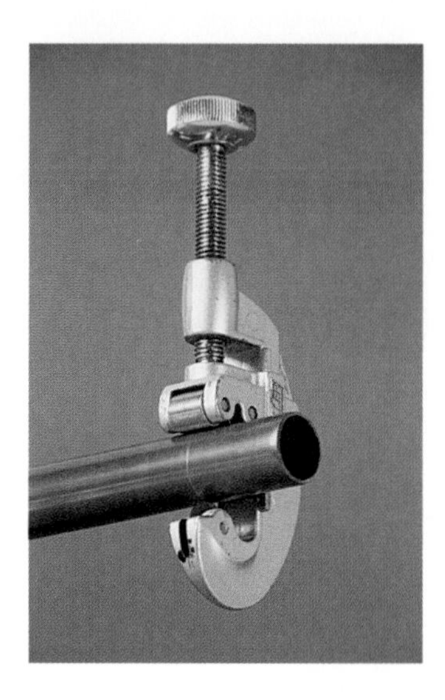

Figure 9-21. A tubing cutter can make a cleaner, more precise cut than a hacksaw. The cutter wheel is tightened by turning the knob-type handle.

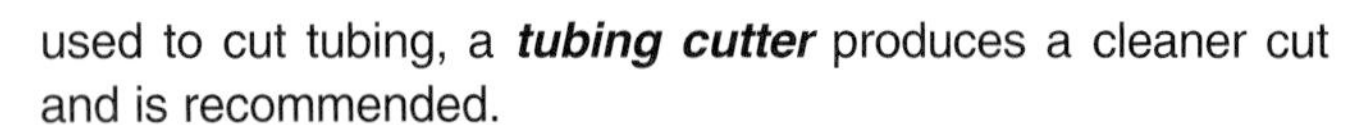

used to cut tubing, a ***tubing cutter*** produces a cleaner cut and is recommended.

The cutter is placed around the tube. The cutting wheel is brought into firm contact and revolved around the tubing. After each complete revolution, the cutter is tightened. Do not overtighten. This process is repeated until the tubing is cut off, **Figure 9-21.**

Removing Burrs

After cutting, there will be a burred edge on the inside of the tubing. Remove the burr using a ***reamer blade,*** which is usually on the cutter tool. Ream only long enough to remove the burr. Excessive reaming will ruin the end for flaring. When reaming, point the end of the tubing downward so the chips will fall free. See **Figure 9-22.**

Cleaning Tubing

Before installation, use compressed air to blow any chips or other foreign material out the tubing. Place the tubing in a clean spot until you are ready to install it. If there is any chance of dirt or grease entering the tubing during installation, cover the ends with masking tape. The slightest amount of foreign material may ruin the job. Keep the tubing spotless.

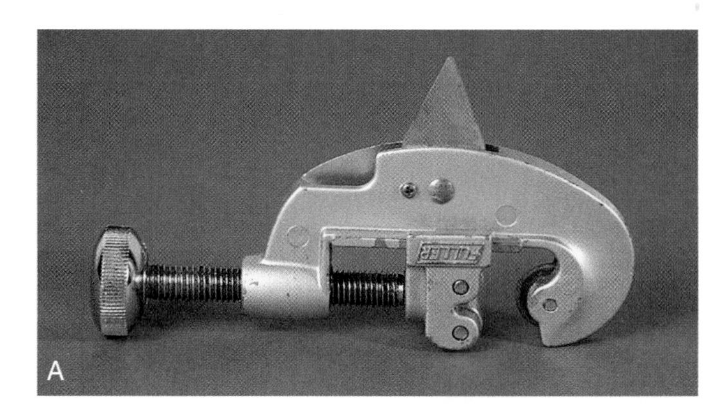

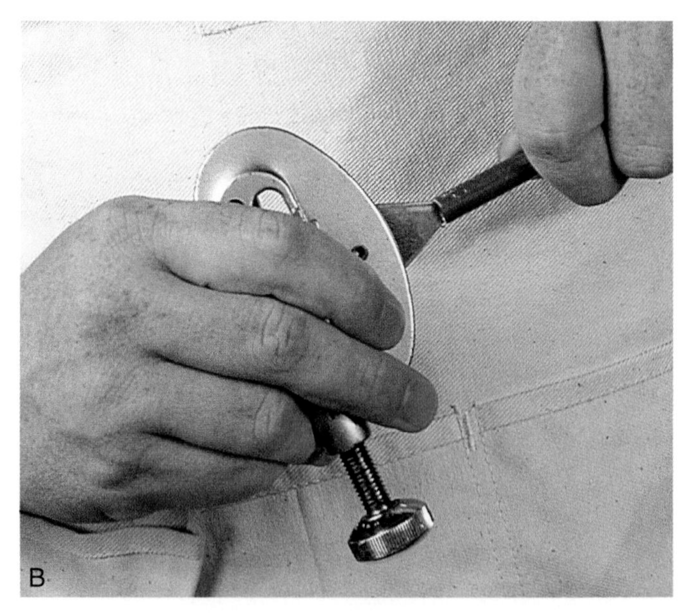

Figure 9-22. Reaming. A—A reamer blade is often built into the tubing cutter. B—Hold the tubing with the open end down while reaming to prevent contamination of the tubing by metal chips from the burrs. Do not ream excessively.

Bending Tubing

Tubing must usually be bent in one or more directions to provide a proper fit. To ensure satisfactory service, it is important to use proper bending techniques.

Soft copper and small-diameter thin-wall steel tubing can be bent by hand. Slip a ***bending spring*** over the tubing and then form the bend with your hands, **Figure 9-23.** When using a bending spring, make sure it is the correct size. Bend the tubing slightly more than needed. When it is bent to the desired shape, remove the spring.

Stiffer and larger-diameter tubing may be handled with a lever-type bender. This tool will make uniform bends. It is often used on softer tubing, as well, when appearance is important. **Figure 9-24** shows tubing inserted in a lever-type bender. Note the tool is marked in degrees to assist in controlling the amount of bend.

It is usually advisable to bend tubing prior to flaring. However, if the bend must be located close to the flare, make the flare first so the bend will not interfere with the flaring tool. Do not locate the bend too close to the flare. To facilitate assembly, allow about twice the length of the flare nut, as shown in **Figure 9-25.**

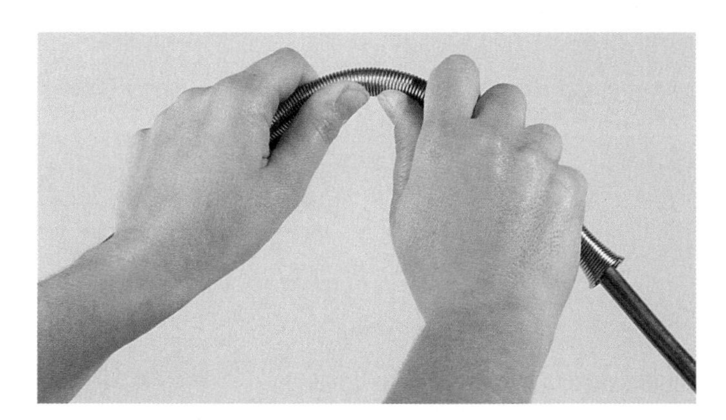

Figure 9-23. Use a spring tube bender to bend tubing without kinking or flattening it. Do not try to bend tubing using only your hands.

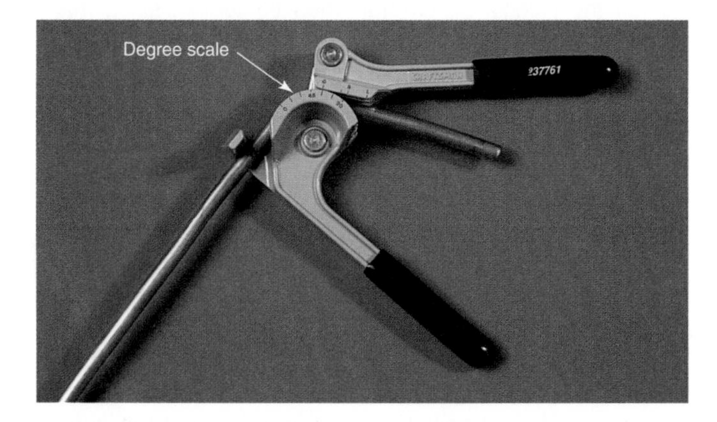

Figure 9-24. Tubing bent in a mechanical bender can be shaped to the desired angle by observing the degree scale on the tool.

Caution: When bending tubing, be very careful to avoid kinks and flat spots, Figure 9-26. Kinked or flattened tubing will restrict flow and lead to trouble. Always use a suitable bending device.

Installing Tubing

Straight tubing runs, especially if short, will not work well. The slightest shifting between the two units will impose a strain on the connections. Straight runs are also difficult to install and remove, **Figure 9-27.** Tubing can fail if subjected to excessive vibration. Secure long runs of tubing with ***mounting clips.*** Junction or distribution blocks and other heavy units must be supported. Never run tubing too close to the exhaust system. Keep it as far away as possible. If necessary, install a ***heat baffle*** or reroute the tubing, **Figure 9-28.**

Tubing Ends Should Align with Fittings

To prevent cross-threading (threads started and turned in a cocked position, ruining the threads), avoid leaks, and facilitate installation, make sure tubing ends are in line with the fitting. See **Figure 9-29**. The tubing should not have to be forced into alignment. Fittings should start easily and turn several revolutions with finger pressure only. If fittings are hard to start, check for damaged threads, alignment, and size. Be careful not to cross-thread fittings.

Assemble Both Ends before Final Tightening

Connect the tubing first at the long leg end, **Figure 9-30.** Leave the fitting loose so the other end can be moved enough

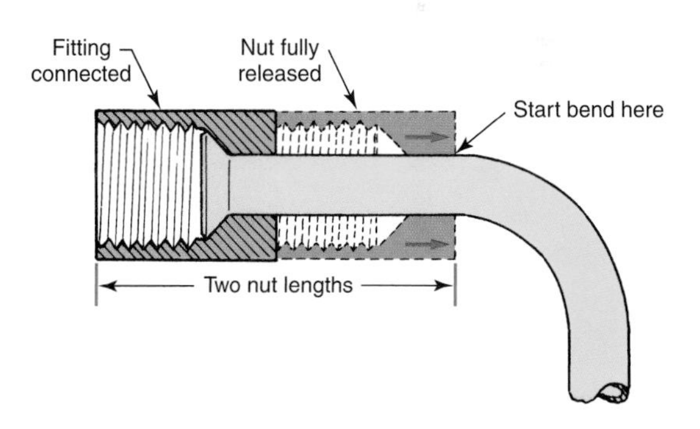

Figure 9-25. When flaring tubing, allow enough space between the fitting and bend so that the nut will slide back as shown.

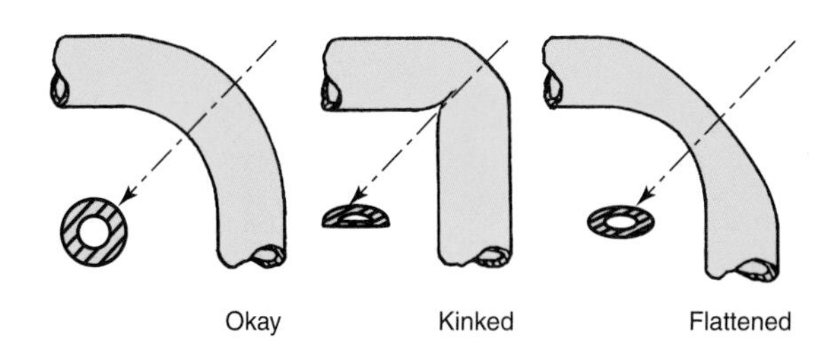

Figure 9-26. When bending tubing, avoid kinking or flattening it. If tubing kinks or flattens, cut it off and form a new flare.

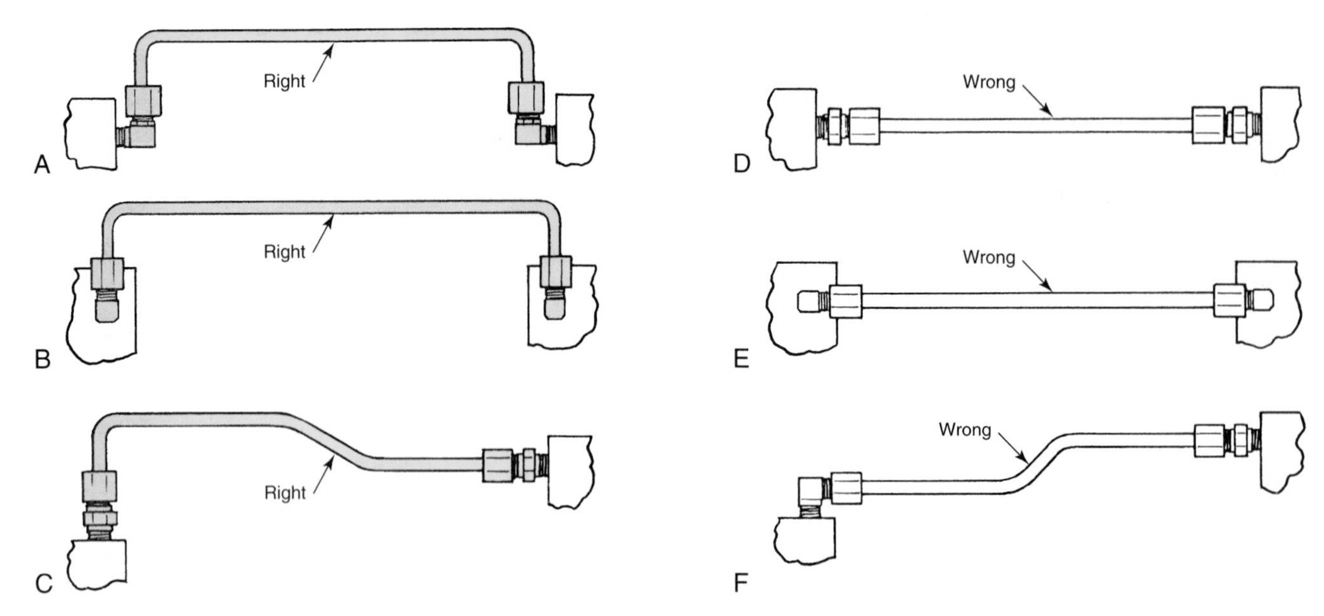

Figure 9-27. Install tubing as shown in A, B, and C to avoid the straight runs shown in D, E, and F.

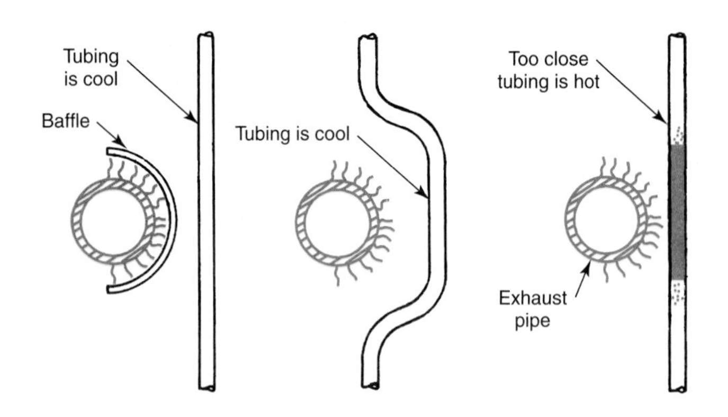

Figure 9-28. Install a heat baffle or reroute tubing to protect it from hot surfaces.

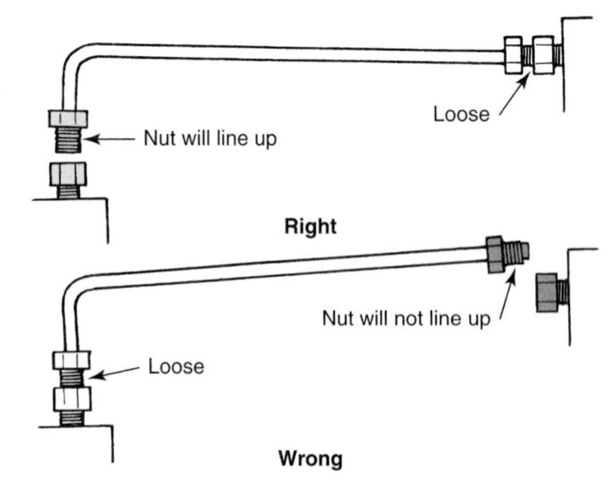

Figure 9-30. Assemble all tubing long leg end first. If short end is assembled first, the long end will be difficult to connect.

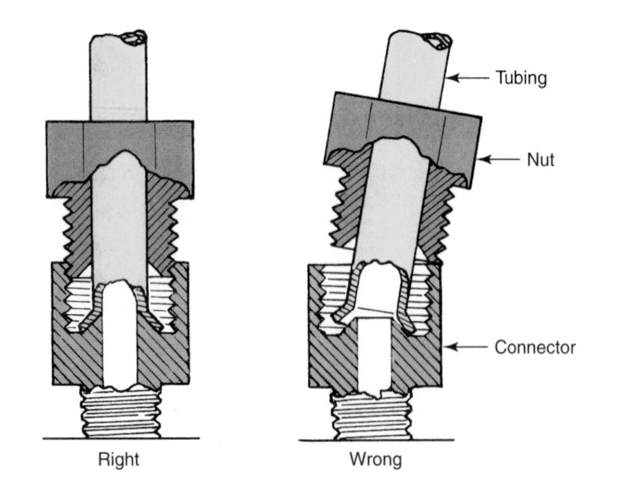

Figure 9-29. Proper fitting alignment is important. Nut on right would cross-thread. Never force a fitting into a thread.

to make the connection. Once both connections are made, tighten the fittings. Be careful when tightening, since many fittings are made of soft material and are easy to twist off. If torque values are available, use them.

Hose

Numerous sections of ***flexible hose*** are used on the modern automobile. Automotive-type hose uses *rubber* or a synthetic compound such as *neoprene* in its construction. They are generally identified by use, pressure capacity, construction, and materials used. Hose will withstand vibration and normal flexing when properly installed.

The cooling, lubrication, fuel, vacuum, steering, and brake systems all use some flexible hose. The technician must know what replacement types are needed and the correct methods of installation.

Cooling System and Heater Hose

Pressures in cooling and heating systems are relatively low. The hose used in such low-pressure applications may be of a ***single-ply*** or ***double-ply construction.*** Heavier hose is available for heavy-duty applications.

Radiator hose is available in straight, curved, molded (usually in a special shape), and flexible (designed to bend without collapsing) types. Radiator hoses often have a built-in spiral of wire to prevent collapse due to the suction of the water pump.

Figure 9-31 illustrates the typical ***molded radiator hose.*** **Figure 9-32** shows the fabric ply and spiral wire construction of the ***flexible radiator hose.***

Hose Clamps

In low-pressure hose installations, such as the heater and radiator, the hose is merely slid over the fitting. A spring or screw-type ***hose clamp*** is then installed. Use a small amount of sealer to ease installation and to provide extra protection against leaks. Locate the clamps so that they may be easily reached for tightening. Tighten them securely.

If the hose fitting has a raised rib, make sure the clamp is installed on the fitting side of the rib. This will prevent the hose from working loose, **Figure 9-33. Figure 9-34** illustrates several methods of attaching a radiator hose using hose clamps. Many types of clamps are reusable; some must be discarded after use. If difficulty is experienced when

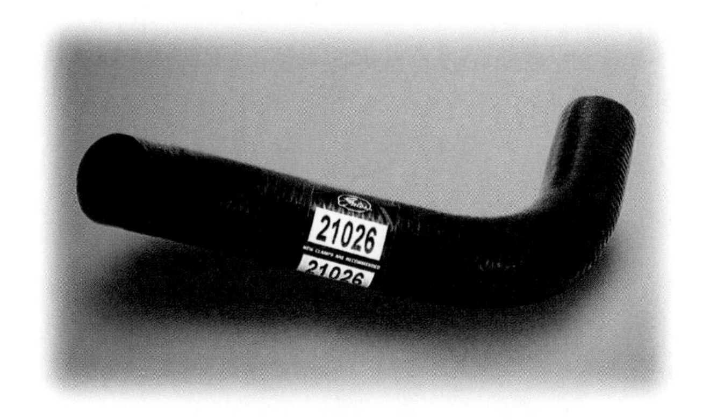

Figure 9-31. A typical molded radiator hose. (Gates Rubber Co.)

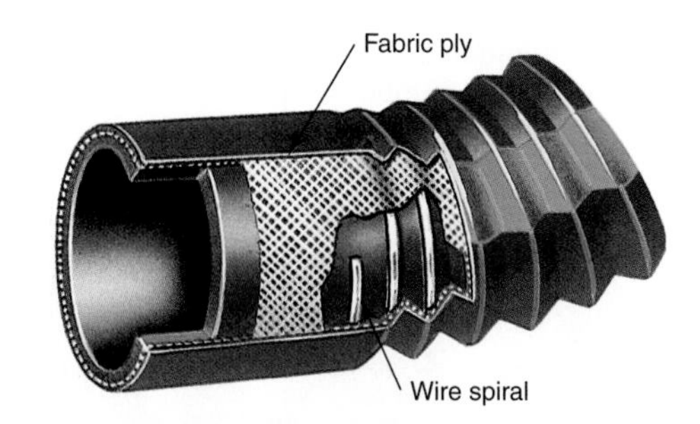

Figure 9-32. Cutaway view of a flexible radiator hose. Note the fabric ply and the built-in wire spiral. (Gates Rubber Co.)

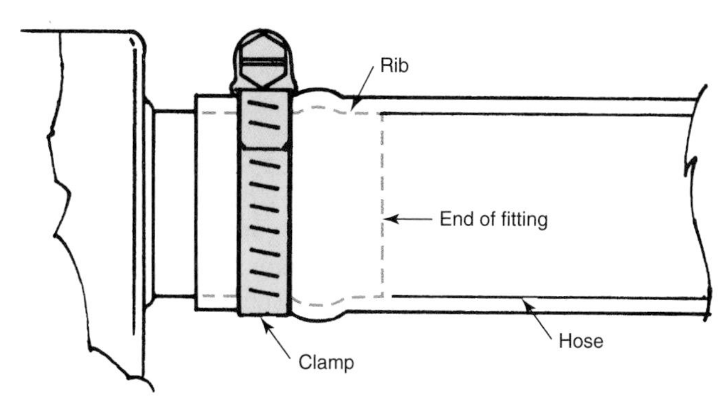

Figure 9-33. Always install hose clamps on the fitting side of the raised rib. (Gates Rubber Co.)

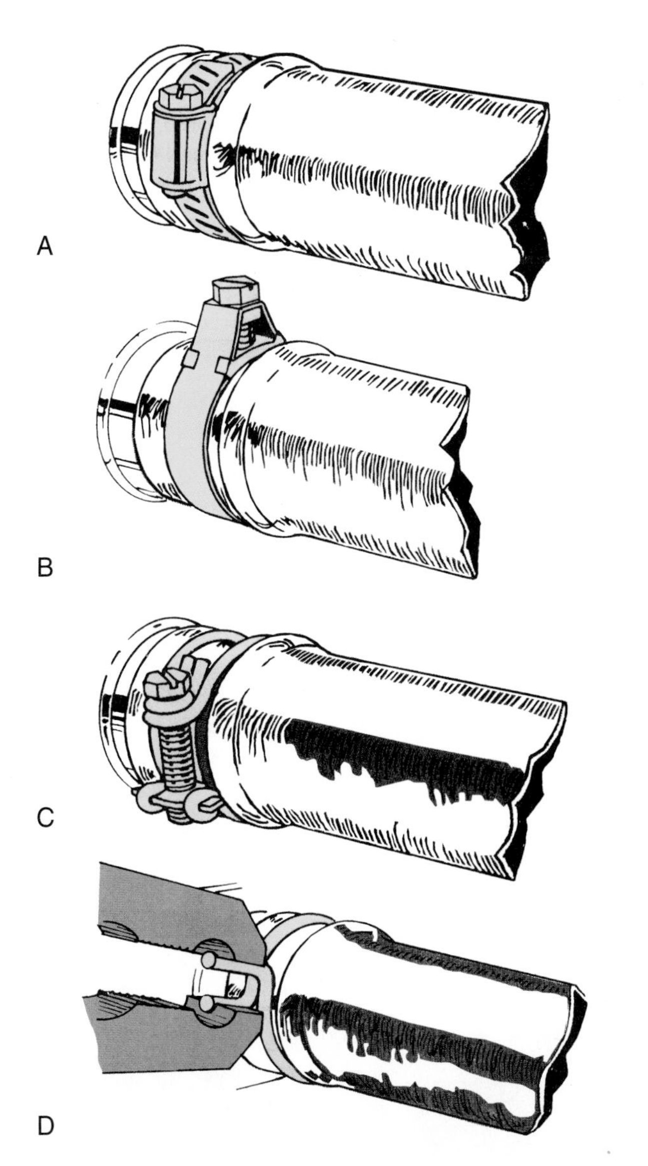

Figure 9-34. Hoses can be secured by various types of clamps. A—Worm drive clamp. B—Screw-tower clamp. C—Twin-wire clamp. D—Spring clamp. (Gates Rubber Co.)

attempting to remove an old hose, use a knife to cut away the portion of the hose over the fitting, as shown in **Figure 9-35.**

Refrigerant Hose

Refrigeration systems on newer vehicles are equipped from the factory with barrier hoses. A ***barrier hose*** has an inner lining, usually made of nylon, **Figure 9-36.** The lining is ***impermeable,*** which means refrigerant molecules cannot pass through it. Barrier hoses are used on all R-134a systems, and must be installed as part of some retrofit operations.

Nonreinforced Hose

Many of the smaller diameter hoses used for vacuum, windshield washer, drain, and overflow applications are made of rubber without reinforcement.

Fuel System Hose

Depending on design, fuel systems operate on either high or low pressure. Therefore, the correct type of fuel system hose must be used. Never use single-ply, synthetic hose in a fuel injection system. The high pressure in these systems can cause a single-ply hose to rupture; a reinforced two-ply hose must be used. See **Figure 9-37.** Never use hose that is not specifically designed for fuel systems.

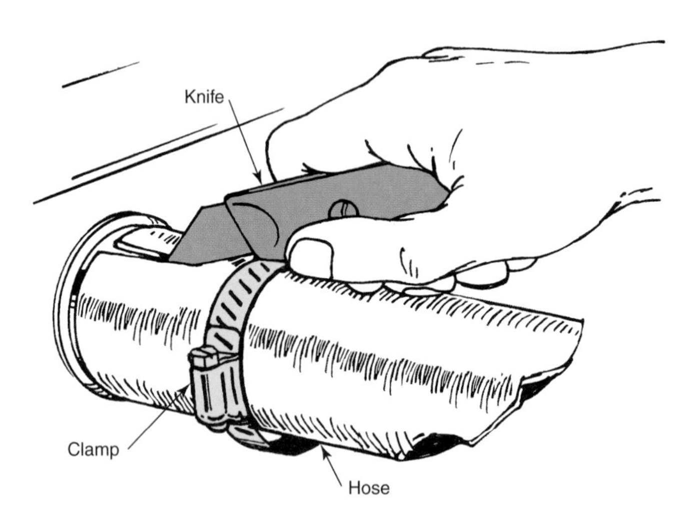

Figure 9-35. A knife can be used to cut away hose that is difficult to remove. (Gates Rubber Co.)

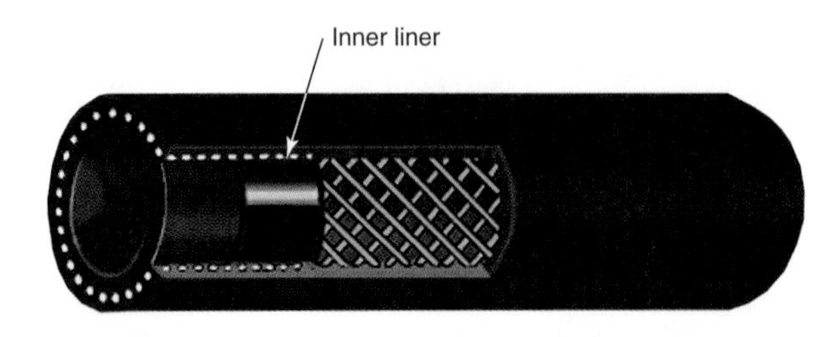

Figure 9-36. Barrier hose includes a lining material that prevents molecules of refrigerant gas from passing through. (Gates Rubber Co.)

Note: The modern fuel system is usually under pressure, even when the engine is off. System pressure must be released before hoses, lines, or fittings are serviced.

Power Steering and Brake Hose

Power steering and brake systems create pressures exceeding 1000 psi (6,895 kPa). The hose used must be of multiple-ply construction. Replacement hoses are readily available. See **Figure 9-38.**

Caution: Do not make up hoses for power steering or brake systems. These systems generate high pressures and must have factory replacement hoses.

Lubrication Hose

Oil filter or oil cooler hoses can either be made up or purchased ready-made. Oil filter hoses utilize a synthetic rubber hose covered with a soft wire braid for pressure strength. Hoses with fabric ply reinforcement are also used. The hose must be oil-resistant.

Hose End Fittings

There are numerous types of end fittings, and many are reusable (they can be taken off and remounted on new hose). Fittings include those designed to connect hose to pipe and to tubing.

The following steps outline installation procedures for different types of hose ends. When assembling hose ends,

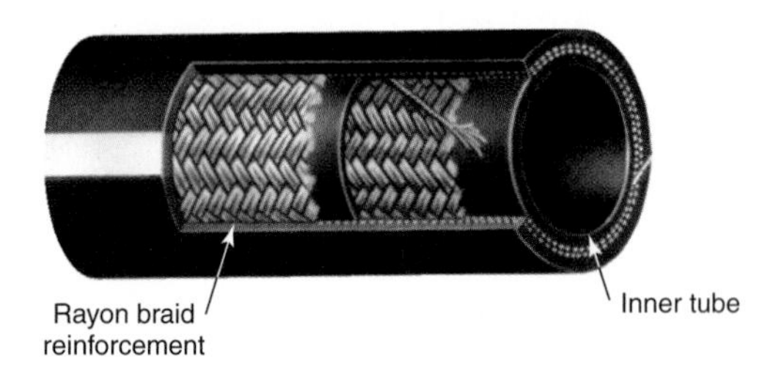

Figure 9-37. Fuel system hose construction. Fuel hose sidewalls are reinforced and relatively thick to prevent collapse under vacuum. (Gates Rubber Co.)

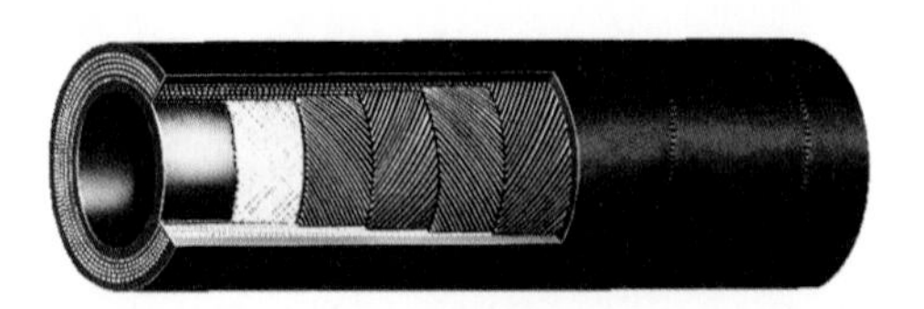

Figure 9-38. Multiple-ply, high-pressure hose. Ply thickness, number of plies, material, and weave must be considered in determining the working pressure. (Gates Rubber Co.)

always lubricate with water, soap, oil, brake fluid, air conditioning compressor oil, or other agent compatible with the system. Refer to the appropriate figures. Directions given for fittings in **Figure 9-39** are general.

Note: Do not use a lubricant that will attack the hose or contaminate the system.

A simple barb-type hose end installation is explained in Figure 9-39A.

1. Lubricate the hose and fitting.
2. Push the hose completely over the barbed end of the fitting.
3. If it is necessary to remove the hose end, cut the hose.

Figure 9-39B illustrates a compression fitting used for wire braid hose.

1. Neck down one end of the braid.
2. Flare the other end.
3. Install the nuts.
4. Install the hose over the nipple to adapt it to size, then remove it.
5. Place the insert over the hose and under the braid.
6. Push the hose against the flat surface to seat the insert fully.
7. Push the nuts over the insert.
8. Push the hose over the nipple.
9. Tighten the nuts.

Figure 9-39C shows a compression fitting on an air brake hose.

1. Slide the air brake hose spring over the hose.
2. Push the hose into the socket.
3. Thread the nipple into the socket, squeezing the hose between nipple and socket.
4. Snap the spring over the socket shoulder.

A different type of compression fitting is shown in Figure 9-39D.

1. Mark and ***skive*** (remove the outer layer of rubber down to first layer of cord) the hose.
2. Push the skived end into the socket.
3. Lubricate the nipple and hose.
4. Thread the nipple into the socket.

In Figure 9-39E, another type of compression fitting is shown.

1. Push the hose into the socket.
2. Lubricate the mandrel (pilot to expand hose and assist in proper seating).
3. Thread in the nipple.
4. Seat the mandrel and then remove it.

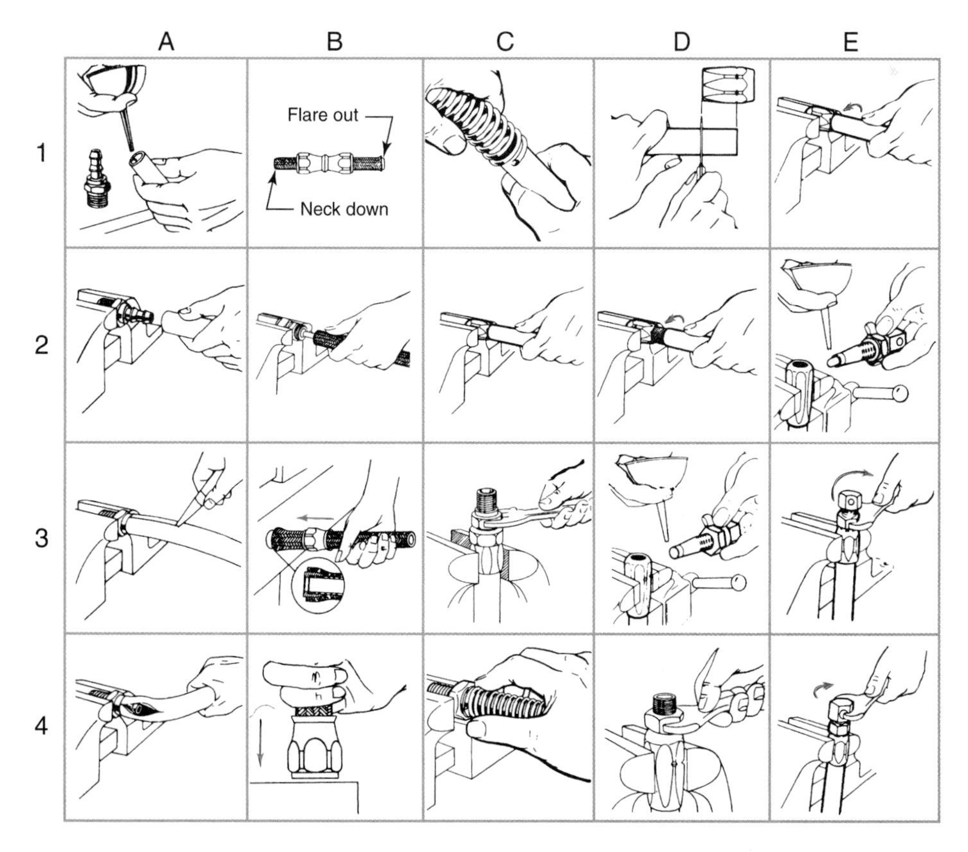

Figure 9-39. General methods of attaching various hose ends are shown. (Imperial Brass Mfg. Co.)

Skived Hose

When instructions call for skiving a hose, be careful not to cut the cord. A skiving knife and mandrel set are shown in **Figure 9-40.** Skive only the portion necessary. The skived portion should not extend out of the fitting. A fitting using a skived section is shown in **Figure 9-41.**

Mounting Hoses

Avoid sharp or double bends and twisting, since this can cause premature failure. In determining how sharp a hose bend may be, figure the radius of the bend should be at least five times the outside diameter of the hose. For example, a hose with an OD of 1/2″ (2.5 mm) should have a minimum bend radius of 2 1/2″ (64 mm). This means if the hose were pulled around to form a circle, the circle would be at least 5″ (127 mm) in diameter. When making straight run connections, allow some slack to avoid stressing the hose from pressure, vibration, or part shifting.

Use flare-nut wrenches when tightening hose fittings. Always support one portion of the fitting with one wrench while tightening with another to prevent twisting the hose. Tighten the swivel end last. **Figure 9-42** illustrates some typical hose installations. The methods shown in the left column are correct. Notice how single, smooth bends, without twisting, are made. Incorrect methods are shown in the right column.

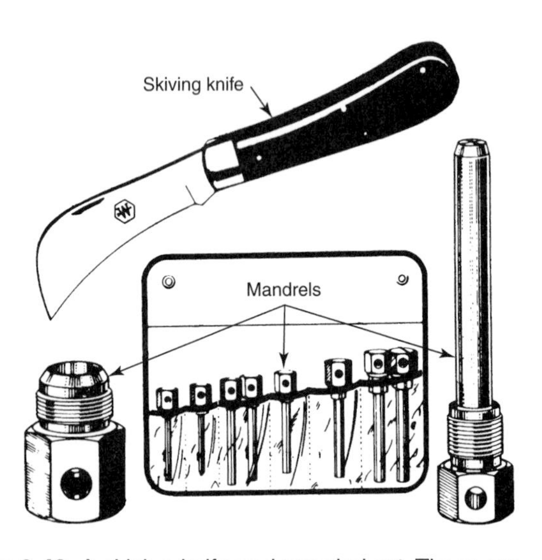

Figure 9-40. A skiving knife and mandrel set. These are essential tools for proper installation of certain types of hose ends.

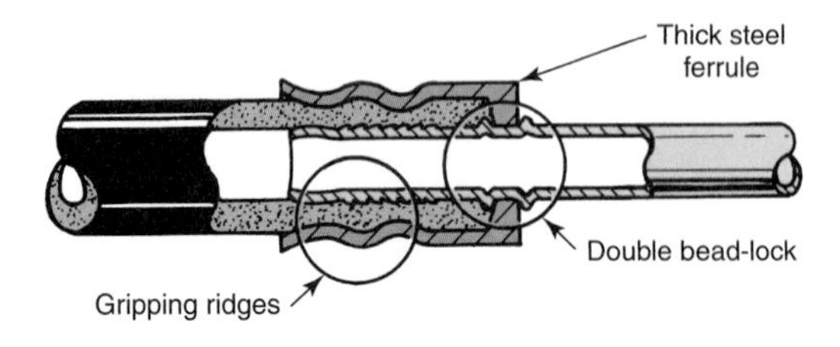

Figure 9-41. A permanent (not reusable) hose end. Note the skived section, which is placed under the steel ferrule. Also note the gripping ridges. (Moog)

Some hoses are installed in the vehicle by the use of special locking devices attached to a flange formed on or fastened to the vehicle body, **Figure 9-43.** These hoses should always be reinstalled on the correct flange opening, using the right retaining device. Keep hoses away from the exhaust system to avoid heat damage. If a hose run is long, use clips to secure it in place. On off-road vehicles, keep hoses and tubing well up within the frame to prevent snagging and to shield them from flying rocks.

Hose Condition

Any hose that shows signs of cracking, undue softness, or swelling should be replaced. Hoses often deteriorate on the inside, causing portions of the hose material to break loose and producing partial or complete blockage. Check hoses carefully; if at all doubtful about their condition, replace them.

Storing Hose

Store hose in a cool spot, preferably in the original container. Avoid exposure to sunlight, fuel, lubricants, and chemical compounds. Do not place heavy objects on the hoses.

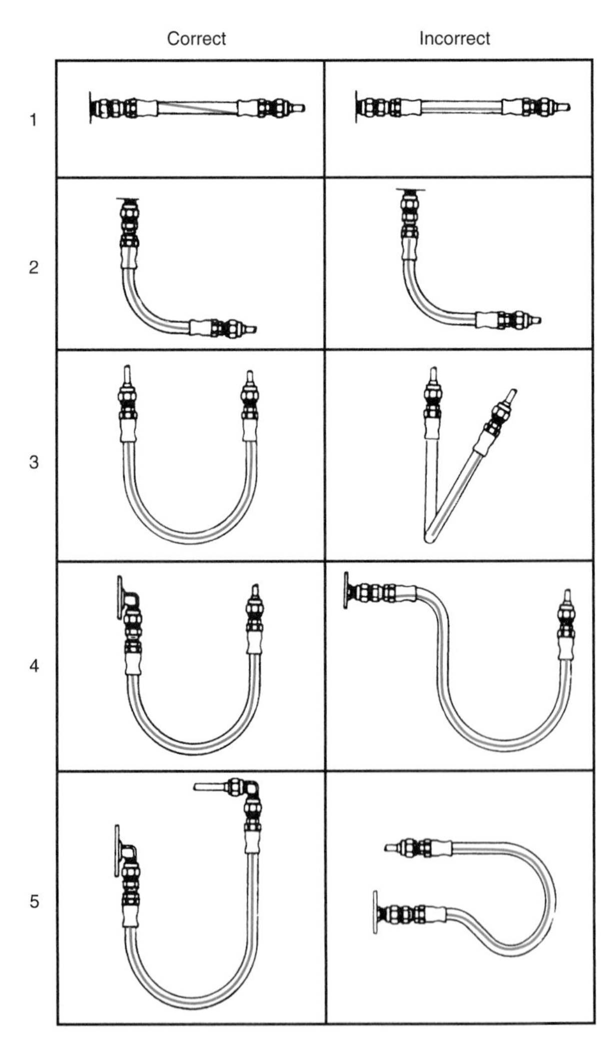

Figure 9-42. Study these correct and incorrect hose installations. Double bends and twisting must be avoided.

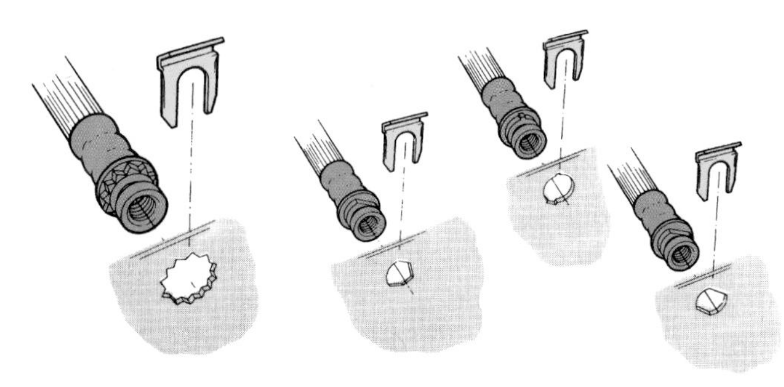

Figure 9-43. Various hose fitting mounting flange shapes. Always use a mounting clip when routing a hose through a flange. (General Motors)

Tech Talk

Flaring tubing is a simple task compared to other automotive jobs, such as reboring an engine or obtaining the proper mesh in a set of differential gears. Since flaring, like many other jobs, is relatively simple, students often tend to overlook its significance and concentrate on what they feel are the important jobs.

The experienced technician, who performs many so-called simple jobs each day, knows that these simple jobs are very important. Many major service jobs have failed due to careless or improper handling of the simple steps. Regarding simple jobs, remember these facts:

- They must be done.
- They must be done correctly.
- Eventually, you will have to learn how to do them.

Keep this in mind as you study this and other texts. Read everything carefully and consider everything you read important. You will be glad you did.

Summary

Copper, steel, aluminum, and plastic tubing are used in automotive work. Brake and steering systems must use double-wrapped, brazed steel tubing. Handle tubing carefully.

Tubing should be cut with a tubing cutter. Bending should be done with either a spring or mechanical bender. Tubing ends must be square and all burrs removed.

Connections are made with either flared, compression, or push-on fittings. Flare fittings must be formed with a flaring tool. Double-flare all double-wrapped brazed steel tubing. Double-flare all high-pressure applications. Both standard and long nuts are available.

Compression, sleeve, and double compression fittings are quick, easy, and suitable for fuel, lubrication, and vacuum lines. When tightening, be sure to hold tubing all the way in the fitting. Compression fittings on soft plastic tubing require a special insert.

Push-on fittings are pushed into place and held by a retaining or locking mechanism. The removal process for some push-on fittings requires special tools.

Connectors, unions, elbows, tees, O-ring, distribution blocks, shutoff, and drain cocks are the most commonly used fitting types.

When installing tubing, avoid straight runs. Support long runs and related parts. Protect from heat. Assemble both ends loosely before final tightening.

Automotive-type hose uses rubber or synthetic compounds in its construction. Nonreinforced, single, and multiple-ply types are used.

Radiator hose is either straight, molded, or flexible. Vacuum wiper, overflow, windshield washer, and other applications often use nonreinforced hose. Fuel line hose must be resistant to gasoline and should have a reinforcing ply. Lubrication system hose must be reinforced and oil-resistant. Power steering and brake hoses use multiple-ply construction. Do not make up these hoses; buy factory replacements.

Hose end fittings can be classed as permanent or reusable. Some hoses are attached with mounting clips. Barb-type fittings, where used, provide sufficient holding power. Threaded hose fittings can be of the flare, compression, or pipe type. Some low-pressure hoses are retained by spring or screw clamps.

When installing hoses, avoid double bends, twisting, and sharp bends. Protect from heat, moving parts, and road-debris damage.

Both hose and tubing must be clean before installation. Where sealant or lubrication is used, it must be compatible with the system involved. Support fittings with a wrench when tightening connections. Tighten swivel ends last. Use flare-nut wrenches. Always test the finished job for leaks or malfunctions. Protect stored tubing and hose from damage.

Review Questions—Chapter 9

Do not write in this book. Write your answers on a separate sheet of paper.

1. The more tubing is worked (bent), the _____ it becomes.
   (A) softer
   (B) harder
2. Always _____ brazed steel tubing.
   (A) double-lap
   (B) single-lap
   (C) use a compression fitting on
   (D) Either A or C.
3. How many flaring steps does it take to make a double-lap flare?
   (A) 1.
   (B) 2.
   (C) 3.
   (D) 4.
4. How many flaring steps does it take to make an ISO flare?
   (A) 1.
   (B) 2.
   (C) 3.
   (D) 4.

5. In a double compression fitting, the sleeve is _____.
   (A) separate
   (B) made from composition material
   (C) part of the nut
   (D) integral
6. When assembling a used 5/16″ compression fitting, how many turns should the nut be tightened after the sleeve just grasps the tubing?
   (A) None.
   (B) One and one-quarter turns.
   (C) One and three-quarter turns.
   (D) Two and one-quarter turns.
7. What type of threads does an O-ring fitting use?
   (A) Straight.
   (B) Pipe.
   (C) Tapered.
   (D) All of the above, depending on design.
8. A union is designed to connect _____.
   (A) tubing and pipe threads inside of a part
   (B) tubing and elbows
   (C) two sections of tubing
   (D) All of the above.
9. Push-on tubing connectors using spring-clips can be removed with _____.
   (A) flare-nut wrenches
   (B) special tools
   (C) impact sockets
   (D) None of the above.
10. Which of the following is the best method of cutting tubing?
   (A) Hacksaw.
   (B) File.
   (C) Tubing cutter.
   (D) Bolt cutter.
11. If you force threaded fittings that start hard, you will _____ them.
   (A) cross-thread
   (B) jam
   (C) split
   (D) Either B or C.
12. In all _____ systems, barrier hoses must be used.
   (A) fuel
   (B) R12
   (C) brake
   (D) R134a
13. Which application would use small-diameter, nonreinforced hose?
   (A) Power steering lines.
   (B) Brake lines.
   (C) Fuel lines.
   (D) Vacuum lines.
14. Removing the outer layer of rubber down to the first layer of cord is called _____.
   (A) necking
   (B) expanding
   (C) skiving
   (D) reducing
15. The radius of a hose bend should be at least _____ times the outside diameter of the hose.
   (A) 2
   (B) 3
   (C) 4
   (D) 5

ASE-Type Questions

1. Technician A says copper tubing should never be used on brake systems. Technician B says copper tubing should never be used on power steering systems. Who is right?
   (A) A only.
   (B) B only.
   (C) Both A and B.
   (D) Neither A nor B.
2. Technician A says as long as the flare is the correct angle, it can be slightly cocked to one side. Technician B says the nut should be placed on the tube before flaring. Who is right?
   (A) A only.
   (B) B only.
   (C) Both A and B.
   (D) Neither A nor B.
3. Tubing fittings should be tightened with a _____.
   (A) pipe wrench
   (B) flare-nut wrench
   (C) crescent wrench
   (D) box-end wrench
4. Technician A says the tubing end should be pointed upward when removing burrs. Technician B says a reamer is the best tool for removing burrs. Who is right?
   (A) A only.
   (B) B only.
   (C) Both A and B.
   (D) Neither A nor B.
5. Technician A says tubing bends that are close to the fitting should be made before the flaring operation. Technician B says bends that are close to the fitting should start at least the distance of the fitting nut from the actual connection. Who is right?
   (A) A only.
   (B) B only.
   (C) Both A and B.
   (D) Neither A nor B.

6. Technician A says a long nut will support tubing a greater distance from the actual connection than a short nut. Technician B says straight runs of tubing should be made whenever possible. Who is right?
   (A) A only.
   (B) B only.
   (C) Both A and B.
   (D) Neither A nor B.
7. Tubing should be arranged to avoid all of the following, *except:*
   (A) exhaust system parts.
   (B) moving parts.
   (C) straight runs.
   (D) the use of mounting clips.
8. Spring and screw clamps are used with _____.
   (A) power steering hoses
   (B) heater hoses
   (C) brake lines
   (D) All of the above.
9. Instead of making up tubing lines, factory replacements should be used whenever _____ system lines are replaced.
   (A) fuel
   (B) power steering
   (C) brake
   (D) Both B & C.
10. Technician A says a hose that looks good on the outside will be good inside. Technician B says hoses should be stored in a very warm area to drive out moisture. Who is right?
   (A) A only.
   (B) B only.
   (C) Both A and B.
   (D) Neither A nor B.

Suggested Activities

1. Practice bending tubing using various bending tools.
2. Inspect the diameter markings on various types of hoses and determine the most important diameter (inside or outside) for matching hoses to other vehicle parts.
3. Write a report on the different types of metals used in automotive tubing. Explain why different types of metal are used in brake systems, fuel lines, air conditioning lines, transmission cooler lines, and oil lines.
4. Cut off a piece of metal tubing using a hacksaw. Cut another piece using a tubing cutter. Is there a difference in the appearance? Which one made the best cut?
5. Try to make a tight 90° bend in a piece of metal tubing with your hands. Did the tubing remain round? Try it with both a spring and mechanical bender. Demonstrate proper bending techniques to your classmates.

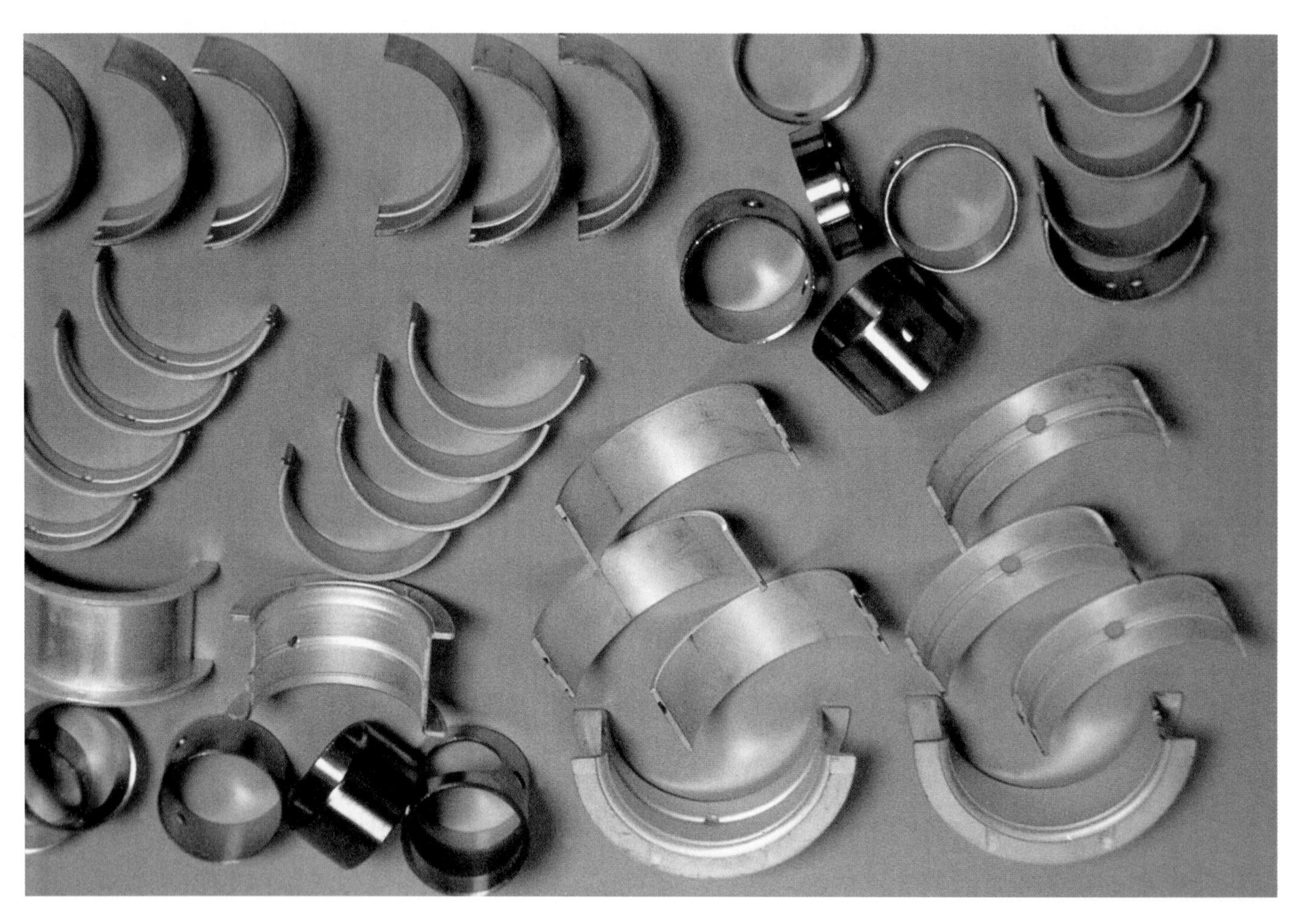

Bearings are critical to the operation of many vehicle components. A variety of friction bearings and bushings are shown here. (Federal-Mogul)

10

Friction and Antifriction Bearings

After studying this chapter, you will be able to:

- Compare the differences between friction and antifriction bearings.
- Explain the application of different bearing designs.
- Properly install friction bearings.
- Diagnose common reasons for bearing failures.
- List the different kinds of antifriction bearings.
- Explain the advantages of each type of antifriction bearing.
- Describe service procedures for antifriction bearings.

Technical Terms

Friction bearing	Races
Antifriction bearing	Cage
Bushings	Ball bearings
Radial loads	Roller bearings
Thrust loads	Needle bearings
Precision insert bearing	Self-aligning bearings
Oil holes	Sealed bearings
Oil grooves	Spalling
Journal	Brinelling
Thrust flange	Overheating
Bearing spread	Electrical pitting
Bearing crush	Bearing packer
Rolling elements	

Bearings are critical to the operation of many vehicle components. This chapter will cover the design, construction, application, and servicing of the bearings used in automobiles and light trucks. Common reasons for bearing failure will also be presented.

Major Classes of Bearings

Bearings can be grouped into two major classifications: friction bearings and antifriction bearings, **Figure 10-1.** The contact area of the ***friction bearing*** slides against the portion of a shaft designed to accept the bearing. This area is usually called the *bearing journal.* The ***antifriction bearing*** utilizes ball or roller elements that roll against the contact area. This reduces but does not eliminate friction.

Both types are used in automobiles and trucks. Major use of the friction bearing is in the engine and transmission. The camshaft, crankshaft, and connecting rods of vehicle engines all use friction-type bearings, **Figure 10-2.** Friction bearings used in transmissions and transfer cases are often called ***bushings.*** Antifriction bearings are used in some transmissions and transfer cases. They are also found in various places in the drive line, wheels, steering system, and belt-driven engine accessories. Antifriction bearing use in engines is largely confined to small, high speed engines used for motorcycles, outboard motors, and chain saws.

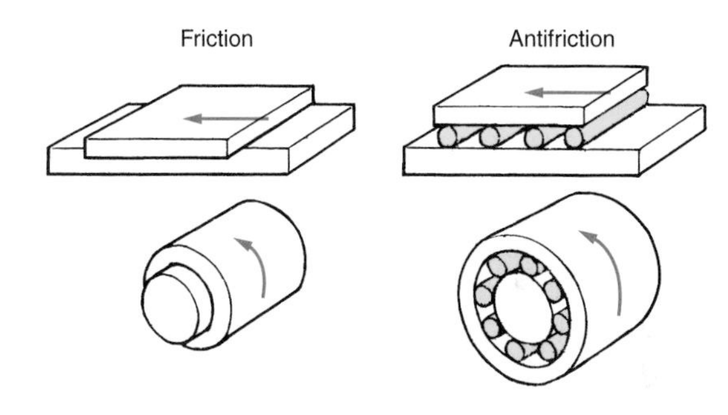

Figure 10-1. A friction bearing uses a sliding contact, while an antifriction bearing utilizes a rolling contact.

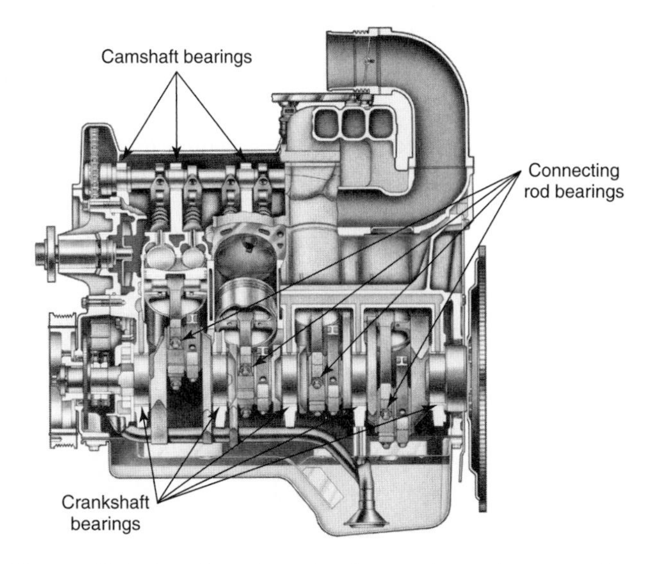

Figure 10-2. An internal combustion engine makes use of a great many friction bearings. (Ford)

The subjects of checking bearing clearance, determining bearing size requirements, prestart lubrication, bearing installation, and torquing will be discussed in detail in the chapters on engine overhaul.

Bearing Load

Any bearing used with a rotating shaft is subjected to two major types of loads, radial and thrust. ***Radial loads*** occur at right angles to the axis of the bearing. An example would be the sideways load placed on a connecting rod bearing as the piston pushes the rod to turn the crankshaft. ***Thrust loads*** are placed parallel to the axis. An example would be the outer pull on a wheel bearing when the vehicle is turned. See **Figure 10-3.**

Friction Bearings

Engine connecting rods, crankshafts, and camshafts use a type of friction bearing called a ***precision insert bearing.*** The precision insert bearing is light and strong. However, it does demand care in handling and installation. Precision insert bearings are made in one- and two-piece designs and in a wide range of sizes. Crankshaft and connecting rod inserts are usually two-piece design.

A precision insert bearing has one or more layers of lead, tin, copper, aluminum, or alloys of these metals commonly referred to as *babbitt metal.* These layers are bonded to a steel core, sometimes called a back. **Figure 10-4** shows the layers of soft metal (lead-copper alloy, copper, tin-lead alloy, and pure tin) in a typical insert bearing.

Camshaft Bearings

A camshaft bearing, **Figure 10-5,** is constructed like the connecting rod and crankshaft inserts. However, it is a one-piece design. The bearing material (usually babbitt metal) is affixed to steel strip stock that is rolled into a full circle and fastened with either a butt or butt-and-clinch joint. The camshaft bearing must be installed by being *pressed* into place. In addition to the standard sizes, camshaft bearings are available in large undersizes to permit *line boring.* Line boring is done by attaching a cutter to a long, rigid steel bar. The bar is passed through the bearings, boring them in line with each other.

Bushings

Bushings are full-round bearings, usually made of solid *bearing bronze,* which is a mixture of copper, lead, tin, and zinc. Some applications use rubber or steel back precision bushings, **Figure 10-6.** Bushings are used where it would be impractical to use two-piece bearings, or where the fit does not have to be as precise. The bushing is pressed into place and either bored, reamed, or honed to size. Steel and rubber suspension system bushings will be covered in the suspension and steering chapters.

Friction Bearing Failure

A properly selected and installed bearing, under normal operating conditions, will last in excess of 50,000 miles (80,000 km), and usually over 100,000 miles (160,000 km).

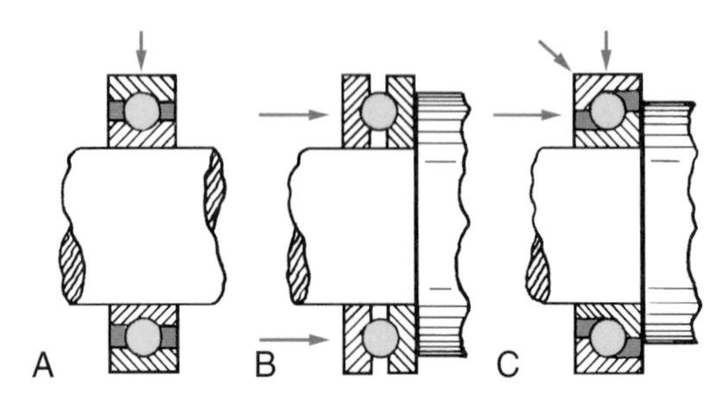

Figure 10-3. Loading designs. A—Radial loading. B—Thrust loading. C—Combination radial and thrust loading. Arrows indicate the direction in which the load is applied.

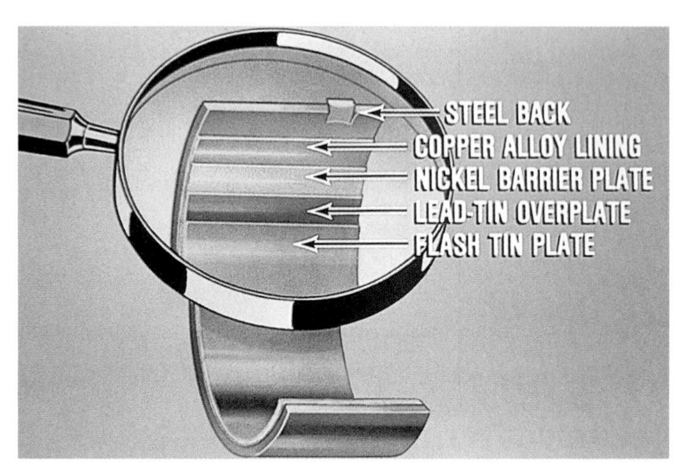

Figure 10-4. This insert bearing has five layers, counting the steel back. (Federal-Mogul)

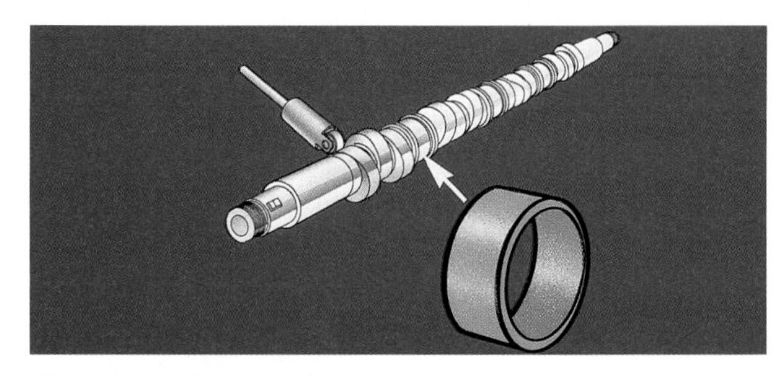

Figure 10-5. A camshaft bearing is a one-piece design. (Federal-Mogul)

Figure 10-6. Typical examples of steel-backed precision bushings. (AE Clevite)

There are many causes or combinations of causes that will result in premature failure. The technician must be familiar with the most significant ones and the effects they have on the bearing insert. Bearing failure is generally preceded by a lowering of oil pressure due to increased clearance. As bearing clearance increases, engine oil consumption will rise from excessive oil throw-off. Eventually, the bearings will start to knock. Whenever an engine is torn down, bearings should be cleaned and carefully inspected. A close study of a damaged bearing will often reveal the cause of the bearing failure.

Dirt—The Number One Cause of Bearing Failure

Field and laboratory studies have shown that dirt is the most frequent cause of bearing failure, **Figure 10-7.** The word *dirt* is used to describe foreign particles that cause damage to moving parts. This can include sand, cast iron and steel chips, pieces of bronze, grinding stone grit, and other materials. Normal engine wear will produce fine particles from the various moving parts. Most of these are normally removed by the oil filtration system. Abnormal engine wear will produce large bits of dirt that will greatly accelerate the wear process. Dirt is a bearing's worst enemy. Get it out of the unit and use every precaution to keep it out. The bearings shown in **Figure 10-8** were damaged by dirt.

Dirt from Reconditioning and Cleaning

Valve grinding, cylinder boring and honing, and shaft grinding deposit metal and other abrasive particles on parts. These must be removed from the reconditioned part by thorough cleaning. Particles from machining are sometimes found in new engines due to poor cleaning.

A sloppy job of cleaning often loosens carbon and other deposits but fails to completely remove them. Rinsing in dirty solvents often contaminates parts, as well. Once the engine is assembled and put into operation, the washing and cleaning action of the oil will cause this debris to reach the bearings. If the engine is very dirty, oil filters will not protect the bearings. The filter will very quickly become completely clogged, forcing the bypass open. This channels dirt directly into the bearings.

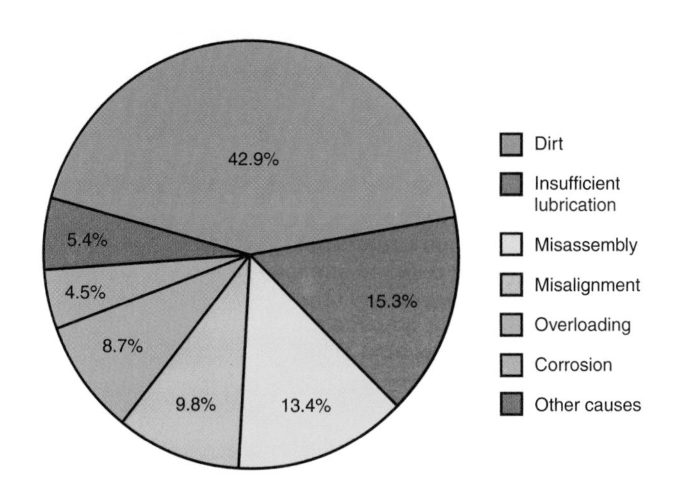

Figure 10-7. Causes of bearing failure and the percentage of occurrence. Note that dirt far exceeds the others.

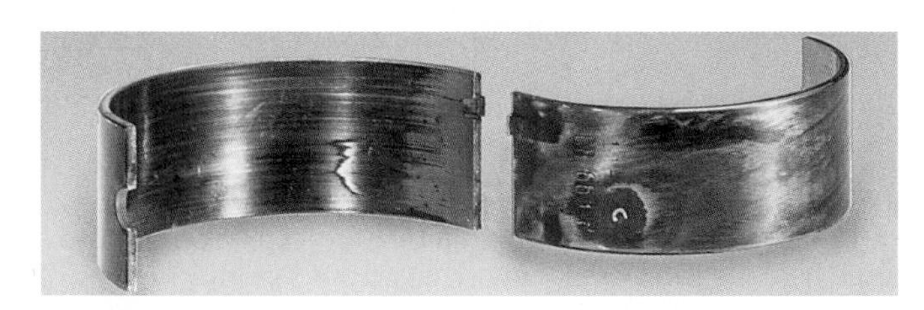

Figure 10-8. Dirt can ruin bearings rapidly, as shown by these inserts. (AE Clevite)

Dirt from External Sources

An engine may be contaminated by working under dusty conditions or by careless handling of parts. Keep clean parts covered until ready for installation. Work in a clean area, protected from windborne dust. When not working on a part, even for a few minutes, cover it. Keep hands and tools (especially wrench sockets) free of dirt when assembling parts. Avoid the use of compressed air blowguns, sandblasters, or steam cleaners near open engines or other sensitive units. Once the engine is assembled and placed in service, dirt can still enter. The most common entrances for dirt are through the air cleaner, breather system, fuel system, cooling system, dipstick, vacuum lines, and lubrication system.

Cover carburetors and throttle body openings when the air cleaner is removed. Keep air cleaners clean and properly serviced. Clean and properly service crankcase breather systems. Maintain a clean filter in the fuel system. Check for coolant leaks into the cylinders (ethylene glycol antifreeze forms a gummy residue in bearings and will cause serious problems). Never lay a dipstick on a dirty surface. Wipe both the stick and the area around the entry hole before replacing the stick. Before adding oil, check the filler tube for dirt. When changing oil filters, wipe the contact area thoroughly. Keep bulk oil tanks clean. Oil filler cans and spouts should be cleaned and stored to prevent contamination. When you remove drain plugs, clean them thoroughly before reinstalling them.

Bearing Lubrication Failure

Low oil pressure caused by worn bearings, a faulty pump, a clogged pickup screen, or an insufficient oil supply will cause rapid failure, **Figure 10-9.** A *dry start* (starting an overhauled engine without initially charging the oil system)

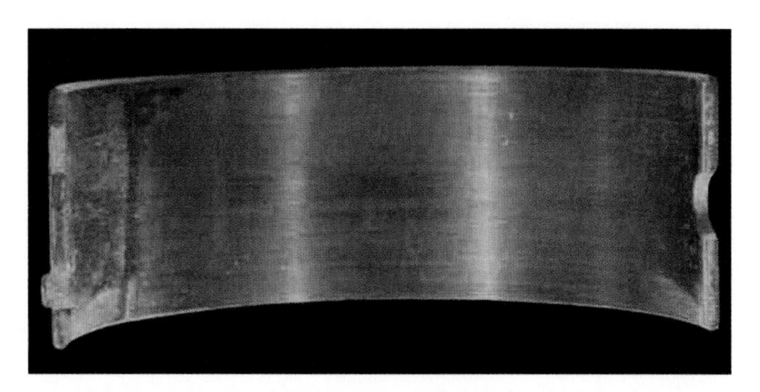

Figure 10-9. Lack of lubrication caused this bearing to wear prematurely.

can cause damage that will reduce the life expectancy of the bearings. Loss of oil through damage to the pan, broken pump or line, leaking gasket, poorly installed oil filter, or failure to replace plug after draining will cause sudden failure.

Bearing Failure from Improper Assembly

Bearing failure can be caused by dirt on the insert back, insufficient clearance, reversing the caps, placing a lower insert in the upper position, a bowed (warped) crankcase, or a bent crankshaft or rods. **Figure 10-10** illustrates various types of damage caused by improper assembly.

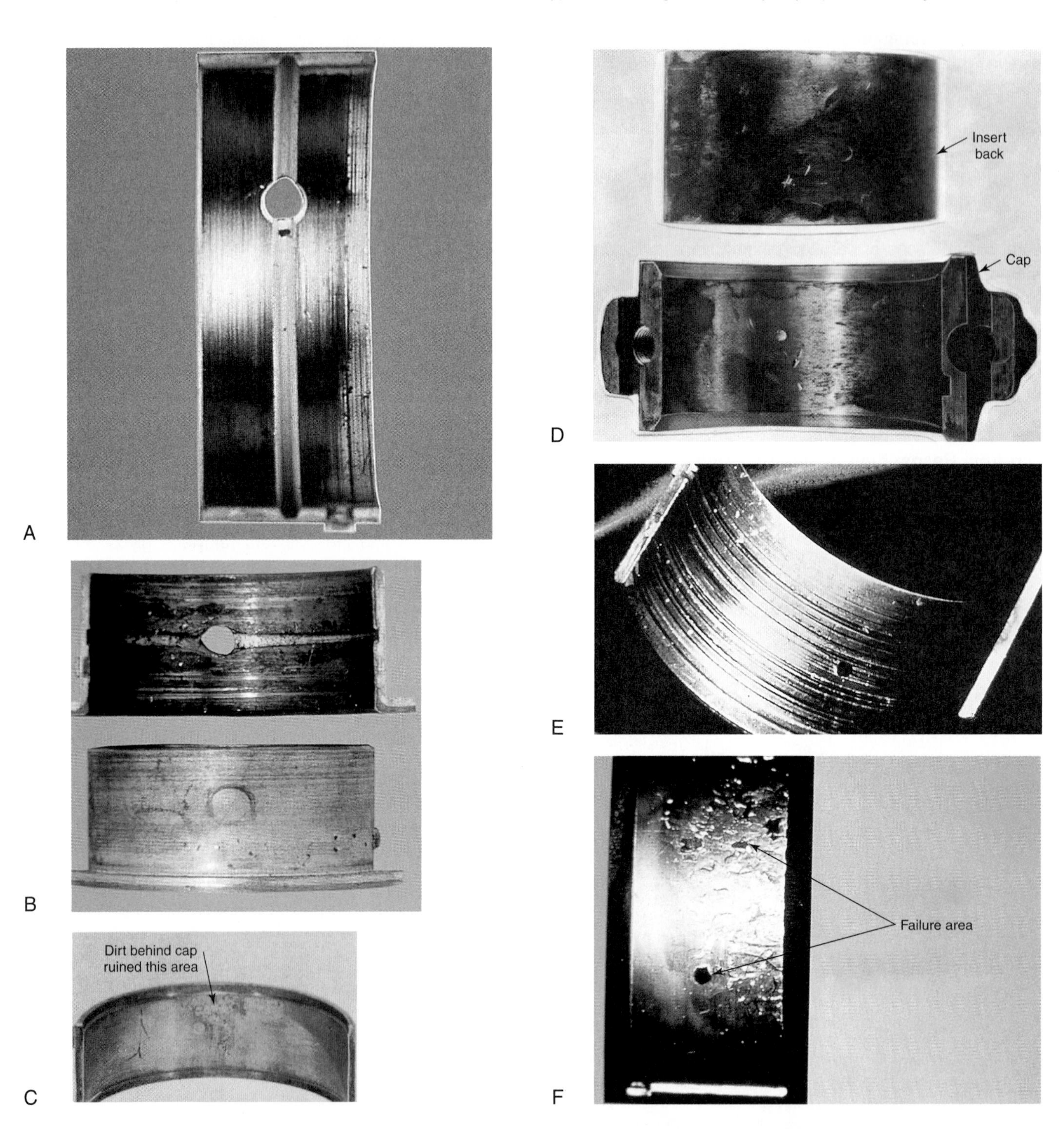

Figure 10-10. Damage caused by improper assembly. A—Bearing damage caused from a tapered housing bore. B—The upper insert, with oil hole, was installed in bottom position. The lower insert (see oil passageway impression on back) then blocked flow of oil to bearing. C—A particle of dirt between insert and bore caused a high pressure area that damaged this bearing. D—A nicked and dented cap bore transferred marks to back of insert, causing localized high pressure areas. E—A rough and scored journal caused this bearing to fail. F—A misaligned connecting rod placed one side of this insert under pressure. Note the failure area. (Federal-Mogul)

Operational Faults

Lugging (pulling hard at low engine rpm), detonation (rapid burning of fuel charge caused by a secondary flame front), preignition (fuel charge firing before plug fires), prolonged slow idling, and excessive rpms will all place the bearings under a heavy load. This can easily lead to premature bearing failure. When bearing condition indicates such problems, the vehicle driver should be informed of the cause to prevent a recurrence. **Figure 10-11** shows damage caused by operational faults.

Friction Bearing Installation

Precision insert bearings are what the name states: precision units that should be handled with utmost care. Do not mix bearing halves. Protect them from dirt and physical damage. Keep your fingers off the bearing surfaces; skin oils and acids can cause corrosion. When installing an insert, never force or pound it into place. Use the proper installation tools and always check for proper clearance. Make certain the bore and insert are spotless and locating lugs are in place. After installation, coat the bearing surface with clean engine oil. Never file an insert.

Insert and Housing Bore

An engine block can become distorted through the effects of heating and cooling. This will throw the camshaft and crankshaft bearing bores out of alignment. This, in turn, will force the camshaft and crankshaft out of alignment, creating heavy bearing loading and uneven stresses. See **Figure 10-12.**

The heavy stresses within the engine can cause the housing bores to elongate. If an insert is installed in such a bore, it will conform to the bore elongation, creating an egg-shaped bearing surface. Clearance in one direction will be excessive while clearance in the other will be insufficient, causing extreme friction and wear, **Figure 10-13.** Such bores must be reconditioned or the part replaced.

The housing bores, insert backs, and parting surfaces must be free of nicks, burrs, or foreign materials. If an insert is prevented from making perfect contact, pressure spots, misalignment, and overheating will result. Always check the housing bores and insert backs carefully to make certain they are smooth and clean. Do not oil these surfaces.

Bearing Oil Grooves and Holes

The insert often will have ***oil holes*** and ***oil grooves*** to permit oil to enter freely. Holes are used to allow oil passage

Figure 10-11. Damage caused by operational faults. A— Excessive idling will produce bearings like this. B—Riding the clutch (holding foot on the clutch all the time) places the main bearing thrust flange under prolonged loading. (Federal-Mogul)

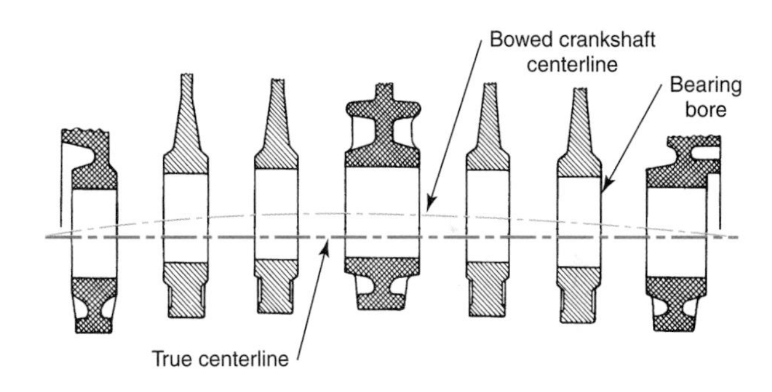

Figure 10-12. A bowed crankshaft will shift main bearing bores out of alignment with their true centerline.

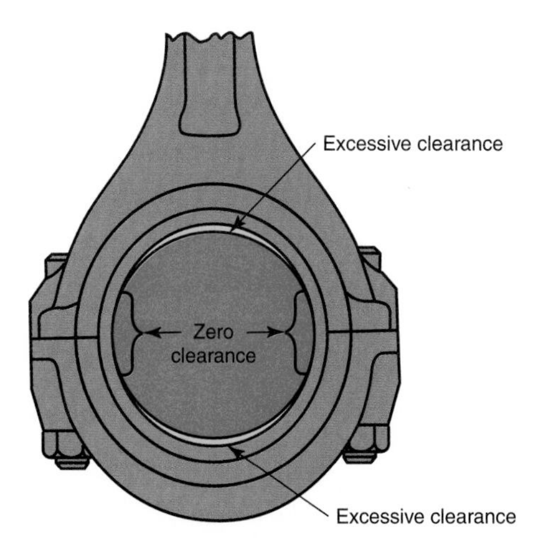

Figure 10-13. Elongated rod bearing bore. Note excessive clearance at top and bottom while zero clearance exists at sides. Insert life would be short. (AE Clevite)

to other areas. Annular, thumbnail, and distribution or spreader grooves are often incorporated, as well. See **Figure 10-14.** Not all inserts are drilled or grooved.

Bearing Journals

The section of a shaft that contacts the bearing surface is called a ***journal.*** It must be round, smooth, and straight, **Figure 10-15.** Nicks and scratches will ruin the bearing material. Manufacturers have established a minimum surface finish of 16 microinches or smoother. The microinch (one millionth of an inch, or 0.000001″) is used as a measurement of surface finish. To measure a surface finish in microinches, tests are made to determine the depths of all grooves or scratches. From this information, the average depth is calculated. This is usually about one-third of the maximum scratch depth. In **Figure 10-16,** you will note the solid line indicates one-third the maximum depth. If the maximum depth is 90 microinches, the finish would be 30 microinches.

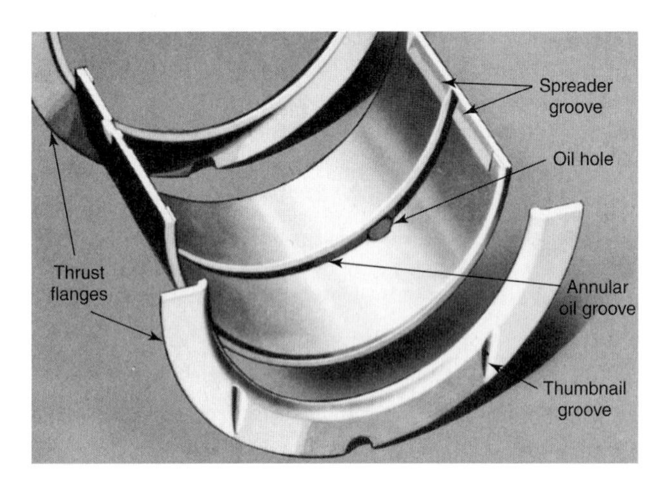

Figure 10-14. Typical bearing insert oil grooves. This main bearing uses separate thrust flanges. (AE Clevite)

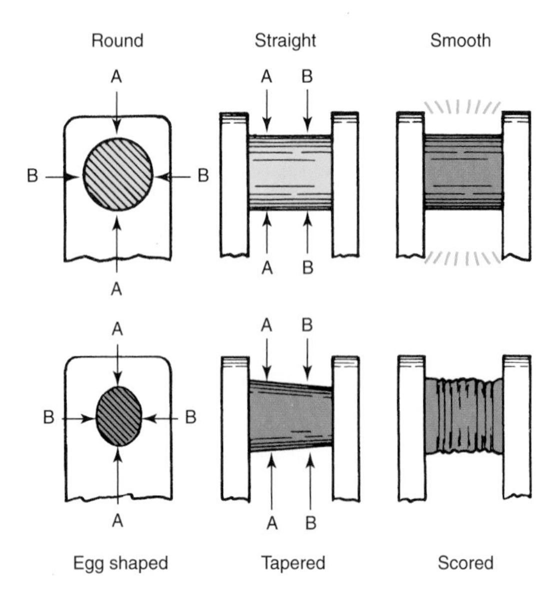

Figure 10-15. Bearing journals must be round, straight, and smooth.

Thrust Flange

Whenever an insert bearing must control thrust forces (pressure exerted parallel to the shaft centerline), a ***thrust flange*** is incorporated on one or both sides of the bearing, **Figure 10-17.** The thrust faces are lined with bearing material. Thumbnail oil grooves allow oil to enter the thrust surfaces. Some thrust flanges are not part of the bearings, but are inserted as separate pieces.

Installing the Insert

To provide adequate support, heat transfer, and alignment, it is essential that the insert properly contact the housing or cap. Inserts are manufactured to produce proper fit by using ***bearing spread*** and ***bearing crush.***

Bearing Spread

The insert diameter across the parting edges is slightly larger (0.005″–0.030″ or 0.13 mm–0.76 mm) than the bore. This makes it necessary to force or snap the insert into the bore by applying thumb pressure to the parting edges. Spread also helps hold the bearing in place during assembly, **Figure 10-18.**

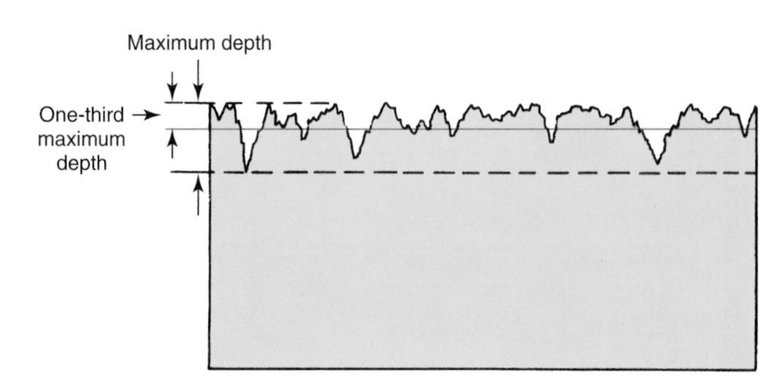

Figure 10-16. Determining surface finish in microinches.

Figure 10-17. Thrust flanges on a crankshaft main bearing. (Federal-Mogul)

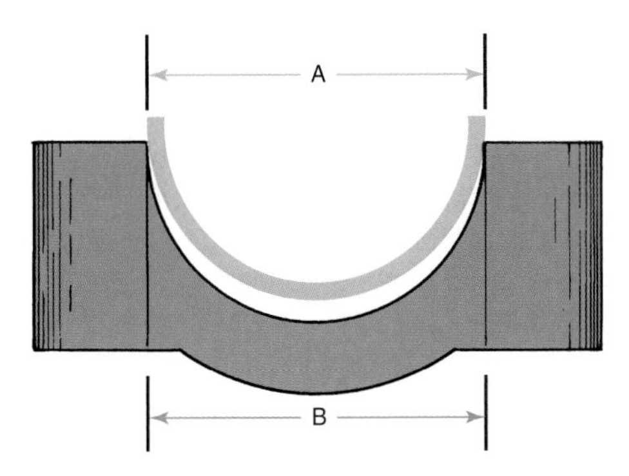

Figure 10-18. Positive bearing spread. Note that insert diameter A across the parting surface is slightly larger than bore diameter B.

Caution: Do not force the insert into place by pressing on its center. This could warp the insert.

Bearing Crush

The insert is also designed so that the parting edges will protrude a slight amount above the bore parting edge after it is snapped into place. In effect, each insert half is slightly larger than a true half circle, **Figure 10-19.** When the bearing is bolted together, the crush area touches first. As tightening progresses, the crush area is forced beneath the bore parting edges. This creates a tight insert-to-bore contact through radial pressure.

Caution: Never file bearing caps or crush. Doing so will ruin the bearing. The insert must not turn.

Inserts have locating lugs (sometimes called tangs) or dowels to prevent the insert from turning. When installing inserts, be certain the lugs are properly aligned with the slots in the housing, **Figure 10-20.** Dowels, when used, must enter their holes.

Align Housing Bore Halves

Even though the bore and insert are clean and the insert spread and crush are correct, the bearing will still be ruined (in case of split bearings) if the upper and lower bore halves are not properly aligned. It is possible to reverse the lower halves of some bearing caps**, Figure 10-21.** This will shift the upper and lower bores out of alignment. When disassembling bearing caps, always mark the upper and lower halves before removal. Use numbers so that you replace the cap in its original position. Used inserts should always be saved for study. If they appear usable, or you cannot obtain replacements, mark them on the back with a fine scribe. If you are sure that they will be replaced, mark them on the bearing surface.

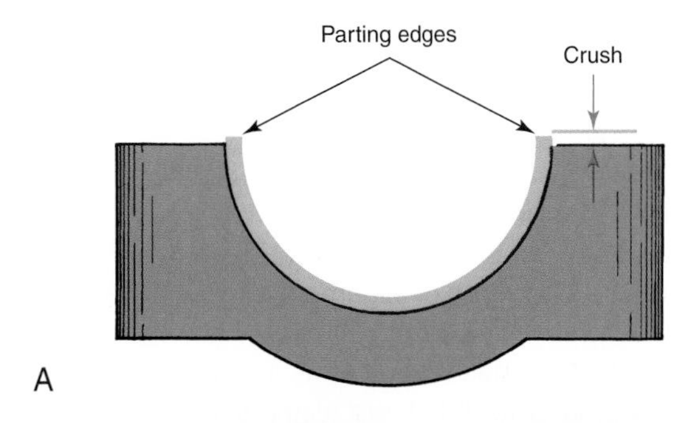

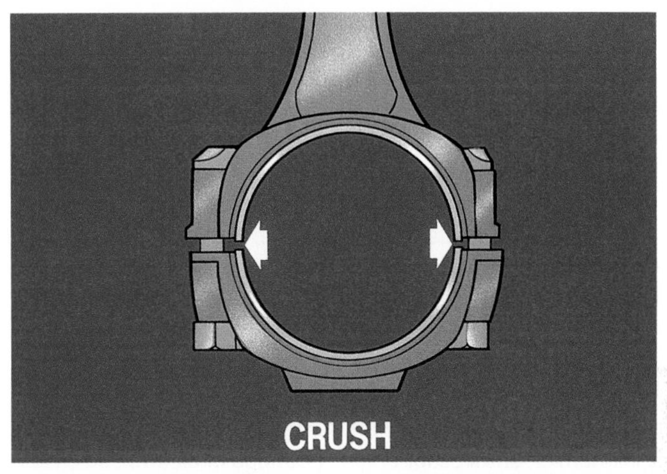

Figure 10-19. Bearing crush. A—Both insert parting edges (exaggerated for emphasis) protrude slightly above cap. B—When rod and cap are drawn together in bearing crush, radial pressure is produced, forcing insert tightly against the bore. (Federal-Mogul)

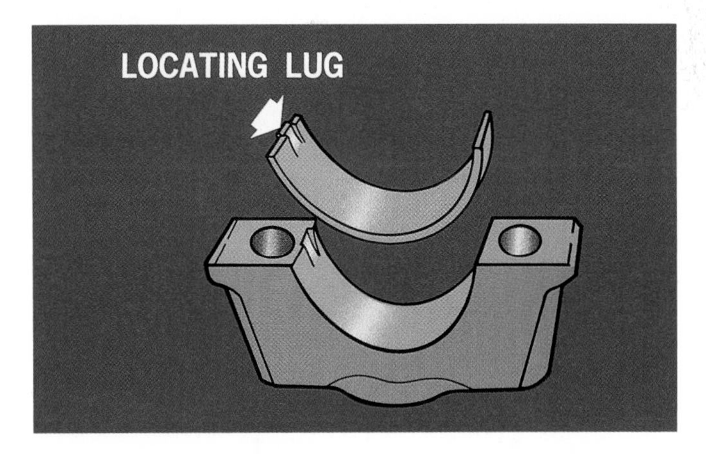

Figure 10-20. Locating lugs keep an insert from turning. Dowels are used for the same purpose on some bearings. (Federal-Mogul)

Do Not Mix Bearing Halves

Insert halves are made in matched pairs, but they are not identical. It is important that they are not mixed. If one of the insert halves is drilled and the other is not, be certain to place the drilled half in the drilled bore. When installing full-round inserts, such as camshaft bearings, make sure the oil holes are aligned, **Figure 10-22.** Neglecting to do this will result in a blocked oil hole, causing immediate bearing failure. Many split bearings are manufactured with both halves drilled to prevent improper assembly.

Wrench Side Pressure

Thick wrenches or sockets can create enough pressure against the cap to shift it out of alignment, **Figure 10-23.** Use the correct size wrench or socket. Tighten by alternating from one bolt or nut to the other. When cap is just snug, tap lightly with a plastic hammer to assist cap alignment. Use a torque wrench to tighten the fasteners to the recommended value.

Bearing Oil Clearance

The precision insert bearing must have enough clearance to allow oil to penetrate and form a lubricating film. The clearance must provide proper flow through the bearing to aid in cooling and passage to other critical areas needing lubrication.

Too much clearance will allow so much oil to flow that it can lower oil pressure. It can also cause excessive throw-off (oil running from bearings thrown off the crankshaft at high velocity). This can flood the cylinder walls with oil beyond the capacity of the piston rings to control. Excessive clearance can also allow sufficient movement between parts to literally pound the bearing. The chart in **Figure 10-24** shows average minimum clearances for engine bearings of different sizes and types. The chart indicates average clearances only, and should not be used when engine manufacturers' recommendations are available.

Checking Bearing Clearance

Approximate clearance of engine bearings can be determined by attaching an engine prelubricator (air-pressure-operated oil tank). Observe the amount of oil dripping from the bearings. This is often done after the pan is removed, but before disconnecting any bearings. It gives the technician an approximate idea of bearing condition. The prelubricator is used again after engine assembly to charge the lubrication system with oil. At the same time, it will provide a final visual check on bearing clearances.

One of the most common methods of obtaining precise clearance measurements is a special plastic wire (trade name Plastigage®). A section is placed either on the journal or on the insert. The bearing is tightened, then removed. The plastic will be flattened. By using a paper gauge supplied with the wire, the width of the wire can be accurately related to clearance in thousandths of an inch, **Figure 10-25.** Complete instruction on the use of the prelubricator and Plastigage will be given in the chapters on engine overhaul.

Undersize Bearings

To compensate for wear, inserts are available in a series of undersizes. If journal wear is slight, the recommended clearance can often be obtained through the use of inserts 0.001″ or 0.002″ (0.025 mm or 0.05 mm) undersize. The shaft

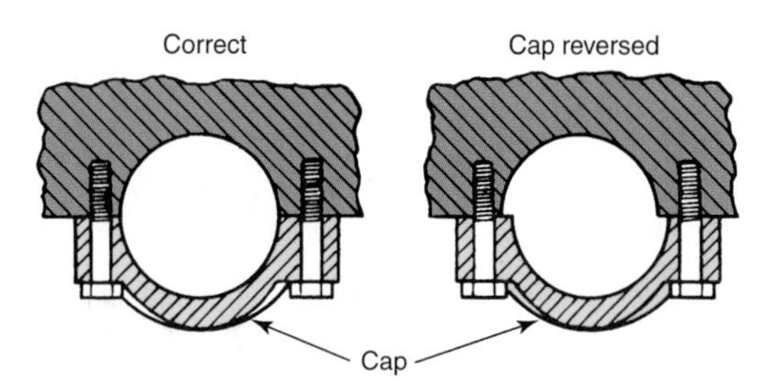

Figure 10-21. If bearing caps are reversed, the upper and lower bore halves will shift out of alignment.

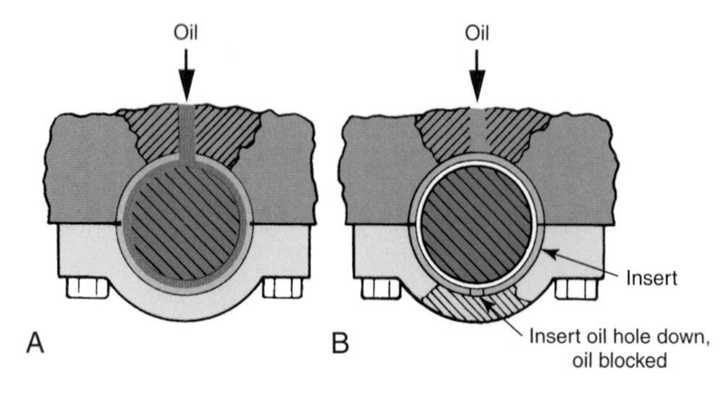

Figure 10-22. Align insert oil hole with the bearing's oil passage. A—Insert oil hole aligned with passageway. Proper lubrication will result. B—The insert oil hole has been placed at the bottom; oil passage is blocked.

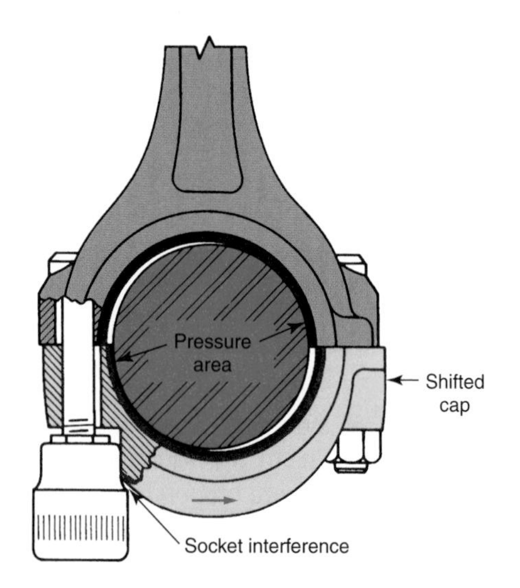

Figure 10-23. A thick wall socket has exerted side pressure, thus shifting cap to one side.

Recommended Oil Clearances for Engine Bearings

Shaft-size	SB (High lead or tin base)	CA (Copper alloy)	AP & CP (Over plated bearing)	AT (Aluminum alloy)
2 –2 1/4	0.0010	0.0020	0.0010	0.0025
2 13/16–3 1/2	0.0015	0.0025	0.0015	0.0030
3 9/16 –4 1/2	0.0020	0.0030	0.0020	0.0037

Note: Chart above indicates minimum diametral clearances. For maximum permissible clearance, add 0.001″.

Figure 10-24. Average minimum oil clearances for various types of engine bearings. (Federal-Mogul)

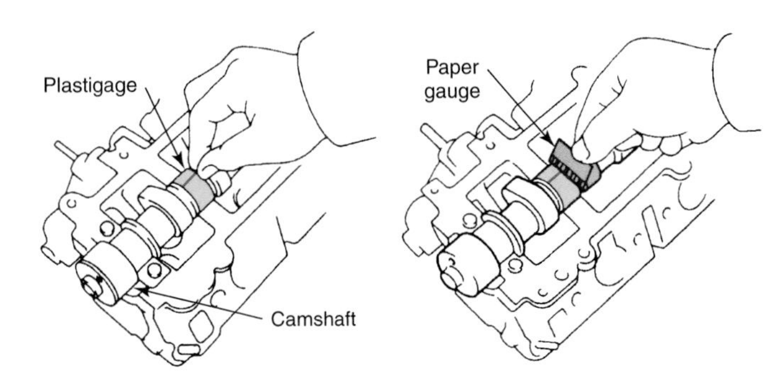

Figure 10-25. Checking bearing clearance with Plastigage®. (Toyota)

must be carefully measured and the largest diameter compared to the original size to determine the correct undersize.

Inserts are available 0.010″, 0.020″, and 0.030″ (0.25 mm, 0.51 mm, and 0.76 mm) undersize. When journal wear is severe or when journals are scored or egg-shaped, the shaft is ground to fit one of the bearing undersizes. This brings the bearing-to-journal clearance up to acceptable standards. Occasionally semifinished (greatly undersize) inserts are bored out to a specified size.

Antifriction Bearings

The antifriction bearing utilizes ***rolling elements*** (balls, rollers, or thin rollers referred to as *needles*) to reduce friction through rolling contact. In most applications, the rolling elements are placed between inner and outer rings, usually called ***races.*** The rolling elements are separated by a ***cage,*** generally made of stamped steel. The cage, or *separator,* prevents the elements from bunching and sliding against each other. In the case of separable bearings (those that can be taken apart), the cage prevents the loss of the elements. The rolling elements and the inner and outer races are hardened and ground to assure proper contact and clearance.

Antifriction bearings are used in many places in modern vehicles. They can be divided into four types: ***ball bearings, roller bearings, needle bearings,*** and ***self-aligning bearings.*** Each type has certain applications it serves best. For instance, the ball bearing produces the least amount of friction, but does not have the load carrying ability of the roller bearing. **Figure 10-26** illustrates ball, roller, and needle bearings used in automotive applications.

There are many variations of the basic antifriction bearing types. Some of the more common are the deep-groove ball, angular contact ball, straight roller, spherical roller, tapered roller, multiple row, and self-aligning. Each design attempts to meet a specific demand. The installation may call for the bearing to handle light or heavy loads, high or low speeds, radial or thrust loading, or a combination of loads. Most antifriction bearing applications involve a combination radial and thrust loads.

Many bearings will sustain thrust in one direction only. Thrust in the opposite direction would force the races apart. By using two or more bearings, facing in opposite directions, thrust in either direction can be handled. See **Figure 10-27.** Bearings can also be designed to provide for thrust loads in both directions, **Figure 10-28.** Understanding the problems

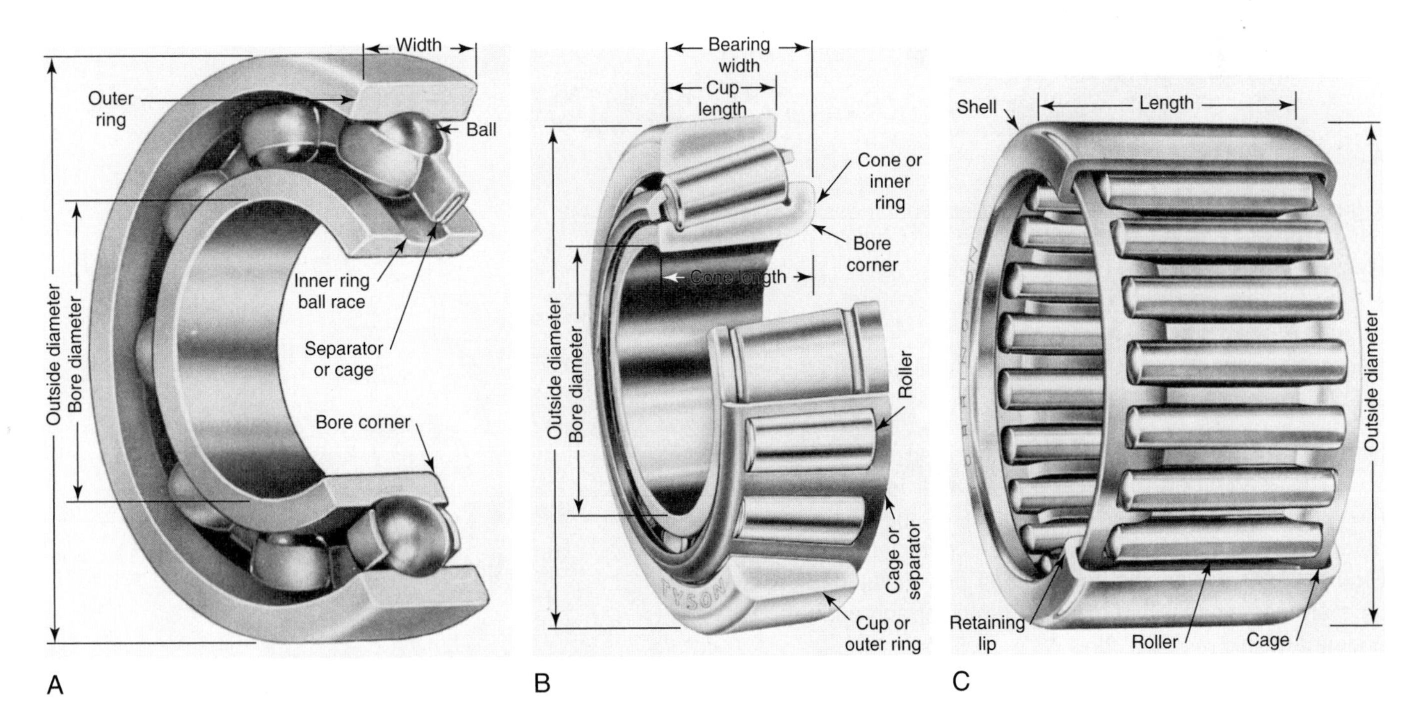

Figure 10-26. Three types of antifriction bearings. A—Typical ball bearing construction. Note how cage keeps balls evenly spaced. (Nice) B—Roller bearing. This bearing uses tapered roller design. Outer race is separate. (SKF) C—Caged needle bearing. Rollers in this bearing operate against the outer shell and in direct contact with the hardened, ground shaft surface. (Torrington)

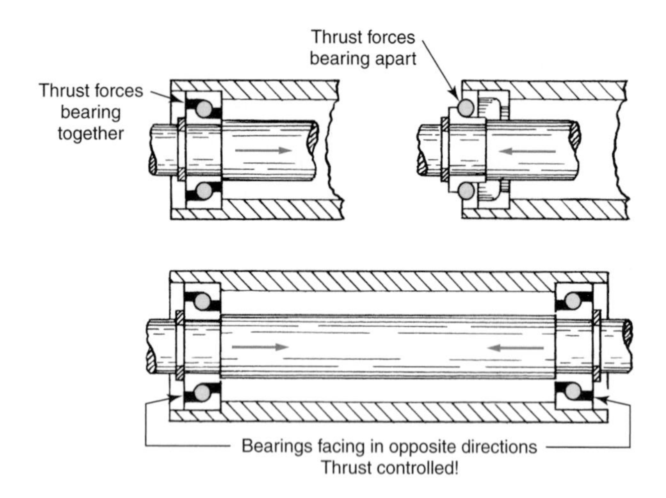

Figure 10-27. By using two bearings, thrust in either direction is controlled. Arrows indicate thrust direction.

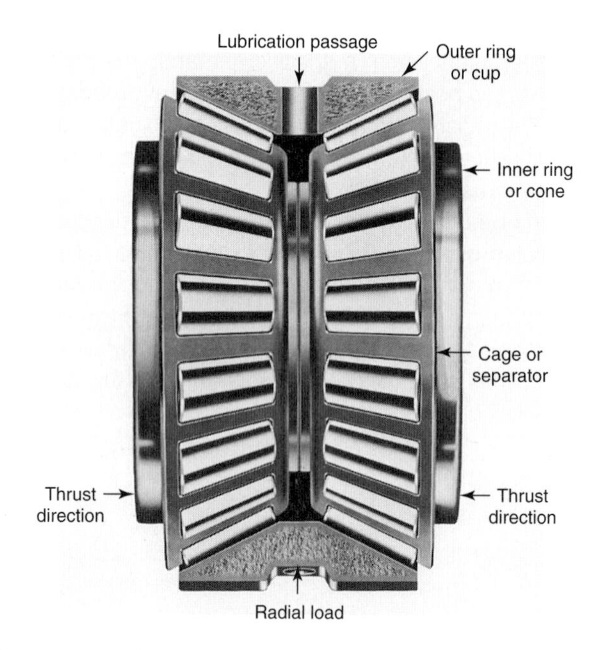

Figure 10-28. A double-row, tapered roller bearing designed to handle thrust loads in both directions. The outer race is one piece, the inner races are separate. (Timken)

involved and the type of bearing needed will help the technician to properly service bearings.

Ball Bearings

The deep-groove ball bearing, **Figure 10-29,** will handle heavy radial and moderate thrust loads. Neither the inner or outer race is separable. The angular contact ball bearing, **Figure 10-30,** will handle heavy thrust and radial loads. The balls are contained within a cage. Both inner and outer races are separable.

Roller Bearings

The straight roller bearing, **Figure 10-31,** is designed to handle heavy radial loads. Most straight roller designs will handle little or no thrust loading. The rollers in the spherical

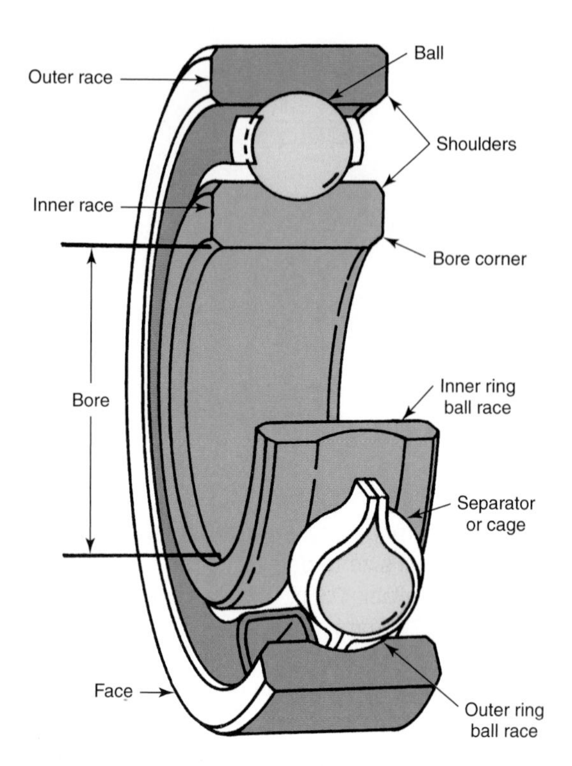

Figure 10-29. Cutaway of a ball bearing assembly. (Deere & Co.)

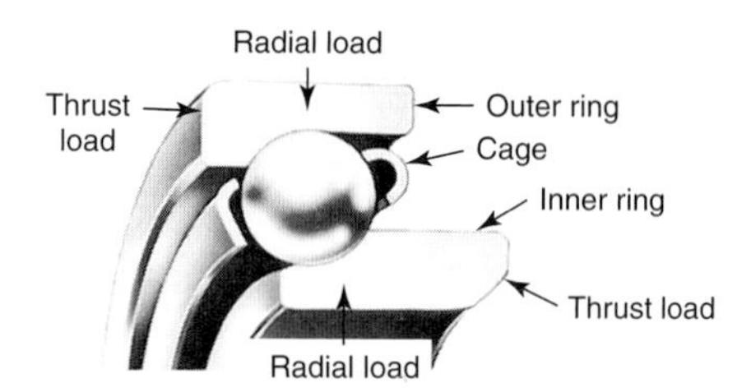

Figure 10-30. An angular contact ball bearing. This type is often used for front wheel bearings on automobiles.

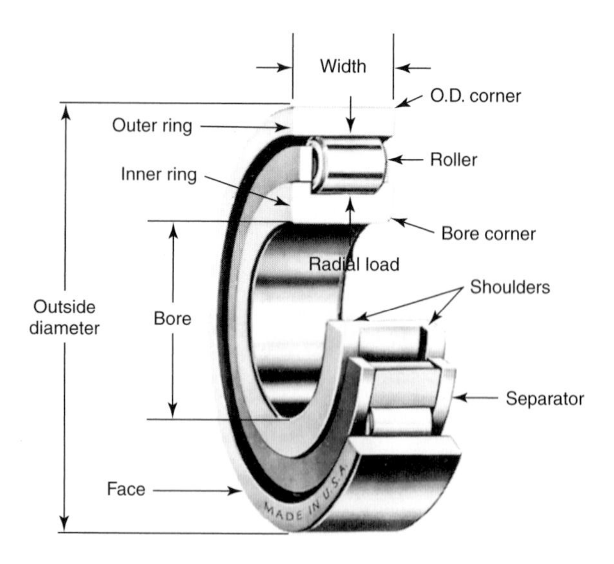

Figure 10-31. A straight roller bearing designed for radial load only. (AFBMA)

roller bearing, **Figure 10-32,** have a curved or spherical shape. This bearing will handle heavy radial loads and moderate thrust loads, and is self-aligning to a degree. The tapered roller, **Figure 10-33,** is the most widely used of the roller bearings, since it will carry both heavy thrust and radial loads. The apex of the angles formed by the rollers and raceways, if extended, would meet on a common axis. This allows the roller to follow the tapered raceways with no bind or skidding. The rollers are secured to the cone with a steel cage. The cone raceway is indented to form a lip that keeps the rollers centered. The cup is then separable.

Figure 10-32. A spherical roller bearing. Note the "barrel" shape of rollers. (SKF)

Needle Bearing

Needle bearings (long, thin rollers) often use only an outer shell. In some needle roller applications, the bore and shaft are hardened, then ground and placed in direct contact with the rollers. A variation of the needle bearing is the Torrington bearing. They are often used in place of thrust washers where the axial loads would quickly wear out a washer. Torrington bearings, **Figure 10-34,** consist of a set of needle bearings that rotate around a central opening. The bearing slips over a shaft and is held in place by the parts that it separates.

Self-Aligning Bearings

A self-aligning bearing is used when there is a possibility or a desirability of either housing or shaft misalignment during operation. This bearing will allow a degree of tilt without distorting the bearing elements. Both internal and external self-aligning bearings are shown in **Figure 10-35.**

Bearing Identification

All bearings are marked with a part number for ease of replacement. The number is usually on the face of the races. If necessary, replacement bearing size can be checked by careful measurement.

Bearing Seals

Bearings can be sealed on one or both sides, **Figure 10-36.** Sealing on one side is often used to help confine lubricant and to prevent the entry of dirt. When both sides are sealed, the bearing is *prelubricated* (lubricated during assembly); the technician cannot add lubricant during service. These bearings are sometimes called ***sealed bearings.***

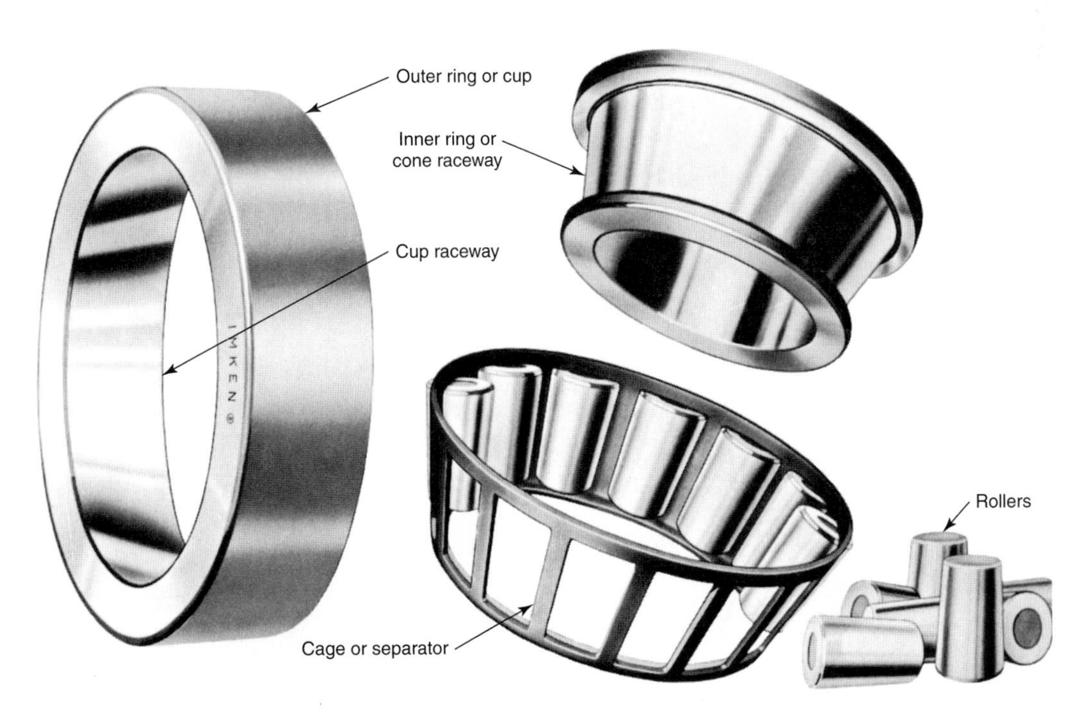

Figure 10-33. The component parts of a tapered roller bearing. Once assembled, this bearing will have a separable outer race but the rollers, cage, and inner race will be one unit.

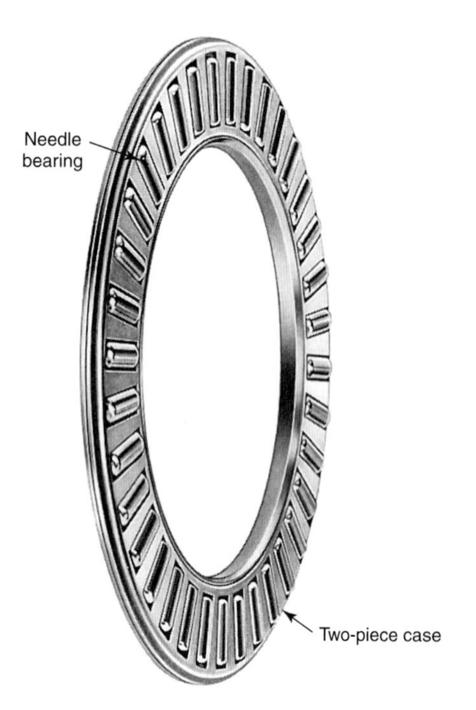

Figure 10-34. A needle thrust bearing. The two-piece case also acts as needle separator. (Torrington)

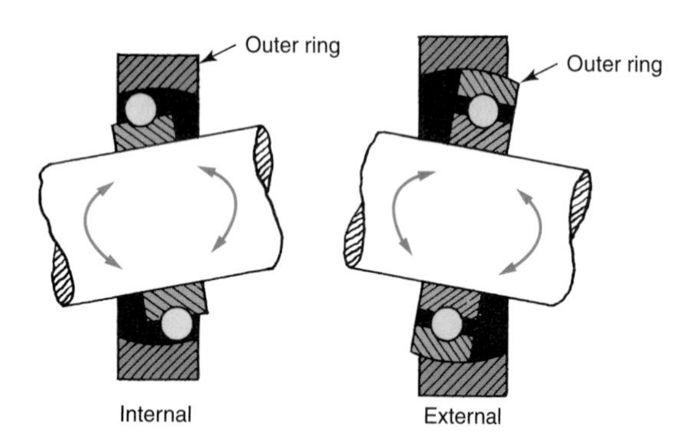

Figure 10-35. Internal and external self-aligning bearings. Note how the shaft is free to tilt. The external design will handle heavier loads, since the ball has a wider contact area with the outer race.

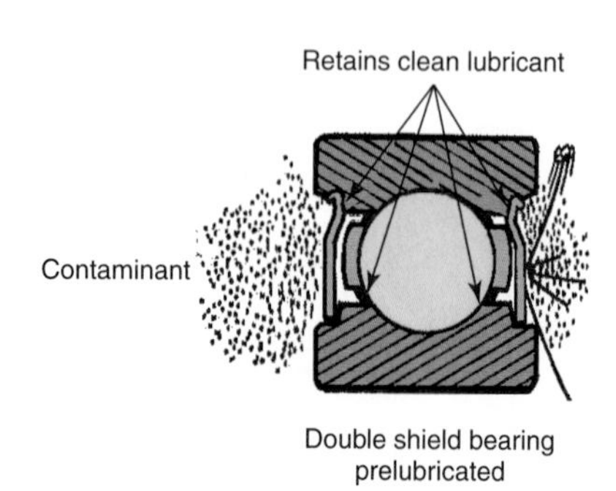

Figure 10-36. Bearing seals hold the lubricant in place, and prevent the entry of contaminants. (Federal-Mogul)

Removing Bearings

The best way to remove a bearing is to use mechanical or hydraulic pushing or pulling tools. These tools exert a constant, heavy force, **Figure 10-37.** A suitable hammer (brass, lead, or plastic) and soft steel drifts, sleeves, and cup drivers can be used when pullers are not available or where their use is not possible or undesirable.

Do not attempt to pull or install a bearing by exerting force on the free (not tight) race. This may chip the balls or rollers. The race itself could crack and fly apart. There are some instances that require force on either the free race or rolling elements. Whenever possible, however, exert force on the tight race only.

Figure 10-38 shows correct and incorrect methods of applying pulling force. In A, the plate supports the inner race only, thus avoiding damage to the outer race and rolling elements. Note that in B, the supporting puller plate rests on the free outer race.

Note: Before pulling bearings, clean the surrounding area to prevent contamination. When a separable bearing is removed, keep the parts together. Do not mix used bearing elements for any reason.

When Inner Race or Bearing Cannot Be Grasped

Occasionally, the bearing inner race is pressed against a shoulder that is as wide or wider than the race. In the case of the tapered roller bearing, a special segmented (made in parts) adapter can be used. It applies the pulling force to the ends of the rollers while forcing them against the cone. This allows the bearing to be removed without damage, **Figure 10-39.**

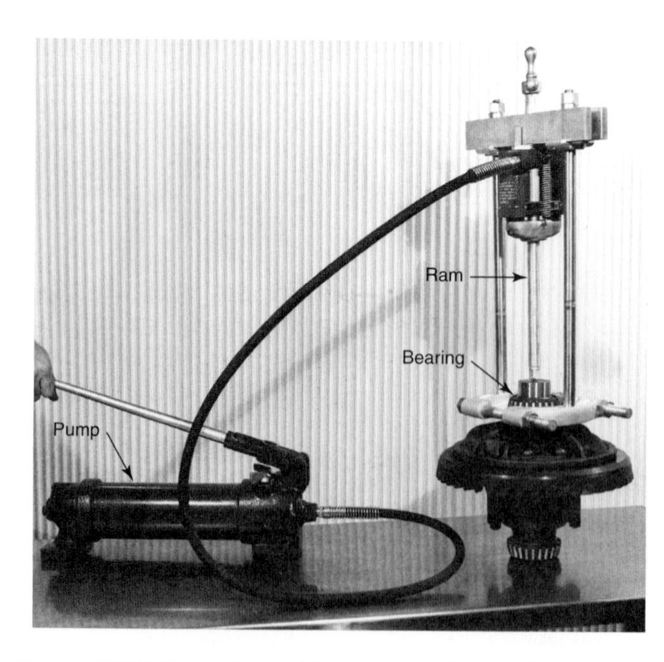

Figure 10-37. Removing a differential carrier bearing with a hydraulic puller. (Timken)

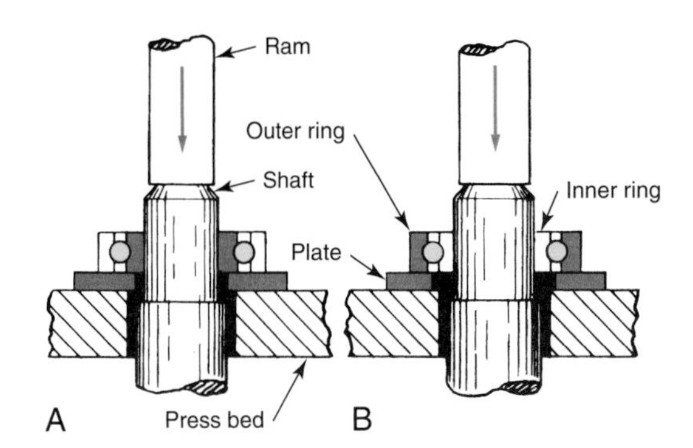

Figure 10-38. Pulling setups. A—Correct. Force is exerted through the tight inner race only. B—Wrong. Force is applied through free outer race and rolling elements.

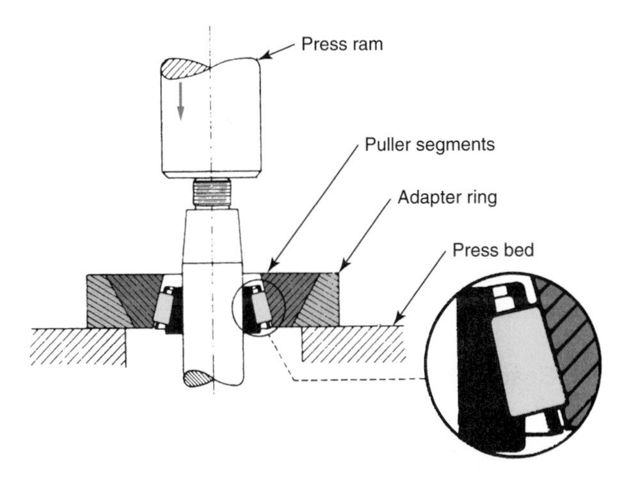

Figure 10-39. Pulling a bearing by applying pressure through rollers. Magnified portion at lower right shows how end of roller is grasped by puller segments. (Timken)

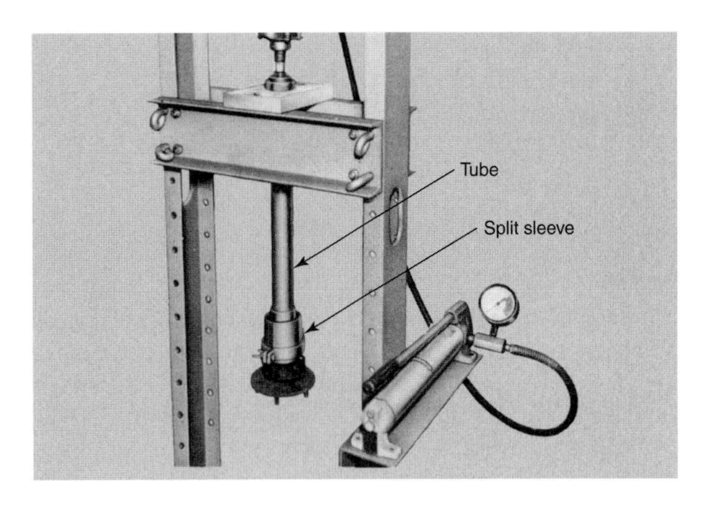

Figure 10-40. Removing an axle shaft bearing with a special puller.

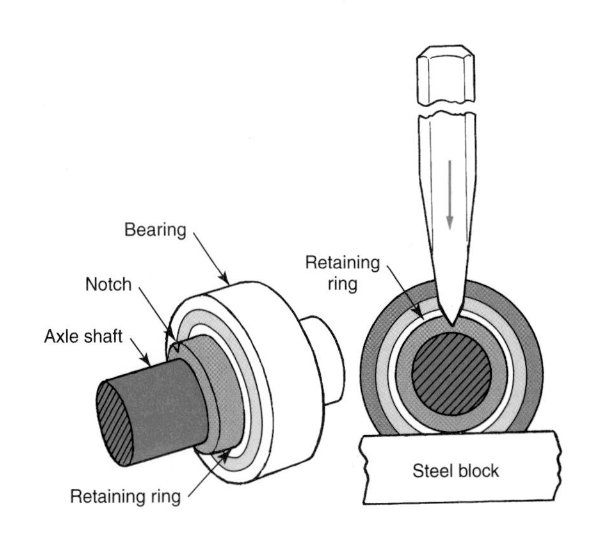

Figure 10-41. Removing a bearing retaining ring by notching it with a chisel.

A special puller for axle shaft bearing work is pictured in **Figure 10-40.** A split sleeve with pulling races is used. The axle shaft passes up through a section of tubing. The puller sleeve grasps both bearing and tubing. The top section of the tubing is fastened to a heavy plate on the bed of the press. As pressure is applied to the shaft end, it is forced through the tube to pull the bearing. The entire bearing is shrouded or shielded to protect the operator from flying parts if the bearing should explode. This puller will remove both tapered roller and ball bearings.

Sometimes, a retaining plate or dust shield is so close to the bearing that it is impossible to grasp. In these cases, you must grind away a portion of the inner race (be careful to protect the shaft with a metal sleeve). Cut out the cage and remove the elements. The outer race can then be removed, exposing the inner race for grasping.

Unhardened retaining rings are sometimes used to hold bearings in place. They are best removed by notching them with a sharp chisel. This will loosen them enough for easy removal. See **Figure 10-41.**

Inner bearing races can also be removed by partial grinding or by cutting with an acetylene torch. Wrap the shaft on both sides of the bearing with wet cloths to prevent heating. Cut only partway through the race, **Figure 10-42.** The race is then squeezed tightly in a vise and struck with a hammer. This will crack the race and allow it to be pulled.

STOP **Warning: Bearing materials are extremely brittle, and may fly apart with great force when broken. Always wear safety goggles when striking bearing parts. Keep other personnel away from work area. Always pull bearings whenever possible. Avoid grinding and cutting with a torch, unless absolutely necessary. Whenever possible, shield the bearing while working on it.**

General Rules for Removing Antifriction Bearings

- Exert force on the tight race when possible.
- Use a puller of the correct size and shape.
- Mount the puller to exert force in a line parallel to the bearing axis.

- Use unhardened, mild steel drifts and sleeves.
- Never strike the outer or free race.
- Do not damage the shaft or housing.
- If it is necessary to hammer a shaft, use a brass, lead, or plastic hammer.
- Keep all bearing parts together.

Antifriction Bearing Service

When the bearing is removed, it must be cleaned before being placed back in service. First, wipe off all surplus grease or oil, then soak in a nonflammable cleaning solvent. A cleaning tank with tray and solvent hose is ideal, **Figure 10-43.** If a tank is not available, a clean bucket will suffice. Never use gasoline, kerosene, or other volatile fluids for cleaning. They cause skin irritation and will ignite readily. Never use carbon tetrachloride—it produces poisonous fumes.

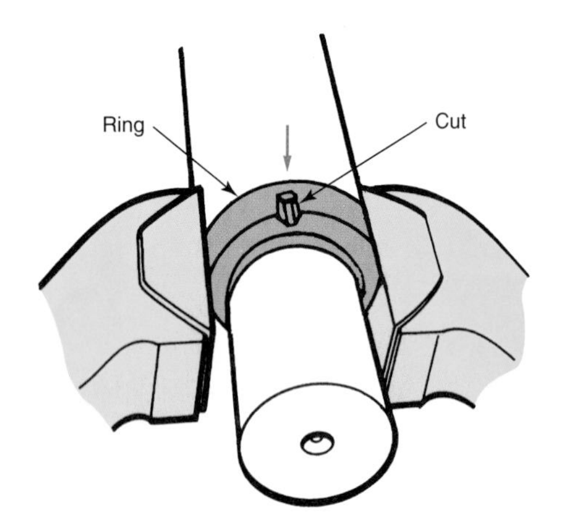

Figure 10-42. Bearing inner race partially cut and then squeezed in a vise. Striking the race with a hammer will crack it, allowing removal. (Federal-Mogul)

Figure 10-43. A cleaning tank works best for removing oil and grease from parts, such as these bearings being lowered into solvent.

While the bearings are soaking, brush each in turn with a nylon bristle brush Carefully use compressed air to blow out the worst of the grease. Continue soaking and brushing until the bearing looks clean. Blow the bearing out again. If any sign of grease is visible, soak, brush, and blow out once more.

Warning: Never spin a bearing with air pressure. Not only will it damage the bearing, it can also be dangerous. When the outer race of a separable bearing is removed, only the sheet metal cage holds the rolling elements to the center race. If the cage and rollers are spun, tremendous centrifugal force is generated that can cause the elements to fly outward with violent force.

When the bearing is clean, rinse in a container of clean solvent and blow dry, **Figure 10-44.** Once the bearings are dry, take them to a clean work area. It is a good idea to reserve an area for bearing work that will be free of dust, dirt, and moisture.

Note: Use clean, dry air. Most air compressor systems are equipped with a filter and moisture trap. Service them often. Directing a stream of air into a white cloth will show if dirt, moisture, or oil is present.

Sealed Bearing Service

When a bearing is factory-packed and completely sealed on both sides, it must not be washed. There is no satisfactory way to relubricate such a bearing, so washing will dilute the lubricant and lead to early failure. Wipe off the outside with a clean, dry cloth.

Antifriction Bearing Failure

Before discussing bearing inspection procedures, it is wise to familiarize yourself with some of the most common

Figure 10-44. Use clean, dry air to blow a bearing dry. Do not allow the bearing to spin. (Federal-Mogul)

bearing defects that can result in failure. As is the case with friction bearings, the number one enemy of antifriction bearings is *dirt.* It will cause scratching, pitting, and rapid wear. Other common defects include corrosion, spalling, brinelling, damage from overheating, physical damage, electrical pitting, and damaged seals.

To inspect a nonseparable bearing, place the fingers of one hand through the center race, **Figure 10-45** Rotate the outer race with the other. The bearing should revolve smoothly with no catching or roughness. If either condition is present, rinse and blow dry again. If the symptoms still persist, discard the bearing. Also check for signs of overheating and wear on the outer surfaces of both races. A bearing that has been loose in the bore or on the shaft will exhibit highly polished areas.

If the bearing can be separated, carefully inspect the raceways and rolling elements. They should be absolutely smooth and free of heat discoloration. Inspect each ball or roller since only one or two may be damaged. If the bearing passes the visual check, place the elements together. While forcing them together, rotate the bearing. The operation should be smooth.

When revolving bearings, do so a number of times. A single damaged ball or roller may not "catch" the first few times around. When checking thrust bearings, place one side on a solid surface. Press down on the other with the heel of your hand and rotate while maintaining pressure. Your hands should be clean and dry. Keep them away from the races and rolling elements.

Do not assume looseness is a sign of wear in a bearing. A new bearing often feels loose before installation. When either the races or the rolling elements are worn enough to produce looseness, the wear will be evident by examining the surfaces. One or more of the following conditions will be visible.

Figure 10-45. Holding a bearing for inspection. Hold the inner race with one hand while rotating the outer race with the other. (Federal-Mogul)

Dirt and Corrosion Damage

If the dirt is very fine, it will have a *lapping* effect (removal of surface metal through fine abrasive action). This will leave the rolling elements and raceways with a dull, matte (nonreflecting) finish. Larger dirt particles will produce scratches and pits. The entry of moisture, use of an incorrect or contaminated lubricant, or storage near corrosive vapors can produce corrosion in the bearing. A bearing that remains *static* (is not rotated) for an extended time often will corrode, **Figure 10-46.**

Spalling

Foreign particles, overloading, and normal wear over an extended period can lead to ***spalling.*** Spalling begins when tiny areas of metal fracture and flake off. These small flakes are carried around in the bearing causing more flaking. Advanced flaking or spalling will produce large craters, **Figure 10-47.**

Brinelling

Brinelling is the term used to describe a series of dents or grooves worn in one or both races. The grooves run across the raceway and are usually spaced at regular intervals. Once brinelling starts (often from inadequate lubrication) a fine reddish iron oxide powder is formed. As the powder is carried around, it increases the wear rate. **Figure 10-48** shows a badly brinelled outer shell.

Figure 10-46. A badly corroded bearing race. (Federal-Mogul)

Figure 10-47. A badly spalled inner bearing race. (AFBMA)

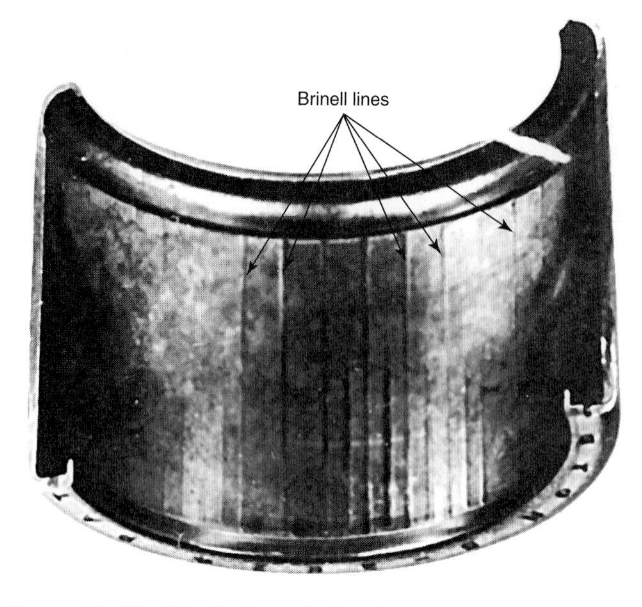

Figure 10-48. Brinelled needle bearing shell.

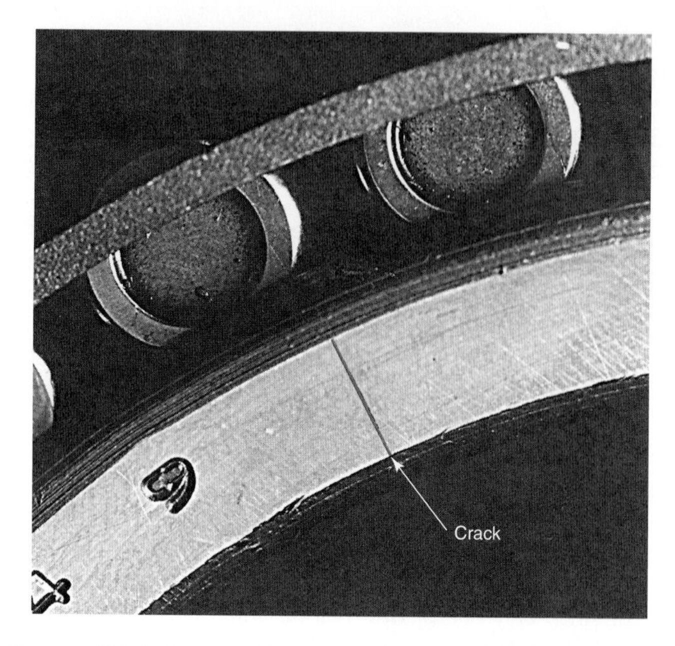

Figure 10-50. A cracked inner bearing race. (CR Industries)

Overheating

Overheating will break down the physical properties of the bearing and cause rapid failure. The principal causes of overheating are inadequate or improper lubrication and poor adjustment. Bearing races and rolling elements that have been overheated will have a blue or brownish-blue discoloration. See **Figure 10-49.**

Physical Damage

One or both bearing races may be cracked, **Figure 10-50.** Improper removal or assembly techniques and wrong bore or shaft size are common causes. Improper removal and assembly procedures will often result in a dented or broken cage, **Figure 10-51.** Pieces of dirt and metal chips will also cause cage breakage. As with a broken cage, careless assembly often produces dented shields, **Figure 10-52.** This could damage the cage as well as cause binding and lubricant loss.

Electrical Pitting

Electric motor or generator bearings are sometimes pitted by the passage of current through the bearing from an

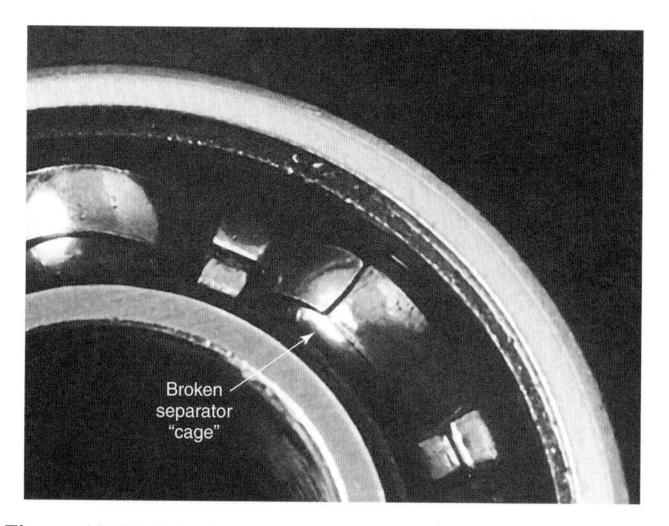

Figure 10-51. A broken bearing cage. (CR Industries)

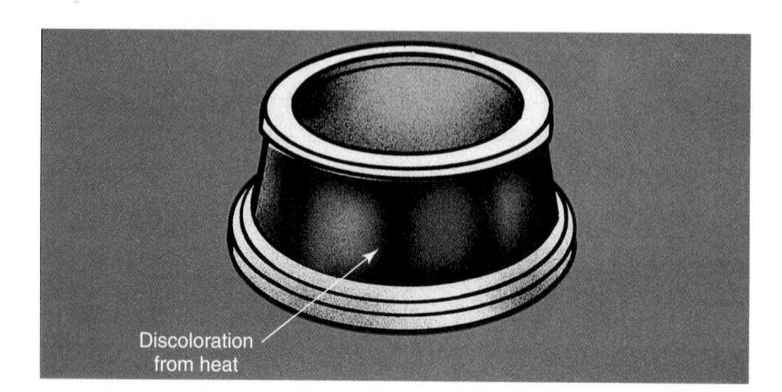

Figure 10-49. A bearing that has been overheated. Note the discoloration on the bearing. (Federal-Mogul)

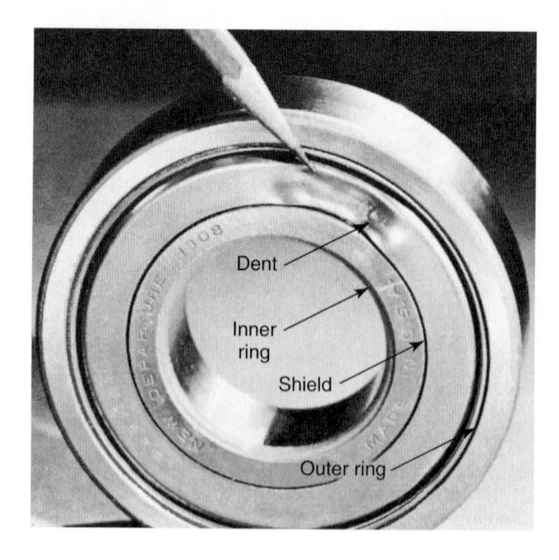

Figure 10-52. This bearing shield or seal has been badly dented. (New Departure)

internal short or from static electricity. The electrical arcing produces numerous tiny pits. Engine and transmission bearings are also frequently pitted when the engine-to-body ground straps are removed or broken. **Figure 10-53** illustrates the effect of ***electrical pitting,*** dirt, corrosion, and poor lubrication on rollers.

Replace the Entire Bearing

If any part of a bearing—the outer or inner race or the rolling elements—is damaged, discard the *entire* bearing. Never replace part of a bearing. Before discarding, write down the part number. It is a good idea to wire the parts together and keep for comparison with the replacement bearing. Mark the bearing as defective.

Bearing Lubrication

If the bearing will be placed into service at once, it may be oiled or packed with grease, depending upon the application. Cover it with a clean cloth until ready to install. If the bearing will be stored for a few days, coat it with oil and place it in a clean box or container.

If the bearing will be stored for an extended period, coat it with light grease and wrap it in oilproof paper, **Figure 10-54.** Place the wrapped bearing in a clean box for storage. Be sure to identify each bearing to prevent unwrapping a number of them when looking for a specific one.

Packing Bearings with Grease

When a bearing must be packed in grease before installation, use a ***bearing packer,*** **Figure 10-55.** If no packer is available, place some grease on the palm of one hand. Your hands must be clean and dry. With the other, press the edge of the bearing into the grease. Repeat this until grease flows out the top. Move around to different sections until the bearing is fully packed. Separable races should be coated also. All grease and oil in the shop should be kept in clean containers and kept tightly covered when not in use. When opening the container, first wipe any dirt off the lid. Avoid dusty areas: an open can of grease near a grinder or cutting torch is an invitation to disaster. If any oil or grease seals are related to the job at hand, inspect them. If necessary, replace the seals at this time. In some instances, seals must be installed after the bearings.

Bearing Installation

Bearings are often similar in type and size, but not exactly the same. Before attempting installation, make certain you are installing the correct one. Be especially careful with new replacement bearings. Check numbers and measurements. Bearing installation calls for care and intelligent use of tools. Many otherwise good jobs have been ruined by careless installation.

Clean and Inspect Bores and Shafts

Clean bearing housing bores and shafts thoroughly. Remove any nicks or burrs with a fine-tooth file, **Figure 10-56.** Be careful not to file a flat spot. After filing, polish with very fine

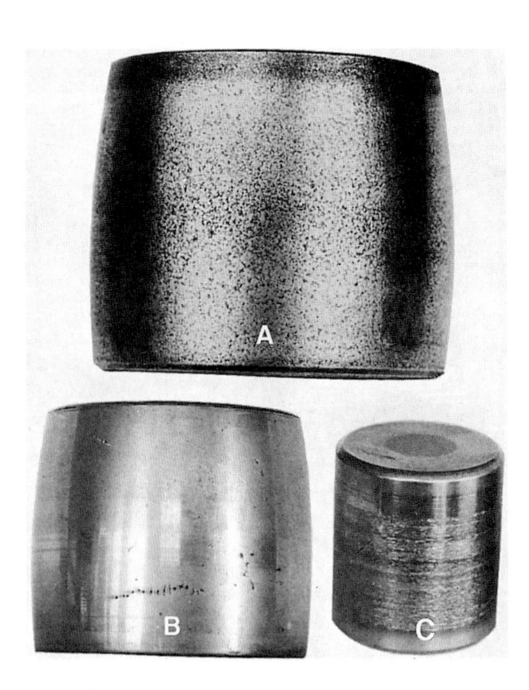

Figure 10-53. Some examples of roller damage. A—Corrosion. B—Electrical pitting. C—Poor lubrication and dirt. (SKF)

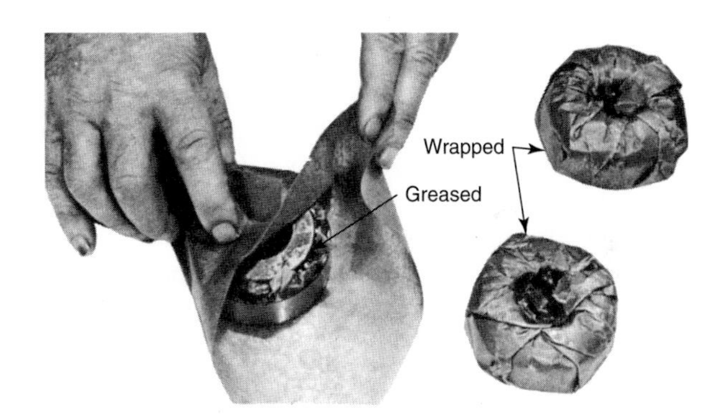

Figure 10-54. Bearings should be greased and wrapped for extended storage.

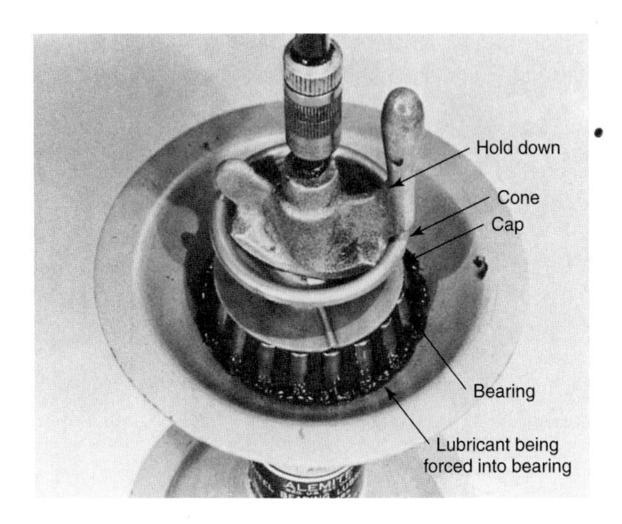

Figure 10-55. A bearing packer is fast and efficient. (Timken)

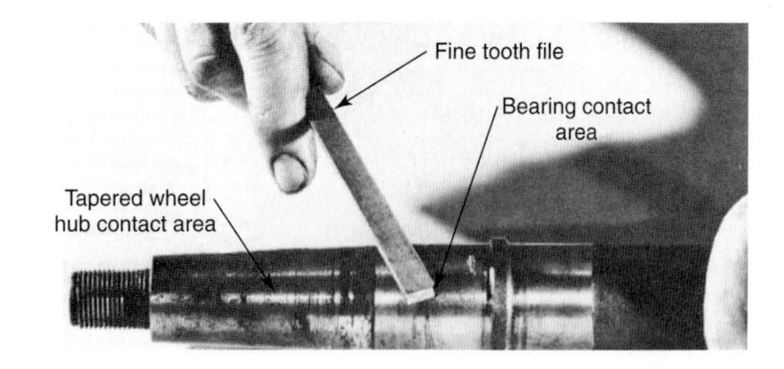

Figure 10-56. Removing burrs from axle shaft bearing with a fine-tooth file.

emery or crocus cloth. On a shaft where the inner race is designed to *walk* (a creeping movement around the shaft), inspect the shaft carefully. Polish if necessary.

If the counterbores or press-fit shaft areas are worn from race slippage, do not center punch or knurl (crosshatch pattern pressed into metal) the bearing as an attempt to increase size. Such procedures will only result in failure, since the bearing, under load, will quickly flatten these raised areas. The area should be built up by metallizing (spraying molten metal onto the shaft) and then grinding to the correct size.

Watch for dirt in threads, splines, and other areas. A sprung shaft or bent housing will cause the bearing to operate in a distorted position, greatly shortening its life. For those jobs in which the bearing failed in a short time, despite proper installation, lubrication, and adjustment, always check shaft and housing for any warpage or other misalignment. The use of a thin film of oil or micronized (finely powdered) graphite will ease installation, prevent corrosion around race contact area, and facilitate future removal. See **Figure 10-57.**

Be careful to determine correct installation sequence and position — install any retainers or snap races that must go on first, and do not press on any element backward. Start the bearing or race with your fingers, then attach a puller or set up in a press and force the bearing into place. Make certain it goes on squarely and to the full distance required. Apply pressure, whenever possible, on the tight race. Observe all safety precautions.

If pressing tools are not available, simple driving tools will handle many installation jobs in a satisfactory manner. Brass tools tend to mushroom and chip, which can contaminate the bearings. Use soft steel tools. Make sure they are clean and in good condition. Strike the tight race only.

Figure 10-57. Use a lubricant to facilitate assembly.

Using Heat and Cold

In difficult assembly jobs, primarily those involving large bearings, place the outer race in dry ice or in a deep freeze. This will reduce the diameter and help installation. Inner races can be heated in clean oil or a special electric oven, **Figure 10-58.** Use a thermometer. Never heat bearings with a torch, or allow the bearing temperature to exceed 275°F (135°C). Follow the manufacturer's instructions.

Bearing Adjustment

Some bearings require adjustment after installation. Proper adjustment depends on the application. Some require a specific amount of free play. Others require preloading (placing bearing under pressure so that when a driving force is applied to parts, they will not spring out of alignment). As the various service operations are described throughout the text, general bearing measurement and adjustment recommendations will be given.

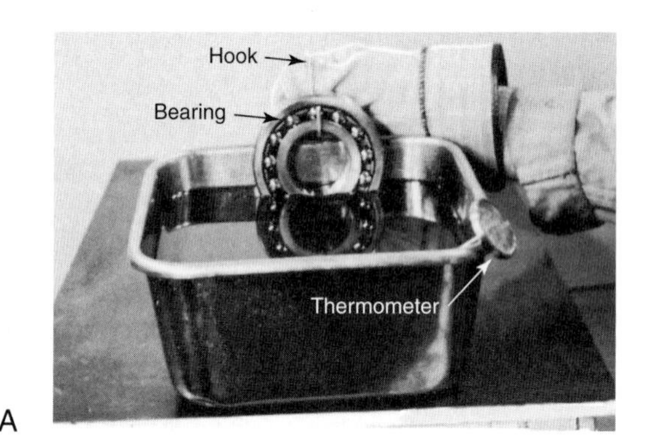

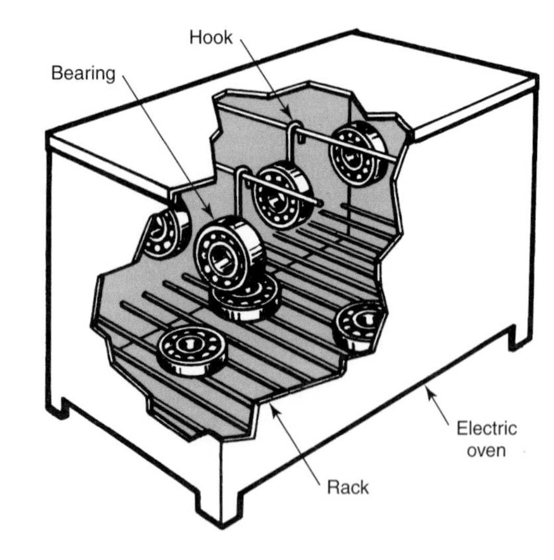

Figure 10-58. Heating bearings. A—Heating a bearing in oil. Hook keeps bearing from touching bottom of container. B—An electric oven used to heat bearings. Bearings are placed on racks or suspended with hooks. Do not put them in direct contact with the heat source. Follow the manufacturer's instructions when operating. (Federal-Mogul)

General Rules for Installing Antifriction Bearings

- Clean all contact surfaces and remove any burrs and nicks.
- Install all parts in the proper order.
- Lubricate for easy installation.
- If heat is required, do not exceed 275°F (135°C).
- Start the bearing squarely.
- Align tools so that bearing will be forced on squarely.
- Press on the full distance required.
- Use only soft steel driving tools.
- Whenever possible, avoid applying pressure through the rotating parts.
- If a vise is needed, use protective jaw covers.
- Driving tools must have smooth, square-cut ends.
- Do not mar shaft or bore surfaces.
- Use safety precautions.

Figure 10-59 illustrates a few bearing installation hints.

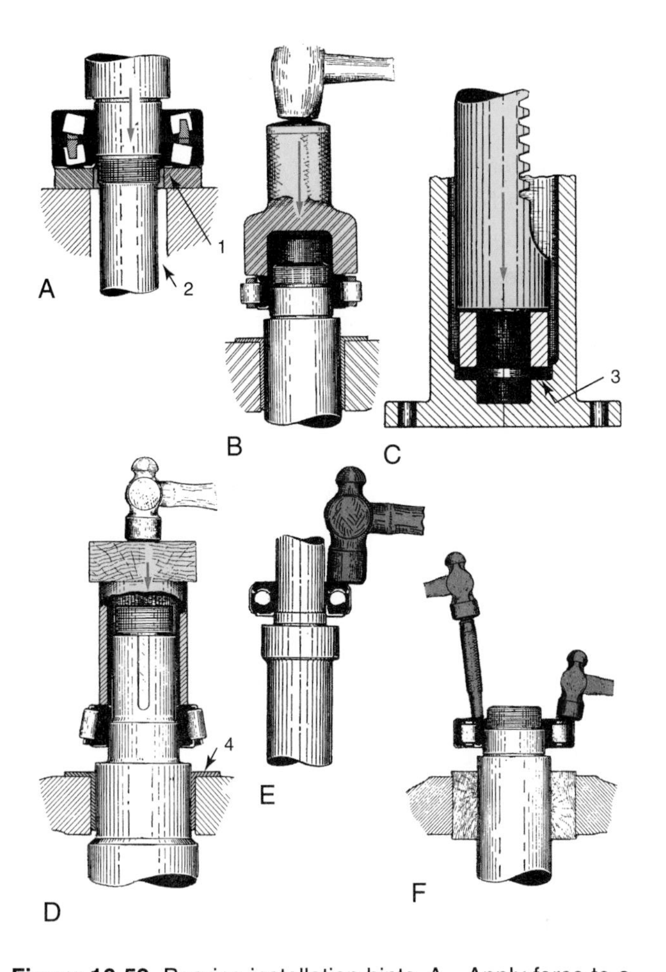

Figure 10-59. Bearing installation hints. A—Apply force to a tight race (1) and provide clearance (2) for the shaft. B—Use a driver with smooth, square cut ends that strike the tight race. C—Clean bearing race recess (3) and force race to full depth. D—Block placed on open pipe driver allows driving force to be centralized. Use protective vise jaw covers (4). E—Do not strike bearing with a hammer. F—Do not use wide punches on bearings. (AFBMA)

Tech Talk

A common mistake when dealing with a bearing (or any moving part) is to assume that penetrating oils or silicone sprays make good lubricants. Even some experienced technicians believe that these products can take the place of the specified lubricants. While penetrating oil and silicone sprays are excellent when used to loosen bolts or drive out water, they have no lubrication value. A bearing assembled with one of these products instead of the proper lubricant will be destroyed in a few minutes of operation.

When lubricating a bearing, or any moving part, make sure that you use the correct lubricant. Also, do not wrap bearings in newspaper for storage. Although newspaper appears clean, it is full of fine paper particles that will flake off and enter the bearing. In addition, newspaper will absorb some of the additives in the lubricant, reducing the amount of lubrication available to the bearing. Always wrap clean, lubricated bearings in oilproof paper.

Summary

Bearings can be classified as friction or antifriction. The *friction bearing* contact area slides against the bearing journal surface. Friction is reduced to acceptable limits by a film of oil. Most modern bearings are of the precision insert type. They can be of the full-round or split-halves type. They utilize steel backs that can be faced with lead-tin babbitt, copper alloys, or aluminum alloy. End thrust is controlled by incorporating thrust flanges on one or more bearings. Bushings are usually bronze or bronze-faced steel, and are bored, reamed, or honed to size.

Low oil pressure, excessive oil consumption, and knocking are danger signals that indicate excessive bearing wear. Bearing failures are usually caused by dirt. Dirt enters the engine from normal wear, poor cleaning and reassembly, or improper storage conditions. Inadequate lubrication, improper assembly, and poor driving habits also cause bearing failures.

Journals must be round, straight, and smooth. Handle bearings carefully. The insert must closely contact the housing bore. Bearing spread, crush, and cleanliness assure a proper fit. Never file bearing inserts or caps. Locating lugs should be in the proper slots. Bearing back and bore must be clean and free of nicks or foreign material. Never reverse or mix bearing caps. Tighten properly using a torque wrench. Check bores for alignment.

Oil grooves and holes are vital. They must be located properly when installing inserts. Bearing clearance is critical. Worn or reground journals must be fitted with undersize bearings. Semifinished inserts may be bored to a specified size. Bearing clearance is best checked with plastic wire (Plastigage).

The *antifriction bearing* utilizes ball or roller elements that roll against the contact area. Antifriction bearings can be divided into four types: ball, roller, needle, and self-aligning. The ball and roller bearings usually consist of an inner and outer race with the rolling elements placed between them and positioned with a cage or separator. The needle bearing can use an outer shell, or can be placed in direct contact with a hardened and ground bore and shaft. Bearings are designed to carry either straight thrust, radial, or combination loads.

The deep-groove ball, angular contact ball, straight, spherical, and tapered roller, needle bearing, and self-aligning bearings are the common variations. Bearings are marked with a part number. Bearings are often sealed on one or both sides. Never wash bearings sealed on both sides.

Hydraulic and mechanical pulling or striking tools can be used to remove bearings. If available, hydraulic and mechanical pullers are recommended. Pull bearings, whenever possible, by the tight race. Special tools are available for pulling by exerting pressure through the balls or rollers. Avoid the use of heat. Do not mar bore or shaft surfaces. Keep all bearing parts together.

Clean bearings in solvent. Blow dry. Rinse in fresh solvent and blow dry again. Do not spin the bearing. Inspect the cleaned bearing. If satisfactory, oil or pack with grease at once. A bearing packer is handy for greasing. Keep bearings covered until ready to install. Rejected bearings may be kept for size comparison with replacements, but mark them as rejects. Never replace one part of a bearing. Replace defective grease or oil seals. Clean bore and shaft. Remove burrs and install any parts that must precede bearing.

Lubricate bearing seat area. Position bearing correctly and start by hand. Pull, press, or drive bearing fully into place, keeping square at all times. Do not damage shaft or bore. Installation tools must be spotlessly clean. In difficult assembly jobs, the use of both heat and cold will ease installation. If necessary, carefully adjust bearing.

Review Questions—Chapter 10

Do not write in this book. Write your answers on a separate sheet of paper.

1. All of the following can cause rapid friction bearing wear, *except:*
   (A) detonation.
   (B) low RPM.
   (C) lugging.
   (D) preignition.
2. Fingerprints on insert bearing surfaces can cause _____.
   (A) warping
   (B) spalling
   (C) brinelling
   (D) corrosion
3. To keep the insert from turning, all of the following are used, *except:*
   (A) lugs.
   (B) keys.
   (C) tangs.
   (D) dowels.
4. A bearing cap can be shifted out of alignment by using a _____ wrench.
   (A) thick
   (B) socket
   (C) 12-point
   (D) All of the above.
5. What is the recommended oil clearance for an aluminum alloy engine bearing installed on a 3″ shaft?
   (A) 0.0015″
   (B) 0.0020″
   (C) 0.0025″
   (D) 0.0030″
6. When journals are worn or reground, _____ bearing inserts are required.
   (A) standard
   (B) oversize
   (C) undersize
   (D) None of the above.
7. Which type of roller bearing can handle heavy radial and thrust loads?
   (A) Straight roller.
   (B) Curved roller.
   (C) Tapered roller.
   (D) Needle.
8. Needle bearings make use of long, thin _____.
   (A) races
   (B) housings
   (C) rollers
   (D) Both A and C.
9. List five general rules for antifriction bearing removal.
10. When a separable bearing is being inspected, every _____ should be checked.
    (A) race
    (B) seal
    (C) ball or roller
    (D) All of the above.
11. A bearing showing some looseness _____.
    (A) should always be rejected
    (B) should be repacked with heavier grease
    (C) may be good
    (D) Both B and C.
12. Describe how a bearing should be stored for later use.
13. All striking type bearing pulling tools should be made of _____.
    (A) soft steel
    (B) brass
    (C) hardened steel
    (D) wood

14. What is the maximum temperature to which a bearing can be heated?
    (A) 212°F (100°C).
    (B) 275°F (135°C).
    (C) 350°F (177°C).
    (D) 400°F (204°C).
15. List five general rules for antifriction bearing installation.

ASE-Type Questions

1. All of the following statements apply to insert type friction bearings, *except:*
    (A) insert bearings rely on sliding friction to reduce wear.
    (B) most insert bearings consist of steel bonded to a soft metal core.
    (C) most insert bearings must be snapped into place.
    (D) lugs hold the insert bearing in place.
2. Technician A says oil filters will always catch all foreign particles in the engine. Technician B says dirt can enter the engine through a vacuum leak. Who is right?
    (A) A only.
    (B) B only.
    (C) Both A and B.
    (D) Neither A nor B.
3. All of the following will cause early bearing failure, *except:*
    (A) reversing insert caps.
    (B) prelubrication.
    (C) bent crankshaft.
    (D) dirt on the insert back.
4. Technician A says reversing or mixing bearing caps will cause the bores to become misaligned. Technician B says a few nicks in the insert housing bore are not harmful. Who is right?
    (A) A only.
    (B) B only.
    (C) Both A and B.
    (D) Neither A nor B.
5. Technician A says thrust flanges are used to control movement of the bearing in its housing. Technician B says one should not force a bearing insert into place by pressing on the center. Who is right?
    (A) A only.
    (B) B only.
    (C) Both A and B.
    (D) Neither A nor B.
6. Technician A says bearing spread helps to hold the bearing in place during assembly. Technician B says the crush area creates a tight insert-to-bore contact. Who is right?
    (A) A only.
    (B) B only.
    (C) Both A and B.
    (D) Neither A nor B.
7. Technician A says hydraulic or mechanical pullers are superior to striking tools for bearing removal and installation. Technician B says you should always apply pulling force to the free race. Who is right?
    (A) A only.
    (B) B only.
    (C) Both A and B.
    (D) Neither A nor B.
8. Technician A says bearings, under pulling pressure, can fly apart with great force. Technician B says if a bearing is started in a "cocked" position, it will line up under pressure. Who is right?
    (A) A only.
    (B) B only.
    (C) Both A and B.
    (D) Neither A nor B.
9. Never _____ a bearing sealed on both sides.
    (A) grease
    (B) wash
    (C) spin
    (D) Both A and B.
10. All of the following statements about servicing antifriction bearings are true, *except:*
    (A) bearing parts should never be mixed.
    (B) when blowing dry, never spin a bearing.
    (C) it is always safer to pull a bearing from its housing or shaft.
    (D) since bearings are hardened, a small amount of dirt will not hurt them.

Suggested Activities

1. Check the clearance of various engine bearings using Plastigage. Compare your measurements to the manufacturer's specifications. Make a chart showing the manufacturer's specifications, the actual clearance, and the difference between these two values.
2. Use a micrometer to measure a used crankshaft (both main and rod journals). Using the manufacturer's specifications, determine the amount of wear. Would the shaft accept a standard undersize bearing? Check the journals for nicks and scoring.
3. Secure a number of damaged friction and antifriction bearings. Clean and inspect each one, then identify the cause of rejection. Write a report summarizing your findings.

Engine oil must provide good lubrication under a wide range of engine operating conditions. Both single-grade and multi-grade oils are shown here. The selection of the proper oil is critical to engine life.

11

Preventive Maintenance

After studying this chapter, you will be able to:

- Describe the purpose of preventive maintenance.
- Describe oil classifications.
- Explain how to change engine oil and filter.
- Explain how to lubricate suspension fittings.
- Explain how to check air filters.
- Identify body parts that require lubrication.
- Explain how to check vehicle fluid levels.
- Explain how to inspect tires.
- Describe tire rotation procedures.
- Explain how to inspect the brake system for defects.
- Explain how to inspect steering and suspension parts.
- Describe engine coolant replacement procedures.
- Explain the procedure for changing transmission/transaxle fluid.
- Identify battery maintenance procedures.

Technical Terms

Maintenance schedule
Impurities
Load up
Viscosity
Multi-grade oils
Oil service grades
API (American Petroleum Institute)
Non-detergent oils
Detergent oils
American Petroleum Institute's certification mark
Oil change intervals
Air cleaner elements
Grease fittings
EP (extreme pressure) greases
Hydrometer
Refractometer
Coolant test strips
Chemical cooling system cleaner
Long-life antifreeze
Bleed valve
Drive belts
Belt tension gauge
Coolant hoses
Electrochemical degradation (ECD)
Aeration
Tread separation
Rotation
Directional tires
Unloaded
Hydrogen gas
Memory saver

This chapter will cover basic preventive maintenance procedures. Most of the material will be concerned with changing the engine oil and filter and lubricating the suspension fittings, but other vital maintenance procedures will also be covered.

Preventive Maintenance

Most automotive service operations are performed to correct vehicle problems. Preventive maintenance, on the other hand, is performed to keep problems from occurring. The most common maintenance operations include changing the engine oil and filter, checking the air filter and various fluid levels, and inspecting tires, brakes and suspension parts. Preventive maintenance also includes changing the engine coolant and transmission fluid.

Maintenance Schedules

Most vehicle maintenance is performed at regular intervals. Intervals can be given as mileage the vehicle has been driven or as a time period. A list of intervals for various maintenance operations is called a ***maintenance schedule.*** A typical maintenance schedule may, for instance, call for changing the engine oil every 6000 miles, (9600 km) and for changing the oil filter at every second oil change. In addition to mileage, most schedules call for performing a service after a certain time interval, no matter what the mileage. For example, the schedule will call for changing the engine oil every 6 months or 6000 miles, whichever comes first.

Most manufacturers publish at least two maintenance schedules, one for normal operation and another for severe or heavy-duty service. Vehicle operating conditions determine which schedule should be followed. Typical severe operating conditions include operation at very low or high outside temperatures, driving for short trips only, or operating primarily in stop-and-go traffic. Heavy-duty service includes conditions that would place a higher-than-normal load on the engine, such as trailer towing. If the normal operating schedule calls for oil changes every 6000 miles (9600 km), the severe or heavy-duty schedule will generally require an oil change every 3000 miles (4800 km). **Figure 11-1** is a typical maintenance schedule.

Engine Lubrication System Service

The importance of a properly functioning engine lubrication system cannot be overemphasized. An ample supply of

MAINTENANCE SCHEDULE B—NORMAL DRIVING CONDITIONS
Ranger/Explorer 2.3L 4-Cylinder, 2.9L, 3.0L and 4.0L 6-Cylinder Engines
B —Required for all vehicles.
b —Required for 49 States vehicles (all States except California); recommended, but not required, for California and Canada vehicles.
(b) —This item not required to be performed. However, Ford recommends that you also perform maintenance on items designated by a "(b)" in order to achieve best vehicle operation. Failure to perform this recommended maintenance will not invalidate the vehicle emissions warranty or manufacturer recall liability.

NORMAL DRIVING SERVICE INTERVALS—PERFORM AT THE MONTHS OR DISTANCES SHOWN, WHICHEVER OCCURS FIRST.

MAINTENANCE OPERATION — MILES (Thousands)	7.5	15	22.5	30	37.5	45	52.5	60	67.5	75	82.5	90	97.5	105	112.5	120
KILOMETERS (Thousands)	12	24	36	48	60	72	84	96	108	120	132	144	156	168	180	192
EMISSION CONTROL SYSTEMS																
Change engine oil and oil filter—every 6 months OR	B	B	B	B	B	B	B	B	B	B	B	B	B	B	B	b
Replace spark plugs—standard				B				B				B				b
platinum type (3.0L)								B								b
Replace coolant—every 36 months OR				B				B				B				b
Check cooling system, hoses and clamps (1)	ANNUALLY															
Replace air cleaner filter				B				b				b				b
Check/clean idle speed control air bypass valve (2.3L) (1)								(b)								(b)
Check/clean throttle body (1)								(b)								(b)
Replace PCV valve (3)								b								b
Replace spark plug wires								b								b
Inspect drive belt condition and tension—2.3L								b								b
OTHER SYSTEMS																
Check wheel lug nut torque (2)	B	B	B	B	B	B	B	B	B	B	B	B	B	B	B	B
Rotate tires	B		B		B		B		B		B		B		B	
Check clutch reservoir fluid level	B	B	B	B	B	B	B	B	B	B	B	B	B	B	B	B
Inspect and lubricate automatic transmission shift linkage (cable system)	B	B	B	B	B	B	B	B	B	B	B	B	B	B	B	B
Inspect and lubricate front wheel bearings				B				B				B				B
Inspect disc brake system and lubricate caliper pins		B		B		B		B		B		B		B		B
Inspect drum brake linings, lines and hoses		B		B		B		B		B		B		B		B
Inspect exhaust system for leaks, damage or loose parts				B				B				B				B
Inspect and remove any foreign material trapped by exhaust system shielding	B	B	B	B	B	B	B	B	B	B	B	B	B	B	B	B
Lubricate driveshaft U-joints if equipped with grease fitting	B	B	B	B	B	B	B	B	B	B	B	B	B	B	B	B
Inspect parking brake system for damage and operation		B		B		B		B		B		B		B		B
Lubricate throttle/kickdown lever ball stud				B				B				B				B
Lubricate rear driveshaft double cardan joint centering ball (Ranger SWB 4×4)	B	B	B	B	B	B	B	B	B	B	B	B	B	B	B	B
Lubricate front drive axle R.H. axle shaft slip yoke (4×4)				B				B				B				B
Inspect spindle needle bearing spindle thrust bearing lubrication (4×4)				B				B				B				B
Inspect hub lock lubrication (4×4)				B				B				B				B
Change transfer case fluid (4×4)								B								B
Lubricate steering linkage joints if equipped with grease fittings	B	B	B	B	B	B	B	B	B	B	B	B	B	B	B	B

(1) Check means a function measurement of system's operation (performance, leaks or conditions of parts). Correct as required.
(2) Wheel lug nuts must be retightened to proper torque specifications at 500 miles/800 km of new vehicle operation. See your Owner Guide for proper torque specifications. Also retighten to proper torque specification at 500 miles/800 km after (1) any wheel change or (2) any other time the wheel lug nuts have been loosened.
(3) At 60,000 miles/96 000 km, your dealer will replace the PCV valve at no cost on 2.3L, 2.9L, 3.0L and 4.0L engines except California and Canada vehicles.
Note: Change rear axle lubricant at 100,000 miles (160 000 km) or if the rear axle has been submerged in water. Otherwise, the rear axle lubricant should not be checked or changed unless a leak is suspected or repair is required.

Figure 11-1. A maintenance schedule for one particular vehicle. This schedule is to be followed when the vehicle is operated under normal driving conditions. (Ford)

clean oil of the correct type must reach all bearing surfaces. The best way to protect the lubrication system is to regularly change the engine oil.

Not even the best filters will remove all ***impurities*** (water, acids, and microscopic dirt and metal particles) from the oil. In addition, as filters begin to ***load up*** (clog), they restrict oil flow. The better the job of filtration an oil filter does, the sooner it will need replacement. Therefore, oil filters should be changed whenever the oil is changed.

Engine Oils

Engine oil must provide good lubrication under a large range of engine temperatures and speeds. It must be thin enough to allow the starter to crank the engine and heavy enough to protect the engine under heavy loads and high temperatures. Oil must also carry away heat and keep impurities in suspension until they can be trapped by the oil filter. In addition, oil must seal the gap between the piston ring and the cylinder wall to prevent compression loss. The selection of the proper oil to do all these jobs is critical. The following sections discuss some of the standards for modern engine oils.

Viscosity

Modern engine oils are classified according to several criteria. The most commonly recognized classification is viscosity, or weight. ***Viscosity*** is an oil's resistance to flow. An oil that flows readily is commonly referred to as thin or light. One that flows slowly is said to be thick or heavy. Engine oils are graded according to their viscosity by SAE (Society of Automotive Engineers) numbers ranging typically from 5 (lightest) to 50 (heaviest). Most modern oils are multi-grade oils. All vehicle manufacturers recommend multi-grade oils for normal use. Single weight oils should be used only when temperatures will be above freezing.

Multi-grade oils meet the viscosity requirements of two or more SAE grades. Typical multi-grade oils are marked 5W-30, 10W-30, 10W-40, 20W-40, or 20W-50. The W stands for winter, and the number before the W is the viscosity of the oil at low temperatures. The number without a W is the oil's weight at normal engine operating temperature.

Always consult the owner's manual for information on the proper oil weight. While heavier oils provide some extra protection for worn bearings and other loose engine parts, lighter oils are better for engines in good condition. Lighter oils actually reduce friction because they can flow more easily. This improves both fuel efficiency and engine life.

Oil Grades

At the present time, there are nine ***oil service grades*** for automotive gasoline engines. The ***API (American Petroleum***

Institute) has classified these oils as SA, SB, SC, SD, SE, SF, SG, SH, and SL. The letter "S" indicates the oil is for use in spark ignition (gasoline) engines.

SA and SB oils were the first engine oils to be classified by the API. They are recommended for light loads, moderate speeds, and clean conditions. They generally contain no additives and, therefore, are known as ***non-detergent oils.*** These oils are still available for use in very old engines. Additionally, they are sometimes called for in gearboxes and two-cycle engines. They should never be used in late-model automotive engines.

In 1964, detergent oils were introduced. ***Detergent oils*** have special additives that hold contaminants, such as dirt, water, and acids, in suspension until they can be trapped by the oil filter. The first detergent oil was grade SC, followed over the years by grades SD, SE, SF, SG, SH, and SL. The latest oil designation is SL. This oil can also be used in place of all former oil classifications. Most manufacturers do not bother to put the earlier grade designations on their oils, as these designations have been superseded by grade SL.

Note: If non-detergent oils (SA or SB) have been used in an engine for a long time, it may be inadvisable to switch to a detergent oil. The detergent oil may loosen internal engine deposits, clogging the filter screen or engine oil passages.

The API has classified oils for use in diesel engines as CA, CB, CC, CD, CE, CF, CF-2, CF-4, CG-4, CH-4. The "C" indicates that the oil is for use in compression ignition (diesel) engines. The second letter indicates the amount of anti-wear additives, oxidation stabilizers, and detergents in the oil. The number following the letters indicates whether the oil is designed for use in two- or four-stroke engines.

Refer to the chart in **Figure 11-2.** Note the relationship between the previously used API classifications and those in use today.

API service classifications and SAE viscosity ratings are displayed on most oil containers, **Figure 11-3.** The ***American Petroleum Institute's certification mark*** is shown in Figure 11-3A. This starburst symbol identifies the specific

Gasoline Engines		
Category	**Status**	**Service**
SL	**Current**	For all automotive engines presently in use. Introduced July 1, 2001. SL oils are designed to provide better high-temperature deposit control and lower oil consumption. Some of these oils may also meet the latest ILSAC specification and/or qualify as Energy Conserving.
SJ	**Current**	For 2001 and older automotive engines.
SH	Obsolete	For 1996 and older engines. Valid when preceded by current C categories.
SG	Obsolete	For 1993 and older engines.
SF	Obsolete	For 1988 and older engines.
SE	Obsolete	For 1979 and older engines.
SD	Obsolete	For 1971 and older engines.
SC	Obsolete	For 1967 and older engines.
SB	Obsolete	For older engines. Use only when specifically recommended by the manufacturer.
SA	Obsolete	For older engines; no performance requirement. Use only when specifically recommended by the manufacturer.

Diesel Engines		
Category	**Status**	**Service**
CH-4	**Current**	Introduced December 1, 1998. For high-speed, four-stroke engines designed to meet 1998 exhaust emission standards. CH-4 oils are specifically compounded for use with diesel fuels ranging in sulfur content up to 0.5% weight. Can be used in place of CD, CE, CF-4, and CG-4 oils.
CG-4	**Current**	Introduced in 1995. For severe duty, high speed, four-stroke engines using fuel with less than 0.5% weight sulfur. CG-4 oils are required for engines meeting 1994 emission standards. Can be used in place of CD, CE, and CF-4 oils.
CF-4	**Current**	Introduced in 1990. For high-speed, four-stroke, naturally aspirated and turbocharged engines. Can be used in place of CD and CE oils.
CF-2	**Current**	Introduced in 1994. For severe duty, two-stroke-cycle engines. Can be used in place of CD-II oils.
CF	**Current**	Introduced in 1994. For off-road, indirect-injected and other diesel engines including those using fuel with over 0.5% weight sulfur. Can be used in place of CD oils.
CE	Obsolete	Introduced in 1987. For high-speed, four-stroke, naturally aspirated and turbocharged engines. Can be used in place of CC and CD oils.
CD-II	Obsolete	Introduced in 1987. For two-stroke-cycle engines.
CD	Obsolete	Introduced in 1955. For certain naturally aspirated and turbocharged engines.
CC	Obsolete	For engines introduced in 1961.
CB	Obsolete	For moderate duty engines from 1949 to 1960.
CA	Obsolete	For light duty engines (1940s and 1950s)

Figure 11-2. API (American Petroleum Institute) motor oil service classification chart. (API)

application that an oil is designed for (such as gasoline engines). Only oils that satisfy the most current requirements of the International Lubricant Standardization and Approval Committee (ILSAC) are licensed to display this symbol. The API service symbol is shown in Figure 11-3B. This symbol identifies the oil's API service grade, as well as its SAE viscosity rating.

Checking Engine Oil Level

The proper oil level is essential to engine life. Operating with a low oil level can reduce lubrication and overheat the remaining oil. A high oil level can cause oil foaming, which may result in loss of lubrication, and oil loss through seals or the piston rings. In extreme cases, high oil levels can cause oil to be forced out of the oil filler or dipstick tube.

Before checking the oil level, turn off the engine and allow it to sit for several minutes. This allows oil in the top of the engine to drain into the pan. Remove the oil dipstick and wipe it clean. Reinsert the dipstick fully; then remove it and check the oil level. See **Figure 11-4.** A normal oil level is shown in **Figure 11-5.** When checking the oil level, also check the condition of the oil. On a warm engine, you should be able to read the dipstick markings through the oil. If the oil is so dark that the markings cannot be read, it is probably time for an oil change. This method can be helpful when the owner is not certain when the oil was changed last.

While the dipstick is out, check the top just below the handle. Rust indicates that there is water in the oil. This may be caused by condensation due to short trip driving or extremely cold weather, or it may indicate an internal cooling system leak. Heavy sludge buildup on the dipstick usually means that the engine oil has not been changed regularly. If the sludge is light colored (sometimes called mayonnaise sludge) there is water in the oil. A clogged PCV valve or a combination of cold operation and poor maintenance can cause this type of sludge.

The best time to check the oil level is after the engine has been sitting overnight with the vehicle level. This allows all oil to drain from the top of the engine and gives a true reading.

Determining when Oil Should Be Changed

Vehicle manufacturer's recommended ***oil change intervals*** range from 2000–7500 miles (3200–12,000 km), depending on operating conditions. There are so many variables that an exact recommendation would be unwise. Engine type (gasoline, diesel, etc.), engine load, trip length, operating speed, temperature, and dust all affect the useful life of engine oil.

The following conditions can be considered near ideal, and if the engine is operated under them, the oil change interval can be extended to the maximum recommended by car manufacturer:

- Engine is mechanically sound.
- Engine is operated for reasonably long periods each time it is started.
- Engine temperature reaches 180°–200°F (82.3°–93.3°C) or higher.
- Fuel injection or carburetion system is operating properly.
- Intake air is properly filtered (clean, undamaged filter).
- Oil filter is in good condition.
- Positive crankcase ventilation system is functioning properly.
- Engine speeds kept from moderate to reasonably high.
- Engine is operated under clean environmental conditions.
- Engine is not overloaded, such as when pulling a heavy trailer.

Figure 11-3. A typical engine oil container. A—The API certification mark can be found on both conventional and synthetic oils. B—The API service symbol shows both viscosity (10W-30) and service classification (SL).

A

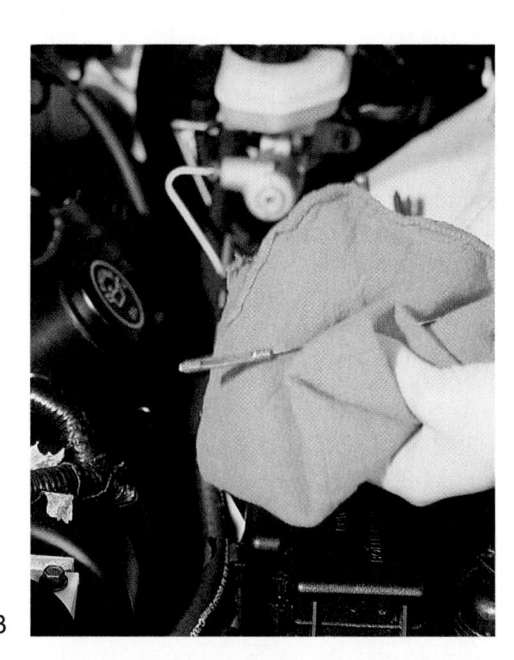

B

C

Figure 11-4. Checking engine oil level. A—Locate the dipstick and remove it from the engine. B—Wipe the oil from the dipstick with a clean cloth and reinstall the dipstick in the engine. C—Remove the dipstick and check oil level.

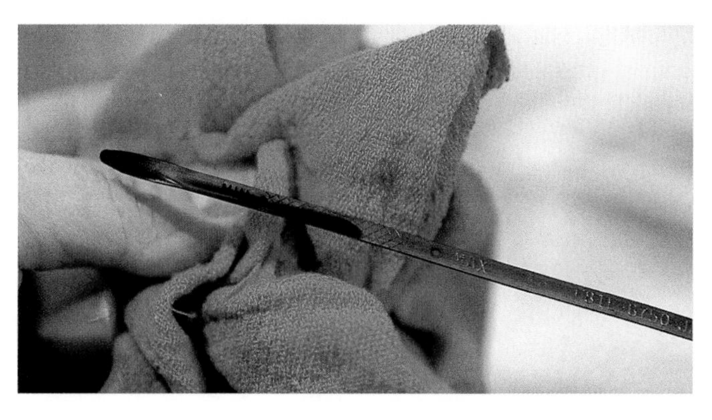

Figure 11-5. The oil level should be between the add and full marks on the dipstick.

If, however, operating conditions are not ideal, the oil change interval will be shortened. In fact, under the worst conditions, such as dust storms, prolonged idling, constant stops and starts, or cold weather operation, the oil change interval can be as little as 500–1000 miles (800–1600 km).

Since few engines are operated under either ideal or extreme conditions, it is important to know the general operating conditions of the engine in question. A reasonable oil change interval is somewhere between the extremes listed earlier. At no time should the manufacturer's recommended maximum interval be extended. Also, keep in mind that diesel engine oil change intervals are shorter than the recommended intervals for gasoline engines.

Changing the Oil and Filter

STOP **Warning: Never pour used oil on the ground or into drains. Refer to Chapter 2, Safety and Environmental Protection, for important information about disposing of used oil and filters.**

Change the oil when the engine is at its normal operating temperature. Place an appropriate container under the oil pan to catch the used oil. Remove the drain plug and allow the oil to drain completely from the oil pan. Clean the drain plug and reinstall it. Use a new drain plug gasket if the old gasket has been damaged.

To change a spin-on oil filter, use a suitable wrench to remove the old filter. See **Figure 11-6.** Wipe the engine's filter base clean, **Figure 11-7A.** Rub a thin film of engine oil (not grease) on the new filter seal ring, **Figure 11-7B.** Install the filter and tighten it until the seal ring engages the base, **Figure 11-7C.** Then give it another half turn, **Figure 11-7D.** When tightening the filter, follow the manufacturer's specifications. Overtightening may split the gasket, distort the filter, and make removal difficult.

Note: Some filters must be filled with oil before installation.

After changing the filter, fill the crankcase with oil of correct grade and viscosity (as recommended by manufacturer). If the filter was changed, add extra oil. Most new filters require anywhere from one-half quart to one quart of additional oil. Before starting the engine, check the oil level on the dipstick.

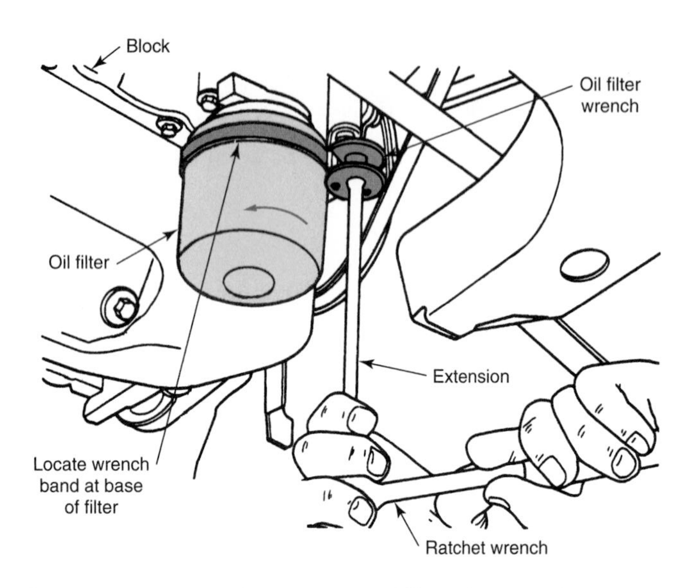

Figure 11-6. Removing a spin-on oil filter element with a strap wrench. (DaimlerChrysler)

Run the engine for a few minutes and then turn it off. Check for leaks around the drain plug and the filter. Recheck the oil level.

Caution: Use care while draining the oil, as it may be hot enough to cause painful burns. Avoid skin contact. Prolonged skin contact with oil has been shown to cause skin disorders and possibly skin cancer.

Checking the Oil Cooler

If an engine uses an oil cooler, check the lines and fittings for leaks. Replace cracked hoses. Check the cooler for the presence of dirt, leaves, or other debris that may interfere with the cooling action. One type of engine oil cooler is shown in **Figure 11-8.** The oil cooler is often connected to the lubrication system through an adapter at the oil filter, **Figure 11-9.**

Other Services during Oil and Filter Change

It is important that several other vehicle systems be lubricated or checked for proper operation when the oil is changed. Some of the most common services are discussed in the following sections. Consult the vehicle service and owner's manuals for other maintenance services.

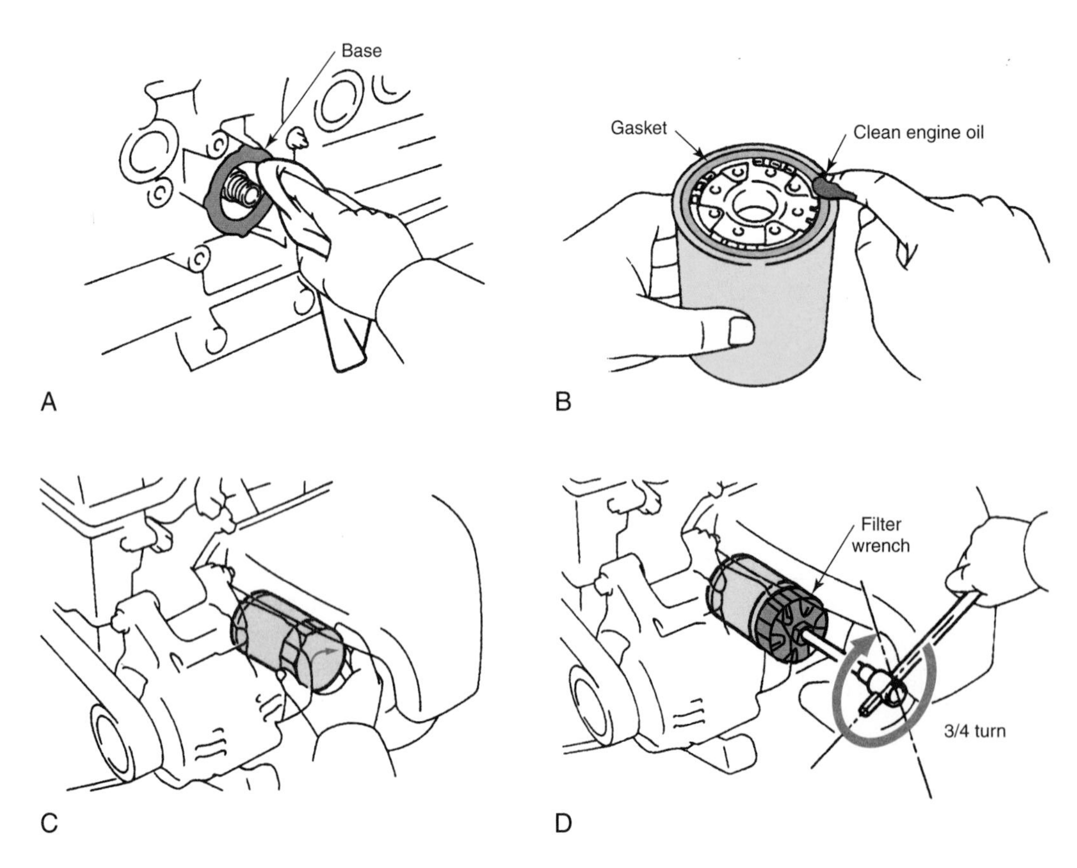

Figure 11-7. Changing a spin-on oil filter. (Toyota)

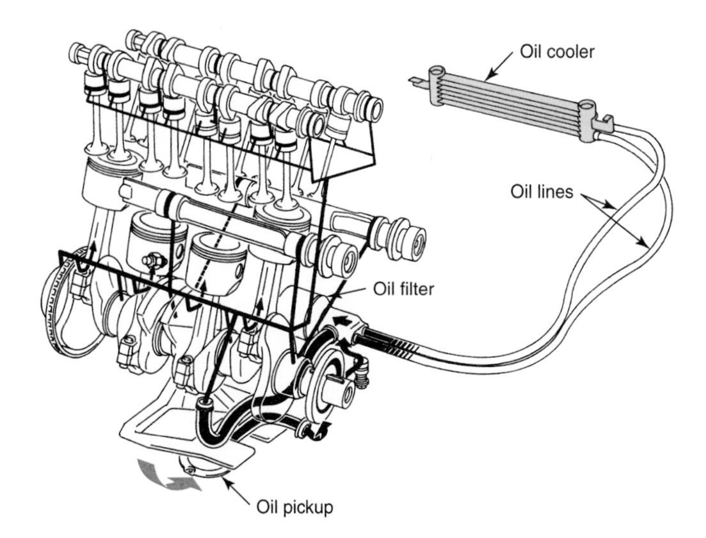

Figure 11-8. Engine oil cooler and lubrication system schematic. (Saab)

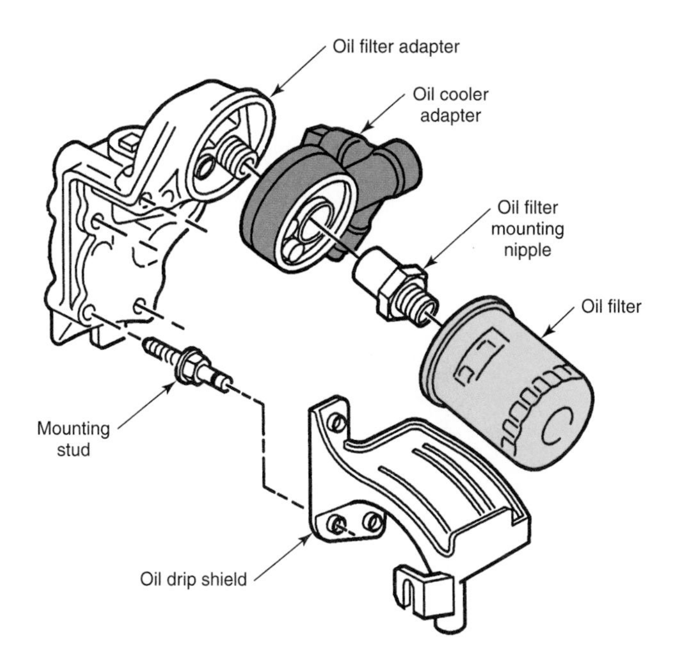

Figure 11-9. Oil cooler adapter location. (General Motors)

Checking the Crankcase Ventilation System

It is pointless to change the oil and filter if the positive crankcase ventilation (PCV) system is not removing crankcase vapors. An inoperative PCV system will permit a rapid buildup of sludge, water, and acids that will shorten the life of the engine. PCV problems will have a negative effect on fuel economy and emissions. Always check the PCV valve and related hoses at every oil change. For complete details on the construction, operation, and service of the PCV system, refer to Chapter 22, Emission System Service.

Checking the Fuel Filters

Fuel filters should be checked at regular intervals and changed when they show any sign of restriction. Fuel filters are discussed in Chapter 17, Fuel Delivery.

Many fuel filters are installed under the vehicle's body, near the fuel tank. These filters are often overlooked and may not be changed according to the manufacturer's schedule. A neglected filter can clog up suddenly, leaving the driver stranded. If a vehicle's owner is not sure when the fuel filter was last changed, recommend that it be changed immediately.

Note: Dirty fuel filters cannot be cleaned. Always replace fuel filters.

Air Cleaner Service

All modern cars and light trucks use paper ***air cleaner elements.*** These elements can be round (radial airflow), as in **Figure 11-10,** or flat (linear airflow), as shown in **Figure 11-11.**

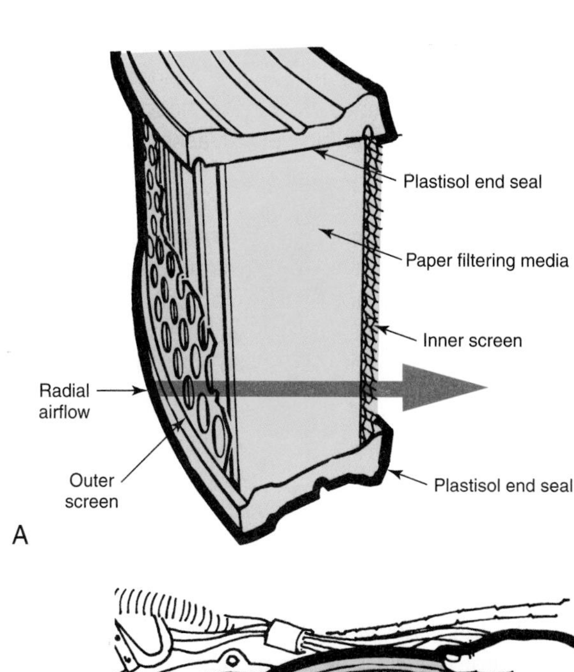

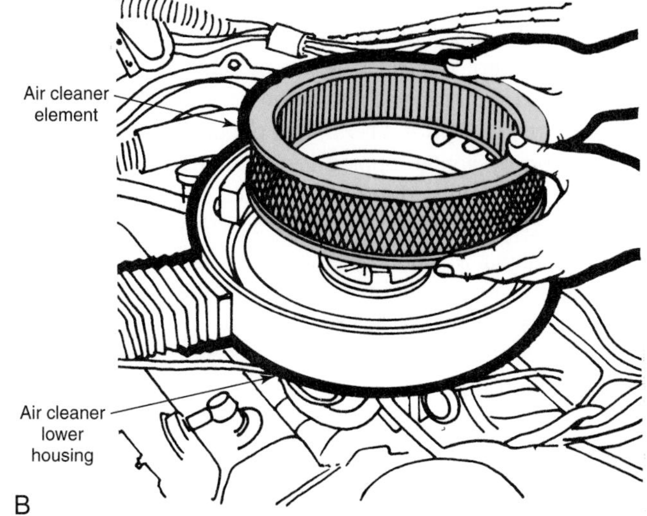

Figure 11-10. A round, radial-airflow air cleaner element. A—Cutaway of a paper element. B—This technician is replacing a dirty air cleaner element with a new one. (Fram)

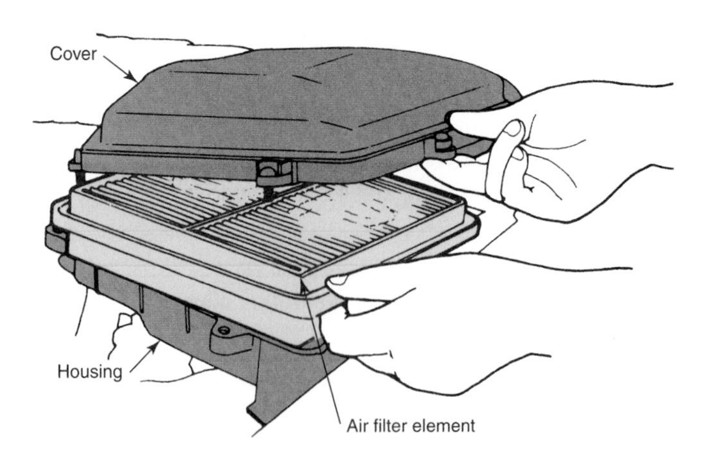

Figure 11-11. This linear flow air filter element is being removed from its housing. Some filter elements are marked to indicate "up" or "airflow" direction. Be sure to install the filter correctly. (General Motors)

The flat air filter element is designed for use in a remotely mounted air cleaner housing. Transverse engines and styling changes that limit clearance under the hood have made it necessary to relocate the air cleaner assembly on some vehicles.

The paper air cleaner element should be replaced at specified intervals. It is important that great care be exercised when removing the element. Do not let dirt fall into the throttle body, carburetor, or intake area. When practical, it is advisable to remove the air cleaner as a unit. When the filter is removed, cover the throttle body or carburetor with a clean cloth.

Caution: If the filter element is wet, especially on vehicles with low mounted air intakes, be sure to caution the driver not to drive through deep standing water.

After removing the filter, inspect it carefully. See **Figure 11-12.** If it is not excessively dirty or oil soaked, the air filter element can be cleaned by directing a stream of compressed air through the element in a direction opposite regular airflow, **Figure 11-13.** The paper element also can be cleaned by lightly tapping it on a flat surface. This will dislodge much of the dirt trapped by the filters.

Before reinstalling a paper air filter element, check the element against a light source, such as a droplight, to make sure there are no ruptured spots. Check the top and bottom gasket surfaces for injury. Install right side up where indicated. The element shown in **Figure 11-14** was badly damaged from careless handling.

When servicing any type of filter element, clean the air cleaner body and cover before replacing element. Make sure all gaskets are in good shape and are in place. Tighten the cover nuts or fasten the cover clips securely. When replacing the air cleaner body, be sure it faces in the correct direction. If a tang or a locating lug is present, see that the tang or lug engages properly.

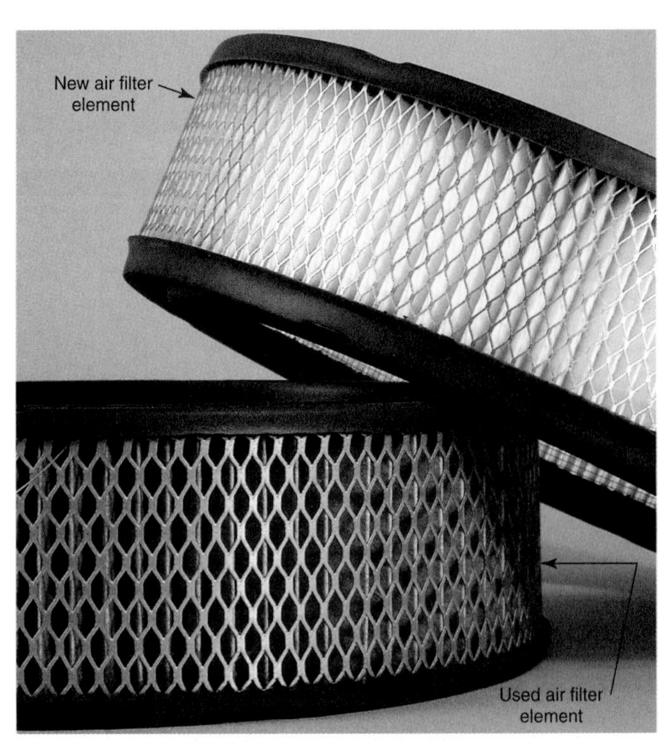

Figure 11-12. New and used paper air filter elements. Note that the used element is relatively dirty. Although a dirty element can sometimes be cleaned, it is often more economical to simply replace it.

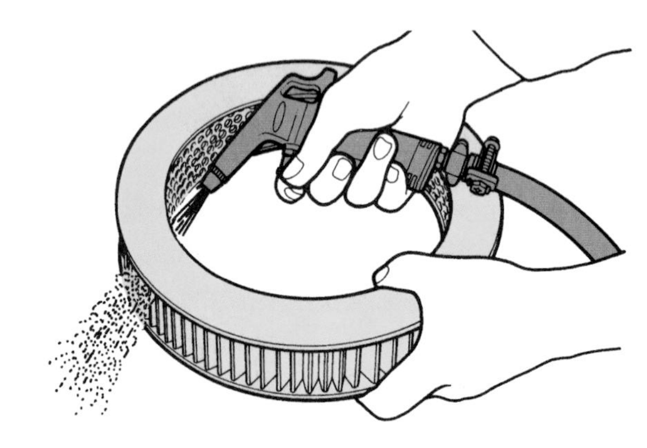

Figure 11-13. Cleaning a paper filter element by blowing air through it from the inside. Use a relatively soft stream of air and keep the nozzle away from the filter surface. Wear safety glasses. (DaimlerChrysler)

Figure 11-14. Physical damage rendered this element unfit for further service. (Perfect Circle)

Lubricating Suspension and Steering Linkage

Steering and suspension parts should be checked for ***grease fittings,*** **Figure 11-15.** Typical parts with grease fittings are the inner and outer tie rod ends, idler arm, pittman arm, drag (center) link, and ball joints. On some vehicles, the control arm bushings have grease fittings. Many companies publish charts showing the lubrication points for specific vehicles, **Figure 11-16.**

If fittings are present, the parts should be lubricated on a regular basis, usually between 2000 and 8000 miles (3218 and 12,872 km). Check the manufacturer's specifications for exact lubrication intervals. Some suspension parts are equipped with plugs rather than fittings. If these components are to be lubricated, the plugs must be replaced with grease fittings, **Figure 11-17.** After greasing is completed, the plugs can be reinstalled or the fittings can be left in place for subsequent lubrication.

There are several grades of grease, and the proper type should be used on the fitting. See **Figure 11-18.** Modern greases usually contain lithium and provide lubrication under extreme pressure. For this reason, they are sometimes called ***EP (extreme pressure) greases.*** These greases also provide protection against corrosion and water entry, which is important to fittings installed under the vehicle. A hand- or an air-operated grease gun can be used to apply these greases.

To lubricate a joint, press the grease gun nozzle straight onto the fitting. Apply pressure slowly to avoid damaging the seals. See **Figure 11-19.** If the joint does not use a sealed boot, apply grease until the old grease is forced from the joint. If the joint has a sealed boot, apply grease until the boot begins to bulge slightly.

Note: In extremely cold weather, allow the vehicle to sit inside long enough to warm the suspension fittings before lubricating a joint. Applying grease to extremely cold joints can damage them. The joint temperature should be above 0°F (-33°C).

If grease will not enter the joint, or if it squirts from around the grease gun and fitting, make sure the fitting is not frozen. In most cases, the fitting has become clogged with dirt or hardened grease and should be replaced. If the fitting is good, check for a damaged or frozen joint. A defective joint should be replaced.

If the joint being greased seems to have excessive movement, check for a worn or loose part. Do not allow the vehicle to leave the shop with an unsafe suspension system. More information on checking and servicing steering and suspension parts is given later in this chapter, as well as in Chapters 36, 37, and 39.

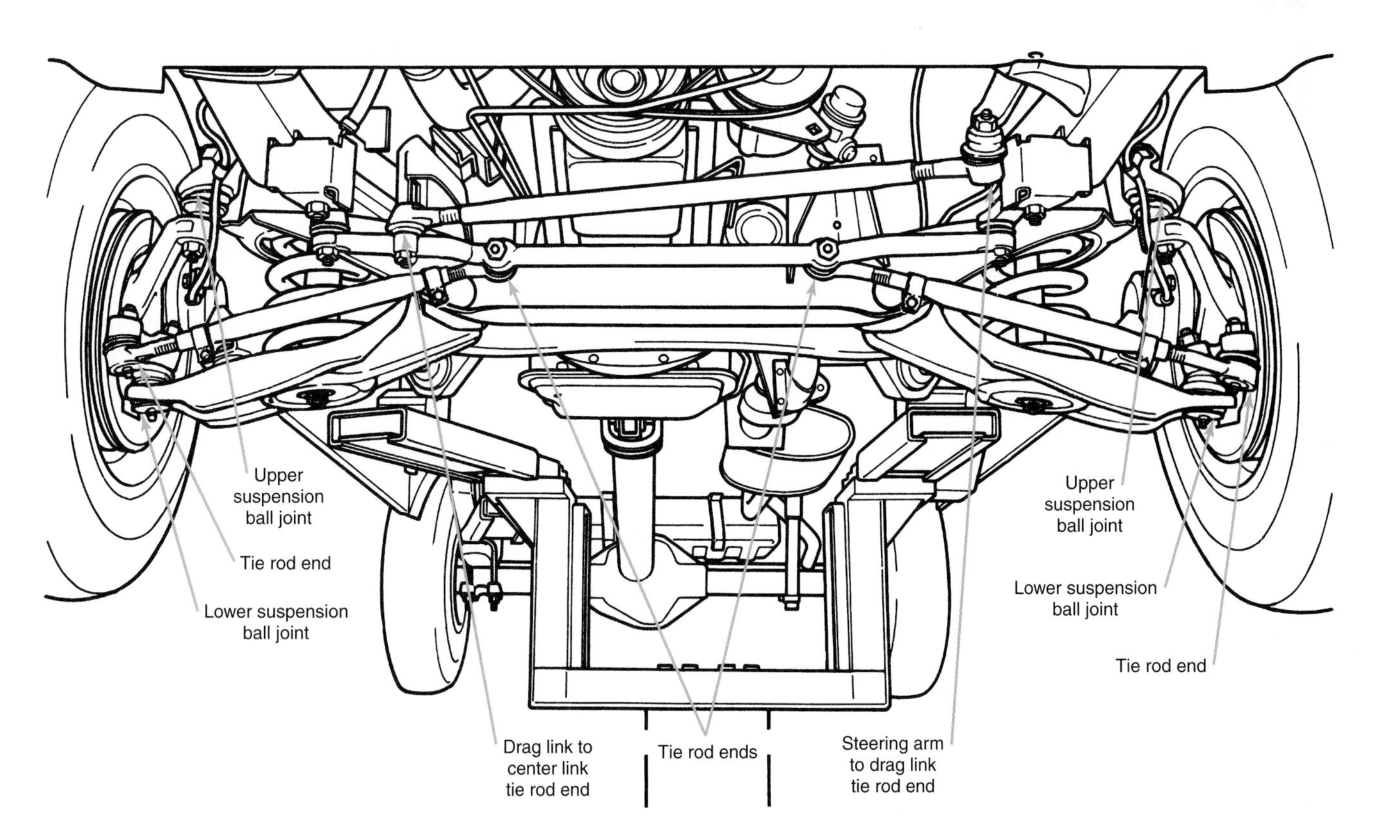

Figure 11-15. Grease fittings on one particular vehicle front suspension and steering assembly. (DaimlerChrysler)

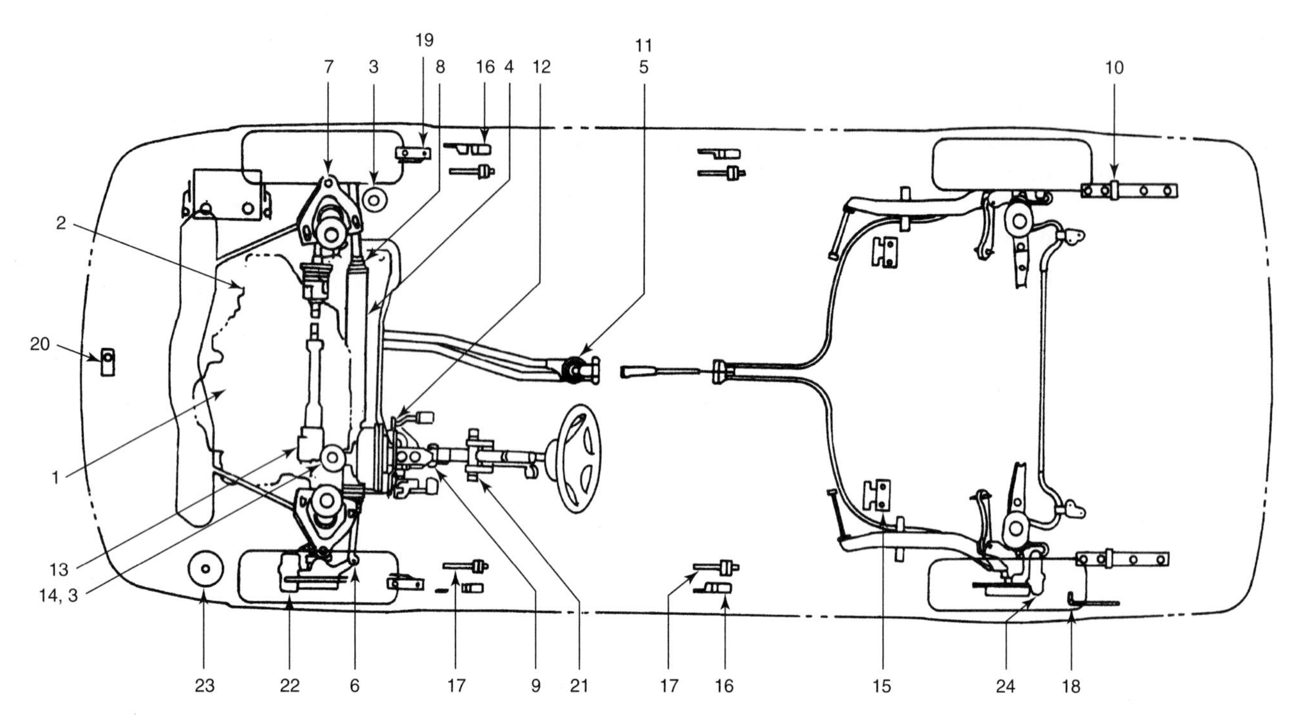

Figure 11-16. Lubrication points on a particular car. 1—Engine. 2—Transmission. 3—Brake parts. 4—Steering gearbox. 5—Transmission shift lever. 6—Ball joints. 7—Ball joints. 8—Steering boots. 9—Steering column bushings. 10—Trunk hinges. 11—Shift lever. 12—Pedal linkage. 13—Intermediate shaft. 14—Master cylinder pushrod. 15—Tailgate hinges. 16—Door hinges. 17—Door opening detents. 18—Fuel filler door. 19—Door hinges. 20—Hood latch. 21—Tilt steering wheel lever. 22—Caliper. 23—Power steering system. 24—Caliper dust seal, pin, and piston. (Honda)

Lubricating the Driveline

When servicing rear-wheel drive vehicles, check for grease fittings on the drive shaft U-joints, **Figure 11-20,** and lubricate them if necessary. U-joints use the same EP lithium grease as suspension and steering parts. It may be necessary to turn the drive shaft to reach the fittings. U-joints require a very small amount of grease, and excess grease can damage the seals. On front-wheel drive vehicles, check the CV axle boots for damage.

Lubricating Body Hinge Pivot Points and Lock Mechanisms

The body hinge pivot points include the door and hood hinges. These points are subject to wear and rust and should be lightly lubricated with oil or white grease whenever the oil and filter are changed. The door locking mechanisms and the hood lock cables should also be lightly lubricated.

Caution: Do not use chassis grease on the door lock mechanisms, as this type of grease will stain clothing if the door lock is accidentally contacted by a person entering the vehicle.

Door and trunk locks can be lubricated with dry graphite. Do not use oil on a lock cylinder, since oil will clog the cylinder mechanism.

Lubricating the Parking Brake Cable

In many cases, the parking brake is never used. Instead, the vehicle driver uses the transmission park position to hold the vehicle in place. The brake cable often rusts and becomes sticky in its sheath. When the parking brake finally is used, it often sticks in the *on* position, preventing vehicle movement or severely damaging the brakes.

The parking brake cable should be cleaned and lubricated whenever the oil and filter are changed. Use a combination of lubricant and penetrating oil on the cable sheath under the vehicle and a small amount of engine oil on portion of the cable that extends out of its sheath under the dashboard. Operate the parking brake a few times to distribute the lubricant.

Checking Fluids

Whenever the oil is changed, be sure to check the fluid levels in the coolant recovery tank, power steering reservoir, windshield washer, and brake master cylinder. Refill as needed and, except for the windshield washer fluid, determine the reason for the low level.

Also check the fluid levels in the automatic transmission/transaxle, manual transmission or transaxle, transfer case, and rear axle assembly, if equipped. Remember that some front-wheel drive vehicles have a separate differential assembly.

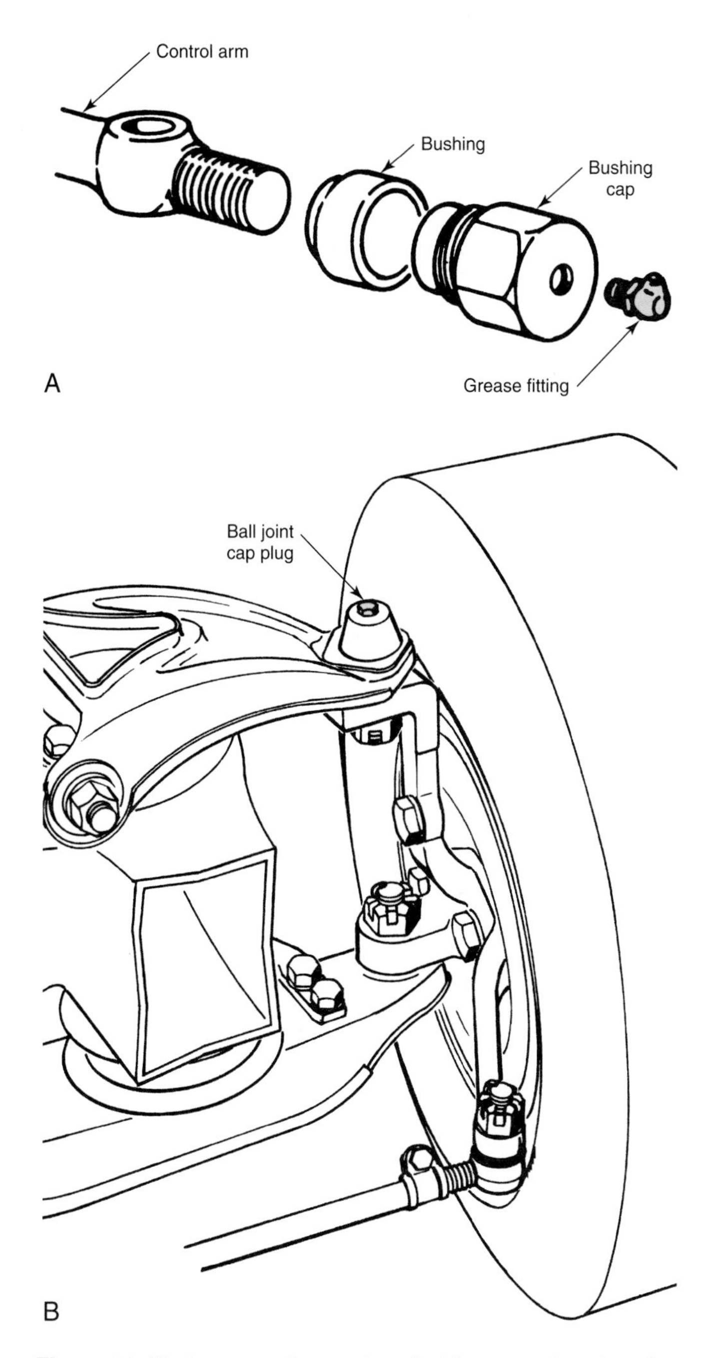

Figure 11-17. A suspension system that incorporates plugs in place of grease fittings. The plugs must be removed and replaced with grease fittings for lubrication. (Moog)

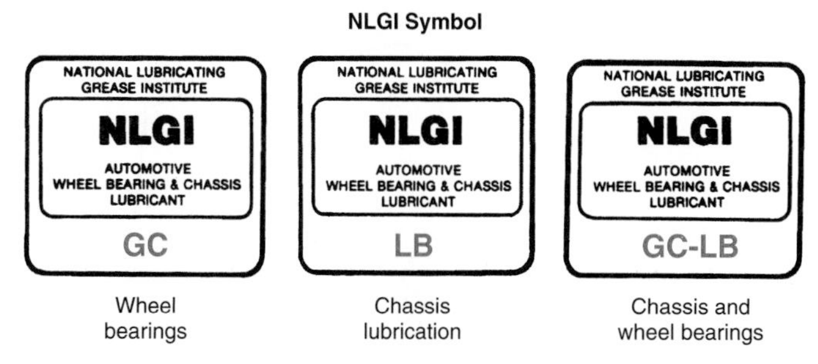

Figure 11-18. National Lubrication Grease Institute (NLGI) symbols. The letter "G" indicates wheel bearing grease. The letter "L" indicates chassis grease. These symbols appear on all approved grease. (DaimlerChrysler)

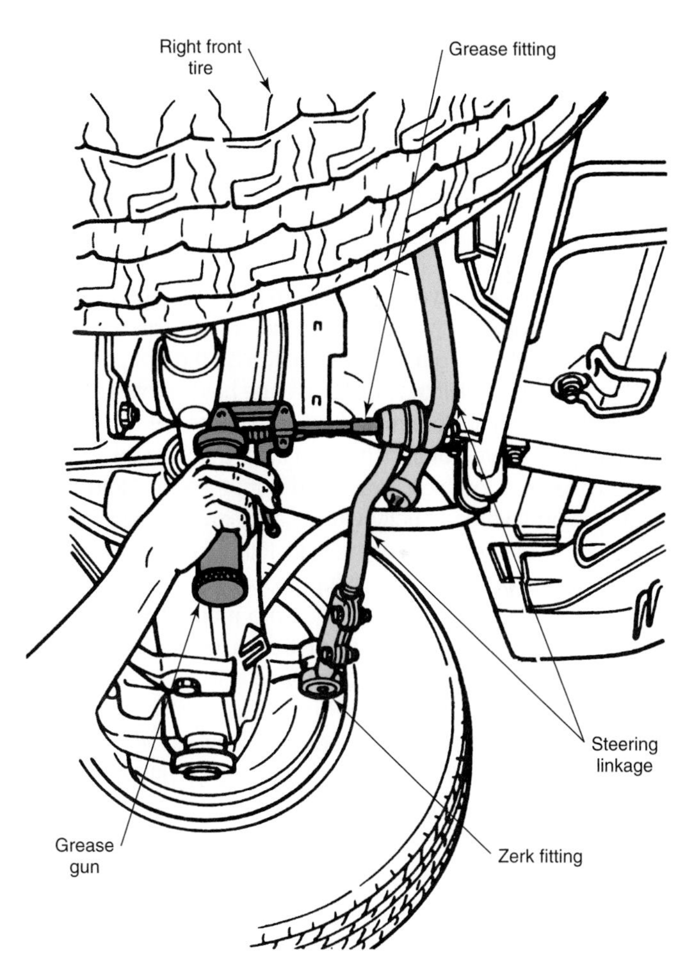

Figure 11-19. Lubricating the steering linkage with a hand-operated grease gun. Always clean the grease fittings before lubricating. This will help prevent dirt from being forced into the fitting and joint. (Ford)

Checking Automatic Transmission/Transaxle Fluid

Transmission and transaxle fluid level is critical to proper operation. A low level will cause air to enter the hydraulic system, causing slippage. If the level is too low, the vehicle will not move. If the level is too high, the fluid will be whipped into foam. The foaming fluid will enter the hydraulic system, causing slippage.

To check the fluid level in an automatic transmission or transaxle, start the engine and move the shifter through all the gear positions. Place the shifter in *Neutral* or *Park* as required by the manufacturer. Then remove the transmission or transaxle dipstick and wipe it with a clean cloth. Reinsert the dipstick and withdraw it. Read the level on the dipstick. Most dipsticks have both hot and cold level positions.

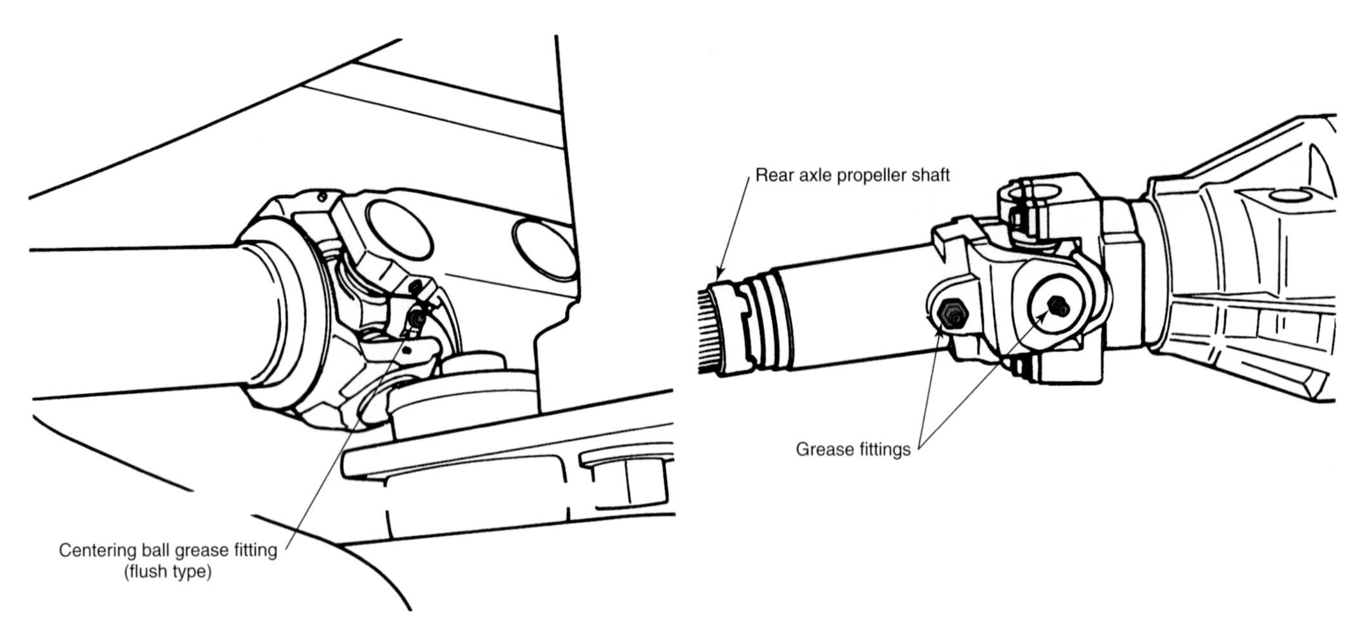

Figure 11-20. Several driveline joint grease fittings and their locations. (DaimlerChrysler)

See **Figure 11-21.** If you are not sure that the transmission is fully heated up, lightly touch the end of the dipstick. If the dipstick is too hot to grasp comfortably, use the hot position to determine fluid level. If the dipstick can be held comfortably, use the cold position to determine fluid level. If the level is too low, add the proper type of transmission fluid. See **Figure 11-22.** If the fluid level is too high, drain fluid from the unit. Note that a slightly high level is usually OK. In addition to checking fluid level, check the condition of the fluid. If the fluid smells burned and is orange or dark brown, it has probably been overheated. The clutch and band friction linings are probably burned and glazed, impairing transmission or transaxle operation. If the fluid is milky looking, it is contaminated with water or coolant. Water can enter the transmission or transaxle when the vehicle is operated in deep water. A leak in the oil cooler will allow coolant to leak in from the engine cooling-system radiator. Water or coolant will ruin the friction materials inside the unit.

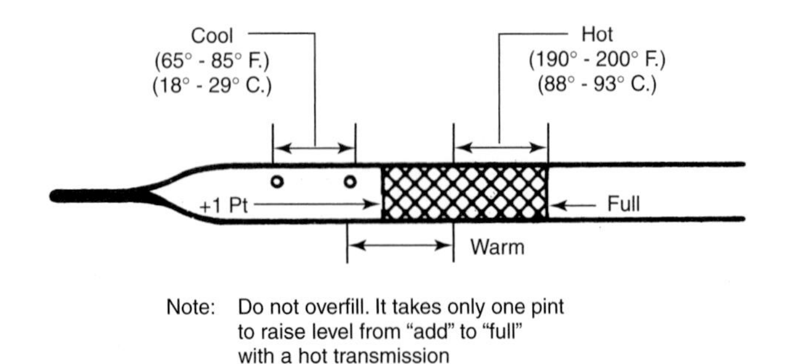

Figure 11-21. The fluid level is important to the proper operation of an automatic transmission or transaxle. Both hot and cold levels are given on most dipsticks. On the dipstick shown, for example, fluid is at an acceptable level if it falls within the ranges shown at the respective temperatures. (General Motors)

Fluids other than Dextron III®	
Vehicle	***Fluid***
Accura	Honda Gen ATF
Audi	Audi/VW Synthetic ATF #G002000
BMW w/5-speed automatic only	BMW LA 2634
Chrysler '88-'98	ATF+3
Chrysler '99-'01	ATF+4
Dodge '88-'98	ATF+3
Dodge '99-'01	ATF+4
Eagle	ATF+3
Ford	Mercon V
Honda	Honda Gen ATF
Hundai	ATF+3
Isuzu	Honda Gen ATF
Jeep	ATF+3
Lexus	Toyota type T-IV ATF
Lincoln	Mercon V
Mazda	Mercon V
Mercury	Mercon V
Mitsubishi	Mitsubishi Diamond ATF
Plymouth '88-'00	ATF+3
Saturn	Saturn Transaxle Fluid #21005966+
Toyota	Toyota type T-IV ATF
Volkswagen without filter tube	Audi/VW Synthetic ATF #G002000

Figure 11-22. Many vehicles use Dexron automatic transmission fluid in their transmissions or transaxles. A considerable number of vehicles, however, use a fluid other than Dexron.

Note: The procedure for changing automatic transmission/transaxle fluid and filter will be covered later in this chapter.

Checking Gearbox Fluid Levels

Check manual transmissions and transaxles, rear axle assemblies, and transfer cases by removing the filler plug and observing the fluid level. See **Figure 11-23.** Slight dripping from the hole is OK. A few manual transaxles have a dipstick, which is read in the same way as an engine oil dipstick. If the oil level is low, add the proper type of oil to bring the level up. Also check the case to determine whether the unit is leaking. If the oil level is very high and the oil is milky looking, water has entered the case. The unit must be thoroughly drained and new oil installed.

Caution: Be sure to use the correct lubricant. Transfer cases and limited slip (positraction) rear axles contain internal friction clutches. The use of the wrong lubricant can cause these clutches to slip and chatter.

Checking Coolant Level and Condition

Coolant recovery tanks are made of clear plastic and the coolant level can be observed without removing the reservoir or radiator caps, **Figure 11-24.** The coolant recovery tank will have hot and cold level markings. Coolant should be at or slightly above the level mark for the particular engine temperature. If the level is low, carefully add a 50-50 mixture of the proper antifreeze. Some newer vehicles use long life coolant, and only this type of coolant should be added to the reservoir.

Caution: On some vehicles, the coolant recovery reservoir is pressurized. Use extreme caution when removing the recovery tank cap. Always turn the cap slightly and allow pressure to escape before removing it entirely. See Figure 11-25.

If the coolant level is extremely low, locate the cooling system leak and correct it as explained in Chapter 23, Cooling System Service.

In addition to checking coolant level, coolant condition and freezing point should also be checked. Visually inspect the coolant. If the coolant is rust colored, it must be changed and the system must be flushed. If oil is floating on top of the

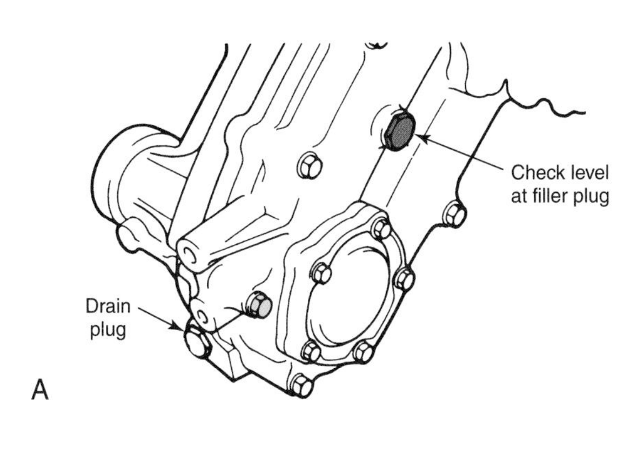

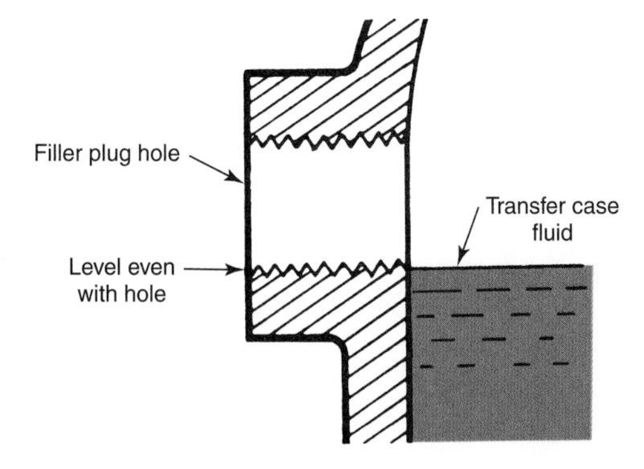

Figure 11-23. Checking fluid level on one particular transfer case. A—Note the location of the drain plug and the filler plug. Make sure you remove the filler plug when checking fluid level. B—In most cases, normal fluid level is at the bottom of the filler hole threads. (Toyota)

Figure 11-24. Coolant level on late-model vehicles can be checked by observing coolant in the reservoir tank.

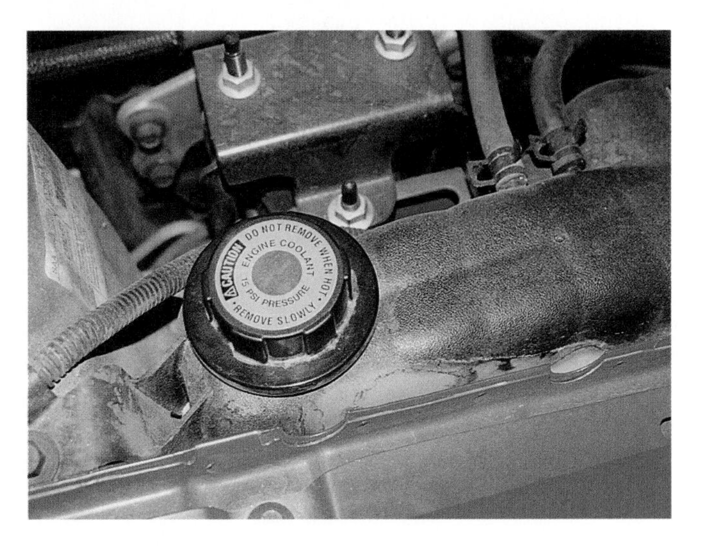

Figure 11-25. Many vehicles have pressurized filler caps on the coolant reservoir instead of the radiator. Remove these caps as carefully as you would a radiator pressure cap.

coolant, engine oil or transmission fluid may be leaking into the cooling system. Even if coolant looks good, there may not be enough antifreeze present to keep the system from freezing in cold weather. The only sure way to check the freezing point of the coolant is to use a hydrometer, a refractometer, or a test strip.

To use a ***hydrometer,*** place the tool's hose into the overflow reservoir. Squeeze and release the bulb to draw a sample of coolant into the hydrometer. Allow the pointer, floats, or balls in the hydrometer to stabilize, and then read the temperature on the hydrometer. See **Figure 11-26.** Some hydrometers have a temperature correcting feature, and it may be necessary to add or subtract from the reading, depending on the temperature of the coolant.

To use a ***refractometer,*** take a small sample of coolant from the radiator. Place a few drops on the refractometer lens. Look through the refractometer to get the freeze point reading.

Some shops now use ***coolant test strips*** to quickly check engine coolant. Chemical test strips can check coolant for freeze protection, as well as alkalinity (acidity) or pH level. To use test strips, remove one strip from the container. Dip the test strip into the engine coolant, making sure both test spots are immersed. Remove the test strip and wait approximately 30 seconds. Compare the two test sections on the strip to the color chart that comes with the strips. See **Figure 11-27.**

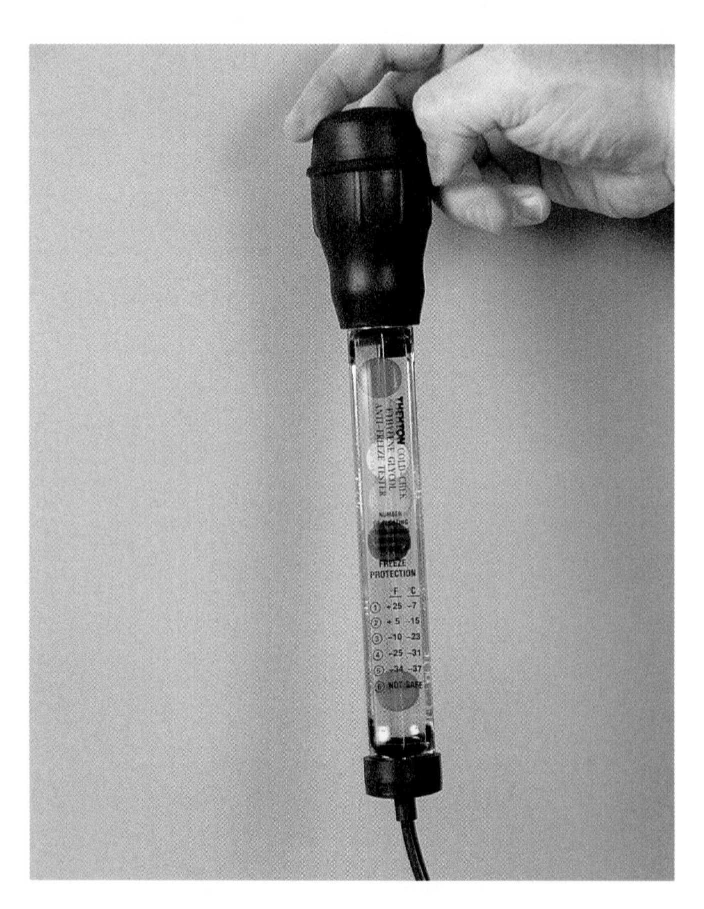

Figure 11-26. A hydrometer can be used to check the coolant's ability to protect against freezing. Most cooling system hydrometers use floating balls or discs.

Note: The procedure for changing coolant is covered later in this chapter.

Checking Power Steering Fluid Levels

Power steering reservoirs usually have a combination filler cap and dipstick, **Figure 11-28.** To check the fluid level, remove the cap with the engine off and observe the level on the dipstick. Some power steering reservoirs are made of

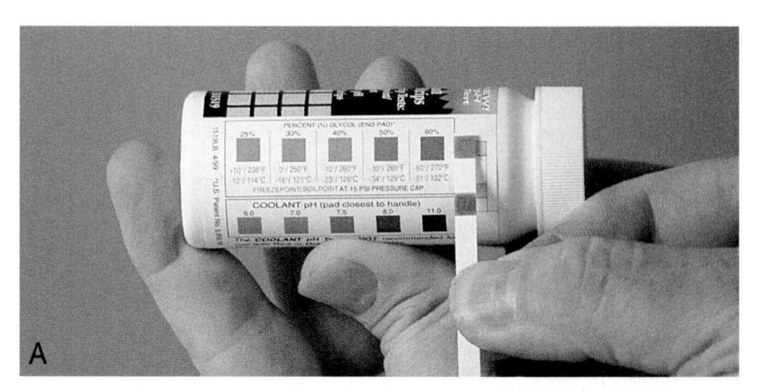

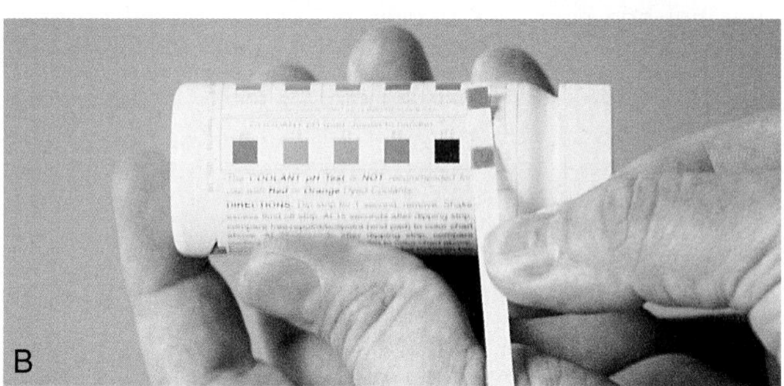

Figure 11-27. Chemical test strips are now available and can quickly determine the cooling system's freeze protection capability, as well as its pH level. A—Test for freeze protection. B—Test for alkalinity, or pH.

Figure 11-28. Periodically check the level and condition of the power steering fluid. This particular reservoir has a combination cap and dipstick. Always use the correct type of fluid to top off the system.

translucent (see-through) plastic, and the fluid level in these reservoirs can be determined without removing the cap. See **Figure 11-29.** Add the proper fluid as needed. Most modern power steering systems require special power steering fluid. Older power steering systems use transmission fluid.

Checking Brake Fluid Levels

To check brake fluid level, remove the master cylinder cap and observe the level in the reservoir. See **Figure 11-30.** On some vehicles, the fluid level can be checked without removing the cap, **Figure 11-31.** On a few vehicles, the reservoir is located away from the master cylinder.

Add the proper type of brake fluid if necessary. Do not leave the reservoir cover off any longer than is necessary to check and add fluid. Brake fluid will absorb water from the air if the cap is left off for long periods. If the brake fluid level is extremely low, immediately locate and correct the cause of the leak. Brake system service is covered in Chapter 34.

Figure 11-29. Some power steering fluid reservoirs are made of translucent plastic. Simply compare fluid level to the markings on the side of the reservoir.

Figure 11-30. In many cases, the cap must be removed to check the fluid level in the master cylinder reservoir. Clean off the top of the reservoir before removing the cap. The fluid should be approximately 1/4″ (6 mm) below the top of the reservoir opening.

Checking Washer Solution

Most windshield washer fluid reservoirs are made of clear plastic, and the fluid level can be observed. If the level is low, remove the cap and add washer fluid as necessary. Do not add plain water. It will freeze in cold weather, breaking the reservoir.

Changing Fluids

In addition to changing oil and checking fluid levels, other vehicle fluids must be changed periodically. These fluids include the engine coolant, automatic transmission/transaxle fluid, gearbox fluid, and brake fluid. Refer to the maintenance schedule for the vehicle at hand to determine the exact fluid change intervals.

Changing Engine Coolant

Most manufacturers recommend changing the coolant at certain intervals. Begin by setting the heater control to the maximum heat position and loosening the radiator cap to prevent pressure buildup. Start the engine and allow it to run until it reaches its normal operating temperature. If the engine is not allowed to reach operating temperature, the thermostat will not open. With the thermostat closed, the old coolant will not drain from the engine block. When the engine reaches normal operating temperature, shut it off. Cover the radiator cap with a thick rag and remove the cap carefully. Open the radiator drain cock and remove the drain plugs from the engine block. Some vehicles do not have engine block drains. Check the vehicle service manual to determine whether the engine block can be drained.

If the coolant is rusty or the interior of the radiator appears to be dirty, clean the system using a ***chemical cooling system cleaner*** before the new antifreeze is installed. If the system is clean, simply flush it out with clean water.

After cleaning, tighten the radiator and heater hose clamps and look carefully for signs of coolant leakage (rust

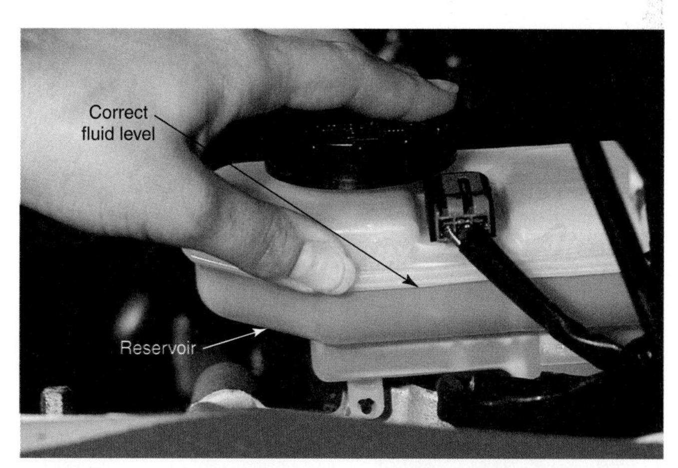

Figure 11-31. If the master cylinder reservoir is made of translucent plastic, compare the fluid level to the markings on the reservoir. If there are no markings, the fluid level should be just below the top of the reservoir.

streaks, discoloration from antifreeze, dampness). Repair leaks as necessary. Close the drain cock and reinstall the block drains.

Most manufacturers specify that coolant should be a 50-50 mixture of antifreeze and water. When refilling the system, add equal amounts of antifreeze and water, or mix the water and antifreeze in equal proportions and then add the solution to the radiator. Make sure you are using the correct antifreeze. Some modern vehicles use ***long-life antifreeze.*** This antifreeze should not be mixed with conventional antifreeze. Add the specified amount of antifreeze and water.

Caution: On some modern vehicles in which the radiator is lower than the engine, the cooling system must be bled to remove air from the engine cooling passages before the engine is restarted. This type of system can usually be identified by a small *bleed valve* on the thermostat housing. See Figure 11-32. Refer to the service manual for exact procedures.

After refilling the radiator, start the engine and allow it to run until it reaches normal operating temperature. The heater control should be set to maximum heat. When the engine is warm, the thermostat will open and release any air trapped in the system. Running the engine until it becomes warm also mixes the antifreeze and water. Add coolant as needed. If the cooling system is open (no reservoir or recovery tank), check the coolant level in the radiator. It should be within about 2″ (50.8 mm) of the bottom of the filler neck. On a closed system, the radiator should be full and coolant in the reservoir should be at the correct level for a warm engine. If the level in the reservoir has risen to the warm engine level, the cooling system is properly filled.

STOP **Warning: Do not remove the pressure cap from a closed system to check the coolant level.**

Changing Automatic Transmission/Transaxle Fluid and Filter

Most vehicle manufacturers recommend periodic automatic transmission or transaxle fluid and filter changes. Intervals vary from around 12,000–100,000 miles (19,300–160,000 km), depending on the type of service and transmission being serviced. Some manufacturers do not recommend changing the fluid and filter.

Note: Some manufacturers recommend draining the torque converter and cooler lines. This is to ensure that all the fluid is replaced. Check the service manual.

Some manufacturers recommend changing the fluid and filter more often in vehicles operated under severe conditions, such vehicles used to tow a trailer and vehicles driven in heavy traffic, or vehicles subjected to severe service, such as police cars and taxicabs. Under these conditions, it is better to change the fluid and filter too often than to take a chance on transmission or transaxle damage.

Before draining the fluid, make sure the transmission or transaxle is at normal operating temperature. This will help ensure proper draining. Some transmissions and transaxles require draining both the torque converter and the oil pan. Other transmissions require draining the oil pan only. Since most modern transmission and transaxle oil pans do not contain a drain plug, you must loosen the pan fasteners to permit draining. Usually, removing all but the last four fasteners at the rear of the pan will allow the fluid to drain out into a container with a minimum of spilled fluid. If the fill tube attaches to the side of the pan, remove the tube to permit draining. See **Figure 11-33.**

Be careful when draining. The fluid may be hot enough to cause serious burns. After the fluid has drained, remove the

Figure 11-32. The cooling system bleed valve is usually located in the thermostat housing, as shown here.

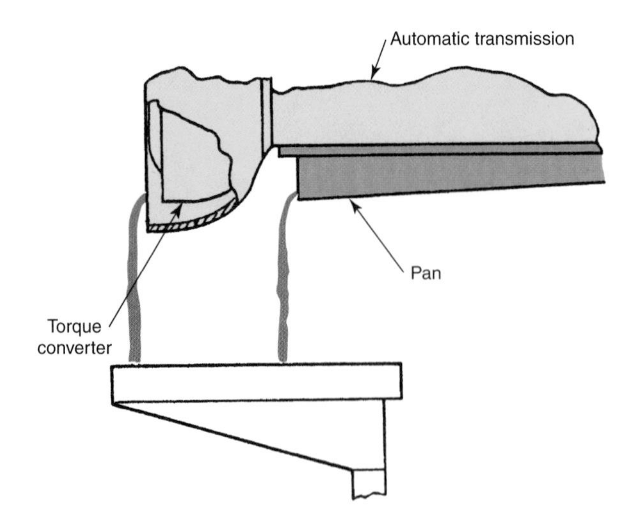

Figure 11-33. Drain both the converter and the transmission pan when required.

oil pan and clean it thoroughly. Remove the old fluid filter and any filter gaskets or seals. Some filters are used to retain check balls, springs, or valves, which can drop out of the valve body when the filter is removed. Note the position of these components and reinstall them when the filter is replaced. Install a new filter using a new gasket or seal, as necessary. Clean the pan and case sealing surfaces thoroughly. Some pans contain a small magnet that aids the filter in removing metal particles from the fluid. Be sure to clean and reinstall the magnet.

Note: If the oil pan contains heavy deposits of sludge and burned material, be sure to inform the owner that the unit will probably need an overhaul soon. Often, the new fluid installed in a transmission or transaxle that is in poor shape will loosen sludge deposits and cause the unit to fail quickly.

If the oil pan must be removed to make band adjustments, make the adjustments before reinstalling the pan. Refer to Chapter 31, Automatic Transmission and Transaxle Service, for more information on band adjustment procedures.

Reinstall the pan using a new gasket or RTV sealant, **Figure 11-34.** Replace the fill pipe, if removed. Add the amount of fluid specified by the manufacturer to the transmission or transaxle. Start the engine and check the fluid level. Add more fluid as needed.

Note: Some transmission pan gaskets can be reused. If a reusable pan gasket is damaged, it can be replaced with a standard cork or rubber gasket.

Caution: Work quickly to add sufficient fluid. The pump or converter can be damaged if they are operated for any period without fluid.

Changing Gearbox Fluid

Most manufacturers do not recommend periodic changes of the fluids used in manual transmissions, rear axles, and transfer cases. Occasionally it becomes necessary to change the oil in one of these units. A common reason is that the original fluid has been contaminated with water. To drain any of these units, raise the vehicle and locate the drain and fill plugs. Some cases do not have a drain plug. These cases can sometimes be drained by removing one of the case-to-extension-housing bolts, **Figure 11-35.** The rear cover must be removed to drain some rear axles. On other vehicles, a suction pump must be used to remove the old fluid.

Once the drain plug has been located, place a pan under the case and remove both the drain and fill plugs. Allow the fluid to drain until only a few drops continue to fall from the opening. If the fluid is contaminated with water, pour a small amount of a petroleum solvent through the fill plug and allow it to drain for at least 30 minutes. This will remove most of the water from the bottom of the case.

After the old fluid has drained, replace the drain plug and add the proper type of fluid through the fill plug. Some

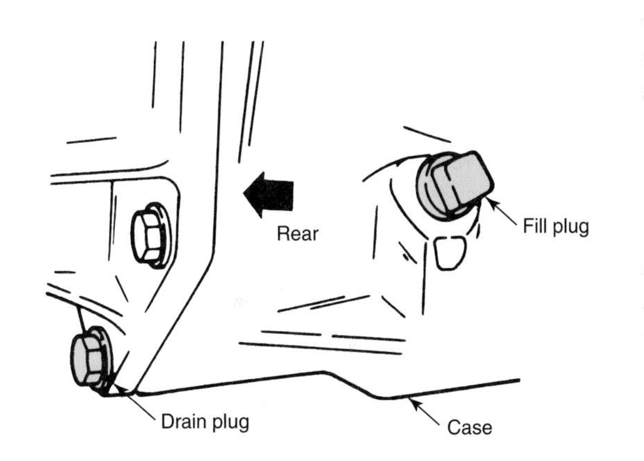

Figure 11-35. Drain and fill plugs for one manual transmission. Note that the drain plug in this transmission is actually the lower case-to-extension-housing bolt. (General Motors)

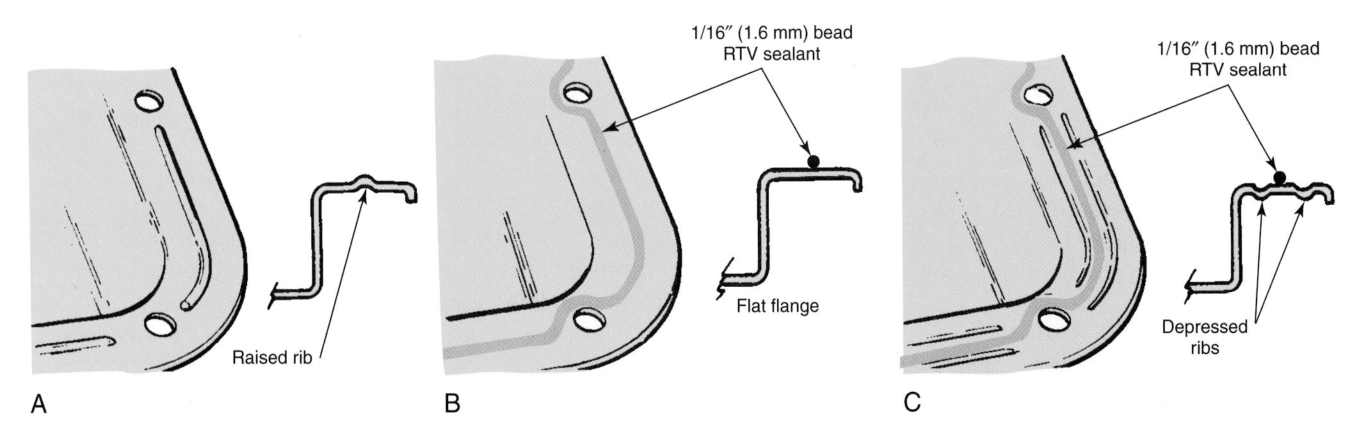

Figure 11-34. Three different transmission pan mounting flange designs. A—Raised center rib. It should only be used with a conventional gasket. B—Flat flange. Use RTV sealer or a gasket. C—Depressed ribs. Use RTV sealer or a gasket. Note the position of the RTV sealer on the inside of the holes to prevent fluid leakage past the fasteners. (General Motors)

manufacturers specify automatic transmission fluid. Transfer cases and limited slip rear axles require a special lubricant that is compatible with the internal friction clutches. Once the fluid has been added, replace the fill plug and lower the vehicle.

Changing Brake Fluid

Some manufacturer's recommend that brake fluid be changed at regular intervals. The easiest way to remove old fluid is to have an assistant press on the brake pedal as you loosen the brake bleeder screw at the right rear wheel. When fluid stops flowing from the bleeder, retighten the bleeder and let up on the brake pedal. Repeat this process two more times at the right rear wheel. Add new fluid to the master cylinder and repeat the process on the left rear, right front, and left front wheels. Be sure to monitor the brake fluid level in the master cylinder and to add fluid after each wheel is bled.

Figure 11-36. Inspect all drive belts for defects. Turn the belt so that the bottom and one side are visible. Then turn the belt the other way and inspect the top and the other side.

General Vehicle Inspection

In addition to checking and changing various fluids, preventive maintenance involves inspecting various vehicle parts and systems. Parts that are usually inspected on a regular basis include drive belts, coolant hoses, tires, brakes, and suspension parts. Inspection procedures for these components are discussed below.

Drive Belt Inspection

Flexible ***drive belts*** drive the water pump, as well as the alternator, power steering pump, air injection pump, and air conditioning compressor. Most belts consist of a woven fiberglass core surrounded by rubber. Many late-model vehicles are equipped with a single belt that drives all accessories. V-belts are still used on some engines. In some designs, several belts are used, each driving a different accessory unit. In other systems, the belts are arranged so the air conditioner and alternator belts will both drive the coolant pump.

To check a belt, grasp it and roll it around so the bottom and one side are clearly visible. See **Figure 11-36.** Look for signs of cracking, oil soaking, glazing, splitting, and fraying, **Figure 11-37.** Replace belts showing these signs and belts that are more than four years old. Modern belts may look OK, but they often fail suddenly from aging. Broken belts can be both inconvenient and dangerous.

Figure 11-37. A—There are pieces of rubber missing from this serpentine belt. B—This V-belt is cracked and frayed. Both should be replaced.

While checking V-belts, also check belt tension. Ideally, this check should be made with a ***belt tension gauge.*** See **Figure 11-38.** If a tension gauge is not available, make sure the belt does not deflect (bend) more than 1/4″ (0.75 mm) under light hand pressure. Serpentine belts use a self-tensioning device called an automatic belt tensioner and, therefore, there is no specification for adjustment.

Coolant Hose Inspection

The ***coolant hoses*** have a simple job: they direct water between the engine, radiator, and heater core. Nevertheless, a hose failure can be catastrophic. Some hose damage may not be evident to the naked eye. Hoses are weakened over time by ***electrochemical degradation (ECD),*** which is a reaction between the chemicals in the coolant and the metals in the engine and radiator. Often, this reaction can cause hose defects. See **Figure 11-39.** ECD has the greatest effect on heater hoses, bypass hoses, and the upper radiator hose. Because of this, coolant hoses should be changed if they are more than three years old.

Visually check all hoses for signs of deterioration. Squeeze each hose. A hose should not be hard and brittle or soft and swollen. See **Figure 11-40** and **11-41.** Bend heater hoses to check for surface cracking, abnormal swelling, and hardness. If any show signs of deterioration, they should be replaced.

Pay careful attention to the bottom radiator hose. It is subject to pump suction whenever the engine is running. If the bottom hose is soft, it will collapse and cut off the coolant circulation. If the bottom hose is loose or cracked, it can admit air into the system. ***Aeration*** (air bubbles in the system) can cause rust to form faster than normal.

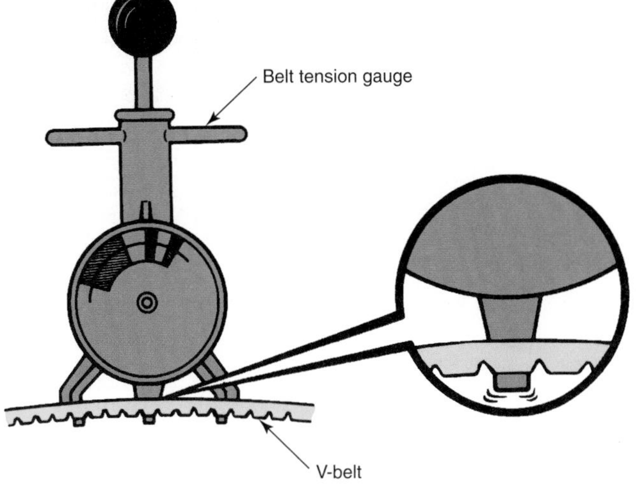

Figure 11-38. Using a belt tension tester to check a V-belt for proper tension. There will sometimes be two specified tensions: one for new belts and one for used belts. (Toyota)

Note: If there is the slightest doubt as to the condition of a belt or hose, replace it.

Checking Tire Condition

Tire condition is critical to safety and proper vehicle operation. Always check tire pressure when performing any kind of maintenance. Tire pressures are listed on the driver's side door frame. Also check tires for excessive or unusual wear patterns, **Figure 11-42.** Also check for ***tread separation,*** a condition in which the tread separates from the tire body. Tread separation will often appear as a raised spot on the tire tread. Check the sidewalls for damage or cracking, and check

Figure 11-39. Hoses are subject to electrochemical degradation, which causes them to deteriorate from the inside. (Gates)

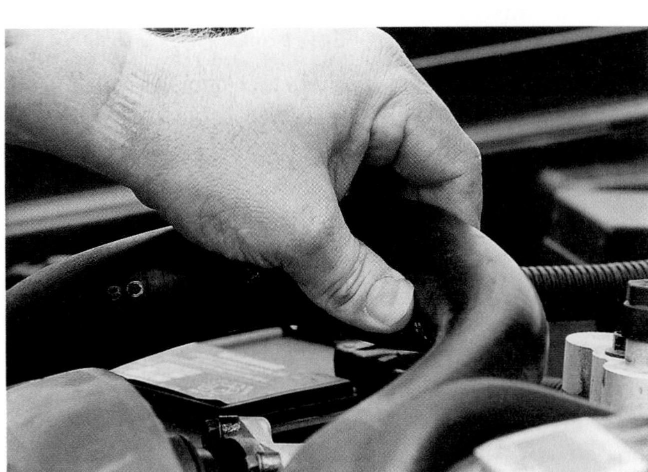

Figure 11-40. Squeeze hoses to check for hardness or for swelling and softness. Allow the engine to cool before making this test.

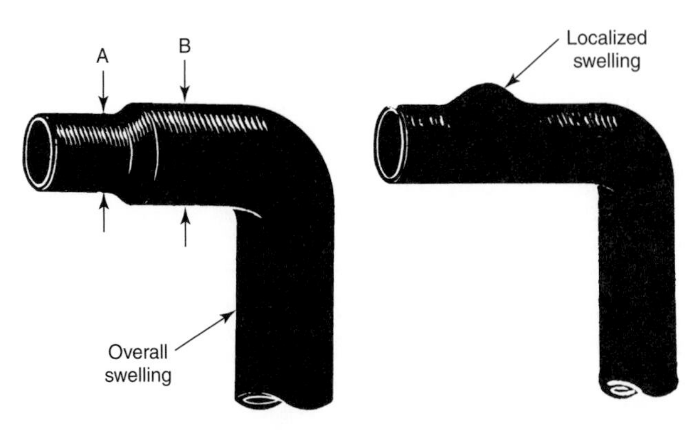

Figure 11-41. Overall and localized swelling of a radiator hose. Note dimensions. A—Normal diameter. B—Swollen diameter. A hose with any abnormal swelling should be replaced. (Toyota)

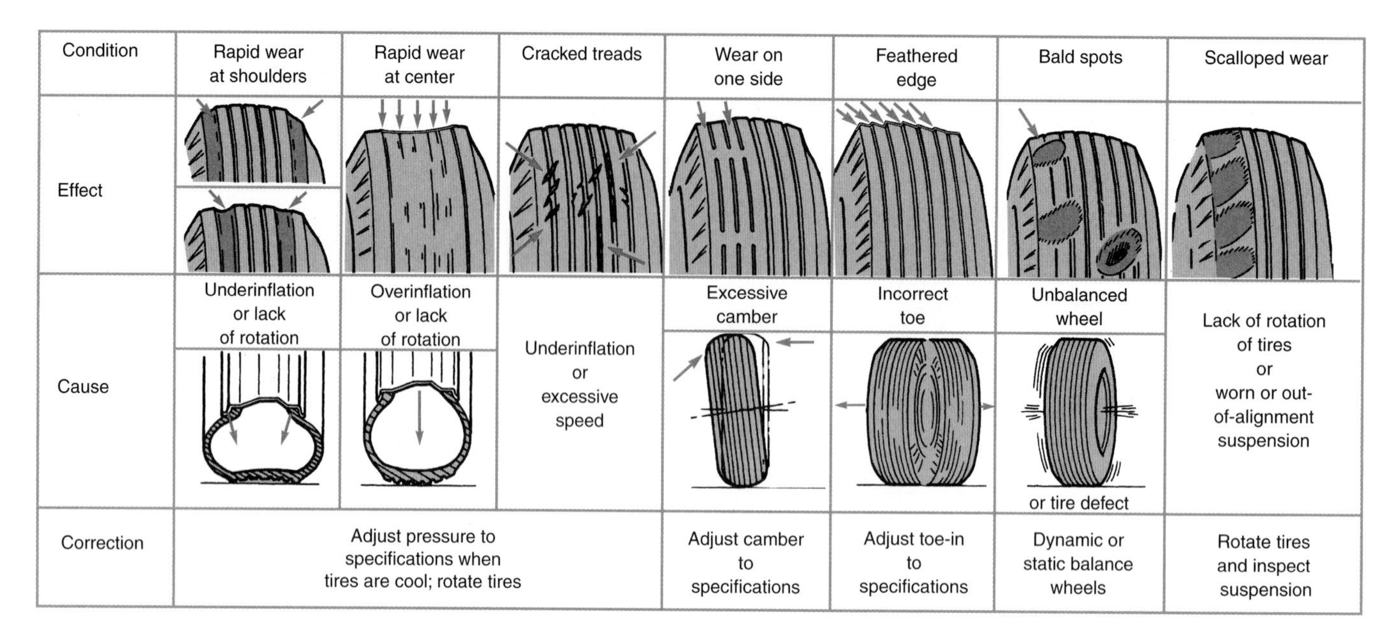

Figure 11-42. Tires showing various wear patterns. (DaimlerChrysler)

the rims to ensure that they are not damaged at the tire sealing edge. Also check the tire valves for cracked rubber and missing caps.

Tire Rotation

Most tire manufacturers call for rotating tires at certain intervals. A common rotation interval is every 5,000 miles (8,000 km). ***Rotation*** helps the tires to wear evenly, increasing their life. Most modern radial tires are rotated between front and back on the same side. A few manufacturers call for rotating tires in a cross pattern. See **Figure 11-43.** Removing the tires to rotate them also allows the technician to inspect the brake system.

To rotate tires, safely raise the vehicle on a lift or jack stands so the wheels are free to turn. Remove the wheel covers and wheel nuts. If any wheel studs are broken or stripped, replace them. Move the tires to the proper axles. Reinstall the wheel nuts and torque them to the proper values. Reinstall the wheel covers and lower the vehicle.

Note: *Directional tires* are designed to rotate in one direction only and, therefore, cannot be rotated in a cross pattern. Directional tires are always identified as such on their sidewalls.

Some maintenance procedures call for balancing the tires when they are rotated. Refer to Chapter 36 for tire balancing procedures.

Inspecting Brakes

Brakes should be inspected every 10,000 miles (16,000 km). This ensures that worn pads and shoes can be replaced before they damage the rotors or drums.

To check disc brakes, remove the tires and observe the condition of the pads and rotors. In most cases, the pads can be observed through openings in the caliper, **Figure 11-44.** If

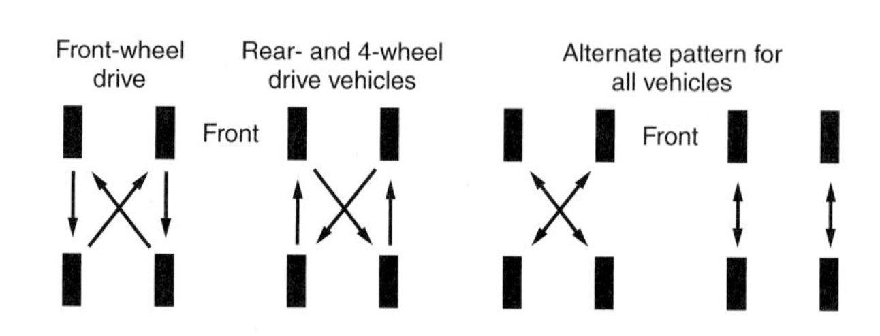

Figure 11-43. Common tire rotation patterns. (Yokohama Tire)

Figure 11-44. To inspect brake pads, remove the tire and wheel, and check the pad thickness through the inspection openings in the caliper.

the pads are worn beyond their minimum thickness, they should be replaced. Always consult the necessary service literature before deciding that the pads should be replaced, since minimum thickness varies from one vehicle type to another.

Carefully inspect the rotor braking surface. A normal rotor will have a shiny braking surface. Also check the rotors for scoring (deep groves in the rotor surface) and signs of overheating (discolored spots in the rotor braking surface). If you find any scoring or overheating, the rotor should be either turned or replaced.

To check drum brakes, remove the tires and the brake drum. Observe the condition of the shoes, **Figure 11-45.** Most shoes should be replaced when the lining thickness is less than 0.06″ (1.6 mm) above the shoe assembly or 0.03″ (0.8 mm) above the rivet heads (if the lining is fastened with rivets). Check the linings for cracks or evidence of oil soaking. If oil is present in the drum and shoes assembly, check for leaking axle seals. Also note whether the wheel cylinders show any signs of leaking. Leaking wheel cylinders will have an oily appearance, and the backing plate may also be oily. Leaking wheel cylinders must be rebuilt or replaced immediately. Inspect the drums for scoring and overheating. Scored or overheated drums should be turned or replaced.

Inspecting Suspension Parts

To check the suspension, the ball joints must be ***unloaded*** (have their spring tension removed). To unload the ball joints place a jack under either the lower control arm or the frame, depending on the suspension type. Then raise the jack until the wheel is off the ground. Chapter 36, Suspension System Service, contains more information about suspension identification and checking. Once the ball joints are unloaded, attempt to move the top of each front wheel in and out. See **Figure 11-46.** The wheels should move very little, less than 1/16″ (0.15 mm). If the wheel moves excessively, either a ball joint is loose or the wheel bearings are worn. Make a visual check for damaged or missing control arm and strut rod bushings, bent parts, loose fasteners, and torn ball joint boots.

Figure 11-45. In drum brake systems, the wheel and brake drum must be removed to check the condition and thickness of the brake shoes.

Inspecting Steering Parts

To check the steering linkage, attempt to move a front wheel from side to side with the wheels off the ground, **Figure 11-47.** The wheels should move slightly, causing the steering wheel to move. If the wheels move very easily or if wheel movement is not transferred to the steering wheel, a steering linkage part is loose. Repeat the test on the other wheel. Also visually check the steering linkage for torn or missing steering socket boots, bent linkage, damaged or missing bushings, and loose fasteners. Refer to Chapter 37,

Figure 11-46. To check suspension parts, unload the ball joints and try to move the top of each front wheel in and out. If movement is excessive, look for damaged or worn suspension parts.

Figure 11-47. To check the condition of the steering linkage, move the wheel from side to side. If movement is excessive or if it is not transferred to the steering wheel, check for worn or damaged steering linkage.

Steering System Service, for information about making further steering system checks.

Battery Maintenance

Typical battery maintenance includes cleaning the cable terminals, battery posts, and battery top, as well as tightening the battery hold-down. Other maintenance procedures, such as charging or replacing the battery, will be covered in Chapter 14, Charging and Starting Systems Service.

The battery hold-down should be tight enough to hold the battery securely, but not so tight that it places excessive pressure on the case. Undue hold-down pressure can cause case failure.

Cleaning Battery Terminals

A battery produces ***hydrogen gas*** as part of its normal operation. This hydrogen gas is highly corrosive. Eventually, the hydrogen attacks the terminals, causing corrosion. Unless the corrosion is removed, the terminals will be eaten away. Corrosion also causes high resistance between the battery terminals and battery posts, making electrical flow difficult. A dirty battery top will attract electrolyte and become conductive, allowing a small but steady flow of current from one post to the other. Such a condition will cause slow discharge of the battery.

Note: Before removing the battery terminals, install a *memory saver* in the cigarette lighter socket. The memory saver provides enough voltage to the electrical system to prevent loss of the radio station settings and computer memory information.

Use a wire brush to clean away the bulk of the corrosion. Remove the battery cable terminals. Plug the cap vent holes (if used) with toothpicks or masking tape. Brush a solution of baking soda and water over the terminals, posts, and battery top. Do not allow the baking soda mixture to enter the battery cells, **Figure 11-48.**

Let the solution stand until the foaming action stops. Apply fresh solution to areas as needed. Then, rinse the battery thoroughly with clean water. Wipe the posts and terminals dry. Brighten the posts and the inside of the terminals with sandpaper or a steel brush. Install the terminals and tighten them securely. If used, install the protective terminal boot. Finally, remove the plugs from the cap vents. Coat terminals, bolts, and nuts with grease to protect against corrosion. Several efficient spray products that prevent corrosion are also available.

Other Maintenance Procedures

Many other preventive maintenance procedures should be performed as needed. These tasks will vary, depending on the manufacturer's recommendations and the condition of the system being serviced. **Figure 11-49** identifies some common maintenance tasks and the chapters in which they are discussed.

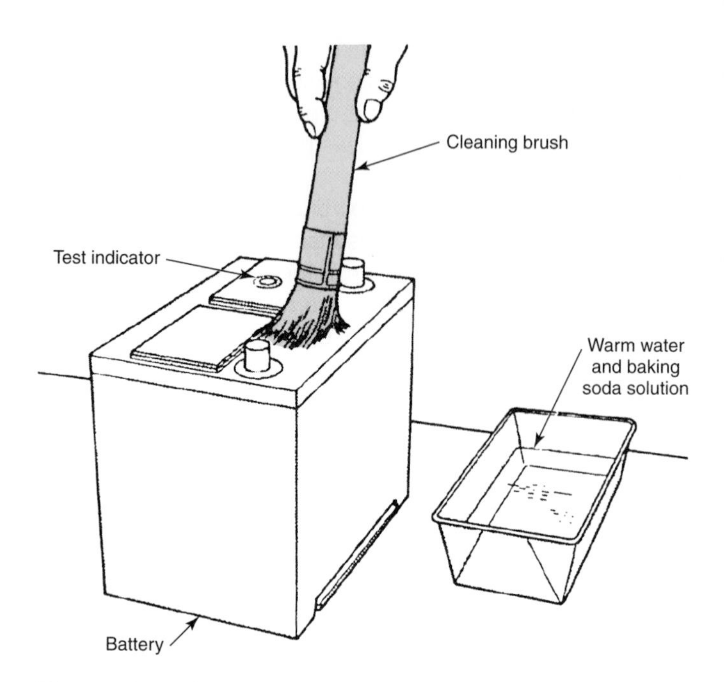

Figure 11-48. A solution of baking soda and water can be used to clean corrosion from a battery. Do not allow the solution to enter the battery cells. (DaimlerChrysler)

Maintenance Task	Chapter
Check cooling system	23
Check exhaust system condition	19
Check emission control device operation	22
Check battery charge and electrolyte level	14
Check clutch adjustment	29
Change automatic transmission fluid	31
Check brake operation and brake fluid level. Inspect brake system for leaks	34
Repack wheel bearings	10, 38
Check tire pressure and condition	38
Check air conditioner/heater operation	40

Figure 11-49. Some common maintenance tasks and the chapters in which they are discussed.

Tech Talk

It is almost impossible to overemphasize the importance of proper vehicle maintenance. However, some technicians develop pet maintenance procedures and services that may not really accomplish anything. While these pet procedures will not cause harm, they may lead the technician to forget about other necessary maintenance procedures. There is the story of the technician who put the vehicle on the lift, drained the oil, and got so interested in cleaning out the body drain holes that he forgot to put the oil drain plug back in.

The only way to avoid this sort of problem is to develop a set maintenance routine, beginning with consulting the factory service manual for the exact maintenance procedures needed. Then follow the procedures exactly. Obviously, the engine oil will be changed, and other fluid levels checked. The other procedures should be performed in order so that nothing is overlooked.

This does not mean that you ignore other, obvious maintenance needs. What it means is that you must always perform the required factory service procedures as a bare minimum. After these procedures are performed, you can investigate the need for other vehicle maintenance services.

Summary

The correct grade and viscosity of engine oil must be supplied in ample quantities to all moving engine surfaces. Oil filtration systems can be full flow or partial flow types.

Filter elements are of the surface type or the depth type. Filters should be changed before they become clogged. Replacement intervals vary with engine wear, vehicle use, and driving conditions.

Use the recommended oil for all late-model engines. It will provide the type of lubrication needed by the engine. The use of quality oil of the correct viscosity and changing it at regular intervals is a must.

Replacement filters must be the correct size and type. Spin-on filters must be replaced at recommended intervals. After removing the old filter, clean the filter mounting base. Apply clean oil to the new filter seal, fill the filter with oil (if recommended), and install the filter. Tighten the filter to the manufacturer's specifications. Fill the crankcase with the correct type and amount of oil. Start the engine and check for oil leakage.

At each oil change, PCV system operation should be checked. An inoperative PCV system can cause sludging, rapid wear, poor mileage, and stalling. Lubricate the front-end components if specified. Check fluid levels and grease the driveline, parking brake cable, and body hinges where indicated.

Fuel and air filters are often neglected and should be checked whenever a vehicle is brought in for other maintenance. Vehicle fluid levels should be checked during any oil change. Levels in the transmission or transaxle and transfer case should always be checked. Other fluid reservoirs to be checked are those for the coolant, power steering, brakes, and windshield washer.

Tire air pressure and tread condition should be checked. Tires should be rotated as necessary to increase tread life. The brake system should be inspected to ensure that the pads or shoes are not worn out and that no other defects are present. Suspension and steering parts should also be checked for looseness and other defects. This is easily performed by shaking the tires either side to side or at the top and bottom.

A common preventive maintenance procedure is to replace the engine coolant. Open the radiator drain (and engine block drains when used) and allow the old coolant to drain. If the cooling system is very dirty, it should be flushed out before the new coolant is installed. Tighten the drains and install the proper mixture of antifreeze and water. If the radiator is lower than the engine, bleed air from the system using the bleed fitting on the engine.

Another common maintenance procedure is to change the fluid in an automatic transmission or transaxle. Begin by raising the vehicle and draining the old fluid. Then remove the oil pan and filter. Clean the oil pan and install a new filter. Install the pan using a new gasket. Add the proper type of fluid and recheck the level with the engine idling.

The fluid in manual transmissions, rear axles, and transfer cases can be replaced by draining the old fluid by removing the drain plug and adding new fluid through the fill plug.

Batteries can be cleaned with a mixture of baking soda and water. Baking soda will also clean the terminals and battery clamps. Cleaned parts should be coated with grease to prevent future corrosion. Before replacing a battery, install a memory saver in the cigarette lighter to prevent loss of the ECM and radio settings.

Review Questions—Chapter 11

Do not write in this book. Write your answers on a separate sheet of paper.

1. Oil viscosity refers to the oil's _____.
   (A) color
   (B) flow rate, or resistance to flow
   (C) quality
   (D) number of additives
2. When discussing oil classification, what does the letters S and C denote?
3. Sludge on the top of the dipstick could indicate:
   (A) short trip driving.
   (B) engine overheating.
   (C) water in the oil.
   (D) lack of oil changes.
4. Oil should be drained when the engine is _____.
5. A spin-on filter should be tightened about _____ turn(s) after contacting the base.
   (A) 2
   (B) 1 1/2
   (C) 1/2
   (D) 3
6. A new filter usually requires about _____ additional quart(s) of oil.
   (A) 1
   (B) 2
   (C) 3
   (D) 5
7. Many fuel filters are installed under a vehicle's _____.

8. Which of the following is a method of cleaning a paper air cleaner element?
   (A) Blow it out with compressed air.
   (B) Wash and blow dry.
   (C) Wash and squeeze dry.
   (D) Shine a light through the element.
9. Name five steering and suspension parts that can be equipped with grease fittings.
10. Door hinges should be lubricated with what type of lubricant?
11. What type of lubricant should be used on lock cylinders?
12. Parking brakes usually stick in the _____ position.
13. Name 8 vehicle units that require a fluid level check during an oil change.
14. You are checking the fluid level in a manual transmission. The fluid should just touch the bottom of the _____.
15. Do not mix conventional antifreeze and _____ _____ antifreeze.
16. To change a transmission filter, the technician must first remove the transmission _____ _____.
17. Belts drive all of the following *except:*
    (A) water pump.
    (B) power steering pump.
    (C) oil pump.
    (D) air injection pump.
18. List five common belt defects.
19. The bottom radiator hose can _____.
    (A) collapse
    (B) leak
    (C) let air into the cooling system
    (D) All of the above.
20. Directional tires should not be rotated in a _____ pattern.

ASE-Type Questions

1. All of the following statements about API service grades for engine oil are true *except:*
   (A) there are now eight oil grades for gasoline engines.
   (B) the "C" in the service grade means the oil can be used in a diesel engine.
   (C) SH oil should always be used in place of non-detergent (SA or SB) oil.
   (D) SL oil is a detergent oil.
2. All the following statements about front-end lubrication are true *except:*
   (A) some suspension parts have plugs instead of grease fittings.
   (B) pressure should be applied slowly to avoid damaging seals.
   (C) if plugs are installed, the part cannot be greased.
   (D) if the joint has a sealed boot, grease until the boot begins to bulge slightly.
3. Technician A says grease should be added to the universal joints until it seeps out from the seals. Technician B says universal joints can use the same grease as other suspension parts. Who is right?
   (A) A only.
   (B) B only.
   (C) Both A and B.
   (D) Neither A nor B.
4. All of the following can be used to lubricate door hinges *except:*
   (A) white lithium grease.
   (B) lightweight oil.
   (C) wheel bearing grease.
   (D) dry graphite.
5. Which of the following maintenance services should be performed whenever the engine oil and filter are changed?
   (A) The transmission filter should be changed.
   (B) Engine coolant should be changed.
   (C) The parking brake should be lubricated.
   (D) The fuel filter should be cleaned.
6. Technician A says the parking brake cables should be lubricated with a combination or lubricant and penetrating oil. Technician B says the parking brake should be actuated a few times after lubrication. Who is right?
   (A) A only.
   (B) B only.
   (C) Both A and B.
   (D) Neither A nor B.
7. Technician A says a low oil level in a transfer case indicates a leak. Technician B says very dark oil indicates that water has entered the transfer case. Who is right?
   (A) A only.
   (B) B only.
   (C) Both A and B.
   (D) Neither A nor B.
8. What type of fluid is used in most modern power steering systems?
   (A) Type A transmission fluid
   (B) Dextron® transmission fluid
   (C) Special power steering fluid
   (D) Brake fluid

9. Brake fluid will absorb _____ from the air if the cover is removed for long periods.
   (A) dust
   (B) water
   (C) nitrogen
   (D) argon
10. Technician A says any coolant hose that shows signs of deterioration should be replaced. Technician B says engine coolant should be replaced only when the engine is cold. Who is right?
    (A) A only.
    (B) B only.
    (C) Both A and B.
    (D) Neither A nor B.
11. All the following statements about changing automatic transmission and transaxle oil and filters are true *except:*
    (A) most modern oil pans do not have a drain plug.
    (B) the filter may be held to the valve body with screws.
    (C) some pans contain a magnet.
    (D) the fluid never becomes hot enough to cause injury.
12. A brake rotor must be serviced if it shows any of the following conditions *except:*
    (A) scoring.
    (B) a shiny surface.
    (C) evidence of heat damage.
    (D) discoloration on the braking surface.
13. Which of the following is the *best* definition of ball joint unloading?
    (A) Removing the ball joint from the vehicle.
    (B) Compressing the ball joint to measure looseness.
    (C) Compressing the ball joint to measure tightness.
    (D) Removing spring tension from the ball joint.
14. All of the following steering parts can be checked visually *except:*
    (A) linkage.
    (B) steering sockets.
    (C) bushings.
    (D) fasteners.
15. Technician A says that the battery surface can conduct electricity if it is dirty. Technician B says that baking soda should not be allowed to enter the battery cells. Who is right?
    (A) A only.
    (B) B only.
    (C) Both A and B.
    (D) Neither A nor B.

Suggested Activities

1. Write a short report listing the consequences of poor maintenance on a modern car or truck.
2. Check the fluid levels in a vehicle's engine, cooling system, transmission, brake master cylinder, power steering reservoir, windshield washer reservoir, and rear differential (if applicable).
3. Perform preventive maintenance on a vehicle, including changing the engine oil and filter, greasing the suspension, and checking fluid levels in the radiator, transmission, master cylinder, and power steering reservoir. Also check the air filter, tire pressure, exterior lights, and other vehicle systems.

A multimeter, such as the one shown here, can be used to diagnose many wiring problems in late-model vehicles. (OTC)

12

Wire and Wiring

After studying this chapter, you will be able to:

- Identify different types of automotive wiring.
- Select the correct type of wiring for the job.
- Make basic wiring repairs.
- Read wiring diagrams.
- Perform basic circuit tests.

Technical Terms

Conductor	Acid core
Insulation	Rosin core
AWG (American Wire Gage)	Resin core
Resistance	Soldering gun
Amperes	Cold solder joint
Watts	Heat-shrink tubing
Candela	Secondary wire
Wiring harness	Corona
Color coded	Resistance wire
Wiring diagram	Printed circuit
Terminals	Voltage drop
Plug-in connectors	Short circuits
Rubber grommets	Near-short circuits
Inline fuse	Ohmmeter
Fusible link	Voltmeter
Circuit breaker	Ammeter
Crimping	Test light
Soldering	Continuity
Crimping tool	

New wiring, when properly installed, is relatively trouble free. As the vehicle ages, the wires and connectors begin to deteriorate from exposure to heat, oil, gas, fumes, acid, moisture, dirt, salt, and vibration. Vehicles damaged by collision or fire often require extensive rewiring. All wiring used on vehicles can be divided into two major classes: primary wire and secondary wire. The use of printed circuits in modern vehicles is increasing and is replacing wire in many systems. A well-trained technician should be familiar with the various types of wire, wire sizes, and insulation. The technician should also be familiar with the various wire terminals, connections, and general installation procedures.

Primary Wire

Primary wire handles battery voltage. On modern vehicles it is usually 12 volts direct current. Older vehicles have 6-volt systems, some tractors and other agricultural machines have 8-volt systems, and some commercial vehicles have 24-volt systems. In the near future, some manufacturers will introduce 42-volt systems. The primary wiring has sufficient insulation to prevent current loss at these voltages. All wiring circuits in the vehicle, with the exception of the ignition high-voltage circuit, are primary wires.

Primary Wire Construction

Primary wiring uses a stranded ***conductor*** made of soft copper. It is an excellent conductor, bends easily, and solders readily. Aluminum also is used to some extent. Stranded wires are made up of a number of small wires twisted together, instead of a single solid wire.

Most modern automotive wire ***insulation*** is made of plastic. Plastic is highly resistant to heat, cold, fumes, and aging. It strips (peels off) easily and offers excellent dielectric (nonconducting) properties. Primary wire insulation is designed to resist engine heat. The insulation on secondary wiring is much thicker than primary wire because of the high voltages used in the secondary system, **Figure 12-1.**

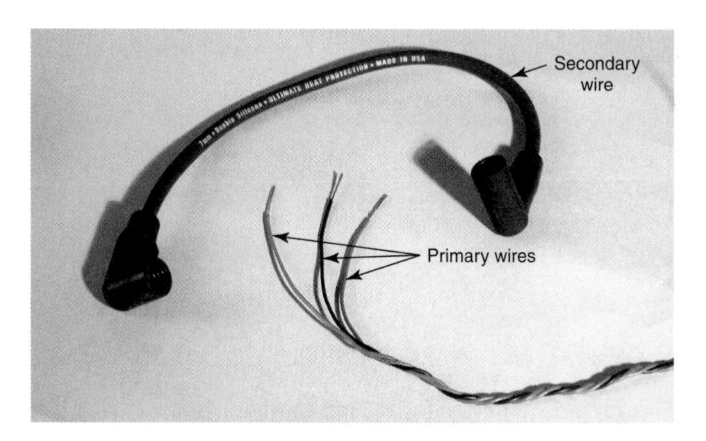

Figure 12-1. Compare primary and secondary wire. More insulation is required on secondary wires due to the high voltage that flows through these conductors.

Wire Size

Every wire is assigned a number. The larger the number, the smaller the wire. This number applies to the size of the actual conductor and does not measure the thickness of the wire insulation. The ***AWG (American Wire Gage)*** is the commonly used standard for wire size. To find the gage of a solid wire, simply measure it with a micrometer. Then locate the measurement, or nearest size, on a wire gage chart. **Figure 12-2** shows a portion of an AWG chart. **Figure 12-3** shows a metric wire size conversion chart.

To find the gage of a stranded conductor, count the number of strands. With a micrometer, measure the diameter of one strand. Square this answer and multiply by the number of strands. This will give you the cross-sectional area of the conductor in circular mils, Figure 12-2. Locate this number (or the nearest one) on the chart. Move horizontally across the chart to the wire gage column to determine the gage. Special steel gauges are also available for quickly checking wire gage.

To avoid blown fuses, overheating, and possible fires, it is vital the technician use a wire that can handle the current to be carried. Electrical load, line voltage, and wire length are the three important factors in determining correct wire gage, or size.

Effect of Wire Size and Length

The electrical load on a wire is the sum of the individual loads of each unit serviced by that wire. As wire size increases, ***resistance*** decreases. Also, as wire length increases, resistance and the resulting voltage drop increases. Resistance causes the conductor to heat. Excessive resistance can cause a wire to heat to the point where the insulation will melt and burn. To prevent high resistance and voltage drop, wire size must be increased as length is increased. With a given voltage and load, a 20 ft. long wire must use larger diameter conductor than a 2 ft. long wire.

Selecting Correct Gage Wire

Most wire manufacturers furnish charts to assist the technician in proper gage selection, **Figure 12-4.** To use the chart shown, determine the total length of the wire needed. The wire lengths shown in the chart are for a single wire ground return. This means that a separate ground wire to the vehicle's battery is not needed, since the vehicle's frame or metal parts act as a return ground wire. If you are installing a two-wire circuit (one wire to a unit and another from the unit to ground), count the length of both wires.

Next compute the total electrical load of the circuit. Be certain to figure the load of all units in the circuit. If the load will fluctuate, use the peak load figure. The load may be figured in ***amperes, watts,*** or ***candela*** (the international term for candlepower). When the load is determined, look on the chart under the appropriate voltage column. Select the listed load that most closely matches your needs. Move across the row of numbers horizontally to the column that most closely matches the length of wire you need. The number listed at the intersection of the load row and the length column is the recommended gage.

Looking on the chart in Figure 12-4, if the vehicle has a 12-volt system, a computed electrical load of 20 amperes, and a wire length of 15 ft., you will find the recommended gage to be No. 14. For the same load and length in a 6-volt system, the recommended gage is No. 10. A 12-volt system uses a smaller gage wire than the 6-volt system used on very old cars. Using a larger gage than necessary will cause no particular harm unless the wire being replaced was originally designed to produce a specific resistance in the circuit.

Wiring Harness

A ***wiring harness*** is made up of various sections of system wiring with common wires (located in same area) pulled through a loom (soft woven or plastic insulation tube),

American Wire Gage	Wire diameter in inches	Cross-sectional area in circular mils
0000	.4600	211600
000	.40964	167800
00	.3648	133100
0	.32486	105500
1	.2893	83690
2	.25763	66370
3	.22942	52640
4	.20431	41740
5	.18194	33102
6	.16202	26250
8	.12849	16510
10	.10189	10380
12	.080808	6530
14	.064084	4107
16	.05082	2583
18	.040303	1624
20	.031961	1022
22	.025347	642.4
24	.0201	404.0
26	.01594	254.1
28	.012641	159.8
30	.010025	100.5

Figure 12-2. American Wire Gage chart (not all sizes are shown). A technician can use this chart to find the proper gage of a wire if the wire diameter or the cross-sectional area is known.

Metric wire sizes (mm²)	AWG sizes (American Wire Gage)
.22	24
.35	22
.5	20
.8	18
1.0	16
2.0	14
3.0	12
5.0	10
8.0	8
13.0	6
19.0	4
32.0	2

Figure 12-3. Note the relationship between AWG and metric wire sizes. (General Motors)

Total approx. circuit amperes	Total circuit watts	Total candle power	Wire gage (for length in feet)											
12V	12V	12V	3′	5′	7′	10′	15′	20′	25′	30′	40′	50′	75′	100′
1.0	12	6	18	18	18	18	18	18	18	18	18	18	18	18
1.5		10	18	18	18	18	18	18	18	18	18	18	18	18
2	24	16	18	18	18	18	18	18	18	18	18	18	16	16
3		24	18	18	18	18	18	18	18	18	18	18	14	14
4	48	30	18	18	18	18	18	18	18	18	16	16	12	12
5		40	18	18	18	18	18	18	18	18	16	14	12	12
6	72	50	18	18	18	18	18	18	16	16	16	14	12	10
7		60	18	18	18	18	18	18	16	16	14	14	10	10
8	96	70	18	18	18	18	18	16	16	16	14	12	10	10
10	120	80	18	18	18	18	16	16	16	14	12	12	10	10
11		90	18	18	18	18	16	16	14	14	12	12	10	8
12	144	100	18	18	18	18	16	16	14	14	12	12	10	8
15		120	18	18	18	18	14	14	12	12	12	10	8	8
18	216	140	18	18	16	16	14	14	12	12	10	10	8	8
20	240	160	18	18	16	16	14	12	10	10	10	10	8	6
22	264	180	18	18	16	16	12	12	10	10	10	8	6	6
24	288	200	18	18	16	16	12	12	10	10	10	8	6	6
30			18	16	16	14	10	10	10	10	10	6	4	4
40			18	16	14	12	10	10	8	8	6	6	4	2
50			16	14	12	12	10	10	8	8	6	6	2	2
100			12	12	10	10	6	6	4	4	4	2	1	1/0
150			10	10	8	8	4	4	2	2	2	1	2/0	2/0
200			10	8	8	6	4	4	2	2	1	1/0	4/0	4/0

Figure 12-4. Wire gage selection chart. Wire lengths shown are for a single-wire ground return. (Belden Mfg. Co.)

taped, or tied together. This speeds installation, makes a neat package, and secures the wire with a greatly reduced number of clamps or clips. **Figure 12-5** shows a typical wiring harness.

Color Coding

All automotive wiring is ***color coded*** (each circuit is given a specific color or set of colors) to assist the technician in tracing various circuits. Manufacturers publish wiring diagrams that show all wires and their colors. Wires can have a solid color or can be striped or banded. This makes it easier to identify and trace a wire.

After aging or exposure to dirt and oil, some wires are difficult to identify by color. In this case, trace the wire back to where it enters the harness. Then, cut away a small portion of the harness covering. This will expose a clean portion of the wire, so the color may be determined.

Wiring Diagrams

A ***wiring diagram*** is a drawing showing electrical units and the wires connecting them. Most diagrams also list the wire colors and terminal designations. Such a diagram is helpful when working on the wiring system. Wiring diagrams are available in various shop manuals and in some automotive reference books. **Figure 12-6** shows a typical wiring diagram.

Since the auto electrical system is becoming more complicated each year, many manufacturers break down the circuits into separate diagrams. This allows the technician to easily see the circuit's purpose and trace wiring specific to that circuit, **Figure 12-7.**

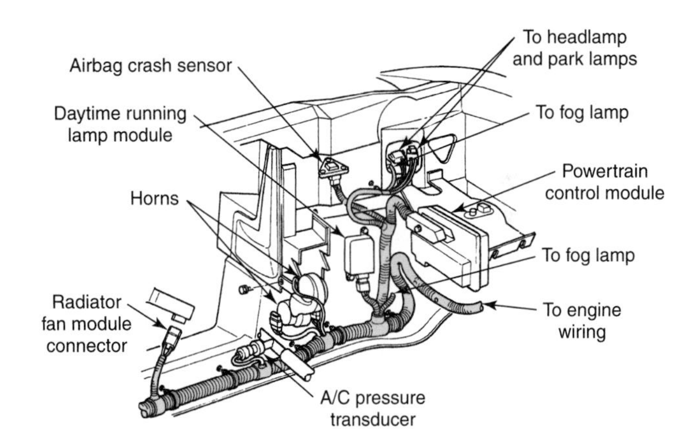

Figure 12-5. One type of automotive wiring harness. Note the use of looms and clips to secure the harness and fuse block. Also, note the plug-in type connectors.

Electrical Wiring Symbols

There is a wide variation in the use of automotive electrical symbols. Some companies use their own drawings for some system components and standard symbols for others. The component's basic internal circuit is sometimes shown. In other diagrams, symbols are used for all components. **Figure 12-8** illustrates a number of typical symbols widely used in automotive electrical diagrams.

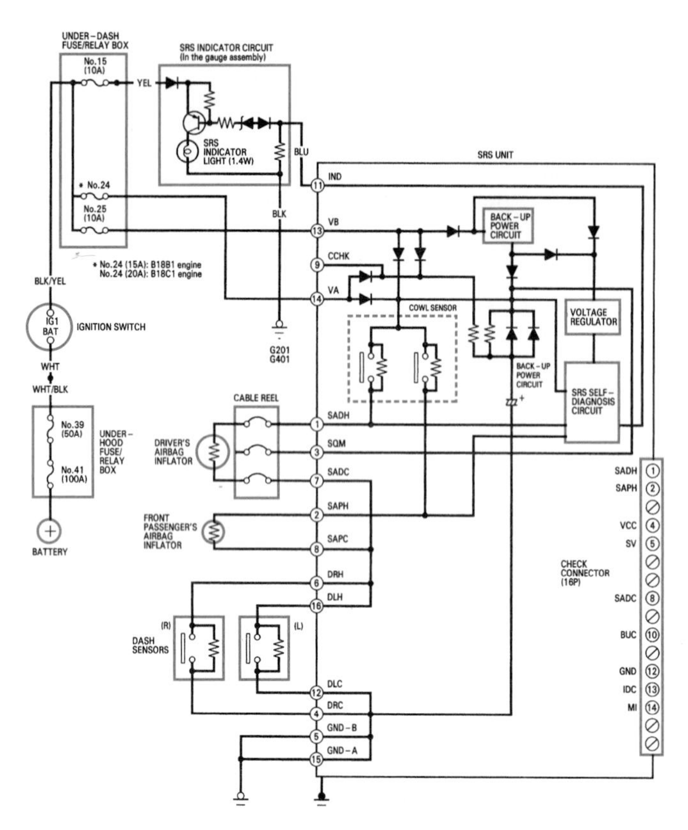

Figure 12-6. Note how each component is explained in an electrical service manual. Failure to use and understand these wiring diagrams can cause considerable electrical damage. (Honda)

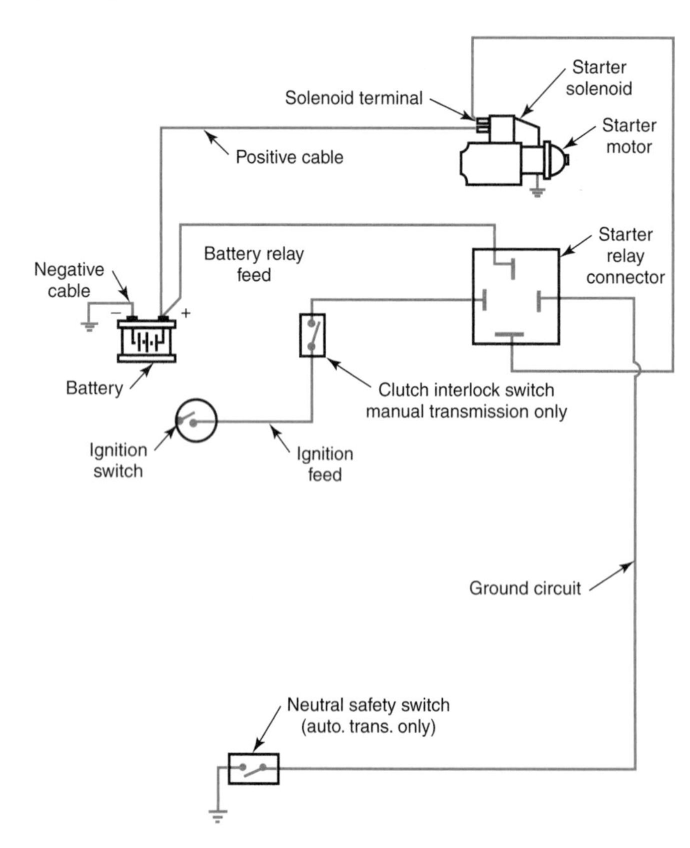

Figure 12-7. Wiring diagram for starter system. (DaimlerChrysler)

Symbol	Meaning	Symbol	Meaning
+	Positive		Zener diode
–	Negative		Motor
	Ground	C100	Connector identification
	Fuse		Male connector
	Circuit breaker		Female connector
	Capacitor		Denotes wire continues elsewhere
	Resistor		Wire goes to one of two circuits
	Rheostat	S100	Splice identification
	Coil		Thermal element
	Step up coil		Timer
	Open contact		Multiple connector
	Closed contact		Optional Wiring with Wiring without
	Closed switch	88:88	Digital readout
	Open switch		Singe filament lamp
	Two-pole single-throw switch		Dual filament lamp
	Pressure switch		L.E.D.–Light emitting diode
	Solenoid switch		Sensor
	Diode or rectifier		Fuel injector

Figure 12-8. Study this chart of electrical symbols commonly used in automotive wiring diagrams. Learn these symbols; this is one of the marks of a well-trained technician. (DaimlerChrysler)

Primary Wire Terminals

Primary wire end ***terminals*** (connecting devices) are available in various shapes and sizes. Primary terminals may be classified as spade, lug, flag, roll, slide, blade, ring, and bullet types. They may either be solderable or solderless and are generally made of tin-plated copper. See **Figure 12-9.**

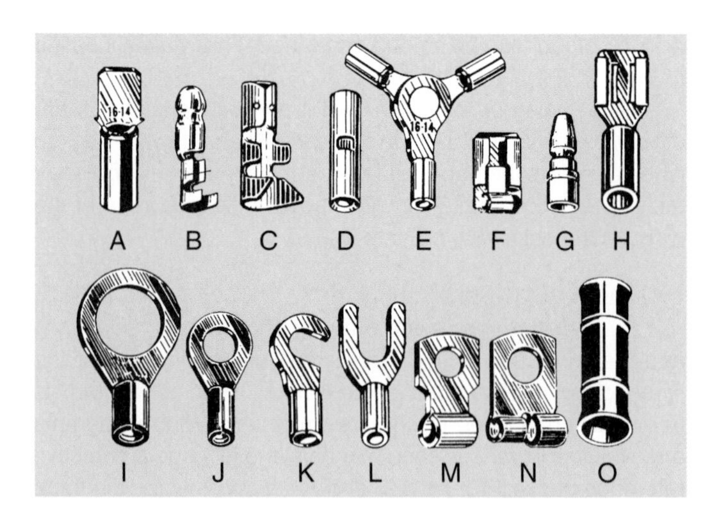

Figure 12-9. Common primary wire terminal types. A—Male slide. B—Bullet or snap-in. C—Female snap-on. D—Butt connector (must be crimped). E—Three-way connector. F—Female slide. G—Bullet. H—Female insulated slide. I—Lug. J—Ring. K—Hook. L—Spade. M—Roll. N—Flag. O—Female bullet connector. (Belden Mfg. Co.)

Plug-in Connectors

Modern vehicles use many ***plug-in connectors,*** sometimes called inline connectors. Several plug-in connectors are shown in **Figure 12-10.** These connectors consist of a male and female connector, which are plugged into each other. This simplifies removal and replacement, since there are no fasteners to loosen or tighten.

Many of these connectors are used to connect several wires at one junction. These are called multiple connectors or they are referred to by the number of wires in the connector, such as the three-wire or four-wire connector. It is almost impossible to plug the connectors in improperly, due to the connector's shape or the presence of special aligning lugs and slots in the connector. If a plug-in connector becomes damaged, it can be replaced with combinations of the terminals shown in Figure 12-9, depending on the usage needed.

The plug-in connectors used with computer-controlled vehicles are often protected with ***rubber grommets*** that have as many as three sealing rings, **Figure 12-11.** This protects the connection from moisture and corrosion, since the voltages used in many computer-related components are very low. The slightest increase in resistance due to moisture and corrosion can affect computer operation. If a plug-in connector on a computer-controlled vehicle becomes damaged, a replacement connector from the manufacturer should be used.

Battery Cable Terminals

New battery cables (with factory installed terminals) are generally used to replace a cable with a corroded, useless terminal. However, it is occasionally desirable to replace only the terminal. See **Figure 12-12.**

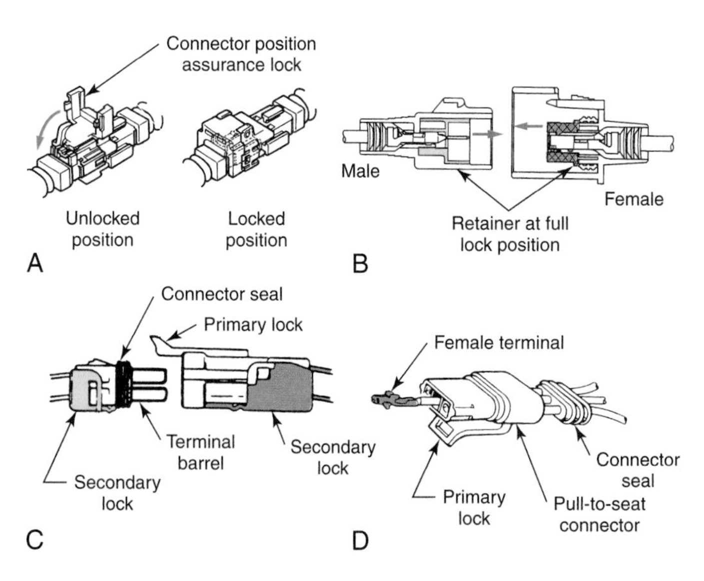

Figure 12-10. Several types of wire connectors. A—One type of airbag harness connector. It uses a connector position assurance (CPA) lock. This ensures the connector is fully seated and will not vibrate or pull apart. B—Single wire connector shown with retainer in full lock position. Note the seals. C—Weather pack connector. D—Metri-pack connector. (Toyota and General Motors)

Fuse Block

The fuse block is a plastic holder that serves two purposes. The first purpose is to serve as a junction or distribution assembly for the wiring harness. Heavy wires deliver current to the fuse block. The fuse block then distributes this current to smaller wires that feed various electrical components. The second purpose of the fuse block is to protect the circuits from electrical overloads. Fuses are installed in each circuit between the power source and the electrical device to be protected. **Figure 12-13A** illustrates a fuse block utilizing the compact miniature fuse. **Figure 12-13B** shows three different types of fuse elements. Newer vehicles have several fuse blocks mounted in various locations. Modern fuse blocks often contain relays and other electrical components.

Adding Fuses

When adding accessory units, such as fog lights, cellular telephones, and CD players, and no provision was made for them in the original wiring, place a fuse in the circuit. Fuse as close as possible to the electrical source to reduce the possibility of a short between the fuse and source. A small fuse

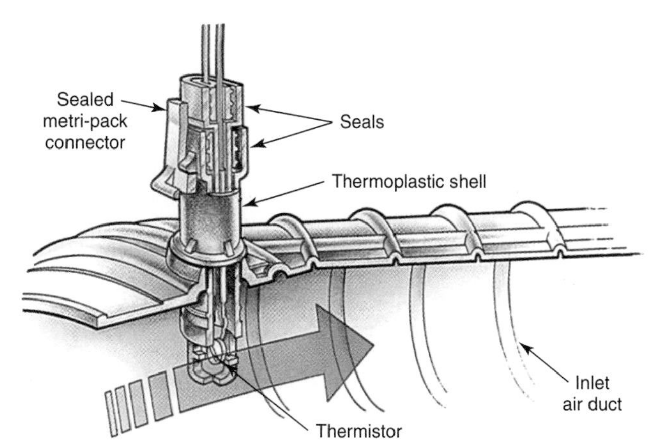

Figure 12-11. Cutaway view of one type of plug-in connector that uses rubber grommets for sealing the connection against dirt, water, etc. (Packard Electric)

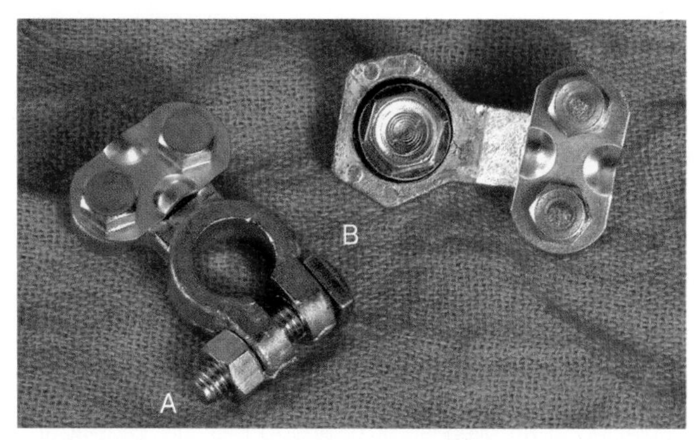

Figure 12-12. Typical replacement battery cable terminals. A—Top-post terminal. B—Side terminal.

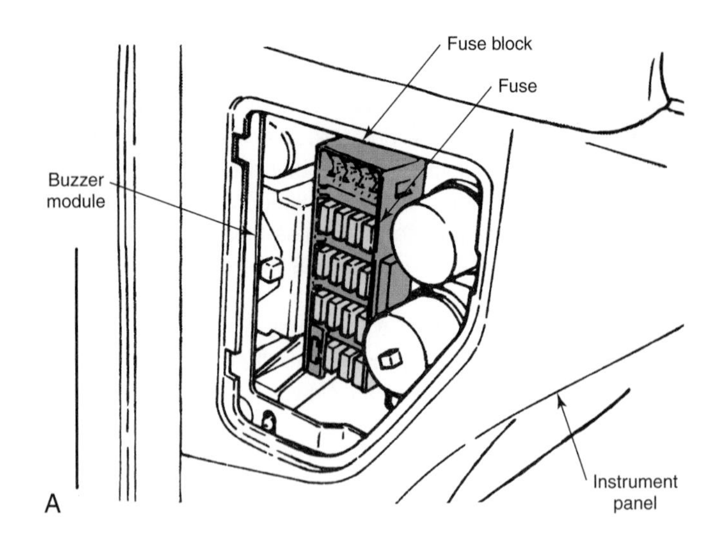

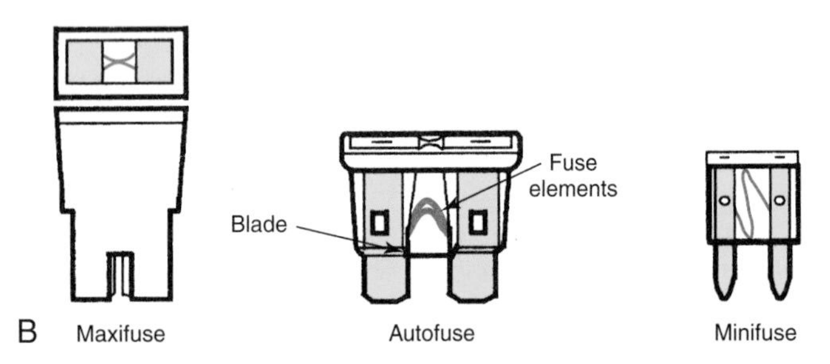

Figure 12-13. A—Fuse block incorporating a number of miniaturized fuses. B—Several different types of fuses used in the automobile. The maxifuse is usually found in junction blocks. Some maxifuses are a larger version of the autofuse. Autofuses and minifuses are typically found in the fuse block, but are sometimes used in junction blocks. (DaimlerChrysler, General Motors)

block or an ***inline fuse*** can be installed. Be sure to inform the owner of the location of the new fuse, **Figure 12-14.** To make the accessory inoperative when the ignition key is *off,* it must be connected so that it can be turned off by the key-switch circuit. Always follow the installation instructions for the accessory.

Caution: Never tap (connect) into an existing circuit to power an accessory. This could overload the system's fuse or circuit breaker and could cause damage to the system. Connect into a terminal block or install a battery cable with an auxiliary lead to power an accessory.

Fusible Link

In some cases, a circuit is protected by a ***fusible link.*** A fusible link is a special wire that acts like a fuse. The wire is covered with Hypalon insulation. When the fusible link is subjected to an overload, it begins to heat up, causing the insulation to blister and smoke. If the overload continues, the fusible link will melt, or blow, breaking the circuit. Fusible link

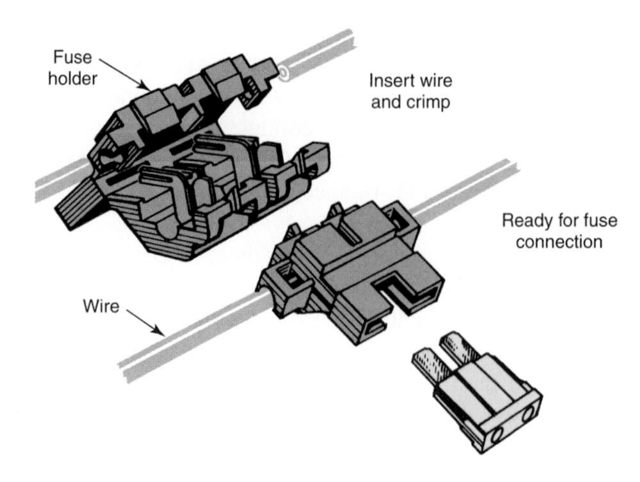

Figure 12-14. Installing an inline fuse. Always try to fuse as close to the source as practical.

wire should be used to repair a blown fusible link. Never replace a fusible link with ordinary wire. Fusible link wire is available at most parts stores. **Figure 12-17** shows how a melted fusible link should be replaced.

Shielded Wiring

Many circuits in the modern vehicle electrical system use very low voltages. Stray electrical impulses or magnetic fields must be kept away from these low voltage circuits. To accomplish this, many wires are *shielded.* Surrounding high voltage or amperage wiring with grounded metal covers or sheaths keeps magnetic fields from escaping. The shielding may be around the wire or device that produces a magnetic field. More commonly, the nearby wiring that could be affected is

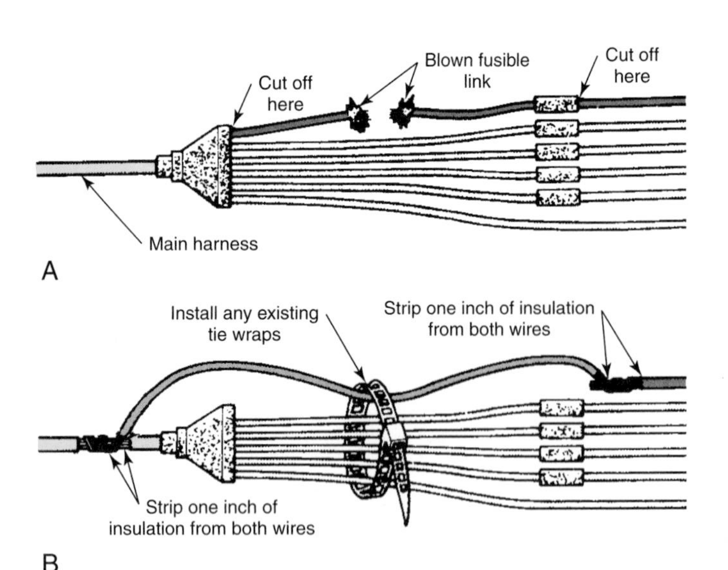

Figure 12-15. A fusible link helps prevent electrical system damage due to overloads, shorts, etc. A—Note how the link has melted. B—Repair requires cutting off the damaged area and installing a new fusible link section. Follow the manufacturer's repair procedures. (DaimlerChrysler)

shielded. Shielding the low voltage wires ensures that they are not affected by magnetic fields from any source.

Repairing shielded wiring is similar to repairing other types of wire. The shield, however, must be kept intact during repairs. Some manufacturers recommend replacing the entire wire when either the wire or shield is damaged.

Circuit Breaker

A ***circuit breaker*** feeds the current through a bimetallic strip, through a set of points, and on to the remainder of the circuit.

When amperage exceeds the breaker's limit, the bimetallic strip heats and bends to separate the points. When the strip cools, it will straighten out, closing the points and reestablishing the circuit. If the overload condition still exists, the strip will heat and reopen the points, **Figure 12-16.**

The circuit breaker will open and close quite rapidly. This gives it an advantage over a fuse when used on a headlamp circuit. With the fuse, an overload will burn out the fuse and the headlights will go out until the fuse is replaced. This could cause an accident. The circuit breaker will cause the lights to flicker on and off rapidly but will still produce enough light to allow the driver to pull safely off the road. A defective circuit breaker should be replaced with one of the same construction and electrical load capacity. See **Figure 12-17.**

Joining Primary Terminals and Wires

After the wire gage is determined, select the connecting terminal. The terminal must be of proper size and type for the unit connecting post or prongs. It must have sufficient current capacity and should be heavy enough to prevent breakage through normal wire flexing and vibration.

Arrange terminals so they have clearance from metal parts that could ground or short them out. On critical applications or where heavy vibration is present, use a terminal such as the ring type that completely encircles the post. If the terminal retaining nut loosens, the wire will not fall off.

Terminals and wires may be connected by either ***crimping*** or ***soldering*** in place. Crimping is fast and forms a good connection. Soldering, if properly done, forms an excellent connection and is recommended in some cases. It is possible to both solder and crimp a connection. Solder forms an electrical path and is not depended on for strength.

Slide and bullet connectors are used where the wires must be separated at some future time. The appropriate slide or bullet terminals are crimped or soldered to the wires. They are then snapped into the connector body, and the two halves are plugged together. If the wire ends are being joined permanently, soldering or butt connectors work very well, **Figure 12-18.** Aluminum wire requires crimped terminals.

Crimping Terminals

A ***crimping tool*** is shown in **Figure 12-19.** It will cut and strip the wire as well as form a proper crimp. To use the crimping tool, first strip the insulation back for a distance equal to the length of the terminal barrel. Then push the wire into the barrel. While being held in, place the crimping tool over the

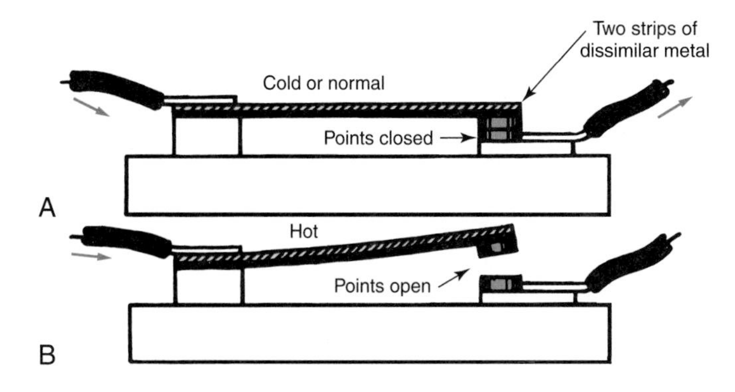

Figure 12-16. Operation of a simple circuit breaker. A—The breaker is carrying a normal system load. The contacts are closed, and the circuit is complete. B—The circuit breaker is overloaded, heating the bimetallic strip. The strip bends upward, separating the points and breaking the circuit.

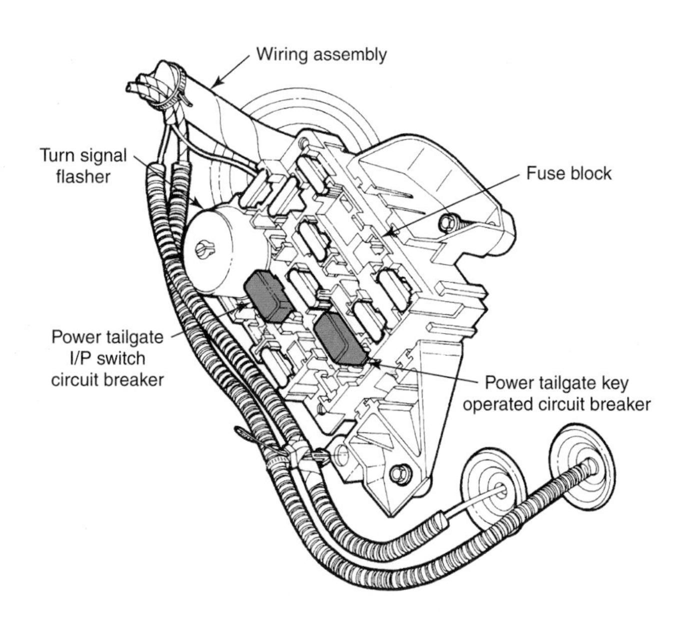

Figure 12-17. This fuse panel contains two power tailgate circuit breakers.

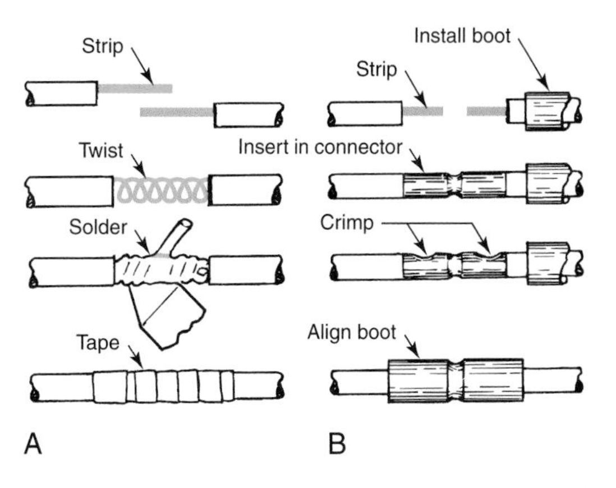

Figure 12-18. Wires can be joined by soldering (A) or using a crimp-type connector (B).

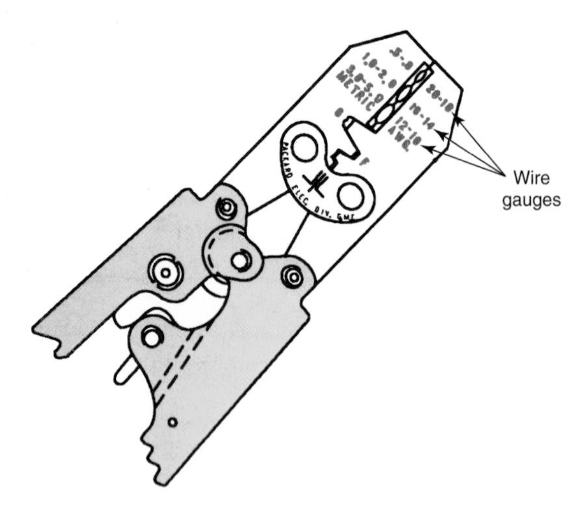

Figure 12-19. Hand-operated crimping tool. Note the jaws provide both metric and AWG size information.

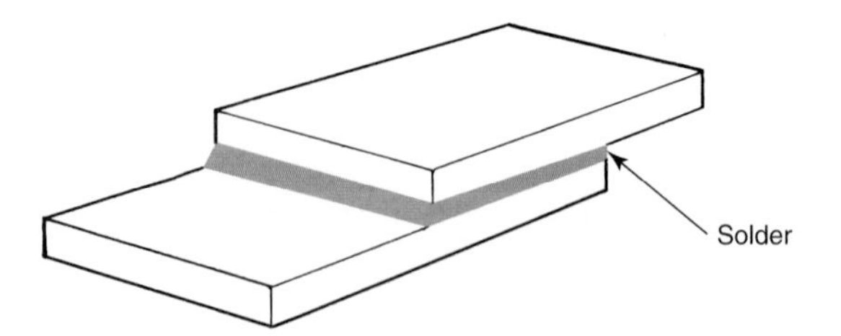

Figure 12-21. The solder and metal "skin" form a tight bond when they cool.

spot to be crimped. Be sure to use the proper crimping edge. Squeeze the handles together to firmly crimp the terminal to the wire, **Figure 12-20.** Follow the tool manufacturer's instructions. Use the correct barrel size for the wire used.

Caution: Never crimp a wire with the cutting edge of a pair of pliers. This would crimp the barrel but weaken it.

Soldering

Soldering can be defined as the act of joining two pieces of metal through the use of lead, tin, and other alloys. There is no actual fusion (melting together) involved. When the base metal is heated to the correct temperature, the solder dissolves a minute "skin" on the metal. Upon cooling, the solder and "skin" amalgamate (mix together), forming a tight bond. See **Figure 12-21.** The pieces to be soldered should fit together as closely as possible. The less solder separating the parts, the stronger the joint.

Solder

As mentioned, solder is a mixture of tin and lead with small amounts of zinc, copper, aluminum, and other substances. The percentage of lead to tin affects both the melting point (point at which solder becomes a full liquid) and the plastic range (temperature span from lowest point at which solder becomes mushy, or plastic, to the highest point just before solder liquefies).

You will note from **Figure 12-22** that pure lead melts at 621°F (328°C). Pure tin melts at 450°F (232°C). A mixture of about 63% tin to 37% lead will melt at 361°F (183°C). Study the chart in Figure 12-22. Melting point and plastic range are affected by alloying in different proportions. Commonly used solders are 40/60 (40% tin, 60% lead), 50/50, and 60/40. Flux core wire solder (wire solder with a hollow center filled with flux), solid wire solder, and solder ground into fine grains and mixed with flux, are used for general soldering.

Acid core solder is excellent for use on radiators and other applications where a corrosive and electrical conductive residue (flux remaining on work after soldering) is not harmful. However, ***rosin core*** solder or ***resin core*** solder must be used for all electrical work. The residue will not cause corrosion or conduct electricity. A special flux is required for soldering aluminum wires.

Soldering obviously heats the metal, which accelerates oxidization (surface of metal combining with oxygen in air). This leaves a thin film of oxide on the surface that tends to

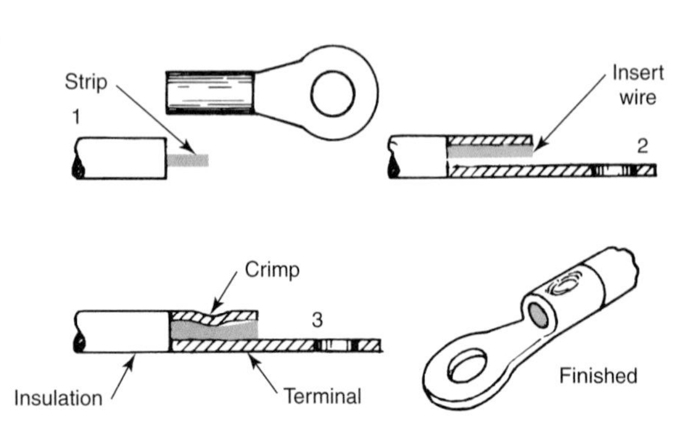

Figure 12-20. Crimping a terminal. Follow the tool manufacturer's instructions.

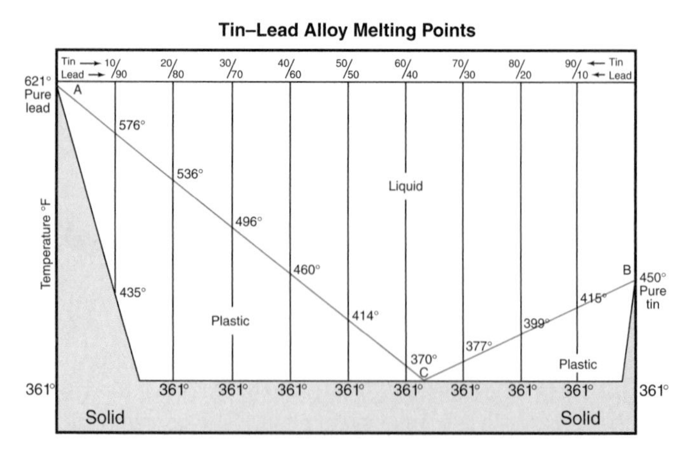

Figure 12-22. Tin-lead alloy plastic range and melting point chart. (Kester)

reject solder. It is the job of the flux to remove this oxide and prevent the reoccurrence during the soldering process.

Soldering Irons

The **soldering iron**, sometimes called a copper, should be of ample size for the job. An iron that is too small will require excessive time to heat the work and may never heat it properly. The proper size iron will bring the metal up to the correct soldering heat (around 525°-575°F or 274°-302°C) quickly and will produce a good solder joint. Electric irons are fast and efficient. A 35- to 100-watt size will handle most wire soldering jobs. See **Figure 12-23.** For electrical wiring, a ***soldering gun*** as shown in **Figure 12-24** is ideal. The tip reaches soldering heat in a matter of seconds.

Preparing to Solder

All traces of insulation, corrosion, and rust must be removed. Removing the insulation will usually expose enough clean wire for soldering. Remember that good soldering requires clean, well-fitted surfaces. The soldering iron tip is made of copper. Through the solvent action of solder and prolonged heating, it will pit and corrode. An oxidized or corroded tip will not satisfactorily transfer heat from the iron to the work. It should be cleaned and tinned. Use a file and dress the tip down to the bare copper. File the surfaces smooth and flat. See **Figure 12-25.**

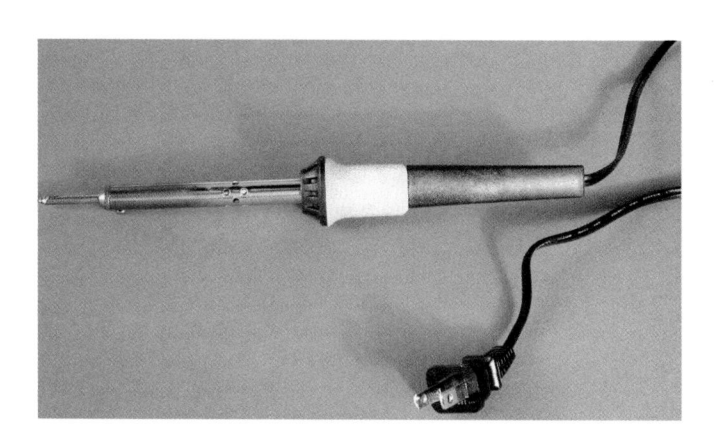

Figure 12-23. An electric soldering iron is commonly used to make wiring repairs.

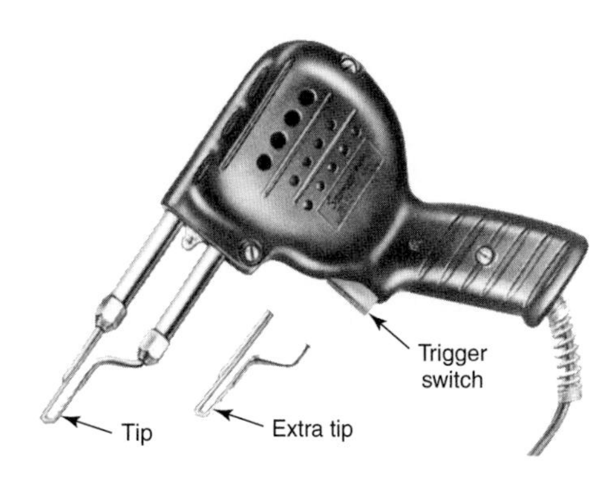

Figure 12-24. A soldering gun such as this works fast. (Snap-On Tools)

Plug in the iron. When the tip color begins to change to brown and light purple, dip the tip in and out of a can of soldering flux (rosin core). Quickly apply rosin core wire solder to all surfaces. If no paste flux is available, use rosin core wire solder. However, dipping the tip in flux provides a faster and better tinning job.

The iron must be at operating temperature to tin properly. When the iron is at the proper temperature, solder will melt quickly and flow freely. Never try to solder until the iron is properly tinned. See **Figure 12-26.** If a surplus of solder adheres to the tip during tinning, wipe off the excess with a rough textured cotton rag.

Soldering Technique

The iron must be held so the flat surface of the tip is in full contact with the work. This will permit a maximum transfer of heat. Look at **Figure 12-27.** Apply the wire solder at the edge of the iron where it contacts the work. This will release

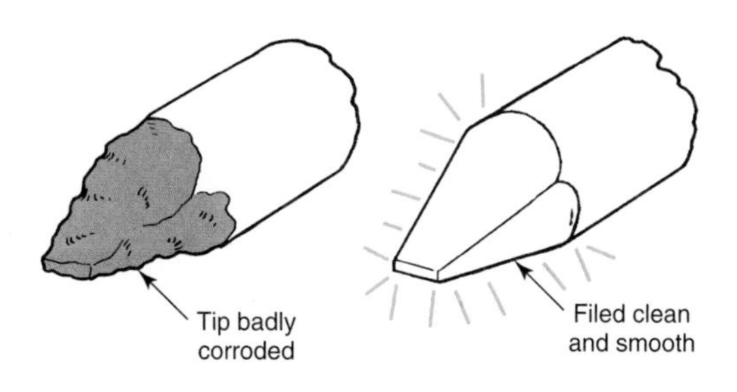

Figure 12-25. File soldering tip surfaces flat and smooth before tinning.

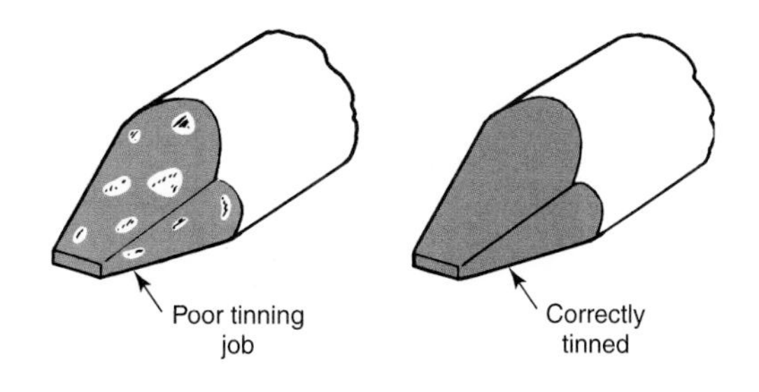

Figure 12-26. Tip must be properly tinned before soldering.

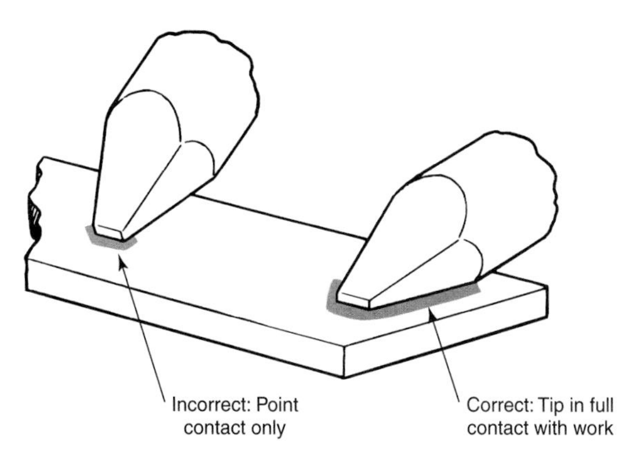

Figure 12-27. Hold the soldering tip flat against the workpiece.

the flux where it will do the most good. Flowing solder at this point will also provide a mechanical bond between the iron and the work. This will speed up heat transfer, **Figure 12-28.**

The workpiece to be joined should be heated so the solder is melted in the metals to be soldered together. When this is done, solder will flow readily and a good solder joint will result. If the solder melts slowly and is pasty looking, the work is *not* hot enough. This can result in a ***cold solder joint.*** If using a gas flame to heat the parts, be careful to avoid overheating.

Soldering Wire Splices

To solder wire splices, first make sure the wires are twisted together firmly. Apply the tip flat against the splice. Apply rosin core wire solder to the flat of the iron where it contacts the wire. As the wire heats, the solder will flow through the splice, **Figure 12-29.**

Soldering Terminals

Terminals do not have to be specially designed for soldering, but the lip-type terminal tang lends itself to soldering better than the closed or open barrel tang, **Figure 12-30.**

To solder the lip type, strip the wire back as shown in **Figure 12-31A.** Insert the wire as shown in **Figure 12-31B.** Crimp the wire holding lips, one after the other, tightly over the wire. Then, carefully fold the insulation tang around the insulated portion of the wire, **Figures 12-31C** and **12-31D.** Using rosin core wire solder, place a drop of solder on the holding lips. Hold the iron in contact with the drop until it flows into the lips and wire. Do not hold the iron in contact with the terminal any longer than necessary. This tends to melt the insulation.

When soldering the open barrel type terminal, strip the wire as for crimping. Tin the exposed wire end (coat with a thin layer of solder) and insert the wire in the barrel. If both crimping and soldering the barrel connector is desired, crimp before soldering. While holding the exposed end upright, heat the socket with the iron. While heating, keep wire solder against socket end. When the solder melts, allow it to flow into the barrel. Hold the iron in place for a few seconds to allow the solder to bond to both the barrel and wire. See **Figure 12-32.** The closed barrel type terminal should be heated, and a small amount of solder should be allowed to flow into the hole. While keeping the barrel hot, press the tinned wire into the hole. Hold the iron in place for several seconds to ensure bonding.

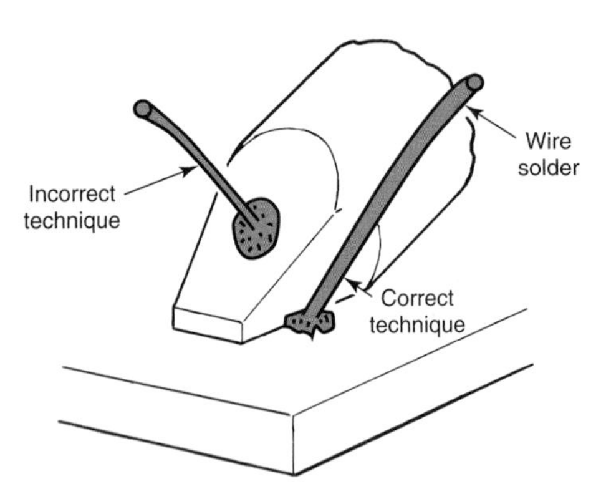

Figure 12-28. Apply solder to the edge of the iron where it contacts the workpiece.

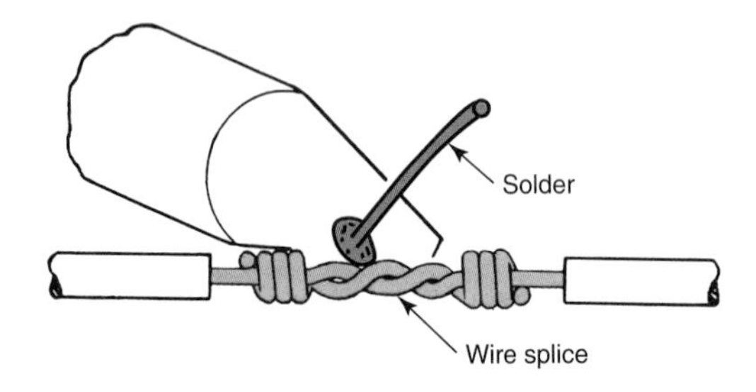

Figure 12-29. Soldering a wire splice. Be sure to heat the wire, not the solder.

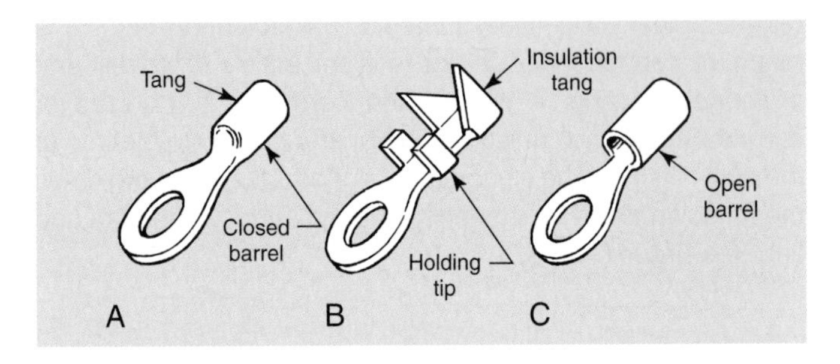

Figure 12-30. Study these different terminal tangs. A—Closed barrel. B—Lip-type. C—Open barrel.

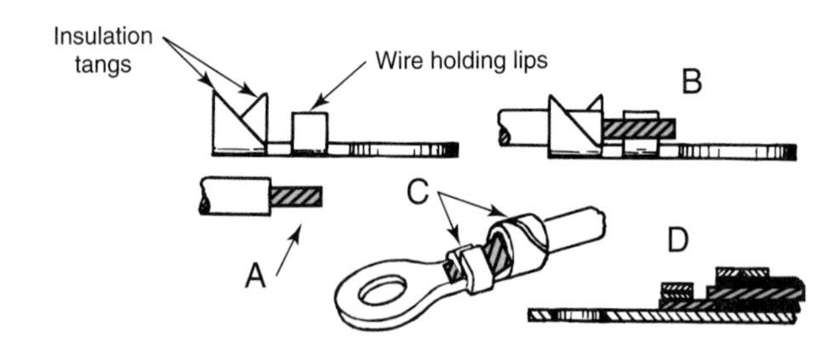

Figure 12-31. The lip-type terminal design allows for easy soldering.

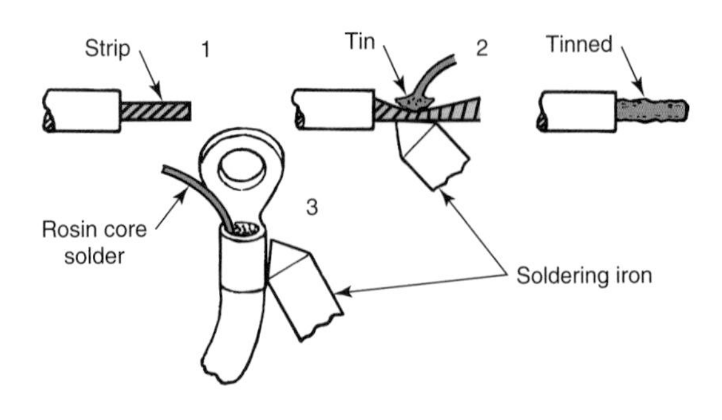

Figure 12-32. When soldering a barrel-type terminal, make sure the wire is tinned before installing. The terminal may be crimped along with soldering, but crimping should be done before soldering.

After Soldering

When joining two wires or terminals by soldering, be careful not to disturb them until the solder has set (cooled until it is solid). If they are moved while the solder is still in a pasty state, it can cause fracture lines that will result in a weak joint.

General Rules for Good Soldering

- Clean the area to be soldered.
- Make a good mechanical connection with the wires before soldering.
- Soldering iron must be of sufficient size and must be hot.
- Soldering iron tip must be tinned.
- Apply full surface of tip flat to work.
- Heat the wires to be joined until solder flows readily.
- Use the proper solder and flux for the job at hand.
- Apply enough solder to form a secure bond but do not waste.
- Do not move wires or connectors until solder sets.
- Place the hot iron in a stand or on a protective pad.
- Unplug electric iron as soon as you are finished.

Use ***heat-shrink tubing*** or tape to insulate soldered wires and noninsulated connectors. When an insulator boot is to cover the terminal tang or when attaching slide-type terminals that will be snapped back into a housing, always slide the boot, housing, etc., on the wire before soldering.

Installing Primary Wire

When installing primary wire, use the following guidelines. Make certain all terminals and posts are clean. Connect all terminals and tighten securely. Lock washers should be used on all screw and post connections. Slip insulator boots, where used, over exposed terminal tangs. Shove slide or bullet-type connectors together tightly and check to see the connection is secure.

Keep all wiring away from the exhaust system, oily areas, and moving parts. Secure all wire with mounting clips or clamps. Fasten wire in enough spots to prevent excessive vibration and chafing.

Where the wire must pass through a hole in sheet metal, install a rubber grommet. When a wire must pass from the fender well or splash shield to the engine, leave enough slack to allow the engine to rock on the mounts without pulling the wire tight. If a number of primary wires travel in a common path, pull them through a loom (woven fiber or plastic conduit) or tape them together. See **Figure 12-33.**

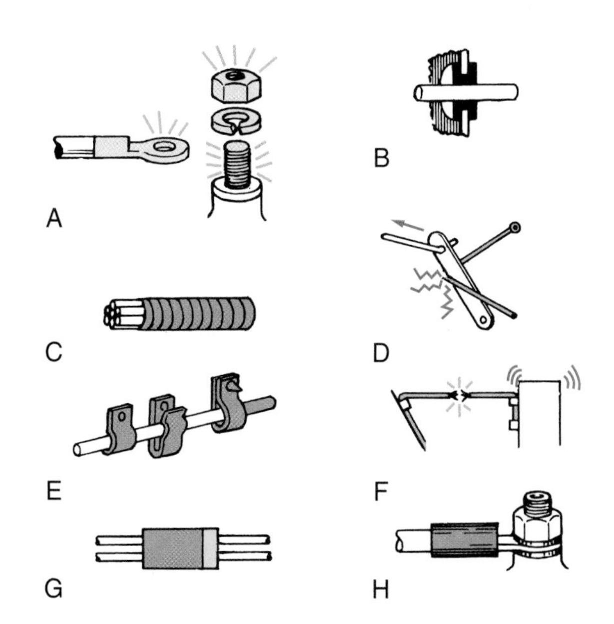

Figure 12-33. Wiring installation hints. A—Connections must be clean and bright. B—Use grommets to protect wire passing through thick metal. C—Tape common wires together. D—Avoid moving parts when locating wires. E—Support with suitable clamps. F—Allow some slack when wire runs to a unit that moves. G—Connectors must be pushed together tightly. H—Use boots on terminal tangs and select terminals heavy enough for job.

Secondary Wire

Secondary wire is used in the ignition system high voltage circuit—coil to distributor and distributor to spark plugs. Since voltages in secondary wire can exceed 50,000 volts, secondary wire has a heavy layer of insulation to protect against shorting to the engine block or another vehicle part. Insulation also limits ***corona,*** which is the loss of electrons to the surrounding air. Excessive corona could impart sufficient current into an adjacent wire to cause it to fire a plug. This is known as crossfiring.

Even with good insulation it is important to arrange spark plug leads so that leads to cylinders that fire consecutively (one after the other) are separated. Modern secondary wire uses a conductor, or core, made of carbon-impregnated string or fiberglass, and elastomer (plastic). The carbon impregnated string and elastomer type have a controlled resistance (between 4000 and 20,000 ohms per foot) in the secondary circuit to reduce radio interference.

Caution: When working on the ignition system, handle resistance wires carefully. Sharp bending and pulling can separate the conductor, ruining the wire. When removing or installing such leads, grip the insulation boot, not the wire.

The secondary wires on older vehicles often had a metal core, usually copper. Copper core secondary wires are sometimes available as an aftermarket (nonfactory) item. ***Resistance wire*** should always be used on modern vehicles. They may be identified by such letters as IRS and TVRS stamped on the insulating material. Secondary wiring will be discussed in more detail in *Chapter 16, Ignition System Service.*

Installing Secondary Wire

When installing secondary wires, avoid sharp bends. If the wires pass through a metal conduit (tube), the conduit should be securely grounded. Install or remove the plug wires by grasping the insulation boots and not the wire. Make sure the terminals snap tightly on the plugs and the distributor ends are all the way in the housing towers. Follow the manufacturer's instructions in arranging the plug wires. If two leads going to consecutively firing cylinders are side by side, there is a danger of crossfiring, especially as the wires age.

Printed Circuit

The ***printed circuit*** uses a nonconducting panel upon which the electrical components are affixed. The components are then connected by either thin conductor strips cemented to the panel or by a special electrically conductive material "printed" in the desired circuit patterns. This eliminates the maze of wires otherwise needed to connect numerous small, complex components. Printed circuits permit a great number of individual circuits in a very small area. The modern auto uses printed circuits in such areas as engine and body control computers, audio accessories, and dash instrumentation. **Figure 12-34** illustrates the use of a printed circuit in a dash instrument cluster.

Troubleshooting Wiring

Many problems throughout the car can be traced to faulty wiring. Loose or corroded terminals; frayed, pinched, bare, or broken wires; and cracked, oil-soaked, or porous insulation are the most frequent causes.

Caution: Do not pierce a wire to check a circuit. Piercing wires allows moisture and dirt to enter the wire, which can change the circuit's resistance. Doing this on a computer-controlled vehicle could cause major damage to the vehicle's computer.

Checking Wires and Connections

When troubleshooting a problem, visually check the wires, fuses, and connections carefully. Wires can separate with no break in the insulation (especially resistance-type secondary wire). A terminal may be tight and still be corroded. A fusible link may burn out at one end instead of in the center. Use a wiring diagram. An experienced technician is patient and thorough in any troubleshooting.

Wires and connections must occasionally be checked for resistance, ***voltage drop, short circuits*** or ***near-short circuits.*** These checks can be made with an ***ohmmeter, voltmeter,*** or ***ammeter.*** This will be discussed in the chapters where these tests pertain.

Checking for Continuity

A small ***test light*** (battery operated) may be used to test wires for internal breaks. Hold one prod against one end of the wire and place the other prod against the other end. If the test lamp burns, the wire is continuous or said to have ***continuity.*** This simple test light is also handy for checking fuses, finding shorted field windings, and for tracing wires where there are no color codes. **Figure 12-35** illustrates several checks.

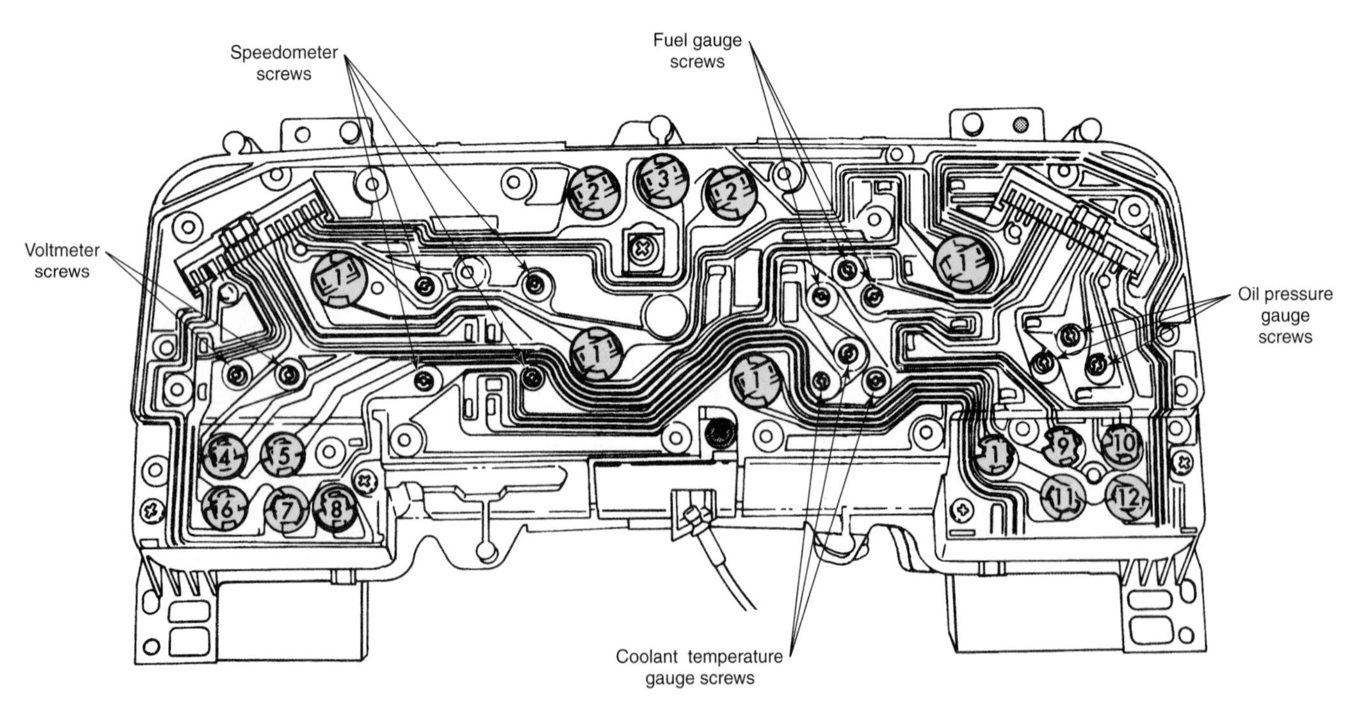

Figure 12-34. Rear view of a printed circuit instrument panel. (DaimlerChrysler)

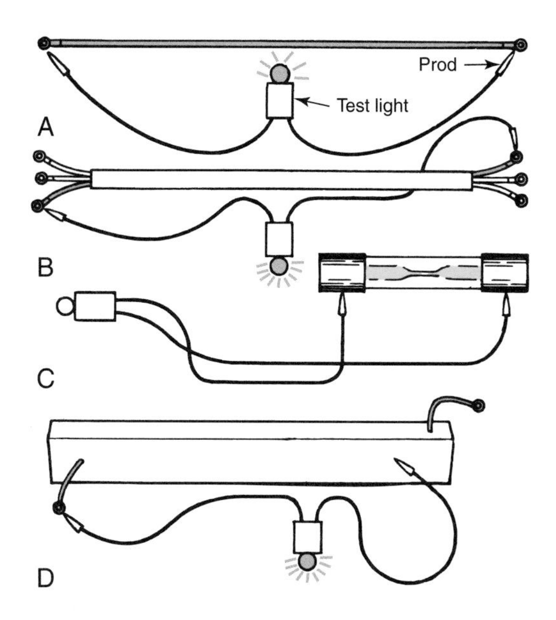

Figure 12-35. Some wiring checks using a simple powered test light. A—Prods on ends of wire. Lamp lights indicating wire is continuous. B—Prod held on the end of one wire and the other prod touched to various wire ends. When lamp lights, proper wire end is identified. C—Checking a fuse. With the prods in place, the lamp does not light. This indicates a faulty fuse. D—One prod touched to a wire end and the other prod to ground. If the lamp lights, the wire is shorted out.

Caution: Refer to the appropriate service manual before using a test light to check any circuit. Using a test light on some circuits can damage electronic equipment.

Tech Talk

One of the most common vehicle electrical problems is a bad ground connection. Current must go through the vehicle's frame, body, or engine to return to the battery. If a ground is bad, the circuit will be open. Even slight resistance will often be enough to affect the performance of electrical units throughout the vehicle, especially on-board computer systems, which operate from very low voltages. Sometimes, a bad ground will ruin transmission or engine bearings, shift or brake cables, or other nonelectrical parts. This is because the electricity tries to return to the battery through these parts, causing them to overheat or transfer metal. The electricity may even try to ground through the engine cooling system, causing severe corrosion.

If any electrical problem is evident, start your trouble-shooting process by checking the grounds for clean, tight connections. Often, someone has attached the negative battery cable to an exhaust manifold bolt. The heat from the exhaust gas causes a bad connection almost immediately. Make sure that there is a ground cable between the engine and body or between the body and the negative battery terminal. These grounds are often overlooked when a battery cable is changed. Ground cables between the engine and the body are sometimes pulled loose if one of the engine mounts breaks.

Methods for checking for bad grounds are covered in later chapters. If you suspect that a bad ground exists between any part of the vehicle and the battery negative, you can run an extra ground wire. Simply connect a heavy gage wire (at least 8 or 10 gage) between the negative battery terminal and the body or between the body and an unpainted spot on the engine. An extra ground will not hurt any part of the electrical system.

Summary

Primary wire (copper stranding, relatively thin insulation) is used for circuits handling battery voltage. Plastic is widely used for insulation. All automotive wire uses a stranded wire conductor.

The AWG (American Wire Gage) is determined by the cross-sectional area in circular mils. The larger the AWG number, the smaller the size. A micrometer or wire gauge can be used to determine wire size.

Automotive electrical systems are color coded. Use an accurate wiring diagram for troubleshooting or replacing wires. Electrical load, line voltage, and wire length must be taken into consideration when choosing wire gage. A wire gage chart will assist in making the right selection. Remember that undersize wires increase resistance, reduce unit efficiency, and can overheat or burn. On two-wire circuits (one wire for ground) count the length of both wires.

Spade, lug, flag, roll, slide, ring, bullet, and push-on terminals are used on vehicle electrical systems. Terminal blocks allow one feeder wire to service a number of other wires. These can be of the screw, bullet, or slide type. Junction blocks provide a central connecting point for a number of wires. Fuse blocks give protection against circuit overloads. A wiring harness contains a number of wires either taped together or pulled through a loom. This keeps common wires neatly arranged and facilitates installation.

Wire ends may be joined by soldering or crimping. Be certain that terminals are of the correct style and size. They may be soldered or crimped to the wire. When crimping, use a suitable crimping tool. Solder is a mixture of lead and tin in varying amounts. Joints to be soldered must fit well. Solder has little strength if the parts are far apart. Wire solder, with flux-filled center core, is desirable.

Flux helps remove oxides and also prevents the formation of oxides while soldering. Be sure to use solder with rosin core only on electrical work. The joint to be soldered must be clean and dry. Lay the flat tip of the iron against the work and apply wire solder where the iron and work contact. Solder must run and tin freely. Do not move work while it is cooling.

Secondary wire (fiberglass, carbon impregnated thread, and elastomer stranding with very heavy insulation) is used on the ignition high-tension circuit. Handle secondary resistance wire carefully.

When installing wires, keep them away from heat, oily areas, and moving parts. Terminals must be clean and tight. Use clips to prevent chafing and excessive vibration. When adding accessories, fuse the circuit as close to the source as possible. Do not tap into an existing circuit to power an accessory. Printed circuits are increasingly used on modern vehicles.

Clean, tight connections, with proper size wire and good insulation, are imperative. When troubleshooting, always check connections and insulation. Replace cracked, spongy, or frayed wires. Many wiring checks can be made with a simple test light.

Review Questions—Chapter 12

Do not write in this book. Write your answers on a separate sheet of paper.

1. Most primary wire used in automobiles is made of _____.
   (A) solid copper
   (B) solid aluminum
   (C) stranded copper
   (D) stranded aluminum
2. No. 16 wire is smaller than No. _____ wire.
   (A) 14
   (B) 18
   (C) 20
   (D) depends on the insulation thickness
3. The higher the system voltage, the _____ the wiring gage can be.
   (A) smaller
   (B) larger
   (C) longer
   (D) None of the above.
4. Which of the following wire arrangements requires the largest wire gage?
   (A) 20 foot long wire, 12 volt system
   (B) 20 foot long wire, 6 volt system
   (C) 6 foot long wire, 12 volt system
   (D) 6 foot long wire, 6 volt system
5. Automotive wiring is _____-coded.
   (A) size
   (B) color
   (C) length
   (D) number
6. The _____ protects a circuit from an overload.
   (A) wire size
   (B) wire length
   (C) wire material
   (D) fuse
7. When soldering electrical work, _____ flux should be used.
   (A) acid
   (B) rosin
   (C) aluminum
   (D) organic
8. When applying wire solder, touch the wire to the _____.
   (A) top of the iron
   (B) work away from the iron
   (C) iron where it contacts the work
   (D) side of the iron
9. The conductor of most modern secondary wire is made of _____.
   (A) copper
   (B) aluminum
   (C) carbon impregnated string
   (D) plastic
10. Match the electrical symbols in the left-hand column by placing the letter of the description in the right-hand column beside the number of the matching symbol.

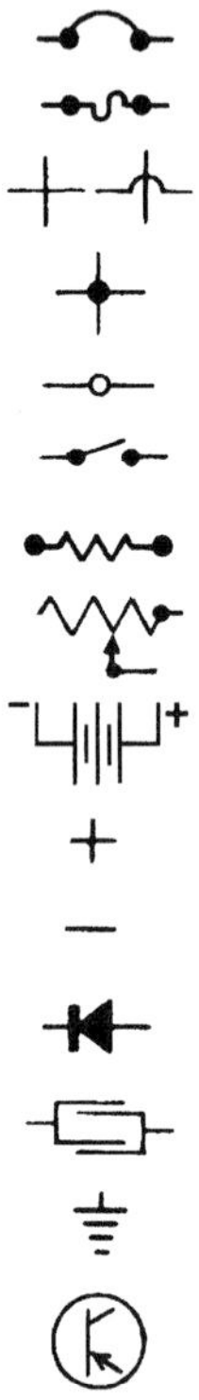

1. _____
2. _____
3. _____
4. _____
5. _____
6. _____
7. _____
8. _____
9. _____
10. _____
11. _____
12. _____
13. _____
14. _____
15. _____

A. Resistor
B. Circuit breaker
C. Wires crossing - not connected
D. Fuse
E. Diode or rectifier
F. Wires crossing - connected
G. Positive
H. Terminal
I. Open switch
J. Rheostat
K. Transistor
L. Battery
M. Negative
N. Condenser
O. Ground

ASE-Type Questions

1. The most commonly used insulation material in modern vehicles is _____.
   (A) rubber
   (B) aluminum
   (C) plastic
   (D) treated paper
2. Technician A says the larger the wire number, the thicker the wire. Technician B says cross-sectional area in square mils determines the wire size. Who is right?
   (A) A only.
   (B) B only.
   (C) Both A and B.
   (D) Neither A nor B.

3. All of the following are considerations when selecting the correct wire gage for a circuit, *except:*
   (A) wire length.
   (B) wire thickness.
   (C) wire insulation.
   (D) electrical load.
4. Technician A says an undersize wire will overheat. Technician B says an undersize wire will increase electrical resistance in the circuit. Who is right?
   (A) A only.
   (B) B only.
   (C) Both A and B.
   (D) Neither A nor B.
5. A number of common wires, taped together, with leads leaving at various spots, is referred to as a wiring _____.
   (A) set
   (B) harness
   (C) package
   (D) herd
6. Technician A says soldering involves fusion. Technician B says the wires should have a good mechanical connection before soldering. Who is right?
   (A) A only.
   (B) B only.
   (C) Both A and B.
   (D) Neither A nor B.
7. Flux is used in soldering to _____.
   (A) clean the metal
   (B) prevent overheating of metal
   (C) cement parts together
   (D) prevent rusting
8. Technician A says resistance-type spark plug wires are used to provide a hotter spark. Technician B says resistor spark plug wires are easily damaged by sharp bends. Who is right?
   (A) A only.
   (B) B only.
   (C) Both A and B.
   (D) Neither A nor B.
9. Technician A says using copper secondary wire may cause radio interference. Technician B says that if spark plug wires pass through a metal conduit, the conduit should be grounded. Who is right?
   (A) A only.
   (B) B only.
   (C) Both A and B.
   (D) Neither A nor B.
10. All the following statements about wires are true *except:*
    (A) spark plug wires can crossfire if wires are too close together when they serve cylinders that fire consecutively.
    (B) as long as the insulation is all right, a wire can be considered OK.
    (C) a frayed wire can cause a short circuit.
    (D) a corroded connection will increase resistance to electrical flow.

Suggested Activities

1. Using the primary wire size selection chart in Figure 12-4, determine the correct size wire for the following:
   A. Load—100 candela; Wire length—11 feet; Voltage—12 volts.
   B. Load—50 amperes; Wire length—20 feet; Voltage—12 volts.
   C. Load—72 watts; Wire length—15 feet; Voltage—12 volts.
2. Use a schematic to trace out a simple wiring circuit, such as tail and stoplights, or the horn. List the switches and relays used in the circuit.
3. Compare wire sizes as used on a modern vehicle. Note variations in wire sizes and determine why different sizes must be used. Write a short report summarizing your findings.
4. Check several electrical connectors on a vehicle, under the hood, in the passenger compartment, and under the chassis. Explain why certain ones are sealed, while others are not.
5. Obtain spark plug firing orders for various engines. Demonstrate how to rewire an engine using the firing order. Also demonstrate how to rewire an engine when the firing order is not available.

Many late-model vehicles are equipped with side air bags to prevent the occupants from hitting their heads on the side windows or roof pillars during a severe side impact. (Saab)

13

Chassis Electrical Service

After studying this chapter, you will be able to:

- Identify and define chassis wiring and related components.
- Explain the differences between chassis wiring and engine wiring.
- Identify lighting systems and components.
- Troubleshoot and replace system components.
- Identify solenoids and relays.
- Troubleshoot and replace solenoids and relays.
- Identify motor-operated components.
- Troubleshoot and replace motor-operated components.
- Identify vehicle security systems and components.
- Troubleshoot and replace vehicle security system components.
- Work safely on vehicles equipped with air bag systems.
- Troubleshoot and service air bag systems.

Technical Terms

Wiring Harness
Printed Circuit
Color Coded
Schematics
Fuses
Fuse Block
Fusible Links
Circuit Breakers
Rheostat
Sealed Beams
Halogen
Daytime Running Lamps
Brake Light Switches
Flasher Unit
Emergency Flasher Switch
Fiber Optic
Light Emitting Diodes
Vacuum Fluorescent Displays
Gauges
Voltage Limiter
Relays
Armed
Disarmed
Pass-Key
Remote Keyless Entry System
Body Computer Systems
Air Bag Systems
Supplemental Inflatable Restraints (SIR)
Side Air Bags
Deploy
Impact Sensors
Diagnostic Control Module
Coil Assembly
Inflator Module
Cruise Control
Vehicle Speed Sensor

Introduction

Although many vehicle electrical systems are directly related to engine operation, they are not the only systems that you will be called on to service. Many vehicle electrical components are not directly connected with the vehicle engine or drive train operation. Examples include safety devices such as vehicle lights, horns, and windshield wipers, as well as entertainment and comfort items such as radios, electronic compasses, power windows, and door locks. This chapter covers electrical components that are used on the vehicle body and chassis and their associated wiring.

Chassis Wiring

Chassis wiring consists of all of the wiring not directly connected with the engine and drive train. This wiring extends throughout the vehicle. Wiring to devices that are operated except when the vehicle is being driven, such as the windshield wipers and radio, is routed through the ignition switch. Wiring to the vehicle lights or other systems that can be operated at any time is powered directly from the vehicle battery.

Almost all vehicle wiring is installed as part of a ***wiring harness.*** Wiring harnesses are groups of wires wrapped together for ease of installation. A typical harness is shown in **Figure 13-1.** Wiring harnesses have molded electrical connectors that can be attached to other harnesses or directly to electrical components. Some vehicles use a ***printed circuit*** for at least part of their wiring needs, usually at the instrument panel. The printed circuit shown in **Figure 13-2** is a plastic sheet with the circuits etched on it.

All chassis wires are ***color coded*** to make identification of individual wires easy. Color codes can be single wire colors, or colors with contrasting stripes or bands. The use of stripes and bands makes hundreds of color combinations possible. The technician should carefully check wire colors, since many wires in a harness may have similar colors.

Using Wiring Schematics

Wiring ***schematics*** are used to trace out vehicle wiring systems. A schematic is a pictorial diagram of electrical wiring throughout the vehicle. Wire colors are given on the schematic, making identification of needed wires easy. Some

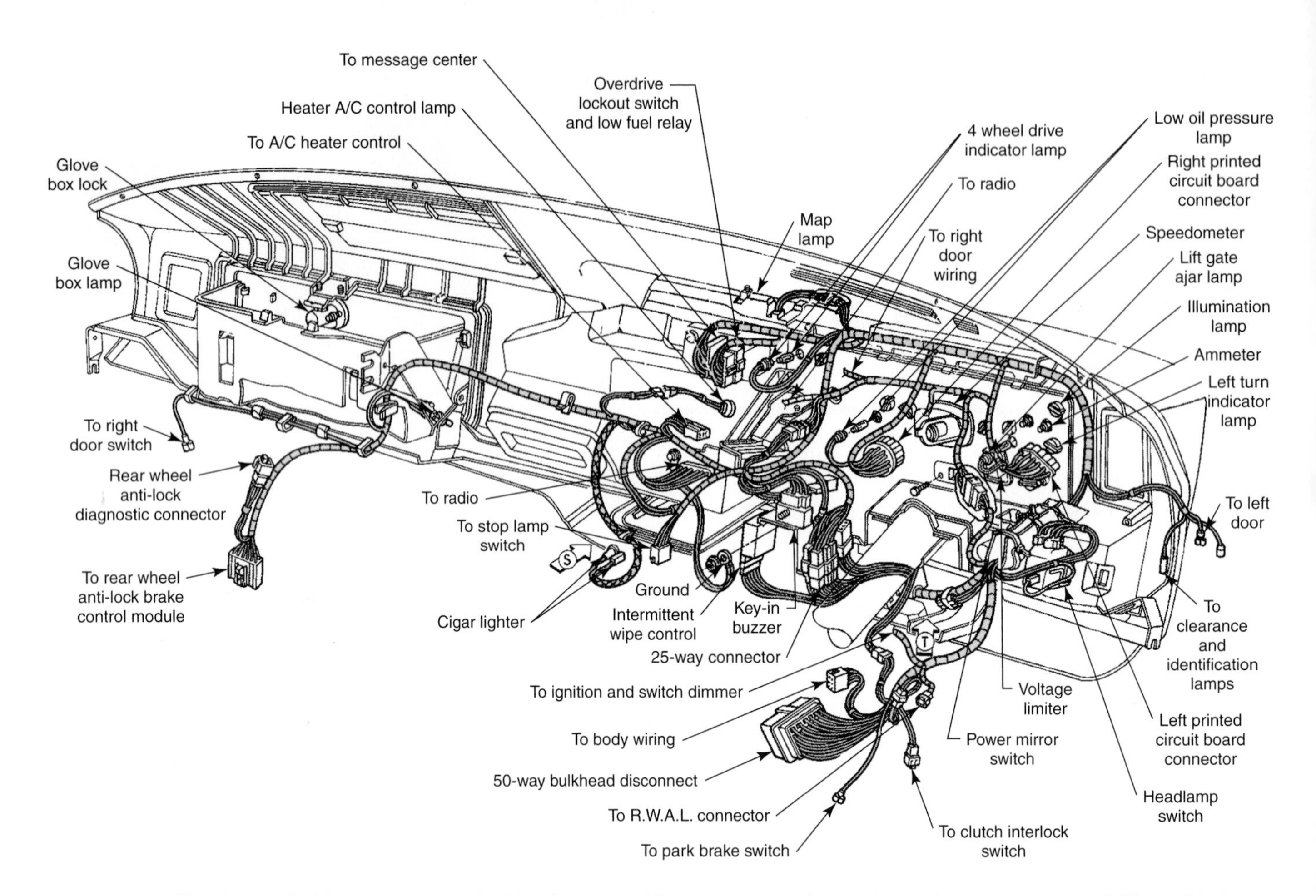

Figure 13-1. This is a typical instrument panel wiring harness with connectors going to the various components. Without this type of "road map," you will quickly become lost in a maze of wires. (DaimlerChrysler)

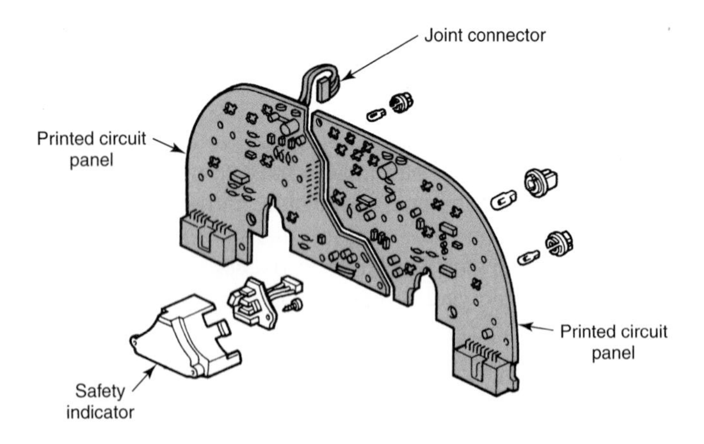

Figure 13-2. This figure shows a rear view of an instrument panel gauge assembly using printed circuit panels (boards). Handle terminals and circuit panels with care. (Honda)

schematics show the exact flow of electricity, such as **Figure 13-3,** while others show the general process of a particular system, such as the computer control schematic shown in **Figure 13-4.** Note the schematic does not show the return flow of electricity. This is because the current flow is assumed to return to the battery's negative terminal through the vehicle's body and frame.

Schematics are often included in the vehicle service manual or may be supplied separately. Tracing the flow of electricity through a schematic is similar to reading a road map. The flow of current can be traced out from its starting point to the component experiencing problems. Conversely, the flow of current can be traced backwards from the inoperable component through switches, control relays, and other components back to the battery.

Fuses, Fusible Links, and Circuit Breakers

To protect the chassis wiring from damage, ***fuses*** are used in the electrical system. Almost all factory-installed fuses are installed in the ***fuse block,*** usually located under the dashboard. A few fuses may be installed at the device that they protect. When an electrical accessory, such as foglights is installed, an ***in-line fuse*** may be installed in the wiring leading to the device.

Fusible links, **Figure 13-5,** which are lengths of wire calibrated to melt when current exceeds a certain value, are also used. Most fusible links are located near the battery positive cable. They can be attached to the positive terminal clamp, a body-mounted junction block, or at the starter solenoid.

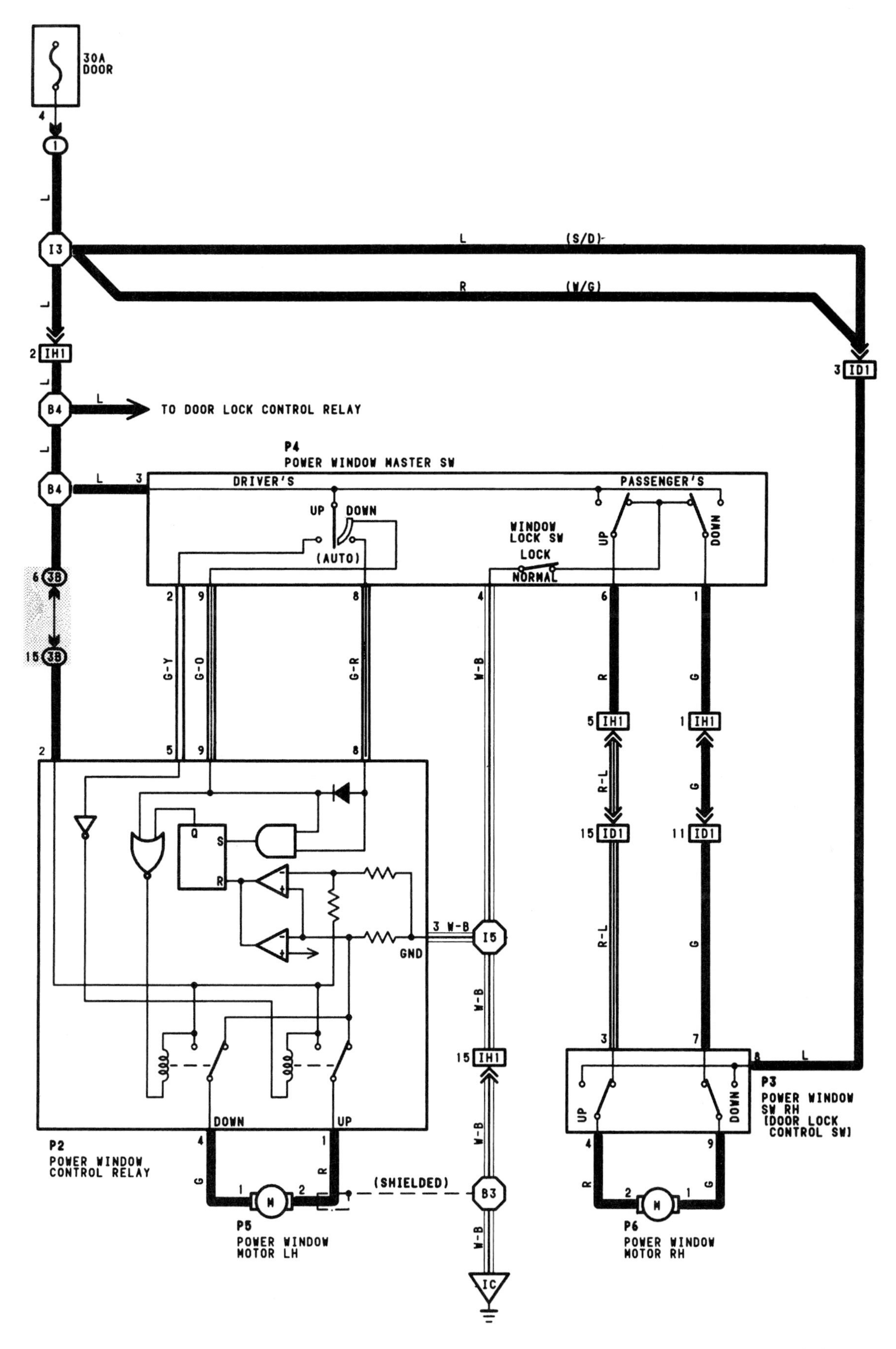

Figure 13-3. A wiring schematic for a power window circuit. Trace the circuit from the switches to the power window motors. (Toyota)

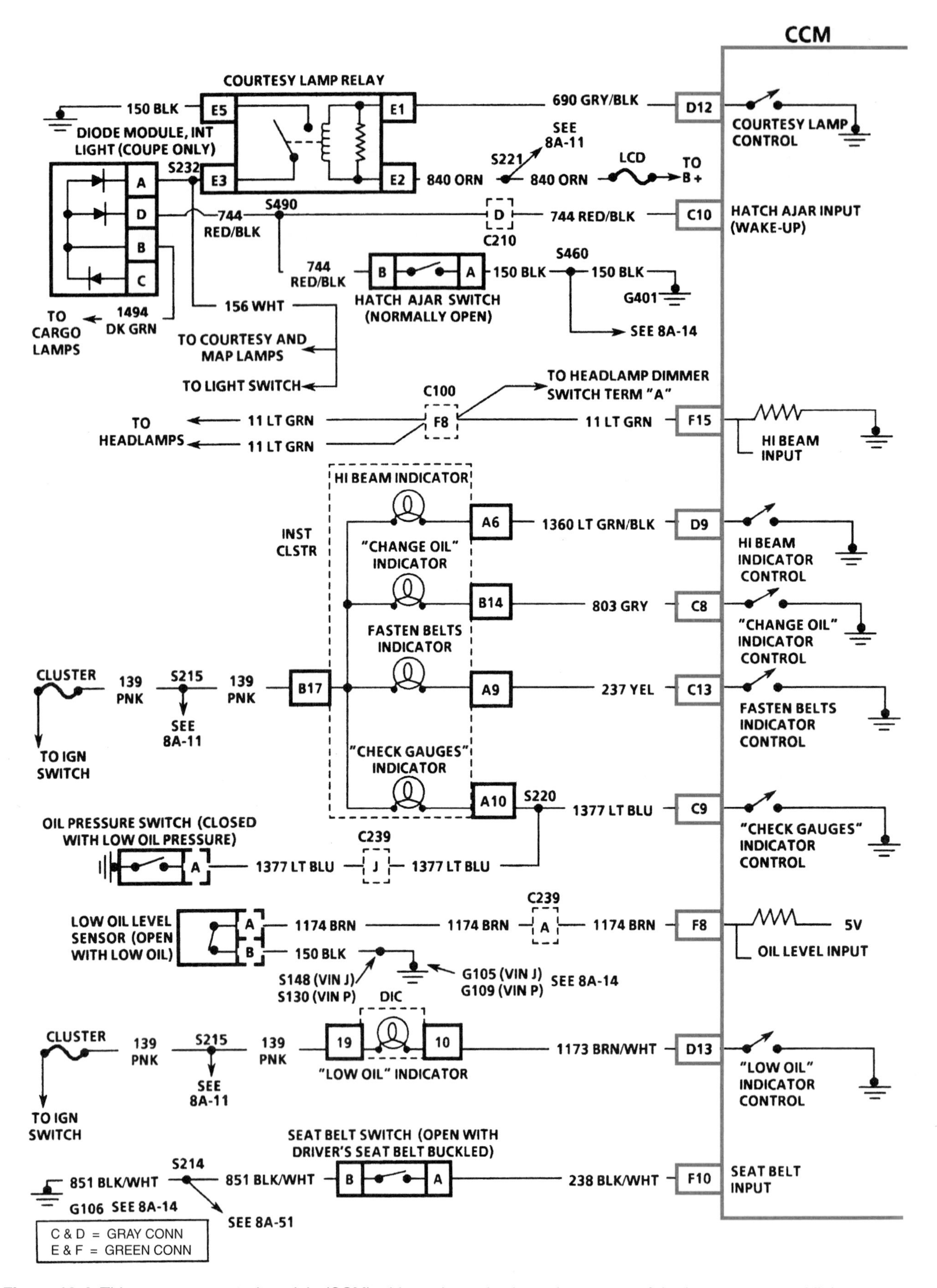

Figure 13-4. This computer control module (CCM) wiring schematic shows how many of the instrument panel lights are operated from the computer. (General Motors)

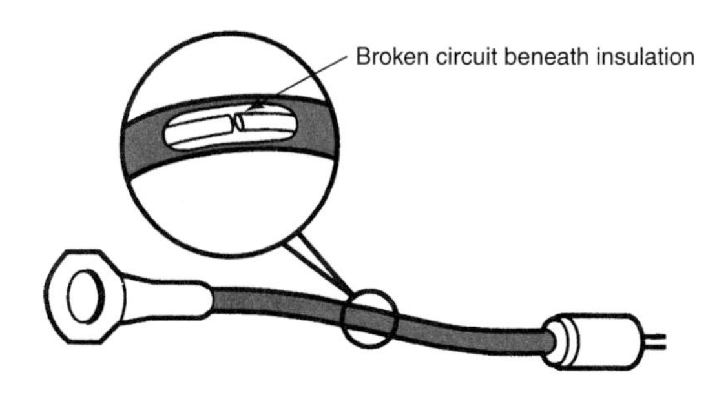

Figure 13-5. This fusible link has melted apart after a short circuit. When replacing fusible link, always use the correct length and gauge. (General Motors)

Circuit breakers are excess-current protection devices that automatically reset. The advantage of the circuit breaker is that it automatically resets when current stops and it has time to cool off. Circuit breakers are used on systems vital to vehicle safety, such as the headlights, and in high-current motors used in tailgate windows and convertible tops. The headlight circuit breaker is located in the headlight switch, while other circuit breakers are installed in the fuse block or at the protected motor.

Chassis Wiring Service

The most common problem encountered in electrical service is *blown (*melted) fuses. Sometimes fuses overheat or vibrate apart after long service. However, the most common cause of blown fuses is a short in the wiring or an electrical device, or an overload in an electric motor. Fusible links can also melt and stop carrying current. A melted link can be spotted by insulation that is discolored or burned off completely. Pull gently on the ends of the fusible link, since they often melt at the end connections where the break is not visible. If the fusible link wire can be pulled apart or feels springy, the wire has probably come apart inside of the insulation. Simply replacing a blown fuse or fusible link may not solve the problem. Always find out why the fuse or fusible link has blown or circuit breaker has opened and correct the problem.

Caution: Do not replace any fuse or circuit breaker with a metal slug or other non-fused connection. This can melt wires, destroy components, or cause a vehicle fire. For the same reason, do not replace a fusible link with regular wire.

Other wiring problems include bad connections at plug-in connectors caused by dirt, grease, corrosion, overheating, or short circuits. Wires can also short inside of the harness due to vibration or from being pinched between vehicle parts. In many cases, careful inspection of the wiring will reveal the damaged section. To check wiring problems that are not obvious, start the engine. With the affected electrical unit operating, wiggle, twist, tap, and generally flex all underhood electrical connectors as you observe the operation of the device. If the device quits working, works intermittently, or experiences drastic changes in its operation when a connector is flexed, the connector should be taken apart for further inspection.

Vehicle Lights and Switches

Vehicle lights consist of two major groups, safety lighting and convenience lighting. Safety lighting devices are the headlights, taillights, marker lights, instrument panel lights, brake lights, turn signal lights, and backup lights. Convenience lighting consists of interior lights and courtesy lights as well as trunk lights, underhood lights, and glove compartment lights.

Safety Lights

Headlights are operated by the driver through the headlight switch. **Figure 13-6** shows an older knob type headlight

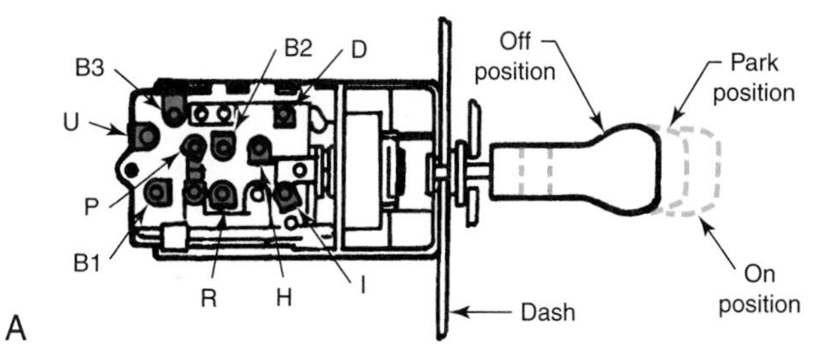

Switch position	Continuity between
Off	B1 to P optical horn
Park	B1 to P optical horn B2 to R park lamps B3 to U headlamps on warning circuit
On	B1 to P optical horn B1 to H headlamps B2 to R park lamps B3 to U headlamps on warning circuit

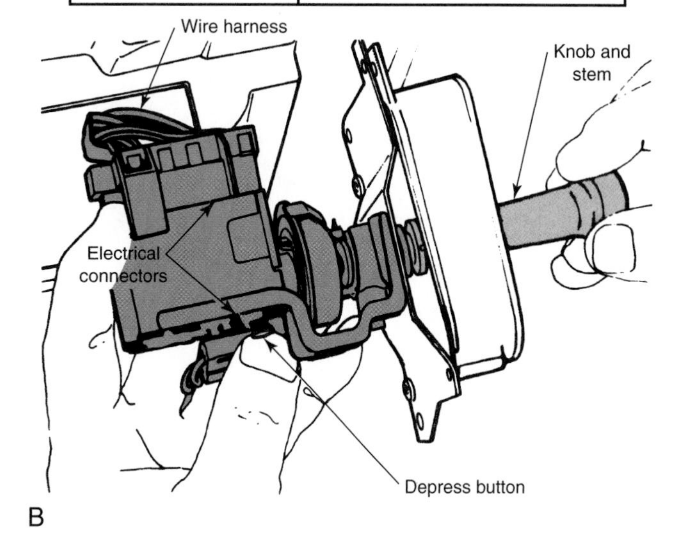

Figure 13-6. This is a push-pull style headlight switch, still used on some vehicles. A—Switch position and continuity. B—Complete switch and wiring harness. Note that depressing the button with your finger allows the knob and stem to pull out of the switch itself for removal from the dash. (DaimlerChrysler)

switch, while **Figure 13-7** shows the newer rocker headlight switch. The headlight switch also operates the taillights and marker lights. The headlight switch also contains a ***rheostat,*** which is used to adjust the brightness of the instrument panel lights and may also be used to operate the interior lights when the doors are closed. Most headlight switches contain a circuit breaker for the headlights instead of a fuse. After passing through the headlight switch, the headlight wiring passes through a dimmer switch, which allows the driver to select high or low headlight beams. The dimmer switches on older vehicles were located on the floorboard by the driver's left foot, but late-model vehicles have the dimmer installed on the turn signal lever.

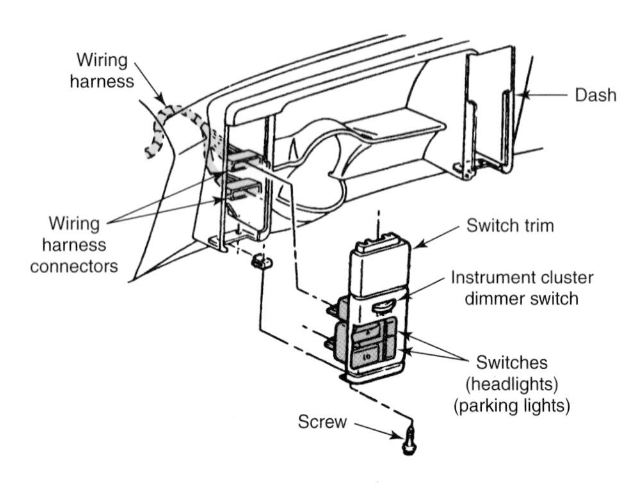

Figure 13-7. This figure illustrates a rocker style, dash-mounted headlight/parking light switch with wiring harness and connectors. (General Motors)

Headlights on older vehicles are known as ***sealed beams.*** Sealed beams are a complete sealed headlight assembly with a heavy glass lens and mirrored interior surfaces to reflect the light produced by the ***filament*** (light-producing element). The sealed beams used on vehicles with two headlights and the outer or top lights on four headlight systems contain two filaments, one each for low and high beams. The inner or bottom lights on four headlight systems contain only a high beam filament. **Figure 13-8** shows the older round-type sealed beam and the more current rectangular sealed beam.

Newer vehicles use ***halogen*** headlights. Some vehicles are equipped with halogen headlights in the form of a sealed beam unit. However, most late-model vehicles are equipped with composite headlights. Instead of a complete sealed beam unit, the composite uses a halogen bulb with a separate filament, **Figure 13-9.** This halogen gas-filled bulb can be replaced without removing the entire headlight assembly. The bulb is installed from the rear of the capsule headlight assembly. The halogen gas allows the filament to be much

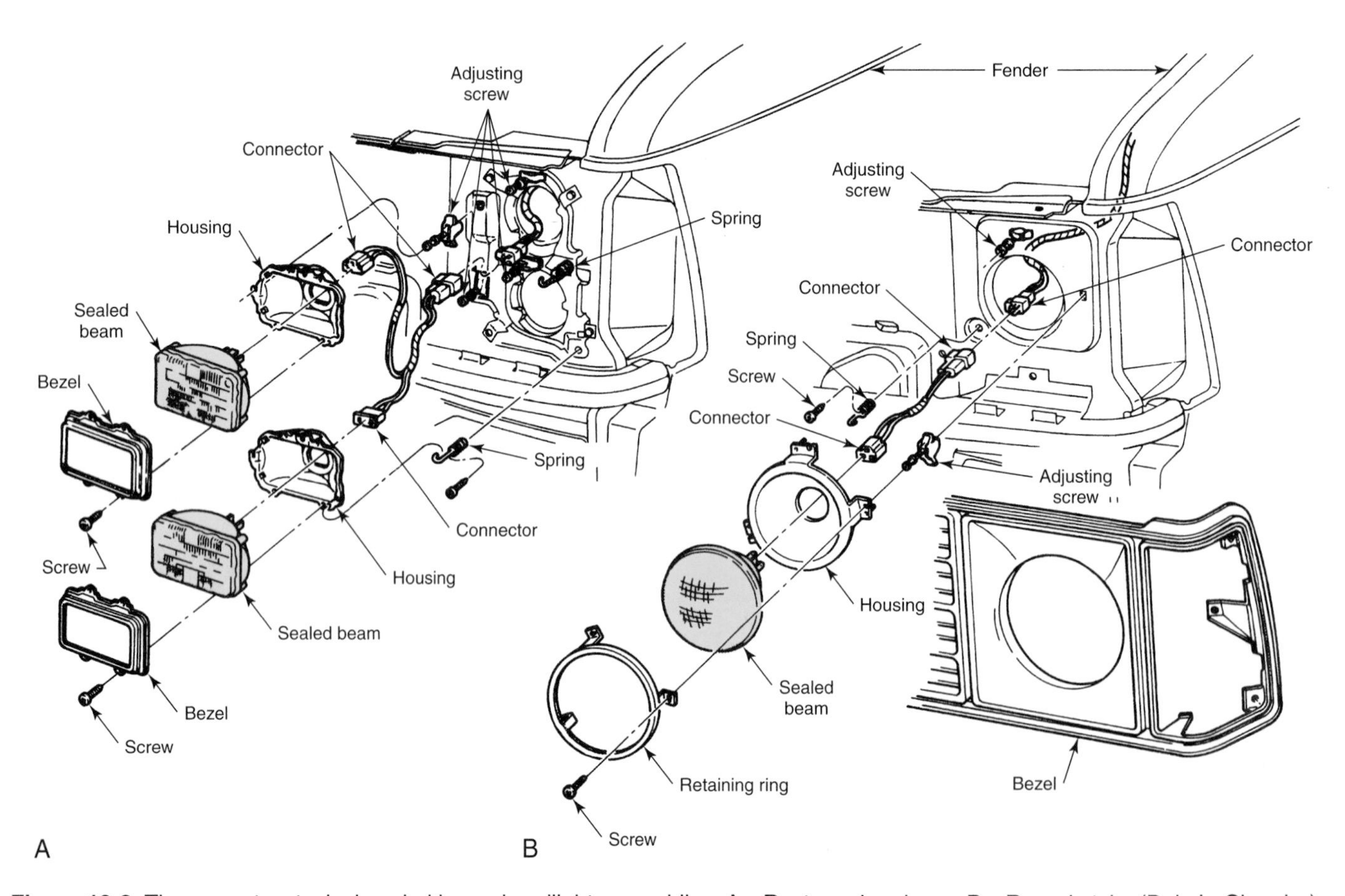

Figure 13-8. These are two typical sealed beam headlight assemblies. A—Rectangular shape. B—Round style. (DaimlerChrysler)

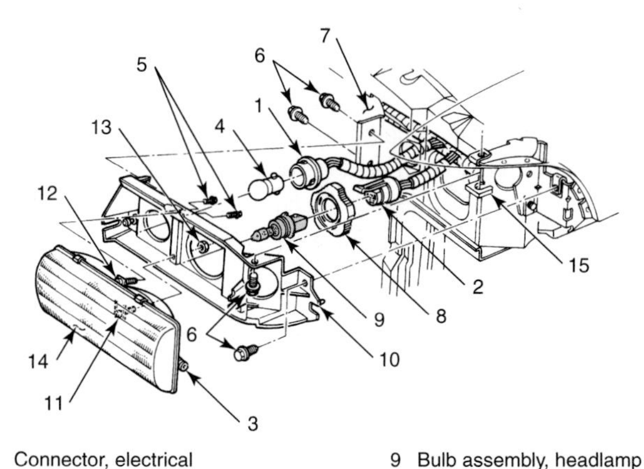

1 Connector, electrical
2 Connector, electrical
3 Pivot, slide
4 Bulb, front parking and turn signal lamp
5 Bolt/screw, headlamp bracket
6 Bolt/screw, headlamp capsule
7 Support assembly, radiator
8 Retainer, headlamp bulb
9 Bulb assembly, headlamp
10 Bracket, headlamp
11 Bolt/screw, headlamp bracket
12 Nut, headlamp bracket
13 Nut, headlamp bracket
14 Capsule assembly, headlamp
15 Shim, headlamp

Figure 13-9. This headlight composite assembly uses a removable halogen headlight bulb. When the bulb burns out, the entire assembly does not need to be replaced. (General Motors)

brighter than older sealed beam designs with about the same service life.

Daytime Running Lamps

Some modern vehicles have ***daytime running lamps.*** These lamps are a safety feature, and make the vehicle more visible during daylight hours. Daytime running lamps are part of the headlight assembly but are wired separately from the other front lights. They are connected to the ignition switch and illuminate whenever the ignition is on.

To diagnose daytime running lights, determine whether they are on whenever the ignition switch is on. If the light on only one side of the vehicle is illuminated, the most likely problem is a burned-out bulb. If both lights are off, the fuse may be melted or there may be a problem with the system wiring or switches.

Tail and marker lights are simple filament lightbulbs, similar to the lightbulbs found in homes. Some taillight bulbs contain two filaments, one for the taillights and one brighter filament for the brake and turn signal lights. Many front marker bulbs also contain two filaments, one for the marker lights and a brighter filament for the turn signals. Modern practice however, is to have separate bulbs for each function. Backup light switches are usually located on the transmission linkage and operate when the vehicle is placed in reverse gear. Backup lights are single filament bulbs, with no connection to the other light assemblies.

Brake Lights and Switches

Brake lights are either single or dual filament lightbulbs operated by a switch attached to the brake pedal. The switch is a simple on-off type. Older ***brake light switches*** are pressure switches, operated by hydraulic pressure in the brake system. Newer brake light switches are mechanical. These switches are usually installed on the brake pedal and energized when the brake pedal is pressed. On some older vehicles, the brake light wiring passes through the turn signal assembly so the brake lights are bypassed when the turn signal is being used. Modern vehicles usually have separate bulbs for the brake and turn signals.

Turn Signals, Flasher Lights, and Switches

Turn signals are operated by the vehicle driver. The selector lever contains a set of contacts that send current to the turn signal lights on the left or right side of the vehicle, as well as left-right indicators on the instrument panel. The turn signal wiring contains a ***flasher unit,*** which causes the lights to turn on and off rapidly, **Figure 13-10.** The turn signal flasher unit is usually installed on the fuse block or in a special holder under the vehicle dashboard.

The turn signal filament is usually in the same bulb as the brake light filament. At one time, however, the turn signals and brake lights used the same filament, usually in the same bulb as the taillight filament. The ***emergency flasher switch*** contains contacts that send current to the turn signal lights on both sides of the vehicle. The emergency flasher is separate from the turn signal flasher and is usually installed on the fuse block or under the vehicle dashboard.

Instrument Panel Lights

There are two classes of instrument panel lights. The first, illuminating lights, light up the speedometer, gauges, and the wiper, headlight, and heater/air conditioner controls. The second group, indicator lights, warn of engine problems such as charging system problems, overheating, and low oil pressure.

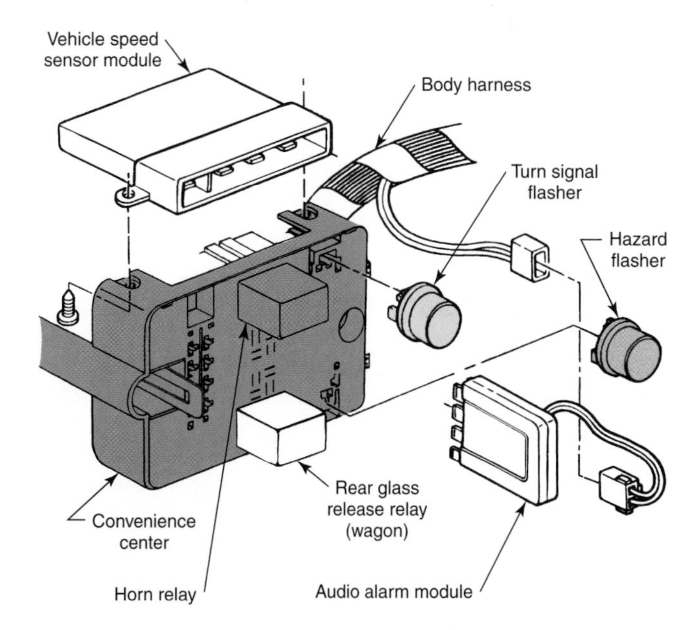

Figure 13-10. Turn signal flashers are usually installed in one of the vehicle fuse boxes. The fuse box shown here is called a convenience center, a common manufacturer's term for fuse boxes. The flasher is normally pushed in or pulled out of the socket for testing or replacement. (General Motors)

The illuminating lights are controlled through the headlight switch, which uses a rheostat to vary their brightness. There are usually several lights, although some modern vehicles use only a few bulbs and distribute their light through a ***fiber optic*** (light-carrying) harness. Indicator lights are installed on the instrument panel and operated by senders. Typical indicator lights monitor engine temperature, oil pressure, and charging system condition. Indicator light senders are simple on-off switches, operated by temperature or pressure changes. The charging system light is usually operated by the alternator voltage regulator. Some indicator lamps, such as the check engine and anti-lock brake indicators, are operated by an external control module.

Digital Instrument Panel Lights

Most digital instrument panel lights are ***light emitting diodes*** (LEDs) or ***vacuum fluorescent displays.*** **Figure 13-11** shows a typical digital instrument panel, sometimes called a digital dashboard. On many vehicles, the instrument panel contains both digital and conventional lights. The digital components are usually serviced as a unit. If a digital instrument panel fails to light up, check the instrument panel fuses and ground wires. If after making these checks, the entire instrument panel is dark, it may require replacement.

Convenience Lights

This group of lights includes all of the lights that are not required for vehicle safety, but are installed to illuminate the passenger sections of the vehicle. The vast majority of these lights are single filament lights operated by on-off switches installed in the door jams or glove compartment, or auxiliary switches in the vehicle's interior or trunk.

Vehicle Light Service

The most common problem in the vehicle light circuit is a burned out bulb. However, other problems can cause a light to fail to illuminate. Many problems are caused by defective switches. On older vehicles, a wire often becomes disconnected, or a connector plug develops high resistance. This is especially common with tail and marker lights where the connector is exposed to water and dirt.

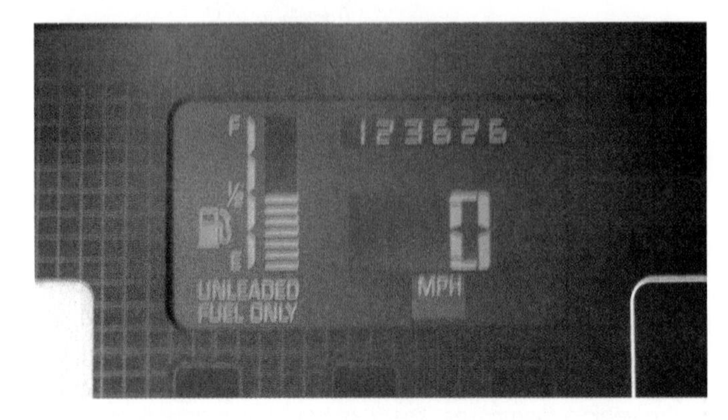

Figure 13-11. A digital instrument panel usually contains some nondigital lights and gauges. On this panel, the speedometer, odometer, and fuel gauge readouts are digital.

On older systems that use two-filament bulbs, one filament can burn out while the other continues to operate. It is also possible for the burned out filament to contact the other filament, causing electrical feedback. This problem can cause unusual symptoms, such as the front marker lights coming on when the brakes are applied. If a lightbulb is broken, the base can usually be removed with pliers.

Caution: When changing halogen lightbulbs (composite), be extremely careful not to touch the bulb surface with your hands. Halogen bulbs are coated with a special substance, and the oil from your hands can cause premature bulb failure. In some cases, the bulb can shatter with considerable force.

Adjusting Brake Light Switches

Pressure switches are not adjustable and should be replaced when they are defective. After the switch is replaced, the brakes should be bled, as explained in Chapter 34, Brake Service. Mechanical switches are adjusted so that only a small amount of brake pedal movement is required to operate the switch. **Figure 13-12** shows the adjustment distance for one specific installation. After installing the switch on most vehicles, pull back on the brake pedal. This ensures that the stoplight switch does not prevent the brake pedal from returning to the fully released position. Some brake light switches adjust automatically.

Instrument Panel Gauges

On modern vehicle instrument panels, ***gauges*** are often used instead of warning lights. Gauges commonly record oil

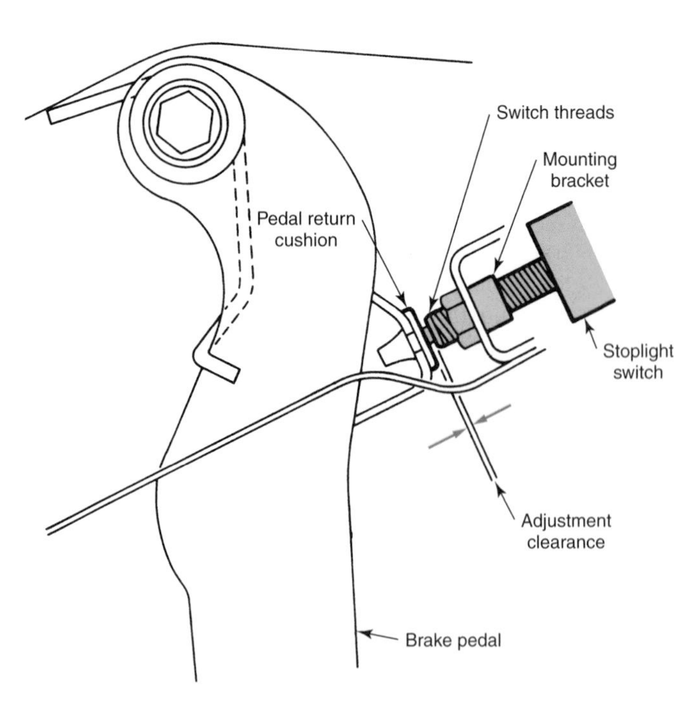

Figure 13-12. One stoplight switch arrangement is shown here. (General Motors)

pressure, engine temperature, charging system operation, vehicle speed, and fuel level. Some vehicles have gauges for engine RPM, intake manifold vacuum, or turbocharger boost pressure. Most instrument panels use both gauges and lights, **Figure 13-13.** Gauges can be electro-mechanical or electronic. Electro-mechanical gauges rely on current flow through a small coil to register such engine operating conditions as temperature, oil pressure, alternator output, and fuel level. These gauges are wired to sensors (sometimes called senders) that vary the amount of current that flows through the gauge coil. Variations in current flow affect the magnetic field in the gauge coil. The field strength of the gauge coil determines the position of the indicator needle. Charging system gauges are usually voltmeters, which measure the alternator output voltage.

The electronic gauges on newer vehicles also use sensors. However, the sensor output is usually sent to an electronic module or an on-board computer. The module or computer analyzes the signal and delivers an output to the gauge. In addition to the needle-type gauge, many modern vehicles use light-emitting diodes (LEDs) or vacuum fluorescent displays to indicate engine operating conditions.

Many modern vehicles have a combination of electronic and electro-mechanical gauges. On many late-model vehicles, the speedometer is a mechanical device that relies on the rotation of a small magnet in the speedometer head to move the speedometer needle. Some very old vehicles use direct mechanical oil pressure and temperature gauges.

To check the operation of electro-mechanical gauges, first check the gauge fuse. If the fuse is OK, unplug the electrical connection from the related sensor. If this causes a change in the gauge reading, you can assume that the gauge is operating and that the problem is in the sensor. To double-check the condition of the gauge and sensor, ground the sensor connection to the engine or vehicle body. If disconnecting or grounding the sensor wire has no effect on gauge operation, the gauge itself is probably defective.

Figure 13-13. This instrument panel has a combination of gauges and warning lights. The vehicle engine and drive train computer operates all of the gauges, including the speedometer and tachometer. The computer bases it outputs on sensor inputs.

Some electro-mechanical gauges use a ***voltage limiter*** to ensure that the voltage to the gauge is the same regardless of the electrical system voltage. A voltage limiter reduces the amount of gauge fluctuation when other electrical devices are being used. If the voltage limiter fails, all gauges in series with the limiter will fail to operate. A defective limiter may also cause gauge readings to appear to be higher or lower than normal.

Electronic gauges cannot be checked by the methods used for electro-mechanical gauges. Unplugging the sensor will sometimes indicate whether the problem is in the sensor or the gauges. Under no circumstances, however, should the technician ground an electronic sensor connection. This will damage the electronic circuitry. Often the only way to check an electronic gauge is by using a special, dedicated tester or substituting a sensor that is known to be good.

If an electronic gauge fails to light up, the gauge itself may be defective. Before condemning the gauge, check the system fuse. Also check the instrument panel ground wires. If all of the gauges fail to light up, or if the entire instrument panel is dark, the instrument panel may require replacement.

Chassis-Mounted Motors

Chassis-mounted motors include the windshield wiper motor, power windows and vent windows, tailgate windows, and sunroof motors, as well as motors that operate convertible tops and power antennas. The modern vehicle may have as many as ten small dc motors to operate various systems. Except for the windshield wiper motor, switches used to control these motors are spring-loaded so they return to the off position when the operator's hand is removed. This prevents accidental operation of the motor, which could result in overheating and damage.

Windshield Wiper Motors and Controls

Windshield wiper motors are high torque motors, capable of overcoming the binding effects of ice on the wiper blades and mechanisms, **Figure 13-14.** For this reason, wiper motors are durable in normal service, requiring relatively little maintenance. Windshield wiper motors have special ***park mechanisms,*** which cause the wiper blades to return to the fully down position when the wipers are turned off. Some older wiper motors were equipped to operate the windshield washer pump, but most modern designs use a separate pump located in the washer reservoir.

Power Window, Tailgate, Convertible Top, and Sunroof Motors

These are generally high-torque motors that are able to overcome the drag of dry or tight operating mechanisms and glass channels. Most of these motors have built-in overload switches or circuit breakers to prevent overheating damage. The convertible top motor is unique in that, instead of working directly to raise and lower the top, it is used to drive a hydraulic pump, which powers hydraulic cylinders that operate the top. The majority of these motors are operated by on-off switches located on the vehicle dashboard, door, or console. See **Figure 13-15.**

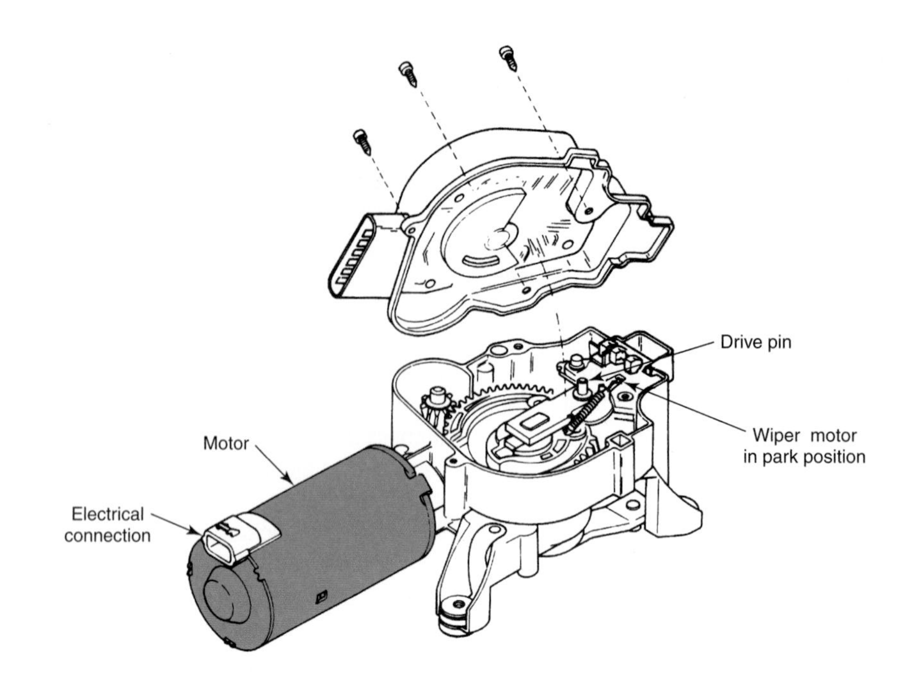

Figure 13-14. A—Wiper motor and gear assembly. B—Wiring schematic for an intermittent (pulse) wiper system. Trace the various circuits until you are familiar with each. (General Motors)

Power Antenna Motors and Switches

Power antenna motors are simple in construction and are usually operated through an automatic relay or by a manual three-position switch. The relay is energized when the radio is turned on or when the ignition switch is turned on when the radio is already on. The manual switch allows for positioning the antenna in any position that best suits the driver.

Electric Motor Service

Electrical motor service in the average vehicle is confined to ensuring the motor is at fault and replacing it. Most of the small electric motors used in these applications are exchanged for rebuilt units. In some cases, as with power antennas, the entire assembly is replaced when the motor goes bad.

Before deciding that a motor-operated device has an electrical problem, check that the mechanical parts are clean and lubricated. Many apparent motor failures can be fixed by cleaning and lubricating the mechanical linkage. Cleaning and lubrication can often restore windshield wipers, power seats, antennas, and sunroofs. Also make sure that movement is not being blocked by debris or, in the case of windows, dry glass

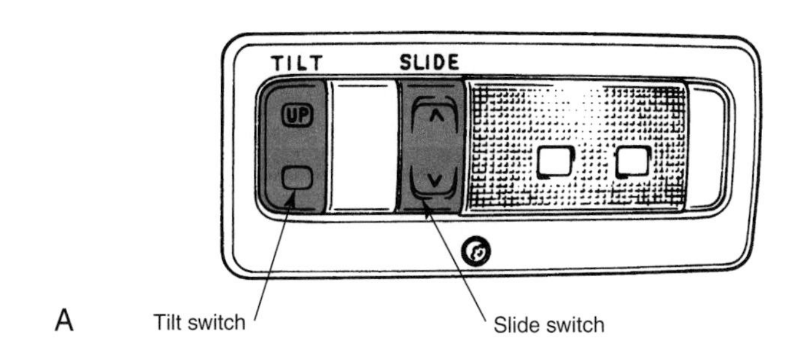

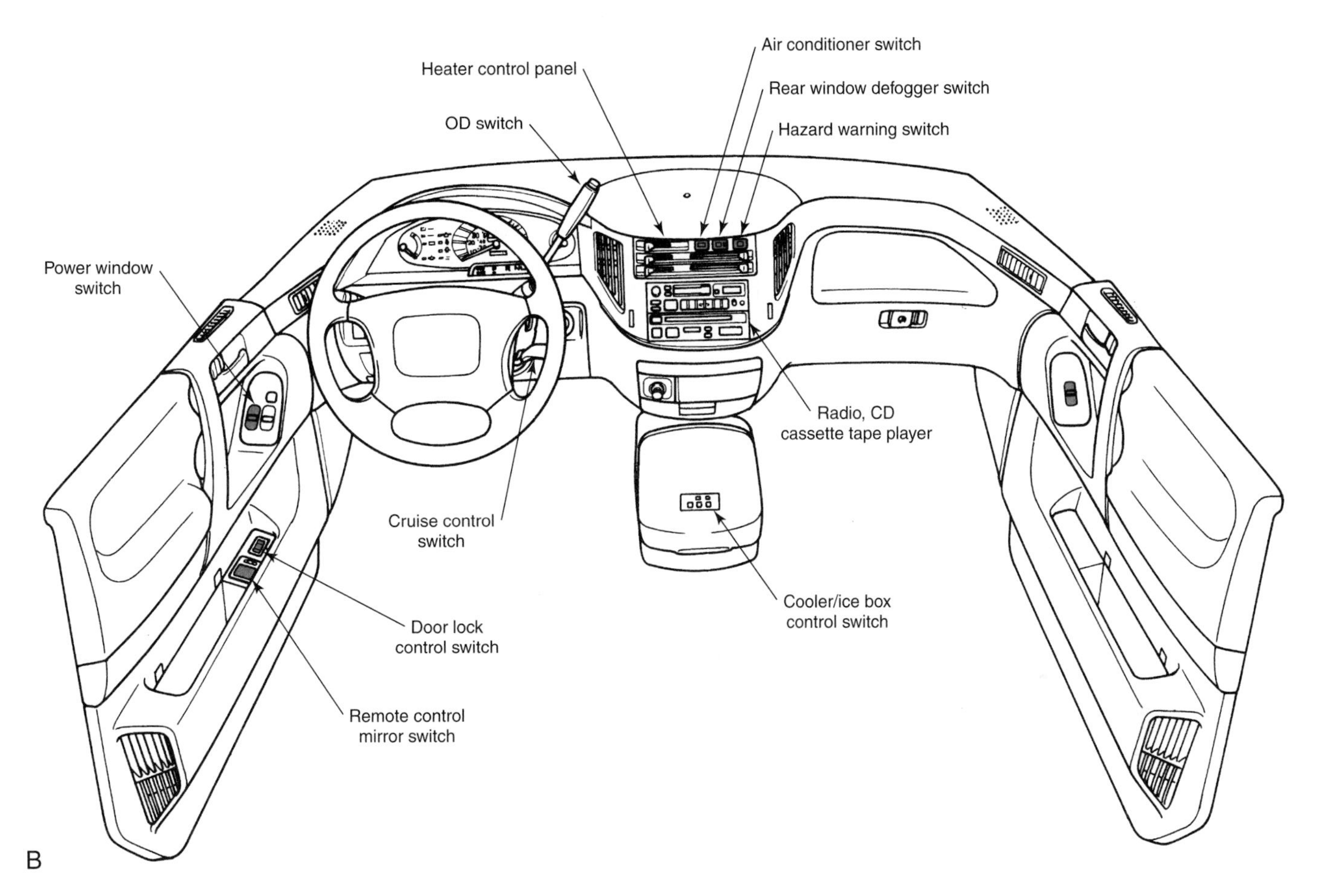

Figure 13-15. A—Sunroof tilt and slide switches. B—Vehicle control switches in their various locations as used by one manufacturer. (Toyota)

channels. If an inoperative convertible top uses a hydraulic system, check the fluid reservoir level.

To check out a vehicle electric motor, bypass the controls (switches, relays) with a fused jumper wire. If the motor operates, the problem is in the control switches or wiring. If the motor does not work, it is defective. Most electric motors are replaced easily. Remove the electrical connectors first, and then remove the mounting bolts or other fasteners. Some power window motors require that holes be drilled in the door-frame to remove the motor-attaching hardware. **Figure 13-16** shows a typical window motor installation.

Caution: Some motor-operated devices, such as door or tailgate glass, may drop when the motor is removed. Always consult the manufacturer's service manual, and block the mechanism in place if necessary.

Some motors, such as those used on windshield wipers and door motors, may require special positioning or aligning when they are reinstalled to ensure the mechanism stops in the correct place. When reinstalling any motor, make sure its ground straps, if it has any, are in place.

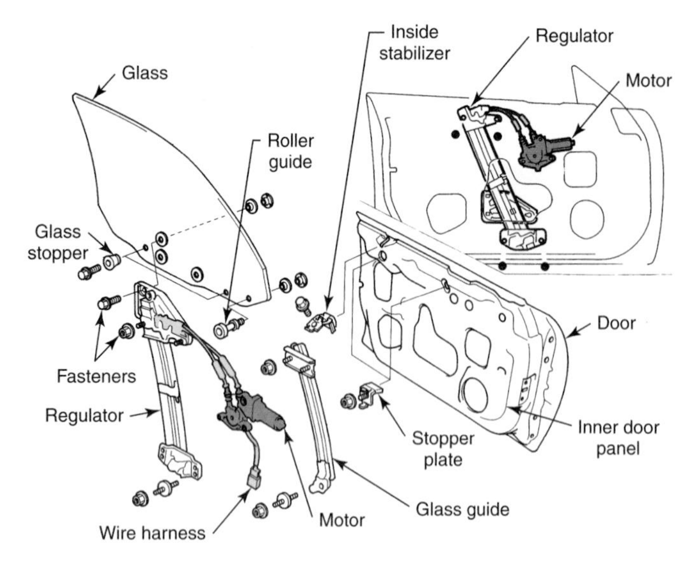

Figure 13-16. A typical power window assembly using an electric motor to cycle the window glass up and down. The motor and regulator mechanism are held to the inner door panel with nuts and bolts. (Honda)

Chassis-Mounted Solenoids and Relays

Chassis-mounted solenoids are similar in construction to the solenoids used to energize the vehicle starter. They are used to operate door locks and trunk releases. ***Relays*** are a type of solenoid that closes electrical contacts, instead of operating a mechanical device. They are used to direct large current flows that might damage a switch. Switches used to control most solenoids are spring-loaded to return to the off position, preventing accidental operation of the solenoid.

Power Door Lock Solenoids and Switches

Power door lock solenoids are two-position solenoids. The solenoids have two windings, allowing them to move a control rod in two directions. When the control switch is moved in one direction, the solenoid moves the door lock to the unlocked position. When the switch is moved in the other direction, the solenoid moves the door lock to the locked position. Wiring is usually straightforward, with separate circuits for the locked and unlocked solenoid positions. See **Figure 13-17.**

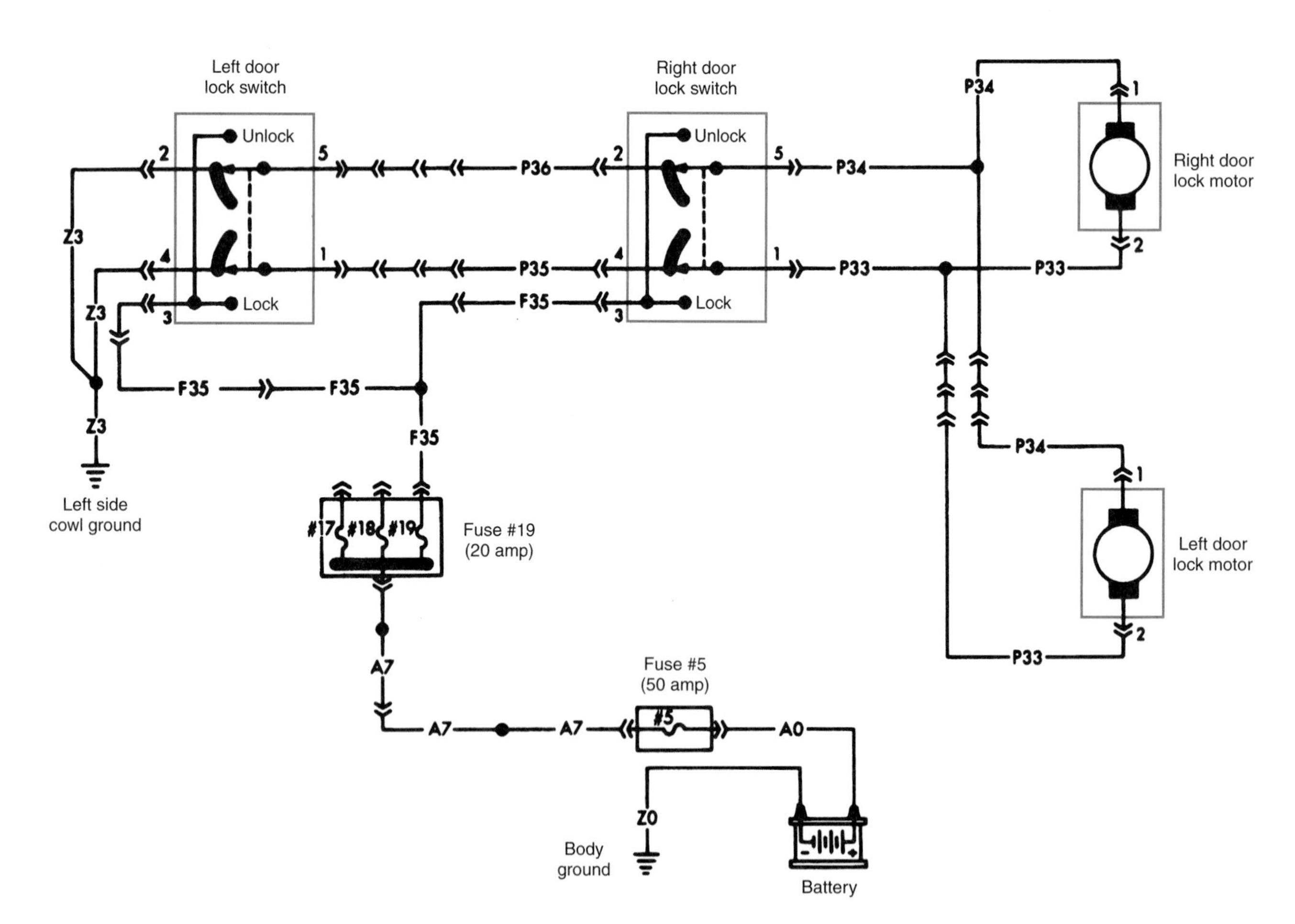

Figure 13-17. This is a wiring schematic for one power door lock circuit. This system uses reversible motors to actuate the linkage. (DaimlerChrysler)

Trunk Release Solenoids

Trunk release solenoids are single position solenoids operated by a switch installed in the vehicle glove compartment, **Figure 13-18.** Power to the trunk release switch passes through the ignition switch. This keeps unauthorized persons from opening the trunk.

Solenoid Service

Although an ohmmeter can be used to check the solenoid windings, the simplest way to check out a solenoid is to bypass the control switch with a jumper wire. If the solenoid operates, the problem is in the control switch or wiring. If the solenoid does not work, it is defective. Occasionally an inoperative solenoid is caused by sticking, misaligned, or bent linkage on the device to which it is attached. Always check this possibility before replacing the solenoid. The easiest method is to disconnect the solenoid and then check its operation. Solenoids cannot be repaired and are replaced after ensuring the solenoid is the cause of the problem.

Control Relays

Control relays are similar to starter solenoids. A small current flow energizes a magnetic winding, closing a set of contact points, which send a heavy current to another unit, such as a motor or heating coil. Use of the relay eliminates having to pass heavy current through the dashboard controls, possibly overheating them and consuming extra electricity fighting the resistance of the extra wiring. Many relays used on late-model vehicles are power transistors with no moving parts. This cuts down on electricity consumption and is more reliable. Relays are used to operate many heavy current-consuming electrical devices on the modern vehicle. Relay service is confined to verifying the relay is not functioning and replacing it.

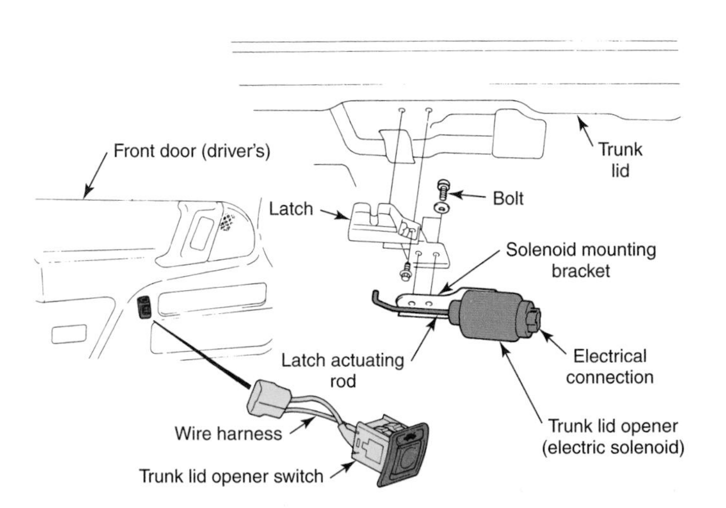

Figure 13-18. This illustration shows a trunk release solenoid and the solenoid's operating switch. (Hyundai)

Security Systems

Security systems protect the vehicle from illegal entry. The body control module (BCM) operates the security system on newer vehicles. When the security system senses an attempt at forced vehicle entry, it sounds an alarm and disables the vehicle's fuel or ignition system. When the security system is ***armed,*** it is able to detect and react to illegal entry attempts. When the security system is ***disarmed,*** it is inactive. The security system may be armed and disarmed by a battery-powered remote control or by a Pass-Key system, described below.

While security systems are complex, some simple tests can be made. The easiest way to test a security system is to simulate an illegal entry. A general procedure is to roll down the driver's side window and then arm the system. Next, reach through the open window and unlock the door. The alarm should sound when the door is opened

On late-model vehicles, a defect in the security system will set a trouble code. Use a scan tool to access the diagnostic information and proceed to check the components identified by the trouble codes. Once the defective security system part has been identified, it can be located and replaced.

Pass Key Systems

Some late-model cars and trucks have what is known as a ***Pass-Key*** system. Pass-Key systems keep the vehicle from being started if the wrong key is inserted in the ignition switch. Some systems are part of a larger security system that will also sound an alarm if the wrong key is used. A Pass-Key system can be recognized by the small resistor built into the ignition key, **Figure 13-19.** In addition to the special key, the system has a small Pass-Key module. When the key is

Figure 13-19. The key of a Pass-Key system can be identified by the built-in resistor. Pass-Key systems were updated over several model years, and the technician should be sure of exactly which system is being serviced.

inserted into the switch and turned, an electrical circuit is completed between the resistor and the Pass-Key module. The current tells the module whether the key is the correct one for the vehicle. If the resistance is correct, the module allows the vehicle to be started. If the key resistance is incorrect, the module disables the starting and fuel delivery system for two to three minutes. The two-three minute time-out prevents potential car thieves from quickly trying different ignition keys until one with the correct resistance is located.

A later version of the Pass-Key system has a small transponder (radio frequency transmitter) in the ignition key. When the key is turned in the ignition switch, the transponder sends a radio signal to the control module. The module uses this signal to determine whether the key being used is the correct one. This system can be identified by a small "p3" stamped on the ignition key near the resistor.

Another version of the Pass-Key system is called the Pass-Lock system. This system can be identified easily since it is the only one that uses a standard ignition key. A Hall-effect sensor reads the key teeth and produces a signal that is read by the control module. All Pass-Key systems use an indicator light. This light usually illuminates the word "security" on the instrument panel. The security light activates during startup and remains on for about five seconds showing the driver that the system is working.

Pass-Key Problems

The security light may or may not illuminate when a Pass-Key system problem is encountered. Most Pass-Key problems prevent the vehicle from starting. A common problem is loss of tension at the ignition-switch resistor contacts. This causes an open circuit, which the module reads as use of an incorrect key. If the open circuit occurs after the engine has been started, the engine will remain running and the security light will come on. With early systems, the engine does not restart if it is switched off. On later systems, a malfunction that occurs while the engine is running disables the Pass-Key system, allowing the vehicle to be restarted. The security light will remain on until the problem is corrected.

The earliest Pass-Key system can only be diagnosed by checking all electrical connections with an ohmmeter. Disconnect the Pass-Key wires at the base of the steering column. If the ohmmeter shows no continuity (infinite resistance) or intermittent continuity, the switch contacts are probably bad. Later Pass-Key systems are diagnosed by attaching a scan tool to the vehicle diagnostic connector and accessing trouble codes. The Pass-Lock system can only be diagnosed with a scan tool.

Pass-Key Reprogramming

If the ignition switch, ignition key, or Pass-Key module is replaced for any reason, the system must be reprogrammed. The system can be reprogrammed using a scan tool, or by using the manual method described below. To reprogram the system manually, attempt to start the vehicle. If it does not start or stalls almost immediately, leave the ignition switch in the run position until the security light stops flashing. This may take up to 10 minutes. Then, attempt to restart the vehicle a second and third time, leaving the key in the run position each time. After the third sequence, the Pass-Key system should be reprogrammed. If the vehicle still will not start, perform the sequence again. The latest Pass-Key systems with a transponder can be reprogrammed more quickly, but you must have the original ignition key to reprogram without a scan tool.

Installing a Security System

It is possible to install a security system on a vehicle that was not originally equipped with one. Most aftermarket security systems will sound the horn and flash the headlights if an illegal entry is attempted. Most systems consist of a control module, door, trunk and hood switches, and a remote device for energizing the system.

To install an aftermarket security system, closely follow the maker's instructions. Often, holes have to be drilled to install the switches. The module should be installed inside of the vehicle to minimize damage from water and dirt. After all of the components have been installed, run wires to connect the components. After installation, make sure the system operates correctly.

Remote Keyless Entry System

The ***remote keyless entry system*** permits locking and unlocking of the vehicle's doors and/or trunk lid using a hand-held radio transmitter. These battery-operated transmitters are electronically coded to each vehicle. Keyless entry systems consist of a battery-operated transmitter and a vehicle-mounted control module (with a receiver). Almost all vehicles using a keyless entry system are equipped with two transmitters. Pressing the proper button on the transponder sends a signal to the module. Based on the transponder input, the module energizes the door lock solenoids to lock or unlock the vehicle doors as requested. The radio signal used is a low frequency wave that is unlikely to interfere with other electrical devices. If the system fails, the vehicle can still be unlocked with the key. The door key should always be carried with the transmitter in case the keyless system fails to work.

Keyless Entry System Problems

If a keyless entry system will not unlock the vehicle doors, first check that the power door lock system operates properly from the door controls. If the power door locks are working properly from the door switches, try to operate the system using both transmitters. If the system will not operate with either transmitter, and the door locks are working properly, the control module is probably defective. If the system operates with one transmitter, but not the other, check the battery of the suspect transmitter. When the system has only one transmitter, or one transmitter is lost, it may be necessary to substitute a known good transmitter or control module and retry. Other causes of keyless entry problems are radio interference from nearby electrical devices, and signal blockage from other vehicles, trees, shrubbery, or buildings.

To prevent part or system damage, follow the manufacturer's instructions when replacing the transmitter's battery or working on the system. If any part of a keyless entry system is replaced, the system will need to be reprogrammed. Reprogramming can be accomplished using a scan tool, or by following a manual procedure outlined in the service manual. Pressing the manual door lock buttons for a set period of time will reprogram most keyless entry systems. When the locks cycle (lock and unlock) once, the system has been reprogrammed.

Radios and Sound Systems

The repair of radios and related equipment, such as cassette tape and compact disc players is beyond the scope of this text. These units are best serviced by a qualified electronics shop. However, some simple checks that can be made before removing the unit for service are explained below.

Diagnosing Radio Problems

If the unit does not come on at all, check the fuse. On aftermarket units, check for the presence of in-line fuses. Also check for broken or disconnected input and ground wires. Make sure the antenna cable is plugged into the radio and the antenna. Check for disconnected wires to the speakers. If the radio has static, the source can be difficult to find. Often, the problem is simple to correct once it is located.

Check that there is no defect in the ignition system or charging system by operating the radio with the key on, but without starting the engine. If the static is only present when the engine is running, unplug the alternator with the engine running. If the noise stops, the stator or diodes are causing the static. Check the plug wires, rotor, cap, and other sources of ignition noise. Make sure the plugs and wires are radio suppression types. Check for a crack in one of the speaker diaphragms. Check all ground straps on the radio, dashboard, antenna, and engine. Also, make sure that shielded wires are intact and that the shielding has not become disconnected. Repair or replace the shielding as necessary. It is sometimes necessary to add additional ground wires to cancel stray magnetic fields. Some manufacturers recommend installing extra filters or capacitors at the radio antenna or sound system's power wire to eliminate static. If these checks do not locate the problem, the radio/sound system must be removed for repair.

Horns

The horn uses a diaphragm with a make and break contact. When energized, the horn diaphragm flexes, or vibrates, creating noise. As it flexes, it breaks the electrical contact inside of the horn body, causing current flow to stop. The loss of current causes the diaphragm to return to its original position, where it is reenergized. This flexing of the diaphragm occurs many times per second, creating a continuous noise.

Horns are operated by steering wheel contacts, which energize a relay. The relay directs heavy current to the horns themselves. A few horn circuits ground the horn circuit directly through the steering column, eliminating the relay. Typical steering wheel contacts are shown in **Figure 13-20.** In some vehicles equipped with air bags, the horn contacts are incorporated with the air bag inflator module.

Most horn problems are caused by defects in the steering wheel contacts. If they do not close, the horn will not blow. If they stick closed, the horn will blow continuously. Servicing the steering wheel contacts usually means at least partially disassembling the steering column. If pressing the horn contacts causes the relay to click, but the horn does not operate, the problem is in the horn, the horn ground, or the relay contacts.

Rear Window and Mirror Defrosters

These devices consist of a heater ***grid*** (a series of fine wires) applied to the glass surface of the rear window or mirror, **Figure 13-21.** The grid material has a calibrated resistance that opposes current flow. As current flows through the grid, it produces heat, causing frost or dew to evaporate. Several components allow the defrosting system to defrost the glass without overheating it or overloading the electrical system. A control switch is located on the instrument panel, and an indicator light is installed in the control switch or nearby.

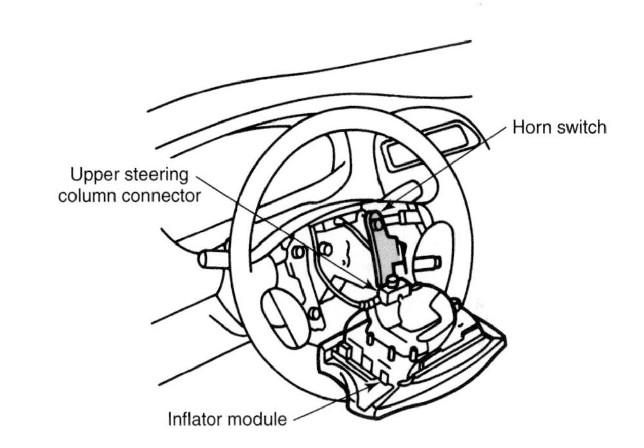

Figure 13-20. Horn switches on the steering wheel. (General Motors)

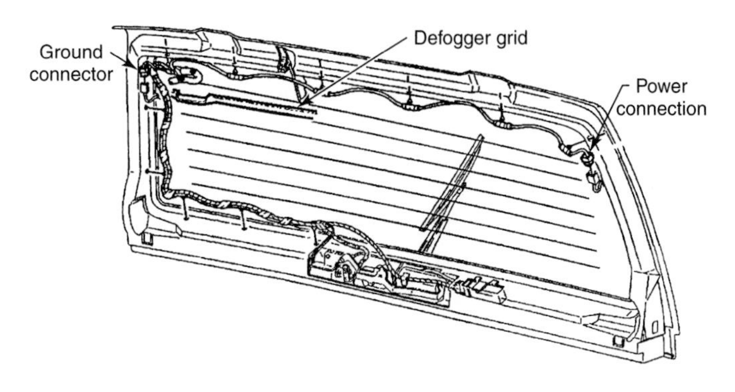

Figure 13-21. This illustration shows a rear window defogger. Note the grid, power connection, and ground connector. (General Motors)

From the control switch, current passes through a timer. The timer shuts off the current flow after enough time has passed to defrost the glass, usually about 10 minutes. The driver can use the control switch to reactivate the defroster if the glass still needs defrosting. From the timer, current passes through the wire grid. The wire grid is made of a metal with relatively high resistance. The current passing through the grid wires causes them to heat up and evaporate dew or frost.

To diagnose a glass defroster, turn the switch on. The indicator light should come on. If the light does not come on, go to the next paragraph. If the light does come on, determine whether the defroster grid is drawing current. Attach a voltmeter to the vehicle's battery terminals. Turn the ignition switch to the on position, with the engine and defroster off, and observe the voltage. Then, turn the defroster on and observe the voltage. Voltage should drop at least 1 volt when the defroster is turned on. On many systems, the voltage drop may be up to 2 volts. If the voltage drop is small or if there is no change in voltage, the defroster is not operating.

If there is any doubt about whether the system is operating, feel the glass at a grid wire. Compare its temperature with that of the glass away from the grid. The glass should be warmer at the grid wire. The same procedure can be used to check heated mirrors. If the rear glass grid does not become hot, make a close visual inspection of the grid wires. A broken wire may sometimes be seen as a gap or discolored spot. If only one grid of a multiple grid system (such as a rear window and mirror heating system) is not working, the problem is in that grid's circuit and not the main control switch or relay.

If the grids are not heating, begin by checking the system fuses or circuit breakers. Some systems use separate relays for the rear window and mirror grids. If the fuses and circuit breaker(s) are intact and not tripped, use a voltmeter or non-powered test light to make sure electricity is reaching the switch. If no voltage is present at the switch and the fuses and relays are good, the problem may be in the wire harness. If voltage is present at the input terminal but not the output terminal, the switch is defective.

If power is leaving the switch and the relay does not operate, the relay may be defective. If possible, bypass the relay by disconnecting the electrical connector and bridging the power and grid connectors with a jumper wire. Then, turn on the defroster control and determine whether the grid is working. Immediately disconnect the jumper wires once you have determined whether the grid is heating up.

If power appears to be entering the grid, but the grid is not warming, use an ohmmeter to check the grid wires, **Figure 13-22.** If an open (infinite resistance) or short (zero resistance) circuit is detected, the wire is defective. Some grids can be checked at their electrical connector. If the rear window grid has a broken section, it can sometimes be repaired by painting conductive material over the broken part, **Figure 13-23.** Other defective parts are simply replaced.

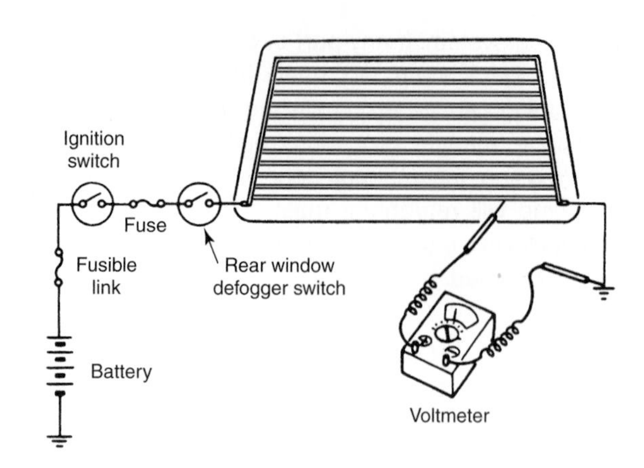

Figure 13-22. This illustration shows an ohmmeter being used to check for a broken defroster grid wire. Infinite resistance indicates that the wire has broken. (Ford)

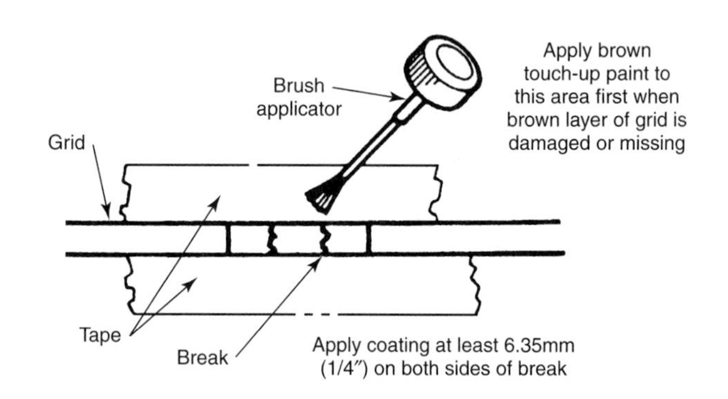

Figure 13-23. Sometimes it is possible to repair a broken wire in the rear defroster grid spot by painting it with conductive paint. Mask off the broken wire to make a neat repair. (Ford)

Body Computer Systems

The remainder of this chapter will deal with ***body computer systems.*** A body computer system controls any function that does not affect engine operating conditions. The sensors, control units, and actuators used in body computer systems are similar to those found in computerized engine control systems. Typical systems controlled by body computers include air bag systems and cruise control systems, which are covered here. In addition, body computers also control many heating and air conditioning, radio, anti-lock brake, traction control, and active suspension systems. These topics will be discussed in the chapters where they apply. Other body computers can control trip odometers, compasses, and other displays.

Air Bag Systems

Air bag systems, or ***supplemental inflatable restraints (SIR),*** are becoming standard on most modern vehicles. These systems are designed to deploy when a vehicle is involved in a frontal collision of reasonable force.

Driver and Passenger Air Bags

On early systems, only a driver-side air bag was used. Most late-model vehicles have both driver- and passenger-side air bags. The driver-side air bag is installed in the center of the steering wheel, and the passenger-side air bag is most commonly installed in the dashboard.

Side Air Bags

Some late-model vehicles use ***side air bags*** installed in the door panels. These operate in the same manner as steering wheel and dashboard air bags. However, they are designed to deploy during a side impact collision. Side air bag impact sensors operate in the same manner as other sensors, but are installed to sense a right angle (side) impact to the vehicle. It is possible for both the front and side air bags to deploy during a crash. Service procedures for side air bags are similar to those for conventional air bags.

Air Bag System Components

Most air bag systems consist of several common components, **Figure 13-24,** which include:

- Front impact sensors.
- Seat sensor
- Diagnostic control module.
- Coil assembly.
- Inflator module.

These components work together to fully ***deploy*** (inflate) the air bag within 50 milliseconds after an impact. After deployment, the air bag will deflate in approximately 100 milliseconds.

Impact Sensors

The ***impact sensors*** are usually located on the front section of a vehicle's frame or on each side of the radiator housing. An impact sensor is an open switch, which is designed to close when an impact provides a velocity change severe enough to warrant air bag deployment. See **Figure 13-25.**

Passenger Seat Sensor

The passenger seat sensor is installed under the passenger-side seat, and tells the diagnostic control module when someone is sitting in the passenger seat. This keeps the module from deploying the passenger-side air bag when no one is sitting in the passenger seat.

Diagnostic Control Module

The ***diagnostic control module*** performs several functions in the air bag system, **Figure 13-26.** The module serves as a monitoring system for the air bag components, warns the driver of system malfunctions by activating the air bag indicator light, and stores trouble codes that are used during system diagnosis. If battery voltage is lost in an accident, most diagnostic control modules provide the air bag system with an alternate source of power. The diagnostic control module also

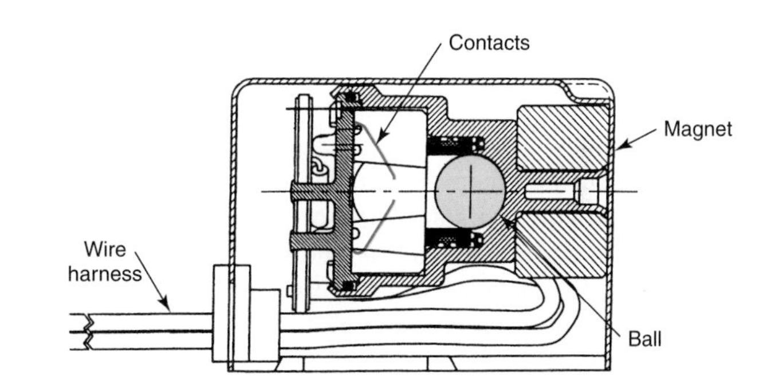

Figure 13-25. This illustration shows a typical air bag diagnostic module. Note that this unit is mounted under the dash and near the center console. (DaimlerChrysler)

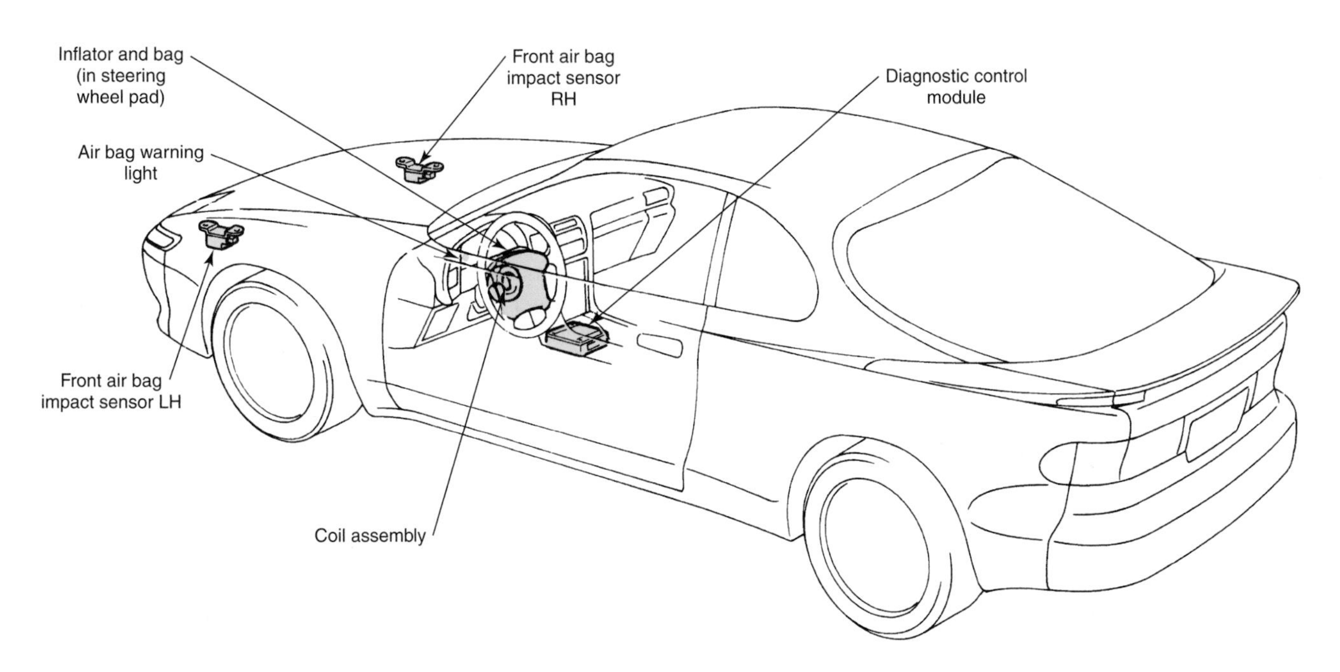

Figure 13-24. Air bag systems typically have these components. Note the location of the front impact sensors and inflator module. (Toyota)

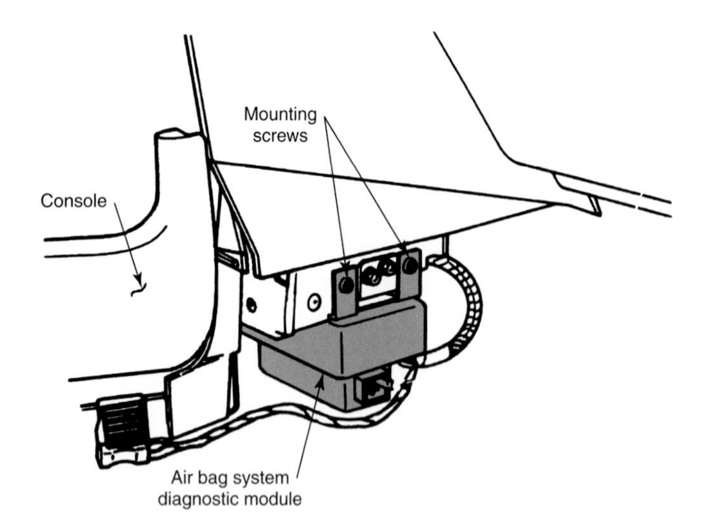

Figure 13-26. This figure shows one type of impact sensor. If the vehicle is involved in a collision, the ball will break free of the magnet field and close the contacts, triggering the air bag system. (Breed Technologies, Inc.)

contains a special safing sensor designed to prevent accidental deployment. The safing sensor responds to a vehicle collision in the same way as the externally mounted impact sensors. This sensor must close simultaneously with one of the impact sensors to trigger the air bag system. The safing sensor prevents a faulty front impact sensor from activating the air bag system.

Coil Assembly

The ***coil assembly*** consists of two current-carrying coils, which are attached to the vehicle's steering column. As the steering wheel is rotated, these coils maintain a continuous electrical connection between the inflator module and diagnostic module. The coil assembly is sometimes referred to as a clock spring. A typical coil assembly is illustrated in **Figure 13-27.**

Inflator Module

The ***inflator module*** is located in the center of the steering wheel in driver-side air bag systems, **Figure 13-28.**

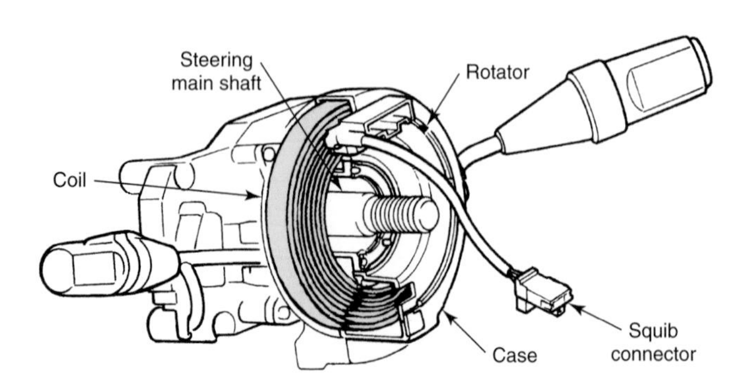

Figure 13-27. Shown is a cutaway view of a typical coil assembly. Note the squib connector, which connects the coil assembly to the inflator module. (Toyota)

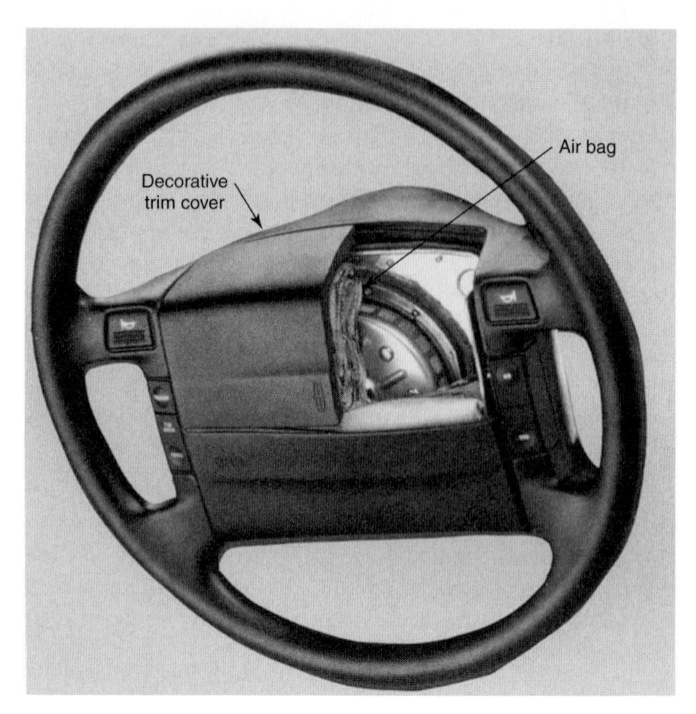

Figure 13-28. This inflator module is housed in a steering wheel assembly. Note the air bag folded under the decorative trim. (Ford)

This module is usually covered with a decorative trim piece. The inflator module contains three main parts:

- An inflatable fabric air bag, **Figure 13-29.**
- A gas-generating material, such as a sodium azide or sodium hydroxide propellant.
- An initiator squib, which provides the ignition source for the inflator.

In addition to a driver's side air bag, some vehicles are equipped with a passenger side air bag. Passenger side inflator modules are usually mounted above the glove box on the passenger side of the vehicle. A passenger side inflator module is shown in **Figure 13-30.**

Air Bag Operation

When a vehicle is involved in a collision severe enough to warrant air bag deployment, the impact sensors close. If at least one impact sensor and the safing sensor in the diagnostic module close, a signal is sent to the initiator squib. When signaled, the squib generates a thermal reaction, which ignites the gas-generating material (sodium azide-based propellant) in the inflator, **Figure 13-31.** This material releases a large amount of harmless gas, which quickly fills the air bag. See **Figure 13-32.**

Disabling and Enabling Air Bag Systems

Many service procedures call for disabling, or making inoperative, the air bag system. Disabling the air bags ensures that they will not deploy during service operations. The following is a general method of disabling air bags.

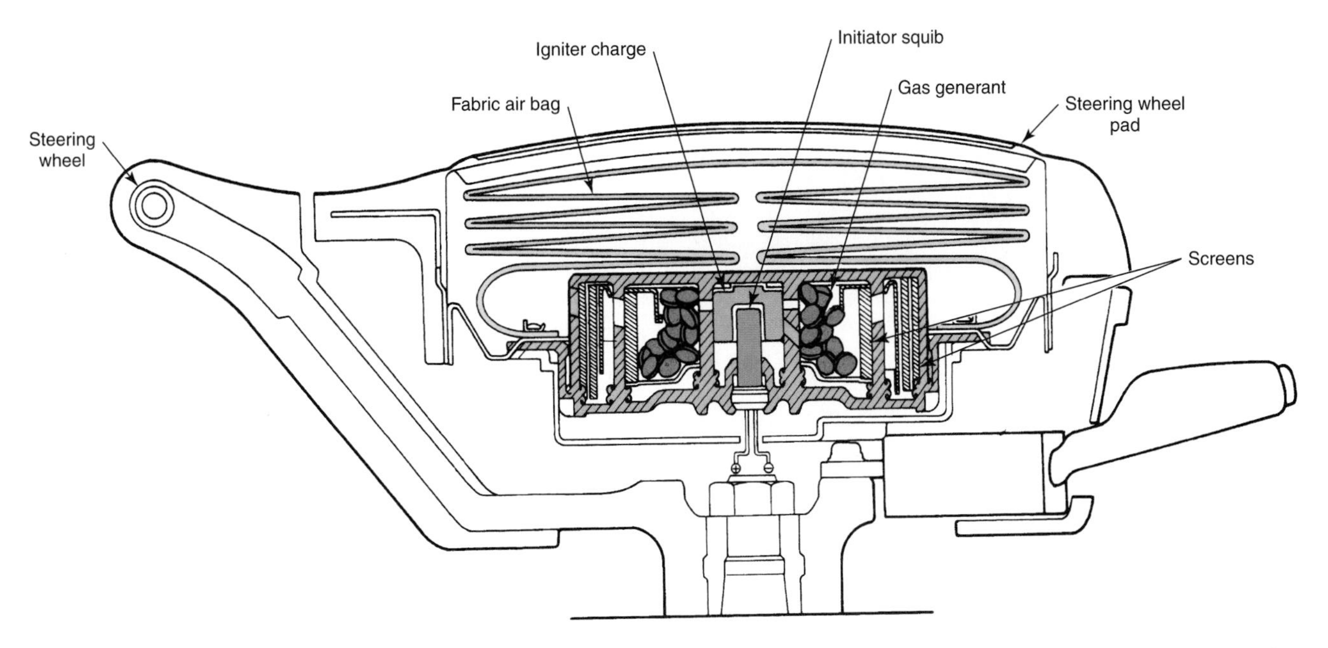

Figure 13-29. This cutaway view of an air bag inflator module shows the major parts of the module. The screens allow gas from the generant to enter the bag. (Toyota)

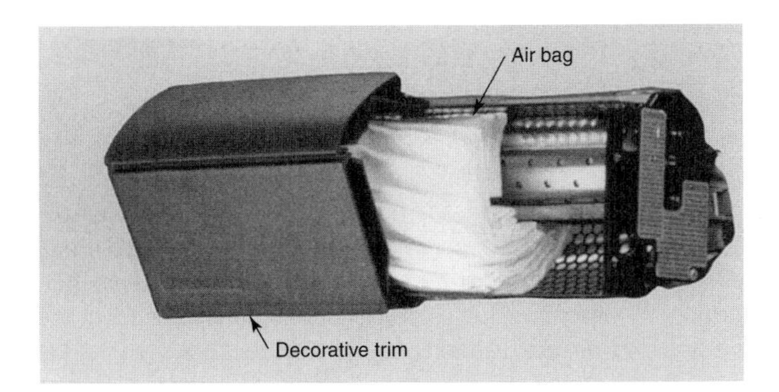

Figure 13-30. The passenger side inflator module is simpler than the drivers' side module. This cutaway view shows how the bag is installed under the trim cover. (Ford)

Warning: Always obtain the proper service literature and follow manufacturer's procedures exactly before attempting to disable any air bag system.

Begin the disabling procedure by pointing the steering wheel in the straight-ahead position. Next, lock the ignition and remove the ignition key. Locate and remove the air bag fuse(s). Next, locate and disconnect the air bag connectors under the dashboard. Check the service manual for exact connector locations. It may be necessary to remove trim parts to access the connectors. Many air bag connectors are yellow, but this should be confirmed by reading the service literature. This disconnects the air bags from the power source so repairs can be performed safely.

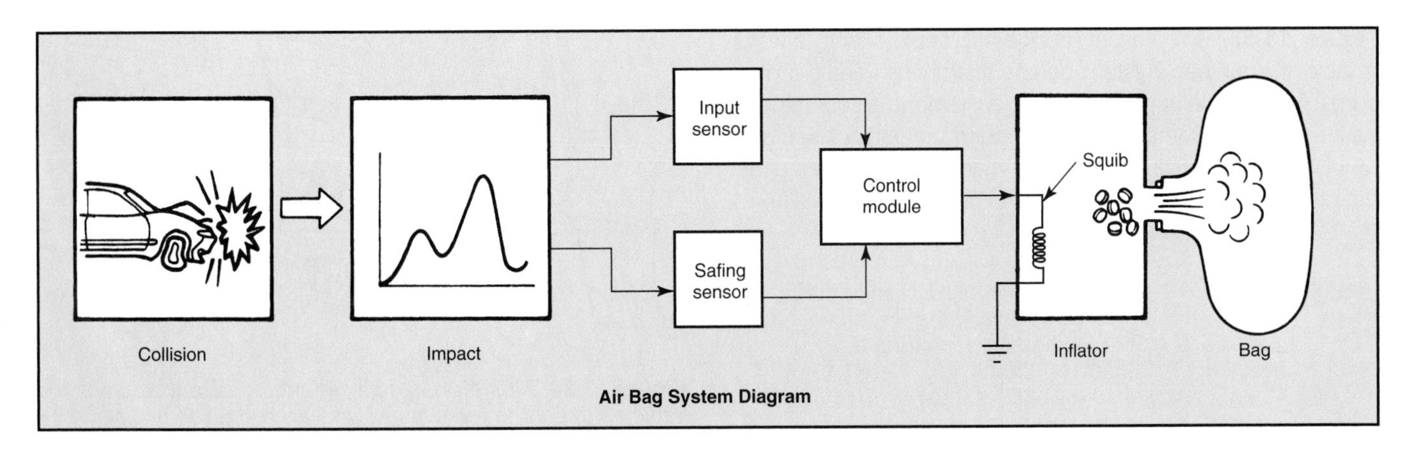

Figure 13-31. Study this air bag activation sequence. Note that the safing sensor and front impact sensors must both be triggered to cause deployment. (Toyota)

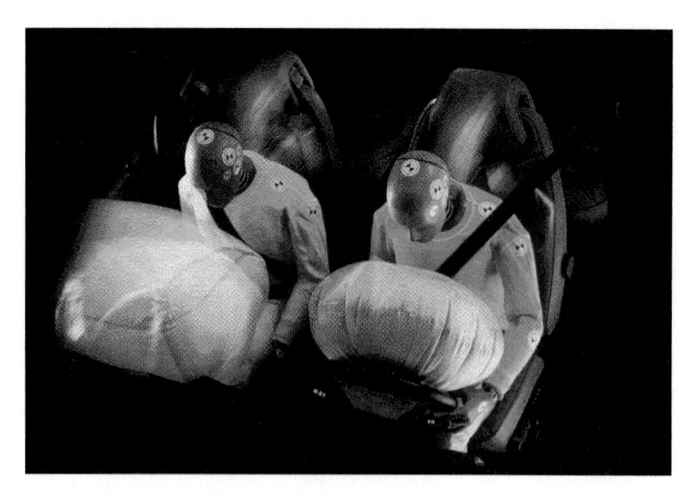

Figure 13-32. A signal from control module causes the squib to ignite a gas-generating material in the inflator module, rapidly filling the airbags with harmless gas. (Saab)

To enable the air bag system, make sure that the key is removed from the ignition switch. Next, reconnect the air bag connectors. Reinstall any trim pieces that were removed. Reinstall the system fuses, turn the ignition switch to the run position, and run the air bag system's self-diagnosis program to ensure that the system is operating correctly.

Troubleshooting Air Bag Systems

Warning: To avoid possible deployment, consult the proper service manual before attempting to troubleshoot air bag systems. Most manufacturers require the use of special testing equipment when servicing an air bag system. Using incorrect testing procedures can cause an accidental deployment.

The self-diagnostic system should be used when troubleshooting air bag systems. Since procedures for accessing trouble codes vary, consult an appropriate manual before attempting to trigger the diagnostic system. Most manufacturers produce a scan tool to retrieve codes. Many components in an air bag system are not serviceable. If the trouble codes indicate a faulty component, it must be replaced, **Figure 13-33.** To prevent accidental deployment, always disable the air bag system before attempting repair procedures. Most manufacturers simply recommend disconnecting battery voltage. Some, however, require the initiator squib to be disconnected. This disables the diagnostic module's alternate power source.

Caution: The diagnostic control module in some systems can retain power well after the battery has been disconnected. Follow the manufacturer's recommendations for disabling the air bag system. Static electricity can also cause an accidental deployment. Make sure you are well grounded before touching any part of the air bag system.

Diagnostic Codes

Code No.	Diagnosis	Trouble Area
11	Short circuit in squib wire harness (to ground) Front air bag sensor turned on at all times Center air bag sensor system turned on at all times	• Wire harness • Front air bag sensor • Wire harness • Center air bag sensor assembly
12	Short circuit in squib wire harness (to +B)	• Wire harness
13	Short in squib circuit	• Squib • Wire harness
14	Open in squib circuit	• Squib • Wire harness
15	Open in front air bag sensor wire harness	• Wire harness
22	Open in air bag warning light system	• Bulb • Wire harness
31	Internal malfunction of center air air bag sensor assembly	• Center air bag sensor assembly
41	Malfunction stored in memory	

Figure 13-33. This illustration shows a trouble code chart for a specific air bag system. Always check for diagnostic (trouble) codes before attempting to service an air bag system. (Toyota)

Servicing a Deployed Air Bag

When an air bag is deployed, sodium hydroxide powder is released into the vehicle's interior. This powder can cause irritation of the skin, eyes, nose, and throat. Wear rubber gloves and safety glasses when removing the deployed bag. After the bag has been removed, vacuum the interior thoroughly to remove residual powder. Make sure to vacuum the heater and air conditioner outlets, **Figure 13-34.** If residue

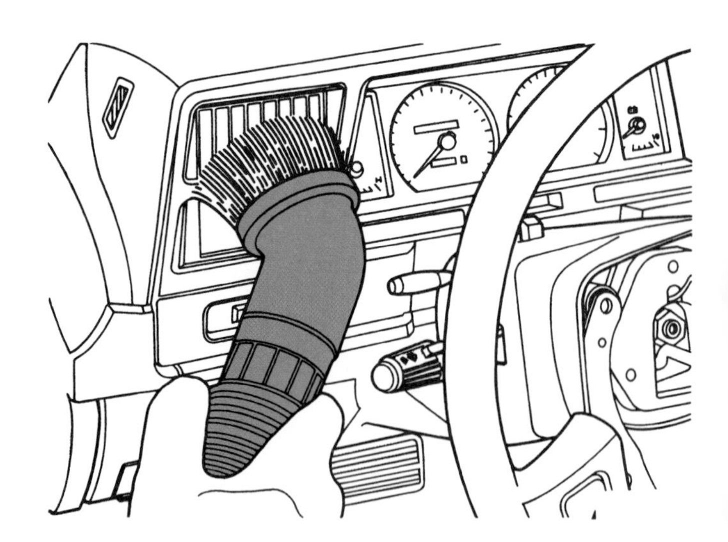

Figure 13-34. After air bag deployment, be sure to vacuum the vehicle's interior to remove any sodium hydroxide powder, which can irritate the skin, eyes, nose, and throat. (DaimlerChrysler)

remains, wipe down the interior surfaces with a damp cloth. Follow manufacturer's recommendations for proper disposal.

Following air bag deployment, most manufacturers recommend replacement of all related parts. Some only require the replacement of the inflator module and the initiator squib. Always replace all items listed in the service manual. In some systems, reusing sensors and diagnostic modules can cause unprovoked deployment of the air bag system. Some air bag systems require the use of a scan tool to reset the system after a deployment.

Air Bag Service Precautions

When working on air bag systems, there are certain precautions that should be taken to reduce the risk of damage to the system and injury to the passengers. These precautions include the following:

- Always disable the air bag system before attempting to service any component on or near the system (steering column, dashboard, or the front of the engine compartment).
- Never subject the inflator module to temperatures greater than 175°F (79.4°C).
- If any part of the air bag system is accidentally dropped, the component should be replaced.
- Never test air bag components with electrical test equipment unless instructed by manufacturer.

Care should be taken when handling live (not deployed) inflator modules. When carrying a live module, point the bag and trim cover away from you. This will minimize the chance of injury in the case of accidental deployment. When placing a live inflator module on a bench, always face the bag and trim cover up. This will allow the air bag to expand freely in the event of accidental deployment.

Note: If you replace a live inflator module for any reason, it is recommended the old inflator module be deployed before it is discarded. Consult the manufacturer's service manual for the recommended disposal procedure.

Cruise Control

Cruise control systems were formerly operated by various electromechanical devices. However, the latest cruise control systems are operated by body computers. Cruise control systems consist of several common components, **Figure 13-35.** These components include:

- Vehicle speed sensor.
- Operator controls.
- Control module.
- Throttle actuator.

Vehicle Speed Sensor

The ***vehicle speed sensor*** uses a rotating magnet to generate a small electrical signal. It is usually mounted on the drive shaft or is driven by the transmission or transaxle governor assembly. It sends a speed signal to the control module. The speed sensor may also send a signal to the engine control computer.

Operator Controls and Control Module

The operator controls are used to set the desired speed. Control location varies between manufacturers, but is usually mounted on the turn signal lever. The operator can set the speed, and can also make minor adjustments to speed as necessary. Another operator control is the brake pedal release switch, usually mounted on the brake pedal bracket. When the operator presses on the brake pedal, the switch sends a signal to the cruise control module, which disengages the cruise control. The control module processes the inputs from the operator controls and the speed sensor and produces an output signal to the throttle actuator.

Throttle Actuator

The throttle actuator opens and closes the vehicle throttle to maintain the operator-set speed. On new vehicles, throttle actuators are electric motors directly operated by the control module. See **Figure 13-36.** On older systems a vacuum diaphragm moves the throttle. A vacuum controller determines the amount of vacuum applied to the diaphragm. The module operates the vacuum controller, which in turn operates the vacuum diaphragm.

Cruise Control Service

When confronted with a cruise control problem, the first step is to determine whether there truly is a problem. Most cruise controls will not operate under 25 or 30 mph. To test the cruise control, you must take the vehicle to a stretch of highway where a speed of 30+ mph can be sustained for a few miles. Engage the cruise control and observe whether it maintains the set speed. Press on the brake pedal to disengage the cruise control.

Note: If the brake fails to deactivate the cruise control, use the manual controls to deactivate the system.

Late-model cruise controls are operated by the body control module, and store trouble codes. The codes can be retrieved with a scan tool. The operation of many newer cruise controls can be tested with a scan tool. One test with the scan tool requires entering the test mode and commanding the cruise control to raise engine speed with the transmission in *Park.* If the cruise control responds to commands and opens the engine throttle, it can be assumed to be operating properly. In such cases, the technician should look for defects in the speed sensors or the brake pedal switch.

Out-of-adjustment throttle linkage and vacuum leaks cause most nonelectronic cruise control problems. Check the throttle cable adjustment and determine whether the vacuum

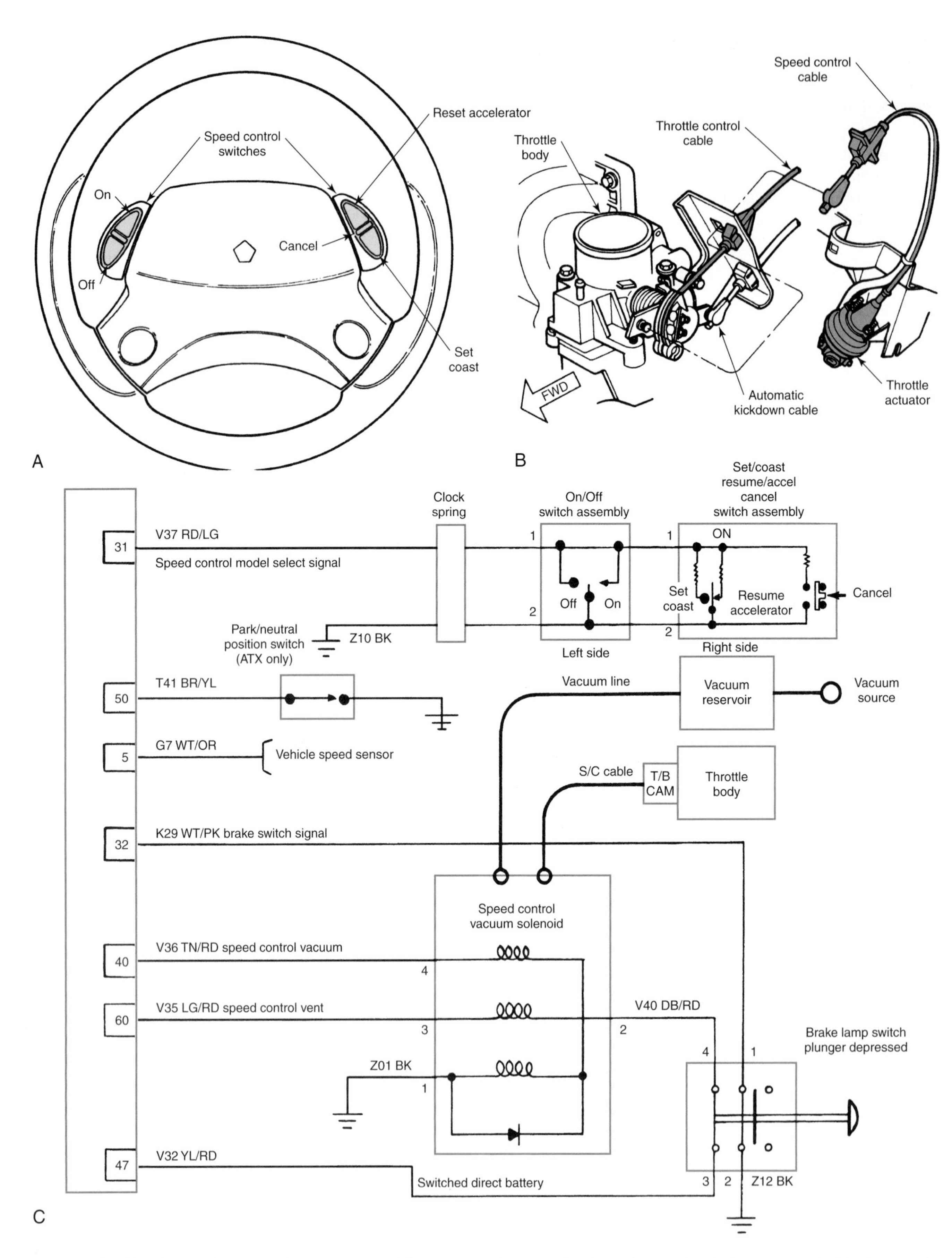

Figure 13-35. This is a cruise control system as used by one manufacturer. A—Speed control switches are located in the steering wheel. B—Mechanical schematic of the system. C—Cruise control circuit schematic. (DaimlerChrysler)

Figure 13-36. This motor opens the throttle on a late-model cruise control. The body control module operates the motor.

diaphragm or hoses are leaking. If other vacuum-operated devices are not working, the vacuum hose has probably come off at the intake manifold. If the problem appears to be in the electrical system of the cruise control, check the fuse first. Then, suspect the speed sensor or the brake pedal switch. Speed sensors on older systems are usually located in the speedometer head, although a few are installed on the drive axle. If the vehicle speedometer is also inoperative, the problem is in the speed sensor.

A common cruise control problem on both electronic and nonelectronic systems is an out of adjustment brake switch. If the brake switch contacts are always closed, the cruise control will not engage. If the brake switch contacts cannot be closed, the cruise control will not disengage when the brakes are applied. Check the brake switch adjustment and adjust if necessary. Brake switch adjustment was discussed earlier in this chapter.

Once the defective part has been identified, replacement is usually simple. If a sensor or body control module is defective, it must be replaced. Refer to the manufacturer's service manual for additional information on cruise control troubleshooting and parts replacement.

Tech Talk

Often, the technician will be asked to add gauges to a vehicle. Although warning lights are sufficient for most driving situations, some special cases make gauges useful. For instance, if a vehicle is being used to tow a trailer or boat, it is a good idea to add an engine or transmission temperature gauge. Oil pressure gauges are often added to give the driver a better idea of engine condition when the vehicle is operated at high RPM. If a vehicle has extra electrical equipment, an ammeter or voltmeter can be added to let the driver know exactly what is going on in the electrical system. Tachometers are used to warn of engine overspeeding or to indicate shift points. Sometimes, a vacuum gauge is added to let the driver know about the state of engine tune and for driving at the most economical speeds and engine loads. In many cases, the owner just wants to know what is going on in the engine.

Gauges are available individually or in sets of two, three, or four. It is a relatively simple job to add engine oil pressure, temperature, RPM, vacuum, or amperage/voltage gauges to a vehicle. The major tasks are:

- Installing the sender unit on the engine.
- Running a wire from the sender to the gauge.
- Running a wire from the gauge to a power source that is hot only when the ignition is on.
- Finding a convenient place to install the gauge, in a location that will be visible to the driver.
- Connecting the gauge illumination light to the instrument panel lighting.

Instead of using a sender, vacuum gauges are installed by running a vacuum line directly to the gauge. Some older oil pressure and temperature gauges are mechanical types, with tubes to direct oil pressure into the gauge. However, most modern gauges have an engine-mounted sender electrically connected to the gauge. A sender is not used with tachometers, ammeters, or voltmeters. These gauges receive a signal directly from the vehicle's electrical system. Installing a transmission temperature gauge is similar to installing an engine temperature gauge, with the obvious difference the sensor is installed on the transmission.

Summary

Chassis wiring consists of all of the wiring and components not directly connected with the engine and drive train. Some devices are routed through the ignition switch, while other devices are powered directly from the vehicle battery. Wiring is wrapped together into wiring harnesses.

Color coding and schematics make wire tracing easy. The most common problem in wire service is blown fuses and fusible links. Overload relays can open when the current draw becomes excessive. Vehicle light systems are divided into safety lights and convenience lights. Safety lights include the headlights, taillights, marker lights, instrument panel lights, brake lights, turn signal lights, and backup lights. Convenience lights include interior lights and courtesy lights, trunk lights, underhood lights, and glove compartment lights. Most lights are serviced by replacing them. The brake light switch may require adjustment.

Gauges are often used instead of warning lights. Disconnect the related sensor to check electromechanical and some electronic gauges. Also check the power supply fuses and instrument panel grounds.

Chassis-mounted motors include the windshield wiper motor, power window, tailgate motors, sunroof motors, convertible top motors, and power antenna motors. Check a motor by bypassing the control switches or relays. Sometimes the problem is not in the motor but the associated linkage.

Solenoids are used to operate door locks and trunk releases. Relays are a type of solenoid used to direct large current flows that might damage a switch.

Security systems are often complex, but can be tested by simulating an attempt at illegal entry. The alarm should go off when this is done. Pass-Key systems can also be checked by some simple methods.

Radios, cassette tape, and CD players should be repaired by a qualified electronics shop. Some simple tests can be made before removing a radio for service. Many other vehicle systems can cause static, and should be checked before the unit is removed for service.

The horn uses a diaphragm with a make and break contact. They are operated by steering wheel contacts, which energize a relay. Rear window and mirror defrosters consist of a heater grid applied to the glass surface of the rear window or mirror. A body computer system controls any function that does not affect engine operating conditions. Typical body computer systems include air bag systems and cruise control systems. Most body computer systems have self-diagnostic capabilities. Service procedures will vary among manufacturers. Consult the appropriate service manual before attempting to service body computer systems.

Glass-heating systems are simple, and use electrical resistance to produce a heating effect on rear windows and mirrors. Most glass-heating system problems can be isolated by simple electrical checks. Some defroster grid problems can be corrected by painting a conductive solution onto the broken grid wire.

Special precautions should be taken when working on vehicles equipped with an air bag system. To reduce the risk of injury, always disable the system before working on or near any air bag system components. After an air bag has been deployed, the interior of the vehicle must be thoroughly cleaned to remove the remaining propellant. Most air bag systems can be diagnosed with a scan tool.

Modern cruise control systems are operated by body computers. Cruise control systems consist of speed sensors and a control module, which operates a vacuum or electrical throttle actuator. Before diagnosing the cruise control system, make sure that there is an actual problem. Replacement of defective cruise control parts is usually simple.

Review Questions—Chapter 13

Do not write in this book. Write your answers on a separate sheet of paper.

1. Wiring _____ are groups of wires wrapped together for ease of installation.
2. Most fusible links are mounted near the battery _____ cable.
3. The rheostat is used to adjust the brightness of _____ lights.
   (A) tail
   (B) marker
   (C) instrument panel
   (D) driving
4. The sealed beams used on older vehicles have two filaments in which of the following vehicle headlight arrangements?
   (A) Two headlights only.
   (B) The outer or top lights on four headlight systems.
   (C) The inner or bottom lights on four headlight systems.
   (D) Both A and B.
5. Touching a(n) _____ lightbulb with oily hands can ruin it.
6. The backup lights are connected to which other light assemblies?
   (A) Taillights.
   (B) Turn signal lights.
   (C) Dashboard lights.
   (D) None of the above.
7. List the two classes of instrument panel lights.
8. Some modern vehicles use only a few bulbs and distribute their light through a _____ _____ harness.
9. Many modern vehicles use both electronic and electromechanical _____. Many of these can be tested by removing the connector to the related _____.
10. What is the purpose of the windshield wiper motor park mechanism?
11. Many motor-operated devices can be restored to proper operating condition by cleaning and lubricating the associated _____ _____.
12. Door or tailgate glass may drop when the _____ is removed.
    (A) motor
    (B) solenoid
    (C) relay
    (D) wiring
13. When a vehicle security system is inactive, it is said to be _____.
14. What is the easiest way to identify a vehicle that has a Pass-Key security system?
15. When the battery of a keyless entry system transmitter is replaced, the system must be_____.
16. The horn relay clicks when the steering wheel horn contacts are closed, but the horn does not sound. The problem is most likely in the _____.
    (A) horn relay
    (B) horn
    (C) Both A and B.
    (D) Neither A nor B.
17. Explain briefly how a broken section of a rear window defroster grid can be repaired.

Match the air bag system component with its description.

18. Safing sensor
19. Impact sensor
20. Diagnostic control module
21. Squib
22. Inflator
    (A) Provides an ignition source.
    (B) Located in the diagnostic control module.
    (C) Located at front of vehicle.
    (D) A type of propellant.
    (E) Detects the presence of a passenger.
    (F) Controls the operation of other components
23. Always _____ the air bag system before attempting to service any component on or near the system's components.
24. Cruise control throttle actuators can be _____ or _____.
25. A common cruise control problem is an out-of-adjustment _____ switch.

ASE-Type Questions

1. All of the following statements about body wiring are true, *except:*
   (A) Chassis wiring extends throughout the vehicle.
   (B) Wires in a wiring harness are not color-coded.
   (C) Some wiring is attached directly to battery voltage.
   (D) Some wiring is routed through the ignition switch.
2. Technician A says some schematics show the exact flow of electricity. Technician B says some schematics show the general process of a system. Who is right?
   (A) A only.
   (B) B only.
   (C) Both A and B.
   (D) Neither A nor B.
3. Which of the following will automatically reset when it cools off?
   (A) Fuse block.
   (B) Fusible link.
   (C) Circuit breaker.
   (D) In-line fuse.
4. Technician A says replacing a blown fuse may not solve the electrical problem. Technician B says fusible links can be replaced with regular wire. Who is right?
   (A) A only.
   (B) B only.
   (C) Both A and B.
   (D) Neither A nor B.
5. Replacing a fuse with a non-fused connection could cause all of the following, *except:*
   (A) melted wires.
   (B) battery overcharging.
   (C) ruined electrical components.
   (D) a vehicle fire.
6. Technician A says the rheostat in the headlight switch circuit is used to control headlight brightness. Technician B says the dimmer switch in the headlight switch circuit is used to control instrument panel illumination. Who is right?
   (A) A only.
   (B) B only.
   (C) Both A and B.
   (D) Neither A nor B.
7. Technician A says the windshield washer can be part of the wiper motor assembly. Technician B says the windshield washer may be located in the washer reservoir. Who is right?
   (A) A only.
   (B) B only.
   (C) Both A and B.
   (D) Neither A nor B.
8. All of the following could cause an inoperative solenoid, *except:*
   (A) defective solenoid winding.
   (B) excessive voltage.
   (C) sticking linkage.
   (D) blown fuse.
9. Technician A says body computer systems control the ignition timing and the air-fuel ratio. Technician B says the impact sensors in an air bag system are located on the vehicle's firewall. Who is right?
   (A) A only.
   (B) B only.
   (C) Both A and B.
   (D) Neither A nor B.
10. All of the following statements about the propellant powder used in air bags are true, *except:*
    (A) when an air bag is deployed, powder is released into the vehicle's interior.
    (B) the propellant powder is harmless.
    (C) wear rubber gloves and safety glasses when removing a deployed air bag.
    (D) if any powder residue remains, wipe down the interior surfaces with a damp cloth.

Suggested Activities

1. Draw an electrical schematic showing a vehicle horn, horn relay, battery, and horn button (on steering wheel). Explain how the components work together to operate the horn.
2. Without referring to a service manual, write a step-by-step procedure for testing the operation of the parking brake warning light from the driver's seat. Compare your procedure with the procedure outlined in the service manual.
3. Use a test light to test a chassis electrical component and/or wiring. Determine what is happening when the light does or does not come on and make a chart of your findings. See if you can develop this chart into a troubleshooting chart for the component and wiring.
4. Use a multimeter or VOM to test a chassis electrical component and/or wiring. Tests should include measuring voltage drop across the components in a live circuit, checking the resistance of deenergized components, and measuring amperage draw of a working circuit.
5. Measure the resistance in a circuit and use Ohm's law to find the amperage draw of the circuit. For instance, if battery voltage is 12 volts, and the circuit resistance is 4 ohms, the amperage draw will be 3 amps. After making the calculation, measure amperage draw with the shop equipment. Do the actual and calculated resistances agree? If they do not, can you think of a reason why? Write a report summarizing your conclusions.

14

Charging and Starting System Service

After studying this chapter, you will be able to:

- Install, test, and service a battery.
- Use jumper cables correctly.
- Test, service, and repair a charging system.
- List the major internal parts of an alternator.
- Test, service, and repair a starting system.
- List the major internal parts of a starter.
- List the major internal parts of a starter solenoid.

Technical Terms

Battery
Electrolyte
Sulfuric Acid
Open-Circuit Voltage Test
Percentage of Charge
Electrical Load Test
Surface Charge
Specific Gravity Test
Hydrometer
Battery Charger
Slow Charging
Fast Charging
Trickle Charger
Dry-Charged Batteries
Battery Groups
Size
Terminal Type
Battery Rating
Cold Cranking Amps (CCA)
Cranking Amps (CA)
Reserve Capacity (RC)
Battery Temperature Sensor
Polarity
Shorts
Parasitic Loads
Jump Started
Booster Battery
Jumper Cables
Alternator
Voltage Regulator
Undercharging
No Charging
Overcharging
Amperage
Voltage
Electronic Regulators
Electro-Mechanical Regulators
Resistance
Voltage Drop Test
Brushes
Diodes
Neutral Safety Switch
Load Test
Starter Drives
Growler
Commutator Bar
Field Coils

This chapter covers the service of the starting and charging systems, including the battery, starter, starter solenoid, alternator, voltage regulator, and associated cables and wiring. Effective operation of a vehicle's electrical system depends on a good battery, a charging system that can keep the battery charged, and a starting system that can properly utilize the battery power.

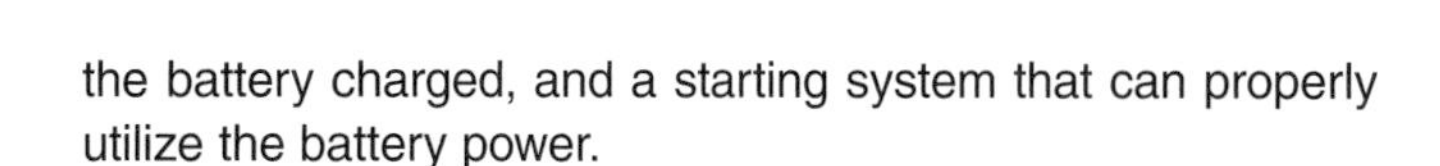

Caution: This chapter describes procedures for handling many typical starting and charging system jobs. When testing batteries, starters, or alternators, the instructions provided by the manufacturer of the test equipment should be carefully followed. Be sure to use the correct electrical specifications for the exact make and model being tested.

Battery Service

The effectiveness of the starting system depends on the vehicle ***battery.*** The battery plays a key roll in the overall functioning of the electrical system. To ensure reliability and extend useful service life, the battery should receive periodic inspection and maintenance. The level of ***electrolyte*** (the liquid in each battery cell) must be correct. The battery must be fully charged. The posts must be clean. Battery cables and terminals must be in good condition and firmly attached. The charging system should also be functioning correctly to recharge the battery during operation. A 12-volt side terminal battery is shown in **Figure 14-1.**

Batteries Can Be Dangerous

Battery service involves two dangerous substances: corrosive electrolyte and explosive gases. Battery electrolyte contains about 38% ***sulfuric acid*** and can cause serious skin and eye burns. If the electrolyte contacts your skin, flush the area with large quantities of cold water. If it contacts your eyes, flush them with cold water and then consult a physician. Battery acid on the vehicle or on clothing should be flushed with cold water. Follow with a mixture of baking soda and water to neutralize the acid.

Battery charging creates a mixture of hydrogen and oxygen gases, which is extremely flammable. The slightest source of ignition can ignite the mixture, causing it to burn with explosive force. This could rupture the battery case and throw acid over a wide area. Never strike a spark, light a match, or bring other open flames near a battery.

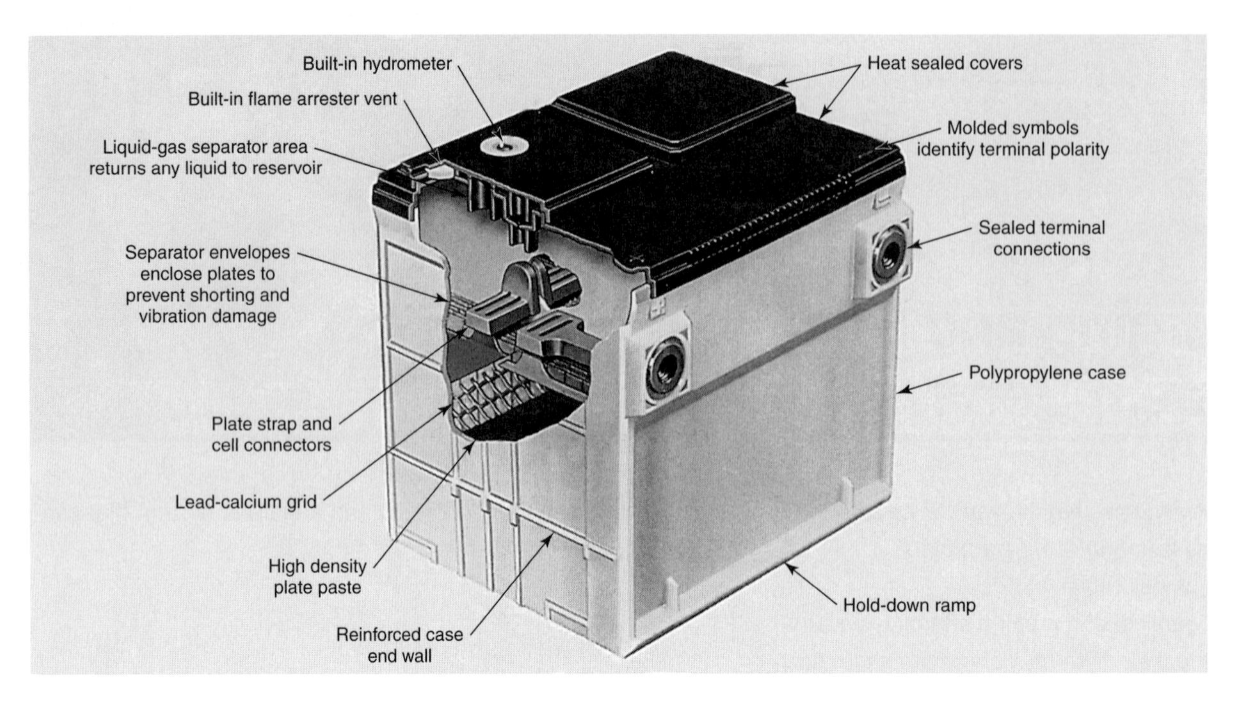

Figure 14-1. This is a cutaway view of a typical 12-volt side-terminal battery. Note the sealed construction and the use of a built-in hydrometer. (General Motors)

The following safety rules should be observed at all times:

- Use a properly fitted lift strap to move batteries.
- When a battery is removed from the vehicle, place it where it will not be knocked over, dropped, or exposed to sparks or flame.
- Do not use a battery as a step.

Check Electrolyte Level

You may occasionally encounter a non-sealed battery, usually on a motorcycle or lawn tractor. Non-sealed batteries can be identified by the presence of removable filler caps. Check the electrolyte level in each cell. Most batteries have a correct level indicator (slot, notch, lip). Add distilled water to bring the electrolyte up to the mark. If no mark is used, raise the level to 3/8″ (9.53 mm) above the top of the separators. To prevent electrolyte leakage caused by expansion pressure, avoid overfilling. When a battery uses an excessive amount of water, check the system for overcharging. Prolonged overcharging will reduce battery life.

Checking Battery Condition

The condition of a given battery is determined by the state of charge, the temperature, and the mechanical condition of the plates, separators, and connectors. Although it is difficult to determine the exact amount of useful life remaining in any battery, several tests will give an indication of its ability to perform satisfactorily for a period of time.

Cold weather reduces the efficiency of batteries. **Figure 14-2** illustrates how battery capacity (amount of electricity that can be drawn from a fully charged battery in a specified length of time) is reduced in cold temperatures. A battery that cranks a vehicle during moderate weather may crank slowly or fail completely during the first cold snap.

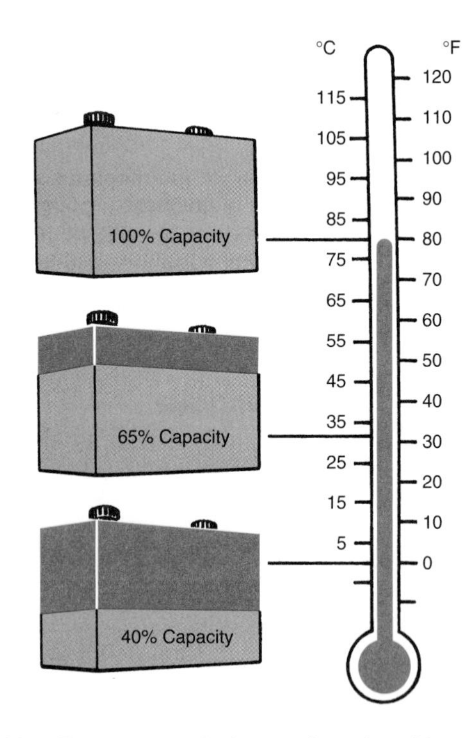

Figure 14-2. Battery capacity is greatly reduced by cold weather. Capacities at indicated temperatures are for sound, fully charged batteries. (Gulf Oil)

A suspect battery should be given an electrical load test with the proper electrical test equipment. If tests indicate a borderline condition, replace the battery. The little additional use that can be squeezed out by leaving the battery in service will be offset by the cost of charging, the possibility of a failure in the field, and the cost of a service call.

Note: If a battery is frozen, it cannot be tested. Only badly discharged batteries will freeze. In addition, freezing generally damages the battery plates. Therefore, frozen batteries should be replaced. If a frozen battery must be tested, allow it to thaw and recharge it before testing.

Checking Open-Circuit Voltage before Testing

The ***open-circuit voltage test*** (no load on the battery) is performed with a voltmeter. The battery cables must be disconnected for this test. Always remove the negative cable first. If the battery was recently charged, boosted, or load tested, allow it to stabilize for 15 minutes before starting the test. Connect the meter leads to the battery terminals. Make sure the positive meter lead is connected to the positive battery terminal and the negative lead is connected to the negative terminal. The voltage reading across the terminals corresponds to the ***percentage of charge*** in the battery. See **Figure 14-3.** Open-circuit voltage should be approximately 12.6 volts. Note the relationship between open-circuit voltage and the percentage of charge.

Warning: A charging battery or a recently charged battery will have a great deal of hydrogen and oxygen gas in its cells. A spark at the voltmeter prods could ignite it.

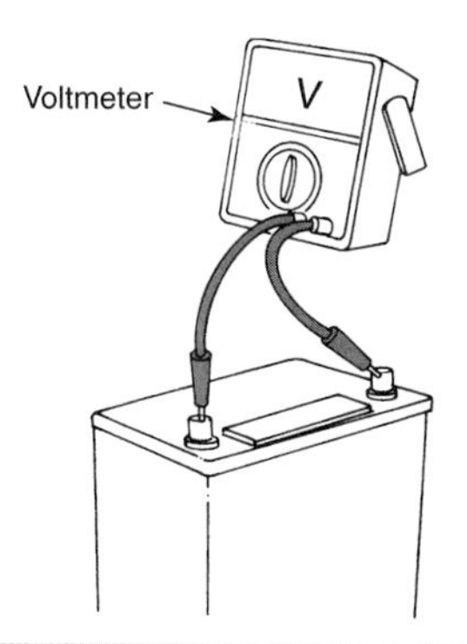

Battery Open-Circuit Voltage	
Open-circuit volts	**Percent Charge**
11.7 volts or less	0%
12.0	25%
12.2	50%
12.4	75%
12.6 or more	100%

Figure 14-3. When checking open-circuit voltage with a voltmeter, follow the test instrument operating instructions carefully. (DaimlerChrysler)

If open-circuit voltage is below specifications, the battery must be charged before further testing can be done.

Electrical Load Test

Open-circuit voltage readings provide a general indication of battery condition. Battery condition can be checked more accurately with an ***electrical load test.***

Note: Do not attempt the electrical load test if the open-circuit voltage reading indicates that the battery is discharged. Recharge the battery before testing.

Attach the electrical load tester as directed by the manufacturer. Connect the positive leads to the battery positive post. Connect the negative leads to the negative post. All electrical load testers have a device used to place an adjustable electrical load on the battery. Therefore, the tester should be switched off to prevent sparking when attaching the leads. If necessary, also attach a voltmeter to the battery.

Note: Many newer battery testers will automatically go through the battery testing procedure once the tester cables are attached. Follow the tester manufacturer's directions carefully.

If the battery was just charged, crank the engine for about 15 seconds or use the tester to apply a load of approximately 300 amps on the battery to reduce the ***surface charge*** (electrolyte in the battery temporarily having a higher-than-average charge). Then, wait for about 15 seconds to allow the battery electrolyte to stabilize.

Next, apply the load specified on the battery label, usually as the "test load." Wait for 15 seconds and then read the battery voltage. Immediately remove the electrical load. At a battery temperature of 70°F (21°C), the battery voltage should not drop below 9.6 volts. If the battery temperature is lower, the voltage will also be lower on a good battery. As a general rule, voltage can be at 9.1 volts at 32°F (0°C) and 8.5 volts at 0°F (–17.8°C) for a good battery. Check the manufacturer's specifications for exact temperature and voltage relationships. Battery temperature can be estimated by determining the temperature the battery has been exposed to for the preceding three hours.

If the battery voltage was at or above the minimum during the load test, the battery is good. If the voltage is below the minimum specification, the battery should be replaced.

Specific Gravity Test

If the battery has removable filler caps, a ***specific gravity test*** can be made. Specific gravity is a comparison between the density of electrolyte and the density of water. The state of charge can be determined by measuring the specific gravity of the electrolyte with a ***hydrometer***.

To check specific gravity with a hydrometer, remove the fill caps. If the electrolyte level is low, add water and charge the battery for about 15 minutes. Never check specific gravity just after adding water. If the battery was just charged, crank the engine for several seconds to reduce the surface charge. Hold the hydrometer in a vertical position and place the hose in the battery cell. Squeeze and release the hydrometer bulb to draw in electrolyte, and then squirt the electrolyte back into the cell. Repeat this sequence several times to bring the float temperature to that of the electrolyte. Next, draw in just enough fluid so the float does not touch either the bottom or top of the float barrel. Allow the gas bubbles to rise to the surface before taking a reading. Then hold the hydrometer at eye level. Note the scale reading at the exact point the float scale emerges from the electrolyte. **Figure 14-4** shows a hydrometer being used.

The chart in **Figure 14-5** shows the relationship between electrolyte specific gravity and state of charge. Specific gravity for a fully charged battery will vary slightly from battery to battery.

Correcting Float Reading for Temperature

To correct for temperature, you must have a hydrometer with a built-in thermometer. Note the temperature of the electrolyte and add 0.004 to the specific gravity reading for each 10°F (5.55°C) above 80°F (26.7°C). Subtract 0.004 from the specific gravity for each 10°F (5.55°C) below 80°F (26.7°C).

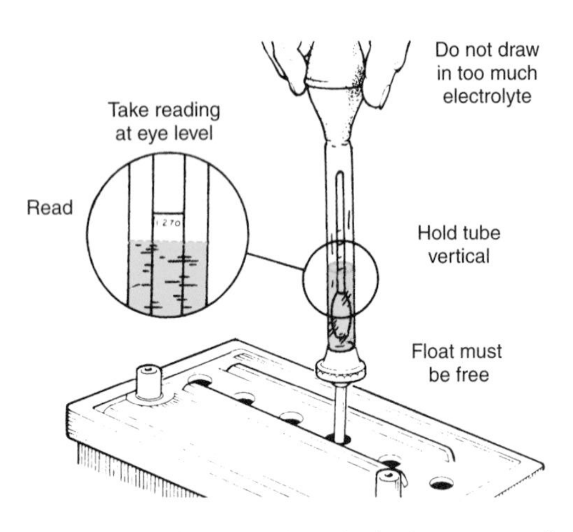

Figure 14-4. Draw electrolyte into the hydrometer until the float is suspended in the fluid and not resting on either the top or bottom of glass barrel. Read the hydrometer at eye level. (British Leyland)

State of Charge*	Specific Gravities as Used in Cold and Temperate Climates		Specific Gravity as Used in Tropical Climates
Fully Charged	1.280	1.260	1.225
75% Charged	1.230	1.215	1.180
50% Charged	1.180	1.170	1.135
25% Charged	1.130	1.120	1.090
Discharged	1.080	1.070	1.045

*State of charge as indicated by specific gravity when discharged at 20 hour rate.

The above are more or less typical specific gravity ranges. Gravity ranges will vary somewhat, depending on battery construction and ratio of electrolyte volume to active material.

Figure 14-5. This figure shows the relationship between specific gravity and a battery's state of charge. (AABM)

Interpreting Hydrometer Readings

When a hydrometer reading indicates the battery is less than 75% charged the battery should be charged and retested. Determine the reason for the low state of charge. A difference of more than 25 points (0.025) between individual cell readings indicates that the cell is bad. Perform a load test for a more accurate picture of battery condition.

Built-In Hydrometer

Many batteries are sealed and have a built-in hydrometer. Refer to **Figure 14-6.** Battery state of charge is indicated by the color of the built-in hydrometer, or "eye" indicator. When the battery is charged, the indicator will show as a darkened area with a bright colored dot in the center. A sound but discharged battery is indicated by a darkened indicator with no center dot. A battery that needs replacing will often be indicated by a bright indicator with no dot.

Battery Charging

Batteries in sound mechanical condition can be brought up to full charge with a ***battery charger.*** The battery charger is used to pass a metered amount of dc (direct current) electricity through the battery. Fast charging or slow charging may be used. As the charging rate is increased, the time necessary to completely recharge the battery is reduced.

Slow Charging

Slow charging passes a relatively small amount of current, usually around 5–7 amps, through the battery for a fairly long period (14–16 hours or longer).

Slow charging is preferred to fast charging if time is available. Heavily sulphated batteries (plate active materials changed to lead sulphate, which resists essential chemical reactions) can overheat during fast charging. Because battery cell condition is not always known, slow charging minimizes

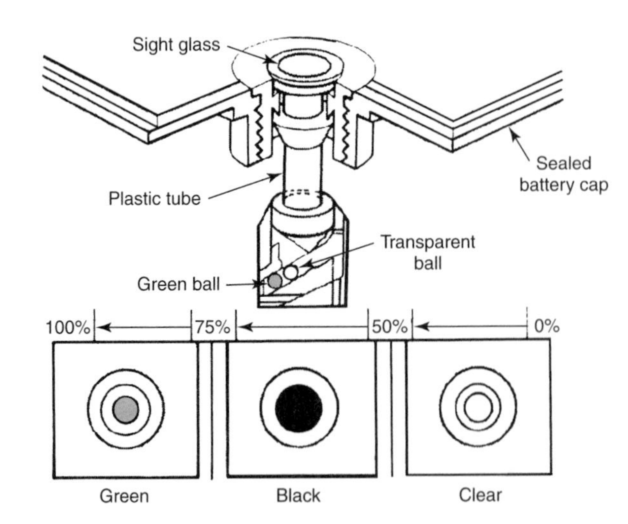

Figure 14-6. A battery indicator "eye" (built-in hydrometer) provides a fast check on battery condition. (DaimlerChrysler)

the risk of possible damage caused by charging. However, a sound battery will not be damaged by proper fast charging.

Begin by cleaning the battery and filling it to the recommended level (if it has filler caps). Replace the caps where used. If the battery will remain in the vehicle, disconnect the cables to prevent damage to electronic components.

Attach the positive charger lead to the positive battery post. Connect the negative lead to the negative post. Switch the charger to either 6 or 12 volts as needed. Set the rate of charge to the lowest possible setting that will recharge the battery in the time allowed. The battery will generally be fully charged within 12–16 hours, although a sulphated battery may require longer. Leave the battery on the charger until the specific gravity or no-load voltage test indicates that it is fully charged.

Watch the battery temperature during charging. If the temperature exceeds 125°F (51.7°C), lower the charge rate. Temperatures in excess of 125°F (51.7°C) will cause serious battery damage. Be certain to remove the battery when charged. Overcharging is harmful.

Fast Charging

Fast charging sends a relatively high initial current, usually about 50–60 amps, through the battery. This charges most batteries in a reasonably short time (one to two hours).

Before fast charging, clean the battery and add water if necessary. Disconnect the battery cables to protect electronic devices. Then, attach the fast charger positive lead to battery positive post and the negative lead to negative post. Set the current control as directed. Switch the charger to either 6 or 12 volts as needed and turn it on. As the battery charge begins to rise, many chargers will automatically reduce the charging rate. Watch battery temperature. If it reaches 125°F (51.7°C), lower the charging rate at once.

If the specific gravity or no-load voltage tests do not show a considerable increase within an hour, try slow charging the battery.

Trickle Charging

Wet batteries (batteries containing electrolyte) that must be kept for any length of time are often placed on a ***trickle charger.*** The trickle charger passes a very low current, often less than one ampere, through the batteries. Despite the small current, batteries can be damaged from trickle charging. Many shops shut off the trickle charger during the night to help prevent overcharging.

Battery Storage

If a wet battery is not to be trickle charged, it should be stored in a cool, dry area. A battery stored at 0°F (–17.8 °C) will retain a charge for nearly a year, while the same battery stored at 125°F (51.7°C) will lose its charge within a month.

Dry-charged batteries (batteries with plates charged but contain no electrolyte) must be stored in a cool, dry area with as even a temperature as possible. Although the dry-charged battery will retain its charge over a long period of time, it is wise to activate the battery by the end of the third year of storage.

Activating a Dry-Charged Battery

Most modern automotive batteries are wet (shipped with the electrolyte installed), but the technician may encounter some dry-charged batteries. These batteries are shipped without electrolyte, and the technician must add electrolyte at the time of installation. Dry-charged batteries are used in farm equipment, lawn equipment, and motorcycles. When activating a dry-charged battery, observe the following instructions. Take the time needed to do the job right. This will prevent unnecessary comebacks and complaints.

Observe the following safety precautions when activating a dry-charged battery:

- When handling battery electrolyte, wear goggles, rubber gloves, and a rubber apron.
- Store battery acid or dry-charge battery electrolyte where the containers will not be broken.
- If it is ever necessary to mix sulfuric acid and water to make an electrolyte mixture, pour the acid into the water. Do not pour water into the acid. Add acid slowly and stir constantly with a glass rod.

Remove the cell caps and the cell cap vent plugs. Carefully open the electrolyte container. In most cases, battery electrolyte is premixed.

Using a glass or plastic funnel, add the specified electrolyte to each cell until the separators are just covered (this allows room for expansion during charging). To prevent a chemical reaction, never use a metal funnel when adding electrolyte.

After filling, replace the caps. Place the battery on the charger and apply a moderate charge to the battery, depending on the instructions supplied with the battery. If excessive gassing occurs, lower the charge rate. Continue charging the battery until the specific gravity reaches at least 1.240 and the electrolyte temperature is 80°F (26.5°C) or above. It is important that both specific gravity and temperature reach the levels indicated.

After charging, add electrolyte (not just water) to bring the level up to the mark. If using a disposable electrolyte container, wash it out with water and discard. Finally, install the battery properly. See the Battery Removal and Installation section in this chapter.

Battery Selection

Never replace an original equipment battery with one that has less electrical capacity. When additional electrical devices have been added to a vehicle, or if the vehicle is operated at low speeds or stopped and started frequently, select a battery of greater capacity.

Battery Groups

Battery manufacturers divide their products into ***battery groups*** based on three criteria:

- ***Size:*** Size is the physical dimensions of the battery. Size varies greatly due to the variety of vehicle types and sizes on the road.
- ***Voltage:*** Although most vehicles made today use 12-volt batteries, some older vehicles had 6-volt

systems. Some modern agricultural and construction equipment and many large trucks have 24-volt systems. 42-volt batteries may be used in the near future.

- ***Terminal type:*** All vehicle batteries have either top or side terminals. Some off-road and marine electrical systems use threaded studs as terminals, often with wing nuts as fasteners.

If a larger battery (physical size) is desired, be certain the battery hold-down will fit properly. Be especially careful if the battery is installed near moving parts. Also check vertical height of the battery in case the hood comes relatively close to the battery when it is closed.

Battery Ratings

A battery's electrical size has almost no relationship to its physical size. Often, a small battery will have a higher cranking-amp rating than a large battery. This is due to variations in battery materials and construction standards. Below are the latest ***battery rating*** measurements:

- ***Cold cranking amps (CCA):*** The maximum amount of current that flows for 30 seconds at 7.2 volts with the battery temperature at 0°F (–17.8°C). This measurement indicates how much current the battery can produce when cold and is the standard measurement for modern batteries.
- ***Cranking amps (CA):*** The maximum amount of current that flows for 30 seconds at 7.2 volts with the battery temperature at 32°F (0°C). This measurement may also be called hot cranking amps (HCA), or marine cranking amps (MCA).
- ***Reserve capacity (RC):*** The number of minutes the battery can produce 25 amps at 10.5 volts with battery temperature at 80°F (26.5°C). Reserve capacity indicates how long the battery can operate the vehicle electrical system in the event of a charging system failure.

Battery Additives

It is recommended that nothing but pure water be added to a battery. Do not add acid unless the battery has been overturned or the electrolyte has leaked out. The use of additives to supposedly improve capacity or prolong battery life, unless specifically approved by the battery manufacturer, can void the guarantee.

Battery Temperature Sensor

Some vehicles use a ***battery temperature sensor*** to help control battery overcharging. See **Figure 14-7.** Temperature data from this sensor is sent to the power train control module (PCM), which controls the battery's charging rate. Excessive overcharging will shorten the battery's useful life. Most temperature sensors are tested using an ohmmeter. The sensor in Figure 14-7 will provide an ohmmeter reading of 9000–11,000 ohms at 75–80°F (23.9°–26.7°C). If it does not, the sensor must be replaced. Follow the manufacturer's recommendations for proper testing.

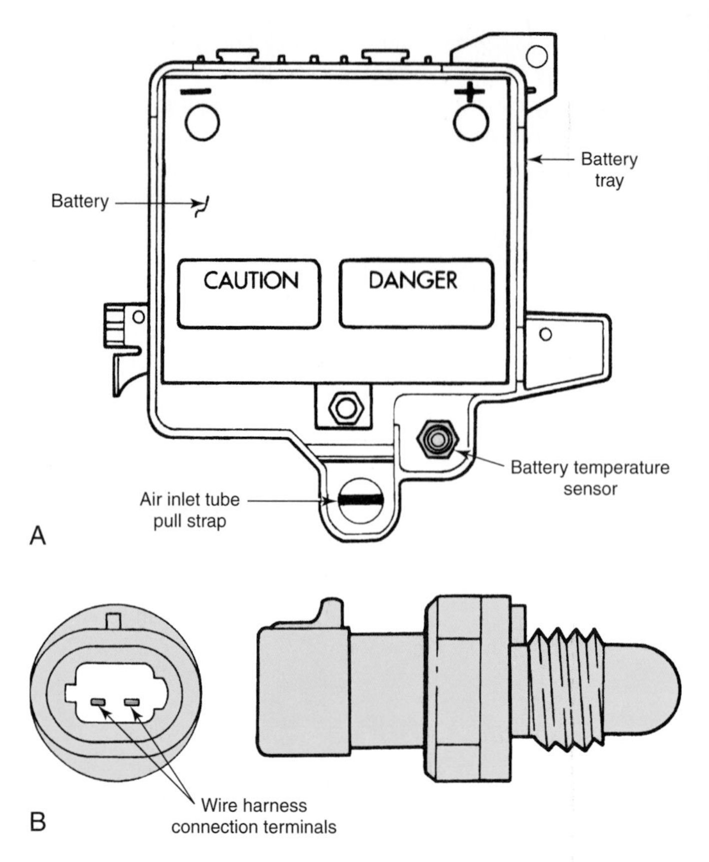

Figure 14-7. Some vehicles have battery temperature sensors. A—Location of the battery temperature sensor on the battery tray. B—The sensor has been removed for testing. The ohmmeter leads are connected to the two wire harness terminals on the sensor.

Battery Removal and Installation

Battery removal is relatively simple, but precautions must be taken to ensure the vehicle is not damaged by acid or electrical shorts. Begin by covering the fender with a protective pad. Before removing the battery cable terminals, note which battery post is grounded. On almost all modern vehicles, the negative post is grounded.

Put on safety glasses and loosen the terminal fasteners. If the terminals are difficult to remove, use a battery terminal puller. Never pound or twist the post or side terminals in an attempt to loosen them. Always remove the ground terminal first, and when installing the battery, replace it last.

Remove the hold-down plate. Using a battery lift strap or an approved removal tool, remove the battery from the vehicle. Make sure the lift strap is securely attached. See **Figure 14-8.**

Before installing a battery, check the battery holder for signs of corrosion. If necessary, remove corrosion with baking soda and water. Make sure the holder is structurally sound enough to support the battery.

Place the battery in the holder so the terminals are properly positioned with respect to the cables. Replace the battery hold-down if it is badly corroded. If the hold-down is corroded but still usable, clean it with baking soda and water. Dry the

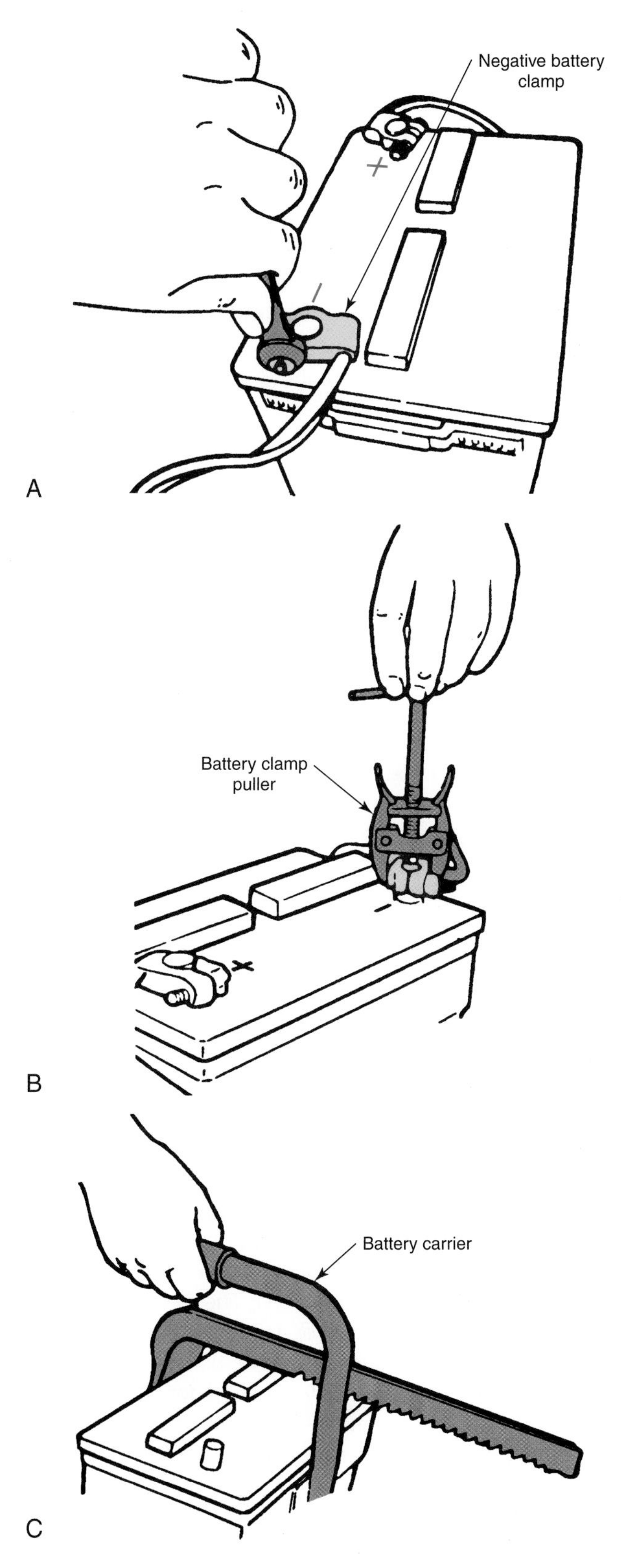

Figure 14-8. Follow these steps when removing a battery. A—Remove the battery cable terminals, ground cable first. B—Use a puller if the terminal is stuck. C—Attach a battery carrier. (Ford)

hold-down and paint it with acid-proof paint. Finally, install the hold-down. Do not overtighten. Clean the battery terminals until bright and coat them with nonmetallic grease.

Battery Polarity Must Be Correct

Before attaching the battery terminals, make certain the correct ***polarity*** (direction of current flow) will be maintained. If the polarity is reversed by accidentally reversing the terminals, the diodes in the alternator and the transistors in the radio, ignition system, or electronic control unit will be ruined when the engine is started.

The positive post is wider in diameter than the negative post. The positive post may be painted red and may have a (+), P, or POS stamped on the top.

The negative post may be painted black and may be marked with (–), N, or NEG on the top. If in doubt as to whether to ground the positive or negative post, refer to the manufacturer's manual.

If the battery has filler caps, attach the terminals so they will not prevent cap removal. Tighten the caps securely. Make sure the cables are in good condition and that they have enough slack to prevent a strain on the connections.

Check the connection between the ground cable and the frame. The connection must be clean and tight. Tighten the starter and solenoid connections.

Battery Drains

Two things can cause the battery to discharge over a period of time: shorts and parasitic loads.

Shorts are unwanted electrical connections. They drain battery power by allowing a constant flow of current through the battery. Typical sources of shorts are frayed or pinched wires; internal defects in switches, solenoids, or motors; and stuck relays. Another type of short is an internal short between the positive and negative plates of the battery. Shorts can run down a battery overnight and must be found and corrected.

Parasitic loads are small drains on the battery that continue after the ignition is turned off. Parasitic loads can be caused by a variety of electronic components. Refer to **Figure 14-9.** The current drain from these loads is extremely small and, therefore, does not discharge the battery, Figure 14-9A. Nevertheless, parasitic loads must be taken into account when testing for excessive battery drain, Figure 14-9B.

Jump Starting with a Booster Battery

When a battery has become discharged to the point it will not crank the engine, the engine can be ***jump started*** with a ***booster battery.*** A booster battery is an additional, properly charged battery of the same voltage as the discharged battery. It is connected to the vehicle's discharged battery with two ***jumper cables.***

Follow These Safety Precautions

When performed correctly, jump starting is safe. However, when safety precautions are ignored, the battery can explode, causing serious injury and vehicle damage. Even discharged batteries contain hydrogen gas that, if ignited, will explode, scattering battery parts and acid in all directions. Follow these safety precautions:

- Make certain that the vehicles are not touching each other.
- Wear protective glasses, avoid contact with battery electrolyte, and do not lean over the battery when making connections.
- In very cold weather, check for frozen electrolyte or no visible signs of electrolyte. If either condition exists, warm the battery until it reaches a temperature of at least 40°F (4.4°C) before attaching the booster. This prevents battery rupture or explosion.
- Keep open flames or sparks away from the battery.
- The last jumper cable connection should be the negative cable to the engine block of the vehicle with the dead battery, not to the negative battery terminal. This reduces the chance of a spark igniting the hydrogen gas buildup.
- Remove metal watchbands, rings, and other jewelry when working on or around batteries. Use proper tools.
- If electrolyte (sulfuric acid) is splashed on you or the vehicle, flush it immediately with water. A solution of baking soda and water can be poured on clothing to neutralize the acid.
- When using a portable starting unit, do not exceed 16 volts to prevent starter, battery, or other electrical system damage.

Jump Starting—Sequence of Operations

When using jumper cables, it is important to connect them in the proper sequence. Turn off all switches on both vehicles. Make certain the vehicles are not touching. Then, proceed as follows:

1. Connect one jumper clamp to the booster positive terminal.
2. Connect the other end of the same cable to the dead battery positive terminal.
3. Connect one end of the other jumper cable to the booster negative terminal.
4. Clamp the other end of the same cable to the disabled vehicle's engine or frame.
5. Start the engine in the booster vehicle. Run the booster vehicle for a few minutes.
6. Start the disabled vehicle.
7. With both vehicles running, disconnect the jumper cables in exact reverse order: negative jumper from the vehicle engine, negative jumper from the booster negative, positive jumper from the discharged battery, and positive jumper from the booster. Never let the clamp ends touch during the operation.

A

Component	Typical (mA) Parasitic	Maximum (mA) Parasitic
BCM	3.6	12.4
ECM	5.6	10.0
Radio	3.0	6.0
Regulator	1.4	2.0
ELC	2.0	3.3
CPS	1.6	2.7
Illuminated entry	1.0	1.0
Theft	0.4	1.0
Auto door locks	1.0	1.0
Chime	1.0	1.0
HVAC power mod	1.0	1.0

B

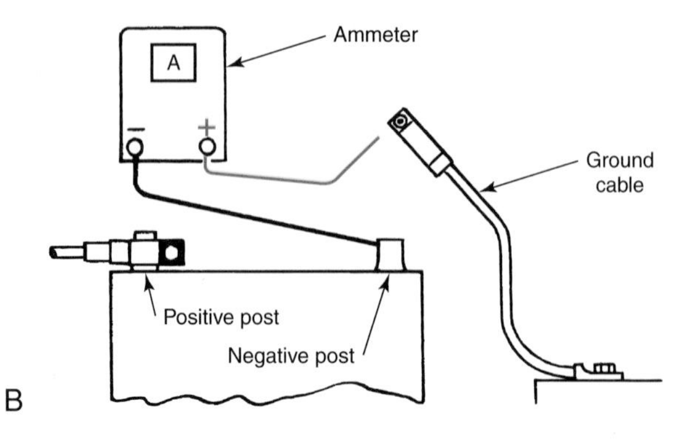

Figure 14-9. Some parasitic loads are normal, since the on-board computers require voltage to retain their programming. A—Typical and maximum parasitic loads (measured in milliamps) caused by various automotive electrical devices. B—An ammeter is used to check for excessive current drain at battery. A small amount of current drain is considered normal.

Caution: Always turn the ignition switch off when working with jumper cables. Failure to do so can damage on-board computers and other electronic parts. Cellular telephones can also be damaged during jump-starting operations. If necessary, disconnect the phone's power lead from the cigarette lighter or remove the power lead fuse. Check with the phone manufacturer for specific instructions.

Charging System Service

All modern vehicles use a charging system that contains an ***alternator*** and a ***voltage regulator.*** The charging system keeps the battery charged and supplies the electrical needs of various units during engine operation. The function and service of the alternator and regulator are covered in this section.

Periodic servicing, such as drive belt tensioning and lubrication, is all that is usually required for thousands of miles. Many late-model alternators use sealed bearings, which eliminate the need for periodic lubrication. In many cases, replacement parts are not available and the entire unit must be replaced if defective.

Charging System Problems

Normal wear, part failure, and service-incurred damage may cause charging system malfunction. Many problems in the charging system start out as minor problems and gradually become worse. Problems in the charging system may produce one or more of the symptoms detailed in the following section.

Undercharging or No Charging

Undercharging and ***No Charging*** are the most common charging system problems. Low or no charging causes slow cranking speed, failure to crank, and dim headlights. If the vehicle has an ammeter, it will indicate a low charging rate. The charge indicator light (when used) may be on or flicker when the engine is running. Usual causes of undercharging are a slipping or missing drive belt, a defect in the alternator or regulator, or high resistance in one or more charging system wires.

Overcharging

Excessively bright lights and frequent bulb replacement indicate that the system is ***overcharging.*** If the vehicle uses an ammeter, it will indicate a high charging rate with a charged battery. A defective voltage regulator is almost always the cause of overcharging. On very rare occasions, disconnected or high-resistance regulator wiring causes this problem.

Noisy Alternator

A light whining noise from the alternator is normal. A loud whine at idle speed can indicate a faulty diode or stator. Diode and stator defects can also cause noise in the AM band of the radio. A loose or glazed belt can cause belt squeal. Dry or worn bearings can cause roaring or squealing noises. A loose pulley can cause clicking or rattling noises.

Ammeter or Indicator Light Problems

If the ammeter needle does not move in either direction when the engine is started and operated, the ammeter is defective. If the charge indicator does not light when the key is on, the bulb is burned out or the regulator is disconnected or defective. If the light is on any time the engine is running, the regulator is defective or idle speed is very low. If the light is on when the key is off, the vehicle wiring is shorted.

A Quick Initial Check Can Save Time and Trouble

Before making any tests or replacing parts, give the system a quick initial inspection. Such a check will often turn up the source of the trouble. Check the drive belt condition and tension. Inspect all connections for tightness and signs of overheating. Examine wires for signs of burning or fraying. Check the regulator and alternator for evidence of physical damage.

A Sound Battery Is a Must

A worn out or badly sulphated battery will produce numerous problems that cannot be corrected until the battery is replaced. Always check battery condition as explained earlier in this chapter before condemning the charging system. A fully charged battery is a must for conducting accurate tests.

Units That May Cause Charging Problems

Charging system malfunctions can originate at the battery, alternator, regulator, ammeter, indicator lights, or wiring. Troubles may involve one unit or, in some cases, all units. Never replace a defective unit without determining what caused the failure. If the failure was brought on by another unit, the unit that caused the failure must be repaired or replaced or the new unit will fail.

Alternator Service Precautions

There are a number of important precautions that are vital when working on the charging system. Failure to observe these rules can result in serious system damage. Learn and observe all the following rules:

- When installing the battery, make sure the correct terminal is grounded.
- When using a booster battery or charger, connect it in parallel—positive to positive and negative to negative.
- Disconnect battery cables when charging the battery.
- When soldering a diode lead, take precautions to protect the rectifier from overheating.
- Never operate the alternator with an open circuit (output lead or battery terminal disconnected). This will allow the alternator to build up very high voltage that can damage diodes and can be very dangerous to anyone touching the alternator battery terminal.
- Do not ground the alternator field circuit, except for brief testing.
- To prevent accidental shorts, remove the battery ground lead before removing any system wires or connecting test equipment other than a voltmeter.
- When adjusting an electro-mechanical regulator, use an insulated tool to prevent accidental grounding.
- Do not ground the alternator output terminal or any regulator terminals.
- Make sure the ignition switch is off before removing or installing a regulator cover.
- Disconnect the connector plug from the regulator before removing the mount screws. Pulling the connector from an ungrounded regulator can ruin it.
- Use the correct alternator and regulator for the vehicle.
- Alternator testing procedures vary. Always follow the manufacturer's recommendations for the vehicle at hand.

Checking Alternator Output

Two electrical properties must be checked to determine whether the charging system is operating properly. These are output ***amperage*** and ***voltage.*** If the alternator and regulator are doing their jobs, both of these will be within specifications. To make this test, connect a voltmeter across the battery terminals and connect an ammeter to the output terminal of the alternator. If you are using a charging system tester, follow the manufacturer's instructions.

Once all connections have been made, start the engine and run it at fast idle. Observe the ammeter and voltmeter. Amperage should be within specifications, and voltage should be between 12.5–14.5 volts, depending on the voltage regulator setting and state of battery charge. Be sure to check the manufacturer's specifications before deciding on the condition of the charging system.

If the amperage and voltage are within specifications, disconnect the tester. If the readings are incorrect, the problem must be isolated to either the alternator or regulator. To make this test, stop the engine and locate the voltage regulator and field lead.

If the vehicle has an external **regulator**, disconnect the field lead and run a jumper wire with an on-off switch from the alternator field terminal to the battery positive terminal. Make sure the switch is set to the off position. Then, restart the engine. Throw the switch to the on position and recheck the charging system output. If the alternator starts charging with the regulator bypassed, the regulator or its associated wiring is defective. If the alternator still does not charge, it is defective. Throw the jumper-wire switch to off, stop the engine, and remove the jumper wire.

Caution: Make this check quickly (a few seconds maximum), as the unregulated alternator can produce enough voltage to damage the vehicle's electrical and electronic equipment.

If the regulator is built into the alternator, follow manufacturer's instructions to bypass the regulator. Some alternators, such as the one shown in **Figure 14-10** allow the regulator to be bypassed by inserting a screwdriver in a hole in the rear of the alternator. Some alternators with built-in regulators have no provision for checking the regulator. These alternators are usually replaced as a unit.

Voltage Regulator Service

If the problem has been isolated to the voltage regulator, it can be serviced or replaced. Most ***electronic regulators*** are not serviceable and must be replaced. Some regulators are part of a non-serviceable alternator assembly. If this type of regulator is faulty, the entire alternator must be replaced. ***Electro-mechanical regulators*** can be serviced, but many technicians prefer to replace them with new units.

Caution: When servicing voltage regulators, observe all the cautions regarding alternator system service.

Electronic Voltage Regulator Service

Modern vehicles are equipped with electronic voltage regulators. These units regulate the alternator's output using transistors, diodes, resistors, and capacitors, **Figure 14-11.** Although electronic voltage regulators handle heavy loads and provide excellent service, they can still malfunction under certain conditions. The bypass test previously described is the simplest way to check the regulator for proper operation.

Most electronic voltage regulators cannot be repaired. If they are faulty, they must be replaced. Replacement is simple when the regulator is installed on the engine firewall or inner fender. Simply remove the battery negative cable, the regulator wiring, and the attaching bolts. To install a new regulator, reverse the removal process. If the regulator is installed in the alternator, the alternator must be disassembled. Alternator disassembly is explained later in this chapter. A few regulators are installed on the back of the alternator and can be replaced without disassembling the alternator.

Electro-Mechanical Voltage Regulator Service

Some older vehicles are equipped with electro-mechanical voltage regulators. These units use electric coils and contact points to regulate alternator output. Always refer to an appropriate service manual before servicing these regulators. Most electro-mechanical voltage regulators can be replaced by unplugging the regulator wiring and removing the fasteners. Install the new regulator. Install and tighten the fasteners, and plug the wire connector into the regulator.

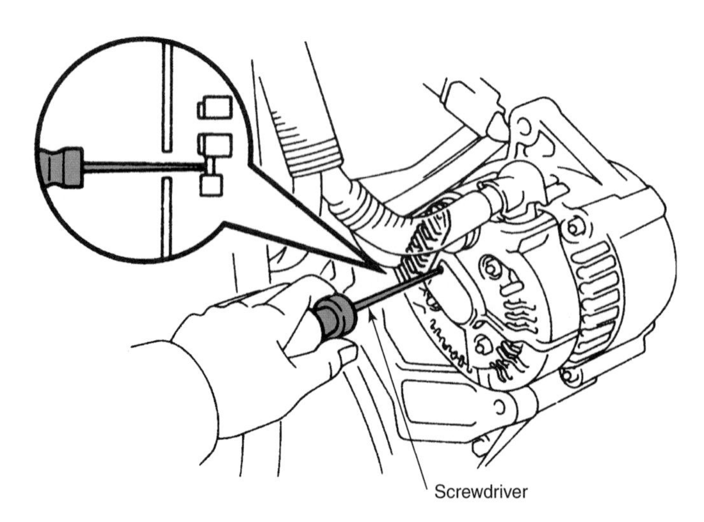

Figure 14-10. This alternator's voltage regulator is being bypassed by inserting a screwdriver in the back of the alternator. (Toyota)

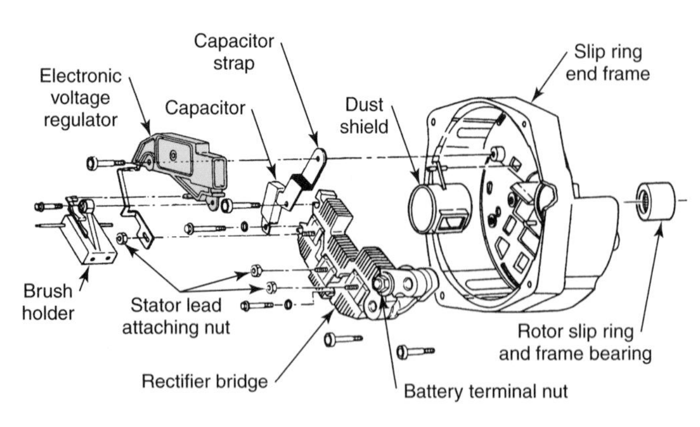

Figure 14-11. An exploded view of an alternator slip ring end frame and components illustrates the various internal parts. Note the electronic voltage regulator. This unit is not adjustable and must be replaced if defective. (General Motors)

Charge Indicator Light

If the indicator lamp fails to light when the ignition key is turned on (engine stopped), check for a burned out bulb; corrosion or looseness in the lamp socket; and loose, corroded, or open connections in the circuit. If the indicator light stays on when the ignition key is turned off, check for a shorted positive diode in the alternator.

When the engine is idling, the indicator light should go out. If it continues to burn, check for slow idle speed. If this does not correct the problem, check the alternator drive belt adjustment, field relay operation, and alternator output. Conduct other system tests as required to pinpoint the bad unit or units.

Computerized Charging System

In many vehicles, an on-board computer controls the current to the alternator's field terminal. This arrangement eliminates the need for a voltage regulator. Most computerized charging systems have self-diagnostic capabilities. When the computer detects a malfunction in the charging system, it stores a trouble code in its memory and activates the check engine light. A typical computerized charging system is illustrated in **Figure 14-12** For more information on trouble codes and computerized systems, see Chapter 15, Computer System Service.

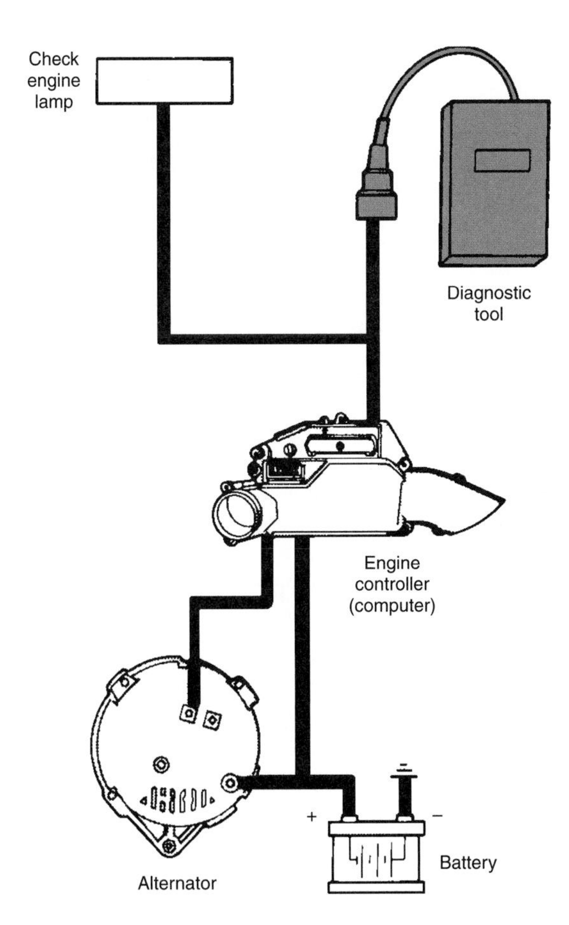

Figure 14-12. Typical computerized charging systems are energized through the computer. The computer is the voltage regulator. Note hookup for diagnostic tool. (General Motors)

Check Alternator Charging Circuit for Excessive Resistance

The charging system cannot function properly when excessive electrical ***resistance*** is present. When system malfunctions are evident and the quick check does not reveal the exact cause, check the system resistance before conducting more exhaustive tests. This is best done by a ***voltage drop test,*** in which a voltmeter is connected across a connection. The circuit is then energized and current flows through the connection. If resistance is high, current flows through the meter, creating a voltage reading. Specifications vary between manufacturers, but the voltage reading generally should not be more than 0.02–0.05 volts. When high resistance is found, replace the wire or clean and tighten connections.

Alternator Disassembly and Overhaul

Most shops prefer to replace defective alternators with new or rebuilt units. Many newer alternators must be serviced as a unit and no replacement parts are available. Some alternators can be disassembled, and faulty parts can be replaced. In many cases, however, the needed parts will cost more than a new or rebuilt unit. Before deciding to rebuild an alternator determine the parts and labor involved.

Alternator Disassembly

Specific disassembly varies with the design of the alternator. The following steps are typical for many alternators.

1. Remove the pulley nut.
2. Using a suitable puller, remove the pulley.
3. Mark the units with a scriber before disassembly. Scribe a single line across the seam between the slip ring end frame and the drive end frame before continuing the disassembly. This will allow you to properly align the parts during the reassembly.
4. Remove the through-bolts.
5. Pry the slip ring end frame and drive end frame apart, being careful not to damage the frames or internal parts. Separate the units. See **Figure 14-13.** On some models, the brush assembly must be removed before dismantling the alternator.
6. When needed, use pullers to remove end plates and bearings.
7. Remove stator and rectifier frames and voltage regulator if used. An exploded view of a typical alternator is shown in **Figure 14-14.**

Clean all parts in solvent, except the rotor, stator, slip ring, and brush assemblies, and then wipe them with a clean cloth. If the bearings are the sealed type, do not place them in solvent. Where applicable, clean and inspect the bearings. If the bearings are the sealed type, inspect them for wear, roughness, and loss or hardening of lubricant. Replace any defective bearings.

Check Brushes

The ***brushes*** should be checked for proper length and replaced if they show any wear. During an overhaul, it is good

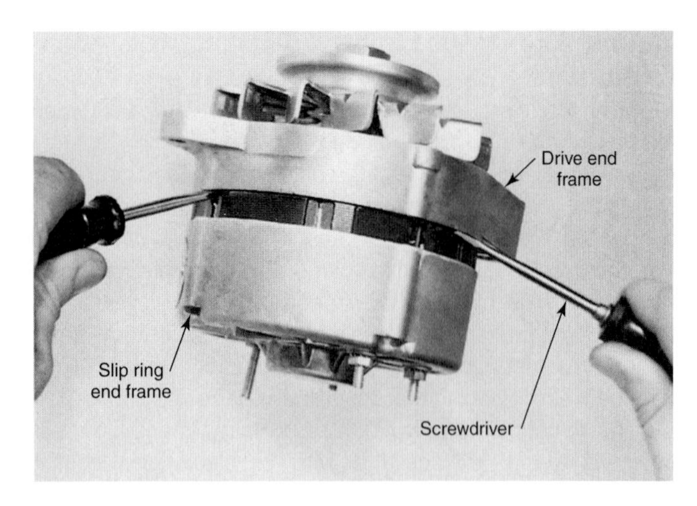

Figure 14-13. Carefully separate the alternator drive and slip ring end frames. (DaimlerChrysler)

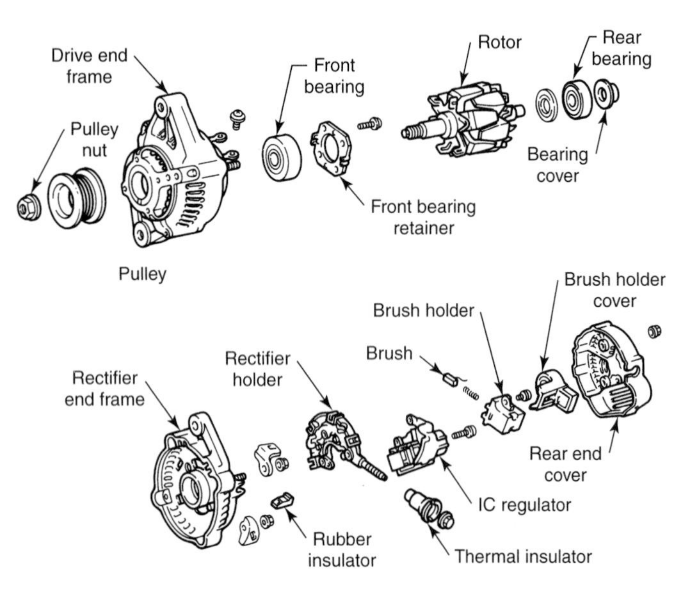

Figure 14-14. This exploded view shows typical alternator construction. (General Motors)

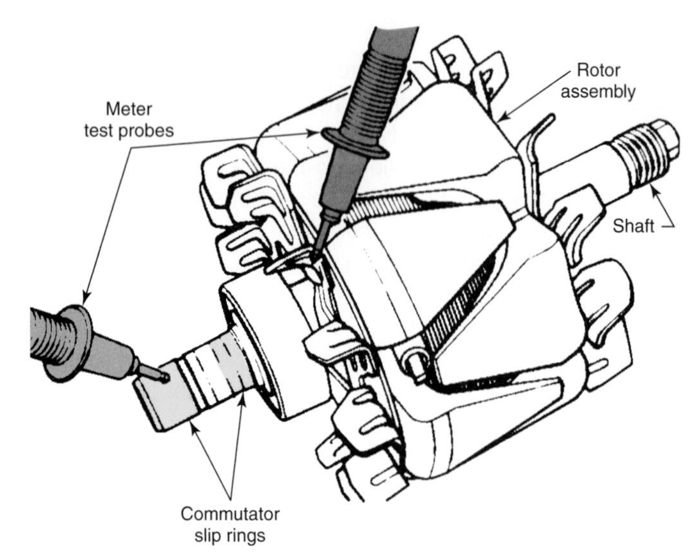

Figure 14-15. The alternator field circuit can be tested for grounding with an ohmmeter. (DaimlerChrysler)

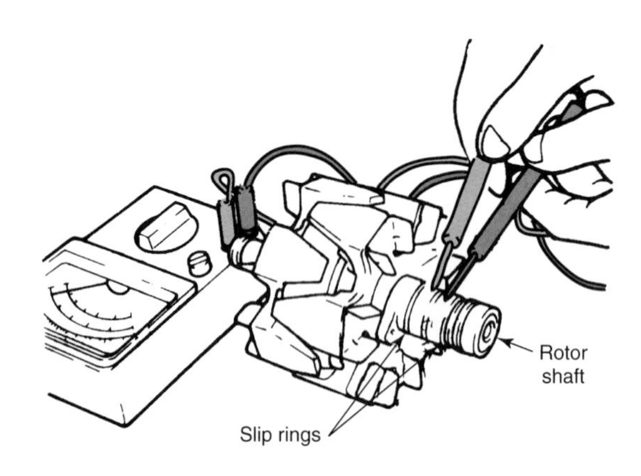

Figure 14-16. In this illustration, the alternator rotor is being tested for shorts and opens. (DaimlerChrysler)

practice to replace the brushes regardless of how much wear they show. Also check the brush springs. Brushes must be absolutely free of oil or grease. If the brushes come in contact with grease during alternator disassembly, clean them at once.

Test Rotor Windings for Opens, Grounds, and Shorts

Use an ohmmeter to check the rotor for grounds, opens, and shorts. To check for grounds, place one ohmmeter lead on one slip ring and the other lead on the rotor shaft or other rotor component that is normally insulated from the commutator. See **Figure 14-15.** If the ohmmeter gives a low reading, the windings are grounded. A powered test light may also be used for this test. The lamp will light if the windings are grounded.

To test for opens, **Figure 14-16,** place one ohmmeter lead on each slip ring. If the windings are open, a high (infinite) reading will occur. If a powered test light is used, the lamp will not light if the windings are open. To check for shorts, attach an ohmmeter lead to each slip ring, Figure 14-16. Check the ohmmeter reading against specifications. If the reading is below specifications, the windings are shorted. When using a test light, never place the prods on the portion of the slip ring contacted by the brush or on the portion of the rotor shaft contacted by the bearing. This protects the vital surfaces from pitting due to arcing.

The slip rings must be smooth and round. If the slip rings are dirty, they should be cleaned by turning the rotor while holding 400-grain polishing cloth against them. Do not use emery cloth. If the rings are scored or out-of-round, the rotor should be replaced.

Check Diodes

Quite often, the ***diodes*** are responsible for charging system problems. Anything that reverses the system polarity, such as reversing battery charger or booster battery cables can ruin the diodes. When the alternator is disassembled, always test the diodes.

Testing Diodes

Some alternators require disassembly in order to test the diodes. Others do not. Follow manufacturer's instructions. Diodes may be tested with an ohmmeter. The diode leads must be disconnected to make this test.

Place one ohmmeter test prod on the diode case. Place the other test prod on the diode lead. Again, make sure the connections are good.

If the diode is good, a high reading will be given with the test prods in one position and a low reading will be given when the prods are reversed. An open diode will give high readings in both positions. A shorted diode will give low readings in both directions.

Prod tips must be sharp. Make certain they penetrate any varnish coating present on the terminals. Be sure to test all diodes. Positive diodes will test the same as negative diodes, but the readings will be opposite of the negative readings. See **Figure 14-17.**

Test Capacitor

Be sure to test the radio suppression capacitor (sometimes called the condenser), at the same time that you test the diodes. A faulty capacitor is often the cause when the diodes test open or shorted. The radio suppression capacitor is always installed in the alternator housing. Some capacitors are pressed into a hole in the rear case half, while others are attached to the rear case half by a bracket.

To test the capacitor, disconnect it and place an ohmmeter across its leads. An infinite reading indicates good condition. A low reading indicates a defective condenser. To test condenser capacity in microfarads, use a condenser tester.

Removing and Replacing Diodes

Diodes can be removed or replaced with a press and suitable tools. **Figure 14-18** shows diodes that have been pressed into place in the heat sink. The pressing tool must bear against the outer edge of the diode case. Follow the manufacturer's recommendations for diode removal and installation.

When soldering diode leads, always grip the lead between the diode and the soldered area with a pair of pliers. The pliers will protect the diode from excessive heat.

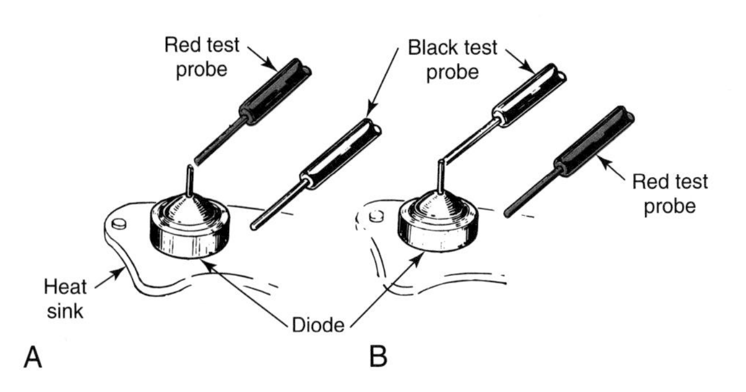

Figure 14-17. This illustration shows a method of testing a diode. A—First test. B—Second test. Note the leads are reversed in B. If the diode is good, the test light should come on in one test and go out in the other. (Nissan)

Some alternators use three diodes (sometimes called a "diode trio") in combination with the internal voltage regulator. These diodes are separate from the output diodes. They are enclosed as a group and in a dielectric (nonconducting) material. These must be serviced as a unit. See **Figure 14-19.**

Test Stator Windings for Opens and Shorts

Figure 14-20 shows a setup for testing stator windings for opens. The lamp should light, and the ohmmeter should show a specified resistance. If the lamp does not light, or if the ohmmeter shows an infinite resistance, the windings are open.

Checking for winding shorts is sometimes difficult and requires more test equipment. Visually inspect the windings for signs of overheating. When the alternator does not produce its specified output and all other electrical checks are OK, you may assume the stator windings are shorted.

Alternator Reassembly

When specified by the manufacturer, bearings must be packed with grease. Assemble the alternator in the reverse

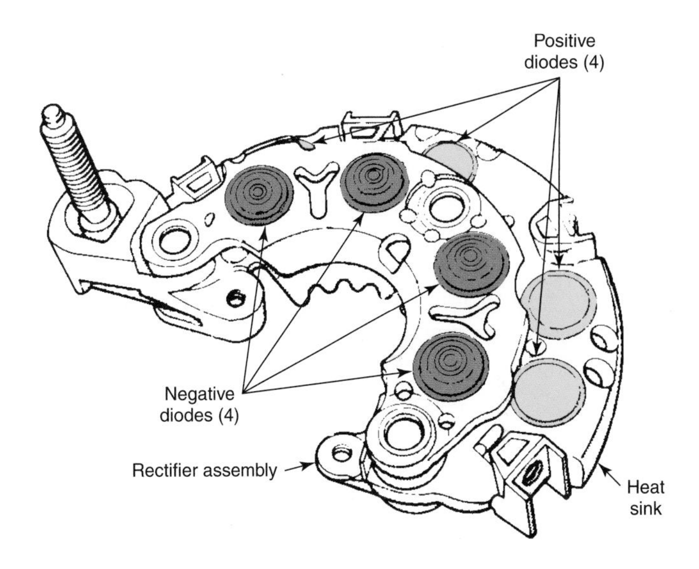

Figure 14-18. These diodes have been pressed into the heat sink. Never hammer a diode into place: it will be ruined.

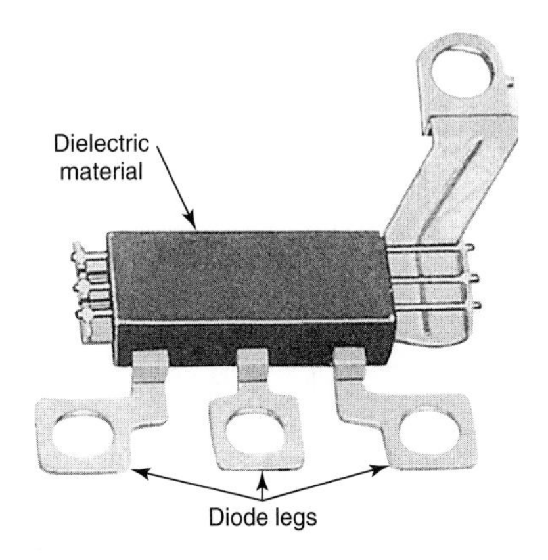

Figure 14-19. Diode trios are used on some alternators. Three diodes are held within a molded dielectric material. (Deere & Co.)

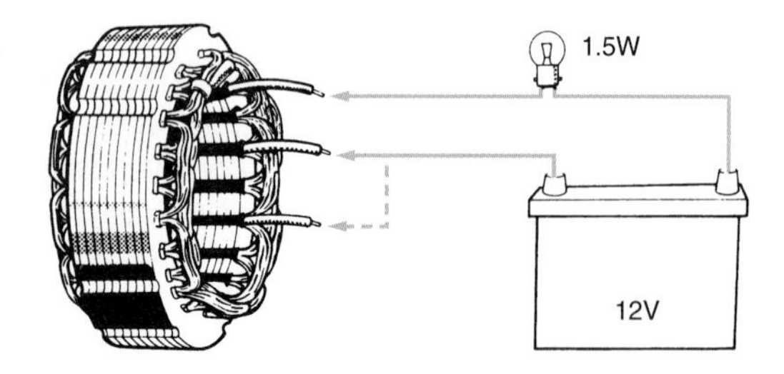

Figure 14-20. Testing the stator windings for opens. (British Leyland)

order of disassembly. Work carefully. Use a press for parts that require a press fit. Align scribe marks. Torque the pulley's retaining nut. Never clamp the rotor to hold it while torquing the pulley nut. Doing so may deform the rotor. Instead, clamp the pulley.

Some alternator brush assembly designs permit the use of a pin to hold the brushes in the holder during assembly. Some require hooking the brush leads over a section of the brush holder for installation. Another type uses simple brush holders that permit installation after the alternator is assembled.

If a brush holder pin is used, or if the leads are hooked, remove the pin or straighten the leads after assembly. See **Figures 14-21** and **14-22.** Make sure all parts are correctly located. Spin the rotor to check for free operation.

When assembling the alternator to the engine, adjust belt tension by prying on the heavy drive end frame edge. Never pry against the center or at the slip ring frame end. Connect all leads to the alternator. Connect the battery leads. Start the engine and test the alternator.

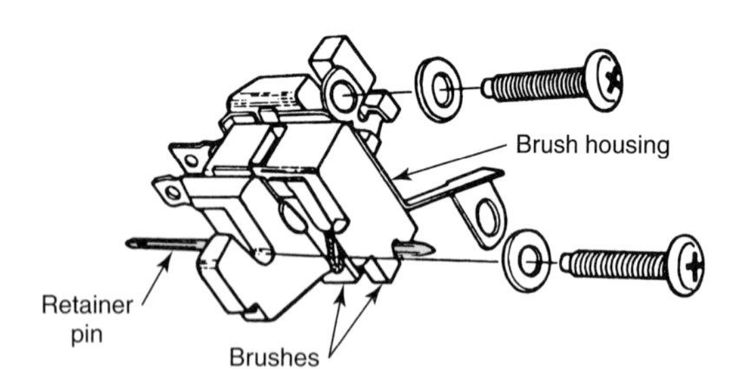

Figure 14-21. A pin holds the brushes in a holder for easy assembly. (General Motors)

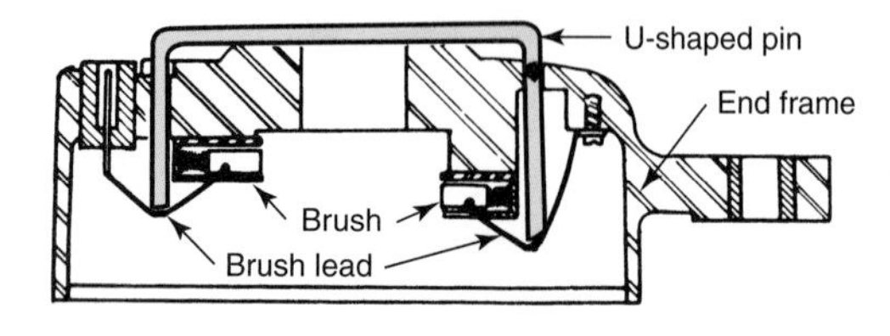

Figure 14-22. A stiff, U-shaped pin keeps these brushes retracted to allow installation of the rotor. (Motorcraft)

Starting System Service

The starting system consists of the battery, starter motor, solenoid, ignition switch, neutral safety switch, and related wiring, **Figure 14-23.** Often, determining which part is at fault when the engine will not crank is more difficult than replacing the defective parts.

Always Check the Battery First

Many complaints of poor starter performance are traced to a discharged or defective battery. Give the battery a thorough check, as explained earlier in this chapter. Remember that proper starter motor performance demands a charged, sound battery.

Checking Starter Circuit

Give the starter circuit wiring a quick visual check. Remove and clean corroded battery terminals. Clean and tighten other loose, burned, or corroded connections. Look for frayed, broken, or shorted wires.

After the visual check of the starter circuit, test the circuit for excessive resistance. Use an accurate low-reading voltmeter.

Make the voltmeter connections as illustrated in **Figure 14-24.** Remove the ignition coil high-tension lead from the distributor. Ground it or ground the primary distributor terminal of the coil. The technique you should use must be compatible with the type of ignition system you are working on.

Crank the engine with the voltmeter leads connected across the connection. If the voltage drop exceeds 0.2 volt, the resistance is excessive. Repeat the procedure for each of the locations indicated in Figure 14-24.

Caution: Do not operate the starter motor for extended periods, as it will be damaged from overheating. Operate it for a maximum of 20 or 30 seconds. Then, allow it to cool for at least two minutes before resuming cranking.

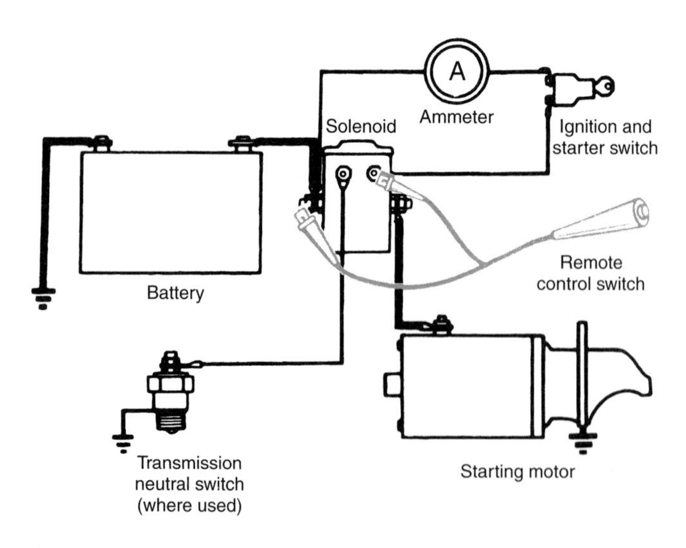

Figure 14-23. This figure shows the component parts of a common starting system or circuit. Note the remote starting (control) switch. (Ignition Mfg's Inst.)

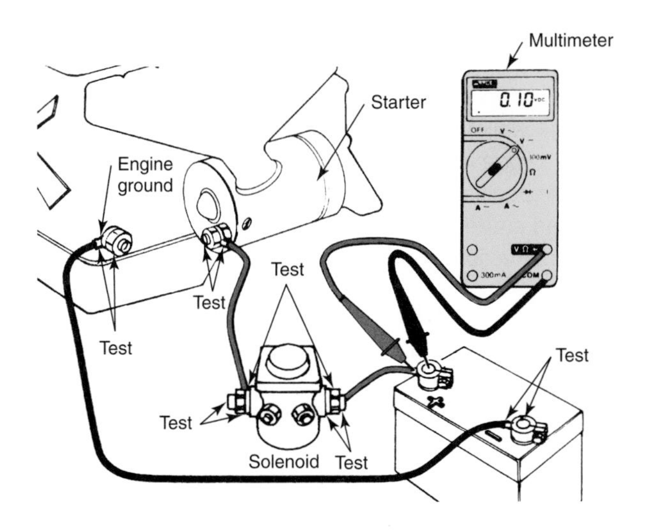

Figure 14-24. This illustration shows various points where starting circuit resistance should be measured.

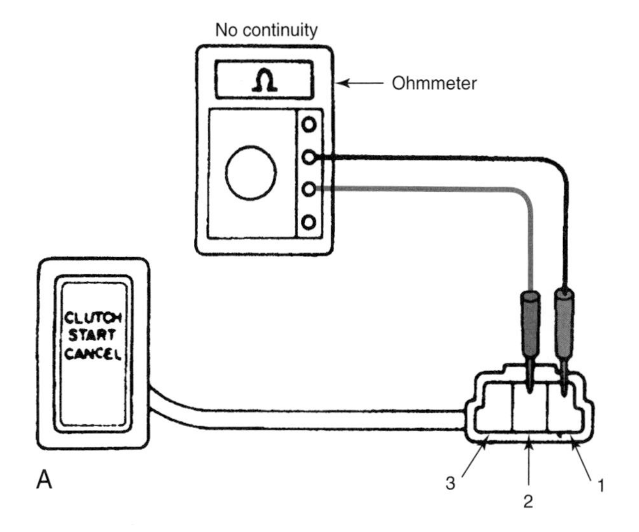

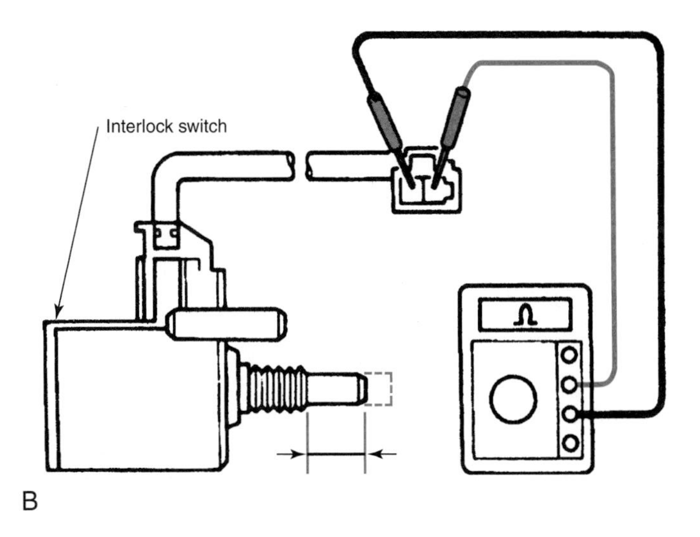

Figure 14-25. Testing procedures for a clutch safety switch are simple, but must be performed carefully. A—Determine which terminals should be used. B—Connect leads between terminals, and depress the plunger. Resistance should change from 0 to infinity ohms as the plunger is moved.

Solenoids, Relays, and Switches

In the event of starter circuit difficulties, do not overlook the ignition switch and starter solenoid. These units occasionally fail. Check each one for correct operation. This can be done by bypassing a switch with a jumper wire or by replacing a suspected relay or solenoid with one known to be good. If the solenoid clicks when energized but the starter does not operate, the problem is in the starter or the internal solenoid contacts. On some vehicles, if the starter turns but does not engage the flywheel, the solenoid may be failing to engage the starter drive mechanism.

All vehicles with automatic transmissions have a safety lockout switch, which is usually called the ***neutral safety switch.*** This switch will not allow the ignition switch to energize the starter solenoid unless the transmission is in park or neutral. Vehicles with manual transmissions have a safety switch installed on the clutch linkage. This switch prevents starting unless the clutch pedal is depressed.

Safety switches can fail or move out of adjustment. Check the correct service manual for exact testing, adjustment, and replacement procedures. A typical test procedure for a clutch safety switch is shown in **Figure 14-25.** Use an ohmmeter for this test procedure. Start by removing the switch connector. Next, place the ohmmeter leads on the connector terminals. If the switch has three terminals, as in Figure 14-25A, determine which terminals connect through the switch electrical contacts and place the leads on these terminals. If the switch has only two terminals, Figure 14-25B, place the leads on the terminals as shown. Then, depress the switch plunger. When the plunger is depressed, the ohmmeter should read zero resistance. Release the plunger and observe the ohmmeter. Resistance should now be infinite. If the switch does not pass these tests, replace it.

Starter Load Test

The starter ***load test*** indicates the current draw in amperes during cranking. It provides an indication of starter motor condition. Excessive engine friction will also be disclosed by this test. Excessive engine friction can be caused by new engine parts (such as rings, bearings, or pistons) that are fitted too tightly. If a good starter draws a high current but will not crank an engine that has been in service (used), there is probably severe internal engine damage. If the starter will not engage the flywheel, the starter drive is defective or the flywheel teeth are stripped.

Instruments for the load test may be connected as illustrated in **Figure 14-26.** If necessary, the technician can use a remote control starter switch to energize the starter without getting into the vehicle to operate the regular starter switch. The method of attaching this remote starter varies according to circuit design.

Ground the coil secondary lead, the coil distributor lead, or the ignition control module to prevent the engine from starting. While cranking the engine, note the exact voltmeter and ammeter readings. Compare the readings with the manufacturer's specifications.

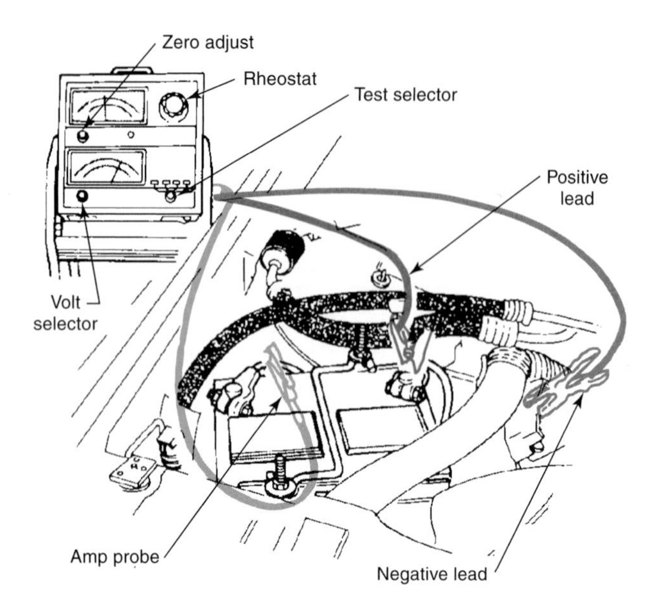

Figure 14-26. Typical starter load test connections are shown here. Three connections allow the tester to measure voltage and amperage.

Starter No-Load and Stall Torque (Locked Armature) Tests

The starter *no-load test* is done by allowing the starter to spin freely while amperage and voltage readings are taken. The *stall torque* test indicates both starter torque and current draw at a specified voltage. Both of these procedures require specialized starter test equipment. Follow the test equipment maker's instructions exactly to perform these tests.

Starter Disassembly

Most modern starters are replaced with new or rebuilt units rather than being rebuilt. Some starters can be disassembled and repaired. Always disassemble a starter carefully, noting the location of all parts. Scribe a line on the pinion housing, frame, and end plate to assist in correct alignment during reassembly. Starters are generally equipped with a locating dowel to guarantee proper alignment. If a vise is used to support the starter during disassembly, do not clamp it tightly. In some cases, it is necessary to unsolder a lead to remove a part.

A cutaway view of a typical solenoid-actuated, overrunning-clutch, drive-type starter is shown in **Figure 14-27. Figure 14-28** shows an exploded view of a starter assembly.

A somewhat different starter drive actuating setup is pictured in **Figure 14-29.** This arrangement employs a

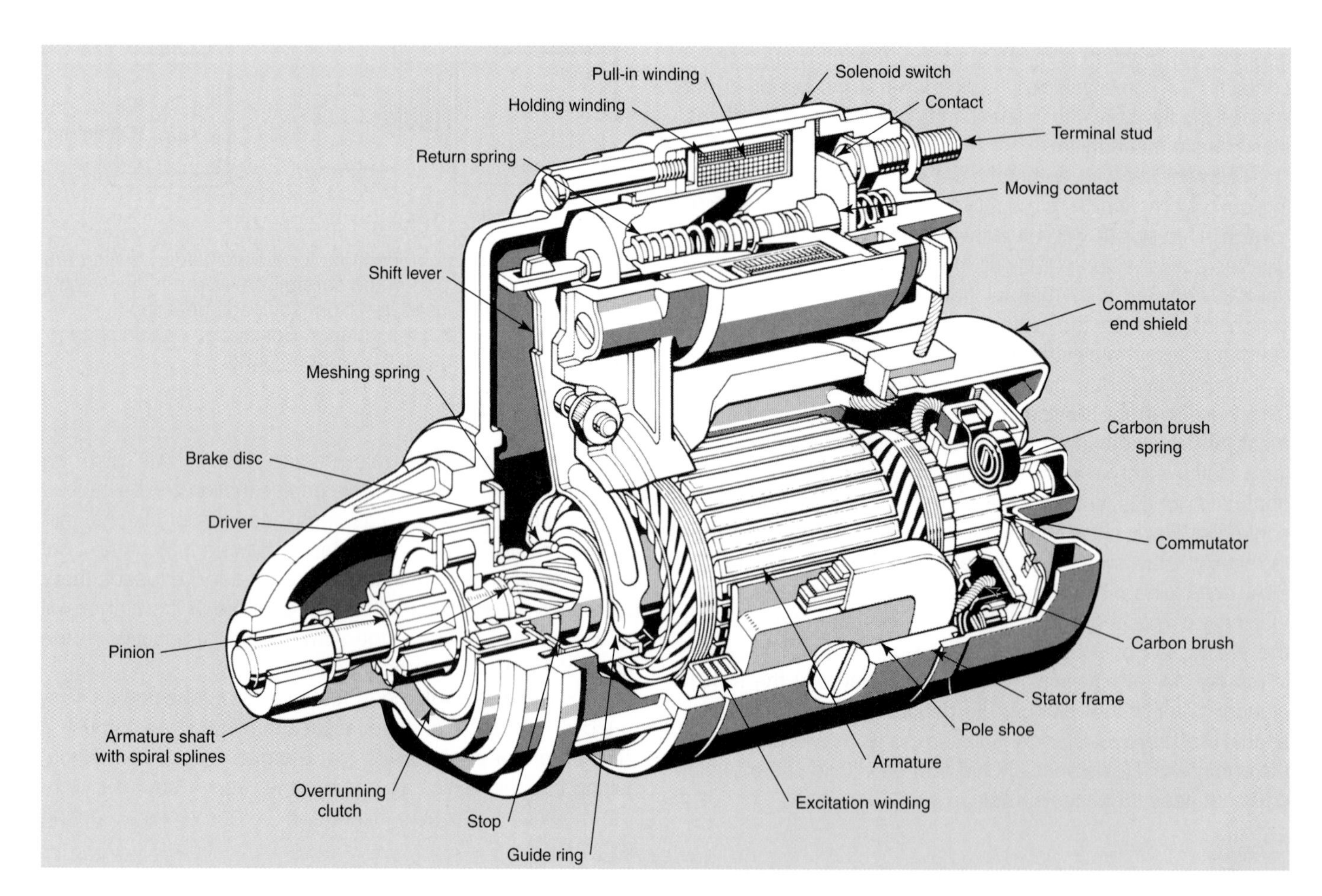

Figure 14-27. This starter uses a solenoid-engaged overrunning clutch drive. The solenoid also actuates the starter motor. (Bosch)

movable pole shoe connected to the starter drive. When the starter is energized, magnetic action causes the pole shoe to slide sideways, thus engaging the starter drive pinion.

Cleaning the Starter

Clean the starter housing, cover plate, and drive housing in solvent. Use a rag slightly dampened with clean solvent to wipe off the armature, the field coils, and the outside of the overrunning clutch unit if they are oily. If no oil is present on these parts, use a soft brush, mild air pressure, or a clean, dry rag.

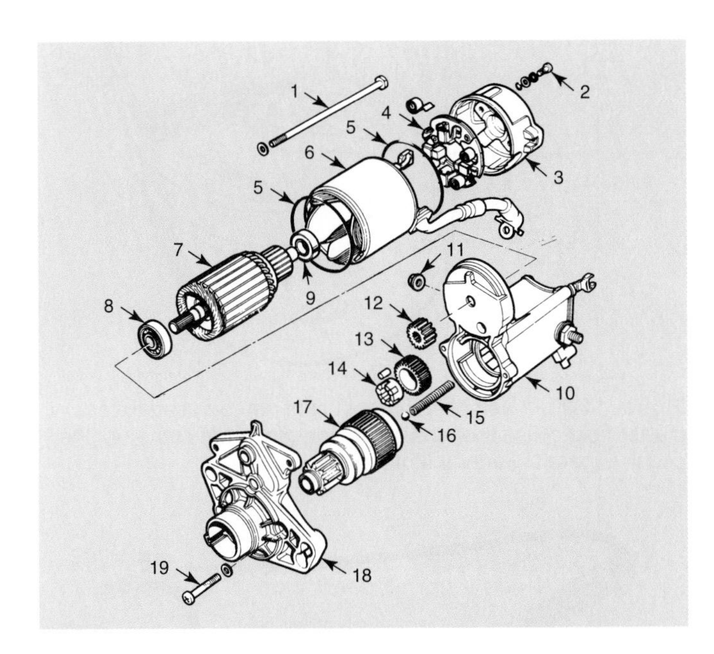

Figure 14-28. This is an exploded view of a starter with a solenoid-actuated overrunning clutch drive. 1—Motor through-bolt. 2—Brush holder screw. 3—End cover. 4—Brush holder. 5—Ring seal. 6—Yoke. 7—Armature. 8—Gear housing bearing. 9—End cover bearing. 10—Solenoid housing. 11—Terminal nut. 12—Pinion gear. 13—Idler gear. 14—Roller bearings and cage. 15—Plunger spring. 16—Steel ball. 17—Overrunning clutch. 18—Gear housing cover. 19—Solenoid screw. (Sterling)

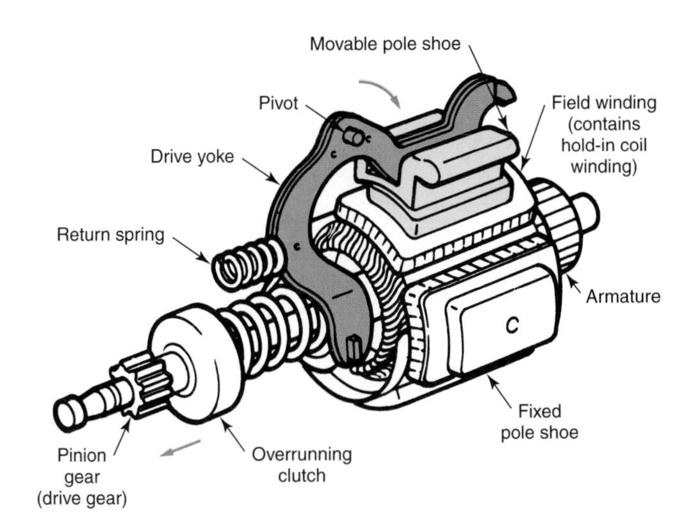

Figure 14-29. This starter uses a movable pole shoe device to actuate the starter drive gear. (DaimlerChrysler)

Cleaning the Starter Drive

Starter drives should be carefully cleaned. A Bendix drive may be cleaned in solvent. Overrunning clutch drives are factory packed with lubricant and must never be placed in solvent.

Testing the Armature for Short Circuits

To test the armature for internal short circuits, a test device called a ***growler*** must be used. A growler is a device that holds the armature where it can be affected by an electromagnet, **Figure 14-30.** Energizing the growler creates a magnetic field in the armature windings. To use the growler place the armature between the magnets as shown in Figure 14-30, then turn on the growler. Do not operate the growler when the armature is not in place.

Hold a thin steel strip (hacksaw blade is fine) loosely on the top of the armature (tip one edge up), Figure 14-30. Turn on the growler. Turn the armature until the entire armature core has passed beneath the strip. If a short circuit exists, the strip will vibrate when the shorted section passes under it.

Checking the Armature for Grounds

Place one test prod on the armature core, **Figure 14-31.** Touch each commutator segment with the other prod. If the armature is grounded, the test lamp will light. Discard grounded or shorted armatures.

Inspecting the Commutator Bars for Opens

Prolonged cranking may overheat the starter and cause the solder on the ***commutator bar*** (commutator segment) connections to melt and be thrown off. Inspect the joints for missing solder. Also, inspect the segments (especially trailing edges) for signs of burning that could indicate an open circuit. If the commutator bars do not show any obvious signs of damage, test them for continuity with a test light. There should be continuity (low resistance) between each bar and *one* other bar, but no additional bars. Refer to **Figure 14-32.** If the commutator fails this test, replace it.

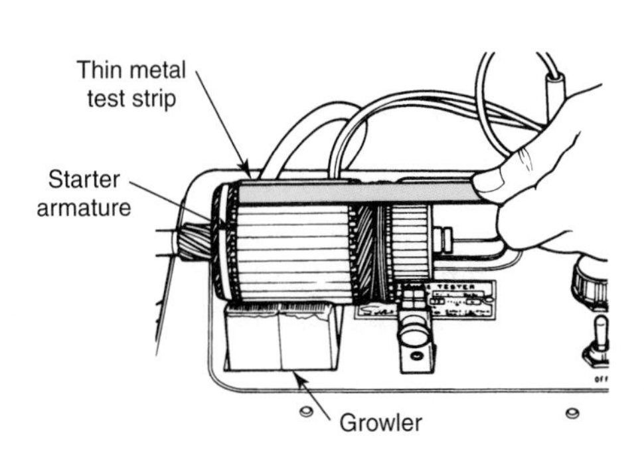

Figure 14-30. In this figure, the technician is using a growler to test the starter armature for short circuits. (General Motors)

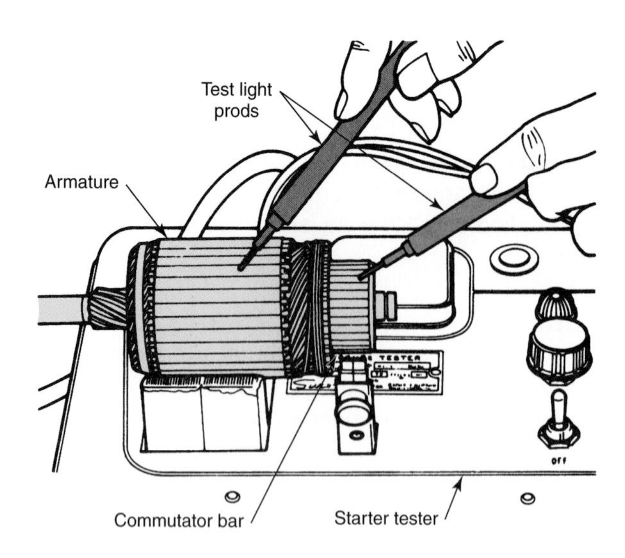

Figure 14-31. Checking a starter armature for grounds. (General Motors)

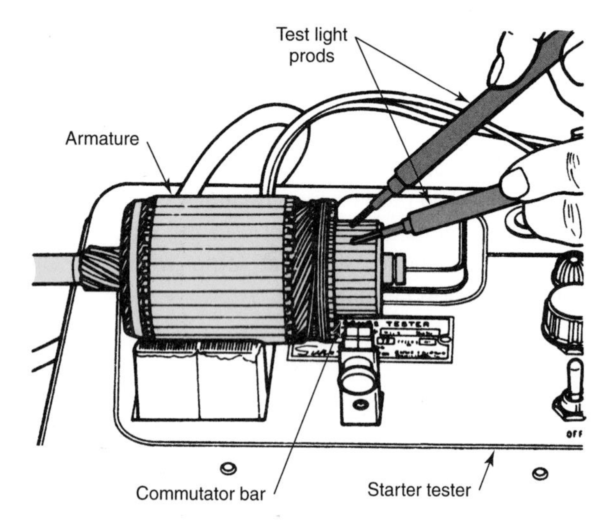

Figure 14-32. Testing armature coils for continuity (opens). (General Motors)

Test Field Coils for Opens

The starter ***field coils*** should be checked with an ohmmeter. Be careful to avoid shocks. Place one test prod on the coil terminal stud or on the connector from the solenoid, **Figure 14-33.** Place the other prod on the insulated brush lead. This, in effect, places a prod on each end of the field coils. The test lamp should light. No light indicates an open circuit.

Test Field Coils for Grounds

Place one prod of the ohmmeter on the field coil connector and the other on the frame or other grounded component. The lamp should not light. A lighted lamp indicates a ground. Be certain that no parts of the field circuit, brushes, or connectors are touching the frame. If the starter uses a shunt coil, disconnect the shunt before making the test. See **Figure 14-34.**

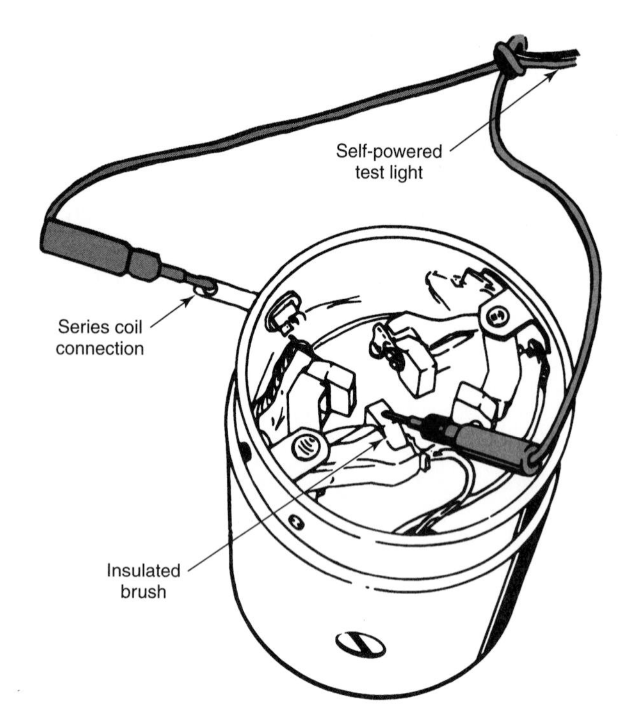

Figure 14-33. A self-powered test light can be used to test starter field coils for opens. If the light does not come on, the circuit is open. (General Motors)

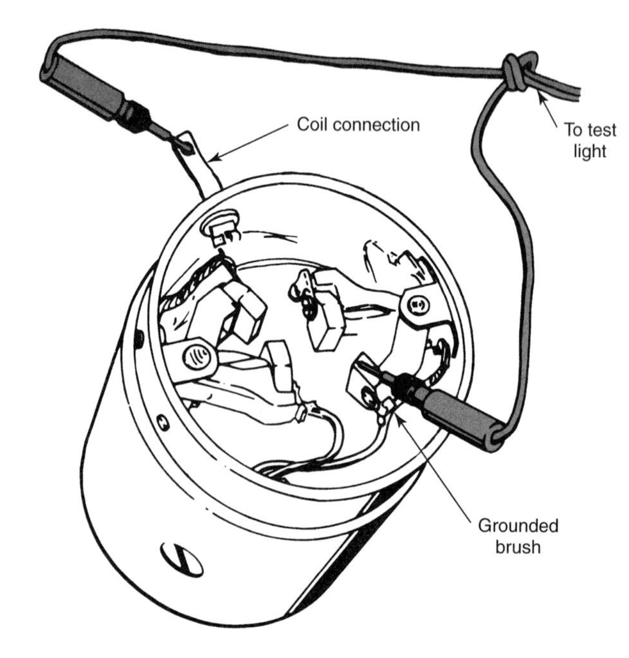

Figure 14-34. The starter field coils can be checked for grounds with a test light. (General Motors)

Check Brushes

Brushes should be replaced if they are oil soaked or worn to one-half of their original length or less. If the starter is dismantled, replace the brushes regardless of their length. When soldering is required, make certain the solder joint is sound. Use high-temperature solder.

Check the insulated brush holders for grounds, **Figure 14-35.** Make sure the brushes slide freely. If fastened to movable arms, the arms should be free and in alignment.

Testing the Starter Solenoid

Most starter solenoids are installed on the starter body. These solenoids operate the pinion gear in addition to sending current into the starter. A few solenoids are installed on the vehicle firewall. Solenoids have internal electrical contacts. These contacts are applied by the solenoid windings. Battery current flows through these contacts to operate the starter. Current flow is very high, and eventually causes the contacts to overheat and fail. Most solenoid problems are caused by damaged internal contacts. If a solenoid clicks when it is energized but does not send current to the starter, it has defective contacts. Some older solenoids could be disassembled to clean or replace the contacts, but most new solenoids are replaced as an assembly.

Note: Sometimes a solenoid will chatter (produce a series of fast clicks) when the ignition switch is turned to the start position. This problem is probably not in the solenoid. More likely causes are a discharged battery or excessive starter current draw.

In addition to a clean, functional set of contacts, proper solenoid operation requires sound pull-in and hold-in windings. Study the typical starter solenoid winding circuit in **Figure 14-36.**

The test in **Figure 14-37** tests the pull-in winding, and the test in **Figure 14-38** tests the hold-in winding. If the solenoid is mounted on the starter, it must be installed for this test. Remove all electrical leads. Depending on which winding is to be tested, make the hookups as indicated. With the switch on, adjust the rheostat to produce the manufacturer's specified voltage. Check the ammeter reading.

If the ammeter reading exceeds specified levels, it indicates a grounded or shorted circuit. A low reading is caused by excessive resistance. If no ammeter reading is apparent, the circuit is open.

Caution: When testing a pull-in winding, do not leave the circuit energized for longer than 15 seconds. This will prevent overheating. Keep in mind that as the circuit heats up, resistance increases and current draw decreases.

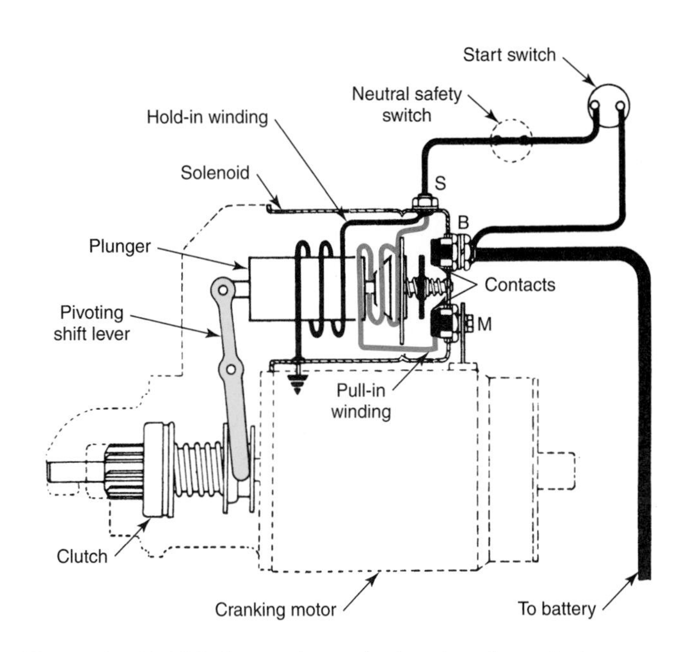

Figure 14-36. This figure shows the interior of a typical starter solenoid. Note the winding arrangement. (General Motors)

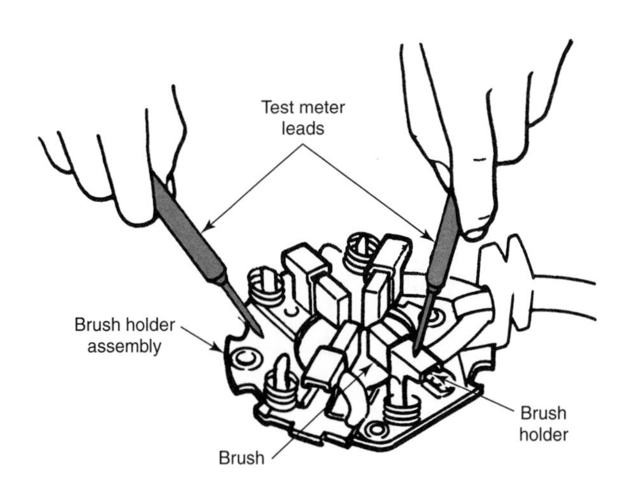

Figure 14-35. Always check the starter insulated brush holders for grounds with a test meter. (General Motors)

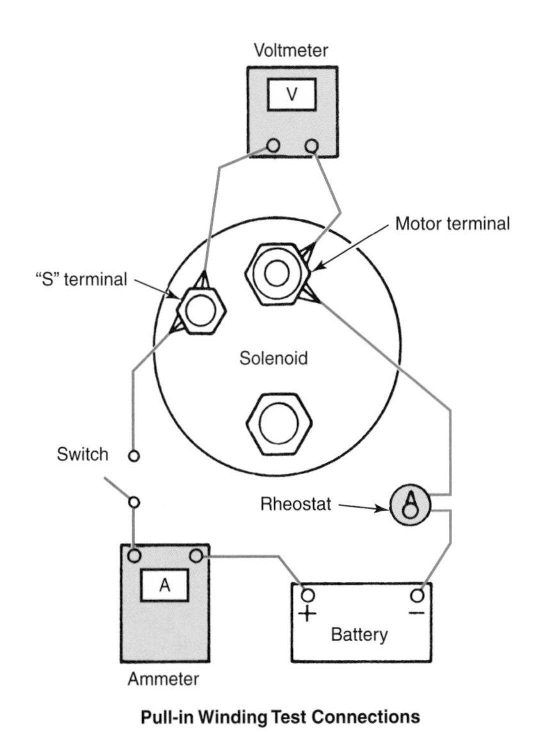

Figure 14-37. This procedure will test starter solenoid pull-in winding circuits. (General Motors)

As with all tests, use a test procedure and specifications for the exact unit.

Other Starter Checks

Inspect the bushings and replace them if excessive wear is present. Check brush spring tension, **Figure 14-39.** Inspect both pinion and flywheel gear teeth for any evidence of chipping, milling, cracking, or improper engagement pattern. The overrunning drive must turn freely in the coast direction and lock tightly in the drive direction. Thrust washers and spacers must be in good condition. The Bendix drive, where used, must be clean and in good condition. Check all soldered connections. Check the pinion housing for signs of cracking.

Starter Assembly

Assemble the starter in the reverse order of disassembly. If necessary, lubricate the two bushings with several drops of recommended lubricant. Wipe a thin coat of light lubricant on the armature shaft splines. Never lubricate unless specified by the manufacturer. Where soldering is required, use rosin-core solder only.

Some starters use a sealer in some locations to prevent the entry of dust and water. When assembling the parts, apply a nonhardening sealer. Where grommets are used, they must be in good condition and properly inserted.

Torque the through-bolts to specifications. Check the overrunning clutch pinion clearance by either forcing (where possible) the solenoid arm to the fully applied position or by energizing the starter solenoid. When energizing the solenoid, ensure the starter motor cannot turn. The solenoid motor terminal can be connected to ground to help prevent starter rotation.

When the solenoid has moved the pinion to the fully applied position, measure the clearance between pinion and pinion stop. Look at **Figure 14-40.** Pinion clearance is adjustable on some starters. On many units, however, clearance cannot be adjusted. When it is wrong, it must be corrected by installing new parts.

Testing the Starter

When assembly is complete, give the starter a no load and stall torque test before attaching it to the engine. If the solenoid is installed on the engine, make sure the solenoid linkage moves the pinion gear.

Installing the Starter

Disconnect the battery ground strap. Clean the mounting flange or pad until it is spotless. This is necessary for proper electrical ground and for accurate mechanical alignment. Install the starter and torque the fasteners.

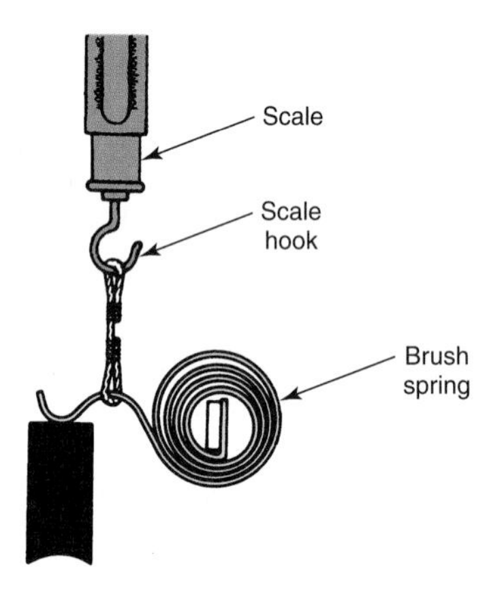

Figure 14-39. Use a spring scale to test starter brush spring tension. (General Motors)

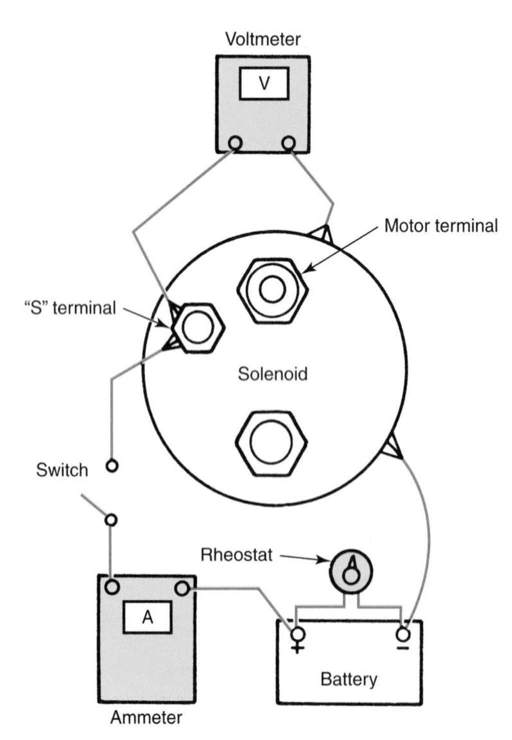

Figure 14-38. To test the starter solenoid hold-in winding circuits, use the procedure shown above. (General Motors)

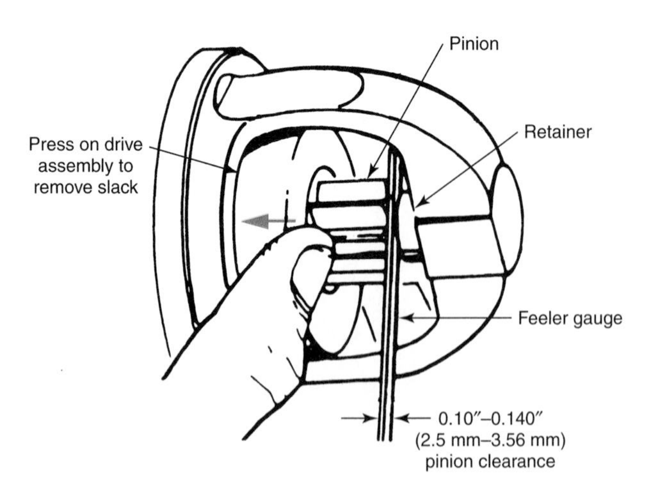

Figure 14-40. Check pinion gear clearance with a feeler gauge. (General Motors)

Some starter setups permit measuring pinion-to-flywheel clearance. **Figure 14-41** illustrates the use of a wire gauge to determine such clearance on one setup.

If the pinion-to-flywheel clearance is incorrect, it is often possible to bring it within the specified range by using mounting shims to tip the starter slightly. This must be done carefully. Make a final clearance check after tightening. Connect the wiring. Connect the battery ground strap. Try the starter. Remember that all splash and/or heat shields, sealing gaskets, and brackets must be in good condition and in place.

Tech Talk

Suppose you are in New York City and want to get to Chicago in a hurry. You go to the airport and are told that to get to Chicago, you must go first to Boston, Atlanta, Miami, New Orleans, and St. Louis. How would you feel about such a roundabout method of getting from one place to another? Do you think an airline that routes its planes in this way would be in business very long?

Many automotive technicians take a similarly inefficient route when trying to figure out what is wrong with a vehicle. They go all over the map to get from the problem to the solution, trying one thing after another instead of isolating the problem and investigating the causes.

For instance, when confronted with a car that will not start, many technicians simply change the battery, only to have the car back in a few days with a brand new, dead battery. When they determine the new battery is dead, they charge the battery, replace the alternator, and return the vehicle to its owner. A few days later, the car is back in the shop and the owner is very angry. It must be embarrassing to find out the problem is a grounded circuit, draining the battery.

Technicians are always being asked to get from New York to Chicago by the most efficient route. In other words, they are asked to find out what is really wrong with a vehicle and fix it. It is up to you to learn the most efficient way to determine what is wrong and what to do about it.

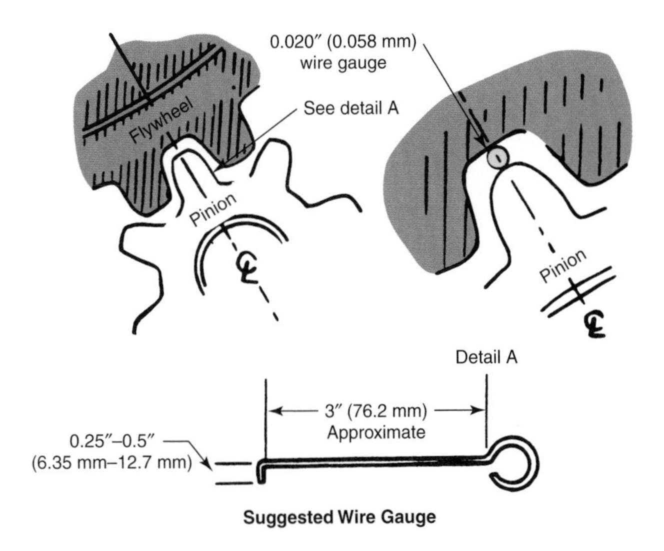

Figure 14-41. Check flywheel-to-pinion clearance with a wire gauge. If the clearance is incorrect, the starter will be noisy, and may damage the flywheel teeth. (General Motors)

Summary

A good battery is vital to the proper functioning of the electrical system. Batteries can be dangerous. Always be careful when servicing any battery. Wear gloves and goggles when handling electrolyte. Avoid creating sparks or open flames around batteries. A stray spark can cause a battery to explode. Keep the battery hold-down snug, cable connections clean and tight, and electrolyte up to the proper level. Modern batteries should be load tested with electrical test equipment. Battery open circuit voltage can be determined with a voltmeter. If possible, check specific gravity with a hydrometer. Some batteries have a built-in hydrometer "eye."

Charge batteries that are less than 75% charged. Batteries may be recharged either by slow charging or by fast charging. Slow charging is preferable when time permits. Batteries may be kept charged in storage by trickle charging. Be careful to avoid overcharging. Store batteries in a cool area.

To activate a dry-charged battery, fill it with the specified electrolyte. Charge the battery before using. A replacement battery should equal or exceed the original in ampere-hour capacity. Three popular battery electrical load ratings are cold cranking amps, cranking amps, and reserve capacity.

When installing the battery, be sure to ground the correct terminal. Current U.S. vehicles ground the negative terminal. The negative post is smaller in diameter than the positive post and may have a NEG, N, or (–) stamped on the top. A voltmeter may be used to check for battery drain.

When using jumper cables to connect a booster battery, connect the booster positive to vehicle battery positive and booster negative to the vehicle frame or engine.

Charging system problems can involve the alternator, regulator, wiring, connections, battery, or any combination of these items. When repairing or replacing one unit, check the others also.

Check the alternator output against specifications. Never disconnect either battery cable when the alternator is charging. This may ruin other electrical components. Always disconnect the battery ground strap before making test connections or removing any part of the charging system. When no other problems are found, check the charging system wiring for excessive resistance.

When testing alternator regulators, observe all alternator system cautions. Some internal alternators can be checked, but others must be replaced along with the alternator when the system is not charging. Electronic regulators are nonadjustable and must be replaced if malfunctioning.

Some alternators can be disassembled and repaired. Clean all alternator parts except rotor, stator, and brush assemblies in solvent. Do not wash sealed bearings. Replace alternator brushes during an overhaul.

Test the rotor for opens, shorts, and grounds. Test the alternator diodes with an ohmmeter. Protect diodes from excessive heat and from mechanical shock. Use a press to remove and replace the diodes. Check stator windings for opens, shorts, and grounds. Inspect the slip rings for signs of scoring and burning. If needed, turn and polish with No. 400 sandpaper. Test the alternator radio suppression capacitor.

Before condemning the starter, check the battery condition. Check the battery cables, the connections, and the entire starter system wiring for high resistance. Clean and tighten connections and replace wiring where needed.

When conducting starter tests, do not operate the starter for more than 20–30 seconds. Allow a two minute cooling off period before resuming cranking. Check solenoids, switches, and relays for excessive resistance, shorting, and mechanical faults.

Sometimes the starter can be repaired, but most shops prefer to replace the entire unit. After disassembling the starter, make tests of the internal components. Test the armature for short circuits and grounds. Check the commutator bars for indication of an open circuit. Check the commutator for run out, burning, and scoring. Test the field coils for opens and grounds. Check the brush holders for grounds. Inspect soldered connections. Replace the brushes when overhauling a starter.

Check the starter bushings and armature shaft bearing area. Inspect the pinion gear and the Bendix or overrunning clutch drive unit. Check the pinion housing for cracks.

During starter reassembly, lubricate the starter bushings and armature shaft splines as recommended by the manufacturer. If a Bendix drive is used, oil it lightly. Check pinion clearance. Install all grommets snugly. Use nonhardening sealer where required. Torque all through-bolts. Clean the starter's mounting pad and torque the mounting fasteners. Replace the heat shields and brace brackets. Test the starter function.

Review Questions—Chapter 14

Do not write in this book. Write your answers on a separate sheet of paper.

1. Battery electrolyte is made up of _____ and _____.
2. When mixing or adding electrolyte to batteries, always protect your _____.
   (A) hands
   (B) eyes
   (C) skin
   (D) All of the above.
3. Wash skin with _____ to remove battery acid.
4. Battery corrosion may be readily removed with a solution of _____ and water.
   (A) baking soda
   (B) sulfuric acid
   (C) chalk
   (D) milk
5. A frozen battery should be _____.
6. A battery should be recharged when the state of charge is reduced to _____.
   (A) 50%
   (B) 25%
   (C) 75%
   (D) Any of the above.
7. All batteries have either top or side _____.
8. The 30-second current flow of a battery at 32°F (0°C) and 7.2 volts is called the _____ amps.
9. The 30-second current flow of a battery at 0°F (–17.8°C) and 7.2 volts is called the ____ ____ amps.
10. The number of minutes that a battery can produce 25 amps at 10.5 volts with battery temperature at 80°F (26.5°C) is called the _____ _____.
11. When installing a battery, always connect the _____ cable last.
12. How is the battery positive post or terminal identified?
    (A) It is wider than the negative post.
    (B) It is painted red.
    (C) A (+) or POS is stamped on the post.
    (D) Any of the above, depending on manufacturer.
13. When connecting a booster battery to a dead battery in a vehicle, hook the booster positive to the vehicle battery _____ and booster negative to the vehicle battery _____.
14. List the three things that determine a battery group.
15. Normal charging voltage is between _____ and _____ volts.
16. Name four alternator parts that must not be cleaned in solvent.
17. When overhauling an alternator, new _____ should be installed if they show any signs of wear or damage.
    (A) brushes
    (B) bearings
    (C) stator windings
    (D) All of the above.
18. List ten precautions regarding service work on alternator charging systems.
19. Always _____ the battery before connecting or disconnecting any unit in the alternator system.
    (A) recharge
    (B) discharge
    (C) disconnect
    (D) remove
20. Diodes may be tested with a(n) _____.
    (A) ohmmeter
    (B) special diode tester
    (C) test lamp
    (D) All of the above.

21. Check the starter circuit for _____, and if it is discovered, clean and tighten connections as required.
22. Never operate starter continuously for more than _____.
    (A) 1 minute
    (B) 20–30 minutes
    (C) 20–30 seconds
    (D) 5 seconds
23. If the starter solenoid clicks when energized but the starter does not operate, the problem is in the _____.
    (A) starter
    (B) internal solenoid contacts
    (C) ignition switch
    (D) Either A or B.
24. List three starter parts that should not be cleaned in solvent.
25. Test the starter field coils for _____ and _____.

ASE-Type Questions

1. Technician A says battery efficiency is improved by cold weather. Technician B says battery specific gravity indicates cold cranking amps (CCA). Who is right?
    (A) A only.
    (B) B only.
    (C) Both A and B.
    (D) Neither A nor B.
2. Technician A says when time is available, slow charging is better than fast charging. Technician B says open-circuit voltage should be checked immediately after charging a battery. Who is right?
    (A) A only.
    (B) B only.
    (C) Both A and B.
    (D) Neither A nor B.
3. To jump start a vehicle with a dead battery, perform all of the following steps, *except*:
    (A) Connect the booster and dead batteries' positive terminals with the same cable.
    (B) Connect the booster battery negative terminal and disabled vehicle engine or frame with the same cable.
    (C) Remove the dead battery negative cable.
    (D) Start the engine in the booster vehicle before starting the disabled vehicle.
4. Technician A says a discharged battery will not affect charging system readings. Technician B says when a defective battery is found, it can be assumed the rest of the starting and charging system is satisfactory. Who is right?
    (A) A only.
    (B) B only.
    (C) Both A and B.
    (D) Neither A nor B.
5. All of the following can cause a short in the electrical system, *except:*
    (A) pinched wiring.
    (B) defective solenoid.
    (C) high battery cable resistance.
    (D) stuck relay.
6. Technician A says electronic regulators are not adjustable. Technician B says alternator rotor and stator windings should be checked for opens, shorts, and grounds. Who is right?
    (A) A only.
    (B) B only.
    (C) Both A and B.
    (D) Neither A nor B.
7. Efficient starter operation is most dependent upon _____.
    (A) a fully charged battery
    (B) proper engine oil
    (C) a well-oiled starter drive
    (D) a dust-free armature
8. Technician A says starter brushes should be replaced when they are worn to within one-half of their original length. Technician B says starter brushes should be replaced if they are oil soaked or any time the starter is dismantled for another reason. Who is right?
    (A) A only.
    (B) B only.
    (C) Both A and B.
    (D) Neither A nor B.
9. All of the following statements about starter brush replacement are true, *except:*
    (A) Replace brushes that are oil soaked.
    (B) Replace brushes that are worn to one-half of their original length.
    (C) Replace brushes every 50,000 miles (80,000 km).
    (D) Replace the brushes anytime that the starter is dismantled.
10. Technician A says starter pinion-gear-to-pinion-stop clearance is automatic and cannot be adjusted. Technician B says starter pinion-to-flywheel clearance is automatic and cannot be adjusted. Who is right?
    (A) A only.
    (B) B only.
    (C) Both A and B.
    (D) Neither A nor B.

Suggested Activities

1. Draw a simple electrical schematic showing a workable starting and charging system. The schematic should include the battery, starter, alternator, ignition switch, solenoid, and related wiring (external voltage regulator optional).

2. Remove, clean, and reinstall a battery. List the electronic devices (such as the clock) that must be reset when the battery cables are removed.
3. Use a battery charger to recharge a battery in a vehicle. Demonstrate the proper method of isolating the battery and charger from the rest of the vehicle electrical system and the correct way of connecting the charger cables to the battery.
4. Using shop test equipment, test the battery in a vehicle. Also make a hydrometer test if possible. Compare the two readings to determine the state of charge of the battery. If the battery is not fully charged, perform Activity 5 to determine the reason.
5. Using the shop test equipment, check the electrical values (amperage and voltage) of a vehicle's starting and charging systems. Perform tests during cranking and as the running engine charges the battery. If the readings are not correct, try to find the reason.

Starting System Problem Diagnosis	
Problem: Starter will not crank engine	
Possible cause	**Correction**
1. Dead battery.	1. Charge or replace battery.
2. Loose or dirty battery connections.	2. Clean and tighten connections.
3. Defective starter switch.	3. Replace starter switch.
4. Defective starter solenoid.	4. Replace solenoid.
5. Defective or improperly adjusted neutral safety switch.	5. Replace or adjust switch.
6. Starter terminal post shorted.	6. Replace insulation.
7. Defective starter.	7. Rebuild or replace starter.
8. Engine bearings seized.	8. Grind crankshaft. Replace bearings.
9. Engine bearings too tight.	9. Install correct bearings.
10. Piston-to-cylinder wall clearance too small.	10. Fit pistons correctly.
11. Water pump frozen.	11. Thaw. Place antifreeze in cooling system.
12. Insufficient ring clearance.	12. Install correct rings.
13. Hydrostatic lock (water in combustion chamber).	13. Remove water and repair leak.
14. Starter drive pinion jammed into flywheel teeth.	14. Remove starter. Install new pinion and replace starter ring gear if needed.
15. Starter armature seized.	15. Rebuild or replace starter.
Problem: Starter cranks engine slowly	
Possible cause	**Correction**
1. Low battery state of charge.	1. Charge battery.
2. Loose or dirty battery cable connections.	2. Clean and tighten connections.
3. Battery capacity too small.	3. Install larger capacity battery.
4. Dirty or burned switch contacts.	4. Replace switch.
5. Excessively heavy engine oil.	5. Drain and install lighter oil.
6. Starter motor defective.	6. Rebuild or replace starter.
7. Engine bearings, pistons, or rings fitted too close.	7. Provide proper clearance.
8. Cold, heavy oil in manual transmission.	8. Hold clutch in while cranking.
9. Extreme cold weather.	9. Preheat engine prior to cranking.
Problem: Starter makes excessive noise	
Possible cause	**Correction**
1. Starter-to-flywheel housing mounting fasteners loose.	1. Tighten mounting fasteners.
2. Dragging armature.	2. Replace armature and/or bushings.
3. Dragging field pole shoes.	3. Tighten pole shoes.
4. Dry bushings.	4. Lubricate bushings.
5. Chipped pinion teeth.	5. Replace pinion.
6. Chipped flywheel ring gear teeth.	6. Replace ring gear.
7. Bent armature shaft.	7. Replace armature.
8. Worn drive unit.	8. Replace starter drive unit.
9. Loose starter through-bolts. Loose end frame bolts.	9. Tighten all starter and frame cap bolts.
10. Flywheel ring gear misaligned.	10. Install new ring gear.
Problem: Starter cranks but will not engage flywheel ring gear	
Possible cause	**Correction**
1. Broken spring or bolt (Bendix type).	1. Replace spring or bolt.
2. Dirty drive unit.	2. Clean or replace unit.
3. Sheared drive key.	3. Replace drive key.
4. Stripped or sheared pinion teeth.	4. Replace drive unit.
5. Section of ring gear teeth stripped.	5. Install new ring gear.
6. Defective or dry overrunning clutch drive unit.	6. Replace drive unit.
7. Snapped armature shaft.	7. Replace armature.
8. Broken spring (overrunning clutch type).	8. Replace drive unit.

(*continued*)

Starting System Problem Diagnosis (*continued)*	
Problem: Starter drive pinion releases slowly or not at all	
Possible cause	**Correction**
1. Dirty Bendix drive sleeve.	1. Clean drive sleeve.
2. Drive pinion binds on drive sleeve splines (mechanical bind).	2. Replace drive unit.
3. Starter switch defective.	3. Replace starter switch.
4. Dirty pinion sleeve.	4. Clean pinion sleeve.
5. Disengagement linkage (overrunning clutch type) binding.	5. Clean, align, and adjust linkage.
6. Linkage retracting spring weak or broken.	6. Install new spring.
7. Actuating solenoid sticking.	7. Clean solenoid.
8. Centrifugal pinion release pin sticking.	8. Replace drive unit.
9. Insufficient drive pinion-to-ring gear clearance.	9. Adjust clearance or replace linkage.

Charging System Problem Diagnosis

Problem: No charge

Possible cause	Correction
1. Alternator drive belt loose or broken.	1. Tighten or replace belt.
2. Voltage regulator fusible link blown.	2. Install new fusible link.
3. Sticking or worn commutator brushes.	3. Free or replace brushes.
4. Loose or corroded connection.	4. Clean and solder connections.
5. Rectifiers open.	5. Correct cause and replace rectifiers.
6. Charging circuit open.	6. Correct as needed.
7. Open circuit in stator winding.	7. Replace stator.
8. Field circuit open.	8. Test and correct as required.
9. Defective field relay.	9. Replace relay.
10. Defective voltage regulator.	10. Replace voltage regulator.
11. Open isolation diode.	11. Replace diode.
12. Open resistor wire.	12. Replace resistor wire.
13. Drive pulley slipping.	13. Install new key and tighten.
14. Brushes oil soaked.	14. Replace brushes.
15. Corroded or loose brush connections.	15. Clean and tighten connections.
16. Seized bearings.	16. Replace bearings. Check shaft for damage.

Problem: Low or erratic rate of charge

Possible cause	Correction
1. Loose drive belt.	1. Tighten belt.
2. Open stator; grounded or shorted turns in stator windings.	2. Replace stator.
3. High resistance in battery terminals.	3. Clean and tighten terminals.
4. Charging circuit resistance excessive.	4. Repair cause of high resistance.
5. Engine ground strap loose or broken.	5. Tighten or replace strap.
6. Loose connections.	6. Tighten connections.
7. Voltage regulator points oxidized.	7. Clean and adjust or replace regulator if required.
8. Voltage regulator setting too low.	8. Increase regulator setting.
9. Defective rectifier.	9. Replace rectifier.
10. Dirty, burned slip rings.	10. Turn slip rings.
11. Grounded or shorted turns in rotor.	11. Replace rotor.
12. Brushes worn. Brush springs weak.	12. Replace brushes and/or springs.

Problem: Excessive rate of charge

Possible cause	Correction
1. Voltage regulator setting too high.	1. Lower regulator setting.
2. Voltage regulator ground defective.	2. Ground properly.
3. Defective voltage regulator.	3. Replace regulator.
4. Alternator field winding grounded.	4. Repair grounded field winding.
5. Open rectifier.	5. Replace rectifier.
6. Loose connections.	6. Tighten connections.

Problem: Noise

Possible cause	Correction
1. Drive belt slipping.	1. Tighten belt.
2. Drive pulley loose.	2. Tighten pulley.
3. Drive pulley misaligned.	3. Align pulley.
4. Mounting bolts loose.	4. Tighten mounting bolts.
5. Worn bearings.	5. Replace bearings.
6. Dry bearing.	6. Lubricate or replace as required.
7. Open or shorted rectifier.	7. Replace rectifier.
8. Sprung rotor shaft.	8. Install new rotor.
9. Open or shorted stator winding.	9. Test. Replace stator as needed.
10. Alternator fan dragging.	10. Adjust fan clearance.

(continued)

Charging System Problem Diagnosis *(continued)*	
11. Excessive rotor end play. 12. Out-of-round or rough slip rings. 13. Hardened brushes.	11. Adjust for correct end play. 12. Turn slip rings. 13. Replace brushes.
Problem: Regulator points oxidized, pitted or burned	
Possible cause	**Correction**
1. Incorrect regulator connections. 2. Rotor coil windings shorted. 3. Regulator setting too high. 4. Poor ground. 5. Brush leads touching each other. 6. Air gap incorrect. 7. Point gap incorrect. 8. Oil on points. 9. Filings or other abrasive particles between points. 10. Use of emery cloth.	1. Replace regulator. Connect properly. 2. Replace rotor. 3. Reduce regulator setting. 4. Correct ground. 5. Separate leads. 6. Adjust air gap. 7. Adjust point gap. 8. Clean points. Replace if needed. 9. File or sand. Clean thoroughly. 10. Replace regulator. Never use emery cloth to clean points.
Problem: Undercharged battery	
Possible cause	**Correction**
1. No charge or low charge rate. 2. Excessive use of starter. 3. Defective battery. 4. Excessive resistance in charging circuit. 5. Defective alternator. 6. Defective regulator. 7. Low regulator setting. 8. Electrical load exceeds alternator rating. 9. Electrical draw in system. 10. Excessive starter motor draw. 11. Water level low in battery cells.	1. See *No charge* or *Low or erratic rate of charge*. 2. Tune engine for faster starting. 3. Replace battery. 4. Test and remove resistance. 5. Rebuild or replace alternator. 6. Replace regulator. 7. Raise regulator setting. 8. Reduce load or install higher capacity alternator. 9. Test. Remove source of electrical draw. 10. Rebuild or replace starter motor. 11. Bring electrolyte up to proper level.
Problem: Overcharged battery	
Possible cause	**Correction**
1. Excessive resistance in voltage regulator circuit. 2. Voltage regulator setting too high. 3. Upper (double-contact) voltage regulator points stuck. 4. Regulator-alternator ground wire loose or open. 5. Defective battery. 6. Voltage regulator coil open. 7. Current regulator setting too high. 8. Other defective regulator parts.	1. Clean and tighten connections. 2. Lower voltage regulator setting. 3. Replace regulator. 4. Tighten or replace wire. 5. Replace battery. 6. Replace regulator. 7. Reduce current regulator setting. 8. Replace regulator.
Problem: Excessive use of water or loss of electrolyte	
Possible cause	**Correction**
1. Battery case cracked. 2. Voltage regulator setting too high. 3. Excessive charge rate from other causes. 4. Battery subjected to excessive heat. 5. Battery sealing compound loose.	1. Replace battery. 2. Lower voltage regulator setting. 3. See *Excessive rate of charge* and *Overcharged battery*. 4. Change battery location or insulate battery against heat. 5. Replace battery.

15

Computer System Diagnosis and Repair

After studying this chapter, you will be able to:

- Retrieve trouble codes.
- Identify trouble code formats.
- Interpret trouble codes.
- Use a scan tool to check components and systems.
- Use a multimeter to check computer system components.
- Identify computer system waveforms.
- Explain how to adjust throttle and crankshaft position sensors.
- Explain how to replace computer control system input sensors.
- Explain how to replace a computer control system ECM.
- Explain how to replace an ECM memory chip.
- Explain how to program an ECM with an erasable PROM.
- Explain how to replace computer control system actuators.

Technical Terms

Electronic control module (ECM)	Titania-type sensor
Maintenance indicator light (MIL)	Artificial lean mixture
Flex test	Artificial rich mixture
Data link connector (DLC)	Forcing the actuator
Fault designators	Anti-seize compound
Hard codes	EPROM
Intermittent codes	FEPROM
Datastream values	Flash programming
Snapshot	Direct programming
Zirconia-type sensor	Indirect programming
	Remote programming
	Drive-cycle test

The on-board computer controls many vehicle systems. The level of control ranges from simply monitoring system operation to aggressively adjusting components or systems to compensate for changes in vehicle operating conditions. Therefore, even a minor computer control system problem can have a dramatic effect on one or more vehicle systems. Be sure to obtain the vehicle's service manual or equivalent service information before you begin any diagnostic operation.

This chapter discusses the diagnosis and repair of the computer control system. This includes the three main sections of any computer control system: the input sensors, the computer, and the actuators. Studying this chapter will help you understand the basic computer system testing procedures.

Note: There are many names for the on-board computers that control engine and vehicle operation. For simplicity, the term *electronic control module (ECM)* will be used when referring to these computers in this chapter. There are also many names for the dashboard light that illuminates to indicate computer system problems. In this chapter, however, this light will be called the *maintenance indicator light (MIL).*

Diagnosing Computer Control System Problems

Often, a vehicle is brought in for service when the MIL is illuminated, when fuel economy drops, or when the vehicle fails an emissions test or has a driveability problem. As a technician, your first impulse may be to blame the computer control system for these problems. The computer controls, however, should be suspected only after you have checked for more obvious—and more common—problems.

Remember that many noncomputer problems can cause the ECM to set trouble codes, illuminate the MIL, or cause malfunctions that might be blamed on the computer control system. Typical noncomputer-related problems that might be blamed on the computer control system include an engine miss caused by a burned valve or a fouled spark plug, a rough idle caused by a vacuum leak, stalling caused by a plugged fuel filter; and overheating caused by a defective fan clutch. Similarly, the ECM can be fooled into sending improper commands to the actuators by such conditions as a dirty cooling system that results in incorrect temperature sensor readings, or a leaking injector that causes the oxygen sensor to read a rich air-fuel mixture. It is a good idea to check overall engine condition before proceeding to the computer control system. Typical computer control problems and some of the systems they can affect are shown in **Figure 15-1.**

The first step in diagnosing computer control system problems is to check for obvious system defects. Always

Figure 15-1. The computer monitors and controls a variety of vehicle systems. The problems listed here are only of few of the possible malfunctions that can occur in computer-controlled systems. (Oldsmobile)

obtain the correct system schematic and determine which circuits and devices directly affect the ECM. Also check the ECM fuses. The ECM may be powered through more than one fuse, fusible link, or relay. If one of the ECM fuses is blown, the MIL may not illuminate at any time.

Note: Never remove fuses before attempting to retrieve trouble codes. Removing fuses can erase the trouble codes from the ECM's memory. If the MIL is not on when the ignition switch is on and the engine is not running, check the bulb before checking any fuses.

After checking the ECM fuses, check all other vehicle fuses. A blown fuse in an unrelated circuit may cause current to attempt to ground through the ECM, affecting vehicle operation or illuminating the MIL.

Remember that computer input sensors operate on very low voltages and slight wiring problems can cause inaccurate readings. Any arcing or sparking caused by loose electrical connections will affect computer system operation.

Carefully check all electrical connections. Look for disconnected or corroded ground wire connections, **Figure 15-2.** Connector problems can also be caused by water entering the connector and by overheating due to high-resistance connections. One way to check for problems in the connectors and wiring is to perform a ***flex test.*** To make a flex test, pull on the various system electrical connectors and wiggle the related wiring as the engine operates. If the engine begins running erratically or the MIL illuminates when the connectors and wires are moved, the connector or the wiring is defective.

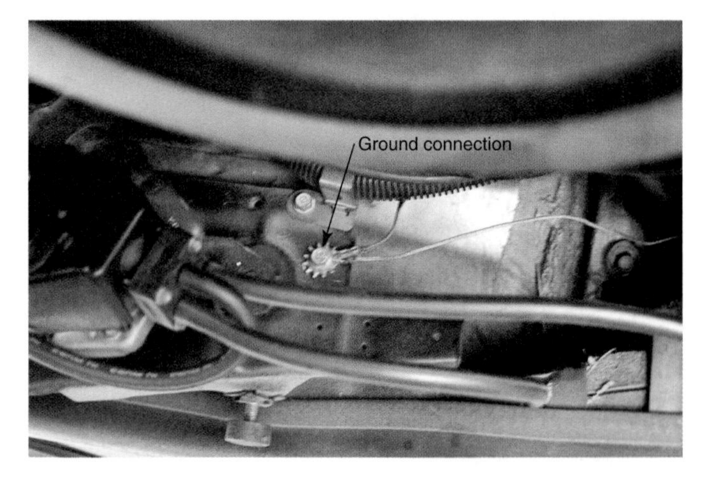

Figure 15-2. Bad grounds are a common source of trouble. Always check grounds by visual inspection and by pulling on the connections. Also try to tighten the attaching bolt when applicable.

Also look for problems in the charging system. Low voltage caused by a defect in the charging system will confuse the ECM and cause erratic operation. Check charging voltage with a voltmeter as the engine idles. Normal charging voltage should be between 12 and 14.5 volts. If charging voltage is low, check the alternator, voltage regulator, battery, and related connections. Also check for a loose or glazed alternator drive belt.

If the voltage at idle is within specifications, increase engine speed to about 2500 RPM and recheck the voltage. If the voltage is above 14.5 volts, suspect a defective voltage regulator. Correct charging system problems before attempting further diagnosis.

Check for any sign of modification or tampering. This includes sensors and output devices that have been removed, unplugged, or disabled in some manner. If a visual inspection does not reveal problems, check for and retrieve any diagnostic trouble codes present in the ECM as outlined in the following section.

Retrieving Trouble Codes

Before performing involved diagnostic routines on computer-controlled vehicles, always check for and retrieve trouble codes from the ECM. If the MIL is on when the engine is running, trouble codes are present in the ECM's memory, **Figure 15-3.** However, it is important to check for trouble codes any time a computer system problem is suspected, since trouble codes may be stored even if the MIL is not on.

To retrieve trouble codes from the ECM, follow the procedures outlined in the factory service manual. Basic procedures for code retrieval are discussed later in this chapter. On most older vehicles, trouble codes can be retrieved without special equipment, although using a scan tool makes the process easier.

For many years, each manufacturer used a different style of diagnostic connector, which is sometimes called a data link connector or data terminal connector. This made accessing diagnostic trouble codes difficult. These connectors could be located almost anywhere, depending on the manufacturer and vehicle model. However, the OBD II diagnostic protocol required manufacturers to use a standardized 16-pin connector located inside the vehicle's passenger compartment. A 16-pin OBD II connector is shown in **Figure 15-4.** In this chapter, we will refer to the diagnostic connector as a ***data link connector (DLC).***

Note: The standard 16-pin DLC is used on some non-OBD II vehicles.

Some older computer systems do not have a DLC, and other methods must be used to access the diagnostic information in the ECM.

The Trouble Code Retrieval Process

Trouble codes are retrieved from the ECM using a two-step process. The first step is to ground a DLC terminal to place the ECM in the diagnostic mode. The second step involves reading the codes that the ECM displays. All methods of trouble code retrieval are variations of these two steps, regardless of whether they are performed manually by the technician or automatically by a scan tool.

Retrieving Trouble Codes without a Scan Tool

In OBD I systems, trouble codes can be retrieved without using a scan tool. In these systems, the most common retrieval method is to ground one terminal of the DLC and then observe a series of flashes from a diagnostic light. The light used may be the MIL, or it may be an LED (light-emitting diode) located on the ECM's case. The sequence of light flashes represents a numerical code. The factory service manual contains instructions for correctly interpreting the light flashes. See **Figure 15-5.**

Another common method of retrieving trouble codes involves connecting an analog voltmeter to specified DLC terminals. Another DLC terminal is grounded and the trouble code output is displayed as a series of deflections of the voltmeter's needle, **Figure 15-6.** Always follow the manufacturer's procedure exactly, since the wrong sequence of steps can erase the trouble codes or, in some cases, cause false codes. Some digital multimeters can display codes as electronic pulses.

The trouble codes on some vehicles can be accessed by turning the ignition switch on and off three times within five

Figure 15-3. The check engine or malfunction indicator light (MIL) is the first place to look when diagnosing any problem. Check the status of this light while test driving the vehicle. (Toyota)

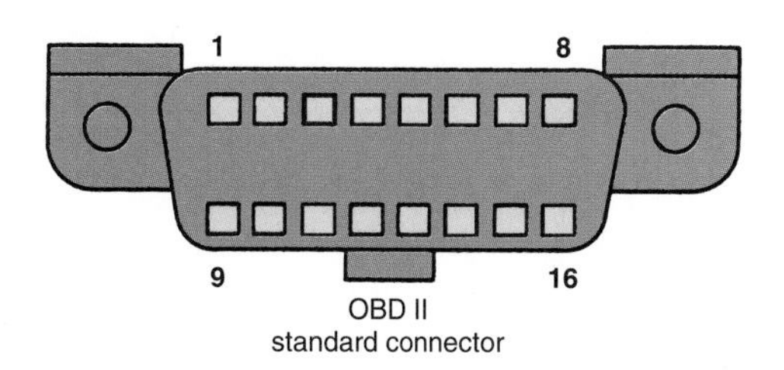

Figure 15-4. This figure shows the standardized data link connector used in OBD II systems. All vehicles built within the last several years use this connector.

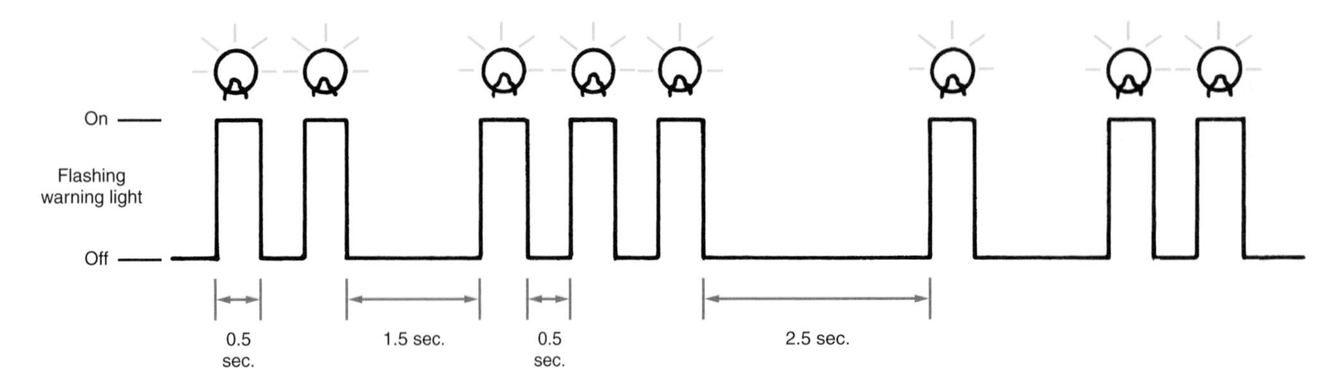

Figure 15-5. Older diagnostic systems made use of the MIL to display codes. The series of flashes is interpreted as numbers. The technician then looks up the numbers in the manufacturer's service literature to determine the problem area.

seconds. The codes will then be displayed by the MIL. Some trouble codes can be accessed by pressing and holding down a certain sequence of buttons on the radio and/or air conditioner control head for a specified period of time. The codes will appear as numbers on the control head display. This feature is found on only a few vehicles with electronic entertainment and climate control systems.

Reading Trouble Codes

To read the MIL flashes or meter pulses, access the codes as described above and observe the light or the needle. If a code 26 is present in a system that displays codes through the MIL, the light will flash two times, pause momentarily, and then flash six times. If a code of 26 is present in a system that uses an anolog meter, the needle on the meter will move twice, pause, and then move six times. Some systems repeat the same code up to three times before moving on to the next code. Some systems repeat the first code as an indication that there are no more codes stored. Write down the numbers of all codes displayed.

Retrieving Trouble Codes with a Scan Tool

On OBD II–equipped vehicles, the proper scan tool *must* be used to retrieve trouble codes. **Figure 15-7** shows a commonly used scan tool. Never attempt to retrieve trouble codes from an OBD II system by grounding a DLC terminal, as this will damage the ECM.

On OBD I systems, it is far easier to use a scan tool to retrieve codes than to perform the procedures explained earlier. Scan tools display the code numbers on a screen and

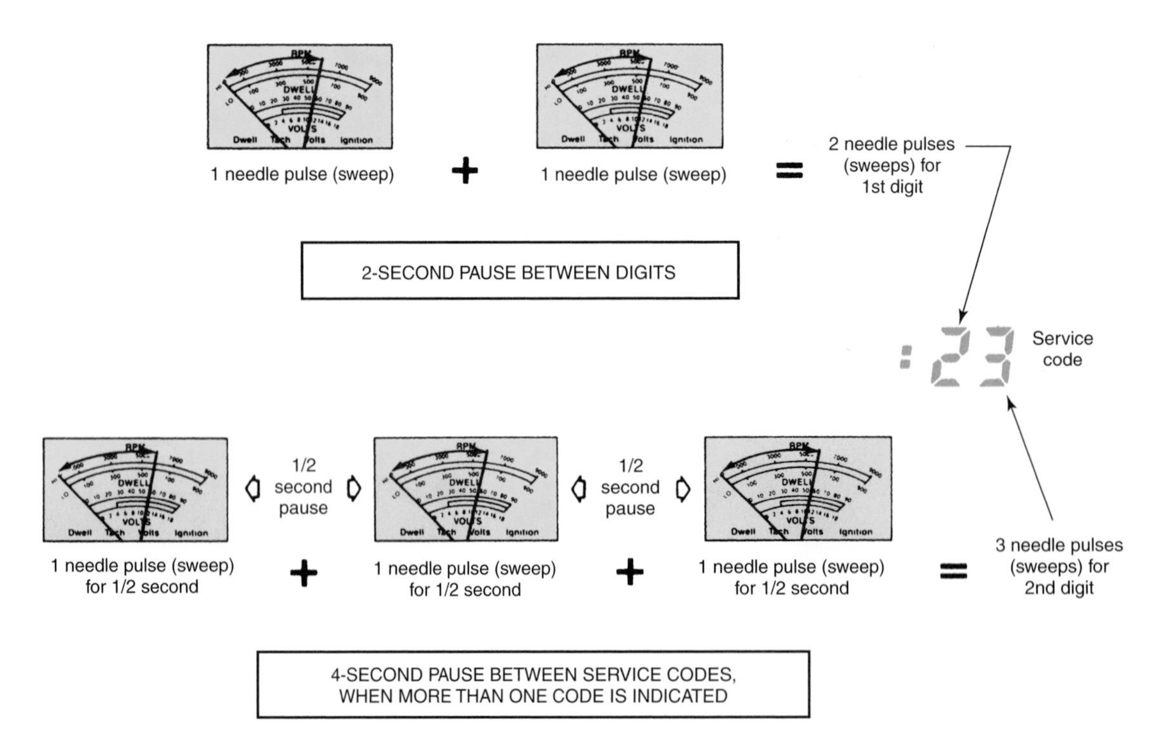

Figure 15-6. An analog meter can be used to read the codes on most OBD I systems. Some digital meters can read codes as digital pulses. (Ford)

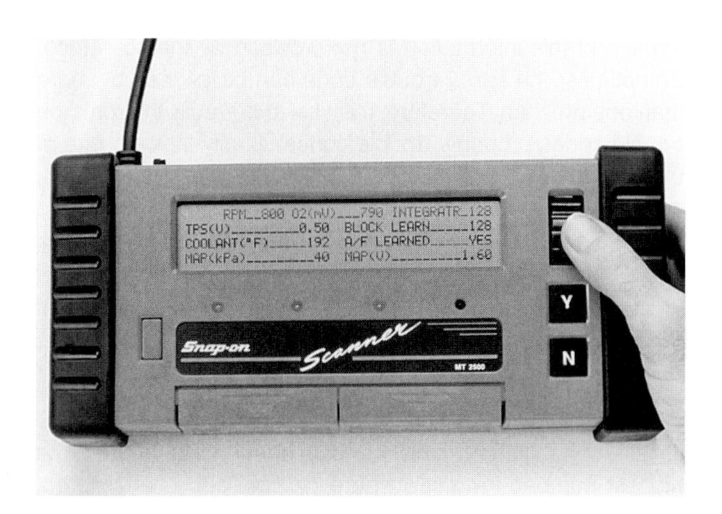

Figure 15-7. This scan tool is widely used. It retrieves trouble codes and displays other engine operating data. This tool also provides information about possible causes of the particular trouble code. (Snap-On)

store them for future reference. They also perform the grounding step, placing the ECM in the diagnostic mode. Typical scan tool connectors are shown being attached to vehicle DLCs in **Figure 15-8.** Always follow the manufacturer's instructions when installing the scan tool and retrieving the trouble codes.

In addition to retrieving trouble codes, scan tools often display sensor and output data and test certain output devices. Some scan tools have the capability to identify the type of ECM and/or PROM in use. Others can be used for reprogramming the ECM.

Any trouble code number(s) present should be written down for later comparison with the corresponding numbers in the service manual. If you are unable to retrieve trouble codes or access sensor operating information, make sure that the DLC wiring is properly connected to the ECM and that the scan tool is properly connected to the DLC. If all the connections are good and you still cannot retrieve codes or access operating information, the ECM is most likely defective. Checking for a defective ECM is covered later in this chapter.

Diagnostic Trouble Code Format

The diagnostic trouble codes for most computer-controlled vehicles are generated as two- or five-character codes that correspond to specific problem areas. Older scan tools displayed trouble codes as numbers and letters. The technician then matched the code to specific problems by looking up the code in the service literature. Newer scan tools are able to process the code letters and numbers, and produce readouts that describe the actual problem. Some scan tool software can also suggest possible solutions and tell the technician if the problem is common on the vehicle being tested.

Note: Remember that trouble code does not necessarily mean that a computer system part or circuit is defective. For example, a constant rich reading from an oxygen sensor may mean the sensor itself is defective. However, it may also mean that a fuel system defect is causing the engine to run rich at all times, triggering the rich sensor reading.

OBD I and OBD II Code Formats

OBD I systems normally display two-digit trouble codes. These codes are correlated to a specific sensor or circuit. In a few cases, the code will specify a certain problem, such as low or high voltage readings. In most cases, however, the code will simply indicate that there is a problem in the system.

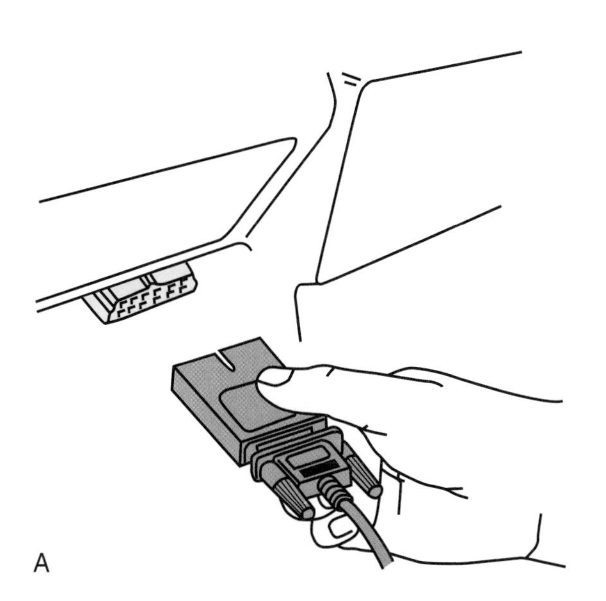

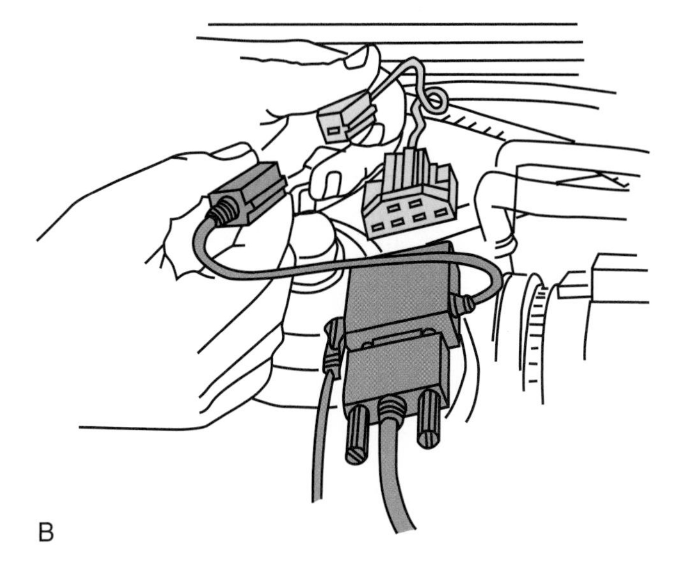

Figure 15-8. The best way to read codes on any computer system is to use a scan tool. A—Connecting a scan tool to a GM diagnostic connector. B—Some diagnostic connectors require a separate power lead, such as this Ford connector. (Snap-On Tools)

OBD II systems display five-character alphanumeric codes consisting of one letter and four numbers. The letter identifies the general function of the vehicle system causing the problem. The first number indicates whether the code is a standardized code (code assigned by SAE) or a manufacturer-specific code. The second number indicates the specific system causing the problem. The last two numbers, which are called the ***fault designators,*** identify the specific component (sensor, output device, or ECM) that might be at fault, as well as the type of fault detected. OBD I and OBD II trouble code formats for a coolant temperature sensor problem are shown in **Figure 15-9.**

In OBD I systems, the two-digit format limits the number of trouble codes to 100. In the five-character OBD II format, there are over 8000 potential codes. OBD II trouble codes do not correspond with OBD I codes. Never try to correlate the older trouble codes with OBD II fault designators. For example, there are only two or three codes used by most older computer control systems to indicate a possible oxygen sensor problem. OBD II systems have *thirty* or more standardized codes for the oxygen sensor and may have additional manufacturer-specific codes, depending on the system. A complete list of the SAE-designated, OBD II trouble codes is located in Appendix B of this text.

Interpreting Trouble Codes

Once trouble codes have been retrieved, they can be compared to the trouble code information in the service manual (if this information is not provided by the scan tool). Sometimes, the same trouble code can be caused by more than one problem. Therefore, it is vital to correctly interpret the trouble codes properly. Trouble codes usually indicate one of three things:

- An engine, drive train, or other vehicle problem is causing one of the sensors to transmit a voltage signal that is out-of-range (too high or too low).
- A sensor's ability to perform is starting to deteriorate.
- Another part or circuit in the computer control system is defective.

Multiple codes should be addressed starting with the lowest code number and working sequentially toward the highest number. In cases where multiple codes are stored, one or more of the codes can often be eliminated using this process. For example, a fuel injector defect on the #4 cylinder (P0204) could be the cause of a misfire on cylinder #4 (P0304). If the technician determines the cause of code P0204, it is likely that the cause of code P0304 will also be determined.

Once the problem area has been identified, concentrate your troubleshooting efforts on that particular area. If the trouble code indicates that the problem is a defective output device or the ECM, you will be able to go to the specific component and determine whether the problem is a disconnected or defective part, a mechanical problem with the part, or a wiring problem. In the case of an out-of-range sensor

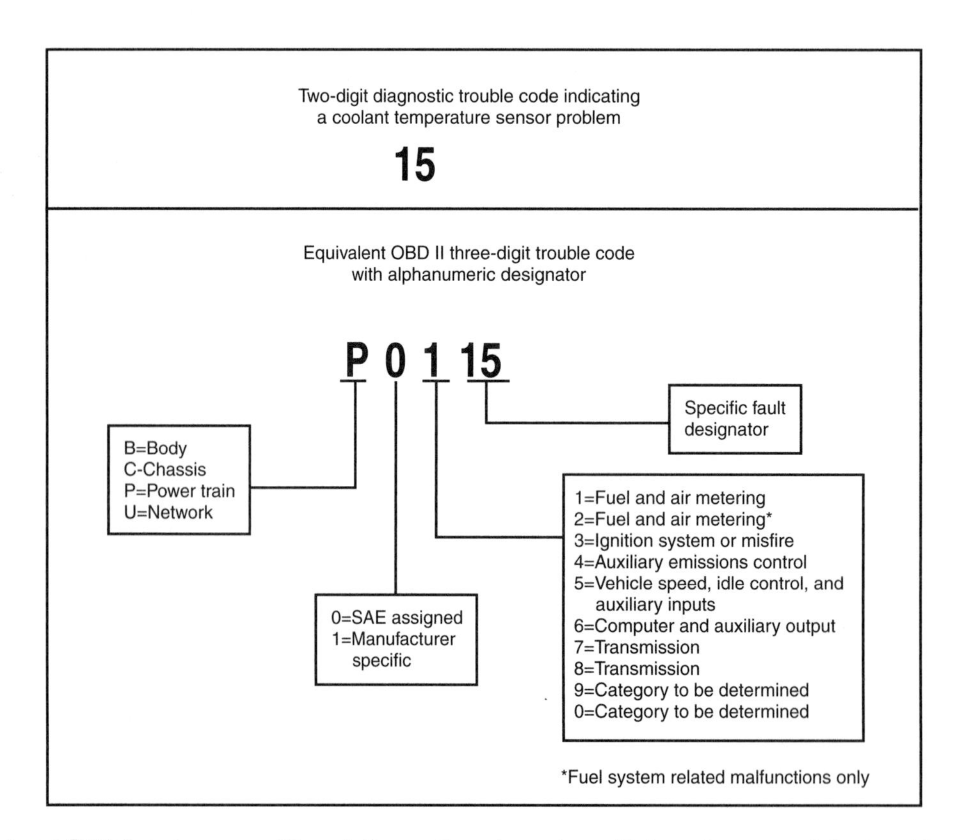

Figure 15-9. OBD I and OBD II systems use different diagnostic code systems. Both systems are distinct and trouble codes from one system should not be correlated with those from the other system.

reading, you must determine whether the problem is a defective sensor or if an abnormal operating condition is causing the out-of-range reading.

Hard and Intermittent Codes

Two types of codes are stored in the ECM memory: ***hard codes*** (permanent) and ***intermittent codes*** (temporary). Hard codes indicate an ongoing problem. These problems are generally easy to track down. Intermittent codes are set by problems that occur occasionally or, in some cases, only once. These problems are sometimes difficult to isolate.

A hard code will illuminate or flash the MIL whenever the engine is running. Intermittent codes will illuminate the MIL only when the problem is occurring, after which time the light will go out.

To differentiate between hard and intermittent codes, record all the trouble codes stored in the ECM's memory. Then erase the stored codes. Most scan tools have the capability to erase trouble codes. On some vehicles, removing electrical power to the ECM for the period of time specified in the service manual will erase the codes. This can be accomplished by disconnecting the negative battery cable or removing the fuse that supplies current to the ECM, **Figure 15-10.** However, the ECM in some OBD II–equipped vehicles can retain trouble codes for several days without battery power. Removing battery power from the ECM on these vehicles may not erase stored trouble codes. Therefore, you must use a scan tool to clear the trouble codes in OBD II systems.

After erasing the codes, restart the engine and allow it to run for the period of time specified by the manufacturer. After the engine has run long enough, stop it and re-enter the diagnostic mode. Hard codes will usually reset almost immediately after the engine is started—when the ECM requests information from a defective sensor or tries to activate a faulty output device. Intermittent codes, on the other hand, will generally not reset immediately.

Hard codes are usually caused by a defective component, an open or shorted circuit, or a faulty connector. In many cases, an intermittent code does not indicate that a sensor or output device is defective, only that it is responding to another problem. An example is an intermittent trouble code for a rich oxygen sensor reading. Although the oxygen sensor may be defective, it is more common for another component, such as a leaking fuel injector or a defective airflow sensor, to be causing the rich mixture problem.

Remember that many sensors in older computer systems can go slightly out-of-range without setting a trouble code. The ECM in an older system often will respond to the incorrect input in the same way it responds to the correct input. OBD II systems are designed to monitor sensor performance and will usually set a diagnostic trouble code if a sensor is slightly out-of-range, even if there is no apparent driveability problem. However, the absence of a trouble code in any computer control system is not absolute proof that a sensor is good.

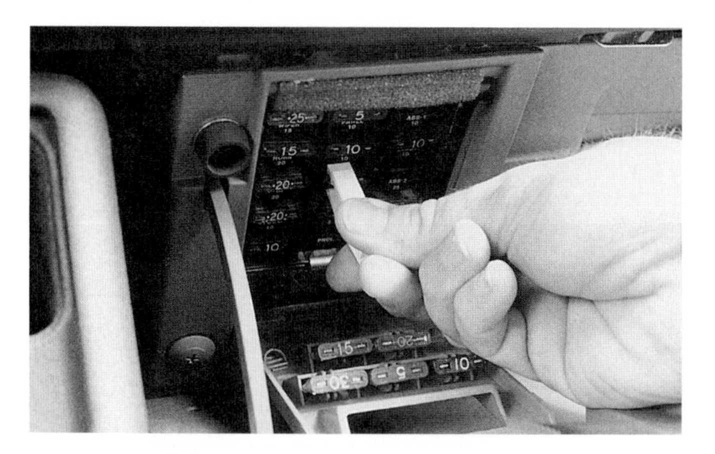

Figure 15-10. The fuse that supplies power to the ECM is usually located in the main fuse block. Pull this fuse to clear stored codes in OBD I systems. Remember that this will not work with most OBD II systems.

Other Scan Tool Information

In addition to retrieving diagnostic trouble codes, some late-model scan tools can display ***datastream values.*** For example, the scan tool can access the ECM and monitor sensors inputs to provide the technician with readings of engine idle RPM, as well as the desired (factory set) idle speed. From this information, the technician can instantly determine whether idle speed is correct. Another scan tool feature will give the technician a series of numbers or percentages indicating whether the fuel trim (air-fuel ratio) is correct for both short- and long-term engine operation, **Figure 15-11.** Other inputs provide exact engine timing in degrees, actual temperature reading from the engine temperature sensors, and voltage or duty cycle readings that indicate manifold vacuum, throttle position, EGR position, and many other operating conditions. See **Figure 15-12.**

Some scan tools can interface with the ECM to provide *snapshots* of engine operating parameters. The ***snapshot*** is a method of recording the engine operating conditions present when a malfunction occurs. This allows the technician to determine exactly what is happening to cause the problem. To record the snapshot information, the technician must drive the vehicle with the scan tool attached until the malfunction occurs. The technician can then retrieve the snapshot readings from the scan tool. Some vehicle ECMs save snapshot information when a malfunction occurs, and the technician can access it once the scan tool is attached. **Figure 15-13,** which shows information obtained while looking for the cause

Short-term fuel trim	Long-term fuel trim
126	130

Figure 15-11. The ECM uses sensor inputs to determine fuel trim for short- and long-term operation. The fuel trim is broken down into numbers or counts that can be read by a scan tool.

Data Scanned from Vehicle			
Coolant temperature sensor 198°F/92°C	Intake air temperature sensor 77°F/25°C	Mass airflow sensor 2.7 volts	Throttle position sensor .78 volts
Engine speed sensor 930 rpm	Oxygen sensor .52 volts	Vehicle speed sensor 0 mph	Battery voltage 13.6 volts
Idle air control valve 38 percent	Evaporative emission canister solenoid Off	Short-term fuel trim 125	Long-term fuel trim 131
Malfunction indicator lamp Off	Diagnostic trouble codes	Open/closed loop Closed	Fuel pump relay On
PROM ID 5248C	Cruise control Off	AC compressor clutch On	Knock signal No
Ignition timing (°BTDC)	Base timing:	6	Actual timing: 19

Figure 15-12. Typical datastream values. The PROM carries an identification code in its programming. A scan tool can be used to access this code. The programming in OBD II systems can be identified by the date code in the program.

of a rough idle, is an example of the type of data obtained using the snapshot feature.

Perform Additional Tests as Needed

A variety of test equipment can be used to check individual computer system components. Many tests can be made using jumper wires, multimeters, or waveform meters. The following sections will quickly review sensor and output device operation and discuss typical testing procedures.

Testing Sensors

Sensors are the inputs to the computer system. If a sensor is defective, the ECM cannot correctly control the related vehicle system(s). The following section covers general test procedures for the most common types of sensors.

Note: What appears to be a sensor problem may be caused by a defect in the engine or another vehicle system. Always make sure that the vehicle has no mechanical or electrical defects that could cause out-of-range sensor readings before condemning a sensor.

Testing Oxygen Sensors

Oxygen sensors measure the amount of oxygen in the exhaust gases and send a signal that represents this amount

Snapshot Data Captured from Vehicle			
Coolant temperature sensor 212°F/100°C	Intake air temperature sensor 79°F/26°C	Mass airflow sensor 3.4 volts	Throttle position sensor 1.99 volts
Engine speed 2476 rpm	Oxygen sensor .31 volts	Vehicle speed sensor 40 mph	Battery voltage 13.2 volts
Idle air control valve 5 percent	Evaporative emission canister solenoid Off	Cooling fan On	EGR pintle position 11 percent
Malfunction indicator lamp On	Diagnostic trouble codes P0272, P0304	Open/closed loop Open	Fuel pump relay On
Transmission gear Drive	Cruise control Off	AC compressor clutch On	Knock signal Yes
Ignition timing (°BTDC)	Base timing:	7	Actual timing: 56

Figure 15-13. Data captured when a malfunction occurs can help you find the cause of a driveability problem. Some ECMs will capture this data automatically when a problem occurs. What type of problem does the data in this chart indicate?

to the ECM. The ECM determines the air-fuel ratio based on the oxygen sensor input. Oxygen sensors are sometimes called O_2 sensors.

There are two basic types of oxygen sensors, the electrically heated and the nonheated. Most vehicles made within the last several years are equipped with heated oxygen sensors. Oxygen sensors with three or more lead wires are heated types. Oxygen sensors with one or two lead wires are nonheated types.

If the MIL is on and a trouble code indicates an oxygen sensor problem, check the sensor's wiring harness for a proper connection. Then make a quick check of the oxygen sensor by tapping on the exhaust manifold or exhaust pipe near the sensor with the engine running. If the MIL goes out, the sensor is defective.

Warning: Exercise caution when working around exhaust manifolds. Exhaust systems can reach extremely high temperatures very quickly.

If the MIL remains on after tapping on the manifold, use a scan tool to access the data from the oxygen sensor in question. Late-model vehicles use more than one oxygen sensor, so make sure you are testing the right one. Connect the scan tool to the DLC and enter the necessary vehicle information. Make sure the engine is warm enough for the computer to operate in closed loop before monitoring O_2 sensor output. If the engine is not in closed loop mode, run it at fast idle for a few minutes to raise engine and exhaust temperature. Then allow the engine to stabilize. After the engine stabilizes, monitor the O_2 sensor readings, **Figure 15-14.**

Oxygen sensors use either Zirconia or Titania elements. The signal from either type of sensor will be displayed as a voltage by the scan tool. The sensor's output voltage should oscillate between 100 and 900 millivolts when the ECM is in the closed loop mode. The voltage readout should change very quickly as the amount of oxygen in the exhaust changes. The meter should show a minimum voltage of less than 300 mV, a maximum of more than 600 mV, and should average about 500 mV.

Note: Some manufacturers allow the oxygen sensor to be bypassed by grounding the sensor input to the ECM. Grounding the oxygen sensor input wire should cause the ECM to produce a set voltage reading. If grounding the wire does not produce this reading, the ECM is defective. The technician should only ground the O_2 sensor wire when specifically instructed by the service manual.

Checking Oxygen Sensors using a Multimeter

Remember that a ***Zirconia-type sensor*** will produce a voltage reading, while a ***Titania-type sensor*** will produce a resistance reading. Most oxygen sensors are Zirconia types. To measure the voltage output of a Zirconia-type sensor, you must have a voltmeter or multimeter that can read very low voltages (in the 100–900 millivolt (mV) range). To measure the resistance of a Titania-type sensor, you will need an ohmmeter or a multimeter that can measure resistance. For this test to be accurate, the engine must be warm enough for the ECM to be in closed loop mode. Unless the oxygen sensor is a heated type, the exhaust system temperature must be at least 600°F (350°C) before starting the test. When tested on the engine, the sensor must be heated either by its heating element (when used) or by exhaust heat. Heated oxygen sensors can be tested within seconds of start-up if the engine is warm. The service manual will tell you whether the sensor is the heated type. In all cases, the manufacturer's instructions must be followed closely.

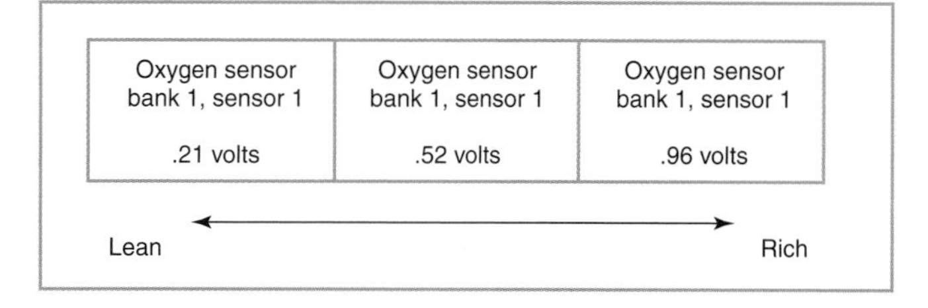

Figure 15-14. Oxygen sensor voltage (or resistance) varies with the content of the oxygen in the exhaust gas. Be sure that you are monitoring the correct sensor on vehicles with multiple oxygen sensors.

Caution: Oxygen sensors are sensitive to excess current flow. They can be destroyed by improper grounding or testing with test lights or low-impedance multimeters. Some manufacturers do not recommend meter tests of the oxygen sensor. Always consult the manufacturer's instructions before testing an oxygen sensor.

After ensuring that the engine is warm enough, connect the meter to the oxygen sensor output wire as shown in **Figure 15-15.** Make sure the meter is connected to the output wire and not to the sensor input or ground wires. One- and three-wire sensors are grounded through the sensor housing. On two- and four-wire sensors, one of the wires is a ground wire. Be very careful to make the proper connections, as slight voltage surges can ruin an oxygen sensor. Observe the meter for several minutes after the ECM enters the closed loop mode. Zirconia oxygen sensors should show a minimum voltage of less than 300 mV and a maximum of more than 600 mV. They should average about 500 mV. Titania oxygen sensors should show low resistance when the air-fuel mixture is rich and high resistance when the mixture is lean.

Testing the Oxygen Sensor Heater Circuit

The heater circuit of a heated oxygen sensor must be checked for proper resistance to ensure that the heater resistor is not burned out or shorted. To test the heater, set the multimeter to the ohms range. Then connect the multimeter

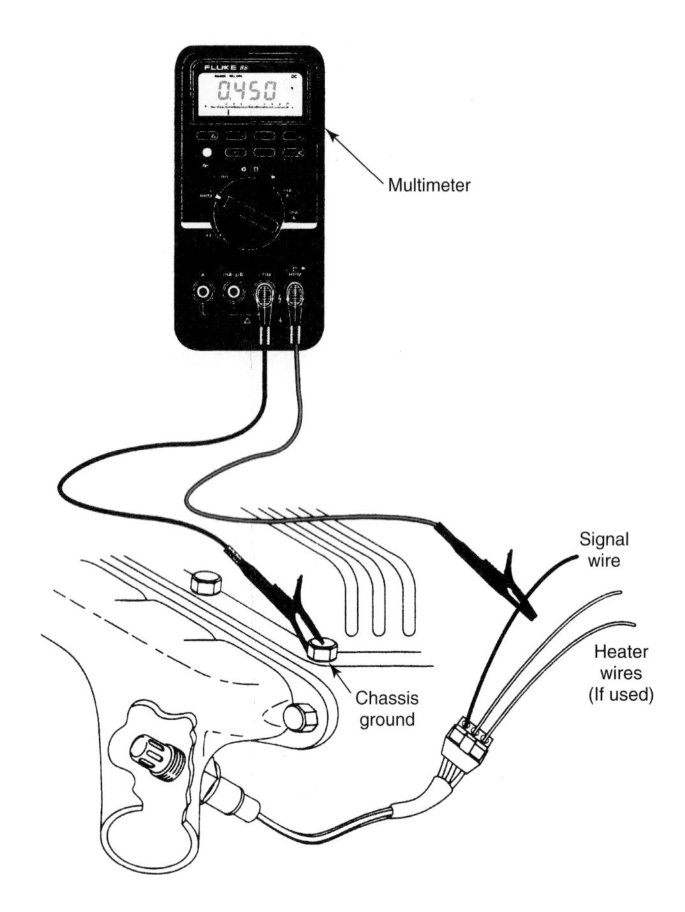

Figure 15-15. A multimeter set to the voltage scale (or resistance) can check an oxygen sensor for proper operation. Be careful when working around hot exhaust parts. (Fluke)

leads to the heater terminals of the oxygen sensor. Polarity is not important, but the leads should not contact the sensor signal terminal. If the resistance is within specifications, the heater circuit is usually good. Set the meter to the voltage scale and test the heater power terminals on the wiring harness. If 12 volts are present, the circuit is good.

Creating an Artificial Lean and Rich Mixture Condition

To test oxygen sensor response time, it is necessary to create artificial lean and rich conditions. To create an ***artificial lean mixture,*** disconnect the PCV hose or another large vacuum hose. Cover the hose opening with your thumb and start the engine. As you remove your thumb from the hose, a lean condition is created due to the vacuum leak. Partially cover the vacuum hose opening if the engine begins to stall. Then observe the meter or scan tool for several minutes. It should show a lower than average voltage or a high resistance as the oxygen sensor reacts to the lean mixture.

Note: The ECM on some fuel-injected engines will react too fast for this test to be effective; the engine speed will just rise and fall.

After making this test, reconnect the vacuum hose and create an ***artificial rich mixture*** with a propane enrichment device, such as that used to adjust carburetors. Do not create a rich mixture with carburetor cleaner or gasoline, as this will simply overload the oxygen sensor. On some vehicles, it is possible to create an artificial rich mixture by blocking the air inlet ahead of the throttle plate. On newer vehicles with idle air control (IAC) solenoids, the IAC will simply open to allow more air into the engine. Blocking the air intake on these engines may not produce definite results. After the enrichment device is in place and operating, check the meter or scan tool. Voltage should be higher than average (or resistance should be low), indicating that the rich condition is being read by the oxygen sensor. If the oxygen sensor does not react, takes longer than two seconds to react, or indicates an out-of-range reading, the sensor is defective and should be replaced.

Checking Oxygen Sensor Waveforms

A waveform meter can be used to detect problems that a voltmeter or scan tool will not reveal. The waveform meter displays a picture of what is happening in the circuit. Some waveform meters require that the wires leading to a particular component be pierced to access the component waveform. Other waveform meters can access needed waveforms through the diagnostic connector.

To check the waveform of an oxygen sensor, attach the waveform meter leads to the proper sensor lead and ground, or to the diagnostic connector. Set the meter to the proper voltage range. Then start the engine and allow the sensor readings to stabilize. Some waveform meter makers recommend running the engine at 2500 RPM for 2 minutes to heat the sensor. Compare the actual waveform with a known good waveform to determine whether the sensor is operating properly. See **Figure 15-16.**

Modern waveform meters can record the operation of two oxygen sensors at the same time. This allows the technician to check the operation of both oxygen sensors and the catalytic converter on OBD II systems. To make this test, attach the meter leads to the upstream oxygen sensor (sensor located ahead of the catalytic converter) and to the downstream oxygen sensor (sensor located after the converter). If the waveform meter uses the vehicle diagnostic connector, attach it now. Set the meter to record the readings of both oxygen sensors. Once the engine is started and allowed to stabilize, compare the two readings. See **Figure 15-17.** The flattened shape of the downstream sensor waveform in Figure 15-17A shows that the converter is operating to clean up the exhaust. In Figure 15-17B, both waveforms have a similar shape. This indicates that the converter is faulty.

Road Testing Oxygen Sensors

If the oxygen sensor tests at idle are inconclusive, perform a road test while monitoring oxygen sensor operation with a scan tool, voltmeter, or waveform meter. A road test will usually uncover a problem that is causing the oxygen sensor to read too lean or too rich, rather than a defective sensor. The

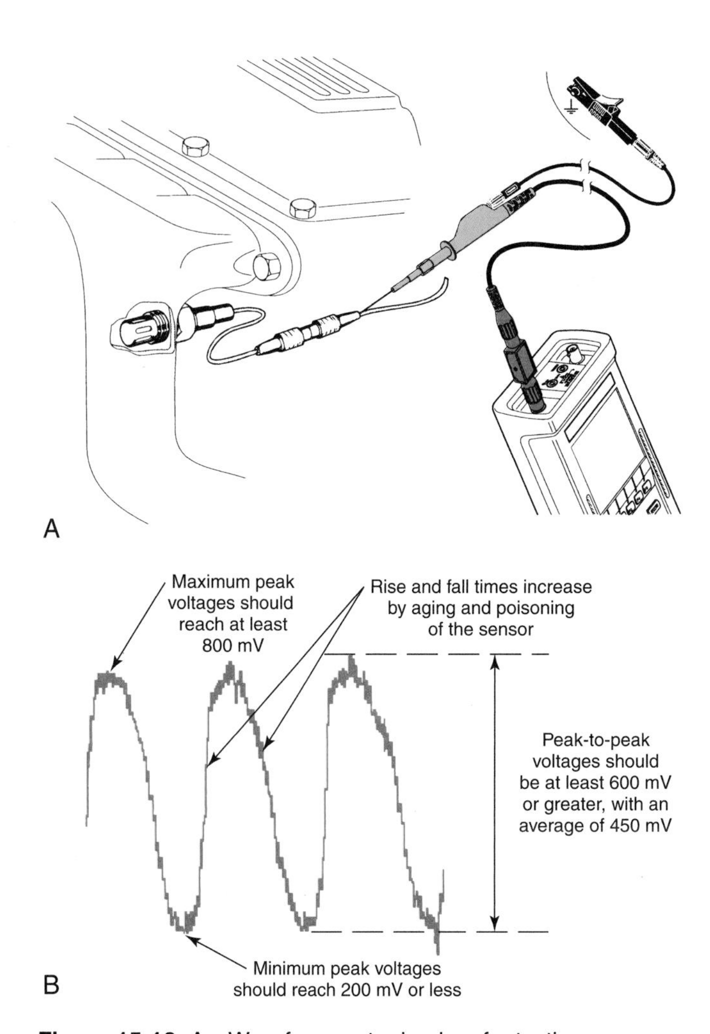

Figure 15-16. A—Waveform meter hookup for testing an oxygen sensor. B—This waveform shows the oscillations of a properly operating oxygen sensor. Lack of oscillations or a pattern that is not consistent indicates a problem. (Fluke)

engine must be completely warmed up before the road test. If a scan tool is used, connect the tool to the DLC. Use an extension lead if the DLC is under the hood. If equipped, use the scan tool's snapshot feature to take a picture of sensor and vehicle operation when the malfunction occurs.

If a multimeter or waveform meter is used, start by connecting the meter to the oxygen sensor through leads that allow the meter to be placed inside the passenger compartment. If the meter you are using has memory and averaging capabilities, follow the manufacturer's instructions to set the meter to record the average reading. Have someone drive the vehicle while you observe oxygen sensor operation. A high average voltage or low resistance reading means that the engine is running too rich, such as when the ECM is leaning out the mixture but cannot compensate for a leaking injector. A low average reading or high resistance indicates too little fuel, such as when the ECM is richening the mixture but cannot compensate for a vacuum leak. After the road test is complete, enter the meter's averaging function. This will allow you to get an average reading of the voltage levels.

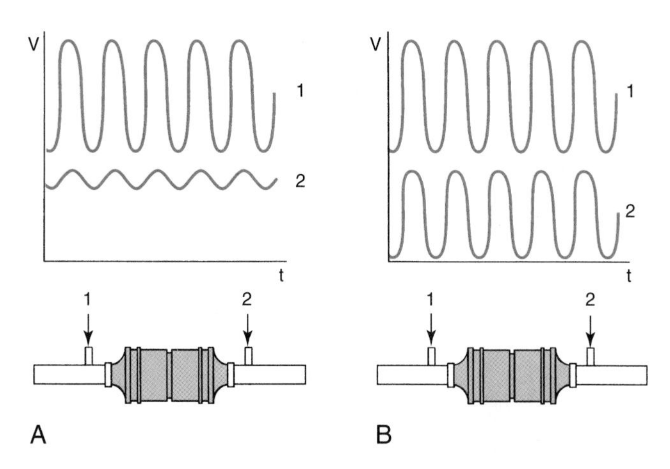

Figure 15-17. Comparing the waveforms of upstream and downstream oxygen sensors allows the technician to determine the condition of the sensors and the converter. 1—Signal from upstream oxygen sensor. 2—Signal from downstream oxygen sensor. Note that the that the signal amplitude from the downstream sensor increases when the efficiency of the catalytic converter declines. (Fluke)

Note: If the above tests are made with the engine running and the oxygen sensor disconnected, false trouble codes will generally be set. After all tests are complete, be sure to clear the ECM's memory.

Testing Mass Airflow (MAF) Sensors

Mass airflow, or MAF, sensors produce either analog or digital signals. Older MAF sensors are usually analog types, while many newer designs are digital types.

Note: Always check carefully to determine which type of MAF sensor you are dealing with, since analog and digital sensors closely resemble each other. If you are not sure which type of sensor you are testing, check the service manual.

Testing Analog MAF Sensors

A quick test using either a scan tool or a multimeter can be performed on analog MAF sensors. Check the proper service manual to determine if the sensor will respond to this test. If a scan tool is used, connect it to the DLC and set it to monitor MAF sensor input. If a multimeter is used, set it to measure dc volts and attach it to the MAF sensor as shown in **Figure 15-18.** With the ignition key in the *on* position and the engine off, voltage will be about 1 volt. Start the engine and observe the scan tool or voltmeter. Voltage should rise to about 2.5 volts, **Figure 15-19.** Tap on the sensor with a small

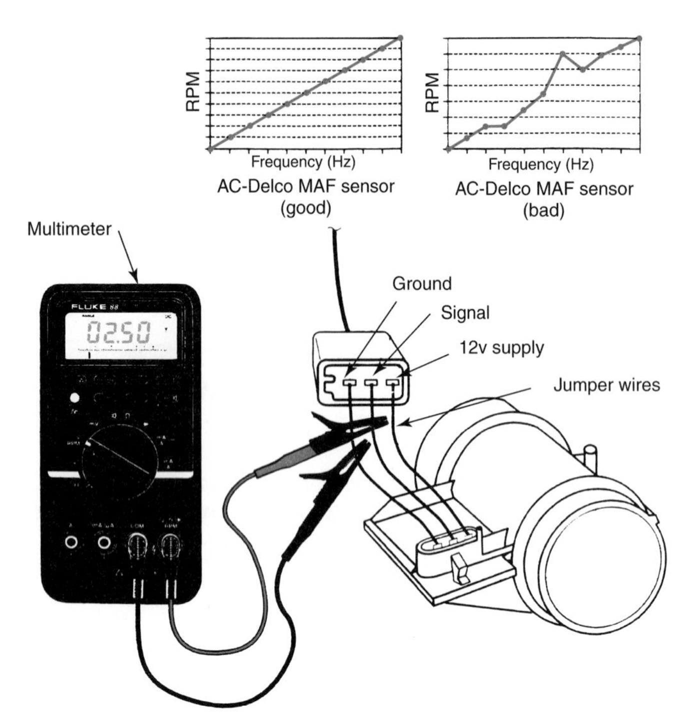

Figure 15-18. A digital meter can be used to check mass airflow sensor operation. Do not use an analog meter to perform this test. (Fluke)

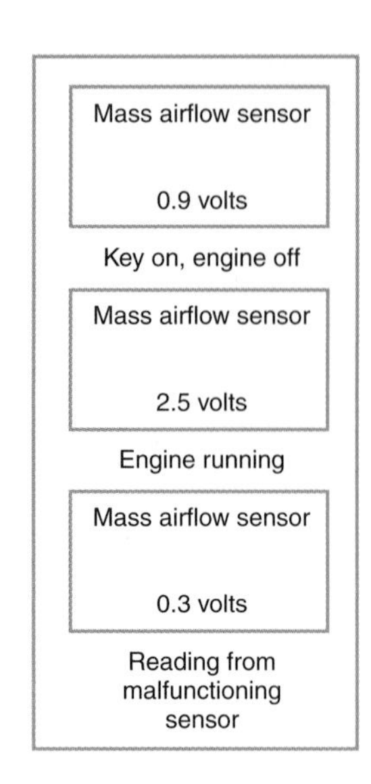

Figure 15-19. Scan tool reading for mass airflow sensor signal. Scan tool readings should be verified using a digital multimeter. (Fluke)

screwdriver handle. There should be no voltage fluctuation or engine misfire. If needed, repeat the test while warming the sensor with a heat lamp. If the voltage fluctuates or engine operation changes dramatically (stalls or idles rough) as the sensor is heated, the sensor is defective and should be replaced.

In a variation of this procedure, the technician can blow through the MAF sensor with the ignition in the *on* position. Voltage should rise, indicating that the sensor is responding to air movement. On some vane-type MAF sensors, the sensor output voltage is determined by inserting an unsharpened pencil into the sensor. The pencil holds the vane in a certain position to produce a specified voltage reading. The sensor voltage output is measured with the ignition *on* and the engine *off.* If the output voltage is not as specified with the pencil inserted, the sensor is defective.

A waveform meter can be used to check the operation of an analog MAF sensor. Attach the waveform meter as instructed by the manufacturer. Start the engine and compare the waveform with a known good waveform for proper shape, frequency, and amplitude (height). See **Figure 15-20.**

Testing Digital MAF Sensors

The output of a digital MAF sensor can be measured by most scan tools, waveform meters, or multimeters capable of reading RPM or duty cycles. Consult the multimeter manual to determine the exact meter capabilities. The test procedure is similar to that for analog MAF sensors. The frequency will be at a set value with the ignition *on* and the engine *off,*

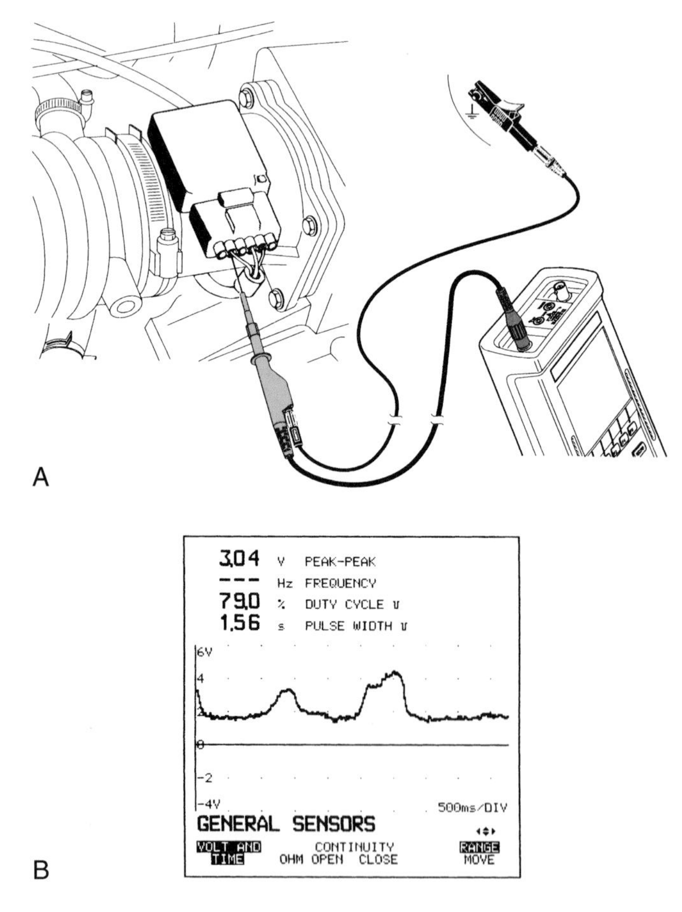

Figure 15-20. A—Connections for testing an analog MAF sensor. B—This waveform indicates correct analog mass airflow (MAF) sensor operation. (Fluke)

Figure 15-21A. When the engine is started, increased airflow through the MAF sensor will cause an increase in the frequency reading, **Figure 15-21B.** If the frequency reading does not change, or if it decreases, the sensor is probably defective.

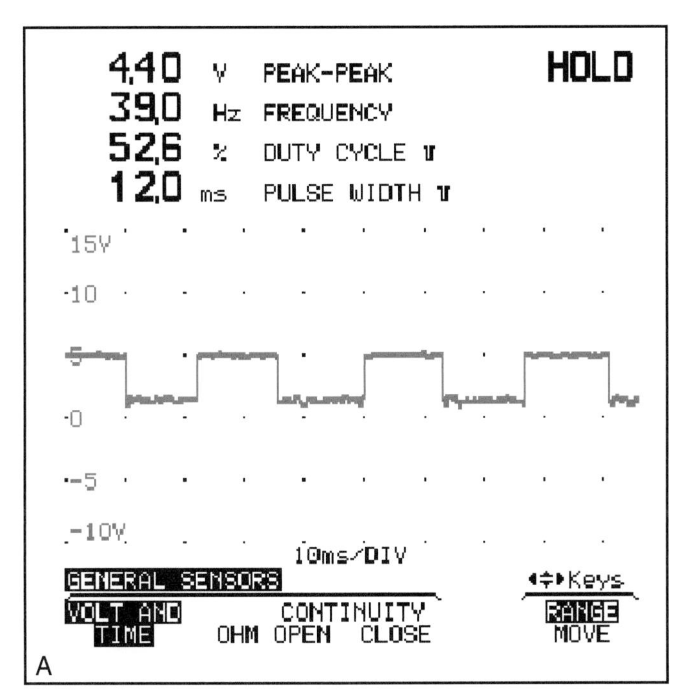

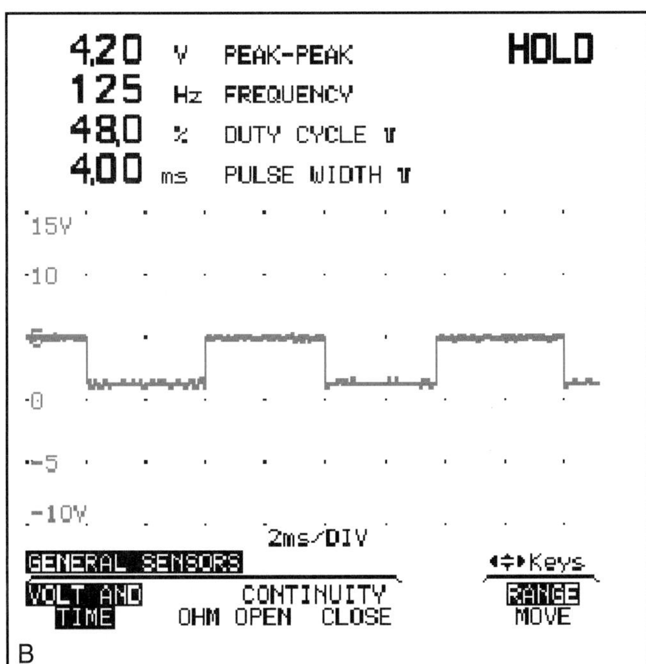

Figure 15-21. A—The frequency of a digital MAF sensor signal can be measured, along with the voltage output. B—Note the increased frequency and pulse width as the engine is started. (Fluke)

Testing Manifold Absolute Pressure (MAP) Sensors

Manifold absolute pressure and similar vacuum and barometric pressure sensors allow the ECM to compensate for manifold vacuum and high altitudes. This sensor is very important in vehicles that use speed density to calculate airflow. Before testing a MAP sensor, check the vacuum hoses for splits or obstructions. Also make sure the engine is providing sufficient manifold vacuum. Consult the service manual to determine whether this test can be made and for the exact procedures.

Begin testing by turning the ignition key *on* without starting the engine. Measure the dc voltage at the ECM and compare it to service manual specifications. Tap the sensor with a small screwdriver while watching for voltage jumps that could signal intermittent problems. Repeat the tapping procedure while warming the sensor with a heat lamp. Next, apply vacuum to the MAP sensor, **Figure 15-22,** while observing the voltage reading. If the voltage reading increases with increases in vacuum, the MAP sensor is probably good. If the MAP sensor fails any of these tests, replace it.

A similar test can be made to the barometric pressure (BARO) sensor on some vehicles. Check the sensor's output voltage at the proper ECM terminals (ignition on, engine not running) and compare it to the specifications for the altitude in your area. A typical altitude compensation chart is shown in **Figure 15-23.** If the voltage is not within specifications, replace the sensor.

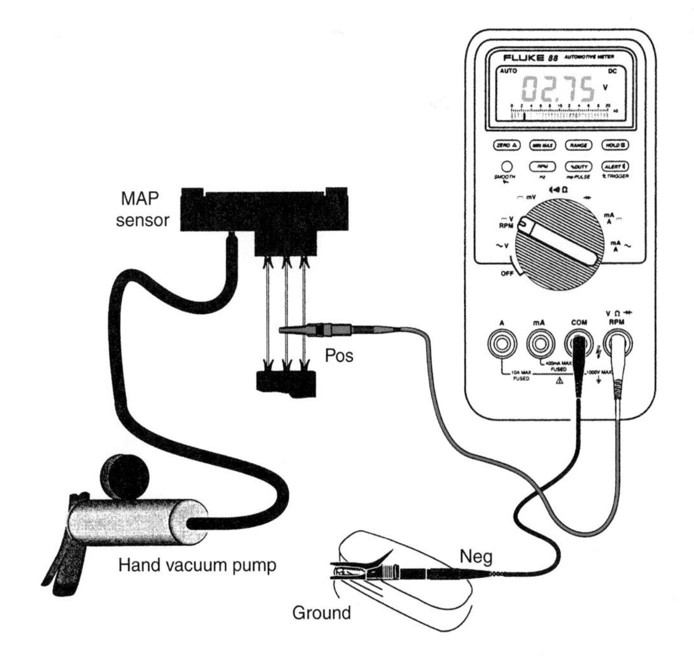

Figure 15-22. Setup for measuring dc voltage from the MAP sensor. Apply vacuum to the sensor to simulate engine vacuum. The readings should be similar to normal readings while the engine is at idle. Compare all readings to service manual specifications. (Fluke)

Altitude		Voltage range
Meters	Feet	
Below 305	Below 1000	3.8–5.5v
305–610	1000–2000	3.6–5.3v
610–914	2000–3000	3.5–5.1v
914–1219	3000–4000	3.3–5.0v
1219–1524	4000–5000	3.2–4.8v
1524–1829	5000–6000	3.0–4.6v
1829–2133	6000–7000	2.9–4.5v
2133–2438	7000–8000	2.8–4.3v
2438–2743	8000–9000	2.6–4.2v
2743–3048	9000–10,000	2.5–4.0v

Low altitude = High pressure = High voltage

Figure 15-23. MAP sensor output voltage varies with altitude. The lower the altitude, the higher the voltage. (General Motors)

A scan tool can be used to check MAP sensors. Attach the scan tool to the DLC and set the tool to read MAP sensor output. Turn the ignition switch to the *on* position without starting the engine and note the reading. Some scan tools will display the actual vacuum reading, which should be 0 in hg when the engine is off. Other scan tools will display voltage, which will be near the reference voltage with the engine off. This reading will vary with altitude and must be compensated for. Next attach a vacuum pump to the MAP sensor fitting. Apply vacuum to the sensor with the ignition *on* and the engine *off.* Scan tool readings should vary smoothly with changes in vacuum. Voltage readings should drop or vacuum readings should increase as vacuum is applied. If the MAP sensor fails these tests, replace it.

MAP sensor signals can also be checked with a waveform meter. Begin testing by determining whether the MAF sensor signal is digital or analog. Then make the proper waveform meter connections and start the engine. See **Figure 15-24.** Check the waveform against a known good waveform. **Figure 15-25** illustrates a good analog MAP sensor waveform. Opening the throttle plate should cause the sensor's voltage signal to rise smoothly. **Figure 15-26** is an example of a good digital MAP sensor waveform. The sensor's output frequency should increase as the throttle plate is opened. Always check the waveform for proper shape, frequency, and amplitude.

It is also possible to test MAP sensors with a waveform meter and a vacuum pump. Make the hookups in the same manner as was done for the voltmeter test. The ignition switch must be in the *on* position. While slowly applying vacuum, make sure the waveform changes smoothly, with no jumping or spikes.

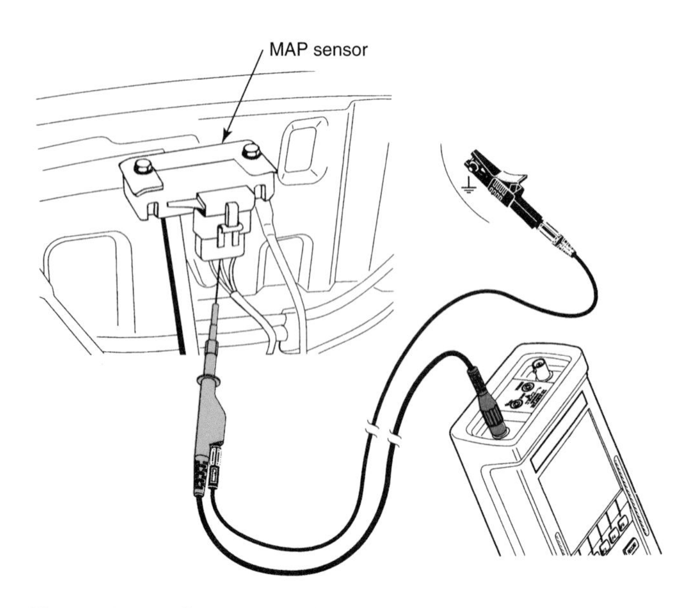

Figure 15-24. Connections for checking a MAP sensor with a waveform meter.(Fluke)

Testing Temperature Sensors

Temperature sensors allow the ECM to compensate for changes in external air temperatures and internal engine temperatures. If a temperature sensor malfunctions, it will usually cause problems that occur when the engine is either

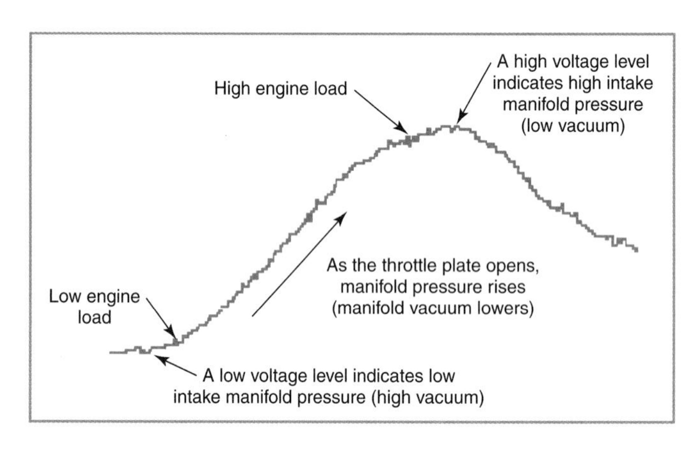

Figure 15-25. Waveform produced by an analog MAP sensor. (Fluke)

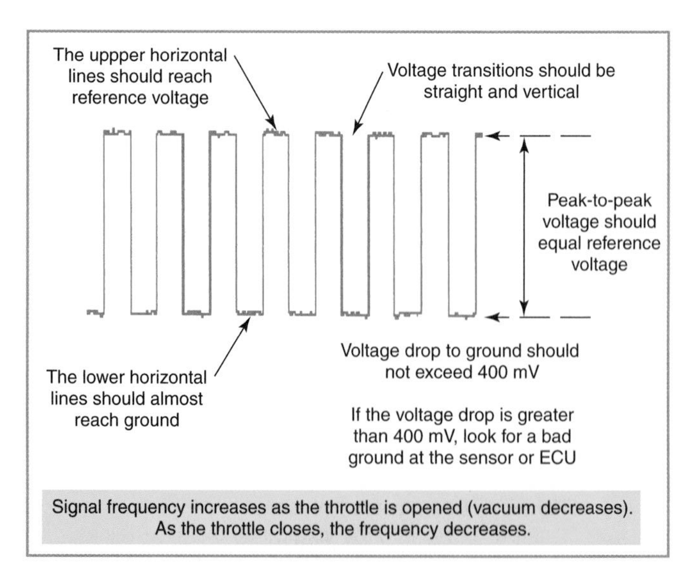

Figure 15-26. Waveform produced by a digital MAP sensor. (Fluke)

hot or cold. The following tests will work with engine coolant, intake air, exhaust gas, and automatic transmission/transaxle fluid temperature sensors.

Note: Before performing tests on an engine coolant temperature sensor, make sure the coolant level in the radiator is correct. The sensor must be in contact with the coolant to operate properly. Also make sure that the sensor electrical connector is attached and that the wiring shows no obvious damage.

Tests Using a Scan Tool

Temperature sensor operation can also be checked using a scan tool. To measure temperature sensor output using this procedure, you must have a scan tool that can convert sensor voltage inputs to temperature readings. This is possible on many scan tools. Follow the manufacturer's instructions.

Begin the test procedure by setting the scan tool to read the temperature as measured by the sensor. Remove the sensor's connector and note the temperature indicated on the scan tool. The scan tool readings should indicate the sensor's lowest reading, usually between -30° and -40°F (-34° and -40°C), **Figure 15-27.** Then, jumper the connector leads to simulate the sensor's highest possible reading, which can be 260°F (127°C) and higher. If the scan tool data shows the output to be correct, the sensor is the likely cause of the problem. If not, there is high resistance in a sensor connection, a shorted or grounded wire, or a problem in the ECM.

Tests Using a Ohmmeter

Most temperature sensors can be checked for a specified resistance at a certain temperature. Unlike most resistors, temperature sensor resistance decreases as temperature rises. To test temperature sensors, you must measure the output resistance in relation to the temperature at the sensor's tip. This will determine whether the sensor is correctly converting the temperature into resistance (which varies the voltage signal) during engine operation.

A temperature sensor can be checked with a digital multimeter or an ohmmeter. If a multimeter is used, set the meter to measure resistance in ohms. Next, disconnect the sensor from the wiring harness and measure the resistance across the sensor leads. Compare the meter readings with the manufacturer's specifications. The readings can be correlated to specific temperatures, **Figure 15-28.** If necessary, a heat gun can be used to raise the temperature to check for resistance variations with temperature. If the resistance measurements are incorrect, the sensor is defective.

If the resistance measurements are correct, the defect is in either the ECM or the wiring. To check the wiring, reinstall the sensor connector, unplug the ECM, and measure the sensor's resistance at the ECM harness. If the resistance is now incorrect, check for a wiring problem. Remember that corroded or loose terminals will cause higher-than-normal resistance readings, while shorts to ground will cause lower-than-normal readings. If the resistance is correct at the harness, the problem is in the ECM.

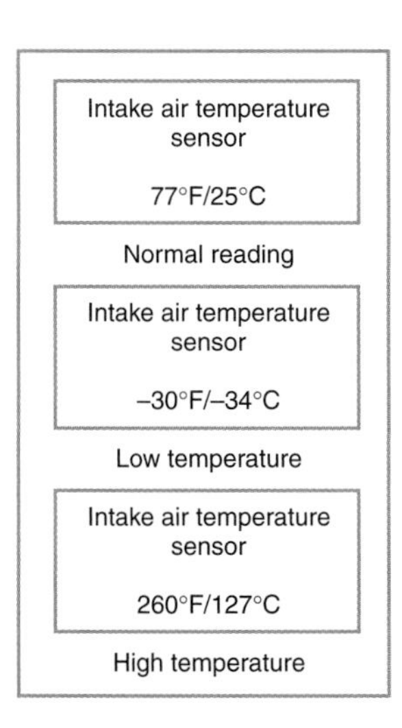

Figure 15-27. Scan tool readings for the intake air temperature sensor. The high temperature reading may be lower than shown here, but it will be much higher than ambient air temperature.

Temperature vs Resistance Valve (Approximate)

°C	°F	Ohms
100	212	177
90	194	241
80	176	332
70	158	467
60	140	667
50	122	973
45	113	1188
40	104	1459
35	95	1802
30	86	2238
25	77	2796
20	68	3520
15	59	4450
10	50	5670
5	41	7280
0	32	9420
–5	23	12,300
–10	14	16,180
–15	5	21,450
–20	–4	28,680
–30	–22	527,000
–40	–40	100,700

Figure 15-28. Most temperature sensors are thermistors. The higher the temperature, the lower the resistance. (General Motors)

Tests Using a Waveform Meter

Coolant and air temperature sensors can be tested with a waveform meter. To check a coolant temperature sensor, connect the meter, start the engine, and then observe the waveform, **Figure 15-29.** Coolant temperature sensors should produce a smooth voltage transition as the coolant temperature increases. There should be no sharp spikes in the waveform. Upward spikes indicate an open in the circuit. Downward spikes indicate a short.

To test an air temperature sensor, remove the sensor from the intake duct. Connect the waveform meter and place the ignition switch in the *on* position. Spray a small amount of water on the sensor and observe the waveform, **Figure 15-30.** A temperature drop should register in an upward movement of voltage as resistance increases. As with the coolant temperature sensor, spikes indicate an open or a short.

Testing Pressure Sensors

Most pressure sensors are simple on-off switches. They can be tested with a scan tool or an ohmmeter. If using a scan tool, connect it to the DLC as specified by the tool manufacturer. Before checking a pressure sensor with an ohmmeter, remove the electrical connector. **Figure 15-31** shows a pressure sensor being tested with an ohmmeter. Note that the placement of the test leads varies, depending on the number of sensor terminals.

If the sensor is normally closed, pressurizing it should cause it to open. The scan tool will indicate that it is open or the ohmmeter will read infinite resistance. The sensor should revert to the closed position, or zero ohms, when the pressure is removed.

If the sensor is normally open, pressurizing it should cause it to close. The scan tool should indicate that it is closed, or the ohmmeter should read zero resistance. The sensor readings should go to the open position, or infinite ohms, when the pressure is removed.

Testing Throttle Position Sensors

The throttle position sensor (TPS) monitors throttle opening and closing so the ECM can adjust fuel and ignition spark timing to match driver demand. Typical TPS voltage can range from 0.2 volts at fully closed throttle to 5.0 volts at wide-open throttle. However, this sensor is adjustable on most vehicles, so depending on the manufacturer and the engine, anything between 0.2 and 1.25 volts might be considered acceptable at closed throttle. Throttle position sensors modify a reference voltage; therefore, an ohmmeter can be used to test these sensors. TPS resistance varies by manufacturer and application.

To test a throttle position sensor, the internal resistance of the sensor must be measured at different throttle openings. Start the test procedure by disconnecting the throttle position sensor's electrical connector. Connect an ohmmeter to the sensor's input and output leads. Next, slowly open the throttle while watching the ohmmeter. The ohmmeter reading should increase smoothly without jumping or skipping. If it does jump or skip, the sensor is defective.

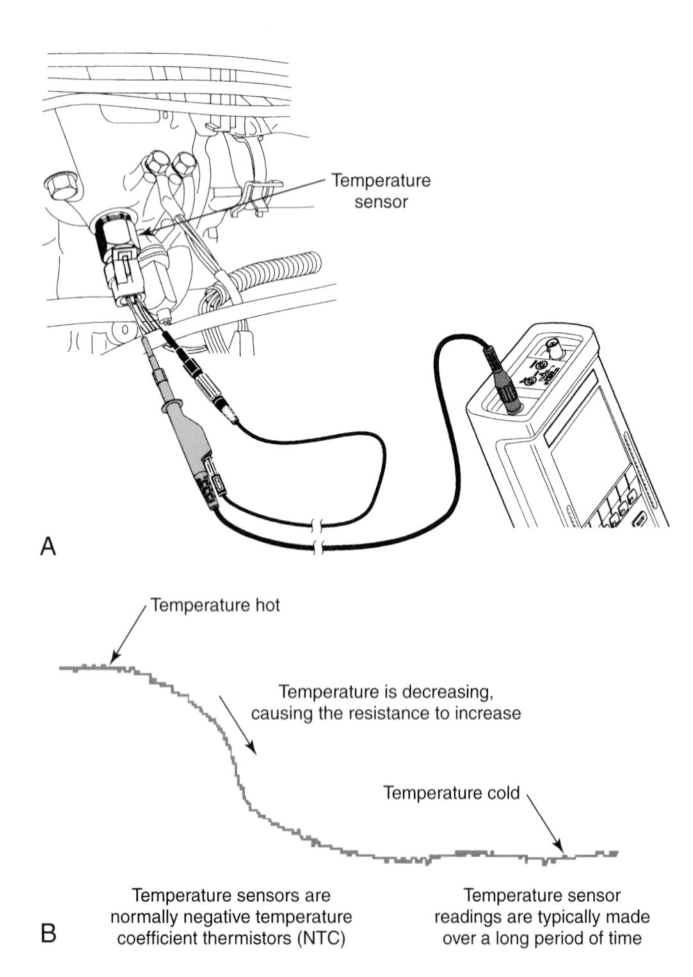

Figure 15-29. A—Setup for testing a temperature sensor with a waveform meter. B—Voltage changes in a temperature sensor as temperature increases. The waveform should show a smooth change with no spikes. (Ferret Instruments)

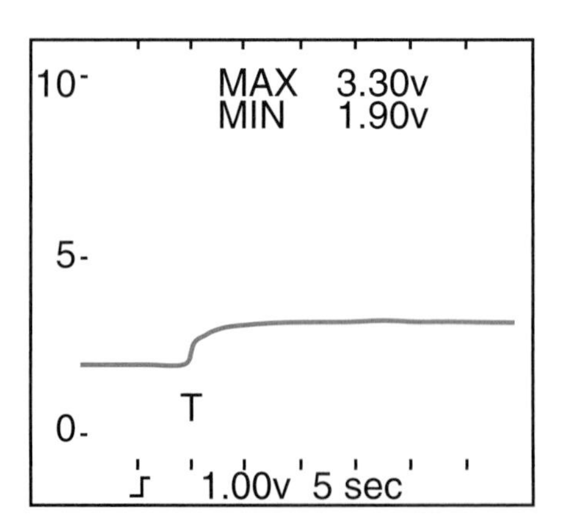

Figure 15-30. Note how the voltage changes in an air temperature sensor. The waveform should rise and fall with changes in temperature, with no spikes or loss of the pattern. (Ferret Instruments)

Using a Scan Tool to Check Operation

The throttle position sensor is another sensor that can be checked with a scan tool. Attach the tool and select the proper settings. With the ignition *on* and the engine *off,* open and close the throttle while observing the TPS voltage. Voltage should rise smoothly as the throttle is opened. The general range of voltage is from 0.5–5 volts. Always check the service literature for exact specifications. Some scan tools read the TPS input as a percentage of throttle opening. If TPS readings are not correct, the sensor should be adjusted or replaced. If the voltage reading fluctuates as the throttle is opened, the TPS is defective and should be replaced.

Testing with a Waveform Meter

The TPS can also be checked with a waveform meter. Connect the meter as specified and place the ignition switch in the *on* position. See **Figure 15-32.** Then observe the waveform while slowly opening the throttle. The meter should show a smooth rise in voltage as the throttle is opened. **Figure 15-33** illustrates the waveform pattern produced by one TPS as the throttle is slowly opened and closed. If there are spikes or breaks in the waveform, the TPS is defective. See **Figure 15-34.**

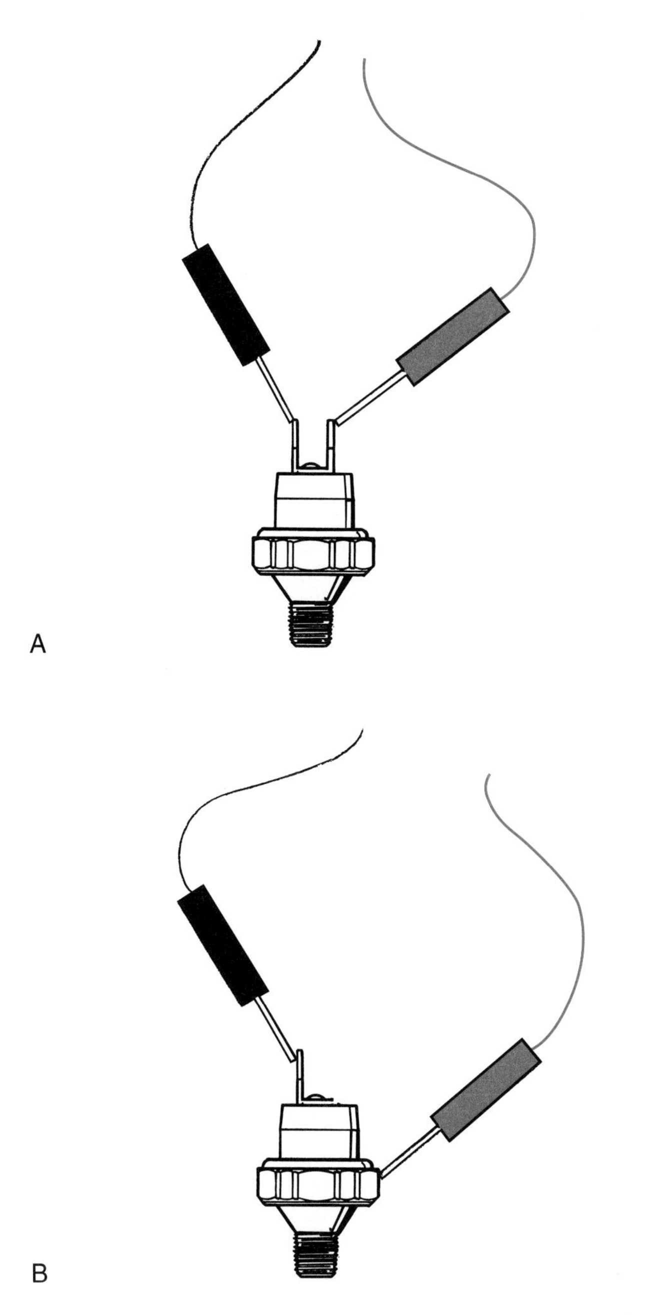

Figure 15-31. Methods of checking a pressure sensor. A—If the sensor has two terminals, check the resistance between the terminals. B—If the sensor has one terminal, check the resistance between that terminal and ground.

Testing Crankshaft and Camshaft Position and Speed Sensors

On most vehicles, the crankshaft and camshaft position and speed sensors provide signals that influence the control of both the fuel and ignition systems. A faulty crankshaft or

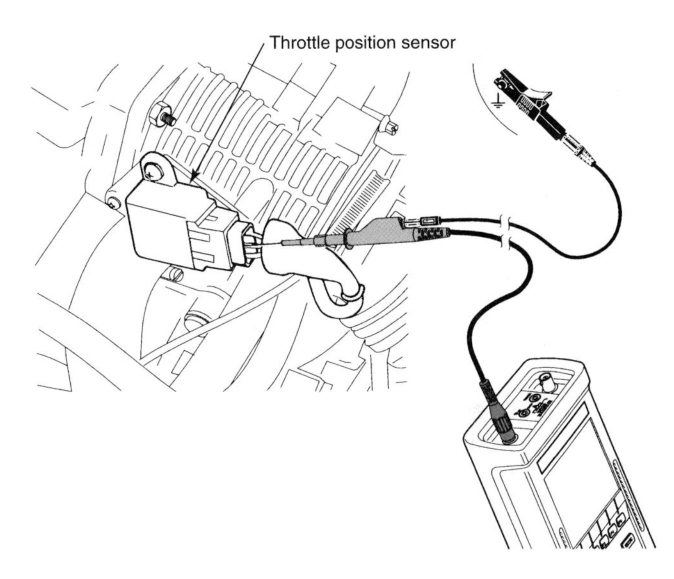

Figure 15-32. Testing a throttle position sensor with a waveform meter. (Fluke)

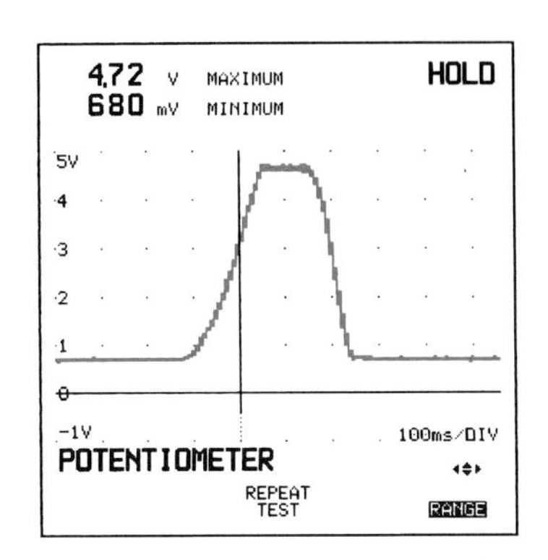

Figure 15-33. This waveform indicates a properly operating throttle position sensor. Note how the pattern rises and falls as the throttle is opened and closed. (Fluke)

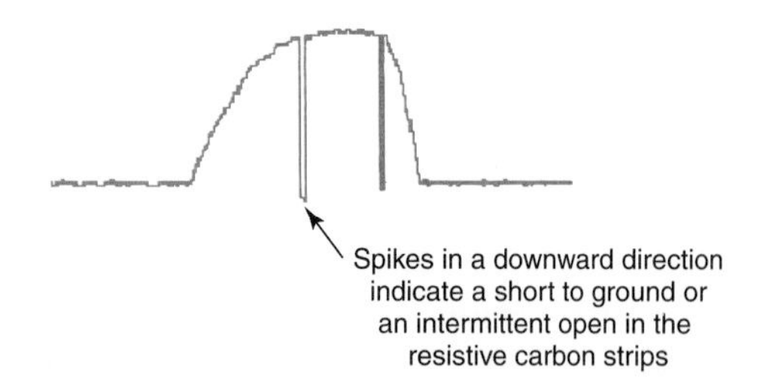

Figure 15-34. This pattern was produced by a defective throttle position sensor. The downward spikes indicate a short to ground or an intermittent open in the sensor. (Fluke)

camshaft sensor may produce either a no-start condition or intermittent stalling. The test procedures for crankshaft and camshaft sensors are similar to those for pickup coils, Hall-effect sensors, and magnetic pickup switches mounted in the distributor. However, their location on or in the engine can make testing difficult.

Note: A problem that appears to be caused by a position sensor can also be caused by problems in the ignition module, ignition coils, fuel injectors, harness wiring, or ECM. The following generalized tests should only be used to narrow down a possible crankshaft or camshaft position sensor problem. If a problem in the position sensor circuits is suspected, it should be verified using tests outlined in the vehicle's service manual.

One test that can be used to check position sensor operation is to crank the engine while monitoring engine rpm using a scan tool. A low rpm reading (or no rpm reading) when the engine is cranking indicates a possible problem in the crankshaft or camshaft sensor circuit, **Figure 15-35.**

To test a position sensor for ac voltage output, disconnect the sensor connector. Obtain a multimeter and connect it to the speed sensor. Set the multimeter to the ac volts range. Crank the engine. The multimeter should read ac voltage as the engine is cranked.

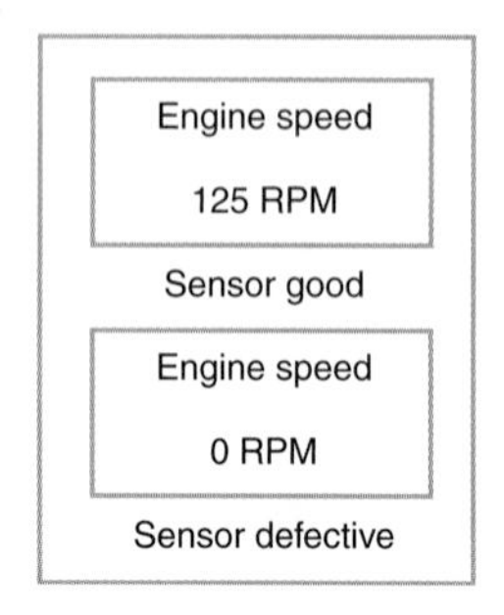

Figure 15-35. A scan tool can be used to quickly check the crankshaft position sensor for proper operation. If the engine has a camshaft sensor, a scan tool reading of engine speed is not reliable, as most ECMs take readings from both sensors to determine engine speed.

Warning: Some engines may start even though the position sensor is disconnected. Keep all tester wires and your hands away from moving parts while making this test.

Check the service literature for the correct specifications. Replace the sensor if it does not produce the proper ac voltage.

A second test involves using a spark tester to check for spark at two or three adjacent spark plug wires. If there is no spark at any of the plug wires, a position sensor circuit problem is possible. If some of the plug wires produce a spark, the problem is most likely in the ignition module or coils.

Another test uses an injector harness "noid" light and a test light. Remove the injector harness connector from one injector, install the light, crank the engine, and note whether the noid light flashes. Repeat this test on two or three adjacent injectors. If injector pulse is present at all the injectors tested, the position sensors are most likely working and the problem is in another circuit. If there is no injector pulse at any of the injectors, either the position sensors or the ECM is defective, or there is a wiring problem. If one or two injectors are not working, use the test light at each injector harness terminal to check for power and ground as instructed by the vehicle's service manual. Injector service will be covered in more detail in Chapter 18, Fuel Injection System Service.

A final test requires a waveform meter. Connect the meter as recommended by the manufacturer. **Figure 15-36** shows some typical crankshaft and camshaft position waveforms. Due to the many variations, always check the manufacturer's service literature for the standard waveform. Any waveform variation indicates that the position sensor is either defective or out of adjustment.

Testing Knock Sensor

The knock sensor allows the ECM to retard ignition timing when spark knock occurs. The knock sensor produces a voltage that is sent to the ECM or modifies a reference voltage from the ECM. Check the service manual for the type of knock sensor you are testing.

Note: Before performing the knock sensor test, make sure that an internal engine problem is not the cause of the knocking condition.

Testing a knock sensor is very easy. All that is needed is a scan tool, voltmeter, or waveform meter and a wrench or other metal object.

To test the sensor using a scan tool, connect the tool to the DLC and set it to monitor knock sensor output. Turn the ignition switch to the *on* position, but do not start the engine. Lightly tap on the engine block with the wrench.

Caution: Tap only on cast iron parts or the block itself. Do not tap on aluminum, plastic, or sheet metal parts.

If the sensor is working properly, the scan tool reading will change, indicating that the ECM is detecting the artificial engine knock, **Figure 15-37.** If the scan tool reading does not change, disconnect the knock sensor connector, connect the voltmeter to the sensor, and repeat the tap test. If the sensor produces a voltage signal, the problem is in the wiring or the ECM. A similar test can be made using a waveform meter. Connect the meter as recommended by the manufacturer. When the engine is tapped, the waveform produced should resemble the pattern in **Figure 15-38.**

A timing light can be used to check the knock sensor circuit on a running engine. Direct the light on the timing marks and tap on the engine. Tapping on the engine should make the timing marks move in the retard direction.

Testing Vehicle Speed Sensors

Most speed sensors produce an alternating current (ac) as they rotate. Therefore, the operation of many vehicle speed sensors can be checked by measuring their output in ac volts. The simplest way to check a speed sensor is to use the proper scan tool. Raise the drive wheels off the ground and shift the vehicle into drive. Modern scan tools can convert the speed sensor output into a speed (miles per hour) reading. If there is no reading, check the speedometer. If the speedometer is indicating speed, a defect exists in the wiring to the ECM or in the ECM itself. If the speedometer is not working, proceed to test the sensor directly with a multimeter

To test the sensor using a multimeter, stop the engine and disconnect the speed sensor connector. Then set the multimeter to the ac volts range and connect it to the speed sensor terminals. Restart the engine and, with the drive wheels off the ground, shift the vehicle into drive and accelerate. The multimeter should start to read ac voltage as the engine begins turning.

This test can also be made using a waveform meter. Look for a pattern similar to the one in **Figure 15-39.** The frequency of the oscillations should increase with vehicle speed.

Some manufacturers call for checking the resistance of the speed sensor winding, but this is done only after other tests have indicated a sensor problem.

Checking the ECM

Diagnosing a defective ECM is sometimes difficult, since it may be too damaged to assist in the diagnostic process. If a scan tool cannot access the ECM's self-diagnostic mode, the ECM is probably faulty. A damaged ECM will sometimes produce false trouble codes. If the ECM produces a trouble code that cannot be confirmed by testing the suspected part,

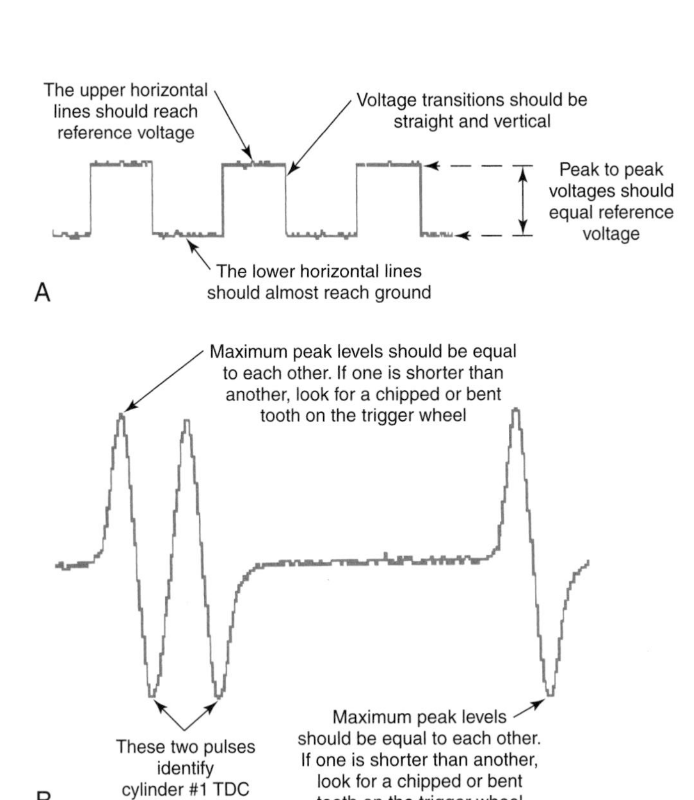

Figure 15-36. Position and speed sensor waveforms vary between types of sensors and manufacturers. Always look up the proper waveform to avoid an incorrect diagnosis. A—Waveform produced by a Hall-effect sensor. B—Waveform produced by a magnetic position sensor. (Fluke)

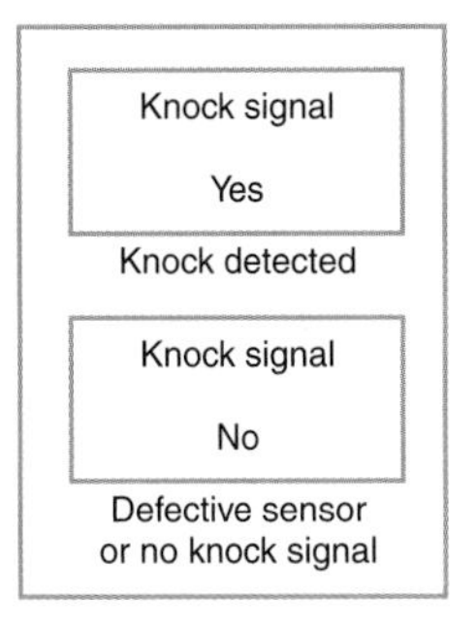

Figure 15-37. The knock sensor signal will show up on most scan tool readouts as a yes (knock present) or no (knock not present).

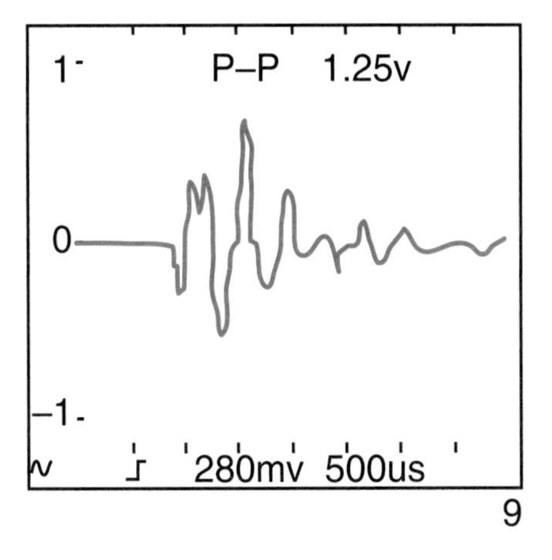

Figure 15-38. Most knock sensor waveforms will resemble this one. (Ferret Instruments)

or if the ECM produces a trouble code number that does exist, it is usually defective.

Often the only way to pinpoint a faulty ECM is through the process of elimination or through substitution. The process of elimination involves checking every other possible cause of a problem and eliminating all of them. If all the sensors, actuators, fuses, wiring, and connectors have been checked and are okay, the problem is most likely in the ECM. Before you determine that the ECM is defective, be sure to thoroughly check every other component and its related wiring.

Substitution involves replacing the suspect ECM with a unit that is known to be good. Before checking by substitution, make sure every related part has been checked. If not, a faulty component could damage the replacement ECM. A shorted actuator, for instance, could destroy the driver in the new ECM.

Note: An ECM-related problem can sometimes be corrected by updating the ECM with new information. This is called reprogramming and is covered later in this chapter.

Checking Actuators

Actuators are the output devices of the computer control system. The following section provides a general overview of common actuator tests.

Testing Solenoids and Relays

Most actuators operated by the computer control system are solenoids or relays. Solenoids are simply wire coils that are energized to move a plunger. Relays are wire coils that are energized to open or close a set of electrical contacts. Some relays are electronic and use transistors in place of the wire coil and contacts.

Many solenoids and relays will make a clicking noise when they are energized. It is often possible to energize the winding and listen for the click. Most solenoids and relays are operated by battery voltage and can be energized with jumper

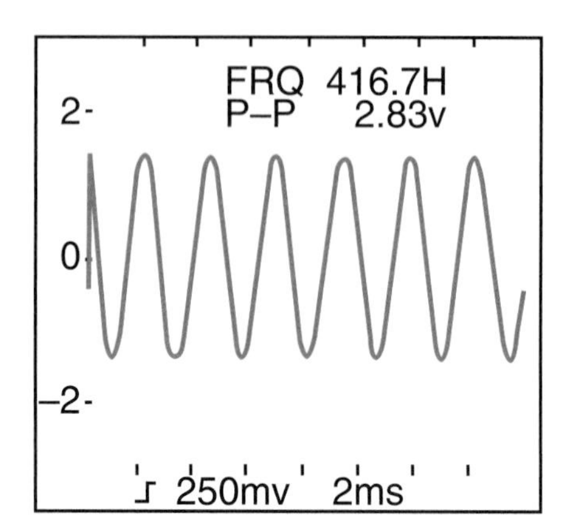

Figure 15-39. Speed sensor waveforms will be a series of oscillations similar to the ones shown here. (Ferret Instruments)

wires from the battery terminals. Be sure to disconnect the actuator's electrical connector before energizing the winding. If the connector is attached, current flow may damage the ECM. Listen for a click as you make and break the circuit with the jumper wires.

It is often possible to determine whether a solenoid or relay clicks without using jumper wires. Park the vehicle in a quiet place and turn the ignition switch to the *off* position. Turn off the radio, air conditioner, and any other noise-producing equipment. Open the hood and locate the actuator. On some vehicles, the actuator may be located under the wheel well. Place your ear close to the actuator and have someone turn the ignition switch to the *on* position without starting the engine. Then listen for a click.

If a solenoid or relay does not click when energized, it may be defective. Make sure the device is being supplied with electricity, usually a reference voltage from the ECM, before deciding that it is defective. Some solenoid plungers, such as those used in the idle speed control or the transmission, may become clogged with carbon or sludge and fail to operate, even when the solenoid is good. Cleaning can sometimes restore these solenoids to proper operation. Electronic relays have no moving parts and will not click. If the solenoid or relay appears to be defective, always follow up by performing the tests outlined in the following sections.

Scan Tool Tests

Most scan tools can be used to check solenoid and relay operation. Examples of these types of solenoids are the idle air control (IAC) solenoid, some transmission solenoids, and mixture solenoids on older vehicles with carburetors. Check the output voltage pulses (usually called counts) as the device operates. The counts can be measured to determine whether the solenoid is operating properly. If readings are within the normal range, the solenoid or relay is working properly. If the counts are not as specified, either the device or ECM is defective, or the ECM is receiving an incorrect input from a sensor.

Some scan tools allow the technician to diagnose actuators by bypassing the ECM and directly operating the devices. This is sometimes called ***forcing the actuator.*** The scan tool then monitors operating voltage during the forcing procedure to determine whether the device is operating properly. Some scan tools can simulate engine and vehicle conditions to check the interaction among various system actuators. Follow the manufacturer's instructions to test solenoids and relays using this method.

Ohmmeter Tests

Solenoid and relay windings can be checked with an ohmmeter as shown in **Figure 15-40.** Solenoid and relay windings have a small range of acceptable resistance. Any reading outside this range means that the actuator should be replaced. Always check the service manual for exact resistance specifications. A reading of zero ohms means the winding is shorted, while an infinite reading indicates an open winding. Either reading means the winding is defective and must be replaced. The circuit in a relay that passes through the internal relay contacts should read zero ohms or infinity, depending on whether the contacts are normally open or normally closed. Energizing the relay winding should cause the reading to change to the opposite value.

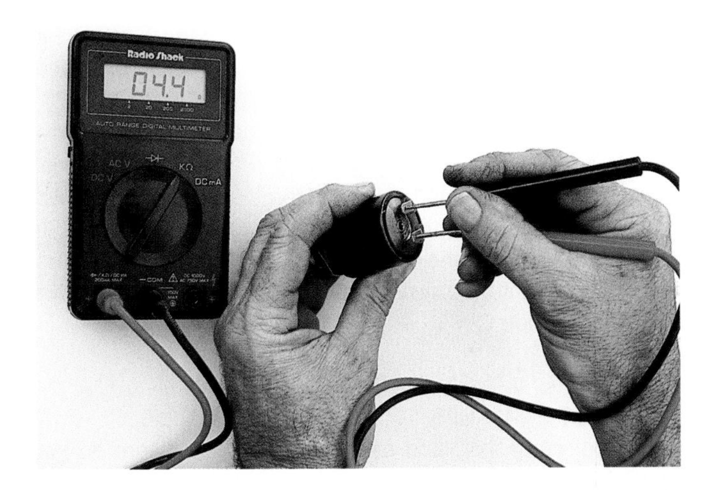

Figure 15-40. Use an ohmmeter to test solenoid and relay windings. Any reading other than the one specified indicates a defective winding.

Operation of solenoids and relays can also be checked with a waveform meter. Connect the meter and start the engine. Operate the engine as needed to energize the actuator. Observe the waveform. **Figure 15-41** shows some possible waveforms. Always consult the manufacturer's specifications to determine what the waveform should look like for the exact component being tested. If the pattern does not look like it should, or shows spikes or breaks, the component is defective.

Testing Motors

A few vehicles use stepper motors to control idle speed, operate the EGR valve, or provide pressure for the anti-lock brake system. These motors operate in small increments, or steps, to move the throttle plate or provide braking pressure. Some anti-lock brake and automatic level control systems have electric motors that operate pumps. In addition, the radiator fan motors in most late-model vehicles are controlled by the ECM.

Note: Before checking any motor, determine whether the ECM operates it through a relay. If so, check the relay as explained previously.

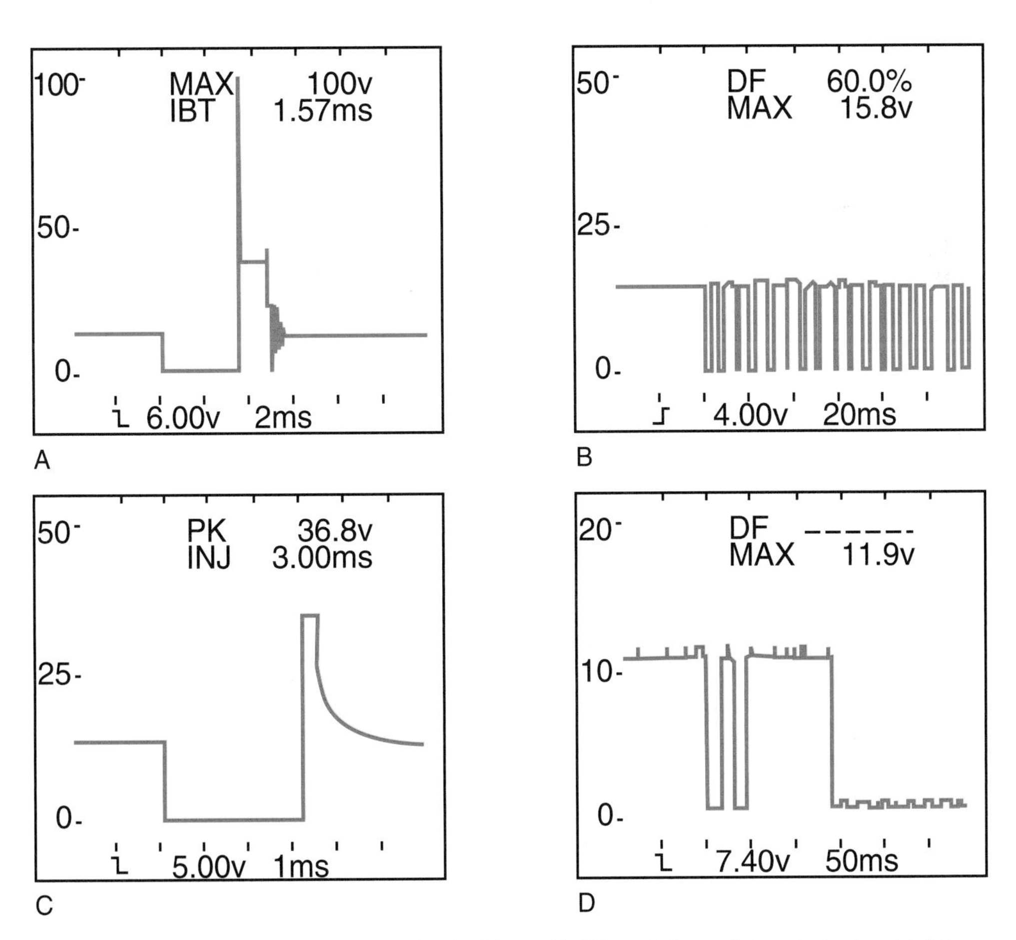

Figure 15-41. Waveforms for various types of actuators. A—Ignition coil. Although the coil is not a solenoid, it is an output device and creates a common pattern. B—EGR valve solenoid. EGR valve solenoids are pulsed on and off rapidly to maintain the proper EGR valve opening. C—Fuel injector. There are many variations of the pattern shown here. D—Idle air control solenoid. (Ferret Instruments)

If the motor is visible, it can be visually checked to determine whether it operates. Some motors can be heard or felt when operating. In some cases, the ECM can be bypassed by using jumper wires to apply voltage directly to the motor. Check the service manual to determine whether the ECM can be bypassed without damaging the motor or the ECM itself.

Motors can also be checked with an ohmmeter. Winding resistance is usually about 0.5–7 ohms for most 12-volt automotive motors. A zero ohms reading indicates that the winding is shorted. If the winding has infinite resistance, it is open. If the winding is defective, the motor must be replaced.

Adjusting Sensors

Some sensors can be adjusted. Typical adjustment procedures for throttle and crankshaft/camshaft position sensors are covered in the following sections. The procedures for adjusting other sensors, such as distributor pickups and ABS wheel speed sensors, are covered in the appropriate chapters.

Throttle Position Sensor Adjustment

A misadjusted throttle position sensor (TPS) can affect ECM operation. A throttle position sensor can go out of adjustment due to linkage wear or changes in the sensor's electrical material. The sensor may also require adjustment when the throttle body is replaced. On most engines, the throttle position sensor is simply mounted on the throttle body, with little or no provision for adjustment. However, some throttle position sensors have slotted mounting holes. The slotted holes allow the sensor to be moved for adjustment. See **Figure 15-42.** With the throttle plate completely closed, rotate the sensor until a specified voltage output is obtained. Specific adjustment procedures are required for some throttle position sensors.

Note: Adjusting some throttle position sensors requires special tools or adjustment procedures. Check the vehicle's service manual for the correct procedure.

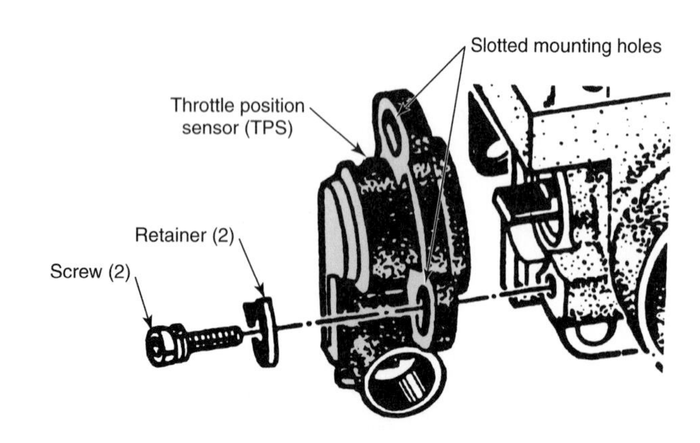

Figure 15-42. In most cases, the throttle position sensor mounts on the throttle shaft and is simply bolted on the throttle body. Note the slotted mounting holes on this TPS. The slotted holes allow the sensor to be turned for adjustment. (General Motors)

To adjust the TPS with a multimeter, obtain the correct voltage specifications and set the meter to a range that will measure these voltages. Using jumper wires to the correct terminals, measure the input and output voltages.

If the input voltage is incorrect, check the wiring and the ECM, and make repairs as necessary. If the output voltage is incorrect, make sure the throttle plate is in the fully closed position. Then loosen the throttle position sensor attaching screws and rotate the sensor until the proper value is shown on the voltmeter.

A scan tool can also be used to adjust the TPS. Plug the tool into the DLC and select the TPS reading from the tool menu. Check the TPS input with the throttle plate in the closed position and compare the input reading to specifications. If necessary, loosen the throttle position sensor attaching screws and rotate the sensor until the proper value is shown on the scan tool.

Adjusting Crankshaft and Camshaft Position Sensors

Most crankshaft and camshaft position sensors do not require adjustment. However, some of these sensors must be adjusted to provide the proper signal to the ECM and prevent damaging contact with rotating engine parts. The adjustment process usually involves the use of a special alignment gauge, as shown in **Figure 15-43.** In this illustration, a special tool is used to adjust a single pickup assembly in relation to the shutter assembly. This positions the shutter precisely between the two parts of the pickup assembly.

If the alignment gauge is not used when specified, damage to the sensor may occur. Consult the vehicle's service manual to ensure that the sensor does not require the use of an alignment gauge.

Some crankshaft sensors use a small piece of cardboard as an adjusting shim between the sensor tip and the toothed wheel. The shim is placed over the sensor's tip before the sensor is installed in the opening. The sensor is tightened into place with the shim holding it in the proper position. When the engine is started, the shim slides from between the sensor and wheel.

Caution: Always check the appropriate service manual for exact procedures before replacing or adjusting any position sensor or pickup. Severe damage can result if the pickup contacts rotating engine parts.

Replacing Computer System Components

This section outlines the general procedures for replacing common computer system components. Many computer system components can be replaced by simply unplugging the electrical connectors and removing the attaching screws. Some devices, such as oxygen sensors, engine temperature sensors, and knock sensors, are

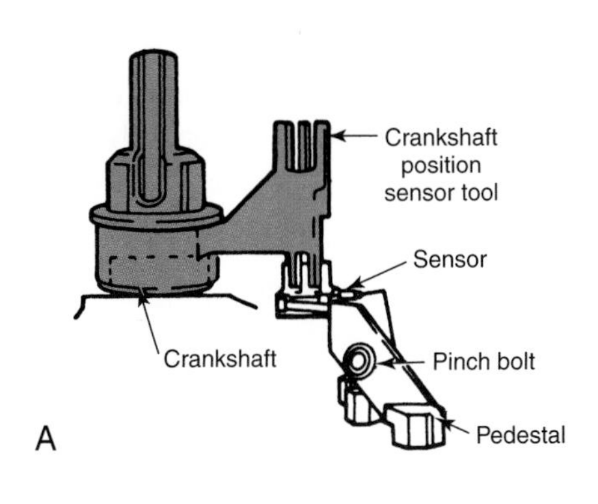

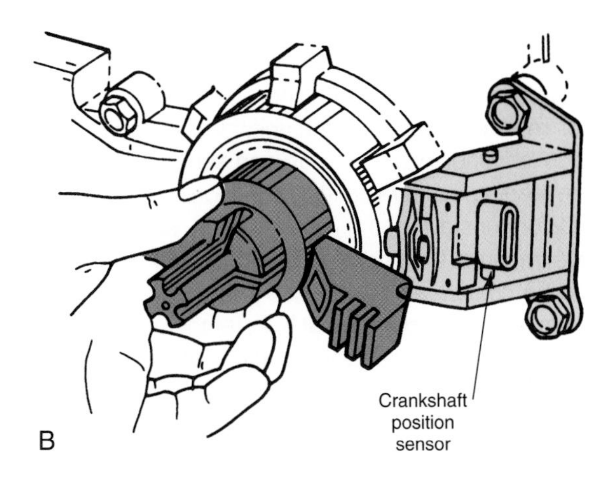

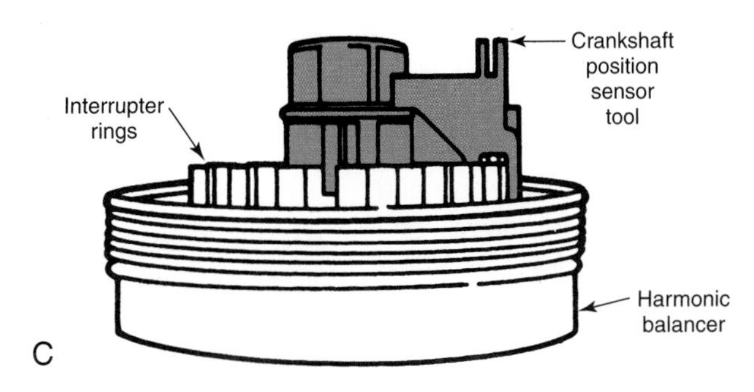

Figure 15-43. Some crankshaft position sensors require adjustment after installation. A and B—Using a position sensor tool to adjust the crankshaft position sensor. C—Using the tool to check the interrupter rings on the crankshaft pulley. (General Motors)

threaded directly into the related component. They must be removed carefully to avoid stripping the threads. Before replacing any computer control system component, always turn the ignition switch to the *off* position and remove the negative battery cable. This will prevent damage to the ECM, the sensors, and other electronic devices from stray electrical charges.

Input Sensor Replacement

The following sections deal with replacing input sensors. There are many different types of input sensors. This makes it necessary to consult the vehicle's service manual for the proper replacement procedures for specific sensors.

Note: The procedures presented in this chapter are general in nature and are meant to describe typical procedures. Always refer to the vehicle's service manual for the exact replacement procedure.

Oxygen Sensor Replacement

The oxygen sensor is probably the most delicate of all the input sensors and should always be handled carefully. Since it is installed in the exhaust manifold or the exhaust pipe, it is subject to high temperatures and corrosion that can cause the threads to seize. Many oxygen sensors require a special socket or tool for removal, **Figure 15-44.** If the oxygen sensor will be reused, always use a special tool to remove it. If the sensor will be replaced and does not require a special tool for removal, loosen it from the manifold using a socket or box end wrench that will contact all sides of the hex.

Figure 15-44. Remove oxygen sensors with the appropriate tool. Be sure you are removing the correct sensor on vehicles with multiple sensors.

Caution: Late-model vehicles have multiple oxygen sensors. Make sure you are removing the correct sensor on these vehicles.

Once the sensor is out, check the manifold or exhaust pipe around the sensor fitting for cracks or pinholes. These will draw in outside air during engine operation, causing a false sensor reading. Replace the manifold or exhaust pipe if any damage is found. Inspect the oxygen sensor for indications of carbon loading, contamination, or poisoning.

Before installing the new oxygen sensor, check the service manual to determine whether the sensor threads should be coated with ***anti-seize compound.*** Some manufacturers recommend a special high-temperature sealant. Use only a light coating of sealant to avoid plugging the external air vents, **Figure 15-45.** Do not overtighten the oxygen sensor, since this may damage the brass shell. If anti-seize compound or sealant is used, wipe off any excess from the exhaust manifold or exhaust pipe and the sensor after tightening.

MAF and MAP Sensor Replacement

The mass airflow and manifold absolute pressure sensors on most vehicles are retained by one or two bolts or clamps and are relatively easy to replace. Start by disconnecting the negative battery cable. Remove any wiring and hoses from the sensor. Loosen the bolts or clamps, and remove the sensor from the vehicle, **Figure 15-46.** Install the new sensor in the reverse order of removal. When installing a MAF sensor, be sure to face the sensor's inlet in the correct direction. See **Figure 15-47.** Tighten all the retaining bolts or clamps. Then reconnect all wiring and hoses. Finally, reconnect the negative battery cable.

Throttle Position Sensor Replacement

To replace a throttle position sensor, make sure the ignition switch is in the *off* position. Then remove the TPS electrical connector. Remove the fasteners holding the throttle position sensor to the throttle body and slide the TPS from the throttle shaft. Slide the new TPS into position; then reinstall the fasteners and electrical connector. **Figure 15-48** shows the sequence used to replace a typical throttle position sensor. After the new sensor is installed, it may be necessary to adjust it using the procedure explained earlier in this chapter.

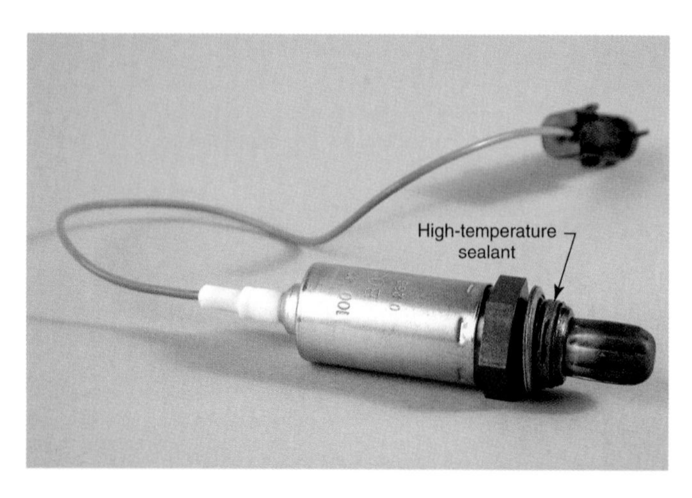

Figure 15-45. If recommended, apply a thin coat of specified sealant to the oxygen sensor threads.

Temperature Sensor Replacement

Temperature sensors are used in many places on the engine and other vehicle components. With the exception of the coolant temperature sensor, they are relatively easy to replace.

Engine Coolant Temperature Sensor

Engine coolant temperature sensors are threaded into a coolant passage in the engine block or the radiator. They should not be removed until the engine cooling system is depressurized and drained below the level of the sensor. Always leave the radiator cap loose while changing any cooling system part to prevent pressure buildup, **Figure 15-49.**

Coolant temperature sensors have *pipe threads*, which are similar to those in household plumbing systems, and are usually installed in the block. Most coolant temperature sensors can be removed by loosening them from the engine using the proper deep-well socket and a ratchet handle. In some cases, the sensor can be removed with a box-end

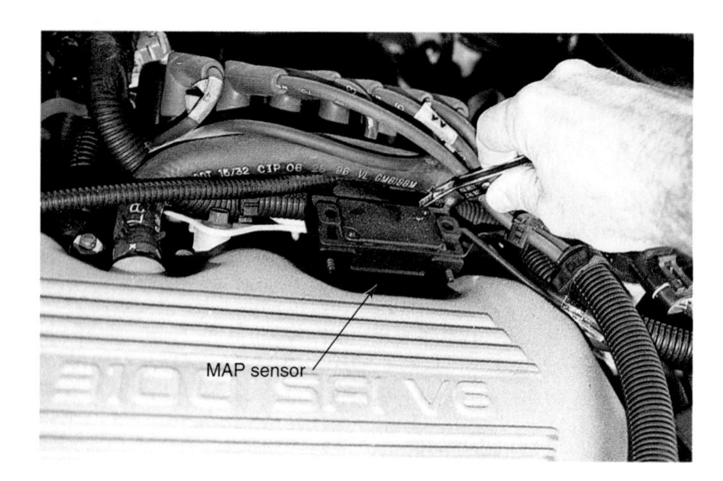

Figure 15-46. When replacing a MAP sensor, disconnect the vacuum line and the electrical connector. Then remove the fasteners and remove the sensor from its mounting.

Figure 15-47. When reinstalling an airflow sensor, make sure it is installed in the correct direction. Note the arrow on this sensor, which indicates the direction of airflow

wrench. A few sensors can only be removed with a special socket. Always disconnect the electrical connector from the wiring harness before installing the socket over the sensor.

Caution: Sensors installed in aluminum parts should not be removed until the engine has cooled for several hours. This cooling period is necessary to prevent damage to the threads in the aluminum part.

A

B

C

Figure 15-48. This figure shows the sequence of steps necessary to remove a throttle position sensor. A—Remove the electrical connector. B—Loosen and remove the attaching bolts. C—Remove the sensor from the shaft. Note the position of the throttle shaft for reinstallation. (General Motors)

Before installing the new sensor, coat the threads with the proper sealant. When installing the sensor, do not over-tighten it, as this may damage the threads or distort the sensor shell. If the cooling system was drained, refill it before restarting the engine. Leave the radiator cap loose and recheck coolant level after allowing the engine to run for about 10 minutes. A few front-wheel drive engines require a special fill procedure. Follow the manufacturer's recommendations. The bleed valve shown in **Figure 15-50** can be loosened to remove air from the cooling system.

Note: When bleeding air from a cooling system, turn on the heater with the blower fan on high. This will provide additional cooling and fluid circulation during the bleeding process.

Intake Air and Exhaust Gas Temperature Sensors

Intake air and exhaust gas temperature sensors are not under pressure when the engine is not running. Intake air

Figure 15-49. Loosen the radiator cap before servicing any part of the cooling system.

Figure 15-50. After the coolant temperature sensor has been replaced, it may be necessary to bleed air from the cooling system. Some engines are equipped with bleed valves. This one is located on the thermostat housing.

temperature sensors can be removed and replaced without special precautions. To replace a temperature sensor installed in the intake manifold or plenum, disconnect the electrical connector. Then use the proper socket to remove the sensor from the manifold. Coat the new sensor with sealing compound (if necessary), install it in the manifold, and reattach the electrical connector. Check for vacuum leaks and sensor operation.

To replace a temperature sensor installed in the air cleaner housing, remove the electrical connector, pull the attaching clip from the sensor, and remove the sensor from the housing. Position the new sensor in the housing, install the clip, and reconnect the electrical connector.

To replace an exhaust gas temperature sensor, allow the exhaust system to cool; then disconnect the electrical connector. Use a special tool or a six-point socket to remove the sensor without damaging its shell. Place anti-seize compound on the threads of the new sensor. Install the sensor and reconnect the electrical connector. Start the engine and check for exhaust leaks and proper sensor operation.

Knock Sensor Replacement

Knock sensors are threaded directly into the engine block, cylinder head, or intake manifold, **Figure 15-51.** This allows them to easily detect and transmit the sound of engine detonation. Knock sensors can be removed and replaced without special precautions. Since the knock sensor shell is usually made of brass, never try to remove this type of sensor with an open-end wrench. Install and tighten the new sensor carefully. Use a socket or box-end wrench to avoid distorting the shell. After tightening, reinstall the electrical connector and check the operation of the knock sensor as explained earlier.

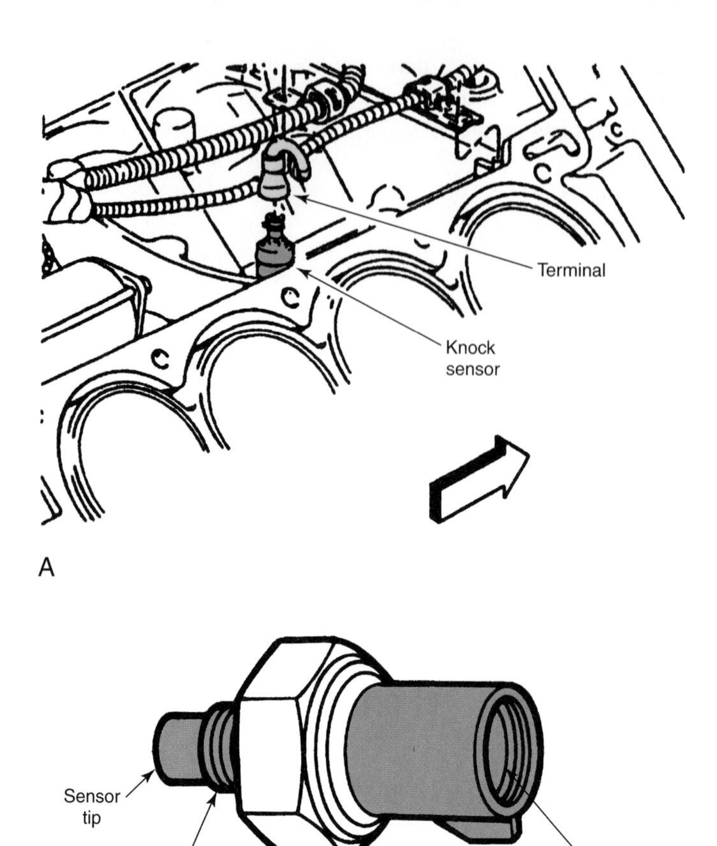

Figure 15-51. A—Knock sensors are threaded into the engine. This one is located under the intake manifold. B—External parts of a knock sensor. (General Motors and Ford)

Crankshaft/Camshaft Position Sensors

Most crankshaft and camshaft position sensors are installed on the engine block, timing cover, camshaft housing, or transmission bell housing. These sensors are secured by a clamp and a single bolt or attached to a bracket held by one or more bolts. After removing the electrical connector and bolt(s), pull the position sensor from the engine.

Check the end of the sensor to ensure that it has not contacted the crankshaft or camshaft sensor ring, **Figure 15-52.** If the end of the sensor has been damaged by contact with the sensor ring or other moving parts, determine the cause and make the necessary corrections before installing a new sensor.

To install the new sensor, make sure replacement O-rings or gaskets are in place. Then lightly lubricate the sensor's tip with engine oil and push the sensor into position. Reinstall the clamp and bolt; then install the electrical connector.

Figure 15-52. After removing a crankshaft or camshaft position sensor, check its tip for signs of damage. If damage is found, be sure to determine and correct the cause before installing a new sensor.

ECM Service

Before replacing an ECM, make sure the ignition switch is turned to the *off* position. Disconnect the negative battery cable, if recommended by the manufacturer. After locating the ECM, disconnect the wiring connector(s) and remove the attaching fasteners. Most ECMs are attached to the vehicle with bolts or screws, but a few are held in place with clips or plastic rivets.

To replace the ECM, set the unit in place and reinstall the fasteners, being sure to reattach any ground straps. Then reconnect the electrical connectors. On some vehicles, it may be necessary to reconnect the electrical connectors before placing the ECM in position on the vehicle.

Caution: To prevent static discharge from damaging the new ECM, follow the manufacturer's instructions concerning static discharge and unit grounding.

Updating ECM Information

Many ECMs can be restored to proper performance by providing them with updated information. On some older vehicles, the PROM chip (sometimes called a mem-cal chip) can be replaced. On many late-model vehicles, the ECM can be reprogrammed from an outside source. These processes will be outlined in the following sections.

Replacing a PROM

To replace a PROM, make sure the ignition key is in the *off* position. Remove the ECM fuse and remove the access cover over the PROM. See **Figure 15-53.** Special tools are available to help remove and install the PROM safely. Remove the old PROM from the ECM. Then install the new PROM, carefully pushing it into position. Finally, replace the access cover and reinstall the ECM fuse.

Replacing a Knock Sensor Module

In some vehicles, the ECM contains a knock sensor (KS) module. This module can be replaced if it is defective. Additionally, it can be removed from a defective ECM and reinstalled in a new unit, ensuring that the proper timing retard is maintained for engine knock protection.

To replace the knock sensor module, make sure the ignition key is in the *off* position. Remove the ECM fuse and then remove the access cover over the KS module. Pinch the module locking tabs to release them and pull the module from the ECM.

Install the new module carefully, making sure the mounting pins are not bent. When pushing the KS module into position, make sure the module's locking tabs snap into place. Finally, replace the access cover and reinstall the ECM fuse.

Figure 15-53. The PROM or mem-cal can be replaced after removing the cover that protects it from water and dirt.

Flash Programming a Computer

As mentioned, the ECMs in late-model vehicles can be updated by erasing old information from the computer's memory and reprogramming the unit with new information. The new information often cures engine driveability and transmission shifting problems. The computer memory section that can be updated is generally called the ***EPROM*** (electronically erasable programmable read-only memory) or the ***FEPROM*** (flash erasable programmable read-only memory). Providing the existing memory with updated information is usually referred to as ***flash programming***.

Actual reprogramming details vary between manufacturers, but the basic procedures are the same for all vehicles. One of three methods can be used to reprogram the ECM:

- Direct programming.
- Indirect programming.
- Remote programming.

Direct Programming

Direct programming is the fastest and simplest method of reprogramming an ECM. The new information is downloaded by attaching a shop recalibration device (usually a computerized analyzer) or a programming computer directly to the vehicle's DLC. See **Figure 15-54A.** The erasure and reprogramming is done by accessing the proper menu and

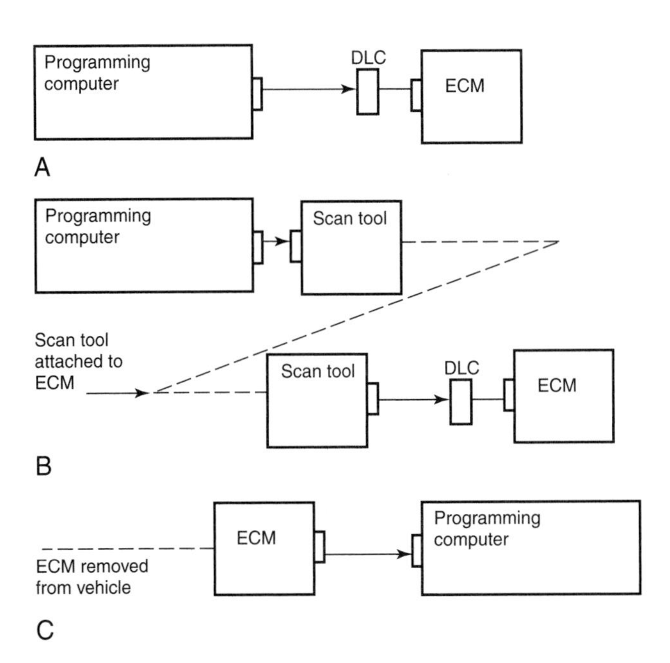

Figure 15-54. Flash programming methods. A—Direct programming. B—Indirect programming. C—Remote programming.

following the instructions as prompted by the recalibration device. Then, the information (often stored in the device's memory, contained on a CD-ROM, or accessed through a connection to the manufacturer's database) is entered into the ECM through the DLC. The shop recalibration device is not a scan tool, and a scan tool is not needed for this programming procedure.

Indirect Programming

To perform ***indirect programming,*** the proper scan tool is used to transfer information from a separate *programming computer* to the ECM. See **Figure 15-54B.** The scan tool can also be used to reset some computer-controlled vehicle systems after programming is complete. The programming computer may resemble a personal computer used in the home, or it may be a computerized analyzer that is similar to the one used for direct programming.

In indirect programming, the scan tool is connected to the programming computer and programming information is downloaded from the computer to the scan tool. Most scan tools use a high-capacity memory cartridge to store the programming information. Some newer scan tools have enough fixed memory to hold the programming information and do not use a separate memory cartridge. Once the programming information has been downloaded, the scan tool can be disconnected from the programming computer, taken to the vehicle, and connected to the DLC. Programming information is then downloaded from the scan tool to the ECM through the data link connector.

Remote Programming

Remote programming is done with the ECM removed from the vehicle. See **Figure 15-54C**.This procedure is used when changes must be made through a direct connection to a manufacturer's database. Remote programming can also be done in cases where normal direct or indirect programming is not practical or possible. Special connectors and tools are required for remote programming. In most cases, this procedure is done only at new-vehicle dealerships.

To perform remote programming, remove the ECM from the vehicle as described earlier in this chapter. Once the ECM has been removed, take it to the programming device. The programming device is generally a computer located in the shop. This device may contain the new ECM information, or it may be used with a modem to connect to a remote database. Attach the programming device's electrical connectors to the ECM. Access the device's programming menu and follow instructions given in the menu to program the ECM. Programming normally takes only a few minutes. When programming is complete (as indicated on the menu), remove the programming device's electrical connectors from the ECM and reinstall the ECM in the vehicle.

Actuator Replacement

Replacing an actuator is often easier than diagnosing the original problem. As you learned in earlier sections, the ECM operates a wide variety of actuators. The following procedures are general examples of actuator replacement. The replacement of specific actuators is covered in the applicable chapters. Always consult the proper service literature when removing and replacing any actuator.

Replacing a Solenoid

Several solenoids are used on modern vehicles, including fuel injectors, idle controls, EGR solenoids, and transmission solenoids. To replace a solenoid, make sure the ignition switch is in the *off* position. Then, remove any parts that block access to the solenoid. Disconnect the electrical connector, remove the fasteners holding the solenoid, and remove the solenoid. Inspect the solenoid. Sometimes solenoid problems are the result of carbon or sludge buildup in the valve and passages. It may be possible to clean the solenoid and restore it to service. If the solenoid is definitely bad, it must be replaced.

Compare the new solenoid to the old one to ensure that the replacement part is correct. Install any needed gaskets or seals on the new solenoid and place it in position. Install the fasteners and the electrical connector. Finally, install any other parts that were removed, start the engine, and check solenoid operation.

Replacing a Relay

Relays are usually installed in one of the vehicle fuse boxes, **Figure 15-55.** To replace a relay, make sure the ignition switch is in the *off* position. Then locate the relay in the fuse box and pull it from its socket. Compare the old and new relays; then install the new relay. Start the engine and check relay operation.

Replacing a Motor

Some vehicles make use of motors as actuators. For example, a motor is sometimes used to control idle speed on fuel injected engines, **Figure 15-56.** To replace a motor, make sure the ignition switch is in the *off* position. Then disconnect

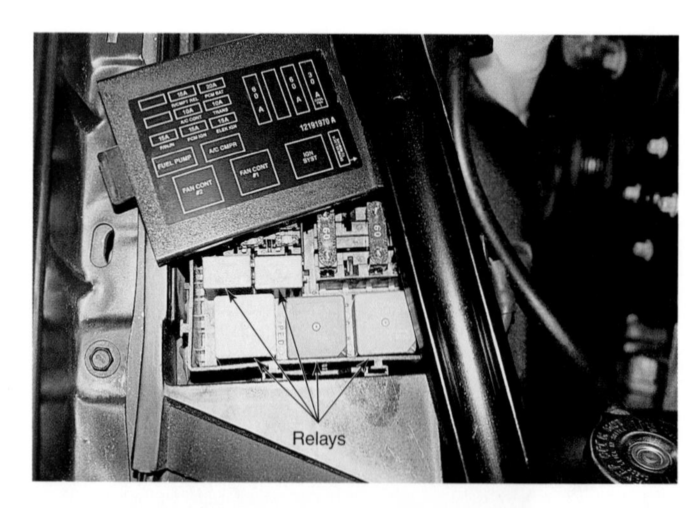

Figure 15-55. Relays are installed in one of several fuse boxes. Be sure that you are replacing the correct relay.

the electrical connector and remove the fasteners holding the motor in position. Transfer any brackets or other hardware from the old motor to the new one; then install the new motor. Install and tighten the fasteners and replace the electrical connector. Start the engine and check motor operation.

Follow-up for Computer Control System Repairs

After replacing computer system parts or reprogramming the ECM, recheck system operation to ensure that the repairs have been successful. After making computer system repairs, erase any trouble codes from the ECM's memory. Then road test the vehicle to ensure that the MIL does not come on. Finally, make sure that none of the trouble codes has been reset. If the repair involved correcting a lean or rich condition, verify that the exhaust emissions output is within specifications.

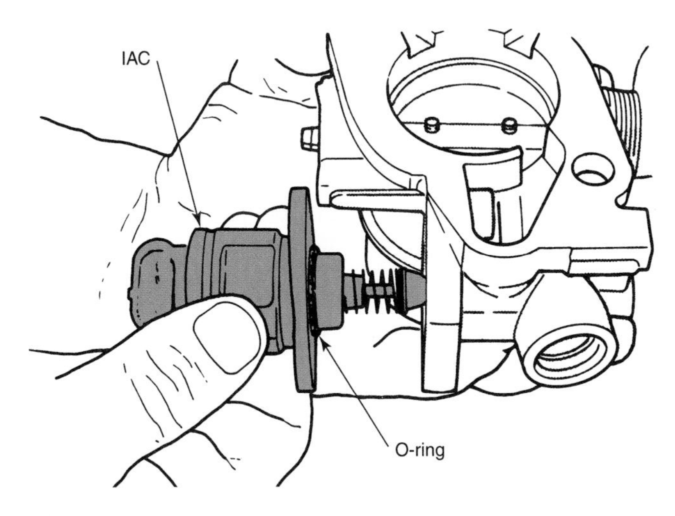

Figure 15-56. The IAC is a commonly replaced motor located on the throttle body, near the throttle plate. It can usually be removed and replaced easily. (General Motors)

ECM Relearn Procedures

After service, the computer system may require a **relearn procedure,** which is a period of vehicle operation that allows the system to adapt to new components and updated programming information. The relearn procedure can often be accomplished by driving the vehicle at various speeds for about ten minutes. Some vehicles require a specific relearn procedure, which may include idling in drive for a specified amount of time or until the engine reaches its normal operating temperature. Always check the manufacturer's service literature for specific relearn procedures. Ignore any unusual engine and transmission conditions until the relearn procedure is complete.

OBD II Drive-Cycle Test

OBD II–equipped vehicles must pass a ***drive-cycle test*** when the ECM or the battery has been replaced, or after trouble codes have been erased. The drive cycle test is also part of some states' I/M emissions testing procedures. The test is performed before the actual emission testing begins.

The drive-cycle test involves attaching a scan tool to the vehicle and driving the vehicle for a set time. The drive cycle consists of specific acceleration, cruising, and deceleration steps. The drive-cycle test is designed to tell the technician whether the OBD II system is operating and whether the vehicle is operating efficiently enough to have a reasonable chance of passing an emissions test. **Figure 15-57** is a chart showing a typical drive cycle and the OBD II systems monitored. Note that the entire test takes 12–15 minutes, starting with a cold engine. If the vehicle fails the drive-cycle test, the technician can use the data gathered by the scan tool to quickly isolate the defective system or component. Once the system has been repaired, the drive-cycle test can be repeated to check OBD II system operation. Emissions testing procedures are covered in more detail in Chapter 20, Emission System Testing and Service.

Typical OBD II Drive Cycle

Diagnostic Time Schedule for I/M Readiness	
Vehicle Drive Status	What Is Monitored?
Cold start, coolant temperature less than 50°C (122°F)	—
Idle 2.5 minutes in drive (auto) neutral (man), A/C and rear defogger ON	HO_2S heater, misfire, secondary air, fuel trim, EVAP purge
A/C off, accelerate to 90 km/h (55 mph), 1/2 throttle	Misfire, fuel trim, purge
3 minutes of steady state—cruise at 90 km/h (55 mph)	MIsfire, EGR, secondary air, fuel trim, HO_2S, EVAP purge
Clutch engages (man), no braking, decelerate to 32 km/h (20 mph)	EGR, fuel trim, EVAP purge
Accelerate to 90-97 km/h (55-60 mph), 3/4 throttle	Misfire, fuel trim, EVAP purge
5 minutes of steady state cruise at 90-97 km/h (55-60 mph)	Catalyst monitor, misfire, EGR, fuel trim, HO_2S, EVAP purge
Decelerate, no braking. End of drive cycle	EGR, EVAP, purge
Total time of OBD II drive cycle 12 minutes	-

Figure 15-57. The OBD II drive-cycle test should be performed to prepare the vehicle for emissions inspection and to reset systems after the battery or ECM has been disconnected. (General Motors)

Tech Talk

About 35 years ago, the only electronic part on an automobile was the radio. The first electronic ignition systems and electronic voltage regulators were introduced in the mid 1960s and did not become common until almost ten years later. Technicians who understood point-type ignition systems and five or six circuits in the carburetor were considered current. Technicians in those days could keep up with things by keeping their eyes open and gaining some trial-and-error experience as newer cars came into the shop.

Of course, all of this has changed. The modern technician must understand electronic ignitions, fuel injection, emission controls, on-board computers, and all sorts of government rules and regulations. Just keeping up from year to year takes considerable reading and study.

So what does this mean to you, the future technician? For starters, trial-and-error learning is out. One error on a computer-controlled system will destroy several hundred dollars worth of computer equipment.

The only way to troubleshoot and repair late-model vehicles properly and efficiently is to get your hands on every bit of new information you can and study it thoroughly. Now is the time to begin collecting information about modern vehicles and taking time to read about the latest developments.

Summary

The first step in diagnosing computer control system problems is to check for obvious defects. Look for problems in the wiring, damaged parts, and other vehicle problems that may be confused with the computer control system malfunctions.

Before performing any other diagnostic steps, retrieve trouble codes from the ECM memory. Trouble codes may be stored even if the MIL is not on. To retrieve trouble codes, follow the procedure outlined in the factory service manual. On OBD I vehicles, trouble codes can be obtained without a scan tool. On OBD II vehicles, a scan tool must be used to retrieve the codes. OBD I and OBD II code formats are different. The next step is to separate the hard codes from the intermittent codes. Then interpret the codes to arrive at a defective system or part.

After the trouble codes have been retrieved and interpreted, check the computer control system devices. Sensors to be tested include the oxygen sensor, mass airflow (MAF) sensor, manifold absolute pressure (MAP) sensor, temperature sensor, throttle position sensor, crankshaft and camshaft position and speed sensors, knock sensor, and vehicle speed sensor. Testing can be done with a multimeter, a scan tool, or a waveform meter.

A faulty ECM can be identified by substituting a known good ECM for the questionable unit or through the process of elimination. If every sensor, actuator, and system fuse, as well as all wiring and connectors, have been checked and are okay, the problem is most likely in the ECM. The ECM will occasionally produce a trouble code that cannot be confirmed by testing or a trouble code that does not exist. This indicates that the ECM is defective.

Actuators include solenoids and relays, and motors. Actuators can be tested with scan tools, ohmmeters, or waveform meters. Sometimes the actuator can be tested by observing whether or not it operates.

Some sensors, such as throttle position and crankshaft/camshaft position sensors, can be adjusted. Check the vehicle's service manual for correct procedures.

Replacing sensors is relatively easy. Some are threaded into the engine. Others are installed on brackets near the engine. Most vehicles have multiple oxygen, temperature, knock, and pressure sensors. Be sure you are removing the correct sensor. Once the sensor has been replaced, it will be necessary to operate the engine to allow the ECM to recalibrate itself.

ECM service involves replacing defective parts or reprogramming the ECM itself. PROMS must be replaced carefully to avoid damaging the ECM. Reprogramming installs new information to cure driveability and other engine problems.

Solenoids, relays, and motors can be replaced without major disassembly. Once the new parts have been installed, start the engine and allow the ECM go through its relearn procedure

Always recheck system operation after performing computer control system service. If possible, road test the vehicle to ensure that it operates properly. After completing the road test, recheck for trouble codes.

Review Questions—Chapter 15

Please do not write in this text. Write your answers on a separate sheet of paper.

1. A vehicle is brought to your shop with a suspected computer system problem. Which of the following should be done *first?*
   (A) Obtain the vehicle's service manual.
   (B) Check for obvious problems.
   (C) Retrieve trouble codes.
   (D) Road test the vehicle.
2. A vehicle has a driveability problem and the charging system voltage is low. Which should be diagnosed and corrected first, the driveability problem or the low charging system voltage?
3. An OBD II computer system will have all of the following, *except:*
   (A) a 16-pin diagnostic connector.
   (B) an oxygen sensor mounted behind the catalytic converter.
   (C) a two-digit trouble code format.
   (D) hard and intermittent trouble codes.
4. Trouble codes in most OBD II systems should be cleared by _____.
   (A) disconnecting the negative battery cable
   (B) using a scan tool
   (C) removing a fuse to disconnect battery power from the ECM
   (D) All of the above.

5. Which of the following sensors can be checked with a scan tool?
   (A) Oxygen sensor.
   (B) MAP sensor.
   (C) MAF sensor.
   (D) All of the above.
6. The waveform meter provides the technician with a _____ of what is happening in the circuit.
7. The two types of MAF sensors are the _____ type and the _____ type.
8. A MAP sensor measures engine _____.
9. A temperature sensor can be checked with a(n) _____.
   (A) scan tool
   (B) ohmmeter
   (C) waveform meter
   (D) All of the above.
10. A pressure sensor can be checked with a(n) _____.
   (A) voltmeter
   (B) ohmmeter
   (C) ammeter
   (D) All of the above.
11. A noid light is used to check which of the following sensors?
   (A) Crankshaft position sensor.
   (B) Throttle position sensor.
   (C) Temperature sensor.
   (D) Knock sensor.
12. Throttle position sensors can go out of adjustment because of _____.
   (A) linkage wear
   (B) changes in the ECM
   (C) changes in the sensor material
   (D) Both A and C.
13. Throttle position sensors are adjusted by attaching what type of meter to the sensor?
   (A) Voltmeter.
   (B) Ammeter.
   (C) Ohmmeter.
   (D) All of the above.
14. What two methods are often the only ways to check an ECM that is suspected of being defective?
15. To check a solenoid, energize it and listen for a _____.
16. Define the process of forcing an actuator.
17. The service manual gives a solenoid winding resistance specification of 0.5–0.7 ohms. Which of the following readings indicates that the winding has the correct resistance?
   (A) 0 ohms.
   (B) 0.5 ohms.
   (C) 7 ohms.
   (D) Infinity.
18. Removing and installing an oxygen sensor usually requires a special _____.
19. Which of the following can cause a false oxygen sensor reading?
   (A) A crack in the exhaust manifold.
   (B) Excessive anti-seize compound on the sensor threads.
   (C) Tight electrical connectors.
   (D) A discolored sensor shell.
20. To install a coolant temperature sensor, the cooling system must be _____.
21. The cooling system bleed valve is used to remove _____ from the cooling system.
   (A) coolant
   (B) air
   (C) sealant
   (D) rust
22. Knock sensors are threaded directly into the _____.
   (A) engine block
   (B) cylinder head
   (C) intake manifold
   (D) All of the above.
23. When changing an ECM, the first thing to remove is the _____.
   (A) negative battery cable
   (B) ECM retaining bolts
   (C) ECM electrical harness
   (D) ROM or PROM
24. When an ECM must be flash reprogrammed, which of the following should be removed?
   (A) PROM.
   (B) ECM.
   (C) ECM wiring harness.
   (D) All of the above.
25. Relays are usually installed in one of the vehicle _____ _____.

ASE-Type Questions

1. Trouble codes on most modern vehicles can be read only by using which of the following?
   (A) The MIL.
   (B) A voltmeter.
   (C) A scan tool.
   (B) A waveform meter.

2. Technician A says that grounding one of the diagnostic terminals on an OBD II system will cause the engine to go into closed loop operation. Technician B says that a scan tool is needed to retrieve trouble codes from an OBD II system. Who is right?
   (A) A only.
   (B) B only.
   (C) Both A and B.
   (D) Neither A nor B.
3. Hard trouble codes are being discussed. Technician A says that hard codes will cause the MIL to illuminate only when the key is on and the engine is off. Technician B says that hard codes will cause the MIL to be on whenever the engine is running. Who is right?
   (A) A only.
   (B) B only.
   (C) Both A and B.
   (D) Neither A nor B.
4. An excessively rich reading from the oxygen sensor can mean all of the following, *except:*
   (A) the oxygen sensor is defective.
   (B) a leaking fuel injector.
   (C) the gasoline is the wrong octane rating.
   (D) a defective airflow sensor.
5. The heater circuit of an oxygen sensor can be checked with which of the following?
   (A) Ohmmeter.
   (B) Voltmeter.
   (C) Ammeter.
   (D) Jumper wires.
6. The output of a frequency-type MAF sensor can be measured by using a _____.
   (A) scan tool
   (B) multimeter that can read RPM
   (C) multimeter that can read duty cycles
   (D) All of the above.
7. Technician A says that testing a BARO sensor requires a vacuum tester. Technician B says that testing a MAP sensor requires a vacuum tester. Who is right?
   (A) A only.
   (B) B only.
   (C) Both A and B.
   (D) Neither A nor B.
8. Technician A says that most pressure sensors are on-off switches. Technician B says that most pressure sensors can be tested with a voltmeter. Who is right?
   (A) A only.
   (B) B only.
   (C) Both A and B.
   (D) Neither A nor B.
9. Technician A says that tapping an engine block with a metal tool will cause the ignition timing to retard. Technician B says that tapping the engine block with a metal tool will cause the knock sensor circuit to operate. Who is right?
   (A) A only.
   (B) B only.
   (C) Both A and B.
   (D) Neither A nor B.
10. All the following statements about speed sensors are true, *except:*
    (A) the output of some vehicle speed sensors is measured in ac volts.
    (B) some vehicle speed sensors are checked by measuring resistance.
    (C) the output of some speed sensors is measured as a pulsed dc signal.
    (D) some speed sensors can be checked with a scan tool.
11. Which of the following computer control system components can be adjusted?
    (A) MAF sensor.
    (B) MAP sensor.
    (C) TPS.
    (D) IAC.
12. Some ECMs have internal memory devices that can be replaced to update the ECM. Which of the following internal parts *cannot* be replaced?
    (A) PROM.
    (B) ROM.
    (C) Mem-cal.
    (D) Knock sensor.
13. All the following are methods of reprogramming the ECM, *except:*
    (A) direct programming.
    (B) indirect programming.
    (C) remote programming.
    (D) mechanical programming.
14. An engine runs rough during the first few minutes of the ECM relearn procedure. Technician A says this indicates another system problem. Technician B says this indicates that the ECM is adjusting to new engine operating conditions. Who is right?
    (A) A only.
    (B) B only.
    (C) Both A and B.
    (D) Neither A nor B.

15. Technician A says that a drive-cycle test is performed before the emissions are checked. Technician B says that a drive-cycle test is performed after the ECM or battery has been disconnected. Who is right?
    (A) A only.
    (B) B only.
    (C) Both A and B.
    (D) Neither A nor B.

Suggested Activities

1. Using factory service information, locate the computer diagnostic test connectors on several different makes and years of vehicles. Make a chart comparing the connector locations by manufacturer and the type of equipment needed to access the trouble codes.
2. Obtain a vehicle with an on-board computer and retrieve trouble codes. Use the procedure outlined in the service manual. If your instructor okays it, create trouble codes by disconnecting a computer sensor or an output device and briefly running the engine. What codes were found? What do the codes indicate?
3. Based on the preceding activities, consult the appropriate factory service manual to determine the next steps to be taken to isolate a computer problem.
4. Determine the effect on the computer system of disconnecting a non-computer device. For instance, what happens to the fuel mixture when the air injector (smog) pump belt is removed? What happens to the timing advance if a vacuum leak develops on an engine with MAP sensor? What happens to the computer open and closed loop cycle if the cooling system thermostat is removed? Discuss you findings with the other members of the class and try to figure out how one system affects another.

A coil pack, such as the one shown here, is commonly used on vehicles with distributorless igniton. This particular coil pack is for a 6-cylinder engine. Each coil fires two spark plugs.

16

Ignition System Service

After studying this chapter, you will be able to:

- Inspect, test, and repair ignition systems.
- Describe the purpose of firing order information.
- Adjust ignition timing.
- Remove, test, and replace distributorless ignition components.
- Explain the use of an ignition oscilloscope.
- Remove, test, and replace a distributor assembly.
- Clean, inspect, test, and replace spark plugs.

Technical Terms

Contact-point ignition systems
Electronic ignition systems
Computer-controlled ignition systems
Distributorless ignition system (DIS)
Direct ignition systems
Ignition switch
Ignition resistor
Ignition coils
Available voltage
Flashover
Resistance-type secondary wiring
Crossfiring
Firing order
Number one wire tower
Number one cylinder
Distributor cap
Rotor
Magnetic pickup coil
Hall-effect pickup coil
Crankshaft position sensor
Camshaft position sensor
Air gap
Ignition control modules
ECM
Oscilloscope
Ignition timing
Distributors
Spark plugs
Spark plug size
Spark gap

All ignition systems are designed to accomplish the same functions:

- To change battery voltage into high voltage capable of jumping the spark plug gap.
- To distribute this voltage to the right plug at the right time.

This chapter covers the diagnosis and service of automotive ignition systems. In some cases, you will be referred to other chapters that cover certain aspects of the ignition system in more detail.

Types of Ignition Systems

For many years, all ignition systems used a distributor to transfer high voltage from the ignition coil to the spark plugs. Early vehicles used ***contact-point ignition systems.*** The distributors in these systems contained a set of contact points, as well as centrifugal and vacuum advance units. The contact points controlled the primary current to the coil. The advance units varied spark timing as operating conditions changed.

As vehicles evolved, ***electronic ignition systems,*** which used an electronic pickup and an ignition module instead of the contact points, became common. Eventually, ***computer-controlled ignition systems*** were introduced. These systems used an ECM to control ignition timing, eliminating the need for the centrifugal and vacuum advance units.

The engines in many late-model vehicles do not have a distributor. Instead, they use ***distributorless ignition systems (DIS).*** A distributorless ignition system contains two or more coils that are connected directly to the spark plugs through secondary wires. This arrangement eliminates the need for a distributor. The DIS coils are usually mounted together in a *coil pack.* The ECM or the ignition module triggers the coils to fire the spark plugs based on signals from position sensors mounted on the engine. **Figure 16-1** shows a typical distributorless ignition system. Some distributorless ignition systems use one coil to fire two plugs, **Figure 16-2.** Others use one coil for each plug, **Figure 16-3.**

Some late-model engines use ***direct ignition systems*** or ***coil-on-plug ignition systems.*** These systems do not use secondary (spark plug) wires. Instead, the coil towers are installed directly over the spark plugs, **Figure 16-4.** Spark plug boots prevent arcing and keep water from entering the coil-to-spark plug connection. The direct ignition system operates in the same manner as distributorless ignition systems.

Service procedures vary from one type of ignition system to another. Contact-point systems are subject to wear and require regular maintenance to maintain top engine performance. The points must be checked and adjusted or replaced at regular intervals. In addition, advance units must be checked and secondary wires and spark plugs replaced periodically. In an electronic ignition system, the spark plugs, secondary wires, distributor cap, and rotor (if used) must be replaced periodically, and the ignition timing must be checked at recommended intervals. If the system uses vacuum and centrifugal advance units, these units must be checked, also.

Distributorless ignition systems require relatively little maintenance other than periodic plug and secondary wire replacement. Direct ignition/coil-on-plug ignition systems only require plug replacement at extended intervals.

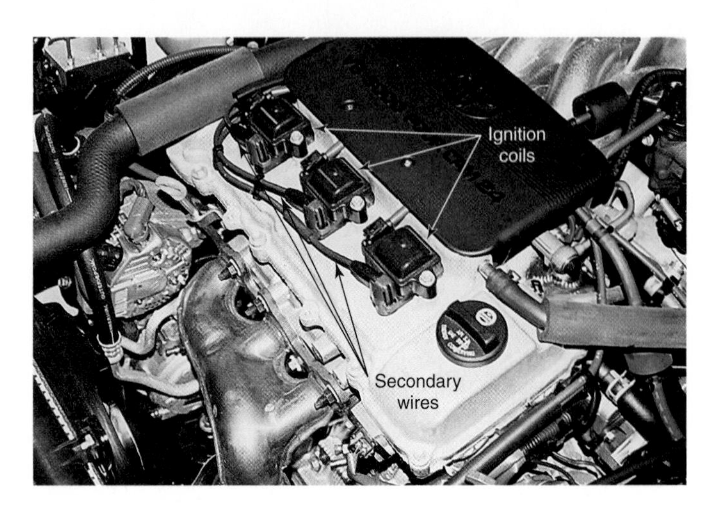

Figure 16-3. This DIS system uses one coil per cylinder.

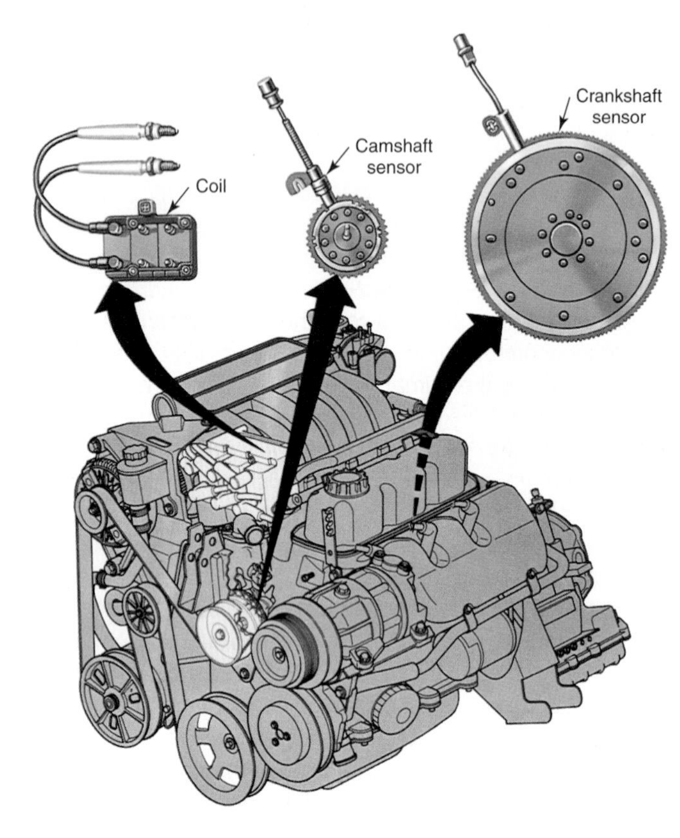

Figure 16-1. Components of one distributorless ignition system. This system uses various sensors to trigger the ignition coils. (DaimlerChrysler)

Figure 16-2. This illustration shows a coil pack used on a common distributorless ignition system. Notice that three coils fire six spark plugs.

Figure 16-4. A—Direct ignition coil assembly. B—Direct ignition coil assemblies installed on an inline 6-cylinder engine. The wiring harness is not shown. (General Motors)

Although the components of modern ignition systems are more reliable than those used in older systems, problems still occur. Remember that secondary (high-voltage) system problems are just as common on new systems as they were on older ignitions.

Ignition System Problem Diagnosis

The first step in ignition system problem diagnosis is to visually inspect the ignition components. Study the primary wires for signs of cracking, burning, and corrosion. All connections must be clean and tight. If the resistance in the *primary circuit* is excessive, the primary voltage may be reduced to a level that can seriously affect the *secondary circuit* available voltage. Check the coil(s) and distributor cap (when used) for signs of flashover, burning, and corrosion. Coil polarity must be correct. Examine the secondary wires for evidence of swelling and deterioration. Wire ends should have undamaged boots and must be inserted fully into or over the correct towers.

On all modern vehicles, the ECM controls the ignition system. After making visual checks, retrieve any trouble codes stored in the ECM's memory. Trouble code retrieval was covered in Chapter 15.

Troubleshooting a No-Start Condition: Spark Test

Note: Remember that a no-start condition can also be caused by a fuel system problem, a defective ECM, or a faulty input sensor. See Chapter 15, Computer System Diagnosis and Repair, for more on ECM and sensor testing.

If an engine will crank but not start, a spark test can be used to help pinpoint the source of the problem. To perform this test on an engine with a distributor, remove the coil secondary wire from the distributor. Attach a spark tester to the secondary wire and ground the tester to the engine, **Figure 16-5.** Crank the engine while watching the tester. An arc should jump between the center and ground electrodes of the tester as the engine is cranked. If this does not occur, check the primary side of the ignition system. Components that may be at fault include the coil, ignition switch, ignition module, ECM, points and condenser (if used), and ignition resistor. Methods for testing these components will be covered later in this chapter.

If a spark is produced at the coil secondary wire, reconnect the wire and remove a secondary wire from a spark plug. Again, attach a spark tester to the spark plug wire, ground the tester, and crank the engine. If there is no spark between the tester electrodes, check the spark plug wire, distributor, and rotor. If a spark is produced, study the spark plugs for signs of fouling and other abnormalities.

When checking a DIS coil for spark, remove the secondary wire from the spark plug and attach the wire to the spark tester. On coils that fire two spark plugs, leave the second wire connected to its plug. If both secondary wires are removed from their plugs on a coil of this type, the resistance

Figure 16-5. Attach the spark tester as shown; then look for a spark as the engine is cranked.

will be too high for the coil to fire, even if it is good. If there is any doubt about the condition of the spark plug that remains connected to the coil, remove the secondary wire from the plug and ground the wire before testing for spark.

Ground the spark tester to the engine. Then crank the engine and check for spark. If a strong spark is produced, reconnect the secondary wire and check for spark at the other secondary wires. If there is no spark at a secondary wire, the coil or the wire itself may be faulty. If there is no spark at two secondary wires served by the same coil, suspect the coil. Procedures for checking secondary wires and ignition coils are presented later in this chapter.

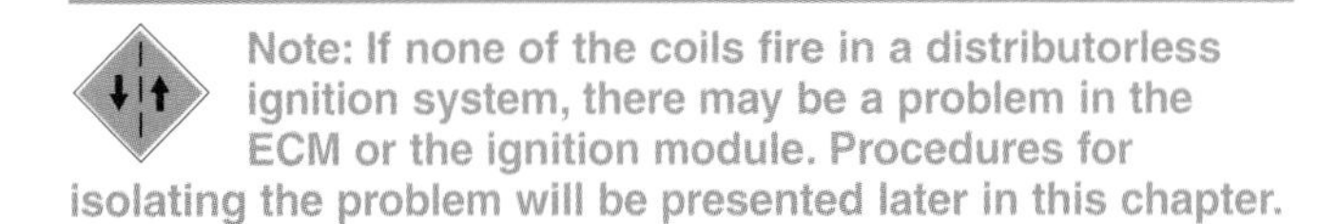
Note: If none of the coils fire in a distributorless ignition system, there may be a problem in the ECM or the ignition module. Procedures for isolating the problem will be presented later in this chapter.

The ignition system can also be checked by attaching an oscilloscope and checking for a secondary pattern as the engine is cranked. Some manufacturers recommend this method instead of checking for a spark by removing secondary wires. Information on using an oscilloscope to diagnose ignition system problems is presented later in this chapter.

Testing, Replacing, and Adjusting Ignition System Components

When an individual part of the ignition system is suspected of being faulty, it should be inspected, tested, adjusted, repaired, or replaced. Always recheck the system after any adjustment or component replacement to make sure the repaired circuit meets specifications.

Ignition switch

The ***ignition switch*** can fail to deliver battery voltage to the ignition system during cranking or when it is in the *on* position. If the engine starts but immediately dies when the ignition switch is returned to the *on* position, the run terminal is defective. If the engine will not crank or cranks but will not start, the start terminal may be defective.

Ignition switches rarely fail, so check all other possibilities before suspecting a bad switch. To check the ignition switch, touch the probe of a nonpowered test light to the battery terminal of the switch. If the light illuminates, the ignition switch is receiving power. If the light does not illuminate, check the wiring between the positive battery terminal and the switch.

If the switch is receiving power, touch the probe of the test light to the *on* terminal (sometimes called the *ign* terminal) with the switch in the *on* position. If the light illuminates, the switch is delivering current to the ignition wire and the problem is in the wiring between the switch and the ignition system or in the system itself. If the light does not illuminate, the switch is defective.

Next, touch the test light probe to the start terminal of the switch as you attempt to crank the engine. If the light illuminates, the switch is good. The problem is farther along in the wiring or in the starting or ignition system. If the light does not illuminate, the problem is in the switch itself.

Ignition Resistor

The ***ignition resistor,*** where used, must provide a specified amount of resistance to the battery-to-coil circuit. An ignition resistor can be a calibrated resistance wire built into the wiring harness or a separate ballast resistor.

Note: Many modern ignition systems operate on full battery voltage at all times and, therefore, do not use an ignition resistor. Consult the proper service manual for exact input voltage specifications.

Since an ignition resistor reduces the amount of electricity in the circuit, the simplest way to check a resistor is to determine whether the voltage at the coil is less than battery voltage. Disconnect the battery wire at the coil and attach a voltmeter. With the ignition switch in the *on* position, voltage should be about 9 volts, assuming the battery is at 12 volts or higher. If the reading exceeds 9 volts, or if there is no voltage at the wire, replace the resistance wire or ballast resistor. Be careful to install a correct service replacement. Recheck voltage after the replacement is installed.

An ohmmeter can be used to check the resistance imparted by the ballast resistor or resistance wire. Compare the reading to the manufacturer's specifications.

Caution: Always use a voltmeter or ohmmeter to test resistors. Never make a test hookup that will connect the resistor directly across the battery. This will overheat and destroy the resistor.

Testing Ignition Coils

Ignition coils should be checked for ***available voltage.*** A simple test for available voltage is to disconnect a plug wire and attach a spark tester. With the engine running, note whether the spark will jump at least 1/4″ (0.6 mm) to ground. Another way to make this test is to attach an oscilloscope as explained later in this chapter. Then disconnect a plug wire and attach a spark tester. When the engine is started, the coil should produce about 20,000 volts on an older ignition system. On newer systems, the coil may produce between 50,000 and 100,000 volts.

If the coil does not produce the specified voltage (or an acceptable spark), check the input voltage to the coil. The coil should receive battery voltage on most newer engines. If the ignition uses a resistor, voltage should be between 9 and 10 volts. If the voltage is below specifications, inspect the wiring and connectors. If input voltage is correct, the coil may be defective.

STOP Warning: Do not remove a plug wire when the engine is running unless specifically recommended by the manufacturer. Removing a plug wire may damage electronic ignition components. Always consult the proper service manual before removing any plug wires.

Other coil checks involve testing both primary and secondary circuit resistance. These tests usually expose internal shorts, grounds, opens, insulation breakdown, and loose or corroded connections. Use reliable test equipment and follow the manufacturer's instructions.

Figure 16-6 shows a conventional coil being tested for both primary and secondary circuit resistance. The ohmmeter readings should be compared to the manufacturer's specifications to determine whether the coil is good. As a general rule, a coil should read about 1.5 ohms across the primary terminals and about 8000–9000 ohms across the secondary (center) tower and either primary terminal. There should be a very high resistance (at least 500,000 ohms) between any terminal and the coil housing.

Figure 16-7 shows typical test connections for a DIS coil pack. To check primary resistance, connect the voltmeter between the battery terminal and the terminal corresponding to the coil in question. To check secondary resistance, connect the ohmmeter to the towers of each coil. Values will vary, depending upon design. Always use specifications that pertain to the exact unit being tested.

Note: Not all the coils in a DIS coil pack will go bad at once. To confirm that a coil is defective, swap it with another coil on the engine. If the problem moves with the coil, the coil is defective. If the problem does not move with the coil, the ignition module, the wires, or the plugs are defective.

Remember that the coil is a vital part of the ignition system. When not in excellent condition, it should be replaced.

Checking the Coil Tower for Corrosion and Flashover

To check for corrosion and flashover, pull the secondary wire from the coil tower. Check the tower for signs of corrosion

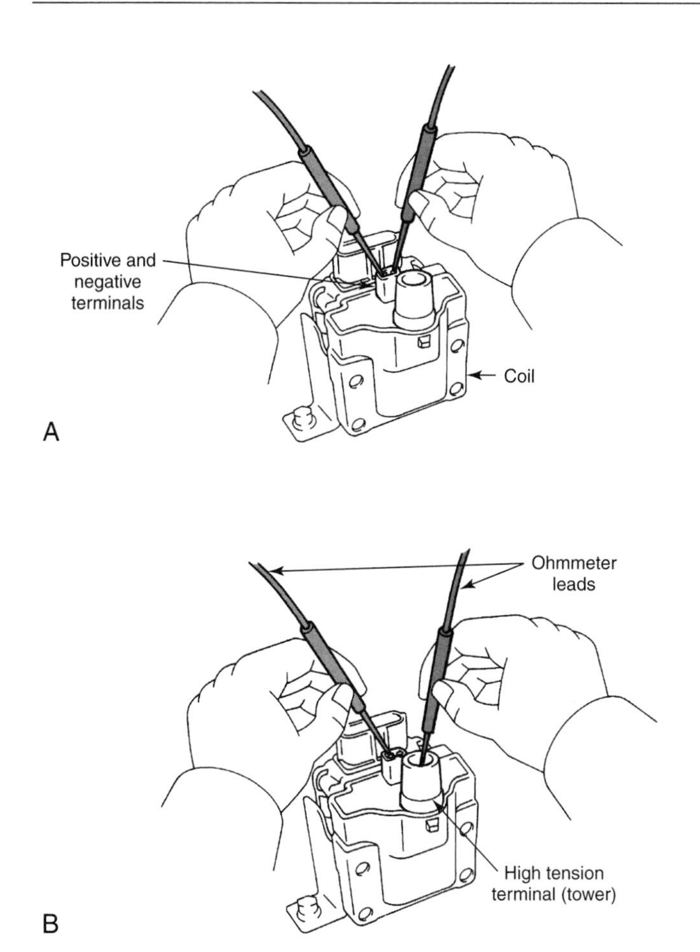

Figure 16-6. Coil secondary and primary resistance test points. A—Primary test connections. B—Secondary coil resistance connections. (Toyota)

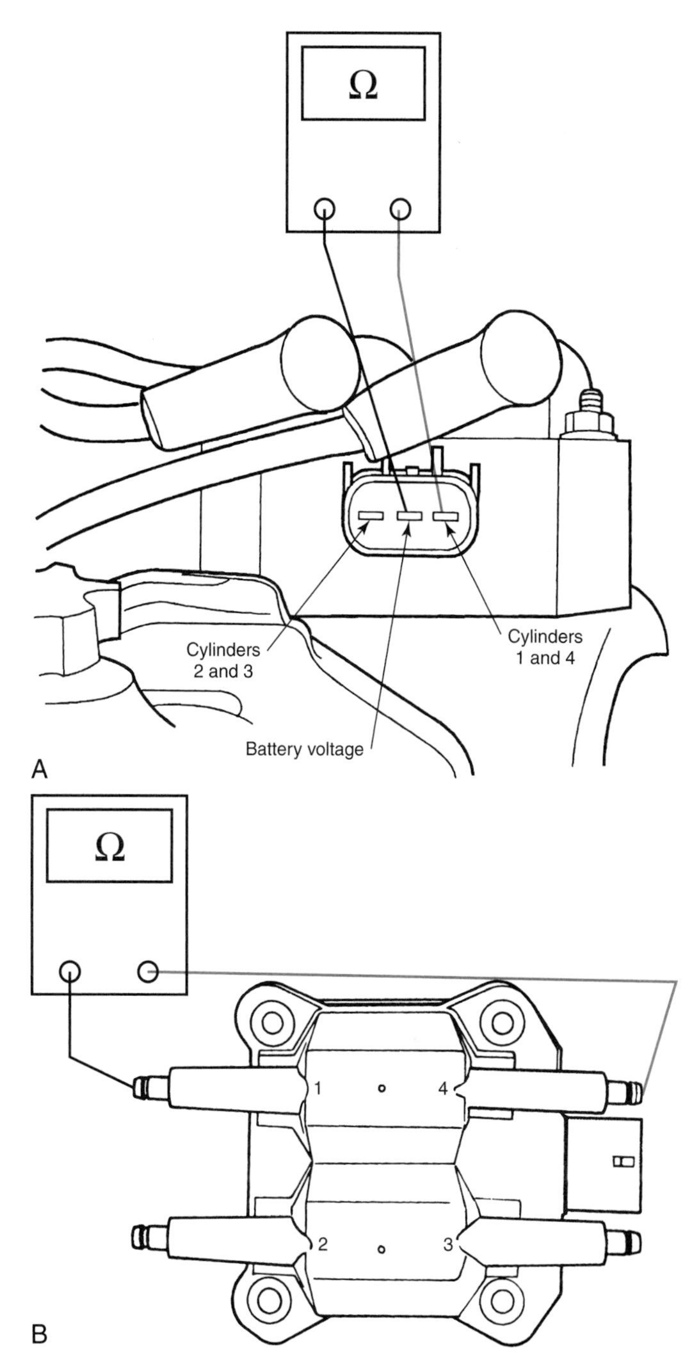

Figure 16-7. DIS coil test connections. A—Checking primary resistance. B—Checking secondary resistance. (DaimlerChrysler)

and burning. Examine the tower for signs of ***flashover*** (high voltage leaving its intended path and leaping down, around, or across the tower and directly to ground). Flashover can be caused by moisture or dirt on the tower surface, a corroded tower, or failure to push the secondary wire fully into the tower.

Continued flashover burns the surface of the tower material, forming a path of carbon tracks. This path promotes additional flashover, and the tower may be severely damaged. Note the carbon tracks (path) left by flashover on the coil tower in **Figure 16-8.** When flashover has cracked the tower or has left a burned path, replace the coil. When replacing a coil damaged by flashover, also replace the boot on the secondary wire. It probably has carbon tracks that will cause flashover with the new coil.

Prevent flashover by having a clean, tight tower wire connection, by having a good boot in place, and by keeping the tower and coil top free of dirt and moisture.

Checking Coil Polarity

The coil must be connected into the primary circuit so the coil polarity marks (+ or -) correspond to those of the battery. If the negative battery post is grounded, the negative terminal on the coil must be grounded. The easiest way to check coil polarity is to visually observe the coil connections. On some older vehicles, the coil connections can be reversed. Many late-model vehicles have molded plastic connectors that cannot be reversed, **Figure 16-9.**

Isolating a No-Spark Problem on a DIS Coil

As mentioned previously, if none of the coils in a DIS coil pack produce spark, the problem is most likely in the ECM or the ignition module. In some cases, the problem can be isolated using a scan tool and test light. With the ignition off, attach the scan tool and remove the primary wire connector at

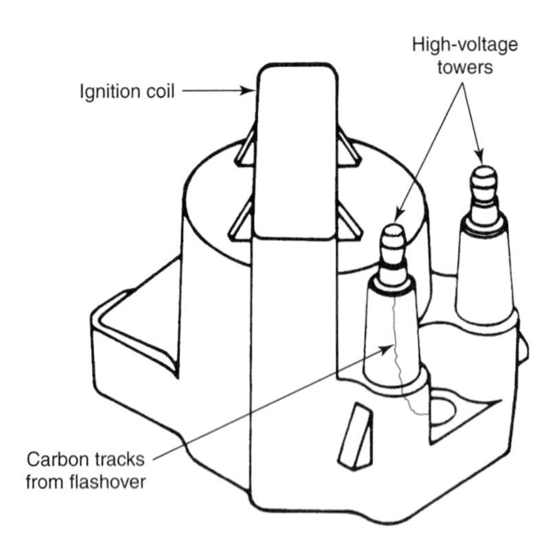

Figure 16-8. Flashover has ruined this coil. Note the carbon tracks. (General Motors)

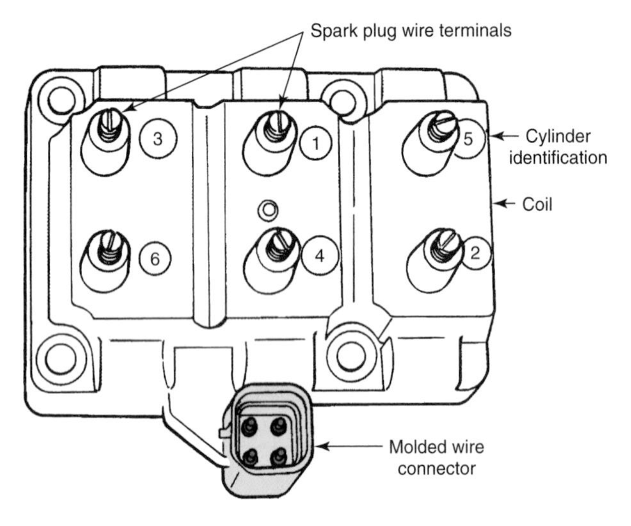

Figure 16-9. This coil pack uses molded plastic connectors for the wiring. Note the cylinder identification number by each spark plug wire tower. (DaimlerChrysler)

the DIS module. Use the service manual to locate the terminal from the ignition module and the ECM speed signal. Ground the test light to the engine block; then set the scan tool to measure engine RPM. Turn the ignition switch to the *on* position and touch the speed signal terminal with the test light probe. The scan tool should briefly indicate RPM when the test light probe is touched and removed from the terminal. If the scan tool shows an RPM signal, the ECM is good and the module may be defective. If the scan tool does not register an RPM reading, the ECM may be defective.

Note: Before condemning the module or ECM, make sure the wiring between the two units is not shorted or cut. Also check that all connectors are properly assembled and show no signs of corrosion or overheating.

Caution: To prevent ECM damage, be sure to probe only the terminal indicated in the service manual.

Checking Secondary Wires

All vehicles use ***resistance-type secondary wiring.*** In the event of ignition trouble, it is often wise to check each wire with an ohmmeter to determine if the resistance is within specified limits. This test will also show if there are breaks in the conductor or poor conductor-to-terminal connections, **Figure 16-10.**

The secondary wire insulation should also be tested to ensure that the voltage is not leaking away before it reaches the plug. One test can be made using a nonpowered test light. Connect the test light clip to a good ground. Then start the engine and move the test light tip over the wires as the engine idles, **Figure 16-11.** If the bulb illuminates at any point, or if an arc is observed jumping from the wire to the probe tip, the insulation is defective.

Secondary wire insulation can also be tested by spraying the wires with a mist of water as the engine idles. If arcing can

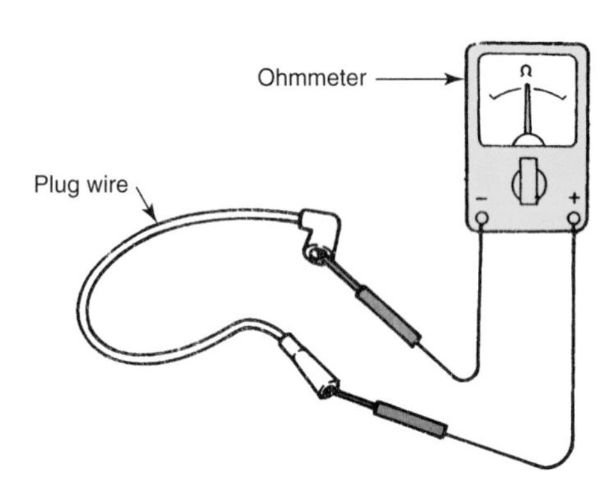

Figure 16-10. Checking plug wire resistance with an ohmmeter. (DaimlerChrysler)

Figure 16-11. A test light placed near a leaking spark plug wire will provide a better ground path than the wire itself, especially if the wire has a high internal resistance.

be seen or heard, or if the engine begins running roughly, the wire insulation has broken down.

Replace wires that are not within specifications or that show signs of faulty insulation. Use wires with specified resistance. Secondary wires can be made up as needed or secured as a factory set. When changing secondary wires, remove one old wire at a time and replace it with a new one. Use new boots. Make sure the wire is snapped firmly on the plug and is fully seated in the distributor or coil tower. Repeat this procedure until the entire set is replaced.

Remember that resistance wire is easily damaged. When removing a wire, grasp the boot (not the wire) and twist it one-half turn to break the seal. Then, pull the boot from the plug or tower. Special gripping tools are available to aid in the removal of old wires and boots, **Figure 16-12.** Never kink or jerk resistance wires. Avoid piercing insulation with test probes.

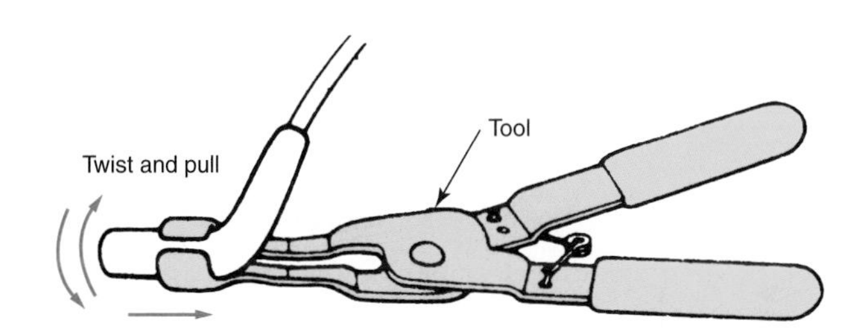

Figure 16-12. A special tool can be used to prevent boot or wire damage when removing the secondary wire from the spark plug. (Ford)

Caution: Some plug wires have distributor cap electrodes that are permanently attached to the wire and lock into the cap. These wires cannot be pulled until the positive locking terminal electrode is compressed from inside the cap.

Arrange Wires to Prevent Problems

When installing secondary wires, be sure to arrange wires in the holders as recommended by the manufacturer. See **Figure 16-13.** Coat the inside of the plug boots with silicone compound, if so suggested. If no directions are available, keep the following points in mind:

- Avoid bunching wires together and running them parallel to each other. Keep them as far apart as practical.
- When two adjacent cylinders (plugs) on the same bank (side) fire in succession, keep their plug wires separated by another wire, **Figure 16-14.** This will help prevent ***crossfiring*** (one wire imparting enough voltage into an adjacent wire to cause it to fire its plug). Crossfiring can cause serious mechanical damage to the engine. Improper wire routing can also cause short circuiting and significant radio interference.
- Route wires so they are not pinched or frayed by other engine parts.
- Support wires properly and keep them away from heat and oil.
- Make sure that all wire ends are clean and firmly in place. All boots and wires must be in excellent condition.

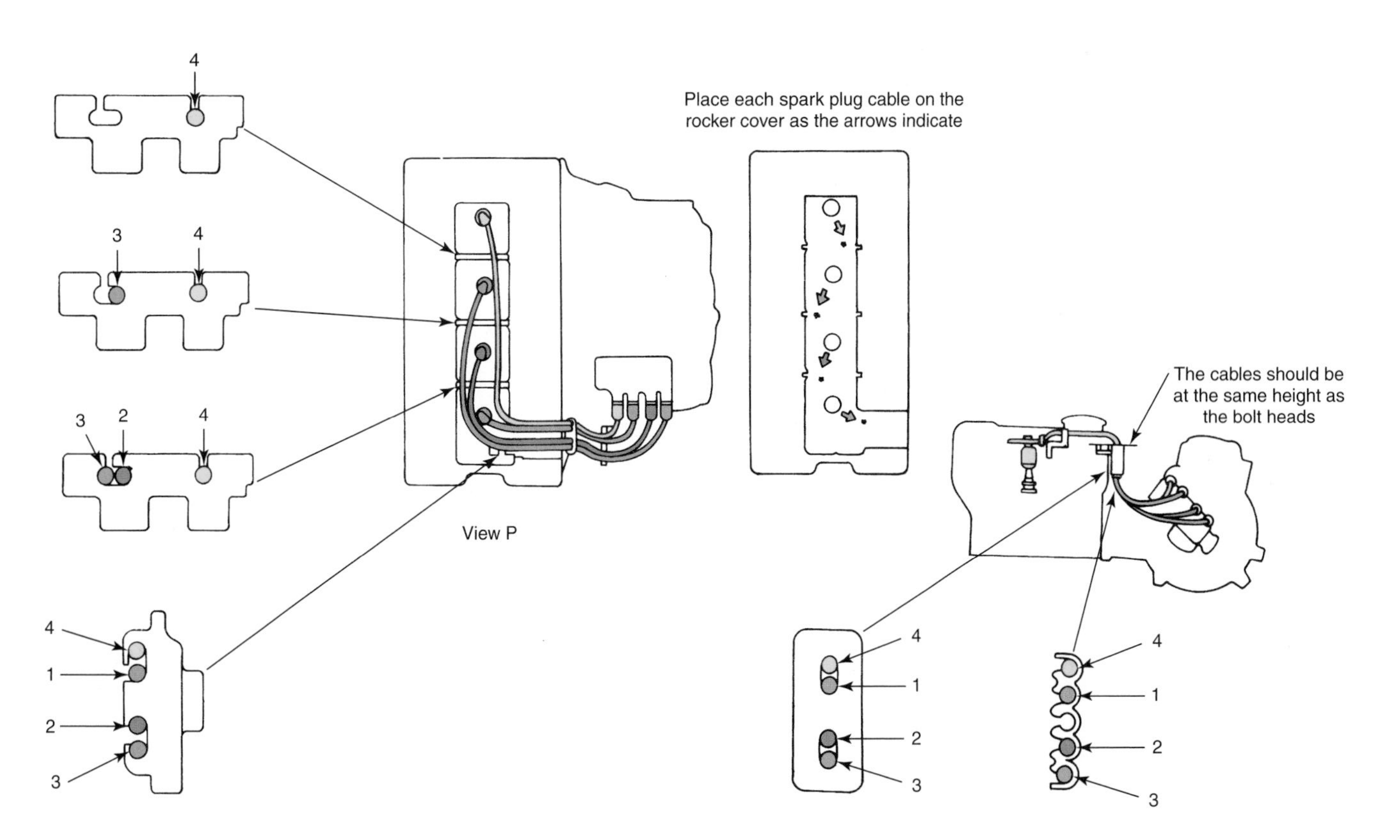

Figure 16-13. Spark plug wires must be routed through the holders properly. (Hyundai)

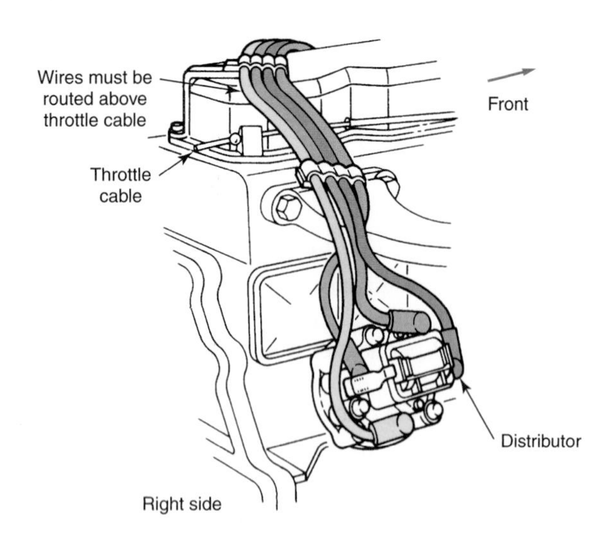

Figure 16-14. To prevent crossfiring, radio interference, grounding, and damage, arrange the spark plug wires as specified by the manufacturer. (DaimlerChrysler)

When replacing plug wires, be sure to maintain the correct firing order. ***Firing order*** is the numerical order in which cylinders fire the fuel charge, starting with the number one cylinder.

Finding the Number One Wire Tower on a Distributor Cap

To find the ***number one wire tower*** on the distributor cap when the manufacturer's shop manual is not available, remove the distributor cap and then remove the spark plug from the ***number one cylinder.*** On inline engines, the front cylinder (nearest the timing cover) is the number one cylinder. On V-type engines, the number one cylinder will be at the front of the engine (timing cover end), but it can be on either cylinder bank. Usually the bank that extends farthest toward the front of the engine contains the number one cylinder. Always check the manufacturer's specifications.

Crank the engine until the number one piston is coming up on the compression stroke. Bump the engine over in small steps until the ignition timing mark is aligned with the pointer. Mark the outside of the distributor housing with chalk directly in line with the front of the rotor tip. Line up the distributor cap and snap it into place, making certain the aligning tang is in place. Because the engine is now positioned to fire the number one cylinder, the rotor (chalk mark) will be aligned with the number one tower. Place the number one plug wire in the number one tower. Insert the remaining wires in the correct firing order, going around the cap in the direction of distributor rotation.

Distributor Cap Inspection

Pull each wire (one at a time) from the distributor cap wire towers. Inspect the towers for signs of corrosion, burning, or flashover. Mild corrosion can be removed with a special wire brush or with sandpaper. Replace the cap if signs of flashover are present.

When reinstalling the wires, make sure the terminals are clean and are pushed fully into the towers. If the distributor cap has been replaced, use new boots when reinstalling the wires.

Remove the ***distributor cap*** by unsnapping the spring arms, removing the attaching screws, or using a screwdriver to press down and turn the latch arms, **Figure 16-15.** Inspect inside of distributor cap for signs of flashover. Check the cap for cracking and check the central rotor contact button for burning or cracking. See **Figures 16-16** and **16-17.** If terminal posts are burned or grooved, replace the cap. Mild scaling caused by sparks leaping from rotor tip to terminal post can be scratched off with a sharp knife. When replacing the cap, also replace the rotor. Make sure the rotor is in place and the cap is properly aligned with distributor housing.

Testing a Rotor

Inspect the ***rotor*** for excessive burning on the tip. Check the contact spring and the resistance rod (if used). **Figure 16-18**

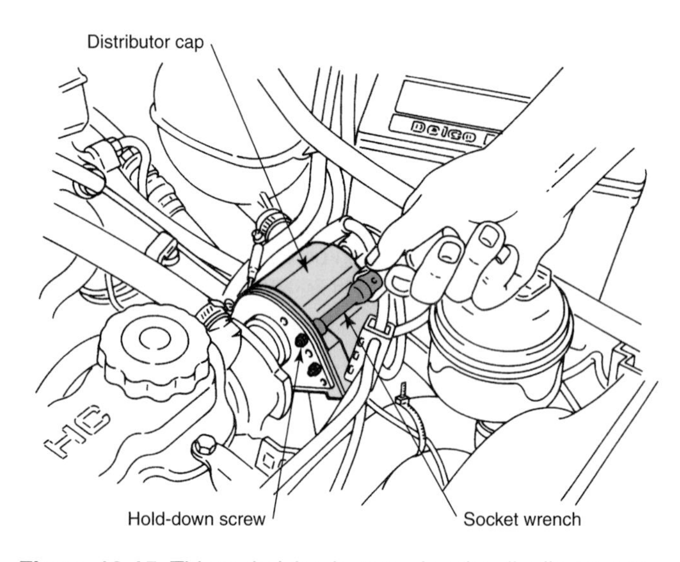

Figure 16-15. This technician is removing the distributor cap hold-down screws with a socket wrench. Some caps are secured with spring clips or spring-loaded latch arms. (General Motors)

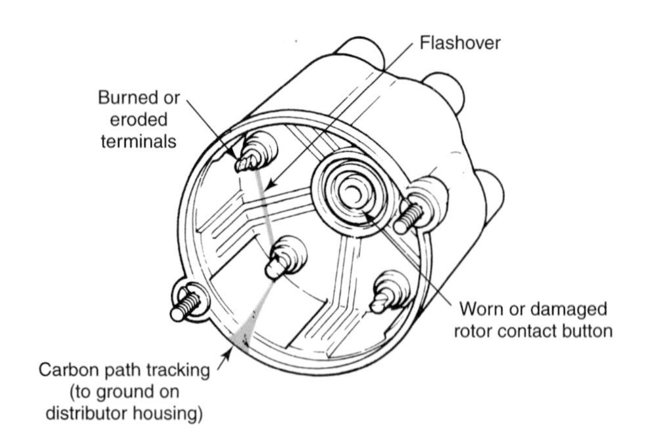

Figure 16-16. Cap cracks, broken towers, and carbon tracking from flashover can cause serious ignition troubles. (DaimlerChrylser)

shows a rotor in good condition. **Figure 16-19** shows a rotor with its tip completely burned off.

Some rotors are removed by simply pulling them straight up from the distributor shaft. Other rotors are attached with screws. When installing a rotor, make sure that it is aligned with the distributor shaft and that it is pressed down fully. Carelessly installing the rotor may damage the rotor and/or the distributor cap. Some manufacturers recommend applying a silicone grease compound to the end of the rotor tip. This helps to suppress radio interference. Apply silicone to the rotor tip only when specified by the manufacturer. Use the recommended compound.

Pickup Coil and Position Sensor Service

Electronic ignition systems use one or several methods to produce and time the output of the coil(s). Many systems use a distributor-mounted trigger mechanism, usually a ***magnetic pickup coil*** or a ***Hall-effect pickup coil.*** Some systems use an optical, or photoelectric, sensor in the distributor. Other systems use a ***crankshaft position sensor*** or a ***camshaft position sensor***. All electronic systems have an ignition module to interpret the signal from the triggering mechanism and to operate the ignition coil(s). The module may be located in the ECM.

The trigger wheel in magnetic pickups and the shutter blades in Hall-effect pickups should be checked for damage, **Figure 16-20.** If the system uses a remotely located crankshaft position sensor, such as the one in **Figure 16-21,** the sensor can often be checked with an ohmmeter. Pickup coils should be inspected for external damage. On some systems, the air gap between the trigger wheel and the pickup coil can be adjusted. If an optical distributor system is used, make sure the slots in the rotor disk are unobstructed.

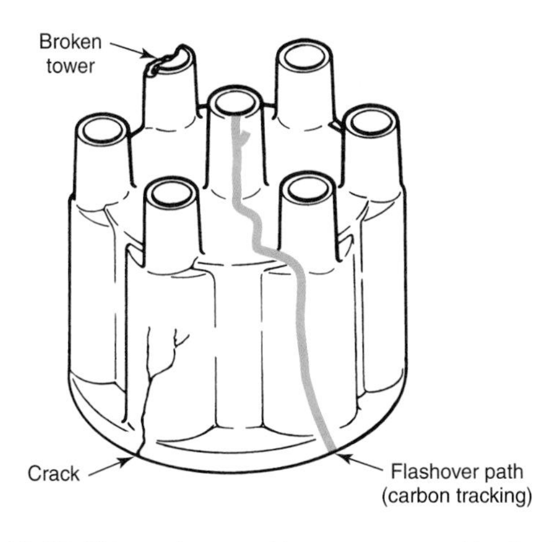

Figure 16-17. This carbon tracking was caused by flashover from the distributor's central coil terminal traveling down one side of the cap to ground on distributor housing. Note other cap damage. (DaimlerChrysler)

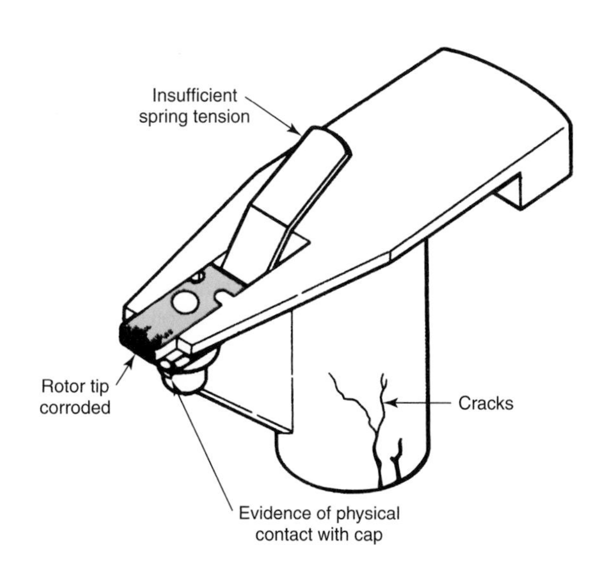

Figure 16-19. The tip has been burned off this rotor. Note other damage. This rotor must be replaced. (DaimlerChrysler)

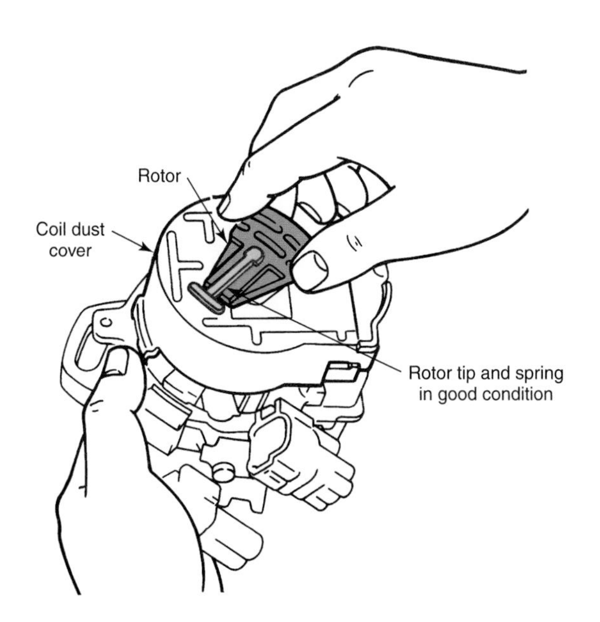

Figure 16-18. Check the rotor for burning, cracking, a broken spring, etc. (General Motors)

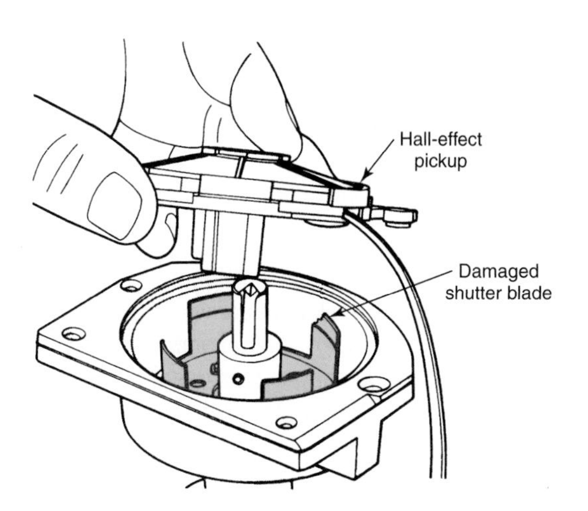

Figure 16-20. Pickup assemblies should be inspected for damage. Note the broken shutter blade on this Hall-effect assembly. (DaimlerChrysler)

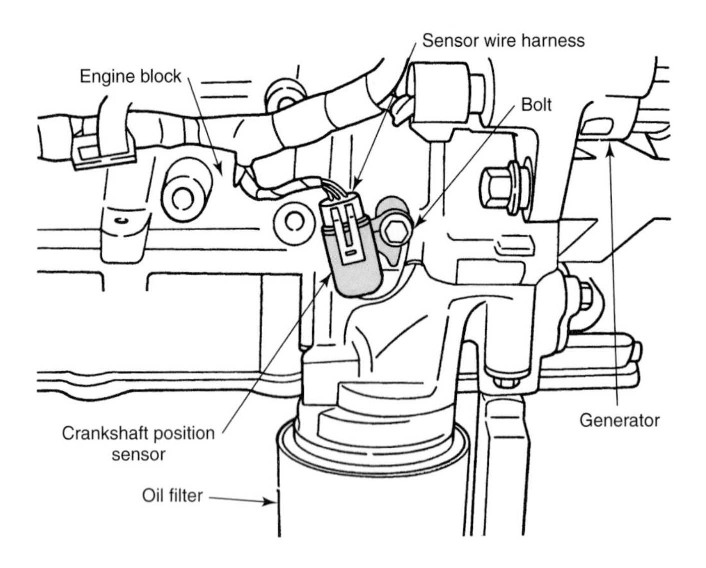

Figure 16-21. A remotely located crankshaft position sensor. (DaimlerChrysler)

Methods for checking magnetic pickups, Hall-effect pickups, and optical sensors vary from manufacturer to manufacturer. Always consult an appropriate service manual before testing these units.

An ohmmeter can be used to check the condition of most magnetic pickup coils and Hall-effect pickups. The pickups do not have to be removed from the distributor to make this test. Connect the ohmmeter leads to the pickup coil leads and read the resistance. Compare the reading to the manufacturer's specifications. Resistance specifications are usually between 300 and 1500 ohms. With the ohmmeter leads still connected, wiggle the pickup coil leads. The reading should not change. If the pickup coil fails either of these tests, replace it. If the coil passes these tests, crank the engine (or turn the distributor by hand if it is removed from the engine). The ohmmeter reading should jump slightly. The jump is caused by a small amount of voltage passing through the ohmmeter circuits and indicates that the pickup coil is producing a voltage signal.

Some electronic ignition distributors contain an additional pickup for a speed sensor used in conjunction with the ECM or cruise control. Make sure you are checking the correct pickup. Many distributor components can be replaced without removing the distributor from the engine. Always refer to the manufacturer's instructions before replacing any part.

Adjusting the Air Gap

On some electronic ignition distributors, it is possible to adjust the ***air gap*** between the trigger wheel and the magnetic pickup. This is done using a nonmagnetic (usually brass) feeler gauge. If a steel feeler gauge is used, the magnet in the pickup will attract the gauge, resulting in drag and a false reading.

To adjust the air gap, loosen the screw holding the pickup coil to the distributor base plate. Then, bump the engine until one tooth of the trigger wheel is directly across from the pickup coil. Using the correct thickness feeler gauge, move the pickup assembly until a light drag is felt on the gauge, **Figure 16-22.** Tighten the pickup screw and recheck the clearance. Finally, crank the engine and recheck the clearance on several teeth.

Note: Refer to Chapter 15 for information on checking and adjusting the gap on crankshaft and camshaft positions sensors.

Control Module Service

Many ***ignition control modules*** can be tested with a voltmeter or an ohmmeter. Some ignition control modules must be checked with special equipment, while substituting a known good unit for the suspect module is the only way to check others.

Testing procedures vary from manufacturer to manufacturer. Always consult an appropriate service manual before testing the module. Improper equipment or incorrect connections can destroy the unit. Always make sure the module is mounted tightly. A tight connection helps the ignition module dissipate heat. Excessive heat can cause serious damage to the unit's internal components.

Note: Some computer-controlled ignition systems do not have a separate ignition module.

To aid in locating intermittent problems, many technicians heat the ignition control module as the engine operates. This can be done with a heat lamp or a heat gun. Heat the unit only until it reaches its normal-operating temperature. Excessive heat can destroy the ignition module's internal

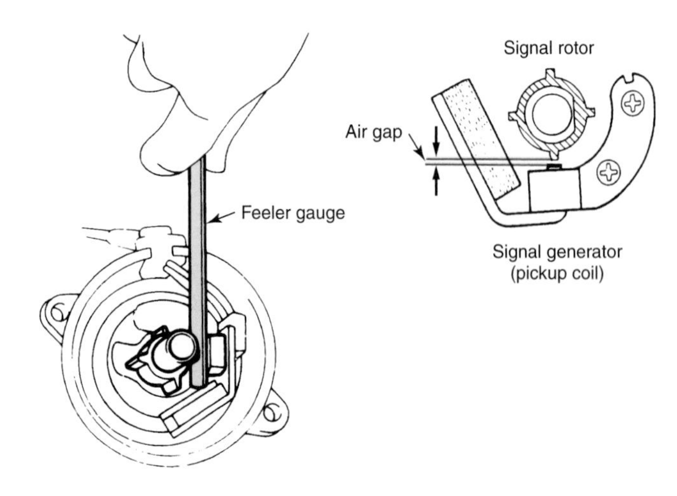

Figure 16-22. Adjusting the air gap on one particular electronic ignition distributor. Note the use of the brass (nonmagnetic) feeler gauge. (Toyota)

components. In some cases, tapping the module with a screwdriver handle as the engine operates can reveal a problem. If the engine begins running rough as the module is tapped, a loose internal connection may be causing the problem.

ECM and Related Components

As mentioned previously, the ***ECM*** controls the ignition system on all late-model vehicles. For more information on diagnosing and repairing the ECM and related components, refer to Chapter 15, Computer System Diagnosis and Repair.

Checking the Ignition System with an Oscilloscope

An ***oscilloscope*** produces a visual pattern, or waveform, on a screen (much like a TV set). See **Figure 16-23.** When attached to the vehicle's ignition system, the oscilloscope will produce a visual record of what is happening in the system. The oscilloscope is often used to diagnose problems in both the primary and secondary sides of the ignition system. Properly used, it will provide fast and accurate information on all ignition system components.

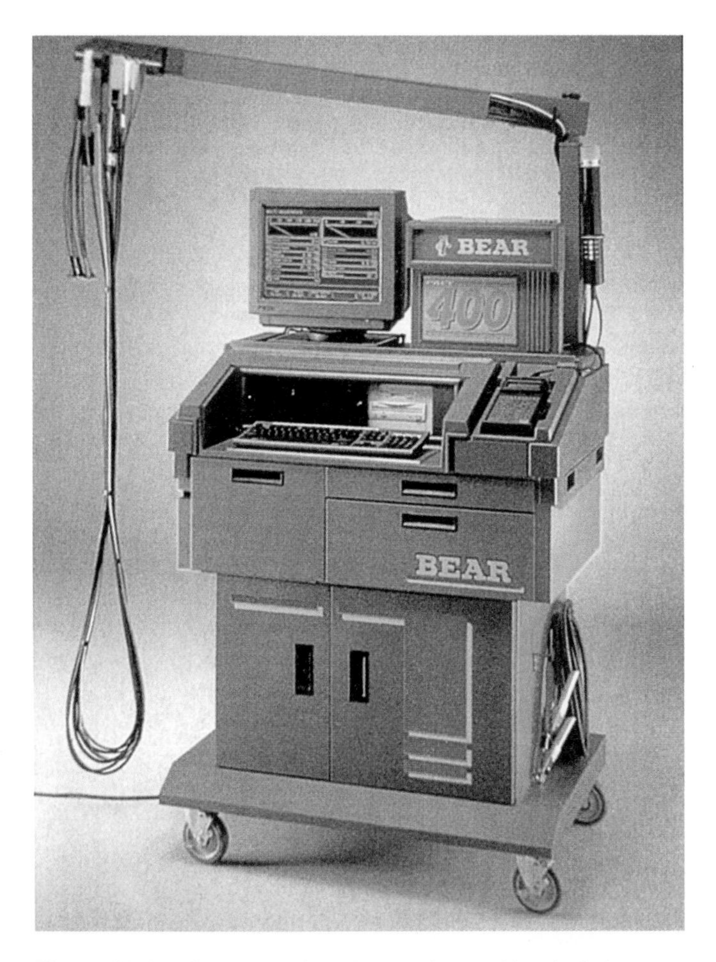

Figure 16-23. One type of engine analyzer with a built-in oscilloscope, computer and screen, and various engine connectors. Note the movable arm, which swings out toward the engine. (Bear)

Secondary Pattern

An ignition system in good order will create a normal waveform on the oscilloscope screen. **Figure 16-24** shows the normal secondary pattern for an ignition system. Point-type and electronic systems produce similar patterns.

Pattern-Firing Section

Refer to Figure 16-24. At point A, the power transistor shuts off (electronic ignition) or the points open (point-type ignition). Primary current stops flowing in the coil, creating a secondary voltage that fires the plug (firing line). The amount of voltage required to fire the plug is indicated by the height of the firing line at B.

As soon as the plug fires, required voltage drops to C, where it remains fairly constant (with the spark still jumping the plug gap) until point D (where spark stops).

Pattern-Intermediate Section

The spark goes out at D in Figure 16-24. Starting at D, unused energy bounces between the coil and the electronic module or condenser, and is dissipated (used up) in a series of gradually reduced oscillations until point E is reached.

Pattern-Dwell Section

At E, the power transistor turns on (electronic ignition) or the points close (conventional ignition). Primary current begins flowing through the coil, producing a series of small oscillations. Power flows through the coil from E to F. This is called the dwell period.

Pattern Presentations

By means of adjustments, it is possible to view the pattern as produced by a single cylinder, **Figure 16-25;** as cylinder patterns superimposed on each other, **Figure 16-26;** or as all cylinders displayed on the screen at once but as separate patterns (parade), **Figure 16-27.**

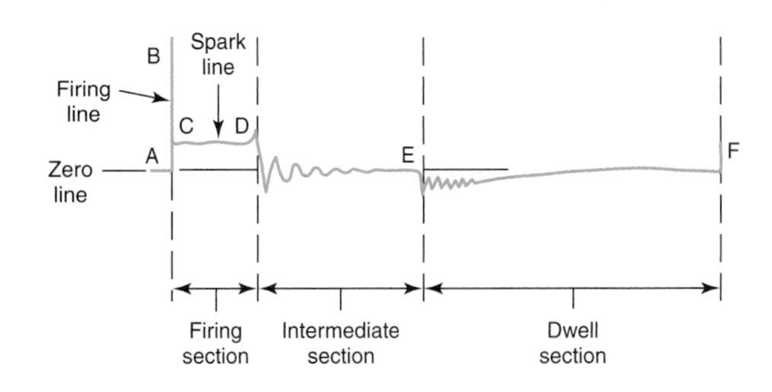

Figure 16-24. Normal oscilloscope pattern (system OK).

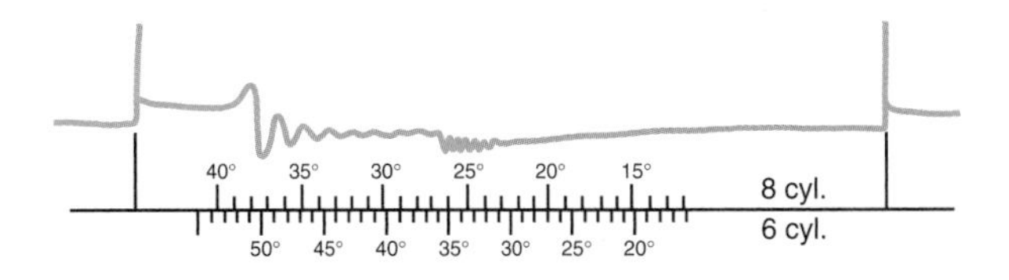

Figure 16-25. Single cylinder oscilloscope pattern presentation.

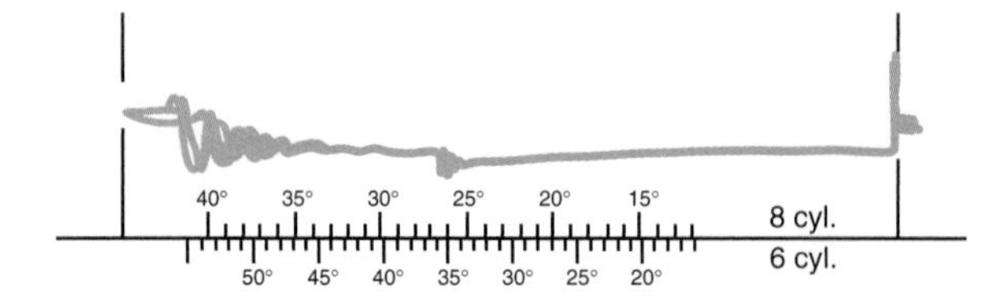

Figure 16-26. Superimposed oscilloscope pattern presentation.

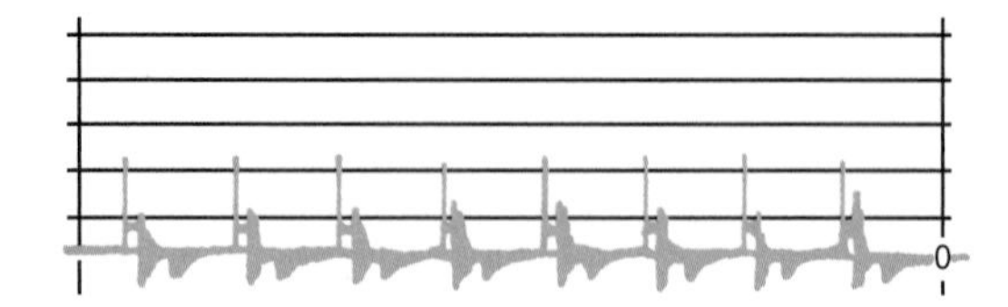

Figure 16-27. Oscilloscope parade pattern.

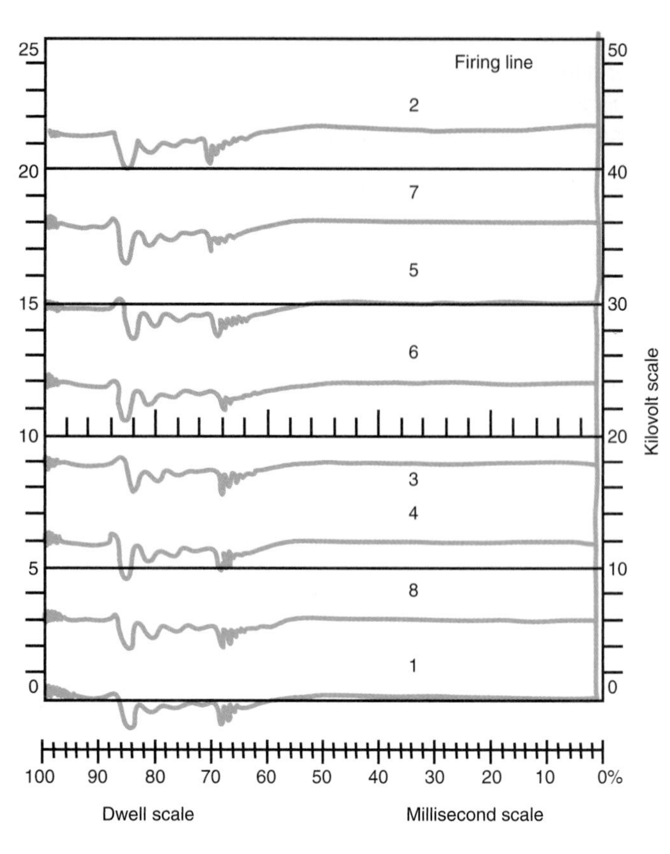

Figure 16-28. Oscilloscope raster pattern. Cylinder patterns are displayed in the order of firing. (Sun Electric)

A raster pattern displays the individual patterns formed by each cylinder. These individual patterns are arranged in firing sequence, one above the other. This allows easy comparison for uniformity. **Figure 16-28** shows a secondary raster pattern.

Primary Pattern

The primary pattern resembles the secondary pattern, except for the extra oscillations in the firing section and the absence of oscillations at the beginning of the dwell section. See **Figure 16-29.**

Typical Oscilloscope Patterns and Their Meanings

Anything that does not look like the normal pattern is evidence of an ignition system problem. Pattern irregularities provide clues as to what unit or units in the system are faulty and to what extent.

Accurate analysis of the various pattern irregularities requires that the technician be familiar with the scope being used. The technician must also be experienced in interpreting the patterns themselves. To become thoroughly skilled in the use of an oscilloscope as a diagnostic tool, be sure to practice whenever possible. Once the tool is understood and pattern

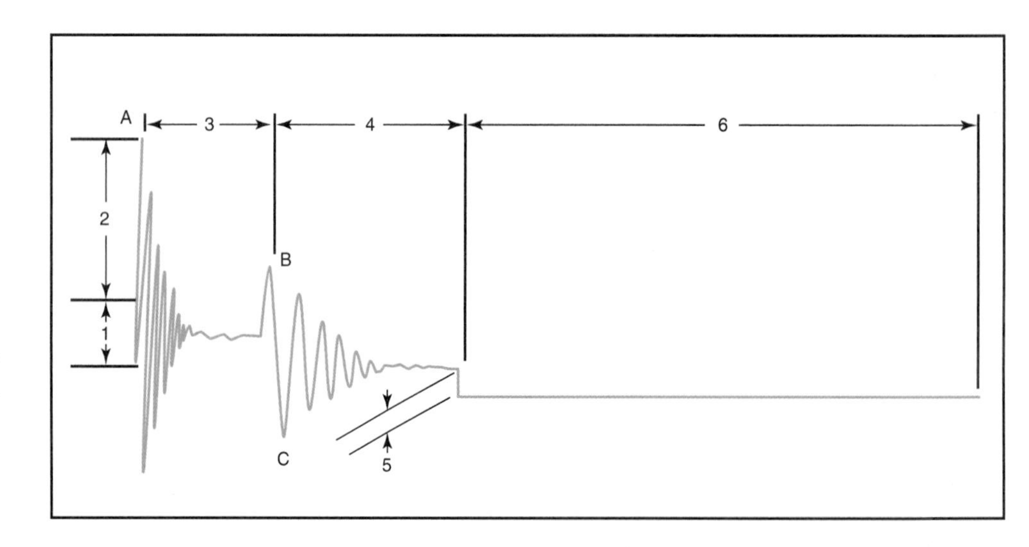

Figure 16-29. The primary pattern was important when diagnosing contact point ignition systems. It is now used to double-check the secondary waveform.

irregularity meaning is clear, the technician will find this a very fast and efficient aid.

Figure 16-30 illustrates 10 different patterns. Each has specific irregularities that indicate a certain problem:

A— Normal pattern. The firing voltages are uniform but too high. Generally caused by worn electrodes, a lean fuel mixture, or improperly gapped plugs.

B— Pattern normal but inverted. Caused by wrong coil polarity.

C— Normal pattern but firing voltages are uneven. Often caused by defective plug wires, defective plugs, a fuel mixture imbalance, or a worn distributor cap.

D— High resistance affecting all cylinders. Generally caused by high resistance in the coil tower, coil wire, or rotor.

E— One plug line is much higher than the others. Caused by a defective or disconnected plug wire.

F— Short spark line, usually accompanied by a high firing line. Caused by a defective or disconnected plug wire.

G— Long spark line, usually accompanied by a low firing line. Caused by a fouled plug, a carbon track in the distributor or coil, or a grounded plug wire.

H— Spark line drops. Caused by high resistance in the wires or terminal ends.

I— Spark line rises. Caused by low compression or a lean mixture.

J— Lack of oscillations. Caused by shorted coil windings.

These patterns represent only a few of the possible pattern irregularities. The manual supplied by the scope manufacturer should be studied carefully.

Checking DIS and Direct Ignition Systems with an Oscilloscope

Oscilloscopes can display DIS and direct ignition system patterns. To read the DIS pattern, the oscilloscope has leads that are attached to each plug wire. Since direct systems have no plug wires, connections must be made to the primary side of the ignition system. If the scope was originally designed for distributor ignitions, an adapter must be used to sort out the

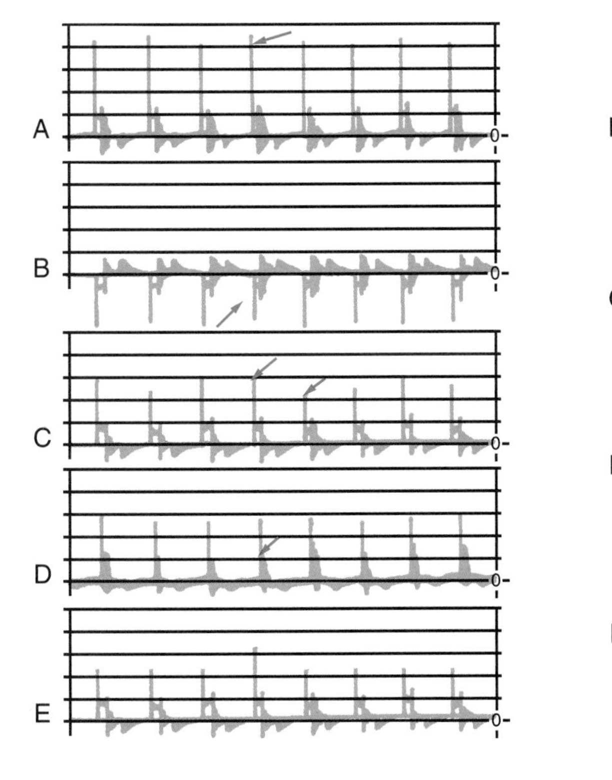

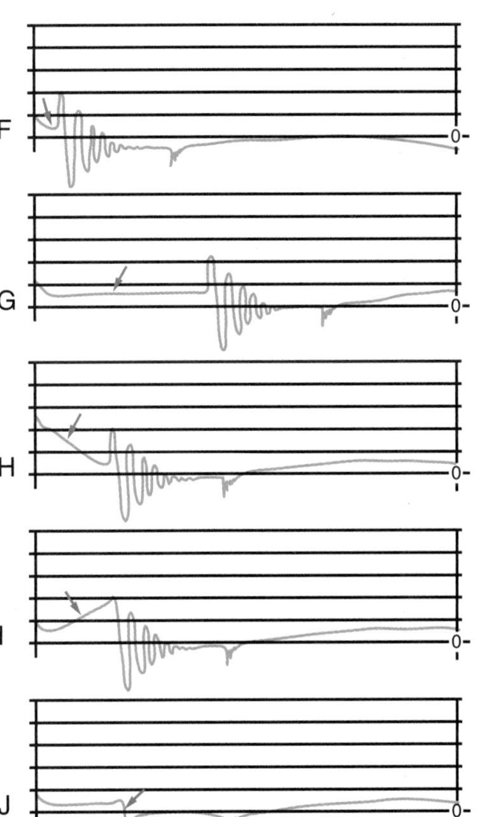

Figure 16-30. Several pattern irregularities and their meanings. A—Firing voltage too high in all cylinders. B—Inverted pattern. Incorrect coil polarity. C—Firing voltages uneven. D—High resistance in all cylinders. E—One firing line is much higher than the others. This is caused by a defective plug wire. F—A short spark line indicates high resistance, usually from a defective wire. G—A long spark line usually indicates a fouled plug or carbon track. H—When the spark line slopes downward, there is high resistance somewhere in the secondary circuit. I—When the spark line slopes upward, look for low compression or a lean air-fuel ratio. J—A lack of oscillations in the pattern indicates a shorted coil winding.

patterns developed by the individual coils. Newer oscilloscopes are able to read the direction of current flow in each wire. They use the current direction information to determine the firing order and which coils fire which plugs. Most modern scopes can also measure the voltage and amperage of individual coils.

Oscilloscope setup for DIS and direct systems is similar to that for a system with a distributor. To check a DIS ignition, make the primary and secondary wire connections. Once the connections are made, set the oscilloscope controls to the DIS settings. Then start the engine and observe the secondary pattern. The scope pattern will resemble the pattern produced by a distributor ignition. Look for high or low firing lines that indicate fouled plugs or faulty secondary wires. As with a distributor system, a lack of oscillations often indicates a shorted coil winding.

Before deciding that the system is OK, check the DIS primary pattern. The primary pattern may show problems that are not visible in the secondary pattern. On a direct, only the primary pattern may be displayed. **Figure 16-31** shows the voltage and amperage flows in a DIS coil during two firing cycles. Comparing this pattern to a known good pattern enables the technician to locate a problem in the coil, module, or primary side connections.

One advantage of testing a DIS or direct system with multiple coils is that the operation of the coils can be compared to determine if one of them is not working properly. Compare the amperage draws for the V8 engine in **Figure 16-32.** All the coils are drawing about the same amount of current, indicating that all are performing at the same level. If one coil draws much more or less current than the others, there is a problem. The problem could be in the coil itself or in the associated plugs or wires. Sometimes the ignition module driver for the coil in question has failed, but this is rare.

Contact Point Service

The ***contact point set*** used on older vehicles should be replaced whenever the ignition system is serviced. Note the location and positioning of the primary lead wire and condenser terminals before removing the old points. It may not be necessary to completely remove the attaching screws to remove the points. Many points will slide out of the breaker plate when the screws are loosened. Remove the old point set, clean the point mounting area, and install the new set.

Note: In an emergency, the used points can be cleaned by lightly filing them with a clean, fine-cut point file. When you are finished filing, pull a clean piece of smooth paper through the points to remove any metal particles.

Setting Contact Points

Point gap (the space between the points when they are fully opened by the cam lobe) is critical. When the gap is too small, the points will arc and burn. Excessive gap will cause missing at high speeds. To adjust the points, turn the distributor

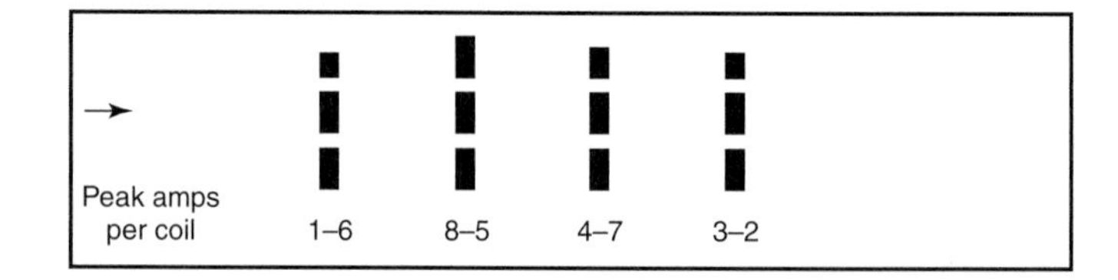

Figure 16-32. The amperage draw of each coil in a DIS system can be observed to determine whether all coils are operating properly.

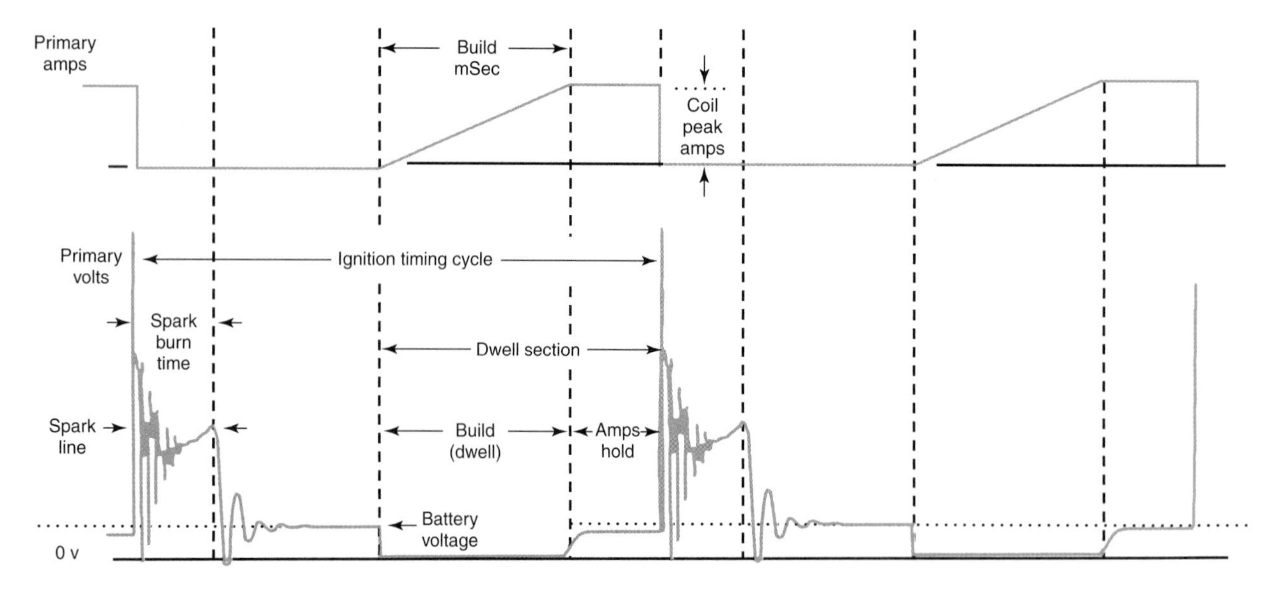

Figure 16-31. This illustration shows the primary pattern of a DIS system. Observing the primary pattern will often reveal problems that are not evident on the secondary pattern.

cam until the contact arm rubbing block is on the highest tip of one of the lobes. Look at **Figure 16-33.** Loosen the attaching screw to allow the point set to be moved. Move the point set until the gap is at the specified dimension, as measured by a feeler gauge. Retighten the attaching screws and recheck the gap. Always recheck the point setting with a dwell meter, which will be discussed later in this section.

Ignition Condenser

The condenser is usually changed with the points. To change the condenser, remove the electrical lead and attaching screw, and then remove the condenser. Install the new condenser and attach the electrical lead to the points.

Lubricate Distributor Cam

When new points are installed, the distributor cam should be cleaned and given a thin coat of high-temperature grease. Do not use engine oil or low-temperature grease. They will be thrown outward into the points. Some point sets include cam lubricant.

Setting Dwell Angle

To set the ***dwell angle,*** attach a dwell meter to the coil. Always follow the test instrument manufacturer's directions. While idling the engine, check the dwell. Adjust point gap as required. The points on some vehicles are adjusted by turning the adjusting screw as the engine idles. Others require that the distributor cap be removed to make the adjustment. If point adjustment fails to give the specified dwell, the distributor cam may be worn or the incorrect point set may be installed on the engine.

Run the engine up to around 1500 RPM. The dwell should not vary more than 6 degrees. Variations in excess of this indicate a worn breaker plate or a faulty distributor shaft. **Figure 16-34** illustrates how a tach-dwell (combination tachometer and dwell meter) is set up to check dwell.

Dwell on Electronic Ignition Systems

The dwell period in electronic ignition systems is controlled electronically, and no adjustment can be made. On many electronic systems, dwell is automatically increased as engine RPM rises and decreases with a speed reduction. This provides the required spark intensity without overheating the coil.

Setting Ignition Timing

On many modern vehicles, the ECM determines ***ignition timing*** based on inputs from various sensors. On these systems, timing is not adjustable. It is still possible to check and, if needed, adjust the timing on most engines that use distributors.

On old contact point distributors, timing would gradually retard as the points wore, increasing the dwell. On electronic distributor engines, the initial timing does not vary unless the distributor body is moved, the distributor shaft bushings become worn, or the advance mechanisms fail (if used).

Checking Ignition Timing

To begin checking timing, perform any recommended preliminary steps. These steps are usually listed on the emissions sticker in the engine compartment, **Figure 16-35.** This could include removing the vacuum advance line and plugging the vacuum fitting, disconnecting the ignition module from the ECM, or placing the ECM in the test or self-diagnostic mode. If the ignition system has contact points, the dwell must be set before the timing is adjusted. Warm up the engine and shut it off. Attach a power (stroboscopic) timing light and tachometer as directed by the manufacturer, **Figure 16-36.** The ideal timing light will have provisions for measuring advance (in degrees) and engine speed. Connect the timing light to the number one cylinder using either a jumper wire from the plug to the plug wire or an inductive-type pickup, **Figure 16-37.** Under no circumstances should the plug wires be pierced with a probe. If the engine has provisions for a magnetic probe, **Figure 16-38,** a "monolithic" (electronic timing) light may be used.

Clean off the timing scale and index mark. Paint or chalk the index mark to make it more legible. See **Figure 16-39.** Some engines have a pointer on the block with the scale stamped on the vibration damper. A few engines have a scale stamped on the flywheel. It is visible through an access hole in the bell housing. Note the timing pointer.

Check the underhood area to make certain all wires and tools are free of the fan and the exhaust system. Then, start

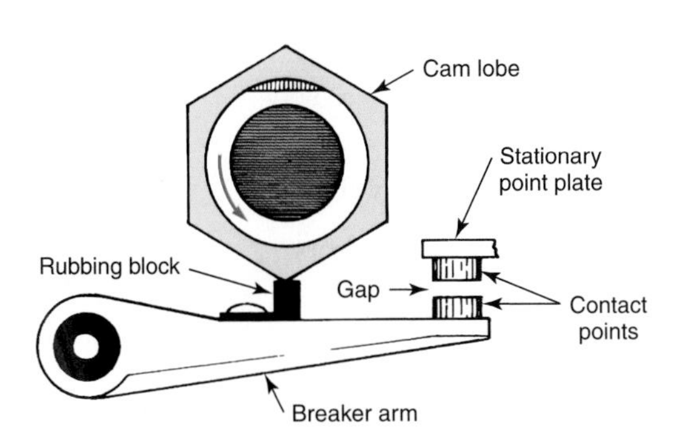

Figure 16-33. Turn the distributor cam until the contact arm rubbing block bears against the highest portion of the cam lobe.

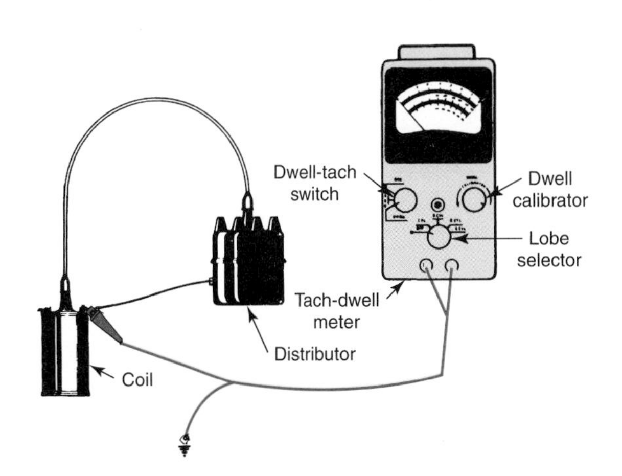

Figure 16-34. Checking dwell angle with a tach-dwell meter. (Sun Electric)

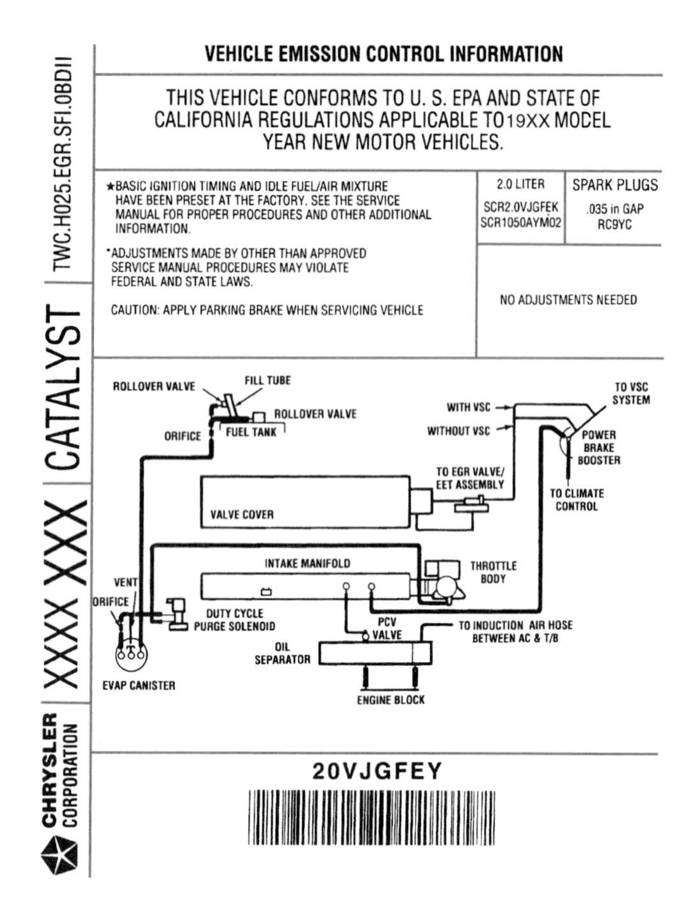

Figure 16-35. One particular vehicle emission control information (VECI) label. If the label is missing, obtain a replacement and follow the recommended procedures and adjustments. (DaimlerChrysler)

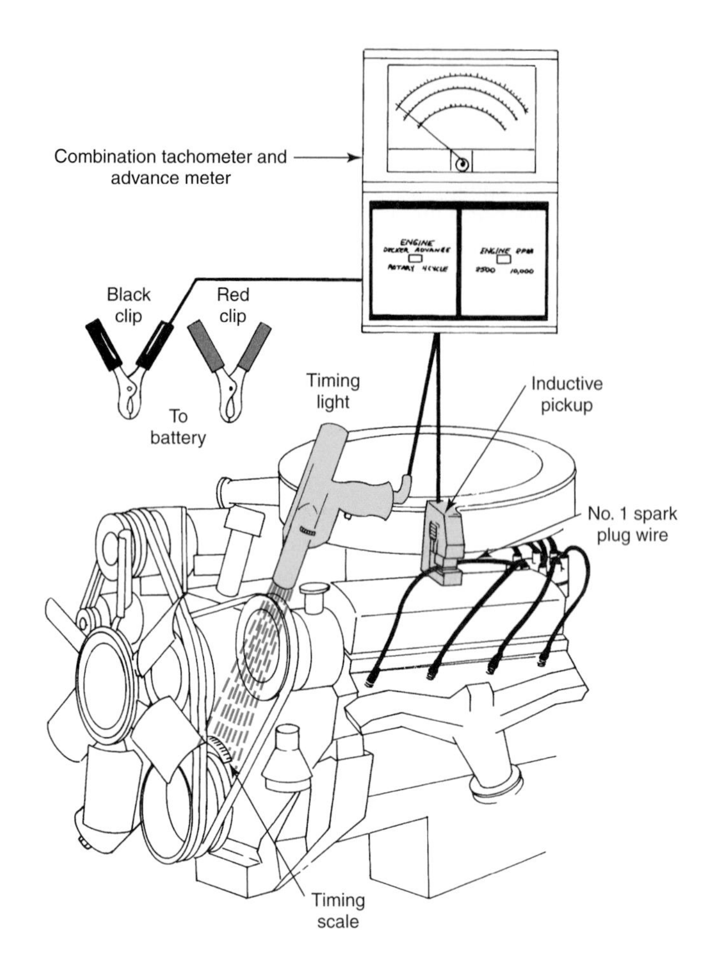

Figure 16-36. Using a timing light to set ignition timing. Note the use of the inductive pickup. The battery hookups are not shown. (Snap-on Tools)

the engine. Operate the engine at idle to avoid bringing the centrifugal advance into play. Direct the timing light beam on the timing marks.

The timing light will make the index mark appear to stand still. Turn the timing light control dial until the index mark lines up with the TDC (zero degrees) mark on the scale. Then read the amount of advance or retard on the timing light's advance meter. If the timing light does not have an advance meter, note the position of the index mark in relation to the marks on the scale to determine timing. Compare your finding to specifications.

Warning: Do not stand in line with the fan. Keep wires and fingers away from the moving fan and belts.

Setting Ignition Timing

If timing is not within specifications on a vehicle with a distributor, loosen the distributor clamp bolts just enough to allow the distributor to be turned without undue force. Look at **Figure 16-40.**

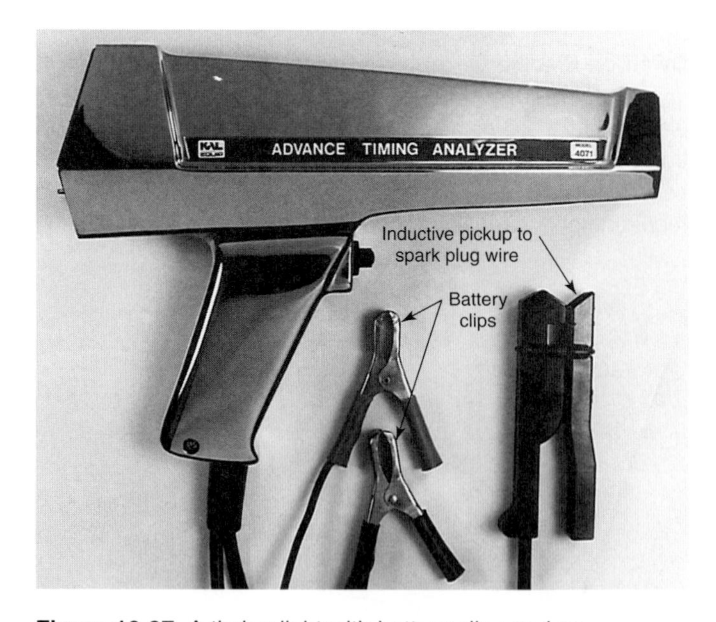

Figure 16-37. A timing light with battery clips and an inductive-type spark plug wire pickup that does not pierce the wire insulation. (Ken Tool)

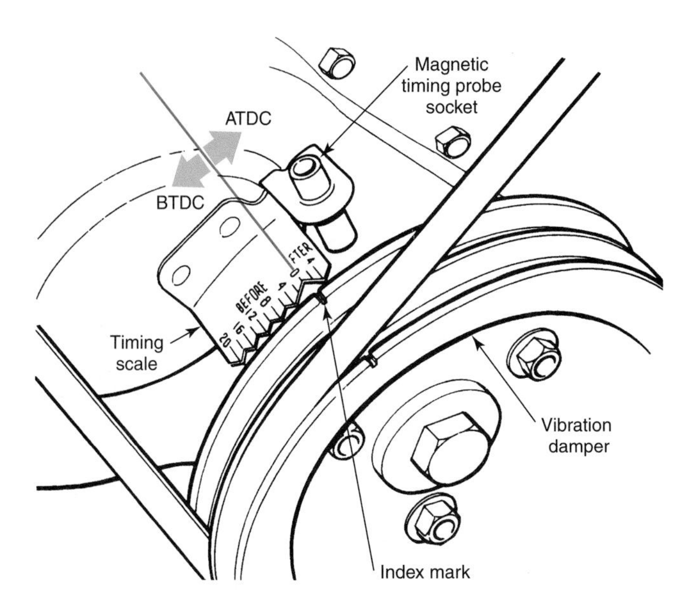

Figure 16-38. This setup has a magnetic timing probe socket to allow the use of electronic timing equipment. (DaimlerChrysler)

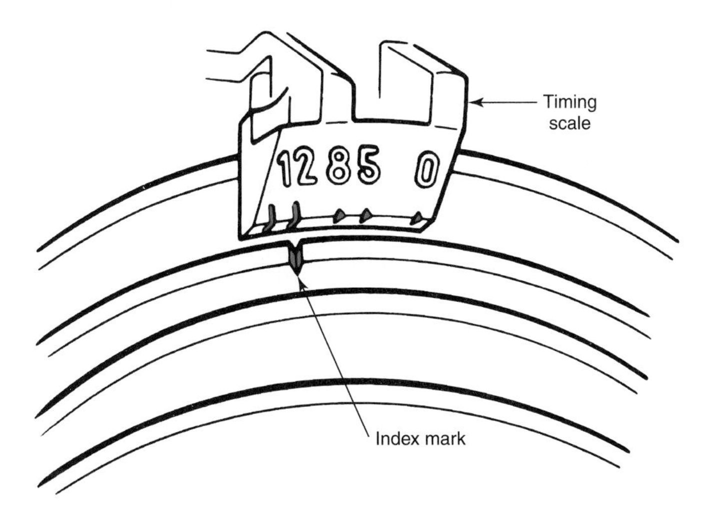

Figure 16-39. Placing paint or chalk on the timing marks to make them more legible. (Toyota)

Turn the distributor as required to bring the painted index mark in line with the scale or pointer. When the mark is exactly in line, tighten distributor clamp. After tightening, recheck timing to make sure the mark is still aligned. Replace vacuum hoses and other connectors, as necessary. After setting timing, the operation of the spark advance devices should be checked, if necessary.

Spark Advance Mechanisms

To provide the proper spark advance or retard to fit all speeds, throttle openings, and engine loading, older distributors are equipped with both a ***vacuum advance*** and a ***centrifugal advance.*** **Figure 16-41** illustrates centrifugal and vacuum advance units. Modern distributors have no spark

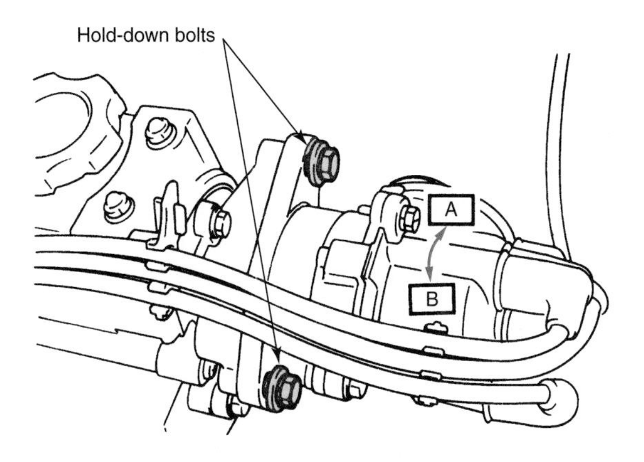

Figure 16-40. Distributor assembly with two hold-down bolts. Turning the distributor toward "A" will advance the timing. The hold-down bolts must be torqued to specifications after timing is adjusted. (General Motors)

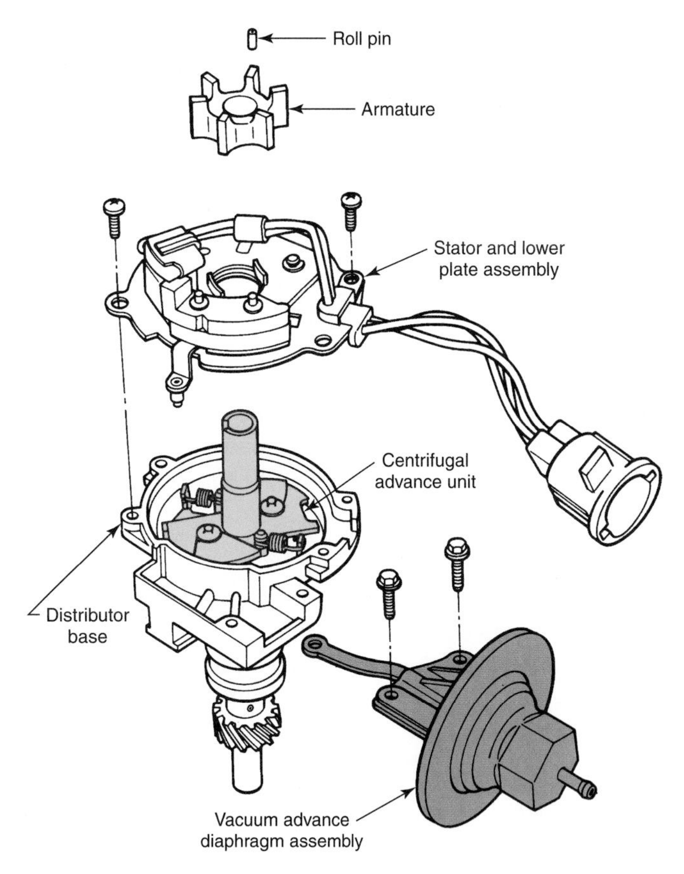

Figure 16-41. This distributor contains both vacuum and centrifugal advance assemblies. (Ford)

advance control units, and the ECM controls the advance and retard functions.

Testing Centrifugal Advance

If the distributor is on the car, disconnect the vacuum line to the vacuum advance and attach a timing light to the vehicle as outlined in the previous section. Slowly increase engine speed to about 2500 RPM. The timing mark should advance smoothly against the direction of engine rotation. If the action is jerky, stop the engine. Remove the distributor cap and twist the rotor in the direction of rotation. When the rotor is released, it should snap back to the original position. A slow return or no movement indicates a gummy or corroded advance unit. Clean and oil the unit. A loose, rattle-type condition indicates broken or stretched return springs. Replace parts as needed. The centrifugal advance should then be rechecked on the engine or with a distributor tester. On some distributors, the advance curve can be adjusted to increase performance or economy. This is done by changing springs or weights. Follow the manufacturer's instructions to modify the advance curve.

Testing Vacuum Advance

With the distributor installed on the engine, disconnect the vacuum advance line. Connect a timing light as described earlier. Start engine and run it at a steady 1200 RPM. Note the position of the timing mark in relation to the pointer. While holding the RPM at 1200, reconnect the vacuum advance line. When line is connected, the timing mark should immediately advance against the direction of engine rotation. If the mark does not move, make sure the vacuum line connection is receiving vacuum. If it is, the problem is in the distributor. Suspect a leaking vacuum unit or, less commonly, a sticking breaker plate.

Overhauling Distributors

Distributors must be overhauled when bushing, shaft, or gear wear becomes excessive, or when the advance mechanisms fail to function correctly. Note the areas of the typical distributor in **Figure 16-42** that could require overhaul. Some newer distributors cannot be overhauled and are replaced as a unit when faulty.

Removing the Distributor

Before removing the distributor from the engine, mark the position of the distributor housing and rotor relative to the engine. Then, remove the electrical connections and the hold-down bolt, and pull the distributor from the engine. If the distributor housing sticks in the engine, a light tap with a plastic hammer will usually free it.

Disassembling the Distributor

Before removing the gear from the shaft, measure from the top of the gear to some reference point on the distributor housing. Then remove the gear from the shaft. See **Figure 16-43.**

Replace the vacuum advance diaphragm unit if it is faulty. Clean and oil the centrifugal advance. Replace rusty or stretched advance springs. Clean, examine, and oil the movable breaker bushing or bearings as required. Replace the shaft if it is worn, rough, or bent. Check the housing for cracks or other damage. Check the drive gear for chipping or wear.

Assembling the Distributor

Assemble the distributor, and check both shaft side and end play. Make sure the pin is through the gear. If the pin is of the solid (not roll pin) type, check that it is securely peened. Lubricate the distributor as required. Install various components. After assembly, check out the distributor on a distributor tester.

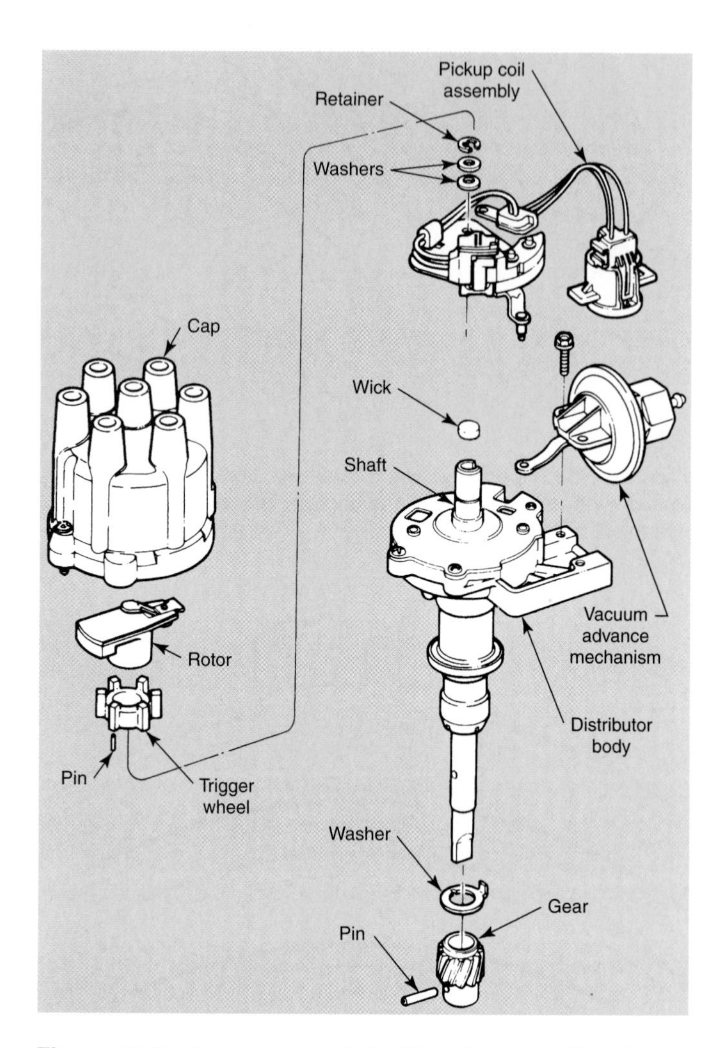

Figure 16-42. Typical electronic ignition distributor. Note the areas that would need service if worn, bent, or corroded. (DaimlerChrysler)

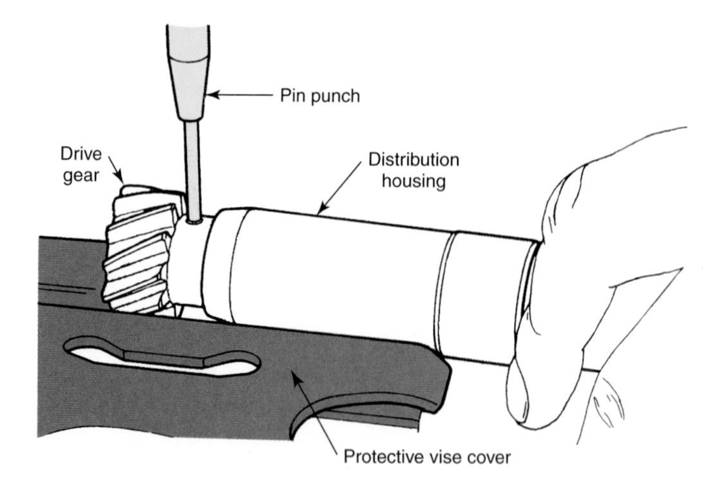

Figure 16-43. Support the gear and shaft when removing or installing the retaining pin. (DaimlerChrysler)

Installing the Distributor (Engine Undisturbed)

If the distributor was marked to indicate the position of the rotor and scribed to show housing-to-engine relationship, installation is simple. Align the rotor with the chalk or scribe mark. See **Figure 16-44.** Then align the housing-to-engine block scribe lines and push the distributor into place. As the distributor is moved down, you will notice the rotor will turn a small amount as the distributor gear meshes. Pull the distributor up far enough to disengage the gear, move the rotor back far enough to compensate for the turning, and press the distributor down again. When the housing flange is flush against the block, the housing-to-engine and rotor housing scribe lines should all be aligned. Lock distributor into place.

Caution: If the distributor will not bottom, do not attempt to force it down by using the hold-down clamp to draw it in. The distributor shaft is probably not aligned with the oil pump shaft slot or tang. Push the distributor downward by hand while cranking the engine. When the two shafts are aligned, the distributor will drop into place.

Installing the Distributor (Engine Disturbed)

If the engine was turned over after the distributor was removed, crank the engine until the number one piston is starting up on the compression stroke. Turn the engine over until the timing marks are aligned. The engine is now ready to fire the number one cylinder. Align the housing-to-engine scribe marks. Turn the rotor to face the number one cap tower. Push the distributor into place. Pull up and adjust for rotor movement. When correct, the distributor will be fully bottomed. The points (if used) will just be opening, and the rotor will point at number one cap tower.

Timing the Distributor

The distributor initial setting (timing mark aligned with the pointer or scale and the rotor facing the number one plug wire tower) will suffice for starting the engine. Try for a slightly retarded setting (spark occurs later than specified) to prevent "kicking back" (engine attempting to rotate backwards due to plug being fired too early), which can damage the starter mechanism. After starting the engine, set the timing as explained earlier in this chapter.

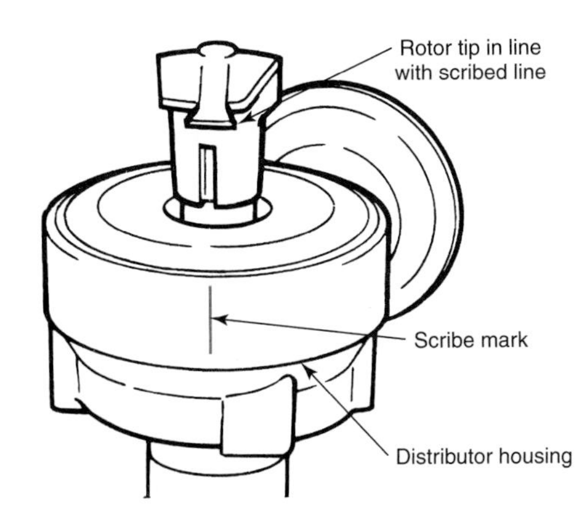

Figure 16-44. Align the rotor tip with the scribe mark on the distributor housing. (DaimlerChrysler)

Servicing Spark Plugs

Almost everyone knows what ***spark plugs*** are, but their familiarity often leads to problems. It is not enough to remove the old plugs and install new ones. The following section explains the proper selection and service of spark plugs.

Spark Plug Service Life

Spark plug service life varies a great deal, depending on such factors as engine design, type of ignition system (contact point or electronic), type of service, driver habits, and type of fuel used. Older vehicles with contact point ignitions and operating on leaded gas required that plugs be cleaned and inspected every 10,000–15,000 miles (16,000–24,000 km). In extreme cases, some plugs required replacement at 5000 miles (8000 km). Plugs in newer vehicles with electronic ignition systems may last beyond the often-recommended replacement interval of between 15,000 and 30,000 miles (24,000 and 48,000 km). Some of the newest vehicles are intended to travel over 100,000 miles (160,000 km) on one set of spark plugs.

In deciding whether to clean and reinstall or replace the plugs, the technician should weigh the cost of services for cleaning, filing, and gapping in light of the remaining useful life. Unless the plugs are in almost perfect condition, it usually pays to replace them.

Removing Spark Plugs

Pull the plug boot and wire assemblies from the plugs carefully. Direct a stream of compressed air around the base of each plug to blow out any foreign material.

Caution: Do not remove the spark plugs from a hot aluminum cylinder head. Damage to the head will result if it is not allowed to cool before spark plug removal.

Using a plug socket (rubber lined), remove the plugs. Make sure the gaskets (where used) are also removed. Keep plugs in order so that any peculiar plug conditions can be related to the cylinder concerned.

Examining Spark Plugs

A careful study of the spark plugs is helpful in determining engine condition, plug heat range selection, and trouble resulting from operational conditions, **Figure 16-45** illustrates typical spark plug conditions.

Spark Plug Problems

Study **Figure 16-46.** Note that in A, the plug fires normally. In B, a dirty insulator caused flashover. In C, a cracked insulator allowed the current to travel to ground. In D, conductive deposits on the insulator allowed the current to

Typical Spark Plug Conditions

Normal Plug Appearance

A spark plug operating in a sound engine and at the correct temperature will have some deposits. The color of these deposits should range from tan to gray. The electrode gap will show growth of about .001 in. (0.025 mm) per 1000 miles (1600 km), but there should be no evidence of burning.

Overheating

Overheating (dull, white insulator and eroded electrodes) can occur when the spark plugs are too hot for the engine. Cooling system problems, advanced ignition timing, detonation, sticking valves, and excessive high speed driving can also cause spark plug overheating.

Fuel Fouling

Fuel fouling (dry, fluffy, black carbon deposits) can be caused by plugs that are too cold for the engine, a high fuel level in the carburetor, a stuck heat riser, a clogged air cleaner, or excessive choking. If only one or two plugs show evidence of fuel fouling, inspect the plug wires for those cylinders. Sticking valves can also cause fuel fouling.

Preignition

Preignition (fuel charge ignited by an overheated plug, piece of glowing carbon, or hot valve edge before the spark plug fires) will cause extensive plug damage. When plugs show evidence of preignition, check the heat range of the plugs, the condition of the plug wires, and the condition of the cooling system. The engine should be checked for physical damage because it has been subjected to excessive combustion chamber pressure.

Oil Fouling

Oil fouling (wet, black deposits) is caused by an excessive amount of oil reaching the cylinders. Check for worn rings, valve guides, or valve seals. A ruptured vacuum pump diaphragm can also cause oil fouling. Switching to a hotter spark plug may temporarily relieve the symptoms, but will not correct the problem.

Detonation

Detonation can cause the insulator nose on a spark plug to crack and chip away. The explosion that occurs during heavy detonation creates extreme pressure in the cylinder. Detonation can be caused by an excessively lean fuel mixture, low octane fuel, advanced ignition timing, or extremely high engine temperatures.

Splashed Fouling

Splashed fouling (plugs coated with splashes of deposits) can occur when new plugs are installed in an engine with heavy piston and combustion chamber deposits. The new plugs restore regular firing impulses and raise the operating temperature. As this occurs, accumulated engine deposits flake off and stick to the hot plug insulator.

High Speed Glazing

High speed glazing (hard, shiny, yellowish-tan, electrically conductive deposits) can be caused by a sudden increase in plug temperature during hard acceleration or loading. This condition often causes misfiring at speeds above 50 mph (81 km/h). If high speed glazing reoccurs, cooler plugs should be used.

Gap Bridging

Gap bridging (carbon-lead deposit connecting the center and ground electrodes) is not often encountered in automotive engines. Prolonged low speed operation followed by a sudden burst of high speed operation can form gap bridging. It can also be caused by excessive fuel additives.

Ash Fouling

Ash fouling (heavy white and yellowish deposits) is caused by the buildup of combustion deposits. The deposits may be caused by burning oil or fuel additives. Although ash fouling is not conductive, excessive deposits can cause spark plugs to misfire.

Mechanical Damage

Mechanical damage can be caused by a foreign object in the combustion chamber. When a plug shows evidence of mechanical damage, all cylinders should be inspected. Valve overlap may allow small objects to travel from one cylinder to another.

Worn Out

Extended use will cause the spark plug's center electrode to erode. When the electrode is too worn to be filed flat, the plug must be replaced. Typical symptoms of worn spark plugs include a drop in fuel economy and poor engine performance.

Figure 16-45. Spark plug appearance can help determine engine condition. (Champion Spark Plug)

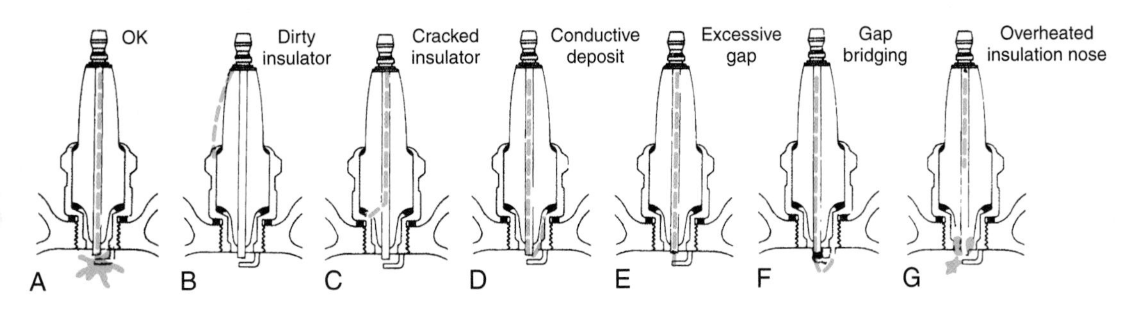

Figure 16-46. Typical spark plug problems.

travel to ground. In E, excessive gap raised the required voltage above that available. In F, conductive deposits between the electrodes allowed current to travel to ground. In G, an overheated insulator nose fired the fuel charge (preignition) before the voltage could build up high enough to force the current to jump the gap.

Cleaning Spark Plugs

Occasionally, plugs must be reused. When this is the case, remove all oil with solvent and blow the plugs dry. If a spark plug cleaning machine is available, install a plug in the machine and apply a stream of abrasive to clean away all deposits. While applying the abrasive blast, rock the plug back and forth to assist in cleaning. Apply the abrasive blast only long enough to clean the insulator and shell. Repeat this procedure for all plugs. Blow each plug free of all abrasive.

Plug Size, Reach, Heat Range, and Type

Spark plug size (18 mm, 14 mm, 10 mm) is determined by the diameter of the threaded section. ***Reach*** is determined by the length of the threaded section. Excessive reach can cause preignition, poor fuel charge ignition, difficult plug removal, and mechanical damage from striking a piston or valve.

When selecting new plugs, choose plugs with the specified ***heat range.*** The plugs in **Figure 16-47** are of the same size and reach, but they have different heat ranges. The heat range is controlled by the length of the insulator, from the tip to the sealing ring. The longer the insulator, the hotter the plug. A plug that is too hot will suffer from burning and preignition. A plug that is too cold will quickly foul out with heavy deposits.

Type indicates resistor, nonresistor, projected core nose, single ground electrode, or multiple ground electrode. See **Figure 16-48.**

Gapping Plugs

Whether the plugs are new or used, the ***spark gap*** should be checked before the plugs are installed in an engine. To adjust gap, bend the side electrode until the proper gap is set between the electrode surfaces. Gaps vary widely, so always check and follow the manufacturer's specifications for the plug and engine. Check for correct gap using a special feeler gauge.

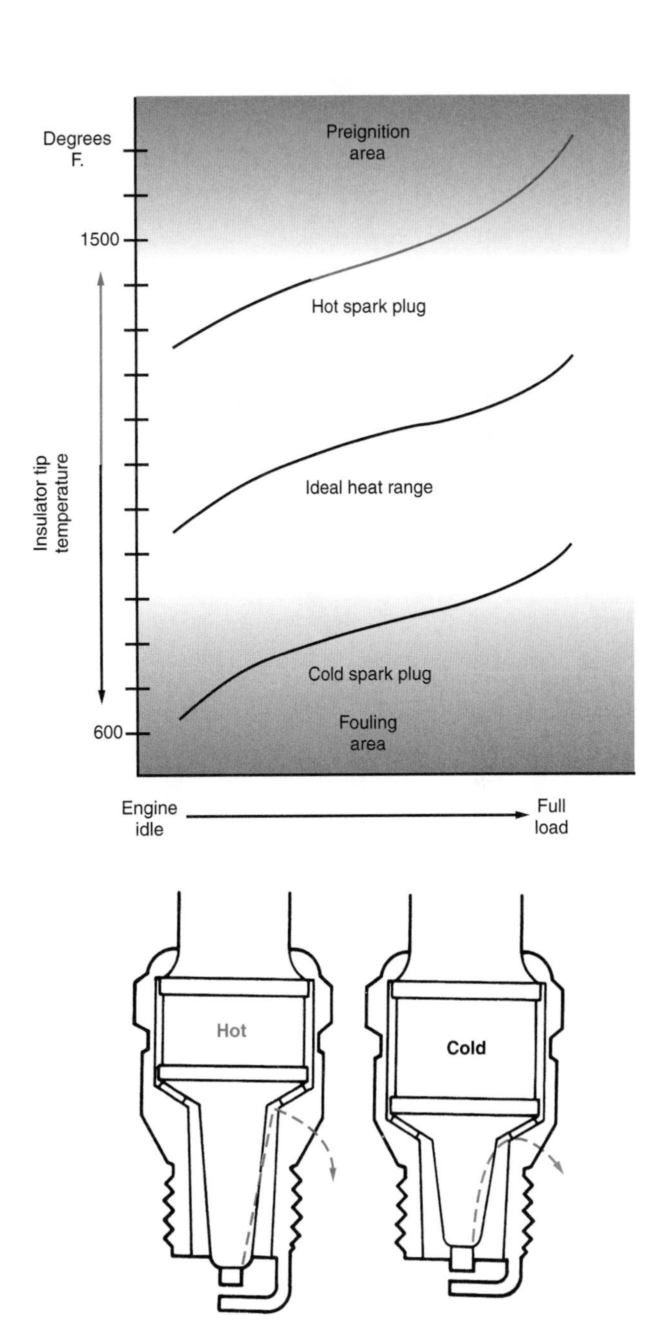

Figure 16-47. Spark plug heat range is controlled by the amount of the insulator exposed to the heat of combustion. The hot plug takes longer to dissipate the combustion heat than a cold plug. (AC-Delco)

Caution: Bend only the side electrode. If any attempt is made to bend the center electrode, the insulator will be cracked. Refer to Figure 16-49.

Installing Plugs

Note: When reusing plugs that have been cleaned and gapped, use new gaskets (where used) if possible. Make sure only one gasket is installed on each plug.

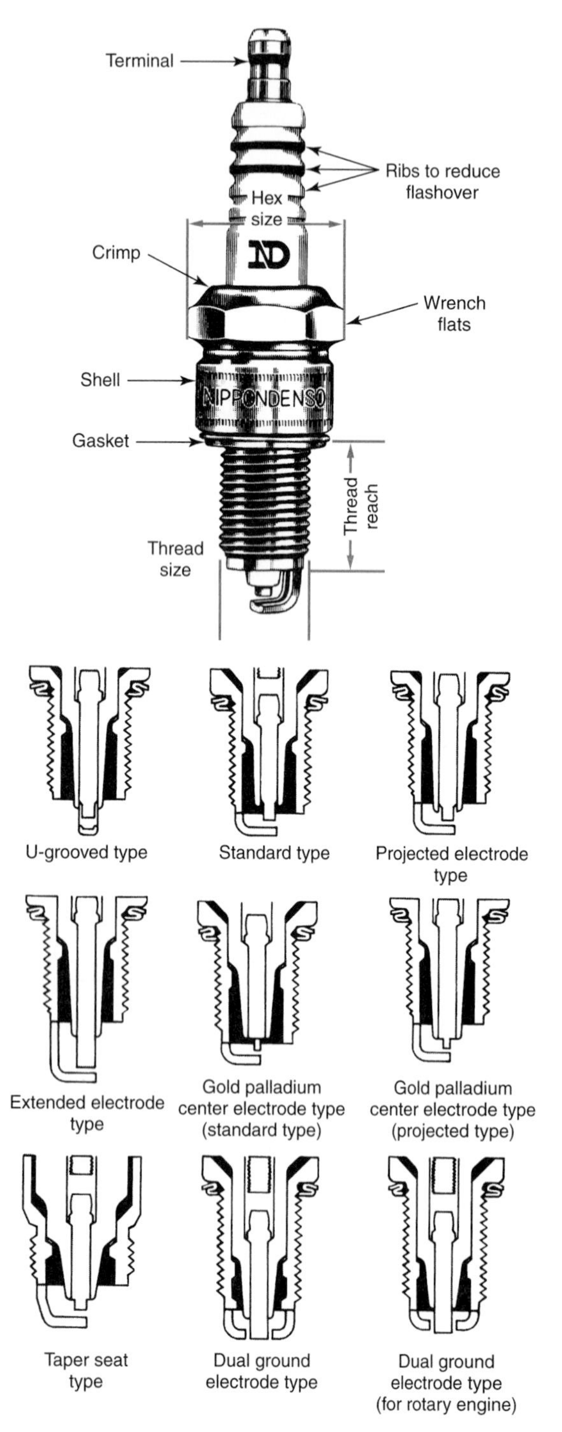

Figure 16-48. Spark plug terminology. Note the various types of plugs offered by this manufacturer. (Nippondenso)

Before installing a spark plug, wipe off the plug gasket seat in the cylinder head. Then install the plug and tighten it by hand until it is snug. Use a torque wrench to bring it up to the specified torque. Be careful to avoid placing side pressure on the wrench. This could tip the socket and crack the plug insulator.

If torque specifications are not available, a good rule of thumb for a plug with a new gasket is to tighten the plug until it is snug (finger tight). Then, give it another quarter turn. For a tapered seat plug, tighten it until snug and then give it another one-sixteenth turn.

Caution: Overtightening plugs can change the gap and cause thread damage to the plug or block. Low torque will cause the plug to overheat and possibly cause preignition. Leakage and loosening can also result.

After installing the plugs, wipe the insulators clean and attach the plug wires. Be careful to maintain the correct firing order.

Tech Talk

Electronic and ECM-controlled ignition systems produce extremely high voltages. Therefore, it is vital to route the spark plug wires properly. Often, the spark will jump through plastic parts, especially if they are damp. After a while, a carbon track is built up through the plastic part, causing a mysterious miss. The high voltages produced by these systems also make it doubly important to check the nonelectronic components. It is common for high secondary voltages to punch through rotors and distributor caps on late-model vehicles.

When checking any ignition system for a hard-start condition, remember to check the voltage available at the coil during cranking.

Before removing a distributor, make sure that you line up the rotor with a spot on the distributor body. In many cases, you can stick a cotter pin on the distributor body directly under the rotor before pulling the distributor. This will locate the rotor when you are ready to reinstall the distributor.

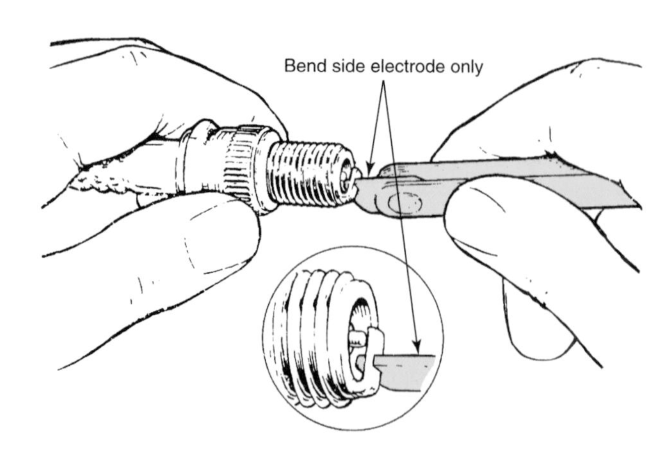

Figure 16-49. When adjusting spark gap, bend only the side electrode. (AC-Delco)

Summary

The first step in ignition system problem diagnosis is to visually inspect all system components. Primary wiring must be in good condition. A quick check for spark will quickly isolate the problem to the ignition system. Check the coil and distributor cap for flashover, burning, or corrosion. Coil polarity must be correct. Secondary wiring must be in good condition. All terminals should be properly connected. Wire ends should have good boots and should be fully inserted into the correct towers. On late-model vehicles, trouble codes should be retrieved from the ECM's memory.

Problems in the primary circuit can be caused by the coil, ignition switch, ignition module, ECM, points, condenser, or ignition resistor. The distributor, rotor, and secondary plug wires can cause trouble in the secondary circuit. These components should be tested carefully.

An oscilloscope can be used for quick and accurate ignition system diagnosis. Use the oscilloscope as recommended by the manufacturer. Any pattern that does not look like the standard pattern for the ignition system being tested indicates a defect. Be sure to look at both the secondary and primary patterns.

The timing can be set on some engines. A timing light is used to set ignition timing. When used, vacuum and centrifugal advance units should be tested for correct operation. On late-model vehicles, the ECM controls ignition timing.

Some worn distributors can be overhauled. When installing a distributor, make sure the timing is correct, the oil pump shaft is engaged, and the distributor is locked into position.

Clean around spark plugs before removing them. Examine the plugs for abnormal wear and deposits. Most plugs are replaced, but occasionally a set of plugs may be cleaned and reused. Before installing plugs, always check the plug gap.

Review Questions—Chapter 16

Do not write in this book. Write your answers on a separate sheet of paper.

1. Excessive circuit resistance lowers _____.
   (A) voltage
   (B) temperature
   (C) firing strength
   (D) Both A and C.
2. Firing order means _____.
   (A) the order in which cylinders fire
   (B) that a piston is at top dead center on the firing stroke
   (C) the direction the distributor turns
   (D) the point at which the timing marks line up
3. Which type of ignition system contains a rotor?
   (A) DIS.
   (B) Direct.
   (C) Distributorless
   (D) Distributor.
4. ECM-controlled ignition systems eliminate the need for _____ and _____ advance systems.
5. When checking a DIS system coil for spark, you should attach one plug wire to the spark tester. What do you do with the other plug wire?
6. When timing the engine with a timing light, it is often necessary to disconnect the distributor _____.
   (A) vacuum lines
   (B) electrical connector
   (C) coil wire
   (D) Both A and B.
7. Which of the following adjustments can be made to an ECM-controlled ignition distributor system but *not* to a DIS system?
   (A) Setting timing.
   (B) Setting dwell.
   (C) Setting timing advance.
   (D) Setting plug gap.
8. Spark plugs in modern engines may operate for as many as _____ before requiring replacement.
   (A) 5000 miles (8000 km)
   (B) 20,000 miles (32,000 km)
   (C) 100,000 miles (160,000 km)
   (D) 250,000 miles (400,000 km)

Match the plug insulator color with its probable cause.

Plug Color	Probable Cause
9. Tan or gray.	(A) Oil fouling.
10. Dry black.	(B) Overheating.
11. Wet black.	(C) Detonation.
12. Dull white.	(D) Normal used plug.
13. Shiny yellow.	(E) Fuel fouling.
	(F) High-speed glazing.

14. When gapping plugs, bend the _____ electrode only.
15. Name the factors to be considered when selecting a new set of plugs for any engine.

ASE-Type Questions

1. Electronic ignition systems are equipped with all the following components, *except:*
   (A) spark plugs.
   (B) ignition module.
   (C) contact points.
   (D) coil.
2. Technician A says a spark test can be used to help pinpoint ignition system problems. Technician B says that when performing a spark test on a DIS coil that serves two spark plugs, both secondary wires should be removed from the plugs during the test. Who is right?
   (A) A only.
   (B) B only.
   (C) Both A and B.
   (D) Neither A nor B.

3. A defective ignition switch can cause _____.
   (A) a no-start condition
   (B) missing at high speeds
   (C) spark plug damage
   (D) Both A and B.
4. All the following statements about distributorless ignition systems are true, *except:*
   (A) some DIS systems use one coil to fire two spark plugs.
   (B) a DIS coil is located on top of the spark plug.
   (C) DIS coils can be swapped to locate a defective coil.
   (D) DIS coils rarely fail all at once.
5. Secondary wire towers should be checked for all the following, *except:*
   (A) corrosion.
   (B) cracks.
   (C) carbon tracks.
   (D) discoloration.
6. Technician A says plug wires can have excessive resistance. Technician B says crossfiring can occur when any two spark plug wires run too close together. Who is right?
   (A) A only.
   (B) B only.
   (C) Both A and B.
   (D) Neither A nor B.
7. All of the following statements about oscilloscopes are true, *except:*
   (A) oscilloscopes produce a visual record of what is happening in the ignition system.
   (B) anything that does not look like the standard pattern indicates a problem.
   (C) using the oscilloscope is quicker than using an ohmmeter or removing the spark plugs.
   (D) the oscilloscope can diagnose problems in the secondary ignition system only.
8. Reversed coil polarity can be easily detected on the oscilloscope because the waveform will appear _____.
   (A) normal but backward
   (B) right side up but with broken lines
   (C) normal but upside down
   (D) on the primary pattern only
9. Technician A says an oscilloscope can be used to pinpoint a defective plug wire. Technician B says an oscilloscope can be used to pinpoint a defective distributor cap. Who is right?
   (A) A only.
   (B) B only.
   (C) Both A and B.
   (D) Neither A nor B.
10 Technician A says a stream of compressed air should be used to clean the area around the base of a spark plug before plug removal. Technician B says spark plugs in aluminum cylinder heads should be removed when the engine is hot to prevent seizing. Who is right?
   (A) A only.
   (B) B only.
   (C) Both A and B.
   (D) Neither A nor B.

Suggested Activities

1. Draw a schematic of an ignition system. Show the secondary and primary systems. Show how high voltage is developed in the system.
2. On a sheet of graph paper, draw the following oscilloscope patterns:
   - Secondary pattern—electronic ignition system using a distributor.
   - Secondary pattern — DIS electronic ignition system.
   - Primary pattern—electronic ignition system using a distributor.
   - Primary pattern— DIS electronic ignition system.
   - Secondary pattern—point-type ignition system.
3. Using the information in this chapter as a reference, demonstrate how to perform a spark test.
4. Examine one or more ignition systems and determine which of the following classes it fits into:
   - Point-type.
   - Electronic with vacuum and centrifugal advance mechanism.
   - ECM-controlled.
   - Distributorless.
   - Direct.
5. Visually inspect the secondary ignition components on one vehicle for signs of arcing and flashover. Discuss your findings with your instructor or the other members of your class.

Ignition System Problem Diagnosis	
Problem: No spark	
Possible cause	**Correction**
1. Distributor pickup defective or misadjusted.	1. Replace or adjust as needed.
2. Discharged battery.	2. Charge battery.
3. Faulty coil or primary circuit resistor.	3. Replace coil or resistor.
4. No primary current to system.	4. Check ignition switch and wiring.
5. Defective coil high tension lead.	5. Replace lead.
6. Defective rotor and/or distributor cap.	6. Replace cap and rotor.
7. Defective plug wires.	7. Replace plug wires.
8. Moisture in distributor cap.	8. Dry cap.
9. Breaker plate not grounded.	9. Replace or tighten ground wire.
10. Defective ignition control module.	10. Replace ignition module.
11. Loose, corroded, or open electronic control module ground lead.	11. Tighten, clean, or connect as needed.
12. Loose, corroded, or disconnected primary connections.	12. Clean, cover with special, protective grease and shove firmly into distributor.
13. Defective distributor electronic pickup.	13. Replace pickup unit.
14. Trigger wheel positioned too high.	14. Reposition correctly.
15. Incorrect trigger wheel-to-pickup air gap.	15. Set correctly. Use nonmagnetic feeler gauge.
16. Defective cam or crankshaft sensor.	16. Replace defective sensor.
17. Breaker points defective or misadjusted (point ignition).	17. Install new points.
18. Defective condenser.	18. Replace condenser.
Problem: Weak or intermittent spark	
Possible cause	**Correction**
1. Discharged battery.	1. Charge or replace battery.
2. Loose or dirty primary wiring connections.	2. Clean and tighten connections.
3. Weak coil.	3. Replace coil.
4. Defective primary circuit resistor.	4. Replace resistor.
5. Burned rotor and cap contacts.	5. Replace cap and rotor.
6. Defective spark plug wires.	6. Replace wires.
7. Insufficient system voltage.	7. Replace regulator.
8. Worn distributor bushings or bent shaft.	8. Replace bushings or shaft.
9. Loose spark plug wires.	9. Clean and tighten connections.
10. Defective distributor electronic pickup.	10. Replace pickup.
11. Trigger wheel pin sheared.	11. Replace pin.
12. Shorted primary wiring.	12. Replace wire and relocate.
13. Loose wiring harness connectors.	13. Tighten connections.
14. Breaker points defective (where used).	14. Install new points.
15. Defective condenser (where used).	15. Install new condenser.
16. Point dwell set incorrectly.	16. Set dwell correctly.
Problem: Missing at idle or low speed	
Possible cause	**Correction**
1. Weak or intermittent spark at plugs.	1. See *Weak or intermittent spark.*
2. Fouled spark plugs.	2. Clean or replace plugs.
3. Spark plug gaps too narrow.	3. Adjust gaps to specifications.
4. Improper plug heat range.	4. Install proper heat range.
5. Damaged plugs.	5. Replace plug or plugs.
6. Defective distributor electronic pickup.	6. Replace pickup unit.
7. Loose harness connections.	7. Clean and tighten connections.
8. Discharged battery.	8. Charge or replace battery.
9. Coil polarity incorrect.	9. Reverse coil primary leads.

(continued)

Ignition System Problem Diagnosis *(continued)*	
Problem: Missing during acceleration	
Possible cause	**Correction**
1. Weak spark.	1. See *Weak or intermittent spark.*
2. Plugs damp.	2. Dry plugs.
3. Fouled plugs.	3. Clean or replace plugs.
4. Plug gap too wide.	4. Gap as specified.
5. Damaged plug.	5. Replace plug.
6. Incorrect trigger wheel-to-pickup air gap.	6. Set gap to specifications.
7. Defective vacuum advance.	7. Repair or replace vacuum advance.
8. Incorrect coil polarity.	8. Reverse coil primary leads.
9. Crossfiring.	9. Rearrange ignition secondary wires.
10. Weak ignition coil.	10. Replace ignition coil.
Problem: Missing during cruising and high-speed operation	
Possible cause	**Correction**
1. Weak spark.	1. See *Weak or intermittent spark.*
2. Improper heat range plug (too hot).	2. Install proper heat range plug.
3. Crossfiring.	3. Arrange wires properly. If needed, install new wires.
4. Fouled plug.	4. Clean or replace plugs.
5. Plug gap incorrect.	5. Gap to specifications.
6. Damaged plug.	6. Replace plug.
7. Ignition timing incorrect.	7. Reset timing.
8. Defective distributor.	8. Repair or replace distributor.
9. Loose or corroded wire connections.	9. Clean and tighten connections.
10. Defective spark plug wires.	10. Replace wires.
11. Weak ignition coil.	11. Replace ignition coil.
Problem: Missing at all speeds	
Possible cause	**Correction**
1. Weak spark.	1. See *Weak or intermittent spark.*
2. Fouled spark plugs.	2. Clean or replace plugs.
3. Damaged plug.	3. Replace plug.
4. Crossfiring.	4. Arrange wiring correctly and if needed, install new wires.
5. Defective coil (DIS systems).	5. Replace coil.
6. Plug gap too wide or too narrow.	6. Adjust gap as needed.
7. Plugs and/or distributor damp.	7. Dry distributor and plugs.
8. Improper plug heat range.	8. Change to correct heat range.
9. Defective distributor.	9. Repair or replace distributor.
10. Defective spark plug wires.	10. Replace plug wires.
11. Distributor trigger wheel pin sheared or missing.	11. Install new pin.
12. Burned or corroded breaker points.	12. Replace and gap points.
13. Defective condenser.	13. Replace condenser.
Problem: Coil failure	
Possible cause	**Correction**
1. Carbon tracking on tower.	1. Replace coil and wire.
2. Oil leak in coil.	2. Replace coil.
3. Engine heat damage.	3. Replace coil. Relocate or baffle against heat.
4. Physical damage.	4. Replace coil.
5. Open plug wire.	5. Replace wire.

(continued)

Ignition System Problem Diagnosis *(continued)*

Problem: Short spark plug life

Possible cause	Correction
1. Incorrect plug heat range (too hot—burns).	1. Install correct (cooler) heat range.
2. Incorrect plug heat range (too cold—fouls).	2. Install correct (hotter) heat range.
3. Mechanical damage during installation.	3. Install correctly.
4. Loose spark plug (overheats and burns).	4. Tighten plugs to proper torque.
5. Incorrect plug reach (too short—fouls).	5. Install plugs with correct reach.
6. Incorrect plug reach (too long—strikes piston).	6. Install plugs with correct reach.
7. Worn engine—oil fouling.	7. Switch to hotter plugs or overhaul engine.
8. Bending center electrode.	8. Bend side electrode only.
9. Detonation.	9. Adjust timing. Change to higher octane gas and/or remove carbon buildup.
10. Preignition.	10. Remove carbon buildup, install valves with full margin, install cooler plugs.
11. Lean mixture.	11. Correct fuel system problem.

Problem: Preignition

Possible cause	Correction
1. Overheated engine.	1. Check cooling system.
2. Glowing pieces of carbon.	2. Remove carbon.
3. Spark plugs overheating.	3. Change to cooler plugs.
4. Sharp valve edges.	4. Install valves with full margin.
5. Glowing exhaust valve.	5. Check for proper tappet clearance, for sticking, and air leaks.

Problem: Detonation

Possible cause	Correction
1. Ignition timing advanced.	1. Retard timing.
2. Engine temperature too high.	2. Check cooling system.
3. Carbon buildup is raising compression ratio.	3. Remove carbon.
4. Low octane fuel.	4. Switch to high octane fuel.
5. Exhaust heat control valve stuck.	5. Free valve.
6. Excessive block or head metal removed to increase compression.	6. Use thicker gasket, change head, or true warped head or block surface.

Problem: Backfiring in intake manifold

Possible cause	Correction
1. Intake valve not properly seating.	1. Check for broken spring, valve clearance, sticking, seat condition.
2. Lean mixture.	2. Adjust mixture.
3. Crossfiring.	3. Arrange plug wires or install new wires if needed.
4. Plug wires installed wrong.	4. Connect wires to proper plugs.
5. Carbon tracking in distributor cap.	5. Replace cap and rotor.
6. Incorrect ignition timing.	6. Set timing to specifications.

Problem: Backfiring in exhaust system

Possible cause	Correction
1. Turning key off and on while vehicle is in motion.	1. Advise driver to avoid this practice.
2. Current flow interruption in primary circuit.	2. Check circuit for loose connections and shorts.
3. Coil-to-distributor cap secondary wire shorting or coil itself shorting.	3. Replace wire or coil.
4. Weak or intermittent spark.	4. See *Weak or intermittent spark.*
5. Incorrect valve timing.	5. Correct timing.
6. Air injection system diverter or diverter valve inoperative.	6. Replace valve.

(continued)

Ignition System Problem Diagnosis *(continued)*	
Problem: Engine kicks (attempts to run backward) during cranking	
Possible cause	**Correction**
1. Ignition timing too far advanced. 2. Plug wires installed incorrectly. 3. Carbon tracking. 4. Moisture in the distributor cap.	1. Retard ignition timing. 2. Attach wires to proper plugs. 3. Replace distributor cap and rotor. 4. Dry or replace cap.

17

Fuel Delivery

After studying this chapter, you will be able to:

- Describe the cleaning, removal, repair, and replacement of fuel tanks.
- Clean, repair, and install fuel lines.
- Test, remove, repair, and replace mechanical fuel pumps.
- Test, remove, repair, and replace electric fuel pumps.
- Service fuel filters.
- Explain vapor lock.
- List the safety rules involved in fuel delivery system service.

Technical Terms

Fuel tanks	Inlet vacuum
Fuel gauge sender	Electric fuel pumps
Cam lock	Fuel pump relay
Cold patching	Cam lock
Bulged	Inertia switch
Fuel lines	Fuel filters
Fuel pump	Water trap
Mechanical fuel pump	Water sensor
Fluid pressure	Vapor lock
Static pressure	Volatility
Pump volume	

This chapter will cover the operation and service of fuel delivery system components, including the fuel tank, fuel lines, fuel pump, and fuel filter. This will prepare you for the upcoming chapter on fuel injection system service.

Be careful of fire when working with gasoline or diesel fuel. Do not smoke while working on any part of the fuel system. Do not leave open containers of fuel in the shop. Before starting work on any part of the fuel system, disconnect the negative battery cable. Gasoline vapors can travel great distances and can ignite if they come into contact with the slightest spark.

Relieving System Pressure

Fuel injection systems can maintain considerable pressure for some time after the engine is turned off. A special pressure relief valve (Schrader valve) is generally provided to relieve system pressure. If the fuel injection system has a pressure relief valve, obtain a hose with a fitting that will depress the valve. Place one end of the hose in a gasoline container and then attach the other end of the hose to the pressure relief valve. This will release system pressure into the gasoline container. See **Figure 17-1.**

Another method is to attach a fuel pressure gauge to the relief valve and place the end of the gauge's pressure relief hose in a gasoline container. Then push the gauge's pressure relief button to release pressure into the container.

If no mechanical provision is made for pressure relief, it may be possible to remove the fuel pump fuse (or disconnect the fuel pump's electrical connector) and operate the engine until it stops from lack of fuel. Consult the vehicle's service manual for the proper relief procedure. Loosen all connections slowly and use an absorbent cloth or an approved container to catch any spillage.

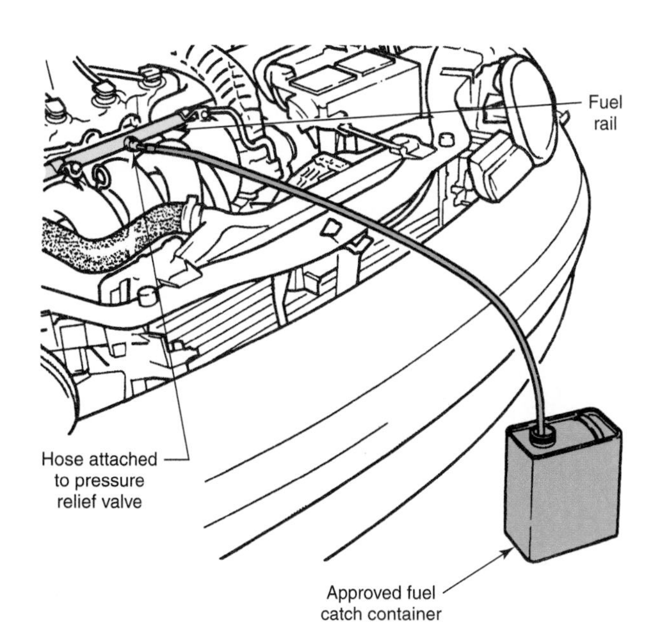

Figure 17-1. Using an approved fuel container to catch fuel during the pressure release procedure. Always relieve fuel pressure before working on the system.

Note: Properly dispose of gasoline-soaked rags and spilled gasoline.

Fuel Tanks

Fuel tanks are designed to safely carry enough fuel to allow the vehicle to travel several hundred miles before refueling is necessary. At the same time, the tank is not so large that it overloads the vehicle when completely filled. A typical fuel tank, along with other fuel system components, is shown in **Figure 17-2.** Older fuel tanks were made of steel. Fuel tanks on late-model vehicles are made of nylon or plastic, such as polyethylene. Some fuel tanks are attached to the vehicle with straps, while others are bolted to the body through rubber mounts. Insulation is placed between the tank and the body to reduce noise transfer. Internal baffles in the tank reduce fuel sloshing.

Draining the Tank

A fuel tank will be much easier to handle if it is empty. Whenever possible, a tool such as the one shown in **Figure 17-3** should be used to siphon fuel from the tank. This tool is equipped with a fuel pump, filter, and storage tank. Hose siphoning should be used only when no commercial siphoning tool is available. Siphoning must be done very carefully to reduce the danger of fire. Some fuel tanks are equipped with a baffle that makes siphoning impossible. Others tanks contain check valves, which could be damaged by conventional siphoning.

Some fuel tanks are equipped with a drain plug. Other tanks can be drained by disconnecting the fuel line at the engine and using the fuel pump to drain the fuel. Check the vehicle's service manual for the proper fuel removal method.

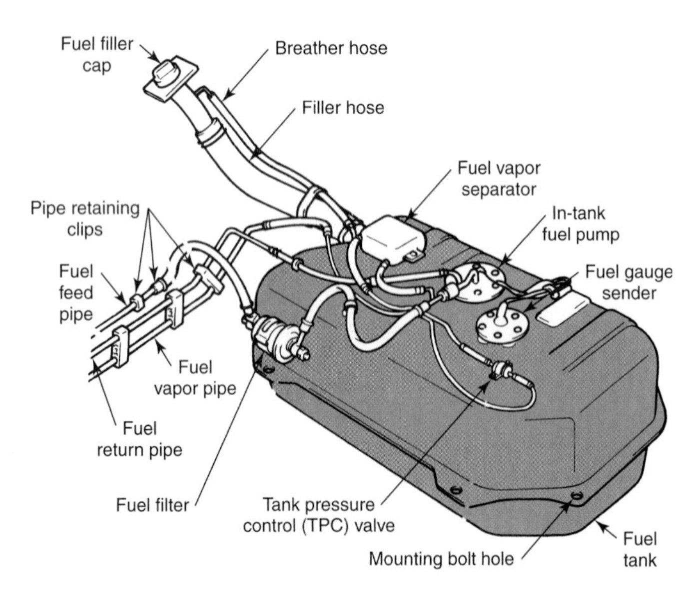

Figure 17-2. A molded steel fuel tank with various lines and fittings. This tank is secured to the vehicle with five bolts. (General Motors)

Warning: Most tanks will contain some fuel even after siphoning. The weight of the fuel and the physical dimensions of the tank can make it awkward to handle. Get someone to help you before you begin removing the tank from the vehicle.

Tank Removal

Raise the vehicle and disconnect the fuel lines. Cover all fuel line ends with masking tape. Disconnect the filler pipe and any external vents. Also remove the wires to the ***fuel gauge sender*** and the fuel pump (if the fuel pump is installed in the tank). Removal of other components may also be necessary on some vehicles. With the help of an assistant, remove the tank support straps or bolts and lower the tank.

Note: The fuel tanks of most late-model vehicles are sealed from the outside air. The tank vent tubes are routed through a carbon canister, which is part of the emissions control system. The carbon canister stores excess fuel vapors until they can be burned in the engine. All tank vents should be carefully noted and reinstalled in their original positions.

Loosen the ***cam lock*** that holds the fuel gauge sender and pickup tube assembly and remove the assembly from the

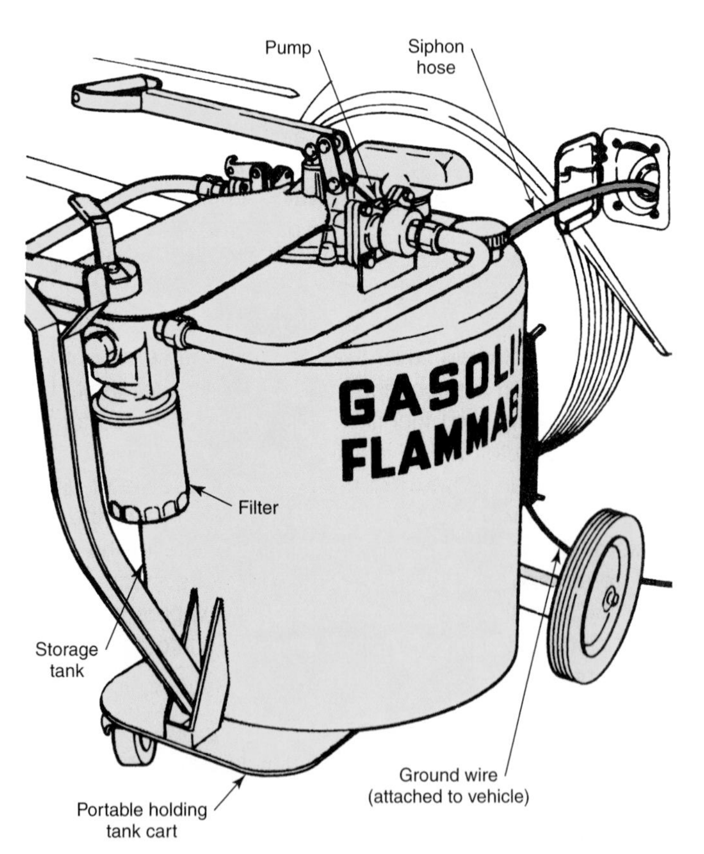

Figure 17-3. Portable fuel storage tank. This unit can drain and then refill the tank. Note ground wire. (DaimlerChrysler)

tank, **Figure 17-4.** Use care to avoid bending the float arm or the pickup pipe. Some fuel pumps are part of the fuel gauge sender assembly, **Figure 17-5.** If the fuel pump is installed in the tank, it will be removed along with the sender. With the help of an assistant, tilt the tank and drain any remaining fuel into an approved container.

Cleaning the Fuel Tank

Begin by inspecting the interior of the tank. A fuel tank that is contaminated with excessive quantities of water or foreign material should be cleaned. If the tank is not cleaned, the contaminants will enter the rest of the fuel system and can cause serious problems if not removed. Replace the tank if the interior is corroded or if any of the baffles are broken. Proceed with cleaning if the tank is not corroded or damaged internally. Place a quart of clean, nonflammable solvent in the tank. You may have to add more or less solvent, depending on the size of the tank. Do not use gasoline to clean a fuel tank. While holding a clean rag over the filler neck and sender hole, shake the tank vigorously to move the solvent around. Drain the tank and repeat the process. Blow the tank dry with compressed air and inspect for any remaining contaminants. Stubborn dirt may require cleaning the interior of the tank with steam. Clean the pickup pipe and filter screen by directing a gentle blast of air down through the pickup pipe. If the pickup pipe is connected to an in-tank fuel pump, remove the pump before cleaning. Replace the pickup filter screen if it is badly clogged.

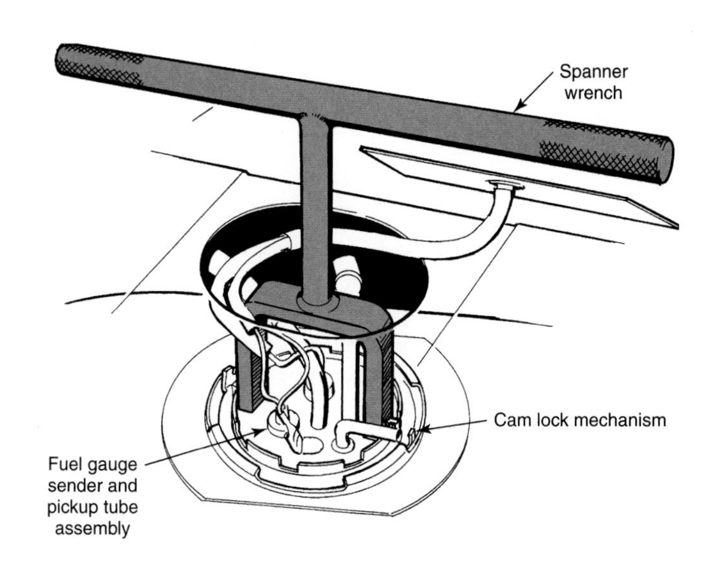

Figure 17-4. Use a special spanner tool to remove the fuel tank pickup unit. Always disconnect the battery before working on any part of the fuel system. (Volvo)

STOP Warning: An empty or partially empty fuel tank will contain vapors. If ignited, these vapors will produce a violent explosion. Never clean or work on the tank anywhere near a source of sparks or flames.

Tank Repair

Note: The following techniques for fuel tank repair apply only to automobile fuel tanks. Heavy steel tanks, high-pressure containers, and oil drums require additional safeguards and different techniques.

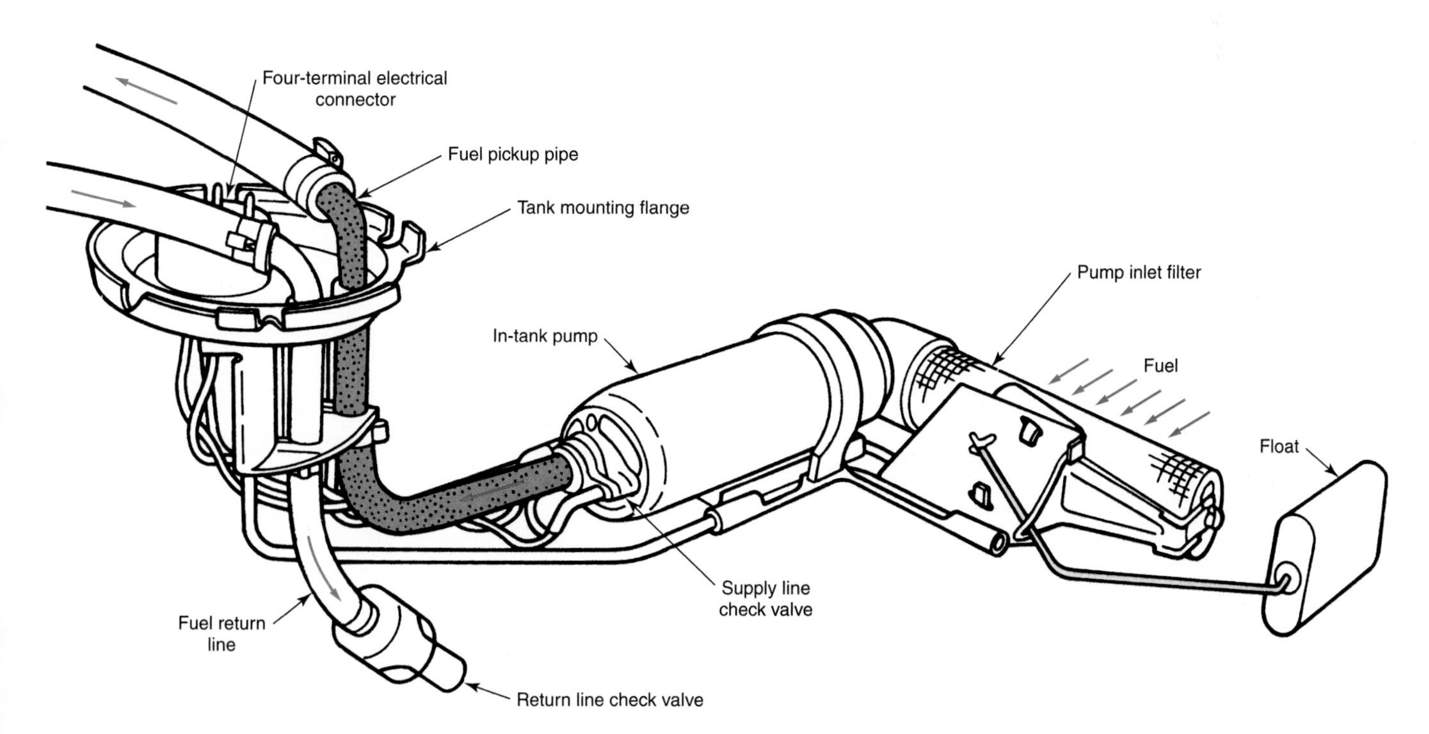

Figure 17-5. Some vehicles have the fuel tank pickup pipe, filter, float, and return pipe combined in one unit. Note the fuel return line. (DaimlerChrysler)

A leaking fuel tank should generally be replaced. However, plastic tanks can sometimes be repaired by patching. Similarly, a metal fuel tank can sometimes be repaired by soldering or brazing *if* adequate safeguards are taken. The fuel tank must be thoroughly steam cleaned inside and out. Following cleaning, the tank should be filled with an nonexplosive gas, such as carbon dioxide or nitrogen, or completely filled with water. See **Figure 17-6.**

STOP Warning: Use the utmost care in all tank cleaning and repair procedures. Fuel tanks can become lethal bombs. Have a fire extinguisher handy and keep other persons away from the operation.

Cold Patching

There are various epoxy sealants and special patching cloths that can be used to repair both metal and plastic tanks. If done properly, ***cold patching*** works very well for some leak repairs. Follow the manufacturer's recommendations. Test all tank repairs by covering the repaired area with a wet, soapy lather. Place an air hose in the tank and apply air. A mild pressure can be applied to the tank by holding a rag around the hose where it enters the tank. If the repair is sound, air bubbles will not appear in the repaired area. After checking the repair, blow the tank dry and reinstall it. Regardless of the type of repair chosen, work carefully. Remember that fuel tank integrity is vitally important. Even a small leak can cause a fire. If there is any doubt as to the success of the repair, *replace* the tank.

Expanded (Bulged) Fuel Tank

In the modern sealed fuel system, considerable in-tank pressure can be generated by such items as a defective fuel cap, pinched or clogged vapor lines, or a plugged vapor canister filter. Fuel tanks can become expanded, or ***bulged,*** from excessive in-tank pressure. This can cause tank damage and fuel leaks. Whenever a bulged tank is encountered, replace it and determine the cause. Make certain that the excess pressure condition is no longer present following repairs. In many systems, the pressure relief valve in the filler cap should open at around 1 1/2 to 2 psi (10 to 17 kPa).

Removing a Dent in the Fuel Tank

Occasionally, the bottom of a fuel tank is pushed inward by striking an object. If the internal baffles are not damaged, the tank can often be straightened. Remove the tank from the vehicle, fill it with water, plug the vent tube, and place a nonvented cap on the filler neck. Remove the in-tank fuel pump if used. Apply air through the pickup tube. The air will exert pressure on the water and will usually cause the dented area to bulge outward. Be sure to apply only enough air pressure to pop the dent out. See **Figure 17-7.**

Caution: Always fill the tank with water before applying air pressure. The use of air alone can cause the tank to rupture and fly apart with great force.

Installing the Fuel Tank

Before reinstalling the tank, replace the sending unit or pickup assembly, if needed. Use a new gasket whenever the sending unit is removed. Make sure the tank insulation strips are in place. With the help of an assistant, carefully raise the tank into position and attach the tank retaining straps or bolts. If the tank uses an airtight, or ***nonvented,*** cap, make sure the vent tube is open. Torque the tank bolts to specifications. Do not overtighten the bolts. Install the fuel lines and other components as needed and reconnect all wiring. Reconnect the battery as the final step. Do not refill the tank until installation is complete.

Fuel Line Service

The ***fuel lines*** are normally trouble free, but they may collect water and dirt from the tank. In other cases, external

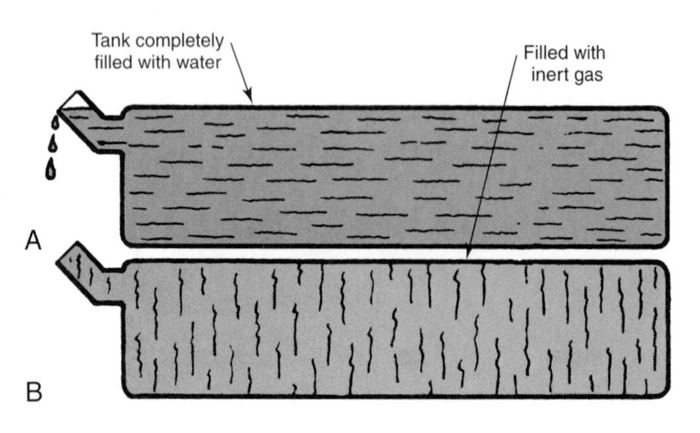

Figure 17-6. Solder or braze fuel tanks only after thorough steam cleaning and preparing the tank as shown in either A or B.

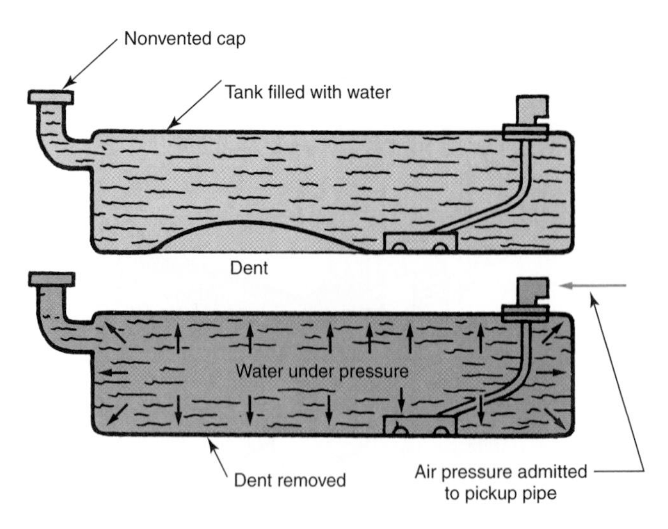

Figure 17-7. Remove a dent in a fuel tank by filling the tank with water and applying low air pressure.

damage, corrosion, or vibration may restrict fuel flow or cause the lines to leak. The following section explains how to service fuel lines. **Figure 17-8** illustrates a typical fuel line arrangement.

Cleaning Fuel Lines

Water or dirt in the fuel lines can contaminate the fuel system and cause serious damage. To clean out a fuel line, disconnect the line at the tank and at the fuel pump. Also, disconnect the line between the fuel pump and the fuel rail or carburetor. Some electric fuel pumps are located in the tank. In these cases, simply disconnect the line from the fuel rail, carburetor, or filter, as applicable. Use tubing wrenches to loosen the line fittings. If push-on fittings are used, a special tool may be needed to release them. Do not kink plastic fuel lines, as this can cause a permanent restriction. Next, remove any inline filters. Direct an air blast through the fuel line until the line is clean. Always blow in the direction opposite fuel flow. If the line remains restricted, check for a dented or kinked section.

After cleaning is complete, install a new fuel filter and reconnect all line fittings. Carefully align and hand thread the fittings for at least two turns to prevent cross threading. Hold the pump and/or the filter fitting with one wrench while tightening the flare nut with another wrench. See **Figure 17-9.** When using push-on type fittings, make sure the tube and the inside of the fitting are clean. Align the fitting and tube and push the fitting on the tube. Be sure to replace the retaining clip, if removed, **Figure 17-10.**

Repairing Damaged Fuel Line

Other than the rubber flex hose used on some carbureted systems, fuel lines rarely need repair. Any rubber hose used must be in good condition. If a metal fuel line is dented, severed, or corroded, the damaged section can be replaced with a new section of tubing. Make sure all connections are tight. Refer to Chapter 8,Tubing and Hose, for full instructions. Plastic fuel lines are not repairable and must be replaced if they are kinked or damaged.

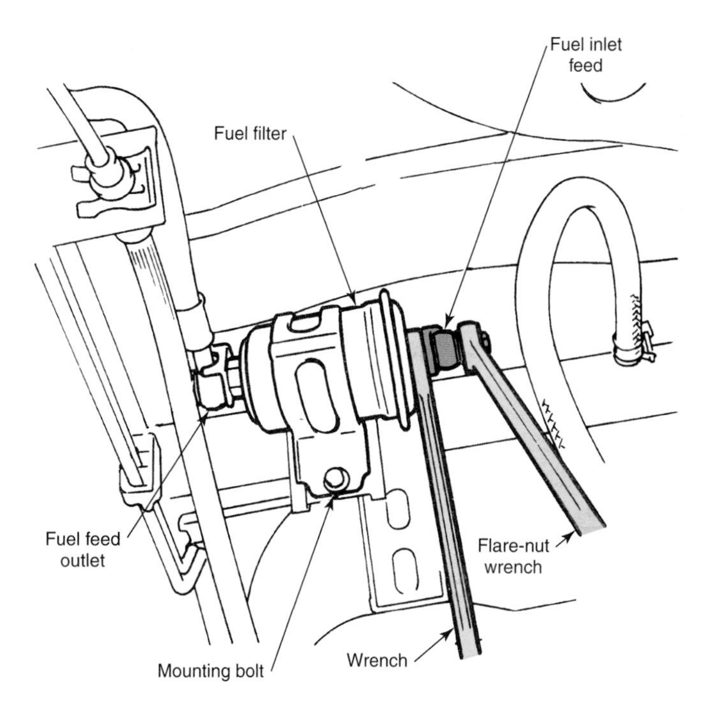

Figure 17-9. Tighten fittings securely. Do not cross thread. Note how the filter fitting is being held while the flare nut is tightened. (General Motors)

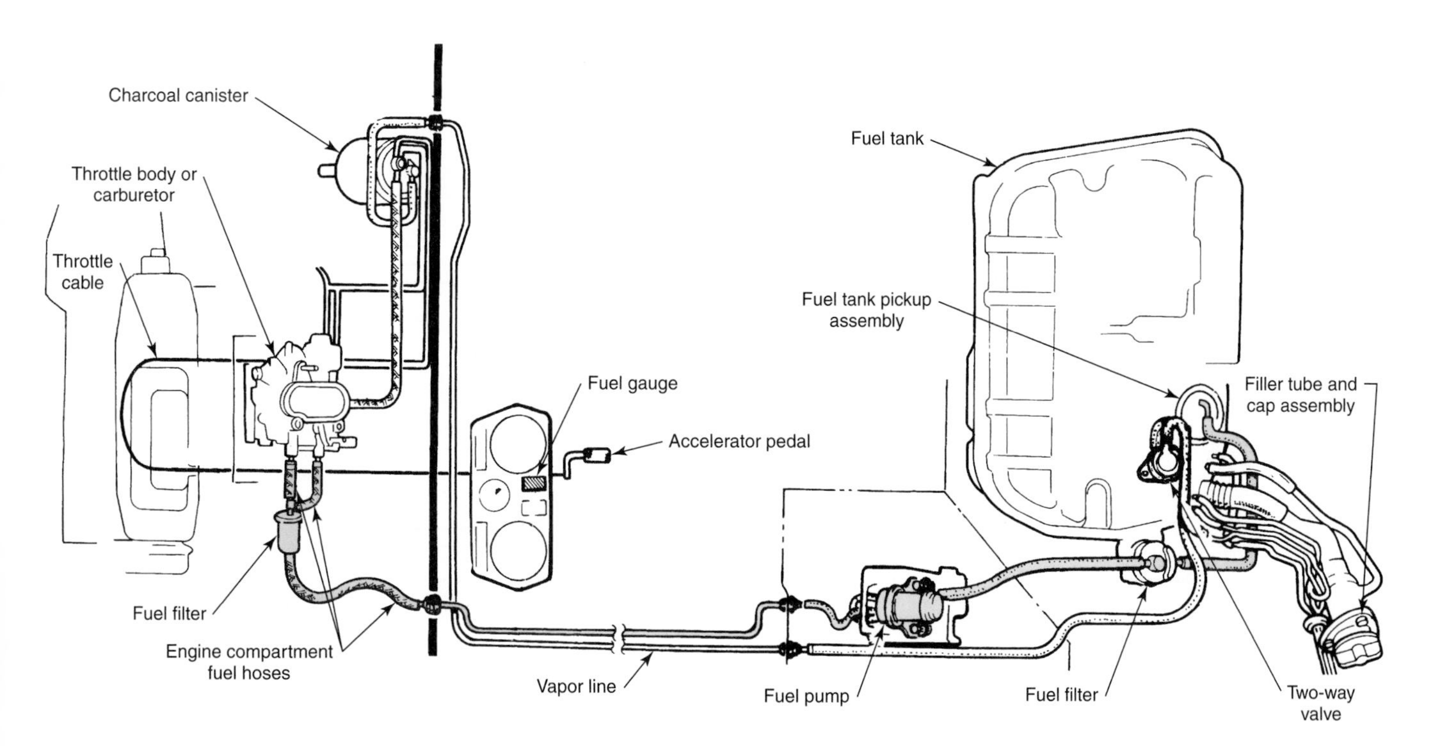

Figure 17-8. Study this fuel line setup from the fuel tank to the engine. Learn the name of each part. (Honda)

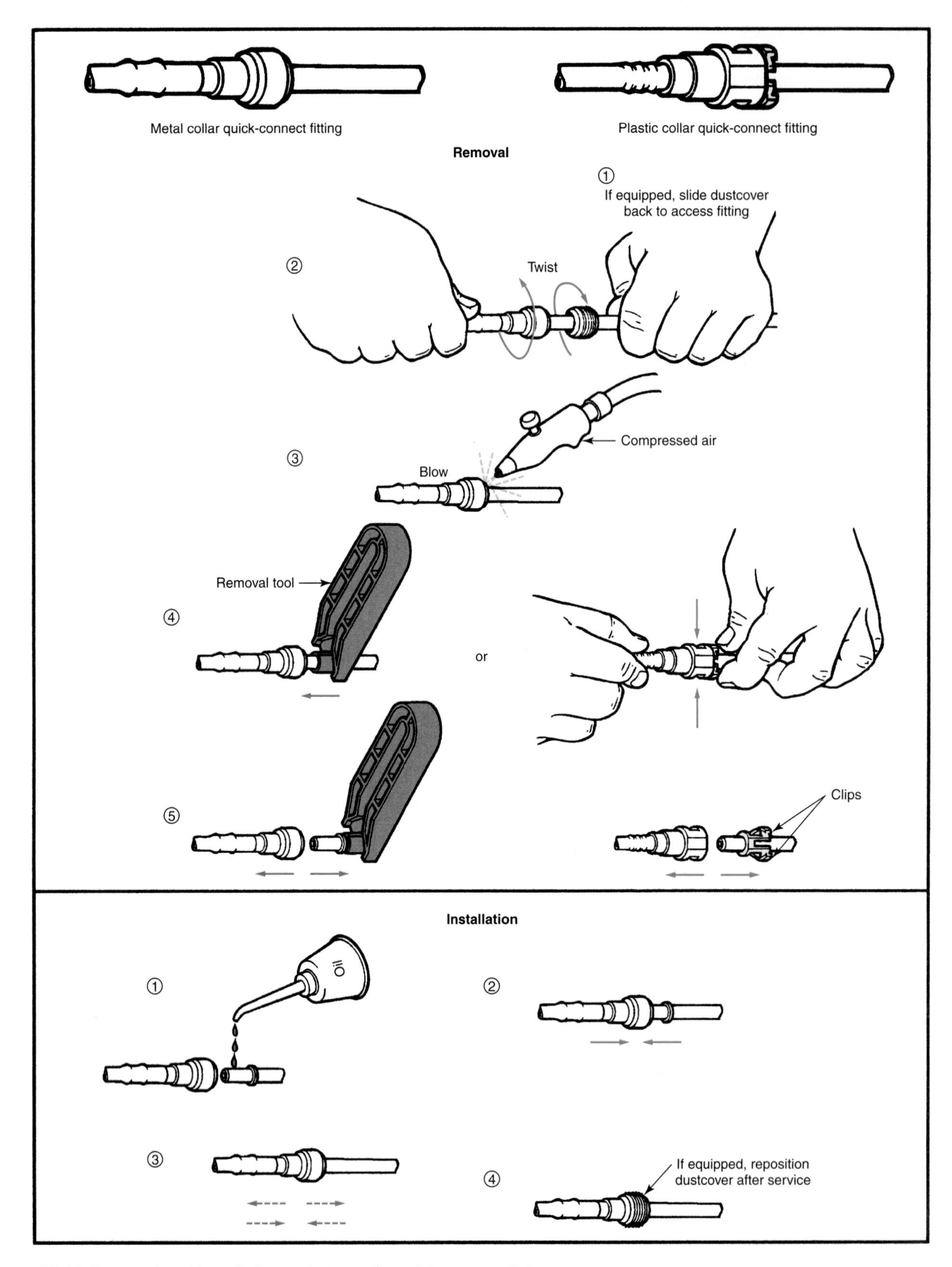

Figure 17-10. Removal and installation techniques for quick-connect fittings.

Fuel Pump Service

The ***fuel pump*** is vital to the proper operation of the vehicle. Without the fuel pump, the other components of the fuel system would be useless. The following sections explain how to test and service both mechanical and electric fuel pumps. Before testing the pump, make sure there is fuel in the tank. Inspect the fuel and vent lines for kinks, dents, and leaks. Check that the pump mounting screws are tight. Clean or replace the fuel filter.

Mechanical Fuel Pump Repair

The following section briefly explains how to test, remove, install, and rebuild a ***mechanical fuel pump.*** There are two tests that must be done to properly evaluate fuel pump pressure. The first is a test of pump ***fluid pressure;*** the second is a test of pump ***static pressure*** (not supplying fuel to carburetor, hence no fuel flow pressure).

Pump Fluid Pressure Test

To perform a pump fluid pressure test, disconnect the fuel line at the carburetor or a recommended point and install a suitable fuel pressure gauge. The gauge illustrated in **Figure 17-11** can be used for both pressure and volume tests. Regardless of the type of gauge used, it should be held at or near carburetor level. The gauge should not be more than 6″ (152 mm) above or below the carburetor to prevent false readings. If you are using a pressure gauge similar to the one shown in Figure 17-11, place the flow volume hose in a container and pinch off the hose before starting the engine. Some manufacturers specify that pump pressure be tested at normal operating speed, while others recommend testing pressure at cranking speed.

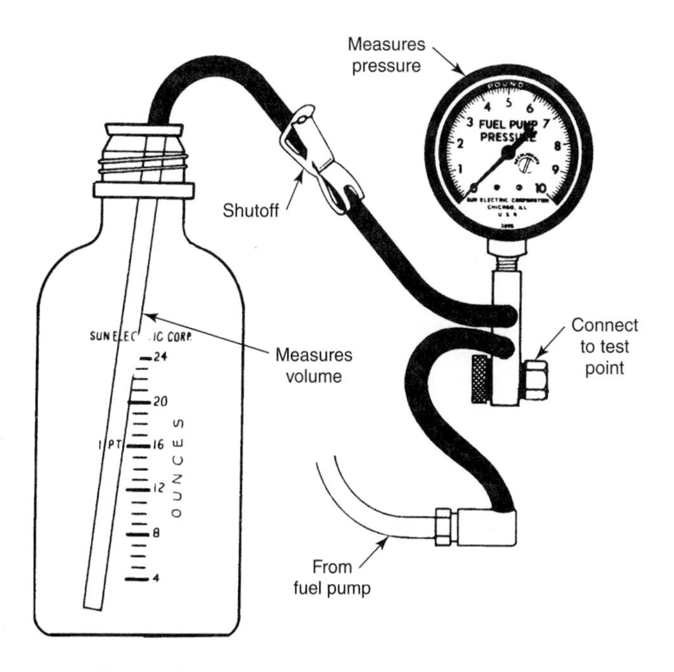

Figure 17-11. This gauge setup can be used to check fuel pump pressure and volume. (Sun Electric)

With the engine cranking or running, open the hose shutoff and draw in about 4 ounces (118 ml) of fuel. This will vent the pump and remove any trapped air that could cause a false reading. Stop the engine and dispose of the fuel in the container. When the container is empty, replace the volume hose and start the engine. With the engine cranking or idling, note the pressure on the gauge. Average fuel pressure should range from 4–6 psi (28–41 kPa) and should stay relatively constant.

Stop the engine and watch the gauge to check pump static pressure. The fuel pressure should either remain constant or fall slowly. A rapid loss of pressure indicates a leaky fitting, a faulty pump outlet valve, or a leaky carburetor float valve. To check pump static pressure, attach the gauge as shown in **Figure 17-12.** Note that the pressure gauge is attached directly to the end of the pressure line. The engine is operated on the fuel remaining in the fuel bowl.

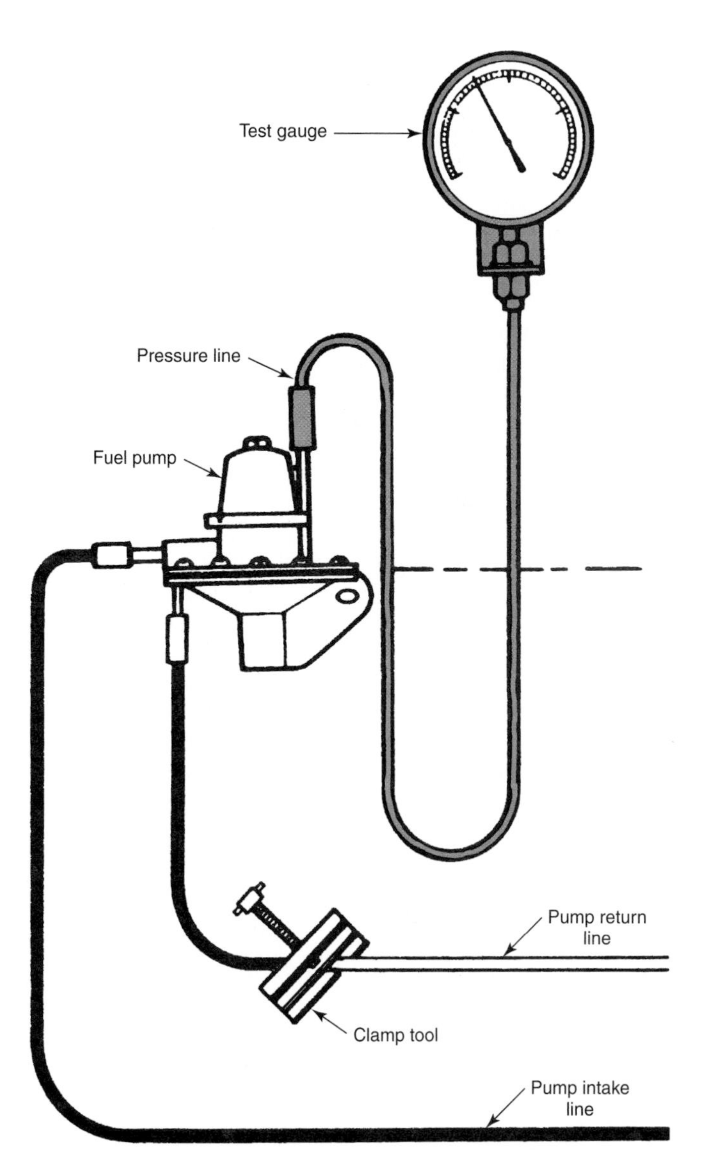

Figure 17-12. Checking pump static pressure.

When making pressure tests on a fuel pump equipped with a vapor return line, the line must be pinched off. An open valve will result in a false pressure reading. Start the engine and remove the return line at the valve. Hold a container under the valve outlet to catch any fuel being discharged. A fuel discharge rate of about 2 1/2 ounces (74 ml) per minute is normal and indicates that the vapor valve is closed. Fuel discharge in appreciable amounts indicates the vapor valve is either open from heat or is stuck open. In the case of heat, cool the pump with wet rags. If the valve is stuck, is should be cleaned or replaced.

Pump Volume Test

Note: If the pump has a vapor return line, close it off or the static pressure test cannot be performed.

After evaluating pump pressure, a ***pump volume*** test must be performed to make sure an adequate supply of fuel is reaching the diesel injection pump or carburetor. Use the gauge setup shown in Figure 17-11 to perform the volume test. Noting the exact time (in seconds), open the flow volume hose shutoff with the engine idling. As soon as there are about 4 ounces (118 ml) of fuel in the container, firmly push the tube into the fuel. Watch for bubbles in the fuel that would indicate an air leak in the intake line. As soon as approximately 16 ounces (473 ml) have been drawn, note the time in seconds, close the flow shutoff, and stop the engine. Manufacturer's specifications will generally call for a flow equivalent to 1 quart in 1 minute at 500 rpm. See **Figure 17-13.**

Caution: Use extreme care when conducting both pressure and volume tests. A fuel pump can spray fuel a long distance. Make certain all connections are tight and that the flow volume hose is in the container. The container should be made of clear plastic.

Pump Inlet Vacuum Test

When volume or pressure does not meet specifications, the pump ***inlet vacuum*** should be determined before condemning the fuel pump. If the suction line (line from the tank pickup tube to the pump) is restricted or leaking air, the pump cannot be expected to perform as required. Disconnect the inlet fuel line from the pump. If gas drips from the open fuel line, cap off the line. Attach the vacuum gauge to the inlet fitting or the pump inlet flex line. Disconnect the output fuel line from the pump at the carburetor. Attach a hose to the carburetor line end and place it in a container to catch the fuel.

Note: Work quickly. The carburetor contains only a small amount of gasoline and will run dry within 30 seconds.

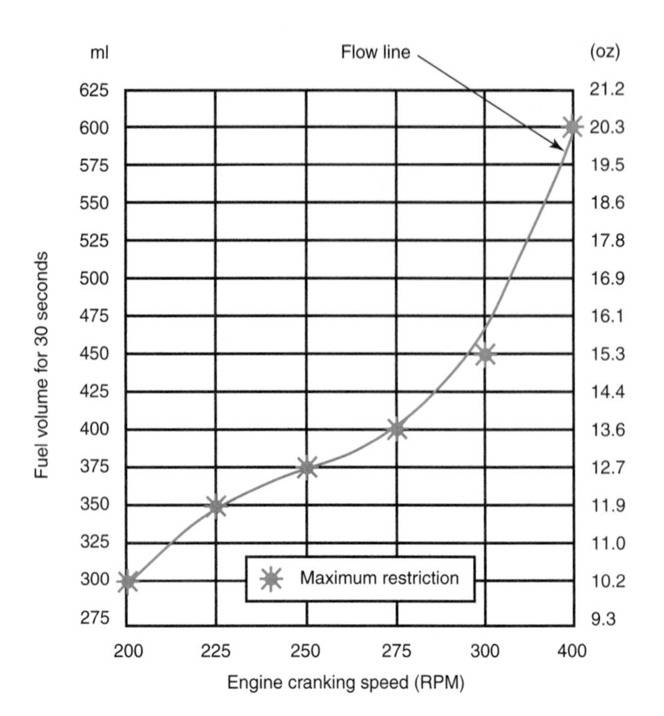

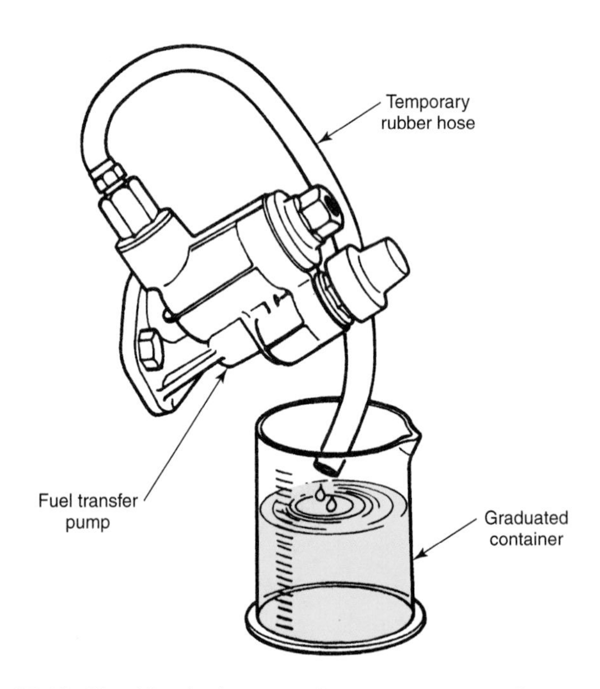

Figure 17-13. Checking fuel pump volume on one type of diesel engine. A—Fuel volume versus engine cranking speed graph. B—Fuel pump volume being measured in a graduated container. (Dodge)

Start the engine and allow it to idle until the gauge reads a vacuum. In general, a minimum vacuum reading of 10″ (34 kPa) should be obtained. When the engine is stopped, the reading should hold steady. A reading of 10″ (34 kPa) or more indicates that the pump valves, diaphragm, flex line, and bowl gasket (where used) are airtight. See **Figure 17-14.** If the reading is below specs or vacuum falls off rapidly when engine is stopped, a leak in the flex line or a leak between the pump

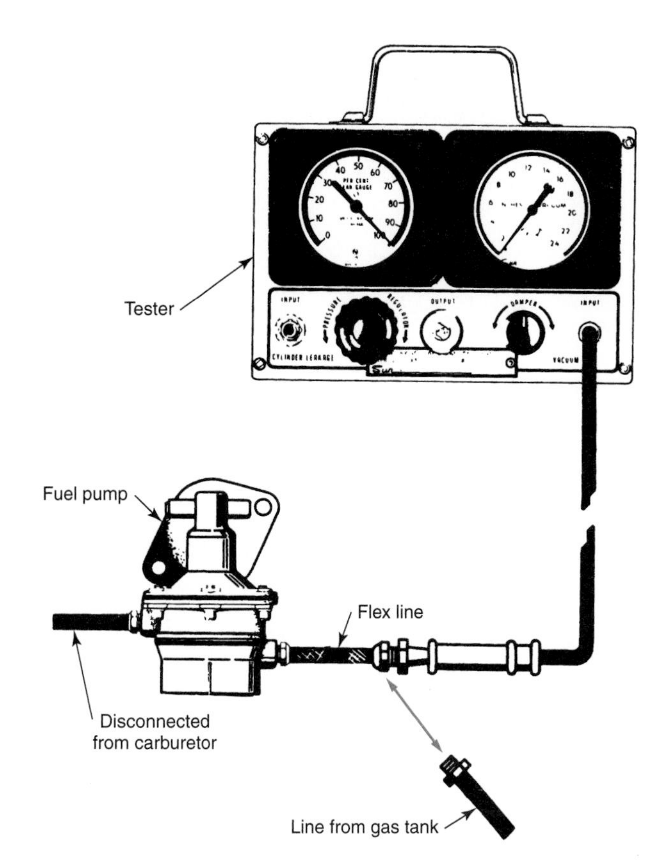

Figure 17-14. This gauge is used to check a fuel pump for vacuum leaks. (Sun Electric)

and the carburetor is indicated. If a flex line is used, remove the line and attach the gauge directly to the pump inlet fitting. If the low vacuum reading or fall-off continues, the pump is defective. If the vacuum reading is now normal, the flex line is leaking. When the vacuum test indicates that the pump and flex line are not leaking, test the entire inlet system by removing the line at the fuel tank and attaching the vacuum gauge at this point. Connect the flex line to the fuel line and operate the engine. If the vacuum reading drops below specifications, or if it falls off rapidly, there is an air leak in the inlet system.

Pump Removal

Before removing a fuel pump, clean all dirt and oil from around the pump line connections and mounting flange. Remove the fuel and vapor lines, and then remove the pump fasteners. Cover the line ends and stuff a clean rag into the engine pump rocker arm opening. If a push rod is used, it should be removed for cleaning and inspection. Brush the outside of the pump with solvent and rinse it off. When ordering a rebuild kit or a new fuel pump, give the vehicle make, year, model, and engine size. If possible, provide the pump number.

Rebuilding Mechanical Fuel Pumps

Today, most mechanical fuel pumps are simply replaced when they are faulty. However, some pumps can be rebuilt. Disassembly procedures will vary somewhat, depending on the pump design. Begin by scribing a line on the pump so that the parts may be reassembled in their correct positions in relation to each other. Remove the valve body cover, noting the relationship of the diaphragm to the valve body. Then remove the valve body-to-pump body screws. Remove the stake marks (places where the metal is dented) from the valve assemblies and pry the valves out of the pump body. Note the location and position, either up or down, of each valve so that the new valves can be reinstalled correctly. Remove the rocker arm pin and pull the rocker arm and spring assembly from the pump body. Note the relative position of the spring and rocker arm assembly. Finally, pull the diaphragm assembly from the pump body. A cutaway view of a mechanical fuel pump is shown in **Figure 17-15.**

Soak all metal parts of the pump in carburetor cleaner for no more than 15 minutes. After rinsing the parts and blowing them dry, lay them on a clean surface. Check all parts for nicks, excessive wear, cracks, and warping. Open the rebuild kit and lay out all the parts. Assemble the pump using all the new parts in the correct relationship to each other. If the pump contains a filter, it must be cleaned or replaced. Realign the scribe marks and start all fasteners. Finally, tighten all the fasteners until they just start to tighten against the lock washers.

Pump Installation

Before installing a new or rebuilt fuel pump, scrape the old mounting gasket from the engine block mounting pad. Lubricate the rocker arm on the pump, and install the pump with a new mounting gasket coated with gasket cement. Check the location of the cam or push rod and install the pump to make correct contact. Make sure the rocker arm rubbing pad (contact surface) bears against the eccentric cam

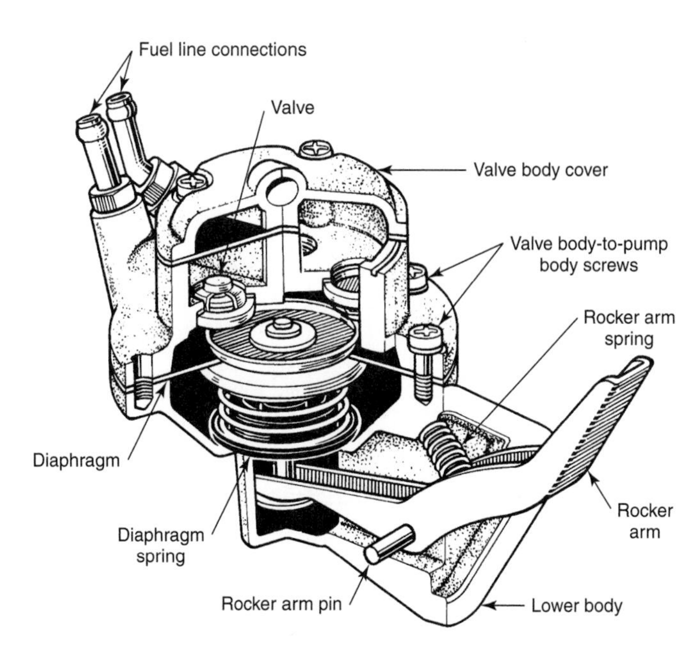

Figure 17-15. Study this serviceable fuel pump. Most mechanical fuel pumps are now simply replaced. (Toyota)

or push rod, where used. **Figure 17-16** illustrates one type of rocker arm pad-to-eccentric contact arrangement. Mounting the rocker arm rubbing pad to one side of the eccentric cam or off the push rod can cause pump damage and possible engine damage. Push the fuel pump inward until the pump is against the mounting pad. Install all pump retaining fasteners and torque them to specifications. See **Figure 17-17.** Never force the pump to the engine by using the fasteners to pull it in. Attach the fuel lines or hoses as explained earlier. Start the engine and check the pump for leaks. Test pump pressure and volume to ensure that there are no other problems.

Electric Fuel Pump Repair

There are four types of ***electric fuel pumps,*** including ***diaphragm, bellows, impeller,*** and ***roller-vane*** types. Accurate pressure and volume tests depend upon a properly charged battery and a mechanically sound pump motor or solenoid. The impeller-type pump shown in **Figure 17-18** is placed inside the gas tank. Other electric pumps are installed on the engine or on the frame, under the vehicle, **Figure 17-19.**

Make sure electrical power is reaching the pump before condemning it. Many pumps are operated by a ***fuel pump relay.*** Check the relay by bypassing it to determine whether the pump operates. If the pump begins working, the relay is defective. Also check the pump fuse and all wires for good connections. The fuel pump circuit shown in **Figure 17-20** contains a pump relay and a fuse. An often-overlooked problem is a faulty fuel pump ground. Many electric fuel

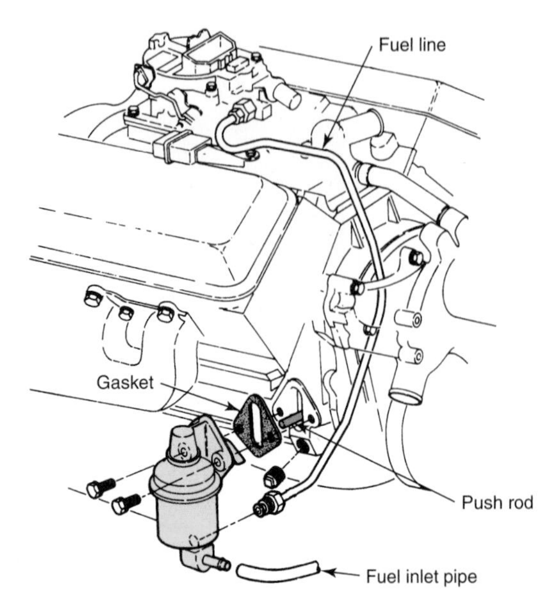

Figure 17-17. When installing a fuel pump, make sure the rocker arm contacts the cam or, as shown here, the push rod, correctly. Do not use fasteners to force pump into place. (AC)

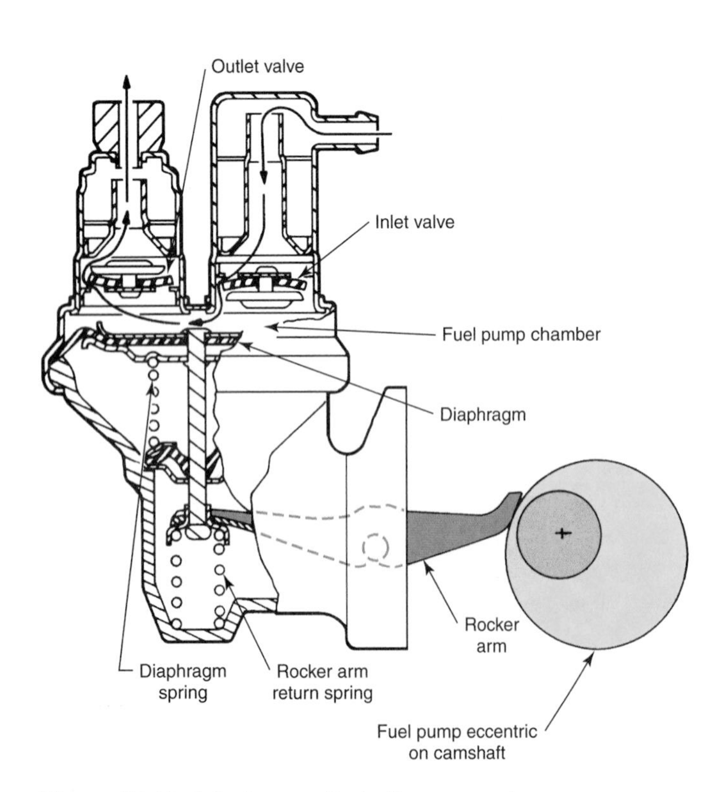

Figure 17-16. A fuel pump illustrating one rocker arm (actuating lever) to fuel pump eccentric on the camshaft. There are a number of different styles. (DaimlerChrysler)

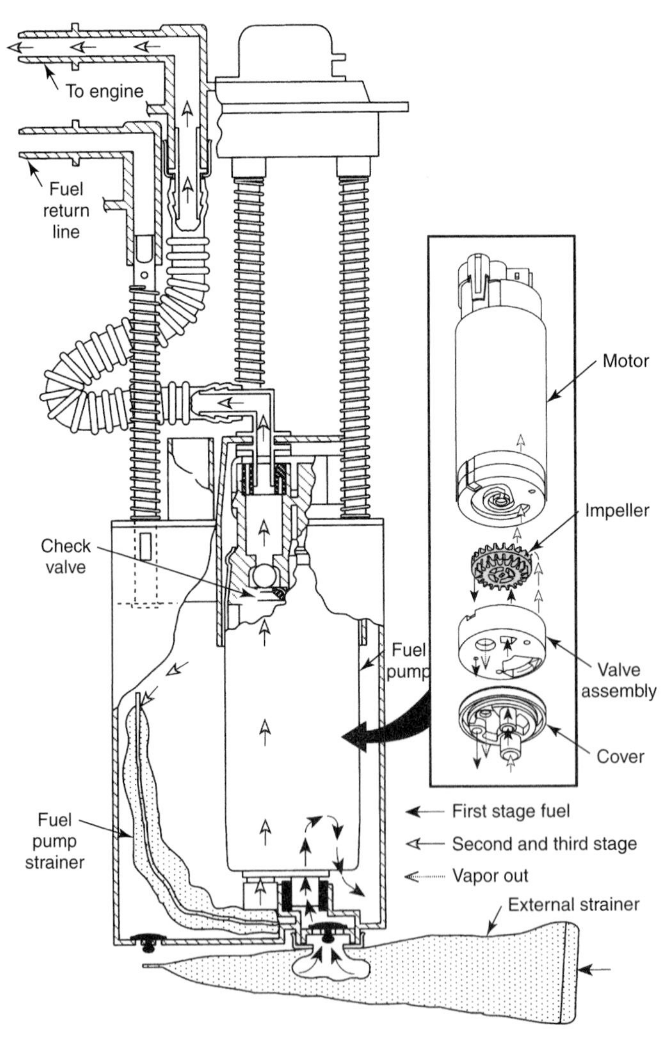

Figure 17-18. A cutaway view of an impeller-type electric fuel pump and related parts. Handle these units carefully to prevent damage. (General Motors)

pumps are grounded through the mounting bracket. Check the bracket mounting screws or ground circuit for tightness.

Testing Electric Fuel Pumps

The procedure for testing an electric fuel pump is generally similar to that for testing a mechanical fuel pump. Since electric fuel pump pressures are higher than those produced by mechanical pumps, the pressure gauge should be designed to read higher pressures. Refer to **Figure 17-21.** Since pressures produced by electric fuel pumps are relatively high, the volume is assumed to be sufficient if the pressure is correct. When stopped, there is no residual line pressure. Never run electric pumps without fuel. When performing electrical checks, follow the manufacturer's procedures to avoid damage to parts and to obtain correct test readings.

Some fuel pumps can be flow tested. To perform a flow test, disconnect the pump outlet at the fuel rail and run a hose to a graduated (marked) cylinder. Have an assistant energize the pump and note the flow rate. As a general rule, about 1/2 pint (1/4 liter) in 30 seconds is acceptable.

If a fuel pump appears to be stuck, try *lightly* tapping it with a wrench or a small hammer. If the pump begins working, replace it. If the pump was stuck, it will most likely stick again.

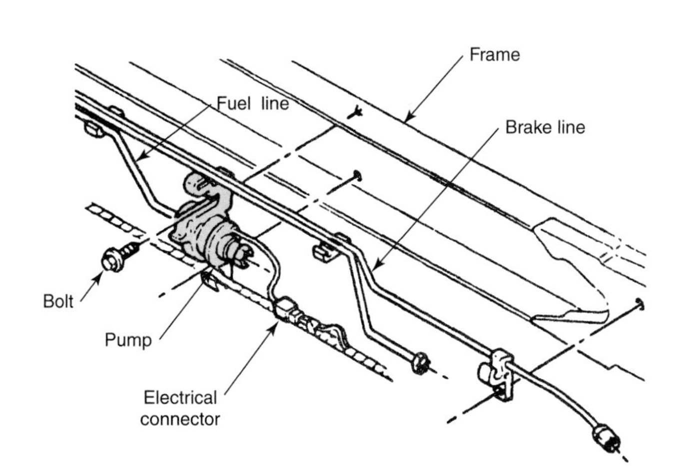

Figure 17-19. This electric fuel pump is mounted on the inside of the vehicle's frame.

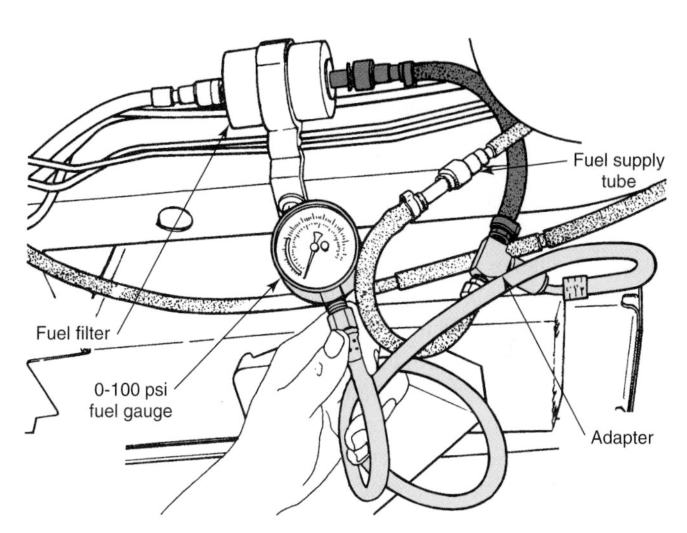

Figure 17-21. This technician is obtaining a fuel pump pressure reading with a fuel pressure gauge that has been attached to the fuel supply line coming from the pump.

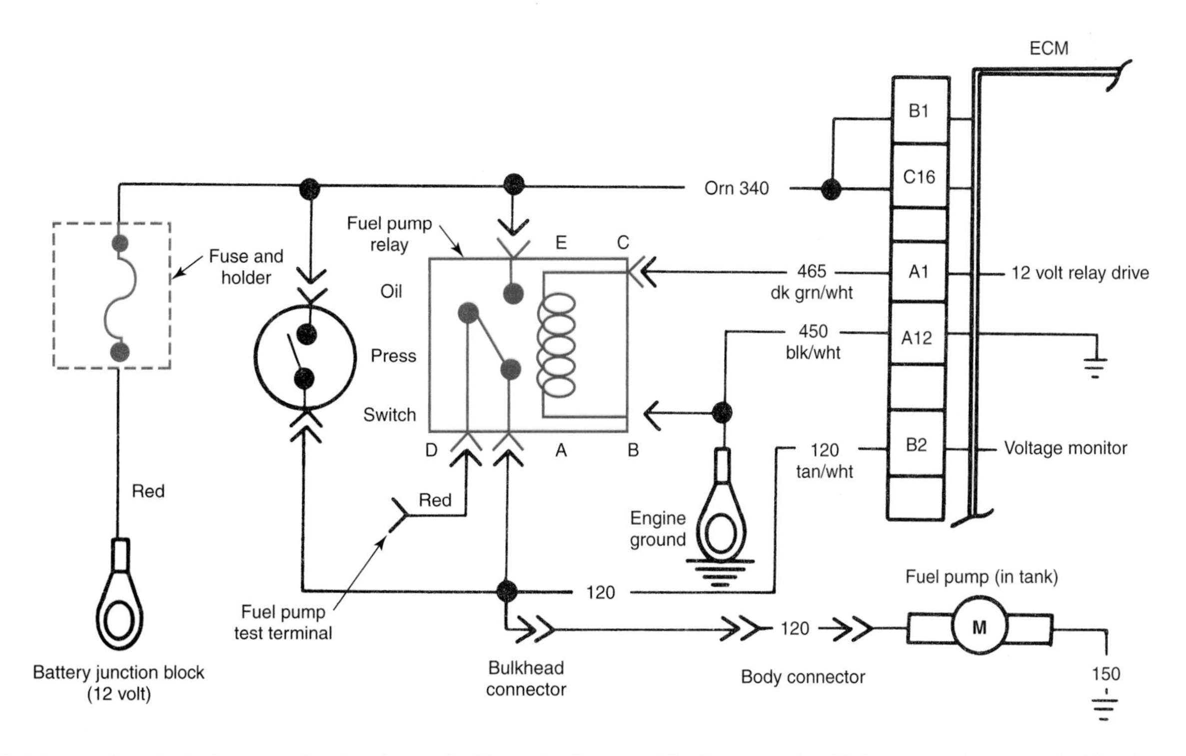

Figure 17-20. An electric fuel pump circuit schematic. Note the fuse and fuel pump relay. This system is controlled by the electronic control module. (AC-Delco)

Electric Fuel Pump Service

Most electric fuel pumps are replaced when they are faulty, and no service is possible. A few pumps can be at least partially serviced. The manufacturer's service manual should be checked before discarding the pump. Before replacing any electric pump, disconnect the negative battery cable.

Replacing an Inline Fuel Pump

The first step in replacing an inline pump, is locate it on the vehicle. Some pumps are placed under the vehicle in the fuel line between the fuel tank and engine. Other pumps are installed under the hood. The easiest way to find the pump is to look under the hood first. If the fuel line leads toward the tank with no sign of the pump, raise the vehicle and trace the fuel line back from the engine or forward from the tank.

After locating the pump, bleed off pressure as explained earlier in this chapter. Then remove the electrical connector. Place a pan under the pump and loosen the fittings. Allow excess fuel to drip into the pan, and then remove the fittings. If fuel continues to drip from the lines, plug them as necessary. Once the fittings are removed, remove the pump fasteners and remove the pump from the vehicle.

Compare the old and new pumps to ensure that the new pump is correct. Place the new pump in position on the vehicle. Install but do not tighten the pump fasteners. Then install and tighten the fuel line fittings. Finally, tighten the pump fasteners and install the electrical connector.

Turn the ignition switch to the *on* position and check pump operation. Make sure the pump's fittings do not leak.

Replacing an In-Tank Fuel Pump

In-tank fuel pumps are usually part of an assembly that also contains the fuel pickup tube and gauge sender. Sometimes, a pulsation damper is also part of the assembly. To remove the assembly, remove the fuel tank as explained earlier in this chapter. Once the tank is out of the vehicle, locate the cam lock. The ***cam lock*** is a large ring that slides under projections on the tank to hold the assembly in place. Locate any tabs that hold the cam lock in place and bend them out enough to allow the cam lock to be turned. Then turn the cam lock using a special tool. If a special tool is not available, the cam lock can be carefully tapped in the unlocking direction to loosen and remove it.

With the cam lock removed, lift the assembly from the tank. Remove the tank-to-assembly O-ring and discard it. Wipe up all spilled fuel, and then place the assembly on a clean workbench. Remove the pump from the pickup tube and sender as needed. Remove the strainer from the pump assembly and inspect it closely. If the strainer appears to be clogged or has any tears or holes, replace it.

Compare the old and new pumps to ensure that the new pump is correct. Then install the pump strainer and reassemble the pump to the pickup tube and sender, as necessary. Place the assembly in the tank, using a new O-ring. Some manufacturers recommend lubricating the O-ring before installation. Watch carefully as you lower the pump into the tank opening to ensure that the strainer is not damaged or folded over enough to restrict fuel flow. Once the pump is in place, install the cam lock. Be sure to tap the lock fully into position and reposition the tabs that hold the cam lock. Reinstall the tank in the vehicle. Add fuel as necessary; then turn the ignition switch to the *on* position and check that the pump operates correctly.

Electric Fuel Pump Inertia Switch

Some vehicles use an ***inertia switch*** to shut off the electric fuel pump during an accident. In a typical inertia switch, a steel ball is held in place by a magnet. See **Figure 17-22.** On impact, the ball breaks away, rolls up a ramp, and strikes a target plate. The target plate opens an electrical switch, cutting off power to the fuel pump. The inertia switch must be manually reset before the vehicle can be restarted. To reset the inertia switch, locate it in the trunk or rear storage compartment. Then push the button on the top of the switch. Start the vehicle and recheck pump operation.

A digital volt/ohm meter can be used to test the inertia switch. If the voltage reading across the electrical contacts exceeds 0.3 volts when the switch is closed and the circuit is energized, the switch must be replaced. To replace the inertia switch, remove the electrical connector and fasteners and remove the switch from the vehicle. Reverse the procedure to install the new switch.

Electric Fuel Pump Oil Pressure Switch

On some vehicles, the fuel pump is wired through an oil pressure switch. When the switch is closed by oil pressure, it allows current to flow to the fuel pump motor. This oil pressure switch may be the same one that operates the instrument panel gauge or light, or it may be a separate switch that operates the fuel pump only. The actual purpose of the oil pressure switch varies from manufacturer to manufacturer. On some vehicles, the switch will not allow the pump to operate unless the engine has oil pressure. Cranking the engine will provide enough pressure to close the switch and energize the pump. On other systems, the oil pressure switch is a backup device for the fuel pump relay. If the relay fails, the oil pressure switch will direct current to the pump as long as the engine has a

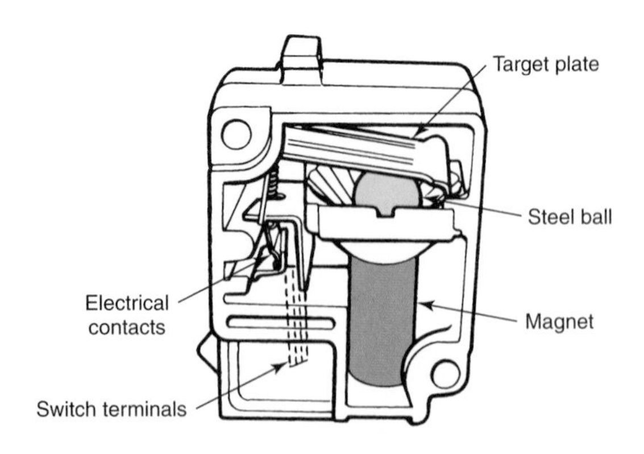

Figure 17-22. Cutaway view of an inertia switch. Caution: If you see or smell fuel, do not reset the switch until the fuel leak is corrected. (Ford)

minimum amount of oil pressure. Usually the needed oil pressure is low, about 5–7 psi (35–48 kPa). Before diagnosing the oil pressure switch, you must know its exact function. Consult the manufacturer's service literature and schematics, and follow all test procedures exactly.

Most oil pressure switches can be tested with a non-powered test light. Connect the test light lead to the positive battery terminal or another source of power. Disconnect the switch's electrical connector and place the test light probe on the switch terminal. Have an assistant start the engine. The test light should come on when the engine starts and oil pressure rises. If it does not, the switch is defective.

To replace an oil pressure switch, allow the engine to cool and then remove the electrical connector. Using the appropriate socket, remove the pressure switch from the engine block. Coat the new switch with sealer if required. Then install the new switch in the block and tighten it to specifications. Start the engine and ensure that the switch responds to rising oil pressure.

Fuel Filter Service

Filters play an important part in maintaining a properly running engine. Clean or change the filters at recommended intervals or as needed. Most modern ***fuel filters*** are designed to be thrown away rather than being cleaned and reused. Some diesel engines and older gasoline engines have an internal paper element that can be changed. Others require disposal of the entire unit, **Figure 17-23**. Some carburetor designs incorporate a filter in the fuel inlet, as shown in **Figure 17-24.**

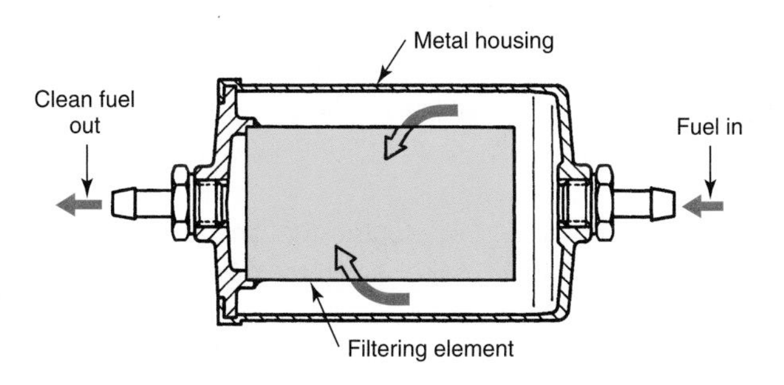

Figure 17-23. A cross section of a gasoline filter. The entire filter assembly must be discarded when clogged. Be sure to install the new filter with the flow arrows pointing in the right direction. (Mercedes-Benz)

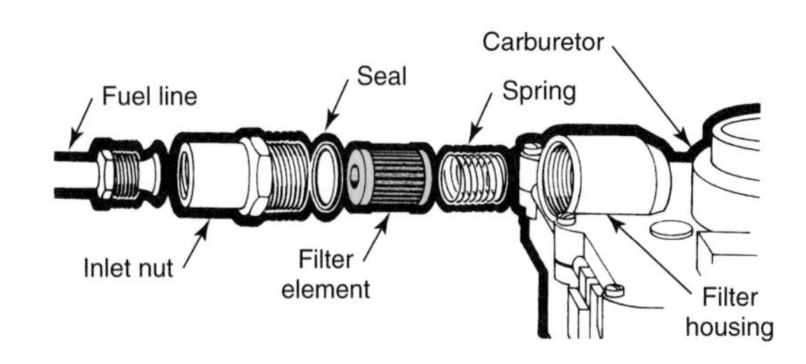

Figure 17-24. One type of carburetor inlet fuel filter. (Fram)

Many diesel fuel systems utilize a filter that not only removes rust and dirt particles but also contains a ***water trap*** that protects against the entry of water, which can cause serious damage if it gets into the injector pump. One such filter is pictured in **Figure 17-25.** This particular filter incorporates a ***water sensor*** to actuate a dash light when water is present. It also has a provision for draining off any water that may accumulate. When apparent fuel pump troubles occur, check filters to make certain they are clean. Some require presoaking to remove filter particles, which can clog injectors. Always check for leaks after servicing a fuel filter.

Vapor Lock

Vapor lock is a condition in which the fuel pump or the lines become heated to the point that the fuel inside begins to vaporize. This vaporization causes the formation of tiny air

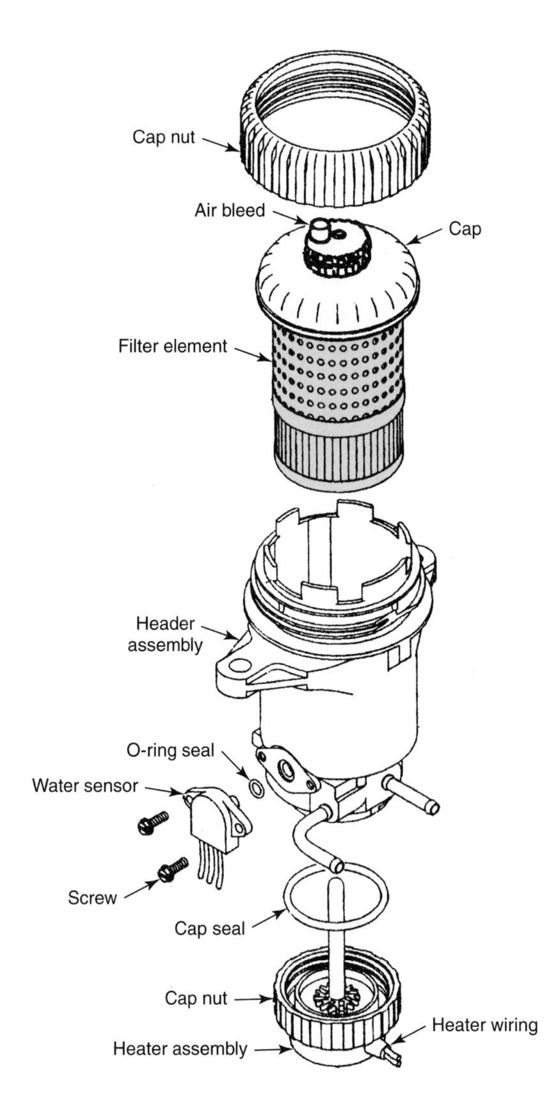

Figure 17-25. Exploded view of a diesel fuel filter assembly. This setup uses a water sensor, which triggers a dash light to warn the driver of water in the fuel. Note the fuel heater, which is used to help thin cold fuel. The heater also helps prevent wax buildup, which can clog the filter. (General Motors)

bubbles. If enough air bubbles are formed, fuel flow to the carburetor can be reduced, or stopped in some cases. Some causes of vapor lock include routing fuel lines too close to the exhaust manifold, excessive looping and bending of the fuel line, or failure to reinstall a heat shield. Vapor lock is seldom a problem on fuel injected vehicles, since pressurizing the fuel raises its boiling point.

When vapor lock occurs, it can be temporarily cured by stopping the engine and placing cold wet rags on the lines and the fuel pump. As soon as the fuel cools, the vapor condenses and the car should start. To correct the situation, determine the cause for the vapor lock condition and make any necessary repairs. Sometimes, gasoline manufacturers will increase the ***volatility*** (ability to vaporize) of gasoline during cold weather. If there is a sudden hot spell or if the vehicle is driven through a hot area, the gasoline can vapor lock even if the fuel lines are correctly mounted. This can sometimes be cured by using high-octane gasoline, which has a lower volatility.

Tech Talk

In the past, automotive components such as fuel pumps, water pumps, alternators, and carburetors were always rebuilt when they began causing problems. Today, however, it is sometimes cheaper and easier to replace components than to rebuild them. Major components that are commonly replaced instead of being rebuilt include alternators, starters, water and fuel pumps, carburetors, fuel injectors, air conditioner compressors, and power steering pumps. Sometimes, complete engines and transmissions are replaced instead of being rebuilt.

Nevertheless, it is still possible to repair some defective major components by replacing a few simple parts. Many alternators, for instance, can be restored to service by replacing the bearings and brushes. It is still common to rebuild brake system wheel cylinders. Carburetors can often be rebuilt successfully.

Some units used on modern vehicles, such as master cylinders and water pumps, should not be rebuilt unless a new part is not available. A good rule of thumb that will save time in the long run is to rebuild a component once only. If it begins to give you trouble again, replace it.

Summary

Fuel tanks are constructed of treated steel or plastic. Use a special siphoning device to remove fuel from tank and store the fuel in an approved container. Try to siphon as much fuel as possible from the tank prior to removal. Metal fuel tanks can sometimes be repaired by soldering, brazing, or cold patching. Remember that fuel tanks can explode with great force.

Fuel lines can be cleaned by forcing air through them. Use the proper size and type replacement tubing or hose. Route fuel lines away from hot areas or use heat baffles. Attach all fuel lines securely and avoid cross threading.

Fuel pumps are either mechanically or electrically operated. Some electric fuel pumps are located in the tank. Test fuel pumps for pressure, flow volume, and inlet vacuum. Use the proper tools and take care to prevent a fire from spraying or leaking fuel. When replacing a pump, obtain as much information as possible to select the correct pump. Support all fittings while tightening the flare nut.

Keep all fuel filters clean. Service the filter or replace if of the disposable type. Vapor lock can slow or completely stop delivery of fuel to the engine. Protect fuel lines from heat to minimize the chance for vapor lock.

Review Questions—Chapter 17

Do not write in this book. Write your answers on a separate sheet of paper.

1. When servicing a fuel injection system, it is important to relieve the residual _____ in the system.
2. What is the most important thing to remember about removing and repairing a fuel tank?
   (A) Some tanks are sealed.
   (B) Tanks are held by straps or bolts.
   (C) Gasoline is extremely dangerous.
   (D) Tanks have electrical connections.
3. What should be done to remedy an expanded fuel tank?
4. When cleaning a fuel line, what direction should air pressure be directed?
5. Mechanical fuel pumps should always be tested for _____ pressure and _____ pressure.
6. Many electric fuel pumps are operated by a pump _____.
7. When installing a mechanical fuel pump, be sure the _____ arm contact surface bears against the cam or eccentric.
8. What is the first step that should be taken when replacing an inline fuel filter?
9. In-tank electric fuel pumps are often part of an assembly that also contains the fuel _____ _____ and the gauge _____.
10. When reinstalling an electric fuel pump in a tank, the technician must make sure that the _____ is not damaged.
11. Inertia switches will turn off the electric fuel pump under what conditions?
    (A) High fuel system pressure.
    (B) Low fuel system pressure.
    (C) An accident.
    (D) All of the above.
12. When used, an oil pressure switch allows current to flow to the fuel pump when the switch is _____ by oil pressure.
13. Diesel fuel systems often employ a special sensor to alert the driver that there is _____ in the fuel system.
14. What causes vapor lock?
15. Pressurizing fuel raises its _____ _____, so vapor lock is seldom a problem on fuel injected vehicles.

ASE–Type Questions

1. Fuel system pressure can be removed by all of the following methods *except:*
   (A) using the pressure relief feature on a pressure tester.
   (B) disconnecting the fuel pump and starting the engine.
   (C) disconnecting the ignition coil and cranking the engine.
   (D) using a hose with fitting that will depress the pressure relief valve.
2. Fuel tanks are usually attached to the vehicle with _____.
   (A) metal straps
   (B) bolts
   (C) an interference fit between the frame and body
   (D) Either A or B.
3. Today, most leaks in fuel tanks are fixed by a method called _____.
   (A) brazing
   (B) cold patching
   (C) soldering
   (D) welding
4. When fuel pump trouble is suspected, all of the following should be checked before condemning the pump *except:*
   (A) fuel tank vent.
   (B) fuel lines.
   (C) fuel filters.
   (D) fuel octane.
5. If a mechanical fuel pump has a vapor return line, what must be done before testing?
   (A) Remove the pump from the engine.
   (B) Pinch off the vapor return line.
   (C) Drain the fuel tank.
   (D) Disconnect the inlet line from the pump.
6. Technician A says that all mechanical fuel pumps can be rebuilt. Technician B says that the pump fasteners should never be tightened to force a mechanical fuel pump into place. Who is right?
   (A) A only.
   (B) B only.
   (C) Both A and B.
   (D) Neither A nor B.
7. Electric fuel pumps can be mounted _____.
   (A) on the engine
   (B) in the fuel tank
   (C) on the vehicle frame
   (D) All of the above.
8. Inertia switch operation is being discussed. Technician A says that the voltage reading across the switch's electrical contacts should not exceed 0.3 volts when the switch is closed and the circuit is energized. Technician B says that once the inertia switch has been triggered (opened), it must be replaced. Who is right?
   (A) A only.
   (B) B only.
   (C) Both A and B.
   (D) Neither A nor B.
9. Technician A says that on some vehicles, the fuel pump will not operate unless the engine has oil pressure. Technician B says the oil pressure switch is sometimes used as a backup device for the fuel pump relay. Who is right?
   (A) A only.
   (B) B only.
   (C) Both A and B.
   (D) Neither A nor B.
10. Volatility is the ability of gasoline to _____.
   (A) burn
   (B) vaporize
   (C) condense
   (D) absorb water

Suggested Activities

1. Draw a schematic of fuel flow from the fuel tank to the engine on a **fuel injected** engine. Include all filters, pumps, and lines.
2. Check fuel pump pressure on a vehicle selected by your instructor and write down your readings.
3. Discuss the fuel pump pressure readings obtained in Activity 2 with your classmates. What do the readings reveal about the condition of the fuel pump?
4. Write step-by-step instructions for checking mechanical fuel pumps and electric fuel pumps.
5. Inspect a fuel filter for clogging. Discuss your findings with your instructor or the class.

This fuel-injected V-8 engine is used in a high-performance sports car. Note the location of the fuel rail and the injectors. (Ford)

18

Fuel Injection System Service

After completing this chapter, you will be able to:

- Identify types of fuel injection systems.
- Identify fuel injection system components.
- Describe the operation of mechanical gasoline fuel injection systems.
- Identify safety rules to be followed when servicing fuel system components.
- Explain how to check fuel injection system pressure.
- Explain how to check fuel injector solenoid and fuel pump operation.
- Explain how to use waveform meters to check fuel injection component condition.
- Explain how to remove and replace fuel system components.
- List the major steps to installing an aftermarket fuel injection system.
- Identify diesel fuel injection system components.
- Explain how to check diesel fuel injection system operation.
- Explain how to remove and replace diesel fuel injection components.

Technical Terms

Electronic fuel injection
Fuel injectors
Solenoid winding
Pulse width
Electric fuel pump
Pressure regulator
Returnless fuel system
Throttle body
Throttle valve
Throttle positioner
Idle air control valve (IAC)
ECM
Drivers
Input sensors
Fuel trim
Fuel injection system pressure
Fuel pressure fitting
Noid lights
Injector balance test
Schrader bleed valve
On-vehicle injector cleaner
Calculated charge system
Mechanical fuel injection system
Cold start injector
Aftermarket fuel injection systems
Surge tank
Diesel fuel injection system
Runaway diesel engine
Test stand
Test liquid
Nozzle tightness
Purged
Glow plugs

Introduction

All modern cars, trucks, and SUVs use electronic fuel injection systems. The onboard computer that governs the ignition and emission control systems also controls the fuel injection system. To service modern fuel injection systems, the technician must understand the electronic and mechanical parts involved in their operation. The technician must also be able to identify and use specialized fuel injection test and service equipment. This chapter covers the components and operating principles of electronic fuel injection, as well as the use of service equipment. Older mechanical gasoline injection systems are also discussed. Additionally, the operating principles of and the service procedures for automotive diesel engines are explained. Studying this chapter will give you the knowledge needed to test and repair modern fuel injection systems.

Electronic Fuel Injection Systems

There are three basic types of ***electronic fuel injection:*** throttle body fuel injection, multiport fuel injection, and central point fuel injection.

Throttle Body Fuel Injection

The ***throttle body fuel injection*** system uses one or two fuel injectors located in a throttle body. The throttle body is mounted on the intake manifold, in approximately the same place that carburetors were formerly mounted. Throttle body fuel injection systems are called by various names, including single-point and central fuel injection (CFI). These systems inject fuel ahead of the throttle valve. A schematic of a typical throttle body injection system with two injectors is shown in **Figure 18-1.** Throttle body fuel injection has been replaced by multiport or central point fuel injection in all late-model vehicles.

Multiport Fuel Injection

Most injection systems found in late-model vehicles are called ***multiport fuel injection systems.*** These systems use one injector per cylinder. The injectors are located inside the intake manifold or in the cylinder head, close to the intake valves. The injectors are connected to the fuel system through a rigid steel tube called a fuel rail. See **Figure 18-2.** Some

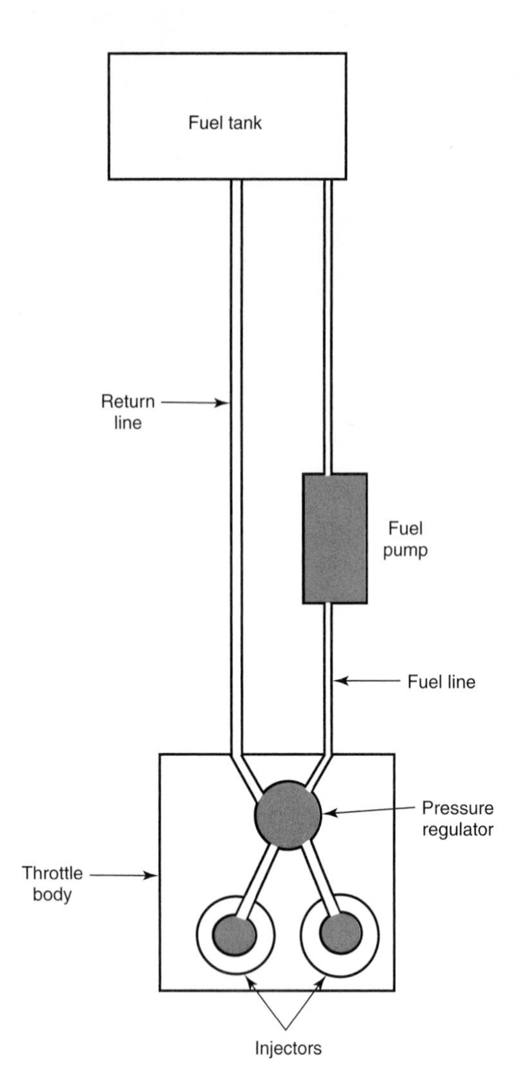

Figure 18-1. Throttle body fuel injection system. The throttle valve, injectors, and pressure regulator are installed in single unit on top of the intake manifold.

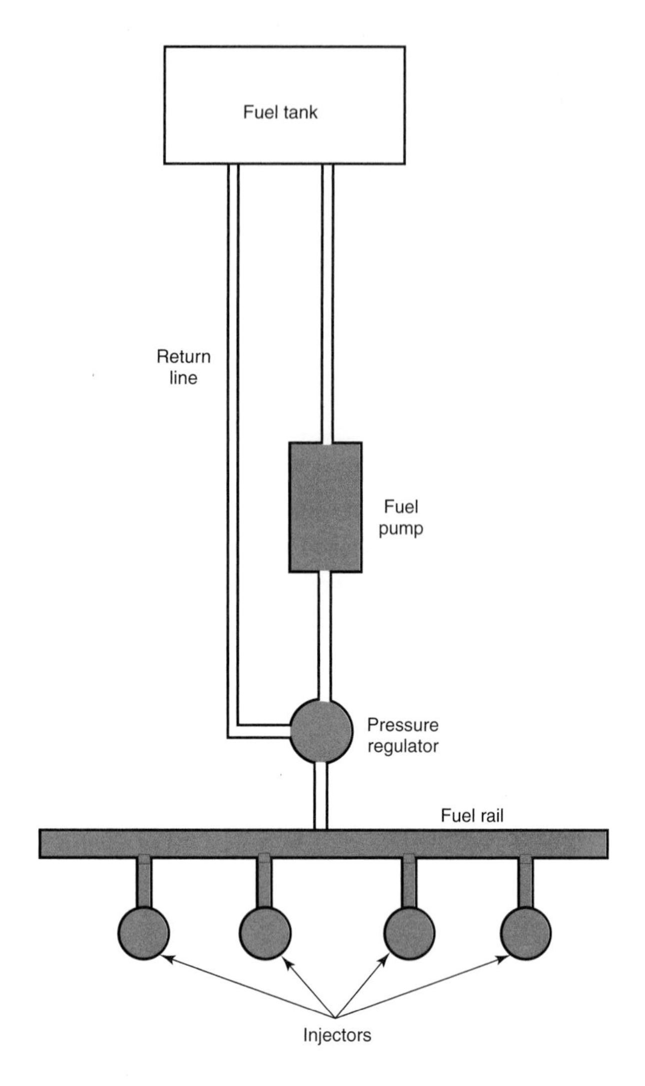

Figure 18-2. Multiport fuel injection system. This particular system is for a four-cylinder engine.

systems use flexible hoses to connect the injectors to the fuel rail.

Central Point Fuel Injection

A variation of the multiport injection system uses one centrally located fuel injector with lines connecting it to outlet nozzles at each intake passage. This system is called ***central point fuel injection,*** or ***CPI.*** The CPI assembly is mounted on the top of the intake manifold. A CPI system combines the simplicity of the single injector with the precise fuel control of the multiport system. See **Figure 18-3.**

Fuel Injection System Components

One of the advantages of fuel injection is its simplicity. A fuel injection system contains surprisingly few components. These components are discussed in the following sections.

Fuel Injectors

The ***fuel injectors*** receive pressurized gasoline and spray it into the intake manifold. When an injector is energized, its ***solenoid winding*** forms a strong magnetic field that pulls the injector's valve upward against spring pressure. This opens the injector, allowing fuel to spray out. When the solenoid is de-energized, spring pressure forces the valve closed. The amount of time the injector remains open, often called the ***pulse width,*** is measured in thousandths of a second.

Fuel Pumps

All electronic fuel injection systems have at least one ***electric fuel pump.*** The pump is energized through a relay that is controlled by the ECM. Some pumps are energized through the oil pressure switch.

A few systems have two fuel pumps. One pump supplies the system with large quantities of fuel at low pressures. The

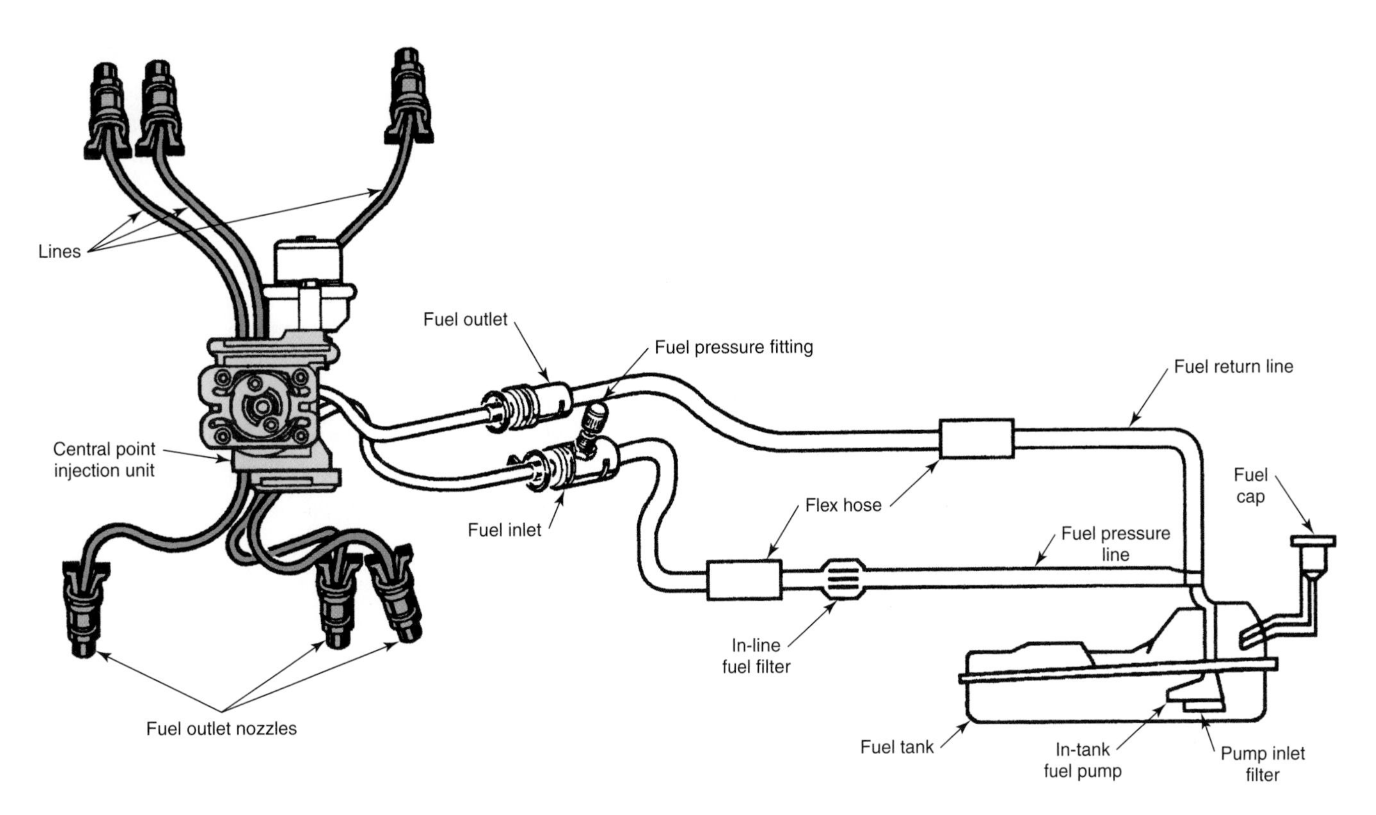

Figure 18-3. The central port injection system uses one injector with individual lines to the outlet nozzles. (General Motors)

other pump creates the high pressures needed to force fuel from the injectors. Electric pumps are covered in detail in Chapter 17, Fuel Delivery.

Pressure Regulators

To control injection system pressures, a ***pressure regulator*** bleeds fuel back into the fuel inlet line or fuel tank. The pressure regulators used on most throttle body fuel injection systems are located in the throttle body and operated by spring pressure only. On multiport and central point fuel injection systems, the regulator is attached to the fuel rail. In multiport regulators, the spring pressure is assisted by intake manifold vacuum. Changes in engine load affect manifold vacuum. The pressure regulator uses the change in vacuum to modify the fuel pressure setting.

Returnless Fuel Systems

Some late-model fuel injection systems do not use a pressure regulator. Instead, the vehicle's computer regulates fuel pressure by varying electric fuel pump output. Since there is no return line, this system is called a ***returnless fuel system.*** A fuel pressure sensor is installed on the injector fuel rail. The computer evaluates the pressure input from this sensor, as well as other sensor information, and adjusts the current to the fuel pump motor as needed.

Some returnless fuel systems have a conventional fuel pump that operates at all times. A pressure regulator mounted on the pump relieves excess pressure directly back to the fuel tank.

Throttle Bodies and Related Components

The ***throttle body*** contains the ***throttle valve,*** which opens and closes to control the amount of air entering the intake manifold or plenum. On throttle body fuel injection systems, the throttle body also contains the fuel injector(s). A throttle positioner is used on throttle body injection systems. The ***throttle positioner*** is a motor or solenoid that opens or closes the throttle plate. On multiport systems, the throttle body contains an ***idle air control valve (IAC).*** The IAC controls idle speed by bypassing the throttle plate, allowing air to flow in response to ECM commands. A multipoint throttle body and related parts are pictured in **Figure 18-4.**

ECM

The ***ECM*** energizes the fuel injectors and varies the injector pulse width based on the inputs from various sensors. The ECM contains internal power transistors called ***drivers,*** which deliver current to the injectors. The ECM also controls the operation of the idle air control valve (IAC). Some ECMs energize the electric fuel pump(s), usually through a relay. The ECM is covered in detail in Chapter 15, Computer System Diagnosis and Repair.

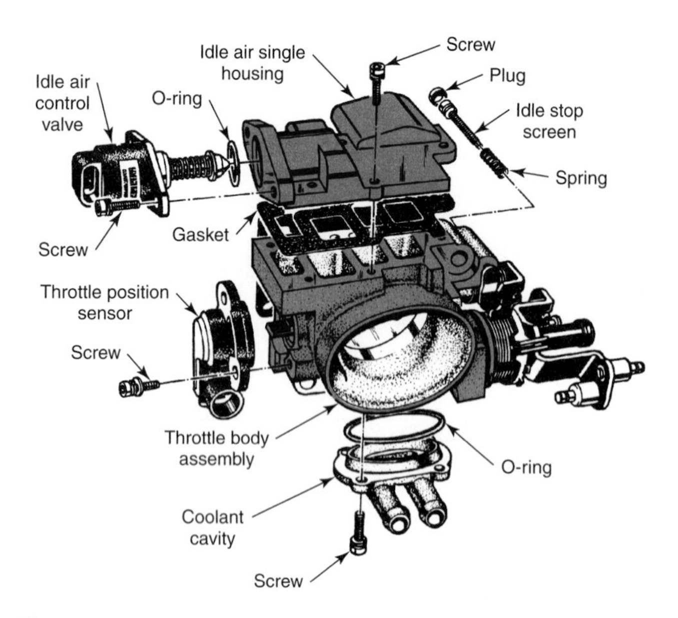

Figure 18-4. This illustration shows a multiport throttle body and related parts. Note the throttle position sensor and the idle air control valve. (General Motors)

Input Sensors

The ***input sensors*** monitor various engine functions and feed this information to the ECM. The number and types of sensors vary with the system. Sensors that have the most effect on the operation of the fuel injection system include the following:

- Oxygen sensor(s).
- Engine speed sensor.
- Throttle position sensor.
- Manifold vacuum (MAP) sensor.
- Atmospheric pressure (BARO) sensor.
- Temperature sensors.
- Airflow sensor.

Sensors are covered in detail in Chapter 15, Computer System Diagnosis and Repair.

Fuel System Safety

Fuel system service can be extremely dangerous. Gasoline and diesel fuels are extremely flammable and explosive. Study the following precautions before reading further in this chapter. Other cautions will be given as they are needed.

- Before working on fuel lines or removing any part of the fuel injection system, always disconnect the battery ground cable.
- Carefully bleed off line pressure by opening a fitting a small amount or by using the bleed valve of a fuel pressure gauge. Cover the fitting or bleeder with a cloth to catch any fuel that may spew out.
- Clean up any fuel spills immediately. Dispose of fuel-soaked cloth properly.
- Never allow sources of ignition near the fuel system or spilled gasoline.

Electronic Fuel Injection Diagnosis

The following portion of this chapter covers the diagnosis of electronic fuel injection systems. The causes and corrections of fuel injection system problems are similar for all systems. Typical injection system problems include the following:

- Engine will not start.
- Engine starts, but stalls.
- Engine is hard to start (excessive cranking) when hot or cold.
- Engine idles rough.
- Engine stays on fast idle.
- Engine has no fast idle when cold.
- Engine hesitates on acceleration.
- Engine cuts out or misfires at all speeds.
- Engine performs poorly at high speeds.
- Engine performs poorly at low speeds.
- Engine has poor gas mileage.
- Engine backfires.
- Engine lacks power.
- Engine surges (speed varies with accelerator pedal steady).
- Engine is hydraulically locked. (One cylinder fills completely with fuel and the engine cannot turn.)
- Black smoke emanates from exhaust.

Initial System Inspection

Before performing diagnostic tests, give the system an initial inspection. This often pinpoints the problem without further checking. A vacuum leak or something as simple as the wrong PCV valve can cause the same symptoms as a severe fuel injection problem. Question the vehicle's owner about the problem. Determine the conditions under which the problem occurs, how it sounds, and how it affects engine performance. If necessary, road test the vehicle to verify the problem.

Perform the following steps as part of the initial inspection:

- Obtain trouble codes. Retrieving the trouble codes sometimes pinpoints the problem, eliminating the need for further checking. For more information on retrieving trouble codes, refer to Chapter 15, Computer System Diagnosis and Repair.
- Check all the fuel lines for loose fittings, cracks, and pinched or collapsed sections.
- Check the vacuum lines for poor connections, improper arrangements, kinks, and leaks.
- Check the wiring for loose connectors, improper connections, shorting, frayed or broken wires, and blown fuses. Carefully check the ground wires.
- Check for debris in the inline filter and gas tank sock filter. Make sure the filter is installed in the proper flow direction.
- Check for carbon on the throttle plates or intake valves.

- Inspect mechanical linkage for freedom of operation and proper adjustment.
- Check the air intake and air cleaner for clogs and obstructions.
- Check the level, type, and octane rating of the fuel being used.
- Inspect related emission controls for possible malfunctions.
- Check the exhaust system for kinks or clogs.
- Make sure the problem is not in the engine ignition or compression systems.

Using a Scan Tool

Before going further with diagnosis, attach a scan tool to the vehicle and observe the condition of the electronic components that affect fuel system operation. Sensors, output devices, and air-fuel ratio can be checked with the scan tool. A scan tool can pinpoint many problems without the need for further checking. The use of scan tools was covered in Chapter 15, Computer System Diagnosis and Repair.

Checking Fuel Trim

Fuel trim is another way of describing the mixture, or air-fuel ratio, of the fuel injection system. Measuring fuel trim allows the technician to determine how well the fuel injection system is maintaining the air-fuel ratio and reacting to other engine conditions. Fuel trim is measured in counts, which are a series of numbers. The counts are a measure of injector pulse width, and indicate a rich or lean condition. Counts can be checked with a scan tool. Comparing the actual counts with the counts specified by the manufacturer will give the technician an overall idea of how the fuel system is performing. The chart in **Figure 18-5** shows how the counts relate to the air-fuel ratio on one manufacturer's fuel injection system. Fuel trim is covered in more detail in Chapter 20, Drivability Diagnosis.

Checking Fuel Injection System Pressure

Checking ***fuel injection system pressure*** is a way to determine the overall condition of the system, including the pump(s), pressure regulator, injectors, filters, and lines.

Short-Term Fuel Trim Operation

Oxygen sensor indication	Fuel control response	Short-term fuel trim (percent)	Short-term fuel trim (counts)
Lean mixture	Increase fuel	1% to 30%	Greater than 128 counts
Correct mixture	No change	0%	128 counts
Rich mixture	Decrease fuel	–1% to –30%	Less than 128 counts

Figure 18-5. Fuel trim shows the relationship between mixture and injector pulse width. In this particular system, a fuel count of 128 indicates the correct mixture and pulse width. A higher fuel count indicates that the ECM is increasing the pulse width to compensate for a lean mixture. A lower count tells the technician that the ECM is reducing the pulse width to correct a rich mixture. (General Motors)

Note: Before beginning the fuel injection system pressure test, obtain the proper pressure specifications. Test results must be compared with the specifications.

After obtaining the specifications, locate the ***fuel pressure fitting,*** **Figure 18-6.** Remove the cap from the pressure fitting and attach the pressure gauge to the fitting. If necessary, clean up any spilled gasoline before proceeding. Turn the ignition switch to the *on* position and note the pressure indicated on the gauge, **Figure 18-7.** If the pressure is within specifications, proceed to the next step. If no pressure registers, the fuel pump is not working or the pressure gauge is not installed correctly. Before deciding that the pump is not operating, recheck the gauge connection and make sure the hose connector is opening the Schrader valve.

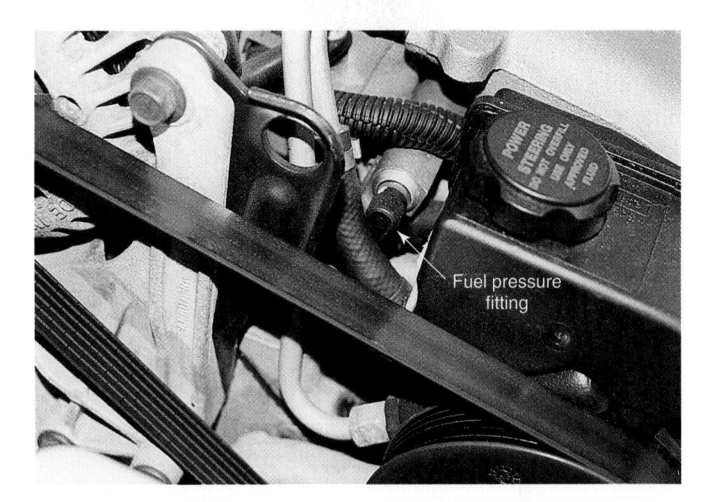

Figure 18-6. Fuel pressure test fittings can be located in many places on the fuel rail. This test fitting is located on the side of the rail, behind the drive belt. A 90° adapter must be used to attach the fuel pressure gauge to the fitting without contacting the drive belt.

Figure 18-7. The gauge reading shown here indicates static pressure, which is the pressure provided by the fuel pump and pressure regulator when the engine is not running.

Note: If the vehicle was driven into the shop, it had at least some fuel pressure. A zero reading on a vehicle that was driven into the shop is most likely the result of an incorrect gauge connection.

A zero pressure reading can also be the result of a defective pump, a faulty relay, or a blown fuse. Refer to the proper schematic and check the condition of the related fuse. If the fuses are good, locate the pump relay. Most relays are in one of the engine compartment fuse boxes, **Figure 18-8.** If the relay is mechanical, place your fingers on the relay and have an assistant turn the ignition switch to the *on* position. The relay should click and vibrate slightly when it is energized. The only way to test a solid-state relay is to substitute a known good unit for the suspect relay.

As a further check, use jumper wires to energize the pump directly. Before applying power to the pump, disconnect the pump wiring harness to ensure that the ECM or relays are not damaged. If the pump does not run when supplied with current, tap the pump with a wrench or screwdriver. Tapping will often free up a stuck pump. If the pump begins working after it is tapped, replace it, because it is likely to stick again.

Note: Occasionally, a hard start may be caused by a defective or disconnected oil pressure switch. Some fuel pumps do not receive power until oil pressure reaches 4 psi–7 psi (24 kPa–42 kPa) and the switch closes.

If the pump is working, record the pressures with the key on. Then, start the engine and recheck the pressure. On a vehicle with a vacuum-assisted pressure regulator, pressure should drop several pounds when the engine is started. Compare the readings in **Figure 18-9** with the reading in Figure 18-7. On a system with a vacuum-assisted pressure regulator, pressure should drop slightly when the engine is started. If pressure does not drop when the engine is started, check the pressure regulator as described in the following section. If pressures are excessively high, suspect a restricted fuel return line or a stuck pressure regulator check ball.

Figure 18-8. Pump relays are usually located under the hood. They can be mounted in a separate relay box, as shown here, or in the underhood fuse box.

Figure 18-9. This gauge reading shows a decrease in fuel pressure once the engine starts. A pressure drop indicates that the vacuum fuel pressure regulator is working.

If the pressure gauge needle rises or falls slightly at regular intervals, an injector may be failing to open. Such a failure can be caused by a disconnected wiring harness, a defective ECM driver, an internal failure in the injector, or a plugged injector screen. Further injector checks are explained later in this chapter.

Checking the Fuel Pressure Regulator

To check the pressure regulator, check the system pressure with key on and the engine off as described earlier. Next, start the engine. Pressure should be lower with the engine running than with engine off. If the pressure regulator has a vacuum connection, remove the vacuum line with the engine running, **Figure 18-10.** If pressure rises, the pressure regulator is probably working properly.

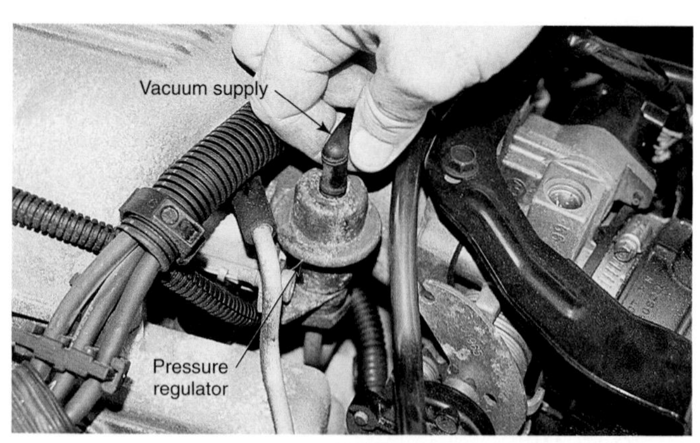

Figure 18-10. To check the operation of a vacuum fuel pressure regulator, remove the vacuum line with the engine running. Fuel pressure should increase when the line is removed.

A leaking pressure regulator check ball sends excessive fuel back to the tank, causing low pressures. The check ball may bleed off all pressure with the engine off, causing hard starting. A stuck check ball causes excessively high pressures, although this is rare. A leaking pressure regulator diaphragm allows fuel to leak directly into the intake passages, causing a rich mixture. If the pressure leaks with the engine off, pinch off the return line. If pressure now holds, the pressure regulator check ball may be defective. If the pressure continues to drop, an injector may be leaking or the pressure regulator diaphragm may be ruptured. A leaking injector causes an excess of gasoline in one cylinder, while a leaking diaphragm causes all cylinders to be rich.

Checking a Returnless Fuel System

Diagnosing a returnless fuel system is similar to servicing a system with a fuel pressure regulator. Returnless fuel systems have a pressure fitting for checking system pressure and pressure changes in response to changes in engine operation. All returnless fuel systems are operated by the on-board computer. The computer will store a trouble code when a malfunction occurs in the system.

Checking Fuel Injector Operation

Fuel injectors are relatively trouble free, but they can cause problems that are difficult to track down. If one cylinder seems to be misfiring, check the ignition system and make sure the cylinder has compression before checking for an injector problem.

An injector that sticks closed may cause the ECM to create a rich condition. This is because the oxygen in the air passes through that cylinder without combining with fuel (burning). The uncombined oxygen causes the oxygen sensor to read a lean mixture. The sensor will tell the ECM to richen the mixture, causing increased emissions and possibly overheating the catalytic converter. One way to check for such a condition is to add propane to the intake manifold as the engine is running. If the rich condition goes away when propane is added, suspect a sticking injector or a plugged injector screen.

Checking Injectors on the Vehicle

The easiest way to check injectors is to place a screwdriver or a long rod on the injector as the engine operates. If you hear a clicking sound, the injector is operating. If no clicks can be heard, the injector is defective or the control system is not energizing it. On throttle body fuel injection systems, the spray can be observed as the engine operates. To make the spray pattern more visible, connect a timing light to the engine and point the light at the injector as the engine runs. The light illuminates the spray every time the spark plug fires, **Figure 18-11.**

Using Noid Lights

Noid lights can be used to check the output of fuel injector drivers. To use a noid light, disconnect an electrical connector lead at an injector and attach the light to the connector, **Figure 18-12.** Then crank the engine. If the light

Figure 18-11. To observe the spray pattern on a throttle body injection system, point a timing light at the injector as the engine operates. The light illuminates the spray, making the spray pattern clearly visible.

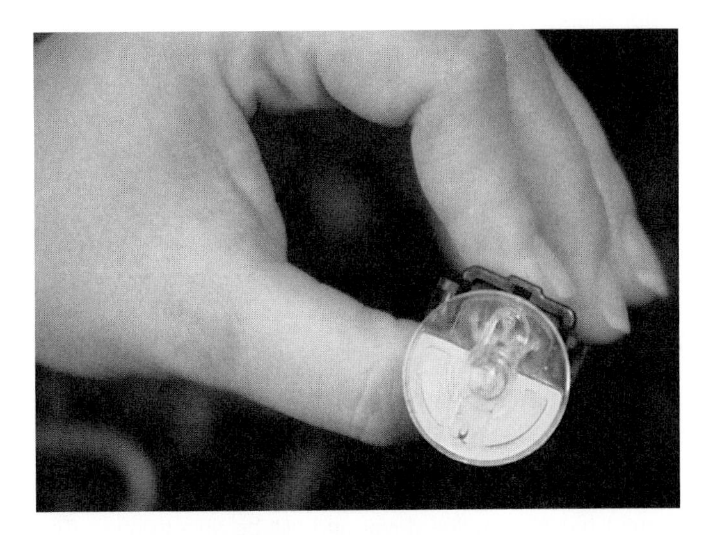

Figure 18-12. A noid light provides a quick way to determine whether voltage is reaching the fuel injector. It can be used on throttle body and multiport injection systems.

comes on, the injector driver is delivering power to the injector. This indicates that the problem is in the injector or fuel supply, or that it is not related to the fuel system. If the light does not come on when the engine is cranked, the computer is defective or is not receiving power.

Note: In some systems, voltage at the connector will illuminate the noid light but be too weak to open the fuel injector.

If there is any doubt about the amount of voltage reaching the injector, attach a voltmeter to the connector lead and measure the exact voltage. If the voltage available at the injector is lower than specified, suspect the ECM, wiring harness, or connectors.

Checking Injector Windings

When the windings in an injector are only partially shorted (some of the windings are bypassed when the internal insulation fails), the injector may continue to operate, but it cannot supply the proper amount of fuel. This often causes a slight miss without setting a trouble code. On some systems, the technician can disable the alternator to check for a weak injector. Injector solenoid windings use relatively large wire gauge with relatively few turns of wire in the winding. Therefore, the magnetic field built up in the winding is more sensitive to changes in amperage than changes in voltage. When the alternator is disconnected, the injectors operate from battery voltage instead of charging system voltage. When voltage decreases, the ECM drivers increase amperage flow to compensate. The extra amperage may cause the injector solenoid to operate properly. Therefore, if the injector begins working with the alternator disconnected, the injector winding is probably bad. To make this test, make sure that the ignition switch is in the off position, then remove the alternator field connection. Start the engine and recheck fuel injector operation. If the miss goes away when the alternator is disconnected, one of the injectors is probably defective. Further testing will be needed to pinpoint the faulty injector. See the procedures for checking individual injectors presented later in this chapter.

Note: Do not forget to reconnect the alternator after making this test.

Isolating the Source of the Problem

As mentioned earlier, the ECM can energize several injectors through one driver. Therefore, a defect in one injector can cause the others operated by the same driver to malfunction. For example, a shorted injector may absorb all the current provided by the driver and still be able to open. The other injectors would be bypassed and would not open, even if they were good. In this case, the other injectors may appear to be defective and the technician could waste a lot of time replacing the wrong injectors.

One way to isolate an open or shorted injector is to check it with an ohmmeter. Remove the electrical connector and measure across the injector's terminals. A very low reading or an infinite reading indicates a defective injector winding. Since injectors are often installed under the plenum or other engine parts, it may be easier to measure the resistance of a group of injectors at a convenient harness connector. Remember that the injectors are wired in parallel, and the total resistance reading should be less than that of any individual injector. Another way to isolate a defective injector is to perform an injector balance test, which is discussed in the following section. A waveform meter can be a big help in isolating problems in injector windings. The use of the waveform meter is covered later in this chapter.

Performing an Injector Balance Test

An ***injector balance test*** isolates an injector that is allowing too much or too little fuel to enter the cylinder. On older fuel injection systems, each injector must be connected to the tester to make the balance test. On OBD II systems, the scan tool can open individual injectors without removing the wiring harness.

Begin the test by making sure the ignition switch is in the *off* position. Next, connect a fuel pressure gauge to the fuel rail. Attach the tester to the first injector, or install the scan tool in the diagnostic connector. Turn the ignition switch to the *on* position. Bleed all air from the gauge lines; removing the air is vital to obtaining correct readings. After the bleeding is finished, wait ten seconds and then record the gauge pressure. Next, turn the injector on using the tester switch or the scan tool. As the gauge needle reaches its lowest point, record the pressure. Repeat the procedure for all injectors, and retest injectors that appear to be defective. Once all injectors have been tested, use the following process to average the readings:

1. Subtract the low reading (injector open) from the high reading (injector closed). Do this for all injectors. This gives the pressure drop for each injector.
2. Add all the injector pressure drop readings.
3. Divide the figure from step 2 by the number of injectors. This gives the average injector drop.

Compare the drop of each injector with the average drop. Any injectors are not close to the average drop are probably defective. The table in **Figure 18-13** shows typical readings for good and defective injectors.

Using Waveform Meters to Check Fuel Injection Components

A waveform meter, sometimes called a lab scope, can be used to isolate fuel injection system problems that cannot be located by other means. Waveform meters were discussed in Chapter 4, Precision Measuring Tools and Equipment.

A fuel injection circuit waveform is a graphic representation of what is happening when the injector is operating.

Injector Balance Test Results

Cylinder	1st Reading	2nd Reading	Amount of Drop	Test Results
1	434 kPa (63 psi)	351 kPa (51 psi)	83 kPa (12 psi)	Injector OK
2	434 kPa (63 psi)	331 kPa (48 psi)	103 kPa (15 psi)	Faulty injector—too much fuel drop
3	434 kPa (63 psi)	351 kPa (51 psi)	83 kPa (12 psi)	Injector OK
4	434 kPa (63 psi)	365 kPa (53 psi)	69 kPa (10 psi)	Faulty injector—too little fuel drop
5	434 kPa (63 psi)	344 kPa (50 psi)	90 kPa (13 psi)	Injector OK
6	434 kPa (63 psi)	344 kPa (50 psi)	90 kPa (13 psi)	Injector OK

Figure 18-13. Injector balance test results. Note that the drop can be too high or too low. (General Motors)

Measuring voltage in relation to time indicates the condition of the circuit. The fuel injection circuit waveform can be thought of as the fuel injection equivalent of an ignition oscilloscope pattern. As with an ignition system waveform, an injector waveform that does not match the standard waveform indicates a problem with the injector.

To check an injector waveform, connect the waveform meter test leads to the harness of the suspect injector. See **Figure 18-14.**

Warning: Never damage the wiring when attaching the meter connectors to the wiring harness. When a wire must be pierced to make a connection, use a special wire-piercing tool.

Once the meter connections have been made, turn on the waveform meter and select the proper range. Then, start the engine and observe the waveform. All fuel injection systems produce one of the four basic waveforms shown in **Figure 18-15.** These waveforms are classified not by the type of injector but by the type of circuit formed by the injector and ECM. Be sure that you have correctly identified the type of injector circuit before deciding whether the waveform indicates a defective injector.

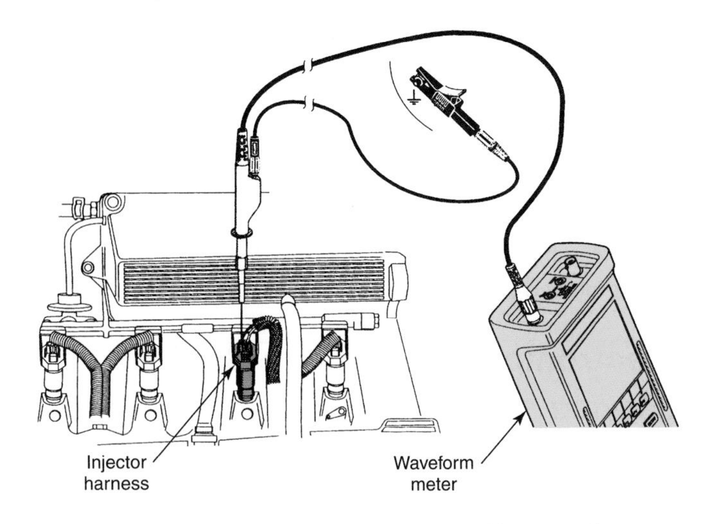

Figure 18-14. When testing a fuel injector with a waveform meter, always follow the meter manufacturer's instructions. Incorrect meter connections can damage the ECM. (Fluke)

The ***saturated switch injector circuit,*** Figure 18-15A, is the most commonly used type. Injectors in this type of circuit are often energized in groups. The injector's on time is shown as a voltage drop when the winding is energized. The inductive spike is caused by unused energy oscillations in the injector winding when the driver shuts off current flow. A shorted winding often causes the inductive spike to be low.

Figure 18-15B shows the waveform produced by a ***peak-and-hold injector circuit.*** This circuit is often used on throttle body fuel injection systems. Since throttle body fuel injectors are usually open for relatively long periods, the peak-and-hold circuit begins to reduce the current once the injector is open. Approximately four times as much current is needed to open the injector as is needed to hold it open. Therefore, once the injector is open, the circuit cuts back on current, causing the current-limiting oscillations shown. A final large spike signals that the current has been turned off. A lack of oscillations is a sign that the winding is shorted.

The Bosch-type peak-and-hold injector circuit is often used on European multiport fuel injection systems. See Figure 18-15C. This injector circuit also limits current flow after the injector is open. However, instead of reducing current, it rapidly pulses the current on and off. This is seen as a series of sharp oscillations just before the current turns off. Once again, a lack of oscillations is a sign that the winding is shorted.

In the ***PNP injector circuit,*** the driver energizes the injector winding by grounding the circuit through the ECM. The PNP injector circuit is shown in Figure 18-15D. Note that the waveform appears to be upside down. As with the saturated switch circuit, a shorted winding often causes the inductive spike to be less prominent.

If the waveform meter has a method of checking amperage, use it to compare current draw to specifications. Average current draw is about 1 amp, but actual specifications should always be used. Excessively high current draw indicates a shorted injector winding, while no current flow means the injector is open.

After the waveform meter tests are complete, disconnect the test leads. If any wires were pierced to make the connections, place a small amount of silicone sealant over the damaged insulation.

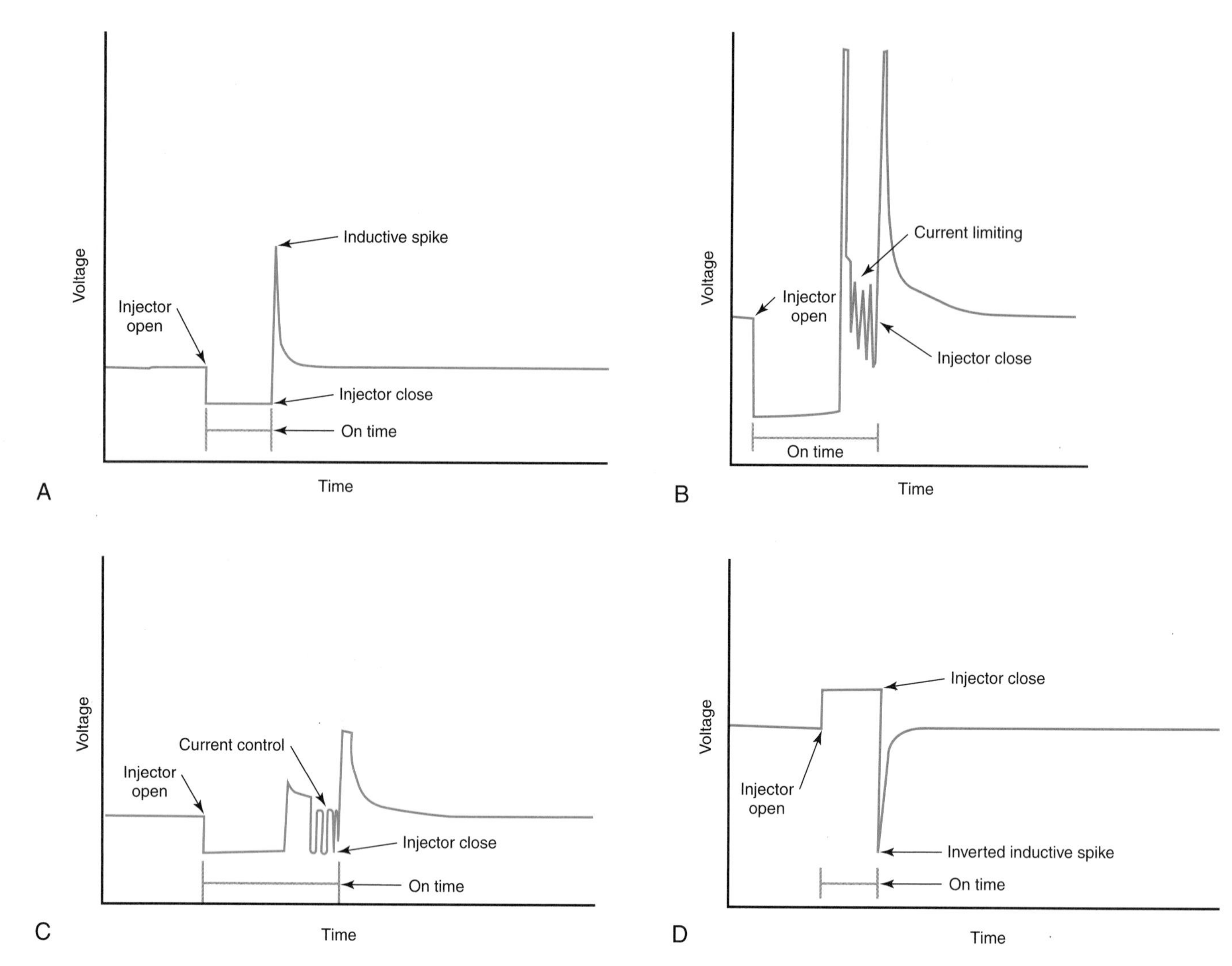

Figure 18-15. Injector circuit waveform patterns. A—The saturated switch injector circuit waveform. Note how voltage drops when the injector is energized and then spikes when the injector is turned off. This type of circuit is probably the most common. B—The peak-and-hold injector circuit limits current flow to the injector to reduce power consumption and heat. Note the oscillations as the ECM limits current flow. C—The Bosch-type peak-and-hold injector circuit limits current flow by rapidly turning the current on and off. This is indicated by the sharp oscillations near the end of the injector on time. D—The PNP injector circuit waveform is similar to the saturated switch injector waveform. Since the ECM sends power to the injector rather than grounding it, the pattern is an inverted version of the saturated switch waveform.

Checking for Deposits in the Throttle Body

Carbon on throttle plates can cause difficult starting when the engine is cold and may result in a rough or erratic idle or stalling. Carbon deposits can be spotted visually, **Figure 18-16.** Simply remove the intake duct, open the throttle valve, and look for deposits.

Checking Idle Speed Control Devices

Before checking any type of idle speed control, look for disconnected wiring or obvious damage. Also make sure no other factor, such as a dirty air filter, vacuum leak, or collapsed intake duct, is causing the problem.

To check a throttle positioner, note the movement of the throttle linkage as the engine is placed in gear or the air conditioner is turned on. Some throttle positioners can be checked with a scan tool. An idle air control valve (IAC) can also be checked with a scan tool. Operation of the idle air control valve is shown on the scan tool as a series of numbers called counts. The counts should match the specifications for the engine. If necessary, remove the IAC from the throttle body and check the valve for damage or deposits. Also check the throttle body passages for carbon buildup.

Electronic Fuel Injection System Service

The following sections detail various service procedures for cleaning or replacing electronic fuel injection components. Always refer to the proper service manual for exact procedures. If any procedure in this chapter conflicts with the manufacturer's service manual, follow the manufacturer's procedure.

Figure 18-16. To check the throttle body for carbon deposits, remove the intake duct and open the throttle valve. Black soot and baked-on carbon are a sign that the throttle body needs cleaning.

Note: Removal and replacement of fuel pumps was covered in Chapter 17, Fuel Delivery. Replacement of the ECM and related sensors was covered in Chapter15, Computer System Diagnosis and Repair.

Warning: Do not allow gasoline spills to remain on the engine or the floor. Clean up gasoline spills immediately.

Removing Pressure

It is important to release fuel system pressure before working on the fuel injection system. Some systems use a special ***Schrader bleed valve,*** which is similar to a tire valve. Disconnect the battery and connect a pressure gauge to the valve. Use the pressure gauge's bleed valve to release system pressure.

A few manufacturers recommend that you disable the fuel pump and then run the engine until it stops to release fuel system pressure. After the engine stops, it should be cranked over a few times. Next, disconnect the battery and then crack (barely loosen) the lines to check for any possible pressure.

Another way to release system pressure is to disconnect the battery and then slowly crack (loosen) a fitting to release pressure. When cracking fittings or using a Schrader valve, cover the area with a shop cloth to catch and contain any fuel that may spray out. Immediately dispose of the fuel-soaked cloth. Wear protective goggles when releasing system pressure and keep an approved fire extinguisher handy.

Cleaning Fuel Injectors

The openings inside the fuel injectors are extremely small, and it is often difficult to clean them successfully. Never soak an injector in solvent of any kind. Injectors are electrically operated, and solvent can damage the solenoid windings, as well as any internal O-rings or seals. The best way to clean the injection system is to use an ***on-vehicle injector cleaner.*** This device attaches to the fuel injection system and cleans the injectors as the engine operates. An on-vehicle injector cleaner also removes deposits from the intake valves.

To use an on-vehicle injector cleaner, assemble it and add the correct solvent. Some systems use a liquid solvent that is poured into a tank, which is pressurized with shop air. Other systems use pressurized cans that are similar to cans of carburetor cleaner. Next, attach the injector cleaner hose to the injection system. If the system has a pressure test fitting, attach the hose to the fitting. A few older systems do not have test fittings. In these systems, the injector cleaner can only be attached by removing the fuel inlet line and connecting the cleaner to the system inlet. **Figure 18-17** is a diagram of an injector cleaner installation.

Disable the fuel pump. This forces the injection system to operate on the cleaner. The easiest way to disable the fuel pump is to disconnect the pump relay, which is located in the engine compartment. If the relay is inaccessible, remove the fuel pump relay fuse. Pressurize the cleaner, if necessary, and start the engine. Allow the engine to operate on the injector cleaner solvent until the solvent is gone and the engine dies. Then, remove the injector cleaner and reconnect the fuel pump. Allow the engine to sit for several minutes. This allows the solvent to remove additional deposits. Finally, start the engine and operate it for several minutes to clear out all the remaining solvent.

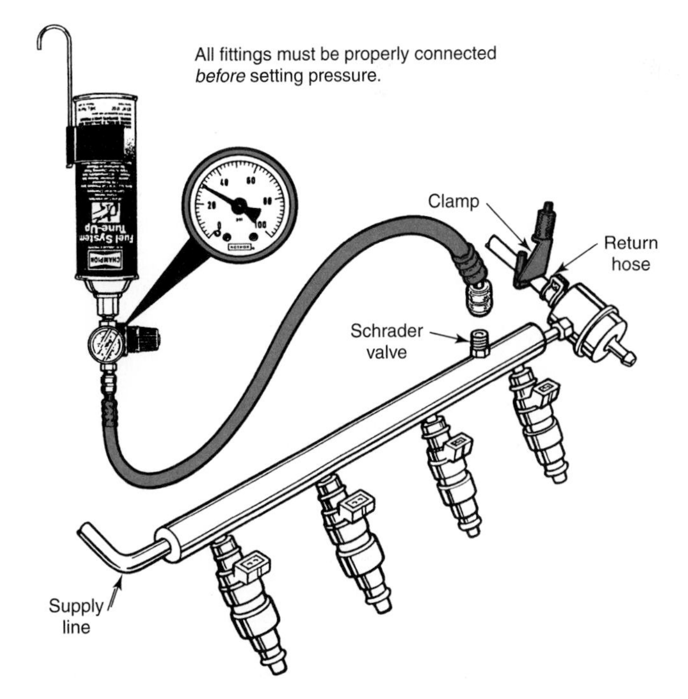

Figure 18-17. This tool set can be used to clean the fuel injectors and check system pressure. The engine operates on the injector cleaner solvent as the solvent dissolves deposits. (OTC)

Cleaning Throttle Valves and Bodies

Carbon and gasoline deposits can be removed from the throttle valve and body by cleaning with carburetor cleaner and a brush. Remove the intake duct at the throttle body and saturate the throttle valve area with carburetor cleaner. Allow the cleaner to soak for a few minutes, and then scrub the throttle valve and intake area with a small brush. See **Figure 18-18.** An old toothbrush can be used for this purpose, but make sure the bristles do not fall off and get into the plenum. After scrubbing, rinse the area with more carburetor cleaner. Reattach the intake duct and start the engine. Allow the engine to operate for a few minutes to clear out all remaining carburetor cleaner.

Removing and Replacing Fuel Injectors

Before attempting to remove injectors, obtain the proper service manual. It may be necessary to remove the upper intake manifold or plenum to gain access to multiport injectors.

Note: Some V-type engines with multiport injectors use different types of injectors on each side of the engine. This is done to obtain exact fuel ratios when the intake plenum is not the same length on both sides of the engine. Do not mix injectors between sides.

Injector Removal

Remove the negative battery cable and then remove the injector's electrical connector. **Figure 22-19** shows the method used to remove one injector from a throttle body. Do not bend or flex the fuel system tubing, since most fuel injector tubing has an internal coating that can flake off if the tubing is bent or flexed. **Figure 18-20** shows removal of a multiport injector fuel rail. First remove any brackets holding the fuel rail to the engine. Next, remove the fuel rail and injector assembly

Figure 18-18. Carburetor cleaner and an old toothbrush are all that is needed to clean a throttle body. Thoroughly clean the valve body and inspect it carefully after cleaning. Make sure that no debris gets into the engine.

by pulling the rail straight up so that the injectors come straight out of the intake ports. If you plan to reuse an injector, place the injector in a secure place so that the tip cannot be damaged before reinstallation.

Checking Individual Injectors

The in-vehicle injector-testing procedures covered in the diagnosis section above should identify most injector problems. Additional tests can be performed on some injectors after they have been removed, including spray pattern, fuel output, and leaks. Use the proper equipment and

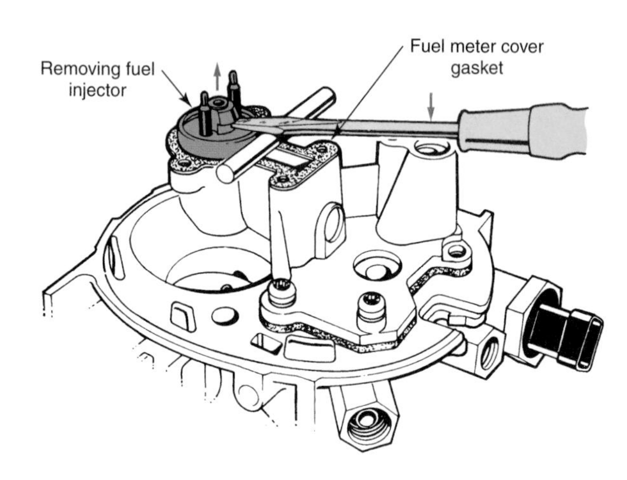

Figure 18-19. When removing a fuel injector from a throttle body, be careful not to damage either unit. (General Motors)

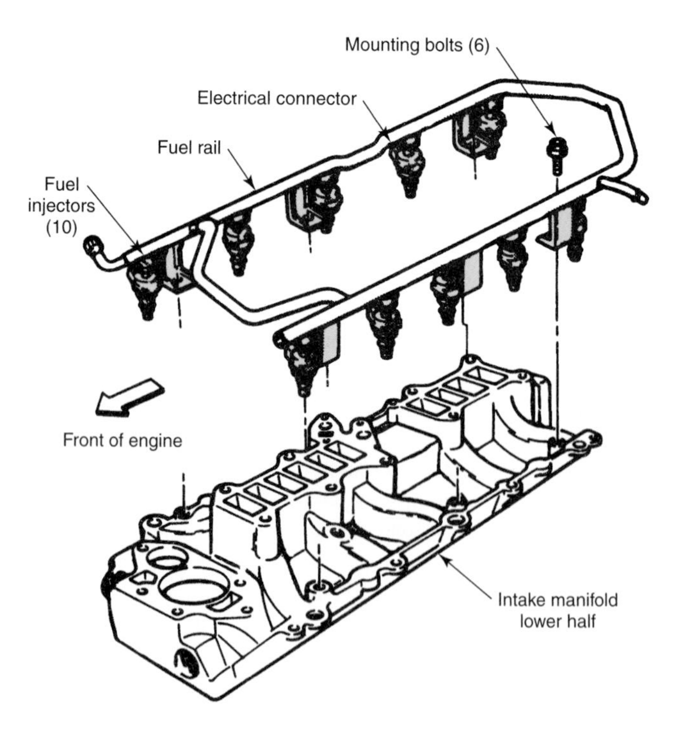

Figure 18-20. Removal of a fuel rail from a V-10 engine intake manifold. Be sure to cover or plug the injector holes in the manifold to prevent the entry of dirt, small parts, etc. (DaimlerChrysler)

recommended test procedures. One test procedure is shown in **Figure 18-21.** If an injector is defective, replace it with a new one. Make certain the replacement injector is correct.

Reinstalling Injectors

Begin injector installation by lubricating the injector's O-rings with transmission fluid or clean motor oil. Carefully install the injector into the intake port or the throttle body, as applicable. Some multiport injectors are installed in the fuel rail before being placed in the engine. If this is necessary, carefully maneuver the rail and injector assembly into place. When all the injectors are properly installed, reinstall the injector electrical connectors, pressure regulator vacuum hose, fuel supply hose, and all other components. Reconnect the negative battery cable and make a final check to ensure that all fittings, fuel lines, ground straps, and wiring harnesses are correctly reinstalled.

Turn the ignition on and allow the fuel pump to fill the fuel lines. This will take about 30 seconds. Then, start the vehicle and check the fuel injection system for leaks. If no leaks are found, check fuel system pressure and general injector system operation.

Removing and Replacing Fuel Pressure Regulators

Pressure regulators are installed on the fuel rail of multiport systems. On throttle body systems, the pressure regulator is mounted on the throttle body assembly. Most pressure regulators are non-adjustable and must be replaced when they are unable to properly control the fuel system pressure. If an adjustable pressure regulator cannot be successfully adjusted, it should also be replaced.

To replace a fuel pressure regulator, begin by removing the air cleaner, hoses, or other components that obstruct access to the regulator. Then, remove the vacuum line to the regulator, if applicable. If the pressure regulator is mounted on the fuel rail, release the fuel pressure, and then remove the fittings and fasteners that hold the pressure regulator to the rail. Finally, remove the pressure regulator from the rail. See **Figure 18-22.**

Install all the required gaskets and seals on the new regulator and place the regulator in position on the fuel rail. Install the fuel line fittings and attaching screws. Reinstall the regulator's vacuum line and all other components that were removed. Start the engine and check for leaks and proper fuel system pressure.

To replace pressure regulators installed on the throttle body, first bleed off fuel pressure. Then, remove the fasteners holding the pressure regulator to the throttle body, **Figure 18-23.** Remove the old gaskets and check the fuel passages for deposits. Install the new pressure regulator and new gaskets. Start the engine and check the fuel system pressure. While checking the pressure, closely observe the area around the pressure regulator mount for leaks.

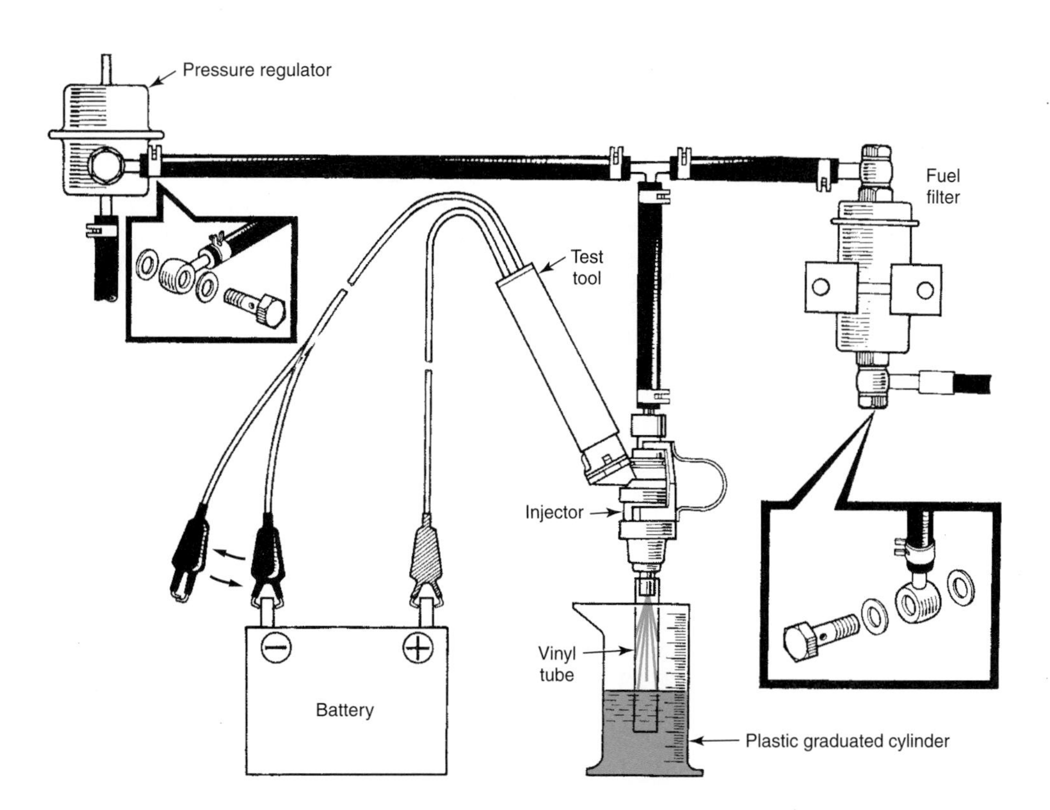

Figure 18-21. When testing a fuel injector's flow rate, place a vinyl tube around the injector to contain the fuel spray in the measuring cylinder. (Toyota)

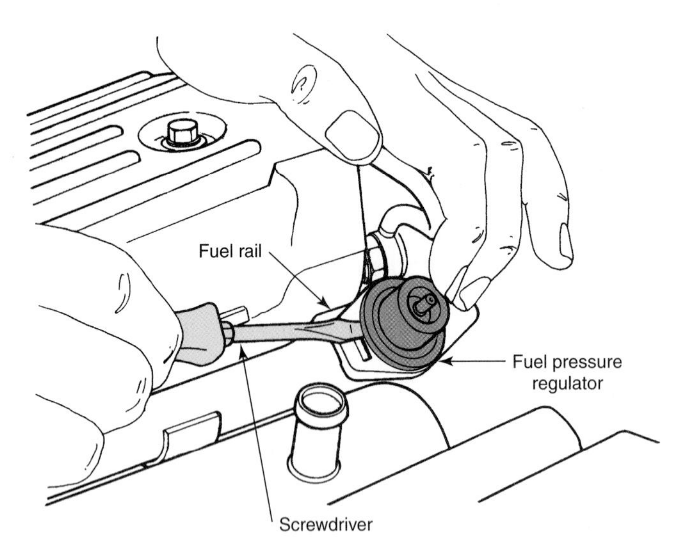

Figure 18-22. Carefully remove the fuel pressure regulator from the fuel rail. Be sure to relieve fuel pressure first. (DaimlerChrysler)

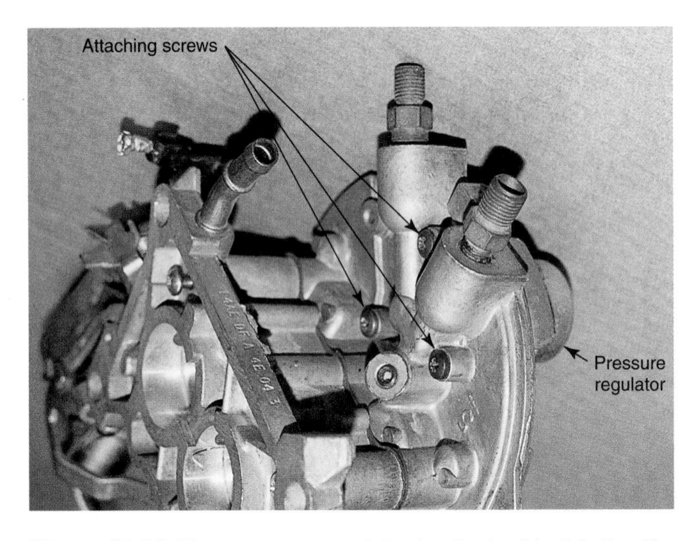

Figure 18-23. The pressure regulator is attached to this throttle body with screws that are mounted under the throttle body's air horn. Use an offset screwdriver to remove the screws without removing the valve body. Other throttle body pressure regulator fasteners are located on the top of the air horn.

Overhauling Throttle Bodies

Throttle body overhaul is relatively simple. The main variation is whether or not the fuel injectors and pressure regulator are installed in the throttle body. The following section gives a general overview of throttle body overhaul.

Note: Central point injection throttle bodies cannot be serviced in the field and must be replaced as a unit.

Throttle Body Removal

Begin throttle body removal by removing the air cleaner or intake ducts, the electrical connections, and the vacuum lines at the throttle body. Then, remove the throttle linkage or cable, the automatic transmission linkage, and the fuel lines. If used, remove the engine coolant lines at the throttle body. Remove the throttle body attaching bolts, and remove the throttle body from the intake manifold or plenum. When working on a central point injection system, remove the throttle body and fuel lines as an assembly. Remove all gasket material and cover the opening in the intake manifold or plenum with a clean shop towel.

Overhauling the Throttle Body in Multiport Injection Systems

To overhaul a multiport throttle body, remove the idle air control, throttle position sensor, and other components as necessary. Clean the throttle body and carefully inspect it for cracks, warping, and wear at the throttle shaft. If the throttle body can be reused, reassemble it using new gaskets. Before reinstalling the idle air control and the throttle position sensor, test them to make sure they function properly. Some technicians prefer to change these components as part of the overhaul.

Overhauling the Throttle Body in Throttle Body Injection System

To overhaul a throttle body used in a throttle body injection system, remove the throttle positioner and the throttle position sensor, if used. If necessary, the throttle body halves can be split. Next, remove all old gaskets, the pressure regulator assembly, and the injector(s), **Figure 18-24.** The throttle body can then be cleaned and rebuilt or replaced. Closely check the pressure regulator for proper operation, and obtain a new regulator if there is any doubt about the old one. Obtain new parts as needed, and closely follow the manufacturer's instructions to rebuild the throttle body. Be sure to use new gaskets and seals where needed.

Throttle Body Installation

Remove the shop towel and install a new gasket on the intake manifold or plenum opening. Place the throttle body in position and install the attaching bolts. Next, attach the linkage, wiring, vacuum lines, and fuel lines as necessary. Install the air cleaner or intake ducts. Finally, start the engine and make any necessary adjustments.

Converting MAF System to Calculated Charge

Early mass airflow (MAF) sensors were unreliable and often failed. Some manufacturers provide a way to eliminate the MAF sensor. This is known as converting the MAF system to a ***calculated charge system.*** The major step in converting a MAF system to a calculated charge system is the installation of a replacement PROM. The new PROM has a modified *algorithm.* An algorithm is a series of mathematical steps used to calculate an answer. The replacement PROM provides an algorithm that allows the ECM to ignore the MAF input and base air-fuel decisions on other sensor inputs.

To install the new PROM, remove the negative battery cable or ECM fuse. Then, remove the ECM fasteners and

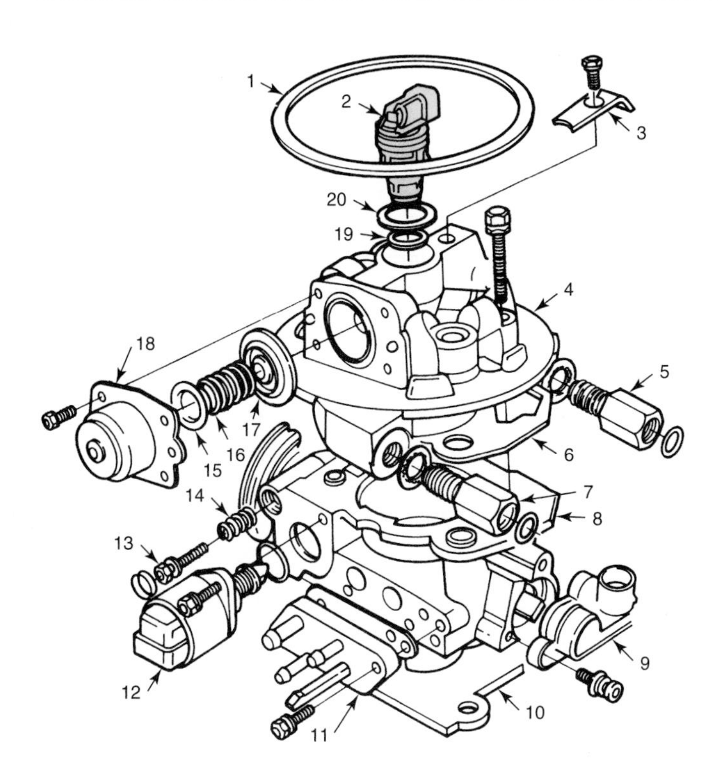

Figure 18-24. Exploded view of a single-injector throttle body arrangement. 1—Air filter gasket. 2—Fuel injector. 3—Injector retainer. 4—Fuel meter assembly. 5—Fuel inlet nut. 6—Fuel meter body-to-throttle body gasket. 7—Fuel outlet nut. 8—Throttle body assembly. 9—Throttle position sensor (TPS). 10—Gasket flange. 11—Tube module assembly. 12—Idle air control valve (IACV). 13—Idle stop screw and washer assembly. 14—Idle stop screw spring. 15—Spring seat. 16—Pressure regulator spring. 17—Pressure regulator diaphragm assembly. 18—Pressure regulator cover assembly. 19—Fuel injector lower O-ring. 20—Fuel injector upper O-ring. (AC-Delco)

remove the ECM from its brackets so that the PROM cover can be removed. It is usually not necessary to remove the electrical connectors from the ECM. Remove the PROM cover, and then unsnap the plastic retainers holding the original PROM in place. See **Figure 18-25.** Pull the PROM from the ECM. Install the replacement PROM, carefully pressing it down until the retainers snap into place. Reinstall the cover and then reinstall the ECM. Replace the fuse or battery cable, and follow the manufacturer's idle relearn procedures.

Mechanical Fuel Injection System

Some older engines use a ***mechanical fuel injection system.*** Mechanical fuel injectors are always open, and fuel pressure is varied to match the fuel flow with the airflow. To keep fuel from dripping when the engine is not running, the fuel injectors have check valves that close when fuel pressure is removed. A mechanically operated pressure control valve regulates fuel flow based on the amount of air flowing into the engine. The fuel pressure on later mechanical systems is modified by an electronic control system.

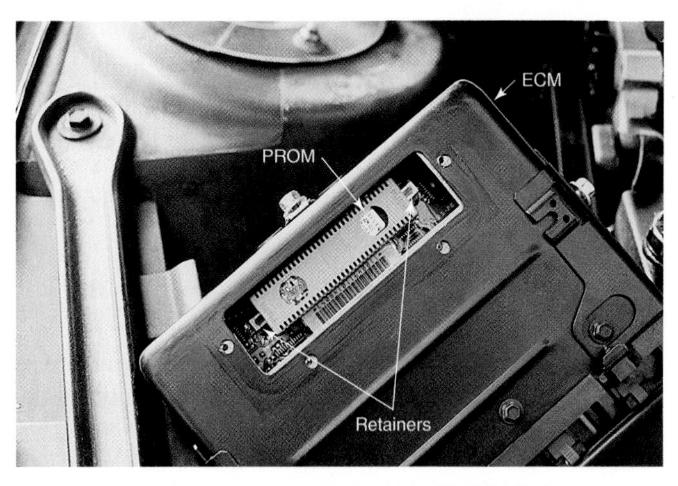

Figure 18-25. An updated PROM can be installed to convert a MAF system to a calculated charge system. The PROM is located under a cover on the ECM. Once the cover has been removed, the PROM can be carefully removed from the ECM. A special tool must be used to remove some PROMs.

Mechanical Injection Fuel Flow Control

Air passing into the intake manifold flows through a mechanical airflow sensor, **Figure 18-26.** This sensor contains an airflow sensor plate that is connected to a fuel distributor control valve. The sensor plate is located in the center of the air venturi and moves according to the amount of air flowing through the venturi. The more air flowing through the venturi (controlled by the driver moving the separate throttle valve), the more the sensor plate rises. Sensor plate movement causes the fuel distributor control valve to move. This causes the control valve to allow more fuel to the injectors as airflow increases. Most mechanical systems have a control pressure regulator to maintain a constant fuel pressure at the fuel distributor control valve.

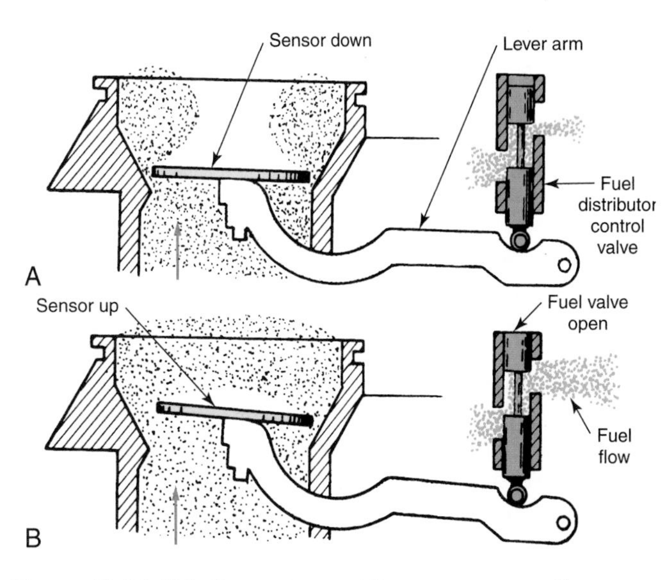

Figure 18-26. This figure shows airflow sensor operation. A—Part load operation. B—Full load operation.

Cold Start Injector

A ***cold start injector*** is mounted near the throttle valve and injects extra fuel into the intake manifold when the engine is cold. On newer systems, the ECM operates the cold start injector. On the original design, a temperature sensor and relay combination called a thermoswitch operated the injector.

Mechanical Fuel Injection System Service

Mechanical gasoline injection is still used on a few vehicles. The following sections cover common mechanical fuel injection system checks and adjustments. Obtain the proper service manual to make other checks. If the system has an electronic control system, refer to the appropriate portions of the electronic fuel injection service section presented previously.

Fuel cleanliness is critical on mechanical injection systems. The internal parts of the injector pump are fitted to extremely close tolerances, and any foreign particles can cause problems. Clean or replace all filters and water traps as recommended.

Mechanical Fuel Injection Diagnosis

Mechanical injection pump pressure and volume may be checked. Fuel pump checks were covered in Chapter 17, Fuel Delivery. Check electrical connections for looseness and corrosion. If the pump seems noisy, check the rubber mountings (where used) and make sure the pump and lines are not touching other vehicle parts. Make sure all hoses are in good condition and that clamps and fittings are tight. Other injection pump checks can be found in the proper service manual.

Since the injectors are mechanically operated, they can be checked while they are off the vehicle. Refer to the procedures for checking diesel injectors, in the diesel section of this chapter, for information on testing these components.

Cold Start Injector Service

If an engine with a mechanical injection system runs poorly when it is cold but operates properly when warm, suspect the cold start injector. To check the cold start injector, begin by allowing the engine to cool off for at least six hours. Allowing the vehicle to sit overnight is best. Then, place the ignition switch in the *on* position and use a test light or multimeter to determine whether the cold start injector is receiving electrical power. If the injector is not receiving power, test the thermoswitch or computer control system to pinpoint the problem. Follow the manufacturer's directions closely to avoid damaging other components.

If the injector is receiving power, check the resistance between the injector terminals and compare it to the specifications. If the resistance is much higher or lower than specified, the cold start injector is defective. Remove the cold start injector by loosening and unthreading it from the intake manifold. Reverse the procedure to install the replacement injector. It is not necessary to release fuel pressure when replacing this component.

Mechanical Fuel Injection System Adjustments

Unlike electronic systems, mechanical fuel injection systems have many adjustment points. Adjustments must be precise to ensure correct system operation. A manufacturer's manual covering the exact system at hand must be used.

When service of internal pump components or a pump overhaul is needed, send the pump to a shop with specialized equipment. Proper injector pump overhaul requires the use of a test stand and appropriate tools. Clearances are precise, and the pumps require absolute cleanliness and care. Do not attempt to overhaul the pump unless the proper tools and specifications are available. When removing or installing the pump, pay particular attention to spacers, adjustment washers, and index marks.

Linkage Adjustments

Linkage adjustments must be made exactly as specified, especially those between the throttle valve and injection pump. Linkage should move freely, without binding or interference with other parts. Lubricate as needed. When adjustment is necessary, follow maker's instructions. When necessary, use the recommended adjustment tools and double-check to ensure that all adjustments are correct. **Figure 18-27** shows the typical detailed linkage adjustment specifications for one section of the linkage system.

Fuel Mixture Adjustment

Adjusting devices, such as those shown in **Figure 18-28,** are used to provide proper air-fuel mixture adjustments at idle, part load, and full load. Some systems have two idle-adjusting

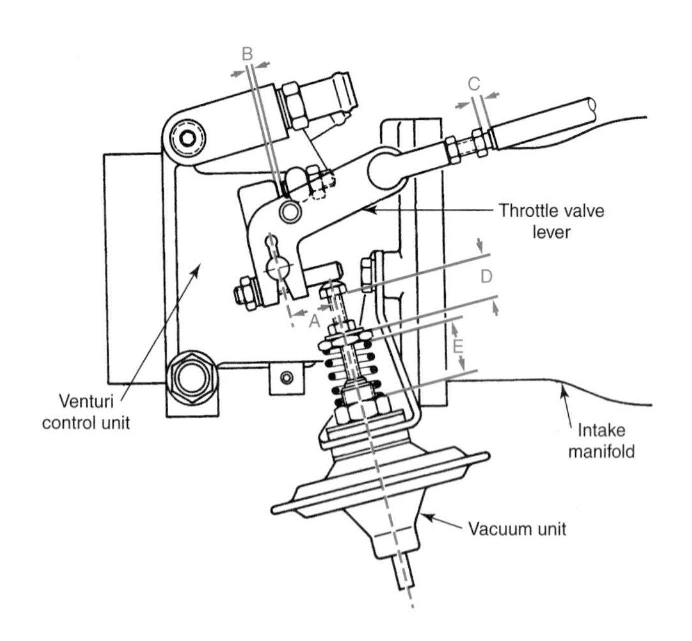

Figure 18-27. Mechanical fuel injection linkage adjustment must be precise. Follow the manufacturer's instructions. A—Shaft distance. B—Throttle valve opening. C—Idle travel of sliding rod. D—Length of thrust bolt. E—Spring height. (Mercedes-Benz)

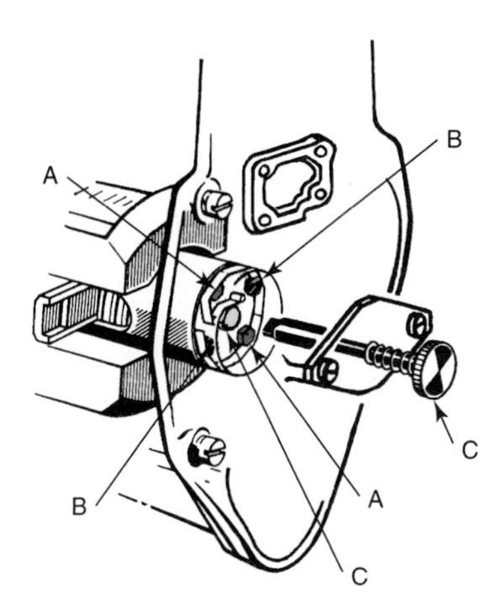

Figure 18-28. One type of idle adjustment for a mechanical gasoline injection system. A—Top part load adjusting screw. B—Bottom part load adjusting screw. C—Idle speed adjusting screw.

screws. One screw on the injector pump controls fuel flow, and is used to adjust mixture. The other adjusting screw, located on the intake manifold, is used to adjust idle speed by varying airflow. Both screws may have to be used in conjunction with each other to secure the proper fuel-air mixture. After setting the air-fuel mixture with the mixture (pump) screw, use the manifold screw to set the idle speed. If setting idle speed changes the air-fuel mixture, readjust the mixture screw. Repeat both adjustments until both the idle speed and air-fuel mixture are correct.

Airflow Sensor/Pressure Regulator and Injector Replacement

Many mechanical gasoline injection systems are designed so that the airflow sensor/pressure regulator assembly and the injectors are removed as an assembly. This section explains how to remove the airflow sensor/pressure regulator and injectors as a unit.

Begin by removing any fuel pressure from the system and disconnecting any electrical connectors. Next, remove the fasteners holding the injectors to the intake manifold. Once the injector fasteners have been removed, remove the ductwork connecting the airflow sensor to the air cleaner and throttle valve. Remove the fasteners holding the pressure regulator to the engine and lift the entire assembly from the engine. Be careful not to bend any of the injector lines.

After repairs are made or a new assembly is obtained, install new O-rings or washers on the injectors, as applicable. Lubricate the new O-rings with transmission fluid or engine oil. Reinstall the assembly on the engine in the reverse order of removal, being careful not to bend or kink any of the injector lines.

Installing an Aftermarket Fuel Injection System

Systems are available that make it possible to replace the carburetor on an older engine with a fuel injection system or to replace an older throttle body fuel injection system with a multiport system. These systems are called ***aftermarket fuel injection systems,*** since they are installed after the vehicle is manufactured. Aftermarket injection systems consist of various sensors, an electronic control unit, and associated wiring harnesses. Some manufacturers also provide injectors, fuel rails, and other fuel system parts.

To begin installation of an aftermarket fuel injection system, carefully read the instructions. In some cases, it is necessary to drill and tap holes in the intake manifold to accept the injectors. Some manufacturers recommend installing two fuel pumps, a high-pressure pump to pressurize the injectors and a low-pressure pump to keep the high-pressure pump supplied with fuel. Sometimes a ***surge tank*** is supplied to damp out fuel pulsations (variations in pressure). Most systems have a fuel pressure regulator. **Figure 18-29** is a schematic of one type of aftermarket fuel injection system.

Some new sensors may have to be installed on the engine. The technician may be able to use some of the existing sensors. Some systems require a speed sensor, such as a Hall-effect switch. The Hall-effect switch is usually installed at the engine crankshaft pulley, and magnets are installed on the pulley to trigger the switch.

For protection against moisture, heat, and stray magnetic fields, install the electronic control module inside the passenger compartment. If this is not possible, install the control unit in a protected place in the engine compartment, as far away from the engine and ignition system as possible. Once the fuel injection parts are in place, install the wiring harness.

Route the power supply wire from a source that has power only when the ignition switch is in the *on* position. The power supply wire must have a fuse. Since the system does not draw much current, a one-amp fuse should be sufficient. After everything has been installed and connected, start the engine and check for fuel leaks.

The injection system manufacturer provides a small programmer unit that connects to the control module. This programmer is used to set the air-fuel ratio. In some states,

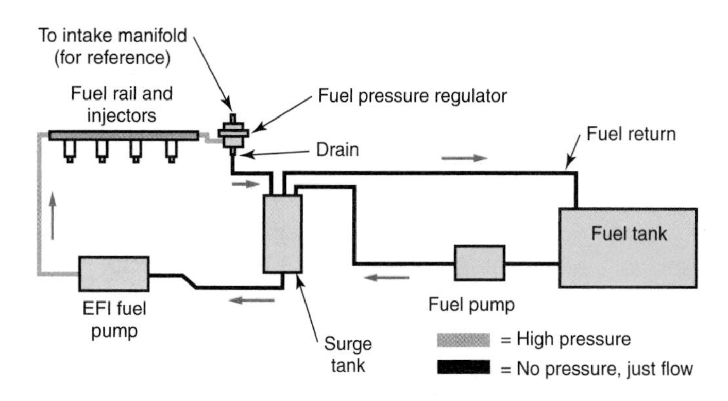

Figure 18-29. Schematic of an aftermarket fuel injection system that can be installed in place of a carburetor. Note the major components. The control system is not shown. (Racetech/SDS)

air-fuel ratio must be set to obtain legal emissions levels. This is not necessary if the vehicle will be used off-road. When this is the case, the system can be set for maximum power.

Diesel Fuel Injection Systems

Diesel engines have fuel injection systems. The ignition source for diesel fuel is the heat caused by high compression ratios. The fuel cannot be compressed along with the air inside the diesel engine cylinder because it would ignite before the piston reaches the top of its compression stroke. Therefore, the ***diesel fuel injection system*** differs from the mechanical gasoline fuel injection in four respects:

- Diesel fuel is injected directly into the cylinder combustion area.
- Injection must be timed to occur at the top of the compression stroke.
- Fuel must be injected into the combustion chamber area when compression is highest. This requires tremendous pressure.
- The diesel engine must have a glow plug to heat the combustion chamber only when the engine is cold. The glow plug has no relation to a gasoline engine's spark plug.

To overcome compression pressure, diesel injection systems pressurize the fuel from several hundred to several thousand psi. Although diesel engines pollute less than gasoline engines, many late-model injection systems are at least partially controlled by the ECM.

Diesel injection systems have an engine-driven injector pump. The engine drives the pump through a chain, belt, or gear on the camshaft or crankshaft. An overall view of a diesel fuel injection system is shown in **Figure 18-30.** This diesel engine is commonly used in large pickup trucks. **Figure 18-31** shows a cutaway view of a four cylinder automotive diesel engine. Note the fuel injection pump, which is driven by a cogged belt. A cross-sectional view of a fuel injector in place in a cylinder head is illustrated in **Figure 18-32.** Also, note the

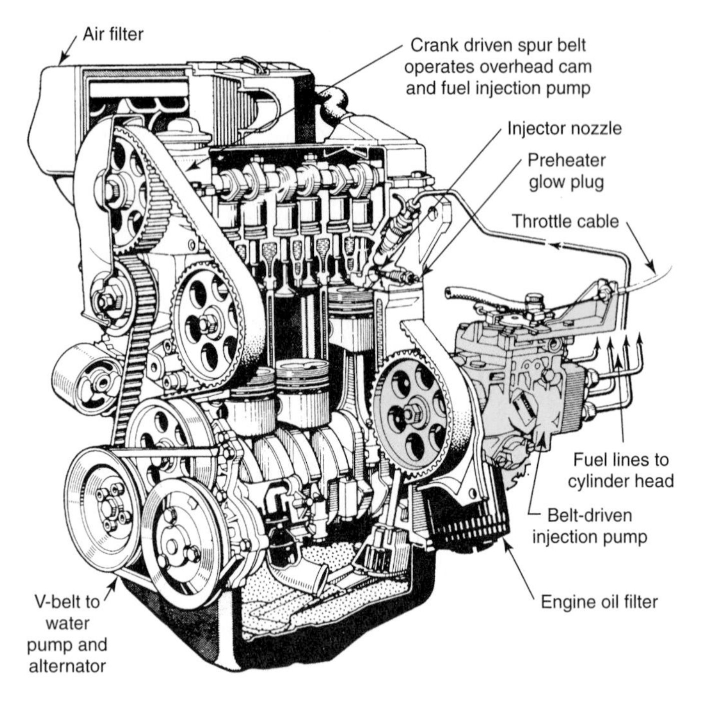

Figure 18-31. Cutaway of a four-cylinder, 1.5-liter diesel engine. This engine develops 48 hp (SAE Net) at 5000 RPM. (Volkswagen)

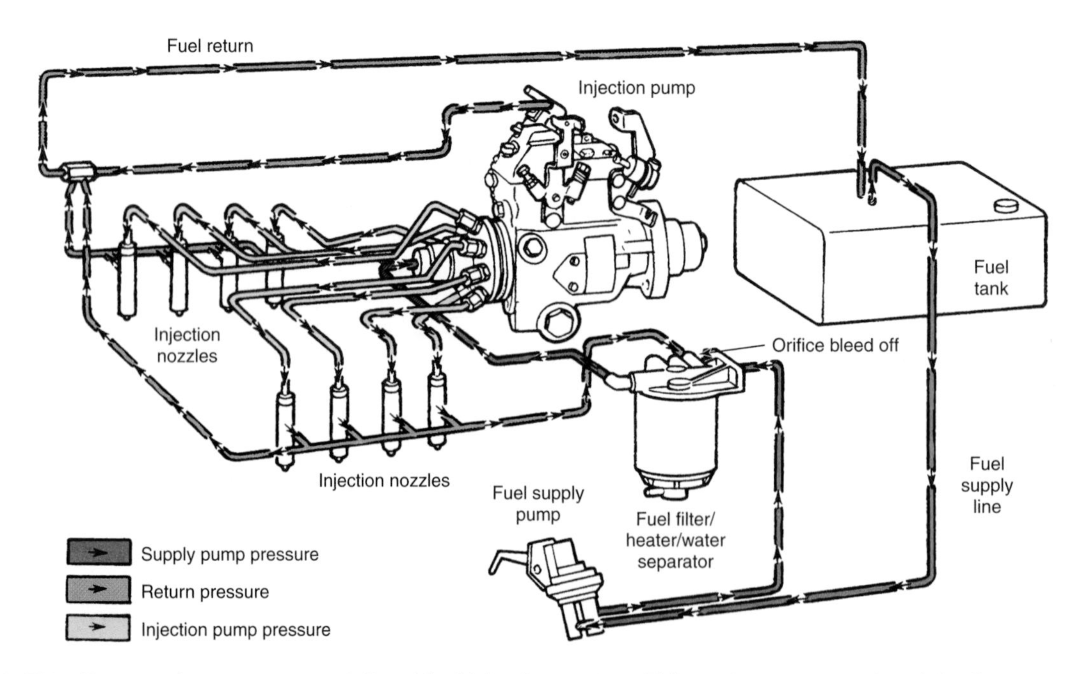

Figure 18-30. This diagram shows one type of diesel fuel injection system. This system uses a rotary injection pump. (Ford)

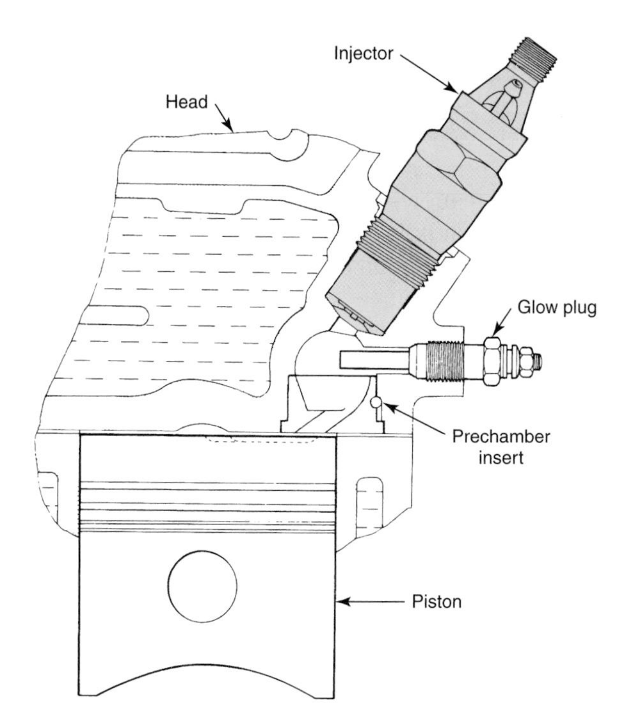

Figure 18-32. The relative arrangement of a diesel fuel injector and glow plug. The precombustion chamber is used on many diesel engines to begin the burning process. (Champion)

glow plug, which is used to heat the combustion chamber when the engine is cold.

The internal parts of an injector pump are fitted to extremely close tolerances, and any foreign material in the pump can cause problems. Since diesel fuel must be perfectly clean, diesel engines have one or more filters. Many diesels have water and sediment traps. The injection system in Figure 18-32 has a combination filter and water trap. One of the most vital maintenance procedures for diesel engine is to frequently check and replace all filters.

Diesel Injection System Service

Before beginning any diesel injection system service, obtain a service manual that covers the exact system being serviced. Diesel injection system checks and adjustments must be precise to ensure correct operation. Closely follow instructions and recommended clearances, adjustments, and pressures. Use the proper tools and test equipment.

When disassembling system lines and parts, always cap lines, nozzles, and pump fittings to prevent the entry of dirt. Before reassembly, always wipe fittings with a clean, lint-free cloth. Some of the more common service adjustments, inspections, and repairs are covered in the following sections.

Make sure you understand the system at hand and that you have the proper tools and specifications. Never steam clean or wash injection pumps with the engine running or with the pump warm. Pump parts are fit very closely. A sudden temperature change can warp the unit and cause its parts to bind.

Diesel Engine Electronics

An ECM controls the injection system of most late-model diesel engines. A typical control method involves using an ECM-operated solenoid to vary the pressure at the injector. The solenoid pulses on, blocking return flow to the fuel tank. Fuel pressure at the injector rises, overcoming spring pressure that normally holds the injector closed. The injector opens, spraying fuel into the combustion chamber. When the solenoid pulses off, fuel begins to return to the fuel tank. Fuel pressure at the injector drops, and the injector closes.

For closer control of engine speed and fuel mixture, some diesel engines have an electronic speed and mixture governor. The decisions of the ECM are based on input from various sensors. When diagnosing an electronically controlled diesel engine, always begin by retrieving trouble codes. After retrieving the codes, make additional checks based on the codes.

Stopping a Runaway Diesel Engine

It is possible for a diesel engine to continue running even after the key is turned off. This condition is called a ***runaway diesel engine*** and can result from wiring problems, vacuum leaks, incorrect vacuum connections, a damaged vacuum control unit, a defective vacuum pump, or a fuel solenoid not returning the fuel valve to the closed position. The correct method of stopping a runaway engine varies with the injection system. Many injection systems stop functioning if the wire to the fuel solenoid is disconnected. If the engine continues to run after the wire is disconnected, interrupt the flow of fuel from the tank to the injection pump by loosening the inlet fitting to the supply pump or by disconnecting the wire to the supply pump.

Diesel Injector Service

To replace diesel fuel injectors, release the fuel pressure by cracking open a fuel line. Next, loosen and remove the fuel line fitting from the injector(s) to be replaced. If the injector has a return line, remove it also. Pull the lines away from the injector, being careful not to bend or kink the tubing. Loosen and unscrew a threaded injector from the cylinder head. Remove a pressed-in injector by removing the bracket and pulling the injector from the head. Thread or press the new injector into the head. Be careful not to damage the sealing washer during installation, and line up any alignment marks on a pressed-in injector before pushing the injector to its fully installed position.

Install new gaskets and tighten all fasteners to the specified torque. For injectors similar to the one shown in **Figure 18-33,** place a wrench on the large hex when tightening the injector. Install the fuel line fittings and torque them to specifications. Purge the injectors when necessary, following the instructions given later in this chapter. Check for leaks following engine startup.

Testing Injectors

Dirty, damaged, or sticking diesel injectors can cause rough running, loss of power, knocking, and smoking. Proper tools must be used to test diesel injectors. Injectors are tested for opening pressure, spray pressure, chatter, spray jet shape, and leakage.

When testing injectors, use a proper ***test stand*** and a special ***test liquid*** instead of diesel fuel. The test liquid minimizes skin problems and inhibits corrosion. It is advisable to shroud the injector nozzle with a clear, protective plastic shield during the test. See **Figure 18-34.** The spray pattern should meet manufacturer's recommendations.

Injectors should also be checked for ***nozzle tightness*** (nozzle sealing ability). Using a test stand, build the pressure in the injector (nozzle tip dry) to slightly below the actual opening pressure (around 150 psi or 1 034 kPa). Hold the pressure for the recommended time (usually a few seconds) and check the nozzle tip. One manufacturer's indicators are pictured in **Figure 18-35.**

Warning: Do not point an injection nozzle toward your body during testing procedures. Never get your hands in front of the nozzle. When spraying, the fuel leaves the nozzle with fearsome force. It can literally drill through flesh, creating a severe injury with chances of blood poisoning. Use care when working on injection pump or injector line fittings. Wear protective goggles! Always crack a fitting and bleed off pressure before working on the system.

Cleaning Injectors

Diesel injectors can be disassembled for cleaning. Tolerances are extremely close, and rough handling can ruin an injector. Keep all parts of each injector nozzle together, as they are carefully matched. Do not mix parts. Thoroughly clean the parts in a sonic bath cleaner or some other suitable cleaner. **Figure 18-36** illustrates the use of a special cleaning tray designed to keep the parts of each injector separate during cleaning. Once the tray is filled with the parts to be cleaned, lower it into the cleaning bath. When the parts have been thoroughly cleaned, reassemble the injectors as recommended. For manual cleaning, use a wooden scraper or a soft brass brush to prevent damaging nozzle parts.

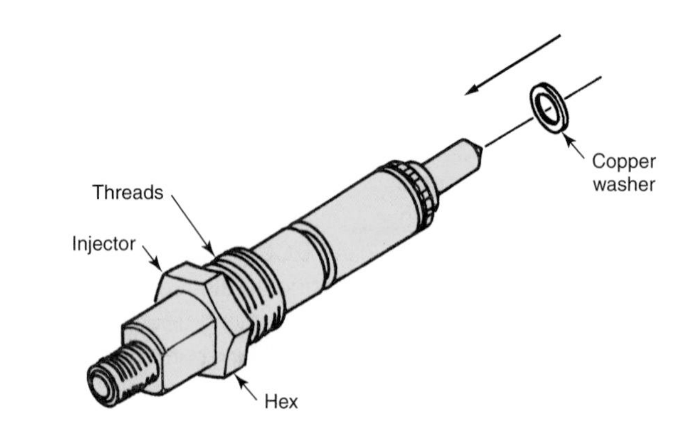

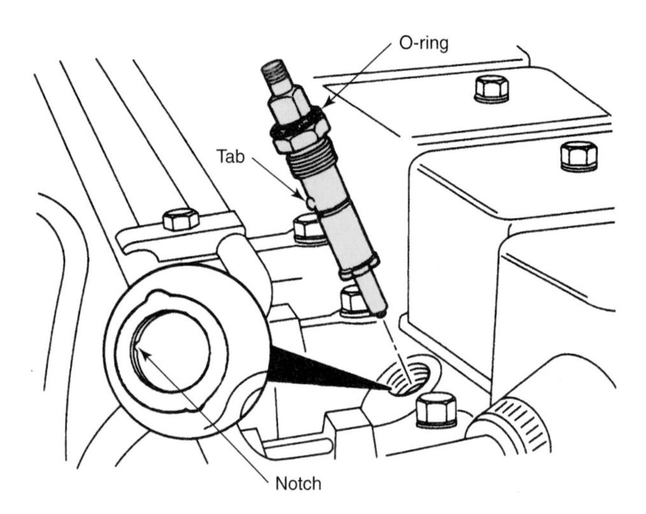

Figure 18-33. This diesel fuel injector uses a copper sealing washer and is threaded into place on the engine. Use a wrench on the hex top to tighten the injector. Notice the alignment tab and notch. (DaimlerChrysler)

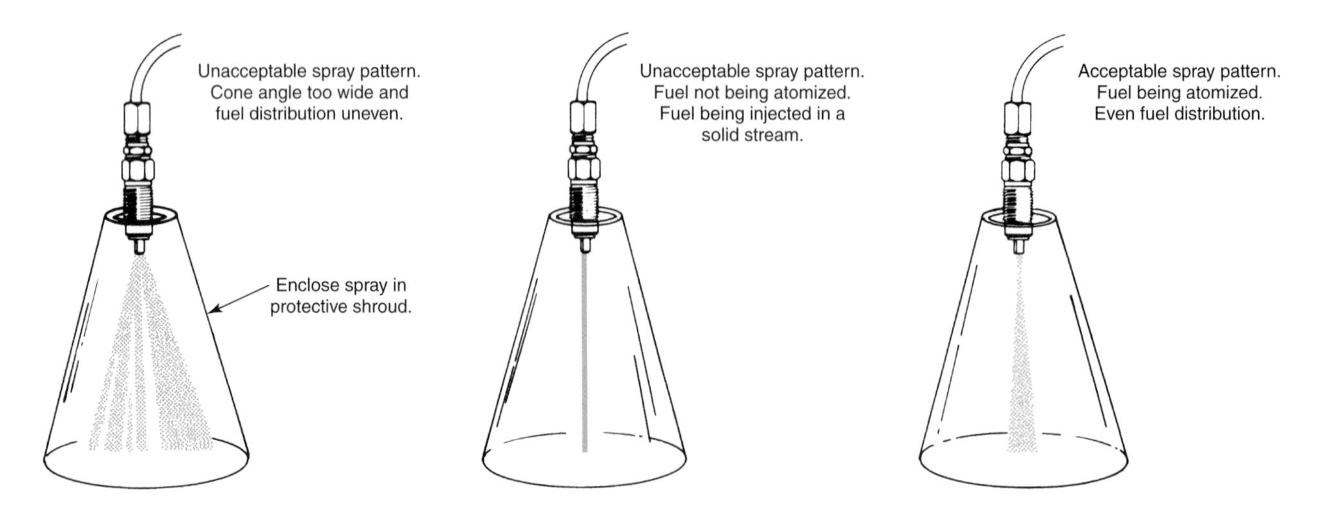

Figure 18-34. Testing diesel injector nozzle spray pattern. Note the use of a protective shroud. (General Motors)

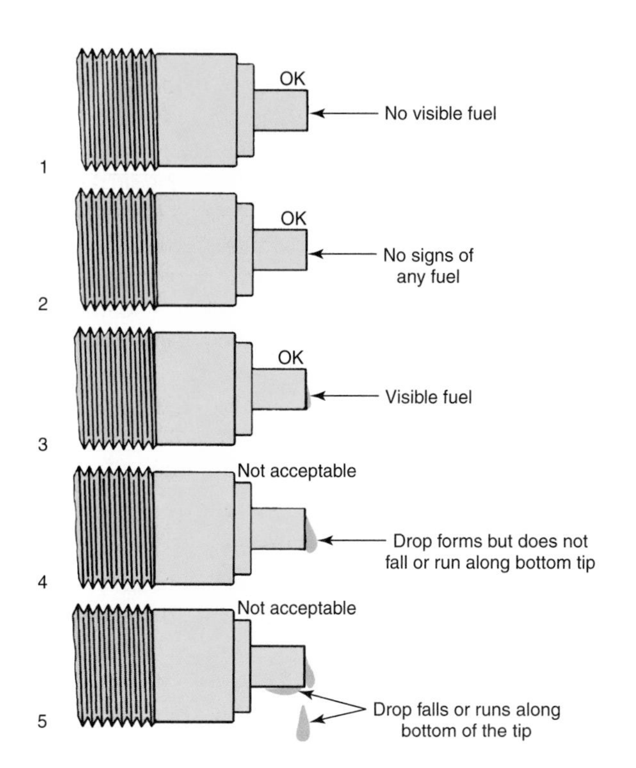

Figure 18-35. Check injector nozzles for tightness. After maintaining a recommended pressure for a few seconds, the nozzle should not show signs of excessive leakage. (General Motors)

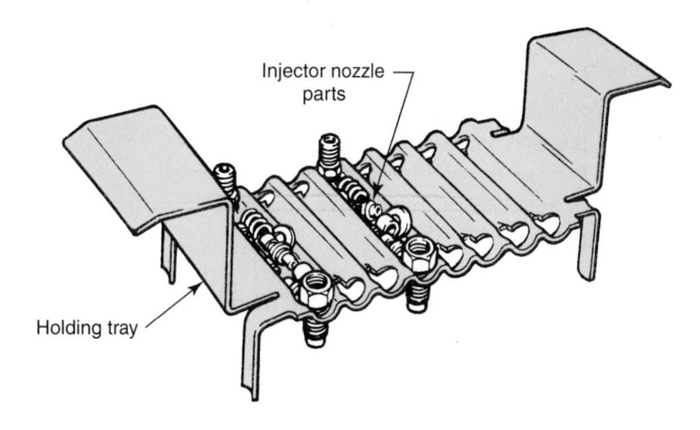

Figure 18-36. This handy tray keeps parts of individual injectors separate during cleaning. Never mix injector parts. (General Motors)

Injector Pump Service

Injector pump problems can cause heavy exhaust smoking, surging, rough running, and noise. Binding or improperly adjusted linkage can affect idle speed range. In some instances, the injection pump timing can be off, resulting in very noisy engine operation or a no-start condition.

An inline injection pump is shown in **Figure 18-37.** In this type of pump, each injector is served by a separate pump

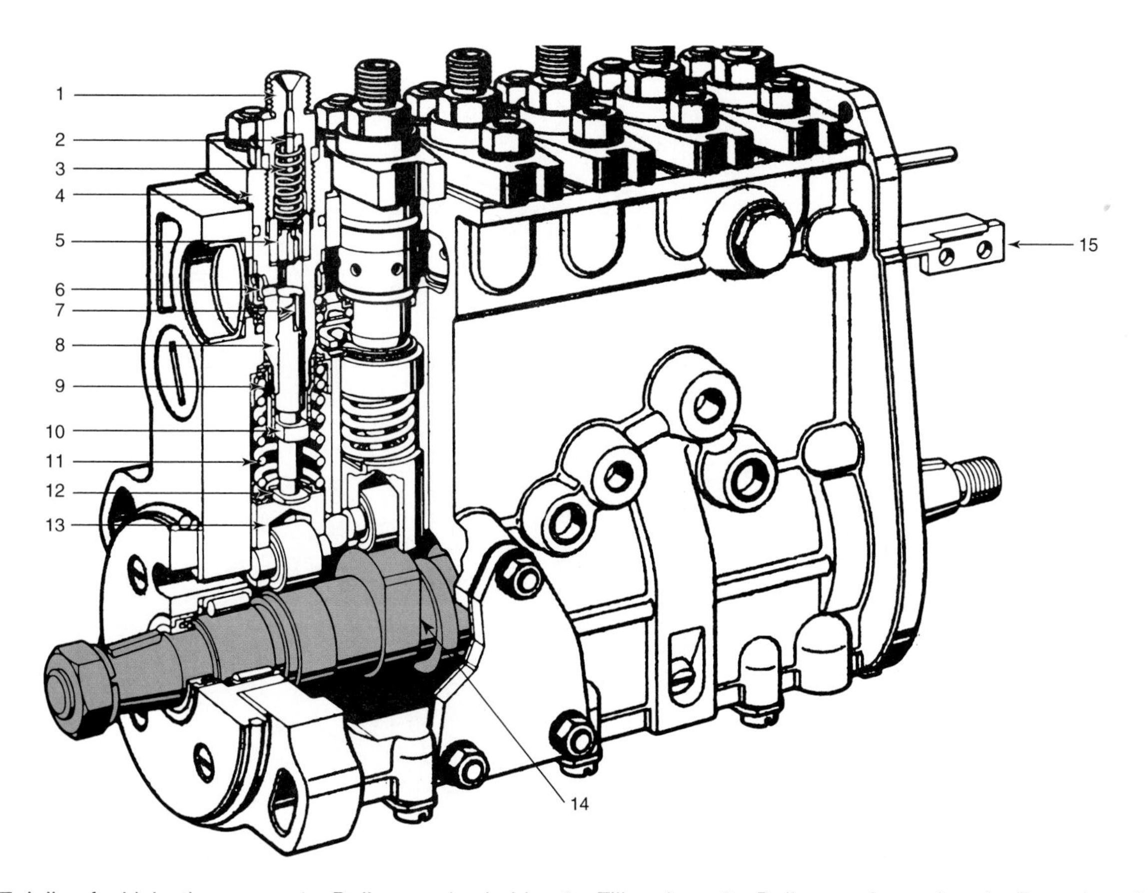

Figure 18-37. Inline fuel injection pump. 1—Delivery valve holder. 2—Filler piece. 3—Delivery valve spring. 4—Pump barrel. 5—Delivery valve. 6—Inlet and spill port. 7—Helix. 8—Pump plunger. 9—Control sleeve. 10—Plunger control arm. 11—Plunger return spring. 12—Spring seat. 13—Roller tappet. 14—Cam. 15—Control rod. (Bosch)

cylinder. The camshaft serves the same purpose as an engine camshaft.

A distributor-type pump is shown in **Figure 18-38.** Note that a single set of pump plungers generates the required pressure. Fuel is fed from the pump to a fuel distributor rotor. As the rotor revolves, it distributes fuel to the various outlets as required.

Injection Pump Removal and Replacement

Before removing a fuel injection pump, mark the position of the pump housing on the engine. This helps ensure that the fuel injection pump is reinstalled in the proper position. Some pumps are removed with the injectors and lines attached, while other pump designs require that the lines and injectors be removed from the pump first. To reinstall an injection pump, line up the marks made before removal. Then carefully put the pump into place and install the mounting screws.

Injection Pump Overhaul

Special injection pump test tools and training in injection pump repair are required to successfully overhaul a diesel injection pump. Great care and cleanliness is absolutely necessary. Take the pump to a shop that specializes in pump rebuilding and testing. Never attempt to make internal repairs to a diesel injection pump without the proper tools and training.

Purging the Injection System

When injectors are removed and replaced, air is admitted to the diesel injection system. If all the injectors were removed and replaced, the injection system would contain so much air that the pump would not be able to develop sufficient pressure to start the engine. The air must be ***purged*** (removed) before the engine can start.

To purge the injectors, reinstall the tubing fittings but do not tighten them. With all the injector fittings loose, crank the engine for approximately one minute or until fuel begins squirting from the fittings. Next, start the engine, following the usual diesel-starting procedure, and allow it to run briefly. Tighten the tubing fittings and restart the engine. Finally check the injector system for proper operation.

If only a few injectors were removed and replaced, the engine can be started without purging, but it will misfire at first. The replacement injectors will be purged of air as the engine runs.

Servicing Diesel Glow Plugs

Glow plugs are used for cold starting because compression heat is not high enough to properly ignite the fuel charge. A glow plug is a low-voltage heating element that is located in the combustion chamber. When the ignition is turned on and the engine coolant is below a certain temperature, the glow plugs are energized and quickly reach full heat. Plug temperature can range from 1832°F–2192°F (1000°C–1200°C). Many glow plug systems reach peak heat in just a few seconds. The engine can then be started. Some glow plugs cycle on and off until the engine reaches normal operating temperature. Three types of glow plugs are illustrated in **Figure 18-39.**

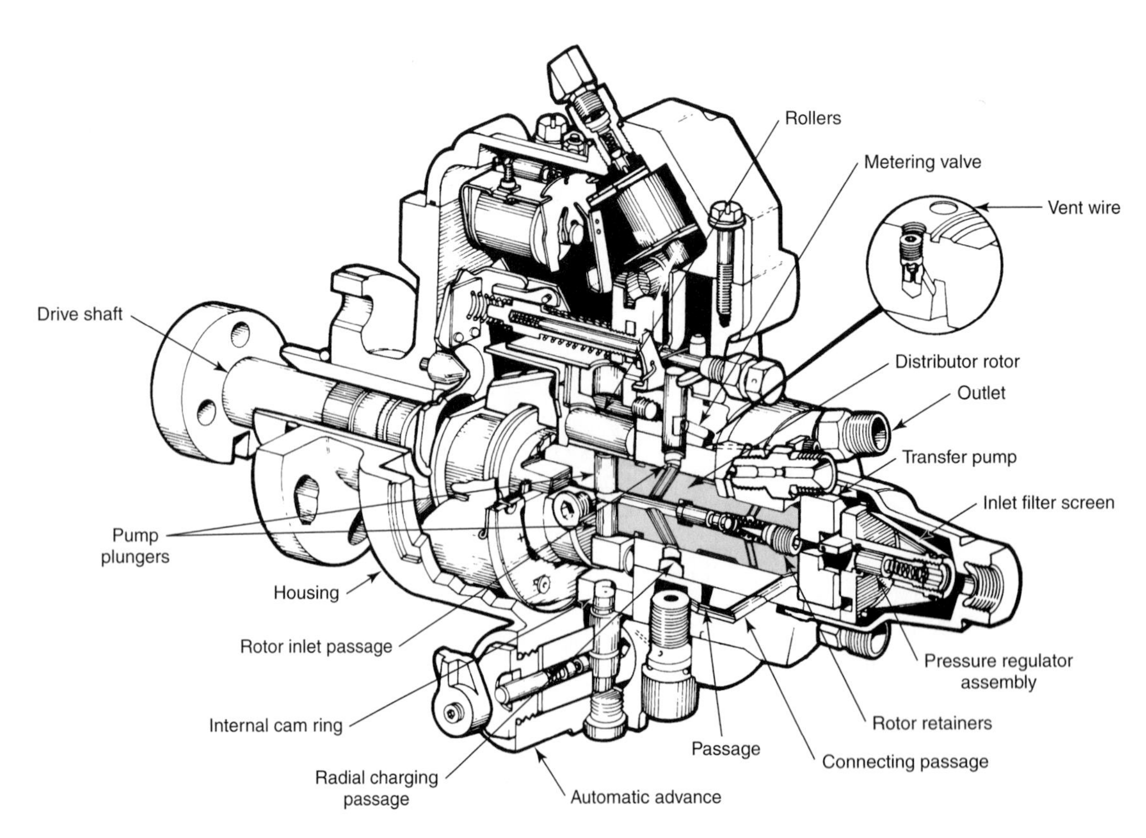

Figure 18-38. Distributor-type diesel injection pump. Note the double pump plungers. As the rotor turns, it passes fuel to the various injectors. (Ford)

A glow plug wiring circuit is shown in **Figure 18-40.** Never bypass the power relay to glow plugs because a constant power source will cause many plugs to quickly overheat and burn out.

Warning: Some glow plug systems use heavy, uninsulated connecting wires. Under certain conditions, current can flow through the connecting resistor wires for almost two minutes after the control light goes out. In addition, the wires can be very hot. Keep hands and arms away from these wires.

Defective glow plugs or glow plug system problems can cause cold start problems. To isolate a defective glow plug, check the glow plug with an ohmmeter. If the reading is infinity or zero, the plug has probably burned out. To test the glow plug system, place a test light on each glow plug terminal with a cold engine and the key in the *on* position. If the light does not come on, the system is not sending current to the glow plugs. Check for defective current relays or other components of the control system.

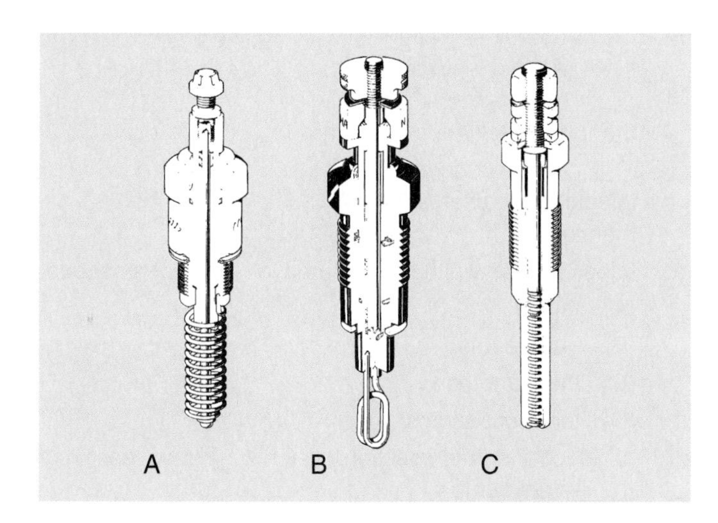

Figure 18-39. Three types of glow plugs are shown here. A—Manifold heater type. B—Open element type. C—Sheathed element type. (Champion)

Tech Talk

Check for vacuum leaks before getting too involved in diagnosing a suspected fuel injection system problem, such as a rough idle. The most common cause of rough idle, surging, and poor acceleration is a vacuum leak. Typical causes of vacuum leaks are disconnected or cracked hoses. If a defective hose connects to a vacuum-operated component, that component may not operate.

To check for vacuum leaks, begin with the vacuum hoses. Start by carefully checking for split and cracked hoses. Unplug each hose with the engine running, plug the end of the hose, and see if idle improves. If the idle does not improve, check the part that is operated by that hose for a defect. If all hoses and vacuum-operated accessories check out, check the intake manifold bolts and throttle body fasteners. Also, closely check the EGR valve, as it often sticks partially open.

If no leaks found, squirt light oil around gaskets that are suspected of leaking, and then race the engine. If the exhaust smokes, the oil is being drawn into the engine through the leak. If all gaskets and hoses are in good condition, check for a cracked manifold or head.

Summary

All modern vehicles use fuel injection systems controlled by the ECM. The three basic types of electronic fuel injection systems are throttle body, multiport, and central point. Major fuel injection components include the fuel injectors, fuel pump, pressure regulator, throttle body, ECM, and input sensors. Some older engines use a mechanical fuel injection system.

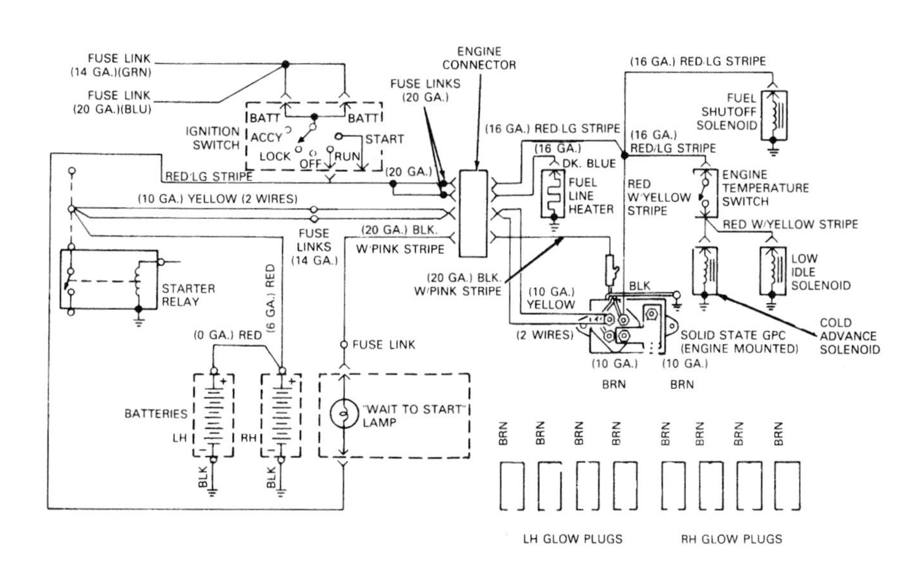

Figure 18-40. The glow plug electrical circuit of a four-stroke, V-8 diesel engine. (International/Navistar)

Safety is important when working with any kind of fuel. Gasoline can ignite or cause skin damage. Before working on the fuel system disconnect the battery ground cable and release fuel system pressure. Clean up spilled gasoline immediately.

All fuel injection systems develop similar problems. Most troubleshooting procedures work on all fuel injection systems. Make a visual check for problems and retrieve trouble codes before making other tests. Check fuel injection system pressure to determine the condition of the fuel pump and pressure regulator. Returnless fuel system diagnosis is similar to that for systems with fuel pressure regulators. Quickly check fuel injectors by listening for a clicking sound as the engine operates. The spray pattern of a throttle body fuel injector can be observed as the engine operates. Noid lights can be used to check the output of fuel injector drivers. Sometimes, the technician can disconnect the alternator to test injector windings. Injectors can also be checked with an ohmmeter and a waveform meter.

Some fuel injectors and throttle bodies can be cleaned rather than replaced. Defective injectors must be replaced. After replacing an injector, turn the ignition on and allow the fuel pump to fill the fuel lines. Then, start the vehicle and check the fuel injection system for leaks, correct pressure, and injector operation.

Pressure regulators are installed on the fuel rail of multiport systems, and on the air horn of throttle body systems. Bleed off fuel pressure before replacing a pressure regulator. To overhaul a throttle body, remove it from the engine. Change all gaskets and any necessary parts. After reinstalling the throttle body, start the engine and check for leaks and proper fuel system pressure.

Mechanical injection systems can be pressure tested. A defective cold start injector causes the engine to run poorly when cold. On many mechanical injection systems, the airflow sensor, pressure regulator, and injectors are removed as a complete assembly.

It is possible to replace a carburetor or a throttle body fuel injection system with an aftermarket multiport fuel injection system. Follow the manufacturers' instructions closely when installing an aftermarket fuel injection system.

To overcome compression pressure, diesel injection systems pressurize the fuel from several hundred to several thousand psi. Diesel injection systems use engine-driven injector pumps. Since diesel fuel must be perfectly clean, diesel engines have one or more filters, which may include a water trap. Diesel injection system checks and adjustments must be precise. The ECM controls the injection system of most late-model diesel engines.

Diesel injectors can be disassembled for cleaning. After cleaning, reassemble the injectors and test them using the proper test equipment. To replace diesel fuel injectors, release the fuel pressure, and then loosen and remove the fuel line fitting from the injector to be replaced. Use a wrench on the hex top to loosen and tighten the injectors. Problems in the glow plug system may prevent the engine from starting or cause rough operation when the engine is cold.

Some diesel fuel injection pumps are removed with the injectors and lines attached, others require that the injectors and lines be detached from the pump before removal. Special injection pump test tools and training in injection pump repair are required to overhaul diesel injection pumps. After repairs, the diesel injection system must be bled to remove air from the pump, lines, and injectors.

Review Questions—Chapter 18

Do not write in this book. Write your answers on a separate sheet of paper.

In Questions 1–5, match the type of fuel injection system with the following statements.

1. _____ Has a fuel rail.
2. _____ Combines the features of two other injection types.
3. _____ Has been replaced by other types of fuel injection.
4. _____ Injects fuel ahead of the throttle valve.
5. _____ Lines connect injector to nozzles at intake manifold.

(A) Throttle body injection.
(B) Multiport injection.
(C) Central point injection.

6. Injector pulse width is measured in _____ of a second.
   (A) tenths
   (B) hundredths
   (C) thousandths
   (D) ten thousandths
7. All electronic fuel injection systems have at least one _____ fuel pump.
8. The pressure regulators on _____ systems are operated by spring pressure only.
9. Pressure regulators on _____ and _____ fuel injection systems are operated by a combination of spring pressure and engine vacuum.
10. If the system has no pressure regulator, it is called a _____ system.
11. The idle air control valve bypasses airflow around the _____ _____.
12. Pulse width is controlled by the _____, which contains internal parts called _____ to energize the injectors.
13. List four fuel system safety rules.
14. As part of the initial fuel system inspection, the fuel lines should be checked for which three defects?
15. As part of the initial fuel system inspection, the vacuum lines should be checked for which four defects?
16. As part of the initial fuel system inspection, the electrical wiring should be checked for which five defects?
17. Fuel trim is another way of describing the _____, or _____, of the fuel injection system.
18. Describe a method of temporarily freeing a stuck fuel pump.

19. A clicking sound from an injector indicates that the injector is _____.
    (A) receiving current
    (B) sticking
    (C) plugged
    (D) disconnected
20. Throttle body spray patterns can be observed by using a _____ light.
21. Injector driver condition can be observed by using a _____ light.
22. If an injector begins working when the alternator is disconnected, the _____ is defective.
    (A) alternator
    (B) ECM
    (C) injector
    (D) pressure regulator

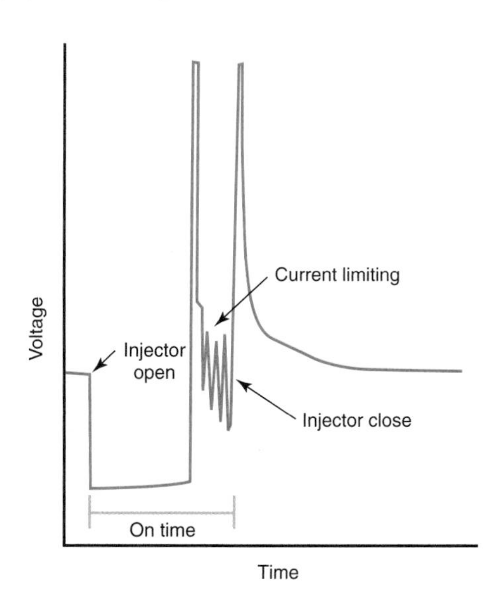

23. The above waveform shows what kind of injector circuit?
24. To use an on-vehicle injector cleaner, the _____ must be disabled.
25. The mass airflow sensor can be eliminated on some engines by replacing the _____.
26. An airflow sensor plate is found on _____ injection systems.
27. Fuel _____ is critical on mechanical injection systems.
28. If a mechanical injection system operates poorly only when cold, suspect the _____.
29. Some carburetors and throttle body injection systems can be replaced with a(n) _____ injection system.
30. Diesel injection pressures are much _____ than gasoline injection system pressures.
31. Diesel injectors can be checked for _____ pattern on a test stand.
32. What should you do to the diesel injector pump housing before removing it?
33. What is the purpose of purging a diesel engine?

ASE Type Questions—Chapter 18

1. Fuel injection systems are being discussed. Technician A says that fuel injection is complex and has many components. Technician B says that modern gasoline fuel injection systems are electronically operated. Who is right?
    (A) A only.
    (B) B only.
    (C) Both A and B.
    (D) Neither A nor B.
2. The ECM controls all of the following fuel injection system components, *except:*
    (A) fuel pump.
    (B) pressure regulator.
    (C) fuel injector.
    (D) idle air control valve.
3. Technician A says that a common fuel injection problem is carbon buildup on the throttle plate. Technician B says that disconnected ground wire could cause fuel injection system problems. Who is right?
    (A) A only
    (B) B only
    (C) Both A and B.
    (D) Neither A nor B.
4. Which of the following *cannot* be checked with a scan tool?
    (A) Injectors.
    (B) Fuel trim.
    (C) Fuel pressure.
    (D) Input sensors.
5. A vehicle is driven into the shop with a complaint of a miss. A fuel pressure gauge is attached and the engine is started. The pressure reading is zero. Which of the following is the *most* likely reason?
    (A) Faulty relay.
    (B) Defective pump.
    (C) Blown fuse.
    (D) Incorrect gauge connection.
6. The fuel pressure on a multiport fuel injection system drops four pounds when the engine is started. Technician A says that this indicates a leaking pressure regulator. Technician B says that this indicates normal pressure regulator operation. Who is right?
    (A) A only.
    (B) B only.
    (C) Both A and B.
    (D) Neither A nor B.

7. All of the following statements about injector balance tests are true, *except:*
   (A) an injector balance test identifies an injector that is allowing too much fuel to enter the cylinder.
   (B) an injector balance test identifies an injector that is not allowing enough fuel to enter the cylinder.
   (C) a scan tool can disable individual injectors on an OBD II system without removing the injector harnesses.
   (D) the fuel pressure gauge should be disconnected before making a balance test.
8. Which of the following determines the type of fuel injection waveform?
   (A) The type of circuit formed by the injector and ECM.
   (B) The size of the internal injector winding.
   (C) The number of injectors used on the engine.
   (D) The position of the injectors in relation to the throttle body.
9. All of the following statements about diesel fuel injection are true, *except:*
   (A) fuel is injected directly into the cylinder.
   (B) injection occurs at the top of the compression stroke.
   (C) fuel is injected into the combustion chamber area when compression is low.
   (D) the glow plugs heat the combustion chamber when the engine is cold.
10. A diesel engine runs roughly when cold. Which of the following is the *least* likely cause?
   (A) glow plugs energized at all times.
   (B) glow plugs not energized at any time.
   (C) one glow plug is defective.
   (D) two glow plugs are disconnected.

Suggested Activities

1. Draw a schematic of a multiport fuel injection system from the fuel tank to the injectors. Identify all of the parts shown on your drawing.
2. Check the engines in the classroom and shop and determine whether they have throttle body, multiport, or central point fuel injection systems. List the types of injection systems on a sheet of paper.
3. Use service literature to determine the pressures for various fuel injection systems. Determine whether you can draw any conclusions about what kind of injection system has the highest and lowest pressures.

Fuel System Problem Diagnosis	
Problem: No fuel delivery	
Possible cause	**Correction**
1. No fuel.	1. Add fuel to the tank.
2. Tank vents clogged.	2. Open fuel tank vents.
3. Tank filter clogged.	3. Replace tank filter.
4. Fuel lines kinked or clogged.	4. Straighten or clean fuel lines.
5. Vapor lock.	5. Cool fuel lines. Change to less volatile gas. Protect fuel lines from heat.
6. Fuel pump inoperative.	6. Rebuild or replace pump.
7. Fuel pump relay or fuse defective	7. Replace as needed.
8. Fuel filter or filters clogged.	8. Clean or replace filters.
9. Frozen fuel line.	9. Thaw and remove water from the fuel system.
10. Air leak between fuel pump and tank.	10. Repair air leak.
11. Clogged injectors.	11. Clean or replace injectors, locate source of deposits.
12. Inoperative injection pump (diesel).	12. Repair or replace pump.
Problem: Insufficient fuel delivery	
Possible cause	**Correction**
1. Tank vent partially clogged.	1. Open tank vent.
2. Tank filter partially clogged.	2. Replace tank filter.
3. Gas lines kinked or clogged.	3. Straighten or clean fuel lines.
4. Vapor lock.	4. Cool fuel lines. Change to less volatile gas. Protect fuel lines from heat.
5. Air leak between fuel pump and tank.	5. Repair air leak.
6. Fuel filter partially clogged.	6. Clean or replace filter.
7. Fuel pump defective.	7. Install new pump.
8. Pressure regulator spring weak.	8. Replace pressure regulator.
9. Clogged injection pipes.	9. Clean or replace pipes.
10. Defective injectors.	10. Clean or replace injectors.
11. Faulty injection pump (diesel).	11. Rebuild or replace pump.
12. Fuel injection pulse width incorrect.	12. Check ECM and sensors. Replace as needed.
13. Mass airflow sensor malfunction.	13. Replace mass airflow sensor.
14. Defective oxygen sensor.	14. Replace oxygen sensor.
Problem: Stalling and/or rough idling	
Possible cause	**Correction**
1. Idle speed too slow.	1. Increase idle speed.
2. Fast idle speed too slow.	2. Increase fast idle speed.
3. Idle speed solenoid or motor defective.	3. Replace as needed.
4. Computer control system defect.	4. Test and replace as needed.
5. Clogged air cleaner.	5. Clean or replace air cleaner.
6. Vacuum leak.	6. Repair leak.
7. PCV system clogged.	7. Clean PCV system.
8. Exhaust heat valve stuck open.	8. Free heat valve or replace.
9. Inoperative canister purge valve.	9. Replace valve or canister.
10. Restricted air cleaner and/or exhaust.	10. Repair restriction or replace restricted part.
11. Defective glow plug system (diesel).	11. Repair glow plug system.
12. Injection pump timing off (diesel).	12. Correct timing.
13. Air in injection lines (diesel).	13. Bleed off air.
14. Low compression.	14. Determine reason and repair.
15. Faulty injection pump (diesel).	15. Repair or replace pump.
16. Malfunctioning injection nozzles.	16. Repair or replace nozzles.
17. Contaminated fuel.	17. Flush system. Add fresh, clean fuel.
18. EGR vacuum hoses misrouted.	18. Correct as required.
19. Defective EGR valve.	19. Replace EGR valve.

(continued)

Fuel System Problem Diagnosis *(continued)*	
Problem: Idle speed varies	
Possible cause	**Correction**
1. Throttle linkage dirty.	1. Clean linkage.
2. Throttle return spring weak.	2. Replace with stronger spring.
3. Accelerator pedal sticking.	3. Clean and lubricate accelerator cable and/or linkage.
4. PCV valve sticking.	4. Replace PCV valve.
5. Dirty or malfunctioning fuel injectors.	5. Clean or replace fuel injectors.
6. Injector control malfunction.	6. Replace injector control or control module.
7. Defective idle air control.	7. Diagnose and repair idle air control.
Problem: Poor acceleration	
Possible cause	**Correction**
1. Exhaust manifold heat control valve stuck.	1. Free heat control valve.
2. Low fuel pump pressure.	2. Replace fuel pump.
3. Air leaks.	3. Repair air leaks.
4. Clogged air cleaner.	4. Clean or replace air filter.
5. Faulty fuel injectors.	5. Clean or replace fuel injectors.
6. Engine control module defective.	6. Replace control module.
7. Plugged catalytic converter.	7. Replace catalytic converter.
8. Clogged muffler.	8. Replace muffler.
9. Bent tail pipe.	9. Replace tail pipe.
10. Clogged fuel filters.	10. Clean or replace filters.
11. Defective injection pump (diesel).	11. Rebuild or replace pump.
12. Defective injection nozzles (diesel).	12. Replace nozzles.
13. Timing not advancing.	13. Test and repair.
Problem: Lean mixture at cruising speeds	
Possible cause	**Correction**
1. Air leaks.	1. Repair air leaks.
2. Defective injection pump (diesel).	2. Repair or replace pump.
3. Fuel injection system control malfunction.	3. Adjust, repair, or replace.
4. Dirty or defective injector nozzles (diesel).	4. Clean or replace nozzles.
Problem: Rich mixture at cruising speeds	
Possible cause	**Correction**
1. Air cleaner clogged.	1. Clean or replace air filter.
2. Excessive fuel pressure.	2. Reduce fuel pressure.
3. Engine running too cold.	3. Check cooling system.
4. Faulty fuel injection pump.	4. Rebuild or replace pump.
5. Malfunctioning fuel injection control system.	5. Adjust, repair, or replace affected parts.
Problem: Low top speed	
Possible cause	**Correction**
1. Incorrect throttle linkage adjustment.	1. Adjust linkage correctly.
2. Clogged air cleaner.	2. Clean or replace air cleaner.
3. Air leak.	3. Repair air leak.
4. Obstruction under accelerator pedal.	4. Remove obstruction.
5. Engine operating too cold or hot.	5. Check cooling system.
6. Clogged catalytic converter.	6. Repair or replace converter.
7. Pinched exhaust pipe.	7. Replace pipe.
8. Clogged muffler.	8. Replace muffler.
9. Faulty injection pump (diesel).	9. Adjust, repair, or replace pump.
10. Plugged injection nozzles (diesel).	10. Clean or replace nozzles.
11. Fuel injection system control malfunction.	11. Repair, adjust, or replace affected unit(s).

(continued)

Fuel System Problem Diagnosis *(continued)*	
12. Low compression. 13. Incorrect (small) tire size. 14. Incorrect final drive ratio. 15. Timing not advancing.	12. Repair engine. 13. Install correct size tires. 14. Install correct ratio gears. 15. Test and repair.
Problem: Hard starting when cold	
Possible cause	**Correction**
1. Air leak. 2. Clogged air cleaner. 3. No fuel delivery. 4. Stale or contaminated fuel. 5. Gasoline not sufficiently volatile. 6. Incorrect grade of diesel fuel. 7. Vacuum leaks. 8. Defective ignition components. 9. Incorrect timing. 10. Discharged or defective battery. 11. Corroded or loose battery and starter connections. 12. Defective starter motor.	1. Repair air leak. 2. Clean or replace air cleaner. 3. Check tank and delivery system. 4. Drain tank. Flush system. Fill with fresh fuel. 5. Change to more volatile fuel. 6. Use correct grade fuel. 7. Locate and repair. 8. Locate, clean, adjust, or replace. 9. Set to specifications. 10. Charge or replace battery. 11. Clean and tighten connections. 12. Rebuild or replace starter.
Problem: Hard starting when hot	
Possible cause	**Correction**
1. Vapor lock. 2. Air leak. 3. No fuel delivery. 4. Stale or contaminated fuel. 5. Overheated engine. 6. Exhaust heat control valve stuck. 7. High elevation. 8. Vacuum hoses split, kinked, or loose. 9. Incorrect ignition timing. 10. Malfunctioning fuel injection system control.	1. Cool lines. Change to less volatile fuel and protect lines from heat. 2. Repair air leak. 3. Check delivery system. 4. Drain tank and fuel system. Fill with fresh fuel. 5. Check cooling system. 6. Free control valve. 7. Change to less volatile fuel. 8. Replace and secure connections. 9. Set to specifications. 10. Repair, adjust, or replace affected parts.
Problem: Excessive fuel consumption	
Possible cause	**Correction**
1. Excessive speed. 2. Rapid acceleration. 3. Heavy loads or trailer towing. 4. Low tire pressure. 5. Dragging brakes. 6. Stop and start driving. 7. Fuel leaks (external). 8. Clogged air cleaner. 9. Low grade or stale gasoline. 10. Exhaust heat control valve stuck. 11. Fuel pressure excessive. 12. Front wheel alignment out. 13. Exhaust system clogged. 14. Transmission slipping. 15. Wrong axle gear ratio. 16. Incorrect tire size. 17. Malfunctioning fuel injection system control. 18. Faulty fuel injector pump (diesel). 19. Injector pump timing incorrect (diesel). 20. Incorrect grade of diesel fuel. 21. Torque converter lockup inoperative.	1. Caution owner to reduce speed. 2. Accelerate moderately. 3. Normal. 4. Inflate tires to proper level. 5. Adjust brakes. 6. Normal. 7. Repair leaks. 8. Clean or replace cleaner. 9. Use higher grade fuel. Use fresh fuel. 10. Free valve. 11. Lower fuel pressure. 12. Align front wheels. 13. Replace muffler and/or tail pipe. 14. Adjust or overhaul transmission. 15. Change to factory ratio. 16. Install proper size tires. 17. Repair, adjust, or replace affected control. 18. Repair, adjust, or replace pump. 19. Set to specifications. 20. Use correct grade. 21. Repair or replace lockup solenoid, converter, or control module.

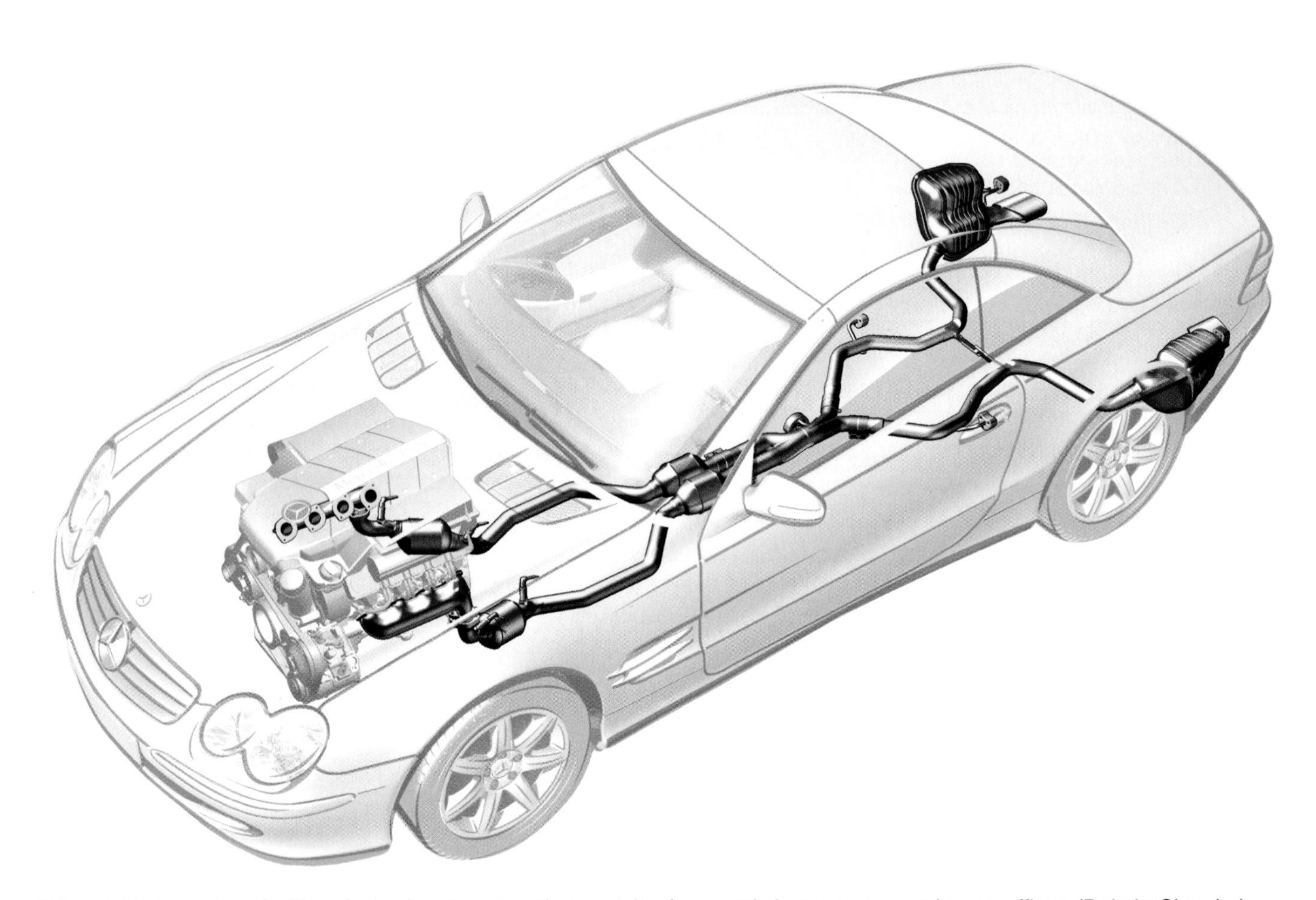

This vehicle is equipped with a dual exhaust system that contains four catalytic converters and two mufflers. (DaimlerChrysler)

19

Exhaust System Service

After studying this chapter, you will be able to:

- Describe exhaust manifold removal and installation.
- Explain heat control valve service, and repair.
- Identify and use special exhaust system service tools.
- Remove and install a muffler.
- Remove and install an exhaust pipe and tailpipe.
- Remove and install a catalytic converter.
- Remove and install a turbocharger waste gate diaphragm.
- Remove and install a turbocharger.
- Remove and install a supercharger.

Technical Terms

Exhaust manifold
Heat control valve
Heat riser
Early fuel evaporation (EFE) device
Mufflers
Back pressure
Resonator
Catalytic converter
Catalysts
Exhaust pipes
Y-pipe
Tailpipe
Joint clamps
Support brackets
Pipe expander
Straightening cone
Exhaust system alignment
Turbochargers
Superchargers
Waste gate
Bypass valve
Intercooler

Although modern exhaust systems last longer than ever, almost every technician will do some exhaust repairs during his or her career. This chapter covers the service of exhaust system parts, including the exhaust manifolds, heat control devices, exhaust pipes, mufflers, catalytic converters, and resonators.

Exhaust System Service

The exhaust system removes the spent exhaust gases from the engine. The exhaust system must do this while keeping noise levels to a minimum and preventing the poisonous gases from entering the passenger compartment. Other devices are placed on the exhaust system to assist in heating the fuel mixture and reducing emissions. In most cases, service is limited to replacing parts and making sure that the system is not leaking.

Warning: The exhaust system parts can become extremely hot. Allow exhaust parts to cool off completely before servicing.

Exhaust Manifold

As the burned gases leave the engine cylinders, they pass into the ***exhaust manifold.*** The manifold is usually made of cast iron or stainless steel and is attached to the cylinder head with a series of fasteners. It is designed with a minimum of sharp bends, allowing it to route the exhaust gases while keeping back pressure at a minimum. Once the exhaust manifold is affixed to the engine, it requires no service.

To remove the manifold, disconnect the exhaust pipe and any braces or tubing that may be connected to the manifold. Next, remove the manifold fasteners. If the fasteners are stuck, apply penetrating oil and allow it to soak for 15 to 20 minutes before attempting to loosen the fasteners.

Note: Exhaust manifold bolts frequently break or strip during removal. You may want to review Chapter 7 for rethreading and bolt removal information.

When installing a manifold, all mounting surfaces must be clean. Use a file to remove burrs and bits of hardened gasket material. Install new gaskets where needed. Torque fasteners in proper sequence. If the new manifold-to-head gaskets are made of a composition material (not steel), the fasteners should be retorqued after the engine has been operated. This compensates for the torque lost when the gasket flattens out after heating.

Use fastener locks, when required, to prevent the fasteners (especially the end ones) from loosening. Connect the exhaust pipe (use a new gasket) and hook up the parts originally attached to the manifold.

Figure 19-1 illustrates a typical exhaust manifold. Note that the intake manifold is connected to the exhaust manifold by a metal tube. The tube provides exhaust gases for EGR (exhaust gas recirculation) valve operation. Exhaust manifolds for V-type engines use similar construction. The cutaway of the exhaust manifold in **Figure 19-2** shows the construction of a modern stainless steel exhaust manifold. The interior stainless steel lining helps absorb noise. The air layer between the inner liner and the manifold reduces heat transfer, allowing more heat to reach the catalytic converter. This assists converter operation.

Heat Control Valve

On older engines, a ***heat control valve,*** or ***heat riser,*** may be installed in the exhaust manifold or between the exhaust manifold and the exhaust pipe. The heat riser provides heat to help vaporize the fuel during engine warm-up. On early engines, the valve is actuated by a bimetallic spring. On later vehicles, a vacuum motor (diaphragm unit) operates the heat riser. The vacuum-operated heat control valve is often called an ***early fuel evaporation (EFE) device.***

The heat control valve will often stick due to carbon buildup. If it sticks in the open position, the engine may hesitate and stall during warm up. Some carburetor and throttle bodies may form ice around the throttle plate, causing repeated stalling. If the valve sticks closed, overheating, detonation, burned valves, and a warped intake manifold may result. It is important to have the heat control valve free in its bushings. The thermostatic spring should not be distorted.

To check an older spring-operated heat control valve, accelerate the engine quickly while watching the heat valve. The counterweight should move, indicating that the shaft is free. To double-check, allow the manifold to cool off thoroughly and then try to move the weight by hand to make sure the valve has full travel.

To check a vacuum-operated heat control valve, allow the manifold to cool, and then attach a vacuum pump to the vacuum motor's vacuum inlet. Apply vacuum while observing the linkage. If the linkage has not moved by the time that 10 inches (33.7 kPa) of vacuum have been applied, the valve is stuck, the linkage is binding or disconnected, or the vacuum motor is leaking.

If the heat control valve is stuck, allow the manifold to cool off and apply several drops of penetrating oil to both ends of the shaft where it passes through the manifold. Work the valve back and forth until it is free. Add more penetrating oil as needed. When the valve is stuck so tight that it cannot be moved by hand, tap the ends of the shaft after applying the penetrating oil. After the valve is free, lubricate the shaft with a special heat-resistant graphite mixture or leave it dry and clean. Never use engine oil, as it will burn and form more carbon.

Caution: Do not use carburetor cleaner or oil in or on a vacuum motor assembly.

If the heat control valve vacuum motor is leaking, it can be replaced without removing any other exhaust system components. Soak the attaching bolts in penetrating oil, and then loosen and remove them. Remove the vacuum line and linkage, and remove the vacuum motor from the manifold. Place the new vacuum motor in position on the manifold and install the attaching bolts. Reattach the linkage and apply vacuum to the new vacuum motor and make sure that the valve linkage moves. If the linkage moves when vacuum is applied, reattach the vacuum line, making sure that it does not touch the manifold.

Figure 19-1. The purposes of exhaust manifolds are the same on all engines. This exhaust manifold is used on an inline engine. (DaimlerChrysler)

Figure 19-2. This cutaway view of an exhaust manifold used on a modern V8 engine shows the double wall construction used on many modern manifolds.

Exhaust System Components

The exhaust system is the series of pipes and other components that exhaust gases travel through on their way from the exhaust manifold to the open air. Exhaust systems may be of the single type, **Figure 19-3,** or the dual type, **Figure 19-4.** As the exhaust gases move through the exhaust system, exhaust sound is silenced. In addition, exhaust gases are chemically altered to be more environmentally safe, and then routed to either the end or the side of the vehicle, where they are discharged into the atmosphere. The parts of the exhaust system are discussed in the following sections.

Mufflers

Mufflers are designed to reduce, or muffle, the sound of the exhaust system. They should silence the exhaust effectively while keeping ***back pressure*** at a minimum. Back pressure occurs when the exhaust gases cannot pass through the muffler fast enough and thus build up pressure, which reduces the ability of the engine piston to push exhaust gases out of the cylinder.

Resonators

In addition to a muffler, some exhaust systems have a ***resonator*** to further dampen the exhaust pulsations. The resonator is a small version of the muffler and is always installed after the muffler in the exhaust system.

Catalytic Converters

Vehicles are equipped with a ***catalytic converter*** to help reduce exhaust emissions. Catalytic converters are always located between the exhaust manifold and the muffler. The catalytic converter contains substances called ***catalysts,*** which cause chemical changes in other substances without themselves being changed.

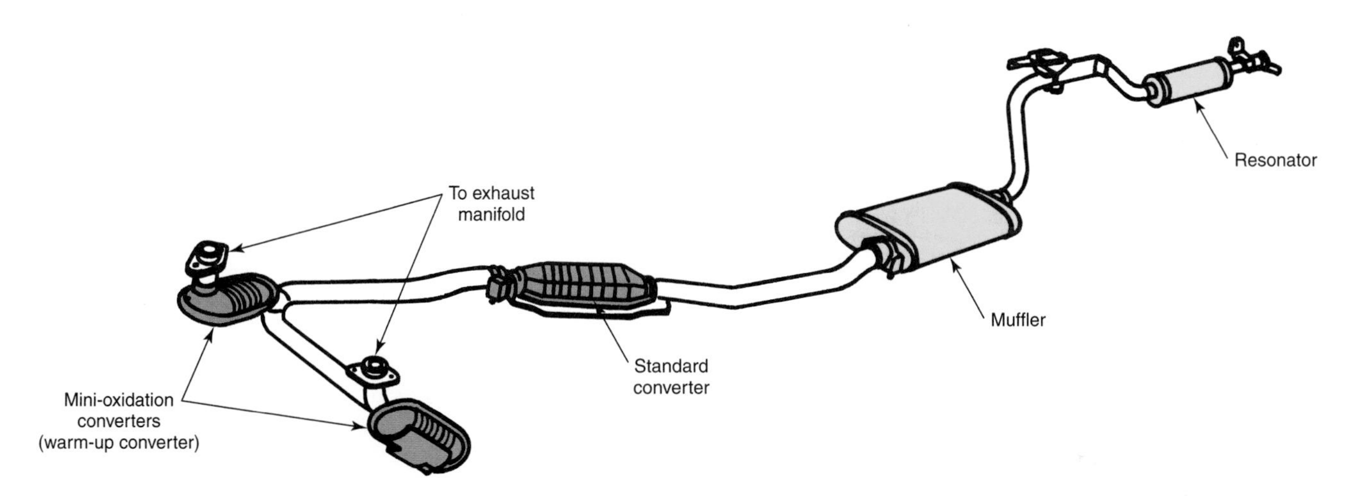

Figure 19-3. This illustration shows the components of a single exhaust system used by one manufacturer. (Sun Electric Corp.)

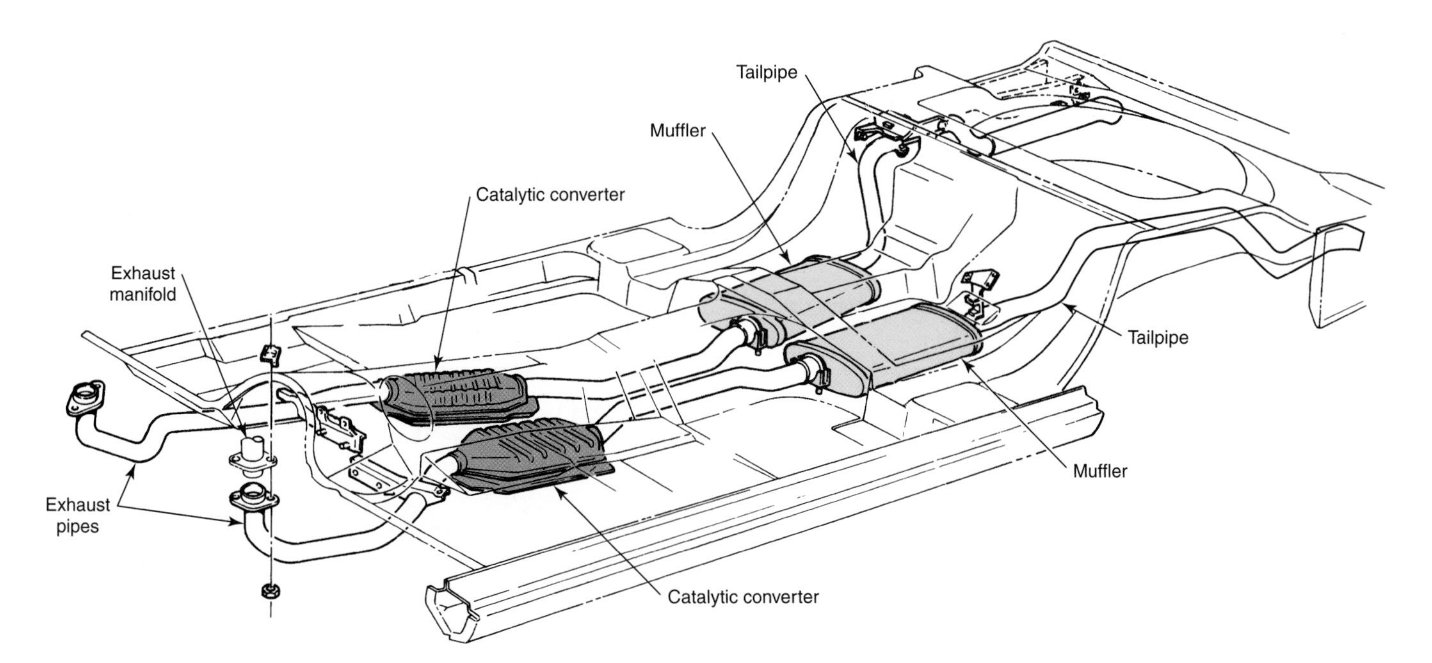

Figure 19-4. This figure shows a typical dual exhaust system. Note the catalytic converters. (DaimlerChrysler)

Some vehicles use a single converter while others employ an additional smaller unit close to the exhaust manifold, **Figure 19-5.** This unit starts the reaction and heats up the exhaust. See **Figure 19-6.** A larger, downstream converter continues and completes the burning and reduction process. This type is referred to as a two-stage converter. Air may be injected into the exhaust between the two converters or into the converter itself. See **Figure 19-7.**

Pipes

The pipes connecting the engine to the muffler (and converter) and the muffler to the resonator are called ***exhaust pipes.*** The first pipe on a V-type engine, which connects the two exhaust manifolds to a single pipe, is called the ***Y-pipe.*** The last pipe in the system, which allows the exhaust gases to exit to the atmosphere, is called the ***tailpipe.*** Exhaust pipes and tailpipes may be aluminized steel and may use single-wrap or double-wrap construction. The double wrap is more effective in reducing exhaust noise resulting from system pulsations.

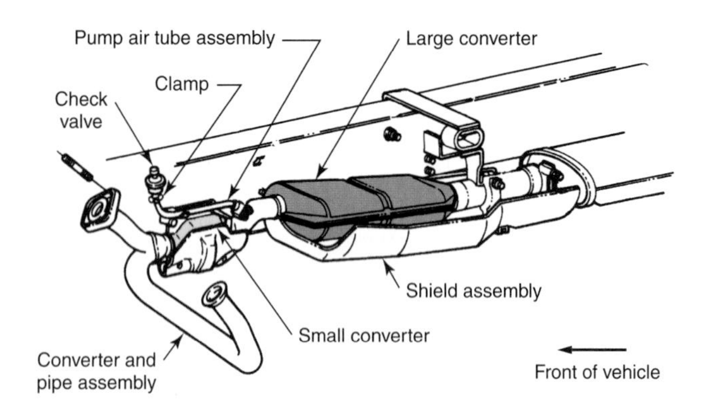

Figure 19-5. Some vehicles use dual converters for maximum efficiency. Note the small converter close to exhaust manifold connecting pipes. Air is pumped into the exhaust pipe connecting the small and large converters. (Ford)

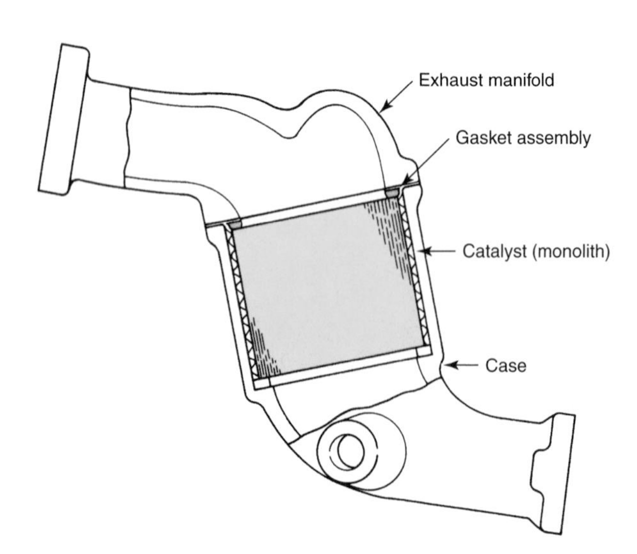

Figure 19-6. This mini-type converter is mounted near the end of the exhaust manifold. A larger converter is used further downstream. (DaimlerChrysler)

Clamps and Support Brackets

The various parts of the exhaust system must be properly joined and supported. ***Joint clamps*** and ***support brackets*** of many types are used, **Figure 19-8.** Figure 19-7A shows a random sampling. Always use clamps of the proper size to ensure a good joint. Check support bracket flexible straps for breakage and replace them as needed. Some catalytic converters and other exhaust system parts are attached through flanges, **Figure 19-9.** Flanges provide a flat sealing surface between the two parts and are always used with a gasket.

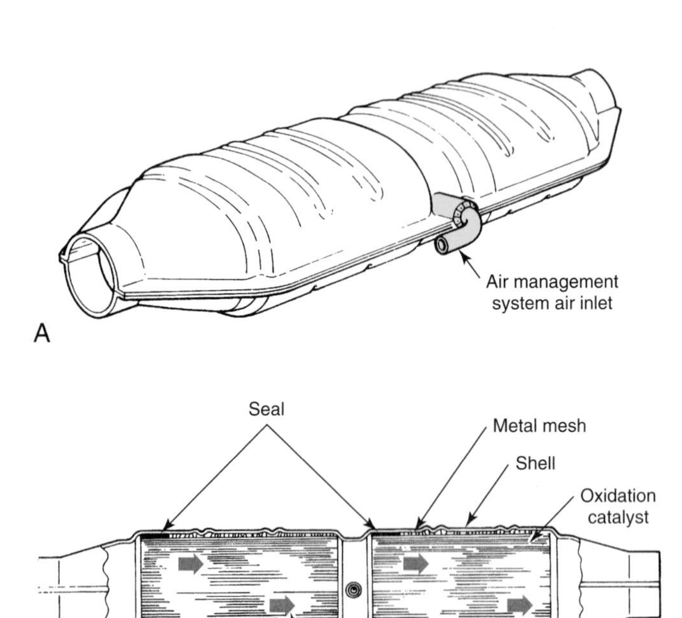

Figure 19-7. As the exhaust gases pass through the converter catalyst sections, hydrocarbons and carbon monoxide are converted into water and carbon dioxide. A—Air entering the mixing chamber provides additional air for the oxidizing (burning) process. B—This converter uses two separate catalyst beds that are separated by the mixing chamber. (General Motors)

Figure 19-8. Many types of clamps and support brackets are used on the modern exhaust system. This illustration shows some commonly used clamps and brackets. (McCord)

Exhaust System Tools

Exhaust system service is highly competitive. To show a profit, the work must be done swiftly. To accomplish this, it is essential that proper tools be used. The power chisel, with suitable cutter heads, is useful for cutting welded mufflers free and for removing tailpipes and exhaust pipes, **Figure 19-10.** Note the assortment of cutting heads with the chisel. A hand pipe cutter, illustrated in **Figure 19-11,** can also be very helpful.

Often, pipes must be expanded to provide a proper joint fit. This is done easily with a ***pipe expander.*** See **Figure 19-12.** The pipe end is often crimped or otherwise distorted. It can be readily brought back to a round shape by using a ***straightening cone,*** **Figure 19-13.** A chain wrench provides a way of both pulling and twisting a pipe, either to free the joint or to provide proper alignment. See **Figure 19-14.**

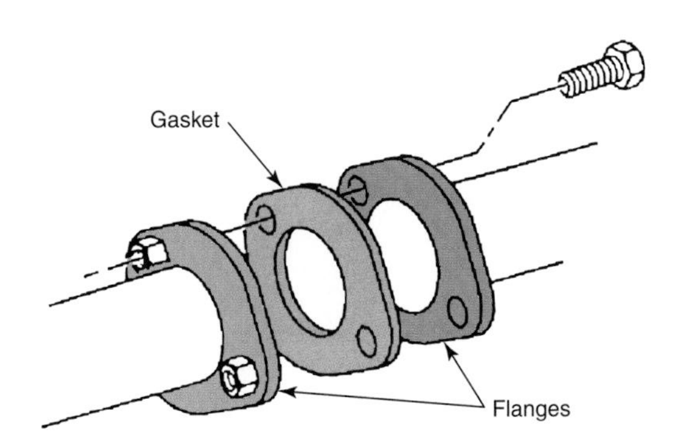

Figure 19-9. The flanged connector shown here is used on many vehicles. A gasket must be used between the flanges. (General Motors)

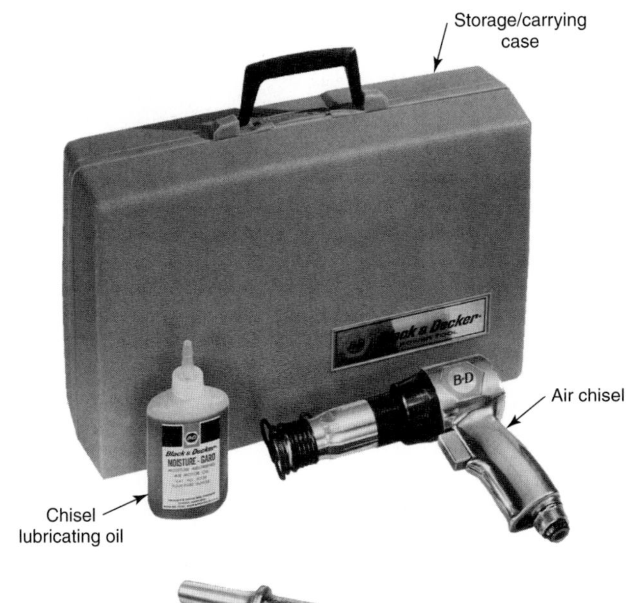

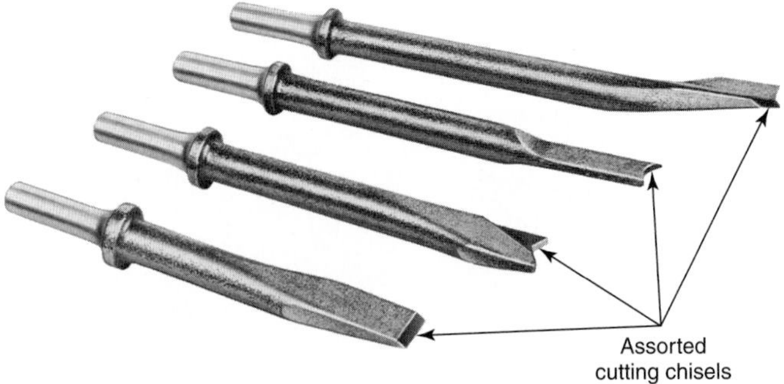

Figure 19-10. Power chisel with assorted cutting heads. (Black and Decker, Walker)

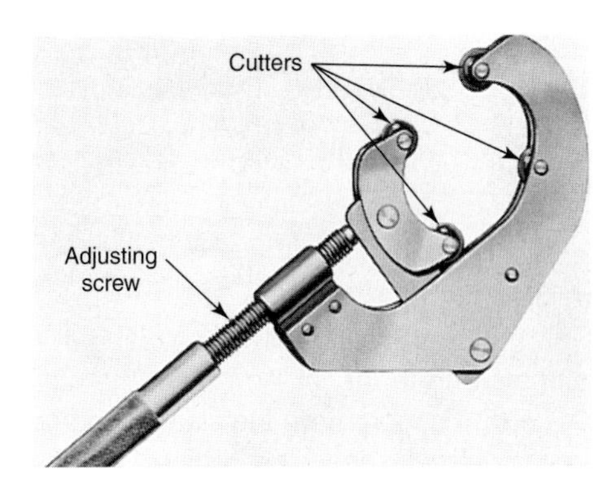

Figure 19-11. This illustration shows a hand-operated pipe cutter. (Walker)

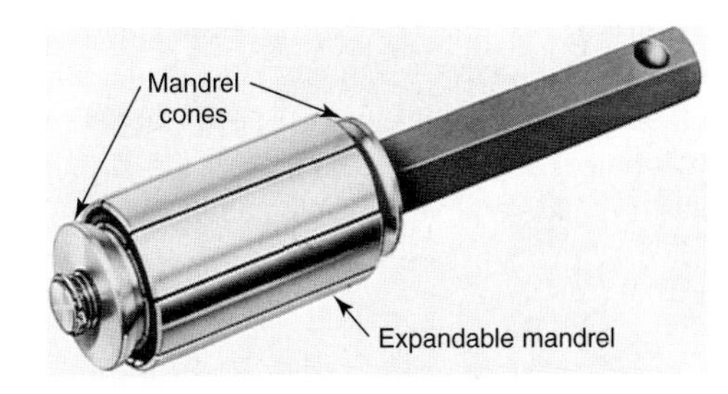

Figure 19-12. The pipe-expanding tool in this figure is used to enlarge pipes or to restore a crushed pipe to its original shape.

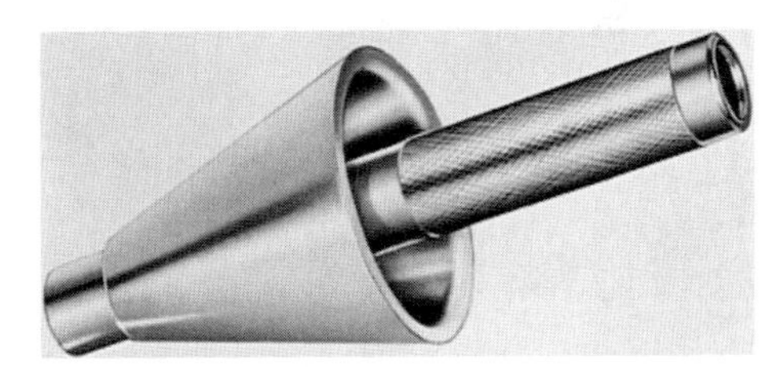

Figure 19-13. The straightening cone shown here is placed in the deformed pipe end and tapped until the pipe is round.

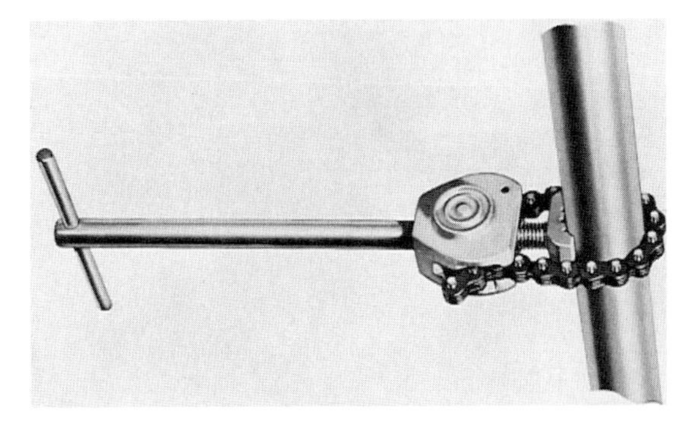

Figure 19-14. The chain wrench provides a good grip on the pipe to facilitate removal or alignment. (Walker)

Exhaust System Service

To perform any exhaust system repair, begin by raising the vehicle. It is very difficult to change exhaust system parts without the clearance provided by raising the vehicle. A frame contact hoist works well for exhaust system service because it allows the rear axle to hang down, providing ample room for part removal.

Warning: Wear eye protection when servicing an exhaust system. Before touching any exhaust system part, allow the exhaust system to cool sufficiently.

Part Removal

After raising the vehicle, apply penetrating oil to the pipe bracket fasteners and to the joint clamp at the muffler outlet. See **Figure 19-15.** Then, remove the clamps and bracket fasteners. An impact wrench speeds up this job. See **Figure 19-16.** If the clamps will not be reused, they can be cut off with a hacksaw or cutting torch. Apply penetrating oil and tap the outlet joint. Apply heat if needed. Pull the pipe free with a chain wrench, **Figure 19-17.** If a pipe is to be reinstalled, use care during its removal. If the pipe remains stuck, cut it off just clear of the muffler or pipe outlet nipple. See **Figure 19-18.** Next, use a power chisel and split the section of the pipe remaining in the nipple. Finally, remove the split section with pliers. If the catalytic converter is held to the exhaust manifold or other exhaust system parts by bolted flanges, remove the bolts and separate the flanges.

Note: Some exhaust parts are welded together. In such cases, cut the pipe where necessary, Figure 19-19. After cutting, straighten the pipe end. The replacement part will have a connection nipple that will engage the existing pipe.

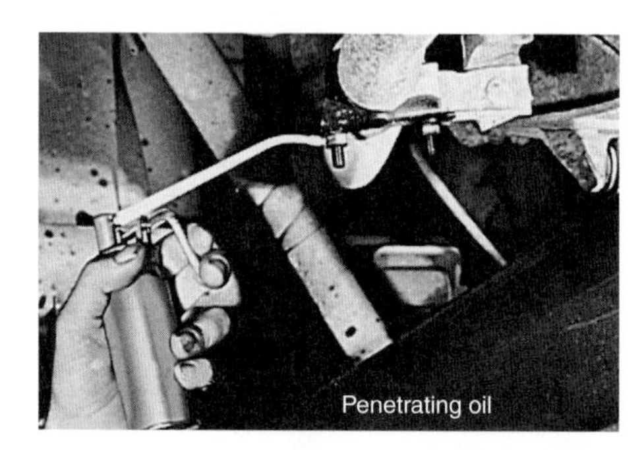

Figure 19-15. Use penetrating oil to facilitate fastener removal. Allow the penetrating oil to soak in for several minutes before loosening the fastener.

Figure 19-16. Muffler clamp nuts can be quickly removed with an impact wrench.

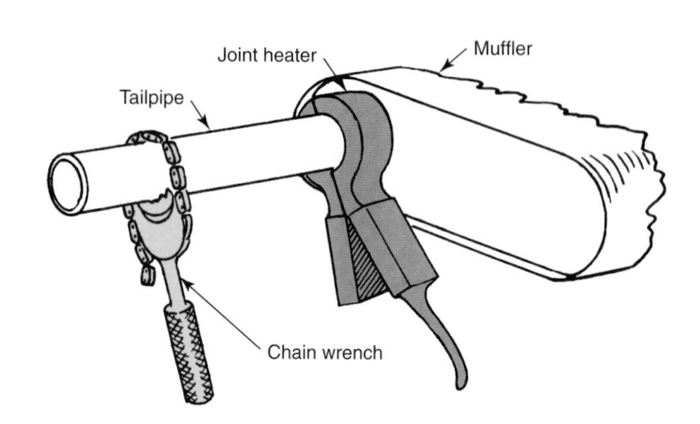

Figure 19-17. As shown here, the tailpipe can often be freed from the muffler by using heat and a chain wrench.

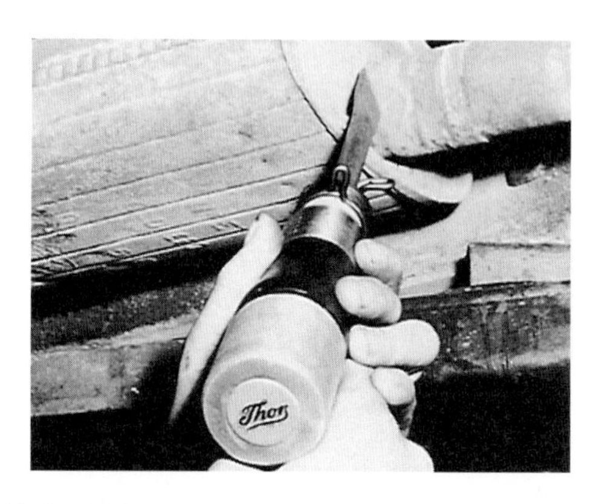

Figure 19-18. Using a power chisel to cut a muffler from an exhaust pipe.

Figure 19-19. Welded tailpipes can be removed with a hand pipe cutter.

Part Cleaning

Use a coarse emery cloth to clean the inside or outside of the nipples that will be reused. If a nipple is distorted, use the pipe-straightening cone to restore it to a perfectly round state.

Obtain and Check New Parts

To ensure that you get the correct exhaust system parts, order new parts by vehicle model and engine size. Always use mufflers and pipes designed for the vehicle at hand. Do not use undersized pipes or mufflers.

Some shops are equipped with pipe benders that can create any exhaust pipe configuration from straight pipe stock. When making pipes on a bender, always follow the manufacturer's instructions. Accuracy is extremely important when making pipes.

Exhaust System Sealant

Before reassembly, apply a liberal coating of exhaust system sealant to the section of the tailpipe that will be in contact with the muffler or converter nipple. Refer to **Figure 19-20.** Use exhaust sealant on all exhaust system joints. It makes the joints slide together easily, assists in alignment, and prevents dangerous exhaust leaks.

Some makers have designed special joints that are first assembled and then sealed. On setups requiring this technique, you must inject a special sealant into an annular ring, or bead, formed in the pipe. After injecting the sealant, idle the engine for about ten minutes to harden the compound. **Figure 19-21** illustrates this procedure. If an exhaust system part is attached with a flange, clean all of the old gasket material from the flanges of the components that will be reused. Use new gaskets when reinstalling the exhaust system.

Assembling Parts

Slide the pipes and nipples together. Make certain that the depth is correct, **Figure 19-22.** Slide the clamp into position, **Figure 19-23,** and tighten the clamps lightly. Make sure the pipes, catalytic converter, muffler, and resonator are installed in the correct position and in the proper direction. Finally, secure all connections.

Setting Pipe Depth in the Nipple

When inserting one pipe into another, insert the pipes so they engage to the proper depth. If the pipe enters too deeply,

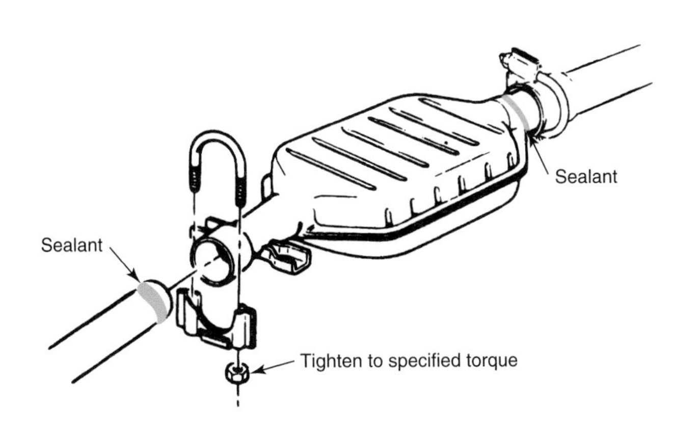

Figure 19-20. Exhaust system sealant helps eliminate leaking joints. Use proper sealant. (General Motors)

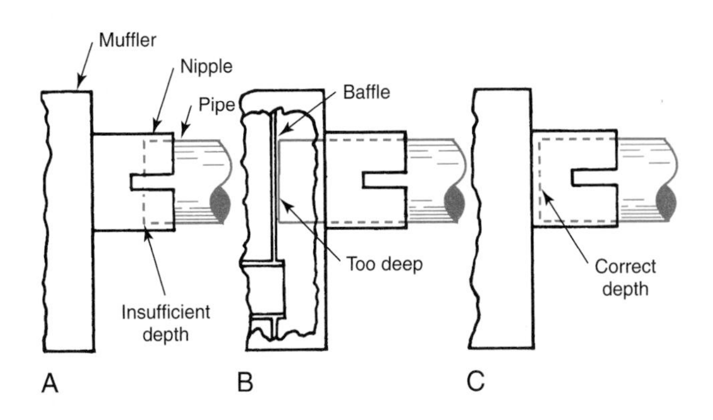

Figure 19-22. A and B—Incorrect depths. C—Pipe must enter the muffler nipple to the correct depth.

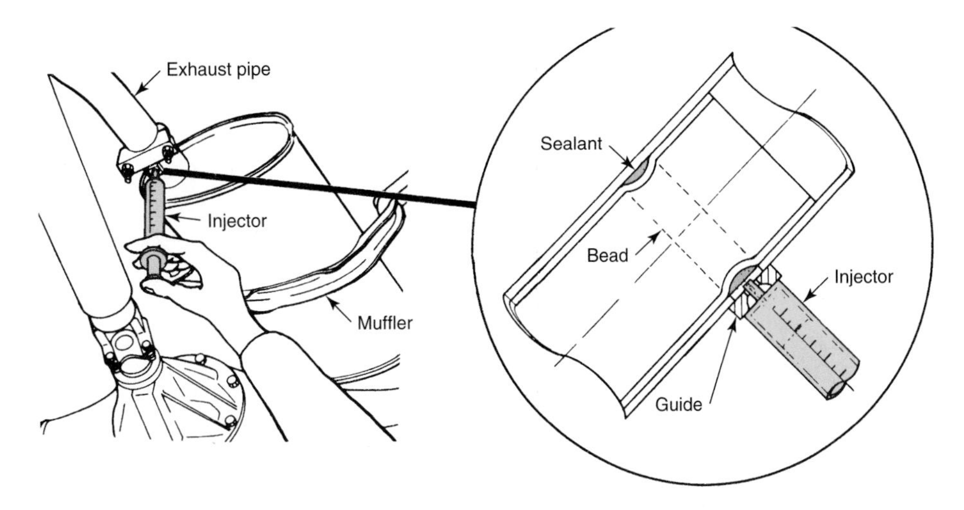

Figure 19-21. This type of exhaust system joint is assembled, tightened, and then sealed by injecting sealant into an annular groove. (Nissan)

it can cause back pressure. Insufficient contact does not allow proper clamping and may permit leakage or pipe separation.

Installing Joint Clamp

Install the clamp so it is about 1/8″ (3.18 mm) from the nipple end (or pipe end when a pipe slips over a nipple). Look at **Figure 19-24.** After aligning all mufflers, resonators, and pipes, tighten the clamps and brackets.

Figure 19-23. A new muffler installed on an exhaust pipe.

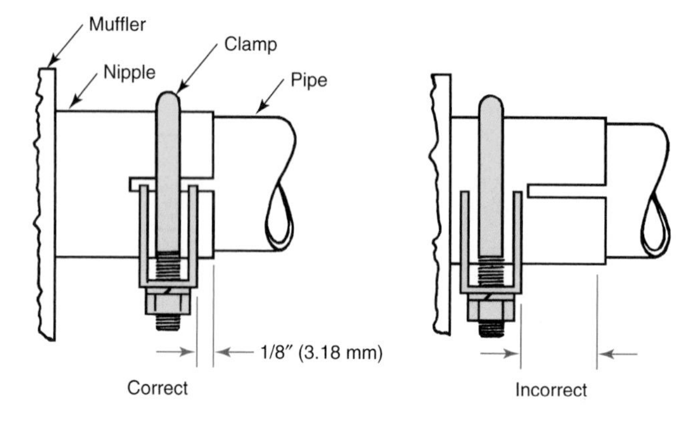

Figure 19-24. The retaining clamp should be positioned about 1/8″ (3.18 mm) from the end of the nipple or pipe.

Note: Do not tighten the clamp to the point that the joint starts to collapse.

Catalytic Converter Service

Theoretically, a catalytic converter could last the life of the vehicle, since the catalyst is not consumed. Converters are designed to last for a minimum of 50,000 miles (80,000 km) and can usually last much longer, if the catalyst is not exposed to lead, overly rich mixtures, or other damage. Initial converter failure can be difficult for the driver to detect. Engine performance is sometimes not affected until the converter becomes clogged. To test converter efficiency, measure the exhaust gas temperature or use an exhaust gas analyzer, which is explained in Chapter 21, Emissions Testing and Service. Most catalytic converters require total replacement. When the converter is replaced, turn the old converter in for recycling; do not discard it as scrap.

Checking Exhaust System Alignment

Exhaust system alignment is critical. It affects the clearances between various parts of the vehicle and the exhaust system. Make a careful check of the entire system to make certain that all parts have sufficient operating clearance. Pay particular attention to the area where the tailpipe crosses over the rear axle. Make sure that the pipe will clear the springs, shocks, and axle when the springs bottom under a heavy load or when hitting bumps. The pipes must also clear the drive shaft, brake lines, and gas lines.

Checking System for Leaks

As a final step, always operate the engine and check each joint for signs of exhaust leaks. See **Figure 19-25.** The system must be airtight to prevent the escape of exhaust gases. If a leak is found, repair the joint.

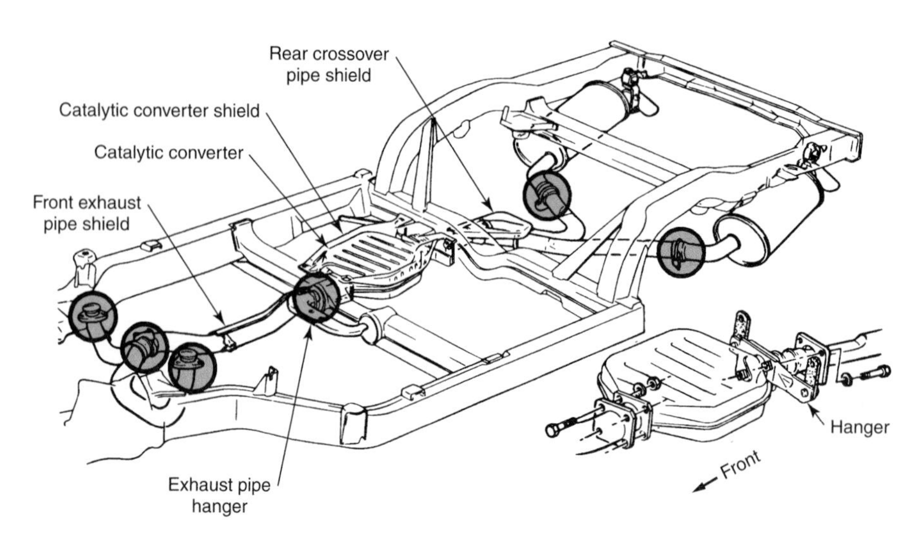

Figure 19-25. Check connections (circled areas) for leaks. (General Motors)

Warning: Exhaust gases contain carbon monoxide (a deadly poison) and, therefore, must not escape under the vehicle. They could flow into the vehicle's passenger compartment, causing serious injury or death.

Turbochargers and Superchargers

To obtain extra power from an engine, devices are sometimes used to pressurize the intake manifold. This forces more air into the engine cylinders. Forcing more air into the engine, in effect, increases the engine's compression ratio, resulting in more horsepower. Two kinds of devices are used: ***turbochargers,*** operated by exhaust gases; and ***superchargers,*** driven by the engine crankshaft. Turbocharger service and supercharger service are explained below.

Turbochargers

Engine exhaust gases drive the turbocharger. **Figure 19-26** is a schematic showing the principle of turbocharging. To keep the turbocharger from developing too much pressure, a ***waste gate*** is installed in the exhaust passages. A pressure operated diaphragm opens the waste gate. The diaphragm is connected to intake manifold through a hose. When the turbocharger overpressurizes the intake manifold, this pressure is transmitted to the diaphragm, which opens the waste gate. Some waste gates are operated by the ECM, which controls a solenoid that allows engine pressure into the diaphragm. When the waste gate opens, exhaust gases bypass the turbocharger, causing it to slow down and reducing manifold pressure. Turbocharger bearings are lubricated by pressurized oil from the engine lubricating system.

Superchargers

Superchargers are driven by the engine crankshaft through belts, chains, or gears. **Figure 19-27** shows the general layout of the supercharger and its related parts. When the supercharger output becomes too great, the ***bypass valve*** opens. The bypass valve diverts air back into the supercharger intake. The bypass valve is operated by mechanical linkage or manifold vacuum. The ECM usually controls bypass valve operation. The supercharger is located between the throttle body and the engine.

Many turbochargers and superchargers have an ***intercooler*** installed between the charger output and the engine. The intercooler cools the pressurized air mixture. Cooling the air reduces engine knocking and improves driveability and performance. Intercoolers are shown in Figures 19-26 and 19-27.

Turbocharger Service

Although turbochargers can spin at speeds over 100,000 rpm, they generally require little service during their normal service life. Gaskets must be in good condition, fasteners must be torqued, and the oil supply system must be open and functioning. Auxiliary controls, such as the waste gate, must operate correctly. It is usually not necessary to adjust boost pressure. On some turbochargers, the linkage between the waste gate and diaphragm has no adjustment provision. The most common turbocharger problem is oil seal failure, resulting in oil being pumped into the exhaust system. This results in excessive oil smoke from the exhaust.

Checking Turbocharger Boost Pressure

A general check of turbocharger operation can be made with the dashboard boost gauge. Start the engine and observe the gauge as you gradually accelerate the engine. The gauge should show the turbocharger boost pressure gradually rising to the maximum range and then remaining steady as the engine is accelerated further. If the boost pressure is within the proper range, and no other problem, such as pinging or low power are encountered, the turbocharger is probably okay.

However, most dash gauges are usually not accurate enough to make a precise check of the boost pressure and should not be used when adjusting the boost pressure. A

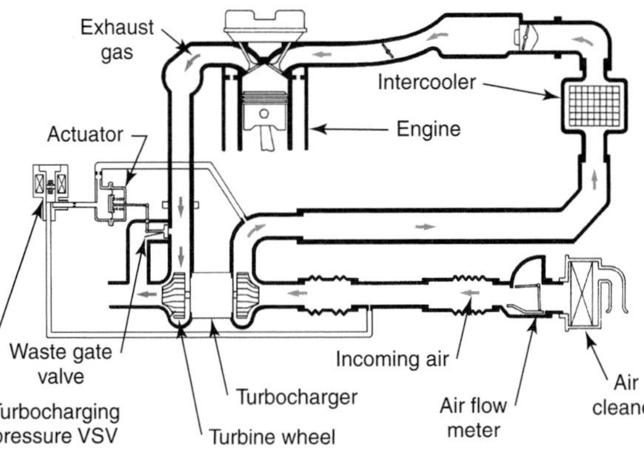

Figure 19-26. This schematic illustrates an exhaust-driven turbocharger. Note the use of an intercooler to drop the mixture temperature. Study the flow of the intake air and exhaust gas. (Toyota)

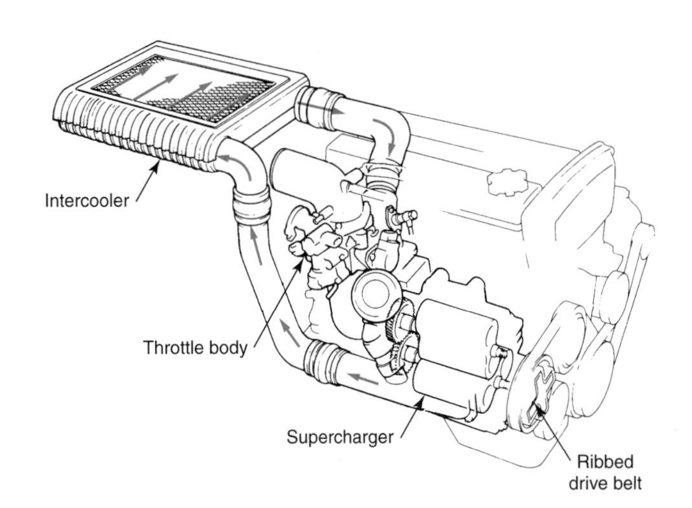

Figure 19-27. This supercharger is mounted on the side of the engine and driven by a belt. The intercooler removes some of the heat from the compressed air. (Toyota)

pressure gauge with a range of 0 lbs–20 lbs is necessary to accurately check boost pressure. The pressure gauge must be installed on an intake manifold fitting or a T-fitting at the waste gate, **Figure 19-28.**

Note: Do not attach the pressure gauge to the turbocharger waste gate fitting in such a way that manifold pressure cannot reach the waste gate. Loss of control to the waste gate diaphragm can cause severe damage.

Install a tachometer on the engine and make sure that the engine is fully warmed up. After the pressure gauge and tachometer are installed, start the engine and increase engine speed until the turbocharger begins to operate. Watch the tachometer carefully and make sure that the engine's speed does not exceed its redline. Most engines in good condition can safely reach 4000 rpm. The safe limit for the engine that you are working on should be determined before starting this test.

Increase the engine speed until the limit of boost pressure is reached. The waste gate should open when boost pressure is within the maximum range. If the boost pressures are below normal, either the turbocharger is defective, or the waste gate is opening too soon or is stuck partially open. If boost pressures are higher than normal, the waste gate is stuck closed, the linkage is disconnected or jammed, or the actuator diaphragm or manifold hose is leaking. After correcting any problems and rechecking the boost pressure, remove the tachometer and pressure fitting.

Caution: Do not attempt to gain more power from the engine by adjusting the boost pressure to be higher than the specifications. This can cause severe engine damage.

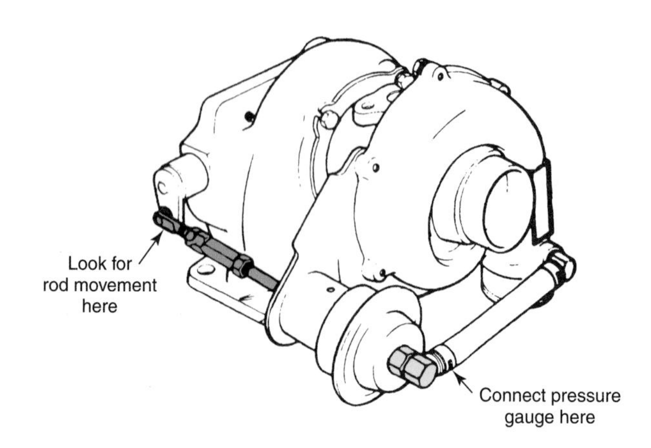

Figure 19-28. Turbochargers should be checked for external oil leaks and unusual noises. The waste gate rod should move freely. Be sure to connect the pressure gauge to the right fitting. (Ford)

If boost pressure is too low, check the exhaust system for restrictions. A restricted exhaust lowers the exhaust gas speed, preventing the turbocharger from turning at sufficient speed. If the exhaust is not restricted, make sure the turbocharger is in good operating condition. If the boost pressure is low because of a damaged turbocharger, the turbocharger should be replaced.

Replacing a Waste Gate Diaphragm

The pressure diaphragms in most turbocharger waste gates can be replaced without replacing the entire turbocharger unit. To remove the diaphragm, allow the engine to cool, and then remove the line from the intake manifold. This line is usually attached to the diaphragm assembly with pipe fittings. These can be removed with the proper-size line wrench. Next, remove the linkage to the waste gate and remove the brackets holding the diaphragm to the turbocharger. To install the new diaphragm, reverse the order of removal. Make sure that the line to the diaphragm assembly does not leak; such a leak could cause excess manifold pressure and possible engine damage. Adjust the linkage as necessary to obtain the proper boost pressure.

Removing, Checking, and Replacing a Turbocharger

Clean the turbocharger and its surrounding area before removing it. To remove the turbocharger, allow the engine to cool off for at least one hour. Next, remove the line from the intake manifold. Loosen the fasteners holding the turbocharger to the exhaust pipes and any brackets holding the turbocharger to the engine. Place a pan under the turbocharger and remove the pressure and return oil lines. Finish removing the exhaust pipe and bracket fasteners and lift the turbocharger from the engine. **Figure 19-29** shows some steps in removing the turbocharger. After removing the turbocharger, cover the exhaust pipe openings.

Caution: Use extreme care during handling and repair to avoid nicking or bending of any part of the turbine or compressor blades. Blade damage can drastically reduce a turbocharger's life. In some cases, it can cause the unit to literally self-destruct.

Remove the turbocharger's end housings to gain access to the turbine and compressor blades. Check the blades for nicks, bends and cracks. Check passages for carbon buildup. The unit should spin freely and quietly. Check endplay (axial play) of the turbocharger shaft with a dial indicator. Inspect boost control (bypass valve) for proper operation. Follow the manufacturer's recommendations. If the turbocharger blades are damaged, it is usually cheaper to replace the entire assembly rather than attempt to rebuild it.

To reinstall the turbocharger, remove all gasket material from the exhaust pipes. Using new gaskets, place the turbocharger in position on the exhaust pipes and loosely install the attaching fasteners. Reinstall the bracket fasteners and reattach the oil lines. Tighten the exhaust pipe fasteners. Make sure that the engine oil level is at the full mark, and then crank the engine over several times with the ignition disconnected. This delivers oil to the turbocharger bearings. Finally, start the engine, check the turbocharger's operation, and look for oil and exhaust leaks.

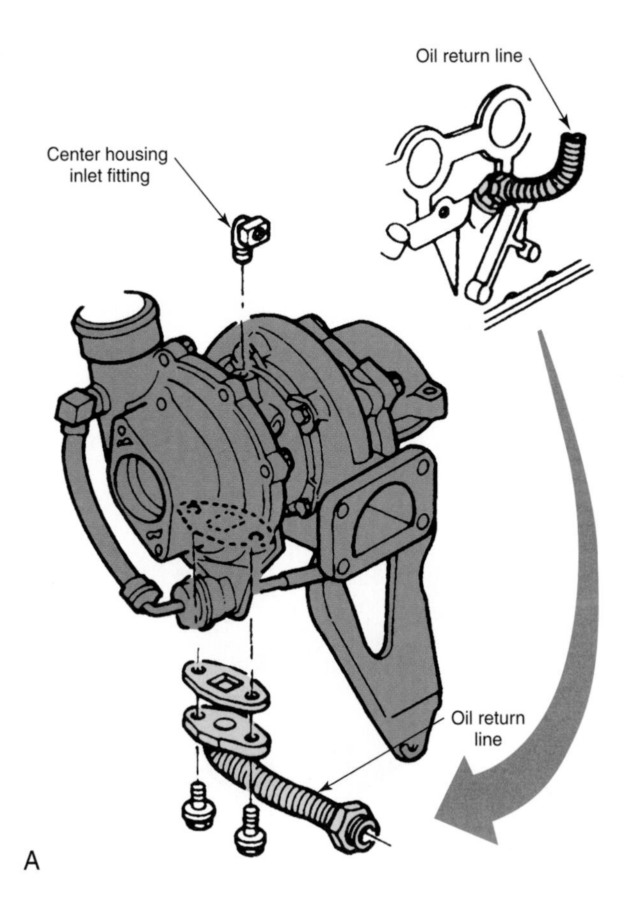

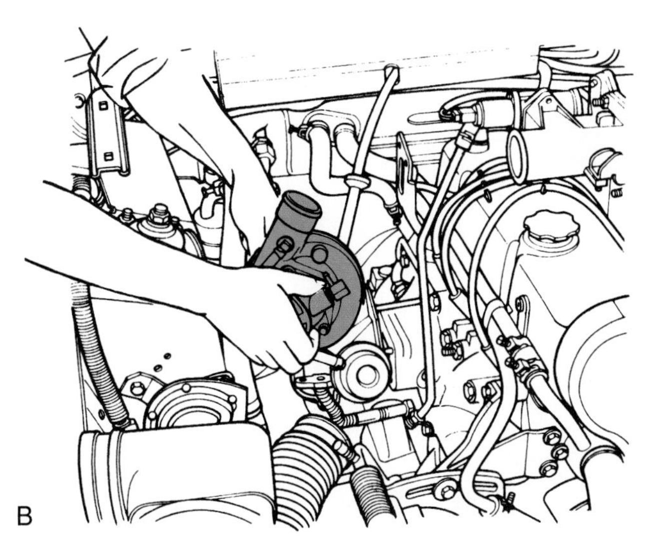

Figure 19-29. Turbocharger removal is a complex procedure on most vehicles. Make sure the engine is completely cool before beginning removal. A—Remove the oil return line and all connections. B—Carefully remove the turbocharger. The turbocharger is heavier than it looks; do not underestimate its weight. (Ford)

Supercharger Service

Supercharger service consists of making sure that the unit is clean, well oiled, and in proper condition. All hoses, clamps, and fittings must be in good condition. The air cleaner element should be checked for restrictions. Small vacuum leaks on the intake side of the supercharger can allow dust to enter the system. Excessive dust can prematurely destroy the supercharger. Some superchargers are pressure lubricated by the engine oil pump. Other superchargers have a separate reservoir that is checked by removing a pipe plug on the supercharger body. Both standard and synthetic oils can be used in an engine with a supercharger.

Checking Supercharger Operation

Begin checking a supercharger by inspecting it for obvious damage. The drive belt should be in good condition, have the proper tension, and be free of cracks and breaks. Check the supercharger pulley area and seams for evidence of leaks. Since many superchargers contain oil for lubrication, an oil leak can cause exhaust smoking.

Note: Some superchargers may seep some oil from the pulley seal. This is normal.

Check for intake restrictions such as a clogged air cleaner or kinked intake hose. Check that all hoses and manifold gaskets are in good condition and that all bolts and clamps are tight. If the supercharger has an intercooler, make sure that it is not clogged with debris or leaking air. Detonation at high engine speeds is one symptom of a clogged intercooler.

Start the engine, if possible, and listen to the supercharger. Some superchargers will have a low-pitched whine, which is normal. If the whine is excessively loud or there are other unusual noises, locate the source. Replace the supercharger if the noise is coming from it.

The best way to test the supercharger is to determine if it is developing boost pressure. Install a pressure gauge between the supercharger and the cylinder heads or intercooler. Start the engine and check the gauge. Normal boost pressure is lower than turbocharger pressure, ranging from 2 psi–15 psi (13.8 kPa–103.4 kPa). Since superchargers are directly driven by the engine, pressure may be present at idle. If the gauge does not show boost pressure as the engine speed is increased, the supercharger or the bypass valve is defective. If the pressures are excessive, especially at higher engine speeds, the bypass valve may be stuck closed or misadjusted. Excess pressure causes severe engine detonation at higher engine speed.

Note: A supercharged engine should develop vacuum in the intake system between the throttle valves and supercharger intake, just as on a naturally aspirated engine.

If the supercharger is defective, it should be replaced as soon as possible. Since superchargers are expensive, make sure the supercharger is truly the cause of the problem. Do not operate a defective supercharger for long periods. It could come apart, allowing metal to enter the engine cylinders, destroying the engine.

Removing and Replacing a Supercharger

Defective supercharger units usually need to be replaced. Some bypass valves can be replaced without removing the supercharger. To replace a supercharger, remove the throttle body assembly and the oil lines. Next, remove the attaching bolts holding the supercharger to the intake manifold and lift the supercharger from the engine. **Figure 19-30** shows the relative position of one particular supercharger in relation to the throttle body and engine. Cover the intake manifold openings to keep debris from entering the engine. Replacement is the opposite of removal. Supercharger remanufacturers use specialized equipment to rebuild superchargers. It is usually not possible to repair a supercharger in the shop.

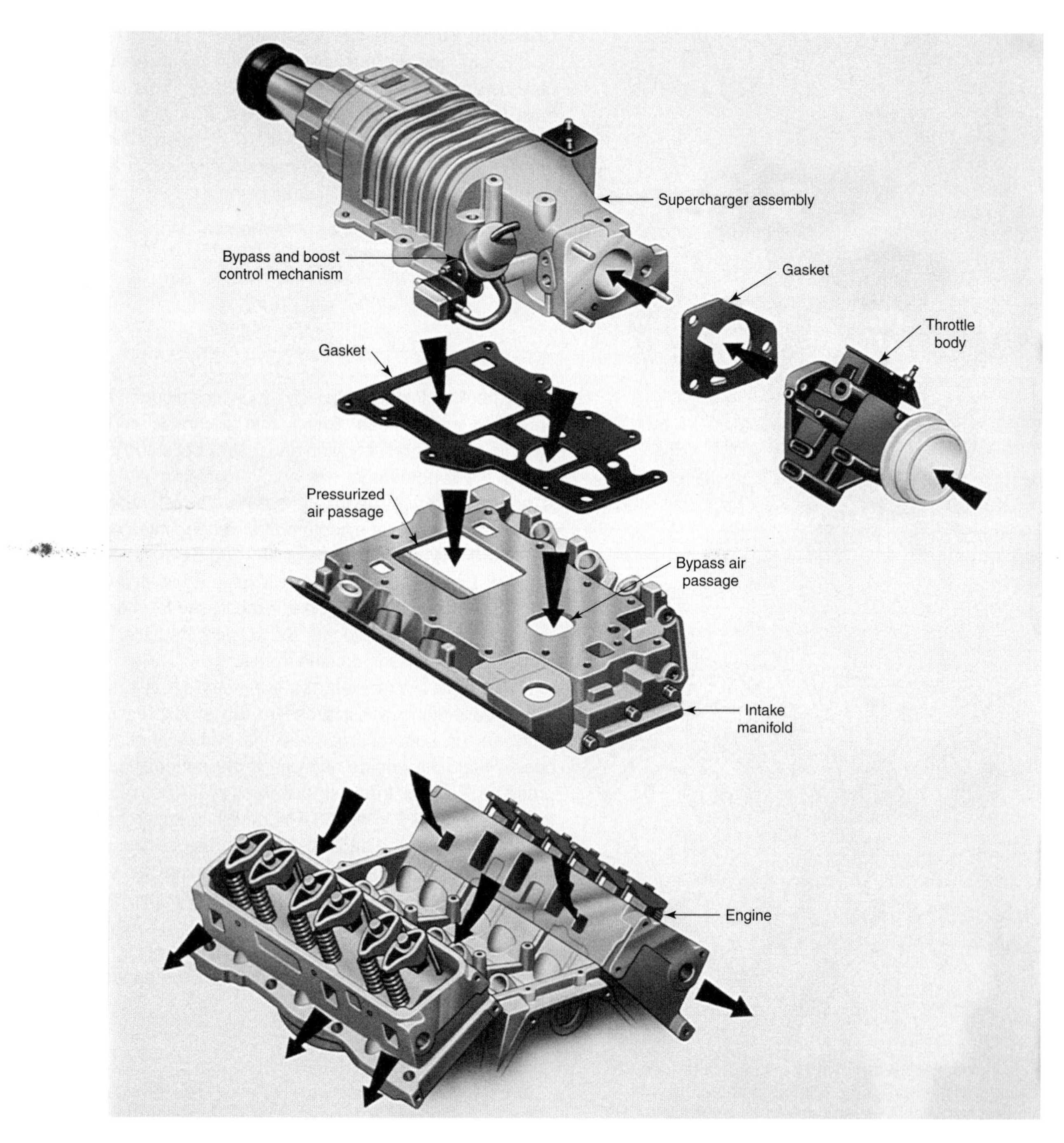

Figure 19-30. Superchargers are simply replaced as a unit. Due to the cleanliness needed, service in the field is not recommended. (General Motors)

Tech Talk

It always pays to use new clamps when installing an exhaust system part, even if you can get the old clamps off without breaking them. The old clamps are usually corroded and bent, and they may not seal properly. In addition, old clamps may break after the new exhaust system has heated and cooled a few times. New clamps are very inexpensive, especially when compared with the cost of follow-up service due to a broken clamp.

Summary

Exhaust gases enter the exhaust manifold from the cylinder head. Exhaust manifold replacement is relatively easy. Use penetrating oil to help loosen the exhaust manifold fasteners. Before installing an exhaust manifold, clean the mounting surfaces. Use a gasket where needed and torque the fasteners to specifications. Retorque the fasteners after engine operation if required.

The exhaust manifold heat control valve on some older engines diverts some of the exhaust gases to warm the incoming fuel-air mixture during engine warm-up. Heat control valves may be operated by a thermostatic spring or a vacuum diaphragm. Check either type by looking for movement of the external linkage.

The exhaust pipe carries the exhaust gases from the exhaust manifold to the converter and/or muffler. Exhaust systems can be either single or dual. Mufflers are designed to reduce noise without causing excessive back pressure. Some systems use a second smaller muffler called a resonator. The catalytic converter reduces exhaust emissions. The final pipe of the exhaust system is called the tailpipe.

Proper tools speed up exhaust system work. Some of the handy tools are the power chisel, hand pipe cutter, joint heater, pipe expander, straightening cone, chain wrench, and power wrench.

If the exhaust system clamps will be reused, soak the threads with penetrating oil before attempting to remove them. Slit the joint of the pipe to be removed. Cut off pipes that are welded. Check the old and new parts to make sure that the new parts are correct. Clean all connections and use sealant. Use new gaskets when reassembling flanged joints. Make sure that the pipes fit snugly, and then tighten clamps securely, but not enough to crush the pipes. Before lowering the vehicle, make sure that all exhaust system parts are correctly aligned and do not contact any other vehicle parts.

Forcing more air into the engine cylinders is a way to obtain extra power from the engine. Turbochargers use exhaust gases to pressurize the intake manifold. Superchargers are driven by the engine crankshaft. Turbochargers can be removed and overhauled, but it is often easier to replace the entire turbocharger. The waste gate diaphragm can often be replaced without replacing the entire turbocharger. Supercharger service is limited to replacement of the entire unit.

Review Questions—Chapter 18

Do not write in this book. Write your answers on a separate sheet of paper.

1. Exhaust systems are classified as _____ or dual exhausts.
2. Older engines sometimes used a _____ to help vaporize fuel.
3. List the five major parts of a typical exhaust system.
4. If a pipe end is crimped or kinked, it can be brought back to form by using a_____.
5. Some pipes can be formed into the needed shape by using what special tool?
6. If exhaust system parts are not to be saved, the quickest way of freeing a stuck joint is to _____ the joint with a _____ _____.
7. To provide additional leak protection, it is wise to coat each joint with _____.
8. When installing a clamp on a joint, allow about _____ between the clamp and the end of the pipe (or nipple) that it surrounds.
9. An exhaust system must be properly _____ in order to prevent noises and damage.
10. What is the last step in any exhaust system repair?
11. Turbochargers are driven by _____.
12. A turbocharger waste gate can be replaced without removing the _____.
13. Turbochargers have both pressure and return _____ lines.
14. Superchargers can be driven from the crankshaft by _____.
    (A) a belt
    (B) a chain
    (C) meshing gears
    (D) All of the above.
15. If a supercharger is defective, what should the technician do?

ASE-Type Questions

1. Each of the following are primary functions performed by the exhaust system *except:*
    (A) reduce noise.
    (B) provide freedom from excessive back pressure.
    (C) prevent poisonous gases from entering the passenger compartment.
    (D) decrease exhaust emissions.

2. Technician A says that composition exhaust manifold gaskets should be retorqued after the first engine warm-up. Technician B says that if exhaust manifold fasteners are properly tightened, locking devices are never needed. Who is right?
   (A) A only.
   (B) B only.
   (C) Both A & B.
   (D) Neither A nor B.
3. The manifold heat control valve _____.
   (A) warms the gasoline coming to the carburetor or fuel injectors
   (B) warms the fuel charge in the intake manifold
   (C) warms the air before it enters the carburetor or fuel injectors
   (D) warms both air and gasoline before it enters the carburetor or fuel injectors
4. Technician A says that a power chisel can be used to remove exhaust system parts that are not to be reused. Technician B says that heat can be used to free exhaust system parts. Who is right?
   (A) A only.
   (B) B only.
   (C) Both A & B.
   (D) Neither A nor B.
5. An exhaust part must be cleaned. Technician A says to use a coarse emery cloth. Technician B says to use a crocus cloth. Who is right?
   (A) A only.
   (B) B only.
   (C) Both A & B.
   (D) Neither A nor B.
6. Exhaust system sealant should _____.
   (A) be applied liberally to the sections to be joined
   (B) be used only when needed
   (C) never be used
   (D) be mixed with a catalyst before using
7. When inserting a tailpipe or exhaust pipe into or over a muffler nipple, the pipe should be shoved into or over the nipple _____.
   (A) one third of the way
   (B) halfway
   (C) the full nipple depth
   (D) Any of the above.
8. Technician A says that the diameters of the exhaust pipe, tailpipe, or muffler are not important as long as they fit. Technician B says that the exhaust system should be aligned before the final tightening of exhaust system clamps and brackets. Who is right?
   (A) A only.
   (B) B only.
   (C) Both A and B.
   (D) Neither A nor B.
9. All of the following statements about exhaust system leaks are true, *except:*
   (A) leaks under the vehicle cannot enter the passenger compartment.
   (B) after completing any work on the exhaust system, the system should be checked for leaks.
   (C) reusing old gaskets can cause leaks.
   (D) failure to tighten joint clamps can cause leaks.
10. An engine with a turbocharger and electronic ignition has blue smoke coming from the exhaust when the engine is running. Technician A says that the problem may be a lack of oil at the turbocharger bearings. Technician B says that the problem may be a leaking turbocharger oil seal. Who is right?
    (A) A only.
    (B) B only.
    (C) Both A and B.
    (D) Neither A nor B.

Suggested Activities

1. Check the exhaust system components on a specific vehicle for damage. Make a sketch of the system and label any defects that you found. Determine whether the defective parts can be fixed or if new parts are required.
2. Use a pressure gauge to check the boost on a turbocharger or supercharger. Discuss your findings with your instructor or class.

Exhaust System Diagnosis	
Problem: Exhaust odor enters vehicle during highway operation	
Possible cause	**Correction**
1. Leaking exhaust system connections. 2. Holes in muffler or pipe system. 3. Tail pipe does not protrude far enough to rear or side. 4. Oil drips on hot exhaust system. 5. Holes in body or fire wall. 6. Operating vehicle with back window down.	1. Tighten connections. Repair or, if needed, replace units. 2. Replace defective units. 3. Install correct length of pipe or an extension. 4. Repair oil leaks. 5. Locate and seal holes. 6. Inform owner.
Problem: Engine lacks power	
Possible cause	**Correction**
1. Clogged muffler. 2. Clogged or kinked exhaust or tail pipe. 3. Muffler or pipes too small for vehicle. 4. Catalytic converter clogged or crushed shut. 5. Manifold heat control stuck.	1. Replace muffler. 2. Replace pipe. 3. Install muffler and pipes of the correct size and type. 4. Replace with new converter. 5. Free or replace.
Problem: Excessive exhaust system noise	
Possible cause	**Correction**
1. Holes in muffler or pipes. 2. System connections leaking. 3. Exhaust manifold or pipe gaskets blown. 4. Muffler of incorrect design. 5. Muffler burned inside. 6. Carbon build-up in straight-through design muffler. 7. Hole in catalytic converter.	1. Replace defective units. 2. Repair connections. 3. Replace gaskets. 4. Replace with correct muffler. 5. Replace muffler. 6. Replace muffler. 7. Replace converter.
Problem: Exhaust system mechanical noise	
Possible cause	**Correction**
1. System improperly aligned. 2. Support brackets loose, bent, or broken. 3. Incorrect muffler or pipes. 4. Baffle loose in muffler. 5. Manifold heat control valve rattles. 6. Engine mounts worn. 7. Damaged or defective catalytic converter.	1. Align system. 2. Tighten. 3. Install correct muffler or pipes. 4. Replace muffler. 5. Replace thermostatic spring. 6. Replace engine mounts. 7. Repair or replace converter.

State-of-the-art diagnostic analyzers, such as the one shown here, are commonly used to diagnose driveability problems in late-model vehicles. (Ford)

20

Driveability Diagnosis

After studying this chapter, you will be able to:

- Define driveability diagnosis.
- Explain the differences between driveability diagnosis and tune-ups.
- Explain the principles of driveability diagnosis.
- List the seven steps in driveability diagnosis.
- Explain the modern use of the term tune-up.
- List the basic steps for a maintenance tune-up.
- Explain the need for road and dynamometer tests.

Technical Terms

Driveability diagnosis
Troubleshooting
Educated guess
Seven-step diagnostic process
Road test
Fuel trim
Misfire monitoring
Power balance test
Maintenance tune-up
Emissions tune-up
Specifications
Chassis dynamometer

Introduction

This chapter explains the process of diagnosing driveability problems. It also explains the difference between driveability diagnosis and tune-ups. Understanding this difference will help you explain to your customers what type of service is really needed. This chapter details the troubleshooting procedures that should be performed when diagnosing a driveability problem. Checks and measurements of specific *internal engine parts* will be covered in the next few chapters. Other chapters deal with the diagnosis and repair of specific engine-related systems, such as fuel, ignition, emissions, exhaust, and cooling systems.

Driveability Diagnosis versus Tune-up

Years ago, restoring performance and smoothness required a *tune-up*. Because many of the engine's systems were controlled mechanically, numerous critical parts would periodically wear out. Regular adjustment or replacement of these parts was required to keep an engine running smoothly.

Because of the increased durability of modern parts, as well as the shift from mechanical controls to electronic controls, tune-ups for late-model engines are simpler and far less critical to engine performance than they were in the past. Today, a tune-up is considered a preventive measure. Restoring performance and smoothness to a modern engine involves troubleshooting the affected system instead of replacing all the related parts.

The modern tune-up involves changing spark plugs, inspecting ignition components, and changing fuel and air filters. If required, the ignition timing and idle speed can be checked and adjusted. However, a tune-up should not be expected to solve a driveability problem. When confronted with the type of problem that used to be corrected with a tune-up, the technician must now determine which parts, if any, are defective and repair the vehicle accordingly.

Driveability Diagnosis

It is easy to change a part or make a simple adjustment to restore vehicle driveability. However, it is often difficult to determine which part requires adjustment or replacement. The process or identifying the faulty part is at the heart of ***driveability diagnosis,*** or ***troubleshooting.*** The ability to find the cause of a problem and its solution separates the professional automotive technician from the parts changer.

To become an expert at driveability diagnosis, the technician must know how internal combustion engines and their related systems operate. In the other chapters in this book, you learned how the various engine systems work together to create an efficient engine. You also learned how to test the engine, ignition, and fuel systems and how to repair or replace defective components. Now that you have this knowledge, you must develop the ability to approach a problem in a calm and logical manner, no matter how mysterious or frustrating it may seem. You must determine what can cause the problem, and just as important, what cannot cause it. You can then proceed from the simplest tests to the most complex. Do not guess at possible solutions, and do not panic when the problem is difficult to find. If you remember these points, you will be able to diagnose most driveability problems with a minimum of trouble. Remember, the engine systems are not the only source of driveability trouble. Problems in the manual clutch, the automatic transmission shift system, the lockup torque converter, the air conditioner, or the starting and charging systems can be mistaken for driveability problems. Occasionally a problem in the suspension, steering, or brakes will cause what seems to be a driveability problem.

Avoid Guessing

Avoid making snap judgments about the cause of a problem. Sometimes you will have to make an ***educated guess*** about the nature of a problem. An educated guess is really not a guess at all, but a reasonable conclusion based on experience, testing, and the process of elimination. Once a reasonable conclusion is reached, it can be verified by a visual inspection and testing. Jumping at the first possible cause that comes to mind is a dangerous way to diagnose problems. Unfortunately, it can quickly become a habit, regardless of how much unnecessary work and aggravation it causes.

Stay Calm

One of the hardest principles of diagnosis is to remain calm during the troubleshooting process, no matter how much you would like to scream and throw things. Controlling your emotions is often difficult, especially if you meet a series of dead ends while looking for a problem. However, staying calm is a necessary skill that must be developed and maintained. Nothing is accomplished by losing your composure. If you have a tendency to overreact to situations, teach yourself to remain calm. In the meantime, try to present an outward appearance of calm to avoid upsetting anyone else, especially the customer.

The Seven-Step Diagnostic Process

As previously mentioned, driveability diagnosis is a process of taking logical steps to reach a solution to a problem. It involves reasoning through a problem in a series of logical steps. The ***seven-step diagnostic process*** that follows will, in most cases, be the quickest way to isolate and correct a problem. Refer to **Figure 20-1** as you read the following sections.

Step 1—Verify the Complaint

Often the vehicle's driver will describe a problem in very general terms, such as "it doesn't run right" or "something is wrong with the engine." The first step in fixing any problem is to verify the problem or complaint. This means determining the cause of the complaint, what its symptoms are, and how it affects the operation of the vehicle. This process involves talking to the driver and road testing the vehicle.

Talking to the Driver

Question the vehicle driver as to the exact nature of the complaint. Interview the driver and try to translate his or her comments into commonly accepted automotive diagnostic terms. Try to determine what is going on by asking a series of basic questions:

- Under which of the following driving conditions does the problem occur?
  - Idling.
  - Accelerating from a stop.
  - Accelerating while moving.
  - Cruising.
  - Decelerating.
  - All the time.
  - Other times.
- Is the engine cold or warm when the problem occurs?
- Does the engine stop running?
- Do you hear any noises when the problem occurs?
- Does the problem occur only during wet weather?
- Has the vehicle been serviced recently?
- Did the problem start suddenly or develop gradually?

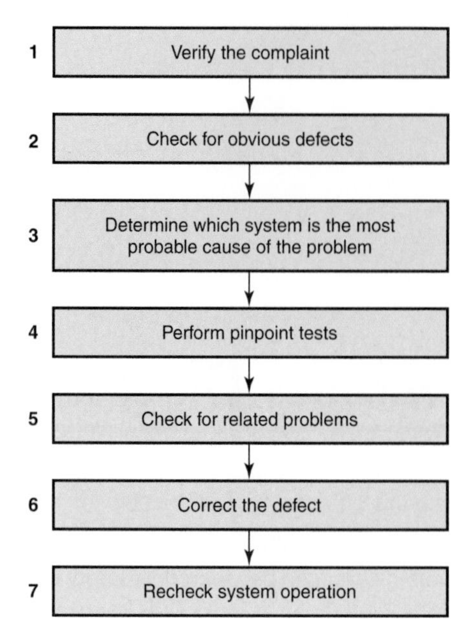

Figure 20-1 The seven-step diagnostic process is the quickest way to isolate the cause of a driveability problem.

You may think of other questions, depending on the answers you get to the above questions. Write down the driver's comments on an inspection form, **Figure 20-2**, or a sheet of paper. Before going on to the road test, make sure you fully understand the driver's complaint.

Performing a Road Test

It may be necessary to perform a ***road test*** to confirm the driver's complaint. In many cases, a short test drive will confirm the problem. In other cases, the road test will reveal that the problem is much different from the malfunction the driver described.

Caution: Do not attempt to road test the vehicle if it has any brake or steering defects.

Perform the road test on a level, lightly traveled road. Be sure to wear your seatbelt and observe all traffic laws when performing the road test. Try to duplicate the problem. This may take several attempts. Once you have determined the exact nature of the problem, proceed to Step 2.

Driveability Worksheet

Name: ______________________________ Date: __________

VIN: ____________________ Year: ____ Make: ______ Model: ________

Style: ______ Color: ____ Engine: _____ Trans: ____ A/C: _____ P/S: ______

Describe the vehicle's problem as accurately as you can: ____________________

____

____

____

Check the boxes that best describe when the vehicle's problem occurs.

☐ Hot ☐ Cold ☐ Constant ☐ Intermittent ☐ Recurring

Check the boxes that best describe the type of symptom(s) that occur.

☐ Stalling	☐ Lack of power	☐ Overheating
☐ Hard start	☐ Surge/Chuggle	☐ Noise
☐ Does not start	☐ Pinging	☐ Vibration/Harshness
☐ Rough idle	☐ Miss/Cuts out	☐ Fluid leak
☐ Incorrect idle	☐ Poor fuel economy	☐ Unusual odors
☐ Hesitation	☐ Indicator light(s) on	☐ Smoke

Other: ____

____

Background information (any service or repair work performed recently, other problems, etc.): ________

____

____

Figure 20-2. A driveability worksheet, such as the one shown here, gives the technician a starting point when diagnosing a problem.

Step 2—Check for Obvious Defects

Step 2 involves checking for obvious problems and performing very basic tests. The problem can often be identified during this step.

As part of Step 2, you should retrieve trouble codes. Do this before removing any fuses or electrical connections. In some cases, the trouble code will identify a defective component. In other cases, the code indicates that there is a problem in a vehicle subsystem, and the subsystem must be diagnosed further to pinpoint the problem. Chapter 15 contains more information on retrieving and interpreting trouble codes.

Once the trouble codes have been retrieved and recorded, proceed to make visual checks of the engine and other underhood components. Visual checks and simple tests take only a little time and might save time in the long run. Things to check for include:

- An illuminated MIL.
- Charging system voltage.
- Loose, overheated, or corroded electrical connections.
- Loose or corroded ground connections.
- Battery terminal corrosion.
- Disconnected vacuum hoses.
- Blown fuses or fusible links.
- Missing or slipping drive belt(s).
- Low oil, coolant, or transmission fluid level.
- Restricted air filter.
- Disconnected or damaged spark plug wires.
- Loose or torn air intake hoses.
- Loose manifold or plenum fasteners.
- Overheating or physical damage to sensors or output devices.
- Evidence of lack of maintenance.

Figure 20-3 illustrates some of the basic areas to check during this step. These simple checks will often uncover the problem.

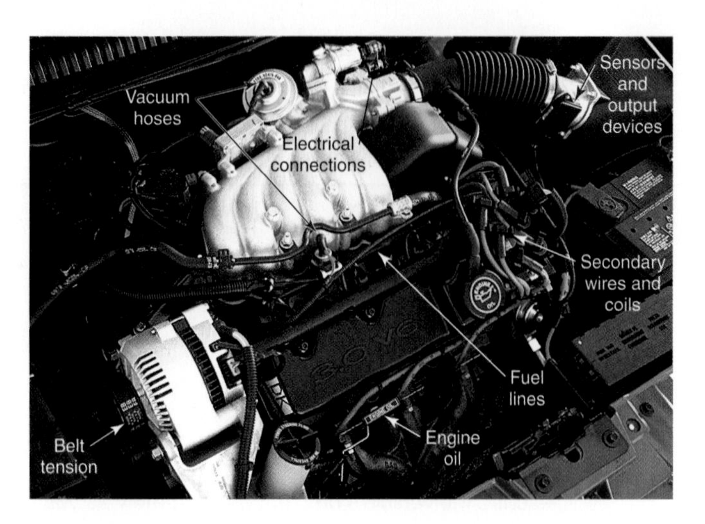

Figure 20-3. Many driveability and performance problems are found during the visual inspection. A quick visual inspection can often save hours performing needless tests.

Step 3—Determine Which System Is the Most Probable Cause of the Problem

Step 3 involves determining which vehicle component or system is the most probable cause of the problem. The engine is composed of compression, ignition, and fuel subsystems. To determine which of these is the most likely source of the problem, you must combine the information that you obtained in Step 1 with your knowledge of which engine parts and systems cause what symptoms. This will help you eliminate many areas that could not cause the problems, so in Step 4, you can concentrate on the areas that could cause the problem. This would be a good time to refer to any available diagnostic and troubleshooting charts. A typical troubleshooting chart is shown in **Figure 20-4.**

Using Scan Tool Information to Diagnose Driveability Problems

In addition to retrieving trouble codes, a scan tool can be used for other diagnostic procedures. On OBD II vehicles, a scan tool must be used to obtain any significant information. Two areas that commonly cause driveability complaints are incorrect fuel trim and misfiring. Using a scan tool to diagnose these problems is discussed in detail in the following sections.

Checking Fuel Trim

The ECM controls air-fuel ratio by adjusting the length of time the injector solenoids are open. This is called ***fuel trim.*** Short-term fuel trim adjusts for short-term effects on the fuel system, such as acceleration, deceleration, and changes in road grade. Long-term fuel trim adjusts for altitude, outside temperature, and other changes that last for a relatively long period. Fuel trim is measured in counts. On many vehicles, counts range from 0 to 256, with 128 considered ideal, **Figure 20-5.** On other vehicles the fuel trim is read as a percentage, with 0 degrees the ideal setting.

Observing the short- and long-term fuel trim allows the technician to determine whether the system is properly controlling fuel mixtures. When the short-term fuel trim drops below 128, the ECM is compensating for a rich condition. When the counts rise above 128, the ECM is compensating for a lean condition.

It is normal for fuel trim to oscillate around 128, going a few counts lower and then a few counts higher. However, if the fuel trim is consistently above or below 128, something is causing the ECM to compensate for an excessively rich or lean mixture. If the system measures fuel trim in percentages, positive percentages indicate that the ECM is compensating for a rich condition, while negative percentages indicate that the ECM is compensating for a lean condition. If long-term fuel trim is off, suspect that the ECM is compensating for a constant condition. This condition may not be caused by an engine or vehicle problem. Instead, it may be the result of operating the vehicle at unusual altitudes or temperatures. Sometimes, the ECM may be compensating for a slight problem, such as a small vacuum leak that has been present for a long time. If the ECM is properly controlling the mixture,

Step	Action	Value(s)	Yes	No
1	**Important** • Check for service bulletins. • DO NOT turn OFF the ignition during this diagnostic unless instructed. • DO NOT clear the DTCs unless instructed to do so. Check for one or more of the following conditions: • The malfunction indicator lamp illuminates with the engine running. • A customer concern of engine performance or driveability. • A suspected fault in a PCM controlled component or system. Did any of the above conditions apply?		Go to Step 2	System OK–Go to Test Descriptions
2	1 Turn ON the ignition, leaving the engine OFF. 2 Observe the malfunction indicator lamp (MIL). Is the MIL illuminated?		Go to Step 3	Go to No Malfunction Indicator Lamp
3	1 Turn OFF the ignition. 2 Install a scan tool. 3 Turn ON the ignition, leaving the engine OFF. Does the scan tool display any PCM data?		Go to Step 4	Go to Data link Connector Diagnosis
4	Start the engine. Did the engine start and run?		Go to Step 5	Go to Engine Cranks but Does Not Run
5	1 Observe the DTC information. 2 Save any DTC and Freeze Frame/Failure Records information. Were any DTCs set?		Go to applicable DTC table	Go to Step 6
6	Is the MIL ON with no DTCs set?		Go to Malfunction Indicator Lamp On	Go to Step 7
7	Compare the scan tool data display with the data values shown in the Engine Scan Tool Data List. Are the displayed values normal or within the typical values range?		Go to Symptoms	Go to the Diagnostic Aids of the DTCs related to that component

Figure 20-4. This troubleshooting chart makes diagnosis easier by outlining a logical series of troubleshooting steps.

the short-term fuel trim will be within the accepted range even when the long-term fuel trim is incorrect.

If the scan tool has capability of producing waveforms, the fuel trim can be compared against oxygen sensor readings to confirm that the ECM is adjusting the fuel trim to compensate for changes in the air-fuel ratio. A comparison of an O_2 sensor reading and short-term fuel trim is shown in **Figure 20-6.** Notice that the fuel trim is shown as a percentage. Observing fuel trim often goes a long way toward determining the source of a driveability problem.

Misfire Monitoring

All OBD II systems have a diagnostic capability called ***misfire monitoring.*** The OBD II misfire monitor checks the engine for misfire at all times and illuminates the MIL if the number of misfires exceeds a preset amount. A scan tool must then be used to isolate the misfiring cylinder or cylinders. When the scan tool misfire test screen is accessed, the tool will display a code that indicates which cylinder or cylinders (if any) are misfiring. The scan tool also can display the number of misfires that have occurred on a particular cylinder in the

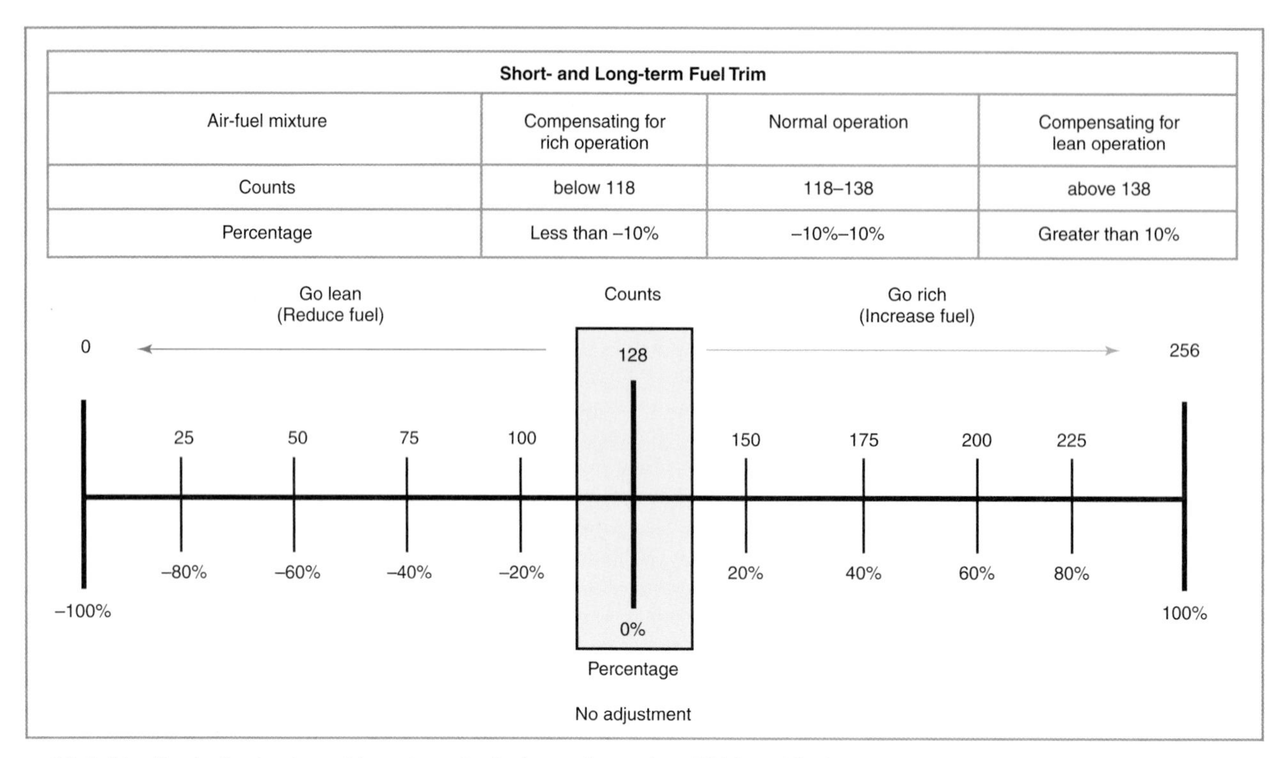

Short- and Long-term Fuel Trim			
Air-fuel mixture	Compensating for rich operation	Normal operation	Compensating for lean operation
Counts	below 118	118–138	above 138
Percentage	Less than –10%	–10%–10%	Greater than 10%

Figure 20-5. Monitor both short- and long-term fuel trim to determine ECM and fuel system condition, as well as the possibility of other engine problems.

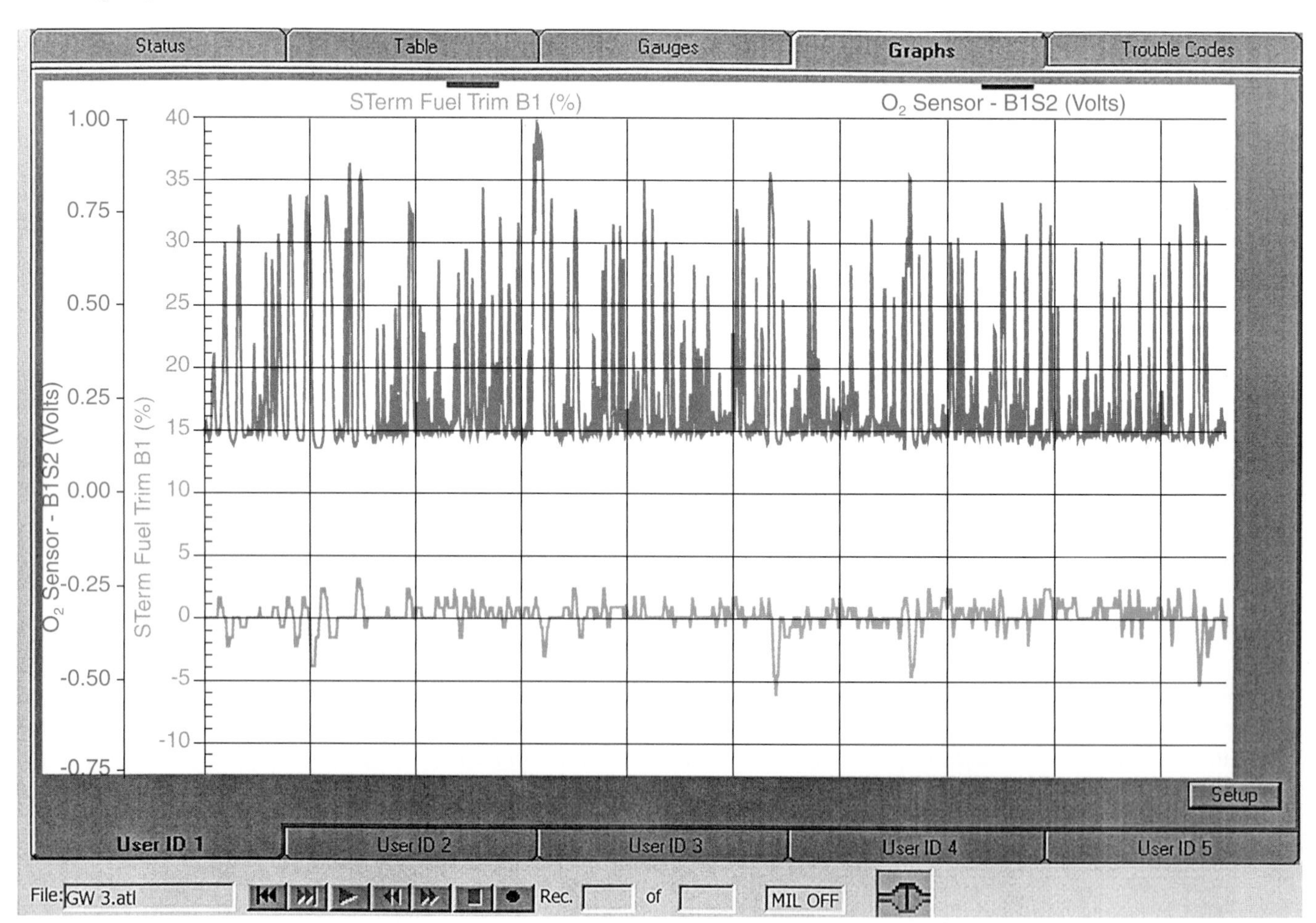

Figure 20-6. O_2 sensor readings and fuel trim can be compared to determine whether the computer control system is operating properly. Note how the fuel trim reacts quickly to changes in O_2 sensor readings. (B&B Electronics)

recent past. See **Figure 20-7.** After the misfire has been isolated to a particular cylinder or cylinders, it will be much easier to determine the cause of the misfire.

Performing a Power Balance Test

To isolate a weak or dead cylinder, modern oscilloscopes and other testers have a feature called a ***power balance test.*** To perform a power balance test, attach the tester as recommended by the manufacturer and operate the engine at a fast idle. Disable one cylinder at a time and record the rpm drops for each cylinder. If disabling one cylinder does not cause as much of a drop as disabling the others, that cylinder is not producing as much power as the other cylinders. The power balance test will isolate the problem to a particular cylinder, reducing diagnosis time.

Step 4—Perform Pinpoint Tests

In the last step, you found the system or general area that is causing the driveability problem. In this step, you begin to check the components of that system, eliminating the causes of the driveability problem, one by one. Always begin this step by checking the components that are the most likely sources of the problem, as well as those parts that are the easiest to reach and test. Then proceed to check the less likely sources or those that are more difficult to test. Do not skip checks because the parts involved "never go bad." Any part can go bad.

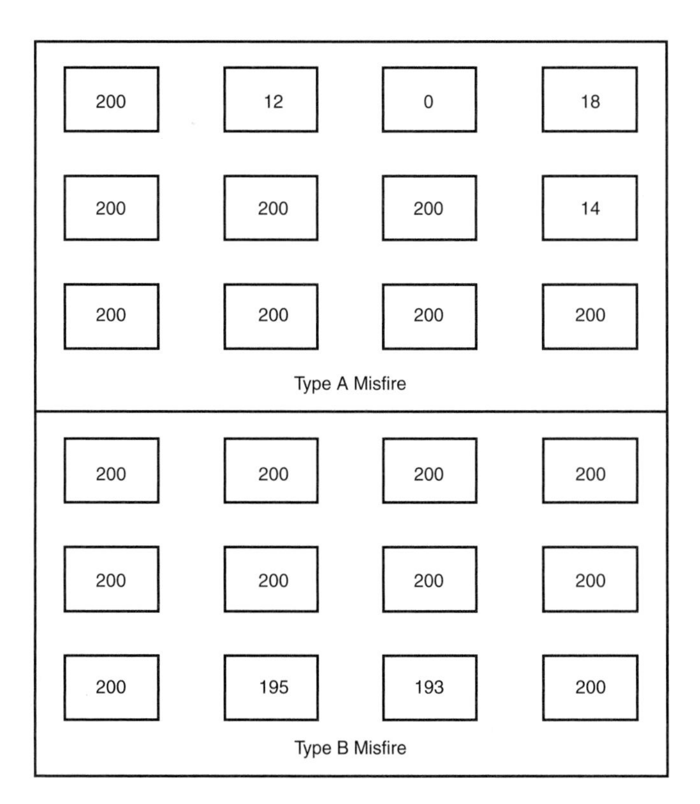

Figure 20-7. The OBD II misfire monitor separates crankshaft rotation into even samples of 200 rpms. If a sample drops significantly (A-type misfire), it will cause the ECM to set a trouble code and flash the MIL. If the misfire is relatively minor (B-type misfire), the MIL will go out when the condition goes away, but the code will remain stored in the ECM.

Typical tests that you might make in Step 4 include:

- Fuel pressure.
- Fuel filter condition.
- Coil output voltage.
- Sensor resistance and voltage.
- Output device operation.
- Available voltage at various electrical devices.
- Available vacuum-to-vacuum devices.
- Operation of vacuum devices.
- Plug condition.
- Cap/rotor condition (when applicable).

Testing the suspected components of modern vehicles may require a scan tool, a multimeter, waveform meter, or other special testers. Step 4 takes the most time and may involve more than one type of testing procedure. Testing in Step 4 may require the removal of parts for further inspection. For example, if Step 3 uncovered a weak cylinder, it may be necessary to remove the fuel injector or spark plug for further checking. It also may be necessary to make a compression or cylinder leakage test of that cylinder. Compression and cylinder leakage testing is discussed in Chapter 22. Other testing procedures were covered in previous chapters.

Step 5—Check for Related Problems

In Step 5, the most likely cause of the problem is isolated and rechecked. This step requires reviewing the various test procedures that were performed in the last step and determining whether the suspect component is likely to be the source of the problem. If is often helpful to take a short break to consider all possible causes of the problem and determine whether the problem you found in Step 4 is the only thing that could be defective. Review how the particular system works, and how the defect could cause the system problem.

Before moving on to the next step, closely recheck the condition of the suspect part, as well as other related parts. This will ensure that you have not condemned the wrong part or overlooked another defect. For instance, if the engine is missing and you have located a clogged fuel injector, do not assume that the other injectors are good. Check all the injectors and related fuel system components. It could be that a torn fuel filter is allowing dirt to enter the injection system.

Step 6—Correct the Defect

In Step 6, you correct the defect by repairing the vehicle as necessary. This repair may be as simple as reconnecting a vacuum hose or it may require that a cylinder head be removed to replace a burned valve. For this step, refer to the repair procedures in other chapters as necessary.

Before beginning repairs, be sure that the customer is informed of what defects you have found, the needed parts and service, and the cost. Do not assume that the customer will want the work done.

Be sure to fix the problem properly. Do not, for instance, clean and regap worn spark plugs if they are actually ready for replacement. When performing repair procedures, you may uncover other problems. Be sure to inform the customer about additional needed work and get permission before proceeding.

Step 7—Recheck System Operation

The final step in the diagnostic process involves rechecking system operation to determine whether the problem has been corrected. The simplest way to do this is to conduct another road test, driving the vehicle at various speeds. Try to duplicate the conditions under which the original problem occurred.

Another method of determining whether the problem has been corrected is to use a scan tool. The tool will confirm proper operation and can also be used to clear trouble codes after repairs are complete. Once the codes are cleared, operate the engine for a few minutes and recheck the diagnostic system to make sure that codes did not reset.

Do not skip this step, since it allows you to determine whether the previous steps corrected the problem. If necessary, repeat Steps 1 through 6 until you are sure that the problem has been fixed.

The Importance of Follow-Up

Once the seven-step checking process has isolated and cured the immediate problem, your first impulse is park the vehicle and get on to the next job. However, it is worth your time to think for a minute to decide whether the defect that you found is really the ultimate cause of the problem, or whether there is a hidden defect that will cause the problem to reoccur. This process is known as *follow-up*.

For example, a customer brings a vehicle into the shop complaining of a miss. You immediately locate a disconnected plug wire. Once the wire is reinstalled, the miss disappears. However, do not assume this is the end of the story. Something caused the wire to fall off the plug. Make sure that the boot is able to hold the wire on the plug and that the wire is not contacting some other underhood part that pulls it loose. In rare cases, a combustion leak through the plug shell can blow the boot from the plug. If you do not locate an underlying problem, the vehicle will be back soon with a dissatisfied customer.

Hidden defects are common and may cause a vehicle to be returned to the shop with the same symptoms. Do not return the vehicle to its owner until you are reasonably sure that the observed defect is the real source of the problem. Some hidden problems can be tricky, such as a vacuum hose that looks good when inspected but collapses under engine vacuum or an electrical connector that allows full current flow when cold but develops resistance as it heats up. This is where good observation skills and customer feedback can be helpful. Whenever you diagnose a driveability problem, be sure to determine the ultimate cause of a failure, even when the problem appears to be simple.

Maintenance Tune-up

This section covers ***maintenance tune-up*** procedures. The maintenance tune-up, sometimes called an ***emissions tune-up,*** consists of changing the spark plugs and filters, as well as the distributor cap, rotor, and secondary wires (if used). Occasionally a fuel system part is also replaced. When replacement is required, use top quality parts that will fit properly, function correctly, and provide adequate service life.

These tune-ups are often done in response to an ***emission flag*** or ***light,*** which appear on the instrument clusters of some cars and light trucks at certain time or mileage intervals.

Specifications

Before beginning a maintenance tune-up, you must obtain ***specifications,*** such as spark plug gap, engine idle speed, and ignition timing, for the exact engine and year model being serviced. Even though an engine may have remained basically unchanged for years, there may be minor changes that can affect specifications. **Figure 20-8** shows some common tune-up specifications.

Checklist for Maintenance Tune-ups

The following checklist shows the tests, adjustments, and other operations commonly performed during a maintenance tune-up. The steps are listed in the general order in which they should be performed. Always use a checklist to avoid duplication and to make sure nothing of importance is overlooked.

- Check the vehicle's prior service record and question the owner regarding vehicle performance.
- Clean and tighten the battery terminals. Tighten the battery hold-down.
- Operate the starter to be certain that the engine cranks over rapidly enough to perform an accurate compression test. If not, check the starting and charging systems and recommend repairs.
- Check oil, coolant, and all other fluid levels.
- Remove the spark plugs (gasoline) or the nozzles and glow plugs (diesel).
- Check engine compression. If compression readings indicate an internal engine problem, conduct further tests and discuss results with the customer before proceeding.
- Gap and install new spark plugs.
- Check drive belts, PCV valve, valve clearance, and heat control valve as needed.
- Inspect all emission control devices.

Note: Some emission control systems are closely tied in with both the fuel and ignition systems. They should be checked and adjusted in the order specified by the manufacturer.

- Replace the air, fuel, PCV, and vapor canister filters as needed.
- Check the distributor wiring (primary and secondary) for condition, connections, and routing.
- If the vehicle has a distributor, inspect the cap and rotor. If either the cap or rotor is defective, replace both. Lubricate the wick under the rotor.
- Reconnect all fittings and reinstall the air cleaner.
- Start the engine.

Engine Tune-up Specifications

Engine Identification					Fuel System		Spark Plug		Engine Timing			Idle Speeds					Fuel Pump[9]	
										Timing Light		Curb (slow) RPM		Fast (cold) RPM				
Displacement Litres/C.I.D.	Cyl.	VIN Code	Produced In^2	Division Usage	Type Carb./Fuel Inj.	Bbls.	AC Type Number	Gap Inches	Magnetic Probe[1]	MT	AT	MT	AT	MT	AT	Cam Step	Pressure P.S.I.	Volume Pint/Sec.
General Motors																		
2.0/121 OHC	4	K	B	2	EFT(TBI-700)	1	**R44XLS**	.045	9.5°	8°	8°	13	13	13	13	–	9-13	0.5/15
2.0/121 OHC Turbo	4	M	B	2	Multi-Port EFI	–	**R42XLS**	.035	9.5°	8	5	13	13	13	13	–	40.5-47[6]	0.5/15
2.0/121 OHV H.O.	4	1	U	1, 4	EFI (TBI-700)	1	**R44LTSM**	.035	9.5°	17	17	13	13	13	13	–	9-13	0.5/15
2.3/138 Quad-4 H.O.	4	A	U	2, 3	Multi-Port EFI	–	**FR3LS**	.035	9.5°	17	17	13	13	13	13	–	40.5-47[6]	0.5/15
2.3/138 Quad-4	4	D	U	2, 3, 4	Multi-Port EFI	–	**FR3LS**	.035	9.5°	17	17	13	13	13	13	–	40.5-47[6]	0.5/15
2.5/151	4	R	U	1, 2, 3, 4	EFT(TBI-700)	1	**R43TS6**	.060	9.5°	17	17	13	13	13	13	–	26-32	0.5/15
2.5/151	4	U	U	2, 3, 4	EFT(TBI-700)	1	**R43TS6**	.060	9.5°	17	17	13	13	13	13	–	9-13	0.5/15
2.8/173	V6	S	M	1, 2	Multi-Port EFI	–	**R43TSK**	.045	9.5°	17	17	13	13	13	13	–	40.5-47[6]	0.5/15
2.8/173	V6	W	U, C, M	1, 2, 3, 4	Multi-Port EFI	–	**R43LTSE**	.045	9.5°	17	17	13	13	13	13	–	40.5-47[6]	0.5/15
3.1/192	V6	T	C, M	2, 3, 4	Multi-Port EFI	–	**R43LTSE**	.045	9.5°	17	17	–	13	–	13	–	40.5-47[6]	0.5/15
3.1/192 Turbo	V6	V	U	2	Multi-Port EFI	–	**R42LTS**	.045	9.5°	17	17	–	13	–	13	–	40.5-47[6]	0.5/15
3.3/204 "3300"	V6	N	U	3, 4	Sequential Port EFI	–	**R44TS6**	.060	9.5°	–	17	–	13	–	13	–	40.5-47[6]	0.5/15
3.8/238 "3800"	V6	C	U	2, 3, 4	Sequential Port EFI	–	**R44TS6**	.060	9.5°	–	17	–	13	–	13	–	40.5-47[6]	0.5/15
3.8/238	V6	7	U	2	Sequential Port EFI	–	**R42LTS**	.035	9.5°	–	17	–	13	–	13	–	40.5-47[6]	0.5/15
4.3/262	V6	Z	U	1	EFI (TBI-220)	2	**R45TS**	.035	9.5°	–	TDC	–	13	–	13	–	9-13	0.5/15

Figure 20-8. This is a chart of common tune-up specifications used by one vehicle manufacturer. These specifications cover four- and six-cylinder engines. (AC Delco)

- Test the operation of the following when applicable:
  - Thermostatic air cleaner.
  - Ignition timing.
  - Distributor vacuum and centrifugal advance.
  - Automatic choke operation.
  - Carburetor idle speed and mixture.

Note: Follow the procedure on the underhood emission control label to set the timing.

- Check the charging voltage and isolate any charging system problems.
- Analyze the combustion gases.

Note: Make certain vehicle's emission levels comply with federal and state requirements.

- Check the radiator, coolant hoses, freeze plugs, and transmission cooling lines for leakage.
- Road test the vehicle.
- Perform a final inspection for leaks.
- Remove grease from the vehicle's interior and body.
- Fill out the vehicle's service record.
- Deliver the vehicle to its owner.
- Answer any questions the owner may have.

Solicit Owner Comments

Before starting a tune-up, ask the owner for any comments regarding engine starting and performance. A set of prepared questions can be used to ensure that all of the important areas are discussed. By doing this, you will be able to provide a proper diagnosis of the vehicle's problem. If the answers to your questions indicate that a maintenance tune-up will not solve the problem, explain this to the customer before proceeding.

Service Records

On all service jobs, document the customer's complaint, the cause, and the corrective actions that you took. Keep an accurate record of all repairs, the parts that were repaired or replaced, and the date and mileage at the time of installation. These records can be helpful in diagnosing trouble at some future date. The life span of many parts may be gauged in time or in miles. Accurate documentation is essential in handling guarantee work and for service billing. When the customer returns for service, check the records to determine the length of time or mileage since the last service. **Figure 20-9** shows a typical service repair order.

13879

Since... ...1912

HARPER MOTORS

4800 NORTH HIGHWAY 101 • EUREKA, CALIFORNIA 95501

"CROSS OVER THE BRIDGE"

BAR # AA001724
EPA # CAD 983595489
P & A CODE: 07755-0

Hazardous Waste Disposal Fee:
Used Motor Oil and/or Used Anti-freeze are regulated **Hazardous Waste** & are being legally transported, stored and recycled.

SERVICE HRS. MON-FRI
7:30 A.M. - 5:30 P.M.

(CHECK (✓) APPROPRIATE BOX)

TERMS	ORIGINAL ESTIMATE (PARTS & LABOR)	AUTHORIZED ADD'L REPAIRS	ADD'L REPAIRS OK'D BY	PHONE
☐ CASH ☐ CREDIT CARD	$	$	DATE	TIME
ALL PARTS ARE NEW UNLESS SPECIFIED OTHERWISE	REVISED ESTIMATE (PARTS & LABOR)	AUTHORIZED ADD'L REPAIRS	ADD'L REPAIRS OK'D BY	PHONE
	$	$	DATE	TIME
ALL PARTS WILL BE DISCARDED UNLESS INSTRUCTED OTHERWISE ☐ SAVE	I ACKNOWLEDGE NOTICE AND ORAL APPROVAL OF ANY INCREASE IN THE ORIGINAL ESTIMATED PRICE AND RECEIPT OF A COPY HEREOF. PLEASE READ REVERSE SIDE. CUSTOMER SIGNATURE X Rebecca Stockel			

☐ CLAIMS REVIEW ☐ AUTHORIZATION TO SUBMIT CLAIM ☐ PARTS SCRAP OUT

$ ______ PARTS $ ______ LABOR $ ______ TOTAL

______ Authorized Signature And Date

ON BEHALF OF SERVICING DEALER, I HEREBY CERTIFY THAT THE INFORMATION CONTAINED HEREON IS ACCURATE UNLESS OTHERWISE SHOWN. SERVICES DESCRIBED WERE PERFORMED AT NO CHARGE TO OWNER. THERE WAS NO INDICATION FROM THE APPEARANCE OF THE VEHICLE OR OTHERWISE THAT ANY PART REPAIRED OR REPLACED UNDER THIS CLAIM HAD BEEN CONNECTED IN ANY WAY WITH ANY ACCIDENT, NEGLIGENCE OR MISUSE RECORDS SUPPORTING THIS CLAIM ARE AVAILABLE FOR (1) YEAR FROM THE DATE OF PAYMENT NOTIFICATION AT THE SERVICING DEALER FOR INSPECTION BY REPRESENTATIVES OF FORD.

(SIGNED) DEALER, GENERAL MANAGER, OR AUTHORIZED PERSON (DATE)

```
RO: 024594-9              ADV: BOB      7325   TAG:                          RO: 024594
CUST:     18992                                VIN:  1FMCU24X5NUB17586
                                                 F 91   91 EXPLORER
                                                 RED
                             CA 95501          STK:                    SLD:
HOME:                                          LICENSE:
                                               DATE/TIME IN:                     14:03
PAY METHOD: CASH                               MILES IN:      26670
PRINTED ON
PROGRAM CODE: 93B15   COMMITMENT CODE:         APPROVAL CODE: 0
------------------------------***CUSTOMER INVOICE***------------------------------
REPAIR: 1   TYPE: W   CONCERN CODE:
CUSTOMER CONCERN :  PROGRAM #93B15             BATTERY CONCERN
         PART                DESCRIPTION   / CC    LIST     PRICE   QTY    TOTAL
,65650                                       79               .00    0
F3TZ,10A687,A          COVER ASY-BATTERY                    15.61    1
                                   PARTS AMOUNT - REPAIR 1:              WARRANTY
 OPERATION            DESCRIPTION          TECH      SSN     TIME           TOTAL
B15A            PROGRAM 93B19               23       4835     1.0
                                   LABOR AMOUNT - REPAIR 1:              WARRANTY
  TECHNICIAN DIAGNOSIS/REPAIR COMMENTS
      PROGRAM COMPLETED
      INSTALLED BATTERY COVER
                                   TOTAL AMOUNT - REPAIR 1:              WARRANTY
----------------------------------------------------------------------------------
GAS: QTY      SALE          OIL: QTY      SALE          GREASE: QTY      SALE
----------------------------------------------------------------------------------
                        WARRANTY
                      LABOR SALES           52.15
                      PARTS W&P CLAIM       15.61
                      W&P RECEIVE           67.76
```

GPD-103821 **LIMITED WARRANTY-ALL WORK GUARANTEED 90 DAYS OR 4,000 MILES WHICHEVER COMES FIRST.**

(END OF REPAIR ORDER) PAGE 1

Figure 20-9. Note this vehicle service repair order. Always document any repairs made on a vehicle, no matter how minor they may seem. This repair order is for a manufacturer's warranty recall. (Harper Motors)

Road Testing

No tune-up is complete until the vehicle has been road tested. Use the road test to check for smooth, responsive, ping-free acceleration. The engine's operation must be smooth (no missing, bucking, or loping) at low, medium, and high speed. The engine should not stall following a fast stop. Check the dash instruments for proper operation. After returning to the shop, inspect the vehicle for water, fuel, and oil leaks.

In addition to checking all engine systems, the road test offers an opportunity to check the brakes, transmission, steering, clutch, and rear axle. Although repairs to these areas are not part of a tune-up, you should call the owner's attention to any such work needed. Many shops also test the horn, headlights, brake lights, backup lights, and signal lights before delivery. This takes little time and helps build customer goodwill. For road testing, select an area where the vehicle can be driven at various speeds, accelerated rapidly, and stopped abruptly. Be careful and obey all traffic laws. Treat the customer's vehicle as though it were your own—remember that you are responsible for it until it is delivered.

Chassis Dynamometer Road Test

The use of a ***chassis dynamometer*** instead of a regular road test has some advantages. It saves time, reduces the possibility of accidents, and allows you to check torque and speedometer accuracy. These checks may not be possible during a road test. A chassis dynamometer also permits the use of diagnostic instruments while the engine is driving the vehicle.

In making a dynamometer test, the vehicle is parked on the dynamometer with the drive wheels resting on rollers. See **Figure 20-10.** The controls are connected to the engine. The engine is started, the transmission is placed in gear, and the drive wheels begin to spin the rollers. The rollers place a load on the engine as though the vehicle were actually on the road.

Figure 20-10. This vehicle is being tested on a dynamometer. Note the drive wheels are on the rollers. An engine-cooling fan (not shown) is sometimes placed in front of the vehicle to help aid in engine cooling. (DaimlerChrysler)

The chassis dynamometer provides a handy and accurate method of quickly and thoroughly road testing a vehicle right in the shop.

STOP **Warning: When using a dynamometer, make sure the area is well ventilated. This will prevent the buildup of carbon monoxide. Also, be sure the vehicle sits securely on the dynamometer and that safety stops and other restraint devices are used. Failure to do so could allow the vehicle to accidentally leave the dynamometer, which could result in injuries and property damage.**

Tech Talk

The top technician arrives at work on time and is absent only for good reasons. If absence is necessary, the technician immediately notifies the employer so that customer commitments and workloads can be adjusted.

Good technicians invariably are hard workers. During periods when they have no work assignments, they clean tools, maintain equipment, sweep the workstation, or help fellow technicians. They take pride in their work and pride in the business. They know that by helping the business to prosper, they too will prosper.

Top technicians devote time and energy toward the betterment of the trade. They gladly share their knowledge with apprentices. They conduct themselves in a way that brings credit to the automotive service profession.

Summary

In the past, restoring performance and smoothness required periodic tune-ups to compensate for wear. Today's tune-ups involve fewer components and are strictly preventive in nature. Restoring performance and smoothness to a modern engine requires troubleshooting the systems affecting driveability.

Isolating and correcting a driveability problem involves diagnostic procedures, as well as repair and service operations. Complete diagnosis includes following every step of the seven-step troubleshooting process. Always remember to check the simple things first, avoid guessing, and remain calm. After checking the operation of vehicle systems to determine problem areas, determine which parts could be the cause of the problem, check the condition and operation of the suspected parts, repair or replace suspected parts, and recheck vehicle operation. Remember to follow up your diagnosis and repair by making sure you have found the ultimate cause of the problem.

The parts usually replaced during a maintenance tune-up are the spark plugs and various filters. Other parts that may be replaced are plug wires, distributor cap and rotor, and occasionally a fuel injector or carburetor vacuum diaphragm. Follow a tune-up checklist to avoid duplication and possible omission of important steps. Keep records showing the parts installed, date, mileage, etc. Use prior service records to help you before starting the tune-up. Be sure to document all repairs and any problems that remain.

The engine's mechanical condition should be checked before performing the tune-up. If the engine checks out OK, perform the preliminary checks on the drive belts, PCV and EGR valves, valve clearance, electrical connections, fasteners, and fittings before starting the tune-up.

Always road test a vehicle following a tune-up. If available, a chassis dynamometer may be used instead of actual road testing. Check out all emission control devices for proper operation and the exhaust emission levels for compliance with emission regulations.

Review Questions—Chapter 20

Do not write in this book. Write your answers on a separate sheet of paper.

1. The hardest part of driveability service is _____.
   (A) diagnosing the problem
   (B) repairing a part
   (C) adjusting a part
   (D) replacing a part
2. What is an educated guess?
3. List the seven steps in driveability diagnosis in order.
4. If the fuel trim is at 158 counts, the ECM is compensating for a _____ condition.
5. Explain how misfire monitoring can help in driveability diagnosis.
6. The power balance test can be used to help isolate a weak _____.
   (A) fuel injector
   (B) spark plug
   (C) cylinder
   (D) All of the above.
7. Owner comments or answers to questions are often helpful for _____.
   (A) locating problems
   (B) determining the nature of the service needed
   (C) deciding how much to charge
   (D) Both A and B.
8. Service records are useful when _____.
   (A) diagnosing trouble at a future date
   (B) handling guarantee work
   (C) service billing
   (D) All of the above.
9. After a tune-up, the vehicle should always be _____ _____.
10. What are the advantages of using a chassis dynamometer instead of performing a road test?

ASE-Type Questions

1. Technician A says that an owner's comments usually mislead the technician and therefore should be ignored. Technician B says that it may be necessary to conduct a road test to confirm the owner's complaint. Who is right?
   (A) A only.
   (B) B only.
   (C) Both A & B.
   (D) Neither A nor B.
2. Which of the following would normally be checked in Step two of the seven-step process?
   (A) Fuel pressure
   (B) Coil output voltage
   (C) Battery terminal condition
   (D) Plug condition
3. Technician A says that the OBD II misfire monitor can be used to determine whether the computer control system is properly advancing the ignition timing. Technician B says that checking fuel trim will isolate a bad spark plug wire. Who is right?
   (A) A only.
   (B) B only.
   (C) Both A and B.
   (D) Neither A nor B.
4. At idle the short-term fuel trim reading on an engine is 128. The long-term fuel trim reading is 134. Which of the following statements about the above readings is *most* likely to be true?
   (A) The ECM is trying to compensate for an ongoing rich mixture.
   (B) The ECM is trying to compensate for an ongoing lean mixture.
   (C) ECM is attempting to compensate for a short-term lean mixture.
   (D) ECM is attempting to compensate for a short-term rich mixture.
5. Technician A says a nonelectronic part is never the source of a driveability problem. Technician B says when a part is visually observed to be defective, it should be replaced without further checking. Who is right?
   (A) A only.
   (B) B only.
   (C) Both A and B.
   (D) Neither A nor B.

6. All of the following statements about driveability diagnosis are true *except:*
   (A) the technician must know how internal combustion engines and their related systems operate.
   (B) engine systems are the only source of driveability trouble.
   (C) do not assume that the customer will want the work done.
   (D) any part can go bad.
7. Technician A says a maintenance tune-up, if properly done, will guarantee satisfactory engine performance. Technician B says specifications for a certain model's engine are good for other years, as long as the engine remains basically the same. Who is right?
   (A) A only.
   (B) B only.
   (C) Both A and B.
   (D) Neither A nor B.
8. A compression test is used to determine _____ condition.
   (A) fuel system
   (B) ignition system
   (C) engine mechanical
   (D) overall vehicle
9. Technician A says a road test enables the technician to determine whether the tune-up was satisfactory. Technician B says while road testing, the technician should ignore any other vehicle problem and concentrate on the engine. Who is right?
   (A) A only.
   (B) B only.
   (C) Both A and B.
   (D) Neither A nor B.
10. Technician A says a chassis dynamometer is preferable to driving the vehicle on the streets. Technician B says the engine should be cold for accurate tune-up work. Who is right?
   (A) A only.
   (B) B only.
   (C) Both A and B.
   (D) Neither A nor B.

Suggested Activities

1. Make a list of all of the engine and drive train systems that could cause driveability problems, such as fuel, ignition, emission controls, transmission, and air conditioning. See how many individual components you can list under each system. Compare your list with those of the other class members.
2. Using the factory service manual, obtain the procedure for retrieving computer trouble codes for an assigned vehicle with computer engine controls. Retrieve the trouble codes and use the service manual's troubleshooting charts to determine the problem. If your instructor permits, create trouble codes by disconnecting wires or hoses. Retrieve these codes and compare them with the created problem.
3. Diagnose a vehicle driveability problem using logical diagnosis procedures and one or more of the following pieces of test equipment:
   - scan tool (code retrieval tool).
   - multimeter.
   - vacuum gauge.
   - compression tester.
   - oscilloscope.
4. Discuss the procedures used in the above activity and your conclusions with your instructor and the other members of your class.

This hybrid vehicle is powered by a gasoline engine and an electric motor. This arrangement improves efficiency and minimizes emissions. (Toyota)

21

Emission System Testing and Service

After studying this chapter, you will be able to:

- Identify the two major types of emissions tests.
- Explain general procedures for conducting emissions test.
- Prepare a vehicle for emissions testing
- Use and exhaust gas analyzer to check vehicle emissions.
- Explain how OBD II systems monitor the operation of various emission control devices.
- Diagnose emission control problems.
- Service and repair the emission control devices used on late-model vehicles.

Technical Terms

Enhanced emissions test
Static test
Two-step idle test
Cost waiver program
Exhaust gas analyzer
Emission information label
OBD II emissions system monitoring
Icing
Thermostatic air cleaner
Early fuel evaporation system
Exhaust gas recirculation (EGR) valve
Ported vacuum
Vacuum override switch
Back pressure transducer valve
Negative back pressure valve
Integrated electronic EGR valve
Digital EGR valve
Air injection system
Air pump
Diverter valve
Check valve
Air control valve
Pulse air injection system
Catalytic converters
Catalysts
Positive crankcase ventilation system
Evaporation control system
Enhanced evaporation system

The air around us may seem inexhaustible, but there is not as much of it as we think. The layer of breathable air around the earth is proportionally as thick as the skin on an apple. Therefore, steps must be taken to ensure that our air does not become contaminated by the emissions from industrial facilities and motor vehicles.

To keep our air as clean as possible, the United States and many other countries have passed laws restricting the production of harmful gases. These laws have prompted programs for testing motor vehicle emissions. See **Figure 21-1.**

If a vehicle fails an emissions test, it must be repaired or the vehicle cannot be driven. On modern vehicles, the emission controls are integrated with all other engine systems. Even when emissions tests are not required, emission control system problems will affect vehicle drivability and fuel economy, or illuminate the MIL.

This chapter will cover the general procedures for conducting the two main classes of state-mandated emissions tests: enhanced tests and static tests. The service of the emission control devices installed on many vehicles is also covered. Because of the interaction of modern emission controls with other engine systems, you will be directed to other chapters for additional information about certain systems.

State Emissions Tests

Emissions tests are performed by various states to comply with federal emissions laws. At least 50% of current automotive technicians will be involved in emissions testing, either as part of a state-mandated emissions testing program, or by diagnosing and repairing vehicles to correct problems uncovered by an emissions test. Therefore, the technician must know how to perform emissions tests, prepare vehicles

Figure 21-1. The growing number of vehicles on the road contributes greatly to air pollution.

for emissions testing, and interpret test results. The test procedures given here can be applied to all testing programs with minor modifications

Types of Emissions Tests

All emissions tests can be roughly divided into two main categories:

- Enhanced emissions tests.
- Static emissions tests.

The most elaborate test is the ***enhanced emissions test,*** sometimes called an ***IM test*** or a ***transient test.*** During this type of test, the vehicle is operated on a dynamometer at different speeds and loads to simulate engine and drivetrain operation during actual driving. As the vehicle is operated on the dynamometer, emission levels are measured. Enhanced emissions tests often include a test of the evaporative emissions controls.

During a ***static test,*** sometimes called the ***basic test*** or an ***idle only test,*** tailpipe emissions are measured as the engine idles. The vehicle is not operated on a dynamometer during a static test. Some states use a static test that includes operation at higher engine speeds, often 2500 rpm. This is sometimes called a ***two-step idle test.*** Vehicles are not tested under load in any static test. In many states, static test procedures include a check of the vehicle's gas cap. Checking the gas cap is a simple test of the most failure prone part of the evaporative emissions system.

Equipment Used to Perform Tests

Various types of equipment are used to perform vehicle emissions tests, including dynamometers (sometimes referred to as treadmills), RPM testers, and automotive analyzers. If an OBD II drive cycle test is performed as part of an enhanced test, a scan tool will be needed. If the particular state requires testing of the evaporative emissions system, the facility must have a pressure-testing device.

Modern emissions test facilities use computers to perform the test procedures, record data, and provide target emissions levels. Many computers can determine the target emissions levels when the technician enters the vehicle identification number (VIN). The test facility computers may be attached to a central computer at the headquarters of the state emissions inspections authority.

Performing Emissions Tests

Emissions test procedures vary greatly between states and sometimes between different areas of the same state. Therefore, the following information should be used as a general guide to emissions testing procedures. Consult your local emissions inspections authority for exact procedures.

Preparing the Vehicle for Testing

Some preparation is needed before the vehicle can be tested. Enter the vehicle data (usually the VIN number) into the computer. The computer will use the data to determine the emissions requirements for the specific type and year vehicle. Next, start the vehicle's engine and let it run until it reaches its normal operating temperature. Cold or overheating engines will give false emissions readings. As the engine warms up, look for obvious problems, such as excessive exhaust smoke or missing. In many states, a visibly smoking engine cannot be tested. Check general engine operation and note whether the MIL is on. If the MIL is on, the vehicle obviously has a problem and should not be tested. Also check for missing, damaged, or disconnected emissions control devices or other parts that could affect engine operation. If required, perform the OBD II drive cycle. The OBD II drive cycle was covered in Chapter 15.

If an enhanced test will be performed, make sure the vehicle can be operated on a dynamometer. Look for a leaking cooling system, inoperative cooling fans, worn or defective tires, and low engine oil or transmission fluid. These defects could cause severe engine or vehicle damage if the enhanced test is attempted. Also check filler neck condition, look for the proper gas cap, and make sure the filler neck unleaded gas restriction is in place. Turn off the air conditioner and any other accessory that would put an extra load on the engine during the test.

Placing the Vehicle on the Dynamometer

Note: If a static test is being performed, the dynamometer will not be used. Nevertheless, many technicians prefer to place the vehicle on the dynamometer during a static test.

Drive the vehicle onto the dynamometer and place it in park. Leave the engine running to maintain temperature and prevent buildup of vapors in the engine.

Once the vehicle is on the dynamometer, install the securing devices. These may be snubbing chains, wheel chocks, or another means of keeping the vehicle from coming off the dynamometer rollers.

Caution: Do not attempt to operate the vehicle on the dynamometer without installing the securing devices.

After securing the vehicle on the dynamometer, position the cooling fans in front of the vehicle radiator. The cooling fans should be operated during the test to prevent engine overheating. A vehicle properly positioned on a dynamometer is shown in **Figure 21-2.**

Attaching the Test Equipment

Once the vehicle is properly placed on the dynamometer and secured, install the analyzer probe 10–16″ into the tailpipe. A collection cone is sometimes used in place of the tailpipe probe. Next, install the engine tachometer. Some tachometers can simply be placed on the vehicle hood. Others must be attached to the engine. If necessary, attach the evaporative emissions pressure tester to the vehicle test fitting. Also attach the OBD II tester to the vehicle diagnostic connector, if required.

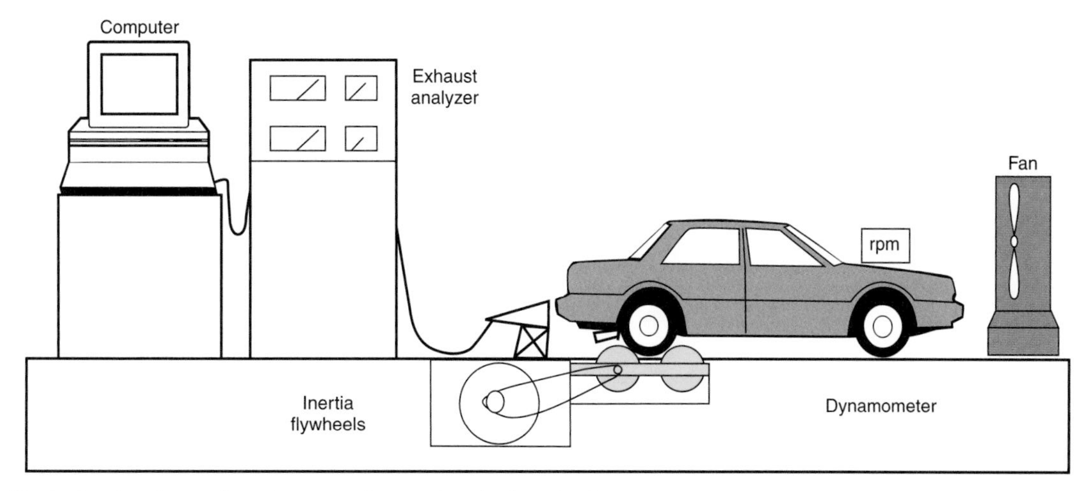

Figure 21-2. Typical setup for performing an enhanced emissions test.

Performing the Emissions Test

Once all preliminary steps have been taken, the technician can perform the emissions test. Both static and enhanced test procedures are outlined in the following sections. Many test sequences have been automated, and the inspector simply responds to prompts from the computer screen. Additionally, exact procedures vary greatly between states. Contact your local vehicle emission inspection authority to determine the exact sequence.

Enhanced Test

If an enhanced test is being performed, enter the vehicle and make sure the dynamometer hand controls are within easy reach. Begin by lowering the dynamometer drive-on plates so the drive wheels are resting on the dynamometer rollers. Place the vehicle in drive and accelerate to the specified speed. The test may require that several sequences of acceleration and braking be performed. **Figure 21-3** shows the sequence of acceleration and deceleration required for one specific enhanced test.

Note: Maximum simulated speeds vary between states. Some test speeds are as low as 15 mph (24 kph), while others can reach 58 mph (93 kph). The simulated driving distance can vary between one half mile (0.8 km) and two miles (3.2 km).

Most test programs will automatically begin the test when the vehicle reaches the desired speed. Many test programs stop the test when the vehicle emissions are very low or very high. After the test is stopped, the vehicle will be passed or failed as applicable.

Other test systems require operator input to begin and end the test. The entire test sequence will have to be performed to obtain final emissions readings. Most test sequences last no longer than four minutes. If the inspector does not maintain the required speeds during the test, the computer will direct that the test be performed again.

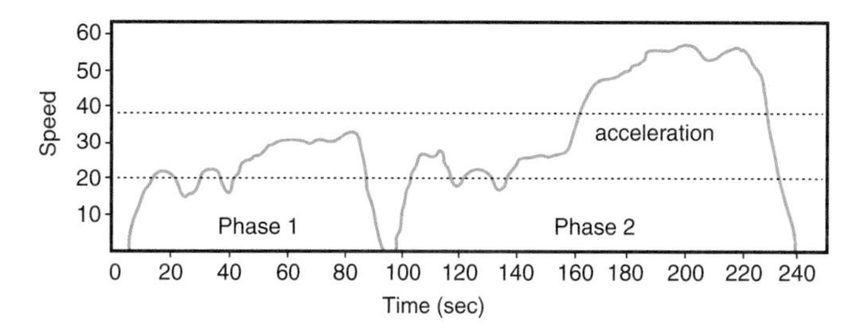

Figure 21-3. Typical acceleration trace for performing one particular enhanced emissions test.

Static Test

If a static test is being performed, allow the vehicle to idle as the tailpipe probe or cone picks up exhaust gases. The gases will be drawn into the analyzer and read. If necessary, accelerate the engine to the desired RPM for the second phase of the static test. As with the enhanced test, the procedure may be stopped when the vehicle passes or fails the test.

Evaporative System Test

Many enhanced and some static tests include testing the evaporative emissions system. Some tests on older vehicles measure flow into the engine from the evaporative system as other tests are being performed. See **Figure 21-4.** Flow should be roughly one liter (slightly more than a quart) during the test. Most vehicles with an operating evaporative emissions system will have much more flow, often as much as 25 liters (28 quarts) during the test.

Vehicles with the OBD II evaporative emissions system can be pressure checked using nitrogen, **Figure 21-5.** Once the nitrogen has been used to pressurize the system, the operator must wait a specified time to determine whether the system loses pressure. A pressure drop indicates a leak in the system. Some evaporative emissions systems can be tested by pressurizing the fuel tank using a scan tool. These tests are discussed later in this chapter.

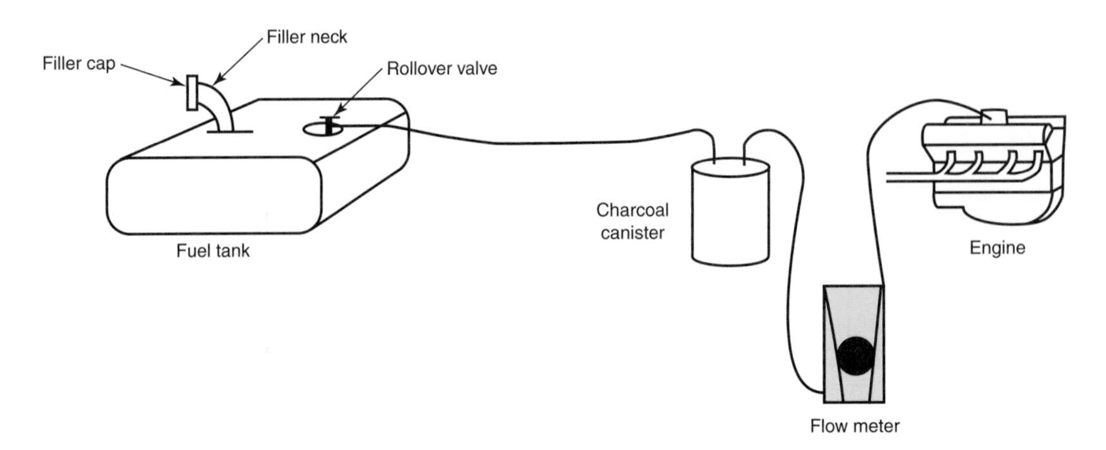

Figure 21-4. In some vehicles, evaporative emission system condition can be determined by connecting a flow meter in the line between the engine and the charcoal canister.

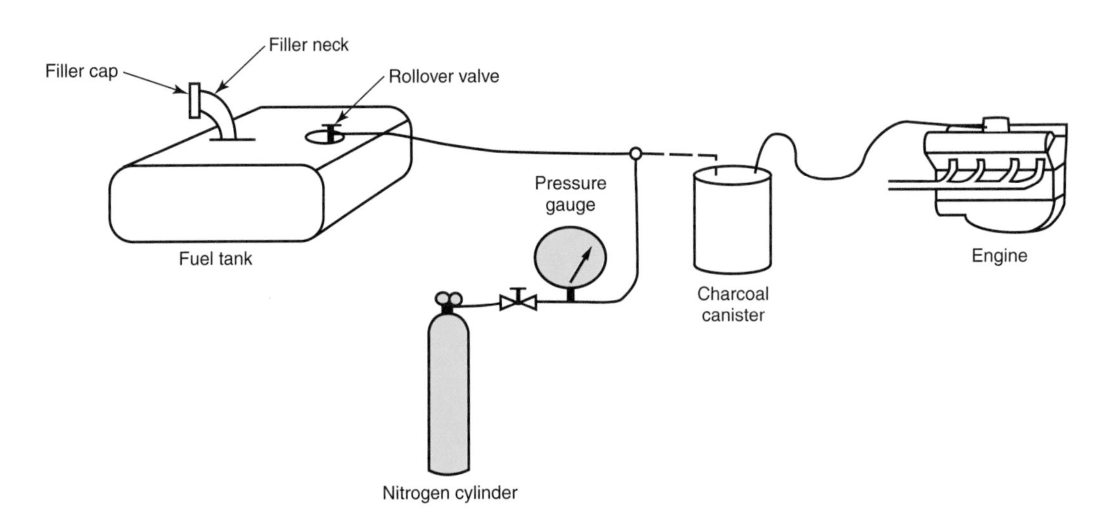

Figure 21-5. Setup for using nitrogen to pressurize an evaporative emissions system.

OBD II System Test

Some states are downloading emissions information from the OBD II system used on 1996 and later vehicles. At this time, the information is being used to study computer operating information and compare it to emissions failure rates. In the future, OBD II information may be an integral part of the emissions testing process.

Remove the Vehicle from the Test Area

After the testing process is completed, raise the dynamometer drive-on plates to release the drive wheels from the rollers. Next, remove the vehicle securing devices and all other test equipment. Then drive the vehicle off the dynamometer and to the customer waiting area of the test facility.

Complete Test Paperwork

At the end of the test, the vehicle receives a pass/fail report that compares the actual readings to the desired values. In many cases, the computer automatically generates the report. Typical exhaust gases checked are:

- Unburned hydrocarbons (HC).
- Carbon monoxide (CO).
- Oxides of nitrogen (NO_x).
- Free oxygen (O_2).
- Carbon dioxide (CO_2).

Evaporative control system flow or pressure results are also given when applicable. In some states, the pass/fail status and readings will be transmitted directly to the central computer at the state motor vehicle department or emissions inspection headquarters.

If the vehicle has passed the emissions test, inform the driver or owner and give the completed test results and other paperwork to the owner. If an emissions windshield sticker is used in the particular state, apply it at this time.

If the vehicle failed the emissions test, the data computer will usually print out a repair diagnostics report in addition to the pass/fail report. The repair diagnostics report lists possible causes of the particular failure(s). This report can be used as an aid to diagnosis and repair of the vehicle.

Inform the driver or owner of the problem and answer any questions. When applicable, inform the vehicle owner of the repair ***cost waiver program.*** The purpose of the repair cost waiver program is to grant an emissions waiver after the owner has spent a certain amount of money, even if the vehicle has not been restored to the desired emissions levels.

Preparing a Vehicle for Emissions Testing

The technician may be called on to get a vehicle ready for an emissions test. Many owners want their vehicles checked out before going to the official inspection station. In addition, the technician must check for proper emissions levels on vehicles that were repaired after failing a state-mandated emissions test. Below are a few methods for improving the chances of passing an emissions test.

- Make sure the engine and vehicle are in good condition. Diagnose and repair any obvious engine missing, rough idle, or other performance problems. Always locate and correct the cause of an illuminated check engine light.
- Change the engine oil and filter. This removes unburned hydrocarbons that are present in the old oil. New oil also tends to do a better job of sealing the piston rings, resulting in a cleaner running engine.
- Try to take the test on the warmest day possible. This will ensure that the engine stays at its normal operating temperature during the test.
- Do not fill the gas tank before getting the test. A full tank may overload the evaporative control system and cause the appearance of a system defect.
- Take a highway drive of about 5-10 miles before the test. A long high-speed drive will tend to remove deposits from the engine internals.
- Do not turn off the engine while waiting for the test. A running engine removes unburned fuel from the crankcase and evaporative control system.

Checking Emissions in the Shop

The technician will often check emissions in the shop. Using an ***exhaust gas analyzer*** is the only sure method of verifying the overall operation of the emission control systems. There are several types of exhaust gas analyzers available. Some are extremely simple and will only measure CO. Other analyzers can measure CO, HC, NO_x, O_2, and CO_2. See **Figure 21-6.** Therefore, the following information is only a general guide. Always follow the equipment and vehicle manufacturers' instructions.

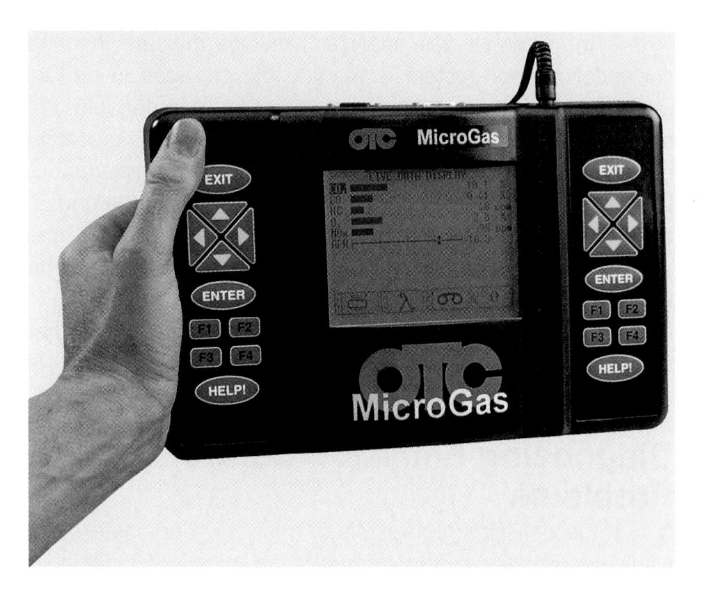

Figure 21-6. An example of an exhaust gas analyzer. Follow the equipment manufacturer's operating instructions. (OTC)

First, start and, if necessary, calibrate the exhaust gas analyzer. Start the engine and run it long enough to reach normal operating temperature. Place the analyzer probe in the vehicle's tailpipe and allow the engine to idle while the analyzer readings stabilize. Then check emissions at idle. If the idle readings are incorrect, check for problems in the fuel, ignition, and compression systems, as well as the emission controls. On carburetor-equipped engines, adjust the idle mixture screws as explained in Appendix A. Once the vehicle air-fuel ratio at idle is correct, allow the exhaust gas analyzer's readings to stabilize.

On vehicles with a carburetor, briefly open the throttle and then allow the engine to return to idle. Observe the exhaust gas analyzer. The mixture should briefly become much richer and then return to the original settings. This will confirm that the carburetor accelerator pump is working properly. Check emissions levels as necessary at different engine speeds.

If the vehicle is equipped with a computer control system and oxygen sensor, the following checks can be made. Start by creating a rich mixture. This can be done by partially closing the choke plate on a carbureted engine, or briefly throttling (restricting) the airflow into a fuel injected engine. The throttle plate should remain closed, restricting the airflow just enough to create a rich mixture but not enough to stall the engine.

When the airflow is restricted, the exhaust gas analyzer should show a momentarily rich mixture. Then the mixture will begin to lean out as the ECM compensates for the rich mixture signal from the oxygen sensor. When this check is complete, remove the airflow restriction and operate the engine at high idle for one minute to clear any extra gasoline from the intake passages. Next, allow the exhaust gas analyzer readings to stabilize and then create a lean mixture by removing the PCV hose or another vacuum hose. The exhaust gas analyzer should show a momentarily lean mixture and then begin to move to the rich position as the ECM begins to compensate for the lean signal from the oxygen sensor. When the test is complete, reconnect the hose.

If the exhaust gas analyzer readings indicate that the computer control system is reading the changes in air-fuel ratio and is changing fuel system settings to compensate, the system is in good condition. If the system does not appear to be reacting to air-fuel ratio changes, refer to the vehicle manufacturer's troubleshooting procedures to determine the cause. After completing all needed tests with the exhaust gas analyzer, turn off the engine and remove the analyzer probe from the tailpipe.

If necessary, the fuel trim can be checked as a further diagnostic step. The procedure for checking fuel trim is covered in Chapter 20.

Diagnosing Emission Control System Problems

If the emission test results or analyzer readings are not within specifications, check for problems in the fuel, ignition, or emission control systems. **Figure 21-7** illustrates the test results for a vehicle with excessive emissions levels. Note how the levels exceed the specifications. Not all incorrect exhaust gas readings are caused by the same problems. The chart in **Figure 21-8** lists common causes for excessive emissions. Notice how some defects cause an increase in certain readings without affecting other readings.

When diagnosing engine or emission problems, remember that many of the vehicle's systems are interrelated. A properly running engine requires a careful balancing of all systems. A failure or misadjustment in one system can affect the operation of other systems. Proper emission control diagnosis is impossible unless the engine is in sound mechanical condition and is properly tuned. Work carefully and follow manufacturer's specifications. Always test vehicle emission levels for compliance with established standards.

Emission Information Label

The ***emission information label*** is located under the vehicle's hood, **Figure 21-9.** This label contains information about the types of emission devices used on the vehicle. The label may also contain service information, such as the proper spark plug gap specification. Read this label before beginning any emissions system diagnosis and use the information provided to determine which emissions systems are used. Remember that the label is intended for use only with the vehicle on which it is attached. Do not remove this label.

OBD II Emissions Systems Monitoring

OBD II diagnostic systems perform a series of self-checks to test the operation of several emission-related systems. This capability is known as ***OBD II emissions system monitoring.*** In addition to constantly monitoring the engine sensors and output devices, OBD II systems monitor all of the following:

- Engine misfires.
- Catalyst efficiency.
- EGR operation.
- Air injection system operation.
- Evaporation control system operation.

Gas	Standards (gpm)	Test Readings
HC	.41	1.13
CO	3.4	2.8
NO_x	.08	0.2

Figure 21-7. Emissions test results. Comparing test results to set standards will give you a place to start when looking for an emissions-related problem.

Information on the OBD II monitoring procedures for specific systems will be covered later in this chapter.

Retrieving Trouble Codes

When diagnosing an emissions-related problem on a computer-controlled vehicle, always check for trouble codes before performing any involved diagnostic routines. For more information on trouble code retrieval, refer to Chapter 15.

Emission Control System Service

To minimize the level of air pollution from vehicle emissions, a number of emission-reducing methods have been developed. They can be divided into three basic classifications:

- Engine modifications and engine controls.
- External cleaning systems.
- Fuel vapor controls.

Engine modifications and engine controls are designed to allow the fuel charge to burn more completely within the combustion chamber. As a result, the gases leaving the chamber are as environmentally safe as possible. Engine modifications include the basic design of internal engine parts such as camshaft lobe shape, combustion chamber shape, and intake manifold design. Control systems are used to alter ignition timing, fuel mixture, and combustion chamber temperature.

External cleaning systems are used to ensure continued burning of the exhaust between the combustion chamber and the tailpipe. These systems include the air injection system and the catalytic converter(s). Fuel vapor controls are designed to prevent the escape of gasoline vapor from the tank filler cap, tank, fuel lines, and other vehicle components. These are a common source of unburned hydrocarbons. Vapor control systems also control engine crankcase fumes caused by unburned fuel leaking past the piston rings.

Note: Several emissions-related devices were covered in previous chapters. Many of these devices, such as the fuel injectors and ignition coils, have another primary purpose, but are controlled by the ECM to reduce emissions. Internal engine components are more conveniently covered in other chapters. The emissions devices covered in the remainder of this chapter can be discussed as separate systems.

Excessive Hydrocarbon (HC) Reading: Excessive HC is usually caused by a problem that results in an incomplete burning of fuel. Sometimes accompanied by a "rotten egg" smell.

- Poor cylinder compression
- Leaking head gasket
- Ignition misfire
- Poor ignition timing
- Defective input sensor
- Defective output device
- Defective ECM
- Open EGR valve
- Sticking or leaking injector
- Improper fuel pressure
- Leaking fuel pressure regulator
- Oxygen sensor contaminated or responding to artificial lean or rich condition
- Fuel filler cap improperly installed

Excessive Carbon Monoxide Reading: Excessive CO is caused by a problem that results in a rich air-fuel mixture. However, excessive CO is often created by an insufficient amount of air or too much fuel reaching the cylinder. Will sometimes coincide with a high HC and/or low O_2 reading.

- Plugged air filter
- Engine carbon loaded
- Defective input sensor
- Defective ECM
- Sticking or leaking injector
- Higher than normal fuel pressure
- Leaking fuel pressure regulator
- Oxygen sensor contaminated or responding to artificial lean condition

Excessive Hydrocarbon (HC) and Carbon Monoxide (CO) Readings: When both HC and CO are excessive, this often indicates a problem with the emissions control system or an on-going problem, usually indicated by a rich air-fuel mixture, that has damaged an emissions control component. You should check all of the systems previously mentioned along with these listed below.

- Plugged PVC valve or hose
- Fuel contaminated oil
- Heat riser stuck open
- AIR pump disconnected or defective
- Evaporative emissions canister saturated
- Evaporative emissions purge valve stuck open
- Defective throttle position sensor

Excessive Oxides of Nitrogen (NO_X): Excessive oxides of nitrogen (NO_X) are created when combustion chamber temperatures become too hot or by an excessively lean air-fuel mixture.

- Vacuum leak
- Leaking head gasket
- Engine carbon loading
- EGR valve not opening
- Injector not opening
- Low fuel pressure
- Low coolant level
- Defective cooling fan or fan circuit
- Oxygen sensor grounded or responding to an artificial rich condition
- Fuel contaminated with excess water

Excessively Low Carbon Dioxide (CO_2) Reading: A low CO_2 reading is usually caused by a rich air-fuel mixture or a dilution of the exhaust gas sample. You should check all of the possibilities mentioned earlier, starting with the ones listed below.

- Exhaust system leak
- Defective input sensor
- Defective ECM
- Sticking or leaking injector
- Higher than normal fuel pressure
- Leaking fuel pressure regulator

Low Oxygen (O_2) Reading: Low oxygen readings are usually caused by a lack of air or rich air-fuel mixture, the same factors that can create an excessive CO reading. Note: Not all analyzers check the exhaust gas for O_2 content.

- Plugged air filter
- Engine carbon loaded
- Defective input sensor
- Defective ECM
- Sticking or leaking injector
- Higher than normal fuel pressure
- Leaking fuel pressure regulator
- Oxygen sensor contaminated or responding to artificial lean conditioning
- Evaporative emissions system valve defective

High Oxygen (O_2) Reading: A high oxygen level is an indication of a lean air-fuel mixture, dilution of the air-fuel mixture, or dilution of the exhaust gas sample by outside air. When a high oxygen reading is present, the CO reading is usually very low or does not register.

- Vacuum leak
- Low fuel pressure
- Defective input sensor
- Exhaust system leak near the tailpipe

Figure 21-8. These are the most common causes of high emissions readings. Each section explains how the excessive reading is created.

Devices for Heating Intake Air

To allow a cold engine to run properly with lean mixtures, the incoming air must be heated to prevent fuel condensation in the intake manifold. Heating the air also helps to prevent ***icing.*** Icing occurs when reduced air pressure in the intake system lowers the freezing point of water vapor. The water vapor freezes around the throttle valve. Ice formation can cause the engine to run roughly or stall in cold, damp weather. To reduce these problems, various devices are used to warm the incoming air.

Thermostatic Air Cleaner

Some older vehicles with throttle body fuel injection or carburetors use a ***thermostatic air cleaner*** to heat the incoming air. A vacuum-operated motor in the air inlet section directs heated air into the intake system. A thermal sensor bleed valve directs vacuum to the motor. **Figure 21-10** illustrates the operation of a thermostatic air cleaner. When the engine is started cold, the bleed valve directs full vacuum to the vacuum motor. This pulls the diaphragm up against spring pressure and opens the air valve. When the valve is opened,

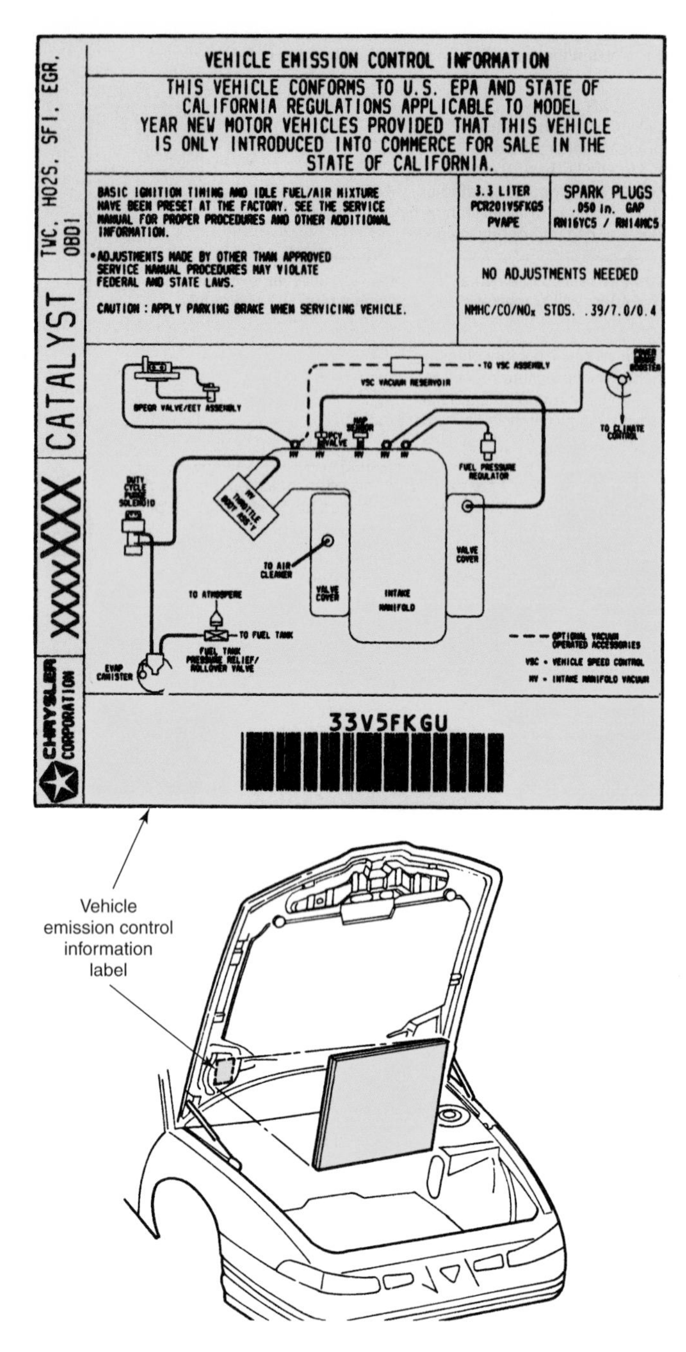

Figure 21-9. An example of an emission control label, which is found under the vehicle's hood. The information on these labels is very specific to the vehicle on which they are mounted. Do not use this information for any other vehicle, even one of same year and model. Labels will vary depending upon vehicle destination, engine size and type, transmission, fuel system, and other factors. (DaimlerChrysler)

heated air is drawn from a shroud around the exhaust manifold. When the engine warms up, the bleed valve reduces the vacuum to the vacuum motor, and it closes the air valve. With the valve closed, cooler outside air is drawn in. The cold engine start position is shown in Figure 21-10A. When the thermal sensor warms, it opens the air bleed. This decreases vacuum to the vacuum motor, allowing the spring to force the diaphragm down and close the valve, Figure 21-10B.

Thermostatic Air Cleaner Service

Make certain that system hoses are in good condition, properly connected, and free of kinks. Following the manufacturer's specifications, make certain that the valve starts to open and goes to the full open position at the correct temperature. Check the operation of both the vacuum bleed sensor and the air motor. **Figure 21-11** illustrates the use of a hand vacuum pump to test the air motor vacuum diaphragm. Check for vacuum leakdown, starting door lift vacuum, and full open vacuum measurement.

Early Fuel Evaporation System

The ***early fuel evaporation (EFE) system*** warms the incoming air by using exhaust heat to warm an exhaust crossover passage under the intake manifold. A vacuum-operated valve partially blocks the flow of exhaust gases, forcing them to exit through the crossover passage before reaching the exhaust pipe. A thermal vacuum switch controls the EFE valve. When the coolant temperature reaches a certain point, the vacuum switch triggers the EFE valve, which diverts the gases through the manifold and on to the exhaust pipe. See **Figure 21-12.** Older engines used a similar device called a heat riser, which was operated by a thermostatic spring. Heat risers, however, were intended only to reduce cold weather warm-up problems, not to control emissions.

Early Fuel Evaporation System Service

Check the operation of the thermal vacuum switch, EFE valve, and hose connections. For full details on the operation of the EFE system, see the heat control valve section in Chapter 19.

Throttle Body Coolant Passages

Many modern vehicles prevent throttle valve icing by diverting some of the engine coolant through the throttle body. **Figure 21-13** shows a typical coolant chamber located under the throttle body. Since the cylinder heads warm more quickly that the rest of the engine, coolant is usually pumped directly from the cylinder head passages to the throttle body.

Throttle Body Coolant Passage Service

If the cooling system is in good condition, the coolant passages in the throttle body will normally be okay. If the cooling system has been neglected, the passages may become filled with sludge, blocking the coolant flow. Sometimes flushing the cooling system will remove deposits in the passages, allowing coolant to flow. If the coolant passages are completely blocked, drain the cooling system and remove the hoses to the passages. Use short bursts of air pressure to clean out the passages. If necessary, scrape deposits from the inside of the passages. Be sure to flush the cooling system after cleaning the passages.

If the throttle body coolant passages seem to be slow to warm up, check to make sure that the thermostat has not been removed or is not stuck open. Install a new thermostat if necessary.

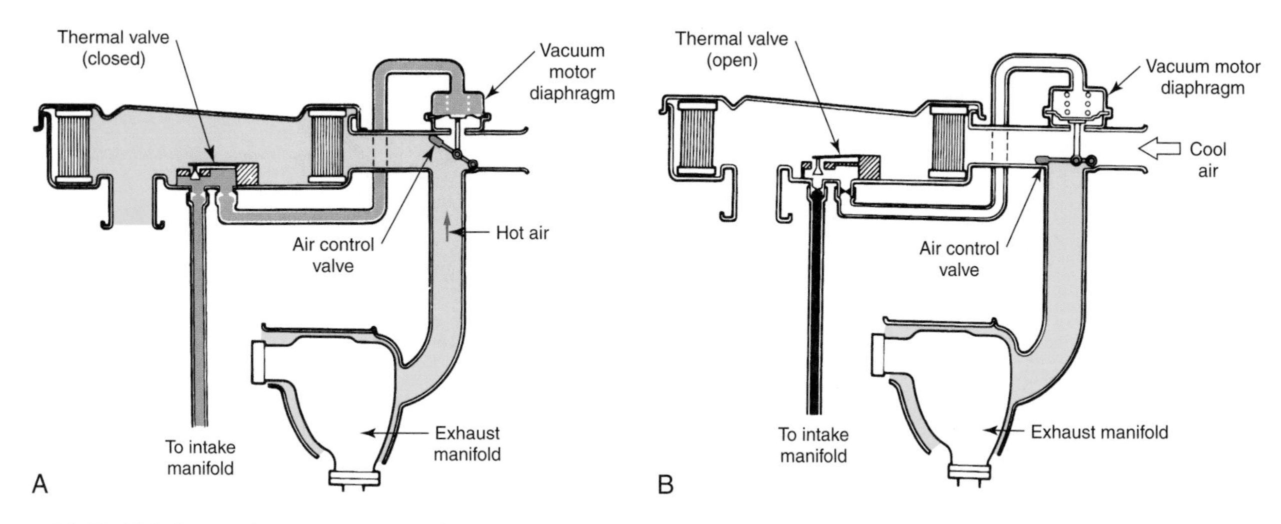

Figure 21-10. This figure shows motor actuation on a thermostatically controlled air cleaner. A—Cold engine, vacuum holds valve open. B—Engine at normal operating temperature, vacuum reduced; valve is closed. (Toyota)

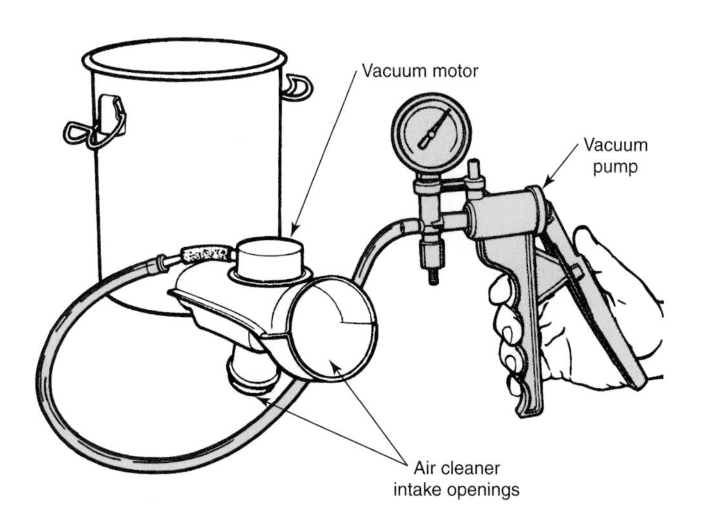

Figure 21-11. Using a hand vacuum pump to test operation of a thermostatic air cleaner vacuum motor. (DaimlerChrysler)

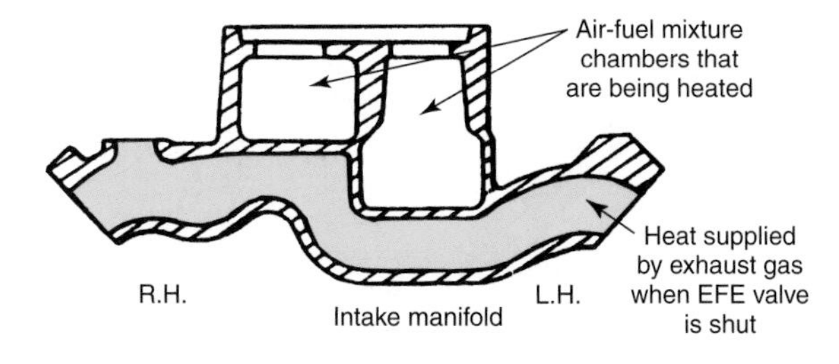

Figure 21-12. In an early fuel evaporation system, exhaust gases warm the intake manifold during cold engine operation. (General Motors)

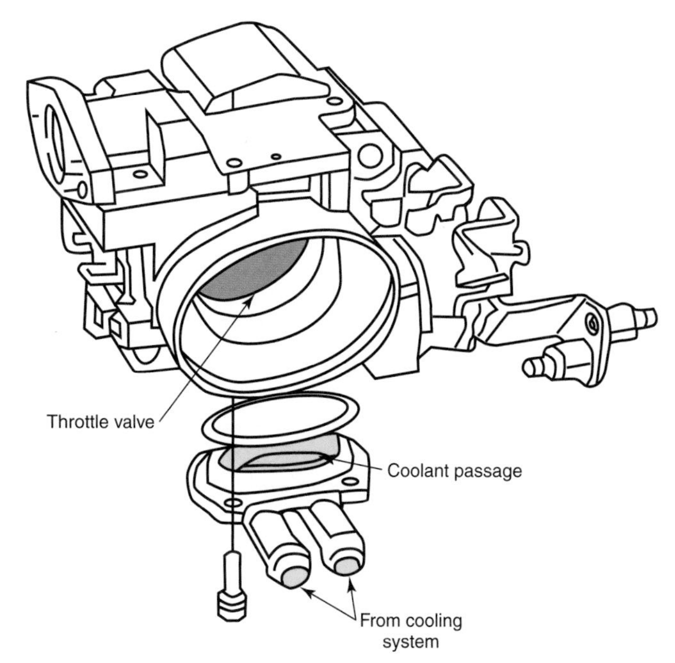

Figure 21-13. Most late-model engines warm the throttle valve area with engine coolant. The coolant circulates through the chamber under the valve whenever the engine is running.

Exhaust Gas Recirculation

When the temperature of the burning fuel mixture exceeds 2500°F (1372°C), nitrogen in the air tends to chemically combine with oxygen. This forms oxides of nitrogen, usually called NO_x. By recirculating a portion of the burned exhaust gas back into the intake manifold, the peak flame temperature in the cylinders is lowered. This significantly reduces the amount of NO_x produced.

The amount of exhaust gas fed into the intake manifold is controlled by the ***exhaust gas recirculation (EGR) valve.*** The EGR valve is open when the engine is under light and moderate loads. During idle and wide-open throttle conditions, the EGR valve is closed. EGR valves are operated by engine vacuum, electrical solenoids, or a combination of both.

Ported Vacuum EGR Valves

The simplest type of EGR valve is operated by ***ported vacuum*** from the throttle body, **Figure 21-14.** A vacuum port above the closed throttle plate cannot receive vacuum at idle,

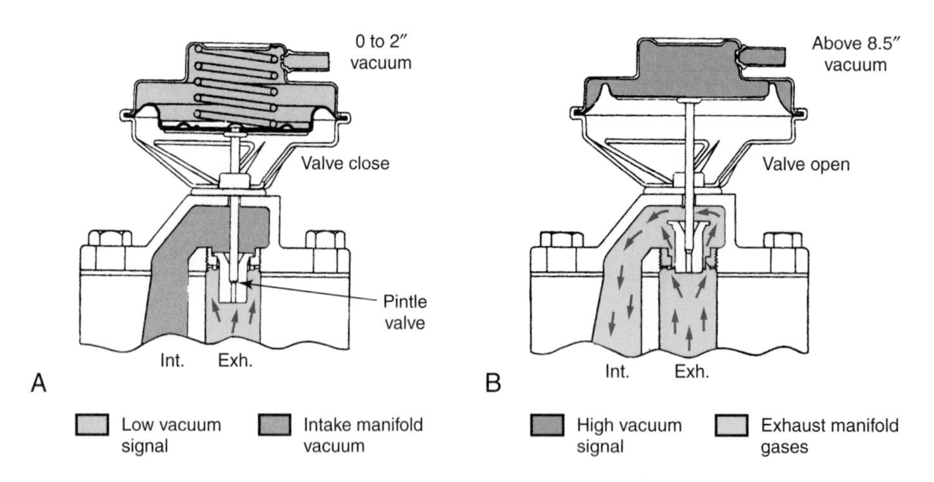

Figure 21-14. Ported EGR valve action. A—Low vacuum, valve is closed. B—High vacuum, diaphragm opens valve. (General Motors)

but it can receive vacuum as soon as the throttle is opened slightly. A spring-loaded diaphragm is used to keep the valve closed. In Figure 21-14A, vacuum is low and the spring keeps the valve closed. In Figure 21-14B, increased vacuum has pulled the valve upward to pass burned exhaust gases back into the intake manifold. Under heavy loads, the manifold vacuum drops off and the spring closes the valve.

Testing Ported Vacuum Operated EGR Valves

When possible, start the engine and depress the EGR diaphragm with the tips of your fingers. If the engine was idling smoothly, it should immediately lose about 200 rpm and show signs of roughness. If this occurs, the EGR valve is all right so far. If rpm does not decrease and engine operation remains smooth, the passage between the EGR valve and the intake manifold may be plugged. Remove and clean the EGR valve and passageway.

If no loss in rpm is evident and the engine idle is rough, the valve may be admitting exhaust gases all the time. This indicates a faulty valve or improper hose routing. Place a T fitting in the vacuum signal line and attach a vacuum gauge. Start the engine and observe the vacuum gauge. There should be no vacuum present. Increase engine rpm and note the vacuum reading when the diaphragm starts to move. Valve opening should fall within specified limits. Also observe the EGR valve stem. If the stem does not move, the valve is stuck or the diaphragm is leaking. If the vacuum gauge does not register vacuum when the throttle is opened, the hoses are misrouted, plugged, or leaking.

Warning: The EGR valve can become very hot. Be careful to avoid burns on your hands and fingers.

EGR Coolant Temperature Override Switch

A coolant temperature-controlled ***vacuum override switch*** is used on some models. The EGR vacuum hose from the carburetor or throttle body port is connected to one side of the switch. The other side goes to the EGR valve. When engine temperature is below a specified point, the switch cuts off vacuum to the EGR to improve cold engine operation. See **Figure 21-15.**

Testing Temperature Override Switches

To test the switch, remove the vacuum line from the EGR valve or from the transducer (if the EGR valve is not of the integral type). With the engine cold, connect a vacuum gauge to the line and run the engine at 1500 rpm. No vacuum should be indicated. If vacuum is present, replace the switch. Operate the engine at 1500 rpm until the coolant temperature reaches the specified level. Vacuum should then register on the gauge. If there is no vacuum, replace the switch.

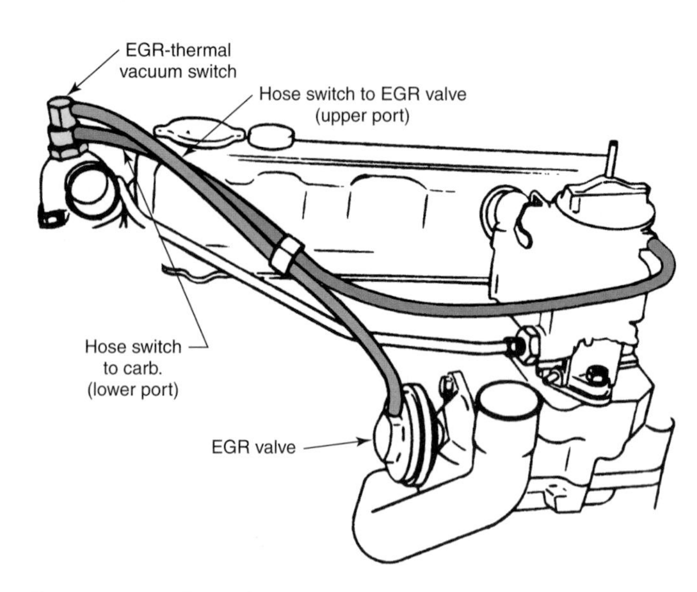

Figure 21-15. The EGR coolant temperature override switch (thermal vacuum switch) will cut off the vacuum signal to the EGR valve until the engine reaches normal operating temperature. (General Motors)

Positive Back Pressure Transducer Valve

Some EGR valves use an exhaust ***back pressure transducer valve*** (BPV) to modulate the amount of vacuum acting on the EGR valve diaphragm, **Figure 21-16.** The BPV is a device that utilizes power from one source to provide control of power or action. The transducer is incorporated into the EGR valve. In Figure 21-16A, exhaust gas cannot flow to the intake manifold, but can create a pressure on the transducer diaphragm. Pressure is low and the spring control valve remains open. This allows air to flow through the bleed holes in the diaphragm plate, past the diaphragm, and through the spring control valve. This weakens the vacuum in the vacuum chamber and the main diaphragm will not pull the exhaust gas recirculation valve open.

In Figure 21-16B, exhaust back pressure has built up, pushing the transducer diaphragm upward. This closes the control valve and stops airflow into the vacuum chamber. Since vacuum is no longer weakened by bleed air, the main diaphragm rises and opens the valve. This permits flow of gases to the intake manifold.

Testing a Positive Back Pressure Transducer Valve

An EGR valve with a back pressure transducer valve should not open if test vacuum is applied with the engine off. If the EGR does open under these circumstances, replace the valve.

Negative Back Pressure Valve

An EGR system controlled by a ***negative back pressure valve*** is pictured in **Figure 21-17.** In this setup, the bleed valve spring is located below the diaphragm. In this position, the spring holds the diaphragm against the air bleed hole, keeping it closed. As soon as engine vacuum is applied to the EGR valve, the diaphragm will be forced upward, opening the valve. The valve then passes a metered quantity of exhaust into the intake manifold. With negative exhaust back pressure, the central diaphragm area is pulled down against the small

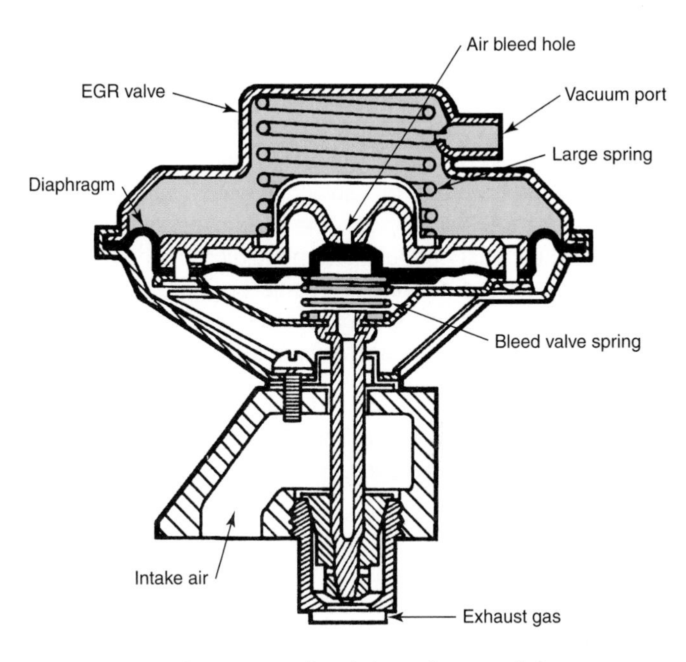

Figure 21-17. A cross-sectional view of a negative backpressure EGR valve. (General Motors)

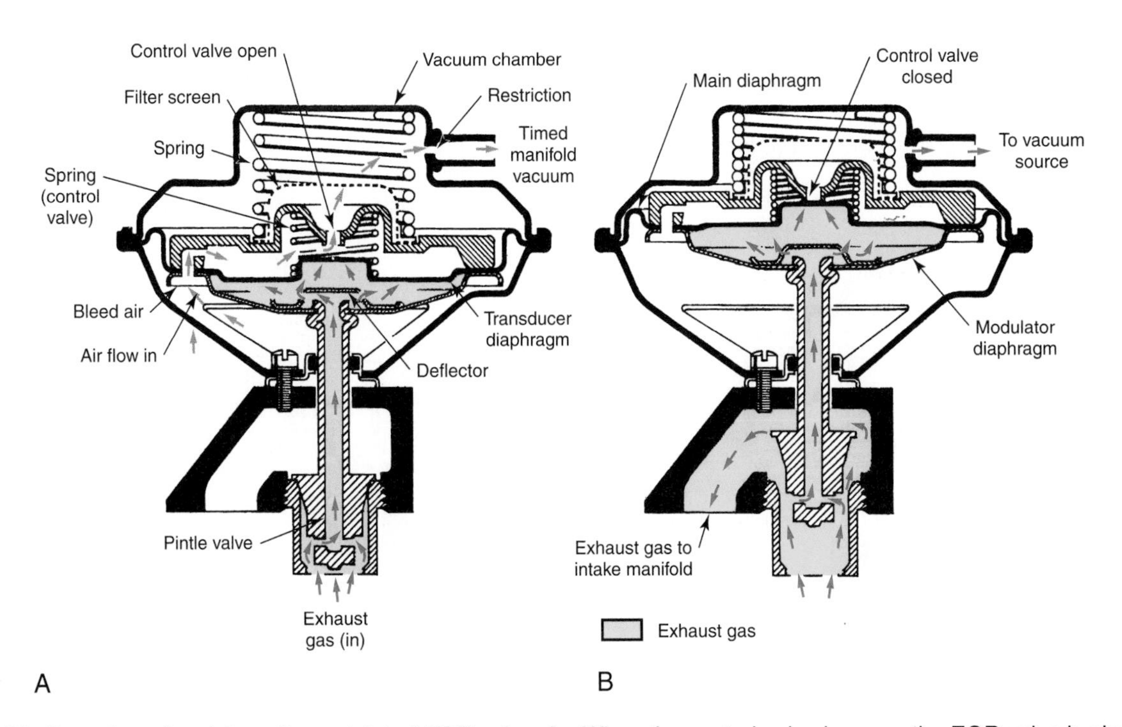

Figure 21-16. Operation of an internally modulated EGR valve. A—When the control valve is open, the EGR valve is closed. B—Exhaust back pressure closes control valve, causing the EGR valve to open. (General Motors)

spring. The bleed hole is then exposed, reducing the effect of engine vacuum. This causes the heavy top spring to force the main diaphragm downward, closing the valve.

Testing a Negative Back Pressure Transducer Valve

An EGR valve with a negative back pressure valve will open if test vacuum is applied with the engine off, **Figure 21-18.** If the EGR does not open under these circumstances, replace the valve.

Integrated Electronic EGR Valve

The ECM controls vacuum flow to the ***integrated electronic EGR valve,*** **Figure 21-19.** The ECM sends an electrical signal to a voltage regulator in the EGR valve. This regulator controls current to a solenoid, which in turn controls vacuum to the EGR diaphragm. The ECM controls the flow of recirculated exhaust gas, using a modulated (modified pulse width) signal based on airflow, throttle position, and engine speed. A pintle position sensor is also used. As EGR flow increases, the position sensor's output increases.

Testing Integrated Electronic EGR Valves

Since the ECM controls the integrated electronic EGR valve, operation must be tested with a scan tool. The scan tool can check the valve circuits and force the valve open to test its mechanical condition.

Digital EGR Valve

The ***digital EGR valve*** functions without engine vacuum. EGR flow passes through one or more orifices, which are opened and closed by electric solenoids. The solenoids are operated by the engine control module. When the solenoids are energized, the armature rises and the pintle is lifted from its seat. The computer uses input signals from the throttle position sensor (TPS), coolant temperature sensor, and the mass airflow sensor (MAF) to calculate the proper orifice openings. This valve usually operates above idle and during warm engine operating conditions. A digital EGR valve is shown in **Figure 21-20.**

Testing Digital EGR Valves

There are three kinds of digital EGR valves: single solenoid, multiple solenoid (usually involving two or three solenoids), and stepper motor valves. All recycle exhaust gases to reduce oxides of nitrogen. Failure of a digital EGR valve will cause pinging and high NO_x levels. Rough idle and stalling can occur if the EGR valve fails while open.

When engine operation or trouble codes indicate that the digital EGR valve is defective, it can be checked by momentarily grounding the solenoid ground wire.

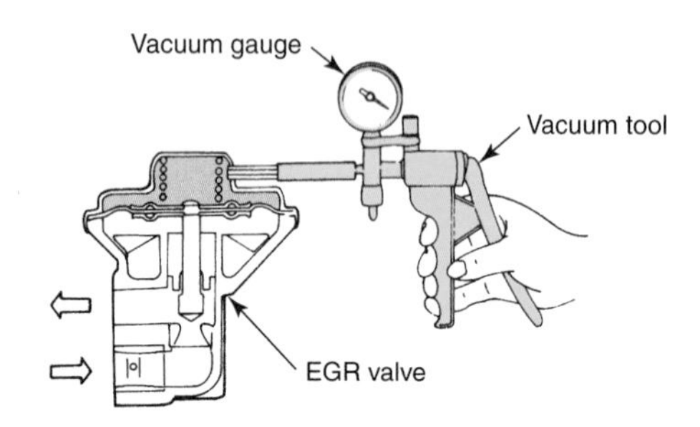

Figure 21-18. An EGR valve with a vacuum pump tool installed onto the vacuum port to check valve for proper operation. (Hyundai)

Caution: Do not ground any EGR wire without first consulting the proper service manual. Grounding the wrong wire can destroy the ECM.

Begin by making sure electrical power is available to the EGR valve. Then, with the engine idling, locate the solenoid ground wire and ground it to the engine block by back probing into the EGR electrical connector. The engine should momentarily idle roughly as the EGR valve opens and allows excess exhaust gas into the intake manifold. If the EGR has multiple solenoids, test each one in turn. This test can also be performed with some scan tools. If the engine does not idle

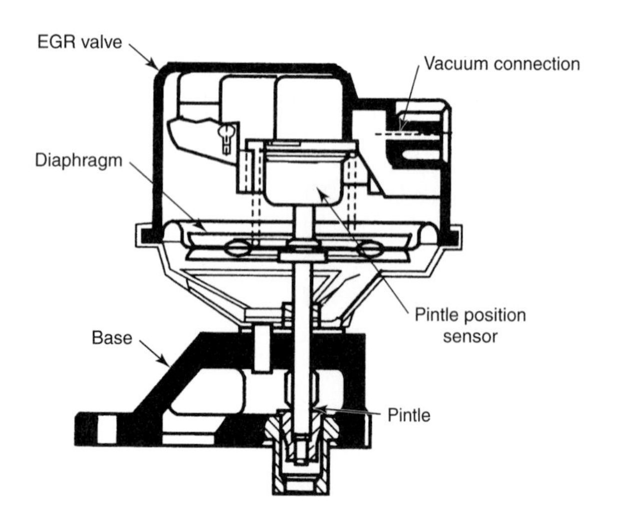

Figure 21-19. Note the pintle position sensor in this cutaway view of an integrated electronic EGR valve. (Sun Electric)

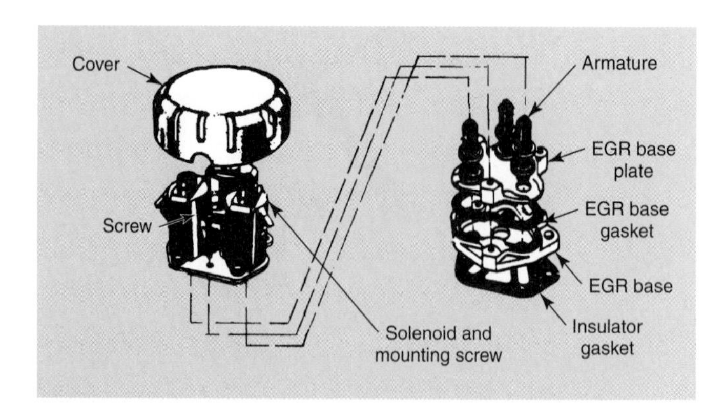

Figure 21-20. Exploded view of a digital EGR valve. Follow manufacturer's recommendations when servicing this unit. (Sun Electric)

roughly, the solenoid winding is defective or the EGR passage is clogged with carbon. Sometimes, the ECM will compensate for an open EGR valve by increasing the idle speed and richening the mixture. In this case, energizing the solenoid will have no effect upon the engine idle.

If you determine that the EGR is defective, remove it for further testing. Begin by visually inspecting the pintle valve for damage or carbon deposits as you would on a vacuum-operated EGR. If the pintle and passages are clear, check the solenoid windings. Some digital EGR valves can be tested by energizing them with a 9V battery and jumper wires. The solenoid windings of some digital EGR valves can be checked with an ohmmeter.

Caution: Never permanently disconnect or render an EGR valve inoperative. To do so is a violation of federal and state law and will cause vehicle emission levels to rise. In addition, many engines will run hotter and in some cases, detonate to the point of serious damage (burned valves and pistons).

OBD II EGR System Monitoring

On vehicles with OBD II systems, the operation of the EGR valve is checked at set intervals of engine operation. Some systems check EGR operation by monitoring intake manifold vacuum through the MAP sensor while opening and closing the EGR valve. If opening and closing the EGR valve does not cause variations in the MAP sensor readings, the system sets a trouble code and illuminates the MIL.

Most OBD II systems check the EGR valve by monitoring oxygen sensor readings while operating the EGR valve. If opening and closing the EGR valve does not result in changes in the oxygen sensor readings, the ECM sets a trouble code and illuminates the MIL.

Cleaning EGR Valves

Some EGR valves may be disassembled for cleaning. On those that cannot be disassembled, the open end of the pintle valve can be tapped lightly with a plastic hammer to break loose any carbon accumulation. A brush can also be used to assist cleaning. Do not sandblast an EGR valve.

Caution: Never soak or wash an EGR valve in cleaning solvent. This could seriously damage the diaphragm.

One type of serviceable EGR valve is shown in **Figure 21-21.** This valve should be cleaned once every three years or 30,000 miles (48,000 km). **Figure 21-22** shows the general steps in disassembly and cleaning. Clean off any gasket material and check for the two punch marks. These marks will ensure proper alignment upon reassembly. Measure the distance from the base to the pintle seat shoulder as shown in Figure 21-22A. Note this distance, because it must be established again during reassembly. Remove the seat with a wrench of the proper size, Figure 21-22B. A small amount of

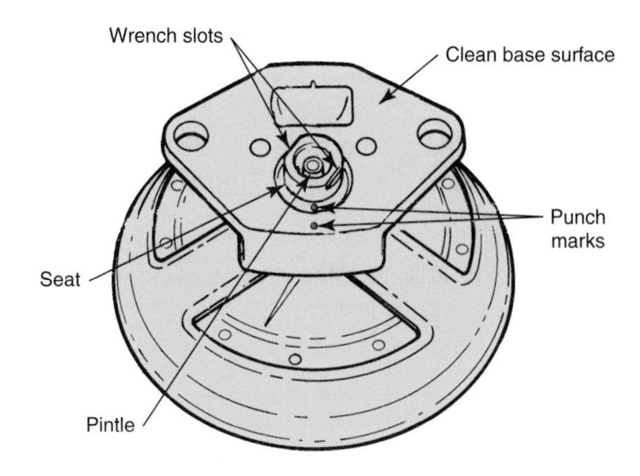

Figure 21-21. Before disassembling a serviceable EGR valve such as this one, make certain there are punch marks and that they are aligned. (General Motors)

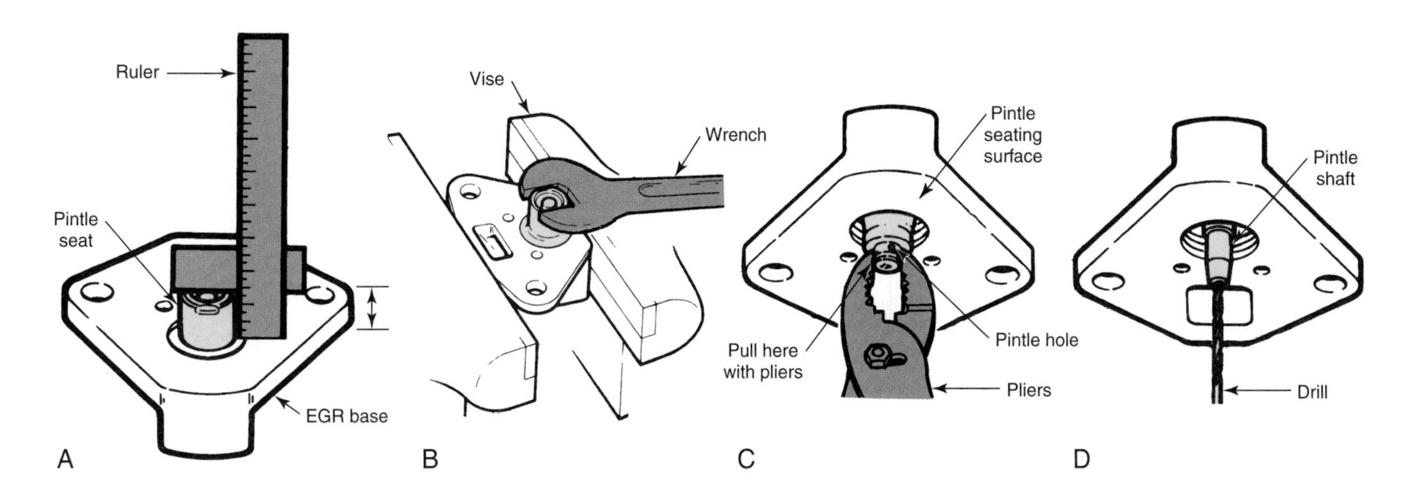

Figure 21-22. Steps involved in disassembly of this EGR valve. A—Check height of pintle seat shoulder-to-base. B—Remove pintle seat. C—Pull pintle from shaft. D—Clean carbon from hollow shaft with appropriate drill bit. (General Motors)

penetrating oil on the threads will ease removal. Do not allow penetrating oil to foul the diaphragm.

After the seat is removed, the pintle can be disassembled. Holding the valve in an upright position, Figure 21-22C, grasp the pintle (not on seating surface) with pliers and pull it downward. Clear carbon from the pintle shaft using a drill bit of the specified size, Figure 21-22D. Turn the drill bit clockwise up into the shaft for a short distance (about an inch); then withdraw it to clear the bit of carbon. Continue the procedure until the drill passes through the shaft hole. Tap the base to jar out any loose particles. Also, brush and shake the valve. If compressed air is used, do not direct air into the shaft opening. Reinstall the pintle on the shaft, applying pressure until it snaps into place. Reinstall the pintle seat and thread it in to its original height. Make sure the punch marks are aligned, and then restake the seat using a prick punch. Reinstall the EGR valve using a new gasket. If cleaning does not restore proper EGR operation, replace the valve.

Air Injection System

The ***air injection system*** helps clean the exhaust by prolonging the combustion process in the exhaust system. These systems are also called *air injection reactors (AIR), thermactors, air guards,* or *smog pumps.* A belt-driven ***air pump*** delivers air to the system. The system uses a system of tubes or passages to route a stream of fresh air into the exhaust just as it passes the exhaust valve. A schematic of one air injection system is shown in **Figure 21-23.**

Air Distribution

The pump intake air is filtered by drawing it from the air cleaner, through a special filter, or by using a centrifugal filter. The pump produces a low-pressure, high-volume air stream. The air flows from the pump through the diverter valve. The ***diverter valve*** reduces the chance of a backfire by diverting airflow from the exhaust manifold to the atmosphere. Diverter valve action is shown in **Figure 21-24.** In Figure 21-24A, airflow is passing through the diverter valve and on to the air injectors. When the throttle is suddenly released, heavy intake manifold vacuum is applied to the metering valve control diaphragm. The diaphragm is drawn downward, forcing the metering valve to block off passage to the air injection manifold. This downward movement opens the diverter passage and pump air is momentarily discharged into the atmosphere. See Figure 21-24B.

A ***check valve,*** **Figure 21-25,** is installed between the diverter valve and the exhaust manifold. The check valve is forced off its seat by pump air pressure, thus allowing air to enter the distribution manifold. If the pump or drive belt fails, or if exhaust pressure exceeds pump pressure, the check valve returns to its seat. This prevents the exhaust gases from flowing through the hose to the pump. From the check valve, the air enters the distribution manifold and air injection tubes. It then enters the exhaust gas near the exhaust valve.

Air Control Valve Operation

Some air injection systems use an ***air control valve.*** The air control valve can be used to perform the normal diverter valve functions. It can also be used for channeling pump air to the intake manifold, exhaust manifold, or the catalytic converter, as determined by system needs. **Figure 21-26** shows an air control valve that channels pump air to the exhaust port area during engine warm-up. Then, when normal operating temperature is reached, it diverts the air to the catalytic converter.

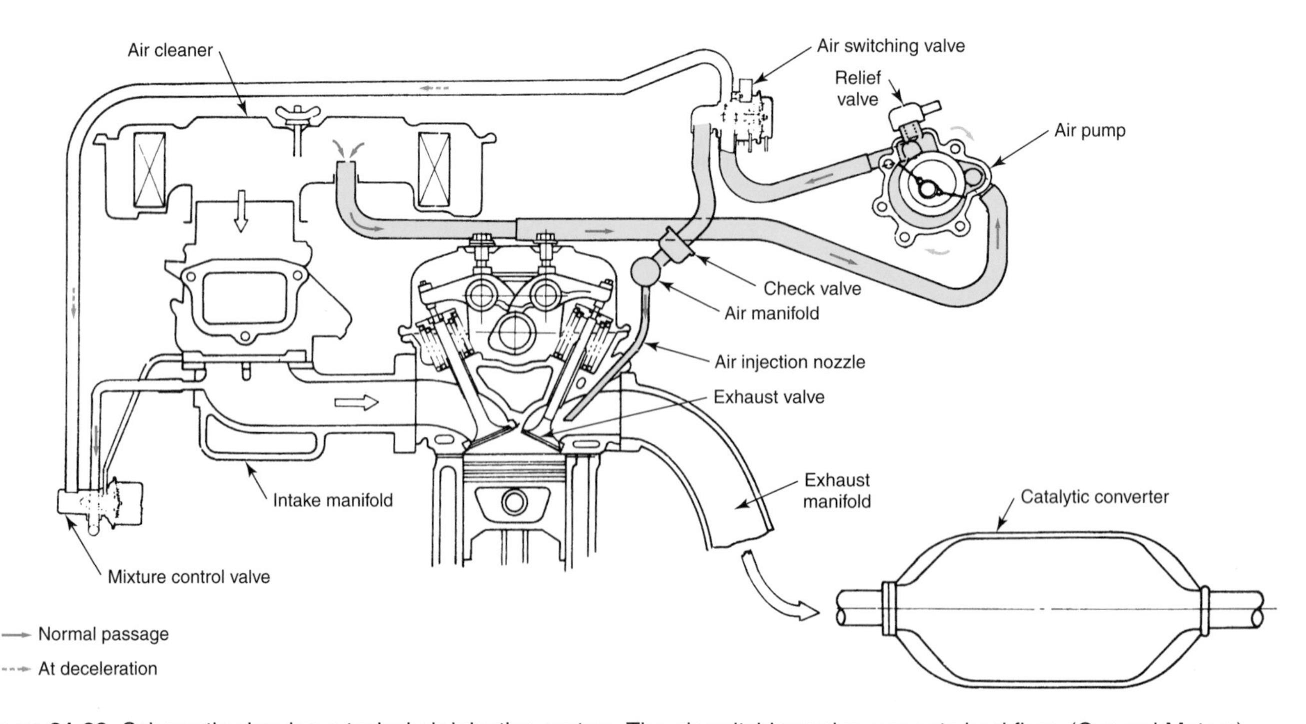

Figure 21-23. Schematic showing a typical air injection system. The air switching valve prevents backfires. (General Motors)

Pulse Air Injection System

The ***pulse air injection system*** makes use of the pulse effect of the engine exhaust to pull air into the exhaust manifold. Small pulses, or gulps, of air are drawn in through the check valves. The check valves operate in the same manner as those discussed previously. See **Figure 21-27.**

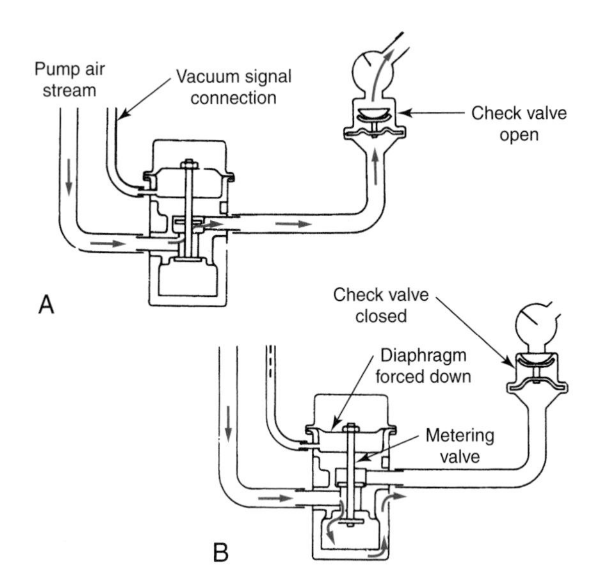

Figure 21-24. Diverter valve action. A—Valve in normal position. B—During deceleration, air is diverted into atmosphere. (Toyota)

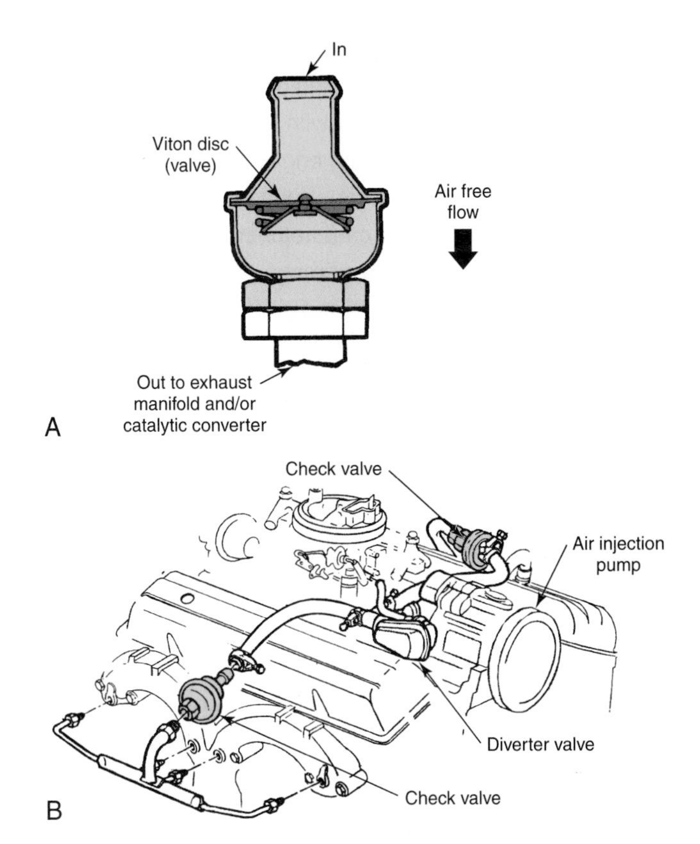

Figure 21-25. Check valve. A—Cutaway of an air check valve. (General Motors) B—Check valves in an air injection reaction (AIR) system. (Ford)

OBD II Air Injection System Monitoring

OBD II systems monitor O_2 sensor readings to verify the proper operation of the air injection system. The ECM compares the sensor readings with the commands that it sends to the air injection system. If the O_2 sensor readings do not match the parameters set by the ECM programming, the ECM sets a trouble code and lights the MIL.

Air Injection System Service

The internal parts of the air injection system are designed to be maintenance-free. When the air pump intake is through the air cleaner, normal air cleaner maintenance

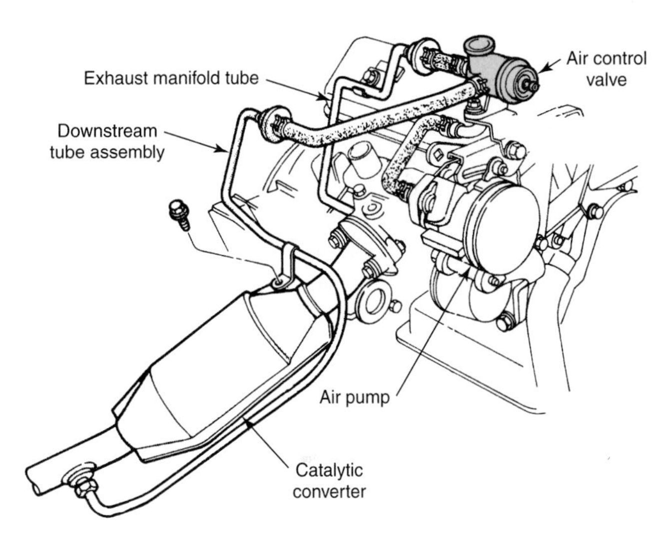

Figure 21-26. The air control valve on this setup directs pump air to the exhaust manifold when the engine is cold and to the exhaust pipe through the downstream tube when hot. (DaimlerChrysler)

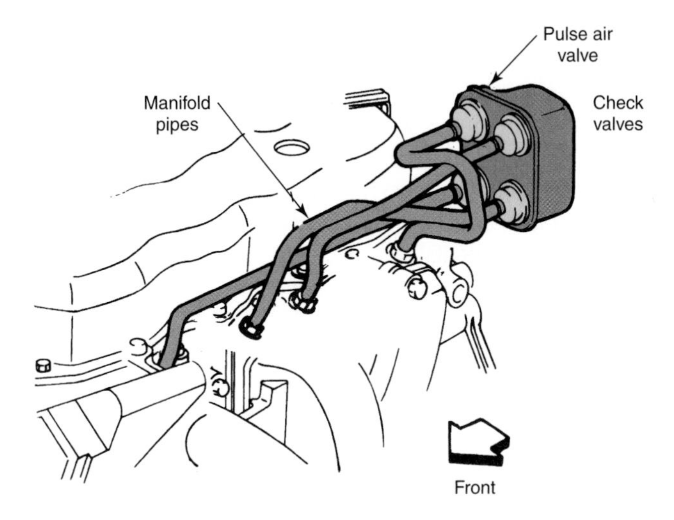

Figure 21-27. An air injection system using the pulse principle. Pulses in exhaust pull air in through check valves in small gulps. (General Motors)

schedules should be followed. When a separate filter is used, replace the filter at recommended intervals. If the vehicle is operated under severe conditions, change filters more often. When installing a new filter, wipe dust and grit off the filter body and air horn assembly.

Drive belt tension is important. A loose belt will reduce pump efficiency, while a tight belt will cause premature wear of the pump bearings. When adjusting the belt, avoid prying on the pump housing. Pull on the pump with your hand only or use a built-in adjustment slot or pin, where provided. Use a gauge to tension the belt properly.

Note: The engine should be brought to operating temperature before conducting the following air injection system tests.

Checking Pump Operation

To check pump operation, remove the outlet hose at the pump and start the engine. You should be able to feel air coming from the pump outlet. If a special low-pressure gauge is available, measure the air pressure produced by the pump. Pressure should be about 1 psi (6.9 kPa). If the pump is not producing sufficient pressure, check the air filter for clogging. Also check the pump relief valve. If the relief valve is bypassing air at all times, remove the pump and install a new valve. If the pump seems noisy, replace the drive belt and recheck the pump. If the pump is still noisy, replace it. Keep in mind that air pumps are not completely silent, even when in perfect condition.

Testing the Check and Diverter Valves

Remove the hose from the diverter valve. Start the engine and operate it at 1500 rpm. There should be no sign of exhaust leakage from the valve. The valve may flutter when the engine is idling, but this condition is normal. With the engine off, use a thin tool to press against the valve plate. The plate should open readily and return to its original position when released. Check both valves on V-type engines and replace them as needed. Check lines (especially the vacuum signal line) for kinks, pinches, or leaks. Remove the vacuum signal line at the diverter valve. A vacuum signal must be present when the engine is running. With the engine running at idle, no air should be diverted. When the throttle is opened up and then quickly released, a sudden gust of air should be discharged into the atmosphere. If the diverter valve is defective, replace it.

Caution: Do not use compressed air to clean diverter valves or check valves. This can ruin the valve. If the valve is dirty, flush it with solvent and shake it dry.

Checking Distribution Manifold, Air Injection Tubes and Hoses

The distribution manifold and air injection tubes do not require periodic maintenance. They are usually made of stainless steel. If the injection tubes become burned, they can be replaced. If the tubes are clogged, they can be cleaned with a wire brush. If the distribution manifold needs cleaning, use regular cleaning methods and solvents. Check the hose system for loose connections, kinks, or other damage. Hoses should also be checked for leaks and restrictions. Repair or replace as needed. Typical hose and tube routing is pictured in **Figure 21-28.** To detect an air leak, soapy water can be placed on the hose or hose connections. Escaping air will cause bubbles to form.

Catalytic Converter

Catalytic converters contain substances called ***catalysts,*** which cause chemical changes in other substances without themselves being changed. Some vehicles use a single converter, while others use a second, smaller unit close to the exhaust manifold, **Figure 21-29.** This unit starts the reaction and heats up the exhaust. The larger downstream converter continues and completes the burning and reduction process. This setup is referred to as a two-stage converter. Air may be injected into the exhaust between the two converters or into the converter itself, **Figure 21-30.**

Converter Cautions

Catalytic converters can overheat from a rich mixture. Allowing unburned fuel to enter the converter can cause serious overheating, resulting in a damaged converter, melted floor insulation, vehicle fires, and other problems. Causes of a rich mixture include:

- A long-term engine miss due to fouled plugs, loose or defective wires, or a cracked distributor cap.
- Using carburetor cleaner with the engine running.
- Shorting out plugs during engine testing.
- Excessive cranking tests with the ignition disabled.
- Excessive use of the carburetor choke.

Converters can also be damaged by backfiring in the exhaust system. Locate and fix all sources of backfiring, such as crossed wires, sticking or burned valves, a carbon-tracked distributor cap, or a defective diverter valve. Do not operate the vehicle until it runs out of fuel, since this can cause backfiring.

OBD II Catalyst Monitoring

All OBD II systems monitor the signal from an oxygen sensor installed behind the catalytic converter to check catalyst efficiency. The ECM compares the readings of the O_2 sensor mounted ahead of the converter with the one behind it. If the sensor readings indicate that the converter is not cleaning the exhaust gases, the OBD II system sets a trouble code. If converter operation returns to normal, the trouble code will be cleared after a set number of engine starts. The ECM will illuminate the MIL when a trouble code is set.

Converter Service

Theoretically, a catalytic converter could last the life of the vehicle since the catalyst is not consumed. Converters are designed to last for a minimum of 50,000 miles (80 000 km), and they usually last much longer. This is provided the

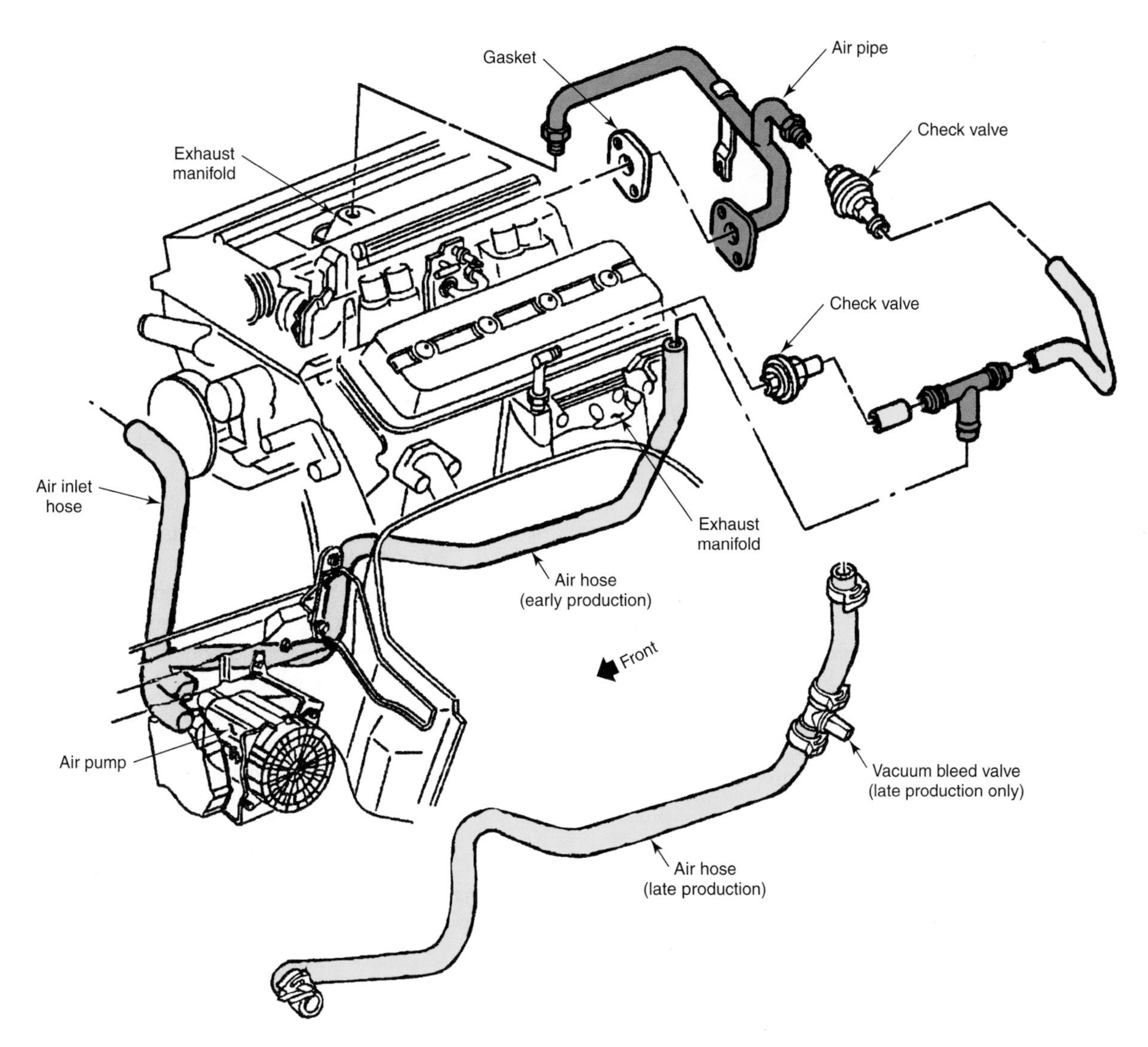

Figure 21-28. Hose and tubing routing used by one AIR system. This system uses an electric air pump that is turned on and off by the electronic control module (ECM), which is not shown. (General Motors)

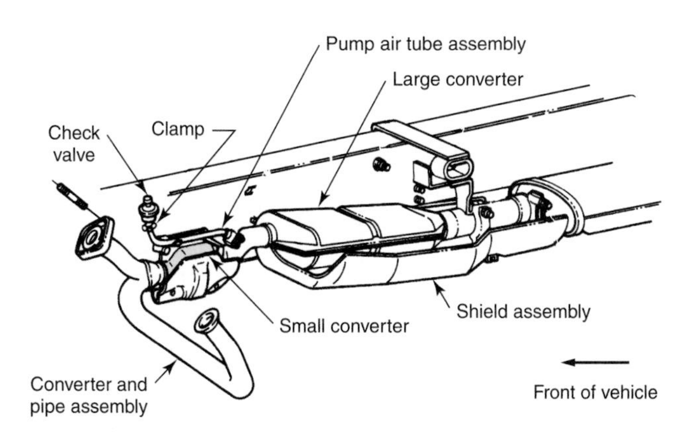

Figure 21-29. A dual converter setup. Note small converter close to exhaust manifold connecting pipes. Pump air is fed into exhaust pipe connecting small and large converters. (Ford)

catalyst is not exposed to overly rich mixtures or suffers other damage. Initial converter failure can be difficult for the driver to detect. Engine performance is sometimes not affected until the converter becomes clogged. To test converter efficiency, measure the exhaust gas temperature or use an exhaust gas analyzer. In OBD II system, converter efficiency can be checked by observing signals from pre- and post converter O_2 sensors. This was discussed in Chapter 15, Computer System Diagnosis and Repair.

Make sure all heat shielding is in good condition and in its proper place. Never spray undercoating on any part of the exhaust system or any portion of the heat shielding. In the event of catalytic converter failure, the unit should be replaced. When a converter is replaced, do not discard it as scrap. Instead, turn it in to be recycled.

Diesel Engine Catalytic Converters

Several diesel engine manufacturers use an oxidizing-type catalytic converter. This emission control unit is placed in

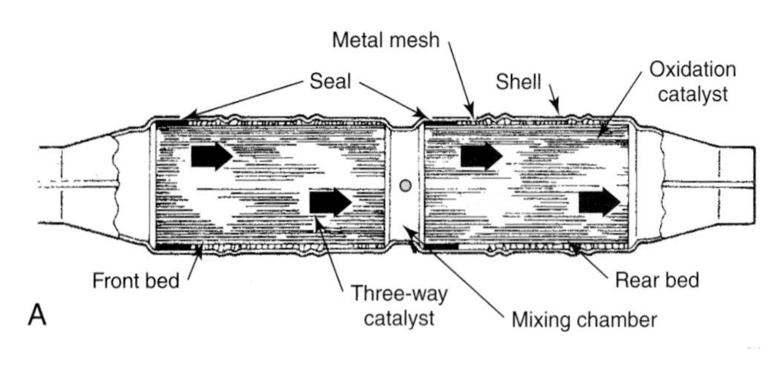

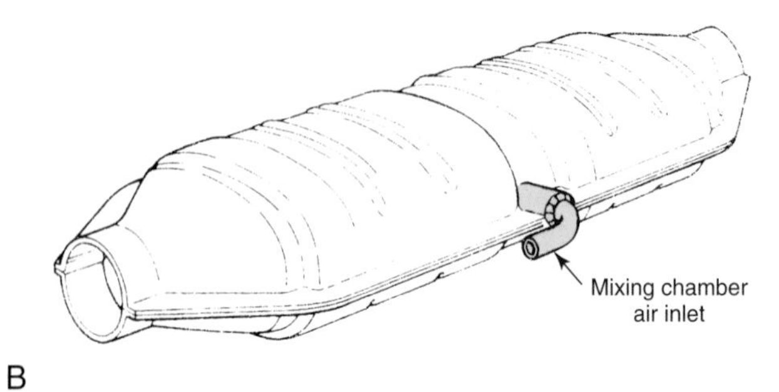

Figure 21-30. As exhaust gases pass through the converter catalyst sections, the sudden increase in temperature changes the hydrocarbons and carbon monoxide into water and carbon dioxide. A—This converter uses two separate catalyst beds that are separated by the mixing chamber. B—Air entering the mixing chamber provides additional air for the oxidizing (burning) process. (General Motors)

the exhaust system to help reduce diesel smoke particulates (tiny particles). The oxidizing converter functions at a normal exhaust system temperature. The oxidation of particles inside the converter does not cause a chemical reaction that raises the temperature. These units are not serviceable. If defective, they must be replaced.

Vapor Control Devices

One of the most obvious and easily controlled emission sources is the evaporation of unburned fuel from the gas tank, fuel supply system, and engine crankcase. Controls to reduce the emission of vapors are explained in the following sections

Positive Crankcase Ventilation System Operation

The ***positive crankcase ventilation system,*** or ***PCV system,*** uses engine vacuum to draw the crankcase fumes back into the cylinders for burning. A malfunctioning PCV system can produce serious engine problems, such as rough idle, oil loss, blown engine gaskets, and increased emissions.

Servicing the Positive Crankcase Ventilation System

Proper PCV valve operation is shown in **Figure 21-31.** PCV system operation should be checked at every oil change. To check system operation, remove the PCV valve from the engine and shake it vigorously. A valve that is not stuck will rattle when shaken. Another way to check is to use a suitable gauge to determine whether air is flowing through the crankcase. Manufacturers' recommendations for cleaning or changing the PCV valve range from once each year or 12,000 miles (19,200 km), whichever comes first, to over 100,000 miles (160,000 km). More frequent cleaning or replacement may be required if the engine is badly worn or if it is operating under severe conditions.

To service a PCV system, remove the connecting hose or pipe and the valve, **Figure 21-32.** Rinse out the PCV system connecting hoses. If the valve can be disassembled for cleaning, carefully take it apart. Sealed valves can often be cleaned by soaking them in solvent, rinsing, and blowing dry. Some valves contain internal parts that can be ruined by solvents and must be replaced if stuck. Follow the manufacturer's recommendations.

Caution: Do not interchange PCV valves. PCV valves are matched to the vacuum and flow needs of specific engines. Always use a valve designed for the engine being serviced.

Clean the hose from the breather or other opening to the air cleaner. If a special mesh is used in the air cleaner for

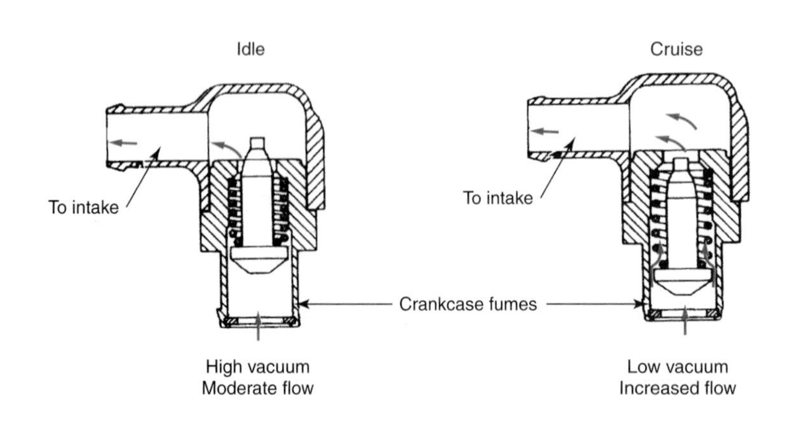

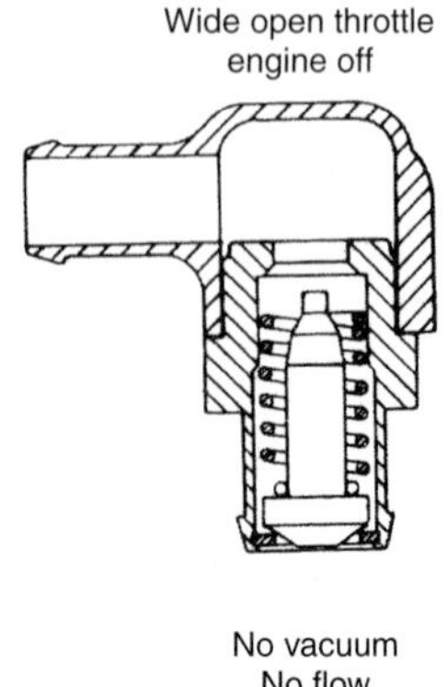

Figure 21-31. This figure shows the operation of a positive crankcase ventilation (PCV) valve during engine idle, cruise, and wide open throttle. (Sun Electric Corporation)

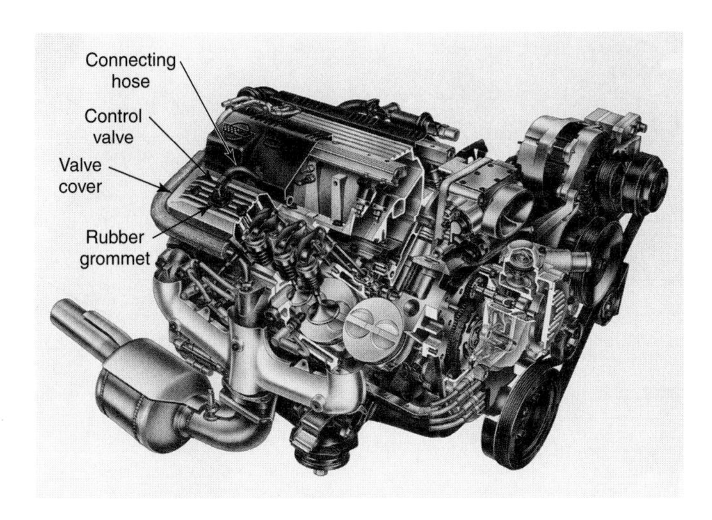

Figure 21-32. To service a PCV system, remove the connecting hose and the valve. (General Motors)

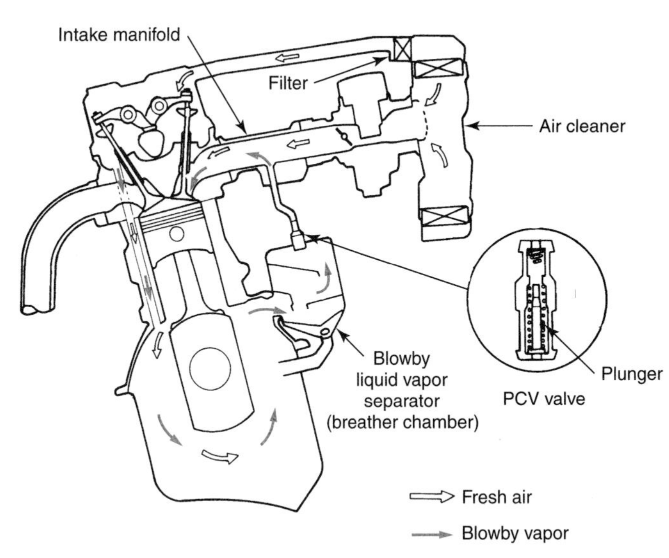

Figure 21-33. A typical closed PCV system. Note how PCV valve allows vacuum from intake manifold to draw fresh air through the air filter and into the engine crankcase. This effectively removes harmful blowby gases from the crankcase and mixes them with intake air for burning in combustion chamber. Note use of liquid-vapor separator to prevent engine oil from being drawn into the intake manifold. (Honda)

filtering the incoming air to the crankcase, wash and re-oil or replace as recommended. A modern closed PCV system is illustrated in **Figure 21-33.** Reinstall the valve so it faces the correct direction (check for directional marks on the valve). Connect all hoses, making sure they are open. Start the engine and bring it to operating temperature before system checking operation.

Evaporation Control System

The major function of any ***evaporation control system*** is to prevent unburned fuel vapors from entering the atmosphere. The fuel tank, fuel lines, and other fuel system components are sealed. Any vapors that develop in the fuel system are channeled to a charcoal-filled canister. The canister holds the vapors until they are drawn into the intake manifold and burned as the engine operates. Early systems used vacuum-operated purge controls to direct vapors into the intake manifold at the proper time. Newer systems use purge solenoids operated by the ECM.

All evaporation control systems have a pressure-vacuum gas filler cap, a liquid-vapor separator, vent and return lines, a charcoal canister, and a purge control. Some systems contain a rollover valve to prevent fuel leakage if the vehicle rolls over in an accident.

Enhanced Evaporation System

An ***enhanced evaporation system*** is used in OBD II vehicles. In addition to the components listed above, the enhanced evaporation system has fuel tank pressure and level sensors, and a service port, called an EVAP service port, for pressure testing. **Figure 21-34** shows a typical enhanced evaporation control system.

OBD II Evaporation Control System Monitoring

OBD II systems monitor the operation of the evaporation control system. The ECM uses engine vacuum to create a slight vacuum in the entire evaporation control system. The ECM then closes the purge valve and monitors the signal from the pressure sensor for a specific amount of time. If the pressure sensor indicates a loss of vacuum, the system is leaking. If the ECM detects a leak, it sets a trouble code and illuminates the MIL.

Evaporation Control System Service

Older evaporation control systems require some maintenance, such as replacing the carbon canister filter. Check vacuum-operated purge valves for proper operation. Carefully inspect all vacuum hoses for leaks.

An enhanced evaporation system does not require periodic maintenance. A problem in an enhanced system will illuminate the MIL. If an evaporation system trouble code has been set, check for a loose or leaking gas cap first. Sometimes, overfilling the gas tank will cause the MIL to come on. If the MIL comes on immediately after the gas tank is filled, recheck the gas cap and then drive the vehicle for a few minutes. The MIL will usually go off after the vehicle is driven a short distance.

To diagnose an enhanced evaporation control system, allow the engine to cool off to the temperature of the outside air. Then start the engine and disconnect the purge hose from the canister. Check for vacuum at the hose, **Figure 21-35.** There should be no vacuum at the hose. Next, reconnect the hose and allow the engine to warm to its normal operating temperature. The engine should go into closed loop operation (O_2 sensor thoroughly warm). Disconnect the purge hose and check for vacuum. Vacuum should be present with the engine in closed loop. If the evaporation control system does not pass this test, make further checks as outlined in the service literature.

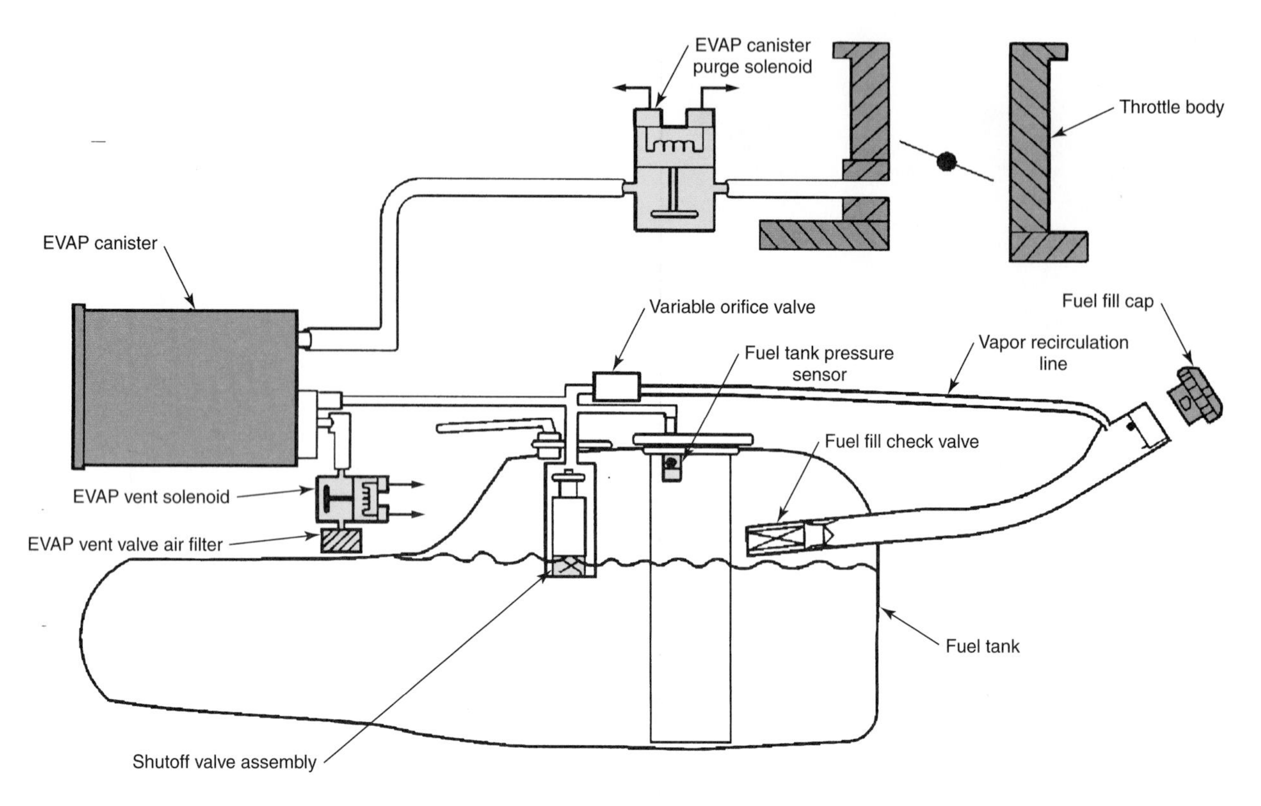

Figure 21-34. Enhanced evaporation control system. (General Motors)

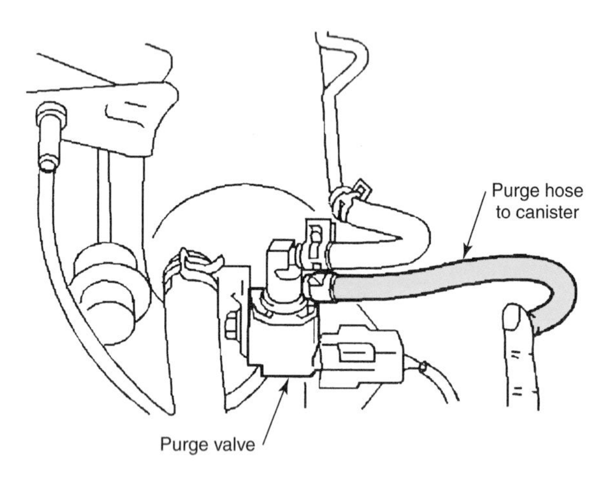

Figure 21-35. Check for vacuum at the hose. There should be no vacuum with the engine cold, and full vacuum with the engine warm. (General Motors)

Evaporation control systems can be pressurizing with nitrogen to check the system for leaks. To make this test, attach the nitrogen tank hose to the EVAP service port. Pressurize the evaporative control system to the specified pressure. Do not exceed this pressure. Then wait the specified time while observing whether the system loses pressure. A pressure drop that is greater than that specified for the time interval indicates a leak in the system. The nitrogen can also be used to flush loose carbon from the system. Consult the proper service literature for this procedure.

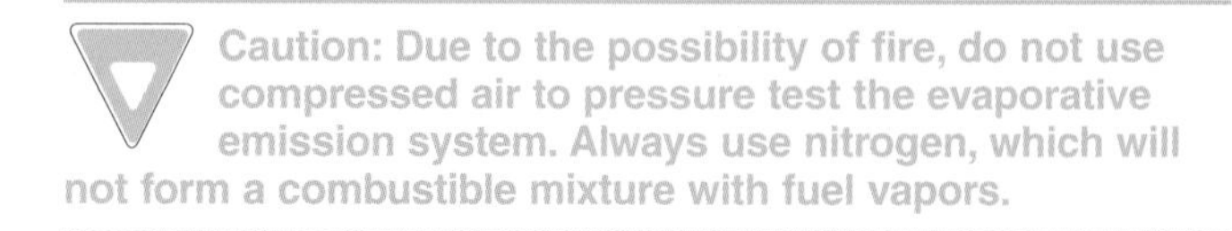

Caution: Due to the possibility of fire, do not use compressed air to pressure test the evaporative emission system. Always use nitrogen, which will not form a combustible mixture with fuel vapors.

Tech Talk

At one time, it was popular to disconnect emission controls. It did not help vehicle performance much, but it was the "in" thing to do. In fact, before emission control laws were tightened, it was possible to advertise that you were in the business of removing emission controls.

Today, however, emission controls are an integral part of the operating system and are controlled by the on-board computer. Bypassing the emission controls is not only illegal, but it is usually impossible. No one with any sense would want to anyway. Thanks to the precise control of all engine systems by the computer, new vehicles run smoother, have more power, and get better gas mileage.

What today's technician must learn is not how to bypass emission controls, but how to work on and with them to restore the vehicle to peak performance.

Summary

Emissions tests are performed by various states to comply with federal emissions laws. Normal vehicle use causes an increase in engine emissions. The main purpose of vehicle emissions testing is to determine whether the vehicle meets federal and state standards. Emissions testing also can be used as a diagnostic tool to correct vehicle defects. Exact emission test procedures vary greatly between states.

Emissions tests can be divided into two categories: enhanced emissions tests and static emissions tests. The enhanced emissions test is sometimes called an IM test or a transient test. Enhanced emissions tests check vehicle emissions as the vehicle is operated on a dynamometer and often include a test of the evaporative emissions controls. Static tests check tailpipe emissions as the engine idles. Vehicles are not tested under load in any static test. In many states, static test procedures include a check of the vehicle gas cap. Another test is the OBD II computer cycle test, which tests engine emissions levels by monitoring the operation of the on-board computer when the OBD II system is used. Various types of test equipment are needed to perform emissions tests.

Emissions can also be checked in the shop using an exhaust gas analyzer. Always consult the emissions label before beginning diagnosis.

The control valve of a thermostatic air cleaner or early fuel evaporation system can be tested with a hand vacuum pump. Many modern vehicles prevent throttle valve icing by diverting some of the engine coolant through the throttle valve body. The coolant passage can be removed and checked for deposits.

Exhaust gas recirculation (EGR) valves reduce oxides of nitrogen (NO_X). Older EGR valves are operated by engine vacuum. Newer valves are solenoid or motor operated. Opening the EGR valve at idle should cause the engine to run roughly. Integrated electronic EGR valves must be tested with a scan tool. Some EGR valves may be cleaned. Do not sandblast an EGR valve. If cleaning the valve does not restore proper EGR operation, replace it.

The air injection system pumps outside air into the exhaust system. A belt-driven pump delivers air through a diverter valve. A check valve keeps exhaust gases from reaching the air injection system. Some systems alternate airflow between the exhaust manifold and catalytic converter. The pulse air injection system uses a check valve that allows air to enter the exhaust system on each exhaust pulse. Air injection system service consists of periodically checking the drive belt and filter, as well as checking overall system operation.

The catalytic converter cleans up the exhaust gases after they leave the engine. A defective converter can be serviced by replacing the entire unit.

Vacuum should flow through the PCV valve when the engine is idling. Covering the PCV valve should cause the engine idle to decrease. If the PCV valve does not rattle when it is shaken, it is defective. Do not attempt to clean a PCV valve and do not interchange PCV valves between engines.

Early evaporation control systems were operated by vacuum valves. Newer systems are operated by solenoids controlled by the ECM. Enhanced evaporation systems are used on OBD II vehicles. Vacuum-operated purge valves should be checked for proper operation. The OBD II evaporation control system has a self-monitoring feature. OBD II evaporation control systems can be checked by pressurizing the system with nitrogen.

Review Questions—Chapter 21

Do not write in this book. Write your answers on a separate sheet of paper.

1. An enhanced emissions test is sometimes called a(n) _____ test.
2. During a(n) _____ emissions test, the engine is not placed under load.
3. The _____ is the most failure-prone part of the evaporative emissions system.
4. In many states, a visibly _____ engine cannot be tested.
5. List the five gases checked in a typical emissions test.
6. What is the purpose of a repair cost waiver program?

Matching

Match the emission control device with the problem it is designed to prevent.

7. Thermostatic air cleaner _____
8. Air injection system_________
9. EGR valve_____
10. PCV valve_____
11. Evaporative emission control _____

(A) Vapor loss from the fuel system
(B) Unburned fuel in the exhaust gases
(C) Throttle valve icing
(D) Knocking at idle
(E) NO_X formation
(F) Release of crankcase blowby

12. The EGR valve is operated by _____.
    (A) intake manifold vacuum
    (B) solenoids and motors
    (C) exhaust back pressure
    (D) Both A and B.
13. In the air injection system, the diverter valve is used to prevent _____.
    (A) NO_X formation
    (B) backfiring
    (C) knocking
    (D) dieseling
14. A(n) _____ mixture can overheat the catalytic converter.
15. Explain why you should never use compressed air to pressurize an evaporation system.

ASE-Type Questions

1. Technician A says that airflow into the engine from the evaporative system should be at least 25 liters (28 quarts) during the test period. Technician B says that airflow during the nitrogen pressure test should be at least one liter. Who is right?
    (A) A only.
    (B) B only.
    (C) Both A and B.
    (D) Neither A nor B.

2. All of the following statements about preparing a vehicle for an emissions test are true *except:*
   (A) cold or overheating engines will give false emissions readings.
   (B) if the MIL is off, the vehicle cannot be tested.
   (C) make visual checks of the engine before entering the testing area.
   (D) check the gas cap before entering the testing area.
3. Technician A says that the simplest exhaust gas analyzer can measure HC levels only. Technician B says that newer exhaust gas analyzers can measure O_2 levels. Who is right?
   (A) A only.
   (B) B only.
   (C) Both A and B.
   (D) Neither A nor B.
4. Air entering the intake manifold is heated to prevent which of the following?
   (A) Fuel condensation in the manifold.
   (B) Water condensation in the engine.
   (C) Preignition.
   (D) Detonation.
5. Which of the following is not a method of cleaning an EGR valve and EGR passages?
   (A) Tapping with a hammer.
   (B) Blowing out with compressed air.
   (C) Sandblasting.
   (D) Using a wire brush.
6. Technician A says that the air injection system diverter valve reduces the chance of backfiring. Technician B says that the air injection system check valve prevents pump air from entering the exhaust manifold. Who is right?
   (A) A only.
   (B) B only.
   (C) Both A and B.
   (D) Neither A nor B.
7. Technician A says that a defective diverter valve can cause backfiring. Technician B says that an excessively lean mixture can cause overheating of a catalytic converter. Who is right?
   (A) A only.
   (B) B only.
   (C) Both A and B.
   (D) Neither A nor B.
8. The PCV system draws fumes from the _____ back into the cylinders for burning.
   (A) exhaust manifold
   (B) crankcase
   (C) intake manifold
   (D) carbon canister
9. Technician A says that all PCV valves can be disassembled and cleaned. Technician B says that some PCV valves can be soaked in solvent to clean them. Who is right?
   (A) A only.
   (B) B only.
   (C) Both A and B.
   (D) Neither A nor B.
10. All the following statements about enhanced emission control system testing are true *except*:
   (A) the purge valve should allow vacuum to the canister when the engine is cold.
   (B) OBD II systems self-check the enhanced emission control system.
   (C) the purge valve should allow vacuum to the canister when the system is in closed loop mode.
   (D) the purge valve should block vacuum flow when the engine is cold.

Suggested Activities

1. Inspect a vehicle and locate the emissions systems components. List all components and identify the function of each. Determine whether the vehicle has any internal emission controls, such as engine modifications.
2. Locate an emissions label on a late-model vehicle. Identify and study the instructions for setting timing or idle. Explain to your classmates what the information means.
3. Obtain a copy of the latest Environmental Protection Agency (EPA) publications on air pollution, fuel mileage, and oxygenated gasoline. The EPA's phone number is listed in your local telephone book. Review the materials and write a short report summarizing them.
4. Check the operation of a vehicle's EGR valve, air injection system, or thermostatic air cleaner by following the instructions given in the appropriate service manual.

22

Engine Mechanical Troubleshooting

After studying this chapter, you will be able to:

- Summarize preliminary test steps.
- Perform a compression test.
- Use a cylinder leakage detector.
- Perform a vacuum test.
- Check engine oil pressure.
- Diagnose engine mechanical problems.

Technical Terms

Preliminary checks	Vacuum gauge
Compression test	Cranking vacuum test
Compression pressure	Oil pressure sender
Cylinder leakage detector	

This chapter explores and explains the process of engine mechanical troubleshooting. Mechanical troubleshooting involves checking the overall condition of the engine. Checks and measurements of specific internal engine parts are covered in the next few chapters. Other chapters deal with the diagnosis and repair of specific engine-related systems, such as fuel, ignition, emissions, exhaust, and cooling systems.

Steps for Diagnosing Engine Problems

The following sections outline the procedures for isolating the causes of engine problems. These procedures help the technician determine which engine component is defective.

Work Logically

Always proceed logically when attempting to locate engine problems. This is true even when the problem seems obvious. For example, a locked-up engine may seem to require a complete engine teardown. However, the actual problem may be that an excessively long flywheel bolt has been installed, jamming the flywheel to the engine block. Jumping to an immediate conclusion or replacing the first component that comes to mind rarely solves a problem. To properly diagnose an engine problem, concentrate on the probable causes and make simple checks first. Proceeding logically and testing everything before replacing anything will pay big dividends in the long run.

Preliminary Checks

Before checking an engine's mechanical condition, several important nonengine systems must be checked, depending on the symptoms. These checks may uncover a problem cause that is unrelated to the engine's mechanical condition. The units tested by these ***preliminary checks,*** if faulty, can produce a variety of apparent faults in other related systems. Failure to make these preliminary checks can lead the technician to disassemble an engine for no reason.

Note: These procedures are covered in later chapters.

Solicit the Owner's Comments

The most important preliminary check is to find out the exact nature of the problem. Before starting diagnostic procedures, ask the vehicle's owner or driver to describe the problem. An owner's comments and answers to specific questions can provide valuable clues. It may also be helpful to take a short test-drive to experience the problem firsthand.

Try to relate the driver's comments to a specific problem, such as hard starting, poor acceleration, or missing. Next, determine the engine and environmental conditions related to the problem. Determine whether the problems occur occasionally or all the time. Ask the owner about recent changes to the car, including new parts, repair work, or a change in fuel. Identify any sounds related to the problem, such as a backfire, howl, grind, or whistle. You may choose to use a set of prepared questions to ensure that all important areas are discussed.

Make Preliminary Checks

The following checks may reveal the problem without further testing. Depending on the nature of the problem, it may be logical to skip certain checks.

- Ignition system—Check the condition of the spark plugs, wires, coil rotor, distributor cap, primary wiring, and other ignition-related components.
- Fuel system—Check the condition of the fuel injection system or carburetor, the air and fuel filters, and the fuel lines. Check for leaks around all lines and fittings.

- Computer control system—Retrieve any onboard computer trouble codes, and track down any computer-related defects.
- Emission control systems—Check the operation of the PCV and EGR valves, smog pump, evaporative emission controls, and other emission components.
- Exhaust system—Study the exhaust smoke for signs of abnormal engine operation. Black exhaust smoke indicates a rich mixture. White smoke (after vehicle has warmed up) may indicate that coolant is leaking into the combustion chambers. Blue-gray smoke indicates that the engine is burning oil.
- Battery and electrical system—Check battery voltage and perform a capacity test. Check the starter, battery, coil, and alternator leads for good connections. Check starter and alternator/regulator operation.
- Oil and coolant levels—Check the dipstick for the amount and condition of engine oil. Look for leaking gaskets or seals. Check the radiator coolant level. Check the radiator and all hoses for leaks. Look for signs of oil in the radiator or coolant in the engine oil.
- Fasteners—Tighten the intake manifold and throttle body fasteners. If needed, torque the head bolts.
- Fittings—Check and tighten fuel line and vacuum fittings.

Check the Engine's Mechanical Condition

If the checks listed above do not pinpoint the problem, check the mechanical condition of the engine. Directions for conducting a compression test, a cylinder leakage test, a vacuum test, and an oil pressure test are presented in the following sections.

The technician should check compression in all cylinders. Checking one or two cylinders where the plugs are easy to remove is not sufficient. If the compression (or cylinder leakage) test indicates a problem, be sure to inform the owner of the need for internal repairs. If the engine fails in the compression test, run a cylinder leakage test to determine just where the trouble lies.

Performing a Compression Test

Before performing a ***compression test,*** operate the engine until normal operating temperature is reached. To perform the compression test, proceed as follows:

1. Remove the spark plugs (this is covered in more detail in Chapter 16, Ignition System Service) and the air cleaner. On a diesel engine, remove the glow plug.
2. Block the throttle valve in the wide-open position. The choke must also be fully opened if the vehicle has a carburetor.
3. Make sure the battery is fully charged.
4. Disable the ignition system by disconnecting the ignition primary lead or by grounding the distributor end of the coil secondary wire.
5. Attach a remote starter control. If the key switch will be damaged by remote cranking when in the lock or off position, turn the key to the on position.
6. Insert a compression gauge tightly in the spark plug or glow plug hole.
7. Crank the engine until the gauge shows no further rise in pressure. This will require at least 4 or 5 compression strokes. See **Figure 22-1.**

On a cylinder with good compression, the first compression stroke forces the gauge indicator needle a considerable distance up the scale. Succeeding strokes continue to raise it, until the cylinder's maximum compression is registered on the gauge. Record the highest reading for each cylinder. For engines with plug holes that are difficult to reach, a compression gauge having an offset tip or a flex hose is useful. See **Figure 22-2.**

When checking compression on diesel engines, remember that compression levels are considerably higher than those produced in a gasoline engine. Note that the gauge in **Figure 22-3** is not hand-held, but is securely attached. In

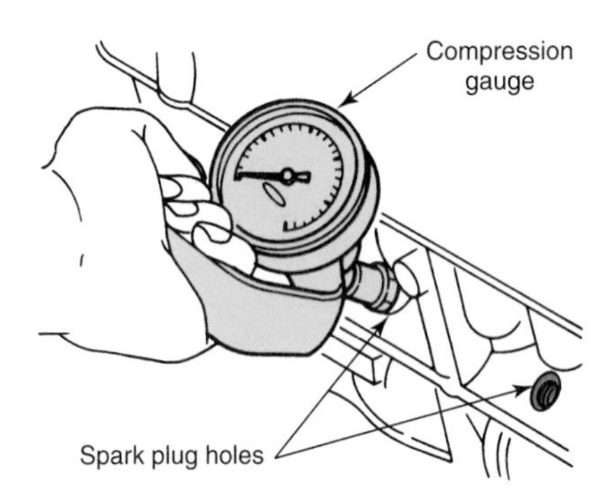

Figure 22-1. A compression gauge can be used to check a gasoline engine's mechanical condition (valves, rings). (Nissan)

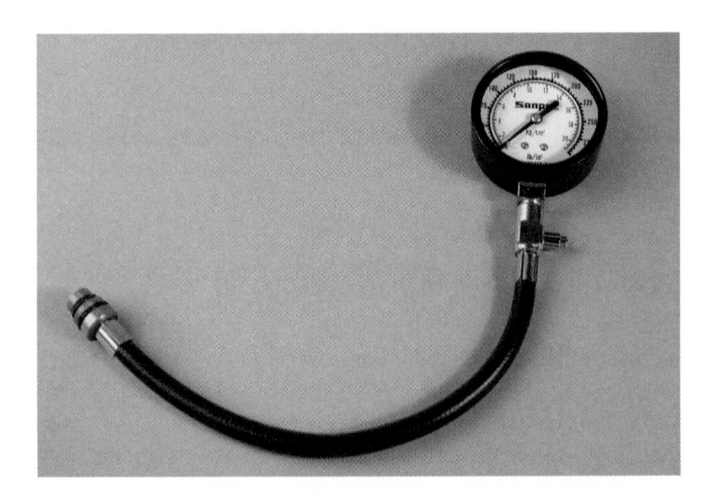

Figure 22-2. This compression gauge has a flexible hose, which allows easier access to spark plug holes and injector holes.

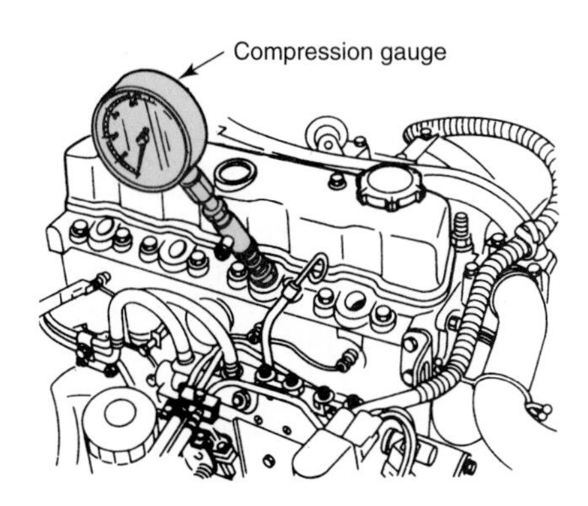

Figure 22-3. This diesel-engine compression gauge is firmly attached to provide a positive, leakproof seal. (Nissan)

some setups, the gauge adapter is bolted in place. In other applications, the gauge is threaded into a glow plug opening. Follow the manufacturer's recommendations.

Interpreting Compression Readings

Examine the readings for all cylinders and compare them to specifications. Also, check for variation between cylinders. The lowest cylinder reading should not be more than 10%–15% below the highest cylinder reading. Some manufacturers' specifications permit greater variation. Variations between cylinders have a more adverse effect on engine performance than overall readings that are even but slightly below specifications.

Low Compression Readings

When taking compression readings, watch the action of the gauge needle. When the needle raises a small amount on the first stroke and only a little more on succeeding strokes, one of the valves is probably not sealing because it is burned, warped, or sticking.

A low buildup on the first stroke with a gradual buildup to a moderate reading on succeeding strokes may indicate worn, stuck, or scored rings. If two adjacent cylinders are low, a blown head gasket or warped head-to-block surface could be responsible.

To help pinpoint the cause of low compression in a gasoline engine, add one tablespoon of heavy engine oil (30W minimum) to the cylinder with a low reading.

Caution: Diesel engines have very high compression. At top dead center, there is a minimum of space between the piston and cylinder head. An excess amount of oil can cause serious engine damage because the oil does not compress. Some diesel makers forbid the use of any oil. Oil should only be used in a diesel engine when recommended by the manufacturer.

After adding the oil, reattach the compression gauge. Crank the engine for a few extra compression strokes and watch the gauge. If the compression goes up a noticeable amount, worn rings are indicated. If the addition of the oil produces no significant rise, valve trouble, a broken piston, or a blown gasket is probably causing the low reading.

Sometimes, sticking valves or rings may free up if special oil treatments are added to the crankcase and the engine is operated for a period of time. Some treatment products can be poured into the throttle body with the engine running. Carefully follow the instructions on the treatment container to prevent damage to the catalytic converter. Some manufacturers do not recommend the use of these types of additives.

High Compression Readings

If the ***compression pressure*** exceeds specifications, the engine may have been modified to increase the compression. There could also be a buildup of carbon on the head of the piston and on the combustion chamber walls. If carbon buildup is causing pinging that cannot be stopped by retarding the timing or by switching to a higher-octane gasoline, the engine should be disassembled and the carbon should be removed.

Another sign of excessive carbon is "dieseling" (engine continues to run after ignition is turned off). Dieseling action can be caused by glowing bits of carbon in the combustion chambers. Hard or slow cranking can also indicate excessive compression from carbon buildup.

Using a Cylinder Leakage Detector

When a cylinder produces a low compression reading, it is often helpful to perform a cylinder leakage test. Unlike a compression gauge, the ***cylinder leakage detector*** can be used to pinpoint the exact cause of the problem. The leakage detector is inserted in the spark plug hole. The piston is then brought up to dead center on the compression stroke, and compressed air is admitted to the cylinder. Once the combustion chamber is pressurized, a special gauge will read the percentage of leakage. Refer to **Figure 22-4.** Leakage exceeding 20% is considered excessive.

While the air pressure is retained in the cylinder, listen for the hiss of escaping air. A leak by the intake valve is audible in the throttle body or carburetor. A leak by the exhaust valve can be heard at the tailpipe. Leakage past the rings is audible at the oil filler hole or the PCV (positive crankcase ventilation) connection. If air passes through a blown gasket to an adjacent cylinder, the noise will be evident at the spark plug hole of the cylinder into which the air is leaking. A stream of bubbles in the radiator may indicate a crack in the block or gasket leakage into the cooling system.

Using a Vacuum Gauge

A ***vacuum gauge,*** **Figure 22-5,** is a useful diagnostic tool. It can be used to detect vacuum leaks, sticking valves, worn rings, clogged exhaust, and incorrect valve or ignition timing. However, great care must be used in interpreting the readings and actions of the gauge indicator needle. In many instances, the readings point to several possible problems. Further checking is required to isolate the exact cause.

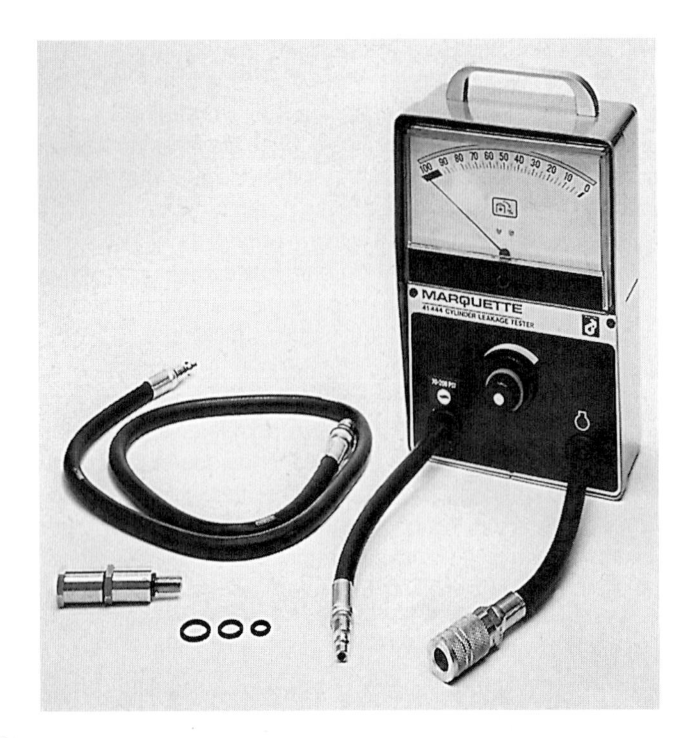

Figure 22-4. One type of cylinder leakage test tool is shown here. (Marquette)

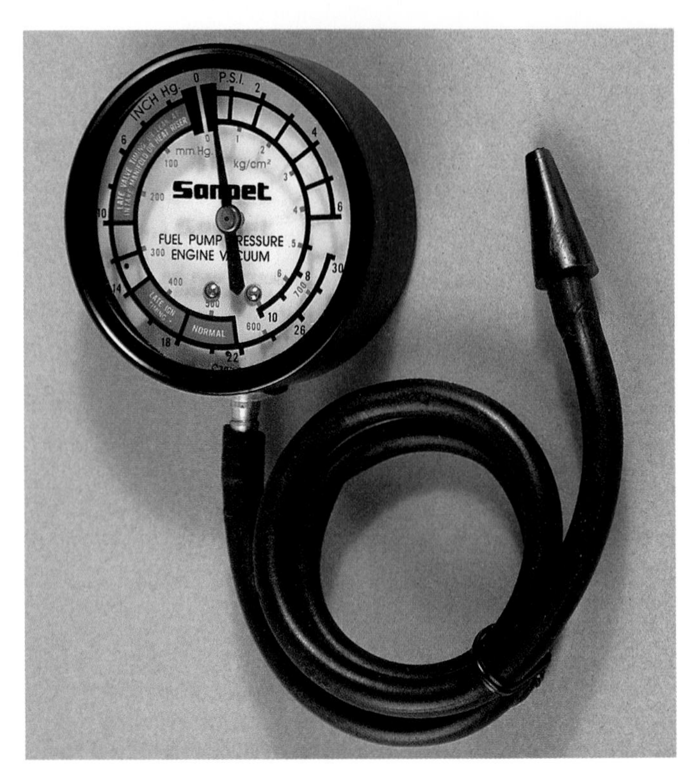

Figure 22-5. Typical vacuum gauge.

Vacuum gauges are calibrated in inches of mercury, which is a common way to express vacuum, or negative pressure. Mercury is usually abbreviated as its chemical symbol, Hg. The vacuum gauge reads the difference in pressure between the air in the intake manifold and the outside air. The vacuum reading for a given engine can be affected by fuel system adjustment and condition, ignition timing, valve timing, valve and valve guide condition, cylinder wear, piston and ring condition, vacuum leaks, PCV condition, exhaust restrictions, and spark plug adjustment.

Attaching a Vacuum Gauge

When possible, connect the vacuum gauge to the intake manifold. Some manifolds incorporate a plug that may be removed so a vacuum line adapter can be installed. If no opening is provided, use a T-fitting to connect the gauge hose to the hose of another vacuum-operated unit. See **Figure 22-6.**

Cranking Vacuum Test

When performing the ***cranking vacuum test,*** the engine must be at normal operating temperature. Connect the vacuum gauge as shown in Figure 22-6. Make sure the throttle valve is fully closed. Ground the coil high-tension wire or unplug the ignition module if recommended. Crank the engine **several times** and average the readings. Cranking speed must be up to specifications.

A relatively steady (some pulsation is normal), high vacuum reading indicates an absence of vacuum leaks and good ring and valve action. A low but fairly steady reading can mean vacuum leaks, worn intake valve guides, poor compression, improper valve timing, and other problems.

An erratic (uneven) reading can point to burned or sticky valves, a damaged piston, or a blown gasket. If the cranking vacuum test indicates problems, conduct a cylinder leakage test. If cylinder leakage test equipment is not available, conduct a compression test.

Vacuum Test with Engine Running

Bring the engine to normal operating temperature and connect the vacuum gauge to the intake manifold. Run the engine at the specified idle speed.

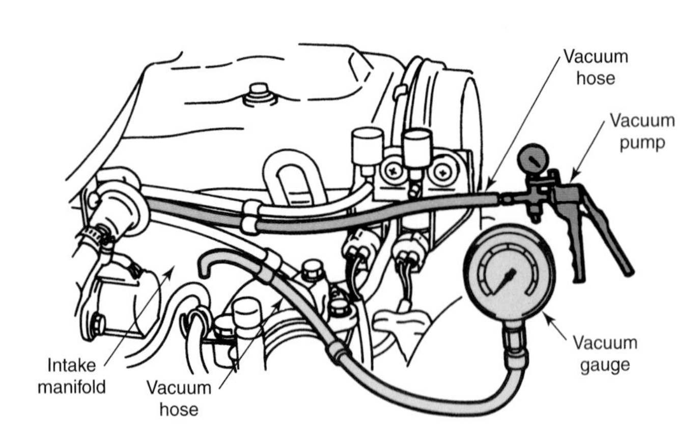

Figure 22-6. The vacuum gauge should be connected to the intake manifold. Note the vacuum pump, which is used to supply a vacuum to certain parts during some vacuum tests. Follow the manufacturer's recommendations for accurate readings. (General Motors)

The vacuum gauge should read from 15–22 inches of mercury, depending on the engine and the altitude at which the test is performed. As altitude increases, vacuum readings decrease. Therefore, subtract one inch Hg from the specified reading for every 1000 feet of elevation above sea level. The reading should be steady. High-performance engines with considerable valve overlap tend to produce a lower, more erratic vacuum reading, especially at idle.

Caution: If the engine is equipped with a turbocharger, do not run the engine at a speed high enough to cause the turbocharger to begin boosting manifold pressure. This positive pressure would be transmitted to the vacuum gauge, ruining it.

Interpreting Vacuum Gauge Readings

Observing the vacuum gauge while the engine idles helps pinpoint trouble areas. Remember, however, that many vacuum gauge readings can be caused by more than one problem. Always conduct all other appropriate tests before arriving at a final diagnostic decision.

Figure 22-7 shows a number of vacuum gauge readings. Note that the gauge area from 15 to 22 in. Hg is marked in color. This is the normal range for most vehicles. Check with the manufacturer for a vehicle's actual specifications.

The typical problems listed below are numbered 1 through 16. These numbers correspond to numbers in Figure 22-7. Study the gauge readings and then read the corresponding descriptions.

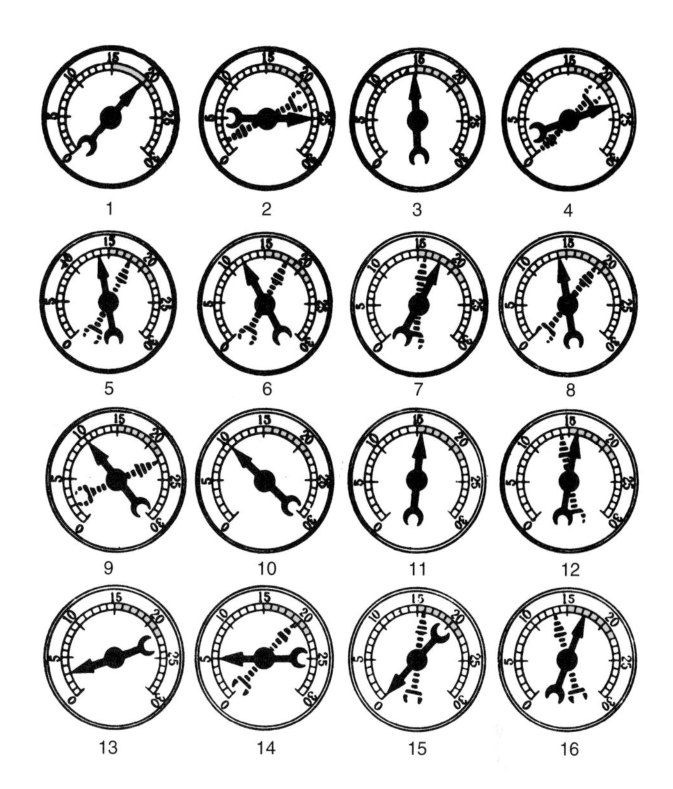

Figure 22-7. A variety of typical vacuum gauge readings are shown here. (Nissan)

1. The needle holds steady between 15 and 22 in. Hg as engine idles: *normal reading.*
2. When the throttle is rapidly opened and closed, the needle (dotted) drops to a low (but not zero) reading. When the throttle is suddenly released, the needle snaps back up to a higher-than-normal figure: *normal reading during rapid acceleration and deceleration.*
3. Needle registers as low as 15 in. Hg, but is relatively steady at engine idle: *normal for high lift cam with large overlap.*
4. When the engine is accelerated the needle (dotted) drops to 0 in. Hg. During deceleration, the needle snaps back to a higher-than-normal reading for a very brief time: *worn rings or diluted oil.*
5. The needle (dotted) remains steady at a normal vacuum as the engine idles, but it occasionally drops about 4 in. Hg sharply, and then quickly returns to normal: *one or more valves may be sticking.*
6. A regular, evenly spaced drop of the needle at idle: *one or more burned or warped valves, or insufficient tappet clearance.*
7. The needle indicates a small but regular vacuum drop at idle: *one or more valves are not seating.*
8. The needle oscillates (swings back and forth) over about a 4 in. Hg range at idle speed. As engine speed is increased, the needle becomes steady: *worn valve guides.*
9. The reading at idle is relatively steady, but the needle oscillation becomes more pronounced as engine rpm is increased: *weak valve springs.*
10. The needle holds a steady but low reading at idle: *late valve timing.*
11. The needle maintains a steady but low reading at idle: *retarded ignition timing.*
12. The needle indicates regular, small pulsations at idle: *spark plug gaps too small or the ignition primary system is not producing enough voltage.*
13. The needle maintains a low, steady reading: *intake manifold or throttle mounting flange gasket leak.*
14. The needle experiences a regular drop of four to eight inches: *blown head gasket or warped head-to-block surface.*
15. When the engine is first started and idled, the reading is normal. However, as the engine rpm is increased, the needle slowly drops to a low reading, sometimes as low as 0 in. Hg: *exhaust back pressure caused by a clogged muffler or kinked tailpipe.*

Note: Excessive exhaust clogging may cause the needle to drop to a low reading at idle.

16. The needle moves slowly back and forth: *improper fuel mixture.*

Note: Improper fuel mixture may be caused by problems in the fuel injection or computer control system or by a maladjusted or defective carburetor.

These examples are typical. The information, if used with care, should be helpful in locating trouble on actual jobs.

Checking Oil Pressure

Oil pressure is critical to the operation of the engine. It prevents premature engine wear by delivering oil to pressure points. It also fills the hydraulic lifters or hydraulic lash adjusters, allowing them to properly operate the valves. An early sign of low oil pressure is often noisy valves. If an engine is operated for any time without oil pressure, it will be destroyed.

Before testing oil pressure, make sure that the engine's oil level and idle speed are satisfactory. If these are OK, obtain an oil pressure gauge of the proper range. Most engine lubrication systems develop no more than 80 psi (552 kPa), so a 0 psi–100 psi (0 kPa–689.5 kPa) gauge is satisfactory for most oil pressure testing.

Note: If the oil pressure specifications are given in kilopascals (kPa) and your gauge reads only pounds-per-square-inch (psi), multiply the psi reading by 6.895 to obtain the kPa figure.

Next, locate the engine ***oil pressure sender.*** This device sends the pressure signal to the dashboard indicator. It is often located near the oil filter or on the side of the engine block. Carefully remove the sender and install the gauge using the proper threaded adapter, **Figure 22-8.** Use of an incorrect adapter will cause oil to spray when the engine is started.

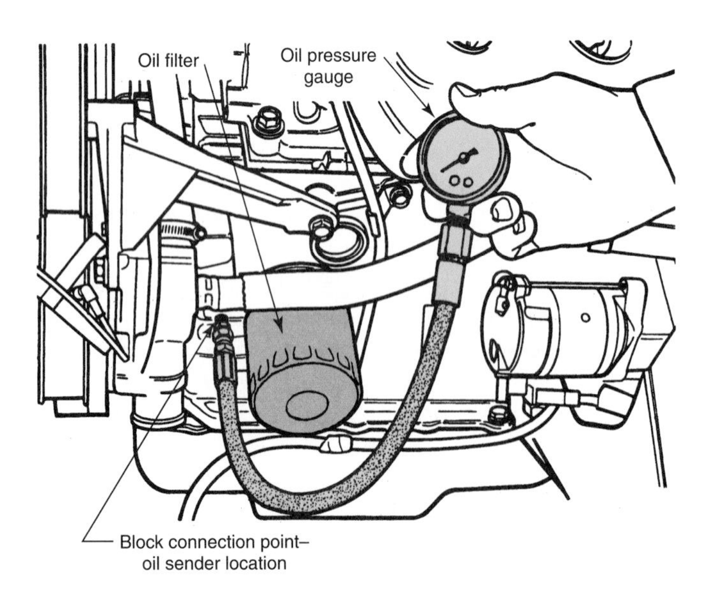

Figure 22-8. An oil pressure gauge should be connected to the oil pressure sender socket. (DaimlerChrysler)

Start the engine and observe the oil pressure. Compare the pressure readings to the specifications for the particular engine. As a general rule, pressure should be at least 20 psi (138 kPa) at idle on a warm engine. Oil pressure should increase as the engine speed increases.

If the oil pressure is low, the oil pump could be defective, the pressure regulator could be stuck open, or there could be internal leaks in the engine's oiling system. In addition, the oil pickup system could be clogged or leaking air or the oil could be too thin. If oil pressure is too high, the pressure regulator could be stuck or the oil could be too heavy.

Engine Noise Diagnosis

The noises that can be produced by the engine and engine accessory systems are numerous. Whines, squeals, knocks, rattles, and many other sounds can come from the engine. They can be loud or soft, sharp or indistinct, metallic or nonmetallic. Accurate diagnosis of engine noises takes a great deal of practice, and even the experienced technician can be puzzled. Try to locate the source and cause of any unusual engine noises. A technician's stethoscope is very helpful in tracking down the source of noises. When the engine is torn down, check to see how accurate your diagnosis was.

Be careful when making a diagnosis based on sounds. Do not recommend an engine teardown unless you are positive that internal engine parts are causing the noise. Remember that other vehicle parts, such as the water pump, fan, alternator, and air conditioning compressor can make a great number of noises. These components can be quickly checked by disconnecting their drive belt(s). The belts themselves can be the source of squeaks and ticking noises. Other noises can come from the flywheel, torque converter, or other parts of the driveline. Broken engine mounts can cause heavy metallic noises and can cause the fan or pulleys to contact stationary vehicle parts. Exhaust pipes, mufflers, and tailpipes can cause various thumps, clangs, and rattles. Exhaust leaks are often the source of ticking noises during acceleration, which can be mistaken for valve train noises.

Tech Talk

Some of the most overlooked tools in your possession are the cheapest: a pencil and paper. Why is it important to have a pencil and paper in your toolbox? They can be used to write down the readings that you get when using your expensive measuring tools. Do not try to remember readings. In addition, writing the readings down gives you a record if the owner wants to see what is wrong with his or her vehicle.

Summary

Always proceed logically when attempting to locate engine problems. Before starting, ask the vehicle's owner or driver for any comments regarding engine performance. Also, make the necessary preliminary checks of nonengine systems.

The technician should check the compression in all cylinders. Variations between cylinders affects engine performance more than low overall readings. Possible causes of low compression are defects in the valves, rings, pistons, head, block, or head gasket. Excessively high compression is usually caused by carbon buildup in the cylinder head. When a cylinder produces a low reading, a cylinder leakage detector can pinpoint the exact problem.

A vacuum gauge can be used to detect vacuum leaks, sticking valves, worn rings, clogged exhaust, and incorrect valve or ignition timing. Incorrect vacuum gauge readings may be due to several possible problems, and further checking is required to isolate the exact problem.

Proper oil pressure prevents premature engine wear by delivering oil to pressure points. Oil also fills the hydraulic lifters or hydraulic lash adjusters. If an engine is operated for any time without oil pressure, it will be destroyed. The oil pressure gauge allows the technician to determine whether the engine has an oil pressure problem. Oil pressure problems are caused by low oil levels, worn oil pumps, clogged oil filters or intake strainers, or internal pressure leaks.

Review Questions—Chapter 22

Do not write in this book. Write your answers on a separate sheet of paper.

1. When taking a compression reading, the technician should crank the engine through at least _____ compression strokes.
   (A) 2
   (B) 3
   (C) 4
   (D) 6
2. Excessive compression pressure could indicate a buildup of carbon in the _____.
   (A) combustion chamber
   (B) oil galleries
   (C) ring groves
   (D) oil pickup tube
3. A steady vacuum gauge reading of 12 in. Hg at idle could indicate _____.
   (A) normal engine operation
   (B) a burned valve
   (C) an over-rich mixture
   (D) late valve timing
4. Installing a high-lift cam with more valve overlap would have what effect on the engine vacuum at idle?
   (A) Vacuum would be lower but steadier.
   (B) Vacuum would be higher and steadier.
   (C) Vacuum would be lower and more erratic.
   (D) Vacuum would be higher and more erratic.
5. The technician should subtract one inch of vacuum from vacuum specifications for every _____ feet of elevation above sea level.
   (A) 500
   (B) 1000
   (C) 2000
   (D) 10,000
6. If an engine comes into the shop with low oil pressure, what is the first thing that the technician should check?
   (A) Oil pump condition.
   (B) Pickup screen condition.
   (C) Engine idle speed.
   (D) Engine oil level.
7. List five possible causes of low cylinder compression.
8. List four engine defects that can be identified with a vacuum gauge.
9. List four possible causes of low oil pressure.
10. Which of the following could cause high oil pressure?
   (A) Internal oil passage leaks.
   (B) Plugged oil filter.
   (C) Low oil level.
   (D) Stuck pressure regulator valve.

ASE-Type Questions

1. Technician A says that owners' comments or answers to questions usually mislead the technician. Technician B says the fuel and ignition systems should be eliminated as sources of problems before the engine's mechanical condition is checked. Who is right?
   (A) A only.
   (B) B only.
   (C) Both A & B.
   (D) Neither A nor B.
2. The compression test and/or cylinder leakage test can be used to determine all of the following engine problems, *except:*
   (A) a defective oil pump.
   (B) a burned valve.
   (C) worn rings.
   (D) a blown head gasket.
3. Technician A says that it is more important for compression readings to be the same in all cylinders than for all cylinders to be up to the specified pressure. Technician B says that adding oil to the cylinder will raise the compression reading somewhat if the valves are burned. Who is right?
   (A) A only.
   (B) B only.
   (C) Both A & B.
   (D) Neither A nor B.

4. When two adjacent cylinders give similar but low compression readings, what is the most likely cause?
   (A) Sticking valves.
   (B) Burned valves.
   (C) Worn rings.
   (D) Blown head gasket.
5. When testing compression on a diesel engine, adding a large amount of oil to a cylinder will _____.
   (A) damage the engine
   (B) help the engine crank over faster
   (C) give a more accurate gauge reading
   (D) produce a low reading
6. Technician A says that low compression pressure on two adjacent cylinders could indicate carbon buildup on those cylinders. Technician B says that carbon buildup can cause pinging. Who is right?
   (A) A only.
   (B) B only.
   (C) Both A & B.
   (D) Neither A nor B.
7. An engine is idling. The vacuum gauge reads a steady 20 in. Hg. Which of the following could be the cause.
   (A) retarded ignition timing.
   (B) retarded valve timing.
   (C) severe vacuum leak.
   (D) normal engine operation.
8. Technician A says that the hydraulic lifters are kept filled by engine oil pressure. Technician B says that low oil pressure could cause the valves to be noisy. Who is right?
   (A) A only.
   (B) B only.
   (C) Both A & B.
   (D) Neither A nor B.
9. Which of the following noises is greatly reduced when the spark plug wire is removed?
   (A) Connecting rod knock.
   (B) Valve tapping.
   (C) Piston slap.
   (D) Both A & C.
10. Technician A says that belt-driven accessories can cause noises that may be mistaken for internal engine noises. Technician B says that many exhaust system problems can be mistaken for internal engine problems. Who is right?
   (A) A only.
   (B) B only.
   (C) Both A & B.
   (D) Neither A nor B.

Suggested Activities

1. Perform a compression test. Write down the readings for each cylinder. Calculate the percentage difference between the highest reading and the lowest reading.
2. Check and record engine oil pressure at various engine speeds.
3. Convert pressure gauge readings from psi to kPa or from kPa to psi.
4. Use a vacuum gauge to diagnose engine problems. Write down the readings at various engine speeds. Discuss the readings from the above tests with other members of your class and decide what they indicate about engine condition.

Engine and Engine System Problem Diagnosis Charts

Problem: Low Compression	
Possible Cause	**Correction**
1. Worn or stuck piston rings.	1. Replace rings, clean piston oil drain holes.
2. Broken rings or pistons.	2. Replace rings and pistons.
3. Bent connecting rod.	3. Replace rod(s).
4. Incorrect connecting rods or pistons.	4. Install new rods or pistons.
5. Blown head gasket.	5. Replace gasket, check for warped head/block.
6. Cracked head or block.	6. Replace head or block.
7. Valves burned or stuck, eccentric (out-of-round) seats.	7. Replace valves, repair seats.
8. Broken timing chain or belt.	8. Replace chain or belt.
9. Broken camshaft.	9. Replace camshaft.
10. Incorrect valve timing.	10. Retime valve train.
11. Valves set too tight.	11. Adjust valves.

Problem: Low Oil Pressure	
Possible Cause	**Correction**
1. Low oil level.	1. Add oil.
2. Improper oil viscosity.	2. Use proper viscosity oil.
3. Diluted oil.	3. Change oil, check for cause of dilution.
4. Camshaft, main, or connecting rod bearings worn.	4. Install new bearings.
5. Crankshaft or camshaft journals worn.	5. Grind journals.
6. Oil pump worn.	6. Replace or rebuild oil pump.
7. Pressure relief valve spring weak.	7. Replace spring or add washers.
8. Oil pump intake clogged.	8. Clean screen and pipe; tighten connection.
9. Hole in oil pickup pipe.	9. Replace pipe.
10. Oil line connection leak.	10. Tighten connection.
11. Defective gauge (direct pressure type).	11. Replace gauge.
12. Defective sender or gauge (electric type).	12. Replace sender or gauge.
13. Plugged oil filter.	13. Replace filter and oil.
14. Improperly installed bypass oil filter.	14. Install correctly.
15. Low idle speed.	15. Set speed to specs.
16. Oil galleys clogged.	16. Clean out or replace block.
17. Loose or missing oil galley plugs.	17. Tighten and/or install plugs.

Problem: Excessive Oil Pressure	
Possible Cause	**Correction**
1. Oil too viscous (heavy).	1. Change to lighter oil.
2. Pressure relief valve spring under too much tension.	2. Reduce spring pressure.
3. Pressure relief valve stuck.	3. Clean valve.
4. Main oil line from pump clogged.	4. Clean line.
5. Defective gauge (direct pressure type).	5. Replace gauge.
6. Defective sender or gauge (electric type).	6. Replace sender or gauge.

Problem: Engine Oil Contamination	
Possible Cause	**Correction**
1. Blowby caused by worn piston rings.	1. Install new rings.
2. Blowby—excessive piston or cylinder wear.	2. Rebore and install new pistons.
3. Coolant entering oil—cracked block or head.	3. Seal leak or replace part. Drain oil, flush, and refill.
4. Coolant entering oil—blown head gasket.	4. Replace gasket. Check head and block surface for warpage. Drain oil, flush, and refill.
5. Fuel entering oil—excessive choking (carburetor only).	5. Adjust choke.
6. Fuel entering oil—float level too high (carburetor only).	6. Adjust float level.
7. Fuel entering oil—float valve leaks (carburetor only).	7. Replace float valve needle and seat.
8. Fuel entering oil—fuel pump diaphragm cracked.	8. Replace diaphragm or pump.

(continued)

Engine and Engine System Problem Diagnosis Charts *(continued)*	
9. Water entering oil—crankcase condensation. 10. Rapid formation of sludge.	9. Clean or replace PCV valve. 10. Clean or replace PCV valve; use detergent oil; raise engine operating temperature if too cold; check for missing thermostat, short trip driving.
Problem: No Oil Pressure	
Possible Cause	**Correction**
1. Oil level too low.	1. Add oil.
2. Oil pump inoperative.	2. Repair or replace pump, check pump drive mechanism.
3. Defective gauge (direct pressure type).	3. Replace gauge.
4. Defective sender or gauge (electric).	4. Replace sender or gauge.
5. Wire between sender and gauge disconnected.	5. Connect wire.
6. Pump intake screen or tube clogged.	6. Clean screen and tube.
7. Pressure relief valve stuck.	7. Clean relief valve, check for free operation.
8. Line to sender or gauge clogged.	8. Clean line.
Problem: Excessive Oil Consumption	
Possible Cause	**Correction**
1. Oil too light.	1. Install heavier oil.
2. Diluted oil.	2. Change oil, refer to engine oil contamination chart.
3. Oil level too high.	3. Lower oil level.
4. Worn or clogged rings.	4. Install new rings.
5. Excessive piston and cylinder wear.	5. Rebore and install new pistons.
6. Worn valve guides.	6. Replace guides or ream to next oversize stem.
7. Worn valve stems	7. Replace valves.
8. Excessive speed.	8. Advise driver to reduce speed.
9. Cylinder torque distortion.	9. Torque head fasteners correctly.
10. Worn bearings—excess oil throw-off.	10. Replace bearings.
11. Clogged PCV system.	11. Clean PCV system.
12. Excessive oil pressure.	12. Reduce pressure.
13. Engine running too hot.	13. Reduce operating temperature.
14. Rear main seal leak.	14. Replace seal.
15. Crankshaft front seal leak.	15. Replace seal.
16. Pan gasket leak.	16. Replace gasket or tighten fasteners.
17. Valve cover gasket leak.	17. Tighten fasteners or replace gasket.
18. Timing gear cover leak.	18. Tighten fasteners or replace gasket.
19. Fuel pump mounting flange loose.	19. Tighten fasteners.
20. Oil filter cover leak.	20. Tighten cover or replace gasket.
21. External line leak.	21. Repair or replace line.
22. Oil pan drain plug leak.	22. Tighten or replace gasket.
23. Oil gallery plug loose.	23. Tighten plug.
24. Oil gauge or sender leak.	24. Tighten or replace.
25. Rear camshaft plug leak.	25. Replace plug.
26. Wrong oil ring design.	26. Install correct rings.
27. Rings installed wrong.	27. Install correctly.
28. Glazed cylinder walls—rings will not seat.	28. Hone walls; install new rings.
29. External leaks.	29. Locate and repair.
30. Piston(s) improperly installed.	30. Install correctly.
31. Improper reading of dipstick.	31. Check with vehicle on level surface; allow sufficient drain-down time; check for possible incorrect dipstick.
32. Damaged turbocharger seals.	32. Replace seals.
33. Loose or broken turbocharger oil feed line(s).	33. Tighten or replace as necessary.
34. Leaking oil cooler.	34. Repair or replace. *(continued)*

Engine and Engine System Problem Diagnosis Charts *(continued)*

Problem: Backfiring in Intake Manifold

Possible Cause	Correction
1. Intake valve not properly seating.	1. Check for broken spring, valve clearance, sticking, seat condition.
2. Lean mixture.	2. Adjust mixture when possible; check fuel injection and computer control system for cause of lean mixture.
3. Cross-firing.	3. Arrange plug wires or install new wires, if needed.
4. Plug wires installed wrong.	4. Connect wires to proper plugs.
5. Carbon tracking in distributor cap.	5. Replace cap and rotor.
6. Choke not closing fully (carburetor only).	6. Adjust choke.
7. Incorrect ignition timing.	7. Set timing to specifications.
8. Incorrect valve timing.	8. Check for jumped timing chain or belt, reset valve timing to specifications.

Problem: Backfiring in Exhaust System

Possible Cause	Correction
1. Turning key off and on while car is in motion.	1. Advise driver to avoid this practice.
2. Coil-to-distributor cap secondary wire or coil shorting.	2. Check wire and coil.
3. Current flow interruption in primary circuit.	3. Check circuit for loose connections and intermittent shorts; check points and condenser where used; check reluctor air gap where adjustable.
4. Incorrect valve timing.	4. Correct valve timing.
5. Air injection system anti-backfire or diverter valve inoperative.	5. Replace valve.

Problem: Starter Will Not Crank Engine (Electrical System OK)

Possible Cause	Correction
1. Defective starter armature or drive.	1. Rebuild or replace starter.
2. Starter drive pinion jammed into flywheel teeth.	2. Remove starter. Install new pinion and replace starter ring gear if needed.
3. Engine crankshaft and/or bearings seized.	3. Grind shaft and replace bearings.
4. Engine bearings too tight.	4. Install correct bearings.
5. Piston-to-cylinder wall clearance too small.	5. Fit pistons correctly.
6. Water pump frozen.	6. Thaw, install antifreeze in cooling system.
7. Insufficient piston ring clearance.	7. Install correct rings.
8. Hydrostatic lock (water in combustion chamber).	8. Remove water. Repair leak.
9. Flywheel or vibration damper bolt too long, contacting engine block.	9. Replace with correct bolt.

Engine Noise Diagnosis

Problem: Noisy Valve Train

Sound Identification: Noisy valves may be identified by a sharp clicking or tapping sound. If excessive tappet clearance exists, the clicking will be very regular and the frequency will increase with engine rpm. Sticking valves and faulty lifters cause intermittent clicking of varying intensity.

Possible Cause	Correction
1. Insufficient lubrication.	1. Provide ample lubrication.
2. Insufficient stem-to-guide clearance.	2. Ream guides to correct size.
3. Warped valve stem.	3. Replace valve.
4. Carboned stem and guide.	4. Clean stem and guide. Replace both if excessive wear is present.
5. Broken valve spring.	5. Replace spring.
6. Weak, corroded, or incorrect spring.	6. Replace with correct spring.
7. Sticking hydraulic lifters.	7. Clean or replace lifters.
8. Improper valve train clearance.	8. Set clearance as specified.
9. Valve seat not concentric with guide or stem.	9. Regrind valve and/or seat.
10. Sticking rocker arm.	10. Clean or replace.
11. Valve lifter loose in bore or chipped.	11. Replace lifter.
12. Valve spring installed incorrectly .	12. Reverse spring position.
13. Excessively low or high oil level in pan.	13. Bring oil level to correct height.
14. Loose or broken rocker arm.	14. Tighten or replace arm.
15. Loose or broken valve seat.	15. Replace insert and stake as recommended.

Problem: Crankshaft Bearing and Flywheel Knocks

Sound Identification: Dull, heavy pound or thud, especially noticeable during periods of heavy engine loading. Frequency is related to crankshaft rpm. The loose bearing may generally be isolated by shorting spark plugs. When the plugs in line with the bearing are shorted, the sound will change. If all the main bearings are quite loose, a great deal more noise will be evident and the frequency will increase. When excessive endplay causes knocking, holding in the clutch usually alters the sound. To test for a loose flywheel, turn off the key and just before the engine stops, turn the key on again. If the flywheel is loose, a distinct knock will occur when the key is turned on.

Possible Cause	Correction
1. Shaft worn.	1. Regrind or replace shaft.
2. Bearing worn.	2. Replace bearing.
3. Thin oil.	3. Change to correct viscosity.
4. Low oil pressure.	4. Correct as required.
5. Excessive endplay.	5. Correct endplay.
6. Sprung crankshaft.	6. Straighten or replace crankshaft.
7. Loose flywheel.	7. Tighten flywheel fasteners.

Problem: Connecting Rod Knock

Sound Identification: The connecting rod knock is usually more evident when the engine is floating (not accelerating or holding back on compression) at speed around 30 mph (48 km/h) in direct drive. The knock, a regular light metallic rap, can either be eliminated or greatly subdued by shorting out the cylinder concerned.

Possible Cause	Correction
1. Crankshaft rod journal worn.	1. Grind journal or replace shaft.
2. Connecting rod bearings worn.	2. Install new bearings.
3. Diluted oil.	3. Change to correct viscosity.
4. Low oil pressure.	4. Correct as required.
5. Bent or twisted rod.	5. Straighten rod. Replace insert bearing. Replace rod if bend or twist is excessive.

(continued)

Engine Noise Diagnosis *(continued)*

Problem: Piston Slap

Sound Identification: Piston slap is caused by the piston tipping from side to side in the cylinder. This tipping produces sounds that can range from a regular clicking to a very distinct hollow clatter, depending on the severity of wear. Piston slap will be more noticeable when the engine is cold. In mild cases, it may actually disappear after the engine is warmed up. The noise may also disappear if the plug wire on the affected cylinder is removed and the vehicle driven. Adding a tablespoon of heavy oil to each cylinder should temporarily quiet the noise. Do not add oil to a diesel engine.

Possible Cause	**Correction**
1. Cylinder worn.	1. Rebore and fit oversize pistons.
2. Pistons badly worn.	2. Rebore and fit oversize pistons.
3. Pistons mildly worn.	3. May be expanded by knurling or peening.
4. Piston pin fitted too tight.	4. Fit pins as specified.
5. Insufficient lubrication.	5. Correct as required.

Problem: Loose Piston Pins

Sound Identification: Loose pins will cause a sharp, double-knock, especially at idle speeds. If only one pin is loose, the knocking will become more distinct when the spark plug in the affected cylinder is shorted. If all pins are loose, shorting one plug will not alter the sounds.

Possible Cause	**Correction**
1. Piston pin worn.	1. Fit new pins of correct size.
2. Piston pin hole worn.	2. Fit oversize pin.
3. Connecting rod bushing worn.	3. Replace bushing or fit oversize pin.
4. Insufficient lubrication.	4. Correct as required.
5. Piston pin locks missing.	5. Install locks.
6. Piston pin lock loose.	6. Tighten lock.

Problem: Timing Gear, Chain, and Belt Noise

Sound Identification: Timing gears, chains, and belts can produce noises varying from a high-pitched howl (fitted too tight) to a low-level chatter or growl (badly worn). The timing chain or belt can slap against the cover, producing a thumping or scraping sound. A missing chain tooth will cause a regular and distinct knock. A tight belt can also produce a whirring sound.

Possible Cause	**Correction**
1. Worn chain.	1. Replace chain and sprockets.
2. Worn sprockets.	2. Replace chain and sprockets.
3. Loose gear or sprocket.	3. Replace gear.
4. Excessive endplay.	4. Correct as needed.
5. Gear misalignment.	5. Align properly.
6. Worn gear.	6. Replace both camshaft and crankshaft gears.
7. Excessive front camshaft or crankshaft bearing clearance.	7. Replace bearings.
8. Gear tooth missing.	8. Replace both gears.
9. Worn belt.	9. Replace belt.
10. Worn or defective belt adjuster.	10. Replace adjuster.

Problem: Combustion Knocks

Sound Identification: Combustion knocks are divided into two classes: preignition and detonation. Preignition occurs when the fuel-air mixture ignites before the spark plug fires. Detonation occurs when an extra flame front is produced in the combustion chamber, creating an explosion of the fuel charge before it can finish burning evenly. The result is the same: a sharp metallic pinging sound. This pinging is most noticeable during acceleration.

(continued)

Engine Noise Diagnosis *(continued)*	
Problem: Preignition	
Possible Cause	**Correction**
1. Overheated engine. 2. Glowing pieces of carbon. 3. Spark plugs overheating. 4. Sharp valve edges. 5. Glowing exhaust valve.	1. Check cooling system. 2. Remove carbon. 3. Change to cooler plugs, check plug tightness. 4. Install valves with full margin. 5. Check for proper tappet clearance, sticking valve, air leaks, overheating.
Problem: Detonation	
Possible Cause	**Correction**
1. Ignition timing advanced. 2. Engine temperature too high. 3. Carbon buildup raising compression ratio. 4. Low octane fuel. 5. Exhaust heat control valve stuck, or vacuum line improperly connected. 6. Block or head shaved to increase compression. 7. EGR valve disconnected. 8. Thermostatic air cleaner valve stuck in cold position.	1. Retard timing. 2. Check cooling system. 3. Remove carbon. 4. Switch to high octane fuel. 5. Free valve; check vacuum line connections. 6. Use thicker gasket or change head. 7. Reconnect EGR and check operation. 8. Repair as needed.
Problem: Fuel Pump Noise (Mechanical Fuel Pump)	
Possible Cause	**Correction**
1. Pump-to-block fasteners loose. 2. Rocker arm or eccentric worn. 3. Rocker arm spring weak or broken. 4. Pushrod or bore worn.	1. Tighten fasteners. 2. Replace. 3. Replace rocker arm spring. 4. Replace pushrod; check for bore wear.
Problem: Other Engine Noises	
Sound Identification: Various accessory units, such as the water pump, alternator, power steering pump, and air conditioning compressor, can produce a variety of squealing, grinding, thumping, and howling noises. They may be quickly checked by disconnecting the accessory drive belts. Engine mounts can cause heavy metallic noises if they are too tight or too loose. Exhaust pipes, mufflers, and tailpipes can also be responsible for various thumps, clangs, and rattles.	

23

Cooling System Service

After studying this chapter, you will be able to:

- Explain the role of antifreeze in an engine cooling system.
- Properly clean a cooling system.
- Detect leaks in a cooling system.
- Test a radiator pressure cap.
- List the safety rules dealing with cooling systems.
- Inspect and replace cooling system hoses.
- Inspect, replace, and adjust drive belts.
- Test and replace a thermostat.
- Inspect, repair, and replace a coolant pump.

Technical Terms

Depressurized	Cooling system cleaner
Overheated	Neutralizer
Overheating protection	Reverse flushing
Antifreeze	Pressure tester
Ethylene glycol	Pressure cap
Propylene glycol	Electrochemical degradation (ECD)
Long-life coolant	Belt tension
Organic acid technology	Electric fan assembly
Hard water	Thermoswitch
Electrolysis	Fluid clutch fan
Hydrometer	Thermostat
Refractometer	Coolant pump
Test strips	Impeller
Alkalinity	Freeze plugs
Ph level	Air-cooled engines
Bleeder valve	
Reverse fill	

This chapter will cover the service requirements of modern cooling systems, including diagnosis, coolant replacement, system flushing, and repair operations. Although most of the information in this chapter relates to liquid-cooled systems, air-cooled systems are also discussed.

Caution: Use care when doing cooling system work on vehicles equipped with airbags. Some sensors are located near the radiator.

The Need for a Cooling System

Automobile engines generate a large amount of heat. About one-third of the heat energy developed by the fuel burning in the cylinders is converted into power to drive the automobile. Another third is wasted and goes out the exhaust. The remaining third is absorbed by the metal of the engine and must be disposed of by the cooling system to prevent overheating.

Figure 23-1 shows the approximate temperatures of the various engine parts during operation. Properly designed and maintained automobile cooling systems are good at keeping these temperatures within the normal range. The failure or malfunction of one or more cooling system components, however, can lead to serious overheating. Under certain conditions, cooling system problems can also cause *overcooling.*

Cooling System Service

Modern cooling systems are relatively trouble free. However, they can develop problems, especially if they are neglected. Routine checks and periodic coolant replacement will usually reveal cooling system problems before they reach the serious stage. It is important to become familiar with problems associated with the cooling system, **Figure 23-2.** You must know which units are responsible for specific problems, how they can be checked, and if faulty, how they are repaired.

Dealing with an Overheated Engine

Warning: Hot coolant can scald or blind you. Before performing *any* service on the cooling system, make sure the cooling system has been *depressurized.* The system is depressurized when all pressure has been released.

If the engine is greatly ***overheated*** (indicated by steam spurting from the overflow), shut it down at once. If an engine is moderately overheated, however, it is best to run the engine at high idle for a minute or two before shutting it down. Flow of the coolant helps to carry excess heat from the cylinders and valves, and there is less possibility of cylinder distortion and warped valves.

Never pour cold water into the radiator of an overheated engine. Once the engine is shut down, open the hood and use

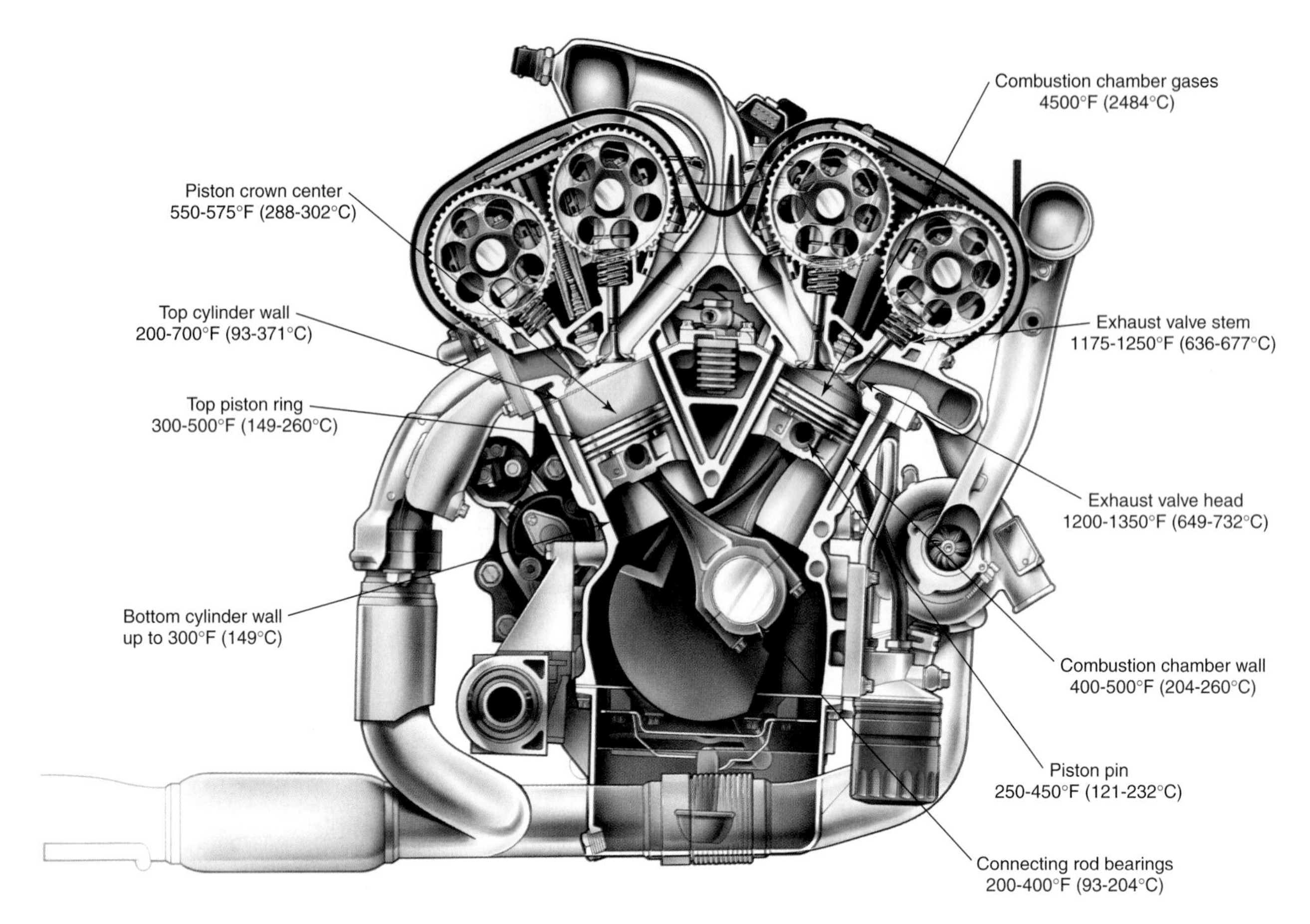

Figure 23-1. Approximate temperatures of various engine components and areas. Temperatures vary depending upon engine design and application. (Saab)

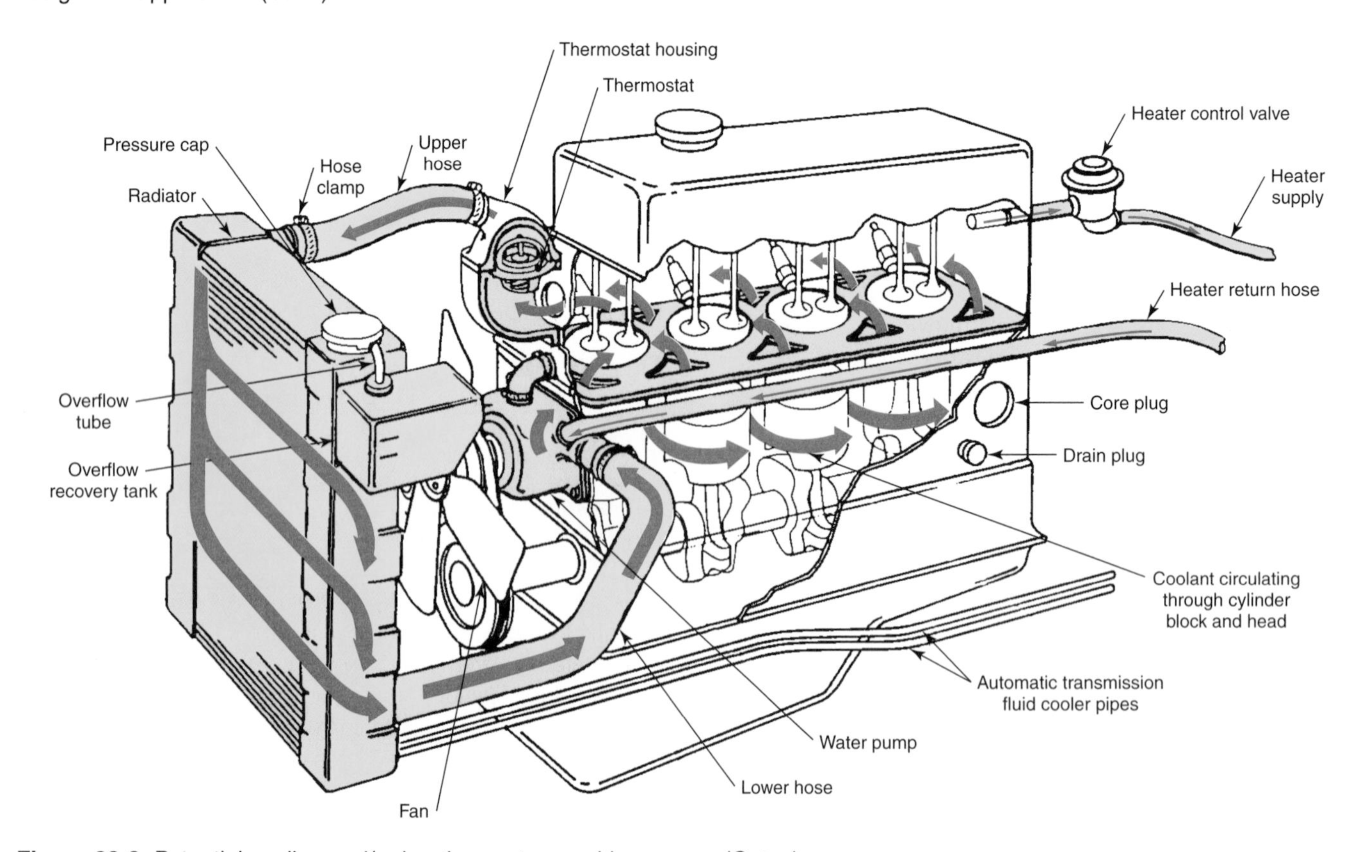

Figure 23-2. Potential cooling and/or heating system problem areas. (Gates)

a shop fan to blow cool air through the radiator and across the engine surface. Allow the engine to cool down until the coolant is no longer boiling (at least thirty minutes). Slowly open the radiator cap, then start the engine. While running the engine at a fast idle, slowly add water to the radiator. The pump will mix the hot coolant and the cold water. Run the engine until the temperature is normal. If the engine begins to overheat again, shut it off. Then, recheck the radiator coolant level. If the coolant level is correct, find out what is causing the engine to overheat and correct the problem.

Overheating Protection

On some newer cars, a special type of ***overheating protection*** is used. If the engine begins to overheat, the computer will shut down some cylinders by turning off the fuel injectors to those cylinders. The usual procedure is to turn off alternating cylinders in the firing order. If the firing order is 1, 3, 6, 2, 5, 4, for instance, the computer will shut down cylinders 1, 6, and 5. This allows the engine to operate as smoothly as possible while the cylinders cool down. After they cool down, their fuel injectors are turned back on, and the fuel injectors for cylinders 3, 2, and 4 are turned off. The cycle repeats as necessary. By causing the engine to run on alternating cylinders, this method ensures no one cylinder is allowed to become too hot.

Antifreeze

In all areas where temperatures may drop below freezing (32°F or 0°C), it is necessary to keep ***antifreeze,* Figure 23-3,** in the cooling system to prevent engine damage. In addition, an antifreeze and water mixture is better than plain water at transferring heat in hot weather. Antifreeze contains rust inhibitors, which prevent damage to the engine, radiator, and heater core, as well as small amounts of water-soluble oils, which lubricate the coolant pump seals and heater shutoff valves.

Figure 23-3. Conventional and long-life antifreeze comes in one gallon containers. Be sure you know which type is used in a vehicle before adding. (Jack Klasey)

Ethylene and Propylene Glycol Antifreeze

In the past, many liquids were used as antifreeze, including kerosene, denatured ethyl alcohol, and methanol (wood alcohol). There are several compounds used to make antifreeze solutions. The first is called ***ethylene glycol.*** This antifreeze is sometimes called EG coolant.

Ethylene glycol will freeze at roughly 9°F (–13°C). When ethylene glycol is mixed with water, which has a freezing point of 32°F (0°C), the resulting mixture has a freezing point lower than either liquid by itself. A half-and-half (50-50) mixture of antifreeze and water has a freezing point of about –35°F (–37°C). A mixture of 70% antifreeze and 30% water will not freeze until the temperature reaches –67°F (–55°C).

Ethylene glycol is not a fire hazard, and it does not harm paint finishes. It does not readily evaporate at normal system temperatures, and it can be used with high-temperature thermostats. Additionally, it can be left in the cooling system for up to a year without causing problems.

Warning: Ethylene glycol antifreeze is poisonous and must not be taken internally. Keep antifreeze away from children and animals. Never put beverages or drinking water in empty antifreeze containers.

The second type of antifreeze compound is ***propylene glycol*** or PG coolant. Propylene glycol coolants are nontoxic and are considered to be "environmentally safe." However, since all used coolants contain some levels of heavy metals from the cooling system, used antifreeze must be disposed of properly.

A 50-50 mixture of propylene glycol antifreeze and water will freeze at –26°F (–32°C). A 60-40 mixture will freeze at –54°F (–48°C). Increasing the antifreeze concentration above 70%, however, will cause the mixture's freezing point to rise. Most manufacturers recommend a 50-50 mixture of water and antifreeze. EG and PG coolants should not be mixed together.

Note: Dyes are used to give antifreeze its color. The color of a coolant can vary (green, blue, red, yellow, orange, or pink) and has no bearing on its composition or service life. Check a service manual for the proper type to use in a given application.

Long-life Coolants

Some manufacturers use ***long-life coolant*** in the cooling system. This coolant can be used for 100,000 miles (160 000 km). Long-life coolant may be referred to as ***organic acid technology*** **(OAT)** coolant. This coolant is made without silicates and other minerals that cause cooling system deposits. Since this coolant is made with ethylene glycol, it has the same freezing and boiling points as a conventional coolant. Do not mix this antifreeze with other types, since this will reduce the long-life properties of the antifreeze.

Long-life coolants have approximately the same freeze points as conventional ethylene glycol coolant. A 50-50 mixture will protect the cooling system down to –34°F (–36°C). The boiling point is raised about the same amount as with conventional coolant. Also, like regular coolants, mixtures greater than 70% coolant are not recommended. Some long-life coolants come prediluted 50-50 from the manufacturer, eliminating the need to add water.

Mixing Antifreeze

All vehicle manufacturers recommend keeping at least a 50-50 mixture of water and antifreeze in the cooling system at all times. The correct amount of antifreeze to use is determined by the capacity of the cooling system. For example, a 20-quart cooling system should have 10 quarts of antifreeze and 10 quarts of water. Never try to save money by adding just enough antifreeze to get by.

Always use clean water to top off the antifreeze solution. In most areas, tap water is acceptable for use in cooling systems. However, the water supply in some localities has high concentrations of lime, minerals, and acids. Water containing these minerals is known as ***hard water.*** Hard water can cause layers of chemicals to build up in the cooling system, reducing the transfer of heat from the metal to the coolant. If your area has this type of water, use another source of water, such as bottled distilled water.

Modern engines and cooling systems contain cast iron, aluminum, steel, brass, copper, and various types of solder. These dissimilar metals can cause ***electrolysis*** (creation of an electric current) in the system, with resulting damage to the metals. An increasing cause of electrolysis is defective or missing ground wires, as well as poorly grounded aftermarket accessory systems. Electrolysis due to poor grounds can destroy a radiator, heater core, or an entire engine.

Useful Life of Antifreeze

The cooling systems of most new cars are filled with a mixture of antifreeze, rust inhibitors, pump lubricant, and clean water. Unless excessive coolant is lost for some reason, coolant can be left in the system for the time period recommended by the manufacturer. This time period for conventional coolant is usually about 24 months. Long-life coolants can be left in for a much longer period. It is suggested the recommendations of the manufacturer be followed regarding the type of antifreeze and length of use. Failure to comply with the manufacturer's recommendations may void the vehicle's warranty.

Checking Antifreeze

To test the coolant protection level at operating temperature, use an antifreeze ***hydrometer,*** **Figure 23-4.** Most antifreeze hydrometers have some form of temperature correction, so they can be used with hot or cold coolant. Draw the coolant in and out of the hydrometer several times to bring hydrometer temperature up to coolant temperature. When using the test unit, follow manufacturer's instructions.

Note: A battery hydrometer cannot be used to check antifreeze.

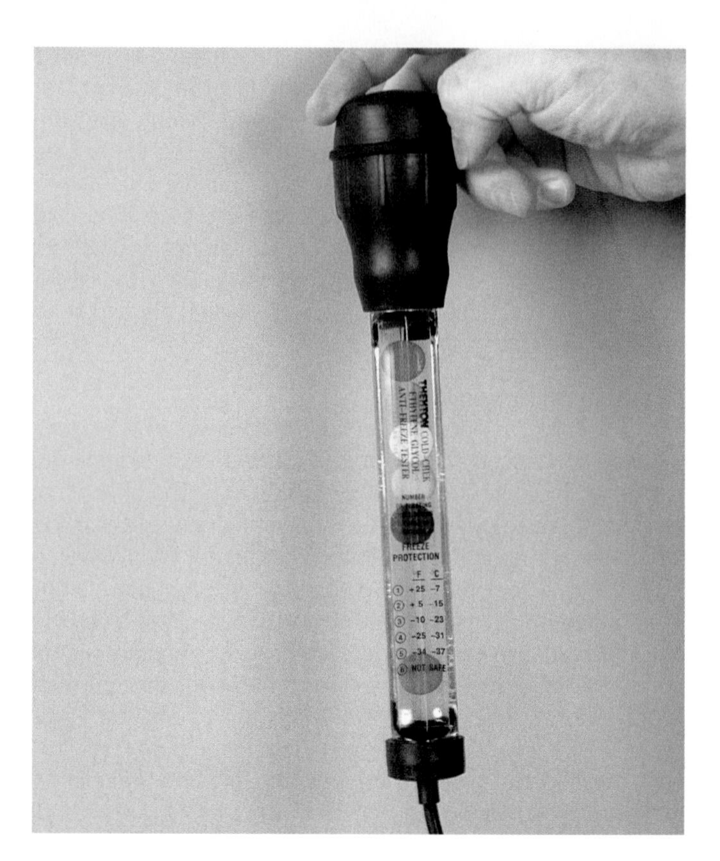

Figure 23-4. One type of antifreeze hydrometer being used to check the concentration (percentage) of antifreeze in the system. (Jack Klasey)

Another tool that can be used to check antifreeze is a ***refractometer.*** To use a refractometer, take a small sample of coolant from the radiator. Place a few drops on the refractometer lens. Look through the refractometer to get the freeze point reading.

Some shops now use ***test strips*** to quickly check engine coolant. Chemical test strips can check coolant for freeze protection, as well as its ***alkalinity*** or ***pH level.*** However, most test strips can only be used for conventional coolants.

To perform the test, remove one strip from the container. Be sure to close the container, since the color chart is often on the side. Dip the test strip in the engine coolant, making sure both test spots are immersed. Remove the test strip and wait approximately 30 seconds. Compare the two test sections to the color chart that comes with the strips, **Figure 23-5.**

Cooling System Maintenance

Cooling system maintenance includes coolant replacement, locating leaks, and checking for proper fan operation. If a cooling system is properly maintained, flushing is usually not needed.

Coolant Replacement

To perform a simple coolant change, operate the engine until the thermostat is open (the top radiator hose is hot). Shut off the engine and place a drain pan under the petcock, then open the petcock. Allow the coolant to drain into the pan.

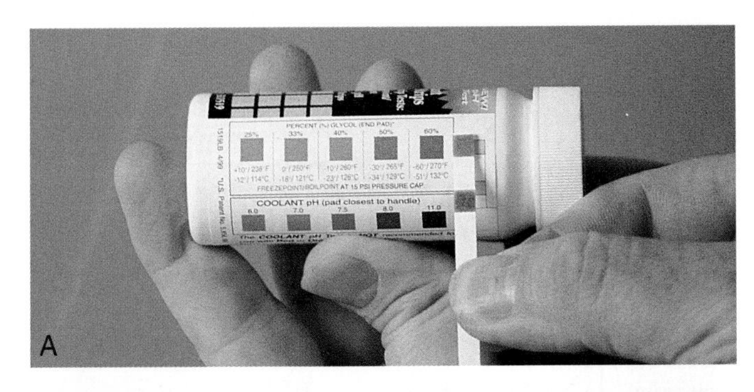

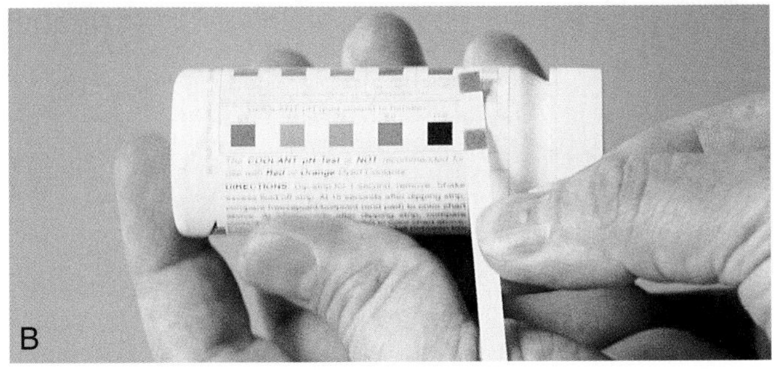

Figure 23-5. Chemically coated test strips can be used to check concentration as well as acidity or pH. A—Checking concentration. B—Checking pH level. (Jack Klasey)

When coolant stops flowing from the petcock, the system pressure has been removed. Remove the pressure cap and allow the coolant to drain completely. Perform other service operations as necessary. Allow the old coolant to cool, then dispose of it properly.

Close the petcock. Using a coolant mixture of 50% antifreeze and 50% water, refill the system. Do not completely fill the radiator, since there must be room for the hot coolant to expand. Then bleed the cooling system as explained in the next section.

Caution: Do not install long-life coolant in a vehicle that uses conventional coolant after replacing a major cooling system part, such as a heater core, water pump, or radiator. A possible loss in corrosion protection may result.

Bleeding the Cooling System

Air pockets often form in the engine when the coolant is drained. These air pockets are hard to remove on many modern vehicles since the radiator filler neck is not the highest point in the system. Since air is lighter than coolant, it rises to the highest point in the engine. On older rear-wheel drive vehicles, the filler neck is above the engine block and air pockets usually remove themselves from the system as the coolant circulates. The following paragraphs explain bleeding procedures according to filler neck placement.

System with Filler Neck above Engine Block

Start the engine and allow it to warm up. Turn the heater on full (mode and temperatures switches). Closely monitor the coolant level, keeping it to within about 3″ (76mm) of the filler neck. On most vehicles, the coolant level will slowly rise until the thermostat opens. The coolant may surge as the engine warms up. When the thermostat opens, the level of coolant in the radiator will drop suddenly. Add coolant to bring the system up to its normal full level. Monitor the level for a few more minutes to be sure all air has been removed. Then, install the pressure cap.

System with Filler Neck below Engine Block

When the filler neck is below the engine block, the engine may be equipped with a ***bleeder valve.*** Locate the valve on the engine—it is usually at or near the thermostat housing, **Figure 23-6.** Open the bleeder valve and add coolant to the radiator until it begins to flow from the valve. Lightly close the bleeder valve. Start the engine, turn the heater on full, and allow the engine to warm up. Monitor the coolant level, keeping it to within about 3″ (76mm) of the filler neck. Open the bleeder valve occasionally to allow air to escape. When only coolant escapes from the bleeder valve, top off the radiator and install the cap. Allow the engine to cool, then recheck the level.

Some older vehicles with the filler neck below the engine block do not have a bleeder valve. To bleed these vehicles, begin by filling the cooling system to the top of the filler neck. Some technicians choose to ***reverse fill*** the engine with coolant to minimize the amount of trapped air in the water jackets. This is done by disconnecting the upper radiator hose from the radiator and adding coolant to the engine through the upper hose.

Start the engine and turn the heater on full. As the engine warms up, monitor both the coolant level and engine temperature. Do not allow the engine to overheat. If the engine begins to overheat, shut it off and run a shop fan over the engine to cool it down. Once the air bubble comes out of the engine, the coolant level will drop significantly. When this occurs, fill the radiator to the top of the filler neck and run the engine to ensure the cooling system has been completely bled. It may

Figure 23-6. Newer engines have one or more bleeder valves to help remove air from the cooling system.

be necessary, in some cases, to raise the front of the vehicle to remove all the air from the cooling system.

System Flushing

Mild rust and scale buildup in the cooling system can usually be corrected by flushing the system with a chemical ***cooling system cleaner.*** When the vehicle has an aluminum engine, cylinder head, or radiator, you must use a cleaner that is harmless to aluminum. Avoid splashing cleaner on the paint finish. Carefully follow the instructions supplied by the manufacturer of the cleaning product. Some cleaners are of a one-step type, while others require two or more separate operations.

To perform a simple flush, drain the old antifreeze solution from the system. Close all drains. Fill the system with clean water, reinstall the radiator cap, and run the engine until normal operating temperature is attained. Add cleaner to the system and start the engine. Set the heater control to maximum heat, allowing coolant and cleaner to circulate through the heater lines and core. After allowing the cleaner to circulate for the recommended period of time, stop the engine and drain the system again. Remove radiator cap while draining. Do not let the solution of water and cleaner boil. If the vehicle is outside and the weather is very cold, be careful of slush ice forming in the radiator.

Warning: When coolant is hot, use extreme caution while removing system pressure caps

Neutralizing and Flushing

If an acid-type cleaner is used, it must be neutralized. A ***neutralizer*** reacts with cleaning chemical that is left in the cooling system, transforming it into a harmless substance. Failure to neutralize and flush properly may leave acids in the system. These acids will attack the system and destroy the protective properties of the inhibitors and the antifreeze. Pour the neutralizer into the system, run the engine for the specified length of time, and drain.

Heavy-Duty System Cleaning

Long periods of neglect often result in a cooling system literally choked with rust and scale. This can often be detected by feeling the radiator surface when the engine is moderately warm. Cold spots on the radiator surface are evidence of clogged tubes. Ultimately, the radiator tubes will become blocked enough to cause the coolant in the system to boil. The boiling action breaks loose large quantities of scale that may plug the radiator completely. Correction requires reverse flushing, **Figure 23-7.**

Radiator Reverse Flushing

Reverse flushing forces water through the radiator in a direction opposite of normal flow, Figure 23-7A. This helps to remove particles that are jammed into openings. Severe radiator clogging may require removing the radiator from the vehicle and boiling it in a hot tank.

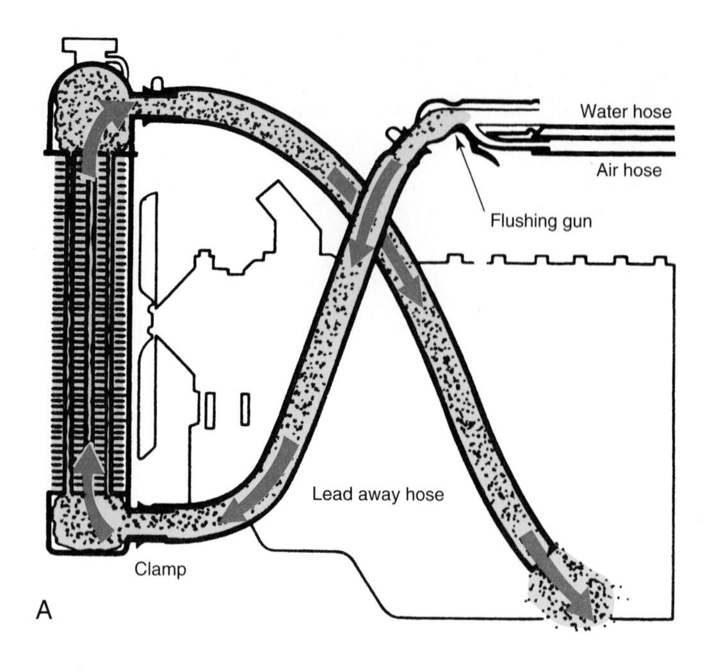

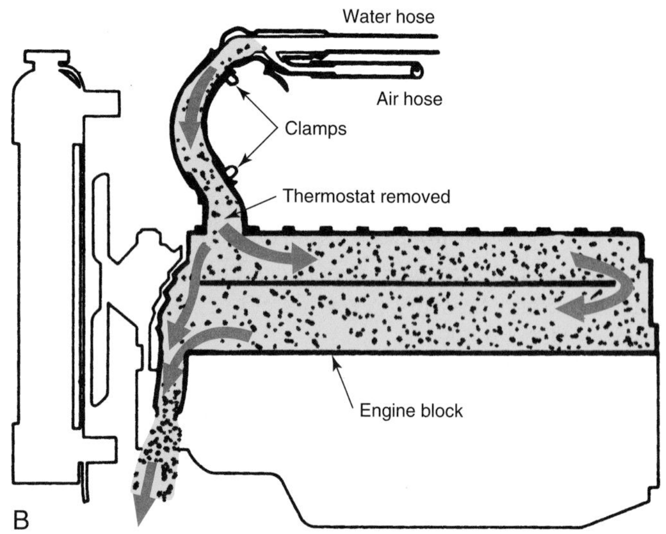

Figure 23-7. Reverse flushing procedures. A—Radiator. B—Engine block.

To begin reverse flushing, remove the upper and lower radiator hoses. Disconnect the heater hoses. Attach the flushing gun to the lower radiator outlet and a lead-away hose to the top radiator outlet. Replace the radiator cap.

Run a stream of water through the radiator and periodically release blasts of air to agitate and loosen particles so they can be flushed out. Do not exceed an air pressure of 20 psi (138 kPa). Pressures higher than this may rupture the radiator. Continue the flushing and air blasting until clear water flows out of the radiator.

Reverse Flushing the Engine Block

Be sure to remove the thermostat before flushing the block. Some vehicles require removing the coolant pump and heater hoses, since pressure flushing can damage the seal.

Attach the flushing gun to the top water outlet of the block. Attach a lead-away hose to the bottom outlet, if necessary. Reverse-flush the block using the procedure described for the radiator. See Figure 23-7B. After flushing, reattach all hoses securely. Fill the system with water and add antifreeze or inhibitor as needed. Finally, test the system for leaks.

Reverse Flushing the Heater Core

After determining the direction of coolant flow, remove the heater hoses from the engine block. Be sure to remove the heater shut-off valve (if used) from the system before beginning the flushing procedure. To reverse flush the heater core, follow the procedures presented in the section on reverse flushing a radiator.

Note: Some heater systems cannot be reverse flushed. Check the manufacturer's recommendations.

After flushing, reconnect the heater hoses and reinstall the heater shutoff valve. Refill the cooling system and test it for leaks.

Finding Leaks in the Cooling System

Cooling system leaks can develop over time. They can be caused by corrosion, vibration, or the gradual loosening of system fittings. Sometimes, cleaning may open cracks or other tiny openings that had been sealed with rust. Removing of the rust allows leaking to start. **Figure 23-8** shows typical examples of radiator damage.

Coolant leakage can cause engine overheating and damage to the pistons, valves, and cylinder head. If coolant leaks into the cylinders, it may plug the rings and cause hard starting, excessive oil consumption, piston corrosion, and bearing failure. Combustion gas leakage into the cooling system can cause overheating and severe cooling system corrosion. Any time the cooling system loses water and an obvious leak cannot be found, the system should be pressure tested, **Figure 23-9,** or checked with special chemicals or a black light.

Pressure Testing the Cooling System

Pressurizing the cooling system raises the boiling point of the coolant mixture. This allows the engine to operate at higher temperatures without causing the coolant to boil. Hotter temperatures increase combustion efficiency, resulting in more power, better mileage, and lower emissions.

Caution: Some of the newest systems use pressures up to 25 psi (172 kPa). Special pressure testers must be used on these systems. Do not pressurize older systems past 20 psi (137 kPa). Higher pressure could cause the radiator, heater core, or other components to rupture.

Fill the radiator to within 1/2" (12.7mm) of the filler neck. Attach the ***pressure tester*** to the filler neck, Figure 23-9A. Build up pressure in the system carefully; do not exceed the pressure for which the system is designed. Check the pressure marking on the radiator cap or obtain information from the manufacturer's manual.

When the system is pressurized, watch the gauge. If the pressure holds steady, the system is probably all right. If the pressure drops, carefully check all areas for leaks. Even a small amount of dampness indicates enough loss of coolant to cause trouble. If no leak can be found, check the tester connection to the filler neck to make certain it is not leaking.

Occasionally, you will find a system that leaks only when cold or when hot. By checking the system before and after engine warm-up, both types of leaks will be exposed. When finished with the pressure test, adjust coolant to the proper level in the radiator.

Radiator Pressure Caps

Cooling systems are pressurized by using a radiator ***pressure cap.*** The pressure cap contains a spring that will not allow pressure to escape from the system until a certain level is reached.

The boiling point of the coolant is raised about 3°F (1.6°C) for each pound of pressure added. If pressure buildup permitted by the radiator cap is 15 psi (103 kPa), the boiling point of the coolant under pressure will be increased by 45°F (15 x 3°F). At sea level, water boils at 212°F (100°C). Adding 45°F to 212°F gives us 257°F, the temperature at which radiator coolant will boil when under 15 psi (103 kPa) of pressure. To maintain specified pressure, the cap pressure valve spring and seal surface must be in good condition.

Pressure Cap Testing

Cap operation can be checked by using a pressure tester, Figures 23-9B and 23-9C. Always place a protective rag over the cap during removal. Stand to one side. Open the cap to the safety stop and wait for steam pressure to subside. The cap may then be safely removed.

STOP **Warning: Removing a pressure cap from a hot radiator can be very dangerous. Sudden release of the pressure may cause the water to turn to steam and literally "explode" into the face of the person removing the cap.**

Install the adapter on the tester, then install the cap and perform the pressure test. The cap should retain a pressure within 1 1/2 psi (10 kPa) of its specified pressure rating. If the cap fails to pass the test, replace it with a new cap of the correct pressure rating. Do not use a nonpressure cap on a pressurized system. Do not use a cap designed for an open system (no coolant reservoir) on a closed system (coolant recovery reservoir).

If the cap tests OK, remove the tester. Loosely install the radiator cap and run the engine until normal temperature is reached. Remove the cap, attach the pressure pump,

Solder bloom—Solder corrosion caused by degradation of antifreeze rust inhibitors. Tube-to-header joints are weakened and corrosion can restrict coolant flow.

Internal deposits—Rust and leak inhibitors (Stop-leak) can form solids that collect in the cooling system and restrict flow.

Tube-to-header-leaks—Solder joint failure resulting in coolant loss.

Leaky tank-to-header seam—Solder joint failure or a cracked header, usually the result of pressure-cycle fatigue.

Leaky oil cooler—Coolant shows traces of oil. Transmission/transaxle or engine damage can result.

Leaky inlet/outlet fitting—Leaks in this area can be caused by fatigue or solder joint corrosion.

Electrolysis—Electrical current created by the chemical reaction between coolant and two dissimilar metals. Causes corrosion of metal components.

Electrolysis—Electrical current created by the chemical reaction between coolant and two dissimilar metals. Will produce voids in tubes.

Figure 23-8. Radiator failures are common. These are some you are likely to see. (Modine)

Fin deterioration—Chemical deterioration of the fins, caused by road salt or seawater.

Fin bond failure—A loss of solder bond between fins and tubes. Fins will be loose in the core.

Blown tank-to-header seam—Caused by extreme cooling system pressure, usually as a result of exhaust leaking into the cooling system.

Loose side piece—Can lead to flexing of the core and radiator-tube failure.

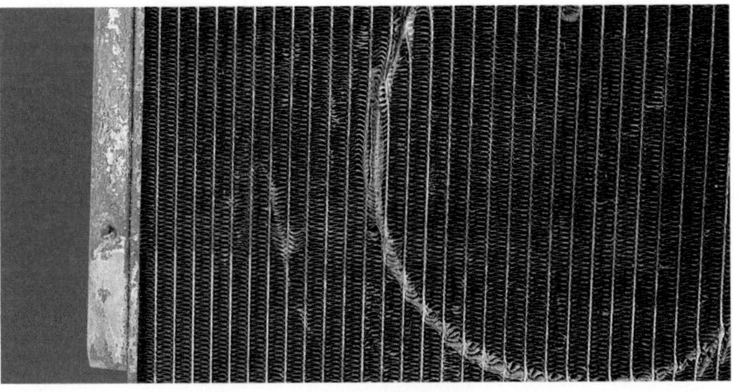

Fan damage—A minor collision, failed water pump, or loose fan support can result in radiator damage.

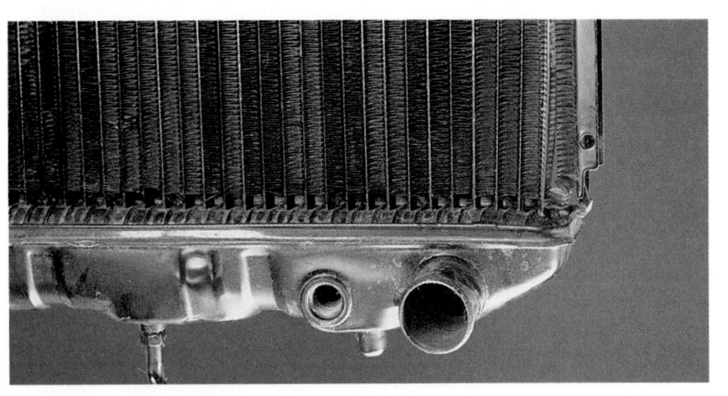

Over pressurization—Excessive pressure in the radiator caused by a defective pressure cap or engine exhaust leak.

Cracked plastic tanks—High stress in the radiator can cause premature plastic tank failure.

Steam erosion—Steam can break down plastic tanks, which will produce thinning and eventually, holes in the tanks. White deposits are often found.

and once again pressurize the system. Recheck the system for leaks.

Checking for Internal Leaks with a Pressure Gauge

If the system loses coolant and no external source is found, check for internal leakage. Apply 6–8 psi (41–55 kPa) of pressure to the system. Run the engine at a slow speed and watch the pressure pump gauge. Pressure buildup indicates a combustion leak, such as a cracked head or blown gasket.

Warning: Do not allow pressure to build up beyond the pressure cap rating.

To determine which bank on a V-type engine is leaking, disconnect all the spark plug wires on one bank and run the engine. If the pressure buildup stops, the leak is in that bank. If not, the leak is in the firing bank. Repeat the test on both banks.

Checking for Internal Leaks with Test Chemicals

By drawing air from the top of the radiator through a special test chemical, it is possible to detect combustion leaks. The test chemical will change color if combustion gases are present in the cooling system.

On V-type engines, operate the engine on one bank and test the system for leakage. Then, repeat the test while operating the engine on the other bank. In this way, you can determine if one or both banks are leaking. **Figure 23-10** illustrates a leak detector tool.

Other Checks for Internal Leaks

Another method of checking for internal leaks requires no special tools. Drain the system down to the level of the engine outlet hose. Then, remove the hose. If necessary, add coolant to bring the level up to the hose fitting neck. Disconnect the coolant pump drive belt. Start the engine and accelerate rapidly several times. Watch for bubbles or for a surge in the water level, either of which indicates a combustion leak.

Pull the engine oil dipstick. Tiny water droplets in the oil clinging to the stick indicate internal leakage. Coolant

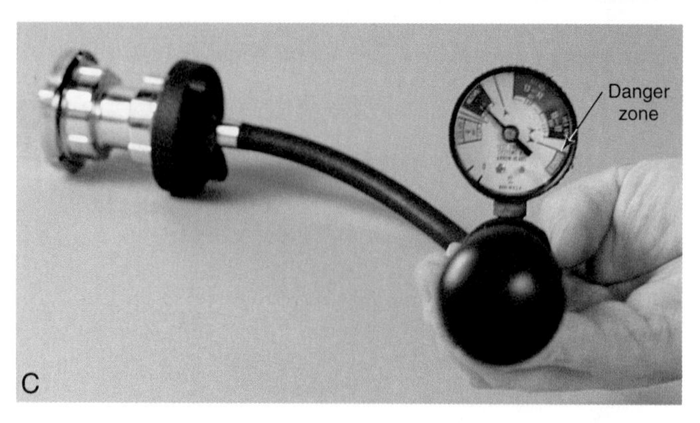

Figure 23-9. Making pressure tests. A—Pressurizing pump attached to a radiator. Pressure is being built up in the system to check for possible leaks. B—Using a pressure tester to check radiator pressure cap function. C—Note the "danger zone" on the gauge. *Never* raise pressure to this level! (Jack Klasey)

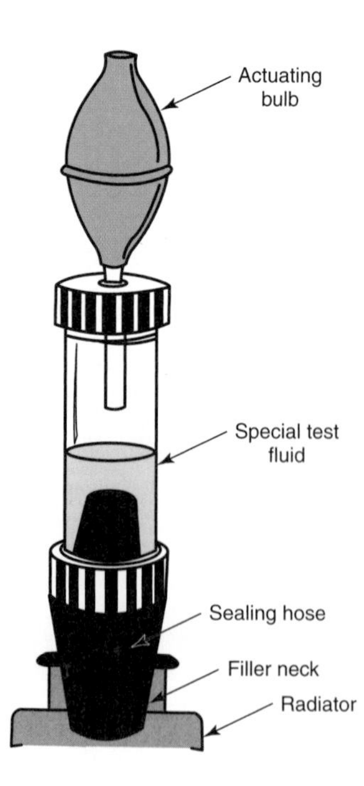

Figure 23-10. Combustion leak detector. Note the special fluid. (P and G Mfg. Co.)

contamination will give oil the appearance of a vanilla milkshake. Oil in the radiator may indicate a combustion leak or a leak in the transmission oil cooler. When draining oil, always watch for water contamination.

Water or steam discharge from the tailpipe when the engine is operating at a normal temperature can mean internal leakage. Water discharge is normal during warm-up. Another way to check for leakage is to use a black light and dye indicator, **Figure 23-11.**

Servicing Internal Leaks

External and internal coolant leaks should be repaired immediately, or they may cause severe engine damage. Sealing compounds should not be poured into the radiator to stop leaks. These sealers can plug up coolant passages and will not stop major leaks.

If internal engine leaks are discovered, the oil should be drained. If the lubrication system has been contaminated with antifreeze, it must be thoroughly flushed. In cases of severe contamination, proper cleaning may require immersing the engine block in a solvent tank.

Checking for Electrolysis

To check for excess electrolysis, connect the negative lead of a digital multimeter (set to read dc volts) to the negative battery terminal. Immerse the meter's positive probe into the coolant at the filler neck. The reading should be no more than 0.3 volt on cars with all cast iron engine parts and 0.1 volt on vehicles with aluminum engine parts. Get a helper to switch vehicle components on and off during the test to ensure no electrical component is causing the electrolysis.

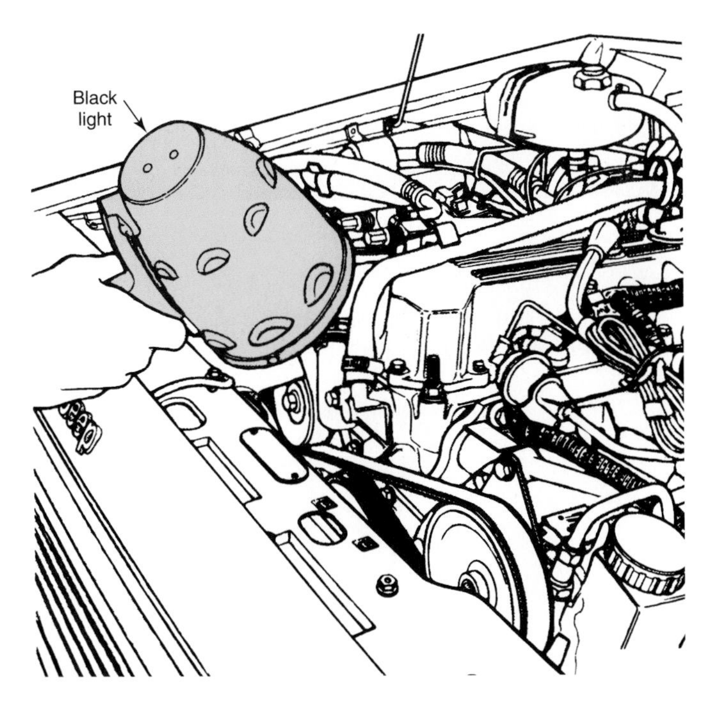

Figure 23-11. Using a black light to check for coolant leaks. Dye is placed in the coolant. Leaking coolant will be illuminated by the black light. Keep tools and hands away from the fan, pulleys, etc. (DaimlerChrysler)

Inspecting the Radiator Filler Neck

Check the condition of the inside sealing seat (surface that contacts the cap) in the filler neck, **Figure 23-12.** It must be smooth and clean. Moderate roughness can be removed with a special reaming tool. Cam edges must be true and the overflow tube must be clean and free of dents. A pressure cap cannot function properly unless the filler neck is in good condition.

Inspecting Hoses

Observe the radiator and heater hose condition. Look for bulges and swelled spots, cuts and abrasions, cracks, and leaks. Observe the hoses as you allow the engine to cool. Any hose that collapses as the engine cools is soft and should be replaced. As a final check, release system pressure and squeeze the hoses near the clamps to check for soft spots, **Figure 23-13.**

Some hose damage may not be evident to the naked eye. Hoses are weakened over time by ***electrochemical degradation (ECD),*** which is a reaction between the chemicals in the coolant and the metals in the engine and radiator, **Figure 23-14.** Because of this, coolant hoses should be changed if they are more than three years old. ECD has the greatest effect on heater hoses, bypass hoses, and the upper radiator hose.

Figure 23-12. Inspect the radiator filler neck for signs of damage. (Jack Klasey)

Hose Replacement

Begin by depressurizing and draining the cooling system below the hose fittings. Then loosen the hose clamps, **Figure 23-15.** Gently twist the hose on the fitting to break it loose. If the hose is stuck to the fitting, use a hose cutter or other knife to slice the hose. Carefully peel the split hose away from the fitting.

Obtain the new hose and compare it with the original. Some hoses are made to fit several engines, and it may be necessary to cut off a portion of the hose to allow it to fit

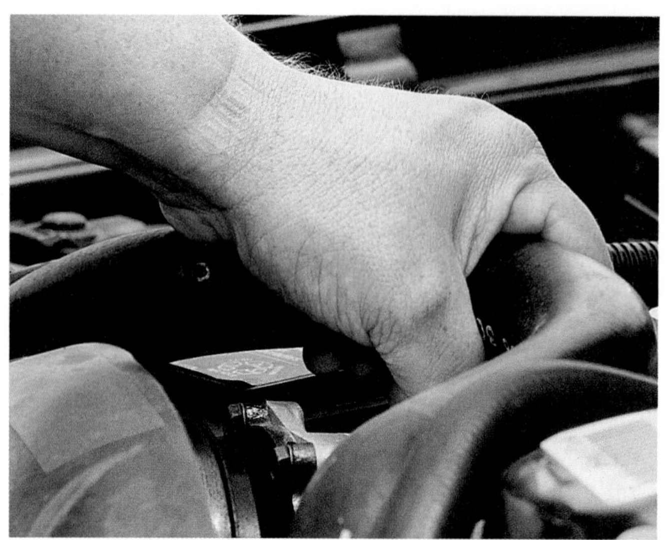

Figure 23-13. Squeeze the hoses to check for hardness or for swelling and softness.

Figure 23-14. This hose should have been replaced long before reaching this advanced state of deterioration. (Gates)

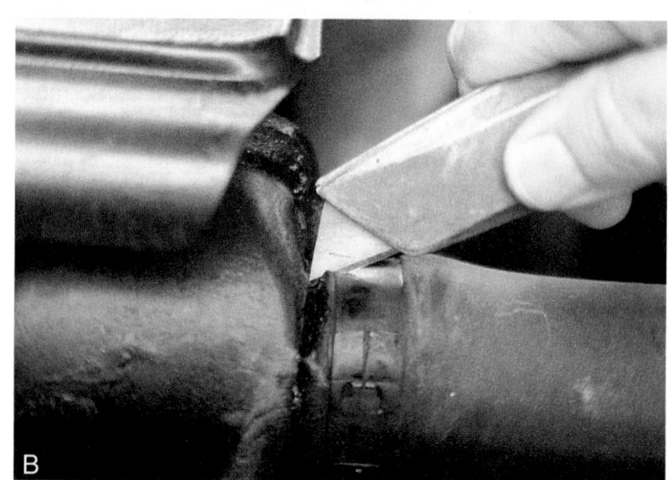

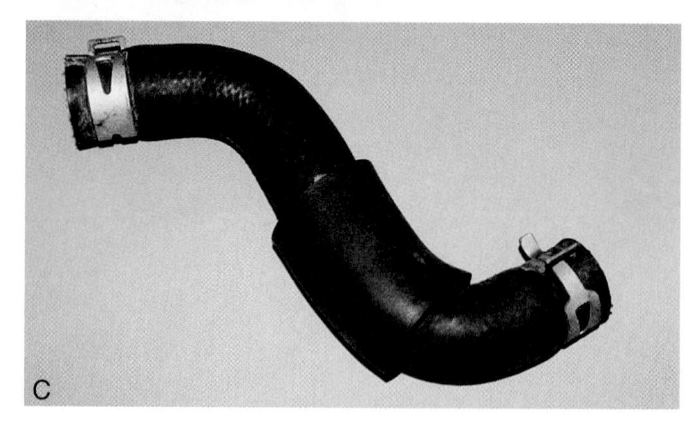

Figure 23-15. Radiator hose replacement. A—After depressurizing and draining the cooling system, use a screwdriver or pliers to remove the clamps, depending on the design. B—A knife is sometimes needed to cut the old hose from the engine or radiator. C—Place the clamps on the hose before installation. Use new clamps whenever possible.

without kinking. Before installing the new hose, slip the clamps over each end. It is best to use new clamps, especially if the original clamps were spring types.

When installing a new hose, thoroughly clean the metal hose fitting. Coat the fitting with a thin layer of nonhardening sealer. Do not coat the inside of the hose, since the sealer may be scraped off into the system.

Make certain the hose ends pass over the raised sections of the fittings far enough to properly position the clamps. Before securing the clamps, make certain any factory alignment marks used on the hoses and radiator are lined up, **Figure 23-16.** If the hose must bend, use either a specially shaped molded hose or a flexible hose.

Inspecting Belts

Observe belt condition and tension. Check belts for cracks, splits, frayed edges, glazing, and oil-soaking, **Figure 23-17.** If the vehicle uses a serpentine belt, check the pulleys for debris or foreign material. The belt should make no noise when the engine is operating.

Use a tension gauge to check belt tightness. If a tension gauge is not available, a general rule is the belt should not deflect more than 1/4″ (6mm) under light thumb pressure. Vehicles with serpentine belts usually have self-tensioning devices. The tensioner should be checked to be sure it is supplying the correct amount of belt tension.

Belt Replacement

Disconnect the battery ground clamp before removing the belt. Slack off the alternator, power steering pump and any other units; then remove the belt. Make sure the replacement belt is of the correct width, length, and construction. See **Figure 23-18.**

Clean oil and grease from the pulley surfaces and install the new belt. If the belt has a directional arrow, install the belt so the arrow faces in the direction of belt travel. Adjust belt tension. When matched belts are being replaced, make sure new belts of exactly the same length are used. Do not force a belt over the pulley edges with a screwdriver or pry bar.

Adjusting Belt Tension

There are several methods that can be used to adjust ***belt tension***—the amount the belt is tightened. One method involves pushing the belt inward and measuring the amount of deflection under a certain pressure. Keep in mind the specifications will vary with different engines. Always follow the manufacturer's directions.

Another way to check belt tension is to utilize a special belt strand tension gauge, **Figure 23-19.** The gauge deflects the belt and indicates belt tension on a gauge. Be sure all belts

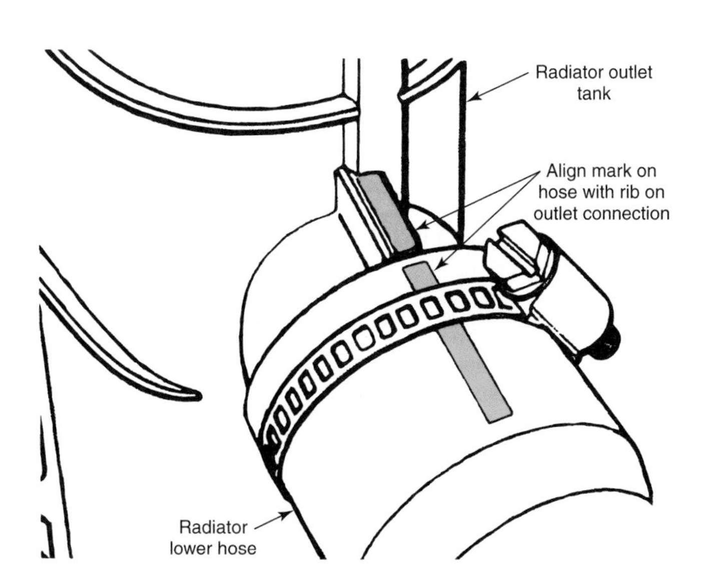

Figure 23-16. This illustrates the proper alignment of the lower radiator hose to the outlet tank. A misaligned (twisted) hose would place a strain on the outlet tank that could cause it to crack or break prematurely. (Ford)

Figure 23-17. Checking belt condition. A—Check for cracks, splits, frayed edges, and other damage. Replace the belt if wear is evident. B—This serpentine belt is cracked and missing pieces. C—This V-belt is worn and should be replaced.

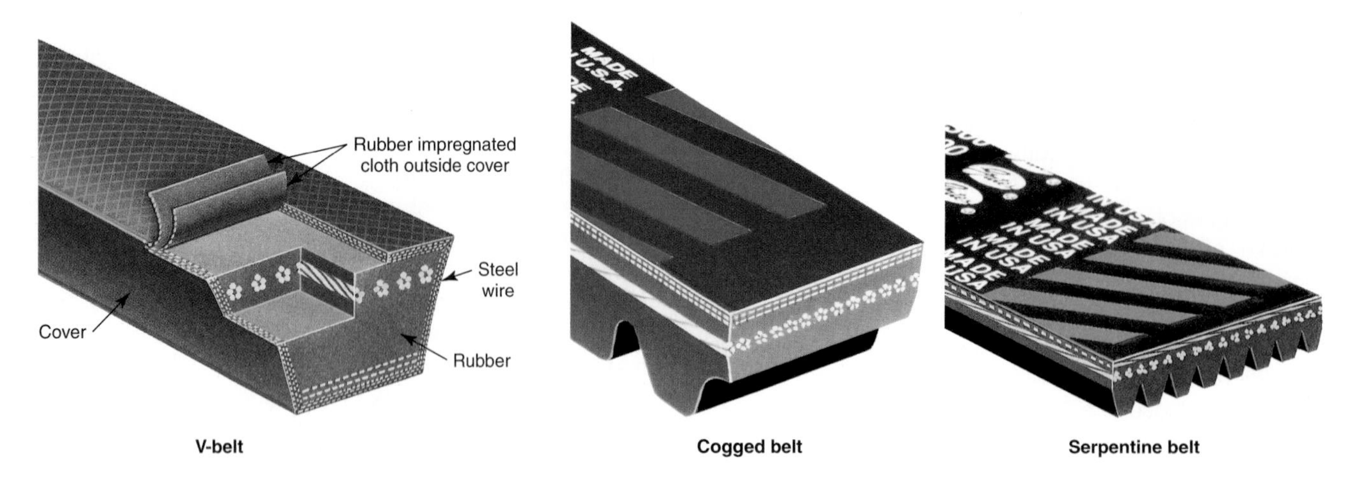

Figure 23-18. The types of belts used on automotive engines. (Gates)

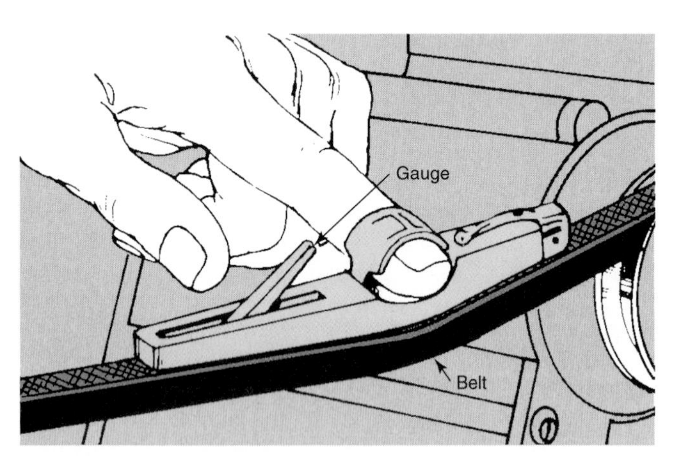

Figure 23-19. Checking belt tension with a tension gauge. (Gates)

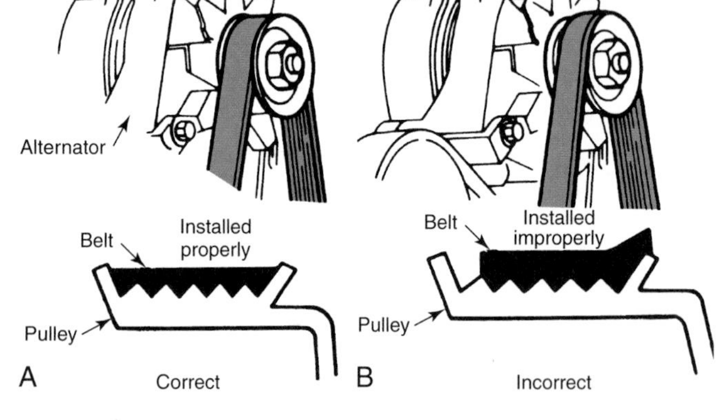

Figure 23-20. Belt seating on pulley. A—Ribbed belt is properly seated. B—The belt has been installed on the pulley incorrectly. This belt will give a false tension reading. (Ford)

are properly seated in their pulleys, **Figure 23-20,** before measuring tension, On many engines with serpentine belts, an automatic belt tensioner is used to maintain proper tension, **Figure 23-21.** If a serpentine belt does not have the correct tension, the belt tensioner is worn and should be replaced.

Belt tension can also be determined by measuring how much torque must be applied to the alternator pulley, power steering pump pulley, or other accessory pulley before it begins to slip on the belt.

Tensioning Specifications

Specifications for tensioning a new belt will be somewhat different from those for tensioning a used belt. Any belt that has been tensioned and placed in operation for a period of 10–15 minutes should be considered a used belt for purposes of retensioning.

Proper Belt Tensioning is Important

A properly tensioned belt will run quietly and will provide maximum service life. Power steering pump action, alternator output, and compressor and coolant pump efficiency will be maintained. A loose belt will squeal, flap, and reduce the efficiency of the unit being driven. Belt life will be greatly reduced.

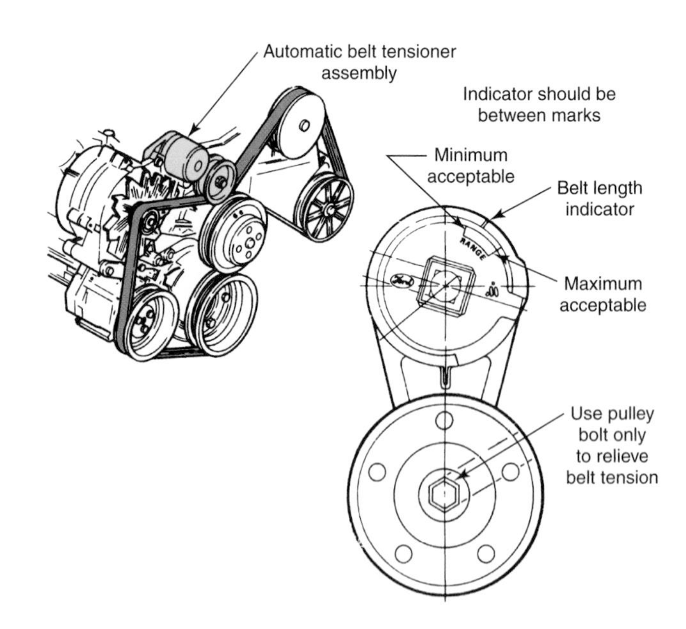

Figure 23-21. Spring-loaded belt tensioner. Note the acceptable belt length indicator. (Ford)

A belt that is too tight will place the alternator, coolant pump, or other unit bearings under a heavy strain and will cause premature bearing wear. Constant strain on the belt will also cause belt breakage. Always tension belts carefully. Use the manufacturer's specifications.

Fan Service

The function of the fan is to draw air through the radiator whenever the vehicle is not moving fast enough to push air through. In most cases, the fan is driven by an electric motor. In some cases, the fan is installed on the front of the coolant pump shaft and driven by the pump belt. Fan blade assemblies are carefully balanced and should be replaced whenever blades are bent or cracked. Do not glue, weld, or braze fan blades. Never attempt to straighten fan blades that are badly bent. When installing a fan, use a spacer where required and torque the fasteners to specifications. Never stand in line with a revolving fan. Keep your fingers away from the blades. Remove the battery ground before working on a fan.

Servicing Electric Fans

Most late-model vehicles use a cooling fan driven by an electric motor. One type of ***electric fan assembly*** is shown in **Figure 23-22.** On most vehicles, the on-board computer controls the fan motor through one or two relays. The computer turns the fan motor on and off using commands from one of two inputs. One input is the coolant temperature sensor. When the coolant temperature reaches a certain level, the computer commands the relays to close, sending power to start the fan motor. When the coolant temperature drops sufficiently, the computer will open the relay and break the circuit. This causes the fan to stop. On vehicles equipped with air conditioning, the computer turns the cooling fan on whenever the air conditioning compressor is activated. On some vehicles, a ***thermoswitch*** is used to turn on the fan independent of the on-board computer.

STOP **Warning: If the coolant temperature sensor is hot enough, the fan can begin to operate at any time, even when the ignition switch is *off*. Always disconnect the battery before working on or near the fan assembly.**

To check fan operation, the engine should be thoroughly warmed up. The fan should come on when the coolant temperature reaches a certain value, as listed in the appropriate service manual. Remember most electric fans will also operate whenever the air conditioner is on, regardless of engine temperature. You can use this to quickly check fan operation by simply turning on the compressor. If the fan does not come on when it is supposed to, check for a defective relay, a blown fuse, or a defective fan motor. Use a scan tool or waveform meter to check computer output and operation of the coolant temperature sensor. One way to isolate the problem is to use jumper wires to directly energize the fan motor. If the motor will not turn, it is defective. If the motor operates, the problem is in the control system.

Electric Fan Replacement

To replace the cooling fan and motor, first disconnect the battery negative cable. Then, remove the electrical connector at the fan motor, **Figure 23-23.** Remove the bolts holding the

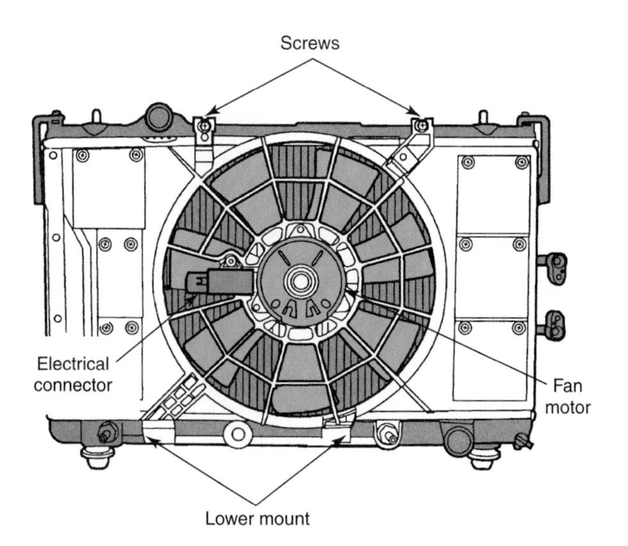

Figure 23-22. One type of electrically driven fan assembly. The fan motor only runs when the coolant temperature reaches 193°–207°F (89°–97°C) or when the air conditioning is on. (DaimlerChrysler)

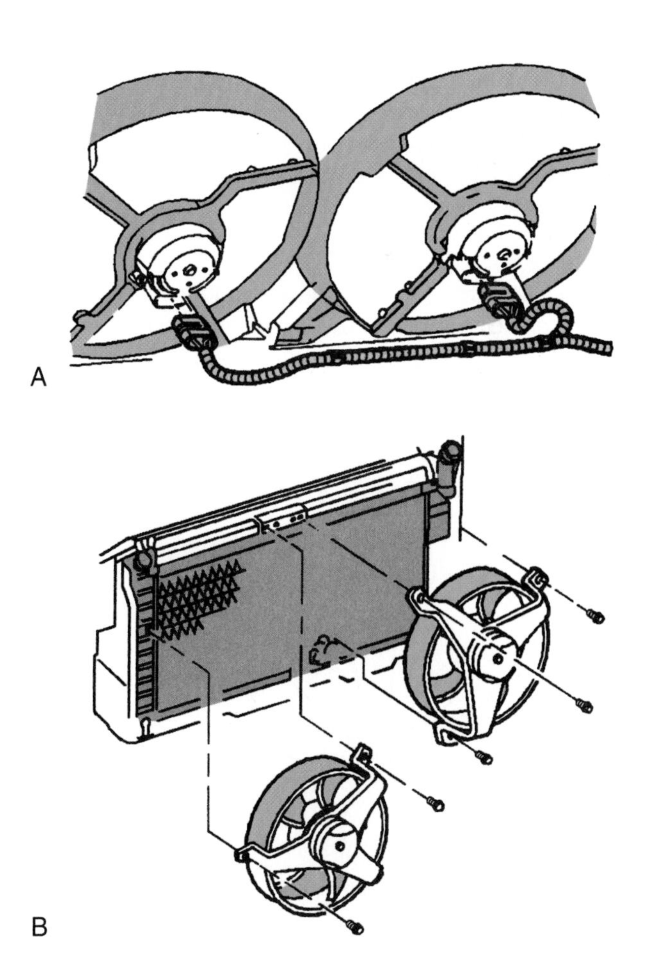

Figure 23-23. Cooling fan replacement. A—Remove electrical connectors. B—Remove fasteners and then, the defective fan. Additional parts may need to be relocated or removed to access the fans. (DaimlerChrysler)

fan assembly to the radiator support, and lift the assembly out of the vehicle. Place the fan on a bench and remove parts as necessary.

Note: In some cases, you may need to replace both the fan and motor.

If the motor is defective, it should be replaced with a unit specifically designed for the vehicle. Install replacement parts and put the assembly into position on the radiator support. Install and tighten the bolts. Reattach the motor electrical connector and battery negative cable. Finally, start the engine and make sure the fan motor energizes at the proper temperature. Always follow the manufacturer's recommendations.

Servicing Fluid Clutch Fans

The ***fluid clutch fan*** is a sealed drive clutch assembly filled with silicone. The two sides of the clutch separate the fan blades from the fan pulley. As engine speed increases, the torque required to turn the fan also increases. At a predetermined speed (about 2500–3200 rpm), the silicone driving fluid in the clutch allows enough slippage to limit maximum fan blade rpm.

Some models of fluid clutch fans use either a bimetallic strip or a bimetallic spring that senses radiator temperature. The bimetallic spring operates a valve designed to either start the fan turning when the temperature indicates the need for cooling or to alter the maximum rpm in accordance with cooling needs.

Checking Fluid Clutch Fan Operation

The fluid clutch fan unit is sealed, and must be replaced when it is defective. Silicone oil leaking from the unit, a broken or stuck spring, or a faulty valve can render the unit inoperative. The fan illustrated in **Figure 23-24** can be checked by running the engine until normal operating temperature is reached. When the engine is stopped, the fan should revolve no more than 1/2 turn. If the fan continues to turn, the fluid clutch unit is defective and should be replaced.

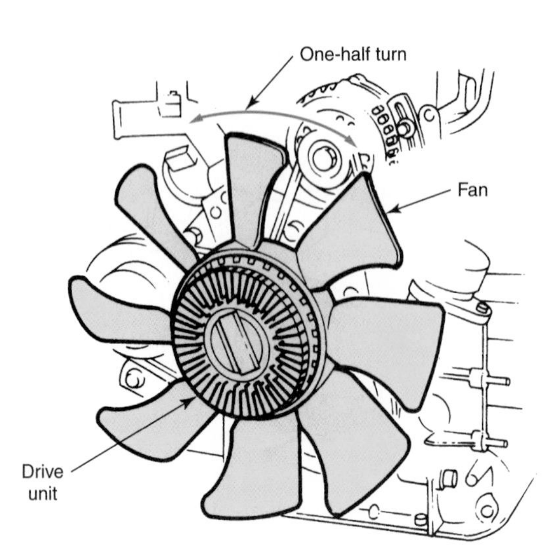

Figure 23-24. Checking the fan fluid drive unit. (Mazda)

Replacing Engine-Driven Fan Clutch

To replace an engine-driven fan or fan clutch, first disconnect the battery negative cable to prevent accidental engine starting. If needed, remove or relocate the fan shroud to allow clearance to take out the fan and clutch. Next, remove the bolts holding the fan and fan clutch to the drive pulley.

Note: If possible, do not remove the belt(s). Leaving the belts in place makes fan removal and installation easier.

Remove the fan and fan clutch. Remove the bolts holding the fan to the fan clutch and separate the two parts. See **Figure 23-25.** Reassemble with new parts as necessary,

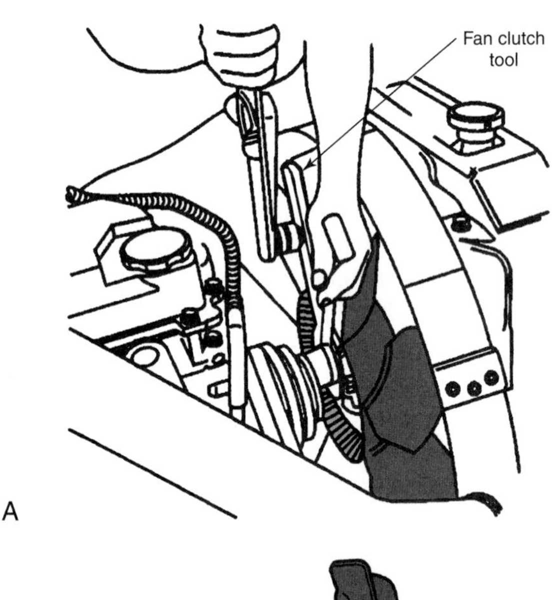

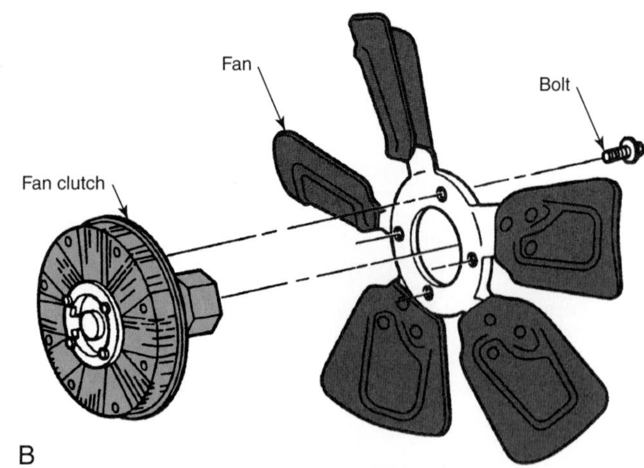

Figure 23-25. Fluid clutch fan replacement. A—Leave the belts in place when loosening the fan's fasteners. B—Fan clutches are usually bolted to the fans. (General Motors, DaimlerChrysler)

being sure the fan blades point in the proper direction. Then place the assembly on the pulley, and install the bolts. Start all bolts before tightening any bolt. Ensure the drive belts are tight, then reinstall the battery negative cable. Start the engine and check fan operation.

Thermostat Service

The ***thermostat*** is used to prevent coolant circulation when the engine is cold. This allows the engine to warm up quickly. Thermostats in most vehicles are set to open at 190°F (88°C). Thermostats can cause overheating by either failing to open or by not opening far enough. The result can be a cracked or warped block or head, blown head gaskets, extra carbon formation, detonation, burned valves, or damaged bearings. Overcooling (engine running too cool) can result from a thermostat sticking in the open position. Damage can include crankcase sludging, poor fuel vaporization, sluggish performance, poor gas mileage, and oil dilution.

OBD II Thermostat Monitor

For the vehicle emissions controls to operate properly, the engine should reach its normal operating temperature as quickly as possible. If the thermostat sticks open, the engine will warm up very slowly. During cold weather, the engine may never reach normal operating temperature. Cold engine operation keeps the emission control system from operating properly, resulting in high emissions and poor fuel mileage.

All vehicles manufactured for the 2000 model year and later have a thermostat monitor to alert the vehicle driver when the thermostat sticks open. The thermostat monitor is part of the OBD II emission control system. The monitor consists of the coolant temperature sensor and an internal timer in the vehicle computer. When the engine is first started, the timer begins counting. If the coolant has not warmed to approximately 80% of the thermostat opening temperature within 5–14 minutes, the computer sets a trouble code and turns on the malfunction indicator lamp (MIL).

Note: If the thermostat sticks closed, the engine will quickly overheat. The temperature gauge or warning light will quickly warn the driver of overheating, even if steam is not visible from under the hood.

Checking Thermostats

When a defective thermostat is suspected, drain coolant from the system until the coolant level is below the thermostat housing. Remove the housing or the housing cap. Remove and rinse the thermostat. In **Figure 23-26,** the housing is off and the thermostat is removed.

Inspect the thermostat valve. It should be closed snugly. Hold it against the light to determine how well the valve contacts the seat. A spot or two of light showing is not cause for rejection. If light shows all around the valve, discard the thermostat.

To check opening temperature, suspend the thermostat with the pellet facing downward in a container of water. The thermostat must be completely submerged and must not touch the container sides or bottom.

Suspend an accurate thermometer (it must not touch container sides or bottom) in the water, **Figure 23-27.** Place the container over a source of heat and gradually raise the water temperature. Stir the water gently as the temperature increases.

Watch the thermostat closely. As the thermostat begins to open, note the water temperature. It should be within 5–10°F (3–6°C) of the temperature rating stamped on the thermostat.

Continue heating the water until the valve is fully opened and note the water temperature. In general, the thermostat should be wide open at a temperature around 20–24°F (11–13°C) above opening temperature. Discard any thermostat that does not meet specifications.

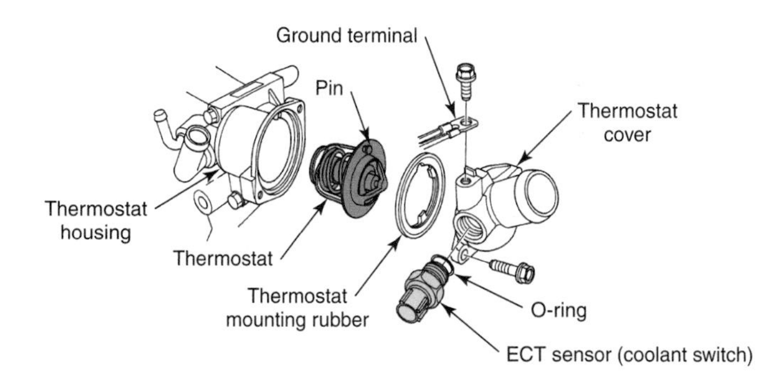

Figure 23-26. Thermostat setup. Note the coolant temperature sensor, which monitors coolant temperature and sends an electrical signal to the electronic control unit. (Honda)

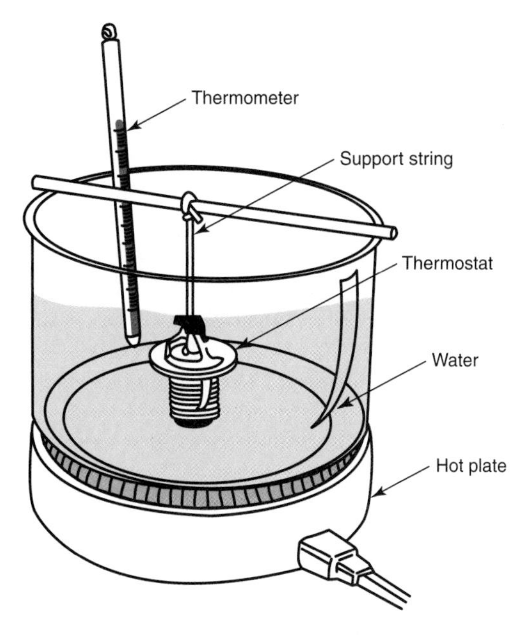

Figure 23-27. Checking thermostat opening temperature. Note that both the thermometer and the thermostat are kept clear of container sides and bottom. (Honda)

Thermostat Installation

When replacement is necessary, select a thermostat of the correct temperature range. Never leave the thermostat out of the engine to try to cure overheating. The thermostat is essential. Clean out the thermostat pocket and housing. If the system is rusty, it should be cleaned and flushed. Always install the device so the thermostatic element will be in contact with the coolant in the engine block. If you cannot determine which way is correct, refer to the manufacturer's service literature. Reversing the thermostat so the pellet faces away from the engine will cause serious overheating. As the coolant in the block heats up, it cannot contact the pellet. The coolant in the engine may begin to boil before the thermostat can open enough to allow sufficient flow through the radiator.

When reinstalling the thermostat housing, be sure to use a new gasket, O-ring, or silicone sealer. Follow the manufacturer's recommendations. Torque the housing fasteners to specifications.

Radiator Removal and Replacement

Before removing the radiator, **Figure 23-28,** drain the cooling system. Then, disconnect hoses and oil cooler lines (where used). Remove the fan(s) and shroud, if necessary. Remove the radiator support fasteners and carefully lift out the radiator. Do not dent the cooling fins or tubes.

If the radiator contains a transmission oil cooler, plug the entry holes so foreign matter cannot enter. Also plug the lines from the transmission. Place the radiator in a spot where it will be protected from physical damage.

In some cases, leaks can be repaired by a radiator repair shop. Leaks may be repaired by careful soldering (brass and copper radiators) and, in some radiators, by replacing gaskets. Special repair materials are available for use on aluminum radiators.

However, in most cases, leaking radiators are replaced. Once the old radiator is removed, compare it to the new radiator, and transfer parts as necessary. Many replacement radiators use pipe plugs to seal the openings. These pipe plugs are left in any fitting not used, so make sure they are tight before radiator installation. Place the new radiator in position and install the fasteners. Reinstall the fan(s), shroud, hoses, and transmission cooler lines. Then add coolant, start the engine and bleed the system. After all air has been removed, check the radiator and hose connections for leaks. Check and add transmission fluid, if necessary.

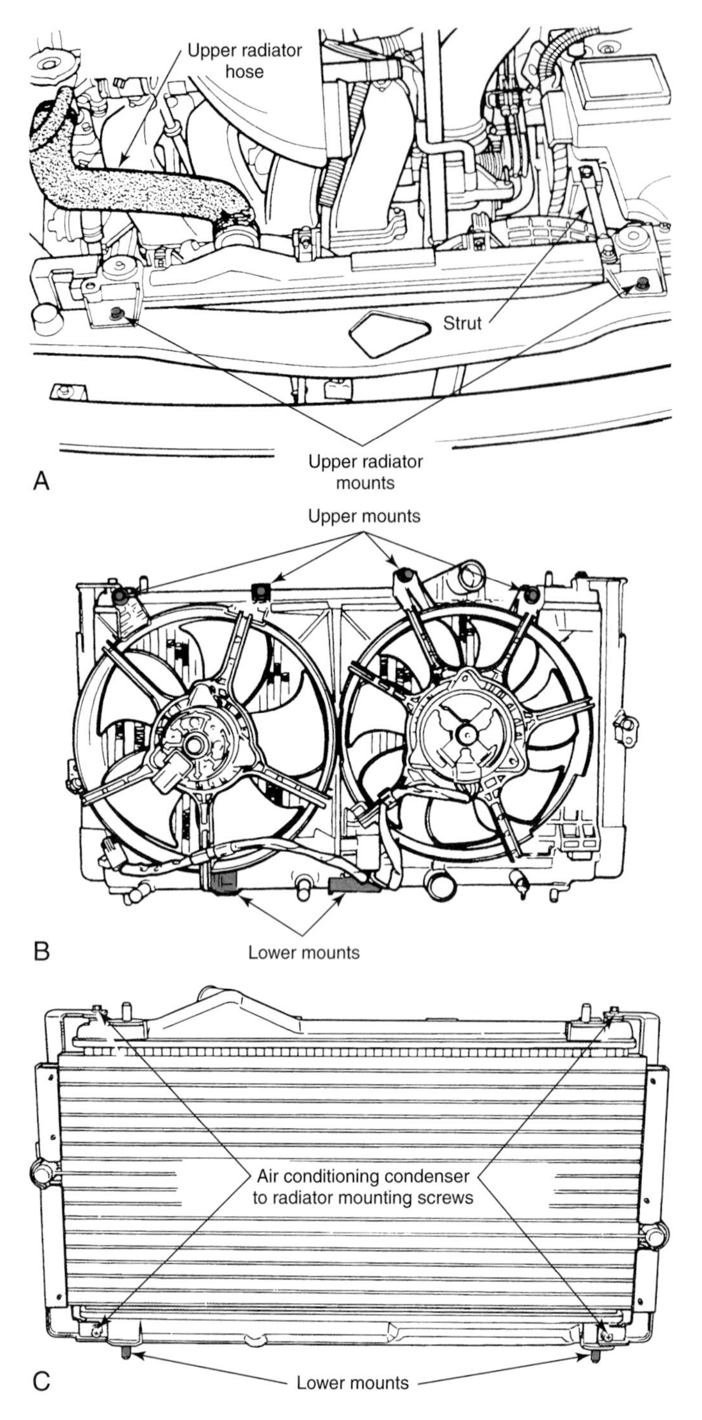

Figure 23-28. Radiator replacement. A—Remove all fasteners and hoses after draining the coolant. B—Remove cooling fans, if needed. C—Some air conditioning system condensers are bolted to the same supports as the radiator. Exercise care in handling such setups. (DaimlerChrysler)

Caution: Proper radiator flow on newer cars is very critical. Make sure any replacement radiator comes from a reputable manufacturer.

Coolant Pump Service

The ***coolant pump,*** also known as the *water pump,* circulates the coolant through the engine and radiator. The pump illustrated in **Figure 23-29** utilizes the block or the engine front cover to form a housing around the ***impeller*** (pumping unit). Other construction methods locate the pump away from the block and connect it to the block by cast passages or hoses. Most pumps are driven by a belt from the engine crankshaft.

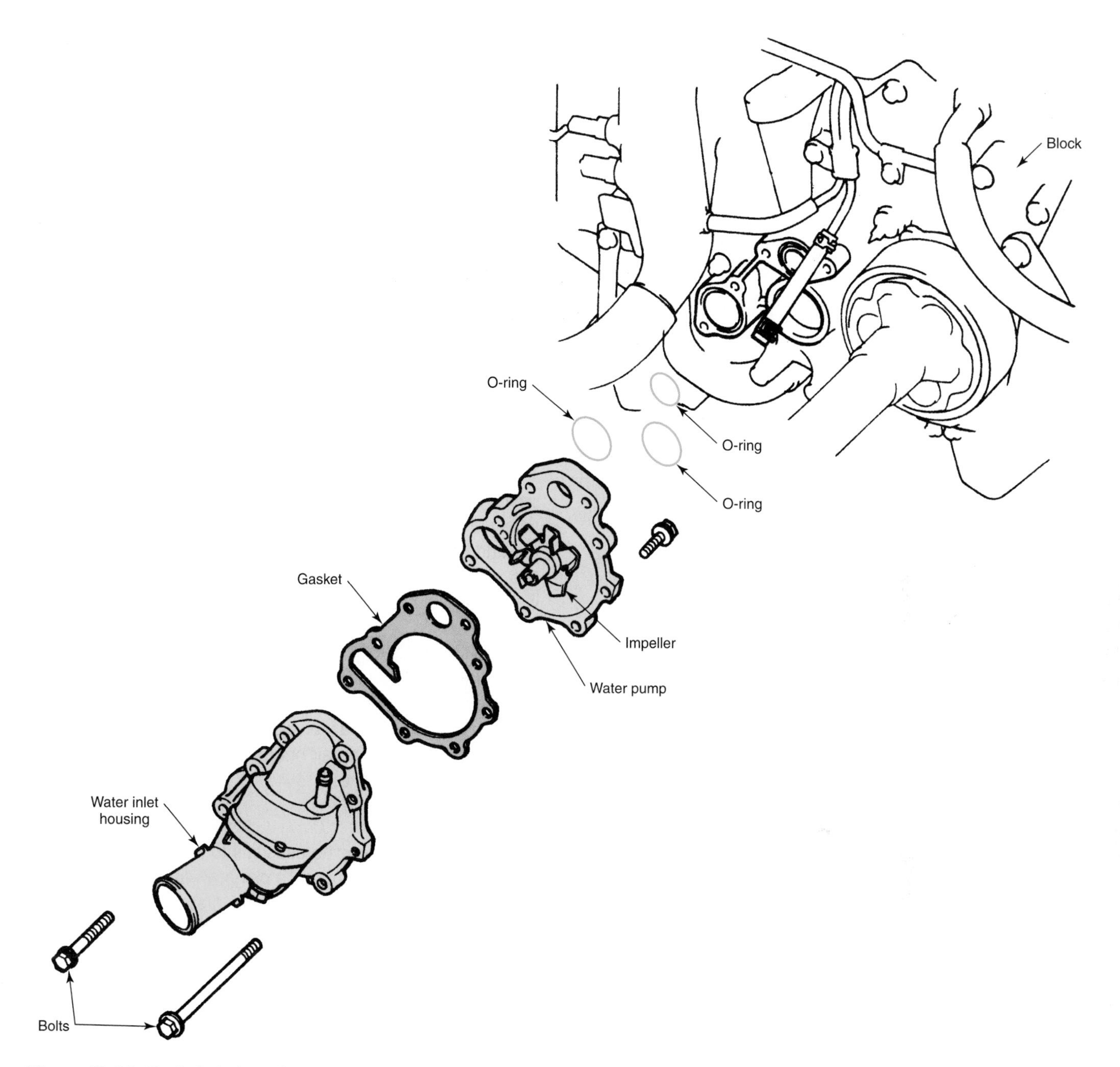

Figure 23-29. Exploded view of a self-contained water pump. Note the O-rings, which are used to seal the pump-to-block connections. New O-rings are generally used if the pump is removed for service or replacement. (Toyota)

Pump Inspection

The most common coolant pump problems are worn bearings and leaking seals. In a few cases, the impeller blades will be worn down by corrosion or broken by debris in the cooling system. Begin coolant pump inspection by checking the pump for signs of coolant leakage at the seal drain hole and gasket area. Also, check the housing for cracks.

Loosen the drive belt to remove pressure from the pump bearings. Grasp the hub and attempt to move the shaft up and down. Little or no play should be present. In many cases, the pump bearings will be so worn they will have play even with the belt tightened. Next, spin the shaft to detect any bearing roughness.

Pump Removal

Coolant pumps are replaced rather than rebuilt. Begin pump replacement by depressurizing and draining the cooling system. Then, remove the fan and fan clutch if they are attached to the pump pulley. If necessary, remove the fan shroud to gain additional clearance. Next, remove the drive belt(s) as necessary and remove the pump pulley. Remove any accessories and brackets as needed to access the pump.

Note: On some engines where the pulley is bolted to the pump hub, it is easier to loosen the pulley-to-hub bolts while the belt(s) still hold the pulley stationary.

If any hoses are attached to the pump, remove them. If the pump uses bypass hoses, now is a good time to replace them. Remove the bolts and nuts holding the pump to the engine and remove the pump, **Figure 23-30.**

Pump Installation

Scrape all old gasket material from the pump mating surfaces on the engine. Inspect any baffles or channel plates in the pump cavity. Replace any plate that is bent or corroded. In some cases, you may need to transfer studs or fittings from the old pump to the new one.

Note: Compare the impellers on the old and new pumps. Both sets of impeller blades should be about the same size, have the same pitch direction, and number of blades. If the impellers do not match, do not use the pump.

Use a new gasket or gaskets, O-rings, or silicone sealer as required. Place the gasket on the pump using a small amount of sealer or spray adhesive. Use only enough to keep the gasket in place. Then place the pump on the engine. Install and tighten the fasteners and install any hoses that were removed. Install the pulley and start the attaching bolts.

Replace the belt(s) and install the fan assembly if necessary. Pulley misalignment will cause rapid belt wear. Sight across the pulleys to make certain the drive hub is positioned to bring the pump pulley in line with the crankshaft drive pulley and any other pulley involved. Refill the cooling system and start the engine. Add more coolant, bleed the system, and check pump operation.

Freeze Plug Repair

Freeze plugs, sometimes called *core plugs,* are installed to close holes left in the engine block and head after casting. Even though these plugs are called freeze plugs, they will not protect the block from freeze damage. Leaking freeze plugs should always be replaced.

Drive a sharp-nosed pry bar through the center of the freeze plug. By prying sideways, the plug should pop out. Another technique involves drilling a small hole in the center of the plug. Punching or drilling near an edge can damage the plug seat ledge in the block. A hook-shaped rod is then inserted in the hole, and the other end of the rod is attached to a slide hammer puller. A few taps should pull the plug out.

Clean the seating area of the plug hole thoroughly. Coat both the plug and the hole seat with nonhardening sealer and drive the plug into place. Special tools can be used to seat both cup and expansion-type core plugs, **Figure 23-31.**

Where driving space is limited, special plugs can be used. These plugs are pulled into place and retained by a screw fastener. Another type is made of rubber with a central nut and bolt arrangement. After the plug is put into place, the bolt and nut are tightened, expanding the rubber against the walls of the block.

Air-Cooled Systems

A few modern vehicles use ***air-cooled engines.*** Although vehicles with air-cooled engines are rare, you may eventually encounter a vehicle so equipped. The following paragraphs describe repair operations that can be performed on an air-cooled engine.

Cleaning Fins on an Air-cooled Engine

The air-cooled engine uses fins to remove heat from the engine. These fins can become plugged with dirt, plant material, or other debris. To clean the fins, first allow the engine to cool completely. Then remove the engine shrouds as necessary and use compressed air to blow loose material away

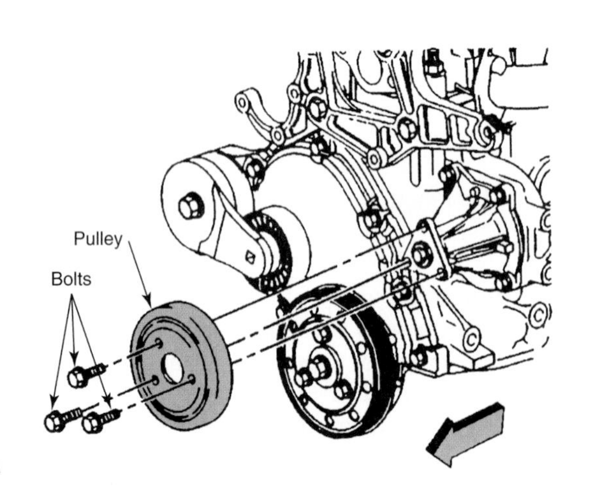

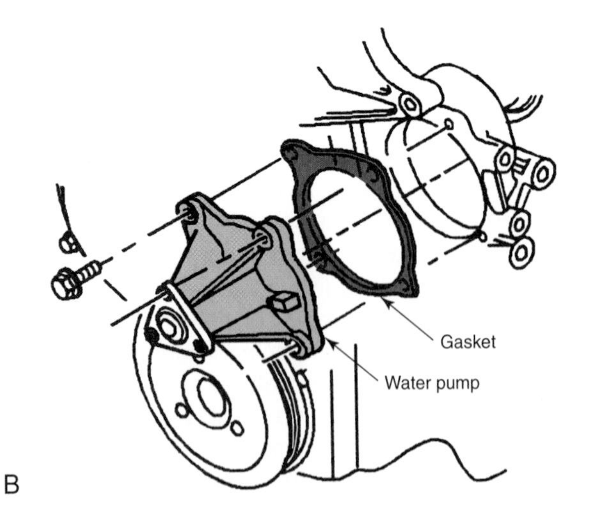

Figure 23-30. Water pump service. A—Remove the drive belt and pulley. B—Remove and replace the water pump. Use a new gasket, O-ring, or sealer when installing the new pump. (General Motors)

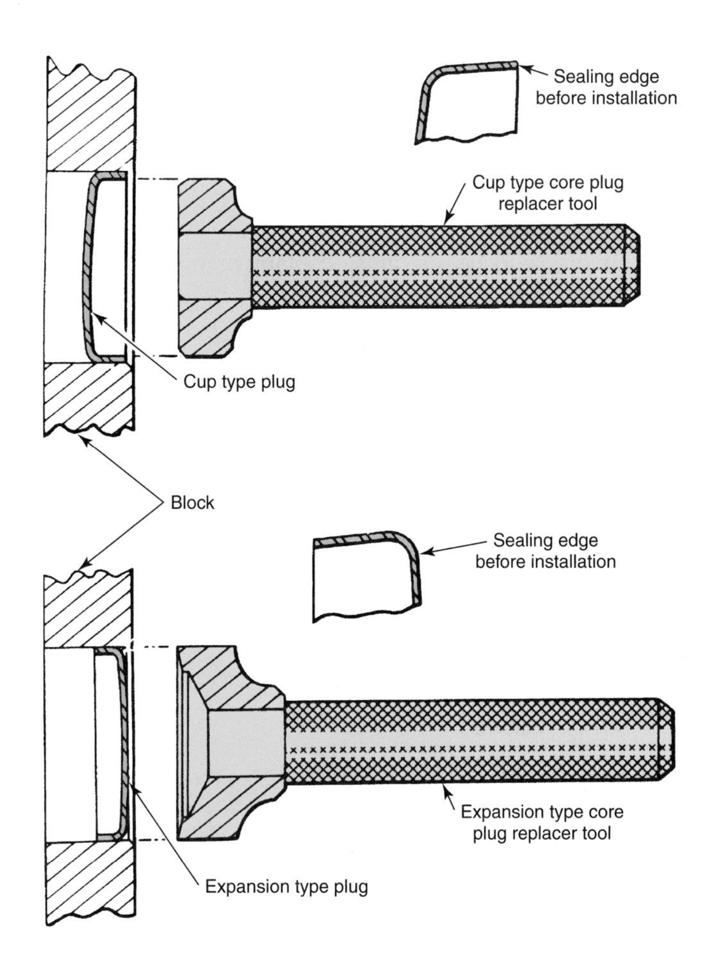

Figure 23-31. Typical core plug installation tools. These tools and plug types are not interchangeable. If the wrong tool or core plug is used, a damaging leak or plug "blow out" could occur. (Ford)

from the fins. If the fins are extremely dirty or oil soaked, spray a solution of detergent and water on them.

Caution: The engine must be allowed to thoroughly cool before spraying any water or detergent on the fins.

Allow the detergent to sit on the fins for about 15 minutes, then spray the fins with a strong stream of water. Do not spray water on ignition or fuel system components. Once the fins are clean, reinstall the shrouds and start the engine to blow out any remaining water.

Replacing Blower Belt

To replace the blower drive belt, determine which of the belt-driven accessories can be moved to remove the belt. Loosen the fasteners at the accessory and push it inward. Once the belt has enough slack, slide it over the pulley and remove it from the engine. To install the new belt, place it over the pulleys and pull the movable accessory outward to tension it. Check belt tightness with a tension gauge before fully tightening the bolts.

Replacing Blower and Bearing

To replace the blower and blower bearing, remove the belt as explained earlier. Then, remove the portion of the shroud holding the bearing assembly. Remove the bearing and blower from the shroud. Some bearings must be pressed from the shroud assembly bracket. Install the new parts on the shroud and reinstall the shroud on the engine. Reinstall and tighten the belt and recheck blower operation.

Replacing Thermostat

To replace an air-cooled engine thermostat, locate the thermostat and remove the surrounding shrouds. Remove the link from the thermostat door and remove the fasteners holding it to the engine. Install the new thermostat and reattach it to the door. Reinstall shrouds as necessary. Start the engine and check to be sure the thermostat opens at the right temperature.

Tech Talk

One of the most common mistakes that beginning technicians make is to overtighten drive belts. Overtightening a belt can ruin bearings in the alternator, water pump, air conditioner compressor, power steering pump, or air injector pump. Many times, overtightening occurs while trying to cure a noisy belt. Usually, when a belt starts squealing, it is glazed (shiny) and should be replaced. If an almost-new belt begins to act up, check for a grooved pulley. Grooved pulleys are especially common on alternators and air conditioner compressors.

Use a belt tension gauge, if one is available. If a gauge is not available, tighten the belt so that there is about 1/4″ (6mm) of deflection when you lightly press on the belt between the pulleys. Another way to check a belt used to drive the alternator is to try turning the alternator fan by hand. If you can turn the fan the belt is too loose. Tighten the belt until it is just past the point where you can no longer turn the fan.

Summary

When working properly, the cooling system is quite efficient. The technician should understand the function of all parts in the system.

Ethylene glycol and propylene glycol are used to make antifreeze. Long-life antifreeze may use organic acid technology to minimize corrosion and extend coolant life. Antifreeze solutions include inhibitors and coolant pump seal lubricants. A hydrometer, dye, or special test strips may be used to check the strength of the system antifreeze solution.

Always use clean, soft water when filling cooling system. Drain and flush the system before filling it with fresh water and antifreeze. Tighten hose clamps and check for leaks.

Moderately contaminated systems may be cleaned by using chemical cleaners, running the engine, and draining. Heavy rusting or scaling requires the use of stronger chemicals and reverse flushing of the radiator and block. Pressure test the system to detect external leaks. Do not exceed the pressure stamped on the pressure cap. Check the system while cold and while hot. A special tool and fluid can be used to detect combustion leaks in the cooling system.

The pressure cap should be tested. Inspect the radiator filler neck cap seat and locking cams. The overflow tube must be open. Use great care when removing a pressure cap while the engine is hot. Never add water to an overheated engine. Allow the engine to cool somewhat. Then, start the engine, run it at fast idle, and add water slowly.

Check hose condition. Look for hardening, cracking, swelling, and softening. Tighten the hose clamps. Use nonhardening cement on hose fittings when replacing a hose. Use hose of the correct shape and size. Avoid forcing bends in hose unless of the flexible type.

Check the drive belts for proper condition. Cracked, frayed, split, glazed, or oil-soaked belts should be replaced. Set belt tension to specifications by using the belt deflection method or the belt strand tension gauge.

Never try to straighten, weld, or glue damaged fan blades. Fans driven by electric motors are used in most installations. They are controlled by coolant temperature sensors. Motors are generally not serviceable and should be replaced when defective. Fluid clutch fans must generally be serviced as a unit, although some models permit the replacement of a bimetallic spring or bimetallic strip and operating piston.

Always remove the ground clamp from the battery before working on a coolant pump, fan, radiator, hoses, or belts. Test thermostats for condition and for initial opening and full opening temperature points. Install thermostats with the thermostatic unit (pellet) facing the engine so the unit will be contacted by the coolant in the block.

In some instances, the radiator must be removed from the vehicle for cleaning, testing, and repair. Some coolant pumps can be rebuilt. When rebuilding, assemble so that part positioning is correct. Use care when installing new freeze plugs.

Review Questions—Chapter 23

Do not write in this book. Write your answers on a separate sheet of paper.

1. Ethylene glycol antifreeze contains _____.
   (A) rust inhibitors
   (B) stop leak
   (C) methanol
   (D) ethanol
2. During the cooling system cleaning process, the heater control should be set to the _____ position.
3. Rust and scale can be removed from a badly clogged radiator by _____.
   (A) reverse flushing
   (B) chemical treatment
   (C) power flushing
   (D) compressed air blasting
4. When reverse flushing the engine block, the thermostat should be _____.
   (A) in place
   (B) removed
   (C) installed upside down
   (D) wired to prevent opening
5. When pressure testing the cooling system, limit the maximum pressure to that stamped on the _____.
6. Name three hose conditions that call for hose replacement.
7. List three drive belt conditions that call for belt replacement.
8. Before working on fans, coolant pumps, or V-belts, always disconnect the _____.
9. Thermostats on most vehicles are set to open at _____.
   (A) 160°F (71°C)
   (B) 180°F (82°C)
   (C) 190°F (88°C)
   (D) 230°F (110°C)
10. List three possible coolant pump problems.

ASE-Type Questions

1. Cooling systems must be protected from rust and corrosion by using _____.
   (A) clean, soft water
   (B) inhibitors
   (C) alcohol antifreeze
   (D) a 15 psi (103 kPa) pressure cap
2. Ethylene glycol is _____.
   (A) poisonous
   (B) a fire hazard
   (C) harmful to paint finishes
   (D) wood alcohol
3. On some modern vehicles, the cooling passages in the engine block must be _____ to properly install new antifreeze.
   (A) cold
   (B) hot
   (C) bled
   (D) Both A and C.
4. Technician A says it is best to reverse flush the radiator and block separately. Technician B says one should always reverse flush the vehicle heater. Who is right?
   (A) A only.
   (B) B only.
   (C) Both A and B.
   (D) Neither A nor B.
5. Technician A says combustion leaks can be detected only by removing the cylinder heads for a visual inspection. Technician B says ethylene glycol leaking into the cylinders can clog rings, bearings, and other parts. Who is right?
   (A) A only.
   (B) B only.
   (C) Both A and B.
   (D) Neither A nor B.

6. Completely removing a pressure cap when an engine is hot can cause a _____.
   (A) cracked block
   (B) warped valve
   (C) sudden, violent flash of steam
   (D) bulged radiator
7. Technician A says thermostats should be installed with the thermostatic element contacting the coolant in the engine. Technician B says leaving the thermostat out of the engine is a good way to cure overheating. Who is right?
   (A) A only.
   (B) B only.
   (C) Both A and B.
   (D) Neither A nor B.
8. All of the following statements about coolant pumps are true, *except:*
   (A) all pumps can be replaced.
   (B) all pumps contain an impeller.
   (C) all pumps can be repaired.
   (D) all pumps contain bearings and seals.
9. Core plugs can be sources of _____.
   (A) leaks
   (B) overheating
   (C) corrosion
   (D) Both A and C.
10. Air cooled engines use all of the following parts, *except:*
   (A) drive belt.
   (B) blower.
   (C) core plugs.
   (D) cooling fins.

Suggested Activities

1. Find the lowest outside temperature that typically occurs in your area. Use an antifreeze chart to determine the ratio of antifreeze and water that will protect against freezing. Then find the cooling system capacity of your vehicle in quarts. Determine how many quarts of antifreeze are needed to protect your cooling system.
2. Obtain a cooling system pressure tester, properly install it on the radiator filler neck, and pressurize the cooling system to the pressure listed on the radiator cap. Wait about 5 minutes and then answer the following questions.

Caution: Allow the engine to cool off before removing the radiator cap.

   • Has the pressure dropped?
   • Are there any cooling system leaks?
   • What was done to correct them?
3. Obtain a cooling system pressure tester and properly install it on the radiator cap. Pressurize it to the listed pressure. Wait about 5 minutes and then answer the following questions.
   • Has the pressure dropped?
   • What do you think is the cause?
4. Obtain an antifreeze tester and test the antifreeze in your vehicle or one assigned by your instructor. Discuss whether the vehicle has sufficient freezing and corrosion protection.
5. Check belt tension using a tension gauge. Adjust the belt(s) if necessary. While adjusting, check belt condition. If any belts are glazed, frayed, or oil-soaked, consult your instructor.

Cooling System Problem Diagnosis

Problem: Overheating	
Possible cause	**Correction**
1. Coolant level low.	1. Add coolant and check for leaks.
2. Drive belt loose.	2. Adjust belt tension.
3. Drive belt broken.	3. Replace belt.
4. Drive belt glazed or oil-soaked.	4. Replace belt.
5. Thermostat stuck closed.	5. Replace thermostat.
6. Radiator pressure cap inoperative.	6. Replace pressure cap.
7. Bugs, leaves, other debris on radiator core.	7. Flush with water from back to front.
8. Rust scale clogging radiator.	8. Flush radiator and install rust inhibitor.
9. Rust scale clogging in block.	9. Flush cooling system and install rust inhibitor.
10. Valve timing off.	10. Reset valve timing.
11. Air leaks into system.	11. Pressure test and repair leaks.
12. Coolant hoses clogged.	12. Replace hoses.
13. Coolant hose collapsed.	13. Replace hose.
14. Low antifreeze boiling point.	14. Change antifreeze or thermostat.
15. Late ignition timing.	15. Adjust timing.
16. Leaking cylinder head gasket.	16. Replace gasket. Check block and head surfaces.
17. Water pump impeller slipping or broken.	17. Replace water pump.
18. Brakes dragging.	18. Adjust brakes.
19. Vehicle overloaded.	19. Advise driver.
20. Manifold heat valve stuck or broken.	20. Loosen or repair.
21. Fan speed slow—improper pulley size.	21. Install larger diameter pulley.
22. Low engine oil level.	22. Add oil to full mark.
23. Frozen coolant.	23. Thaw and add antifreeze.
24. Exhaust system back pressure.	24. Change muffler or open up dented pipe.
25. Lean carburetor mixture.	25. Clean carburetor. Install proper size jets.
26. Wrong cylinder head gasket.	26. Install correct head gasket.
27. Ignition timing retarded.	27. Advance ignition timing.
28. Defective electric fan motor or control.	28. Replace as necessary.
29. Defective spark delay valve.	29. Replace valve.
30. Core sand in head or block.	30. Clean or replace.
31. Inoperative fan fluid coupling (fan clutch).	31. Replace fan clutch.
32. Aftermarket (add-on) air conditioner on vehicle equipped with a standard cooling system.	32. Install heavy-duty cooling system parts.
33. Defective electric fan ambient temperature switch.	33. Replace fan switch.
34. Clogged catalytic converter(s).	34. Replace catalytic converter(s).
35. Lean fuel mixture.	35. Adjust fuel mixture.
36. Inoperative electric fan.	36. Test and repair.
Problem: Overcooling and/or slow warmup	
Possible cause	**Correction**
1. Thermostat stuck open.	1. Replace thermostat.
2. Weather extremely cold.	2. Cover a portion of the radiator.
3. No thermostat.	3. Install a thermostat.
4. Low-temperature thermostat.	4. Install correct thermostat for engine.
Problem: Apparent overheating or overcooling	
Possible cause	**Correction**
1. Faulty temperature sender.	1. Replace gauge sender.
2. Faulty temperature gauge.	2. Replace gauge.
3. Faulty gauge wiring.	3. Check and repair any breaks or loose connections.
4. Complete unit faulty (bulb type).	4. Replace entire unit—gauge, tubing, and bulb.
5. Improper fan size, type, or speed.	5. Replace with correct unit.

(continued)

Cooling System Problem Diagnosis *(continued)*	
Problem: Belt squeal upon acceleration	
Possible cause	**Correction**
1. Belt loose. 2. Belt glazed. 3. Excessive friction in belt-driven accessory.	1. Adjust belt tension. 2. Replace belt. 3. Repair defective unit.
Problem: Belt squeal at idle	
Possible cause	**Correction**
1. Belt loose. 2. Pulleys misaligned. 3. Uneven pulley groove. 4. Foreign material on belt. 5. Belt width not uniform. 6. Belt tensioner loose or broken.	1. Adjust belt tension. 2. Align all pulleys. 3. Replace pulley. 4. Clean or replace belt. 5. Replace belt. 6. Tighten or replace.
Problem: Belt jumps from pulley or rolls over in pulley groove	
Possible cause	**Correction**
1. Belt loose. 2. Pulleys misaligned. 3. Broken cords (internal). 4. Mismatched belts. 5. Eccentric pulley. 6. Loose pulley.	1. Adjust belt tension. 2. Align all pulleys. 3. Replace belt. 4. Install matched set of belts. 5. Replace. 6. Tighten or replace.
Problem: Noisy water pump	
Possible cause	**Correction**
1. Bearing worn and rough. 2. Seal noisy. 3. Loose impeller.	1. Repair or replace pump. 2. Add inhibitor-water pump lube mixture to system. 3. Rebuild or replace pump.
Problem: Buzzing radiator cap	
Possible cause	**Correction**
1. Coolant boiling.	1. Shut engine off and correct cause of overheating.
Problem: Coolant loss	
Possible cause	**Correction**
1. Leaking radiator. 2. Leaking hose. 3. Cracked hose. 4. Overheating. 5. Overfilling. 6. Air leak at bottom hose. 7. Blown head gasket. 8. Water pump seal leaking. 9. Heater core leaking. 10. Cracked block or head. 11. Radiator pressure cap inoperative. 12. Leaking block freeze plugs. 13. Improper cylinder head tightening. 14. Leak at temperature sender. 15. Leaking surge tank. 16. Leak at fasteners that enter water jacket. 17. Cracked water jacket or thermostat housing. 18. Damaged coolant recovery bottle. 19. Leaking petcock.	1. Repair leak or replace radiator. 2. Tighten clamp or replace hose. 3. Replace hose. 4. Correct cause. 5. Fill to correct level. 6. Tighten clamps or replace hose. 7. Replace gasket and check mating surfaces. 8. Replace seal or entire pump. 9. Replace heater core. 10. Repair or replace. 11. Replace radiator cap. 12. Replace freeze plugs. 13. Torque as recommended. 14. Tighten or replace sender. 15. Repair leak or replace tank. 16. Remove fasteners, seal, and reinstall. 17. Repair or replace. 18. Replace recovery bottle. 19. Tighten or replace petcock.

The engine compartment in most late-model vehicles is crowded. This can complicate engine removal. (Mitsubishi)

24

Engine Removal, Disassembly, and Inspection

After studying this chapter, you will be able to:

- Describe general procedures for removing an engine from a car.
- Explain the use of engine removal equipment.
- List safety rules that apply to engine removal.
- Describe general procedures for disassembling an engine.
- Explain how to make visual checks of major engine parts.

Technical Terms

Cable	Split keepers spring
Strap	Spring retainer
Bar	Valve spring compressor
Attachment points	Valve guide cleaner
Balance point	Vibration damper
Pull point	Harmonic balancer
Lifting angle	Front cover
Short block	Ring ridge
Long block	Ridge reamer
Engine stand	Oil galleries
Variable valve timing	Freeze plugs
Valve springs	Resurfacing

Introduction

This chapter explains how to remove, disassemble, and check the internal parts of a typical internal combustion engine used in modern vehicles. These procedures apply to any piston engine built within the last 50 years. The engine removal technique will vary depending on whether the vehicle is front- or rear-wheel drive. The proper disassembly and inspection procedures depend on whether the engine has overhead camshafts or pushrod-type valve mechanisms and a cast iron or aluminum block and cylinder heads.

General Removal Procedures

There are many ways that an engine can be removed from a car. When removing an engine, you must consider whether the vehicle is front- or rear-wheel drive, placement of the engine in the vehicle, frame and body clearance, accessory equipment, and the type of transmission or transaxle. Manufacturers' shop manuals will be helpful in determining the exact steps for removing specific engines.

On most rear-wheel drive vehicles and some front-wheel drive vehicles, the engine is pulled upward out of the engine compartment. A different procedure is required when the engine must be removed from beneath the vehicle.

Some installations allow the removal of the engine with the transmission or transaxle attached. Others require it to be separated and the engine pulled by itself.

Support the Transmission or Transaxle

If the engine alone is to be pulled, be certain to provide proper support for the transmission or transaxle. The drive plate (provides drive from crankshaft to torque converter) will not support a load. If the transmission or transaxle is not properly supported, serious damage can be done. An adjustable stand or a special frame cross-member support may be used.

Protect the Vehicle

Before removing the engine, cover the vehicle's fenders with protective pads. If the attaching points for the hood hinges are adjustable, scribe around the hinges with a sharply pointed tool or a marking pencil. The scribe lines will speed up hood alignment when replacing the hood. See **Figure 24-1.** Remove hinge fasteners. Lift off hood and store it upright in a

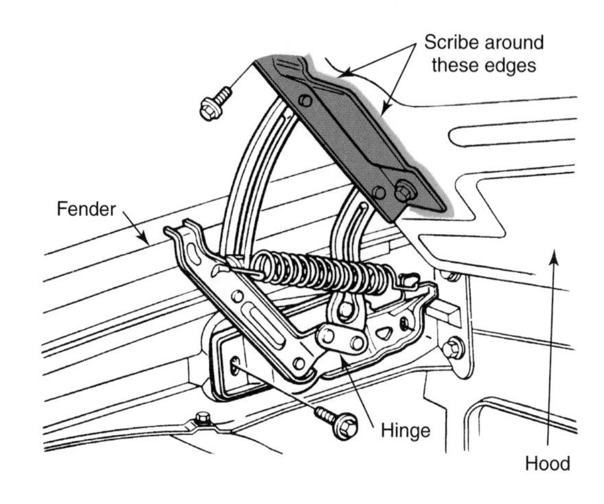

Figure 24-1. Scribing around edges of hood hinge attaching plate will make hood alignment easy during reassembly. (General Motors)

protected area. Replace all fasteners in their original holes so they will not be lost.

Disconnect All Attached Wiring, Tubing, Hoses, and Controls

Remove the battery cables and battery. Disconnect the coil primary lead, starter, computer control, and alternator wires, oil pressure and temperature indicator wires, engine ground strap, and any other accessory wires. As the parts and wires are removed, they should be marked with masking tape for correct installation, **Figure 24-2.**

Caution: Avoid part damage. When pulling tubes and hoses out of the way, be careful not to kink or damage them. Cover the ends of hoses and tubes with tape to prevent the entry of dirt.

Disconnect the fuel and emissions lines, vacuum lines, oil pressure gauge line (if used), and any other lines attached to the engine. Remove the air intake hose or air cleaner and cover the throttle body with a plastic bag.

Warning: Correctly release fuel pressure before disconnecting the fuel lines. Failure to release pressure can result in serious injury. Refer to Chapter 17, Air and Fuel Delivery, for information concerning the release of fuel system pressure.

Disconnect throttle body linkage and transmission or transaxle throttle valve cable where used. Disconnect the exhaust pipe at the exhaust manifold. Disconnect the clutch linkage, speedometer cable, and transmission or transaxle control rods and cables (if the transmission will be pulled with the engine).

Note: Once a wire, control rod, or other part has been removed, put the fasteners back into place. This will speed up reassembly and avoid improper placing of fasteners.

Drain the oil from the engine and remove the oil filter. Drain the cooling system. Remove the radiator hoses. On vehicles with an automatic transmission or transaxle, remove the fluid cooler lines. Plug the lines to prevent the entry of dirt. Remove the radiator. Handle the radiator carefully and protect it during storage. Disconnect the propeller shaft or drive axles and wire them out of the way.

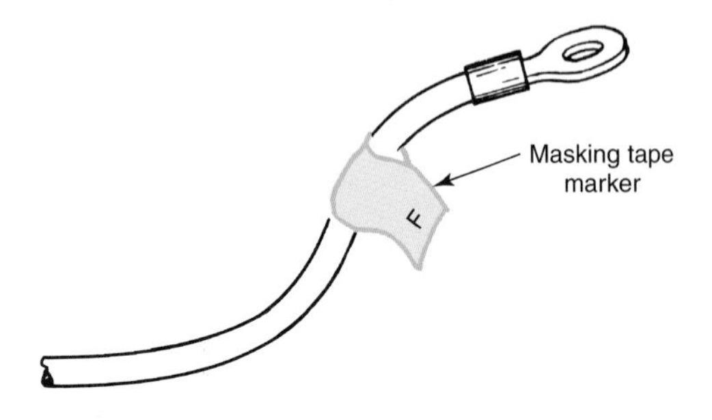

Figure 24-2. Marking wires with tape will facilitate installation.

Remove the starter and alternator, if necessary. The power steering pump, air conditioning compressor, and smog pump may be moved to one side or removed.

Note: It may be necessary to recover the refrigerant from the air conditioning system before removing the engine. Refer to Chapter 40, Air Conditioning and Heater Service, for more information on air conditioning system service.

Loosen the engine mounting (motor mount) bolts, but do not remove them. Make a thorough final check to make certain all necessary items have been removed before attempting to remove the engine.

If the vehicle will be removed from the bottom of the vehicle, remove the CV axles. Refer to Chapter 32, Axle and Driveline Service, for more information on CV axle removal. On many vehicles, the engine cradle and related suspension parts must be removed where they attach to the frame.

Warning: Do not remove any suspension part without first consulting the appropriate service manual. Many suspension parts are under spring tension and can fly apart with great force if the fasteners are removed.

Engine Removal from Top

Most engines are removed upward from the engine compartment. The following section explains how this is done.

Attach a Lifting Device

Attach the puller ***cable, strap,*** or ***bar*** to a suitable spot on the engine. Eyebolts may be used or cylinder head capscrews may be removed, placed through the puller brackets, and reinstalled. Some engines have specific ***attachment points.*** Consult the correct vehicle manual. Regardless of the attachment point, make certain the eyebolt, capscrew, or bolt is threaded into the hole to a depth at least one and one-half times its diameter. This will ensure proper holding strength. See **Figure 24-3.**

Puller Bracket Must Be Snug against the Engine

Occasionally, the head or heads have been removed from the block before engine removal. Never use the head capscrews or studs to attach the puller brackets unless they are shimmed to force the bracket against the block. Failure to do this will place a strong side pull on the fastener, causing it to fail. This principle also applies to any fastener that is too long, **Figure 24-4.** When attaching puller brackets, select fasteners of sufficient strength. Thread them into areas that will withstand the pressure of lifting. Finally, make sure the lifting brackets are firmly sandwiched between the block and the fastener caps, either directly or through the use of sufficiently strong shims.

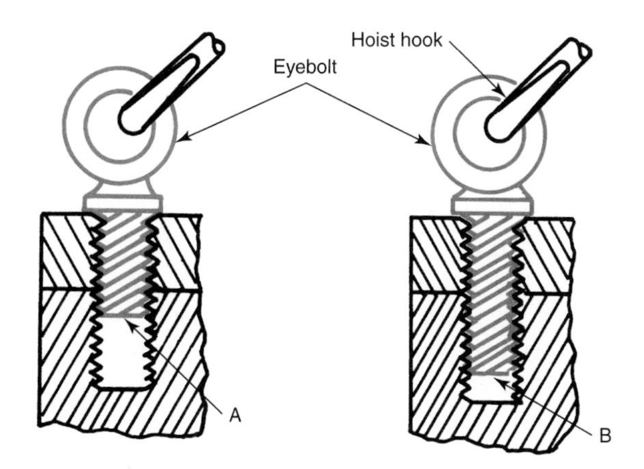

Figure 24-3. Puller fastener must have ample thread. A—Eyebolt threads a very short distance into hole and will very likely rip out under pulling pressure. B—Using a longer eyebolt ensures that enough thread is used.

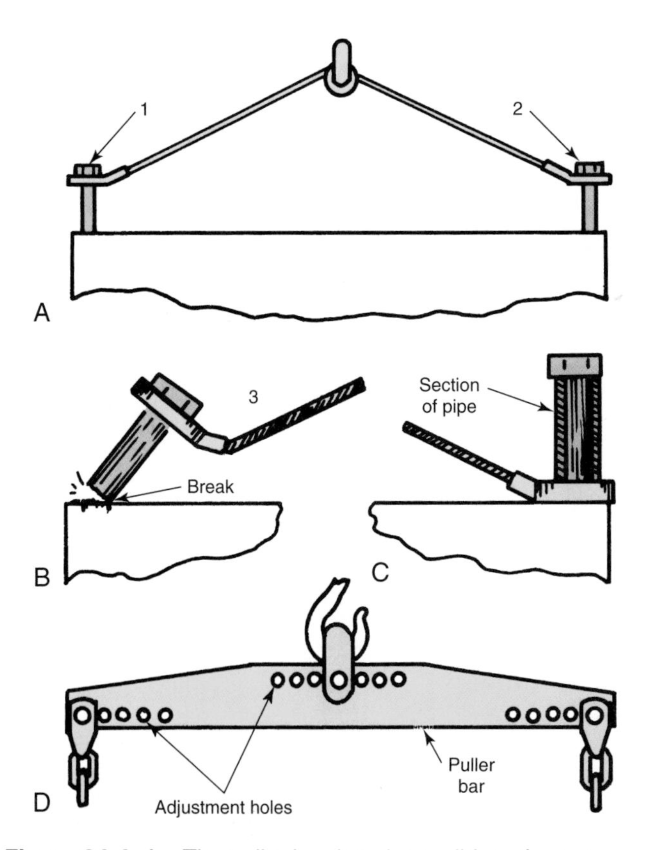

Figure 24-4. A—The puller brackets have slid up the capscrews. B—When the hoist exerts a force on the puller cable, the puller bracket will force the capscrew sideways, causing it to break or bend. C—The bracket is held against the block by a short section of pipe to prevent capscrew damage. D—Typical puller bar. Note the adjustment holes.

Select the Proper Balance Point

Attach the puller at the appropriate ***balance point*** so the engine (or engine and transmission) will be balanced at the angle desired. Failure to do this will cause tipping that could damage parts and make removal difficult. See **Figure 24-5.**

Make certain the ***pull point*** (point of attachment on puller) cannot slip under pressure. **Figure 24-6** shows what can happen when a chain hook is placed on a plain-cable pulling strap.

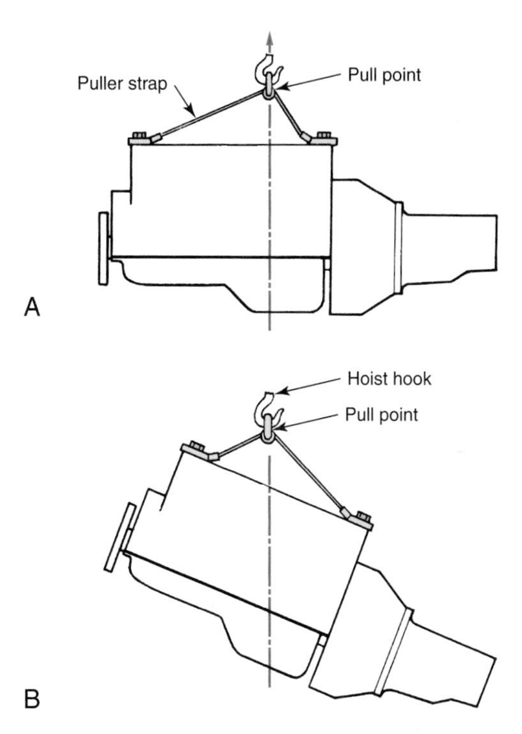

Figure 24-5. A—The engine can be lifted in a level position by arranging the pull point properly. B—Moving the pull point toward the front of the engine alters the lifting angle. Any number of angles is possible.

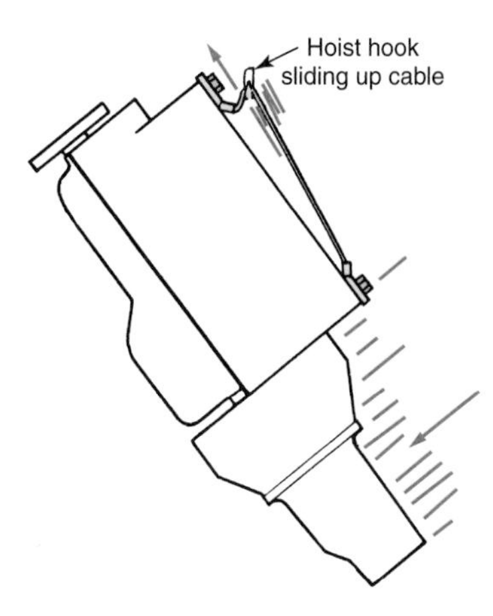

Figure 24-6. This engine was being lifted by placing a hoist hook around a plain cable puller strap. The rear of the engine tipped down, and the hoist hook slid to the front end of the cable. The rear of the engine is now falling downward with dangerous force. Make certain the pull point cannot slip.

Position the Lift

After the pulling device is firmly attached, move the lift into a position that will raise the engine without causing any undesirable side or fore-and-aft pressures. Insert the lift hook into the puller. Place a light lifting strain on the engine. Finish removing the engine mounting bolts.

Lift the Engine

Start raising the engine while checking for proper clearance. Be careful of the ***lifting angle.*** If the engine assumes the wrong balance angle, lower it back into position and change either the pull point on the puller or the location of the puller brackets.

As the engine begins to rise, pull it forward until it is free of the transmission or transaxle (if the transmission or transaxle will be left in the vehicle). As lifting progresses, make sure the flywheel ring gear or drive plate (automatic transmission or transaxle) does not hang up. Also watch carefully for any wires or hoses that you may have forgotten to remove.

If you are removing the transmission with the engine on a rear-wheel drive vehicle, the unit often has to assume a relatively steep angle to clear. See **Figure 24-7.** Be sure to install a plug on the transmission output shaft to prevent loss of fluid.

As the pulling continues, gently rock the engine on occasion. This will check whether the engine is free. If the engine stops moving at one point and continues at another, stop and check for an obstruction. Continue raising and guiding the engine, moving the lift as needed.

Raise the engine to a height sufficient to clear the vehicle. Back the lift and engine away from the vehicle. Once the engine has cleared the vehicle, immediately lower it until it is just above the floor. Move the engine to the cleaning area and clean the exterior. Remove the transmission or transaxle, if attached.

Engine Removal from Under the Vehicle

When the engine is removed from beneath the vehicle, the engine and transaxle are removed as a unit. To avoid unnecessary part removal, check the appropriate service manual before removing components from under the vehicle. Once the appropriate parts have been removed, raise the engine removal jack under the engine and secure it with safety straps. It is very important to attach the safety straps properly. The straps will prevent the engine from falling off the jack as the engine is lowered or moved to the workbench.

Once the engine removal jack is in place, raise the engine slightly to remove the tension from the engine mounts. Then remove the mount fasteners and other parts as needed. Once all parts are removed and placed out of the way of the engine, slowly lower the engine. As the engine is lowered, watch for wires or hoses that may not have been removed.

Continue lowering the engine until the jack is at its lowest point. See **Figure 24-8.** Do not attempt to move the jack and

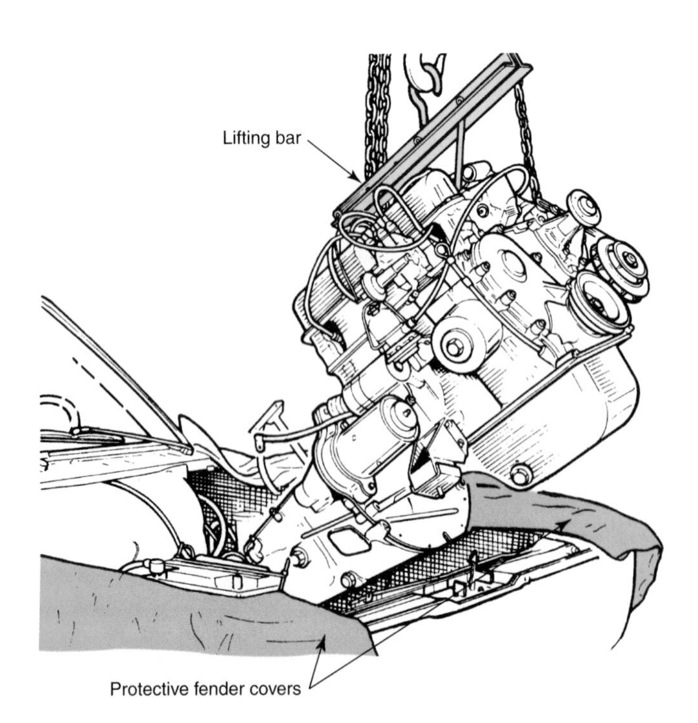

Figure 24-7. Using a chain hoist unit to pull an engine. Be very careful. (Honda)

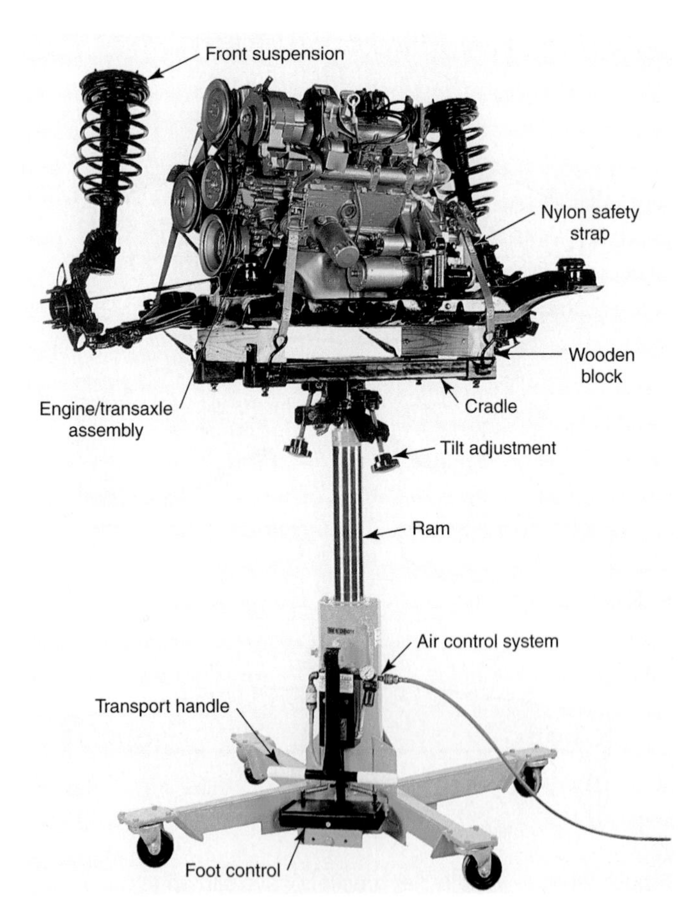

Figure 24-8. Engine, transaxle, and front suspension that have been removed from under a vehicle. Note the nylon safety straps that secure the assembly to the lift. Follow the manufacturer's recommendations. (Meyer Hydraulics)

engine until the jack is completely lowered. If necessary, remove the transaxle from the engine and then position the engine on an engine stand.

Safety Rules for Pulling Engines

- Attach a lift strap or bar at the correct balance point.
- Lift strap fasteners must have ample thread and strap brackets should be in contact with both the engine and the ends of the lift strap fasteners.
- Watch your hands and keep clear of the engine at all times.
- Lower the engine as soon as it has been removed.
- Do not use a rope as an engine sling.
- Do not depend on a knot in a chain. Bolt it together.
- If a chain is used as a strap, use heavy, wide washers under the head of the fastener to prevent the fastener head from pulling through the link.
- If the lift setup fails at any point, allow the engine to drop.
- Make sure the pulling point cannot slip.

Rebuild or Replace?

When servicing engines, you can choose to either rebuild the engine or replace the engine. Each has its advantages and disadvantages. Rebuilding an engine is more expensive and time consuming than replacement. However, engine rebuilding allows for custom building for increased horsepower or torque.

Replacement is done on high mileage engines and in cases where a major component, such as the block, crankshaft, or heads are cracked or distorted. These engines are either rebuilt in a clean, controlled environment or, in some cases, are new engines. Buying a quality OEM or aftermarket replacement engine is generally cheaper than the parts and labor of rebuilding. There are two types of replacement engines. The ***short block*** consists of the block, crankshaft, and piston assemblies. The ***long block*** comes with the cylinder heads, oil pump, and other components, depending on the supplier.

Engine Disassembly and Inspection

The following sections cover engine disassembly and the visual inspections used to check for obvious engine damage. Other chapters will cover further disassembly and making detailed measurements with the appropriate measuring tools.

Place the Engine on a Stand

The engine should be placed on an ***engine stand.*** This allows the technician to easily reach all parts of the engine and to turn the engine when necessary to reach the bottom components. See **Figure 24-9.**

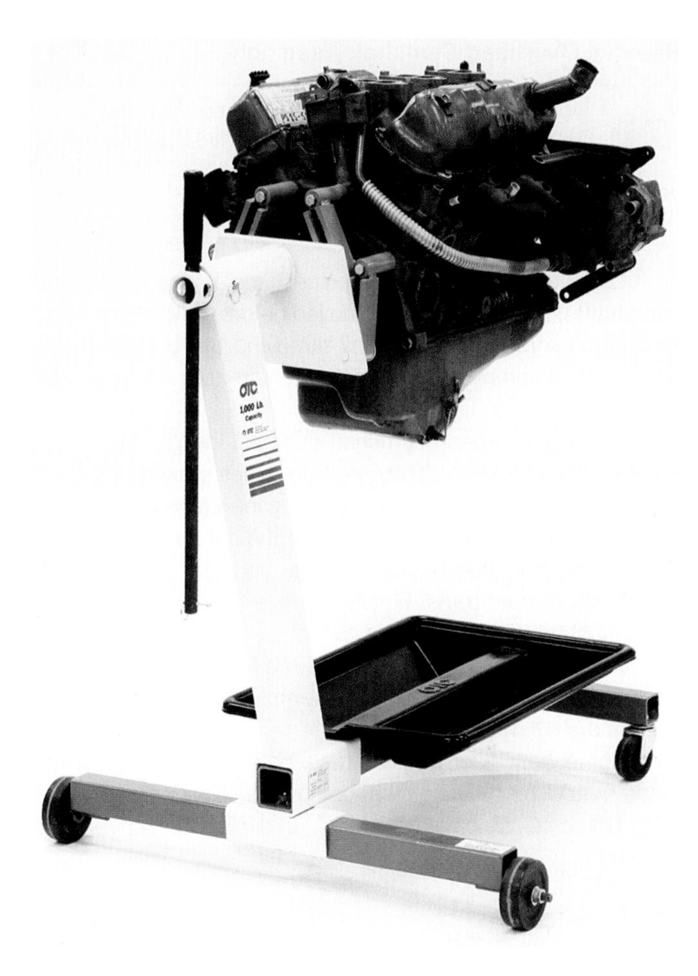

Figure 24-9. After removal, the engine should be mounted in a repair stand. (OTC)

Intake and Plenum Removal

Many modern engines have an upper and lower intake manifold or plenum. In most cases, the upper manifold must be removed to gain access to the lower manifold fasteners. To remove the intake or plenum, first remove the throttle body. Then remove the attaching bolts and any fittings. Gently pry on the manifold to remove it. Be careful not to damage the sealing surfaces of the manifold or head cylinder head. If the manifold does not come off easily, check for any overlooked fasteners before applying heavy pressure.

Note: Some engines have plastic intake manifolds. Service procedures are similar to those for cast iron or aluminum manifolds.

Cylinder Head Removal

Never remove a cylinder head until the engine has cooled. Removing a cylinder head while it is still hot can cause the head to warp as it cools. In hot weather, it may take as long as 6 hours for an engine to cool down completely. Begin by removing the intake and exhaust manifolds (when necessary), spark plugs, wires, rocker arm cover, and any accessory units attached to the head.

Remove Overhead Camshaft Assembly

Note: Consult the manufacturer's service manual for information on removing the camshaft(s) and timing belt or chain from an overhead camshaft engine before proceeding. Chapter 27 has more information on overhead camshafts.

On overhead camshaft engines, the timing chain and camshaft(s) may need to be removed before the cylinder head bolts can be reached. Begin by removing the camshaft and front cover. Then remove the belt or chain from the camshaft sprockets.

On dual overhead camshaft engines, identify the intake and exhaust camshafts. They are sometimes marked with an *I* for intake and an *E* for exhaust. Mark the camshaft bearing caps and remove them. Then remove the camshafts from the engine, **Figure 24-10.** Make sure you do not mix the camshafts or their parts. Finally, remove the followers, lifters, and adjustment shims, if used.

Some camshafts are enclosed in a housing that contains the bearing journals. The entire assembly must be removed to access the cylinder head bolts.

Some 4-cylinder engines have balance shafts mounted in the head. The balance shaft compensates for engine vibration. Removal of balance shafts is similar to removal of camshafts.

Variable Valve Timing

A few overhead camshaft engines have a camshaft sprocket with a hydraulic device that can vary the valve timing. Engines with this type of sprocket are said to have ***variable valve timing.*** Variable valve timing sprockets are operated by engine oil pressure. The variable valve timing assembly retards the valve opening at low speeds for smooth operation, and advances the valve timing at high speeds for more power. Some engines use variable valve timing at the exhaust valves only. On these engines, the valve timing mechanism keeps the exhaust valves open for part of the intake stroke at certain times. Keeping the exhaust valve open allows exhaust gases to enter the cylinder on the intake stroke. This lowers emissions and allows elimination of some add-on emission control devices.

Most variable valve timing devices are operated by engine oil pressure. The flow of oil pressure into the timing devices is controlled by electrical solenoids that are operated in turn by the engine control computer. The computer memory is accessed to diagnose system problems. Computer operation and service will be explained in later chapters.

The variable valve timing mechanism is installed at the front of the camshaft. Engines with dual overhead camshafts may have a timing sprocket at each camshaft. Most variable valve timing components are replaced as units rather than being rebuilt. When replacing these parts, closely follow manufacturer's instructions to mark the part position for reassembly. Setting timing is as important on these engines as on engines with nonvariable valve timing.

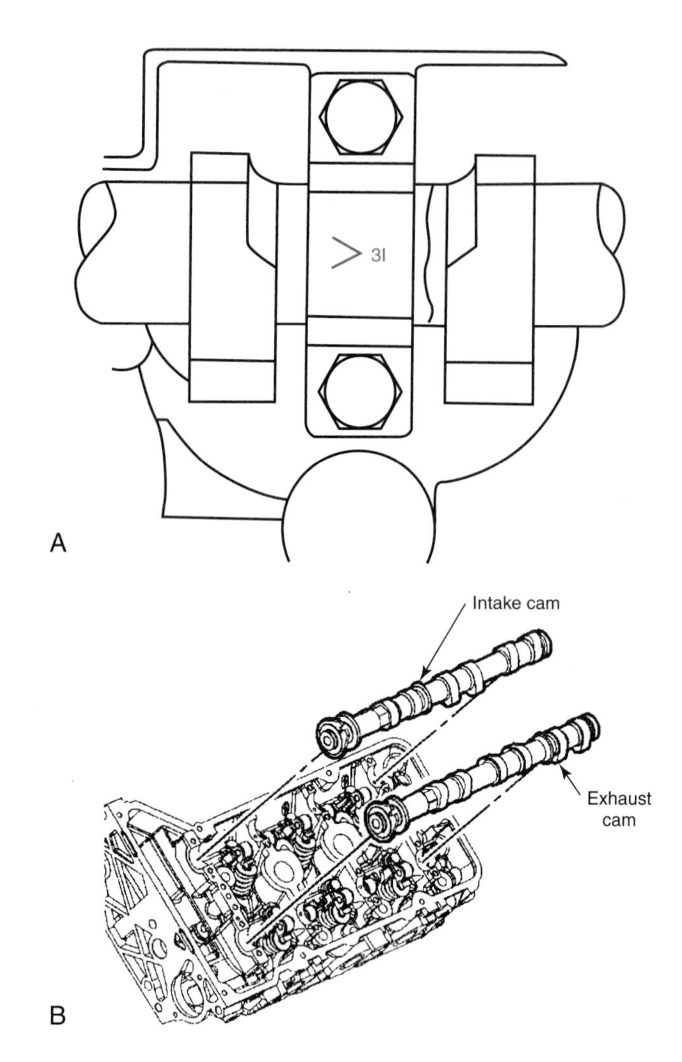

Figure 24-10. Camshafts and related components must be reinstalled in the correct positions. A—Marking the camshaft bearing caps will facilitate reassembly. B—When removing dual overhead camshafts, be sure to identify the exhaust and intake camshafts. (General Motors)

Remove the Rocker Arm Assembly

On most OHV engines, it will be necessary to remove the rocker arms to gain access to the head capscrews, or head bolts.

On engines using a rocker arm shaft, remove the rocker arm assembly by starting at one end. Loosen each support bracket bolt a couple of turns. Repeat this step until the assembly is free. If each bracket bolt is completely removed before moving to the next, the last bracket could be damaged. Valve spring pressure could push the free portion of the shaft upward. See **Figure 24-11.**

On engines using ball stud-type rocker arms, loosen each ball nut until the rocker arm can be swiveled sideways to clear the push rod. Look at **Figure 24-12.** In cases where rocker arm shaft support brackets are an integral part of the head, the head may be pulled before sliding the rocker shaft out of the brackets.

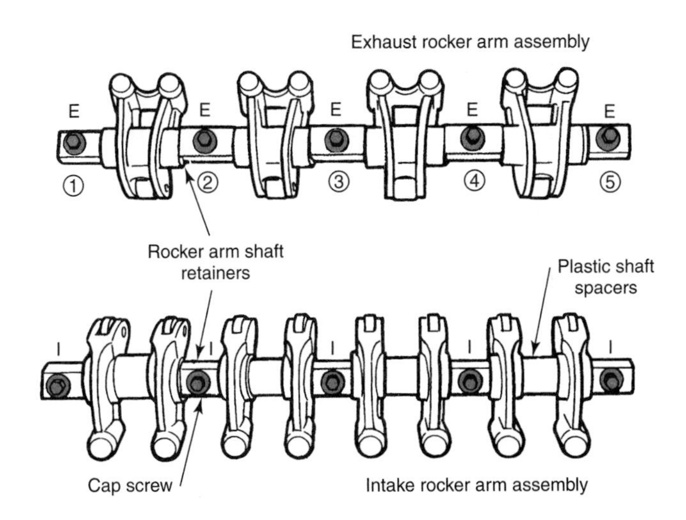

Figure 24-11. The proper sequence must be used when unbolting the rocker arm assembly. Keep all parts in order. (DaimlerChrysler)

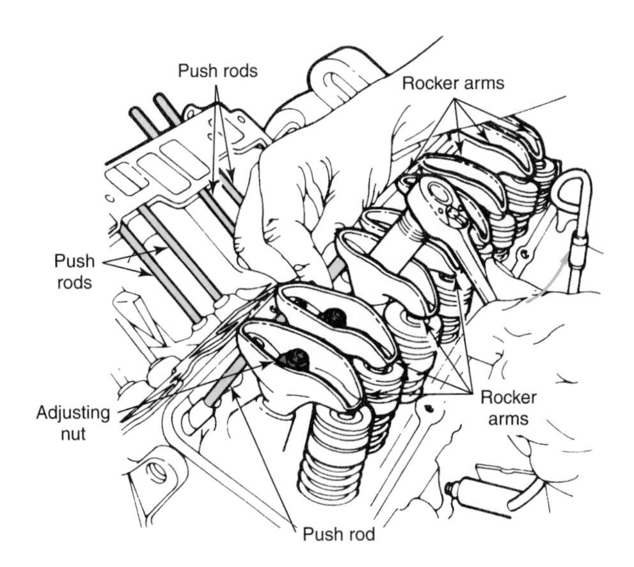

Figure 24-12. Loosen the adjusting nut so the rocker arm will move, allowing push rod removal. (General Motors)

Remove the Push Rods

If push rods are used, remove and place each push rod in a marked holder so it can be replaced in its original position. See **Figure 24-13.** Skip this step if the engine does not use push rods.

Loosen the Cylinder Head Fasteners

Using the recommended tightening sequence, reverse the order and crack (just break loose) each head capscrew. Once all have been loosened, they may be removed. If length varies or if a capscrew is drilled or machined for oil passage, note the correct location. See **Figure 24-14.**

If the cylinder head is stuck, carefully position a pry bar between the head and block and pry upward to loosen the head as in **Figure 24-15.** Be careful not to damage the head.

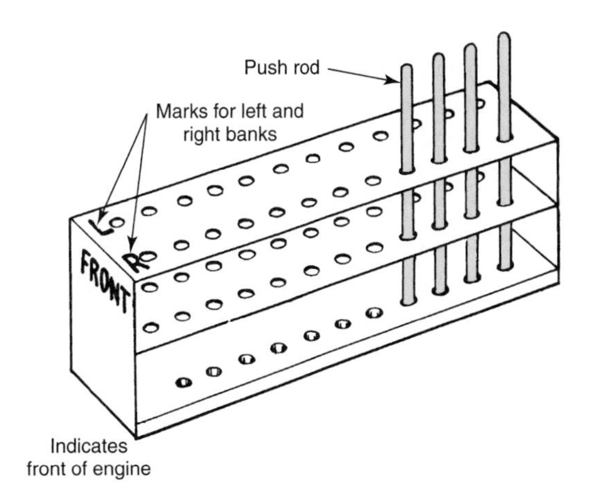

Figure 24-13. Keep the individual push rods in order by placing them in a marked holder.

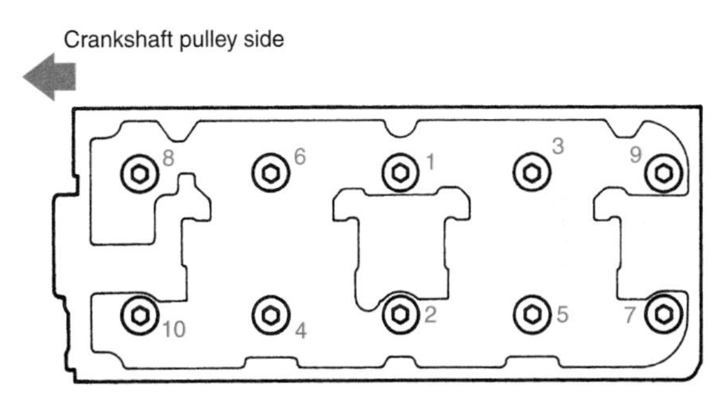

Figure 24-14. "Crack" the cylinder head bolts one at a time. Remove them in the reverse order of tightening. (DaimlerChrysler)

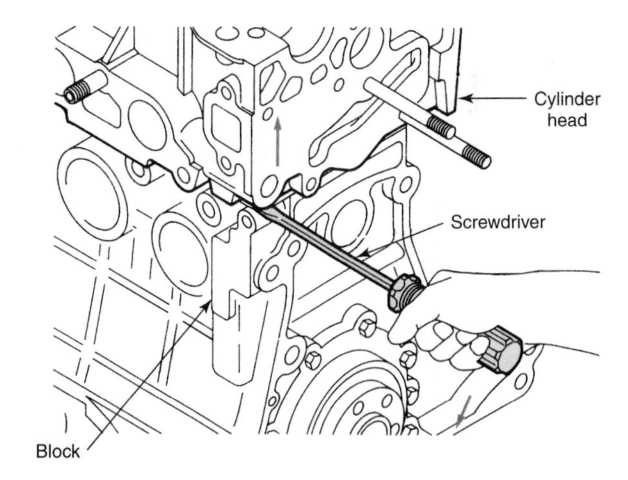

Figure 24-15. Using a large screwdriver to loosen the cylinder head. Be careful not to damage the head and block mating surfaces. (General Motors)

Do not hammer or force a tapered object between the head and block mating surfaces. The slightest nick or dent may cause serious damage. Instead, check for a missed head bolt. When the head is loose, lift it from the block.

Cylinder Head Disassembly

After removing the cylinder head from the block, place the head in a suitable repair stand. See **Figure 24-16.** If the head has overhead camshafts or balance shafts and they have not already been removed, remove the fasteners holding the bearing caps and remove the caps from the head. Mark the caps or place them in order for reinstallation in the same position. Then remove the camshafts or balance shafts.

Next, compress the ***valve springs*** and remove the ***split keepers, spring,*** and ***spring retainer*** assemblies. A typical ***valve spring compressor*** is shown in **Figure 24-17.** This compressor uses a clamping action to compress the spring and retainer so the keepers may be removed.

Another method of removing the valves is to place a socket over the valve as shown in **Figure 24-18.** When the socket is struck with a hammer, the spring will be compressed and the shock of the hammer blow will cause the retainers to pop out. As the valves are removed, place them in a rack so they can be replaced in their original guides. Use a rack similar to that shown in **Figure 24-19.** Determine which cylinder head parts can be reconditioned and which must be discarded. Discard all burned, cracked, or warped valves. Also discard any broken or worn springs, keepers, and related parts.

Figure 26-16. After removing a cylinder head from the block, place the head on a stand. This will make further disassembly much easier.

Figure 24-17. Compressing a valve spring with a pneumatic valve spring compressor. (Fel-Pro)

Cleaning the Cylinder Head

Begin cleaning the head with a wire brush to remove any carbon from the combustion chambers and valve ports, **Figure 24-20.** Clean the head-to-block surface with a scraper. Be careful not to scratch the surface. If the cylinder head coolant passages are badly clogged, give the head an additional cleaning in a hot tank. Do not hot tank an aluminum head.

A spring-type ***valve guide cleaner*** can be used to remove the carbon in each guide. Look at **Figure 24-21.** Follow the valve guide cleaner with a valve guide bristle brush to remove all loosened carbon, **Figure 24-22.** Blow all dust and carbon from the combustion chambers, ports, and guides. Push a cloth, moistened with solvent, through all the valve guides to remove any remaining foreign material. The stem clearance check will not be accurate if any foreign material is left in the guide. Additionally, if the valve guides are not clean when the pilot for valve seat grinder is inserted, the pilot may be tipped, throwing the seat out of alignment.

Camshaft Timing Mechanism Removal

The manufacturer's service manual instructions must be followed when removing the camshaft timing gears, chain, or belt. This section covers camshaft-timing removal on OHV engines. In almost all cases, the timing mechanism must be removed before the crankshaft can be removed.

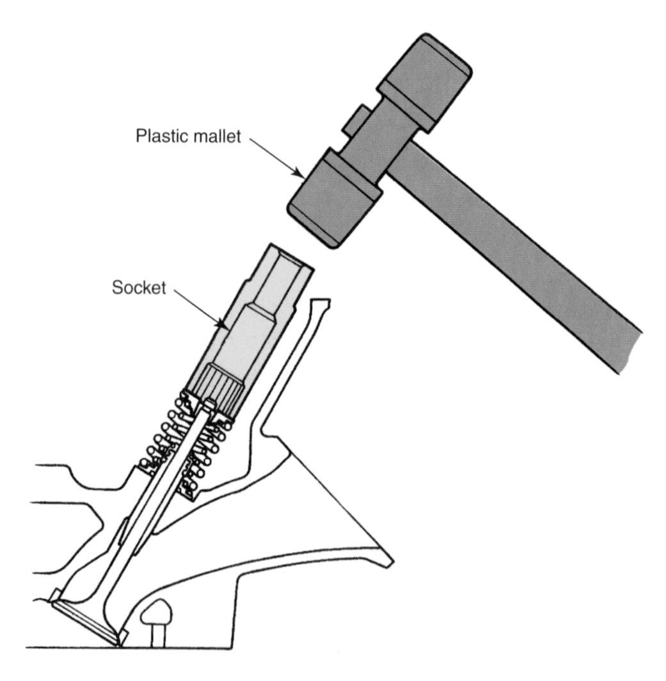

Figure 24-18. A plastic mallet and a socket can be used to loosen keepers before using the valve spring compressor. (Acura)

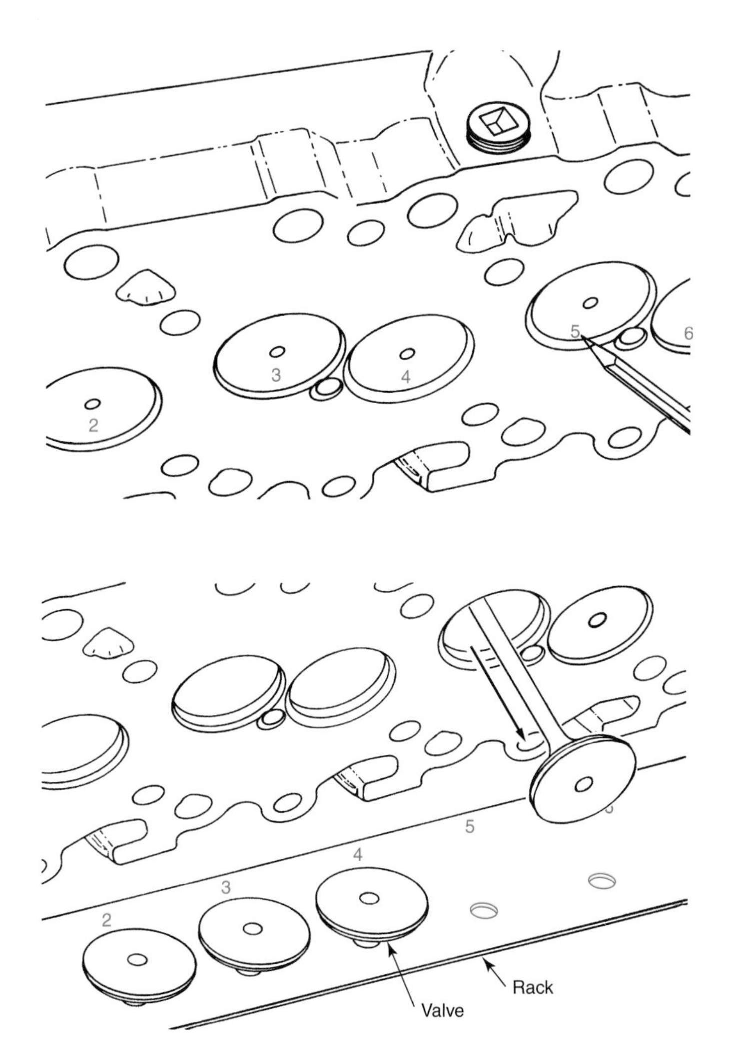

Figure 24-19. Place valves in a rack and number them in proper order during removal and installation. (DaimlerChrysler)

Note: If the engine has overhead camshafts, the timing mechanism was removed as part of head removal.

Remove the Vibration Damper

The first step in servicing the camshaft drive components is to remove the ***vibration damper,*** or ***harmonic balancer.*** To remove the vibration damper, remove the retaining capscrew in the end of the crankshaft, if used. Remove any bolt-on pulleys. Attach a suitable puller to the damper *hub* (do not pull on the outer rim) and withdraw the damper, **Figure 24-23.**

When tightening the puller screw against the hollow end of the crankshaft, be careful not to damage the crankshaft threads. If the threaded hole in the crankshaft is larger than the puller screw, place a cap over the end of the crankshaft, Figure 24-23B. Never allow the puller screw to enter the threaded hole in the crankshaft.

Remove the Front Cover

To remove the ***front cover,*** remove the screws that hold the cover to the engine block or the oil pan. Watch for variations in capscrew lengths. If possible, leave the water pump in place. Once the capscrews have been removed, gently pry on the cover to break the gasket seal. Then, pull the cover away from the engine. See **Figure 24-24.**

Figure 24-20. Removing carbon from cylinder head combustion chamber and ports using a wire brush. Wear safety goggles! (General Motors)

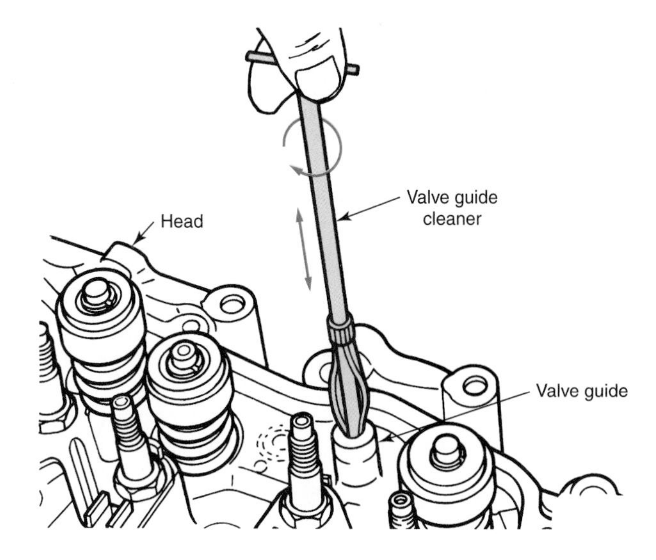

Figure 24-21. Using a spring blade valve guide cleaner to remove carbon from a valve guide. (General Motors)

Remove the Timing Gear Assembly

Visually inspect both camshaft and crankshaft gears or sprockets and the timing chain or belt, if used. Worn, chipped, or galled parts must be replaced. If a timing chain is used, the timing gears and the chain can generally be removed as an assembly after the attaching bolts are removed, **Figure 24-25.** When timing belts are used, the belts can be removed first, and then the timing gears can be removed. If the engine uses two timing gears with no chain or belt, each gear can usually be removed without removing the other.

Note: When it is necessary to replace any of the camshaft timing parts, both the gears and the timing chain or belt should be replaced.

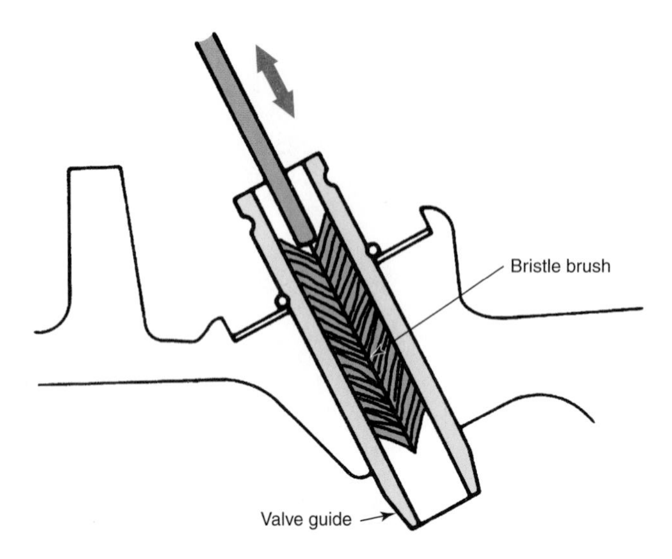

Figure 24-22. Use a brush to remove loosened carbon from a valve guide. (Toyota)

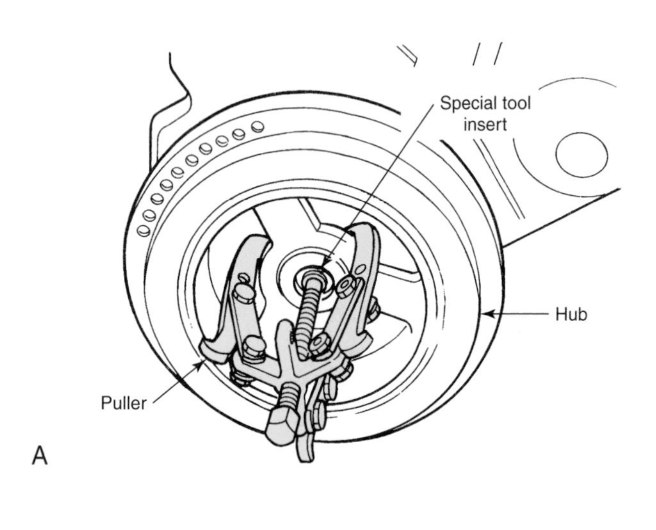

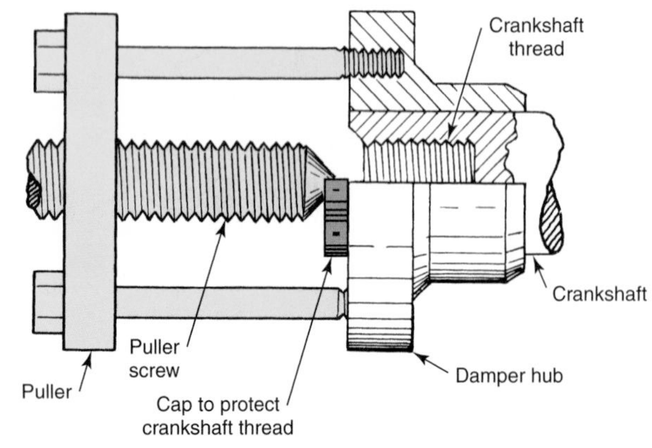

Figure 24-23. Pulling a vibration damper. A—The pulling force is applied to the center of the hub. B—If necessary, a plug can be used to protect the crankshaft threads.

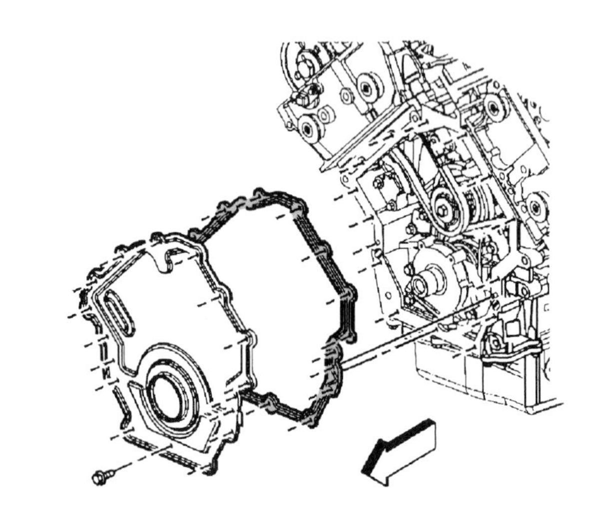

Figure 24-24. Front cover shapes and materials vary widely. This illustration shows a typical sheet metal front cover. (General Motors)

Figure 24-25. This timing chain uses a bolt-on camshaft gear and a pressed-on crankshaft gear. (DaimlerChrysler)

Remove the Camshaft

Some camshaft gears are bolted in place and are simple to remove. Others, however, are force fitted on the camshaft and are best pressed off. Turn the cam gear so the thrust-plate-retaining capscrews are accessible. Remove the capscrews. After the retainers are removed, pull the camshaft from the engine, being careful to avoid damage to the camshaft bearings.

Note: Before the camshaft can be pulled, the distributor or oil pump (depending on which one has a gear that engages a gear on the camshaft), fuel pump, and valve lifters must be removed.

Oil Pan Removal

To gain access to the crankshaft, connecting rods, oil pump, and oil pickup, the oil pan must be removed. Most oil pans are held to the block by a series of small capscrews. Loosen and remove these fasteners and carefully pry on the oil pan lip to break the gasket seal. Sometimes, it is possible to break the oil pan loose from the engine by tapping on the pan body with a plastic or wooden hammer. Once the gasket seal has been broken, remove the pan from the engine.

Oil Pump Removal

To remove the oil pump from the engine, remove the fasteners holding the pump to the engine block and lift the pump from the block. A rod driven by the camshaft is used to turn some oil pumps. Meshing gears on the rod and camshaft transmit power to the pump. The camshaft gear may also drive the distributor, when used.

Note: Some oil pumps are built into the engine block and are turned directly by the crankshaft. This type of oil pump can be removed by removing the cover and removing the gears from the crankshaft.

Disassemble the Oil Pump

After removing the pump cover, but before pulling either the rotors or gears, lightly mark them with a sharp scribe. Then when reassembled, the same ends of the rotors or gears can be positioned to face the cover plate. Both units should mesh with each other in the same position. Pump disassembly and inspection procedures are covered in more detail in Chapter 26, Engine Block, Crankshaft, and Lubrication System Service.

Removing the Crankshaft

At this point, you should be ready to remove the crankshaft. So far, you should have removed the vibration damper, the front cover, the timing belt or chain, and the timing gears. If you have not done this, go back and remove them. On a few 4-cylinder engines, balance shafts are installed in the oil pan. Remove any balance shafts at this time.

Mark Caps before Removal

Before removing the main bearing caps, mark each cap and the corresponding crankcase web with a prick punch or a number stamp. This is shown in **Figure 24-26.** Place the cap and web marks on the same side of the engine. Never mark on the top of the cap. Use a heavy section near one side to prevent distortion.

Rotate the crankshaft to bring each rod journal near bottom dead center (BDC), one at a time. Remove the rod caps. If working on an older model car in which shims are used, remove the shims and mark the position on the rod. Install protectors over the rod bolts to prevent them from nicking the journals.

Caution: The ring ridge must be removed before pulling pistons. This will prevent the piston rings from catching on the ridge and possibly breaking the piston lands.

Main Bearing Cap Removal

To remove the main bearing caps, remove any locking devices and crack each cap bolt loose. Remove all capscrews while watching for variations in diameter and length. If main bearing cap cross bolts are used, remove them *before* removing the cap bolts, **Figure 24-27.** Do not misplace the cross bolt crankshaft-to-cap spacers. Keep them in the proper order. After all bolts are removed, carefully pry the caps free. Light tapping with a plastic or rawhide hammer will help. Some caps must be removed with a special puller.

Caution: Never pound on the caps or use a pry bar between the crank journal and the cap bore. The caps may look sturdy, but they can be damaged easily.

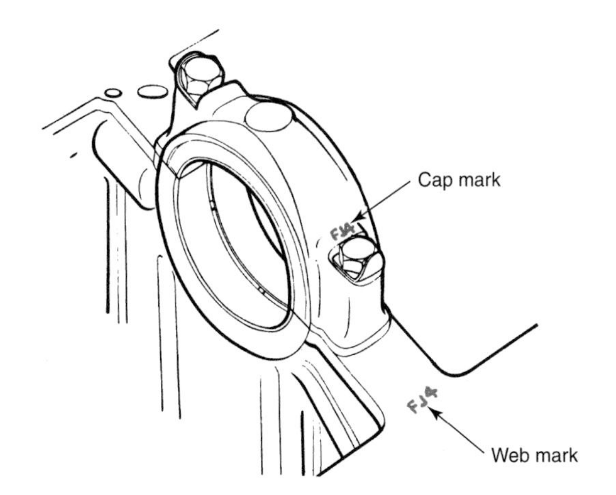

Figure 24-26. Mark the bearing caps before they are removed. They must be reassembled in the proper position and direction. (DaimlerChrysler)

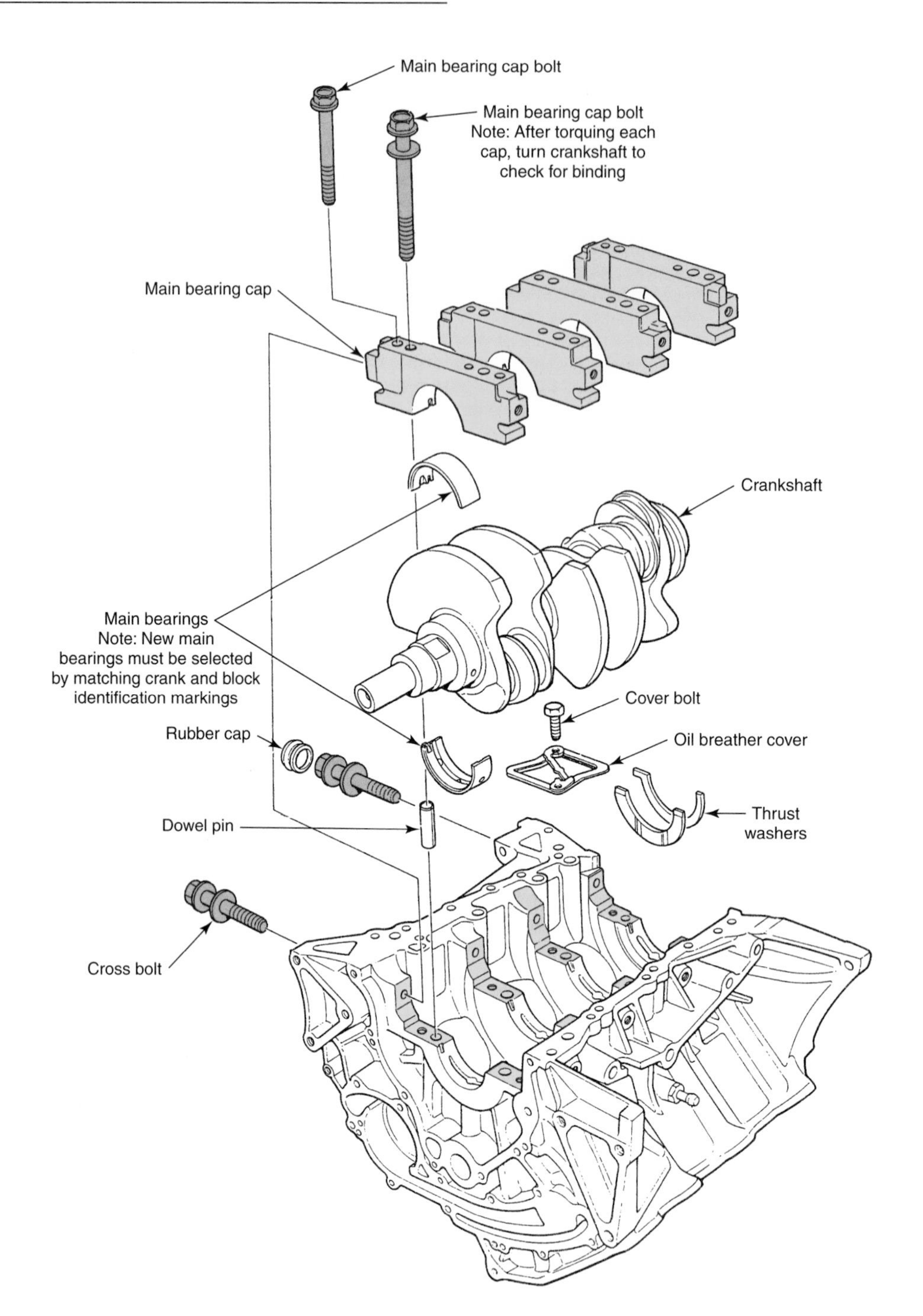

Figure 24-27. Note the use of main bearing cross bolts in this engine. (Honda)

Lifting the Crankshaft from the Crankcase

After removing the bearing caps, lift the crankshaft straight up, being careful to avoid damage to the journal or thrust surfaces. If the crank is too heavy to handle by hand, use a sling and a lift. **Figure 24-28** shows a crankshaft being lifted from a crankcase. After removing the crankshaft, remove all the bearing inserts and mark them for study. **Figure 24-29** shows the effects of a spun bearing. The bearing locked to the crankshaft and turned inside the block and cap. A machining process called line boring may be able to fix this block. If the damage is too extensive, the block must be discarded. Once the block has been thoroughly checked, clean the web, cap bore, and parting surfaces. Then reinstall and torque all bearing caps.

Cleaning and Storing the Crankshaft

Before performing any service, the crankshaft must be thoroughly cleaned. Use a rifle brush to clean the oil channels, **Figure 24-30.** Flush all passages and blow them dry with air. Lightly oil all journal surfaces immediately after drying. It should then be covered until it is ready to use. If the shaft is to be stored for some time, stand it on end or support it in several

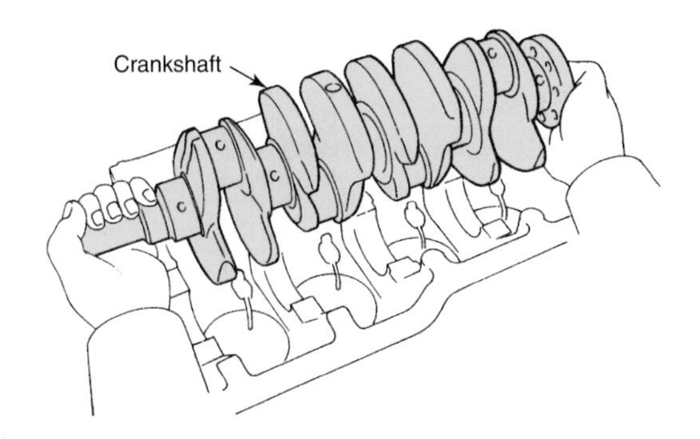

Figure 24-28. Removing a crankshaft by hand. Be careful! If the crankshaft is too heavy, use a lift and a sling. (Toyota)

Figure 24-29. When a bearing spins in the block, it causes the type of damage shown here. This block must be repaired by a machine shop or discarded.

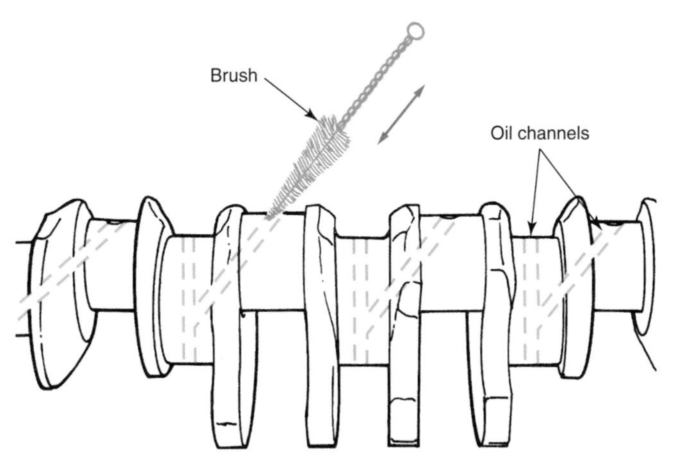

Figure 24-30. Using a rifle-type brush to clean out the oil channel in a crankshaft. (Cummins Engine Co.)

Figure 24-31. Removing the cylinder ring ridge. The reamer is supported with lips on the top of the guide fingers. (DaimlerChrysler)

spots. If the shaft is allowed to sag during storage, it can take on a permanent set (bend), throwing it out of alignment.

Piston Removal

With the crankshaft removed, the pistons can be pushed out of the engine block. However, most engines develop a ***ring ridge*** at the top of the block. The top ring does not reach the very top of the cylinder, so the top of the cylinder remains at its original diameter while the rest of the cylinder is worn to a wider diameter. The rings expand to fit the worn part of the cylinder. Pushing the piston out before removing the ring ridge would cause the rings to strike the ridge, breaking the rings, the piston lands, or both. This is the reason ring ridges must be removed.

Remove the Ring Ridge

Before attempting to remove the ring ridge, run the piston down in the cylinder. Wipe the cylinder and block surfaces with an oily rag and insert a suitable ***ridge reamer.***

The ridge reamer should be expanded tightly in the cylinder. If it is the type shown in **Figure 24-31,** a downward pressure should be exerted to keep the guide finger lips against the block surface. Turn the tool with smooth strokes. After each revolution, adjust the pressure to keep the cutter tight against the cylinder. This helps prevent the cutter from catching and making chatter marks. For cylinders that terminate in a tapered block surface, use a ridge reamer supported and aligned by the cylinder walls. Stop cutting when the ridge has been removed. Be very careful to avoid cutting below the ridge and into the cylinder. The ridge area should blend smoothly into the cylinder proper, **Figure 24-32.**

With a driving and installing tool, drive the rod and piston up and out of the cylinder. If a driver is not available, a clean hammer handle can be used to tap the rod. Following the removal of the piston assemblies from the engine, remove the rings and soak each piston and rod assembly in a good carbon-removing solvent. Rinse and dry the piston and rod assemblies and place them in a clean work area.

Inspect the Cylinders

Following piston removal, wipe out the cylinders. Using a bright light, carefully inspect each cylinder for cracks and score marks. Minor scoring and heavy scratches will require reboring to a suitable oversize. If the cylinder is smooth and wear is within limits, new rings will function correctly after the

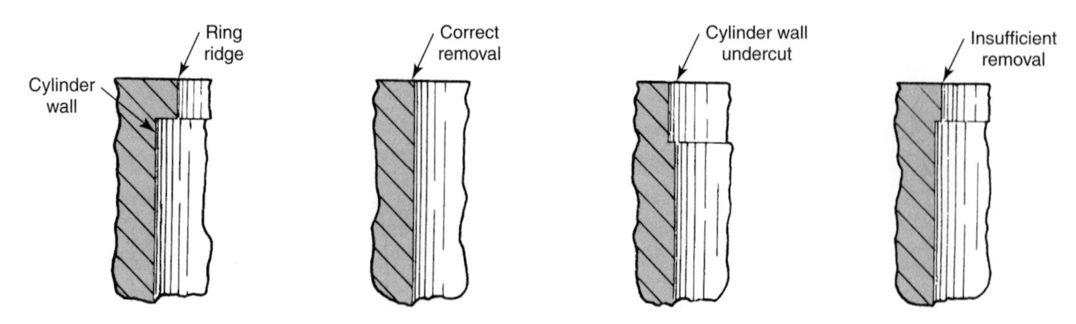

Figure 24-32. The ring ridge should be removed until the ridge area is cut just flush with the cylinder wall.

cylinder walls are properly prepared. Measuring cylinder wear limits and cylinder repair are covered in Chapter 26, Engine Block, Crankshaft, and Lubrication Service. Heavy scoring (scoring that a maximum rebore will not clean up) will require the installation of a sleeve or the replacement of the block. A crack in the cylinder, or anywhere else on the block, usually means the block must be replaced. Depending on its location and severity, a cracked block may be saved by installing a sleeve.

Remove the Piston Rings

If the rings will not be reused, they can be expanded over the piston lands and removed. Special tools are available to remove the old rings and install new rings. Be sure to carefully remove all dirt and carbon from the ring lands after the rings are removed.

Remove the Piston Pins

Before removing a piston pin, make certain the piston is marked so that it may be reassembled to the same rod in the same position. Prick punch marks will suffice if no factory identification has been provided. Exact removal procedures are covered in Chapter 26, Engine Block, Crankshaft, and Lubrication Service.

Engine Block Cleaning

In preparing to check the cylinder block for warpage or distortion, the first steps are dismantling and thorough cleaning. All parts should be cleaned down to the bare metal. This allows close inspection and accurate measurement.

During block reconditioning, it is imperative that all ***oil galleries*** be cleaned. The best method to do this is to remove the gallery end plugs before cleaning the block in a hot tank solution. Do not hot tank an aluminum block. All passageways should be cleaned out with a stiff wire brush, rinsed, and blown dry. The cooling system must be free of rust and scale. Any dirt, sludge, or particles left in the distribution passages can circulate and cause immediate and severe wear. Replace and tighten all end plugs.

Replace all defective ***freeze plugs.*** The main bearing bores and camshaft should be checked for alignment. These procedures are covered in Chapter 26, Engine Block, Crankshaft, and Lubrication Service, and Chapter 27, Engine Timing and Intake Service. Inspect all threaded holes for evidence of dirt, rust, scale, and stripped or galled threads. Thread repair is discussed in Chapter 8, Fasteners, Gaskets, and Sealants.

Checking for Cracks and Warpage

If the block or cylinder heads are to be reused, they should be carefully checked for cracks and warpage. A block or cylinder head with a distorted mating surface will cause problems and can ruin an otherwise good rebuild.

Many cracks can be seen through simple visual inspection. It is also wise to check for cracks that cannot be seen by the naked eye, especially if the engine is known to have been subjected to freezing, severe overheating, or mechanical damage. Many other parts of the automobile, such as transmission cases, gears, axles, steering gears, and wheel spindles, can crack during service. Any part can develop one of three main types of cracks:

- Cracks that are plainly visible to the eye.
- Cracks that are so fine they are invisible without detection equipment.
- Internal cracks that do not reach the surface.

A part may develop one, two, or all three types of cracks during service. Locating and repairing one cracked area does not mean that you have located all of the cracks in a part, especially in a large part such as an engine block. Any cracked part should be discarded.

Crack Detection Methods

There are a number of techniques used to check for the presence of cracking, including *magnetic, fluorescent, dye penetrants,* and combinations of these techniques. The X-ray technique requires expensive equipment and is only used in large specialty shops. The following crack detection methods can also be used to find cracks in cylinder heads, connecting rods, and crankshafts.

Magnetic Field with Iron Powder

A powerful permanent magnet or an electromagnet is placed across a suspected area. The metal under the magnet's feet becomes heavily magnetized. A fine iron powder is then dusted over the area. A crack will interrupt or break the magnetic field enough to cause the iron powder to collect along the crack. Because this process works best when the crack is at right angles to the magnetic field, the magnet should be moved into different positions. This process works only on cracked parts made of iron or other ferrous (iron-containing) metals, such as steel. Parts made of aluminum or other nonferrous material cannot be checked by this method. **Figure 24-33** illustrates the use of a powerful permanent magnet. Note the crack that has been exposed by iron powder collecting along the entire length. The poles of the magnet are at right angles to the crack.

Note: If you are unsure whether a part contains iron, apply a magnet to the part. If the magnet sticks, the part contains iron and can be checked magnetically.

Magnetic Field with Fluorescent Ferromagnetic Particles

A test with ferromagnetic particles requires that a strong magnetic field be established through the part to be tested. This method will work only on ferrous metals. A special solution that contains the fluorescent particles is sprayed on the area to be tested. As with iron powder, the ferromagnetic particles are attracted to and held along the crack line. When exposed to a black light, the particles packed along the crack line will glow white while the remainder of the part will remain blue-black.

Crack Detection Using Fluorescent and Dye Penetrants

Crack detection using penetrants involves coating a part with a special fluorescent chemical or dye that can readily enter even the smallest cracks. If a crack is present, it is brought out by a fluorescent light or by the use of a special developer. See **Figure 24-34.** These penetrating methods will work on both ferrous and nonferrous materials.

Penetrants are ideal for use on aluminum blocks or cylinder heads. The area to be checked is first cleaned with a special cleaner. See Figure 24-34A. Then, the penetrant is sprayed over the area, Figure 24-34B. A small amount of cleaner is sprayed on the gear, and the excess penetrant is wiped off with a clean cloth, Figure 24-34C. The part is then sprayed with a developing solution. The developer will draw the penetrant to the surface of any cracks, Figure 24-34D. The gear is examined under a lamp that emits black light. If any

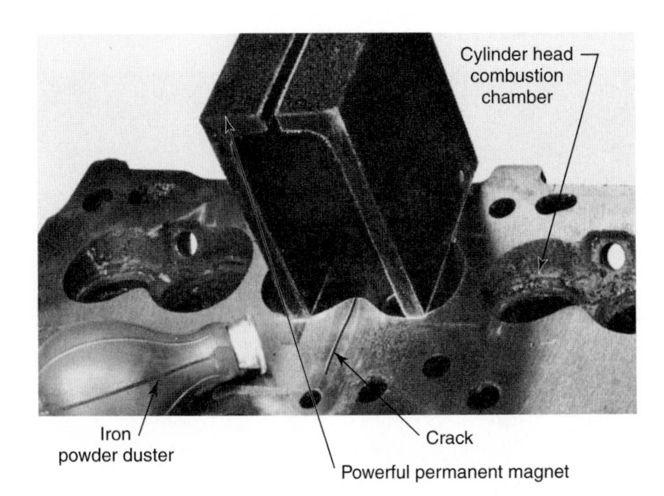

Figure 24-33. A crack in this cylinder head is exposed by using a powerful magnet and iron powder. Follow the manufacturer's recommendations. (Storm-Vulcan, Toyota)

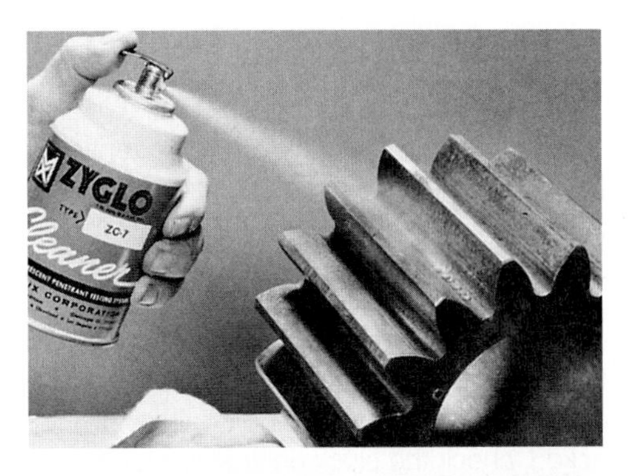

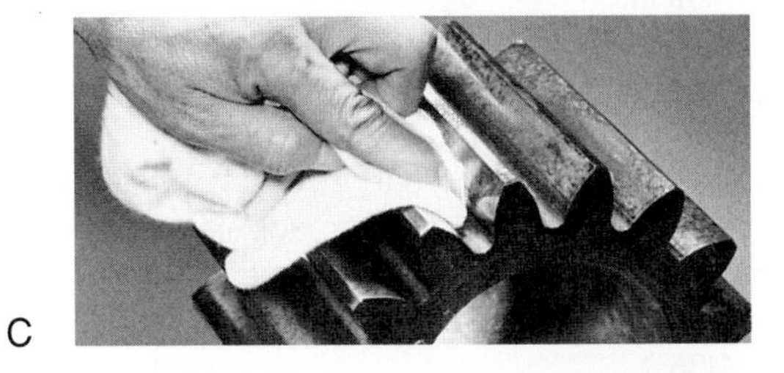

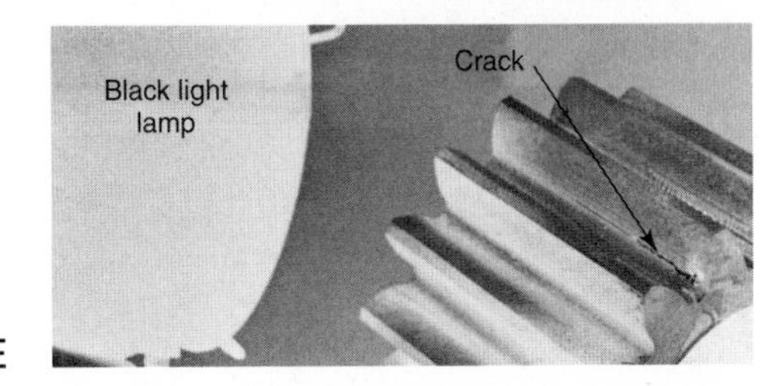

Figure 24-34. Make sure that any part to be tested is clean before applying the fluorescent penetrant. A—Cleaning section of a large gear before fluorescent penetrant is applied. B—Apply the fluorescent penetrant. C—Remove any excess penetrant. D—Apply the developer solution. E—The crack will show when the part is examined under a black light. (Magnaflux)

cracks are present, the developed fluorescent penetrant will glow visibly, Figure 24-34E. Penetrating dye will show as a bright line against a whitish background when exposed to a developer.

Checking Block for Cracks

The cylinder block is the engine part that most commonly develops cracks. Block cracks can be caused by general engine overheating, localized overheating due to clogged cooling passages, lack of antifreeze in the cooling system during below freezing weather, and mechanical damage from a foreign object in a cylinder. Pay particular attention to the cylinder walls, lifter galleys, water jackets, and the main bearing saddles.

Checking Cylinder Head for Cracks

Check the head for obvious cracks. Most cylinder head cracks occur in or close to the combustion chambers. Some cracks can occur between water and oil galleys. If the head surface is cracked, check the block deck at the location where the crack on the head meets the block. The cylinder head surface must also be free of nicks and scratches.

Checking Crankshaft for Cracking

It is good practice to visually inspect the crankshaft for signs of cracking. Many shops use a special penetrating dye or a magnetic process to check for cracking. See Chapter 26, Engine Block, Crankshaft, and Lubrication System Service, for more information on crack detection.

Checking Block and Head Surfaces for Warpage

To ensure the close fit necessary between the head and block, both should be checked for warpage. A straightedge and feeler gauge should be used. Checking engine block and cylinder head surfaces for distortion is shown in **Figure 24-35** and **Figure 24-36.**

Place the straightedge across the part as shown. Make a cursory examination of the surface by sighting along the straightedge. Slide a feeler gauge between the straightedge and the block or head surface to determine the amount of distortion. Some warpage, around 0.003″ (0.08 mm) in any 6″ span (152.4 mm) or 0.006″ (0.15 mm) overall, is permissible.

If the distortion is not within these limits, the head and/or block require ***resurfacing,*** **Figure 24-37.** A minimum amount of metal should be removed.

Removal of metal from the head or block will raise the compression ratio in most engines by reducing the size of the combustion chamber. It will also change the effective distance from the lifters to the rocker arms. On an overhead camshaft engine, the timing chain length will be altered. Special head gaskets, which are thicker than standard, are available to maintain compression and working dimensions when metal has been removed from either the head or block.

Tech Talk

Big factories make money by doing something called production planning. They figure out how to make the most of their materials and time by carefully planning when to do certain jobs, when to send out for more parts, and when to send some work out to be done by other companies. This allows the factory to make its finished product with as little

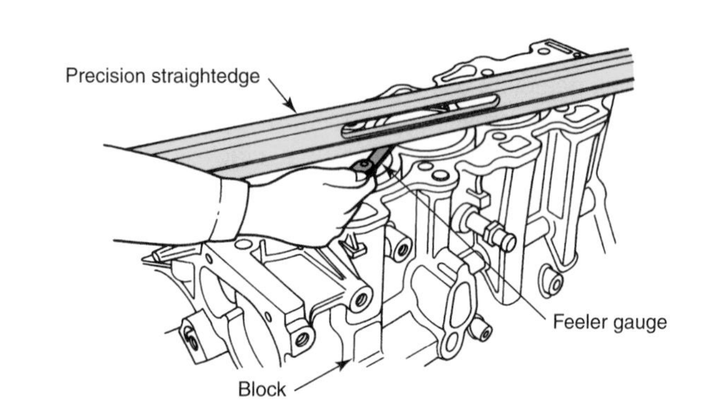

Figure 24-36. A straightedge and a feeler gauge can also be used to check block warpage. (Acura)

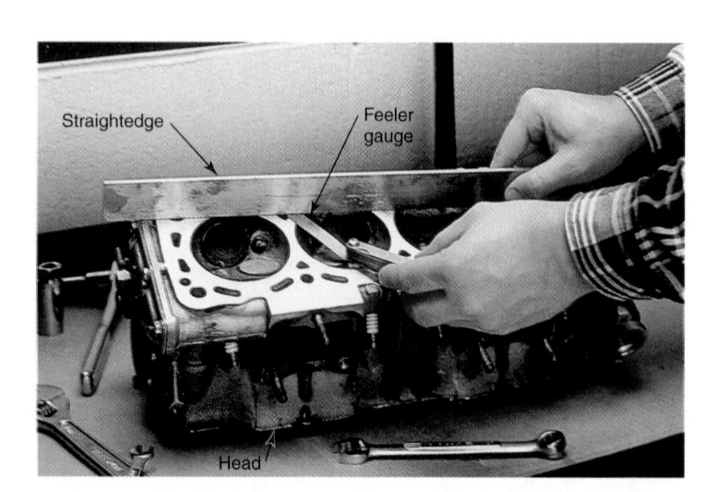

Figure 24-35. Use a straightedge and feeler gauge to check a cylinder head surface for warpage. (Fel-Pro)

Figure 24-37. Resurfacing a cylinder head with a grinder. Remove only a minimum amount of metal. (Fel-Pro)

downtime (time spent doing nothing) as possible and to make the most of its time, parts, and employees.

The automotive technician can also use production planning. Think of a big repair job, such as overhauling an engine. You do not want to be sitting around a half-assembled engine waiting for parts to be delivered or the heads to come back from the machine shop.

Try to schedule part delivery and outside services. For example, as you tear down the engine, find out exactly what parts and outside services you will need to make a complete repair. You can then order the parts and send the components that need machining to the machine shop. While you are waiting for the parts and machining, you can clean the internal engine parts, hone the cylinders, and make other repairs. When the new parts and machined components are delivered, you will be ready to put the engine back together.

Compare this process with cleaning the internal parts, honing the cylinders, preparing to put the engine back together, and then ordering parts. You will be in for a long wait, sitting around and watching dust settle on your clean engine parts.

Before beginning any job, take a few minutes to decide how it can be done most efficiently. Pay particular attention to one of the most common problem areas: not ordering all the needed parts and services. These few minutes spent ordering parts and services will save you hours later. Production planning will pay big dividends in time saved, but only if you do it.

Summary

Before removing an engine, determine if the transmission or transaxle will be pulled with the engine. Cover the fenders. Scribe around the hinges and remove the hood. Drain the coolant; then remove the radiator hoses and radiator. If desired, drain the engine oil and transmission fluid. Disconnect all wiring, tubing, hoses, and controls attached to the engine and, if necessary, to the transmission or transaxle. Attach the puller strap securely. Position the pull point to properly balance the assembly. Pull the engine slowly, constantly checking that all parts are free. When the engine is high enough to clear the vehicle, remove and lower it. Steam clean the engine and place it in a repair stand. Be very careful. Stay out from under the engine at all times.

Begin engine disassembly by removing the cylinder heads. Next, remove the oil pan, crankshaft, timing cover, camshaft timing mechanism, and camshaft. Then, remove the cylinder ridge and remove the pistons. Remove the oil pump, noting whether timing is affected. Check all parts for obvious wear.

Review Questions—Chapter 24

Do not write in this book. Write your answers on a separate sheet of paper.

1. Engines must be pulled with the _____.
   (A) transmission and transaxle attached
   (B) transmission and transaxle removed
   (C) Either A or B, depending on the application.
   (D) Neither A nor B.
2. _____ around adjustable hood hinges before removal.
3. _____ wire ends after removal to facilitate reassembly.
4. When pulling hoses and tubes free of the engine, be careful to avoid _____.
5. Whenever practical, always _____ fasteners after a part is removed.
6. When removing an engine from a vehicle, make sure that the _____.
   (A) fasteners securing the engine to the puller are threaded into areas that will withstand the pressure of lifting
   (B) puller is attached at an appropriate balance point
   (C) pull point cannot slip under pressure
   (D) All of the above.
7. Engine angle during lifting should be _____.
   (A) level
   (B) back tipped down
   (C) front tipped down
   (D) Depends on situation
8. A gentle _____ motion will help to determine if the engine is clear during pulling.
9. List eight safety rules for engine pulling.
10. Name three defects to look for in an engine cylinder.

ASE-Type Questions

1. As soon as the engine assembly clears the vehicle, it should be _____.
   (A) lowered to just above the floor
   (B) steam cleaned
   (C) moved to the bench
   (D) disassembled
2. Technician A says the cylinder head should always be removed before it has time to cool. Technician B says that, on most engines, it is not necessary to remove the overhead-camshaft-to-head capscrews. Who is right?
   (A) A only.
   (B) B only.
   (C) Both A and B.
   (D) Neither A nor B.
3. Which of the following is not removed during an overhead-valve cylinder head removal?
   (A) Rocker arm cover.
   (B) Push rods.
   (C) Camshaft.
   (D) Head capscrews.
4. Technician A says the cylinder head capscrews may be of different lengths. Technician B says head capscrews may be drilled for oil passages. Who is right?
   (A) A only.
   (B) B only.
   (C) Both A and B.
   (D) Neither A nor B.

5. If improperly used, a vibration damper puller can damage the _____.
   (A) damper
   (B) crankshaft threads
   (C) timing cover
   (D) Both A and B.
6. When it is necessary to replace a timing chain, what else should be replaced?
   (A) Camshaft sprocket.
   (B) Crankshaft sprocket.
   (C) Both A and B.
   (D) Neither A nor B.
7. Technician A says piston rings should be removed with a special tool. Technician B says carbon deposits in the ring lands should be left in place to assist the seating of new rings. Who is right?
   (A) A only.
   (B) B only.
   (C) Both A and B.
   (D) Neither A nor B.
8. Removing the ring ridge will prevent damage to all of the following, *except:*
   (A) rings.
   (B) piston lands.
   (C) engine block.
   (D) None of the above.
9. Which of the following can be used on to check aluminum cylinder heads for cracks?
   (A) Iron powder.
   (B) Fluorescent ferromagnetic particles.
   (C) Fluorescent chemicals and dye penetrants.
   (D) None of the above.
10. Technician A says that a straightedge and a feeler gauge can be used to check engine block and cylinder head mating surfaces. Technician B says that if head or block surface distortion is greater than specifications, the head or block can be resurfaced. Who is right?
   (A) A only.
   (B) B only.
   (C) Both A and B.
   (D) Neither A nor B.

Suggested Activities

1. Tear down an engine and list all the parts needed to overhaul it.
2. Calculate the cost of overhauling the above engine by the following process:
   a. Using a flat rate manual, find the prices of the needed parts.
   b. Using a flat rate manual, find the labor times for replacing needed parts and other labor needed to overhaul the engine.
   c. Total the labor times and multiply them by the average labor rate for your area.
   d. Add the total cost of parts to the total labor.
   e. Add other charges, such as supplies (rags, cleaners, and oil absorbent) and outside services (machine shop work) for a grand total of what it would cost to overhaul the engine.
3. Discuss your price calculation with the other members of your class. Ask if there is anything that you missed. Determine whether the cost of an overhaul would be greater or less than the cost of a replacement engine. Also, decide what a fair profit on this job would be.

25

Cylinder Head and Valve Service

After studying this chapter, you will be able to:

- Inspect the cylinder head and parts for defects.
- Properly grind valve seats and valves.
- Test valve springs.
- Service valve guides.
- Service valve seats.
- Reassemble a cylinder head.
- Install a cylinder head.

Technical Terms

Burning	Valve seat inserts
Pitting	Peened
Carbon deposits	Swaged
Valve grinding machine	Refaced
Cutting stones	Valve seat reamers
Dressed	Concentricity
Dressing tool	Lapping
Runout	Prussian blue
Valve face angle	Rocker arm studs
Interference angle	Spring tension
Sodium-filled valve	Free length
Chamfer	Squareness
Valve stem clearance	Installed height
Removable Valve guides	Torque-to-yield
Valve guide reamer	Torque-plus-angle

This chapter covers cylinder head and valve service. Although some of these service procedures can be performed without removing the head from the engine, you should thoroughly study Chapter 24, Engine Removal, Disassembly, and Inspection, before beginning this chapter. Study this chapter carefully and practice the various repair procedures, especially valve and valve seat resurfacing. Proper valve and head service are critical for maximum engine life and performance.

Cylinder Head Reconditioning

The following sections explain the steps in servicing the cylinder head. When disassembling a cylinder head, keep all valves, springs, keepers, and retainers in order so that they may be replaced in the same location. Pay particular attention to the sections on checking valve guide clearance and refinishing the valve seats. These operations have a great effect on head durability.

This chapter is a continuation of the engine rebuild started in Chapter 24. The chapter information assumes the head has already been removed from the engine, cleaned, and checked for cracks and warpage. Head removal is covered in Chapter 24, Engine Removal, Disassembly, and Inspection. Check the service manual for any specific procedures.

Valve Reconditioning

Inspect each valve for signs of ***burning, pitting,*** and heavy ***carbon deposits.*** Insufficient valve lash (clearance), weak springs, clogged coolant passages, sticking, a warped valve stem, and improper ignition or valve timing can cause valves to be burned or pitted. See **Figure 25-1.**

Heavy carbon deposits, especially under the head of the intake valve, indicate worn valve guides, damaged seals, worn rocker arm bushings, clogged oil drain holes in the head, or rocker arm shaft oil holes facing the wrong direction. Look at **Figure 25-2.** Discard all burned, cracked, or warped valves. The grinding necessary to clean them will leave insufficient valve margin. Using a power wire wheel, brush all traces of carbon from the valve head and stem, **Figure 25-3.** Following wire brushing, rinse the valve in solvent and blow it dry. Place the valve back in the rack so that it may be reinstalled in the original guide.

Valve Grinding

A typical ***valve grinding machine*** is shown in **Figure 25-4.** Study the names of the parts. Note the

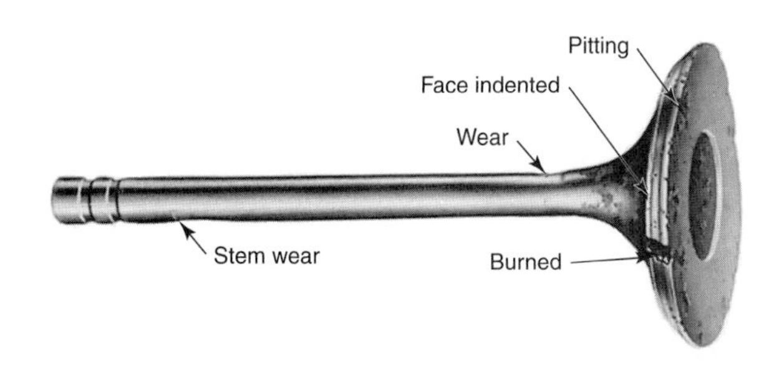

Figure 25-1. A burned valve indicates cylinder head or valve train problems.

Figure 25-2. Heavy carbon deposits under the valve heads indicate excess oil consumption through the valve guides.

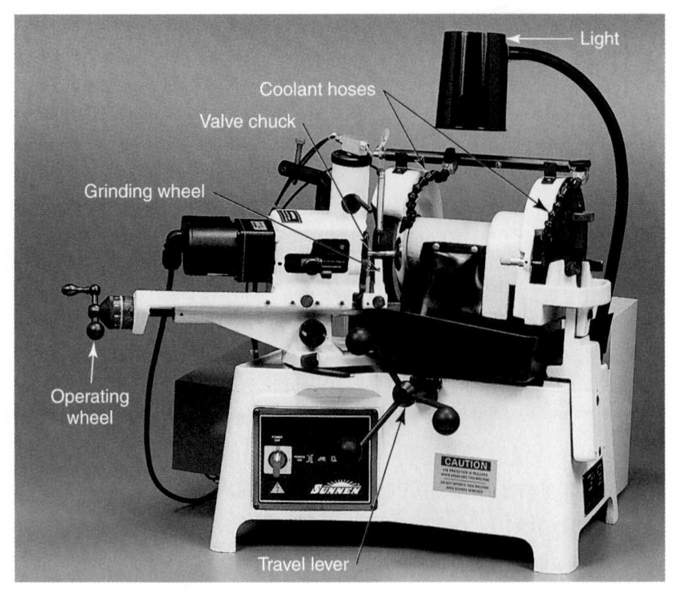

Figure 25-4. A typical valve-grinding machine. Learn the name of each part. (Sunnen)

provisions for setting the valve cutting angles and cooling the valve as it is ground. The quality of the job is directly related to the condition of the valve grinder's ***cutting stones.*** The cutting stones must be trued up, or ***dressed,*** and kept in that condition. A properly dressed stone works better and faster.

Put the diamond-tipped ***dressing tool*** into position and tighten it securely. Start the machine and *slowly* advance the stone toward the diamond. When the diamond just touches the stone, turn on the coolant. Move the diamond back and forth slowly across the stone until the stone is smooth, clean, and true. Several very fine cuts may be required.

After dressing the cutting stones, loosen the chuck swivel nut and swing the chuck to the desired angle. Carefully adjust the chuck aligning edge to the desired angle marking. Lock the swivel and recheck the angle setting. Place the valve in the chuck. Various gripping devices are used, so follow manufacturer's recommendations. Make sure that an excessive amount of the valve stem does not protrude. This can cause chatter (valve vibration during grinding), **Figure 25-5.** Close the chuck tightly. Turn on the machine and watch the valve rotate. If a noticeable amount of wobble or ***runout*** is present, stop the chuck and reposition the valve. If excessive runout is still present, a warped stem is indicated, **Figure 25-6.** If warped to the point that grinding will leave insufficient margin, discard the valve.

Grinding the Valve Face

Determine the correct ***valve face angle.*** On some engines, both intake and exhaust angles are the same. On

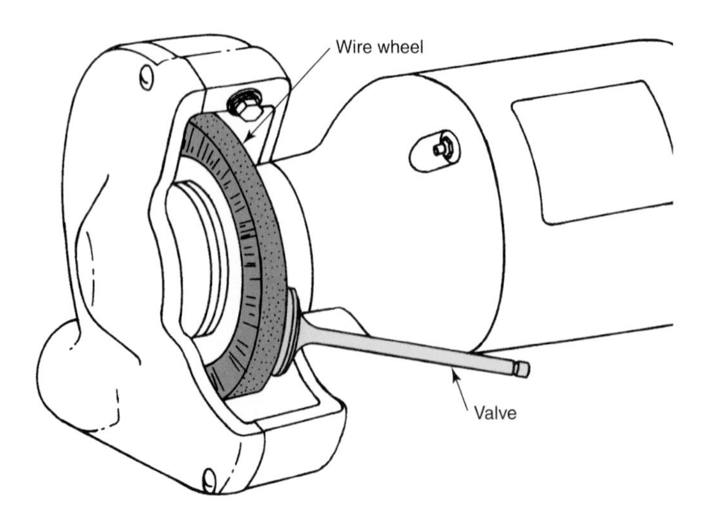

Figure 25-3. Using a wire wheel to clean carbon from a valve. Valves must be free of carbon and gum before reconditioning can begin. For illustration purposes, the safety guard is not shown. (DaimlerChrysler)

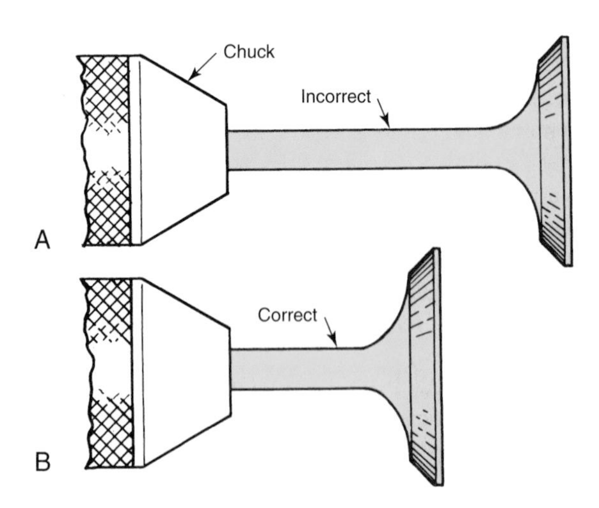

Figure 25-5. A—A valve that is protruding too far out of the chuck will chatter. B—Valve depth is correct.

others, they are different. Common angles are 30° and 45°. To provide fast initial seating, it is often recommended practice to grind the 30° valve to 29° and the 45° valve to 44°. This provides an ***interference angle*** that produces a hairline contact between the valve face and the top of the valve seat. Some manufacturers feel that due to valve design and material, the interference will allow the valve to form a perfect fit when heated, **Figure 25-7.** Other manufacturers don't recommend grinding valves to an interference fit. Always consult the manufacturer's literature for your specific application.

Move the chuck until the valve is in front of but not touching the stone. Turn on the valve-grinding machine and engage the chuck drive to spin the valve. Turn on the coolant and direct the stream toward the valve. If you have selected the proper angle, the valve face and stone should be parallel. See **Figures 25-8** and **25-9.** Once you are sure the valve face and stone are parallel, make sure the valve is turning, and then slowly advance the stone until it just starts to cut. Move the valve face back and forth across the stone. Never run the valve off the stone, **Figure 25-10.**

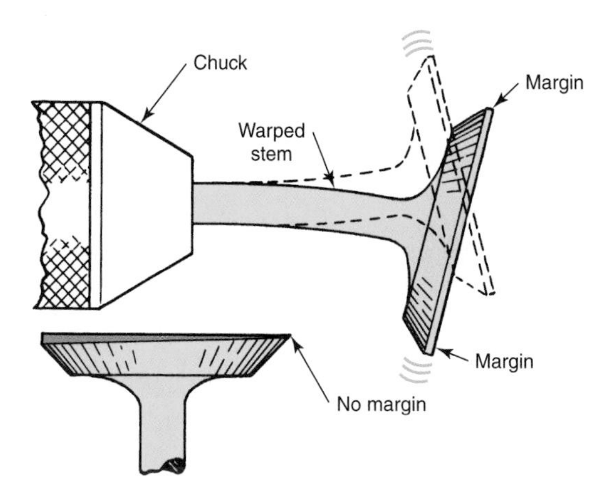

Figure 25-6. Excessive valve wobble causes the valve margin to be removed on one edge.

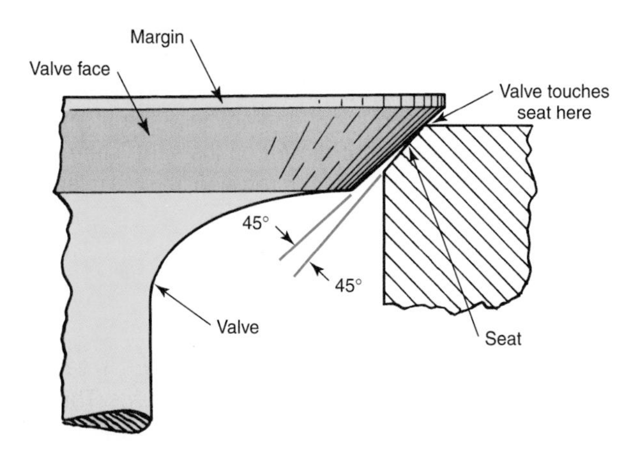

Figure 25-7. Interference angle. Note the 1° difference in angles and how the valve face contacts the top edge of the seat. One manufacturer recommends a 2° difference.

Caution: Some valves use a special coating, such as nickel-chrome, on the face area. Only a limited amount of this coating can be removed.

If your machine has a micrometer feed, set it to zero at the point where the stone just starts to cut. Advance the stone

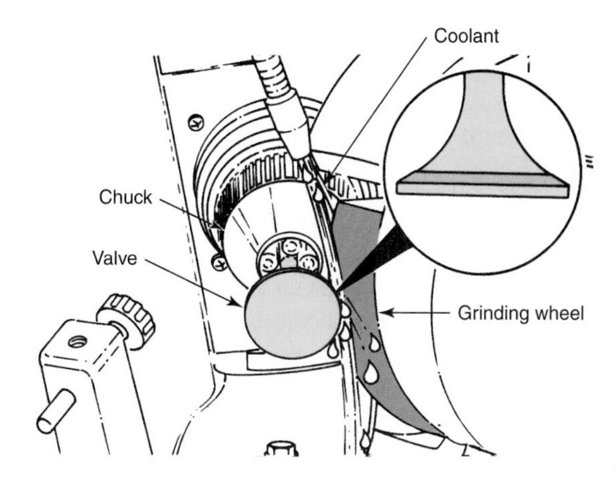

Figure 25-8. Grinding a valve face. Proceed slowly and grind a little at a time. (Cummins Engine Co.)

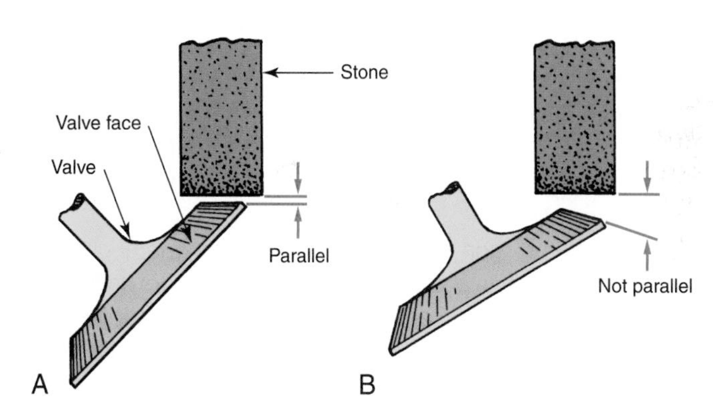

Figure 25-9. If the chuck is set at the proper angle, the valve face and stone will be parallel. A—Correct angle setting. B—Wrong angle adjustment.

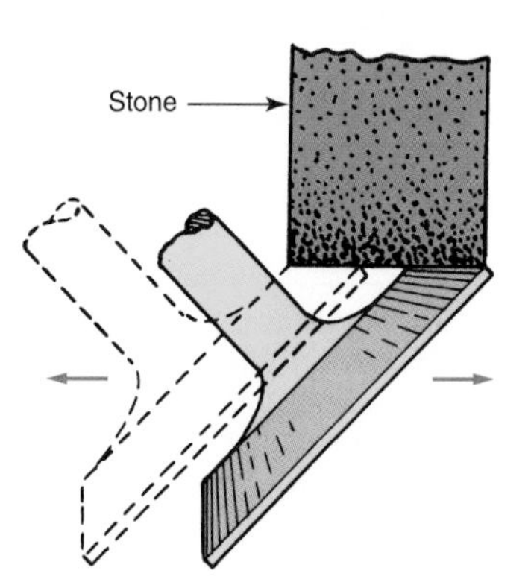

Figure 25-10. When grinding, move the valve back and forth while keeping the valve face in full contact with the stone.

against the valve 0.001″–0.002″ (0.025 mm–0.050 mm) at a time. Watch the valve face carefully. As soon as all dark spots disappear, center the valve face on the stone. Allow the stone to run a few seconds without advancing it. Then, carefully back the stone away from the valve.

Disengage the chuck drive and rotate the valve by hand while examining it closely for any remaining pits or burns. The valve face should be bright, smooth, and free of *all* defects. The margin should be ample, 1/32″ (0.8 mm) or more. If the valve is not cleaned up, repeat the process. When finished, inspect the micrometer feed dial and mark down the amount of material removed from the valve. Return the valve to the rack.

Note: If you have marked down the amount removed from each valve face, it is recommended that you remove a comparable amount from the stem. Valve stem grinding procedures are explained below.

Grind the remaining valves using the same procedure. Do not forget to change angles if intake and exhaust face angles are different. When using a valve grinder, proceed slowly. Many beginners inadvertently turn the feed wheel the wrong way or too fast and jam the stone against the valve. If the cut is suddenly too heavy, do not panic and crank the wheel. You may turn it the wrong way. Instead, shut the machine off. When it has stopped, move the stone away.

Grinding Valve Stem Ends

The valve stem end should always be trued up and smoothed by grinding. This will help maintain original tappet clearance. Never remove an excessive amount (more than 0.010″ or 0.25 mm), as the surface hardening is not deep on some valves. If the valve stem is ground below the hardening, rapid wear will result.

Dress the side of the wheel used for stem grinding. Chuck the valve in the V-block holder. Run it in until it just touches the stone. If so equipped, set the micrometer feed dial to zero. Back off the valve and start the wheel. As with valve-face grinding, direct a good stream of coolant on the portion of the valve being ground. Advance the stem against the wheel. Continue advancing with light cuts until the micrometer dial indicates that you have removed the same amount as was taken from the face. If the machine has no micrometer feed, remove enough to produce a smooth square end. The operator in **Figure 25-11** is grinding a valve stem end. Notice how the grinding stone's feed lever is grasped. Once the stone is close to the valve, hold it as shown. This method permits smoother and more accurate adjustments.

Warning: Some exhaust valve stems are partially filled with metallic sodium, which aids in valve cooling. A *sodium-filled valve* has a thicker stem than a normal valve, Figure 25-12. Sodium is extremely toxic and reacts violently with any form of water, even the moisture in the air. Never cut, drill, melt, or burn a sodium-filled valve. When the valve is worn out, dispose of it properly. Always follow manufacturer's recommendations for proper disposal.

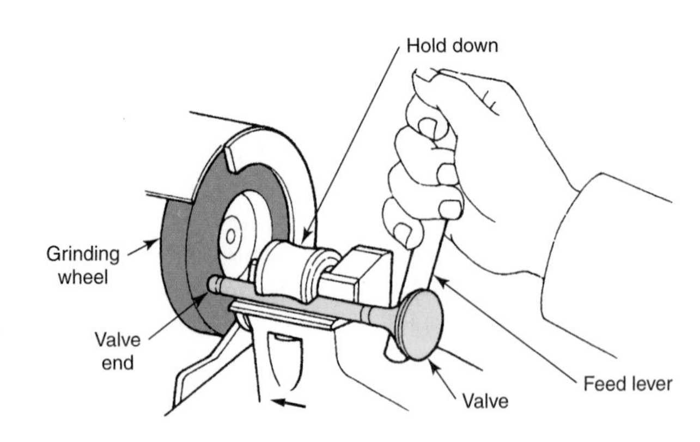

Figure 25-11. Truing a valve stem end. Note the position of the operator's hand. (Lexus)

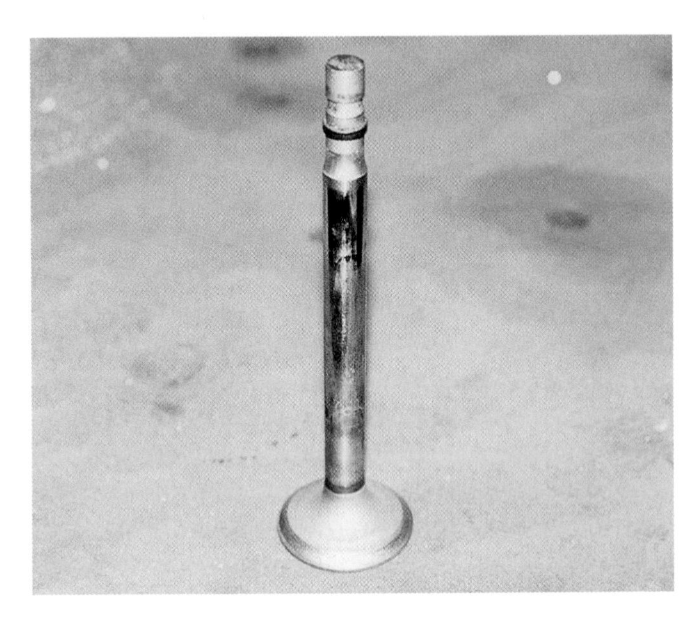

Figure 25-12. A sodium-filled valve has a much thicker stem than other valves. Do not attempt to grind a sodium-filled valve. Sodium begins burning when it contacts the moisture in the air.

When using a plain V-block in which the valve must be hand held, make certain the block is close to the wheel. This will prevent the valve stem from catching and pulling the valve between the block and the wheel. Position the valve stem in the block. Hold it down firmly and advance the stem against the wheel. If some of the ***chamfer*** on the valve stem end has been removed through wear and refacing, the chamfer may be renewed by grinding. Place the valve in the V-block. Set the holder at 45° and adjust the stop to grind about a 1/32″ (0.80 mm) chamfer, **Figure 25-13.**

Valve Inspection

Perform a final inspection of each valve. Each valve must be smooth and free of pits, scratches, and burns. There must be ample margin remaining to prevent the valve from burning. Valve stem wear must not be excessive. The stem should be free of nicks and scratches that could cause eventual breakage or sticking. Keeper grooves must be undamaged.

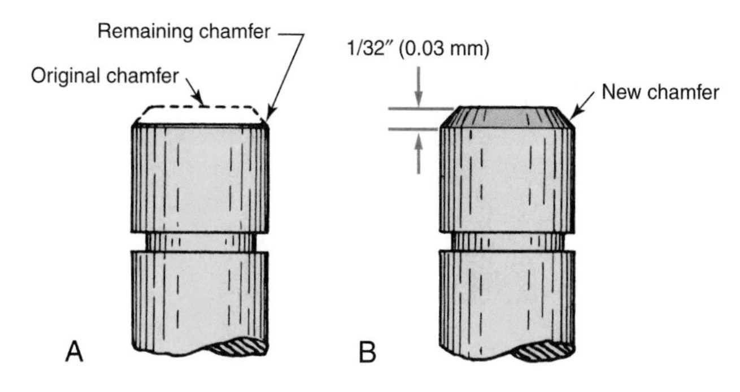

Figure 25-13. A—The chamfer on this valve stem end has worn off. B—The same end after renewing chamfer.

Valve stem ends must be smooth, squared, and lightly chamfered. Check the manufacturer's specifications. **Figure 25-14** illustrates two valves. One is acceptable; the other is not. Following the final inspection, each valve must be thoroughly washed and blown dry. Place the valves in a clean rack and cover them until you are ready to use them.

Valve Guides

When deciding whether or not to use the old valve guides, do not be concerned about too little clearance unless new valves with oversize stems are being installed. However, excessive clearance is a concern. Excessive clearance causes greater oil consumption, poor seating, and possible valve breakage, **Figure 25-15.**

Two methods are commonly used to check for excessive ***valve stem clearance.*** In one method, a small hole gauge is carefully fitted to the largest valve guide diameter (do not measure exhaust guide counterbores). Some valve seat grinder pilots can be used if a hole gauge is not available. The hole gauge is read directly or removed and measured with an outside micrometer, **Figure 25-16.** The valve stem is then miked at a corresponding wear area. The difference between the measurements is the stem clearance.

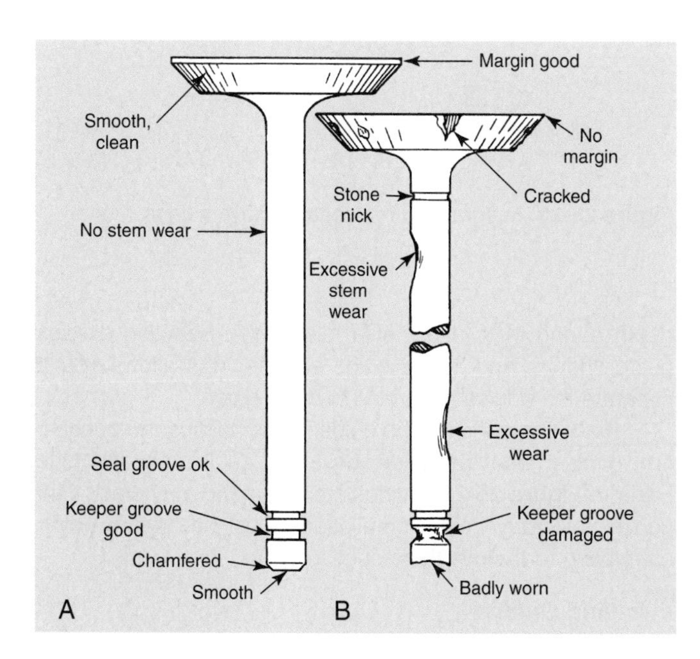

Figure 25-14. A—This valve is acceptable. B—This valve is not acceptable. Note the characteristics of each.

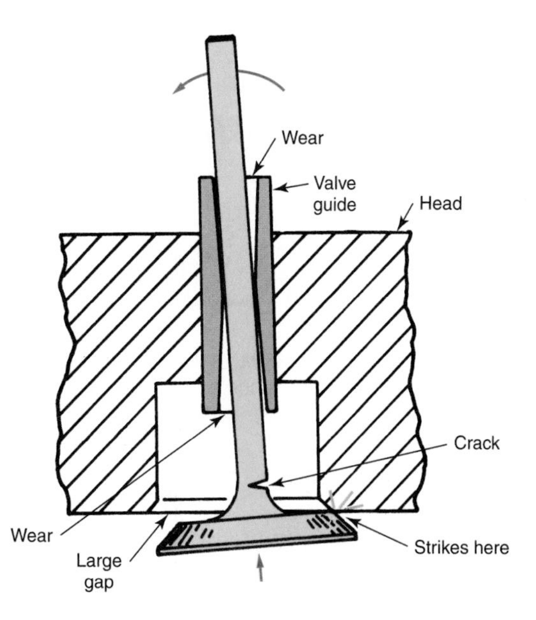

Figure 25-15. Excessive valve guide wear causes trouble and can ruin an otherwise good valve job.

Another method is to drop the valve into position with the head just free of the seat. It can be held in this position by a special tool or by slipping a piece of rubber tubing over the valve stem as shown in **Figure 25-17.** A dial indicator is then clamped to the head. The indicator stem is placed against the valve margin. Without raising the valve, move it back and forth against the stem. Watch the indicator to determine the travel. Remember the reading is not the actual clearance, because the measuring point is above the guide. The extra length magnifies the reading. Follow the manufacturer's recommendations for maximum allowable movement, **Figure 25-18.**

Engine design, type of oil seal, and amount of lubrication all determine acceptable clearance. Follow the manufacturer's

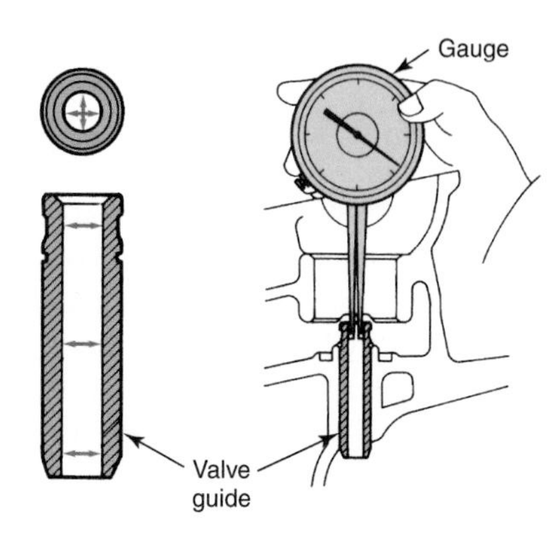

Figure 25-16. Measuring valve guide inside diameter using a hole gauge. Note the depths at which measurements are obtained. (Lexus)

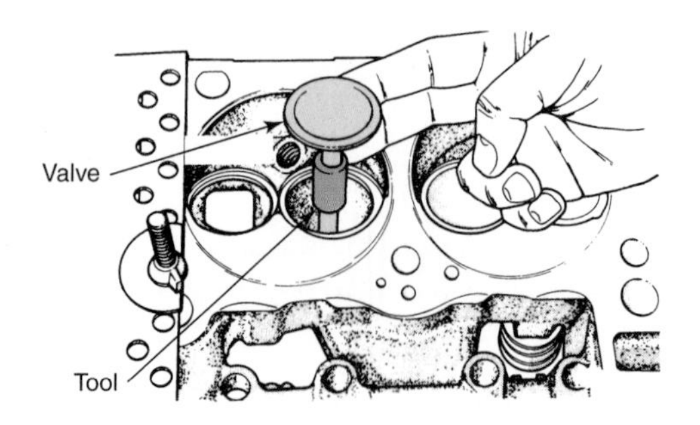

Figure 25-17. Positioning the valve with a special tool prior to checking stem-to-guide clearance with a dial indicator. (DaimlerChrysler)

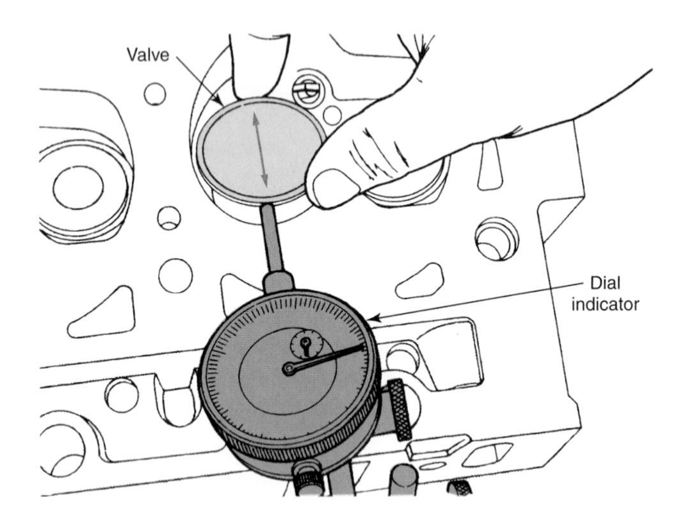

Figure 25-18. Checking valve stem play to determine stem-to-guide clearance. (DaimlerChrysler)

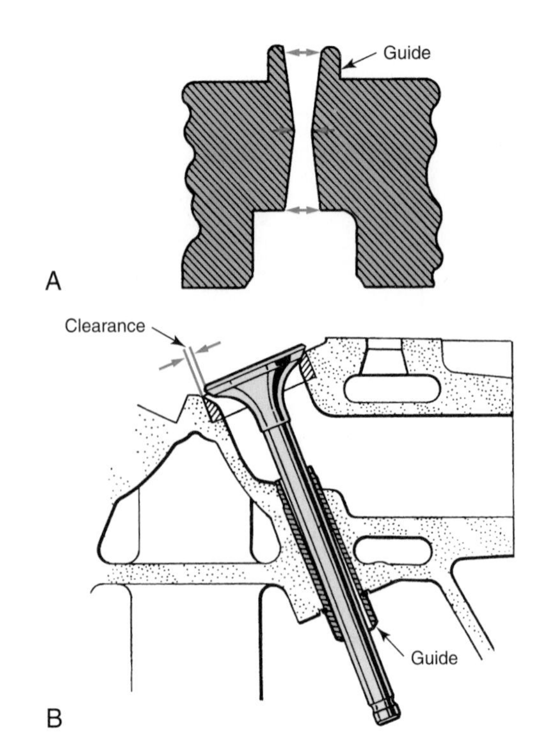

Figure 25-19. A—Stem-to-guide clearance near end of guide must be within limits. B—Note that correct clearance in guide center will not prevent tipping. (Jaguar & DaimlerChrysler)

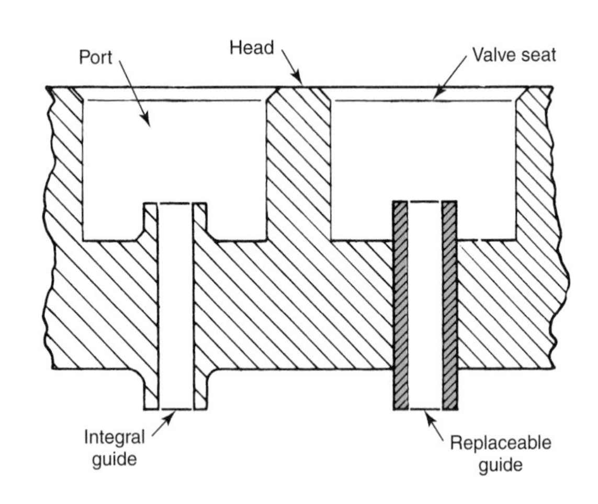

Figure 25-20. Integral and replaceable valve guides.

specifications. Generally, when the valve stem clearance exceeds 0.006″ (0.15 mm), it is considered excessive. Remember that both the guide and stem wear less in the center. Even though a stem-to-guide clearance at the center is correct, the clearance at the ends may be excessive and cause tipping, **Figure 25-19.** Some valve guides are removable. When an excessive stem-to-guide clearance is present, the valve guides may be replaced. Integral guides cannot be removed. When worn, they can be reamed to an oversize and new valves with oversize stems can be installed. See **Figure 25-20.**

Replacing Valve Guides

Removable valve guides may either be driven or pressed out. The punch should have a pilot section extending into the guide. The pilot should be a few thousandths of an inch smaller than the guide hole to prevent binding after the guide is cooled for installation. The main body of the punch should be slightly smaller than the guide so it will follow the guide through the hole. The contact edge should be smooth and square with the punch centerline.

Before driving out the guides, make a note of the distance from the surface of the head to the face of the guide. Also, note the shape of the end that extends into the combustion chamber. The proper end must be driven in to the correct depth during installation. Also, distinguish between exhaust guide shapes and intake guide shapes. This allows you to separate the guides according to valve type.

Place the punch in the guide. While holding the punch in firm contact with the guide, drive the guide from the hole. Refer to **Figure 25-21.** Guides are brittle and may crack if the punch is loosely held. Some heads must be heated before attempting to remove the guides.

Installing Guides

Before installing guides, make sure guide holes are spotlessly clean. If a refrigerator or freezer is handy, the guides may be placed in the freezer compartment long enough to thoroughly chill them. Dry ice or liquid nitrogen may also be used. The resulting reduction in diameter aids in their installation. Give the guide and hole a thin coat of hypoid lubricant (Lubriplate or similar type). Insert the proper end of the guide into the correct guide hole. Drive the guide to the specified depth; do not drive it past the required depth. If needed, a mark may be used to provide a means of measuring valve guide height from a given surface. See **Figure 25-22.** Some guide installation tools have a stop that will contact the head surface when the guide is at the correct height.

Reaming Guides after Installation

Some guides are factory reamed and do not require additional reaming following installation. If the guides must be reamed, use a ***valve guide reamer*** of the exact size. Start the reamer carefully and turn it clockwise both while entering and leaving the guide. Ream dry, while being careful to avoid any side pressure on the tool. Allow the pilot portion of the reamer to guide it through. A properly reamed guide provides approximately 0.002″ (0.05 mm) valve stem clearance, **Figure 25-23.** Consult the manufacturer's specifications for each vehicle.

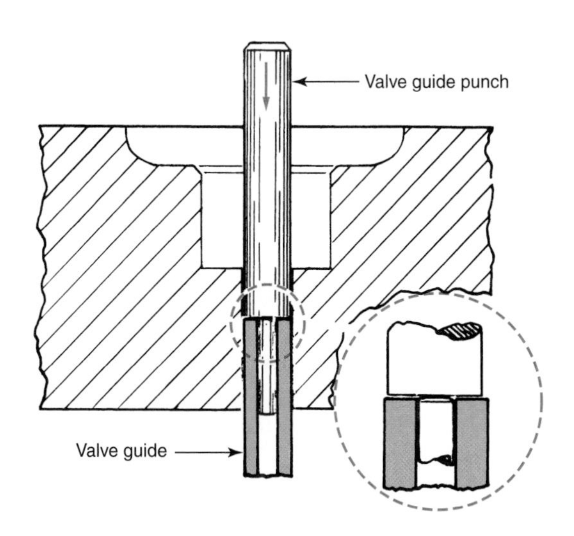

Figure 25-21. One form of a valve guide punch. Use a punch mark on tool to determine when guide has been driven to the proper depth.

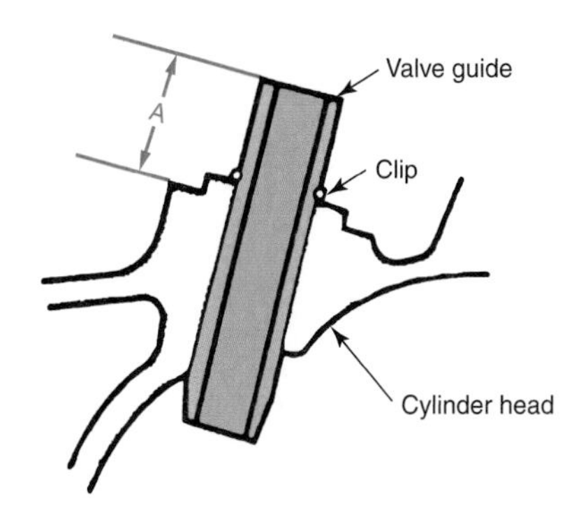

Figure 25-22. This valve guide has been installed to the proper depth. (Mazda)

Servicing Integral Guides

When the guide is cast as part of the head, it is necessary to determine the extent of the wear. If wear is excessive, select a new valve with a suitable oversize stem. Stem oversizes generally are available in 0.003″, 0.015″, and 0.030″ (0.08 mm, 0.38 mm, and 0.76 mm). Ream the worn guide to fit the valve stem. As with removable guides, use a sharp reamer of the correct size. Following reaming, wash the guides and blow them dry.

Valve Seats

Inspect each valve seat for signs of excessive burning or cracking. ***Valve seat inserts*** are steel rings pressed into the head. These inserts are often used on aluminum heads. If the seat is an insert type and shows any signs of damage, it must be removed and replaced. If an integral valve seat is damaged, it must be cut out and a replacement insert must be installed. In some cases, replacement inserts cannot be used and the head must be replaced. Typical seat damage includes looseness, burning, or cracking.

Servicing Valve Seat Insert

A special chisel or mechanical puller may be used to remove valve seats, **Figure 25-24.** When removing, be careful not to damage the seat recess. Make certain you have the correct size insert. ***Outside diameter*** (called ***OD***), depth, and ***inside diameter*** (called ***ID***) should match that of the insert being replaced.

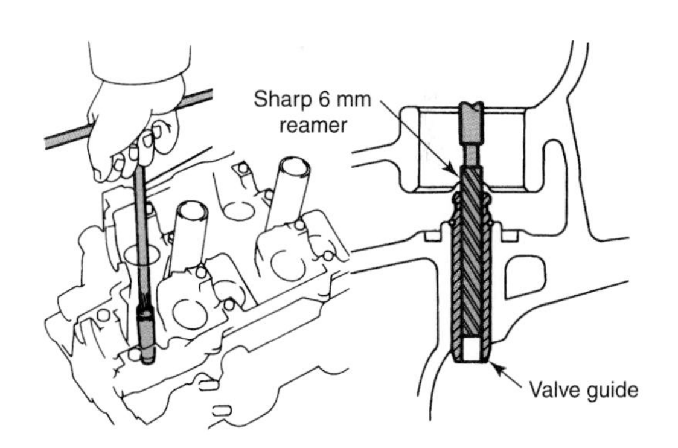

Figure 25-23. Reaming a valve guide. (Lexus)

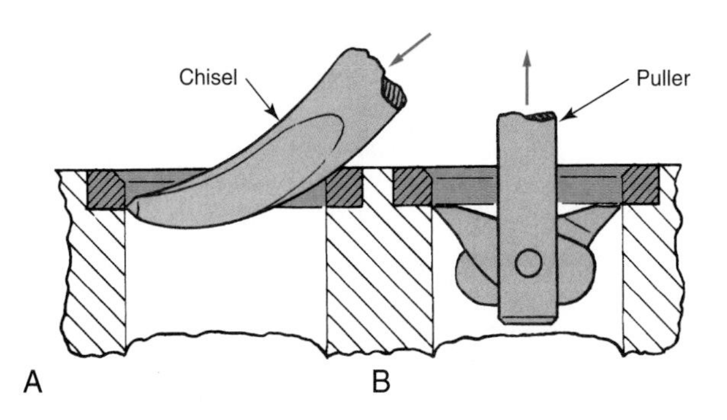

Figure 25-24. Removing valve seat inserts. A—Special chisel. B—Mechanical puller.

If the original inserts were cast iron, cast iron replacements can be used. If a hard valve seat insert is removed, replace with a similar type. Hard seat inserts are usually made of special heat-resistant steel such as Stellite. The recess must be clean and free of nicks and dents. Place a special driver pilot in the valve guide. Install a driving head on the driver. The head should be just a little smaller than the insert OD.

The insert's outside diameter will be 0.001″–0.002″ larger than the recess. This produces an interference fit to assist in securing the insert. It also produces good heat transfer from the insert to the head. If the inserts have been chilled in dry ice or in the freezer, remove and install them one at a time. If all the inserts are removed from the freezer at the same time, some of them will warm up before they can be installed. Keeping the inserts cold reduces their ODs, making them easier to install. Lay the insert over the recess in the head, making sure the beveled outside edge is down. Slide the driver over the pilot and start the insert with several firm blows. As the insert nears the bottom, reduce the strength of the hammer blows. Do not continue to pound the insert after it is fully seated. **Figure 25-25** shows a cross section of a typical insert driver setup.

Peening or Swaging Insert

The head metal around the outside of the valve seat insert may be either ***peened*** (upsetting head metal around the insert outside diameter to hold it in place) or ***swaged*** (upset by a rolling or rubbing action). All hard inserts and inserts set in an aluminum head must be peened or swaged. The insert will have a small chamfer on the upper OD into which the head metal is forced.

For peening, a pilot is placed in the valve guide and a special peening tool body is dropped over the pilot. The peen is adjusted so that it contacts the head metal along the edge of the insert. By turning and hammering the peening tool, the metal is upset (bulged), **Figure 25-26.** Soft cast iron inserts have the same coefficient of expansion as cast iron head metal. If cast iron inserts are properly fitted, they will not have to be peened. Many technicians peen all types of inserts to provide an extra measure of safety. Other tools apply a rolling pressure to swage the metal into the chamfer.

Cutting Insert Seat Recess

Where no insert is used and the integral seat is damaged beyond repair, a recess may be cut and an insert seat can be installed. In cases where an insert is used but is loose, a recess may be cut for an insert of slightly larger outside diameter. If the guide is in good shape, select a pilot that fits the guide as recommended by the tool manufacturer. Choose a cutter of the correct size and install.

Fit the guide to the pilot assembly and drop the tool body over the pilot. All alignment screws must be loose. Place the anchor bolt slot over a convenient head bolt hole and install the anchor bolt. Align the tool body with the pilot by shaking the tool slightly. Lock the anchor bolt and alignment screws securely. The cutter should revolve with finger pressure when all screws are secured. If binding is present, loosen the alignment screws, readjust, and retighten. The object is to have the tool body and drive mechanism secure without binding the pilot and cutter assembly. **Figure 25-27** shows two types of insert recess tools.

With the cutter just touching the work, place the insert ring on the stop block. Run the stop collar down until it touches the ring. Lock the feed screw to the cutter sleeve and remove the ring. The tool will then cut to the exact depth of the ring. Make certain all alignment screws are tight. Use either a ratchet handle or a power drive mechanism to rotate the cutter.

With the cutter just clearing the work, start the cutter. Feed the cutter into the work by turning the knurled stop collar. Do not force the cutter. Give the stop collar several turns, and then feed the cutter down lightly. Repeat this process until the

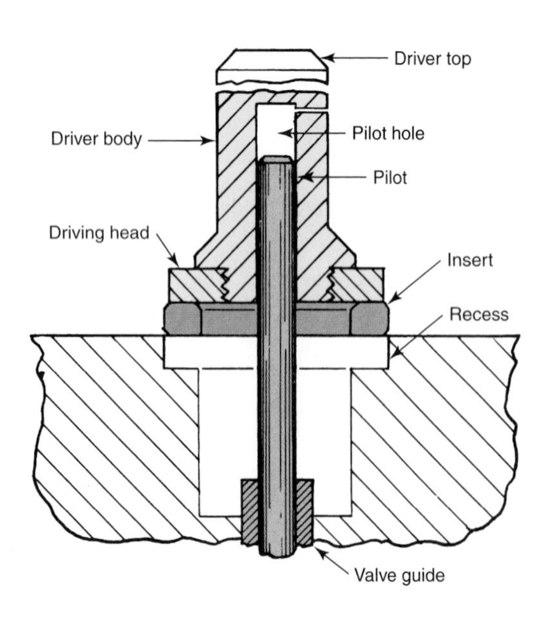

Figure 25-25. Installing a valve seat insert with a special pilot driver combination.

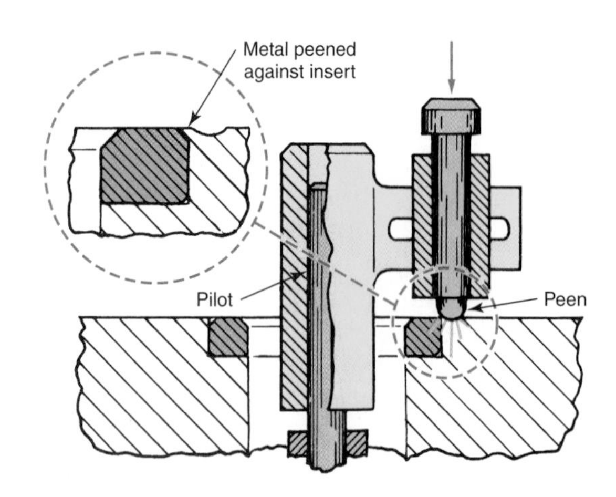

Figure 25-26. Using a special tool to peen the metal around an insert's edge. Note how the metal is forced against the insert.

stop collar engages the stop block. At this point, give the tool a few additional turns to produce a smooth seat for the insert. Turn the cutter out. Figure 25-27B shows the technician moving the cutter into the work by turning the stop collar.

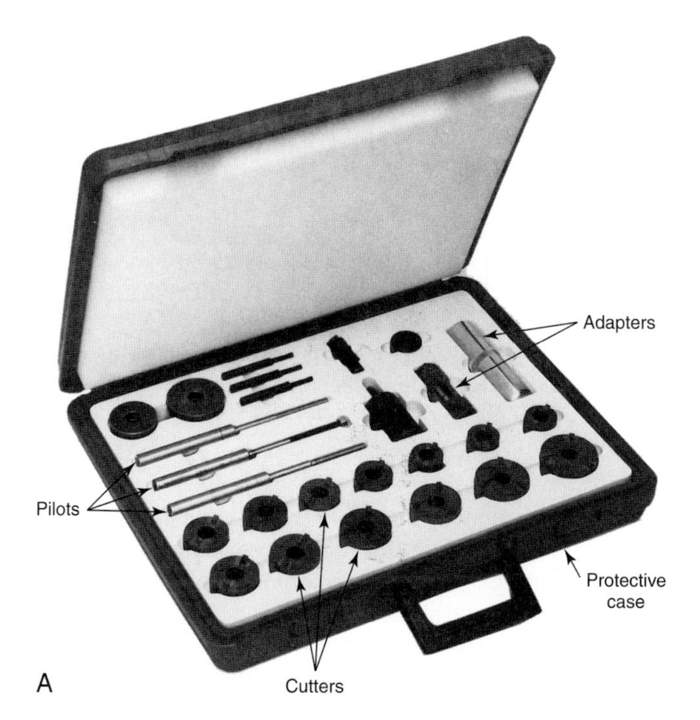

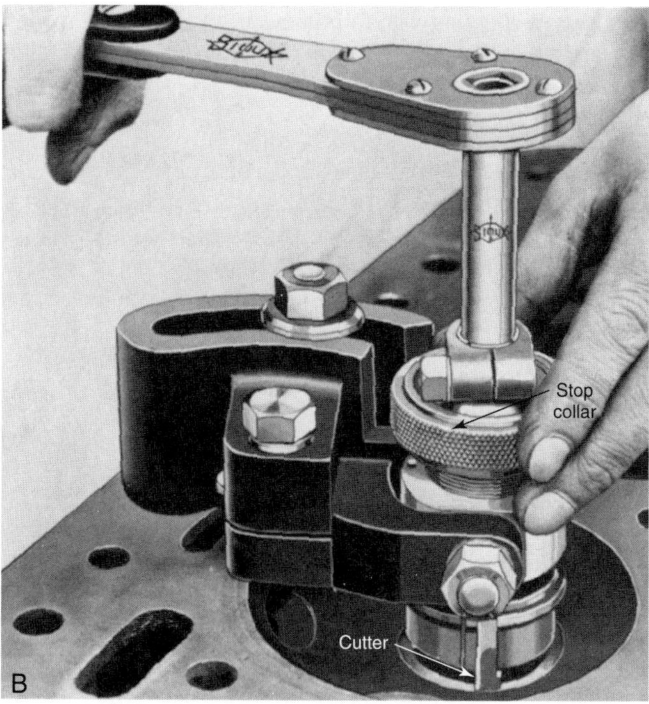

Figure 25-27. A—A set of valve seat recess cutters with various pilots and adapters stored in a protective carrying case. Avoid dropping these cutters. They can be easily damaged. B—A different style of cutter being used to cut a new insert recess. Follow the toolmaker's operating instructions. (Lisle Tools and Sioux Tools)

Valve Seat Service

After all valve guide and insert work is complete, the valve seats are ready to be ***refaced.*** The seat must be free of carbon, oil, and dirt. The cutting device will quickly fill with debris if the seat is not clean, ruining the cutting action. The valve guides must be clean to allow the pilot to properly align with the guide hole. Valve seat refacing can be done using grinding stones or hardened cutters.

The valve seat must be cut at the correct angle and be clean, smooth, and free of cracks, nicks, and pits. It must be of the correct width and engage the face of the valve near its central portion. Common seat angles are 30° and 45°. If an interference fit is desired, the interference angle may be ground on either the seat or the valve. Follow manufacturer's specifications.

Seat width varies with each manufacturer, but average around 1/16″ (1.58 mm) for both intake and exhaust seats. A seat that is too narrow will pound out of shape easily. It will also fail to dissipate enough heat from the valve face. A seat that is too wide will tend to collect carbon, eventually preventing a good seal. This can lead to valve overheating and burning. See **Figure 25-28.**

When refacing a seat, the removal of stock widens the seat beyond its original specifications. The seat must be narrowed by removing metal from its upper portion. In cases where the valve port walls narrow or are uneven, metal must be removed from the bottom of the seat. If the walls are smooth and of constant diameter, only a very light cut with a 60°–70° stone should be taken. If inserts are used, the bottom cut is not necessary, **Figure 25-29.** The light bottom cut produces a seat that is the same width at all spots.

Preparing to Grind Valve Seats

The following sections explain how to use grinding stones to reface valve seats. The information concerning grinding stone selection and preparation should be carefully

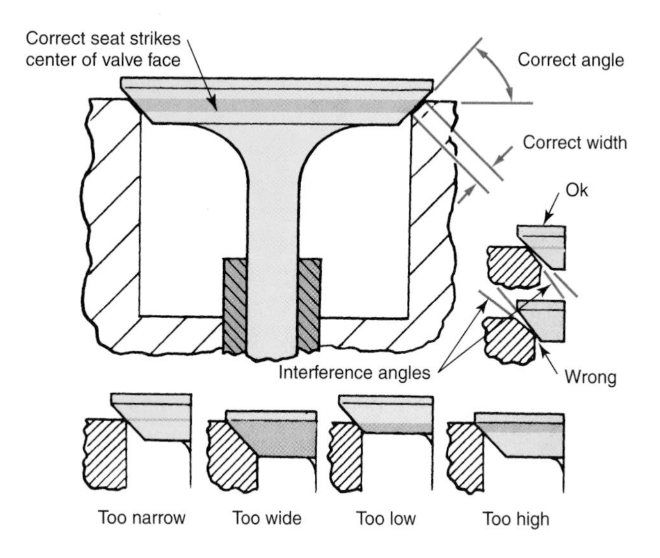

Figure 25-28. Correct and incorrect valve seats. Note the interference angles.

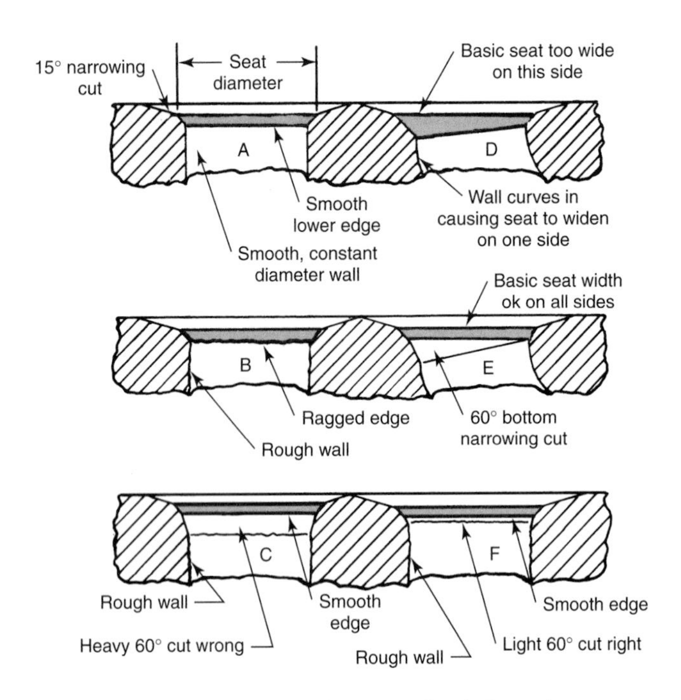

Figure 25-29. Narrowing a valve seat. A—15° cut from top makes a good seat when the port walls are smooth and of constant diameter. B—A rough wall leaves a ragged lower edge on the seat. C—A heavy bottom cut produces a smooth lower seat edge, but widens seat diameter. D—Curved port walls produce an uneven seat width. E—A bottom cut produces an even width. F—Very light bottom cut smoothes seat edge without appreciable increase in seat diameter.

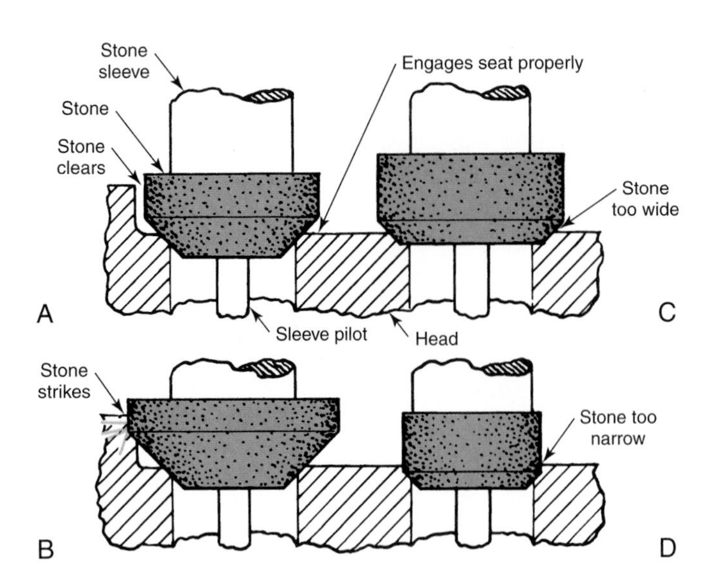

Figure 25-30. Stone must be of correct width. A—Stone OK. B—Too wide. C—Too wide, makes a horizontal step at bottom of seat. D—Too narrow, makes a vertical step at top of seat.

followed. Many heads have been ruined by the use of a worn stone or a stone of the wrong angle.

Seat stones are available in various widths. Coarse-textured roughing stones are used for the initial, or roughing, cut on steel seats. The fine-textured finishing stone is used for the last cut on steel seats. The cast iron block or head requires only the use of the finishing stone. Special stones are available for grinding Stellite and other hard seat inserts. The stone must be a little wider than the finished seat in order to prevent counterboring. It must not be so wide as to strike other parts of the combustion chamber. **Figure 25-30** illustrates how various stone widths affect the job.

Select a cutting stone with the correct angle for the job. This saves time in dressing and prolongs the life of the stone. Many stones are constructed so that an angle may be ground on both ends. After selecting a cutting stone of the correct size and texture, screw it snugly onto the stone holder or sleeve. Place the stone sleeve on the dressing stand pilot. Adjust the diamond holder to the correct angle and lock all adjustments. Back the diamond away from the stone. Engage the stone drive motor and run the diamond tip fully across the stone's face. Use care not to make the cuts too heavy when dressing the stone. Take light cuts on the stone until the angle is correct and the full stone face is clean and true. The full stone angle must also be dressed to prevent injury to the diamond.

There are basically two types of stone pilots in use. One is the adjustable type that is slipped into the guide and then expanded. The other is of tapered construction that is secured through friction between the guide and a tapered section. Make sure the guide is clean, regardless of the type used. Wipe the pilot off with a clean, lightly oiled rag before inserting it into the guide. The pilot must be rigid. See **Figure 25-31.** Mount the correct seat angle stone on one sleeve and the 15° and 70° stones on two other sleeves. This allows you to grind and narrow the seat without removing and changing stones. Once the pilot is inserted, finish the complete seat operation before moving to the next one.

Grinding the Valve Seat

Once the stone is dressed and clean, place the sleeve on the pilot, Figure 25-31B. The stone should contact the seat. Insert the motor drive head into the sleeve. Tilt the motor up, down, and sideways to feel for a non-binding, central position. While supporting the motor, engage the switch. Allow the stone to grind for a few seconds. Then, stop and remove the motor. Raise the sleeve and examine the seat. Repeat this procedure until the seat is smooth, clean, and free of burns and pits.

Caution: Remove only enough stock to clean the seat. If an integral seat is hardened, excessive grinding can cut through the hardened area. Check manufacturer's specifications.

On hard inserts, dress the stone several times for each seat. Never continue grinding when the stone surface needs dressing. If using a roughing stone, stop when the seat is cleaned up. Switch to a finishing stone and polish the seat. The finished seat will be only as accurate as the stone.

Using a 60°–70° stone, grind until the 60°–70° angle touches the basic 30°–45° seat surface all the way around.

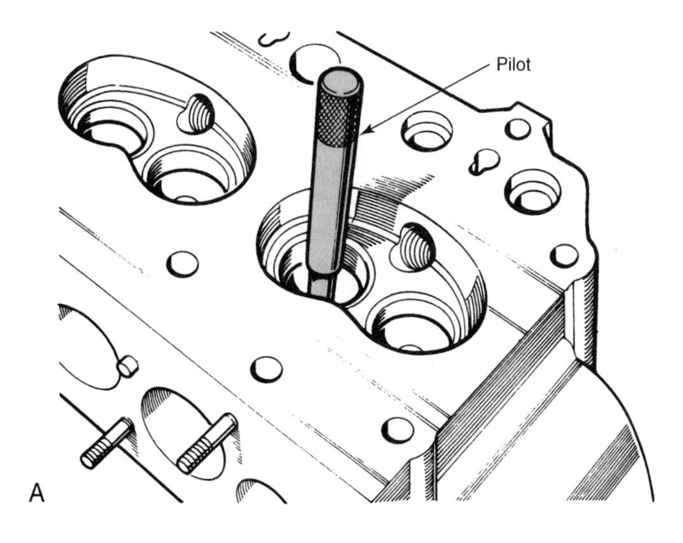

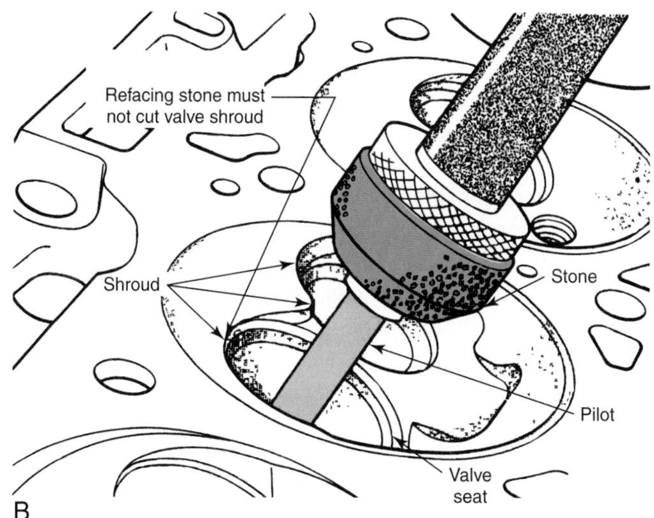

Figure 25-31. A—Insert the pilot for the valve-seat stone into the valve guide. The pilot must be tight. B— Refacing (grinding) a valve seat. Wear safety glasses. (DaimlerChrysler)

This 60°–70° stone cuts very quickly. Do not apply downward pressure and cut for only about two seconds before checking. With the 15°–30° stone, (see manufacturer's specifications) remove stock until the seat is down to the specified width. A small measuring tool will assist in a careful measurement of the seat width. A trick often used for seat grinding is to mark the seat (after grinding the basic angle) with a series of soft pencil marks across the width. When removing stock from above and below the seat, the pencil marks clearly show what remains of the base seat angle, **Figure 25-32.**

Using Valve Seat Reamers

Valve seat reamers, sometimes called *seat cutters,* are replacing grinding stones in many shops. The seat reamer consists of a cutting head with several blades attached. The cutter blades resemble small files. Cutting heads are available in all of the common valve seat angles, as well as extra wide and narrow versions for narrowing and positioning the seat.

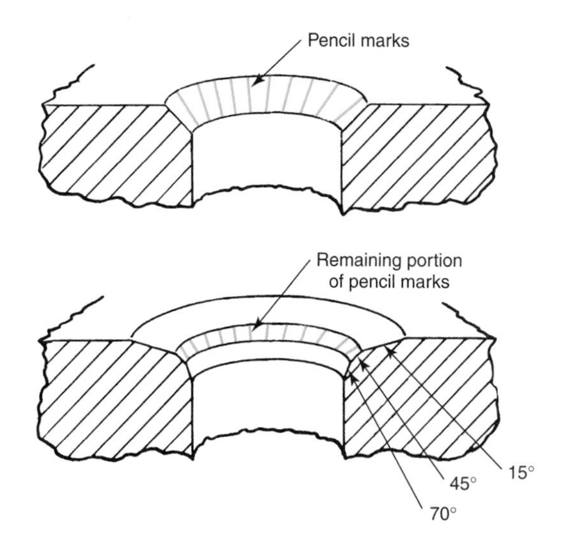

Figure 25-32. Using pencil marks to determine the width of a basic 30° or 45° seat.

The advantage of the valve seat reamer is that it is hand-operated. The reamer teeth do not require dressing and are replaced when the blades become dull. A typical valve seat cutter is shown in **Figure 25-33.** As was the case when using a grinding stone, a pilot shaft must be installed in the valve guide. The reamer must always be turned in the proper direction, usually clockwise. The finished cut is checked in the same way as with a grinding stone.

Note: Valve seat reamers cannot be used on some hardened valve seats. Check the appropriate service manual and use a grinding stone where indicated.

Testing Valve Seats

To test a valve seat for ***concentricity*** (true roundness), place a special valve seat dial indicator on the pilot. Adjust the indicator bar so that it contacts the center of the valve seat. The dial needle should travel about a half turn when the bar length is correct. Set the dial to zero. Hold the upper dial section and slowly turn the bottom section around so the bar travels completely around the seat. The dial needle will indicate any runout. The entire seat should be within 0.002″ (0.05 mm). If runout exceeds 0.002″ (0.05 mm), check the setup carefully and try again. If runout is still excessive, regrind the seat.

Lapping Valves

Valve ***lapping*** consists of turning a valve against a seat coated with lapping compound, which is a fine abrasive paste. Lapping was often performed in the past, sometimes as a substitute for grinding the valves and seats. Today, lapping is sometimes performed to make a final fit between the valve and seat. Some technicians feel that it produces a more accurate seal between valve and seat, while others contend that it is of no value. Many manufacturers say that when modern valve-grinding equipment is properly used, lapping is not necessary. Lapping when an interference fit is desired can actually damage the seal.

Final Check of Seat and Valve Face

Before placing the valves in the guides, the head and the guides should be thoroughly washed, flushed, and blown dry. Rub a very thin film of ***Prussian blue*** on the valve face. Place the valve in position. While pressing in the center of the valve, rotate the valve about one-quarter turn and back. Remove the valve and examine the seat. It should be marked with blue around its entire circumference. The seat should mark the valve face near the center. Pencil marks about 1/4″ (6.35 mm) apart around the valve face will also provide a check, **Figure 25-34.** Turning the valve should wipe away all graphite where the valve seats.

Rocker Arm Stud Service

If the cylinder head uses individual rocker arm studs, check them for signs of damage or looseness. If a replacement is necessary due to breakage, a standard size replacement will suffice. If the stud is loose, the hole will have to be reamed to one of several available oversizes. Most rocker arm studs are pressed in, however, some heads use screw-in studs.

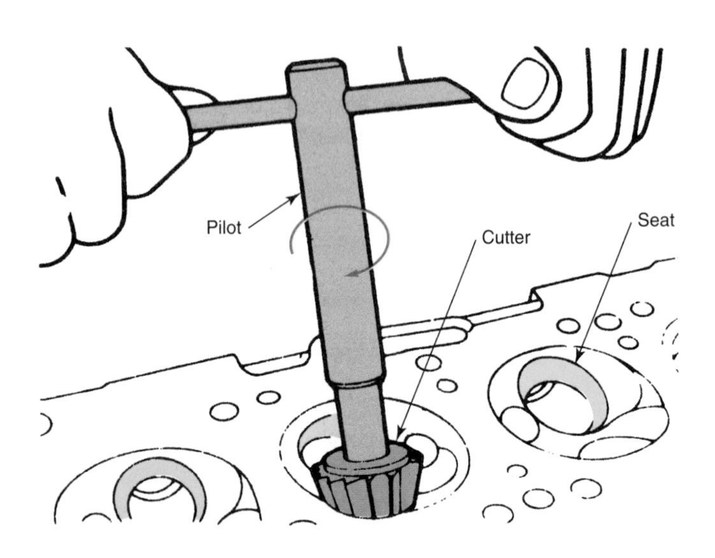

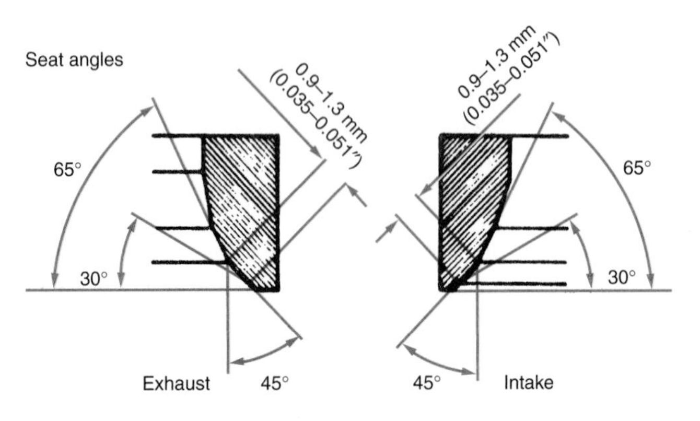

Figure 25-33. A valve seat cutter is used to clean and true seat. Always check the valve guide for wear before cutting the seat. (Hyundai)

To remove the stud, place the pulling sleeve over the stud. Tighten the nut against the sleeve and turn the nut to pull the stud, **Figure 25-35.** If the stud is broken off at the boss, drill into it and remove it with a stud extractor. If an oversize stud is required, ream the hole with a special reamer of the correct size. If a larger oversize stud is needed, ream the hole in two steps. Use the smaller oversize reamer first, and then finish with the desired size. See **Figure 25-36.** Thread the replacement stud in the driver. Coat the plain end with hypoid lubricant or Lubriplate. Place the stud over the hole and drive it down until the driver body touches the stud boss. This will be the correct depth. Remove the driver tool.

Valve Spring Service

Valve springs should be soaked in solvent, brushed, and thoroughly rinsed. Never clean painted springs in strong cleaners. This will remove the paint and other coatings that prevent rust. Power wire wheels will also remove this protective coating, shortening spring life. After extended service, valve springs tend to lose tension. Since correct ***spring tension*** is important to proper valve action, each spring must be tested to make certain it meets minimum requirements. Manufacturers provide specifications listing the amount of pressure, in pounds or kilograms that a given spring should exert when compressed to a specific length. The spring is placed in an appropriate measuring device and compressed to the specified length. The pressure is then determined, **Figure 25-37.**

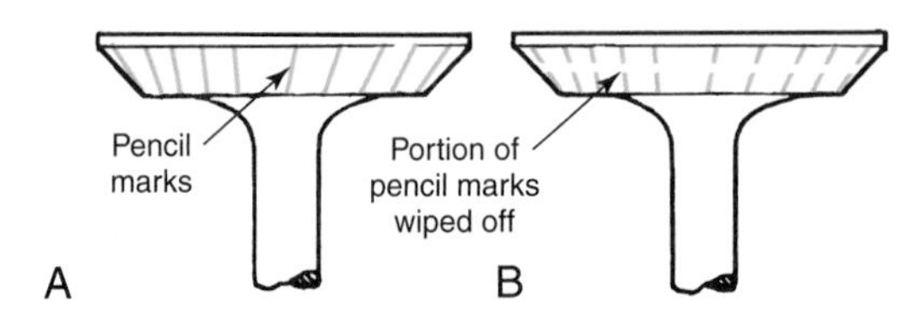

Figure 25-34. Pencil marks on the valve face indicate the valve's face-to-seat accuracy. A—Marks applied. B—Portion of marks wiped off by placing the valve in the seat and giving it a quarter turn.

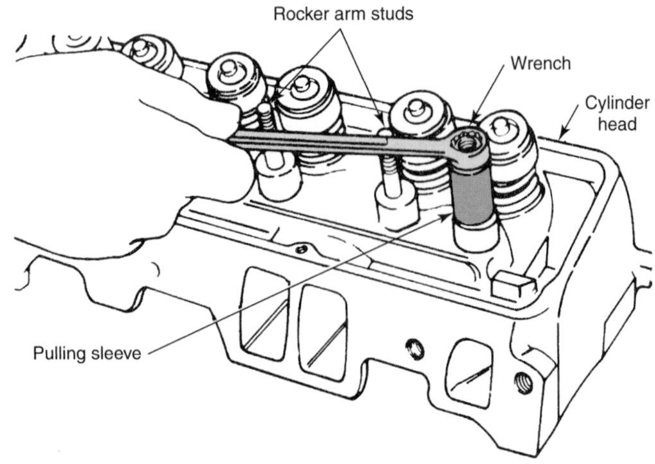

Figure 25-35. Pulling a rocker arm stud. (General Motors)

Place the spring on a flat surface. Slide a combination square up to the spring (do not tip spring). Using the scale on the blade, measure the spring's ***free length*** (length when spring is not under pressure). It should meet the manufacturer's specifications. Check the spring for ***squareness*** by carefully sighting the spring between its edge and the blade. Give the spring a partial turn and check again. Place the spring on its opposite end and check again. The spring should be parallel to the blade in all cases. If both sightings indicate the spring is parallel (not more than 1/16″ (1.59 mm) difference between top and bottom), you can assume the spring is square, **Figure 25-38.**

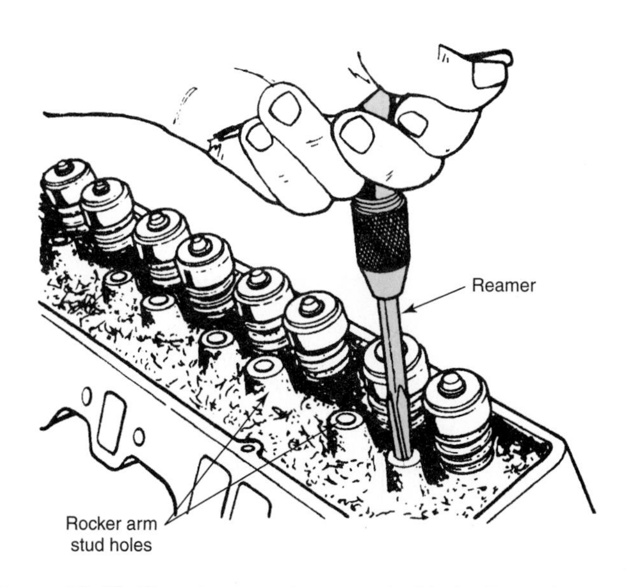

Figure 25-36. Reaming a rocker arm stud hole. Ream in steps if installing an oversize stud. (General Motors)

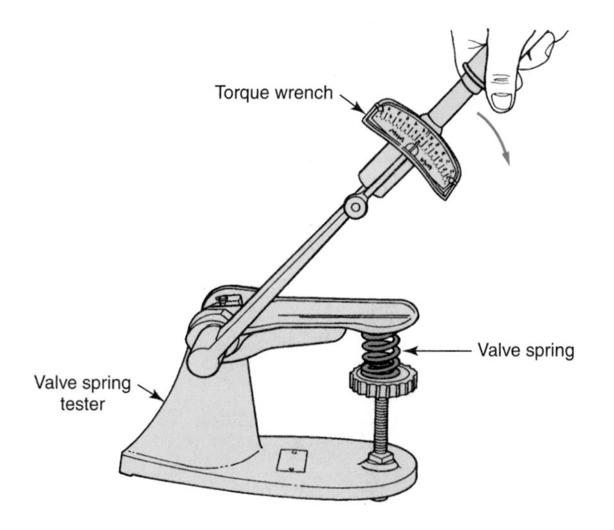

Figure 25-37. Testing valve spring tension. The spring is placed on the base. When the lever is pulled down, the pointer pad compresses the spring to the specified distance on scale. The tension is then read on dial. (DaimlerChrysler)

Note: While inspecting the valve springs, check for any signs of rust, corrosive etching, scratches, and nicks.

Replace springs that fail to meet specified compressed pressure, free length, squareness, or that show other problems. A weak spring will cause valve float (valve closing so slowly the camshaft lobe starts to open it again before it has fully seated). Valves may start sticking in the guides, causing heavy tappet noise, missing, burning, and breakage. Remember that using poor valve springs can be expensive. New springs are inexpensive and will certainly raise the level of reliability and performance. Inspect damper springs (used inside regular spring to reduce spring vibration) and damper clips if used. Discard any that are worn or fail to meet specifications.

Cylinder Head Assembly

With the cylinder head back in a suitable fixture and spotlessly clean, oil the valve guides. Select the proper valve, oil the stem, and insert the valve into the guide. Each valve should be installed in the port that it was removed from. On engines that do not have provisions for adjusting the clearance between a rocker arm and its push rod, the height of the valve stem from the head should be checked. The amount of metal removed from the valve face and seat might allow the stem to protrude farther. If this happens, the rocker arm will be tipped down on the push rod side, forcing the hydraulic lifter plunger near the bottom of its travel. Malfunctions can result if provisions are not built into the lifter to adjust to this change. If the stem height is excessive, the valve must be removed and the stem end ground down the proper amount. Check all valves, **Figure 25-39.**

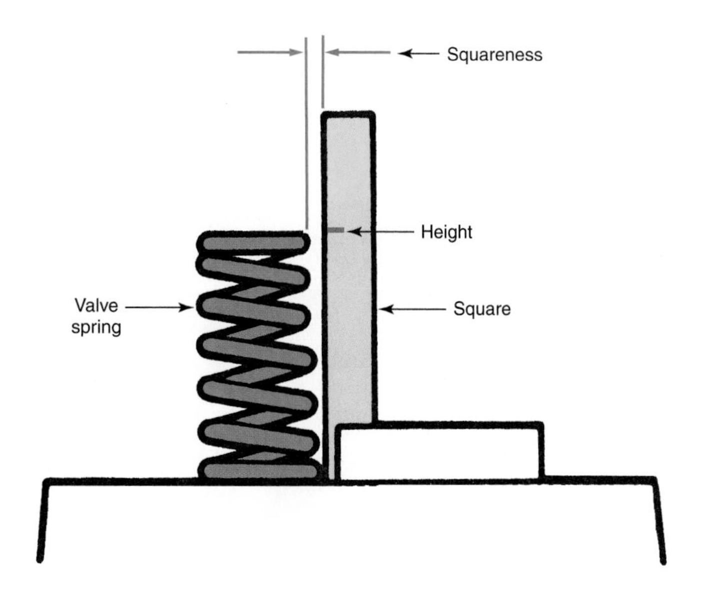

Figure 25-38. Checking the spring's free length and squareness. (Suzuki)

Note: Checking and service of camshafts installed in the cylinder head is covered in Chapter 27, Engine Timing and Intake Service.

Valve Stem Seals

Some guides are designed to accept valve stem seals. **Figure 25-40** shows several types of valve stem seals. Both integral and removable guides may be prepared for a seal. To install one type of seal, the valve stem end is covered with a protective plastic cap, **Figure 25-41.** The seal is then pressed over the end and down the stem. The seal is forced over the machined section of the guide as far as possible with the fingers. A special tool is used to complete the seating in some cases by grasping the seal and forcing it fully down, **Figure 25-42.**

In addition to the special guide seal shown, umbrella seals and neoprene rings are often used on the valve stem ends to prevent oil from flowing down the stem to the guide, **Figure 25-43.** Valve guides are often cut at an angle to prevent oil from puddling on the top. When installing the valve assembly, be careful to avoid damage to the seals. Occasionally, only the intake valves are protected with guide seals, tapered guide heads, and stem end shields. Make sure they are correctly installed.

If required, place a steel washer around the guide and in contact with the head. Check springs for a closed coil end (end of the spring in which the coils are spaced closer together). Place closed end toward the head, over the stem, and in contact with the head, **Figure 25-44.** On some engines, there are differences between intake and exhaust springs and retainers. Be careful to assemble them in the proper locations. If dual coils or a damper spring are used, space the coil ends according to the manufacturer's instructions, usually about 180° apart.

Using a spring compressor, compress the spring just far enough to expose the stem oil seal groove. Install a stem-to-guide oil seal if used. Slip the seal into the groove. Make sure it is positioned properly and is not twisted. Insert the split keepers, or locks, and slowly release the spring. See **Figure 25-45.** As the spring rises, guide the retainer so it is centered around the keepers. When fully released, check keepers to make certain they are fully engaged.

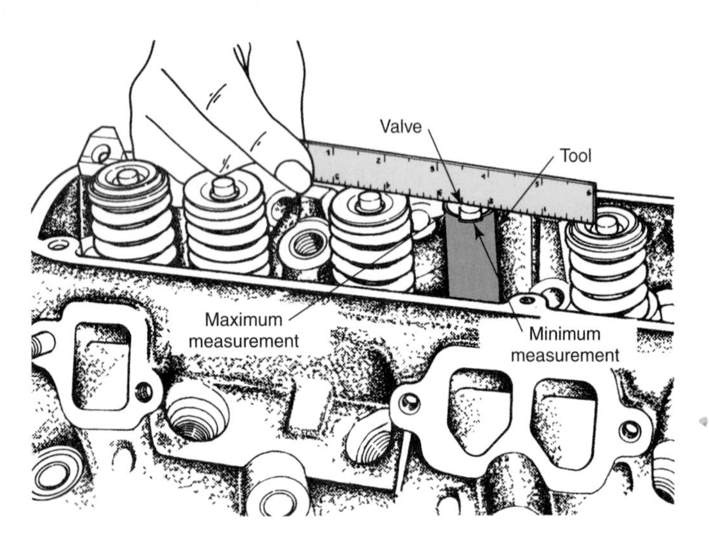

Figure 25-39. Checking valve stem height. Incorrect valve stem height can cause serious problems. (DaimlerChrysler)

Figure 25-41. Placing a protective cap on the valve stem end. (Fel-Pro)

Figure 25-40. Various types of valve stem seals are shown here.

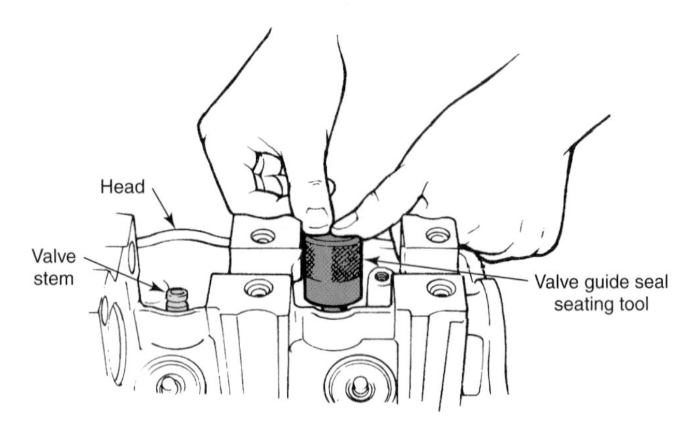

Figure 25-42. Seating valve guide seal using a special tool. (General Motors)

Warning: If keepers are not locked into position, they can slip and fly out with dangerous force. Stay to one side of the valve and spring assembly. Always wear safety goggles when compressing springs.

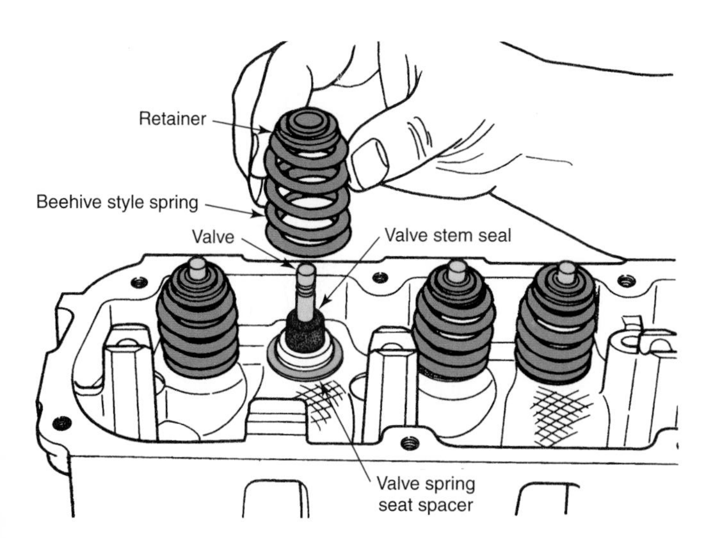

Figure 25-43. Valve stem seal installation. A—Umbrella-type seal. B—Ring-type seal. (Fel-Pro)

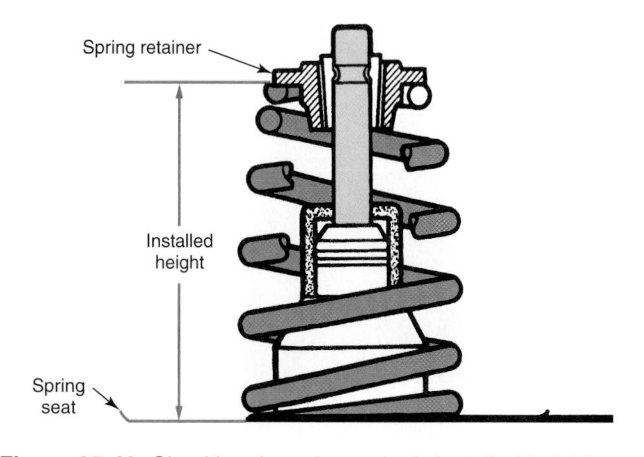

Figure 25-44. After the valve guide seal is installed, the spring and retainer can be placed in position. (DaimlerChrysler)

When a stem seal is employed, it can be tested with a vacuum pump and a small vacuum cup adapter. Place the vacuum cup over the retainer and squeeze the vacuum pump. When the pump is released, the reading on the gauge should hold steady, indicating an airtight seal.

Checking Installed Height of Valve Spring

As with the valve stem end, removal of stock from valve face and seat will allow the keeper grooves to protrude higher above the head. This will increase the installed height of the spring and reduce spring tension. Using manufacturer's specifications, measure the ***installed height*** of each spring, **Figure 25-46.**

If the height is excessive, it must be corrected by removing the spring and placing a special steel washer or insert between the spring and the head. These washers are available in different thicknesses. Do not install washers that are too thick, as the spring pressure can be increased to the point of causing rapid lifter and camshaft wear.

Figure 25-45. As the spring and retainer are compressed, position the keepers so they are centered in the retainer. (Fel-Pro)

Figure 25-46. Checking the valve spring's installed height as measured from the spring seat to the bottom of the spring retainer. (DaimlerChrysler)

Reinstalling the Head on the Engine

Before installing the head, make sure that the block surface is clean and not warped. Check the block for warping before installing the head. If necessary, remove carbon from the top of the piston. Clean the bolt holes in the block and make sure the threads are not damaged.

Clean the head bolts, and check for damaged threads. If required, coat the bolt threads with thread compound. Carefully check the new head gasket to ensure that it is the right one for the engine, and then place the head gasket on the block. Ensure that the correct side is facing up and that the front of the gasket is at the front of the block, **Figure 25-47.** Install guide pins into several holes in the block. Lower the head into position over the guide pins. Install several head bolts; then remove the guide pins. Install all head bolts, and then tighten them in the sequence called out in the manufacturers' service literature. Finish tightening the head bolts to the proper torque. Some engine blocks have integral guide pins that do not have to be removed after cylinder head installation. See **Figure 25-48.**

Caution: Modern head bolts are often of the *torque-to-yield* type, designed for one-time use. Do not reuse these fasteners, since they are permanently stretched and will not properly seal, even if torqued correctly.

Torque-Plus-Angle Tightening

The cylinder head specifications on many modern engines call for a special bolt-tightening method called ***torque-plus-angle.*** This method is often used to torque aluminum heads on cast iron blocks. Begin by torquing the bolts to a relatively low torque, as listed in the manufacturer's specifications. Then, tighten the bolt the specified additional fraction of a turn (usually 90° or 1/4 turn). Special angle gauges are available to accurately measure the fraction of a turn needed. Each bolt is tightened one time only. Manufacturers claim that this method is more accurate, since it stretches the bolt the exact amount needed to allow for the different expansion rates of cast iron and aluminum. Always check the manufacturer's specifications, and never substitute traditional torquing methods for the torque-plus-angle method.

Valve Service with Head on Engine

It is possible to service valve springs, retainers, and seals with the head on the engine. Bring the piston to top dead center (TDC) on the compression stroke (both valves closed) and remove the rocker arms for that cylinder. Remove the spark plug and insert an air hose adapter. Connect an air hose to the adapter, **Figure 25-49,** and apply full air pressure to the cylinder. With the rocker arm out of the way, the spring may be compressed and the keepers removed, **Figure 25-50.** A new spring or valve guide seal may be installed. Keep air pressure to the cylinder until the valve spring is replaced and the keepers installed.

Tech Talk

Did you ever duck into a fast-food place, order a cheeseburger, and get a fish sandwich instead? When you sent it back, did you get a hamburger instead of a cheeseburger? Annoying, wasn't it? Chances are you avoid that place now.

Figure 25-47. The cylinder head gasket must be installed with the correct orientation. It must fit properly and provide openings for all passages. (Fel-Pro)

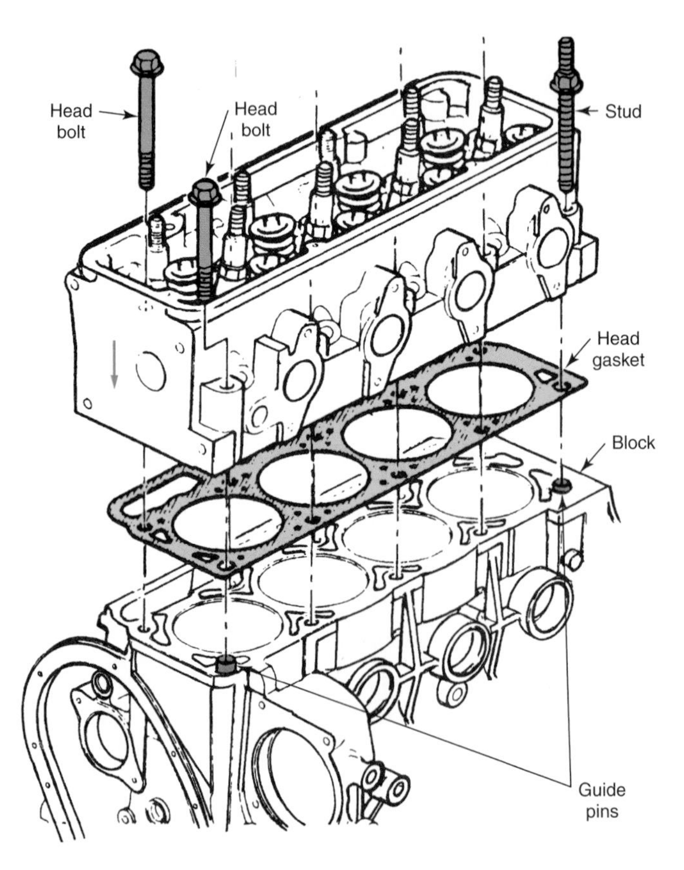

Figure 25-48. Positioning the cylinder head for installation. Note the short, permanent guide pins on the block. (General Motors)

Figure 25-49. An air hose adapter allows you to apply air pressure to the cylinder. With the rocker arm out of the way, the spring may be compressed and the keepers removed. (Fel-Pro)

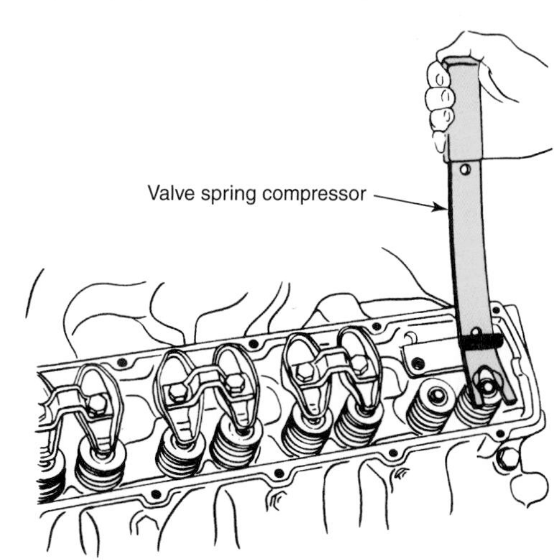

B

Figure 25-50. After applying air pressure to a cylinder, a new spring or valve guide seal may be installed. A—Using a screw-type compressor to remove a valve spring. B— Another type of valve spring compressor. This one is particularly handy for removing a spring with the head on the engine. (Acura)

The restaurants that you go to regularly are not like that. You get what you want the first time. That is why you go back.

The automotive service business works the same way. Your customers will come back if they get what they want. Customers expect something for their money—service. Service consists of giving them what they asked for. There are so many places that do not give good service that a place that does is almost guaranteed success.

A person who wants to get a vehicle fixed is not very different from a person who wants a cheeseburger. Always do your best possible work to guarantee your success.

Summary

When disassembling a cylinder head, keep all valves, springs, keepers, and retainers in order so that they may be replaced in the same location. The head, valves, guides, and other parts must be thoroughly cleaned. Do not scratch the head surface. Using an accurate straightedge, check the cylinder head for warpage. Reface all valves and reject any that will not clean up or that have insufficient margin. An interference angle may be used in some engines. Dress all cutting stones. Smooth and chamfer valve stem ends. Stems must not be worn beyond limits.

Be sure to check valve stem clearance. If clearance is excessive, replace the guides or ream for an oversize stem. When replacing valve guides, be certain to get the proper guide and install it right side up in the correct hole. Drive the guide in the specified distance. Some guides require reaming after installation. Seals are often used on the guides and on the stems of both intake and exhaust valves to prevent excessive oil consumption.

A cracked or burned valve seat can be repaired by installing a valve seat insert. Grind the valve seat at the correct angle until it is cleaned up. Narrow the seat to the specified width by using a 15°–30° stone on the top and, in some cases, a 60°–70° stone on the bottom. Keep all cutting stones properly dressed and remove no more metal than is necessary. Test the valve seat for concentricity.

Replace any broken, loose, or damaged rocker arm studs. If loose, ream and install an oversize stud. Check valve springs for squareness, tension, rust, or nicks. Replace any that show the slightest defect. Lubricate and install the valves. Check the stem height above the head. Install the springs with the closed coil end against the head. Check the installed spring height. Add an insert between the spring and the head, if needed. Check the stem seal with a vacuum cup.

Before reinstalling the head, make sure that the block surface is clean and accurate. Place the head gasket right side up with the correct end forward on the block. Using guide pins, lower head into position. Head bolts and holes in block must be clean and coated with thread compound. Torque head to manufacturer's specifications. Always use the proper method to torque the heads.

Review Questions—Chapter 25

Do not write in this book. Write your answers on a separate sheet of paper.

1. Clogged coolant passages or insufficient valve clearance could cause the valves to be _____ or _____.
2. Too much clearance between the valve stem and valve guide can cause _____ deposits on the valve.
3. Valve grinding stones are dressed with:
   (A) a file.
   (B) another stone.
   (C) a diamond.
   (D) a hardened steel rod.
4. When the valve is ground at a slightly different angle (about 1°) than the seat, a(n) _____ fit is produced.
5. To control stem height above the head, it is necessary to grind the _____ end.
6. Where excessive valve stem clearance is present, it may be corrected by _____ guides or by _____ for an _____ valve stem.
7. Valve seat runout should be kept within approximately:
   (A) 0.002″ (0.05 mm).
   (B) 0.006″ (0.15 mm).
   (C) 0.020″ (0.51 mm).
   (D) 0.0003″ (0.0762 mm).
8. The valve seat should engage the valve face near the _____.
9. Valve springs should be tested for _____ and _____.
10. To facilitate accurate head, gasket, and block alignment, _____ should be used when installing the head.
11. By _____, valve springs and seals can be replaced while the head is on the engine.
    (A) applying air pressure to the cylinder
    (B) bringing the piston in the affected cylinder to top dead center on the compression stroke
    (C) bringing the piston in the affected cylinder to bottom dead center
    (D) Both A and B.

ASE-Type Questions

1. It is necessary to keep all cylinder head parts in order because:
   (A) they may be lost.
   (B) they can be kept in a smaller area.
   (C) it is important that they be returned to their original positions.
   (D) it is just a good habit.
2. Technician A says valves should be closely inspected to determine what caused them to fail. Technician B says that damaged rocker arm studs can be replaced instead of replacing the head. Who is right?
   (A) A only.
   (B) B only.
   (C) Both A and B.
   (D) Neither A nor B.
3. The most important reason to keep wheels dressed is:
   (A) they cut faster.
   (B) they will produce accurate angles.
   (C) they wear longer.
   (D) they look better.
4. When grinding the valve face:
   (A) keep the valve in the center of the stone.
   (B) move the valve back and forth—staying on the stone.
   (C) move the valve back and forth—off both sides of the stone.
   (D) keep the valve on the right-hand side of the stone.
5. What tool would be used with a hole gauge when checking valve stem clearance?
   (A) Micrometer.
   (B) Dial indicator.
   (C) Feeler gauge.
   (D) Caliper.
6. Technician A says a cracked intake valve seal causes excessive oil consumption. Technician B says excessive exhaust valve stem clearance causes excessive oil consumption. Who is right?
   (A) A only.
   (B) B only.
   (C) Both A and B.
   (D) Neither A nor B.
7. A valve seat that is too wide will:
   (A) pack with carbon and start to leak and burn.
   (B) run too cold.
   (C) break the valve stem.
   (D) be loose.
8. Technician A says the valve seat must be concentric with the guide hole. Technician B says a dressed valve seat stone can grind approximately twelve seats before it must be dressed again. Who is right?
   (A) A only.
   (B) B only.
   (C) Both A and B.
   (D) Neither A nor B.

9. Excessive valve spring installed height can cause:
   (A) heavy spring tension.
   (B) valve float.
   (C) slow valve timing.
   (D) seal damage.
10. Technician A says all valves should always be installed in the spot from which they were removed. Technician B says gasket cement must always be applied to the head gasket. Who is right?
   (A) A only.
   (B) B only.
   (C) Both A and B.
   (D) Neither A nor B.

Suggested Activities

1. Disassemble a cylinder head and determine the repairs needed. Prepare a list of all needed parts and services (seat reconditioning, guide reaming, surface machining).
2. Using a flat rate manual, calculate the parts and labor needed to restore the head to service.
3. Discuss your price calculation with other members of your class. Ask if there is anything that you missed. What would be a fair profit on this job?
4. Determine which parts of the cylinder head overhaul process can be done in your shop and make a list of the jobs to be done by an outside machine shop. Write a short report outlining how you will handle having the outside work performed.

Cutaway of a late-model, 5.7 liter V-8 engine. Note the location of the pistons, crankshaft, and related components. (General Motors)

26

Engine Block, Crankshaft, and Lubrication System Service

After studying this chapter, you will be able to:

- Measure cylinder wear.
- Hone a cylinder wall.
- Describe cylinder reboring.
- Check crankshaft and main bearing bores for problems.
- Install new main bearing inserts and rear seal.
- Measure main bearing clearance and crankshaft endplay.
- Service connecting rods.
- Service automotive pistons.
- Properly install piston rings.
- Correctly install a piston and rod assembly in its cylinder.
- Measure bearing clearance with Plastigage.
- Service engine oil pumps.

Technical Terms

Distortion
Aligning bar
Ring ridge
Cylinder diameter
Taper
Out-of-round
Ring float
Tipping
Scuffing
Piston slap
Deglazing
Honing
Oversize
Reboring
Boring bars
Cylinder sleeve
Camshaft rear bearing oil seal plug
Undersize
Crankshaft regrinding
Wick seal
Split-type (two piece) synthetic rear main bearing seal
Crankshaft endplay
Twist
Bend
Ring grooves
Ring groove cleaner
Ring groove spacer
Piston pin fit
Compression rings
Oil control rings
Ring gap
Groove depths
Ring side clearance
Ring expander
Ring compressor
Bearing clearance
Side clearance
Rotor pump
Gear pump

This chapter covers engine block, crankshaft, and lubrication system service and repair. Block reconditioning, ring ridge removal, boring, honing, cylinder sleeve repair, piston service, and connecting rod service are discussed. The installation and service procedures for various types of engine oil pumps are also covered.

Engine Block Service

The engine block is the foundation on which all of the other engine parts are assembled, **Figure 26-1.** Part alignment and wear demand the block be free of ***distortion*** and cracks. Engine block inspection was covered in Chapter 24, Engine Removal, Disassembly, and Inspection.

Check Main Bearing Bores

Using an inside micrometer, carefully check the bores for distortion (out-of-roundness). Distortion greater than approximately 0.0015″ (0.0381 mm) requires correction in one of two ways. In the first correction method, undersize, semifinished inserts are installed and align bored. In the second method, material is ground from the caps' parting edges. The caps are

Figure 26-1. The engine block is the foundation of the engine.

then replaced, torqued, and align bored to their original specifications. Grounding material from the caps' parting edges is preferred because it allows standard size precision inserts to be used. Any future work on the crankshaft bearings will also be facilitated.

If the individual main bearing bores are acceptable, they must then be checked for alignment. A study of the bearing inserts will indicate bore misalignment and other problems, **Figure 26-2.** Wiping wear of the bearing insert surfaces is usually more evident at the center mains and diminishes toward both ends of the crank.

An ***aligning bar*** that is about 0.001″ (0.025 mm) smaller than bore specifications may be placed in the main bores. The parting surfaces and bores must be clean, and the caps must be torqued. The bar should turn by hand using a bar or wrench with a handle of 12″ (304.8 mm) or less in length. If you cannot turn the bar, the bores are out of alignment. Make sure no caps were reversed.

Alignment can also be checked by placing an accurate straightedge across the bores. Keep the straightedge parallel to the bore centerline. Check in several different positions, **Figure 26-3.** If a 0.0015″ (0.0381 mm) feeler can be inserted between the straightedge and any bore, alignment must be corrected. Removal of material from the cap parting surfaces and align boring is the preferred method of correction. Remember, proper clearance between the inserts and journals requires round bores in proper alignment.

Cylinder Service

High-mileage engine cylinders may be out-of-round and tapered to the extent that machining is required. Before pulling the pistons, the cylinder ***ring ridge*** must be removed. This prevents damage to the rings, piston lands, or both, during piston removal. After the pistons are removed, the cylinder must be inspected for cracks and scoring. This was covered in detail in Chapter 24, Engine Removal, Disassembly, and Inspection.

Before any other measurements and repairs are performed, the ***cylinder diameter*** must be determined. This measurement forms the basis upon which new rings are ordered and rebore sizes figured. Carefully measure the bottom of each cylinder with an inside micrometer. Write down these measurements. They will help in determining the original factory cylinder diameter.

If your measurements are not more than 0.009″ (0.23 mm) larger than factory specifications, the cylinders are standard size. If your measurements are 0.010″–0.019″ (0.25 mm–0.48 mm) larger than specs, the engine has already been rebored to 0.010″ (0.25 mm) oversize. If the measurements are 0.020″–0.029″ (0.51 mm–0.74 mm) larger than specs, the cylinders have been rebored to 0.020″ (0.51 mm) oversize, etc.

For example, assume that your measurements show a cylinder diameter of 3.924″ (99.67 mm) and the specifications list 3.910″ (99.31 mm) as standard. The cylinder measurements are 0.014″ (0.36 mm) larger than standard. This indicates the engine was bored 0.010″ (0.25 mm) oversize. The additional 0.004″ (0.10 mm) would be wear in the lower area. See **Figure 26-4.**

Checking Cylinder for Wear

Cylinder wear is usually heaviest in the upper portion of the cylinder. This is due to the pressure of the rings, intense heat, poor lubrication, combustion pressure, and abrasive material introduced to the combustion area through the air inlet system. The unworn area above the ring travel is referred to as the ring ridge. The bottom of the cylinder generally shows little wear. The cylinder bottom is usually free of ring wear because it is lubricated better and is subjected to less piston thrust pressure than the upper portion. Wear in the upper portion of the cylinder above normal ring travel is also minor.

From **Figure 26-5,** you will note the heaviest cylinder wear is at the top of the ring travel. The heavy wear area extends downward about 0.75″–1.0″ (19.05 mm–25.4 mm). Below this, there is a steady reduction in wear. This condition is known as ***taper.*** Another ridge often exists at the bottom of the ring travel. However, it is much less pronounced.

The cylinder may have an oval shape, considered worn ***out-of-round.*** The greatest wear will be at right angles to the engine or crankshaft centerline. Out-of-roundness is primarily caused by the side thrust forces generated during the compression and firing strokes. Since the crankshaft connecting rod journal is offset to the line of piston travel, the piston attempts to move sideways in addition to up and down. See **Figure 26-6.**

Measuring Cylinder Taper and Out-of-Roundness

It is important that each cylinder be checked for taper and out-of-roundness. Rings must follow the cylinder wall if they are to function properly. As taper increases, ring efficiency decreases. ***Ring float*** and ***tipping*** will destroy the seal and can cause rings to break or scuff the cylinder wall. When taper is excessive, the rings are compressed at the bottom of their travel. As the piston enters the enlarged upper cylinder area, the piston tips back and forth. This causes both the upper and lower ring edges to round off. As the piston travels to the wider cylinder top at high rpm, it can begin to literally float at the top of the cylinder before the rings have a chance to expand outward. See **Figure 26-7.** Sealing efficiency is lost and ***scuffing*** can occur. In addition to tipping, the piston is slammed from one side of the cylinder to the other as the crankshaft rod journal passes over TDC (top dead center). This produces a noise called ***piston slap.*** In addition to ring damage, the piston can fatigue and literally disintegrate.

A quick and accurate method of measuring taper and out-of-roundness is to use a cylinder dial gauge. Slide the gauge near the bottom of the bore and zero the indicator. While keeping the guide feet in firm contact with the wall, slowly pull the indicator up through the cylinder. Slide the gauge up and down in several different sections and note the total indicator reading. This will determine the amount of taper.

Slide the gauge into the bore so the indicator stem is located in the area of ring wear. While holding the guide feet in firm contact, slide the gauge around in the bore until the

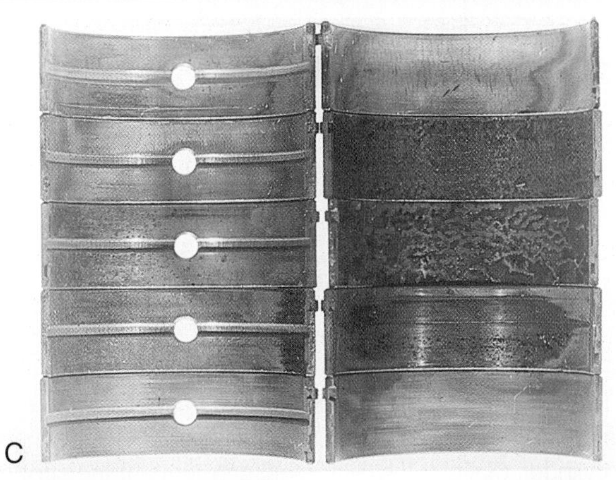

Figure 26-2. A—These aluminum bearings were ruined from lack of lubrication. B—This bearing damage was caused by a tapered housing bore. C— A bent crankshaft ruined this set of main bearings. D— A rough and scored journal caused this bearing to fail. E— Excessive idling produces bearings like this. F— Riding clutch (holding foot on the clutch all the time) places the main bearing thrust flange under prolonged loading. Note the ruined thrust surface. G— Antifreeze leaking into an engine contaminates bearings. As shown here, deposits can build up and eliminate oil clearance, with disastrous results. (Federal-Mogul, AE Clevite)

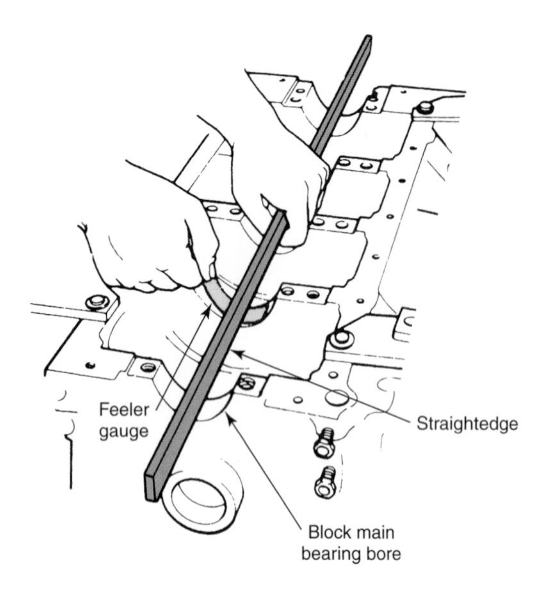

Figure 26-3. A straightedge and a feeler gauge can be used to check bearing bore alignment.

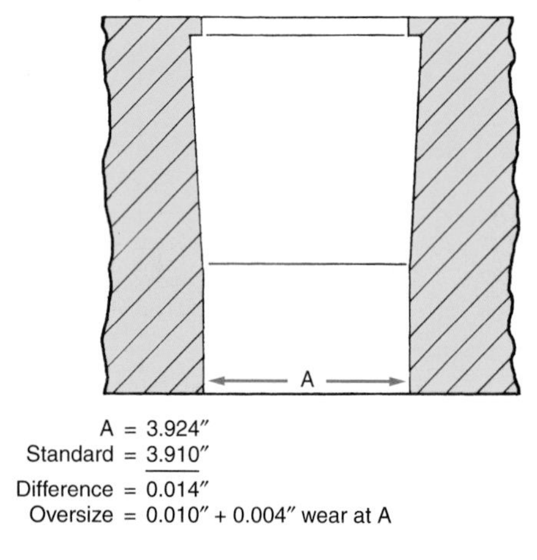

Figure 26-4. Compare measurement A with the manufacturer's specifications to determine cylinder oversize.

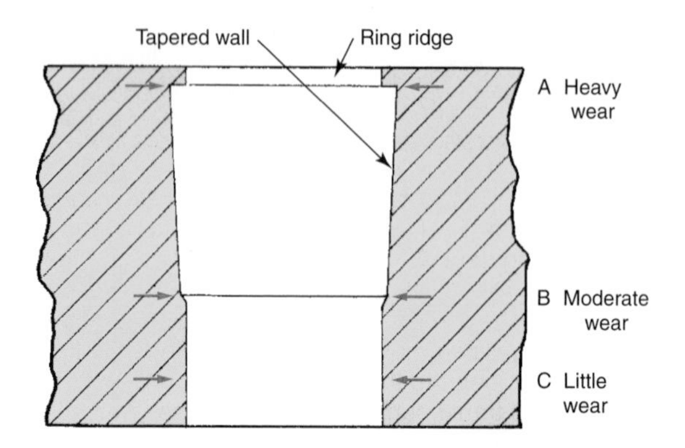

Figure 26-5. Study this cylinder's wear pattern. The diameter through A (top of ring travel), minus the diameter through B (bottom of ring travel) indicates the amount of taper. Note the sharp edge formed by upper ring ridge while the ridge at bottom of ring travel is much less pronounced.

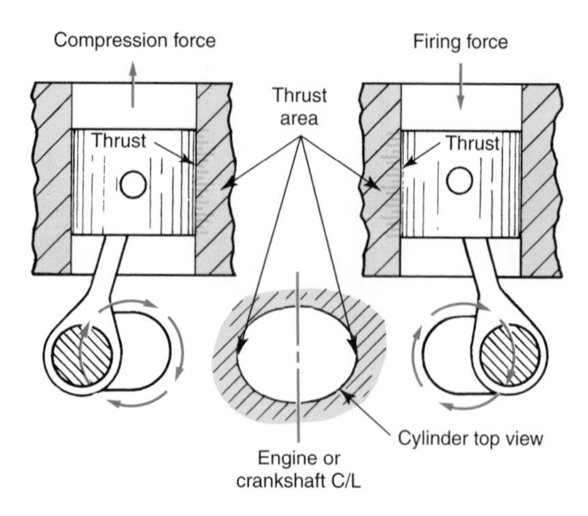

Figure 26-6. Thrust forces can wear a cylinder out-of-round at right angles to the engine centerline.

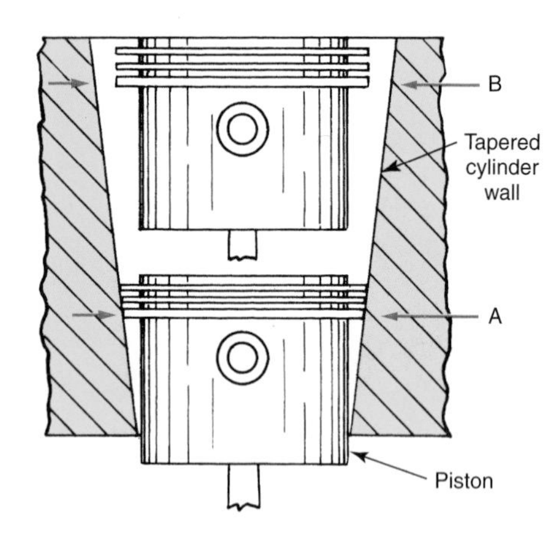

Figure 26-7. Excessive cylinder taper can cause ring float. The rings contact the cylinder wall at A, but before they can expand outward at B, the piston starts down again.

indicator reads maximum out-of-roundness. Write down the readings for each cylinder, **Figure 26-8.** The bore may also be measured with an inside micrometer, **Figure 26-9.** To do this, make two measurements near the bottom of the cylinder, one parallel to the engine centerline and the other at a right angle to the cylinder. Make two similar measurements at the spot of greatest wear at the top of the cylinder. See Figure 26-9B. Write all of the measurements down.

Note: This is a critical step. Be sure to make accurate readings.

The various readings may be listed as shown in Figure 26-9C to facilitate taper and out-of-roundness computations. Be certain to indicate the number of each cylinder concerned. If the cylinder is not scored, cracked, or scuffed, taper up to a maximum of 0.012″ (0.30 mm) is usually permissible. Taper beyond this point will prevent the new rings from working properly. Excessively tapered cylinders must be rebored. Cylinder out-of-roundness should not exceed 0.005″ (0.13 mm). The rings cannot conform to the cylinder wall beyond this point, resulting in heavy oil consumption. Some manufacturers specify a maximum of less than 0.005″ (0.13 mm) due to differences in engine design and application. Always follow the manufacturer's specifications.

Cylinder Reconditioning

A shiny, glazed cylinder wall surface causes the time required for ring break-in to become excessive. In some cases, the rings may never seat properly. Cylinders with minor taper and out-of-roundness can be deglazed using a spring-loaded hone or an abrasive-tipped wire deglazer. Excessively worn cylinders or cylinders that are scored may need to be honed or bored, followed by a final honing to impart the correct microinch finish.

New rings have minute tool thread marks around the ring-to-cylinder edge. The cylinder wall should also have

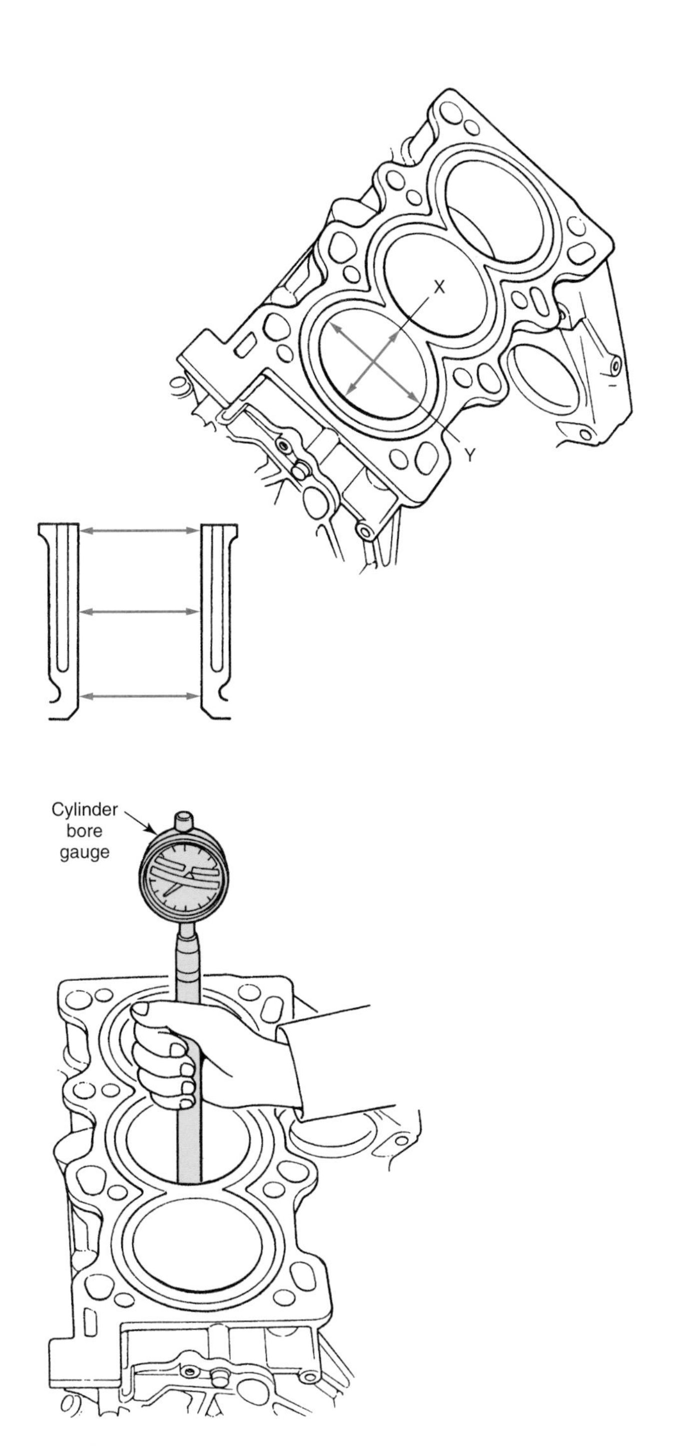

Figure 26-8. Using a dial indicator cylinder bore gauge to check for cylinder taper and out-of-roundness. Note the recommended measuring areas (X and Y) and the three measuring levels. (Acura)

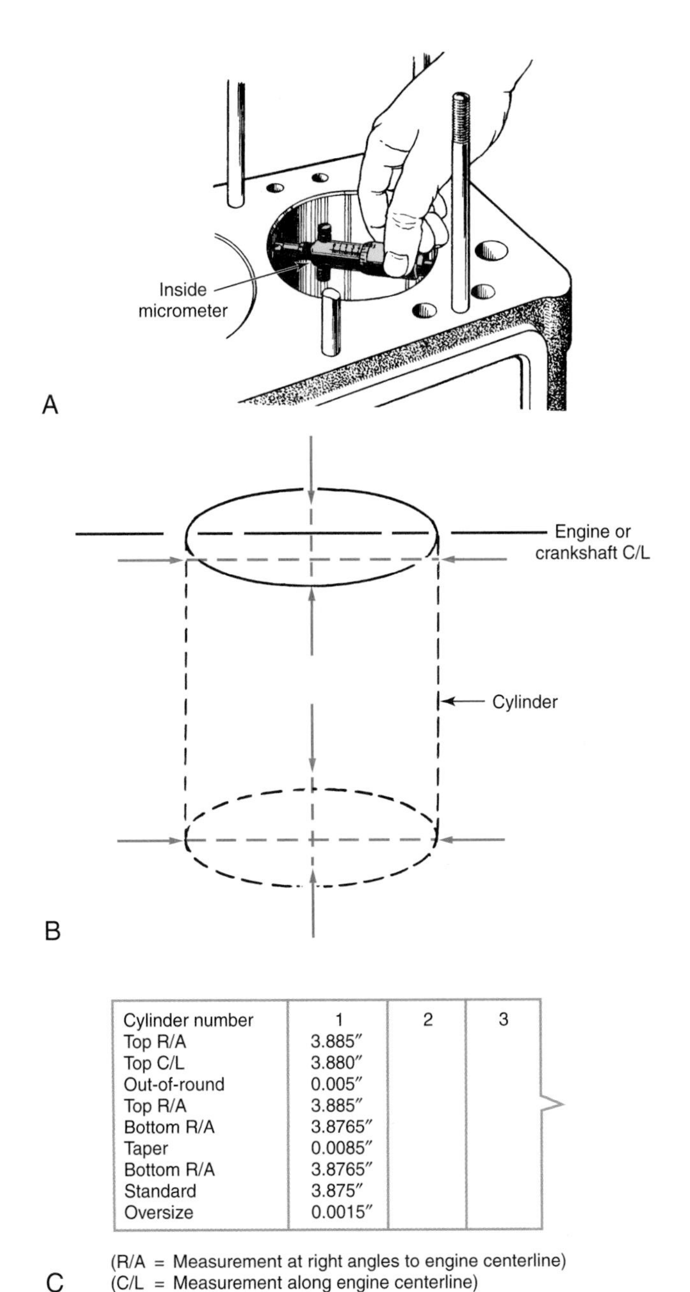

Cylinder number	1	2	3
Top R/A	3.885″		
Top C/L	3.880″		
Out-of-round	0.005″		
Top R/A	3.885″		
Bottom R/A	3.8765″		
Taper	0.0085″		
Bottom R/A	3.8765″		
Standard	3.875″		
Oversize	0.0015″		

(R/A = Measurement at right angles to engine centerline)
(C/L = Measurement along engine centerline)

Figure 26-9. A—Checking cylinder diameter with an inside micrometer. B—Measure each cylinder at these points. C—Cylinder measurements may be listed in this manner to facilitate taper, out-of-roundness, and oversize computations.

minute scratches imparted by the hone. As the new rings travel up and down, both the cylinder wall and the rings wear a tiny amount. High spots on the rings tend to wear off, allowing the low spots to contact the wall. The correct cylinder wall microinch finish allows the rings to seat at about the same time the cylinder wall returns to its glazed condition. Finishing stones with grit sizes of 180 and 220 produce finishes of around 20–30 and 15–25 microinches (millionths of an inch) respectively. Equipment for checking in microinches is not available in most shops, however. A finish that is too rough will wear the rings excessively.

A finish that is too smooth can prolong or even prevent seating. Properly seated rings retain some thread marks for thousands of miles. The microinch finish for any given stone is controlled by varying the pressure of the stone against the wall. A light pressure produces a finer finish. Heavy pressure causes the abrasive particles to cut deeply, producing a rougher surface. Light to medium pressure is recommended.

Deglazing Cylinder Walls

Although discouraged by a few manufacturers, ***deglazing*** is commonly accepted as good practice if:

- Cylinder honing is not excessive.
- The correct microinch surface is imparted.
- The cylinders are thoroughly cleaned after honing.

Deglazing hones are available in a variety of sizes. Cover the crankshaft with rags and swab honing oil on the walls. Using a suitable drill for power, insert the hone and start it spinning, **Figure 26-10.** Move the hone up and down in the cylinder. Do not let stones protrude more than 0.5″ (12.7 mm) on the top or bottom. Move the hone rapidly enough to produce a crosshatch finish similar to that shown in **Figure 26-11.** Note the cross lines form an included angle of about 50°–60°. Do not be concerned about an exact angle. An angle from 20°–60° is usually acceptable. Make about twelve complete strokes with the hone. Wipe the bore and inspect the walls. If a hone pattern is visible over most of the ring travel area, consider the cylinder finished. If the pattern is not visible, repeat the deglazing procedure and check the walls again.

Honing Cylinders

When cylinder wear has almost reached maximum acceptable taper and out-of-roundness, an adjustable rigid hone should be used. This type of hone does not flex to fit the wall taper and helps to remove both taper and out-of-roundness. Ideally, all ***honing*** operations should take place with the crankshaft removed from the block. If this is not possible, cover the crankshaft.

Apply honing oil and insert the hone in the bottom of the bore. Using 180 or 220 grit stones, adjust the stones outward until firm contact with the cylinder wall is obtained. Ensure that the hone is at the bottom of the cylinder, and note the exact position on the hone assembly in the cylinder. This will assist in determining when the proper depth has been reached. **Figure 26-12** illustrates the honing sequence. Drive the hone with a 1/2″ or 3/4″ drill.

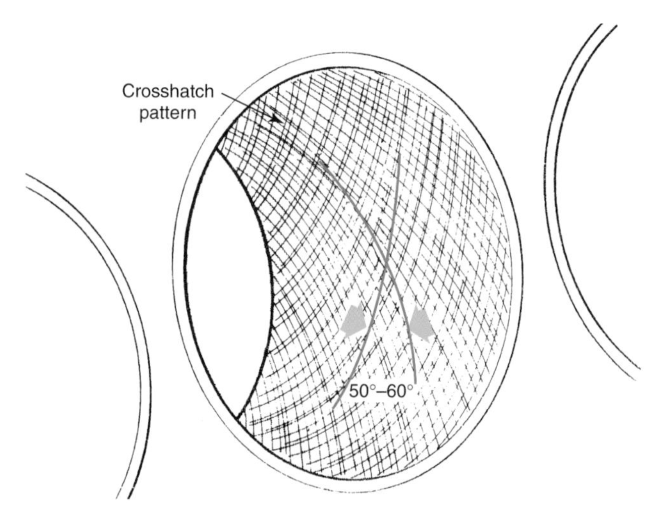

Figure 26-11. A deglazing hone was used to create this desirable crosshatch pattern. (DaimlerChrysler)

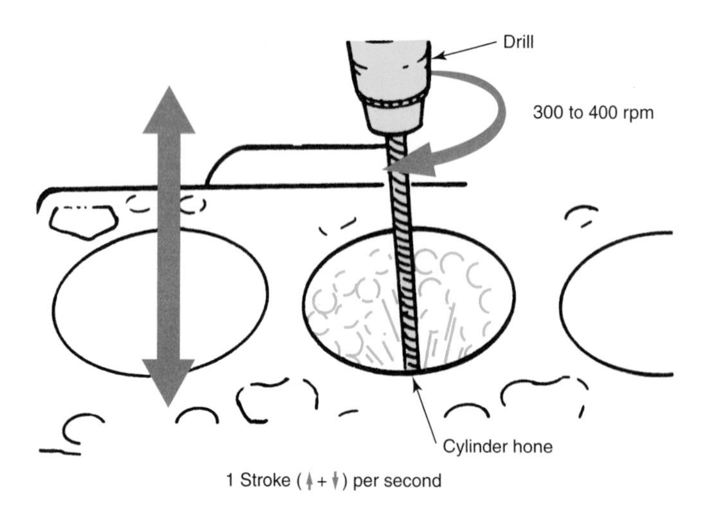

Figure 26-10. This illustration shows a cylinder bore being honed with an abrasive-tipped brush. (DaimlerChrysler)

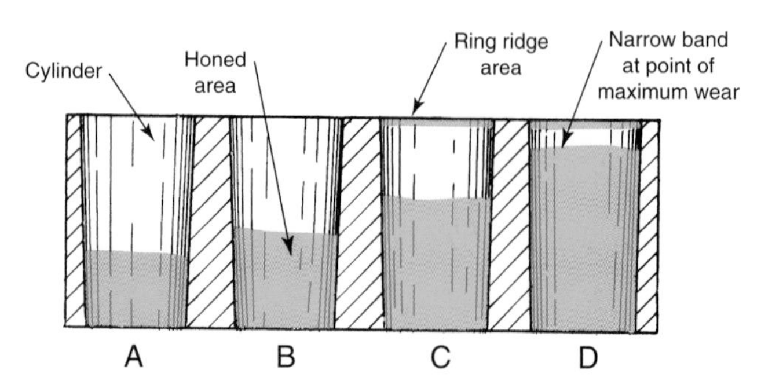

Figure 26-12. Note this honing sequence. The hone was started in the bottom of the cylinder (A). As honing progresses, the crosshatch pattern covers more and more of the bore (B, C, D).

Warning: This type of hone requires considerable torque to start. Grasp the drill handles tightly. Keep your clothing away from the spinning hone. Never pull the hone out of the cylinder while it is spinning. The hone parts are rotating at high speeds and can fly apart with great force.

Start honing at the bottom of the cylinder, using short up and down strokes. Since the cylinder walls are the least worn at the bottom, they will keep the hone properly aligned. Keep adjusting the hone to ensure firm stone-to-wall contact. After approximately twenty short, fast strokes, loosen the stones, withdraw the hone, and inspect the cylinder. The crosshatch pattern shown in Figure 26-11 should be visible over the bottom part of the cylinder. If no crosshatch pattern can be seen, continue honing until it is visible. If the crosshatch pattern is visible, repeat the honing process, moving slightly higher in the cylinder. When the hone marks cover about 70% of the cylinder's length, begin moving the hone over the full length of the cylinder.

Allow the stones to protrude about 0.5″ (12.7 mm) above and below the cylinder. If the crankshaft was left in the block, be careful not to hit it. Try for a crosshatch pattern as mentioned. The use of the rigid hone as described necessitates the use of oversize pistons and rings in order to maintain a suitable working clearance between the piston and cylinder. This is discussed later in this chapter.

Reboring Cylinders

When cylinders have exceeded wear limits or when heavy scoring is present, they should be rebored to a suitable ***oversize.*** Ideally, all cylinders should be bored to the same size. After careful measurement of all cylinders, compute the amount of oversize necessary to recondition the worst one. ***Reboring*** is done in multiples of 0.010″ (0.25 mm). Oversizes of 0.010″, 0.020″, 0.030″, and 0.040″ (0.25 mm, 0.51 mm, 0.76 mm, and 1.01 mm) are most commonly used.

Do not try to rebore the cylinders with an oversize that is just barely larger than the poorest cylinder. Small variations in boring bar centering may leave areas untouched. However, reboring at the smallest practical oversize permits future correction for wear. Do not use excessive oversizes unless raising engine displacement for performance purposes.

Note: Some aluminum blocks use cast iron sleeves that are relatively thin. On these engines, it is especially important to remove a minimum amount of metal.

There are numerous types of ***boring bars,*** ranging from fairly portable to massive production-type units. Always follow the manufacturer's directions when using this equipment. The block face should be draw filed by laying a large, flat file across the top of the block (cylinder deck) and pulling it toward you. This removes any burrs and high spots. Center the bar carefully in the cylinder. Because the bottom is the least worn, it is advisable to center in this area. Following centering, the bar should be firmly clamped to the block. The cutter must be kept sharp. It should be carefully set to cut a hole that will be 0.0025″ (0.064 mm) smaller than the finished size. This allows the final cylinder honing to remove boring tool marks and fractured metal, and to impart the correct microinch finish.

Note: While some boring bar manufacturers insist no further finishing is needed after boring, it is a good precaution to hone the cylinder after boring. Boring the cylinder sometimes leaves the cylinder finish too rough for proper ring and cylinder operation and durability.

The boring cut will heat up the cylinder. There is a chance the cylinder will be distorted upon cooling if the adjacent cylinder is bored next. Always bore alternate cylinders. This allows each cylinder to cool before boring the one next to it. Back off the cutter before withdrawing it from the finished cylinder. The sharp edge at the top of the cylinder must be chamfered by hand-feeding the bar. Remember that accurate centering, firm clamping, sharp tools, and exact cutter settings are essential. Mike each bore for size and use a cylinder dial gauge to check for taper or out-of-roundness. A typical boring bar is shown in **Figure 26-13.**

Figure 26-13. A machine used to bore cylinders.

Honing after Boring

Use a hone with 180 or 220 grit stones. Hone the full length of the bore to produce a correct crosshatch finish. Hone the cylinder until the new piston is correctly fitted. If the desired 0.0025″ (0.064 mm) of material was removed following boring, a satisfactory base metal (final finish that is produced in solid, unfractured block metal) finish will result.

Installing a Cylinder Sleeve

A badly scored or cracked cylinder can often be salvaged by cutting the block to accept a ***cylinder sleeve.*** **Figure 26-14** shows the steps for installing a cylinder sleeve. After the sleeve is installed, it is bored and honed to size. Maximum bar cuts should be around 0.050″ (1.27 mm). Take several cuts. Use a light finishing cut when nearing the proper size. Follow specifications for sleeve-to-block fit.

Cleaning Cylinders Following Honing

A thorough cleaning following honing is of vital importance. Do not use gasoline, kerosene, cleaning solvent, or oil. Use hot, soapy water and a stiff bristle brush. Scrub the cylinders and rinse twice with a hot, clear water rinse. This procedure removes excess honing oil and minute, abrasive metal particles. Immediately wipe the cylinders dry using a clean shop rag. Check the rag after drying to make sure the cylinders are completely clean. See **Figure 26-15.** When the cylinders are clean and dry, swab them with clean engine oil. If the crankshaft was left in place, it must be cleaned, dried, and the journals must be oiled. Check the crankcase area under the cylinders carefully. Despite a fine job of cylinder reconditioning, the whole job may be destroyed if cleaning is not thorough.

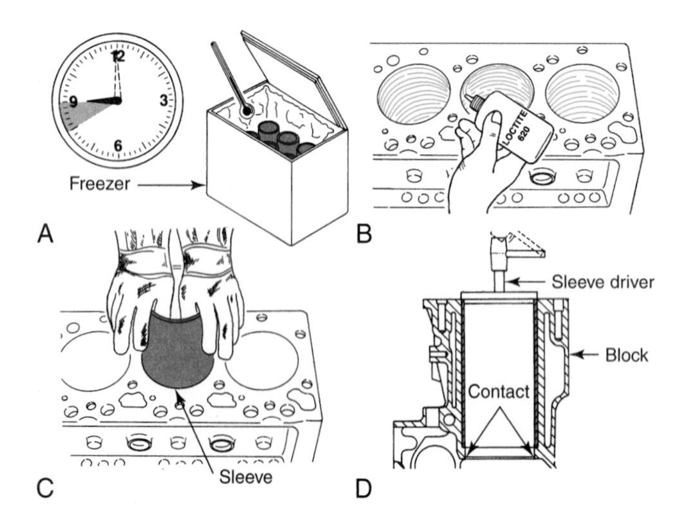

Figure 26-14. One manufacturer's recommended procedure for installing a cylinder sleeve. A—Cool sleeves to 10°F (–12°C) or below for one hour in a freezer. B—Apply Loctite 620 to the bore that is to be sleeved. C—Push the sleeve into the bore as far as possible by hand. Wear gloves. D—Drive the sleeve into its bore until it contacts the step at the bottom of the bore. (DaimlerChrysler)

Installing In-Block Camshaft Bearings

With one-piece bearing inserts, the old inserts are either driven or pulled out. The bore may be slightly chamfered on the front side so that material is not shaved from the new bearing as it is forced into place. Look at **Figure 26-16.** Clean the bores and oil the delivery holes. Then check the bores for size and alignment.

There are a number of camshaft bearing removal and installation tools. In one type, the proper size mandrel is selected and fitted to a drive bar. The drive bar is passed through the bores until the mandrel is positioned properly. The insert is oiled and slipped on the mandrel. Rotate the bar until the mandrel expands snugly inside the insert. Align the oil hole in the bearing with the oil hole in the bore and start the bearing by hand.

At this point, rotate the bar one-eighth turn counterclockwise. This reduces the mandrel diameter about 0.004″ (0.10 mm) to allow the bearing ID to reduce as it is driven into

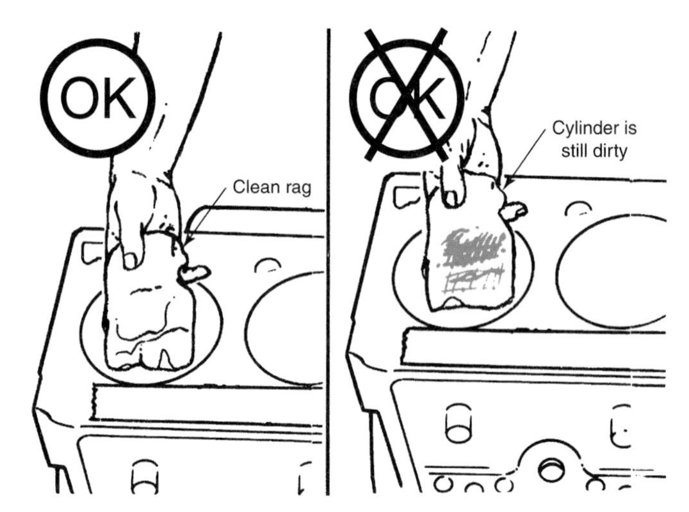

Figure 26-15. Cylinders must be thoroughly cleaned after honing. After washing, dry the cylinders with a clean, white shop rag. Continue wiping until the rag comes out clean. (DaimlerChrysler)

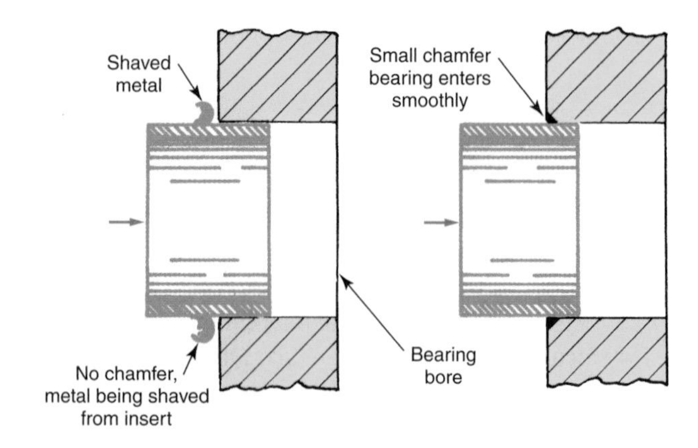

Figure 26-16. Chamfering a bearing bore on the front side prevents the insert from being shaved during installation. A small chamfer is ample.

place. Using a hammer, drive the bearing into the bore. Stop driving when the drive face of the mandrel is flush with the bore, **Figure 26-17.** Remove the bar. Check to make certain the oil hole in the bearing is correctly aligned with the oil hole in the block, **Figure 26-18.** If the oil hole is not aligned, remove the bearing and reinsert it.

When installing cam bearings, be sure they are started properly. The oil holes must be aligned, and material must not shave as the bearing enters. If the bearing is cocked, it will be distorted. It can even become loose enough to rotate in the bore, cutting off the oil supply.

If the front bearing is specially designed to provide lubrication for the timing chain or gears, make sure the correct end of the bearing faces forward and that it is installed to the proper depth. If the bearing shells are chamfered on one end, start them so the chamfer enters the bore first. When possible, install the split seam toward the top of the engine (away from the high load area). If the original inserts were staked in place (a portion of shell dented into a recess with a punch), the new inserts should also be staked to prevent the insert from turning.

Figure 26-17. This illustration shows a camshaft bearing that has been driven into position with a drive bar and a proper mandrel. (Federal-Mogul)

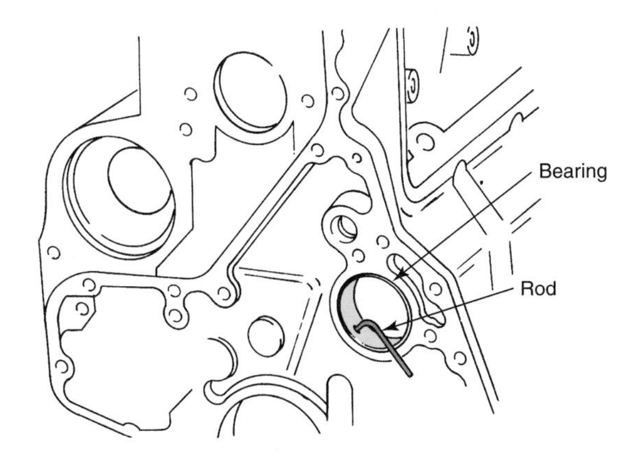

Figure 26-18. Checking the camshaft bearing oil hole alignment after installing the bearing. This engine calls for a 0.128″ (3.25 mm) diameter rod to check for proper clearance. The rod must pass through the bearing hole and into the oil hole. (DaimlerChrysler)

Camshaft Rear Bearing Core Plug or Oil Seal Plug

When replacing the ***camshaft rear bearing oil seal plug*** (often called *Welch plug* or *core hole plug*), clean out the plug counterbore. Coat both the OD of the new plug and the counterbore sides with sealer. Drive the plug in squarely. Do not drive the plug below recommended depth. In some cases, a stop ledge is provided. Use a suitable driver that contacts the plug's outer edge.

In some designs, striking the plug in the center (after fully seating it with a driver) causes it to expand tightly against the walls of the counterbore. This type of plug is crowned. Be sure to install it with the crown facing outward, **Figure 26-19.** As with all plug installations, make sure you have a strong, permanent seal. If a leak develops after engine installation, the repair can be very expensive.

Crankshaft Service

This section discusses inspecting, measuring, and servicing the crankshaft. To begin, place the crankshaft on a pair of smooth, clean, oiled V-blocks. If the blocks are not absolutely smooth, cover the V's with thin, hard paper.

Check Main and Rod Journals for Finish and Accuracy

Turn the shaft slowly and visually check each journal for signs of scratching, ridging, scoring, or nicks. You will often note a dark line around the journal. This is caused by the oil groove in the insert and is not harmful unless it protrudes more than 0.0003″–0.0004″ (0.0076 mm–0.0102 mm) from the surface of the journal. This dark line should be polished

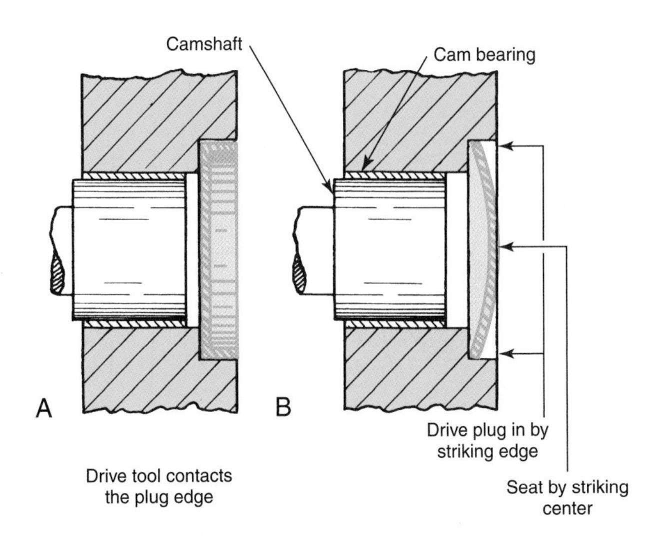

Figure 26-19. This illustration depicts typical camshaft rear journal oil seal plugs. A—This plug is driven in on edge. B—This plug is driven in on edge and seated with a blow in its center.

with crocus cloth (extremely fine abrasive) to remove any accumulated carbon or gum.

Crocus cloth can also be used to remove any small burrs caused by tiny nicks or scratches. Pull the cloth around the journal in a “shoe shine” motion. By keeping the cloth pulled around half of the journal, the polishing will not produce flat spots. All journals must be absolutely smooth. Any roughness, ridging, or scoring must be ground smooth.

If the journal surfaces are in good shape, they must then be checked for out-of-roundness, taper, and wear. Measurements should be taken near one end of the journal in several spots around the diameter. Be careful to keep off the corner fillet radius. Write down each measurement. Repeat this process near the other end of the journal, **Figure 26-20.**

Out-of-Roundness, Taper, and Undersize

The measurements performed in Figure 26-20 determine three important points: out-of-roundness, taper, and undersize.

Out-of-roundness is computed by figuring the difference in diameter measurements at various points. Out-of-roundness must not exceed 0.001″ (0.025 mm). **Figure 26-21** illustrates three journals—one within limits (would accept a standard size insert), one requiring grinding, and one within limits but worn 0.001″ (0.025 mm) undersize.

You will note the journal in Figure 26-21A is only 0.0003″ (0.0076 mm) out-of-round and has worn a mere 0.0001″ (0.0025 mm). Providing the taper is within limits, this journal will be satisfactory. If a new insert is required (it pays to use new inserts), a standard size is required.

In Figure 26-21B, the journal is 0.003″ (0.08 mm) out-of-round and has worn a maximum of 0.001″ (0.025 mm). This journal is unfit for service and should be reground.

In Figure 26-21C, the journal is only 0.0002″ (0.0058 mm) out-of-round, but has worn an even 0.001″ (0.025 mm). If the taper is satisfactory, this journal can still be used but will require a 0.001″ (0.025 mm) ***undersize*** insert.

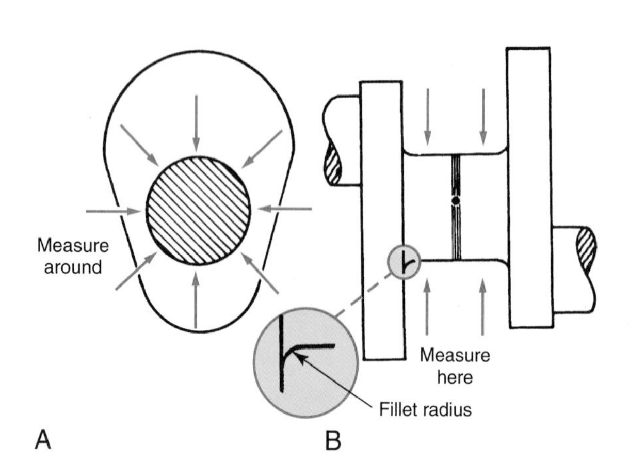

Figure 26-20. A—This is an end cross-sectional view of a journal. B—This is a side view of a journal. Make measurements as indicated by the red arrows.

Taper

The amount of taper can be determined by computing the difference in diameter readings between both ends of the journal. Taper should not exceed 0.001″ (0.025 mm). In **Figure 26-22,** journal #1 taper is 0.0002″ (0.0051 mm) and thus is acceptable. Journal #2 has a taper of 0.003″ (0.08 mm) and therefore must be reground.

Connecting rod journals tend to wear more out-of-round and tapered than the main bearing journals. This is basically due to the fluctuating load, which places certain areas of the journal under heavy, sudden stresses. Rod twist and bend exert uneven edge loading, which tends to taper the journal.

Selecting Correct Undersize Insert

When journal wear is minor and the journal out-of-roundness and taper are within limits, the proper oil clearance can often be maintained by installing 0.001″ or 0.002″ (0.025 mm or 0.050 mm) undersize inserts. When determining the correct undersize, always use the largest journal measurement. If the smallest measurement or an average of all measurements is used, there could be insufficient oil clearance and the bearings will quickly fail. If the largest measurement indicates the journal has worn 0.0005″ (0.0127 mm) below standard, no undersize is required. For wear from

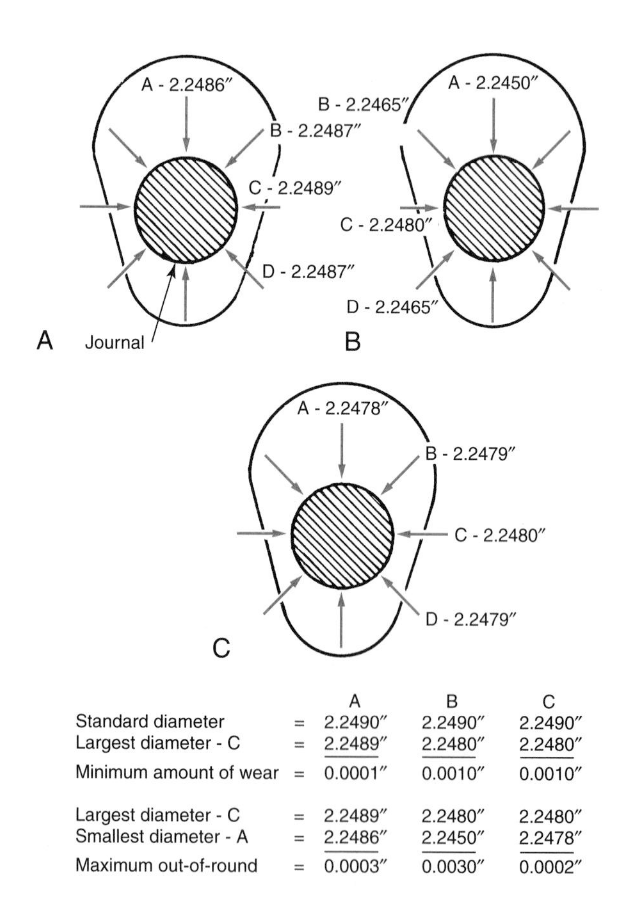

		A	B	C
Standard diameter	=	2.2490″	2.2490″	2.2490″
Largest diameter - C	=	2.2489″	2.2480″	2.2480″
Minimum amount of wear	=	0.0001″	0.0010″	0.0010″
Largest diameter - C	=	2.2489″	2.2480″	2.2480″
Smallest diameter - A	=	2.2486″	2.2450″	2.2478″
Maximum out-of-round	=	0.0003″	0.0030″	0.0002″

Figure 26-21. This illustration shows end cross-sectional views of three journals. A—This journal is acceptable for a standard insert. B—This journal must be reground. C—This journal is acceptable, but requires a 0.001″ (0.025 mm) undersize insert.

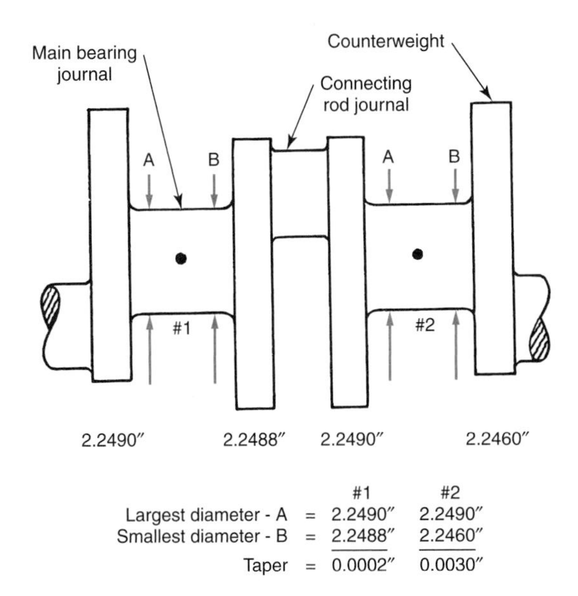

Figure 26-22. Journal #1 has only 0.0002″ (0.0051 mm) taper and is acceptable. Journal #2 shows a taper of 0.003″ (0.08 mm) and must be reground.

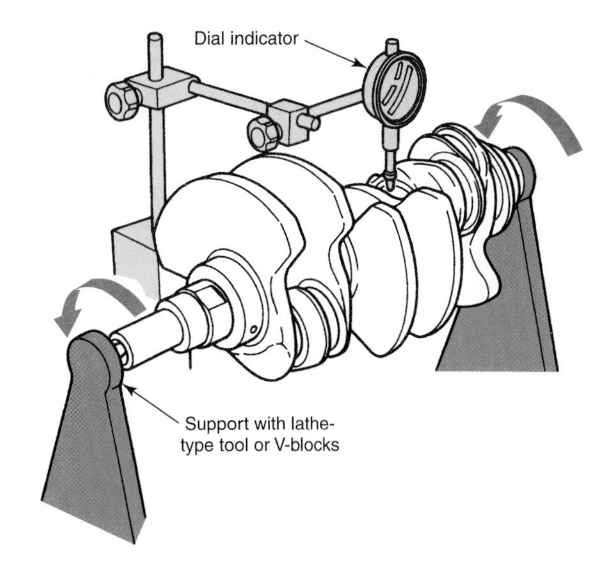

Figure 26-23. This illustration shows a dial indicator being used to check crankshaft alignment.

0.001″–0.0019″ (0.025 mm–0.048 mm), a 0.001″ (0.025 mm) undersize is needed. For wear from 0.002″–0.003″ (0.050 mm–0.080 mm), a 0.002″ (0.050 mm) undersize will suffice. For wear much above 0.003″ (0.076 mm), a 0.002″ (0.050 mm) undersize will not bring the oil clearance within limits. An average oil clearance would be about 0.001″ (0.025 mm) for each 1″ (25.4 mm) of journal diameter in pressure-lubricated systems. Bearing material, engine design, and rpm all affect the amount of clearance required, so *always* check the manufacturer's specs.

Crankshaft Alignment

Even though the journals are smooth and within limits, bearing life will be greatly shortened if the crankshaft is out of alignment.

Check the shaft by placing both end journals on a lathe-type tool or a set of V-blocks, **Figure 26-23.** Adjust a dial indicator to the center or intermediate mains. Turn the shaft and record the amount of runout. Check each main journal. To record the end journals, move a V-block in to the intermediate main.

On long shafts, it may be necessary to support the shaft at the intermediate journals to prevent sag. *Average* maximum misalignment (consult manufacturer's specs) should be held to 0.001″ (0.025 mm) between journals and 0.002″ (0.050 mm) overall. Remember that any journal out-of-roundness must be taken into consideration, as it will affect the indicator reading.

If the crankshaft is only slightly out of alignment, it may be possible to straighten it. This can be done by either cold bending or bending with pressure and heat. In some cases, the application of heat alone causes the crankshaft to straighten. In general, forged steel crankshafts allow a greater degree of straightening than the cast iron or nodular iron types.

Regrinding a Crankshaft

Crankshaft regrinding is a specialty operation. Most auto shops do not have the necessary heavy equipment. Crankshaft grinding requires accurate machinery and a skilled operator. They send, or "farm out," such jobs as crankshaft and camshaft grinding, reboring, align boring, and shaft straightening to machine shops. **Figure 26-24** shows a crankshaft set up for grinding.

Polishing the Journals

A properly ground shaft may look smooth, but in reality, the surface has thousands of tiny, sharp edges. Following grinding, the journals should be polished to remove these abrasive edges. Each fillet radius and shaft thrust surface should also be polished. A good polishing operation produces an extremely smooth surface—7 microinches or smoother (a 16 microinch finish is satisfactory).

Main Bearing Insert Installation

Bearing inserts must be of the correct size, design, and material. The block, oil galleries, and bearing bores must be spotlessly clean. On some crankshafts, all mains are the

Figure 26-24. A crankshaft being reground. (Federal-Mogul)

same size. Other shafts are designed so the front main is the smallest, with a gradual increase in size from front to back. If an engine uses the graduated type, the inserts are not interchangeable. See **Figure 26-25.**

Working on one bearing at a time, install the proper inserts. Bearing bores and insert backs must be clean and dry. The inserts should snap into place. The *locating lugs* should fit the recesses properly, and the correct amount of crush should be evident. The bearing half that is drilled for oil entry must be placed in the upper, or crankcase, bore. Check each insert to be sure the oil hole aligns with the oil passageway. See **Figure 26-26.** Some front main inserts are designed to facilitate timing chain lubrication. In such cases, make certain they are properly located.

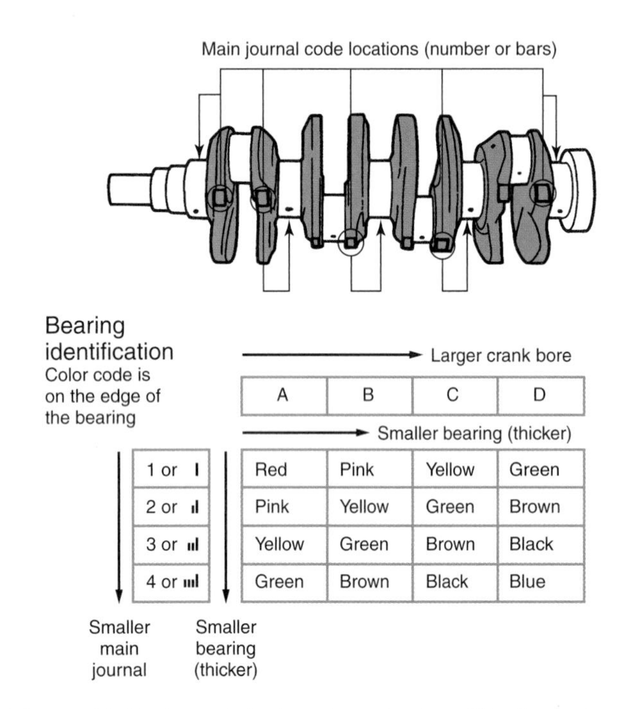

	A	B	C	D
1 or I	Red	Pink	Yellow	Green
2 or II	Pink	Yellow	Green	Brown
3 or III	Yellow	Green	Brown	Black
4 or IIII	Green	Brown	Black	Blue

Figure 26-25. A crankshaft and bearing identification color code and size chart is shown here. (Honda)

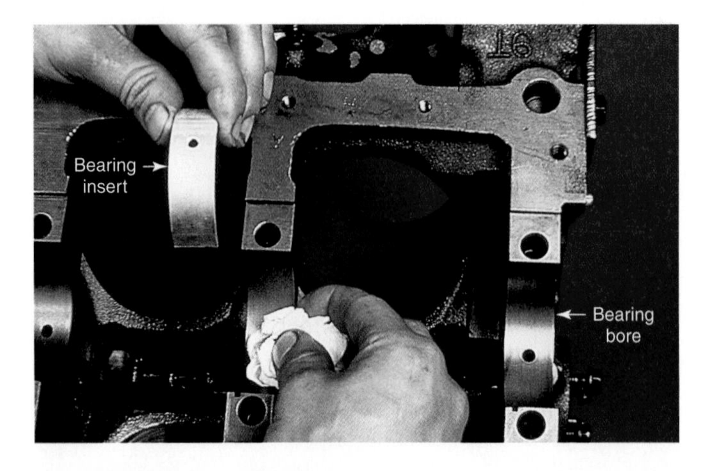

Figure 26-26. In this illustration, an upper main bearing insert is being installed. (Federal-Mogul)

Installing the Crankshaft

Before installing the crankshaft, wipe a heavy film of oil on all bearing surfaces. Lower the crankshaft into place. Place caps in position. Do not reverse them. Insert the cap bolts and tighten them snugly. However, leave the thrust bearing cap loose. Tap the caps lightly with a plastic mallet to assist in alignment. Torque all caps except the one used for the thrust bearing.

Pry the shaft forward against the lower thrust flange. While holding the shaft forward, pry the cap back to force the cap thrust flange against the shaft thrust surface. Maintain forward pressure on shaft and torque the cap. This ensures an even contact between crank and thrust bearing flanges.

Checking Bearing Clearance

Wipe a thin film of clean engine oil over the surface of all upper inserts. The rear main oil seal is left out at this time. It tends to hold the shaft away from the bearing surface when checking clearance. Carefully lower the crankshaft into position. Be careful to avoid damage to the thrust bearing flange surfaces. Rotate the shaft several times to seat the journals.

Using a clean cloth, wipe off the exposed surface of each journal. Place a length of Plastigage across each journal, about 0.25″ (6.35 mm) from top center. The Plastigage should span the width of the insert. Do not place it across an oil hole. After the Plastigage is in place, install the caps and torque them to specifications.

Caution: Do not rotate the shaft until you are finished with the clearance check or the Plastigage will be smeared.

Remove each cap. Check the width of the Plastigage with the paper scale. Slide the scale over the flattened Plastigage until you find the marked band that is closest to the width of the Plastigage, **Figure 26-27.** The number on the band indicates the bearing clearance in thousandths of an inch. All bearings should show specified clearance.

If there is a variation in the width of the flattened Plastigage (from one side of the journal to the other), taper is present. The difference between the widest and narrowest section would determine the amount of taper. After determining clearance and taper, wipe off the Plastigage and remove the shaft.

Rear Main Bearing Oil Seals

There are two common materials used for main bearing oil seals: *graphite impregnated wick* and *synthetic rubber.* A number of different designs are used to seal the rear main bearing against the passage of oil. Regardless of design, this is one operation that must be done with care. Far too often, this job is done in a haphazard manner. Basically, oil can find its way into the flywheel housing by passing along the shaft, through the rear main parting surfaces, along the rear main

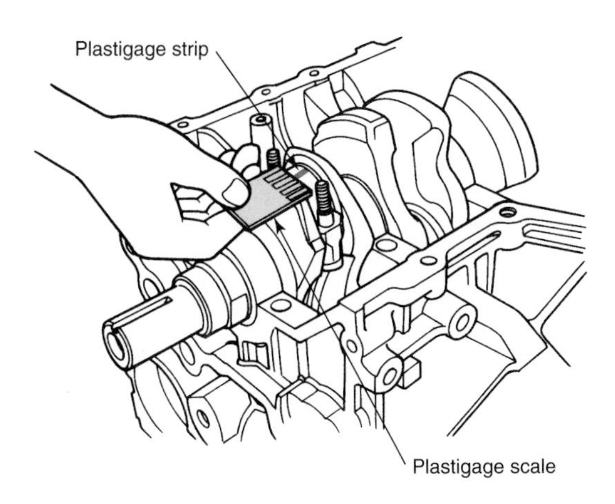

Figure 26-27. Check flattened Plastigage to determine bearing clearance. (Honda)

cap edges (in some applications), and in one design, through the crankshaft flywheel flange bolt holes. When installing the seal, give some thought to the basic setup and how it prevents oil leakage.

Wick Seals

If properly installed, the ***wick seal*** does a good job of preventing oil leakage. It is good practice to soak the seal in engine oil for 30 minutes before installation. Clean out the seal groove in the cap and crankcase bore. Lay one section of wicking over the groove in the upper bore. Starting in the center, press the seal fully into the groove. Work up each side by pressing *in* and *down* to bottom the seal. When almost seated (an equal amount should protrude above each parting surface), place an installation tool against the seal and tap the seal into place. Refer to **Figure 26-28.** If an installation tool is not available, a smooth, round bar or a large socket may be used.

When the seal is fully seated, keep the installation tool in place. Using a sharp knife or razor blade, cut off the protruding seal ends flush with the parting surfaces. Make certain no loose ends of seal material are left that could jam between the parting edges and prevent the cap from seating. See **Figure 26-29.**

Install the seal wicking in the cap. Seat the seal in the center of the cap and work the seal down and in toward the bottom. When the seal is fully seated, hold the installation tool against the seal opposite the parting edge and trim the seal flush.

Synthetic Seals

Instead of the wick-type seal just discussed, a number of engines employ synthetic (neoprene) main bearing seals. **Figure 26-30** illustrates the procedure for removing a ***split-type (two piece) synthetic rear main bearing seal.***

Carefully install the split seal according to the manufacturer's instructions. Use engine oil on the seal lips and apply sealer where needed. The rear main cap used in the setup shown in **Figure 26-31** uses side seals and cap screw plugs to prevent leakage. These must be carefully installed also.

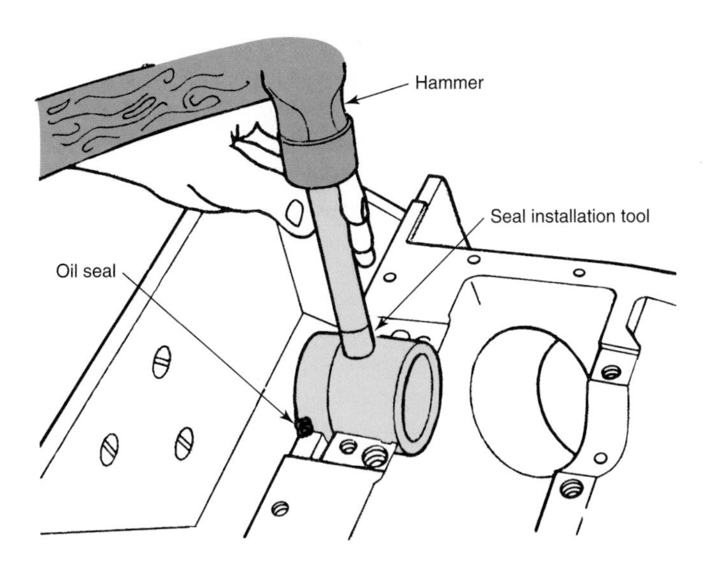

Figure 26-28. In this illustration, a special installation tool is being used to seat a rear main upper oil seal. (DaimlerChrysler)

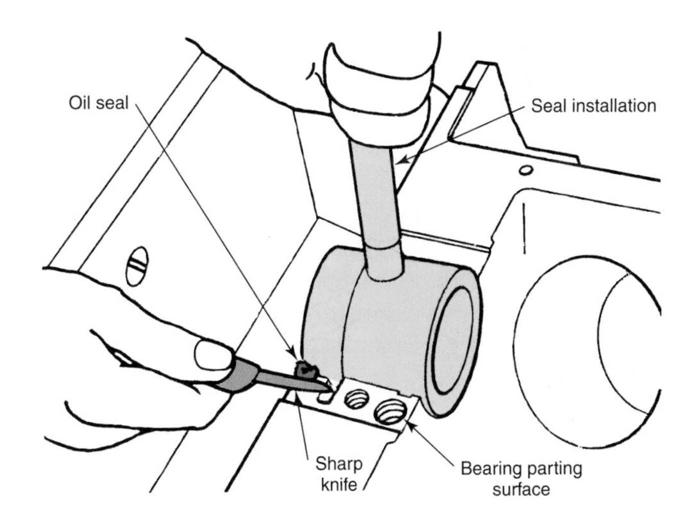

Figure 26-29. Trim seal ends flush with the parting surface. The soft bumper protects the knife edge as it passes through the seal. (DaimlerChrysler)

With parting edges clean, apply a thin layer of sealer on the cap. Some technicians coat only the last portion past the oil slinger. Some synthetic seals have a special glue on their ends. In these cases, be careful to avoid getting oil or sealer on the ends. Do not get sealer on the journal. Be sure the coating on the parting surface is not so heavy that it will squeeze out onto the journal.

Other engines utilize a one-piece (continuous) rear main bearing seal, **Figure 26-32.** Remove the seal carefully, Figure 26-32A. Clean the recess. Coat the seal as needed. Using a seal guide to prevent seal damage, Figure 26-32B, drive the seal into place. Make certain the seal lip faces inward and the shaft seal contact area is smooth. The use of a special seal driver is recommended. See Figure 26-32C.

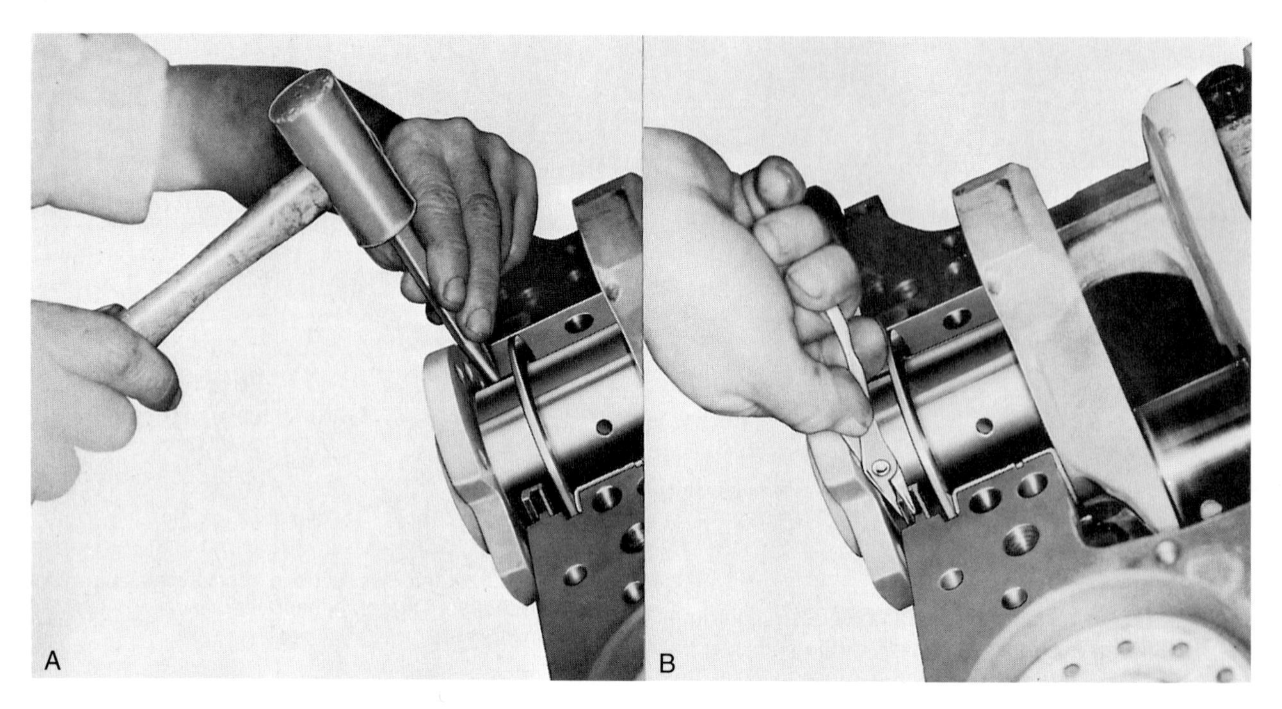

Figure 26-30. Removing a synthetic upper rear main bearing seal. A—One end of the seal is tapped down with a punch until the opposite end protrudes far enough to grasp it with pliers. B—The seal is then pulled from the groove. (General Motors)

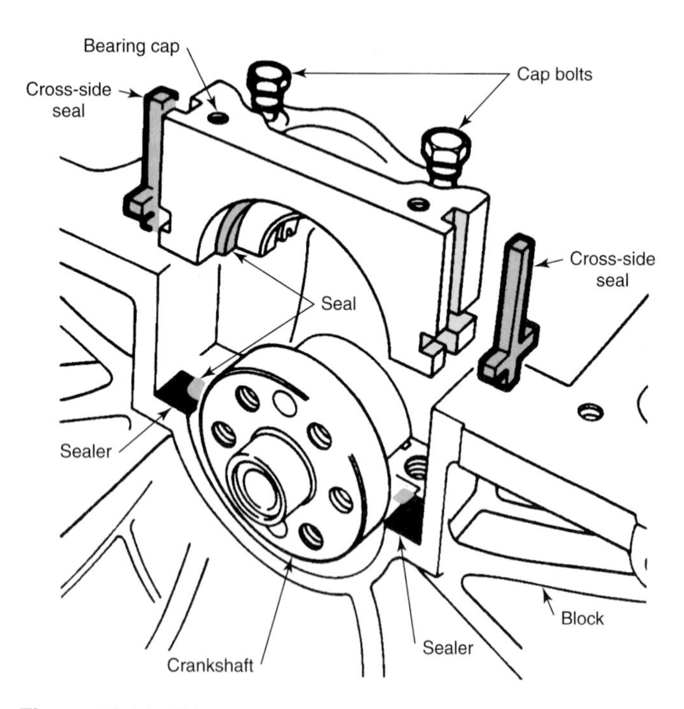

Figure 26-31. This setup uses a synthetic, two-piece rear main bearing seal. (Land Rover)

Checking Crankshaft Endplay

A dial indicator or a feeler gauge can be used to check ***crankshaft endplay,*** **Figure 26-33.** If a feeler gauge is used, force the shaft to its limit of travel in one direction. Slip the feeler gauge between the insert thrust flange and the crank thrust face on the free side. This lets you determine clearance.

Crankshaft endplay should be checked against the manufacturer's specifications, but generally ranges from 0.004″–0.008″ (0.10 mm–0.20 mm). Do not jam the feeler blade into the clearance area or the thrust bearing flange could be marred.

Most engines have thrust flanges built into one of the main bearings, and the entire bearing must be changed to adjust endplay. Some engines have separate thrust shims (washers). Other engines use a thrust plate that allows endplay to be adjusted by adding or removing shims. This setup is used on the front of the shaft, **Figure 26-34.**

Measuring Crankshaft Main Journals—Engine in Car

At times, it may be desirable to install new main bearings without removing the engine from the car. To select the proper replacements, it is necessary to measure each journal.

There are several different types and styles of measuring devices available. One is a gauge that allows the shaft to be measured without rolling out the upper insert. To use this gauge, the journal and gauge contact pads are wiped clean. The central plunger is locked down, and the gauge is placed against the shaft. The plunger is then released. When the plunger is in full contact with the shaft, the plunger is locked with the thumbscrew. A regular outside micrometer is used to measure across the length of the plunger. This distance is the radius (one-half of the diameter) of the shaft and must be doubled to determine shaft diameter. As with all precision measuring tools, this gauge must be used with extreme care, **Figure 26-35.**

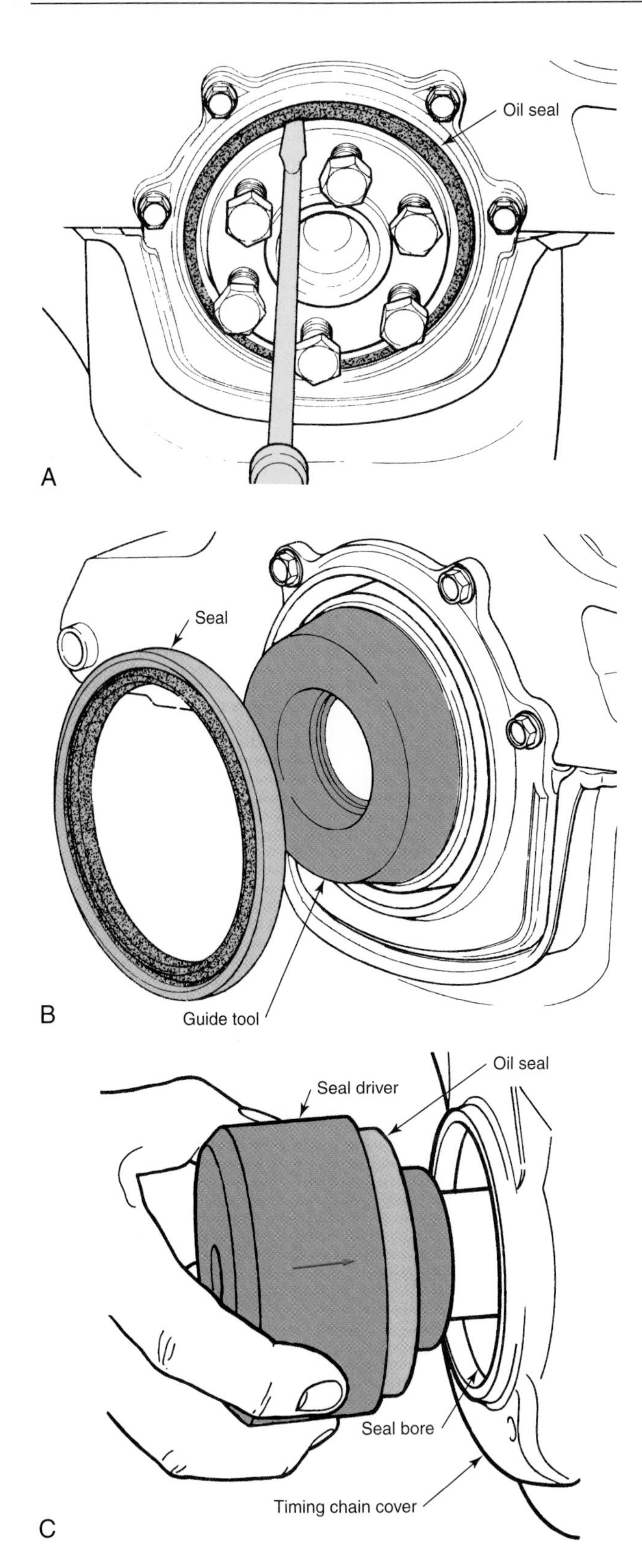

Figure 26-32. A—Do not scratch or nick the seal lip contact surface while removing a continuous (one piece) synthetic rear main bearing oil seal. B— When installing a one-piece rear main bearing seal, use a properly designed seal installation guide tool to prevent seal damage. C—One type of seal driver, used for installing one-piece rear main bearing oil seals, is shown here. (DaimlerChrysler)

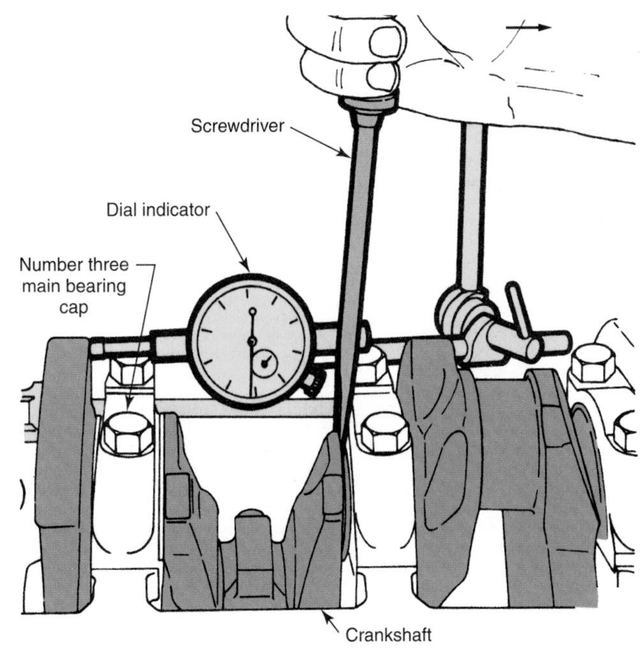

Figure 26-33. Check crankshaft endplay with a dial indicator. (DaimlerChrysler)

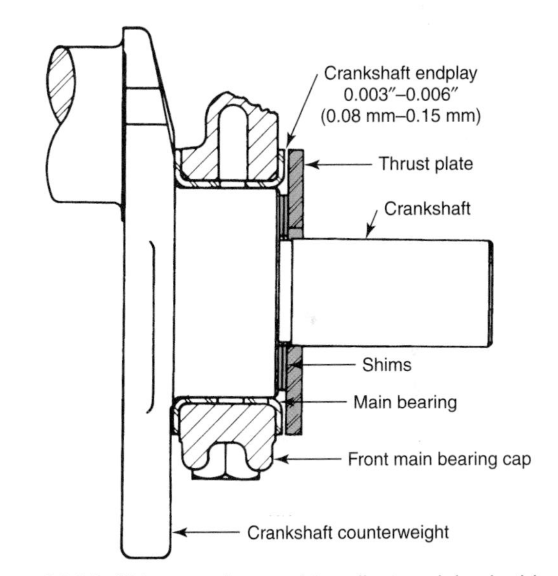

Figure 26-34. Shims can be used to adjust endplay in this shaft.

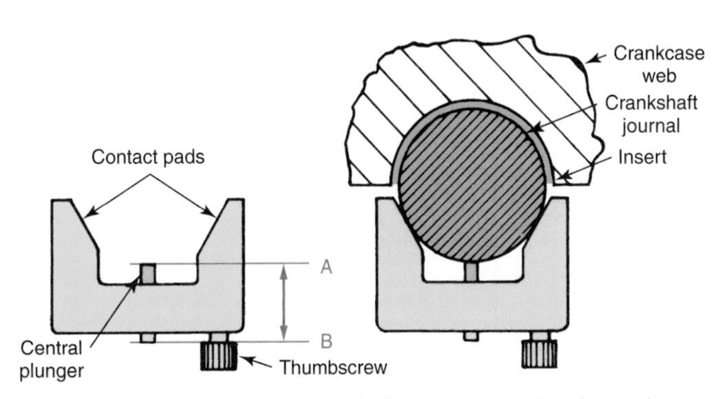

Figure 26-35. A crankshaft main journal gauge is shown here. Measure distance A–B and double it. Note that the gauge is placed against the journal and that insert may be left in place.

Another gauge consists of an outside micrometer with special caliper-type jaws. It requires the upper insert be rolled out to allow room for jaw entrance. Look at **Figure 26-36.**

Whatever type of measuring tool is used, use care, as it is easy to make a poor reading. Rotate the shaft so you can take measurements at several points.

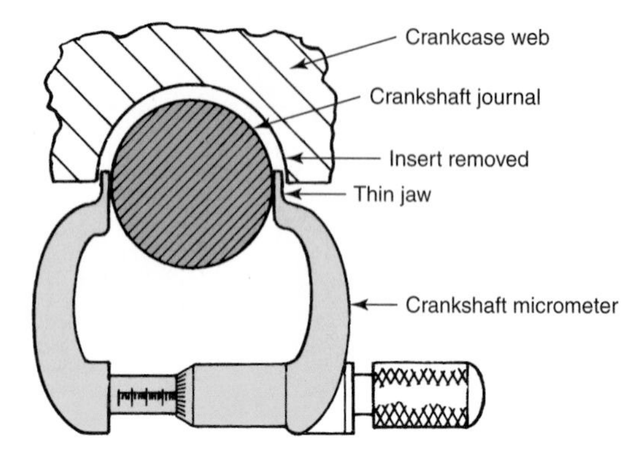

Figure 26-36. The thin jaws of this special crankshaft micrometer enter between the journal and bore after the insert is removed.

Removing Crankshaft Main Bearing Upper Inserts—Engine in Car

On many engines, it is possible to remove and replace the upper main bearing inserts with the engine in the car. Loosen the front and rear caps but do not remove them. This will assist in the removal of the intermediate bearing inserts. Remove the intermediate caps. Insert a suitable removal plug into the journal oil hole, **Figure 26-37.** The flat section of the plug must be a little narrower and thinner than the insert to prevent binding in the bore. It should contact the insert end squarely, Figure 26-37A. In an emergency, a plug can be made from a cotter pin, Figure 26-37B.

Rotate the shaft to bring the flat section of the plug against the insert end *opposite* the locating lug. Make sure it will clear the bore. Slowly rotate the shaft and the insert will be forced around and out of the upper bore, Figure 26-37C.

Rolling in a New Bearing Insert—Engine in Car

Before installing a new insert, the upper bore and crank journal should be thoroughly cleaned. Do not leave any loose threads of cloth in the bore. Lubricate the journal and the new insert. Place the insert against the journal, plain end toward

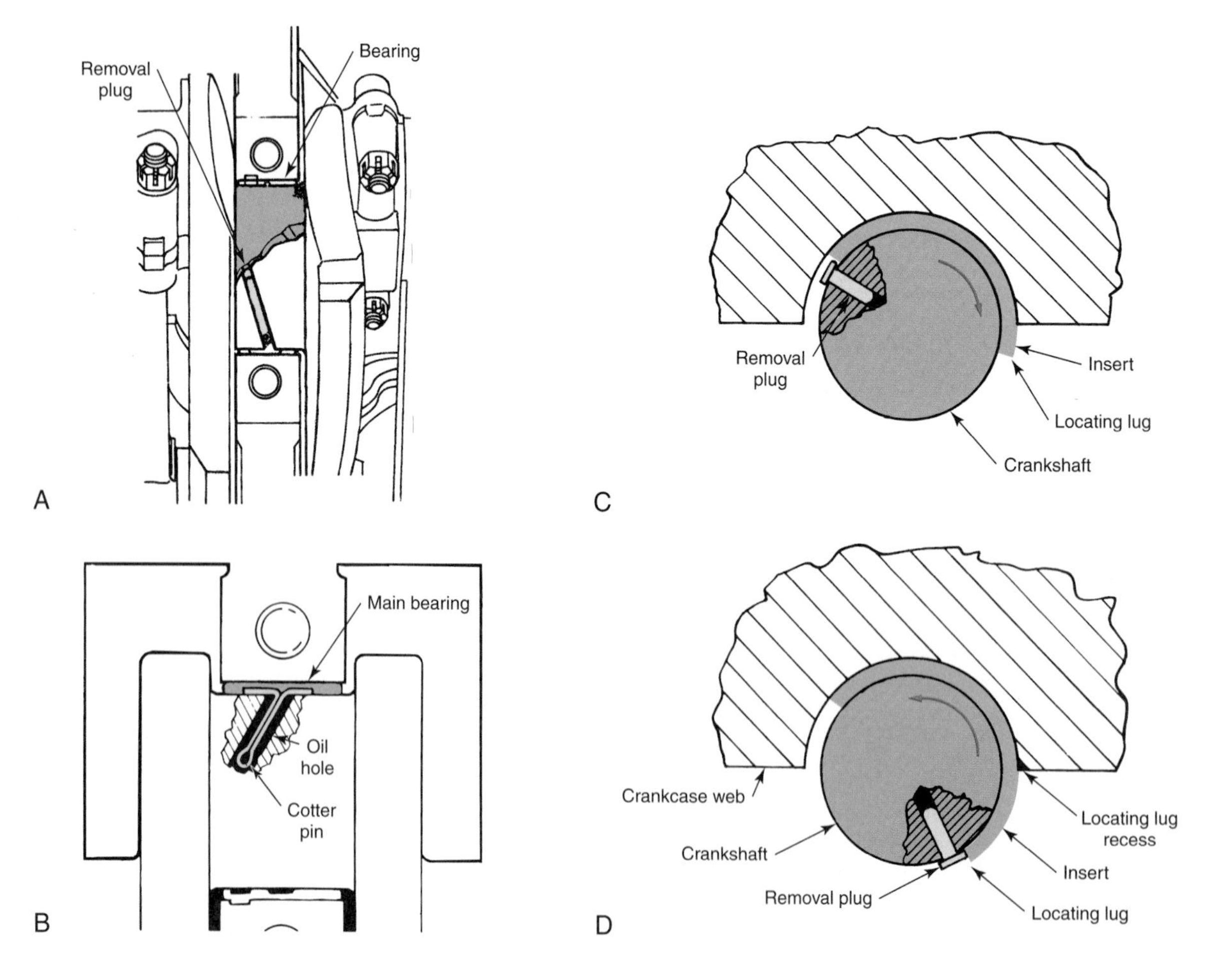

Figure 26-37. A—A typical insert removal plug. B—A cotter pin can be used as a removal plug. C—As the shaft is rotated, the removal plug forces out the insert. D—The plug can be used to roll a new insert into position. (DaimlerChrysler)

the bore locating lug recess. Slip the insert into the bore as far as possible. Place the removal plug in the journal's oil hole and rotate it against the locating lug end of the insert. Continue rotation until the insert is properly seated. Make certain the insert locating lug engages the lug recess in the bore. See Figure 26-37D.

Snap the lower insert into the cap. Lubricate the insert and install the cap. Install the cap bolts and tighten them snugly. Back them off a few thousandths and remove the front and rear main inserts. When installing the rear main upper insert, make sure the oil seal does not rotate out of position. When all inserts are in, torque all caps but the one with the thrust bearing. Align the upper and lower flanges with the crank thrust surface and torque the cap. Bearing clearance should be checked with a Plastigage.

Checking Main Bearing Clearance—Engine in Car

Working on *one* bearing at a time, remove the cap. Wipe the journal and the cap insert clean. With an extension jack, apply upward pressure to the crankshaft adjacent to the journal being checked. This keeps the shaft in contact with the upper insert so the Plastigage will provide a true reading. If the shaft is allowed to sag downward, the Plastigage material will be flattened when the cap is torqued, giving a false reading.

Place a strip of Plastigage across the lower insert, install the cap, and torque the cap bolts to specifications. Finally, remove the cap and determine clearance by measuring the flattened Plastigage.

Installing a Rear Main Bearing Oil Seal—Engine in Car

Some engines are designed so the rear main oil seal can be replaced without removing the crankshaft. Replacing these seals is often tricky and requires care.

Remove the cap. Pry out the old seal and clean the groove. If a synthetic rubber seal is used, install it in the cap at this time. Using a clean brass drift punch, tap on the end of the upper seal (it may help to rotate the shaft also) to push the other end out. As soon as the end is clear, grasp it with pliers and pull the seal from the bore. If difficulty is encountered when removing a wick-type seal, it will help to drop the shaft a few thousandths of an inch. A corkscrew-type attachment can be screwed into the end of the wick to help pull it out to a point where pliers can be used. Use care not to nick the journal seal surface.

Lubricate the new upper rubber seal and insert it into the groove. To seat the seal, push it into the groove while rotating the crankshaft in the same direction.

In cases where a wick seal or packing is used, remove it as previously described. To insert the new upper wick seal, attach a soft wire to one end of the wick. Using the cap, preform the wick by installing it in the cap groove. Do not cut off the ends. Remove the wick from the cap and lubricate the wick thoroughly. Pass the wire through and around the upper groove. Force the end of the wick into the groove. Tuck all fibers in and pull on the wire. Wiggle the wick where it enters the groove, making sure the wick does not bulge out and hang up. Pull the wick through the groove until an equal amount protrudes from each side. Rotating the crankshaft will help. Cut the wick off flush with the parting surface. To prevent nicking the journal surface, a thin piece of shim stock can be slid between the end of the wick and the journal.

Checklist for Crankshaft and Bearing Service

The answer to all of the following questions should be yes.

- Were the crankshaft journals within specifications for size, out-of-roundness, and taper? If not, were they reconditioned to meet specifications?
- Was the crankshaft stored properly?
- Were all parts thoroughly cleaned before reassembly?
- Were the bearing inserts' spread and crush correct?
- Were the oil holes and aligning lugs properly installed?
- Were bearing clearances checked with Plastigage and within specifications?
- Were the bearing caps placed in their original positions?
- Were all bearing cap alignment marks in their original positions?
- Were the bearing caps properly torqued?
- Was crankshaft endplay checked? If it was not within specifications, was it corrected?
- Were seals installed properly?

Connecting Rod Service

When pistons must be removed for ring work, the rods usually need reconditioning, too. Modern high-rpm and high-horsepower engines impose tremendous loads on the rods. In order to reduce reciprocating weight, the rods are made as light as possible. This reduction in weight, while beneficial, does tend to allow distortion of the large end bearing bore. It can also allow the rod to twist and bend. High rpm, heavy loads, centrifugal and inertial forces, and the effects of heating and cooling are primarily responsible. Connecting rods should always be checked for twist, bend, and bearing bore distortion.

Make sure all rods are marked and that both upper and lower bearing halves are marked with the same number. They should be marked in order starting with the number one cylinder. Note the relationship between the numbers and the block so the rods can be installed without reversing them. If the rods are not numbered or if one or more rods show the wrong number, renumber them in the proper order.

Note: Correct numbering of all rods is important. The rod and piston must be in the correct relationship to each other and replaced in the proper cylinder with the marks facing in the original direction.

Connecting Rod Twist and Bend

Rods can have twist and bend distortion. ***Twist*** is a condition in which the centerlines of the upper and lower rod bearing bores are out of alignment in a horizontal plane. ***Bend*** is present when the centerlines are misaligned in a vertical direction. See **Figure 26-38.** If a rod is misaligned, it causes a diagonal wear pattern that extends down to the lower portion of the skirt.

One type of connecting rod alignment gauge is shown in **Figure 26-39.** To use this gauge, the piston pin must be reinstalled in the connecting rod. The crankshaft bearing end of the connecting rod is installed on the mounting rod so that it is parallel to the steel plate. An adapter is placed over the piston pin as shown in the figure. Once this is done, the technician can use a feeler gauge to measure the amount of bend and twist.

Do not confuse a misaligned rod with a regular offset connecting rod. Rods are occasionally offset to provide proper alignment between the cylinder and crank journal. Some normally offset rods are shown in **Figure 26-40.** Notice how the web centerline (C/L) intersects the lower bore to one side. Also note how the offset on each rod faces different sides. It is possible to straighten a bent or twisted rod. This is normally done in a machine shop. Ideally, a bent or twisted rod should be discarded.

Checking Connecting Rod Large End Bore

The connecting rod *large end bearing bore* should be checked for roundness, bore size, straightness, and surface condition, **Figure 26-41.** Checking and reconditioning the upper connecting rod bore and bushings is covered in this chapter in the section on piston and piston pin service. After the cap is aligned and torqued to the rod, the connecting rod large end bore should be checked for out-of-roundness by either measuring with a telescoping gauge and micrometer or with an inside micrometer as shown in **Figure 26-42.**

The rod bore should be checked in several positions to give an accurate reading. Out-of-roundness should not exceed 0.001″ (0.025 mm). If the crankshaft journal is at its maximum out-of-round limits, the bore out-of-roundness

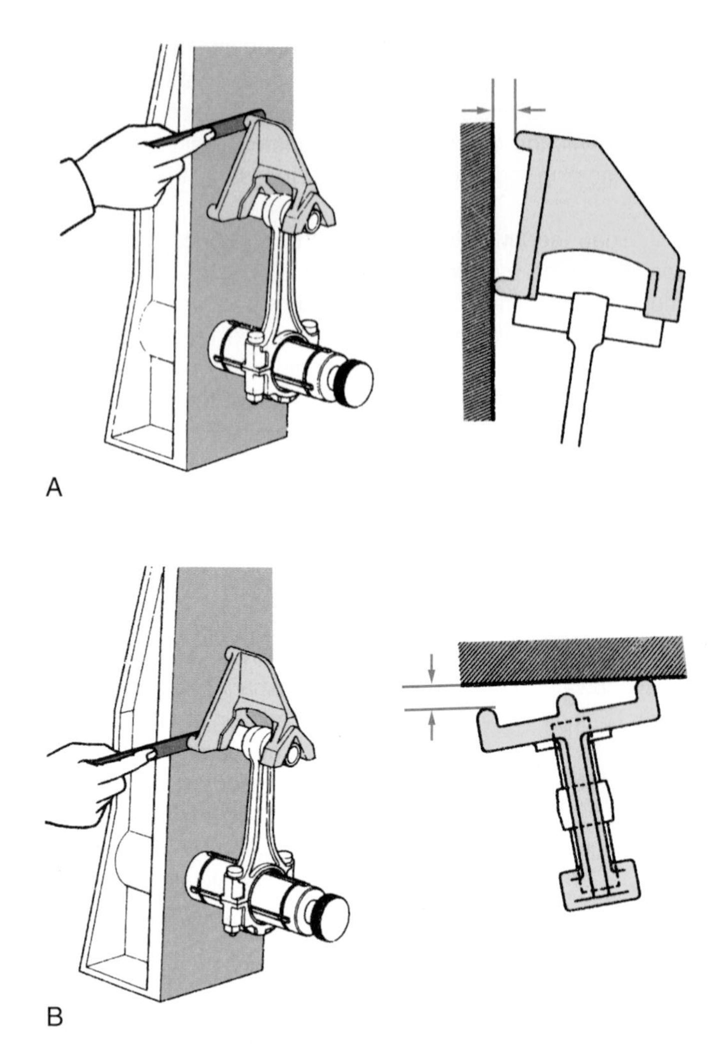

Figure 26-39. This is a special tool for checking connecting rods. A—Checking for rod bend. B—Checking for rod twist. (Toyota)

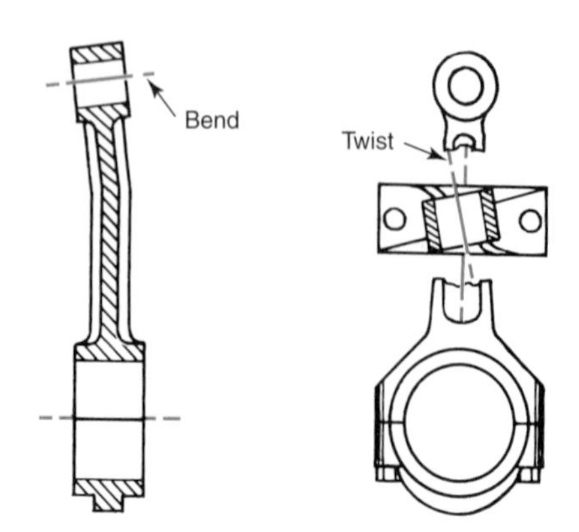

Figure 26-38. Bend, twist, or a combination of both can be present in a connecting rod.

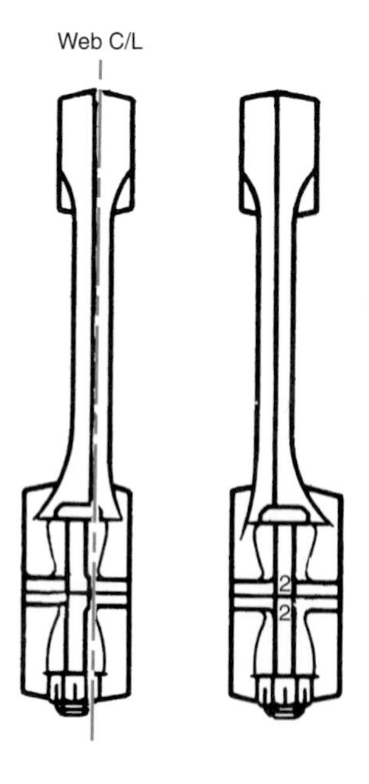

Figure 26-40. A pair of regular offset connecting rods. Note the numbering on the rods. (Fiat)

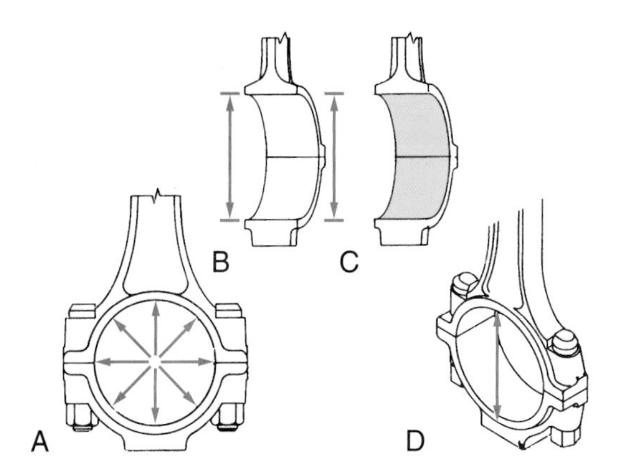

Figure 26-41. These diagrams illustrate several important checks for connecting rods' large end bores. A—Out-of-roundness. B—Straightness. C—Surface condition. D—Bore size.

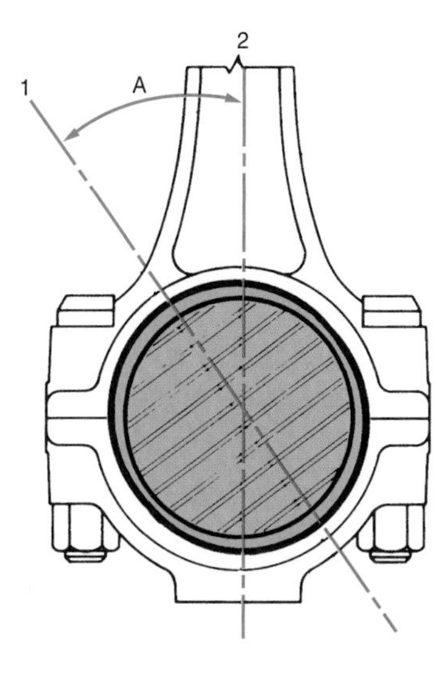

Figure 26-43. A connecting rod's bore elongations are generally in area A, between centerlines 1 and 2.

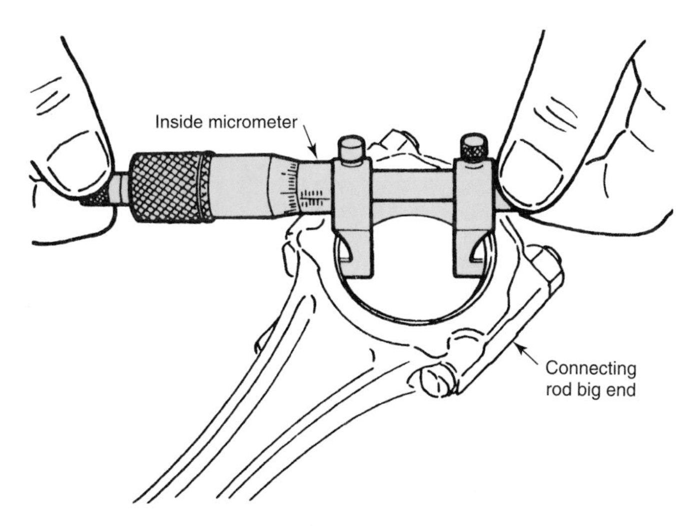

Figure 26-42. The out-of-roundness of a connecting rod's large end bore can be quickly determined with an inside micrometer. (Nissan)

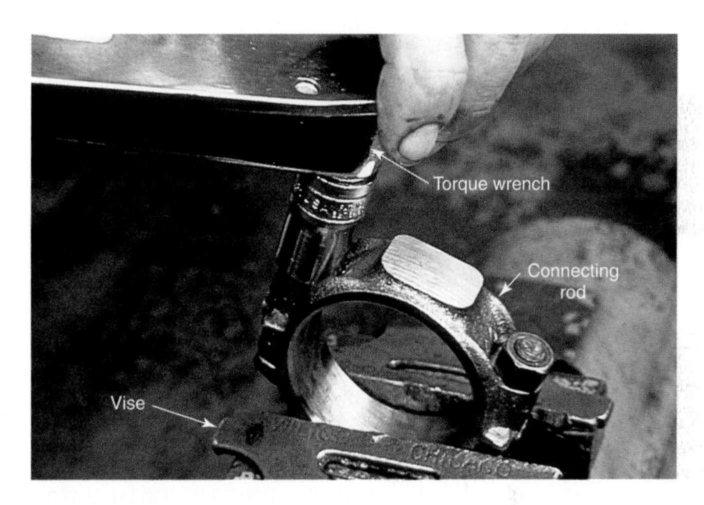

Figure 26-44. Holding connecting rod halves in a vise with protective jaw covers helps relieve strain on the rod when torquing the fasteners. (Federal-Mogul)

should be less than 0.001″ (0.025 mm). As a general rule, bore elongation direction varies from vertical to around 30° from vertical, **Figure 26-43.**

Correcting Out-of-Roundness

After determining the amount of out-of-roundness, remove half of the amount of material required for correction from the sides of the connecting rod at the crankshaft bore, and half from the rod cap. This requires a special grinder. Following material removal, assemble the cap to the rod and torque it to the specified amount, **Figure 26-44.**

Since material was removed from the sides of the bearing bore, the bore must be honed to make it completely round again. When honing, direct a stream of honing oil so an ample amount enters the bore. Move the rods back and forth over the stones, being careful to avoid exerting any side pressure that would cause the rods to tip. Keep the stones snug in the bores to reduce chatter. The stones must be of the correct grit size to produce a 30–40 microinch finish and kept dressed to ensure straight, smooth bores.

After the initial removal of rough stock with the hone, check the bore size often to avoid honing oversize. With careful work, it is possible to bring the bore back to the specified size and limit the out-of-roundness to the generally recommended 0.0003″ (0.0076 mm). The rod should always be carefully checked for bend and twist after reconditioning either the wrist pin or large end bore.

Piston Service

All pistons must be of the correct type, weight, and size. They must be free of cracks, scoring, burning, and damaged ring grooves. They must be properly fitted to the cylinders, rings, and piston pins. If the cylinders have been rebored, the original pistons must be discarded and new pistons of the proper oversize obtained.

Inspect Pistons

After cleaning and drying, clamp the connecting rod lightly in a vise with the piston skirt just clearing the jaws. Clean any remaining carbon from the ***ring grooves.*** Be careful to avoid cutting any metal from either the side or the bottom of the grooves. **Figure 26-45** illustrates one type of ***ring groove cleaner.*** Check each piston and reject pistons that show any signs of cracking, scuffing, scoring, burning, or corrosion. Pay particular attention to the piston pin bosses and skirt. Also note damaged snap ring grooves.

Caution: The piston should be discarded if it displays any cracking, no matter how small. Do not attempt to repair a cracked piston.

Scoring is often caused by insufficient clearances. Scuffing is caused by metal-to-metal contact. Excessive heat builds up and particles are torn from one surface and deposited on another. Scuffing and scoring are closely related. Frequently, the heat generated will discolor the scuffed areas. Burning is sometimes severe, as shown in **Figure 26-46.** When pistons show signs of corrosion, look carefully for possible coolant leaks such as a cracked head or cylinder, warped head or block, damaged gasket, or other problems.

Examine all ring grooves for burrs, dented edges, and side wear. Pay particular attention to the top compression ring groove. It is the one most subject to wear, **Figure 26-47.** Groove width can be checked by sliding a new ring into the groove and using a feeler gauge to determine clearance. Check clearance at several spots around the groove. Special gauges are also available for a quick check of groove wear. On some pistons, the ring groove can be reconditioned by cutting the groove wider and installing a steel ***ring groove spacer*** on the top edge.

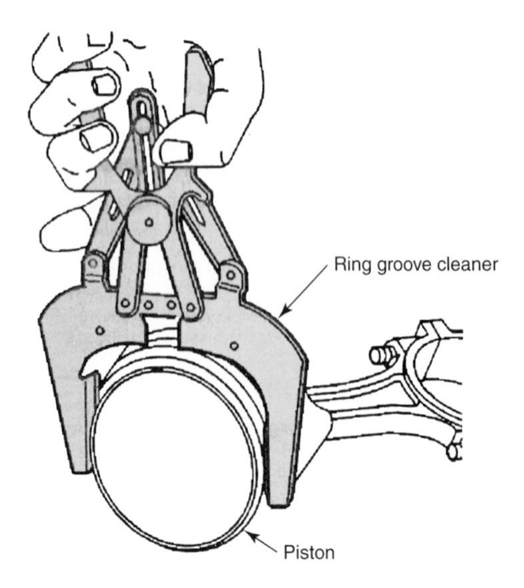

Figure 26-45. This is one type of hand-held tool used to clean piston ring grooves. Note different blade sizes. (Lisle Tool)

Fitting Pistons

New pistons must be checked for proper clearance, or *fit,* in the cylinder. New pistons require the cylinder bore be honed until the piston clearance is correct. An oiled, long feeler gauge strip is placed in the bore. The piston is inverted and shoved down into the bore so the skirt thrust surface bears against the strip. A spring scale is attached to the feeler gauge strip and the strip is withdrawn. The manufacturer's specifications should indicate the correct pull in pounds or kilograms for the specific feeler gauge thickness. In worn cylinders, the check should be made near the bottom.

Piston-to-cylinder clearance may also be determined by careful measurement of both piston and cylinder. Measure the piston at both top and bottom of the skirt across the thrust surfaces. Some manufacturers specify an exact location for piston measurement. Measurements should be taken when metal is at room temperature, 60°F–70°F (16°C–21°C). Once a piston is properly fitted, it should be marked for the cylinder concerned. Clearances with 0.001″–0.0015″ (0.025 mm–0.038 mm) are about average.

Figure 26-46. Preignition has burned a hole through the head of this piston.

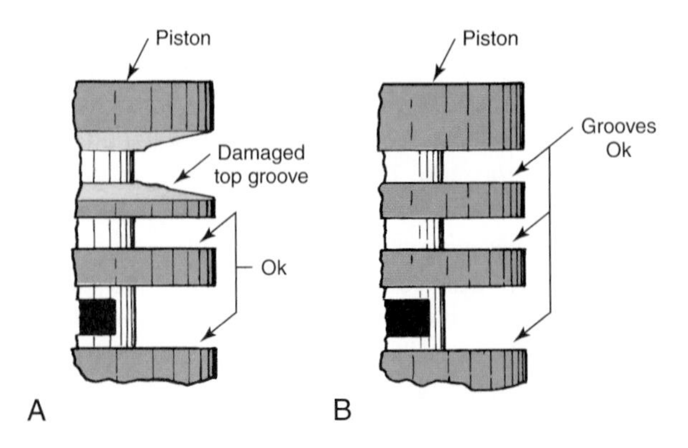

Figure 26-47. Compare these two ring grooves. A—A badly damaged top ring groove. B—A normal groove.

Piston Pins

Most piston pins are designed to oscillate in the connecting rod, in the piston, or in both. Piston pins are sometimes called wrist pins. **Figure 26-48** illustrates five pin arrangements. ***Piston pin fit*** (clearance between pin and bearing surface) is important. A pin must have ample clearance for oil, yet it must not have looseness that will result in pin knock and ultimately pin failure. The bearing surfaces must be round, smooth, straight, and in perfect alignment, **Figure 26-49.**

Clamp the connecting rod lightly in a vise. Attempt to rock the piston on the pin. Any discernible movement between pin and piston or pin and rod is cause for pin rejection. Do not confuse the rod sliding along the pin with up-and-down movement. Careful measurement of pin and boss using a vernier micrometer and a small hole gauge will determine exact clearance. If wear is excessive, it will be necessary to install an oversize pin and, in some instances, new bushings. If the pin is bushed in the rod or piston and the bushing shows only minor wear, the old bushings may be honed to fit the oversize pin.

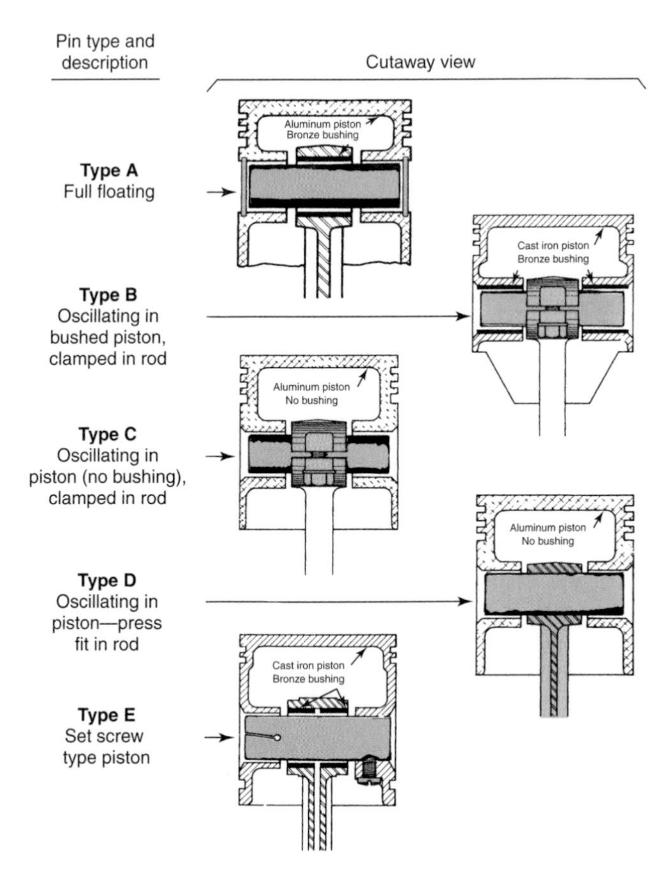

Figure 26-48. Study these various piston pin arrangements. Type D is currently in wide use. (Sunnen)

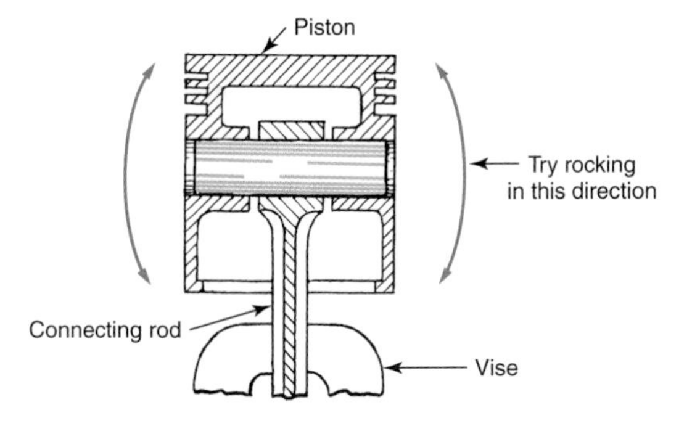

Figure 26-49. There should be no discernible movement between the piston and piston pin or the connecting rod and piston pin.

Piston Pin Removal

Before removing the piston pin, make certain the piston is marked so it can be reassembled in the same position to the same rod. Prick punch marks will suffice if no factory identification has been provided. **Figure 26-50** illustrates one type of factory identification. If the piston pin is the floating type, remove the end locks (snap rings) and tap the pin free. Do not mar the piston pin bearing area. Use care in removing the end locks so as not to distort them. Most engines currently use a pin that oscillates in the piston and is a press fit in the rod. With this type, use a press or puller arrangement to remove the pin, **Figure 26-51.**

Fitting Piston Pins

Since piston-to-pin clearances are extremely small (0.0002″–0.0005″ or 0.005 mm–0.0127 mm typical in aluminum pistons), pin fitting is an exacting job. Modern honing and boring machines do such highly accurate work that a pin can slide freely through the piston with as little as 0.0001″ (0.0025 mm) clearance. Even though the pin feels free, this space would not provide sufficient oil clearance. The pin would probably seize in the piston. Seizure can literally demolish a piston.

Pin clearance in bushed rods is somewhat greater, averaging around 0.0005″ (0.0127 mm). In pressure-fed rod

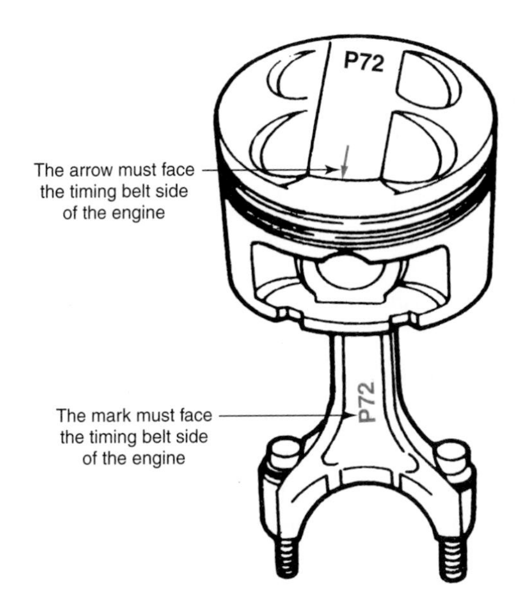

Figure 26-50. Factory identification marks are used to ensure correct relationship between piston, connecting rod, and block for one specific engine. (Honda)

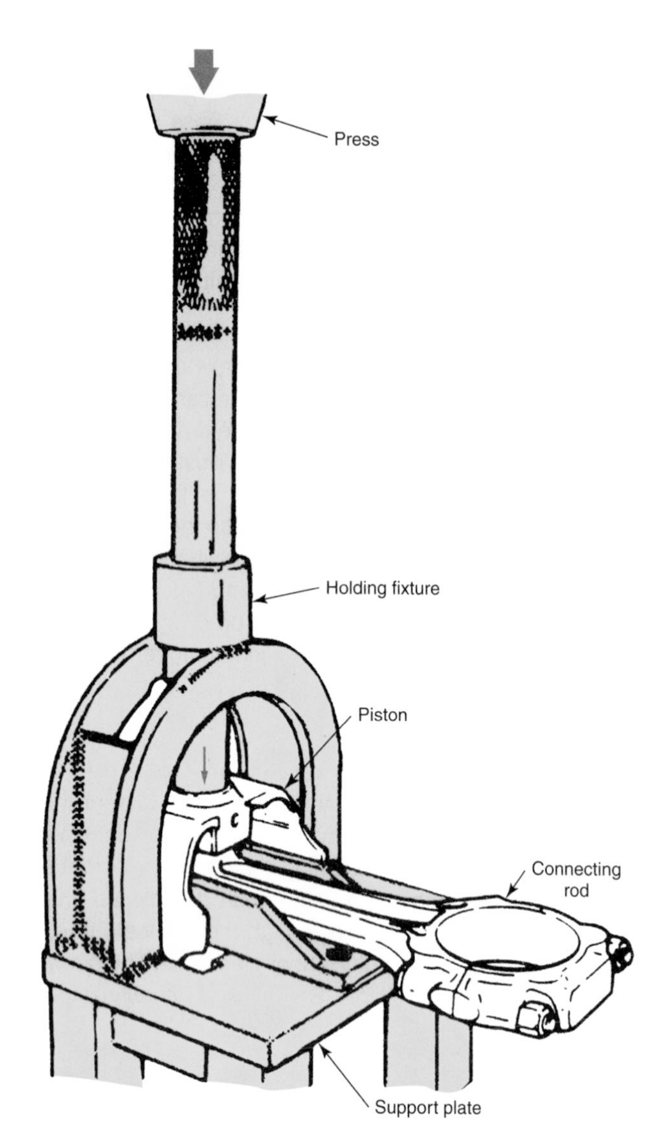

Figure 26-51. A hydraulic press and special holding fixture are used to force out a connecting rod wrist pin. Note how the piston and rod assembly are carefully held on the support plate. Wear safety glasses and use care during this procedure. (General Motors)

bushings, the clearance is about 0.001″ (0.025 mm). Proper pin clearances depend on pin diameter, method of attachment, type of piston material, piston operating temperature, and the use of a bushing in the rod. Many pin-fitting machines are equipped with highly accurate measuring devices that control pin clearance within a few ten-thousandths of an inch. The pin may also be measured with a vernier micrometer. There is no ideal average that works well on all engines. Consult and carefully follow the manufacturer's specifications. **Figure 26-52** demonstrates the difference between a reamed finish and a honed or diamond-bored finish. Bearing bore diameter should be checked with a small hole gauge and vernier micrometer. If the measurements are done carefully, accurate clearances can be determined.

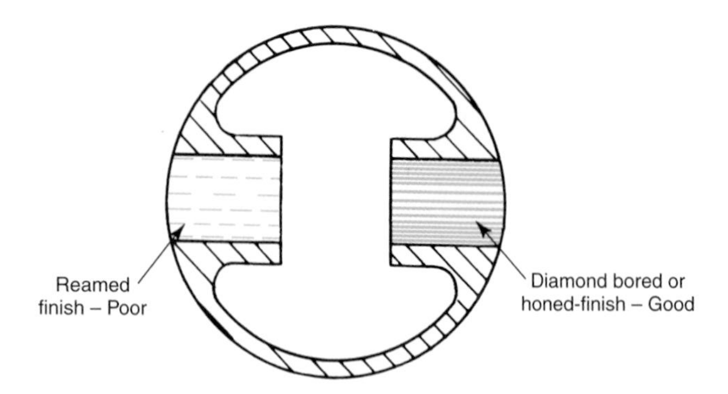

Figure 26-52. Do not use a reamer for a final pin fit. Such a surface allows high initial wear. Fit pins by actual measured clearances and never by thumb or palm push fits.

Assembling Connecting Rod and Piston

Clean pistons, pins, and rods twice with a round bristle brush and hot, soapy water and rinse thoroughly. Pay special attention to pinholes and hollow pins, locking lip recesses, bolt holes, and other areas where grit may be trapped.

Caution: When torquing rod caps out of the engine, pressing piston pins into the rod upper bore, or grasping rods in a vise, use extreme care to avoid bending or twisting the rods.

All parts must be clean and dry, checked for proper clearance, and thoroughly lubricated. While holding the piston and rod in the correct relationship to each other, pass the pin through the units. If an interference fit in the rod is used, a press or puller is required, **Figure 26-53.** Some shops apply controlled heat to the rod bore to ease pin installation and prevent galling. If the pin is a press fit in the rod or held to the rod with a clamp setup, the pin must be carefully centered so that rod side movement will not cause the pin to strike the cylinder wall. See **Figure 26-54.** The interference fit between the pin and rod can be checked by measuring the torque required to pass the pin through the rod. Be sure to check the manufacturer's specifications.

On the floating pin installation, make certain the end locks are not distorted or weak. Check to see that they are fully seated in their grooves. Install the open end toward the bottom of the piston. This tends to cause them to expand into the grooves during the shock period imposed during the firing stroke. Loose pin locks can actually cut through the boss from the inertial force. When removing or installing locks, spring them only as far as necessary. If the lock breaks and moves out of its groove, the cylinder, piston, or both can be damaged.

Ring Service

The two basic ring types, ***compression rings*** and ***oil control rings,*** are available in a multitude of designs. Replacement rings are available in sets designed for either a

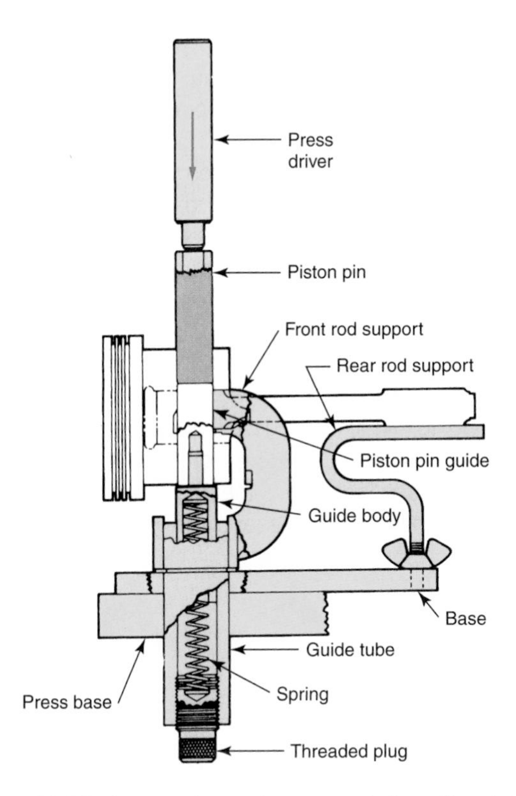

Figure 26-53. A press is used to properly install a piston pin. The pin is press fit in the connecting rod. (General Motors)

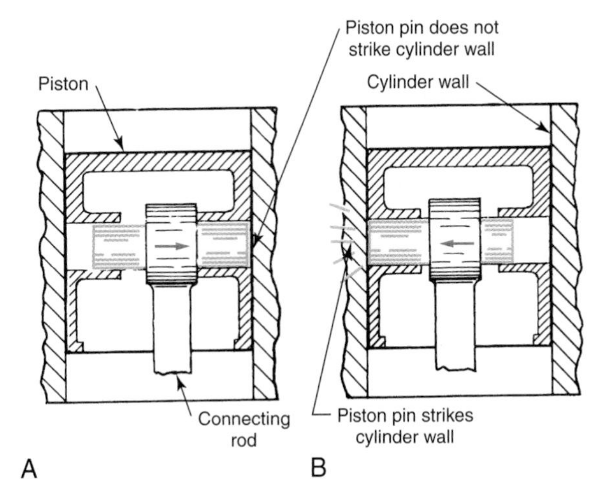

Figure 26-54. A—When the piston pin is centered in the rod, the rod's side movement will not cause the pin to strike the cylinder wall. B—An offset pin causes trouble.

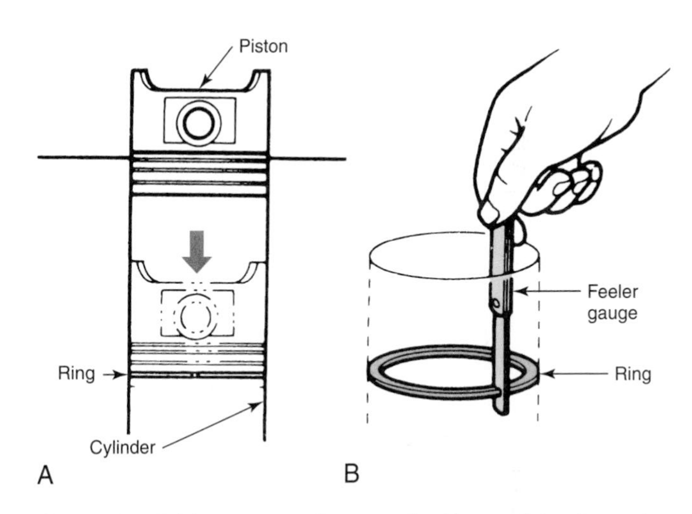

Figure 26-55. To measure ring gap, A—Use a piston to seat ring squarely at bottom of ring travel in cylinder. B—Measure the ring gap with a feeler gauge. (Mazda)

rebored or a worn cylinder. The set for installation in a rebored cylinder is commonly referred to as a factory or rebore set while rings for a worn cylinder is called an engineered or oil control set. Mild taper (up to 0.005″ [0.13 mm]) can usually be handled well by a rebore set. The oil control set is of a somewhat different design with stiffer expander springs to force the rings to follow tapered walls. The oil control set can be used in cylinders with severe taper, but will produce more drag and wear. Specify cylinder diameter (standard, 0.010″, 0.020″, etc., oversize) when ordering ring sets.

Checking Ring Groove Clearance

When rings are installed in a cylinder, a certain amount of clearance, or ***ring gap,*** must exist between the ends. As the rings heat up, they expand. If the ends touch and expansion continues, scuffing, scoring, and damage will occur. It is advisable to allow a minimum of 0.003″ (0.08 mm) gap for each 1″ (25.4 mm) of cylinder diameter. For example, a 3″ (76.2 mm) cylinder would require a gap of 0.009″ (0.23 mm). Start the ring by hand, and then use a piston to push it to the bottom of the ring travel. This squares the ring with the bore. Use a feeler gauge to check compression rings for proper gap, **Figure 26-55.**

A gap of up to 0.008″ per 1″ (0.20 mm per 25.4 mm) of bore diameter is acceptable. Anything above this would indicate a wrong size ring set. If the gap is a few thousandths small, it may be widened by clamping a small, fine-tooth mill file in the vise and rubbing the ends of the ring across the file surface. Hold the ring near the ends and be very careful. Remove the ring from the cylinder by pulling upward on the ring directly opposite the gap. If the side of the ring near the gap is lifted, the ring can be distorted or broken.

Checking Groove Depth and Side Clearance

Rings are made for a variety of ***groove depths.*** The ring must not touch the bottom of the groove when installed in the cylinder. A gauge may be used to check the oil ring groove depth, **Figure 26-56.** The groove should not be too deep because certain types of expanders push outward from the groove bottom. Excessive groove depth reduces expander pressure. Shallow grooves can be deepened in some cases if the correct rings are not available. Some manufacturers supply shims to reduce excessive depth.

Rings should be rolled around in their respective grooves to check for binding. See **Figure 26-57.** Check ***ring side clearance*** by inserting the ring into the groove and passing a feeler gauge between the ring and groove side. Check clearance in several spots around the groove. Clearance should not exceed 0.006″ (0.15 mm) or be less than 0.0015″ (0.038 mm) for top rings. See **Figure 26-58.** Some multiple piece oil rings are designed so the expander-spacer forces the rails not only against the cylinder walls but also against the ring groove sides.

Installing Rings

Grasp the connecting rod in a clean vise. All ring grooves must be clean. Make certain the drain holes in the oil ring grooves are open. Starting with the bottom oil ring, install each ring according to the directions supplied by the ring's manufacturer. Multiple piece oil ring side rails must be spiraled over the piston. This is shown in **Figure 26-59.** Do not try to expand them. The ends of the flexible spacer in Figure 26-59A must be butted together. Do not let them overlap. Butted ends must be located over a solid portion of the groove bottom. Some grooves have little solid area. Special shims are available to prevent the spacer ends from bending inward through the groove.

When spiraling rails into position, Figure 26-59B and Figure 26-59C, be careful to avoid scoring the piston with the sharp rail end. The end can be slid over a piece of stiff feeler stock to prevent piston damage. After both rails are installed, check the spacer ends to make sure they are not overlapped. Use a good quality ***ring expander*** to install the remaining rings. Do not expand any ring more than necessary. See **Figure 26-60.**

Be careful to install the rings with the recommended side up. Many rings are marked with the word *top.* This is to remind the technician to face that side of the ring toward the top of the piston. If no markings are present and no illustrations are

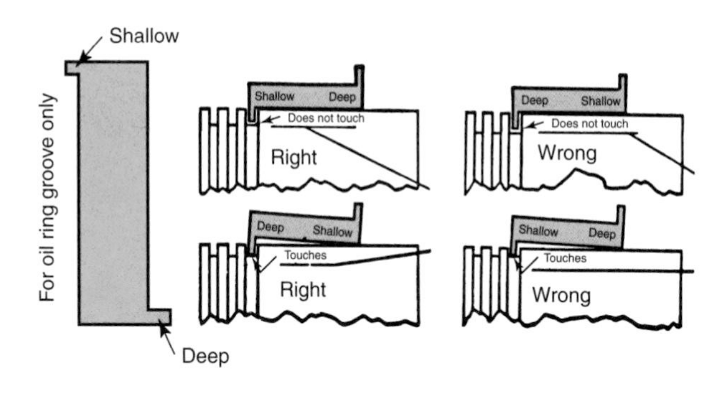

Figure 26-56. When a gauge such as this is enclosed with the ring set, be sure to use it. The shallow tip should not touch. The deep tip should touch. If the deep tip does not touch or if the shallow tip does, the ring set is wrong for the piston in question. (Hastings)

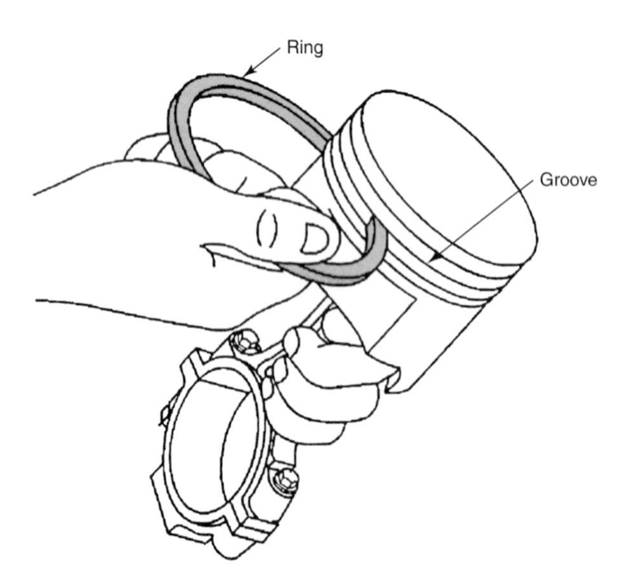

Figure 26-57. Roll the ring completely around the groove to check for binding.

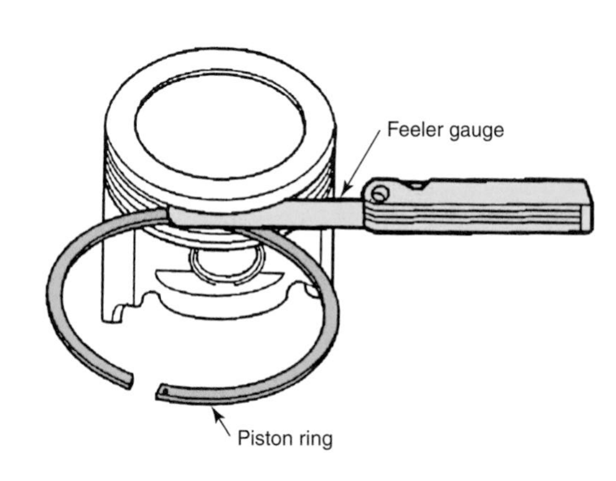

Figure 26-58. Use a feeler gauge to check the ring's side clearance in its groove. (DaimlerChrysler)

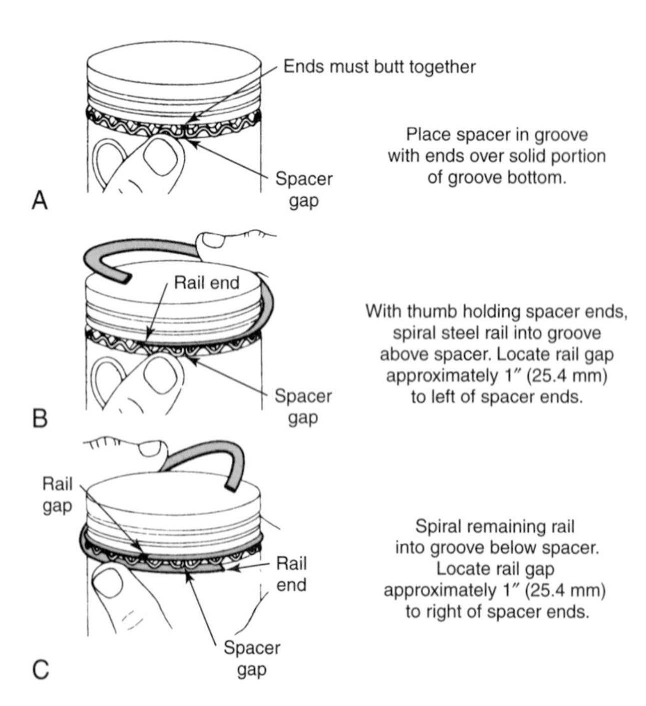

Figure 26-59. The oil ring's side rails must be spiraled into place. Do not cut off the ends of the expander spacer.

provided, a study of the ring's profile will usually determine the top side. Study the typical ring profiles in **Figure 26-61.** Do not forget to install expanders for compression rings when they are so equipped. On multiple piece rings, follow manufacturer's recommendations. Although rings tend to float (move around in their grooves), it is a good idea to space the ring gaps around the piston so they are not in alignment. See **Figure 26-62.**

Installing Rod and Piston Assembly

Give the rod and piston assembly a final check. Check that the piston pin is centered and secure and the piston and rod are correctly assembled in relation to each other. Check the rings are properly installed and that oil holes, where used, are open. See **Figure 26-63.** Squirt a heavy coat of clean engine oil over the rings and piston. Apply plenty to the pin

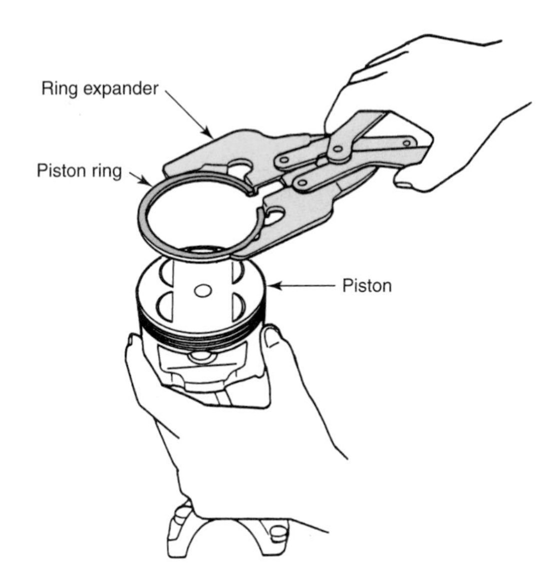

Figure 26-60. Use a good quality ring expander to install rings. (Honda)

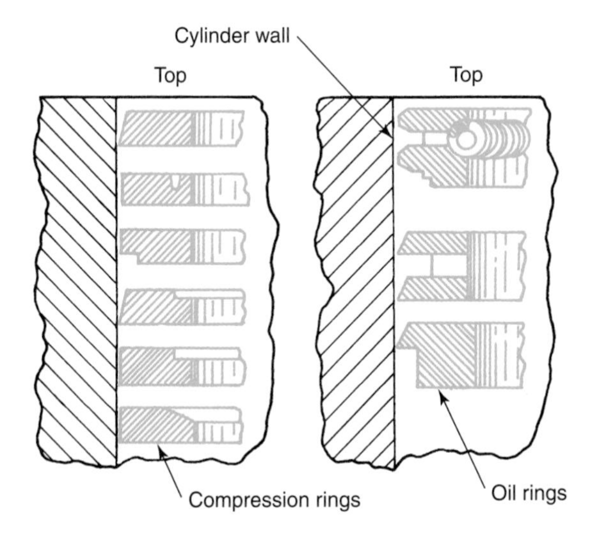

Figure 26-61. This figure shows the various profiles of compression and oil control rings.

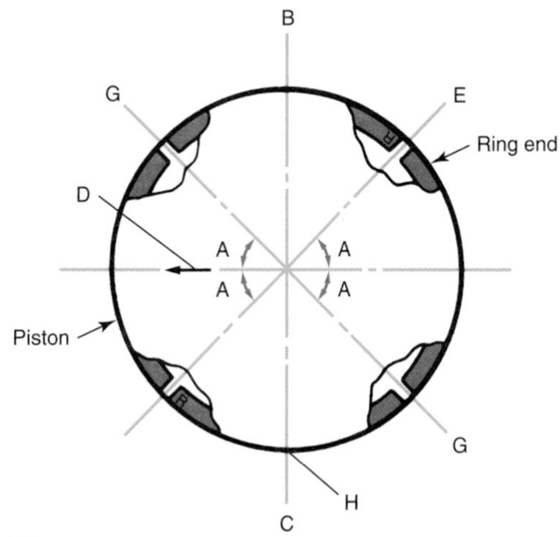

A 45°
B Intake side
C Exhaust side
D Arrow mark
E Upper compression ring end gap
F Lower compression ring end gap
G Oil ring rail gaps
H Oil ring spacer gap

Figure 26-62. This diagram illustrates ring gap arrangement for one particular engine. (General Motors)

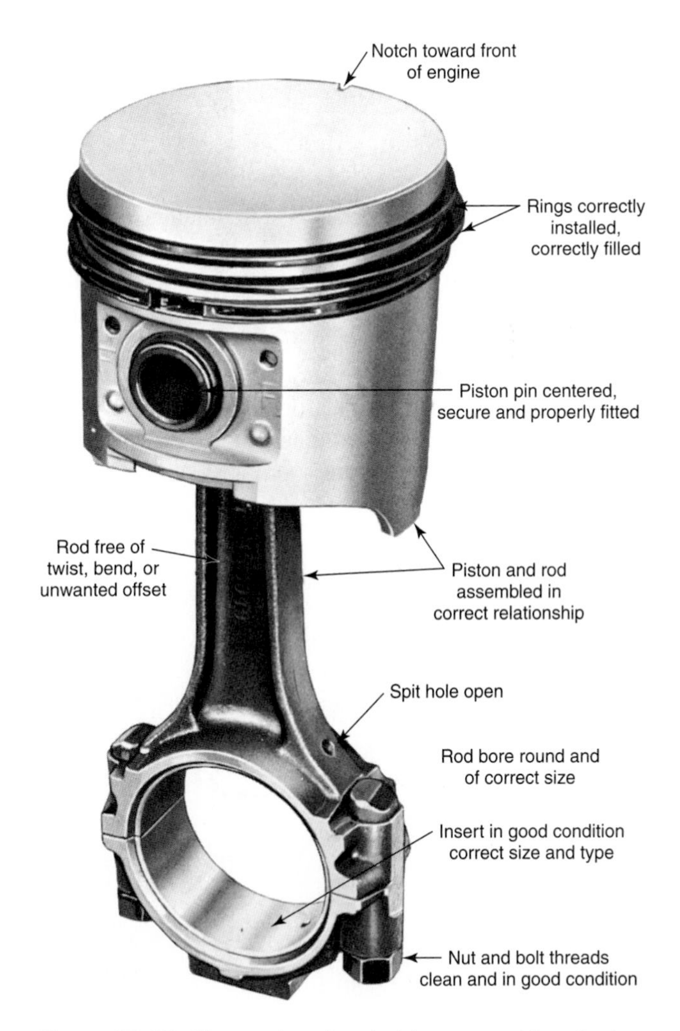

Figure 26-63. Give each rod and piston assembly a final check before installing in the engine. (DaimlerChrysler)

and work the rod back and forth to ensure oil enters the pin bushing. See **Figure 26-64.** Hands and work area must be clean. Grasp the rod in a vise. Ensure the ***ring compressor*** is clean and lightly lubricated. Then slide the ring compressor down over the rings until the lower tightening band is below the lowest ring. Tighten the compressor securely.

Note: Tap lightly around the outside of the compressor using the tightening wrench. Then, retighten. This will ensure the rings are fully compressed.

Snap in the upper rod bearing insert and lubricate it. Install the journal protectors. Turn the crankshaft so the connecting rod journal is at bottom dead center. Slide the exposed piston skirt into the cylinder while keeping the rod aligned with the journal. Make certain the piston identification marks face the correct direction. Using a hammer handle, tap the piston through the compressor and into the cylinder. The piston should enter with light tapping. If the piston catches on the way in, a ring is probably hung up on the cylinder block surface.

Caution: Do not force the piston into the cylinder. Remove and reinstall the piston and ring compressor.

While tapping the piston into the cylinder, it is important to keep the compressor firmly against the block. Failure to do this may allow it to ride up far enough for an oil ring side rail to pop out and hang up. See **Figure 26-65.** A slightly different shaped compressor is required for a block with a slanted top surface. Guide the rod bearing around the journal as the piston is tapped or pulled down through the cylinder.

Journal protectors provide a handle that is very handy for pulling the rod into position on the journal. Rubber hose can also be used if journal protectors are not available. See **Figure 26-66.** Snap in the lower bearing insert. Lubricate and install the cap so the cap number is on the same side as the upper mark. Tighten the cap bolts or nuts until the cap is snug.

Figure 26-64. The cylinder, piston, rings, pin, and rod bearings must be heavily oiled. (Federal-Mogul)

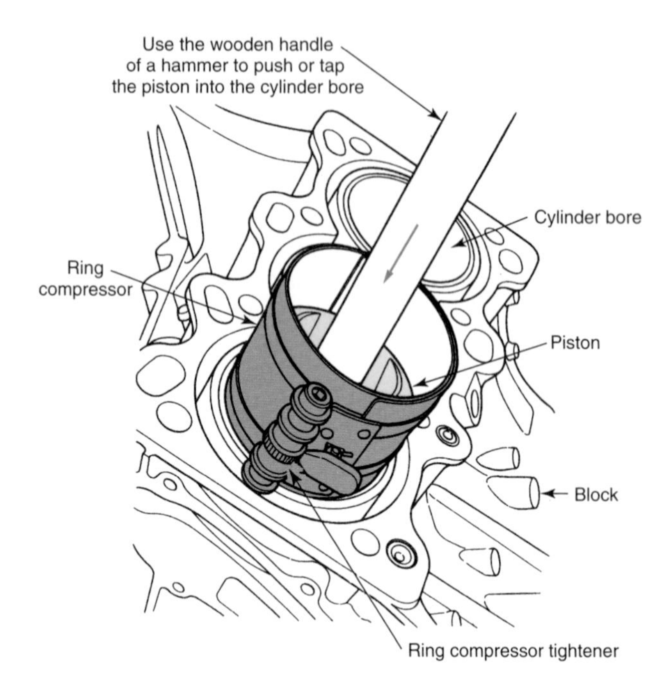

Figure 26-65. When installing a piston with a ring compressor, maintain a downward pressure on the ring compressor to prevent the rings from expanding before they enter the cylinder bore. (Acura)

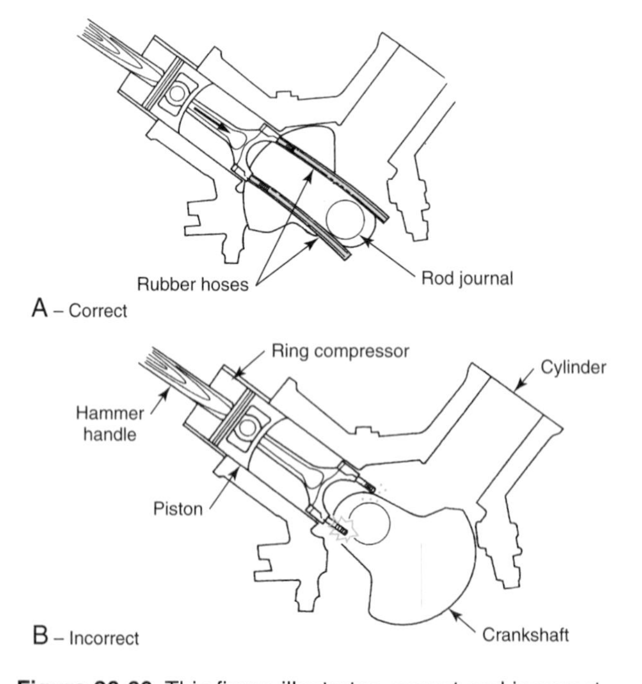

Figure 26-66. This figure illustrates correct and incorrect methods of installing a connecting rod and piston. A—Correct. The rubber hoses guide the bolts past the journal, preventing bearing surface damage. B—Incorrect. The bolts hit the crank journal. (Acura)

If the journal serves a single rod, turn the crankshaft a couple of revolutions to allow the rod to center before the final bearing tightening. If two rods operate on the same journal, install both rods and then revolve the shaft.

Checking Connecting Rod Clearances

After the cap is installed, the rod ***bearing clearance*** must be checked. Rotate the crankshaft to BDC if you are sure the journal is round. If the journal is out-of-round, rotate it downward just far enough to remove the cap. This allows a bearing clearance check more in line with the widest journal diameter. Remove the cap and wipe all oil from the journal and insert. Place a strip of Plastigage across the insert about 0.25″ (6.35 mm) off center. Install the cap and torque it to specs. It is important to make certain the upper rod bearing is held against the top of the journal while torquing. This prevents the lower cap from drawing the rod and piston assembly downward, thus flattening the Plastigage and giving a false reading.

Without turning the crankshaft, remove the cap and check the width of the flattened Plastigage. See **Figure 26-67.** An even width of flattened Plastigage indicates a straight journal. Remove the Plastigage, lubricate the insert, and install cap and torque. Following manufacturer's specifications (0.004″–0.010″ or 0.102 mm–0.254 mm average), check the connecting rod ***side clearance.*** Use a suitable feeler gauge or dial indicator. See **Figure 26-68.** Retorque all rods. Palnuts, if used, should be installed. Turn the crankshaft to make sure all parts are clear and that excessive drag is not present.

Balance Shaft Service

If an engine is equipped with a balance shaft, check the shaft journals for wear. Replace the shaft if the journals are worn excessively. Also check the shaft bearings, when used. If the bearings are worn, they should also be replaced. Some balance shaft journals ride directly on the enclosing parts and do not use bearings. The journal riding surfaces should be carefully checked for wear. If the riding surface is worn excessively, replace the related part.

To reinstall a balance shaft, lubricate the journals and slide the shaft into the engine. Reinstall the drive gear and ensure that the gear is timed to the engine. The gear timing mark should be in a certain position when the engine is at top dead center. Failure to time the shaft gear will result in severe vibration when the engine is started.

Note: On some engines, the balance shafts are located in the cylinder head(s). Service procedures for these shafts are similar to those covered here.

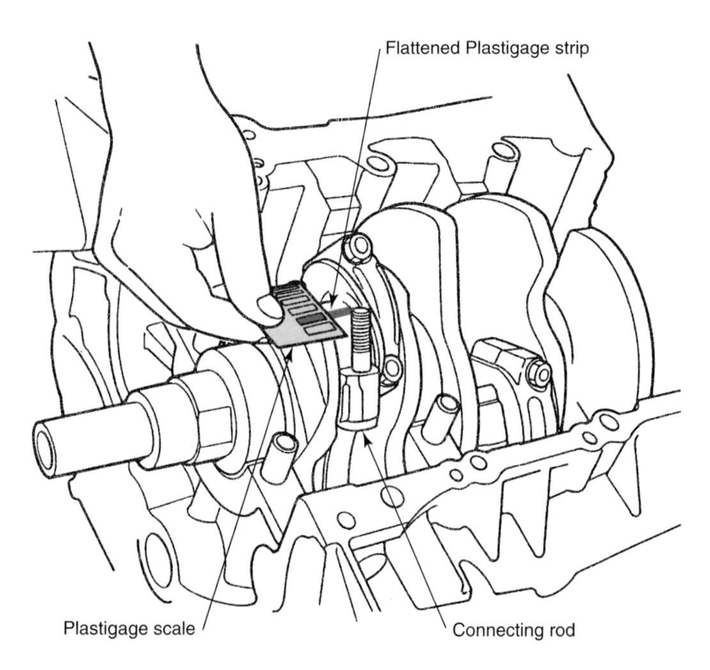

Figure 26-67. The width of flattened Plastigage, measured with a paper scale, indicates the bearing clearance. This engine has a bearing clearance of 0.0009″–0.0018″ (0.022 mm–0.046 mm) and a service limit of 0.002″ (0.05 mm). (Acura)

Oil Pumps

There are two main types of oil pumps, the ***rotor pump*** and the ***gear pump.*** A typical rotor-type oil pump is shown in **Figure 26-69.** This particular pump mounts directly to the engine and is driven by the crankshaft. Note the pump contains an oil pressure relief valve. Check the end clearance of both inner and outer pump rotors by placing a straightedge across the pump body and passing a suitable feeler gauge (0.004″ or 0.10 mm maximum) between the straightedge and rotor surfaces, **Figure 26-70.** The type of gasket, if any, between the pump body and cover must be considered when checking end clearance. The measured clearance between the rotors and the body is actual end clearance. If a thin gasket is used, the thickness of the compressed gasket must be added to the feeler gauge reading.

When the end clearance is excessive, determine if the wear is in the pump body or in the rotors. Measure the length of both the inner and outer rotors. Use manufacturers' wear

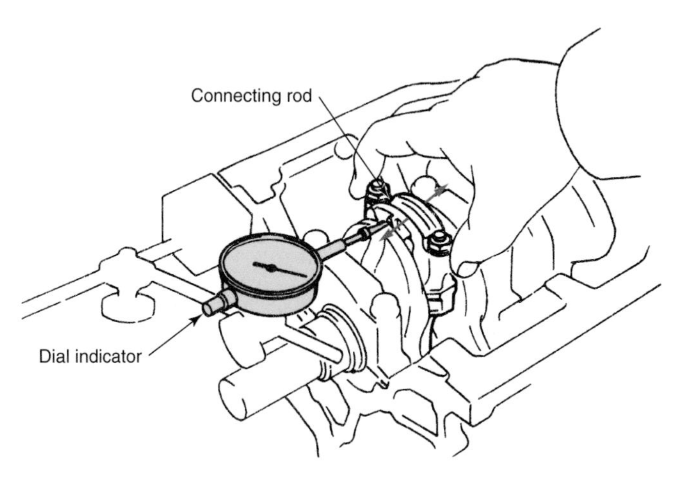

Figure 26-68. Checking the connecting rod's side play (clearance) with a dial indicator. (Toyota)

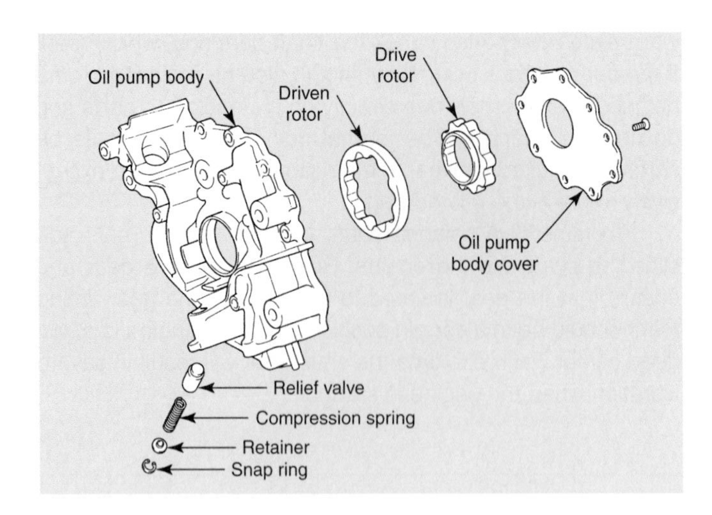

Figure 26-69. The engine crankshaft drives the rotor-type oil pump shown here. (Toyota)

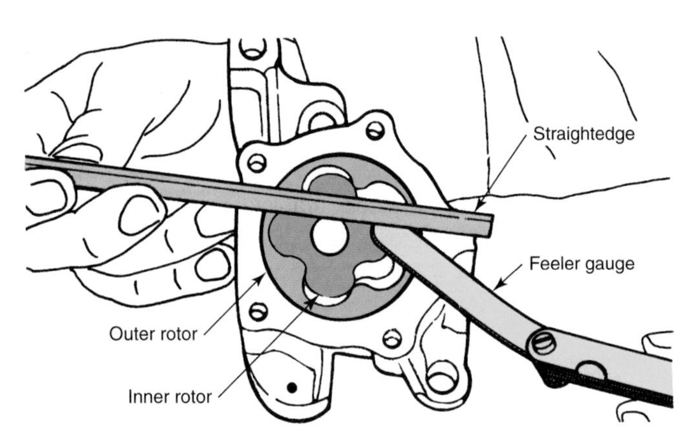

Figure 26-70. A feeler gauge and straightedge are used to check the oil pump rotors' end clearances. (DaimlerChrysler)

limit specifications. See **Figure 26-71.** Use a feeler gauge to check the clearance between the outer rotor and the pump body (0.012″ or 0.30 mm maximum). See **Figure 26-72.** Next, check the tip clearance between the inner and outer rotors (0.010″ or 0.25 mm maximum). See **Figure 26-73.** Place a straightedge on the cover and use a feeler gauge to determine cover wear (0.0015″ or 0.038 mm maximum). See **Figure 26-74.**

The inner rotor shaft-to-body bearing clearance should also be checked (0.001″–0.003″ or 0.025 mm–0.076 mm average range). This can best be done by carefully measuring the shaft and the bearing hole. Inspect the rotors and shaft for scoring, galling, and chipping. Check the pump body for cracks.

Note: Always replace rotors as a pair. Never replace one rotor. Remember that in order to function properly, the rotor pump working clearances between the inner and outer rotor, between the outer rotor and pump body, and between the rotor ends and the pump cover must be within specified limits.

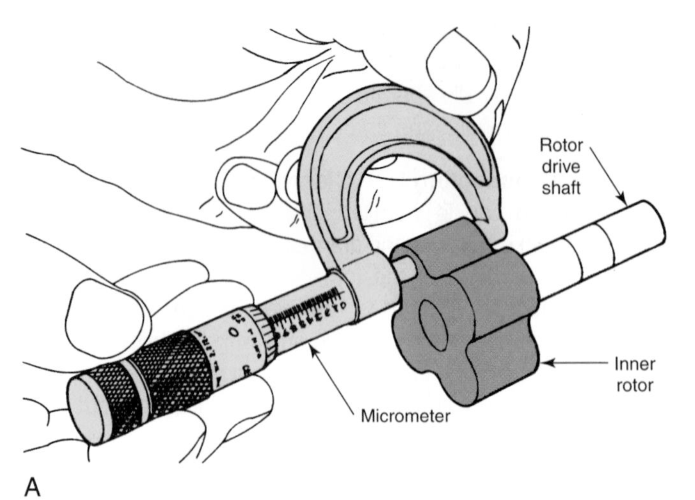

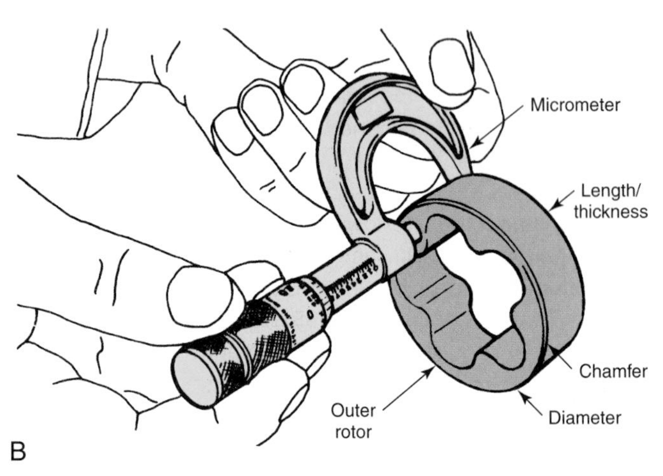

Figure 26-71. A—Use a micrometer to measure the inner rotor length. B—Measure the length of the outer pump rotor with a micrometer. (DaimlerChrysler)

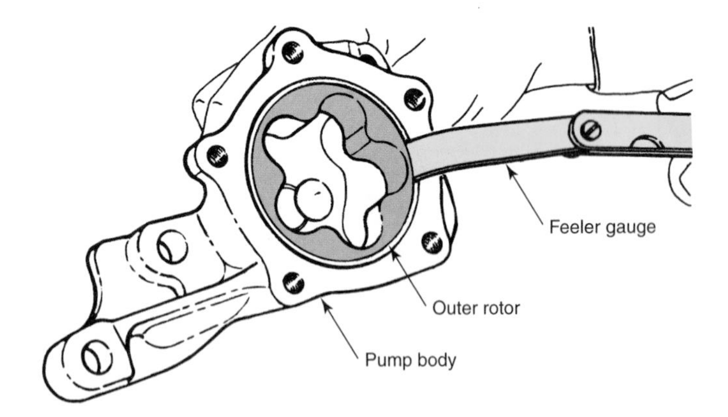

Figure 26-72. Measure the clearance between the pump body and the rotor. If this wear exceeds the manufacturer's specification, the pump assembly must be replaced. (DaimlerChrysler)

Checking Gear-Type Oil Pump

Study the disassembled gear-type oil pump in **Figure 26-75.** This pump also incorporates a pressure relief valve. Use a straightedge and feeler gauge to check the end

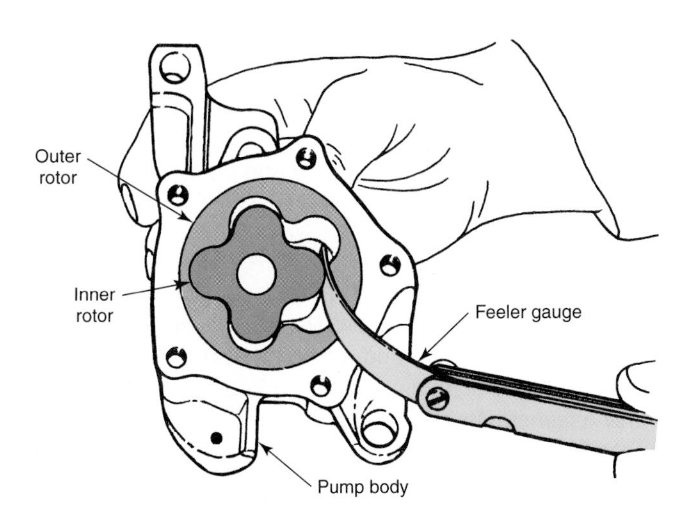

Figure 26-73. Use a feeler gauge to measure the clearance between the widest part of the inner rotor and the narrowest part of the outer rotor. (DaimlerChrysler)

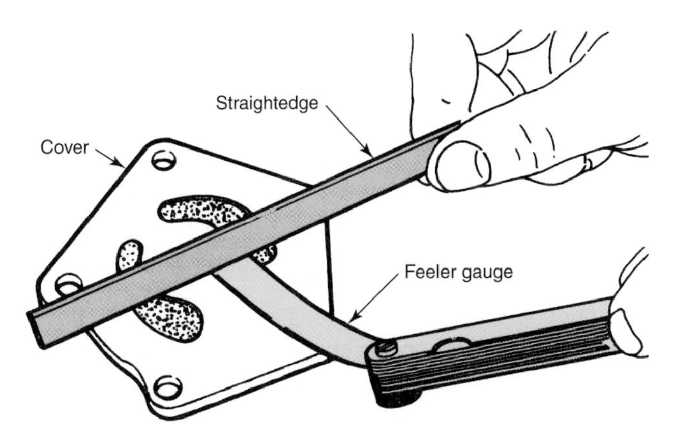

Figure 26-74. Use a straightedge and feeler gauge to check the pump cover for wear. (DaimlerChrysler)

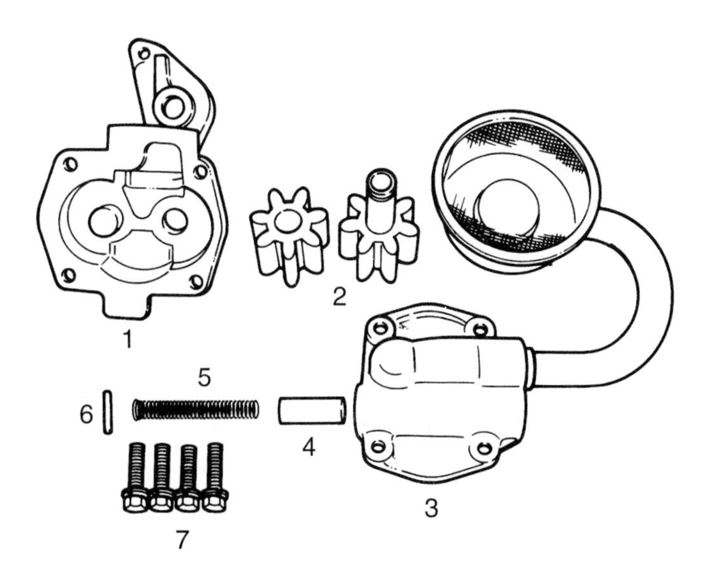

Figure 26-75. A typical gear-type oil pump contains only a few simple components. 1—Pump body. 2—Gears. 3—Cover/pickup assembly. 4—Relief valve. 5—Relief valve spring. 6—Retaining pin. 7—Cover bolts. (General Motors)

clearance between the gears and pump body (0.004″ or 0.10 mm maximum). Remember to consider the gasket when determining clearance. **Figure 26-76** illustrates pump gear end clearance being checked with a straightedge and feeler gauge. Also check the clearance between the pump drive shaft and the bushing installed in the pump body. Rock the pump shaft in the bushing. If the shaft seems loose, measure the clearance with a wire feeler gauge or dial indicator. The clearance should not exceed 0.003″ (0.08 mm).

Use a feeler gauge to check the clearance between the gear teeth (use tip of tooth) and the pump body. See **Figure 26-77.** Check the backlash between the gear teeth

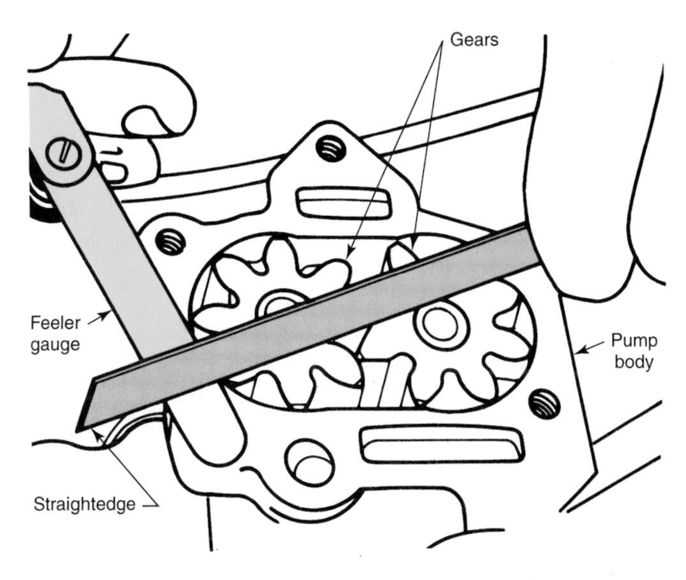

Figure 26-76. Measure the oil pump end clearance with a straightedge and feeler gauge. Some pumps require a gasket to be in position when checking the clearance. (General Motors)

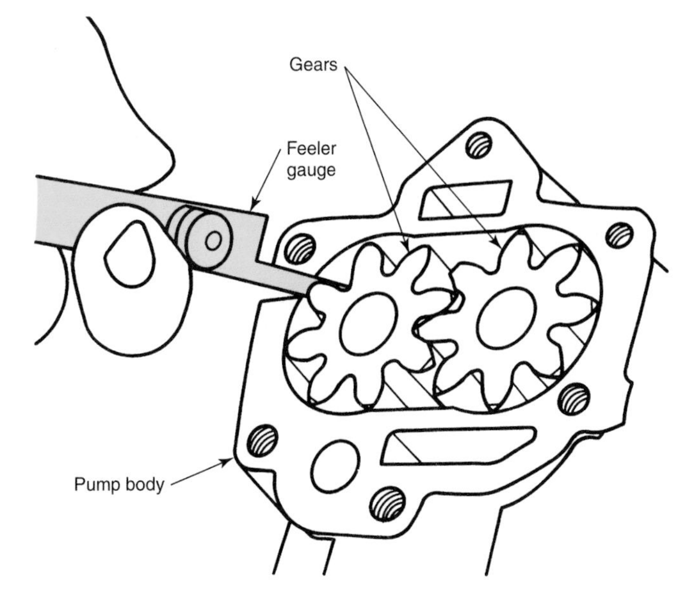

Figure 26-77. Use a feeler gauge to measure the clearance between the tips of the gear teeth and the pump body. (General Motors)

with a feeler gauge. It generally measures between 0.002″–0.008″ (0.05 mm–0.20 mm). If backlash is not within recommended limits, replace both gears as a pair, **Figure 26-78.** Check the pump cover for flatness using a straightedge and feeler gauge as shown in **Figure 26-79.** A wear depth of 0.002″ (0.05 mm) is generally the maximum allowable. If the cover is not within this specification, it can be sanded down on the face with 400 grit or finer sandpaper on a flat surface. Always follow manufacturer's recommendations.

After sanding the cover, polish the surface with crocus cloth to remove scratches and clean it thoroughly. Other gear pump checkpoints include gear tooth wear, scoring, and chipping. Check the pump body and cover for cracks. Also, clean and check the relief valve. End clearance between the gears and the cover plate can be checked by placing a strip of Plastigage across the face of the gears and bolting the cover in place. Without turning the gears, remove the cover and check the flattened Plastigage.

Pump Assembly

All pump parts must be spotlessly clean. Lubricate shafts and gear or rotor. Install parts, gaskets, and tighten all fasteners. Before putting the pump into position, fill the gear or rotor cavities with engine oil. Some manufacturers recommend packing the pump with petroleum jelly (never chassis lube) to make sure it primes. If a drive gear shaft was pinned, make certain a new pin is in place and properly peened. The pump drive shaft should turn freely.

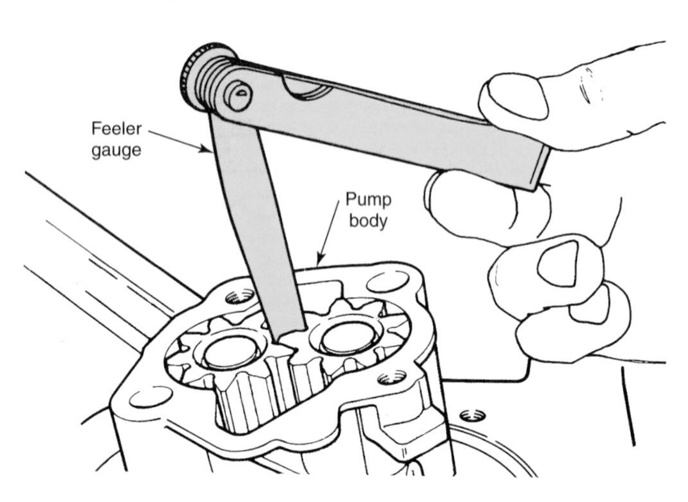

Figure 26-78. Use a feeler gauge to check the backlash between gear teeth. Excessive lash (clearance) lowers the pump output pressure. (DaimlerChrysler)

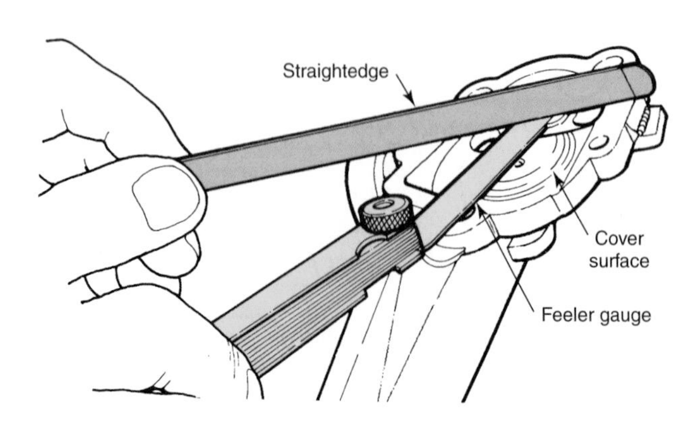

Figure 26-79. Use a steel straightedge and the proper size feeler gauge to check the oil pump cover for flatness. (DaimlerChrysler)

Pump Installation

Install the pump carefully. If ignition timing is affected, it will have to be reset. This is covered in Chapter 16, Ignition System Service. Clean and attach any external lines. Check pickup tube and screen positioning. Make certain the pump is firmly attached. Also check that any gaskets or seal rings are in good condition and properly positioned. See **Figure 26-80.**

Note: Before starting an engine following reconditioning, the lubrication system should be charged, or primed, with oil under pressure. The pressurizing procedure is described in Chapter 28, Engine Assembly, Installation, and Break-In. This procedure assists the pump in priming itself (drawing a vacuum and pulling oil from sump).

A pressure relief valve is incorporated in the lubrication system to limit the maximum pressure. The valve may be part of the pump or can be built into the block. See **Figure 26-81.** The pressure relief valve should be disassembled, cleaned, and checked. Check the spring for free length (the length of the spring when not under pressure). If specified, check the spring's pressure when it is compressed to a specified length. Inspect the fit of the plunger valve in the bore. The plunger and bore must be free of scoring. Crocus cloth may be used to remove carbon and minor scratches from both bore and plunger. The crocus cloth cleans and smoothes the valve and bore while removing very little metal. Afterwards, the valve should be thoroughly rinsed, dried, lubricated, and assembled.

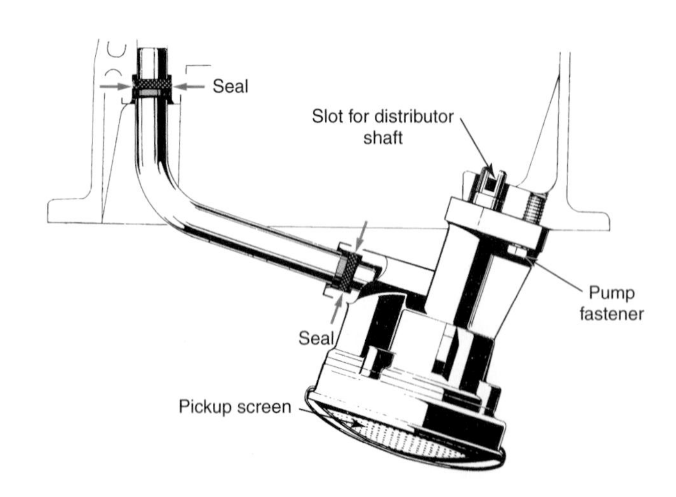

Figure 26-80. Be careful when installing an oil pump. Make certain that all fittings, fasteners, seals, and gaskets are in good condition and properly located. If adjustable, check the pump pickup height. (Volvo)

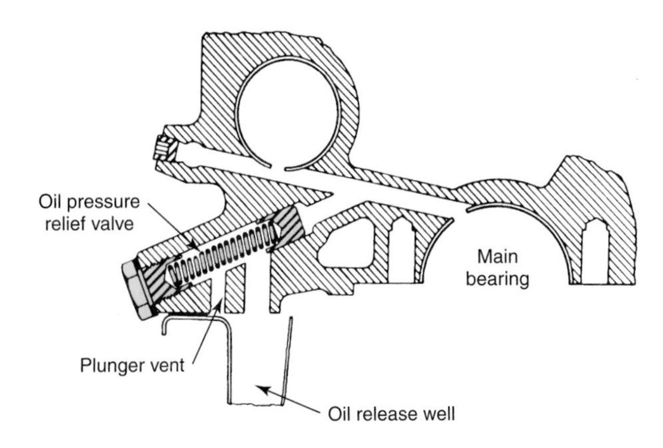

Figure 26-81. This figure shows a pressure relief valve built into the block. (Clevite)

Note: The relief valve's pressure control is altered by any change in the spring length, either by stretching or by adding or subtracting shims. Never stretch the spring. If it does not meet specifications, replace it.

Tech Talk

After boring engine cylinders oversize, you will probably want to know the new engine displacement. To find the displacement of each cylinder, find the area of the cylinder by squaring the bore (multiplying the bore by itself [D^2]) and then multiplying that figure by 0.7854. This gives you the area of the cylinder. Multiply the area of the cylinder by the piston stroke, or travel from TDC to BDC. Multiply this answer by the number of cylinders to get the total engine displacement in cubic inches. The formula is stated below:

$$\text{Total Displacement} = D^2 \times 0.7854 \times \text{Stroke} \times \text{Number of Cylinders}$$

For example, you want to find the displacement of a 318 cubic inch (5.2 liter) V-8 engine that has been bored 0.030″ over. Start by adding 0.030″ to the original bore (3.91″) to get a new bore of 3.94″. The stroke is 3.31″, so you can use the formula:

$$\text{Total Displacement} = D^2 \times 0.7854 \times \text{Stroke} \times \text{Number of Cylinders}$$

$3.94^2 \times .7854 \times 3.31 \times 8 =$

$15.52 \times .7854 \times 3.31 \times 8 = 322.78$ cu. in.

The oversize bore has changed the "318" (5.2 liter) engine into a "323" (5.3 liter) engine.

When the displacement is increased, the compression ratio is also increased. The formula for finding the exact increase in compression involves a complex series of measurements. However, a simple way to determine the approximate increase in compression is to add 2% to the compression ratio for each 0.030″ increase in bore size. To make the calculation, simply add the extra percentage to 100%, which was the original compression ratio, and then multiply the new percentage by the original compression. If you show the percentages as decimals (2% would be 0.02), the formula looks like this:

$$\text{New Compression Percentage} = \left[0.02 \times \left(\frac{\text{bore increase}}{0.030''}\right)\right] + 1.00$$

$$\text{New Compression Ratio} = \text{New Compression Percentage} \times \text{Original Compression Ratio}$$

For instance, if the 318 above had a compression ratio of 9.5 to 1, multiply 9.5 by 102%, or 1.02. This gives an approximate new compression ratio of 9.69 to 1, which can be rounded off to 9.7 to 1. If the bore had been increased by 0.060″, the new compression ratio would be 9.5×1.04, which gives a new compression ratio of 9.88 to 1.

Summary

Cylinder taper up to 0.012″ (0.30 mm) and out-of-roundness not exceeding 0.005″ (0.13 mm) should be considered maximum. Any cylinder wear beyond this point requires reboring. Cylinders should be deglazed to assist ring break-in. Cylinders should be honed with 180 or 220 grit stones to produce a 20–30 microinch finish. When reboring, rebore to the nearest standard oversize. Cylinders should be carefully cleaned with hot soapy water and rinsed with clear, hot water. Dry and oil at once.

Each connecting rod should be numbered on the upper and lower halves on the same side. The cylinder ring ridge must be removed before pulling the piston and rod assembly. Use rod bolt protectors to prevent damage to the crankshaft journals. Each rod must be checked for twist and bend. The large end bore must be round, straight, smooth, and of the correct size. Rods not meeting specifications must be replaced or reconditioned.

Clean pistons by soaking and then scraping them. Do not use a wire brush. Use a suitable ring groove cleaner to remove carbon from the ring grooves. Oil drain holes must be clean and open. Check each piston for wear, scoring, scuffing, and cracks. A worn ring groove can be repaired by cutting it wider and installing a steel spacer.

Proper piston-to-cylinder fit is important. The use of a feeler gauge strip and spring scale is used to check for correct clearance. When measuring pistons, measure across the skirt at right angles to the pin. If replacing one or more pistons, pistons used for replacement should weigh the same as the others in the engine. Piston tops are generally marked so they may be correctly installed. On some engines, pistons are not interchangeable from one cylinder bank to the other.

Before removing piston pins, determine correct relationship between the piston and rod so they are assembled together correctly. Piston pins are retained by snap ring locks, are bolted to the piston or rod, or press-fitted to the rod. New bushings should be burnished before honing to secure them properly. Never fit pins by feel or by using a reamer. A proper fit must be determined by careful measurement. Pinholes must be straight, smooth, round, and in perfect alignment.

Oversize pins should be used to remove excessive clearance from a piston and rod assembly. Honing or diamond boring are recommended. Accurate work when fitting pins is a must.

Following reconditioning, clean piston and rod assemblies in hot, soapy water, rinse, dry, and oil. Install the snap ring end locks with the open ends down. The snap ring lock fits tightly in the groove. When fixing the pin to the rod, center it carefully to prevent scoring the cylinder walls. When handling rods, do not subject them to any twisting or bending forces.

Ring sets are available in rebore (accurate cylinder) and oil control (worn cylinder) types. Rings must be of the correct size and width compatible with the cylinder. Check each ring for proper end gap, groove clearance, and groove depth. Install rings as recommended by the manufacturer. Use a ring expander if needed. Space ring gaps around the piston.

Lubricate piston pin, rings, piston, cylinder, and upper rod insert before installing each rod and piston assembly. Use a ring compressor and tap the piston into the cylinder. Check for proper bearing clearance, lubricate, and torque rod caps. Check for rod bearing and side clearance on the crankshaft journal.

When servicing an oil pump, check clearances and compare them to specifications. When assembling an oil pump, make sure all parts are clean and fill the gear or rotor cavities with engine oil. After assembly, make sure the pump drive shaft turns freely. When installing the pump, be sure to check pickup tube and screen positioning. Also make sure that gaskets and ring seals are in good condition and properly positioned.

Review Questions—Chapter 26

Do not write in this book. Write your answers on a separate sheet of paper.

1. If the ring ridge is not removed _____.
   (A) the rings will not seat properly
   (B) the cylinder walls will be distorted
   (C) the top ring and piston can be broken
   (D) the piston will be hard to install
2. Cylinder wear is greatest at _____.
   (A) the bottom of the ring travel
   (B) the center of the ring travel
   (C) the top of the ring travel
   (D) the ring ridge.
3. When honing worn cylinders, always_____.
   (A) hone at the top
   (B) hone at the bottom
   (C) start honing at bottom then work up the cylinder
   (D) start honing at top then work down into the cylinder
5. If the crankshaft journal is 0.0015″ (0.0381 mm) out-of-round, the rod bore is 0.002″ (0.05 mm) out-of-round, and the bearing is fitted so the minimum clearance is 0.002″ (0.05 mm), what would the maximum clearance be?
6. Heavy score marks on pistons require _____.
   (A) scrapping the pistons
   (B) filing to remove marks
   (C) knurling
   (D) grinding the piston
7. Pistons are often fitted to the cylinders by using a _____ and a spring scale.
8. A typical pin fit in an aluminum piston would have the following clearance_____.
   (A) 0.002″–0.004″ (0.05 mm–0.10 mm)
   (B) 0.0002″–0.0005″ (0.0051 mm–0.0127 mm)
   (C) 0.020″–0.0205″ (0.51 mm–0.5207 mm)
   (D) 0.006″–0.010″ (0.15 mm–0.25 mm)
9. The clearance between the ends of the ring when installed in the cylinder is referred to as _____.
10. Rings should be checked for _____ gap, side _____, and proper size.
11. It is necessary to use a ring _____ to install the piston and rod assemblies.
12. The clearance between a connecting rod bearing and crankshaft journal is best determined using _____.
13. The crankshaft journal can be damaged by the _____ when removing or installing pistons and rods with the crankshaft installed.
    (A) ring ridge
    (B) bearing crush
    (C) bearing cap
    (D) rod bolts
14. The two main types of oil pumps are _____ pumps and _____ pumps.
15. Some manufacturers recommend packing an oil pump with _____ to facilitate priming.
    (A) chassis lube
    (B) petroleum jelly
    (C) high temperature grease
    (D) desiccant

ASE-Type Questions

1. Technician A says that distortion in the main bearing bores can be corrected by installing and align boring undersize semi-finished inserts. Technician B says that distortion in the main bearing bores can be corrected by grinding material from the cap's parting edges and then align boring the caps to their original specifications. Who is right?
   (A) A only.
   (B) B only.
   (C) Both A and B.
   (D) Neither A nor B.

2. Technician A says the ring ridge should not be completely removed. Technician B says that, when removing the ring ridge, one should undercut into the cylinder wall. Who is right?
   (A) A only.
   (B) B only.
   (C) Both A and B.
   (D) Neither A nor B.
3. All of the following statements about cylinder wall taper are true, *except:*
   (A) taper is the difference between the diameter at the top of the cylinder and the diameter at the bottom of the ring travel.
   (B) the ring ridge is still at the original (factory) diameter of the cylinder.
   (C) if the taper is more than 0.006″ (0.15 mm), the cylinder must be rebored.
   (D) the cylinder wears more at the top than the bottom.
4. Technician A says cylinders must be deglazed or honed before installing new piston rings. Technician B says cylinders should not be honed following reboring. Who is right?
   (A) A only.
   (B) B only.
   (C) Both A and B.
   (D) Neither A nor B.
5. Technician A says worn cylinders must be honed until all of the glazed surface is removed. Technician B says a reconditioned cylinder should be cleaned first by wiping with an oily rag. Who is right?
   (A) A only.
   (B) B only.
   (C) Both A and B.
   (D) Neither A nor B.
6. Technician A says rods can become bent or twisted in normal service. Technician B says that bent or twisted rods should be replaced. Who is right?
   (A) A only.
   (B) B only.
   (C) Both A and B.
   (D) Neither A nor B.
7. All of the following statements about worn pistons are true, *except:*
   (A) worn piston pin bores in the rod or piston can be restored by knurling to expand the pin.
   (B) cracked pistons should be discarded.
   (C) scuffing is caused by metal-to-metal contact.
   (D) proper oversize pistons must be used with a rebored cylinder.
8. Technician A says the ends of the oil ring expanders must be lapped over each other. Technician B says the gaps on all rings must be in a straight line. Who is right?
   (A) A only.
   (B) B only.
   (C) Both A and B.
   (D) Neither A nor B.
9. Technician A says that oil control rings can consist of multiple pieces. Technician B says a ring should be removed by pulling on one end of the ring. Who is right?
   (A) A only.
   (B) B only.
   (C) Both A and B.
   (D) Neither A nor B.
10. Technician A says an oil pump pressure relief valve may be adjusted by stretching the control spring to the proper length. Technician B says that after reconditioning an engine, its lubrication system should be primed before starting. Who is right?
   (A) A only.
   (B) B only.
   (C) Both A and B.
   (D) Neither A nor B.

Suggested Activities

1. Obtain a used engine block and measure all of the cylinders for wear. Be sure to record your measurements.
2. Using a flat rate manual, calculate the labor time to bore one cylinder to an oversize. Discuss your price calculation with the members of your class.
3. Using a block that must be bored, set up the boring bar, make the cut(s), and take final measurements. Record your answers and calculate the new displacement of the engine.

Cutaway of an overhead valve engine showing typical valve train components.

27

Valve Train and Intake Service

After studying this chapter, you will be able to:

- Remove, check, and install a block-mounted camshaft.
- Remove, check, and install a cylinder head–mounted camshaft.
- Properly remove and install a vibration damper.
- Remove, check, and install valve lifters.
- Remove, check, and install rocker arms.
- Remove, check, and install push rods.
- Measure timing gear and chain wear.
- Remove and install camshaft gears and sprockets.
- Inspect a camshaft drive mechanism for problems.
- Service a front cover oil seal.
- Remove and install a timing belt.
- Clean and check intake manifolds and plenums.

Technical Terms

Valve timing
Torsional vibration
Interference engine
Noninterference engines
Camshaft
Camshaft bearings
Plastigage
Cam lobe
Thrust
Cam lobe lift
Lifters
Valve lash adjusters
Leakdown rate
Lash adjuster
Rocker arm shaft
Push rods
Crankshaft
Vibration damper
Harmonic balancer
Front cover
Backlash
Thrust plate
Woodruff key
Timing belt
Top dead center (TDC)
Timing chain
Sprocket
Slack
Crankshaft front oil seal
Centering sleeve
Oil nozzles

This chapter covers servicing the engine camshaft, the camshaft drive mechanism, and other associated valve train parts. In addition to the camshaft, valve train parts include valve lifters, push rods, and rocker arms. The information presented in this chapter is applicable to both cam-in-block engines and overhead cam engines. Any difference in service between the two types is described in the text.

Note: This text describes the replacement and timing of the camshaft drive components while the engine is installed in the vehicle. All of the methods described can be used during engine overhaul.

Interference and Noninterference Engines

Camshaft drive service is very important on all engines. Worn gears, chains, belts, and sprockets alter ***valve timing*** and may cause ***torsional vibration*** that can damage the camshaft. Engine overheating, accelerated wear, and sluggish operation can result. Wear makes accurate valve lash settings impossible. Advanced wear causes objectionable noises. Extreme wear will cause the valve to hit the top of the piston. Therefore, gears, chains, belts, and sprockets should receive careful attention.

Because clearance between the valves and pistons is so close, severe engine damage can occur if the timing belt, chain, or gears fail. The type of engine with such close clearance is referred to as an ***interference engine.*** On interference engines, proper valve timing is critical, **Figure 27-1.** If a timing belt, chain, or gear has failed on an interference engine, severe engine damage has probably occurred; in most cases, a complete engine overhaul or replacement will be needed.

Noninterference engines have much more space between the valves and piston heads. If a belt or chain fails on these engines, simply resetting the sprockets to the proper timing marks and installing a new belt or chain is all that is needed.

Camshaft Service

This section explains how to service the ***camshaft.*** Camshaft service is similar for all engines, no matter where the cam is placed or how it is driven.

Checking Journal-to-Bearing Clearance

Worn ***camshaft bearings*** can seriously lower oil pressure and produce excessive throw off, increasing oil consumption. Other vital engine bearings may be starved, and deposits (carbon) in the combustion chambers can build to dangerous levels. Check the bearings for signs of wiping, imbedded dirt, and scoring.

Note: A few overhead cams do not have bearing inserts. The cam journals ride on the head and cap surfaces. If no insert is used, check the riding surfaces of the head and bearing cap for wear or scoring. If damage is found, replace the cap and/or head.

Camshaft-journal-to-bearing clearance can be checked in several ways. Feeler (narrow blade) gauges of varying thickness can be inserted between the camshaft journal and the bearing until the clearance is determined. Do not force the gauge, or you may damage the bearing surface. Maximum clearance should be 0.005″–0.006″ (0.13 mm–0.16 mm).

A dial indicator can be set up so that the clearance (movement) is indicated when the camshaft is forced up and down (right angles to bore).

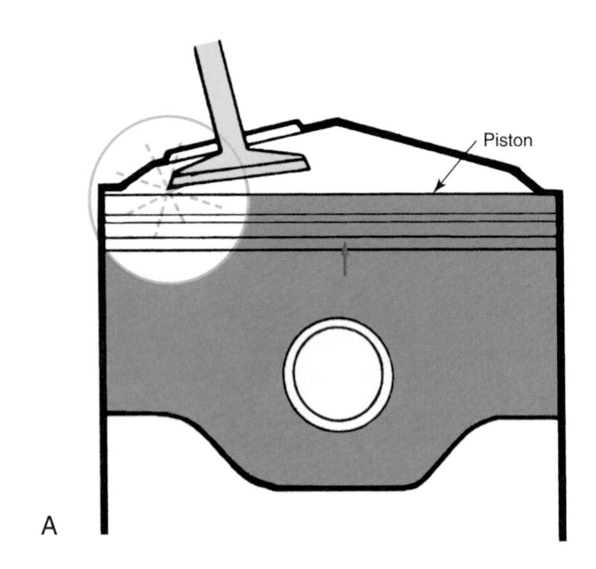

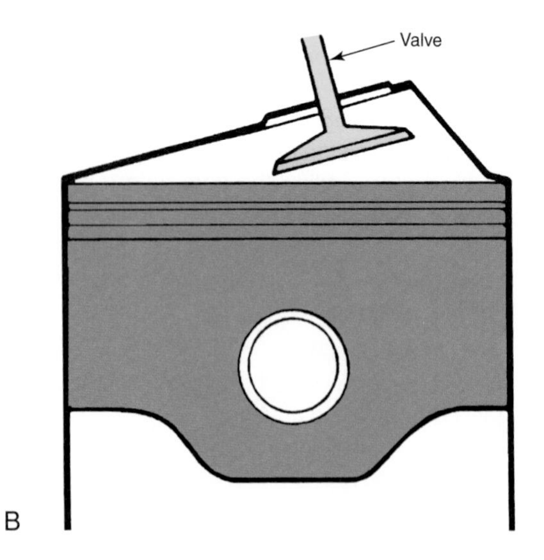

Figure 27-1. Engines with timing belts can be classified as (A) interference or (B) noninterference engines, depending on what happens if the piston and valve synchronization is lost due to a broken or damaged timing belt. Note piston-to-valve contact in A. This can cause severe engine damage.

If the camshaft rides in pressed-in bearings, a hole gauge (camshaft removed) can be used to measure the bearing inside diameter (ID). By subtracting the diameter of the corresponding journal from the bearing ID, an accurate indication of clearance will be given.

Plastigage can be used to check clearance on camshafts with insert-type bearings, as well as those that ride on head and cap surfaces. Place the Plastigage on the camshaft bearing journal, install the bearing cap (and bearing, if applicable), and torque the cap bolts to specifications. Remove the bearing cap carefully and measure the Plastigage. Do not rotate the camshaft with the bearing cap in place. A false reading will result. See **Figure 27-2.**

Inspecting the Camshaft

Clean, rinse, blow dry, and lightly oil the camshaft. Mike each journal in several spots to determine the amount of wear and the extent of out-of-roundness. Overall wear exceeding 0.0015″ per inch (0.038 mm per 25.4 mm) of journal diameter or out-of-roundness beyond 0.001″ (0.025 mm) requires regrinding and the use of undersize inserts. Journals must be smooth.

Check camshaft journal alignment with V-blocks and a dial indicator. Check specs for maximum runout. Depending on design, maximum runout should range from 0.002″–0.005″ (0.05 mm–0.13 mm), depending on design. Refer to **Figure 27-3.**

Inspect each ***cam lobe*** for signs of galling, chipping, or excessive wear, which can reduce valve lift and damage lifters. The wear pattern usually varies in width, being somewhat narrower on the base circle and widening toward the nose of the cam. The pattern may be somewhat off-center, **Figure 27-4.**

The wear pattern should not show across the full width of the lobe. The majority of camshafts are ground so that the

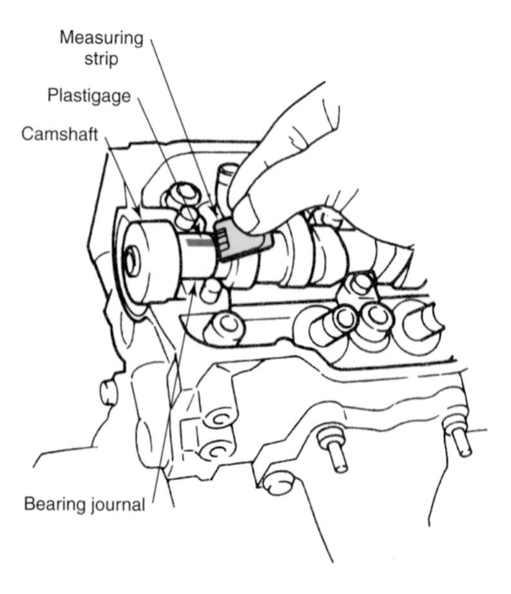

Figure 27-2. Checking camshaft journal bearing clearance with Plastigage. (General Motors)

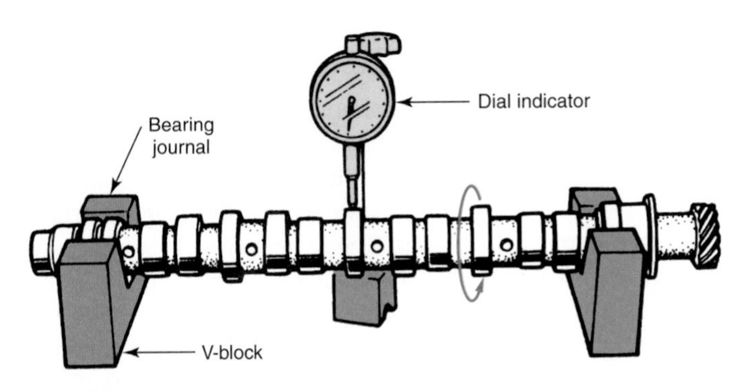

Figure 27-3. Checking camshaft runout. Mount the camshaft in V-blocks and measure runout with a dial indicator. This cam shaft has a runout limit of 0.0039″ (0.10 mm). (Suzuki)

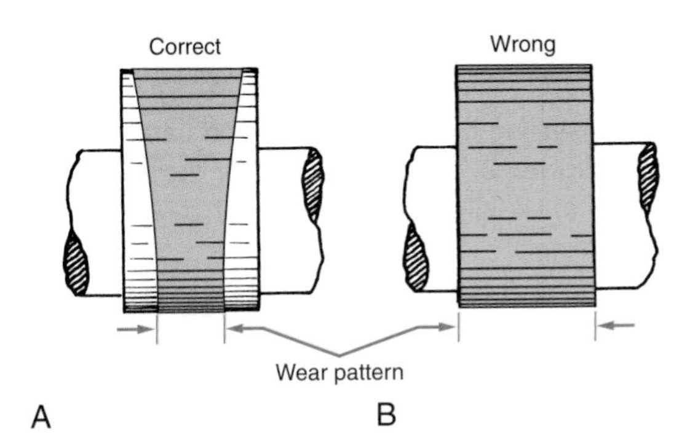

Figure 27-4. Cam lobe wear patterns. A—Correct. B—Incorrect.

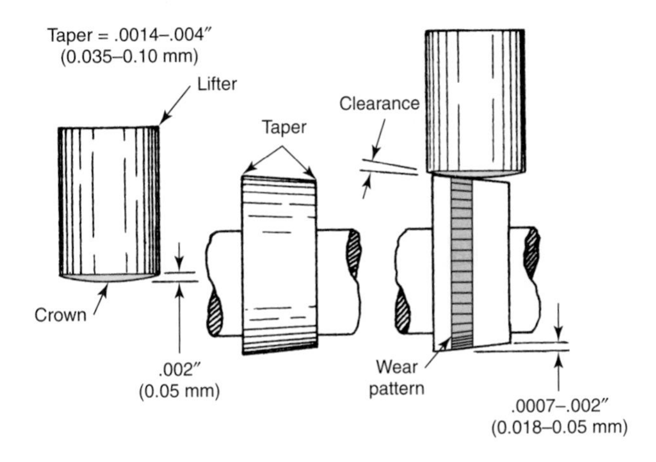

Figure 27-5. Cams are often ground with a taper. The lifter base is crowned.

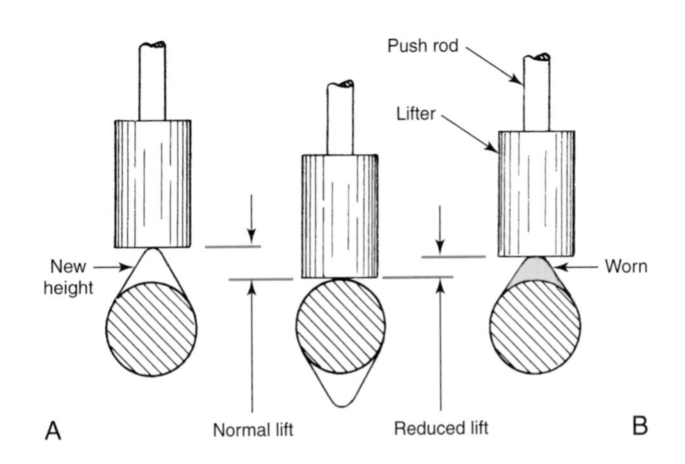

Figure 27-6. Cam wear lowers lift. A—No wear on cam, normal lift. B—Worn cam, reduced lift.

cam lobe surfaces are slightly tapered. Grinding the bottom of the lifter with a slight crown and placing it slightly off-center in relation to the cam tends to cause the lifters to rotate. Also, the loading (pressure) area will not extend to the edge of the lobe, where it could cause damage, **Figure 27-5.**

Camshaft Drive Thrust Direction

During operation, some camshafts tend to exert a ***thrust*** toward the rear of the engine. This type of shaft does not use a bolt-on thrust flange. Camshafts that exert a forward thrust must be restrained. This is the job of the front thrust washer. In addition, some engines use a small spring located between the camshaft gears and the front cover to place a rearward pressure on the gear.

Checking Cam Lobe Lift

When the cam lobes wear, they may not raise the lifters to the specified height. This, in turn, reduces the valve opening, or lift, distance. Refer to **Figure 27-6.** The procedure for measuring ***cam lobe lift*** depends on the engine type. When the camshaft is installed in the block, lift can be checked with a dial indicator. To make this check, the camshaft must be in the block with the lifter and push rod (if used) installed. Mount the dial indicator as shown in **Figure 27-7.** Hold the push rod down and turn the crankshaft until the lifter is riding on the cam base circle. The base circle is the point on the cam lobe directly opposite the maximum lift of the lobe, sometimes called the cam nose. Then zero the dial indicator and turn the crankshaft until the cam nose has pushed the lifter to its maximum height. Do this slowly, so that you can record the highest reading on the indicator as the nose passes under the lifter. The total indicator reading (TIR) will be the amount of lift. Check specifications for maximum lobe lift wear and discard the camshaft if it does not meet the specifications.

On overhead cam engines, camshaft lift is usually measured by checking the lobe lift with a micrometer once the camshaft has been removed from the head. This procedure can also be used to check block-mounted camshafts once they are removed from the block. Refer to **Figure 27-8.** Compare the micrometer reading to the published specifications. If the reading is not within specifications, the camshaft should be replaced.

With the cam out of the engine, lift can be checked with V-blocks and a dial indicator.

Caution: When the camshaft or lifters require replacement, you must replace both. A worn camshaft or lifter will quickly wear the new parts.

Installing the Camshaft

Inspect the cam sprocket before installing the camshaft. In many cases, first installing the sprocket can make camshaft installation much easier.

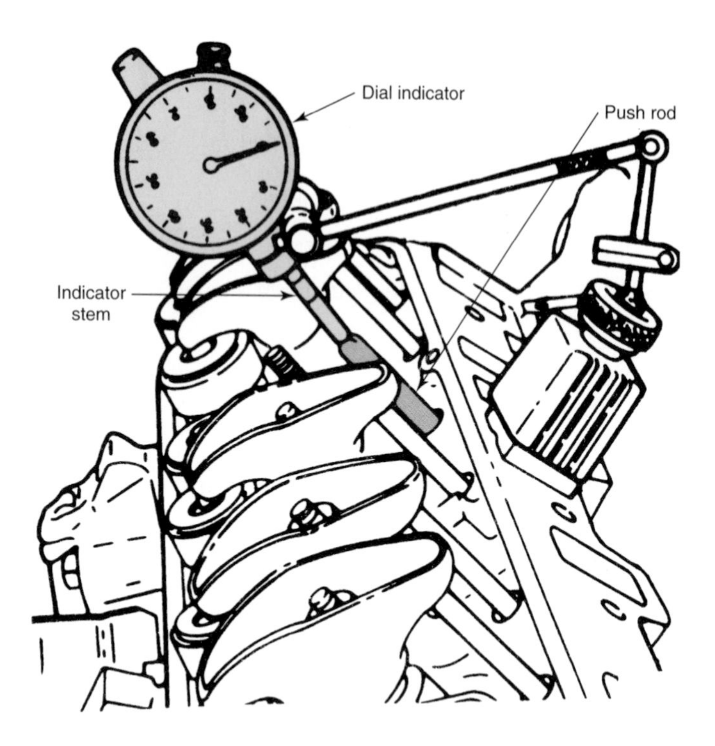

Figure 27-7. Camshaft lobe lift can be measured with a dial indicator. A—Checking lift on overhead cam engine. B—Checking lift on a push rod engine. The indicator stem and push rod must be parallel to obtain the correct measurement. (General Motors)

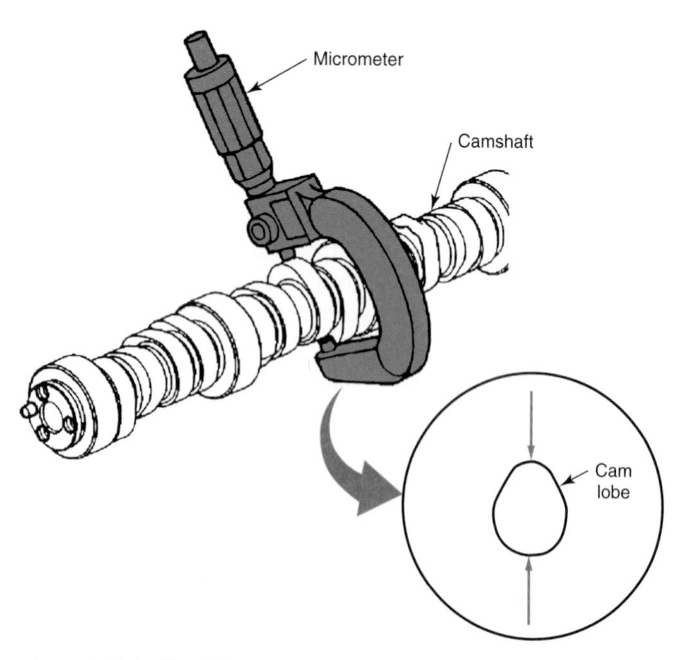

Figure 27-8. Checking camshaft lobe lift. Note that the measurement is taken at the widest section of the cam lobe. (General Motors)

If applicable, always install new cam bearings when the camshaft is replaced. Some engines with overhead camshafts have pressed-in camshaft bearings, **Figure 27-9.** These can be pressed out and new bearings pressed in. Note the position of the oil hole in the front bearing of this photograph. Proper location of the oil hole is critical. Other vehicles have two-piece camshaft bearings. The camshaft bearing caps are removed to allow access to the bearings and the camshaft. The bearings are replaced much like crankshaft main bearings. On some engines, the camshaft rides directly on the camshaft bore and cap, and no bearings are used, **Figure 27-10.**

Note: Replacement of one-piece cam bearings was discussed in Chapter 26, Engine Block, Crankshaft, and Lubrication Service.

After installing the bearings, oil and carefully install the camshaft, **Figure 27-11.** The edges of the cam lobes and gear teeth can easily damage the soft bearing lining. Use extreme care.

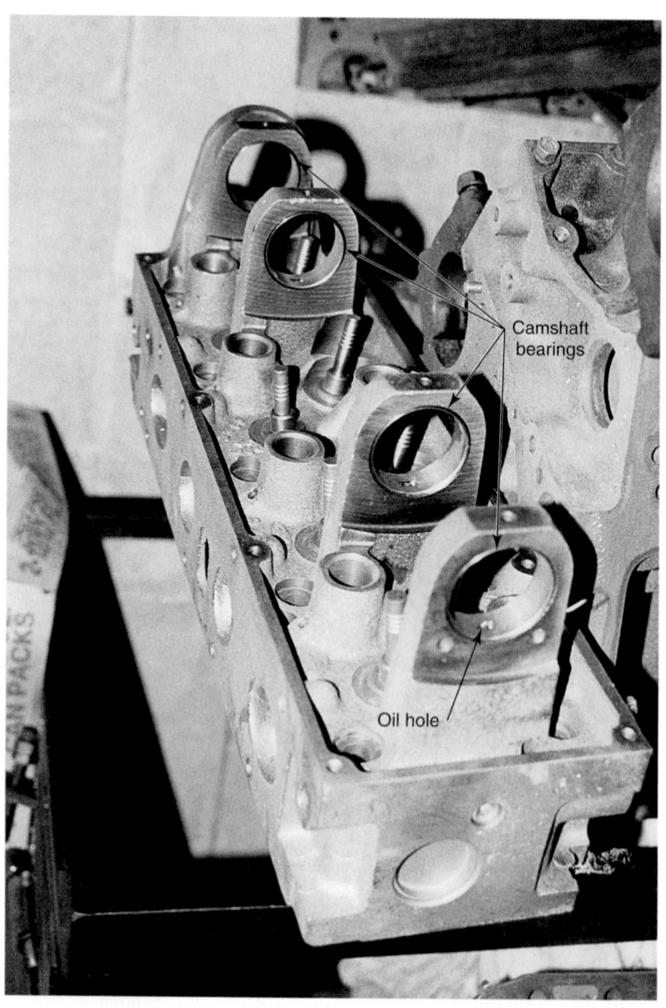

Figure 27-9. This particular cylinder head uses pressed-in camshaft bearings.

If the camshaft rides on two-piece bearings, lubricate and install the lower bearing halves into the bearing saddles in the head. Make sure the bearings are correctly seated in the saddles. Lubricate the camshaft journals and install the camshaft. Next, lubricate the upper bearing halves and install them into the bearing caps. Then install the caps over the camshaft journals. Be sure that the caps are installed in their original positions and are facing in the right direction. If the bore and cap do not use bearings, lubricate them thoroughly before installing the camshaft. Finally, tighten the fasteners to the proper torque.

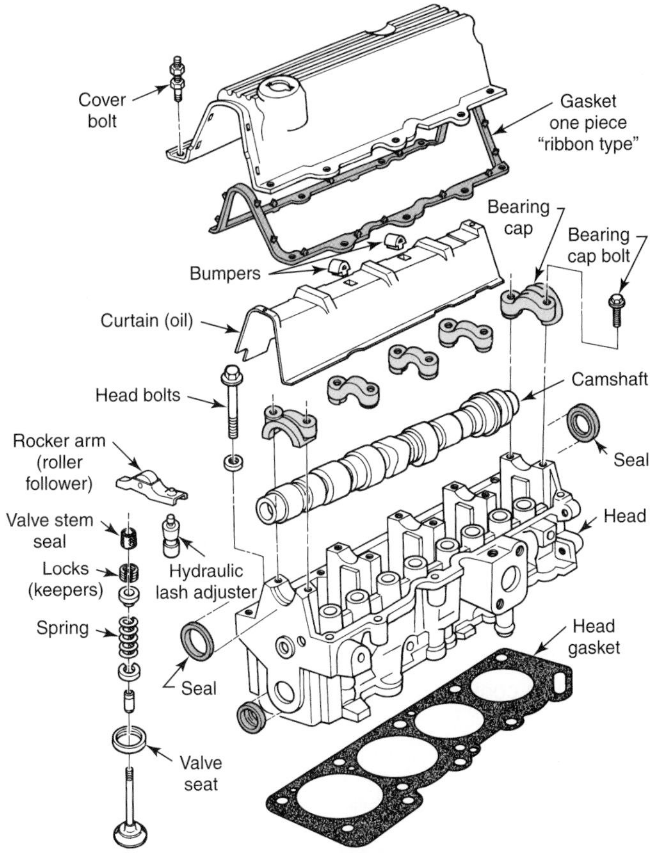

Figure 27-10. One four-cylinder engine using two-piece camshaft bearings. The cylinder head houses the integral bottom-half of the bearings. Removable bearing caps form the top half. Note the use of roller rocker arms. (DaimlerChrysler)

Figure 27-11. When installing a camshaft, be sure to lubricate it and exercise care to avoid damaging the bearing linings. (Federal-Mogul)

Lifter Service

Worn or sticking ***lifters*** can cause missing or noises and contribute to the wear of the other engine parts. Therefore, they should be carefully checked.

Valve lifters can be divided into two main types: hydraulic lifters and mechanical lifters. The hydraulic lifter is by far the most common. The hydraulic valve lifters used on overhead camshaft engines are sometimes called ***valve lash adjusters.*** The basic lifter operating principles remain the same, however.

The portion of the lifter body that protrudes below the guide bore is often coated with gum and varnish. This makes removal difficult unless a special puller is used to grasp the lifter. The tool is engaged, and the lifter is pulled upward with a twisting motion, **Figure 27-12.**

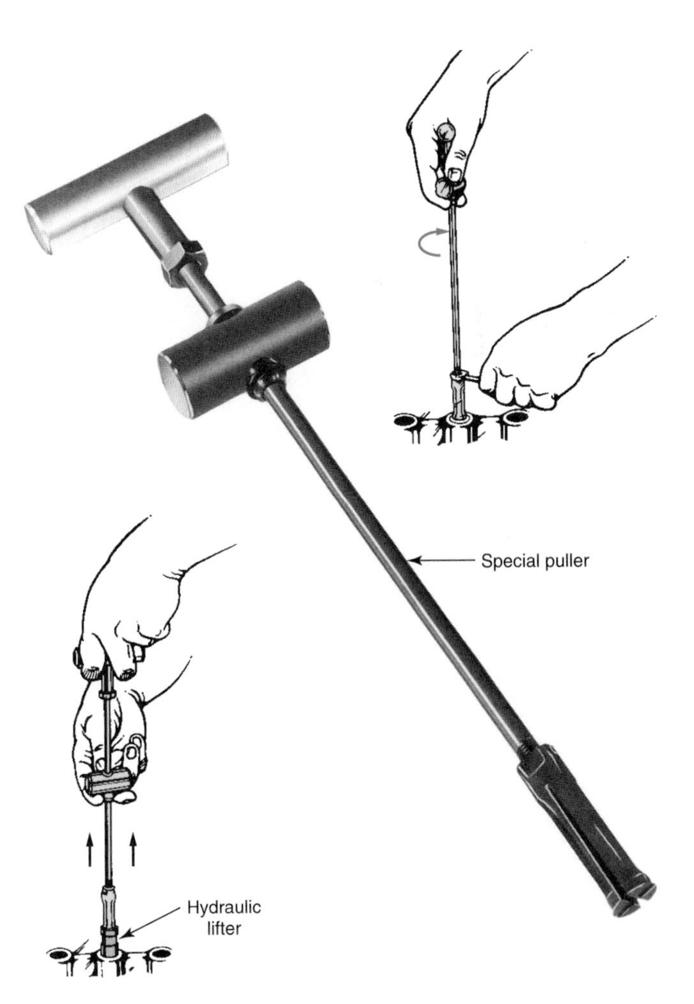

Figure 27-12. Using a special puller to remove a hydraulic lifter. (General Motors)

Servicing Hydraulic Lifters

Hydraulic lifters contain internal parts. While these lifters can be disassembled for cleaning and service, the best solution is to replace any lifter that is worn or shows evidence of leakdown.

Keep Lifters in Order

Each lifter should be placed in a marked holder so that it may be returned to the guide bore from which it was removed. A block of wood with two rows of holes, each row representing one bank of lifters, will do.

Cleaning Lifters

Due to the close working tolerances, lifters must be thoroughly cleaned and assembled in a spotless condition. The slightest trace of grit, dust, or lint will cause faulty operation. Rinse each lifter in clean solvent. You may need to soak the lifters to remove excess gum. Wipe all lifter surfaces with a clean, lint-free cloth. Use a firm wiping action to remove all remaining gum. Finally, blow all parts dry.

Inspecting Lifter Parts

Check the lifter body. It must be smooth and free of scoring. The lifter-to-cam-lobe surface must also be smooth and free of galling, chipping, and excessive wear. A round wear pattern (rotating lifter) or a square wear pattern (nonrotating lifter) is acceptable as long as the pattern is smooth and free of excessive wear.

The outer portion of the lifter body that contacts the lifter guide bore usually shows a distinct wear pattern caused by cam lobe side thrust. It, too, can be considered acceptable unless scoring or pronounced wear is evident. If the seat for the push rod is scored or badly worn, the lifter must be replaced.

Inspect the lifter plunger with a magnifying glass to check for signs of galling. Any scratches on the plunger body that can be felt with the fingernail are cause for rejection. Ignore the slight edge that may occur where the plunger extends beyond the inner working surface of the lifter body. However, if this edge is quite sharp, the plunger must be considered defective.

Checking Leakdown Rate

Each lifter must possess the correct leakdown rate characteristic. ***Leakdown rate*** is the length of time it takes for a specified weight to move the plunger (lifter filled with test fluid) from the top to the bottom. This is usually a specified distance. If a test tool similar to that shown in **Figure 27-13** is available, test the leakdown rate as follows:

1. Raise the weight arm and ram.
2. Place the lifter in the special sleeve inside the test cup. The cup must have sufficient clean test fluid to completely cover the lifter.
3. Lower the ram against the push rod seat.
4. Swing the weight arm down on the ram and depress lifter plunger.
5. Work the weight arm up and down to completely fill the lifter with fluid. After a number of strokes, you will notice a firm resistance on the compression stroke. Give the arm 8–10 additional fast pumps to make certain that all air is expelled.
6. Raise the weight arm and allow the plunger to rise against the stop ring.
7. Place the weight on the ram.
8. Using a watch with a second hand, observe the time the instant the indicator needle begins to move.
9. Give the cup lever a complete turn every two seconds while the plunger is being depressed.
10. When the indicator needle has traveled the prescribed distance, check to see how many seconds have elapsed. See the manufacturer's specifications for the acceptable leakdown rate.

Another leakdown tester is shown in **Figure 27-14.** To use this tester, remove the push rod seat and submerge the lifter in clean solvent. Depress the check valve with a clean, soft rod. This will allow the bottom area to fill. When completely filled, remove the lifter and reinstall the push rod seat. Engage the test pliers as shown and squeeze the handles. The plunger should slowly move downward. If travel is rapid, replace the lifter. Make sure the lifter is completely filled with solvent before testing.

Replacing Parts

Since the cost of new lifters is small compared to that of the possible comeback from premature lifter failure, almost all

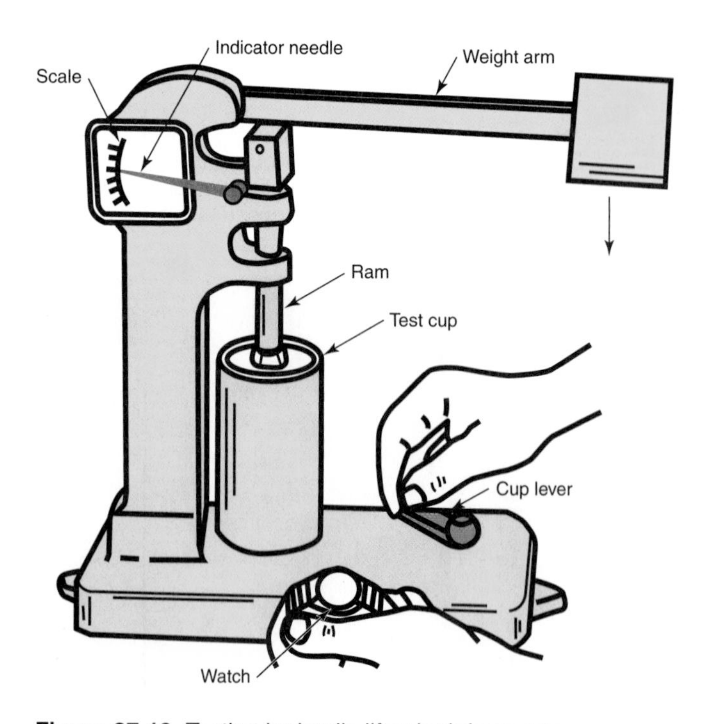

Figure 27-13. Testing hydraulic lifter leakdown rate. (General Motors)

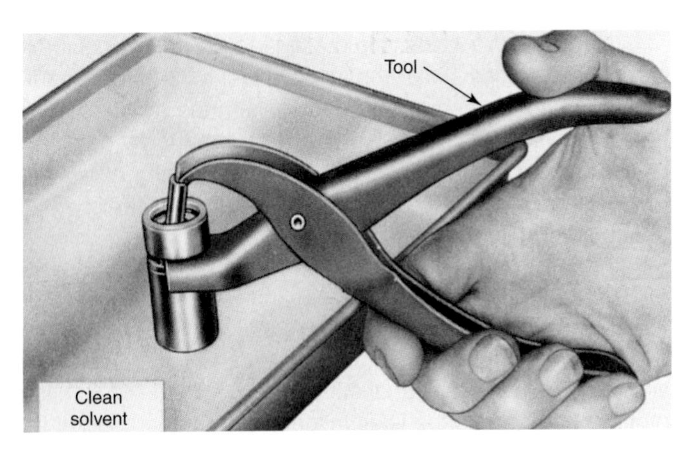

Figure 27-14. Testing leakdown rate with special test pliers. (DaimlerChrysler)

shops discard lifters that are damaged, show wear, or have excessive leakdown. When there is considerable mileage on an engine, most shops will automatically install a new lifter set. Disassembly, cleaning, inspection, reassembly, and testing take time. If the cost of this labor is deducted from the price of new lifters and the increased reliability factor is taken into account, replacement is by far the more desirable option.

Installing Hydraulic Lifters

Before installation, fill the lifters with engine oil. Begin by removing the push rod seat and draining out the solvent. Fill the plunger body with clean oil. Jiggle the check valve open to allow oil to fill the lower compartment. When this compartment is full, fill the plunger body and replace the push rod seat. Lubricate the outside of the lifter body and the lifter guide bore. Rub a small amount of assembly lube or other thick lubricant on both the cam lobe and the push rod ends of the lifter. Install the lifter in the hole from which it was removed.

When lifters have been installed without filling them with oil, the engine RPM should not exceed a fast idle until all lifters are pumped up (filled with oil).

Overhead Camshaft Hydraulic Lifters

Most modern engines with overhead camshafts use hydraulic lifters (followers). The lifters can be incorporated into the overhead valve assembly as shown in **Figure 27-15.** The hydraulic adjusting mechanism is located between the camshaft lobe and the valve stem. To replace the lifters, remove the camshaft. Once the camshaft is out, simply pull the lifters from the head, **Figure 27-16.** Thoroughly lubricate the new lifters, insert them in the head, and then replace the camshaft.

Another type of lifter, sometimes called a ***lash adjuster,*** is shown in **Figure 27-17.** This type incorporates a rocker arm that is directly acted on by the overhead camshaft. One side of the rocker arm opens the valve. The lash adjuster is installed on the opposite side and adjusts the clearance as the engine operates. The internal operation of both of these lifter types is identical to lifters installed in the engine block.

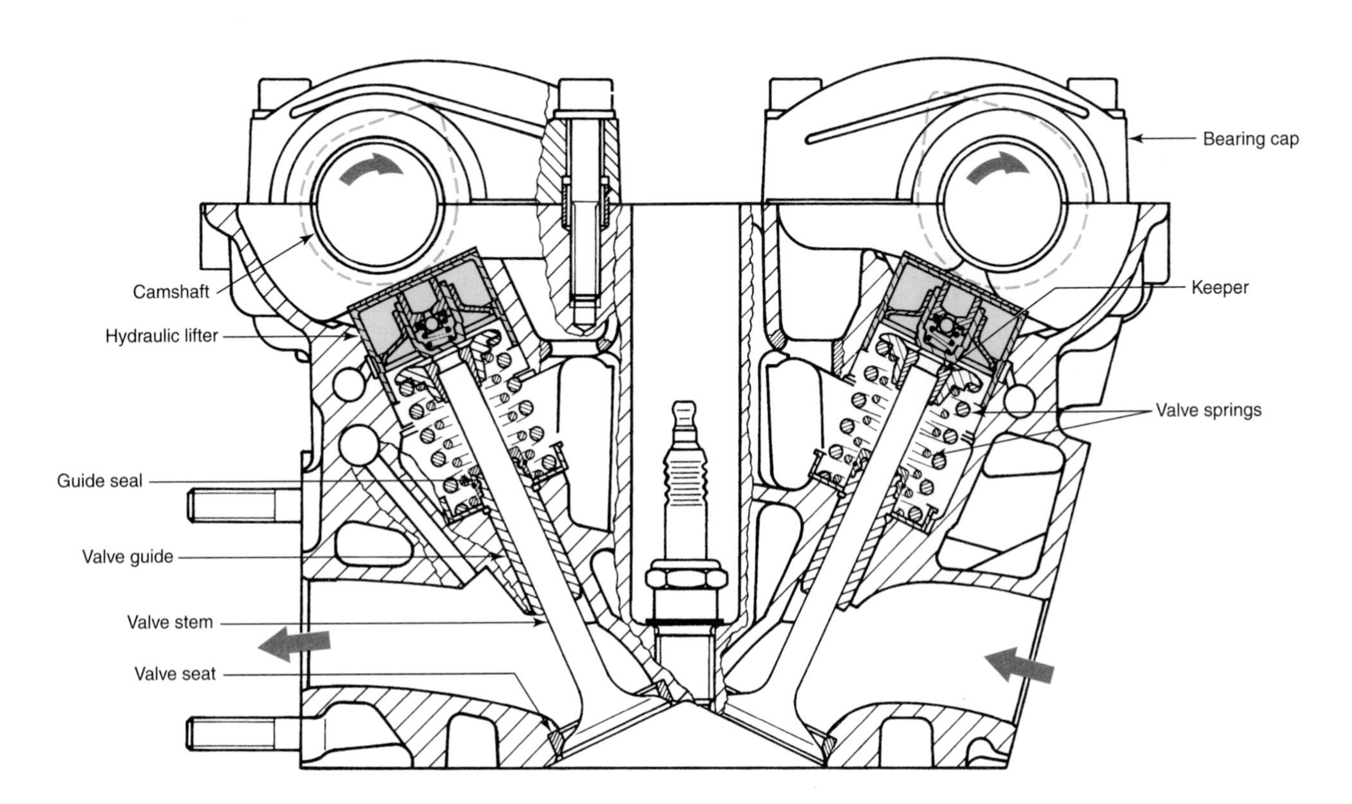

Figure 27-15. An overhead camshaft assembly that uses hydraulic valve lifters. (Mercedes-Benz)

Roller Lifters

Many late-model engines are equipped with roller lifters, which help reduce friction and lessen camshaft and lifter wear. A roller lifter operates in the same manner as a conventional hydraulic lifter with the exception of the roller, which engages the camshaft lobe. Instead of sliding on the cam, the lifter rolls on the cam surface. See **Figure 27-18.** Factory roller lifters are all hydraulic. Some aftermarket high performance roller lifters are solid.

Servicing Roller Lifters

When removing roller lifters, be sure to keep them in proper order. If any part of the lifter needs replacing, discard the entire unit and replace it with a new lifter assembly. Roller lifters are serviced in the same manner as non-roller lifters.

After the lifter has been thoroughly cleaned, check all components for scoring, pitting, galling, varnish buildup, and evidence of other problems. Examine the roller closely. It should be free of pits and roughness, and it must rotate smoothly.

If pitting or roughness is present on the roller, check the corresponding camshaft lobe for damage. If the same condition exists on the camshaft, the camshaft and all lifters must be replaced. Unlike other types of lifters, a single defective roller lifter can be replaced without replacing the camshaft. Always check to be completely sure that the camshaft is undamaged.

Servicing Mechanical Lifters

Mechanical lifters, or tappets, are found on some high-performance engines and some older vehicles. They are simple in construction and operation. Begin the inspection process by cleaning the lifter. Inspect the push rod socket for signs of wear or galling. The lifter-to-camshaft surface should be smooth and free of cam wear, grooving, chipping, and galling. Lifters showing heavy camshaft wear or worn sockets should be replaced. If the wear is minor, the tappet may be resurfaced on the valve-grinding machine. Lifter wear patterns are shown in **Figure 27-19.**

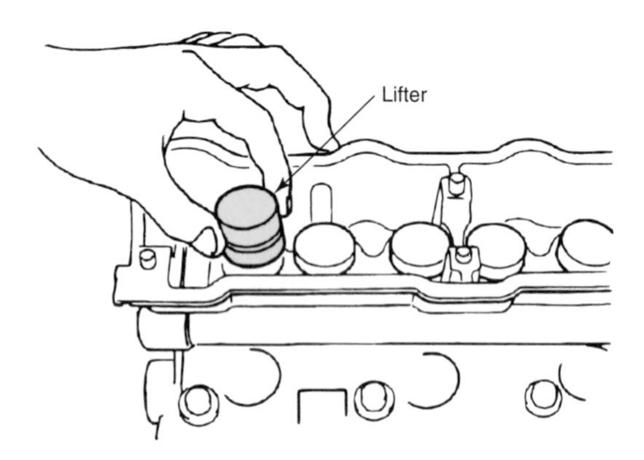

Figure 27-16. Removing a hydraulic lifter from the cylinder head of an overhead cam engine.(General Motors)

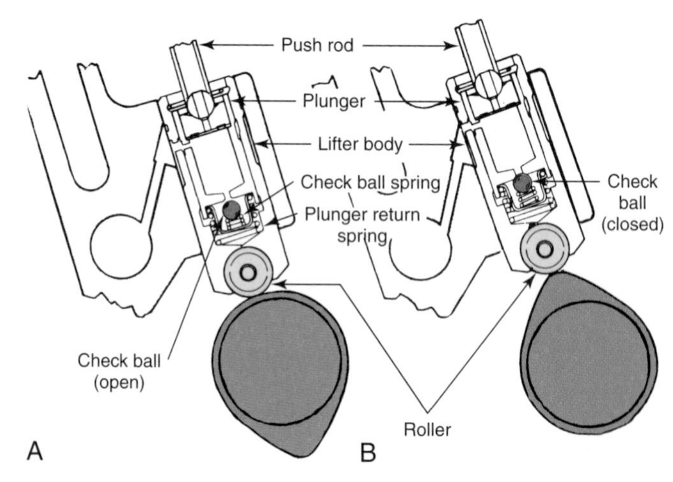

Figure 27-18. Cross section of a hydraulic roller lifter. A—Valve in the closed position. B—Valve open. (General Motors)

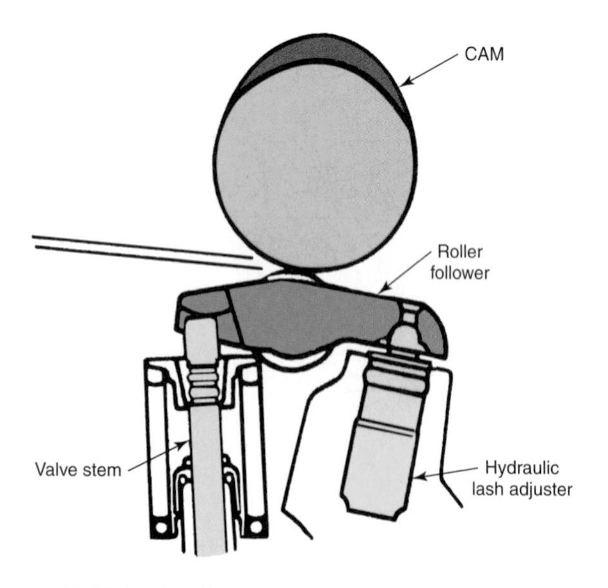

Figure 27-17. Lash adjusters are used on overhead camshaft valve trains. They use the same principle as the hydraulic lifter, but are installed on the rocker arm or on the opposite side of the valve stem. (Ford)

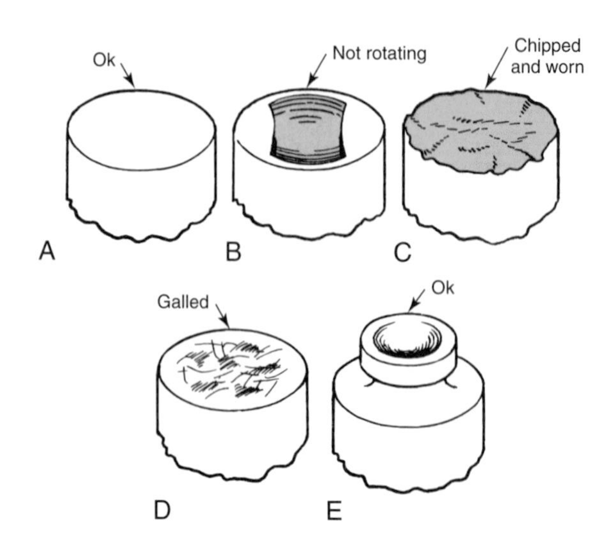

Figure 27-19. Lifter wear patterns. A, B, C, D—Camshaft end of lifter. E—Push rod end of lifter.

Grinding Mechanical Lifters

Before grinding a mechanical lifter, dress the grinding wheel surface. Secure the lifter in the V-block holder. While applying a stream of coolant to the lifter end, advance the lifter against the stone. Cuts should not exceed 0.002″ (0.05 mm). Move the lifter back and forth over the stone surface. Do not remove more stock than is absolutely necessary. At the end of the last cut, continue to move the lifter back and forth until the cutting action stops. This produces a smooth finish. If both ends of the lifter are adaptable to grinding, reverse the lifter and repeat the process.

When lifter wear is pronounced or galling and chipping are present, check the cam lobes carefully for damage. Oversize lifters may be used to correct lifter-to-bore clearance. When clearance exceeds 0.005″–0.006″ (0.13 mm–0.15 mm), replacement is necessary. The bores should be reamed to the exact oversize needed.

Servicing Rocker Arms and Related Components

If the engine uses ball stud–type rocker arms, check the rocker arm studs for wear and replace any that are worn or have damaged threads. Rocker arm stud replacement was discussed in Chapter 25.

If the engine uses rocker arm assemblies, clean each ***rocker arm shaft*** thoroughly. Pay special attention to the hollow center. Examine the shaft for signs of wear and scoring. Replace it if necessary.

Rocker Arm Inspection

Check each rocker arm carefully. Inspect the rocker arm tip that contacts the valve stem. Wear should be even. If deep ridges or uneven wear is evident, it is possible to grind the rocker arm to remove all imperfections. However, most technicians prefer to replace worn rocker arms.

Check the rocker arm surface that contacts the push rod or lash adjuster. If applicable, check the condition of the rocker-arm-to-shaft bearing surface. If bushings are used, wear can be corrected by installing a new bushing and honing it to size. Excessive rocker-arm-to-shaft clearance will permit a heavy flow of oil that can flood valve stems and increase oil consumption.

If the rocker arm is a ball stud–type, check the area that rides against the pivot ball. Also check the pivot balls and adjusting nuts for wear or cracks. Any part showing excessive wear or damage should be replaced.

Inspecting Push Rods

Push rods should be straight and both ends must be smooth. If the push rod is designed to carry oil, be certain to clean the inside and blow it dry. Rod straightness can be checked with V-blocks and a dial indicator, **Figure 27-20.** Maximum allowable runout will vary. Consult the manufacturer's service manual for specifications. A bent push rod should be replaced. Do not attempt to straighten a bent push rod.

Assembling and Installing Rocker Arm Assemblies

After cleaning and inspection, the rocker arms, spacers, springs, and related parts should be lubricated and assembled on the shaft. Be very careful to install the arms in the correct locations and in the right direction. The arms must also be correctly placed in relation to the front of the shaft. **Figure 27-21** shows the installation of rocker arms on a shaft that bolts to the head.

Positioning the Rocker Arm Shaft

The hollow rocker shaft carries a supply of oil to the rockers. Therefore, it is important that the support bracket, which is designed to transfer oil from the cylinder head to the shaft, be properly located. **Figure 27-22** shows one method of carrying oil through the support brackets.

To ensure that the oil supply opening in the shaft indexes with the correct bracket, make sure that the marked end (flat or notch) faces the specified end of the engine. The notch or flat must also be positioned (up, down, to the side) as recommended.

The individual rocker oil passages are generally positioned so they face the head. This provides positive lubrication for the heavily stressed lower rocker bearing area. It also permits less oil flow because of the reduced clearance between the rocker and the bottom of the shaft. If the oil passages were turned upward, an excessive amount of oil would be passed. This would overlubricate the valves, resulting in heavy oil consumption. **Figure 27-23** illustrates the usual positioning of these oil passages. Note that less clearance exists between shaft and the rocker arm at the bottom than at the top.

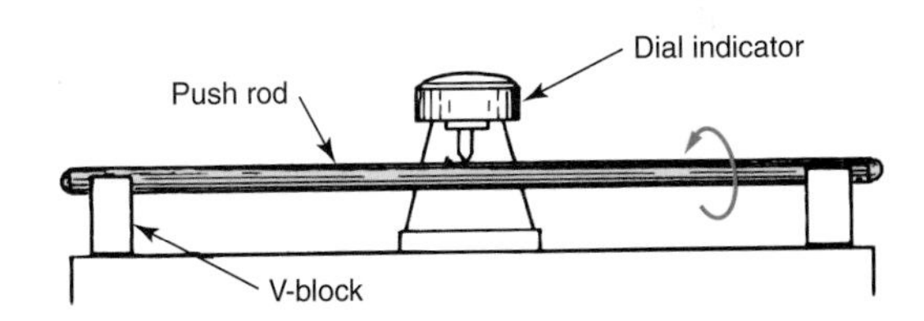

Figure 27-20. Checking push rod straightness with V-blocks and a dial indicator.

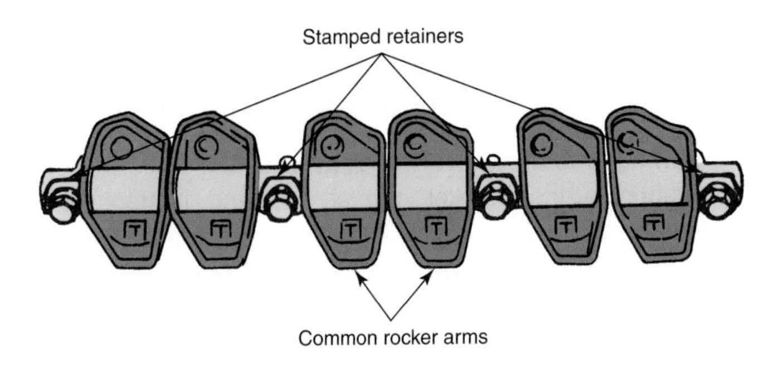

Figure 27-21. Properly installed rocker arms. The shaft is bolted to the integral supports, which are cast with the head. (DaimlerChrysler)

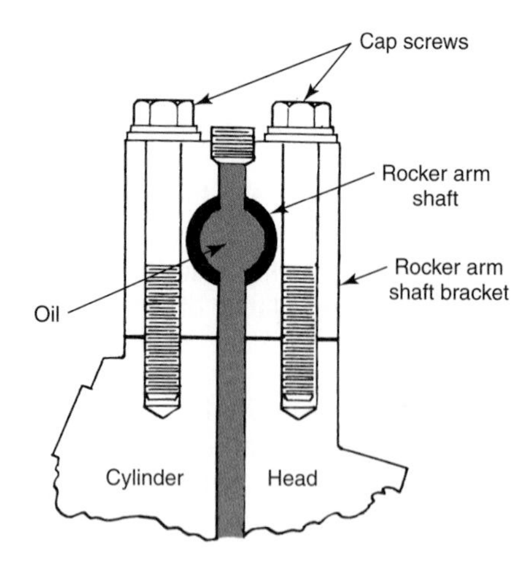

Figure 27-22. One method of furnishing engine oil to the rocker arm shaft via the support bracket.

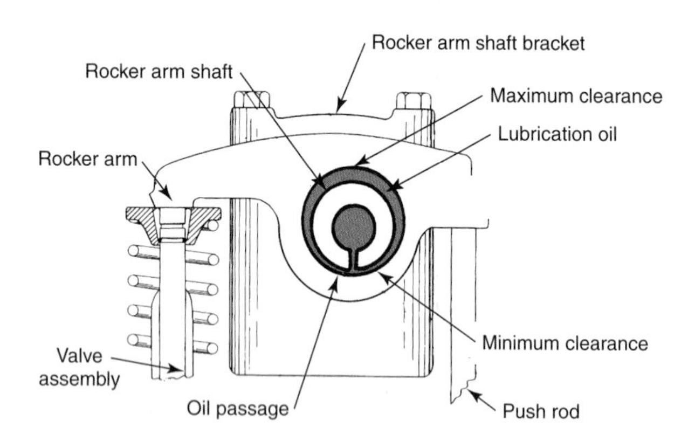

Figure 27-23. Rocker arm oil passages in shaft generally face toward the head.

The individual ball stud rocker arms are lubricated by a metered flow of oil delivered through hollow push rods. Shaft-mounted rocker arms are drilled in various ways to facilitate the flow of oil to both valve stem and push rod ends.

Installing the Rocker Arm Assembly

On some engines, the push rods are installed before the rocker arm assembly. On others, the rocker assembly is installed first. The push rods are placed in the lifters, and the valve springs are compressed. This tips the rocker high enough to place the push rod under the rocker ball end. A small amount of Lubriplate or other suitable lubricant should be applied to each end of the push rod before installation.

Tightening Rocker Shaft Brackets

Lubricate the bracket cap screws and tighten them finger-tight. Give each bracket bolt several turns. Proceed slowly and evenly to prevent damage; if the hydraulic lifters are filled with oil and the shaft assembly is drawn rapidly against the head, the result can be bent push rods, bulged lifters, warped valve stems, and sprung rockers. By drawing the assembly down slowly, the lifters have time to leak down without undue strain on the various parts.

If rocker arm–mounted valve-adjusting screws are used, they should be backed off before tightening the assembly. This also applies to conventional lifter setups. When the brackets are snugged against the head, torque them to specifications, **Figure 27-24.** If an oil overflow line is incorporated in the rocker assembly, make sure it is installed properly.

Installing Ball Stud–Type Rocker Arms

When installing ball stud–type rocker arms, place the rocker arms over the studs in the head(s). Lubricate the pivot balls and place them over the arms. Install the adjusting nuts and draw them down over the pivot balls. While tightening the nuts, make sure that the rocker arms are correctly positioned to contact the push rods and valve stems. Do not tighten the rocker arms down completely. Leave the rocker arms nuts loose by several turns to allow the engine to start. Final adjustment will be made with the engine running.

Servicing the Camshaft Drive Components

The ***crankshaft*** drives the camshaft by one of three methods: two mating gears, a timing chain and sprockets, or a timing belt and sprockets. This section covers the service of the timing gears, chains, belts, and sprockets. Although these parts have a relatively long service life, it is an extremely poor practice to neglect them.

Camshaft Drive Service with Engine in Vehicle

You will be often called on to service a timing belt, chain, or gear set without removing the engine from the vehicle. This next section covers procedures for in-car camshaft drive service. Many of the procedures are similar to those used in engine rebuilding.

Removing the Vibration Damper

The first step in servicing the camshaft drive components is to remove the ***vibration damper,*** or ***harmonic balancer,***

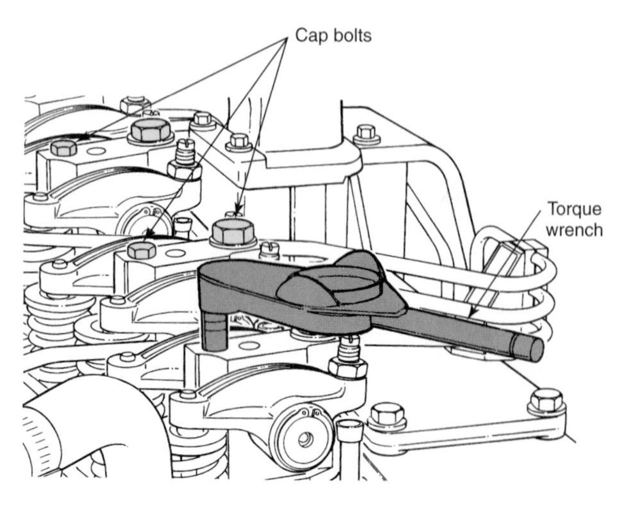

Figure 27-24. Torquing the rocker arm shaft retaining cap bolts.

Figure 27-25. To remove the vibration damper, first unscrew the retaining cap screw in the end of the crankshaft (if used). Remove any bolt-on pulleys. Attach a suitable puller to the damper *hub* (do not pull on the outer rim) and withdraw the damper, Figure 27-25A.

When tightening the puller screw against the hollow end of the crankshaft, be careful not to damage the crankshaft threads. If the threaded hole in the crankshaft is larger than the puller screw, place a cap over the end of the crankshaft, Figure 27-25B**.** Never allow the puller screw to enter the threaded hole in the crankshaft.

Note: Some vibration dampers are a slip fit with the crankshaft and can be pulled off once the crankshaft bolt is removed.

Inspecting the Vibration Damper

Most vibration dampers consist of an outer inertial weight and a hub that fits over the crankshaft. A rubber insert is installed between the weight and hub. After removing the damper, check for deterioration of the rubber insert. Also make sure that the inertial weight has not slipped in relation to the hub. The rubber may be deformed or twisted by slippage between the hub and weight. Sometimes the hub and inertial weight will be out of alignment.

Check the hub and weight for chips, cracks, dents, or other damage. Carefully inspect the keyway in the hub to ensure that it is not worn. When applicable, inspect the surface of the vibration damper that contacts the front cover seal. If you find any damage, replace the vibration damper.

Removing the Front Cover

To dismount the ***front cover,*** remove the screws holding the cover to the engine block or the oil pan. Watch for variations in cap screw lengths. If possible, leave the water pump in place. Once the cap screws have been removed, gently pry on the cover to break the gasket seal. Then, pull the cover away from the engine.

Caution: Some front covers contain the oil pump gears. Consult the service manual before removing this type of cover.

Servicing the Timing Gears

The following procedure applies to engines using two gears to drive the camshaft. Visually inspect both camshaft and crankshaft gears. If any teeth are obviously worn, chipped, or galled, replace both gears. If the gears appear to be in good condition, check the ***backlash*** (distance one gear can rotate without moving the other gear). Backlash can be checked with a dial indicator or with a feeler gauge, **Figure 27-26.** Always consult the manufacturer's service manual for exact procedures.

Caution: If any one of the camshaft drive parts requires replacement, all of the parts should be replaced. A worn gear or chain will not engage the new parts perfectly and will cause wear and premature failure.

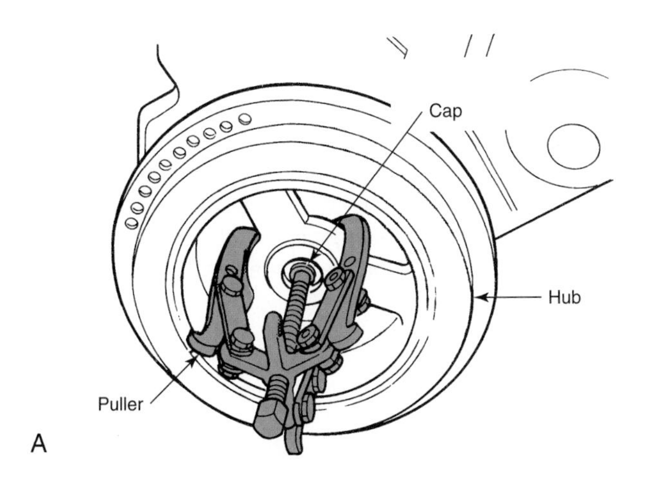

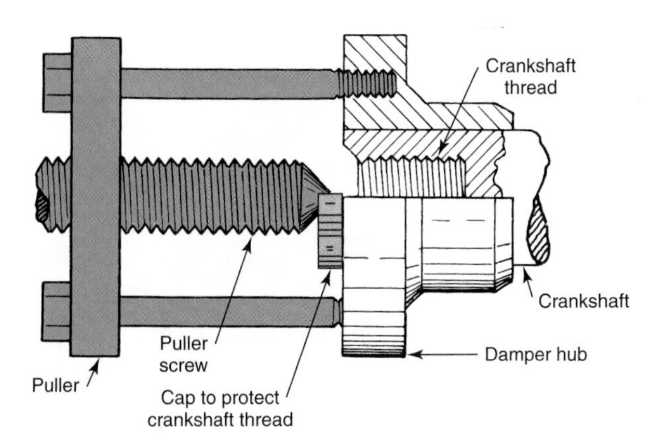

Figure 27-25. Pulling a vibration damper. A—The pulling force is applied to the center of the hub. B—If necessary, a cap can be used to protect the crankshaft threads.

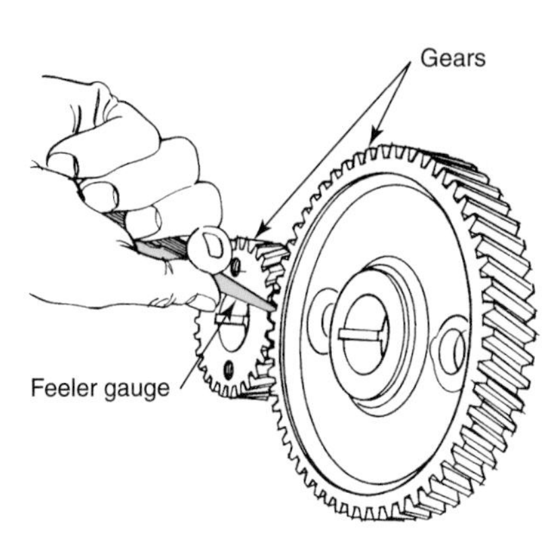

Figure 27-26. Feeler gauge being used to check backlash between the gear teeth. Check backlash in several different spots. (Sealed Power)

Removing the Camshaft Gear

Some camshaft gears are bolted in place, making removal simple. Others, however, are force-fit on the camshaft and must be pressed off. Before removing the camshaft gear, the camshaft must often be removed from the engine. To remove the camshaft, turn the cam gear so the cap screws retaining the ***thrust plate*** (when used) are accessible. Remove the cap screws, **Figure 27-27.**

Note: If the gear is made of plastic or composite material, it may be possible to drill a small hole and then use a punch to split the gear. This can ease removal without requiring removal of the camshaft from the engine.

Before removing the camshaft, remove the distributor and mechanical fuel pump if used, and the oil pump if it is geared to the camshaft. Also remove the valve lifters. Then, pull the camshaft from the engine. Pull the camshaft straight out to avoid damaging the camshaft bearings. After removing the camshaft, the gear can be pressed from the shaft. The camshaft must be held so it will not drop and be damaged.

Installing a New Camshaft Gear

Before installing a new camshaft gear, thoroughly clean the gear engagement area and lubricate it lightly. Slide a new spacer and thrust plate into place. Be sure to face them in the correct direction. Install a new ***Woodruff key.*** Set up the camshaft in a press (supporting the shaft under the front bearing journal edge) and press the timing gear on fully. Remember to apply pressure to the steel hub only.

Note: Camshaft alignment and condition checks are discussed later in this chapter. These checks should be done before installing the camshaft.

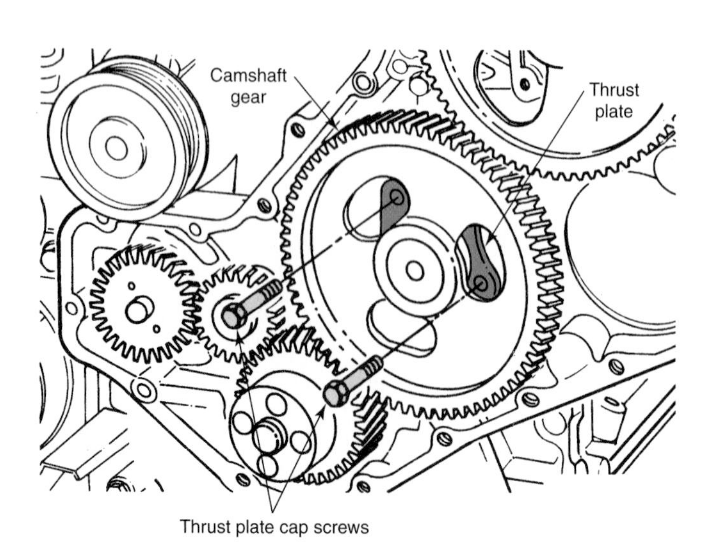

Figure 27-27. Removing camshaft thrust plate cap screws.

Checking End Play at Thrust Plate

After the camshaft gear is pressed on, use a feeler gauge or dial indicator to check clearance (end play) between the thrust plate and the face of the camshaft journal. See **Figure 27-28.**

Removing and Replacing the Crankshaft Timing Gear

If the crankshaft gear is to be replaced, pull it from the crankshaft before installing the camshaft gear, **Figure 27-29.** If there are no threaded holes in the gear for puller bolts, use a puller setup similar to the one shown in Figure 27-29A**.** Be careful not to damage the gear teeth during pulling.

To replace the gear, clean and lubricate the end of the crankshaft. Make certain the Woodruff key is in place. The forward key often has a tapered end to aid in alignment of the crankshaft gear and vibration damper. Make sure the tapered end faces forward.

Note: Reinstall the crankshaft gear before the camshaft and gear are installed.

Install the crankshaft gear with the timing mark facing outward. Drive the gear fully into place with a suitable sleeve and a heavy hammer, Figure 27-29B. Do not mar the gear's teeth.

Installing Camshaft and Gear

Before installing the camshaft, turn the crankshaft so the timing mark faces the center of the front cam bearing. Lubricate the cam bearing journals and lobes. Slide the camshaft carefully into position, making certain the camshaft gear timing mark is aligned with the crank gear mark, **Figure 27-30.**

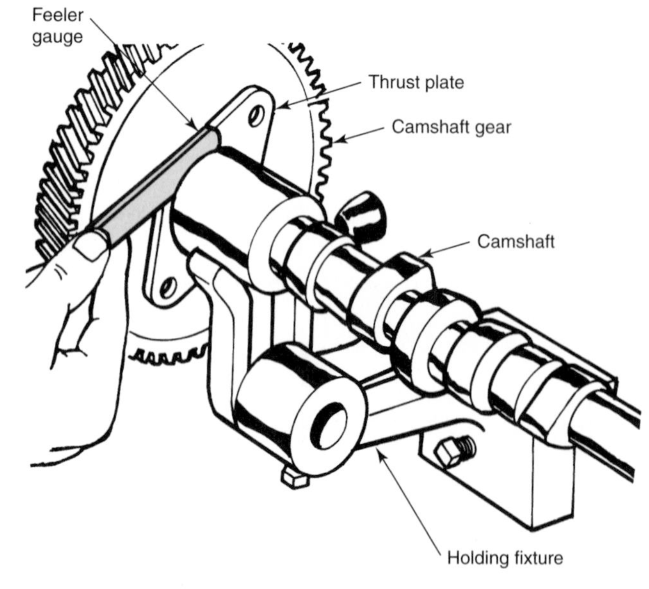

Figure 27-28. Using a feeler gauge to check end play between the thrust plate and the face of the camshaft journal.

Install the thrust plate cap screws (using new locks) and torque them to specifications. Check for correct backlash and runout values and then lube the gears thoroughly. The gears must be properly aligned with each other so that teeth are engaged across full gear width.

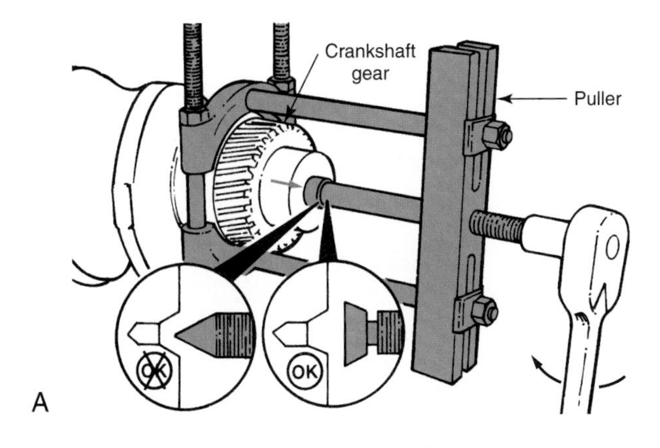

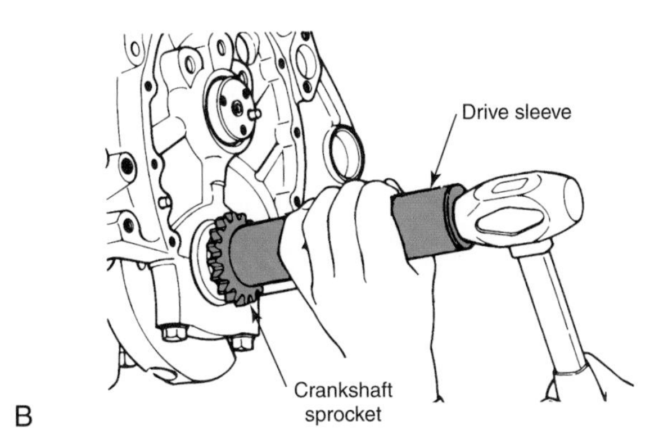

Figure 27-29. Replacing a crankshaft timing gear. A—Crankshaft gear being removed with a puller. B—Driving crankshaft sprocket into place. (DaimlerChrysler and General Motors)

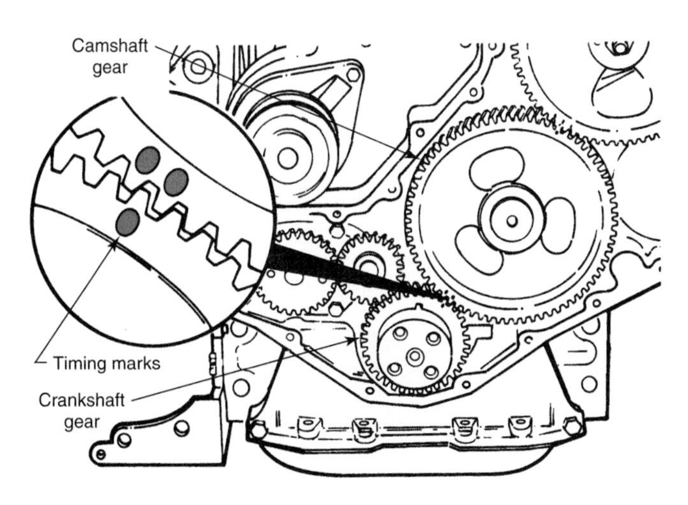

Figure 27-30. Timing marks on the camshaft and crankshaft gears must be correctly aligned. (DaimlerChrysler)

Servicing the Timing Belt and Sprocket

Some engines use a ribbed, fiberglass-reinforced, rubber ***timing belt*** to drive the camshaft, fuel pump, oil pump, and accessory shaft. See **Figure 27-31.** The belt itself is generally driven with sprockets made from aluminum or sintered iron (metal formed by heating and compressing powdered iron and graphite into a solid piece).

Caution: On some OHC engines, you must secure the camshaft(s) using a special tool or bolts before removing the belt. Failure to do this can result in engine damage. Check the service manual.

Removing the Timing Belt

The following is a basic procedure for timing belt removal. Follow the manufacturer's recommendations for the specific engine you are working on.

1. Remove the water pump, air pump, and other belts.
2. Remove the pulley(s).
3. Remove the timing belt cover.
4. Line up the timing marks.
5. Loosen the belt tensioner.
6. Remove the timing belt.

Belt Inspection and Precautions

If a timing belt is to be reused, examine it thoroughly for damage. When examining the belt, do not twist or bend it into

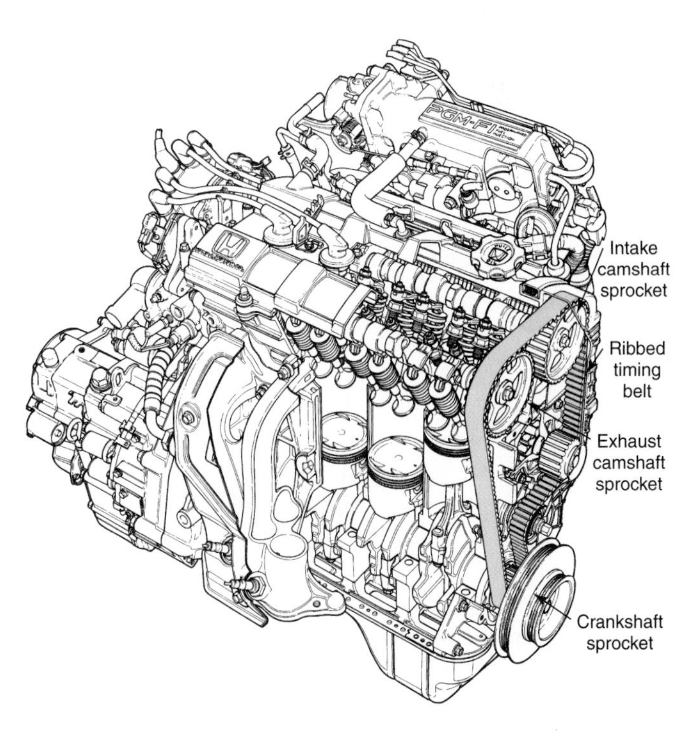

Figure 27-31. This four-cylinder, dual-overhead camshaft, 16-valve engine uses a ribbed timing belt. Note that one camshaft operates all intake valves and one camshaft operates all exhaust valves. (Honda)

a curved diameter of 1″ (25 mm) or less. Also, do not turn it inside out. Such careless handling of the timing belt can cause core damage. Keep the belt away from oil, chemicals, direct sunlight, and excessive heat.

Check the belt for the following problems:

- Hardened surface rubber.
- Cracked back rubber.
- Cracked or peeling fabric.
- Badly worn teeth.
- Cracked tooth bottom.
- Missing teeth.
- Worn or cracked belt sides.
- Oil or water saturation.

If one or more of the listed conditions is present, replace the belt with a new one of the same quality. Some belt conditions are shown in **Figure 27-32.**

Inspecting, Removing, and Installing the Sprockets

After the belt has been examined, check all sprockets for looseness, tooth wear, burrs, or cracks. If any damage is found, replace the sprocket.

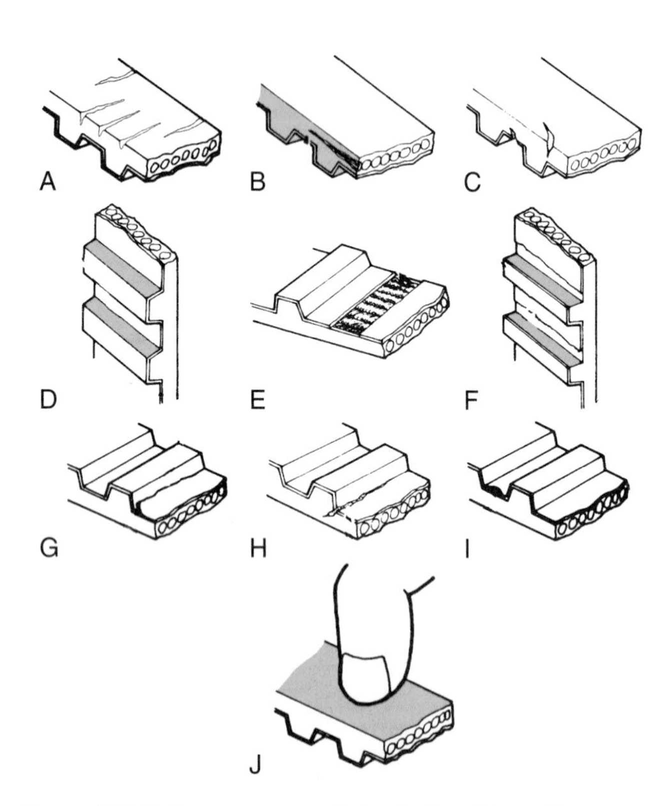

Figure 27-32. Some common timing belt problems. If any are present, discard the belt. A—Cracked belt back surface rubber. B—Side wear. C—Edge cracking. D—Flank on teeth worn (load side). E—Missing tooth. Reinforcing belts exposed. F—Extreme flank wear. Canvas worn. Rubber exposed. G—Cracked and separated canvas. H—Cracked canvas. I—Canvas separating from rubber. J—Back surface rubber hardened. Even firm pressure with fingernail will not leave marks. (DaimlerChrysler)

Sprockets are generally pressed and/or bolted onto the camshaft, crankshaft, or accessory shaft. When removal is necessary, follow the manufacturer's recommendations. **Figure 27-33** shows a crankshaft sprocket being removed with a special puller.

When installing sprockets, make sure they are not turned around (front to back). Install all keys (where used) and line up all timing marks. Replace sprocket fasteners and torque them to specifications.

Installing the Belt

The following is a basic belt installation procedure. Line up all sprocket timing marks and install the belt, **Figure 27-34.**

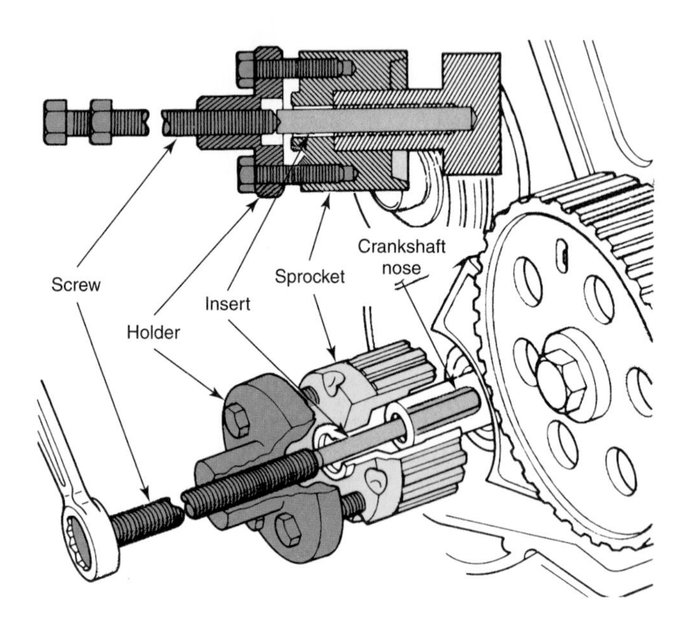

Figure 27-33. The crankshaft sprocket is removed from the crankshaft nose using a special puller. (DaimlerChrysler)

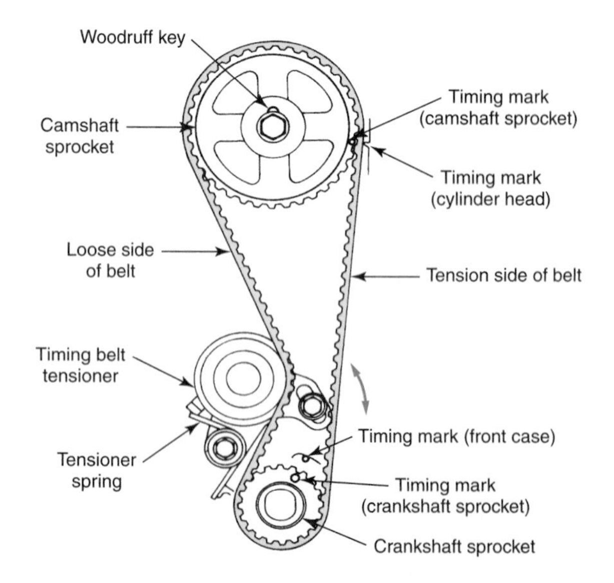

Figure 27-34. Sprockets and belt correctly assembled with timing marks aligned. Note the spring-loaded belt tensioner. (Hyundai)

Be sure the belt is routed correctly. Note that many belt drives have more than one timing mark. Extra timing marks are needed for dual overhead camshafts and timing belt–driven distributors.

Remove all spark plugs and revolve the crankshaft to ***top dead center (TDC)*** in the direction of normal travel (turning the crankshaft backwards could cause the belt to skip teeth, altering the timing). Place the belt tensioner tool horizontally on the tensioner hex and loosen the tensioner locknut. Rotate the crankshaft through two full revolutions from TDC. Secure the locknut on the tensioner while keeping the belt-tensioning tool in position. See **Figure 27-35.**

Caution: If a whirring sound is heard after you start the engine, stop the engine. This sound indicates the belt is too tight. This condition will greatly reduce the belt's service life.

Servicing Timing Chain and Sprocket

The following section explains how to service the ***timing chain*** and ***sprocket*** assembly. This arrangement is the most common type of camshaft drive used on engines without overhead camshafts. It is also used on some overhead cam engines.

Checking Chain for Wear

To check a timing chain for wear, place a steel rule against the block as shown in **Figure 27-36.** Turn the camshaft sprocket clockwise as far as it will go (the crankshaft must not turn). Align one of the chain link pins with a mark on the rule. Without moving the rule or the crankshaft, turn the gear counterclockwise as far as possible. Note the distance the link pin has moved and check this distance against manufacturer's specifications.

Another technique for checking timing chain wear involves turning the crankshaft so that all ***slack*** is taken up on one side. Carefully measure from the tightened chain surface (about midpoint) to a reference point on the block. Rotate the crankshaft in the opposite direction to produce maximum amount of slack on the same side. Pull the slack chain outward toward the reference point as far as possible. While holding the chain out, measure from the surface to the reference point. The difference in the two measurements should, in general, not exceed 1/2″ (12.7 mm), **Figure 27-37.** A new chain on new sprockets will have about 1/4″ (6.35 mm) slack. Note there are also special spring-loaded timing chains that have no slack.

Some overhead camshaft engines use long timing chains or timing belts, **Figure 27-38.** Slack is controlled by idler sprockets, rubbing blocks, or spring-loaded tensioning devices.

Removing the Timing Chain

Crank the engine until the timing marks on both sprockets face each other and are aligned with a line running from the center of the crankshaft to the center of the camshaft, **Figure 27-39.** To facilitate sprocket and chain installation, do not rotate the crankshaft until the chain and sprockets are replaced.

Remove the sprocket-to-camshaft cap screws. These cap screws are often used to retain a fuel pump eccentric and, in some cases, a distributor drive gear.

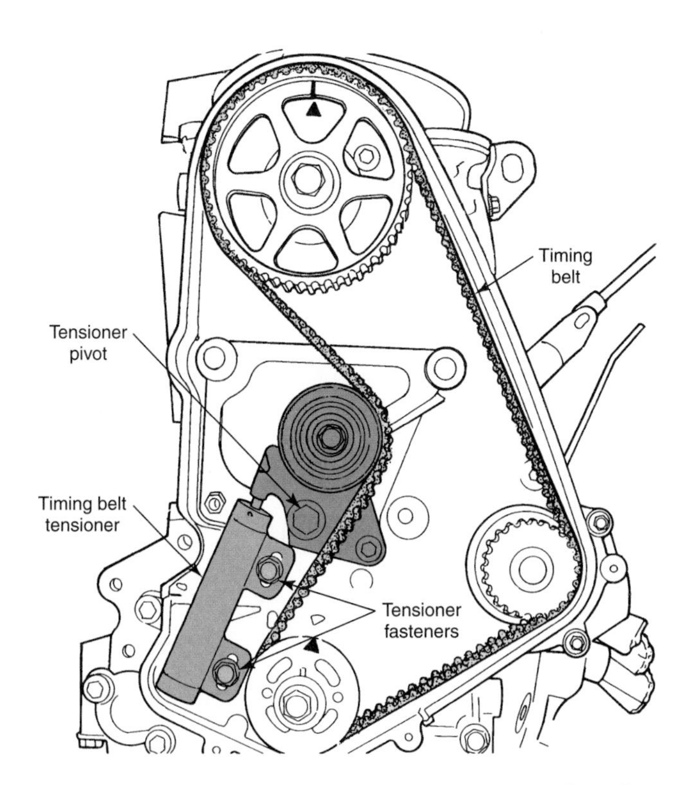

Figure 27-35. Timing belt tension is kept properly adjusted with a belt tensioner. Some of these tensioners may contain silicone oil. Always check for signs of leakage. (DaimlerChrysler)

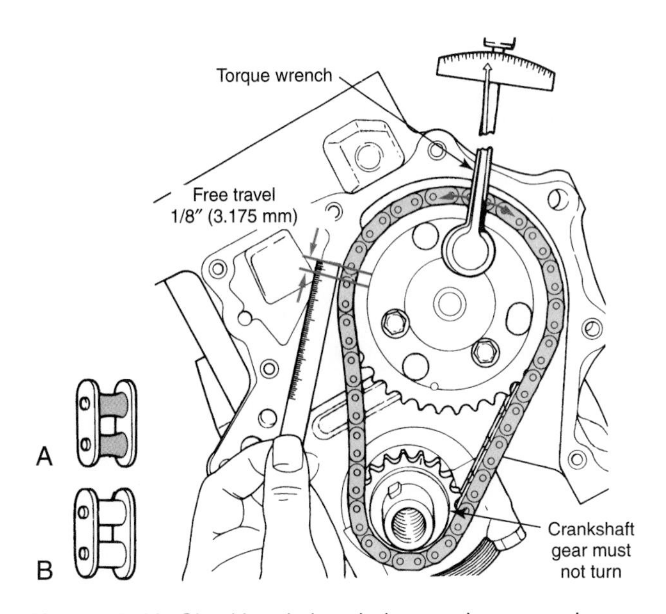

Figure 27-36. Checking timing chain wear by measuring sprocket free travel. A—Wear on link. B—Link cracked. (DaimlerChrysler and Nissan)

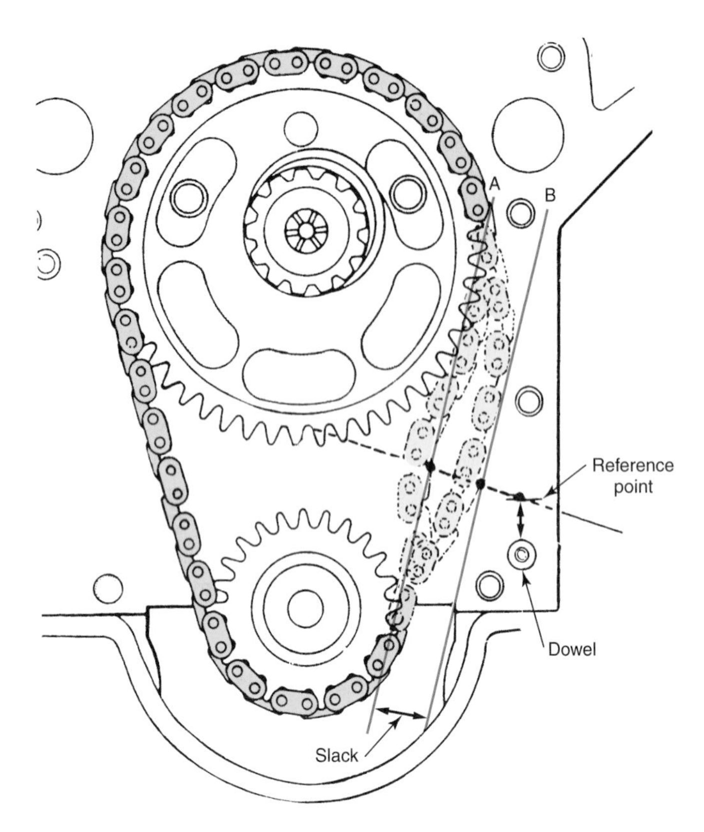

Figure 27-37. Timing chain wear can be checked by measuring slack. The distance between lines "A" and "B" represents the amount of slack in the chain. (DaimlerChrysler)

Figure 27-38. Tension is maintained on this long timing chain through the use of rubbing blocks and a tensioning device.

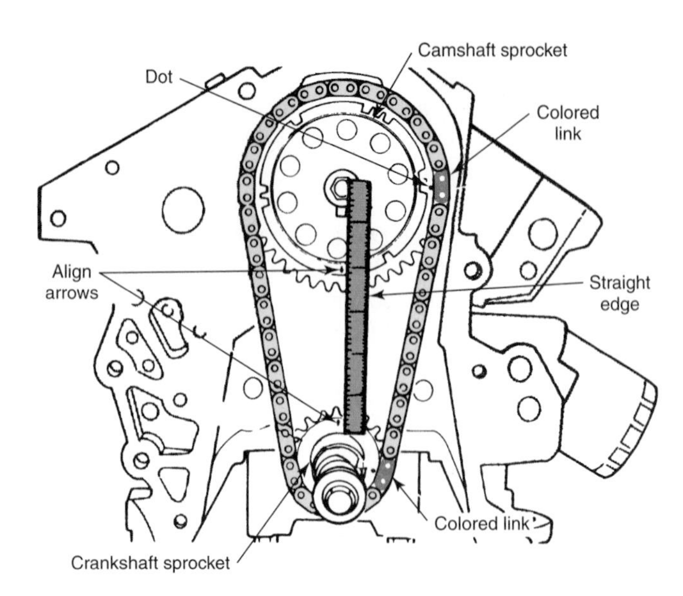

Figure 27-39. Before removing the timing chain, align timing marks with crankshaft and camshaft centers using a straight edge. (DaimlerChrysler)

Work both sprockets forward until the camshaft sprocket is clear of the shaft end. Two large screwdrivers can generally be used for this purpose. In some cases, a puller must be used to remove the sprockets. On some installations, the cam sprocket will clear without removing the crank sprocket. After removing the camshaft sprocket, remove the crankshaft sprocket from the crankshaft.

Inspecting Sprockets

After removal, carefully inspect both sprockets for signs of wear or chipping. Look for shallow chain imprints on the sprocket teeth. The slightest indication of wear is cause for rejection, **Figure 27-40.** If a thrust plate is used, check it for signs of excessive wear and replace it if necessary.

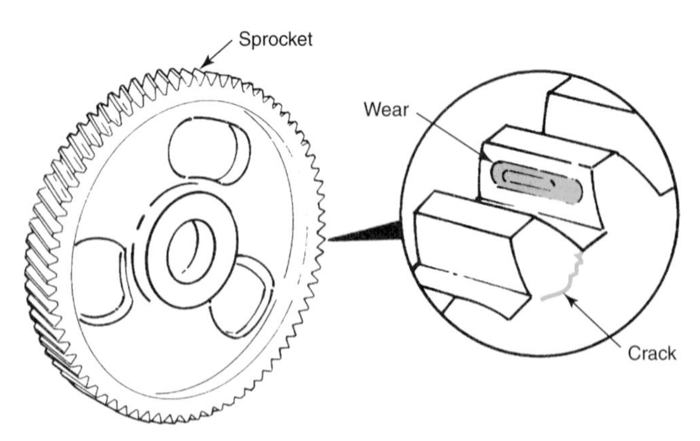

Figure 27-40. Sprocket teeth show wear and cracking. This sprocket must be replaced. (DaimlerChrysler)

Use New Sprockets

When installing a new timing chain, the best practice is to use new sprockets. Worn sprockets will increase the wear rate on the new chain. Used sprockets will also wear faster when a new chain is installed.

Installing Chain and Sprocket

If the camshaft sprocket can be installed without disturbing the crank sprocket, drive the crank sprocket fully into place, as described earlier in this chapter.

When both sprockets must be installed together, align the timing marks and place the chain around the sprockets. With the timing marks facing away from the engine, slip the crankshaft sprocket onto the crank until the cam sprocket touches the camshaft. At this point, make certain the timing marks are together and in line with the center of the crankshaft and the camshaft, **Figure 27-41.** If necessary, rotate the crankshaft and camshaft to obtain proper alignment. Use the attaching cap screws to pull the cam sprocket into place. Do not hammer the cam sprocket onto the camshaft, because this could loosen the rear oil seal plug (also called a Welch plug or core hole plug).

Make sure the fuel pump eccentric or distributor drive, if used, is in place. Torque the cap screws. Make a final timing mark alignment check. Install the crankshaft oil slinger (when used) and check sprocket runout, chain slack, and camshaft end play.

Other Timing Chain Marks

Some manufacturers specify a certain number of links or pins between timing marks. Some timing chains have colored links that must be aligned with the timing marks, **Figure 27-42.**

Removing and Replacing Front Cover Oil Seal

Whenever the vibration damper is removed, the ***crankshaft front oil seal*** should be replaced. Either drive or pull out the oil seal. When driving out a seal, support the cover so that it is not sprung or cracked. **Figure 27-43** illustrates the use of a special seal puller.

In Figure 27-43A, the puller blocks have been expanded outward to grasp the retainer lip. A removal sleeve is then placed over the puller screw, Figure 27-43B. The puller screw is held stationary and the draw nut is tightened to pull the seal out.

Before installing a new seal, clean the cover thoroughly. Coat the OD of the new seal with sealant and start it into the seal recess, **Figure 27-44.** Install the puller and plate. While holding the puller screw stationary, turn the draw nut to force the seal into place.

Remove the installing tool and check the seal to make sure it is fully seated. On the seal shown in Figures 27-43 and 27-44, a 0.001″ (0.025 mm) feeler gauge should not fit between the cover and seal edge when the seal is properly seated, **Figure 27-45.**

Note: Always install the seals so the seal lip faces the engine.

Installing the Front Cover

Before installing the front cover, cement a new cover gasket (if used) into position. If dowel pins are used to properly

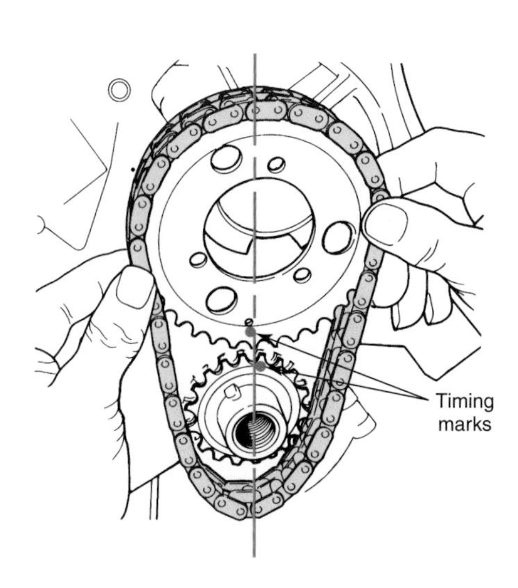

Figure 27-41. Installing chain and camshaft sprocket. Timing marks are together and aligned with the shaft centers. (DaimlerChrysler)

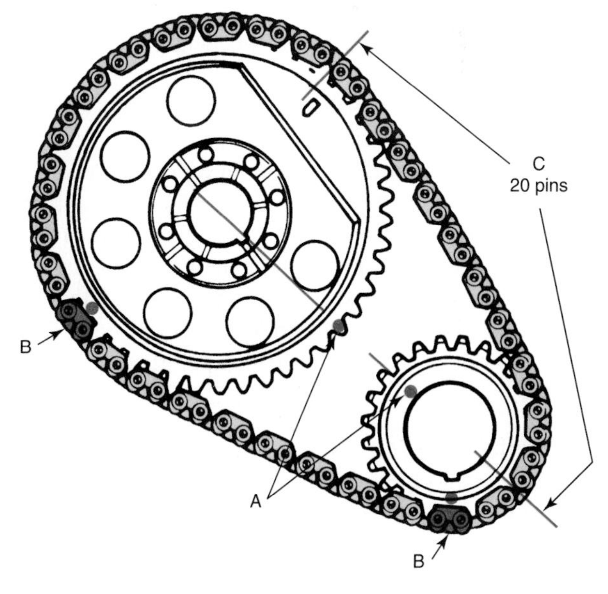

Figure 27-42. Three methods of timing the camshaft. A—Marks together and aligned with shaft centers. B—Colored links aligned with sprocket marks. C—Specific number of pins or links between marks. (DaimlerChrysler)

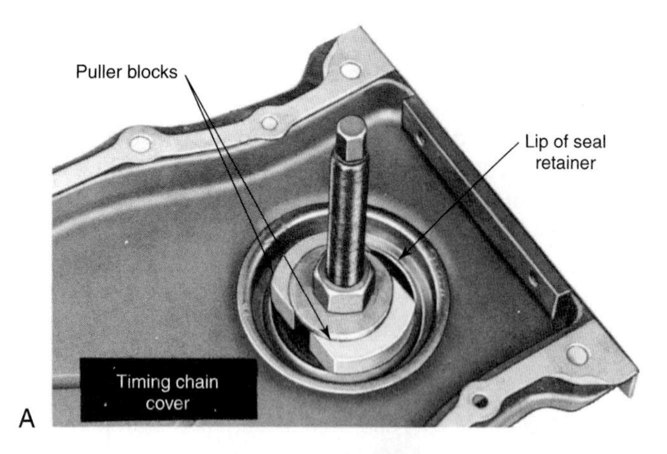

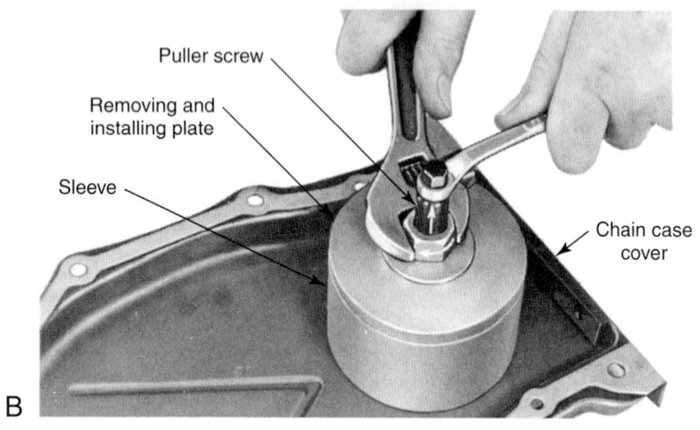

Figure 27-43. A—Puller blocks expanded to grasp seal retainer lip. B—Pulling chain cover oil seal. (DaimlerChrysler)

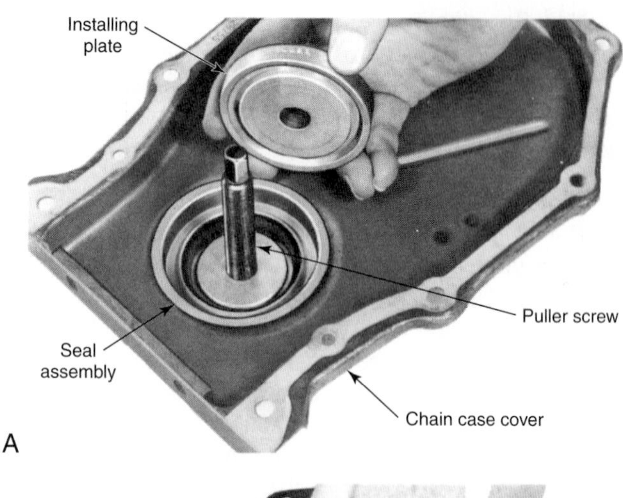

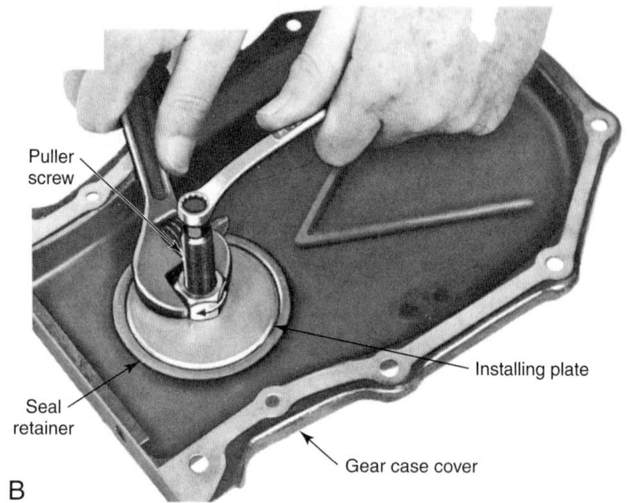

Figure 27-44. A—Setting up the puller for seal installation. B—Pulling the seal into place. (DaimlerChrysler)

locate the front cover, the cover can simply be bolted into place. See **Figure 27-46.**

If the cover is not positioned with dowel pins, it is imperative that a suitable ***centering sleeve*** be used. The centering sleeve aligns the cover crankshaft seal with the crankshaft. If the sleeve is not used, the seal may be off-center, causing it to leak. Slide the sleeve over the crankshaft, **Figure 27-47.** Lubricate the sleeve. Slide the seal over the sleeve and force the cover against the block. Start all screws with finger pressure. With the sleeve in place, tighten screws properly. Finally, withdraw the sleeve, leaving the seal centered perfectly on the crankshaft.

Installing the Vibration Damper

Before installing the vibration damper, lubricate the seal surface of the damper. Make sure it is absolutely smooth to prevent seal failure. Install the damper (do not forget the Woodruff key) with a puller or by driving it into place. Make sure the damper is seated fully, **Figure 27-48.**

Driving can damage some vibration dampers, unless a special driver that supports the outer rim section is used. Install the pulley and, if used, the crankshaft end cap screw.

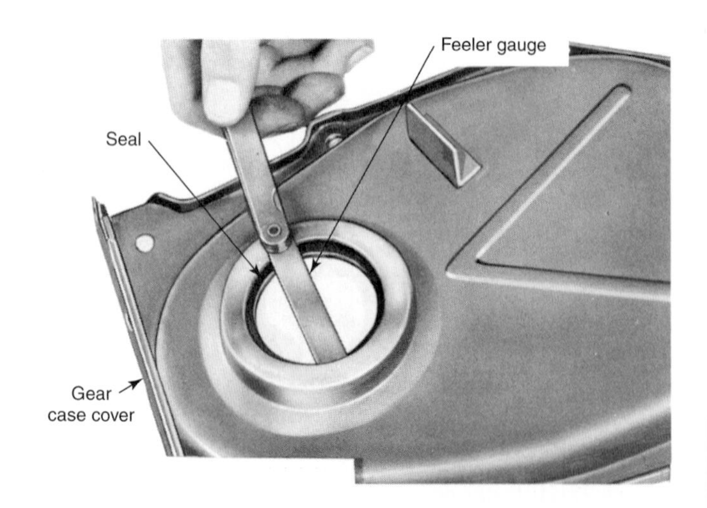

Figure 27-45. Checking the oil seal for proper seating depth. (DaimlerChrysler)

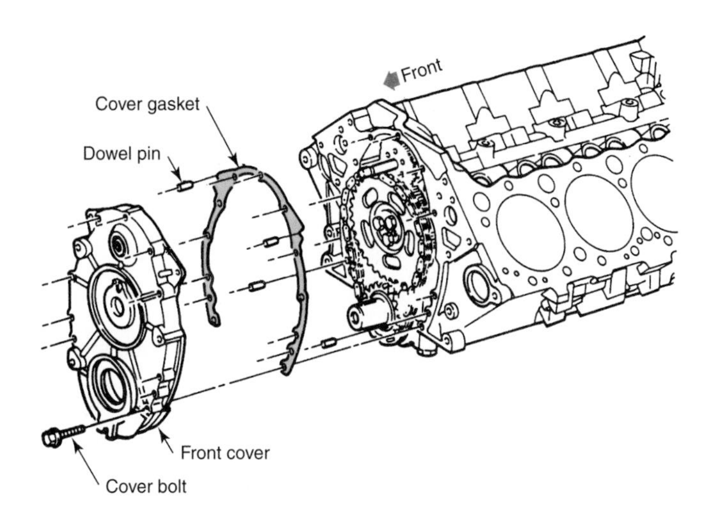

Figure 27-46. Dowel pins are used to align this cover, centering the seal on the crankshaft. (General Motors)

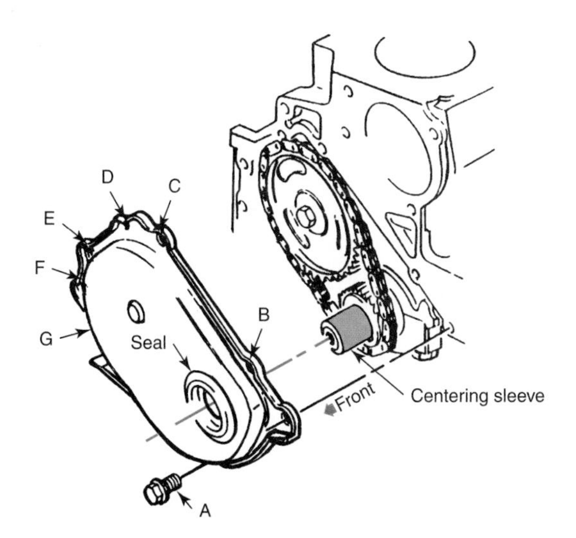

Figure 27-47. Using centering sleeve to align oil seal on crankshaft. Tighten cover and bolts in sequence (A–G) with the sleeve in place. (General Motors)

Chain and Gear Lubrication

Always check the timing chain or gear ***oil nozzles*** to make certain they are open. If feeder troughs are used, they must be clean and properly located. Improper lubrication will cause both the gears and the chain to quickly fail. When you remove a front cover and find that the chain or gears are badly worn or overheated, check out the lubrication system thoroughly. Note the timing gear oil nozzle in **Figure 27-49.**

Handle Gears and Sprockets with Care

Both gears and sprockets can be damaged by careless handling. Never hammer on them except with a proper driver. Make sure the driver is only used on the center hub. Avoid nicking or scratching the teeth. Keep gears and sprockets clean and lubricate them thoroughly upon assembly.

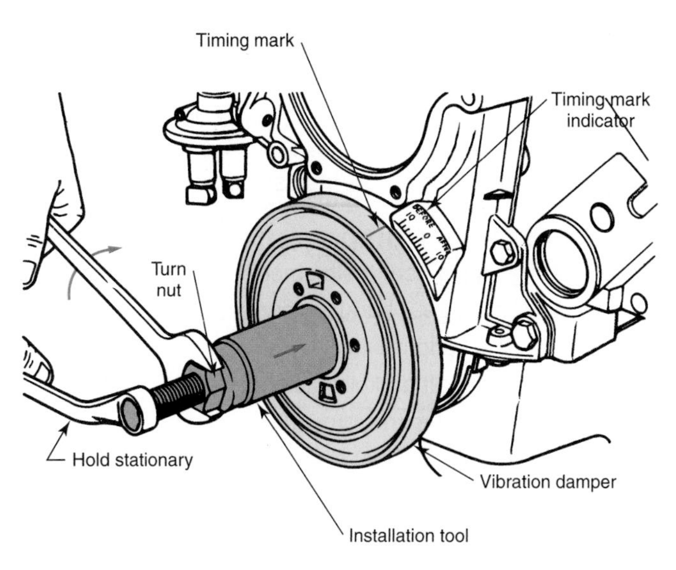

Figure 27-48. Using a special tool to install and seat the vibration damper. (DaimlerChrysler)

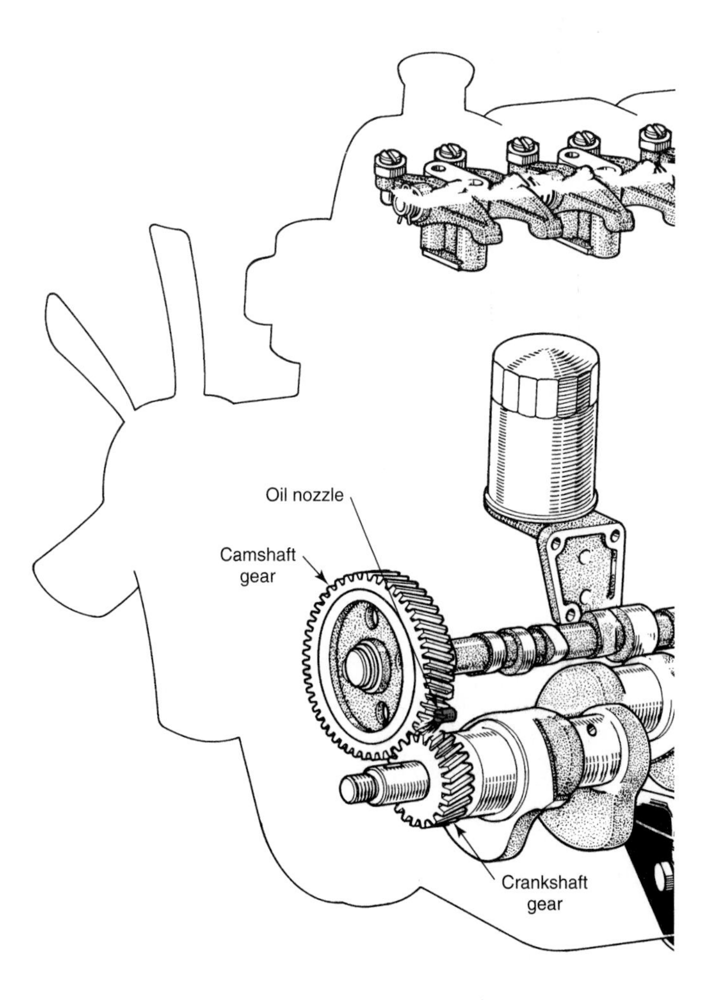

Figure 27-49. The oil nozzle must be open for proper gear lubrication. (Toyota)

Camshaft and Valve Train Service Checklist

If you have performed a thorough job, you will be able to answer "yes" to each of the following important questions relating to camshaft and valve train service.

Camshaft Drive Components

- Are the timing gear teeth smooth and sound?
- Is the timing gear backlash correct?
- Is the gear runout within limits?
- Is the camshaft end play as specified?
- Are the timing marks properly aligned?
- Are the chain (or belt) sprockets in good condition?
- Is the chain (or belt) slack within limits?
- Is the sprocket runout acceptable?
- Are the chain (or belt) and sprockets meshed so the timing marks align properly?
- If rubbing blocks or tensioners are used, are they in good condition and properly placed and adjusted?
- Were all Woodruff keys replaced?
- Is the chain or gear oiling nozzle clean?
- Are the camshaft journals smooth, round, and within maximum wear limits?
- Is the camshaft runout within limits?
- Are the cam lobes in good condition and is the lift as specified?
- Are the cam bearings good and do they provide the correct clearance?
- Are the cam bearing oil holes aligned?
- Are the cam bearings secure in the bore and were they pressed in the proper distance?
- Is the drive gear for the distributor and oil pump in good condition?
- If a thrust plate is used, is it securely bolted into place and is within wear limits?
- Have camshaft gears and sprocket cap screws been torqued?
- Is the fuel pump eccentric and distributor drive gear (if detachable) secured in place?
- Is the rear camshaft bearing oil seal plug sealed and tightly installed?
- Is the crankshaft oil slinger correctly in place?
- Was a new oil seal installed in the gear or chain cover with the lip facing the engine?
- Was the chain cover oil seal properly centered in relation to the crankshaft before tightening the cover?
- Is the damper installed securely and to the proper depth?
- Is the damper seal lip contact surface smooth and oiled?
- If a damper-retaining cap screw is used, is it in place and torqued properly?
- Was the damper carefully installed so it was not damaged?
- Were all gears, chains, journals, and cams thoroughly lubricated before assembly?

Rocker Arms

- Is the end contacting the valve stem smooth and accurately ground?
- Are the oil holes open?
- Is the rocker shaft or ball stud bearing surface smooth and within wear limits?
- Is the push rod ball or socket end smooth and free of wear?
- Is the rocker correctly installed and does it contact the valve properly?
- Is the rocker clean?

Rocker Shaft and Ball Studs

- Is the rocker shaft clean inside and out?
- Are the rocker arm bearing areas smooth and within limits?
- Is the correct end of the shaft forward?
- Do the rocker arm oil holes face in the correct direction?
- Are the shaft brackets in the correct location, torqued, and free of cracks?
- Is ample oil reaching the assembly?
- If an overflow pipe is used, is it correctly located?
- Are the ball studs tight in the head?
- Are the ball stud adjusting nut threads in good shape?
- Are the ball stud nuts within breakaway specifications?
- Are the self-locking rocker arm valve clearance–adjusting screws within breakaway specifications?

Push Rods

- Are the rods straight?
- Are rod ends smooth and free of excessive wear?
- If the rods carry oil, is the hollow section thoroughly clean?
- Is the correct end up?
- Are both ends in proper contact?
- If no clearance adjustment is provided, are the rods the correct length?

Lifters

- Have the mechanical lifters been trued on the grinder?
- Are ends and sides smooth and free of wear and galling?
- Is the lifter-to-lifter-bore clearance correct?
- Are the hydraulic lifters immaculately clean and in good condition?
- Have the hydraulic lifters been checked for leakdown?

General

- When possible, were all parts replaced in the location from which they were removed?
- Were all parts thoroughly cleaned?
- Were all parts properly lubricated before assembly?

Intake Manifold and Intake Plenum Service

The vacuum chamber between the throttle valve and cylinder head can be called an intake manifold or a plenum. Many vehicles have upper and lower intake manifolds. Manifolds and plenums should be thoroughly cleaned and checked for cracks. Many modern intakes are made of aluminum or plastic. These are more easily damaged than cast iron manifolds and should be handled carefully.

Note: Some manufacturers recommend replacing plastic intake manifolds that have been removed from the engine.

Always follow the manufacturer's recommendations for cleaning these types of manifolds. Plastic manifolds may be damaged by common solvents. They should not be placed in a hot tank or soaked in carburetor cleaner. Plastic manifolds should not be sandblasted or glass bead cleaned.

Manifolds and plenums can be checked for cracks using the same methods as were used for checking heads and blocks. Intakes can become warped and develop vacuum leaks. Check any suspect manifold or plenum with a straightedge and feeler gauge. Some warped intakes can be resurfaced, but this must be done carefully since they are lighter and more awkwardly shaped than heads and blocks. Make sure that the manifold is tightly and accurately placed before attempting to resurface it.

Tech Talk

Many technicians spend a lot of time and effort replacing a camshaft and lifters. They thoroughly clean everything, press in new cam bearings, carefully install the new cam and lifters, and then ruin the whole job by not lubing the cam lobes. When the engine first starts with a new camshaft, it takes a few minutes for the oil to reach the cam lobes. If the lobes are dry, the new camshaft can be destroyed in five minutes.

Never install a camshaft (or any engine part) without plenty of lubrication. Use a general-purpose heavy lube or the special camshaft lube supplied with many new camshafts. It is far better to overlubricate a camshaft than to take a chance on underlubricating.

Summary

Camshaft and valve train conditions are of critical importance to proper engine performance. Camshaft service is similar for all engines, regardless of where the cam is placed or how it is driven. Worn camshaft bearings can seriously lower oil pressure and produce excessive throw off. Camshaft-journal-to-bearing clearance should always be checked when servicing the camshaft.

With the camshaft out of the engine, mike each journal in several spots to determine the amount of wear and the extent of out-of-roundness. Make sure the journals are smooth. Inspect each cam lobe for galling, chipping, and excessive wear. The wear pattern should not show across the full width of the lobe. Also check cam lobe lift. Inspect the cam sprocket before installing the camshaft. If applicable, install new cam bearings when the camshaft is replaced. Lubricate the bearings and camshaft journals before installing the camshaft.

Worn or sticking lifters can cause missing or noises and contribute to the wear of other engine parts. Therefore, each lifter should be carefully checked. The lifter body must be smooth and free from scoring. The lifter-to-cam-lobe surface must also be smooth and free of galling, chipping, and excessive wear. Also, check the leakdown rate and compare it to specifications. Most shops simply replace lifters that show damage or have excessive leakdown rates. Before installing hydraulic lifters, fill them with engine oil. Lubricate the outside of the lifter body and the lifter guide bore. Also lubricate the cam lobe and the push rod ends of the lifter.

To replace overhead camshaft hydraulic lifters, the camshaft must generally be removed. Once the camshaft is out, pull the lifters from the head. Lubricate the new lifters, insert them in the head, and replace the camshaft.

Many late-model engines are equipped with roller lifters. If any part of the roller lifter needs replacing, discard the entire unit and replace it with a new lifter assembly. Unlike other types of lifters, a single defective roller lifter can be replaced without replacing the camshaft.

Check rocker arms carefully for wear. Most technicians simply replace worn rocker arms. Push rods should be straight and both ends must be smooth. Rod straightness can be checked with V-blocks and a dial indicator. A bent push rod must be replaced.

Removing the vibration damper is the first step in servicing the camshaft drive components. Pull and install the vibration damper by exerting pressure on the hub portion. Do not damage the crankshaft end threads.

Timing gear teeth must be smooth. Gear runout and backlash must be within limits. Some camshaft timing gears are bolted on; others are pressed on. When replacing a gear or a sprocket, it is good practice to replace the mating gear or sprocket.

When replacing a timing belt, check sprockets for possible damage. Remove and install the camshaft carefully to avoid damage to the bearing surfaces.

When pressing a new timing gear into place, apply pressure to the hub only. Check camshaft endplay at the thrust plate. Install the crankshaft gear before the camshaft gear. Make certain the gear timing marks face outward and that they are aligned properly.

When replacing a timing chain, install new sprockets. Check sprocket runout and chain slack. The chain and sprockets must mesh to properly align the timing marks. Some timing instructions call for sprocket marks to face each other on a line between crank and camshaft centers. Some call for colored links to align with the marks. Others give a specific number of links or pins between marks.

Slack in long chains is often controlled with rubbing blocks and tensioning devices. Check these carefully when servicing the chain. Replace any parts that do not appear to be in good condition.

Place the chain around the sprockets and install them as a unit. Replace all Woodruff keys. Then, install gears, sprockets, or vibration damper. Never force a camshaft back into the engine, as it may loosen the rear bearing core hole plug.

Always install a new chain or gear cover oil seal with its lip facing the engine. Center the seal to the crankshaft before tightening the cover screws. The vibration damper seal contact area must be smooth. Lubricate all gears, chains, and bearings before installation. Never lubricate a timing belt.

Intake manifolds and plenums should be cleaned and checked for cracks. Handle aluminum and plastic manifolds carefully. Plastic manifolds must not be soaked in strong cleaning solutions or sand or bead blasted. Intakes can be checked for cracks and warping. Some warped intakes can be resurfaced.

Review Questions—Chapter 27

Do not write in this book. Write your answers on a separate sheet of paper.

1. Define an interference engine.
2. A cam lobe should be checked for what three defects?
3. Most camshafts are ground so that the camshaft lobe surfaces are slightly _____.
4. Cam bearings can be _____ in, or can be _____ types that are replaced like crankshaft bearings.
5. Before installation, valve lifters should be filled with _____.
6. Which of the following parts should be replaced if they are worn or damaged?
   (A) Push rods.
   (B) Rocker arms.
   (C) Rocker arm shafts.
   (D) All of the above.
7. To remove the vibration damper, a special _____ may be needed. A few vibration dampers are a _____ fit and do not require a special tool for removal.

Matching

For Questions 8–10, match the gear type with the checking method.

8. Timing gears.	(A) Measure slack.
9. Timing gear and chain.	(B) Visual inspection.
10. Timing belt.	(C) Measure backlash.

11. Some timing gears must be _____ off the camshaft.
12. When installing a camshaft and gear, make sure that the _____ marks are aligned with the other gear.
13. When removing the timing belt(s) from some OHC engines, a special tool must be used to hold the _____ in position.
14. List six timing belt defects.
15. Which of the following is the *best* practice when replacing a timing chain?
    (A) Replace the crankshaft gear.
    (B) Replace the camshaft gear.
    (C) Replace both gears.
    (D) Do not replace the gears.
16. When driving a crankshaft oil seal out of the cover, _____ the cover so it will not be damaged.
17. Some vibration dampers can be damaged unless the technician uses a special driver that supports the _____.
18. Do not forget to install the crankshaft _____ after the vibration damper is in place.
19. Timing gears may be lubricated by an oil _____.
20. Intake manifolds should be checked for _____ or warpage.

ASE-Type Questions

1. Technician A says that worn camshaft bearings can reduce oil pressure. Technician B says that worn camshaft lobes can be checked with a micrometer. Who is right?
   (A) A only.
   (B) B only.
   (C) Both A and B.
   (D) Neither A nor B.
2. Plastigauge can be used to check which of the following?
   (A) Insert type cam bearings.
   (B) Pressed-in cam bearings.
   (C) Timing gear backlash.
   (D) Intake manifold warping.
3. Which of the following tools can be used to check camshaft lobe lift with the cam removed from the vehicle?
   (A) Micrometer.
   (B) Dial indicator.
   (C) Feeler gauge.
   (D) Ruler and straightedge.

4. All of the following statements about valve lifters are true, *except:*
   (A) Hydraulic lifters contain internal parts.
   (B) Excessive wear on the lifter means that it should be replaced.
   (C) Both hydraulic and mechanical lifter valve trains can have adjustment devices.
   (D) Overhead camshafts always have mechanical lifters.
5. A camshaft is used with roller lifters. Under which of the following conditions does the camshaft *not* have to be changed when a single lifter is replaced?
   (A) The same push rod is used.
   (B) The same roller is used.
   (C) The camshaft is undamaged.
   (D) The timing gear is undamaged.
6. Technician A says that some timing gears must be removed from the crankshaft with a puller. Technician B says that some timing gears slip over the crankshaft. Who is right?
   (A) A only.
   (B) B only.
   (C) Both A and B.
   (D) Neither A nor B.
7. Technician A says that the sprocket timing marks should be lined up before a timing belt is removed. Technician B says that a whirring sound after the belt is replaced indicates a loose belt fit. Who is right?
   (A) A only.
   (B) B only.
   (C) Both A and B.
   (D) Neither A nor B.
8. Timing chain imprints on the gear teeth indicate which of the following?
   (A) The chain and gears have worn to a good fit.
   (B) The gear was specially made to fit the chain.
   (C) The chain was made of high strength steel.
   (D) The chain and gears should be replaced.
9. A Woodruff key is used on all of the following, *except:*
   (A) crankshaft gears.
   (B) camshaft gears.
   (C) vibration dampers.
   (D) rocker arm shaft
10. Technician A says that a plastic manifold should not be cleaned in a parts washer. Technician B says that a plastic manifold should not be cleaned by bead blasting. Who is right?
    (A) A only.
    (B) B only.
    (C) Both A and B.
    (D) Neither A nor B.

Suggested Activities

1. Obtain a used camshaft and measure the lift of the lobes. Be sure to identify the intake and exhaust lobes before measuring. Record your measurements. How does the lift compare with the factory specifications?
2. Check several lifters for wear on the body and the contact face. Record your findings.
3. Determine whether the camshaft and lifters checked above should be replaced. Present your findings to your classmates.
4. Count the teeth on a camshaft and crankshaft gear. Explain why the cam gear has twice as many teeth as the crank gear.

Cutaway of a late-model V-6 engine. (Lexus)

28

Engine Assembly, Installation, and Break-In

After studying this chapter, you will be able to:

- Summarize a typical sequence for assembling an engine.
- Describe how to install an engine in a car.
- Pressurize the engine lubrication system with oil.
- Make final checks before starting the engine.
- Operate an engine properly for safe break-in.

Technical Terms

Sequence of assembly	Scuffing
Lash	Pressurizer
Adjusting disc	Break-in period
Lift strap	Exhaust disposal lines
Protective pads	Carbon monoxide
Hoist	Retorque
Scoring	Dynamometer

No single method of reassembling engine components is right every time. Technician preference, engine design, and parts availability all help to determine the order of assembly for any given engine.

This chapter provides a general overview for engine reassembly and installation. The sequence of assembly that follows is typical for most engines and may be used as a general guide. However, it is important for the technician to study the construction of each specific engine and the manufacturer's shop manual, if one is available. This advance planning will help the technician anticipate and prevent any problems caused by assembling parts in the wrong sequence.

For example, assume you are working on an engine with a mechanical fuel pump. If you first install the fuel pump, then try to install the camshaft, you will find the fuel pump arm contacts the camshaft journals, preventing you from installing the camshaft. You will have to remove the fuel pump to install the camshaft.

Or, assume that you are working on an overhead cam engine and that you have installed the camshaft before installing the head on the block. On some engines, you would be forced to remove the camshaft to install the head bolts.

In both examples, as a result of your failure to study the engine construction and to "think ahead," you would have to waste valuable time removing parts to permit the installation of others.

Typical Engine Assembly Sequence

The ***sequence of assembly*** described in this chapter covers major units. It is assumed that subassemblies such as valves, springs, bearings, and rings have been correctly installed. It is further assumed that all parts have been cleaned, checked, lubricated, and replaced or repaired as needed. Any particular step in this sequence may be reviewed in detail by referring to the chapter covering the operation or part concerned. If additional information is needed, consult the appropriate service manual.

To assemble an engine, begin by mounting the engine block on a suitable stand. Then, install the following components:

1. Oil gallery and core hole plugs.
2. Crankshaft.
3. Camshaft. (If the camshaft timing gear is pressed on, make sure it is in place before installing the camshaft.)
4. Timing gears, timing chain, and sprockets. (Make sure the timing marks are aligned properly.) On overhead camshaft designs, installation of the timing chain, gear, or belt must be delayed until the cylinder head is installed.
5. Piston and rod assemblies.
6. Oil pump, oil pickup, and connecting lines.
7. Front cover.
8. Vibration damper and pulleys.
9. Oil pan.
10. Cylinder head(s).
11. Water pump.
12. Valve lifters, push rods, and rocker arm assemblies. (Set valve clearance at this time.)
13. Intake manifold.
14. Exhaust manifold(s).
15. Distributor (make certain timing is correct), plugs, coil, and secondary wires.
16. Fuel pump, fuel injectors or carburetor, and fuel and vacuum lines.
17. Wiring connectors, alternator, starter, and the temperature and oil pressure senders.
18. Drive belts.

19. Flywheel. (On models with a standard transmission, the clutch disc, pressure plate, and clutch or flywheel housing can be installed at this time.)
20. Transmission or transaxle. (In some cases, it is desirable to attach the transmission or transaxle to the engine and install them together; in other cases, the engine must be installed alone.)

Do Not Hurry!

Technicians often become anxious as the time to start the engine draws near. This is natural because your knowledge, care, and skill will receive a crucial test when the engine is started.

It is at this point that many fine overhaul jobs are ruined by carelessness brought about by hurrying. Work energetically, but work carefully. Think and plan; avoid a "last minute rush!"

Cover Engines for Protection

When not actively working on an engine, it is advisable to cover it with a fabric or plastic tarp to protect them from dirt and debris. When the heads and the intake manifold are installed, cap off the manifold and install the spark plugs to prevent the entry of dirt or small, loose parts.

Adjusting Valve Lash for Hydraulic Lifters

Hydraulic lifters are used primarily to eliminate the need for ***lash,*** or clearance, between the end of the valve stem and the rocker arm. When the parts heat up and elongate during engine operation, the lifter will leak down. Any shortening will cause the lifter to pump up. In this way, "zero" clearance is constantly maintained. Once set, the hydraulic lifter does not require further adjustment.

The object in adjusting hydraulic lifters is to place the lifter plunger somewhere near the center of its stroke. This will allow changes in both directions. If the plunger is forced to the bottom, it will act as a solid lifter. If allowed to remain at the top, it cannot compensate for wear and temperature contractions.

Some engines have no provision for valve lash adjustment on the rocker arms. Valve stem length above the head, head gasket thickness, and push rod and rocker wear all become critical on an installation of this type. Replacement push rods are available in different lengths to compensate for the small changes needed.

Lifter Must Be on Cam Lobe Base Circle

Rotate the camshaft until the cam lobe nose faces directly away from the lifter. The lifter will then rest on the base circle. See **Figure 28-1.**

There are several ways of determining when the lobe is in this position. On some engines, such as the overhead camshaft type, the lobe is visible. If the engine is in the car and the ignition is properly timed, the engine can be slowly turned over until the plug lead to the affected cylinder fires. At this instant, both valves are closed and the lobes are in the proper position for lash setting.

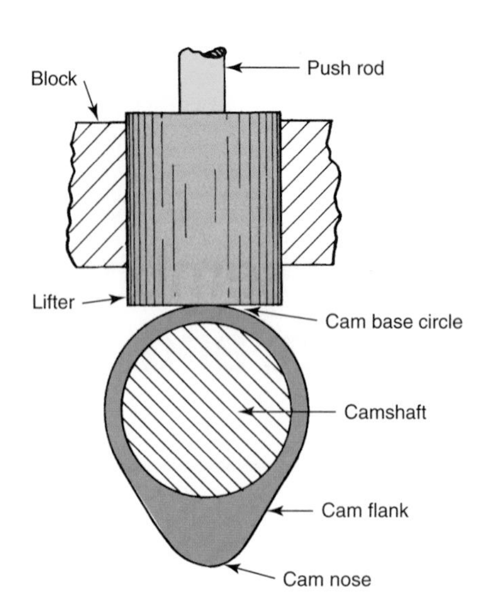

Figure 28-1. To set valve lash or clearance, the lifter must rest on the base circle of the cam lobe.

Slowly crank the engine until a particular valve is fully opened. Then give the crankshaft exactly one full turn (mark the damper with chalk). The cam lobe will be turned one-half revolution, placing the cam lobe nose opposite the lifter.

When a piston is brought to TDC (top dead center) on the compression stroke (both valves closed), the lobes will be in the correct position for that cylinder.

Another technique involves using chalk marks to divide the damper into two 180° sections (four-cylinder), three 120° sections (six-cylinder), or four 90° sections (V-8). One of the marks is on the timing notch, and the others are related in degrees to this mark. By cranking the engine until the marks index with the timing pointer, it is possible to set certain valves and reduce the amount of cranking required.

Lifter Plunger Must Be at the Top of Travel

The rocker arm adjustment should be loosened so the lifter plunger travels to the top of its stroke. At this point, the push rod can be "jiggled" sideways and up and down. See **Figure 28-2.**

Grasp the push rod with the thumb and forefingers. While gently shaking the push rod from side-to-side, slowly tighten the rocker arm adjusting nut, **Figure 28-3.** As the rocker arm push rod end moves downward, the amount of shake will be reduced. Stop at the instant that play or shake disappears. At this point, no lash is present between valve stem and rocker arm or the rocker arm and push rod.

Give the rocker arm's adjusting nut an additional number of turns following manufacturer's specs (typically 1 1/2 turns). This will force the plunger down to the midpoint of its stroke. Repeat this process on all rockers. Where there is no provision for rocker arm adjustment, compress the lifter and check push-rod-to-rocker arm clearance against specifications. If necessary, install a longer or shorter push rod to correct clearance.

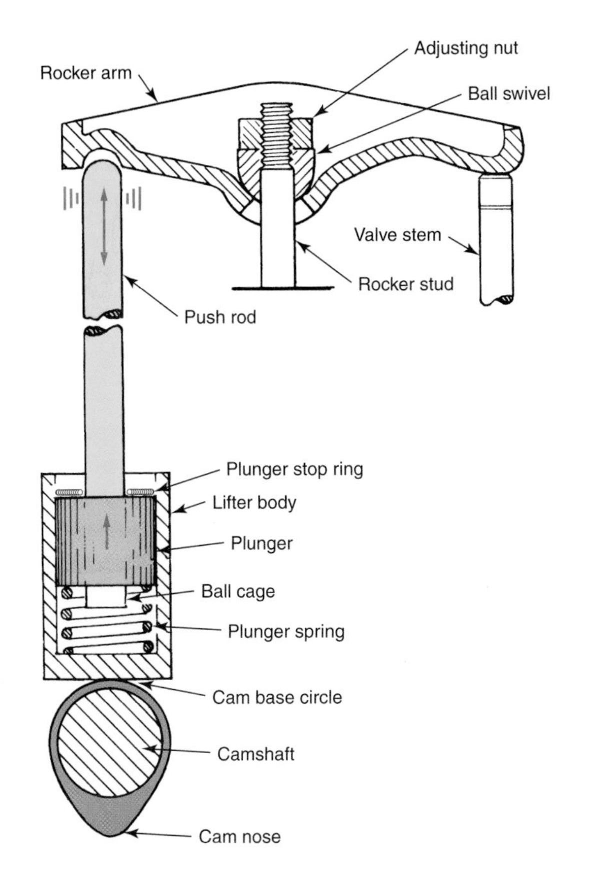

Figure 28-2. With the hydraulic lifter plunger against the stop ring, the rocker arm adjusting nut is backed off until the push rod can be "jiggled" by the technician.

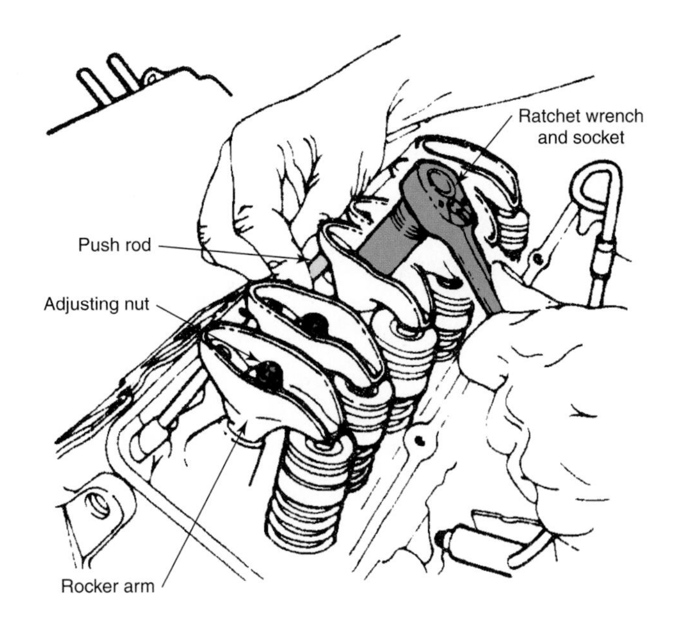

Figure 28-3. The rocker arm adjusting nut is tightened until the push rod no longer moves. The adjustment nut is then tightened to manufacturer's specifications. (General Motors)

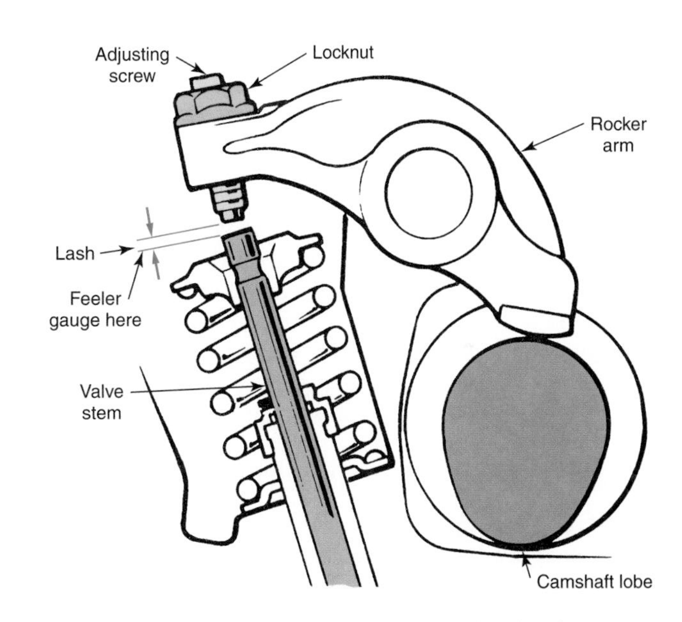

Figure 28-4. A mechanical lifter requires adjusting for clearance between the rocker arm and valve stem. Measuring is done with a feeler gauge. (General Motors)

Adjusting Valve Lash for Mechanical Lifters

A certain amount of lash, or clearance, between the valve stem and the rocker arm is a must when mechanical lifters are employed. The exact amount will vary from engine to engine, depending on use, design, and construction. Always provide the amount of clearance specified by the engine manufacturer.

Too much clearance will cause noisy operation, late valve opening, early valve closing, lowered valve lift, excessive wear, and possible valve breakage. Too little clearance will cause early valve opening, higher lift, late closing, and valve burning.

As with the hydraulic lifter, the mechanical lifter must rest on the cam base circle during adjustment. The rocker arm is carefully adjusted so correct clearance exists between the valve stem and rocker arm.

Clearance can be measured with a feeler gauge of the exact thickness or a stepped *go/no go* blade (go = 0.001″ or 0.025mm below specs; no go = 0.001″ or 0.025mm above specs). To check clearance, hold the push rod end of the rocker arm down, and slip the feeler gauge blade between the rocker arm and valve stem. The blade should pass between them with a slight drag, **Figure 28-4.** Valve clearance can also be checked with a dial indicator, **Figure 28-5.** This device gives highly accurate settings.

Cold and Hot Clearance Settings

When an engine is reassembled, an initial, or cold, setting of the valve clearance is necessary. For a final hot clearance setting, the engine must be up to normal operating temperature (oil and water temperature). This will require about thirty minutes of engine operation. The procedure for setting hot clearance of mechanical lifters is identical to the procedure for setting cold clearance.

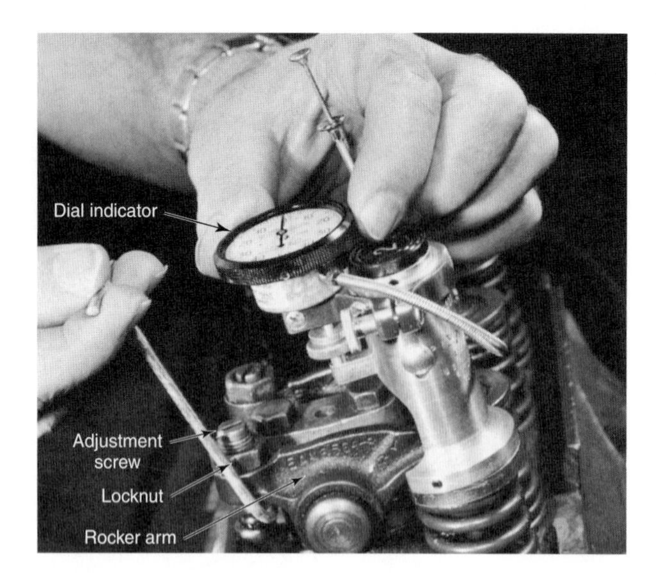

Figure 28-5. A dial indicator can be used to measure rocker-arm-to-valve stem clearance. (P and G Co.)

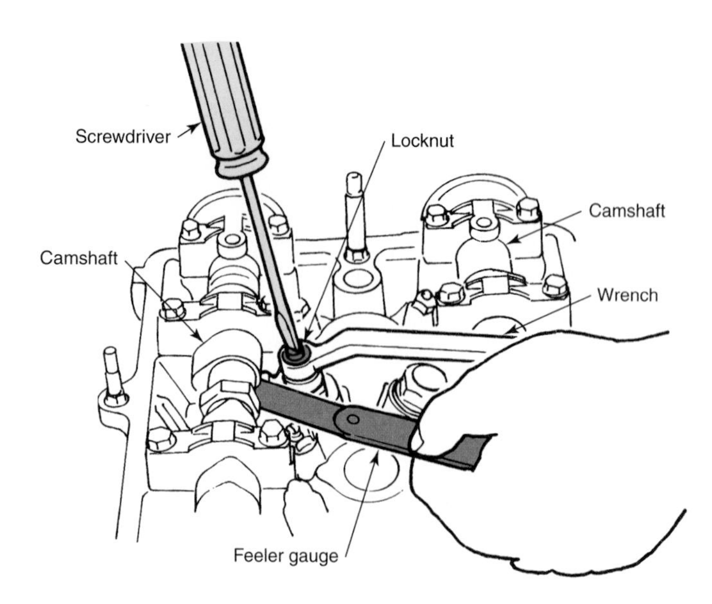

Figure 28-6. Some rocker arm adjusting screws use locknuts. Loosen the locknut, turn the screw until proper clearance (measured with a feeler gauge) is achieved, then tighten the locknut. (Acura)

Note: Accurate valve clearance is important. Make certain the engine is hot and the clearance settings are exact. Follow the manufacturer's suggested adjustment procedures and clearances.

Rocker Arm Adjusting Screws

Some rocker arm adjusting screws are self-locking. A specified amount of torque must be applied to move them. If the "breakaway" torque is below accepted limits, change the screw or the nut.

If a locknut adjusting screw is used, loosen the nut and adjust the screw. While holding the screw, firmly tighten the nut. After tightening, recheck clearance, **Figure 28-6.**

Direct Acting Overhead Camshaft Adjustments

In some overhead camshaft engines, the cam lobes act directly on the valves through a cam follower. No rocker arms are needed. Valve clearance is set by inserting an ***adjusting disc*** on the top of the follower, **Figure 28-7,** or between the follower and the valve stem. Various disc thicknesses are available to alter clearance as needed.

Another setup uses a tapered adjusting screw to set clearance, **Figure 28-8.** The taper produces a wedging effect that alters clearance when the screw is turned.

Engine Installation

Proper engine installation is just as important as proper assembly. Not only can the rebuilt engine be ruined by careless installation, but other parts of the vehicle can be damaged in the process.

Attach Lift Equipment

Begin installation by attaching a ***lift strap*** or fixture to the engine. Review Chapter 24, Engine Removal, Disassembly,

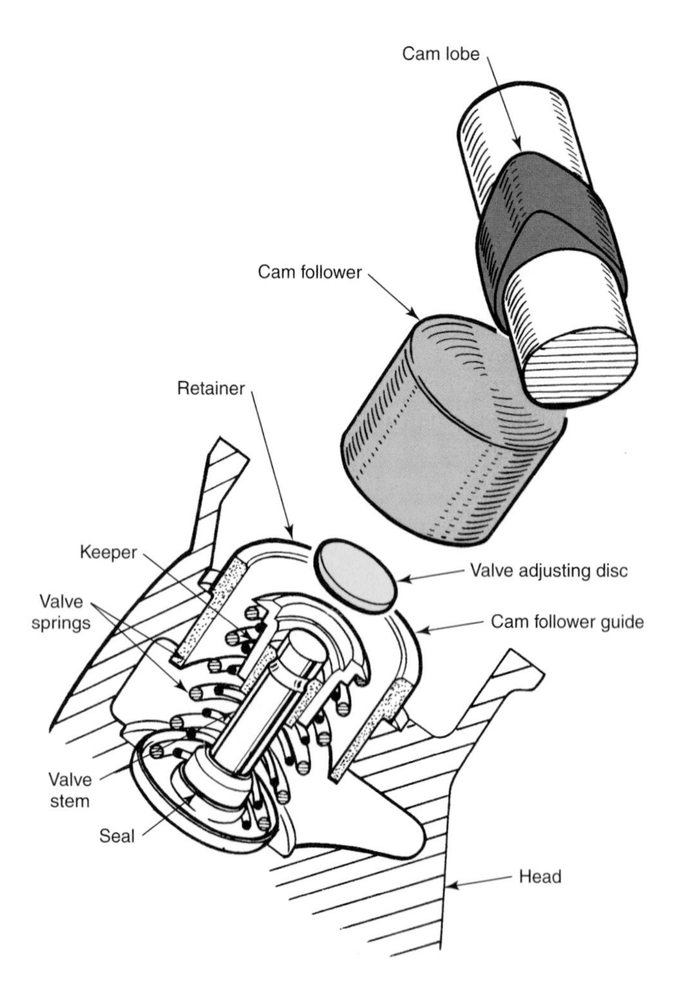

Figure 28-7. Direct-acting cam follower systems use an adjusting disk to alter valve clearance. Disks are available in various thicknesses to achieve the desired clearance. (Jaguar)

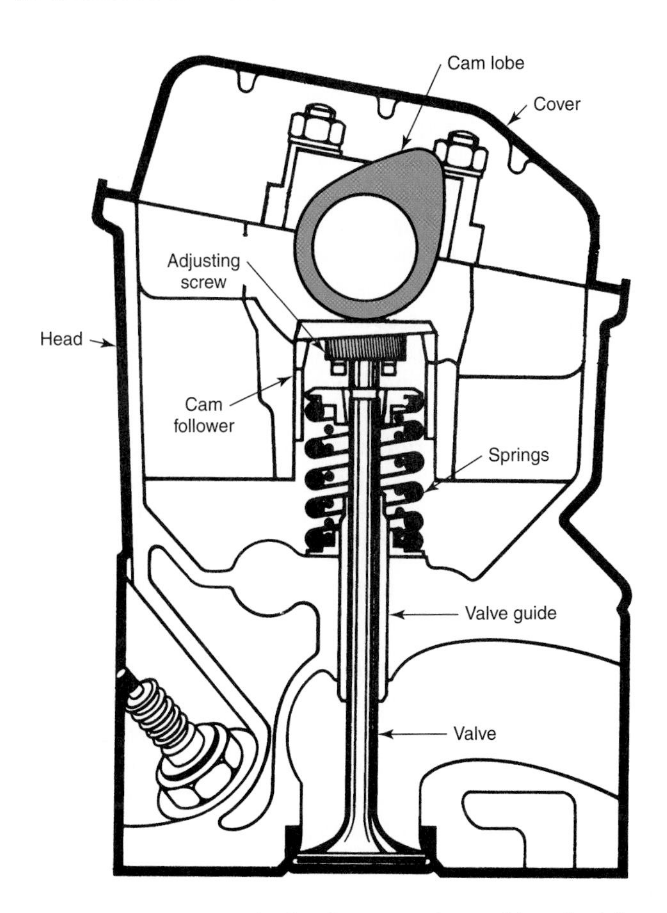

Figure 28-8. A tapered adjusting screw is used to adjust the valve clearance for this overhead camshaft engine. (DaimlerChrysler)

and Inspection, for recommendations on attaching lift devices. Pay particular attention to the safety rules for engine removal—they also apply to installation. It is usually easiest and safest to attach the lifting equipment in the same way that it was attached during removal.

Place ***protective pads*** on the fenders. Raise the engine with a suitable ***hoist,*** **Figure 28-9,** being careful to balance the engine at the angle desired.

Installation with Transmission or Transaxle in Car

To install an engine when the transmission or transaxle is in the vehicle, guide the engine into the engine compartment. When engine crankshaft and transmission/transaxle input shaft centerlines are at the same level and parallel to each other, move the engine backward to engage the transmission or transaxle. For manual shift vehicles, make sure the input shaft passes through the throwout bearing, through the clutch disc, and into the crankshaft pilot bearing. See Chapter 29, Clutch and Flywheel Service, for full details. For automatics, start the converter pilot into the crankshaft. Be careful not to bend the relatively light flex-type flywheel or drive plate. Remove the converter-restraining strap.

Install the fasteners that secure the clutch or converter housing to the engine. On automatics, install fasteners that

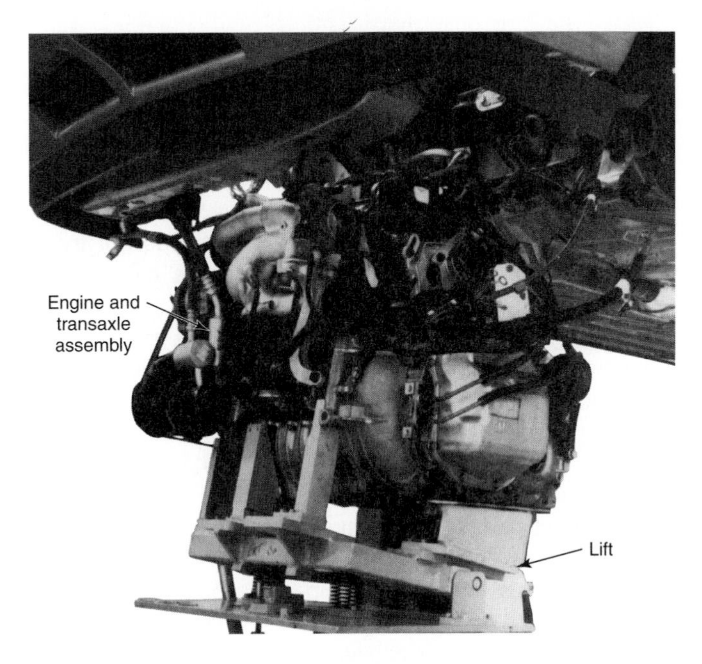

Figure 28-9. Use great care when removing or installing an engine. Make certain the hoist or lift is securely attached. Always correctly support the vehicle to prevent tipping or falling as weight is added or removed. (General Motors)

secure the converter to the flywheel. Torque the engine mount bolts.

Installation with Transmission or Transaxle on Bench

To install an engine when the transmission or transaxle is out of the vehicle, guide the engine into place on the front mounts. Install the mount bolts loosely. Use a strap to support the rear of the engine. Then, remove the engine lift or hoist.

Raise the car and install the transmission or transaxle. Attach the engine rear mount cross member and tighten the engine mounts. Remove the support strap and torque all mount bolts to factory specifications.

Installation with Transmission or Transaxle on Engine

Installing an engine with the transmission or transaxle attached requires care and skillful maneuvering. On a rear-wheel drive vehicle, the engine and transmission form a long unit. The engine will usually have to be tipped at a steep angle (up to 45°) and guided into place. Position the transmission with a jack, using a wooden block to prevent damage to the transmission pan. Install the rear support cross member and attach to the engine rear mount. Install and torque all mount bolts. Remove the jack and install the drive shaft.

On a front-wheel drive vehicle, the extra size and weight of the transaxle will make maneuvering the assembly more difficult. Begin by guiding the engine and transaxle into place in the engine compartment. Install the lower engine and transaxle mounts, and then position and install all side and upper mounts. Finally, install both CV axle shafts.

Position Engine Carefully

When installing the engine, lower it slowly and carefully. Avoid damage to the car body or to underhood accessories. As the lowering progresses, make sure the engine is not catching on some part of the vehicle. Gentle shaking, pressure with a pry bar, and careful use of a jack are all helpful in engine positioning. If difficulty is encountered, do not try to force the engine into place. Find out what is causing the trouble and remedy it. Make sure the mounts are properly assembled and torqued. **Figure 28-10** shows how engine mounts are used on a front-wheel drive vehicle with a transverse-mounted (sideways) engine.

Connect All Wiring, Hoses, Tubing, and Linkages

Connect all fuel and vacuum lines. Attach the radiator hoses, air conditioning lines, and oil cooler lines, if used. Reconnect the exhaust system pipes. Make sure the ignition switch is turned to the off position and then reconnect all wiring. Connect battery cables last, but do not install the battery at this time. Attach transmission or transaxle shift linkage and throttle linkage or wiring. Connect power steering and air conditioning lines. Refer to the chapter on air conditioning for recharging procedures. Check each unit to make sure it is properly connected.

Marking Pays Off

As mentioned in the chapter on engine removal, all wires, lines, and hoses should be clearly marked so they may be reinstalled in a minimum amount of time. If you marked the various items properly when they were removed, you will now appreciate the importance of careful marking.

An Overhauled Engine Needs Instant Lubrication

When an overhauled engine is started, it is vitally important that all moving parts receive adequate lubrication. Even though all parts may have been thoroughly lubricated during assembly, most of this oil will have drained off into the pan. The oil galleries are dry, the filter is dry, and the tappets (hydraulic lifters) will need additional oil.

Upon starting, the oil pump must prime itself. It must then force oil throughout the system, filling the galleries, filter, and other areas before supplying oil to the bearings. This takes time. Do not be fooled by the oil pressure gauge registering immediate pressure. This could be caused by air in the lines.

During this critical period of time, the engine is operating without proper lubrication. New parts are closely fitted and areas such as cylinder, ring, and piston surfaces will quickly heat up and cause ***scoring*** and ***scuffing.*** Bearings and journals can also be damaged by a dry start (lack of oil upon starting). This can be prevented by *pressurizing the lubrication system* before starting the engine.

Pressurizing the Engine Lubrication System

The lubrication system may be pressurized by using a special tank and hose setup, **Figure 28-11.** The tank is filled to the indicated level with the same kind of oil that will be used in the engine. Air pressure up to the normal system pressure (around 40 lb. psi or 276 kPa) is then admitted to the tank. Watch the gauge to avoid excess pressure.

The hose fitting is attached to some external entry point into the lubrication system, such as the oil sender hole. The valve is opened and oil under pressure flows through the system. This primes the oil pump and fills the galleries, filter, lifters, and bearings. This ensures prompt lubrication upon starting. Rotate the engine several times while pressurizing, **Figure 28-12.**

Pressurizer as a Bearing Leak Detector

The ***pressurizer*** (often referred to as a bearing leak detector) can also be used to check bearing clearance. Some technicians prefer to pressurize the lubrication system before installing the oil pan. This technique permits watching the leakage rate at the ends of the bearings. The oil should pass through each bearing and fall in a series of individual drops. A

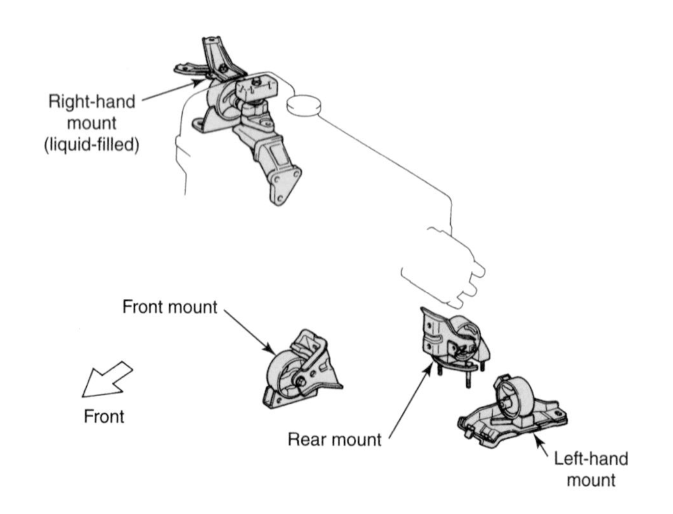

Figure 28-10. A four-point engine/transaxle mount setup. Note the right-hand mount is liquid-filled. This helps to reduce engine noise and vibration. (Toyota)

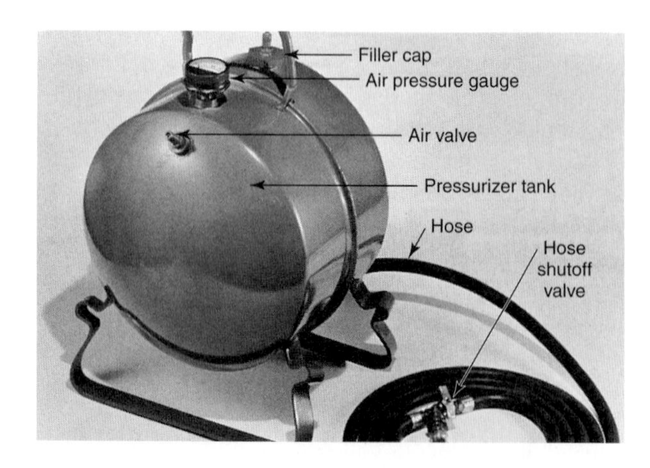

Figure 28-11. Typical lubrication system pressurizer. (Clevite)

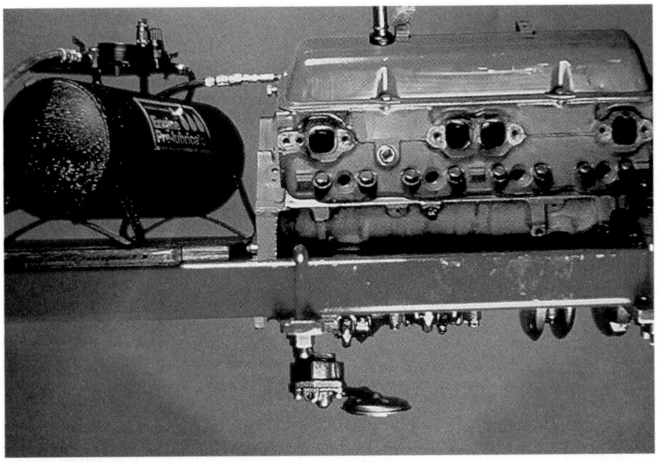

Figure 28-12. Preparing to pressurize the lubrication system with a pressurizer. (Federal-Mogul)

Figure 28-13. The amount of engine oil passing by the bearings will provide some indication as to the bearing clearances. (Federal-Mogul)

steady stream indicates excessive clearance. When a steady stream is found, rotate the crankshaft one-half turn in case registration (alignment) of the oil holes is responsible. A rate of fewer than 20-25 drops per minute indicates insufficient clearance, **Figure 28-13.**

An engine that has been in storage for any length of time should be pressurized upon installation, even though it may have been pressurized upon completion of the overhaul.

Fill with Oil

Bring the oil level in the pan up to the full mark. Do not overfill. Use a top-quality oil of the type and viscosity recommended by the manufacturer. Prestart pressurizing eliminates the need to add additional oil to fill the filter.

Fill the Cooling System

Fill the radiator to the full mark (do not overfill) with the factory-recommended mixture of clean water and antifreeze. This will provide water pump seal lubrication, resistance to leaks, protection from rust and corrosion, and protection from freezing and overheating.

Caution: Do not use water alone in the cooling system. Most carmakers insist on the use of antifreeze. Failure to add antifreeze could damage the engine and void the manufacturer's warranty.

Check the Fuel Tank

Check the fuel level, and if needed, add fuel to the tank. If the vehicle has been in storage for an extended period of time, it is a good idea to drain the tank and refill it with fresh fuel. If the tank is to be drained, refer to Chapter 17, Fuel Delivery, for the necessary safety precautions.

Check the Battery

If possible, check the battery electrolyte level and add distilled water to bring the electrolyte up to the proper level. The battery should be in good condition and fully charged. Install the battery and connect the battery cables.

Double-Check before Starting

Before starting the engine, double-check the following items:

- Engine oil level.
- Coolant level in the radiator.
- Spark plug wire installation order.
- Point gap or reluctor-to-armature air gap, if applicable.
- Ignition timing.
- Valve clearance.
- Automatic choke adjustment, if applicable.
- Transmission shifter position.

Remove tools, extension cords, and wiping cloths from the engine compartment. Have a fire extinguisher ready in case of emergency.

Starting the Engine

A properly overhauled engine should start readily. Fuel injection systems and carburetors on cars equipped with electric pumps will fill if the ignition switch is turned on for about 60 seconds without cranking.

Vehicles with mechanical fuel pumps may have to be cranked for a short time to allow the fuel pump to fill the carburetor. Some technicians fill the carburetor with a gravity feed can or use an electric pump to force gas from a container into the carburetor. This helps eliminate excessive cranking.

If the vehicle has a carburetor, make certain the automatic choke has closed the choke valve. Do not place the palm of your hand over the carburetor air horn to choke the engine while cranking. A backfire through the carburetor could inflict a serious burn.

If the engine cranks slowly because of an old battery, a booster battery may be used to facilitate starting. Hook the booster in parallel to the car battery—positive to positive and negative to negative. Never hook a booster battery in series. A series hookup doubles battery voltage and can cause extensive damage to the electrical system. See Chapter 14 on charging and starting systems for details on the use of booster batteries.

Proper Break-In

Modern design, materials, machines, and repair procedures make it possible to assemble an engine with highly accurate clearances and controlled finishes. This has eliminated the long, old-fashioned ***break-in period.*** Despite the fact that break-in is simplified, proper break-in is still of major importance. The first hour or so of engine operation is extremely critical. Lubrication, rpm, temperature, and loading are all vital. If these are correct, they will produce proper wearing in of the rings, cylinder walls, and bearings until mating parts are smooth enough to provide proper sealing and to reduce friction to a normal level.

Failure to follow accepted break-in rules may result in extensive engine damage from scuffing and scoring. The amount of damage is often hard to determine immediately, but the engine may fail in service thousands of miles sooner than would be normally expected.

Run the Engine at Fast Idle

When the engine is first started, the choke (on older vehicles) or the electronic control system (on newer vehicles) will cause the idle speed to be higher than normal. This is desirable to ensure an oil pressure and throw-off sufficient to adequately lubricate the cylinder walls. Run the engine at this speed until normal operating temperature is reached (usually 15 to 20 minutes). If the cold idle is not approximately 1200 rpm, raise the speed by some means until the engine is warm.

If it was not possible to pressurize the lubrication system before starting the engine, a few squirts of engine oil should be directed into the carburetor air horn during the first minute or two of operation. This will provide cylinder lubrication until bearing throw-off and "spit hole" lubrication take over. Follow the manufacturer's recommendations.

Check oil pressure and coolant temperature occasionally during the warm-up. Also check coolant level, since there may have been air in the system when filling. Inspect for any signs of fuel, oil, or coolant leakage.

Provide Adequate Ventilation

Never operate on an engine in an enclosed area. Open windows and doors to provide fresh air. If special ***exhaust disposal lines*** are available, run the exhaust into these lines. If the weather permits, run the engine outside.

Remember that exhaust gases contain ***carbon monoxide.*** Carbon monoxide is a deadly poison and is cumulative. If you are exposed to exhaust fumes every day, even in relatively small amounts, it will build up in your system until you become physically ill. Avoid all possible exposure to exhaust fumes.

Make Final Adjustments

When normal operating temperature is reached, turn off the engine. ***Retorque*** the head and manifold bolts, and give valve clearance a final adjustment, if required, **Figure 28-14.** If setting mechanical lifter clearance, use the hot setting recommendations. Start and run the engine long enough to check the ignition timing and idle speed if they are not controlled by the computer.

Do a Road Test/Break-In Run

Drive the vehicle to a spot where you can safely reach a speed of 50 mph (80km/h). Accelerate rapidly up to about 50 mph (80km/h). Immediately let up on the accelerator and allow the car to coast down to around 30 mph (48km/h). Drive the vehicle at 30 mph (48km/h) for a block or so and again accelerate rapidly up to 50 mph (80km/h). Once again, coast back to 30 mph (48km/h). Accelerate and coast fifteen to twenty times. When slowing down, watch for cars behind you.

The object of the acceleration is to increase ring loading against the cylinder walls, speeding up break-in. During the coast period, strong vacuum in the cylinders will draw additional oil up around the rings.

Devote as much time as practical to the break-in run. The fifteen to twenty acceleration-coast cycles mentioned are minimum requirements. Observe oil pressure, temperature, steering, braking, engine performance, and shifting during the run. When back at the shop, check again for any possible leakage.

Dynamometer Break-In

If possible, mount the engine or vehicle on a ***dynamometer*** for break-in. The dynamometer allows the technician to make a road test in the shop. See **Figure 28-15.**

Deliver the Vehicle to the Owner

After the initial break-in, final checks, and cleanup, the vehicle is ready for delivery.

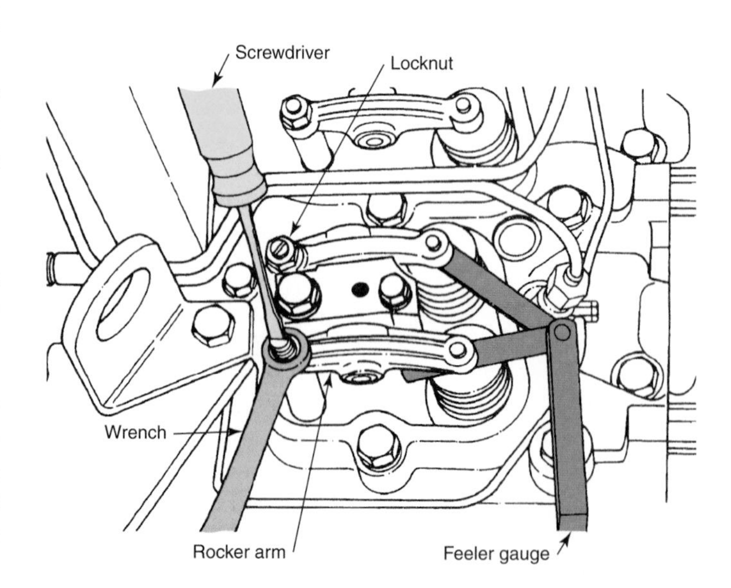

Figure 28-14. After the initial break-in, it is a good idea to recheck valve clearances on engines that do not use self-adjusting valve trains. Be sure to follow the manufacturer's recommendations. (Cummins)

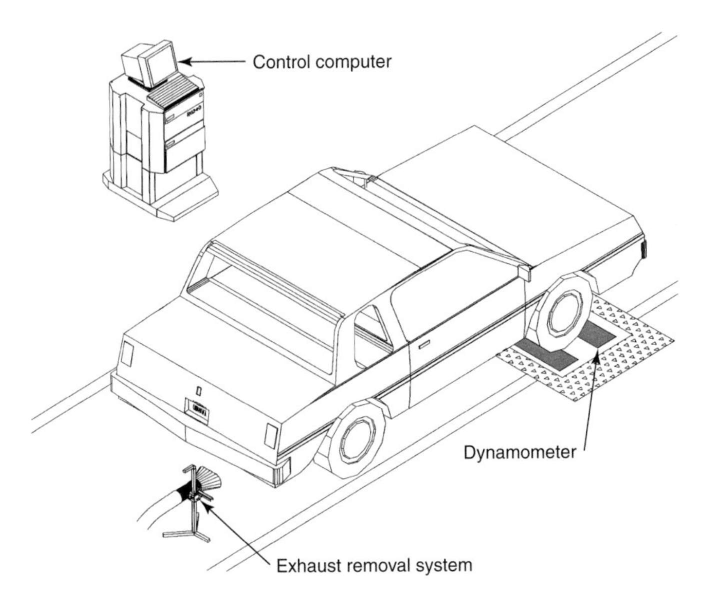

Figure 28-15. A front-wheel drive vehicle having a newly rebuilt engine broken in on a dynamometer. (Clayton Mfg.)

Note: Be sure the owner is told to avoid sustained high-speed driving and heavy loads during the first 200 to 300 miles (320 to 480km). Also inform the owner that oil consumption may be noticeable until the rings are seated.

Ask the owner to bring the car in at the end of the first 500 miles (800km) for an oil change and a checkup. During break-in, metal particles will be dislodged and enter the oil. Changing the oil and the filter at the end of the first 500 miles (800km) eliminates the possibility of engine damage caused by the prolonged use of contaminated oil. At the time of the oil change, be sure to check for leaks. Also, ask the owner how the car has been performing.

Ring Seating Takes Time

The overhauled engine may continue to burn oil for a short period of time. Oil consumption should drop to an acceptable level after about 2500 miles (4000km) or 65 hours of operation. Remember that the normal amount of oil consumption varies with engine condition, design, vehicle use, operating conditions, and driving habits. A general rule of thumb is to consider excessive oil consumption a condition in which a quart of oil is consumed in less than 700 miles (1120km). One quart per 1500 miles (2400km) may be considered good oil mileage. These are approximations and are for normal driving.

The Engine Needs Help

While an engine may be in good mechanical condition, the ignition, fuel, and cooling systems also must be functioning properly or engine performance will be substandard.

Protect your engine overhaul efforts and your reputation by encouraging the owner to have essential work performed in these areas. Point out why the work is important and what can be expected if it is not done. It is a good idea to put the suggestions in writing, so the owner will not later say, "I certainly wish the technician would have told me this work was needed."

Tech Talk

Dirt in an overhauled engine will cause scuffed pistons, worn bearings, clogged ports, sticking and collapsed lifters, and many other problems. Much of the dirt that enters the engine during an overhaul is the result of carelessness. This is especially true if you lay engine parts around the shop where dust, grinding grit, welding smoke, and other contaminants can get to them. The undersides of parts can pick up dirt and metal filings from workbenches. To keep dirt out of the engine, make sure that you do the following:

1. Thoroughly clean each part before and after working on it.
2. Cover all parts that will not be assembled immediately.
3. Carefully inspect all parts before installing them.

When working on all automotive assemblies, think clean!

Summary

Engine component order of assembly varies. Study the engine construction and "think ahead." Do not rush the assembly work. Proceed carefully, and follow safety rules when installing the engine. If a manufacturer's shop manual is available, use it for assistance. Lower the engine into place slowly and avoid damage to the vehicle or accessory units.

Attach all wiring, hoses, and tubing. Connect transmission, clutch, and accelerator linkages. Torque engine mounts and adjust belts.

Pressurize the lubrication system before starting the engine. Bring the oil level up to the full mark. Fill the radiator and check the battery. Clear away tools, cords, and other items. Have a fire extinguisher ready. Never use your hand for a choke.

Start the engine and run it at about 1200 rpm until normal operating temperature is reached. Check for leaks while the engine is running. Also check oil pressure and coolant temperature. Shut off the engine and retorque the heads and manifolds. Drive the car up to 50 mph (80km/h) and then coast to 30 mph (48km/h). Repeat this procedure fifteen to twenty times. If desired, the engine may be run in on a dynamometer. Instruct the owner as to the importance of proper operation during the first 200 miles (240km), a 500-mile (800km) oil change, and taking care of any needed adjustments or repairs in the cooling, fuel, and ignition systems.

Review Questions—Chapter 28

Do not write in this book. Write your answers on a separate sheet of paper.

1. Excessive valve clearance will _____.
   (A) increase horsepower
   (B) cause early valve opening
   (C) prolong the life of the valve
   (D) cause late valve opening and a lower lift

2. When installing the engine, the transmission should be _____.
   (A) on the engine
   (B) in the car
   (C) on the bench
   (D) varies with different makes and models
3. List eight things that should be double-checked before attempting to start an overhauled engine.
4. When hooking up a booster battery, connect _____.
   (A) positive to positive and negative to negative
   (B) positive to negative and negative to positive
   (C) positive and negative both to car battery positive
   (D) positive and negative both to car battery negative
5. Name four things that should be checked during the initial warm-up period.
6. Name five things that should be done after the initial warm-up.
7. Brief, rapid acceleration followed by coasting will help to _____ on a new engine.
   (A) lube the camshaft lobes
   (B) wear in the rod bearings
   (C) seat the rings
   (D) seat the valves
8. Break-in can also be performed with the engine mounted on a _____.
9. Tell the vehicle driver to avoid sustained high-speed driving and heavy loads during the first _____ miles of operation.
   (A) 50 (80km)
   (B) 100–200 (160–320km)
   (C) 200–300 (320–480km)
   (D) 500–1000 (800–1600km)
10. The first oil change on the overhauled engine should occur at _____.
   (A) 500 miles (800km)
   (B) 1000 miles (1600km)
   (C) 2500 miles (4000km)
   (D) 7500 miles (12000km)

ASE-Type Questions

1. Technician A says there is an exact order of assembly that applies to all engines. Technician B says anticipating problems will slow both engine assembly and engine installation. Who is right?
   (A) A only.
   (B) B only.
   (C) Both A and B.
   (D) Neither A nor B.
2. Technician A says that hydraulic lifter plungers should be about in the center of their travel when properly installed and adjusted. Technician B says that the valve lifter should be on the high point of the cam to adjust valve clearance. Who is right?
   (A) A only.
   (B) B only.
   (C) Both A and B.
   (D) Neither A nor B.
3. The lubrication system should be pressurized before starting the engine to prevent immediate engine _____.
   (A) wear
   (B) noise
   (C) overheating
   (D) Both A and B.
4. Technician A says pressurizing the lubrication system will protect against engine damage and can be used to check for excessive bearing leakage. Technician B says the oil placed on parts during assembly will provide adequate lubrication when the engine is first started. Who is right?
   (A) A only.
   (B) B only.
   (C) Both A and B.
   (D) Neither A nor B.
5. Technician A says clean water is all that is needed to fill the cooling system. Technician B says placing a hose in the radiator and allowing the hose to run while the engine is warmed up will help the rings seat. Who is right?
   (A) A only.
   (B) B only.
   (C) Both A and B.
   (D) Neither A nor B.
6. When the engine is first started, it should be operated at _____ until normal operating temperature is reached.
   (A) normal idle
   (B) fast idle
   (C) 3000 RPM
   (D) varying speeds
7. During the road test, you should accelerate to a speed no faster than _____.
   (A) 30 mph (48km/h)
   (B) 50 mph (80km/h)
   (C) 65 mph (104km/h)
   (D) 55 mph (88km/h)

8. A newly rebuilt engine should be road tested for _____.
   (A) 5 to 10 acceleration-coast cycles
   (B) 15 to 20 minutes minimum
   (C) 20 to 30 minutes maximum
   (D) 15 to 20 acceleration-coast cycles
9. Engine _____ may be noticeable until the rings are seated.
   (A) noise
   (B) oil consumption
   (C) leakage
   (D) sluggishness
10. A rebuilt engine burns one quart of oil in the first 500 miles (800km) of operation. Technician A says the cause could be rings that have not yet seated. Technician B says the vehicle must be driven further to determine whether oil consumption is excessive. Who is right?
   (A) A only.
   (B) B only.
   (C) Both A and B.
   (D) Neither A nor B.

Suggested Activities

1. Using a factory service manual as a reference, write step-by-step instructions for reassembling a particular engine.
2. Locate the engine mounts on various vehicles and make sketches of mount locations.
3. Using the shop engine hoist and a junk engine, determine how to attach the lifting fixtures to keep the engine:

   level.

   tilted toward the rear.

   tilted toward the front.

   tilted to the right or left side.

   After deciding where to make the attachments, lift the engine a few inches from the floor and determine whether your calculations are correct.

Caution: Engines are heavy. Your instructor should supervise this activity.

4. Identify all the connections (electrical, fuel, and vacuum lines, brackets, control cables, and linkages) that must be made when reinstalling an engine in a vehicle.

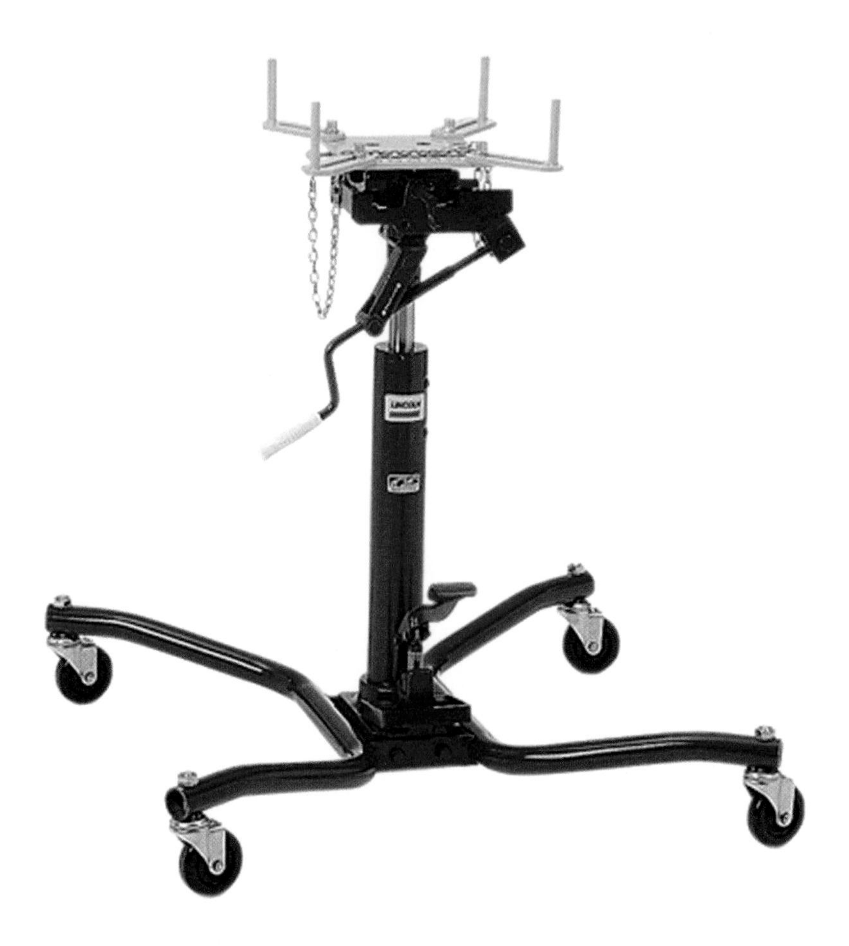

A transmission jack, such as the one shown above, should be used when removing a manual transmission or transaxle during clutch service. (Lincoln Automotive)

29

Clutch and Flywheel Service

After studying this chapter, you will be able to:

- Explain the construction, operation, and service of diaphragm spring clutches.
- Describe the construction, operation, and service of coil spring clutches.
- Adjust different types of clutch linkages.
- Summarize clutch break-in procedures.
- Diagnose clutch problems and suggest possible corrections.

Technical Terms

Friction disc	Pilot shaft
Asbestos	Glaze
Coil springs	Heat checking
Diaphragm spring	Ring gear
Pressure plate	Pilot bushing
Throwout bearing	Pilot bearing
Throwout fork	Free play

Although most new vehicles are sold with automatic transmissions or transaxles, a significant percentage of vehicles on the road have manual transmissions. Every vehicle with a manual transmission or transaxle has a clutch, and every clutch will eventually wear out. Therefore, the technician must be familiar with clutch service. This chapter will cover the service of automotive clutches and flywheels.

Clutch Service

The clutch is a simple friction device that connects and disconnects the engine from the transmission or transaxle. All modern clutches are of the single-plate type—they have one ***friction disc*** installed between the flywheel and pressure plate surfaces. Servicing the clutch is usually confined to making linkage adjustments or replacing clutch components. These procedures are explained below.

Asbestos Warning

Clutch friction materials contain ***asbestos***—a known carcinogen (substance that can cause cancer). Small airborne particles of asbestos when grinding clutch friction materials, cleaning clutch assemblies, etc. These particles, which may cause cancer, are easily inhaled by the technician, unless appropriate procedures are followed. When exposure to asbestos fibers is unavoidable, wear an *approved respirator*. Never use compressed air to blow clutch assemblies clean. Clean the assembly by flushing with water, or using a vacuum device.

Clutch Design and Operation

The modern, single-plate, dry-disc clutch uses either individual, direct-pressure ***coil springs*** or a ***diaphragm spring*** to apply force to the ***pressure plate,*** **Figure 29-1.** A direct-pressure, coil-spring-loaded clutch is illustrated in Figure 29-1A. The diaphragm spring clutch is shown in Figure 29-1B.

Determining the Reason for Clutch Failure

Before repairing a damaged clutch, it is a good idea to study the various parts to determine the cause of the failure. Merely replacing the worn or damaged parts, without identifying the reason for the wear or damage, is an invitation to trouble.

Before condemning a clutch as defective, check for correct pedal free play as explained later in this chapter. Inspect the linkage for wear and binding. Check and tighten the engine mounts. Tighten the rear spring clamp or U-bolt assemblies.

Removing the Clutch

Using suitable tools and equipment, remove the transmission or transaxle. Chapter 30 contains specific instructions for removing manual transmissions and transaxles.

Warning: Before proceeding, disconnect the battery ground strap to prevent the engine from being cranked while working on the clutch.

Mark the Pressure Plate and Flywheel

Use a sharp prick punch to mark both the pressure plate and the flywheel, **Figure 29-2.** If the pressure plate assembly will be used again, it must be reinstalled in exactly the same position. This is necessary so the balance of the flywheel-clutch assembly will not be thrown off.

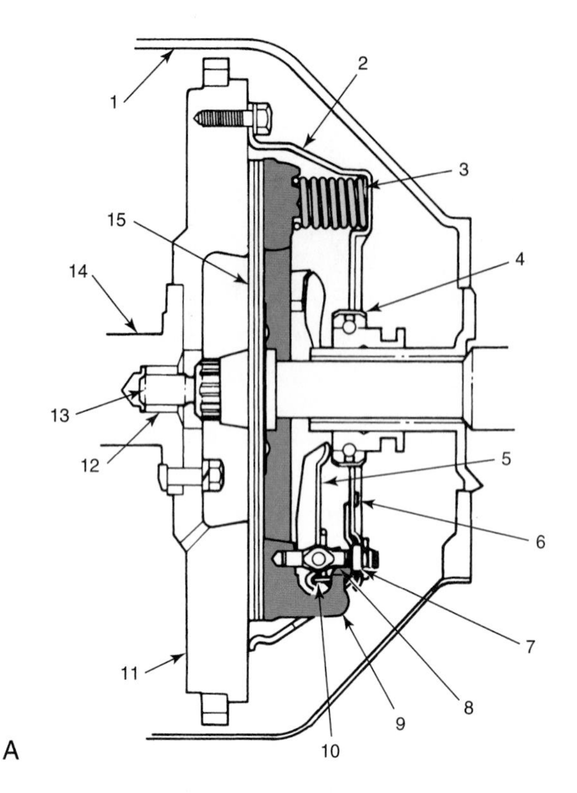

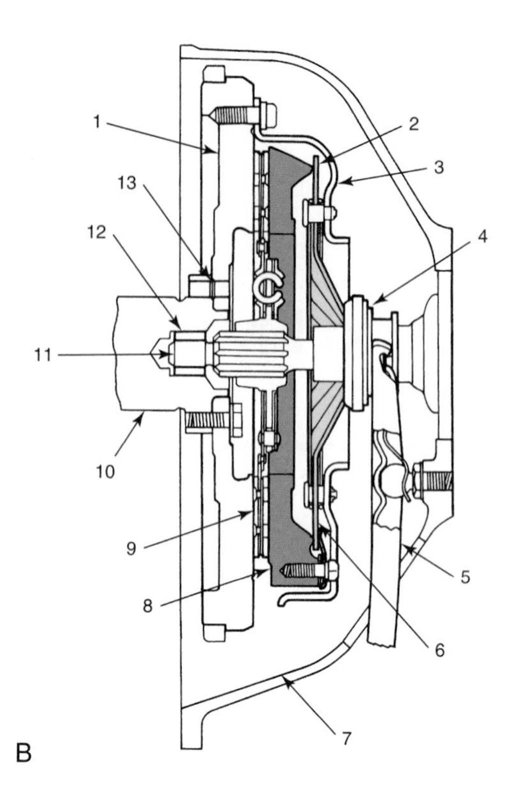

Figure 29-1. Clutch design. A—With this clutch, force is applied to pressure plate by coil springs. 1—Bell housing. 2—Pressure plate clutch cover. 3—Coil spring. 4—Clutch release bearing. 5—Release lever. 6—Anti-rattle spring. 7—Adjustment nut. 8—Eyebolt. 9—Pressure plate. 10—Strut. 11—Flywheel. 12—Pilot bearing. 13—Transmission input shaft. 14—Crankshaft. 15—Drive disc. (Chevrolet) B—Force is supplied to pressure plate by diaphragm spring. 1—Flywheel. 2—Diaphragm. 3—Pressure plate. 4—Clutch release bearing. 5—Throwout fork. 6—Retracting spring. 7—Bell housing. 8—Pressure plate. 9—Drive disc. 10—Crankshaft. 11—Transmission input shaft. 12—Pilot bushing. 13—Dowel hole. (Chevrolet)

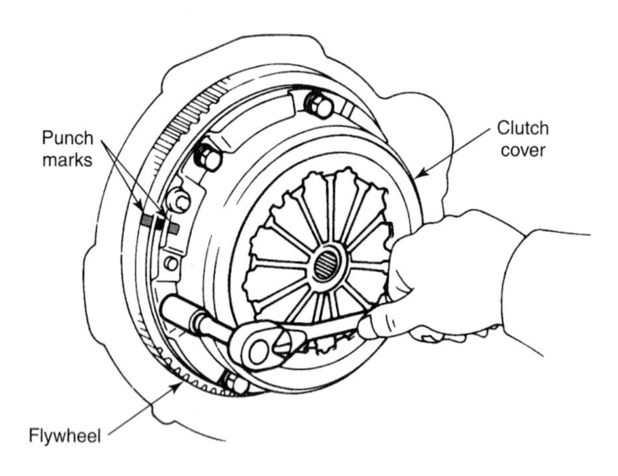

Figure 29-2. Mark the pressure plate and flywheel before removing the pressure plate.

Remove the Pressure Plate

When the transmission or transaxle has been removed, pull the ***throwout bearing*** out of the ***throwout fork.*** If working clearance is needed, disconnect the clutch linkage from the fork. The fork may then be tipped back or removed.

In some instances, the clutch assembly can be exposed for removal by taking off a sheet metal splash pan. Other designs require the removal of the entire flywheel housing. Remove only those parts necessary for access.

Loosen each pressure-plate-to-flywheel fastener one turn at a time. Continue around the fasteners until all have been loosened. This permits releasing the pressure of the coil springs or diaphragm spring evenly and prevents twisting or warping of the pressure plate.

If desired, a clutch disc ***pilot shaft*** can be inserted through the clutch disc hub and into the crankshaft pilot bearing, **Figure 29-3.** This will hold the disc in position until the pressure plate fasteners are removed. A used transmission input (clutch) shaft can be used in place of a pilot shaft.

Remove all but one of the pressure plate fasteners. While pressing the pressure plate assembly up and in, remove the final fastener. Remove the pressure plate and clutch disc, **Figure 29-4.** Do not touch the pressure plate, clutch disc, or flywheel clutch surface with greasy fingers.

Cleaning Parts

Clean the flywheel face and pressure plate assembly using a nonpetroleum-based cleaner. Brake or electrical winding cleaner is ideal for this purpose, **Figure 29-5**. Both units must be spotlessly clean and absolutely free of oil or grease. Once cleaning is completed, use care to prevent contamination.

Never wash the throwout bearing in any kind of solvent. The throwout bearing is packed with grease and sealed. Washing it would remove or dilute the lubricant. The throwout

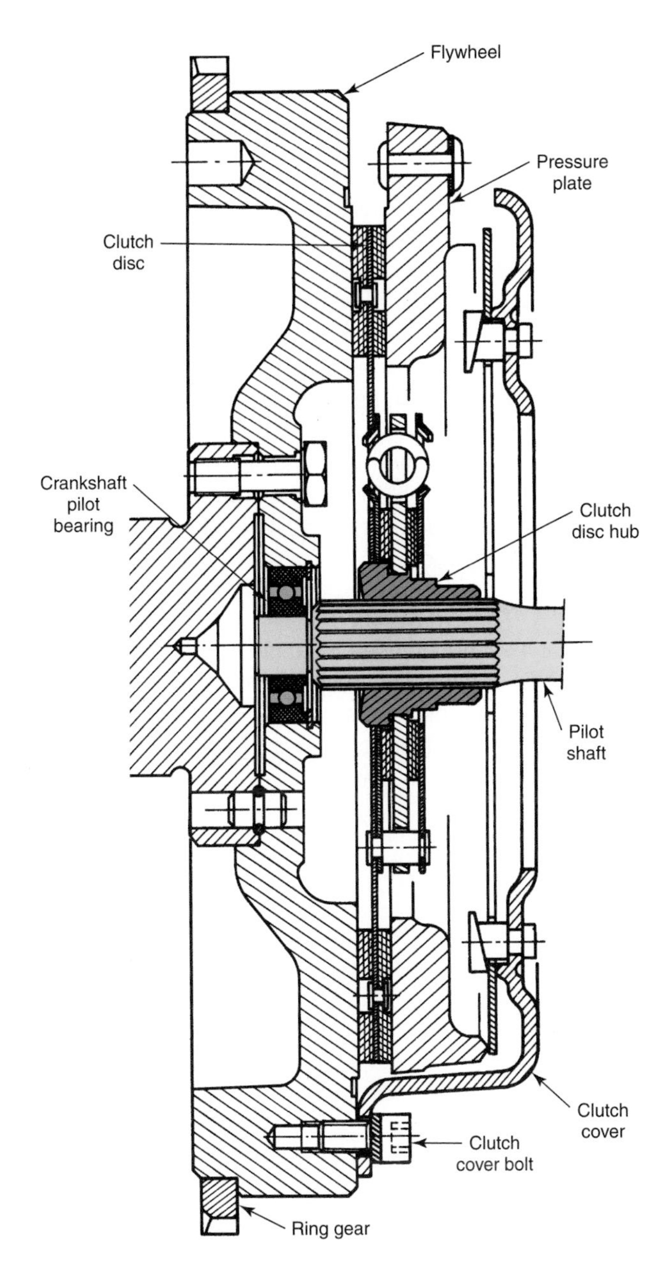

Figure 29-3. Inserting a pilot shaft through the clutch disc hub and into the crankshaft pilot bearing will properly align the parts as the fasteners are secured.

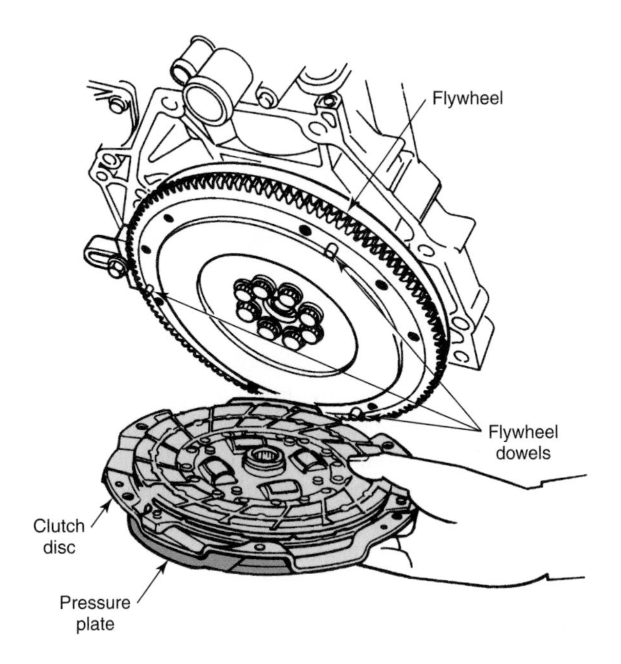

Figure 29-4. Removing the clutch disc and pressure plate. (Honda)

Figure 29-5. A nonpetroleum-based brake cleaner or electrical winding cleaner can be used to degrease the flywheel face and pressure plate assembly.

bearing may be wiped off, using a clean cloth moistened with solvent.

Sanding Clutch Friction Surfaces

Use medium-fine emery cloth or equivalent aluminum oxide paper to sand the friction surfaces of the flywheel and the pressure plate. Sand lightly until the surfaces are covered with fine scratch lines. The sanding scratch marks should run *across* the surface.

This will break the ***glaze*** (coating of shiny, hardened metal) on these surfaces and remove any carbonized oil deposits. The new clutch disc will seat smoothly and quickly against the sanded surfaces.

Note: If the surfaces of the flywheel and pressure plate cannot be restored by sanding, they must be machined (usually called *turning*) with special equipment to remove any heat marks or grooves.

Inspecting the Flywheel-to-Clutch Friction Surface

The flywheel-to-clutch friction surface must be clean, dry, lightly sanded, and free of heavy ***heat checking*** (cracking). It must also be free of scoring and warpage. Do not expect a clutch disc and pressure plate to work properly when assembled on a glazed, dirty, rough, or warped flywheel surface.

Checking Flywheel Runout

To check flywheel runout, set up a dial indicator and rotate the flywheel. Force the flywheel in one direction while turning to prevent end play from affecting the reading. Determine runout by reading the indicator. Check both the clutch disc contact surface and the rim edge. If necessary, replace the flywheel or have the friction surface reground. **Figure 29-6** shows a flywheel clutch surface being checked for warpage with a dial indicator.

Mounting the Flywheel

The flywheel crankshaft flange surface must be clean and true. If dowel pins are used, properly align the flywheel. Install cap screws and torque them to specifications. If required, coat the cap screw threads with oil-resistant sealer. Use lock washers or a lock plate when needed. Some flywheels use a hardened plate that is intended to protect the flywheel against damage from the cap screw heads. If the clutch disc contact face is scored or wavy, it must be reground, using a special grinding machine, **Figure 29-7.**

Removing the Flywheel Ring Gear

The flywheel ***ring gear*** is located on the outside of the flywheel and contacts the starter teeth during cranking. Some ring gears are heated so that they expand, and then they are placed in position. Upon cooling, they shrink and grip the flywheel tightly. Other ring gears are welded or bolted in position.

If the ring gear is welded to the flywheel, begin removal by cutting the welds. Next, heat the old gear and drive it from the flywheel. Ring gears that are shrink-fit can also be heated (keep heat away from flywheel) and removed. If necessary, the ring gear can also be drilled almost through and a chisel used to spread the gear for removal. When driving the gear off, do not strike the flywheel.

Installing the Ring Gear

If a shrink-fit will be used, heat the gear to no more than 450°F (232°C). A controlled temperature oven is ideal for this purpose. If this type of oven is not available, the gear can be heated in oil. Use a thermometer and keep the gear from touching the bottom of the tank.

Caution: To avoid fire, cover the tank while heating.

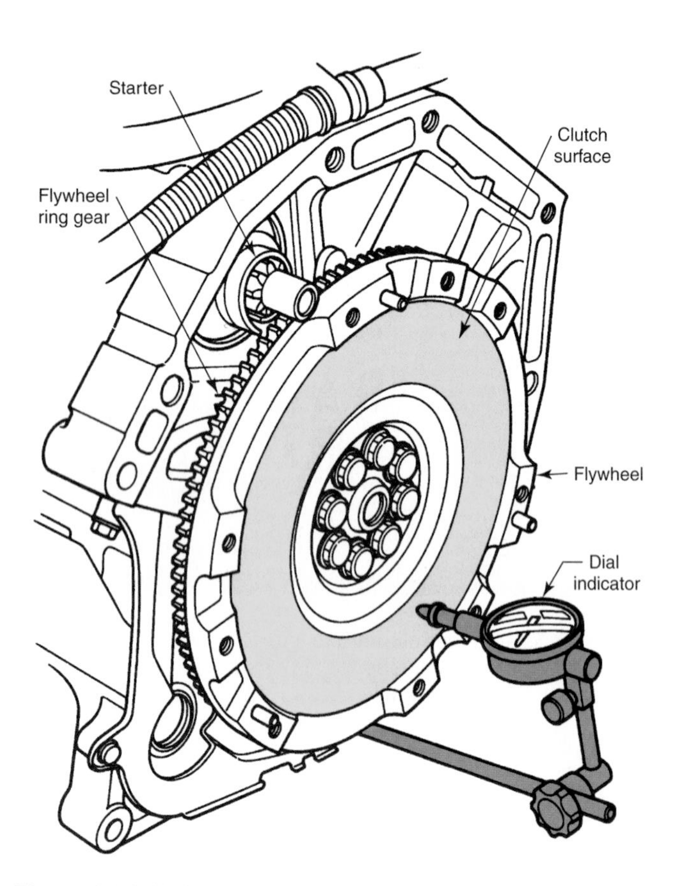

Figure 29-6. Using a dial indicator to check for warping of the flywheel face. (Honda)

Figure 29-7. A special grinder used to resurface the clutch discs and flywheels. (Van Norman Equipment Co.)

When the ring is hot, quickly place it over the flywheel contact surface and drive it into place. Chilling the flywheel to shrink it slightly will help. Be sure the ring is placed so the relieved edge of the teeth (if so designed) face in the desired direction, **Figure 29-8.**

Another installation method involves placing the gear on a firebrick surface and heating it with an acetylene torch. Move the flame around the ring. Wire solder (50/50 or 40/60 type) is touched against the ring frequently as the temperature is raised. When the solder starts to melt as it is pressed firmly against the ring, the temperature is high enough. The 50/50 (50% tin, 50% lead) solder melts at 414°F (212°C). The 40/60 (40% tin, 60% lead) solder melts at 460°F (238°C).

Caution: Do not overheat the ring, or the teeth will be softened.

When installing a ring gear that is arc-welded to the flywheel, add as little weld material as possible to maintain flywheel balance.

Checking the Pressure Plate Assembly

Inspect the pressure plate for excessive burning, heat checking, warpage, and scoring. Check the coils or diaphragm spring for evidence of cracking, loss of temper (overheating), and looseness. Check for wear at the ends of the release levers where they contact the throwout bearing. In the case of the diaphragm spring, check the ends of the release fingers, **Figure 29-9.**

As with the flywheel, the pressure plate friction surface must be clean, dry, lightly sanded, and free of scoring, warpage, and heavy checking. Pressure plate assemblies should be rebuilt or replaced unless they are satisfactory in all respects.

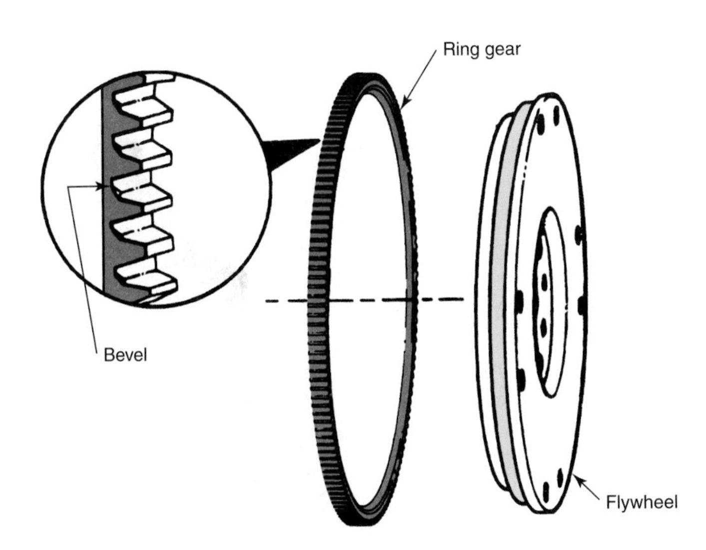

Figure 29-8. A correctly positioned, beveled-edge ring gear that is ready to be installed on the flywheel. (DaimlerChrysler)

Rebuilding a Pressure Plate Assembly

Instead of rebuilding the units themselves, many shops prefer to use rebuilt or reconditioned pressure plate assemblies. Unless proper tools are available and the technician is thoroughly skilled in clutch rebuilding, it is advisable to use a factory-rebuilt unit.

If rebuilding must be done in the shop, mark the pressure plate parts so that all the parts can be reassembled in the same relative positions. Place the assembly on a special base plate fixture or set it up in a hydraulic press. When using a hydraulic press, insert the spacer blocks under the pressure plate so the pressure plate can move downward as pressure is applied.

Apply pressure to the pressure plate while loosening the adjustment nuts. When the adjustment nuts have been removed, release the pressure. This will allow the cover to move upward and off.

Check all pressure plate parts for cracking, wear, overheating, and other damage, **Figure 29-10.** Check coil spring tension. If the pressure plate is not badly damaged, it can be resurfaced.

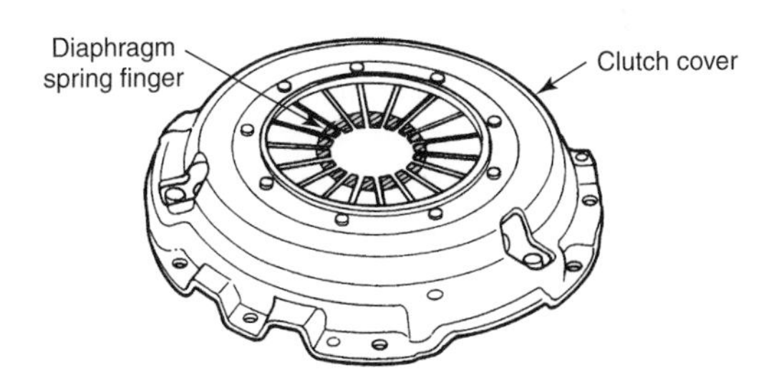

Figure 29-9. Check the pressure plate diaphragm spring fingers for damage. (Honda)

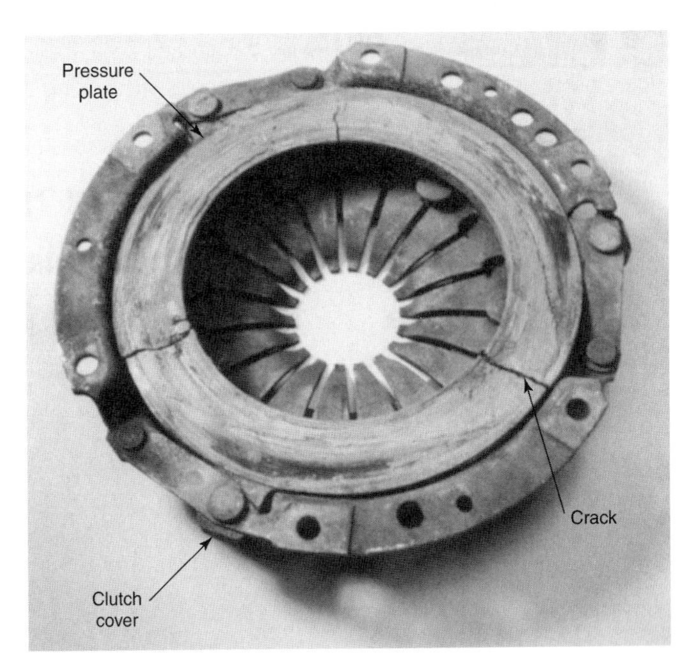

Figure 29-10. This pressure plate is badly cracked and must be replaced. (Luk Automotive)

Reassemble all pressure plate parts in their correct order. Align the marks. Use new or reconditioned parts as required. Check the clearance between the pressure plate drive lugs and the openings in the pressure plate. Apply lithium (high-temperature) grease to the lug-to-pressure-plate contact areas.

Adjusting Clutch Release Fingers

It is important the pressure plate be withdrawn an equal distance around the entire circumference when the release fingers are depressed. This will permit complete disengagement for shifting and smooth engagement when the clutch is reapplied.

After installing the pressure plate and clutch disc on the flywheel, place a straightedge across the pressure plate and measure the distance from each release finger. Carefully adjust each finger to the specified distance by tightening or loosening the adjusting nuts. See **Figure 29-11.**

Caution: Never use an air wrench to install a pressure plate. The air wrench may warp the pressure plate, leaving the actuating fingers at unequal heights.

Use a press or the clutch fixture to actuate the clutch several times. This will allow the parts to seat. Following actuation, check release finger adjustment for the final time. When adjustment is complete, some manufacturers recommend staking the adjusting nuts so they will not loosen in service. Stake the nuts in at least two places.

Replacing the Clutch Pilot Bushing or Bearing

The ***pilot bushing,*** or ***pilot bearing,*** is installed in the rear of the crankshaft and aligns and supports the transmission/transaxle input shaft. It is good practice to always install a new clutch pilot bushing (or bearing, in some cases) when doing a clutch job. Worn pilot bushings can cause clutch chatter, spot burning, and transmission/transaxle input gear damage. The pilots are relatively inexpensive and are easily changed.

An expandable finger or a threaded puller can be used to pull the pilot bushing or bearing, **Figure 29-12.** Sometimes the pilot bushing can be removed by packing the recess with heavy grease and inserting a punch that fits tightly inside the bushing. As shown in **Figure 29-13,** striking the punch will transfer the force through the grease and push the bushing out.

Clean out the pilot bushing recess in the end of the crankshaft. Wipe the outside of the new pilot with a light film of high-temperature grease. Place the pilot on a driver (chamfered inner hole end facing outward) and drive the bushing into place. When driving a bearing, use a driver that contacts the outer race only, **Figure 29-14.** Where retainers are used to secure the bearing or bushing, install them as specified.

When installing a pilot bearing, position it with the open side of the bearing facing inward (seal end facing the transmission/transaxle). Apply a thin film of high-temperature

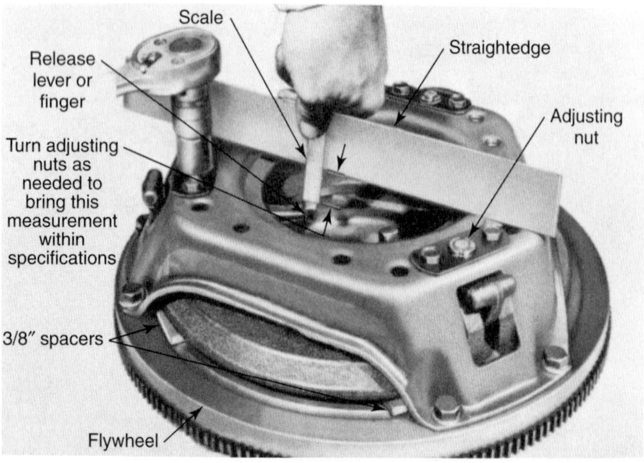

Figure 29-11. Checking the clutch release finger or lever adjustment. Turn the lever adjusting nuts as required. (General Motors)

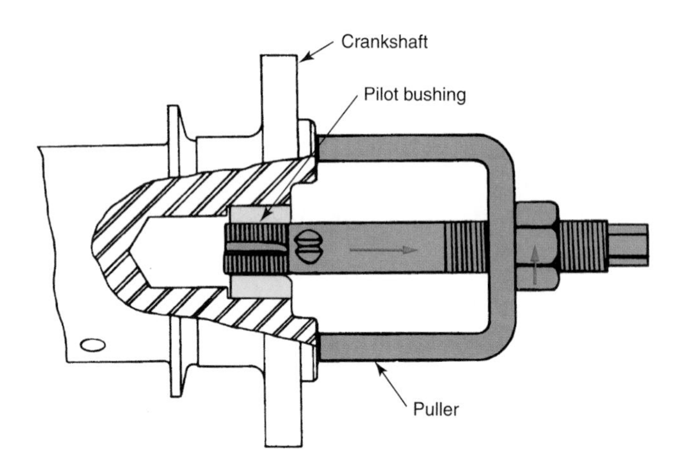

Figure 29-12. Using a threaded tip puller to remove the clutch pilot bushing. If a ball bearing is used instead of a bushing, an expandable finger tip puller will be needed. (Oldsmobile)

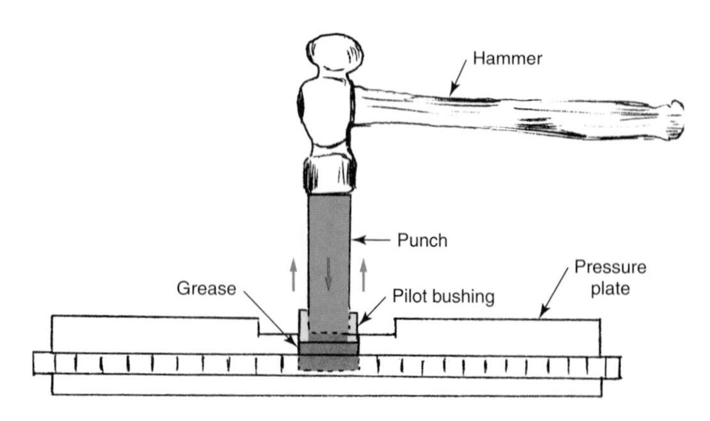

Figure 29-13. Thick grease and a snug fitting punch can be used to force the pilot bushing from its bore housing in the pressure plate. The punch tries to displace the grease, which in turn, forces the pilot bushing up and out.

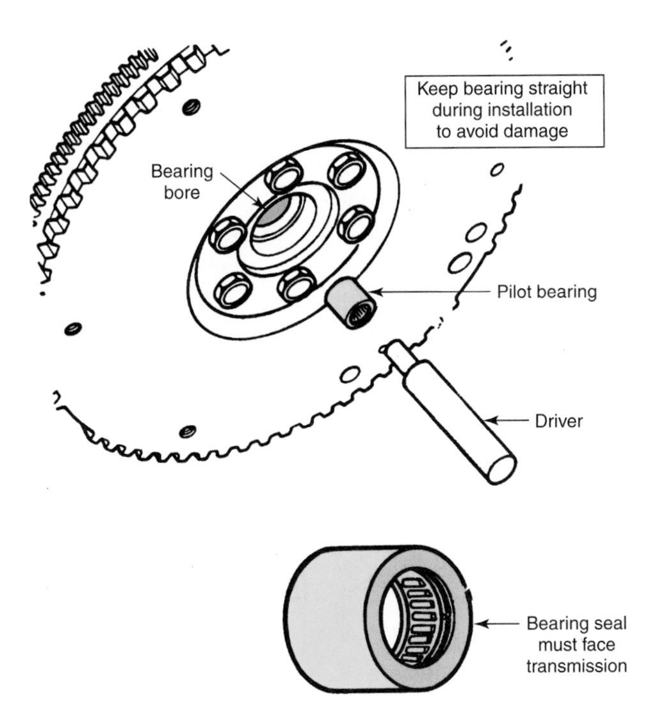

Figure 29-14. Installing a pilot bearing with a special driver tool. Start and drive the bearing squarely. (DaimlerChrysler)

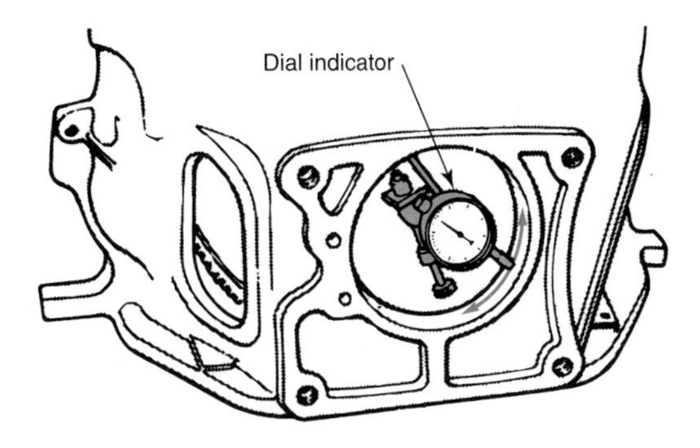

Figure 29-15. Checking clutch housing bore runout. (DaimlerChrysler)

grease to the inside of the bushing. Never over lubricate the bushing — the excess grease will find its way onto the clutch disc facing, causing clutch chatter and grabbing.

Aligning the Clutch Housing as Necessary

The portion of the clutch housing to which the transmission or transaxle is attached must be properly aligned with the crankshaft centerline.

Short throwout bearing life, clutch chatter, early input shaft bearing failure, and jumping out of gear indicate possible housing misalignment. In such instances, both the housing bore and face runout should be checked. If the clutch housing has been removed, it is good practice to check the alignment after installation.

Checking Clutch Housing Face and Bore Runout

Check for clutch housing bore runout by setting up a dial indicator as shown in **Figure 29-15.** Make certain the indicator stem rides against the machined bore surface. The indicator mounting bar must also be firmly affixed to the flywheel. The stem must be at a right angle to the bore.

Zero the indicator and slowly turn the flywheel until the indicator stem has traveled completely around the bore. Watch the indicator needle throughout the stem travel. The total needle travel (amount the needle moved from both sides of the zero mark) represents double the actual runout. For example, if the total needle travel is 0.018″ (0.46 mm), actual runout would be 0.009″ (0.23 mm).

To check housing face runout, adjust the dial indicator so the stem end rides against the machined face of the housing. The stem should be at a right angle to the face. Rotate the housing one complete turn while noting the total needle travel, **Figure 29-16.**

If either bore or face runout exceeds specifications, the situation must be corrected by using thin shims between the housing and the engine block. Some manufacturers provide offset dowel pins that can be used to correct bore runout. When the offset dowels are used, shims must be employed to correct face runout. The use of offset dowel pins is shown in **Figure 29-17.** Whenever a correction for bore runout has been made, face runout should be rechecked. When face runout has been changed, bore runout must be rechecked.

Make careful adjustments and keep checking runout until it is brought within limits. If bore or face runout cannot be brought within limits, or if an excessive shim thickness is required, replace the clutch housing.

Installing the Clutch Disc and Pressure Plate Assembly

Installing the clutch disc and pressure plate assembly is a relatively easy task. However, the job is often ruined or the service life of the assembly seriously shortened by careless

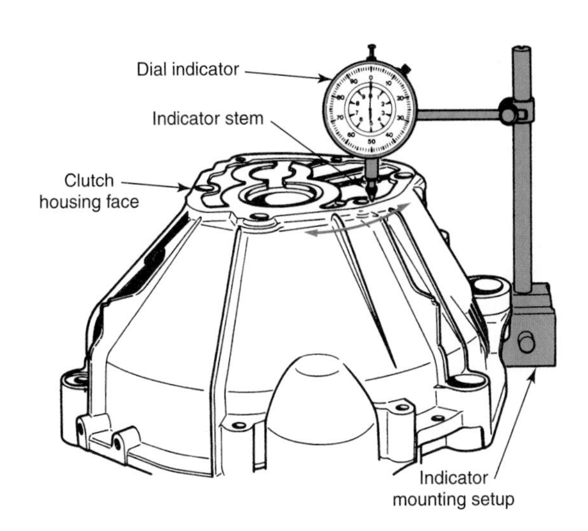

Figure 29-16. Checking clutch housing face runout with a dial indicator. The housing is turned in a circle, keeping the dial indicator stem in contact with the mounting face (surface).

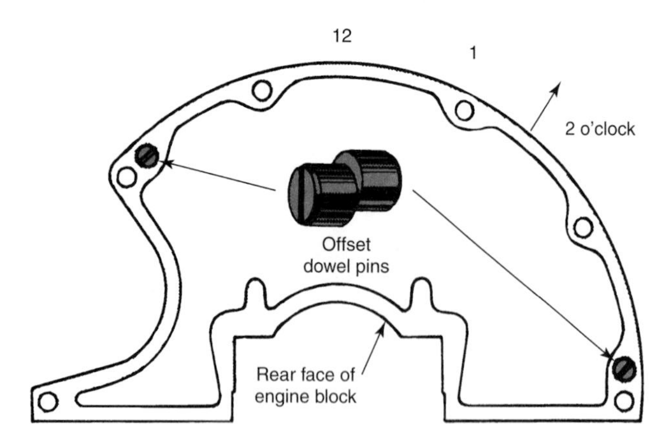

Figure 29-17. Offset dowel pins may be used to correct bore runout. (DaimlerChrysler)

handling of the parts, improper tightening, and disc damage during transmission or transaxle installation. Use care and follow the directions given in this section.

Check for Oil Contamination

Oil leaking from the transmission or transaxle, flywheel bolts, or rear main seal can quickly ruin the friction material on a new clutch disc. Always check these two potential trouble spots carefully before installing the clutch assembly. Repair as needed.

Note: Oil leaks that flow into the clutch area must be stopped.

Use a New Clutch Disc

It is advisable to install a new clutch disc whenever the clutch has been disassembled. When you consider the cost of a new disc compared to the overall costs and the assurance of proper operation and extended service life offered by a new disc, it is obviously a poor practice to reinstall an old disc.

If the old disc must be considered for installation, carefully examine the areas shown in **Figure 29-18.** Check the disc friction facing for signs of looseness, glazing, oil soaking, or wear, Examine the hub torsional coil springs (cushion springs) to make certain they are not broken or loose. Look for signs of warpage or cracking. Hub splines must be free of excessive wear. Generally, 0.03″ (0.8 mm) is the maximum amount of hub spline wear allowable. If the disc is faulty in any way, it must be discarded.

Install the Clutch Disc with the Correct Side Facing the Flywheel

Examine the clutch disc friction facing. If one side is marked "flywheel," place that side toward the flywheel. If neither side is marked, place the disc against the flywheel so the hub and hub cushion or damper spring assembly will clear both the flywheel and the pressure plate.

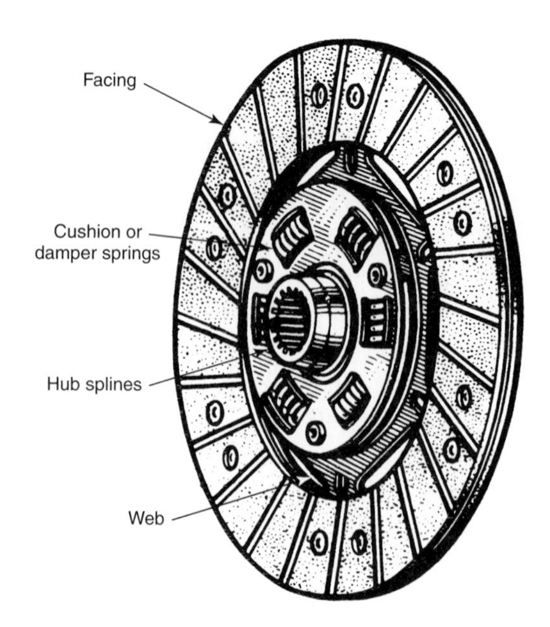

Figure 29-18. If the old clutch disc must be reinstalled, check these areas to determine serviceability. Remember, it is usually good practice to install a new clutch disc.

By careful study of the clutch and disc design, the correct side to place against the flywheel will become obvious, **Figure 29-19**. Note the long side of the hub faces away from the flywheel in Figure 29-19A. It faces toward the flywheel in the setup shown in Figure 29-19B.

Use a Clutch Disc Pilot Shaft

While holding the clutch disc and pressure plate assembly against the flywheel, insert a pilot shaft or a used transmission input shaft through the disc hub and into the pilot bearing. See **Figure 29-20.**

The pilot shaft will align the disc hub with the clutch pilot bearing. This will hold the parts in alignment while the pressure plate is attached.

Start all the pressure plate fasteners, being sure to install lock washers. Tighten each fastener one turn. Continue around the pressure plate, tightening fasteners one turn at a time, until the plate is snug against the flywheel. At this time, use a torque wrench and tighten the fasteners to specifications.

After the pressure plate is torqued, remove the pilot shaft. When the transmission or transaxle is installed, the input shaft will pass through the disc hub and into the pilot bearing without difficulty.

Install a New Throwout Bearing

Properly installed and properly used, a clutch will last a long time. Even though the throwout bearing appears to be sound, it is good practice to install a new bearing whenever the clutch is overhauled. Failure to do so may result in bearing failure long before the clutch is worn out.

If reuse of the old throwout bearing must be considered, for some reason, inspect it carefully. The bearing should spin freely, but with enough drag to indicate the presence of grease. Press the bearing against a flat surface. While maintaining

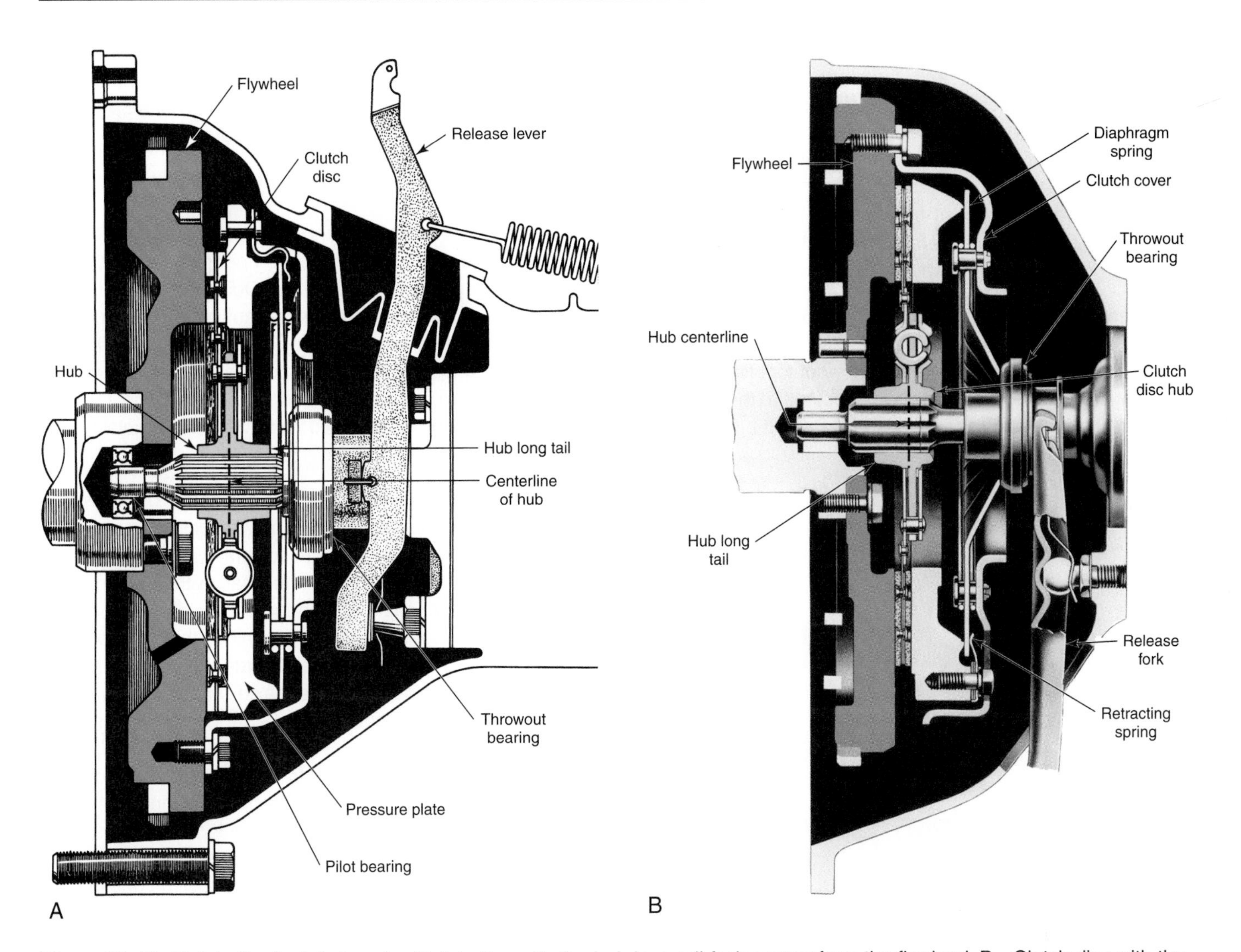

Figure 29-19. Clutch disc installation. A—Clutch disc with the hub long tail facing away from the flywheel. B—Clutch disc with the hub long tail facing the flywheel. (Toyota, Ford)

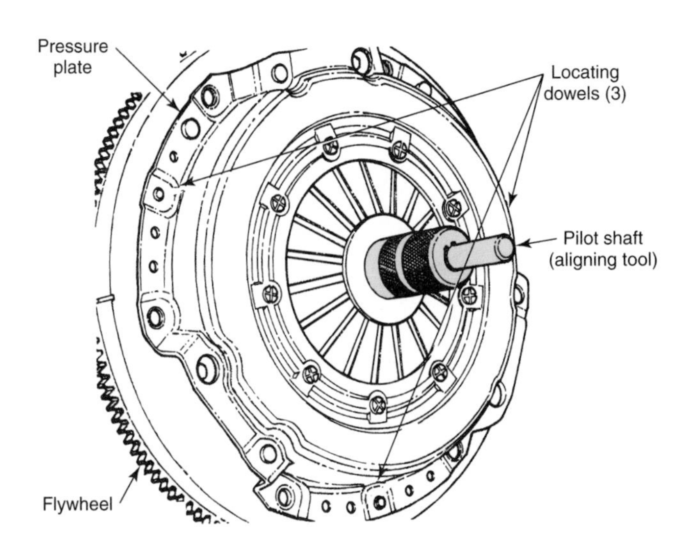

Figure 29-20. A pilot shaft, or clutch disc aligning tool, can be used to hold the pressure plate and clutch disc to the flywheel. A scrapped transmission input shaft also can be used. (DaimlerChrysler)

pressure, revolve the bearing. It should turn smoothly, with no sign of catching or roughness.

In some cases, there are provisions for greasing. If the bearing is mechanically sound, the addition of proper grease will render it fit for further service.

On some throwout bearing assemblies, the bearing may be pressed from the sleeve. This allows the use of the old sleeve by merely pressing a new bearing into place. In other sleeve designs, such as the one shown in **Figure 29-21,** the bearing is an integral part of the sleeve.

If a new throwout bearing is to be installed in the sleeve, use a press or a large vise. Never use a hammer to seat a throwout bearing. Press the bearing on squarely and until fully seated.

Pack the inner groove of the throwout sleeve with high-temperature grease. Also, coat the throwout fork groove with a *thin* coating of the same lubricant.

Hydraulic Throwout Bearing Assembly

Some systems employ a hydraulic throwout bearing assembly, which combines a throwout bearing and a slave cylinder into a single unit. The slave cylinder encircles the transmission input shaft, and the bearing is permanently attached to the cylinder's piston, **Figure 29-22.**

During operation, piston movement causes the throwout bearing to travel in a linear direction, engaging or disengaging the pressure plate.

The hydraulic throwout bearing and the slave cylinder must be serviced as an assembly. Do not try to disassemble the unit. The hydraulic lines will leak if they are disturbed.

Lubricate and Install the Clutch Throwout Fork

Lubricate the throwout fork pivot with Lubriplate or similar grease. Wipe a *thin* coating on the throwout fork fingers.

Install the fork. Make certain it is secured to the pivot and that any internal retracting spring is in place. If the fork fingers are held to the throwout bearing with retaining springs or clips, make certain the fingers are in their proper positions and the clips are in place. Install the dust boot where used. A typical throwout fork and bearing setup is illustrated in **Figure 29-23.**

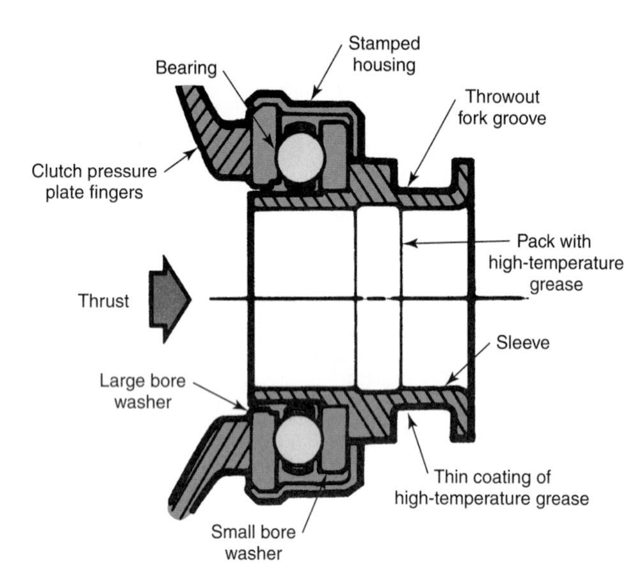

Figure 29-21. In this throwout bearing assembly, the bearing is an integral part of the sleeve. Note the use of high-temperature grease inside the sleeve and on the throwout fork groove. (Federal-Mogul)

Do Not Depress Clutch Pedal until Installation Is Complete

After the throwout fork and throwout bearing have been installed, avoid depressing the clutch pedal before the transmission or transaxle is fully in place. Doing so will exert pressure on the clutch release fingers, causing them to pull the pressure plate away from the disc. This would release the disc and allow it to drop down, preventing the transmission input shaft from passing through the disc hub and into the pilot bearing. Transmission installation would be impossible without disc realignment.

Use Care when Installing the Transmission or Transaxle

Tighten the input shaft bearing retainer bolts, **Figure 29-24.** Using a clean rag dampened with solvent, wipe the input shaft until it is absolutely clean. Dry the shaft with a clean cloth. Apply a very thin coat of high-temperature grease to the portion of the input shaft bearing retainer that supports the throwout bearing sleeve. Do not lubricate the input shaft.

If the transmission or transaxle shows signs of oil leakage through the input shaft bearing, correct the leak before installing the unit.

Mount the transmission or transaxle on a suitable stand and align it with the engine crankshaft centerline. Place the transmission/transaxle in gear. Pass the input shaft and front bearing retainer through the throwout bearing sleeve. When the input shaft splines strike the disc hub, turn the output shaft to rotate the input shaft. This will align the splines on the shaft and the hub. Push inward on the transmission as the output shaft is turned. When the splines are aligned, force the shaft through the hub and into the pilot bearing.

If the shaft resists entering the pilot, carefully move the rear of the transmission or transaxle up and down and

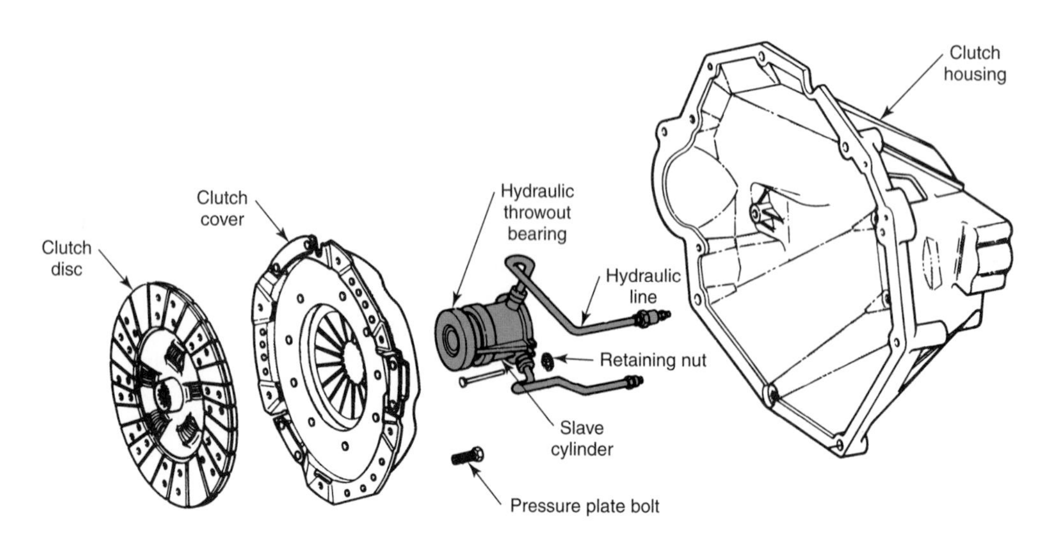

Figure 29-22. Overall view of a hydraulic throwout bearing assembly. (DaimlerChrysler)

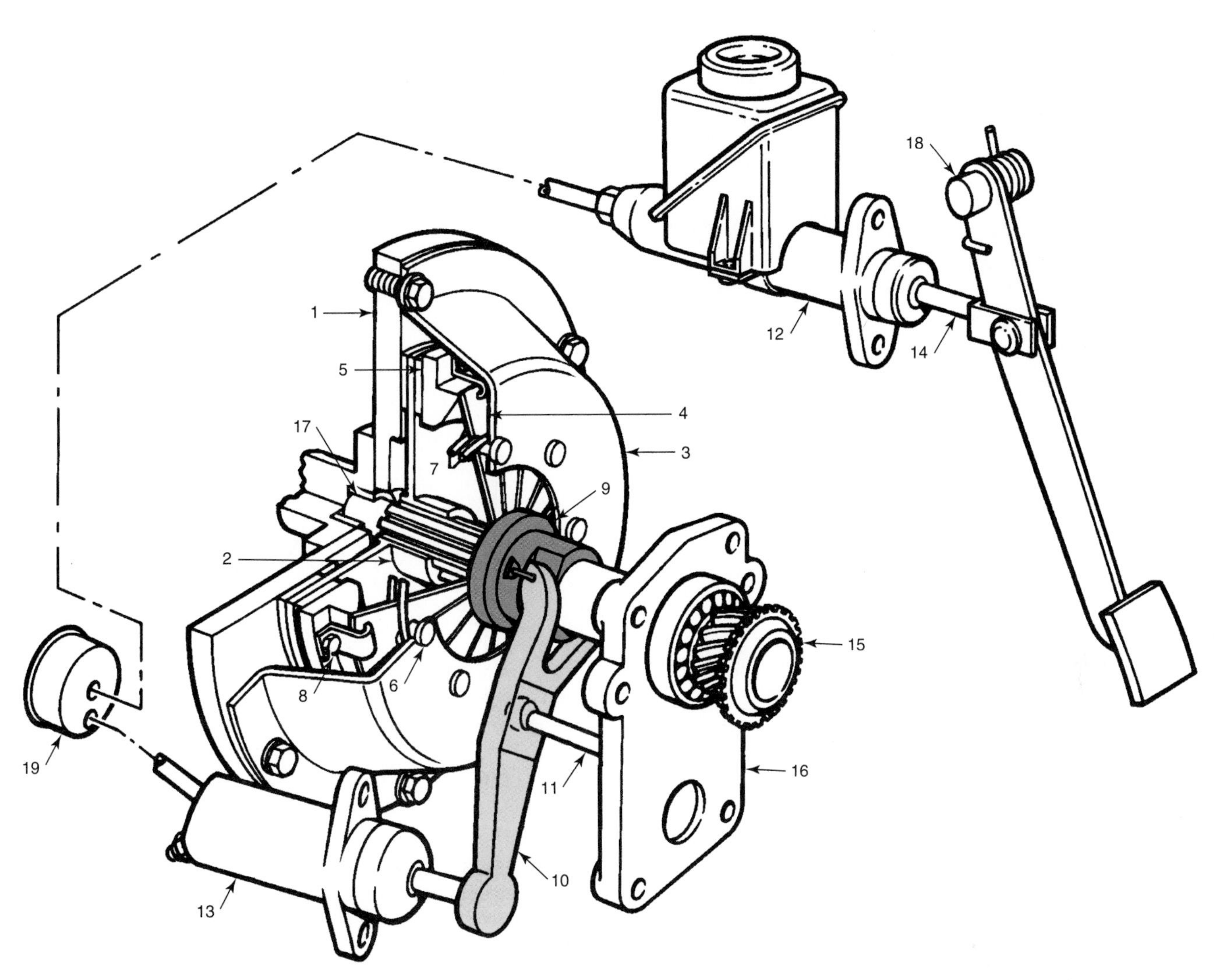

Figure 29-23. A cutaway view of a throwout fork and hydraulic system. 1—Flywheel. 2—Disc hub. 3—Clutch cover. 4—Spring finger. 5—Pressure plate. 6—Rivets. 7—Springs. 8—Bolt. 9—Throwout bearing. 10—Throwout fork. 11—Fork pivot. 12—Master cylinder. 13—Slave cylinder. 14—Push rod. 15—Transmission input shaft. 16—Input shaft retaining plate. 17—Pilot bushing. 18—Clutch pedal. 19—Hydraulic damper. (Land Rover)

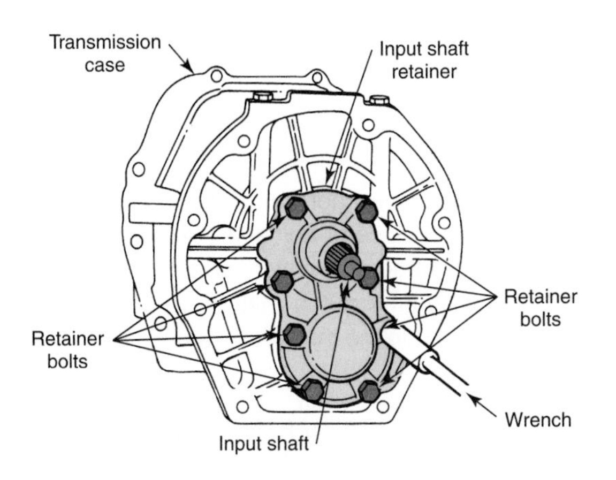

Figure 29-24. Tightening the transmission input shaft bearing retainer bolts. Torque bolts to specifications. (General Motors)

sideways. When the input shaft is fully seated, both sides of the transmission/transaxle front will be touching the clutch housing or engine block. Install fasteners and bring them to proper torque.

Never use the transmission/transaxle fasteners to draw the unit into place. Never allow the weight of the transmission or transaxle to hang on the input shaft and clutch disc; keep it supported until the fasteners are in place. **Figure 29-25** shows a cross-section of a typical clutch assembly with the transmission installed.

Attach the Clutch Linkage

Connect the clutch linkage assembly. Install all springs, washers, cotter pins, and other components. Lubricate where required. Operate the clutch pedal several times to check linkage operation. Adjust pedal free play as explained in the next section. If the dust boot is cracked or torn, replace it.

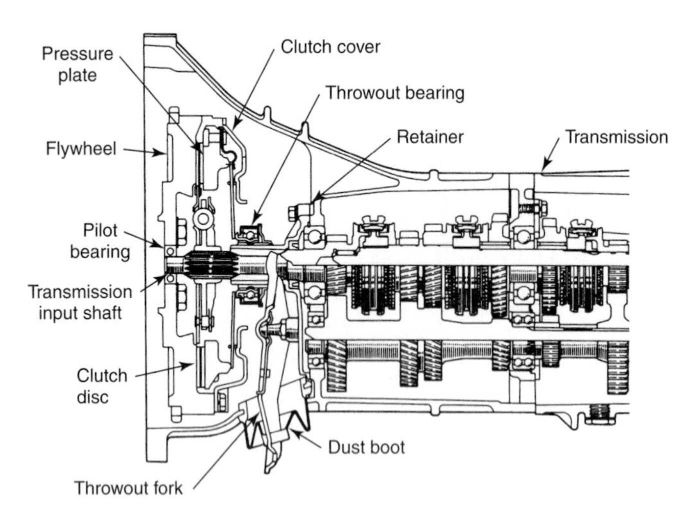

Figure 29-25. A cross-sectional view of a typical clutch assembly with the transmission attached. (Mazda)

Pedal Free Play

The throwout bearing should touch the clutch release fingers only when the clutch pedal is depressed.

Note: One clutch assembly utilizes an automatic self-adjuster that removes pedal free play and keeps the throwout bearing in contact with the release fingers at all times.

Most clutch release assemblies must be adjusted so the throwout bearing moves away from the whirling clutch release fingers when the clutch pedal is released. This allows the throwout bearing to stand still, prolonging its service life.

From the fully released position, the clutch pedal must be depressed a certain distance before the throwout bearing is forced against the clutch release fingers. This distance is called clutch pedal ***free play,*** or free travel, **Figure 29-26.** Failure to properly adjust free play can result in severe damage. If free play is insufficient, the clutch will slip and overheat, resulting in short clutch life. The throwout bearing will be in constant motion, causing it to wear out quickly. If the free play is excessive, the clutch may not release completely. This can cause hard shifting, gear clash, and damage to the gear teeth.

Important Checks before Adjusting Pedal Free Play

Some clutch assemblies provide only for free play adjustment. Others, however, permit adjustment for pedal height and total pedal travel, as well. When this is the case, both pedal height and total travel should be checked before adjusting pedal free play. Clutch return action should also be observed.

Clutch Pedal Height

Where adjustment is possible, check the pedal height against specifications. Height measurement is usually determined by checking the distance from the pedal to a specific spot. It also may be checked by comparing clutch pedal height to the height of the brake pedal. Adjust the pedal stop as required.

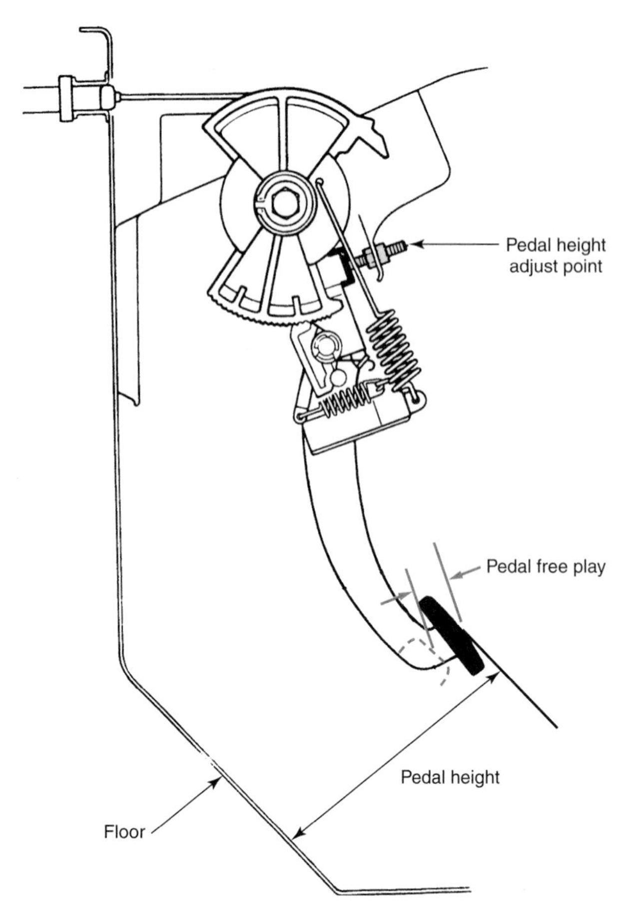

Figure 29-26. Clutch pedal free travel and pedal height. Proper adjustment of free play is important to prevent damage or excessive wear to the clutch. (Toyota)

Clutch Pedal Total Travel

Where required, check pedal total travel. Measure the distance the clutch pedal moves in traveling from the fully released position to the fully extended position. Adjust as needed.

Clutch Pedal Return Action

Check the clutch pedal return action. The pedal should return until it is firmly against the stop. If the pedal sticks or catches, check for binding, interference, or a weak return spring. If the pedal does not fully return, check the return spring or springs. Replace or adjust as required.

Never try to force the clutch pedal to return the full distance by adjusting the clutch linkage. To do so will remove free play, which will ruin the throwout bearing and possibly the entire clutch.

Adjusting Clutch Pedal Free Play

Make sure the clutch pedal is in the fully released position and that it is firmly against the pedal stop. Use two or three fingers to depress the clutch pedal until the throwout bearing engages the clutch release fingers. The pedal should move downward from the fully released position under moderate finger pressure. When the throwout bearing engages the clutch release fingers, you will feel a sharp increase in resistance to movement.

The measured distance the pedal moves from the fully extended position to the point at which the release fingers are engaged is the amount of free play. Average free play is about 1″ (25.4 mm). Check specifications. If pedal free play does not meet specifications, adjust the linkage as needed.

There are many linkage setups. Study the action as the clutch pedal is depressed. Find the adjustment device and move it as required to provide proper free play. Tighten the locknuts and replace the snap rings and cotter pins after the adjustment is made. Check pedal free play and readjust if necessary.

One linkage arrangement is shown in **Figure 29-27.** On this setup, pushing the pedal (A) downward forces the push rod (B) to rotate the cross shaft (C), thus forcing the push rod (D) to actuate the throwout fork (E). Note the adjustment threads on the end of the push rod (D) where it passes through the swivel.

A somewhat similar linkage arrangement is shown in **Figure 29-28.** Note that the adjustment threads are on the upper push rod end. To set pedal free play on this setup, run both nuts (A) and (B) away from the swivel. Force the push rod end toward the vehicle's firewall while pushing the cross-shaft lever in the opposite direction. Move the shaft lever until the throwout bearing engages the release fingers. Hold in this position and run the nut (B) up to within 1/4″ (6.35 mm) of the shaft lever. Release the lever and the push rod. Tighten the nut (A) until the cross-shaft lever is secured between both nuts. Check the pedal for correct free play.

Self-Adjusting Clutch Linkage

When a self-adjusting linkage is used, check the clutch cable for kinks, cuts, and pinched areas, **Figure 29-29.** Be sure the self-adjusting clutch mechanism is clean and lightly lubricated, Check for damaged parts on the self-adjusting mechanism, such as worn pawl and detent (quadrant) teeth or a loose mounting bracket.

Operate the clutch to make sure that it releases fully. If the vehicle is equipped with a neutral start switch, **Figure 28-30,** check to see that it is properly secured and adjusted. Follow the manufacturer's recommendations for the specific vehicle.

Hydraulic Linkage

Where a hydraulic slave cylinder, **Figure 29-31,** is used to actuate the throwout fork, check the master cylinder fluid level. If the fluid level is low, check the master cylinder, slave cylinder, hydraulic line, and connections for leaks. Add fluid as required. If the fluid is old or contaminated, flush the system. (See Chapter 33 on brake service for information on servicing hydraulic units.) Adjust pedal free play to specifications. An overall view of one type of hydraulic clutch linkage is shown in **Figure 29-32.**

Caution: Use the proper type of brake fluid only. Motor oil, transmission fluid, or any kind of petroleum-based oil will damage rubber parts.

Clutch Break-In

It is good practice to subject a newly installed clutch disc to about twenty starts. This will wear off the friction-facing "fuzz" and seat the disc properly. Following this initial break-in, recheck the clutch pedal free play.

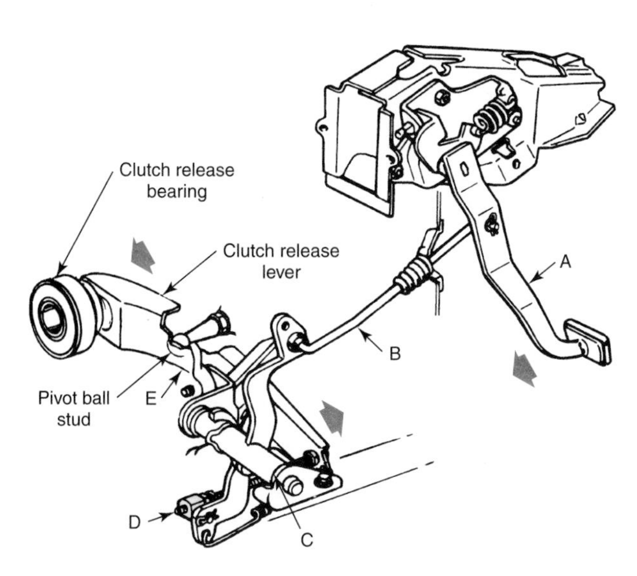

Figure 29-27. Clutch pedal linkage arrangement. (Federal-Mogul)

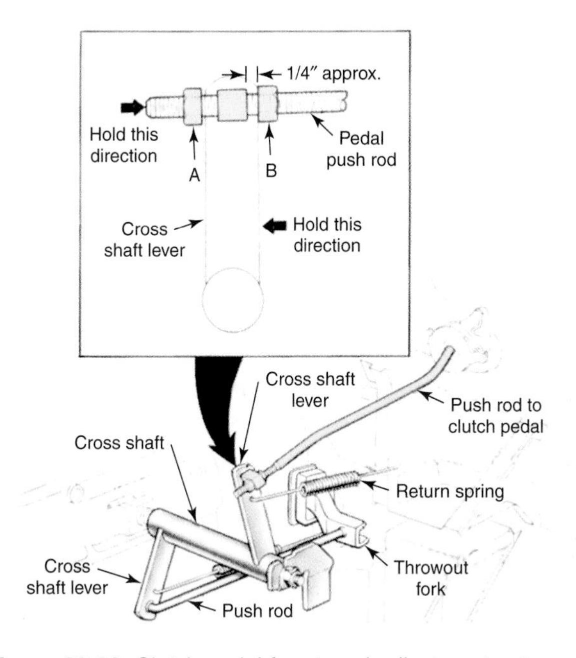

Figure 29-28. Clutch pedal free travel adjustment setup.

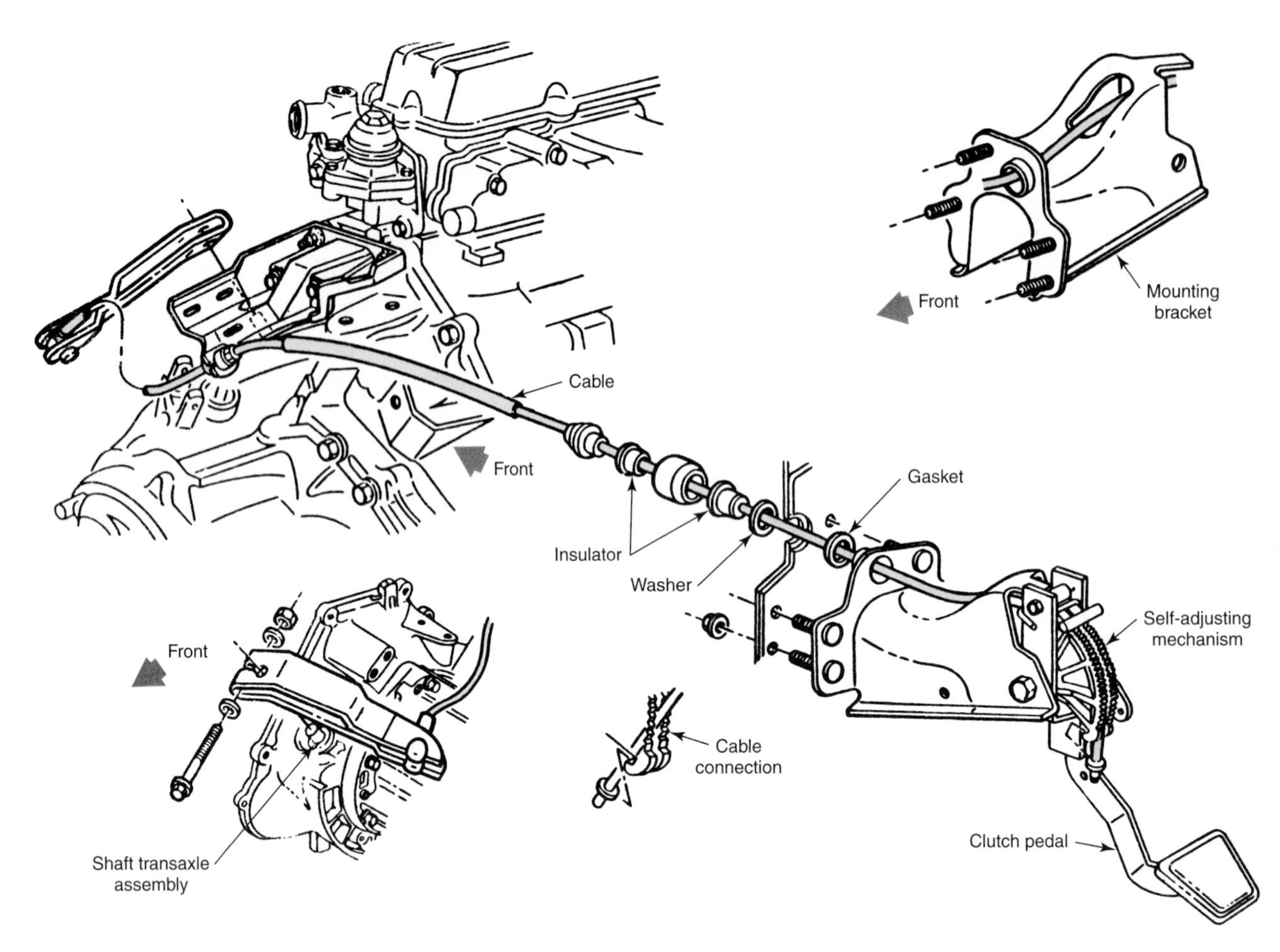

Figure 29-29. Overall view of one type of clutch mounting bracket and cable assembly. The use of a cable greatly simplifies linkage location and routing. Note the self-adjusting mechanism used to take up part wear and cable stretch. (General Motors)

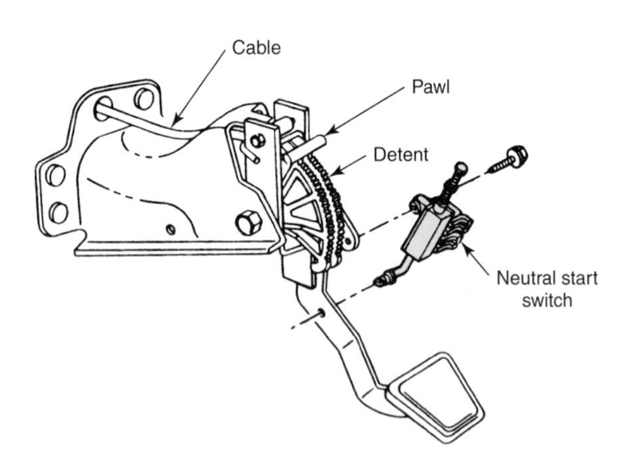

Figure 29-30. Exploded view of a self-adjusting clutch pedal assembly. Note the neutral start switch and its location. (General Motors)

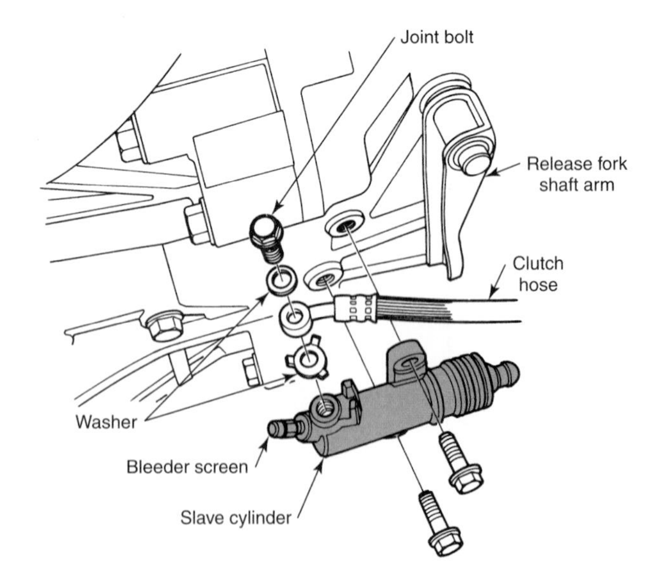

Figure 29-31. A hydraulic slave cylinder with related parts. (Honda)

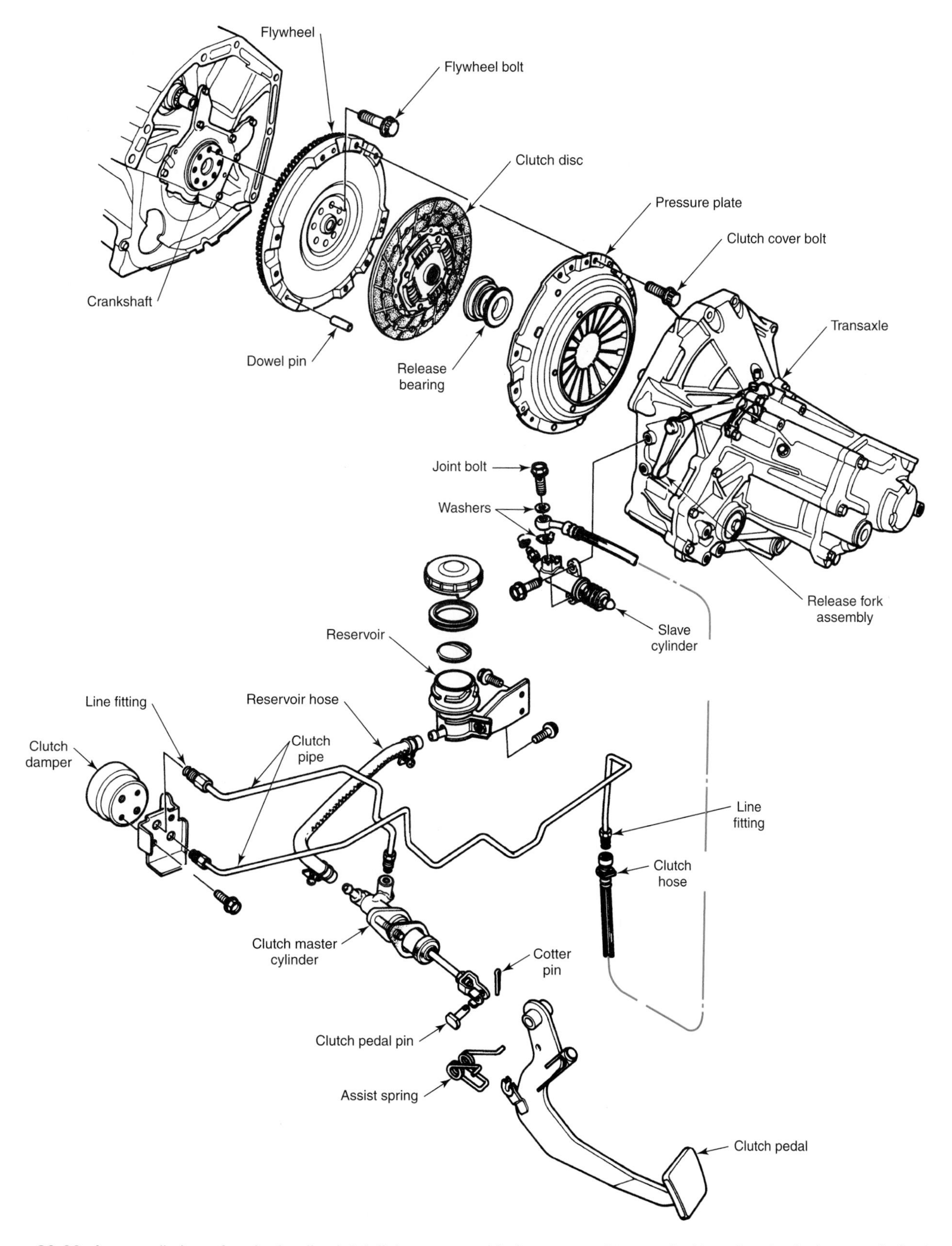

Figure 29-32. An overall view of an hydraulic clutch linkage assembly for a manual transaxle. Note the clutch damper. A diaphragm helps absorb hydraulic fluid shock, smoothing the clutch engagement.

Steam Cleaning Precautions

When the engine and clutch housing are steam cleaned, a certain amount of moisture enters the housing. Rapid heating of the housing by the steam blast will also cause condensation.

If the clutch is not used for some time following steam cleaning, corrosion may form on the clutch unit. The clutch disc facing tends to absorb moisture, which can cause enough corrosion to literally "freeze together" the flywheel, clutch disc, and pressure plate. If this happens, the clutch assembly must be disassembled to separate the units.

To prevent corrosion after steam cleaning, start the engine, set the brakes, and shift the transmission into high gear. Run the engine at a moderate speed and slowly let the clutch pedal out until the engine tries to drive the car forward. Hold the pedal at this point and allow the clutch to "slip" for about five or six seconds. This will heat up the clutch enough to dry it.

Tech Talk

There may come a time in your career as a technician where you are having a difficult day and feel you have insufficient time to perform a job on a customer's vehicle. If a part or adjustment is not in direct sight and will not adversely affect the vehicle's performance, you may be tempted to charge the customer for it anyway—reasoning that he or she might never know the difference. This practice is referred to as "smoking" by technicians, but is simply another name for stealing. If you are caught intentionally "smoking" a repair, you will lose your job and could possibly face criminal prosecution. The best remedy for this problem is not to do it at all, no matter how strong the temptation.

Summary

A diaphragm spring or coil springs are commonly used to apply force to the pressure plate. Always study the disassembled parts in an attempt to determine what caused the clutch trouble. Be sure to disconnect the battery ground strap when working on the clutch.

The pressure plate and flywheel should be marked before removal to ensure correct balance during reassembly. Loosen pressure plate fasteners one turn at a time, in sequence, to prevent warping the pressure plate during removal.

Do not clean the clutch disc or the throwout bearing in solvent. Use a nonpetroleum-based cleaner on the pressure plate and flywheel.

Sand both the flywheel and pressure plate surfaces with medium-fine emery cloth to break the mirror-like glaze. For satisfactory service, flywheel and pressure plate surfaces must not be warped, scored, or badly checked.

The flywheel mounting surface and the crankshaft flange must be clean and free of burrs. Torque flywheel cap screws and use locks where needed. Check flywheel runout. Flywheel ring gears may be removed and installed by heating, but temperatures should not exceed 450°F (232°C). Install the ring with the teeth pointing in the correct direction. Tack weld if required.

If the pressure plate assembly is to be rebuilt, mark the pressure plate release fingers and the pressure plate to ensure assembly in the same relative positions. The pressure plate and flywheel friction surface may be reground, if appropriate.

When reassembling the pressure plate, check the fit of the pressure plate drive lugs in the pressure plate. Adjust the release fingers. Always install a new pilot bearing or bushing during a clutch overhaul. Lubricate the pilot.

Check the clutch housing for misalignment. Shims and offset dowel pins are used to correct the problem.

Be sure the flywheel and pressure plate surfaces are clean. To avoid contamination, do not touch the new clutch disc friction lining or the pressure plate and flywheel friction surfaces.

When installing the clutch, align the pressure plate and flywheel marks. Tighten each pressure plate fastener a little at a time until the pressure plate touches the flywheel. Torque the fasteners. Always use a new clutch disc and install a new throwout bearing during a clutch overhaul.

Once the clutch, throwout bearing, and fork have been installed, do not depress the clutch pedal until the transmission or transaxle is in place. Support the transmission or transaxle while installing—never let it hang on the input shaft.

Attach the clutch linkage, lubricate it, and check alignment and action. Check the clutch pedal height, total travel, and free play. Check the hydraulic linkage master cylinder and slave cylinder for leaks. Flush the hydraulic system if the fluid is dirty.

Give the clutch a quick break-in and recheck pedal free play. Readjust if needed. Slip the clutch for a few seconds following steam cleaning to prevent corrosion.

Review Questions—Chapter 29

Do not write in this book. Write your answers on a separate sheet of paper.

1. List two safety precautions for dealing with asbestos in the clutch assembly.
2. Force is applied to the pressure plate by either _____ springs or a _____ spring.
3. Before removing the pressure plate fasteners, always _____.
   (A) wipe them off
   (B) block the flywheel
   (C) prick punch the pressure plate and flywheel
   (D) check for disc warpage
4. What should you check the flywheel-to-clutch friction surface for?
5. When installing a new flywheel ring gear, never heat it above _____.
   (A) 200°F (93°C)
   (B) 900°F (482°C)
   (C) 650°F (343°C)
   (D) 450°F (232°C)
6. Name the clutch parts which should be replaced whenever the clutch is serviced.

7. Name some indications of a worn clutch pilot bushing.
8. Align the clutch disc hub with the _____ by using a clutch aligning arbor or a used transmission input shaft.
9. Name the three ways in which the clutch linkage can be operated.
10. To prevent clutch corrosion following steam cleaning, the technician should _____.
    (A) allow the car to stand for several hours before running
    (B) start the engine and slip the clutch
    (C) raise the rear end and spin the wheels
    (D) oil the clutch disc facing

ASE-Type Questions

1. What is the first thing that should be removed when replacing a clutch?
    (A) Transmission.
    (B) Inspection cover.
    (C) Battery cable.
    (D) Clutch linkage.
2. Technician A says each pressure plate fastener should be removed completely before proceeding to the next one. Technician B says the pressure plate fasteners should be tightened before installing the pilot shaft. Who is right?
    (A) A only.
    (B) B only.
    (C) Both A and B.
    (D) Neither A nor B.
3. Technician A says the pressure plate or flywheel friction surfaces can be sanded to restore them to service. Technician B says pressure plate assemblies can be rebuilt satisfactorily in some cases. Who is right?
    (A) A only.
    (B) B only.
    (C) Both A and B.
    (D) Neither A nor B.
4. Technician A says flywheel runout is preset and does not require checking. Technician B says some flywheel ring gears are secured by short arc welds. Who is right?
    (A) A only.
    (B) B only.
    (C) Both A and B.
    (D) Neither A nor B.
5. When installing the pressure plate assembly to the flywheel, always _____.
    (A) align the prick punch marks
    (B) install the top fastener first
    (C) use a C-clamp to secure the unit
    (D) tighten each fastener fully before starting on the next one
6. Technician A says when replacing a clutch, it is advisable to install a new pilot bearing or bushing. Technician B says to assist in forcing the input shaft into the pilot bearing, it is permissible to tighten the transmission to clutch housing fasteners. Who is right?
    (A) A only.
    (B) B only.
    (C) Both A and B.
    (D) Neither A nor B.
7. Insufficient pedal free play can cause _____.
    (A) clutch slipping
    (B) excessive throwout bearing wear
    (C) rapid clutch disc wear
    (D) All of the above.
8. Excessive pedal free play can cause all of the following, *except:*
    (A) hard shifting.
    (B) fast throwout bearing wear.
    (C) gear clash when shifting.
    (D) transmission gear damage.
9. If the clutch pedal sticks or catches, the technician should do all of the following, *except:*
    (A) check for linkage binding or interference.
    (B) check for a weak return spring.
    (C) adjust the clutch linkage to remove free play.
    (D) replace worn parts when found.
10. On most vehicles, clutch pedal free play is adjusted by _____.
    (A) placing shims under the pressure plate
    (B) adjusting the clutch linkage
    (C) bending the pedal stop
    (D) aligning the clutch housing

Suggested Activities

1. Inspect several different used clutch parts and determine why each one failed. Be sure to follow the asbestos precautions listed earlier.
2. Obtain the necessary measuring equipment and measure the runout of a flywheel. If it is out of specification, try to determine whether the problem is in the friction surface or the flywheel itself.
3. Take a vehicle with a manual transmission on a road test and perform all of the checks you learned in this chapter.
4. Give complete estimates for replacing the worn clutch parts you studied earlier. Include all applicable parts, reconditioning, and labor prices.

Clutch Problem Diagnosis

Problem: Clutch slips	
Possible cause	**Correction**
1. Insufficient pedal free travel.	1. Adjust free travel.
2. Disc facing soaked with oil or grease.	2. Clean clutch and pressure plate, replace disc. Correct source of oil contamination.
3. Broken or weak pressure plate spring or springs.	3. Rebuild or replace pressure plate.
4. Clutch disc facing worn.	4. Replace clutch disc.
5. Hydraulic or mechanical linkage sticking.	5. Clean, align, and lubricate where needed.

Problem: Clutch chatters and/or grabs	
Possible cause	**Correction**
1. Clutch disc facing oil or grease soaked.	1. Replace clutch disc. Correct source of leak.
2. Burned clutch disc facing.	2. Replace clutch disc.
3. Warped or worn clutch disc.	3. Replace clutch disc.
4. Pressure plate warped.	4. Grind or replace.
5. Pressure plate or flywheel surface scored.	5. Grind or replace.
6. Pressure plate fingers bind.	6. Free fingers.
7. Clutch housing-to-transmission surface out of alignment with crankshaft centerline.	7. Align or replace housing.
8. Sticking linkage.	8. Free linkage.
9. Pilot bearing worn.	9. Install new pilot bearing.
10. Pressure plate release fingers improperly adjusted.	10. Adjust fingers.
11. Engine mounts loose or worn.	11. Tighten or replace mounts.
12. Transmission loose.	12. Tighten fasteners.
13. Rear spring shackles or axle housing control arms loose.	13. Tighten shackles or replace control arm insulators and tighten.
14. Worn splines or transmission input shaft.	14. Replace shaft.
15. Faulty throwout bearing.	15. Replace throwout bearing.

Problem: Clutch will not release properly	
Possible cause	**Correction**
1. Excessive pedal free travel.	1. Adjust pedal travel.
2. Warped clutch disc.	2. Replace clutch disc.
3. Clutch facing torn loose and folded over.	3. Replace clutch disc.
4. Warped pressure plate.	4. Grind or replace.
5. Clutch housing misaligned.	5. Align housing.
6. Clutch disc hub binding on transmission input shaft.	6. Free hub.
7. Pilot bearing worn.	7. Replace pilot bearing.
8. Faulty throwout bearing.	8. Replace throwout bearing.
9. Throwout fork off pivot.	9. Install fork properly.
10. Clutch disc is frozen (corroded) to flywheel and pressure plate.	10. Replace disc and clean flywheel and pressure plate.
11. Excessive idle speed.	11. Adjust idle speed.

Problem: Clutch is noisy when pedal is depressed—engine running	
Possible cause	**Correction**
1. Dry or worn throwout bearing.	1. Replace bearing.
2. Worn pilot bearing.	2. Replace pilot.
3. Excessive total pedal travel.	3. Adjust pedal travel.
4. Throwout fork off pivot.	4. Install fork correctly.
5. Clutch housing misaligned.	5. Align housing.
6. Crankshaft endplay excessive.	6. Correct endplay.

(continued)

Clutch Problem Diagnosis *(continued)*	
Problem: Clutch is noisy when pedal is depressed—engine not running	
Possible cause	**Correction**
1. Dry, sticking linkage. 2. Dry or scored throwout bearing sleeve. 3. Pressure plate drive lugs rubbing clutch cover.	1. Lubricate and align linkage. 2. Lubricate or replace. 3. Lubricate with high temperature grease.
Problem: Clutch noisy when pedal is fully released—engine running	
Possible cause	**Correction**
1. Insufficient pedal free travel. 2. Clutch disc worn. 3. Clutch disc springs broken. 4. Clutch housing misaligned. 5. Worn clutch disc hub splines. 6. Worn input shaft splines. 7. Sprung input shaft. 8. Input shaft transmission bearing worn.	1. Adjust free travel. 2. Replace clutch disc. 3. Replace clutch disc. 4. Align housing. 5. Replace clutch disc. 6. Replace input shaft. 7. Replace input shaft. 8. Replace transmission bearing.
Problem: Excessive pedal pressure	
Possible cause	**Correction**
1. Linkage needs lubrication. 2. Pressure plate release fingers binding. 3. Linkage misaligned. 4. Throwout bearing sleeve binding on transmission bearing retainer. 5. Sticking linkage in master or slave cylinder.	1. Lubricate. 2. Free and lubricate. 3. Align linkage. 4. Free and lubricate retainer. 5. Clean or replace as needed.
Problem: Rapid clutch disc wear	
Possible cause	**Correction**
1. Insufficient pedal free travel. 2. Scored flywheel or pressure plate. 3. Driver "rides" the clutch (rests foot on clutch while driving). 4. Driver races engine and slips clutch excessively during starting. 5. Driver holds vehicle on hill by slipping clutch. 6. Weak pressure plate springs.	1. Adjust free travel. 2. Resurface or replace. 3. Advise driver. 4. Advise driver. 5. Advise driver. 6. Rebuild or replace pressure plate assembly.

Cutaway of a 5-speed manual transmission used in a late-model vehicle. (Ford)

30

Manual Transmission and Transaxle Service

After studying this chapter, you will be able to:

- Explain manual transmission construction and operation.
- Explain manual transaxle construction and operation.
- Disassemble, check parts, and reassemble a manual transmission.
- Disassemble, check parts, and reassemble a manual transaxle.
- Diagnose manual transmission and transaxle problems.

Technical Terms

Transmission	CV axle housings
Transaxle	Output shaft
Sliding gears	Countershaft
Shift linkage	Antilash plate
Synchronized gears	Reverse idler
Pilot bolts	Cone surface
Transmission jack	Drive chain
Shift plate	Drive sprockets
Inspection cover	Differential unit
Alignment marks	Snap rings
Input shaft	Thrust washers
Extension housing	Dummy shaft

This chapter will cover the design of manual, or standard, transmissions and transaxles. The manual transmission or transaxle gears are selected by the driver and are always used with a manual clutch.

Transmission and Transaxle Designs

A ***transmission***, **Figure 30-1,** is used on vehicles with front engines and rear-wheel drive. The parts layout of all manual transmissions is similar. A ***transaxle*** is used on vehicles with front engines and front-wheel drive. In some cases, transaxles are found on vehicles with rear engines and rear-wheel drive.

The gears in all manual transmissions and transaxles are ***sliding gears,*** which are moved in and out of engagement by the driver through ***shift linkage.*** Modern manual transmissions and transaxles usually have ***synchronized gears*** with special internal clutches to prevent gear clash when shifting in forward speeds. The four main classes of transmission/transaxle gears are *reduction, direct drive, overdrive,* and *reverse.* The following sections explain the workings of various types of manual transmissions and transaxles.

Manual Transmission Types

Manual transmissions in modern rear-wheel drive vehicles can have three, four, five, or six forward speeds. The transmission is installed on the clutch housing. A drive shaft connects the transmission to the rear axle and differential assembly.

The three-speed transmission is being replaced by transmissions with more gears. However, many three-speeds are still in use. The three-speed transmission provides drive ratios of approximately 2.79:1 in first gear (the input shaft turns 2.79 times to rotate the output shaft once); 1.70:1 in second gear, and 1:1 in third gear. Gear, or drive, ratios vary, depending on vehicle weight, engine horsepower, and other factors.

Power flow through the gears of a typical three-speed transmission is shown in **Figure 30-2.** Study the positioning of the second and high synchronizer gears and of the low and reverse sliding sleeve and gear. The operation of other rear-wheel drive transmissions is similar.

In four-, five-, and six-speed transmissions, the first gear ratio is similar to that of the three-speed transmission. The extra gears in these transmissions give the driver a wider latitude in selecting an appropriate gear for a given situation.

Typical ratios for a four-speed transmission are around 2.78:1 for first gear, 1.93:1 for second, 1.36:1 for third, and 1:1 for fourth (high), gear. As with three-speed transmissions, these ratios vary. Study the working parts in the four-speed transmission shown in **Figure 30-3.** Note that all forward gears are synchronized.

On many four- and five-speed transmissions, the highest gear (fourth or fifth) is an overdrive gear, with a ratio such as 1:0.7 (for each full turn of the input shaft, the output shaft rotates 7/10 of a turn). On some five- and six-speed transmissions, both upper gears are overdrives. Study the transmissions in **Figures 30-4.** Overdrive transmissions are diagnosed, serviced, and repaired in the same way as manual transmissions in which the top gear is direct (1:1). Refer to the manual transmission section of this chapter for service and repair information.

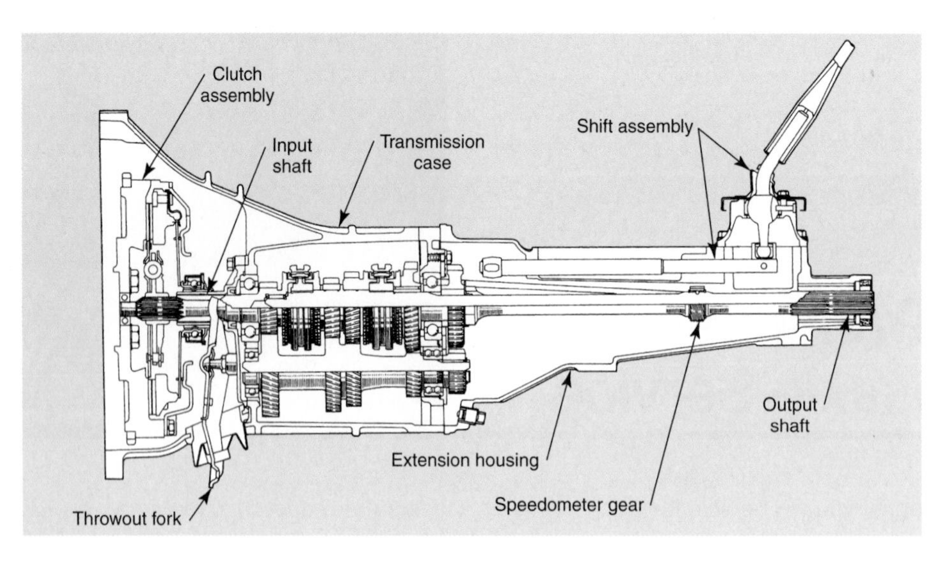

Figure 30-1. Cross-sectional view of a four-speed transmission. (Mazda)

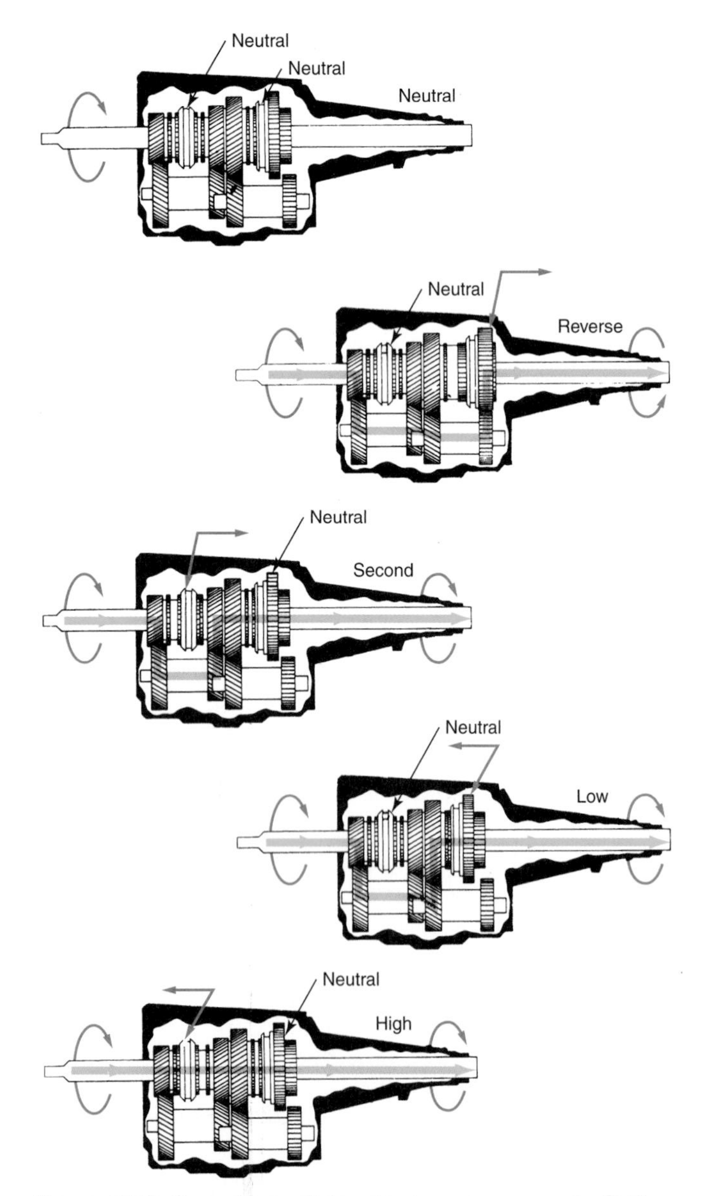

Figure 30-2. Gear drive relationship in various speeds. Shift movement is indicated by the arrows. (General Motors)

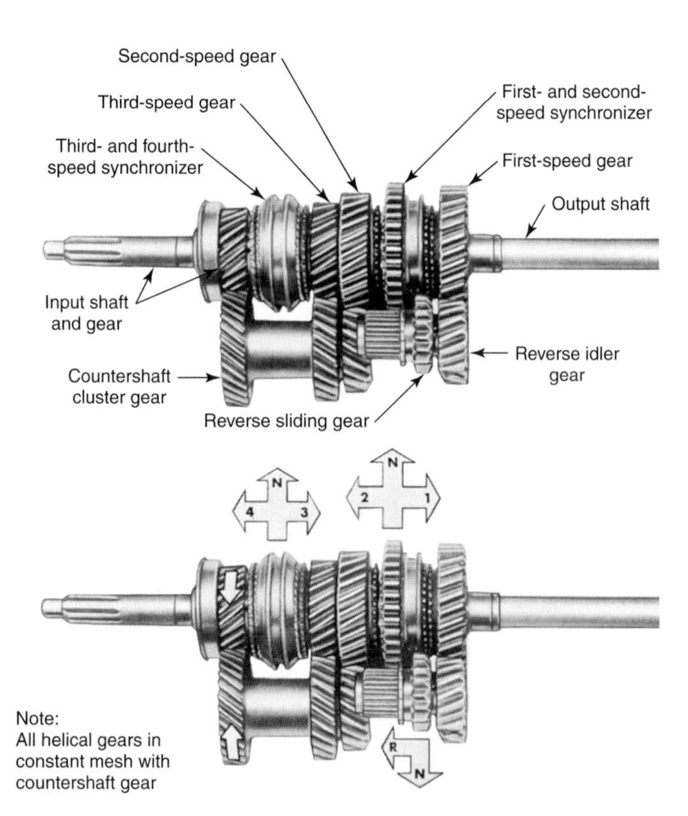

Figure 30-3. Working parts of a typical four-speed, synchronized transmission. (Ford)

Manual Transaxle Types

On front-wheel drive vehicles, the transmission and differential assemblies are installed in a single housing and commonly referred to as a transaxle. Transaxles operate in the same way as rear-drive transmissions and rear axle differentials, but all parts are housed together. The manual transaxle has the same kinds of gears, synchronizers, and shifting mechanisms as a manual transmission. The number of forward gear ratios varies from three to five.

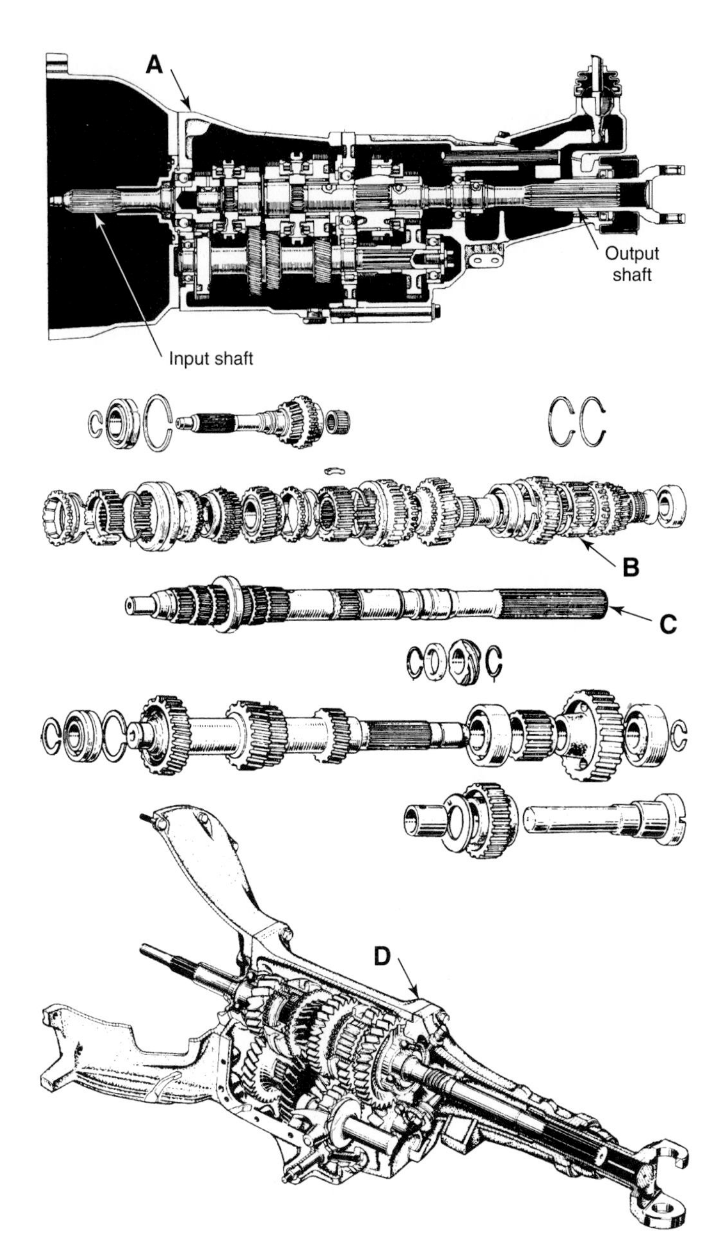

Figure 30-4. Overdrive transmissions. A—A five-speed transmission in which fifth gear is an overdrive gear. The input shaft turns about one-third slower than the output shaft in fifth gear. B—Input shaft, gears, synchronizers, bearing, etc. C—Output shaft and cluster gear and idler assembly. D—Four-speed transmission in which fourth gear is the overdrive gear. (Toyota and DaimlerChrysler)

The major differences between a transaxle and a transmission are the shape of the case and the placement of the parts within the case. Additionally, the transaxle contains the differential gears and may use a drive chain or two gears to transfer power inside of the transaxle. The most obvious external difference is the presence of two axle (output) shafts connecting the transaxle to the front wheels, compared to the single output shaft of a transmission. See **Figure 30-5.**

There are two basic variations of front-wheel drive transaxle construction. The type used depends on engine placement in the vehicle. An engine can face sideways in the vehicle, or it can face forward. An engine mounted sideways is called a *transverse engine*, and an engine mounted forward is called a conventional engine. A manual transaxle with a transverse engine is shown in **Figure 30-6.** Power flow takes place in a straight line, without any change in the angle once it leaves the engine.

Figure 30-7 shows the power flow through a transaxle on a vehicle with conventional engine placement. In this design, the output shaft drives the ring and pinion gear in the differential. This ring and pinion gear is a *hypoid* type, like the ones used on rear-wheel drive vehicles. The design of the ring and pinion causes the power to make a 90° turn. This type of transaxle more closely resembles the rear-wheel drive manual transmission.

Servicing Manual Transmissions and Transaxles

Service procedures are similar for the internal components of manual transmissions and transaxles. The major differences occur in removing the unit from the vehicle. Always obtain and refer to the correct manufacturer's service manual. The service manual procedures should be followed exactly.

Talk with the Owner

Talk with the owner about observations and complaints regarding transmission or transaxle operation. Ask questions to help pinpoint trouble areas. Make a list of possible problems based on the owner's statements.

Check Clutch and Shift Linkage

If the owner complains of hard shifting, gear clash, or jumping out of gear, check the clutch and shift linkage operation and adjustment before road testing.

The clutch pedal free travel must be within specifications. Excessive free travel will prevent full withdrawal of the pressure plate from the clutch disc. This will cause the transmission or transaxle input shaft to continue turning, making shifting difficult and noisy. Clutch adjustments were covered in Chapter 29.

The shift linkage must operate smoothly and must be adjusted so the transmission or transaxle is shifted fully into gear. Failure to provide full shift engagement can cause the transmission or transaxle to jump out of gear. Linkage adjustments are covered later in this chapter.

Road Test when Possible

Whenever possible, the vehicle should be road-tested. Some discretion must be used. Check the transmission or transaxle lubricant level before road testing. If a road test is performed, it should include some heavy acceleration and deceleration. Operate the vehicle at various speeds. If practical, include some bumpy sections and a hill in the route.

Check closely for abnormal noise, jumping out of gear, vibration, hard shifting, gear clash during shifting, and leaks. Note the gear (reduction, direct drive, overdrive) in which any problem was most evident.

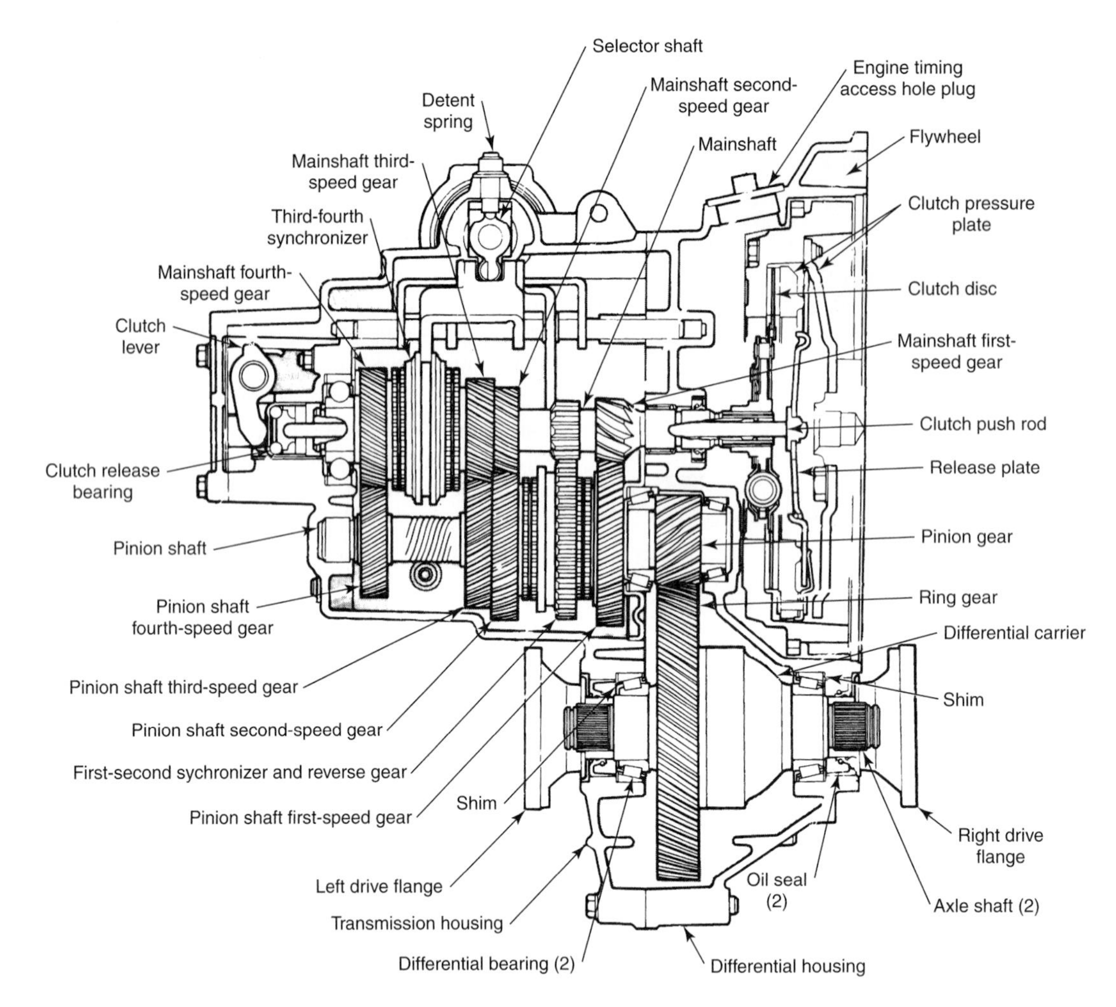

Figure 30-5. Cross section of a four-speed manual transaxle. Note the two axle shafts connected to the differential. (DaimlerChrysler)

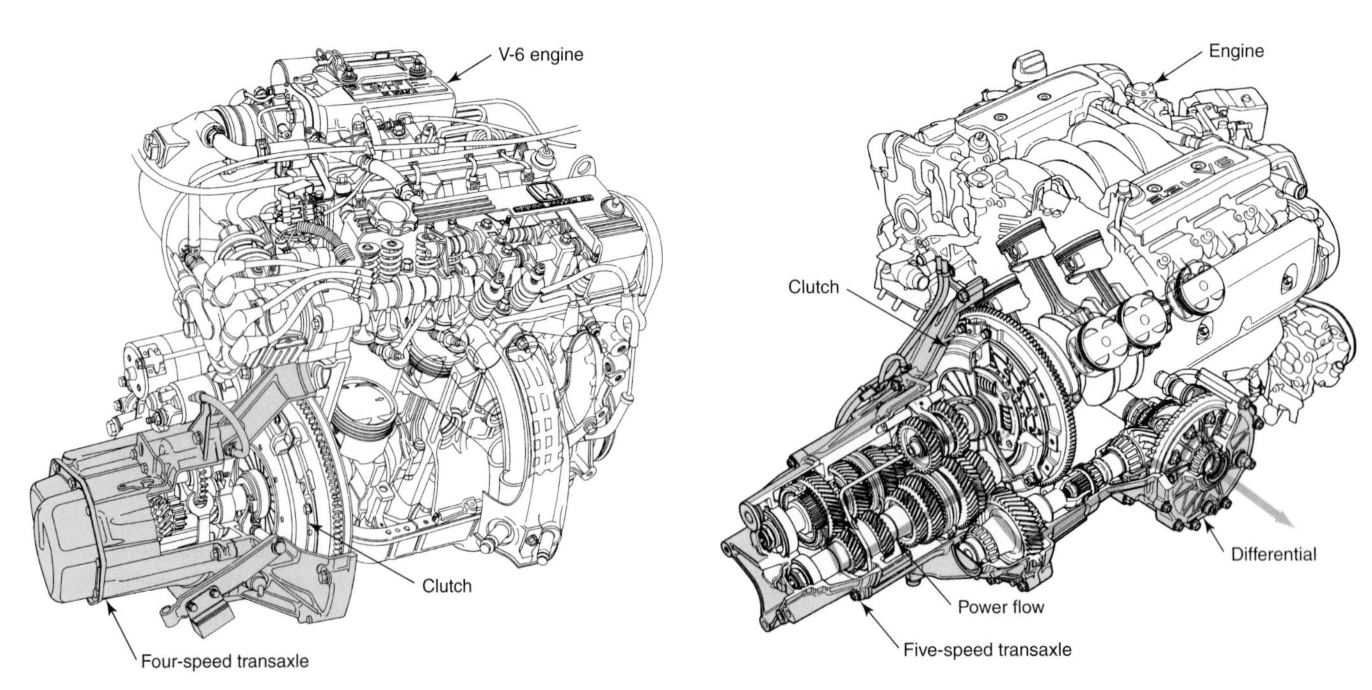

Figure 30-6. Cutaway of a transverse-mounted engine and a four-speed manual transaxle assembly. (Honda)

Figure 30-7. Cutaway of a five-speed transaxle with an inline, V-6 engine. (Honda)

Remove Transmission or Transaxle

Note: Before removing the transmission or transaxle, make sure that removal is necessary. Some repairs can be performed with the transmission or transaxle in the vehicle. Among these are shift linkage adjustment, shift cover overhaul, cover gasket replacement, and output shaft oil seal replacement.

If transmission or transaxle removal is necessary, raise the vehicle and drain the transmission oil. Remove the shift linkage (marking it to facilitate assembly). Remove the speedometer cable.

Note: Once a cable or shift rod has been removed, put the fasteners back into place. This will speed up reassembly and prevent improper placing of fasteners.

Disconnect the drive shaft or CV drive axles and wire them out of the way. Before removing the shaft(s), mark them so they can be reassembled in their original positions. See Chapter 32 for complete details.

If the universal joint is a cross-and-roller type, like the one shown in **Figure 30-8,** tape the loose roller bearings to the cross to prevent them from falling off.

Remove the transmission or transaxle mounts as needed. If a support member must be removed, be sure to provide support for the engine with either a jack stand or an engine support strap.

If the engine must be raised to gain access to the transmission mounts, be careful to avoid damage to any attached parts. When pushing upward on the oil pan, place a wide block of wood between the pan and the jack.

Use a Transmission Jack or Pilot Bolts

If the transmission or transaxle is of a size that can be easily handled, remove either the two upper or two lower transmission-to-clutch-housing fasteners. Replace them with ***pilot bolts*** (which can be made by cutting off the heads of two bolts of the correct size). Pilots should be long enough to provide support until the input shaft is clear of the clutch disc hub. Remove the remaining fasteners. Slide the transmission (or transaxle) away from the clutch housing. When the transmission is free, lower it to the floor.

If desired, a ***transmission jack*** can be used for support during removal. Attach the transmission or transaxle firmly to the jack. Remove the fasteners holding the transmission to the clutch housing. Then, guide the transmission away from the housing and lower it to the floor.

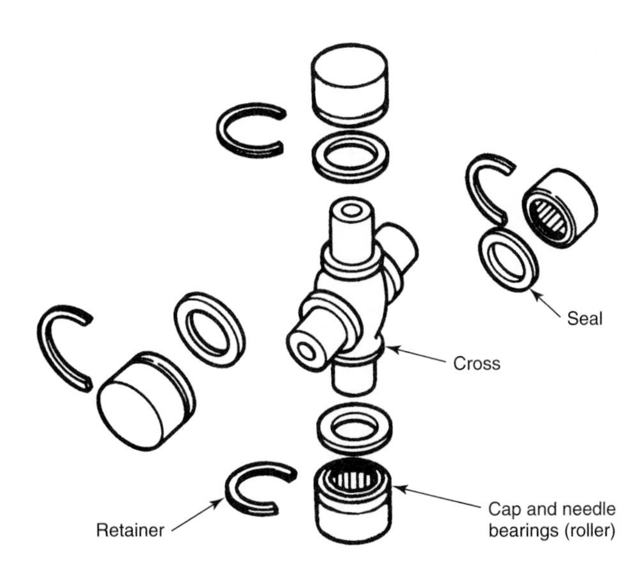

Figure 30-8. Exploded view of a typical cross-and-roller universal joint. (DaimlerChrysler)

Clean Exterior

Before beginning disassembly, clean the outer surfaces of the transmission or transaxle thoroughly. This will permit disassembly of the unit with a minimum amount of contamination.

Flush Interior

Remove the transmission or transaxle ***shift plate*** or ***inspection cover.*** Pour a pint of clean solvent into the case (the oil was drained before transmission or transaxle removal). Spin the input shaft. Continue turning the shaft while rocking the case to provide additional agitation. Drain the case and repeat the process. This flushing will remove enough of the heavy lubricant to permit a visual inspection of the gear teeth.

Inspect Gears, Shafts, Synchronizers, and Other Parts

Before disassembling the transmission or transaxle, remove the top or side covers, if such covers are used. Slowly rotate the gears while carefully inspecting the teeth for chipping, galling, and excessive wear. Rock the gears on the shaft to determine approximate clearance. Check end play of the cluster gear, reverse idler gear, and input and output shafts. Inspect the synchronizer units for excessive looseness. Check the condition of the teeth engaged by the synchronizer clutch sleeve.

Note: This step is not possible on transmissions or transaxles without top or side covers. In these cases, the transmission/transaxle must be entirely or partly disassembled to check internal parts.

This initial inspection will help to indicate the location and extent of the problem. Refer to **Figure 30-9.**

Disassembly Procedure

Although the basic designs of transmissions or transaxles are similar, disassembly procedures vary widely among the different makes and models.

Some transmissions require the extension housing and output shaft to be removed first. Some permit the removal of the input shaft first, while others require the cluster shaft be lowered to the bottom of the case to permit removal of the output shaft and gears. Some transmissions are designed with two aluminum end castings that are attached to a central support. The end castings must be removed to gain access to the gears and shafts.

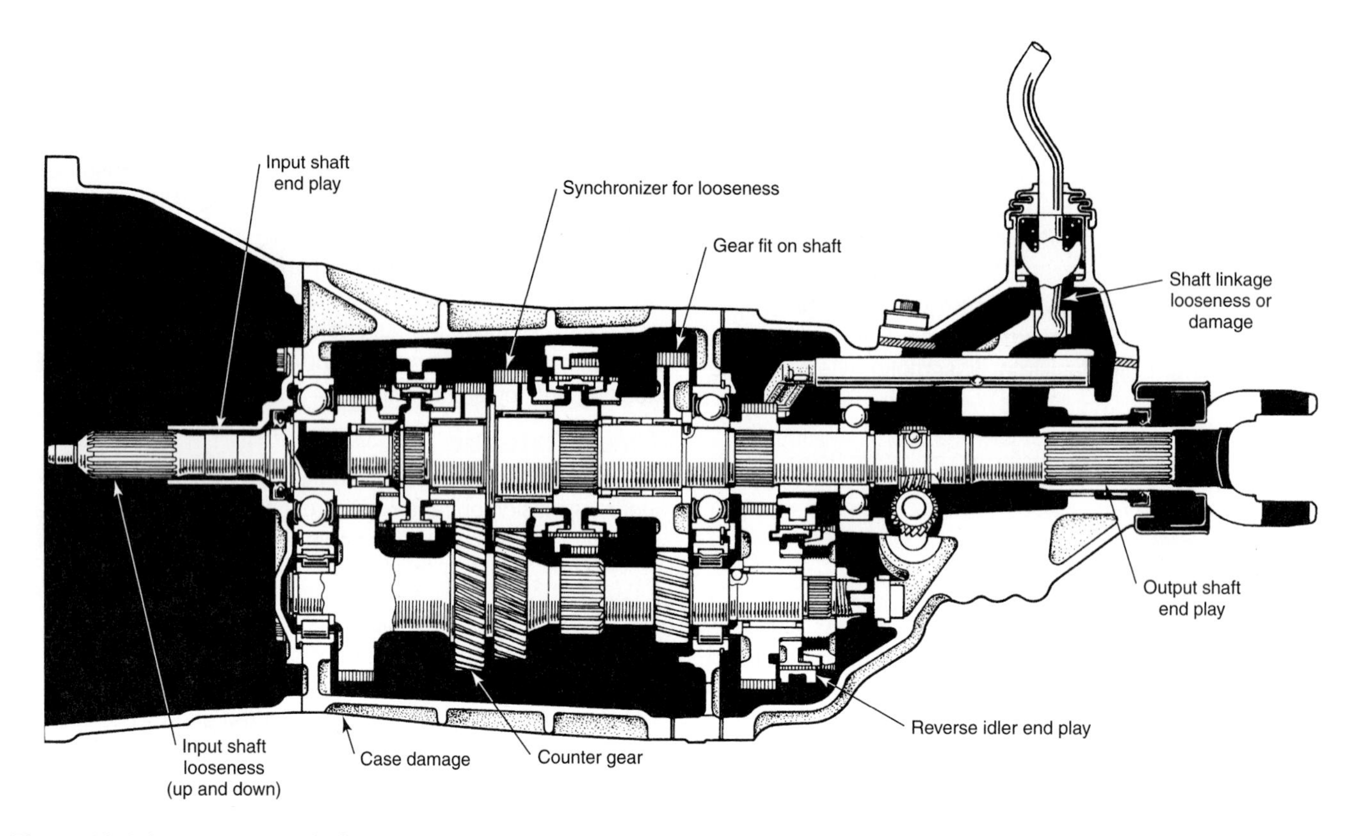

Figure 30-9. Inspect gears, shafts, synchronizers, bearings, and other parts before disassembling the transmission or transaxle. (Toyota)

Some transaxles require removal of the differential assembly before beginning repairs. Other transaxles must be split to expose the internal gears. In still other designs, both CV axle housings must be removed before repairs can be made. Where one design may require input shaft removal by passing the shaft into the case and out the cover hole, another may permit the shaft to be pulled directly from the case. In all cases, the differential assembly must be removed, either as part of the transaxle (transverse engines), or as a separate unit (conventionally positioned engines).

Due to space limitations, disassembly procedures for all types of transmissions and transaxles cannot be covered in this text. Therefore, one typical method of removal and installation will be shown. Instead of dwelling on a specific disassembly procedure, the transmission or transaxle parts will be covered separately. General inspection techniques will also be discussed.

Manufacturer's Service Manuals

The technician should obtain a service manual covering the specific transmission or transaxle to be overhauled. Disassembly and assembly order and technique will be shown. Exploded views will assist in the correct positioning of all parts, and specifications will be given.

General Disassembly Procedures

Place the transmission or transaxle in a suitable stand, **Figure 30-10.** Follow the manufacturer's recommended order of part removal. If no manual is available, study the method of construction. This will provide clues as to which part should be removed first, second, and so on. Careful study will also usually indicate how the parts must be removed. Be sure to place ***alignment marks*** on the transmission or transaxle housing and any extension or CV axle housings before beginning disassembly. See **Figure 30-11.**

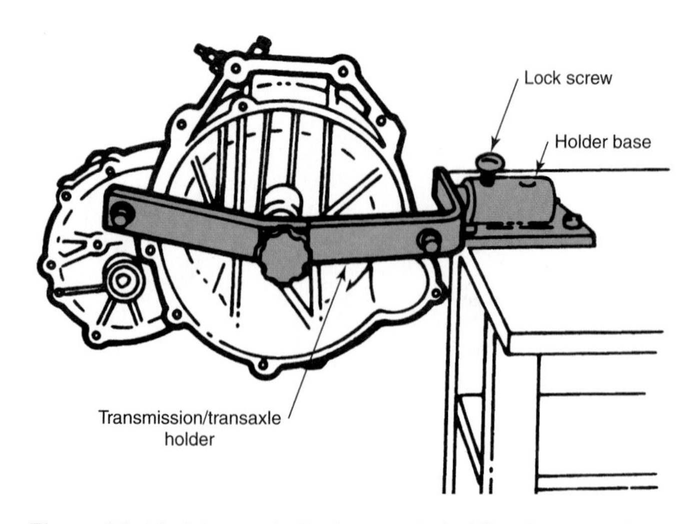

Figure 30-10. A transmission/transaxle holding fixture makes service faster and easier. The holder base allows the holder to be rotated and locked in many convenient working locations. (General Motors)

Remove the input shaft bearing retainer, if one is used. On a rear-drive transmission, remove the extension housing fasteners and pull the housing. This will usually allow either the output shaft or the input shaft to be pulled far enough to determine the exact removal procedure.

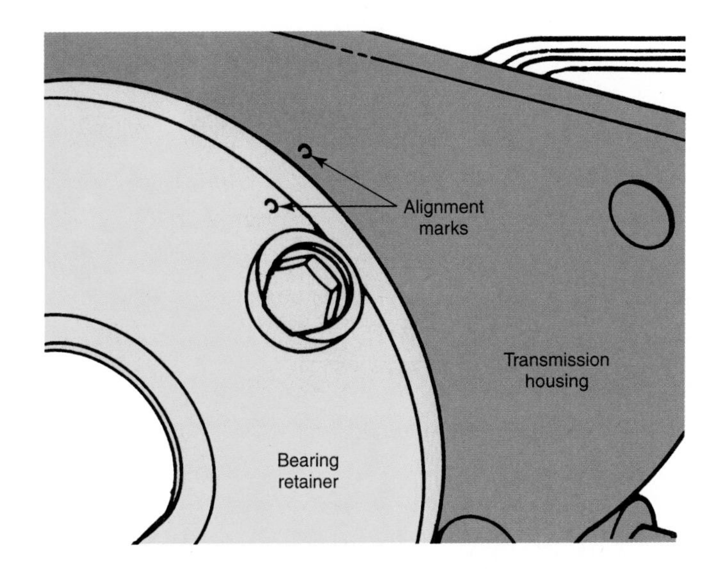

Figure 30-11. Bearing retainer-to-transmission case alignment marks will help assure proper reassembly. (General Motors)

On some transmissions and transaxles, lowering the cluster gear to the bottom of the case is necessary to permit shaft removal. Proceed with the disassembly, being careful to avoid excessive hammering. If a hammer is required, use one with a soft face of lead, plastic, brass, or rawhide. A drift punch should be brass or soft steel.

Note: When disassembling a transmission or transaxle, use care to avoid distorting snap rings in case they must be reused. Whenever possible, use new snap rings during assembly.

Remove each synchronizer unit as one part. Be certain to keep the cones or blocking rings with the unit and on the same side as originally installed. Be careful not to lose any roller or needle bearings used in the transmission or transaxle.

In some transmissions and transaxles, the shift rails operate in the case rather than in a separate shift cover. See **Figure 30-12.** This design generally requires shift fork and rail removal. Be careful to avoid losing any of the detent springs, detents, interlocks, or setscrews. The cluster gear will usually contain a number of needle bearings. Do not lose them.

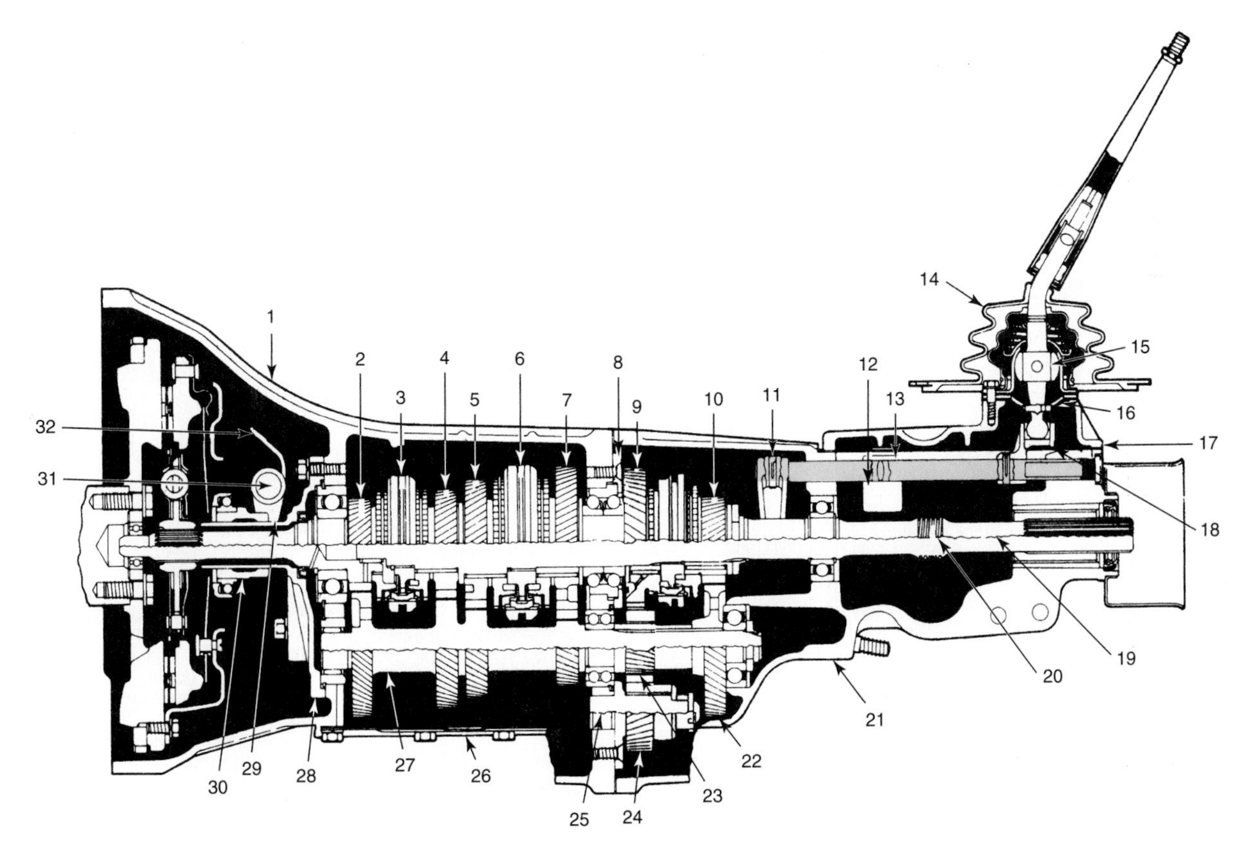

Figure 30-12. A shift rail that operates in the case. 1—Transmission case. 2—Main drive pinion. 3—3-4 synchronizer. 4—Third gear. 5—Second gear. 6—1-2 synchronizer. 7—First gear. 8—Bearing retainer. 9—Overdrive synchronizer assembly. 10—Overdrive gear. 11—Control finger. 12—Neutral return finger. 13—Control shaft. 14—Control lever cover. 15—Control lever assembly. 16—Stopper plate. 17—Control housing. 18—Change shifter. 19—Mainshaft. 20—Speedometer. 21—Extension housing. 22—Counter overdrive gear. 23—Counter reverse gear. 24—Reverse idler gear. 25—Reverse idler gear shaft. 26—Under cover. 27—Counter gear. 28—Front bearing retainer. 29—Clutch shift arm. 30—Release bearing carrier. 31—Clutch control shaft. 32—Return spring. (DaimlerChrysler)

> **Caution:** Avoid excessive hammering to loosen parts. Handle parts carefully to avoid chipping and nicking. Place all small parts (loose needles and rollers, detent springs, detent plugs or balls) in a separate container to prevent loss.

Remove all the parts, including the input and output shaft bearings, from the case. Drive the reverse idler shaft free of the case and remove the reverse idler gear and thrust washers. **Figure 30-13** illustrates a typical four-speed transmission. **Figure 30-14** illustrates a typical transaxle that

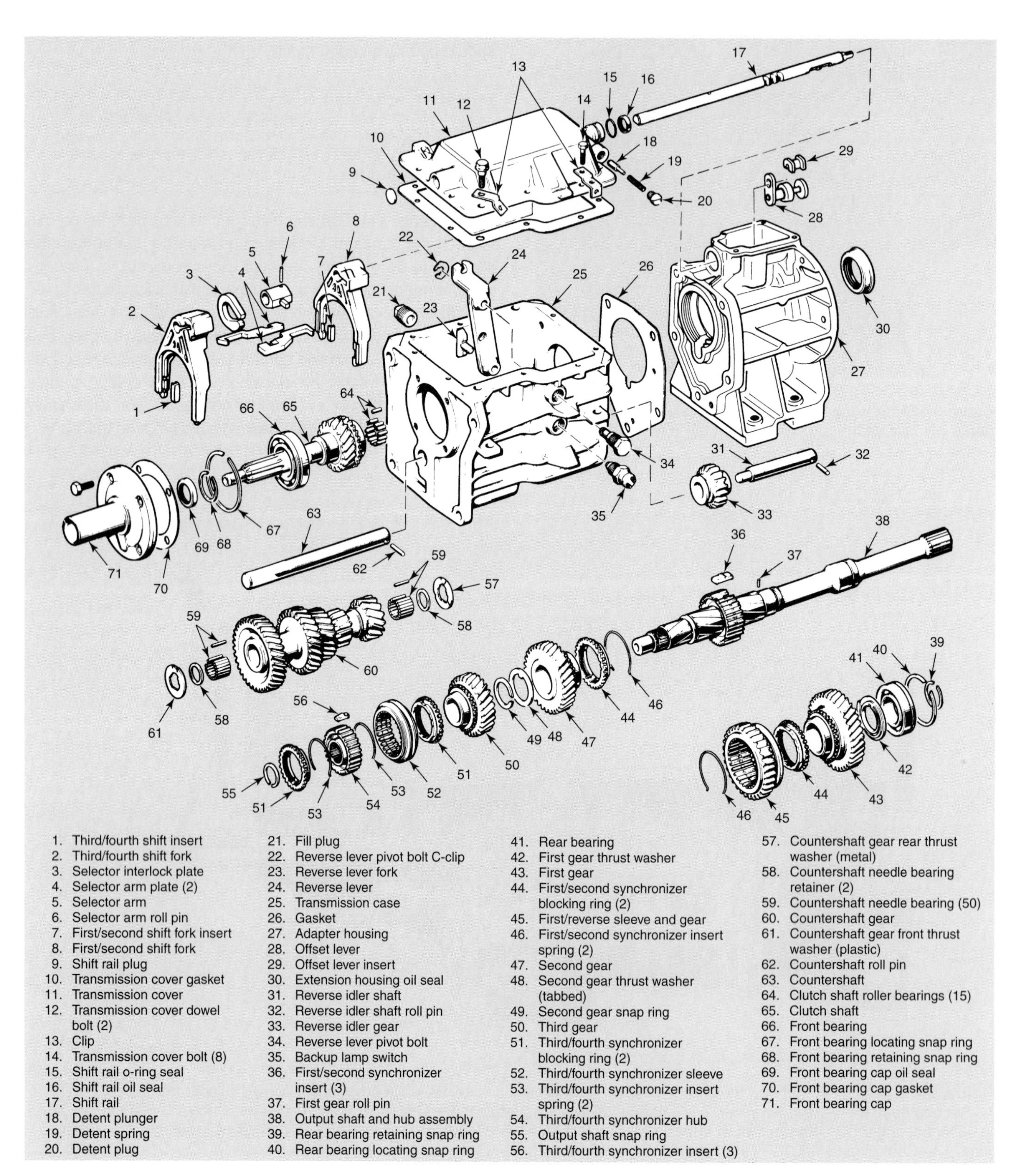

Figure 30-13. An exploded view of a four-speed transmission. Study the parts carefully. (DaimlerChrysler)

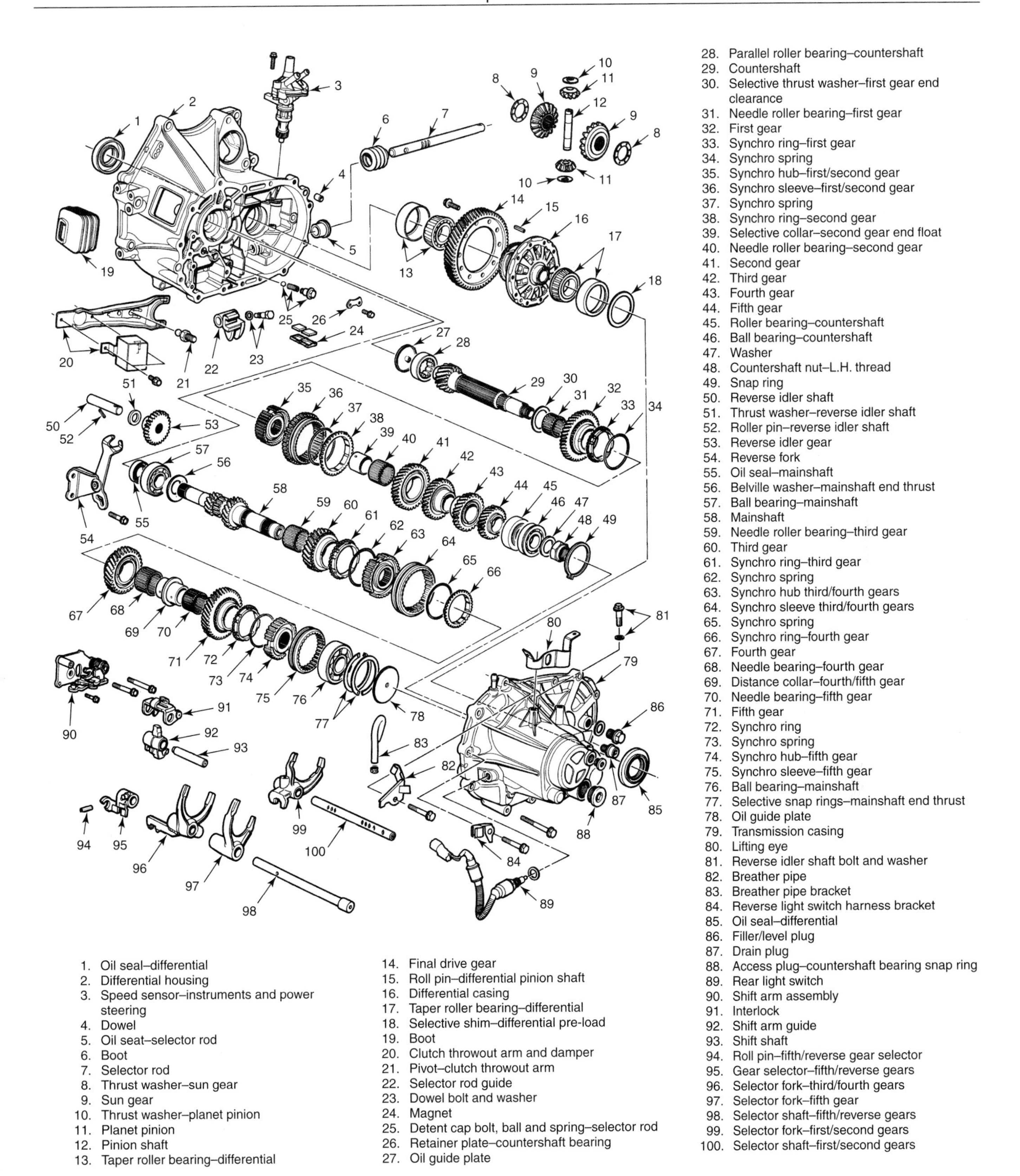

Figure 30-14. A disassembled view of a five-speed manual transaxle assembly. (Sterling Motor Cars)

has been completely disassembled. Study the part names, positioning, and relationships.

Clean All Parts Thoroughly

Clean all gears, bearings, and shafts until spotless. Pay particular attention to the inside of the case and extension housing. Any tiny metal particles (from chipped gear teeth or bearing wear) that remain in a crevice or hard-to-reach spot will be loosened by the new transmission fluid. Such particles will eventually find their way into the moving parts, where they will cause accelerated wear or even broken gear teeth. The quality of automotive repair is closely related to the thoroughness of parts cleaning—be meticulous in all cleaning.

Inspect Internal Parts

Once the transmission or transaxle has been completely disassembled and cleaned, the internal parts can be inspected. In the following sections, internal parts are grouped by major assemblies. Keep in mind that many minor variations are possible.

Inspect Input Shaft

The ***input shaft*** connects the other internal transmission parts to the clutch. Check the clutch pilot bearing end for wear and scoring. Examine the shaft splines where they contact the clutch disc splines. They must be smooth and free of excessive wear. Inspect the drive gear for wear, galling, pitting, and chipping. The drive gear synchronizer clutch teeth also must be free of wear. Check the clutch teeth closely for tapering. The end of each clutch tooth is normally chamfered for easy engagement with the clutch sleeve. The remaining portion of the tooth body, however, must not be tapered. Taper or excessive tooth wear can cause the transmission to jump out of gear. Do not reuse gears if the teeth show excessive wear, chipping, or galling.

Rotate each roller bearing while watching for signs of chipping or flaking. Check the condition of the roller bearing contact surface in the end of the shaft. The blocking ring (shift cone) contact surface must be true and smooth.

Examine the input shaft bearing for wear or other damage. The bearing contact surface on the shaft must be of full diameter, with no sign of wear caused by the inner race turning on the shaft. **Figure 30-15** shows the various areas and parts of the input shaft that require inspection.

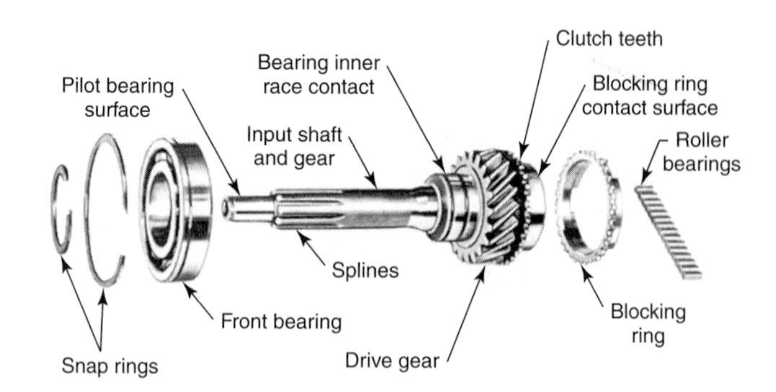

Figure 30-15. Check these areas and parts of the input shaft assembly. (Ford)

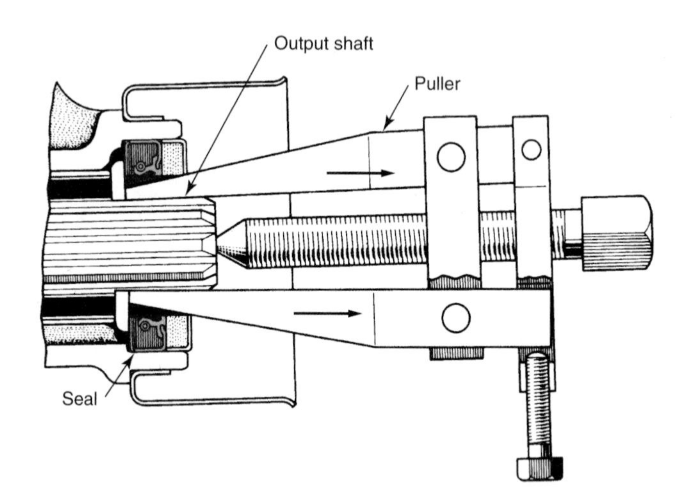

Figure 30-16. Removing a seal with a special two-leg puller. Be careful not to damage the seal bore or output shaft. (Toyota)

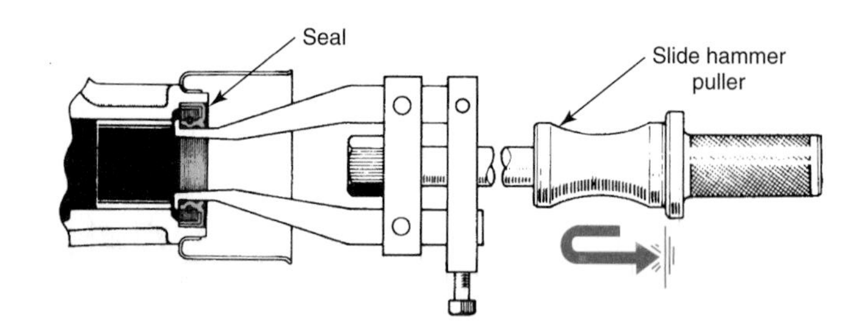

Figure 30-17. Removing an extension housing oil seal with a slide hammer puller. (Toyota)

Replace the Input Shaft Bearing Retainer Oil Seal

Always replace the bearing retainer oil seal, **Figure 30-16.** Wipe a coat of sealer around the outside of the new seal and drive it into place. Make sure the seal is driven squarely to the proper depth. The seal lip must face the transmission or transaxle.

Inspect Extension Housing or CV Axle Housing

Check the bushings in the ***extension housing*** or ***CV axle housings*** (where the CV axles enter the transaxle) and replace them if necessary. Always replace the housing rear oil seal, **Figure 30-17.** Clean the housing and drive in a new bushing if one is needed. Coat the outer edge of the new oil seal with sealer. With the seal lip facing inward, drive the seal squarely into the housing. Make sure the seal is driven to the proper depth. Refer to **Figure 30-18.**

Inspect Output Shaft and Gears

Inspect the ***output shaft*** bearing surfaces, **Figure 30-19.** They should be perfectly smooth, with no evidence of galling. Try the gears on the shaft. They should turn smoothly without excessive rocking. Where gears are splined, check for excessive play.

Examine every tooth on every gear. There must be no signs of chipping, galling, or wear. If the gear has a blocking ring surface, it must be smooth. Look at **Figure 30-20.** Check the inboard pilot bearing surface of the shaft. It, too, must be perfectly smooth.

All snap ring grooves must have sharp, square shoulders. Thrust washers must be smooth, and their thickness must meet specifications. Inspect the output shaft rear bearing. The outboard splines should be in good condition. Synchronizer inspection will be covered later in this chapter.

Check the shafts (input and/or output) for runout with a dial indicator, **Figure 30-21.** The shaft assembly should be supported on each end with V-blocks. The dial indicator is placed against the shaft at a specified point, and the dial is set to zero. The shaft is then rotated (in either direction) while noting the maximum runout reading on the dial. Runout should generally not exceed 0.002" (0.06 mm). Follow the manufacturer's specifications.

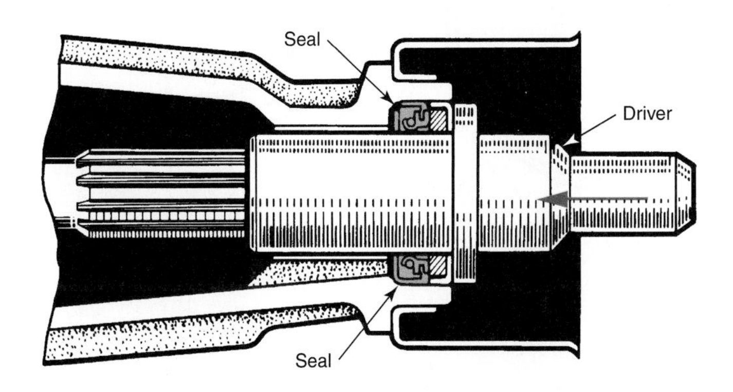

Figure 30-18. Installing a new oil seal into the extension housing. Note the seal lip is facing inward (toward transmission). (Toyota)

Inspect Countershaft and Gears

Lift the ***countershaft*** from the case and retrieve any needle bearings that have fallen to the bottom of the case. Clean the shaft, gears, needle bearings, thrust washers, and other parts. Inspect the teeth of all gears. Examine the needle bearings and the countershaft. Check the thrust washers, spacers, and retainer washers. One type of countershaft assembly is illustrated in **Figure 30-22.**

If the countershaft is equipped with an ***antilash plate*** (a part that prevents rattle caused by normal gear backlash), check the teeth and springs. The antilash plate shown in **Figure 30-23** is riveted to the countershaft and should not be removed. Replace the gear and plate as a unit. Some antilash plates are removable.

Inspect Reverse Idler Gear

Examine the ***reverse idler*** shaft. Inspect the idler gear bushings (some transmissions and transaxles use needle bearings) and thrust washers. Check the gear teeth carefully. Try the gear on the shaft and test it for wear. A reverse idler gear unit is shown in **Figure 30-24.**

Inspect Synchronizers

Since the synchronizers prevent gear clash, they should be carefully checked for damage, and all defective parts replaced. Scribe (mark) each blocking ring and the hub so that rings can be returned to their original side. If the clutch sleeve and hub, **Figure 30-25,** are not marked, scribe them so they

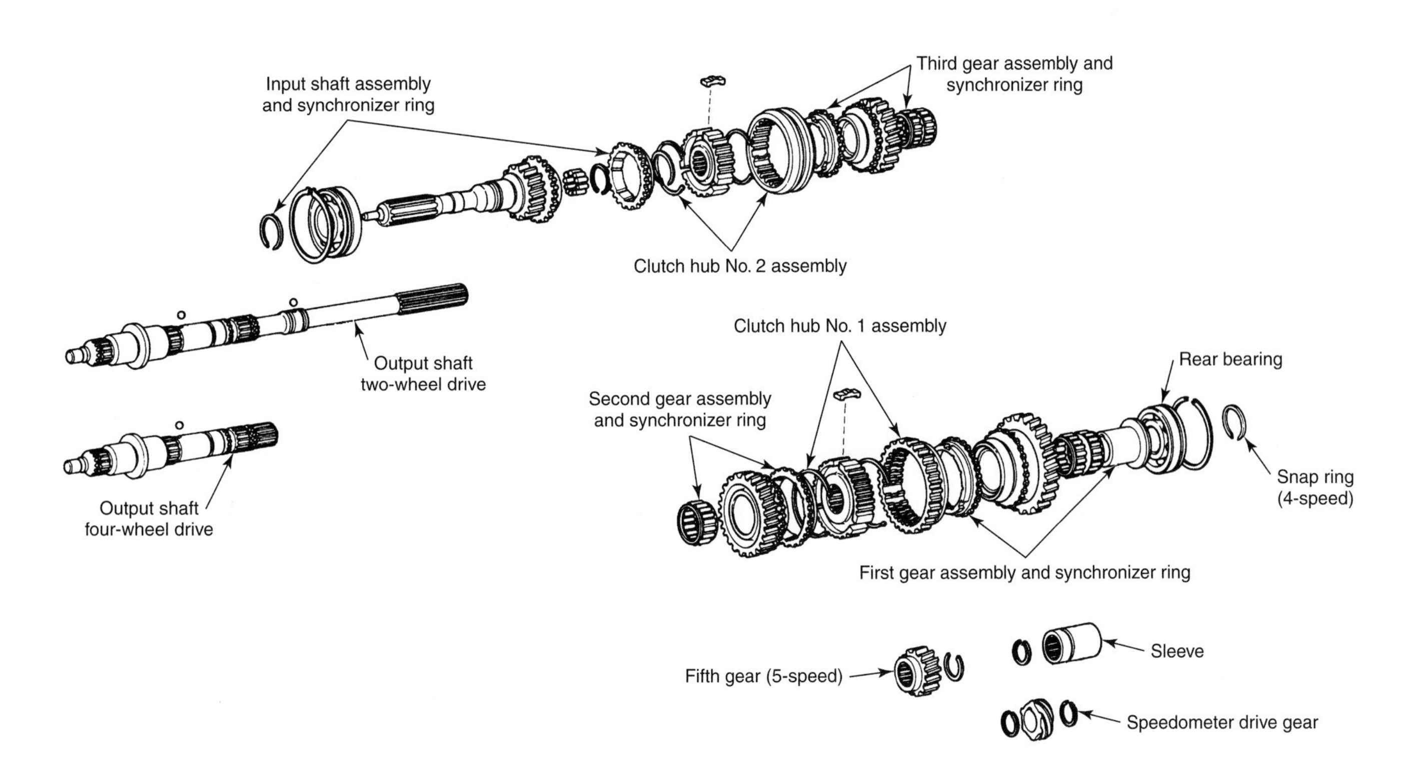

Figure 30-19. Exploded view of an output shaft assembly. Note the difference between the two-wheel and four-wheel drive shafts. (Toyota)

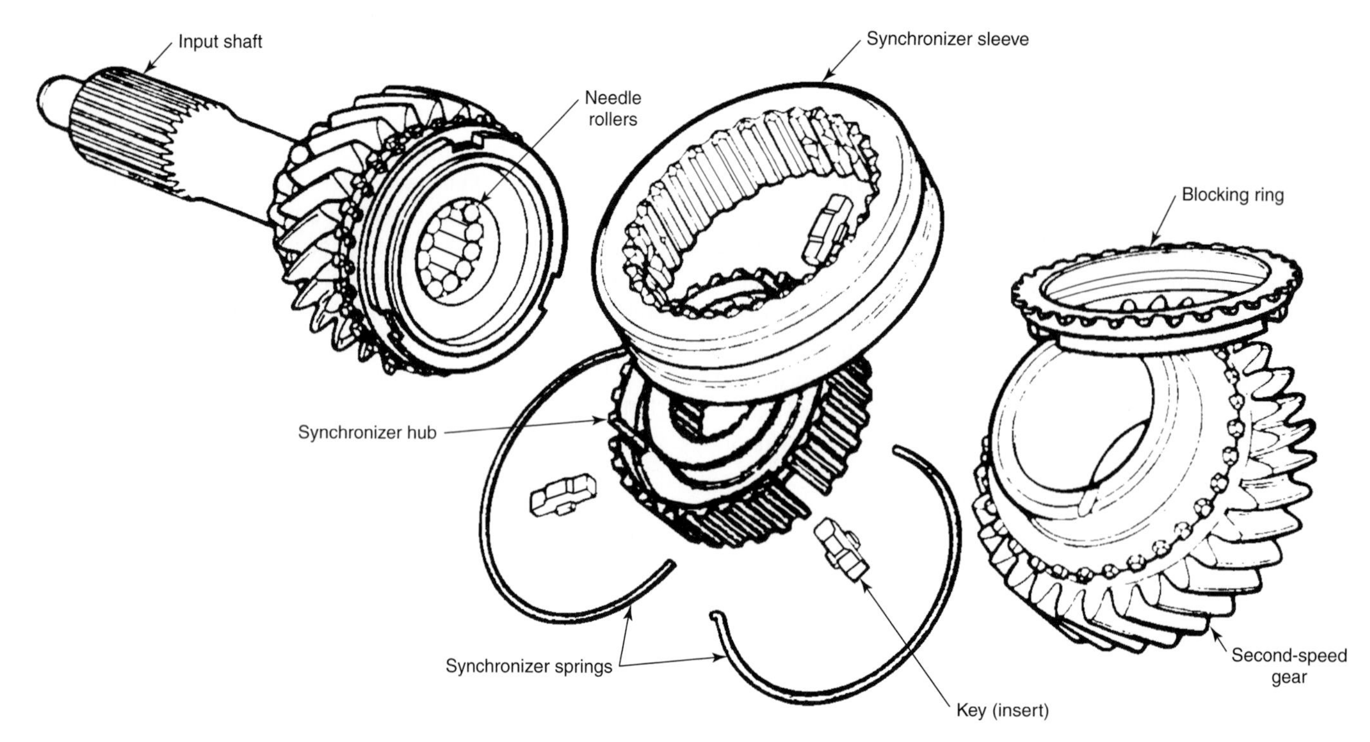

Figure 30-20. A blocking ring and related components. (Borg-Warner)

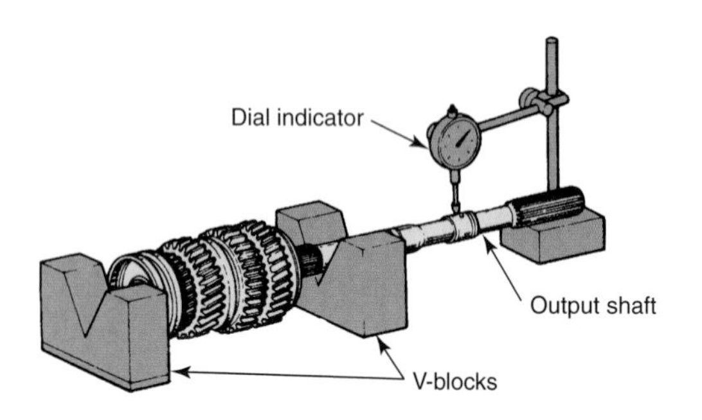

Figure 30-21. Checking an output shaft for runout with a dial indicator. Note the V-blocks, which accurately support the shaft. (Toyota)

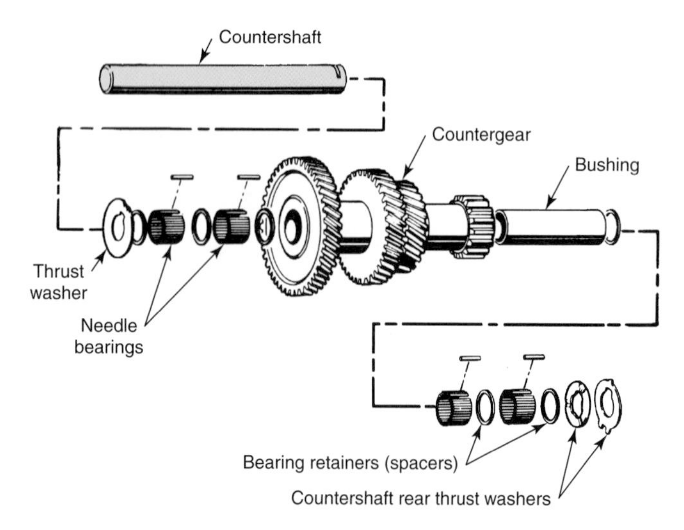

Figure 30-22. Typical countergear (cluster gear), shaft, and bearings. (DaimlerChrysler)

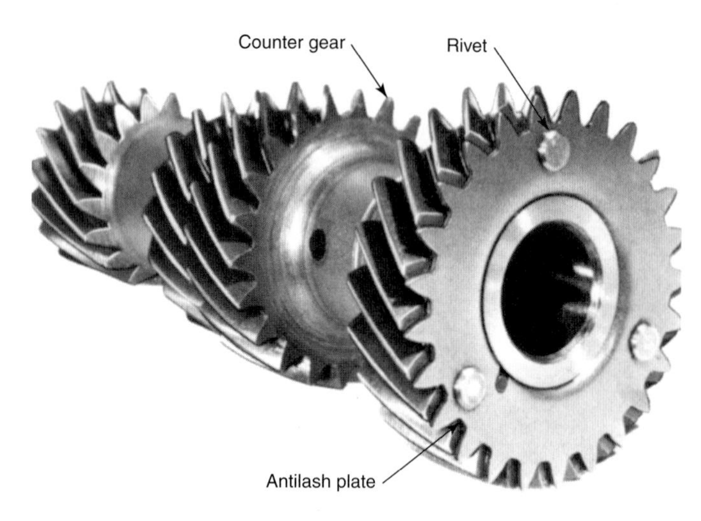

Figure 30-23. Countergear antilash plate. This plate and gear must be replaced as a unit. (General Motors)

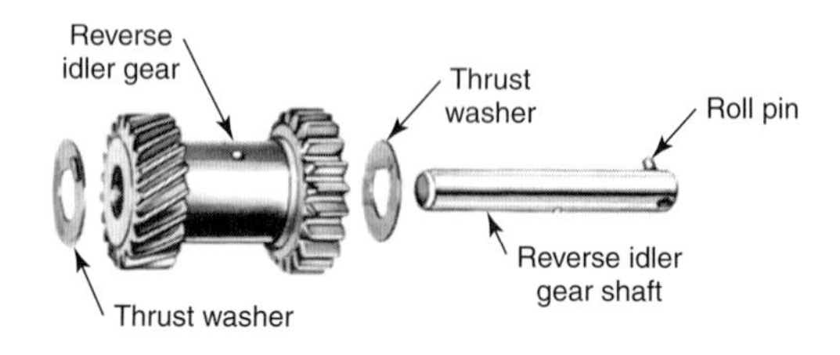

Figure 30-24. One type of reverse idler gear assembly. This gear uses bronze bushings.

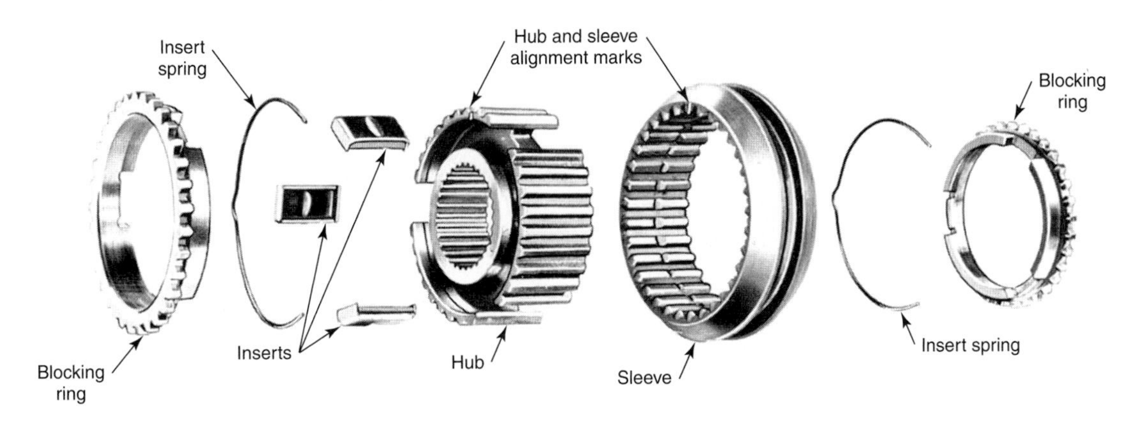

Figure 30-25. Disassembled synchronizer. Note the alignment marks on the sleeve and hub.

may be reassembled in the same position. Slide the clutch sleeve from the clutch hub. Remove the inserts and insert springs. Clean all parts.

Check the inserts and insert springs, **Figure 30-26,** for excessive wear. Slide the sleeve on the hub (with marks aligned) and test play. Inspect the hub inner splines. The sleeve clutch teeth must not be battered or tapered. Pay particular attention to the blocking rings. The inside should still show fine grooves and the teeth should be in good shape. The notched sections that fit over the inserts should not be battered and worn.

The ***cone surface*** of the gears engaged by the synchronizer should be smooth. **Figure 30-27** shows a synchronizer clutch sleeve (inserts and springs are shown, hub is not shown), blocking rings, and the two gears served by the synchronizer. Note the smooth gear cone surfaces and the grooves in the inner section of the blocking rings.

Assemble Synchronizer

To assemble the synchronizer shown in **Figure 30-28,** lubricate parts with transmission or transaxle lube. Place one

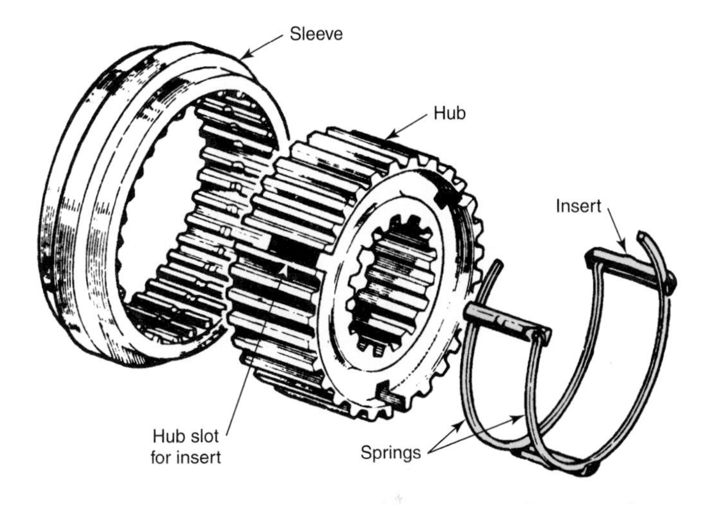

Figure 30-26. Check inserts and insert springs for excessive wear, and replace if necessary. When installed, the springs are inside the hub, one on each end, with the inserts in the hub slots.

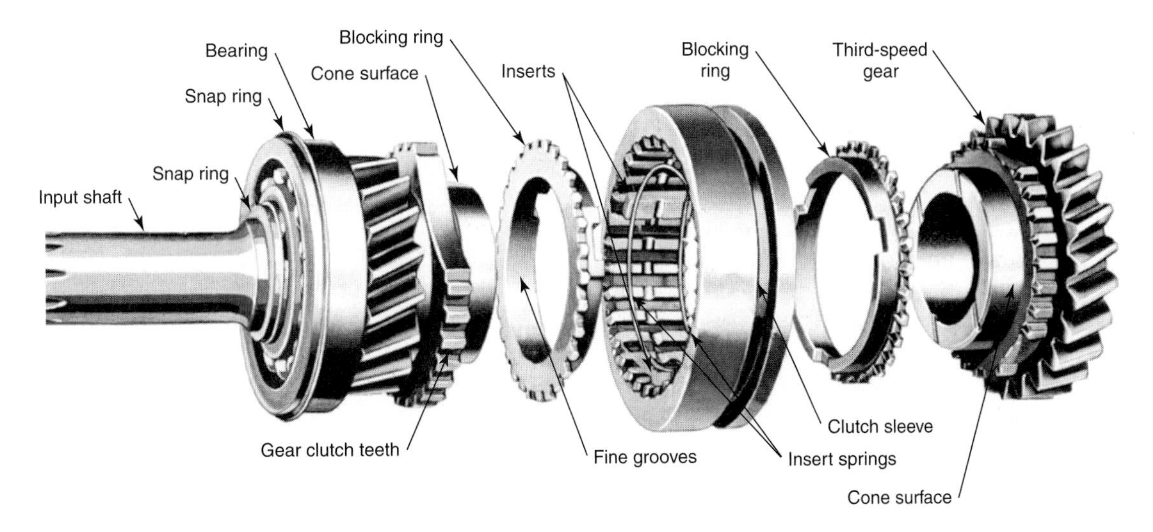

Figure 30-27. Synchronizer (minus hub) and the gears it serves. (GMC)

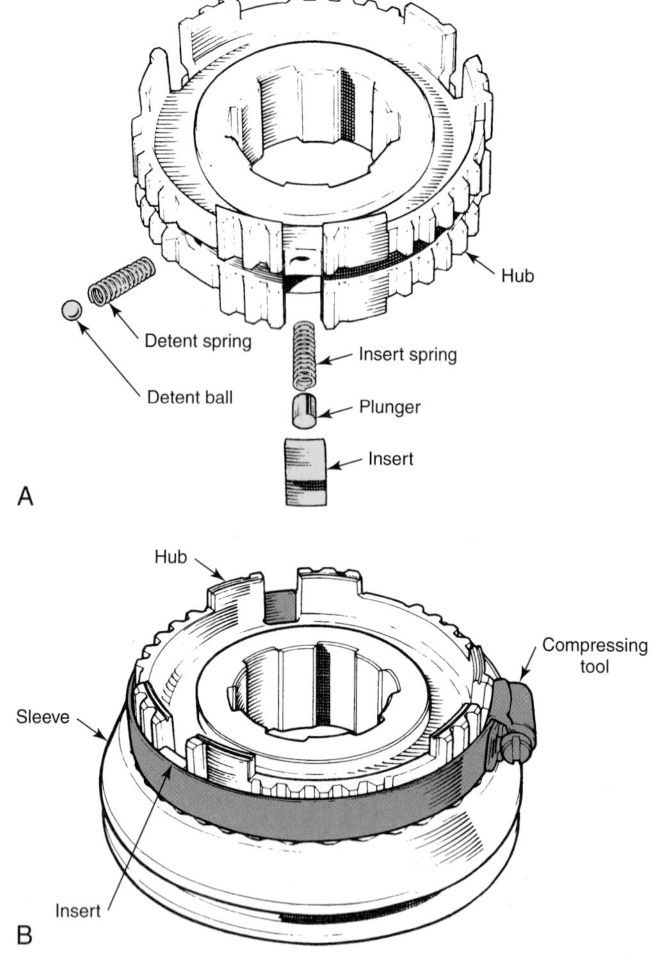

Figure 30-28. Assembling a synchronizer. A—Synchronizer hub with insert spring, insert plunger, and insert. Note the detent ball and spring. B—A compressing tool is being used to secure the insert and detent assemblies while the hub is slipped into the sleeve. (British Leyland)

of the insert springs in the hub so the humped portion rests in one of the hub insert slots. Align the hub and sleeve marks and start the sleeve on the hub. Be sure the sleeve is facing in the correct direction.

Install the three inserts and push the sleeve into place. Make sure the clutch sleeve and hub marks are aligned and the insert springs are securely in place behind the lips or tabs on the ends of the inserts. Install the blocking rings.

Installation of insert springs varies. Follow the manufacturer's instructions. Note how the springs are installed in **Figure 30-29.** The bent tip on each spring is installed in the same insert.

Some synchronizers are easily installed by installing the springs and inserts in the hub, then placing a compressing tool around the hub. This holds the inserts down so the hub can be gently tapped into the sleeve, **Figure 30-30.**

Inspect Transaxle Chain and Sprockets or Gears

Closely examine the transaxle ***drive chain*** and ***drive sprockets*** for wear or damage. Check the chain for slack and worn pins. If a transmission uses two large gears for power transfer, check them for chipping, wear, and loose fit on the shaft splines.

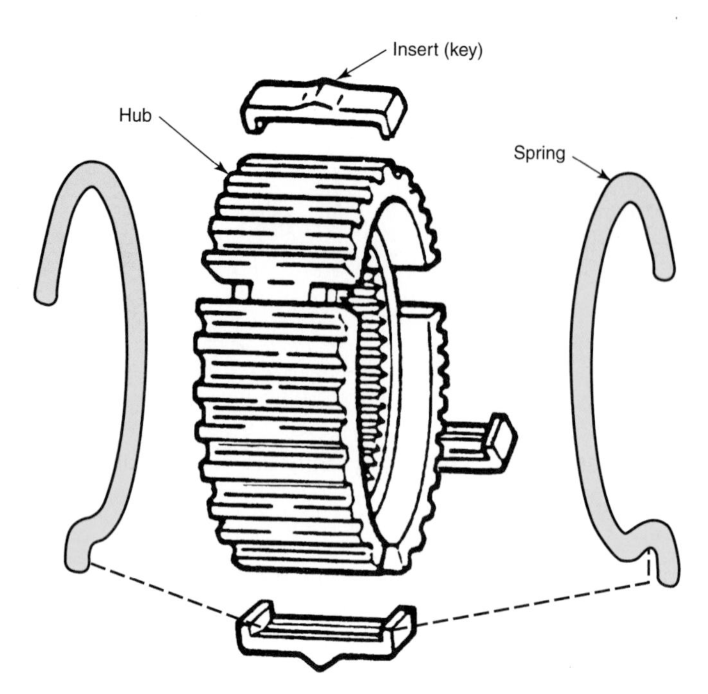

Figure 30-29. Insert springs with bent tips. Note that both bent tips go into the same insert. (Borg-Warner)

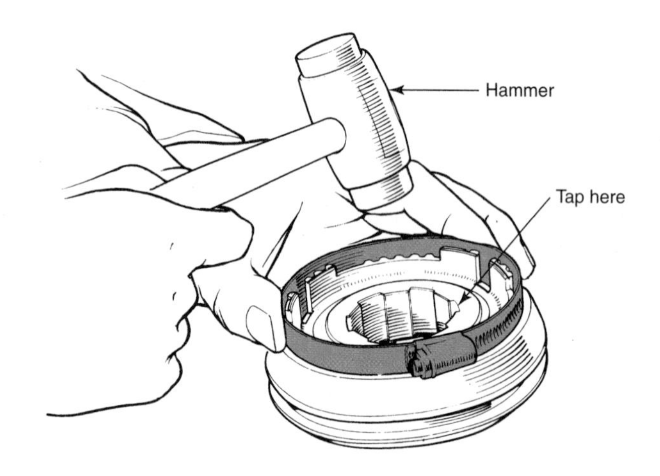

Figure 30-30. With a compressing tool holding the insert and detent units in place, the hub can be gently tapped into position with a plastic hammer.(British Leyland)

Inspect Transaxle Differential Unit

Closely inspect the ***differential unit*** for wear or damage. Differential gears are checked in the same manner as the gears in the transmission. The hypoid-type differential assembly can be checked following the procedures outlined in Chapter 32.

Check Case and Housings for Cracks and Burrs

Inspect the front bearing retainer, transmission or transaxle case, and any extension or CV axle housings,

Figure 30-31, for signs of cracking. Carefully examine the areas around bolt holes and shaft and bearing openings.

On aluminum cases, check for porous areas (slight flaws in the aluminum casting) that allow gear lubricant to seep to the outside. Follow the transmission or transaxle manufacturer's recommendations for repair.

Also check all shift covers, inspection covers, and other sheet metal parts for dents or warping that could prevent them from sealing properly. Carefully straighten any damaged areas or replace the part.

Check the extension-to-case and the case-to-clutch-housing surfaces for any burrs that could cause misalignment. If burrs are found, remove them with a fine mill file. If the transmission or transaxle has a vent, make sure it is open.

Obtain and Check New Parts

When a part is unfit for service, new parts must be obtained. Always check the new part against the old for matching design, size, and shape. Try the part in the transmission or transaxle to make sure it fits properly.

Use New Snap Rings, Thrust Washers, and Gaskets

To prevent future problems, use new ***snap rings*** whenever possible. New ***thrust washers*** will provide proper end play. New gaskets (installed with gasket cement) are a must. Some transmission and transaxle cases use RTV (room temperature vulcanizing) sealer in place of gaskets. Always use the recommended sealer. An inferior seal can cause a complete loss of lubricant, leading to transmission or transaxle failure. If drive-in expansion plugs were removed, install new plugs. Use the recommended sealer on all plugs and pins.

Caution: Parts must be absolutely clean and well lubricated before assembly.

After cleaning and inspection, all parts should be oiled and placed in clean containers. Before installation, every part should be heavily lubricated with transmission oil or transmission fluid as recommended by the manufacturer.

Seal and Stake Pins

Where pins pass through the outer wall of the case, apply sealer to the hole so the pin will not leak oil. Drive the pin slightly below the surface of the case and stake it to prevent loosening.

Check for Cracks

When a transmission or transaxle has suffered heavy gear damage (shattered teeth), use crack detection chemicals on the remaining gears. Check the shafts and shaft openings in the case also. Discard all parts showing the slightest sign of cracking. Replace the bearings, even if they appear to be good.

Transmission or Transaxle Reassembly

Basically, the transmission or transaxle is reassembled in the reverse order of disassembly. All parts must be lubricated

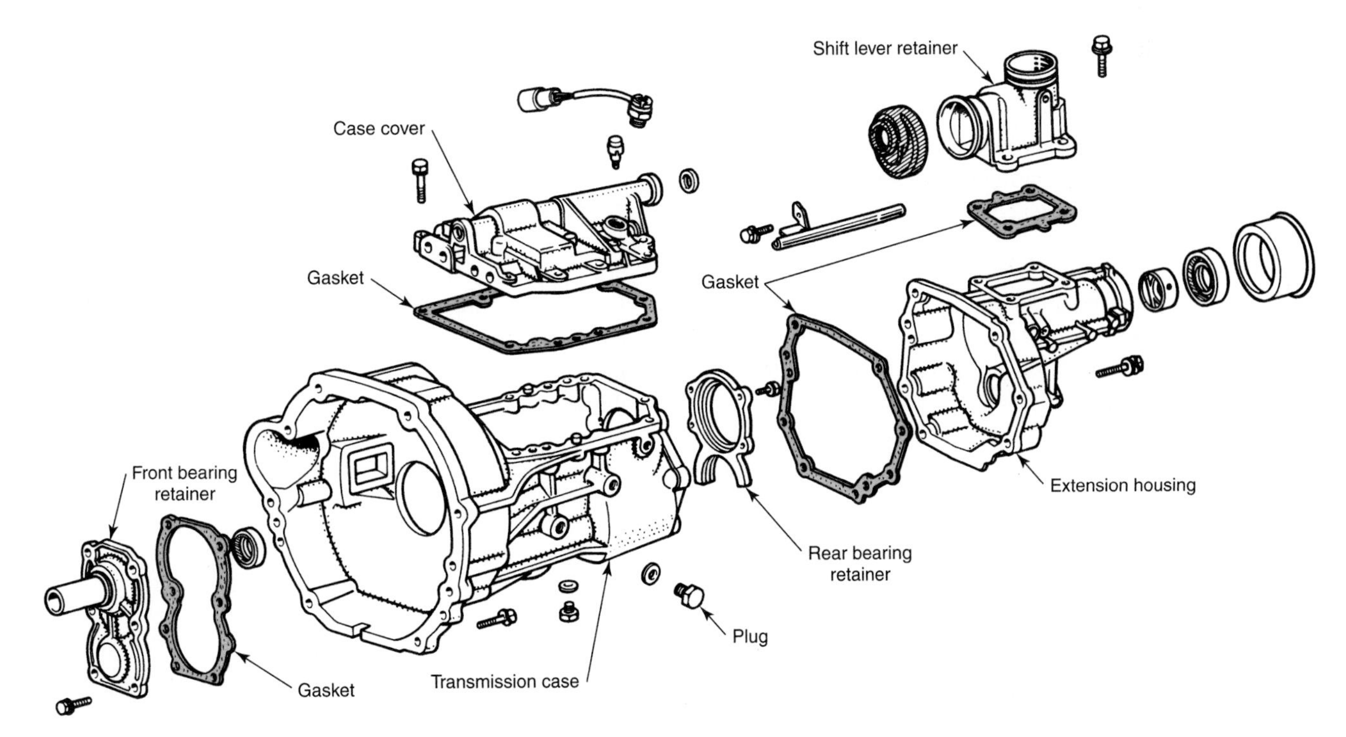

Figure 30-31. Transmission case and gasket assembly. Check the case, extension housing, bearing retainer, and cover for cracking, porosity, and other damage. (Toyota)

and properly installed. Make constant use of the manufacturer's service manual to ensure that all parts are reinstalled correctly.

General Internal Part Installation Procedures

Lubricate all bushings and shafts with transmission oil before beginning assembly. To install thrust washers, place a coating of soft grease on each end of the gear and press the thrust washers into the grease. Make sure they face in the right direction. The grease will hold the washers in place as the gear is installed.

Note: Any assembly lube/grease used to hold internal parts in place must be compatible with, and dissolve readily in, the normal oil or transmission fluid used in the transmission or transaxle.

If any gears, such as the counter gear or the reverse idler gear, use needle bearings, the bearings may be held in place with a ***dummy shaft.*** A dummy shaft is a wooden or metal shaft that is the same diameter as the regular shaft but only as long as the combined thickness of the gear and thrust washers. The gear is then placed in the case with the needle bearings and thrust washers already installed. When the regular shaft is pushed through the case opening and into the gear, the dummy is forced out the other side. This procedure provides proper alignment and holds the needles, spacers, and thrust washers in line.

If a counter gear must be lowered to install the output or input shaft, place it into the case but do not install it until the main shafts are in place.

Never use excessive force to make a part fit. If a part does not slide into position as it should, stop and check for the source of difficulty.

Spin all gears after they have been installed. They must turn freely. Use a dial indicator or feeler gauge to check the end play of all units for which an end play specification is given. **Figure 30-32** shows a typical method for checking end play.

Lubricate Assembled Internal Parts

When the internal parts of the transmission or transaxle are fully assembled, pour fresh gear oil over the gears and shaft through the cover opening. Use oil of the recommended viscosity. Some units use automatic transmission fluid. Turn the input shaft while shifting through the gears. The shafts and gears should turn freely with no catching. Check shaft end play if it was not checked earlier.

Install Shift Cover

To install the shift cover, place the transmission or transaxle in neutral. Place the shift fork levers in the neutral position. Install a new gasket. Hold the cover in line with the cover hole. The shift forks must align with the clutch sleeve and gear fork grooves, **Figure 30-33.** Guide the shift forks into their grooves. Insert the cover fasteners and tighten them to specifications. Try the shift mechanism for proper operation.

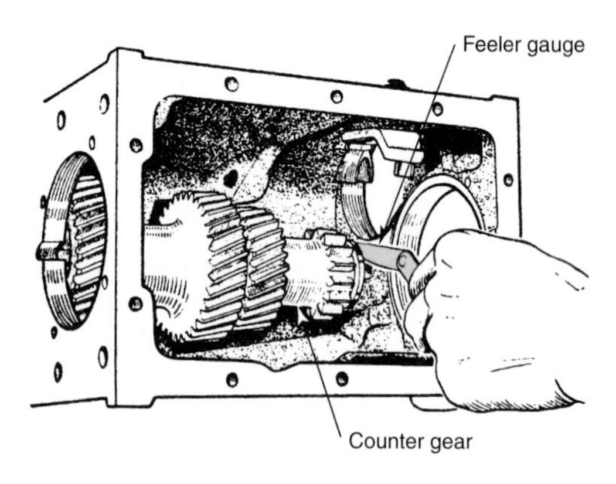

Figure 30-32. Checking countergear end play with a feeler gauge. (British Leyland)

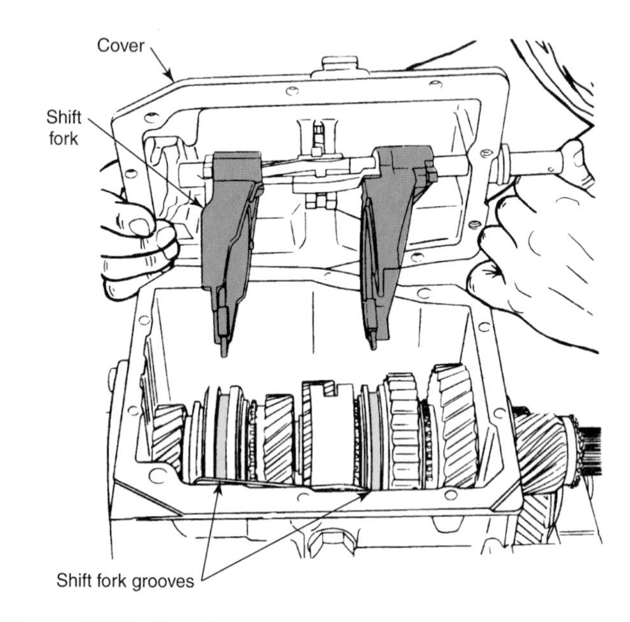

Figure 30-33. Installing the shift cover. Shift forks must align with the shift fork grooves. Do not force the cover into position! (General Motors)

Install Transmission or Transaxle

Wipe the clutch housing and transmission or transaxle face clean. Check for burrs. Using a transmission or transaxle jack, raise the transmission or transaxle into alignment with the engine and insert the input shaft through the clutch throwout bearing, through the disc hub, and into the pilot bearing.

Caution: Never let the transmission or transaxle hang with all of its weight supported by the input shaft.

Install and torque the fasteners holding the transmission or transaxle to the clutch housing or engine. Install the drive line components, speedometer cable, shift linkage, and other parts.

Adjust Shift Linkage

Disconnect the shift rods at the transmission or transaxle (or at the column shift levers, if used). Place the transmission or transaxle shift levers in neutral. If the linkage has slotted adjustment holes, loosen the adjustment nuts and leave the shift rods connected.

With the transmission or transaxle shift levers in neutral, place the shift levers in neutral. Adjust the levers, stop block, etc., **Figure 30-34.** It may be necessary to pass an aligning pin through the levers or to use a special tool as shown in **Figure 30-35.** Tighten the linkage adjustment nuts.

If the linkage is the type that was disconnected, adjust the linkage length so the rods just reach from the transmission or transaxle to the column shift levers. Insert and secure the linkage.

A typical floor shift is illustrated in **Figure 30-36.** To adjust this linkage, loosen the shift linkage adjustment nuts. Place the gear shift lever in neutral and pass the alignment pin

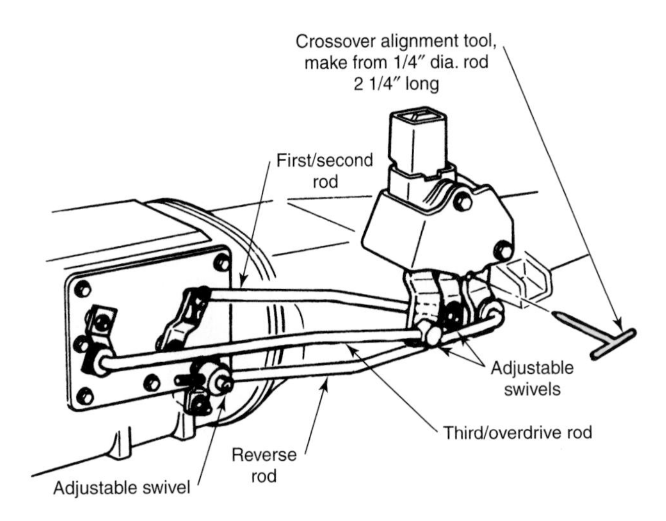

Figure 30-35. Adjusting shift linkage on a floor shift transmission. Note the use of an alignment pin to hold the gearshift in the correct position during adjustment.

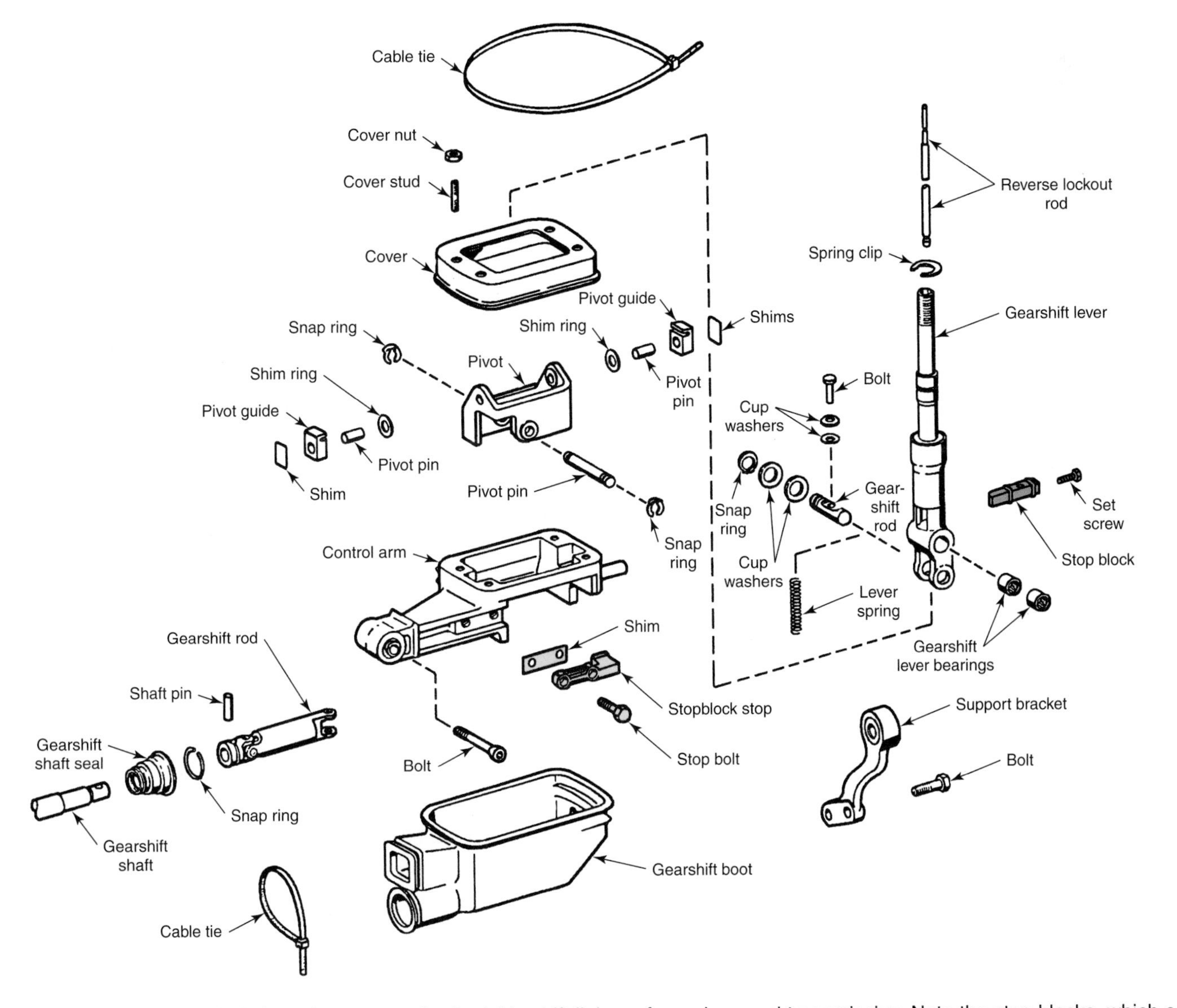

Figure 30-34. Exploded view of one type of adjustable shift linkage for a six-speed transmission. Note the stop blocks, which can be adjusted to control linkage movement. (General Motors)

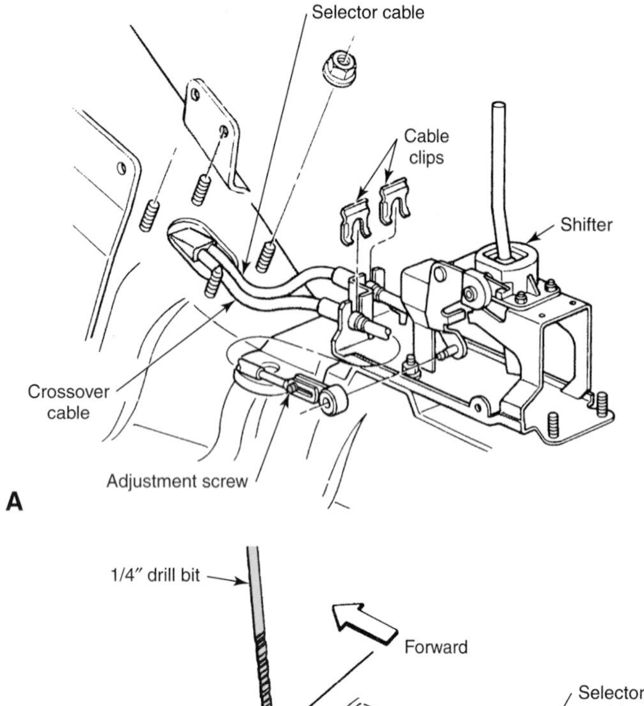

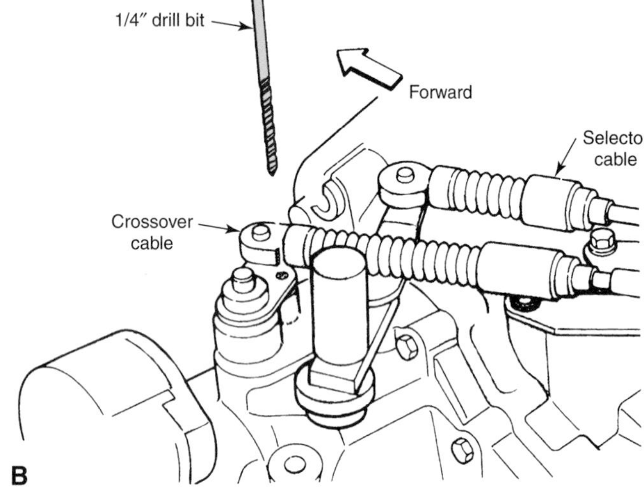

Figure 30-36. Adjusting linkage in a floor shift. A—Shift (crossover) cable assembly with adjustment screw. B—Alignment pin (drill bit used here) being placed into aligning holes to provide the correct linkage alignment before and during tightening of the adjustment screw. (DaimlerChrysler)

through the aligning holes. Place the transmission or transaxle shift levers in neutral. Tighten the adjustment nuts. Finally, remove the pin.

Fill Transmission or Transaxle

Fill the transmission or transaxle to the level of the filler plug with the recommended gear oil or automatic transmission fluid, **Figure 30-37.** Fill slowly so the oil will have time to flow. If the differential assembly of the transaxle is a separate unit, fill it with clean gear oil of the correct viscosity and type.

Road Test

Road test the vehicle. The transmission or transaxle should operate quietly and smoothly. Shifting should be smooth and positive with no jumping out of gear. Shift up and down to test the synchronizers. When back at the shop, check the transmission or transaxle for leaks and recheck the lubricant level.

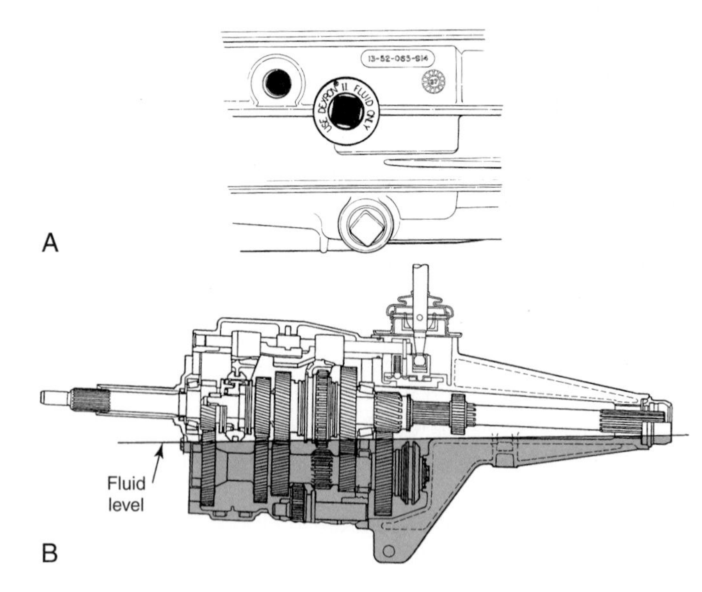

Figure 30-37. Filling the transmission or transaxle. A—Be sure to use the correct type of transmission lubricant, which will be noted in the manufacturer's specifications. It may be noted on or near the filler plug. B—A proper lubricant fluid level for one type of five-speed transmission. (Borg-Warner)

Tech Talk

One of the most frustrating jobs in manual transmission or transaxle work is installing the spring-loaded detent balls that hold the internal linkage pieces in position. It seems that there is never enough room to reach in and place the ball properly. You will probably drop the balls several times before you are finished—if the springs don't launch them across the shop.

An easy way to install detent balls involves using an old hacksaw blade. Bend the blade so that one end will reach into the recesses of the transmission. Place the detent ball on the hole in the end of the blade with a little grease. The hole will hold the ball in place while you position it against its spring. Then, install the levers or rails over the ball and the hacksaw blade. After everything is in position, carefully slide the blade out, leaving the ball in place.

Summary

Vehicle transmissions and transaxles are of three-, four-, five-, and six-speed designs. An overdrive may be incorporated. Transaxles are used on front-wheel drive vehicles.

Remove the transmission or transaxle only when necessary. To remove a transmission or transaxle, drain the transmission oil. Remove all controls, wires, etc. Drop the drive shaft or shafts. Remove the fasteners holding the housing to the transmission or transaxle and pull the transmission or transaxle back. Use pilot bolts or a jack. Do not allow the transmission or transaxle to hang on the input shaft at any time.

Clean the outside of the transmission or transaxle before disassembly. Flush the inside of the transmission or transaxle and examine the gear teeth. Make any needed end play and clearance checks before disassembly.

General disassembly requires dropping the countergear to the bottom of the case (use a dummy shaft) and then removing either the input or output shaft. A manufacturer's service manual should be available for reference.

Place the transmission or transaxle on a stand. If hammering is required, use a soft-face type. Keep all synchronizer parts together. Be careful to avoid losing roller or needle bearings, snap rings, springs, or detent balls. Place them in a separate container. All parts must be thoroughly cleaned.

Inspect all synchronizer parts for excessive wear, chipping, galling, and cracking. Check synchronizer clutch sleeve teeth and the corresponding gear clutch teeth. Teeth should not be chipped, tapered, or excessively worn.

Synchronizer blocking rings should show some sign of fine tooling lines on the inside tapered surface. When reassembling synchronizers, make certain the hub and clutch sleeve marks are aligned and the inserts and springs are correctly installed.

Always install new seals. Use new gaskets and apply gasket cement. In some cases, RTV sealer is used. When lock pins pass through the case, use sealer and stake the pins. Use new thrust washers and snap rings. Lubricate all parts as they are installed. Avoid using heavy force to make parts fit.

Compare new parts with used parts for size, shape, and operation. Where heavy damage was incurred, check the case, gears, and shaft for cracking.

Use soft grease to hold roller and needle bearings in place during assembly. Do not plug lubricant entry holes. Use grease that readily dissolves in oil. Make certain snap rings are properly seated. Use a dummy shaft to install the countergear in the case.

When assembled, check for correct end play on all parts. All moving parts must turn freely and without catching. Check shifting action.

Align the shifter forks carefully when installing the shift cover. Install the transmission or transaxle. Slowly fill with recommended lubricant. Make all necessary connections. Check the shift linkage adjustment. Road test the vehicle and recheck the lubricant level.

Review Questions—Chapter 30

Do not write in this book. Write your answers on a separate sheet of paper.

1. A _____ has only one output shaft, while a _____ has two output shafts.
2. A transverse engine is installed _____ in the vehicle.
3. Before road testing, always check the _____ in the transmission or transaxle.
4. List five defects that you should watch for while road testing a manual transmission or transaxle.
5. List six inspection points for the input shaft assembly.
6. After disassembly and cleaning, check the transmission or transaxle case and extension housing for _____.
7. Always check the case-to-clutch-housing surfaces for any burrs that could cause _____.
8. A dummy shaft is the same _____ as the regular shaft, but it is not as _____.
9. When removing or installing a transmission or transaxle, never allow the weight of the unit to hang on the _____.
10. The transmission or transaxle should be filled to the bottom of the _____ with the recommended type and grade of lubricant.

ASE-Type Questions

1. Modern manual transmissions in rear-wheel drive vehicles can have _____ forward gears.
   (A) four.
   (B) five.
   (C) six.
   (D) All of the above.
2. Technician A says that an average ratio for first gear in a transmission or transaxle is 5.69:1. Technician B says that a four-speed transmission or transaxle has a much lower first gear ratio than does a three-speed transmission. Who is right?
   (A) A only.
   (B) B only.
   (C) Both A and B.
   (D) Neither A nor B.
3. Technician A says that clutch adjustment can affect transmission or transaxle shifting. Technician B says that synchronizer condition can affect transmission or transaxle shifting. Who is right?
   (A) A only.
   (B) B only.
   (C) Both A and B.
   (D) Neither A nor B.
4. A gear should never be reused if the teeth show any signs of _____.
   (A) chipping.
   (B) galling.
   (C) excessive wear.
   (D) Any of the above.
5. All of the following should be replaced any time the transmission or transaxle is disassembled, *except:*
   (A) snap rings.
   (B) seals.
   (C) bearings.
   (D) gaskets.

6. Technician A says that some antilash plates are removable. Technician B says to replace a worn antilash plate and corresponding gear as a unit. Who is right?
   (A) A only.
   (B) B only.
   (C) Both A and B.
   (D) Neither A nor B.
7. When assembling synchronizers, you should do all of the following *except:*
   (A) lubricate the synchronizers.
   (B) align the hub and clutch sleeve marks.
   (C) make sure the sleeve is facing in the correct direction.
   (D) separate the insert springs from the synchronizers.
8. Grease used to hold internal parts in place must dissolve in _____.
   (A) parts cleaning solvent.
   (B) normal grade oil or transmission fluid.
   (C) kerosene.
   (D) alcohol.
9. Some transmissions and transaxles use _____ as a lubricant.
   (A) gear oil.
   (B) automatic transmission fluid.
   (C) brake fluid.
   (D) Both A and B.
10. Before adjusting the length of the shift linkage rods, both the shift lever and the transmission or transaxle must be in _____.
   (A) neutral.
   (B) first.
   (C) third.
   (D) any gear.

Suggested Activities

1. Do a survey at your school to determine what percentage of the faculty, staff, and student body owns vehicles with manual transmissions. How do your figures compare with the national average of about 10%?
2. Using the appropriate service manual, check and (if necessary) adjust the linkage on a manual transmission/transaxle.
3. Tear down a manual transmission or transaxle and list all the parts needed for a rebuild.
4. Calculate the cost of repairing the above transmission by the following process:
   a. Use a flat rate manual to find the prices of the needed parts.
   b. Use a flat rate manual to find the labor times for replacing needed parts.
   c. Decide if the clutch will be replaced and add the clutch parts and labor to the list.
   d. Total the labor times and multiply them by the average labor rate for your area.
   e. Add the total of parts and labor.
   f. Add other charges, such as supplies (rags, cleaners, and oil absorbent) and outside services (machine shop work) for a grand total of what it would cost to repair this transmission.
5. Discuss your price calculations with the other members of your class. Ask if there is anything you missed. Determine whether the cost of the repair is greater or less than the cost of a replacement transmission.

31

Automatic Transmission and Transaxle Service

After studying this chapter, you will be able to:

- Explain automatic transmission and transaxle in-vehicle service and diagnosis.
- Explain towing procedures for vehicles with automatic transmissions and transaxles.
- List common automatic transmission and transaxle problems and corrections.
- Road test an automatic transmission or transaxle.
- Summarize the adjustment of automatic transmission and transaxle linkage.
- Explain automatic transmission and transaxle shift linkage and band adjustment.
- Summarize automatic transmission and transaxle removal and installation.

Technical Terms

Torque converter
Planetary gearsets
Holding members
Hydraulic control system
Electronic control system
Lockup clutch
Bands
Clutch packs
Overrunning clutches
Oil pump
Spool valves
Main pressure regulator
Manual valve
Shift valves
Throttle valve
Governor valve
Servos
Clutch pistons
Accumulators
Filter
Valve body
Transversely
Conventionally
Troubleshooting chart
Band and clutch application charts
Aeration
Shift points
Limp-in mode
Gear skipping
Stall test
Band adjustments
Throttle valve (TV)
Detent rod
Detent cable
Vacuum modulator
Air pressure checks
Add-on transmission oil cooler

Automatic transmission and transaxle operating principles and service procedures are relatively complicated. The increased reliance on electronic controls and overdrive units, as well as frequent design changes, all combine to further complicate service. It is not within the scope of this text to cover complete automatic transmission and transaxle service and repair. Hundreds of pages would be needed to provide complete overhaul, testing, and adjustment details for the various types and models. This chapter covers the in-vehicle service (service that can be performed while the unit is in the vehicle) of modern automatic transmissions and transaxles. Removal and installation is also covered.

Automatic Transmissions and Transaxles

Modern automatic transmissions and transaxles are usually three- or four-speed units. They contain a ***torque converter,*** one or more ***planetary gearsets,*** multiple sets of ***holding members*** (clutches and bands), a ***hydraulic control system,*** and an ***electronic control system.***

Torque Converters

Torque converters transfer power through the movement of transmission fluid. Since the power is transferred through fluid, there is no mechanical connection between the engine and road. This allows the engine to continue running when the transmission is in gear and stopped. Modern torque converters use a hydraulically operated clutch, called a ***lockup clutch,*** **Figure 31-1.** This clutch is applied to lock up internal parts of the converter to prevent slippage. The converter lockup clutch may apply in any gear other than Low and Reverse. The locking action provides direct, no-slip drive for increased fuel economy.

Planetary Gearsets and Holding Members

All transmissions and transaxles use planetary gearsets to provide different gear ratios. By holding and driving different parts of the planetary gearset, many different ratios are possible from the same set of gears. The range of gears available from a typical planetary gearset is shown in **Figure 31-2.** To obtain additional gear ratios, more than one gearset may be used. Gearsets may be combined, sharing some common members. Holding members are used to apply and drive various parts of the planetary gearsets. Typical holding members are ***bands,*** multiple disc ***clutch packs,*** and ***overrunning clutches*** (one-way clutches). The action of

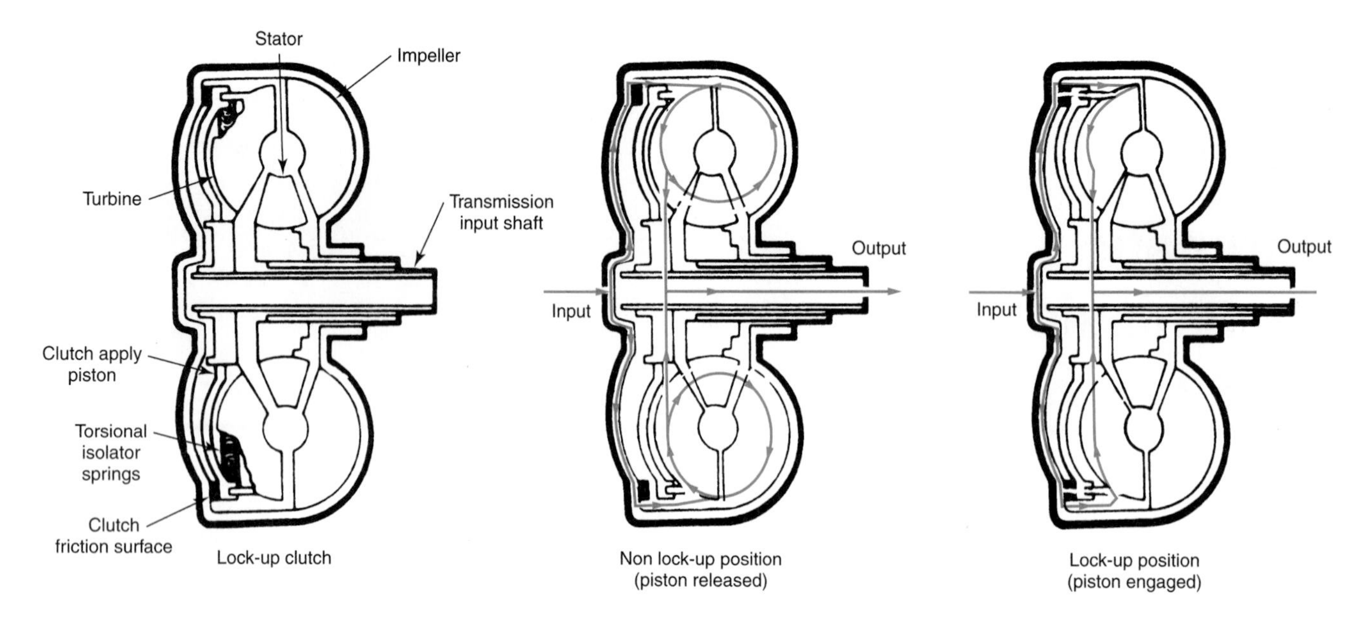

Figure 31-1. This diagram illustrates the operation of one type of lockup converter. When in lockup position, the converter acts as a solid drive unit. (DaimlerChrysler)

the various holding members is controlled by the hydraulic system.

Hydraulic Control System

The hydraulic control system depends on pressure developed by a transmission ***oil pump.*** The pump creates hydraulic pressure, which is controlled by ***spool valves.*** The spool valves control hydraulic pressure and use it to fill the torque converter and obtain different gears as needed. The major valves that are contained in automatic transmissions and transaxles include:

- The ***main pressure regulator,*** which controls the overall system hydraulic pressure by regulating the oil pump's output. The main pressure regulator valve setting is controlled by the spring tension.
- The ***manual valve*** is operated by the driver to engage the transmission in Park, Reverse, Neutral, Drive, Manual Second or Low, and other available gears.
- The ***shift valves*** redirect hydraulic pressure to the holding members to obtain the different gears in a drive range. These valves are controlled by the valves described below.
- The ***throttle valve*** is controlled by linkage from the engine throttle plates, or by a vacuum modulator operated by engine manifold vacuum. The throttle valve tries to move the shift valve to the downshifted (lower gear) position. The more the throttle is depressed, the higher the throttle pressure. Throttle valves are not used in electronically controlled transmissions and transaxles.
- The ***governor valve*** is mounted on or driven by the output shaft. As governor pressure rises, it tries to move the shift valve to the upshifted (higher gear) position. The governor pressure increases as the shaft speed increases. Governor valves are not used in electronically controlled transmissions and transaxles.

Some other components of the hydraulic system include ***servos,*** which operate the bands; ***clutch pistons,*** which operate the multiple disc clutches; and ***accumulators,*** which cushion the application of the holding members. Another important part of the hydraulic system is the ***filter,*** which removes dirt and metal from the transmission fluid. Many of the valves and other components are installed in a ***valve body*** that is attached to the bottom of the transmission or transaxle case. **Figure 31-3** shows the layout of a simple automatic transmission hydraulic system. Note the relationship of the valves just discussed. Other valves are used to cushion shifts, provide detent shift for passing, and control the lockup torque converter.

Electronic Control System

Almost all modern automatic transmissions and transaxles use the ECM to control shift points, system pressures, and torque converter lockup. The basic theory of electronic transmission and transaxle control is similar to that for electronic engine controls. Engine sensors, such as the throttle position sensor and engine temperature sensors, create input signals. Transmission-specific sensors, such as oil pressure and temperature sensors, provide other input to the transmission control computer. The transmission control computer, which may be part of the engine or body control computer, makes decisions based on the sensor input. It then sends command signals to solenoids attached to the valve body. Typical solenoids include the shift solenoids, which move the shift valves, and pressure control solenoids, which position the main pressure regulator valve. Other solenoids control the operation of the torque converter lockup clutch.

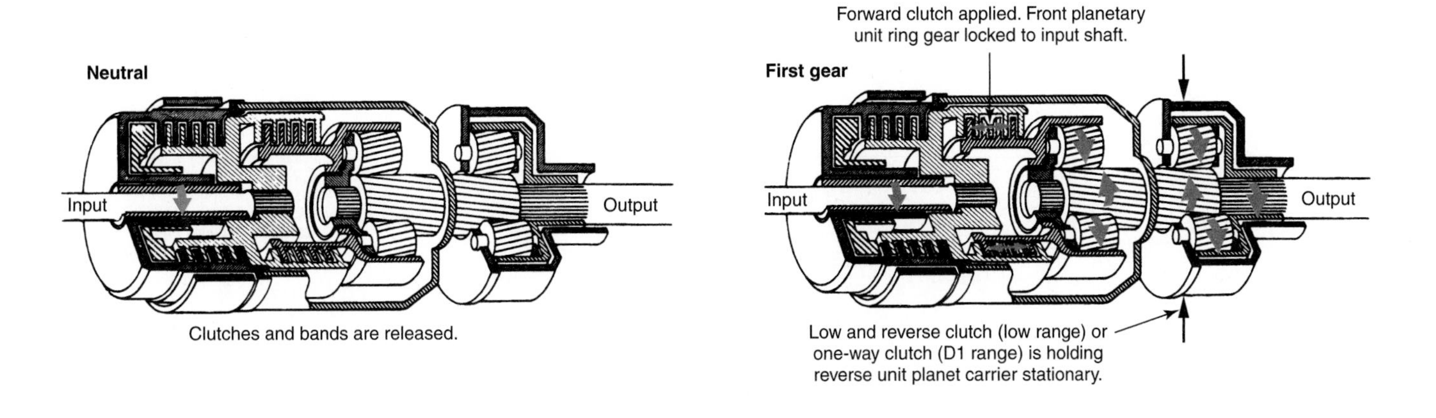

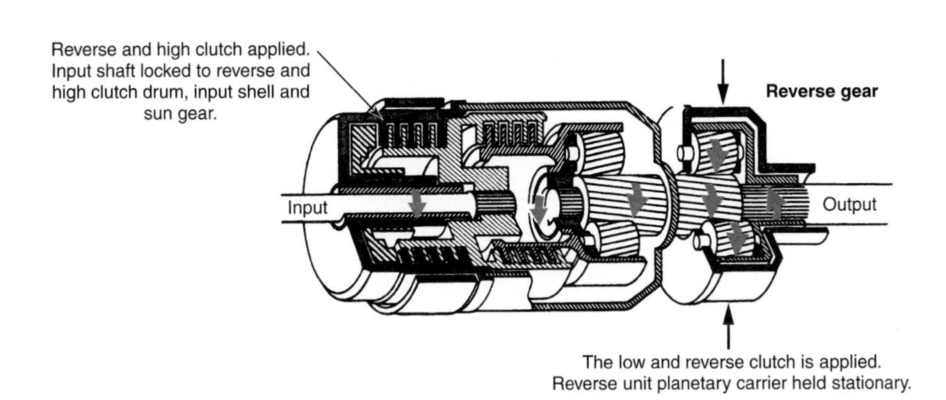

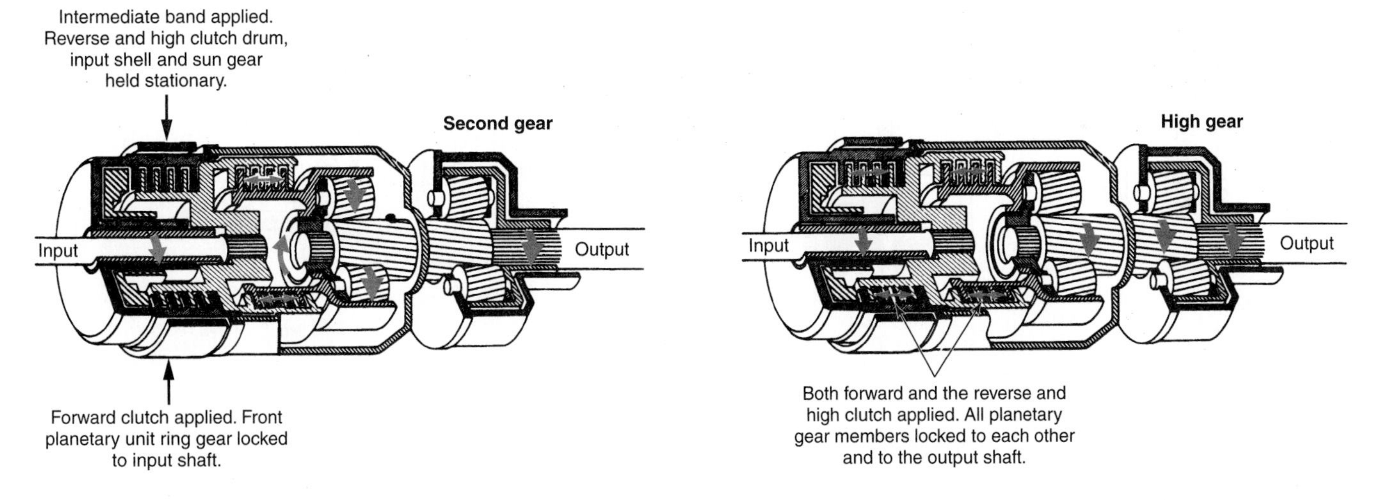

Figure 31-2. Clutch, band, and gearset action during various drive ranges is revealed in this series of diagrams. (Ford)

See **Figure 31-4.** Electronic control systems eliminate the need for throttle valves and governor valves.

A control loop, discussed in Chapter 15, is formed by the interaction of the input sensors, the computer, the output devices, and the transmission operation. **Figure 31-5** shows a simple control loop.

Transmission and Transaxle Designs

Transmissions are always found on vehicles with front engines and rear-wheel drive. The parts layout of all automatic transmissions is similar. From the torque converter, the input shaft passes through the hydraulic pump to mate with the clutch drums, planetary gears, and output shaft, **Figure 31-6.**

Automatic transaxles are used on vehicles with front engines and front-wheel drive, as well as vehicles with rear engines and rear-wheel drive. The parts layout of transaxles may vary slightly, since the power flow must be split. Compare the transaxle in **Figure 31-7** with the transaxle in **Figure 31-8.** In the transaxle shown in Figure 31-7, engine power flows from the torque converter, through a drive chain, and then into the clutch drums, planetary gearsets, and differential assembly. In the unit shown in Figure 31-8, power flows from

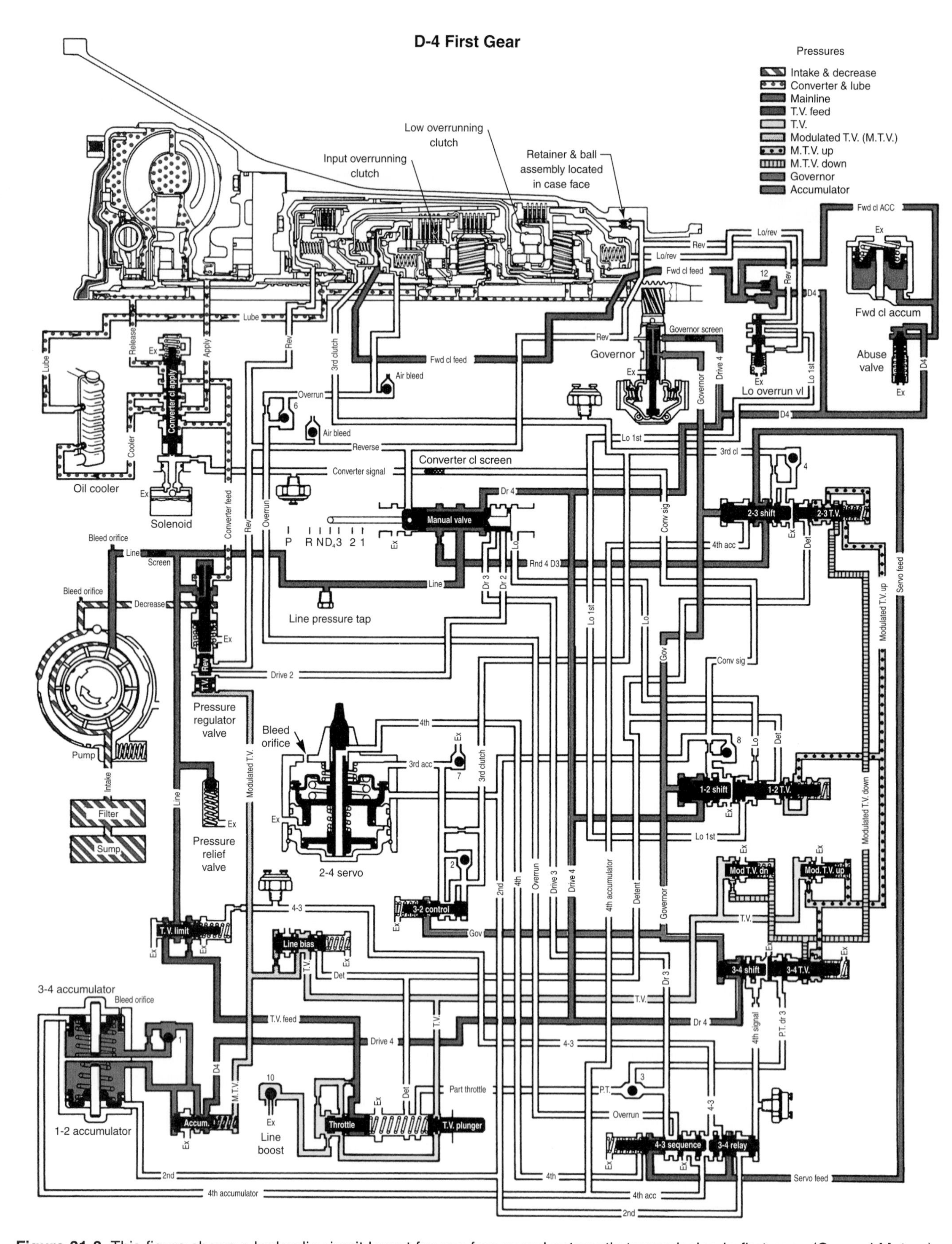

Figure 31-3. This figure shows a hydraulic circuit layout for one four-speed automatic transmission in first gear. (General Motors)

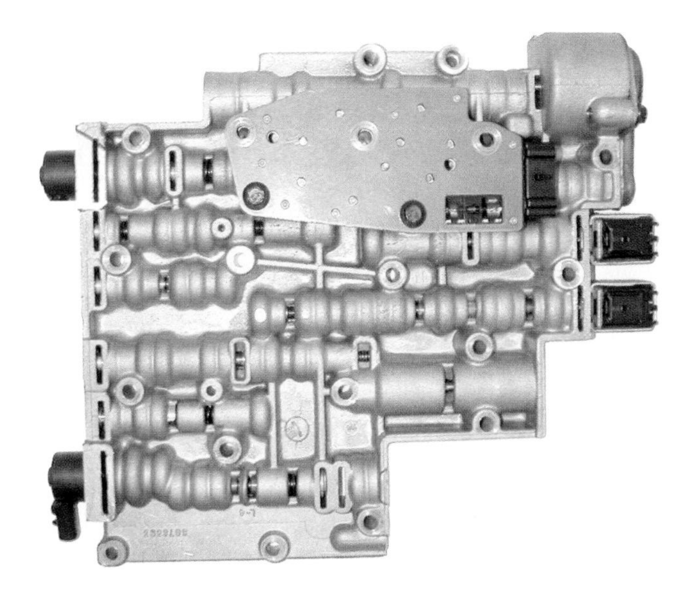

Figure 31-4. Pressure control and shift solenoids on a common valve body. Solenoids may also be located on the outside of the transmission case. (Sonnax)

the torque converter, through the input shaft, and into the clutch drums and planetary gears. From the planetaries, it flows to output shaft. A gear on the output shaft drives a gear on the transfer shaft. The transfer shaft pinion gear drives the differential ring gear and case assembly. Transaxles can be placed

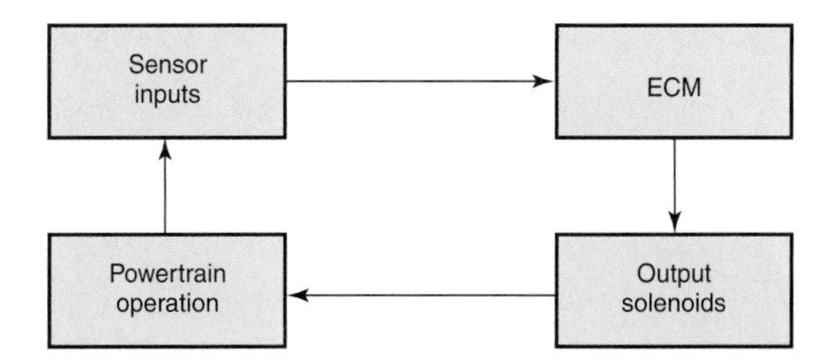

Figure 31-5. This illustration shows the control loop operation of a electronically controlled transmission or transaxle. All electronically controlled transmissions and transaxles use a more complex version of the simple loop shown here.

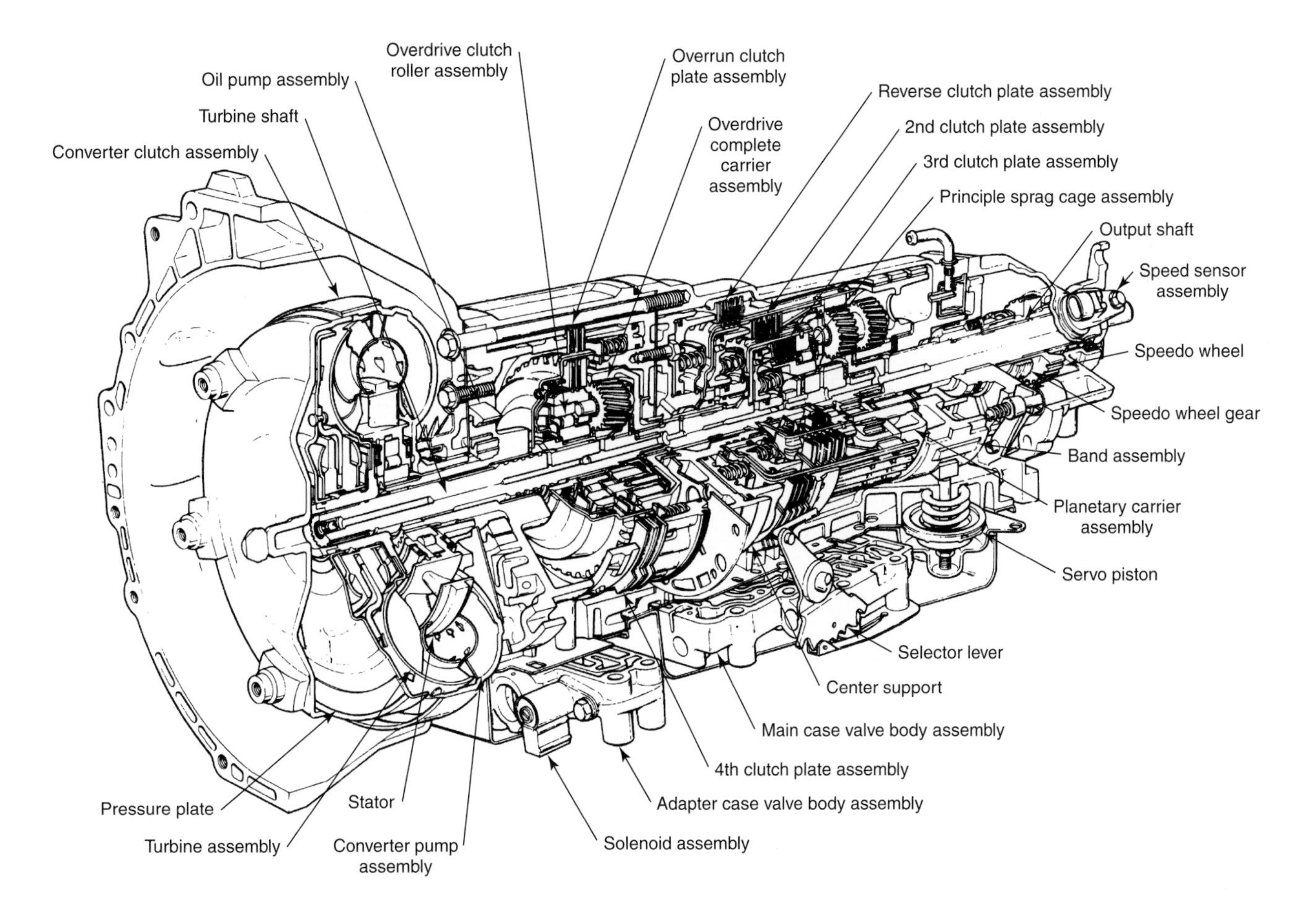

Figure 31-6. Cutaway of an electronically controlled, four-speed automatic transmission. Study the parts closely. (General Motors)

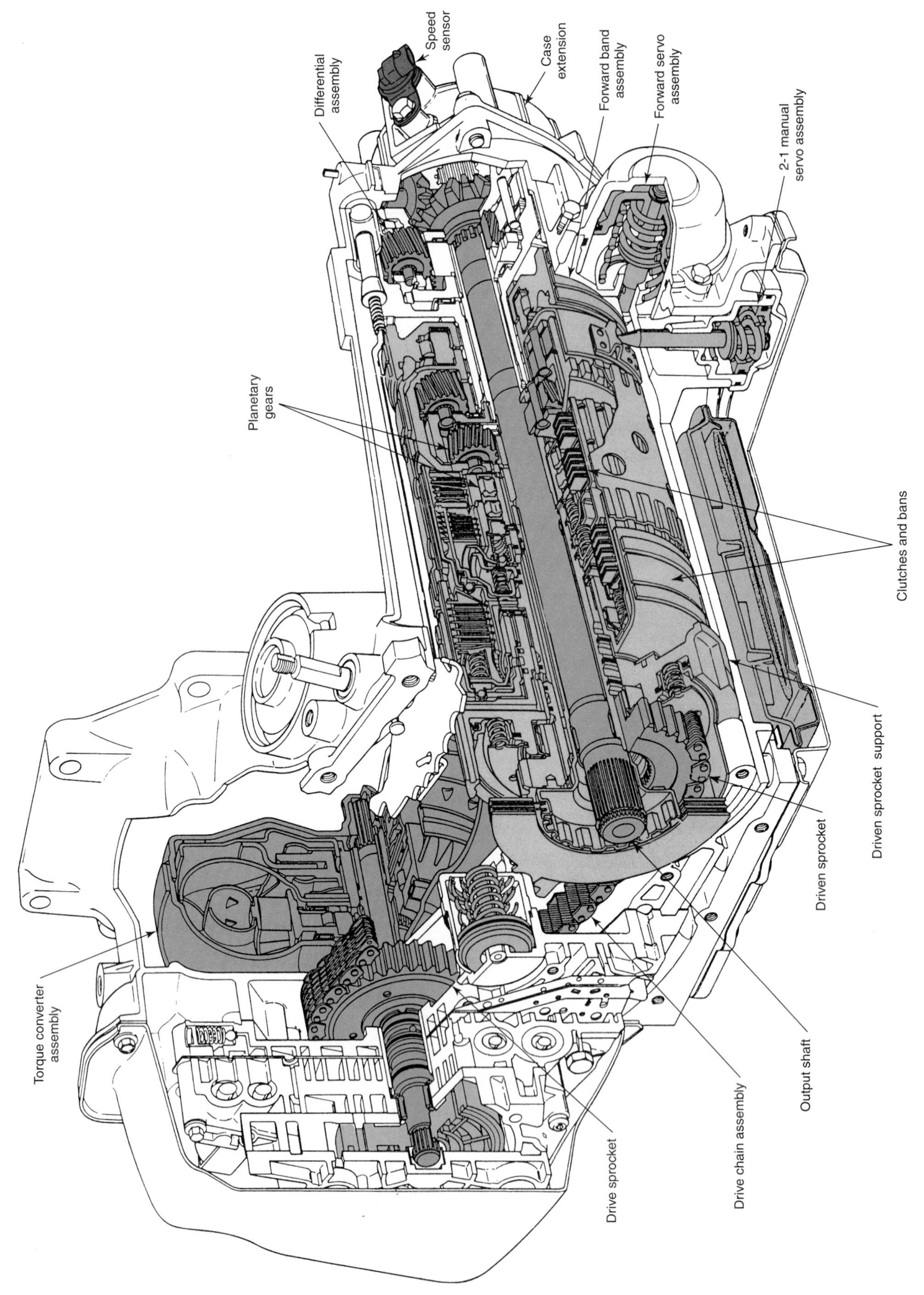

Figure 31-7. One automatic transaxle arrangement. Note that torque from the converter is transmitted to the transmission by a multiple-link chain. (Toyota)

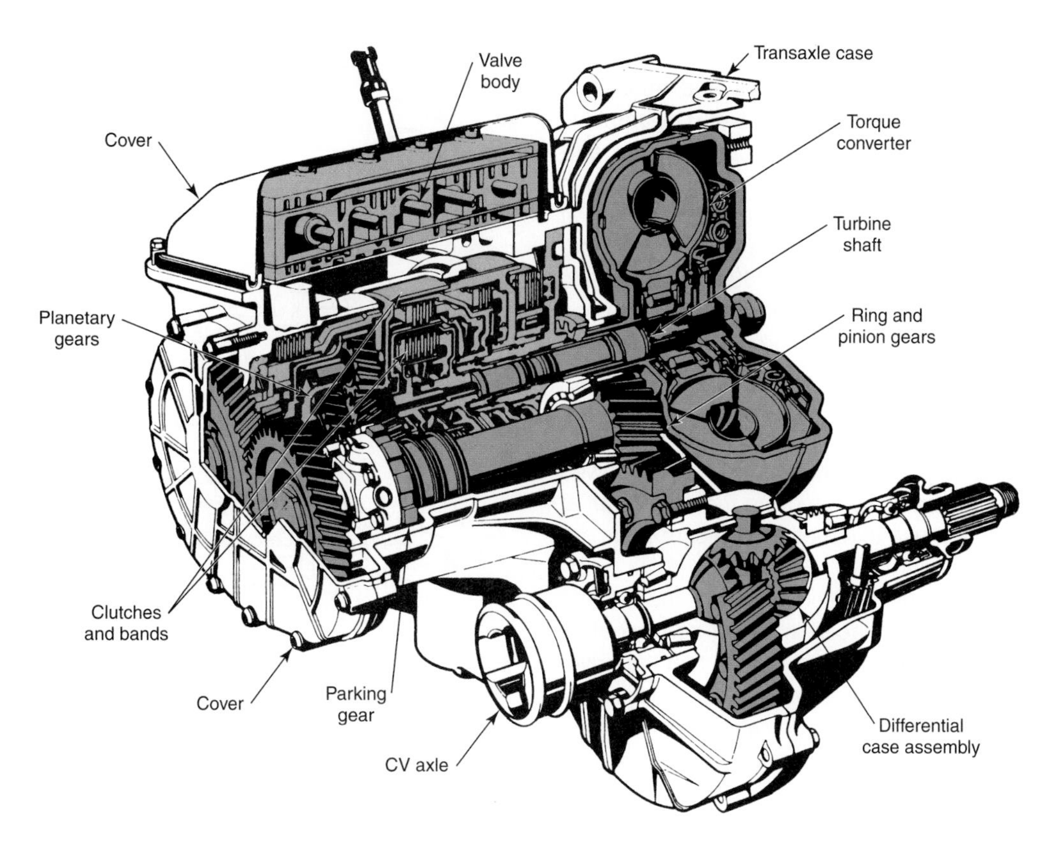

Figure 31-8. This transaxle is obviously different from the one shown in Figure 31-7. However, many of the parts are similar. (Saab)

transversely (sideways in the vehicle) or ***conventionally*** (facing the vehicle front). The transverse mounting is more common, but many conventional placements are found. Compare **Figures 31-9** and **31-10.**

Automatic Transmission and Transaxle Service

Many transmission and transaxle problems can be solved by a simple adjustment or repair. As stated earlier, this chapter discusses typical in-vehicle service and problem diagnosis. For information relating to a specific unit, always refer to the manufacturer's service manual. The manual will provide complete disassembly, inspection, and assembly details. Manufacturers' service manuals contain exploded views and illustrations covering the disassembly, parts inspection, and reassembly of the transmission or transaxle.

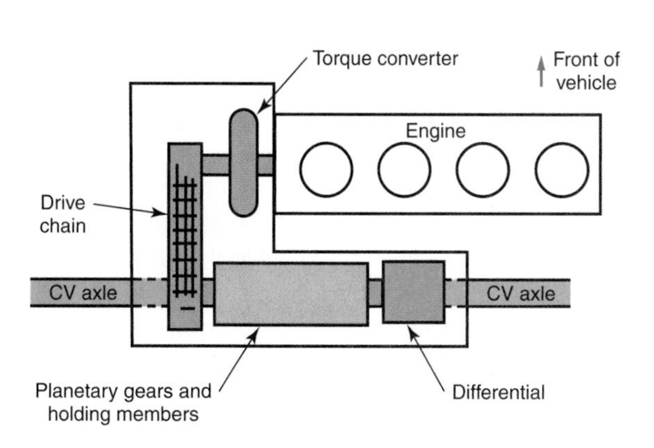

Figure 31-9. This transaxle arrangement is used with a transverse engine.

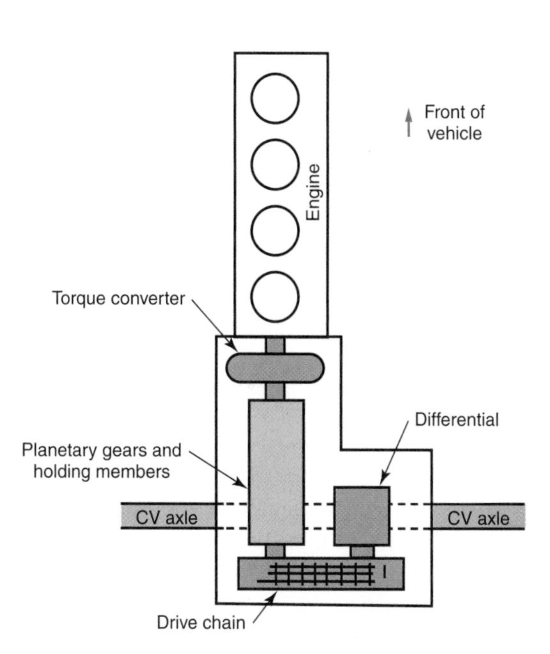

Figure 31-10. The conventional transaxle arrangement is used with a longitudinally mounted engine.

Transmission and Transaxle Service Basics

Vehicles with automatic transmissions and transaxles cannot be push-started. If the vehicle will not start by cranking with the starter, determine the problem and correct it. If a rear-wheel drive vehicle's rear end, drive shaft, and transmission are in sound condition, the vehicle can be towed in Neutral at a nominal speed. Generally, do not exceed 45 mph (72 kph) or travel a distance greater than 50 miles (80 km). Be sure to release the parking brake. Some manufacturers caution against towing for distances exceeding 12–15 miles (24 km–80 km). Check the transmission fluid level before towing. When the transmission or driveline components are inoperative, a rear-wheel drive vehicle must be towed from the rear end (rear wheels off the ground). The vehicle may be towed without raising the rear wheels by removing the drive shaft at the differential end. If you use this method, be sure to tie the shaft up out of the way securely.

Front-wheel drive vehicles should have the driving wheels raised off the road or placed in wheel dollies before towing. Some four-wheel drive vehicles require that the driveline be removed or that the wheels be placed in dollies for towing. Others can be towed with all four wheels on the road by placing the transmission in Park and the transfer case in Neutral. Towing speed is generally around 30 mph (48 kph) with a distance of around 15 miles (24 km). Remember that whenever a vehicle has driveline, wheel bearing, drive shaft, or transmission/transaxle problems, use wheel dollies or lift the affected end so that towing does not cause additional damage. The best method of towing a disabled vehicle is with a platform towing vehicle, often called a skid truck. Always follow the vehicle manufacturer's recommended towing method. Do not use the ignition lock mechanism in the steering column to hold the wheels in a straight-ahead position. Damage to the lock and steering column is likely. Use an approved steering wheel clamping tool. Always use a safety chain or strap. **Figure 31-11** illustrates typical tow truck attaching methods. Do not allow passengers to ride in the towed vehicle.

Figure 31-11. Several common towing methods for vehicles are shown here. Follow all of the vehicle manufacturer's towing recommendations to prevent damage. (General Motors)

Transmission and Transaxle Service Tools

Sets of tools should be available for such tasks as adjusting the bands, checking the throttle linkage, or making other adjustments. In addition to the basic set, special purpose tools must be acquired to service specific transmissions and transaxles. **Figure 31-12** shows some of the special tools needed for in-vehicle adjustments. A 0 psi–300 psi (0 kPa–2068 kPa) pressure gauge should also be included. Major repair jobs require many more standard and specialized tools.

Transmission and Transaxle Identification

Before any service work can be performed, the vehicle's transmission or transaxle must be properly identified. On most vehicles, simply noting the shape of the pan is all that is needed. On others, a code or date is sometimes required. These are usually either on a label in the vehicle or stamped on the transmission/transaxle case. Make certain that you have correctly identified the type of transmission or transaxle in the vehicle. Double-check your identification by consulting the manufacturer's service manual. This ensures that you have the correct information when ordering parts.

Aluminum Parts

Automatic transmission and transaxle cases and other parts are made of aluminum. While quite suitable for this purpose, aluminum parts require careful handling to avoid nicking, scratching, or burring the machined surfaces. Threads are soft and easily stripped. Always use a torque wrench to prevent stripping the threads. To prevent galling, dip the fastener threads in transmission fluid before installation. If a thread is stripped, use a Heli-Coil to make a repair. Thread repair was discussed in Chapter 8, Fasteners, Gaskets, and Sealants.

Diagnosing Transmission and Transaxle Problems

The following section covers methods used to detect some minor transmission or transaxle problems and explains how to determine whether further checking is needed. The variety of transmission and transaxle types, coupled with yearly modifications, make it impractical to formulate a generalized diagnosis chart of significant value. As with service techniques and procedures, the use of the manufacturer's service manual is recommended. It contains a comprehensive diagnosis chart or guide that applies to the specific unit being serviced. Whenever possible, discuss transmission performance with the vehicle's owner. It is a good idea to have the owner present during the road test. This enables the owner to point out exact symptoms.

Do not attempt to diagnose transmission or transaxle problems until the engine has been determined to be in sound

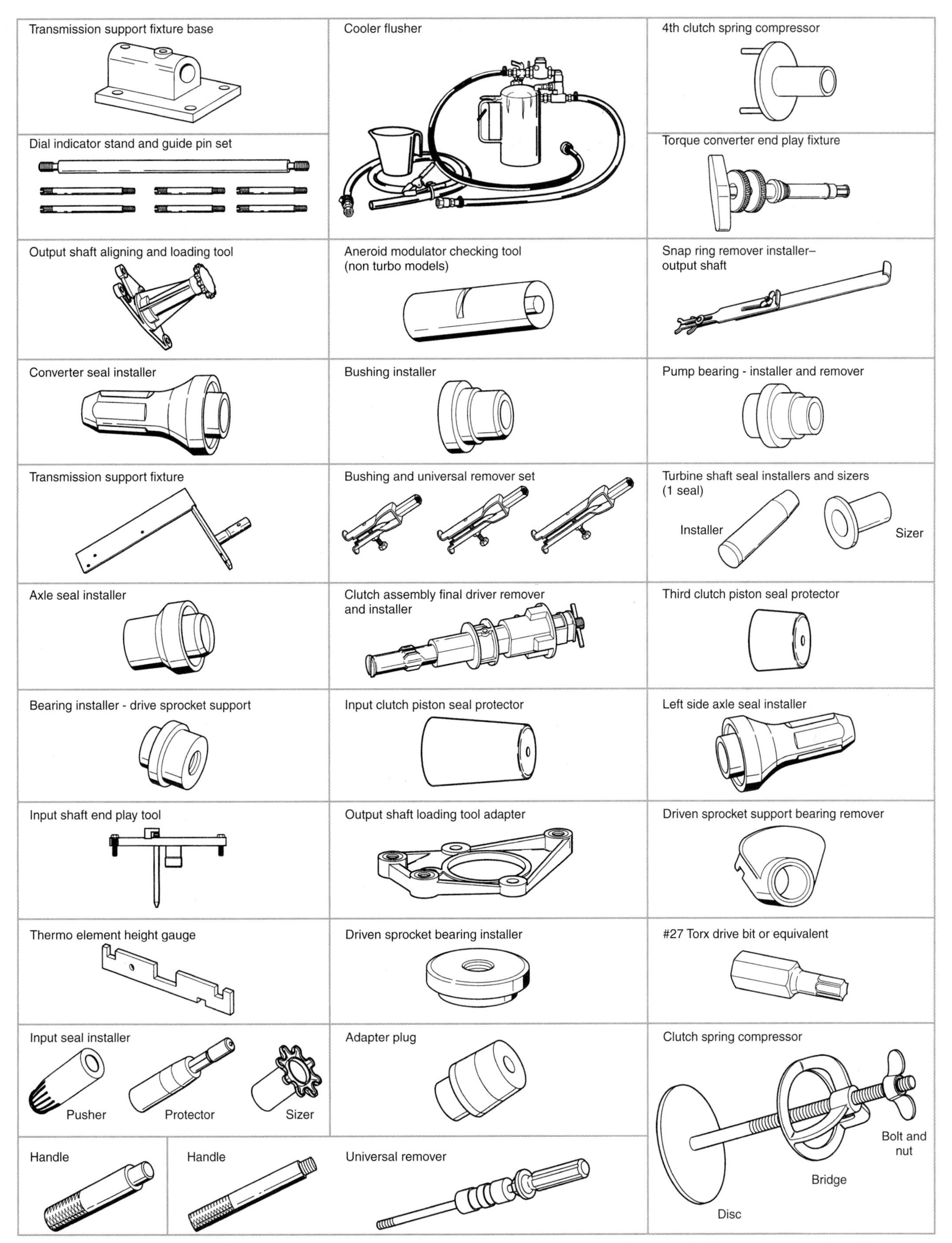

Figure 31-12. Several special tools are needed to properly service and repair automatic transmissions and transaxles. Always use the correct tools. (General Motors)

mechanical condition and properly tuned. The idle speed must be correct. The engine should accelerate from a standing start to any desired speed without missing, hesitation, or lack of power. Locate and correct any engine problems before checking the transmission or transaxle. Both the engine and the transmission/transaxle must be at operating temperature before testing. If the unit is electronically controlled, retrieve any trouble codes before proceeding.

A ***troubleshooting chart,*** such as the one shown in **Figure 31-13,** allows the technician to quickly match a problem with the possible causes and corrections. In many cases, the "problem" is actually a normal condition. This is especially true of newer transmissions and transaxles, which shift more firmly than older units. In such cases, a tactful explanation of transmission operation will be all the repair needed. ***Band and clutch application charts*** indicate which holding members are applied in which gear. This makes it easy to determine which band or clutch is defective when the transmission slips in a particular gear.

Fluid Level

Transmission fluid level is very important. The level should be checked at least once a month or whenever a transmission or transaxle problem is suspected. Always check the fluid level before conducting any other tests. A high or low fluid level can have a noticeable effect on hydraulic system operation. A high fluid level (above the full mark) allows the moving parts to churn the fluid. The churning fills the fluid with tiny air bubbles, resulting in foamy fluid. This is known as fluid ***aeration.*** When aerated fluid is pumped throughout the transmission, faulty clutch and band operation, erratic shifts, overheating, cavitation noise (noise caused by the pump or torque converter operating in aerated fluid), and other problems can result. A low fluid level can cause overheating, slipping in all gears, erratic shifts, suction noises, and other malfunctions.

Checking Transmission Fluid Level

Before checking the transmission fluid level, the fluid must be brought to normal operating temperature (170°F–200°F or 77°C–93°C). Driving four or five miles with

Remedial Steps If Oil Pressure Is Not Normal

Trouble symptom	Probable cause	Remedy
1. *Line pressures are all low (or high). NOTE *"Line pressures" refers to oil pressures 2, 3, 4 and 5 in the "Standard oil pressure table" on the previous page.	a. Clogging on oil filter	a. Visually inspect the oil filter; replace the oil filter if it is clogged.
	b. Improper adjustment of oil pressure (line pressure) regulator valve	b. Measure line pressure ② (kickdown brake pressure); if the pressure is not the standard value, readjust the line pressure, or if necessary, replace the valve body assembly.
	c. Sticking of regulator valve	c. Check the operation of the regulator valve; repair if necessary, or replace the valve body assembly.
	d. Looseness of valve body tightening part	d. Tighten the valve body tightening bolt and installation bolt.
	e. Improper oil pump discharge pressure	e. Check the side clearance of the oil pump gear; replace the oil pump assembly if necessary.
2. Improper reducing pressure	a. Improper line pressure	a. Check the ② kickdown brake pressure (line pressure); if the line pressure is not the standard value, check as described item 1 above.
	b. Clogging of the filter of the reducing-pressure circuit	b. Disassemble the valve body assembly and check the filter; replace the filter if it is clogged.
	c. Improper adjustment of the reducing pressure	c. Measure the ① reducing pressure; if it is not the standard value, readjust, or replace the valve body assembly.
	d. Sticking of the reducing valve	d. Check the operation of the reducing valve; if necessary, repair it, or replace the valve body assembly.
	e. Looseness of valve body tightening part	e. Tighten the valve body tightening bolt and installation bolt.
3. Improper kickdown brake pressure	a. Malfunction of the D-ring or seal ring of the sleeve or kickdown servo piston	a. Disassemble the kickdown servo and check whether the seal ring or D-ring is damaged. If it is cut or has scratches, replace the seal ring or D-ring.
	b. Looseness of valve body tightening part	b. Tighten the valve body tightening bolt and installation bolt.
	c. Functional malfunction of the valve body assembly	c. Replace the valve body assembly.

Figure 31-13. A sample from one manufacturer's troubleshooting chart is shown here. It lists the trouble, probable cause, and remedy (correction) for several improper transmission fluid pressures. (Hyundai)

frequent stops and starts usually produces a normal fluid temperature. Also, operating the vehicle at a fast idle with the gear select lever in Park or Neutral, the wheels blocked, and parking brake set may raise the fluid to normal operating temperature. If the end of the dipstick is almost too hot to touch, the transmission fluid is at operating temperature. Before checking the fluid, shift the transmission or transaxle through all the ranges before returning it to Park or Neutral. Make sure that:

- The vehicle is in a level position.
- The fluid temperature is normal.
- The engine is idling.
- The shift lever is in Neutral or Park.

Note: Never check the oil level with the transmission or transaxle in any drive range unless specifically recommended by the manufacturer.

With the engine idling, wipe off the dipstick cap and the end of the filler tube. Remove the dipstick and wipe it clean. Then, reinsert into the filler tube. Make certain that the dipstick enters to the full depth. Remove the dipstick and observe the fluid level. The fluid level (with the transmission/transaxle at operating temperature and the shift lever in Neutral or Park) should be between the "add" and "full" marks on the dipstick. See **Figure 31-14.** Add fluid if needed. Under no circumstances should the level move above the "full" mark. If the transmission fluid level is above the "full" mark, drain off the required amount. Unless you are positive the fluid is hot, do not add fluid to bring the level to the "full" mark. If the level is just below the maximum, it will not be overfull when the fluid does reach operating temperature. See **Figure 31-15.**

Check Fluid Condition

Factory-installed fluid generally contains a red dye, which makes finding leaks easier. The fluid in these cases normally has a reddish hue. When examining the fluid level on the dipstick, check the oil for discoloration and a burned smell.

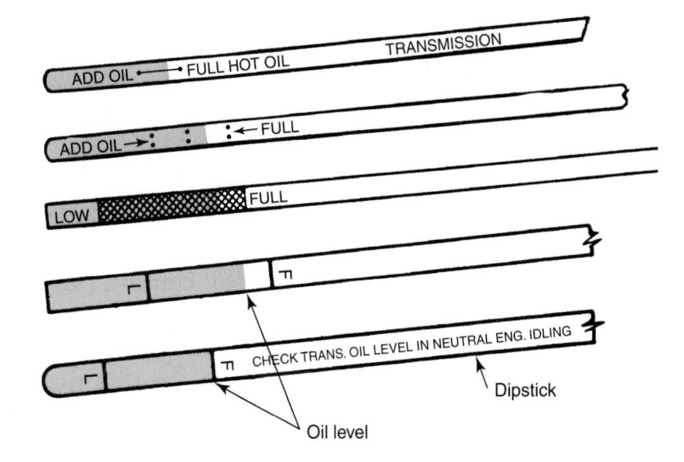

Figure 31-14. Transmission fluid level must be between the "add" and "full" marks on the dipstick. Some dipsticks have a one-quart range from "add" to "full;" others have a one-pint range.

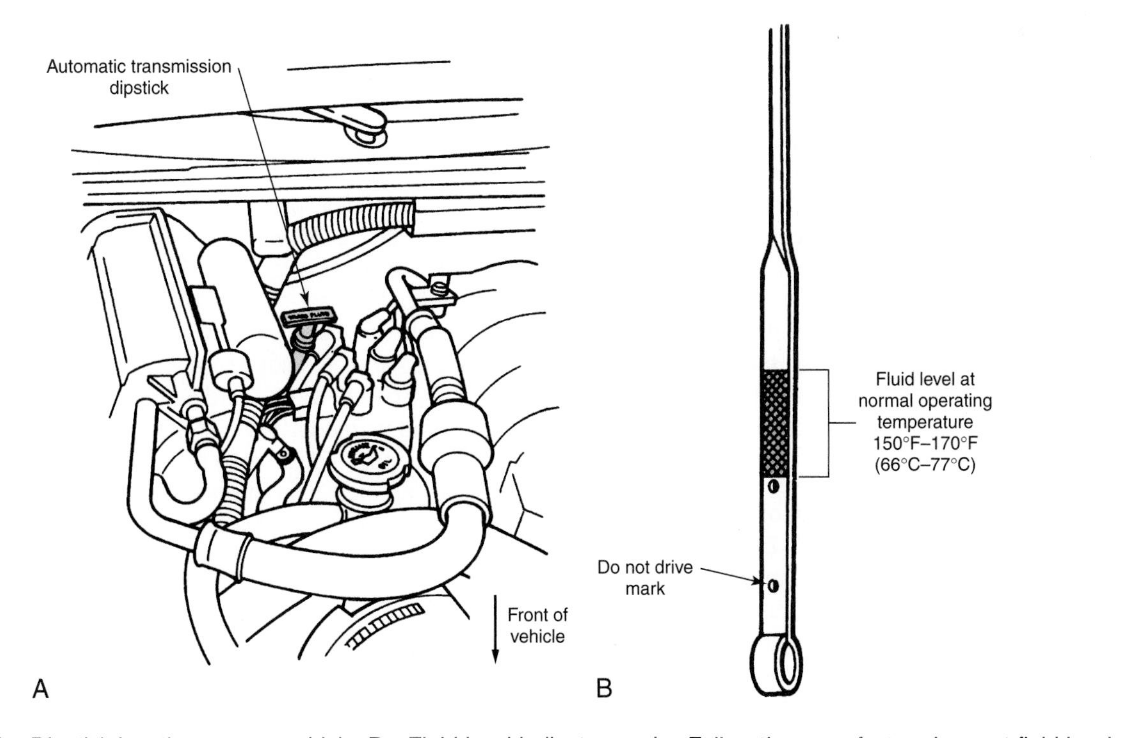

Figure 31-15. A—Dipstick location on one vehicle. B—Fluid level indicator marks. Follow the manufacturer's exact fluid level checking procedures. (Ford)

Such a condition indicates damaged bands or clutches. Also, check the dipstick for deposits of varnish, which indicate that the transmission fluid has been repeatedly overheated.

If any antifreeze or water has entered the transmission through the oil cooler or by driving through very high water, the fluid will look milky. Any appreciable amount of water entry will raise the fluid level. If evidence of antifreeze or water is found in the fluid, check the transmission oil cooler for an internal leak. A cooler leak must be fixed immediately and the transmission flushed, or the antifreeze will severely damage the holding members. Air bubbles indicate aeration or an air leak in the pump suction (usually in the filter).

Fluid Type

When adding fluid to any modern automatic transmission or transaxle, be sure to use the correct fluid. Most General Motors and Ford vehicles use *Dexron III/Mercon* fluid. Chrysler vehicles use ATF +3 or ATF +4. Many imported vehicles take special types of fluids. Sometimes these fluids are available only from the vehicle dealer. Some older vehicles may use Type F fluid. Follow manufacturer's recommendations. Fluid additives are available to convert Dexron III/Mercon fluid for use in imported vehicles. Use care to prevent dirt from entering the transmission/transaxle when checking the fluid level or adding fluid.

Caution: Do not use fluid marked with the letters AQ-ATF. This is obsolete Type A, Suffix A fluid, which is not suitable for modern vehicles. Also, do not use Dexron II in transmissions or transaxles that use Dexron III.

Road Test

A road test can help to determine the exact transmission or transaxle problem. Some shops use road test charts, such as the one shown in **Figure 31-16.** The use of such a chart ensures that all pertinent checks are made. It also provides the technician with a written record that is helpful for diagnostic purposes. Shift points and shift patterns obviously must relate to the unit being tested. Perform each check or adjustment carefully. An error, even though small, can often seriously upset transmission/transaxle performance. Remember that in a series of diagnostic steps, the accuracy of each check and adjustment can be completely dependent on the preceding steps.

Checking Shift Points

Shift points are a good indicator of internal and external transmission or transaxle problems. Bring the unit to the normal operating temperature. Move the selector lever from Neutral through all drive ranges (engine at normal idle rpm). These initial holding member engagements should be smooth. Harsh engagements can indicate either excessive engine idle rpm or incorrect control pressures.

Road test the vehicle. From a standing start with normal acceleration, the transmission or transaxle should make all shifts within a reasonable range. Check the service manual for the exact shift points. Specified shift points (road speed or engine rpm at which the transmission or transaxle shifts) depend upon such things as engine size, tire size, and rear axle ratios. If the unit uses a lockup torque converter, the converter apply might feel like an extra shift. If the converter lockup is hard to detect or to separate from a gearshift, attach a tachometer to the engine. A converter lockup causes a smaller rpm drop than a gear change. On some computer-controlled automatic transmissions and transaxles, a scan tool can be used to detect when lockup occurs.

Automatic shifts should go from lowest to highest. Coasting downshifts should be smooth. Part throttle and detent downshifts should occur when the accelerator pedal is depressed. Engagement at each shift should be smooth, yet positive, with no sign of slippage. If the transmission or transaxle has a lockout feature for first gear (vehicle starts in second to provide better traction for slippery roads), place the selector in the low gear lockout (D2 or 2) range. The vehicle should start out in second gear.

Checking Electronic Transmission and Transaxle Components

Electronically controlled transmission or transaxle problems often resemble the problems that occur in hydraulically controlled units. Therefore, diagnosis should begin with some basic checks, such as fluid level and condition.

Once the basic checks have been made, connect a scan tool to the vehicle's diagnostic connector. Retrieve trouble codes and, if necessary, use the scan tool or other electrical testers to investigate the problem further. **Figure 31-17** shows possible electronic control system problems.

Note: The use of scan tools and other test equipment to check electronic components was discussed in Chapters 3 and 15.

Limp-In Mode

If a major problem occurs in the computer control system, the ECM will place the system in ***limp-in mode.*** When the system is in the limp-in mode, the ECM ignores most input sensor signals and operates the transmission or transaxle devices based on internal settings. The transmission or transaxle solenoids are energized in a way that gives the unit only one or two forward gears. Whenever the system goes into limp-in mode, the ECM will illuminate the dashboard maintenance indicator light, or MIL. An illuminated MIL is always an indication that the computer control system has a problem and that trouble codes should be retrieved.

Input Sensor Problems

Faulty input sensors are the most common cause of electronic transmission or transaxle problems. A defective throttle position sensor (TPS) can cause improper shifting or erratic application of the converter lockup clutch. A failed TPS can also cause rough upshifts or a bump on closed throttle downshifts.

Speed sensors can cause problems by sending incorrect engine or output shaft information to the ECM. This usually affects shift speeds. A speed sensor located inside the case can collect metal filings that affect the signal. Some metal particles are produced as part of normal operation, and they often become stuck on the sensor magnet, affecting the production of the magnetic field. Wheel-mounted speed sensors can be damaged by road debris, or they can become coated with road tar.

Temperature sensors that are defective or produce out-of-range signals can cause hard or soft shifts. Some defective temperature sensors can keep the transmission from

Road Test Symptom Chart

Numbers in chart below correspond with those indicated in troubleshooting charts. (Located in proper service manual for vehicle being tested.)

			Shift quality											
			Rough	Shift timing (Mark km/h (MPH))	No shift	Shift slippage	Vehicle won't move	Cruise slippage	Poor power/ acceleration	Noisy	Engine won't start	Vechicle won't stand still	No engine braking	Comments
Park range	Eng. start										Ⓐ			
	Holding									Ⓑ		Ⓒ		
"R" range	Man. shift (Vehicle at halt)	P-R					Ⓤ			Ⓥ				
	Reverse						Ⓔ-Ⓤ	Ⓔ	Ⓔ	Ⓥ				
"N" range	Man. shift (Vehicle at halt)	R-N								Ⓥ				
	Eng. start										Ⓐ			
	N									Ⓑ		Ⓓ		
"D" range	Man. shift	N-D	Ⓕ				Ⓖ-Ⓤ			Ⓥ				
	1st						Ⓖ-Ⓤ		Ⓘ	Ⓥ				
	Auto shift	1-2	Ⓛ		Ⓙ	Ⓝ				Ⓥ				
	2nd								Ⓟ	Ⓥ				
	Auto shift	2-3	Ⓜ		Ⓚ	Ⓞ				Ⓥ				
	3rd in lock-up "OFF"								Ⓟ	Ⓥ				
	Auto shift lock-up "OFF" (3) → Lock-up "ON" (3)				Ⓐ2	Ⓐ3								
	3rd in lock-up "ON"								Ⓟ	Ⓥ				
	Auto shift lock-up "ON" (3) → Lock-up "OFF" (3)									Ⓥ				
	Decel.	3-2			Ⓠ	Ⓣ				Ⓥ				
	Kickdown	3-2			Ⓠ-Ⓢ	Ⓣ				Ⓥ				
	Decel.	2-1			Ⓡ									
	Kickdown	2-1			Ⓡ									
"2" range	Man. shift (vehicle in operation)	D-2			Ⓦ		Ⓗ-Ⓤ			Ⓥ				
	1st						Ⓗ-Ⓤ		Ⓘ	Ⓥ				
	Auto shift	1-2	Ⓛ		Ⓙ	Ⓝ				Ⓥ				
	2nd								Ⓟ	Ⓥ				
	Decel.	2-1			Ⓡ					Ⓥ				
	Kickdown	2-1			Ⓡ					Ⓥ				
"1" range	Man. shift (vehicle in operation)	2-1	Ⓐ1		Ⓡ-Ⓩ					Ⓥ				
	Man. shift (vehicle in operation)	D-1			Ⓡ-Ⓧ					Ⓥ				
	Acceleration						Ⓗ-Ⓤ		Ⓘ	Ⓥ				
	"1"	Engine braking								Ⓥ			Ⓨ	

Figure 31-16. One form of transmission road test chart. Use one for the exact transmission being tested. (Nissan)

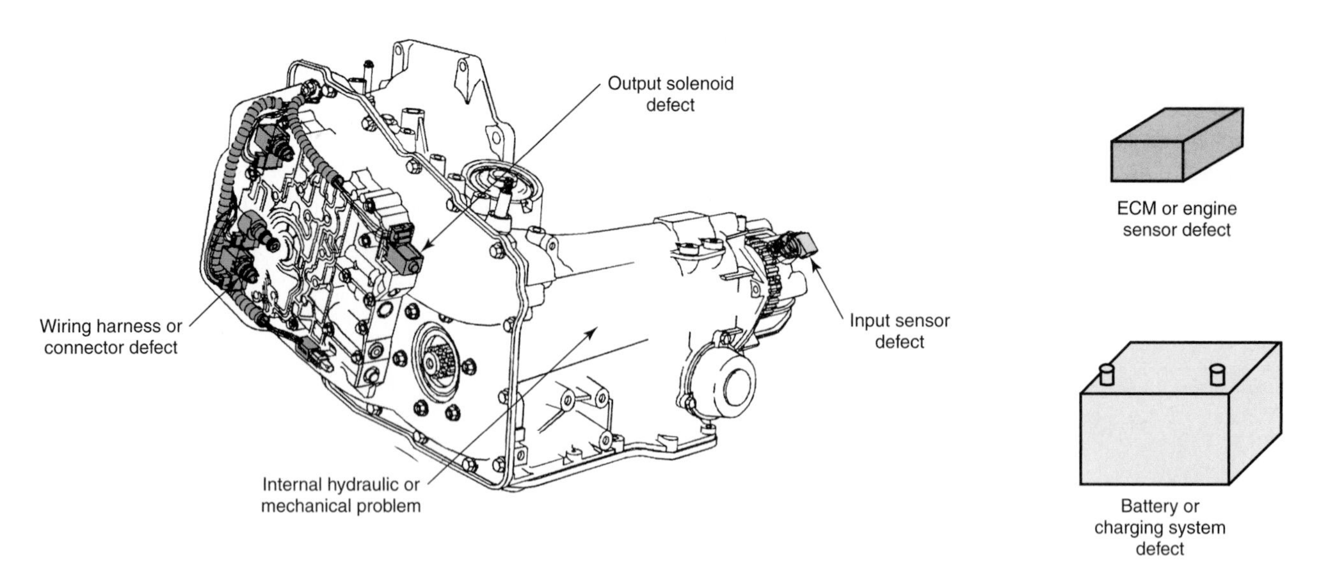

Figure 31-17. Most electronic control system problems involve the input sensors, solenoids, or the related wiring. Always check for hydraulic or mechanical problems before deciding that the electronic control system is defective. (General Motors)

upshifting or prevent converter clutch apply. Other sensors can indirectly affect shift speeds and shift quality.

A loss of voltage to a sensor (usually called reference voltage) will result in an inoperative sensor. Reference voltage problems are usually caused by a defective ECM or a ground problem.

Solenoid Problems

Solenoid problems will vary depending on whether the solenoid is an on-off type or a pulsed type. On-off solenoids can stick open or closed. The usual result is the loss of some gears. In many cases, the transmission or transaxle will take off from a stop in a gear other than first. Other solenoid failures can cause ***gear skipping*** (shifting from first to third, for example), failure to shift into higher gears, or incorrect application of the converter lockup clutch. Occasionally, a solenoid will stick at times and work properly at other times. Typical causes of intermittent sticking include a high-resistance solenoid winding, a bad electrical connection, or buildup of sludge. Erratic shifting is a common symptom of an intermittently sticking solenoid. The transmission or transaxle may work well most of the time, with only occasional shift problems. In some cases, the transmission will shift improperly only when cold or only when hot.

Pulsed solenoids are generally used to control pressures. Therefore, instead of skipping gears, a failed pulsed solenoid can cause slipping or excessively hard or soft shifts. A defective line pressure solenoid will cause problems in any or all gears. Many pulsed solenoids control pressures of a specific operation, such as converter clutch apply feel or part throttle downshift. Defects in the related solenoid will cause problems during that process only.

Almost every solenoid has one or more small filters. If the filters become plugged, oil pressure will not be able to pass through the solenoid valve to the rest of the hydraulic system. Additionally, a torn filter can cause the solenoid to stick.

ECM Problems

Since it contains many complex circuits, a faulty ECM can cause a variety of problems, depending on which internal part or circuit has failed. Sometimes, the ECM will keep the transmission or transaxle from shifting. A defective ECM may cause the unit to stick in one gear. Another common ECM problem is erratic shifting. Examples include downshifting at cruising speeds or occasionally failing to upshift. The operation of the ECM is often affected by heat. A cold ECM may work well when the vehicle is first driven but cause problems as it heats up.

A defective ECM can stick in one mode. Turning the ignition switch off and then back on may temporarily correct this problem. If the ECM controls pressures through a pulsed solenoid, failure may cause slipping, hard shifting, shudder during shifts, and other problems. In many cases, a failed ECM will also cause engine drivability problems.

A defective ECM can set false trouble codes and may set codes that do not exist. A faulty ECM may illuminate the MIL when there is no problem, or it may not illuminate the MIL when a problem is present.

Pressure Checks

Checking the fluid pressure is a good way to determine the condition of the hydraulic system, including the pump, valves, pressure regulator, and internal seals. Since fluid pressures, checkpoints, and methods of checking vary, always consult the manufacturer's service manual. Begin by checking the fluid level with the transmission or transaxle at normal operating temperature. Remove dirt from around the pressure taps (plugs). Remove the tap and connect a suitable pressure gauge. The gauge's hose should be long enough to reach the driver's compartment for road testing. A typical gauge (0 psi–300 psi [0 kPa–2069 kPa]) is shown connected to the checkpoint in **Figure 31-18.** Note the use of a long hose.

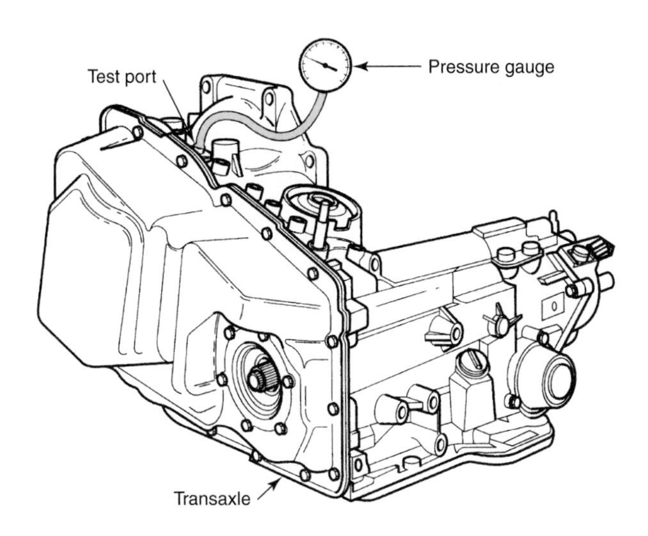

Figure 31-18. A hydraulic fluid test gauge is connected to the automatic transaxle to obtain a pressure check. (General Motors)

Drive the vehicle at the recommended road speeds. Check the pressure in the specified drive range. Compare the gauge readings to the manufacturer's specifications. Excessive pressure readings indicate that a pressure regulator or pressure relief valve is stuck closed. Low pressure indicates a worn front pump, a pressure regulator or pressure relief valve that is stuck open, internal leaks, or a plugged filter. Closely watch the gauge during shifts to determine whether a particular servo or clutch seal is leaking. If pressure is low in only one gear, the related apply piston seals or seal rings are probably leaking.

Other pressure tests can be performed in the shop. Attach a tachometer to provide accurate rpm checkpoints. Be careful to place the unit in the correct drive range and operate at the exact rpm when performing the tests. Compare gauge readings with those specified.

A vacuum gauge (in addition to a pressure gauge and tachometer) is required to check control oil pressure on transmissions or transaxles with a vacuum modulator. The vacuum modulator test shown in **Figure 31-19** uses a hand-held vacuum pump in place of engine vacuum. Manufacturers provide specifications showing the correct control pressure at a given engine vacuum. If an altitude-compensating vacuum unit is used, be sure to make allowance for the barometric pressure during the test. Vacuum gauge readings will be about 1″ lower for every 1000 ft (305 m) rise in elevation. Weather conditions can also affect barometric pressure, but usually not enough to seriously affect pressure readings.

Stall Testing

The ***stall test*** is used on some transmissions and transaxles to determine the condition of the disc clutches, bands, one-way clutch, and other parts. To perform a stall test, bring the engine to its normal operating temperature. Connect a tachometer so that it can be read from the driver's seat. Apply both the parking and service brakes. Place the selector

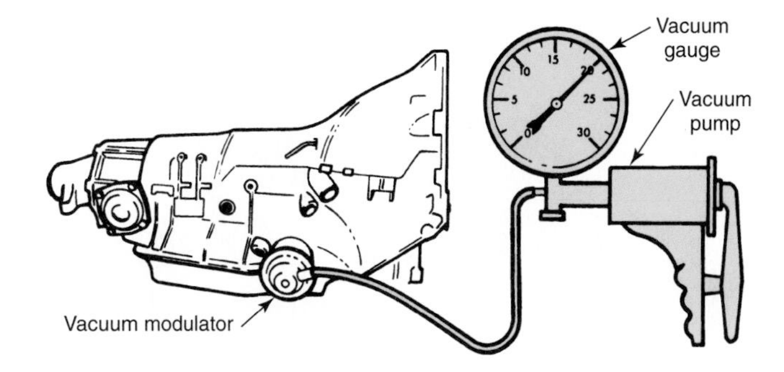

Figure 31-19. Vacuum gauge and pump combination being used to check a vacuum modulator. (General Motors)

lever in the recommended stall test position (usually Drive, Manual Low, or Reverse). While holding the brakes on with great force, push the throttle to the recommended throttle position. Do not go beyond full throttle into the kickdown position unless so recommended.

Note the rpm indicated while operating at full throttle. If engine speed is below specifications, it can indicate engine or converter stator problems. If the rpm is above the manufacturer's specifications, the bands or clutches may be slipping. When stall testing, never keep the accelerator pedal in the full throttle position for more than 5 seconds or you will overheat the transmission or transaxle. Return the selector lever to the Neutral position and operate the engine at around 1200 rpm for a minute or two before stall testing a different drive range. If the engine rpm exceeds stall specifications during the stall test, release the accelerator immediately. Make certain the brakes are firmly applied and keep people out of the way.

Caution: Some transmissions must not be stall tested. Follow manufacturer's instructions.

Checking the Oil Pan.

If other checks indicate that the unit may have slipping holding members, remove the oil pan and check for sludge and metal particles. While some deposits are normal, a coating of more than 1/4″ (0.6 mm) indicates worn and slipping holding members. Most modern transmissions and transaxles contain a small magnet to trap metal particles. Check it for excessive metal deposits. The oil pan in **Figure 31-20** shows an oil pan with a normal amount of deposits.

In-Vehicle Repairs

The following repair operations are designed to be performed with the transmission or transaxle installed in the vehicle. Always consult the manufacturer's service manual for the exact procedures.

Band Adjustment

Some transmissions and transaxles require periodic ***band adjustments*** to compensate for normal wear. Other

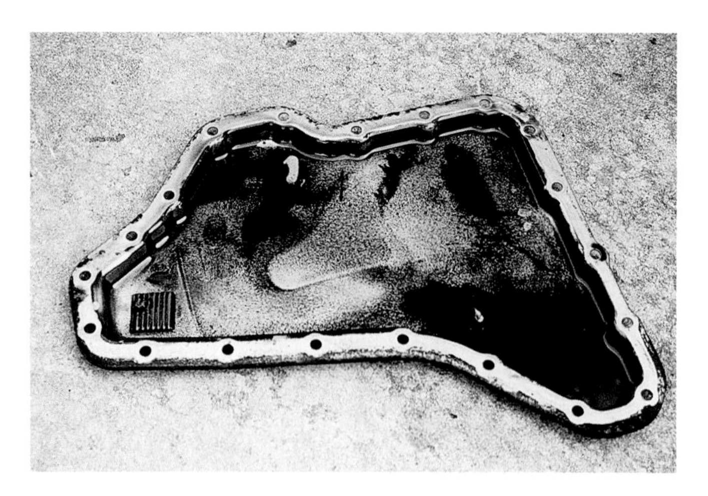

Figure 31-20. This transaxle oil pan has a small amount of sludge, with a few metal particles on the magnet. The clutches and bands in this transmission are probably okay, and fluid has not been overheated.

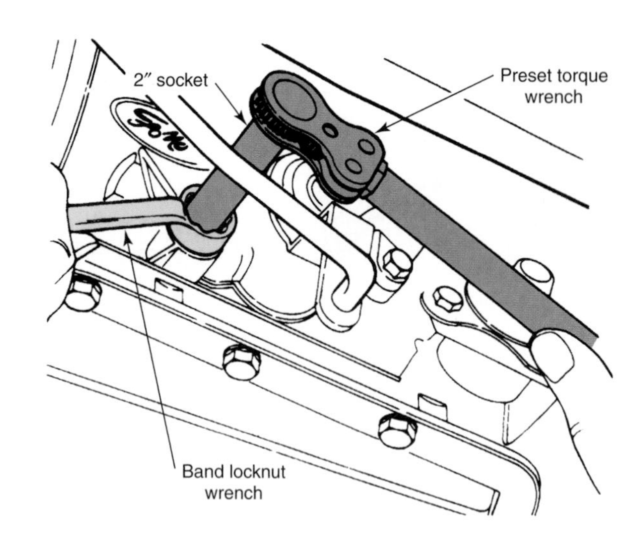

Figure 31-21. A preset torque wrench is being used to tightening a band-adjusting screw from under the vehicle.

designs, in which the bands are subjected to moderate wear, do not require periodic adjustment. In many cases, the bands can be adjusted only when the transmission or transaxle is disassembled for overhaul. The bands in these units are adjusted by adding or subtracting shims inside the servo assembly or by replacing the servo apply pin. When you are making band adjustments, make certain the specifications relate to the transmission or transaxle at hand. Follow the specifications exactly. Failure to do so may cause serious damage.

Some external band adjustment mechanisms on older vehicles are reached through the floor pan after turning the mat to one side. Others are accessible from beneath the vehicle, **Figure 31-21.** Some require removal of the transmission oil pan, **Figure 31-22.** The typical external adjustment consists of an adjusting screw passing through the case and engaging one end of the band. A locking nut is provided to secure the adjusting screw.

General adjusting procedure requires loosening the locknut several turns, tightening the adjustment screw to an exact torque, and then backing off the screw an exact number of turns. The screw is held in this position while the locknut is tightened. Remember this is a critical adjustment and must be done exactly as specified.

When band adjustments are difficult to reach, an extension like the one shown in **Figure 31-23** is helpful. Remember that when an extension is used with the torque wrench, the indicated torque (dial reading) is less than actual torque. Following a band adjustment, road test the vehicle for proper shift operation.

Manual Shift Linkage Adjustment

An internal detent assembly controls the position of the manual valve in the valve body. Even if the external linkage is misadjusted, the internal manual valve will be positioned correctly. However, the exterior linkage can become so

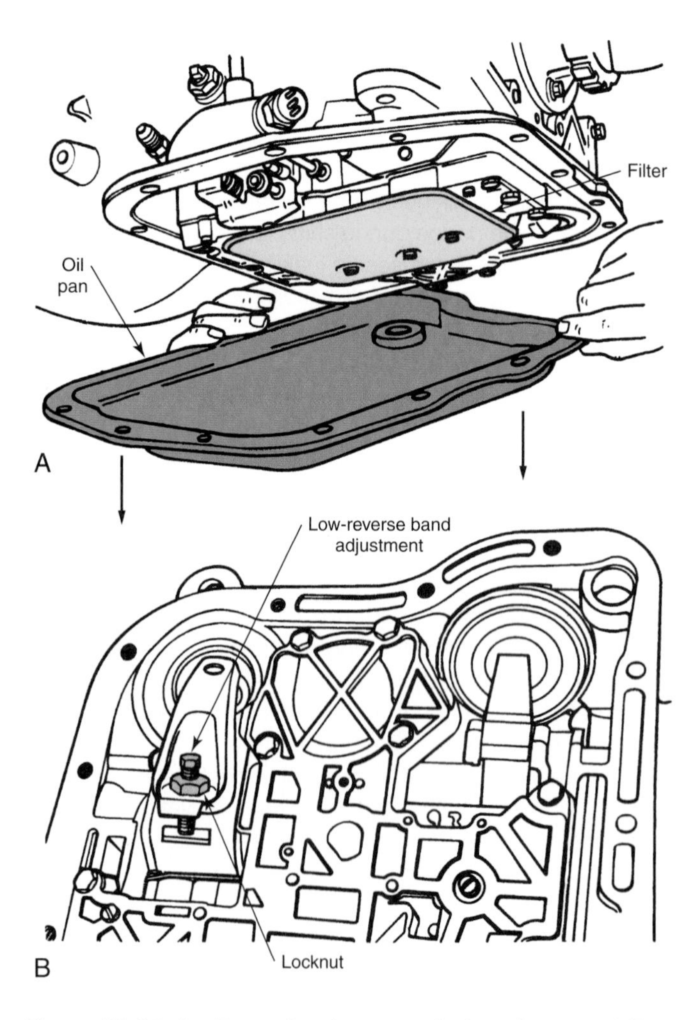

Figure 31-22. A—Removing the transmission oil pan and filter provides access to the low-reverse band adjustment screw and locknut shown in B. (DaimlerChrysler)

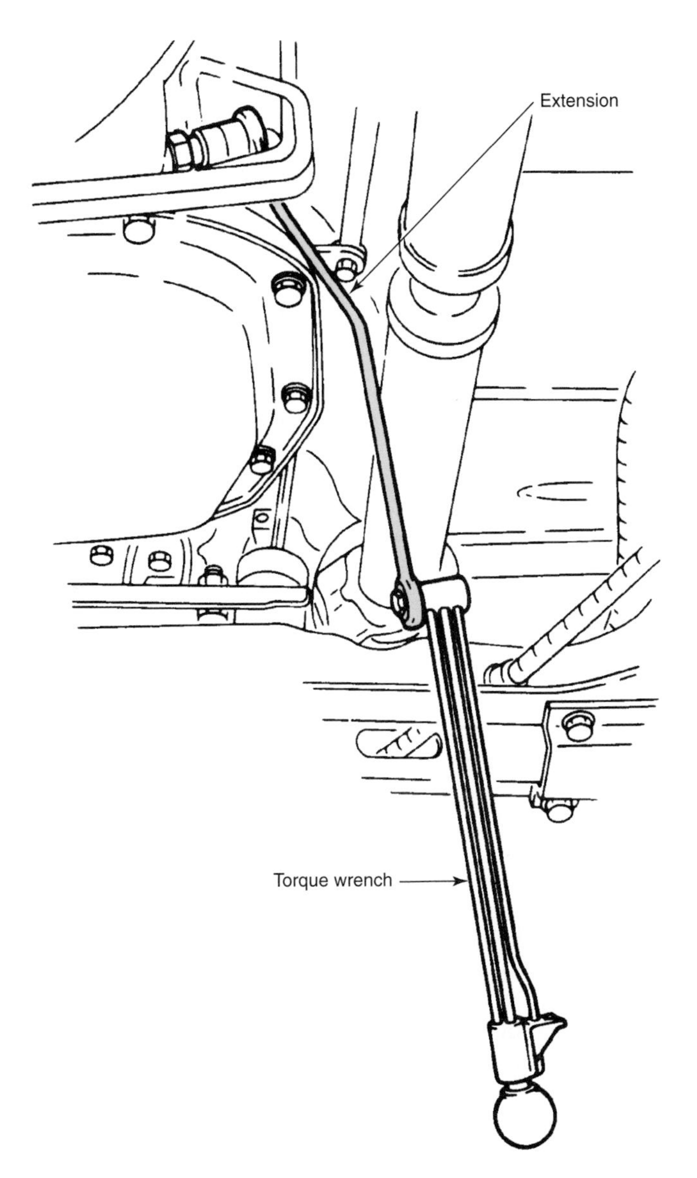

Figure 31-23. An extension makes this adjustment screw easy to torque. (DaimlerChrylser)

misadjusted the driver will be unable to tell what gear is being selected. In such a case, the external linkage can be adjusted so that the manual valve's position matches the drive range indicated by the gearshift lever's position.

Even though the linkage was initially set correctly, wear, loosening of locknuts, and deterioration of engine mounts can alter the setting enough to cause trouble. If the engine mounts are damaged, do not adjust the linkage until new mounts are installed. The old mounts could allow further shifting of the engine and transmission/transaxle, which will throw the linkage out of adjustment again.

The linkage setup shown in **Figure 31-24** is used on some vehicles with column shift levers. To adjust this particular linkage, loosen locknut A until the shift rod slides in the swivel clamp (trunnion) freely. Move the manual lever into the Park position (second from rear). Place the gearshift selector lever

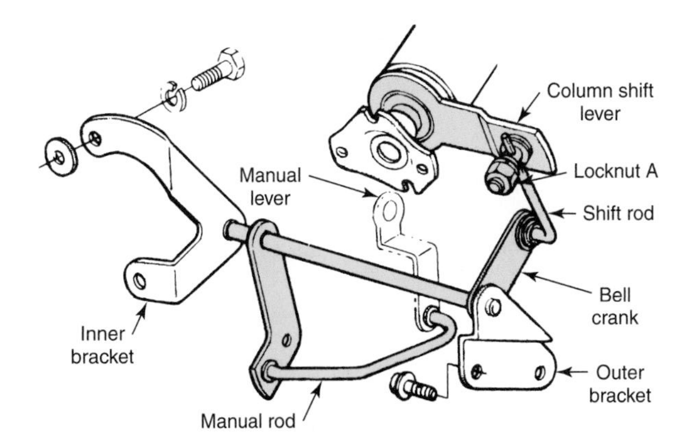

Figure 31-24. One type of column shift arrangement. (Ford)

in the Park position. When both the transmission/transaxle manual lever and the gearshift lever are in the Park position, tighten the locknut. Move the shift lever through all ranges and check for operation and alignment.

Modern vehicles use a cable instead of a linkage system for manual shifting. The floor shift in **Figure 31-25** adjusts in much the same way. Adjust the shift linkage or cable carefully, as accuracy is a must. Make certain the parking pawl functions correctly.

Throttle Linkage (TV) Adjustment

The throttle linkage is usually called the ***throttle valve,*** or ***TV,*** cable or linkage. TV adjustment is critical. The relationship between the throttle opening and the position of the throttle valve controls the shift points. On many transmissions and transaxles, the throttle valve affects the operation of the main pressure regulator and, therefore, affects the overall hydraulic pressures. Faulty adjustment can result in incorrect shift points and internal pressures and can cause transmission or transaxle failure. Any movement of the accelerator pedal must produce a corresponding change in the positioning of the throttle plates and in the throttle valve position. Some automatic transmissions use a TV rod to connect the throttle valve lever and the throttle body or carburetor linkage. Any change in throttle plate positioning causes a corresponding change in the lever. Many transmissions and transaxles use a TV cable to produce the same result. **Figure 31-26** illustrates the TV cable arrangement of a common transmission.

Adjusting Throttle Valve Rods and Cables

Most throttle valve rods can be adjusted by pressing the accelerator pedal to the floor. Next, make sure the throttle plates are completely open and the throttle lever is completely bottomed against its stop inside the transmission or transaxle. Then, loosen the cable stop (located at the top of the engine) and pull the TV cable all the way forward, **Figure 31-27.** Some TV cables are self-adjusting. Pull the cable all the way forward, and then step hard on the accelerator pedal. This adjusts the cable. Recheck the shift operation to ensure that the adjustment is correct.

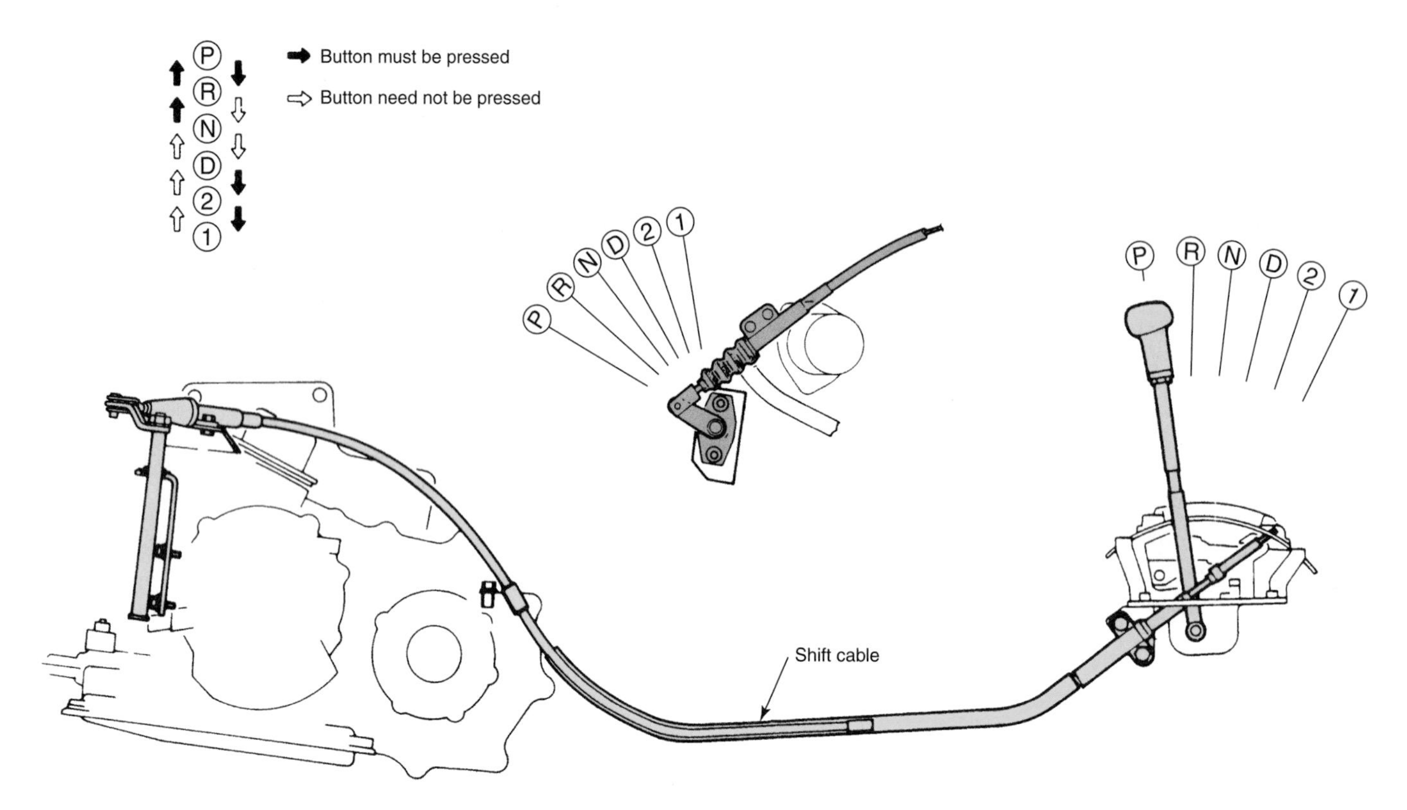

Figure 31-25. A typical floor shift linkage arrangement with shift cable. (Mazda)

Downshift (Detent) Adjustment

Some transmissions and transaxles are equipped with a vacuum modulator to control normal upshifts and downshifts and a mechanical linkage to control forced downshifting, usually called a ***detent rod*** or ***detent cable.*** The detent should be adjusted carefully to ensure the transmission downshifts in accordance with manufacturer's specifications and engine demand. Detent rod linkage is shown in **Figure 31-28.**

Detent downshifts may also be accomplished with a solenoid controlled by a downshift switch. If such an arrangement is used, the switch must be properly adjusted, **Figure 31-29.**

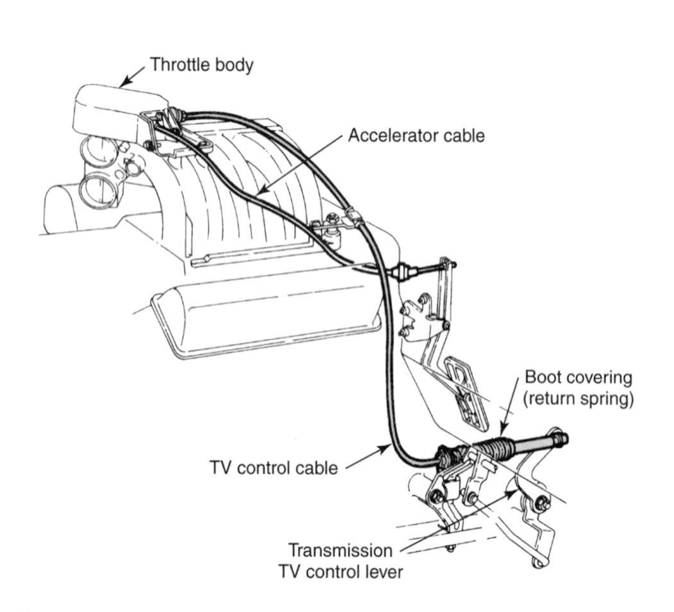

Figure 31-26. A typical throttle valve (TV) cable attaches to throttle body linkage on one end and to the TV lever on other. (Ford)

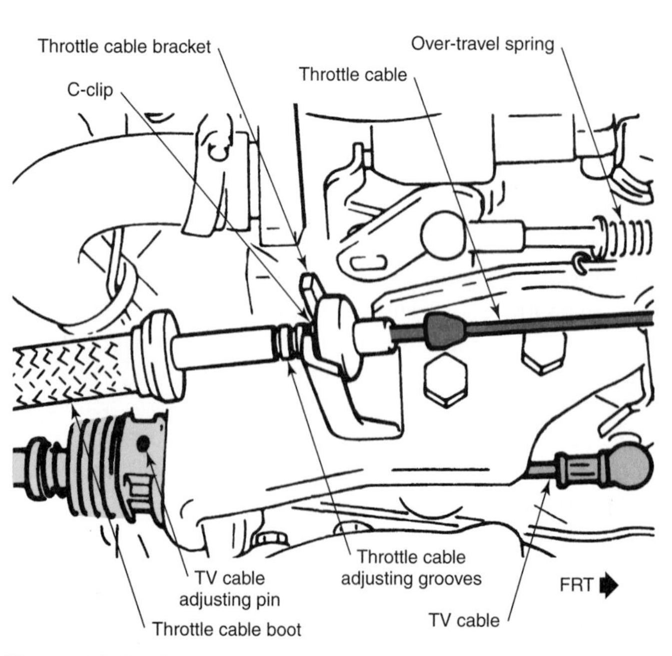

Figure 31-27. One particular TV cable adjustment arrangement. (General Motors)

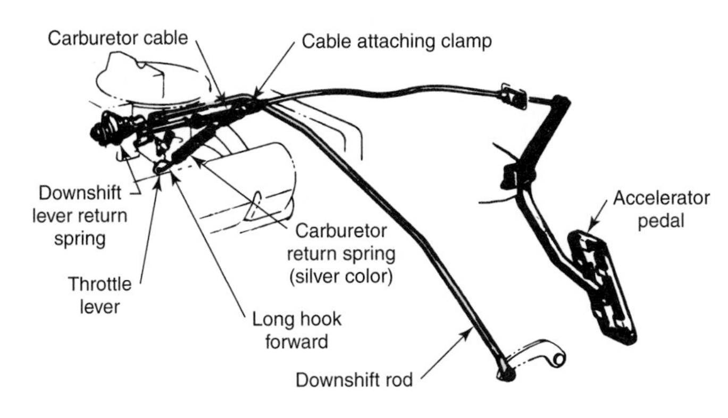

Figure 31-28. A downshift rod assembly and its location on one particular vehicle.

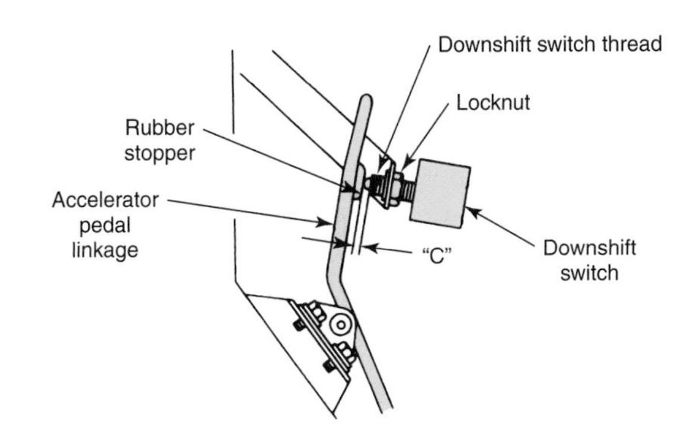

Figure 31-29. Electric downshift switches are usually installed on the firewall inside the passenger compartment. The switch may have a threaded adjustment as shown here, or it may have a slotted mounting bracket. The switch can be adjusted so that it energizes the detent solenoid when the accelerator pedal is depressed to about 90% of wide-open throttle. (Nissan)

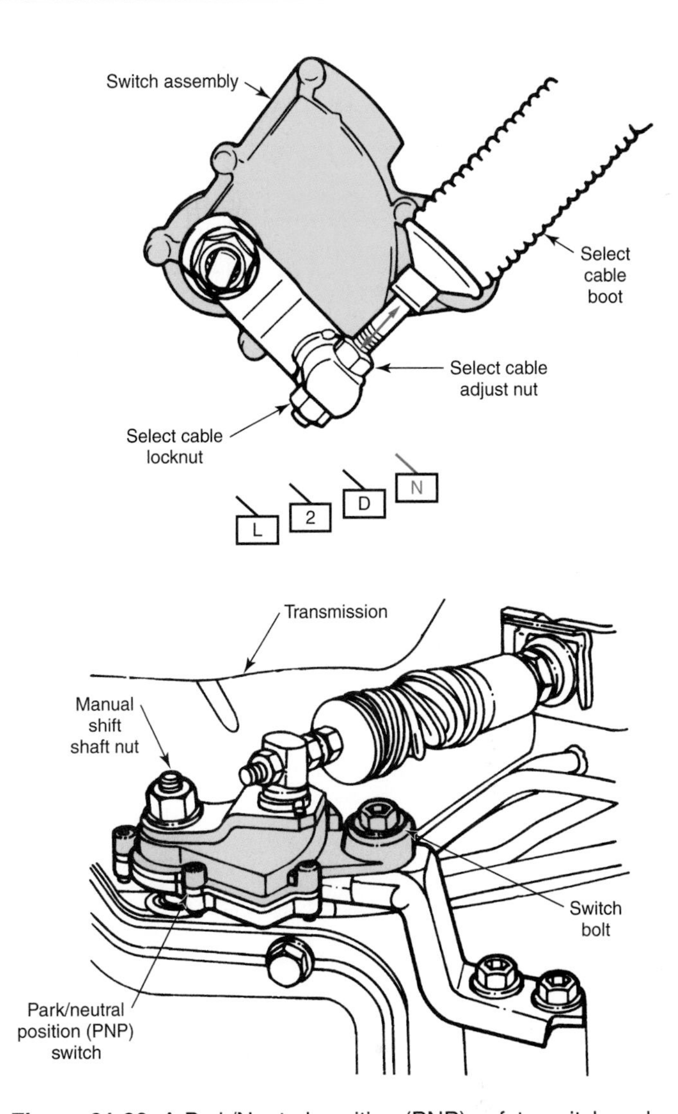

Figure 31-30. A Park/Neutral position (PNP) safety switch and the select cable. Follow the vehicle manufacturer's adjustment recommendations. The switch must be correctly aligned with the various positions so that the engine will start in Park or Neutral, but not in the other positions. (General Motors)

Accelerator Pedal Height and Linkage Action

Many throttle linkage adjustment specifications indicate a definite distance between the bottom of the accelerator pedal and the floor mat. With this distance set to specifications, the engine hot idle speed, the adjustment of the throttle valve linkage or cable, and the adjustment of the detent rod, cable, or electric detent switch must all be correct. When the accelerator is fully depressed, the throttle valve(s) must be in the wide-open position. The linkage action must be smooth and free of binding. Lubricate the linkage as needed.

Neutral Safety Switch

Always check the neutral safety starting switch for proper operation. The engine should crank with the selector lever in either the Neutral or Park position. The engine should not crank with the selector lever in any other position. Adjust or replace the neutral safety switch as needed. The switch may be located on the steering column or on the floor shift console. In some cases, the switch is located on or in the transmission itself. Refer to **Figure 31-30.**

Leak Detection

When you have to add fluid frequently, it indicates a leak somewhere in the system. There are a number of potential leak areas. Typical leak areas include the pan or housing gaskets, oil filler or manual valve seal, front pump seal or O-ring, rear seal, converter neck, drain or pressure plugs, cooling line connections, and cooler. Refer to **Figure 31-31.** In rare cases, a leak may be caused by a crack or porosity in the transmission/transaxle case. At road speed, wind passing under the vehicle body causes fluid to flow back toward the rear of the vehicle. For example, it is quite possible for engine oil from a leaking valve cover to flow back to the end of the transmission extension housing. At first glance, this might appear to be a leaking rear transmission seal. Other leaking areas can be misleading in the same manner.

Note: A leaking vacuum modulator diaphragm may allow transmission fluid to be drawn into the engine and burned. In such cases, there is no external sign of a leak.

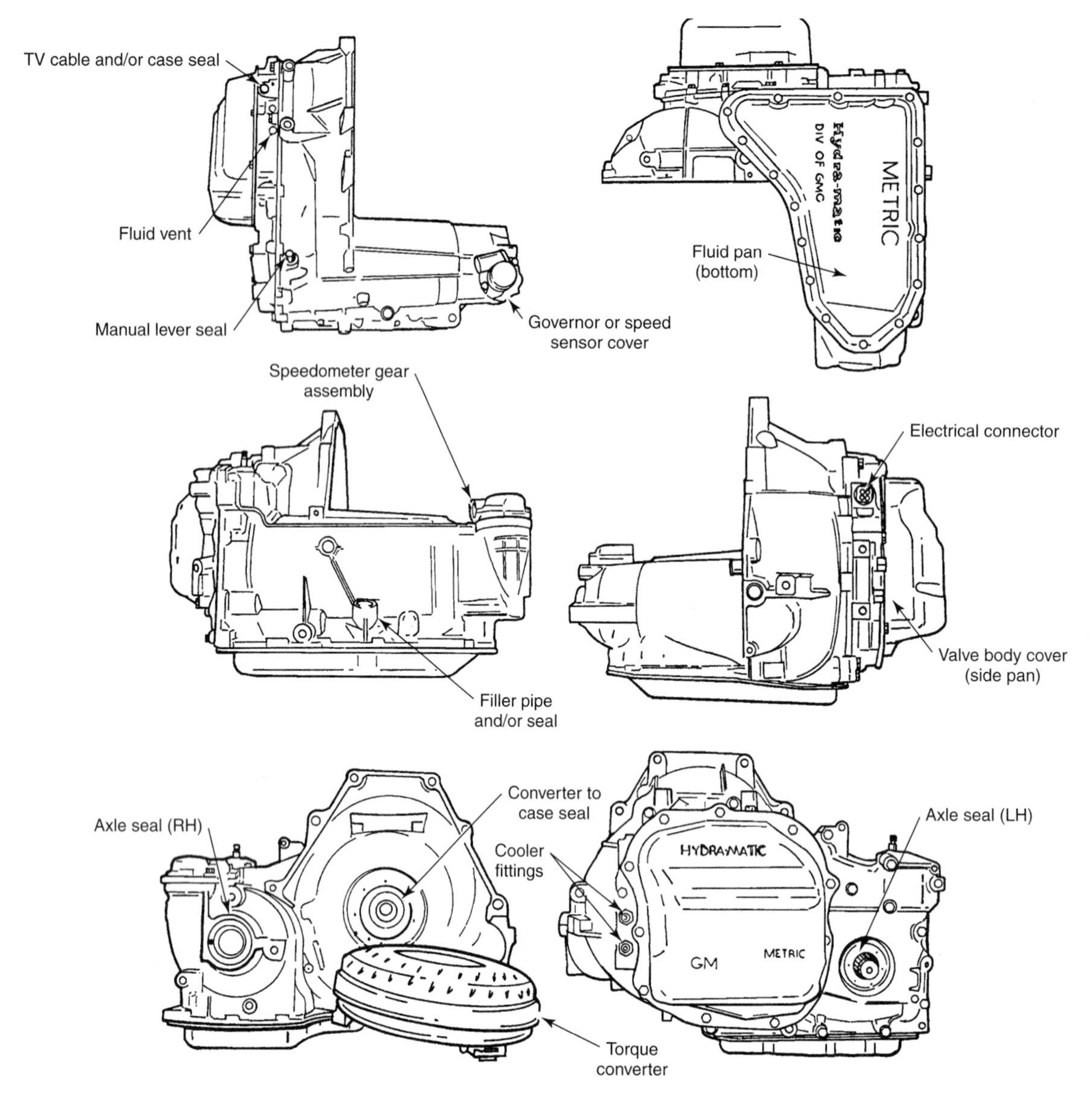

Figure 31-31. This series of diagrams identifies a few possible transmission fluid leak points on one particular automatic transaxle. (General Motors)

Identify Leaking Fluid

Do not assume that fluid dripping from the rear of the transmission or from the converter housing is transmission fluid. It may be engine oil, coolant, brake fluid, or power steering fluid. Transmission fluid is generally dyed red and can usually be identified by the color. A black light is very helpful in determining the type of fluid. Compare the appearance of the leaking fluid to that on the engine, power steering, and transmission/transaxle dipsticks. If needed, a special fluorescent dye may be added to the transmission fluid for positive identification. It is very difficult to pinpoint the source of a leak when the parts are covered with dirt and oil. Clean the engine, the converter housing, and the transmission or transaxle; then blow them dry with compressed air. Remove the converter housing inspection pan and clean the converter. Clean the inside of the housing and blow it dry with air.

Following cleaning, operate the engine at a fast idle to bring the transmission fluid to normal operating temperature. Stop the engine, place the vehicle on a lift, and restart the engine. Examine the engine and the transmission or transaxle for leaks. Shifting the transmission or transaxle through all drive ranges will help you spot existing leaks. Watch carefully for the first sign of a fluid leak. If no leaks are apparent, operate the vehicle on the road for several miles, with frequent starts and stops. Place the vehicle on a lift and recheck for leaks. When fluid flows from the converter housing, it could mean a loose converter drain plug, leaking converter, defective front seal, housing-to-case fastener leak, or similar defect.

Figure 31-32 illustrates possible leakage points and the flow of fluid in a typical converter-housing assembly.

Casting Repair

If a fluid leak is the result of a porous casting or a small crack, a satisfactory repair can often be made with an epoxy resin. Operate the vehicle until the transmission or transaxle reaches full operation temperature. Clean the case thoroughly and mark the leakage area. Apply several coats of nonflammable solvent to the leak and surrounding areas. Blow the area dry after each coat. Give the area a final scrubbing and immediately blow it dry. Mix a batch of epoxy resin following the epoxy manufacturer's directions. Apply a heavy coat to the case. The case should still be hot. Allow the epoxy to cure for a minimum of three to four hours before starting the vehicle. Following the waiting period, road test and recheck for leaks.

External Seal Replacement

There are many seals, O-rings, and gaskets on the average automatic transmission or transaxle. Most of them can be changed without removing the unit from the vehicle. Simply remove any parts that interfere with access. Most seals can be pried out and a new seal carefully driven into place. Note the following precautions:

- Ensure the shaft and seal bore are free of nicks and dents.
- Clean the seal area of the transmission or transaxle.
- Coat the outside of the new seal with sealer.
- Coat the lips with transmission fluid before installation.
- Use a suitable installation tool.
- Ensure the seal lip faces toward the inside of the transmission or transaxle.

Replacing O-rings is similar to replacing seals. The sealing surfaces should be carefully cleaned and checked for damage. The new O-ring should be oiled and placed into the bore or on the separate part as specified. Install all fasteners and check for leaks. When gaskets are replaced, all the old gasket material should be removed. Use gasket cement only if the gasket manufacturer calls for it. After installing the new gasket, carefully torque the fasteners to the proper torque. The table in **Figure 31-33** indicates whether the transmission or transaxle must be removed from the vehicle to service a particular seal.

Electronic Control System Component Replacement

Once a problem has been isolated to the electronic control system, repairs are relatively easy. To replace transmission or transaxle solenoids and sensors, begin by removing the oil pan. Next, remove the electrical connector from the device to be replaced. Remove the old part and install the new part. Reinstall the electrical connector and check the condition of the internal wiring harness. Wiring harnesses often fail, especially at the connectors and in areas where they pass through brackets. If the harnesses are undamaged, reinstall the oil pan. Always recheck the operation of the engine and the transmission or transaxle after replacing an electrical device.

Note: Detailed procedures for replacing computer control input sensors are presented in Chapter 15, Computer System Diagnosis and Repair.

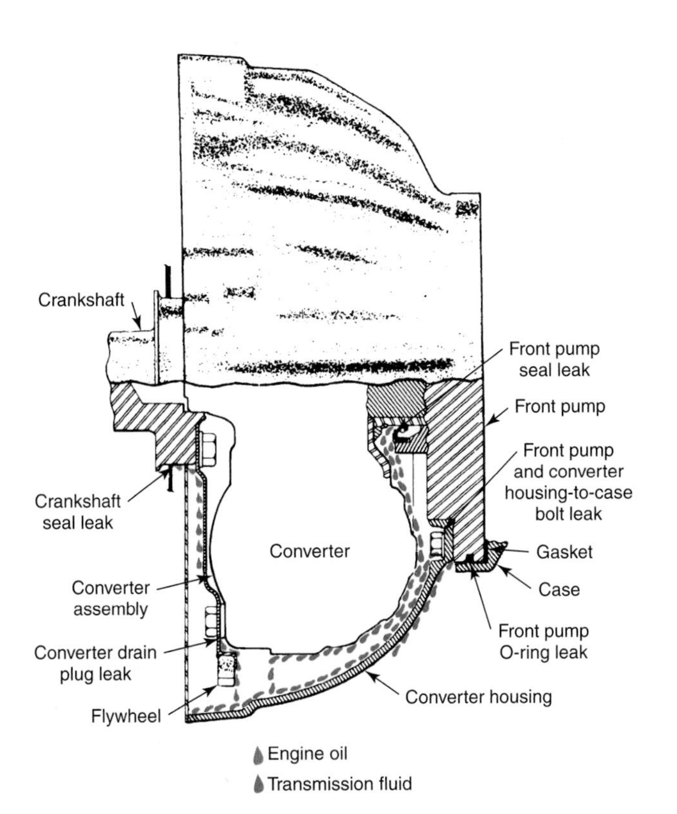

Figure 31-32. This drawing points out possible leakage points and fluid flow in typical converter housing assembly. (Ford)

Transmission in-vehicle	Transmission must be removed
Extension housing seal	Front pump seal
Manual shaft seal	Front pump O-ring
Governor seals	Front pump gasket
Governor cover gasket	Clutch piston seals
Speedometer gear seal and O-ring	Side pan gasket (transaxles)
Accumulator O-rings	
Servo piston O-rings	
Pan gasket	
Valve body gasket	
Extension housing gasket	
Filler tube seal	

Figure 31-33. In-vehicle and out-of-vehicle transmission service and repair procedures.

Vacuum Modulator Replacement

Some transmissions and transaxles have a ***vacuum modulator.*** On many transmissions, the modulator replaces the TV linkage to operate the throttle valve. On other transmissions, the modulator controls transmission pressures, while a separate rod or cable controls upshift speeds. On one popular electronic transaxle, a modulator is used to control pressures, while the ECM controls shift speeds. Modulators contain a vacuum diaphragm assembly with a hose attached to intake manifold vacuum. Manifold vacuum, which corresponds to engine load, operates on the diaphragm to alter the position of the throttle valve. An additional refinement incorporates an evacuated bellows. It provides diaphragm action and adjusts pressure to the valve in accordance with changes in barometric pressure. **Figure 31-34** illustrates an altitude-compensating vacuum modulator.

Adjusting the vacuum modulator alters the transmission control pressure, **Figure 31-35.** Most original equipment modulators do not have adjusting screws. Aftermarket modulators are usually adjustable. Always check shift speeds and shift feel after installing a modulator. Soft, low shifts will destroy the holding members in very short time.

When shift points are too high or low, the vacuum modulator should be checked for a bent neck and for vacuum leakage. To check for leakage apply a controlled vacuum to the unit. The modulator should hold vacuum for several minutes. Also check for a plugged, split, or disconnected modulator vacuum line.

Governor Assembly Replacement

The valves in the governor assembly are operated by centrifugal force and are prone to sticking. Therefore, it is often necessary to remove the governor for cleaning or replacement. Some governors are mounted on the output shaft. These governors can only be accessed by removing the extension housing (transmissions) or the oil pan and valve body (some transaxles). On other transmissions and transaxles, the governor is installed on the side of the case and can be accessed by removing a cover plate.

Once the governor is removed, check it for sticking and dirt. Check all valves for free movement. If the governor is case-mounted, carefully check the drive gear for damage and the governor bore for wear. Carefully check the governor springs. A weak or misplaced spring will affect shift points. Thoroughly clean the governor and oil it before reinstalling. Always recheck shift speeds after changing the governor.

Valve Body Replacement

The valve body assembly is removed by pulling the pan and removing the valve body-to-transmission fasteners. Before removing the fasteners, remove the filter, linkage, and electrical connectors as necessary. When removing the valve body, be sure to remove it slowly in order to catch any valves, servo and accumulator springs or pistons, check balls, or linkage parts that may fall free. An exploded view of a typical valve body is shown in **Figure 31-36.** With the valve body on the bench, check it for metal particles and sludge. Check every valve for free movement. If any valve is sticking, it can be carefully removed and cleaned. Also check the valve bore for sludge or metal. If the valve cannot be made to work properly, replace the valve body.

Check the valve body for proper placement of all valves and springs. Reversed valves or misplaced springs adversely affect valve body operation. Check the condition of the sheet metal spacer plate and any check balls. Thoroughly clean the valve body and oil it before reinstalling. Be sure to install any check balls, accumulator or servo pistons and springs, and linkages. Do not use gasket cement when replacing the valve body. Torque the valve body bolts from the inner bolts outward. Install a new filter and reinstall the pan. Always recheck transmission or transaxle operation after servicing the valve body.

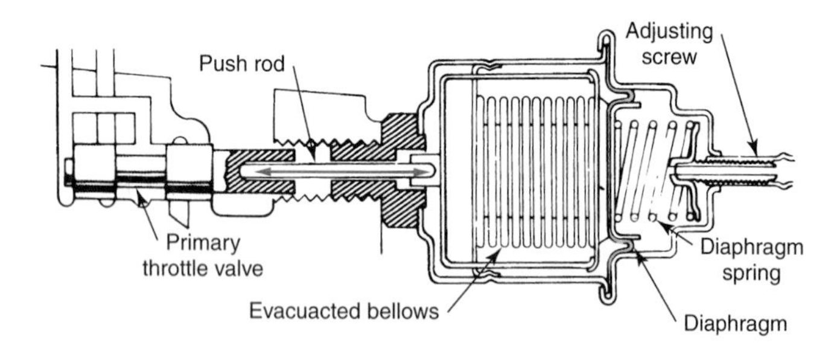

Figure 31-34. Vacuum-controlled primary throttle valve. The bellows is sensitive to barometric pressure. (Ford)

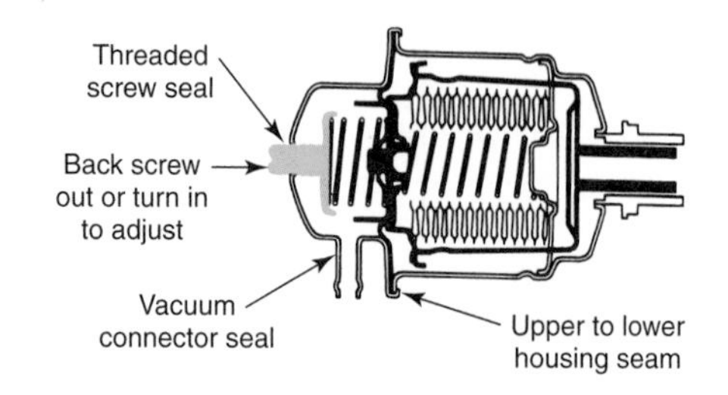

Figure 31-35. Cutaway view of an adjustable vacuum modulator unit. (General Motors)

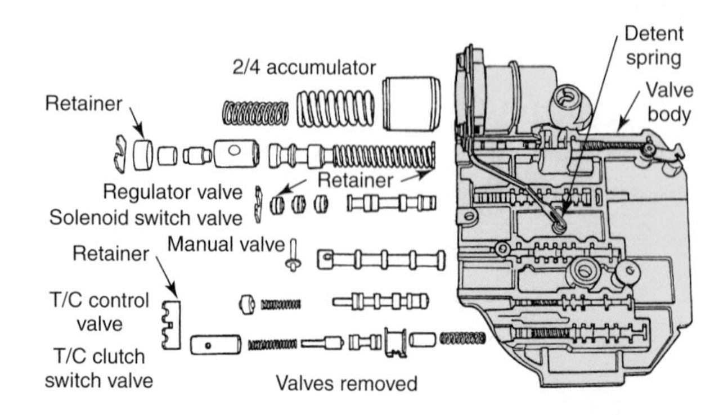

Figure 31-36. An exploded view of one type of valve body. This valve body is used with a four-speed automatic transaxle. (DaimlerChrysler)

Clutch and Band Operation

Once the valve body has been removed, ***air pressure checks*** can be made. This procedure consists of applying air pressure to check clutch and band operation. The transmission in **Figure 31-37** has had the control valve body removed. Note the case's oil passageways. Clutch action is being checked in **Figures 31-38** and **31-39.** When air is applied, a distinct thump can be heard or felt if the clutch piston is functioning. The application of air to the band servo apply passageways causes the band to tighten. Application of air to the servo release passage causes the band to loosen. Use only clean, dry air. The air nozzle must be clean. Be sure to use air on correct passageways and do not exceed the maximum air pressures recommended by the manufacturer.

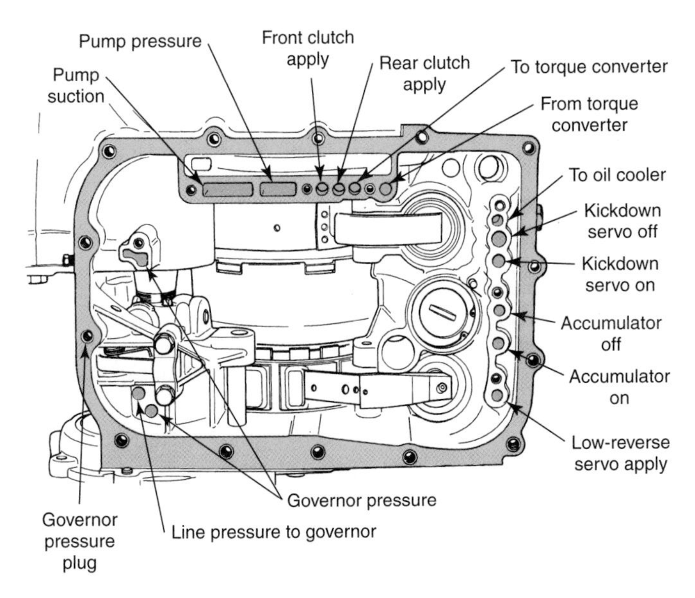

Figure 31-37. Removing the oil pan and valve body assembly exposes the transmission case's internal oil passages. (DaimlerChrysler)

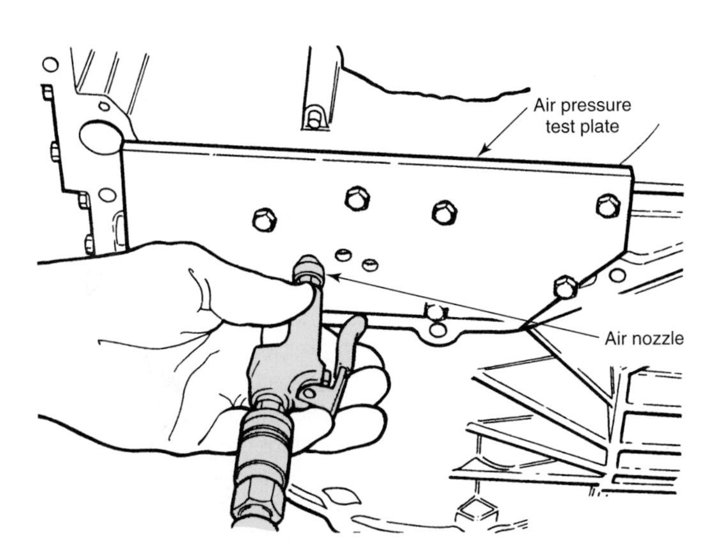

Figure 31-38. Apply air pressure to check clutch action. Note the air pressure test plate, which exposes the correct test holes and seals the holes not used. Use clean, moisture-free air and do not exceed the recommended test pressure. Wear your safety glasses. (DaimlerChrysler)

Other In-Vehicle Checks and Repairs

Depending on the transmission or transaxle design, a number of other tests and repairs can be performed with the unit in the vehicle. The vacuum modulator, modulator valve, servo, and park-lock device can often be removed, inspected, and replaced without pulling the transmission. Follow manufacturer's recommendations.

Add-On Transmission Oil Coolers

On vehicles used for trailer towing, commuting in heavy traffic, or other severe service, an ***add-on transmission oil cooler*** should be installed, **Figure 31-40.** Such a cooler greatly extends fluid and transmission/transaxle life.

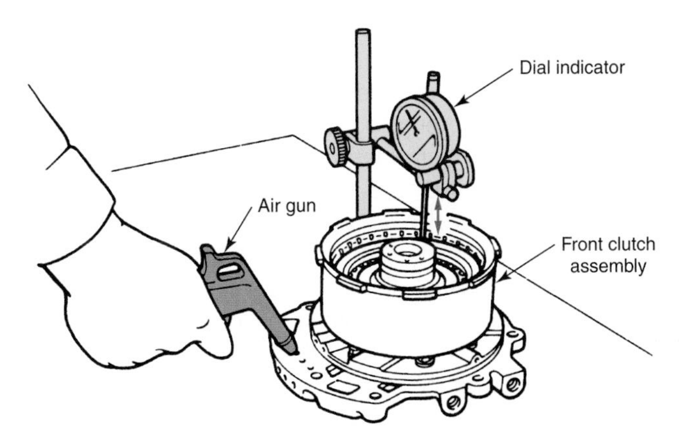

Figure 31-39. This illustration shows a front clutch unit being checked with an air gun. This particular clutch has been removed from transmission for testing. Note the dial indicator being used to measure clutch clearance (travel). (Mazda)

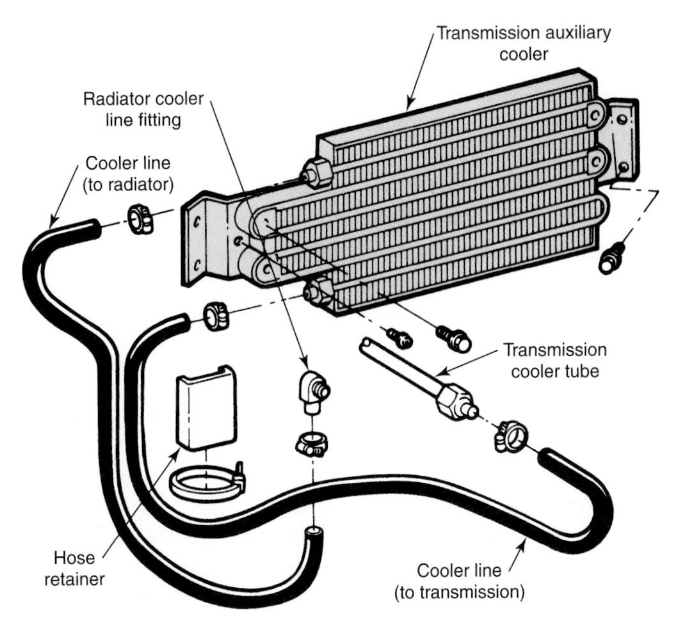

Figure 31-40. An auxiliary transmission-oil cooler is shown here. Note how hoses connect between the radiator cooler line fitting and the transmission cooler tube. (DaimlerChrysler)

Add-on transmission coolers are generally located in front of the radiator. The oil cooler should be installed in the return line to the transmission or transaxle. This will keep the oil from absorbing heat from the coolant after it has been cooled in the add-on cooler. The cooler itself should always be mounted ahead of the radiator and air conditioner condenser.

To install an add-on oil cooler, first note the radiator and/or AC condenser placement. Determine where the cooler can be located and how the hoses should be routed through the radiator support. Then locate the oil cooler lines on the vehicle radiator and determine which line is the return line to the transmission or transaxle. One way to determine the return line is to disconnect one of the lines and attach hoses to the radiator outlet and the line. Place both hoses in a container and start the engine. If fluid is pumped from the hose connected to the disconnected line, it is the inlet line, and the other line should be used. If fluid comes from the hose connected to the radiator, use the disconnected line.

Another method is to start the engine when it is cold and open the hood. Place the transmission in gear with the parking brake applied. As the engine warms up, check the temperature of the inlet and return lines, being careful not to touch any moving parts. The inlet line will become warm before the return line.

Once the return line has been identified, cut it in a spot where the cooler hoses will not contact moving parts. Install hose supplied with the cooler kit to the return line. Installing both ends of the hose will reduce fluid loss while other installation steps are performed.

Install the cooler in front of the radiator and AC condenser. The cooler should be at least 1/4″ (0.6 mm) ahead of the condenser or radiator and should not touch the condenser or radiator at any point.

Cooler manufacturers have various methods of securing the cooler, including metal brackets that are bolted in place and plastic ties that install through the radiator or condenser fins. Check the installation instructions for exact mounting methods.

Once the cooler is installed, cut the hose connected to the return line at the center of its length and route the ends through the radiator support to the cooler. Finally, install the hoses on the cooler. Install rubber grommets to keep the hoses from rubbing on the radiator support.

Start the engine and check the cooler hoses for leaks. Make sure the hoses do not contact moving parts. The cooler and hoses hold anywhere from 1/2–1 quart (0.47–0.95 liter) of fluid, and some oil will probably be lost during installation. Therefore, it is important to check and add transmission fluid as needed.

Transmission or Transaxle Removal

To begin transmission or transaxle removal, disconnect the battery ground strap. Remove the starter unless it is needed to rotate the engine to remove the bolts or nuts holding the converter to the flywheel or drive plate. Some manufacturers recommend using a ratchet and socket to turn the crankshaft balancer nut, **Figure 31-41.**

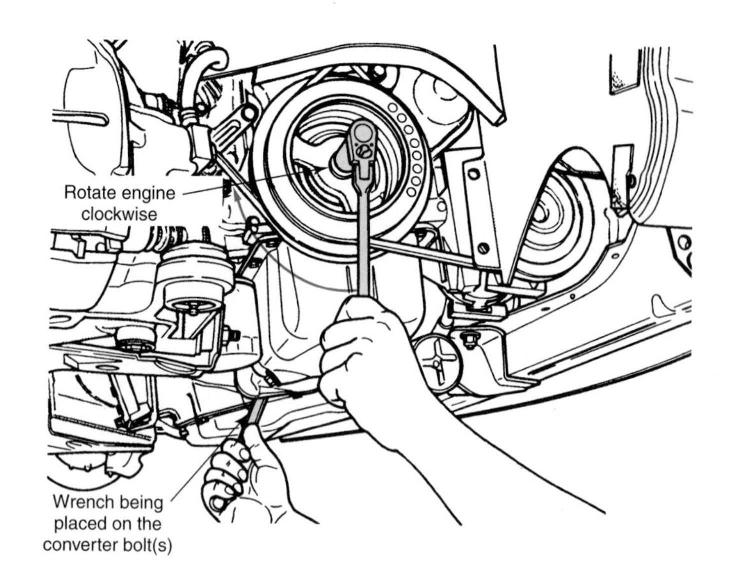

Figure 31-41. An engine is rotated clockwise so the next converter bolt can be accessed. If you are rotating the engine using this method, disconnect the battery to prevent accidental starter engagement. (DaimlerChrysler)

Note: Do not pry on the ring gear teeth to rotate the engine and converter. This can damage the ring gear. Special *flywheel turners* are available.

Remove the converter-to-flywheel fasteners. Also, remove wires, coolant lines, shift rod, vacuum line, downshift rod, filler tube, speedometer cable, and any other parts that would prevent the transmission's removal. Disconnect the drive shaft or CV drive axles and wire them out of the way. Before removing the shafts, mark them so they can be reassembled in their original positions. See Chapter 32, Axle and Driveline Service, for complete details. If the U-joint is a cross-and-roller type, tape the loose roller bearings to the cross. Remove the transmission or transaxle mounts as needed. If a support member must be removed, support the engine with either a jack stand or an engine support strap or fixture. If the engine must be raised to gain access, be careful to avoid damage to any attached parts. If pushing upward on the oil pan, place a wide block of wood between the oil pan and jack.

Use a Transmission Jack

Automatic transmissions and transaxles are heavy, and a transmission jack should be used for removal. Whenever a transmission or transaxle is being removed, get someone to help you stabilize the unit as it is lowered. Attach the transmission or transaxle firmly to the jack. After the unit is securely fastened to the jack, remove the fasteners holding the transmission or transaxle to the engine. Guide the transmission or transaxle away from the engine and lower it to the floor.

Note: The transmission/transaxle and converter should always be removed as an assembly. The retaining bar prevents the converter from dropping off during removal. See Figure 31-42.

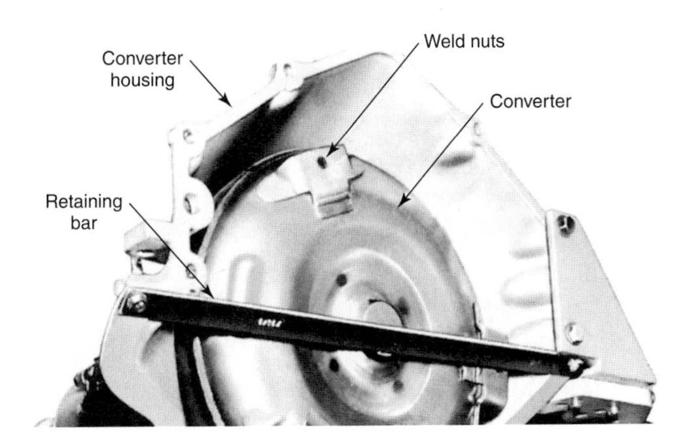

Figure 31-42. Always use a retaining bar or strap to prevent the converter from falling off during transmission removal or installation. (General Motors)

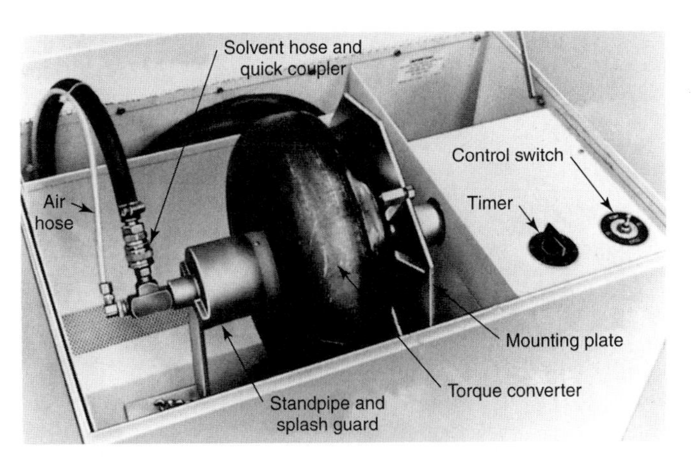

Figure 31-43. This torque converter cleaner helps remove metal particles and sludge that can circulate into a transmission or transaxle, plugging filter screens and valve bodies. This unit can also be used for flushing oil coolers. (Owatonna Tool Co.)

Torque Converter Cleaning

After the torque converter has been removed and thoroughly drained, which usually takes about 15 minutes, purge any remaining fluid with air pressure. Put approximately 2 quarts (2.2 liters) of recommended cleaning solvent into converter and shake vigorously. Drain again and use compressed air to blow dry.

Most converters can be pressure-cleaned and drained with a special cleaning machine. One is shown in **Figure 31-43.** The converter is bolted to a mounting plate and the proper hoses are connected. The converter is rotated at 20 rpm while cleaning solvent is passed through it. Air is injected into the solvent to aid cleaning action through agitation.

Flushing Oil Cooler

To avoid the possibility of damage to a new or rebuilt transmission/transaxle, the oil cooler and lines should be thoroughly ***flushed*** to remove any metal particles or other material. Use clean solvent and a pressure gun. Reverse flush until the solvent comes out clean. Flush out the solvent (in the normal direction of flow) with automatic transmission fluid. Pass fluid through the lines until all the solvent is removed. Cap the lines until you are ready to reconnect them.

Caution: Oil cooler hydraulic pressures are low, usually about 10 psi (68.9 kPa). Do not use air pressure in excess of normal hydraulic pressures.

A flushing unit, such as that shown in **Figure 31-44,** does a good job of flushing the cooler. Follow the manufacturer's recommendations for the correct flushing procedure. After flushing, check all connections. If the oil flow through the lines seems impeded, check for dents or a pinched section. Replace damaged lines. Be extremely careful with plastic cooler lines and connectors. Keep them away from sharp or hot areas. Replace the radiator if the cooler is clogged and cannot be cleaned. **Figure 31-45** shows two typical cooler line arrangements.

Flushing may also be done by installing the transmission or transaxle and filling it with the recommended amount of fluid. Connect the transmission fluid outlet line only. Place the inlet end in a container. Start the engine, move the gear selector through the various ranges and return to Neutral. Allow about one quart of fluid to pump through the lines. Stop the engine and add another quart of fluid. Continue this process until the fluid coming out the return pipe is clear and clean. Reconnect the inlet line and add transmission fluid.

Transmission or Transaxle Installation

Install the transmission or transaxle in the reverse order of removal. Always install the transmission and torque converter as an assembly. Make sure the converter is installed into the front pump properly and to the full depth. See **Figure 31-46.** Use a retaining strap to keep the converter from falling out. Align converter and drive plate or flywheel marks and be certain the converter hub or pilot enters the recess in the crankshaft.

Caution: Do not allow the weight of the transmission to hang on the drive plate or flywheel. Avoid prying on the drive plate.

After installing the transmission or transaxle on the engine, make sure the converter can turn after the engine-to-transmission/transaxle housing bolts are tight. If the converter cannot be turned, find out why and correct the problem before proceeding. If the converter can be turned, install it to the drive plate or flywheel. When the transmission or transaxle is completely installed and all lines, wires, linkage, and other parts are connected, add the amount of fluid specified by the manufacturer. Start the engine and move the selector through the gears, check the fluid level, and add fluid if required. Bring the transmission or transaxle to normal operating temperature, check for leaks, recheck the fluid level, and perform a road test. Make final adjustments and another check for leakage.

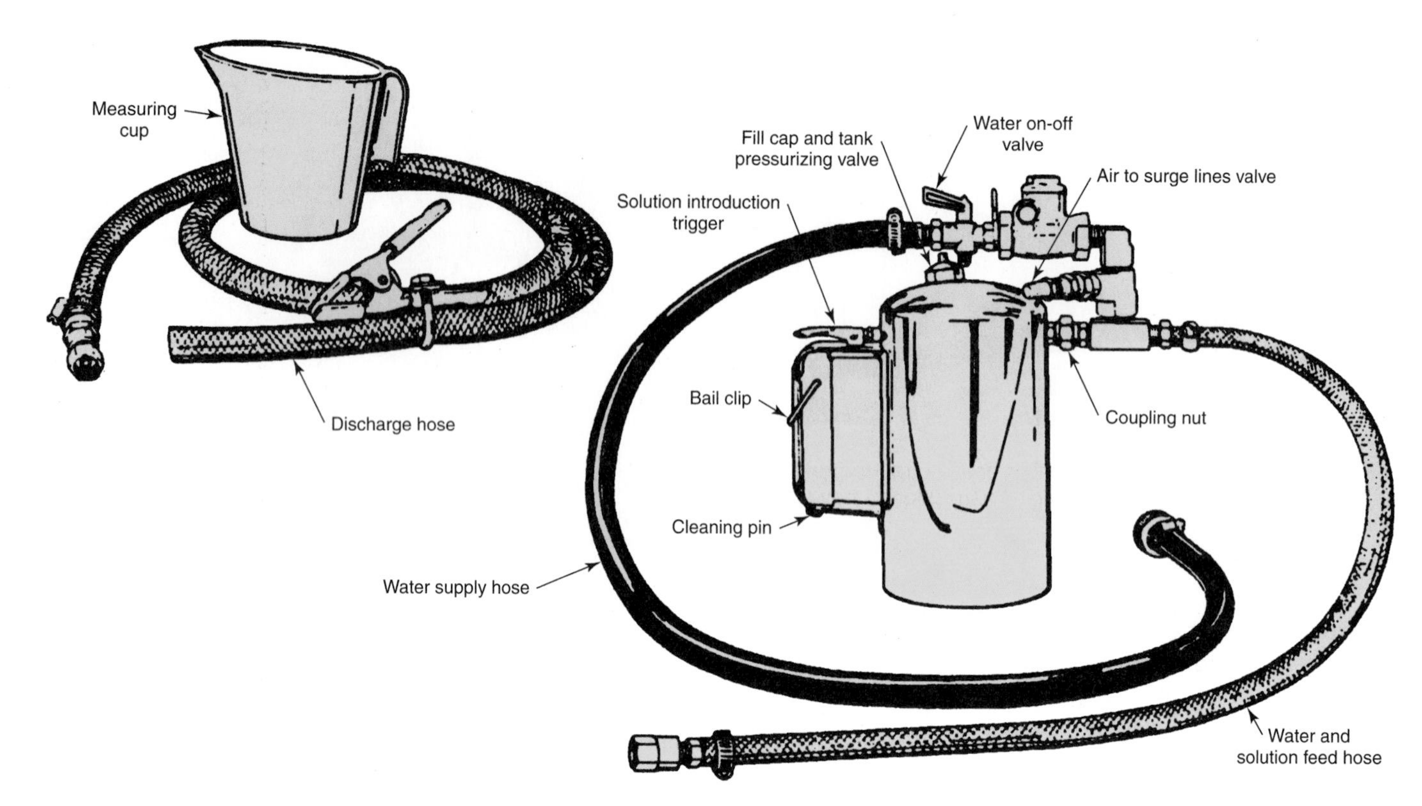

Figure 31-44. When using this oil cooler and line-flushing tool with shop compressed air, the air line should be equipped with a oil/water filter unit. Never exceed the vehicle manufacturer's recommended fluid flushing pressure. (General Motors)

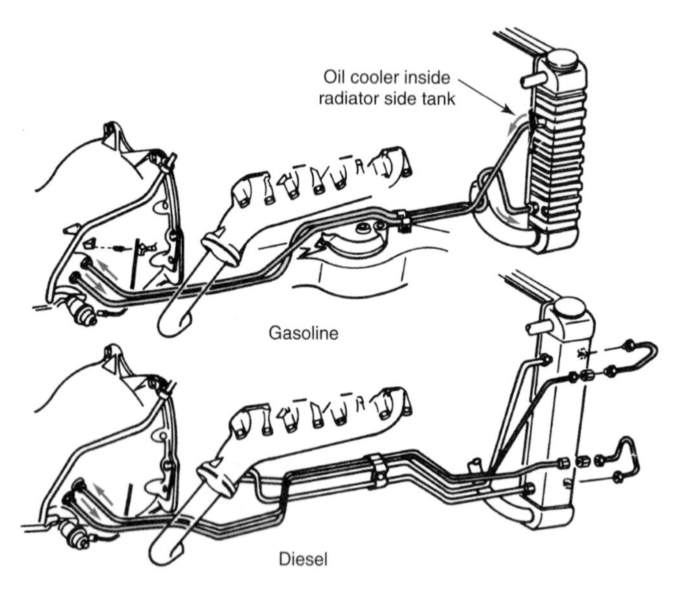

Figure 31-45. Typical transmission oil cooler line arrangements for a gasoline engine and a diesel engine. (General Motors)

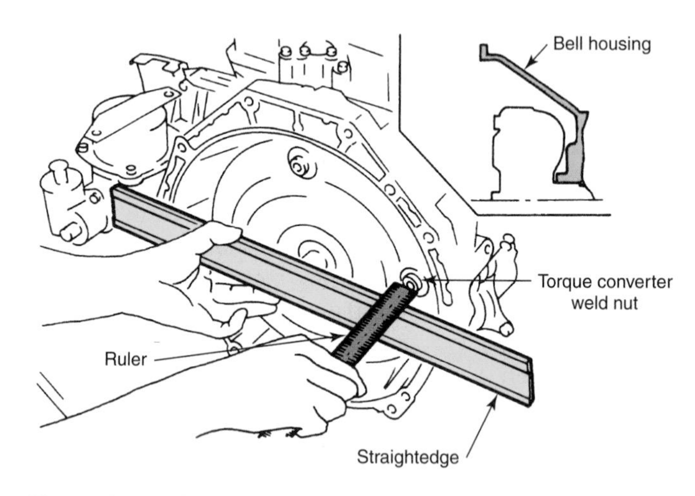

Figure 31-46. By placing a straightedge across the front of the bell housing and measuring to the torque converter weld nut, you can confirm that the converter is fully seated. (DaimlerChrysler)

Tech Talk

One of the most controversial areas of automotive service is whether or not to use chemical additives. There are many of these additives available. Engine oil additives are available to reduce oil burning and noises, clean the engine, reduce wear, or free sticking valve lifters. Cooling system chemicals are often used to clean the system, reduce corrosion, or seal leaks. Other additives clean the fuel injectors, remove water from the gas tank, raise octane, or seal the power steering system. If used as maintenance items, some of these additives may reduce deposit buildup, wear, and corrosion, and possibly seal some minor leaks. Oil viscosity improvers (thickeners) can reduce—although not eliminate—oil burning, blowby, and noises. However, many additives are sold as cures for vehicle problems that can only be solved by major repairs. As cures, it can only be said that they probably will not make the problem worse.

Additives are frequently used in automatic transmissions. Many transmission additives claim to seal leaks, stop slipping, and free sticking valves. Using these products usually results in disappointment. No additive can repair burned clutches or bands or broken parts. Any solvent strong enough to free sticking transmission valves is strong enough to damage the seals, friction materials, and plastic parts. Occasionally, using an additive will seal a minor leak, but in most cases, the only permanent cure is to replace the seal or gasket. Too much of the seal swelling solvent can make the leak worse and cause other external and internal seals to start leaking.

Some transmission additives are advertised as transmission "tune-up" products. While these may be of some help, the best tune-up for a transmission is to change the oil and filter regularly and to make band adjustments when necessary.

If you are tempted to use some kind of additive to solve an automatic transmission problem, do not be disappointed if it does not work. Since additives are relatively cheap, it might be worthwhile to try a can for a minor oil leak. But if the leak does not stop, it is time to change the seal or gasket. Adding more than one can of additive may overexpand other seals, causing additional leaks.

Summary

A large percentage of transmission and transaxle problems can be solved by in-vehicle service procedures. Modern transmissions and transaxles are either three- or four-speed automatic units with a standard or lockup torque converter, compound planetary gearset, holding members, and a hydraulic control system. On many late model transmissions and transaxles, the hydraulic system is computer-controlled.

The torque converter drives the transmission directly. The converter drives the transaxle directly with a multiple link chain or with gears. Transmissions vary from year to year and from model to model. Always use manufacturer's service manual for exact repair and adjustment procedures. The technician should have the tools required to perform accurate transmission diagnosis, adjustment, and repair.

Before towing a vehicle, place its transmission in Neutral. Do not exceed 45 mph (72 kph) or the distance recommended by maker, usually around 50 miles (80 km). If the transmission or drive components are inoperative, raise the drive wheels for towing.

The engine must be in good condition and properly tuned before diagnosing transmission problems. Fluid level is critical to proper transmission/transaxle operation. Overfilling can cause aeration, leaking, and poor operation. A low fluid level can cause overheating and slipping. When checking fluid level, make certain the vehicle is level and the shift lever is in Neutral or Park, the fluid is at operating temperature, the engine is idling, and the dipstick enters to full depth. Move the selector through all ranges and check fluid level again. Do not allow dirt to enter the fill pipe. Check fluid for odor, discoloration, aeration, and signs of water. When adding fluid, be sure to use the correct type.

Shift points must occur at recommended speeds and should be smooth and positive. Transmission fluid must be at operating temperature before conducting pressure tests. The rpm, drive range, and road speed are important in determining pressure. When conducting a stall test, apply both parking and service brakes. Do not operate at full throttle for longer than five seconds. Operate the engine in Neutral to cool the fluid between stall test periods.

Checking fluid pressure is a good way to determine the condition of the hydraulic system. Since procedures for checking pressure vary, consult the manufacturer's service manual. A stall test is used on some units to determine the condition of the disc clutches, bands, one-way clutches, and other parts.

After removing the oil pan, check it for sludge and metal particles. A coating of more than 1/4″ (0.6 mm) indicates worn and slipping holding members.

Some transmission designs require periodic band adjustments. Adjust bands very carefully and exactly as specified by the manufacturer. Improper band adjustment can cause serious damage.

Shift linkage must be adjusted accurately. Engine mounts must be in good condition to prevent altering linkage adjustment. Accelerator pedal height, throttle plate, TV, and downshift linkage adjustment must be exact. Adjust downshift switch if needed. Check operation of neutral safety switch and adjust or replace it as needed. To detect leaks, clean off the transmission and converter housing. Allow the engine to idle and check for evidence of a transmission fluid leak. A black light is helpful in locating leaks. Some leaks may be repaired without transmission removal. Porous castings can often be repaired with epoxy resin.

Some governors are exposed by removing the extension housing; others are covered with a removable plate. The valve body may be removed by draining the fluid and dropping the pan. Band and clutch action can be checked with air pressure on some transmissions. When performing air pressure tests, use clean, dry air.

When removing a transmission, be careful to avoid springing the drive plate. Secure the transmission to the jack. Get someone to help you when removing or installing a transmission. Remove the converter and transmission as an assembly. Some front-wheel drive vehicles require the removal of the engine with the transmission. Others must have the engine properly supported before transmission removal. Use a retaining bar or clip to prevent the converter from dropping off. Mark the drive plate and converter. Always flush the oil cooler and lines before connecting them to a new or repaired transmission. Reverse flush with clean solvent followed with transmission fluid. Use care to align marks and avoid damage to the drive plate during transmission installation. Dip fastener threads in fluid before installing, and repair stripped threads with Heli-Coils. Torque all fasteners to specification.

Review Questions—Chapter 31

Do not write in this book. Write your answers on a separate sheet of paper.

1. Electronic control systems eliminate the need for the _____.
   (A) throttle valve
   (B) shift valves
   (C) oil pump
   (D) valve body
2. When installing aluminum parts, dip the fastener threads in _____ _____ before installation.
3. Before checking transmission fluid level, make sure that _____.
   (A) the vehicle is level
   (B) the shift lever is in Neutral or Park
   (C) the engine is idling
   (D) All of the above.
4. Transmission fluid level should be kept _____.
   (A) on the full mark
   (B) between the add and the full marks
   (C) slightly below the add mark
   (D) slightly above the full mark
5. Transmission fluid installed at the factory is usually _____ in color.
   (A) yellow
   (B) green
   (C) red
   (D) clear
6. Most modern automatic transmissions require a special fluid called _____.
   (A) DEXRON III/MERCON
   (B) TYPE F
   (C) DEXRON II
   (D) AQ-ATF
7. When troubleshooting an electronically controlled transmission or transaxle problem, what should be done after basic checks have been made?
8. The oil pressure gauge used for transmission or transaxle oil pressure checks should read 0 psi–_____ psi or 0 kPa–_____ kPa.
9. When stall testing, never maintain the full throttle position for longer than _____ seconds.
   (A) 5
   (B) 15
   (C) 45
   (D) 60
10. When checking the transmission oil pan for deposits, a coating or more than _____ indicates worn and slipping holding members.
11. A porous casting can often be repaired by cleaning and covering the leaking area with _____.
12. List four precautions that must be taken when replacing transmission or transaxle seals.
13. Name three seals that can be replaced without removing the transmission or transaxle from the vehicle.
14. Before installing a new or repaired transmission, the _____ and _____ should be flushed.
15. Never allow the weight of the transmission and converter to rest on the _____.

ASE-Type Questions

1. Technician A says that most transmissions are of the same design. Technician B says that most transmission problems require transmission removal for correction. Who is right?
   (A) A only.
   (B) B only.
   (C) Both A and B.
   (D) Neither A nor B.
2. All of the following statements about pushing and towing vehicles with automatic transmissions are true, *except:*
   (A) the ignition lock mechanism on the steering column should be used to hold the wheels in the straight-ahead position.
   (B) modern vehicles with automatics cannot be started by pushing.
   (C) front-wheel drive vehicles can be towed with the front wheels off the ground.
   (D) when the transmission is inoperative, the vehicle should be towed with the drive wheels off the ground.
3. Before checking transmission or transaxle operation, which of the following should be checked first?
   (A) Oil level.
   (B) Engine condition.
   (C) Idle speed.
   (D) All of the above.
4. Check fluid level when the fluid temperature is _____.
   (A) cold
   (B) slightly warm
   (C) at normal operating temperature
   (D) overheated
5. Milky fluid indicates the presence of _____ in the transmission or transaxle.
   (A) sludge
   (B) varnish
   (C) water or coolant
   (D) motor oil

6. Linkage adjustments are important to maintain _____.
   (A) proper upshift points
   (B) proper downshift points
   (C) transmission service life
   (D) All of the above.
7. Excessive transmission or transaxle pressures indicate a(an) _____.
   (A) pressure regulator that is stuck closed
   (B) worn front pump
   (C) plugged filter
   (D) leaking clutch seal
8. Technician A says that some bands do not require periodic adjustments. Technician B says that an air pressure test will locate an out-of-adjustment band. Who is right?
   (A) A only.
   (B) B only.
   (C) Both A and B.
   (D) Neither A nor B.
9. When removing the transmission, always remove the converter _____.
   (A) with the transmission
   (B) before the transmission
   (C) after the transmission
   (D) the converter can be left on the engine

Suggested Activities

1. Check the fluid level of at least three automatic transmissions or transaxles. Also, inspect the dipstick for signs of fluid overheating. Write a short report summarizing the process and describing any special precautions that must be taken (such as allowing some transaxles to warm up before checking the level).
2. Draw a simple diagram showing how the governor and throttle valves work on the shift valve in an automatic transmission oil circuit. Explain how extra valves could be added to obtain more gears.
3. Use a set of planetaries and shafts from a scrap transmission to get various gears. Other class members can hold various parts of the gear train to get reduction, direct, overdrive, and reverse. Try to determine which parts to hold and drive to get the same number of gears as the transmission had originally.
4. Change the oil and filter of an automatic transmission or transaxle. Inspect the bottom of the pan for the presence of metal or sludge. After servicing the transmission/transaxle, check the hydraulic pressures according to the service manual procedures. Make a chart showing how the pressure changes in different gearshift positions.
5. Tear down an automatic transmission or transaxle and list all the parts needed to rebuild it. Calculate the cost of repairing the unit by finding the prices of parts and labor using a flat rate manual. Total the labor times and multiply them by the average labor rate for your area. Add the total of parts and labor. After determining other charges, such as supplies and outside services, calculate a grand total of what it would cost to repair this transmission.
6. Discuss your price calculation in Activity 5 with the other members of your class. Ask if there is anything that you missed. Determine whether the cost of repair is greater or less than the cost of a replacement transmission or transaxle.

Automatic Transmission and Transaxle Diagnosis	
Problem: Fluid leaks	
Possible cause	**Correction**
1. Defective gaskets or seals. 2. Loose bolts. 3. Porous or cracked case. 4. Leaking vacuum modulator diaphragm. 5. Overfilled transmission.	1. Replace defective parts. 2. Tighten bolts. 3. Repair or replace case. 4. Replace vacuum modulator. 5. Reduce fluid level.
Problem: Slipping in gear	
Possible cause	**Correction**
1. Low fluid level. 2. Clogged filter. 3. Stuck valve. 4. Burned holding members. 5. Misadjusted bands (when used). 6. Internal leaks.	1. Add fluid and check for leaks. 2. Replace filter. 3. Remove valve body and free sticky valves. 4. Disassemble transmission, replace burned holding members. 5. Adjust bands, recheck operation. 6. Disassemble transmission and correct leaks.
Problem: No up or downshifts	
Possible cause	**Correction**
1. Linkage misadjusted. 2. Governor stuck. 3. Stuck valves. 4. Defective or disconnected vacuum modulator (when used). 5. Internal leaks. 6. Computer control system problem.	1. Readjust linkage. 2. Remove and free sticky governor or replace. 3. Remove valve body and free sticky valves. 4. Replace modulator, check vacuum lines. 5. Disassemble transmission and correct leaks. 6. Test and correct system.
Problem: Noises	
Possible cause	**Correction**
1. Clogged filter. 2. Pump or torque converter defective. 3. Defective gears.	1. Replace filter. 2. Replace pump or torque converter. 3. Replace gears.

32

Axle and Driveline Service

After studying this chapter, you will be able to:

- Explain the construction and operation of one- and two-piece Hotchkiss drive shafts.
- Service one- and two-piece drivelines.
- Describe cross-and-roller universal joints.
- Service cross-and-roller universal joints.
- Diagnose driveline and universal joint problems.
- Explain the construction and operation of front-wheel drive CV axles.
- Explain the construction and operation of CV joints.
- Service CV axles, joints, and boots.
- Explain the construction, operation, and service of axle housings.
- Compare drive axle types.
- Describe the construction, operation, and service of differentials.
- Diagnose differential and axle problems.

Technical Terms

Hotchkiss
Slip yoke
Center support bearing
Cross-and-roller universal joint
Spider
Trunnion
Bearing caps
Constant-velocity universal joint
Drive shaft angle
Constant-velocity joints
Tripod joints
Rzeppa joints
CV boot
Steering knuckle
Semifloating
Flanged end axles
C-lock
Bearing retainer ring
Endplay
Ring gear
Pinion gear
Hypoid gears
Differential gears
Preload
Gear ratio
Runout
Pinion gear depth
Backlash
Contact pattern
Drive pattern
Coast pattern
Limited-slip

This chapter covers the operating principles and service of drive shafts and universal joints found on rear-wheel drive vehicles. The service of front and rear drive axles is also covered. Information on the removal and replacement of rear-wheel drive ring and pinion and differential units is also presented.

Rear-Wheel Drive Shaft Service

Modern drivelines use an open drive shaft. This design is commonly called the ***Hotchkiss*** design. Drive force and axle housing windup are handled by leaf springs or by control arms. A typical Hotchkiss drive assembly is shown in **Figure 32-1.**

The drive shaft may consist of one or more pieces. A one-piece drive shaft is illustrated in **Figure 32-2.** Note the ***slip yoke,*** which allows lengthwise movement between the transmission and the rear axle housing. The slip yoke slides onto the splined transmission output shaft. The end yoke is attached to the differential pinion shaft or to the pinion shaft flange, depending on the design. Modern long-wheelbase pickup trucks and a few older cars utilize two-piece drive shafts. This application requires the use of a ***center support bearing.*** Study the center support bearing arrangement in **Figure 32-3.**

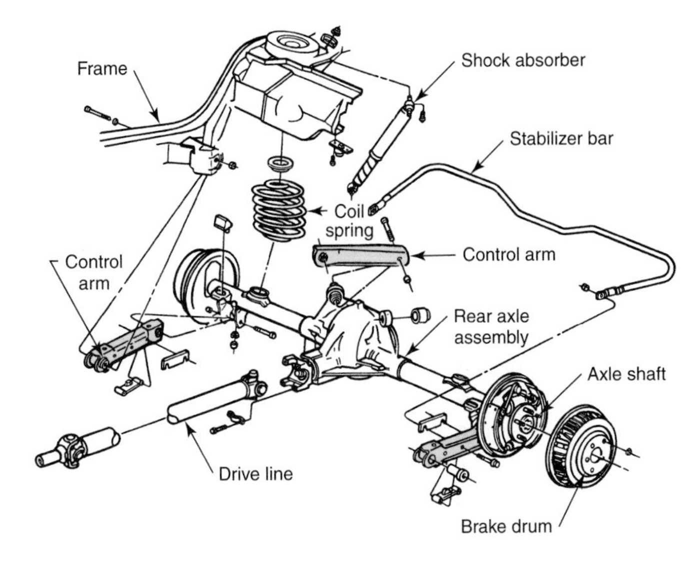

Figure 32-1. Exploded view of a Hotchkiss-type rear axle and driveline assembly. (General Motors)

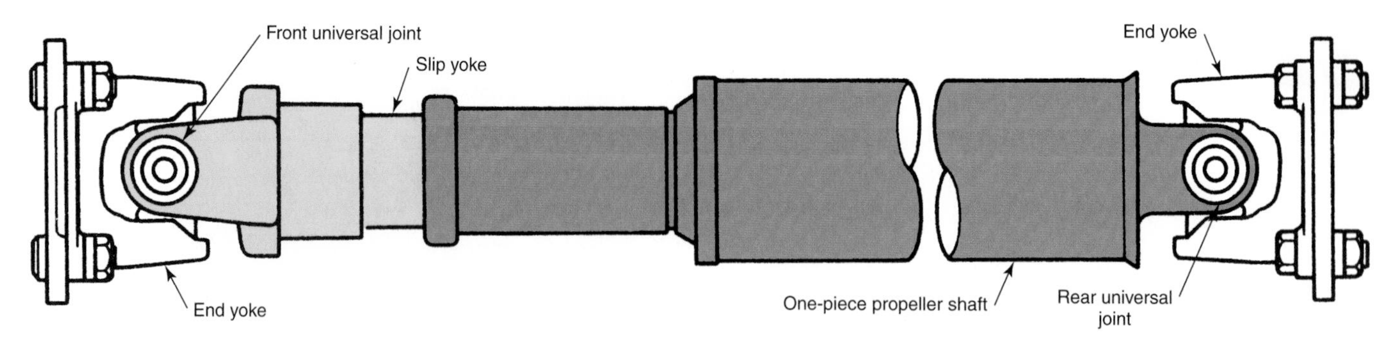

Figure 32-2. A one-piece drive shaft. (Toyota)

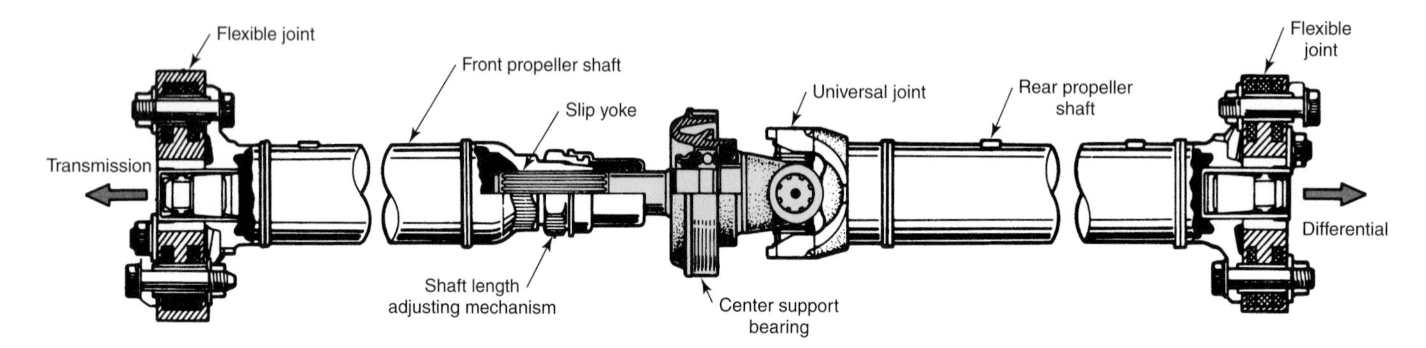

Figure 32-3. A two-piece drive shaft. Note the use of a center support bearing. (Toyota)

Cross-and-Roller Universal Joint

The ***cross-and-roller universal joint*** is used on rear-wheel drive shafts. It is usually referred to as a U-joint. Needle bearings are used to reduce friction. An exploded view of the cross-and-roller joint is shown in **Figure 32-4.** The cross-shaped piece in the center is often called a ***spider.*** Each arm of the spider is a short shaft called a ***trunnion.*** The ***bearing caps*** in this design are retained by snap rings set into the yoke at the outer ends of the bearing caps. Other U-joints are held in place by U-bolts passing around the bearing cap and through the yoke or by injection-molded nylon rings. When this type of joint is disassembled, removing the caps shears the nylon ring. Conventional bearing caps with snap ring retainers are used for replacement.

Constant-Velocity Universal Joint

A ***constant-velocity universal joint*** transfers rotation between input and output shafts smoothly and through a wide range of drive angles. In addition, the driven side of a constant-velocity joint does not experience the cyclic fluctuations in rotational speed that are found at the driven side of a cross-and-roller universal joint.

Some rear-wheel drive vehicles use the constant velocity joint shown in **Figure 32-5.** This constant velocity joint has two U-joints held by a center yoke. The front U-joint connects the slip yoke to the center yoke. The rear U-joint connects the center yoke to the drive shaft. A centering socket and ball inside of the center yoke maintain the relative position of the U-joints and the center yoke. The centerline of rotation passes through the ball. The only part that is subject to cyclic rotational fluctuations is the center yoke. Since the center yoke is relatively small in relation to the rest of the drive shaft, it does not produce noticeable vibrations.

The centering socket between the joints forces each half of the unit to rotate on a plane forming one-half of the total angle between the drive shaft and the transmission or differential pinion shaft. The use of a constant-velocity joint produces a very smooth flow of power, even over fairly acute driving angles. One or more constant-velocity joints may be used.

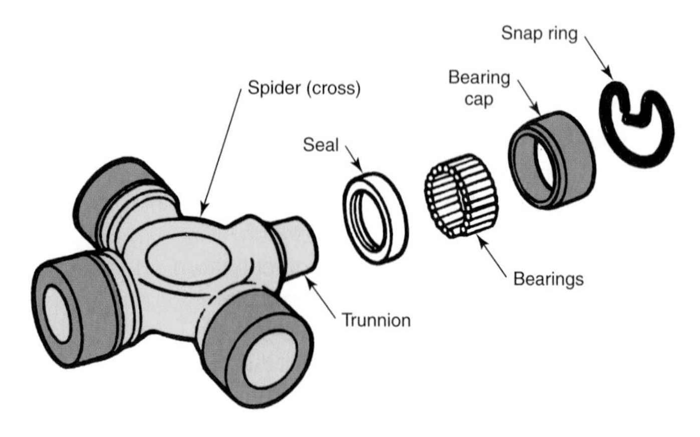

Figure 32-4. A basic cross-and-roller universal joint. Bearing caps are retained by snap rings. (Spicer)

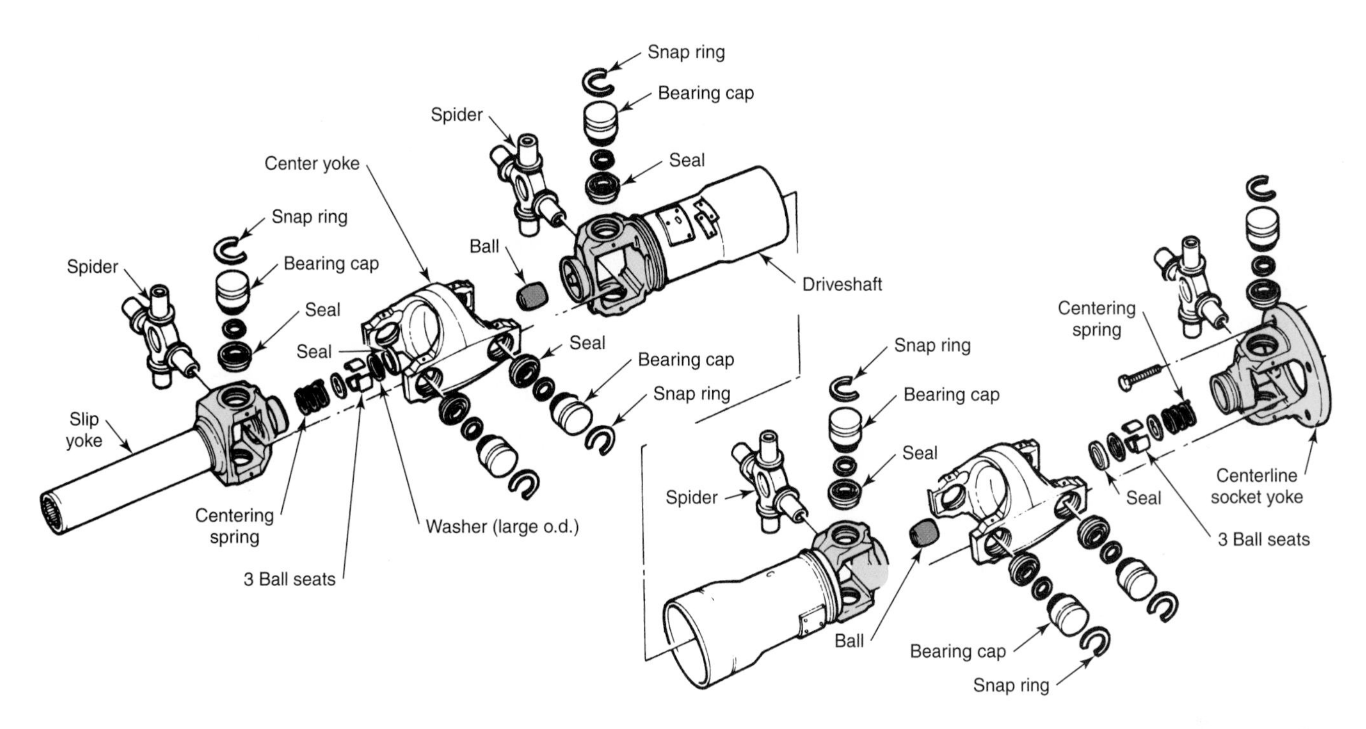

Figure 32-5. Exploded view of a drive shaft and two constant-velocity universal joints. Note that the ball seats in the socket of one yoke; the spring seats in the socket of the other yoke. (Ford)

Removing Rear Drive Shaft

Before disassembling a U-joint and removing the drive shaft, mark the drive shaft, slip yoke, flange yoke, and companion flange. Make sure the marks can be used to reassemble the parts in exactly the same relative positions. The yokes at both ends of the drive shaft must be in the same plane. See **Figure 32-6.** The yokes in Figure 32-6A are correctly aligned so that both will operate in the same plane. The yokes in Figure 32-6B are incorrect. When the yokes are permanently affixed to the ends of the shaft, do not be concerned about yoke alignment.

Begin drive shaft removal by removing the fasteners holding the rear U-joint to the rear axle pinion flange. **Figure 32-7** illustrates removal of the rear drive shaft yoke from the pinion flange. Other designs use straps to hold the U-joint bearing caps in place. Remove the straps to detach the caps from the flange.

Do Not Drop Bearing Caps

If the rear bearing caps are not retained on the cross with a thin strap, tape them on. This will keep them from dropping and possibly losing the needle bearings.

Support Drive Shaft

After disassembling the rear U-joint, carefully lower the shaft end. Do not allow the shaft to fall. Do not allow the shaft to hang supported by one U-joint. Never force the shaft to flex

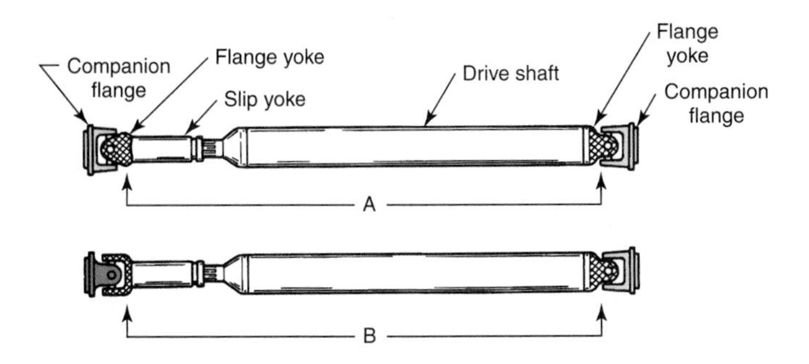

Figure 32-6. Yoke alignment is critical. The yokes in A are aligned. The yokes in B are not, and vibration, damage, etc., will occur.

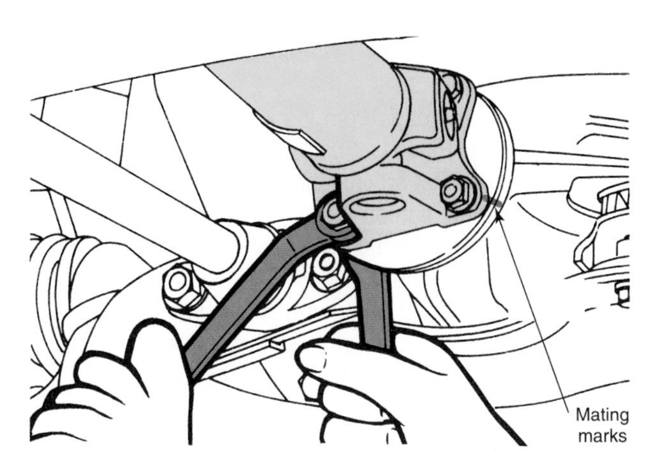

Figure 32-7. This rear universal joint is part of a yoke that is bolted to the pinion flange. To remove the yoke, loosen and remove the attaching bolts. (DaimlerChrysler)

the U-joint beyond its capacity to swivel. Careless handling can cause severe damage to the shaft and joints. If the shaft is long and clumsy, have another technician help during its removal and installation.

Two-Piece Shaft and Center Support Bearing

If a two-piece shaft is used, the center support bearing must be removed to permit shaft withdrawal. Check between the support and the frame for the presence of shims. If shims were used, replace them when you reinstall the center support. Always check the condition of the center support bearing before reinstalling it.

If a slip joint is placed between the shaft and one of the U-joints, it is possible to insert the splined stub shaft into the slip yoke in a number of positions.

Caution: Engage the stub shaft with the slip yoke so that both yokes are in the same plane. If the yokes are misaligned, serious vibration can result.

Protect Slip Yoke Surface

If the front shaft slip yoke engages the transmission output shaft, cover the slip yoke with cardboard or several layers of rags. This will protect the yoke from dirt and nicks. When needed, place a spare yoke or special plug in the transmission to prevent fluid leakage while the shaft is removed.

Repairing Cross-and-Roller U-Joints

Clamp the solid portion of the U-joint in a vise. If the yoke must be clamped, clamp it lightly. Avoid clamping the tube portion of the drive shaft. All drive shafts (aluminum, composite, or steel) are thin and can be easily damaged. This damage can cause shaft failure during operation. Support the drive shaft according to the manufacturer's recommendations.

If the bearing caps are held with snap rings at the outer edge, tap the bearing cap inward a small amount to free the snap ring, and drive the snap rings out with a thin punch or screwdrivers, **Figure 32-8.**

If the caps are held by external snap rings, use pliers to remove them, **Figure 32-9.**

Next, press the caps from the yokes. One method uses a vise as a press. Place a small socket against one bearing cap and a large socket against the yoke on the opposite side, **Figure 32-10.** Then, tighten the vise. As the vise is tightened, the small socket forces the cross to push the opposite bearing cap partially into the large socket. Once one cap is partially out, knock it out the rest of the way with a small punch and a hammer. Then, use the punch and hammer to drive the remaining cap back through the yoke, and remove it.

Note: If the bearing caps are held by injection-molded plastic, heat the yoke with a torch to soften the plastic. While the yoke is still hot, drive the bearing caps from the unit.

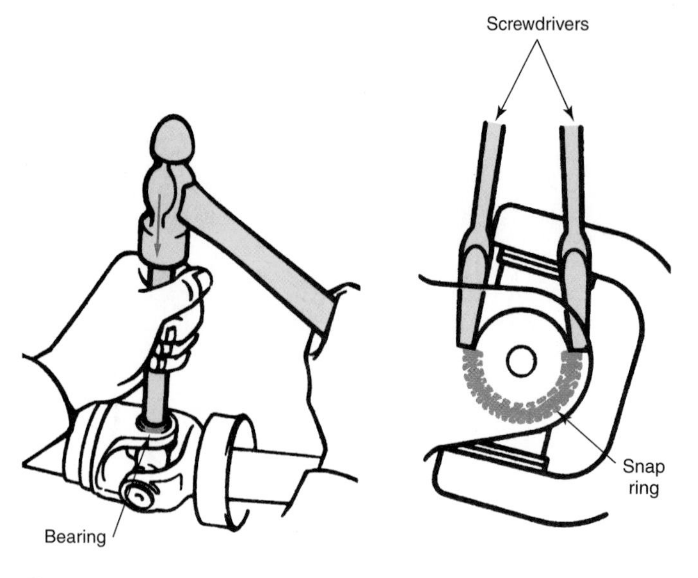

Figure 32-8. Tap the bearing cap inward a small amount to free the snap ring. Then use two screwdrivers as shown to remove the snap rings from the bearing caps. (Toyota)

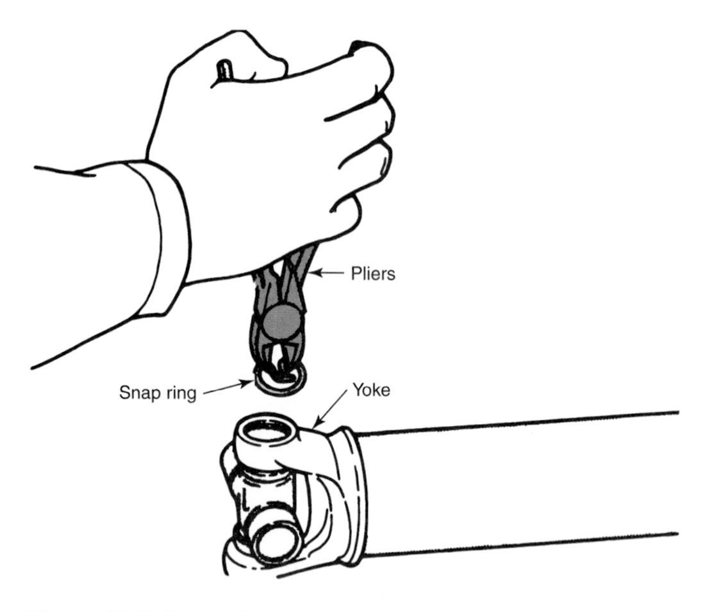

Figure 32-9. Removing a snap ring with pliers after the bearing cap has been tapped inward. (DaimlerChrysler)

Once one yoke has been removed, place the remaining yoke between the jaws of a heavy vise. Adjust the jaws so the yoke is just free to move. Rest the cross trunnions on the top of the jaws. If the U-joint will be reused, use jaw covers to protect the cross trunnions.

Strike the yoke smartly with a lead, brass, or plastic hammer. This drives the yoke downward, causing the cross to force the bearing cap partially out of the yoke lug, **Figure 32-11.**

When the bearing cap is forced out, grasp the cap and strike the yoke to complete the removal. Some bearing caps, after they are loosened, can be removed with pliers. Do not spill the needle bearings after the bearing cap is freed.

Figure 32-10. This illustration shows a method of removing the U-joint caps from the yoke, using two sockets and a vise. Position the two sockets as shown and tighten the vise. This will cause the smaller socket to press the cap into the larger socket. (DaimlerChrysler)

Figure 32-11. Driving the yoke downward to remove the bearing cap. (Dana Corp.)

Force the cross in the opposite direction to remove the other bearing cap. The cross may then be forced against one lug, tipped outward, and removed. Refer to **Figure 32-12.**

Universal Joint Cleaning and Inspection

Wipe off the cross trunnions. If they are worn, discard the cross, bearing caps, and snap rings. A U-joint repair kit will be needed. If the trunnions look good, clean the cross thoroughly. Blow out the grease passages in the cross. Check the condition of the yoke's lug holes.

Wash the bearing caps and needles. Blow them dry. If the inside of the bearing caps and the trunnion surfaces are free of corrosion and grooves, the parts may be reused. Check all of the needle bearings for signs of chips or breaks. Place the bearing caps on the trunnions and check for looseness.

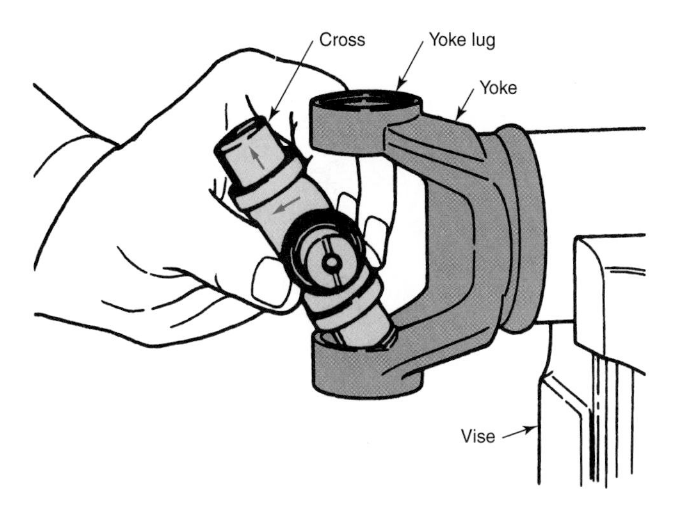

Figure 32-12. Tip the cross and remove it from the yoke. (DaimlerChrysler)

A U-joint repair kit is relatively inexpensive. If the old joint shows even the slightest sign of wear, install a repair kit. If either the bearing caps or the cross is worn, replace both. Never install new bearing caps on an old cross or vice versa.

Assembling a Universal Joint

Begin the reassembly of the universal joint by packing the bearing cap with the recommended lubricant. If the bearing is the sealed type (no grease fitting), pack the grease reservoirs at the ends of the trunnions. Pack the reservoirs carefully to eliminate trapped air. If the new U-joint has a grease fitting, skip this step.

Note: Many universal joints are prepacked with grease and no additional grease is needed.

If necessary, install the seals on the cross trunnions. On many modern universal joints, the seals are part of the cap assembly.

Start one of the bearing caps in a yoke lug hole. Insert it from the bottom with the open side of the bearing cap up to prevent the loss of needles. Make sure that each bearing cap contains the specified number of needles. Insert one of the cross trunnions into the bearing cap. Start the other bearing cap, making certain it slips over the trunnion.

When partially seated, place the two bearing caps between the vise jaws. Close the vice jaws until the bearing caps are flush with the yoke. Stop tightening when the caps are flush; do not over tighten or you may damage the caps, trunnions, and yoke. Tap one of the bearing caps. Use a soft-faced punch that is the full width of the bearing cap and tap the cap until it is slightly below the snap ring groove in the lug. If an inner-side snap ring arrangement is used, tap the lug through until the snap ring can be inserted. Insert a new snap ring, **Figure 32-13.** Make certain the snap ring is seated to its full depth.

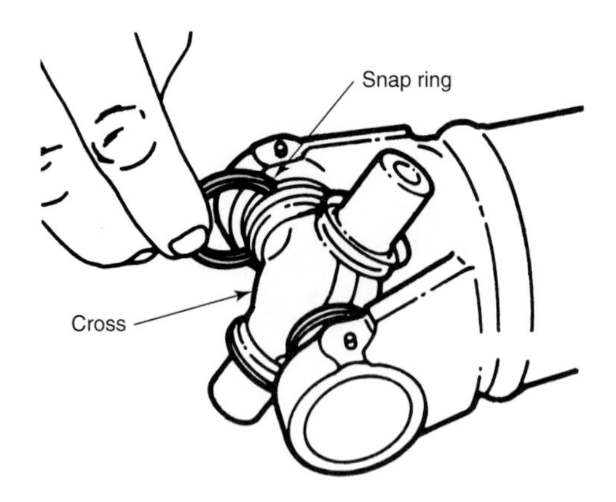

Figure 32-13. Inserting a snap ring into bearing cap groove. (General Motors)

Some joints require measuring snap ring-to-yoke groove clearance to be sure the fit is correct. After assembling the universal joint and the installing snap rings, place the joint in a vise. Force the cross to one side. With a feeler gauge, measure ring-to-yoke groove distance as shown in **Figure 32-14.** If the measurement is not within manufacturer's specs, replace the snap rings with thinner or thicker ones until the recommended clearance is obtained. All snap rings should have the same thickness for proper drive shaft balance.

Caution: Using an excessive amount of grease when rebuilding a U-joint can prevent bearing caps from seating fully. This will give a false snap ring fit reading.

Support the cross and strike the yoke to force the bearing cap into firm contact with the snap ring. With an inner snap ring, this forces the snap ring against the inner face of the yoke. Always seat the bearing cap in this fashion to prevent improper centering of the cross. Refer to **Figure 32-15.** Install the other snap ring, or rings, and seat the bearing caps against rings.

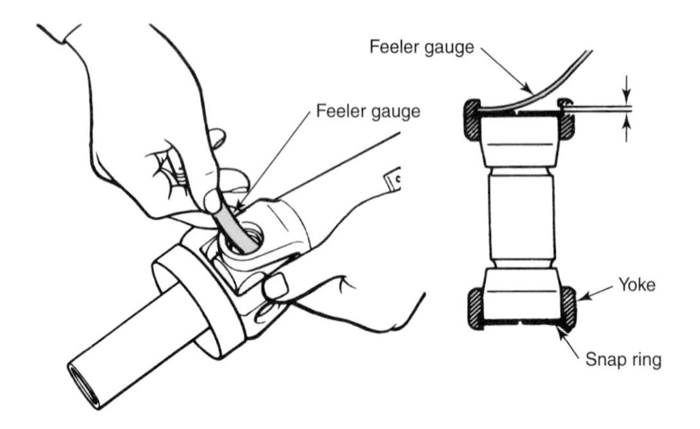

Figure 32-14. Measuring snap ring-to-groove clearance with a feeler gauge. Note how the gauge is inserted between the top of snap ring and the groove. (DaimlerChrysler)

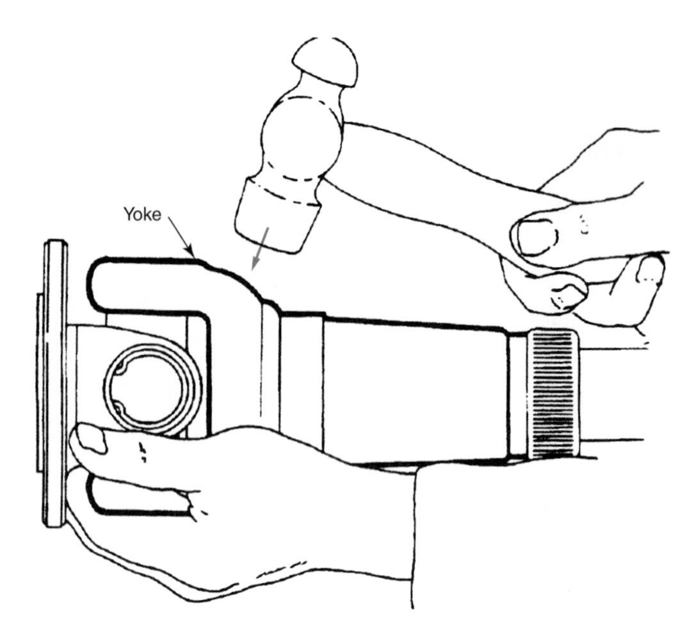

Figure 32-15. Striking the yoke to seat the bearing cap snugly against the snap ring and groove. (British-Leyland)

If a grease fitting is used, force universal joint grease into the joint slowly until it starts to show at the seals. Use a low-pressure hand grease gun or a power gun equipped with a pressure relief valve. Never use a high-pressure gun without this adapter. Excessive pressure can blow the bearing caps out of the yokes with the tremendous force or damage the seals.

Test the action of the assembled joint. It should move throughout its range without binding. If a slight bind exists, rap the yoke lugs with a soft hammer. This usually frees the joint. If it does not, disassemble the joint and check for the source of the bind.

Repairing Rear-Wheel Drive Constant-Velocity Universal Joints

The constant-velocity U-joint is literally two cross-and-bearing cap joints attached by a center yoke. Mark the center yoke, slip yoke, and shaft yoke so that all parts can be reassembled in the same order. Mark crosses, if they will be reused, so that grease fittings will be accessible, **Figure 32-16.**

Remove the snap rings and force the bearing caps from one end. A special bearing cap remover can be used to force the bearing caps partially from the yoke. The bearing cap is then grasped in the vise, and the center yoke is driven upward to complete pulling of the bearing cap. The bearing caps are lifted from the center socket yoke, and the cross is tipped and removed.

The tool is then used to force out the bearing caps in the other end of the center yoke. Avoid forcing the center yoke too far to one side. Stop when the slinger ring just touches. Grasp the bearing caps and remove them.

Reassemble the joint in the reverse order of disassembly. Use new parts where required. Lubricate the centering device. When assembled, remove the grease plugs, lubricate, and

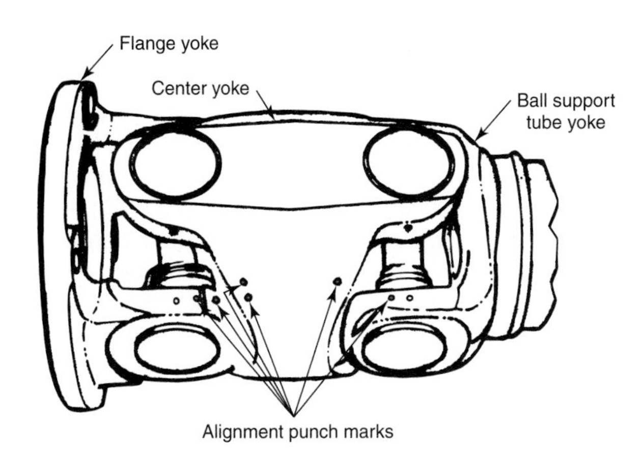

Figure 32-16. A constant-velocity joint marked before disassembly. Note the double marks on the left side. This prevents mixing the ends. (DaimlerChrysler)

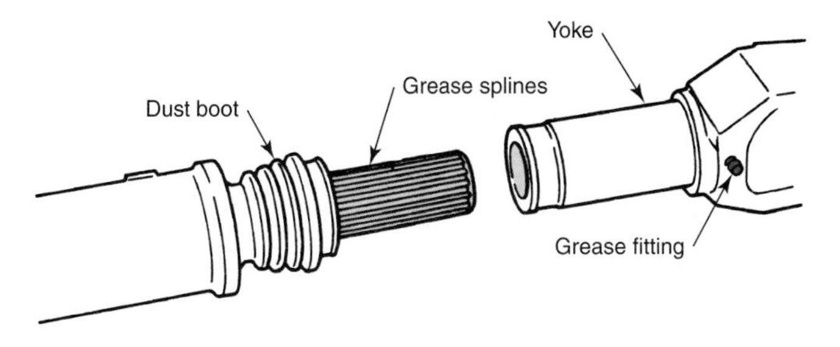

Figure 32-18. This slip yoke setup requires greasing the inner spline surface. Note the grease fitting. (Suzuki)

replace the plugs, **Figure 32-17.** When reassembling a sealed joint, lubricate all parts before reassembly.

Drive Shaft Installation

Check the joints to make sure all marks are aligned. Cover the slip yoke with cardboard or rags. Position the shaft in the vehicle. Support the shaft during installation to prevent damage to the U-joints.

Remove the covering from the slip yoke and lubricate the outside surface as recommended. Some installations use grease; others use transmission fluid. The inner, splined surface may be lubricated with transmission fluid. If the slip yoke has a grease fitting, **Figure 32-18,** it should be greased after installation.

When connecting the U-joint, make certain the marks on the shaft yoke and flange yoke are aligned. Check the flange for nicks or burrs.

In cases where a drive shaft has a splined stub, make certain the arrows (factory balance marks) or punch marks on the shaft and slip yoke are aligned.

If U-bolts are used to connect bearing caps to the flange yoke, torque the U-bolt nuts as specified. Excessive tightening will distort the bearing caps and cause shaft shudder and short life.

Before tightening U-bolts, make certain the bearing cap heads are underneath the locating tang. After the bolts have been torqued to specification, rap the joint with a soft hammer and retorque the fasteners. If a strap is used on the cross, make certain that it fits into the pockets provided in the yoke, **Figure 32-19.**

Inspect the driveline to make sure all fasteners are secured. If a center support bearing is used, torque the mounting fasteners. Where specified, check the clearance between the end of the transmission extension housing and the front slip yoke face. Grasp the shaft and shake it sideways. There should be no discernible movement. Road test the vehicle to check for quiet operation.

Drive Shaft Balance

The drive shaft turns at engine RPM in high gear. This requires the shaft be accurately ***balanced.*** If the shaft is bent or badly dented, it should be replaced. Proper straightening and balance are beyond the capabilities of a regular shop.

If the vehicle is being undercoated, keep the shaft and U-joints covered. Undercoating on the shaft may cause serious vibration.

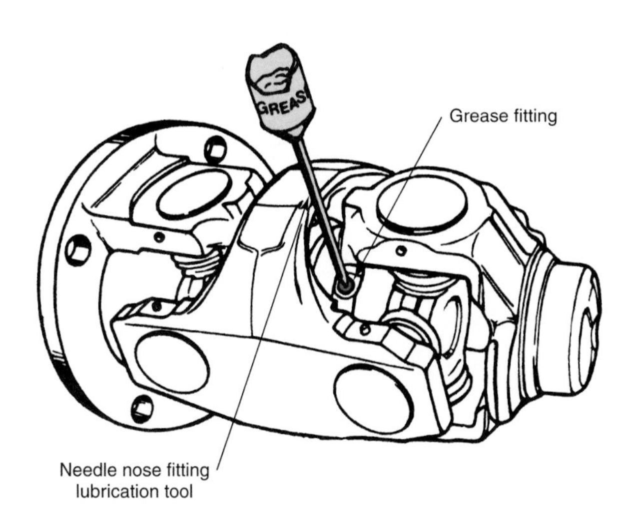

Figure 32-17. Lubricating a constant-velocity universal joint. Note the needle-nose grease fitting. (General Motors)

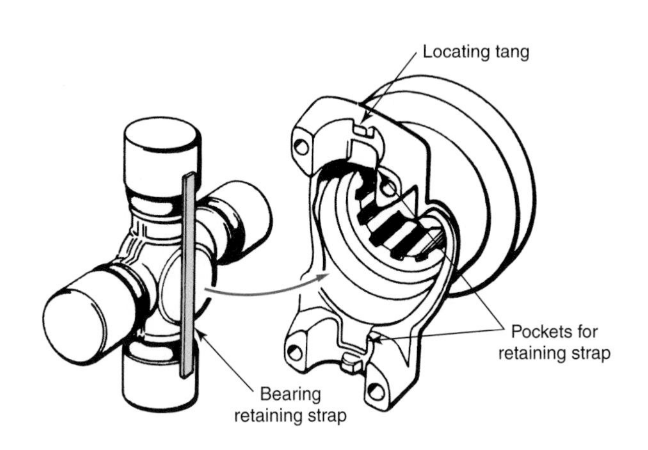

Figure 32-19. The retaining strap, when used, must fit into the yoke pockets. (Ford)

The shaft can be checked for runout in the vehicle by using a dial indicator. Mount the indicator to some rigid spot and place the stem on the driveline near one end. Turn a back wheel to rotate the drive shaft. Note the indicator reading. Move indicator to other end and then to the center.

When noting the indicator reading, do not count sudden changes from a weld, flat spot, or minor tube out-of-roundness. **Figure 32-20** illustrates typical measuring points on one- and two-piece drivelines. Total indicator reading (TIR) at these points should generally not exceed 0.030″ (0.76 mm). Follow the manufacturer's recommendations for your particular driveline.

Minor shaft unbalance can occasionally be corrected by using screw-type hose clamps. The clamps are attached to the shaft and rotated around until the shaft is balanced. See **Figure 32-21.**

If everything checks out all right but vibration persists, disconnect the rear U-joint. Turn the pinion flange 180° and reconnect the joint. If the vibration is still present, disconnect the front slip yoke. Rotate the yoke 180° and reconnect it.

Drive Shaft Angle

If the ***drive shaft angle*** (angle formed between the front and rear U-joints) is too great, driveline vibration is likely to occur at cruising speeds. This is due to the design of the cross-and-roller U-joint, which can cause great variations in drive shaft speed when angles are too great. When shaft vibration is present, check the angle formed between the centerline of the drive shaft and the differential pinion shaft.

A special gauge is shown in **Figure 32-22.** The gauge is placed on the front universal joint and drive shaft. The pointer indicates shaft-to-joint angle. The vehicle must be level and at normal curb weight (no passengers or luggage; gas tank full). Use a drive-on or axle-engagement lift. Do not use a frame contact lift, as this will allow the axle housing to hang down and alter the drive angle. The gauge can also be used to check the rear universal joint angle. The gauge is attached to the yokes and not the crosses.

A spirit level gauge can also be used to check the drive shaft drive angle. The gauge is adjusted by centering the level

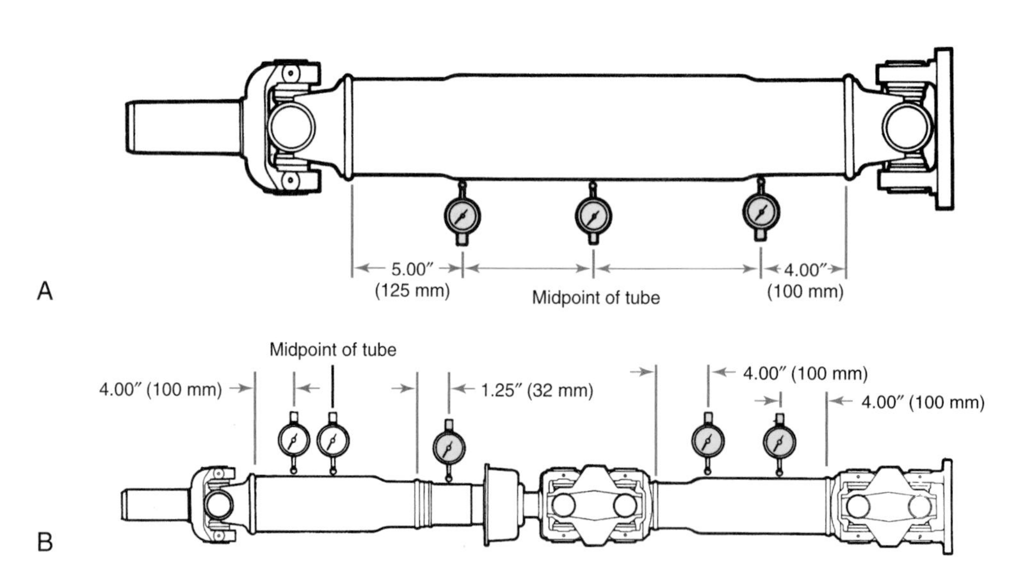

Figure 32-20. Checkpoints to be used for obtaining runout. A—One-piece driveline. B—Two-piece driveline. (General Motors)

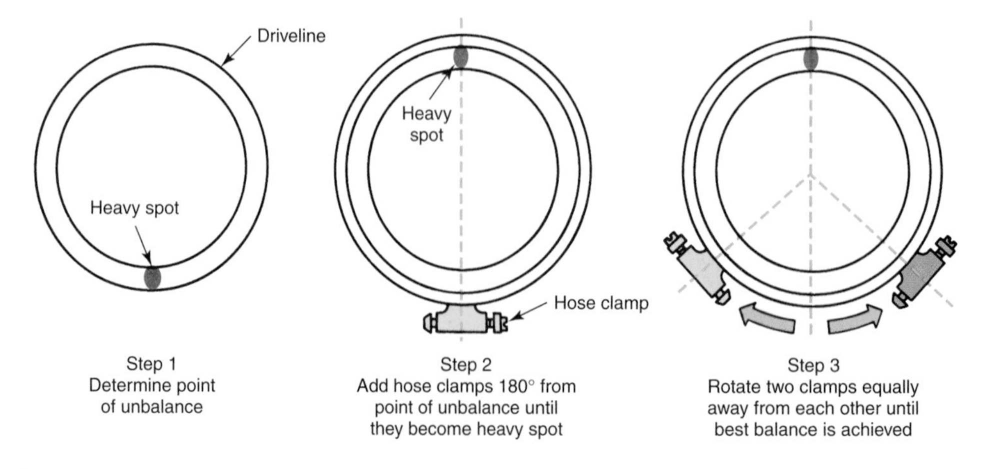

Figure 32-21. Eliminating driveline imbalance with screw-type hose clamps. (General Motors)

bubble while holding it on the differential carrier housing. The gauge is then held against the drive shaft, and the bubble's position is checked against specifications.

A third technique for checking drive angle is illustrated in **Figure 32-23.** This requires the measuring device to be placed at three different locations: the front drive shaft, the rear drive shaft, and the rear axle housing. The measurements at all three places must match the specifications given in the service manual.

If driveline angles are not within the recommended range, the drive shaft drive angle must be adjusted. On some vehicles, the transmission extension can be raised or lowered with shims. However, many vehicles do not provide for transmission mount adjustments, and drive angle adjustment is made to the axle housing only.

The rear axle housing can be tilted by adjusting the control arms. If the arms are not adjustable, new arms must be installed. When rear leaf springs are used, tapered wedges may be inserted between the spring and the axle spring pad.

If the thick portion of the wedge faces the rear of the vehicle, the pinion shaft companion flange will be tilted downward. To raise the flange, insert the wedge so the thick side faces the front. See **Figure 32-24.**

Front-Wheel Drive CV Axle Service

Constant-velocity joints, or CV joints, are used on most modern front-wheel drive vehicles and on some rear-wheel drive vehicles with independent rear suspensions. Their design produces less vibration than the conventional cross-and-roller U-joint. This is important when the possibility of transmitting driveline vibration is great, such as when drive shafts (axles) are located in the front of the vehicle. One particular front-wheel drive CV joint and axle assembly is shown in **Figure 32-25.** The one- or two-piece front-wheel drive axles are usually equipped with inboard (inside) and outboard (outside, nearest to the wheel) CV joints. See **Figure 32-26.**

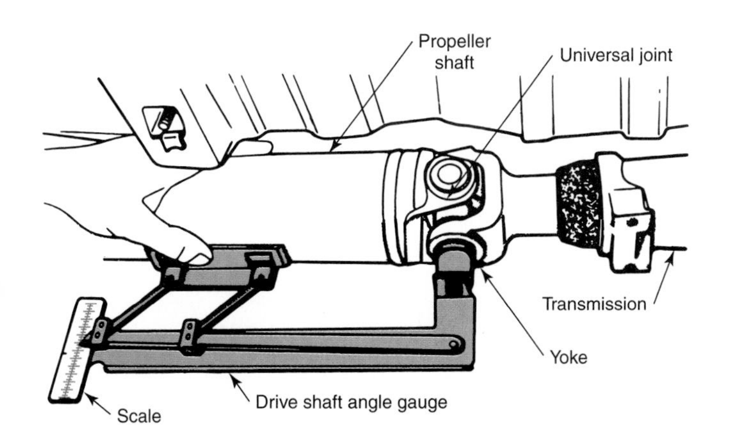

Figure 32-22. Using special gauge to measure the front universal-joint angle. Keep the gauge firmly in position while measuring. (DaimlerChrysler)

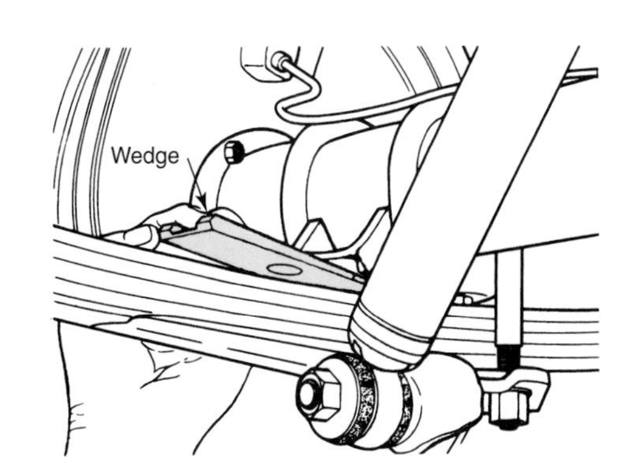

Figure 32-24. Inserting a tapered wedge between the rear spring and the axle housing to change the drive shaft "drive" angle. (DaimlerChrysler)

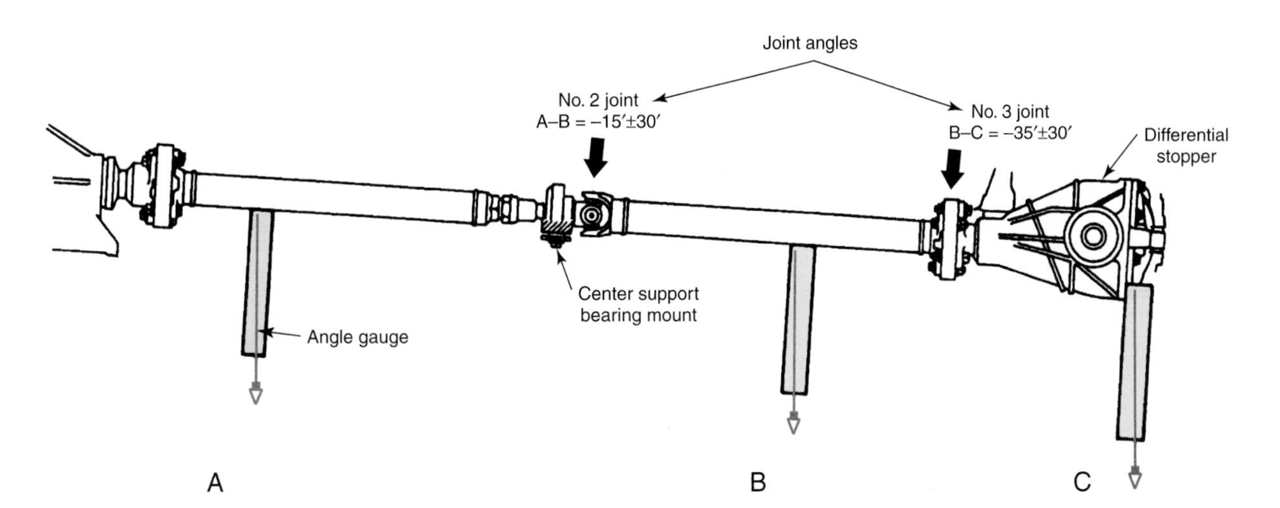

Figure 32-23. Measuring drive shaft angle in three spots with angle gauges. Follow the tool and vehicle manufacturers' specifications. (Toyota)

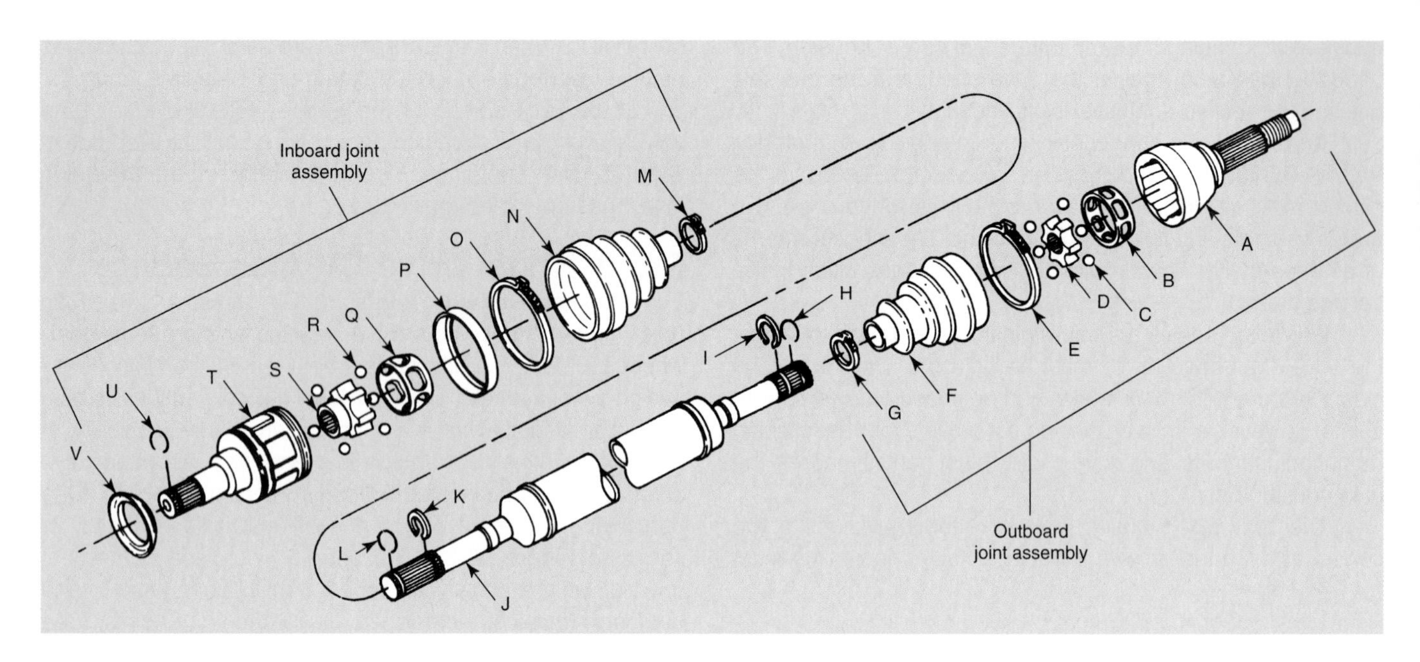

Figure 32-25. Exploded view of one type front-wheel drive axle and universal joint assembly. A—Outer bearing race and stub shaft assembly. B—Bearing cage. C—Ball bearings. D—Inner bearing race. E—Boot clamp (large). F—Boot. G—Boot clamp (small). H—Circlip. I—Stop ring. J—Interconnecting shaft. K—Stop ring. L—Circlip. M—Boot clamp (small). N—Boot. O—Boot-clamp (large). P—Bearing retainer. Q—Bearing cage. R—Ball bearings. S—Inner bearing race. T—Outer bearing race and stub shaft assembly. U—Circlip. V—Dust deflector. (Ford)

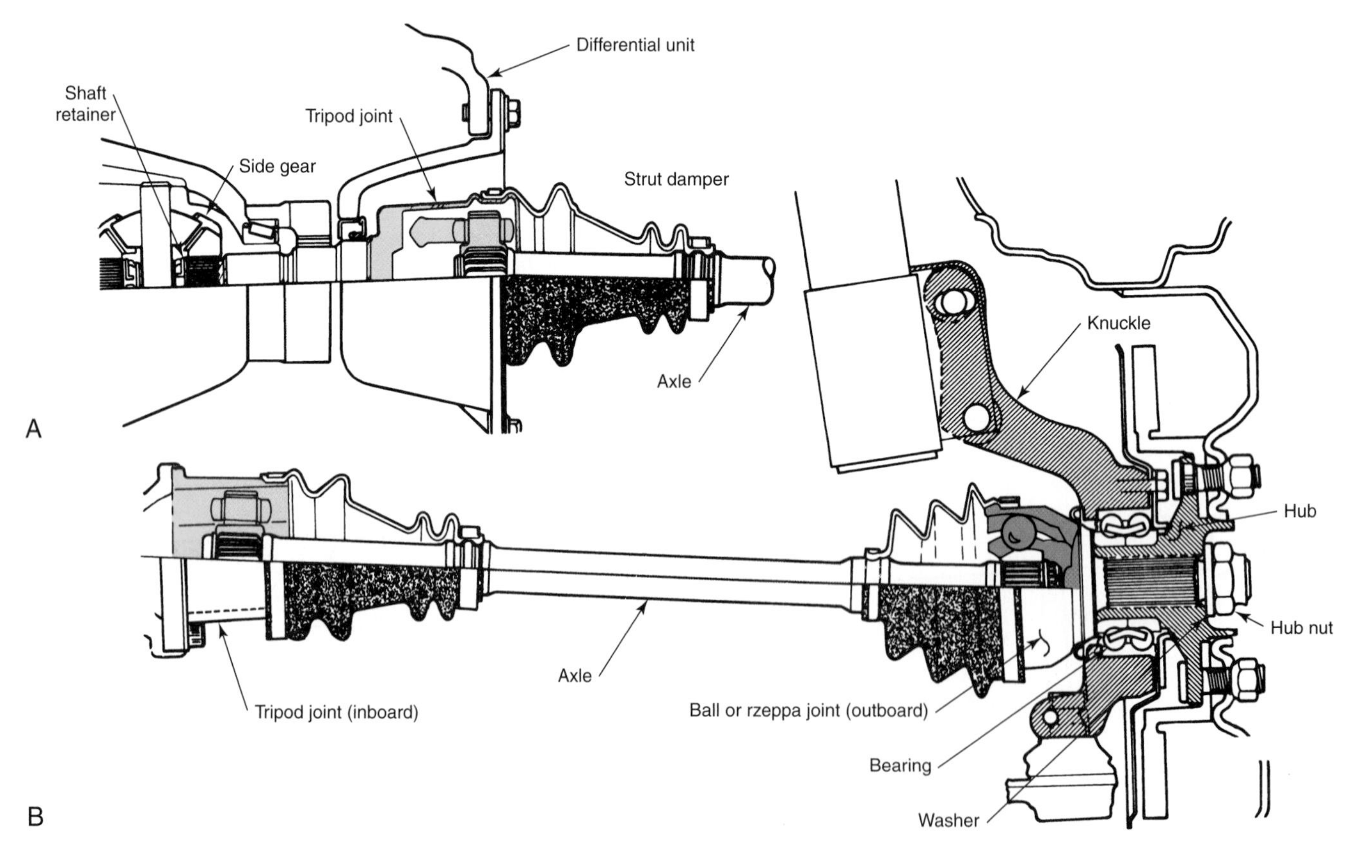

Figure 32-26. One front-wheel drive axle and universal joint setup. Joints used on this unit are constant-velocity tripod and ball types. Note somewhat different inboard joint-attachment methods. A—Automatic transmission. B—Manual transmission. (DaimlerChrysler)

There are two basic CV joint types: ***tripod joints*** and ***Rzeppa joints*** (also called ball and channel). The Rzeppa joint is by far the most common. **Figure 32-27** pictures both the tripod and Rzeppa units.

CV joints are protected by a ***CV boot.*** CV boots are bellows-shaped covers that completely seal the interior parts of the CV joint. Typical boots can be seen in Figure 32-27. CV joints usually last the life of the vehicle if they are not damaged. If the boot becomes torn, all of the CV joint lubricant will be thrown out, and dirt and water will enter, causing the joint to fail quickly. Therefore, CV boots must be inspected frequently. Damaged boots should be replaced immediately.

Removing CV Axles

The following is a general procedure for removing CV axles. Follow manufacturer's recommendations for your vehicle. Remove the battery negative cable. The vehicle should be elevated with a floor lift or with a jack and placed on jack stands. This will give you room to work. Remove the front wheels on the side of the axle to be removed. Remove the hub nut and loosen the axle from the front wheel bearing located in the ***steering knuckle.***

Note: Some axles must be pressed from the wheel bearing and steering knuckle assembly. Consult the manufacturers' service manual.

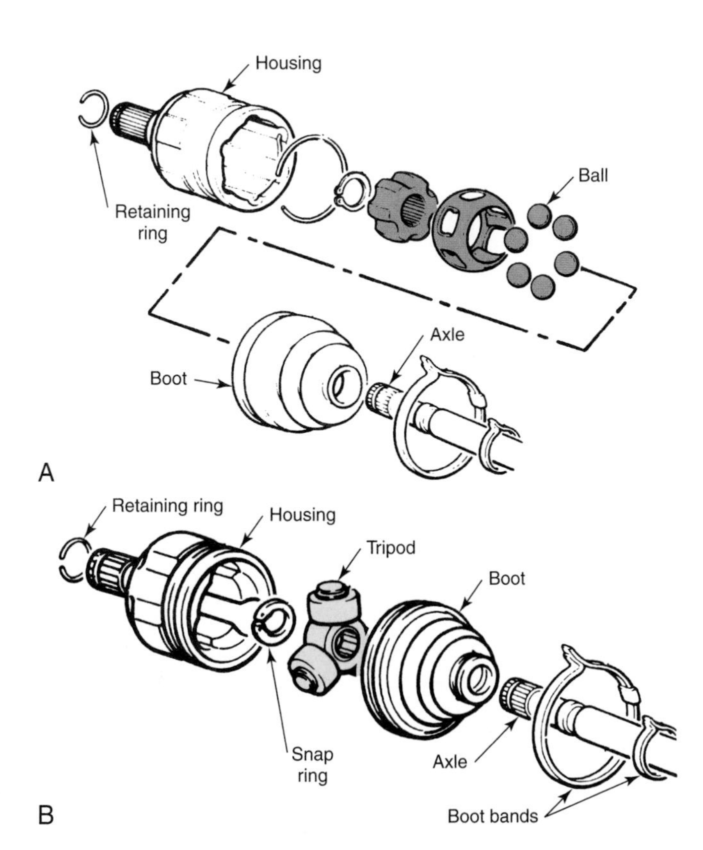

Figure 32-27. A—Exploded view of a ball (Rzeppa) joint. B— Exploded view of a tripod joint. (DaimlerChrysler)

Next, remove any part that restricts access to the CV joint fasteners. Some vehicles require the partial or complete removal of the steering knuckle assembly before the CV axle can be removed. If so, remove the fasteners holding the steering knuckle to the lower control arm, **Figure 32-28.** Remove the tie rod end and strut rod mounting, if necessary. Slide the steering knuckle assembly off the CV axle shaft and wire it out of the way. The inner end of most CV axle shafts is held to the transaxle by an internal snap ring, and the axle can be removed with a sharp pulling motion. Some CV axles must be removed with a special tool. Others are removed by carefully prying between the inner CV joint and the transaxle case, **Figure 32-29.**

Note: A few CV axles are held to the transaxle out put shaft flange by bolts or by studs and nuts. Remove these fasteners before removing the axle.

CV Joint Disassembly

Begin disassembly of either a tripod or Rzeppa joint by removing the ***straps*** holding the CV boot, **Figure 32-30.** The boot can then be either cut off or slid out of the way to gain access to the CV joint.

Next, remove any snap rings, **Figure 32-31,** and carefully separate the joint. If the joint will not come apart easily, tap it lightly with a hammer.

Caution: Be sure that all snap rings have been removed before striking the CV joint with a hammer.

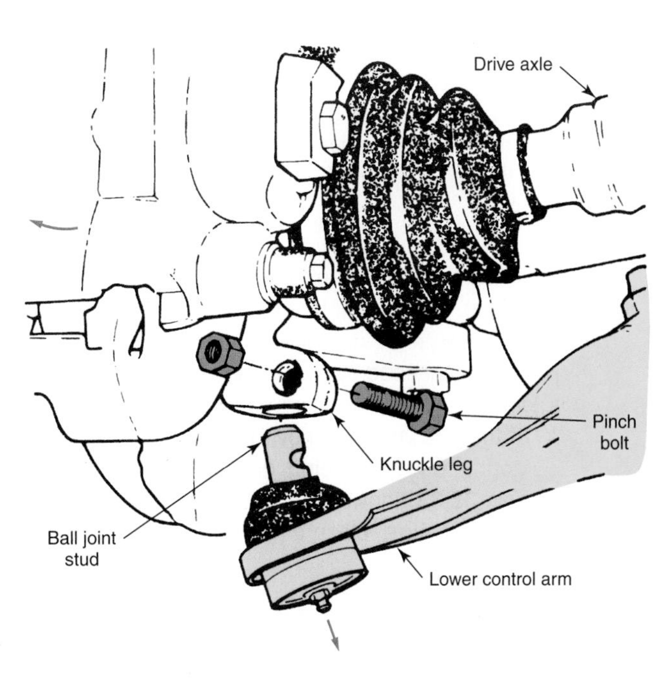

Figure 32-28. Removing the pinch bolt to separate the lower control arm and steering knuckle. Use caution if the parts are under tension. (Driveshaft Technology, Inc.)

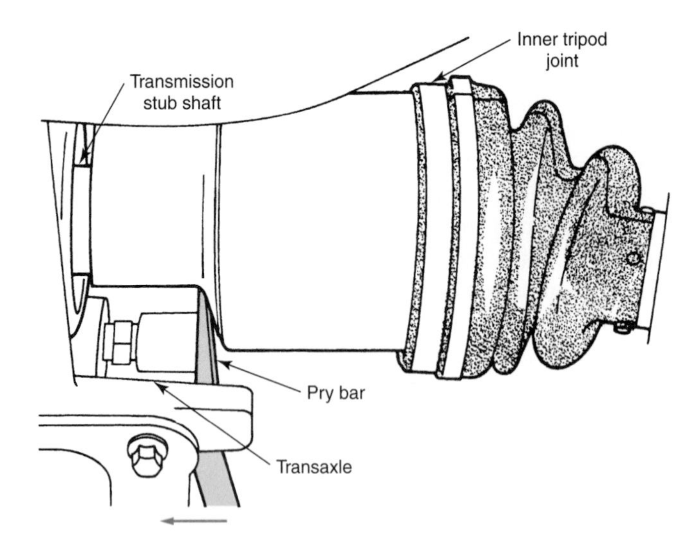

Figure 32-29. Removing the CV axle and joint from the transaxle stub shaft with a pry bar. Be careful not to damage the parts. (DaimlerChrysler)

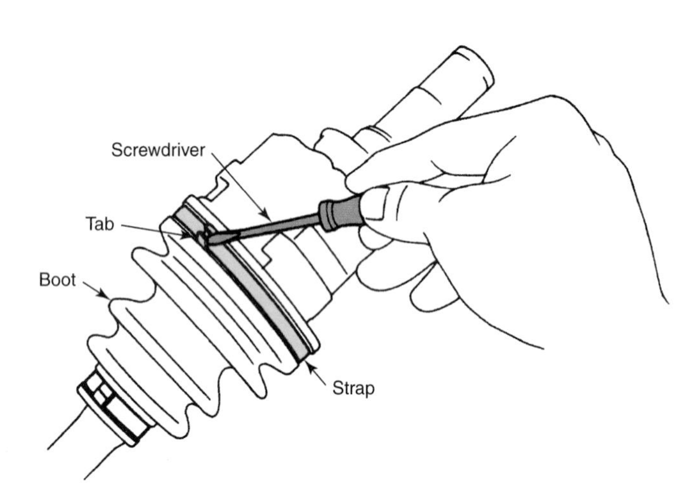

Figure 32-30. Removing boot straps by bending up and releasing the strap tab with a screwdriver. If the boot is going to be reused, do not damage it during removal. (Toyota)

Clean all external parts, except the rubber boot, thoroughly in solvent. Disassemble the joint and inspect the internal parts for wear or damage. The internal parts of a tripod joint can be accessed by sliding the joint apart. The balls in Rzeppa joints are contained in a cage. The cage must be tilted inside of the housing to remove the balls. **Figure 32-32** illustrates a tripod CV joint being disassembled after boot removal. **Figure 32-33** shows a disassembled Rzeppa joint. Once the CV joint is apart, clean it thoroughly.

Note: If the boot only is being changed, install the new boot and reverse the removal steps to complete the job. Be sure to clean the CV joint and check it for wear before reassembly. Thoroughly grease the joint before reinstalling the boot.

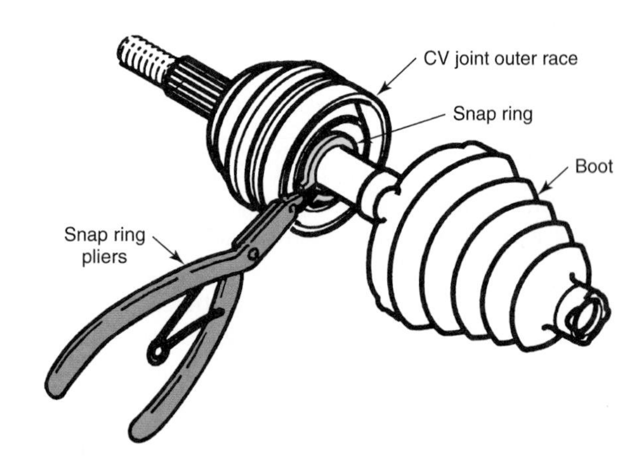

Figure 32-31. Removing a snap ring so the joint may be disassembled. Work in a neat and clean area. (General Motors)

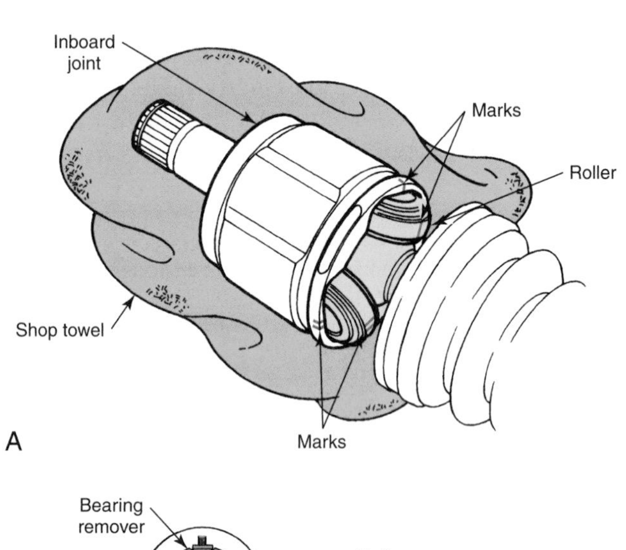

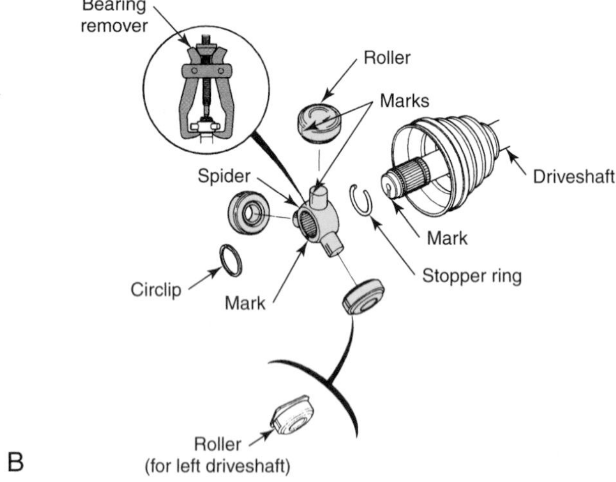

Figure 32-32. Disassembling a tripod CV joint. A—Mark components before disassembly. B—Remove parts as shown. (Honda)

CV Joint Reassembly

Replace internal CV joint parts as required. Although replacement parts are available, many technicians prefer to replace the entire CV joint.

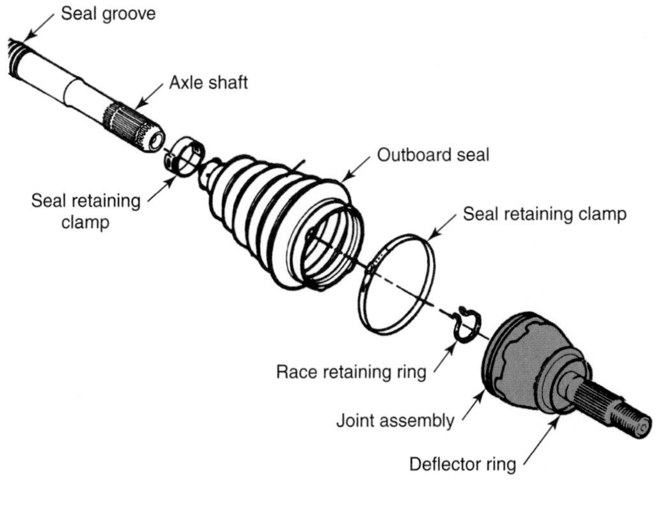

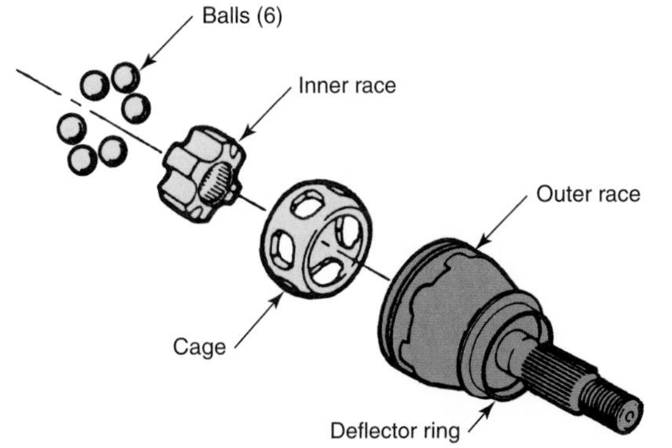

Figure 32-33. Disassembling a Rzeppa CV joint. (Saginaw Division of General Motors)

CV joints are usually assembled in the reverse order of disassembly. Be sure to replace all defective parts. Where required, lubricate the joint using the special CV joint lubricant provided. Align any scribe marks. Follow the manufacturer's procedures for reassembly.

Note: Install a new CV joint boot whenever the CV joint is disassembled. Pack any remaining CV joint lubricant into the boot before replacing the boot seal straps.

There are several methods of retaining the CV boot straps, **Figure 32-34.** If the straps are not properly installed, centrifugal force will throw all of the lubricant out of the CV joint.

CV Axle Installation

Reverse the removal procedure. Align the scribe marks. Lubricate where necessary. Be sure to follow the manufacturer's instructions carefully.

Caution: Be sure to torque all fasteners and double-check all areas. A failure in this area can lead to loss of vehicle control and occupant injury.

Servicing Axle Shafts

This section covers the servicing of the side axles of rear-wheel drive rear axle assemblies. The information presented here also applies to solid axles used on four-wheel drive vehicles.

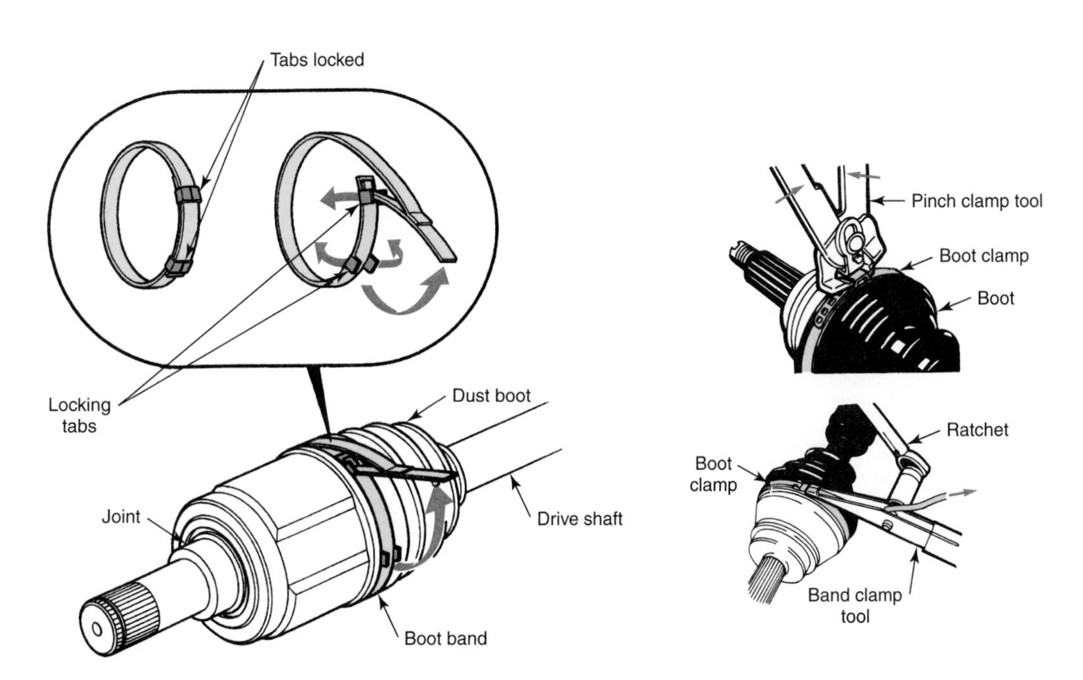

Figure 32-34. Installing and restraining boot straps with a special strap (band) and pinch tools. (Perfect Circle & Honda)

Modern Side Axles

The modern side axle assembly is a ***semifloating*** design, in which the axle drives, retains, and supports the wheel. A single ball or bearing cap wheel bearing is used at the outer end of each axle housing. The axle can be retained in the housing by a retainer plate or by axle shaft locks on the inner ends of the axles. These types of axles are also used on the front axle assembly of a four-wheel drive vehicle.

The differential carrier may be of the integral type (permanent part of the axle housing), **Figure 32-35.** It may also be a removable type, sometimes called a chunk or pumpkin. See **Figure 32-36.**

An exploded view of one type of independent front axle assembly for a four-wheel drive vehicle is shown in **Figure 32-37.** A cutaway of an assembled drive axle is shown in **Figure 32-38.** Study part names and relationships.

Axles for Independent Rear Suspensions

The rear axles used with independent rear suspensions resemble small drive shafts. The axle shafts are attached to the differential assembly output shaft and contain U-joints or CV joints. A typical joint is shown in **Figure 32-39.** Service for these axles is the same as for the Hotchkiss and CV axles discussed earlier in this chapter.

Flanged End Side Axle Removal

Flanged end axles are shown in Figures 32-35 and 32-36. To pull an axle, remove the wheel. Pull the brake drum off after unscrewing the small drum-retaining capscrew or the flat nuts threaded over the lug bolts.

Remove the nuts from the bearing retainer plate. If the design permits, pull the retainer plate outward far enough to reinstall one nut to hold the brake backing plate in place.

Attach a slide hammer puller to the axle flange. With a few sharp blows, pull the axle bearing free of the housing. Remove the tool and slide the axle from the housing. Refer to **Figure 32-40.**

If a nut was not placed on one backing plate bolt before axle removal, make certain the backing plate is not disturbed when the axle is pulled. Place a nut in position as soon as the axle is out. This is very important as the brake line can be bent, kinked, or weakened if the backing plate is moved. A typical housing end, backing plate, and axle assembly are illustrated in **Figure 32-41.**

C-Lock Axle Removal

On some rear axles, the axle is retained with an inner end lock, which is often called a ***C-lock.*** To remove such an axle, pull the wheel and brake drum. Drain the differential housing and remove the inspection plate. Remove the differential pinion

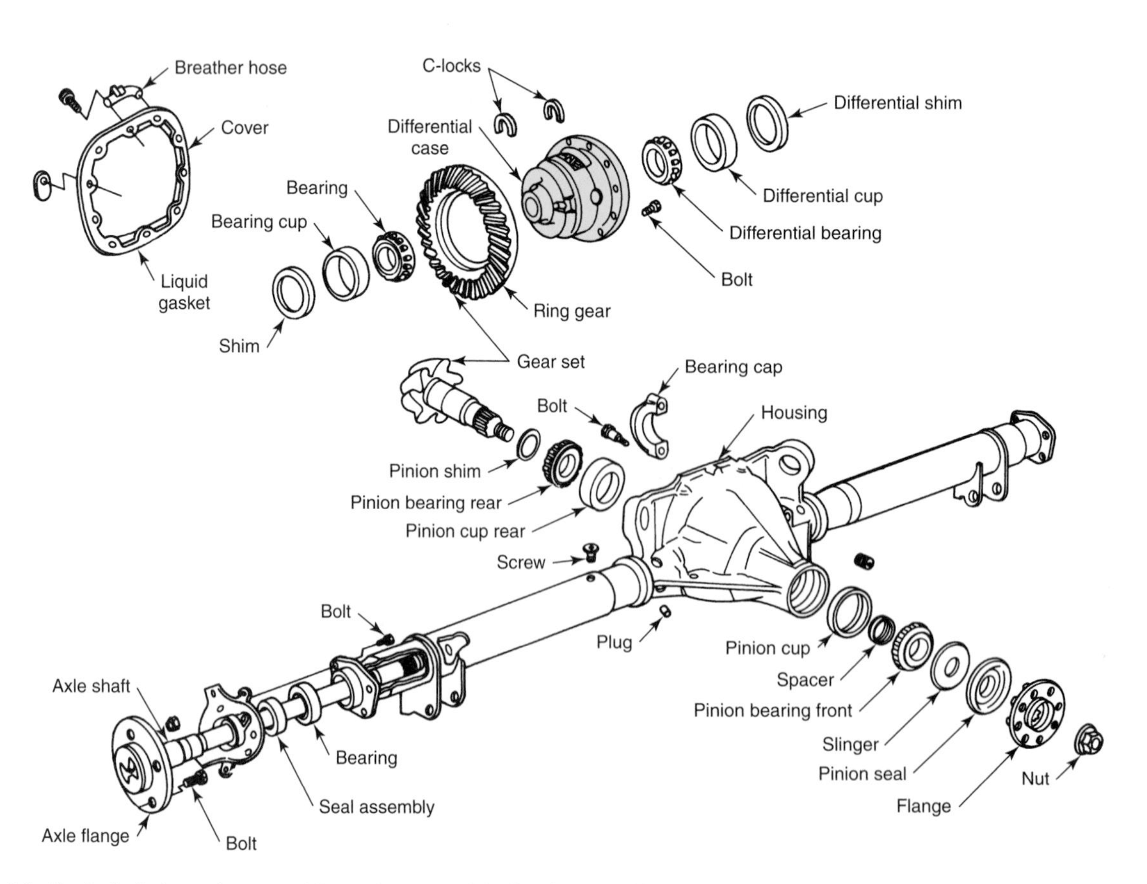

Figure 32-35. Exploded view of a rear-drive axle assembly. Study the different parts and relationships. (Ford)

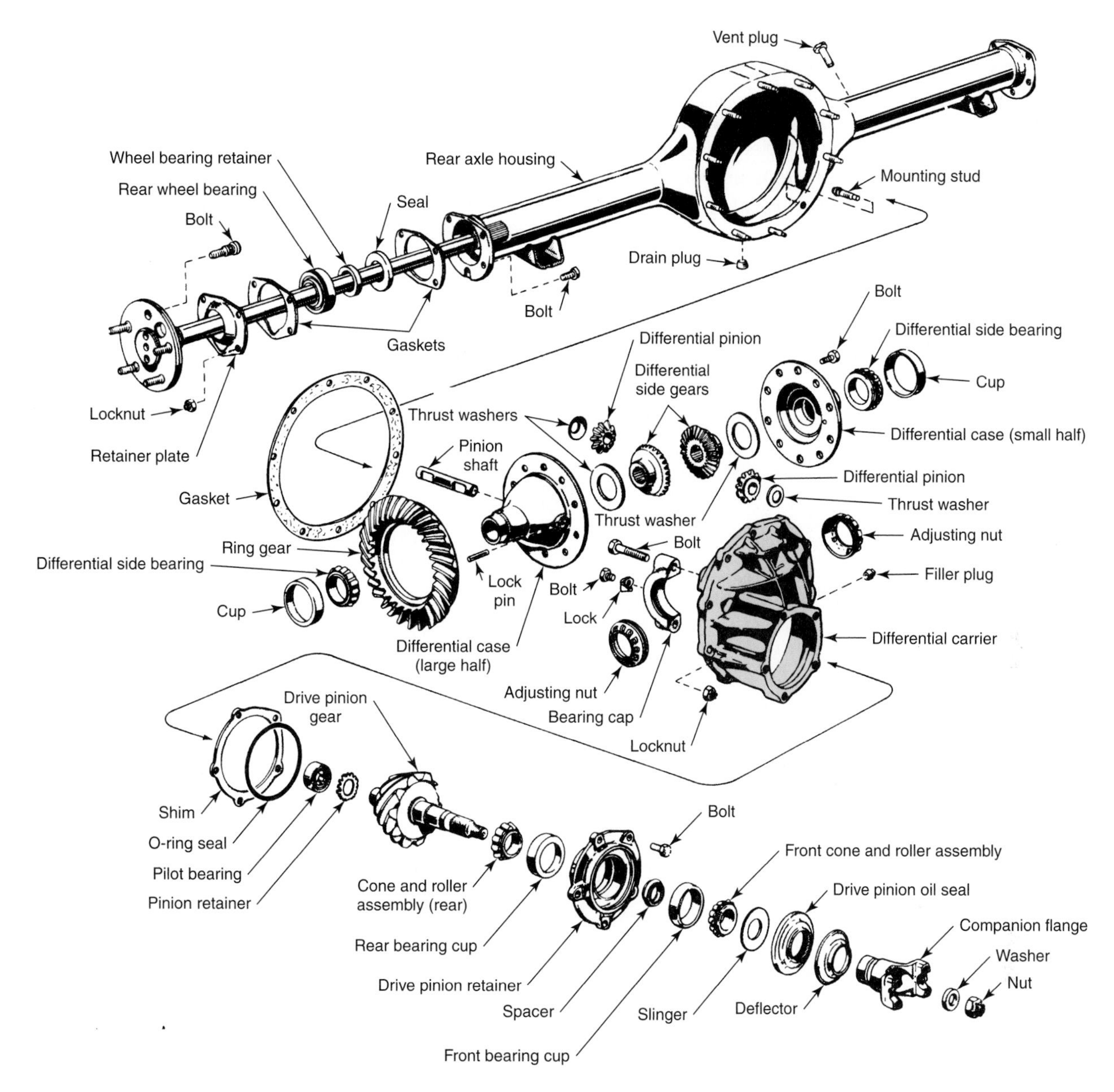

Figure 32-36. Drive axle assembly with a removable differential carrier. (Ford)

shaft (not the drive pinion shaft). Push the axle inward as far as it will go. This will free the C-lock from the recess in the axle side gear. Remove the C-lock. Withdraw the axle. See **Figure 32-42.**

Tapered End Axle Removal

On some older vehicles, the wheel hub or hub-drum assembly is attached to the axle by means of a taper and key, **Figure 32-43,** or by a taper, key, and splines, **Figure 32-44.** This requires the use of a heavy-duty puller to remove the hub.

Broken Axle Removal

If an axle has been broken, a short piece or stub may be left in the housing. You can use a powerful magnet on a long handle, a tapered spring on a handle (the spring grips the broken section when it is turned), or some other special tool to retrieve the segment.

Examine the break carefully by placing the broken ends together. If the break is clean (no missing pieces), drain and flush the housing thoroughly.

If small pieces are missing from the break, clean and flush the housing until all particles are removed. In some cases, complete disassembly of the differential and drive pinion assembly is required. Remember that a very tiny particle of axle shaft metal can ruin bearings and gears.

Axle Shaft Inspection

When an axle is removed, it should be carefully washed and blown dry. Do not wash sealed bearings. Inspect the splines for evidence of wear. Look carefully for indications of twisting. See **Figure 32-45.** Check the oil seal contact surfaces. Polish off any burrs.

The shaft can be placed on V-blocks or between centers to check for runout, bending, or other problems. **Figure 32-46** shows the axles being held between centers while checking for bends and flange runout with a dial indicator.

On flange-type axles, inspect wheel lugs and use a press to replace any that are broken or stripped. See **Figure 32-47.** Discard shafts showing signs of twisting, excessive wear, or runout beyond specifications.

Axle Oil Seal Replacement

Following axle removal, always install new oil seals. Some installations use two oil seals—one on the inside of the

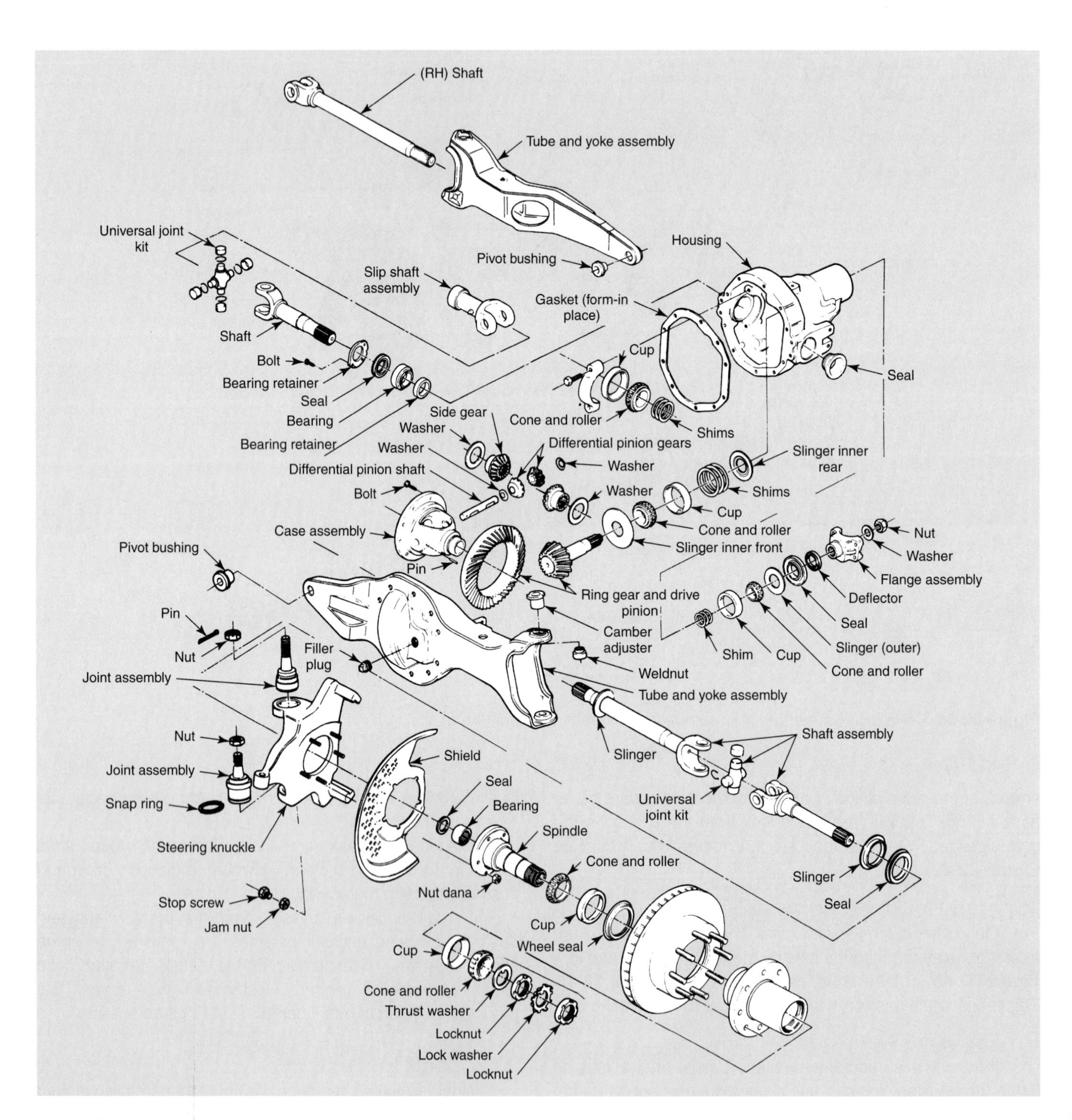

Figure 32-37. An exploded view of an independent suspension, front drive axle assembly used on one four-wheel drive vehicle. (Ford)

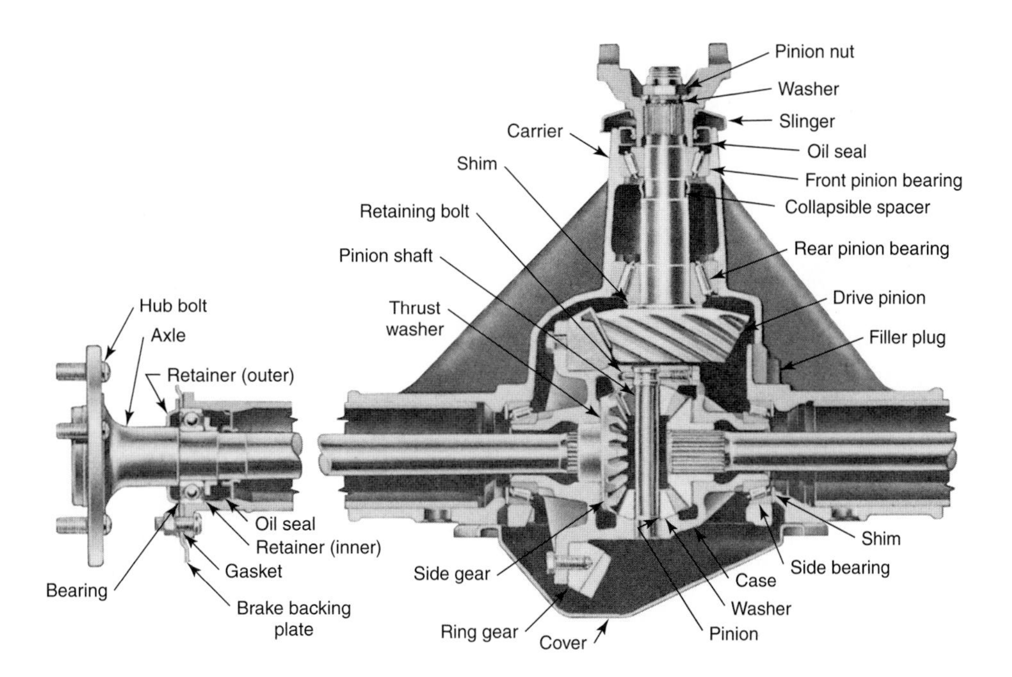

Figure 32-38. Assembled view of a drive axle. (General Motors)

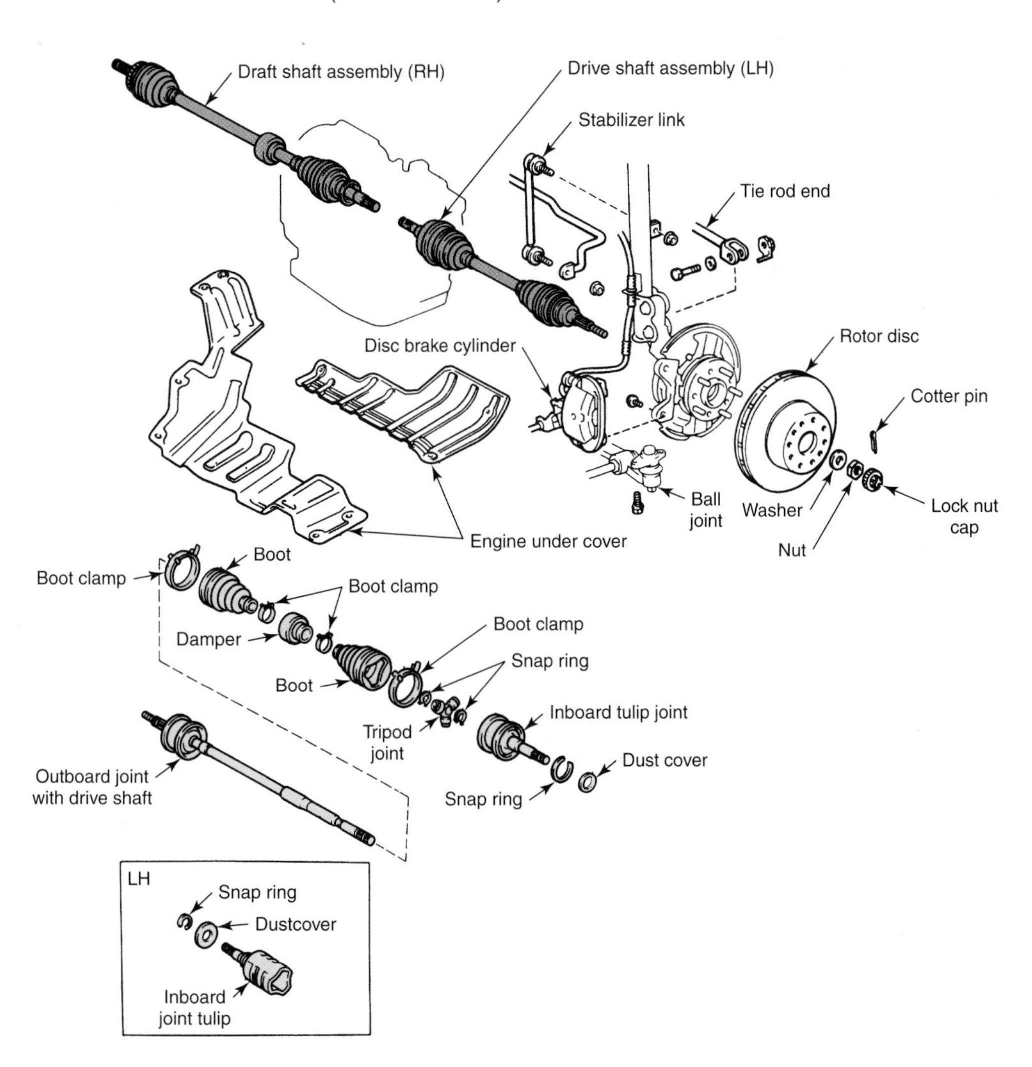

Figure 32-39. Exploded view of a rear-wheel drive, independent-suspension, drive axle arrangement. Study the axle and CV joint construction. (Toyota Motor Corp.)

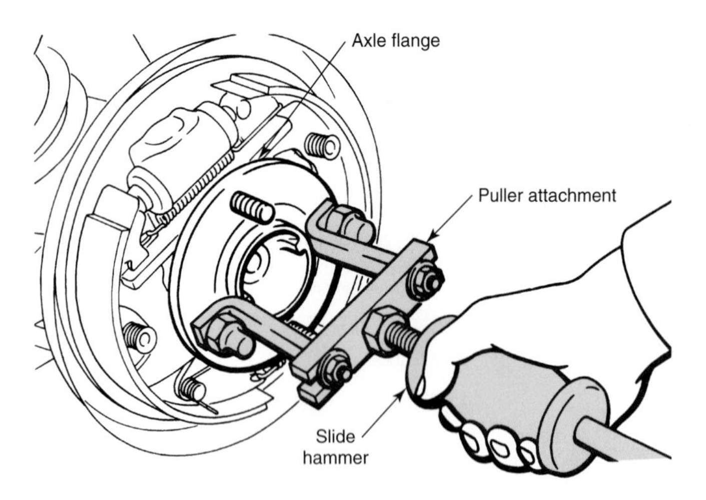

Figure 32-40. Using a slide hammer and puller attachment to remove a rear axle. (DaimlerChrysler)

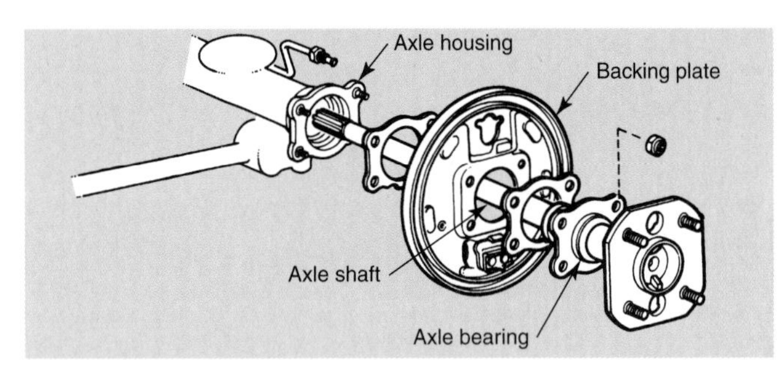

Figure 32-41. Typical rear axle assembly. (Toyota)

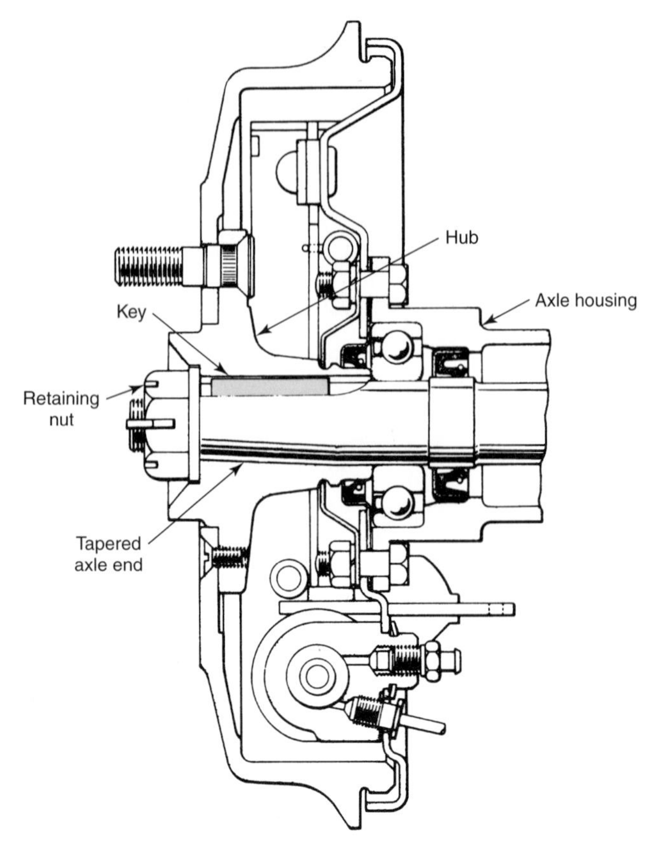

Figure 32-43. This hub is attached to the axle with a tapered axle end and key. The retaining nut holds the hub firmly against the taper and key. (British-Leyland)

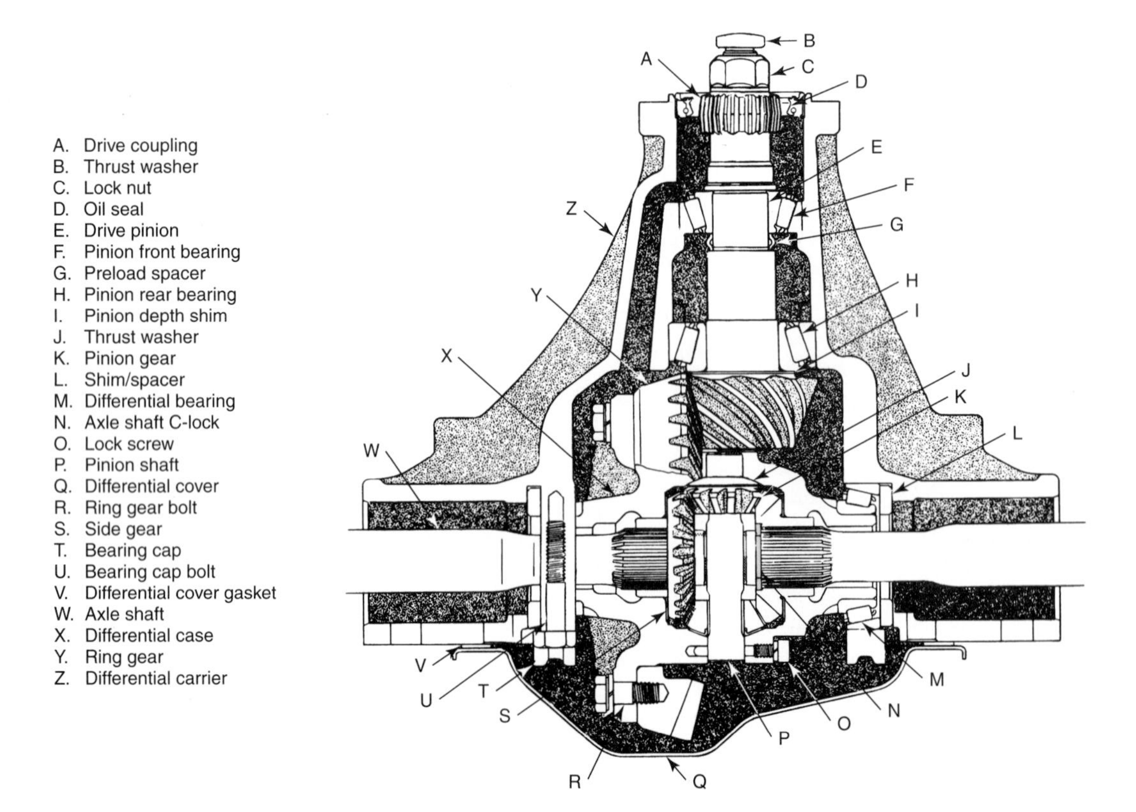

Figure 32-42. These axles are retained by C-locks (N) on the axle shaft inner ends. Note how the differential pinion shaft (P) keeps the axle ends in the outward position, thus forcing the C-locks into the recesses in the side gears (S). Also note the special drive coupling (A) used with the extension housing (torque tube) setup. (General Motors)

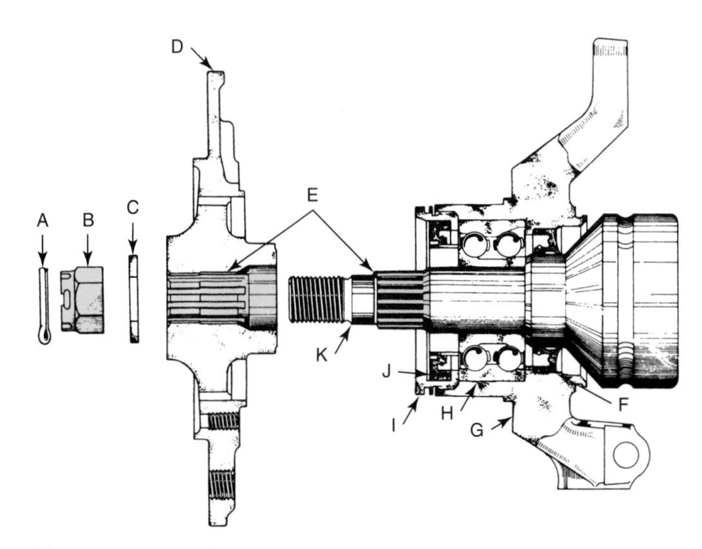

Figure 32-44. This hub is secured to the axle with splines and a retaining nut. A—Cotter pin. B—Castle retaining nut. C—Washer. D—Hub. E—Splines. F—Seal. G—Steering knuckle housing. H—Double row ball bearing. I—Nut. J—Seal. K—Axle (drive). (Saab)

Figure 32-45. Axle damage. A—Axle has been twisted. B—Splines have step wear (worn thin where axle engages gears). (Isuzu)

bearing and the other on the outside. The bearing is either sealed for life or it depends upon periodic applications of wheel bearing grease. If the wheel bearing is lubricated by the differential lubricant, only an outer seal is used.

A slide hammer puller with a hook nose is very handy for removing the housing inner seal. When pulling the seal, avoid scoring the housing seal counterbore, **Figure 32-48.**

Clean the seal counterbore. Remove nicks and burrs. If the seal is made of leather, soak it in light oil for 30 minutes. Coat the outer seal edge with nonhardening sealer. Apply gear oil to the seal lip. Using a suitable tool, drive the seal into

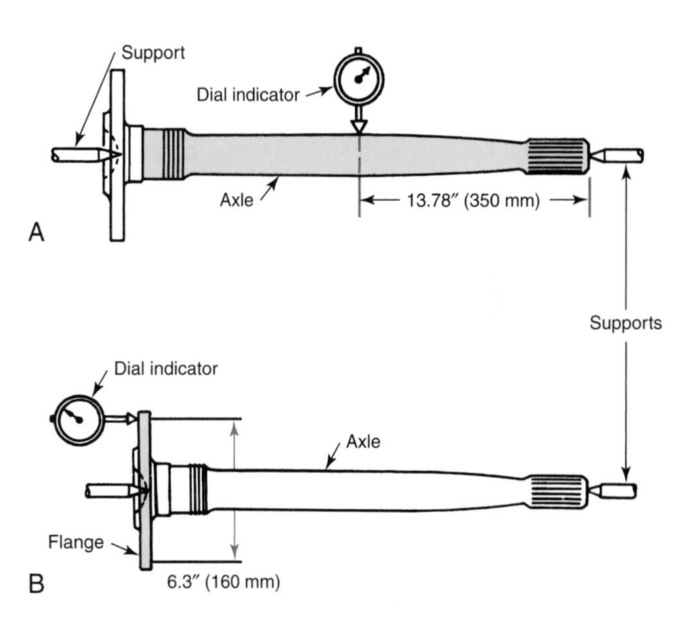

Figure 32-46. A—Checking axle for bend with a dial indicator. B—The indicator is placed against the flange to measure flange runout. Note that the axles are supported on tapered centers. (General Motors)

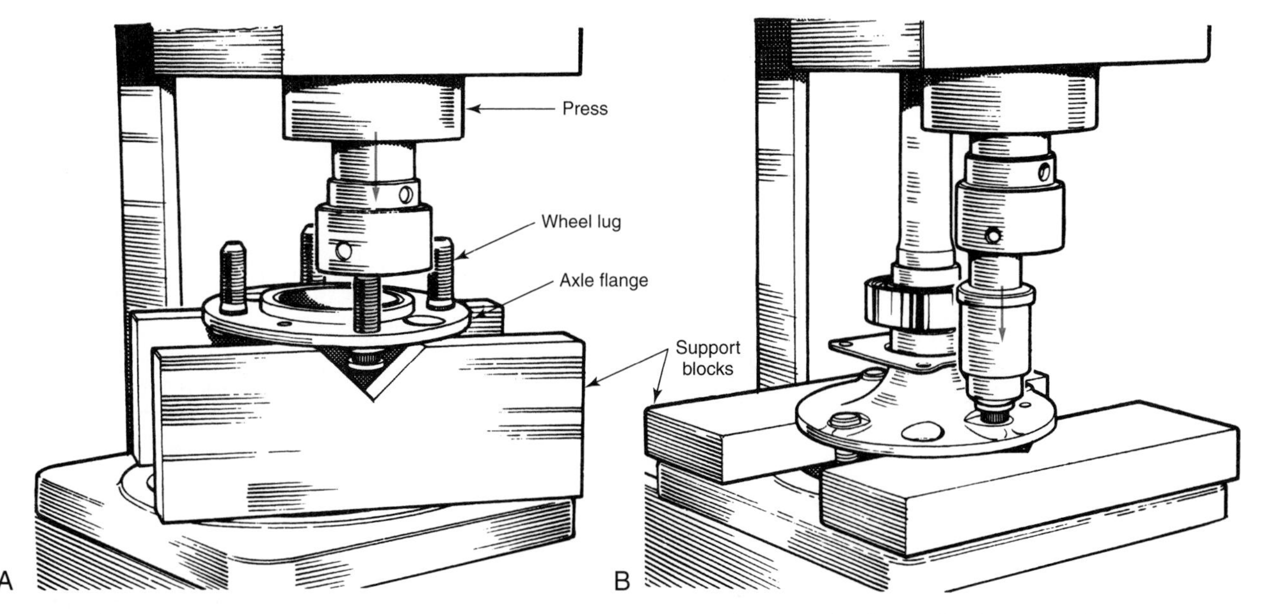

Figure 32-47. Replacing a damaged wheel lug. A—The press is forcing the lug from the flange. B—A new lug is pressed into position from the back of the flange. (DaimlerChrysler)

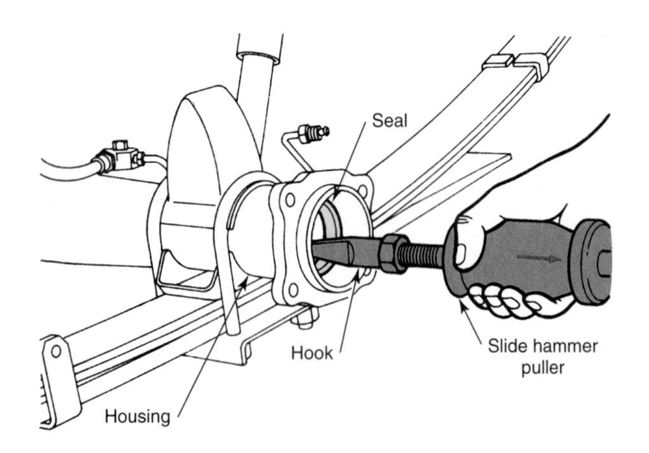

Figure 32-48. Pulling an axle housing oil seal. (DaimlerChrysler)

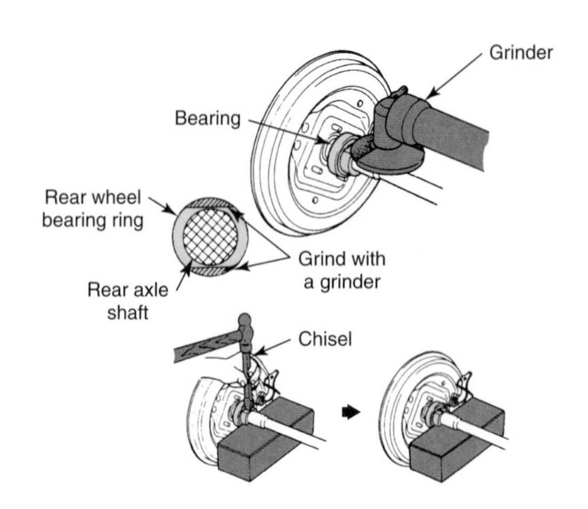

Figure 32-50. Notch the bearing ring with grinder. Then, crack the ring with a hammer and chisel.

place. The seal lip must face inward. Drive the seal squarely and to proper depth. See **Figure 32-49.**

The outer seal may be built into the bearing itself. It may also be incorporated in the oil seal retainer.

Rear Wheel Bearing Replacement

Where a ***bearing retainer ring*** is used, place the axle so the ring rests on a solid support, such as a vise. Slide a protective sleeve up to the ring. Notch the ring with a grinder and break it away with a hammer and chisel. See **Figure 32-50.** Do not grind completely through the ring, as this would damage the axle. Make certain a protective sleeve is used.

Slide the retainer ring from the shaft. Set the axle up in a press (a puller can also be used) so the bearing may be grasped while the axle is forced through the bearing. Grasp the bearing, not the retainer plate. Make certain the axle flange is clear of the puller and press.

Note: Use a bearing cover, Figure 32-51. Also use protective goggles to prevent injury from flying parts in the event the bearing explodes under pulling pressure.

Clean the axle and coat the bearing face and retaining ring contact surface with a film of lubricant. Use the type used for ball joints.

Put the retaining flange into position on the axle. Slide the new bearing on the axle. The bearing must face in the direction specified. Place the axle in a press so the bearing inner race is supported. Do not install the bearing by exerting pressure on the outer race.

Before pressing the bearing into place, check the retainer plate for proper positioning and the bearing for correct installation. Press the bearing on the shaft to the specific distance from the flange surface.

Slide the retaining ring into position. It must face as recommended. Set the axle in the press so the retaining ring is well supported. Support the ring around its full circumference, not in just two spots. Force the axle through the ring until the ring just contacts the bearing. See **Figure 32-52.** Never try to press the bearing and retaining ring on at the same time. Always use a new retaining ring.

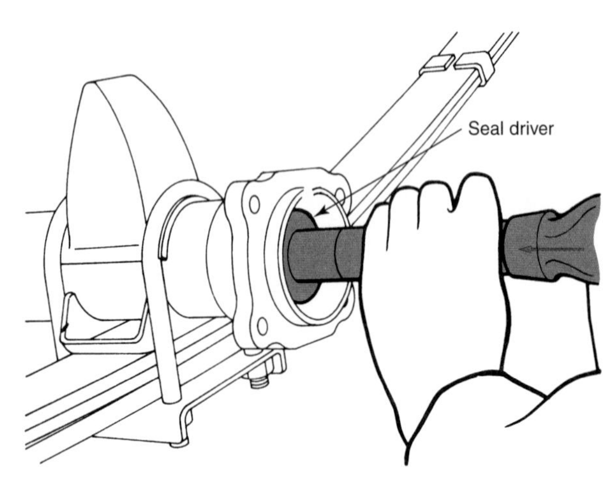

Figure 32-49. Driving an axle seal into position with a seal driver. (DaimlerChrysler)

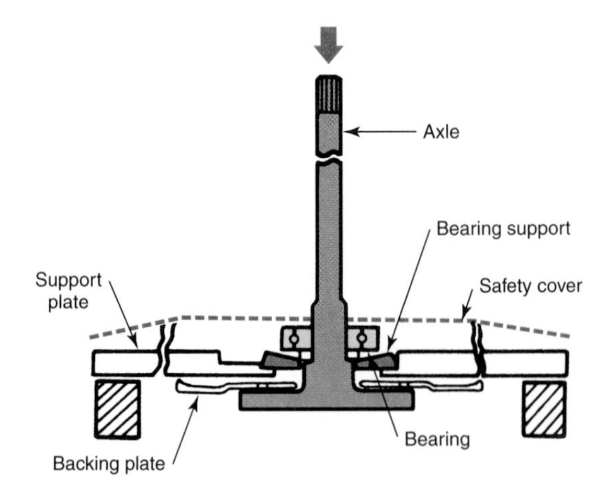

Figure 32-51. Axle bearing being removed. Support the bearing to prevent damage. Place a safety cover or shield over the bearing to protect yourself and others. Bearings under pressure can fracture and come apart with great force! (Mazda)

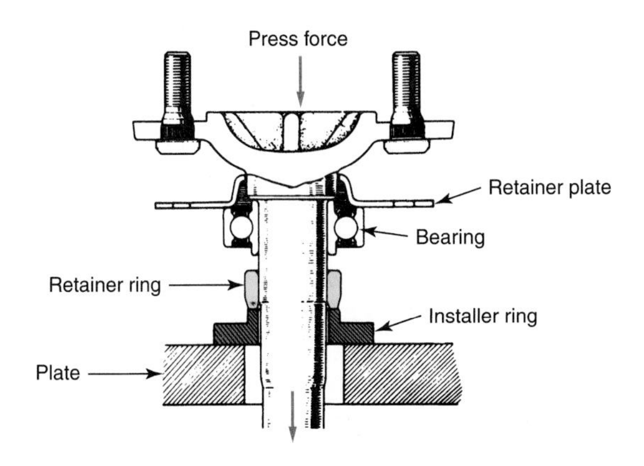

Figure 32-52. Pressing a bearing-retainer ring into position. Note how the installer ring supports the retainer ring. (Toyota)

One axle setup has the retainer plate so close to the bearing the bearing cannot readily be grasped for pulling. When this is the case, cut off the roller cage outer section. Grind off a section of the inner race. Use a sleeve to protect the axle. Then, remove the rollers one at a time.

Cut out the remaining portion of the cage and pull the outer race from the axle. The inner race may now be grasped for pulling. Some bearings are retained with a locknut. Remove the bearing carefully on these to avoid possible thread damage.

Axle Installation

An axle must be clean. If the wheel bearings were damaged, clean and flush the housing. Place a new gasket on the housing end. Place a new O-ring gasket on the bearing outer face.

Coat the length of the axle with clean gear oil. Lubricate the bearing counterbore in the housing. Lubricate the bearing if required. Slide the axle into the housing. Pass the splined end and axle length through the oil seal very carefully to avoid damage.

As the shaft is passed into the housing, support its weight. Engage the splined end in the axle side gear. Align the bearing and start it into the housing counterbore. Sometimes the axle is longer on one side than the other. Make sure the axle is installed on the correct side.

Align the retainer plate so the oil drain section, **Figure 32-53,** is over the drain hole in the backing plate. The gasket must be aligned to expose the oil drain hole. Do not plug the hole with sealer. If the oil drain is not aligned, oil leakage past the bearing will pass into the brake assembly. Oil can then reach the shoes and drum or pads and rotor, ruining the linings.

When installing an axle with a tapered roller bearing and shims, make certain the shims are clean and in the proper position. Use gaskets and sealer where required. On axles with the hub attached to a tapered end, install the shims, backing plate, and an outer seal. Use new gaskets and coat with sealer. Tighten the fasteners. Attach the brake lines and bleed the brakes (see Chapter 34, Brake Service).

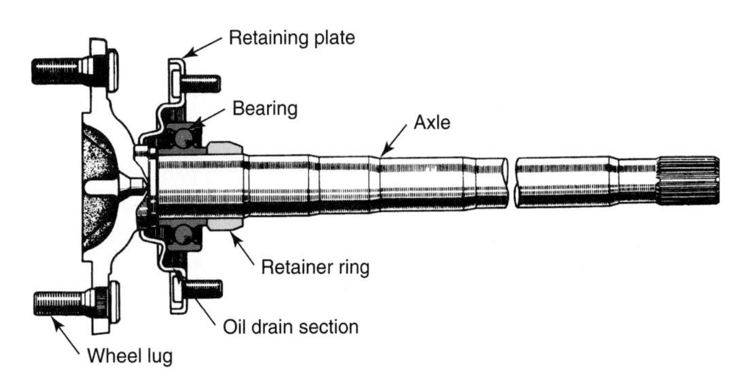

Figure 32-53. Bearing and retainer ring installed properly on the axle. Note the position of the oil drain section. (Toyota)

Check Axle Endplay

After the axle retainer plate fasteners are torqued, check the axle for proper ***endplay.*** Attach a dial indicator firmly to the axle flange and adjust the indicator stem against the backing plate. Pull the axle in and out and note the total reading. This will be the axle endplay.

Some axles are designed so that endplay is controlled by the play in the wheel bearings (ball type), by shims (tapered roller), or by an adjusting nut on one end of the axle housing (tapered roller bearings). Follow the manufacturer's specifications. See **Figure 32-54.**

Ring and Pinion, Differential Unit Service

The ***ring gear*** and ***pinion gear*** used in modern rear axle assemblies are referred to as ***hypoid gears.*** This refers to the placement of the pinion gear below the center of the ring gear. This is done to reduce the height of the driveline and to allow for a lower driveline hump in the floor of the passenger compartment. In addition to the ring and pinion, the rear axle

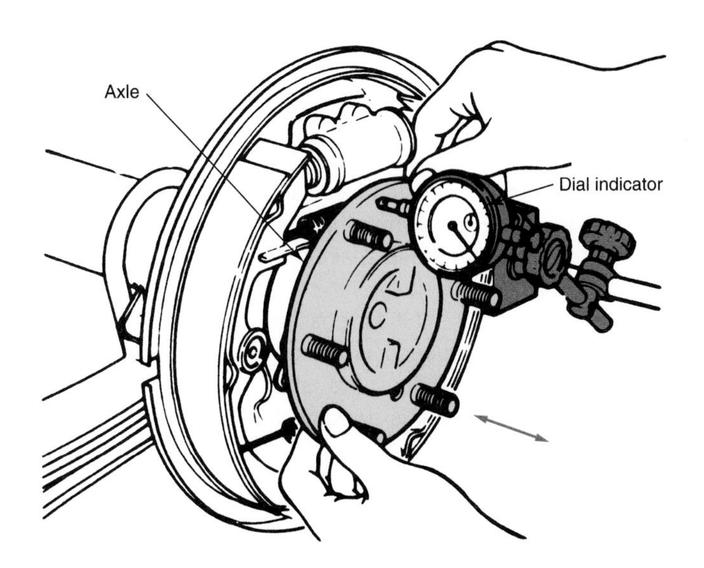

Figure 32-54. Checking axle endplay. Move the axle back and forth while watching the indicator. (Nissan)

assembly contains the ***differential gears.*** These gears allow the drive wheel on one side to rotate at a different speed than the drive wheel on the other side when making turns.

The following sections detail the basic troubleshooting and repair of the rear axle assembly, including ring and pinion removal and replacement and setting of pinion depth, gear backlash, and ring gear-pinion tooth pattern. The information here also applies to the front wheel axles used on four-wheel drive vehicles.

Troubleshooting

Drive axle sounds can be among the hardest to isolate and cure. Many other sounds can be mistaken for noise in the rear axle assembly. Certain body types, such as station wagons, may transmit more noise to the passenger compartment.

Road Testing

Before road testing for drive axle assembly problems, check the lubricant level. Add lubricant if required. Check the tires for a saw-tooth wear pattern or for a mud and snow tread design. Both can produce distinct rumbles, growls, or other noises. Bring the tire pressure to specifications.

Drive the vehicle far enough to warm the lubricant. Then check action during drive (acceleration), cruise (engine driving enough to maintain vehicle speed), float (engine neither driving nor holding back; vehicle speed slowly decreases), and coast (accelerator released; engine braking) conditions.

Noises Often Mistaken for Drive Axle Sounds

If the tires are suspected, inflate both front and back tires to 50 psi (345 kPa). If the tires are responsible, the noise should be noticeably altered. Reduce pressure to the recommended level immediately following road test.

Worn or improperly adjusted front wheel bearings can produce sounds similar to those caused by a defective rear axle. Raise the front end and shake the wheel to detect looseness. Spin to test for roughness. During the road test, noise produced by the front wheel bearings can usually be reduced or altered by pressing on the brake while maintaining vehicle speed.

Certain road surfaces produce distinct sounds. By driving on a different surface, these sounds can be quickly identified. The transmission can also produce noises easily confused with typical drive axle problems.

Engine noise is occasionally mistaken for rear axle sounds. With the vehicle standing still, operate the engine at the approximate RPM at which the sound was noticed during the road test. If the sound is again heard, it will obviously not be the drive axle.

Check the drive shaft for possible unbalance or wear. When road testing, sounds produced by tires, front wheel bearings, road surface, or drive shaft unbalance will not change when the vehicle is switched from drive to coast (or vice versa).

Relating Drive Axle Sounds to Specific Parts

Bearing sounds tend to be low-pitched whines or growls, fairly constant in pitch, that extend over a wide range of road speeds.

Sounds produced by gears are apt to be of variable pitch and most pronounced in specific speed ranges or pull (drive, cruise, float, coast) conditions.

Defective rear wheel bearings produce a continuous growl that is the same in all pull conditions. Sudden turns to the right or left will increase or decrease the load on a given bearing and will alter the sound somewhat. Jack up the rear of the vehicle and turn the wheels slowly while "feeling" for any signs of roughness. Chipped bearings can produce a clicking sound.

Ring and pinion noise will usually be related to a specific pull condition. If it shows during drive, it will probably disappear during coast.

Differential pinion and side gear noises are noticeable on turns, as there is little movement of these gears in straight-ahead driving. Pinion bearing noise is low pitched and continuous.

When drive pinion bearings or differential case side bearings are worn, the ring and pinion backlash and tooth contact pattern is altered. This can produce a compound noise made up of bearing growl and gear whine.

A low-speed squeal can be caused by the pinion oil seal. A clanking sound occurring during acceleration or deceleration may be caused by worn universal joints, worn transmission, excessive ring and pinion backlash, worn pinion and axle side gear teeth, or worn drive pinion shaft.

Service Operations

The following sections cover the servicing of various rear axle and differential assembly parts.

Replacing Pinion Shaft Seal and/or Flange

Disconnect the drive shaft as explained above and move it out of the way. If required, measure the pinion shaft bearing preload by using an inch-pound torque wrench.

With the rear wheels free and the emergency brake released, tap the brake backing plates to free the brake shoes from the drum. Spin each wheel several times to make certain there is no drag. One method of ensuring complete freedom from brake drag is to remove both rear wheels and brake drums.

Use a torque wrench with the proper socket to turn the drive pinion shaft through several complete revolutions. Note the inch pound reading during turning. Write this down. See **Figure 32-55.**

Scribe Shaft, Nut, and Flange

Scribe a line starting on the pinion shaft end and running along the threads up to and partway across the end of the nut. Using a punch, prick the companion flange in line with the scribe mark. Note the number of threads exposed beyond the nut. This procedure will ensure correct reassembly of all parts. If a new companion flange will be installed, it will be unnecessary to scribe and punch the mark.

Use a Special Holding Tool

Use a pinion companion flange holding tool to keep the flange from moving while the retaining nut is removed. Do not

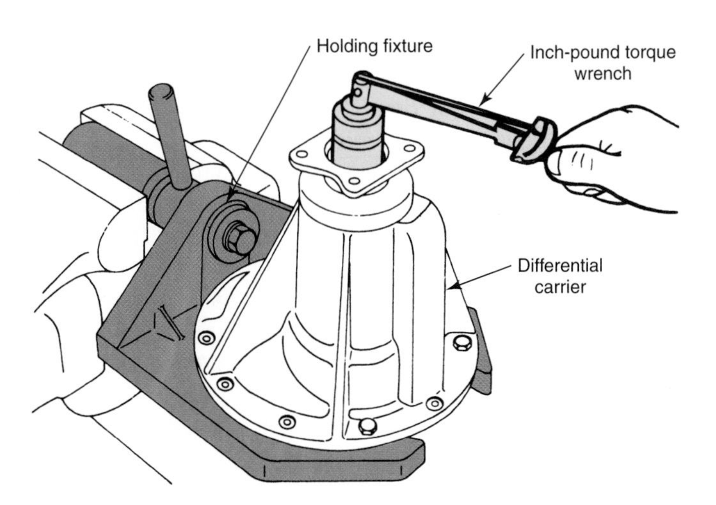

Figure 32-55. Measuring the torque required to turn the pinion shaft. Note the carrier holding fixture. (DaimlerChrysler)

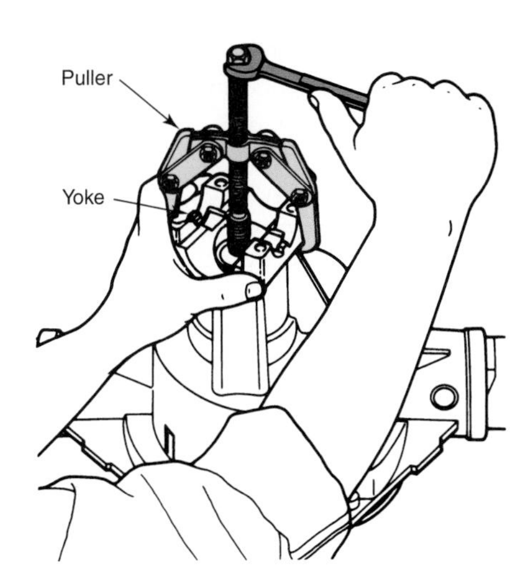

Figure 32-57. Using a puller to remove the differential pinion yoke or flange. (Ford)

try to use a pry bar or a large crescent wrench to hold the flange. To prevent damage, use a tool designed for the job, **Figure 32-56.** Following pinion nut removal, use a pinion flange puller to remove the flange. One such puller is shown in **Figure 32-57.** Do not try to pound the flange off.

Seal Removal and Installation

Note the depth at which the pinion oil seal is seated. Next, remove the pinion oil seal by prying it out of the carrier or by using a threaded puller as in **Figure 32-58.**

Wipe out the seal recess thoroughly. Make certain the new seal is the correct size and type. Wipe off the outer diameter of the seal and coat the seal with a thin layer of nonhardening sealer. With a proper seal installer, drive the seal (lip facing inward) into place. Drive the seal squarely and to the correct depth. Some manufacturers recommend installing the seal with a special installer that forces the seal into position by tightening the pinion nut. This eliminates the need for hammering and the possibility of distorting the seal.

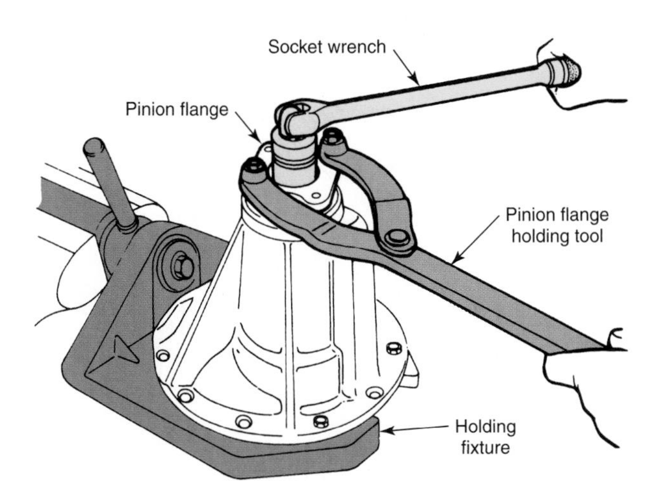

Figure 32-56. Using a pinion flange tool to prevent flange rotation while removing the retaining nut with a socket wrench. (DaimlerChrysler)

Check the Pinion Flange

If the old pinion shaft flange is to be reused, wash it and blow it dry. Inspect the seal contact surface. If necessary, remove any nicks or burrs. If the seal surface is worn or if the splines show evidence of wear, discard the flange. Check the new flange for nicks and burrs. Also, make sure it is the proper size and shape.

Install the Pinion Flange

Lubricate the seal lip, the flange seal contact surface, and the splines with gear oil. Align the punch mark on the flange with the scribe mark on the shaft. Start the flange into place. If a new flange is used, alignment is not required. If it is difficult to get the flange on far enough to start the retaining nut, use a puller. Never pound on the flange, as it can cause serious damage to the bearings and ring-and-pinion gears.

Install the washer and the pinion shaft nut. Place a small amount of lubricant on the face of the washer and on the nut threads. Use a new nut and washer if specified.

Grasp the flange with a holding tool and tighten the nut. When the nut starts to tighten the flange in place, rotate the

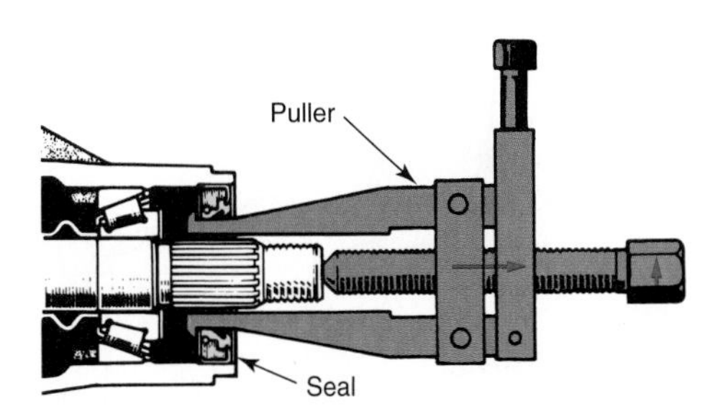

Figure 32-58. Removing the pinion oil seal with a special puller. (Toyota)

pinion shaft a few times to make sure the bearings are seated. Then, check the ***preload*** as outlined in the rear axle overhaul section.

Note: Pinion bearing preload must be correct. Follow the manufacturer's specifications.

Connect the universal joint. Inspect the lubricant level in the axle housing.

Ring and Pinion, Differential Removal

Drain the housing. Remove both axle shafts. Disconnect the drive shaft. Check drive pinion shaft preload. If the differential carrier is the removable type, unscrew the fasteners and remove the carrier. Watch out; it is heavy. Place the carrier in a repair stand, **Figure 32-59.**

If the carrier is the integral type, it may be necessary to remove the entire housing. Check the manufacturer's manual. Repair stands designed to handle the entire housing are available.

Mark Parts before Removal

Before removing the differential case side bearing caps, make certain each cap and adjusting nut (where used) is marked. The caps and carrier are factory marked in **Figure 32-60.** If no marks are visible, mark the caps and carrier with a scribe or prick punch. See **Figure 32-61.**

It is also good practice to check the backlash between the ring and pinion gears before differential case removal (see the section on adjusting backlash later in this chapter). Tooth contact pattern and bearing preload can also be checked.

Remove the bearing-cap fasteners and rap the caps to remove them. Do not mar the cap parting surface. If the differential case uses mechanical adjusters as shown in **Figure 32-62,** the case will lift out readily.

If the shim is the adjusted preload type, use a couple pinch bars or a differential puller to force the case out of the housing while in the repair stand. Tie each shim pack together and identify. Do not shift shims from one side to the other, **Figure 32-63.** Do not drop the side bearing outer races. Identify the races to prevent mixing them.

A special adapter may also be used to remove the case. The adapter is attached with two ring gear cap screws. A slide hammer puller is then connected to the adapter. The case is withdrawn with a series of sharp blows.

Another technique for removing the case involves using a differential housing spreader, **Figure 32-64.** The spreader is installed as directed, and the turnbuckle is expanded. Never expand the housing more than specified. Consider a 0.020″ (0.51 mm) expansion the absolute limit. Following expansion, remove the case and immediately release the spreader.

Differential Disassembly, Inspection, and Repair

Use a suitable puller to remove the differential case side bearings (if required), **Figure 32-65.** Tie the side bearings to the outer races or cups. Free the ring gear by removing the fasteners. Tap the gear free of the case. Drive out the pinion

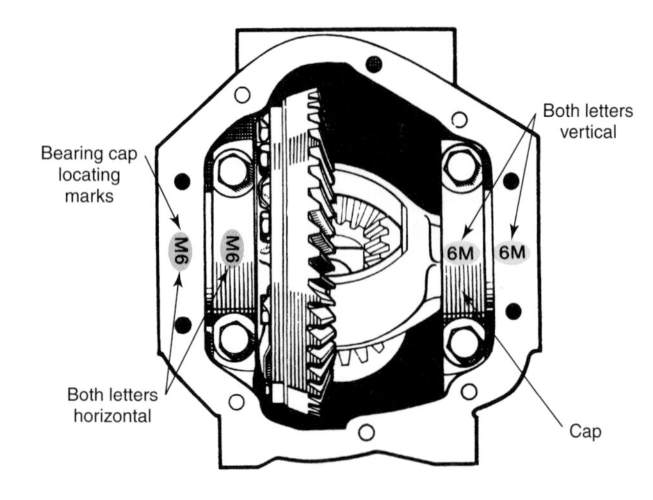

Figure 32-60. These differential case side bearing caps are factory marked for proper positioning.

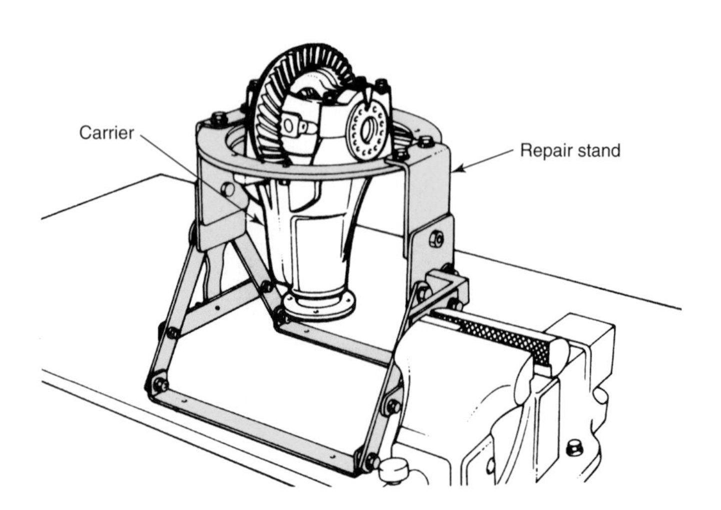

Figure 32-59. Differential carrier bolted to a repair stand.

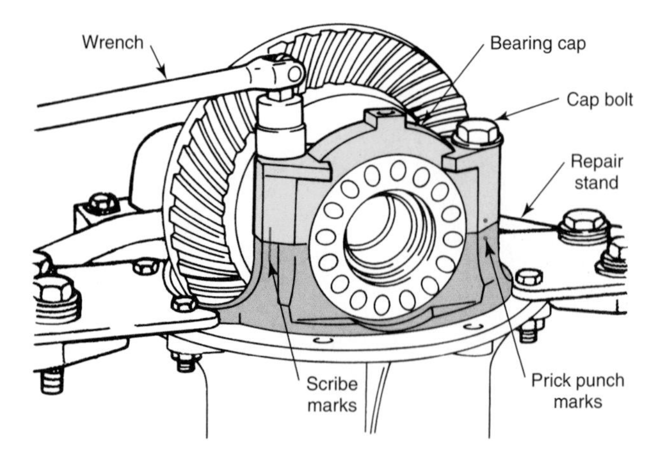

Figure 32-61. Mark the caps or carrier before removing the caps. (General Motors)

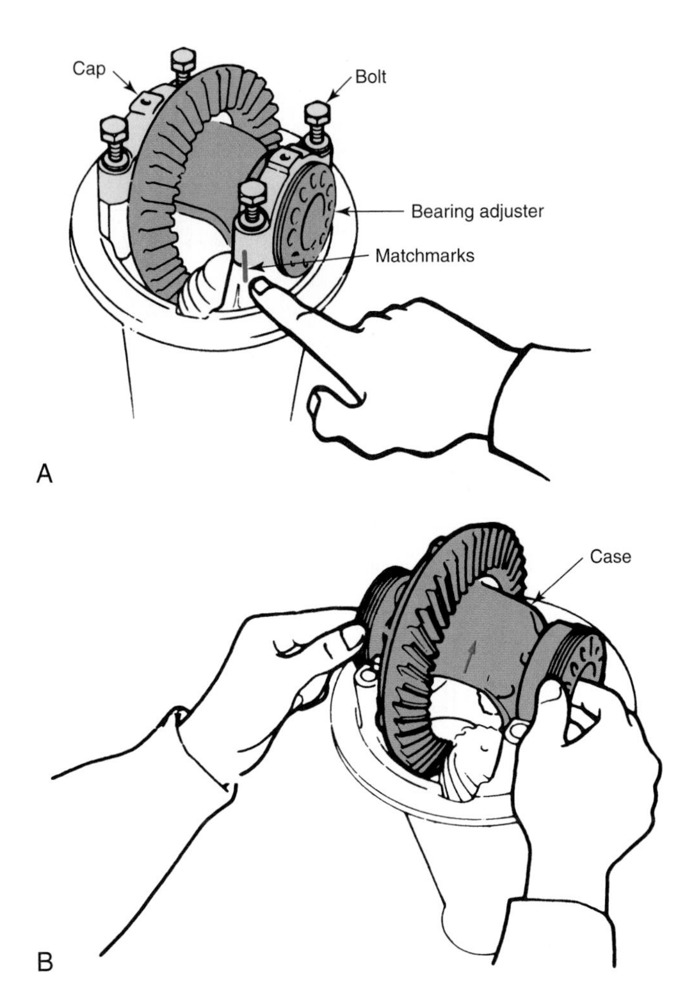

Figure 32-62. A—Placing match marks on the housing and bearing caps before removal. Loosen the bolts and remove with caps. B—The differential case and bearing should lift right out. (Toyota)

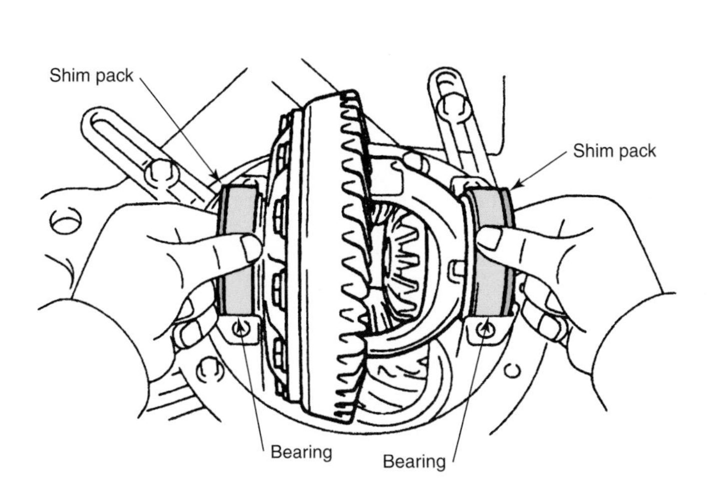

Figure 32-63. Keep the shim packs from each side with the bearings as you remove the case. Tie the shim packs and bearings from each side together. Mark them left and right. (Toyota)

gear lock pin and remove the pinion shaft, **Figure 32-66,** or remove the pinion shaft lock screw, **Figure 32-67.**

Remove the pinion gears. Wash all parts and blow them dry. Inspect the pinion and axle side gears for excessive wear or chipping. Check the ring and drive pinion gear for scoring, chipping, and other damage.

Check Differential Case for Wear and Runout

Examine the case side bearing contact surfaces. They must be in perfect condition, with no sign that the bearing inner races have turned on them. Thrust washer surfaces inside the case must be smooth and free of excessive wear.

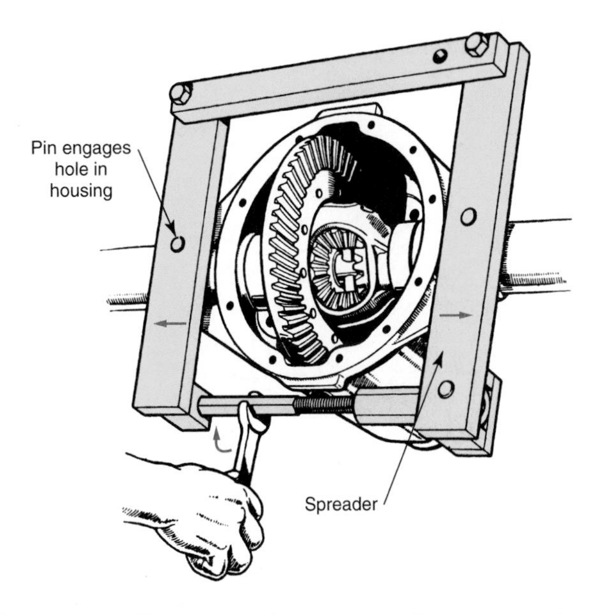

Figure 32-64. Using a housing spreader. Never expand housing more than 0.020″ (0.51 mm).

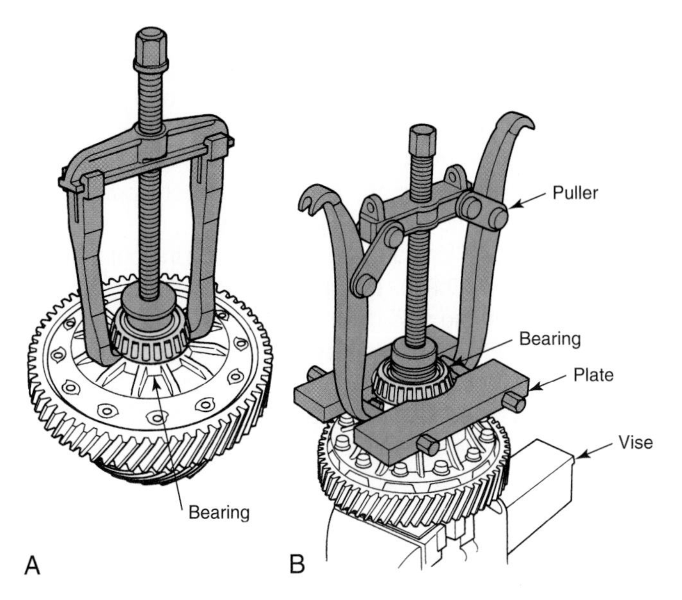

Figure 32-65. Removing the differential case side bearings. A—Using a two-leg puller. B—Using a two-leg puller and plate. (Honda)

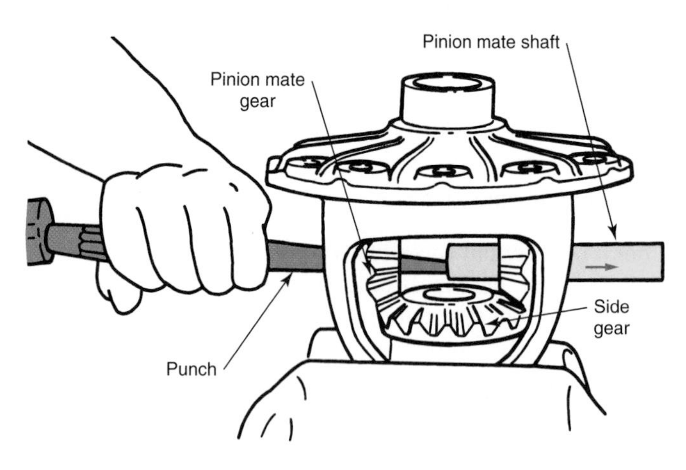

Figure 32-66. Removing the pinion mate shaft with a hammer and punch. (DaimlerChrysler)

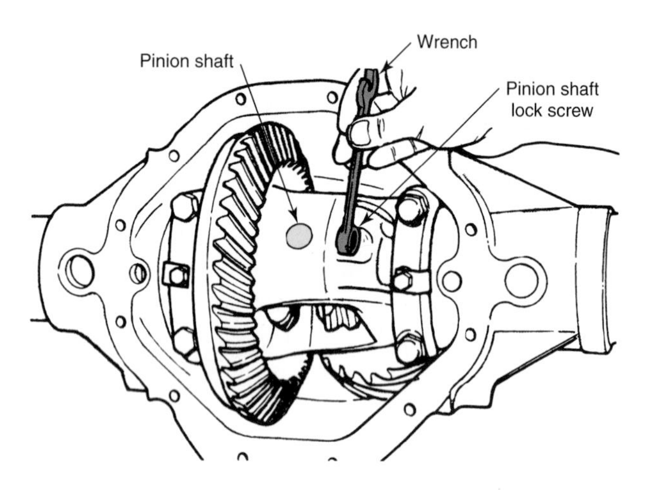

Figure 32-67. Removing the pinion-shaft lock screw with a wrench. (DaimlerChrysler)

Place the case in a set of V-blocks and check the ring gear flange-attaching surface with a dial indicator. Runout must be within specifications, **Figure 32-68.** If new side bearings are required, lubricate the case contact surface. If required, install shims between the case and the bearing. Drive or press the bearings into position. Apply force to the inner cone, not to the rollers, **Figure 32-69.**

If the side bearings are replaced, replace the outer cones also. Lubricate the case, thrust washers, pinions, and axle side gears. Place the side gears and washers into position in the case. "Walk" the pinion gears around the axle side gears until aligned with the shaft hole. Insert the pinion shaft, and install the lock pin. Rotate the gears a few times. Next, check the clearance between the side gear and the thrust washer, **Figure 32-70.**

Note: If inner axle-end C-locks are used, it will be necessary to remove the pinion shaft and spacer block to install the axles.

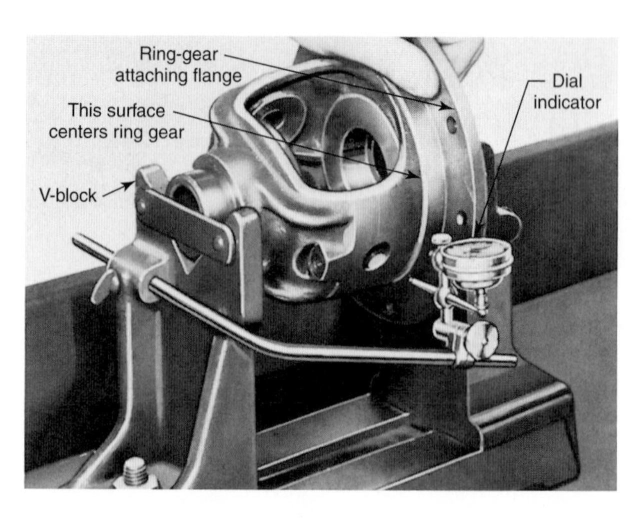

Figure 32-68. Checking ring-gear centering-surface runout. The attaching flange must also be checked. (General Motors)

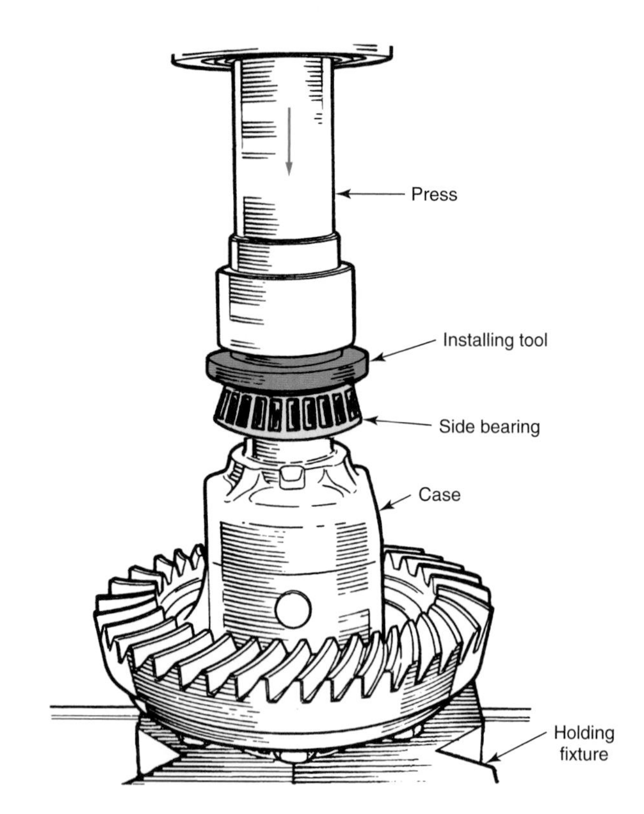

Figure 32-69. Installing a new differential side bearing. The installing tool must apply force to the inner race only! (DaimlerChrysler)

New Ring Gear Installation

If either a new ring gear or a pinion drive gear is required, both gears must be replaced. Never change one without the other. Always check the new ring and pinion to make certain that they are a matched set. The match number on the pinion must be the same as the number on the ring. Note the numbers on the ring and pinion in **Figure 32-71.** The number 4 appears on both the ring and the pinion, thus indicating they

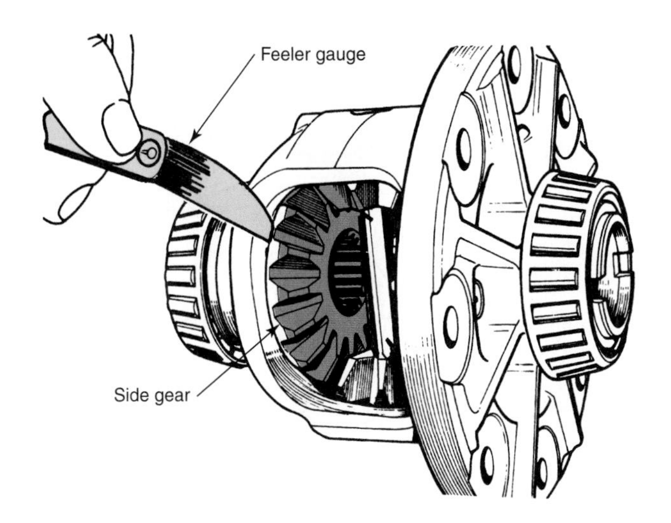

Figure 32-70. A method of checking clearance between the axle side gear and the thrust washer. (DaimlerChrysler)

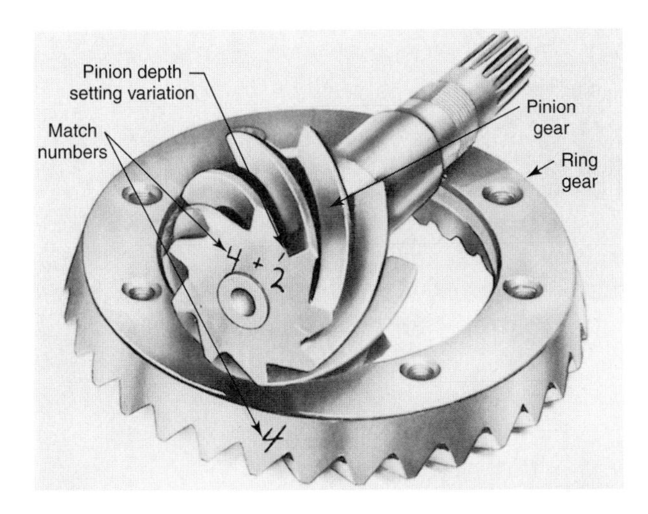

Figure 32-71. Shown here are typical ring and pinion gear markings that indicate a matched set. Notice that the match numbers are identical on both gears.

are a matched set. The +1/2 marking indicates the variation from a standard pinion depth setting. This number is needed to determine the thickness of the pinion shaft shim pack. This topic will be discussed in the pinion drive gear installation section of this chapter.

Check the case flange and the attaching surface of the ring gear for any sign of burring, dirt, and other problems. The two contact surfaces must be spotless.

Some ring gear setups require the gear to be brought to a specified temperature before installation. Use a special heating oven or oil bath heater. Never attempt to heat the ring gear with an acetylene torch, as this can remove the tempering.

Insert several guide studs into the ring gear to provide accurate alignment with the differential case. If guide studs are not available, thread a cap screw into each side of the ring. Lubricate all of the attaching cap screws. Start the cap screws into ring gear and tighten them alternately until the ring gear just touches the case. Remove the guide studs and install the remaining fasteners.

Tighten the ring gear fasteners from alternating sides of the case, in a star pattern. Bring the fasteners to one-half recommended torque the first time around. Go over the fasteners a second time, bringing them to full torque.

Note: Use ring gear fasteners only. Do not substitute regular cap screws for ring gear fasteners. Fasteners must be lubricated. Fasteners must be a snug fit in both the case and the ring gear. Do not use lock washers unless they were used in the original installation. Some fasteners require the use of a liquid adhesive. Follow the manufacturer's recommendations.

Gear Ratio Must Be Correct

The new ring and pinion set must have the correct number of teeth to maintain the desired ***gear ratio*** (number of times the pinion gear turns during each drive ring turn). If the ratio is to be changed, make certain the new set is adaptable to the differential case. On some installations, different cases are used, depending on the gear ratio.

Check Ring Gear Runout

After the differential is mounted in the carrier, check the ***runout*** of the ring gear. If runout is beyond specified limits, remove the ring gear. Check for nicks, burrs, and dirt on the contact surfaces. Reassemble and check runout again. Runout must be within specified limits, **Figure 32-72.**

Pinion Drive Gear Removal

Remove the pinion flange retaining nut. Remove the pinion flange (see the section covering the replacement of the pinion shaft seal and/or pinion flange in first part of chapter). With differential case removed, tap the pinion shaft inward until it is free.

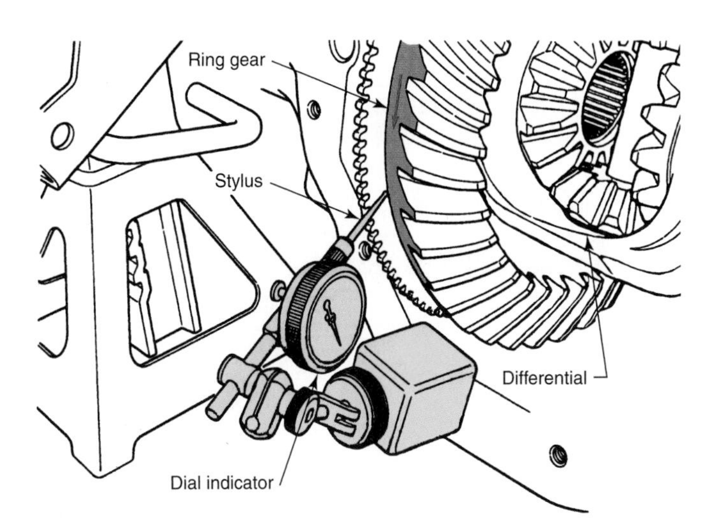

Figure 32-72. Measuring ring gear runout with a dial indicator. (DaimlerChrysler)

On one type of pinion arrangement, the carrier housing can be pulled off the carrier after the fasteners have been removed, **Figure 32-73.**

Be careful to save all shims. Tie individual shims in the pack together and identify their location. Measure the thickness of each shim and write it down, in case shims are lost. Wash all parts of the assembly and blow them dry. Inspect parts for wear, chips, and scoring.

Installing New Pinion Bearings

Remove the front and rear bearing cups from the carrier. Do not mar the surface of the counterbore during removal. **Figure 32-74** shows a puller being used to remove a bearing cup.

Check the bores. Lubricate and install new cups. Drive the cups in squarely to the correct depth. If adjusting shims are used beneath the inner cup, be certain they are spotlessly

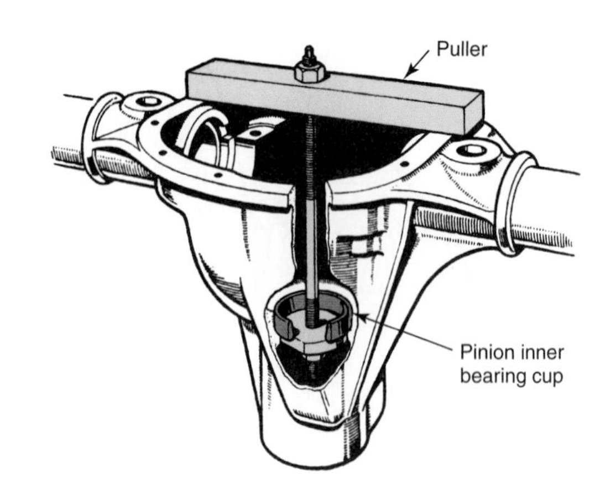

Figure 32-74. Pulling the pinion inner bearing cup from the carrier.

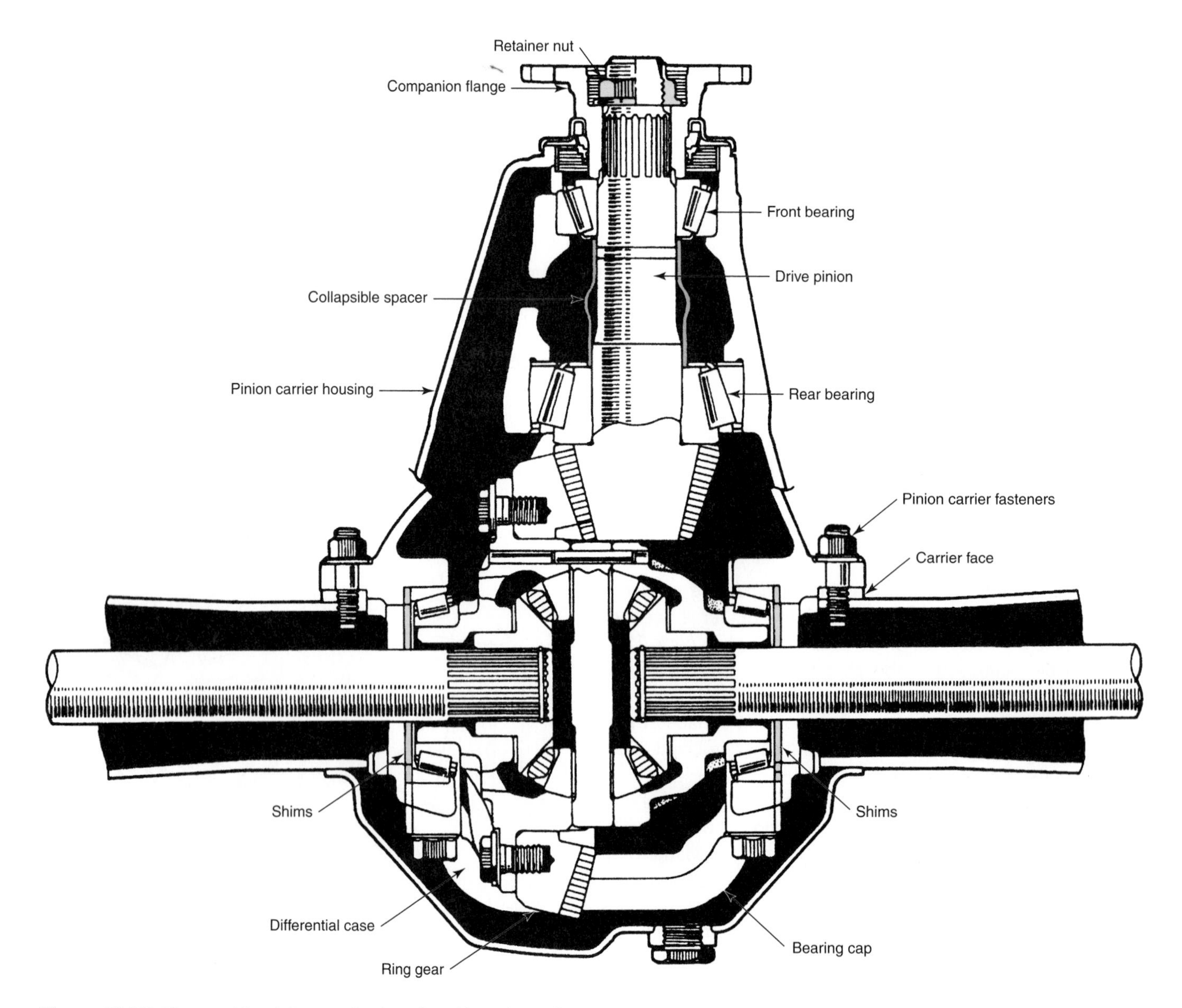

Figure 32-73. Removable pinion carrier housing. Note the collapsible spacer. (Toyota)

clean and in place. If a new pinion gear is being installed, adjust the shim pack as required.

The pinion assembly in **Figure 32-75** uses shims for positioning the pinion gear. A collapsible spacer controls the preload.

Remove the pinion shaft inner bearing. Lubricate the pinion shaft. Install a spacer or shim, if used. Drive the bearing firmly into position. Place the pinion gear against a soft, clean surface. Apply pressure to the inner cone, not to the rollers, **Figure 32-76.** If pinion depth must be checked, do not install

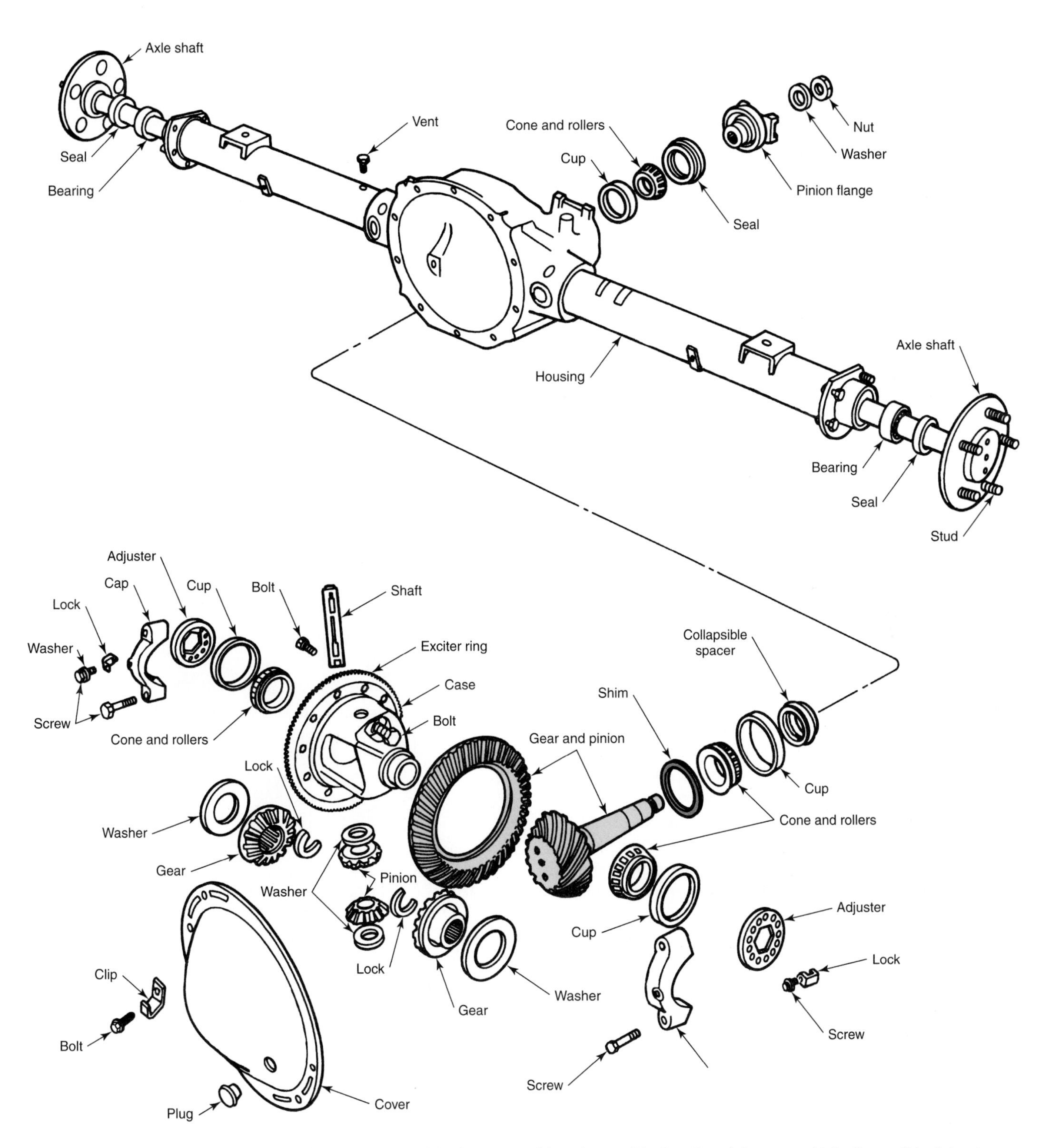

Figure 32-75. Exploded view of a differential assembly that uses shims for positioning the pinion gear. Note the anti-lock brake system sensor (exciter) ring. (DaimlerChrysler)

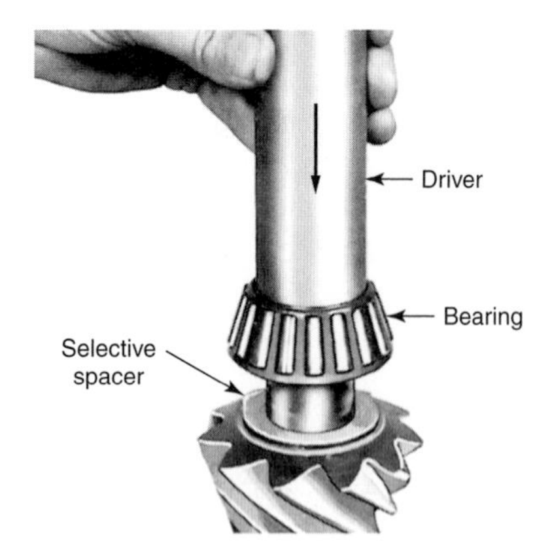

Figure 32-76. Driving a pinion shaft bearing. Note the pinion depth spacer is in position. (DaimlerChrysler)

the collapsible spacer on the pinion shaft until the depth check is made.

Checking and Adjusting Pinion Depth

Pinion gear depth (distance from face of pinion gear to centerline of ring gear) is critical. Each matched ring and pinion set is tested at the factory for the relationship between the ring gear and pinion that will produce the best tooth contact pattern. This relationship (actual pinion depth) is compared to a nominal (standard) pinion depth. The difference is marked on the pinion gear.

If the actual pinion depth is such that the pinion is closer to the ring gear centerline, the amount in thousandths marked on the pinion is preceded by a – (minus) sign. If the pinion depth is such that the pinion is a greater distance from the centerline, the amount is preceded by a + (plus) sign. The position of the pinion in relation to the ring gear centerline for a + (plus) and a – (minus) pinion marking is shown in **Figure 32-77.**

Pinion depth is controlled by placing shims between the pinion gear and the bearing, Figures 32-75, 32-76, 32-77. In other designs, pinion depth is controlled by placing shims between the inner bearing outer race or cup and the carrier. Some designs control pinion depth by placing shims between the pinion retainer and the carrier face, Figure 32-73.

As illustrated in Figure 32-77, shims must be added for pinions marked with a minus and removed for pinions marked with a plus. An exception to this is the Ford pinion illustrated in Figure 32-73. In this design, the addition of shims is required for a plus marking. Pinion depth should be checked with special gauges when a new ring and pinion set, new pinion bearings, or a new carrier is installed.

If the original ring and pinion, inner bearing, and carrier will be reused, the original shim pack will give the proper depth. To set the pinion depth, use a special gauge designed for that purpose. There are numerous types available.

One type of setting gauge is shown in **Figure 32-78.** The bearings are lubricated and held in position while the gauge plate and clamp plate are secured with the clamp screw. Note the gauge body resting in the case side bearing bores.

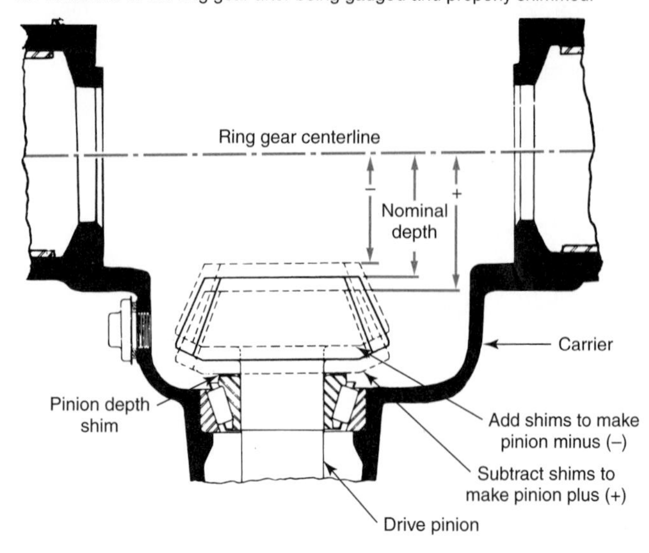

Figure 32-77. Relative distance from the ring gear centerline for a pinion marked with a minus (–), a plus (+), and a nominal (0). (General Motors)

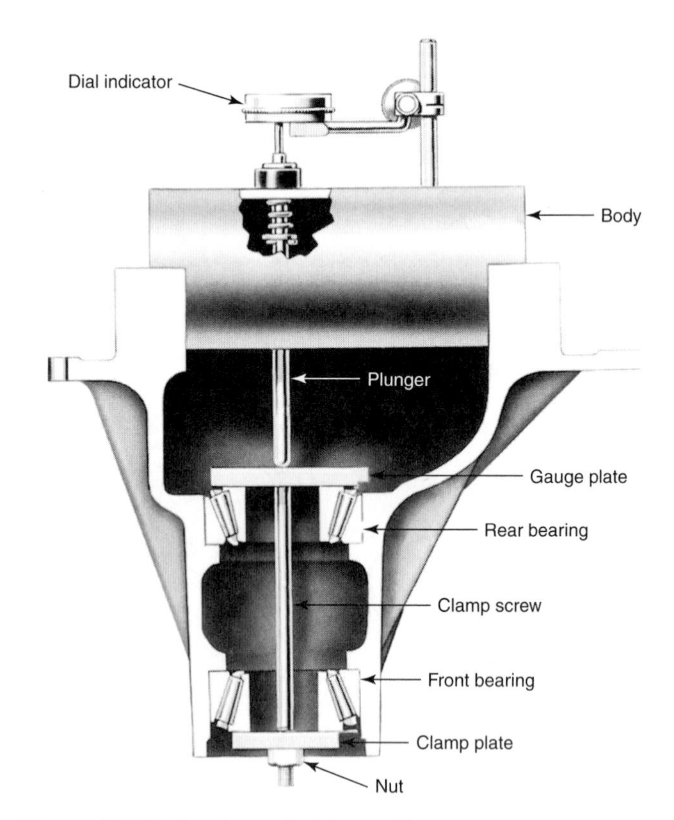

Figure 32-78. One type of pinion setting gauge.

The clamp screw is tightened until the bearings are preloaded to 20 inch pounds (2.26 N•m). Place the gauge body in the case side bearing bores. The bores and the gauge must be clean.

Mount a dial indicator on the gauge body. The indicator stem must contact the body plunger. Swing the gauge body so the plunger is clear of the gauge plate. Set the indicator to zero. Swing the gauge body in the bores so the plunger moves across the gauge plate. Swing it back and forth until the highest reading is noted.

Another type of gauging device is being used in **Figure 32-79.** A special gauge block is inserted in the pinion bearings and brought to specified torque. The arbor is held in place by hand or with bearing cups. Selective thickness spacers are tried between the face of the gauge block and the arbor. When a spacer that will just fit is found, write down the thickness. Examine the pinion for a plus or minus amount. If the chosen spacer is marked 0.094″ and the pinion is marked minus 2, add 0.002″ (0.05 mm) to the thickness of the 0.094″ (2.38 mm) spacer. The correct spacer to place between the pinion gear and bearing will then be 0.094″ + 0.002″ = 0.096″ (2.38 + 0.05 = 2.43 mm).

If a new pinion gear is being used in the same carrier and with the original bearings, adjust the original shim pack by the plus or minus amount marked on the pinion.

> Note: When the pinion is marked with a plus number, subtract this amount from the shim pack. When a minus number is used, add this amount to the thickness of the shim pack. The pinion in Figure 32-73 is an exception. For this specific setup, reverse the procedure and add the amount to the shim pack when the pinion is marked with a plus number, or subtract the amount when the pinion is marked with a minus number.

Install the pinion after determining the correct depth-setting shim pack. Lubricate all parts before assembly. Parts must be clean and free of burrs.

Pinion Preload Must Be Correct

To prevent the pinion gear from moving away from the ring gear under load, the pinion bearings must be properly preloaded. If preload shims are used, add or subtract shims until the preload is correct when the pinion-flange retaining nut is brought to the specified torque.

After the pinion-flange retaining nut is torqued, turn the pinion shaft several turns to allow the bearings to seat before checking preload.

If preload shims and a solid spacer are used, manufacturers may recommend tightening the pinion nut to a specified torque. Follow the manufacturer's instructions. If a collapsible preload spacer is used, the directions may call for bringing the nut up to the original mark (old flange being used) plus an additional amount such as 1/8 turn or 1/32″ (0.79 mm).

Another technique involves gradually tightening the retaining nut until a specific bearing preload is reached. Preload is set with an inch-pound torque wrench. Check the preload frequently during the tightening process to avoid exceeding the recommended preload. A few additional inch-pounds may be recommended. This method works with either a new or used flange.

If the bearing preload is exceeded on designs employing a collapsible preload spacer, a new spacer must be installed. Do not try to correct by backing off the pinion nut to obtain the correct preload. When preload is exceeded, the spacer is collapsed to the point where the ring and pinion contact pattern is changed.

Installing Differential Case in the Carrier (Threaded Adjuster Type)

Lubricate the differential side bearings and place cups on the bearings. Lubricate the side bearing bores in the carrier to allow easy cup side movement.

Place the case in the carrier. If the ring and pinion gearset is the nonhunting type (any one pinion gear tooth contacts only a certain number of ring gear teeth), make certain the marked ring and pinion teeth are meshed. See **Figure 32-80.** Hunting-type gearsets (any one pinion gear contacts all ring gear teeth) will not be marked and may be meshed in any position.

Move the assembly in the bores until a small amount of ***backlash*** (play between ring and pinion gear) is present.

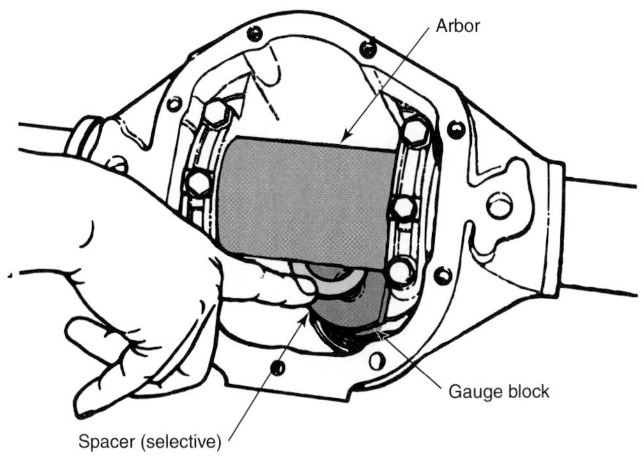

Figure 32-79. Measuring a housing for correct pinion shim thickness with a special gauge block tool. (DaimlerChrysler)

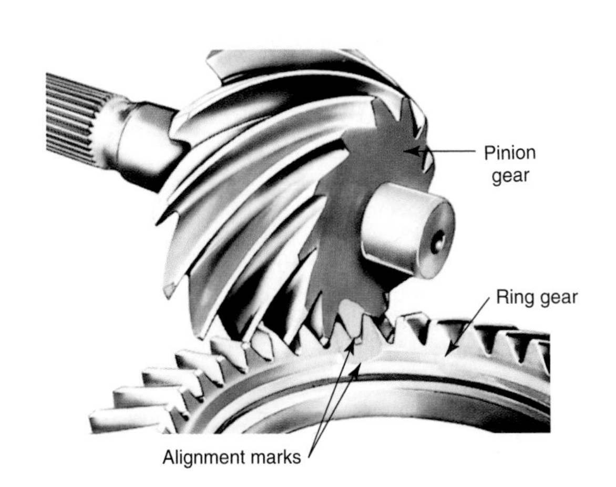

Figure 32-80. The nonhunting gearset marking should be aligned when meshing the ring to the pinion. (Ford)

Then, install the threaded adjusters snugly against the bearing cups. Adjusters must be installed with the same number of threads showing on the outside of each adjuster.

Install the bearing caps so the marks are aligned. Insert the cap fasteners and bring them to the correct torque. Check for smooth operation of the adjusters as the fasteners are torqued. If binding is experienced, remove and check for dirt, burrs, and nicks. Following full torquing, loosen the fasteners and retorque to around 25 foot pounds (33.90 N•m). Some manufacturers recommend loosening the fasteners and then torquing one fastener on each cap to full torque.

Loosen the right-hand adjuster (on the pinion side of the ring gear) and tighten the left-hand adjuster (on the back of the ring gear) until no backlash is present. Rotate the ring gear while tightening the left-hand adjuster. **Figure 32-81** illustrates how the adjusters are moved using spanner wrenches.

Tighten left-hand bearing cap fasteners to full torque. Turn the right-hand adjuster in until the left-hand bearing is in firm contact with the left-hand adjuster. Loosen the right-hand adjuster and retighten. At this point, the case bearings are just snug with no endplay or preload. There is no backlash between the ring and pinion.

Set up a dial indicator to check backlash. The indicator stem must be in line with the direction of tooth travel. Look at **Figure 32-82.**

After bringing the right-hand cap fasteners to full torque, turn the right-hand adjuster inward two or three notches. This preloads the bearings and should give the specified backlash.

Check the backlash at four different spots around the ring. Leave the ring in the position producing the smallest amount of backlash. If backlash is set to specifications (around 0.006″–0.008″ or 0.15 mm–0.20 mm), insert the adjuster locks. Torque the lock fasteners.

If backlash varies at the different points more than the specified allowable variation (around 0.002″ or 0.05 mm), check the ring gear for runout. If necessary, the ring must be removed, cleaned, reassembled, and rechecked. Check case flange runout also, **Figure 32-83.**

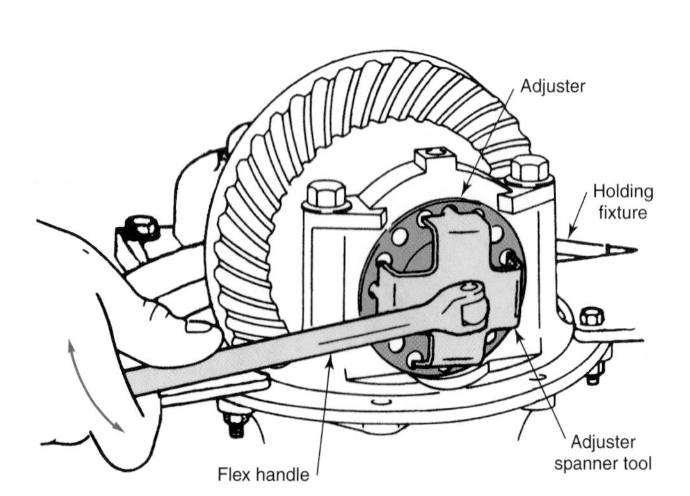

Figure 32-81. Adjusting bearing backlash with an adjuster spanner tool and a flex handle. (General Motors)

If backlash is not quite as specified, loosen one adjuster and tighten the other until backlash is correct. The bearing cap fasteners must be at full torque during these adjustments.

When moving the adjusters, loosen the appropriate adjuster *two* notches. Tighten the other adjuster *one* notch. Finally, tighten the first adjuster one notch. This procedure assures that solid contact is made and prevents loosening in service.

Another method sometimes specified for preloading differential side bearings involves bringing the adjusters up so that no endplay exists in the bearings. With no backlash between the ring and pinion, a dial indicator is used to measure the carrier spread as the right-hand adjuster is tightened. When the carrier spread (distance between side bearing bores being increased due to bearing preloading) reaches the specified amount, the backlash should be close to specifications. If not, adjust by loosening one adjuster and tightening the other.

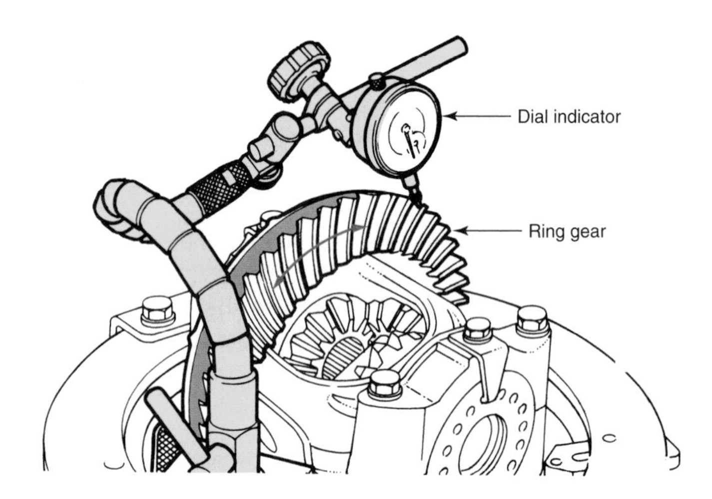

Figure 32-82. A dial indicator set up to measure gear backlash. (DaimlerChrysler)

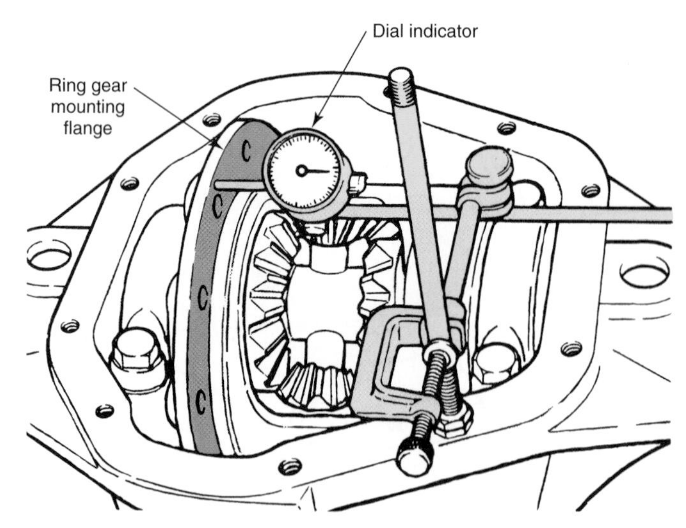

Figure 32-83. Checking the ring gear mounting flange for runout with a dial indicator. (DaimlerChrysler)

Installing the Differential Case in the Carrier—Shim Preload Adjustment

Clean the side bearing carrier bores. Lubricate the bearings and install the cups. If old bearings are being used, use the original shim packs. Place shims on each end of the bearing cups. Start the case into the carrier bores.

If a new case, new carrier, or new bearings are used, the original shim packs cannot guarantee the correct preload. In these instances, it is necessary to determine the proper thickness shim pack.

In general, the selection of shims involves placing enough shims or spacers on each side to remove all side play (just snug, no preload). Refer to **Figure 32-84.**

The manufacturer will usually require that a spacer of a certain thickness be placed on one side to start with. This adjusts the ring and pinion backlash to within workable measurements when the side play is removed.

Shims may then be moved from one side to the other until the backlash is correct. Check in the backlash at 90° intervals around the ring gear. See **Figure 32-85.**

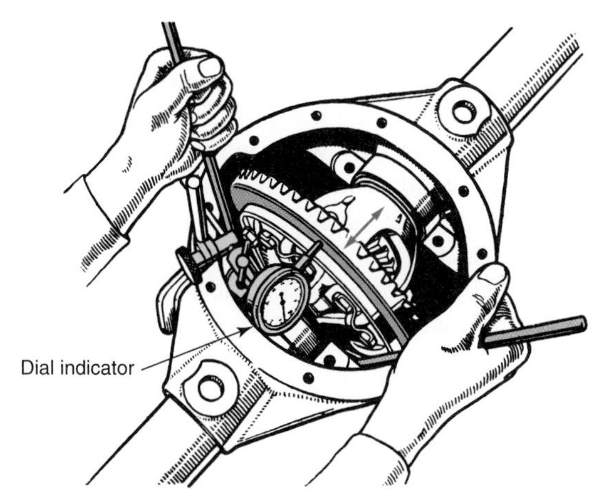

Figure 32-84. Checking side play in the differential case bearings. Add shims until the backlash is eliminated.

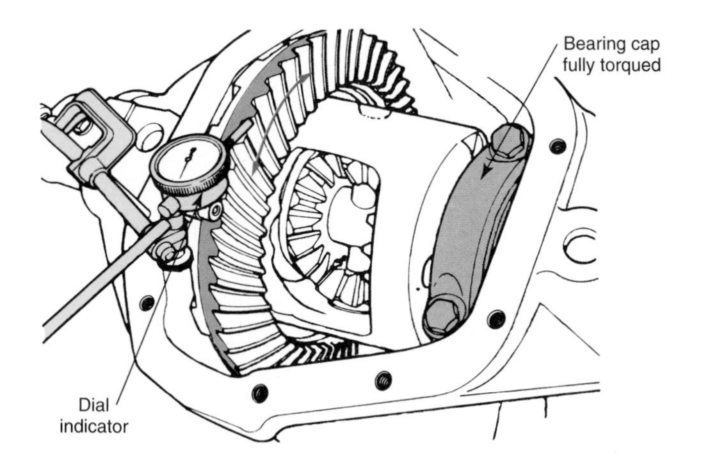

Figure 32-85. Checking backlash between the ring and pinion gears. Caps should be in place and fully torqued. (DaimlerChrysler)

When the backlash is correct and there is no endplay or preload, add a shim to each side. Make sure that both shims are the same thickness, as specified by the manufacturer.

Use a carrier spreader, if needed, to permit the installation of preload shims. Lubricate the shims and avoid undue pounding. Light tapping on the shims with a soft hammer is permissible.

Some differential assemblies place the side bearing shims between the carrier and the bearing inner cone. See **Figure 32-86.** Some production side bearing spacers or shims cannot be reused. They must be replaced with steel service spacers and shims.

Some manufacturers require a specific preload adjustment check to be made with a torque wrench on one side of the ring gear fasteners or on the pinion shaft. When preload and backlash adjustments are set to specifications, recheck the torque on the bearing cap fasteners. Rap the top of the fasteners with a hammer and retorque.

Checking the Tooth Contact Pattern

The relationship between the ring and pinion gears must be adjusted to produce the correct ***contact pattern.***

To check the contact pattern (preload and backlash must be correct), cover all the ring gear teeth with a light coating of hydrated ferric oxide or some other suitable compound. Some new gear sets come with a special compound for this purpose. Apply the compound to the teeth with a reasonably stiff paintbrush. See **Figure 32-87.**

Using a wrench, turn the pinion shaft in the normal forward drive direction while creating a drag on the ring gear. The technician in **Figure 32-88** is using a brass drift to bind the ring gear. Continue turning the pinion until the ring gear has made one full turn. This produces a ***drive pattern.*** The drive pattern is produced by the pinion gear driving the ring gear, **Figure 32-89.**

Turn the pinion in the opposite direction, while binding the ring, to produce a ***coast pattern.*** This is the pattern produced when the ring gear drives the pinion gear.

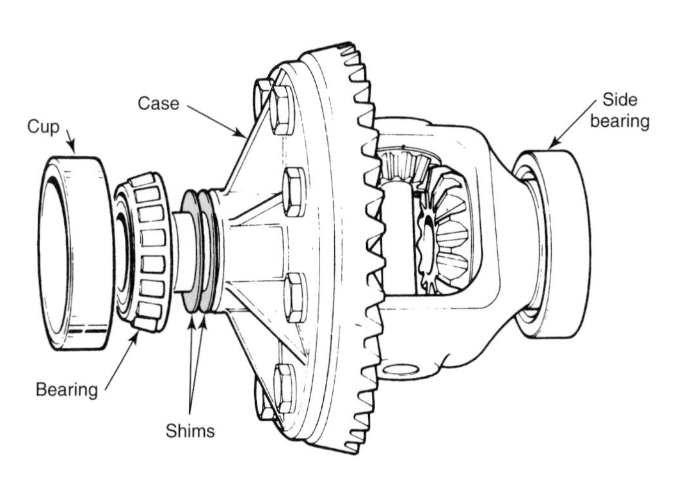

Figure 32-86. This differential carrier uses shims between the case and the side bearing cones.

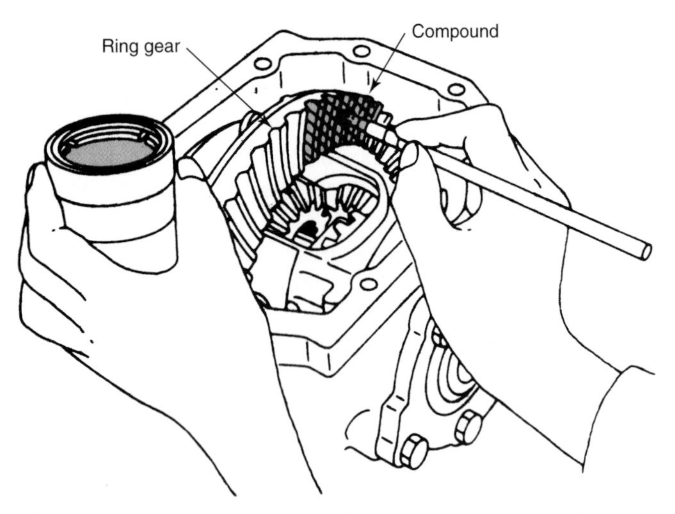

Figure 32-87. Applying a special pattern compound to the gear teeth with a brush. The compound will show the tooth contact pattern when the gears are rotated. Incorrect tooth contact can produce noise and/or a short gear life span. (Infiniti)

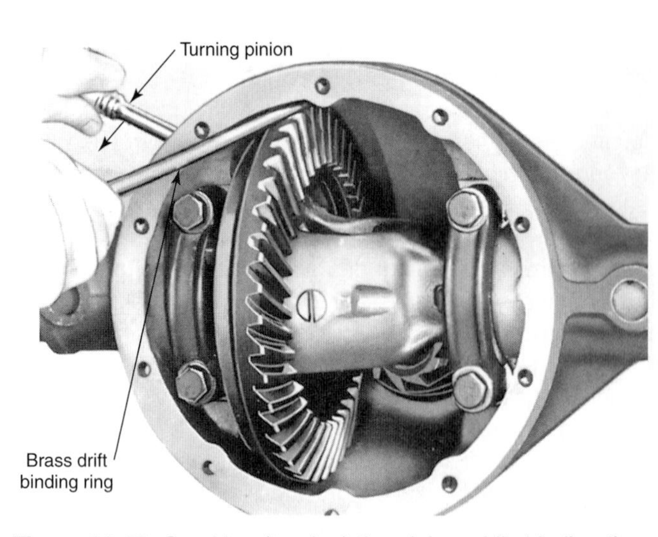

Figure 32-88. Cranking (turning) the pinion while binding the ring gear to produce a drive pattern.

There is no exact pattern that must be formed. The pattern shape varies, depending on gearset design, wear, and load. In general, however, the pattern (contact area) should be even around the ring. Unevenness indicates excessive ring runout.

The drive pattern should be centrally located between the top and bottom of the tooth. It can be somewhat closer to the toe. Under increased loading, the pattern spreads out and tends to move towards the heel of the tooth. When the point of heavy loading is reached (pulling, hills, or rapid acceleration), the pattern may extend almost the full distance from toe to heel. The drive side of a ring gear tooth is the convex side. The coast side of a ring gear tooth is the concave side. Note the gear tooth nomenclature in **Figure 32-90.**

The coast pattern should also be centralized between the top and bottom of the tooth. In some cases, it may be a little longer and closer to the toe. Examine the pattern closely for the presence of thin lines, which indicate an area of narrow contact that will produce unusually high, localized pressure. Such lines should not be present.

A heel contact pattern is caused by excessive backlash. It is corrected by moving the ring toward the pinion. Toe contact indicates insufficient backlash. To correct this problem,

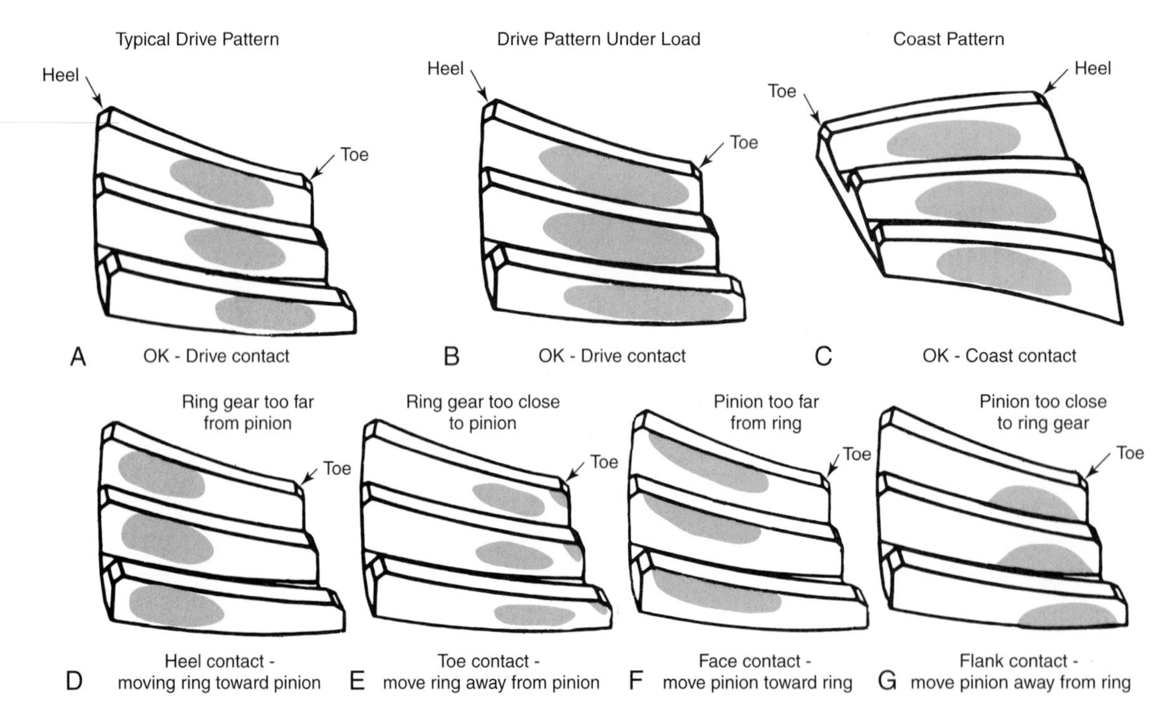

Figure 32-89. Typical ring gear tooth contact patterns.

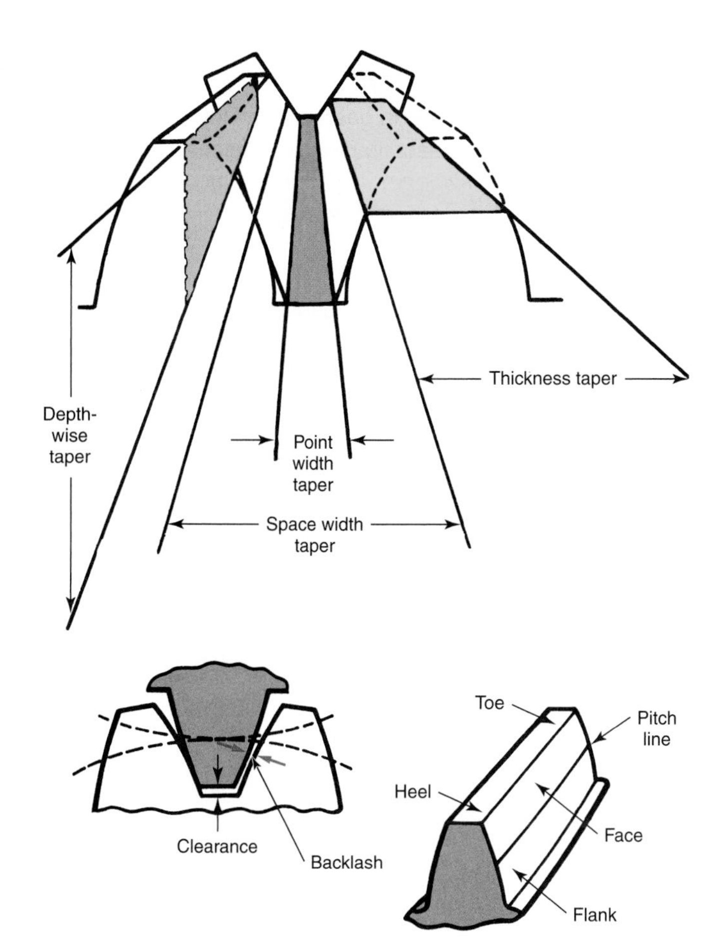

Figure 32-90. Common gear tooth nomenclature. (General Motors)

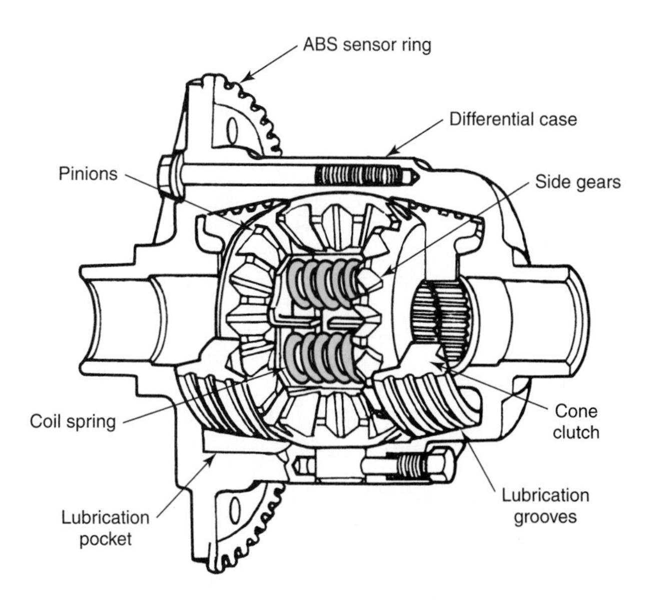

Figure 32-91. Cross-sectional view of a traction-type differential that incorporates coil springs. Note the anti-lock brake system sensor (exciter) ring. (DaimlerChrysler)

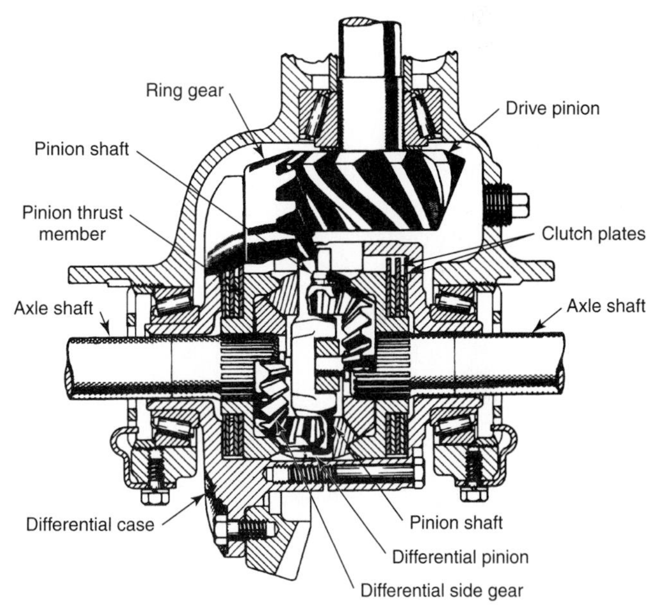

Figure 32-92. Cross-section of a Sure Grip differential.

move the ring away from the pinion. Face contact indicates the pinion is set too far from the ring. To correct, move the pinion toward the ring. Flank contact requires moving the pinion away from the ring gear.

When moving either the ring or the pinion to correct the contact pattern, move in small amounts. Make certain the bearings are properly preloaded, the cap fasteners are fully torqued, and the backlash is within specifications before rechecking the pattern.

Limited Slip Differential Service

The ***limited-slip*** (also called Sure Grip, No-Spin, Anti-Spin, Positive Traction) type of traction differential uses friction members (cone or disc clutch). When one wheel attempts to spin, driving force is still being applied to the nonspinning wheel by the differential. See **Figure 32-91.**

In the conventional, nontraction differential, when one wheel spins, no driving force is applied to the other. A traction differential, which uses a clutch pack on either side of axle side gears, is shown in **Figure 32-92.**

In the Sure Grip design, Figure 32-92, two pinion shafts are used. Note the ends of the shafts are wedge shaped. They operate in V-shaped ramps. When a driving force is applied to the ring gear, the pinion shafts slide up the V-ramps and compress both clutch packs. This results in a mechanical connection between both axles.

When rounding a corner, the outside axle turns faster than the inner axle. This causes the pinion shaft on the outer side to slide down its ramp. This releases the clutch pack on that side and allows normal differential action.

If one wheel spins, the clutch pack on that side is released. The pack on the slower-turning wheel causes it to remain clutched to the differential case. It thus receives turning force despite differential action. **Figure 32-93** depicts a cross-sectional view of a traction differential that has a preload spring instead of V-shaped ramps found in the Sure Grip type.

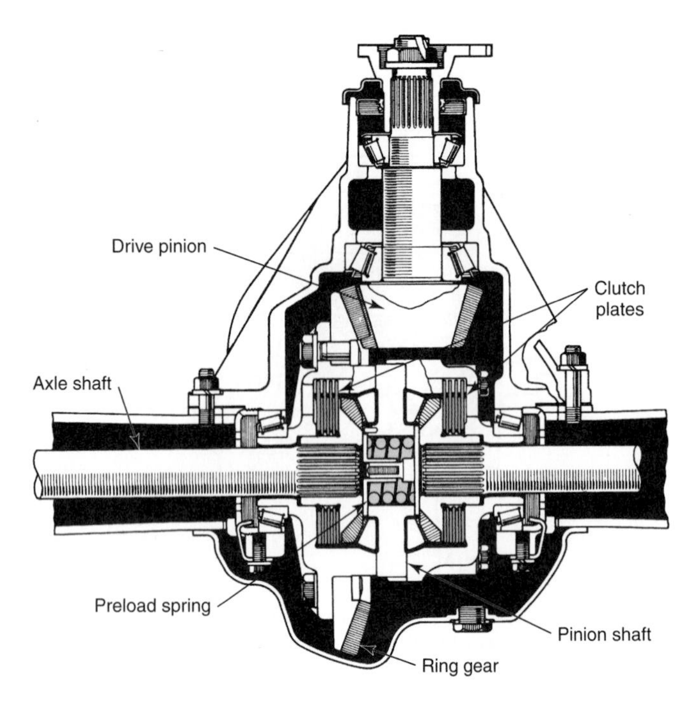

Figure 32-93. This limited-slip traction-type differential uses a single preload spring. Note the clutch plates. (Toyota)

A somewhat different type of traction differential is illustrated in **Figure 32-94.** This unit uses cone clutch brakes, one on each side of the axle side gears, to provide traction.

Normal differential action causes clutches to slip. When one wheel starts spinning, the slower-moving axle still remains clutched to the differential case and therefore receives some driving force.

Cone brakes are kept preloaded to the case by means of preload springs. The normal tendency for the side gears to move away from the pinions under load also increases the cone loading.

One particular positive-action type traction differential uses a governor assembly. The purpose of the governor is to positively lock both rear axles (wheels) together when one wheel spins more than 100 RPM faster than the other during slow vehicle operation. **Figure 32-95** shows an exploded view of this unit.

Lockup takes place through the use of flyweights in the governor, cam system, and the multiple-disc clutch units. The flyweights move outward to engage the latching bracket as one axle turns faster than the other.

This retards the camform side gear. In turn, this compresses the multiple-disc clutch unit, locking one side gear to the case. Equal driving torque is then transmitted to

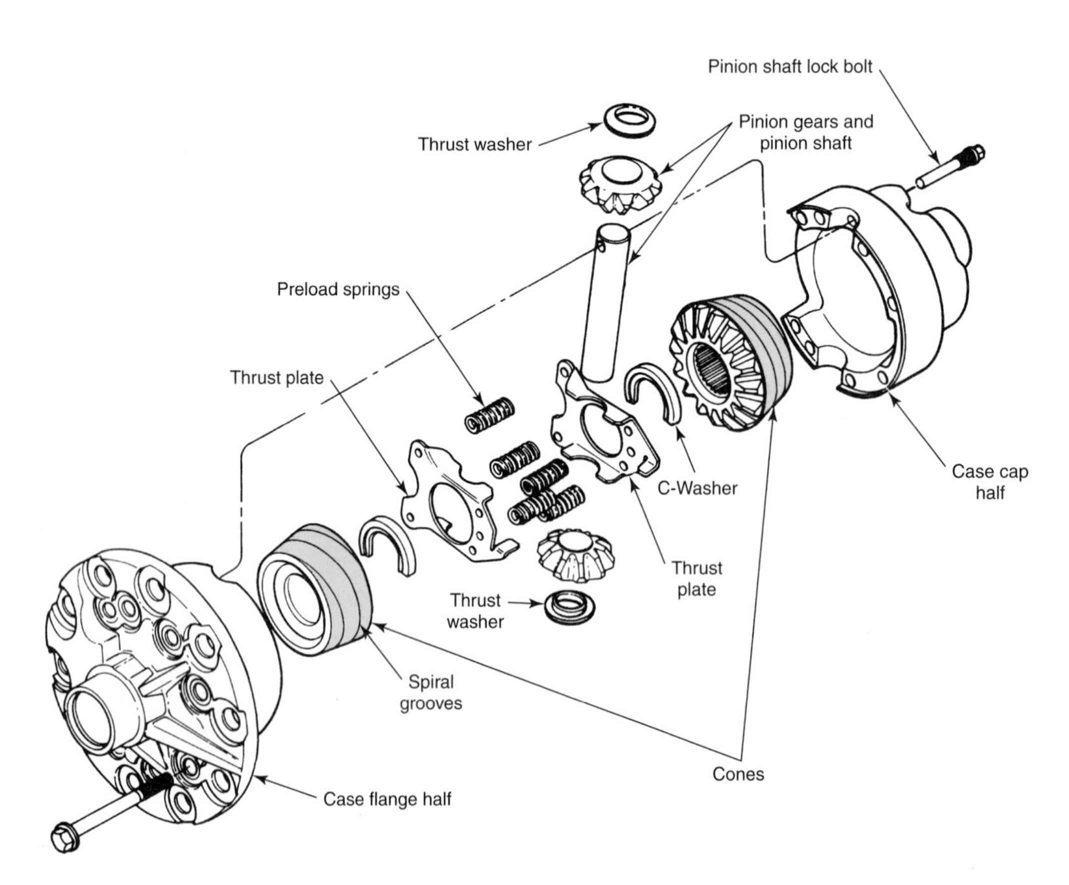

Figure 32-94. Exploded view of a cone-type positive-traction differential. Note the spiral grooves on the cones. These grooves provide lubrication passages. (Ford)

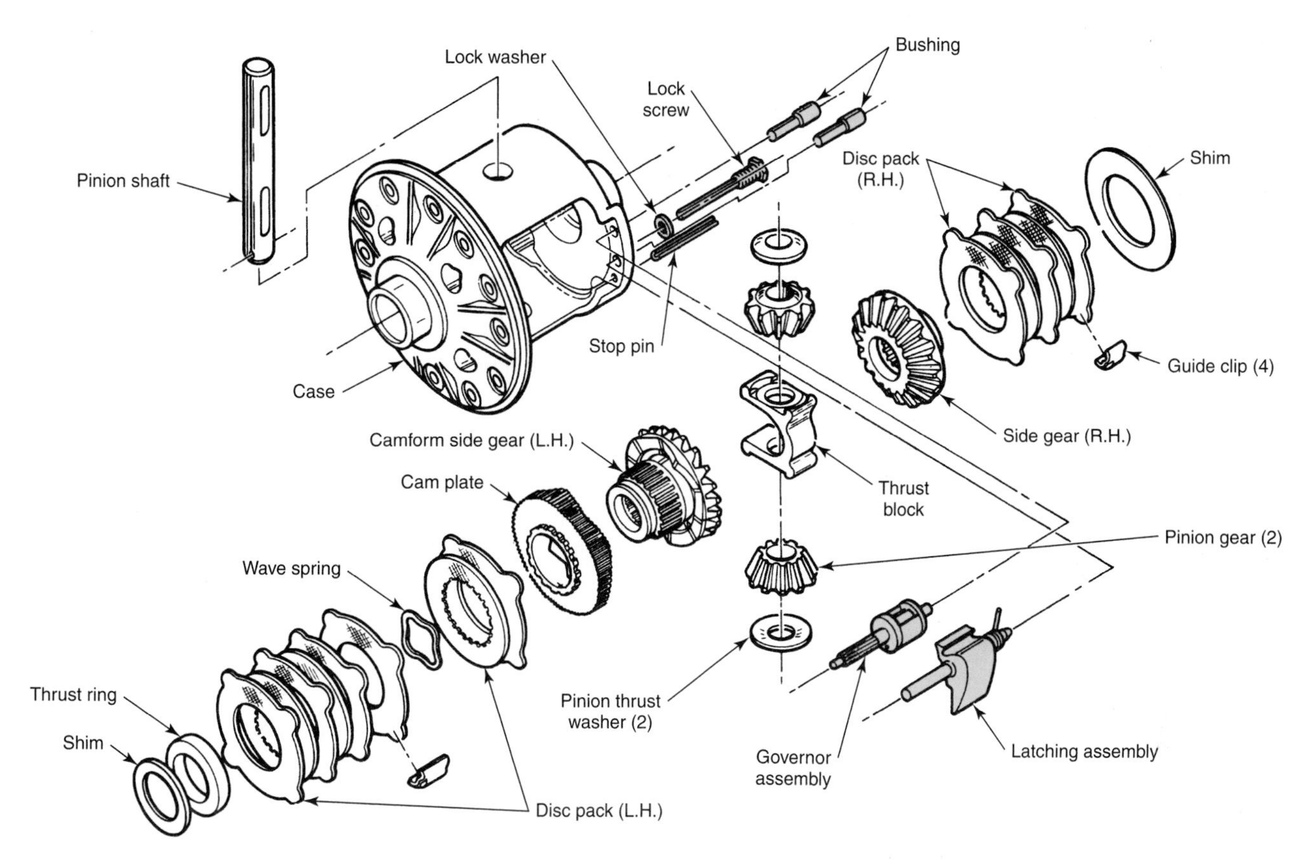

Figure 32-95. Exploded view of a positive-traction differential. Note the governor assembly. (General Motors)

each wheel. When axle (wheel) speed difference drops below 100 RPM or road speed of vehicle is 20 mph (32 km/h) or faster, the differential operates as a standard (nonpositive traction) type. Service of this unit is basically the same as for other types using clutch plates.

Testing Traction Differential Action

To check traction-type differential action, raise *one* wheel off the floor. Next, place the transmission in neutral, block the wheel remaining in contact with the floor, and release the parking brake.

Use an adapter similar to that shown in **Figure 32-96** to provide a means of attaching a torque wrench to the raised wheel. The wrench head must be in the center of the wheel.

Turn the wheel and note the amount of torque required to keep the wheel moving. Check this amount against specifications. Torque specifications will vary, depending on design. Some positive action differential units can be tested while they are out of the axle housing. They can be held in a vise with a special tool and rotated (in the direction of normal travel) with a torque wrench. See **Figure 32-97.** The initial breakaway torque (the force at which the unit begins to slip) is then recorded. Use the manufacturer's torque reading for your specific unit. Some traction differentials use only gears instead of cones or clutch plates. They generally do not require special lubricants.

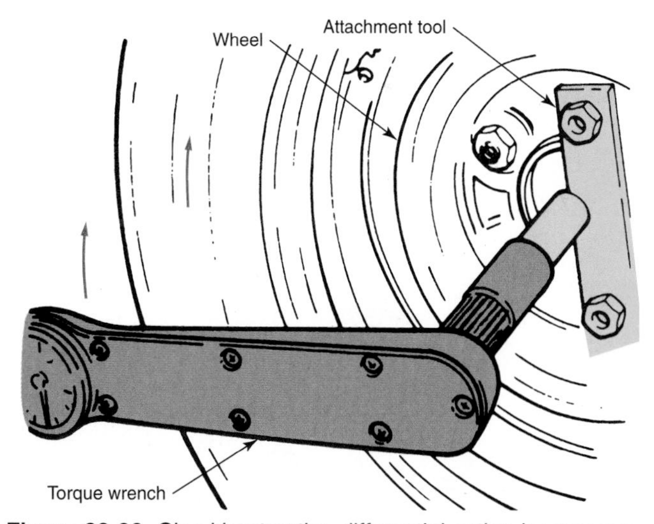

Figure 32-96. Checking traction differential action by measuring torque required to turn the raised (off ground) wheel. Note attachment tool to connect and center torque on wheel. (Ford)

Traction Differential Service

Other than the clutch packs, cone brakes, and springs that make up the limited-slip action, service on the traction differential is much like that for the conventional unit already discussed.

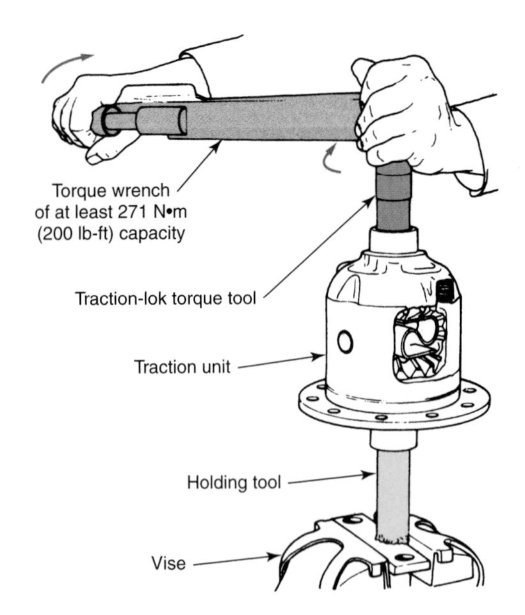

Figure 32-97. Testing a traction differential unit out of the carrier. Use proper tools and mounting procedures for an accurate reading. (Ford)

If the differential case is split (made in two pieces), make certain the two halves are marked before disassembly. See **Figure 32-98.** After disassembly, clean and inspect all parts. If brake cones will be reused, be careful not to reverse their positions. If one-half of the case is damaged, replace both halves.

Before reassembly, lubricate all parts thoroughly with special traction-differential lubricant. Use the axles to align the side gears and cone brakes or pinion thrust blocks while the case is bolted together.

Install the differential in the axle housing. If the axle-end thrust blocks are the type that can fall out of position, check that they are in place before installing the axles.

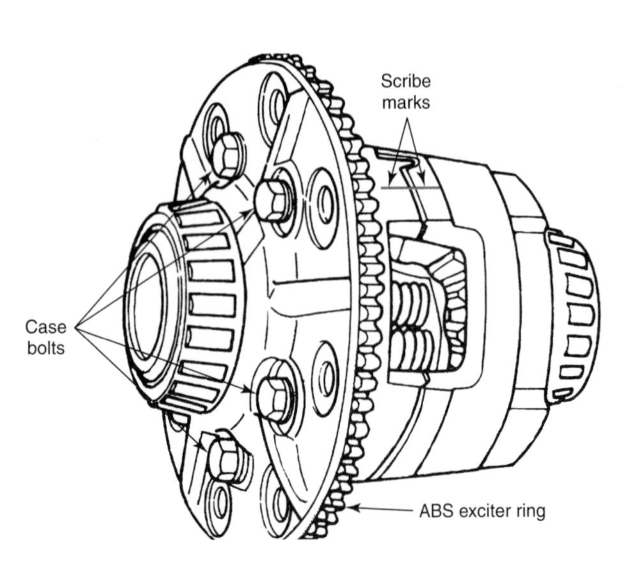

Figure 32-98. A two-piece (split) differential case. Mark both halves before disassembly. Keep all parts in their correct order. (DaimlerChrysler)

Drive Axle Lubrication

For conventional differentials, the use of a hypoid or a multipurpose gear lubricant is generally specified. Use the recommended viscosity. Most traction-type differentials require a special lubricant. Never use anything but the specified oil.

Maintain the oil level in the housing so that if checked warm, the level is in line with the bottom of the filler hole. If cold, the level can be up to 1/2 in. (12.70 mm) below the hole. Wipe off the filler plug and the surrounding area before plug removal. Wipe off the filler gun nozzle. Under no circumstances must any foreign material enter the housing. See **Figure 32-99.**

Keep Vent Open

The axle housing is vented by means of a small hole, a tiny, capped pipe, or a hose. The vent prevents a buildup of pressure within the housing as the lubricant warms up. Make certain the vent is open.

Housing Alignment

One method of checking housing alignment is shown in **Figure 32-100.** The flanges on the axle ends must be clean and free of burrs. Straight edges must be straight and held in firm contact with the flanges during measurement. See the manufacturer's specifications for allowable misalignment. Check in both horizontal and vertical positions.

Tech Talk

CV joints will outlast the rest of the vehicle if they are properly lubricated. The lubricant is good for the life of the CV joint and does not have to be changed. Often, however, CV axles fail due to a lack of lubricant. The prime

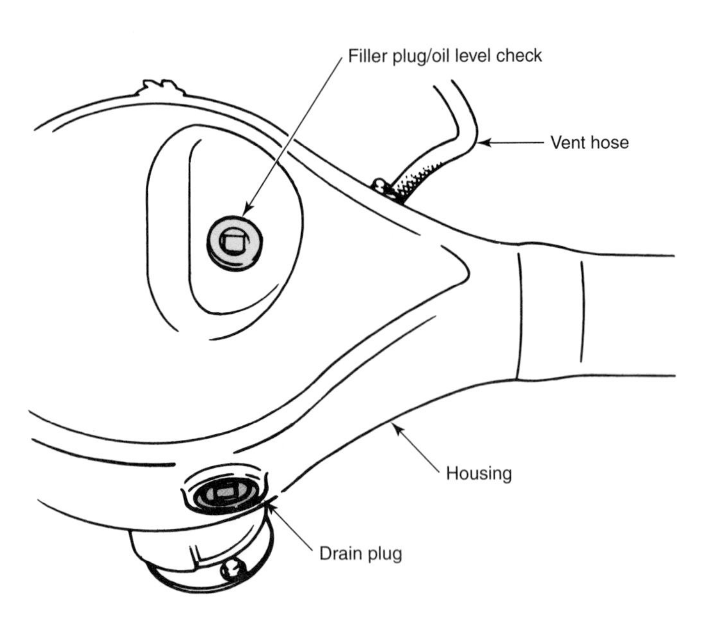

Figure 32-99. The locations of the lubricant drain plug and fill plug in one particular differential housing are shown here. Always refill to the specific level recommended by the manufacturer with the correct lubricant. (General Motors)

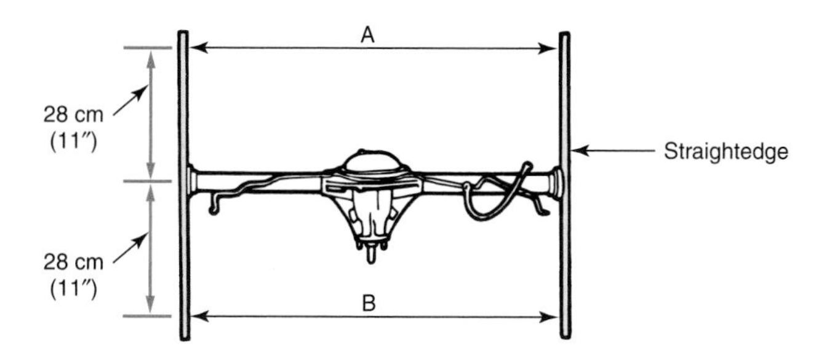

Figure 32-100. Checking housing alignment. For this particular housing, the difference between measurement A and B should not exceed 3/32″ (2.38 mm). Check in a vertical plane and a horizontal plane. (DaimlerChrysler)

reasons for CV joint failure are a damaged boot and failure to use the proper lubricant during service. Both of these problems are easy to prevent.

Whenever a front-wheel drive vehicle is on the rack, check the CV joint boots for damage. A torn boot is usually easy to spot, and the stationary parts near the joint will be covered with grease. If the boots are damaged, advise the owner to replace them immediately and to check the CV joint for wear and corrosion.

Whenever you replace a defective CV joint or boot, always clean the interior of the joint thoroughly and repack it with the correct replacement lubricant. New boots and CV joints generally come with the correct lubricant, which is usually packaged in a plastic bag. Use as much of this lubricant as you can squeeze out of the bag. If a new joint or boot does not come with lubricant, do *not* substitute chassis grease. The CV joint will fail quickly. When you reassemble the joint, make sure the boot retaining straps are correctly positioned and tightened to prevent lubricant loss.

Summary

Modern vehicles use the open shaft, or Hotchkiss, drive shaft. Drive shafts can be made of one or more pieces. The cross-and-roller universal joint is commonly used by modern vehicles. The cross-and-roller joint consists of two yokes connected by a cross. Bearing caps, containing needle bearings, reduce friction. The bearing caps may be retained by snap rings, U-bolts, cap screws, or other means.

Before disconnecting a U-joint to remove the shaft, mark the shaft, joint, and drive flange to preserve balance. Yokes on both ends of the shaft must be in the same plane.

Support the shaft during removal. Cover the slip yoke. Avoid bending sharply at the U-joint. Do not allow the shaft to hang supported by a U-joint. Tape the bearing caps to the cross to prevent them from dropping off during shaft removal. Never tighten vise jaws on the drive shaft tube. Support the free end of the shaft while the other end rests in a vise. Use jaw covers on the vise.

Remove the snap rings and dismantle the joint. Clean and inspect its components. If the cross or bearing caps are worn, replace both. Use new seals and snap rings. Lubricate the joint with universal joint grease. Use a low-pressure gun and fill the joint slowly. On sealed types, lubricate before assembly. After the joint is assembled and the snap rings are installed, strike the yokes to seat the bearing caps against the snap rings. Joint action should be smooth.

When installing a drive shaft, cover the slip yoke to prevent the entry of dirt and damage from contact with vehicle's underbody parts. Support the shaft. Lubricate the slip yoke. Yokes must be in the same plane. All marks must be aligned. Torque the joint U-bolts or cap screws. Tap the joint and retorque. Crimp the locking tabs against the cap screws. Check all fasteners. Shake the shaft to detect looseness. Road test the vehicle; watch for proper shaft operation.

If shaft vibration is present, check the shaft for runout, presence of undercoating, misaligned marks, or improper drive shaft angle. Replace sprung or badly dented shafts. Drive shaft angle may be checked by various methods, including special gauges, levels, protractors, and plumb bobs. The vehicle must be level and at curb weight for drive angle check. Tilt the rear axle housing to provide the correct drive angle by adjusting the control arms (coil springs) or by using tapered wedges (leaf springs). Some vehicles allow adjustment of the transmission height also.

Constant-velocity (CV) joints are used on most modern front-wheel drive vehicles and on some rear-wheel drive vehicles. The two main types of CV joints are the tripod joint and the Rzeppa joint. Torn CV boots are a common source of CV joint failure. Removing the CV axle often requires the steering knuckle to be removed. CV joints can be disassembled and worn parts replaced, but many technicians prefer to replace the entire joint when defective.

The typical drive axle assembly contains the ring and pinion gears and the differential assembly. Axles are retained in the housing with a retainer plate or with C-locks.

Before road testing, check the lubricant level, tire pressure, and tread pattern. Check action during drive, cruise, float, and coast conditions. Road surface, tire tread, front wheel bearings, transmission, and other factors can make noises that are easily mistaken for drive axle problems. Check carefully before pulling the axle apart.

Check bearing preload. Scribe a line on the pinion shaft, retaining nut, and flange before loosening retaining nut. Hold the companion flange with a special wrench while loosening the flange nut. Use a puller to remove the flange. Clean the pinion seal recess. Soak the new seal if required. Coat the outer diameter of the new seal with cement. Install the seal, lip in, to the correct depth.

Inspect the pinion flange for nicks and burrs. Lubricate before installation. Draw the flange on with a nut. Never pound on the pinion flange or pinion shaft. Tighten the pinion flange nut as required. If preload is exceeded, install a new collapsible spacer (where used).

Pull axles by removing retainer plates or inner end C-locks. On front-wheel drive units, remove universal joints and/or the spindle assembly. Use a slide hammer puller on conventional axles. Do not disturb the backing plate unless necessary. If the backing plate must be pulled, disconnect the brake line. If an axle was broken or if other damage occurred, flush the housing thoroughly. Check the axle for twists and bends.

Do not wash sealed wheel bearings. Always install new oil seals following axle removal. Notch the bearing retainer rings to remove them. When pressing new wheel bearings into place, apply pressure to the inner race only. Press a new retaining ring into firm contact with the bearing. Do not press the ring and bearing on at the same time. The bearing must face in the correct direction and must be pressed to the proper position.

When installing axles, do not damage the seals. The retainer plate must be positioned so the oil drain hole can function. Where axle and endplay shims are used, install them in their original positions. Check axle endplay. If a tapered end axle is used, make certain the drive key is in place when the hub is attached to the axle. Use a new cotter pin to secure the hub retaining nut.

Some carriers are removable, others are an integral part of the housing. Mark differential case bearing adjusters and bearing caps before disassembly. Check backlash and tooth contact pattern before removing the case. Use a puller or pry the case out of the carrier. Mark and save all shims. A spreader may be required. Generally, do not spread housings more than 0.020″ (0.51 mm).

Clean and inspect the differential case and parts. Replace as needed. When case side bearings must be replaced, use new outer cups. The ring and pinion must be replaced as a matched pair. The ring and pinion must have matching numbers. If the ring and pinion gears are the nonhunting type, mark them before disassembly.

Check the ring gear flange for runout. Attach the ring gear with special fasteners. The contact surfaces of the ring and pinion must be spotless. Lubricate ring gear fasteners or coat with recommended liquid adhesive before assembly. Fasteners must be a snug fit in the ring and flange. When removing the pinion shaft, mark and save shims. Do not damage the bearing counterbore during the removal of pinion outer races.

Set proper pinion depth with a special setting gauge. When using a new drive pinion, correct the depth setting by allowing for the plus or minus amount marked on the pinion. Shims must be clean and of the proper thickness to provide exact pinion depth. When installing the pinion, preload must meet specifications. Use a new seal.

Before installing the differential case, lubricate the entire assembly thoroughly. If the ring and pinion gearset is the nonhunting type, mesh the marked teeth during case assembly. Adjust the case side bearing preload as specified. Check ring and pinion backlash. Check the ring and pinion tooth contact pattern by coating the ring gear with a special compound. Load the ring gear and turn the pinion until the ring makes one full turn. Do this in both drive and coast directions. The contact pattern should be centralized between bottom and top of the tooth. The pattern may be slightly toward the toe.

To correct a faulty pattern, adjust the pinion in or out by moving the ring toward or away from the pinion. Traction-type differentials use clutch packs or cone brakes to provide power to both axles, even when one wheel is spinning.

Traction differentials may be tested in the vehicle by measuring the torque required to turn the raised wheel. The other wheel must contact the floor. The transmission and/or transfer case must be in neutral, with emergency brake off. Lubricate all traction differential parts with recommended lubricant during assembly. Fill differential housing to within 1/2″ (12.70 mm) of the filler plug or as recommended by the manufacturer. Use multipurpose gear lubricant for conventional differentials and a special lubricant for traction-type units if required. Do not allow dirt to enter the housing when checking lubricant level or when filling. The axle vent must be open.

Review Questions—Chapter 32

Do not write in this book. Write your answers on a separate sheet of paper.

1. The two-piece drive shaft utilizes a _____ bearing.
2. The bearing caps in the cross and bearing cap joint are retained by _____.
   (A) snap rings
   (B) U-bolts
   (C) cap screws
   (D) One or more of the above, depending on manufacturer.
3. To allow driveline length to change, a _____ _____ is installed in the drive shaft.
4. What portion of a drive shaft is easily damaged by clamping in a vise?
5. If the bearing caps are held in place by injection-molded plastic, what should be done before they are removed?
   (A) Heat them in a oil bath.
   (B) Heat them with a torch.
   (C) Chill them with dry ice.
   (D) Strike them hard with a brass hammer.
6. The _____ on each end of the drive shaft should be in the same plane.
7. Before removing a drive shaft, it is important the shaft and pinion flange or yoke be _____ to preserve shaft _____.
8. List the two basic types of CV joints.
9. Some CV axles are held to the transaxle by an internal _____.
10. Most drive axles found on cars and light trucks are of the _____ design.
    (A) full-floating
    (B) three-quarter floating
    (C) semifloating
    (D) non-floating
11. Most axles are retained in the housing by using a _____ plate or by using _____ on the axle inner ends.

12. List three things that can make noises that can be mistaken for drive axle problems.
13. A rear axle noise that varies during sudden turns to the left or right is probably caused by _____.
    (A) axle bearings
    (B) front wheel bearings
    (C) rear axle gears
    (D) worn U-joint(s)
14. Before removing a pinion flange, check the bearing _____.
15. Hold the pinion flange with a _____ while loosening the retaining nut.
    (A) crescent wrench
    (B) pipe wrench
    (C) large screwdriver
    (D) special tool
16. If the differential carrier is the integral type, it may be necessary to remove the _____ to service the internal components.
17. When tightening a pinion-flange retaining nut on installations using a collapsible spacer, what must be done if the preload is exceeded?
    (A) Install a new spacer.
    (B) Back off the nut until preload is correct.
    (C) Back off the nut past correct preload and tighten until preload is correct.
    (D) Leave preload alone.
18. Axle wheel bearing retainer rings may be removed from the axle after _____.
    (A) heating
    (B) chilling
    (C) notching with a chisel
    (D) cutting with a torch
19. The wheel hub on a tapered end axle is kept from turning by a _____.
20. Pinion depth is adjusted by placing shims between the _____ and _____.
    (A) bearing, case
    (B) gear, bearing cone
    (C) case, gear
    (D) either A or B, depending on the design
21. Proper ring and pinion tooth contact pattern is important. The pattern should generally be _____.
    (A) centralized on the tooth
    (B) centralized between top and bottom of the tooth but closer to the toe
    (C) centralized between top and bottom of the tooth but closer to the heel
    (D) near the top of the tooth and extending over the entire length
22. In the hunting-type ring and pinion gear set, any one tooth on the pinion will contact _____ on the ring.
    (A) all teeth
    (B) certain teeth
    (C) one other tooth
    (D) either A or B, depending on the design
23. When checking backlash between the ring and pinion, place the dial indicator stem against one of the _____ gear teeth.
24. Tooth contact pattern is adjusted by moving either the _____ or the _____.
25. When one rear wheel is spinning, the limited-slip differential unit applies power to the wheel that _____.
    (A) is spinning
    (B) is not spinning
    (C) has the most traction
    (D) Both B & C.

ASE-Type Questions

1. The Hotchkiss-type drive shaft is sometimes called a(n) _____ design.
    (A) closed
    (B) open
    (C) hypoid
    (D) semifloating
2. Technician A says if the cross is in good condition but the bearing caps are not, it is permissible to replace the bearing cap only. Technician B says some U-joints are sealed and must be lubricated before assembly. Who is right?
    (A) A only.
    (B) B only.
    (C) Both A and B.
    (D) Neither A nor B.
3. Technician A says minor drive shaft imbalance can sometimes be corrected with hose clamps. Technician B says a newly assembled U-joint should have a rather stiff action until it is broken in. Who is right?
    (A) A only.
    (B) B only.
    (C) Both A and B.
    (D) Neither A nor B.
4. All of the following are the result of a torn CV boot, *except:*
    (A) CV joint grease is thrown out.
    (B) dirt and water enter the CV joint.
    (C) the CV axle is bent.
    (D) the CV joint wears out.

5. Technician A says front-wheel drive axles can be of a one- or two-piece design. Technician B says rear-wheel drive shafts can be of a one- or two-piece design. Who is right?
   (A) A only.
   (B) B only.
   (C) Both A and B.
   (D) Neither A nor B.
6. A rear axle noise that varies between drive and coast is probably caused by _____.
   (A) axle bearings
   (B) front wheel bearings
   (C) rear axle gears
   (D) worn U-joint(s)
7. All of the following statements about axle replacement are true, *except:*
   (A) always install a new oil seal whenever an axle has been removed.
   (B) when removing or installing bearings, use the proper equipment to protect against flying parts, as bearings can explode.
   (C) the brake backing plate must always be removed before pulling an axle.
   (D) whenever an axle has broken or other damage has occurred, always inspect the housing thoroughly.
8. Technician A says differential side bearing shims may be mixed and used on either side. Technician B says ring and pinion gears must be replaced as matched sets. Who is right?
   (A) A only.
   (B) B only.
   (C) Both A and B.
   (D) Neither A nor B.
9. Technician A says pinion depth can be set satisfactorily by measuring from the pinion gear to the axle centerline with an accurate ruler. Technician B says a pinion gear marked –2 must be shimmed so that it will operate 0.002″ closer to the ring gear centerline than a pinion marked 0. Who is right?
   (A) A only.
   (B) B only.
   (C) Both A and B.
   (D) Neither A nor B.
10. Technician A says the limited-slip clutch on the spinning wheel is released. Technician B says limited-slip differentials should always be lubricated with regular multipurpose gear lubricant. Who is right?
   (A) A only.
   (B) B only.
   (C) Both A and B.
   (D) Neither A nor B.

Student Activities

1. Draw a sketch showing the differences between one- and two-piece drivelines used on rear-wheel drive vehicles. Illustrate the center support bearing on the two-piece design.
2. Inspect a rear-wheel drive shaft for dry, loose U-joints and/or damage to the shaft. Make a report for your instructor.
3. Inspect a front CV axle shaft for loose CV joints and leaking boots. Make a report for your instructor.
4. Replace a U-joint or CV joint. Write a report listing the replacement procedures in the proper order, and the tools and supplies needed.
5. Calculate the axle ratio on a vehicle by turning one wheel (other wheel stationary) and counting the revolutions of the drive shaft. Does this agree with the published specifications for the vehicle? Discuss the results with your instructor and the other members of your class.

Rear Axle Problem Diagnosis

Problem: Noise during straight-ahead driving

Possible cause	Correction
1. Insufficient lubricant.	1. Fill housing to correct level.
2. Improper lubricant.	2. Drain. Flush and fill with correct lubricant.
3. Differential case bearings worn.	3. Replace bearings.
4. Drive pinion shaft bearings worn.	4. Replace pinion bearings.
5. Ring and pinion worn.	5. Replace ring and pinion.
6. Excessive backlash.	6. Adjust backlash.
7. Insufficient backlash.	7. Adjust backlash.
8. Excessive ring and pinion backlash.	8. Adjust backlash.
9. Insufficient ring and pinion backlash.	9. Adjust backlash.
10. Pinion shaft or differential case bearings not preloaded.	10. Preload as specified by the manufacturer.
11. Excessive ring gear runout.	11. Remove ring, clean, and check flange runout. Reinstall and check runout. Replace ring or case as needed.
12. Ring gear fasteners loose.	12. Torque fasteners.
13. Ring and pinion not matched.	13. Install a matched ring and pinion set.
14. Differential case bearing cap fasteners loose.	14. Torque fasteners.
15. Warped housing.	15. Replace housing.
16. Pinion shaft companion flange retaining nut loose.	16. Torque flange nut.
17. Tooth (ring and pinion) contact pattern incorrect.	17. Adjust as needed.
18. Loose wheel.	18. Tighten wheel lugs.
19. Wheel hub loose on tapered axle.	19. Inspect, if not damaged, torque retaining nut.
20. Wheel hub key (on tapered axle) sheared.	20. Install new key.
21. Wheel (axle) bearing worn.	21. Replace axle bearing and seal.
22. Bent axle.	22. Replace axle.
23. Wheel hub or axle keyway worn.	23. Replace axle or hub as needed.
24. Dry pinion shaft seal.	24. Replace pinion shaft seal.
25. Loose universal joint retainers.	25. Tighten universal joint retainers.
26. Damaged universal joint.	26. Replace universal joint.
27. Worn or broken front-wheel drive front suspension parts.	27. Repair or replace as necessary.
28. Worn or broken transaxle unit.	28. Repair or replace as needed.

Problem: Noise when rounding a curve

Possible cause	Correction
1. Differential pinion gears worn or broken.	1. Replace gears.
2. Differential pinion shaft worn.	2. Replace pinion shaft.
3. Axle side gears worn or broken.	3. Replace side gears.
4. Excessive axle side gear or pinion gear end play.	4. Install new thrust washers or replace case and/or gears.
5. Excessive axle end play.	5. Adjust end play.
6. Improper type of lubricant.	6. Drain. Flush and fill with correct lubricant.
7. Loose or broken suspension parts (front-wheel drive).	7. Repair or replace parts as needed.
8. Loose or broken universal joints.	8. Tighten or replace universal joints.

Problem: Clunking sound when engaging clutch, accelerating, or decelerating

Possible cause	Correction
1. Excessive ring and pinion backlash.	1. Adjust backlash.
2. Excessive end play in pinion shaft.	2. Preload bearings.
3. Axle side gears and pinions worn.	3. Replace worn gears.
4. Differential bearings worn.	4. Replace bearings.
5. Side gear thrust washers worn.	5. Replace thrust washers.
6. Differential pinion shaft loose in case or pinions.	6. Replace pinion shaft, gears, or differential case.
7. Axle shaft splines worn.	7. Replace axle.
8. Wheel hub or axle keyway worn.	8. Replace hub or axle.
9. Loose wheel or hub.	9. Tighten fasteners.
10. Loose or broken universal joints.	10. Tighten or replace universal joints.

(continued)

Rear Axle Problem Diagnosis *(continued)*	
Problem: Axle leaking lubricant	
Possible cause	**Correction**
1. Breather clogged. 2. Worn seals. 3. Carrier-to-housing or inspection cover loose. 4. Carrier or inspection cover gasket damaged. 5. Lubricant level too high. 6. Wrong type of lubricant. 7. Porous housing (standard and transaxle). 8. Stripped fill plug threads. 9. Cracked housing (standard and transaxle).	1. Open breather. 2. Install new seals. 3. Tighten fasteners. 4. Install new gasket or sealer. 5. Drain lubricant to proper level. 6. Drain. Flush and install correct lubricant. 7. Repair or replace housing. 8. Repair or replace as needed. 9. Repair or replace housing.
Problem: Noises that may be confused with drive axle assembly	
Possible cause	**Correction**
1. Low air pressure in tires. 2. Road surface. 3. Transmission. 4. Bent propeller shaft. 5. Loose U-joints. 6. Engine. 7. Front wheel bearings. 8. Tire tread. 9. Dragging brakes. 10. Excessive front wheel end play.	1. Inflate tires to proper pressure. 2. Test on several different road surfaces. 3. Check transmission. 4. Replace shaft. 5. Tighten or replace U-joints. 6. Check engine. 7. Replace bearings. 8. Inflate temporarily to 50 psi (34.5 kPa) for road test only. 9. Adjust brakes. 10. Adjust wheel bearings.
Problem: Rear axle overheating	
Possible cause	**Correction**
1. Wrong type of lubricant. 2. Insufficient lubricant. 3. Overloading (pulling heavy trailer). 4. Gears worn. 5. Bearing preload too great. 6. Insufficient backlash between ring and pinion.	1. Drain, flush, and fill with correct lubricant. 2. Fill lubricant to proper level. 3. Reduce vehicle load. Advise driver. 4. Replace gears. 5. Adjust preload as specified. 6. Adjust backlash.

33

Four-Wheel Drive Service

After completing this chapter, you will be able to:

- Explain four-wheel drive construction and operation.
- Define part-time and full-time four-wheel drive.
- Diagnose mechanical and electronic transfer case problems.
- Disassemble, check parts, and reassemble a transfer case.
- Disassemble, check parts, and reassemble a typical locking hub assembly.
- Diagnose four-wheel drive problems.

Technical Terms

Transfer case
Part-time
Full-time
Sliding clutch
Windup
Differential assembly
Viscous clutch
Pattern failures
Locking hubs
Overrunning clutches
Vacuum motor

This chapter covers the design and operation of four-wheel drive systems, with an emphasis on the transfer case. This chapter also explains the differences between part-time and full-time four-wheel drive systems. Theory and service principles of electronic transfer cases are also covered. The design, operation, and service of transfer cases and locking hubs are also covered.

Four-Wheel Drive Components

Modern four-wheel drive systems may be used with either a manual or an automatic transmission. All four-wheel drive systems use a ***transfer case*** between the transmission output and the drive shafts. The purpose of the transfer case is to divide the power flow so that both the front and rear wheels are driven by engine power. All four-wheel drive systems have two drive shafts. Systems that are constantly engaged require special components to compensate for the differences in the speeds of the front and rear axles during turns.

Transfer Case

The transfer case is needed to apply transmission output-shaft torque to the front and rear drive shafts. Engine power enters the transfer case from the transmission's output shaft. Power is transmitted to each drive shaft through a set of gears or a chain and sprockets. In some designs, a differential assembly or a viscous coupling is used to compensate for changes in front and rear axle speeds. **Figure 33-1** illustrates a four-wheel drive setup with a transfer case. Note how both drive shafts are attached to the transfer case. Transfer cases can be either ***part-time*** of ***full-time*** units.

Part-Time Transfer Case Operation

On modern part-time transfer cases, the driver can select one of three different drive modes by operating a lever on the vehicle console or floor:

- Two-wheel drive—high range.
- Four-wheel drive—high range.
- Four-wheel drive—low range.

Although it would be possible to operate in the two-wheel drive—low range, the transfer case shift linkage does not allow selecting this mode, **Figure 33-2.** Trace the power flow starting with the transmission output shaft. The output shaft is splined to and drives the transfer case's main drive gear. The main drive gear is in constant mesh with the idler cluster gear. The idler cluster turns freely in needle bearings on the idler shaft. Note the idler cluster is made up of two gears, one large (high range) and one small (low range). These gears are in constant mesh with the output gears. Both output gears turn freely in bushings on the transfer case output shaft.

The ***sliding clutch*** is splined to the output shaft. In the neutral position (shown), it is centered between, but not engaging, the output gears. Low range (high engine RPM, low wheel RPM) is engaged when the sliding clutch is shifted into engagement with the low-range output gear. High range (low engine RPM, high wheel RPM) is engaged when the sliding clutch is moved into engagement with the high-range output gear. The output gear that is engaged with the clutch turns the sliding clutch. Since the clutch is splined to the output shaft, the shaft also turns.

Two-Wheel, Four-Wheel Drive Action

The transfer case output shaft is made of two parts, Figure 33-2. The long section is driven by the sliding splined clutch. This long output shaft section, in turn, drives the rear drive shaft. The shorter front section is connected to the long

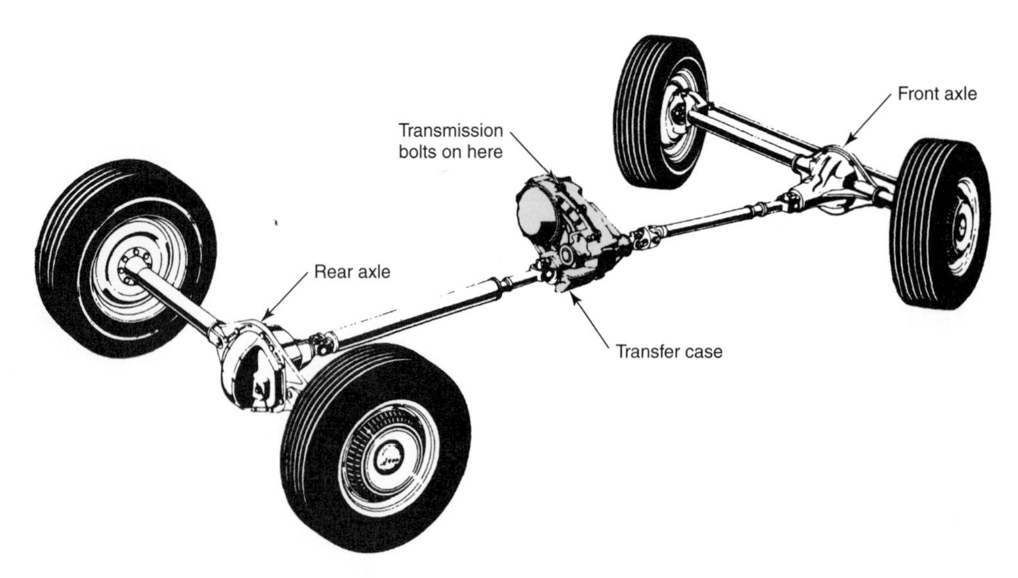

Figure 33-1. A typical four-wheel drive setup is shown here. Note how the transfer case drives both front and rear drive shafts. (DaimlerChrysler)

section only when the four-wheel drive sliding splined clutch (which is splined to and constantly turned by the long section) is moved into engagement with the splined end of the short shaft. When two-wheel drive is selected, the four-wheel sliding clutch disengages from the short output shaft. The long shaft then drives the rear drive shaft, but does not apply torque to the front. Four-wheel drive mode is accomplished by merely moving the four-wheel clutch into engagement with the short shaft. Torque is then applied to both front and rear drive shafts.

Vehicles equipped with this type of transfer case must never be operated in four-wheel drive on dry, hard-surfaced roads. The vehicle's front wheels rotate slightly faster, because they follow a more curved path than the rear. As a result, ***windup,*** or internal stresses between the front and rear axle

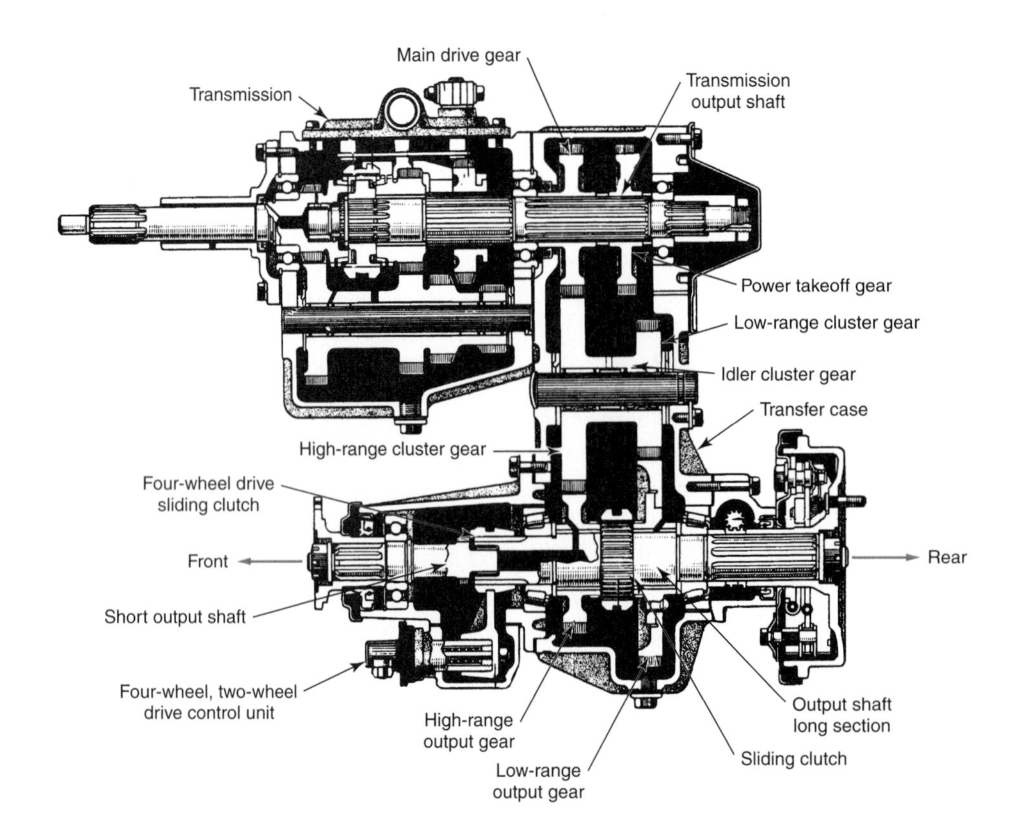

Figure 33-2. This figure shows one type of transfer case used on a part-time four-wheel drive system. (Toyota)

parts builds up in the entire drive train until something breaks. Either the tires will break loose or slip on the pavement or a drive train part will be damaged. Since it is very difficult for the tires to break loose on hard, dry surfaces, the vehicle will probably be damaged.

On another type of transfer case, the front drive shaft and ring gear assembly is driven at all times. A vacuum motor is used to disconnect the front wheels from the front axle, **Figure 33-3.** The vacuum motor operates a shift collar installed in the front drive axle, which is similar to the collar

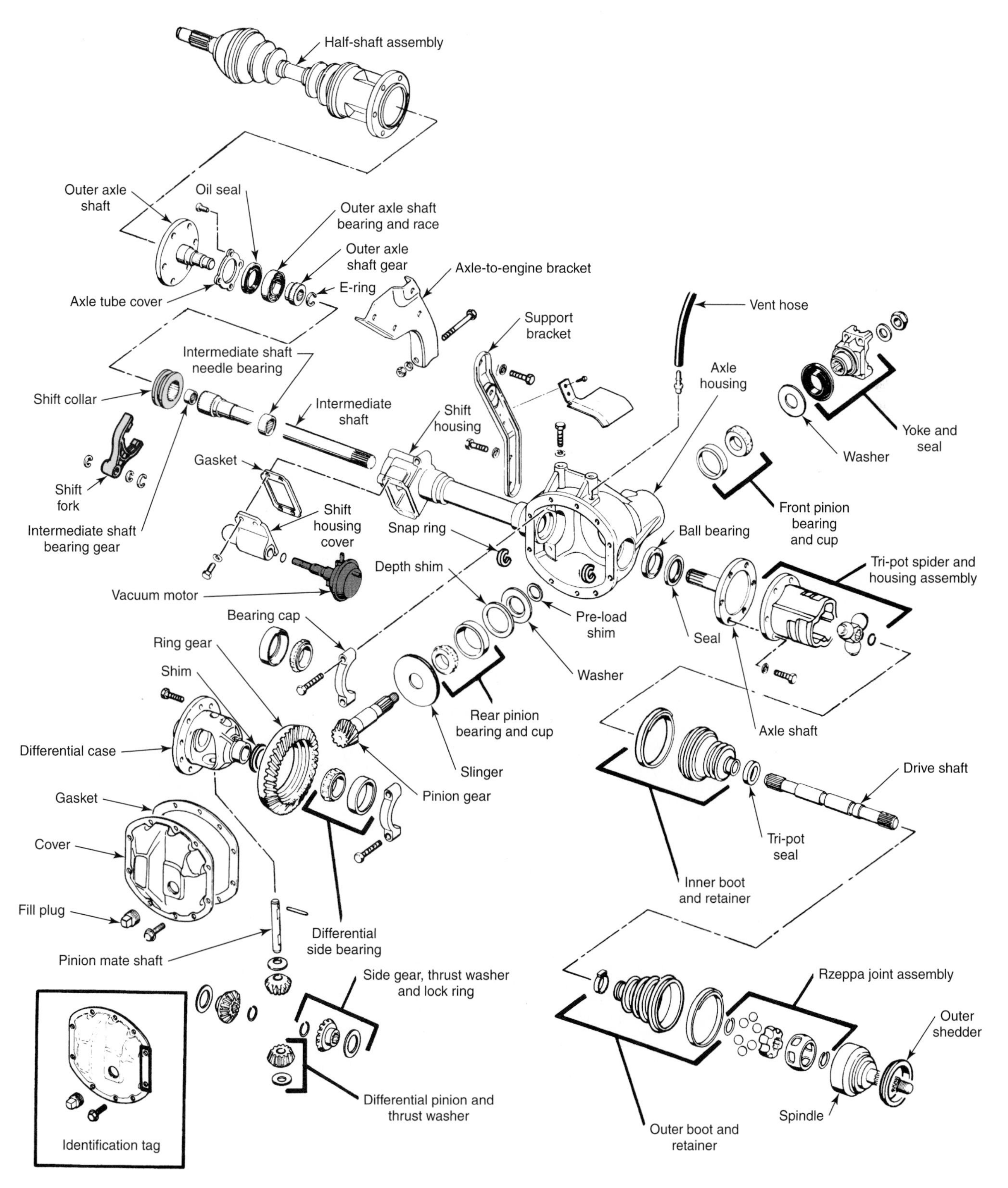

Figure 33-3. This type of front axle assembly is driven continuously by the transfer case. The axle is shifted in and out of four-wheel drive with a vacuum motor and shift fork unit. (DaimlerChrysler)

used with a manual transmission synchronizer. This design eliminates windup by disconnecting the front drive axles from the front differential carrier.

Full-Time Four-Wheel Drive

Some transfer cases permit the vehicle to constantly operate in four-wheel drive on any road surface. These units make use of a ***differential assembly*** or a ***viscous clutch*** to allow for differences in front and rear axle speeds. Although these assemblies work in different ways, they both accomplish the same thing. They are discussed in detail below.

Full-Time Transfer Case with Differential Unit

One manufacturer's full-time transfer case is shown in **Figure 33-4.** Note the drive chain transmits power from the drive sprocket to the differential case sprocket. A differential unit inside the case applies driving torque to both the front and rear transfer case output shafts. The differential allows the front and back shafts to rotate at different speeds while still applying power. This keeps drive train windup from damaging parts and permits constant use of four-wheel drive. This unit employs a limited-slip differential that provides some drive to either the front or rear axle if one wheel is spinning. For very severe traction situations, the unit has a mechanism to lockout (stop) the differential action. This provides drive to both front and rear drive shafts. Never use the lockout control on dry, hard roads. Use it only when stuck or under very poor traction situations.

An exploded view of the transfer case is illustrated in **Figure 33-5.** This transfer case is available with or without a low-range reduction unit. Another type of full-time transfer case is pictured in **Figure 33-6.** It uses a chain drive and a

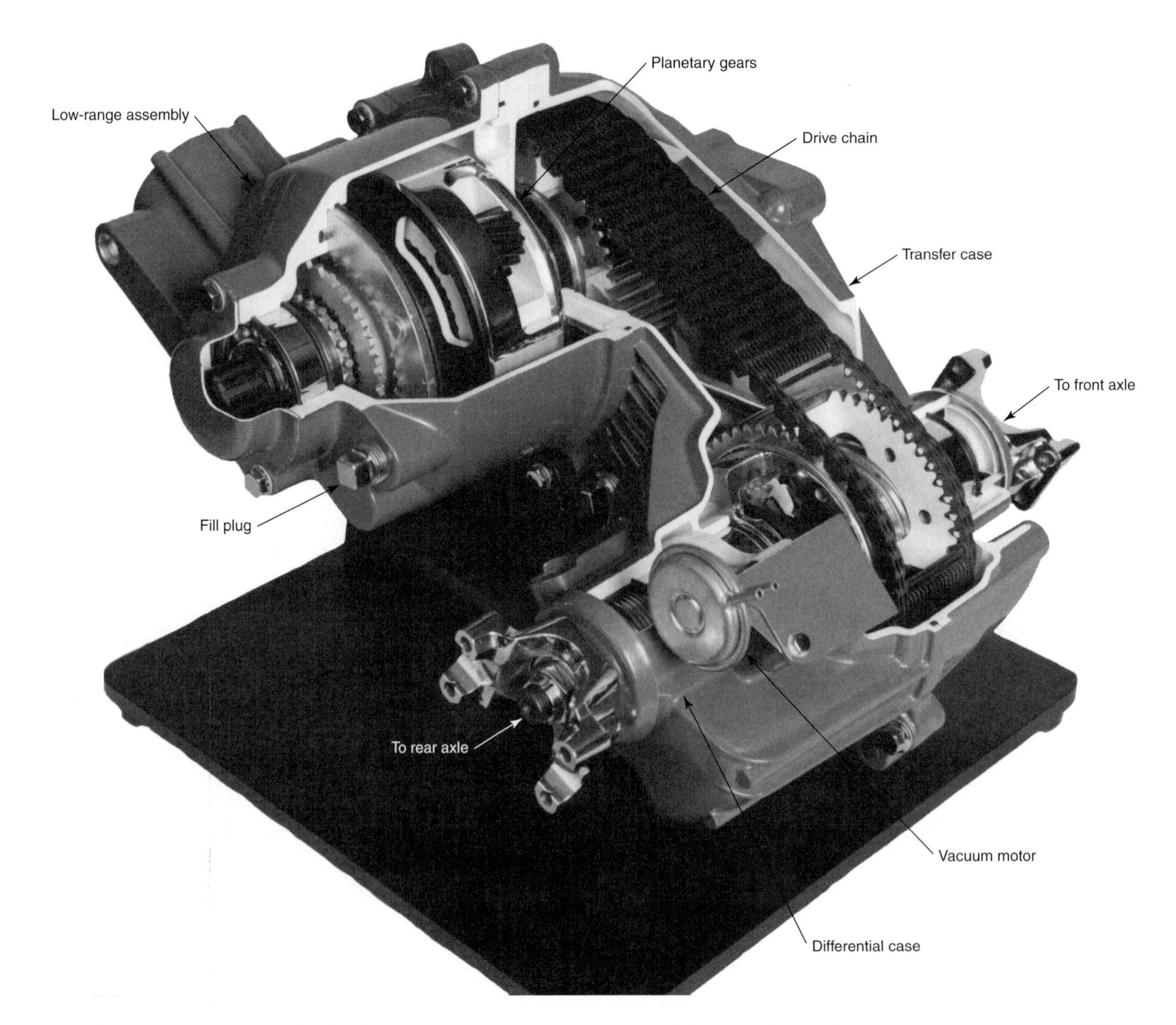

Figure 33-4. A cutaway view of a full-time transfer case is shown here. Note the use of a chain to drive the differential unit. This case is also equipped with a planetary gear, low-range assembly. (Warner Gear)

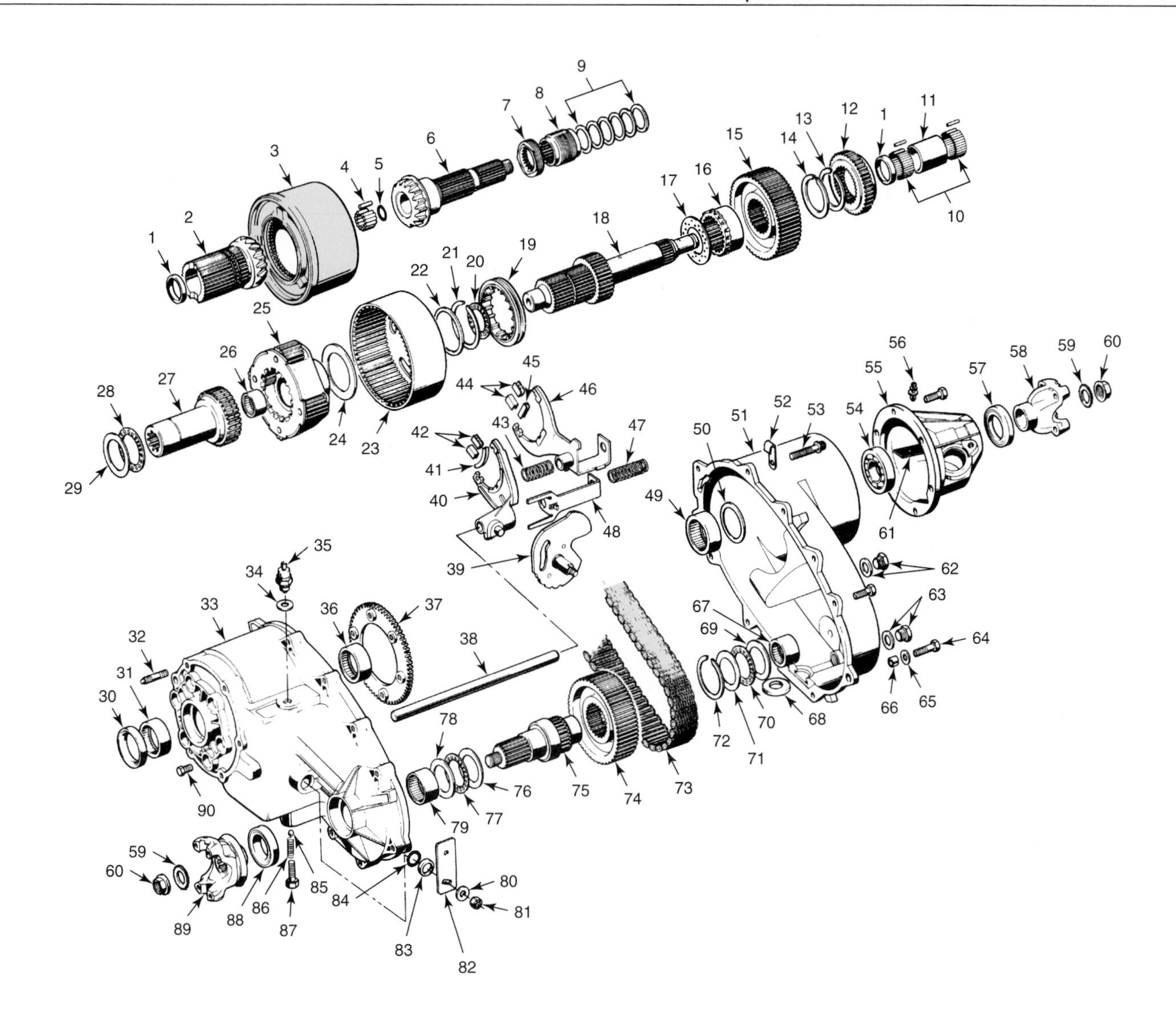

1. Mainshaft rear bearing spacer–short (2)
2. Side gear
3. Viscous coupling and differential assembly
4. Mainshaft rear pilot roller bearings (15)
5. Mainshaft O-ring
6. Rear output shaft
7. Oil pump
8. Speedometer gear
9. Differential end play shims
10. Mainshaft needle bearings (82)
11. Mainshaft rear bearing spacer
12. Clutch gear
13. Clutch gear locating ring
14. Drive sprocket locating ring
15. Drive sprocket
16. Side gear clutch
17. Mainshaft thrust washer
18. Mainshaft
19. Clutch sleeve
20. Mainshaft thrust bearing
21. Annulus gear retaining ring
22. Annulus gear thrust washer
23. Annulus gear
24. Planetary thrust washer
25. Planetary assembly
26. Mainshaft front pilot bearing
27. Input gear
28. Input gear thrust bearing
29. Input gear thrust bearing race
30. Input gear oil seal
31. Input gear front bearing
32. Front case mounting stud (6)
33. Front case
34. Lock mode indicator switch gasket
35. Lock mode indicator switch
36. Input gear rear bearing
37. Low-range lockplate
38. Shift rail
39. Range sector
40. Range fork
41. Range fork insert
42. Range fork pads
43. Mode fork spring
44. Mode fork pads
45. Mode fork insert
46. Mode fork
47. Shift rail spring
48. Mode fork bracket
49. Rear output shaft bearing
50. Rear output shaft bearing seal
51. Rear case
52. Wiring clip
53. Spline bolt
54. Rear output bearing
55. Rear retainer
56. Vent
57. Output shaft oil seal
58. Rear yoke
59. Yoke seal washer
60. Yoke locknut
61. Vent chamber seal
62. Fill plug and gasket
63. Drain plug and gasket
64. Rear case bolt
65. Washer (2)
66. Case alignment dowel
67. Front output shaft rear thrust bearing
68. Magnet
69. Front output shaft rear thrust bearing race (thick)
70. Front output shaft rear thrust bearing
71. Front output shaft rear thrust bearing race (thin)
72. Driven sprocket retaining snap ring
73. Drive chain
74. Driven sprocket
75. Front output shaft
76. Front output shaft front thrust bearing race (thin)
77. Front output shaft front thrust bearing
78. Front output shaft front thrust bearing race (thick)
79. Front output shaft front bearing
80. Washer
81. Locknut
82. Operating lever
83. Range sector shaft seal retainer
84. Range sector shaft seal
85. Detent ball
86. Detent spring
87. Detent retaining bolt
88. Front output shaft seal
89. Front yoke
90. Lockplate bolts

Figure 33-5. This figure shows an exploded view of a full-time transfer case. (DaimlerChrysler) standard (nonlimited-slip) differential unit. Low and high range is accomplished through gearing. This unit may be locked out when wheel spin is experienced. The lockout must be disengaged whenever the vehicle is driven on dry paved roads.

Full-Time Transfer Case with Viscous Coupling

This transfer case operates somewhat like the differential-equipped transfer case. The basic difference being the viscous coupling used in place of the mechanical limited-slip

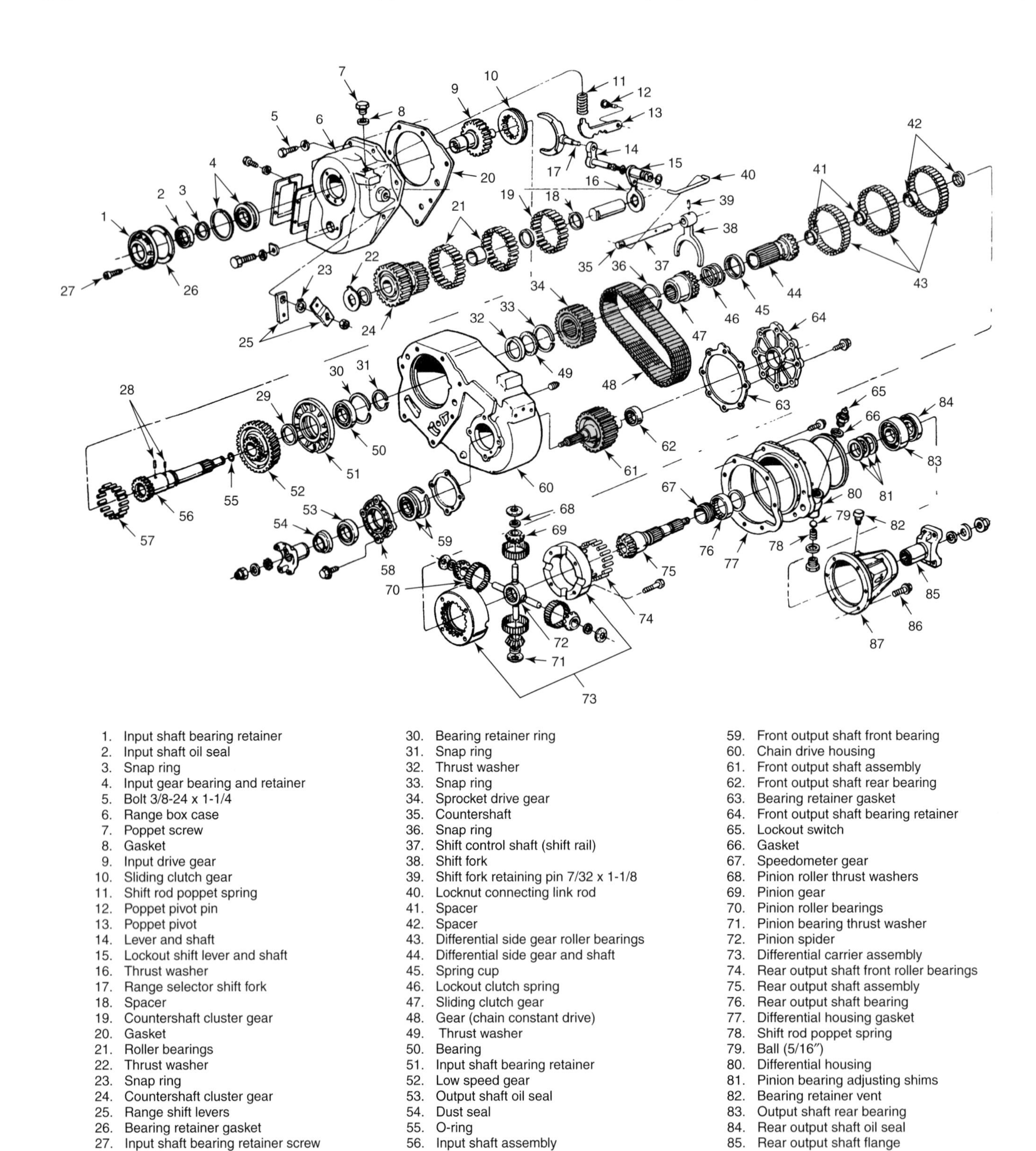

Figure 33-6. This figure shows an exploded view of a transfer case that uses a chain to drive the differential unit. It does not employ a limited-slip feature in the differential. It does not use planetary gears to achieve low range. Conventional gearing is used. (Ford)

differential. See **Figure 33-7.** The viscous coupling contains a special silicone fluid that acts as a torque biasing, limited-slip unit. The coupling is attached to the front drive shaft by the side gear and drive sprocket. These turn the driven sprocket and front output shaft. The two sprockets are connected with a drive chain. The rear drive shaft is attached to the viscous coupling through the teeth on the rear output-shaft side gear. These teeth are meshed with the differential pinions.

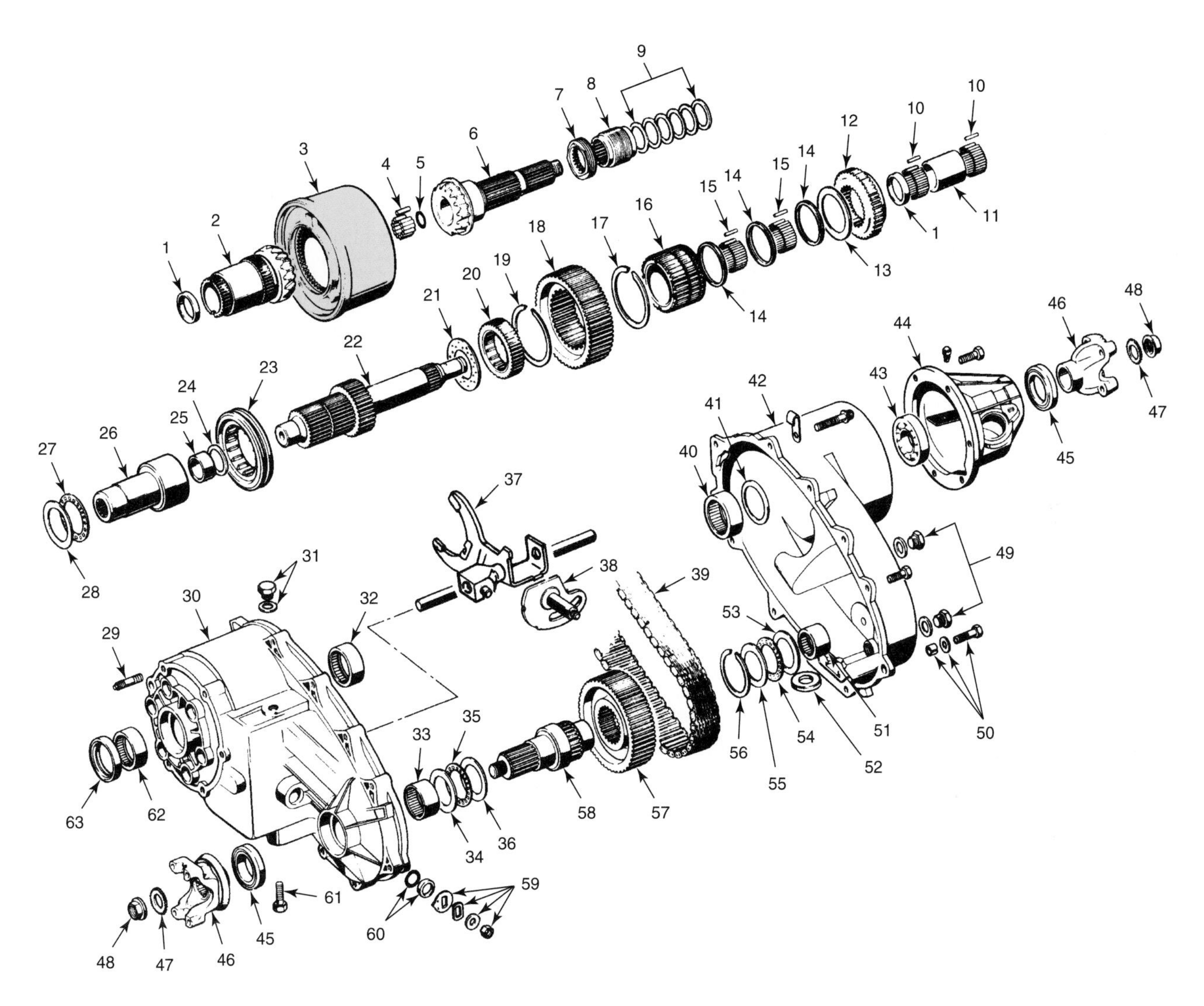

1. Mainshaft bearing spacers (short) (2)
2. Side gear
3. Viscous coupling
4. Mainshaft pilot roller bearings
5. Mainshaft O-ring
6. Rear output shaft
7. Oil pump
8. Speedometer drive gear
9. Differential shims
10. Mainshaft needle bearings (82)
11. Mainshaft needle bearing spacer (long) (1)
12. Clutch gear
13. Clutch gear thrust washer
14. Sprocket carrier needle bearing spacer (3)
15. Sprocket carrier needle bearings (120)
16. Sprocket carrier
17. Sprocket carrier snap ring
18. Drive sprocket
19. Sprocket carrier snap ring
20. Spline gear
21. Mainshaft thrust washer
22. Mainshaft
23. Clutch sleeve
24. Mainshaft thrust washer
25. Mainshaft bushing
26. Input gear
27. Input gear thrust bearing
28. Input gear thrust bearing race
29. Mounting gear
30. Front case
31. Plug and washer
32. Input gear rear bearing
33. Front output shaft front bearing
34. Front output shaft front thrust bearing race (thick)
35. Front output shaft front thrust bearing
36. Front output shaft front thrust bearing race (thin)
37. Range fork and rail
38. Range sector
39. Drive chain
40. Rear output shaft bearing
41. Rear output shaft bearing seal
42. Rear case
43. Rear output bearing
44. Rear retainer
45. Yoke seal
46. Yoke
47. Seal washer
48. Yoke nut
49. Fill and drain plugs
50. Alignment dowel, washer and bolt
51. Front output shaft rear bearing
52. Magnet
53. Front output shaft rear thrust bearing race (thick)
54. Front output shaft rear thrust bearing
55. Front output shaft rear thrust bearing race (thin)
56. Driven sprocket retaining snap ring
57. Driven sprocket
58. Front output shaft
59. Range sector shaft retaining locknut and washers
60. Range sector shaft seal and retainer
61. Positive lock detent bolt
62. Input gear front bearing
63. Input gear seal

Figure 33-7. An exploded view of a viscous coupling transfer case assembly is shown here. This case can be shifted manually or with vacuum motors. (DaimlerChrysler)

During normal operation, the viscous coupling does not operate. When excessive wheel spin is present, the coupling transfers torque to the axle having the best traction, either front or rear. The silicone fluid in the coupling is very thick, or viscous, and will not thin due to heat. As one axle overspeeds because of wheel slippage, the coupling rotational speed also increases. As the coupling rotational speed increases, the fixed clutch plates are forced to rotate in the silicone fluid as they increase in speed. As fluid is forced between the clutch plate and displaced, it expands, creating friction and more resistance to higher input speed. The coupling does not lock the axles together; it controls the amount of slipping. At the same time, it sends maximum torque to the slower-moving axle. Transfer case power flow is shown in **Figure 33-8.**

Caution: The viscous coupling and pinion assembly are not serviceable at a component level. They are sealed together and the silicone fluid is not replaceable. If the coupling and/or pinions are damaged, replace the entire unit with a new one.

Electronic Transfer Cases

On many modern vehicles with four-wheel drive, mechanical shift linkage has been replaced with electronic controls. Vehicles with electronic transfer cases use an on-board computer to select transfer case ranges. As with mechanical transfer cases, the purpose of any electronically controlled transfer case is to select either two- or four-wheel drive to provide the best traction for any road conditions. The electronic transfer case makes selections without requiring the driver to make decisions or operate levers.

The major components of an electronic transfer case are:

- Control computer. The computer may be called an *ECM* (electronic control module), *PCM* (power train control module), or *TCCM* (transfer case control module).
- Input sensors. The vehicle speed sensor is the most important input, but many other sensors affect transfer case operation.
- Output motors. The output motor is installed in the transfer case. It operates the shift fork to change from two- to four-wheel drive.

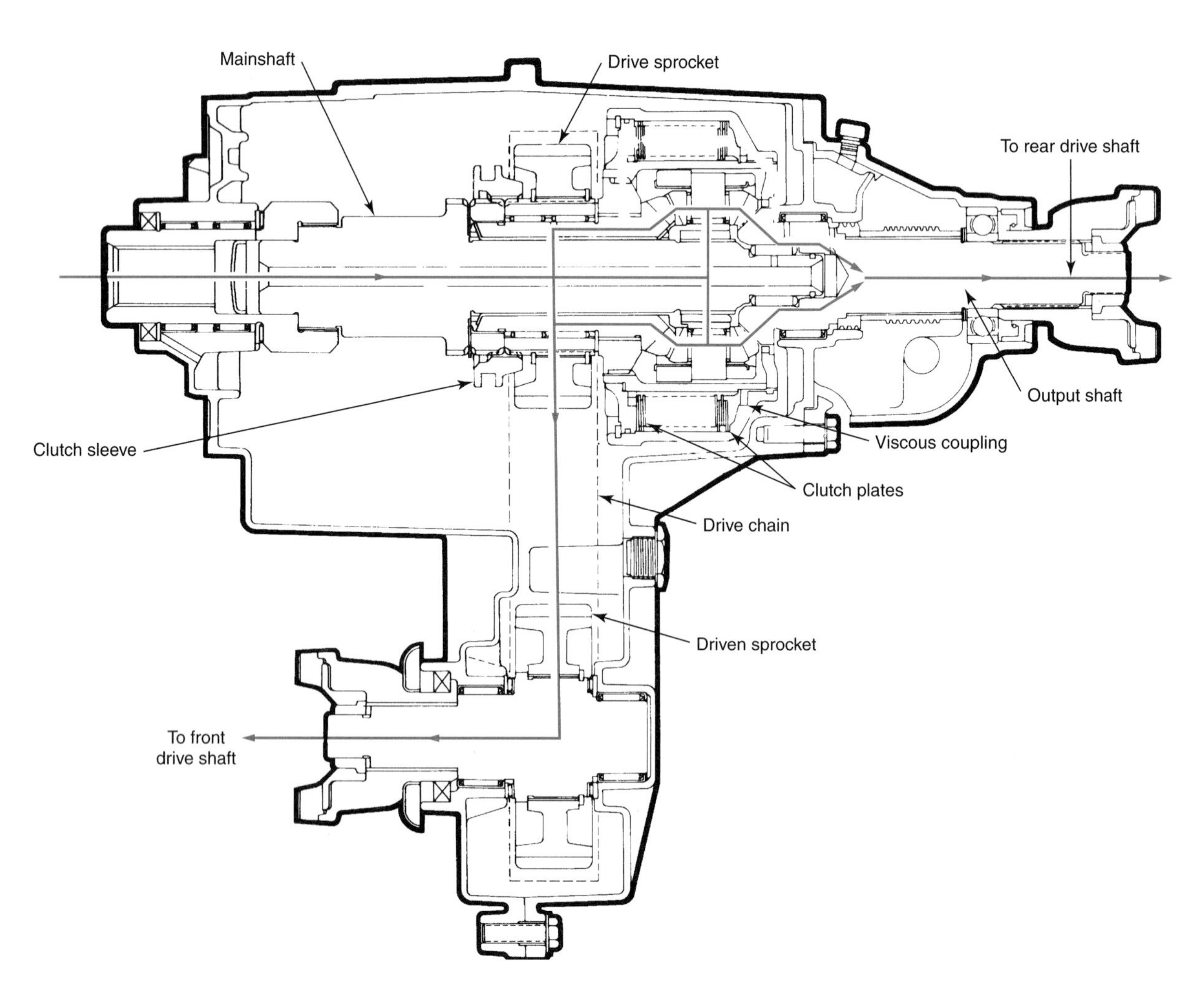

Figure 33-8. This figure shows a power flow schematic of a viscous coupling transfer case. The solid line is the flow of power to the front and rear axles. The dotted line represents power flow through the coupling when excessive wheel spin (loss of traction) is encountered. (DaimlerChrysler)

Figure 33-9 shows a cross section of a popular electronically controlled transfer case. The motor operates an actuator arm that applies the clutch pack. Other than the motor control of clutch pack application, the transfer case operates in the same manner as other transfer cases. A motor is also used on the front axle to engage and disengage the front axles and differential carrier.

Transfer Case Service

The following sections describe the service procedures for typical transfer cases. These procedures are general, due to the many types of four-wheel drive systems in service. Always obtain and use the proper manufacturer's service manual. Follow the manufacturer's directions for transfer case lubricant type and viscosity. Check the lubricant level and do not overfill. If periodic changes are recommended, be sure to use the correct type of fluid. Any transfer case using a limited-slip differential requires a special lubricant or lubricant additive. Talk to the owner about observations and complaints regarding transfer case operation. Ask questions to help pinpoint possible trouble areas. Make a list of possible problems based on the owner's statements.

Whenever possible, the vehicle should be road tested. Check the transfer case lubricant level before road testing. **Figure 33-10.** If the unit has a viscous coupling, check the coupling for leakage by inspecting the transfer case oil. Leaking silicone appears as globules of silicone fluid in the transfer case oil. When performing the road test, include some heavy acceleration and deceleration. Operate the vehicle at various speeds. When possible, the route should include some bumpy sections and a hill. Check closely for abnormal noise, jumping out of gear, vibration, hard shifting, gear clash during shifting, and leaks. Note the transfer case position (two-wheel drive—high range, four-wheel drive—high range, four-wheel drive—low range) in which any problem was most evident.

Check Clutch and Shift Linkage

If the owner complains of hard shifting, gear clash, or jumping out of gear, check the clutch and shift linkage

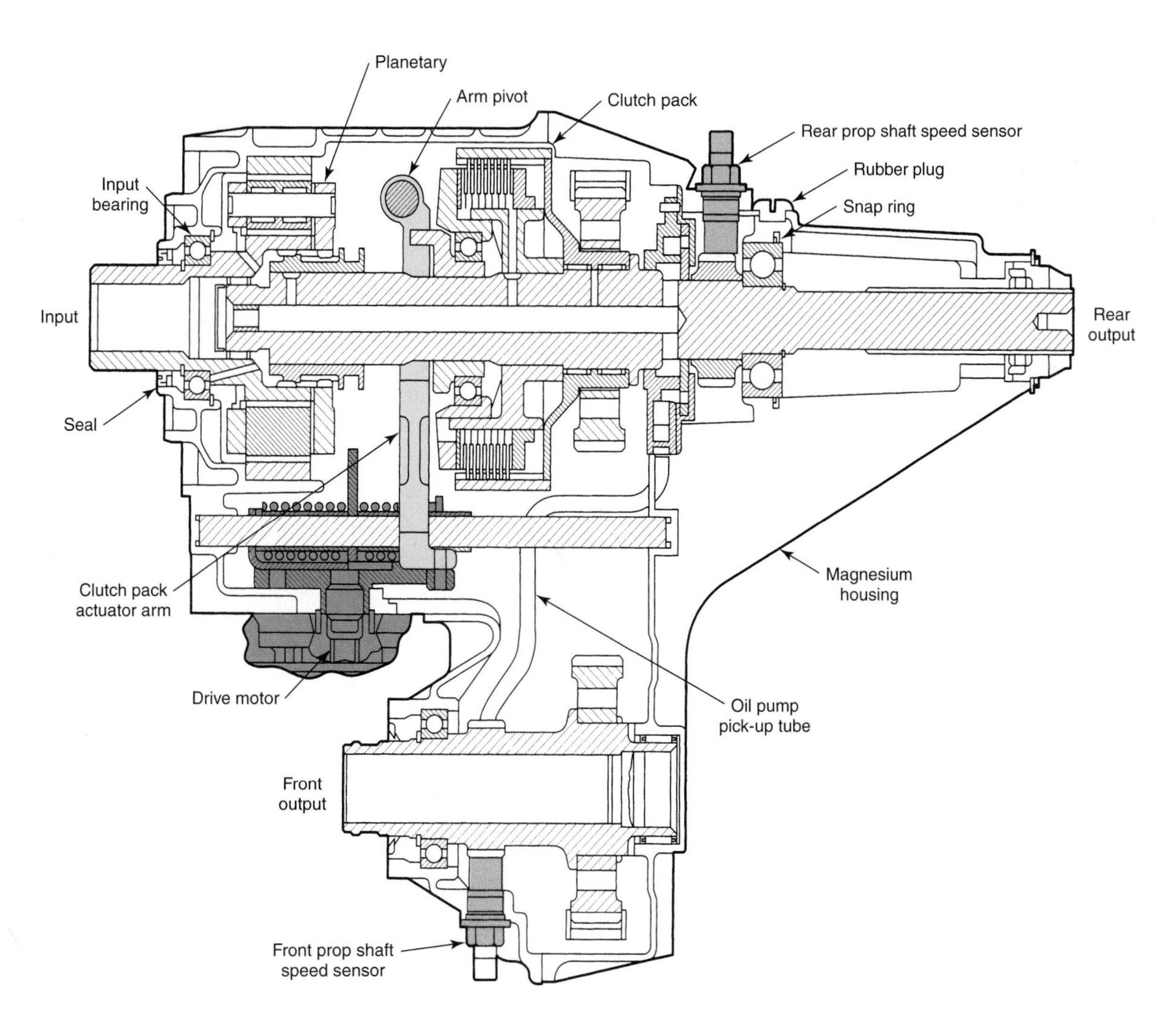

Figure 33-9. This electronically controlled transfer case is similar to a nonelectronic model. Study the cross section and observe the motor assembly and actuator arm. (General Motors)

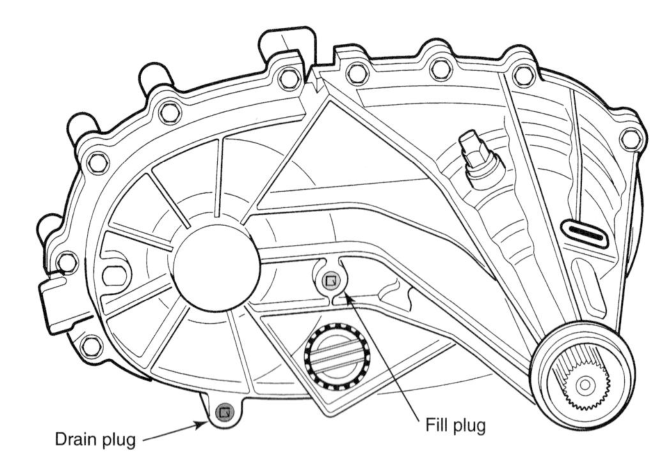

Figure 33-10. This transfer case has oil drain and fill plugs. Other transfer cases may have a fill plug only. (General Motors)

operation and adjustment before road testing. The free travel in the clutch pedal must be within specifications. Excessive free travel prevents full withdrawal of the pressure plate from the clutch disc. This causes the transfer case input shaft to continue turning, making shifting difficult and noisy. Clutch adjustments are covered in Chapter 29, Clutch and Flywheel Service. The shift linkage must operate smoothly and should be adjusted so the transfer case is shifted fully into gear. Failure to provide full shift engagement can result in the transfer case jumping out of gear. Linkage adjustments are covered later in this chapter.

A vacuum diaphragm problem can be caused by low engine vacuum, a disconnected vacuum line, a faulty vacuum control, or a leak in the diaphragm itself. If the diaphragm is receiving vacuum and does not operate, the diaphragm or linkage is probably at fault.

If all external checks are OK, the unit probably has an internal problem. Common problems include a leaking viscous coupling, worn clutch discs, or damaged gear teeth, or a loose drive chain. All of these require that the transfer case be removed from the vehicle for service.

Electronic Transfer Case Problems

The problems of an electronically controlled transfer case often resemble the problems that occur in mechanically operated transfer cases. Other problems are unique to the electronic control system. The following section covers common electronic control system problems.

Many computer control system problems are called ***pattern failures.*** A pattern failure is a problem that is common to a certain type of vehicle. The experienced technician will learn to spot pattern failures and quickly determine the commonly defective part.

Input Sensor Problems

Sometimes symptoms result from the transfer case reacting to problems in the transmission, engine, or vehicle computer. Sensors are the most common cause of electronic transfer case problems. A common problem is improper or erratic range selection caused by a defective speed sensor. Speed sensors can cause problems by sending incorrect engine or output shaft information to the computer. This usually affects transmission shift speeds as well as transfer case operation. A speed sensor inside of the transmission or transfer case can collect metal filings. Some metal particles are always produced during the normal operation of the transfer case. These particles often become stuck on the sensor magnet, affecting the signal. Speed sensors can be removed for inspection and replacement without removing the transfer case, **Figure 33-11.**

Temperature and throttle position sensors can indirectly affect transfer case operation. Most defects in these sensors will show up first as engine performance problems.

Output Motor Problems

Output motor problems will vary depending on the type motor. Motors can stick in any position. This usually results in loss of some ranges. Occasionally a motor will stick sometimes and work properly at other times. Usual causes of intermittent sticking are a high-resistance winding, bad electrical connection, or sludge buildup. Erratic operation is a common symptom of an intermittently sticking motor. The transfer case may work well most of the time, but occasionally shift improperly only when cold or only when hot.

Control Computer Problems

Since it contains many complex circuits, a failed control computer can cause different problems, depending on what internal part has failed. Sometimes the computer will keep the transfer case from shifting. The unit may remain in one range. Another common computer problem is erratic shifting. Examples include range changes when speed and road surface are unchanged, or the occasionally failure to select the appropriate range. The operation of the computer is often affected by heat. A computer may work well when the vehicle is first driven, and cause problems as it heats up.

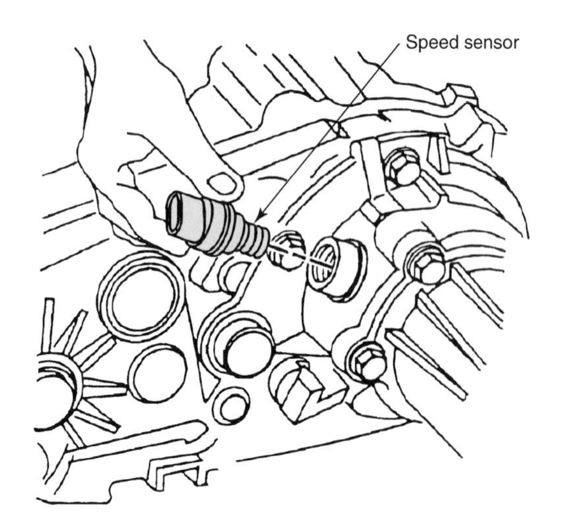

Figure 33-11. Speed sensors can be removed without removing the transfer case. Check the sensor tip for metal particles that can affect sensor operation. (General Motors)

A defective computer can hang up, or stick in one mode. Turning the ignition switch off and then back on may temporarily correct the problem. In many cases, a failed computer also causes engine drivability problems. A defective computer can set false trouble codes, sometimes codes that do not exist. A failed computer may illuminate the MIL when there is no problem, or fail to illuminate the MIL when a problem is present.

Wiring and Connector Problems

The voltages used to operate the sensors are usually much lower than battery voltage. Therefore any wire damage or corrosion at the connectors greatly affects the sensor inputs to the computer. Even a slight increase in resistance can cause incorrect sensor inputs, leading to improper or erratic operation. A commonly overlooked wiring problem is a corroded ground connection or disconnected ground wires. Remember that the return path for the current is as important as the input path. Resistance through a ground circuit can affect several control loops at once. Therefore, when a transfer case control system seems to have several unconnected or intermittent problems, look for a poor ground.

Often a wire has been chafed or broken by movement. This commonly occurs where wires must pass through confined spaces or small openings in the body or other sheet metal. Manufacturers often place electrical connectors near or under the vehicle battery. The battery acid and hydrogen gas can cause corrosion inside of the connector. A very common wire problem occurs when a wire touches an exhaust system part. The insulation melts, and the wire grounds against the exhaust.

Transfer Case Electrical Checks

Before proceeding with electronic tests, make some basic electrical checks. Begin by ensuring that the system fuses are not melted (blown). There may be more than one fuse controlling the system.

Note: Do not remove any fuses until you have retrieved trouble codes from the computer. Removing fuses may erase trouble codes.

Next, make a careful inspection for obviously burned, chafed, and disconnected wires. In some cases, the wiring harness wrapping may have to be pulled back to expose damaged wires. Be sure to look carefully for burned insulation on any fusible links, and pull on them to determine whether the wire has broken internally. Fusible links are usually located at the battery positive terminal, a nearby power relay, or at the starter solenoid.

Check all vehicle ground wires. On many vehicles, several ground wires are attached to one of the bolts on the engine thermostat housing. These wires can become corroded due to slight coolant leaks. Often they are removed during thermostat replacement and not reattached.

Using Scan Tools to Check the Transfer Case

If the preceding checks do not reveal a problem, the next step is to obtain a *scan tool.* The scan tool allows the technician to obtain information directly from the computer. This information would not be available by any other means. Trouble code retrieval was discussed in Chapter 15, Computer System Diagnosis and Repair.

Caution: On OBD II-equipped vehicles, the proper scan tool *must* be used to retrieve trouble codes. Do not attempt to retrieve trouble codes from an OBD II system by grounding one of the diagnostic connector terminals. Grounding any terminal will damage the computer.

Transfer Case Overhaul

Transfer case overhaul is similar to that of a manual transmission. The technician should obtain a service manual covering the specific transfer case to be overhauled and refer to it as the unit is disassembled. Disassembly and assembly order, specifications, and techniques will be shown in such a manual. Exploded views will assist in the correct positioning of all parts. Although basic transfer case designs are similar, the disassembly procedure and order of disassembly vary widely among the different makes and models.

Note: Some transfer case repairs, such as shift linkage adjustment and seal replacement, can be performed with the transfer case in the vehicle.

Once an attaching part has been removed, put the fasteners back into place. This facilitates reassembly and avoids the misplacement of fasteners.

Raise the vehicle and drain the oil from the transfer case and, if necessary, the transmission. Remove the shift linkage and mark it for easier reassembly. Remove any electrical connectors and vacuum lines. Before removing the drive shafts, mark them so they can be reassembled in their original positions. Disconnect the front and rear drive shafts and use wire to secure them out of the way. See Chapter 32, Axle and Driveline Service, for details on drive shaft service. Tape the roller bearing caps to the cross to prevent them from falling off.

Caution: The transfer case is very heavy and awkward. Get someone to help you support and lower the transfer case from the vehicle.

Remove the transfer case mounts as needed. If a support member must be removed, support the engine and transmission assembly with a jack stand or other support. If the jack must push against the engine or transmission oil pan, place a wide block of wood between it and the pan. A transmission jack should be used for removal. Attach the transfer case firmly to the jack. Then, remove the fasteners holding the transfer case to the transmission, slide the transfer case away from the transmission and lower the transmission jack.

If the transfer case is a size and weight that can be easily handled, pilot bolts can be used for easier removal. Remove all but two of the transfer case-to-transmission housing fasteners. Install the pilot bolts in two mounting holes. The pilot bolts should be long enough to provide support until the transfer case input shaft is clear of the transmission output shaft. Remove the final two transfer case-to-transmission housing fasteners. With the help of an assistant, slide the transfer case away from the transmission and lower it to the floor.

General Disassembly Procedures

To aid in locating external fasteners, clean the transfer case's outer surfaces thoroughly before beginning disassembly. This will also minimize the contamination of internal parts. Some transfer cases are designed with two aluminum end castings attached to a central support. The end castings must be removed to gain access to the gears and shafts. Other transfer cases must be split to expose the internal gears. Due to space limitations, it is obvious that a text such as this one cannot cover the disassembly procedure for all transfer cases. Instead of discussing specific disassembly procedure and order, the transfer case parts are covered separately and general inspection techniques are discussed.

Place the transfer case on a clean workbench or in a suitable stand, if available. Follow the manufacturer's recommended order of part removal. If no manual is available, study the method of construction. This often provides clues about which part should be removed first, second, etc. Careful study also usually indicates how the parts must be removed. Be sure to place alignment marks on the transfer case housing before beginning disassembly. Two common transfer cases are shown in **Figures 33-12** and **33-13.** Note that one case is driven by gears, while the other case uses a chain drive. If the transfer case has internal shift forks, remove them first. Do not lose any of the detent springs, detents, interlocks, or setscrews. Next, remove all the parts from the case, including all bearings. The internal gears may use individual needle

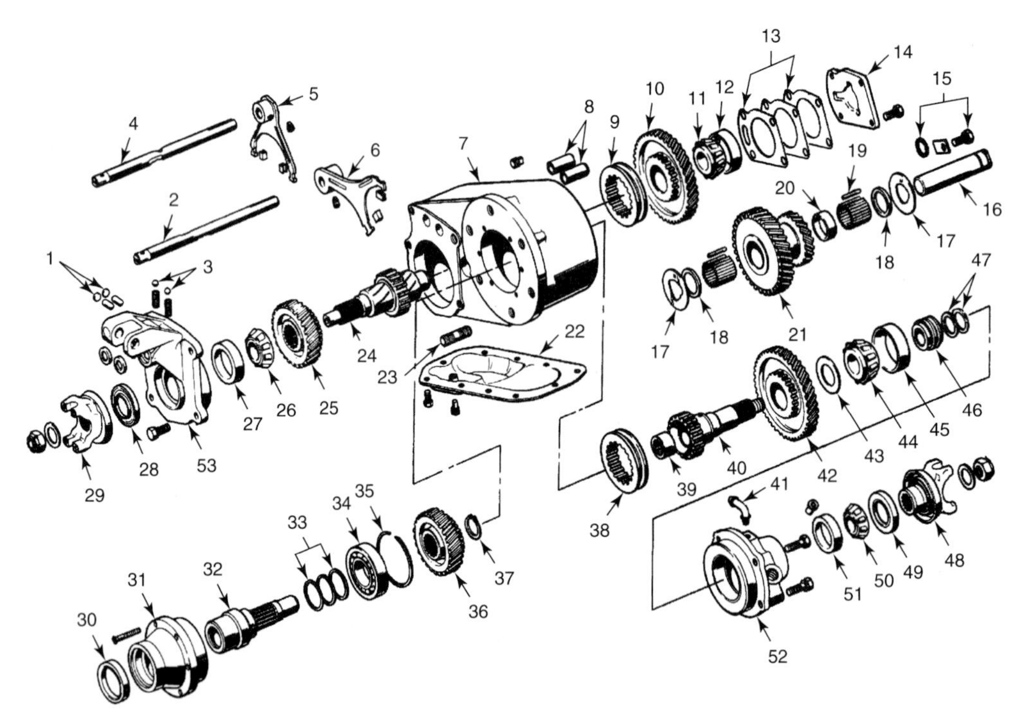

1. Interlock plugs and interlocks
2. Shift rod – rear output shaft fork
3. Poppet balls and springs
4. Shift rod – front output shaft fork
5. Front output shaft shift fork
6. Rear output shaft shift fork
7. Transfer case
8. Thimble covers
9. Clutch sleeve – front output shaft
10. Clutch gear – front output shaft
11. Bearing – front output shaft rear
12. Race – front output shaft bearing
13. End play shims – front output shaft
14. Cover plate
15. Lock plate, bolt and washer
16. Intermediate gear shaft
17. Thrust washer
18. Bearing spacer (thin)
19. Intermediate gear shaft needle bearings
20. Bearing spacer (thick)
21. Intermediate gear
22. Bottom cover
23. Stud (case-to-trans.)
24. Front output shaft
25. Front output shaft gear
26. Front output shaft bearing (front)
27. Front output shaft bearing (race)
28. Oil seal
29. Front yoke
30. Seal
31. Support – input shaft
32. Input shaft
33. Shims
34. Input shaft bearing
35. Input shaft bearing snap ring
36. Rear output shaft gear
37. Snap ring
38. Clutch sleeve – rear output shaft
39. Input shaft rear bearing (needle) (or pilot bearing)
40. Rear output shaft
41. Vent
42. Clutch gear – rear output shaft
43. Thrust washer
44. Bearing – rear output shaft front
45. Race – rear output shaft bearing
46. Speedometer drive gear
47. End play shims
48. Rear yoke
49. Rear output shaft oil seal
50. Bearing – rear output shaft gear
51. Bearing race
52. Rear bearing cap
53. Front bearing cap

Figure 33-12. An exploded view of a part-time, gear-driven transfer case is shown here. Study the parts carefully. (DaimlerChrysler) bearings. Retrieve all of these bearings and check the service manual to ensure that you have the proper count. Place any small parts (loose needles and rollers, springs, fasteners) into a separate container to prevent loss. Avoid distorting snap rings in case they must be reused. Lay out all parts in order, so the transfer case can be reassembled simply by reversing the order of disassembly.

Caution: Avoid excessive use of a hammer to loosen parts. Use a brass hammer if necessary. Handle the parts carefully to avoid chipping and nicking. If possible, use new snap rings for reassembly.

Cleaning and Inspecting Parts

Clean all internal parts, especially gears, bearings, and shafts. Thoroughly clean the inside of the case. Any tiny particles of chipped teeth or bearing material that remain will be loosened by the new lubricant and will eventually find their way into the moving parts, where they will cause damage. Once the transfer case has been completely disassembled and cleaned, the internal parts can be inspected. Keep in mind that many variations in part construction and layout are possible, depending on:

- Transfer case manufacturer.
- Variations in the vehicle body type, engine, and transmission.
- Whether the unit is from a full-time or part-time four-wheel drive system.
- Whether the forward drive mechanism uses gears or a chain and sprockets.

When a transfer case has suffered heavy gear damage such as shattered teeth, use crack detection chemicals on the

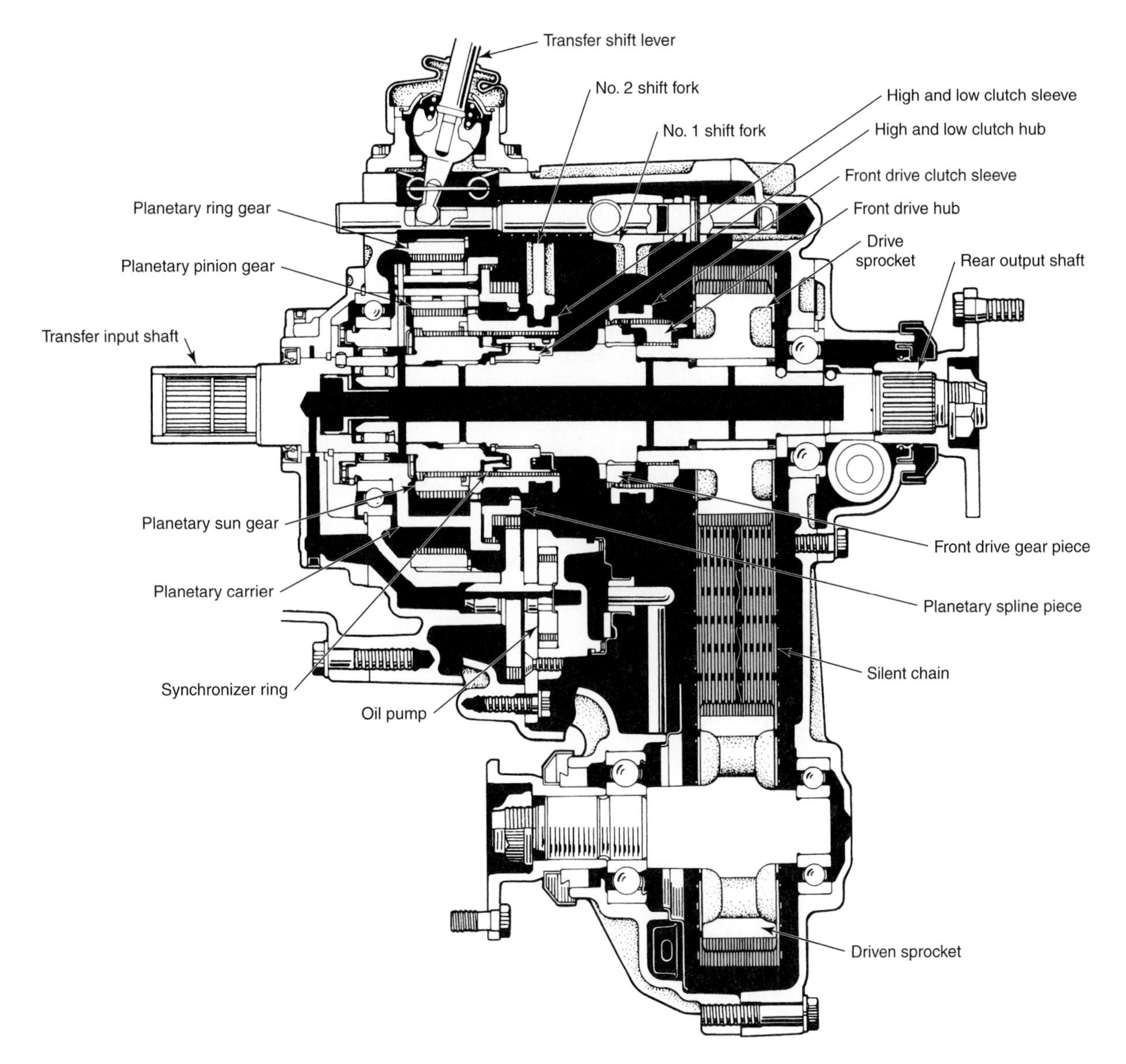

Figure 33-13. This figure shows a cutaway view of a two-speed, part-time transfer case. This case uses a planetary gear reduction setup, along with a drive chain. Low-range reduction is 2.566:1. (Toyota)

remaining gears. Crack detection was covered in Chapter 24, Engine Removal, Disassembly, and Inspection. Check the shafts and shaft openings in the case also. Discard all parts showing any sign of cracking. Replace the bearings, even if they appear to be good.

Inspect Transfer Case Housing

Inspect the transfer case for signs of cracking. Look carefully, especially around bolt holes and shaft and bearing openings. On aluminum cases, check for porous areas or other flaws in the case. Also check all sheet metal covers for dents and warping that would prevent them from sealing properly. Carefully straighten any damaged areas or replace the part. Check the transfer case mating surfaces for any burrs that could cause misalignment or leaks. Remove any burrs with a fine mill file. If the transfer case has a vent, make sure it is open. Check the bushings in the transfer case housing (if used) and replace if necessary. Always replace any external transfer case oil seals, such as those shown in **Figure 33-14.**

Inspect All Shafts and Gears

Inspect all shaft bearing surfaces. They should be smooth with no evidence of wear or damage. Place the gears on the shaft. They should turn smoothly without excessive rocking. If the gears are splined, check for excessive play. Look over every tooth on every gear. There must be no signs of chipping, galling, or wear. All snap ring grooves must have sharp square shoulders. Thrust washers must be smooth and not worn appreciably. Thrust washer thickness should be measured with a micrometer. Inspect all shaft support bearings and internal shaft or gear bushings. Replace any that are not in good condition. Check the synchronizers if the transfer case uses them. A typical transfer case shaft and gear assembly is shown in **Figure 33-15.** Closely inspect the drive chain and drive sprockets of the chain drive assembly for wear or damage. Check the chain for slack and worn pins, **Figure 33-16.** If the transfer case uses two large gears for power transfer, check them for chipping, wear, and loose fit on the shaft splines.

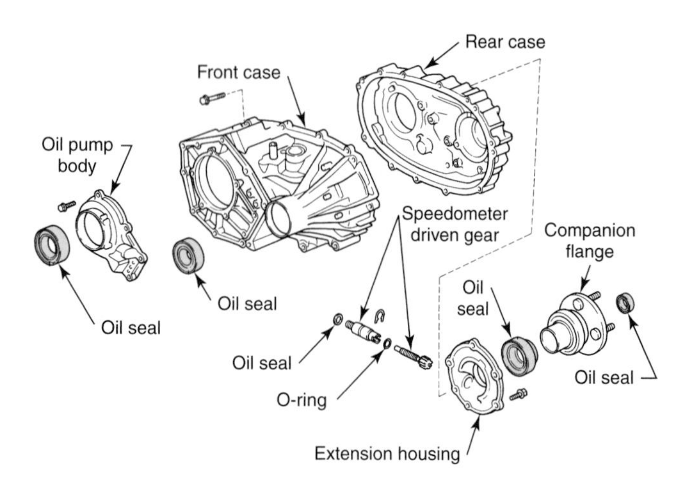

Figure 33-14. Replace transfer case external seals as needed during inspection and repair. (Toyota)

Inspect Transfer Case Pump, Differential Unit, and Viscous Coupling

If the transfer case has an internal lubrication pump, such as the one in **Figure 33-17,** check it for wear and proper clearances. Replace the pump if it shows any signs of wear or if the clearances are excessive. Closely inspect the differential unit or viscous coupling for wear or damage. Differential gears are checked in the same manner as the gears inspected earlier. The viscous coupling should be checked for leakage,

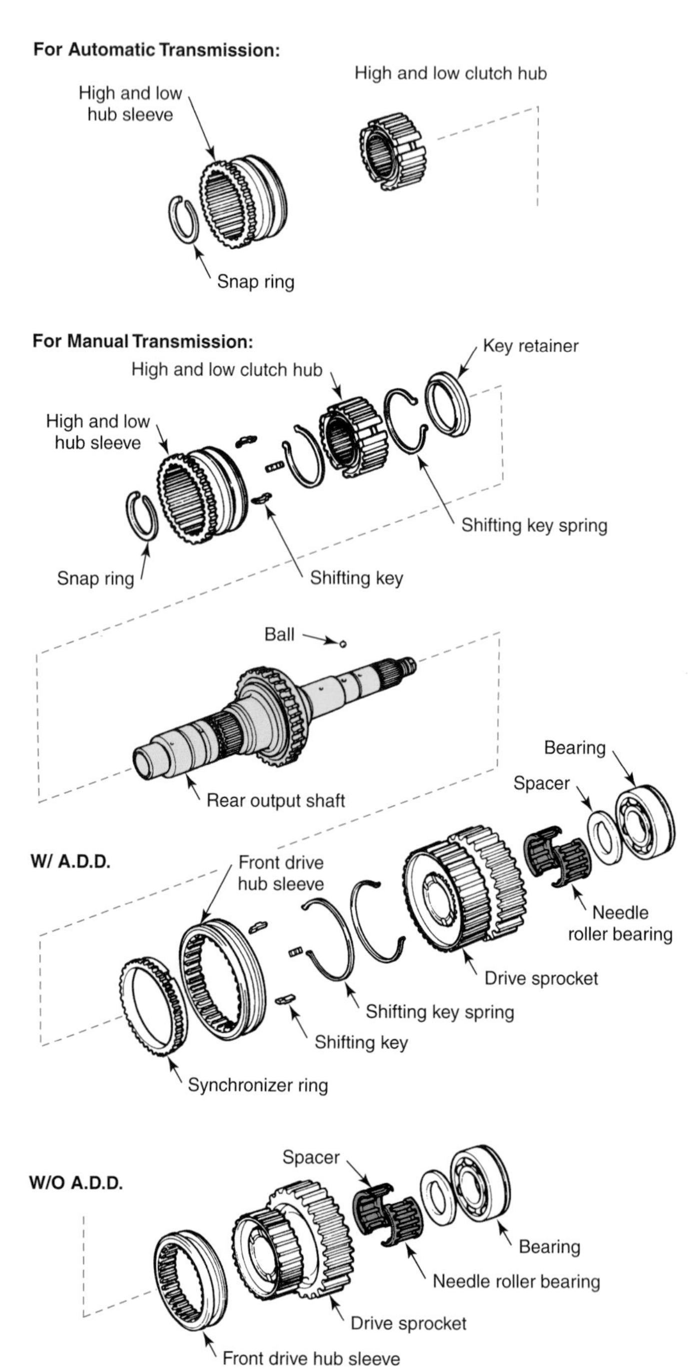

Figure 33-15. This figure shows a typical transfer case shaft and gear assemblies with related components. Note the two-piece needle roller bearings. (Toyota)

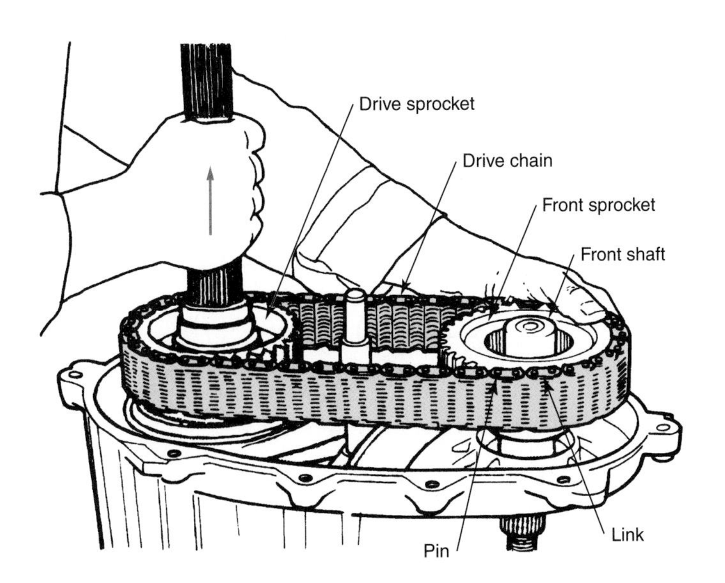

Figure 33-16. Remove the drive chain and sprockets. Check the chain for excessive wear and broken or missing links and pins. (DaimlerChrysler)

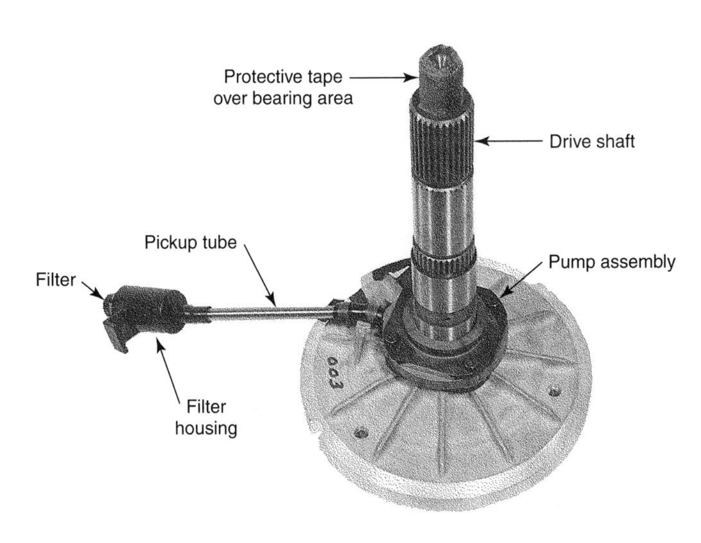

Figure 33-17. This figure shows a transfer case's internal lubrication pump assembly, pickup tube, filter, and filter housing. (Borg-Warner)

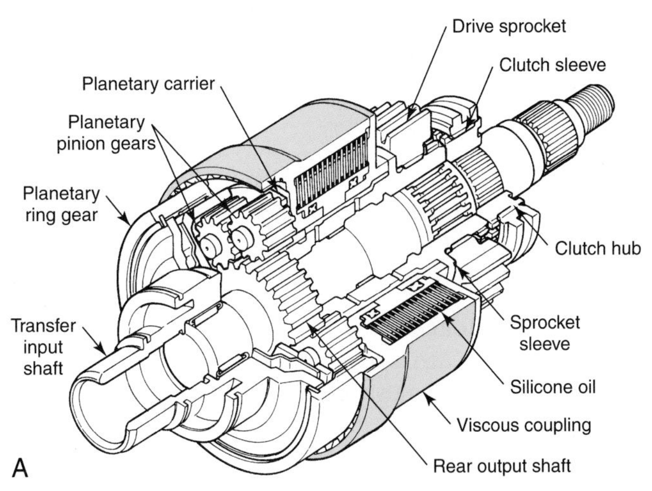

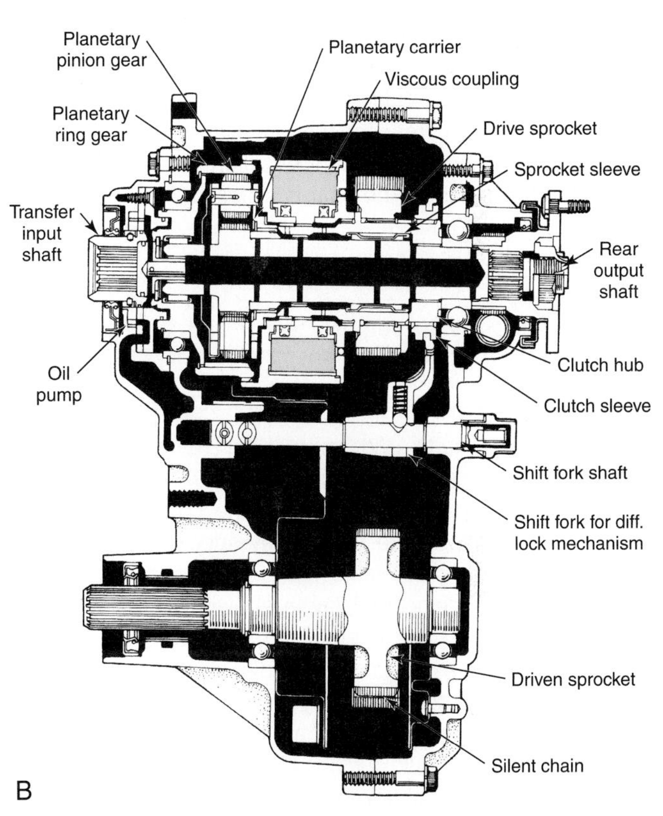

Figure 33-18. A—A cutaway of one particular type of transfer case viscous coupling. The inside of the coupling is filled with silicone oil. The unit slips to allow for torque buildup without damaging the drive train. B—Cross-sectional view of a transfer case that incorporates a viscous clutch. (Toyota)

which appears as globules of silicone in the transfer case oil. Replace any suspected coupling. A viscous coupling is shown in **Figure 33-18.**

Note: The viscous coupling cannot be disassembled for repairs. Replace the entire unit when defective.

Obtain New Parts

When a transfer case component is worn or damaged, replace the defective parts. Considering the difficulty of removing and disassembling the transfer case, any suspect part should be replaced. Always check the new part against the old for design, size, and shape. Try the part on the associated shaft or in the transfer case to make sure it fits properly. Use new snap rings whenever they can be obtained. New thrust washers will provide proper endplay. New gaskets and seals are a must. Some transfer cases use RTV (room temperature vulcanizing) or anaerobic sealer in place of gaskets. Always use the recommended sealer. An inferior seal can cause a complete loss of lubricant, leading to transfer case failure.

Caution: Parts must be absolutely clean and well lubricated before assembly.

Following cleaning and inspection, all parts should be oiled and placed in clean containers. Before installation, every part should be heavily lubricated with gear oil or automatic transmission fluid as recommended by the manufacturer.

Transfer Case Reassembly

The transfer case is reassembled in the reverse order of disassembly. All parts must be lubricated, properly positioned, and secured before installation. Make frequent use of the proper manufacturer's service manual to ensure that all parts are reinstalled correctly. Never use excessive force to make a part fit. If a part does not slide into position as it should, stop and check for the source of difficulty. Lubricate all bushings and shafts with transmission oil before beginning assembly. To install thrust washers, place a coating of soft grease on each end of the gear and press the thrust washers into the grease. Make sure they face in the right direction. The grease holds the washers in place as the gear is installed.

Note: Any grease used to hold internal parts in place must be compatible with and dissolve readily in the normal oil or transmission fluid used in the transfer case.

Check all shafts and gears after reassembly. There should be no binding or hard to turn parts. If any parts are binding, disassemble and determine the cause. Check the endplay of all units where an endplay specification is given, using a dial indicator or feeler gauge as needed. A typical procedure for checking endplay is shown in **Figure 33-19.** When the transfer case internal parts are fully assembled, pour fresh gear oil over the gears and shaft through the cover opening. Use the recommended gear oil and correct viscosity. Turn the input shaft while shifting through the gears. The shafts and gears should turn freely with no binding. Check shaft endplay, if not checked earlier. After all checks have been made, install the shift mechanism, if applicable.

Transfer Case Installation

Wipe the clutch housing and transfer-case face clean and check both for burrs. Using pilot bolts or a transmission jack, raise the transfer case into alignment with the engine and align the case input shaft with the transmission output shaft. Then carefully push the transfer case into position at the rear of the transmission.

Caution: Never let the transfer case hang with all of its weight supported by the shafts. Damage to the transfer case and transmission will result. Also, the transfer case could slip off of the transmission, resulting in damage and/or injuries if it hits someone.

Install and torque the fasteners holding the transfer case to the transmission. Reinstall all transmission and transfer case mounts. Install the selector linkage (if used), electrical and vacuum lines, and other parts. Install the front and rear drive shafts, being careful to match the alignment marks.

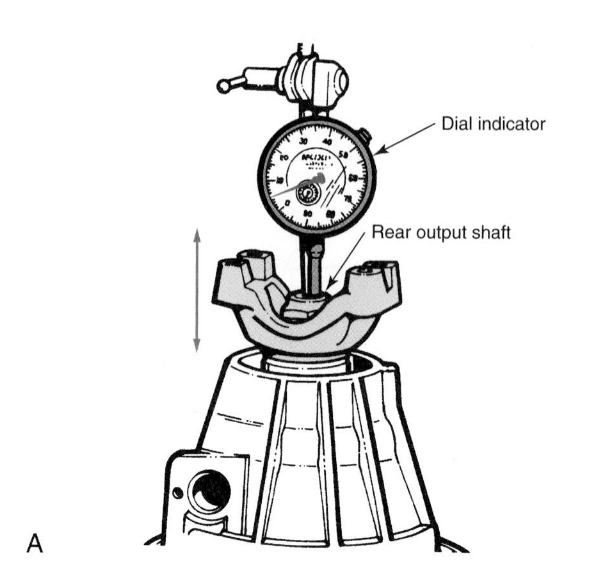

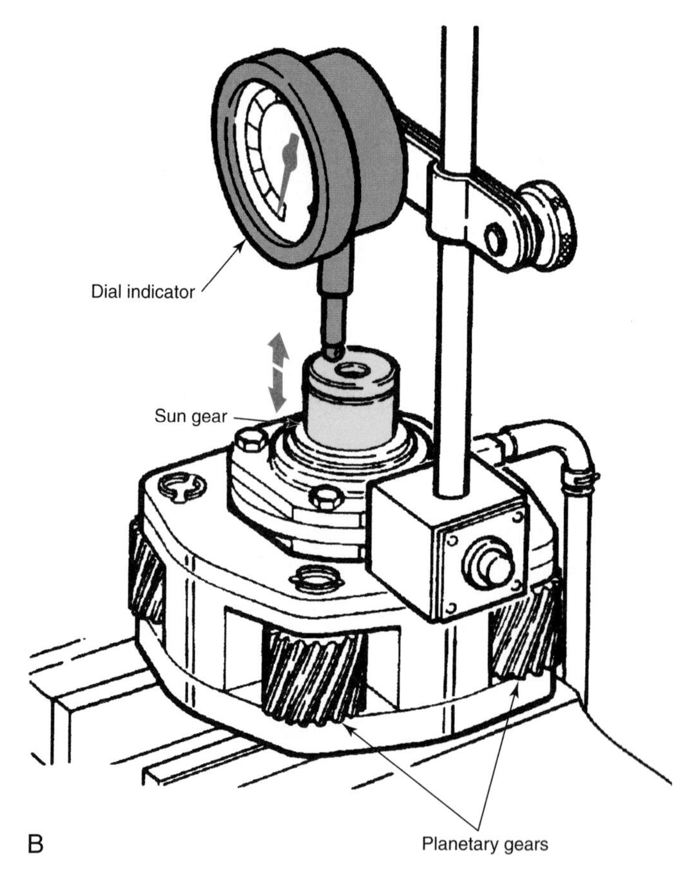

Figure 33-19. A—Checking rear output shaft endplay with a dial indicator. B—Checking the clearance between the sun gear and the planetary gear set with a dial indicator. (Ford and Range Rover)

Adjust the selector linkage as necessary. Fill the transfer case to the level of the filler plug with the recommended lubricant. Fill slowly so the oil has time to flow into all moving parts.

Road Test

Road test the vehicle. The transfer case should operate quietly and smoothly. On part-time units, shifting between all ranges should be should be smooth and positive, with no jumping out of gear. Full-time units should operate quietly and without any unusual feel. After returning to the shop, check for fluid leakage and recheck the lubricant level.

Locking Hubs

Some four-wheel drive vehicles use front ***locking hubs*** to engage and disengage the front wheels and front drive axle. This type of arrangement is a component of a part-time four-wheel drive system, but is not controlled by internal parts in the transfer case. Locking hubs are a simple device for disengaging the front wheels from the front drive axles. The locking hub is a device that locks the front wheels to the front axle at the wheel hub. A typical manual locking hub is shown in **Figure 33-20.**

Some locking hubs are engaged and disengaged automatically by the use of one-way, or ***overrunning clutches*** in the front drive axles. These one-way clutches transfer power in one direction but will freewheel, or overrun, if power is applied in the other direction. These clutches lock up whenever the engine is driving the front wheels through the transfer case and unlock when the transfer case is returned to 2-wheel drive operation. Other locking hubs can be automatically released by moving the transfer case to the two-wheel position and backing up the vehicle for a few feet.

Locking Hub Service

Locking hubs are simple in construction and are relatively simple to repair. However, locking hubs must be carefully cleaned and checked for corrosion since they are much more likely to become contaminated with water and dirt. Seals and gaskets must be in perfect shape to protect the internal parts. See **Figure 33-21.** To begin removal of the locking hub assembly, lift the vehicle at the affected axle and remove the wheel. Most locking hub components are held in place by a snap ring. Remove the front hub cover and the wheel bearing nut first. Then remove the snap ring and slide the hub parts

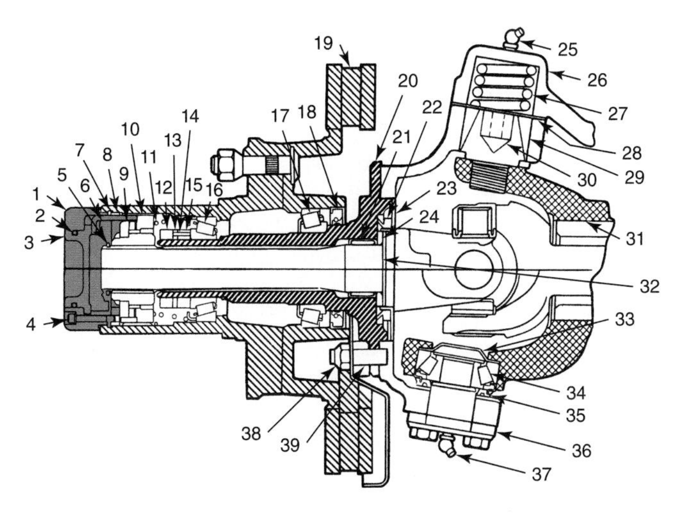

1. Retaining plate
2. O-ring
3. Actuator knob
4. Retaining plate bolt
5. Axle shaft snap ring
6. Actuating cam body
7. Internal snap ring
8. Outer clutch retaining ring
9. Axle shaft sleeve and clutch ring
10. Inner clutch ring
11. Spring
12. Lock nut
13. Lock adjusting nut
14. Pin adjusting nut
15. Adjusting nut
16. Outer wheel bearing
17. Inner wheel bearing
18. Seal
19. Hub and disc
20. Spindle
21. Spindle bearing
22. Seal
23. Deflector
24. Spacer
25. Lube fitting
26. Upper bearing cap
27. Pressure spring
28. Gasket
29. Bushing, king pin
30. King pin
31. Yoke
32. Outer axle shaft
33. Grease retainer
34. Lower bearing
35. Seal
36. Bearing cap
37. Lube fitting
38. Spindle attaching nut
39. Spindle attaching bolt

Figure 33-20. This figure shows a cutaway view of a manually locking hub assembly as used by one manufacturer. (General Motors)

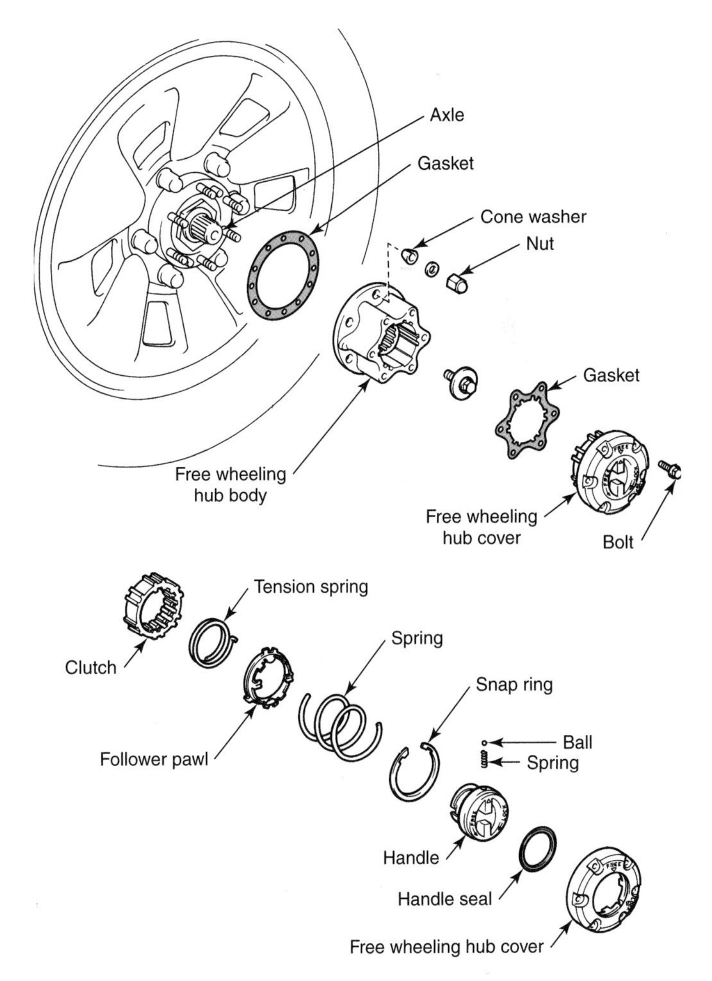

Figure 33-21. An exploded view of a manual locking hub assembly is shown here. Always replace seals, gaskets, and O-rings when servicing the hub and/or bearings. (Toyota)

from the axle shaft. Disassemble the locking hub on a clean workbench. Carefully check all parts for signs of wear, corrosion, and dirt. Typical locking hubs are shown in **Figures 33-22** and **33-23.**

Note: Carefully check the seals and gaskets for any damage that would permit the entry of water or loss of lubricant. This is extremely important when servicing locking hubs, since they are often submerged in water and mud during off-road operation.

Install new parts as needed, being careful to coat all parts with the correct type of lubricant. Ensure that sufficient grease has been installed in the hub cavity. If the front hub assembly has adjustable wheel bearings, adjust them as outlined in Chapter 38, Wheel and Tire Service.

Vacuum Motor Service

On those front-wheel drive assemblies having a ***vacuum motor,*** the front wheels may fail to drive if the vacuum motor is inoperative or the linkage is stuck. If the shift collar in the axle shift collar assembly is stuck, the front axle may be engaged at all times. Check the vacuum motor with a vacuum pump to ensure the diaphragm is not leaking. If the motor is defective, it can be replaced.

To service the vacuum motor and linkage, remove the differential inspection cover and vacuum motor as a unit. The shift fork and linkage is also attached to the cover, **Figure 33-24.** Check all parts for wear and ensure the shift collar is not stuck or damaged. See **Figure 33-25.** Replace any worn parts as necessary. If the shift collar must be removed, refer to

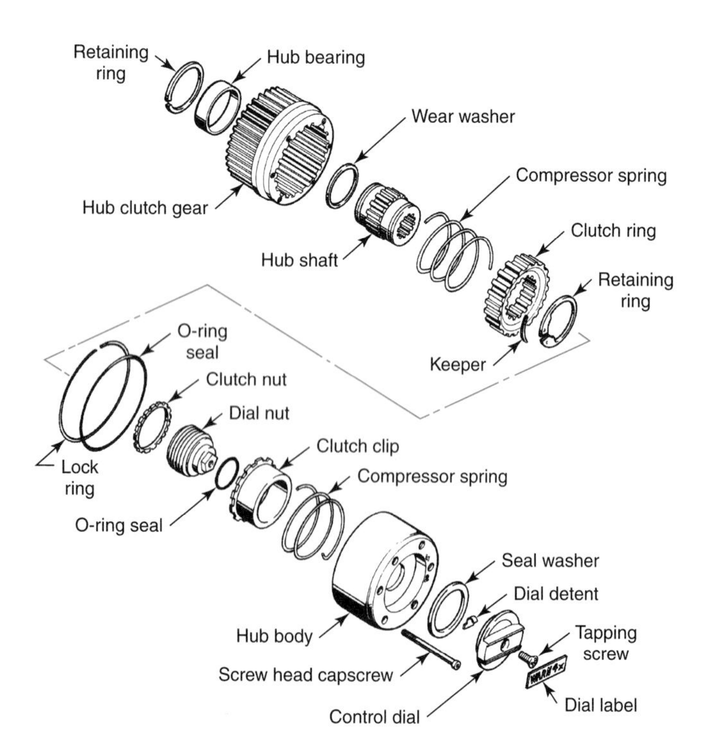

Figure 33-22. This figure shows an exploded view of a manual locking hub. (DaimlerChrysler)

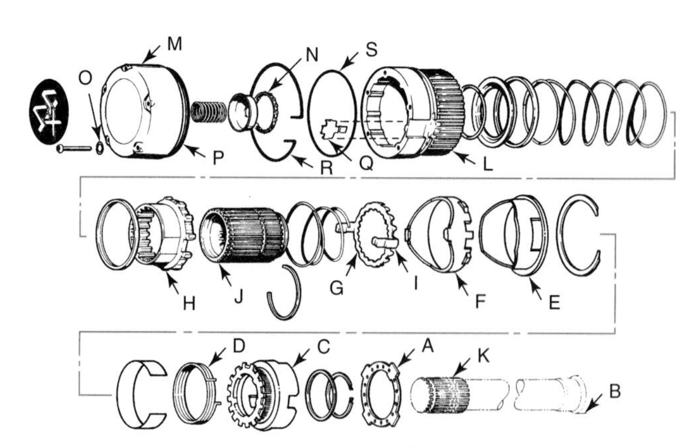

Figure 33-23. Study this exploded view of an automatic locking hub assembly. Note the names of each part. A—Drag sleeve retainer washer. B—Axle housing. C—Drag sleeve. D—Brake band. E—Steel inner cage. F—Plastic outer cage. G—Cam follower. H—Clutch gear. I—Cam follower. J—Hub sleeve. K—Axle shaft. L—Clutch housing. M—Hub outer cover. N—Bearing assembly. O—Cover sealing O-ring. P—Cover sealing O-ring seal groove. Q—Seal bridge retainer. R—Wire retaining ring. S—Cover sealing O-ring. (General Motors)

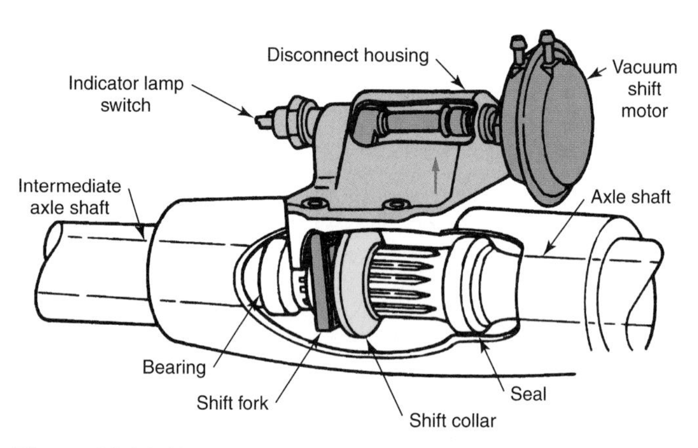

Figure 33-24. Remove the vacuum motor, cover, and shift fork assembly after removing all fasteners, wiring, and vacuum hoses. Cover the opening with a clean shop towel to prevent dirt or small parts from entering the housing. (DaimlerChrysler)

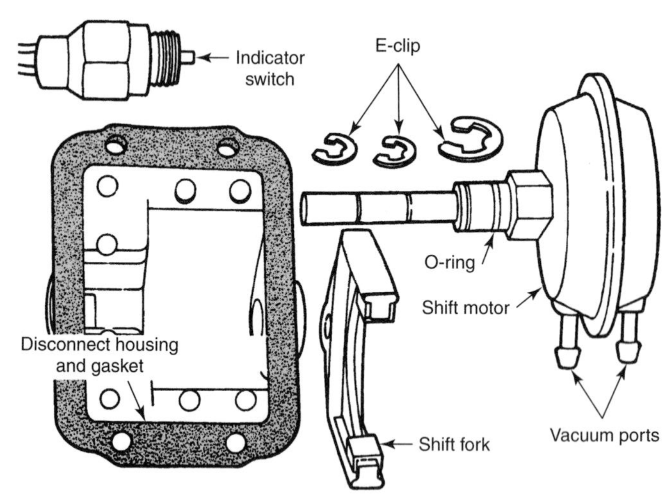

Figure 33-25. Check all shift motor parts carefully for signs of binding, excessive wear, water damage, breakage, etc. Repair or replace as necessary. (DaimlerChrysler)

Chapter 32, Axle and Driveline Service. When reassembling the vacuum motor and cover to the axle, make sure the shift fork engages the shift collar. Refill the axle with the correct lubricant and recheck system operation.

Tech Talk

Transfer cases may not seem very heavy, but they are oddly shaped and awkward to handle. If you try to lift a transfer case into position by yourself, it is likely to slip and fall, crushing your hand or foot. If it does not hurt you, the case will be damaged when it hits the floor.

The best way to install a transfer case or any heavy underbody component is with a transmission jack. Always strap the unit to the jack with the retaining chain. If a transmission jack is not available, make sure that you have at least one other person to help you. Tell the other person exactly how you intend to lift and place the transfer case.

Summary

Transfer cases are either part-time or full-time units. Never operate the part-time unit in four-wheel drive on dry, hard roads. Full-time units are always in four-wheel drive, with a differential assembly or viscous coupling to allow for differences in axle speeds. Full-time units must not be operated in the lockout position on dry, hard road surfaces. Many modern transfer cases are electronically operated. They make use of input sensors, a control computer, and an output motor.

Transfer case problems are sometimes difficult to separate from transmission or clutch problems. Always make sure that the problem is actually in the transfer case before beginning repairs. Electronic transfer case diagnosis may require a scan tool.

Use proper lubricant for the transfer case to prevent noises or other problems. When removing or installing a transfer case, use the proper handling equipment. Remember, transfer cases are heavy. During an overhaul, use the manufacturer's service manual and follow the instructions carefully. Disassembly, checking, and cleaning procedures are similar to those for transmissions and transaxles. The differential and viscous coupling require special attention.

Locking hubs should be serviced carefully, since they are often operated in dirty and wet conditions. Front wheel bearings should be adjusted when necessary. Vacuum motors can be serviced by removing the inspection cover and motor, as a unit, from the axle assembly. Check all parts for damage or wear and make sure the shift collar is not stuck. Recheck four-wheel drive operation after reassembly.

Review Questions—Chapter 33

Do not write in this book. Write your answers on a separate sheet of paper.

1. On a vehicle with four-wheel drive, which of the following receives engine power first?
   (A) Rear axle.
   (B) Front axle.
   (C) Front driveshaft.
   (D) Transmission.
2. Drive train _____ can damage the transfer case, drive shafts, or other drive train parts.
3. Full-time four-wheel drive vehicles should be driven in *lockout* when operating _____.
   (A) on dry paved roads
   (B) in muddy fields
   (C) on snow or ice
   (D) Both B and C.
4. One type of transfer case uses a viscous coupling filled with _____.
5. Name the three major components of an electronic transfer case.
6. Which of the following could *not* cause a vacuum diaphragm problem?
   (A) High engine vacuum.
   (B) Disconnected vacuum line.
   (C) Leaking diaphragm.
   (D) Vacuum control fault.
7. Define a pattern failure.
8. When an electronic transfer case seems to have several intermittent or unrelated problems, suspect a bad _____.
9. On an OBD II system, you will need a _____ to recover transfer case trouble codes.
10. Warped and dented areas in sheet metal covers can be repaired by _____ or replacing the part.
11. Name two transfer case repairs that can be performed with the transfer case installed in the vehicle.
12. Most locking hub parts are held to the axle shaft with _____.
13. To service the vacuum motor and linkage, remove the _____ and vacuum motor as a unit.

ASE-Type Questions

1. All modern four-wheel drive systems have the following parts, *except:*
   (A) two drive shafts.
   (B) a transfer case.
   (C) a viscous coupling.
   (D) front and rear drive axles.
2. Technician A says full-time four-wheel drive systems drive the rear wheels at all times. Technician B says part-time four-wheel drive systems drive the front wheels at all times. Who is right?
   (A) A only.
   (B) B only.
   (C) Both A and B.
   (D) Neither A nor B.

3. Technician A says a part-time four-wheel drive unit uses a differential or viscous clutch. Technician B says a part-time four-wheel drive unit uses a sliding clutch. Who is right?
   (A) A only.
   (B) B only.
   (C) Both A and B.
   (D) Neither A nor B.
4. All of the following statements about a differential unit installed inside of the transfer case are true *except:*
   (A) the differential unit applies driving torque to the front output shaft only.
   (B) the differential allows front and rear drive shafts to be driven at different speeds.
   (C) the differential keeps drive train windup from damaging parts.
   (D) the differential allows constant operation in four-wheel drive.
5. While tearing down a transfer case with a viscous clutch, a few globules of silicone were found. Technician A says to replace the viscous clutch. Technician B says since only a few globules of silicone were found in the old oil, it is OK to reuse the viscous clutch. Who is right?
   (A) A only.
   (B) B only.
   (C) Both A and B.
   (D) Neither A nor B.
6. If the owner complains of hard shifting or gear clash, check the ______ first.
   (A) drive shafts and U-joints
   (B) clutch adjustment
   (C) locking hubs
   (D) oil level
7. If a hammer is being used to loosen parts, be sure to use a ______ hammer.
   (A) large
   (B) steel
   (C) claw
   (D) brass
8. If any transfer case shafts appear to be binding, the transfer case should be ______.
   (A) installed and driven until loose
   (B) disassembled and the problem located
   (C) replaced with a new transfer case
   (D) Either A or B.
9. Technician A says locking hubs may be used to disengage the front axles of a part-time four-wheel drive system. Technician B says the front axles of a part-time four-wheel drive system may be disengaged by a vacuum motor. Who is right?
   (A) A only.
   (B) B only.
   (C) Both A and B.
   (D) Neither A nor B.
10. Which of the following is the *least* likely to become damaged by dirt or water entry?
    (A) Viscous clutch.
    (B) Internal shaft and gear bearings.
    (C) Locking hubs.
    (D) Transmission output shaft.

Suggested Activities

1. Draw a schematic of a typical part-time four-wheel drive unit. Label the major components and show power flow in both two- and four-wheel drive modes.
2. Do a survey at your school to determine what percentage of the faculty, staff, and students own vehicles with four-wheel drive systems. Also, ask whether the systems are full time or part time.
3. Write a short report explaining why you think four-wheel drive vehicles are becoming more popular.
4. Obtain some damaged four-wheel drive parts and discuss the possible causes of the damage with your classmates.

Transfer Case (Part-Time Drive) Problem Diagnosis	
Problem: Jumps out of gear in two-wheel drive	
Possible cause	**Correction**
1. Shift lever detent spring weak or broken. 2. Sliding clutch spline engaging surface worn or tapered.	1. Replace spring. 2. Replace worn parts.
Problem: Jumps out of gear in four-wheel drive	
Possible cause	**Correction**
1. Shift lever interference with floor pan. 2. Excessive transfer case movement. 3. Sliding clutch engaging surfaces tapered or worn. 4. Bent shift fork. 5. Shift rod detent spring weak or broken. 6. Shift lever torsion spring (where used) not holding. 7. Worn bearings, gear teeth, or shafts.	1. Provide proper clearance. 2. Check and replace transfer case mounts. 3. Replace worn parts. 4. Replace shift fork. 5. Replace detent spring. 6. Replace torsion spring. 7. Overhaul unit.
Problem: Noise Note: Transfer cases using a gear drive produce considerable gear whine, which is normal.	
Possible cause	**Correction**
1. Worn bearings, splines, chipped gears, or worn shafts. 2. Low lubrication level. 3. Loose or broken mounts.	1. Rebuild unit. 2. Fill to proper level. 3. Tighten or replace mounts.
Transfer Case (Full-Time) Diagnosis	
Problem: Noisy operation	
Possible cause	**Correction**
1. Low lubrication level. 2. Operating in "lockout" on hard, dry surface roads. 3. Improper lubricant. 4. "Slip-stick" condition ("Quadra-Trac" type). Makes a grunting, pulsating, rasping sound. 5. Excessive wear on gears, chains, or differential unit. 6. Loose or deteriorated mounts.	1. Fill to correct level. 2. Shift out of "lockout." Advise driver. 3. Drain and fill with recommended lubricant. 4. Normal if vehicle has not been driven for a week or two. Should stop after some usage. If it persists, drain fluid and refill. Use special additive if required. Make certain tire sizes are the same and pressures are equal. 5. Rebuild as needed. 6. Tighten or replace.
Problem: Jumps out of low range and/or is hard to shift into or out of low range	
Possible cause	**Correction**
1. Shift linkage improperly adjusted, bent, or broken. 2. Shift rails dry or scored. 3. Improper driver operation. 4. Reduction unit parts worn or damaged.	1. Adjust correctly. Straighten or replace. 2. Clean, polish, or lubricate or replace as needed. 3. Follow shift procedure recommended by manufacturer. 4. Repair as needed.
Problem: Lockout will not engage	
Possible	**Correction**
1. Lockout parts damaged. 2. Defective vacuum control. Loose or damaged vacuum lines. 3. Defective shift linkage. 4. Electronic control system defect.	1. Repair as needed. 2. Replace control. Replace or connect vacuum hoses. ("Quadra-Trac"). 3. Repair or replace. 4. Test and repair as necessary. *(continued)*

Transfer Case (Full-Time) Diagnosis *(continued)*	
Problem: Will not engage in two-wheel drive	
Possible cause	**Correction**
1. No vacuum. Loose or broken hoses. 2. Defective shift motor (axle). 3. Defective shift motor (transfer case). 4. Electronic control system defect.	1. Replace hoses. Secure all loose connections. 2. Replace shift motor. 3. Replace shift motor. 4. Test and repair as necessary.
Problem: Will not engage in four-wheel drive	
Possible cause	**Correction**
1. No vacuum. Loose or broken hoses. 2. Defective axle shift motor. 3. Binding or broken transfer case shift linkage. 4. Defective axle shift linkage. 5. Damaged transfer case. 6. Electronic control system defect.	1. Replace hoses. Secure all loose connections. 2. Replace shift motor. 3. Repair or replace shift linkage. 4. Repair or replace shift linkage. 5. Repair or replace transfer case. 6. Test and repair as necessary.
Problem: Loss of lubricant	
Possible cause	**Correction**
1. Clogged breather vent in transmission and/or transfer case. 2. Lubricant level too high in transmission and/or transfer case. 3. Improper lubricant or viscosity. 4. Defective seals, worn bearings, and/or shaft. 5. Porous case area. 6. Transfer case-to-transmission bolts loose.	1. Open clogged breather vent. 2. Drain lubricant to the correct level. 3. Drain and fill with proper lubricant. 4. Replace worn parts as needed. 5. Clean and repair area per manufacturer's instructions. 6. Tighten. Replace seal if necessary. Check lubricant level.
Problem: Vehicle wanders when driving straight ahead	
Possible cause	**Correction**
1. Improperly matched tire size. 2. Uneven tire pressure.	1. Use a matched set of tires. 2. Adjust air pressure to recommended levels.

34

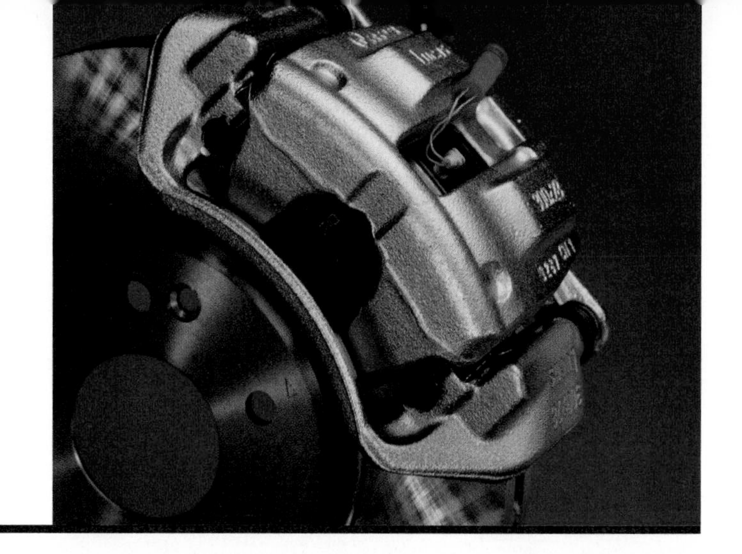

Brake Service

After studying this chapter, you will be able to:

- Explain drum brake construction, operation, and service.
- Summarize disc brake construction, operation, and service.
- Explain the operation and service of power brakes.
- Explain the operation and service of anti-lock brake system.
- Describe master cylinder operation, construction, and service.
- Diagnose brake hydraulic system problems.
- Diagnose brake friction system problems.
- Diagnose power brake system problems.

Technical Terms

Asbestos
Carcinogen
DOT 3
DOT 4
Silicone brake fluid (DOT 5)
Master cylinder
Check valve
Reservoir
Primary cup
Compensating port
Single-piston master cylinders
Double-piston master cylinder
Floating piston
Dust boot
Free travel
Wheel cylinder
Brake caliper
Caliper piston
Fixed calipers
Floating caliper
Proportioning valve
Height-sensing proportioning valve
Metering valve
Pressure differential switch
Combination valves
Bleeding
Manual bleeding
Pressure bleeding
Surge bleeding
Retracting springs
Hold-downs
Self-adjuster mechanism
Disc brake pads
Break-in
Brake drums
Rotors
Turning
Grinding
Parking brake
Power brakes
Vacuum power booster
Hydraulic power booster

Brake service is one of the most commonly performed automotive repairs. During the first 100,000 miles (160,000 km) of its life, the average vehicle will need two or three sets of front brake pads, the rotors turned at least once, and the rear pads or shoes replaced at least once. Every automotive technician will perform brake service at some point in his or her career.

This chapter covers the operation and servicing of modern brake systems. This includes both the hydraulic system (master cylinder, wheel cylinders, calipers, lines and hoses, and valves) and the friction system (brake shoes and drums; disc brake rotors and pads). Power brakes will also be covered.

Safety in the Shop

Before beginning hands-on brake service, note this possible on-the-job hazard to your health. Brake friction materials contain ***asbestos***—a known ***carcinogen*** (a substance that can cause cancer). Removing brake drums, cleaning brake assemblies, and other operations can produce small airborne particles of asbestos. The technician easily inhales these. Breathing these particles may cause cancer.

Observe the Following Rules to Prevent Breathing Asbestos

1. Never use compressed air to blow brake assemblies clean. Use a vacuum source or flush with water.
2. Equip all brake service equipment with an efficient dust removal system. Turn on the system whenever the equipment is in operation.
3. Because some exposure might be unavoidable, wear an approved filter mask.

Closed Cleaning Systems

Brake dust is extremely hazardous to the human respiratory system. Therefore, most brake shops have closed cleaning systems that contain brake dust and ensure that it does not enter the atmosphere. A typical brake dust containment system is shown in **Figure 34-1.** Other systems use a wet cleaning agent that traps the brake dust in a liquid solution for later disposal, **Figure 34-2.**

General Brake Cautions

1. Use care when removing wheels to avoid damage to the calipers' external brake lines.

Figure 34-1. This illustration shows a technician using a typical containment system to reduce the amount of dust that enters the air during brake service. (Nilfisk of America)

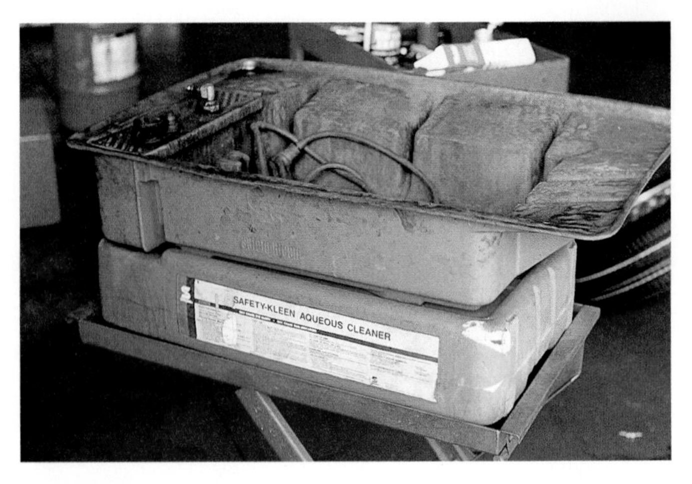

Figure 34-2. A wet cleaning device is shown in this Figure. A wet cleaning agent is sprayed on the brake assembly to trap and remove dust. The dust and cleaner drain into the pan for later disposal.

2. Reject any drums or rotors that do not meet thickness specifications.
3. Turn drums and rotors in axle pairs. Variations in thickness between right and left sides should meet minimum specifications.
4. Replace fluids siphoned from the master cylinder in order to retract pistons.
5. Pump up brakes to bring pads into contact with the rotor before operating the vehicle.
6. Do not separate caliper halves unless so specified.
7. When removing the caliper to pull the rotor, place cardboard or another suitable block between the pads to prevent the pistons from working out of the cylinders.
8. Tap the master cylinder, wheel cylinders, and calipers during bleeding to help dislodge trapped air.
9. Caution the driver that "riding" the brake will quickly ruin the brakes.
10. If the caliper mounting bolts are the prevailing torque type, replace them with new ones.
11. When fitting either oversize or offset wheels, make certain they clear the caliper assembly.
12. Adjust the front wheel bearings to remove play.

Brake Inspection

Periodic brake inspections are a must for safe and efficient brake operation. The inspection should be thorough. It is best to develop a checklist so that no important area is overlooked. Specific instructions concerning various brake system components are presented below.

Brake Pedal and Master Cylinder

Check the brake pedal free play. Check the total pedal travel. There should be ample travel remaining when the brakes are fully applied. The pedal should be firm with no spongy feeling that could indicate air in the system. The pedal, when firmly applied and held, must remain at one point and should not move slowly toward the floor. Pedal action should be smooth and quiet.

Check the fluid level in the reservoir. If the vehicle is equipped with power brakes, exhaust the vacuum reservoir (engine off) by repeated brake applications. Hold the brake pedal down firmly while starting the engine. As soon as the engine starts, the brake pedal will move downward if the vacuum booster is functioning. Check for signs of leakage.

Brake Light Switch and Bulbs

Check switch operation by inspecting brake lights while brake is being applied. Lights should come on quickly, even with very mild brake pressure. If service is needed, refer to Chapter 13, Chassis Electrical Service.

Wheel Cylinders and Brake Shoe Assemblies

A periodic inspection generally involves pulling a wheel and drum. If any trouble is indicated (brake lining wear, fluid leakage, part scoring), all wheels should be removed.

Check the wheel cylinders for leakage by pulling back the lip of the dust boots. Any fluid, other than normal dampness in the boot, indicates a leak. The shoe lining should have ample material remaining and should be free of oil or grease. Inspect retracting springs, shoe hold-downs, automatic adjusting device, and shoe contact pads on the backing plate. Backing plate and shoe anchors must be tight.

Brake Drums

Check drums for out-of-round or tapered condition. The drums must be free of scoring, cracking, grease, and oil.

Disc Brakes

Check brake pads for wear. Check rotor for scoring, cracking, uneven wear, or warping. Caliper pistons should show no signs of leakage.

Seals

Oil and grease seals must be in good condition with no visible signs of leakage. Seals should be changed whenever the bearings are removed.

Parking Brake

The parking brake should hold the vehicle securely. When the brake is firmly applied, there must still be ample pedal travel remaining. Check for missing cotter pins, frayed cables, rust, and other problems.

Brake Lines and Hoses

Inspect all hoses for cracking, softening, or swelling. Check lines for leakage, damage from contact with moving parts, and vibration. All line clips, hold-down assemblies, and other fasteners must be in place.

Chassis

The brake inspection must also include a check for loose wheel bearings, worn ball joints, worn steering parts, defective shock absorbers, broken struts, and worn springs. These parts can affect braking action.

Road Test

Following the above checks, drive the vehicle to test brake action. The vehicle should stop quickly and smoothly. There must be no tendency to dive or to pull to one side or the other. Inspection points for a typical brake system are illustrated in **Figure 34-3.**

Full information on the inspection and servicing of all of the brake components will be given in this chapter.

STOP Warning: The brakes, when needed, must work and must work right. Make every inspection thorough. Perform all brake work carefully and to the highest standards. Use only quality parts. Refuse to do any "half-way" jobs. Remember that every time the pedal is depressed, the lives of a number of people can depend upon your skill, knowledge, and care. Do not let them down.

Brake Hydraulic System Operation

The hydraulic brake system uses a master cylinder to develop hydraulic pressure. The master cylinder may be power (vacuum or hydraulic) assisted. Pressure is transmitted to each caliper and wheel cylinder through steel tubing and flexible hoses.

Brake Fluid

Use only top-quality, DOT (Department of Transportation) approved brake fluid. The latest brake fluid is marked ***DOT 3*** or ***DOT 4.*** Alcohol-based fluid, where available, can be used, but most manufacturers recommend the newer ***silicone brake fluid (DOT 5).*** The fluid must have a boiling point of at least 284°F (140°C). Do not reuse old brake fluid. Keep fluid in clean, tightly sealed, well-marked containers. Protect it from contamination with dust, water, and oils. Do not mix regular and silicone fluids.

Brake Fluid Will Ruin Paint

Do not spill brake fluid on the vehicle's paint. If fluid gets on a painted surface, wipe it off and immediately wash the

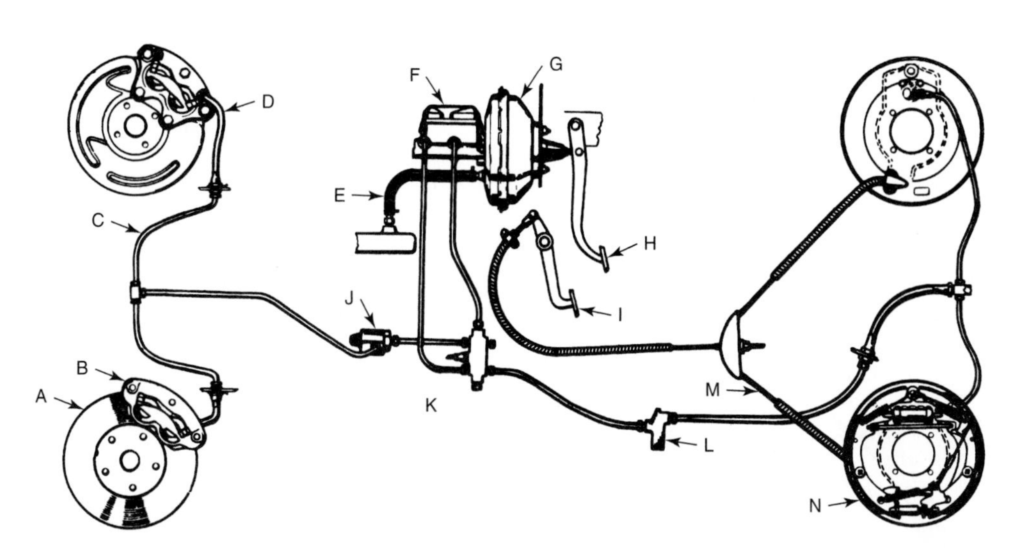

Figure 34-3. This illustration shows common brake system inspection points. A—Rotor. B—Caliper. C—Line. D—Flex hose. E—Vacuum line. F—Master cylinder. G—Power booster. H—Brake pedal. I—Parking brake. J—Metering valve. K—Brake warning light. L—Proportioning valve. M—Parking brake cable system. N—Brake shoe assembly. (Bendix)

surface with mild soap and water. Although the newer silicone brake fluid is harmless to paint, most brake systems contain alcohol-based fluid. Remember that most brake fluids (other than pure silicone) can ruin the finish.

Flushing the System

If necessary, the brake system may be flushed with new brake fluid. It can also be flushed with alcohol or an approved brake-cleaning solvent and rinsed with brake fluid. This procedure should be performed only if the master cylinder, calipers, and wheel cylinders are to be reconditioned. Flushing should always be done prior to repairs.

Master Cylinder Operation

The ***master cylinder*** converts the pressure on the brake pedal into hydraulic pressure in the brake system. When the master cylinder is in the released position, **Figure 34-4,** there is no pressure in the lines.

Note: On drum brakes only, a low static pressure remains in the lines due to a special check valve.

When drum brakes are used, the brake-shoe retracting springs pull the shoes away from the brake drum and force the wheel cylinder pistons inward. This action causes fluid to flow through the lines toward the master cylinder. The fluid lifts the ***check valve*** from its seat and flows into the master cylinder ***reservoir.***

When pressure in the lines drops to the point where it is less than the force developed by the check valve spring, the check valve returns to its seat. This maintains residual line pressure. Systems with disc brakes do not use residual check valves. A dual master cylinder on a system with front disc and rear drum brakes will have a residual check valve on the drum brake side only. A dual master cylinder that operates one front and one rear brake separately will have two check valves, one for each side drum brake.

Figure 34-4 shows the braking system (on one wheel only) when brakes are released. Note the master cylinder piston is released to the point where the ***primary cup*** clears the ***compensating port.*** This allows fluid to move into the reservoir.

When the brake pedal is depressed, the master cylinder piston is forced into the cylinder. As soon as the primary cup passes the compensating port, the fluid ahead of the piston is trapped. Any further piston movement will force the fluid to flow through the brake lines.

Single-Piston Master Cylinder

Older vehicles (before 1967) may be equipped with ***single-piston master cylinders.*** Because they had only one operating piston, a faulty unit or a fluid leak could cause brake system failure. For this reason, single-piston master cylinders are no longer used.

A cross-sectional view of a single-piston master cylinder is shown in **Figure 34-5.** Note that this type of master cylinder has only one brake fluid reservoir. The residual pressure check valve maintains a slight pressure in the system. This pressure is needed to keep the wheel cylinder piston cups expanded when the brakes are not being applied. Residual pressure check valves are used on most vehicles with drum brakes. A few brake systems rely on internal wheel cylinder springs to keep the cups expanded.

Dual-Piston Master Cylinder

The ***double-piston master cylinder*** (also called dual-piston or split system) was developed so the hydraulic systems for the front and rear wheels could be completely separated. With such a separation, a failure or leak in the front system would not affect the rear and vice versa. A simultaneous failure in both systems is unlikely.

One type of dual-piston master cylinder is shown in **Figure 34-6.** When the master cylinder push rod is forced

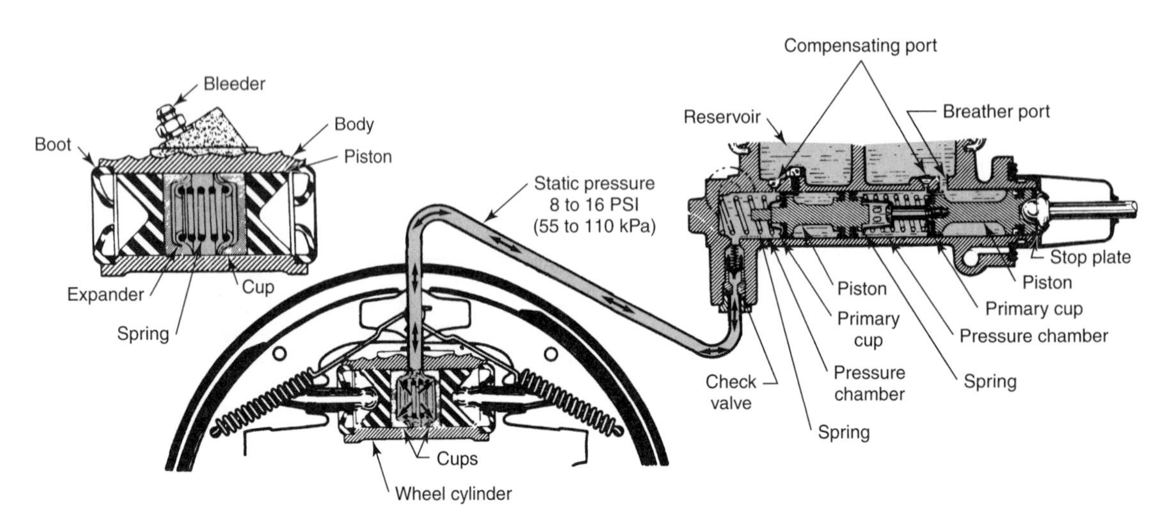

Figure 34-4. This illustration shows the state of a drum brake system when the brakes are released. (General Motors)

inward, the primary piston moves toward the ***floating piston*** (also called secondary piston) until the compensating port is closed. Further movement of the primary piston is then transmitted to the floating piston. As the floating piston is forced forward by the hydraulic fluid separating it from the primary piston, it closes off the other compensating port. Further movement of the primary piston builds up pressure in both outlet systems (one to the front brakes and the other to the rear brakes).

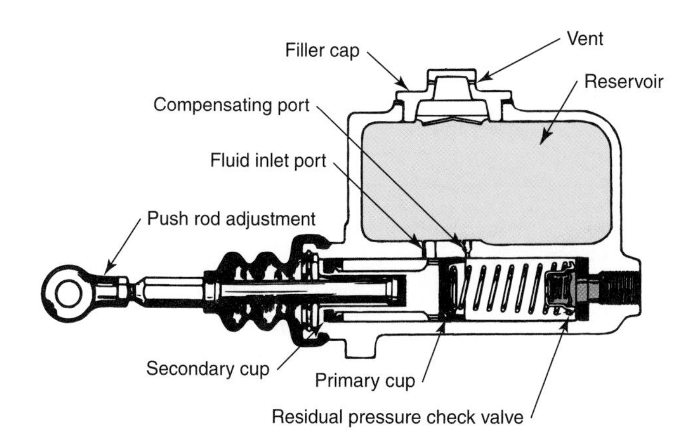

Figure 34-5. Single-piston master cylinders, like this one, are no longer used.

Figure 34-7 illustrates the action of the dual-piston master cylinder in protecting against complete brake failure. In Figure 34-7A, the master cylinder push rod is in the released position. In Figure 34-7B, the push rod has forced the primary and floating pistons to build up pressure in both front and rear hydraulic systems. In Figure 34-7C, the front brake line has ruptured, allowing the primary piston to move in until the floating piston strikes the end of the cylinder. Pressure is maintained to the rear wheels. In Figure 34-7D, the rear wheel brake line has ruptured, allowing the primary piston to move inward until it bumps the floating piston, thus maintaining pressure to the front wheels.

Failure of either the front or rear system causes a sudden increase in the brake pedal travel required to apply the brakes. A pressure differential safety switch is used to activate a

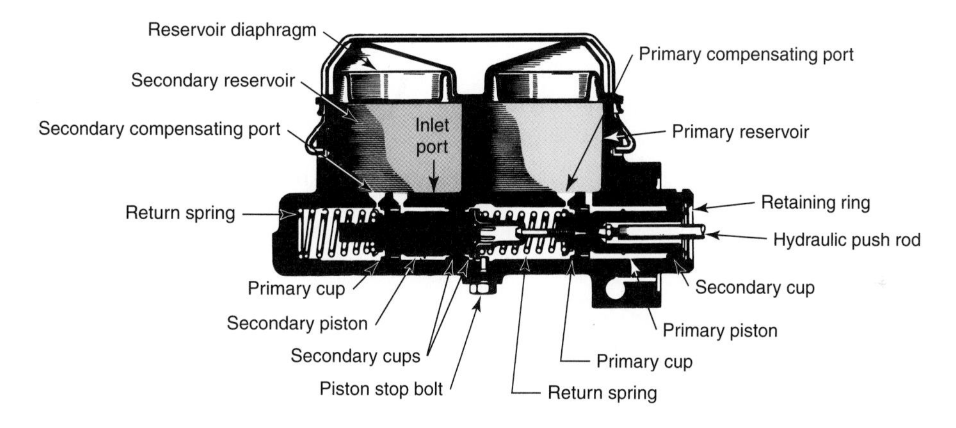

Figure 34-6. This illustration shows a typical dual-piston master cylinder. (Bendix)

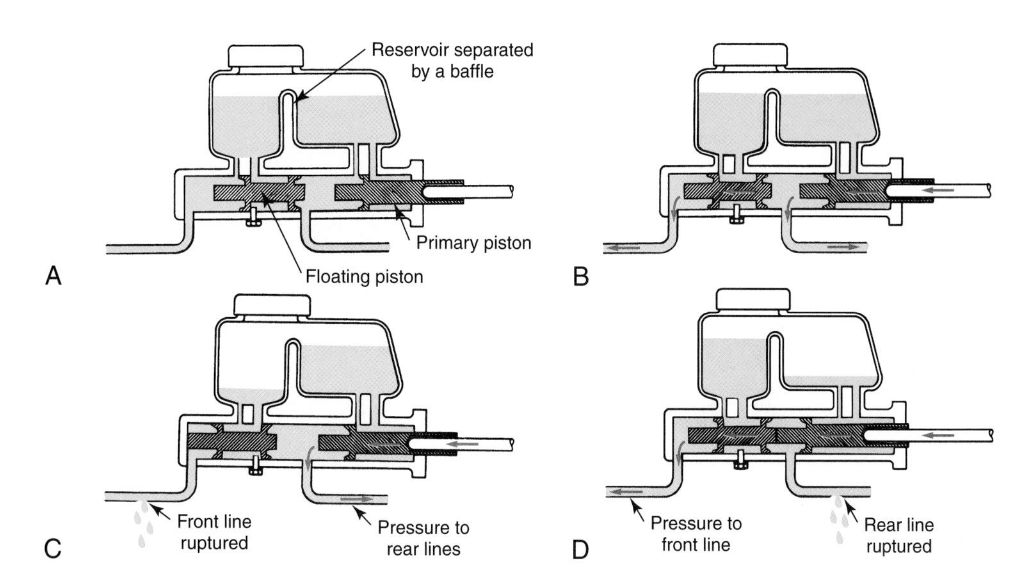

Figure 34-7. Action of a tandem master cylinder in the event of either front or rear brake system failure. (Mercedes-Benz)

warning light. Some dual-piston master cylinders have completely separate reservoirs, **Figure 34-8.** Others merely use a baffle to form two reservoirs. Look at **Figure 34-9.**

Compact Master Cylinder

The compact master cylinder is a composite design (plastic reservoir and aluminum body) that is used in diagonally split brake systems, **Figure 34-10.**

Compact master cylinders incorporate the functions of a standard dual master cylinder, a fluid-level sensor, and integral proportioners. The proportioners are used to improve front-to-rear balance during heavy brake applications.

Quick Take-Up Master Cylinder

The quick take-up master cylinder is used with low-drag disc brake calipers. The quick take-up feature supplies a large volume of low-pressure fluid to the brakes during the start of application. This low-pressure fluid rapidly supplies the increased fluid displacement needs caused by the no-drag caliper pistons.

There are four separate outlet ports on the cylinder housing. Each feeds an individual brake assembly. See **Figure 34-11.** A proportioning valve and warning light switch are housed in an integral bore at the bottom of the cylinder.

During the start of brake application, more fluid is displaced in the low-pressure primary chamber than the high-pressure chamber. See **Figure 34-12.** This displacement is created because the low-pressure chamber has a larger diameter than the high-pressure chamber.

As continued application occurs, brake fluid is forced around the outer edge of the primary piston lip seal. From

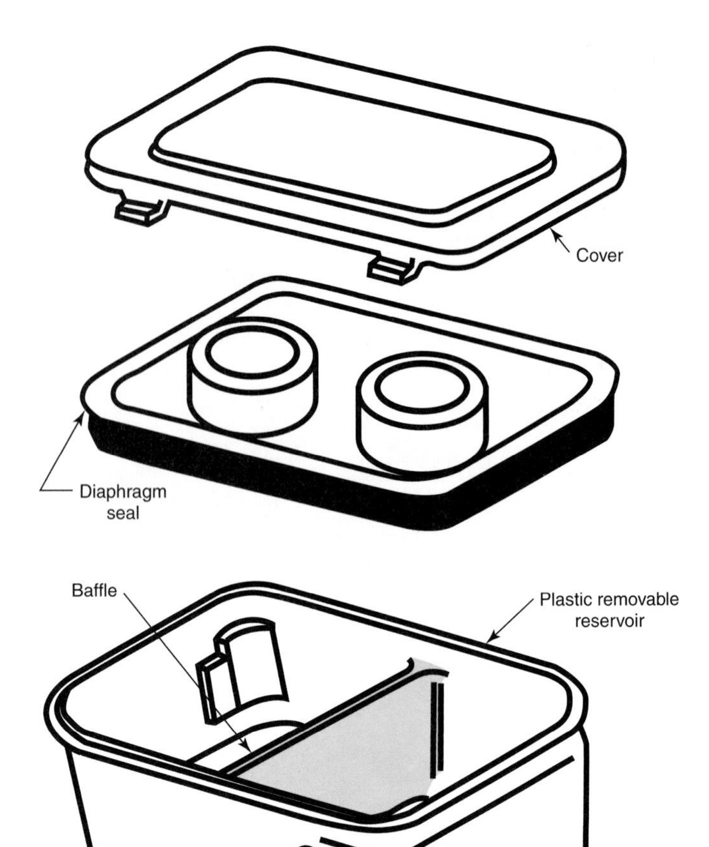

Figure 34-9. This plastic, removable master cylinder reservoir has an internal baffle to create two reservoir chambers. (DaimlerChrysler)

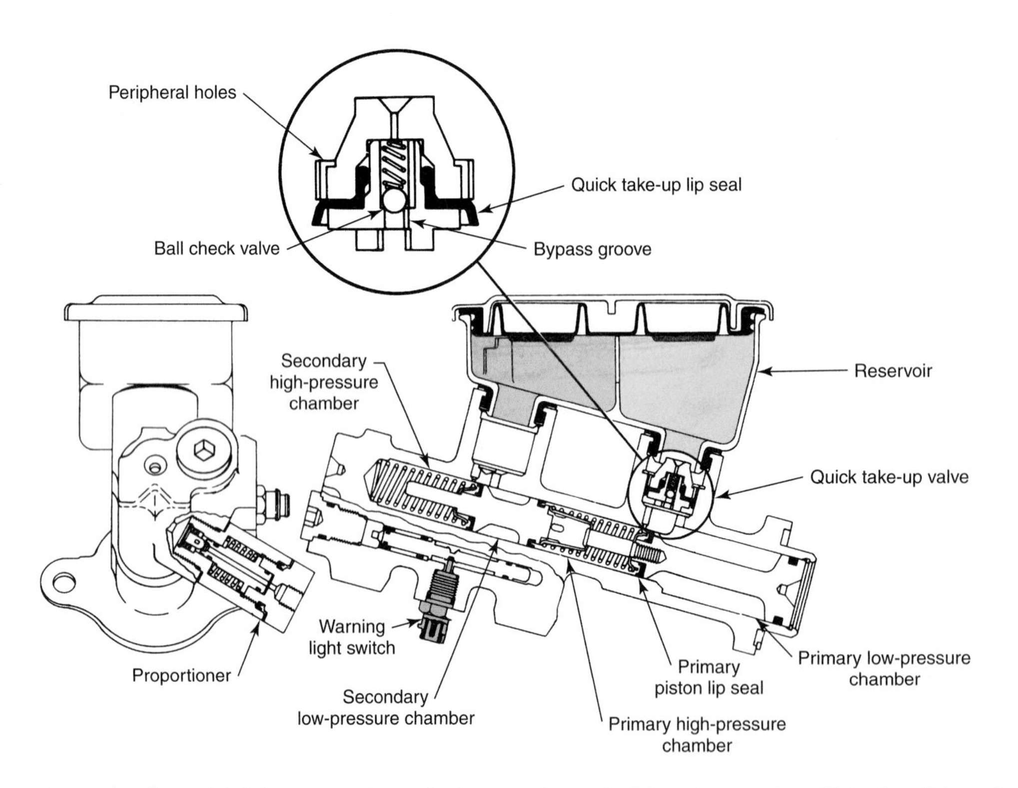

Figure 34-8. The major parts of a quick take-up master cylinder are shown in this cutaway view. Note that this unit also has a warning-light switch and a proportioner attached to the cylinder body. (Delco)

there it goes into the high-pressure chamber and on to the braking units at the wheels. Look at Figure 34-12A.

To create equal pressure and displacement in both systems, the primary piston travels a lesser amount than the secondary piston. At a preset pressure, the quick take-up valve "ball check" opens. This allows excess fluid in the larger primary chamber to flow into the reservoir. See Figure 34-12B. Once the primary piston lip seal has passed the quick take-up valve port, high pressure is built up ahead of each piston. The master cylinder functions like a conventional unit.

When the brakes are released, the master cylinder piston springs return the pistons faster than the returning fluid can travel. A vacuum is then generated in both the high- and low-pressure chambers. If the vacuum were not relieved, fluid would pass around the piston seal cup lips in an effort to eliminate the vacuum, Figure 34-12C.

To stop this vacuum from forming, the primary chamber vacuum is compensated by the bypass groove in the quick take-up valve. The primary piston is compensated by fluid traveling from the reservoir through small-diameter holes in the quick take-up valve. The secondary piston chamber vacuum is relieved by fluid traveling from the reservoir through the compensating port and bypass hole.

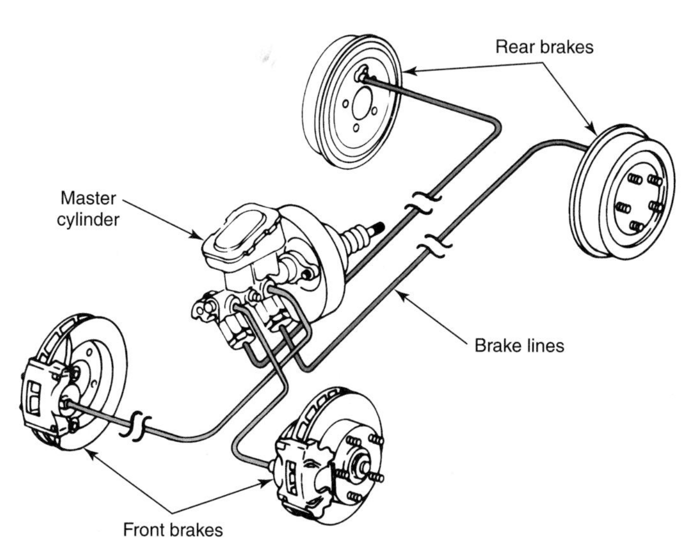

Figure 34-10. This is a sketch of a diagonally split braking system. Each brake line has its own attachment point on the master cylinder. (Bendix)

Checking Master Cylinder

The master cylinder fluid level can be easily checked. Refer to **Figure 34-13.** After thoroughly cleaning the cover(s) and the surrounding area, remove the cover to expose the fluid. Fluid level should be around 1/4″ (6.35 mm) from the top of the reservoirs (G). Fluid must be clean and free of discoloration. Make sure the tiny hole of the cover vent (C) is open to prevent pressure buildup in reservoir. Check the condition and placement of the reservoir seal diaphragm (F). The bail wire (D) must snap on tightly.

Inspect beneath the ***dust boot*** (I), for evidence of fluid leakage past the secondary cup (K). Check primary cups (L) by applying the brakes firmly. Hold pressure. The pedal should not move inward. If it does, it indicates that one or both primary cups could be allowing fluid to escape. A leak elsewhere in the system will also cause the pedal to fall away.

The compensating ports (A) must be open when the pedal is released. If it is closed by dirt, corrosion, a swollen primary cup, or improper linkage adjustment, it can cause a pressure buildup in the lines that will keep the brakes applied.

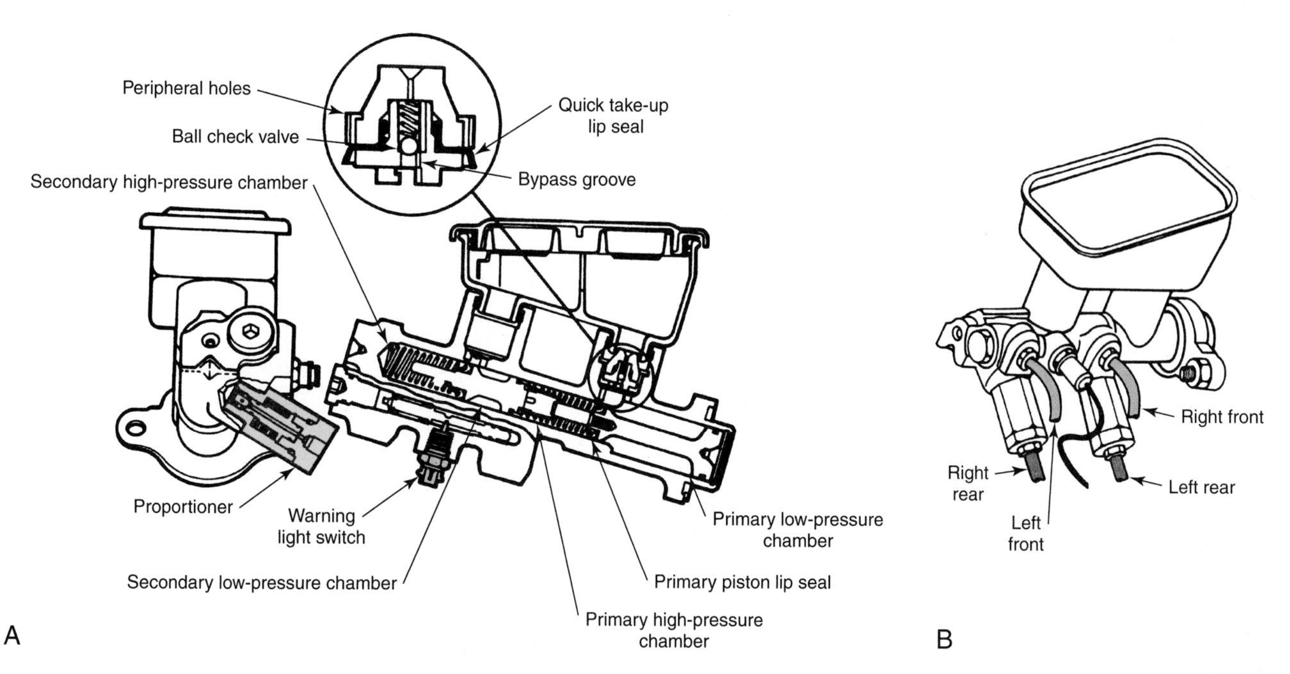

Figure 34-11. A—Cross-sectional view of a quick take-up master cylinder with a proportioning valve and warning-light switch. B—Four brake line connection points. (Bendix)

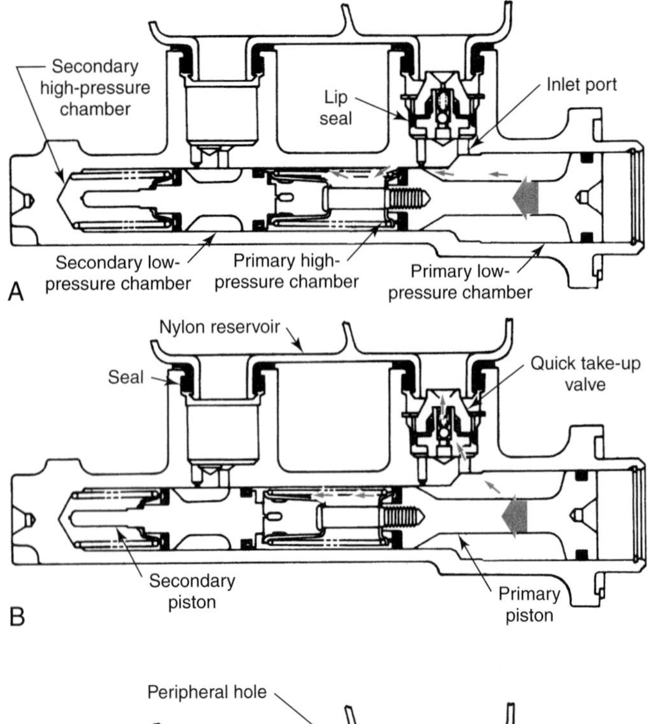

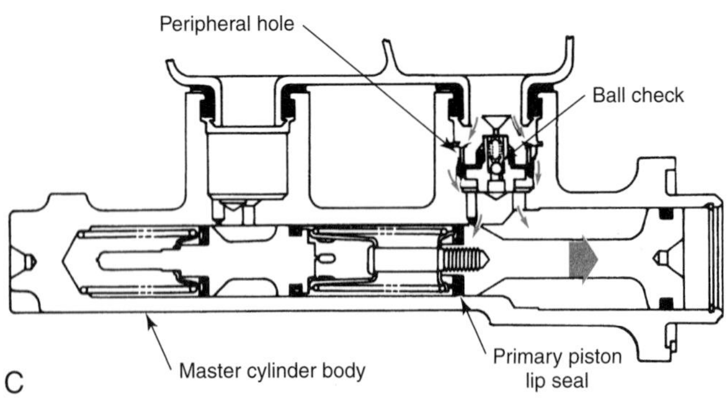

Figure 34-12. Study this illustration to understand the operation sequence of a quick take-up master cylinder. A—The start of quick take-up application. Note fluid flowing around primary piston cup. B—The primary piston and secondary piston now have equal pressure. The cylinder now operates as a conventional unit. Note fluid flow through quick take-up valve into reservoir. C—Releasing. (Delco)

Nylon reservoirs are usually equipped with molded-in windows (thin area) that are used to view the fluid level. See **Figure 34-14.**

Residual check valves, where used (drum brakes only), must function properly to maintain a static pressure in the lines. A faulty check valve can cause excessive pedal travel before the brakes apply. Worn linings or improperly centered shoes can also cause excessive pedal travel.

The check valve can be tested by applying and releasing the brakes and then cracking open a wheel cylinder bleed screw. If residual pressure is present, fluid will briefly squirt out.

Master Cylinder Repair

Remove the master cylinder. Cap the brake lines to prevent the entry of foreign material. Scrape off exterior dirt. Wash the exterior in approved brake system solvent. Remove the push rod where used. (In some designs, the stop plate or retainer must be removed first.)

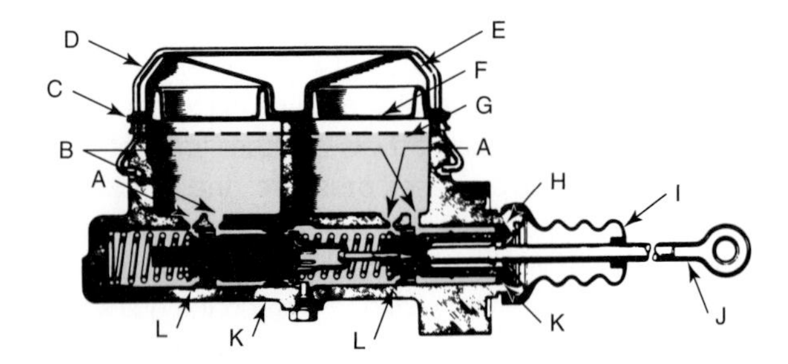

Figure 34-13. This illustration shows common master cylinder checkpoints. A—Compensating ports. B—Inlet ports. C—Vent. D—Bail. E—Reservoir cover. F—Diaphragm reservoir seal. G—Fluid level. H—Push rod retainer. I—Dust boot. J—Push rod. K—Secondary cups. L—Primary cups. (Bendix)

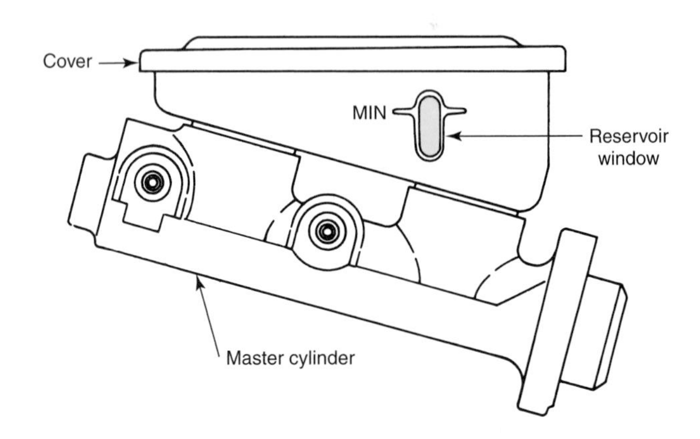

Figure 34-14. When a master cylinder has a reservoir window, the cover does not have to be removed to check fluid level. (General Motors)

Caution: Some stop plate retaining rings can fly out with great force. Hold the piston in until the snap ring is removed. Withdraw pistons, springs, and other parts.

If residual check valves are used, remove the valves by pulling the tube seats. This is done by threading a screw into the seat and prying up on the screw, **Figure 34-15.**

Use Special Solutions for Cleaning

Caution: Clean all parts in a recommended brake-cleaning solution or brake fluid. Never allow a brake system's rubber parts to contact gasoline, kerosene, oil, or any type of petroleum-based cleaner. Never touch rubber parts with oily or gasoline-soaked fingers. Wash your hands with soap and water before handling parts.

If exceptionally dirty, it is permissible to wash the disassembled cylinder body, head nut, push rod, and reservoir cap in fresh petroleum-based solvent provided the parts are blown dry, rinsed in approved brake solvent or alcohol, blown dry

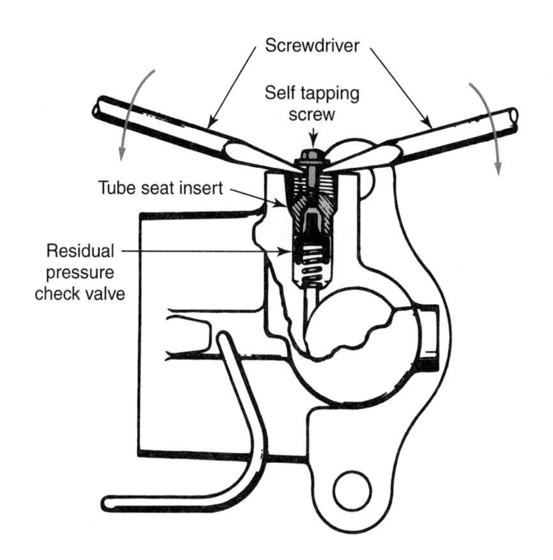

Figure 34-15. Pull the tube seat insert to free the check valve for removal.

again, rinsed again in brake solvent or alcohol, and blown dry a third time.

When all parts are clean, inspect the cylinder surface. Hold it up to a good light. Scoring, pitting, or heavy corrosion will require discarding the entire master cylinder. Minor (very light) scratches and corrosion can be removed with crocus cloth. Do not use emery cloth or sandpaper.

The cylinder, in some cases, can be honed to remove minor scratches. The hone stones must be true, clean, and extremely smooth. Hone as little as possible. Use brake fluid as a hone lubricant. Place crocus cloth over the hone stones for a final smoothing.

Some manufacturers diamond bore the master cylinder to a very smooth finish and then roll the surface to produce a glassy finish. Honing this type of cylinder during service will destroy the glassy finish and, therefore, is not recommended. Do not attempt to remove scratches, pits, or corrosion from aluminum master cylinder bores. This removes the protective anodized surface. If any scratches or pits show or if aluminum can be seen through the anodizing, the cylinder must be replaced.

Note: Do not try to salvage a master cylinder unless the cylinder is in excellent condition. Trying to save a few pennies can cost dearly.

When the cylinder has been cleaned, scrub the bore with a clean, lint-free cloth dipped in approved brake solvent. Repeat the flushing process two more times. Compressed air must be oil free.

Ports Must Be Open and Free of Burrs

Check the compensating and inlet, or breather, ports. They must be clean and open. The compensating port may be cleaned by passing a thin (0.020″ or 0.508 mm), smooth, copper wire through the opening. Do not pass square or rough-tipped steel wires through the port.

If a burr is present at the port, remove it with a deburring tool. The slightest burr will cut the primary cup and cause leakage. Remove all burrs. Give all parts a final rinse with brake-cleaning fluid and blow dry.

Lubricate with Brake Fluid

Coat the cylinder wall with brake fluid. Dip the piston and rubber cups in clean brake fluid. Assemble in the reverse order of disassembly.

Cylinders, cups, and pistons must be coated with brake fluid before assembly. Parts assembled dry can cause sticking and scoring. Use a repair kit if the cylinder is in excellent condition. Never install old parts. Make certain the piston stop plate lock ring is properly engaged. **Figure 34-16** illustrates

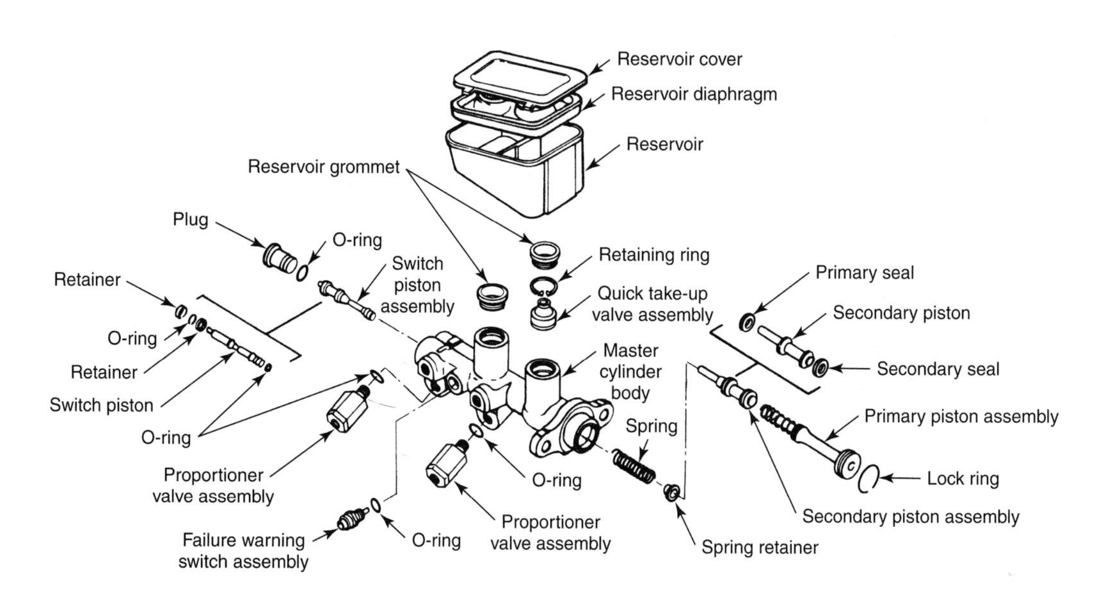

Figure 34-16. Exploded view of one type of master cylinder. Carefully inspect *all* parts during service. Follow the manufacturer's overhaul procedures carefully. (Bendix)

the parts of a typical master cylinder. All parts should be carefully inspected during service.

Bleed Master Cylinder before Installing

Attach bleeder tubes, **Figure 34-17.** Add the recommended brake fluid to the reservoir until the fluid level is above the ends of the bleeder tubes. Force the pistons in and out several times until all air is dispelled from the cylinder. Maintain the reservoir level while bleeding. Remove the bleeder tubes. Install the cover to prevent fluid contamination. Plugs or a special bleeding syringe can be used in place of the bleeder tubes. Follow the manufacturer's recommendations. See **Figures 34-18** and **34-19.**

Caution: Avoid spraying brake fluid. Wear safety goggles.

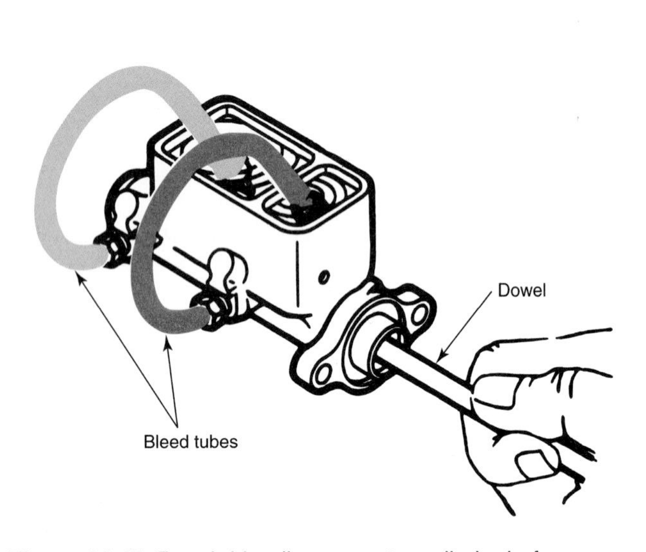

Figure 34-17. Bench bleeding a master cylinder before installation. Note the bleeder tubes. Operate the pistons until all air is expelled. (DaimlerChrysler)

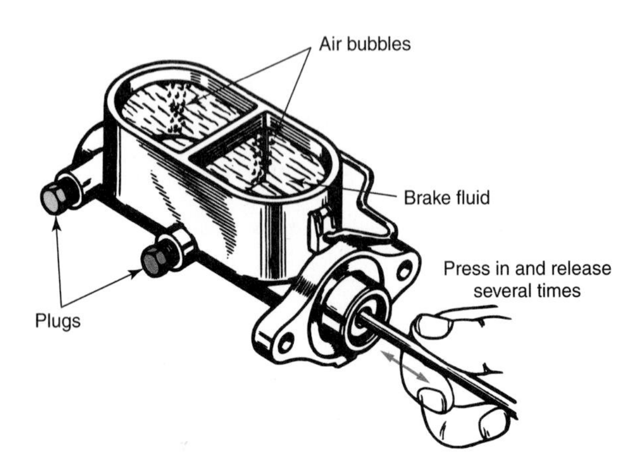

Figure 34-18. Bleeding a master cylinder with the aid of plugs. Force the piston in and out until air bubbles stop. (Bendix)

Install the cylinder and connect the lines. At this time, it is good practice to crack each master cylinder line connection open a small amount. Wrap a rag around the connection. Depress the brake pedal gently to force the remaining air out of the master cylinder. Tighten the connection before releasing. Adjust the brake pedal. Bleed the brakes (discussed later in this chapter). Fill the master cylinder to the proper level and install the cap and seal.

Adjusting Brake Pedal Height

If the pedal has an adjustable stop, measure the distance from the floorboard to the bottom of the pedal. Check against specifications and adjust if needed, **Figure 34-20.**

Adjusting Brake Pedal Free Travel

Brake pedal ***free travel*** is the distance the pedal moves before the push rod engages the cylinder piston. Free travel is

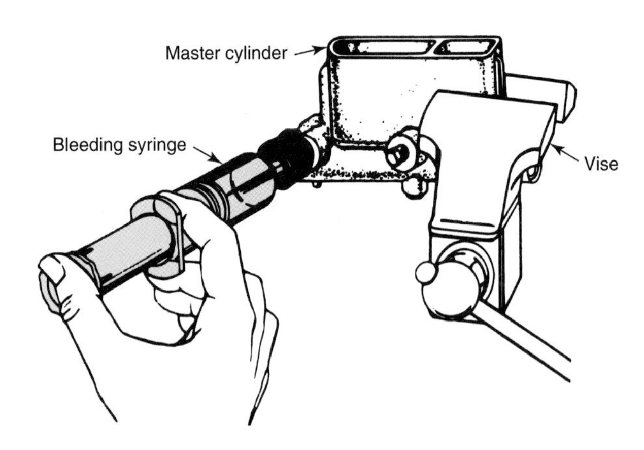

Figure 34-19. Some master cylinders can be bled with a special syringe. Plug the outlet holes. Fill the cylinder with fluid (about half full). Remove a plug. Depress the syringe plunger fully. Place it firmly against an outlet. Pull the plunger out until the syringe is half full. Remove the syringe. Point it straight up. Expel air until fluid squirts out. Put the syringe (with fluid) on the same outlet and depress the plunger. When air bubbles in the reservoir stop, remove the syringe and plug the outlet. Repeat this procedure on the other outlets. Plug the outlets firmly. (EIS)

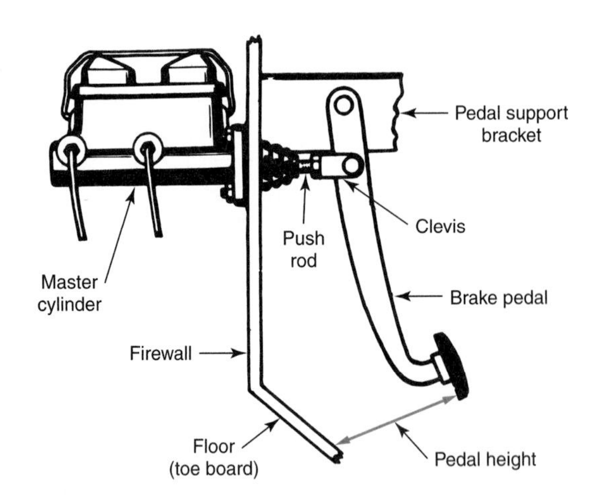

Figure 34-20. Brake pedal height should be as specified.

needed to ensure the piston will not be applied when the pedal is in the released position.

Free travel is automatically set on some cylinders when the brake pedal height is correctly adjusted by lengthening or shortening the push rod clevis. See **Figure 34-21.**

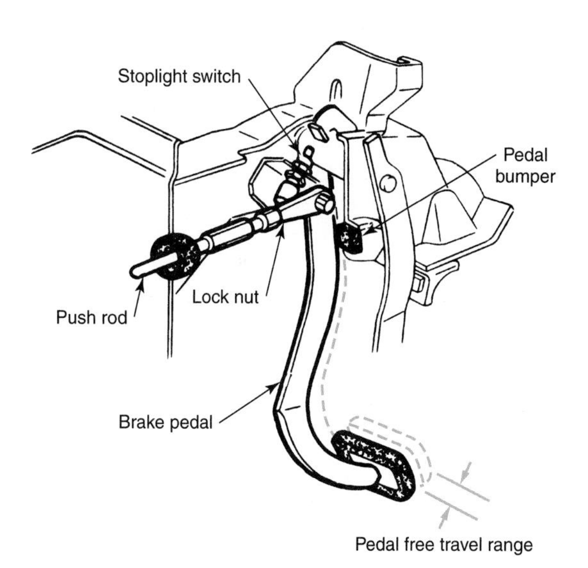

Figure 34-21. Brake pedal free travel is critical to proper brake system operation.

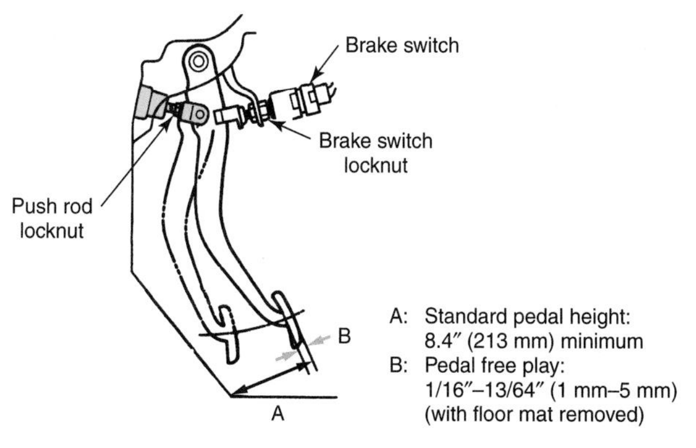

Loosen the push rod locknut and screw the push rod in or out with pliers until the standard pedal height from the floor is 8.4″ (213 mm). After adjustment, tighten the locknut firmly.

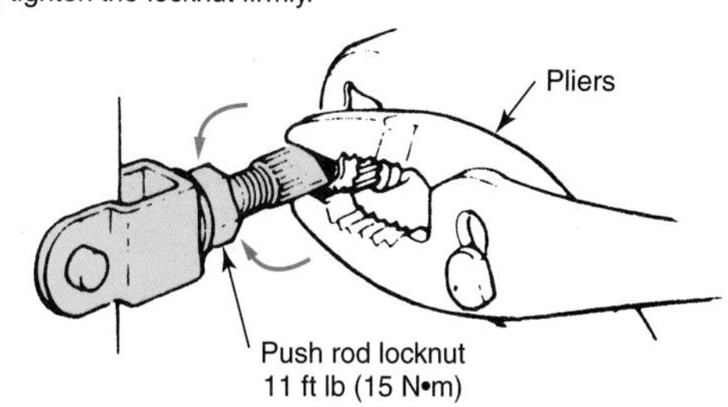

Screw in the brake switch until its plunger if fully depressed (threaded and touching the pad on the pedal arm). Then back off the switch 1/2 turn and tighten the locknut firmly.

Figure 34-22. This illustration shows a way of obtaining brake pedal free travel by adjusting the push rod. It may be necessary to adjust the brake-light switch *after* push rod adjustment. (Honda)

Other types use a separate pedal stop. The push rod is adjusted until the correct free travel exists. **Figure 34-22** illustrates pedal free travel.

On some vehicles with power brakes, the technician must check the distance from the push rod to the face of the vacuum booster. Use a special gauge and adjust as needed before mounting the cylinder, **Figures 34-23** and **34-24.** Proper free travel for manual (non-power) brakes is around 1/4″–1/2″ (6.35 mm–12.70 mm). For power-assisted brakes, free travel ranges from 1/8″–3/8″ (3.18 mm–9.53 mm).

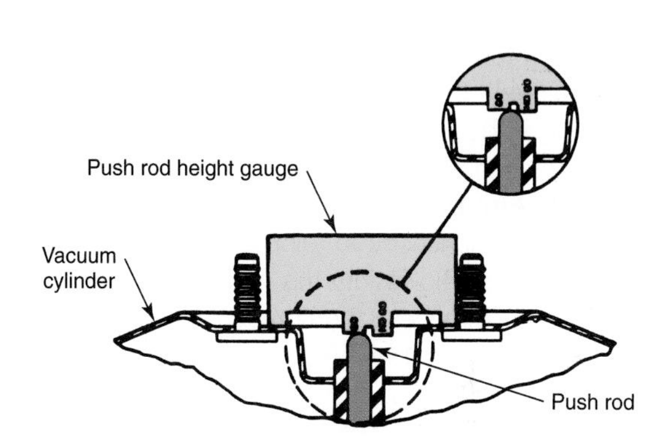

Figure 34-23. Use a special tool to check the distance the master cylinder push rod protrudes from the vacuum booster unit. (General Motors)

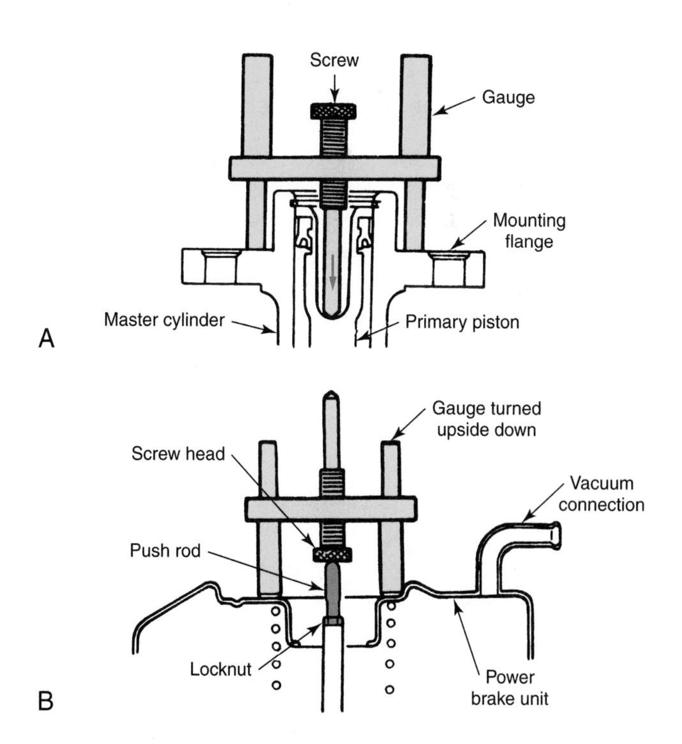

Figure 34-24. A special adjustment gauge is being used to measure and set push rod length. A—The gauge is set on master cylinder mounting flange. The screw is turned in until it just touches the primary piston. B—The gauge is turned upside down and placed on the power brake unit. Loosen the locknut and turn the push rod until it just touches the screw head. Tighten the locknut. Proper push rod clearance is now set. (Mazda)

Wheel Cylinder Operation

Pressure from the master cylinder is delivered to the ***wheel cylinder*** by lines and hoses. This pressure moves the wheel cylinder pistons outward until the brake shoes contact the drum. Any increase in brake pedal pressure beyond this point will cause a corresponding increase in shoe-to-drum contact. **Figure 34-25** shows the system with brakes being applied. Since the brake drum is attached to the axle, contact with the brake shoes causes the drum and axle to stop turning.

When the brake pedal is released, the brake shoe retracting springs force fluid to flow backward into the master cylinder. Tiny holes in the piston prevent vacuum action that could draw air into the cylinder. The holes also permit instant piston release until the check valve can move from its seat. The primary cup lips are forced away from the cylinder wall by the passage of fluid through tiny bleeder holes in the head of the piston. See **Figure 34-26.**

As the piston continues to release, fluid forces the check valve off its seat and flows into the cylinder. When the piston releases to the point where the compensating port is uncovered, excess fluid flows into the reservoir. Look at **Figure 34-27.**

Wheel Cylinder Service

Always remove, disassemble, and clean the wheel cylinders when the brakes are relined. With some designs, relining the brakes can force the wheel cylinder pistons deeper into the cylinder. If corrosion, rust, or gum has formed between the cups, forcing the pistons inward would cause the cups to operate on the rough surface. Scoring and leakage would soon follow. Replace the cylinders if necessary.

Remove Wheel Cylinder

To gain access to the wheel cylinder, pull the wheel and brake drum. Remove the brake shoes (see Brake Shoe Removal in this chapter). Loosen the brake line flare nut at the wheel cylinder. Do not bend the brake line out of the way. This can damage the line and make alignment difficult during installation. Remove the fasteners holding the wheel cylinder to the backing plate. Remove the cylinder. See **Figure 34-28.**

Observe Cautions

When disassembling, repairing, and assembling any wheel cylinder, keep rubber parts away from oil and grease. Wash hands with soap and water before handling rubber parts. Use approved (nonpetroleum-based) brake-cleaning solvents or denatured alcohol for cleaning. Lubricate parts with brake fluid before reassembly.

Wheel Cylinder Repair

Pull the rubber boots free of the cylinder and push the pistons, cups, and spring from the cylinder. See **Figure 34-29.** Use air on single-end cylinders to force the piston free.

Wash all parts in approved brake solvent or denatured alcohol. Some wheel cylinder pistons are made of iron and are impregnated with lubricant. Do not wash these pistons; wipe them off. Pistons should not be reused if pitted or scratched.

Aluminum wheel cylinder bodies should be replaced if they are pitted, corroded, or scratched. Clean cast iron

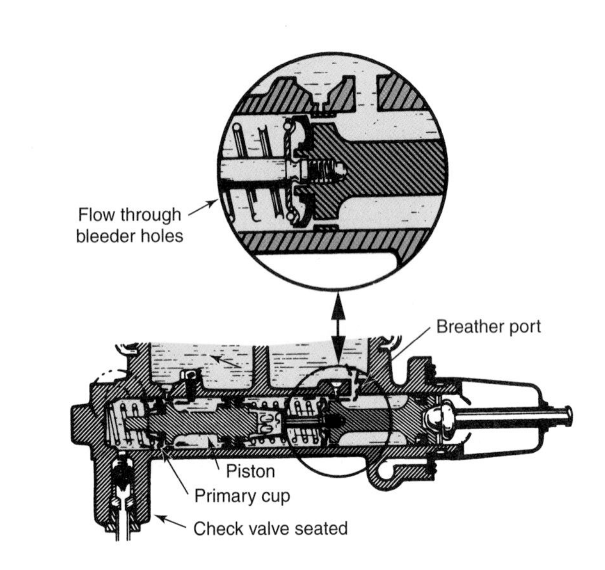

Figure 34-26. Master cylinder operation at the start of a fast release. (General Motors)

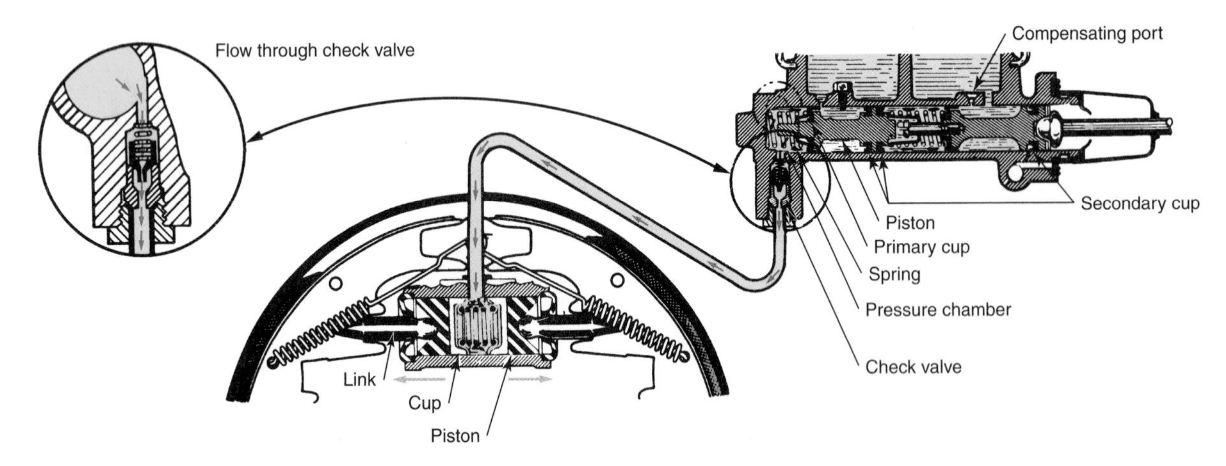

Figure 34-25. Drum brake system with brakes applied. (General Motors)

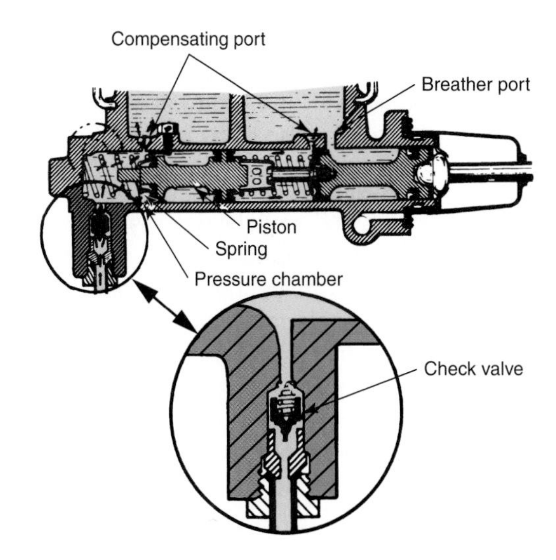

Figure 34-27. Master cylinder operation during the finish of a fast release. (General Motors)

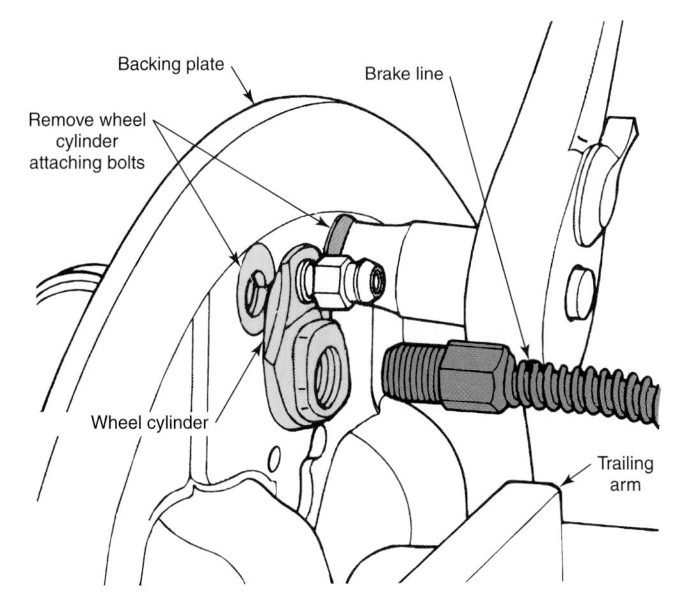

Figure 34-28. Remove the brake line and wheel cylinder attaching bolts so the cylinder can be removed from the backing plate. (DaimlerChrysler)

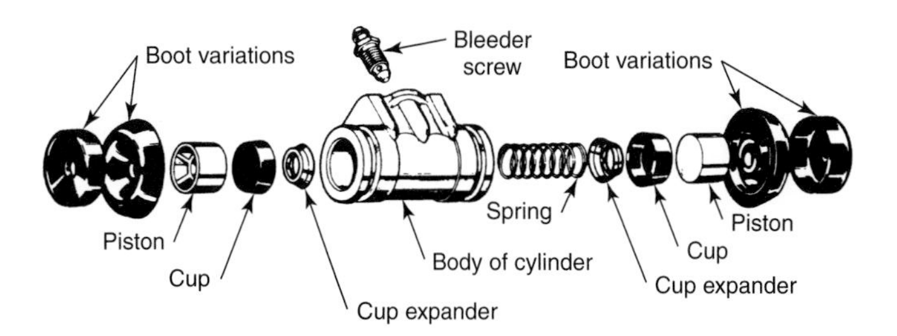

Figure 34-29. Note the parts in this exploded view of a typical wheel cylinder. Most modern wheel cylinders have the same parts as this one. (Bendix)

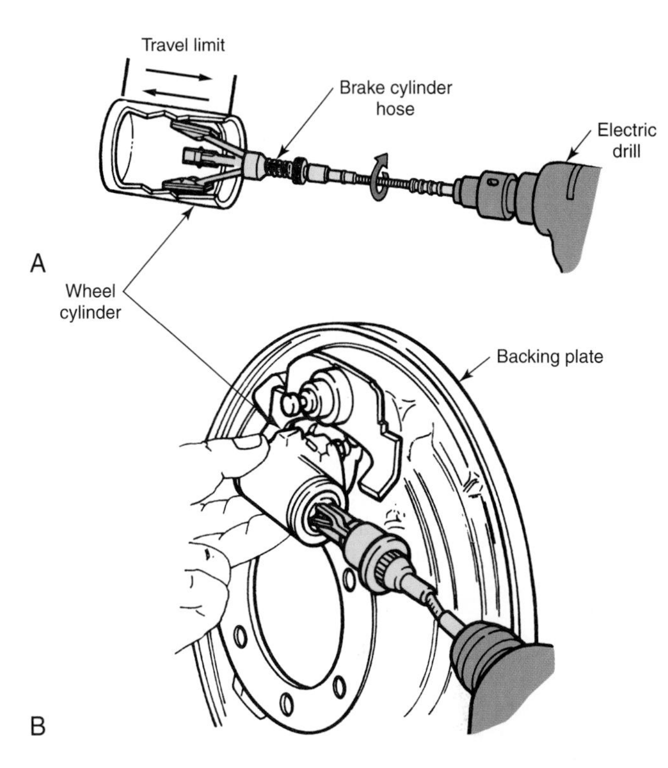

Figure 34-30. This Figure shows the process of honing a wheel cylinder. A—Hone in place. Note stone travel limit. *Do not* push stones all the way through or pull them completely out while they are turning. Compress spring-loaded fingers and insert them before starting to turn the hone. B—Wheel cylinder being honed while mounted on the backing plate. (Niehoff)

cylinders with crocus cloth and inspect. If crocus cloth fails to remove scratches or pitting, light honing (with very fine grit stones) may be used. Use brake fluid for stone lubrication, **Figure 34-30.** If the wheel cylinder has a special glassy, rolled finish, do not attempt honing.

Following honing or the use of crocus cloth, scrub the cylinder with a rag soaked in brake cleaner. Rinse at least two times and blow dry.

Note: The compressed air used to blow cylinders dry must be dry and free of oil. If it is not, allow parts to air dry.

Inspect the cylinder. The finish must be free of scoring and pitting. A slight pitting in the very center of the cylinder is permissible as long as it is not in the area of the cylinder that engages the rubber cups. When in doubt, discard the cylinder.

Many shops will not overhaul wheel cylinders. They feel the cost of the labor involved, along with a possible reduction in the service life of used cylinders, does not warrant such work. When a cylinder is defective, it is replaced with a new one.

Check the clearance between the piston and cylinder wall. It must not exceed 0.005″ (0.127 mm) for cylinders 1″ (25.4 mm) or less in diameter, or 0.007″ (0.178 mm) for cylinders exceeding 1″ (25.4 mm) in diameter. Use a No-Go gauge or a feeler gauge. Pistons must be free of corrosion, scoring, and flat spots. If the pistons pass this test, coat the cylinder, new cups, and pistons with brake fluid and assemble. See **Figure 34-31.**

Note: Make certain the cup lips face inward. If cup expanders are used, they must be in place. Do not damage cups when starting them into the cylinder. Do not force the cups over the center ports.

Snap the end boots into place to hold the pistons in the cylinder. Use a cylinder clamp if needed. Install the cylinder on the backing plate. Connect the brake line. Install the shoes and drum. Bleed the brakes. A typical wheel cylinder installation is illustrated in **Figure 34-32.**

Disc Brake Caliper Service

Disc brakes use a heavy disc, called a rotor, instead of a conventional brake drum. The rotor is bolted to the wheel hub. A ***brake caliper,*** bolted to the spindle, surrounds the rotor. When the brake is applied, hydraulic pressure from the master cylinder forces the ***caliper piston*** outward. The caliper piston presses the friction pads against the revolving rotor, stopping the rotor.

Disc brakes are highly resistant to fade (loss of friction from overheating). A typical disc brake assembly is shown in **Figures 34-33.** The friction pads are self-adjusting and operate with very little clearance between the lining and rotor. One design actually allows the pads to rub the rotor very lightly at all times.

Disc Brake Caliper

A caliper with two hydraulic pistons on each side is illustrated in **Figure 34-34.** This particular caliper may have the two halves separated for repair. New seal rings must be used when bolting the halves back together. Some calipers must not have the halves separated. The technique used to maintain pad clearance in the caliper shown in **Figure 34-35** is simple and effective.

When the piston is forced outward to apply the pads, the piston seal stretches slightly to the side, Figure 34-35A. When the pedal is released, the seal straightens up and draws the piston back far enough for the pad to clear the rotor, Figure 34-35B. In addition to the withdrawal action of the seal, any slight runout in the rotor also helps to push the pad away.

Fixed and Floating Calipers

Some caliper assemblies are held rigidly in place. These are called ***fixed calipers*** and require the use of at least one piston on each side to force the pads against the rotor. **Figure 34-36** illustrates one arrangement.

The ***floating caliper*** uses one piston and is free to slide sideways. As the piston is forced out, it first shoves the brake pad it contacts against the rotor. When all movement in this direction is taken up, the caliper itself slides away from the piston. This draws the other pad (installed in outboard side of caliper) against the rotor. Further piston travel applies pressure to both pads.

Actual movement is very small as the pads barely clear the rotor after the brakes have been released. **Figure 34-37**

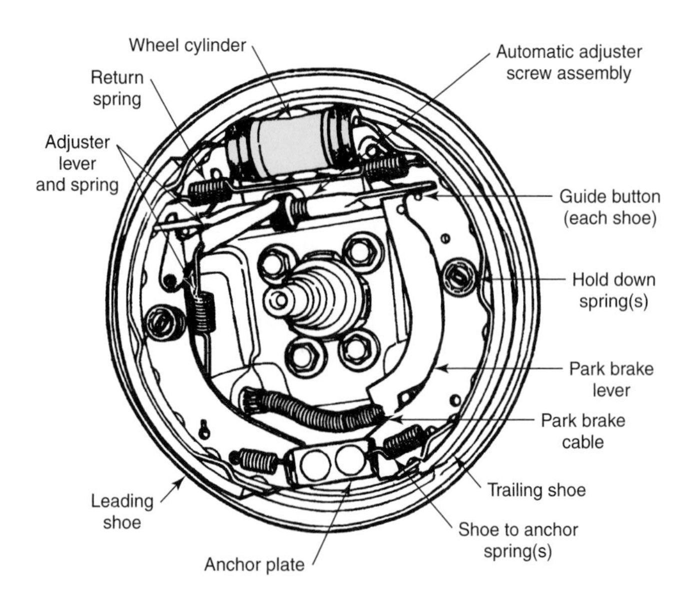

Figure 34-32. A typical wheel cylinder installation. This is from the left side of the vehicle. (DaimlerChrysler)

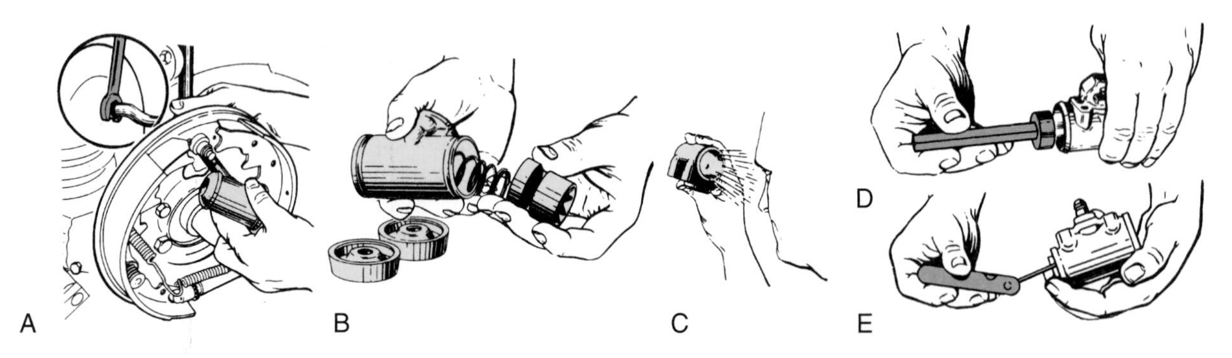

Figure 34-31. Some major steps in wheel cylinder reconditioning. A—Remove cylinder. B—Disassemble and clean. C—Inspect cylinder. D—Checking cylinder with No-Go gauge. E—Checking cylinder-to-piston clearance with a feeler gauge. (Wagner)

shows the various parts of the common single-piston, floating-caliper assembly. Modern vehicles often have disc brakes on the front, with drum brakes on the rear. Some vehicles have disc brakes on both front and rear. The hydraulic system action is essentially the same for all system configurations.

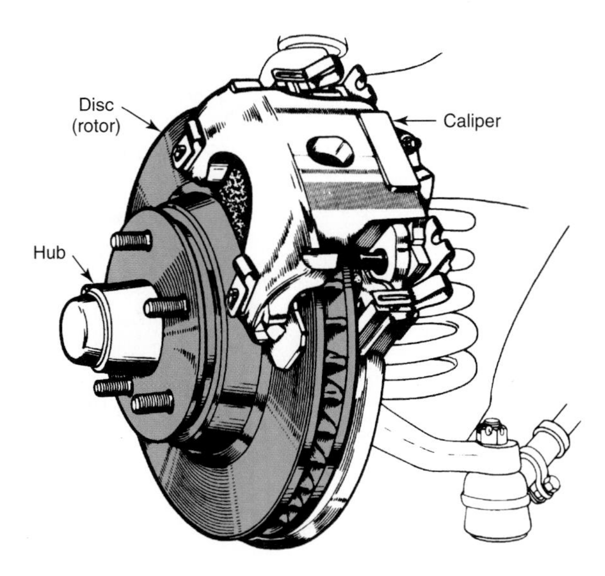

Figure 34-33. This disc brake assembly is commonly used. Most modern disc brakes will resemble the one in this illustration. (Bendix)

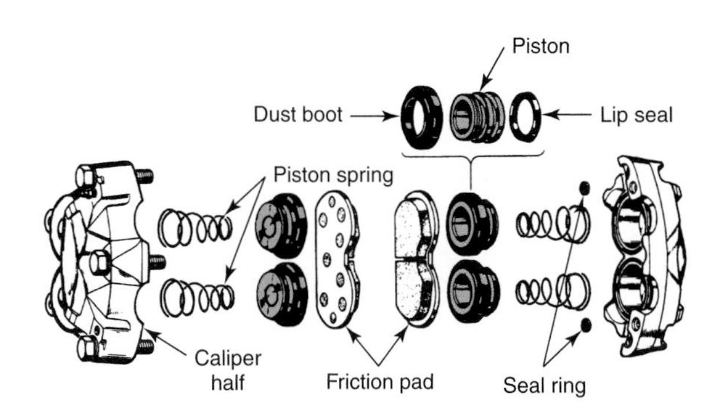

Figure 34-34. Exploded view of a disc brake caliper using two pistons in each half. This type of caliper is rare today. (DaimlerChrysler)

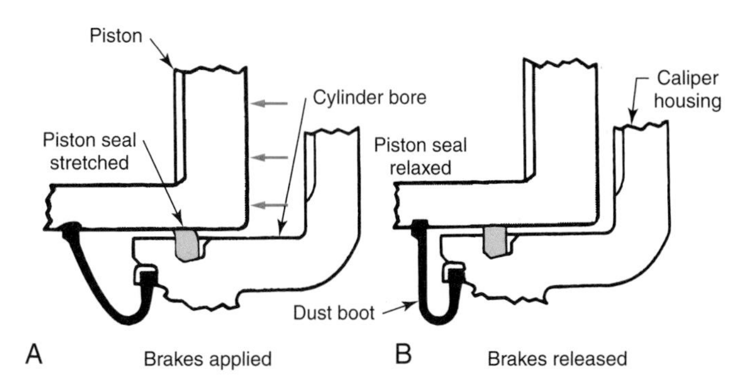

Figure 34-35. Seal action maintains the proper pad-to-rotor clearance. A—With the brakes applied, the seal is forced sideways. B—With the brakes released, the seal pulls the piston back.

Servicing Disc Brake Calipers

When the caliper is removed, check carefully for any aligning shims. Tag the proper location. If the caliper halves can be separated, as in Figure 34-34, do so. Next, continue disassembly, **Figure 34-38.** Remove the piston dust boot(s). Use care to avoid scratching either the piston or the cylinder wall, Figure 34-38A. A fiber or plastic pry stick is useful for safely removing the dust boots.

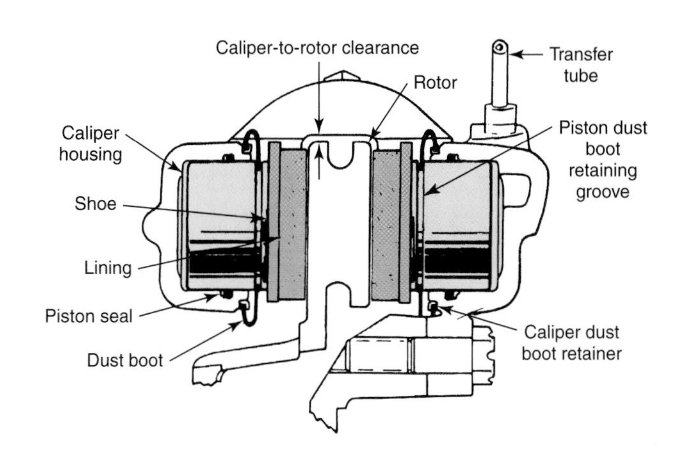

Figure 34-36. One type of caliper pad and piston design.

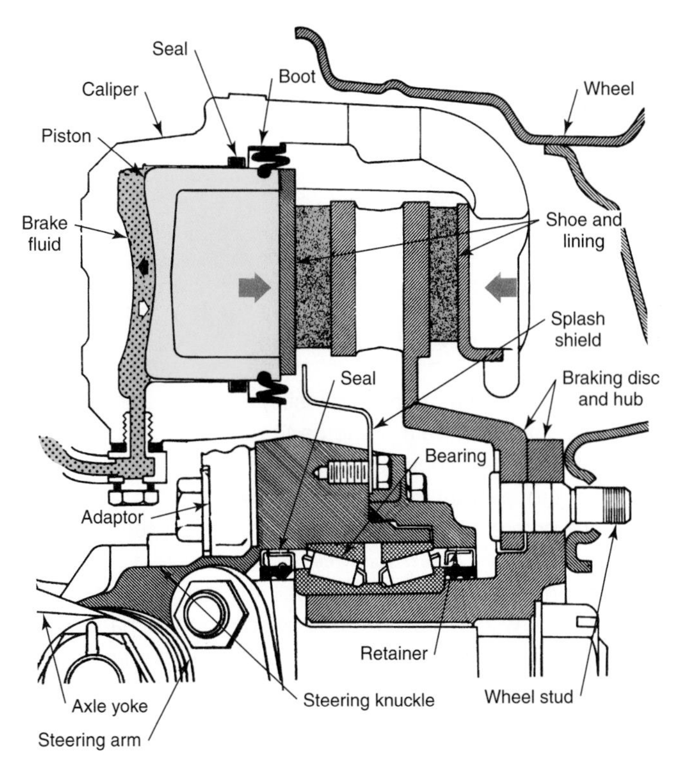

Figure 34-37. Cross section of a single piston, floating caliper disc brake as used on one particular front-wheel drive vehicle. (DaimlerChrysler)

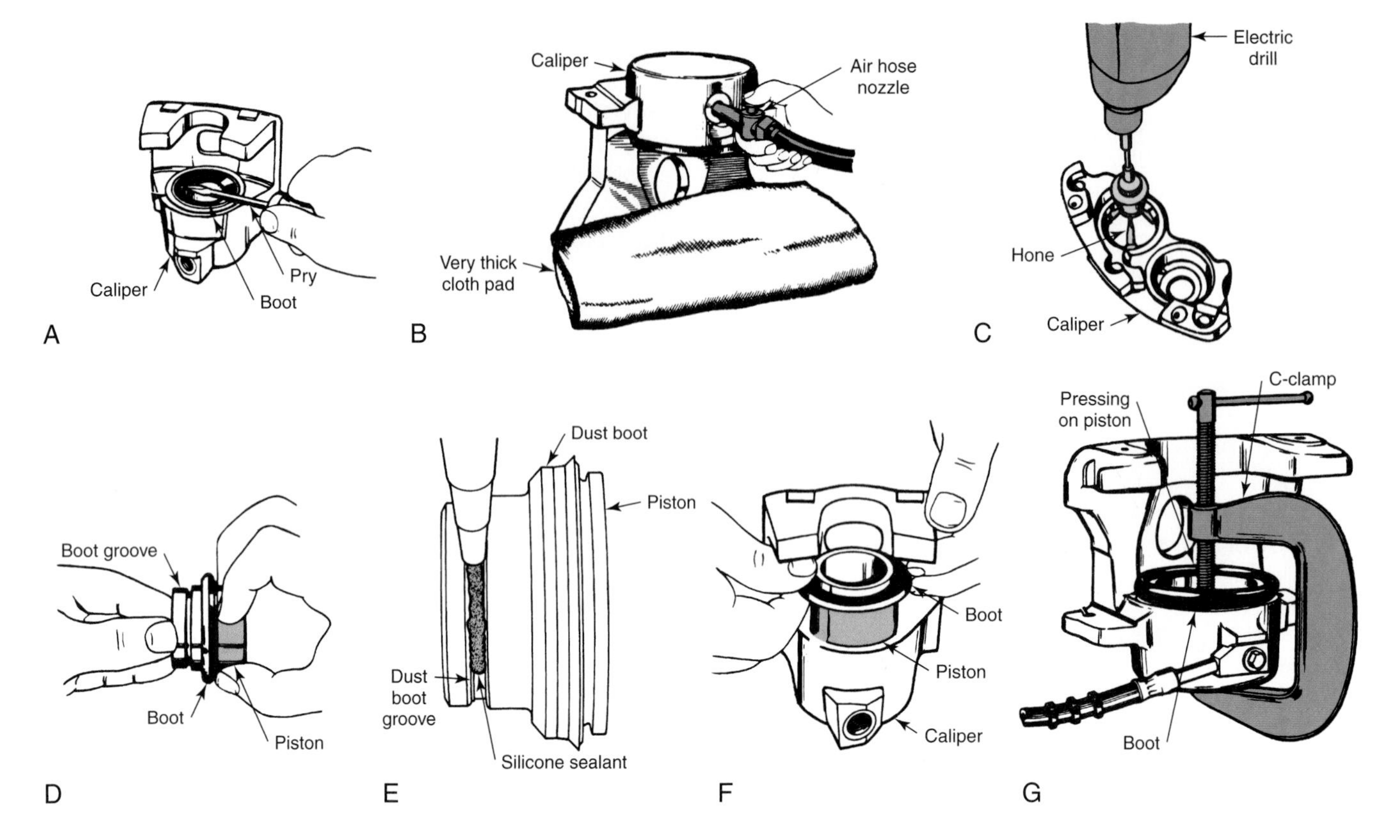

Figure 34-38. Common brake caliper overhaul operations. A—Removing the piston boot. B—Using air to remove the piston. Be careful! C—Honing the cylinder bore. D—Installing the boot on a cleaned and lubricated piston. E—Placing silicone sealer in the dust boot groove. F—Inserting the piston in a clean, lubricated bore. G—Forcing the piston all the way in with a small C-clamp. (Bendix)

Remove the pistons. The careful use of an air hose, as shown in Figure 34-38B, makes removal easy on some calipers.

Warning: Be careful when using air to remove the caliper piston. Place a thick pad of cloth between the piston and the front of the caliper, as shown in Figure 34-38B. Apply air gradually with little pressure. If the piston does not come out, remove the air. Rap the caliper with a soft hammer and try air pressure again.

Do not remove phenolic (plastic) caliper pistons with air pressure. They require more force to move in their bore than the cast iron type. They can come out with enough force to damage the piston. Remove these with system hydraulic pressure before the brake lines are disconnected from the calipers.

When removing phenolic caliper pistons, cover the cylinder bore with a pad of cloth to avoid piston damage. Keep fingers clear of the cylinder, as the piston can come out very fast. Turn the unit away from your face and body. Keep other personnel away and cover the whole assembly with a heavy shield cloth.

Remove the cylinder seal. Use a nonmetallic pry tool. When the caliper assembly is stripped, clean all parts in brake cleaner and blow them dry. Make certain the air source is oil free. If in doubt, blow the parts dry, give them a final rinse in clean brake cleaner or alcohol, and let them air dry. Be careful with alcohol—it is very flammable.

Check caliper cylinders carefully for scoring or corrosion. Some assemblies require that a feeler gauge be used between the piston and bore to check for proper fit. If the bore is in relatively poor condition, discard it. If it is in fair condition, hone as shown in Figure 34-38C. Bore size must not be increased by more than around 0.001″ (0.025 mm). Use brake fluid on the hone. The honing stones must be fine.

A cylinder bore in quite good condition can often be cleaned by hand, using crocus cloth. Following honing (or use of crocus cloth), the entire unit must be thoroughly cleaned in brake cleaner or alcohol. Brush out the boot and seal grooves with a nonmetallic brush. Thoroughly blow out the grooves and all passageways. Repeat cleaning several times.

Coat the new piston seal with special lubricant and place it into the groove in the cylinder. Use your fingers only. Lubricate the bore. Make sure the seal is seated.

Check the condition of the piston(s) and replace them if the plating is scored, corroded, or worn. If the caliper contains more than one piston, always replace them in pairs. Check phenolic (plastic) pistons for cracks, gouges, or chips.

Thoroughly clean the piston and groove. Coat the new boot with special lubricant. Slide the boot over the piston, Figure 34-38D. Some pistons require the use of a special silicone sealer in the dust boot groove. Install it before the boot is placed on the piston. This is shown in Figure 34-38E. Slide the boot into its groove.

Make certain the cylinder bore and piston are coated with brake fluid. Place the piston in the bore and press inward, Figure 34-38F. If needed, a C-clamp may be used against the piston. See Figure 34-38G. The clamp will hold the piston down while the edge of the boot is snapped into place in the caliper boot groove.

When using a C-clamp with plastic pistons, place a smooth piece of metal or wood across the piston. Do not place the C-clamp in direct contact with the piston, or damage to the piston may result. Use a blunt tool. Do not puncture the boot. If you do, replace the boot.

Install the brake pads. Some designs allow installation of the pads after the caliper is mounted. Secure the pads as needed. Lubricate any recommended surfaces. Never get lubricant on the pad or rotor. Install the caliper as recommended.

Note: Do not switch calipers from side to side. Mounting the calipers on the opposite side will often place the bleeder screw in a position that makes it impossible to completely bleed the brakes.

Bleed and actuate the brakes to make sure the caliper pads are in full contact with the rotor. The preceding techniques are common, but variations in calipers might call for slightly different installation methods. Learn and follow the recommended procedures for the brake system being serviced.

Servicing Brake Lines and Hose

Brake lines must be in excellent condition, free of rust, dents, kinks, and abraded areas. They must be supported to prevent vibration. Hoses must be free of cracking, kinking, swelling, and cuts. Hoses must not contact any moving parts. When replacing brake lines, use only double-wrapped, coated steel tubing.

Warning: Never use copper tubing for brake lines.

Use double-lap flares on the tubing ends. See Chapter 8, Fasteners, Gaskets, Sealants, for complete instructions on flaring, cutting, and bending. **Figure 34-39** shows the formation of a double-lap flare. **Figure 34-40** illustrates an I.S.O.-type (International Standards Organization) flare. These are not interchangeable.

Servicing Brake System Hydraulic Valves

To provide even braking and warn of problems, the brake hydraulic system of modern vehicles contains many valves. The most common valves are the proportioning valve, the metering valve, and the pressure differential switch. These valves are sometimes installed in a single housing, called a combination valve.

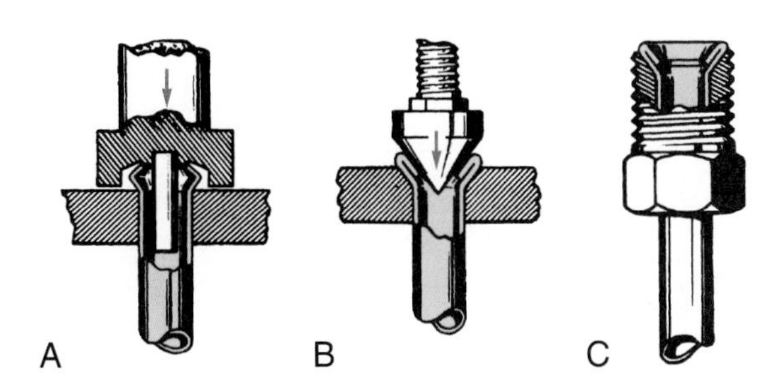

Figure 34-39. Steps in forming a double-lap flare, A—Forming a single lap. B—Forming a double lap. C—The finished double-lap flare with fitting. (Bendix)

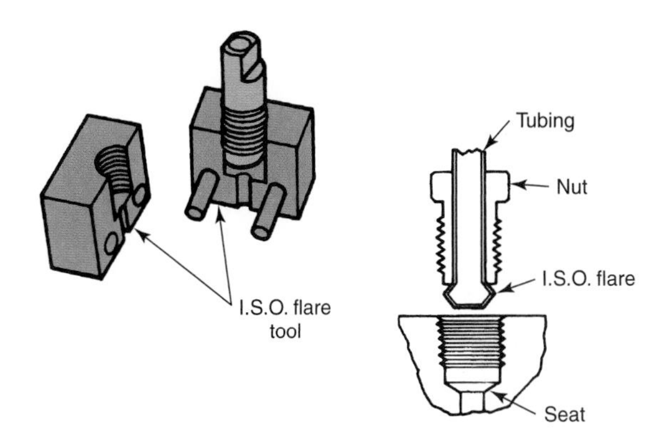

Figure 34-40. Cutaway view of an I.S.O. (International Standards Organization) flare and seat. Note the special tool required to form this flare. Follow the tool manufacturer's instructions. *Do not* use this flare with a single-lap or double-lap flare arrangement. (Bendix)

Proportioning Valve

A ***proportioning valve*** is used in brake systems with disc brakes in the front and drum brakes in the rear. Under mild stops, braking effort is about equal to the front and rear. As pedal pressure is increased, the proportioning valve controls, and finally limits, pressure to the rear wheels. This reduces the possibility of rear-wheel lockup during heavy braking. The proportioning valve can be a separate unit or it can be incorporated into a combination valve.

Dual Proportioning Valve

Some vehicles have a diagonally split brake system with a dual proportioning valve. The master cylinder is connected directly to the valve. From there, the system is divided diagonally.

Height-Sensing Proportioning Valve

The ***height-sensing proportioning valve*** uses a variable pressure range feature, which increases the pressure to the rear brakes as the vehicle's weight (cargo) increases. This pressure diminishes as the vehicle's weight decreases. Most valves are located on the vehicle's chassis and are connected to the rear axle with a calibrated tension spring or a rod-type linkage. See **Figure 34-41.**

Vehicle weight transfer during a stop causes the chassis-to-axle distance to change, changing the spring or rod linkage

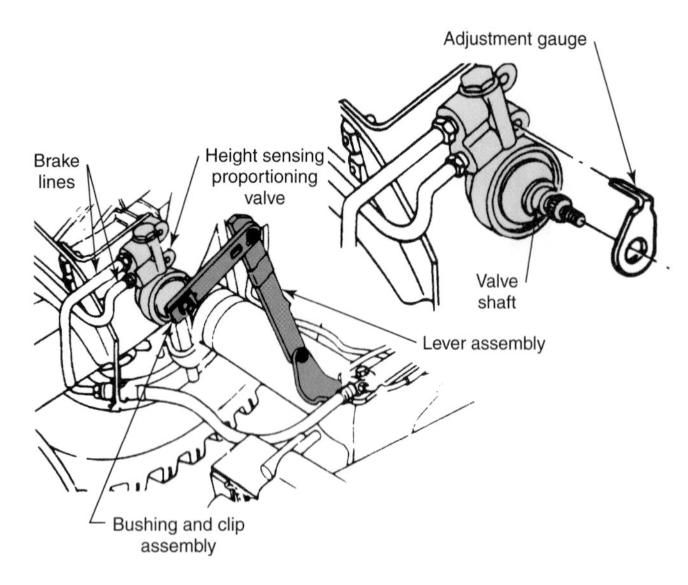

Figure 34-41. A height-sensing proportioning valve assembly is shown here. This unit is adjustable; others are not. Some require special bleeding procedures. Follow the manufacturer's service recommendations. (Fram)

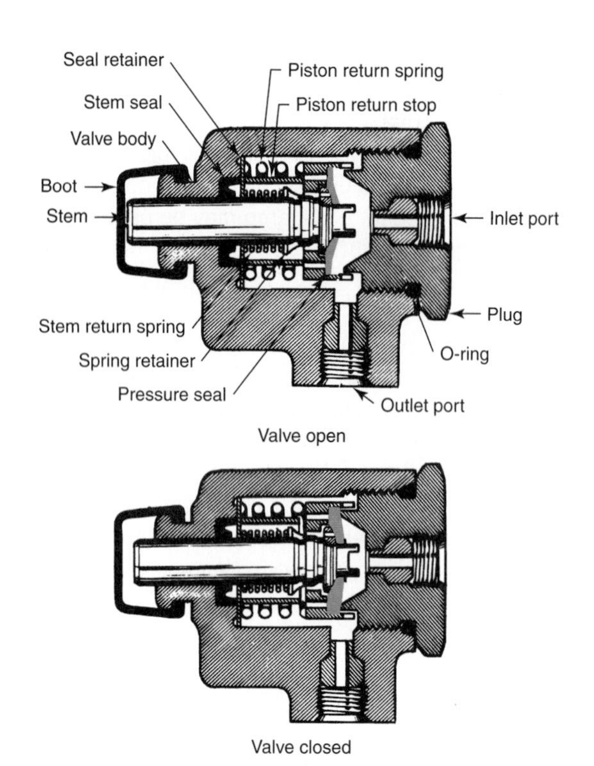

Figure 34-42. Most metering valves resemble the one shown here. Metering valve operation is relatively simple. The valve is held closed by spring pressure and opens when the fluid pressure reaches a certain value. (General Motors)

length. This, in turn, adjusts the valve, limiting pressure to the rear brakes. Loading the vehicle (wood in the bed of a truck, for example) also actuates the valve.

Warning: Do not alter a vehicle's riding height with air shocks or springs when the vehicle is equipped with this type of proportioning valve. Unsafe braking action could result.

Disc Brake Metering Valve

Vehicles with front disc and rear drum brakes require the use of a ***metering valve.*** See **Figure 34-42.** The metering valve closes off pressure to the front disc brakes until a specified pressure level is generated in the master cylinder. This allows pressure to force the back brake shoes to overcome retracting spring pressure and move into contact with the drum. Pressure beyond this opens the metering valve, and both front and rear brakes receive pressure.

Pressure Differential Switch

All dual brake systems use a ***pressure differential switch*** to warn the driver that one-half of the split brake system has failed. A small piston "floats" in a cylinder separating two pressure chambers. One side of each chamber is connected to one side of the master cylinder. The piston is centered by a spring on each end. An electrical switch is placed in the center of the piston. The switch is grounded whenever the piston moves to one side. This completes an electrical circuit through the dashboard-mounted brake warning light.

Figure 34-43 illustrates one type of differential pressure switch. In Figure 34-43A, the switch is in the normal, light-off, position. Each side has equal pressure and the piston remains centered. When one side of the system develops a leak, the pressure drops on that side of the valve. The piston is forced toward the low-pressure side. It then touches the electrical plunger and provides the ground needed to light the warning light, Figure 34-43B.

Combination Valve

Combination valves contain either one or two of the valves discussed previously. See **Figure 34-44.** They are called the two-function valve and the three-function valve. The two-function valve combines the metering valve and the brake warning light switch in one unit, Figure 34-44A. Some units may contain a proportioning valve instead of the metering valve.

The three-function valve houses the metering valve, the proportioning valve, and the brake warning light switch, Figure 34-44B. These valves cannot be adjusted or repaired. If they are defective, the entire unit must be replaced.

Bleeding the Brakes

Bleeding the brakes means removing air from the brake system. Air can enter in a number of ways: a low fluid level in the master cylinder, a disconnected system component, or leaky wheel or master cylinder cups. Air in the system causes a springy or spongy feel when braking.

Bleeding consists of pumping fresh fluid throughout the system. This forces air out through the wheel cylinder or caliper bleeder valves. Brakes can be bled manually or with a pressure bleeder.

Manual Bleeding

Manual bleeding requires an assistant to pressurize the brake system by pressing on the brake pedal. Clean all wheel cylinder or caliper bleeder screws. Remove the cap or plug from the bleeder screw, if used. Attach a bleeder hose to the bleeder screw farthest from the master cylinder. Place the free end in a clear glass jar. See **Figure 34-45.**

Note: If the master cylinder is equipped with a bleeder screw, bleed the master cylinder first.

Clean the master cylinder reservoir cap and the surrounding area. Fill the reservoir almost to the top. Open the bleeder screw 3/4 turn. Place the free end of the hose in the jar of fluid. Have the assistant press the brake pedal slowly to the floor. This will force air and fluid from the wheel cylinder, **Figure 34-46.** Have the assistant hold the pedal in the depressed position until the bleeder screw is closed. The pedal may then be released swiftly. Press the pedal down and reopen the bleeder. When the pedal reaches the floor, shut the

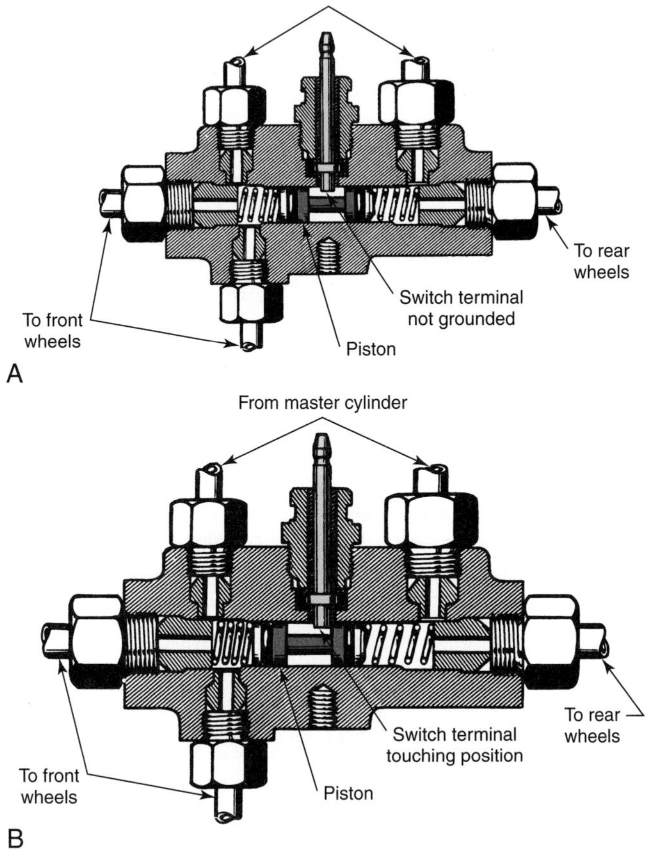

Figure 34-43. A—Pressure differential brake warning switch in normal position (no brake failure). The switch terminal is not touching the piston. B—Pressure differential brake warning switch with one side of system faulty. Note how the piston is forced toward the faulty side, grounding the switch terminal. The warning light would come on. (General Motors)

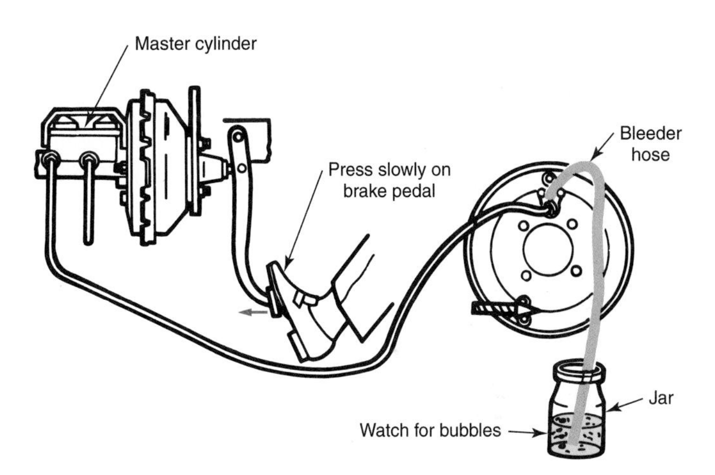

Figure 34-45. Manually bleeding a wheel cylinder requires an assistant. Be sure to place the bleeder hose in a brake fluid jar. (Niehoff)

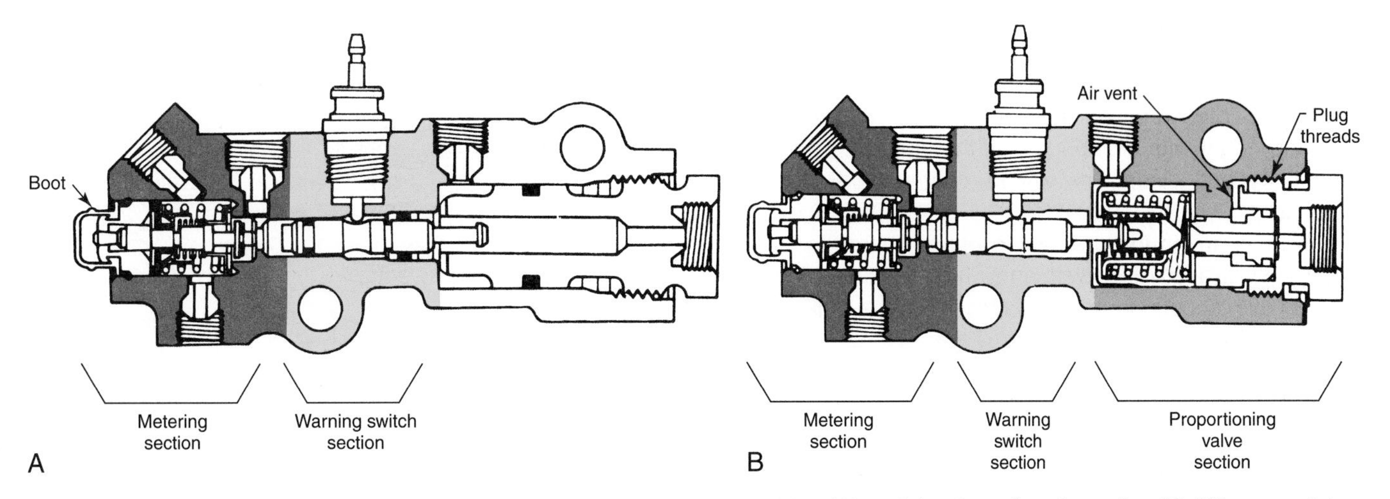

Figure 34-44. The two types of combination valves are the two-function valve (A) and the three-function valve (B). When servicing the brakes, always inspect these valves for leakage inside the boot and around the fittings. (Bendix)

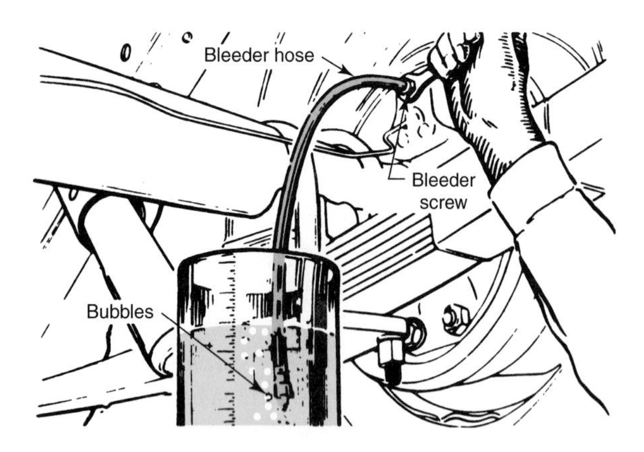

Figure 34-46. Bleed the cylinder until clear fluid, with *no* air bubbles, flows from the hose. Note that bubbles are still being forced from this system.

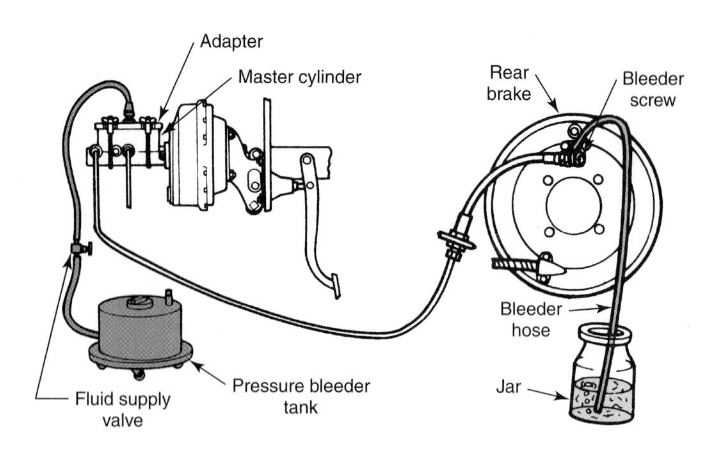

Figure 34-47. Bleeding a rear brake with a pressure bleeder tank and adapter. This method is generally preferred to manual bleeding. (Niehoff)

bleeder again before releasing the brake. Repeat this process (keep the reservoir full at all times) until fresh brake fluid, with no air bubbles, flows into the jar.

Tighten the bleeder screw. Remove the hose and move the setup to the wheel on the opposite side of the vehicle. Bleed the remaining cylinders or calipers in the same manner.

Make certain all bleeder screws are properly seated and tightened. When bleeding disc brakes, rap the caliper with a plastic hammer to dislodge air bubbles clinging to the caliper wall.

Note: Discard the brake fluid in the jar. Never reuse this fluid in a brake system.

Pressure Bleeding

Pressure bleeding requires only one person and is faster than manual bleeding, which requires two people. For pressure bleeding, a pressure tank, partially filled with brake fluid, is attached to the master cylinder reservoir. **Figure 34-47** illustrates a typical pressure-bleeding hookup.

Pressure-Bleeding Tank

Some pressure tanks separate the fluid from the compressed air with a diaphragm. If the tank does not use a diaphragm, only clean, dry, oil-free, compressed air should be used.

Fill the tank to the specified level and charge to 20 psi–30 psi (138 kPa–207 kPa) with an air hose. Avoid shaking the tank, as this tends to form air bubbles. Keep the tank at least 1/3 full. Bleed the tank as required. Then, use an adapter to attach the tank to the filled master cylinder reservoir. Both the adapter and the master cylinder must be clean.

Turn on the tank hose valve to admit fluid pressure to the master cylinder. Attach a bleeder hose. Open the bleeder and allow fluid to flow from the cylinder until it emerges clean and free of air bubbles. Close the bleeder securely. Repeat this process on the remaining wheels. Shut off the pressure tank and remove it. Siphon off enough fluid to lower the master cylinder fluid level to 3/8″ (9.53 mm) from the top.

Bleeding Systems with Dual Master Cylinder

If the master cylinder has two caps but a common reservoir, attach the pressure bleeder tank to one hole. Insert a blind (no vent hole) cap in the other. If separate reservoirs are used, attach the tank to one and bleed that side. Then, attach the tank to the other side. When bleeding tandem master cylinders, bleed the wheels served by the primary piston (not the floating piston) first. See **Figure 34-48.**

Surge Bleeding

In cases where it is difficult to remove the air from the wheel cylinder, try ***surge bleeding.*** Attach a pressure bleeder and admit pressure to the master cylinder. Open a bleeder screw and have a helper depress the brake pedal with a fast

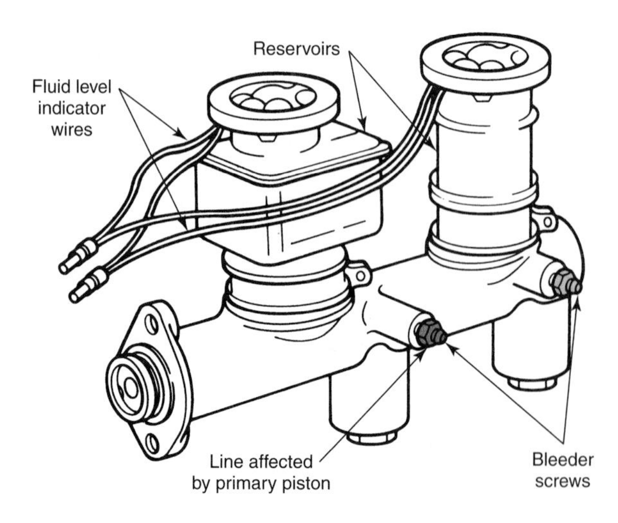

Figure 34-48. Master cylinder bleeder screws. Note the separate reservoirs and the use of fluid level indicators and wires. (Niehoff)

movement. Release the pedal slowly. Wait a few seconds and repeat. Continue until air is expelled. On the last downstroke of the brake pedal, close the bleeder screw quickly.

Bleeding Disc Brake Systems

When front disc brakes are used, the metering valve must be blocked open. A spring-like hold-open tool is recommended to prevent valve damage. Some installations require the removal of the pressure differential warning light switch terminal and plunger to prevent switch damage during bleeding.

Bleeding Power Brakes

Power brakes are bled using the methods already described. However, it is helpful to start the engine and allow the booster to help apply the brakes during the bleeding process.

Bleeding Anti-Lock Brake Systems

Procedures for bleeding anti-lock brake systems vary from manufacturer to manufacturer. Always consult the vehicle's service manual and follow the recommendations carefully.

Warning: The working pressure of some boost systems can reach 2600 psi (1 793 kPa). These systems must be pumped down to relieve pressure before the lines can be loosened. Failure to do so can result in serious injury.

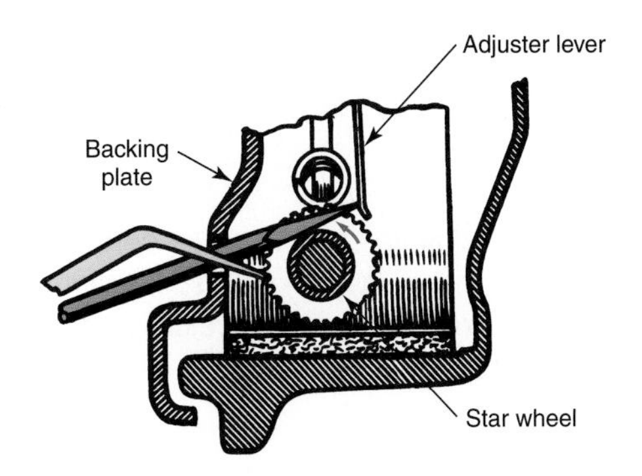

Backing off on adjusting screw (access slot in backing plate)

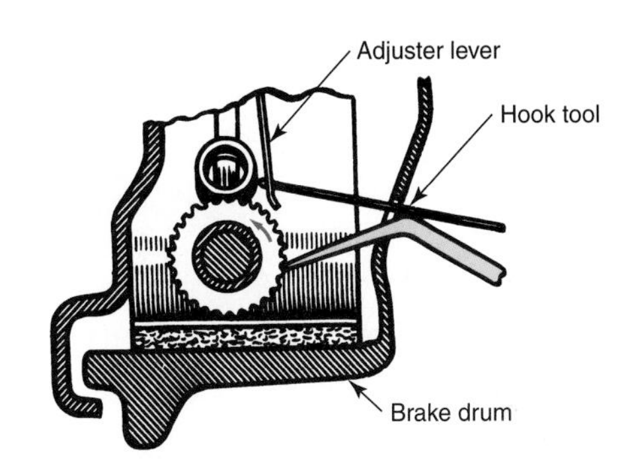

Backing off on adjusting screw (access slot in brake drum)

Figure 34-49. The adjuster lever must be held out of the way before turning the adjuster star wheel.

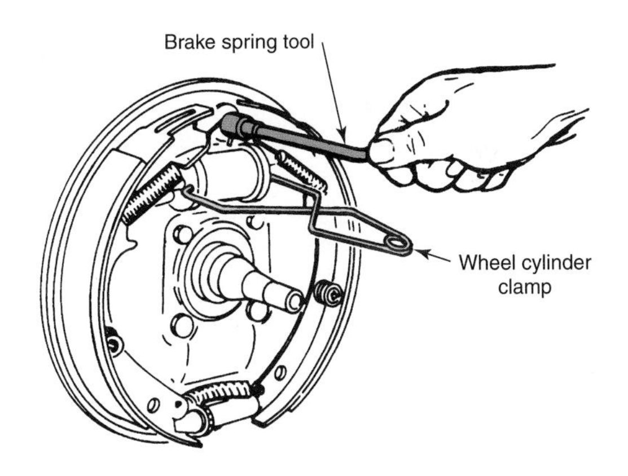

Figure 34-50. A brake spring tool is being used to remove shoe return springs. Note the cylinder clamp. (FMC)

Brake Friction Member Service

Friction members are the devices that use friction to stop the vehicle. Friction members include brake shoes and drums, and disk brake pads and rotors. The service of these units is covered below.

Brake Shoe Removal

Pull the wheel and drum. The parking brake must be off to remove rear drums. If the brake shoes are too tight to allow the drum to be pulled free, back off the adjustment. On self-adjusting brakes, pass a thin screwdriver through the adjustment slot in the backing plate. Hold the adjuster lever free while loosening the adjuster wheel (star wheel). If adjustment is through a slot in the brake drum, a hook may be used to pull the adjuster lever free, **Figure 34-49.**

Study the Parts Arrangement

Before attempting to remove the shoes, study the arrangement of the brake parts. Note the color of the springs, where the springs are connected, in what order the springs are connected, and how hold-downs are installed. This will help during assembly.

Clamp the Wheel Cylinder

Install a wheel cylinder clamp to prevent the pistons from popping out of the cylinder. Leave the clamp in place until the shoes are reinstalled. See **Figure 34-50.**

Remove the Shoes

Use a brake spring tool to remove the ***retracting springs,*** as shown in Figure 34-50. A number of spring arrangements for single-anchor, duo-servo brakes are shown in **Figure 34-51.**

Remove the shoe ***hold-downs,*** **Figure 34-52.** If the shoes are fixed to anchors, remove anchors or fasteners as needed. As parts are removed, lay them out in proper order. Keep the parts for each wheel in one group. Clean and inspect all parts.

Warning: See the cautions about breathing asbestos at the beginning of this chapter.

Check the springs carefully to make certain they are in good condition. Springs that show discoloration, stretched areas, nicks, or deformed end hooks are damaged and should be replaced. See Figure 34-51B.

Installing Brake Shoes

Clean the backing plate and torque plate mounting fasteners. Sand the shoe pads (raised portions of backing plate used to support shoes). Coat the pads with a film of high-temperature grease, **Figure 34-53.** Clean and back off the ***self-adjuster mechanism.*** This allows the drum to clear the new, thicker lining, **Figure 34-54.**

Place a small amount of high-temperature grease on the adjuster screw threads and on the ends where they contact the brake shoes. Lubricate the area between the hold-down and the shoe surface and the area where the wheel cylinder links or push rods contact the shoes. Use special high-temperature brake lubricant and use it sparingly. Never use oil or other greases.

Caution: Keep grease and oil away from lining. Avoid touching the lining with your fingers as much as possible. Keep your hands free of grease and oil. Remember that even the slightest bit of oil on the lining will ruin the brake job.

Install the shoes, being careful to place the primary and secondary shoes in their proper positions. On single-anchor, duo-servo applications, the primary shoe will have the shortest lining and will face the front of the vehicle. The primary shoe will

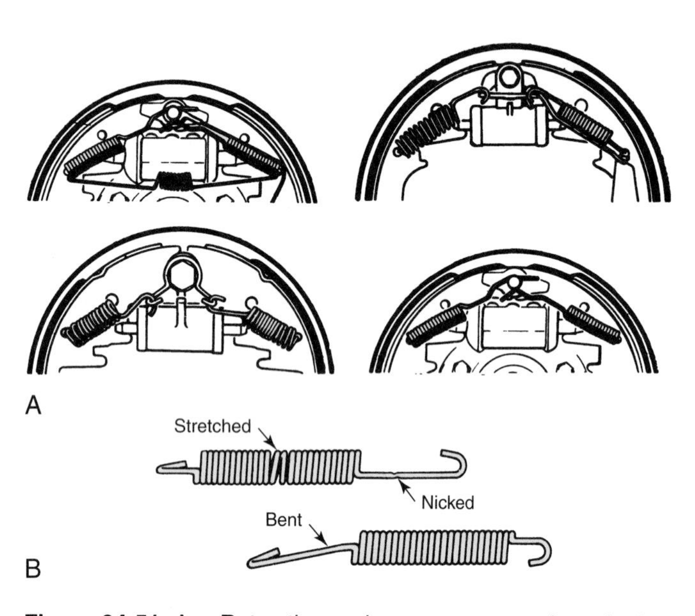

Figure 34-51. A— Retracting-spring arrangements in a single-anchor, duo-servo brake system. B—These retracting springs are unsuitable for further use.

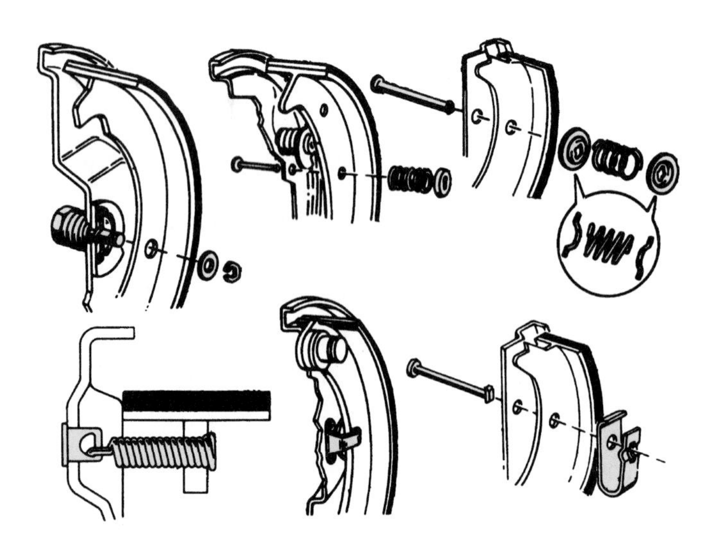

Figure 34-52. Various types of brake shoe hold-downs are used on modern vehicles.

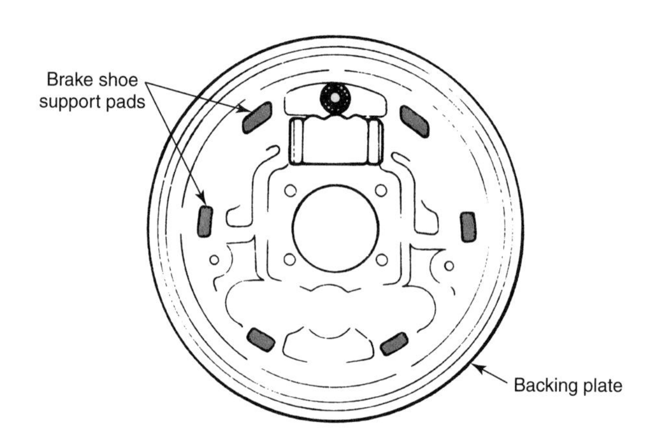

Figure 34-53. Clean and lubricate the brake shoe support pads (red). Lightly grease the anchor if recommended (yellow). (DaimlerChrysler)

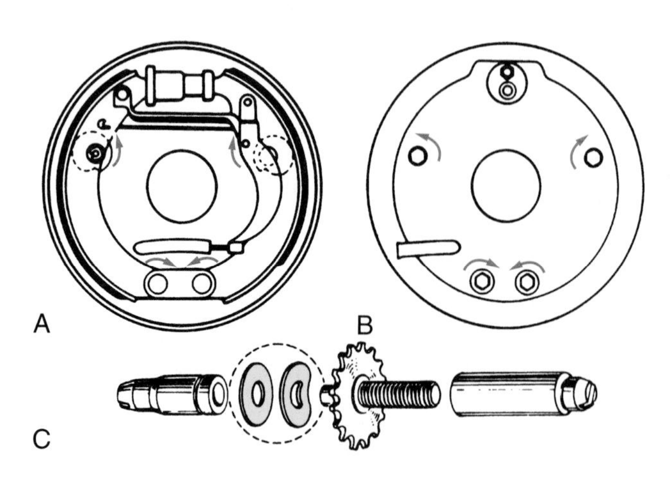

Figure 34-54. Back off the adjuster cams (A and B) to provide the clearance required to install the brake drum. Clean, lubricate, and collapse the adjuster type shown in C. (Wagner)

be the first shoe encountered moving away from the wheel cylinder in the direction of forward wheel rotation.

Install hold-downs and retracting springs. Make certain the springs are in the correct position and hooked in the proper spot. Use a brake spring tool to avoid damage to the spring, **Figure 34-55.**

When installed (parking-brake lever, cable, and automatic adjusters, where used, must be in place), rap the shoe assembly back and forth to check for freedom of movement. Recheck the entire assembly. Install the brake drum and adjust the shoes.

Parts Must Be in Correct Position

The shoes must be in their correct positions. The star wheel must face the adjustment slot in the backing plate. Self-adjusters must be installed on the correct side of the vehicle.

Using Gauge to Make Initial Shoe Adjustment

Brakes with self-adjusting shoes need only an initial adjustment following the installation of new shoes. Normally, the automatic adjuster will then maintain proper lining-to-drum clearance for the life of the lining.

Release the parking brake and slack off the cable so that both shoes are in firm contact with the anchor pin. Adjust the special gauge to match the brake drum's diameter, **Figure 34-56.** Lock the gauge securely.

Expand the shoes outward (hold the lever away from the star wheel to prevent burring the wheel) until they just fit in the opposite side of the adjusting gauge, Figure 34-56. Install the drums and wheels. Start the vehicle and make a series of stops in reverse. This causes the brake shoes to stick to the drum and follow it around far enough to activate the automatic adjuster. Repeat reverse stops until a full pedal is attained.

Making Initial Shoe Adjustment by Hand

If no setting gauge is available, install the drum. With the wheel free of the ground and the parking brake disconnected, remove the cover from the star wheel's adjustment slot. It may be in the backing plate or drum. While holding the adjuster lever out of the way, **Figure 34-57,** back off the star wheel about 30 notches. Make reverse stops until a full pedal is attained.

With a brake-adjusting tool, turn the adjusting star wheel to expand the shoes. While turning the wheel, revolve the tire in the direction of forward travel. Continue turning the star wheel until a firm drag is felt when attempting to rotate the tire. As soon as the brakes are applied, the shoe material will be smoothed down and the drag will be eliminated.

Figure 34-55. This illustration shows a technician installing a retracting spring with a brake spring tool. (DaimlerChrysler)

Note: Some cars do not have an opening in front of the star wheel. Careful inspection will reveal a slug (metal still in opening) that must be punched out to form the opening. These slugs can be in the backing plate or in the face of the brake drum itself. When you are finished with the adjustment, fill the slot with a recommended slot plug.

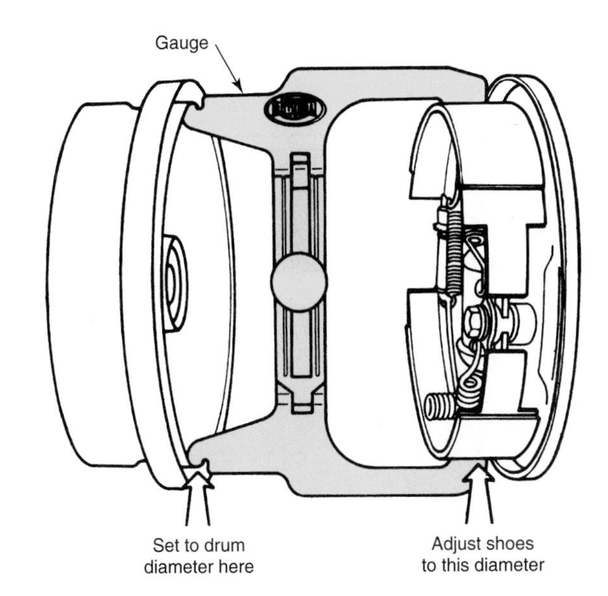

Figure 34-56. To use a brake shoe adjusting gauge, first set the gauge to the drum, and then adjust the shoes to fit. (Ford)

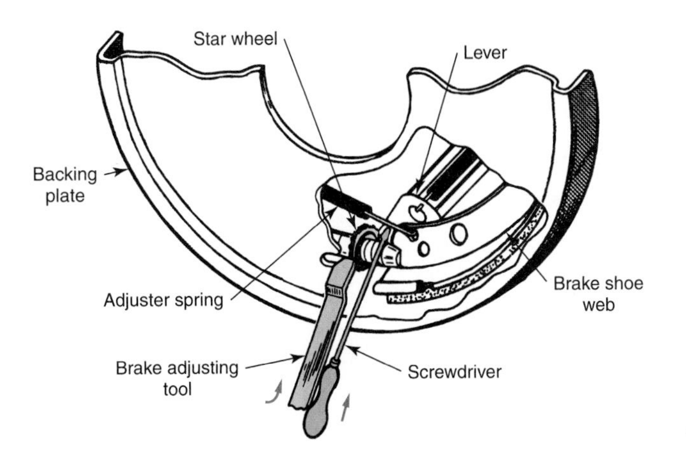

Figure 34-57. Hold the adjuster lever out of the way while turning the star wheel. (DaimlerChrysler)

Caution: If the slug is knocked out with wheel and drum installed, remove the wheel and drum and discard the slug. Never leave it inside.

Changing Disc Brake Pads

In some cases, the ***disc brake pads*** are equipped with small metal tabs that contact the rotor when the pads become excessively worn. The tabs make a shrill noise every time excessively worn brakes are applied. Regardless of the wear-determining device, pads should be changed when worn to within 1/8″ (3.18 mm) of their bases. Refer to **Figure 34-58.**

Before forcing the piston back, siphon off brake fluid from the fluid reservoir (until it is about one third full) to make room for fluid that is displaced when the piston is forced inward. See **Figure 34-59.** Failure to do this will result in flooding the master cylinder.

If necessary, remove the caliper. Do not allow the caliper to hang from the brake hose (or sensor wire). Use a C-clamp to force the pistons back into the bore. This will free the pads from the rotor, **Figure 34-60.** Some pads can be pulled away from the rotor with pliers.

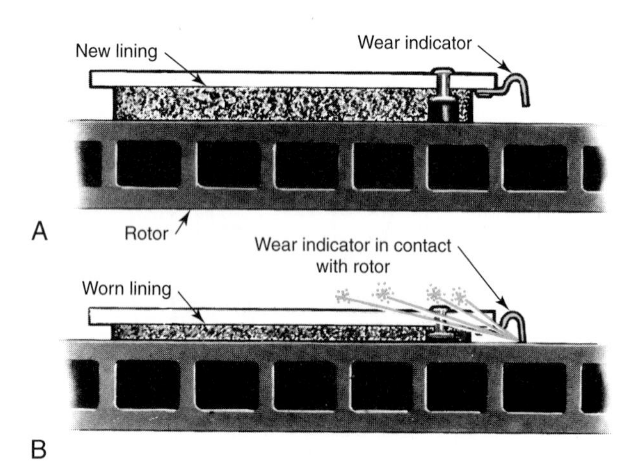

Figure 34-58. Wear-indicator tab action. A—New pad. The indicator does not touch the rotor. B—Pad worn. The indicator strikes the rotor and alerts the driver of a worn pad. (Oldsmobile)

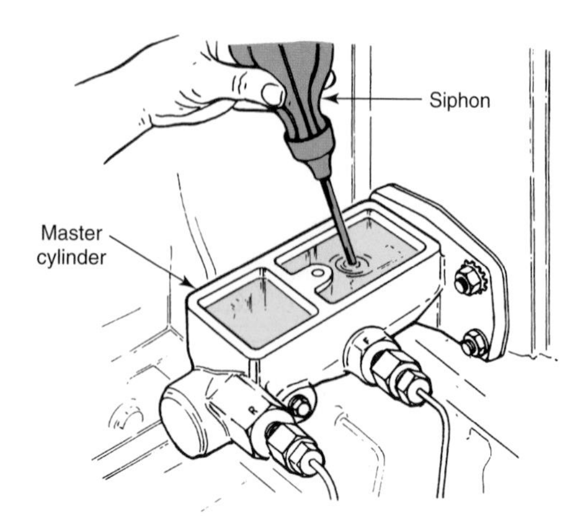

Figure 34-59. Removing brake fluid from the reservoir with a siphon. Do not reuse old fluid. (FMC)

Note: A few pads can be removed without removing the caliper from the wheel.

Once the pads are moved away from the rotor, remove the caliper fasteners and remove the caliper, **Figure 34-61.** Some caliper fasteners have recessed Allen or Torx® heads, and must be removed with a special tool. Once the caliper is removed, wire it to the vehicle. Do not let the caliper hang by the brake hose. If the pads are removed with the caliper, Figure 34-61A, loosen the retainer clips and remove the pads from the caliper. If the pads stay with the rotor when the caliper is removed, Figure 34-61B, slide them from the rotor.

When the pads are removed, check the pistons and boots for leakage, corrosion, and gumming. If necessary, disassemble and clean. (See the Caliper Repair section in this chapter.) Check the rotor for excessive scoring and oil or grease contamination.

Caliper Alignment

If the caliper is an older fixed type, check that it is centrally located over the rotor, **Figure 34-62.** This check is not required with floating calipers.

Make sure the pads are parallel with the rotor surface. See **Figure 34-63.** If the alignment error exceeds specifications, check for missing or wrongly placed shims, front-end damage, or incorrect installation. **Figure 34-64** shows correct and incorrect caliper-to-rotor alignments.

Insert new pads. Lubricate (as recommended) any sliding metal surfaces with special grease (such as molydisulfide). Install the splash shield or any other pad-holding device. Always use the appropriate fasteners.

Pump the brake pedal to force the pads out against the rotor. After all other service is performed, add fluid to bring the master cylinder to the correct level. Do not attempt to drive the vehicle until a full pedal is obtained by pumping the pedal until

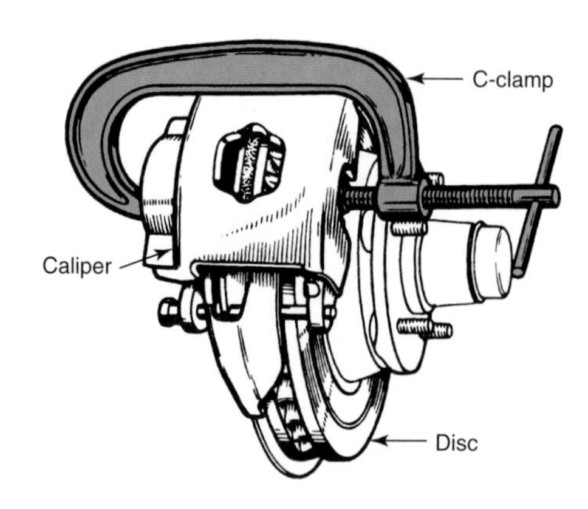

Figure 34-60. Use a C-clamp to force the brake piston back into its bore to free the brake pads. (Bendix)

the pads are against the rotor. If this is not done, the brakes will not function on the first application of the pedal.

Breaking in a New Set of Brake Linings or Brake Pads

Following the installation of new brake linings or pads, it is important that they be given the proper ***break-in.*** When

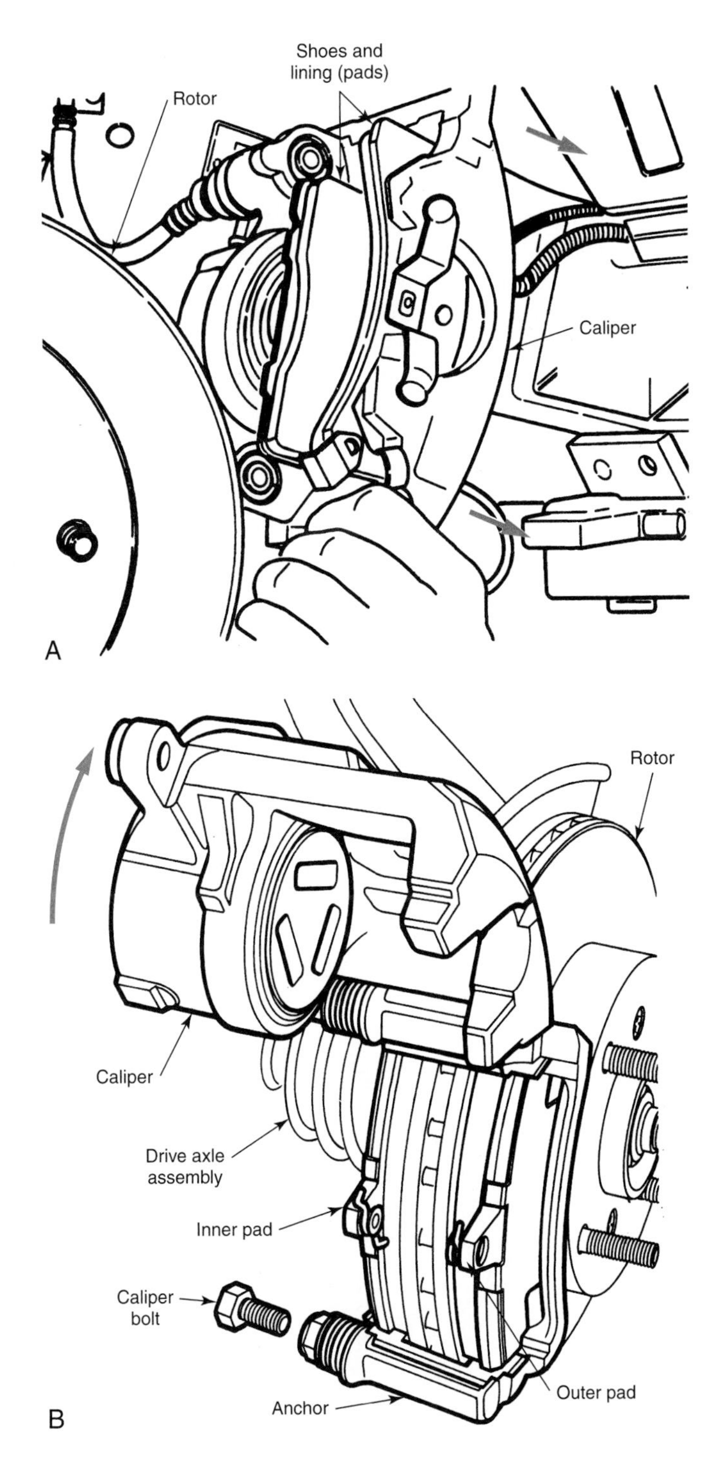

Figure 34-61. A—The pads come off with the caliper in this design. B—In this design, brake pads remain with the rotors and anchor unit when the caliper is removed.(DaimlerChrysler, Sterling)

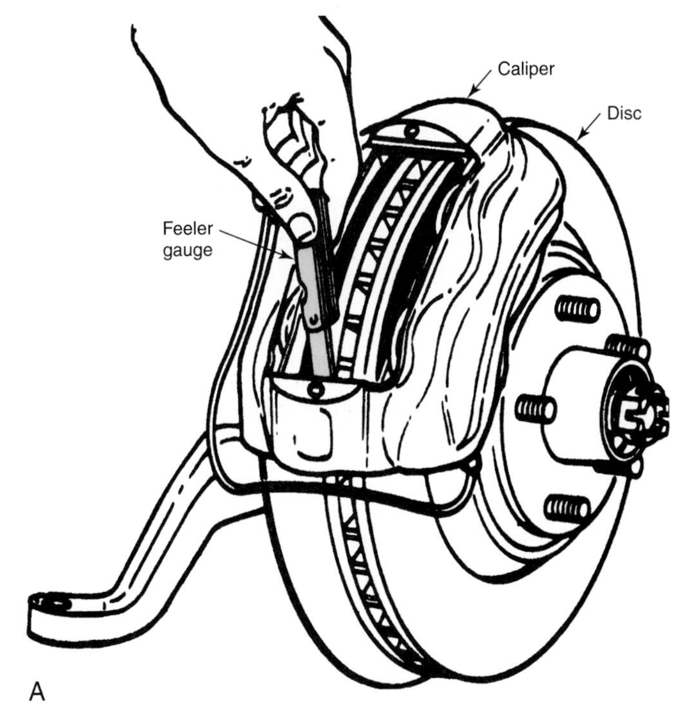

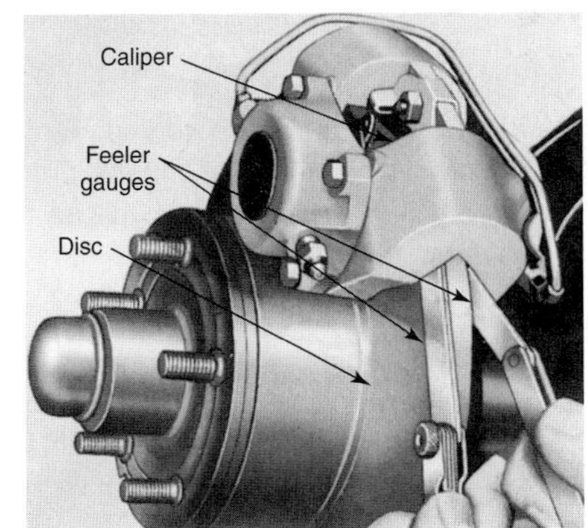

Figure 34-62. A—This procedure is used to check the rotor-to-caliper clearance on some older vehicles with fixed calipers. B—Checking a fixed caliper for proper centering on the brake rotor. Because these calipers cannot move, correct clearance is a *must!* Double-check the clearance by turning the rotor by hand. There should be no interference between the rotor and caliper. (Wagner and Bear)

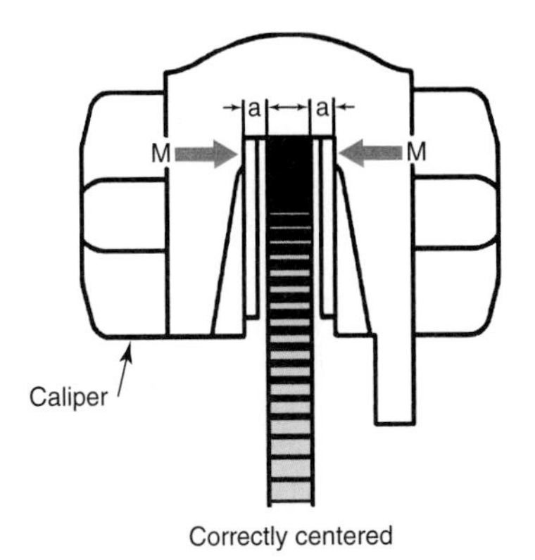

Figure 34-63. Checking a fixed caliper to make certain it is parallel with the rotor. (Mercedes-Benz)

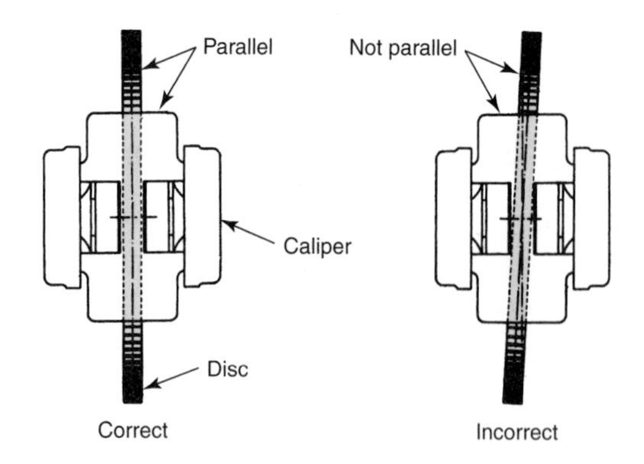

Figure 34-64. Correct and incorrect caliper-to-rotor alignment. (Mercedes-Benz)

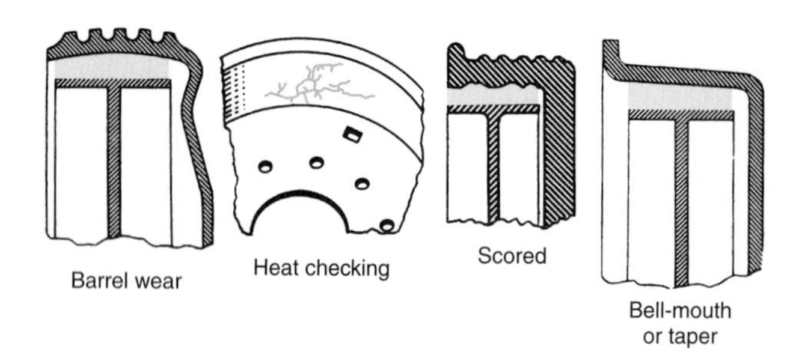

Figure 34-65. These defects can occur on any brake drum. If turning cannot remove defects, replace the drum. Do not reinstall a damaged drum. (Wagner)

road testing following the completion of a brake job, make eight to ten mild stops from around 25 mph (40.2 km/h). Make the same number of stops from around 45 mph (72.4 km/h) at one mile (1.6 km) intervals. Stops must be mild.

Caution the owner to avoid severe use of the brakes for several hundred miles. This will seat the linings properly and lengthen service life.

Brake Drum and Rotor Service

The ***brake drums*** and ***rotors*** are as important as the pads and shoes to proper brake operation. The drums and rotors should be carefully checked for damage as outlined below. Correct repair and refinishing procedures are vital.

Brake Drum Service

Wash the brake dust from the drum. If grease or oil is present in the drum, remove it with cleaning solvent and blow the drum dry. Wipe the drum's braking surface with a clean, alcohol-soaked rag. Wipe again with a dry rag. Repeat this process until the drum is spotless.

Inspect Brake Drum

Inspect the brake drum for scoring, cracking, heat checking, bell-mouth wear, and barrel-shape wear. Refer to **Figure 34-65.** Scoring, bell-mouth, and barrel wear may be removed by turning if the damage is not too deep. Heat checking can also be minimized and often removed by turning the drum. Use a brake drum micrometer to check for out-of-roundness. Measure in 45° increments around the drum.

Warning: Destroy any drum that is cracked. Never try to weld a cracked drum.

Although some specifications call for less runout, any drum measuring more than 0.010″ (0.254 mm) out-of-round or showing more than 0.005″ (0.127 mm) taper should be trued by turning or grinding. Drums that measure over 0.060″ (1.524 mm) above standard should be destroyed. This does not apply to truck drums. Some vehicle drums cannot exceed 0.030″ (0.762 mm). Check the manufacturer's specifications.

If the drum appears serviceable without turning, polish it with fine emery cloth as shown in **Figure 34-66.** This removes the glassy surface that can cause poor brake action with new shoes. Most shops prefer to lightly turn drums, no matter how good they look. **Figure 34-67** shows a special micrometer being used to check for drum out-of-roundness.

Turn Drums in Pairs

Although brake drums can differ slightly in diameter between the front and rear, the front drums must have diameters within 0.010″ (0.254 mm) of each other. The rear drums must also be the same diameter. When a drum on one side needs turning, turn the one on the other side to the same diameter.

Remove as Little Metal as Possible

Because turning makes the drum thinner, remove only enough metal to true the drum. Never increase the standard drum diameter by more than 0.060″ (1.524 mm). The removal of too much drum metal can cause overheating, checking, and failure.

Using a Drum Lathe or Grinder

Drums may be trued by ***turning*** (using a cutter bit) or by ***grinding*** (passing a high-speed grinding wheel across a

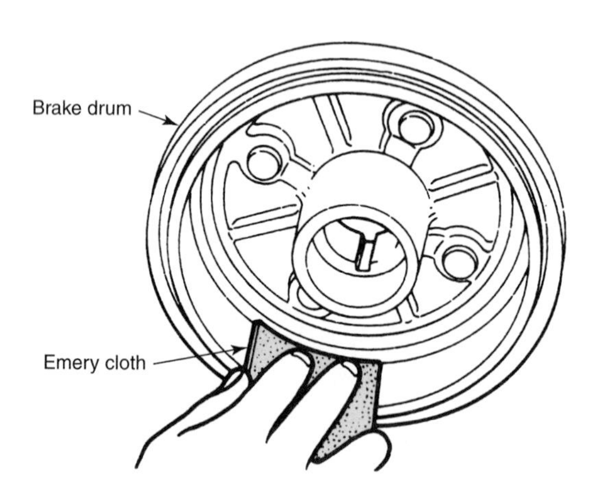

Figure 34-66. Use emery cloth to clean and deglaze the brake drum contact surface. Clean thoroughly when finished. (Mazda)

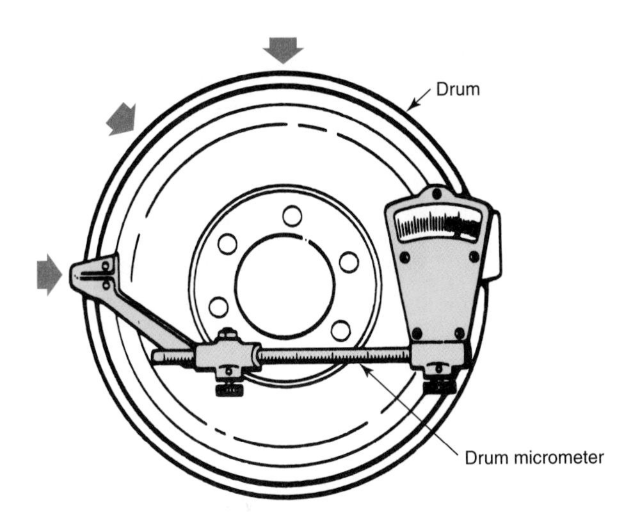

Figure 34-67. A drum micrometer is used to check the brake drum for out-of-roundness. Check in several spots. (Wagner)

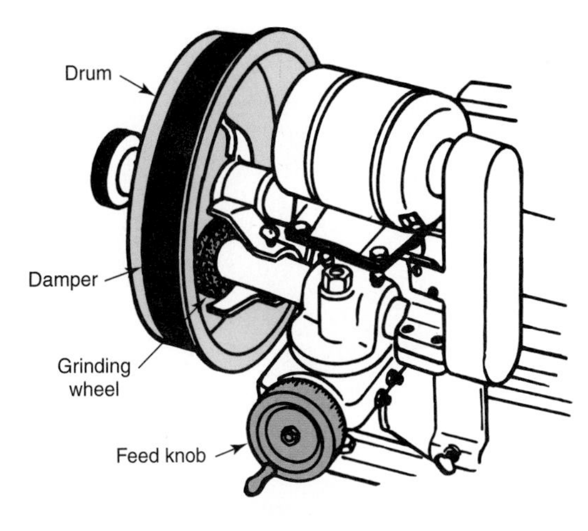

Figure 34-69. Grinding a brake drum with a special grinder. *Wear safety goggles!* (FMC)

revolving drum). Many shops use only turning. Some shops turn the drums and then dress them very lightly with a grinder. Others use only the grinder.

If the tool is sharp, the cut is light, and the feed is slow, the turned finish will be satisfactory. It can be slightly more resistant to squeal and chatter than the ground surface. **Figure 34-68** illustrates a brake drum lathe. Note the cutting bit.

Figure 34-69 shows a drum grinder in action. The drum must be accurately mounted. The cut must be light and the feed slow to produce a satisfactory finish. Be sure to place a damper on the drum to prevent chatter marks. Follow the tool manufacturer's directions.

After truing, clean the drum's braking surface with a scrub brush and hot, soapy water. Rinse it with hot water and dry it immediately. Finally, wipe the drum with a clean, alcohol-soaked rag.

On-Car Lathe

On-car brake lathes perform the same function as bench-mounted lathes. The difference is that the on-car brake lathe is used without removing the rotor from the vehicle, **Figure 34-70**. The on-car lathe saves time when the rotor is extremely difficult to remove from the vehicle. Many rotors are pressed onto the CV axle. If the on-car lathe was not available, the entire CV axle would have to be removed and the rotor pressed off in order to cut the rotor. The procedures for setting up and using an on-car brake lathe differ greatly from those for a bench lathe. The series of photographs in **Figure 34-71** show the procedures for setting up and using a typical on-car brake lathe. The most critical part of the setup procedure is aligning the cutters and turning device so the rotor is cut with a minimum of warping and thickness variations.

Figure 34-68. To turn a drum in a brake lathe, the drum must be installed so that it turns true with no wobbling. The cutting tool must be set to remove only enough metal to clean up defects. (Ammco)

Figure 34-70. An ASE-certified technician is using an on-car brake lathe to turn a rotor. This particular vehicle is front-wheel drive. When using these lathes, carefully follow all of the tool manufacturer's installation and machining procedures. (Hunter Engineering Company)

Figure 34-71. The following series of photographs shows the operating sequence of a vehicle powered on-car brake lathe. The procedures will vary between the various lathe manufacturers. 1—The vehicle has been raised to a comfortable height. 2—The wheel is removed. 3—Lug bolts are reinstalled (if needed) to secure the rotor to the hub. 4—Remove the caliper mounting bolts. 5—Support the caliper on a wire hook. 6—Clean the caliper mounting area. 7—Bolt on the lathe mounting legs. 8—Mount the lathe head on the vehicle. 9—Carefully align the carbide cutter bits to the centerline of the rotor. 10—Place the vibration dampener on the rotor; remove the clip before machining. 11—Adjust the cutter bits. 12—Manually turn the head in as far as the cutter bits will allow. Then, engage the lathe head drive. Repeat steps 11 and 12 as needed. (Kwik-Way Mfg. Co.)

Clean New Drums

New drums are usually given a protective coating to guard against rust. Remove the coating and clean the braking surface with alcohol. Some coatings require lacquer thinner for removal.

Rotor Service

The rotor should be free of excessive or heavy scoring. Because the rotor is not completely protected against the elements, some scoring is natural. Scoring up to 0.015″ (0.38 mm) deep, as long as the rotor is smooth, is permissible. Clean up minor roughness with fine emery cloth. See **Figure 34-72.**

Check the rotor for lateral runout (side-to-side wobble). Before checking, set the front wheel-bearing clearance to just remove any endplay.

Mount a dial indicator to a solid surface. Place the indicator anvil in about 1″ (25.4 mm) from the wear surface's outer edge. Slowly rotate the rotor and read the dial. Maximum runout should not exceed 0.004″ (0.102 mm). Refer to **Figure 34-73.**

Also check the rotor for parallelism (same rotor thickness all the way around). Using a micrometer, check the rotor thickness in six or eight spots around the rotor. Select locations that are in from wear edge about an inch (25.4 mm). Carefully record each measurement.

The maximum difference in readings should not exceed 0.0005″ (0.013 mm). This may seem like a small amount, but anything more than this causes a pulsating brake pedal and possible brake shudder or chatter. A few manufacturers call for even smaller maximums, such as 0.00025″ (0.0064 mm).

Rotor Minimum Thickness

When wear has reduced rotor thickness beyond the recommended minimum, the rotor should be discarded. This minimum thickness is generally marked on the rotor assembly. See **Figure 34-74.** Never reduce to this thickness by

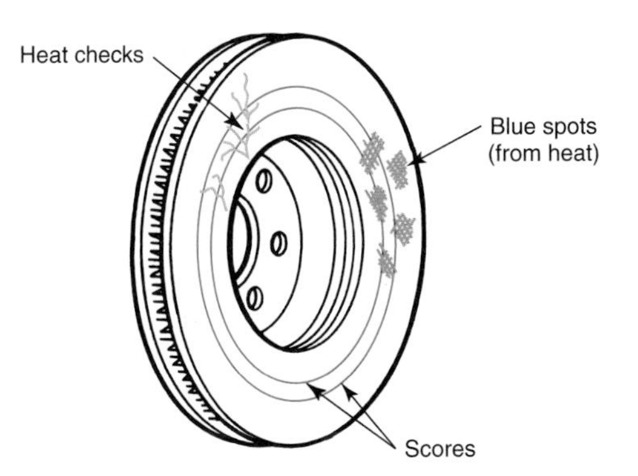

Figure 34-72. Common rotor problems are shown here. Blue spots and minor scoring can be removed. A crack or a series of cracks requires rotor replacement. (Niehoff)

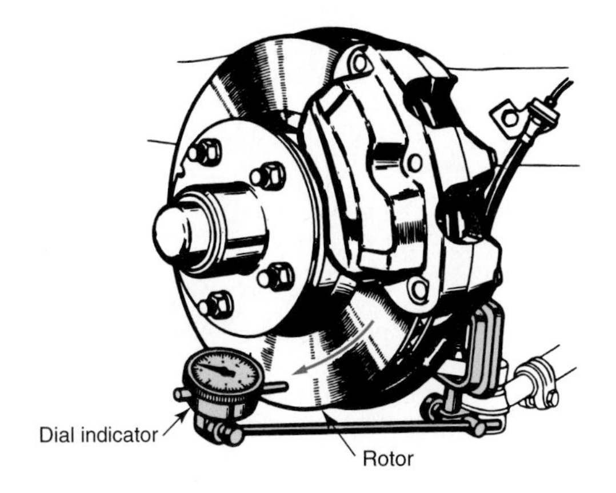

Figure 34-73. A dial indicator is used to check rotor runout. If runout is excessive, check the bearing condition before deciding that the rotor is warped. (Bendix)

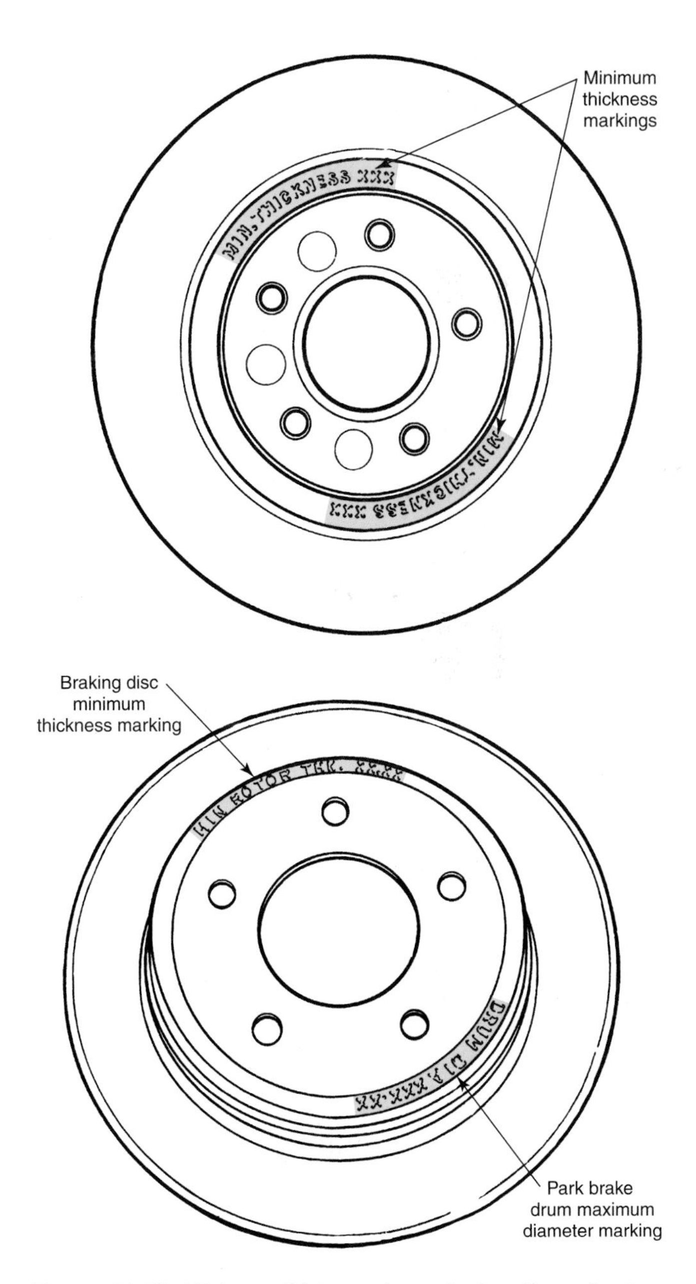

Figure 34-74. Minimum thickness is marked on these front and rear brake rotors. Note the rear rotor also provides a wear limit for the parking brake drum minimum thickness. (DaimlerChrysler)

grinding—this is the wear limit. Minimum grinding thickness must leave more material.

Truing Rotor

Measure the rotor with a micrometer at 6 to 8 different locations around the braking surface to check for parallelism (thickness variations). See **Figure 34-75.** If it is not worn beyond turning or truing thickness (different than minimum-wear thickness), the rotor may be turned on a rotor-refinishing machine, **Figure 34-76.** This tool cuts both sides of the rotor at once, thus reducing the chance of chatter.

Remove only enough stock to clean up the rotor surface on both sides. Do not reduce the thickness beyond recommendations. Follow all of the tool manufacturer's directions. This is a precision operation.

Upon completion of the turning operation, the rotor surfaces should be given a nondirectional, crosshatch finish using the proper grinding attachments. The surface should run between 20–80 microinches with 50 microinches being about average. A typical crosshatching attachment is shown in Figure 34-76B.

Never resurface one rotor. Rotors must be done in pairs to ensure smooth, even braking. When installing a rotor, all contact surfaces must be spotless to prevent runout.

Parking Brake Adjustments

The ***parking brake*** adjustment is more important than most technicians realize. If the parking brake is set too loose, it will not hold properly. If the parking brake is set too tight, the brakes will drag when the vehicle is moving. This will overheat and ruin the brakes.

Adjusting Rear Wheel Parking Brake

Apply the parking brake around 3 notches (about 1 3/4″ or 44.5 mm travel). Adjust the equalizer nut until a slight drag is noticeable at the rear wheels. Refer to **Figure 34-77.**

Release the brake. The wheels should turn freely. Parking brake cables must operate freely. Lubricate the cables if necessary.

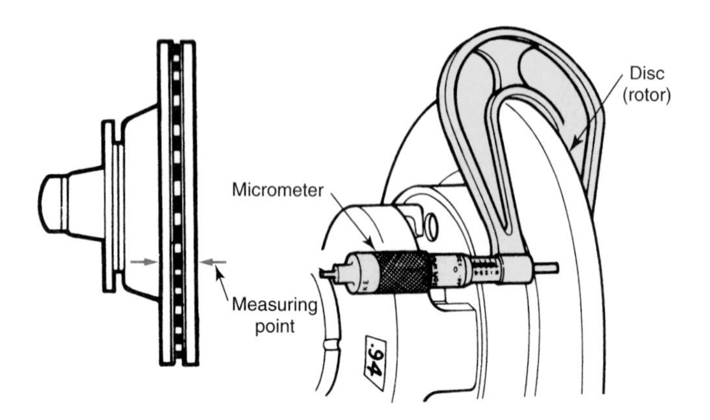

Figure 34-75. Measuring rotor parallelism with a micrometer. Measure at twelve equally spaced locations around the braking surface. (Niehoff)

Note: The parking brake must release fully. If the brake is set too tight, it can cause the automatic adjuster to malfunction. It can also ruin the lining from overheating. The service brakes should be properly adjusted before adjusting the parking brake.

Adjusting Disc Brake–Type Parking Brake

Vehicles with disc brakes in the rear require a somewhat different procedure for adjustment of the parking brake mechanism. There are two basic types: the internal expanding shoe-and-drum type and a screw-actuated unit that is an integral part of the caliper. See **Figures 34-78** and **34-79.**

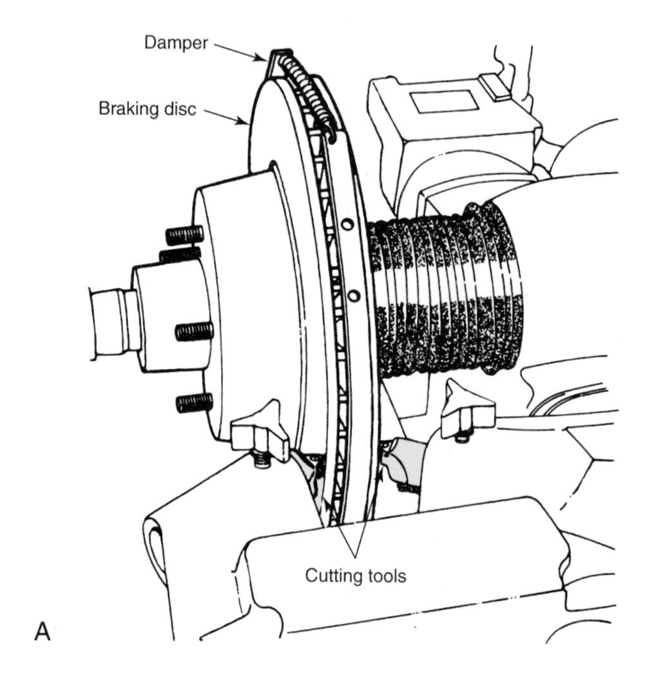

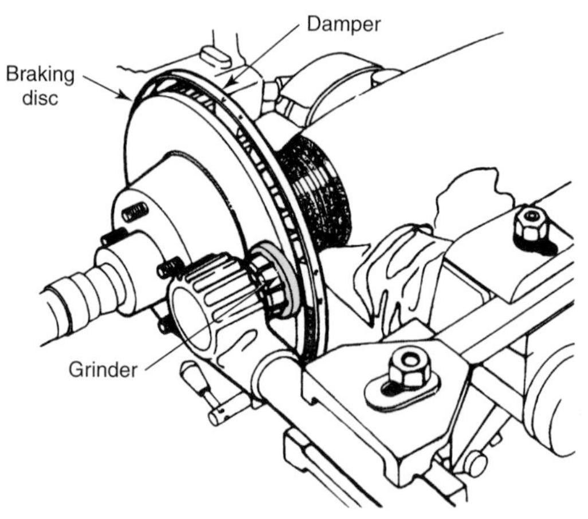

Figure 34-76. A—Rotor-cutting tool. This particular tool has a cutter on each side (straddle cutter) and resurfaces both faces of the rotor at the same time. B—Rotor-refacing (sanding) machine attachment. This sander will give the rotor a 50 microinch surface and nondirectional hatch marks. Use extreme caution and be sure to keep the machine clean. Always wear safety goggles! (DaimlerChrysler)

To adjust the disc (drum) type, **Figure 34-80,** rotate the star wheel by placing an adjusting tool through an opening generally located in the face of the rotor-drum.

If parking brake shoes require replacement, disconnect the caliper and back off the star wheel (if necessary). Remove the rotor-drum unit. The shoes are serviced in the same manner as a servo-type service brake.

To compensate for new shoes or cable wear, engage the parking brake pedal or hand lever one notch. Adjust cable for slack at the equalizer.

For adjustment of the screw-actuated unit, all that is generally necessary is adjustment of the cable. This too, is performed at the equalizer. Check all parts for freedom of movement. Follow the manufacturer's service procedures.

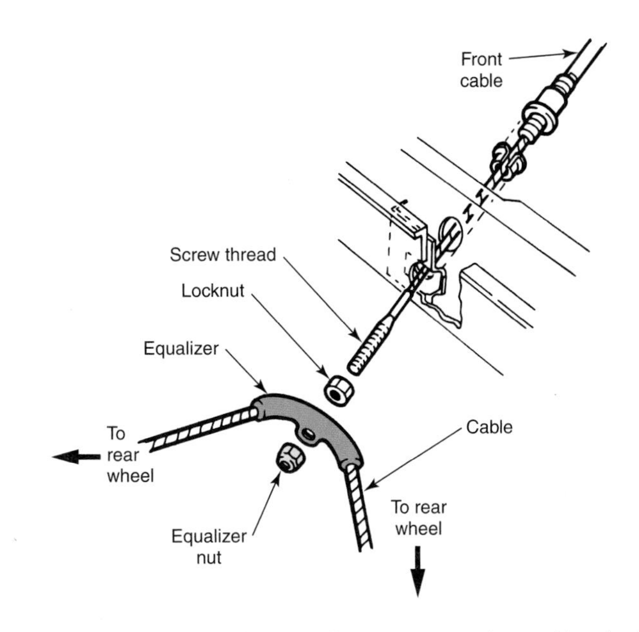

Figure 34-77. Adjust the equalizer nut to set the parking brake drag. (TRW Inc.)

Power Brake Service

Power brakes consist of a power booster. The power booster reduces the amount of pedal and foot pressure required to generate the hydraulic power necessary to stop the vehicle.

Most power brakes operate by engine vacuum and atmospheric pressure acting on a vacuum diaphragm to apply pressure directly to the master cylinder pistons, **Figure 34-81.** A typical ***vacuum power booster*** is shown in **Figure 34-82.**

A booster is shown in the released position in **Figure 34-83.** Note that engine vacuum exists on both sides of the diaphragm. When the brake pedal forces the push rod inward, the vacuum port is closed and the atmospheric port is opened, admitting atmospheric pressure to one side of the diaphragm. This moves the diaphragm assembly, causing the master cylinder push rod to actuate the master cylinder pistons and build up hydraulic pressure in the system. Other boosters, some containing two diaphragms for increased pressure, are also used.

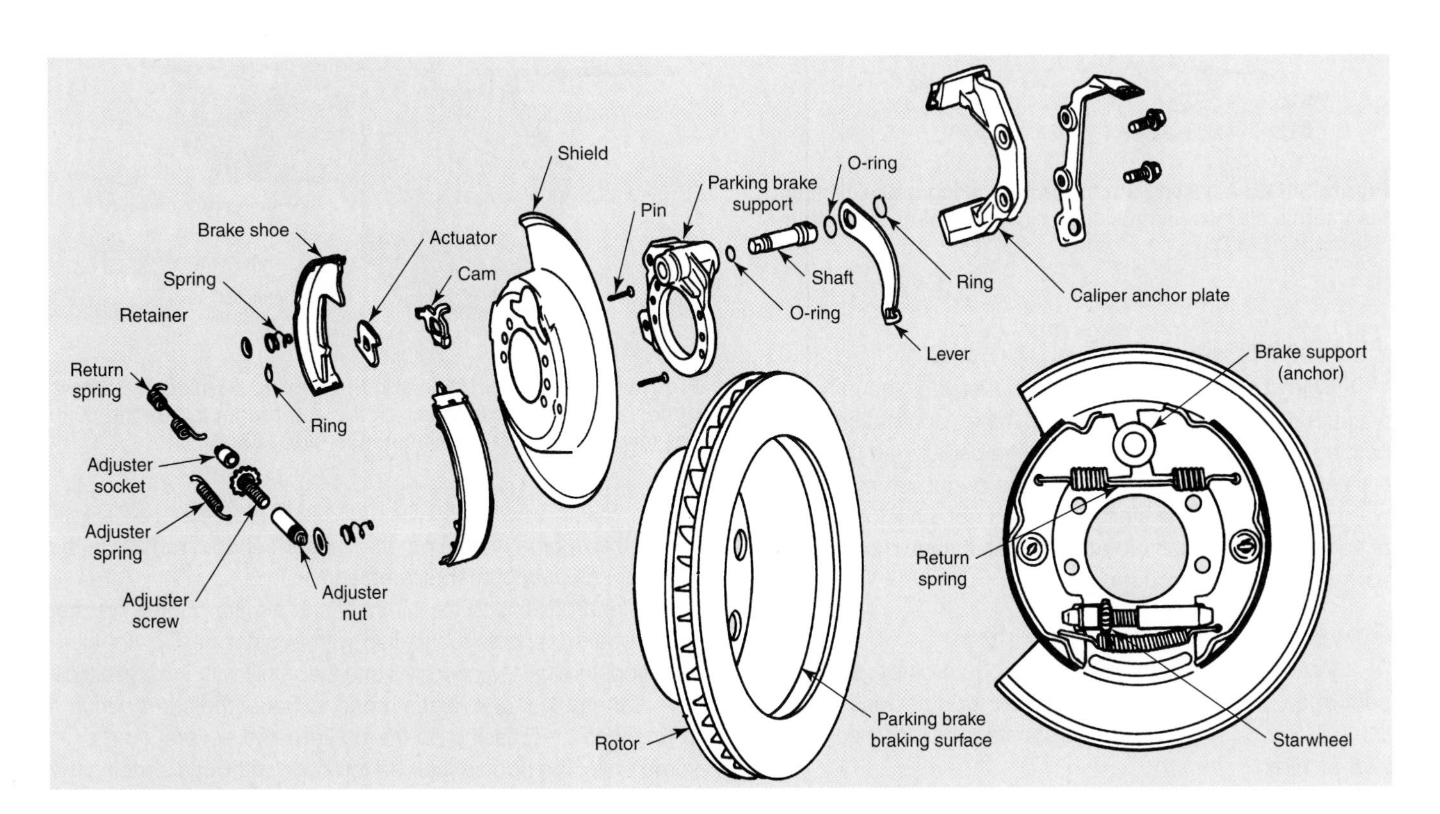

Figure 34-78. This parking brake assembly uses a small set of drum brakes inside of the rotor. This parking brake is used on only a few vehicles. (FMC)

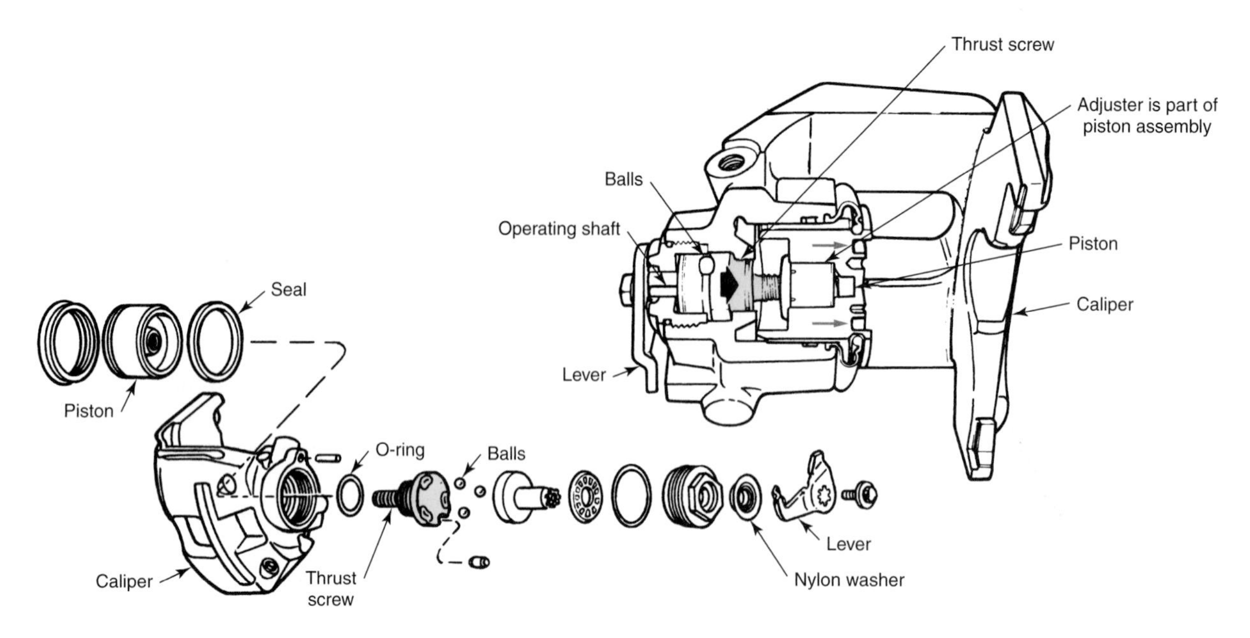

Figure 34-79. Cutaway and exploded views of a screw-actuated parking brake assembly. As the lever turns the thrust screw, the piston firmly clamps the rotor between the pads. (FMC)

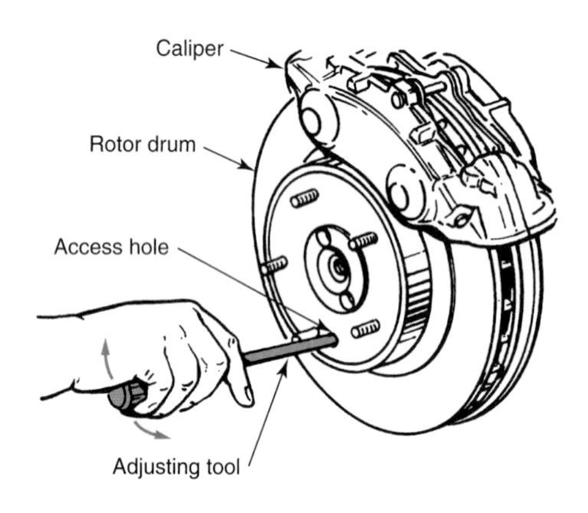

Figure 34-80. Adjusting a rotor-drum parking brake. Rotating the internal star wheel provides adjustment of shoe-to-drum clearance. (FMC)

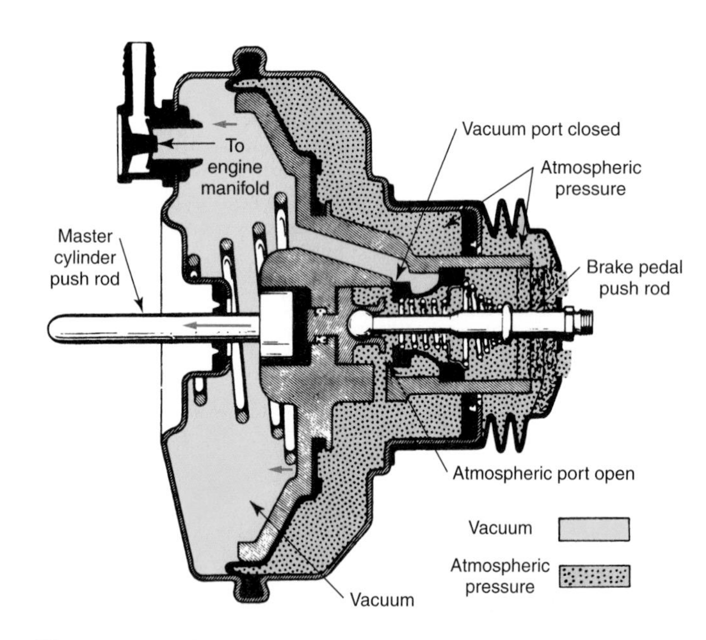

Figure 34-81. Power booster in the applied position. Note how atmospheric pressure acting on the diaphragm forces the push rod toward the master cylinder. (General Motors)

Checking Power Brake Operation

With the engine stopped, exhaust the vacuum in the vacuum tank by making several brake applications. While holding firm foot pressure on the brake pedal, start the engine. If the unit is functioning properly, the pedal will move downward when the engine starts. If it does not, check the amount of vacuum at the booster vacuum inlet. It should be the same as existing engine vacuum.

Power Booster Maintenance

Periodic inspection of the vacuum lines should be performed. Replace any cracked, soft, or otherwise defective lines. A few older power brake boosters call for cleaning the inlet air filter.

The engine should be in sound mechanical condition so that a proper vacuum is created. Check for vacuum leaks. When the engine is shut off, the vacuum should remain in the lines. Loss of vacuum indicates a leak.

Clean or replace the atmospheric air filter. Some booster units require periodic lubrication with booster oil. Do not lubricate boosters unless specified and then only with the correct oil.

If the power unit needs repair, exhaust the static vacuum in the lines by pressing on the brake pedal several times. On some units, the booster can be removed without disturbing the hydraulic lines to the master cylinder. On others, the entire assembly must be removed.

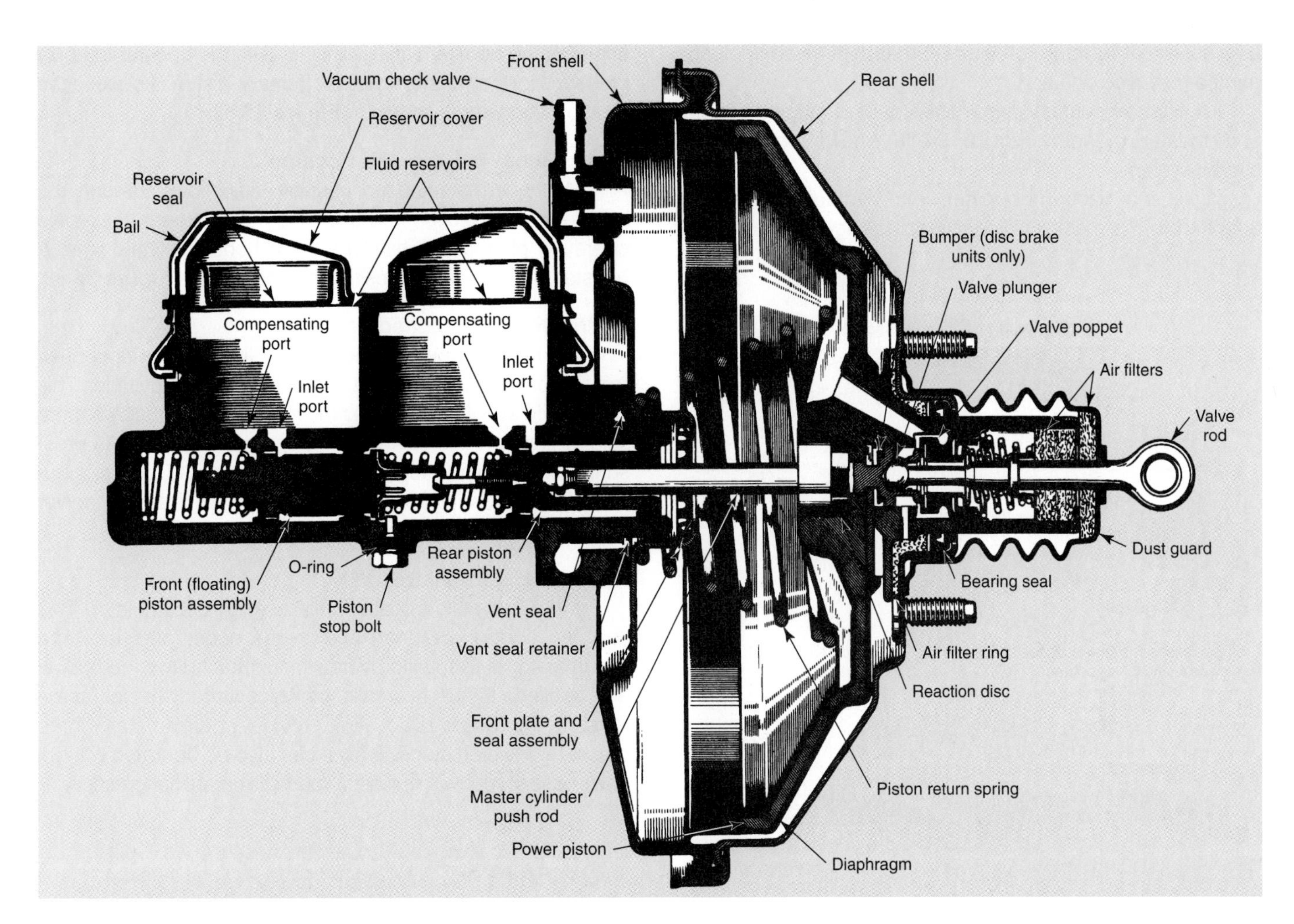

Figure 34-82. This figure is a cutaway view of a power booster and master cylinder. Note how a diaphragm acts as a seal between the housing and the power piston. (General Motors)

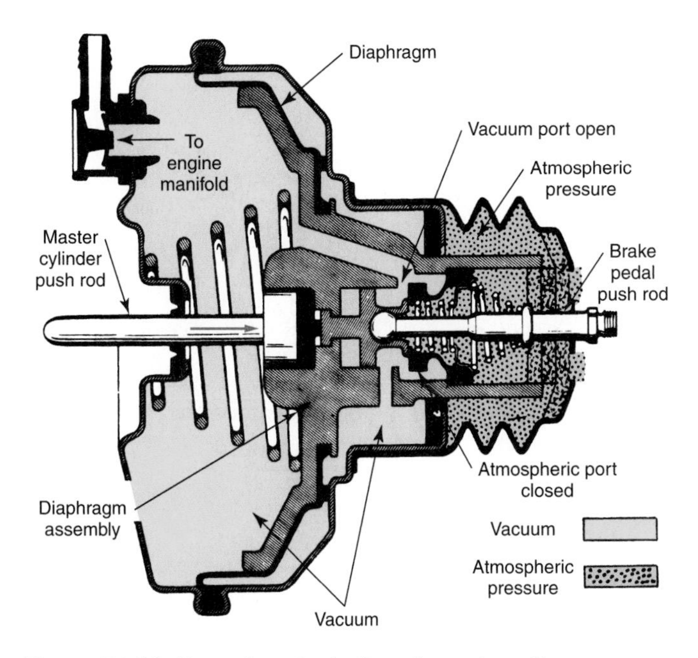

Figure 34-83. Power booster in the released position. (General Motors)

Disassemble the vacuum booster. Clean all rubber parts in alcohol. Check parts for wear and replace as needed. The power piston (sliding-type) cylinder wall must be free of rust and dents.

Assemble the unit, using care to place all parts in the correct position. Use special lubricant where it is called for. Install new gaskets and apply sealer where it is needed. Check the length of the master cylinder push rod. Be sure to stake the tab socket if it is required by the manufacturer. See **Figure 34-84.**

Install the booster and hook up the lines. Bleed the brakes and test the unit. An exploded view of a double-diaphragm power booster is shown in **Figure 34-85.**

Note: Many shops do not perform power booster overhaul. They prefer to install new or rebuilt units.

Hydraulic Power Booster

A ***hydraulic power booster*** (also called "Hydro-Boost") is used on some vehicles. The power steering pump is used as a source of hydraulic pressure. Other belt-driven or electric pumps may also be used.

A Hydro-Boost system consists of a pump, booster, and master cylinder. **Figure 34-86** shows one type of overall system.

Like the vacuum booster, the hydraulic booster is attached to the master cylinder. Depressing the brake pedal actuates the booster valves. This causes the booster to apply pressure to the master cylinder primary piston. An exploded view of a booster is shown in **Figure 34-87.**

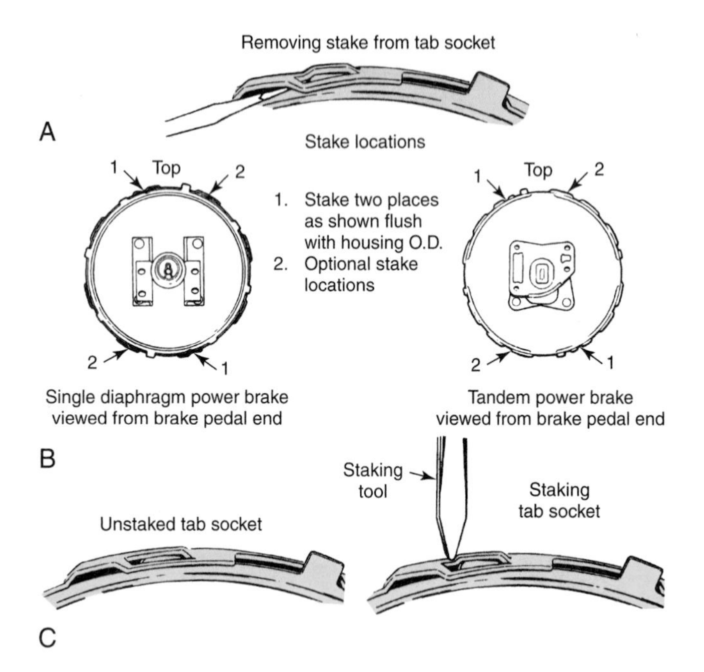

Figure 34-84. A—Removing stake from the tab socket. B—Various stake locations on a single and dual diaphragm power brake unit. C—Unstaked socket being staked. (General Motors)

Hydro-Boost in Released Position

While in the released position, fluid flows through the booster power section and on to the power steering gear. No operating pressure is built up in the booster at this time. A booster in the released position can be seen in **Figure 34-88.**

Hydro-Boost in Applied Position

As the brake pedal is depressed, the booster input push rod and piston move forward, toward the master cylinder. This causes the spool valve to also move forward. This allows more fluid to flow into the cavity behind the power piston. As pressure builds behind the power piston, it moves forward, actuating the master cylinder. **Figure 34-89** illustrates a power booster in the applied position.

Hydro-Booster Failure

Most hydraulic booster units use an accumulator. This unit can contain a spring and/or gas under pressure. The accumulator is filled with hydraulic fluid each time the brakes are applied. If the hydraulic power source fails for some reason, the accumulator will provide approximately three pressure-assisted stops. Brakes can also be operated without power assist, but will require a much firmer pedal pressure.

Caution: Accumulators contain strong springs and pressurized gas. They can come apart with lethal force. Do not apply heat. Follow the manufacturer's recommendations for correct disposal. Do not attempt accumulator service without the proper training and tools.

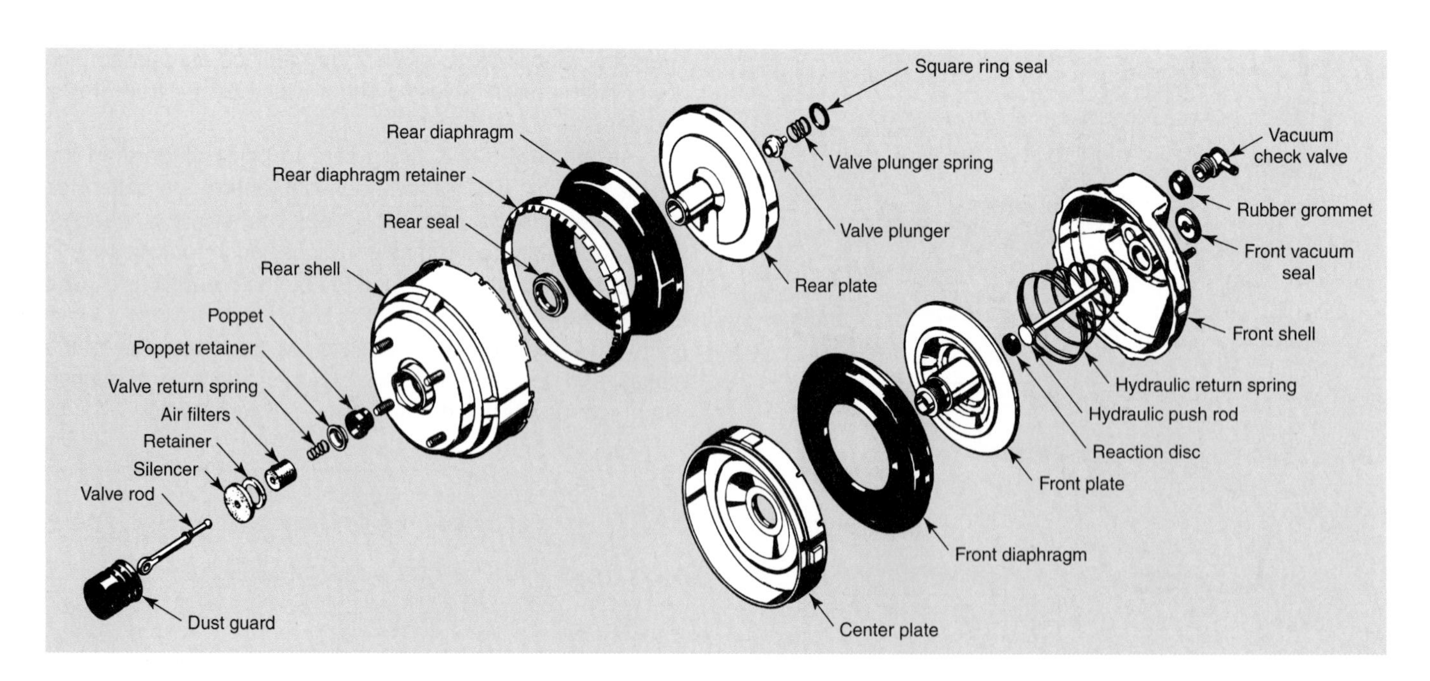

Figure 34-85. The double-diaphragm power booster is used on some vehicles. Note the position of the diaphragms in relation to the other booster parts. (General Motors)

A common problem with hydraulic brake boosters is belt squeal when the brake is applied. This is often caused by a loose belt or a defective accumulator. In some cases, the brake pedal travel is excessive, placing an extra load on the hydraulic system.

Hydro-Boost Service

The following are general service steps for a Hydro-Boost unit. Always follow the manufacturer's procedures.

1. Remove the Hydro-Boost from vehicle—with or without the master cylinder connected. Do not bend or kink steel brake lines.

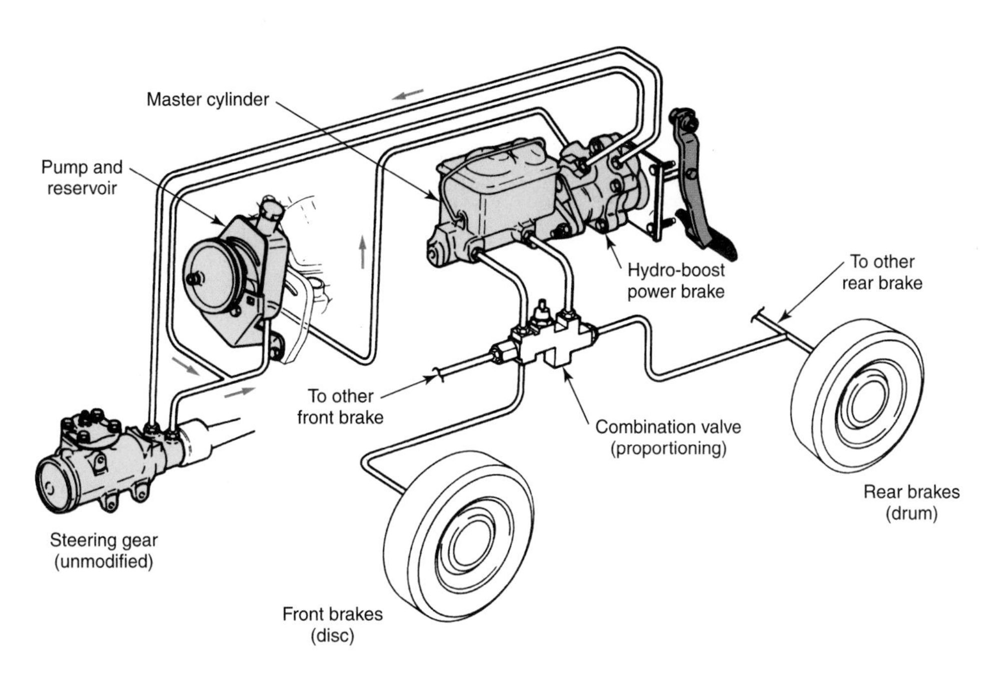

Figure 34-86. The Hydro-Boost system was used on some older vehicles. This Figure illustrates the main components and their relationship. Note arrows (black) indicating hydraulic fluid flow. (Bendix)

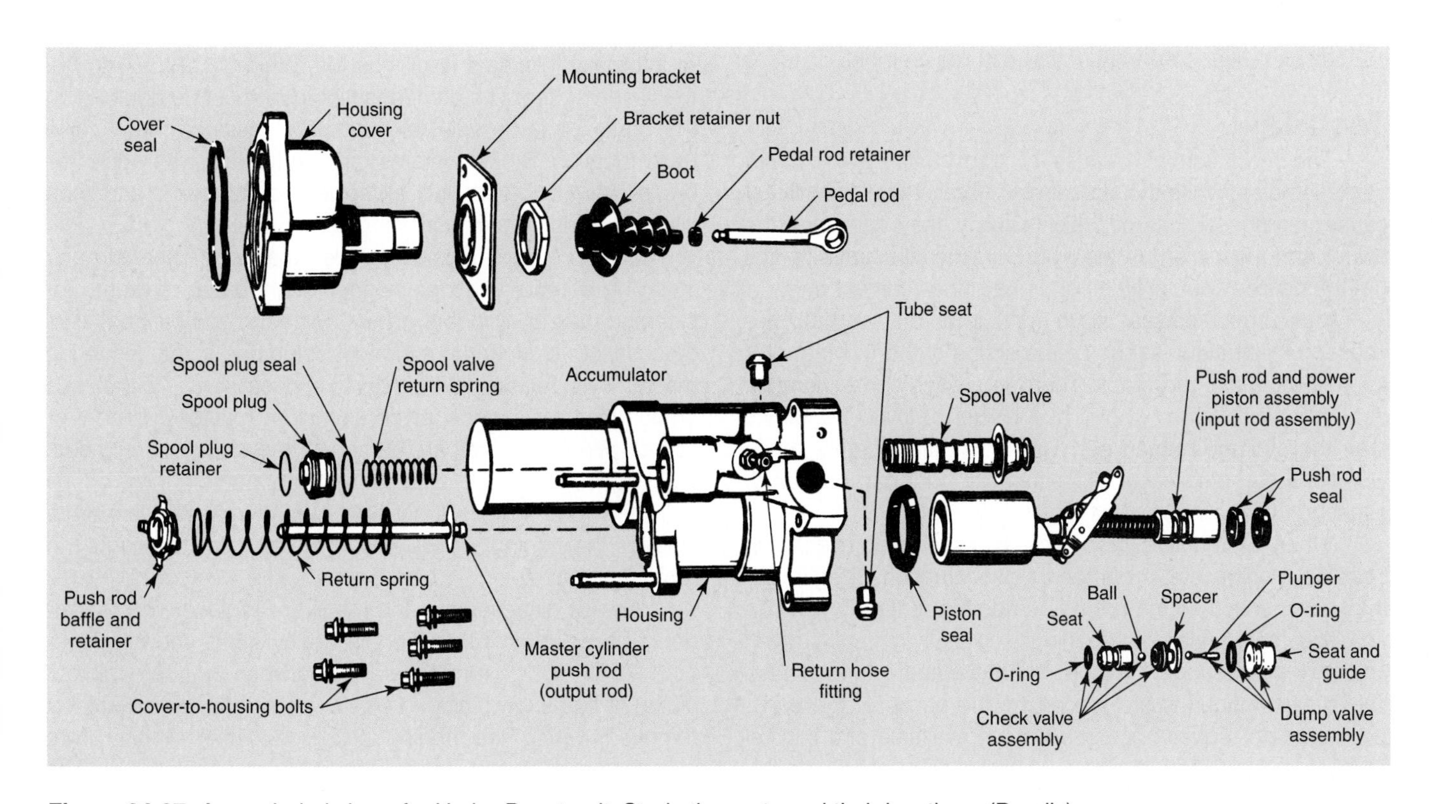

Figure 34-87. An exploded view of a Hydro-Boost unit. Study the parts and their locations. (Bendix)

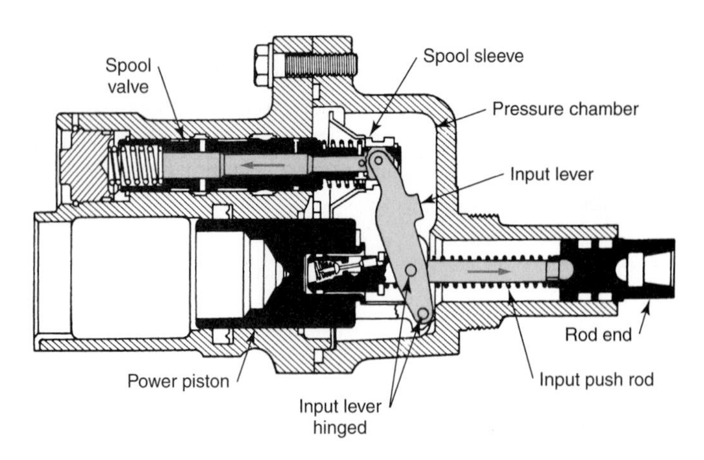

Figure 34-88. Cross-sectional view of a Hydro-Boost unit. The booster is in the fully released position. (FMC)

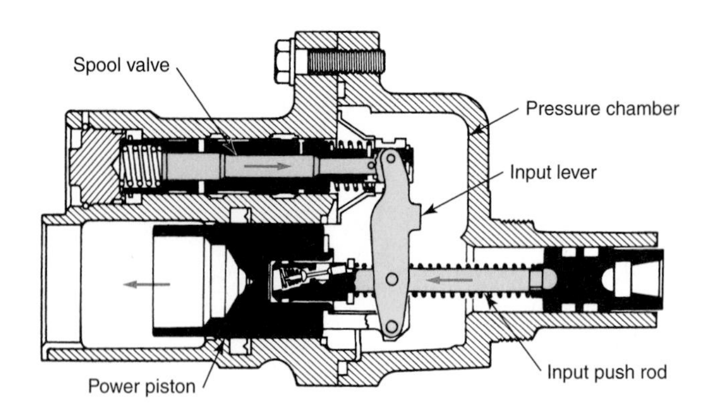

Figure 34-89. A Hydro-Boost unit in the applied position. The input lever has moved the spool valve, allowing more pressure in the pressure chamber. The power piston is forced forward, applying pressure to the master cylinder. (FMC)

2. Mount the booster on a special holding fixture or clamp it in a vise. Usually, the master cylinder end goes up.
3. Remove the spool valve plug and retainer. Take out the spool valve plug, O-ring(s), spring, and other parts.
4. Remove all remaining parts. Be sure to keep them in order.
5. Clean all parts in denatured alcohol or new, clean, power steering fluid. Check the valve and bores for scratches or indications of wear. If scratches can be felt with your fingernail, the booster unit should be replaced. These units maintain extremely close tolerances and parts are not interchangeable.
6. Reassemble in the reverse order of disassembly. Replace all seals and O-rings. Lubricate all parts with clean power steering fluid as you reassemble.

Tech Talk

The most common cause of disc brake repair comebacks is noise. While the squeals from brake systems will not cause damage or loss of braking ability, they can drive the owner and technician up the wall. Semimetallic brake linings are the noisiest, but all types of brake linings can act up.

Most brake noises heard right after an overhaul are caused by vibrations set up by the action of the moving rotor against the stationary pads. There is always some vibration during braking, but most of it at a higher or lower frequency than the average human ear can hear. Unfortunately, these vibrations can also occur at frequencies that can be heard, especially by the vehicle owner.

When overhauling the brakes, you should do everything possible to eliminate or dampen these vibrations. Begin with the basics: machine the rotor as smooth and flat as you can and use high-quality brake linings. Use all the anti-squeak shims or compounds furnished by the manufacturer. These shims or compounds are placed on the backs of the pads. Additional anti-squeak compounds are available and can be used as needed. As a simple final precaution, make sure the wear indicator is not bent and touching the rotor.

Summary

The master cylinder check valve maintains a residual pressure in the lines to drum brakes, even when the brake pedal is released.

Use a complete checklist when doing a brake inspection. Inspect the master cylinder, brake pedal, stoplight switch, wheel cylinders, calipers, brake lining, brake drums, brake rotors, seals, brake lines and hoses, parking brake, switches, valves, and chassis components. Also, road test the vehicle. Do not perform halfway jobs. Use approved brake fluid only. Keep brake fluid away from a car's painted surfaces.

Modern master cylinders are usually the double-piston (tandem) type. The tandem cylinder provides an extra measure of safety. Master cylinders must not leak, externally or internally. The fluid level must be within 3/8″ (9.53 mm) of the reservoir top. The compensating port must be open when the brake pedal is fully released. The residual check valve must function. Brake systems may be flushed with approved brake solvent (bleed until all solvent is removed) and then filled with fresh brake fluid.

Master cylinders can often be repaired by cleaning and installing a repair kit. Use nonpetroleum-based solvents for cleaning master cylinders, wheel cylinders, and calipers. Use crocus cloth to clean up minor corrosion and scratches in the master cylinder, wheel cylinders, and calipers. Do not use crocus cloth or sandpaper on aluminum master cylinders or wheel cylinders. If they are damaged, they must be replaced. Light honing (not on aluminum components), followed by polishing with crocus, is possible in many cases. Use brake fluid on the honing stones. Clean the cylinder thoroughly to remove abrasive.

Always disassemble and rebuild or replace wheel cylinders at every brake reline job. Discard cylinders showing more than 0.005″ (0.127 mm) clearance between cylinder wall and piston or that show pitting, corrosion, or scratches that are not removed during the honing process. Remove burrs from cylinder ports. Before assembling cylinders, coat the pistons and cups with fresh brake fluid. If you are unsure whether a cylinder can be successfully repaired, throw it away. Air used to blow brake cylinders dry must be free of water and oil. If it is not, finish the cylinders with an alcohol rinse and allow them to air dry.

When removing wheel assemblies from disc brakes, use care to avoid damaging the caliper brake lines. The caliper halves on some disc brakes must not be separated. Some calipers require removal to change the friction pads. Others do not. Siphon fluid from the master cylinder reservoir before forcing pad pistons inward.

When replacing a brake line, use double-wrapped, coated steel tubing only. Use double-lap or I.S.O. flares only. Protect tubing from vibration. Avoid sharp bends. Keep the tubing away from moving parts and heat.

Do not attempt an overhaul of the metering valve, differential pressure switch, proportioning valve, or either type of combination valve. If any of the above valves are defective, replace them.

If the master cylinder has a bleeder valve, bleed it first, otherwise, bleed the cylinder farthest from the master cylinder first. On split systems (tandem cylinders), bleed the wheel cylinders served by the primary piston first. Brakes may be bled manually or with a pressure bleeder. Bleed the brakes until clear fluid with no air bubbles appears. Never reuse fluid. Anti-lock brake systems require special bleeding techniques. Brake pedal height and free play must be correct. Check operation of power booster where used.

Install new grease seals at every brake reline. Study the wheel brake shoe assembly before removal. Keep all parts together. Use proper tools. Clamp the wheel cylinder. When installing shoe assemblies, clean and lubricate the shoe support pads. Lubricate the push rod-to-shoe, adjuster, adjuster-to-shoe, and hold-down areas. Lubricate sparingly and only with special high-temperature grease.

Install shoes in their correct locations. Install springs, hold-downs, and other parts. Springs must be in good condition and in the proper position. Keep shoe lining free of oil and grease.

Never adjust brakes too close. Some anchor pins are adjusted by rotating (eccentric type) or by moving up and down (slotted type). Shoes must be properly centered in the drum. Self-adjusting brakes need only an initial adjustment following relining. They will then maintain correct clearance for the life of the lining. Adjustment slots for moving the star wheel may be in the backing plate or in the brake drum.

If a caliper is removed, check for mounting shims. Never replace one-half of a caliper. If caliper cylinders cannot be cleaned with crocus cloth, discard the caliper. When installing a caliper, reinstall shims in their original locations. Check the caliper for proper centering over the rotor. The caliper must also be parallel to the rotor.

Brake drums must be free of cracking, excessive scoring, and heavy heat checking. Drum should not be more than 0.010″ (0.254 mm) out-of-round. Taper should not exceed 0.005″ (0.127 mm).

Drums may be trued by turning or grinding. Remove only enough metal to clean up the drum. Never remove more than 0.060″ (1.524 mm). When a drum on one side needs truing, turn the drum on the other side to the same size.

Drums turned to an oversize of 0.030″ (0.762 mm) or larger require the use of thicker lining or shim stock under the lining. Clean the drums thoroughly after turning or grinding.

Check the rotor for excessive scoring, wear, and runout. After installing new pads, pump up the brakes to force the pads against the rotor before driving the vehicle. The pedal should be firm. Keep the front wheel bearings adjusted to the manufacturer's specifications. Wheel bearing adjustment is especially critical on disc brake jobs.

Keep the parking brake adjusted. Do not set it too tight. Adjust the parking brake only after adjusting the service brakes.

Efficient booster operation depends on a sound unit and a normal vacuum or hydraulic source, depending on the booster type. Check the vacuum lines for leaks.

Review Questions—Chapter 34

Do not write in this book. Write your answers on a separate sheet of paper.

1. Residual pressure in the brake lines is maintained by _____.
   (A) the proportioning valve
   (B) the master cylinder check valve
   (C) the master cylinder compensating port
   (D) a slight pressure on brake pedal
2. List five important items to be checked during a brake inspection.
3. A dual master cylinder _____.
   (A) provides greater braking power
   (B) applies pressure to separate front and rear systems
   (C) works the clutch also
   (D) has two cylinders, one on top of the other
4. When the brake pedal is fully released, the _____ in the master cylinder must be open to relieve pressure buildup in the system.
5. Use _____ cloth to remove slight corrosion and scratches in wheel and master cylinders.
6. The maximum allowable clearance between a brake cylinder piston and its cylinder wall is _____.
   (A) 0.010″ (0.254 mm)
   (B) 0.050″ (1.270 mm)
   (C) 0.005″ (0.127 mm)
   (D) 0.001″ (0.025 mm)
7. Brake fluid containing _____ will ruin the paint on a vehicle.
   (A) alcohol
   (B) silicone
   (C) oil
   (D) Either A or B.

8. When bleeding tandem master cylinders, the brakes on which wheels should be bled first?
9. Even trace amounts of _____ or _____ on brake shoes or pads can render the brakes ineffective.
10. When replacing a brake line, use _____.
    (A) copper tubing
    (B) brass tubing
    (C) double-wrapped steel tubing
    (D) single-wall steel tubing
11. After installing new pads, always pump the brakes to force the _____ out against the _____ before driving the vehicle.
12. The safest way to remove phenolic (plastic) pistons from calipers is by the use of _____.
    (A) a large screwdriver
    (B) special pliers
    (C) air pressure
    (D) brake system pressure
13. The Hydro-Boost unit receives necessary operating pressure from the _____.
14. The _____ proportioning valve incorporates a variable pressure range feature to prevent brake lockup.
15. A dual proportioning valve is used on a _____.
    (A) single master cylinder
    (B) Hydro-boost unit
    (C) diagonal split brake system
    (D) None of the above.

ASE-Type Questions

1. Technician A says wheel cylinders should be rebuilt or replaced when the brake shoes are replaced. Technician B says the master cylinder should be rebuilt or replaced when the brake shoes are replaced. Who is right?
    (A) A only.
    (B) B only.
    (C) Both A and B.
    (D) Neither A nor B.
2. A failure or leak in the front system of a dual piston master cylinder will _____ the rear system.
    (A) cause a gradual leak in
    (B) have no effect on
    (C) cause an immediate loss of pressure in
    (D) increase pressure to
3. Technician A says brake springs are all alike and may be installed at random. Technician B says a cracked brake drum can be salvaged by welding. Who is right?
    (A) A only.
    (B) B only.
    (C) Both A and B.
    (D) Neither A nor B.
4. When a drum on the front left of a vehicle is turned 0.030″ (0.762 mm), the drum on the front right should be turned _____.
    (A) 0.010″ (0.254 mm)
    (B) 0.030″ (0.762 mm)
    (C) 0.060″ (1.524 mm)
    (D) 0.300″ (7.620 mm)
5. Always adjust the parking brake before _____.
    (A) adjusting the brake shoes
    (B) turning the drums
    (C) bleeding the brakes
    (D) None of the above.
6. Start the brake-bleeding process by bleeding the wheel cylinder or caliper that is _____.
    (A) closest to the master cylinder
    (B) on the same side as the master cylinder
    (C) farthest away from the master cylinder
    (D) on the left front side
7. Technician A says pressure bleeding requires two people. Technician B says fluid bled from the system can be used again if it is strained. Who is right?
    (A) A only.
    (B) B only.
    (C) Both A and B.
    (D) Neither A nor B.
8. The disc brake caliper does not have to be removed to perform any of the following tasks, *except:*
    (A) replacing the master cylinder.
    (B) bleeding the caliper.
    (C) replacing caliper piston seals.
    (D) changing a brake hose.
9. The quick take-up master cylinder supplies a large volume of _____ fluid at the start of brake application.
    (A) cool
    (B) reverse-flow
    (C) high-pressure
    (D) low-pressure
10. Technician A says after the brakes are overhauled, the technician should make 8 or 10 stops from high speed to seat the linings. Technician B says aluminum wheel cylinders may be honed. Who is right?
    (A) A only.
    (B) B only.
    (C) Both A and B.
    (D) Neither A nor B.

Student Activities

1. Make a diagram of the hydraulic system needed for front-disc/rear-drum brakes. Explain the operation and interaction of the hydraulic and mechanical components.

2. Draw a wheel cylinder and a single piston caliper on the blackboard and explain how the hydraulic force is turned into movement of the shoes and pads.
3. Draw a pressure differential valve on the board and explain the differences in fluid flow in each of the following modes:
   - Normal operation.
   - Pressure loss at the rear (or right front-left rear of a split system).
   - Pressure loss at the front (or left front-right rear of a split system).
4. Check the fluid level in the master cylinders of at least three vehicles. Explain how to determine what type of fluid should be added.
5. Bleed the brakes of a vehicle without ABS. Write a short report summarizing the bleeding procedures.

Brake Problem Diagnosis	
Problem: No brakes	
Possible cause	**Correction**
1. Broken line, hose, or other leak.	1. Repair source of leak.
2. Air in system.	2. Bleed system. Repair source of air entry.
3. Lining and/or pads worn.	3. Adjust or reline brakes.
4. Master cylinder cups leaking.	4. Rebuild or replace master cylinder.
5. Low fluid level in master cylinder.	5. Fill reservoir and bleed system.
6. Brake pedal linkage disconnected.	6. Reconnect brake pedal.
7. Automatic shoe adjusters not functioning.	7. Repair or replace adjusters. Adjust shoes.
8. Vaporized fluid from excessive braking.	8. Allow to cool. Install new fluid. Advise driver.
9. Caliper seal or piston damage.	9. Repair or replace caliper as needed.
Problem: Spongy pedal	
Possible cause	**Correction**
1. Air in system.	1. Bleed system. Repair source of air entry.
2. Shoes not centered in drum.	2. Adjust anchors to center shoes.
3. Drums worn or too thin.	3. Replace drums.
4. Soft hose.	4. Replace hose.
5. Shoe lining wrong thickness.	5. Install correct lining.
6. Cracked brake drum.	6. Replace drum.
7. Brake shoes distorted.	7. Replace shoes.
8. Insufficient brake fluid.	8. Bleed system. Fill with fluid.
9. Bent pads.	9. Replace pads.
Problem: Hard pedal (excessive foot pressure required)	
Possible cause	**Correction**
1. Incorrect lining.	1. Install proper lining.
2. Linings contaminated with grease or brake fluid.	2. Replace or reline shoes. Repair source of leak.
3. Brake shoes not centered.	3. Center brake shoes.
4. Primary and secondary shoes reversed.	4. Install shoes in correct location.
5. Brake linkage binding.	5. Free and lubricate.
6. Master or wheel cylinder pistons frozen.	6. Rebuild or replace cylinder.
7. Linings hard and glazed.	7. Sand lining with medium grit sandpaper.
8. Lining ground to wrong radius.	8. Grind lining as specified.
9. Brake line or hose clogged or kinked.	9. Repair or replace line or hose.
10. Power booster unit defective.	10. Repair or replace power booster.
11. No vacuum to power booster.	11. Replace clogged, soft lines. Repair vacuum leaks.
12. Engine fails to maintain proper vacuum to booster.	12. Tune or overhaul engine.
13. Pads worn excessively thin.	13. Replace pads.
14. Seized caliper piston(s).	14. Repair or replace.
15. Heat-checked or blued rotor.	15. Replace rotor.
16. Faulty proportioning valve.	16. Replace valve.
17. Quick take-up valve center orifice clogged.	17. Replace master cylinder.
Problem: Brakes grab (one or more wheels)	
Possible cause	**Correction**
1. Grease or brake fluid on lining and/or pads.	1. Replace lining and/or pads.
2. Lining charred.	2. If mild, sand. If severe, replace.
3. Lining loose on shoe.	3. Replace brake shoes.
4. Loose wheel bearings.	4. Adjust wheel bearings.
5. Defective wheel bearings.	5. Replace bearings.
6. Loose brake backing plate.	6. Torque backing plate fasteners.
7. Defective drum.	7. Resurface or replace drum.
8. Sand or dirt in brake shoe assembly.	8. Disassemble and clean linings and drum.
9. Wrong brake lining.	9. Install correct lining.
	(continued)

Brake Problem Diagnosis *(continued)*	
10. Primary and secondary linings or shoes reversed. 11. Loose caliper. 12. Defective power brake booster. 13. Uneven tire pressure.	10. Install correctly. 11. Tighten to specifications. 12. Repair or replace. 13. Inflate to specifications.
Problem: Brakes fade	
Possible cause	**Correction**
1. Poor quality shoes and/or pads. 2. Excessive use of brakes. 3. Overheated brake fluid. 4. Improper lining-to-drum contact. 5. Thin brake drums. 6. Dragging brakes. 7. "Riding" the brake pedal. 8. Excessively thin rotors.	1. Replace shoes and/or pads. 2. Use lower gears, reduce speed or load. 3. Flush. Install super heavy-duty fluid. 4. Adjust shoes or resurface drum. 5. Install new drums. 6. Adjust or repair other cause of dragging. 7. Advise driver to keep foot off brake pedal unless needed. 8. Replace rotors.
Problem: Brakes pull vehicle to one side	
Possible cause	**Correction**
1. One wheel grabbing. 2. Shoes not centered or adjusted properly. 3. Different lining on one side or shoes reversed on one side. 4. Plugged brake line or hose. 5. Uneven tire pressure. 6. Front end misaligned. 7. Sagged, weak, or broken spring. Weak shock absorber or strut. 8. Wheel cylinder bore diameter different on one side. 9. Pads contaminated with grease or brake fluid. 10. Caliper or backing plate loose.	1. See *Brakes grab.* 2. Center and adjust lining-to-drum clearance. 3. Replace lining or install shoes in proper position. 4. Clean or replace brake line. 5. Use same pressure on both sides. 6. Align front end. 7. Install new spring, shocks or struts. 8. Install correct size cylinder. 9. Replace pads. 10. Tighten backing plate retainers to specifications.
Problem: Brakes drag	
Possible cause	**Correction**
1. Parking brake adjusted too tight. 2. Clogged hose or line. 3. Master cylinder reservoir cap vent clogged. 4. Brake pedal not fully releasing. 5. Insufficient pedal free travel. 6. Brakes adjusted too tight. 7. Brakes not centered in drum. 8. Master cylinder or wheel cylinder cups soft and sticky. 9. Loose wheel bearing. 10. Parking brake fails to release. 11. Shoe retracting springs weak or broken. 12. Out-of-round drum. 13. Defective power booster. 14. Seized caliper piston. 15. Sliding caliper bound. 16. Rotor thickness out of specifications. 17. Loose caliper bolts. 18. Bent pads. 19. Improper or contaminated brake fluid.	1. Adjust properly. 2. Clean or replace brake line. 3. Open vent in cap. 4. Adjust brake pedal. 5. Adjust pedal free travel so that compensating port will be open when brake is released. 6. Adjust correctly. 7. Center shoes in drum. 8. Rebuild or replace cylinders. Flush system. 9. Adjust wheel bearings. 10. Clean and lubricate parking brake linkage. 11. Replace shoe springs. 12. Resurface drum. 13. Rebuild or replace booster. 14. Rebuild or replace caliper. 15. Free caliper. Clean sliding surfaces and lubricate if required by manufacturer. 16. Replace rotor. 17. Tighten to specifications. 18. Replace pads. 19. Repair as necessary.

(continued)

Brake Problem Diagnosis *(continued)*	
Problem: "Nervous" pedal (pedal moves rapidly up and down when applying brakes) Note: This is a normal operating condition on vehicles equipped with anti-lock brakes.	
Possible cause	**Correction**
1. Brake drums out-of-round. 2. Excessive disc runout. 3. Loose wheel bearings. 4. Drums loose. 5. Rear axle bent. 6. Brake assembly attachments loose or missing.	1. Resurface drums. 2. Resurface or replace disc. 3. Adjust wheel bearings. 4. Tighten wheel lugs, adjust brakes. 5. Replace axle. 6. Repair as necessary.
Problem: Brakes chatter	
Possible cause	**Correction**
1. Weak or broken shoe retracting springs. 2. Defective power booster. 3. Loose backing plate. 4. Loose or damaged wheel bearings. 5. Drums tapered or barrel shaped. 6. Bent brake shoes. 7. Dust on lining. 8. Lining glazed. 9. Drum damper spring missing. 10. Grease or fluid on linings. 11. Shoes not adjusted properly. 12. Incorrect brake pads. 13. Damaged brake pads. 14. Damaged rotors.	1. Replace springs. 2. Rebuild or replace booster. 3. Tighten fasteners. 4. Adjust or replace bearings. 5. Resurface drums. 6. Replace brake shoes. 7. Sand linings and clean. 8. Sand linings and clean. 9. Install damper spring. 10. Repair source of leak and reline brakes. 11. Center and adjust shoes. 12. Install correct pads. 13. Replace pads. 14. Replace or resurface rotors.
Problem: Brakes squeal	
Possible cause	**Correction**
1. Glazed or charred shoes and/or pads. 2. Dust or metal particles imbedded in lining. 3. Lining rivets loose. 4. Wrong type of lining. 5. Shoe hold-down springs weak or broken. 6. Drum damper spring missing. 7. Shoes improperly adjusted. 8. Shoes bent. 9. Bent backing plate. 10. Shoe retracting springs weak or broken. 11. Drum too thin. 12. Lining saturated with grease or brake fluid. 13. Pad wear sensors contacting rotor. 14. Rotor contacting caliper. 15. Loose outboard pads.	1. Sand or replace shoes and/or pads. 2. Sand lining and clean. 3. Replace pads or shoes. 4. Install pads or shoes with the correct lining. 5. Replace hold-down springs. 6. Install damper spring around drum. 7. Adjust shoes. 8. Replace shoes. 9. Replace backing plate. 10. Replace springs. 11. Replace drum. 12. Replace lining. Repair leak. 13. Replace pads. Resurface rotor if necessary. 14. Check for loose fasteners, missing shims, and other problems. Correct as required. 15. Bend tabs to tighten. Replace if tabs are broken.
Problem: Shoes click	
Possible cause	**Correction**
1. Shoe is pulled from backing plate by following tool marks in drum. 2. Shoe bent. 3. Shoe support pads on backing plate grooved.	1. Smooth drum braking surface. 2. Replace shoes. 3. Smooth and lubricate pads or replace backing plate. *(continued)*

Brake Problem Diagnosis *(continued)*	
Problem: Red brake warning light comes on	
Possible cause	**Correction**
1. Air in system. 2. Malfunctioning master cylinder. 3. Worn out brake lining. 4. Defective shoe adjusters. 5. Contaminated or improper brake fluid.	1. Check for reason and bleed system. 2. Repair or replace as necessary. 3. Replace pads or shoes. 4. Adjust or replace as necessary. 5. Repair as necessary.
Problem: Amber anti-lock brake (ABS) light comes on	
Possible cause	**Correction**
1. Anti-lock brake system malfunction.	1. Refer to Chapter 35.
Problem: Automatic shoe adjusters will not function	
Possible cause	**Correction**
1. Adjuster wheel (star wheel) rusty or dirty. 2. Parts installed wrong. 3. Adjuster lever dirty and sticky. 4. Star wheel notches burred. 5. Adjuster lever bent.	1. Clean threads. Lube with high temperature grease. 2. Install adjuster correctly. 3. Clean and lube adjuster. 4. Install new adjuster. 5. Install new adjuster lever.

A few late-model vehicles are equipped with an adaptive cruise control system, which used radar to help maintain a safe following distance. A pictogram in the center of the gauge cluster depicts the proximity of the leading car. If necessary, the system will slow the vehicle to maintain the desired following distance. (DaimlerChrysler)

35

Anti-Lock Brake and Traction Control System Service

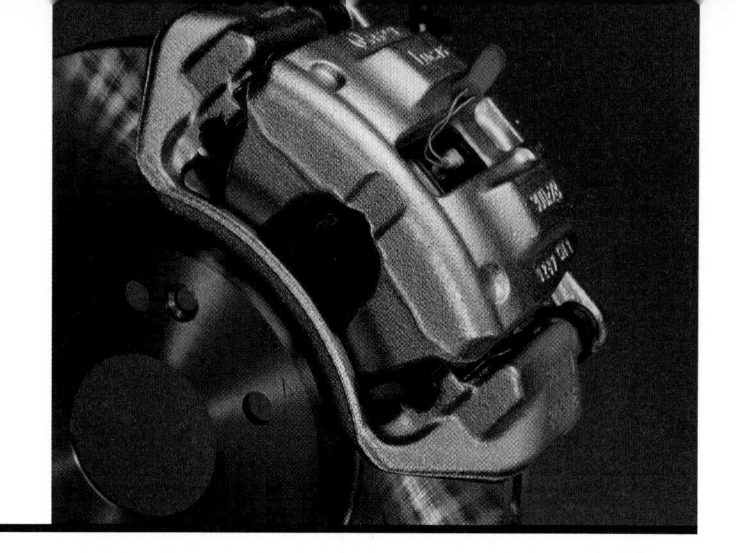

After studying this chapter, you will be able to:

- Explain anti-lock brake system operation.
- Identify anti-lock brake system components.
- Diagnose anti-lock brake system problems.
- Explain traction control system operation.
- Identify traction control system components.
- Diagnose traction control system problems.

Technical Terms

Anti-lock brake systems	Hydraulic actuator
ABS systems	Solenoid valves
Foundation brakes	Hydraulic pump
Base brakes	Accumulator
Wheel speed sensor	Pulsation
G-force sensor	Hydraulic modulator
Lateral accelerometer	Ball screws
Control module	Pedal travel switch
ABS warning light	Traction control systems

This chapter covers the principles of anti-lock brakes and traction controls. Both are similar in that they use electronic components to control the brake hydraulic system. The hydraulic system, in turn, controls the application of the brake's friction elements. The major differences between the systems include the ultimate purpose of each system and the fact the traction control system also affects the engine output on some vehicles. On many new vehicles, the anti-lock brakes and the traction control system are operated by a single electronic/hydraulic control system.

Anti-Lock Brake Systems

Many vehicles are now being equipped with ***anti-lock brake systems,*** often referred to as ***ABS systems.*** Most ABS systems used on automobiles are similar to the one shown in **Figure 35-1** and control all four wheels. The ABS systems used on most light trucks and a few small cars control only the rear wheels, **Figure 35-2.** All ABS systems, whether two- or four-wheel systems, use electronic and hydraulic components to help prevent wheel lockup during hard braking. Anti-lock brakes allow the driver to maintain directional control while providing maximum braking efficiency.

Always Obtain a Service Manual

Many vehicle and parts manufacturers produce anti-lock brake systems. These systems, while similar, require different troubleshooting and service procedures. The same system may be installed on different makes of vehicles, and the same line of vehicles may have two or more brands of anti-lock brake systems. Therefore, it is important to refer to the correct service manual before beginning any repairs on an anti-lock brake system.

The Reason for Anti-Lock Brakes

When a situation calling for a quick stop arises, the average driver tends to apply the brakes as hard as possible. Although this is a natural reaction, it is the wrong thing to do in most situations. Applying the brakes as hard as possible causes one or more wheels to lock up, causing a skid. Skidding a tire greatly reduces its efficiency in slowing the vehicle. In addition, if any one of the tires is skidding, the vehicle cannot be steered effectively.

On wet or icy pavement, the tires slip on the moisture on the road. When the road is dry, locking the wheel causes a

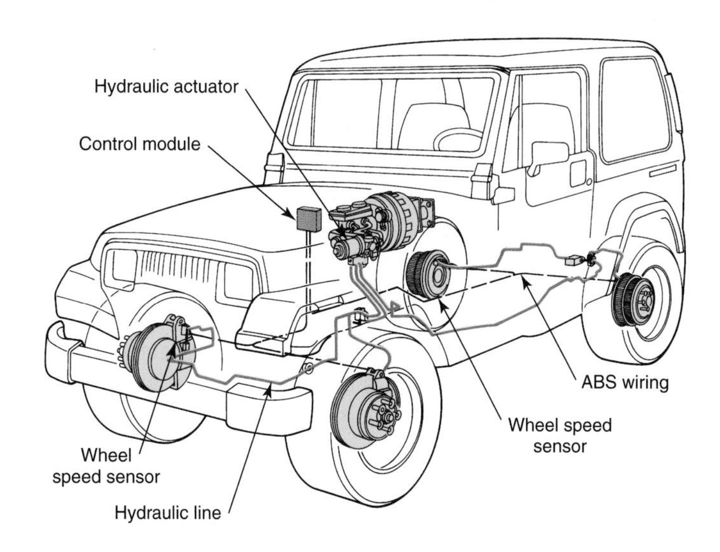

Figure 35-1. A vehicle equipped with a four-wheel anti-lock braking system. Study the various systems and parts. (DaimlerChrysler)

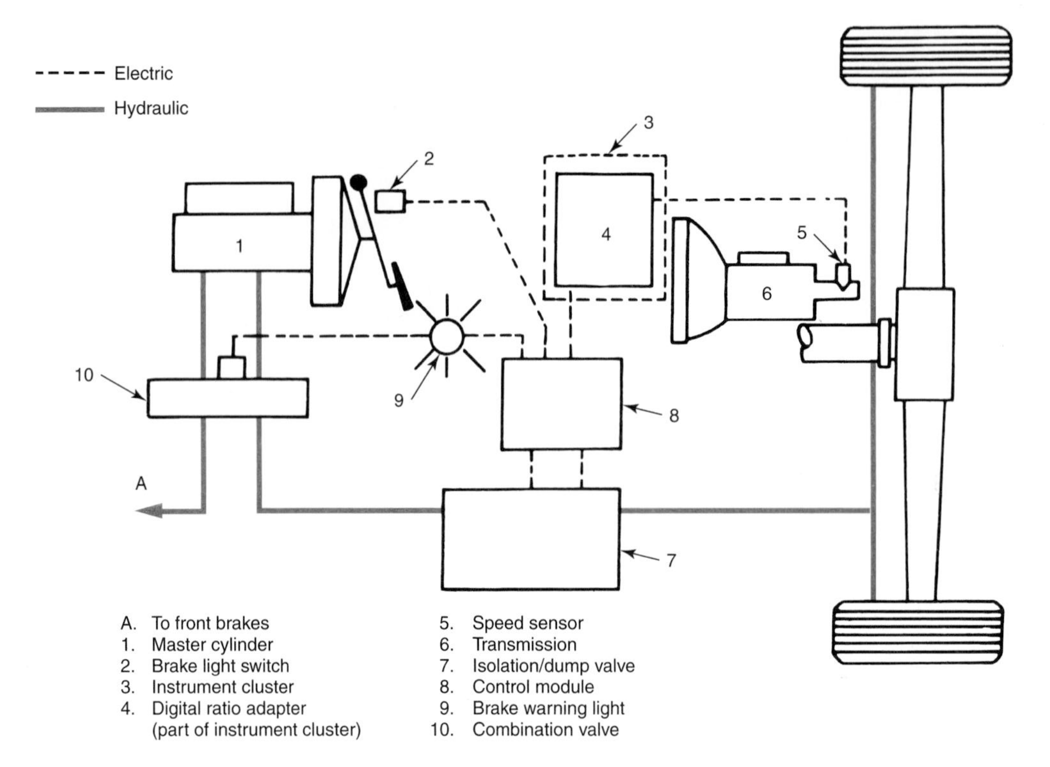

Figure 35-2. A rear-wheel anti-lock brake system. Study the electrical and hydraulic circuits and various components. (General Motors)

single area of the tire to slide across the pavement. The tremendous friction from the tire rubbing against the pavement causes the contact surface of tire to melt. The tire then slips on the melted rubber. Skid marks commonly seen on roads are actually melted rubber from a panic stop. Since the rubber melts only in the area where the skidding tire contacts the road, flat spots are worn into the tire.

Safety experts recommend avoiding skids by pumping the brakes. Pumping the brakes means alternately applying and releasing the brake pedal as quickly as possible during a panic stop. While this will help prevent wheel lockup, it is hard for many drivers to remember. Additionally, it is very difficult to pump the brakes fast enough to have any real effect.

The force of the brake pedal is transmitted to the brakes by hydraulic pressure developed in the master cylinder. Therefore, modulating (alternately reducing and increasing) the brake hydraulic pressure between the master cylinder and the wheel calipers or wheel cylinders can prevent lockup. The ABS system pumps the brakes at a much faster rate than any driver could. Since the anti-lock brake system only operates during very hard stops, normal braking is unaffected by an ABS system. The end result is improved braking capabilities, improved handling during emergency braking, and reduced tire wear.

Anti-Lock Brake System Components

Anti-lock brake systems use an electronic control system to modify the operation of the brake hydraulic system. The electronic and hydraulic components work together to prevent wheel lockup during periods of hard braking. Most anti-lock brake systems, regardless of the manufacturer, contain several common components. These components include wheel speed sensors, an anti-lock control module, and a hydraulic actuator. See Figure 35-1.

The brake friction components (shoes and pads), most of the hydraulic components (wheel cylinders, caliper pistons, master cylinder, and hydraulic lines), and the power brake system components of an ABS system are the same as those used on vehicles without ABS. When discussing ABS systems, the standard friction and hydraulic components are referred to as the ***foundation brakes*** or the ***base brakes.***

Wheel Speed Sensors

Anti-lock brake systems use ***wheel speed sensor*** units to determine the rate of wheel rotation. Most wheel speed sensors consist of a toothed rotor, or wheel, and a sensing unit, **Figure 35-3.** The toothed rotor is attached to the vehicle's axle or brake rotor and rotates at the same speed as the wheel and tire. As the rotor spins, it creates a magnetic field in the sensing unit. The sensing unit converts the magnetic field into a pulsing voltage signal, which is sent to the control module. The strength and frequency of this signal varies in relation to the speed of the wheel.

Some anti-lock brake systems have wheel speed sensors that are mounted at each wheel. Other systems use speed sensors mounted on the rear axle housing or in the transmission, **Figure 35-4.** Axle and transmission-mounted sensors are commonly used on two-wheel ABS systems and are found on some four-wheel ABS systems.

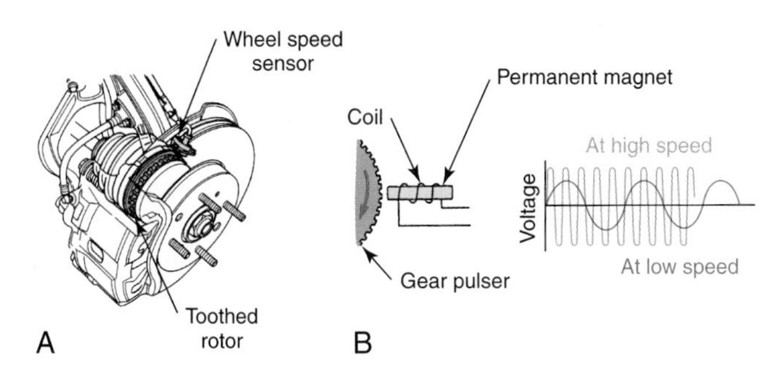

Figure 35-3. A—A wheel speed sensor and toothed rotor. B—When the toothed rotor turns, the magnetic flux around the coil in the wheel sensor alternates, producing voltages with a frequency in proportion to the speed of the rotating wheel. These pulses are sent to the anti-lock control module to identify wheel speed. (Honda)

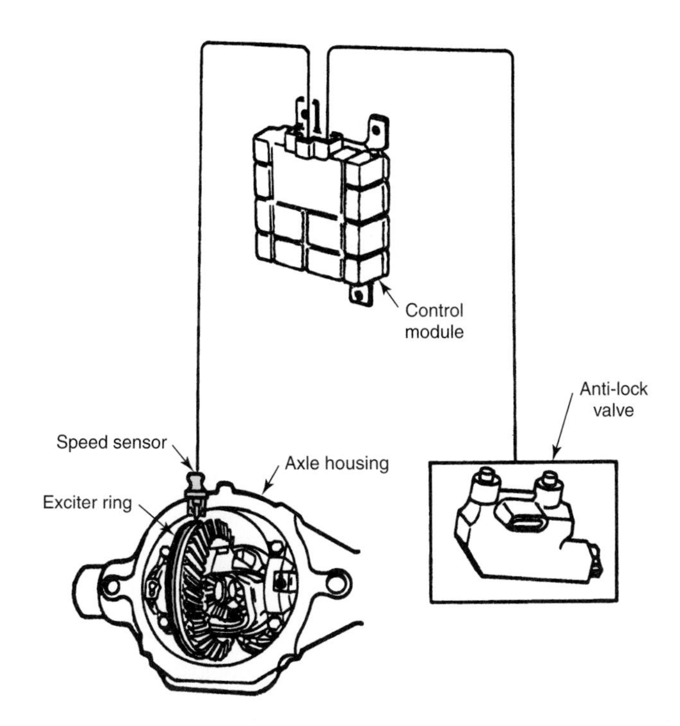

Figure 35-4. A speed sensor that is mounted in the rear axle housing. Note the toothed exciter ring, which is fastened to the ring gear. (DaimlerChrysler)

G-force Sensor

Some systems contain a ***G-force sensor,*** which measures the rate of deceleration by comparing the vehicle tilt during braking to the normal ride position. The G-force sensor is often installed in the passenger compartment. The control module generates outputs to the hydraulic system based on G-force and wheel speed sensor inputs. See **Figure 35-5.** Some ABS systems use a form of the G-force sensor called a ***lateral accelerometer*** to sense cornering speed.

Control Module

The ***control module*** uses the signals produced by the wheel speed sensors to determine when the anti-lock system should be activated. When a wheel is nearing a lockup

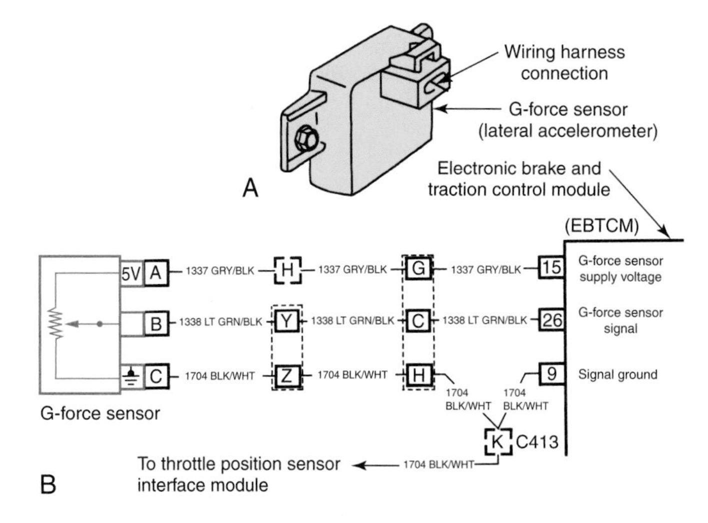

Figure 35-5. A—Typical G-force sensor (lateral accelerometer). B—G-force sensor and wiring going to the electronic brake control module. The G-force sensor cannot be repaired and must be replaced as a complete assembly. Handle these sensors with care; they are easily damaged! (General Motors)

condition, the control module signals the hydraulic actuator to regulate fluid pressure to that wheel. The control module processes inputs and delivers outputs using the same general process as the engine control computers discussed in earlier chapters. A typical anti-lock control module is shown in **Figure 35-6.**

Most control modules are equipped with self-diagnostic capabilities. The control module illuminates an ***ABS warning light*** on the vehicle dashboard if there is a malfunction in the ABS system. The light also comes on during and shortly after starting the vehicle, during the bulb and wiring check. This light can be used on some systems to retrieve trouble codes.

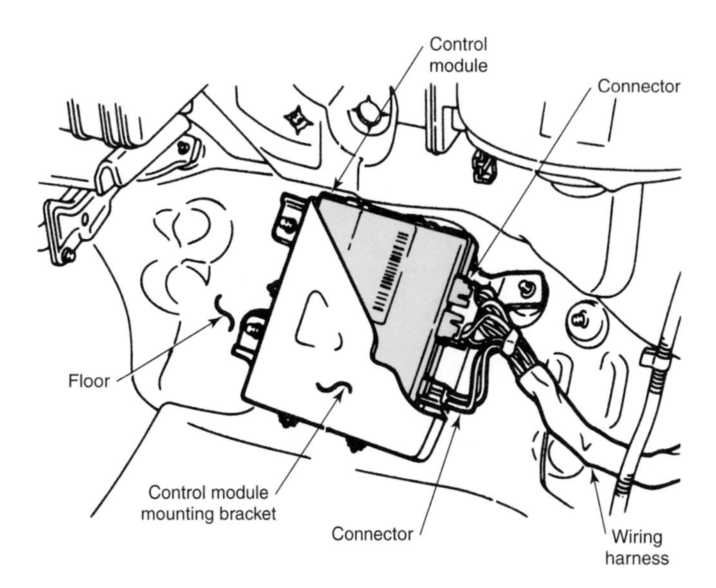

Figure 35-6. Anti-lock brake control module, mounting bracket, and wiring harness. (General Motors)

Note: Do not confuse the ABS warning light with the *brake system warning light* that is used on all vehicles. The ABS light is usually amber. The brake warning light is red. Typical lights are shown in Figure 35-7. In some cases, both lights are operated by the control module.

Hydraulic Actuator

The ***hydraulic actuator*** is the unit that regulates the pressure delivered to the brakes based on commands it receives from the control module. The actuator usually consists of two or four ***solenoid valves,*** a ***hydraulic pump,*** and an ***accumulator.*** All of the above components may be housed in a single casing or may be separately mounted and connected by high-pressure hoses. See **Figure 35-8.** The actuator on a two-wheel ABS system has two solenoid valves, and the actuator on a four-wheel ABS system has four solenoid valves. Some systems use three solenoid valves—one for each front wheel and a single solenoid to operate both rear wheels. No matter how many solenoids there are, they control the hydraulic system in the same way.

The solenoid valves control the hydraulic pressures by two methods, depending on the system and the severity of the lockup position:

- When the lockup condition is slight, the ABS solenoids are positioned to seal the passage between the master cylinder and brakes lines so no additional pressure can reach the affected wheel cylinder or caliper.
- When the lockup condition is severe, the ABS solenoids are positioned to dump small amounts of hydraulic fluid back to the reservoir, the accumulator, or the hydraulic pump intake. This fluid removal reduces, or bleeds off, hydraulic pressure going to the individual wheel cylinder or piston.

These processes occur many times per second, causing the slight pulsation that can be felt in the brake pedal during ABS operation. This ***pulsation*** is caused by the ABS system pumping the brakes to give maximum braking power without wheel lockup. Note the hydraulic actuator controls the brake hydraulic system pressure during ABS operation, no matter what degree of pressure is applied to the brake pedal by the driver.

During ABS operation, hydraulic system pressure is constantly being bled off and then reapplied to provide maximum braking without wheel lock. Therefore, if the ABS is used for long periods, the original pressure applied may not be sufficient to operate the brakes. The hydraulic actuator pump provides makeup pressure to the hydraulic system. The intake of some hydraulic pumps is piped through the solenoids and used to quickly draw off pressure when a particular wheel must be depressurized quickly. This process is controlled by the solenoids.

An accumulator is also installed in the hydraulic actuator or between the master cylinder and the actuator. It absorbs extra fluid when the actuator valves are bleeding off pressure. The accumulator also holds fluid pressure in reserve to allow for brake operation if the pump is unable to keep up with pressure demands.

Piston-Operated Hydraulic Actuator

One variation of the hydraulic actuator uses moveable pistons instead of a hydraulic pump to produce hydraulic pressure. The pistons are operated by small electric motors through reduction gears. This actuator is called a ***hydraulic modulator.*** The hydraulic modulator regulates the pressure delivered to the brakes in a manner similar to the conventional hydraulic actuator. However, the solenoids used in the hydraulic modulator have only one function—to isolate, or seal, the passage between the master cylinder and the affected wheel cylinder or caliper. If additional pressure control is needed, the pistons are moved to increase or reduce the hydraulic pressure to a particular circuit.

A cutaway view of the hydraulic modulator is shown in **Figure 35-9.** Notice the pistons are operated by motor-driven screw assemblies called ***ball screws.*** The ball screws operate by moving up or down on threaded rods. Note that each front wheel piston has a separate ball screw, while the rear pistons are both operated by a single ball screw.

The pistons can be driven forward to increase pressure or backward to reduce pressure. In addition, the pistons can be held in position to maintain a certain pressure. The pistons serve as accumulators for extra pressure. The piston position can be precisely controlled by extremely close tolerances in the ball screws and reduction gears, and by the use of brakes on the drive motors. The brakes can be electromagnetic brakes or expansion spring brakes. A typical drive motor and brake assembly is shown in **Figure 35-10.**

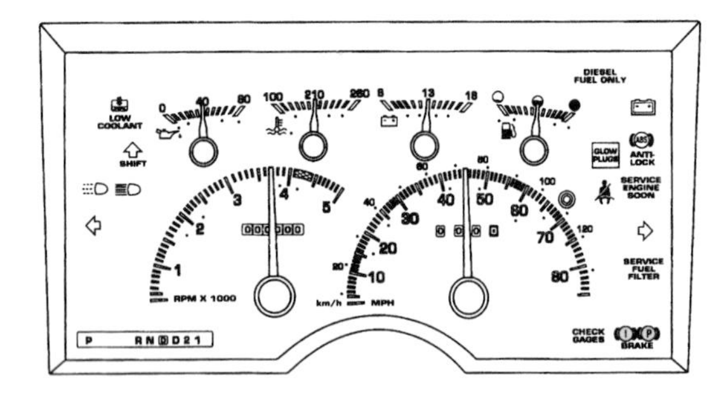

A

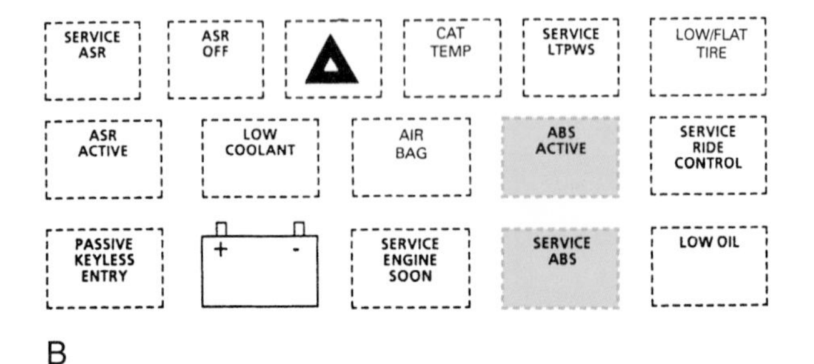

B

Figure 35-7. A—An instrument cluster with ABS and brake warning lights. B—A driver information center with ABS active and service lights. (General Motors)

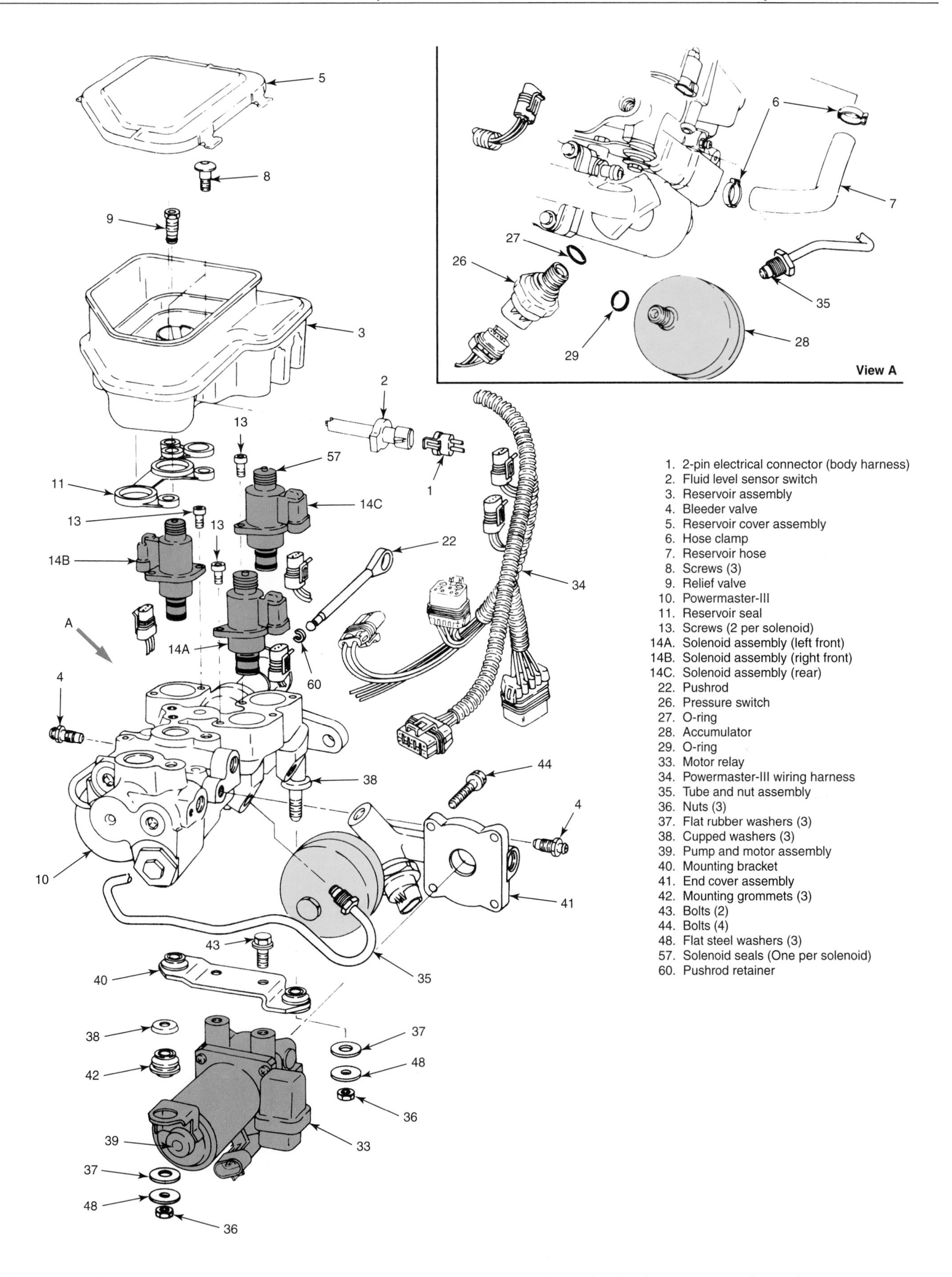

Figure 35-8. A—Hydraulic actuator. B—Anti-lock brake system operation schematic. C—Cutaway view of the electric motor, pump, and the relief valve. (Honda)

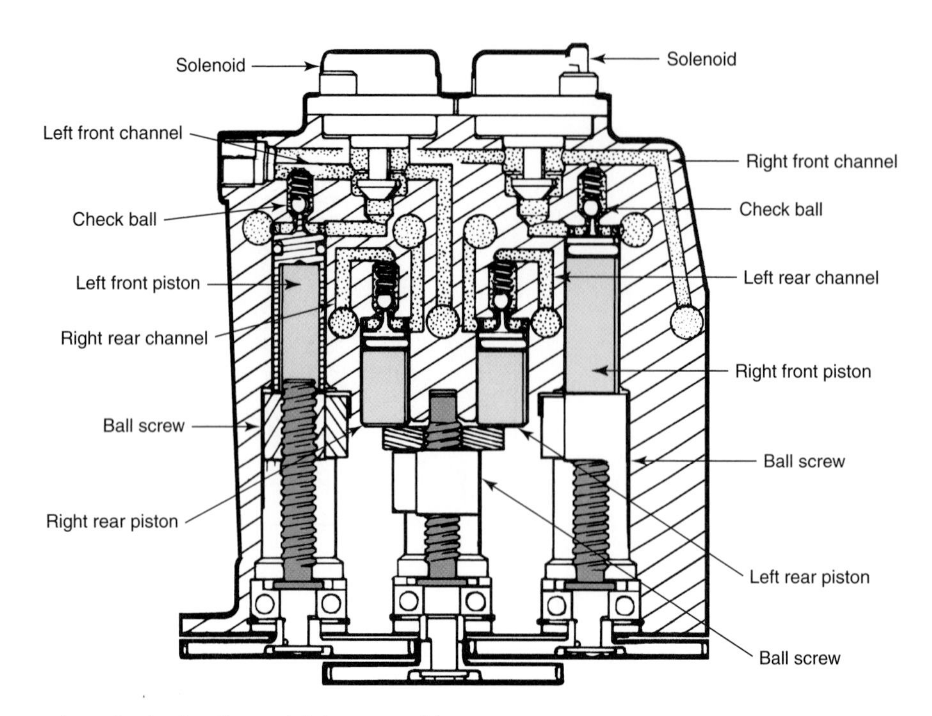

Figure 35-9. A cutaway view of a hydraulic modulator assembly.

Pedal Travel Switch

Some ABS systems are equipped with a ***pedal travel switch,*** which is mounted on the brake pedal assembly. The pedal travel switch is electrically connected to the control module. The signal from the pedal travel switch alerts the control module when pedal pulsation becomes excessive during ABS operation. The control module then modifies the action of the hydraulic actuator solenoid valves to reduce pulsation.

Anti-Lock Brake Operation

Under light braking conditions, the anti-lock portion of the brake system does not operate. The sensors continuously monitor wheel rotation and send signals to the anti-lock control module. When the brake pedal is pressed, fluid flows from the master cylinder, through the hydraulic actuator, and into the wheel cylinder or caliper, **Figure 35-11.** Basic hydraulic brake operation was discussed in Chapter 34, Brake Service.

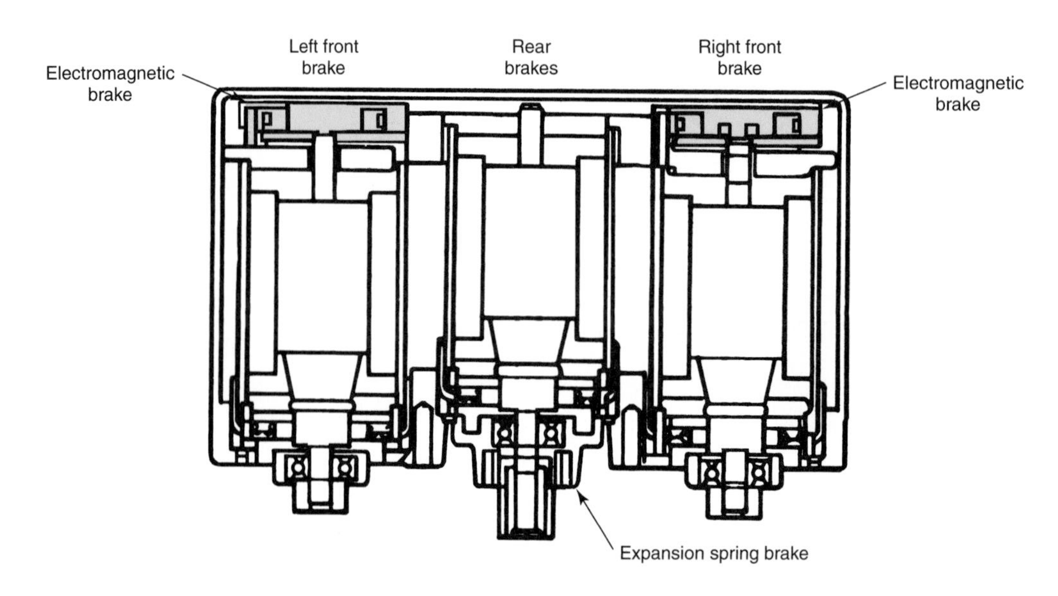

Figure 35-10. Cutaway of an electric drive motor and electromagnetic brake (EMB) assembly.

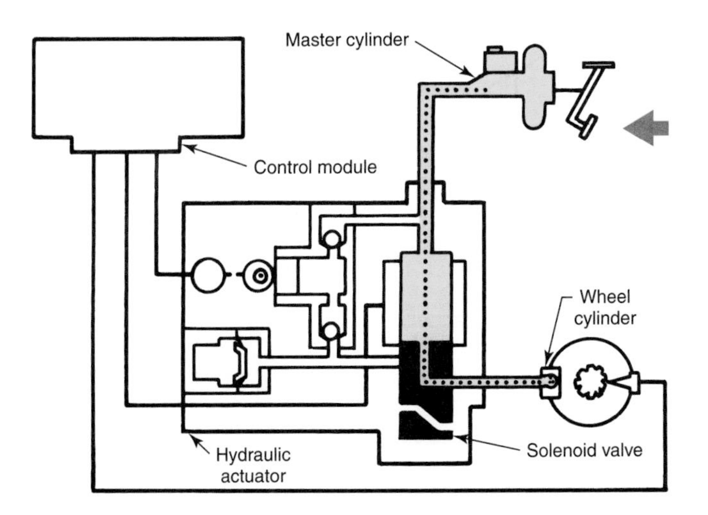

Figure 35-11. Anti-lock brake system operation during normal braking. Hydraulic fluid flows directly through the solenoid into the wheel cylinder.

When the control module senses that a wheel is nearing a lockup condition, it signals the solenoid valve in the hydraulic actuator to block the fluid passage between the master cylinder and the wheel cylinder. When this occurs, pressure is trapped between the wheel cylinder and the hydraulic actuator. Master cylinder fluid pressure cannot flow through the solenoid valve, and the brake pressure at the affected wheel is held constant, **Figure 35-12.**

If the control module detects a complete lockup, it commands the hydraulic actuator to decrease pressure to the affected wheel cylinder. To accomplish this, the solenoid valve in the actuator moves to cut off fluid pressure from the master cylinder and allow brake fluid at the caliper to flow into the accumulator, reservoir, or pump intake, **Figure 35-13.** When this occurs, pressure at the wheel is decreased.

On the piston-type hydraulic modulator, the solenoid valves isolate the circuit from the master cylinder and the pistons are moved to reduce pressure. **Figure 35-14** shows the solenoid and piston operation to control the front brakes.

When all the wheels are rotating normally, the solenoid valves in the actuator return to their original positions and the foundation braking system takes over. At the same time, the actuator pump delivers any excess fluid in the accumulator back to the master cylinder. If necessary, a typical anti-lock system can repeat this cycle up to 15 times a second.

Anti-Lock Brake System Maintenance

No specific ABS maintenance is needed. The wheel speed sensors should be periodically checked for damage. Check the toothed rotors for contamination with dirt, grease, or road debris. Check the level in the brake master cylinder as you would with a non-ABS brake system. Add fluid as necessary. If the fluid level seems excessively low, check for leaks in the system. Other brake system checks can be made according to the information in Chapter 34, Brake Service.

Note: Check the manufacturer's requirements before adding brake fluid. Many ABS systems are not designed to accept silicone brake fluid.

Troubleshooting Anti-Lock Brakes

Under hard braking, the brake pedal pulsates slightly. This is a normal condition when the anti-lock brake system is operating. If no pulsation is felt, the anti-lock system may not

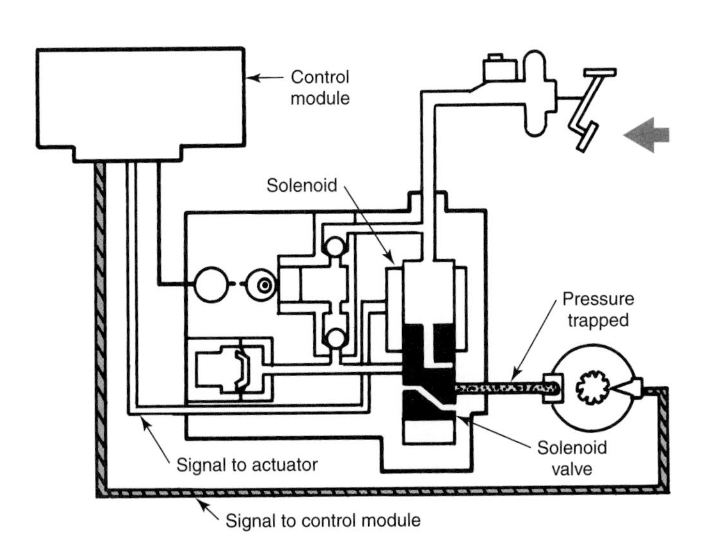

Figure 35-12. When the electronic control unit senses a potential wheel lockup, it instructs the solenoid to block hydraulic fluid to the wheel cylinder.

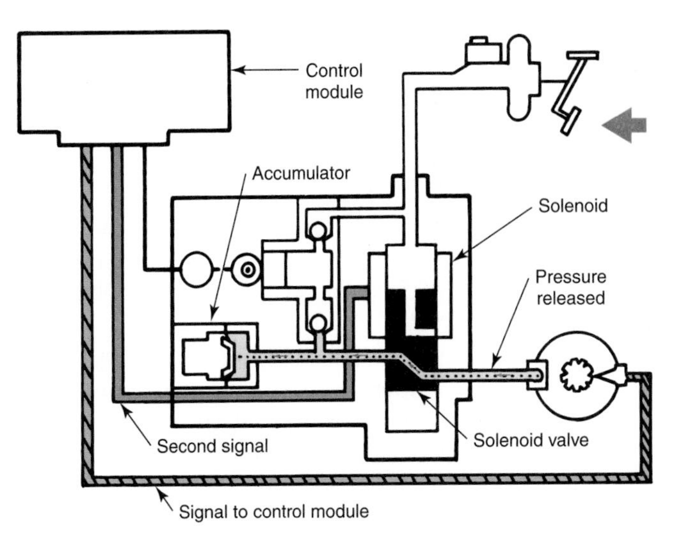

Figure 35-13. If a wheel locks up, the solenoid is instructed to release hydraulic pressure at the affected wheel. Note the fluid flowing back into the accumulator.

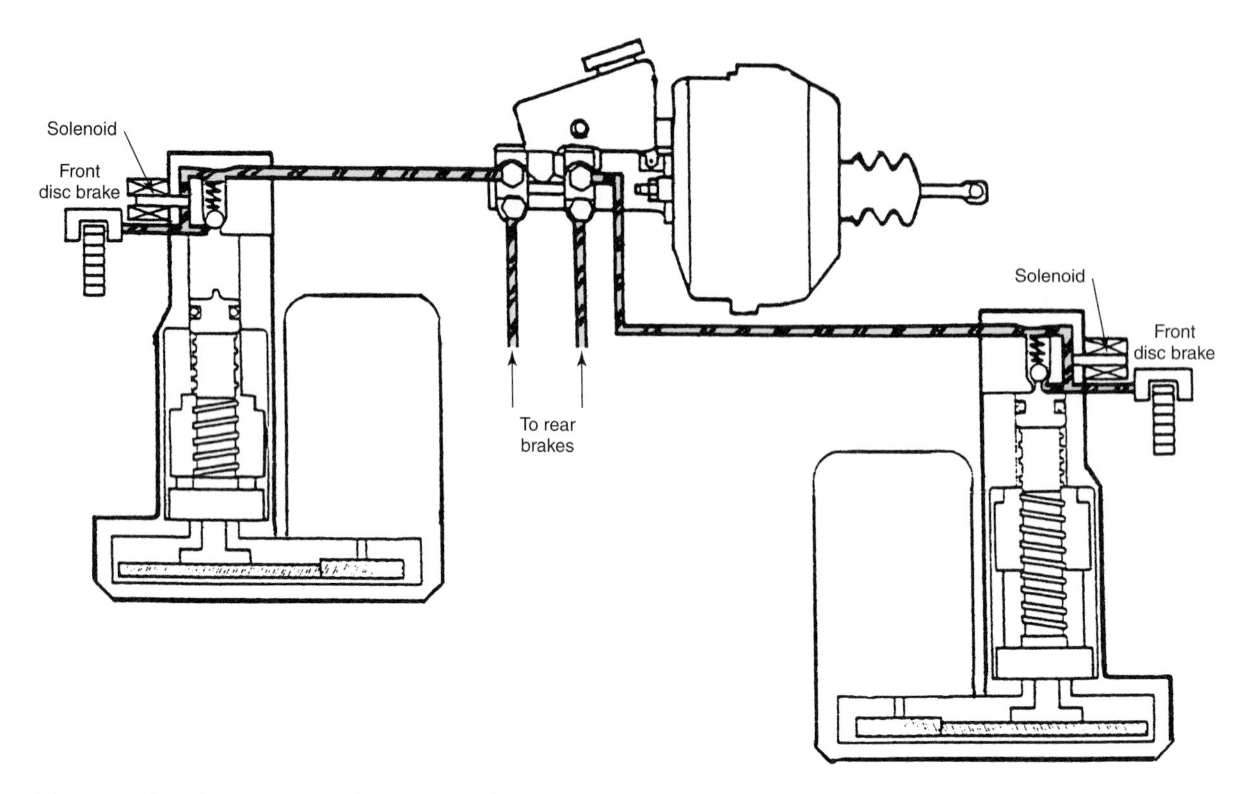

Figure 35-14. Anti-lock brake system decreasing hydraulic pressure to the front brake system to help prevent wheel lockup. The piston and the ball screw have moved down. This permits the check ball to seat, isolating fluid pressure from the front brake.

be operating. Another sign of trouble is the brakes locking up at speeds above 6 mph–10 mph (9.7 km/h–16.1 km/h). Below 6 mph–10 mph, the system is deactivated and lockup may occur under normal operating conditions. An obvious sign of trouble is the ABS warning light staying on after the vehicle has been operating for a few minutes, or the light flickering during vehicle operation. It is normal for the light to be on for several seconds after the vehicle is started.

Caution: Be sure to observe all traffic rules when testing the anti-lock brake system. Perform all ABS road testing in a safe area, away from other vehicles.

Making Preliminary Checks

Many preliminary checks can be made visually or with simple test equipment and hand tools. If ABS trouble is suspected, or if the ABS light is on, start by checking for the following problems:

- Low fluid level in the master cylinder—A low fluid level causes the ABS light to come on and may also trigger the red brake warning light. Add only brake fluid approved for the system being tested. Many ABS systems are not designed to accept silicone brake fluid.
- External fluid leaks—External leaks cause erratic pressure application and a low fluid level in the master cylinder. Leaks may also trigger the ABS light.
- Worn disc brake pads and rotors or worn and maladjusted shoes and drums—Excessive wear or maladjustment forces the hydraulic system to provide more fluid to the caliper piston or wheel cylinder, upsetting hydraulic pressure development.
- Problems in the power brake booster—A defect in the booster can cause a hard pedal or slow brake application, either of which can cause ABS problems.
- Stuck parking brake cable—A stuck parking brake cable causes partial application of the parking brakes, resulting in hydraulic system or sensor malfunction.
- Incorrect charging system voltage—Low voltage causes the ABS computer to operate incorrectly. Charging system service is covered in Chapter 14, Charging and Starting Systems.
- Blown fuses in the ABS input—There may be several fuses supplying power to various circuits in the ABS system, **Figure 35-15.**
- High resistance or disconnected control wiring—Check carefully at connections under the vehicle. Also check for disconnected or corroded ground wires.
- Defective relays—As shown in **Figure 35-16,** there may be several relays operated by or delivering power to the ABS system. Check that these relays are not loose or disconnected. Also check that the sockets are not overheated, corroded, or otherwise damaged.
- Mismatched tires—All tires on a vehicle equipped with ABS should be the same size and type. The tires should always be the same height as the original equipment.

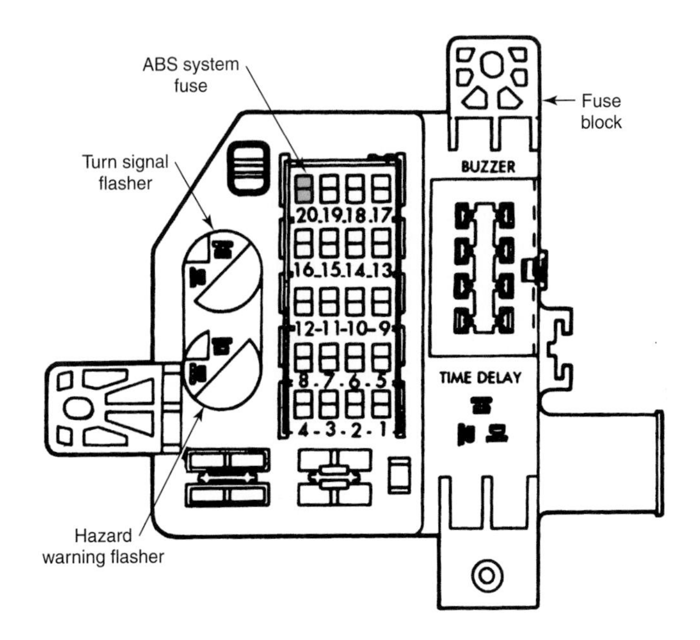

Figure 35-15. One particular fuse block that contains an anti-lock brake system fuse. (DaimlerChrysler)

Retrieving Trouble Codes

Most anti-lock brake systems have self-diagnostic capabilities. If no problems are uncovered during the preliminary checks, retrieve the trouble codes stored in the control module's memory. Recovery procedures are similar to those for the engine control computer. Refer to Chapter 15, Computer System Service, for detailed information on retrieving trouble codes.

In some anti-lock brake systems, trouble codes can be accessed by going through a series of code retrieval steps and watching a series of flashes from the dashboard-mounted ABS trouble light. Another retrieval process involves using a voltmeter to observe voltage pulses created by the control module in the self-diagnosis mode. Other systems require that a special retrieval tool or a breakout box be used to access the codes. On most late-model vehicles, a scan tool must be used to retrieve the codes. See **Figure 35-17.** Always consult the vehicle's service manual for the appropriate method. If necessary, the trouble codes should be compared to the appropriate trouble code chart to determine potential problems, **Figure 35-18.**

Checking Wheel Speed Sensors

Wheel speed sensors are a common source of ABS problems. They should be inspected for signs of physical damage. A buildup of dirt or grease between the rotor teeth can cause the control module to set a trouble code. Many manufacturers recommend checking wheel speed sensors with an ohmmeter to determine their resistance. If the reading does not fall within specifications, the sensor should be replaced.

If possible, check the gap between the sensor and the rotor teeth. A gap that is too wide or too small causes erratic readings. A typical gap-checking procedure is shown in

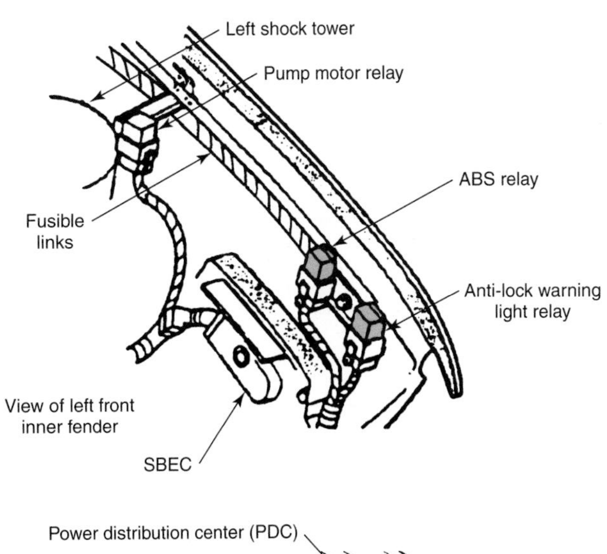

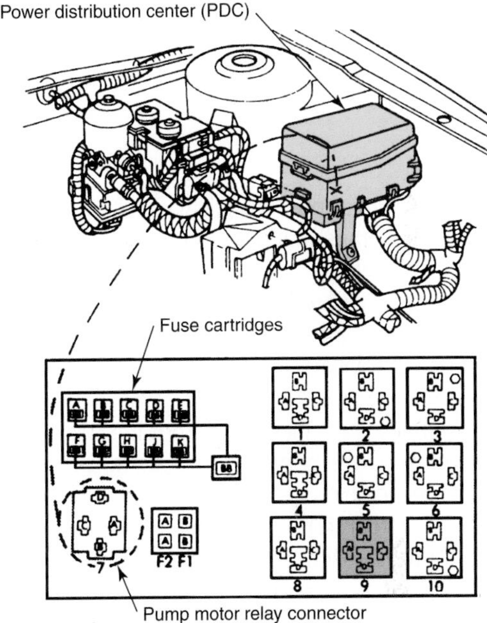

Figure 35-16. ABS relay locations. (DaimlerChrysler)

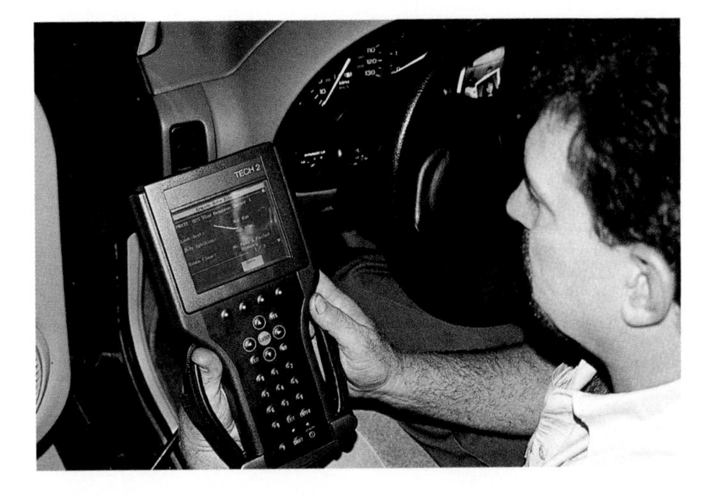

Figure 35-17. A scan tool is invaluable when checking the anti-lock brake system for trouble codes. This tool attaches to the data link connector (DLC). (General Motors)

Diagnostic Code (Sensor Check Mode)

Code No.	Diagnosis	Malfunctioning Area
Normal	All speed sensors and sensor rotors are normal.	
71	Low output voltage of front right speed sensor	• Front right speed sensor • Sensor installation
72	Low output voltage of front left speed sensor	• Front left speed sensor • Sensor installation
73	Low output voltage of rear right speed sensor	• Rear right speed sensor • Sensor installation
74	Low output voltage of rear left speed sensor	• Rear left speed sensor • Sensor installation
75	Abnormal fluctuation in output voltage of front right speed sensor	• Front right sensor rotor
76	Abnormal fluctuation in output voltage of front left speed sensor	• Front left sensor rotor
77	Abnormal fluctuation in output voltage of rear right speed sensor	• Rear right sensor rotor
78	Abnormal fluctuation in output voltage of rear left speed sensor	• Rear left sensor rotor

Figure 35-18. A trouble code chart for one specific anti-lock brake system. This chart only contains trouble code information for the wheel speed sensors. (Toyota)

Figure 35-19. While not specifically called for on some systems, using a brass feeler gauge results in a more accurate reading.

Some systems do not have gap specifications. When this is the case, make sure the sensor assembly is mounted in exactly the same position as the old sensor. Make sure that all fasteners are correctly torqued.

Always make sure the wiring from the sensor is routed correctly. A sensor wire passing too close to a source of heat or strong magnetic fields can cause the wiring to provide false information to the control module. Wires should, therefore, be kept away from the exhaust system and ignition wires. Also, make sure the wiring is retained by the proper clips to reduce the possibility of cutting a wire during turns or braking.

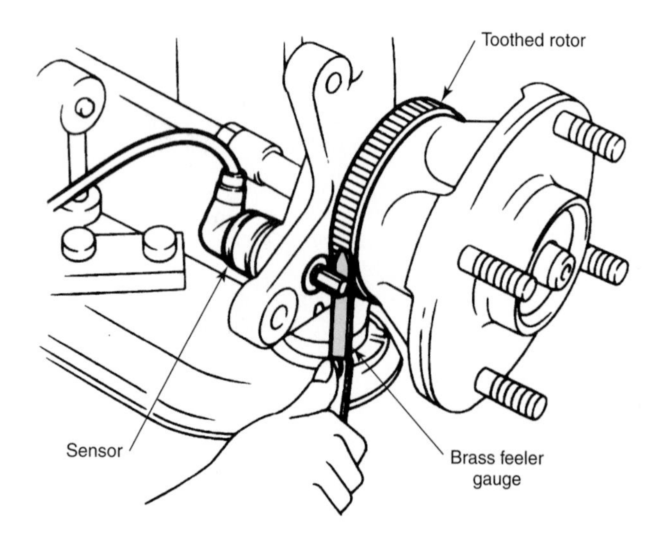

Figure 35-19. Checking the gap between the speed sensor and the toothed rotor. Set this gap (when adjustable) to the vehicle manufacturer's specifications. (Nissan)

Another cause of improper speed sensor readings is a physical problem at the wheel. A common problem is a warped rotor or drum caused by overtorquing the lug nuts. The sensor rotor may also become warped, causing an erratic signal to be produced by a good sensor. If the sensor rotors are overheated, dropped, or hammered into place on the hub, they may lose their magnetic properties. Loose wheel bearings, CV joints, or suspension and steering system parts can also cause erratic speed-sensor readings.

Checking Control Module

If a trouble code indicates the anti-lock control module is faulty, check the input voltage to the unit. If the control module does not receive sufficient voltage, it will not operate properly. Check the wiring to the control module. Make sure all connections are clean and secure. Verify the operation of all related anti-lock system components before condemning the control module.

Caution: Do not try to make ohmmeter checks to the control module unless they are specifically called for by the manufacturer. The electronic circuitry inside the module can be destroyed by ohmmeter current.

Some scan tools can operate the control module by inputting test signals. If the control modulator does not provide the proper outputs based on the inputs of the scan tool, it can be assumed the module is defective. Follow the scan tool manufacturer's instructions to perform these tests. In most cases, the internal components of the control module cannot be serviced. The module must be replaced if it is defective.

Checking Hydraulic Actuator

If a faulty hydraulic actuator is suspected, it should be tested according to the manufacturer's recommendations.

Some manufacturers specify ohmmeter tests to determine the condition of the solenoid valves and motor in the actuator. Another test, shown in **Figure 35-20,** checks for proper hydraulic pressures at the actuator under various conditions. Some hydraulic actuators can be disassembled to check for internal problems.

Replacing ABS Parts

Most ABS parts, such as the wheel speed sensors, control module, and hydraulic actuator, are not field repairable. They should be replaced if they are defective. Note that most of the hydraulic system and friction elements are similar to those on units without ABS. Refer to Chapter 34, Brake Service, for service procedures on all non-ABS components. Before replacing any part of the ABS system, refer to the cautions below.

- Make sure the ignition key is off before disconnecting any ABS electrical connector. Some manufacturers recommend removing the battery cable.
- Depressurize the brake system before beginning any repairs to the hydraulic system. Pumping the brake pedal a minimum of 40 times discharges the accumulator. Also make sure the ignition key is off, since the hydraulic actuator motor may attempt to recharge the accumulator if pressure is relieved while the key is on.
- If repair procedures require the replacement of any hydraulic hoses or lines, make sure the correct replacement lines are used. Hydraulic pressures are high in an ABS hydraulic system, and a standard brake hose may rupture.
- Be sure the ignition switch remains in the off position while bleeding ABS system brakes. The system may attempt to repressurize itself if the ignition is turned on during bleeding.

Replacing a Speed Sensor

To replace a speed sensor mounted on the axle, begin by removing the wheel and tire. Then remove the fasteners holding the sensor to the steering knuckle or axle housing. Pull the speed sensor out of the housing or knuckle and disconnect the electrical connector. See **Figure 35-21.** After reinstalling, be sure to recheck the sensor-to-rotor gap. Reinstall the wheel assembly and check ABS operation.

To replace a speed sensor mounted on the differential assembly or transmission, raise the vehicle and remove the fasteners holding the sensor to the transmission or differential, **Figure 35-22.** Pull the speed sensor out of the mounting bracket and disconnect the electrical connector. Install the new sensor, making sure that all wiring is clipped in place properly. Adjust the gap if possible and recheck ABS operation.

Replacing a Toothed Rotor

The toothed rotor is sometimes bolted to the rear of the brake rotor or axle assembly. In some cases, it is pressed on. Some rotors are not removable, and must be replaced by replacing the entire brake rotor or axle assembly. In most

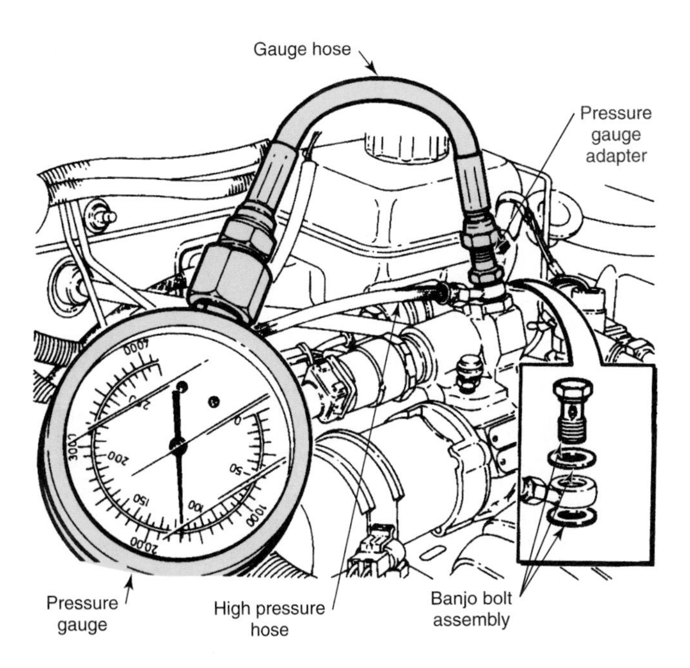

Figure 35-20. Checking hydraulic pressure. Use the recommended type of pressure gauge and follow the vehicle manufacturer's instructions. The system must be depressurized before attaching the gauge. Wear safety glasses.

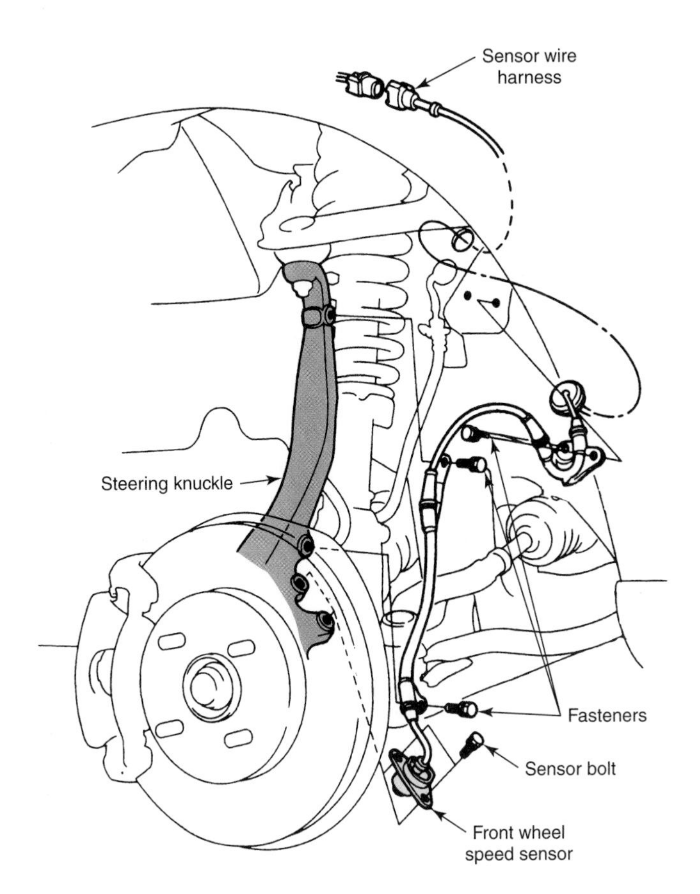

Figure 35-21. Removing a wheel speed sensor that is attached to the steering knuckle. (Honda)

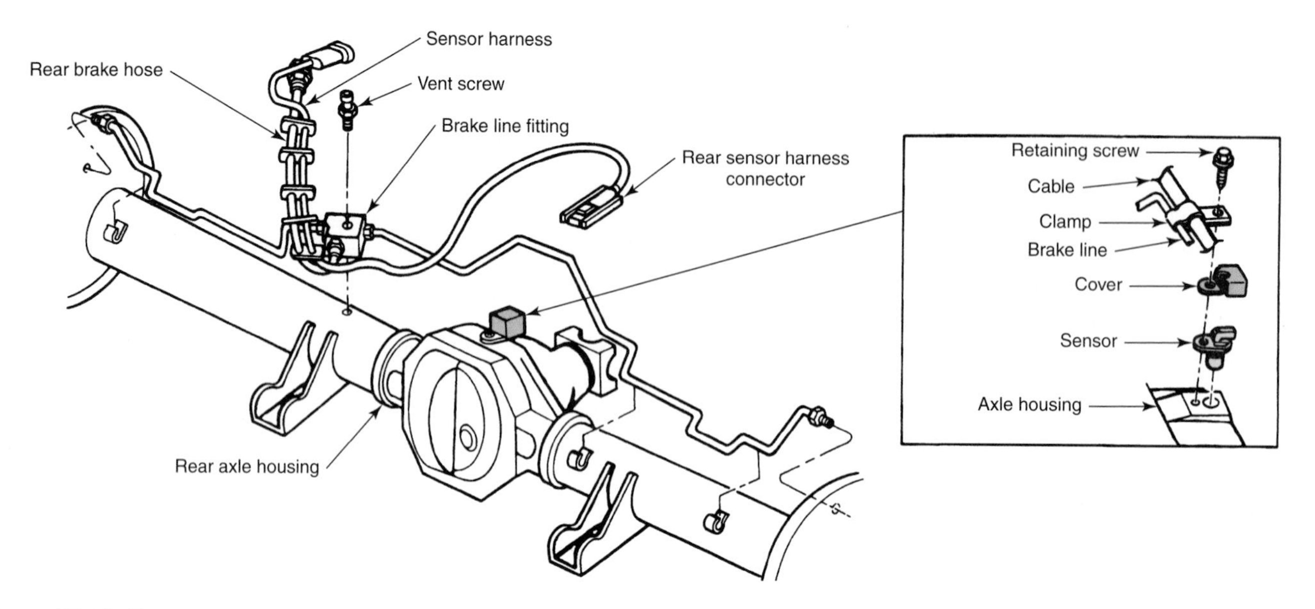

Figure 35-22. Removing a speed sensor from the rear axle housing. Clean around the sensor before removing to prevent dirt from entering the axle housing. (DaimlerChrysler)

cases, the brake rotor or axle must be removed to gain access to the toothed rotor. Follow the manufacturer's directions.

Caution: The replacement rotor must never be hammered into position or dropped. Hammering the rotor may slightly magnetize it, which would result in sensor malfunction.

Replacing a G-Force Sensor

The G-force sensor is usually mounted in the passenger compartment, **Figure 35-23.** To remove the sensor, remove any covering trim or carpet and remove the wiring harness. Loosen the bracket fasteners and remove the sensor. Reverse the removal process to install the new sensor.

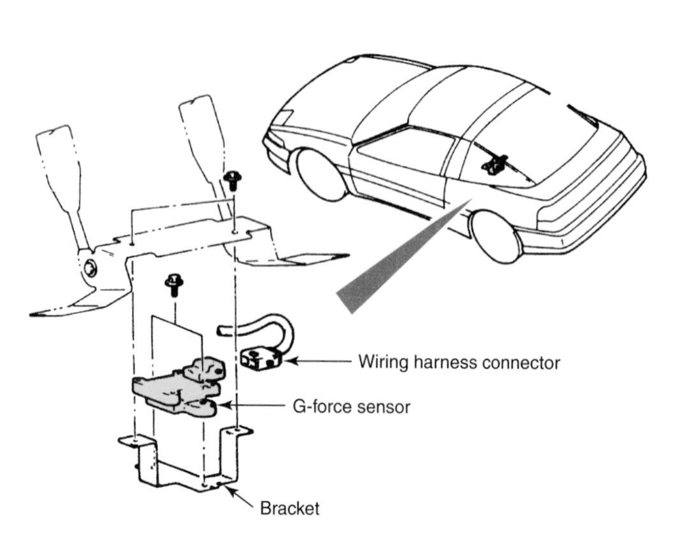

Figure 35-23. A typical G-force sensor and its location inside one particular vehicle. (General Motors)

Caution: The G-force sensor measures vehicle tilt to determine G-forces. The replacement sensor must be mounted in exactly the same position as the old sensor, or the readings will be inaccurate.

Replacing a Control Module

Make sure the ignition switch is off or the battery is disconnected as applicable. Then, remove the control module electrical connectors. Remove the module bracket fasteners and remove the module from the vehicle. The location of a typical module is shown in **Figure 35-24.**

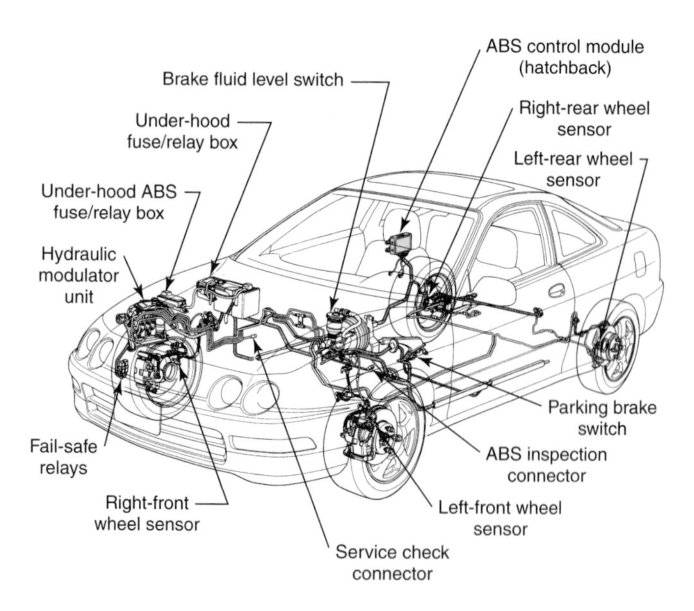

Figure 35-24. Location of the control module in a specific vehicle. Module location varies from vehicle to vehicle. (Honda)

Note: A few modules are attached to the hydraulic actuator and master cylinder. The entire assembly must be removed to gain access to the module.

Install the new module and tighten the bracket fasteners. Reconnect the electrical connectors and check ABS operation.

Replacing Hydraulic System Components

Hydraulic system components include the hydraulic actuator, pump and motor, and accumulator. These components may be housed in a single unit, or they may be separate units connected by high-pressure hoses. **Figures 35-25** and **35-26** show some typical hydraulic system components and locations. Due to the variations in hydraulic system design, only general service procedures are outlined below.

Begin the service of hydraulic system components by depressurizing the hydraulic system according to manufacturer's instructions. If necessary, remove any electrical connections; then remove the defective component. Many of the ABS hydraulic components are attached to the master cylinder. Therefore, the master cylinder may have to be removed. If the defective component must be disassembled, follow the manufacturer's instructions carefully. Clean all parts in the recommended cleaner and lay them out on a clean workbench before beginning reassembly.

Begin reassembly by rebuilding all subassemblies. If a repair kit is used to overhaul a specific component or its subassembly, use all the new parts, even if the old ones seem good. This is especially true of seals and gaskets. Lubricate all new seals with clean brake fluid before installation. Make all possible bench tests before reinstallation to ensure the component will operate properly on the vehicle.

Install the new or overhauled component. Use new gaskets and seals as needed. Tighten all fasteners to the correct torque and use new fasteners where called for by the manufacturer. Reattach the electrical connectors and bleed the brakes according to the manufacturer's instructions.

Note: Do not turn the ignition switch on until the bleeding procedure is completed.

Start the vehicle and check ABS operation. The ABS warning light may remain on for a few minutes, until the system has repressurized. After correct operation has been verified, recheck the hydraulic system for leaks and sufficient reservoir level.

Traction Control Systems

To reduce wheel spin when accelerating on slippery surfaces, some vehicles are equipped with ***traction control systems.*** These systems reduce engine power and operate

Figure 35-25. A hydraulic actuator and its related components. (DaimlerChrysler)

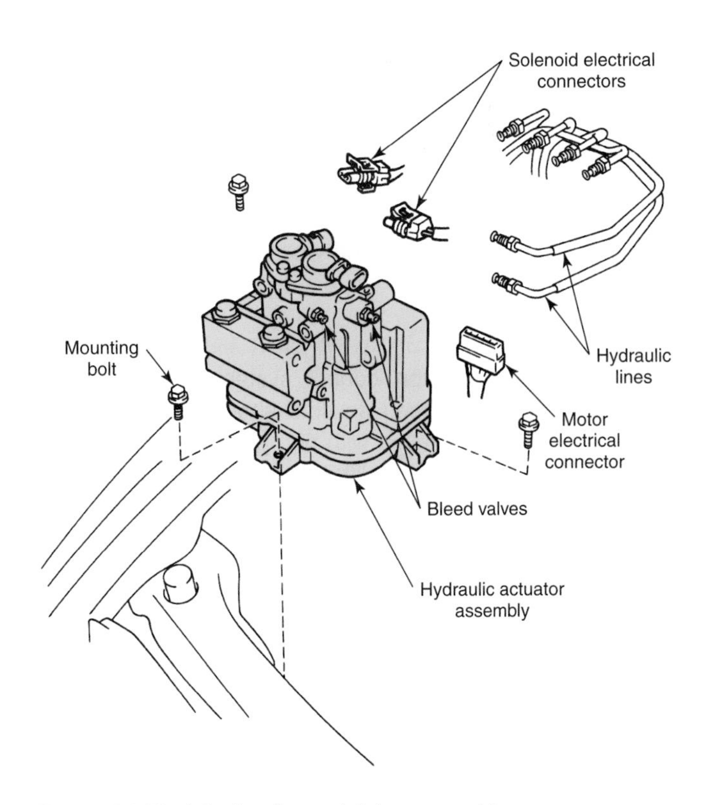

Figure 35-26. A hydraulic modulator assembly. (General Motors)

the brake system to increase vehicle acceleration and stability on low-friction or uneven road surfaces, **Figure 35-27.** Traction control systems also provide higher levels of cornering performance.

The traction control system has the ability to apply the brakes on the drive axles. On a two-wheel drive system, it controls only the two driving wheels. On a full-time four-wheel drive system, it can apply any one of the four brakes. If the system detects one drive wheel spinning at a faster rate than the others, it applies the appropriate amount of braking force to slow the wheel to the correct speed.

If the system determines that both (or all) drive wheels are spinning excessively, it can close the throttle or briefly retard ignition timing to prevent further spinning. Most vehicles with traction control also have an anti-lock brake system. On many new vehicles, the anti-lock brake system and the traction control system are controlled by a single electronic control module. A system with one electronic module may have one or two hydraulic actuators. See **Figure 35-28.**

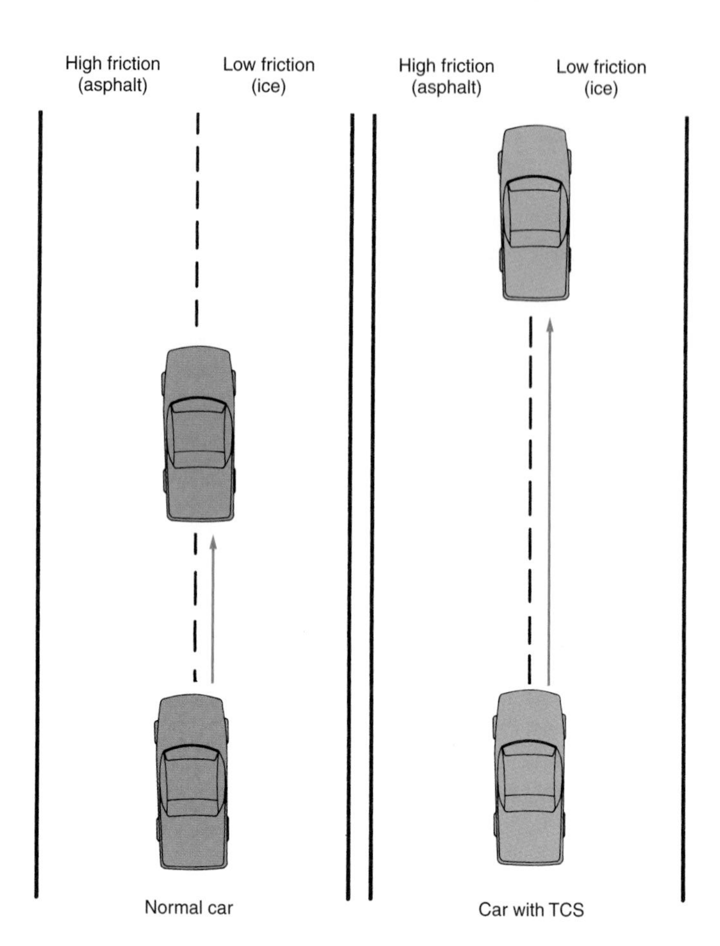

Figure 35-27. Vehicles equipped with a traction control system accelerate efficiently on low-friction surfaces.

Traction Control System Components

Most traction control systems have several common components. Typical systems consist of the following parts:

- *Wheel speed sensors*—Monitor wheel speed and send signals to the control module. Often, they are the same sensors used in the anti-lock brake system.
- *Control module*—Receives signals from the sensors and decides on outputs to control the hydraulic brake actuators and throttle motor or ignition timing control.
- *Throttle motor*—Moves the throttle plate as instructed by the control module.
- *Hydraulic actuator*—Varies the amount of brake pressure at the drive wheels based on signals from the control module.
- *Indicator lamp*—Alerts the driver when the traction control system is operating. On some systems, there are separate lights to indicate when the traction control system is operating and when a system malfunction has occurred.

Traction Control System Maintenance

No specific traction control system maintenance is needed. The wheel speed sensors should be checked for damage and contamination with dirt, grease, or road debris whenever the vehicle is on a lift. The engine control section of the system should only be checked if trouble develops. Periodically check the fluid level in the brake master cylinder.

Troubleshooting Traction Control Systems

Most traction control systems are equipped with self-diagnostic capabilities. Methods for retrieving trouble codes for these systems vary from manufacturer to manufacturer. Refer to the appropriate service manual for proper diagnostic and service techniques.

Replacing Traction Control System Components

Most traction control components are similar to anti-lock brake components, as are the procedures for their replacement. Refer to the engine section of the proper service manual for engine component replacement.

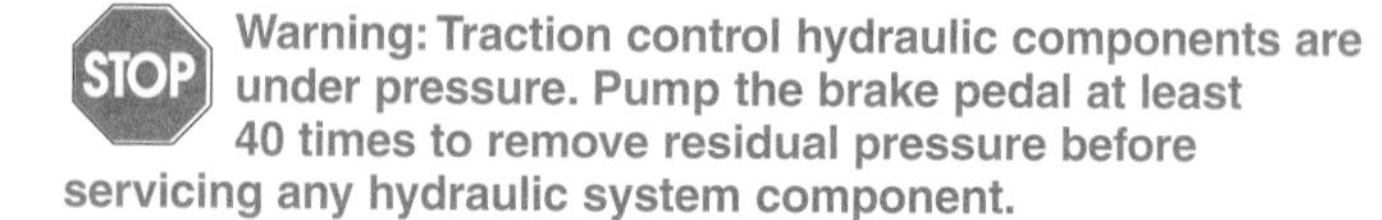

Warning: Traction control hydraulic components are under pressure. Pump the brake pedal at least 40 times to remove residual pressure before servicing any hydraulic system component.

Foundation Brake Service on Vehicles Equipped with ABS or Traction Control.

Many common foundation brake service procedures, such as pad and shoe replacement, rotor and drum service, and wheel bearing replacement, are not affected by the presence of ABS or traction controls. If the replacement procedures involve the wheel speed sensors, treat them gently and

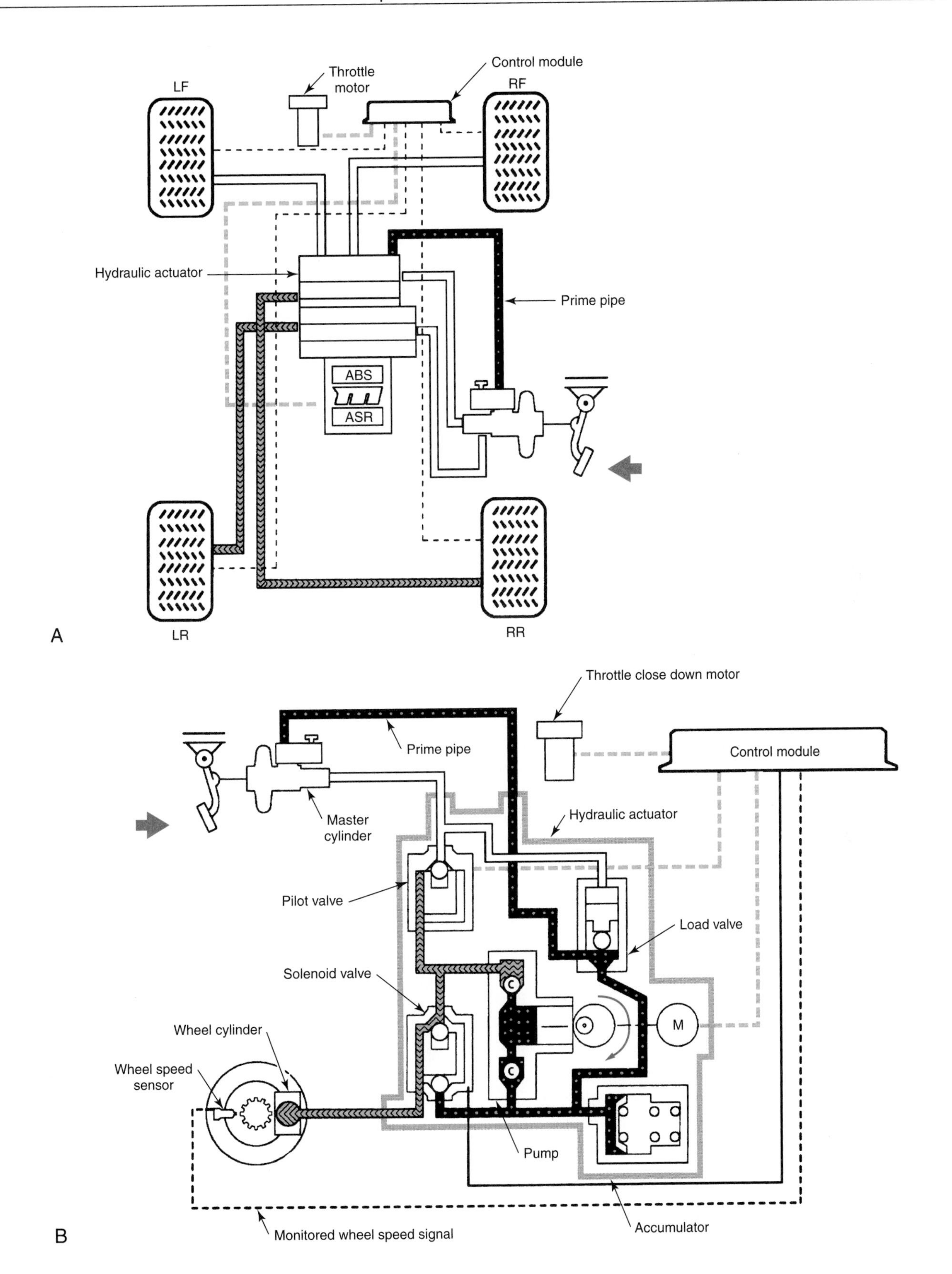

Figure 35-28. A—Traction control system schematic. This particular system is called an automatic slip-regulation (ASR) system. B—Traction control operation schematic. Note that both the anti-lock brake system and the traction control system are governed by the same control module. (General Motors)

recheck the gap where applicable. Remember not to drop or hammer on the sensor rings. Never use the sensor rings to pry on other components.

Hydraulic components on ABS and traction control vehicles are serviced in the same manner as conventional systems. However, all replacement parts must be able to withstand higher pressures than standard components. Make sure that all hoses, lines, fittings, rebuilt components, and wheel cylinder and caliper kits are designed for the system on which you are working.

Make sure the hydraulic system has been depressurized before disconnecting any hydraulic system part. Also, make sure the ignition switch remains off until the hydraulic system is completely reassembled and bled.

Bleeding an Anti-Lock Brake System

Bleeding procedures for anti-lock brake and traction-control systems can vary greatly from the procedures used on standard brake systems. Proper bleeding will remove all air from the hydraulic actuator assembly, including the pump, solenoids, accumulator, and lines. Air must also be removed from the foundation brake parts, including the master cylinder, calipers, and wheel cylinders, as well as the proportioning/metering valve and the pressure differential valve. If the entire system is not bled properly, air may be trapped in the system. This can cause a spongy brake pedal and erratic system operation.

Exact bleeding procedures vary from manufacturer to manufacturer. Some systems can be manually bled. This procedure is similar to the bleeding procedure for standard brake systems. Note that it may be necessary to open lines at the hydraulic actuator or accumulator to remove additional trapped air. After the ABS and/or TCS components are bled, bleed the foundation brakes to ensure that all air has been removed from the system.

STOP **Warning: Follow the bleeding procedure outlined in the service manual or other literature for the system at hand. Failure to follow the proper bleed procedure may result in air becoming trapped in the hydraulic actuator. This trapped air may not become apparent until the first time the ABS system is activated.**

As part of the bleeding procedure for some ABS systems, a scan tool is used to manually operate the hydraulic actuator. If this step is not performed on these systems, a low brake pedal or improper ABS operation may result. Begin by manually bleeding the master cylinder and the lines to the actuator. Follow the manufacturer's recommendations for manually bleeding the actuator. Then, install the scan tool and follow the manufacturer's directions to energize the actuator. As the actuator operates, it will purge itself of air by venting it to the brake calipers, the wheel cylinders, or the fluid reservoir. Repeat as needed to purge all air from the system.

Once the ABS and traction control system circuits have been bled in this manner, the foundation brake system can be pressure or manually bled as specified by the manufacturer.

Caution: While bleeding the foundation brakes, be sure to keep the master cylinder filled to avoid introducing air back into the ABS/TCS components.

Tech Talk

In earlier chapters, you learned to check all non-electronic engine components before investigating a problem in the electronic control unit or other electronic components. The same procedures apply to ABS systems as well. Always keep in mind the basic principles of brake system operation are the same for vehicles with ABS as they are for non-ABS vehicles. The ABS system only operates during very hard braking, and does not affect the normal operation of the wheel brake assemblies. Worn out friction materials or hydraulic leaks cause the same problems on ABS and non-ABS systems.

When beginning to diagnose an ABS problem, make sure that all of the non-ABS components are in good condition. Check the condition of all brake pads, shoes, rotors, and drums. Check the brake calipers and wheel cylinders for leaks or binding, and, if they are used, check the metering, proportioning, and pressure differential valves for proper operation. The only time these steps may be bypassed is when ABS trouble codes indicate a definite problem in the ABS system itself. If your checks reveal the non-ABS brake components are not the source of the problem, only then should you proceed to check the ABS components.

Summary

Anti-lock brake systems use electronic and hydraulic components to help prevent wheel lockup during hard braking. The anti-lock brake systems used on most automobiles control all four wheels. Most light truck ABS systems operate only the rear wheels. The basic principle of all ABS systems is that they control the brake hydraulic system to pulse the brakes on and off, preventing wheel lockup. The components and operation of all anti-lock brake systems are similar, but enough differences exist to make it necessary to carefully read the manufacturer's service manual.

Common ABS components include the wheel speed sensors, control module, and hydraulic actuator. Other components that can be used are the G-force sensor and brake-travel switch. Most anti-lock brake systems have an amber warning light, in addition to the red brake warning light used on all vehicles.

ABS troubleshooting may involve the use of special testers. However, many ABS checks can be made with standard tools and test equipment. Most anti-lock brake service can be performed without special tools. Due to the high pressures created by anti-lock brake systems, safety precautions should always be followed. Always depressurize the hydraulic system before doing any work on the ABS hydraulic components.

Traction control systems increase vehicle acceleration and stability on slippery or uneven road surfaces. They reduce engine power output and apply the brakes on the drive wheels to maximize traction.

Normal brake service operations, such as pad or shoe replacement and caliper or wheel cylinder rebuilding, are similar to those on vehicles without ABS or traction controls. However, the system must be depressurized before doing any work on the hydraulic system.

Review Questions—Chapter 35

1. ABS systems used on most automobiles are used to control lockup on _____ _____ wheels.
   (A) the front
   (B) the rear
   (C) all four
   (D) All of the above, depending on manufacturer.
2. Name the three main components used by both the anti-lock braking and traction control systems.
3. Most anti-lock brake systems have _____ capabilities.
4. Pulsation in the brake pedal during ABS operation is a sign of _____.
   (A) low fluid level
   (B) a defective sensor
   (C) a defective actuator
   (D) normal system operation
5. On a piston-operated hydraulic actuator (hydraulic modulator), the pistons take the place of the _____.
   (A) hydraulic pump
   (B) accumulator
   (C) solenoids
   (D) Both A and B.
6. If an ABS equipped vehicle has a pull to one side when braking, which of the following is the most likely cause?
   (A) Hydraulic modulator.
   (B) Sticking caliper.
   (C) Restricted brake hose.
   (D) All of the above.
7. When the wheel lockup condition is severe, the solenoids in the ABS hydraulic actuator will be positioned to dump small amounts of hydraulic fluid back to the _____.
   (A) reservoir
   (B) accumulator
   (C) hydraulic pump intake
   (D) All of the above, depending on the manufacturer.
8. Name ten things that should be checked before retrieving the ABS system trouble codes.
9. When bleeding the brakes on an ABS system, do not turn the _____ on until bleeding is completed.
10. Most brake service procedures for ABS and traction control brake systems are _____ to normal brake system service.

ASE-Type Questions

1. ABS wheel speed sensors can be installed at all of the following places on the vehicle, *except:*
   (A) flywheel.
   (B) brake rotor.
   (C) transmission.
   (D) differential.
2. Technician A says the amber ABS warning light illuminates only when the system is malfunctioning. Technician B says the traction control system may have two indicator lights. Who is right?
   (A) A only.
   (B) B only.
   (C) Both A and B.
   (D) Neither A nor B.
3. Technician A says some scan tools can read ABS trouble codes from the control module. Technician B says some scan tools can send input signals to the control module to test for the correct outputs. Who is right?
   (A) A only.
   (B) B only.
   (C) Both A and B.
   (D) Neither A nor B.
4. On many ABS systems, a scan tool can be used to _____.
   (A) check voltages
   (B) check hydraulic pressures
   (C) retrieve trouble codes
   (D) bleed the hydraulic system
5. Many manufacturers recommend checking wheel speed sensors with _____.
   (A) a voltmeter
   (B) an ohmmeter
   (C) a test light
   (D) All of the above.
6. All of the following can cause the ABS controller to set a trouble code, *except:*
   (A) leaking accumulator.
   (B) faulty wheel speed sensor.
   (C) low fluid level.
   (D) an out-of-round drum.
7. An ABS-equipped vehicle has a pedal pulsation during light braking. Technician A says the pedal pulsation means the ABS system is working normally. Technician B says the pedal pulsations during light braking indicates the ABS system is defective. Who is right?
   (A) A only.
   (B) B only.
   (C) Both A and B.
   (D) Neither A nor B.

8. Technician A says the ABS accumulator must be depressurized before the electronic control unit can be replaced. Technician B says static electricity can ruin an electronic control unit. Who is right?
   (A) A only.
   (B) B only.
   (C) Both A and B.
   (D) Neither A nor B.
9. Pumping the brake pedal at least _____ times discharges the accumulator.
   (A) 20
   (B) 40
   (C) 60
   (D) 80
10. In operation, the traction control system controls the operation of the brake hydraulic system and the _____.
   (A) engine
   (B) transmission and torque converter
   (C) positraction differential
   (D) four-wheel drive system

Suggested Activities

1. Draw a sketch showing the relationship of the ABS hydraulic and electronic control components to the base brake system. Explain how the ABS hydraulic system operates to control the base brake hydraulic system. Also explain how the electronic components control the operation of the hydraulic components.
2. Locate a vehicle with an anti-lock brake system and study the ABS code retrieval procedure and troubleshooting charts in the appropriate manufacturer's service manual. Retrieve and list any ABS trouble codes that may be stored based on the information obtained.
3. Check the sensor-to-wheel gaps of the wheel speed sensors. If available, check a vehicle that has a drive train speed sensor. Perform other tests listed in the service manual.
4. Bleed an anti-lock brake system according to manufacturer's specifications.

36

Suspension System Service

After studying this chapter, you will be able to:

- Explain the construction, operation, and service of conventional front suspensions.
- Explain the construction, operation, and service of conventional rear suspensions.
- Describe the function of coil springs, torsion bars, and leaf springs.
- Describe the function of load-carrying and following ball joints.
- Explain the construction, operation, and service of MacPherson strut suspensions.
- Describe the function of control arms, strut rods, and stabilizer bars.
- Describe the function of shock absorbers and MacPherson strut dampers.
- Summarize the operating principles and service of front and rear suspensions.
- Diagnose problems in suspension systems.

Technical Terms

Conventional suspension	Tension-loading
MacPherson strut suspension	Compression-loading
Coil springs	Control arm
Torsion bars	Control arm bushings
Spring compressor	Strut rods
Shock absorbers	Stabilizer bar bushings
MacPherson strut dampers	Leaf spring
Gas charged	Automatic level control
Ball joints	Height control sensor
Load-carrying ball joint	Height control valve
Preloaded	Curb height
	Active suspensions

This chapter is intended to familiarize you with the various types of front and rear suspension systems found on modern vehicles. The variety of suspension designs is great, but all have a common purpose of absorbing shocks while keeping the wheels in contact with the road surface. All suspension systems contain many of the same components, such as springs, control arms, and ball joints. Some systems have strut rods, stabilizer (sway) bars, shock absorbers, or MacPherson struts.

Warning: All suspension systems contain parts under spring tension. Be sure to identify all parts under spring tension and remove this tension before disassembling any part. Failure to do so can result in part or equipment damage and/or a severe or even fatal injury.

Front Suspension Systems

This section covers the service of various components on the two major types of front suspension systems, the ***conventional suspension*** and the ***MacPherson strut suspension.*** Differences and similarities between the two types are called out.

Conventional Suspensions

Modern conventional front suspension systems are equipped with ***coil springs*** or ***torsion bars.*** Most coil springs are mounted between the vehicle frame and the lower control arm, as shown in **Figure 36-1.** Some older vehicles, however, used coil springs that were mounted between the upper control arm and vehicle body.

A torsion bar is a long spring steel rod, which takes the place of a conventional spring. An example of a torsion bar front suspension system is shown in **Figure 36-2.** When the lower arm moves upward, it twists the torsion bar. Another torsion bar arrangement, called transverse mounting, is illustrated in **Figure 36-3.** The bars go across the width of the chassis and back to the control arms.

Front Coil Spring Service

Raise the vehicle and place jack stands under the frame. Remove the wheel and tire assembly. Disconnect the stabilizer bar and remove the shock absorber. Disconnect the lower arm tie strut, if used. Attach a safety chain. Place a jack under the lower control arm and align it with the arm. This allows the jack to roll, following the movement of the control arm's free end. Separate the lower ball joint stud from the spindle body as recommended in the ball joint removal section of this chapter. Lower the jack and control arm until pressure is

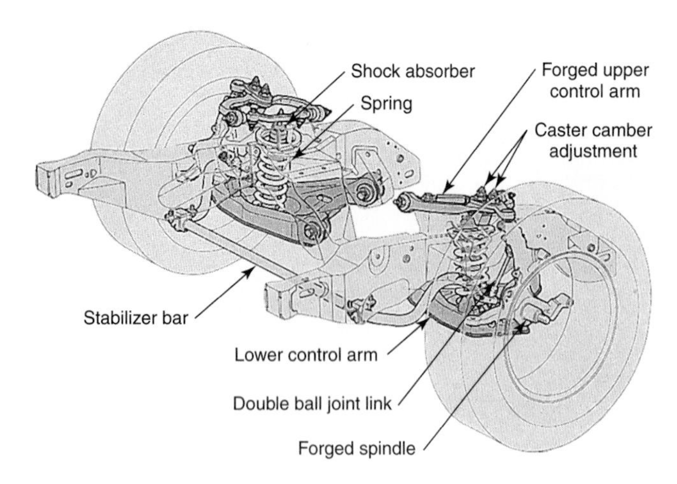

Figure 36-1. This is a conventional front suspension used on a rear-wheel drive car. (Ford)

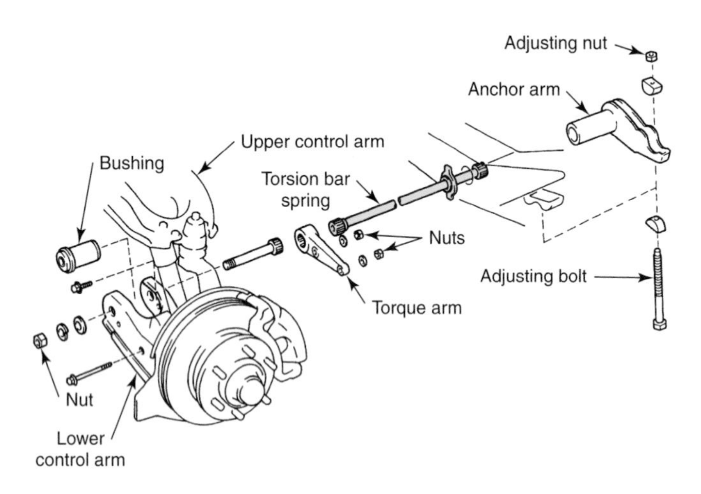

Figure 36-2. A torsion bar front suspension setup is shown here. This is how most torsion bar suspensions are arranged. (Toyota)

removed from the spring. Remove the spring carefully. If it is slightly loaded (under pressure), it could snap out violently. See **Figure 36-4.**

To install a spring, reverse the above procedure. Make certain that any spring insulators used are in place and the spring is correctly installed. Specifications usually call for the spring end to be in a certain location. **Figure 36-5** illustrates the correct spring end location for one vehicle. When the control arm is raised, make sure the spring does not rotate out of position. Raise the arm until the ball joint stud passes through the spindle body. You may have to guide the spindle onto the ball joint stud. Install the stud nut, torque, and insert a cotter pin. Install the shock absorber, stabilizer bar, and tie strut where used. Use extreme care when removing and installing springs. Carelessness can result in a serious accident.

Spring Compressors

Some coil springs cannot be removed or installed without the use of a ***spring compressor.*** The compressor is inserted into the spring, and then the draw bolt is tightened. This pulls

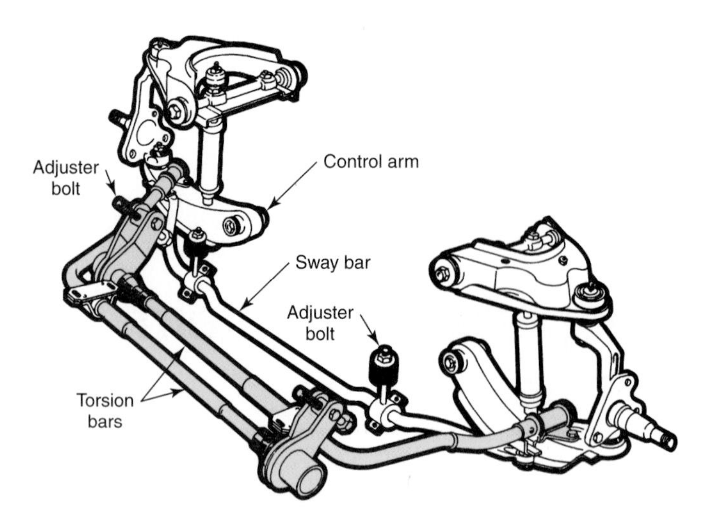

Figure 36-3. These torsion bars are arranged in a transverse manner. Note that each bar has its own adjustment bolt. (Moog)

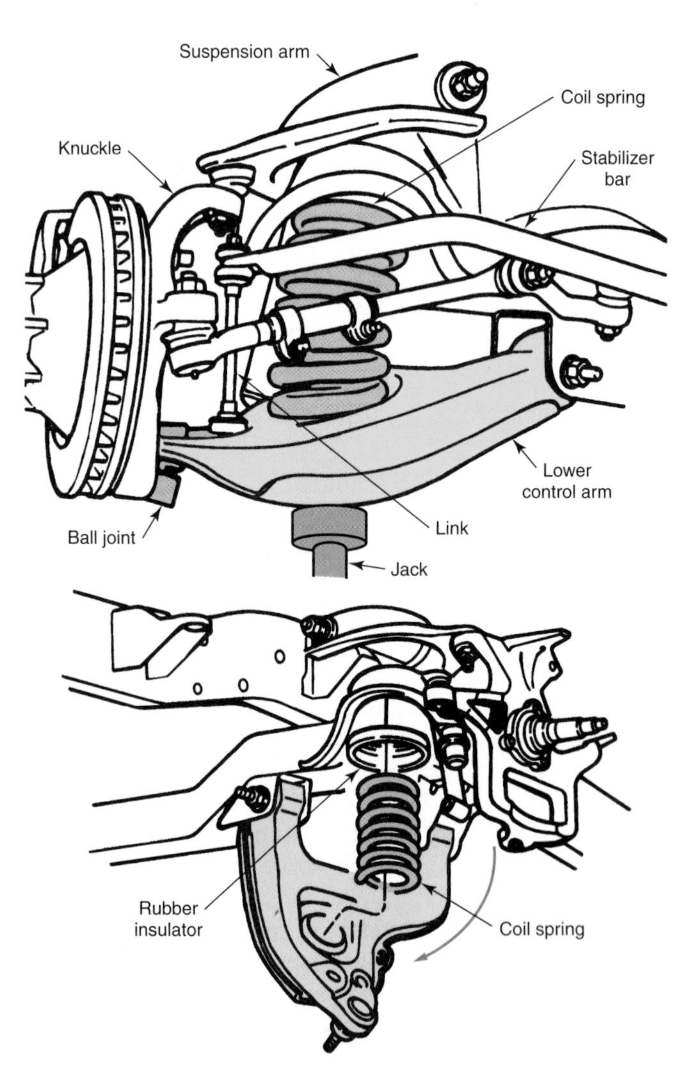

Figure 36-4. In this illustration, a lower control arm, still under coil spring tension, is being lowered with a jack. Be careful during this procedure! (DaimlerChrysler)

the coils together to both shorten and unload the spring, making removal possible. **Figure 36-6** illustrates one type of spring compressor in position. Make sure the compressor is installed correctly and the draw bolt is secure.

Torsion Bar Service

To service a torsion bar, first raise the vehicle by the frame until the front control arms are in the full rebound (down) position. On some vehicles, the rubber rebound bumper must be removed to permit the upper arm to reach the full rebound position. Turn the torsion bar adjusting bolt to unload the torsion bar. Mark the torsion bars to indicate the right and left side in order to avoid confusion and to make installation easier. **Figure 36-7** illustrates a typical adjustable torsion bar rear support. A torsion bar, rear support, upper and lower control arms, and strut rod are shown in **Figure 36-8.**

A detailed view of one method of attaching the torsion bar to the lower control arm is shown in **Figure 36-9.** Disengage the lock ring and plug from the rear anchor or support. Clean the torsion bar ahead of the seal, **Figure 36-10.** Disengage the seal from the anchor and carefully slide the seal partially up the bar. If necessary, remove the front torsion bar clip. Then slide the bar backward far enough to disengage the front hex from the control arm. Pull the bar to one side and then remove from the front. Some installations permit sliding the bar backwards out through the rear anchor.

If the bar is stuck in the anchors, use a special clamp-on striking pad to avoid denting or nicking the surface. The pad is clamped to the bar and struck with a hammer to loosen the bar. Never use heat on the bar or on the front and rear anchors. Do not hammer on the bar or use a pipe wrench. Clean and inspect the bar. Remove any small nicks or burrs by filing and polishing with fine emery cloth. Paint the polished sections with rust-resistant primer. Like all springs, a torsion bar can crack if it is nicked. Clean, inspect, and lubricate the rear anchor swivel and adjusting bolt. Replace the parts if needed. Clean the front and rear anchor hex holes.

Make sure the bars are installed on their designated sides. The bar must be clean before installation. Slide the bar through the rear anchor and slip a new seal over the bar. Coat both ends of the torsion bar with multipurpose grease. Insert the bar front hex end into the lower control arm. Install the rear anchor lock ring and plug. Fill the opening around the front of

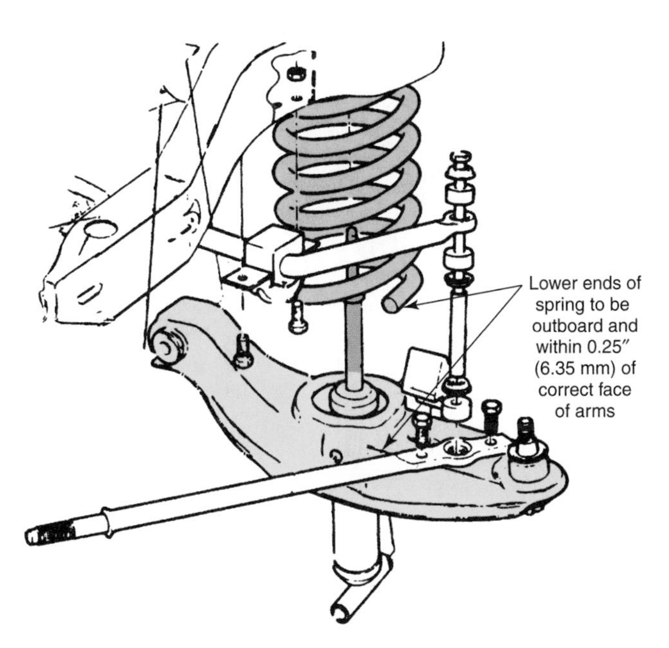

Figure 36-5. This drawing shows the correct spring end location for one vehicle. (General Motors)

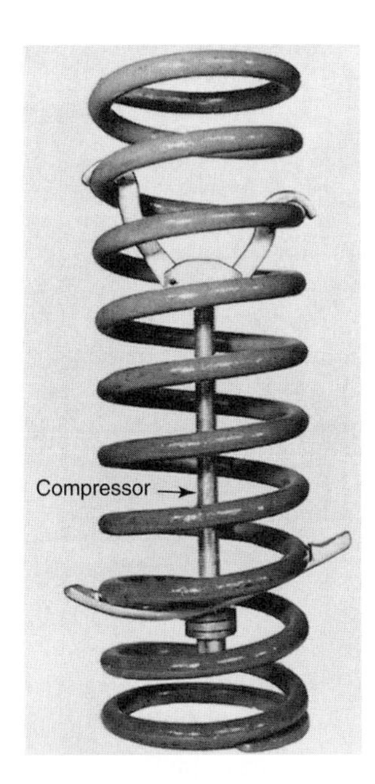

Figure 36-6. The proper positioning for one type of spring compressor is shown here. Always use care when compressing any spring. (Branich)

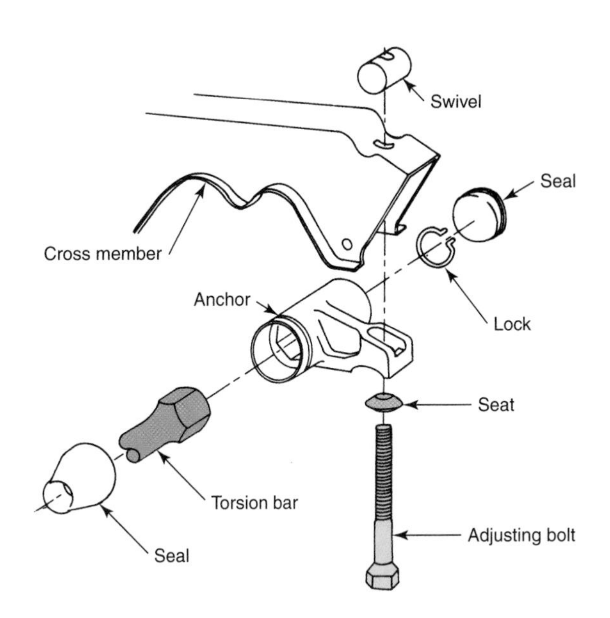

Figure 36-7. This is one type of torsion bar rear support. (DaimlerChrysler)

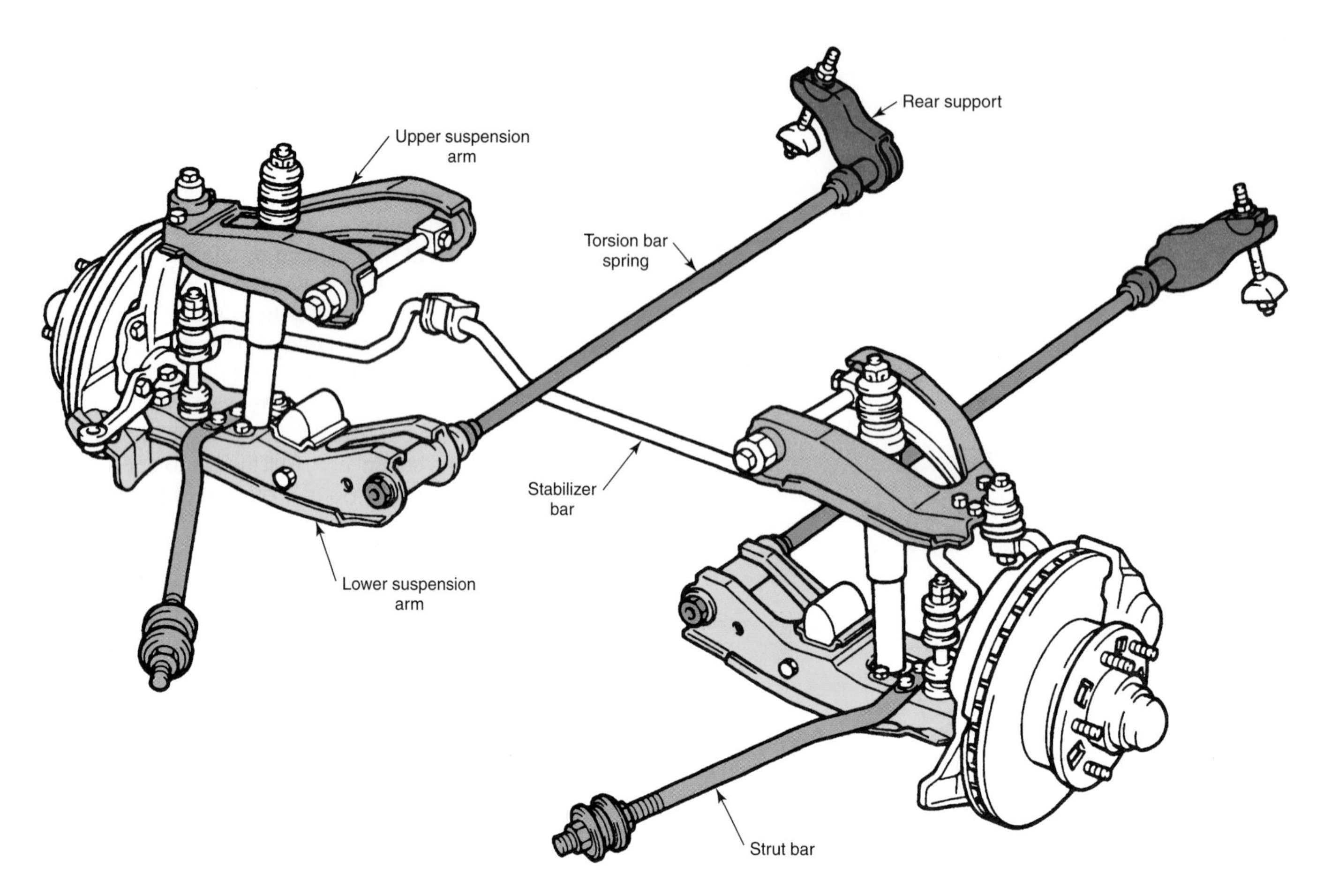

Figure 36-8. A torsion bar arrangement. Note that the torsion bars are connected to the lower control arm. (Toyota)

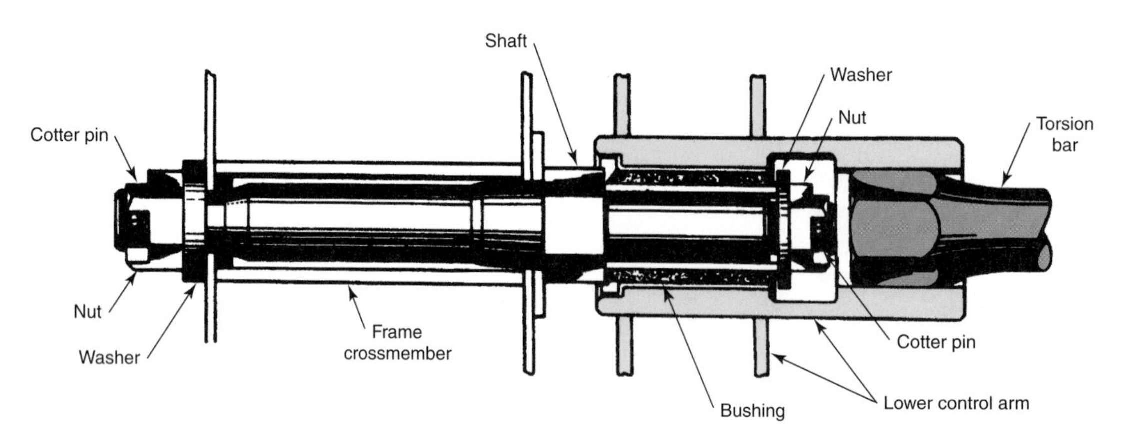

Figure 36-9. This is a cross-section of the torsion bar pivot shaft section in a lower control arm. Note how the hexagonal head of the torsion bar fits into the lower control arm.

the rear anchor with multipurpose grease and carefully position the seal. Use the adjusting screw to place a small load on the bar. Place the vehicle on the floor and adjust front ride height by turning the torsion bar adjusting bolt. Bounce vehicle vigorously and recheck height. **Figure 36-11** illustrates the measuring points for checking the front and rear control height for a specific vehicle equipped with an independent front and rear suspension.

MacPherson Strut Service

A MacPherson strut assembly is shown in **Figure 36-12.** The compact design of the MacPherson strut eliminates the

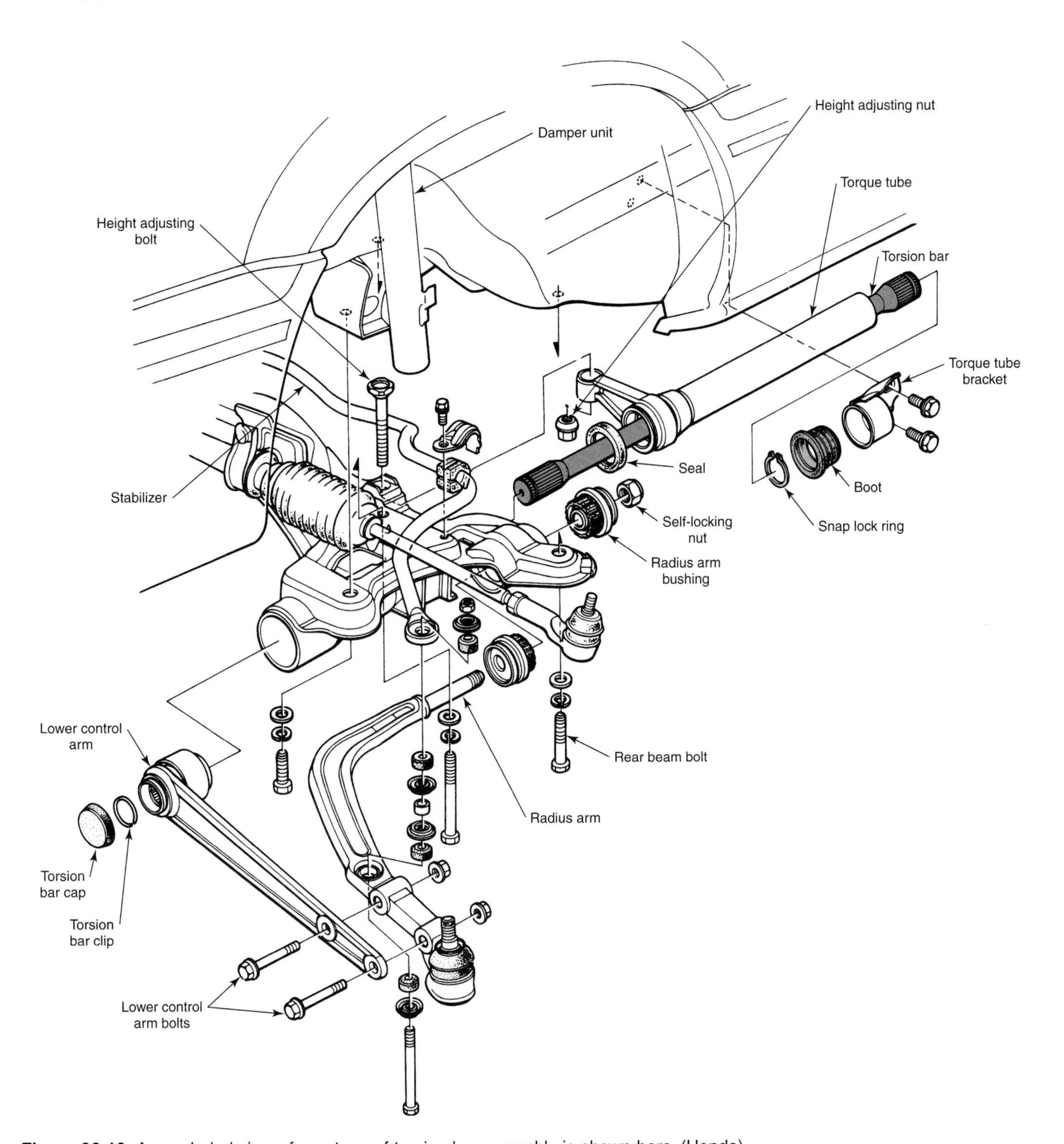

Figure 36-10. An exploded view of one type of torsion bar assembly is shown here. (Honda)

need for an upper control arm and other suspension components. Note the MacPherson strut suspension usually contains a coil spring. There are two basic MacPherson strut coil spring arrangements. One type uses the coil spring mounted on the lower control arm, **Figure 36-13.** The most common type (which is covered here) mounts the coil spring so that it surrounds the strut damper or shock. See **Figure 36-14.**

To remove the MacPherson strut, place the vehicle on a hoist or jack stands. Raise the hood in order to access the strut bolts (generally three) at the top of each inner fender

Figure 36-11. These are the measuring points for checking the control height of a particular minivan. (Toyota)

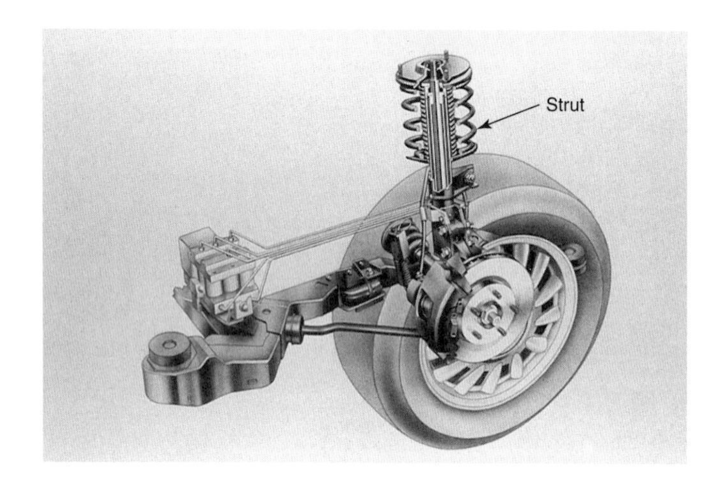

Figure 36-12. This is one type of front-wheel drive, MacPherson strut front suspension. (General Motors)

well. Mark one bolt on each side of the vehicle. Leave at least one bolt or nut attached until the bottom of the strut is disconnected. Remove the tire and wheel. Disconnect the lower mounting bolts or ball joint, depending on the attachment method used. Disconnect the brake assembly and/or line and any wiring or brackets. Support the strut so it will not fall. Remove the nut or bolt that was left connected on the fender well. Remove the strut and mount it in a special holding fixture or vise. See **Figure 36-15.**

Attach a coil spring compressor. Compress the spring until tension is removed from the spring retainer, rebound bumper, and other parts. Loosen the strut rod nut while holding the strut rod, **Figure 36-16.** Pull off the rebound bumper, spring retainer, and other parts. Lift off the coil spring. Be sure to mark the spring top and bottom (if different). Also label which side of the vehicle it was removed from. If the spring requires replacement (sagging, bent, broken), carefully relieve the compressing-tool tension. Remove the spring and discard it. Place the new spring in the tool and compress it.

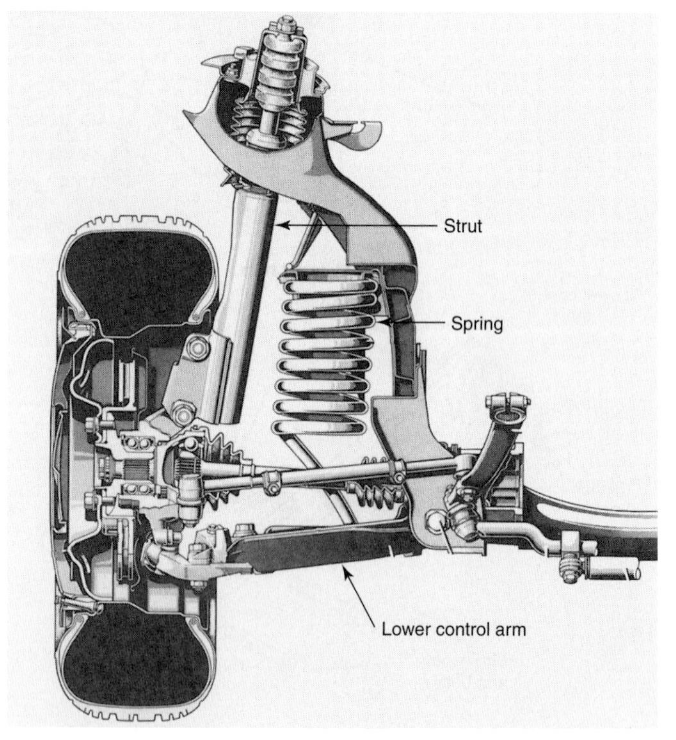

Figure 36-13. This type of MacPherson strut front suspension has springs between the frame and lower control arm. (Mercedes-Benz)

Reinstall the parts in reverse order of disassembly. Torque all fasteners to specifications. Reconnect the brake lines and bleed the brakes if necessary. Align the suspension and road test the vehicle.

Warning: As with all springs under tension, be extremely careful. They can fly off with lethal force.

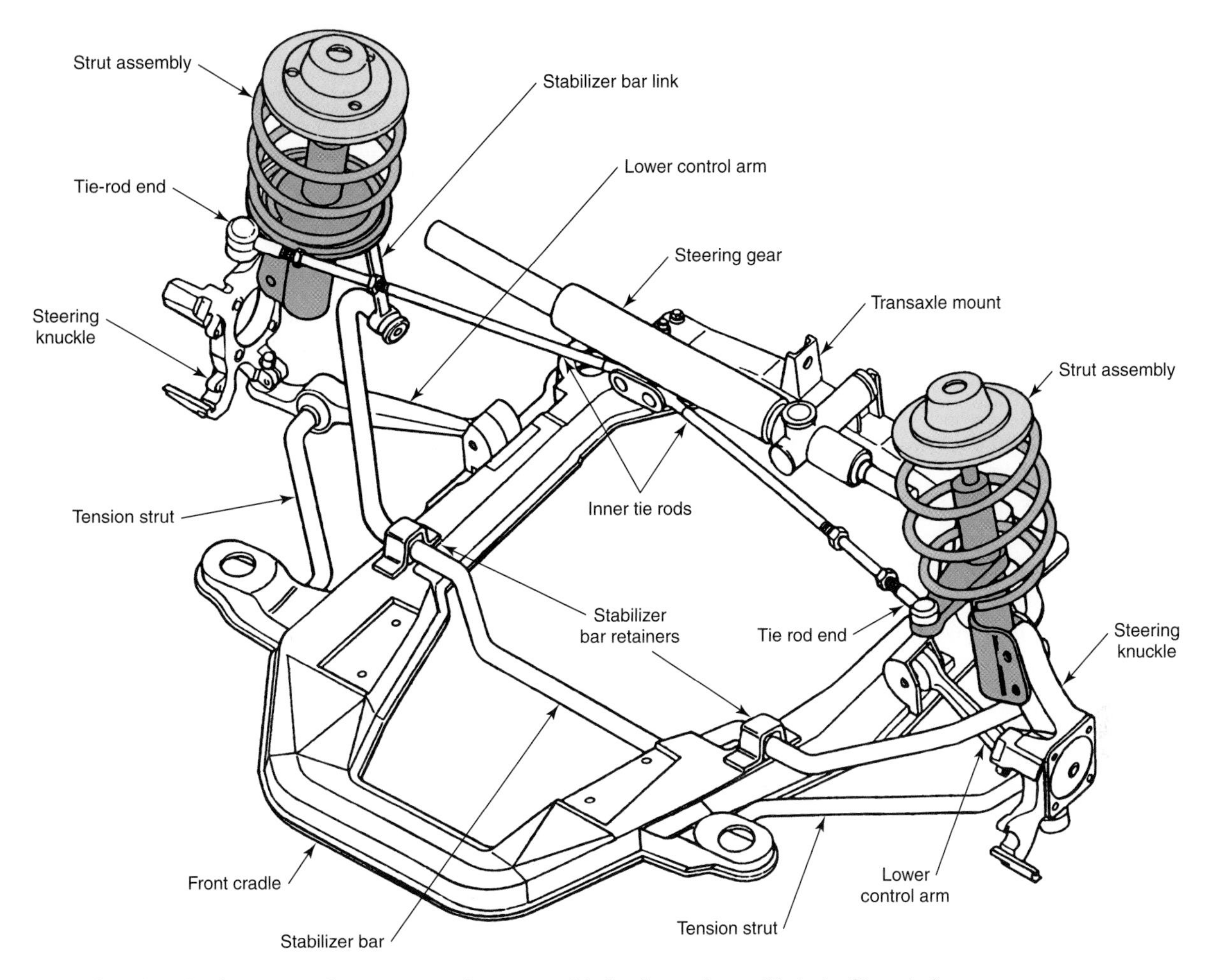

Figure 36-14. A coil spring/strut-type front suspension assembly is shown here. (DaimlerChrysler)

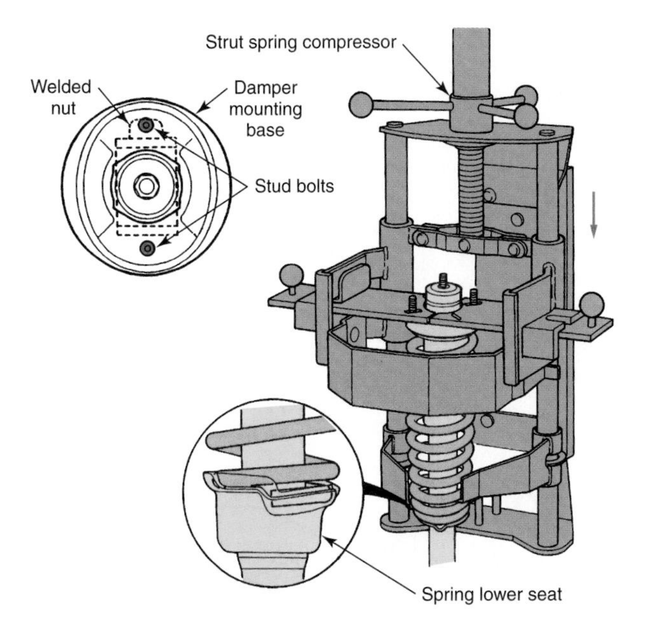

Figure 36-15. A strut spring compressor and holding fixture are shown here. When compressing a strut spring, make sure you follow the same safety precautions that apply to conventional springs. (Honda)

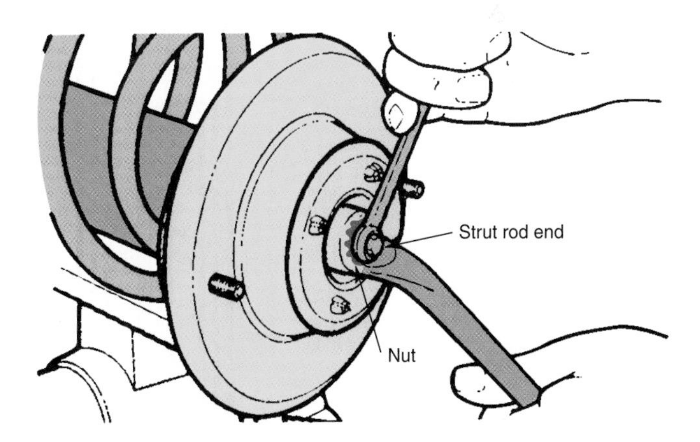

Figure 36-16. Hold the strut rod while removing the strut rod nut. Be sure that spring tension has been relieved before completely removing the nut.

Shock Absorber Service

Shock absorbers are hydraulic units that absorb spring oscillations to provide a steady, smooth ride. The shock absorbers must be in good condition. When a shock is operating properly, it will control spring oscillation, spring rebound, and the rate of spring compression. When mounted at an angle or straddle-mounted, body sway and lean are also minimized. ***MacPherson strut dampers*** are also considered shock absorbers, since they perform the same function that a shock absorber does in a conventional system.

The most commonly used type of shock is the telescopic or airplane design. Typical telescopic shock action is illustrated in **Figure 36-17.** Note the extended shock in Figure 36-17(1). The rebound stroke pulls the piston upward, forcing fluid in A through the valve in the piston to compartment B. Vacuum action draws fluid through the base valve from reservoir C. Figure 36-17(2) shows the action during the compression stroke. The piston is forced downward. This causes fluid to be forced through the piston valve into compartment A, and through the base valve into reservoir C. Area D is air space above the fluid.

The fluid passing through the various valves or orifices slows down the movement of the piston, thus placing a damping action on the movement of the springs. Most shock absorbers made today are ***gas charged.*** This type uses a gas-filled chamber to keep pressure on the shock absorber's hydraulic fluid at all times.

Some shock absorbers have adjusting devices to control the amount of damping action. Some are controlled by a manual valve that may be turned from inside the vehicle.

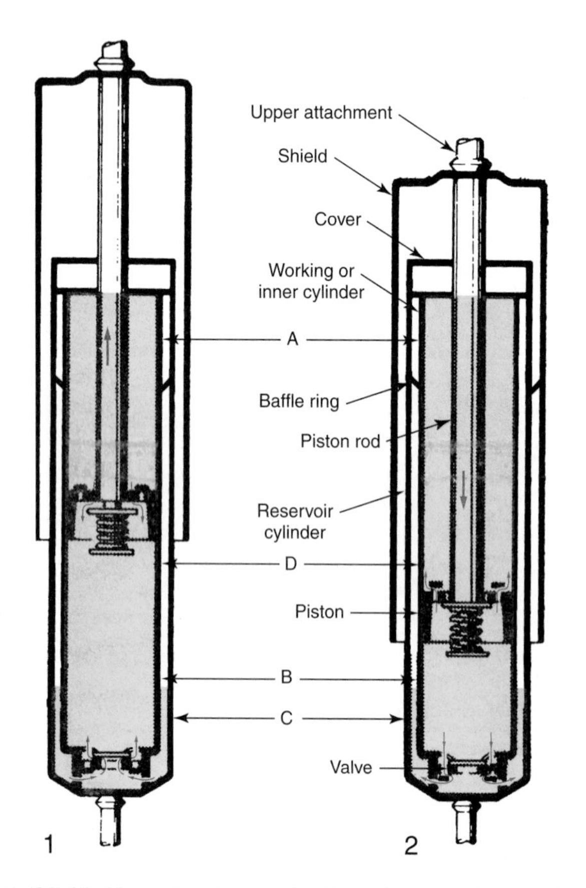

Figure 36-17. Normal action or double-acting shock absorber. (Volvo)

Some have an automatic control, while others must be adjusted at the time of installation. Most shock absorber valves are calibrated for an average load/road condition for a given vehicle and are not adjustable. Most shocks cannot be repaired.

Checking Shock Absorbers and Strut Dampers

Inspect each shock absorber or MacPherson strut damper for signs of fluid leakage, and replace them if leaks are found. Check the condition of the rubber bushings. Shocks or dampers can rattle, pound, and cause poor control if the mounting bushings are worn or missing. Also inspect mounting brackets and fasteners. Grasp shock or damper and try to shake sideways and up and down. Compare shock action by bouncing each corner of the vehicle vigorously and quickly releasing at the bottom of the down stroke. Shocks or dampers in good condition will allow about one free bounce and will then stop any further movement. If shock action is doubtful, service or replace.

Replacing Shock Absorbers

Modern gas-filled shock absorbers expand to their full length when not installed, and may need to be compressed slightly to fit the attachment points. Use new rubber bushings with new shocks. Install bushings and washers in the correct order. Make certain that any stone guards face in the correct direction. Tighten the shock mounting bolts with the vehicle at normal curb height (weight on wheels) to prevent placing the bushing under a strain. Do not overtighten the fasteners. A typical mounting arrangement for a front shock absorber is illustrated in **Figure 36-18.** Follow the shock manufacturer's installation guidelines.

Replacing Strut Dampers

The procedure for replacing the strut damper assembly is identical in most cases to that given for MacPherson strut spring replacement discussed earlier in this chapter. Some

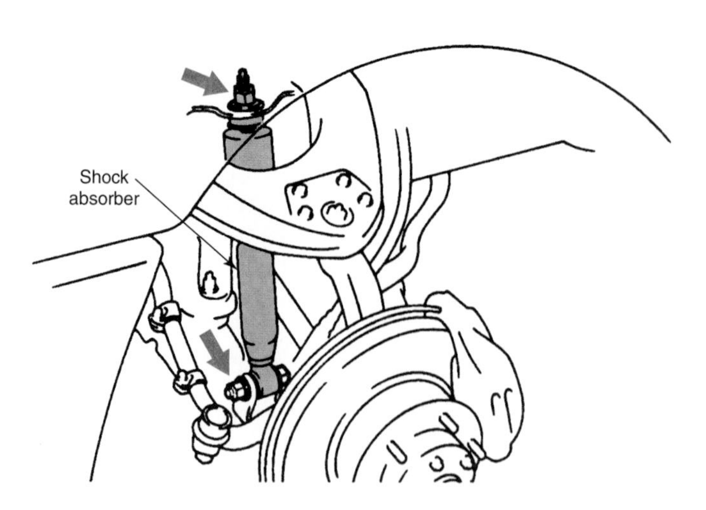

Figure 36-18. Inspect all shock absorbers, dampers, and mounts for wear, leaks, damage, etc. (Toyota)

strut dampers can be replaced as a cartridge while others require replacement of the entire strut. To replace a strut damper cartridge, remove the spring as outlined in the spring replacement section. Be sure to use an adequate spring compressor when needed. Unscrew the nut holding the damper to the housing and pull the cartridge from the housing. If necessary, clean the bottom of the housing and add fresh oil. Usually very little oil is needed. Install the new damper assembly. Finally, reinstall the spring. A cutaway of a cartridge-type strut damper is shown in **Figure 36-19.**

Ball Joint Service

All modern vehicles, whether they are equipped with conventional or MacPherson strut assemblies, use ***ball joints.*** The function of the ball joint is to carry the vehicle load to the wheels while still allowing relative movement between the wheel assembly and vehicle body. The ball joint that carries the majority of the vehicle weight is called the ***load-carrying ball joint.*** **Figure 36-20** shows one ***preloaded*** by a coil spring or rubber pressure ring. This preloading keeps the joint bearing surfaces in constant contact, **Figure 36-21.**

Depending on the arrangement, both ball joints can be placed under either ***tension-loading*** (forces attempt to pull joint apart) or ***compression-loading*** (forces attempt to compress joint). See **Figure 36-22.**

A second arrangement tension-loads the follower joint (upper joint in this case) and compression-loads the main load-carrying joint (lower joint in this case). See **Figure 36-23.** Note the coil spring is between the lower control arm and frame in the previous illustration.

Ball Joint Wear

Excessive ball joint wear will alter wheel alignment and can cause hard steering, shimmy, and tire wear. Although the follower joint does wear, the load-carrying joint usually wears out first, so check both joints. If the follower shows any

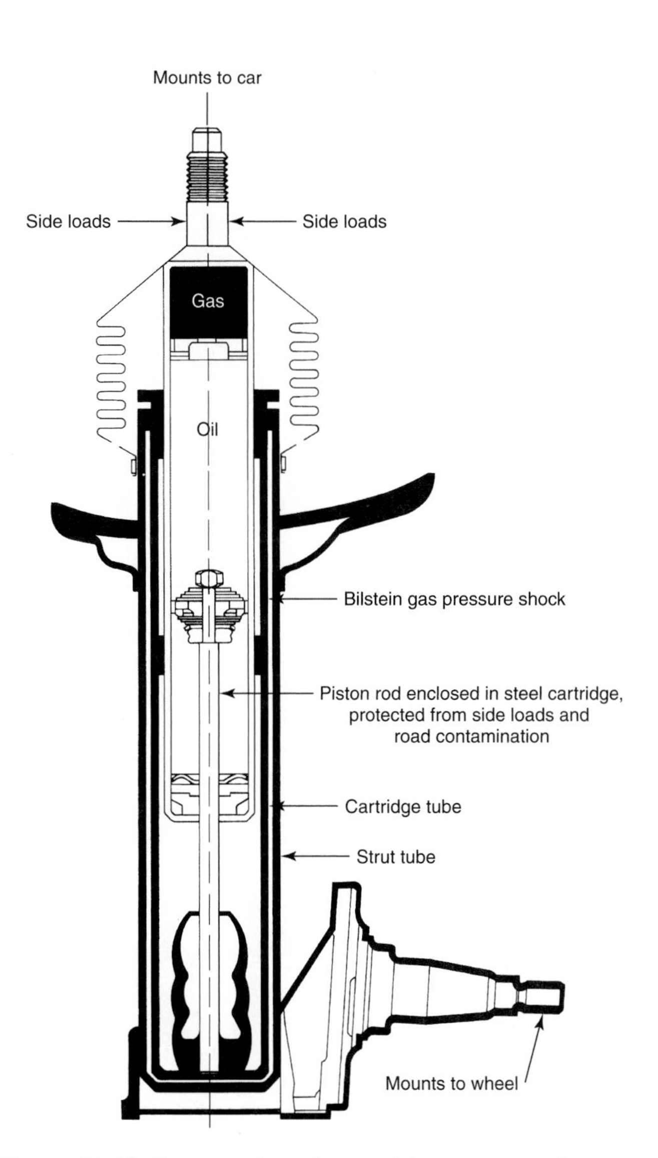

Figure 36-19. Cutaway view of a cartridge-type strut damper. Not all struts use replacement cartridges. (Bilstein)

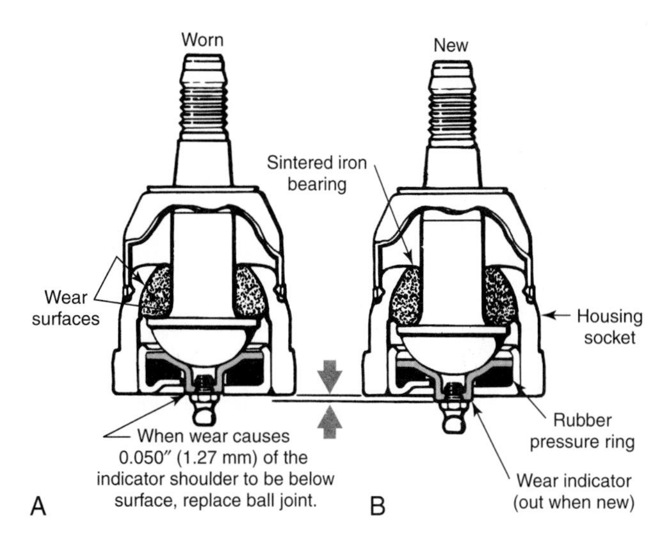

Figure 36-20. One type of load-carrying ball joint is shown here. Note the wear indicator. (General Motors)

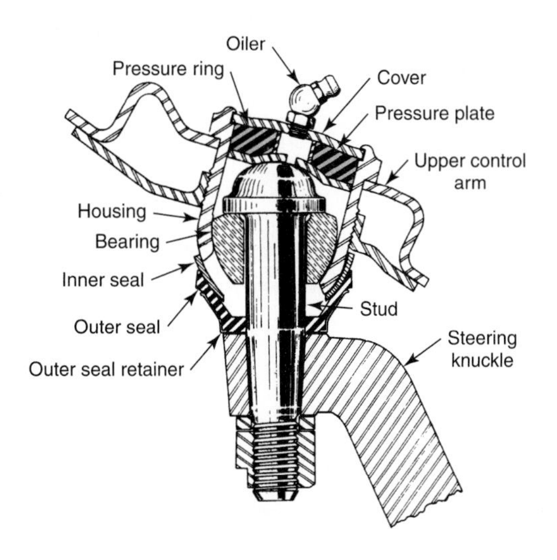

Figure 36-21. This is a cross-sectional view of a typical follower (non-load-carrying) ball joint. Note the rubber pressure ring used to provide constant loading.

discernible looseness, replace it. The main joint clearance or play should not exceed manufacturer's specifications.

To check ball joints for wear, they must be properly unloaded (pressure removed). When the coil spring is located between the lower control arm and the frame, **Figure 36-24,** place the jack under the lower control arm. When the jack is raised high enough to provide clearance between the tire and floor, the ball joints will be unloaded.

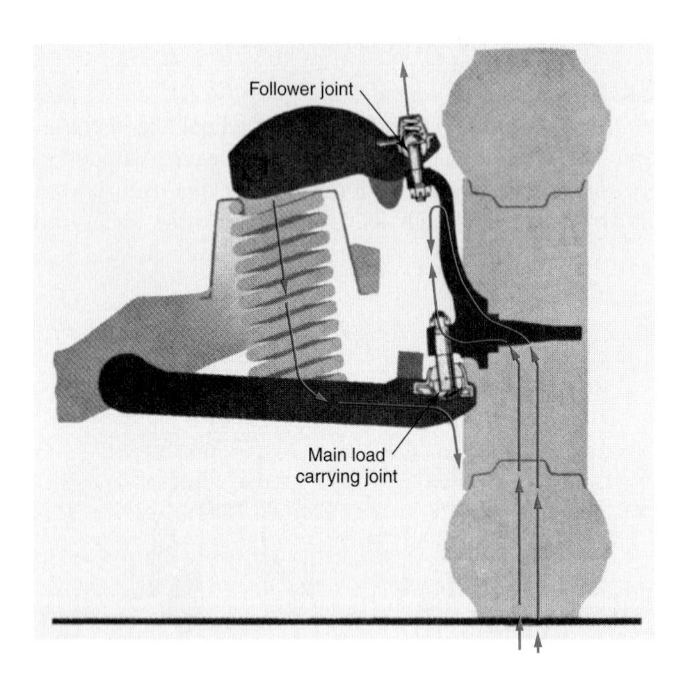

Figure 36-22. This suspension arrangement tension loads either ball joint. The main load-carrying joint is at the bottom. (Moog)

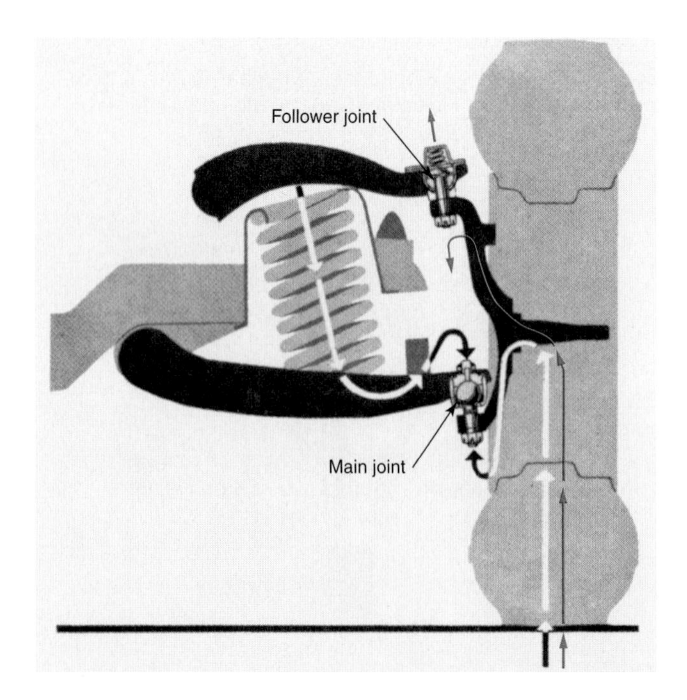

Figure 36-23. This suspension arrangement tension loads the follower joint and compression loads the main joint. (Moog)

After properly unloading the ball joint, the amount of wear may be determined in one of two ways. See **Figure 36-25.** Some manufacturers recommend checking joint play by measuring the axial movement. Axial movement is the up and down movement of the wheel and tire assembly. See Figure 36-25A. Other manufacturers specify testing for excessive clearance by measuring tire sidewall movement. See Figure 36-25B. Because the amount of acceptable ball joint varies from design to design, follow the manufacturer's specifications.

Load-Carrying Ball Joint Service

When replacing a load-carrying ball joint, you will use some of the procedures that were covered earlier. If the coil spring is mounted between the lower control arm and the frame or body, as in **Figure 36-26,** raise the vehicle and place a jack under the lower control arm. Some technicians prefer to use a spring compressor to hold the spring in place, as shown in **Figure 36-27.**

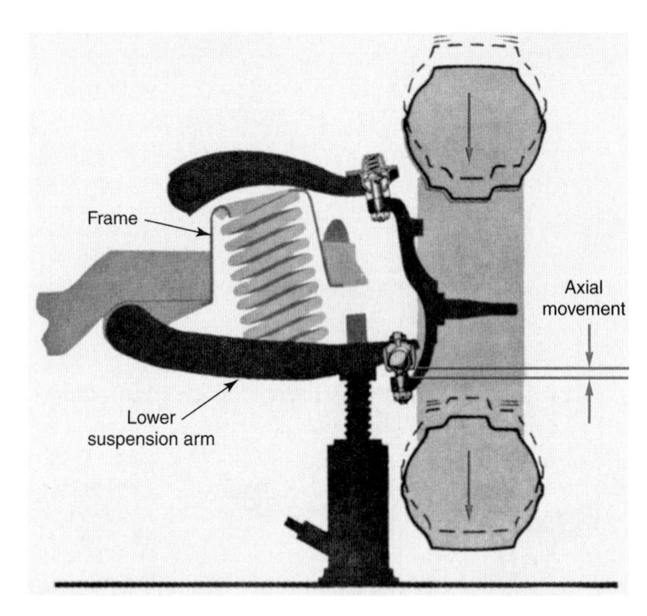

Figure 36-24. To unload ball joints when the coil spring is mounted between the lower suspension arm and the frame, place a jack under the lower suspension arm. Be careful!

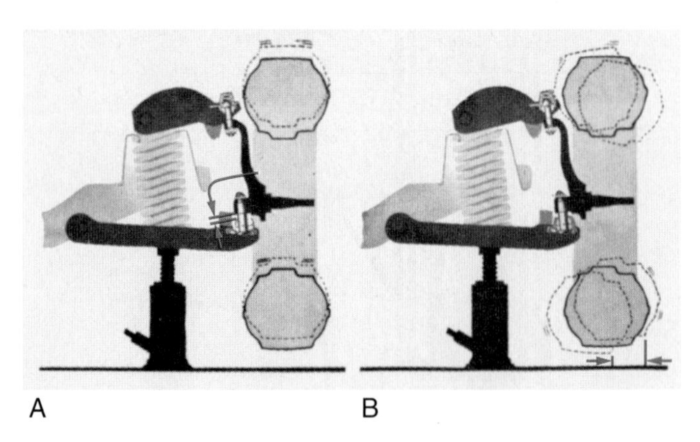

Figure 36-25. Two methods of measuring ball joint wear are shown here. A—Axial movement. B—Tire sidewall movement.

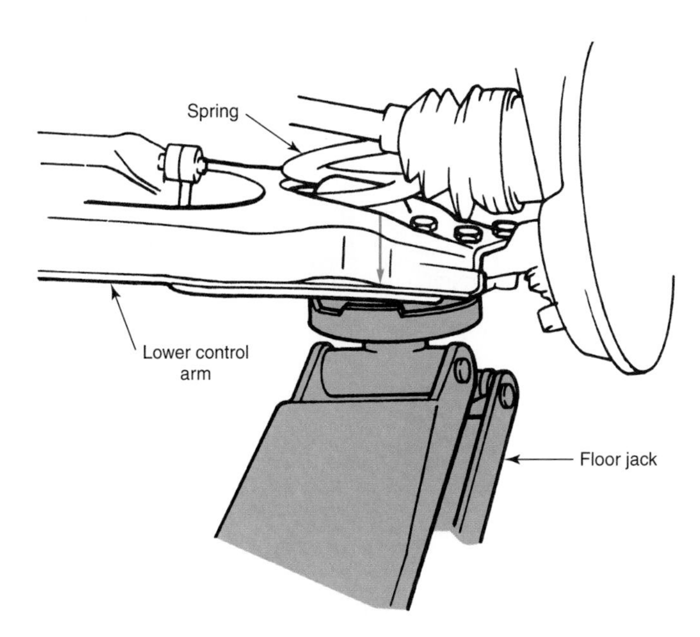

Figure 36-26. Support the lower control arm with a jack. Be sure the jack plate is securely under the control arm before unloading spring pressure. (General Motors)

STOP **Warning: Always place a jack beneath the lower control arm when the coil spring is mounted between the lower arm and the frame. Unless a jack is in place against the lower control arm, removal of either the upper or lower ball joint will allow the spring to propel the lower control arm downward with lethal force.**

Remove the cotter key from the ball joint stud nut. Loosen the stud nut several turns. In some cases, the castle nut is removed and a standard hex nut is placed on the stud to within two or three threads of the steering spindle or knuckle. A special stud removal tool may be placed between the ends of the upper and lower stud ends. A stud removal tool is shown installed in **Figure 36-28.** Turn the tool to apply pressure to the ball stud. When the stud is under pressure, strike the steering spindle sharply with a hammer to free the stud in the spindle or knuckle body. Never try to force the ball stud out of the spindle body using tool pressure only. To do so would distort the spindle. Figure 36-28 illustrates spot on the spindle body that should be struck to loosen the ball joint stud. When the ball stud is loose, remove the tool and stud nut. Make certain the lower arm is securely supported. Lower the control arm slowly until the ball joint is clear. If necessary, disconnect the lower control arm pivot shaft and remove the arm.

Some ball joints are screwed in place while others may be riveted, pressed, bolted, or welded. If the joint is threaded, unscrew and remove it. Clean both the arm and threads completely. Install the new joint and torque to specifications. If the joint is riveted to the arm, drill through the rivets with the specified size drill bit. Cut off the rivet heads and drive the rivets out. A power chisel makes rivet head removal easy. Some manufacturers recommend drilling off the rivet heads that secure the joint. This procedure is illustrated in

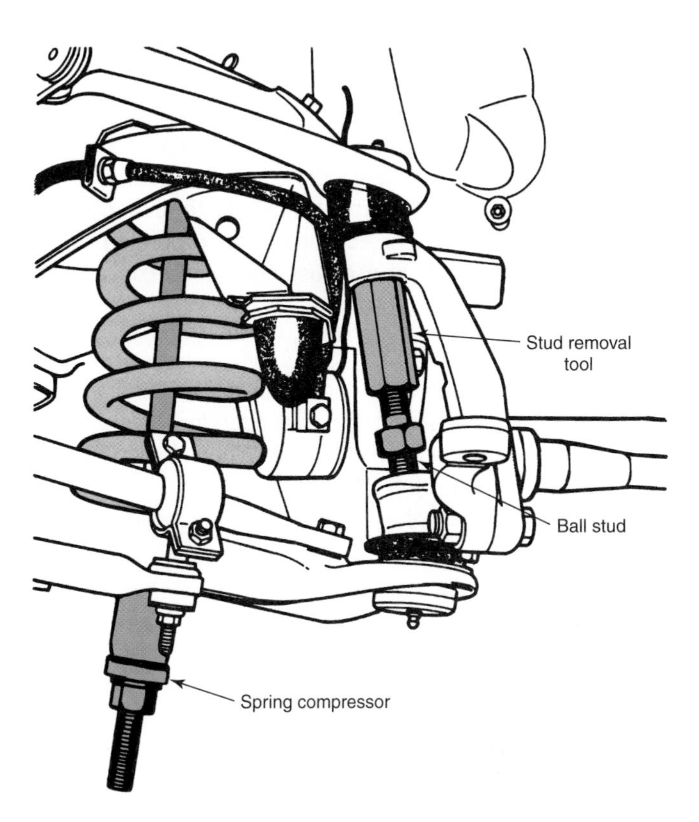

Figure 36-27. Using a spring compressor to hold the spring in place when removing a ball joint. (DaimlerChrysler)

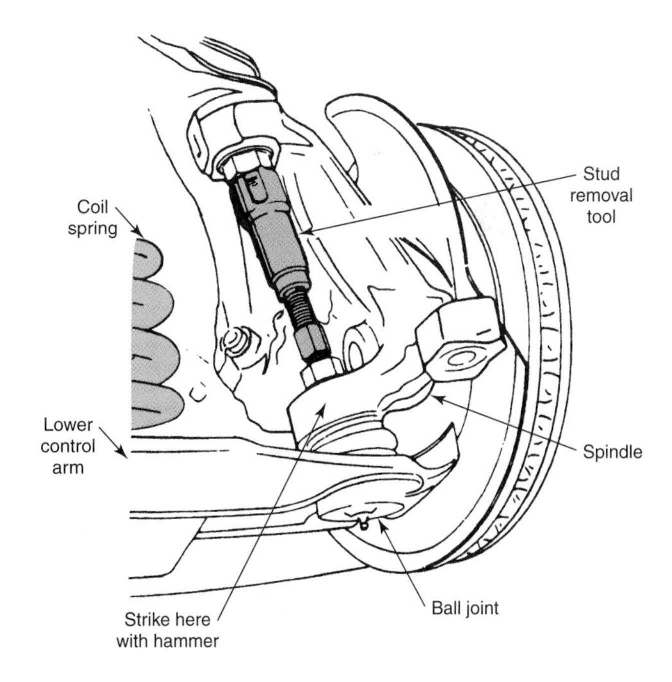

Figure 36-28. Hit the spindle body with a hammer to remove a ball joint stud. Be careful not to damage any drive axles or anti-lock brake sensors that may be present. (General Motors)

Figure 36-29. Clean the arm thoroughly and install the ball joint. Instead of inserting new rivets, use the special bolts supplied specifically for mounting the ball joints. Torque the bolts to specifications.

Figure 36-30 illustrates the removal and installation of a ball joint that is pressed into place. Clean the ball joint stud and the hole it fits through. If removed, replace the control arm. With coil spring properly in place, raise arm with jack until ball stud passes through the spindle body. Install the retaining nut and torque it to specification. Install a cotter pin and lubricate the joint. If a camber eccentric is incorporated in the upper ball stud, check and adjust camber as covered in Chapter 39, Wheel Alignment.

Follower Ball Joint Service

The general procedure for removing and installing a follower ball joint is similar to that described for removing and installing the main load-carrying joint. When the coil spring is mounted between the lower control arm and frame, **Figure 36-31,** support the lower arm. Do not allow spring

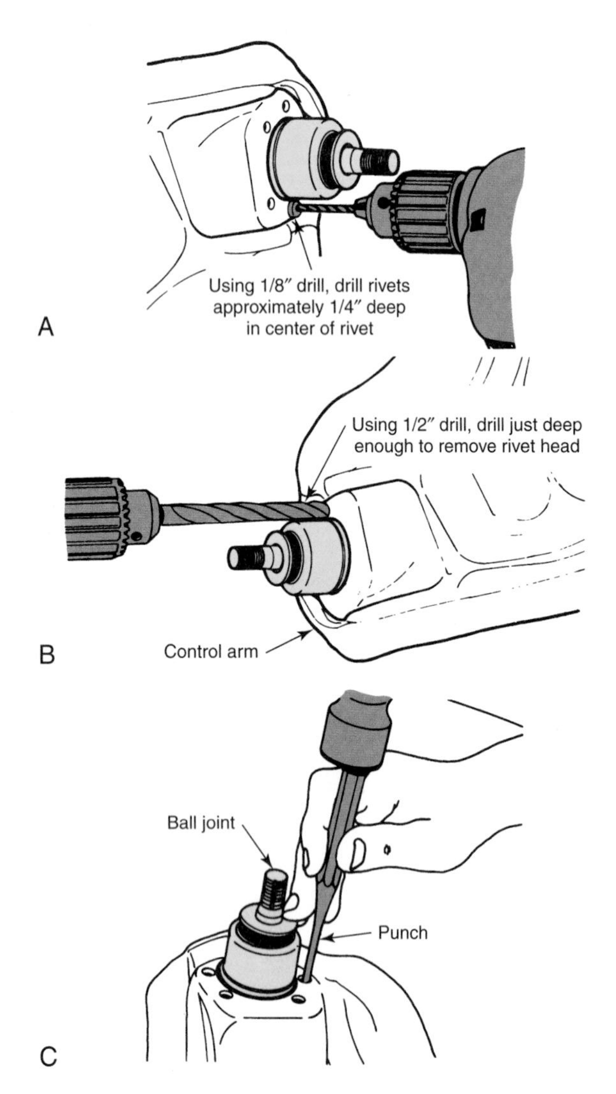

Figure 36-29. Some ball joints require rivets to be drilled out before the joint can be removed. A—Drilling a 1/8″ (3.16 mm) hole 1/4″ (6.35 mm) deep in the rivet's center. B—Rivet heads have just been drilled off using a 1/2″ (12.7 mm) bit. Do not drill into the control arm itself. C—Using a punch to drift out the rivets, freeing the ball joint. (General Motors)

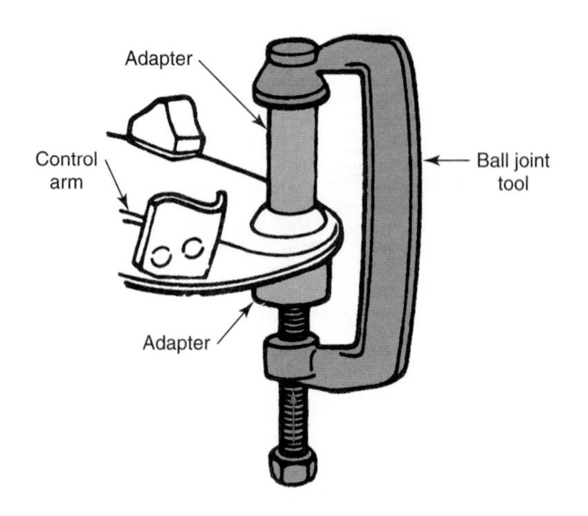

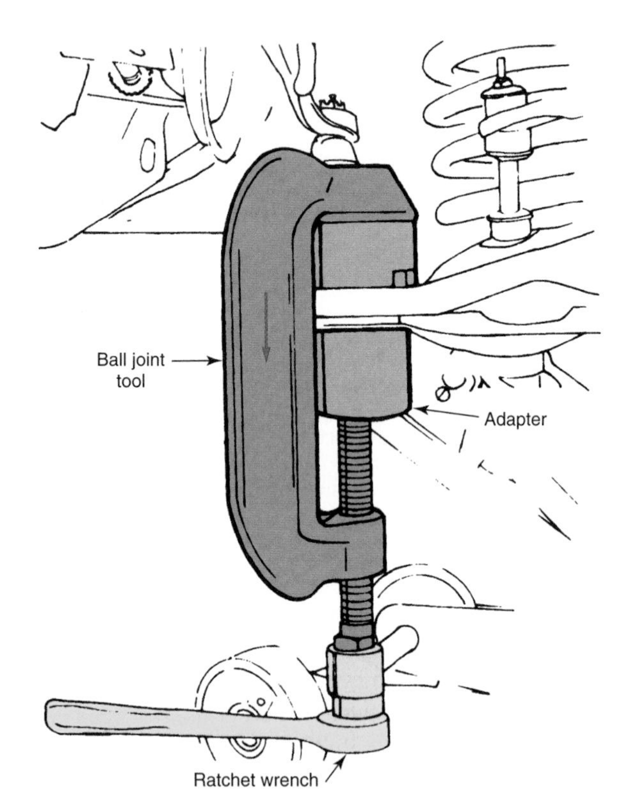

Figure 36-30. A—Removing a pressed-in ball joint. B—Installing a pressed-in ball joint. (General Motors)

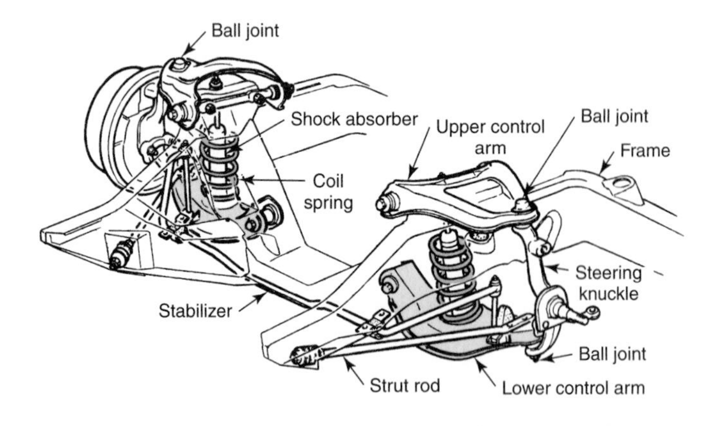

Figure 36-31. This figure shows the ball joint locations in one coil spring suspension setup. (General Motors)

pressure to slam down the control arm when the follower joint stud is removed. Some follower joints are welded to the control arm. When the joint is worn, the entire assembly must be replaced.

Steering Knuckle

The steering knuckle is mounted between the two ball joints in a conventional suspension system and at one end of the damper in a MacPherson strut suspension. The knuckle allows the suspension system to turn the front wheels while retaining the wheel and brake assembly. The steering knuckle on most conventional suspensions has a spindle cast into it. This spindle provides a smooth surface for wheel bearings to turn.

Control Arm Removal

The ***control arm*** must be removed when it becomes damaged or to replace the ***control arm bushings.*** Disconnect the shock absorber, stabilizer bar, tie strut, and other parts as needed. Block the lower control arm with a jack. Remove the ball joint stud from the steering knuckle. When removing the control arm's mounting shaft, mark the location of camber and caster shims, position of bushings, or other alignment components to assist in rough resetting of camber and caster. When removing the tie strut (where used), do not move the rear locknut (nut on lower arm side of the rubber strut bushing). **Figure 36-32** illustrates a typical upper control arm that has been completely disassembled.

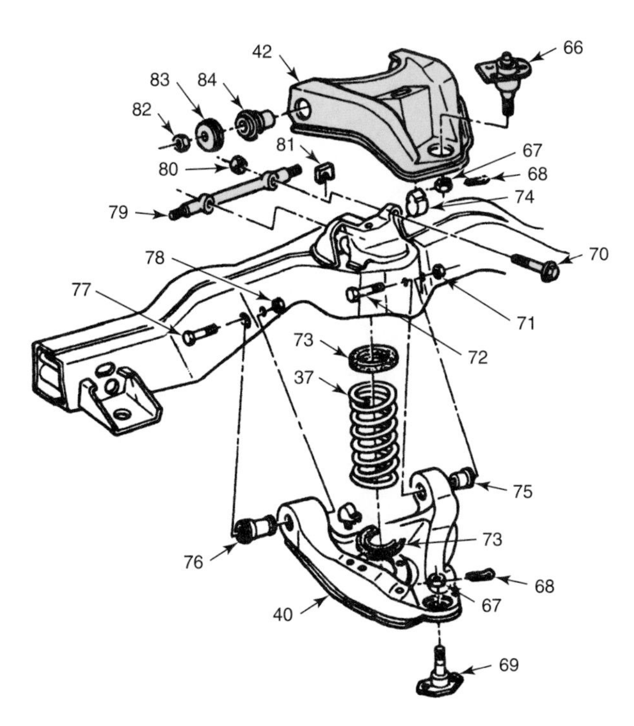

Figure 36-32. Study the parts and their relative locations in this exploded view. 37—Coil spring. 40—Lower control arm. 42—Upper control arm. 66—Upper ball joint. 69—Lower ball joint. 73—Insulator. 74—Bumper. 75—Bushing. 76—Bushing. 79—Shaft. 81—Shim.

Control Arm Bushings

Always replace both control arm bushings even if only one is worn. When installing a new inner shaft, use new bushings. Remove the fasteners and washers from the end of the bushings. Apply penetrating oil around the bushing. To facilitate bushing removal, bolt on a tool similar to that shown in **Figure 36-33.** Note that several tools are used in this setup.

Set the arm up in a vise and press the bushing out by turning the nut on the press while holding the receiving cup on the other end. Reverse the arm and remove the other bushing. If threaded-type bushings are used, they may be removed by unscrewing. Clean the control arm and check for cracks, heavy dents, springing, and other problems. Check the inner shaft for excessive wear.

To install new bushings, place an install tool on the bushing, making sure the bushings are on the correct ends. Install a stiffener tool to prevent the arm from being bent, **Figure 36-34.** Set the arm up in a press and start the new bushings into place. Using the bushing-pressing tools, force the bushings into place, Figure 36-34. Install outer washers and fasteners. Torque to specifications. If threaded bushings are used, make certain the inner shaft is correctly centered in the arm. In a lower control arm with a tie strut, there is only one inner bushing. Note how a small spacer tube is used to prevent arm distortion while removing the lower arm bushing, **Figure 36-35.** The spacer, when required, should also be used when installing the bushing.

Strut Rods and Bushings

If the vehicle is equipped with ***strut rods,*** disconnect them from the control arms and frame, **Figure 36-36.** In most cases it will not be necessary to remove any other parts. If the strut rods are threaded for caster adjustment, count the

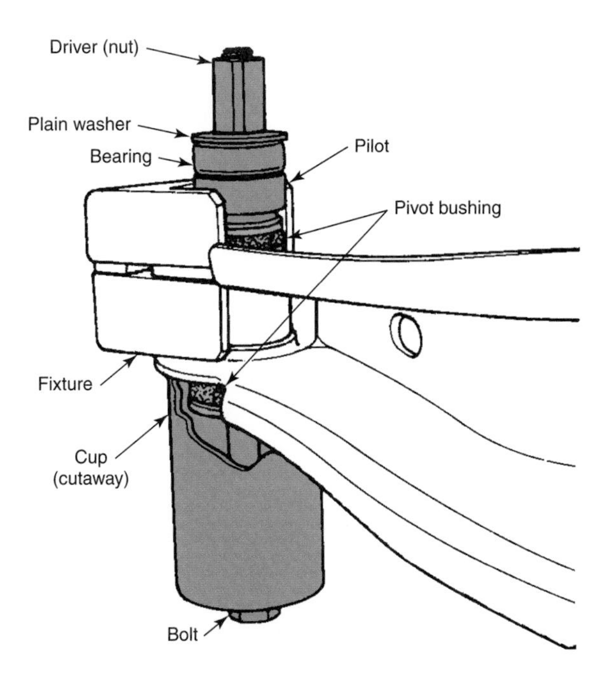

Figure 36-33. Removing a pressed-in control arm bushing. (DaimlerChrysler)

number of threads on one side and reinstall the rod in the same position. This will make alignment easier. Install the replacement rods and torque to specifications.

Stabilizer Bar Bushings

To replace ***stabilizer bar bushings,*** disconnect the stabilizer bar rod at the control arm. Remove the long bolt and remove the bushings. Reinstall the assembly with new bushings, being sure to properly assemble all bushings, washers, and spacers in their original position. Install the long bolt in its original position and tighten the nut. **Figure 36-37** shows a typical stabilizer bar bushing assembly.

The bar bushings can be replaced by unbolting the bar brackets and removing the bushings. The bushings are split, allowing them to be slipped over the bar. Installation is the reverse of removal.

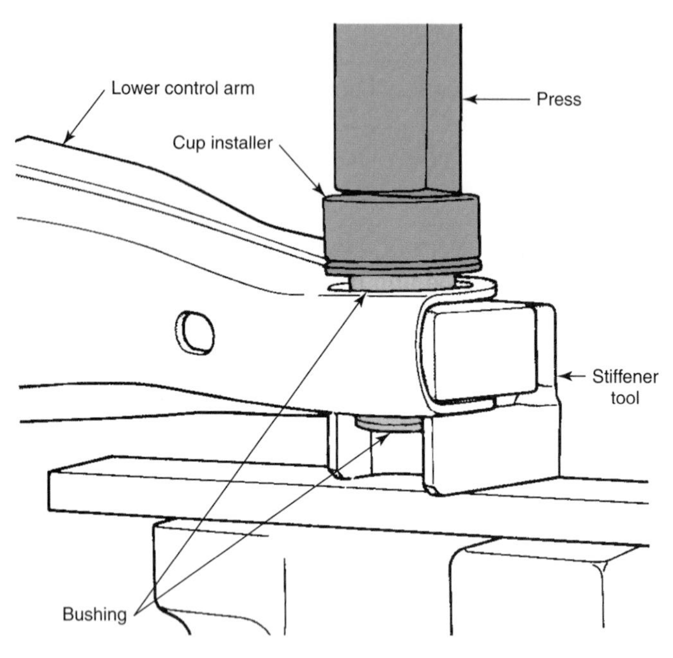

Figure 36-34. Replacing a control arm bushing using a special pressing tool. (DaimlerChrysler)

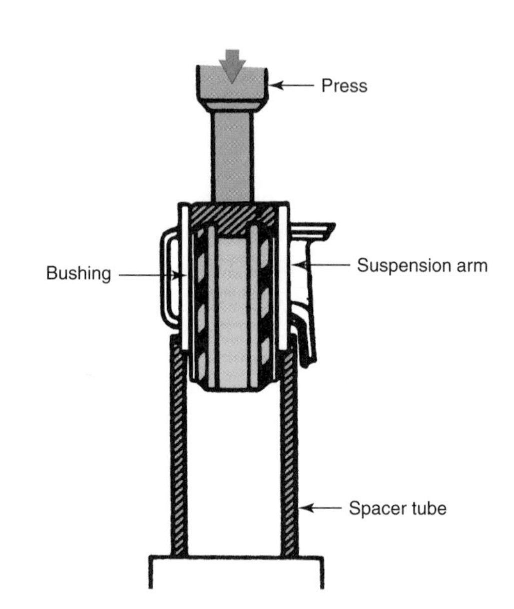

Figure 36-35. A spacer tube prevents control arm distortion as the bushing is pressed out. (Nissan)

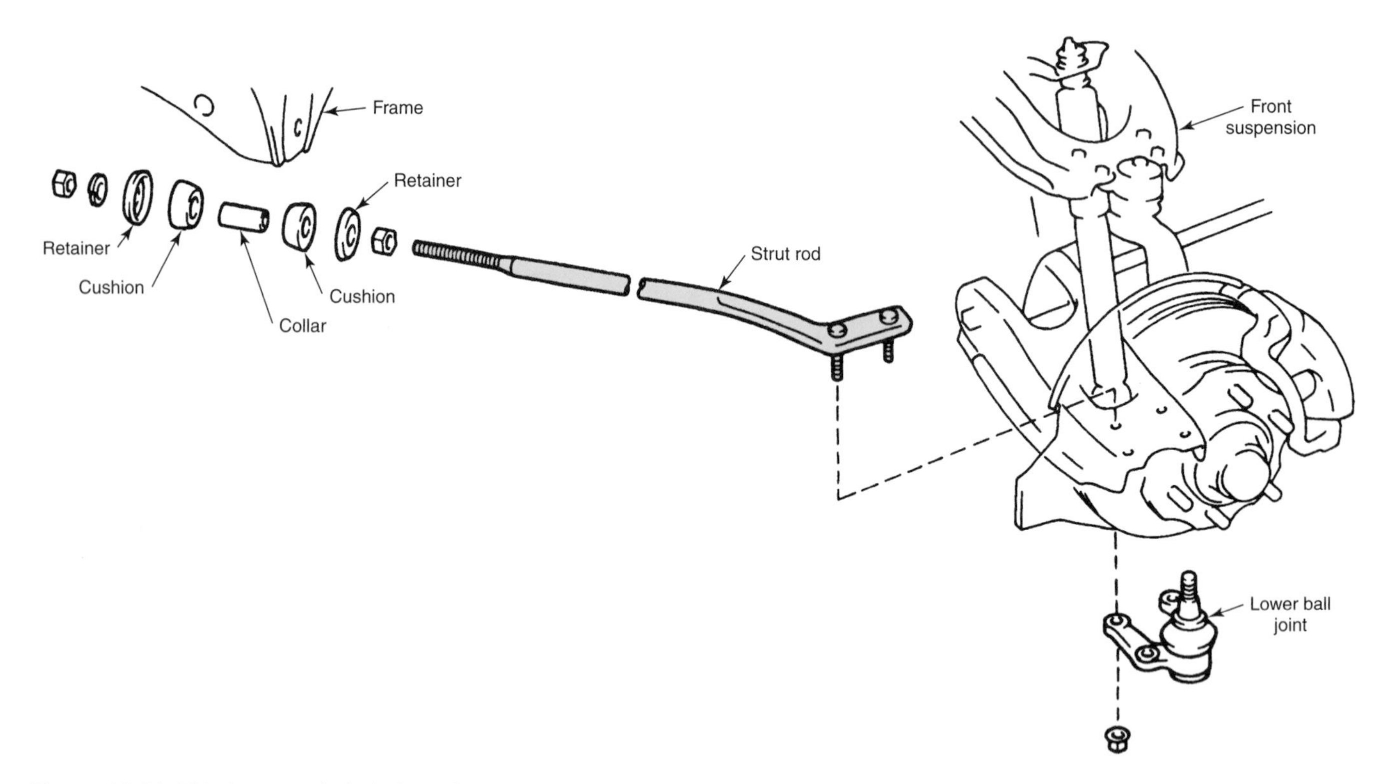

Figure 36-36. This is an exploded view of a strut rod assembly. (Toyota)

If the stabilizer bar is bent, it can be replaced by removing the fasteners at the bushings. Once the bushings are removed, the stabilizer bar can be removed from the vehicle. Some stabilizer bars pass though the front frame rail and must be slid out between the rail openings. To install a new stabilizer bar, place it in position and reinstall the bushings and fasteners. If there is any doubt as to the condition of the bushings, they should be replaced when the stabilizer bar is replaced.

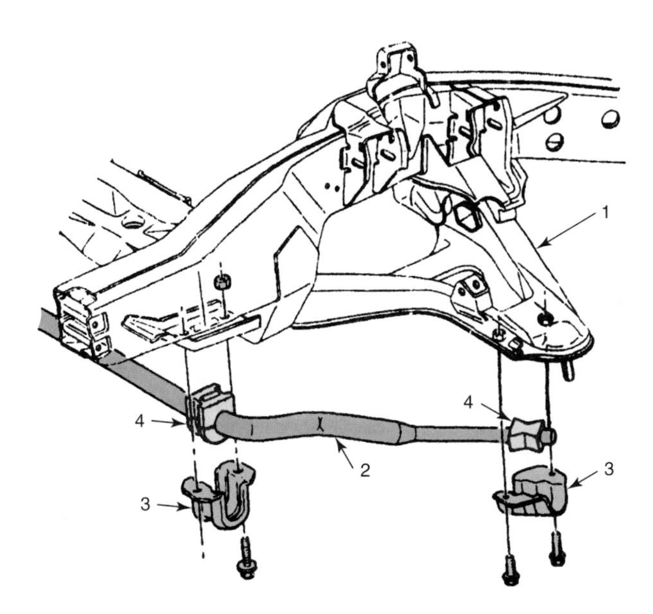

Figure 36-37. Stabilizer bar bushing assemblies. 1—Lower control arm. 2—Stabilizer bar. 3—Bushing clamps. 4—Bushings. (General Motors)

Rear Suspension Systems

Most vehicles generally use coil, leaf spring, or MacPherson strut suspension systems. When coils are used, control arms (also called control links, control struts, or trailing arms) must be used to provide proper rear axle housing alignment. **Figure 36-38** illustrates a typical control arm setup. A coil spring rear suspension system is shown in **Figure 36-39.** Note how the upper control arm is placed at an angle. These arms also function as the conventional track bar. A coil spring rear suspension, as used with one particular front-wheel drive, is illustrated in **Figure 36-40.**

Figure 36-41 shows a different form of rear suspension using leaf springs. The independent rear suspension system

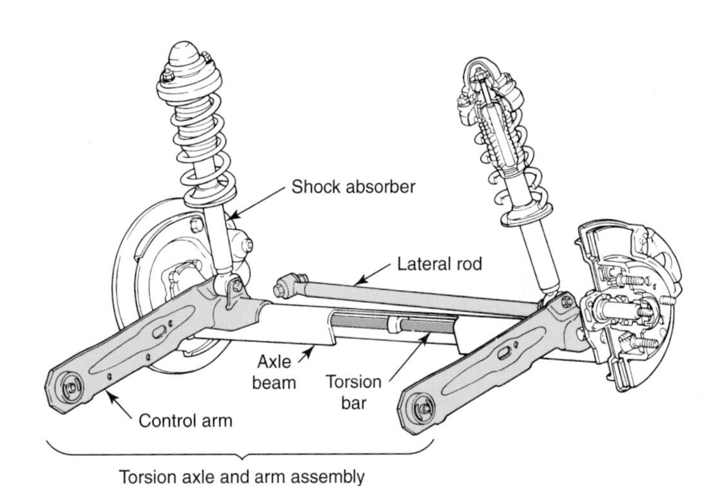

Figure 36-38. This coil spring rear suspension also uses a torsion bar. (Hyundai)

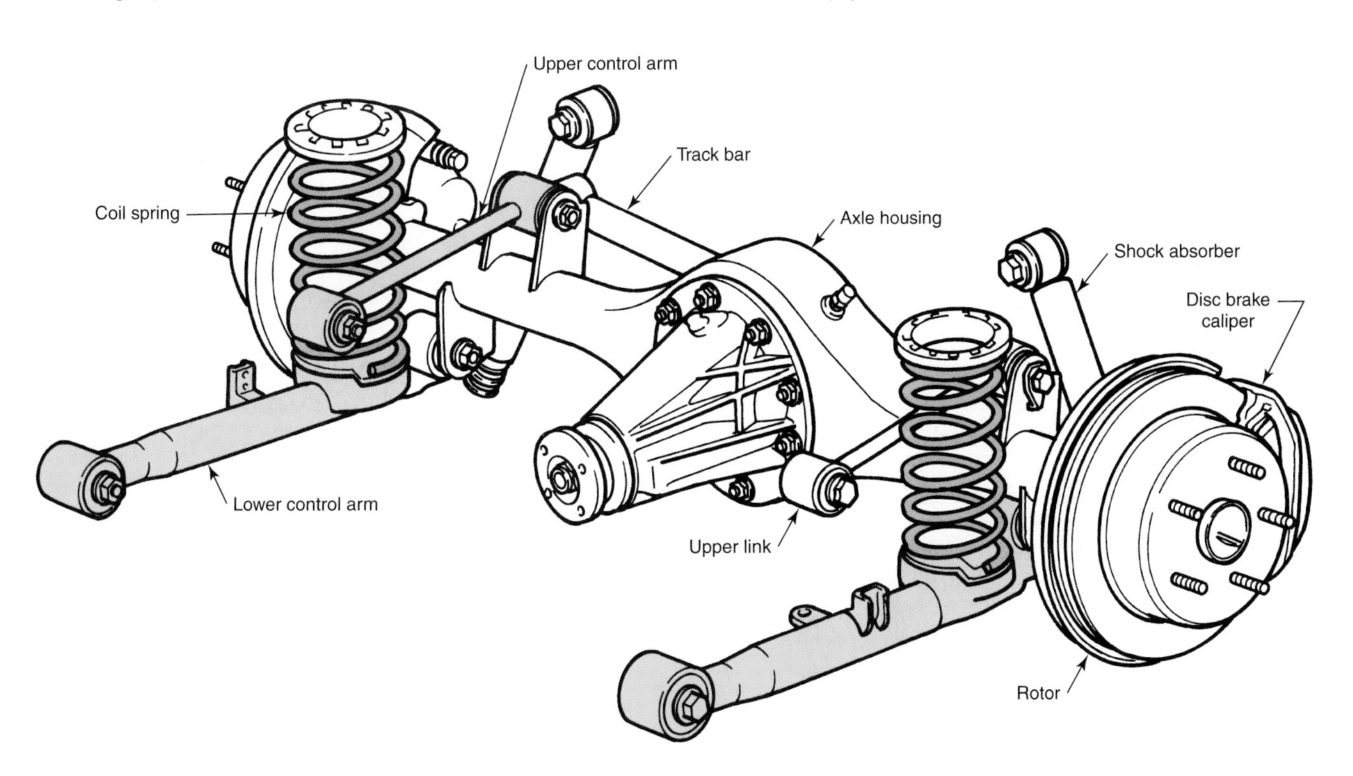

Figure 36-39. This is a coil spring rear suspension used on a rear-wheel drive vehicle. (Toyota)

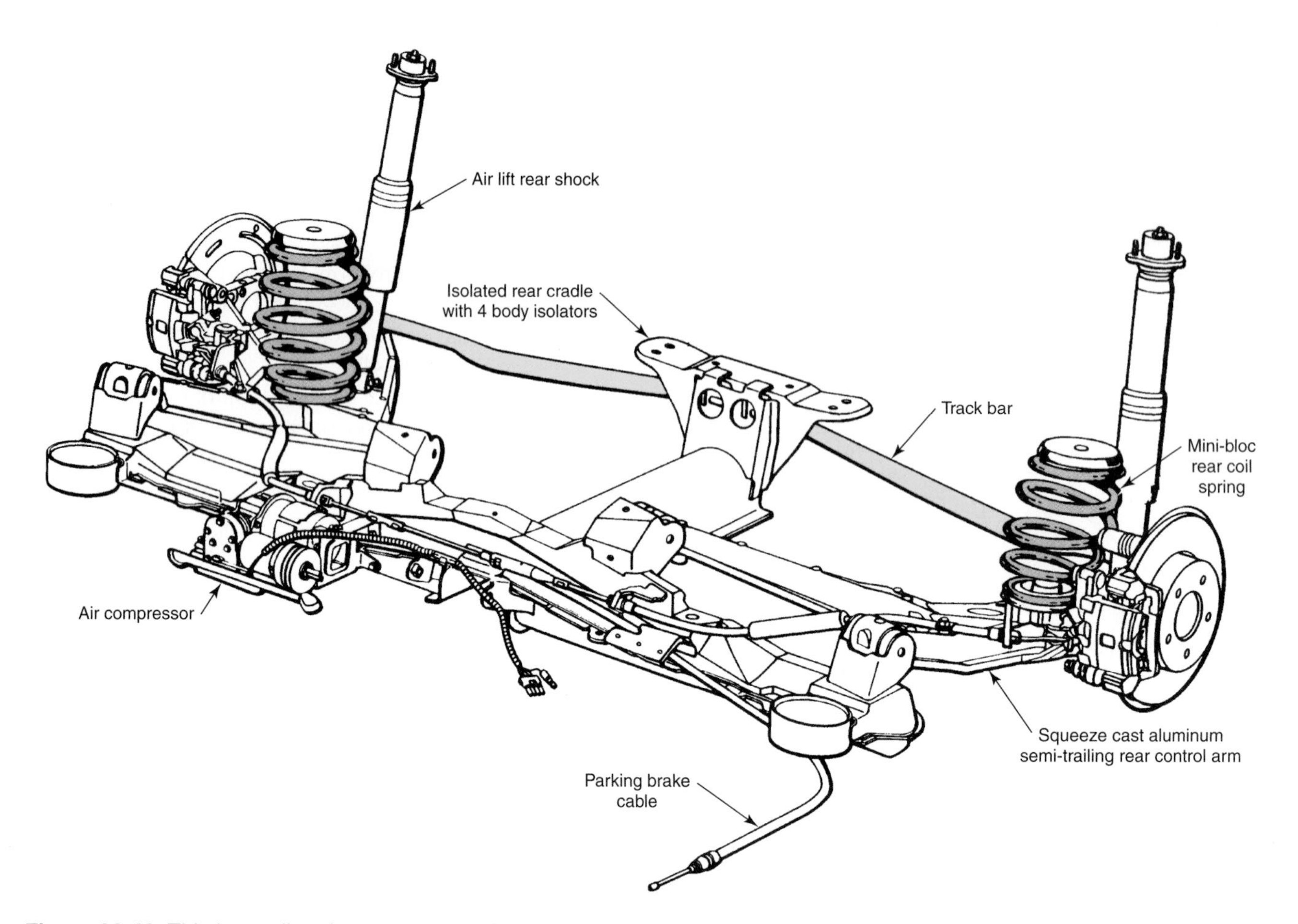

Figure 36-40. This is a coil spring rear suspension with air shock absorbers. (General Motors)

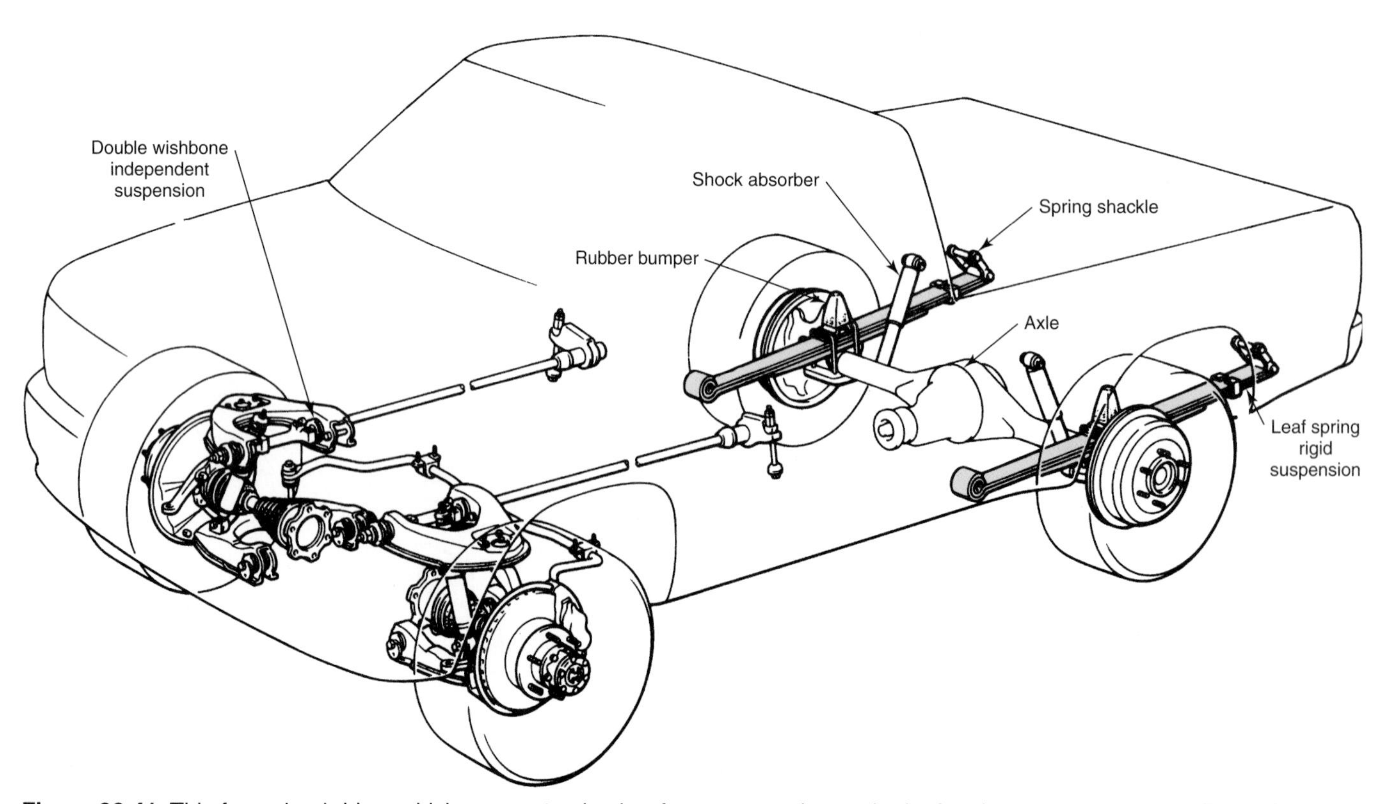

Figure 36-41. This four-wheel drive vehicle uses a torsion bar front suspension and a leaf spring rear suspension. (Toyota)

in **Figure 36-42** uses a single multiple-leaf spring that is mounted in a transverse position. Another similar rear suspension has a single leaf spring made of fiberglass. Removal and replacement procedures for coil springs and MacPherson strut assemblies are similar to those given in the front suspension section.

Rear Leaf Spring Service

The most common ***leaf spring*** failure point is the front and rear bushings. Some leaf spring bushings are best removed and installed by using a suitable puller as illustrated in **Figure 36-43.** Following the removal of the old bushing, clean the spring eye thoroughly.

Coat the new bushing with a suitable lubricant before pulling it into position. Allow the weight of the vehicle to rest on the bushings before torquing the shackle bolts. Replace any broken spring leaves. Use inserts between the spring leaves (where required). The spring center bolt must be tight. Be sure to torque the spring U-bolts. All of the spring rebound clips must be in place. Be careful when removing or replacing a fiberglass spring. Nicks, chips, or other damage from rough handling can lead to spring failure.

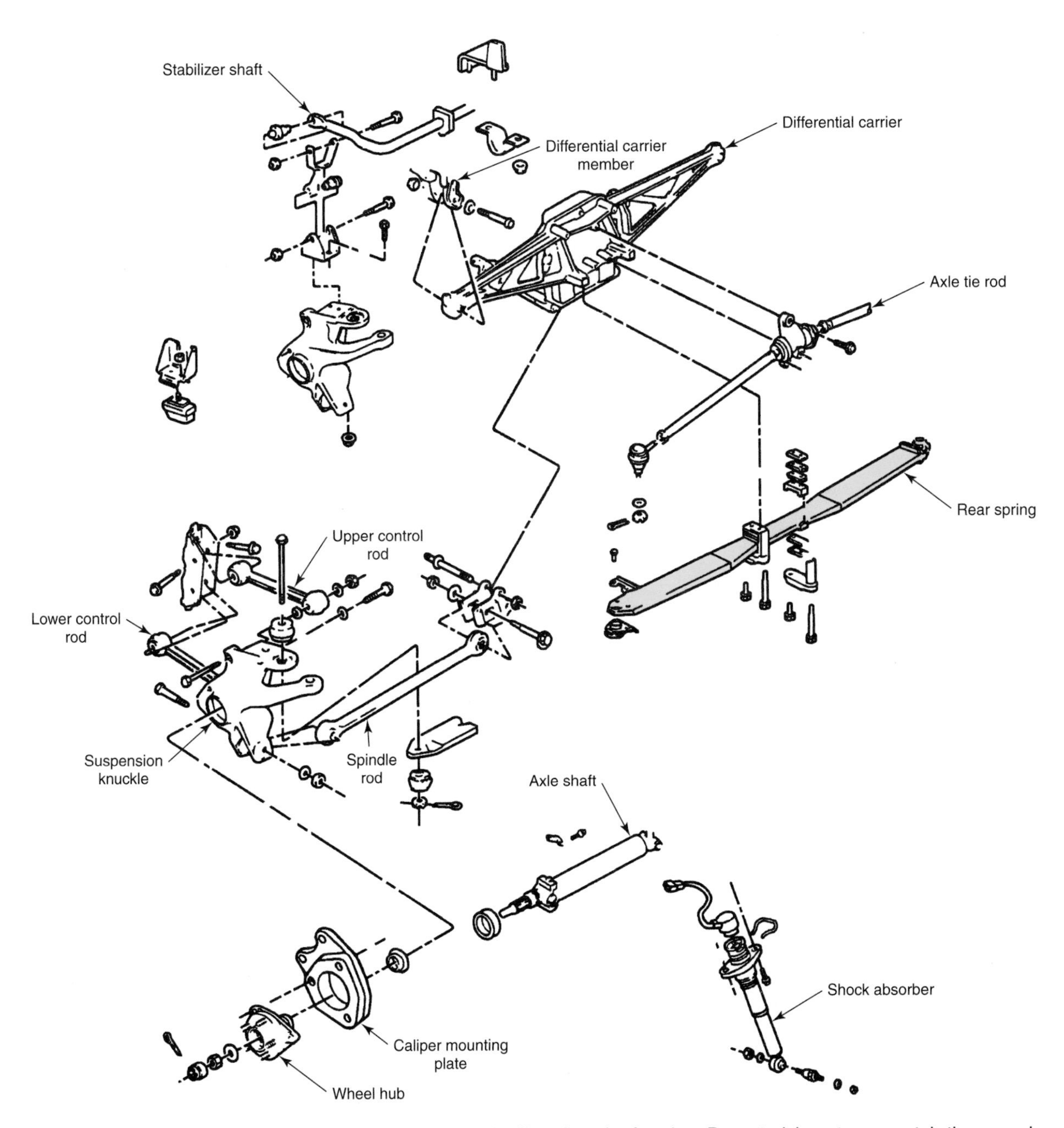

Figure 36-42. This independent rear suspension uses a single fiberglass leaf spring. Do not nick, cut, or scratch these springs, or complete spring failure may result! (General Motors)

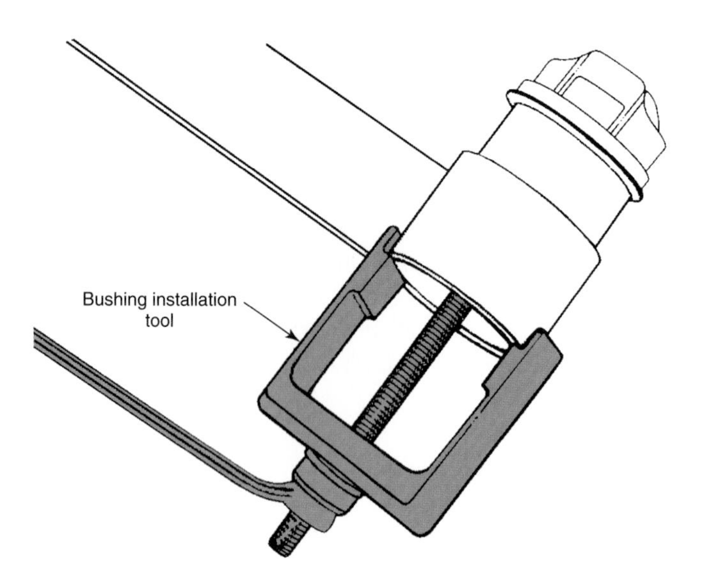

Figure 36-43. Installing a new leaf spring pivot bushing. (DaimlerChrysler)

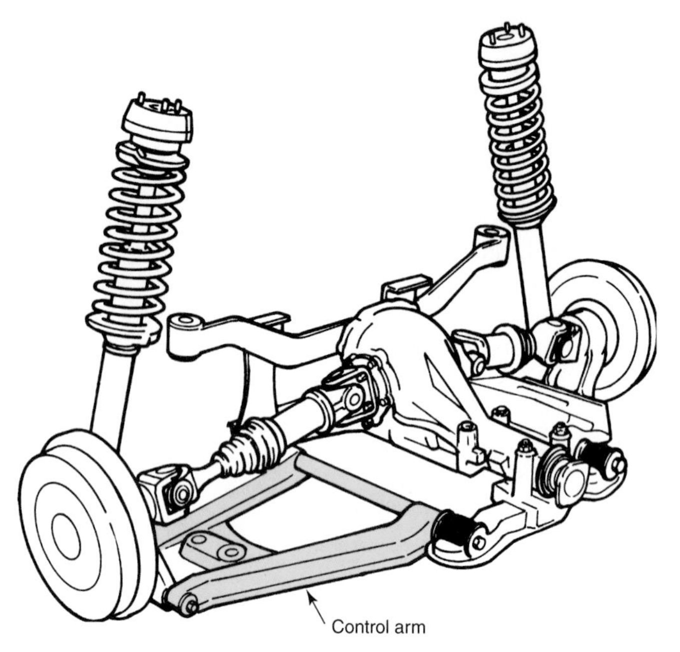

Figure 36-44. Be sure to always replace control arms to the proper side. (Moog)

Replacing Rear Suspension Control Arms and/or Bushings

When replacing rear suspension control arms or control arm bushings, it is important that the vehicle be at its normal standing height. The weight of vehicle must be on the rear wheels before tightening the control arm pivot bolts. This procedure applies to track bars as well. By allowing the vehicle to return to its normal standing height, the rubber bushings will be at rest (no twisting strain on them) while tightening the control arm bolts. This will allow them to flex in either direction from normal without damage to the bushings.

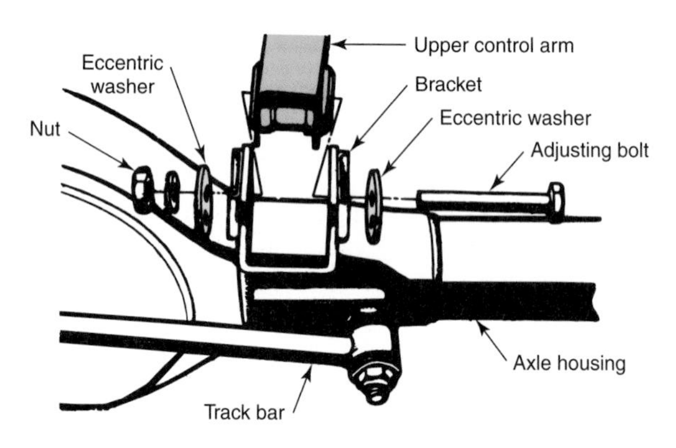

Figure 36-45. This rear axle housing may be tilted to adjust the pinion drive angle by varying the upper control arm length. Length may be altered by the eccentric washer setup.

Caution: Rubber bushings should be lubricated with rubber lube only. Never lubricate with engine oil, grease, or kerosene.

Be very careful to place the control arms on the proper side, **Figure 36-44.** Check the drive angle of the differential pinion and adjust as needed. Adjusting the length of the control arm or arms may change drive angle. Note in **Figure 36-45** the upper control arm frame-to-rear axle housing distance may be varied through the use of an eccentric washer. For complete details on differential pinion-angle adjustment, see Chapter 32, Axle and Driveline Service.

Automatic Level Control

Some vehicles are equipped with an ***automatic level control*** system. This system compensates for the weight of passengers or cargo. It can maintain normal curb height at the rear of the vehicle, as measured from the frame or bumper to the ground. Most systems will level the vehicle with additional weight of up to approximately 500 pounds (227 kg). A basic system will generally include an air compressor (vacuum or electrically-operated air pump), air tank (reservoir), pressure regulator, height control sensor, special shock absorbers or struts, connecting wire, tubing, hose, and an air dryer. See **Figure 36-46**. A hydraulic level control system using an oil pump is shown in **Figure 36-47.**

Height Control Sensor

A ***height control sensor,*** or ***height control valve,*** uses an overtravel lever (actuating arm) to detect distance changes between the frame and suspension. The unit is attached to the vehicle frame and is connected to a suspension arm or the rear axle housing. As the distance changes, the overtravel lever causes the control valve to admit more air to the shocks or exhaust some, either mechanically by moving valves or electrically by starting the compressor. **Figure 36-48** illustrates an electric height control sensor with linkage.

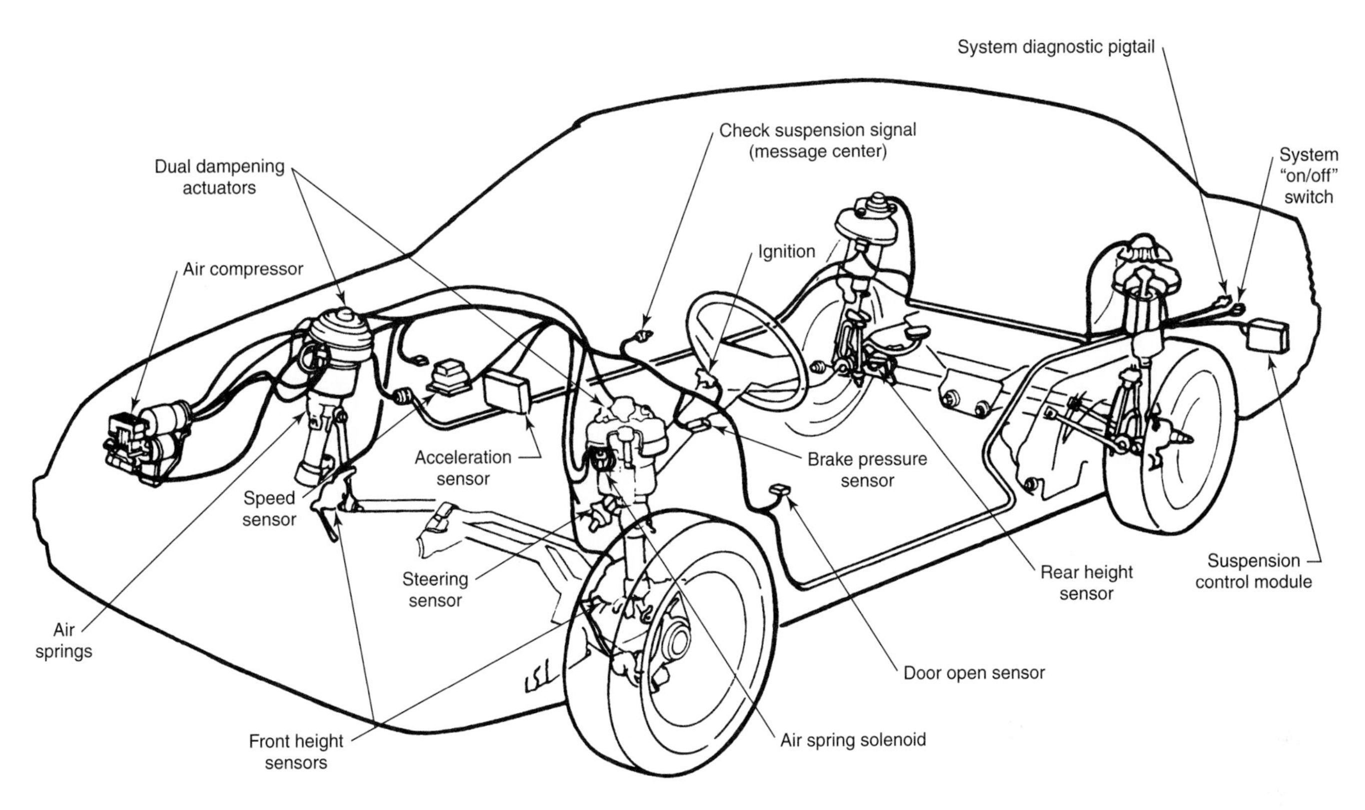

Figure 36-46. This vehicle has an automatic level control system. Carefully study the overall system. (Ford)

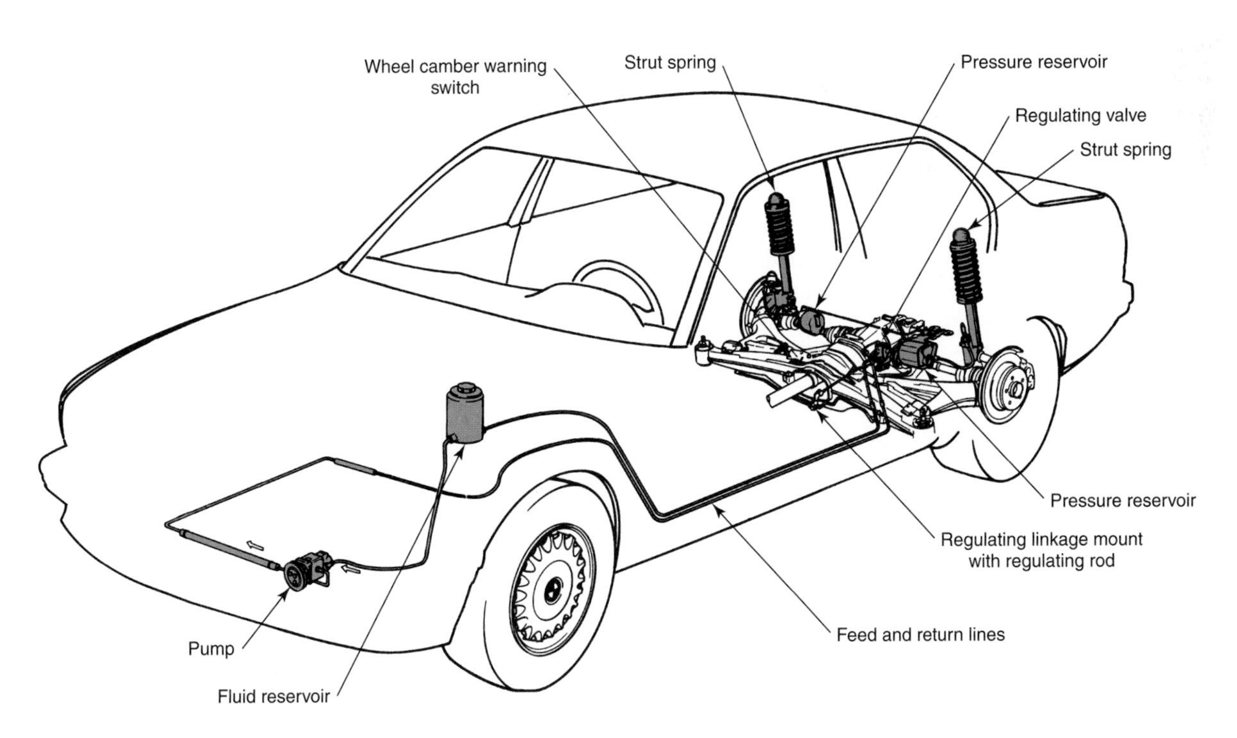

Figure 36-47. This type of load-leveling system uses a hydraulic power steering pump. Note the wheel camber switch, which triggers an instrument cluster light when overloading of the vehicle produces excessive wheel camber. (BMW)

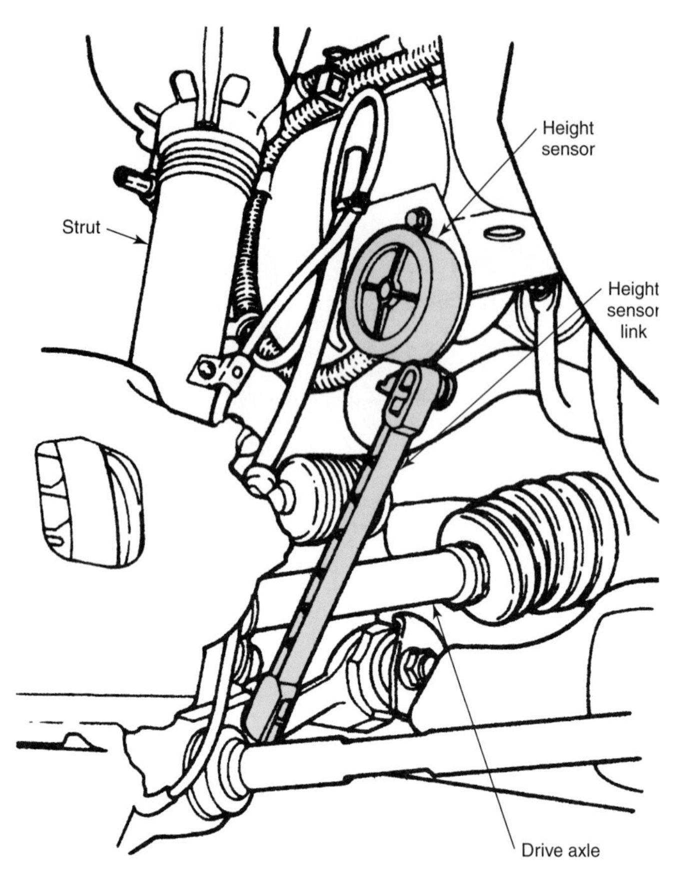

Figure 36-48. An electrically operated height control sensor with linkage. (Ford)

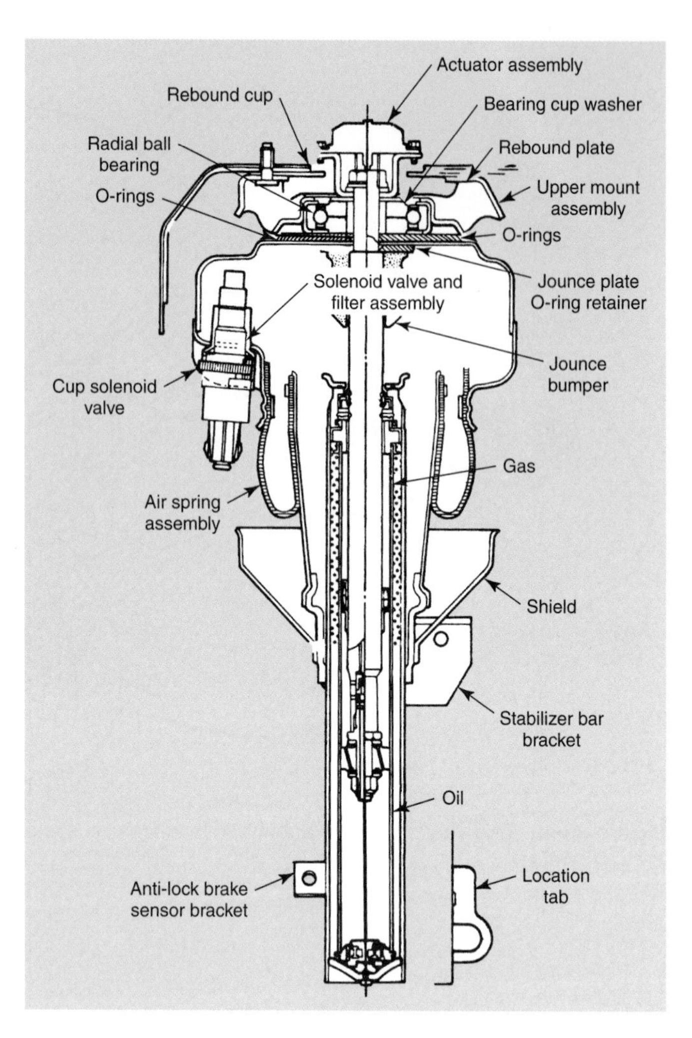

Figure 36-49. This is a cross-sectional view of an electronically controlled, air-operated strut. (Ford)

Air-Operated Struts and Shocks

The automatic level control system generally uses two air-operated struts or shock absorbers. The struts are hydraulically damped and operate in the conventional manner with the added benefit of being able to extend or decrease their operating range length with an air chamber, rubber boot, and fittings. A cutaway of an air-hydraulic strut is illustrated in **Figure 36-49.**

Automatic Level Control Service

The following is a general service procedure to be used before other checks are made. Determine the vehicle recommended ***curb height,*** or trim height. Start the engine and run it for 15–30 seconds. Place two people in the rear seat or the equivalent weight in the trunk and turn the ignition on. There should be a delay of 8–14 seconds with most systems before the vehicle begins to rise.

The height obtained when the compressor shuts off should be within 3/4″ (19.1 mm) of the specified curb height at the beginning of the test. Remove the people or equivalent weight from the vehicle. There should be an 8–14 second delay before the vehicle begins to lower. If the level control system fails this preliminary test, consult the vehicle manufacturer's service manual for any specific tests. Follow all diagnosis and repair steps exactly. Areas of inspection should include the following:

- Compressor.
- Height sensor.
- Tubing, hose, and connections.
- Air shocks.
- Electrical connections and wire.
- Electric relays.
- Fuses.
- Pressure regulator.
- Exhaust solenoid.

Computerized Ride Control Systems

Computerized ride control systems, also called ***active suspensions,*** are computer-controlled electronic suspension systems that can adapt to specific driving conditions, **Figure 36-50.**

Under normal driving conditions, a driver can pick the ride best suited for handling and comfort. For example, if a vehicle is being driven on a winding road, the driver can pick

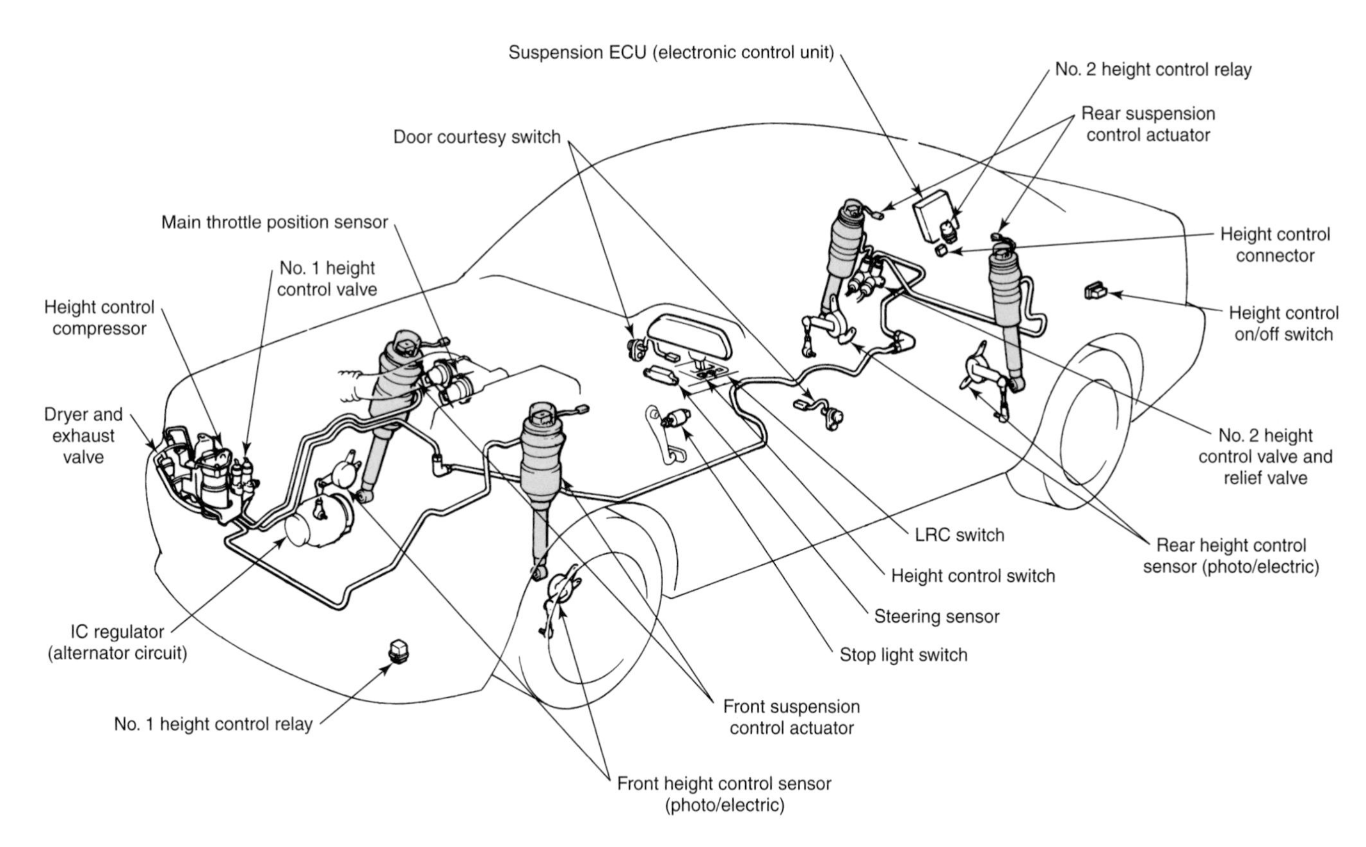

Figure 36-50. This vehicle is equipped with a computerized ride control "active suspension" system. Study all of the parts. (Toyota)

a firm setting. In this mode, the ride control module signals actuators to close valves on the adjustable shocks. This produces a stiff ride, allowing the vehicle to corner better. If the vehicle is being driven on a highway, however, the driver can switch to a soft mode. In this setting, the valves in the adjustable shocks are opened and a soft, comfortable ride is produced.

Under certain driving conditions, the control module will override the driver's selection. During periods of hard braking, a brake sensor signals the control module to close the valves on the adjustable front shocks. This stiffens the front shocks and minimizes front-end diving.

Under heavy acceleration, the acceleration sensor signals the control module to stiffen the adjustable rear shocks, preventing rear-end squatting, **Figure 36-51.** When the driver steers the vehicle into a sharp turn, a steering sensor sends an appropriate signal to the control module. The module signals the actuators to increase shock pressure on the outside of the vehicle. This action decreases the vehicle's leaning during turns. Other systems also provide programmed ride control. Typical ride control systems consist of the following components:

- Control module (uses signals from the steering and brake sensors to control actuators).
- Steering sensor (monitors the direction and speed of steering wheel rotation).
- Brake sensor (monitors brake system applications).
- Acceleration signal sensor (monitors rate of acceleration).
- Mode select switch (allows driver to adjust ride control system).
- Actuators (control the flow of hydraulic fluid in the adjustable shock absorbers).

Ride Control System Troubleshooting

Troubleshooting procedures for ride control systems vary among manufacturers. Consult the appropriate service manual for proper instructions. Most ride control systems have self-diagnostic capabilities, which simplify the troubleshooting process. The ride control module is usually located in the vehicle passenger compartment. It continuously monitors data sent by the various sensors and controls the output circuits of the appropriate systems. This unit also controls diagnostic system functions.

If the computer recognizes a problem in one of the input or output devices, it will alert the driver by triggering a trouble indicator light on the dashboard. A trouble code is then stored to aid the technician in making proper repairs. Always follow the manufacturer's recommendations when accessing diagnostic codes and servicing the system. A suspension control system check sheet is shown in **Figure 36-52.**

Tech Talk

Many automotive students dream of working for a racing team. While this would be a fun way to make a living, the chances that you will actually do this are slim. Out of the millions of automotive technicians, no more than a few thousand have paying jobs in the racing industry. The odds are at least 100,000 to 1 that you will never collect a paycheck from a racing organization. It is far more likely that you will be repairing the vehicles of people who barely drive the speed limit rather than repairing race cars.

There is nothing wrong with having an interest in working on high performance vehicles. However, chances are that you

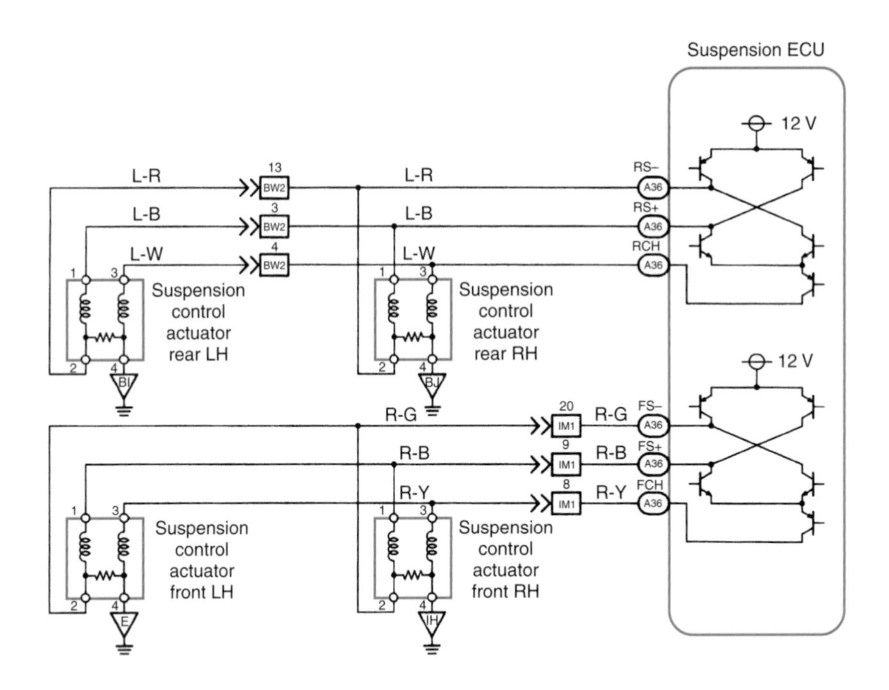

Control		Function
Anti-roll control		Changes spring rate and damping force to "firm" mode. This control suppresses rolling and minimizes change of the vehicle posture improving controllability.
Anti-dive control		Changes spring rate and damping force to "firm" mode. This control suppresses nose diving of the vehicle during braking and minimizes changes of the vehicle posture.
Anti-squat control		Changes spring rate and damping force to "firm" mode. This control suppresses squatting of the vehicle during acceleration and minimizes change of the vehicle posture.
High speed control		Changes spring rate to "firm" and damping force to "medium" modes respectively. This control improves driving stability and controllability at high speeds.
Rough road control		Changes spring rate and damping force to "medium" or "firm" mode as needed to suppress bottoming of the vehicle and thus improves riding comfort when driving on uneven roads.
Pitching control		Changes spring rate and damping force to "medium" or "firm" mode. This suppresses pitching of the vehicle when running over uneven road.
Bouncing control		Changes spring rate and damping force to "medium" or "firm" mode. This suppresses up and down bouncing of the vehicle when running over uneven road.

B

Figure 36-51. A—Active suspension electronic control unit and system electrical schematic. B—This chart illustrates the control and function of the active suspension system in different situations. (Lexus)

will only work on race cars in your spare time, and make a living repairing the engines, brakes, suspensions, automatic transmissions, and air conditioners of standard production vehicles. Therefore, it is to your advantage to know how to service these systems, not just how to build high performance vehicles.

SUSPENSION CONTROL System Check Sheet

TECHNICIAN'S NAME ____________

Customer's Name		Registration No.	
		Registration Year	/ /
		Frame No.	
Date Vehicle Brought In	/ /	Odometer Reading	km Mile

Date of Problem Occurrence		/ /
Frequency of Problem Occurrence		☐ Constant ☐ Sometimes (times per day, month) ☐ Once only
Conditions at Time of Problem Occurrence	Weather	☐ Fine ☐ Cloudy ☐ Rainy ☐ Snowy ☐ Various/Others
	Outdoor Temperature	☐ Hot ☐ Warm ☐ Cool ☐ Cold (Approx. °F(°C))
	Place	☐ Highway ☐ Suburbs ☐ Inner City ☐ Hill (☐ Up, ☐ Down) ☐ Rough Road ☐ Others ( )

Problem Symptom		
	☐ Malfunction in damping force and spring rate control.	☐ Cannot be changed by operating LRC switch. ☐ Anti-roll control does not operate. ☐ Anti-squat control does not operate. ☐ Anti-dive control does not operate. ☐ High speed control does not operate. ☐ Others ( )
	☐ Malfunction in vehicle height control	☐ Vehicle height cannot be changed by operating the height control switch. ☐ High speed control does not operate. ☐ Ignition Switch OFF Control does not operate. ☐ Others ( )
	☐ Others	

Diagnostic Trouble Code Check	1st Time	☐ Normal Code ☐ Malfunction Code (Code)
	2nd Time	☐ Normal Code ☐ Malfunction Code (Code)

Figure 36-52. This is a typical suspension control system check sheet used to record the owner's comments and the technician's diagnostic results. (Lexus)

Summary

Front suspension systems use coil springs, torsion bars, MacPherson struts, or a combination of each. Ball joints allow the up and down and swivel action of the spindle body. One ball joint is loaded (main load-carrying joint) while the other is non-loaded (follower joint). A ball joint may be tension or compression-loaded.

Some coil springs, for both conventional and MacPherson strut front suspensions, must be compressed for removal or installation. Others may be unloaded by slowly lowering the lower control arm. Be careful, loaded coil springs have tremendous force. Use a jack to lower the suspension arm.

Be sure to reinstall spring insulators where required. Position the spring end as specified. Do not nick or dent springs. Sagging springs (broken or weak) may be detected by measuring the vehicle's curb height. Remove the torsion bar by unloading the front suspension arms and then backing off the bar tension anchor. The torsion bar must be free of nicks and burrs. Lubricate both ends of the torsion bar before installing it. Never heat a torsion bar or the bar anchors. Set the vehicle's standing height by turning torsion bar adjusting bolt.

Shock absorbers and strut dampers must function well to provide safe handling and a comfortable ride. Check shock operation by bouncing vehicle or by bench testing shocks. Most modern shocks are gas-filled for better control. Some shocks are adjustable. If a stone shield is used, place it so it faces the correct direction. Do not overtighten the shock bushing fasteners.

To check ball joint wear, the joints must be unloaded. When a coil spring is between lower arm and frame, unload the joint by placing a jack beneath the lower control arm. When a coil spring is between the upper suspension arm and body, place the jack beneath the frame. In some cases, a wedge must be placed between the underside of the upper arm and the frame. Check ball joint wear by measuring either axial (up-and-down) movement or side-to-side movement of the tire.

The use of a suitable puller eases ball joint removal. Striking the spindle with a hammer also helps. Always use great care when removing ball joints to prevent the coil spring from being released violently. When bolting new ball joints to suspension arms, use special bolts supplied for the purpose. When the joint is welded to the arm, both the arm and joint should be replaced as an assembly.

When removing control arms, note the number and location of camber and caster shims or other adjusting devices. When pressing suspension arm bushings out or in, use a stiffener tool to prevent the arms from distorting. Rear suspension systems commonly use either leaf or coil springs. Replace sagged rear coil springs and weak or broken leaf springs. Place the weight of the vehicle on the spring shackle bushings before tightening bushing fasteners. When installing rear axle control arms, check the differential pinion shaft drive angle. Place the vehicle at its normal standing height before tightening the control arm bushing bolts. Use only rubber lubricant on rubber bushings.

Some vehicles are equipped with automatic level control. This system keeps curb height correct when the vehicle has additional weight or passengers. The system uses an air compressor, valves, tubing and hose, height sensor, and air shocks or struts. Some vehicles employ an electronically controlled, air-operated, active suspension at all wheels.

Review Questions—Chapter 36

Do not write in this book. Write your answers on a separate sheet of paper.

1. Torsion bars _____ to absorb road shocks.
   (A) bend
   (B) compress
   (C) twist
   (D) expand
2. Front suspension ball joints can be _____ or _____ loaded.
3. Name three results of excessive ball joint wear.
4. Before checking ball joint wear, the joint must be _____.
5. Allowable wear tolerances for ball joints vary between _____.
   (A) makes of vehicles
   (B) upper and lower joints
   (C) cars and trucks
   (D) All of the above.
6. Place the following ball joint removal steps in order, from first to last.
   (A) Remove the cotter pin from the ball joint stud.
   (B) Strike the spindle with a hammer.
   (C) Loosen the ball joint stud nut.
   (D) Place a jack under the lower control arm.
   (E) Remove the ball joint stud nut completely.
7. To avoid distorting the suspension arm when pressing out bushings, use a _____.
   (A) stiffener tool
   (B) vise
   (C) new bushing
   (D) small hammer
8. What important rear axle angle can be adjusted by altering the length of rear control arms?
9. Bouncing the vehicle is one way to test shock general condition. After releasing the vehicle, if the shocks are good, how many bounces (oscillations) should the vehicle make before stopping?
   (A) 1
   (B) 2
   (C) 3
   (D) 4

10. Which of the following is the best description of a gas-charged shock absorber?
    (A) A shock absorber containing pressurized gas instead of hydraulic fluid.
    (B) A shock absorber containing pressurized gas and hydraulic fluid.
    (C) A shock absorber filled with compressed air from a vehicle-mounted compressor.
    (D) A shock absorber filled with compressed air from an outside source.

ASE-Type Questions

1. Technician A says MacPherson strut suspensions have only one ball joint per wheel. Technician B says on systems having two ball joints per wheel, one of the ball joints is the main load-carrying joint while the other acts as a follower or guide. Who is right?
    (A) A only.
    (B) B only.
    (C) Both A and B.
    (D) Neither A nor B.
2. Technician A says all suspensions contain parts under spring tension. Technician B says all suspensions contain coil springs, either at the front or rear. Who is right?
    (A) A only.
    (B) B only.
    (C) Both A and B.
    (D) Neither A nor B.
3. Technician A says lowering the lower arm to release coil spring tension can be dangerous if done incorrectly. Technician B says a spring compressor is always needed to remove front suspension system coil springs. Who is right?
    (A) A only.
    (B) B only.
    (C) Both A and B.
    (D) Neither A nor B.
4. The standing height of a vehicle equipped with torsion bars can be changed by adjusting the _____.
    (A) torsion bar length
    (B) torsion bar anchor tension
    (C) rear leaf spring height
    (D) rear leaf spring length
5. All of the following statements about hydraulic fluid leakage at the MacPherson strut damper are true, *except:*
    (A) excessive leakage means the damper should be replaced.
    (B) slight leakage is a normal condition.
    (C) leakage is usually the result of seal failure.
    (D) when the damper is replaced, the spring should be replaced also.
6. Technician A says all MacPherson strut assemblies incorporate a coil spring around the strut damper. Technician B says the action of MacPherson strut dampers is the same as that of conventional shock absorbers. Who is right?
    (A) A only.
    (B) B only.
    (C) Both A and B.
    (D) Neither A nor B.
7. The procedure for replacing a MacPherson strut damper assembly and a MacPherson strut spring in most cases is _____.
    (A) identical
    (B) similar
    (C) completely different
    (D) dependent on the manufacturer
8. When replacing a front MacPherson strut assembly, the technician must gain access to the upper strut bolts by _____.
    (A) removing the wheel
    (B) opening the hood
    (C) pulling up the front carpet
    (D) removing the inner fenders
9. Technician A says a common service procedure performed on leaf springs is bushing replacement. Technician B says a common service procedure performed on control arms is bushing replacement. Who is right?
    (A) A only.
    (B) B only.
    (C) Both A and B.
    (D) Neither A nor B.
10. All of the following are true statements about computer ride control systems, *except:*
    (A) under certain driving conditions, the control module will override the driver's selection.
    (B) in the soft mode, the vehicle will corner well.
    (C) most ride control systems have self-diagnostic capabilities.
    (D) during sharp turns, the actuators increase shock pressure on the outside of the vehicle.

Suggested Activities

1. Write a report listing the differences between conventional and MacPherson strut suspensions.
2. Obtain the vehicle's specifications and measure the vehicle's ride height. Determine whether the ride height can be adjusted. If the ride height is adjustable, correct improper ride height. If the ride height is not adjustable, determine which parts must be replaced to restore the correct ride height.

3. Inspect suspension components for wear. The check should include all ball joints, control arm bushings, and the strut rod and stabilizer bushings if used. Demonstrate the checking procedure to other members of the class.
4. Replace shock absorbers on a conventional suspension system. Make a list of the tools that you needed to perform the job.
5. Replace a MacPherson strut assembly, including removal disassembly, parts replacement, reassembly, and reinstallation on the vehicle. Write a short report, listing the tools that you needed to perform the job and explaining why the vehicle must be aligned after strut replacement.

37

Steering System Service

After studying this chapter, you will be able to:

- Explain the differences between conventional and rack-and-pinion steering systems.
- Identify the components of conventional steering systems.
- Identify the components of rack-and-pinion steering systems.
- Diagnose problems in conventional and rack-and-pinion steering systems.
- Summarize the construction, operation, and service of steering linkage components.

Technical Terms

Conventional steering system
Rack-and-pinion steering system
Steering shaft bearings
Recirculating-ball design
Worm bearing preload
Over-center adjustment
Power cylinder
Control valve
Thrust bearing preload
Worm-to-rack-piston preload
Pitman shaft over-center preload
Sector shaft
Vane
Slipper
Roller
Fluid level
Pressure test
Bleeding
Power steering coolers
Pressure-sensing switch
Steering linkage
Pitman arms
Idler arms
Tie rod ends
Adjuster sleeves
Center links
Electronic assist steering

Steering systems, though varying in design, all contain the same basic elements. They consist of a steering wheel, steering shaft, steering gearbox or rack-and-pinion assembly, steering arms, and steering knuckle assemblies. This chapter covers both conventional and rack-and-pinion steering systems.

STOP Warning: When working on a vehicle equipped with a supplemental restraint system (air bag), always follow the manufacturer's service, repair, and handling instructions. Failure to do so can result in personal injury caused by the unintentional activation of the system. Neglecting the manufacturer's precautions can also prevent the air bag from inflating during an accident, resulting in occupant injury or death. For more information on air bag systems, see Chapter 13, Chassis Electrical System.

Types of Steering Systems

The two most common types of steering systems are the ***conventional steering system***, or parallelogram steering system, and the ***rack-and-pinion steering system.*** A typical conventional steering system is shown in **Figure 37-1,** and a rack-and-pinion steering system is illustrated in **Figure 37-2.**

Steering Column and Steering Wheel

Because of the number of steering column designs, only a general discussion of steering columns and steering wheel removal and installation is presented here. Squeaks, roughness, binding, and other steering system problems can be caused by worn, dirty, or dry ***steering shaft bearings*** in the steering column. Column looseness or misalignment also can cause trouble. **Figure 37-3** is an exploded view of a typical steering column. An exploded view of a tilt steering column assembly is illustrated in **Figure 37-4.** Consult the manufacturer's service manual for specific service procedures for each

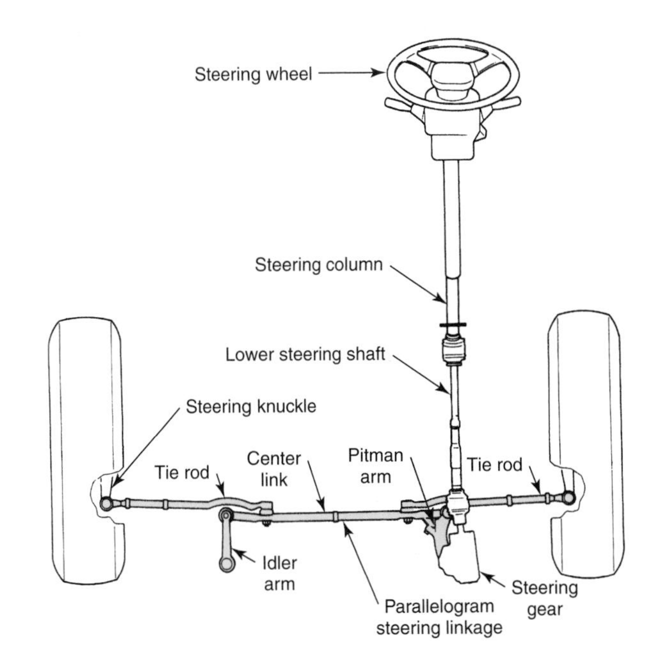

Figure 37-1. Conventional steering system components. (General Motors)

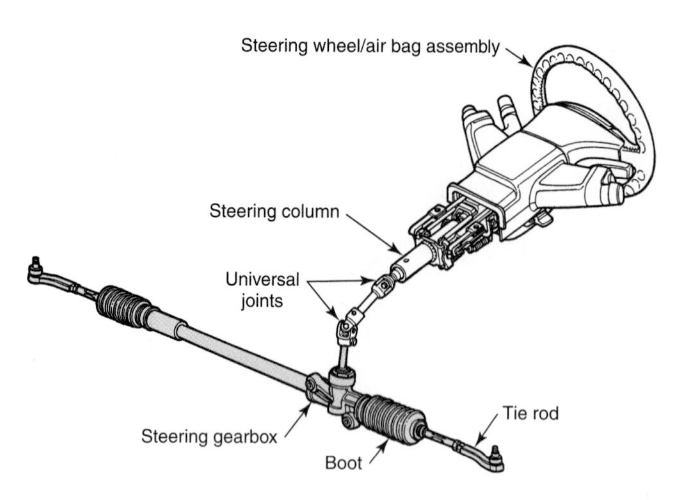

Figure 37-2. Manual rack-and-pinion steering assembly. (Hyundai)

steering column. If the steering wheel is mounted on the steering shaft so that marks are in alignment, the spoke position, if incorrect, must be altered by moving the tie rod adjuster sleeves. For complete details on this operation, see the toe-in and steering wheel position adjustment in Chapter 39, Wheel Alignment.

Steering Wheel Service

If the vehicle is equipped with an air bag system, be sure to disable it according to the manufacturer's recommended procedure. Remove the steering wheel center cap or inflator module and any other parts required to expose the steering wheel retaining nut. As a safety measure, some columns are fitted with a retaining ring that snaps into a groove above the nut. Remove the ring and the nut carefully. Check the steering wheel and shaft for alignment marks. If none are used and if the wheel is to be replaced in the same position, mark the steering wheel and shaft. Use a suitable puller to remove the wheel. Refer to **Figure 37-5.** Never hammer on the end of the steering shaft. Damage to the column, steering shaft, and/or collapsible joint can occur.

Install the parts in the reverse order of their removal. Install the wheel on the steering shaft, aligning the marks to ensure proper positioning. See **Figure 37-6.** If no aligning marks are used, center the steering gear and mount the wheel with the spokes in proper position for straight driving. Make sure the wiring for the horns, air bag, and steering wheel controls, (if used) is not chafed or pinched as the wheel is installed. Coat the steering-shaft threads with locking compound to hold the nut in place. Install the washer and retaining nut. Torque the retaining nut to factory specifications. Reinstall all remaining components and enable the air bag system, if equipped.

Conventional Steering Gears

This section covers the steering gears used in most older vehicles and many current cars and light trucks. The steering gear must convert the rotation of the steering wheel into

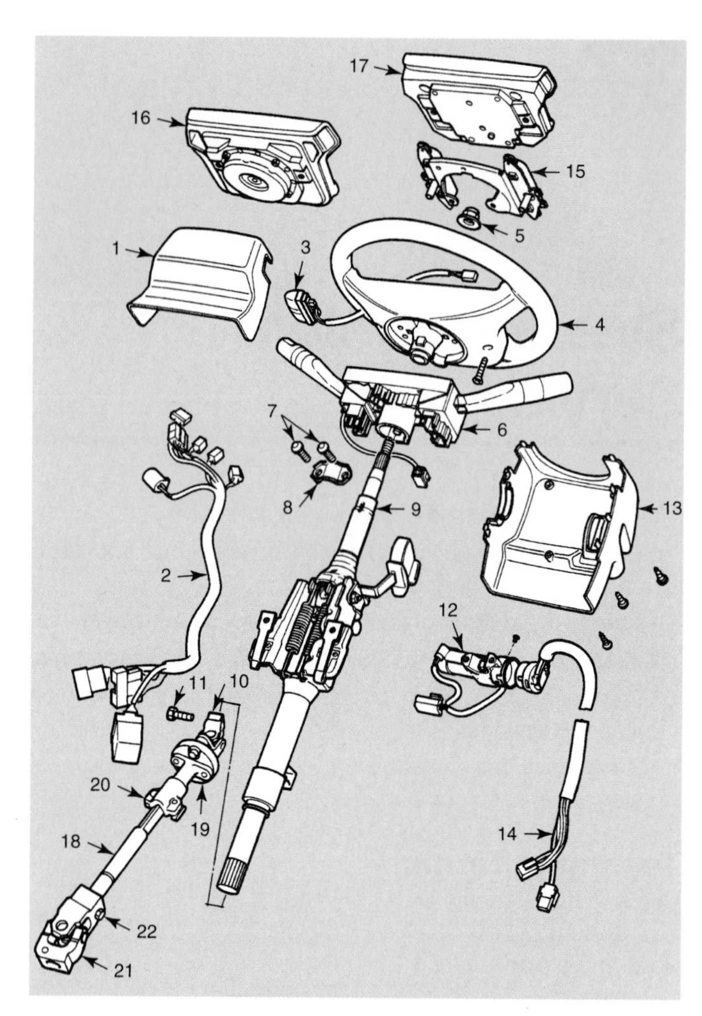

Figure 37-3. An exploded view of a typical steering column. 1—Upper cover. 2—Wiring harness. 3—Cruise control harness. 4—Steering wheel. 5—Nut. 6—Turn signal and wiper control assembly. 7—Bolt. 8—Column bracket. 9—Steering shaft. 10—Pinch collar. 11—Pinch bolt. 12—Ignition lock cylinder. 13—Lower cover. 14—Ignition wiring harness. 15—Support plate. 16—Horn pad. 17—Inflator module. 18—Intermediate steering shaft. 19—Steering collar. 20—Pinch collar. 21—Stub shaft pinch collar. 22—Pivot joint. (Land Rover)

side-to-side motion at the steering linkage. Conventional steering gears use the ***recirculating-ball design.*** The gearbox is attached to the frame and is usually connected to the steering shaft by a shock-absorbing universal joint.

Steering Gear Inspection

A periodic inspection of the steering gear is important to safety as well as ease of handling. Check gear alignment and lubricant level. Turn gear from full left to full right and back to detect any roughness or binding. Shake steering wheel as much as possible without moving the front wheels (engine off) to check for gear wear or misadjustment. If needed, adjust worm shaft thrust preload and pitman shaft over-center preload. Inspect the flexible coupling between the gear housing and steering shaft. See **Figure 37-7.**

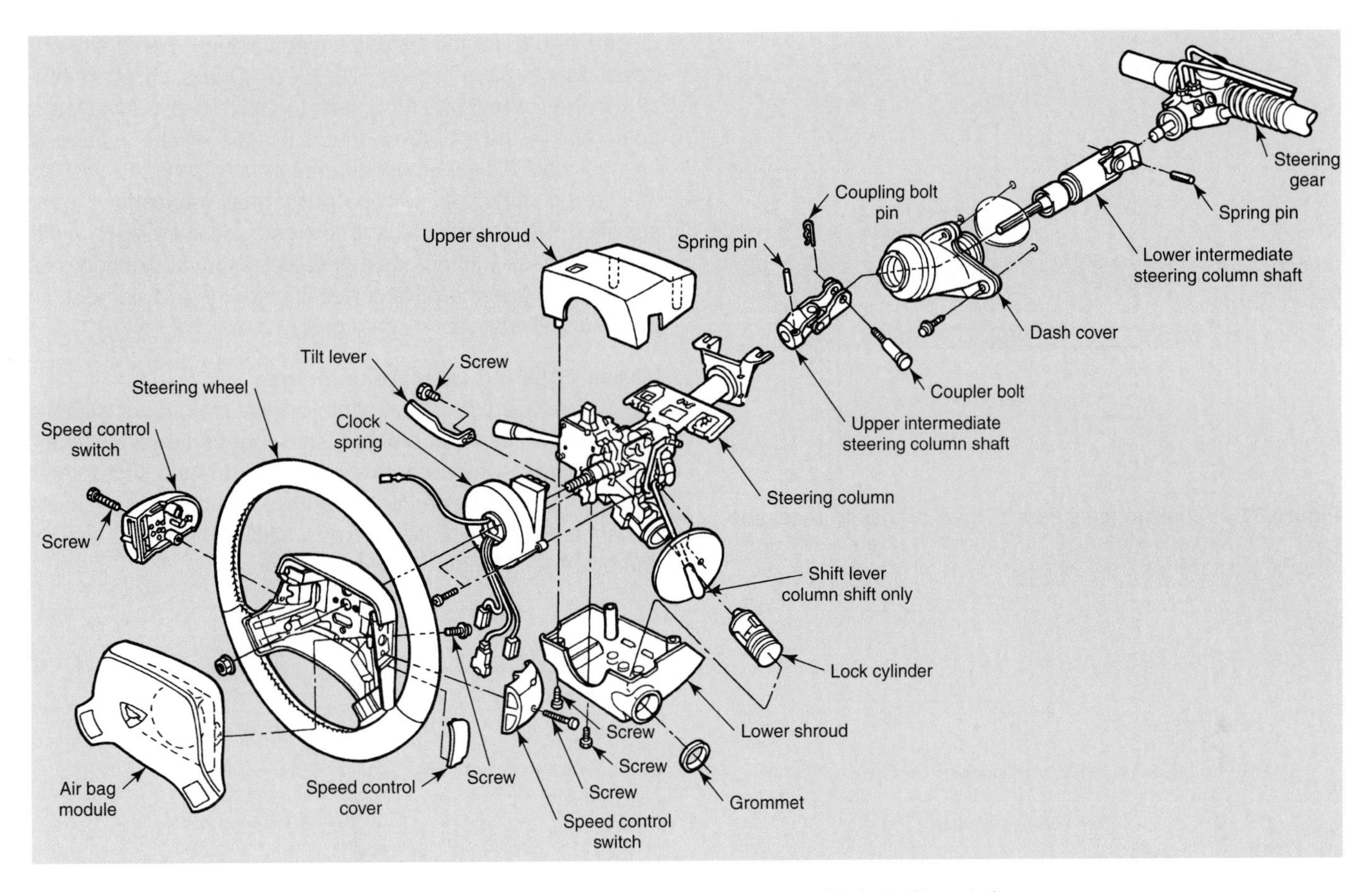

Figure 37-4. Exploded view of a tilt steering column. Note the air bag module. (DaimlerChrysler)

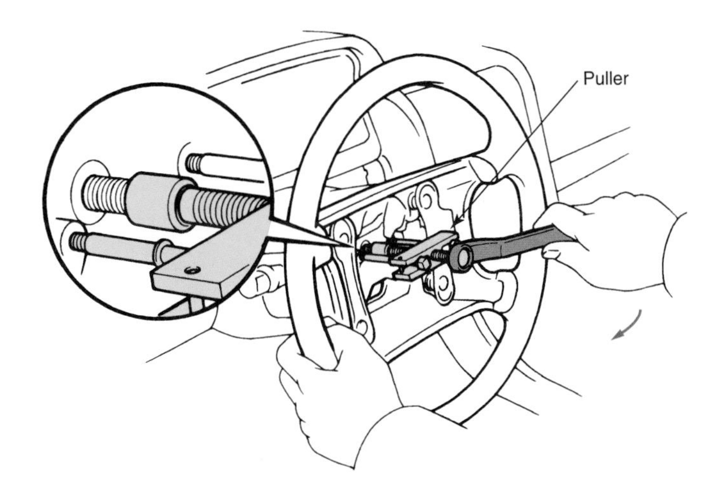

Figure 37-5. Using a puller to remove a steering wheel. Be careful not to damage the steering shaft threads. (Lexus)

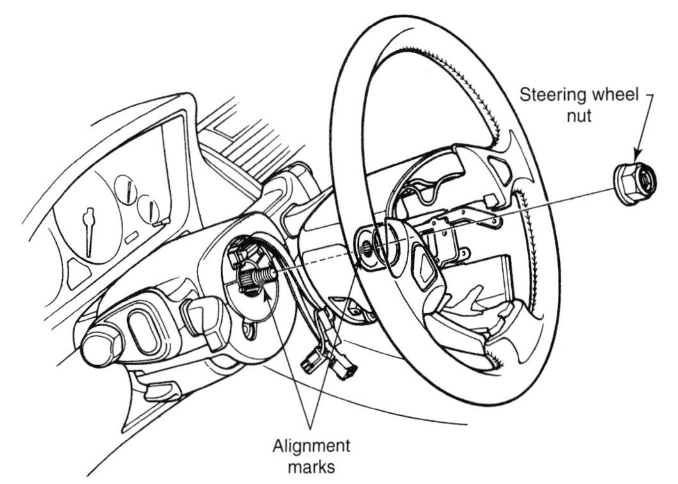

Figure 37-6. When replacing the steering wheel, be sure to align the alignment marks. (Honda)

Manual Steering Gear

Common manual steering gears use the recirculating-ball design. The gearbox is attached to the frame and is usually connected to the steering shaft by a shock-absorbing universal joint. One form of the recirculating-ball, worm-and-nut steering gear is shown in **Figure 37-8.** A different recirculating-ball, worm-and-nut steering gear is illustrated in **Figure 37-9.** Note how the pitman arm is attached to the pitman shaft (cross shaft, sector shaft, etc.). A rack-and-pinion gearbox is illustrated in **Figure 37-10.**

Checking Manual Steering Gear Lubricant Level

Clean off the filler plug and surrounding area. Remove the filler plug to check the level of the gear lubricant. If needed, fill the gearbox to the indicated level using lubricant

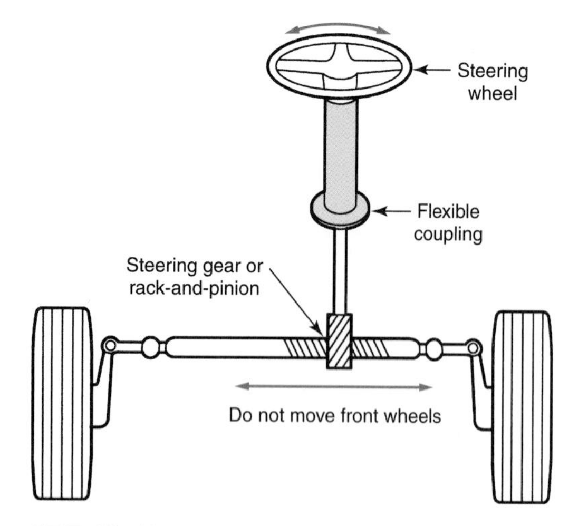

Figure 37-7. Shaking the steering wheel slightly to check for gear wear and adjustment. Do not turn the steering wheel far enough to move the front wheels. (General Motors)

recommended by the vehicle's manufacturer. Some steering gears do not have regular filler plugs. One such gearbox is shown in **Figure 37-11.** To check the fluid level in this type of gear, remove the two cover-attaching bolts. Fill the gearbox at fastener hole A until lubricant appears at hole B. Always use the correct lubricant. Multipurpose gear lubricant is often specified. Occasionally, a soft, extreme pressure (E.P.) multi-purpose chassis lubricant is used as a manual steering gear lubricant. Unless the oil is contaminated, there is no need for periodic steering gear oil changes.

Manual Steering Gear Adjustments

There are two adjustments on the majority of steering gears, worm bearing preload and over-center adjustment, which is the clearance between the ball nut and sector teeth with the gear in the center of its travel. Before making these adjustments, make sure that any binding is not caused by gearbox and steering column misalignment.

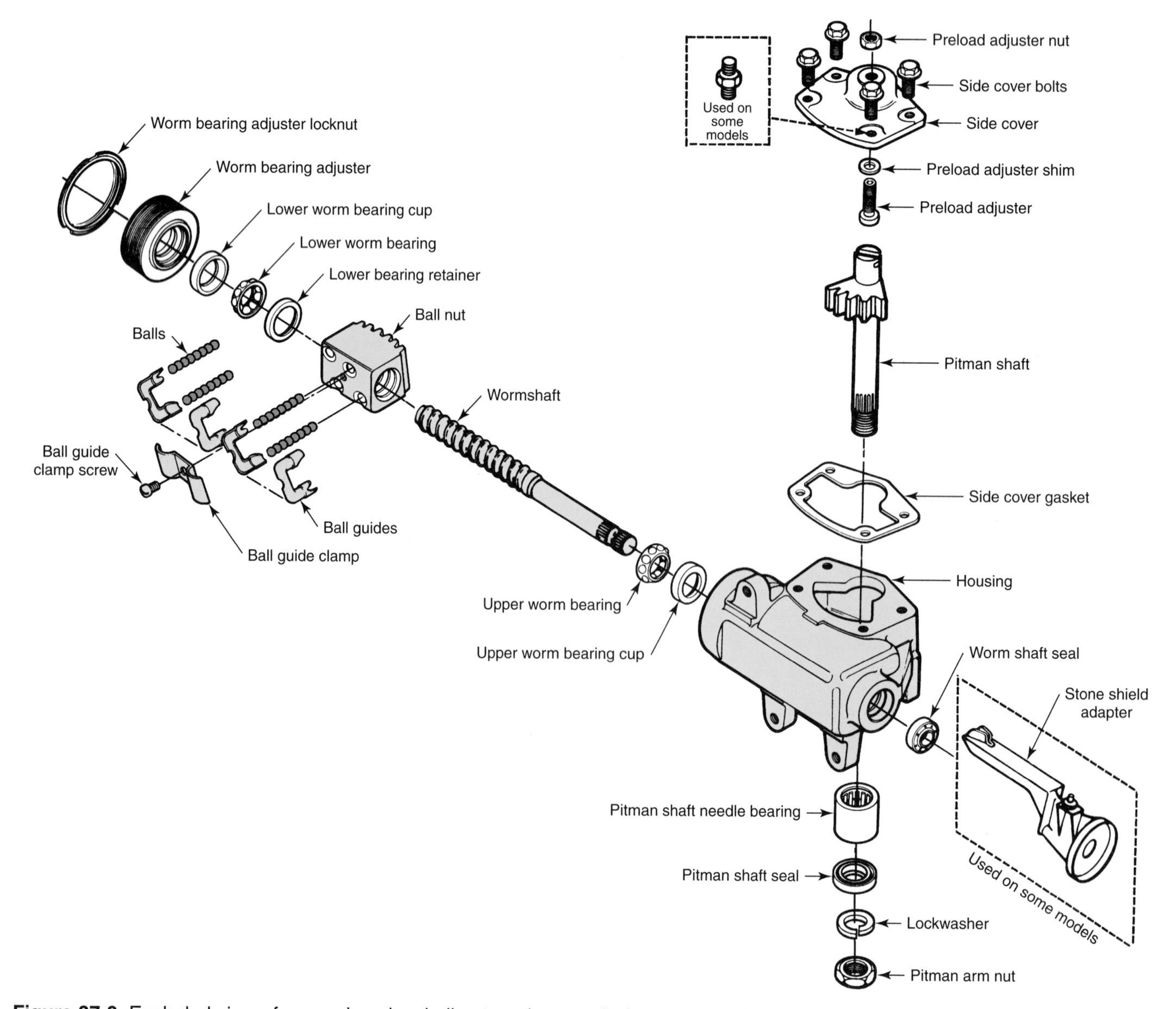

Figure 37-8. Exploded view of a gear housing, ball nut, and worm shaft assembly. (Saginaw Division of General Motors)

Caution: Adjust the worm shaft bearing preload first, with the gear in an off-center position. Make the over-center adjustment last, with the gear in the center of its travel.

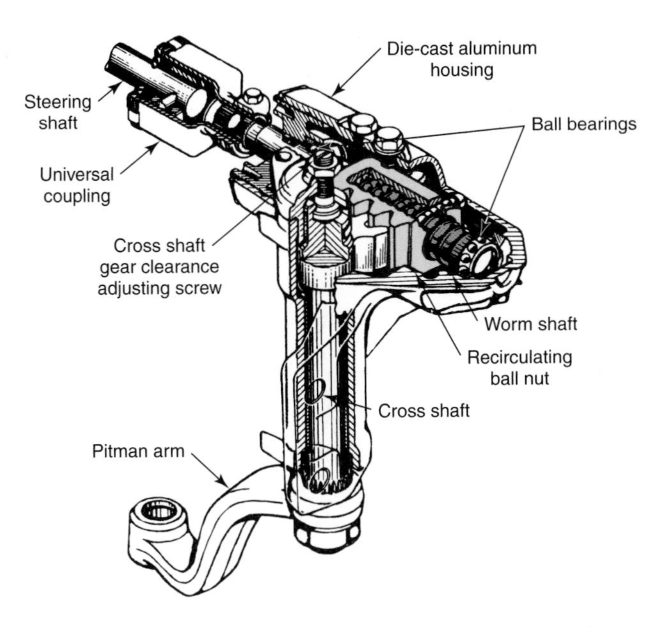

Figure 37-9. Another type of recirculating-ball, worm-and-nut steering gear. (DaimlerChrysler)

Worm Bearing Preload Adjustment

To make the ***worm bearing preload*** adjustment, remove the pitman arm from the shaft, a procedure that is discussed later in this chapter. Loosen the pitman shaft adjusting screw locknut. Back off the adjusting screw two or three turns. See **Figures 37-12** and **37-13.** Remove any steering system load on the worm gear by turning the steering wheel the specified number of turns away from the center position. Do not continue turning the steering wheel until it strikes the stops, as

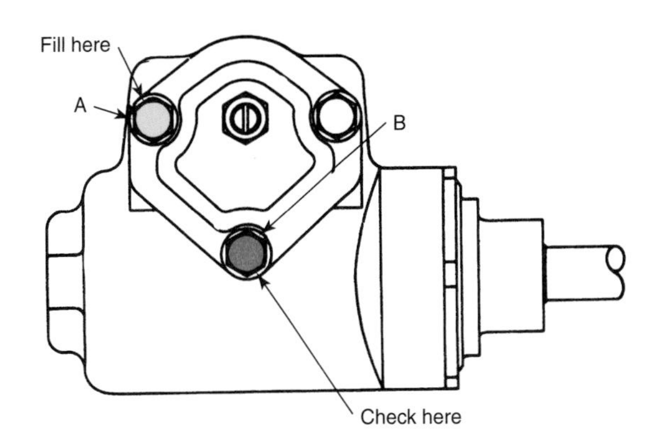

Figure 37-11. Remove the gear cover fasteners to check and adjust the lubricant level. (General Motors)

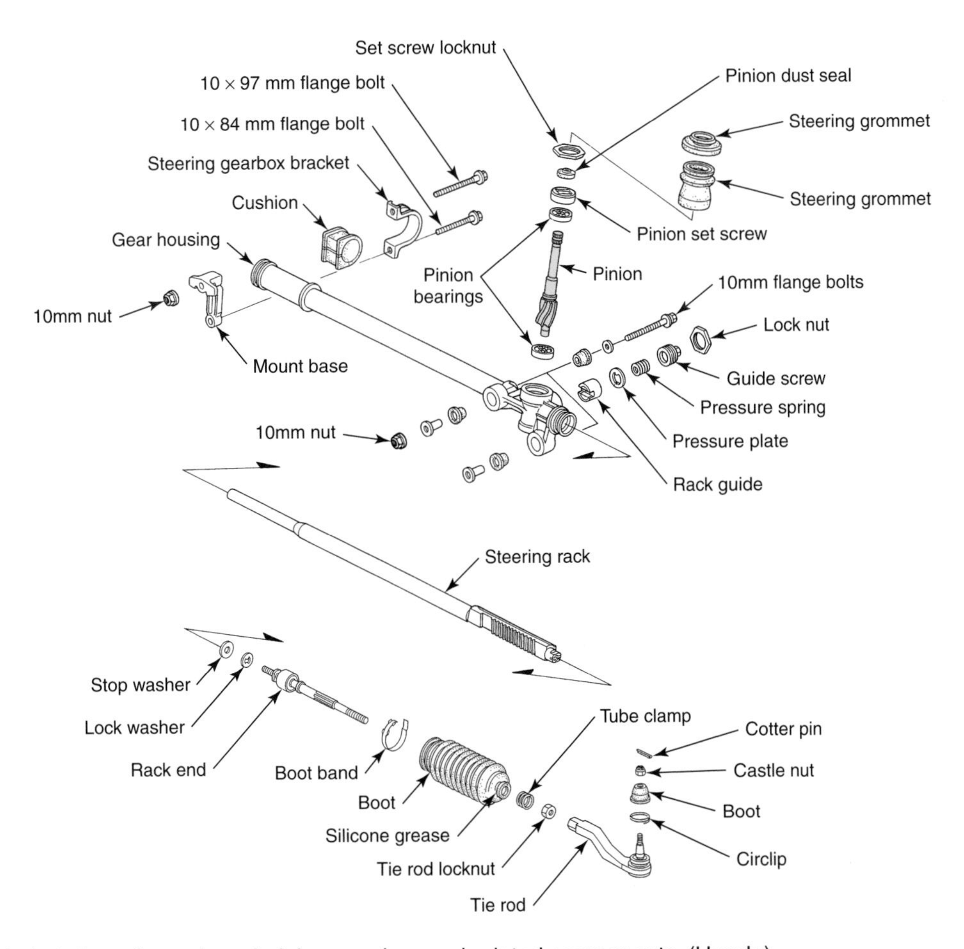

Figure 37-10. Exploded view of a rack-and-pinion gearbox and related components. (Honda)

this would damage the gearbox. Always turn the steering wheel slowly if the pitman arm is disconnected.

Use an inch-pound torque wrench to turn the steering wheel toward the center position, **Figure 37-14.** Note the reading while turning the steering wheel the specified number of turns to either side of center. Compare with factory specifications. If any adjustment is required, loosen the worm shaft bearing-adjuster locknut. Turn the bearing adjuster as required to produce the specified worm shaft bearing preload. See **Figure 37-15.** Tighten the locknut securely and recheck turning torque. The wheel should turn freely from stop to stop without binding or roughness. If roughness is present, the worm bearings may be worn and must be replaced. When worm shaft bearing preload adjustment is complete, proceed with the over-center mesh adjustment.

Pitman Shaft Over-Center Adjustment

To make the ***over-center adjustment,*** which is the depth of gear mesh at the center of its travel, determine the exact center of the steering gearbox travel. Then, turn the pitman shaft's lash-adjuster screw clockwise until all play is removed. Tighten adjuster screw locknut. Check the pull required to move the steering wheel through the center high point with an inch-pound torque wrench. To do this, first center the gear. Move the wheel an additional one-half turn to the left. Then, use the torque wrench to pull the steering wheel one full turn to the right. Note the highest reading and compare with manufacturer's specifications.

Loosen the locknut and move the pitman shaft lash-adjuster screw as required. Tighten the locknut securely and recheck the over-center adjustment. The pull required

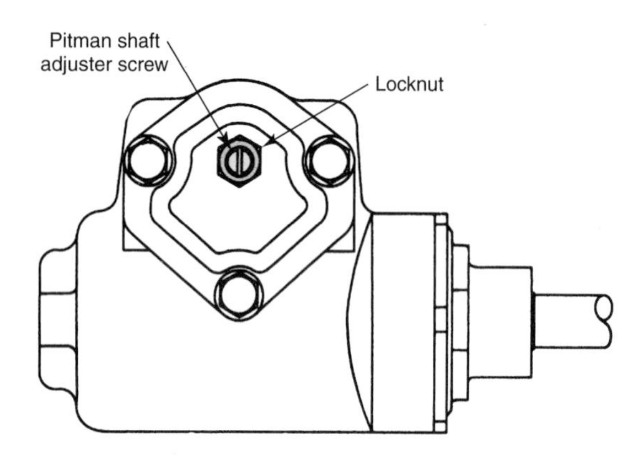

Figure 37-12. Loosen the locknut and unscrew the pitman shaft adjuster screw. (General Motors)

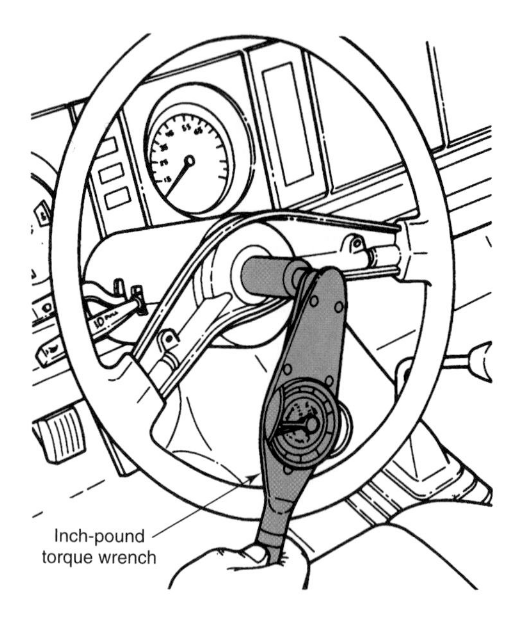

Figure 37-14. Using an inch-pound torque wrench to check worm shaft preload. Try to maintain a constant turning speed with the wrench for the most accurate reading. (Ford)

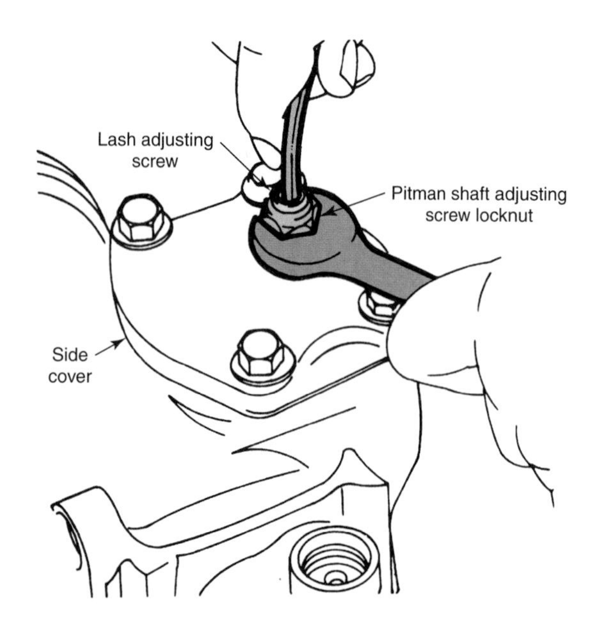

Figure 37-13. Turning the pitman shaft lash-adjusting screw. Note that some locknuts have left-hand threads. (Jeep)

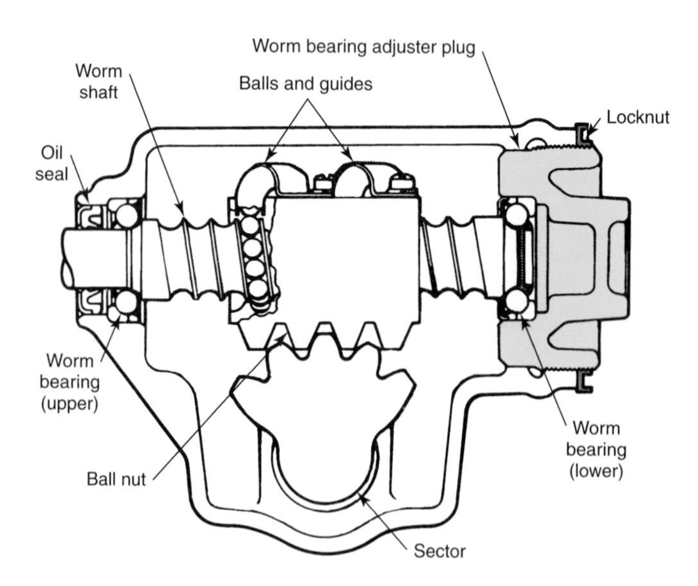

Figure 37-15. After loosening the locknut, the worm shaft's bearing adjuster plug is turned until bearing preload is set to specs. (DaimlerChrysler)

represents the original worm preload plus the over-center load. Turning the lash-adjuster screw to force the pitman shaft inward reduces the clearance between the teeth on the pitman shaft sector gear and the teeth on the ball nut. See **Figure 37-16.**

Some manufacturers specify the use of a spring scale on the steering wheel in place of a torque wrench. If a scale is used, it must be accurate. Attach as shown in **Figure 37-17.** Pull on the scale. Keep it in line with the direction of wheel travel.

Manual Steering Gear Overhaul

Remove the pitman arm. Remove the universal joint between the worm shaft and the steering shaft. Remove the fasteners holding the gear housing to the frame and remove the gearbox. Some cars require removal of the steering wheel or the entire steering column to facilitate gearbox removal. Clean the exterior of the housing thoroughly. Disassemble the gearbox according to the manufacturer's instructions. Generally, the pitman shaft cover should be removed to gain access to the pitman shaft. After the pitman shaft is removed from the housing, remove the steering shaft and ball nut assembly. When withdrawing the worm shaft and ball nut assembly, hold it in a horizontal position.

Disassemble and clean the worm shaft and all other parts as detailed in the manufacturer's service manual. Inspect bearings, cups, and worm shaft bearing surfaces for pitting, galling, and wear. Check the worm shaft, sector teeth, ball nut, and ball nut bearing surfaces. Inspect the pitman shaft for wear, scoring, and galling. Replace parts as needed. Make sure that new gaskets and seals are used.

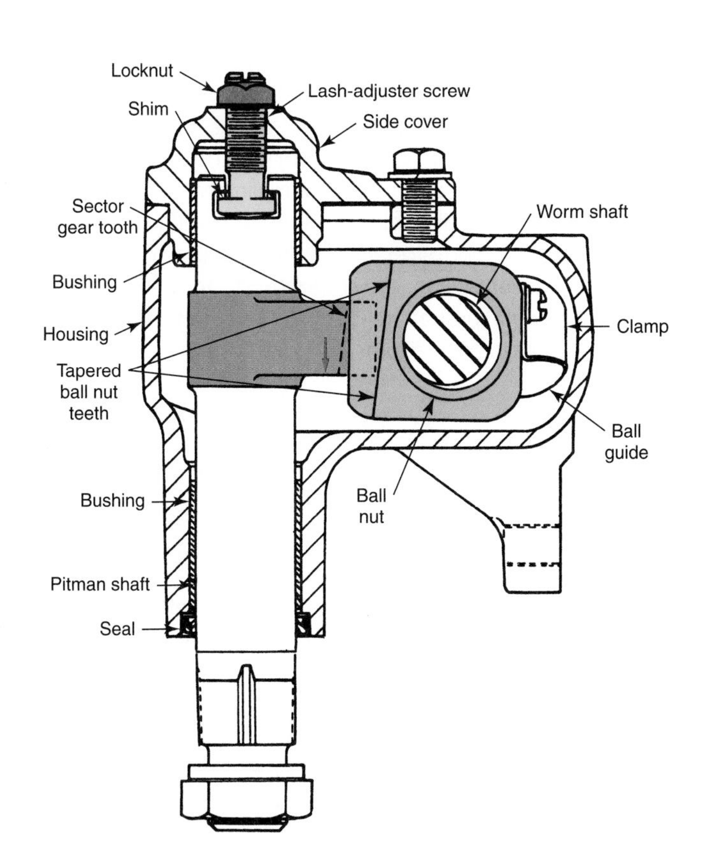

Figure 37-16. When the lash-adjuster screw forces the pitman shaft inward, the pitman sector gear moves into closer mesh with the tapered ball nut teeth. (General Motors)

STOP **Warning: The steering gear condition is very important to the safe operation of the steering system. Replace all parts showing signs of damage or noticeable wear.**

Reassemble the internal parts of the gearbox. Install the ball nut onto the worm shaft so that when the assembly is placed in the gear housing, the deep side of the nut teeth will face the cover. Replace the worm gear in the gearbox and install the pitman shaft and cover. Readjust as outlined earlier.

Conventional Power Steering Service

The conventional power steering gear uses an integral ***power cylinder.*** The power cylinder is part of the steering gear and is geared to the pitman shaft. A ***control valve*** is also used in the housing. Internal power steering gears are of two general types, offset and inline. The offset power steering gear connects the power piston to the pitman shaft as illustrated in **Figure 37-18.**

The inline design, so named because the ball nut, worm shaft, and control valve are all in line, uses the recirculating-ball nut as part of the power piston. See **Figure 37-19.** The action of the inline-type, during a left turn, is illustrated in Figure 37-19A. Note how the control valve has admitted fluid under pressure (red arrows) to one side of the ball nut. This causes the ball nut to move the sector gear. The fluid being displaced by the ball nut (white arrows) flows back through the control valve to the pump reservoir.

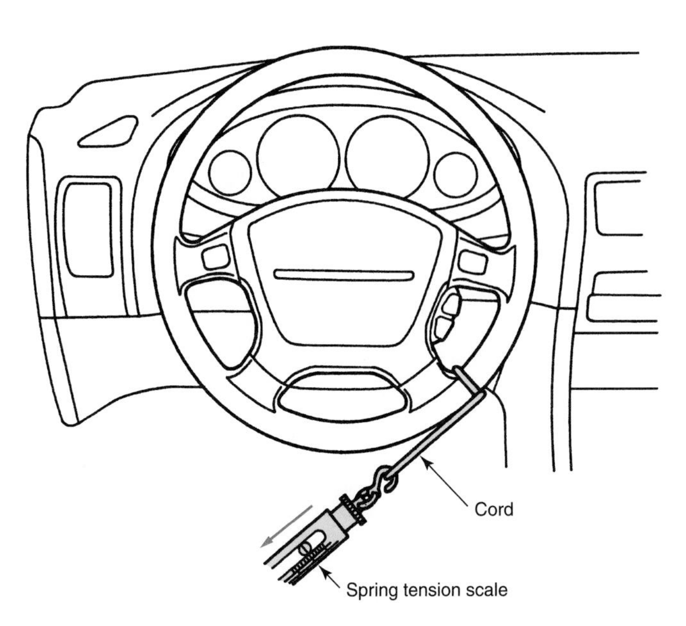

Figure 37-17. Checking steering gear adjustment with an accurate spring tension scale.

When the steering wheel is turned hard enough in the other direction to activate the control valve, fluid under pressure (red arrows) is admitted to the opposite side of the ball nut. This forces the ball nut to reverse the direction of the pitman shaft sector gear. See Figure 37-19B. The steering gear shown in **Figure 37-20** is the inline type and has a torsion-bar-operated spool valve. The ball nut uses the recirculating-ball principle for low-friction operation.

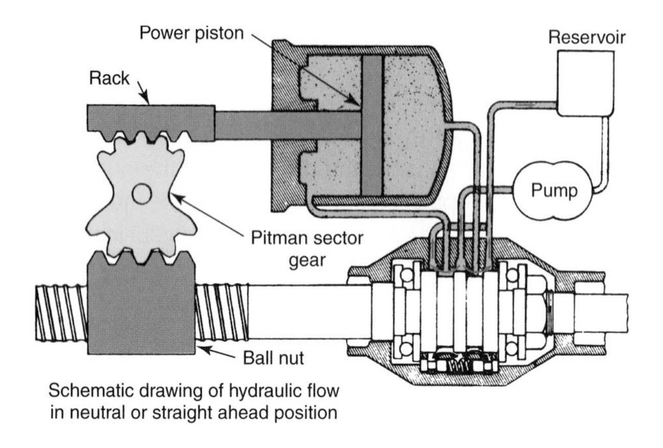

Figure 37-18. Offset-type internal power steering gear. (General Motors)

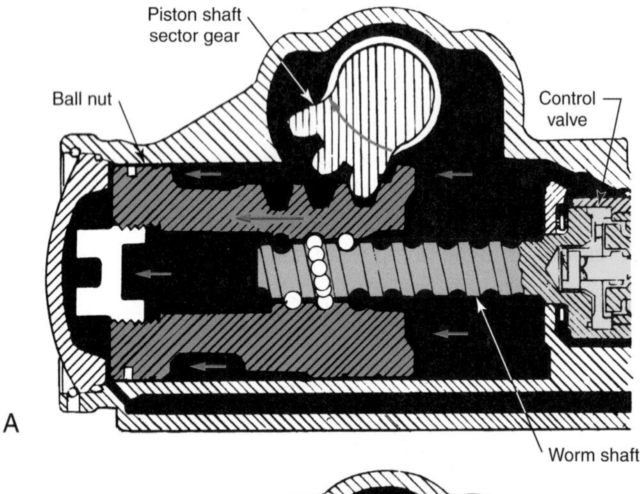

Figure 37-19. A—Action of an inline power steering gear during a left turn. B—Action of an inline power steering gear during a right turn.

Power Steering Gear Adjustments

The inline power steering gear has three basic adjustments, ***thrust bearing preload, worm-to-rack-piston preload,*** and ***pitman shaft over-center preload.*** These adjustments, unless specified otherwise, can be made while the gearbox is in the car. If disassembly is required, consult the manufacturer's service manual.

Thrust Bearing Preload

Thrust bearing preload can be checked with the gearbox in the car. Adjustment, however, may require removal. Remove the pitman arm and back off the pitman shaft's lash adjuster if specified. Also, if required, disconnect the fluid return line from the pump reservoir and drain the gearbox by turning it from right to left several times. Turn the steering wheel to the extreme right or left and back the specified amount. Using a torque wrench or a spring scale, measure the pull required to turn the steering wheel a specified amount. This indicates the thrust bearing preload only. **Figure 37-21** shows how the thrust bearing preload of one power steering gear is adjusted. Turn gear stub shaft the specified amount from the center position. Loosen the pitman shaft lash-adjuster screw. Attach an inch-pound torque wrench and turn the stub shaft through a specified arc. Tighten or loosen adjuster plug as needed to produce the correct preload.

Worm-to-Rack-Piston Preload

With the pitman arm removed and the pitman shaft's lash-adjuster loosened, move the steering wheel to a position about one-half turn from the center. Pull the wheel through a very short arc (about 1″ or 25.4 mm) and note the torque or scale reading. The reading should be slightly higher than the thrust bearing preload reading. The gearbox must be dismantled to adjust the worm-to-rack-piston preload. You may want to consider overhauling the gearbox if an adjustment is needed. A general procedure for power steering gear overhaul is covered later in this chapter.

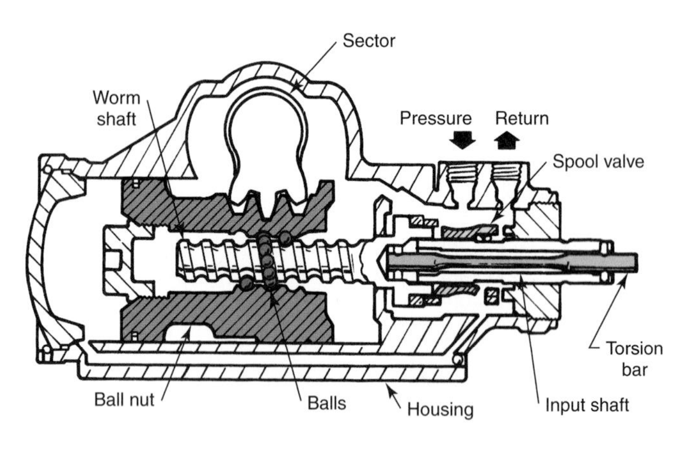

Figure 37-20. Cutaway of a typical inline, torsion bar–operated, spool-valve power steering gear. (Moog)

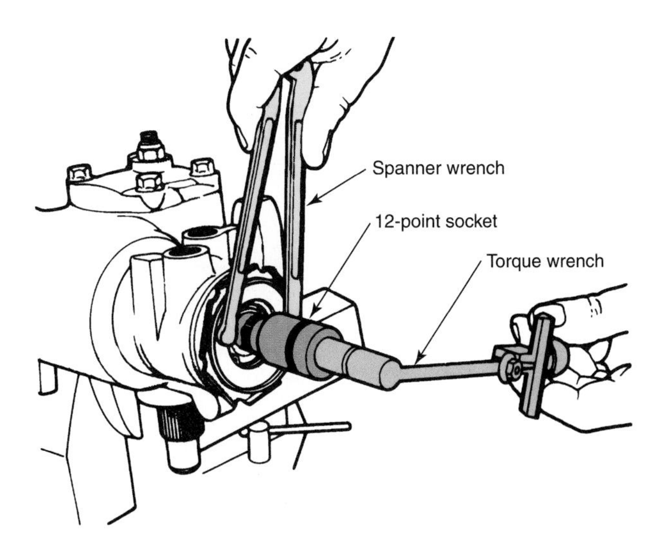

Figure 37-21. Using a spanner wrench to move the adjuster plug until the torque wrench indicates correct thrust bearing preload. (DaimlerChrysler)

The worm groove is ground with a high point in the center. When the rack-piston passes this point, a mild preload is produced. With the steering gear disassembled, clamp the rack-piston lightly in a vise. Make sure to protect the rack-piston with jaw covers. With the valve assembly in place on the worm, rotate the worm until the required distance is obtained from the end of the rack-piston to the thrust bearing face. This locates the rack-piston on the high point of the worm.

With an inch-pound torque wrench, rotate the stub shaft in both directions covering an overall arc of about 60°. Note the torque reading in both directions. Average the two highest readings and compare your results with the manufacturer's specifications. Look at **Figure 37-22.** If the preload is lower than required, it may be brought to specifications by replacing the balls with the next larger size. The sizes of replacement balls offered by one manufacturer are shown in **Figure 37-23.** Note the very small size differences.

Pitman Shaft Over-center Preload

The pitman shaft over-center preload is perhaps the most critical as far as vehicle handling is concerned. This adjustment can usually be made with the steering gear in the vehicle. Center the steering gear. Using either a torque wrench or spring scale, measure the amount of pull required to move the wheel through the over-center position. Perform the test several times and average the readings. If the over-center preload must be adjusted, first center the steering gear. Adjust the pitman shaft's lash-adjuster screw until the torque reading is as specified when the gear is turned through a 20° arc. Tighten the adjuster screw locknut. **Figure 37-24** illustrates how the over-center preload is adjusted on the bench. The procedure is basically the same when the gearbox is on the car. The pitman arm must be disconnected.

Power Steering Gear Overhaul

Begin the overhaul by disconnecting the pressure and return hoses from the steering gear. Elevate and cap the line fittings. Plug the housing openings. Remove the pitman arm and mark the pitman shaft and steering shaft coupling for correct reinstallation. Remove the coupling fastener and steering gear fasteners. Remove and drain the gearbox. Cycle the rack-piston by turning the stub shaft lock-to-lock several times. This assists in draining the gearbox. Plug the hose fitting openings. Clean the exterior of the gear thoroughly and disassemble it as needed. Clean and inspect all parts and replace them if necessary. Install new seals, O-rings, gaskets, rack-piston ring, and other parts as needed. Complete rebuilding kits, such as illustrated in **Figure 37-25,** are available.

Assemble in reverse order of disassembly and make necessary adjustments. **Figure 37-26** shows typical power steering system leakage points. Clean all parts thoroughly and handle parts carefully to avoid nicking, burring, or

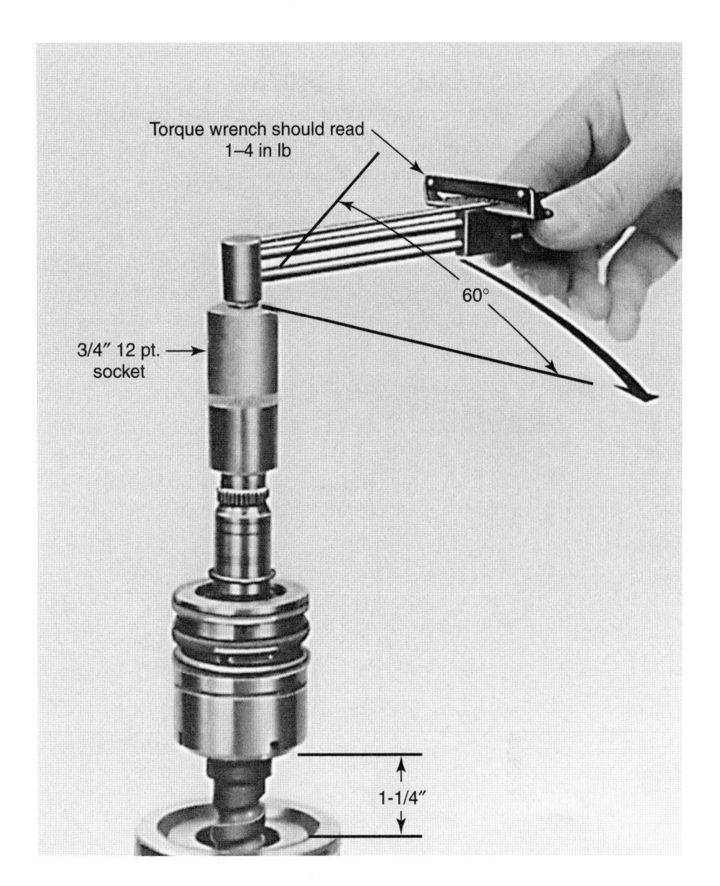

Figure 37-22. Checking worm-to-rack-piston preload with a torque wrench.

Size Code	Mean Diameter	Size Range of Ball
6	0.28117″	0.28112–0.28122″
7	0.28125″	0.28120–0.28130″
8	0.28133″	0.28128–0.28138″
9	0.28141″	0.28136–0.28146″
10	0.28149″	0.28144–0.28154″
11	0.28157″	0.28152–0.28162″

Figure 37-23. Replacement ball size chart from one manufacturer.

distorting. Keep your work area, tools, and hands clean. Remember that dirt, even in very small amounts, can ruin the power steering gear, pump, control valve, power cylinder, and test instruments. Never strike or hammer the housings. Do not steam clean unless specified by the manufacturer.

Steering Gear Alignment

When reinstalling the steering gear, check for proper alignment with the steering shaft. Misalignment can cause binding and premature wear. The steering gear, when in the exact center of travel, should be attached to the steering shaft. Make certain the steering wheel spokes are in the correct position. All alignment marks should be in line. **Figure 37-27** shows the alignment marks on one gear installation. After proper gear housing alignment, the steering wheel should turn through its full travel smoothly without roughness or binding.

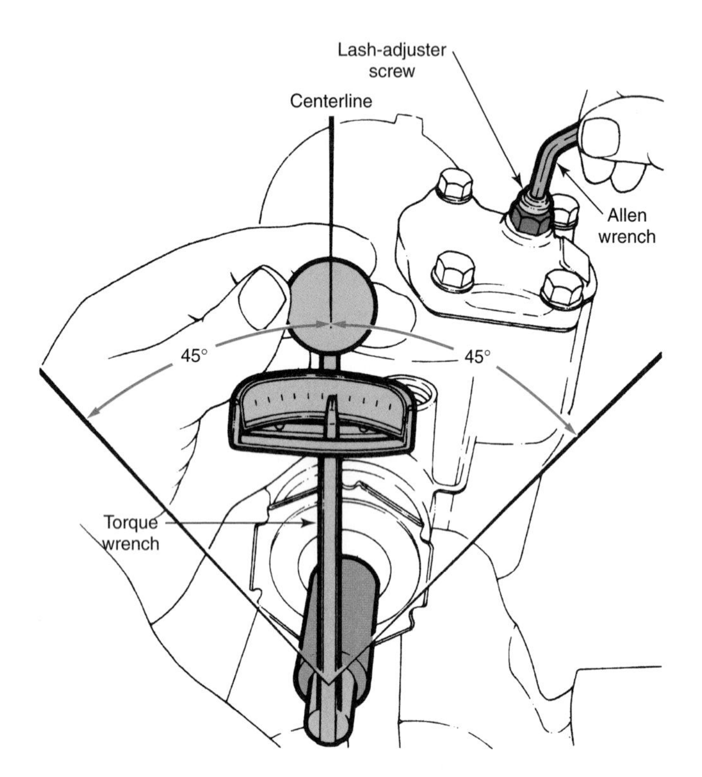

Figure 37-24. Using a torque wrench and an Allen wrench to properly adjust pitman shaft over-center preload. (DaimlerChrysler)

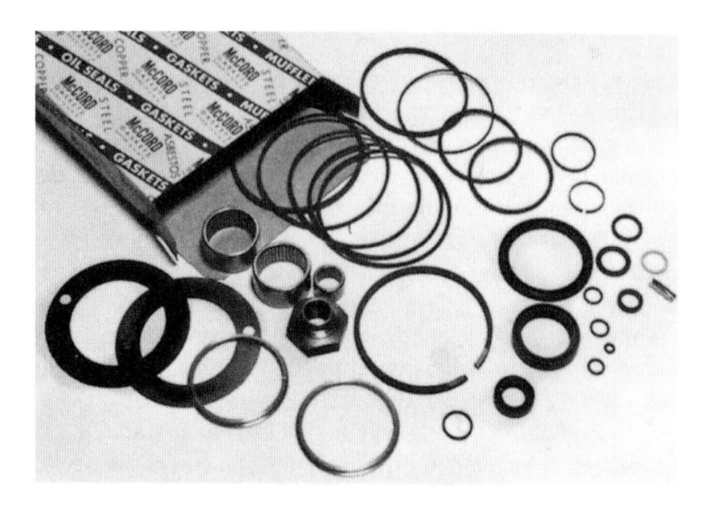

Figure 37-25. A power steering rebuild/service set. (McCord)

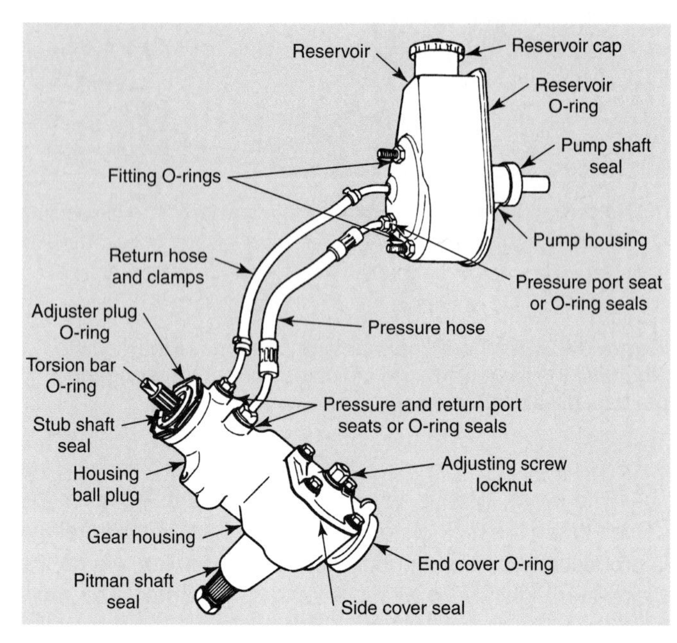

Figure 37-26. Possible power steering fluid leak points on the pump and gear. (DaimlerChrysler)

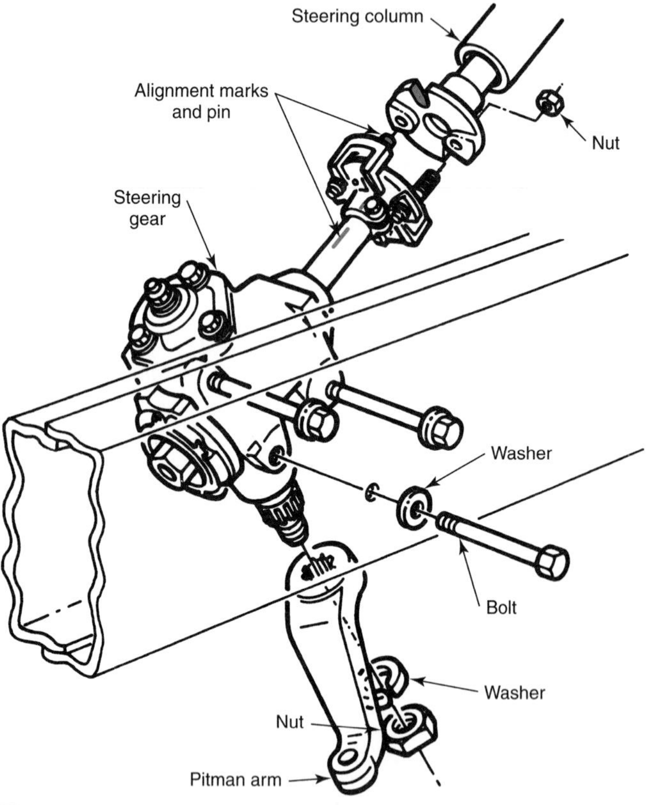

Figure 37-27. Be sure to align all witness (alignment) marks and pins when reassembling the steering gear shaft and coupling. (Ford)

Caution: If the vehicle is elevated, do not grasp a front tire and twist it quickly to move the steering wheel to full lock. Severe steering column damage can occur. This is especially true of vehicles equipped with rack-and-pinion systems.

Gear housing and steering shaft alignment is important. Make sure the gear fasteners and coupling fasteners are torqued to specifications. Attach and torque the pitman arm.

Rack-and-Pinion Steering Gears

This section covers the rack-and-pinion gear used on most newer vehicles. Rack-and-pinion steering is simpler than other systems. A rack-and-pinion steering system is simpler because the steering column shaft connects directly to a ***sector shaft,*** usually called a pinion. The pinion operates the rack section, which operates the tie rods, and therefore the steering knuckles. Many parts, including the linkage system used with conventional gearboxes, are eliminated by this system. **Figure 37-28** shows a typical rack-and-pinion steering system.

Manual Rack-and-Pinion Steering Gear Overhaul

Ideally, a rack-and-pinion steering gear should be replaced when leaking or malfunctioning. However, if an overhaul is preferred, begin by marking the location of the steering shaft coupling on the pinion shaft. Then, remove the bolt securing the coupling to the shaft. Remove the tie rod ends. Unbolt the rack assembly from the frame cross member. On some vehicles, the rack is secured to the vehicle's body. Carefully remove the rack assembly from the vehicle. Do not tear the rubber dust boots. Be sure to secure any shims. See **Figure 37-29.**

Mount the rack in a special holding fixture or gently grip it with a vise. Obtain a reassembly toe setting by measuring from the center of the tie rod end to the boot-retaining groove on both sides. Record these measurements; they will be used during reassembly. Next, remove the tie rods, dust boots, and clamps. Tie rod removal is explained later in this chapter. Some units will require new clamps because the old ones must be cut off. New boots should always be used. Depending on the rack type, remove the retaining pin or the claw washer. See **Figure 37-30.**

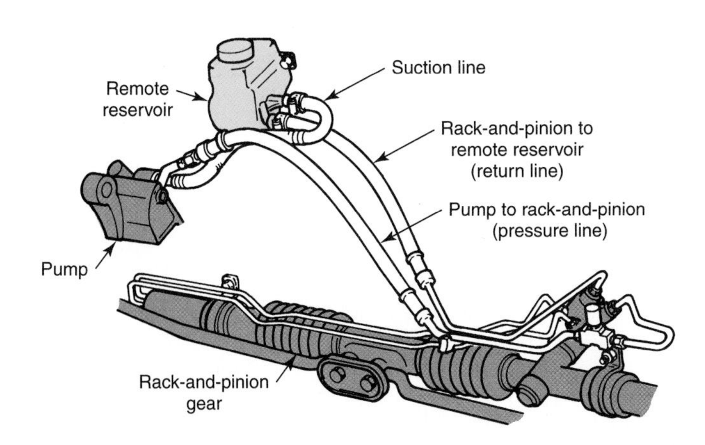

Figure 37-28. Power rack-and-pinion steering system. This pump uses a remote fluid reservoir. (Moog)

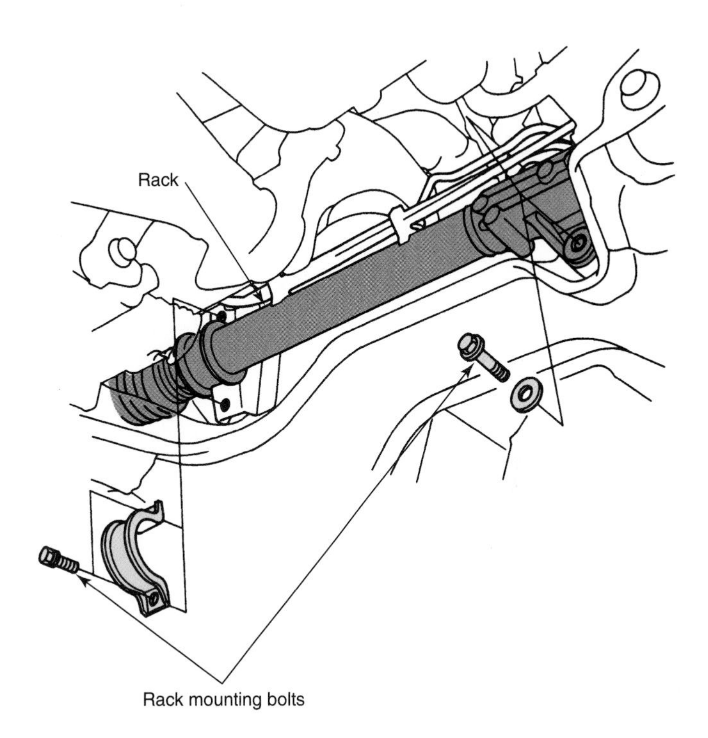

Figure 37-29. Removing the rack assembly from the vehicle. Use care to avoid part damage. Keep shims and other parts in the correct order. (Honda)

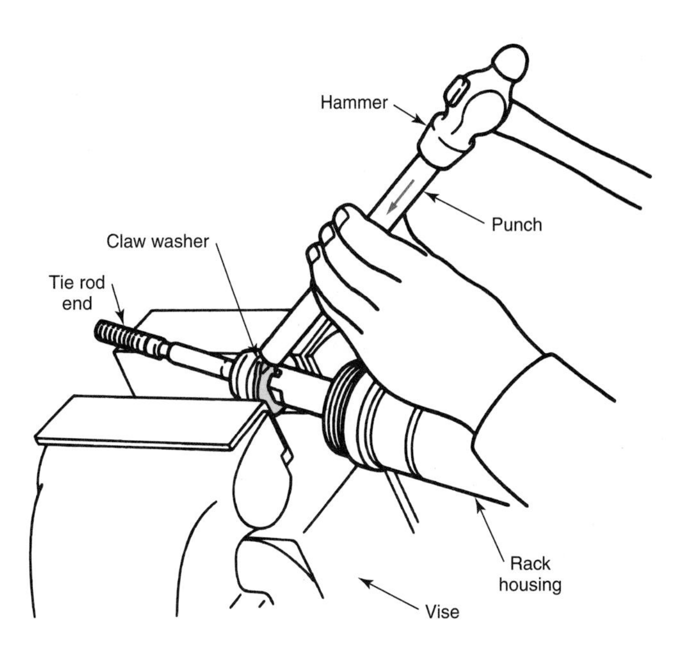

Figure 37-30. Unstaking a claw washer with a hammer and punch before attempting to remove the tie rod end. (General Motors)

Caution: If the rack has worn or damaged threads, scoring in the housing bores (especially in aluminum housings), or more than two retaining pin holes in it, a new steering gear assembly must be installed.

Remove the steering gear locknut, adjuster plug, spring, and rack bearing guide. Take off dust cover. Unscrew the pinion bearing cap and remove the pinion shaft's locknut, snap ring, and pinion. Disconnect any remaining fasteners and pull the rack from its housing. Thoroughly clean all metal parts in solvent. Check for wear, cracks, scoring, missing teeth, and damaged seals. An exploded view of a rack-and-pinion assembly is illustrated in **Figure 37-31.** Reassembly is basically a reversal of disassembly procedures. Follow the

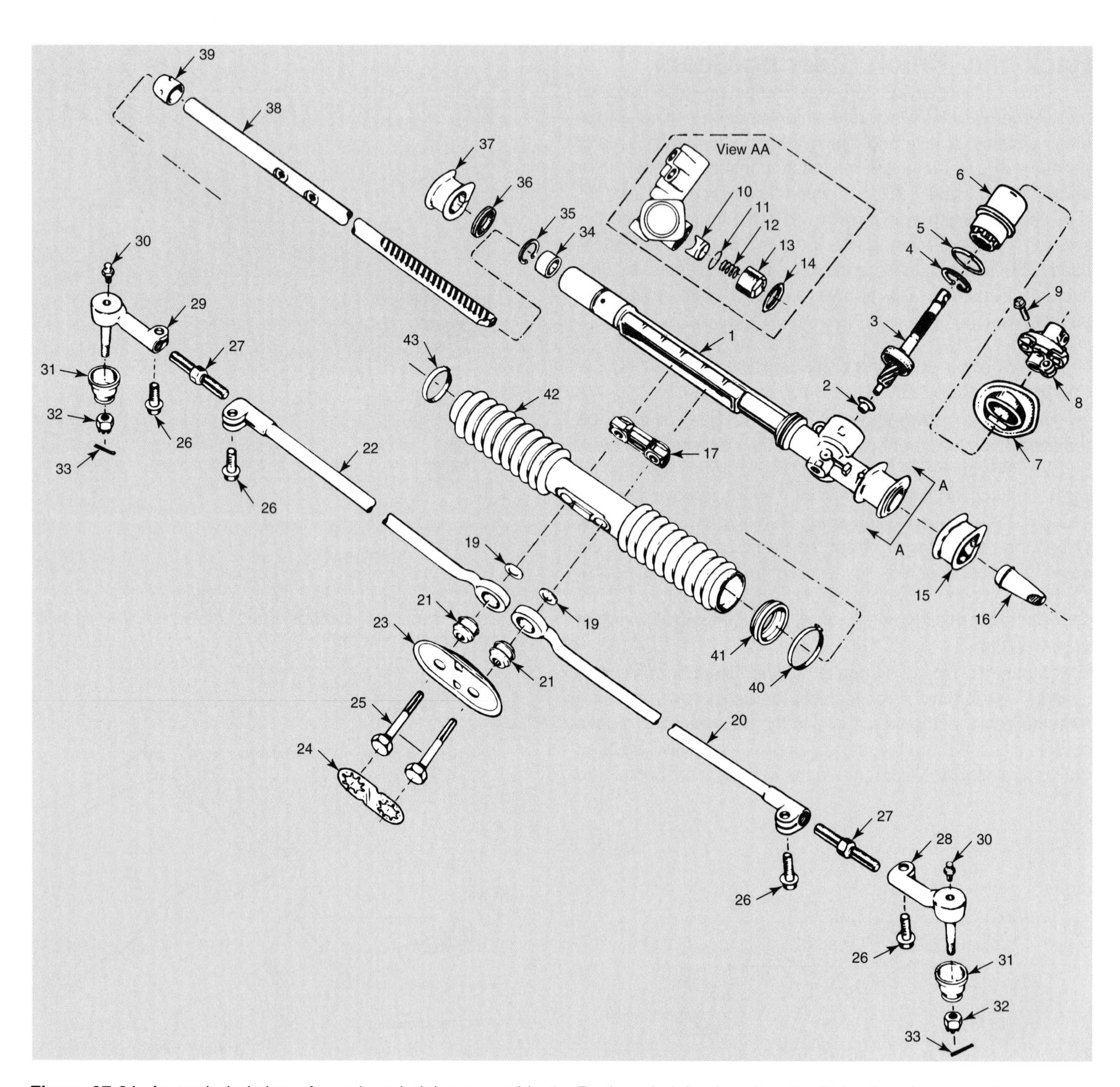

Figure 37-31. An exploded view of a rack-and-pinion assembly. 1—Rack-and-pinion housing. 2—Roller bearing assembly. 3—Bearing and pinion assembly. 4—Retaining ring. 5—Dust seal. 6—Damper assembly. 7—Dash seal. 8—Steering coupling assembly. 9—Pinch bolt. 10—Bearing. 11—O-ring seal. 12—Spring. 13—Adjuster plug. 14—Lock nut. 15—Mounting grommet. 16—Housing end cover. 17—Rack guide. 19—Washer. 20—Inner tie rod. 21—Inner pivot bushing. 22—Inner tie rod. 23—Bolt support plate. 24—Lock plate. 25—Inner tie rod bolt. 26—Bolt. 27—Tie rod adjuster. 28—Outer tie rod assembly. 29—Outer tie rod assembly. 30—Lubrication fitting. 31—Tie rod seal. 32—Hex nut. 33—Cotter pin. 36—Retaining bushing. 37—Mounting grommet. 38—Steering rack. 39—Housing end cover. 40—Boot clamp. 41—Retaining bushing. 42—Rack-and-pinion boot. 43—Boot clamp. (Saginaw Division of General Motors)

manufacturer's procedures when servicing an assembly. Torque all fasteners to specifications and double-check all connections.

Rack-and-pinion Power Steering

The power rack-and-pinion system uses a rotary control valve to regulate the fluid coming from the pump. It routes the fluid to the appropriate side of the rack piston, **Figure 37-32.** The piston converts hydraulic pressure to a linear movement. This movement is transmitted to the steering arms by the tie rods.

Control Valves

The control valve directs the flow of power steering fluid from one side of the rack piston to the other, **Figure 37-33.** As the steering wheel is turned, the resistance created by the vehicle's front tires causes the valve's torsion rod to twist. The twisting action causes a difference in the degree of rotation between the valve and the spool. As this occurs, the passages in the control valve change. Fluid can no longer flow directly through the valve and back into the reservoir. Instead, it must pass through the passages of the power cylinder.

For a right turn, a circuit is opened that allows the fluid to be pumped through the spool's top radial groove and into the right side of the power cylinder. At the same time, fluid is discharged from the left side of the power cylinder by the spool's bottom radial groove. This fluid flows back into the valve chamber above the spool and is routed into the reservoir. For a left turn, the opposite circuit is opened. As long as the torsion rod is twisted, hydraulic pressure acts on the rack. The amount of force on the rod is reduced when the fluid acts on the rack. This reinforces the action of the pinion. The moment the force on the torsion rod is released, the valve's return passage is opened and the fluid flows directly back into the reservoir.

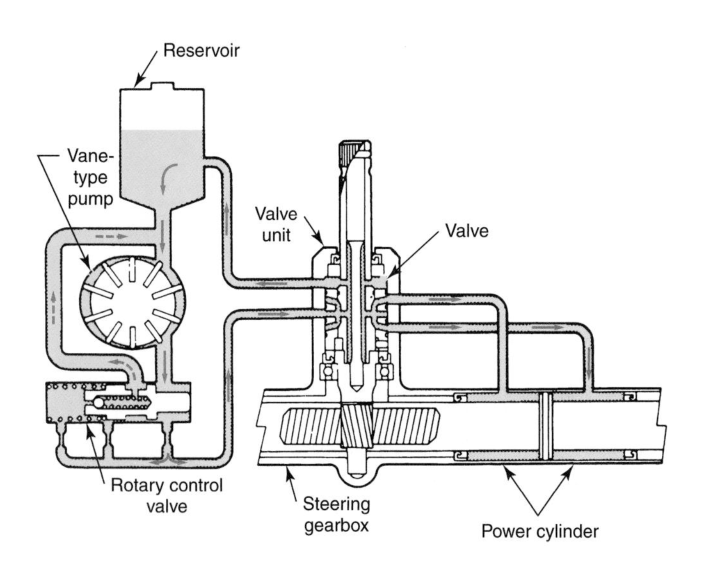

Figure 37-32. Cutaway view of a power rack-and-pinion steering system. Note the rotary flow-control valve, which regulates the fluid traveling from the pump. (Honda)

Power Cylinder

The power cylinder is an integral part of the rack housing, **Figure 37-34.** There are two power steering fluid lines that run between the cylinder and the control valve. The lines are connected on the left and right side of the piston. When turning right, fluid is pumped to the right side of the power cylinder, forcing the piston and rack to the left. The fluid is then discharged from the left section of the power cylinder. Simultaneously, the boot on the left side is distended and the boot on the right side is compressed. This causes air to flow through a capillary tube between the boots, keeping the air pressure constant.

Movement of the rack is transferred to the steering members by the tie rods. When an overhaul of the power rack-and-pinion system is necessary, follow all of the manufacturer's recommendations carefully. Disassembly procedures are similar to those for manual rack-and-pinion systems. Remember, steering is critical. Perform only top-notch work.

Power Steering Pumps

Power steering systems use pressurized fluid to provide most of the turning force to the front wheels. These systems assist both front and rear wheels on vehicles equipped with four-wheel steering. All units use a pump to provide hydraulic pressure to the control valve.

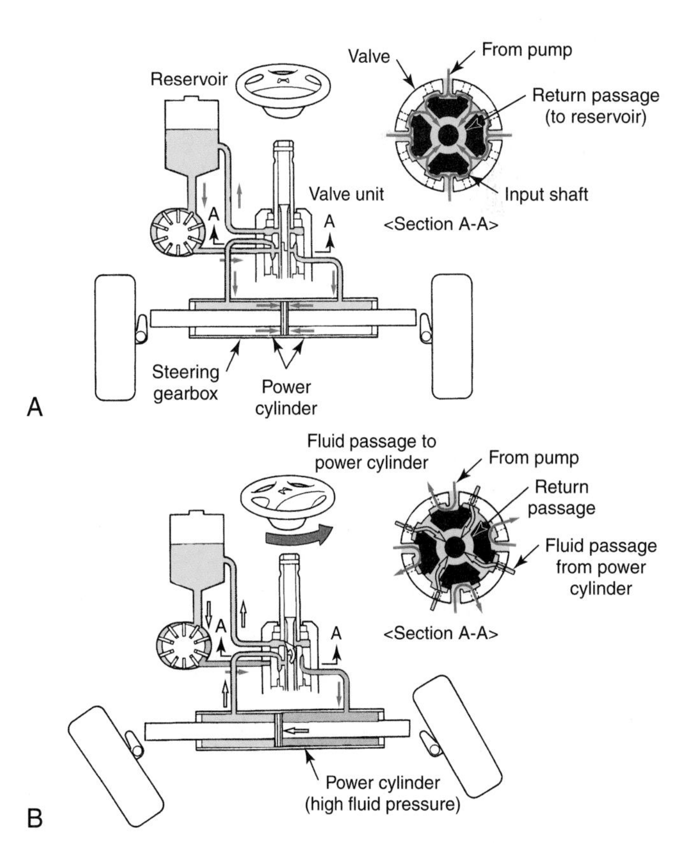

Figure 37-33. Control valve operation. A—Steering straight ahead. B—Turning left. (Honda)

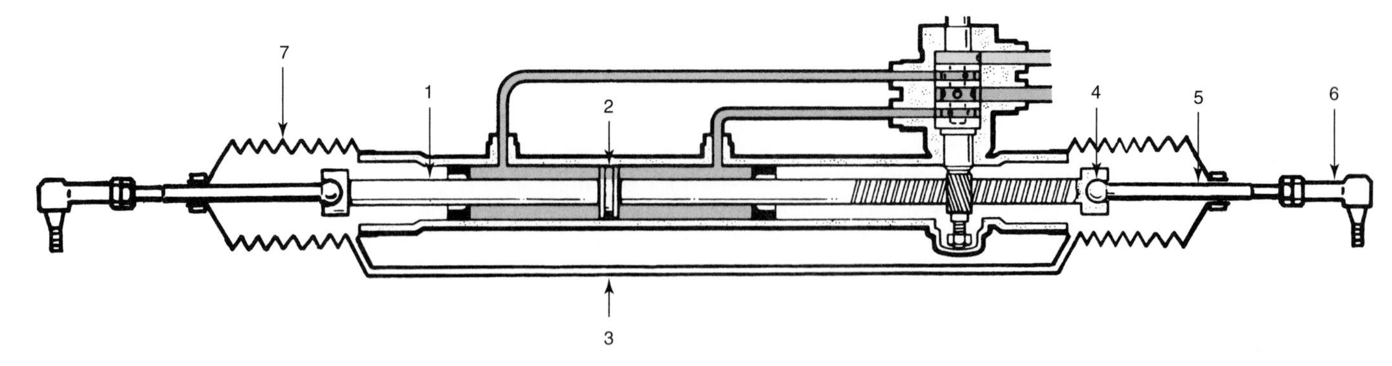

Figure 37-34. A power cylinder cutaway. 1—Rack. 2—Piston. 3—Pressure-equalizing tube. 4—Inner ball joint. 5—Tie rod. 6—Tie rod joint. 7—Boot. (SAAB)

Power steering pumps, usually are ***vane, slipper,*** or ***roller*** types. All use a ring with a cam-shaped inner opening. A rotor turns inside the cam ring. The rotor may employ vanes to form a seal between the rotor and cam. **Figure 37-35** illustrates a typical vane-type power steering pump. Note how the vanes contact the cam-shaped pump ring inside. When the rotor turns, each space formed between the rotor and ring by the vanes grows from a small size to a large size, **Figure 37-36.** This forms a vacuum and fills the space between the vanes with oil.

As the rotor continues to turn, the space reduces in size, Figure 37-36. This compresses the fluid and forces it through the outlet of the power steering pump. The vanes are kept in contact with the ring cam walls by centrifugal force and by fluid pressure fed to the base of each vane. The pump in **Figure 37-37** is a slipper-type pump. It uses spring-loaded slippers to form a seal between the rotor and ring. The roller pump utilizes a series of rollers rather than vanes or slippers, **Figure 37-38.** A gear-type power steering pump is illustrated in **Figure 37-39.**

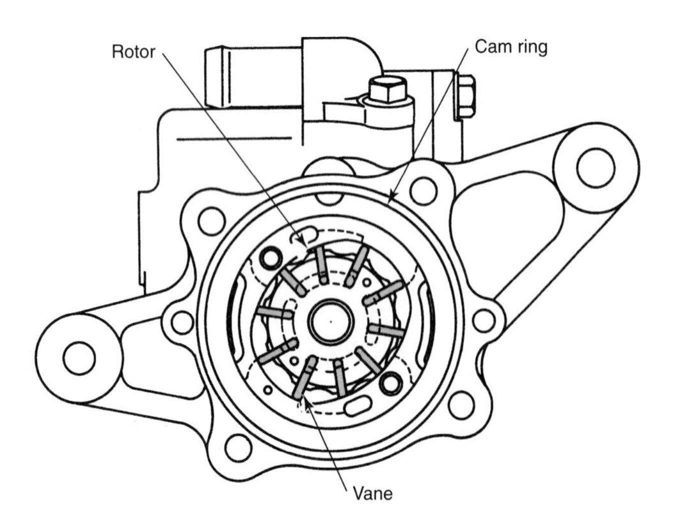

Figure 37-35. A belt-driven, vane-type power steering pump. Note that this pump uses ten vanes to help smooth fluid pulsations. (Honda)

Power Steering Pump Belt

Most installations use one or two V-belts or a serpentine belt to drive the power steering pump. The belt or belts can operate from the engine crankshaft pulley. **Figure 37-40** illustrates a typical power steering arrangement that employs a single belt drive. Note the fluid reservoir is separate from the pump on this setup. Carefully inspect power steering belts for

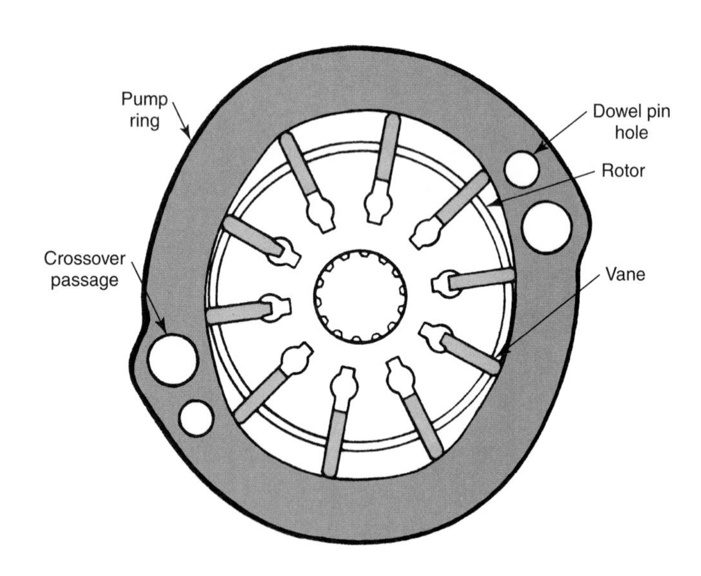

Figure 37-36. Vane-type power steering pump. Note the relationship of the rotor and vanes to the pump ring. (DaimlerChrysler)

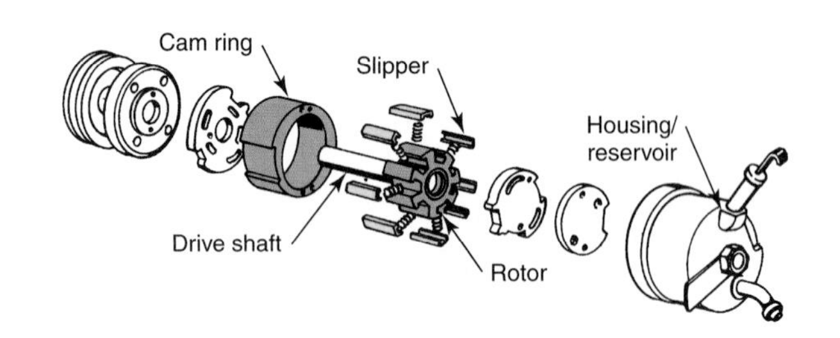

Figure 37-37. Exploded view of a slipper-type power steering pump. (Moog)

glazing, rotting, cracking, or swelling and replace as needed. If excessive belt squeal occurs during sharp turning of the front wheels, the belt is probably loose and should be adjusted. Adjustment may be checked with a belt tension gauge. Replace or adjust the belt as needed.

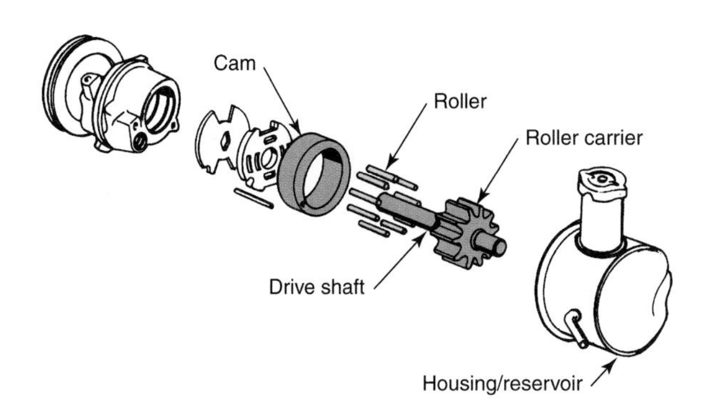

Figure 37-38. Exploded view of a roller-type power steering pump. Note the location of the rollers. (Moog)

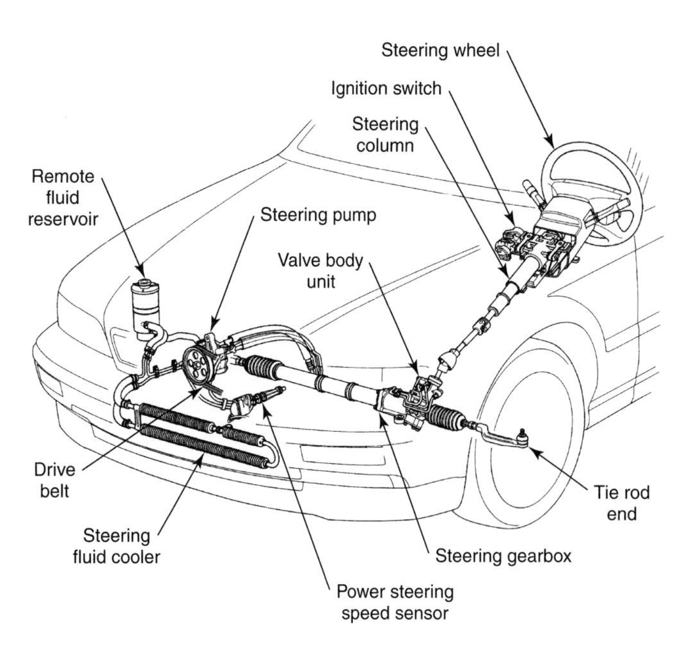

Figure 37-40. Most power steering pumps are belt driven. This power rack-and-pinion steering setup that uses a belt-driven pump and a remote fluid reservoir. (Honda)

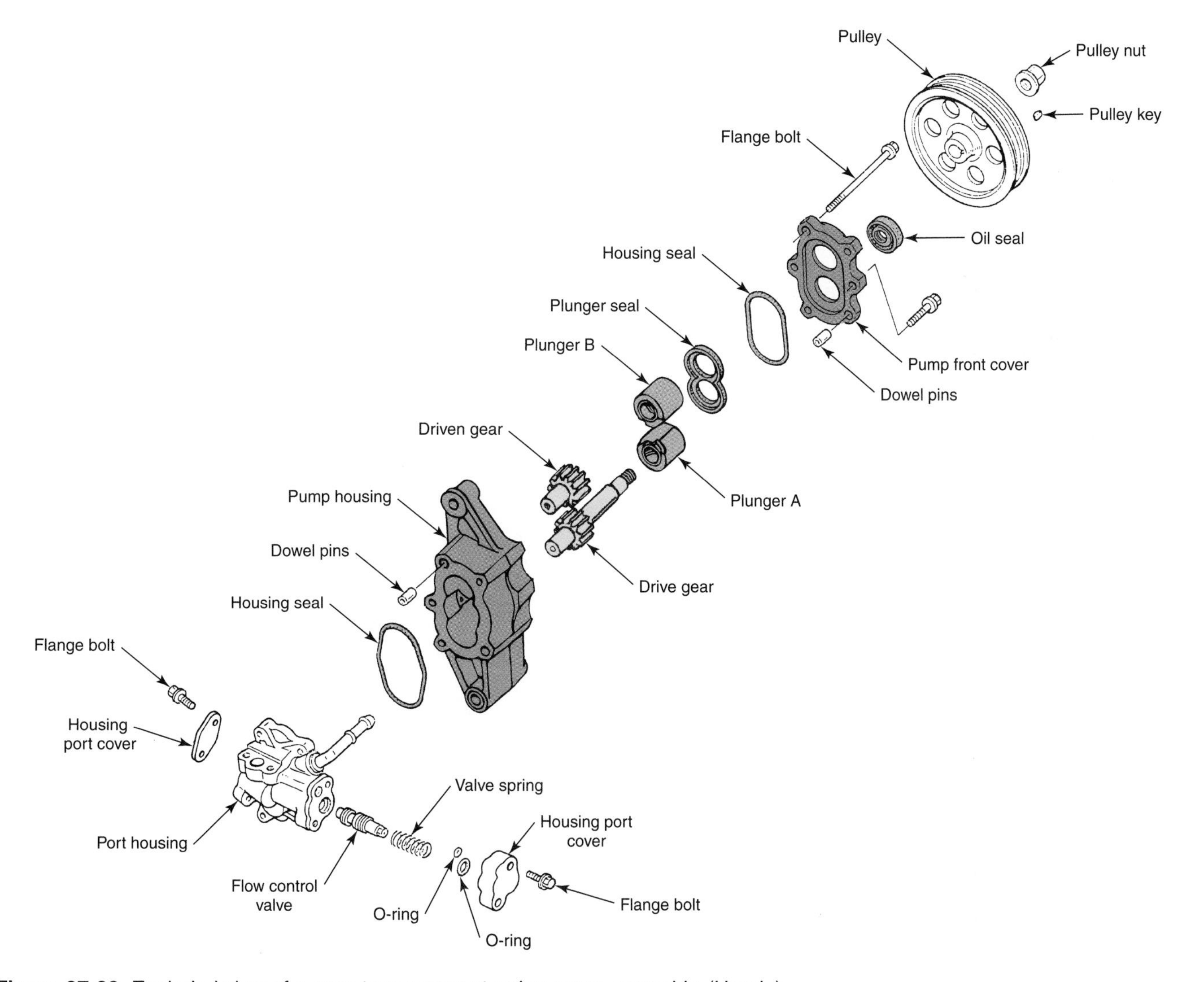

Figure 37-39. Exploded view of a gear-type power steering pump assembly. (Honda)

Power Steering Service

If a vehicle does not use a serpentine belt with an automatic self-adjuster, belt tension may be checked using a torque wrench, a ruler to measure belt deflection, or a strand tension gauge. The torque wrench method calls for adjusting the belt tension until a certain torque is required to turn the power steering pump pulley. The belt deflection method requires pressing inward on a certain spot and measuring the resulting amount of belt deflection. The strand tension gauge is the most accurate method of testing belt tension. The instrument is slipped around the belt, and the pump is moved in or out until the tension gauge reading matches the specifications.

If the belt needs adjustment, loosen the pump's attaching fasteners. Move the pump outward as needed and tighten the fasteners. Be careful when prying on pumps to adjust belt tension. Pry only on the heavy section. If the pump is equipped with a special wrench tab, use it. Never pry on the reservoir or filler neck. Some pumps are equipped with an adjusting bolt, which eliminates the need to pry on the pump body, **Figure 37-41.**

To replace the belt or belts, loosen fasteners. Rotate pump inward to slack and remove belts. Install and tension new belts as specified. See **Figure 37-42.** New belts are tightened somewhat more than used belts to compensate for initial stretching and seating. When removing and installing belts, disconnect the battery's ground terminal to prevent injury.

Power Steering Fluid Level

Check the ***fluid level*** in the power steering pump reservoir. Fluid level should be between the "add" and "full" marks. Check the level with the fluid at operating temperature. The reservoir illustrated in **Figure 37-43** has a dipstick to make checking easy. Clean the reservoir top before removing it to prevent dirt from entering the reservoir. If the fluid level is low, inspect entire system for leaks. See **Figure 37-44.** If fluid must be added, always use power steering fluid intended for the system being worked on. Modern vehicles are designed to use special power steering fluid, not automatic transmission fluid. Always check manufacturer's specifications.

Power Steering System Pressure Test

By conducting a ***pressure test,*** the origin of power steering system problems can be easily identified. Pressure problems may be found in the pump itself or in the hose or gear system. There are three major precautions for making an accurate pressure test, proper test connections, correct engine RPM, and specified fluid temperature. Before conducting a pressure test, belt tension must be checked. If needed, adjust the belts. Fluid must be at the specified level.

Remove the high-pressure outlet hose from the pump. Insert a 0 psi–2000 psi (0 kPa–13,790 kPa) gauge and shutoff

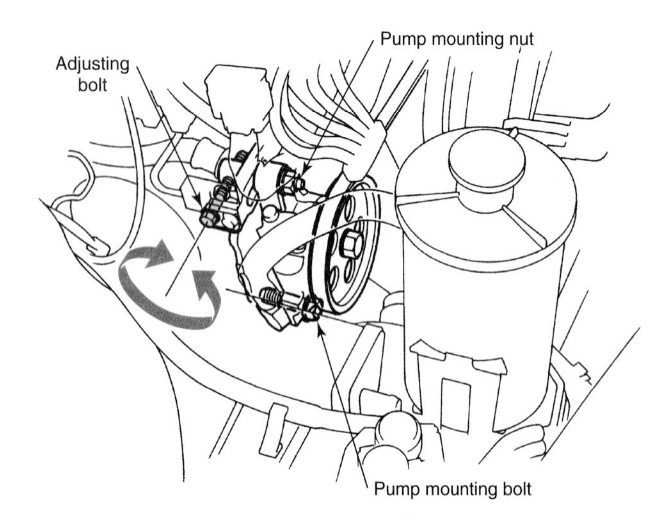

Figure 37-41. Adjusting belt tension. In the setup shown in the illustration, the belt is tightened or loosened with an adjusting bolt built into the pump housing. Follow the manufacturer's recommendations for the correct tension. Too much tension can damage the pump bearings. Too little tension will cause belt slippage and poor overall pump output. (Honda)

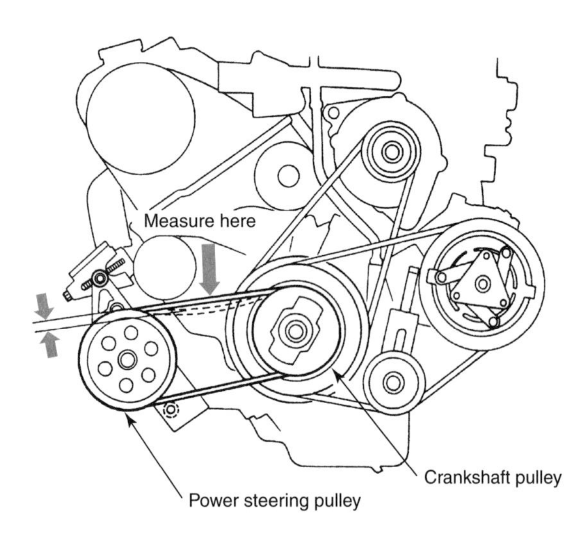

Figure 37-42. A new power steering belt has been installed. (Honda)

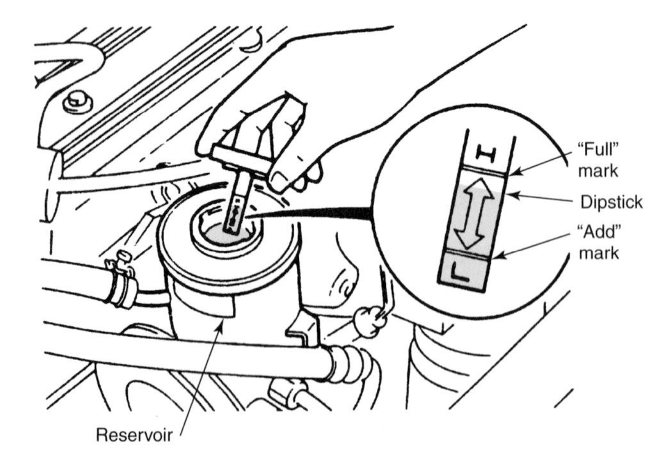

Figure 37-43. Keep the power steering pump filled to the proper level. In this system, a dipstick is used to check fluid level. (Mazda)

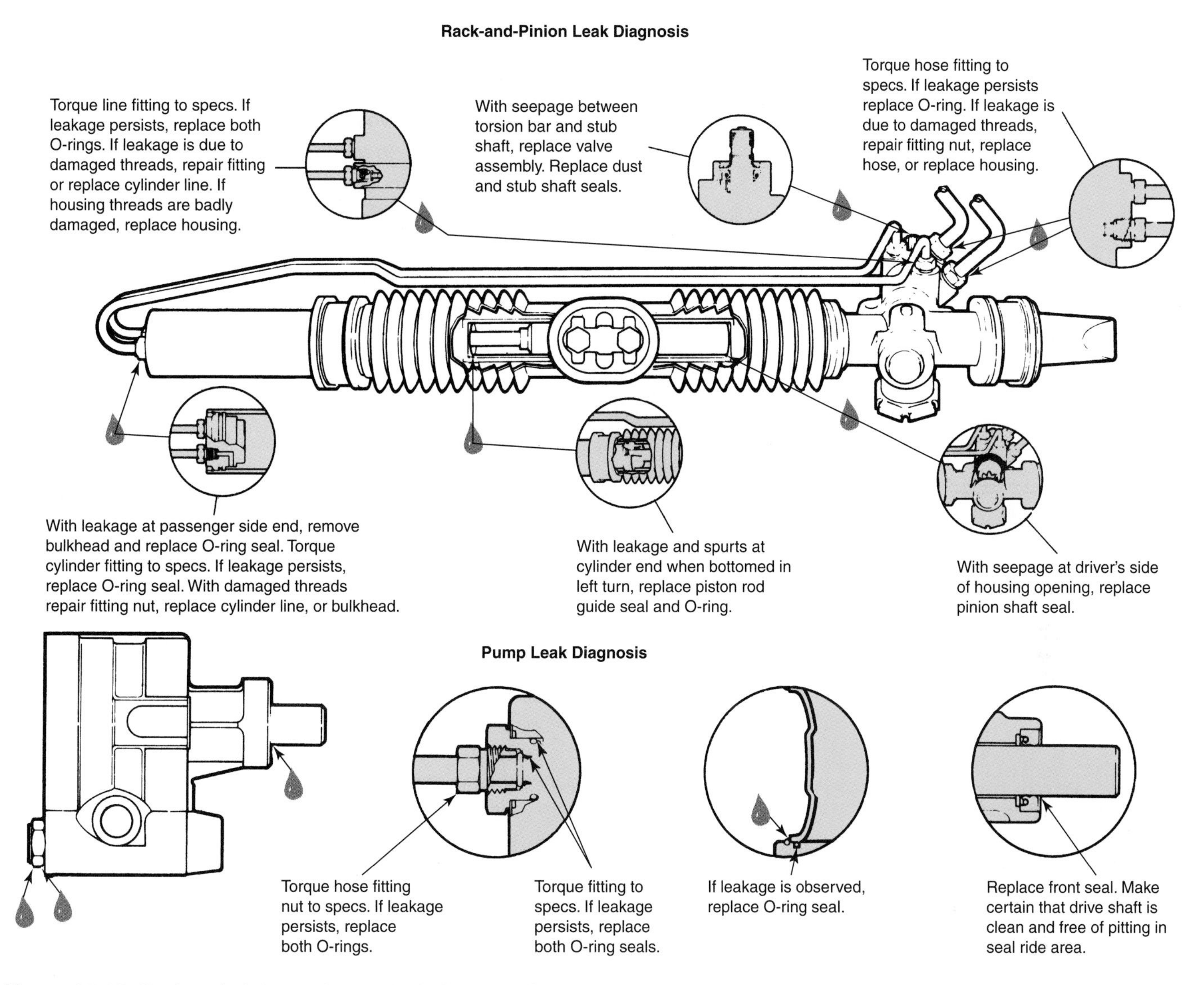

Figure 37-44. Rack-and-pinion and pump leak diagnosis for one particular manufacturer. Note the recommended correction with each leak area. (General Motors)

valve between the disconnected hose and the pump outlet. The gauge must be between the shutoff valve and the pump, **Figure 37-45.** Torque the fittings and open the gauge fully. If the manufacturer's specifications call for an exact RPM and fluid temperature, install a thermometer in the pump reservoir and connect a tachometer, **Figure 37-46.** Bleed the system by starting the engine and turning the steering wheel from full right to left and back several times. Check fluid level and add if needed. Operate the engine until fluid temperature reaches indicated level. The shutoff valve must be open. Raise the engine RPM to factory test specifications and note system pressure with the valve open (wheels in straight ahead position).

With the weight of the car on its wheels, turn the steering wheel to the right or left until the stop is reached. Hold the wheel hard against the stop and read the maximum pressure developed. Never hold the wheels hard against the stops for longer than five seconds. This would cause a rapid rise in fluid temperature and damage the pump. Check pressure against

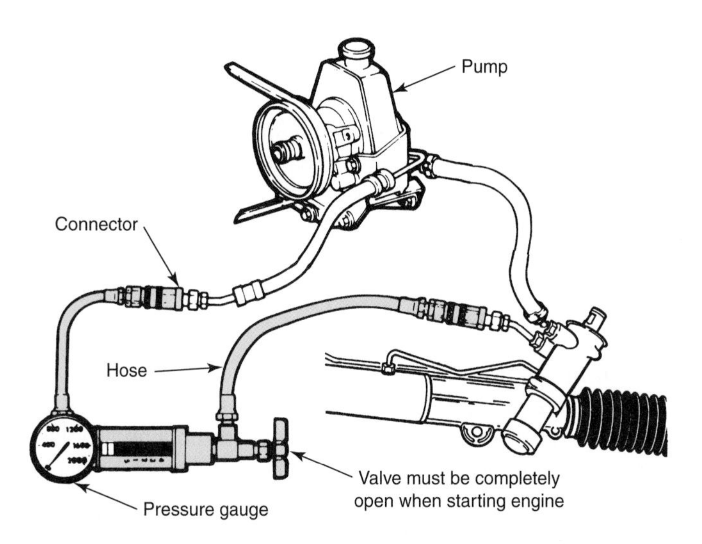

Figure 37-45. Setup used for obtaining power steering pressure. Follow the manufacturer's direction carefully. Wear goggles! (Saginaw Division of General Motors)

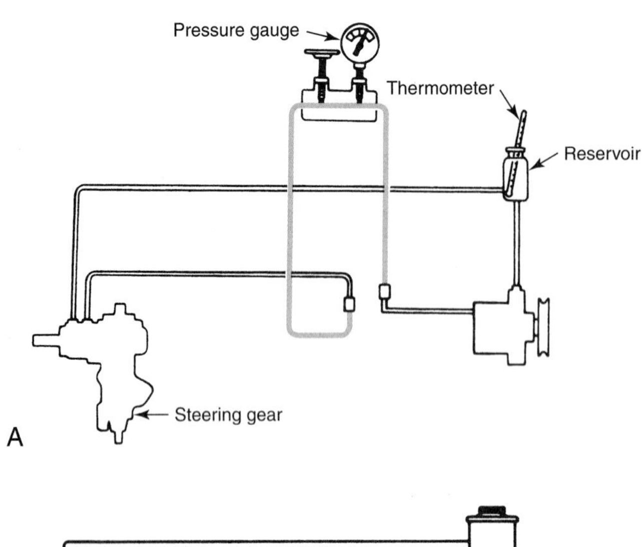

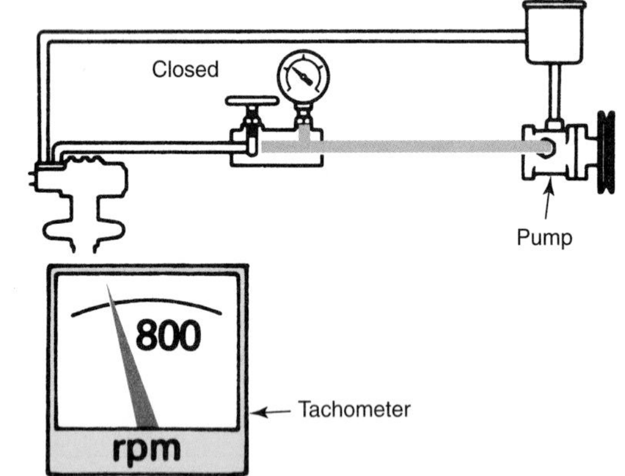

Figure 37-46. A—Some pressure tests call for the use of a thermometer and pressure gauge. B—A tachometer and pressure gauge may also be needed. Do not shut the gauge off (closed) for more than about five seconds to prevent system damage. (DaimlerChrysler, Toyota)

specifications. Pressure specifications generally call for pressures exceeding 1000 psi (6895 kPa). If the pressure is below an acceptable level, slowly close the shutoff valve by returning the steering wheel to the straight-ahead position. If the pressure rises to the proper level with the valve off, the pump is working properly and the pressure drop is in the hoses or gear system. If the system pressure does not rise, the pump is at fault. Never leave the shutoff valve in the closed position longer than five seconds. To do so would cause a rapid fluid temperature rise, resulting in pump damage. Some manufacturers call for checking maximum pressure by using the shutoff valve only.

Bleeding Steering System

When a hydraulic steering system has been repaired, an accurate fluid level reading cannot be obtained unless ***bleeding*** is performed to remove trapped air from the system. A basic bleeding procedure is as follows:

1. Fill the reservoir with the correct type of fluid.
2. Start the engine and allow it to run for about one minute. Do not turn the wheels.
3. Shut off the engine and allow the vehicle to sit for about two minutes. This allows tiny air bubbles and foam to form larger, more easily expelled bubbles.
4. Refill the reservoir.
5. Start the engine and let it idle for about two minutes.
6. Check the fluid level and add fluid if necessary.
7. Check the repaired part of the system for leaks.
8. Road test the vehicle. Check for abnormal steering function, noise, and leaks.
9. Recheck the fluid level and add if necessary.

Note: Do not overfill the reservoir.

Power Steering Pump Overhaul

Leaking seals or worn or damaged parts require pump disassembly and repair. When removing the pump pulley, use a puller. See **Figure 37-47.** Never try to hammer the pulley off. Some pumps have fiberglass or nylon reservoirs, which can be damaged by prying or hammering. Clean pump exterior before disassembly. Drain as much power steering fluid from the pump as possible. Obtain a service manual and refer to it during the overhaul. Disassemble the pump as required, using care to avoid excessive force. Clean all parts and lay out on a clean table. **Figure 37-48** shows a typical vane-type pump completely disassembled. Carefully inspect all parts for excessive wear. Check for galling, nicks, or other physical damage. Check the flow control-valve. Replace damaged or worn parts.

When assembling, lubricate all parts. Most modern vehicles call for special power steering fluid instead of automatic transmission fluid. Always use the exact fluid specified by the manufacturer. The vanes' rounded ends must ride against the cam ring. To facilitate proper installation of parts without damage to the O-rings, use a coating of petroleum jelly. Install new seals, O-rings, and gaskets. Protect the shaft seal when inserting the rotor drive shaft. **Figure 37-49** illustrates the use

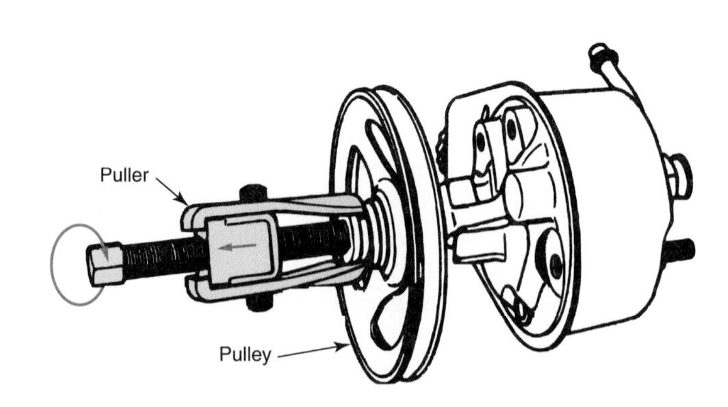

Figure 37-47. Removing the pulley from a power steering pump with a special puller. (DaimlerChrysler)

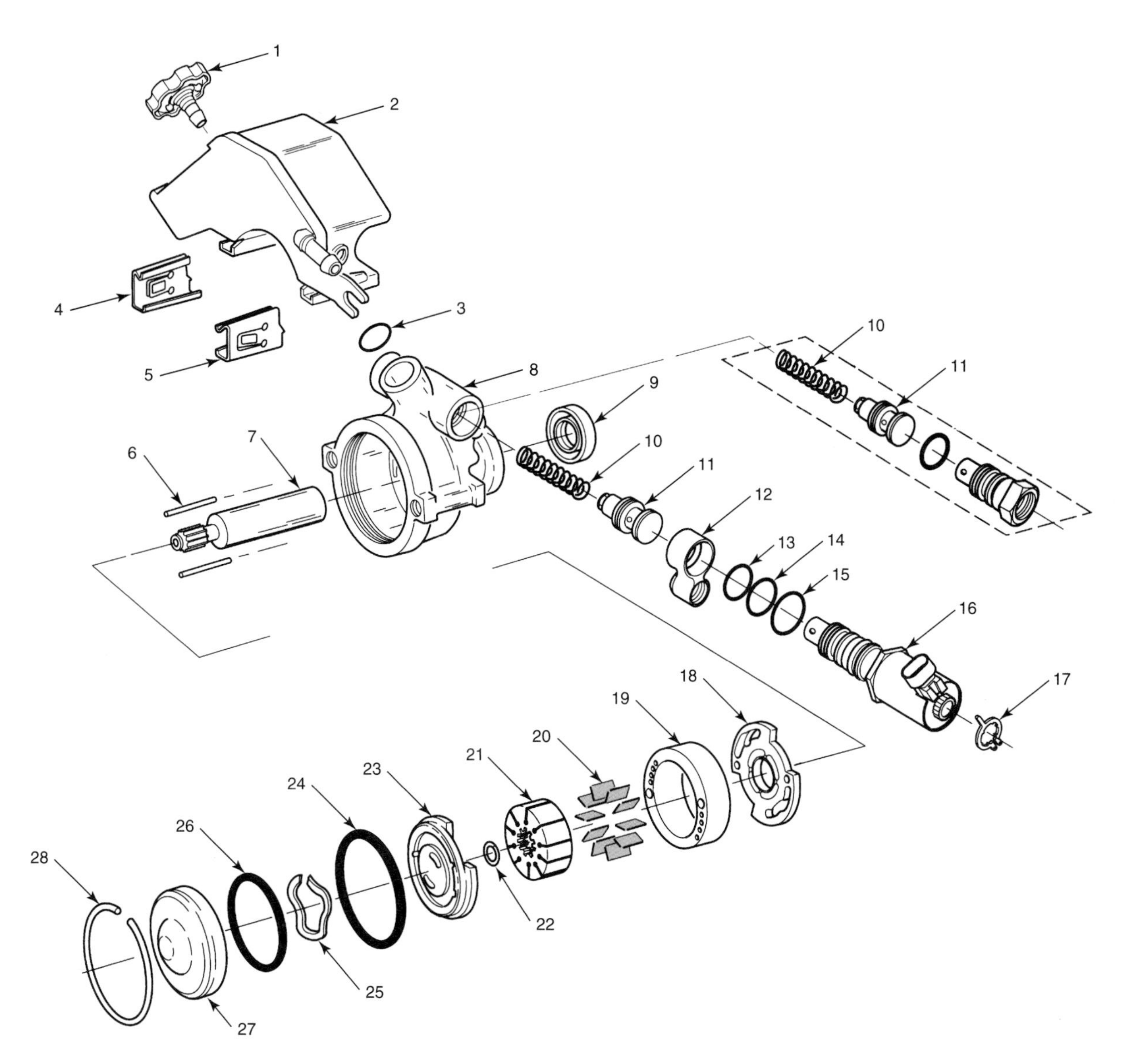

Figure 37-48. A rotor vane steering pump completely disassembled. 1—Reservoir cap. 2—Reservoir. 3—O-ring. 4—Retaining clip. 5—Retaining clip. 6—Dowel pin. 7—Drive shaft. 8—Housing assembly. 9—Drive shaft seal. 10—Spring. 11—Flow control valve. 12—Adaptor housing. 13—O-ring. 14—Seal. 15—Seal. 16—Power steering sensor. 17—Lock ring. 18—Thrust plate. 19—Pump ring. 20—Vanes (10). 21—Pump rotor. 22—Shaft retaining ring. 23—Pressure plate. 24—O-ring. 25—Pressure plate spring. 26—Seal. 27—End cover. 28—Retaining ring. (Saginaw Division of General Motors)

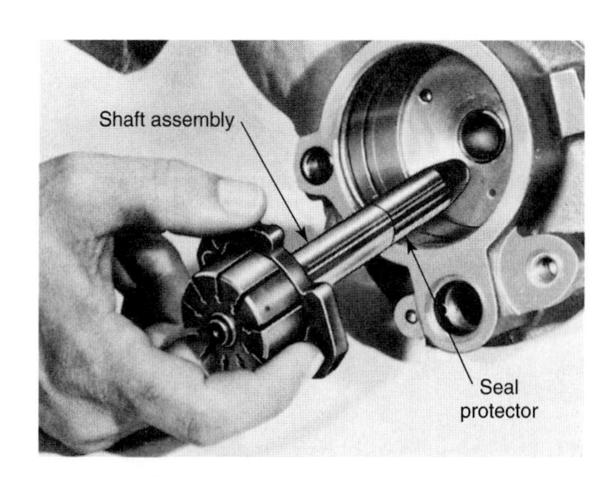

Figure 37-49. By using a seal protector, the shaft can be passed through the seal without damage.

of a shaft seal protector. If the pressure plate is spring-loaded, use a press to hold the end plate while inserting the retaining (snap) ring, **Figure 37-50.** Make certain the flow control valve springs are not distorted, nicked, or damaged. Install springs and valve correctly.

Install the pulley using a special installation tool and mount the pump on the engine. Make certain the pulley's drive key is in place. Adjust belt tension as described earlier in this chapter. If a leaking drive shaft seal is the only trouble, it can often be replaced without removing the pump from the car. The first step in on-the-car seal replacement is to remove the drive belt and pulley. Next, pry out the old seal. Then drive in a new seal using a suitable driving tool, **Figure 37-51.**

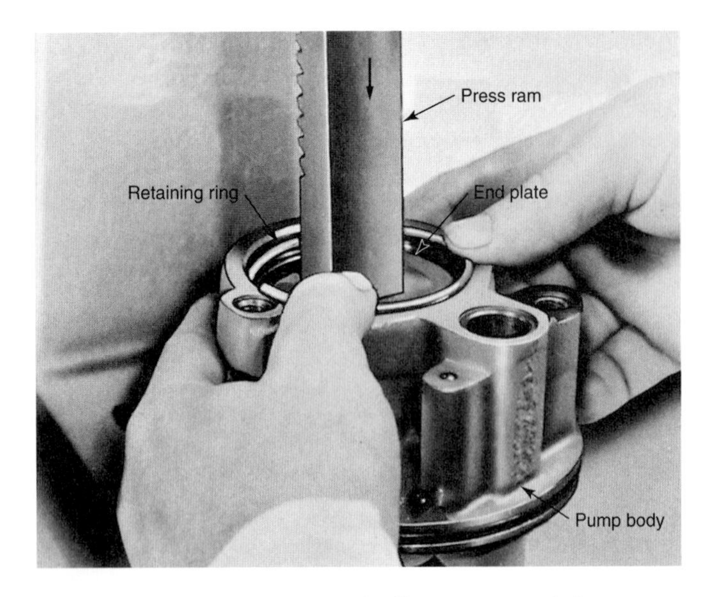

Figure 37-50. Using a press to facilitate pump end plate retaining ring installation.

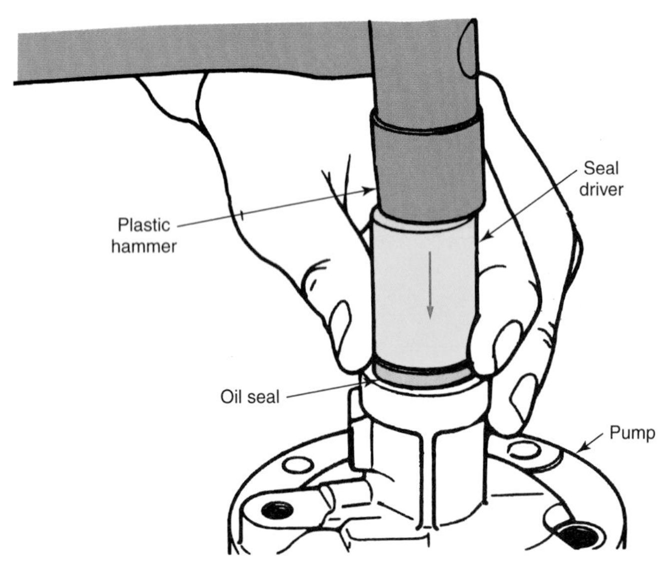

Figure 37-51. Installing a new pump shaft seal. To prevent a leak, be sure the seal is facing the proper direction before installation. (DaimlerChrysler)

Power Steering Hoses

Power steering hose is subjected to very high hydraulic system pressures. Carefully inspect each hose for softness, swelling, cracking, abrasion, etc. Inspect hose fittings for cracking and looseness and replace as needed. Power steering hose replacement is relatively simple. Remove hose or hoses from the steering pump and cap the pump outlets. Tip the hose, end-down, into a container to drain. Remove the fitting from the power steering gear or from the power piston control valve. Install the new hose and tighten fittings to specifications. Use only quality replacement hose, do not use remanufactured hose. Install high-pressure hose on the pump's outlet-to-gear circuit.

Power Steering Coolers and Switches

Most ***power steering coolers*** are simply extra lengths of tubing placed where air can flow over them. They are usually part of the low-pressure return circuit. If the car is equipped with a power steering cooler, check it for leakage. If necessary, clean the cooler surface. See **Figure 37-52.** A leaking cooler should be replaced. Some manufacturers incorporate a ***pressure-sensing switch*** into the power steering pump. This switch protects the engine from overloading by controlling idle speed. It can also cycle the air conditioning compressor on and off. To replace this switch, unscrew it and install the replacement.

Steering Linkage

The ***steering linkage*** consists of the various parts that connect the steering gear to the wheels. Steering linkage is vital to the operation of the overall steering system. There are many places where the steering linkage can become worn, damaged, or bent. Steering linkage arrangements vary depending upon need and basic design. One typical conventional steering system is illustrated in **Figure 37-53.** A rack-and-pinion system is shown in **Figure 37-54.** Note that they contain some of the same parts, such as tie rod ends.

Note: See Chapter 11, Preventive Maintenance, for information on steering system lubrication (greasing) as part of preventive maintenance.

Steering Linkage Service

Raise the car and inspect all steering linkage connections. Check the ball sockets for looseness by shaking the center link and tie rods. Check the idler arm mounting fasteners. Check the pitman arm retaining nut tension. Inspect

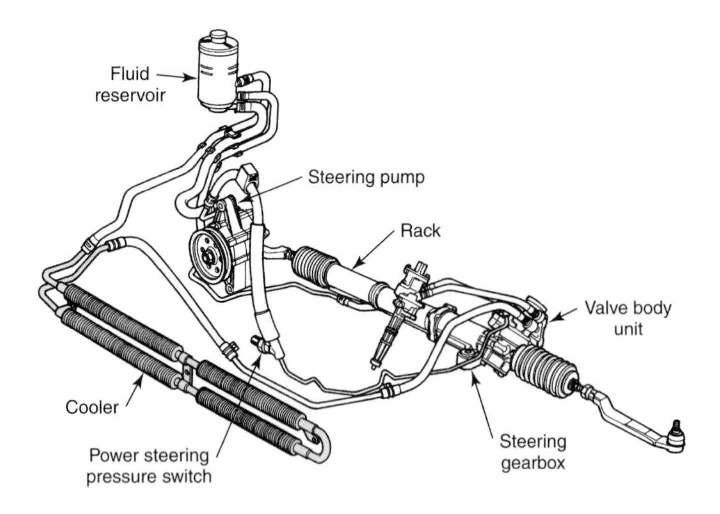

Figure 37-52. One type of power steering pump cooler. Check the cooler pipe for kinks, cracks, loose hose connections, and debris. Air passing over the tubing and the fins reduces the fluid temperature. (Honda)

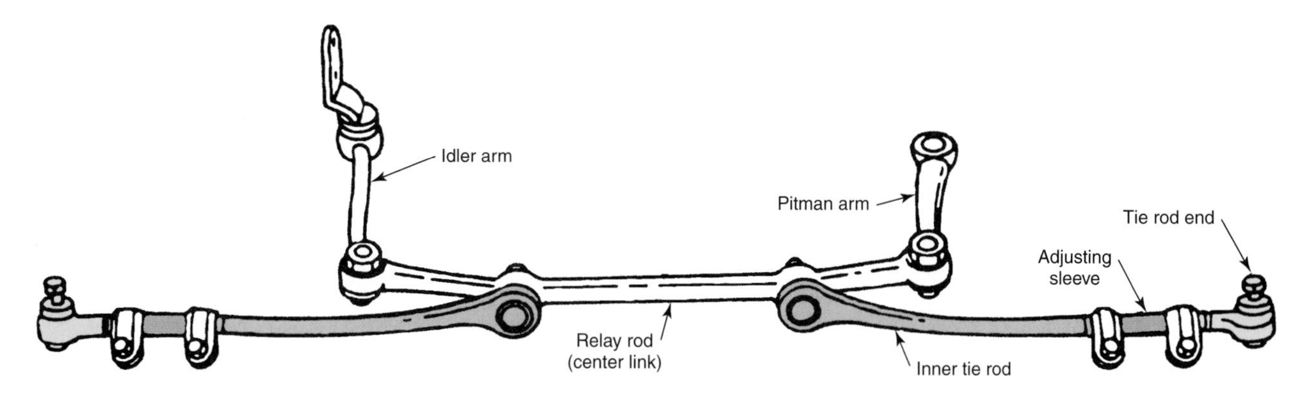

Figure 37-53. A parallelogram steering linkage arrangement. (Monroe Auto Equipment)

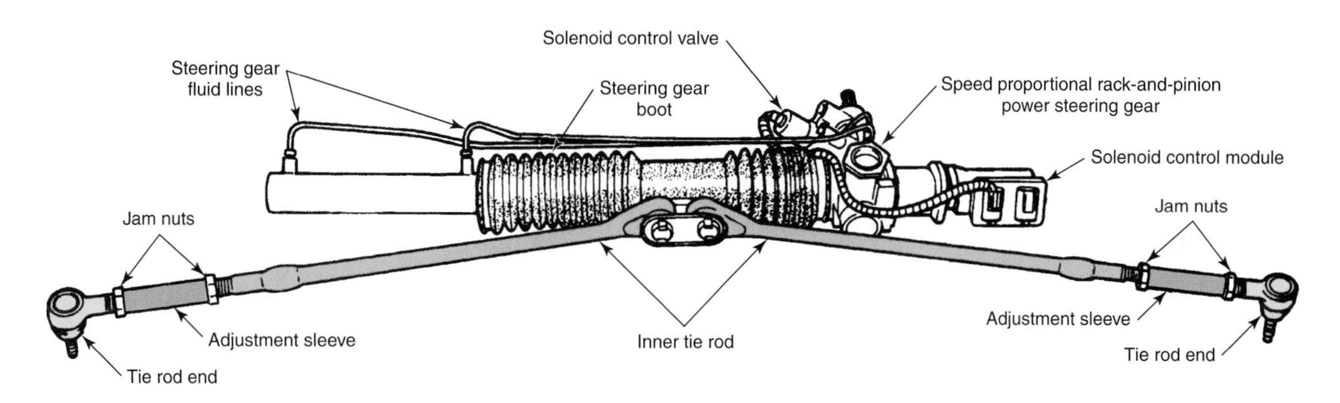

Figure 37-54. A rack-and-pinion, variable assist, speed-sensing (proportional) steering gear setup. (DaimlerChrysler)

the tie rod sleeve adjusters and sleeve clamp bolts. Check the steering arm fasteners. All nuts and cotter pins should be in place. Tighten loose fasteners and replace worn joints and leaking grease seals. Check for bent parts. Review the section on tie rod end replacement later in this chapter. Repair or replace parts as needed.

Pitman Arms

Pitman arms are used on conventional steering systems only. Remove the nut holding the pitman arm to the pitman shaft. Install a suitable puller. Tighten the puller until the pitman arm pulls free. Protect the pitman shaft and threads. Do not hammer on the end of the puller or on the end of the pitman shaft. Hammering could damage the steering gear. See **Figure 37-55.** To remove the pitman arm from the drag link or from the center link ball socket stud, see the section on tie rod end removal.

To install the pitman arm, clean the end of the pitman shaft thoroughly. Lubricate the splined area of the pitman shaft. Install the pitman arm on the shaft. Be certain to align the arm and the shaft correctly, **Figure 37-56.** Install the lock washer and the retaining nut. Torque the nut to specifications. Support one side of the pitman arm. With a heavy hammer, strike the other side with a sharp blow. Never hammer on the retaining nut or against the end of the shaft. Torque the nut

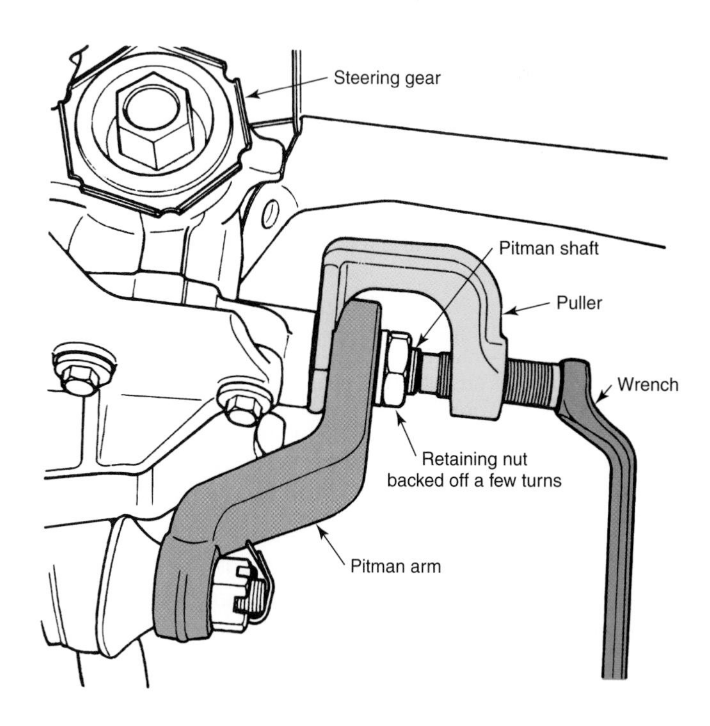

Figure 37-55. Using a special puller to remove the pitman arm from the steering gear. (DaimlerChrysler)

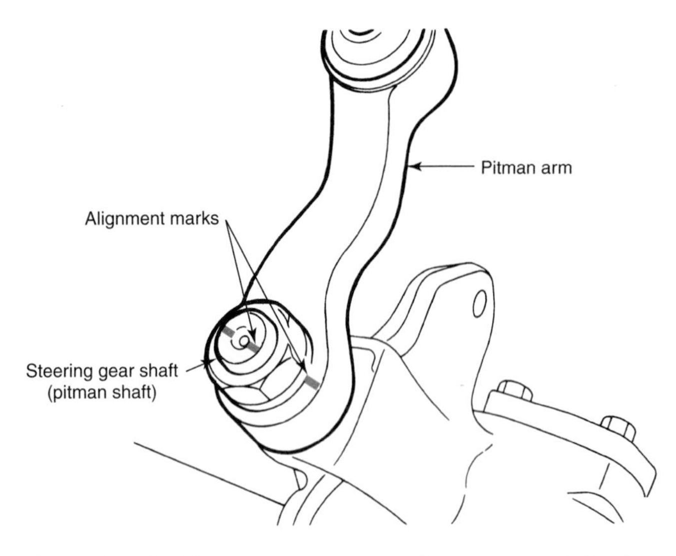

Figure 37-56. Alignment marks on a pitman arm and steering gear shaft. (DaimlerChrysler)

again. The sharp blow assists in proper seating of the pitman arm on the shaft. Stake the nut to prevent it from loosening.

Idler Arms

Idler arms are found only on conventional steering systems. Remove the bolts and nuts holding the idler arm to the frame. Next, remove the idler arm from the center link stud as applicable. To install the new idler arm, reverse the removal process. After all fasteners are in place, tighten the ball socket fastener and install a new cotter pin.

Tie Rod Ends

All steering systems contain ***tie rod ends.*** If a deflection test proves a tie rod end to be worn (loose), the tie rod end should be discarded. See **Figure 37-57. Figure 37-58** illustrates typical tie rod ball socket construction. In Figure 37-58A, the half-ball type is shown, while Figure 37-58B displays the full ball design. Note that a plug is used to allow periodic lubrication. In most tie rod designs, a fitting is normally installed in place of a plug. Removal and replacement procedures are the same for both types.

Note: For ease of reassembly, count the number of exposed threads on the old tie rod where it threads onto the rest of the steering linkage. Install the new part with the same number of threads showing. This simplifies alignment.

Remove the cotter pin and retaining nut. The stud is held in place by a taper fit, which must be broken or loosened. A special removing tool, such as shown in **Figure 37-59,** may be used to break the taper. This tool removes the tie rod end without damaging the seal or the socket. If a puller is not available, the tie rod stud can be removed by striking the steering arm sharply several times with a hammer while supporting it with a heavy block of steel. When installing the tie rod ball

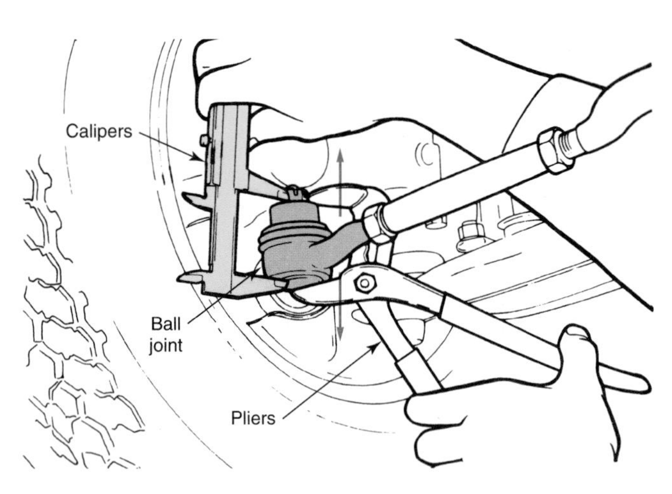

Figure 37-57. Grip the ball joint with pliers, compress it fully, and measure the compressed joint. The service limit (allowable wear) for this joint is 0.060″ (1.5 mm). (DaimlerChrysler)

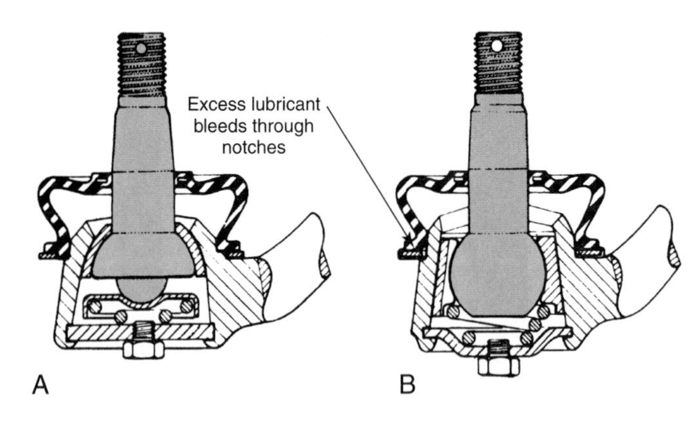

Figure 37-58. Tie rod end (ball socket) construction. (DaimlerChrysler)

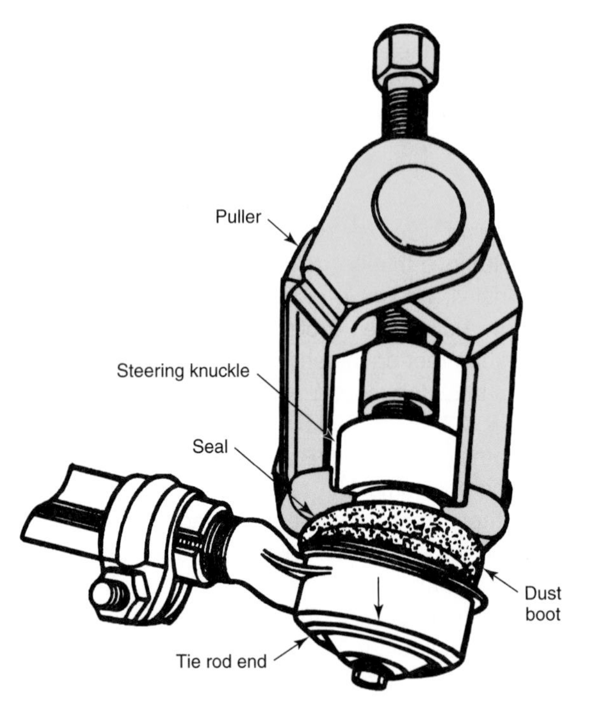

Figure 37-59. Using a puller to remove a tie rod end from the steering knuckle. (DaimlerChrysler)

stud, clean the stud and the tapered hole into which it fits. Wipe stud with a thin coat of oil. Insert stud and install the retaining nut. Torque the nut to specifications and insert a new cotter pin.

Tie Rod Adjuster Sleeve

In conventional systems, ***adjuster sleeves*** are threaded into the inner and outer tie rod ends. Turning the sleeves adjusts the steering toe by lengthening or shortening the overall tie rod assembly. Thoroughly clean the tie rod, the end threads, and the adjuster. Lubricate the adjuster sleeve threads. Force the tie rod and tie rod end against the adjuster sleeve while turning the sleeve. Continue turning the sleeve until toe is correct. The tie rod and tie rod end should have about the same number of threads turned into the adjusting sleeve. Position the adjuster sleeve clamps over the threaded portion of the tie rod in the sleeve. The sleeve slot and clamp opening should be correctly aligned. Torque clamp bolts to specifications. **Figure 37-60** illustrates a typical linkage system. Note tie rod end construction, clamp arrangement, and other parts.

Rack-and-Pinion Tie Rods

Rack-and-pinion tie rod removal is similar to removal on a conventional system. Remove the cotter pin and retaining nut. Then, loosen and remove the stud by breaking the taper with a power removing tool or puller. To remove the tie rod, remove the tie rod end from the shaft of the tie rod. Remove the bellows (boot) over the end of the rack-and-pinion assembly and remove the tie rod retaining pin. Consult the manufacturer's service manual to determine whether the retainer must be unscrewed, drilled, or driven out of the tie rod end. Then, using a special tool such as the one shown in **Figure 37-61,** unscrew the tie rod. Some vehicles have a crimped sleeve to hold the tie rod end in place. This must be ground or chiseled off, and a new sleeve must be crimped into place. Using the removal tool, install the new tie rod and a new retaining pin. Replace the bellows and tie rod end, and align the vehicle.

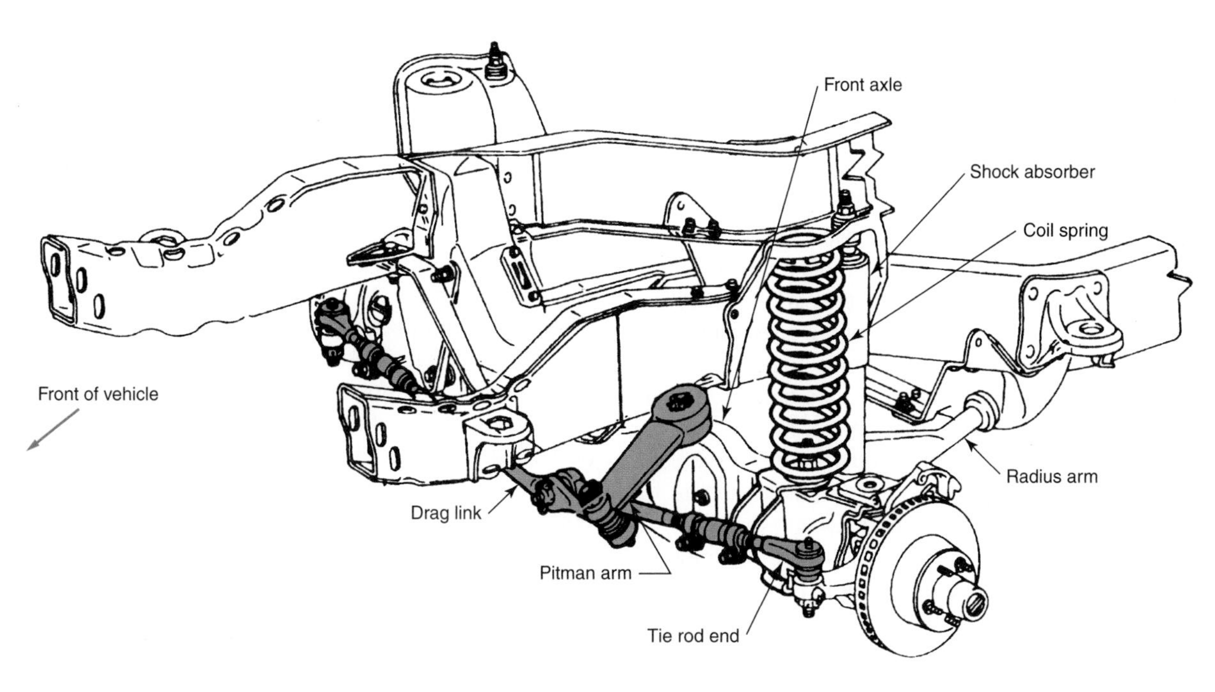

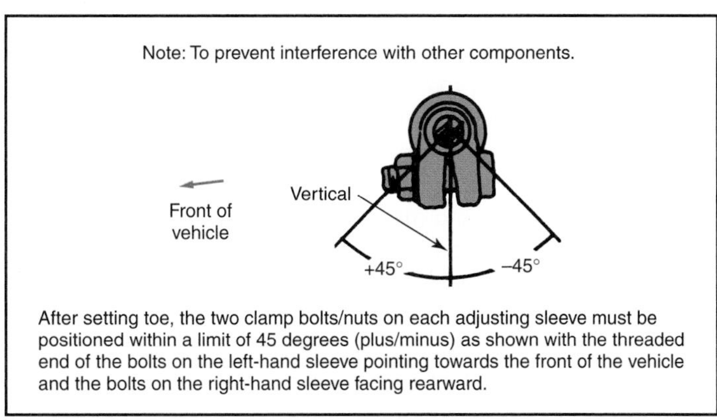

Figure 37-60. A steering linkage arrangement used on a four-wheel drive vehicle. Study this arrangement carefully. The pitman arm transfers the steering gear movement through the drag link and tie rods to the spindles. (Ford)

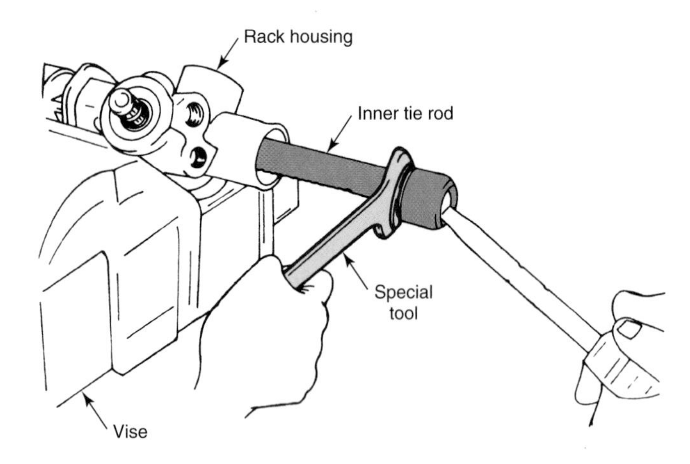

Figure 37-61. Removing a rack-and-pinion tie rod end by unscrewing it with a special wrench. (Hyundai)

Center Links

The procedure for replacing ***center links,*** or drag links, is similar to the procedure for replacing tie rod ends. Remove the cotter pin and nut from any tie rod ends that need removal and break the taper with a power tool or puller. Reverse the removal procedure to install the new link. Be sure to use a new cotter pin.

Electronic Assist Steering

Electronic assist steering is installed on some late-model vehicles with power steering. Electronic assist steering reduces power steering pressure at cruising speeds to increase control and road feel. High pressures are maintained at low speeds for ease of steering. Electronic assist operation is relatively simple. Major components include steering wheel rotation and vehicle speed sensors, a control module, and a flow control solenoid. The control module uses the vehicle speed sensor and steering wheel rotation sensor inputs to decide on the amount of power steering assist provided by a flow control solenoid. The flow control solenoid is mounted in the high-pressure side of the power steering system. If the electronic assist system fails, the power steering system operates normally.

Electronic Assist Steering Service

The most common problem caused by a defective electronic steering assist system is a lack of control and road feel at cruising speeds. On some vehicles, a warning light may come on if the system develops a problem. Before deciding that an electronic assist steering system is defective, check the basic power steering system. Check the fluid level, and make sure the system provides full power assist at low speeds. Once the power steering system has been eliminated as the source of the problem, use a scan tool to test the operation of the electronic assist system. Based on scan tool readings, proceed to test the sensors, the module, and the flow control solenoid.

Once the problem has been located, part replacement is relatively easy. Speed sensors are often the same ones used to provide input to the engine and power train computers, and are located in the same positions. Steering wheel rotation sensor removal may require that the steering wheel be removed. Most flow control solenoids are located in the outlet of the power steering pump. After replacing parts, always recheck system operation.

Summary

Steering systems consist of a steering wheel, steering shaft, steering gearbox or rack-and-pinion assembly, pitman arm, linkage, steering arms, and steering knuckle assemblies. Steering systems can be conventional (parallelogram) systems, or rack-and-pinion systems. All steering system work is critical to the safe operation of the vehicle. Steering wheels should be carefully aligned to center them.

The most common type of steering gear is the recirculating-ball, worm-and-nut design. Check the lubricant level in a manual steering gear periodically and add only the recommended lubricant. Manual steering gears usually have two adjustments, worm bearing preload and over-center adjustment. Make the worm bearing preload adjustment first. The conventional power steering gear has three basic adjustments, thrust bearing preload, worm-to-rack-piston preload, and pitman shaft over-center preload. When repairing power steering gears, use care to keep dirt away from the parts. Avoid denting or nicking the parts. Do not force the parts together. Use new seals and gaskets. Lubricate parts before assembly. The steering gear must be properly aligned with the steering shaft. Adjust the steering wheel spoke position by turning the tie rod adjusting sleeves. Steering wheel hub and shaft marks must be aligned. When installing a steering wheel, align marks and torque the retaining nut.

Power steering pumps can be of the vane, slipper, roller, or gear types. Steering pumps are generally belt driven. When overhauling power steering pumps, use care to avoid nicking parts. Clean all parts thoroughly. Use new seals and gaskets and lubricate parts before assembly. Replace soft, cracked, or abraded power steering hoses. Use quality replacement hose. Torque all connections. With the engine running, turn the steering wheel from one side to the other several times to bleed air from the hoses, gear, and pump.

The steering system should have a thorough inspection at regular intervals. Inspection should cover linkage connections, tie rod adjuster sleeve clamps, all system fasteners, steering arms, pitman arm, steering gear, power steering pump lubricant level, power steering pump belts, power steering hose, and cooler (where used). Road test the vehicle to check for proper steering.

Tension a new belt slightly tighter than a used belt to allow for initial belt stretch. When adjusting, removing, or installing belts, remove the battery's ground strap. Before conducting a power steering pressure test, check belt tension and reservoir fluid level. Use a 0 psi–2000 psi (0 kPa–13,800 kPa) gauge. Never close the pressure test shutoff valve or hold the steering wheel hard against the stops for a period exceeding five seconds. When adjusting pump belt tension, pry on a heavy section (not filler pipe or reservoir) of the pump or use the notch for a wrench in the mounting bracket.

When removing or installing the pitman arm, never hammer on the end of the pitman shaft or the end of the puller. When installing the pitman arm, align it on the shaft correctly. Torque pitman arm retaining nut. Loose idler arms must be replaced. Tie rod adjuster sleeves must be installed with an equal number of threads engaged on each end. Position sleeve clamps over threaded area and turn to specified angle.

Electronic assist steering is installed on some late-model vehicles with power steering. The most common problem caused by a defective electronic steering assist system is a lack of control and road feel at cruising speeds. After eliminating the power steering system as the source of the problem, use a scan tool to check the operation of the electronic assist system. Based on scan tool readings, proceed to test the sensors, the module, and the flow control solenoid. Once the problem has been located, part replacement is relatively easy.

Review Questions—Chapter 37

Do not write in this book. Write your answers on a separate sheet of paper.

1. Worn, dirty, or dry steering column shaft bearings can cause what kind of steering system problem?
   (A) Squeaks.
   (B) Roughness.
   (C) Binding.
   (D) All of the above.
2. Conventional steering gears use the _____ design.
3. When checking manual steering gear lubricant level, the filler plug and surrounding area should be _____.
4. The typical manual steering gear has two basic adjustments, _____ and _____.
5. Which of the two adjustments in question 4 is done first?
6. Name the three types of power steering pumps.
7. A squealing sound when turning the steering wheel against the stops usually indicates a loose power steering _____.
8. Name three important considerations for making an accurate test of power steering system pressure.
9. To disconnect two steering parts that are held by a taper, the taper must be broken (loosened), usually by _____.
   (A) striking it with a hammer
   (B) heating it cherry red with a torch
   (C) cooling it with dry ice
   (D) The taper will come loose when the attaching nut is removed.
10. Some idler arms can be repaired by installing new _____.
    (A) bushings
    (B) seals
    (C) grease fittings
    (D) All of the above.

ASE-Type Questions

1. Technician A says the steering gear must be correctly aligned with the steering shaft when it is reinstalled. Technician B says it is OK to adjust steering wheel spoke position by removing the wheel and relocating it on the steering shaft. Who is right?
   (A) A only.
   (B) B only.
   (C) Both A and B.
   (D) Neither A nor B.
2. The lubricant in a manual steering gearbox should be changed every _____.
   (A) two years
   (B) 30,000 miles
   (C) each spring and fall
   (D) the lubricant does not require periodic changing
3. Technician A says conventional steering gear adjustments are usually checked with a torque wrench or spring scale. Technician B says there are no adjustments to be made on a power steering conventional steering gearbox. Who is right?
   (A) A only.
   (B) B only.
   (C) Both A and B.
   (D) Neither A nor B.
4. When overhauling any hydraulic unit, one of the most important rules is _____.
   (A) working fast
   (B) painting the exterior
   (C) cleanliness
   (D) use of labor saving tools
5. All of the following statements about checking power steering fluid level are correct, *except:*
   (A) The fluid must be cold to make an accurate check.
   (B) Some power steering reservoir caps have dipsticks.
   (C) Some power steering reservoirs have level marks on the reservoir body.
   (D) Always top off with special power steering fluid when recommended.
6. Technician A says the power steering system is bled using bleeder fittings located on the high-pressure side of the gearbox. Technician B says the power steering system is bled using bleeder fittings located on the low-pressure side of the gearbox. Who is right?
   (A) A only.
   (B) B only.
   (C) Both A and B.
   (D) Neither A nor B.

7. All of the following statements about power steering hoses are true, *except:*
   (A) High-pressure hoses are used between the pump outlet and the gearbox.
   (B) Power steering coolers are usually installed in the high-pressure circuit.
   (C) Hoses should be replaced if they show any swelling or cracks.
   (D) Hose replacement is simple, but results in draining some of the fluid.
8. When conducting a power steering pressure test, never leave the shutoff valve closed for longer than _____.
   (A) five minutes
   (B) five seconds
   (C) twenty seconds
   (D) one minute
9. Technician A says the pitman shaft can be heated to remove the pitman arm. Technician B says using a pickle-fork-type tie rod separator can damage the seal. Who is right?
   (A) A only.
   (B) B only.
   (C) Both A and B.
   (D) Neither A nor B.
10. Technician A says the clamps on a tie rod adjuster sleeve can be tightened in any handy position. Technician B says removing the inner tie rod end on a rack-and-pinion steering system requires a special tool. Who is right?
   (A) A only.
   (B) B only.
   (C) Both A and B.
   (D) Neither A nor B.

Suggested Activities

1. Inspect two vehicles, one with a conventional steering system and the other with a rack-and-pinion steering system. Determine the difference between each system and write a short report explaining the differences. Add simple sketches to improve the report.
2. Inspect steering components for wear. The inspection should include all tie rod ends, idler arm, and pitman arm if used. Demonstrate the inspection procedure to other members of the class.
3. Replace a tie rod end, center (drag) link, or idler arm on a conventional steering system. Make a list of the tools that you need to perform the job.
4. Replace a tie rod and/or bellows boot on a rack-and-pinion system. Make a list of the tools that you needed to perform the job. Explain how the lubricant is added to the unit.
5. Replace a conventional steering box and a rack-and-pinion unit. Make a list of the tools that you needed to perform each job.

Manual and Power Steering Systems Diagnosis	
Problem: Hard steering and poor recovery following turns	
Possible cause	**Correction**
1. Tire pressure low.	1. Inflate to correct pressure.
2. Power steering pump defective.	2. Repair or replace pump.
3. Power steering pump fluid level low.	3. Add fluid to reservoir.
4. Manual steering gear lubricant level low.	4. Add lubricant.
5. Incorrect front wheel alignment.	5. Align front wheels.
6. Ball joints dry.	6. Lubricate ball joints.
7. Steering linkage sockets dry.	7. Lubricate linkage.
8. Linkage binding.	8. Relieve binding.
9. Damaged suspension arms.	9. Replace arms.
10. Steering gear adjusted too tight.	10. Adjust gear correctly.
11. Steering shaft bushing dry.	11. Lubricate bushings.
12. Steering shaft bushing or coupling binding.	12. Align shaft or coupling.
13. Excessive caster.	13. Adjust caster.
14. Sagged front springs.	14. Replace springs.
15. Bent spindle body.	15. Replace spindle.
16. Steering wheel rubbing steering column jacket.	16. Adjust jacket, check for steering shaft damage.
17. Steering gear misaligned.	17. Align gear.
18. Sticky valve spool.	18. Clean or replace spool valve.
19. Steering pump belt loose.	19. Adjust belt tension.
20. Power steering hose kinked or clogged.	20. Replace hose.
21. Different size front tires.	21. Install correct size tires on both sides.
22. Malfunctioning steering gear pressure port poppet valve.	22. Repair or replace.
23. Rack-and-pinion adjusted incorrectly.	23. Adjust to specifications.
24. High internal leakage of rack-and-pinion assembly.	24. Repair leaks or replace gear.
25. Rack-and-pinion mountings loose, causing binding.	25. Tighten mountings to specifications.
26. Defective steering stabilizer.	26. Replace stabilizer.
Problem: Vehicle pulls to one side	
Possible cause	**Correction**
1. Uneven tire pressure.	1. Equalize pressure.
2. Brakes dragging.	2. Adjust brakes.
3. Improper front end alignment.	3. Align front end.
4. Wheel bearings improperly adjusted.	4. Adjust bearings.
5. Damaged or worn steering valve assembly.	5. Replace steering valve assembly.
6. Tire sizes not uniform.	6. Install tires of the same size.
7. Broken or sagged spring.	7. Replace spring.
8. Rear axle housing misaligned.	8. Align rear housing.
9. Bent spindle.	9. Replace spindle.
10. Frame sprung.	10. Straighten frame.
11. Radial tire problem.	11. Switch front tires. If vehicle now pulls in the other direction, tires are defective.
Problem: Vehicle wanders from side-to-side	
Possible cause	**Correction**
1. Weak shock absorber or strut.	1. Replace shocks or struts.
2. Loose steering gear.	2. Torque mounting fasteners.
3. Loose rack-and-pinion mountings.	3. Tighten to specifications.
4. Ball joints and steering linkage need lubrication.	4. Lubricate suspension.
5. Steering gear not on high point.	5. Adjust steering gear properly.
6. Broken or missing stabilizer bar or link.	6. Replace stabilizer or link.
7. Rack-and-pinion improperly adjusted.	7. Adjust to specifications.
8. Tie rod end loose.	8. Tighten. Replace if worn.

(continued)

Manual and Power Steering Systems Diagnosis *(continued)*	
Problem: Sudden increase in steering wheel resistance	
Possible cause	**Correction**
1. Pump belt slipping. 2. Internal leakage in gear. 3. Fluid level low in pump. 4. Engine idle too slow. 5. Air in system. 6. Low tire air pressure. 7. Insufficient pump pressure. 8. High internal leakage in rack-and-pinion. 9. Defective steering stabilizer.	1. Adjust belt tension. 2. Overhaul or replace gear. 3. Add fluid. 4. Adjust idle. 5. Bleed system. 6. Inflate to recommended level. 7. Test and repair or replace as required. 8. Repair leaks or replace gear. 9. Replace stabilizer.
Problem: Steering wheel action jerky during parking	
Possible cause	**Correction**
1. Loose pump belt. 2. Oily pump belt. 3. Defective flow control valve. 4. Insufficient pump pressure.	1. Adjust belt tension. 2. Replace belt. Clean pulleys. Repair source of leak. 3. Replace flow control valve. 4. Test and repair.
Problem: No effort required to turn wheel	
Possible cause	**Correction**
1. Steering gear torsion bar broken. 2. Broken tilt column U-joint. 3. Steering wheel hub-to-shaft key missing. Splines stripped. Loose nut.	1. Replace spool valve and shaft assembly. 2. Replace U-joint. 3. Replace key, shaft, or wheel. Tighten nut to specifications.
Problem: Excessive wheel kickback and play	
Possible cause	**Correction**
1. Steering linkage worn. 2. Air in system. 3. Front wheel bearings improperly adjusted. 4. Gear over-center adjustment loose. 5. Worm gear not preloaded. 6. No worm-to-rack-piston preload. 7. Loose pitman arm. 8. Loose steering gear. 9. Steering arms loose on spindle body. 10. Excessive play in ball joints. 11. Defective rotary valve. 12. Worn steering shaft universal joint. 13. Extra large tires. 14. Defective steering stabilizer.	1. Replace worn linkage components. 2. Bleed and add fluid if needed. 3. Adjust front wheel bearings. 4. Make correct over-center adjustment. 5. Preload worm gear. 6. Install larger set of rack piston balls. 7. Torque pitman arm nut. 8. Tighten mounting fasteners. 9. Tighten arm fasteners. 10. Replace ball joints. 11. Replace rotary valve. 12. Replace joint and/or shaft. 13. Advise owner. Install a steering stabilizer or replace with a larger stabilizer. 14. Replace stabilizer.
Problem: No power assist in one direction	
Possible cause	**Correction**
1. Defective steering gear.	1. Overhaul or replace gear as needed. *(continued)*

Manual and Power Steering Systems Diagnosis *(continued)*	
Problem: Steering pump pressure low	
Possible cause	**Correction**
1. Pump belt loose. 2. Belt oily. 3. Pump parts worn. 4. Relief valve springs defective or stuck open. 5. Low fluid level in reservoir. 6. Air in system. 7. Defective hose. 8. Flow control valve stuck open. 9. Pressure plate not seated against cam ring. 10. Scored pressure plate, thrust plate, or rotor. 11. Vanes incorrectly installed. 12. Vanes sticking in rotor. 13. Worn or damaged O-rings.	1. Adjust belt. 2. Clean pulleys. Replace belt. Correct source of leak. 3. Overhaul pump. 4. Repair or replace as needed. 5. Add fluid. 6. Correct source of leak. Bleed system. 7. Replace hose. 8. Clean or replace valve. 9. Repair or replace cam ring and pressure plate. 10. Replace damaged parts and flush system. 11. Install vanes correctly. 12. Free vanes. Clean thoroughly. 13. Replace O-rings.
Problem: Steering pump noise	
Possible cause	**Correction**
1. Air in system. 2. Loose pump pulley. 3. Loose belt. 4. Glazed belt. 5. Hoses touching splash shield. 6. Low fluid level. 7. Clogged or kinked hose. 8. Scored pressure plate. 9. Scored rotor. 10. Vanes installed wrong. 11. Vanes sticking in rotor. 12. Defective flow control valve. 13. Loose pump. 14. Reservoir vent plugged. 15. Dirty fluid. 16. Pump bearing worn. 17. Chirp type noise. 18. Whine or growl.	1. Correct leak and bleed system. 2. Tighten pulley. 3. Tension belt correctly. 4. Replace belt. 5. Reroute hose to prevent contact. 6. Add fluid, check for leaks. 7. Replace hose. 8. Polish. Replace if badly scored. 9. Polish. Replace if badly scored. 10. Install vanes correctly. 11. Free vanes and clean thoroughly. 12. Replace flow control valve. 13. Tighten pump mounting fasteners. 14. Clean vent. 15. Drain, flush, and refill. 16. Overhaul as needed. 17. Tighten loose belt. 18. Low fluid. Fill to proper level.
Problem: Steering gear dull rattle or chuckle	
Possible cause	**Correction**
1. Gear loose on frame. 2. Loose over-center adjustment. 3. No worm shaft preload. 4. Insufficient or improper lubricant (manual gear).	1. Tighten gear mounting fasteners. 2. Make correct over-center adjustment. 3. Adjust preload. 4. Fill with specified lubricant.
Problem: Hissing sound in gear	
Possible cause	**Correction**
1. Normal sound when turning wheel when vehicle is standing still or when holding wheel against stops. 2. Gear loose. 3. Noisy pressure control valve. 4. Intermediate shaft rubber plug missing.	1. Normal condition, advise driver. 2. Tighten mounting fasteners. 3. Replace valve. 4. Replace plug.

(continued)

Manual and Power Steering Systems Diagnosis *(continued)*	
Problem: Tire squeal on turns	
Possible cause	**Correction**
1. Excessive speed.	1. Advise driver.
2. Low air pressure.	2. Inflate to correct pressure.
3. Faulty wheel alignment.	3. Align wheels.
4. Excessive load.	4. Advise driver.
5. Rack-and-pinion mountings loose.	5. Tighten mountings to specifications.
Problem: External fluid leaks	
Possible cause	**Correction**
1. Defective hose.	1. Replace hose.
2. Loose hose connections.	2. Tighten to proper torque.
3. Cracked hose connections.	3. Replace hose.
4. Pitman shaft seal in gear defective.	4. Replace seal. Check bearing for excessive wear.
5. Gear housing end cover O-ring seal leaking.	5. Replace seal.
6. Gear torsion bar seal leaking.	6. Replace valve and shaft assembly.
7. Adjuster plug seals leaking.	7. Replace seals.
8. Side cover gasket.	8. Replace gasket.
9. Pump too full.	9. Reduce fluid level.
10. Pump shaft seal defective.	10. Replace shaft seal.
11. Scored shaft in pump.	11. Replace shaft.
12. Oil leaking out of reservoir from air contamination.	12. Correct source of air leak.
13. Pump assembly fasteners loose.	13. Torque fasteners.
14. Leaking power cylinder (linkage type).	14. Overhaul as needed.
15. Rack-and-pinion housing cracked.	15. Replace steering gear.
16. Extreme cam ring wear.	16. Replace parts. Flush system.
17. Scored pressure plate, rotor, or thrust plate.	17. Replace parts. Flush system.
18. Vanes incorrectly installed.	18. Install pump vanes properly.
19. Reservoir cracked.	19. Replace reservoir.
20. Pump reservoir cap leaking.	20. Repair or replace cap.
21. Defective rack-and-pinion stub shaft seal.	21. Replace stub shaft seal.
22. Defective rack-and-pinion rotary valve.	22. Replace rotary valve.
23. Pinion shaft seal leaking.	23. Replace shaft seal.
24. Rack-and-pinion bulkhead seal defective.	24. Replace seal.

38

Wheel and Tire Service

After studying this chapter, you will be able to:

- Describe the construction, operation, and service of wheel bearings.
- Explain tire and wheel construction and service.
- Summarize tire size, type, and quality ratings.
- Diagnose common wheel bearing- and tire-related problems.

Technical Terms

Rollers	Temperature resistance
Cage	Traction
Races	Tread wear
Spindle	Space-saver
Wheel rims	Plug
Custom wheels	Liquid sealer
Torque stick	Patch
Wheel lugs	Stitcher
Lug nuts	Bead expander
Plies	Wheel tramp
Bias	Wheel shimmy
Bias-belted	Static balance
Radial	Dynamic balance
Uniform tire quality grading	

This chapter covers the design and service procedures for wheel bearings, wheel rims, and tires. This chapter provides basic information about tire and wheel service. Tire repair, wear problems, mounting, and balancing are discussed.

Wheel Bearings

The following sections cover the servicing of wheel bearings found on front- and rear-wheel drive vehicles. These bearings form the connection between the rotating wheel and brake assemblies and the nonrotating wheel spindles. The drive and nondrive axles of a vehicle are supported by bearings. Three main types of bearings are used on modern vehicles:

- Tapered roller bearings, **Figure 38-1.**
- Straight roller bearings.
- Ball bearings, **Figure 38-2.**

Tapered roller bearings can be cleaned, regreased, and adjusted. The straight roller and ball bearings used on most modern vehicles cannot be serviced and are replaced as a unit. A few older vehicles use ball bearings that can be regreased in the front axles. The first part of this section covers the service of tapered roller bearings. The following section covers the service of sealed roller and ball bearings.

Note: Rear wheel bearings used in the axle assemblies of rear-wheel drive vehicles are covered in Chapter 32, Axle and Driveline Service.

Tapered Roller Bearing Service

Rear-wheel drive vehicles have tapered roller bearings in the front wheels, Figure 38-1. Some front-wheel drive vehicles may use tapered roller bearings in the rear axle hub assembly.

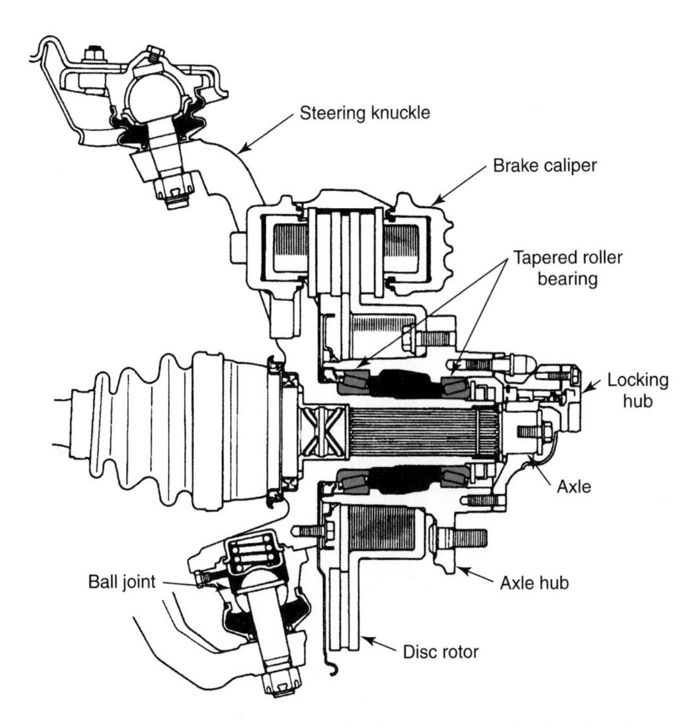

Figure 38-1. A cutaway view of a four-wheel drive front axle and hub assembly that uses tapered roller bearings. (Toyota)

The tapered roller bearing consists of the ***rollers*** and ***cage*** and the inner and outer ***races.*** To remove the bearings, other wheel components must be removed. Begin by prying off the wheel cover, being careful to avoid springing it out of shape. Pry a little at a time, moving around the cap.

Some wheel covers are held on by *wheel cover locks,* as shown in **Figure 38-3.** Other vehicles with custom wheels have *wheel rim locks,* **Figure 38-4.** The key to these locks is in the vehicle and must be located to remove the wheel covers. After using the wheel lock key, replace it in the vehicle. Without the key, the driver will not be able to remove the wheel and tire in an emergency.

Tapered Roller Bearing Removal

After removing the wheel cover, remove the wheel lug nuts and remove the wheel and tire. Remove the bearing dust cap. Straighten the cotter pin and remove it, then unscrew the adjusting nut. Note that some vehicles use a left-hand thread nut on the rear wheels. If the vehicle has front disc brakes, remove the caliper. Caliper removal was discussed in Chapter 34, Brake Service. Also remove the anti-lock brake (ABS) sensor if applicable.

Caution: The drum or rotor may be hot. Use a rag, if necessary, to protect your hands.

Remove the washer from the steering knuckle, or ***spindle,*** and shake the drum or rotor from side to side or pull it outward a short distance. Then, push it back on. This moves the outer bearing so it can be grasped and removed from the hub. Remove the outer bearing and place in a clean container. Then grasp the drum or rotor firmly and pull it straight off the spindle. Be careful that the inner bearing and seal are not dragged across the spindle threads. See **Figure 38-5.** If the

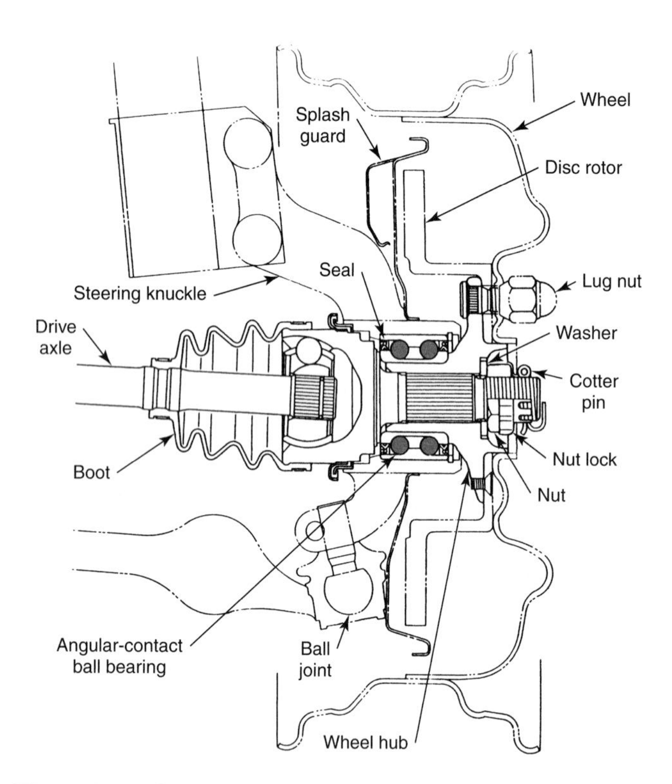

Figure 38-2. One type of front-wheel drive hub and bearing assembly. This particular arrangement uses angular-contact ball bearings. (Daihatsu)

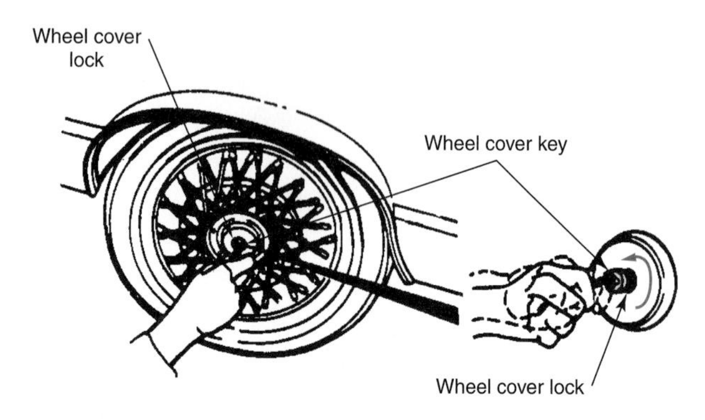

Figure 38-3. One type of wheel cover lock arrangement.

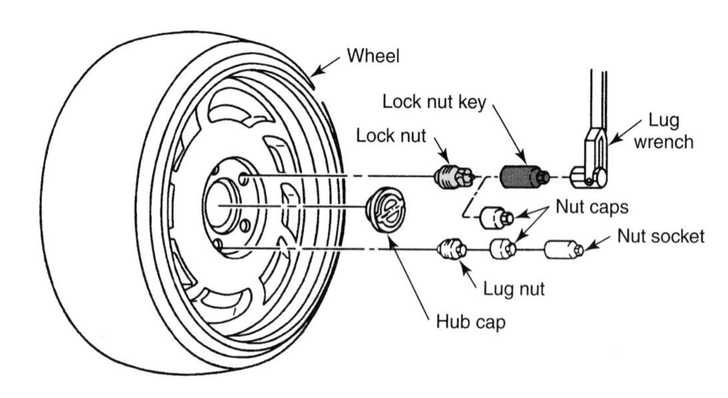

Figure 38-4. Special light alloy (aluminum) wheel locking nut. These replace one standard nut per wheel. Note the unique pattern on the nut head. A special key with a corresponding pattern fits over the nut head for removal. (General Motors)

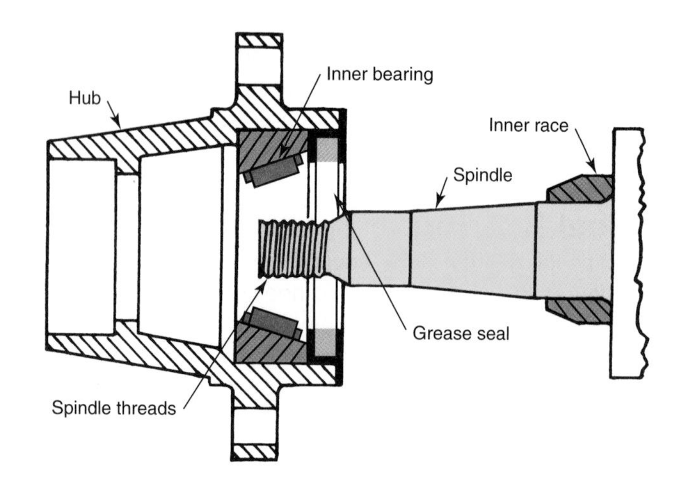

Figure 38-5. Support the wheel when removing or installing it to prevent the inner bearing and grease seal from dragging over the spindle threads.

brake shoes drag and make wheel assembly removal difficult, back off the brake shoe adjustment (see Chapter 34, Brake Service). Lay the drum or rotor over a clean rag or piece of paper. Use a long, soft steel drift to engage the inner bearing race. Do not strike the roller cage. Tap the inner bearing from the hub. Tap a little at a time, moving the drift around the race. This removes the bearing and grease seal. Discard the old grease seal. Always use a new grease seal following bearing or brake service. See Chapter 10, Friction and Antifriction Bearings, for complete instructions on bearing cleaning and checking. Important points to remember are:

- Use a final rinse of clean solvent.
- Never spin the bearing with air pressure.
- Examine each roller and race.
- Pack bearings with lubricant and store in a clean container until ready to use.
- Discard bearings showing the slightest signs of chipping, galling, or wear.
- If any part of the bearing is damaged, replace the entire bearing.
- A used roller assembly must always be installed in the original race.

Thoroughly clean and flush the inside of the hub. Do not allow solvent and grease to contact the brake drum or disc surfaces. Wipe the hub dry with a clean cloth and inspect the bearing races. If damaged, or if either the inner or outer roller bearing assembly is damaged, remove the corresponding race, **Figure 38-6.** Use a brass or soft steel drift. If the hub is slotted, place the drift on the exposed edge of the outer race at the slot. If slots are not provided, place the drift on the thin exposed race edge. Tap the race from the hub. Move the drift around so that the race is forced out without excessive tipping. Tipping the race can distort the hub. Use care to avoid damaging the hub with the drift. A puller, as in **Figure 38-7,** will remove the hub races easily and will avoid tipping.

Wipe the hub recess clean. Lubricate new bearing race and place in position over recess. Make certain that the race faces the correct direction. Using a driver, drive the race inward until it is seated firmly against the stop. Refer to **Figure 38-8.** The race may be pressed into place if such equipment is available. See **Figure 38-9.** If a drift punch is used, tap a little at a time. Move the punch around the race to avoid cocking or tipping. Use a soft steel drift. Never use a hardened punch.

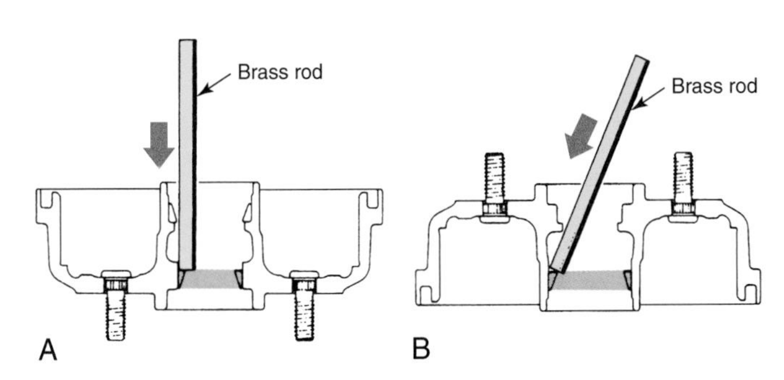

Figure 38-6. A—Driving the outer bearing race from the hub. B—Driving the inner race from the hub. (Mitsubishi)

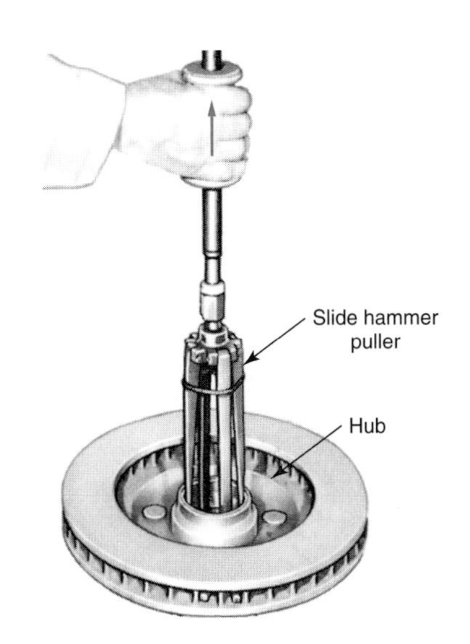

Figure 38-7. Using a slide hammer puller to remove the hub bearing race. (Ford)

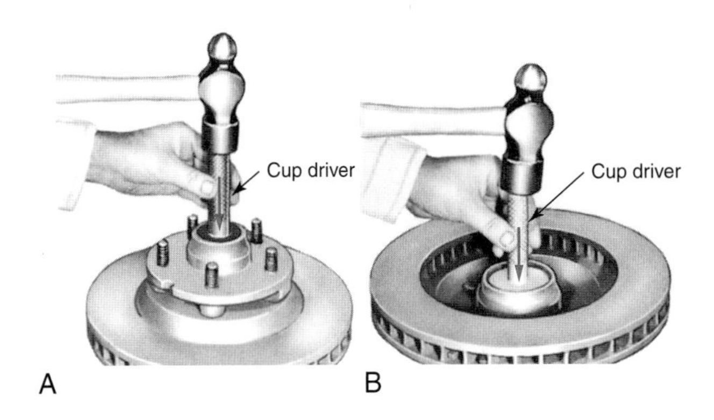

Figure 38-8. A—Installing an outer bearing race. B—Installing an inner bearing race. Note the use of the race driver tool.

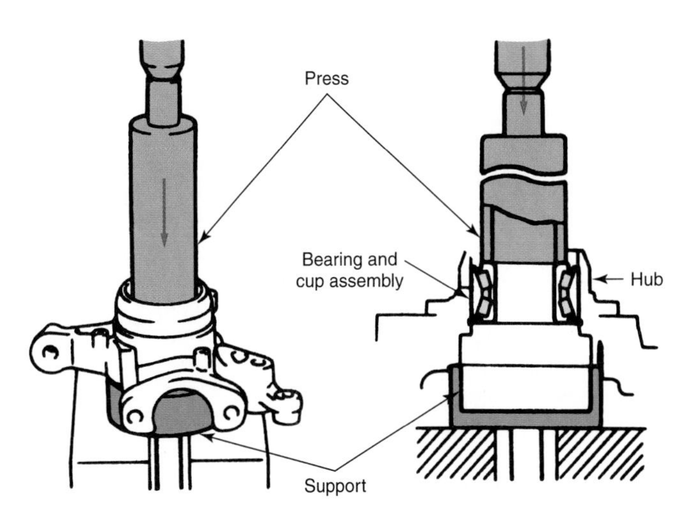

Figure 38-9. Using a press to install the bearing and race assembly in the hub. Wear your safety glasses! (Toyota)

Packing Wheel Bearings

Never repack a bearing with any kind of grease without first removing all of the original grease from both bearings and the hub. Many types of grease are not compatible and should never be mixed. Use only high-temperature grease on a modern vehicle, since all modern brake systems operate at high temperatures. Using a bearing packer or your hands, pack each bearing with the specified wheel bearing grease. Make sure each bearing is fully packed and that a generous amount of grease is applied to the outside of the rollers. Never apply grease to an oily bearing; the grease will not adhere properly. The bearing must be clean and dry before packing.

To prevent rust from moisture condensation and grease runoff from the bearings, the hub, race surfaces, and dust cap interior should be given a coating of grease. The coating in the cap can be relatively light. Coat the hub inner cavity to a depth that will bring the grease level up to the inner edge of the bearing races. Never pack the hub full of grease. Pack as illustrated in **Figure 38-10.**

Insert the packed inner bearing into the hub. The neoprene seal lips should be lubricated with a small amount of wheel bearing grease. Then wipe out the seal recess in the hub and place the seal on the hub with the sealing lip facing inward. Using a seal driver, drive the seal to the proper depth. Do not drive the seal in so deeply that it engages the bearing. Seals are usually driven in until flush with the top surface of the hub. Refer to **Figure 38-11.** If a punch or hammer must be used to seat the seal, strike the seal's outer edge only. The inner portion is unsupported and a blow there will destroy the seal. Wipe off any grease on the outside of the seal or on the hub.

Cup
Roller
Hub
Spindle
Seal
Cage
Cone
Nut lock
Bearing race
Grease
Hub grease cavity
Hub cavity

Pack flush with inner edge of the bearing race

Apply a light film of grease

Figure 38-10. Fill the hub grease cavity to the depth shown. Also, place a light film of grease inside the dust cap. Follow the vehicle manufacturer's recommendations for the type and amount of grease used. (DaimlerChrysler)

Brake Assembly and Spindle

Cover the spindle with a clean cloth. Clean off the brake assembly with approved equipment. Do not apply a direct blast of cleaning fluid, as the dust and dirt could be forced into the wheel cylinders or calipers. Clean the spindle thoroughly with a cloth dampened with solvent. Wipe dry with a clean cloth. Brake cleaning was discussed in Chapter 34, Brake Service.

Warning: Do not use compressed air to clean or dry brake parts as they can contain asbestos. Review the asbestos warning in Chapter 34, Brake Service.

Examine the spindle carefully for signs of cracking, wear, or scoring. If the vehicle has considerable mileage or is used in heavy service, the spindle should be tested for cracks. The inner race is designed to move on the spindle. This movement exposes a constantly changing portion of the race to the heaviest pressure from the rollers, which increases bearing life. To ensure proper race movement, the inside of the race and the area of the spindle that touches the race must be very smooth and coated with a film of grease. If rusty or at all rough, polish the race contact areas of the spindle with crocus cloth. Wipe the spindle clean and lubricate it with a light coat of wheel bearing grease to prevent rusting and to reduce the friction of race motion.

Installing Tapered Wheel Bearings

Check the brake shoes and drum or rotor and pads for grease and fingerprints. Clean them with a nonpetroleum solvent if necessary. Support the drum or rotor and slide straight on spindle, **Figure 38-12.** Do not drag the grease seal

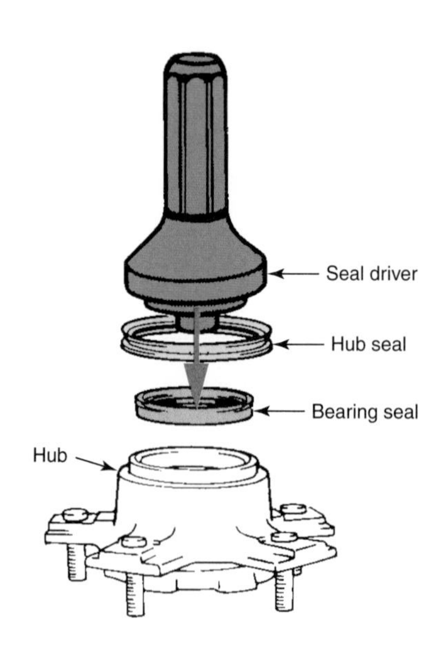

Figure 38-11. Using a seal driver to install the hub and bearing seal. Be sure the seals are facing in the proper direction.

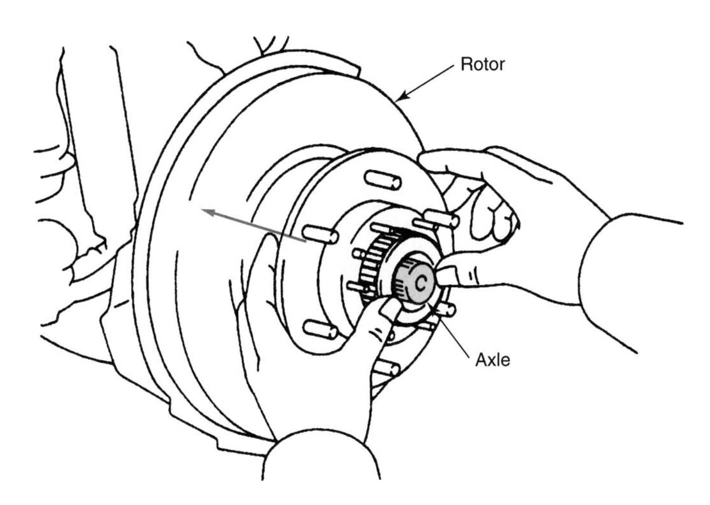

Figure 38-12. Installing a hub on the spindle. Be careful not to damage the grease seal by dragging it over the end of the axle or spindle threads. (Toyota)

over spindle threads. When the drum or rotor is in position, insert the outer bearing, safety thrust washer, and locknut. Do not forget to install the pronged safety washer (if used). To avoid interference by brake friction pads, do not reinstall the caliper until the bearing adjustment is complete.

Wheel Bearing Adjustment

Proper wheel bearing adjustment is very important. Improper adjustment can cause poor brake performance, wheel shake, poor steering, or rapid bearing wear. See **Figure 38-13.** Each manufacturer specifies a particular adjustment procedure. There are several methods that may be used for adjusting wheel bearings. Two of the most accurate and widely recommended are the torque wrench method and the dial indicator method.

Bearing Adjustment—Torque Wrench

In this method, the locknut is tightened to a specified torque while spinning the wheel in the direction of tightening. This heavier initial torque seats the bearings. The nut is then loosened until it can be rotated by hand. Then, the nut is once again torqued, this time to a lower value. The nut is then backed up until the cotter pin will pass through the nearest hole in the spindle and one of the slots in the nut. See

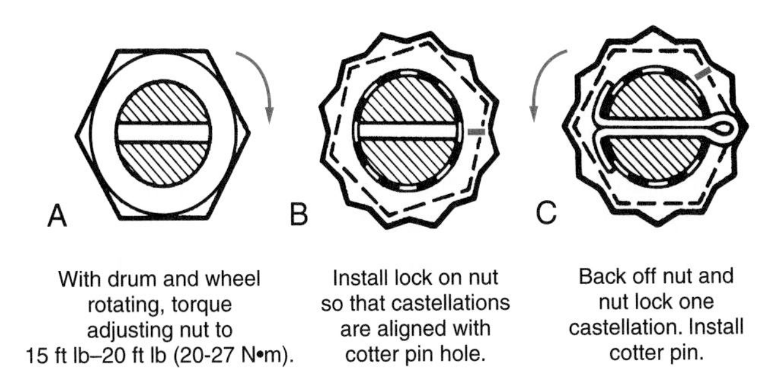

Figure 38-13. Adjusting the wheel bearing using a stamped adjustment nut. (Ford)

Figure 38-14. Some vehicles use a stamped nut lock that fits over the regular nut. This provides a finer adjustment in that more combinations of nut position and alignment with the cotter pin holes are available, **Figure 38-15.**

This special lock cap is compared with other nut types in **Figure 38-16.** Note that the nut may be adjusted as little as 1/24 turn. Another technique employing a torque wrench when the special stamped locking nut is used is illustrated in **Figure 38-17.** The adjusting nut is tightened to around 20 ft lb (27 N·m) while spinning the wheel. The stamped locknut is then tried over the adjusting nut until one of the slots is aligned with the cotter pin hole in the spindle. The adjusting nut and locknut are then carefully loosened by one slot. The cotter pin is then inserted. Some vehicles use a special staked nut in place of a cotter pin. Look at **Figure 38-18.** When the staked nut is removed, it must be discarded. The nut has an integral, thin lip that is forced into a groove cut in the axle stub shaft with a special tool. Do not use a screwdriver or other sharp edged tool for staking.

Bearing Adjustment—Dial Indicator

Tighten the adjusting nut firmly with a medium size adjustable wrench while spinning the wheel. Loosen the

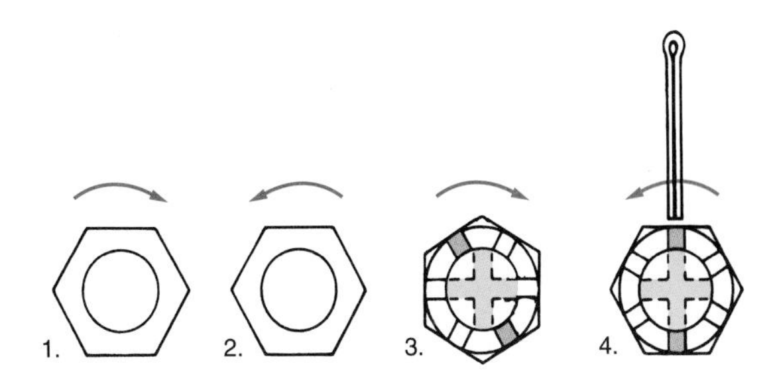

Figure 38-14. One wheel bearing adjustment procedure that involves the use of a torque wrench. (General Motors)

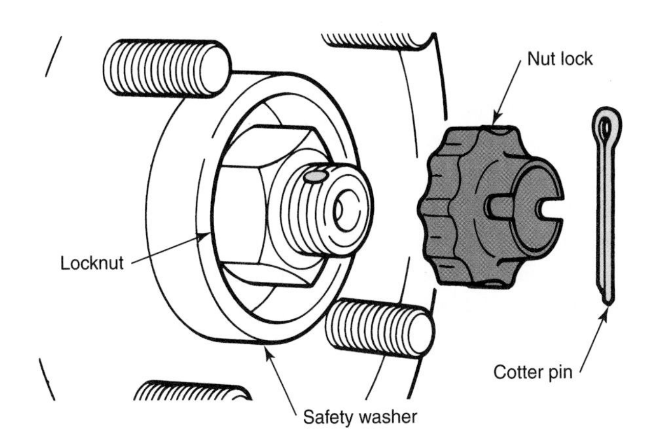

Figure 38-15. By using a stamped adjustment nut lock, fine wheel bearing adjustment is possible. (DaimlerChrysler)

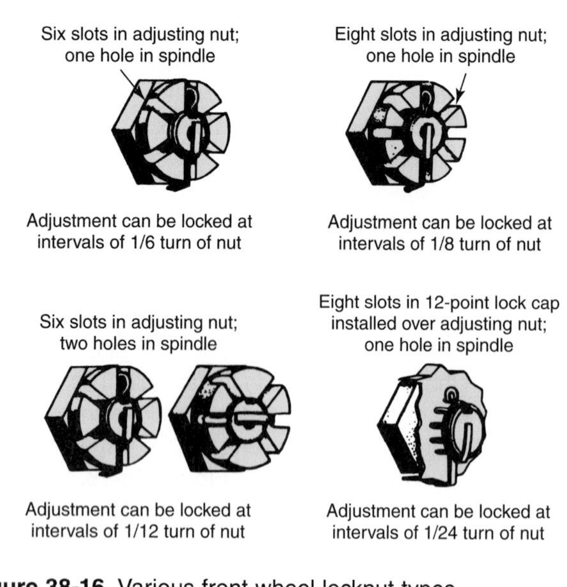

Figure 38-16. Various front wheel locknut types.

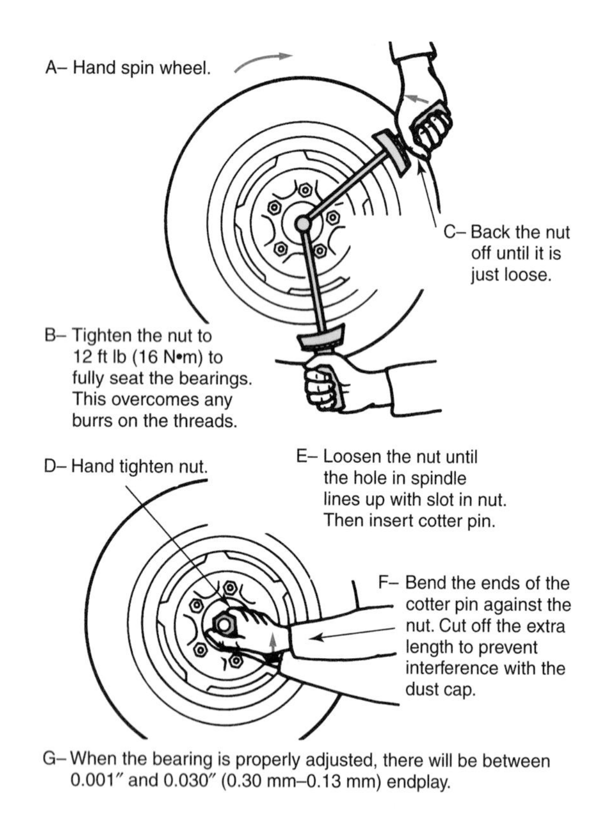

Figure 38-17. One method of wheel bearing adjustment with the tire and wheel on the vehicle. (General Motors)

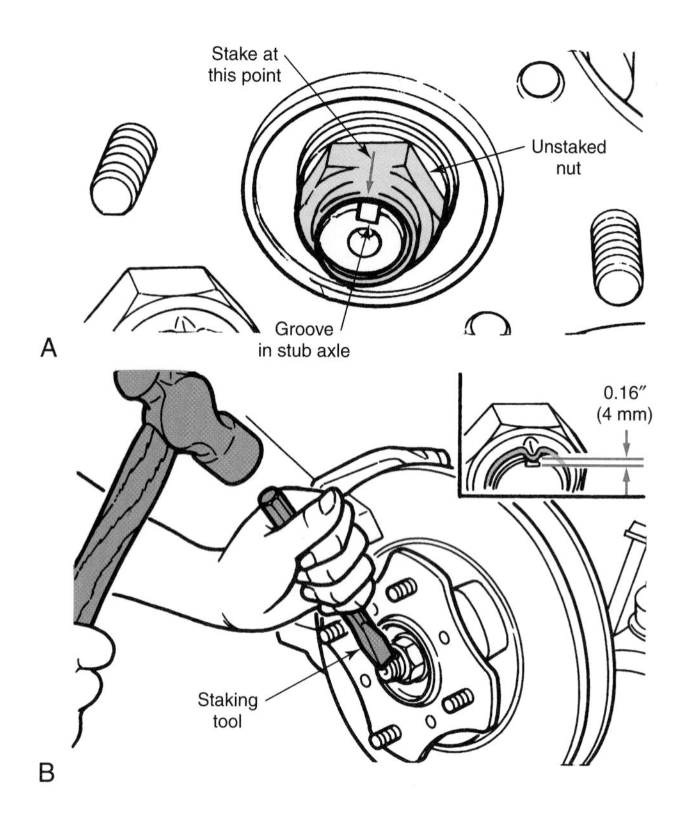

Figure 38-18. Staking the wheel nut to prevent rotation. A—An unstaked nut. Note the groove in the stub axle. B—A staking tool being used to force the nut lip into the axle groove. Note the proper staking depth. (Mazda)

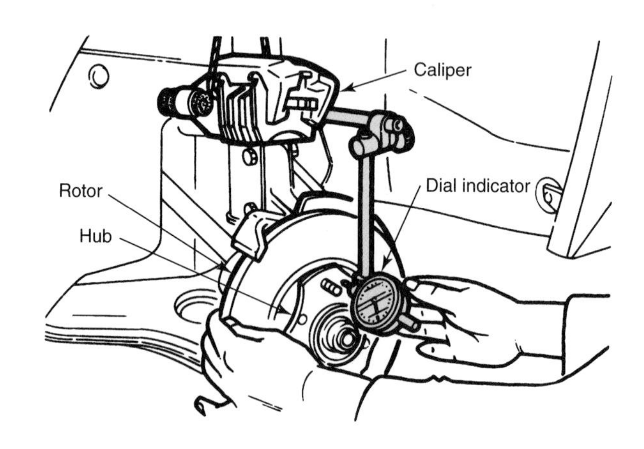

Figure 38-19. A dial indicator being used to check bearing play. Note that the caliper has been removed. This makes accurate measuring easier. (Mazda)

adjusting until it is just finger loose. Attach a dial indicator so that the indicator stem just contacts the machined end (outer side) of the hub, **Figure 38-19.** Grasp the sides of the rotor or the tire and pull straight in and out. Read gauge for amount of endplay. Adjust nut until endplay is within manufacturer's specifications. The dial indicator may be mounted to the spindle nut with the indicator touching the hub end. It may also be mounted to a wheel stud with the indicator stem contacting the end of the spindle.

Regardless of the technique used, the wheel will spin smoothly, freely, and with no appreciable lateral play (side shake) when the adjustment is correct. Do not confuse ball joint or suspension arm bushing wear with wheel bearing looseness. If in doubt, mount a dial indicator as described and check actual endplay. For most tapered roller wheel bearings, adjust so that endplay ranges from 0.001″–0.005″ (0.025 mm–0.127 mm). Bearing service is illustrated in **Figure 38-20**. If disc brakes are used, up to 0.001″ (0.025 mm) endplay is normally acceptable. Excessive endplay allows the disc to wobble.

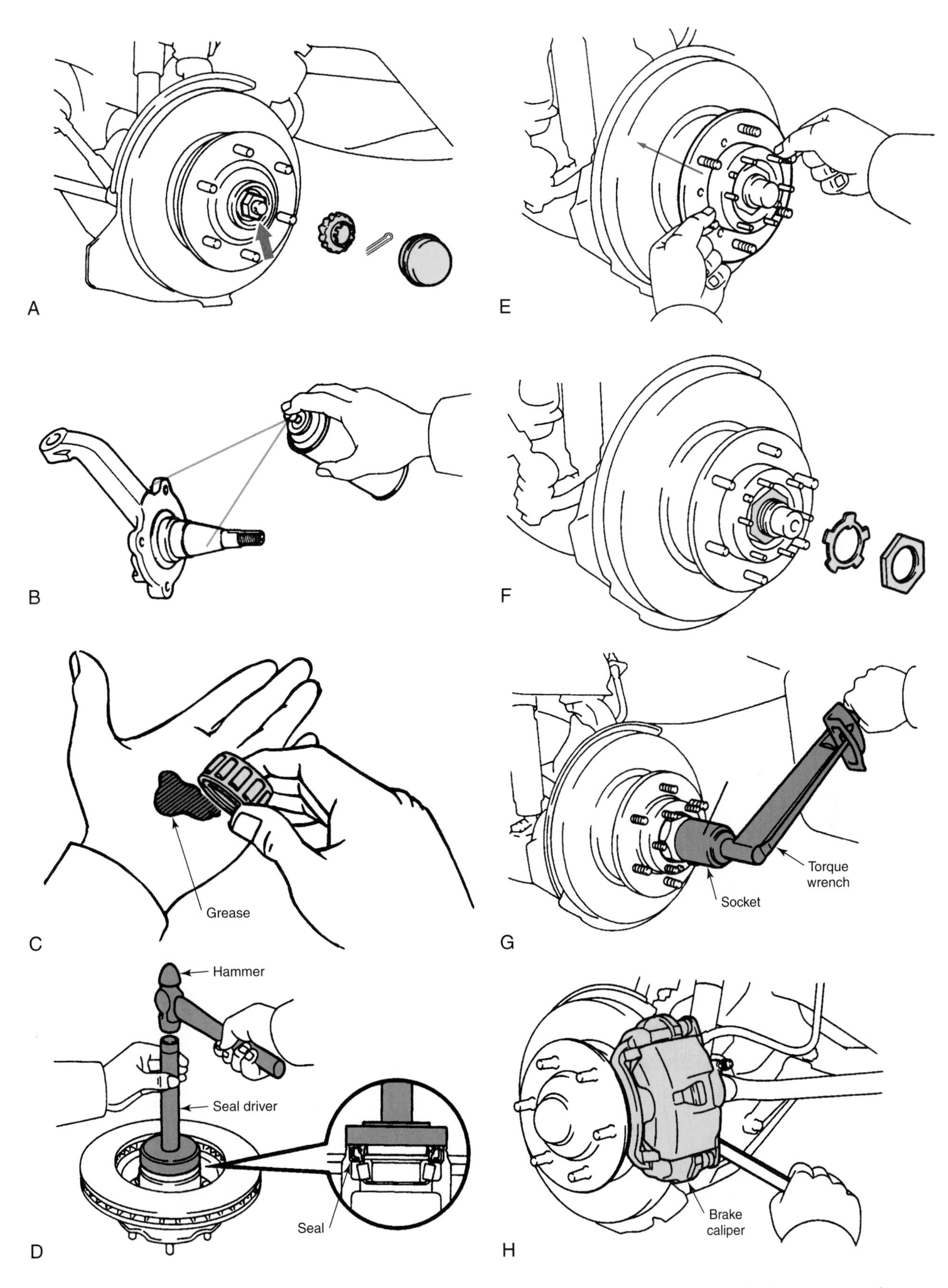

Figure 38-20. Some steps involved in wheel bearing service. A—Remove the dust cap, cotter pin, adjustment cap, and nut. B—Clean the spindle and check for damage. C—Clean the hub and bearings. Inspect for damage. Repack with grease. D—Install new seals. E—Install the hub and bearings onto the spindle. F—Replace lockwashers and nut. G—Torque adjustment nuts to specifications. H—Reinstall the brake caliper. (Toyota)

Tap the head of the cotter pin firmly into the nut slot. Cut the pin to a length that will permit bending the ends, **Figure 38-21.** If a static collector is used in the dust cap, make certain the end of the pin bent over the spindle is short enough so that it will not hit the collector prong. The cotter pin may be installed as shown in **Figure 38-22.** Regardless of the technique used for bending the pin ends, always use a new cotter pin. Use the thickest cotter pin that can be passed easily through the hole. Make certain the pin is tight after the ends are bent. A loose pin can break from vibration. Use new O-ring or sealer where required. Always make a final check before installing the dust cap to make certain that the safety washer, cotter pin, and staked nut (if used) are properly installed.

Sealed Ball and Roller Bearings

Many front-wheel drive vehicles use sealed straight roller or ball bearings, **Figure 38-23.** The rear axle bearings of some front-wheel drive vehicles are sealed. Sealed bearings must be replaced when they are defective or have damaged grease seals.

Sealed Bearings—Front-Wheel Drive Axles

Removal of most sealed front bearings require pressing out the bearing. This requires the removal of the steering knuckle assembly from the vehicle. A few bearings will slide out of the steering knuckle after the knuckle is removed from the vehicle and a snap ring removed from the knuckle. Some bearings are simply bolted to the steering knuckle and are replaced along with the hub as an assembly, **Figure 38-24.** Some bearings can be serviced only by replacing the

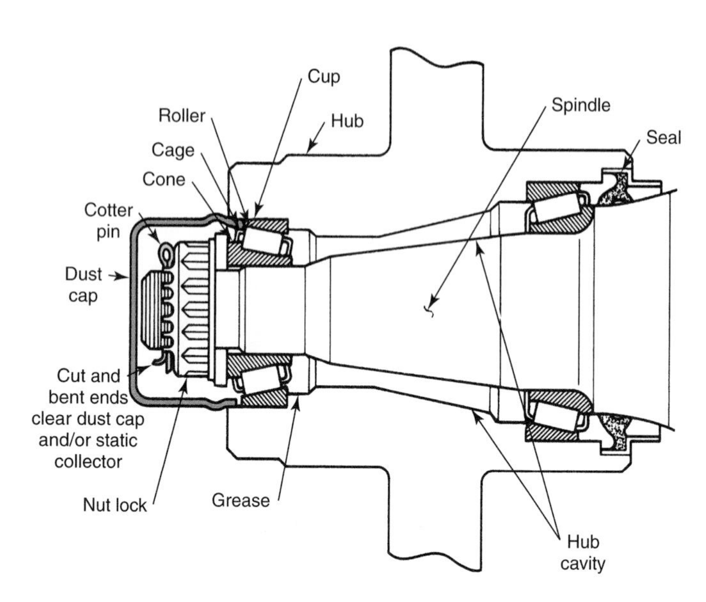

Figure 38-22. A properly installed cotter pin. Note how the ends of the pin are cut and bent to clear the dust cap sides and front. Never reuse the old cotter pin or one that does not fit the hole snugly. (DaimlerChrysler)

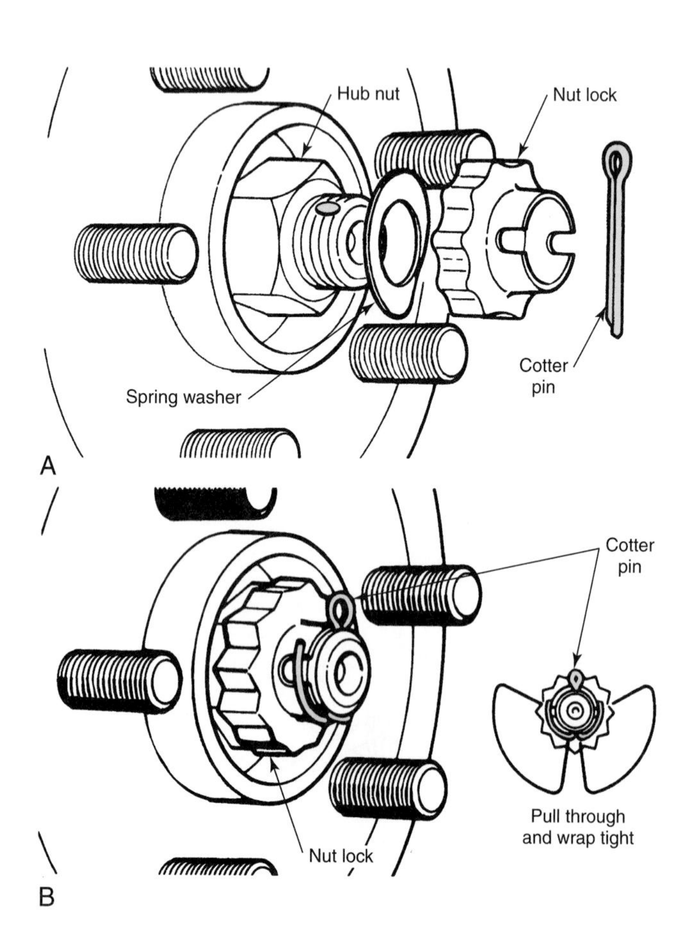

Figure 38-21. A—Installing the spring washer, nut lock, and cotter pin into a hub nut. B—The parts are installed, and the cotter pin is properly positioned and bent. (DaimlerChrysler)

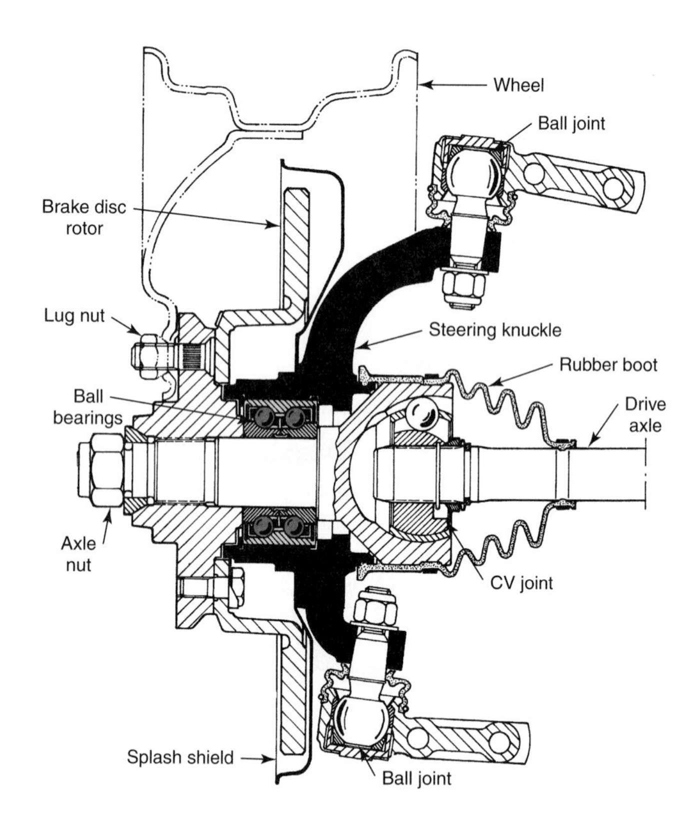

Figure 38-23. Front-wheel drive hub assembly that incorporates sealed ball bearings. (Saab)

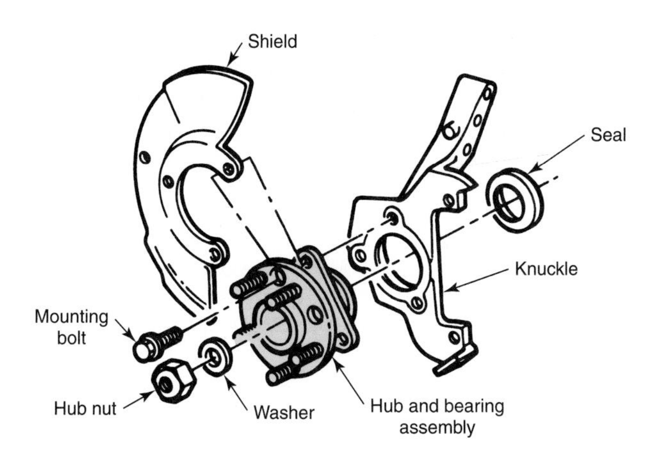

Figure 38-24. Exploded view of one type of wheel hub and bearing assembly. (General Motors)

complete steering knuckle assembly. Therefore, the following removal process is a general procedure. Consult the manufacturer's service manual for the exact procedure for a specific vehicle.

Begin by taking off the wheel cover and removing the wheel and tire as explained earlier. Remove the cotter pin, nut lock (when used), hub nut, and washer.

Note: The staked nut or cotter pin should always be replaced with new units. Also, some sealed bearing assemblies contain wheel speed sensors for anti-lock brakes. Replace the bearing, hub, and sensor as a unit if defective. They are not field serviceable.

Remove the brake caliper and brake rotor. Support the caliper with a wire hook. Never depend on the brake hose to support the caliper. Loosen and remove the anti-lock brake wiring harness and sensor (if equipped), hub mounting bolts, and splash shield. Carefully observe the position of all parts so they can be reassembled in the correct position. An exploded view of hub components is shown in **Figure 38-25.** Some vehicles require that a puller be attached to push the CV drive axle out of the bearing. See **Figure 38-26.** Do not use a hammer or heat the assembly to aid removal. The steering knuckle can be removed from the suspension by removing the outer tie rod end, ball joint, and MacPherson strut attaching bolts. See Chapter 36, Suspension System Service, for more detailed removal procedures.

Pressed-in Bearing Replacement

To remove the pressed-in bearing, first remove any snap rings or dustcovers in the steering knuckle and place the knuckle on a suitable press. Press the bearing out of the steering knuckle using the correct adaptors, **Figure 38-27.** Note the position of any bearing spacers or other internal parts. Refer to Chapter 10, Friction and Antifriction Bearings, for additional information on bearing-pressing techniques. Clean and inspect the steering knuckle and all parts that will be replaced. If the knuckle has any cracks or wear spots,

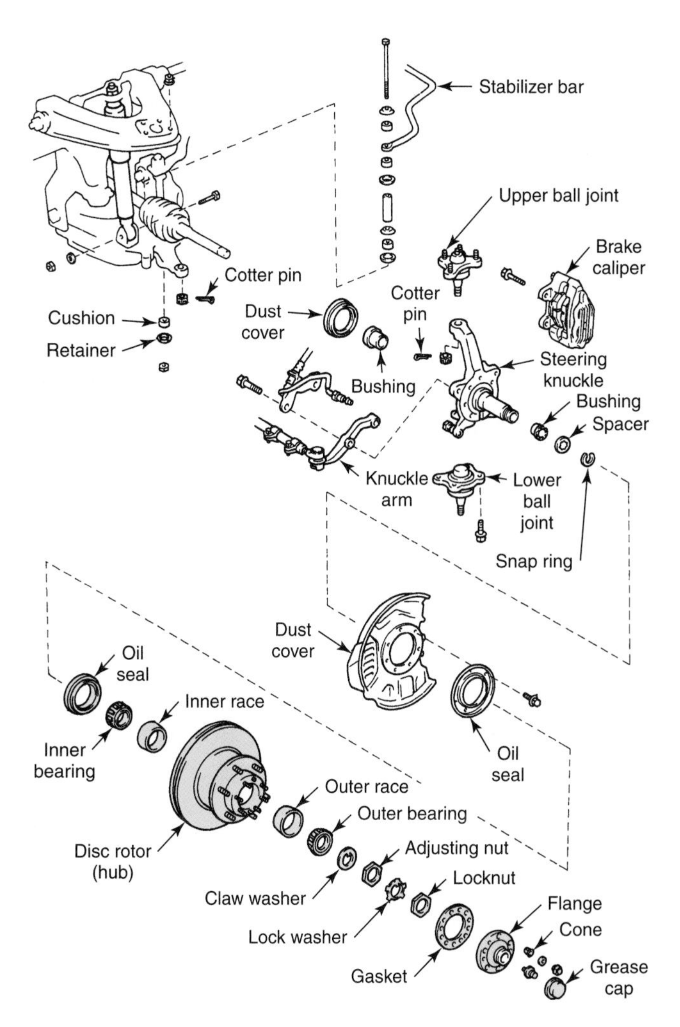

Figure 38-25. Exploded view of a four-wheel drive front hub and its related parts. (Toyota)

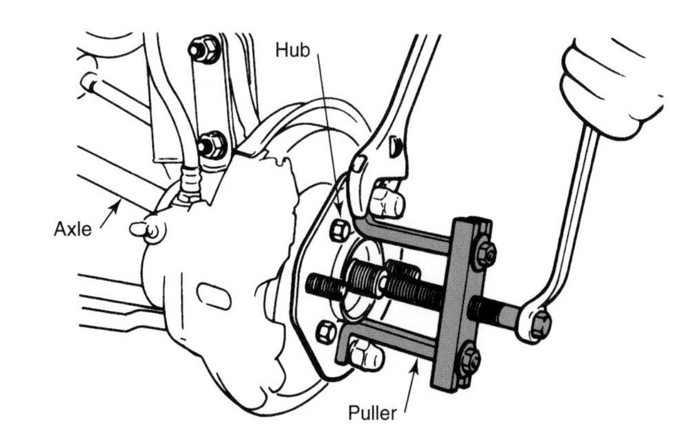

Figure 38-26. Using a puller to remove the hub. The adjustable wrench is used to keep the hub from turning as the puller is tightened. Do not use heat. (DaimlerChrysler)

replace it. Replace any dented or worn dustcovers or wear plates. Carefully position all parts of the bearing assembly over the steering knuckle. Using the correct adaptors, press the new bearing into the steering knuckle. Install any snap rings or dustcovers in their correct positions.

Reinstalling Steering Knuckle

Replace the steering knuckle on the vehicle, sliding the CV axle shaft into the bearing opening. Reinstall and torque all MacPherson strut and ball joint fasteners. Replace the nut holding the CV axle shaft into place and tighten to specifications. Install the cotter pin or deform the staked nut as required, **Figure 38-28.** Replace the brake caliper, ABS sensor and harness (if equipped), wheel, and wheel cover.

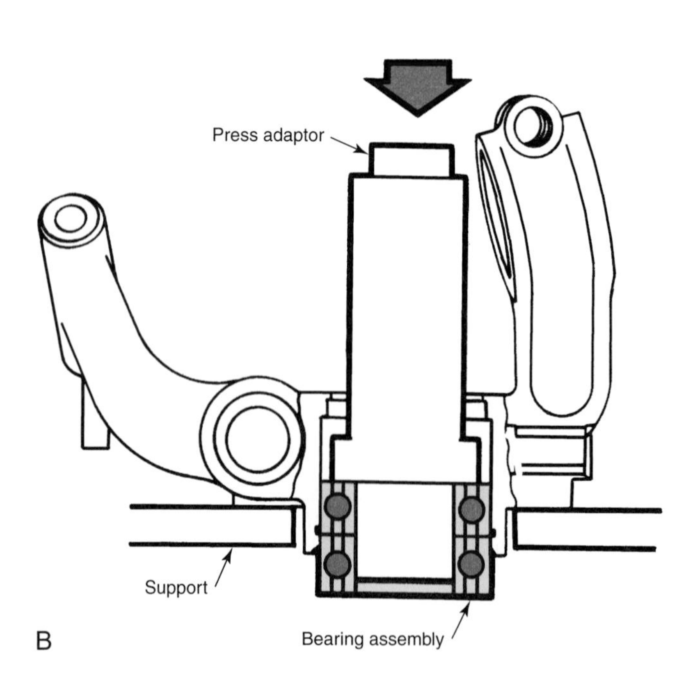

Figure 38-27. A—Remove the snap-ring and place the steering knuckle in a press. Support the knuckle properly. B—Remove the bearing from the steering knuckle with an arbor press and a special bearing adaptor tool. (Sterling)

Note: After removing and reinstalling the steering knuckle on the vehicle, front-end alignment must be checked. See Chapter 39, Wheel Alignment.

Replacing Sealed Bearings—Rear Nondriving Axles

Most sealed bearings used on rear wheel, nondriving axles are replaced as a unit. No other bearing service is possible. Many manufacturers recommend checking bearing play and preload in a method similar to the tapered bearing check discussed previously. Check the manufacturer's service manual for exact methods and specifications. Begin the replacement procedure by removing the wheel cover, then remove the wheel and tire. Remove the cotter pin or nut lock as applicable, hub nut, and washer. Remove and support the brake caliper, and remove the brake rotor. If the rear axle has drum brakes, remove the drum. If the vehicle is equipped with anti-lock brakes, remove the sensor and/or sensor connector.

Caution: Make sure that the brake assembly is supported before removing the bearing hub. Damage could result if the brake assembly is left to hang.

Loosen and remove the bearing hub mounting bolts and remove the hub. Carefully observe the position of all parts for reassembly. If the hub is being reused, install a new bearing in

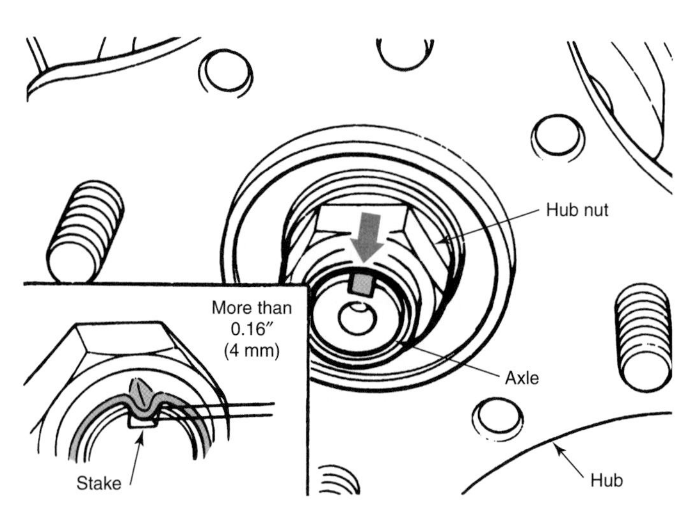

Figure 38-28. Hub nut on an axle showing the proper staking position and stake depth. If staked improperly, the nut may come off, allowing the hub to separate from the axle. (Mazda)

the hub. Reinstall the hub and bearing assembly. If any alignment shims were behind the hub assembly when it was removed, make sure they are reinstalled in their original location. Install the rotor and caliper or drum as applicable. Install and tighten the axle nut. Recheck axle play if necessary. Reinstall the wheel and tire assembly and wheel cover.

Note: Where used, the staked nut or cotter pin should always be replaced with new hub bearing units.

Vehicle Wheel Rims

In the past, factory supplied vehicle ***wheel rims*** (the assembly on which the tire is mounted) were all made of stamped steel, painted to match the car color, or simply painted black. The only way to obtain any other type of wheel in the past was from an aftermarket supplier. Today, however, ***custom wheels*** are available as original equipment from vehicle manufacturers. In addition, aluminum, aluminum alloy, composite (graphite or plastic), and chromed steel wheels are available from many aftermarket manufacturers. Some wheels are solid aluminum, while others have a steel core covered with aluminum. Alloy wheels are usually a mixture of aluminum and magnesium. Wire wheels are also available. **Figure 38-29** shows a custom wheel available on one particular vehicle.

Wheel Rim Size

Rim size is determined by three measurements: rim width, rim diameter, and flange height. Rim width and diameter are measured in inches. Flange height is identified by letters such as J or K. The letters indicate a definite flange height. For example, a K rim flange is 0.77″ (19.5 mm) high. See **Figure 38-30.** These measurements apply to rims made of any material.

Figure 38-29. This custom five-spoke, cast-aluminum wheel is standard equipment on one particular high-performance vehicle. (Ford)

Wheel Torquing

When the wheel rims are made of aluminum, or a composite of aluminum and steel, proper torque is critical. Always torque to the manufacturer's specifications. See **Figure 38-31.** Do not use an impact wrench on any aluminum or alloy wheel unless it is equipped with a ***torque stick,*** Figure 38-31A. Torque sticks are flexible extensions that limit the output of the impact wrench to a preset value. This is accomplished by the flexing action built into the torque stick. When a certain torque value is reached, the stick flexes instead of transmitting further torque. They are available in several common torque settings. Although correct torque is not as important with steel rims, the front disc brake rotors on most modern vehicles can be deformed by over-torquing. If torque sticks are not available, use a torque wrench to tighten the rim, Figure 38-31B.

Checking for Wheel Damage

Check all rims for cracking, elongated mounting holes, and bent mounting flanges. Slight leakage of a steel rim can be corrected by welding or brazing without compromising rim safety. A porous aluminum rim can be repaired with an epoxy sealer. This wheel repair is similar to the automatic transmission porous casing repair described in Chapter 31, Automatic Transmission and Transaxle Service. Aluminum rims should not be welded for any reason. The rim can be tested for excessive runout as outlined in the tire balancing section.

Wheel Lug Service

Wheel lugs are pressed into the bearing hub. ***Lug nuts*** are used with the lugs to hold the wheel to the hub. Lug nuts should be tightened to the proper torque and in the sequence shown in Figure 38-31B. This prevents wheel or hub distortion and protects the lug from damage. If the wheel lugs are

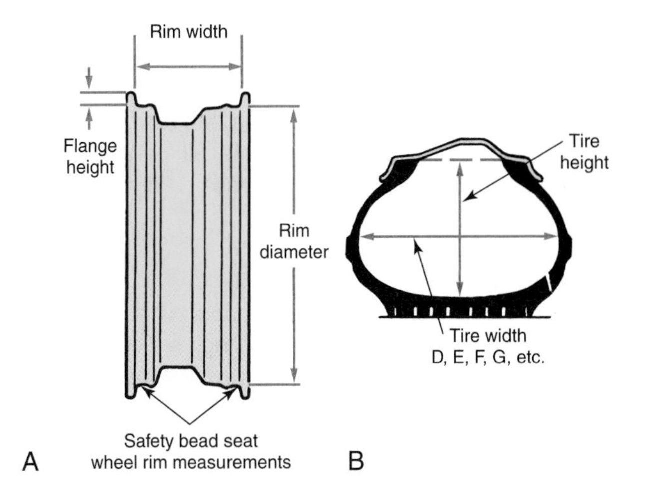

Figure 38-30. A—Rim size is identified by rim width, rim diameter, and flange height. Note the safety bead seats. These are used to keep the tire on the rim in the event of tire failure. B—The relationship between tire height and tire width determines the aspect ratio. If the tire height is 78% of the tire width, the aspect ratio is 78. (DaimlerChrysler, Firestone)

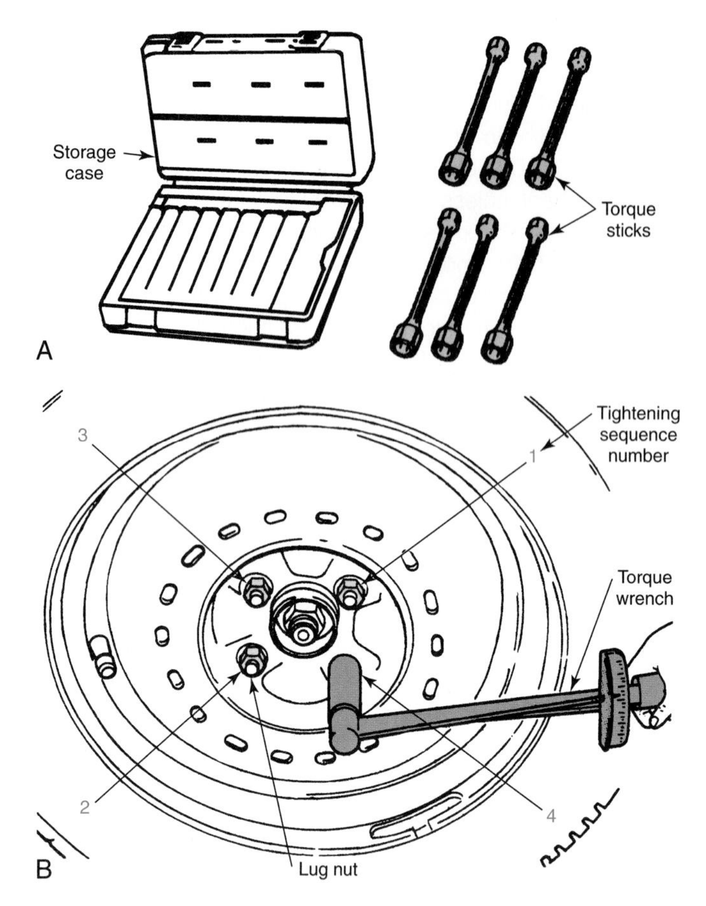

Figure 38-31. A—Torque sticks. B—Torque wrench being used to set the lug nuts to a specific torque. (DaimlerChrysler, General Motors)

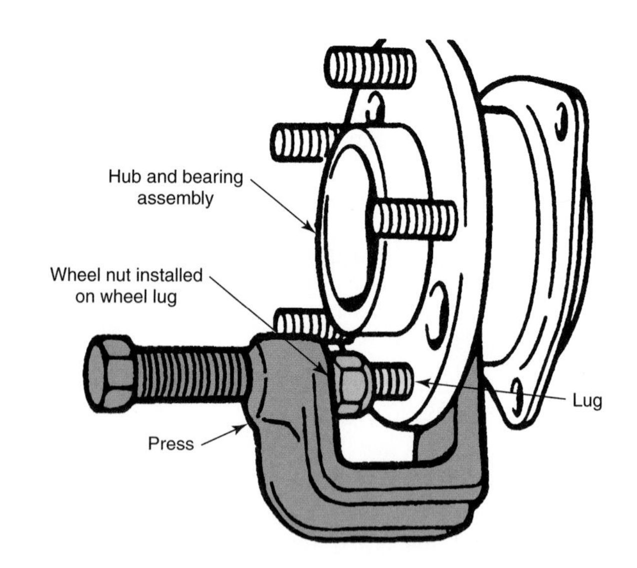

Figure 38-32. Removing a damaged lug bolt with a special press. (General Motors)

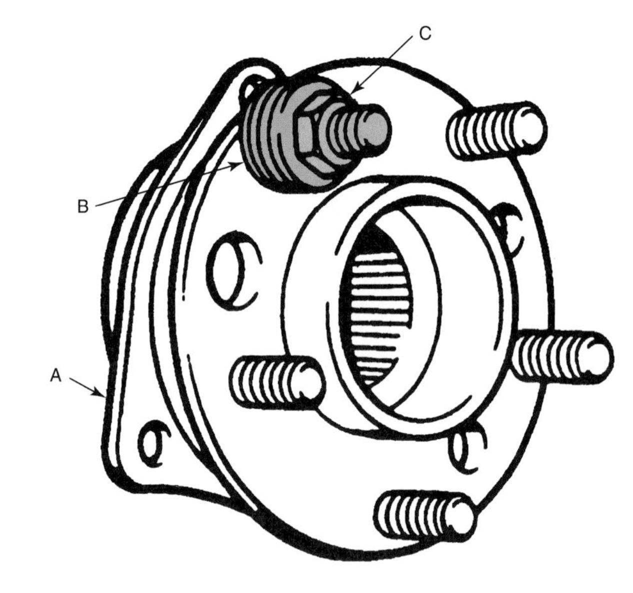

Figure 38-33. Installing a new lug bolt into the hub by using washers and a lug nut. As the lug nut is tightened, the lug bolt is drawn into position. (General Motors)

difficult to remove, they should be coated with anti-seize compound to prevent damage.

If a lug nut is damaged or missing, it should be replaced as soon as possible. Always replace broken lug nuts or bolts with the correct size and type. Most lugs can be knocked out of the wheel hub after the wheel rim and the brake drum or rotor are removed, **Figure 38-32.** The new lug can be installed by using a press or by drawing the lug through the hub using a wheel nut, **Figure 38-33.** Lubricate all parts thoroughly before installation. If the lug will not press solidly into the hub, replace the hub.

Vehicle Tires

Tires are a vital, but often overlooked part of overall vehicle performance. Improper handling, braking, and ride quality may result from defective tires or tires that are wrong for the vehicle. Serious handling problems can be caused by something as simple as underinflation. The following section covers modern tire grading, selection, and installation techniques.

Tire Cord Construction

Tires may be constructed with various cord arrangements, **Figure 38-34.** Many older tires were constructed using two or four ***plies*** (layers of cord material) laid at an angle to the tire centerline. Look at Figure 38-34A. This is termed ***bias*** construction. Another arrangement of the plies is illustrated in Figure 38-34B. This is called a ***bias-belted*** cord arrangement. Note the difference in cord angles between Figures 38-34A and 38-34B.

The third type of construction is ***radial,*** in which the body cords cross the tire centerline almost at right angles to the belts, Figure 38-34C. Today, radial tires are almost universally used for all applications, including retrofitting older vehicles

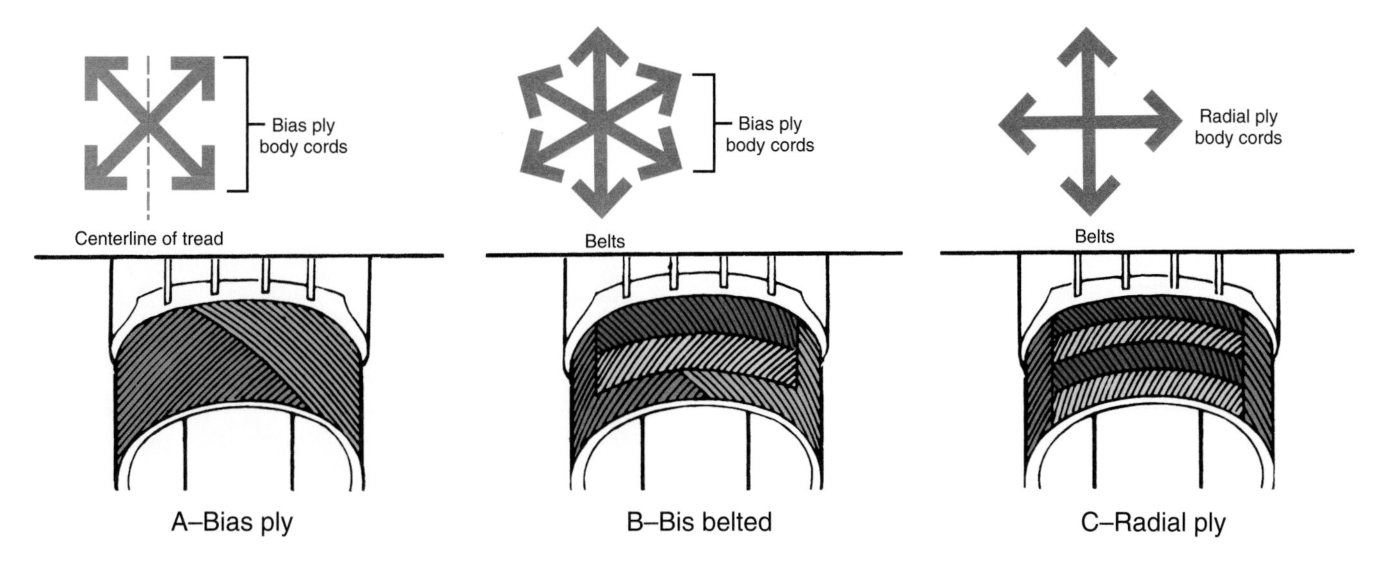

Figure 38-34. Three methods of arranging tire cords. Note the arrows above each ply method. These indicate the direction of the ply cords. (Goodyear Tire and Rubber Co.)

that originally came with bias tires. Rayon, nylon, polyester, aramid, Kevlar, and fiberglass are used as cord materials. Steel wire is often used in the belt section of radial tires. Other materials are also being tested and evaluated.

Caution: Manufacturers do not advise mixing radial and bias tires on the same vehicle. Poor vehicle handling will result. If radial and bias tires must be used on the same vehicle, ensure that the bias tires are placed on the front axle.

Tire Rating Information

Tire ratings are determined by a system of numbers and letters that are molded on the side of the tire or listed in the vehicle owner's manual. The rating system identifies the tire size, rim size, and type of construction, as well as the maximum speed and load handling capabilities. A typical modern tire might have the series of letters and numbers shown in **Figure 38-35** on its sidewall.

The letter *P* indicates the tire is designed for a passenger vehicle, such as a car or light truck. Other possible tire designations are *T* for a tire designed for temporary use, such as a space-saver tire, *LT* for light trucks, and *C* for commercial or large trucks. This letter may be absent on some tires. The number *225* represents the tire's section width in millimeters, measured at the tire's widest point. This is the actual tire size. Tire size can range from 145 to 315, with most sizes from about 185 to 235. The higher the number, the larger the tire. The number *70* is the aspect ratio (relationship of a tire's cross-sectional height to its width). Other common aspect

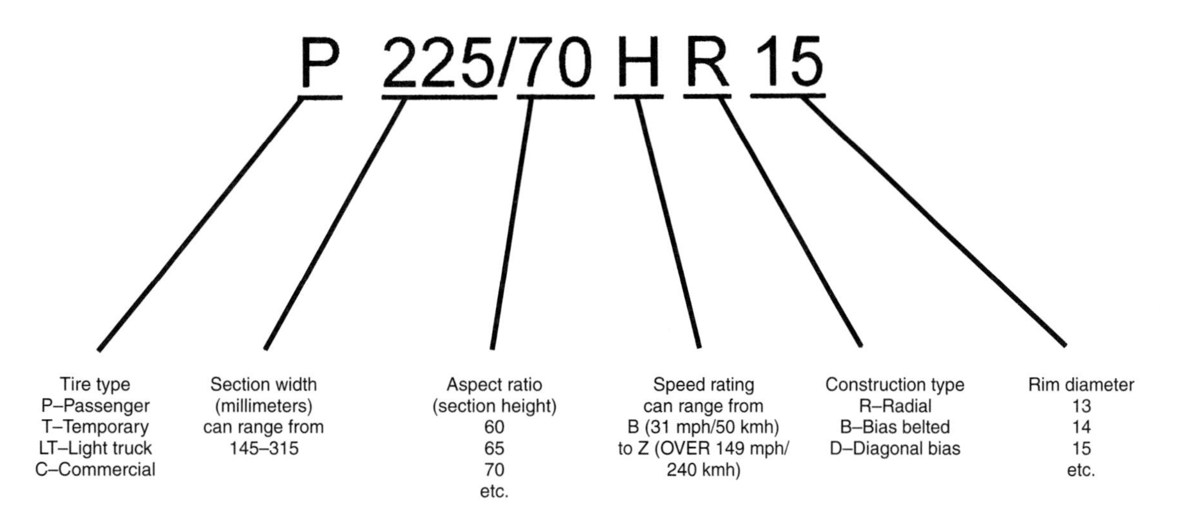

Figure 38-35. Typical tire rating information.

ratios are 60, 65, 75, 78, and 80. The higher the aspect ratio, the taller the tire. Low aspect ratios provide more traction, while high aspect ratios give better mileage. The letter *H* is the speed rating. The speed rating gives the maximum speed at which the tire can be operated. This ranges from *B* (31 mph or 50 kmh) to *Z* (over 149 mph or 240 kmh). The letter *R* indicates radial ply construction. A bias ply tire is marked *B*. Almost all passenger car and light truck tires made today are radial tires. The number *15* is the rim diameter in inches. Modern wheel rims range from 12 inches (rare) to 17 inches (also rare). The most common rim sizes are 13, 14, and 15 inches.

Tire Quality Grading

All passenger and light truck vehicle tires now being produced are graded by the Department of Transportation (DOT) ***uniform tire quality grading*** system. The quality grading system is applicable to the following three areas:

The ***temperature resistance*** rating has three letter-grade levels: A, B, and C. "A" offers the greatest resistance to heat generation. "C" provides the least. All tires must meet the "C" rating. Tire ***traction*** is also graded on three levels: A, B, and C. "A" offers the best traction (wet roads), while "C" offers the smallest amount. Tire ***tread wear*** is graded by using a set of numbers ranging from 100 to about 500. A tire that has a tread wear grade of 150 should supply approximately 50% more mileage than a tire with a tread wear rating of 100.

Special Service Tires

Many safety and high-speed tires are available. High-speed tires generally use steel wires in the outer ply and tread area. Steel resists expansion from centrifugal force, which helps to control distortion. They also resist punctures better than fabric belts. Most vehicles are now equipped with a compact, or ***space-saver*** spare tire. It is designed to take up less cargo space and is lighter than a normal tire. The compact spare tire is for temporary use only. Use it to replace a flat until the normal tire is repaired or replaced. Do not rotate a space saver tire in with the other tires. When replacing this tire, mount only on a special wheel designed for its use. Do not use hub caps or wheel covers on the rim. Possible tire damage could result. Keep inflated to recommended pressure. A compact spare tire is shown in **Figure 38-36.**

Figure 38-36. A compact spare tire compared to a tire of normal size. (Goodyear Tire and Rubber Co.)

Tire Selection

A tire load information label can be found on most vehicles. It is generally located on the driver's door, on the door pillar, or inside the glove box. The load information label contains specific information on the maximum vehicle load, tire size (including the spare tire), and inflation pressures. Older vehicles with obsolete grading systems can be equipped with modern tires by the use of interchange charts, such as the one in **Figure 38-37.**

Tire Pressure

All tires must be inflated to the recommended pressure. See **Figure 38-38.** The sidewalls of an underinflated tire will flex excessively and quickly generate damaging heat. In addition, the center section of the tread will buckle up away from the road, causing rapid wear on the tire edges. See Figure 38-38A. Overinflation will make the tire ride hard and will also make it more susceptible to ply breakage. Note how the overinflated tire in Figure 38-38B bulges in the center, thus pulling the edges of the tire away from the road. This produces rapid wear in the center. To provide proper steering, ride, wear, and dependability, keep the tire pressure within the manufacturer's recommended range. Note how the full tread width of the correctly inflated tire in Figure 38-38C contacts the surface of the road.

Inflate tires to recommended pressure when cold (at prevailing atmospheric temperature). A tire may be considered cold after standing out of direct sunlight for three or four hours. When tires are driven, their temperature rises. The amount of temperature increase depends on speed, load, road smoothness, and prevailing temperature. A cold tire at 24 psi (165 kPa) will build up pressure to about 29 psi (200 kPa) after 4 miles (6.4 km) at speeds over 40 mph (64.4 kmh). When checking hot tires, the pressure measured will be greater than the cold tire pressure. Never let air out of hot tires to get the specified cold pressure. For heavy loads or sustained high speed driving, many manufacturers recommend increasing cold pressure 4 psi (25 kPa).

Tire Rotation

Tire life can be extended by periodic tire rotation. The front tires are placed at the rear before any front wheel misalignment can cause serious tire imbalance or tire wear. Rotate tires about every 5000 miles (8000 km). **Figure 38-39** shows popular methods of rotation, for both conventional and radial tires.

Caution: Some manufacturers do not recommend radial tire rotation. Other radial tire manufacturers recommend that radial tires be rotated so they always turn in the same direction, Figure 38-39. Do not rotate in the compact (space-saver type) spare tire. Space-saver spares are for emergency driving only.

P-metric size **(35 psi max. pressure)	If vehicle tire placard specifies a P-metric tire size, the following are acceptable substitute sizes: Use inflation pressure specified on vehicle tire placard.
P215/60R16	P225/55R16, P235/50R16
P225/60R16	P255/50R16
P205/55R16	P225/50R16
P225/55R16	P215/60R16, P235/50R16
P225/50R16	P215/60R16, P225/55R16, P235/50R16
P235/50R16	P215/60R16, P225/55R16
P245/50R16	P225/60R16
P255/50R16	P265/50R16
P265/50R16	P275/50R16
P275/50R16	(None)
P245/40R17	(None)
P275/40R17	P315/35R17
P315/35R17	P275/40R17

* The letters H, S, V, or Z may be included in the tire size designation of P-metric and other metric sizes preceding the "R"
** For standard load tires

Alphanumeric size on vehicle placard (32 psi max. pressure)	If vehicle tire placard specifies an alphanumeric tire size, the following are acceptable substitute sizes Important: Add 3 PSI above the pressure specified on the vehicle tire placard to assure adequate load capacity.
AR78-13	P165/80R13, P175/75R13, P185/70R13, P195/60R13, P215/50R13, 185/70R13
BR78-13	P175/80R13, P185/75R13, P195/70R13, P195/65R13, P205/60R13
CR78-13	P185/80R12, P195/70R13, P215/60R13, P235/50R13
BR78-14	P175/75R14, P185/70R14, P185/70R14
CR78-14	P185/75R14, P195/70R14, P205/65R14, P215/60R14, P195/70R14
DR78-14	P185/80R14, P195/75R14, P205/70R14, P205/65R14, P215/60R14, P245/50R14,
ER78-14	P195/75R14, P205/70R14, P255/60R14, P245/50R14
FR78-14	P205/75R14, P215/70R14, P235/60R14, P265/50R14
GR78-14	P215/75R14, P225/70R14, P245/60R14 P255/55R14, P265/50R14
HR78-14	P225/75R14, P235/70R14
BR78-15	P165/80R15, P175/75R15, P185/70R15
ER78-15	P195/75R15, P215/65R15, P245/50R15
FR78-15	P205/75R15, P215/70R15, P215/65R15, P235/60R15, P245/50R15
GR78-15	P215/75R15, P215/70R15, P235/60R15 P255/55R15, P265/50R15
HR78-15	P225/75R15, P235/70R15, P255/60R15 P275/50R15
JR78-15	P225/75R15, P235/70R15, P255/60R15
LR78-15	P235/75R15, P245/70R15,P255/65R15, P265/60R15, P295/50R15

Figure 38-37. One type of tire size interchange chart.

Tire Demounting and Mounting

Use the proper equipment to demount and mount tires. Use care to avoid damaging the bead-sealing surfaces.

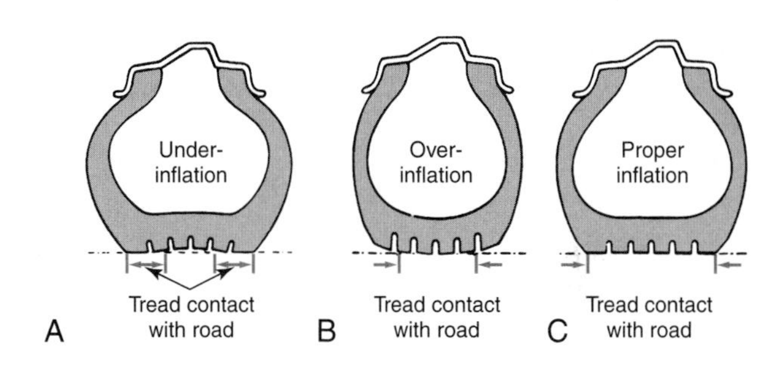

Figure 38-38. The effect of inflation pressure on the tire-to-road contact. (Rubber Mfg. Association)

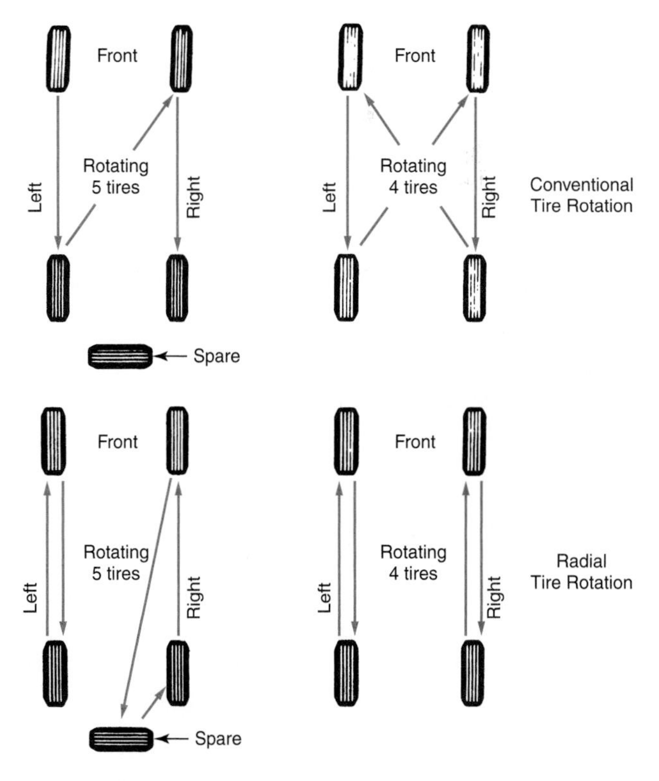

Figure 38-39. Tire rotation chart for both conventional and radial tires. Follow the specific rotation pattern recommended by the vehicle or tire manufacturer. (DaimlerChrysler)

Always use an approved rubber lubricant before demounting and mounting. Otherwise, you may tear the sealing surface and/or cause excessive strain on the bead wire. Refer to **Figure 38-40.** Before mounting a tire on a wheel, clean off the rim sealing area with coarse steel wool. File off nicks and burrs. Straighten any dented area on the sealing flanges (edges). If the rim has any cracks or splits, discard it. See **Figure 38-41.**

Note: When mounting oversize tires, make sure the wheel is approved for their use.

Repairing a Punctured Tire

If a tire is suspected of having a leak, first fill it with air. Place the tire and wheel in a drum filled with water. A hose can also be used. Look for air bubbles forming on the tire surface and at the bead. Also check the wheel rim and valve stem for leakage. When the leak is located, mark it with crayon or chalk. All tires have repairable and nonrepairable areas. Never repair a tire with:

- Ply separation.
- Damaged bead wires.
- Tread separation.
- Loose cords.
- Cuts or cracks that extend into the tire cord (fabric) material.
- Tread wear indicators showing.
- Sidewall punctures, bulges, or damage.
- Punctures larger than 0.5″ (13 mm).

In the past, punctures were repaired by inserting a rubber plug in the puncture hole without demounting the tire. However, due to safety concerns, this method of repair is no longer recommended. The use of plugs without demounting the tire can result in sudden tire failure.

To repair a puncture properly, demount the tire and remove the puncturing object, noting the angle of penetration. Clean the puncture area with a special tool. A correct puncture repair must fill the damage area and patch the inner liner. A ***plug*** or ***liquid sealer*** is used to fill the puncture hole from the inside of the tire. If a plug is used, cut it off slightly above the tire's inside surface. After filling the hole, scuff the puncture area well beyond the actual damage. Then apply cement to the scuffed area. Apply a ***patch*** that covers the puncture well beyond the damaged area according to the manufacturer's directions. A tool called a ***stitcher*** is used to help bond the patch tightly to the tire's inner liner. See **Figure 38-42.** Allow sufficient time for drying.

Tire Inflation Following Mounting

When mounting the tire, use a liberal amount of rubber lubricant. Do not use silicone. Use a clip-on air chuck to

Figure 38-40. Demounting a tire. Use caution and wear safety goggles! (Hunter Engineering Co.)

Figure 38-41. Clean the rim and check for cracks, dents, and old wheel weights before mounting the tire. (Rubber Mfg. Association)

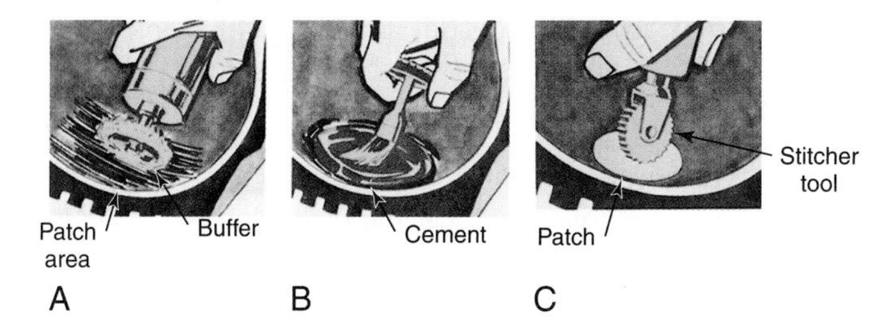

Figure 38-42. Installing a tire patch. A—Buff an area slightly larger than the patch and clean the buffed area thoroughly. B—Apply the cement with a brush (allow for recommended drying time). C—Install the patch. Use the stitcher tool to firmly roll the patch into contact with the cement. Roll over the entire surface of the patch. (Goodyear Tire and Rubber Co.)

attach the air hose to the valve stem. Back away from the tire while inflating. Never exceed a pressure of 40 psi (276 kPa) in an endeavor to force the beads out against the rim flanges. If difficulty is experienced, deflate the tire and check for the source of trouble. Stay away from the tire while seating the beads. Reduce air pressure following bead seating. Look at **Figure 38-43.** On some tires, the use of a ***bead expander*** or seal ring will help trap sufficient air to inflate a tubeless tire. Remove the expander before tire pressure exceeds 10 psi (69 kPa).

The tire and rim can explode with extreme force if too much air is applied. When a tire bead is forced into place, it can cock to one side and bind. If air pressure is allowed to exceed 40 psi (276 kPa), the wire bead can snap. This can allow the bead edge to explode outward against the rim flange with such force that the flange is sheared off. Never exceed maximum bead-seating pressure. Do not inflate tires with air contaminated with oil, gasoline, or cleaning solvents because tire damage or explosion can occur.

Figure 38-43. Inflating a newly mounted tire. (Hunter Engineering Co.)

Tire Wear Patterns

An important part of tire service is inspecting the tire treads for unusual wear patterns. Although worn suspension parts or poor alignment can be responsible for rapid or uneven tire wear, underinflation, overinflation, toe, camber, and cornering wear are the most common causes. Refer to **Figure 38-44.**

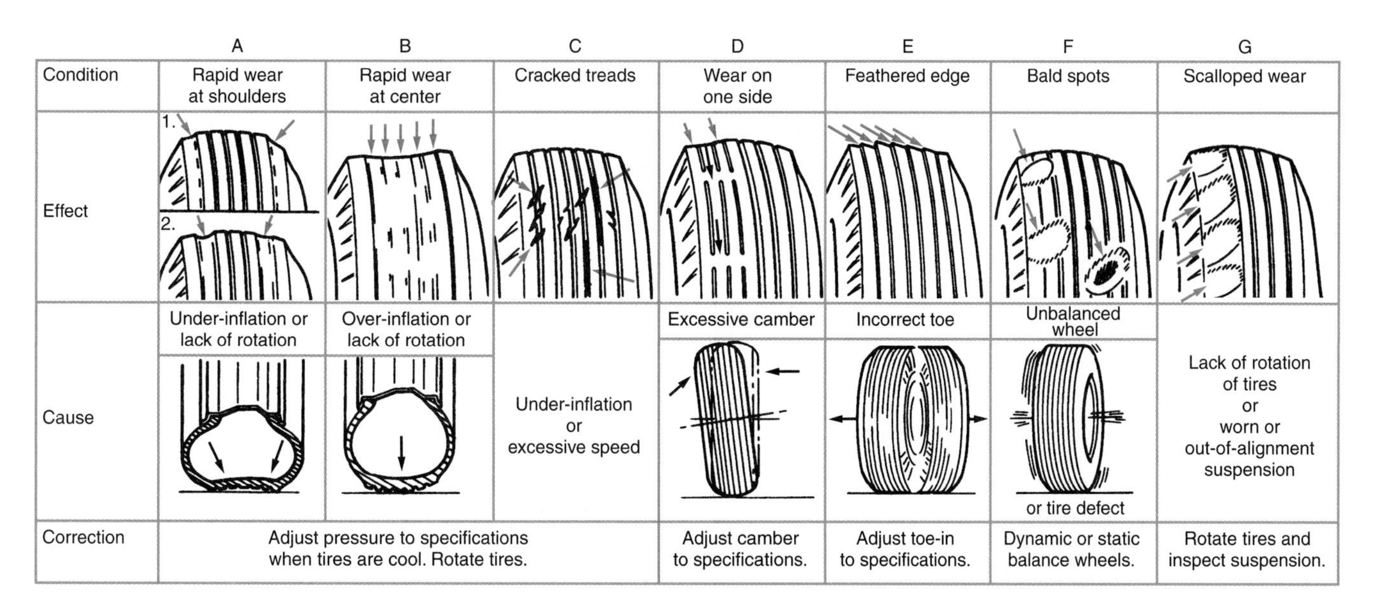

	A	B	C	D	E	F	G
Condition	Rapid wear at shoulders	Rapid wear at center	Cracked treads	Wear on one side	Feathered edge	Bald spots	Scalloped wear
Effect	1. 2.						
Cause	Under-inflation or lack of rotation	Over-inflation or lack of rotation	Under-inflation or excessive speed	Excessive camber	Incorrect toe	Unbalanced wheel or tire defect	Lack of rotation of tires or worn or out-of-alignment suspension
Correction	Adjust pressure to specifications when tires are cool. Rotate tires.			Adjust camber to specifications.	Adjust toe-in to specifications.	Dynamic or static balance wheels.	Rotate tires and inspect suspension.

Figure 38-44. Some abnormal tire wear patterns. Note the cause, the effect, and the correction. (DaimlerChrysler)

Underinflation can cause rapid edge or shoulder wear, Figure 38-44A. Overinflation can wear out the center of the tread, Figure 38-44B. Excessive toe-in or toe-out causes extremely rapid wear and is evidenced by feathered edges as in Figure 38-44E. Improper camber angle wears only one side of the tread, Figure 38-44D. Excessive speed around corners generally causes a rounding of the outside shoulder. There are certain similarities to toe wear in some cases. A combination of causes can produce numerous wear patterns. One common symptom of combination wear is a series of cupped out spots around the tire, Figures 38-44F and 38-44G. This pattern is also common when toe is off on a rear tire. When a vehicle's tires wear in one of the patterns illustrated in Figure 38-44, corrective measures are needed.

Wheel and Tire Balance

Irregularities in construction, shifting of the cords and weight mass from tread wear, and other problems can cause the tire and wheel assembly to become unbalanced. At highway speeds, even a slight imbalance can cause ***wheel tramp*** (tire and wheel hopping up and down) or ***wheel shimmy*** (shaking from side to side). Wheel assemblies must be in both static and dynamic balance.

Caution: If the tire is defective or the rim is damaged, do not try to correct the problem by balancing. Replace the affected tire or rim.

Static Balance

To be in ***static balance,*** the weight mass must be evenly distributed around the axis of rotation. See **Figure 38-45.** A wheel in static balance will remain balanced in any position, whereas static imbalance will cause the heavy side to rotate to the bottom. The wheel is brought into static balance by clipping weights to the rim opposite the heavy side. If more than 2 ounces (56 g) of weight is required, the weight should be split by adding half the required amount to the inside of the rim and the remainder to the outside. By placing half of the weight on each side, dynamic balance is not disturbed. Look at Figure 38-45B.

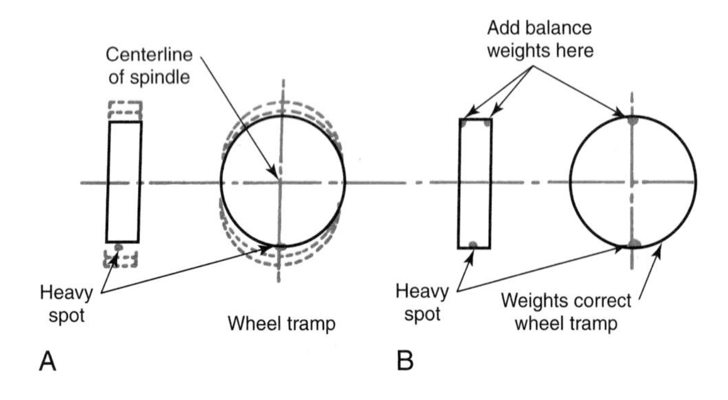

Figure 38-45. A wheel brought into static balance by adding properly sized balance weights to the light side. Note how the weights are split, with one-half placed on each side of the wheel. (General Motors)

Dynamic Balance

To be in ***dynamic balance,*** the centerline of the weight mass must be in the same plane as the centerline of the wheel. **Figure 38-46** shows a wheel and tire assembly that is dynamically imbalanced. Note how the weight mass centerline fails to coincide with the plane of the wheel centerline. When the wheel rotates, centrifugal force attempts to force the weight mass to align with the wheel centerline. Since the direction of this force is one way on one side of the assembly and the other is on the opposite side, Figure 38-46A, it causes the wheel to shimmy, Figure 38-46B. Dynamic imbalance is corrected by adding wheel balance weights in amounts sufficient to bring the weight mass and wheel centerlines into the same plane. To be in dynamic balance, the assembly must also be in static balance. **Figure 38-47** pictures a dynamically balanced wheel and tire assembly. Note how the wheel and weight mass centerlines coincide, Figure 38-47A. When the wheel rotates, there is no shimmy, Figure 38-47B.

Measuring Wheel and Tire Runout

A wheel and tire assembly with excessive lateral or radial runout cannot be balanced properly. Modern tires often

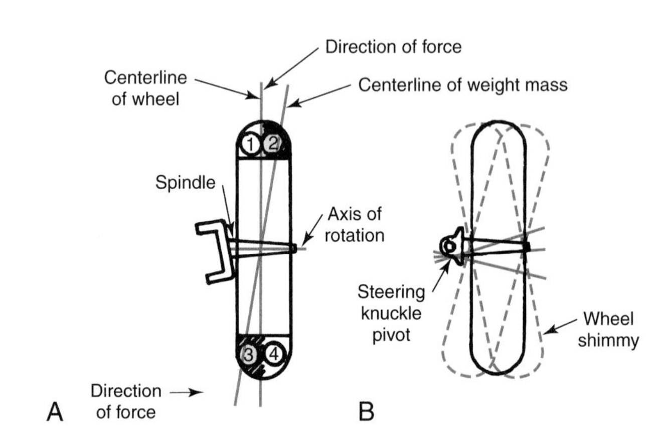

Figure 38-46. A dynamically unbalanced wheel and tire assembly. (DaimlerChrysler)

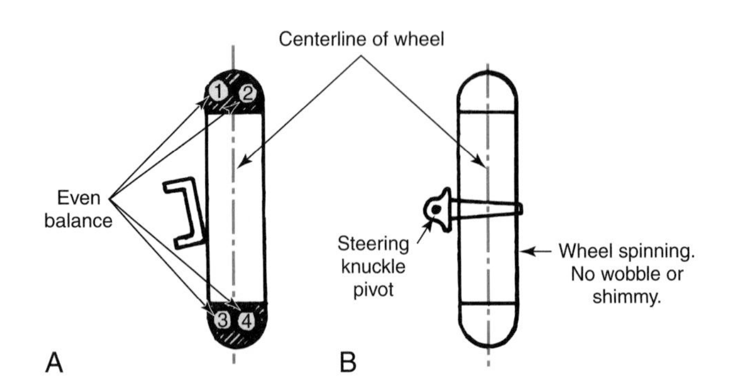

Figure 38-47. A dynamically balanced tire and wheel assembly.

develop tread separations, which result in excessive runout. Such tires cannot be balanced and should be discarded. Always check tire runout before balancing. Check runout at the points indicated in **Figure 38-48.**

Generally, tire radial and lateral runout should be kept within the figures shown in Figure 38-48. When possible, use a dial indicator setup as shown in **Figure 38-49.** Follow manufacturer's specifications. If tire runout is excessive, check the wheel rim runout. It may be possible to bring tire runout within specs by shifting the tire on the rim until the point of maximum tire runout is opposite the point of maximum wheel runout. Runout may be checked by using a dial indicator or a special indicator. This is shown in Figure 38-49. The indicator stand must be heavy enough to hold the instrument in a fixed position. Raise the wheel and position the indicator. Slowly rotate the wheel and note the reading. The tire must be warm to avoid any flat spots from sitting in one place for a long time.

Balancing Techniques

Modern radial tires must have both static and dynamic balance. The tire should be balanced statically first and then balanced dynamically. Tire and wheel assemblies may be balanced on the vehicle or off the vehicle. An off-car balancer is shown in **Figure 38-50.** Follow manufacturer's instructions. An on-car balancer is shown in **Figure 38-51.**

Warning: Front-wheel drive vehicles should have the front wheels spun by the engine, not with an on-car balancer. The suspension should not be allowed to hang unsupported while spinning the wheels. Driving constant-velocity joints at extreme angles can cause excessive vibration and part damage.

Limit wheel speed to around 35 mph (56.3 kmh), as indicated on speedometer. Remember that the speedometer

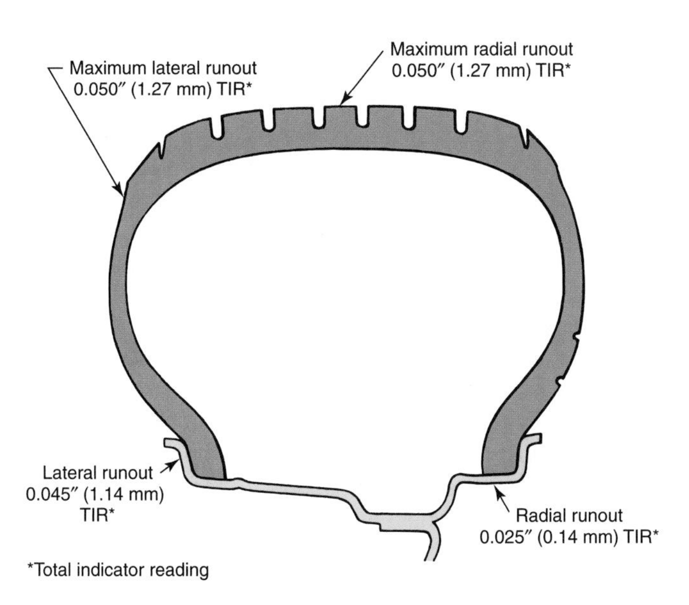

Figure 38-48. Check the wheel and tire at these points with a dial indicator. (General Motors)

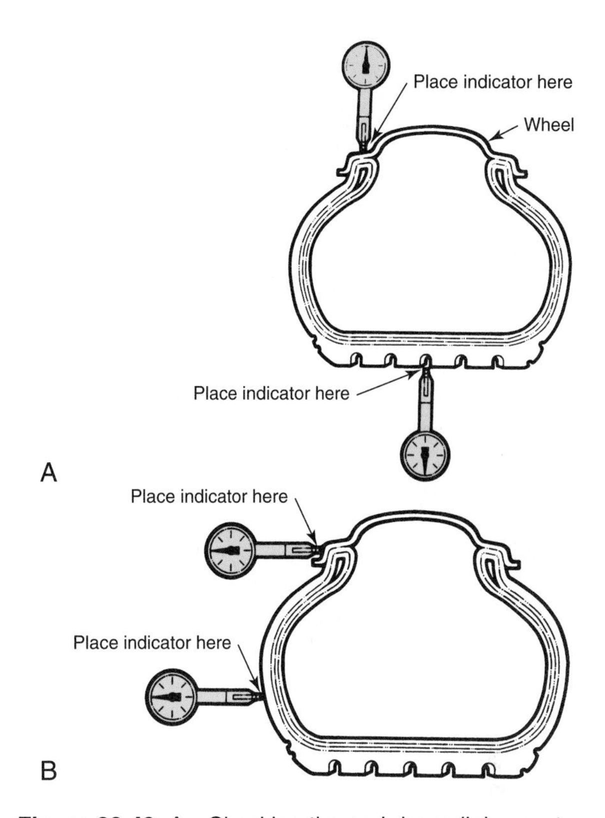

Figure 38-49. A—Checking tire and rim radial runout. B—Checking lateral runout with a dial indicator. (B.F. Goodrich Co.)

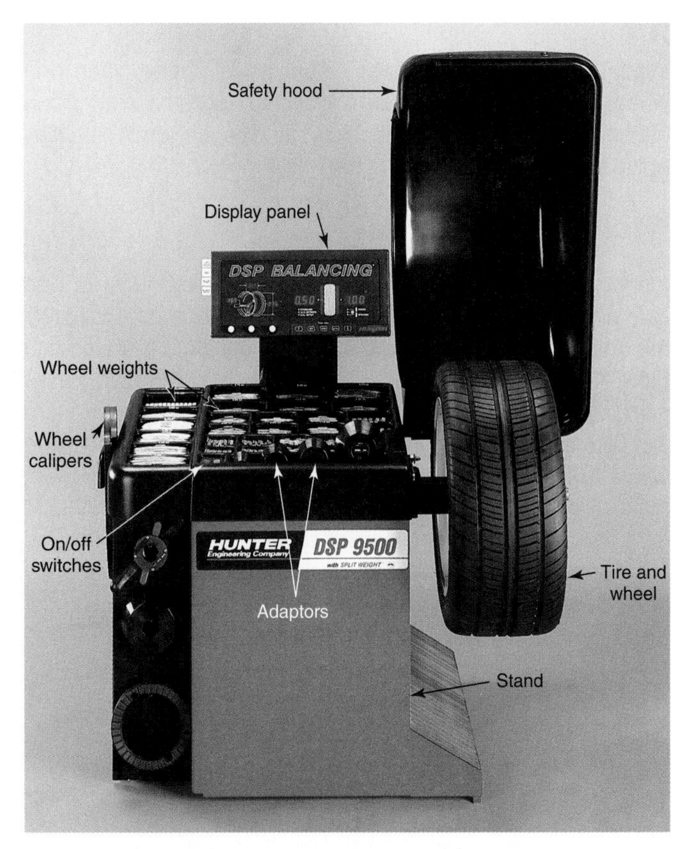

Figure 38-50. One type of electronic, off-vehicle wheel balancer. (Hunter Engineering Co.)

Figure 38-51. An on-vehicle electric wheel balancer. Follow the equipment and/or vehicle manufacturer's specific balancing procedures. (Hunter Engineering Co.)

indicates only one-half of the actual wheel speed when one wheel is spun and the other is stopped. Extreme speed can cause the tire to disintegrate. It can also cause possible differential or transaxle damage.

Balancing Rear Wheels

The rear wheel and tire assemblies should also be balanced. When a standard differential is used, jack up one side only. When balanced, lower the vehicle to the floor. Raise the other wheel and balance. Do not try this on the limited-slip type differential, as it will move the vehicle forward. When using an on-car balancer on vehicles equipped with a limited-slip differential, raise both wheels from the floor. Remove one wheel and replace a couple of the lug nuts to hold the drum in place. Balance the side with the wheel in place. Once that side is balanced, the other wheel should be reinstalled and balanced. It is not necessary to remove the wheel already balanced. Make certain the vehicle is secure on jack stands. Keep personnel away from the front of the vehicle.

Caution: When balancing the rear wheels, if one wheel is on the floor and the other is free to turn, the revolving wheel turns twice as fast as the speedometer indicates. Do not exceed a speedometer reading of around 40 mph (64 kmh). This produces a wheel speed of about 80 mph (129 kmh).

With the limited-slip differential setup (both sides free of floor), the speedometer indicates the wheel speed. Use quality equipment in top shape and follow the manufacturer's instructions. Always clean the wheel of mud and grease and pick out rocks and other debris stuck in the tread before balancing.

Tech Talk

Specialization (concentrating on one area of the vehicle) is the trend in most modern automotive shops. It is common to work for one of the many shops that specialize in exhaust systems, brakes, alignment, transmissions, air conditioning, or driveability. Even shops where you will be expected to perform many different types of work will have specialists in alignment, air conditioning, transmissions, and other areas. This is a natural result of the increasing complexity of modern cars and trucks.

However, you must still have knowledge of all the vehicle systems. If not, you may be stuck on a problem caused by a system outside your area of expertise or destroy a part related to the system you are working on. For instance, a front-end vibration caused by a faulty torque-converter clutch will never be solved by a front-end specialist who does not know anything about automatic transmissions. An air conditioning specialist may never find out that the reason the compressor clutch will not engage is because of a diagnostic code or other problem in the electronic control unit. Even changing a battery can become a nightmare if you do not know that disconnecting the battery will wipe out any stored information in the engine control computer, ABS system, radio, and other systems.

Therefore, it is vital that you have a working knowledge of all of the systems of a modern vehicle, not just those you plan to work on. Try to keep up with all developments in automotive technology, since almost every one will somehow affect your job.

Summary

Most front wheel bearings are tapered roller designs. Some front wheel bearings may be the ball or straight roller design. Clean, pack, and adjust tapered roller wheel bearings on a mileage schedule as recommended by the manufacturer. Pack the hub recess and coat the spindle and dust cap with wheel bearing grease. Adjust roller bearings to produce up to 0.003″ (0.076 mm) endplay. Ball bearings should have a mild preload. Use a new cotter pin or staked nut following adjustment. Be sure to bend cotter pin ends open so that the pin is tight. A loose cotter pin can break. Always check to make sure the safety washer and nut lock (if used) are in place. Use a new grease seal when repacking wheel bearings. Keep grease off of the brake linings, drums, pads, and rotors. When a bearing must be replaced, replace the bearing outer and inner race. Never exchange bearings from one wheel to another. Do not mix wheel bearing lubricants.

Many kinds of wheel rims are available today. Wheel rim size is determined by flange height, width between flanges, and the diameter across the rim. Wheel lugs are critical to vehicle operation and should be replaced when broken. Tire size is determined by the diameter across beads when mounted and cross-sectional width. Rayon, nylon, polyester, fiberglass, aramid, Kevlar, and steel are widely used in tire manufacture. The tire fabric or cord may be applied in diagonal, radial, or a combination of these two patterns. Tires are rated according to a system that determines size, load, and speed ratings. Tires are graded for temperature resistance, traction, and tread wear by the Department of Transportation (DOT). Tire pressure must be correct. Check pressure when cold. Rotate tires about every 5000 miles (8000 km).

Use rubber lubricant when demounting or mounting tires. Clean the rim and check for dents, burrs, and cracks before mounting a tire. Punctures can only be repaired by patching and filling the hole. When seating tire beads against the rim flanges, stand back and do not exceed 40 psi (276 kPa) pressure. Reduce to normal running pressure when seated.

Tire wear patterns can often indicate the cause of wear. Both front and back wheel and tire assemblies must be statically and dynamically balanced. Clean the wheel and tire before balancing. Check tire radial and lateral runout before balancing. Check wheel lug torque and tighten in the proper sequence. Static balance should be corrected first, then dynamic balance.

On-car or off-car balancing equipment may be used. Some front-wheel drive vehicles should use the engine to spin the tires when using an on-car balancer. When balancing back wheels with on-car equipment, do not exceed 35 mph–40 mph (56.3 kmh–64.4 kmh) speedometer reading. In cases where a limited-slip differential is used, on-car balancing requires raising both sides. Remove one wheel and balance the remaining one. Replace the first wheel and then balance it.

Review Questions—Chapter 38

Do not write in this book. Write your answers on a separate sheet of paper.

1. Where are tapered roller bearings usually used?
   (A) The front axles of rear-wheel drive vehicles.
   (B) The rear axles of front-wheel drive vehicles.
   (C) The rear axles of rear-wheel drive vehicles.
   (D) Both A and B.
2. When a vehicle is equipped with wheel cover locks, where is the removal key usually located?
   (A) On the wheel.
   (B) In the vehicle.
   (C) At the manufacturer's headquarters.
   (D) At the dealer's service department.
3. When is it okay to pack a bearing hub with grease?
4. Explain why high-temperature wheel bearing grease should be used to repack wheel bearings on modern vehicles.
5. Rotate tires every _____ miles or _____ km.
6. _____ should always be used to ease bead seating.
7. The aspect ratio refers to the relationship between the tire _____, and _____.
8. Wheel tramp is caused by _____ imbalance.
9. Wheel shimmy is caused by _____ imbalance.
10. List the three grades used in the Uniform Tire Quality Grading System.

ASE-Type Questions

1. Technician A says a torque stick should be used when tightening an aluminum or composite wheel rim with an impact wrench. Technician B says overtorquing a steel wheel could warp the brake rotor. Who is right?
   (A) A only.
   (B) B only.
   (C) Both A and B.
   (D) Neither A nor B.
2. All of the following statements about wheel lugs are true, *except:*
   (A) A damaged lug nut should be replaced as soon as possible.
   (B) Lug nuts should be tightened in the proper sequence.
   (C) A damaged lug can be knocked out of the hub.
   (D) If a new lug will not press solidly into the hub, a larger lug should be used.
3. Technician A says sealed ball bearings can be repacked if care is taken. Technician B says all old grease should be removed from tapered roller bearings before repacking. Who is right?
   (A) A only.
   (B) B only.
   (C) Both A and B.
   (D) Neither A nor B.
4. Before driving the dustcover into place, _____.
   (A) apply a thin coat of grease to cover inside
   (B) make certain cotter pin is in place
   (C) make sure safety washer is in place
   (D) All of the above.
5. An old outer bearing race may be used with the _____.
   (A) the original bearing assembly
   (B) a new bearing assembly
   (C) the old bearing assembly from the other side of the vehicle
   (D) All of the above.
6. Technician A says to always replace the cotter pin when packing or replacing wheel bearings. Technician B says a staked nut may be reused if it is not badly bent. Who is right?
   (A) A only.
   (B) B only.
   (C) Both A and B.
   (D) Neither A nor B.

7. If you drive on wet roads, which of the following traction grades would give the best traction?
   (A) A.
   (B) B.
   (C) C.
   (D) These grades do not apply to wet pavement conditions.
8. Technician A says a leak in a tire's sidewall can be patched. Technician B says a leak in a tire's sidewall should not be repaired under any circumstances and that the tire should be replaced. Who is right?
   (A) A only.
   (B) B only.
   (C) Both A and B.
   (D) Neither A nor B.
9. Technician A says both front and rear tire and wheel assemblies should be balanced. Technician B says a tire in dynamic balance is also in static balance. Who is right?
   (A) A only.
   (B) B only.
   (C) Both A and B.
   (D) Neither A nor B.
10. Extreme wear on the outside on both sides of the tire tread indicates _____.
   (A) overinflation
   (B) underinflation
   (C) incorrect toe
   (D) bent rim

Suggested Activities

1. Obtain the correct measuring tools to check wheel rim runout. After checking several rims, determine the average runout.
2. Service (remove, clean, repack, reinstall, adjust) a tapered roller wheel bearing assembly. Refer to the manufacturer's service manual for the proper type of grease and for bearing tightening procedures.
3. Remove and replace a tire installed on a steel rim.
4. Remove and replace a tire installed on an aluminum or other custom rim. List the differences between servicing this type of rim and a steel rim.
5. List the reasons that a leak in a tire may not be repairable. Consult your instructor as necessary. Patch a tire after determining that it is repairable.

39

Wheel Alignment

After studying this chapter, you will be able to:

- Explain the importance of wheel alignment.
- Describe the purposes of major alignment settings.
- Identify adjustable and nonadjustable alignment settings.
- Summarize wheel alignment procedures.
- Explain the difference between two-wheel and four-wheel alignment.
- Diagnose common alignment-related problems.

Technical Terms

Wheel alignment	Two-wheel alignment
Caster	Four-wheel alignment
Camber	Alignment angles
Road crown	Turning plates
Steering axis inclination	Heads
Included angle	Ride height
Setback	Curb weight
Toe	Runout
Toe-out on turns	Eccentrics
Tire scrubbing	Threaded rods
Wheel tracking	Four-wheel steering
Dog-tracking	Shims
Thrust angle	Eccentric bushings

This chapter covers the purpose and methods of wheel alignment. It presents an overview of the various angles involved in front-end alignment. Different alignment methods for front- and rear-wheel drive vehicles, and the principle of two-wheel and four-wheel alignment are covered also. After completing this chapter, you will have a good basis for a thorough study of the principles of front-end and steering geometry.

Defining Wheel Alignment

Wheel alignment refers to the process of measuring and correcting the various angles formed by the front and rear wheels, spindles, and steering arms. Correct alignment is vital. Improper alignment can cause hard steering, pulling to one side, wandering, noise, and rapid tire wear.

Major Alignment Angles

The various alignment angles, (caster, camber, toe, steering axis inclination, and toe-out on turns) are all related. A change in one can alter the others. Some of the angles, such as caster, camber, and toe-in, are adjustable. Others, such as steering axis inclination and toe-out on turns are built in and can only be adjusted by changing parts or bending the vehicle frame.

Caster

Caster is the tilting of the spindle support centerline from a true vertical line as viewed from the side of the vehicle. The spindle support centerline is an imaginary line drawn through the center of the upper and lower ball joints, or the ball joint and upper strut bearing on MacPherson strut suspensions. When the tire support centerline intersects the roadway at a point ahead of the tire, the caster is positive. Tilting the spindle so that the support centerline strikes the road behind the tire support centerline produces negative caster.

Both positive and negative caster angles are illustrated in **Figure 39-1.** Positive or negative caster is measured in degrees from true vertical. Positive caster tends to assist the wheels in maintaining a straight-ahead position. A negative caster setting can make turning the wheels easier. Incorrect caster can cause hard steering, wandering, high-speed instability, and with less caster, pulling to the side. Improper caster angles will not cause tire wear.

Camber

Camber is the tilting of the wheel centerline away from a true vertical line as viewed from the front of the vehicle. When the top of the wheel is tilted outward from the vehicle, camber is positive. When the top of the wheel is tilted inward, camber is negative. Both positive and negative camber are illustrated in Figure 39-2. Camber angles are small, usually no more than positive or negative 1° from zero. Too much variation in camber causes pulling. Excessive camber, whether positive or negative, causes tire wear.

When adjusting front wheel camber, the left wheel can be set with approximately 0.25°–0.5° more positive camber to compensate for ***road crown.*** Since the vehicle tends to pull toward the side with greater positive camber, setting the left front wheel camber more positive will offset the pull effect of the crowned road. It is not necessary to set the rear wheel camber (when adjustable) to compensate for road crown.

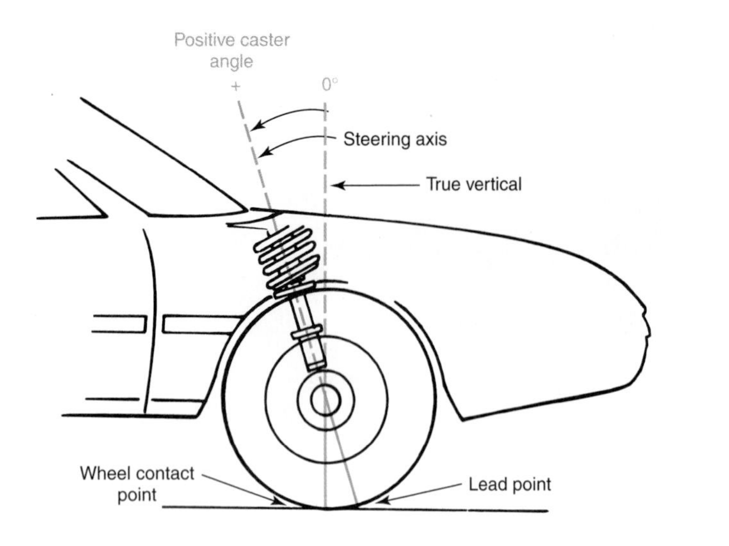

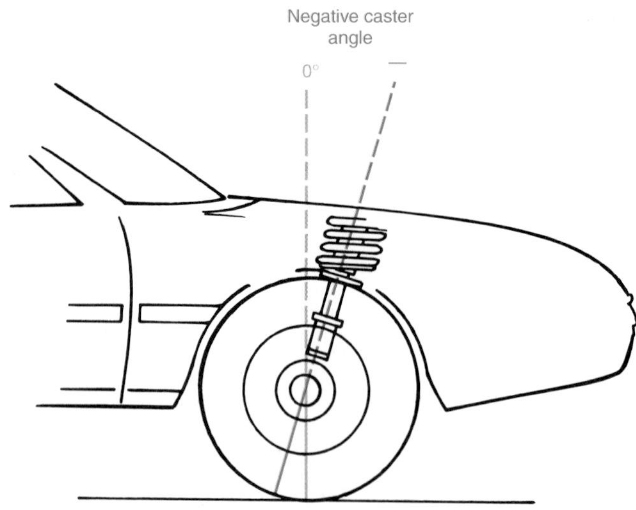

Figure 39-1. Positive and negative caster in relation to true vertical. (Perfect Equipment Corp.)

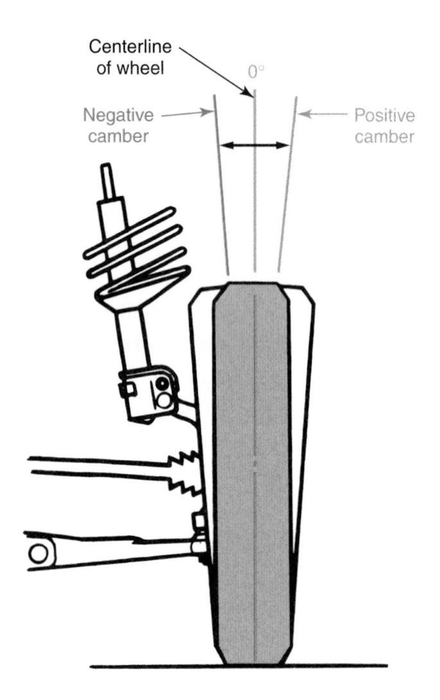

Figure 39-2. This figure illustrates positive and negative camber on a front-wheel drive vehicle. (General Motors)

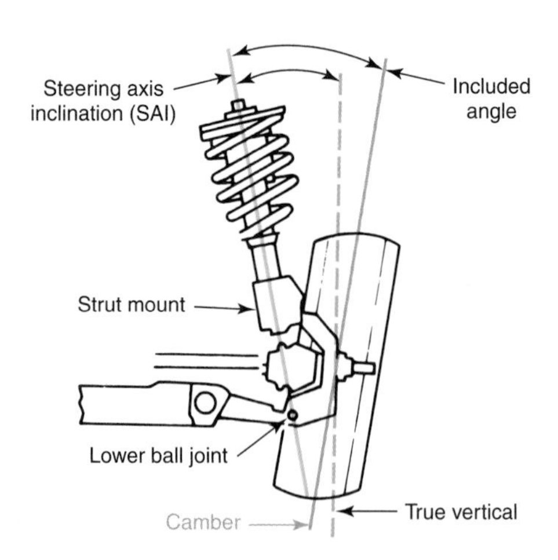

Figure 39-3. Steering axis inclination (SAI) is formed by tilting the upper mount of the strut and the lower ball joint as viewed from the front of the vehicle. This setting is normally negative on front-wheel drive vehicles. Note the included angle. This is the sum of the camber angle and the steering axis inclination. (General Motors)

Steering Axis Inclination

Steering axis inclination, or SAI, is the imaginary line formed by tilting the top ball joint or strut mount inward. This angle is measured in degrees from true vertical, **Figure 39-3.** Front suspensions are designed so that the imaginary lines formed by SAI and camber contact the road very close to each other under the tire. The close relationship of the SAI and camber lines helps the suspension to absorb road shocks. Steering axis inclination is nonadjustable. If the vehicle's SAI is incorrect, check for a bent spindle, frame, strut, or other problem and replace parts as needed. If the incorrect SAI reading is a result of a bent frame, it may be necessary to have the frame straightened at a body shop. The average steering axis inclination varies greatly depending on the manufacturer. The sum of the SAI and the camber on any one wheel is referred to as the ***included angle.***

Setback

Setback is a condition in which one front wheel spindle is positioned behind the spindle on the opposite side. This condition can also be present in rear wheel assemblies. Setback is usually caused by collision damage. One indication of setback is a caster reading that varies by more than 1° from one side of the vehicle to the other. Many types of alignment equipment cannot measure setback. Severe setback can sometimes be detected by measuring between the rear of each tire and the wheel opening.

Toe

Toe is the relative positions of the front and rear of a tire in relation to the tire on the other side of the vehicle. Look at

Figure 39-4. Note that the distance between the backs of the tires is greater than the distance between the fronts. Rear-wheel drive vehicles are toed in to compensate for the natural tendency of road-to-tire friction to force the wheels apart. On some front-wheel drive vehicles, the front tires are toed out. This is done to offset the force created by the drive axles, which tends to drive or throw the tires inward during operation. The toe setting compensates for this and allows the front tires to run parallel to one another while rolling straight down the road.

Toe-Out on Turns

When the vehicle rounds a corner, the front wheel on the inside of the turn is forced to follow a smaller arc than the outer wheel. If the wheels remained parallel during the turn, the tires would be forced to slip. Note that the steering arms in **Figure 39-5** are parallel to the centerline of the vehicle. With such an arrangement, the front wheels remain parallel when making a turn. This can cause tire slip, Figure 39-5B.

Tire slip during turns is avoided by angling the steering arms, as in **Figure 39-6.** Because the steering arms are angled, the inner wheel turns more sharply than the outer wheel during cornering. This arrangement, known as ***toe-out on turns,*** allows all wheels to turn from the same center, thus eliminating tire slip. See **Figure 39-7.**

In effect, the actual toe-out is small. The specification for toe-out on turns may call for the outer wheel to be at a certain angle when the inner wheel is turned to exactly 20°, or a certain angle for the inner wheel when the outer wheel is turned to 20°. Although specifications vary, they usually call for around a 1°–3° difference between the inner- and outer-wheel's turning angles.

Tire Scrubbing

Tire scrubbing (sideslip) is generally a result of either toe-in or toe-out. It may also be caused by incorrect camber, incorrect caster, or damaged parts. A wheel normally tries to run in the direction of its toe angle. However, because the wheel is fastened to the vehicle, it is forced to travel straight ahead (when driving in that direction). A tire-scuffing action occurs, which is similar to dragging the tire sideways on the road surface. A toe problem that is severe will quickly wear a tire.

Wheel Tracking

Wheel tracking is the ability of a vehicle's rear wheels to follow directly behind the front wheels. If the rear wheels are not tracking correctly, the vehicle will not travel in a straight line unless the front wheels are turned to compensate for the misalignment. This condition is often referred to as ***dog-tracking,*** **Figure 39-8.**

Tracking is set by aligning the vehicle's ***thrust angle*** with its geometric centerline, **Figure 39-9.** Ideally, the thrust angle should cover the centerline. This would result in perfect

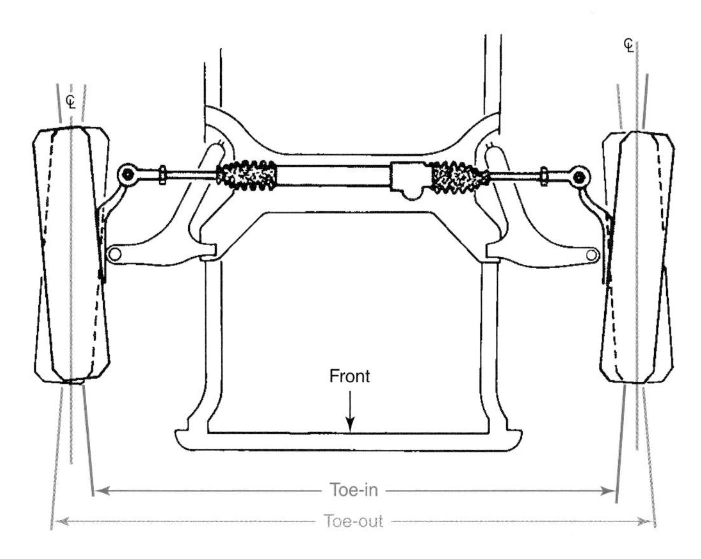

Figure 39-4. This overhead view shows toe-in and toe-out. With toe-in, the wheels are closer together at the front than at the back. (DaimlerChrysler)

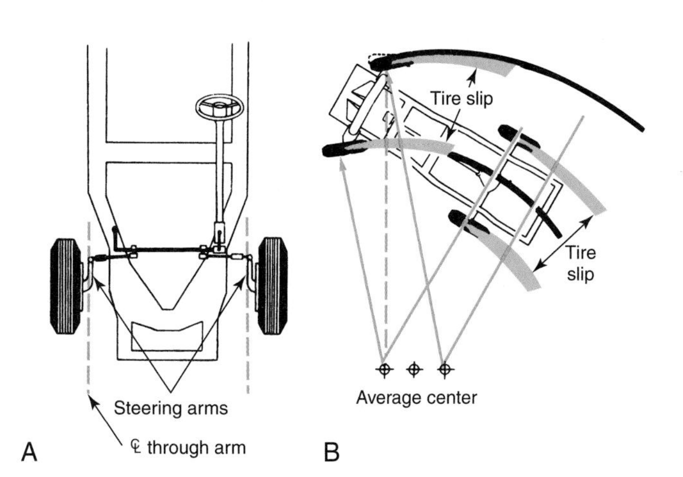

Figure 39-5. When the steering arms are parallel (red lines), as in A, the front wheels remain parallel during turns, thus causing tire slip, B. (Hunter Engineering Co.)

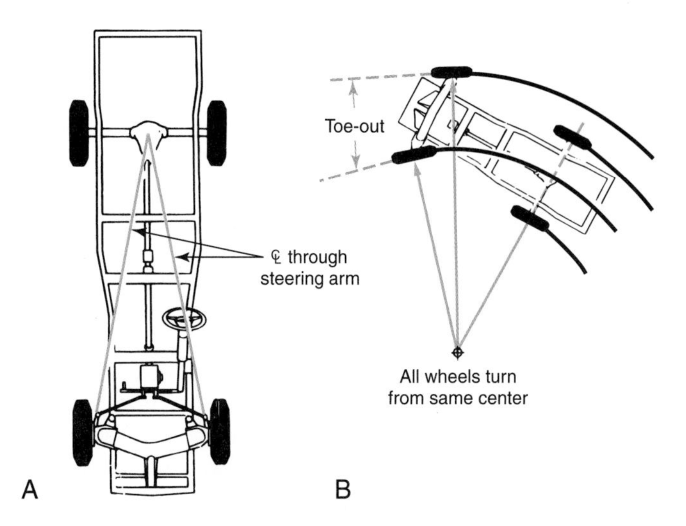

Figure 39-6. A—Angling the steering arms causes toe-out to occur on turns. B—This allows the tires to roll about their respective arcs without slipping.

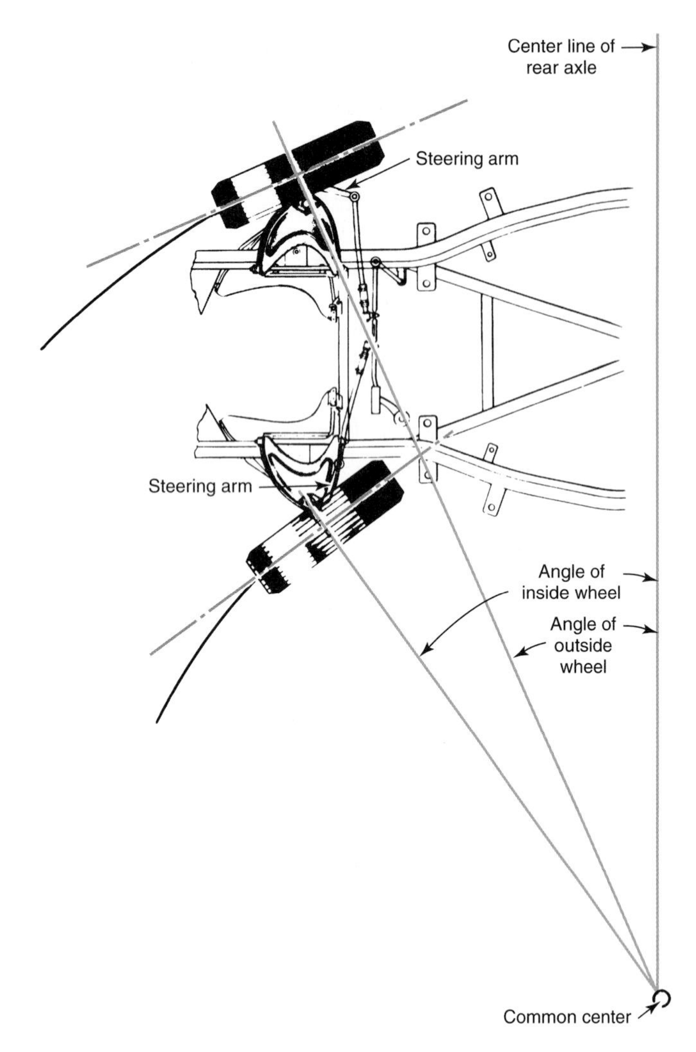

Figure 39-7. Toe-out on turns must be specified. Wheels turn about a common center determined by the wheelbase of the vehicle. With respect to common point, the inside wheel is ahead of the outside wheel and makes a sharper angle than the outer one. (Bear)

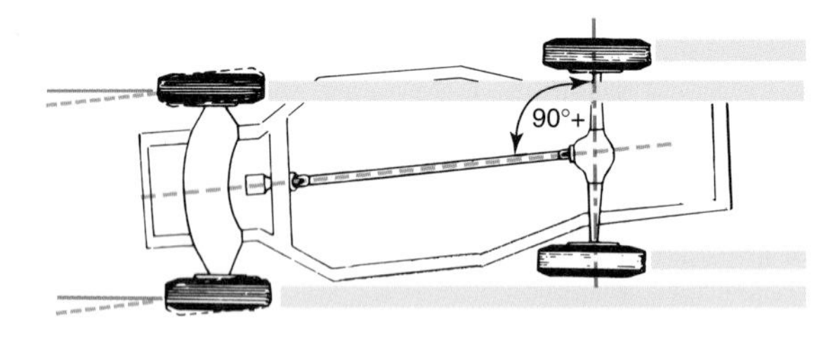

Figure 39-8. Dog-tracking occurs when the rear wheels do not follow directly behind the front wheels. (Hunter Engineering Co.)

tracking. Manufacturer's tolerances, accident damage, and normal wear, however, make perfect alignment rare. Vehicles with front-wheel drive, four-wheel drive, and four-wheel steering also complicate thrust angle alignment. The technician must strive to align the thrust angle and the geometric centerline as closely as possible. Tire wear, poor fuel economy, and incorrect handling can result if the vehicle's thrust angle is not properly set.

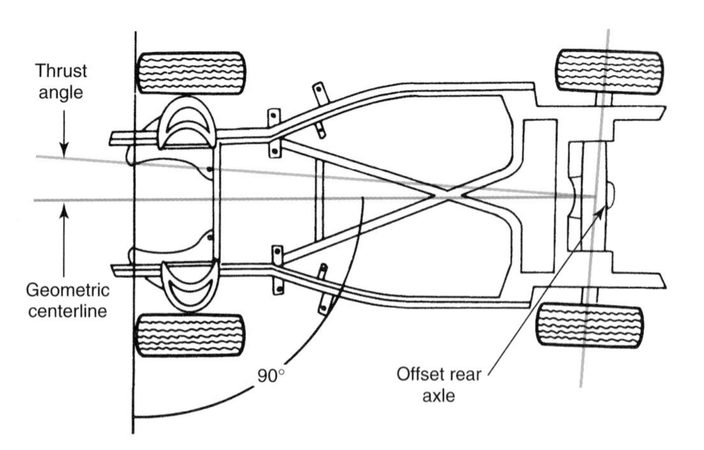

Figure 39-9. Relationship between a vehicle's thrust angle and centerline. The rear axle is offset and the angle is greater than 90°. (Sun Electric)

Types of Wheel Alignment

In the past, the only type of wheel alignment performed was the ***two-wheel alignment,*** or front wheel alignment. The rear wheels were attached to a solid rear axle assembly that could not be adjusted and was rigid enough to stay in reasonable alignment. Today, the ***four-wheel alignment,*** in which the alignment of the front and rear wheels are checked and adjusted, is commonly performed. Most front-wheel drive vehicles have provisions for adjusting the rear wheels. In addition, many rear-wheel drive vehicles are equipped with independent rear suspensions, which also must be adjusted. Modern solid rear axles and suspension systems are lighter than on earlier cars and can be knocked out of adjustment easily.

Adjustable settings on the front wheels are caster, camber, and toe. Nonadjustable settings are steering axis inclination and toe-out on turns. The rear wheel settings that can be made on many vehicles are camber and toe. These settings, both adjustable and nonadjustable, are referred to as ***alignment angles.*** The modern practice is to check both front and rear wheel alignment. Checking all four wheels also makes it possible to set the thrust angle to ensure perfect wheel tracking.

Wheel Alignment Equipment

Wheel alignment equipment varies greatly, from simple mechanical gauges to computerized devices. To do any kind of alignment, the technician must have equipment capable of checking all of the alignment angles listed previously. Simple camber-and caster-checking devices, held by magnets to the wheel hub, allow fairly accurate checking, **Figure 39-10.** Toe can be checked with trammel bars, **Figure 39-11,** a drive-over plate, **Figure 39-12,** or even a tape measure. However, to do an accurate four-wheel alignment on a modern vehicle, more elaborate equipment is needed.

The basic alignment rack is equipped with ***turning plates*** to allow the front wheels to be turned for measuring caster and to allow the vehicle suspension to settle into its normal riding position after being raised. The ideal alignment

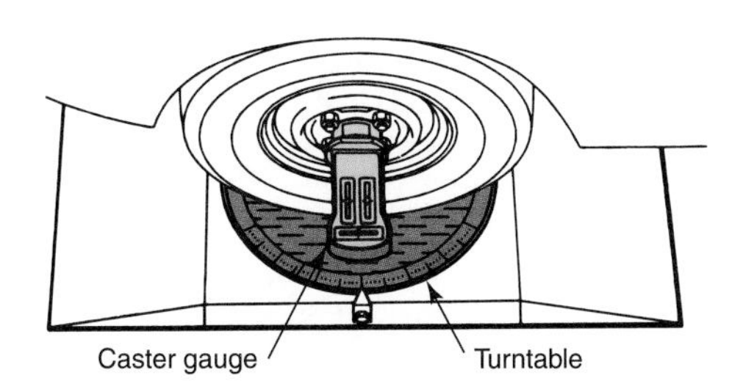

Figure 39-10. This simple caster gauge can be used with a turntable to check caster angle.

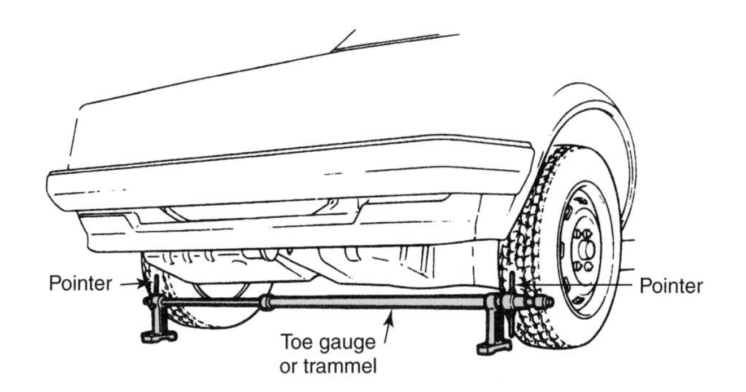

Figure 39-11. A trammel bar can be used to make a rough toe check. (Mazda)

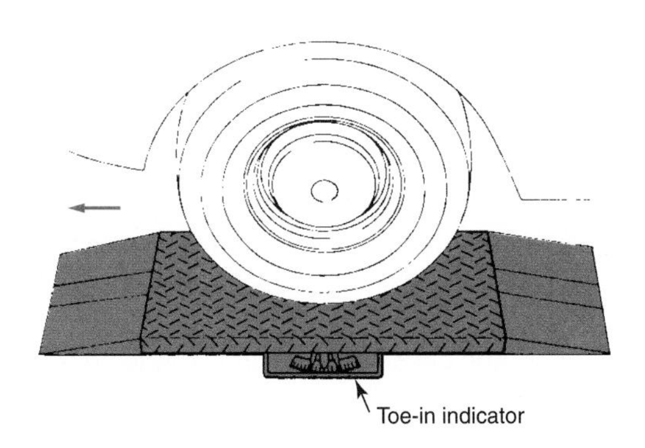

Figure 39-12. In this illustration, a vehicle is being driven over a mechanical toe indicator. (Toyota)

rack is high enough or has a pit deep enough to permit easy access to the underside of the vehicle. The best alignment racks are attached to a hydraulic lift, so they may be lowered to drive the vehicle on and raised to gain access to the underside of the vehicle for extended periods.

The alignment machine should be able to accurately measure all of the alignment angles. Modern alignment machines must be capable of measuring the rear wheel alignment as well as front wheel alignment. This necessitates the use of rear wheel sensing devices. The wheel-mounted sensing devices used on modern alignment machines are usually called ***heads.*** Older alignment machines use wheel-mounted light beam generators to measure the alignment angles, **Figure 39-13.** The latest alignment machines use wheel-mounted electronic sensors and a computer to provide readouts on a screen, **Figure 39-14.**

Figure 39-13. A vehicle set up to have toe-out on turns checked. As the wheel is turned, the light beam moves on the chart. (Hunter Engineering Co.)

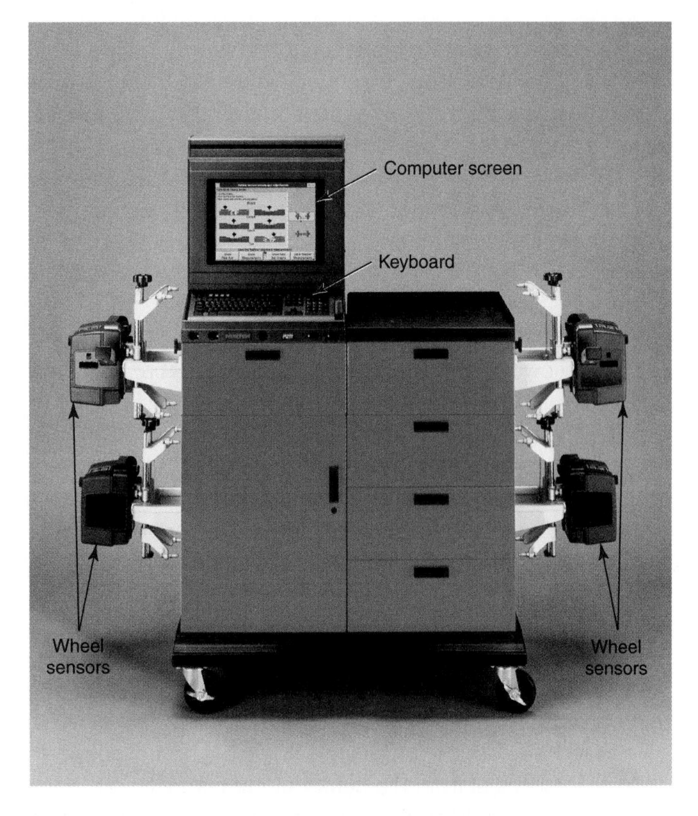

Figure 39-14. Computerized alignment machine. The sensors are mounted on the vehicle's wheels, and alignment readings are displayed on the screen. (Hunter Engineering Co.)

There are many manufacturers of alignment equipment. Obtain and study the user manuals to familiarize yourself with the alignment equipment's operation. When you are checking and adjusting wheel alignment, use quality equipment that is in good condition. Wheel alignment is a precision operation; both the equipment and the techniques must be perfect.

Performing Wheel Alignments

The following procedures apply to both conventional and MacPherson strut assemblies. Alignment angles are designed to produce the same results no matter what type of suspension, steering, or drive train are used. When it is possible, road test the vehicle before beginning the wheel alignment. Check the vehicle for handling problems, noises, and other factors that might be the cause of problems. Talk to the customer and find out why the vehicle is being brought in for an alignment. Then, perform the prealignment checks listed in the following section.

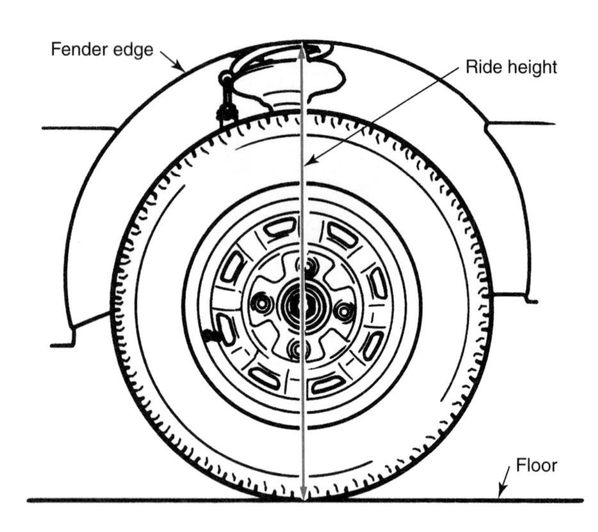

Figure 39-15. Measuring the distance shown in this illustration is a method of checking the trim height on a vehicle with MacPherson strut front suspension. (General Motors)

Note: When road testing part-time four-wheel drive vehicles, make sure the transfer case is in the two-wheel high range.

Prealignment Checks

Begin by driving the vehicle onto the alignment rack. The vehicle must be driven on as straight as possible. If the vehicle is not reasonably straight on the rack, it will be impossible to set the steering wheel in the straight ahead position.

Many complaints of improper wheel alignment can be traced to wear, part damage, problems in the tires, brakes, wheel bearings, or other nonrelated areas or items. Do not attempt to align a vehicle that has worn or damaged parts. Check the suspension and steering parts in the front and rear of the vehicle for wear or damage as explained in Chapter 36, Suspension System Service, and Chapter 37, Steering System Service.

Suspension parts to be checked include: ball joints, upper and lower control arm bushings, stabilizer bar bushings, strut rod bushings, and shock absorbers or struts. In addition, check all suspension parts for bends, kinks, and other damage. Steering parts to check include the gearbox or rack and pinion, pitman arm, idler arm, tie rods, and center link. Be especially careful to check the gearbox play and gearbox-to-frame fasteners. Check for wear, improper adjustment, or loose fasteners. Check power steering operation. Also carefully check the vehicle frame for bends or twists. Correcting a bent frame requires the services of a body shop equipped to perform frame straightening. Correct all problems before proceeding.

Place the vehicle on a level surface and bounce both ends of the vehicle from the center of the bumpers. Allow the vehicle to come to its normal rest position and check the ***ride height,*** or curb height, **Figure 39-15.** If the ride height is incorrect, look for broken or sagging springs. Replace all defective springs. Check the passenger compartment and trunk for excessive weight such as luggage, sales samples, tools, sporting goods, and other items. Remove any excess weight. On some vehicles with torsion bar suspensions, the ride height can be adjusted. Check the manufacturer's service manual.

Inspect the tires for wear, bulging, or other damage. Carefully note any tire conditions (as discussed in Chapter 38, Wheel and Tire Service) that might indicate an alignment problem. Tire size and pressure must be as specified on all four wheels. Never attempt to align a vehicle that has mismatched tires or a combination of good and worn tires. Check the front wheel bearings for damage or excessive play. Spin the wheels and check for bent rims, dragging brakes, and loose wheel lugs.

Final Preparation for Wheel Alignment

After all needed repairs have been made, final preparations for alignment can be made. To ensure accurate alignment checks and adjustments, the vehicle must be at ***curb weight.*** Curb weight is the weight of the vehicle with all normal accessories and a full tank of gas, but without the driver or passengers. Many manufacturers specify bouncing both the front and rear of the vehicle at the center of the bumpers. This allows the vehicle to come to its normal rest position before making alignment checks. See **Figure 39-16.**

Note: Some import vehicle manufacturers specify that weights be placed in the vehicle to simulate an average weight of passengers and luggage. Refer to the manufacturers' specifications for exact weight and placement.

Alignment Specifications

One of the most important parts of performing a wheel alignment is obtaining the correct specifications. All vehicle

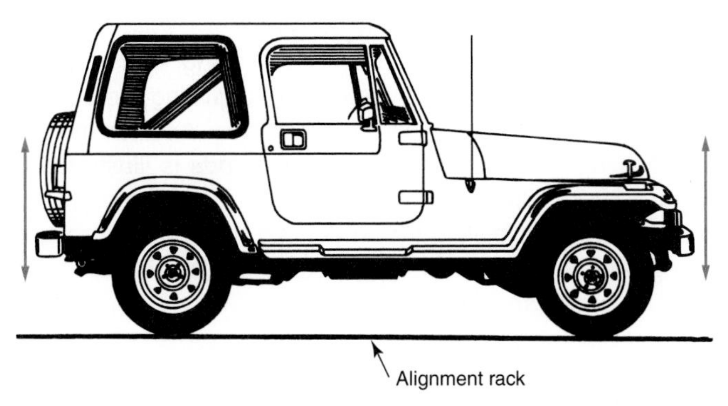

Figure 39-16. Bounce the front and rear of the vehicle. Allow it to come to normal curb height before making alignment checks. With this method, no alignment spacers are used. (DaimlerChrylser)

manufacturers publish and regularly update alignment specifications for current and older models. Always look up the most current alignment data for the vehicle before beginning the alignment. Do not guess at specifications.

Alignment Setup

Be sure to set up the alignment equipment as specified by the manufacturer. If the equipment is attached to the wheel hub or the wheel rim, **Figure 39-17,** the wheel cover must be removed. Many modern wheel covers are held to the rim by wheel cover locks or by the lug nuts themselves. Custom wheels are secured by wheel locks, **Figure 39-18.** The key to these locks is often kept in the vehicle and must be used to remove the wheel covers. If you are unable to locate the key, ask the owner. Many modern wheel rims are made of aluminum or chrome plated steel, and special precautions must be taken to prevent scratching the rim. Refer to the manufacturers' service literature.

Note: After using the wheel lock key, replace it in the vehicle. Without it the driver will not be able to remove the tire in an emergency. Be sure to attach any safety straps to the rim to prevent damage to the head if it falls off the rim.

Runout Compensation

Wheel rims always have some distortion or ***runout.*** Even a new rim has some runout. If the alignment heads are installed on the edges of the rims, this runout must be accounted for, or it will cause false readings. To compensate for runout, attach the alignment equipment to the rim and raise the wheels from the rack. On front-wheel drive vehicles, apply the parking brake and place the transmission in neutral. Perform runout compensation as recommended by the operator's manual for the alignment equipment. Repeat this procedure for each wheel, and then lower the vehicle.

Turn the steering wheel from side to side several times to equalize any play in the steering system. If the vehicle has power steering, the engine should be running. Next, center the

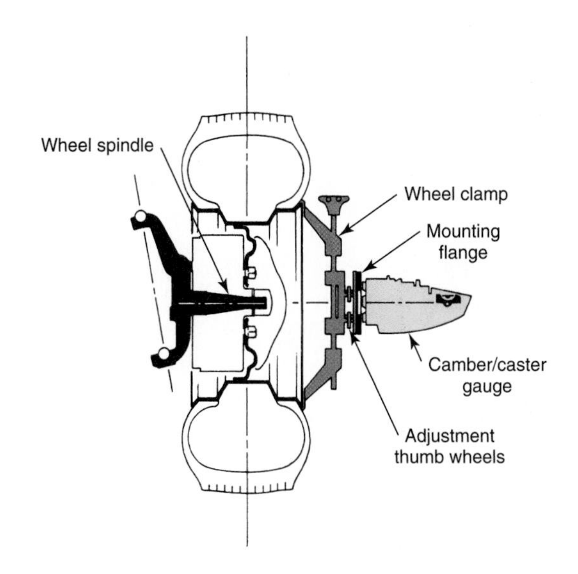

Figure 39-17. This alignment gauge is attached to the wheel rim. Be sure to connect the tool safety strap (not shown) if used. Handle these tools with care. (Ammco Tools)

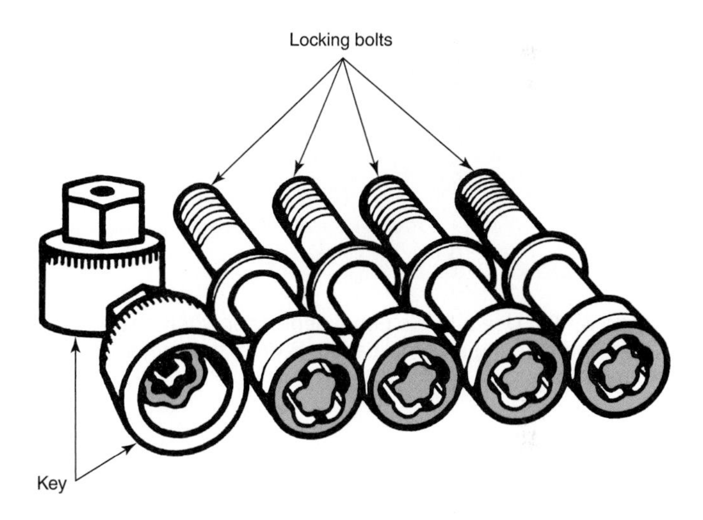

Figure 39-18. Wheel locking bolts and removal key. These bolts are for use with alloy wheels. The styles vary. (Mercedes-Benz)

steering wheel, **Figure 39-19,** and lock it in place with a special wheel-locking tool. Turn the engine off if applicable. To prevent excessive wheel movement during the caster check, lock the brakes by using a brake pedal depressor, as shown in **Figure 39-20.** Apply the parking brake and, if the transmission is an automatic, shift into park.

Measuring, Recording, and Adjusting Alignment

The first step in adjusting alignment is to measure and record all alignment readings. If the vehicle is receiving a four-wheel alignment, start by measuring, recording, and adjusting the rear camber and toe. If the vehicle is receiving a two-wheel alignment, start with the front wheel measurement section later in this chapter. In many cases, it is advisable to check

Figure 39-19. Center the steering wheel by turning it from side to side several times to equalize any play. (Lexus)

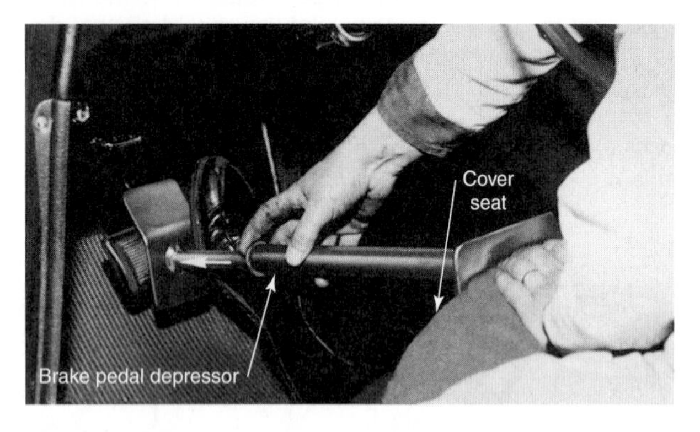

Figure 39-20. Use a brake pedal depressor when checking caster and steering axis inclination. Protect the seat surface. (Ammco Tools)

and record the rear wheel alignment, even if the rear cannot be aligned. For example, rear alignment should be checked and recorded if the steering wheel cannot be centered or rear tire wear is excessive. Specific methods of using the various types of alignment equipment are not discussed, due to the great number and variety of alignment devices available. Always consult the manufacturers' service information for the recommended type of alignment equipment.

Note: Always adjust vehicle toe last, as it is affected by other adjustments.

Rear Wheel Adjustment

Since the rear wheels do not affect steering, the most common effects of improper settings are tire wear and noise. However, if the rear toe is off, the steering wheel may be difficult to center. Note that there is no adjustment for caster on the rear wheels, since this is a factor that affects steering and is not needed on the rear.

Rear Camber

There are several methods of setting rear camber. **Figure 39-21** shows some methods of adjusting the camber using egg-shaped cams called ***eccentrics.*** Another method of adjusting the camber with ***threaded rods*** is illustrated in **Figure 39-22.** Other designs require loosening the rear strut bolts, pushing or pulling the wheel into position, and retightening the bolts. All alignment settings must be within the manufacturer's specified range.

Rear Toe

Rear toe can be set by one of several methods depending on the manufacturer. Like rear camber, rear toe can be set by either eccentric cams or threaded rods. On some vehicles, toe is adjusted by loosening a lock bolt or jam nut and pushing or pulling the suspension part into position, **Figure 39-23.** The lock bolt is then retightened. **Figure 39-24** illustrates a method of setting toe by rotating the tie rod.

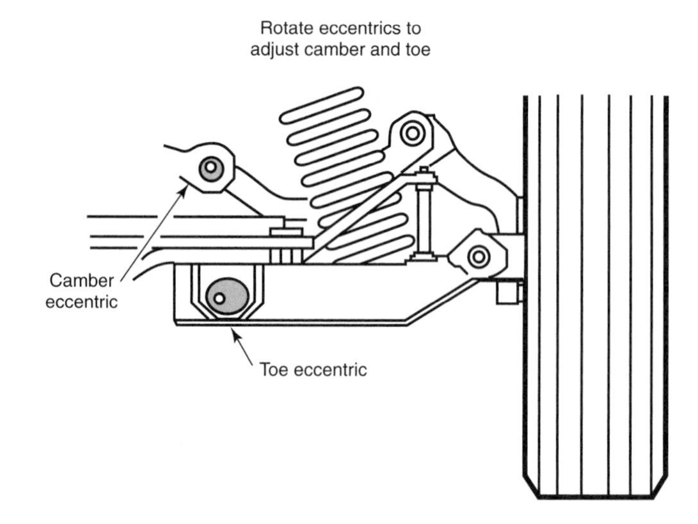

Figure 39-21. Many modern vehicles use eccentric cams to properly adjust rear camber. (Hunter Engineering Co.)

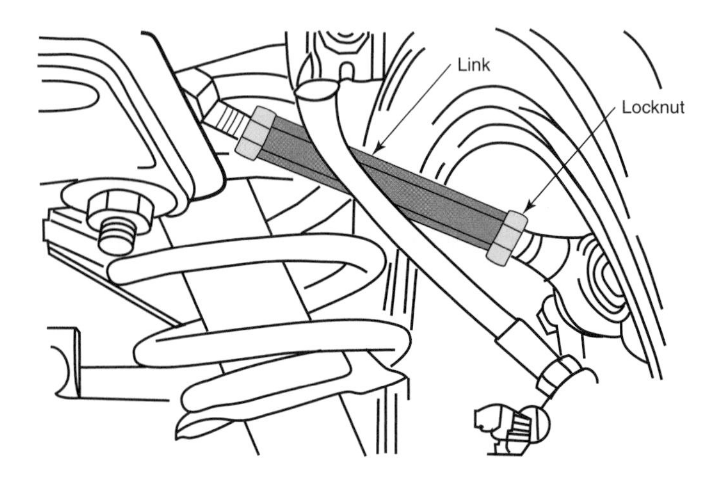

Figure 39-22. A threaded rod is commonly used to adjust rear camber. (Hunter Engineering Co.)

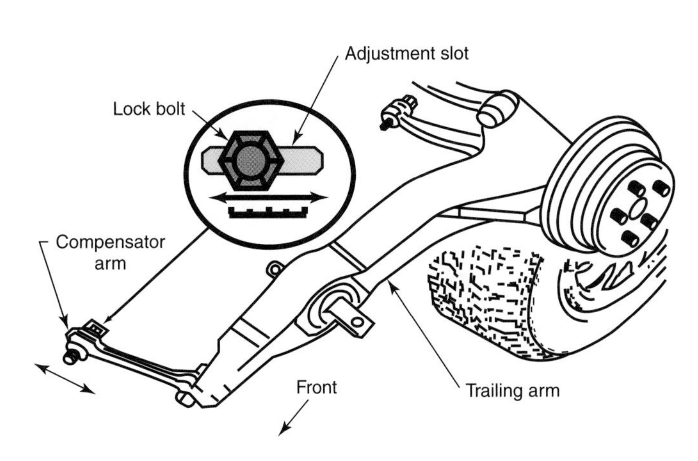

Figure 39-23. Adjusting rear toe by loosening the lock bolt and sliding the suspension into the correct position. (Hunter Engineering Co.)

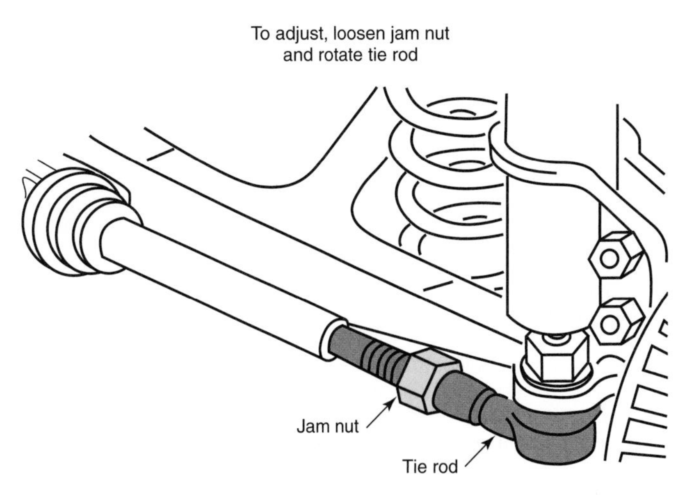

Figure 39-24. Many cars have a tie rod that can be lengthened or shortened to set rear toe. (Hunter Engineering Co.)

Note: On some older alignment machines, toe is measured in inches. This is also true if you are using a trammel bar to set front toe-in.

Some vehicles with ***four-wheel steering*** (the rear wheels turn with the front wheels) must have the rear toe set by a different method than that used by vehicles with two-wheel steering. To make this adjustment, the technician must have a special tool to lock the rear steering gearbox. The tool is installed to lock the gearbox before setting the rear camber and toe. It is removed to set the front caster and camber, and then reinstalled to set the front toe. Refer to the manufacturer's service manual for the exact procedures and specifications.

Setting Rear Alignment Using Shims

On many late model vehicles, there are no adjusting devices for setting camber and toe. A method has been devised for using ***shims*** to adjust the vehicle's alignment. Shims can also be used on the rear axles of some vehicles in which the factory adjustment is insufficient. Rear axle shims are round metal or plastic discs that are thicker on one side than the other, as seen in **Figure 39-25.** The amount of caster and toe change needed is determined by recording the actual readings and comparing them to the factory specifications. The type of shim needed and its exact placement are then calculated, **Figure 39-26.**

The rear brake drum or rotor must be removed and the bearing housing separated from the rear axle. The shim is then placed between the rear bearing and axle, as in **Figure 39-27.** This tilts the bearing, and therefore the wheel, in the correct direction to give the proper camber and toe specifications. This procedure must be done carefully to ensure that the final reading is correct and that all rear axle parts are reinstalled in their original positions. The rear wheel bearings must be properly adjusted and a new cotter pin must be used. This procedure was discussed in Chapter 38, Wheel and Tire Service.

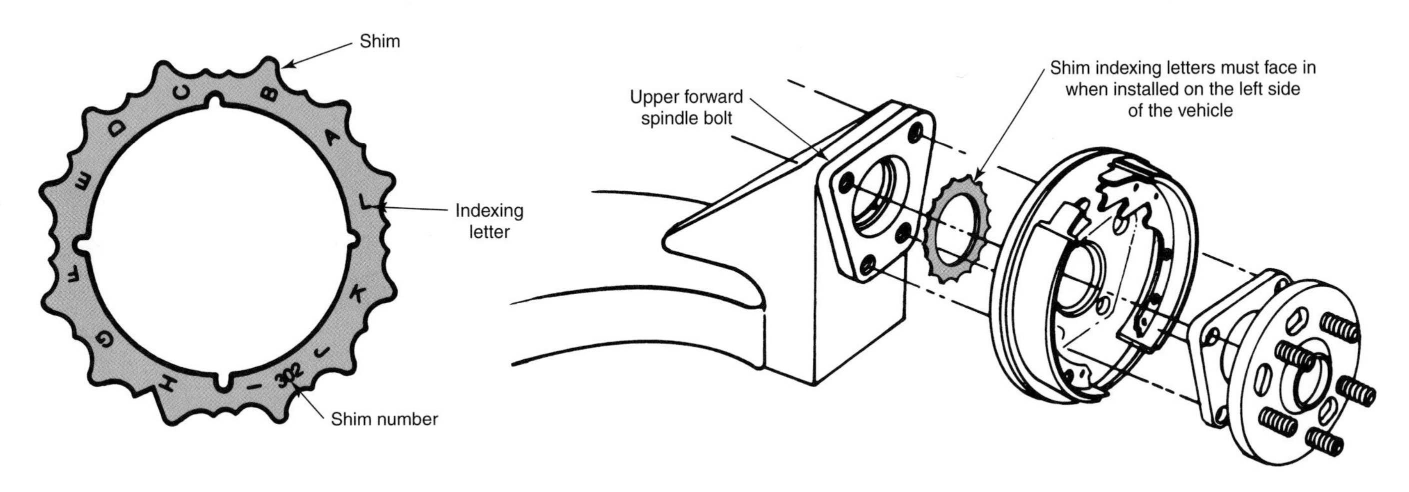

Figure 39-25. This exploded view shows where a round metal shim is installed to set rear alignment. Proper positioning is critical to avoid making the alignment worse instead of better. If needed, two shims may be used together to obtain the most accurate adjustment. (Perfect Equipment Co.)

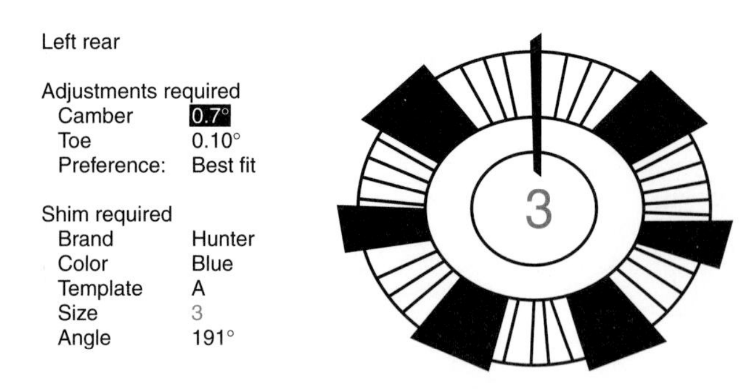

Figure 39-26. This illustration shows where to place a shim to make camber and toe adjustments. Dark areas show where the shim must be cut to provide mounting bolt clearance. (Hunter Engineering Co.)

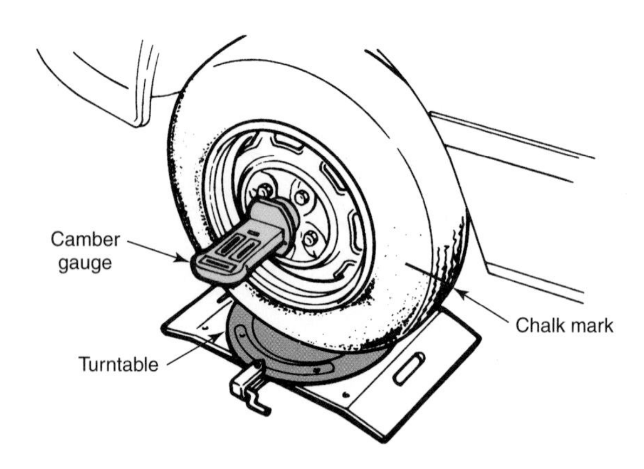

Figure 39-28. Camber gauge attached to the wheel. Be sure the wheel is pointed straight ahead. Lock the turntable. (Mazda)

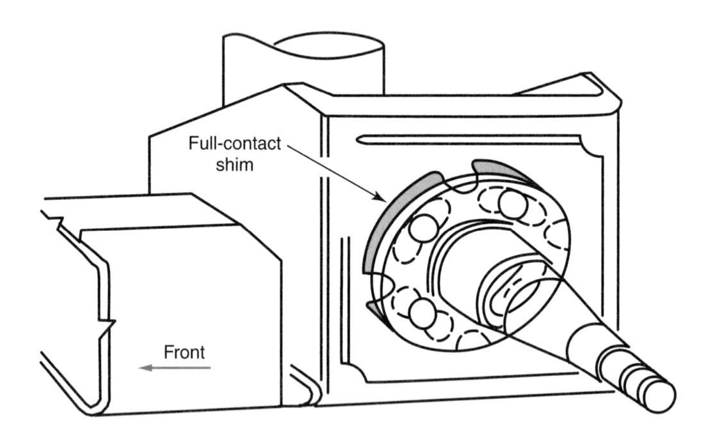

Figure 39-27. Camber and toe being set with a full-contact shim. (Hunter Engineering Co.)

Front Wheel Alignment

After measuring, recording, and setting rear wheel alignment, check the front alignment. On many vehicles, either the caster or camber is not adjustable. On other vehicles, only the toe can be adjusted. Be sure to check the manufacturer's specifications to determine what can be adjusted. On many vehicles, adjusting the caster will affect the camber and vice versa. For this reason, caster and camber are usually adjusted and checked together. Many methods are provided for adjusting caster and camber with the same adjustment.

Caution: Never attempt to make nonfactory adjustments by bending or welding suspension or steering parts. In some cases, aftermarket parts are available to make caster or camber adjustments that cannot be made by following factory procedures. Check with local parts suppliers for the availability of these parts.

Front Caster

Unlock the steering wheel and carefully check the caster angle by following the equipment manufacturer's instructions. Usually the front wheels are turned a certain amount, and then turned the same amount in the opposite direction. Repeat the caster check on the other wheel. **Figure 39-28** illustrates the use of one form of caster gauge. Always set caster exactly as specified. Make certain both sides are within 0.5° of the same setting. Remember that handling is often improved on crowned roads by deliberately setting the driver's side caster up to 0.5° more negative than the passenger's side. Regardless of setting variation to compensate for road crown, settings must be within the manufacturer's specified range and with no more than a 0.5° difference between sides.

Caster is adjusted on many modern vehicles by moving the lower strut rod in or out, **Figure 39-29. Figure 39-30** illustrates how caster is adjusted on some vehicles with MacPherson strut suspensions. On these vehicles, the nuts holding the top of the strut tower are loosened, and the tower is slid forward or backward. Sometimes, the strut tower mounting holes may have to be cut or filed to obtain enough movement.

Front Camber

Before checking the camber, make sure that the wheels are in the straight-ahead position. Use equipment as directed by the manufacturer. **Figure 39-31** shows one type of camber-checking tool. Remember that it is acceptable to set the

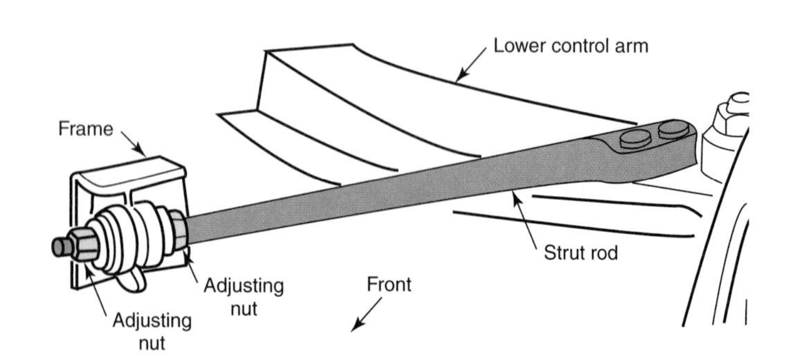

Figure 39-29. Caster adjustment is made by moving the strut rod with adjusting nuts. On this design, the technician can increase caster by shortening the strut rod or decrease caster by lengthening the strut rod. (Hunter Engineering Co.)

camber slightly more positive on the left wheel to compensate for road crown.

Figure 39-32 shows common ways of adjusting the camber by loosening the nuts holding the top of the strut tower and repositioning the tower. Another method, moving an eccentric attached to the lower control arm, is shown in **Figure 39-33.** In **Figure 39-34,** the camber is adjusted by moving an eccentric located on the top or bottom bolt holding the strut assembly to the spindle. On other designs, camber is adjusted by loosening the strut bolts, pushing or pulling the wheel into position, and then retightening the bolts. The strut rod mounting slots may require filing or cutting to allow enough movement.

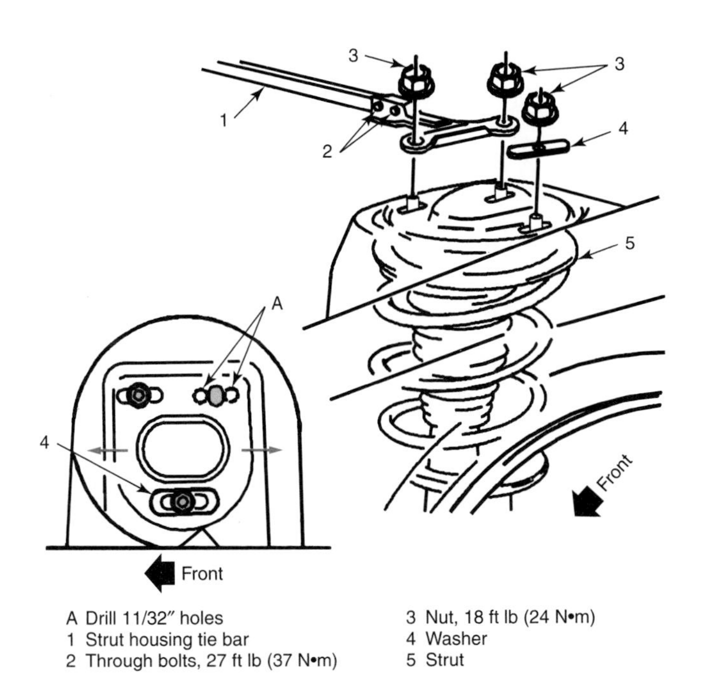

Figure 39-30. One type of caster adjustment on a MacPherson strut suspension uses slotted holes in the body. The strut assembly is moved in or out to set camber. (General Motors)

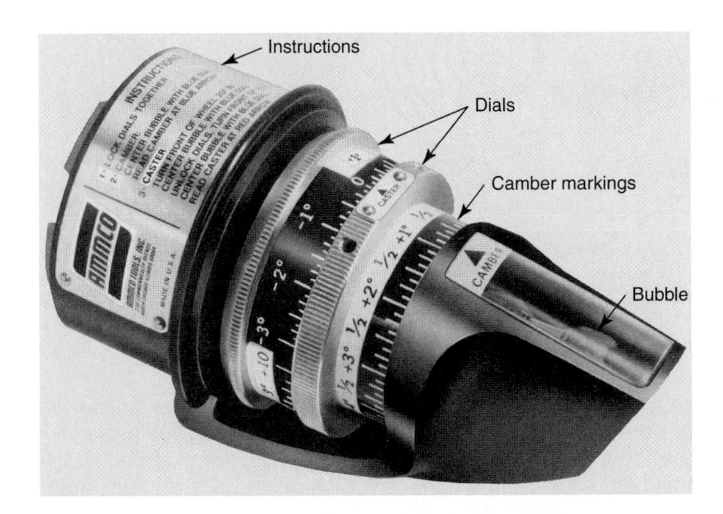

Figure 39-31. One mechanical camber-checking tool. Use the tool as directed by the manufacturer. (Ammco Tools)

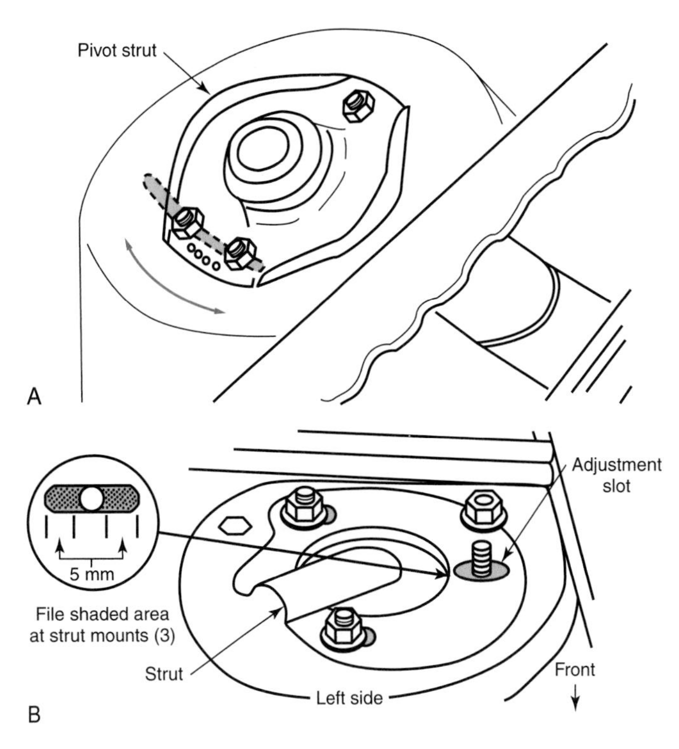

Figure 39-32. These are camber adjustment methods that use slotted holes in the body. A–By pivoting the strut. B–By sliding the strut sideways. (Hunter Engineering Co.)

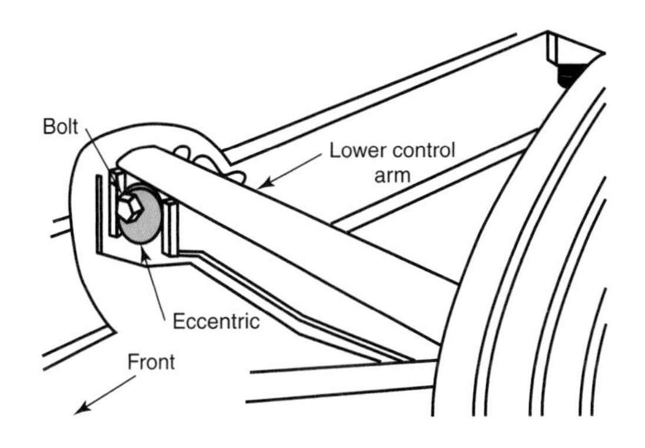

Figure 39-33. Move this lower control arm by rotating the adjusting eccentric. (Hunter Engineering Co.)

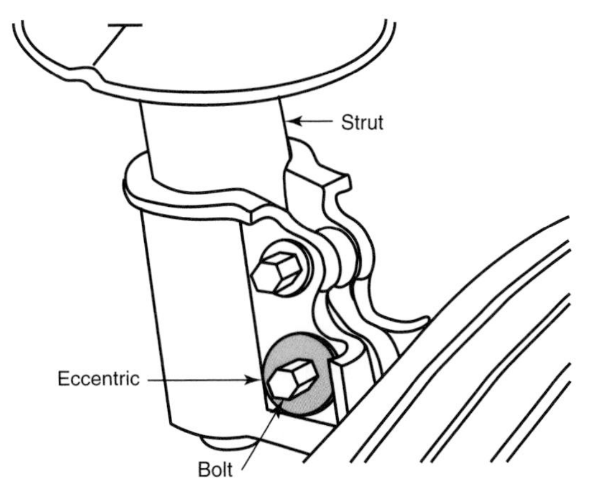

Figure 39-34. Rotate the eccentric cam on the strut to set correct camber. (Hunter Engineering Co.)

Other Methods of Adjusting Caster and Camber

There are a number of other methods used for caster and camber adjustment, **Figure 39-35**. On many older vehicles, current light trucks, and large rear-wheel drive domestic cars, shims are used at both sides of the upper control arms, Figure 39-35A. Another method uses slotted holes in the vehicle frame or control arm, Figure 39-35B. Eccentrics on the upper or lower control arm bushings may be used to make adjustments, Figure 39-35C and Figure 39-35D. In another design, the spot welds or rivets holding the top of the strut

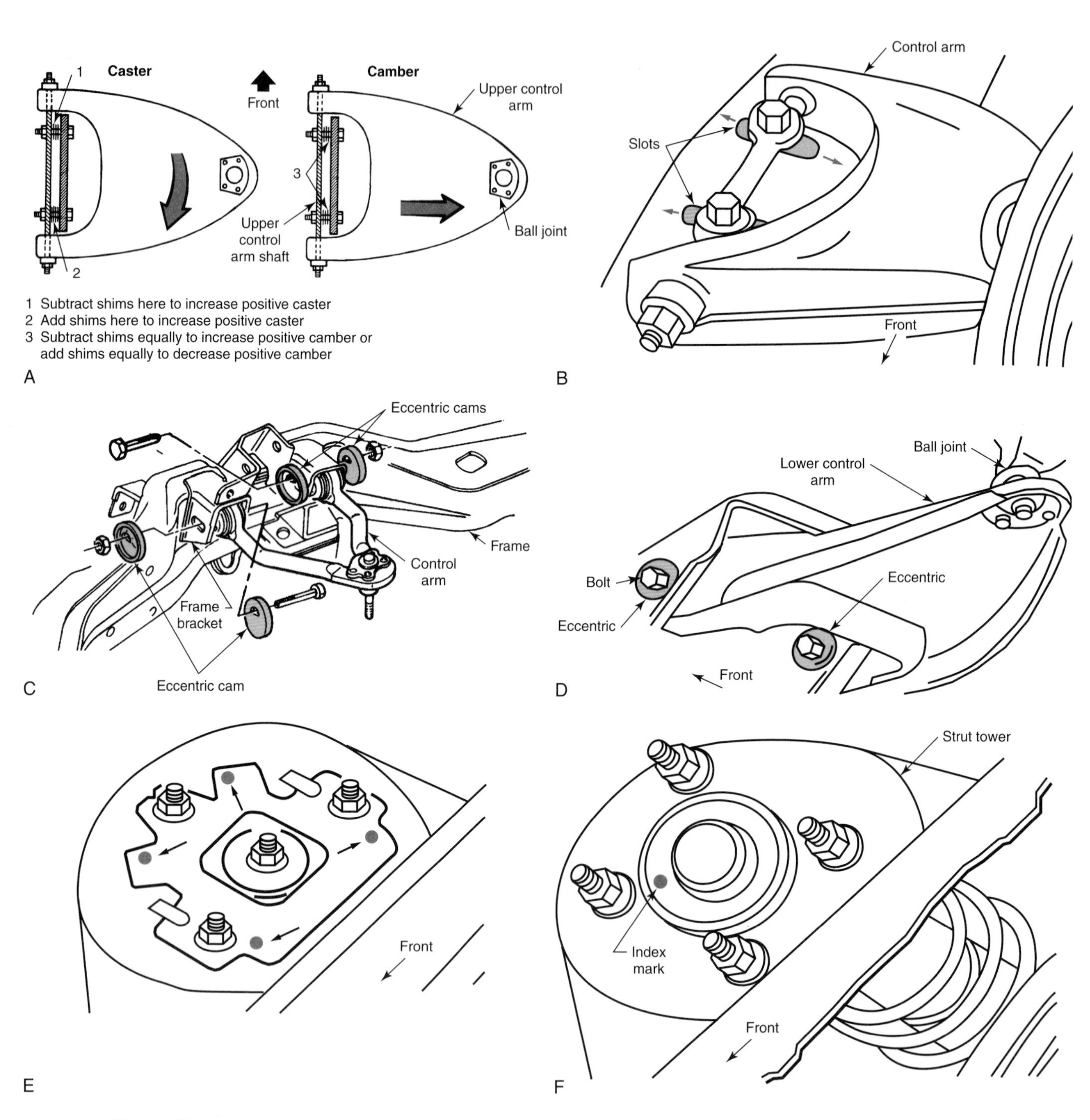

Figure 39-35. This illustration shows some methods of adjusting alignment. A—In this arrangement, caster and camber adjustments are made by adding or removing shims. B—Caster and camber are adjusted by moving the control arm in or out in the provided slots. C—Rotate the eccentric cams to set caster and camber on this particular control arm. D—Adjustment eccentrics located on the lower control arm. E—To adjust caster or camber in this setup, loosen the upper strut mounting bolts and drill out the spot welds (do not drill more than 3/8″ deep). Move the strut to achieve the desired camber and caster, and torque all fasteners to specifications. F—To make adjustments in this arrangement, remove the strut tower nuts; then lower and rotate the tower to align the index mark. Replace the fasteners and torque to specifications. (General Motors, Hunter Engineering Co.)

tower to the body are drilled out or chiseled off, and the tower moved to the desired position, Figure 39-35E. On some vehicles, the strut tower nuts are removed and the strut assembly lowered and rotated to obtain the correct reading, Figure 39-35F.

When using any of these methods, the technician must calculate the change made to both caster and camber. For instance, adding a shim to the rear of the upper control arm in Figure 39-35A will move the upper ball joint backward and increase positive caster, but it will also move the ball joint inward, making camber more negative. Certain control arms are designed with the ball joint off center, as in **Figure 39-36.** Adjusting one end will have more effect on the camber, while adjusting the other end will have more effect on the caster. With experience, you will be able to dictate how much any one adjuster can be moved to get the desired caster and camber. Even after you develop that experience, always recheck both caster and camber when making any adjustments.

Adjusting Caster and Camber Using Eccentric Bushings

Caster and camber are not adjustable using conventional adjusting devices on many light trucks having a solid front axle. To adjust caster and camber on these vehicles, special ***eccentric bushings*** have been developed, **Figure 39-37.** To use one of these bushings, first calculate the amount by which caster and camber must be changed. Then select the proper bushing using a chart similar to the one shown in **Figure 39-38.** To install the bushing, remove the upper ball joint from the axle. Remove the old bushing, which is cylindrical, and install

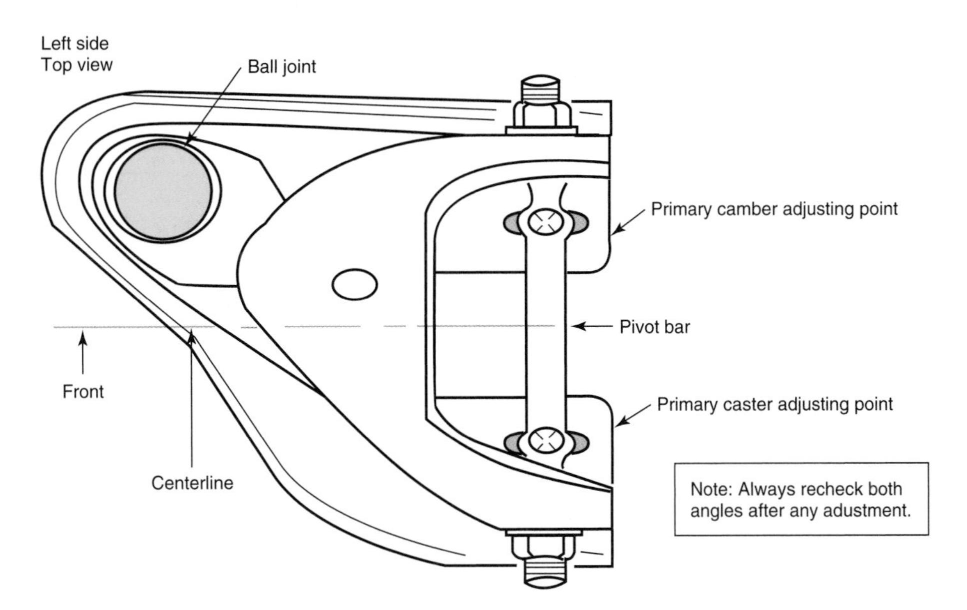

Figure 39-36. This control arm has the ball joint mounted off center to make adjustments. Moving the control arm at the top slot adjusts camber. Moving the control arm at the bottom slot adjusts caster. (Hunter Engineering Co.)

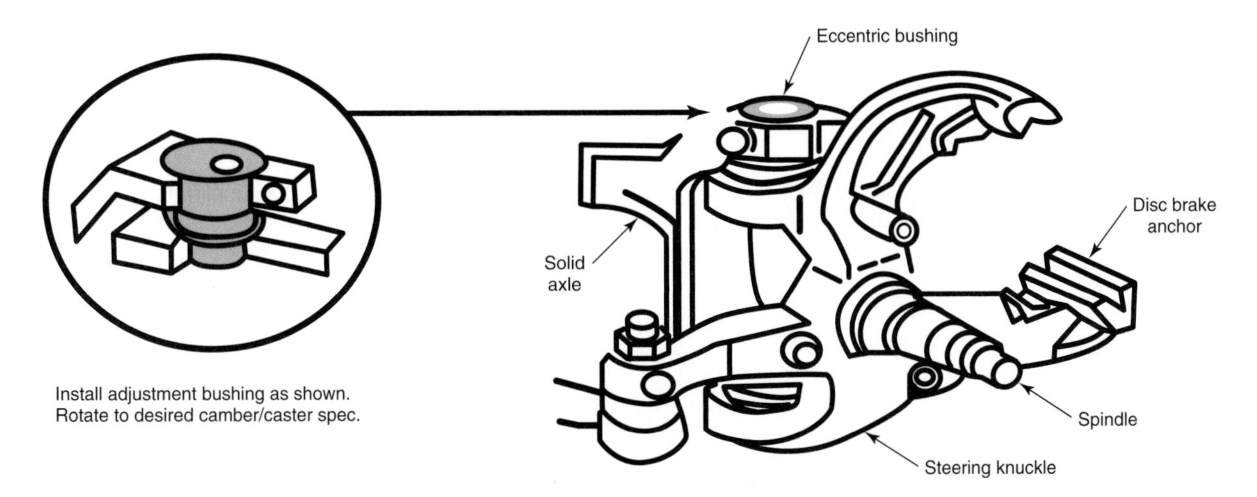

Figure 39-37. On this solid front axle the camber/caster adjustment is made with an eccentric bushing. The bushing is round but the center hole is drilled off-center. The bushing can be rotated to place the hole at a different position in relation to the front suspension. (Hunter Engineering Co.)

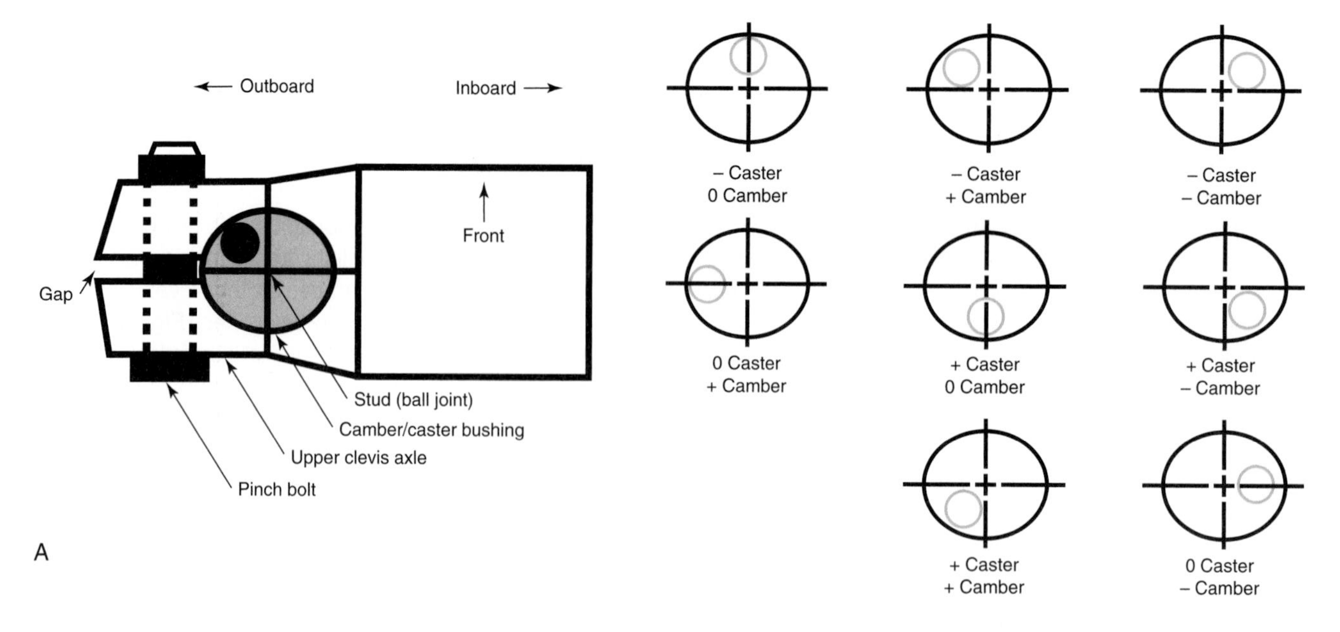

On higher degree bushings, it may be necessary to widen the gap of the pinch bolt assembly slightly for ease of installation.

2WD & 4 WD Applications	Bushing I.D. Number	+ or - Degrees of Adjustment: 2WD Max. Camber or Caster	2WD Combined Camber & Caster	4 WD Max. Camber or Caster	4 WD Combined Camber & Caster
FORD 2WD: 87 & Up F150, F250, & F350; 89-90 Bronco II; 91 & Up Explorer; 89 & Up Ranger	HR-1/4	.25 (1/4)	.2 (3/16)	.2 (3/16)	.1 (1/8)
	HR-1/2	.50 (1/2)	.3 (5/16)	.4 (3/8)	.2 (1/4)
	HR-3/4	.75 (3/4)	.5 (1/2)	.6 (5/8)	.4 (3/8)
	HR-1	1.00 (1)	.7 (11/16)	.7 (3/4)	.5 (1/2)
FORD 4 WD: Mid 90 Bronco II; 90 & Up Explorer; Mid 90 & Up Ranger (All three with pinch bolt style axle)	HR-1-1/4	1.25 (1-1/4)	.9 (7/8)	.9 (7/8)	.6 (5/8)
	HR-1-1/2	1.50 (1-1/2)	1.1 (1-1/16)	1.0 (1)	.7 (3/4)
	HR-1-3/4	1.75 (1-3/4)	1.2 (1-3/16)	1.2 (1-3/16)	.9 (7/8)
	HR-2	2.00 (2)	1.4 (1-3/8)	1.4 (1-3/8)	1.0 (1)
MAZDA 4 WD: 91 & Up Navajo	EX -1-3/4	DO NOT INSTALL		1.75 (1-3/4)	1.1 (1-1/8)
	EX-2	DO NOT INSTALL		2.00 (2)	1.3 (1-1/4)

B

Figure 39-38. A–Caster and camber bushing adjustment positions. B–One type of bushing selection chart for both two- and four-wheel drive vehicles. (Hunter Engineering Co.)

the eccentric bushing in its place, **Figure 39-39.** Be sure that the eccentric bushing is turned to where it will make the proper change. Reinstall the ball joint and wheel and recheck the caster and camber.

Checking Steering Axis Inclination

Although steering axis inclination (SAI) cannot be adjusted, it should be checked whenever another cause cannot be found for a handling or tire wear problem. Keep the brakes locked while checking inclination. Most alignment machines check SAI as the technician turns the front wheels, similar to the method of checking caster. Some alignment machines are able to check caster and SAI at the same time. Camber, caster, and steering axis inclination are illustrated in **Figure 39-40.**

Checking Toe

If the vehicle has power steering, start the engine. Turn the steering wheel from side to side, and then center it. Stop the engine and lock the steering wheel in the centered position. Then loosen the tie rod adjusting-sleeve clamp bolts (conventional steering, **Figure 39-41**) or locknuts (rack and pinion steering, **Figure 39-42**). Adjust toe by turning the sleeves or rods to obtain exactly half of the needed toe on each wheel. Unlock the steering wheel, turn it from side to side, and then recenter it. Recheck the toe on each side and readjust as needed. Some vehicles have only one sleeve, in which case the sleeve must be adjusted to provide the entire amount of toe needed. If the vehicle has only one sleeve, the steering wheel cannot be centered. At least one vehicle has separate sleeves for setting toe and centering the steering wheel.

Caution: On some vehicles with conventional steering, the clamp bolts on the tie rod adjusting sleeve must be in a specific location in relation to the tie rod's top or front surface. This prevents interference with other parts. The toe-in adjustment is critical. An improper toe-in setting can severely wear the tire tread in a very short period of driving.

Follow the manufacturer's recommendations for proper clamp positioning. If none are available, study the steering linkage and be certain no interference is possible. Figure 39-41 illustrates the proper positioning for one particular setup. Sleeve clamps must be over the threaded rods and must be torqued. The sleeve should be centered over the tie rod and tie rod socket shaft.

Warning: A loose tie rod sleeve, an improperly positioned clamp, a loose fastener, or a poorly centered sleeve can cause an accident. Double-check all fasteners and the position of all parts.

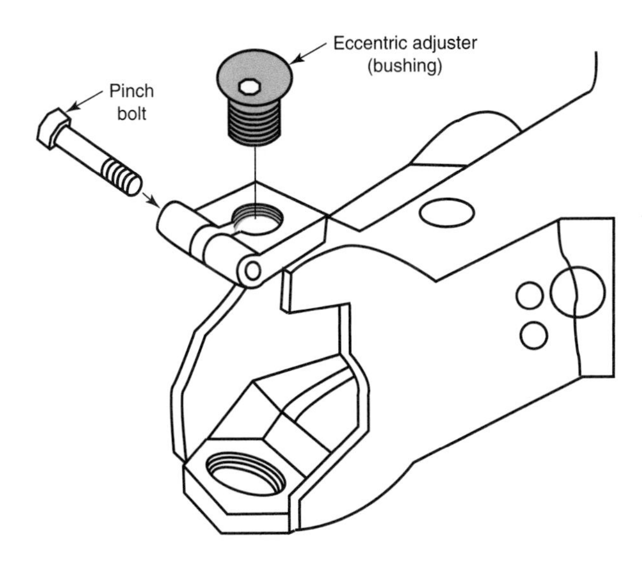

Figure 39-39. Installing an eccentric adjuster bushing into correct position before reinstalling the ball joint. The hole is drilled off-center to change the alignment. (Hunter Engineering Co.)

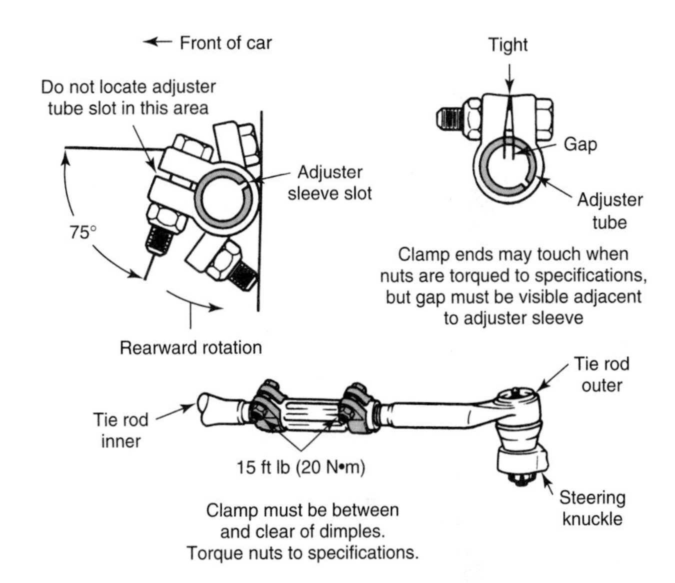

Figure 39-41. Proper positioning for the tie rod adjusting-sleeve clamp is important to keep the clamp from loosening or striking stationary parts. (General Motors)

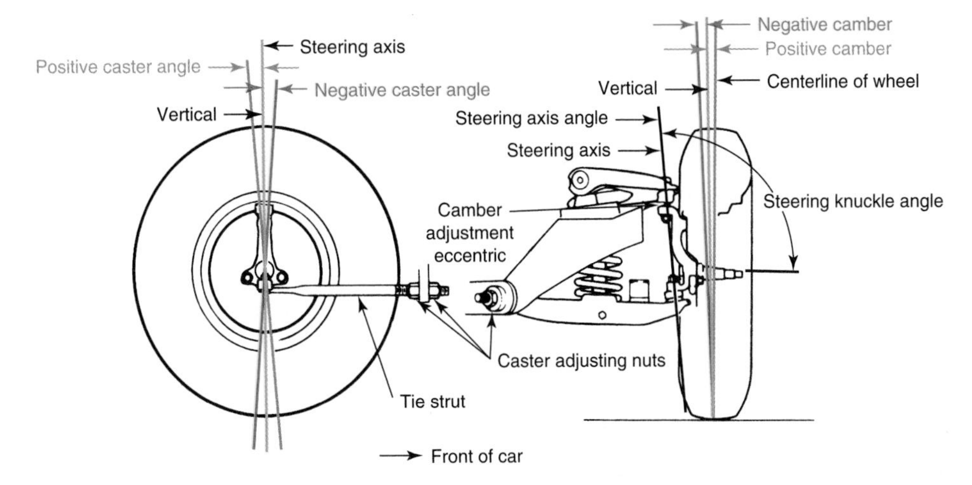

Figure 39-40. This illustration shows camber, caster, and steering axis inclination (angle). Note how the camber and caster are adjusted on this setup. (General Motors)

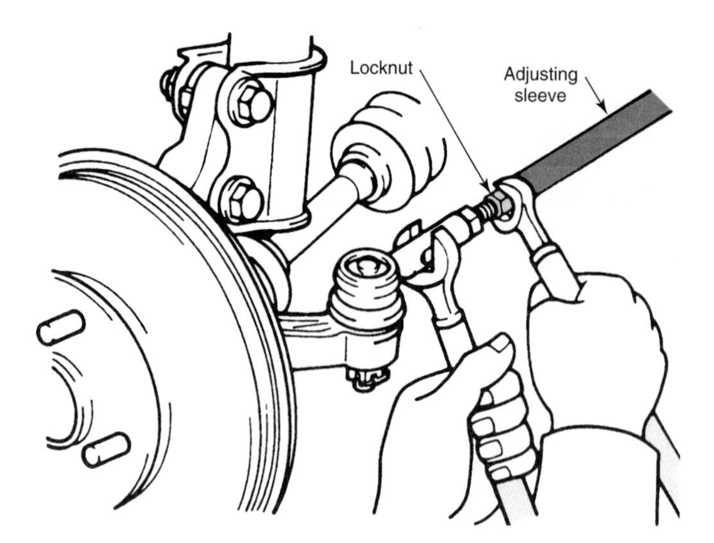

Figure 39-42. This type of front toe adjuster is common on vehicles with rack-and-pinion steering gears. Turn the adjusting sleeve as specified in the service manual. (General Motors)

When toe is correct, turn both connecting rod sleeves downward to adjust spoke position

When toe is correct, turn both connecting rod sleeves upward to adjust spoke position

Steering wheel centerline

When toe is not correct, lengthen left rod to decrease toe-in.

Shorten right rod to increase toe-in.

When toe is not correct, shorten left rod to increase toe-in.

Lengthen right rod to decrease toe-in.

Adjust both rods equally to maintain normal spoke position

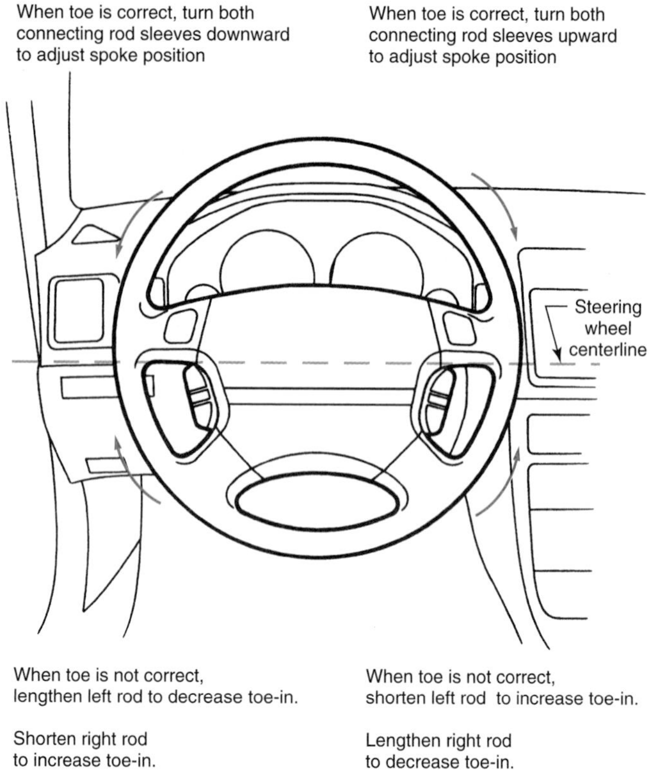

Figure 39-43. Correct the steering position by the adjusting tie rods. The steering wheel should be centered when the wheels are pointing straight ahead. (Honda)

When the toe-in is adjusted correctly, the steering wheel should be straight when the wheels are in the straight-ahead position. If the steering wheel is off-center, it may be straightened by turning both tie rod sleeves in the same direction until the steering wheel is aligned. Spoke positioning for the steering wheel of one vehicle is shown in **Figure 39-43.** When the front-end is properly aligned and the front wheels are in the straight-ahead position, the steering wheel should be straight. After finishing the alignment, confirm the steering wheel position by driving the vehicle.

Checking Toe-Out on Turns

Leave the brake pedal depressor applied. Turntables must be set with the locking pins removed. Wheels should be in the straight-ahead position. Turn the front of the right wheel inward until the turntable dial reads exactly 20°. Read the indicator on the left wheel. This reading is the angle of toe-out for the left wheel. It should be slightly more than 20°.

Turn the front of the left wheel inward 20° and read the indicator on the right wheel turntable. This is the angle of toe-out for the right wheel. Compare with specifications. Turning radius or angle of toe-out is built into the steering arms. If the angles are not according to specifications, the steering arms are bent and must be replaced. **Figure 39-44** illustrates the use of a turntable in checking turning radius. **Figure 39-45** illustrates a setup for checking toe-out on turns with a computerized alignment machine.

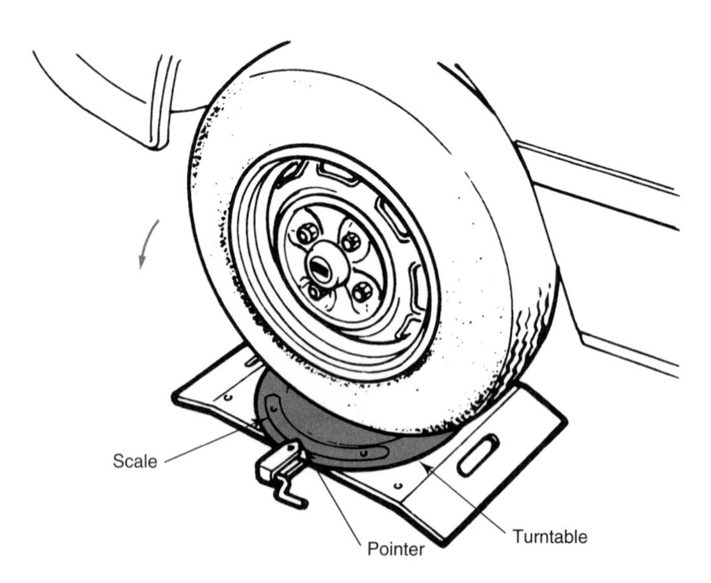

Figure 39-44. Vehicle turning radius must be checked on a turntable. The reading is obtained with the pointer and scale. (Mazda)

Final Alignment Check

After completing the alignment, give the vehicle a final inspection. In addition to repeating the prealignment inspection, make certain that vehicle curb height is as specified. Check that caster, camber, and toe-in are correct. Check that the steering axis inclination is as required and that toe-out on turns meets specifications. Be sure that all fasteners are tight after the alignment is complete. Road test the vehicle after adjusting the wheel alignment. If the vehicle is aligned correctly, it will not pull to the right or left and will not wander excessively. The steering wheel will be straight when driving straight forward.

Figure 39-45. This vehicle is being checked for toe-out with a computerized alignment machine. Follow the manufacturer's operating instructions. Use care when handling the machine. (Hunter Engineering Co.)

Tech Talk

In most cases, it is not obvious that a vehicle has been aligned. The vehicle will probably handle about the same as it did when it was brought in. Sometimes, the customer may not believe that you actually aligned the vehicle. Even worse, you may get blamed for a problem elsewhere on the vehicle or, in a large shop with many alignment technicians, someone else's incorrect alignment.

After making alignment adjustments, always place a mark on the front suspension, always in the same general location. A can of paint in an unusual color or mechanic's crayon works well. This will let you know that you worked on the vehicle the next time that it is in the shop. Also write down all alignment readings for that particular vehicle—before and after alignment adjustments. You may wish to record the information in a small stenographer's notebook, which can be kept out of the way in your toolbox. This information will help you diagnose a comeback problem.

Finally, always check that the steering wheel is straight during the road test, and straighten it if it is crooked. Most drivers have no idea what was done to align the suspension of their vehicle, but they can't help but notice a crooked steering wheel.

Summary

Wheels must be properly aligned to produce good steering, handling, and tire wear. Alignment angles are caster, camber, toe-in, steering axis inclination, and toe-out on turns. Wheels must also track correctly.

Caster (top ball joint forward or backward) may be specified as negative (tipped to front) or positive (tipped to rear). Camber (top of wheel in or out) may be specified as negative (tipped inward) or positive (tipped outward). Toe may be in or out depending on the vehicle. Toe is critical to proper tire wear. It must be carefully checked and corrected. Toe-out on turns allows the inner wheel to turn more sharply than the outer so that it may move about a smaller radius without the tires scrubbing. Toe-out on turns is determined by the angle of the steering arms. Steering axis inclination (tipping top ball joint inward) places the load nearer the tire-road contact area. This makes steering easier and reduces the need for excessive camber. Steering axis inclination is not adjustable.

Many types of alignment-checking equipment are available. A modern alignment machine must be used to perform a four-wheel alignment. Before checking wheel alignment, make the recommended prealignment inspections. Check for worn or damaged parts and replace any that are not up to specifications. Check vehicle ride height and curb weight. Tires must be in good condition and match by type, amount of wear, and size. Air pressure must be correct in all tires.

Vehicle manufacturers have devised many methods to adjust caster and camber. Never heat or bend suspension or steering parts to make changes. Special shims and bushings are available to make extra adjustments. More positive camber may be adjusted into the left-front wheel to compensate for road crown. Caster can also be varied in the left wheel to handle the normal road crown. Use quality alignment equipment in proper condition. Make all checks and adjustments carefully and to exact specifications.

When the alignment is complete, all angles should match factory specifications. The steering gear must be at midpoint in its travel. The steering wheel spokes should be in the proper position. Always road test the vehicle before and after a wheel alignment. If the vehicle is aligned correctly, it will not pull to the right or left and will not wander excessively. The steering wheel will be straight when driving straight forward.

Review Questions—Chapter 39

Do not write in this book. Write your answers on a separate sheet of paper.

1. Camber can be changed by _____.
   (A) tipping the top ball joint in or out
   (B) tipping the top of the tire in or out
   (C) tipping both the ball joints inward
   (D) Both A & B.
2. Performing a four-wheel alignment reduces the chance of _____.
3. Curb _____ and ride _____ should be as specified before aligning front wheels.
4. Vehicles with front-wheel drive or independent rear suspension systems require _____ and _____ checks at the rear wheels.
5. The pull created by road crown is best offset by _____.
   (A) running the left tire with less pressure
   (B) increasing negative camber in the left front wheel
   (C) increasing positive camber in the left front wheel
   (D) increasing toe-in

6. Steering and suspension parts must never be _____.
   (A) bent
   (B) welded
   (C) heated
   (D) All of the above.
7. Steering axis inclination and camber are added together to form the _____.
8. Vehicle toe must be set _____.
   (A) in
   (B) out
   (C) at zero
   (D) Any of the above, depending on the vehicle.
9. When doing a four-wheel alignment, which of the following angles must be set last?
   (A) Front camber.
   (B) Rear camber.
   (C) Front toe.
   (D) Rear toe.
10. Compensating the heads removes the effect of runout in the _____,
    (A) bearings
    (B) control arms
    (C) wheel rim
    (D) All of the above.

ASE-Type Questions

1. Technician A says that caster angle involves tipping the top of the tire in or out. Technician B says that caster is an imaginary line through the wheel spindle. Who is right?
   (A) A only.
   (B) B only.
   (C) Both A & B.
   (D) Neither A nor B.
2. Technician A says that toe-in is required to offset toe-out on turns. Technician B says that toe-out on turns is adjusted by altering the length of the tie rods. Who is right?
   (A) A only.
   (B) B only.
   (C) Both A & B.
   (D) Neither A nor B.
3. When doing a four-wheel alignment, which of the following angles must be set first?
   (A) Front camber.
   (B) Rear camber.
   (C) Front toe.
   (D) Rear toe.
4. Technician A says that toe-out on turns is adjustable. Technician B says that steering axis inclination cannot be adjusted. Who is right?
   (A) A only.
   (B) B only.
   (C) Both A & B.
   (D) Neither A nor B.
5. Technician A says that shims are often used to set rear toe. Technician B says that moving the lower strut rod is a common method of setting front caster. Who is right?
   (A) A only.
   (B) B only.
   (C) Both A & B.
   (D) Neither A nor B.
6. The brake pedal depressor must be applied before checking the _____.
   (A) front caster
   (B) front toe
   (C) rear camber
   (D) rear toe
7. All of the following methods are used to set front caster and camber on modern vehicles, *except:*
   (A) shims at the control arm mounting.
   (B) eccentrics on the control arm bushings.
   (C) bending the control arm.
   (D) rivets removed and strut tower moved into position.
8. When shimming a rear axle, the shim is placed between the rear wheel bearing housing and the _____.
   (A) brake drum or rotor
   (B) axle
   (C) wheel rim
   (D) brake backing plate
9. All of the following statements about a vehicle that has been properly aligned are true, *except:*
   (A) the vehicle will pull slightly to the left.
   (B) the vehicle will not wander.
   (C) the tire wear will be minimal.
   (D) the steering wheel will be straight when driving straight forward.
10. Technician A says that, when centering the steering wheel on a vehicle with power steering, the engine should be running. Technician B says that the brake pedal lock prevents excessive movement when checking caster. Who is right?
    (A) A only.
    (B) B only.
    (C) Both A & B.
    (D) Neither A nor B.

Suggested Activities

1. Inspect at least three vehicles and determine the alignment-adjusting devices. Then, refer to the service manual to determine whether you found all of the adjusters. Write a report to the other members of your class on the things to look for when searching out alignment devices.
2. Look up alignment specifications for a particular make of vehicle over the last 10 years in a manual and determine the range of settings, from highest to lowest. Be sure to include the following:
   - Caster in degrees.
   - Camber in degrees.
   - Toe in fractions of an inch.

   Also include the following, if applicable:
   - Rear camber in degrees.
   - Rear toe in fractions of an inch.
3. Perform a two-wheel alignment on a rear-wheel drive vehicle. Write a short report on how you aligned the vehicle, and what adjustments were needed. Also record the alignment readings before and after the alignment adjustments are made.
4. Perform a complete four-wheel alignment on a front-wheel drive vehicle. Write a short report on how you aligned the vehicle, the alignment readings before and after, and what adjustments were needed.
5. Research aftermarket (non-factory) devices that can provide additional alignment adjustment, and report to the other members of your class on how these devices are used to gain additional adjustment on front and rear wheels.

Wheel Balance and Alignment Problem Diagnosis	
Problem: Wheel tramp	
Possible cause	**Correction**
1. Brake drum, rotor, wheel, or tire out of static balance.	1. Balance assembly statically and dynamically.
2. Wheel or tire out-of-round (excessive radial runout).	2. Change tire position on wheel or discard tire or wheel as needed.
3. Defective shock absorbers.	3. Replace shocks.
4. Bulge on tire.	4. Replace tire.
5. Defective front stabilizer.	5. Replace stabilizer.
6. Loose or worn wheel bearings.	6. Adjust or replace bearings.
7. Defective MacPherson strut.	7. Replace strut.
Problem: Wheel shimmy	
Possible cause	**Correction**
1. Wheel and tire assembly out of dynamic balance.	1. Balance assembly statically and dynamically.
2. Tire pressure uneven.	2. Inflate both front tires to same pressure.
3. Worn or loose front wheel bearings.	3. Adjust or replace bearings.
4. Defective shock absorbers or struts.	4. Replace shocks or struts.
5. Improper or uneven caster.	5. Adjust caster angle.
6. Excessive tire or wheel runout.	6. Correct by moving tire on rim or replace defective parts.
7. Abnormally worn tires.	7. Move to rear if still serviceable.
8. Improper toe-in.	8. Adjust to specifications.
9. Defective stabilizer bar.	9. Replace stabilizer.
10. Tire pressure too low.	10. Inflate tires to correct pressure.
11. Loose wheel lugs.	11. Tighten lugs.
12. Front end misaligned.	12. Align front end.
13. Bent wheel.	13. Replace wheel.
Problem: Poor recovery following turns and/or hard steering	
Possible cause	**Correction**
1. Low tire pressure.	1. Inflate to proper pressure.
2. Lack of lubrication.	2. Lubricate steering system.
3. Front wheels misaligned.	3. Align front wheels properly.
4. Bent spindle assembly.	4. Replace spindle assembly.
Problem: Vehicle pulls to one side	
Possible cause	**Correction**
1. Uneven tire pressure.	1. Inflate both front tires to same pressure.
2. Improper toe-in.	2. Adjust toe-in to specifications.
3. Incorrect or uneven caster.	3. Adjust caster angle.
4. Incorrect or uneven camber.	4. Adjust camber angle.
5. Improper rear wheel tracking.	5. Align rear axle assembly.
6. Tires not same size.	6. Install same size tires on both sides.
7. Bent spindle assembly.	7. Replace spindle.
8. Worn or improperly adjusted wheel bearings.	8. Adjust or replace bearings.
9. Dragging brakes.	9. Adjust brakes.
Problem: Vehicle wanders from side to side	
Possible cause	**Correction**
1. Low or uneven tire pressure.	1. Inflate tires to recommended pressure.
2. Toe-in incorrect.	2. Adjust toe-in.
3. Improper caster.	3. Adjust caster angle.
4. Improper camber.	4. Adjust camber angle.
5. Worn or improperly adjusted front wheel bearings.	5. Replace or adjust wheel bearings.
6. Vehicle overloaded or loaded too much on one side.	6. Advise owner regarding vehicle load limits.
7. Bent spindle assembly.	7. Replace spindle.

(continued)

Wheel Balance and Alignment Problem Diagnosis *(continued)*

Problem: Tire squeal on corners	
Possible cause	**Correction**
1. Low tire pressure. 2. Toe-out on turns incorrect. 3. Excessive cornering speed. 4. Bent spindle assembly. 5. Improper front end alignment.	1. Inflate to recommended pressure. 2. Replace bent steering arm. 3. Advise driver. 4. Replace spindle. 5. Align front wheels.
Problem: Loose, erratic steering	
Possible cause	**Correction**
1. Loose front wheel bearings. 2. Loose wheel lugs. 3. Wheel out of balance.	1. Replace or adjust. 2. Tighten lugs. 3. Balance wheel assembly.
Problem: Hard riding	
Possible cause	**Correction**
1. Excessive tire pressure. 2. Improper tire size. 3. Heavy-duty shock absorbers installed.	1. Reduce pressure to specifications. 2. Install correct size. 3. Advise driver and/or change shocks.
Problem: Improper wheel tracking	
Possible cause	**Correction**
1. Frame sprung. 2. Rear axle housing sprung. 3. Broken leaf spring. 4. Broken spring center bolt; spring shifted on axle housing. 5. Wheels misaligned.	1. Straighten frame. 2. Replace or straighten housing. 3. Replace spring. 4. Install new spring center bolt. 5. Align all wheels.
Problem: Noise from front or rear wheels	
Possible cause	**Correction**
1. Wheel lugs loose. 2. Defective wheel bearings. 3. Loose wheel bearings. 4. Lack of lubrication. 5. Lump or bulge on tire tread. 6. Rock or debris stuck in tire tread. 7. Cracked wheel. 8. Wheel hub loose on axle taper (where used). 9. Wheel bearing worn or defective.	1. Tighten lugs. 2. Replace wheel bearings. 3. Adjust wheel bearings. 4. Lubricate bearings. 5. Replace tire. 6. Remove rock or debris. 7. Replace wheel. 8. Inspect and tighten. 9. Replace wheel bearing.
Problem: Tires lose air	
Possible cause	**Correction**
1. Puncture. 2. Bent, dirty, or rusty rim flanges. 3. Loose wheel-rim rivets. 4. Leaking valve core or stem. 5. Striking curbs with excessive force. 6. Flaw in tire casing. 7. Excessive cornering speed especially with low tire pressure. 8. Porous wheel rim.	1. Repair puncture. 2. Clean or replace wheel. 3. Peen rivets. 4. Replace as needed. 5. Advise driver. 6. Repair or replace tire. 7. Advise driver. 8. Repair porous wheel rim.

(continued)

Wheel Balance and Alignment Problem Diagnosis *(continued)*	
Problem: Tire wears in center	
Possible cause	**Correction**
1. Excessive pressure.	1. Reduce tire pressure to specifications.
Problem: Tire wears on one edge	
Possible cause	**Correction**
1. Improper camber. 2. High speed cornering.	1. Align camber angle. 2. Advise driver.
Problem: Tire wears on both sides	
Possible cause	**Correction**
1. Low pressure. 2. Overloading vehicle.	1. Inflate tires to specifications. 2. Advise driver.
Problem: Tire scuffing or feather edging	
Possible cause	**Correction**
1. Excessive toe-out (inside edges). 2. Excessive toe-in (outside edges). 3. Excessive cornering speed. 4. Improper tire pressure. 5. Wheel shimmy. 6. Improper toe-out on turns. 7. Excessive runout. 8. Improper camber. 9. Bent spindle assembly.	1. Correct toe-out. 2. Correct toe-in. 3. Advise driver. 4. Inflate tires to specifications. 5. Balance wheels statically and dynamically. 6. Replace bent steering arm. 7. Correct or replace tire or wheel. 8. Adjust camber angle. 9. Replace spindle.
Problem: Tire cupping	
Possible cause	**Correction**
1. Uneven camber. 2. Bent spindle assembly. 3. Improper toe-in. 4. Improper tire pressure. 5. Excessive runout. 6. Wheel and tire assembly out of balance. 7. Worn or improperly adjusted wheel bearings. 8. Grabby brakes.	1. Correct camber angle. 2. Replace spindle. 3. Adjust toe-in. 4. Inflate tires to specifications. 5. Correct or replace wheel or tire. 6. Balance both statically and dynamically. 7. Replace or adjust wheel bearings. 8. Repair brakes.
Problem: Heel and toe wear	
Possible cause	**Correction**
1. Grabby brakes. 2. Heavy acceleration.	1. Repair brakes. 2. Advise driver.

40

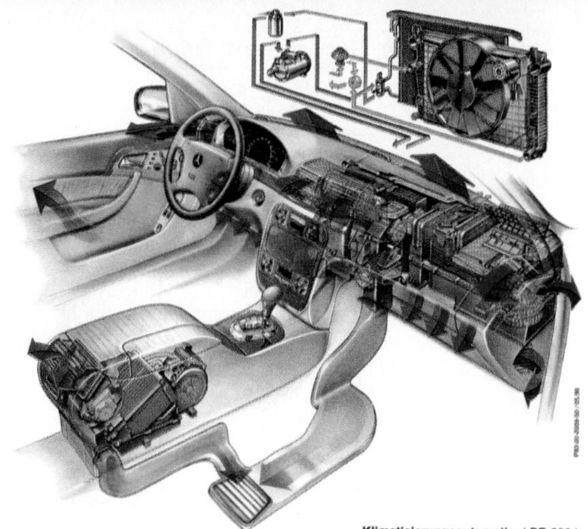

Air Conditioning and Heater Service

After studying this chapter, you will be able to:

- Explain basic refrigeration theory.
- Identify major air conditioning system components.
- List the safety rules for air conditioning service.
- Inspect an air conditioning system for problems.
- Recover, evacuate, and recharge an air conditioning system.
- Service air conditioning and heater parts.
- Install air conditioning system parts.
- Service refrigerant oil.
- Diagnose air conditioning and heating problems.

Technical Terms

Air conditioning
Orifice tube
Cycling clutch orifice tube (CCOT)
Fixed orifice tube (FOT)
Ford orifice tube cycling clutch (FOTCC)
Variable displacement orifice tube (VDOT)
Thermostatic expansion valve (TXV)
Pressure-temperature relationship
Refrigerant
Boiling point
R-134a
Tetrafluoroethane (CF_3CH_2F)
One-pound can
R-12
Freon
Dichlorodifluoromethane (CCl_2F_2)
Blended refrigerants
R-22
Chlorofluorocarbons (CFCs)
Chlorine (Cl)
Montreal Protocol
Clean Air Act
Environmental Protection Agency (EPA)
Significant New Alternatives Policy
SNAP program
Refrigerant service centers
Service hoses
Manifold gauge
Drop-in refrigerant
Refrigerant oil
Miscible
Hygroscopic
Compressor
Compressor clutch
Condenser
Flow restrictor
Fixed orifice tube
Evaporator
Interior air filter
Lines
Hoses
Muffler
Receiver-drier
Sight glass
Accumulator
Service fittings
Push-on service fittings
Screw-on service fittings
Schrader valve
Manual service valves
Evaporator pressure control
Pressure cycling switch
Thermostatic temperature cycling switch
Pressure cutoff switches
Cut-out switches
High-pressure cutoff switch
Low-pressure cutoff switch
Temperature sensors
Thermistors
High temperature sensor
Ambient air sensor
Pressure relief valve
Fusible plug
Phosgene gas
Functional test
Performance test
Refrigerant identifier
Electronic leak detector
Discharging
Recover
Refrigerant recycling
Evacuating
Vacuum pump
Purging
Flushing
Barrier hose
Biocide
Retrofitting
Heater core
Shut-off valve
Blend door
Dual Zone

Originally, automobile air conditioners and heaters were separate from each other and were manually controlled by the vehicle's occupants. On modern vehicles, the air conditioner and heater are combined into one system that provides air at whatever temperature is desired. Some systems automatically respond to changes in outside temperature, humidity, and sun load. This chapter will cover the principles of both air conditioning and heating.

Note: The air conditioning service procedures in this chapter are general in nature. Before starting work on any air conditioning system, be sure to obtain a service manual for the vehicle. You should also review all up-to-date materials regarding tools, refrigerant handling, and service procedures.

Principles of Air Conditioning

Air conditioning is a process in which air entering the vehicle is cooled, cleaned, and dehumidified. The parts of a typical air conditioning system are shown in **Figure 40-1.** The

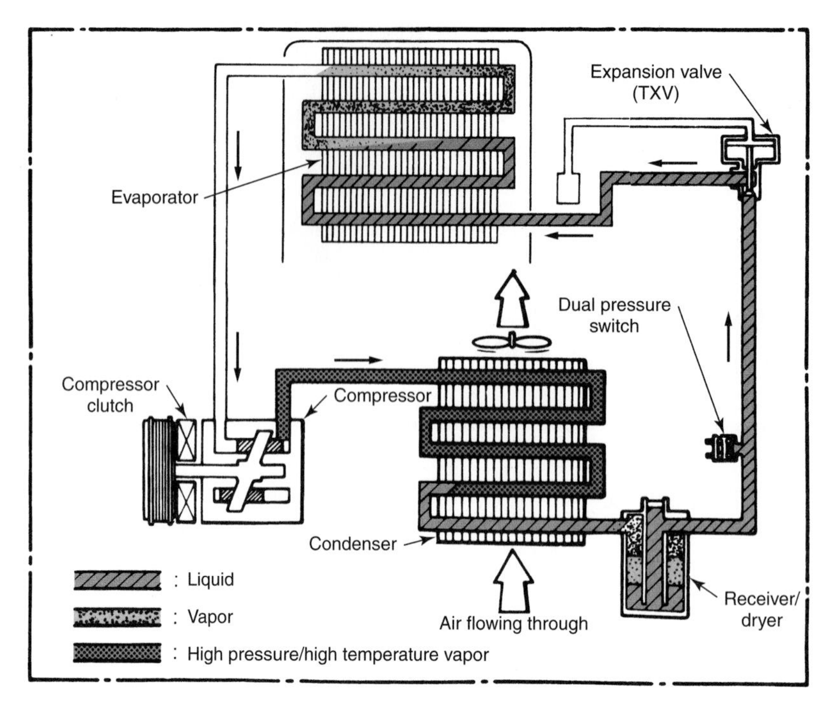

Figure 40-1. The parts of a typical automotive air conditioning system. (General Motors)

various parts of the system are connected by tubing and flexible hose. The system contains a charge of refrigerant, which provides the actual cooling effect. The job of the air conditioning system is to vary pressures so the refrigerant will absorb heat and change from a liquid to a vapor (boil) at one point in the system and condense from a vapor to a liquid, giving up heat, in another part of the system.

Types of Air Conditioning Systems

There are two major types of air conditioning systems. They are classified by the type of restrictor used to control refrigerant flow between the condenser and evaporator. One type of restrictor is called an ***orifice tube.*** The opening in orifice tube systems is not variable, so the flow of refrigerant is varied by controlling the operation of the compressor. On some orifice systems, the compressor is turned on and off. This is called a ***cycling clutch orifice tube (CCOT)*** system. Some systems using orifice tubes are called ***fixed orifice tube (FOT), Ford orifice tube cycling clutch (FOTCC),*** thermostatic switch systems, pressure switch systems, or accumulator systems. On other systems, the pumping capacity of the compressor is varied as needed. These are termed ***variable displacement orifice tube (VDOT)*** systems.

The second type of restrictor is the ***thermostatic expansion valve (TXV).*** In this system, the entrance of liquid refrigerant into the evaporator is controlled by varying the opening of the expansion valve, based on the temperature at the evaporator outlet.

Air Conditioning Cycle

There is a distinct ***pressure-temperature relationship*** for refrigerants. When the system pressure is lowered, the refrigerant boiling point decreases, turning liquid refrigerant into a vapor. As system pressure rises, the refrigerant boiling point is increased, so vaporized refrigerant returns to a liquid state.

Figure 40-2 illustrates the refrigerant state in various parts of the system during the air conditioning cycle. With the system operating, high-pressure liquid refrigerant collects on the *discharge* or *high-pressure* side of the system. The refrigerant moves through a restrictor into the evaporator. The restrictor (orifice tube or TXV) causes the liquid refrigerant to enter the evaporator at low pressure. This lowered pressure decreases the refrigerant's boiling point. This side of the refrigeration system is referred to as the *suction* or *low-pressure* side.

Warm air is forced through the evaporator coils by the blower motor, giving up heat to the refrigerant. As more heat from the air is absorbed, the refrigerant begins to boil, turning into a vapor. By the time the refrigerant reaches the evaporator outlet, it is completely vaporized. The air (which has given up its heat to the refrigerant) cools the passenger compartment. From the evaporator, refrigerant vapor is drawn into the compressor. This results in an increase in vapor pressure and a rapid rise in refrigerant temperature. The vapor is pumped to the condenser under high pressure, sometimes as much as 400 psi (2400 kPa). This raises the boiling point of the refrigerant to a temperature higher than the outside air.

In the condenser, the hot, high-pressure refrigerant vapor gives up its heat to the stream of outside air moving over the condenser fins. Air is forced through the condenser by vehicle movement, or by the radiator cooling fan when the vehicle is stationary. The change in temperature causes the refrigerant to condense, or return to its liquid state. Still under

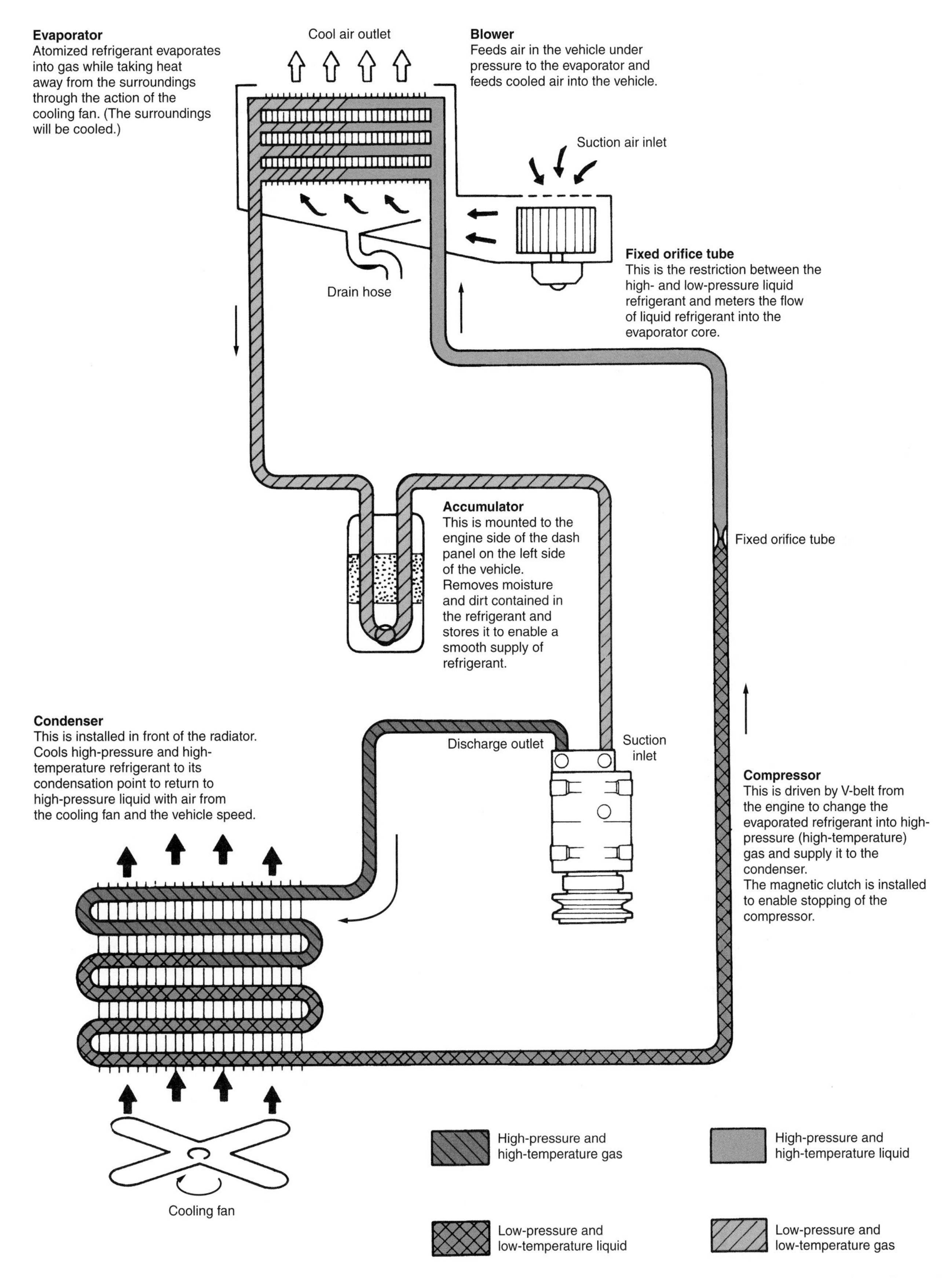

Figure 40-2. The automotive refrigeration cycle. Note how the refrigerant changes state in various parts of the system. (Hyundai)

pressure from the compressor, the liquid refrigerant flows to the restrictor for another cycle through the air conditioning system.

Refrigerants

A ***refrigerant*** is a compound of chemical elements (such as carbon, hydrogen, fluorine, and in some cases, chlorine) that is able to readily change its state from a liquid to a vapor or from a vapor to a liquid. By undergoing these changes in state, a refrigerant can provide a considerable cooling effect by absorbing and releasing large amounts of heat in relation to its volume. The refrigerant's ***boiling point,*** or temperature at which it will change into a vapor, can be varied by pressure changes in the air conditioning system.

Refrigerant is used over and over in the system and will maintain its efficiency indefinitely unless contaminated with dirt, water, or air. It is colorless in both the vapor and liquid state. Refrigerant is nonpoisonous, except when in direct contact with an open flame. Unless combined with moisture, it is noncorrosive. It is heavier than air and will become a vapor when released into the atmosphere.

R-134a

While there are many refrigerants on the market today, only one meets all manufacturers' requirements for motor vehicle use. That refrigerant is ***R-134a,*** sometimes called ***HFC-134a.*** Its chemical name and formula is ***tetrafluoroethane*** **(CF_3CH_2F)**.

R-134a was installed in all OEM vehicle air conditioners beginning with the 1994 model year. The molecular structure of R-134a does not include a chlorine atom, so it does not contribute to ozone layer depletion. While R-134a does have some effect on global warming, it is many times less damaging than R-12. **Figure 40-3** is a pressure-temperature chart for R-134a. This refrigerant is supplied in 30-pound (13.5 kg) cylinders, or in one-pound cans, **Figure 40-4.** All R-134a cylinders are painted light blue for easy identification.

R-134a Temperature-Pressure Chart

Temperature °F (°C)	Pressure psi (kPa)	Temperature °F (°C)	Pressure psi (kPa)
16 (–9)	15 (106)	100 (38)	124 (857)
18 (–8)	17 (115)	102 (39)	129 (887)
20 (–7)	18 (124)	104 (40)	133 (917)
22 (–6)	19 (134)	106 (41)	137 (948)
24 (–4)	21 (144)	108 (42)	142 (980)
26 (–3)	22 (155)	110 (43)	147 (1012)
28 (–2)	24 (166)	112 (44)	152 (1045)
30 (–1)	26 (177)	114 (46)	157 (1079)
32 (0)	27 (188)	116 (47)	162 (1114)
34 (1)	29 (200)	118 (48)	167 (1149)
36 (2)	31 (212)	120 (49)	172 (1185)
38 (3)	33 (225)	122 (50)	177 (1222)
40 (4)	35 (238)	124 (51)	183 (1260)
45 (7)	40 (272)	126 (52)	188 (1298)
50 (10)	45 (310)	128 (53)	194 (1337)
55 (13)	51 (350)	130 (54)	200 (1377)
60 (16)	57 (392)	135 (57)	215 (1481)
65 (18)	64 (438)	140 (60)	231 (1590)
70 (21)	71 (487)	145 (63)	247 (1704)
75 (24)	78 (540)	150 (66)	264 (1823)
80 (27)	88 (609)	155 (68)	283 (1948)
85 (30)	95 (655)	160 (71)	301 (2079)
90 (32)	104 (718)	165 (74)	321 (2215)
95 (35)	114 (786)	170 (77)	342 (2358)

Evaporator range: 19 (134) – 45 (310)
Condenser range: 147 (1012) – 301 (2079)

Figure 40-3. A pressure-temperature relationship chart for R-134a refrigerant. (General Motors)

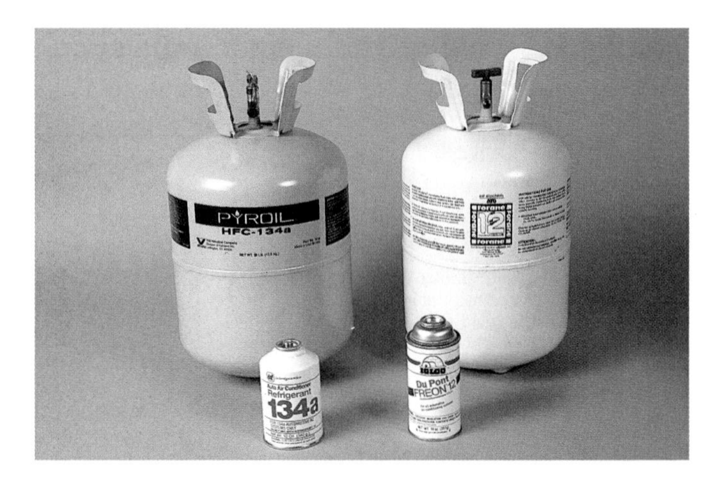

Figure 40-4. Refrigerants are packaged in 30 lb. cylinders and in one-pound cans. Shown are containers for R-134a, at left, and R-12, at right.

Note: The term *one-pound can* is used to describe small cans of refrigerant. However, a one-pound can almost never contains one pound of refrigerant. In terms of actual content, a "one pound can" of R-134a contains 12 ounces (340 g) of refrigerant while R-12 cans contain 14 ounces (397 g).

R-12 and Blended Refrigerants

R-12 or ***CFC-12*** (usually called ***Freon***) was the only automotive refrigerant for many years. Since it contains chlorine, which contributes to ozone depletion, R-12 is no longer used. It was phased out of new vehicles beginning with the 1994 model year. The chemical name and formula for R-12 is ***dichlorodifluoromethane*** **(CCl_2F_2).** R-12 is supplied in 30-pound (13.5 kg) and larger cylinders, painted white. One-pound cans may still be found in some areas.

A pressure-temperature chart for R-12 is shown in **Figure 40-5.** If you compare pressure and temperature figures, you might note R-12 is a little more efficient than R-134a. R-12 evaporates at a slightly higher pressure and condenses at a slightly lower pressure than R-134a. The manufacture and importing of R-12 in the United States and most other countries ceased on December 31, 1995, in accordance with the Montreal Protocol and the Clean Air Act.

Caution: Several developing countries were granted an extension on the R-12 manufacturing ban. This led to the smuggling of R-12 refrigerant. Some confiscated caches of smuggled R-12 have been found to be contaminated with water and other impurities. Be sure any R-12 you use comes from reputable sources.

R-12 Temperature-Pressure Chart			
Temperature °F (°C)	Pressure psi (kPa)	Temperature °F (°C)	Pressure psi (kPa)
16 (–9)	18 (127)	100 (38)	117 (808)
18 (–8)	20 (136)	102 (39)	121 (833)
20 (–7)	21 (145)	104 (40)	125 (859)
22 (–6)	22 (155)	106 (41)	129 (893)
24 (–4)	24 (165)	108 (42)	133 (917)
26 (–3)	25 (175)	110 (43)	136 (940)
28 (–2)	27 (185)	112 (44)	140 (969)
30 (–1)	28 (196)	114 (46)	145 (997)
32 (0)	30 (207)	116 (47)	149 (1027)
34 (1)	32 (219)	118 (48)	153 (1057)
36 (2)	33 (230)	120 (49)	158 (1087)
38 (3)	36 (249)	122 (50)	162 (1118)
40 (4)	37 (255)	124 (51)	167 (1150)
45 (7)	44 (287)	126 (52)	171 (1182)
50 (10)	47 (322)	128 (53)	176 (1215)
55 (13)	52 (359)	130 (54)	181 (1248)
60 (16)	58 (398)	135 (57)	194 (1334)
65 (18)	64 (440)	140 (60)	207 (1425)
70 (21)	70 (484)	145 (63)	220 (1519)
75 (24)	77 (531)	150 (66)	235 (1618)
80 (27)	84 (580)	155 (68)	250 (1721)
85 (30)	92 (633)	160 (71)	265 (1828)
90 (32)	100 (688)	165 (74)	281 (1940)
95 (35)	108 (746)	170 (77)	298 (2057)

Evaporator range (shaded: 24 (165) through 47 (322)); Condenser range (shaded: 136 (940) through 281 (1940))

Figure 40-5. R-12 pressure-temperature relationship chart. (General Motors)

Blended refrigerants are sometimes used as replacement refrigerants in automotive air conditioning systems. A blended refrigerant is a combination of two or more different refrigerants and sometimes other gases to create a new refrigerant. Some blended refrigerants contain significant amounts (often from 59%–80%) of R-134a. Other refrigerant blends contain ***R-22,*** which is used in commercial cooling applications.

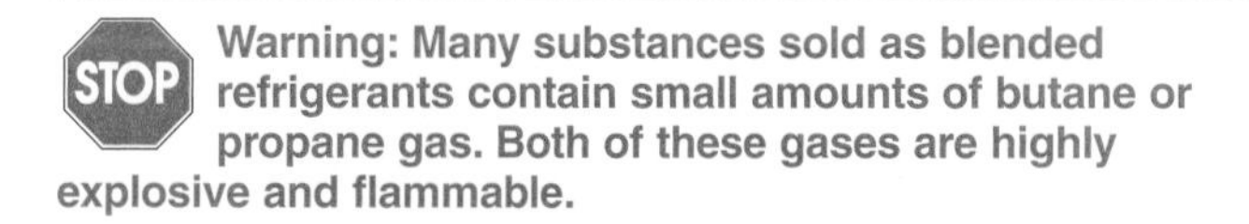
Warning: Many substances sold as blended refrigerants contain small amounts of butane or propane gas. Both of these gases are highly explosive and flammable.

Refrigerants and Governmental Regulation

In the 1970s, scientists discovered that chemicals called ***chlorofluorocarbons*** or ***CFCs*** were damaging the ozone layer, which protects the earth from harmful solar radiation. It was found that CFCs were breaking down into their component parts, including the element ***chlorine (Cl).*** As the chlorine reached the upper atmosphere, it was combining with oxygen atoms, depleting the ozone layer.

While most nations quickly banned CFC usage in spray cans, CFC compounds continued to be used throughout the world in many other applications, including the automotive refrigerant R-12. In 1987, leaders from the major industrialized nations reached and signed an international agreement called the ***Montreal Protocol.*** This agreement called for the reduction and elimination of CFC and all other ozone depleting chemicals.

Clean Air Act

A second document, called the ***Clean Air Act,*** enforces the Montreal Protocol's standards at the national level in the United States. The Clean Air Act regulates more than refrigerants; sections of the Act outline restrictions regarding the manufacture, import, release, and service of refrigerants and refrigerant systems.

A section of the Clean Air Act charges the ***Environmental Protection Agency (EPA)*** with implementation and enforcement of the regulations. It also provides for severe penalties for any infractions as well as substantial rewards, encouraging individuals and businesses to turn in violators. In cases of conflict between the Montreal Protocol, the Clean Air Act, and any other regulatory law regarding refrigerants, the more stringent law is enforced.

Warning: Businesses or technicians who violate any provision of the Clean Air Act may be fined, lose certification, and possibly face criminal charges.

SNAP Program

Under the Clean Air Act, the EPA examines all R-12 substitutes. The procedure used to review potential replacement refrigerants is called the ***Significant New Alternatives Policy*** or ***SNAP program.*** Refrigerants are classified as unacceptable if they are found to have a high global warming potential, or are ozone depleting, flammable, or toxic. The EPA updates the SNAP list regularly.

Acceptable refrigerants are subject to various requirements, but SNAP's evaluation does not determine how the replacement refrigerant will perform. The EPA has determined several refrigerants are acceptable for use as R-12 replacements, **Figure 40-6.** Using refrigerants found unacceptable under SNAP is a violation of the Clean Air Act.

Caution: All vehicle manufacturers and most after market parts suppliers do not allow the use of replacement refrigerants with their systems or parts. Use of any refrigerant other than R-134a or R-12 may void the vehicle or part warranty. EPA approval *does not* mean manufacturer approval.

Handling Refrigerants

The rules in this section for handling refrigerants apply to all mobile vehicle refrigeration systems:

- All refrigerants must be recovered; there are no exceptions. Venting refrigerants to the atmosphere is illegal.
- If your shop is not equipped to handle a particular type of refrigerant, you must treat it as contaminated refrigerant or decline the job.
- Do not mix refrigerants (for example R-134a and R-12).

Alternative Refrigerants to R-12 (From EPA SNAP list)			
Name	**Trade Name**	**Oil Compatible**	**Comments**
HFC-134a	R-134a	PAG/POE	Acceptable for all systems.
R-406A	GHG	MO	Contains R-22, must be used with barrier hoses.
GHG-X4 R-414A	GHG-X4 Autofrost, Chill-it	MO	Contains R-22, must be used with barrier hoses.
Hot Shot R-414B	Hot Shot Kar Kool	MO/POE	Contains R-22, must be used with barrier hoses.
FRIGC FR-12 (HCFC Blend Beta) R-416A	FRIGC FR-12	POE	Contains R-143a and butune, but not flammable.
Free Zone (HCFC Blend Delta)	Free Zone RB-276	MO	Contains a lubricant (approx. 2%).
Freeze 12	Freeze 12	MO	Approx. 80% R-134a.
GHG-X5	GHG-X5	MO	Contains R-22, must be used with barrier hoses.

MO–Mineral oil. PAG–Polyaclkylene glycol. POE–Polyol ester.

Figure 40-6. Replacements for R-12 that are considered acceptable by the Environmental Protection Agency.

- Do not add refrigerant to a system that has a detectable leak.
- Blended refrigerants can be recycled and reused *only* in the *original vehicle.*
- Technicians who service air conditioning systems must be certified in refrigerant recovery and recycling.

Note: In this text, unless reference is made to a specific refrigerant, the term "refrigerant," when used by itself, should be taken to mean R-134a.

Refrigerant Service Center

Every type of refrigerant must have its own dedicated service tools and equipment. Every HVAC service shop must have at least two dedicated ***refrigerant service centers,*** one for R-134a and one for R-12. These may be combined into a dual function machine, **Figure 40-7.** To use another refrigerant, the shop must purchase or convert another refrigerant service center.

Service hoses also must comply with regulations set forth in the Clean Air Act. They must be equipped with shut-off valves to prevent refrigerant blowback. Manifold gauge and service machine hoses must have a unique fitting permanently attached to the hose for each refrigerant type.

Figure 40-7. A refrigerant service center for use with either R-134a or R-12. (Robinair)

Manifold Gauge

An air conditioning ***manifold gauge*** set is used for checking system pressure, charging a system, and recovering contaminated or unknown refrigerants. It consists of two gauges set in a manifold that contains two gauge valves and three outlet connections, **Figure 40-8.** The low-pressure gauge is graduated in pounds per square inch in one direction and in inches of vacuum in the other.

The center manifold connection is common to both valves. It is used for attaching a hose for charging with refrigerant, attaching to a vacuum pump for system evacuation, or for injecting oil into the system. Separate manifold gauge sets should be on hand for R-134a, R-12, and other refrigerants.

Handling Contaminated or Unknown Refrigerants

To recover a blended refrigerant, the shop must have a refrigerant recovery and recycling machine dedicated to that

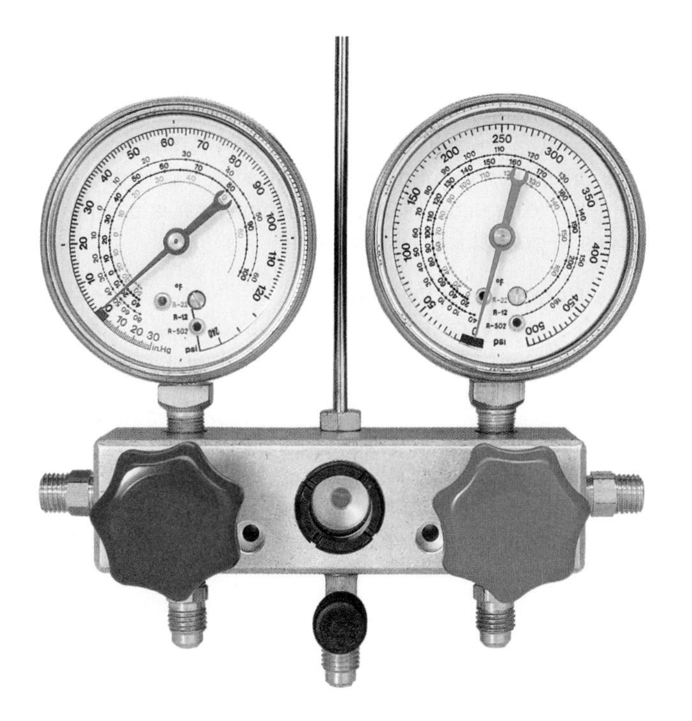

Figure 40-8. The manifold gauge set consists of high-pressure and low-pressure gauges and valves. There are three outlets—one for each valve and a common connection. (TIF Instruments)

blend. If the refrigerant is contaminated or unidentified, or if the shop does not have a machine dedicated to the blend, the refrigerant should be discharged into the shop's contaminated refrigerant container. To minimize the chance of cross-contamination, use a manifold gauge set to remove any unknown or contaminated refrigerants.

Caution: If you have no way to handle blended or contaminated refrigerants, turn down the job.

Blended refrigerants containing flammable gases such as propane or butane can create an explosion or fire hazard. These refrigerants must be handled carefully. Many recovery and recycling machines contain internal arcing and sparking parts, and should not be used to handle possibly flammable refrigerant blends. Recovery stations are available to handle such refrigerants.

The shop must have a special container for contaminated or unknown refrigerants. Contaminated refrigerant containers are painted gray with a yellow top, **Figure 40-9.** The Department of Transportation (DOT) certifies this color for interstate shipment of waste. Contaminated refrigerant should be shipped to a reclaiming facility for recycling or disposal.

Drop-In Refrigerants

The term ***drop-in refrigerant*** has been used to mean a refrigerant substitute that can be added to a partially charged refrigeration system with no modifications. Currently, there is no such thing as a "drop-in" refrigerant, in fact, it is illegal to mix refrigerants.

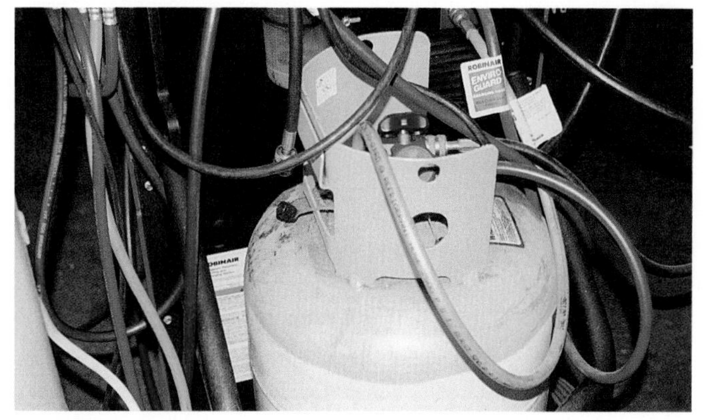

Figure 40-9. Contaminated refrigerant must be placed in a special container painted gray with a yellow top.

A refrigeration system using R-12 must either have R-12 added or be retrofitted. A system with R-134a must have R-134a added. Any system with a blended refrigerant must have that blend added. If the refrigerant is unknown or no longer available, it must be completely reclaimed and destroyed according to EPA guidelines.

Note: The EPA does not use the term "drop-in" to describe any refrigerant.

Refrigerant Oils

Refrigerant oil is needed to lubricate the moving parts. The oil must be compatible with the refrigerant, since it is carried through the system by the refrigerant. If the oil is not compatible with the refrigerant, it may not mix well enough to stay in the refrigerant stream. Any refrigerant oil that will mix well with refrigerant is said to be ***miscible.***

All refrigerant oils are ***hygroscopic.*** A hygroscopic material will absorb water. Some oils are more hygroscopic than others. No matter what type of compressor oil is being used, containers should be kept closed at all times to keep the oil from absorbing water. Three types of refrigerant oil are used today, **Figure 40-10.**

Types of Refrigerant Oil

Systems using R-134a require a type of oil called ***polyalkylene glycol (PAG).*** This oil is synthetic (nonpetroleum) oil similar in chemical makeup to cooling system antifreeze. PAG oil is usually light blue in color. It is intended for R-134a systems, and cannot be used in an R-12 system. It has a viscosity rating of about 150. This viscosity applies only to refrigerant oils, and cannot be compared to motor or gear oils.

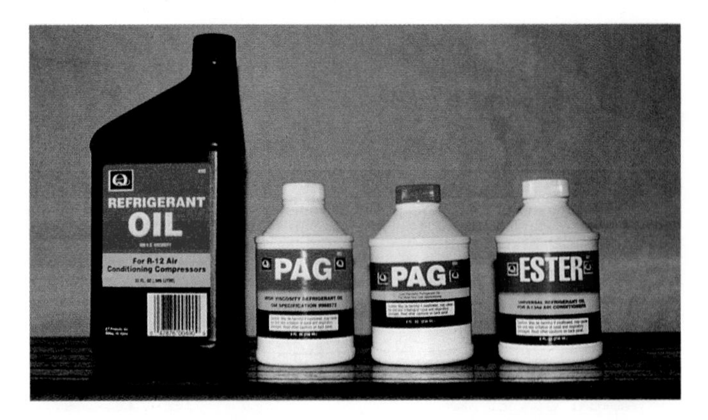

Figure 40-10. Mineral oil, polyalkalene glycol (PAG), and polyol ester (POE) are the three types of refrigerant oil in use. Mineral oil, at left, can be used only in R-12 systems. PAG (the red-labeled container is for use only in General Motors systems) can be used only with R-134a, POE can be used in either R-12 or R-134a systems.

Petroleum based ***mineral oil*** is used in R-12 systems. They are often called 500 or 525 viscosity refrigerant oils. This viscosity is a standard for refrigerant oil only. Mineral oil is usually clear to light yellow in color. Mineral oil is also used with some blended refrigerants. However, mineral oil should never be used in an R-134a system, since it will not mix with the refrigerant.

A third type of oil, called ***polyol ester (POE),*** can be used with both R-134a and R-12. It is an alcohol-based oil compatible with small amounts of PAG or mineral oil. POE is usually clear with no tint. It has a slight odor similar to brake fluid. A feature of POE oil is its ability to change viscosity in relation to temperature changes. Some manufacturers do not permit the use of POE in their R-134a systems.

Components of the Air Conditioning System

All automotive air conditioning systems can be broken down into their basic parts. The parts common to all air conditioning systems include:

- Compressor.
- Condenser.
- Restrictor.
- Evaporator.
- Receiver-drier or accumulator.
- Lines and hoses.
- Pressure switches.

If any one of these parts should fail, the air conditioning system will not work. In addition, some air conditioners contain these components:

- Sight glass.
- Muffler.

Other parts are not vital to overall air conditioner operation. However, they increase air conditioner efficiency, durability, and operational quality.

Compressor

The refrigerant leaves the evaporator as a low-pressure vapor and travels to the ***compressor*** through the flexible low-pressure vapor line. Look at **Figure 40-11.** The compressor draws in the vapor and forces it through the high-pressure vapor line to the condenser. The compressor is used to both move and pressurize the refrigerant.

Basic Compressor Designs

Most automotive refrigerant compressors are piston types that internally resemble small engines. Older air conditioning systems often use two-cylinder compressors. The pistons in these compressors are attached to a rotating crankshaft in the same manner as engine pistons.

Axial Compressors

The most common type of compressor is the ***axial*** type. In an axial compressor, the pistons are parallel to the rotating input shaft. The pistons are forced to move back and forth in an axial direction by the action of a rotating swash or wobble plate. On a typical compressor, there may be three double end pistons operating in six separate cylinders. This would be referred to as a six-cylinder compressor.

Radial Compressors

Other air conditioning systems employ ***radial*** compressors, **Figure 40-12.** The radial compressor also uses a type of crankshaft to operate the pistons. Note these pistons are at right angles to the crankshaft. In addition to the piston compressors, a few units are rotary types, compressing the refrigerant by the use of rotary vanes.

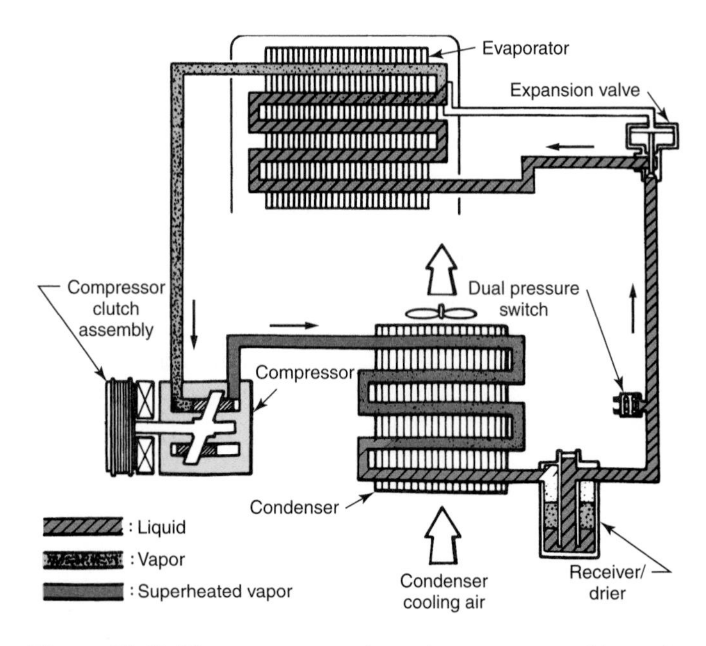

Figure 40-11. The compressor draws low-pressure refrigerant vapor from the evaporator, then sends it onward to the condenser as a superheated high-pressure vapor. (General Motors)

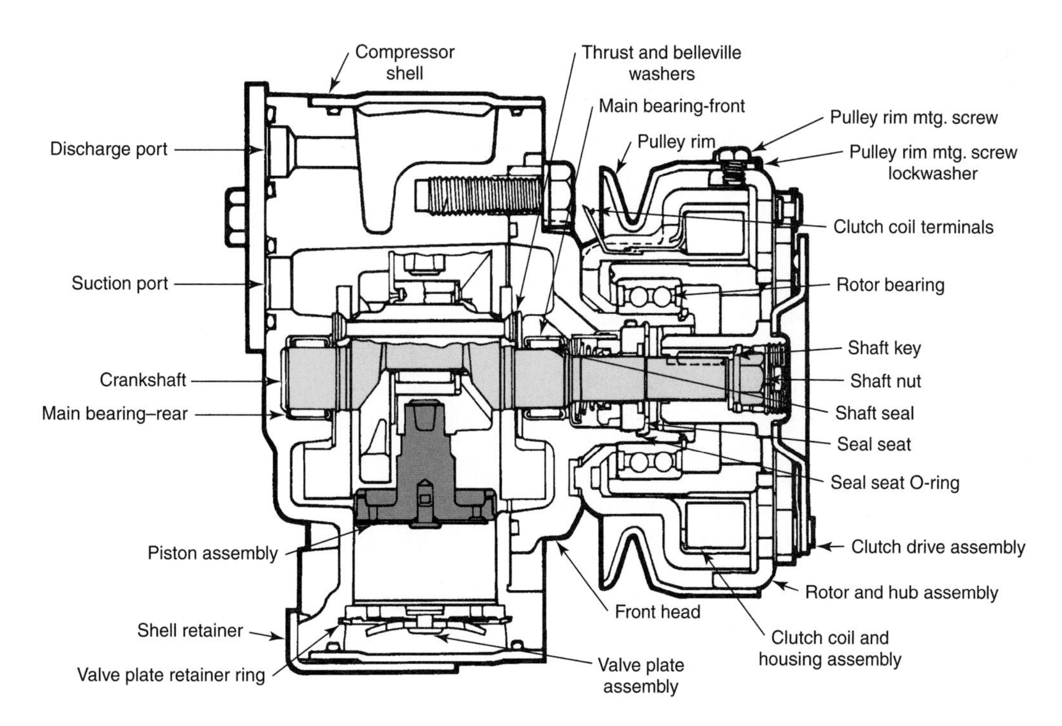

Figure 40-12. A cross-sectional view of a radial-type air conditioning compressor. Study the construction. (Harrison Radiator)

Variable Displacement Compressors

Some compressors are designed with internal valves to allow their pumping capacity to be varied. This variation in compressor output is used to control evaporator temperature to prevent icing. While ***variable displacement compressors*** are usually used with fixed orifice air conditioning systems, they are sometimes used on systems with expansion valves. A variable displacement compressor is shown in **Figure 40-13.** Note the internal valve (bellows control valve) that changes compressor capacity according to system pressure. This compressor can meet all air conditioning needs without cycling on and off.

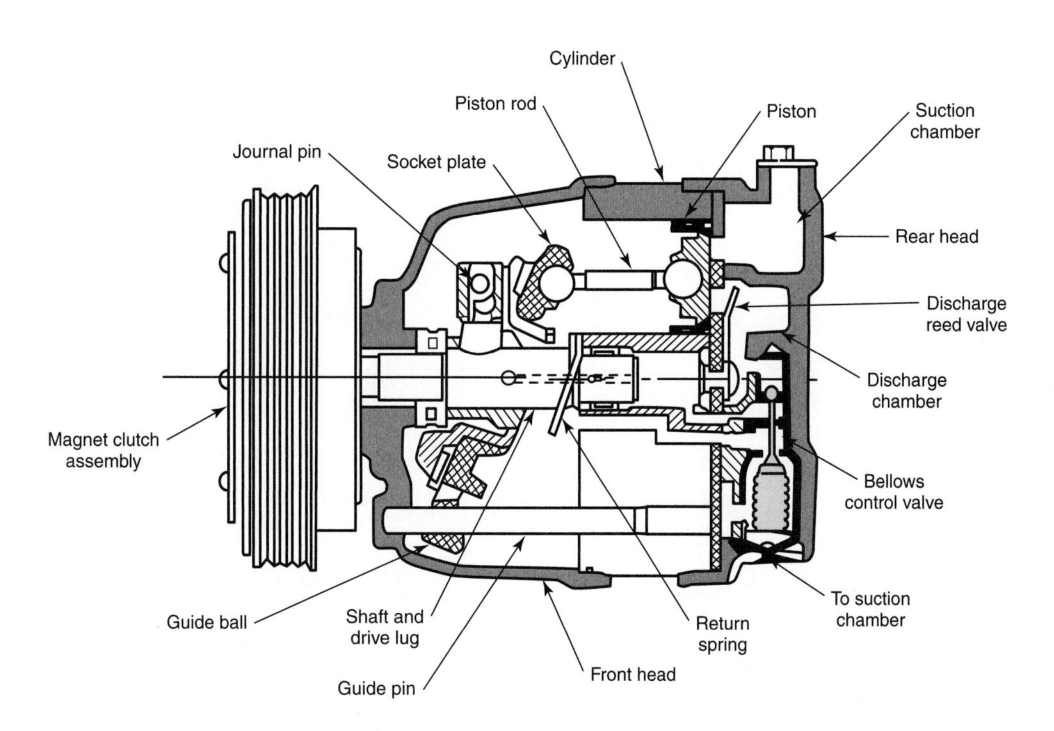

Figure 40-13. A cross-sectional view of a variable displacement compressor. Capacity changes in response to changes in system pressure. (Subaru)

The compressor's internal intake valve reduces the refrigerant output whenever the evaporator pressure becomes too low. This reduces the load on the engine, reducing the amount of fuel used to operate the air conditioning system. The internal valve is moved by pressures in the evaporator, and changes the piston displacement to either reduce or increase refrigerant output.

Scroll Compressors

Some vehicles have ***scroll compressors,*** **Figure 40-14.** Scroll compressors are sometimes called *spiral* or *orbital* compressors. Scroll compressors are smoother than piston compressors and more durable than rotary vane compressors.

The main components of this type of compressor are the scrolls. A ***scroll*** is a length of flat metal, formed into a spiral shape. One scroll is fixed to the compressor body and does not move. The other scroll is attached to the crankshaft in an eccentric (off-center) position. The fixed and moving scrolls are shaped and carefully machined so they do not touch during operation. Compressed refrigerant exits at the center of the scroll assembly.

Compressor Clutch

The ***compressor clutch*** is energized to place the air conditioning system into operation. The clutch coil is an electromagnet that draws the clutch hub inward, locking the revolving pulley to the compressor shaft. On some systems, the compressor clutch is applied whenever the air conditioner is on. On cycling clutch systems, the compressor is turned on and off to control evaporator pressure. Various designs are used for magnetic clutches. One type is shown on the compressor in **Figure 40-15.**

Condenser

Condensers, **Figure 40-16,** are often made of aluminum. During system operation, condenser pressures and temperatures are both high. The hot, high-pressure vapor from the compressor is forced into the ***condenser,*** where heat is given up to cooler air passing over the condenser fins. Follow the path of the high-pressure vapor from the compressor as it travels through the condenser, **Figure 40-17.** As the vaporized refrigerant travels through the condenser coils, it gives up enough heat to the passing air stream to return to its liquid state. As the liquid leaves the condenser, it is stored in the receiver-drier. The condenser is mounted in front of the radiator so it will be exposed to the stream of incoming air.

Figure 40-14. Cutaway view of a scroll compressor.

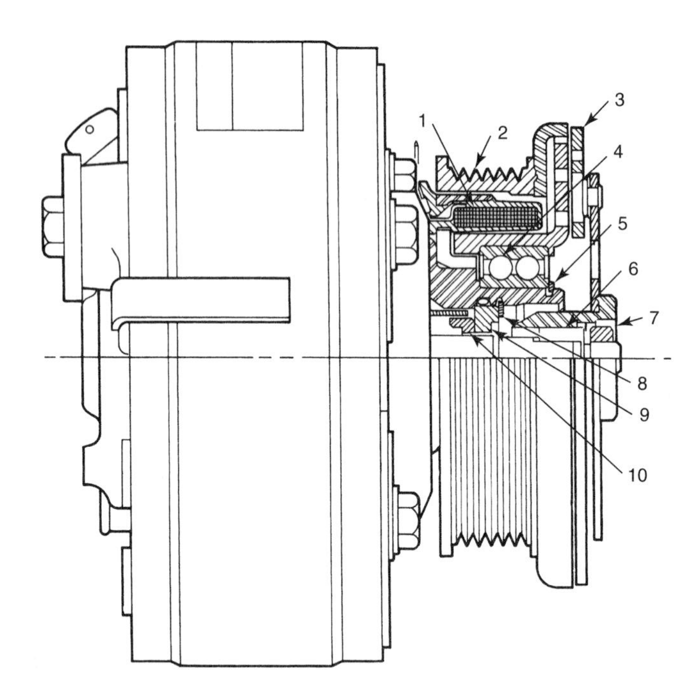

Figure 40-15. Cutaway of a magnetic clutch assembly as used on one radial compressor. 1—Clutch coil. 2—Pulley rotor. 3—Clutch drive assembly. 4—Rotor bearing. 5—Bearing retainer. 6—Shaft key. 7—Shaft nut. 8—Seal seat retainer. 9—Shaft seal seat. 10—Shaft seal. (General Motors)

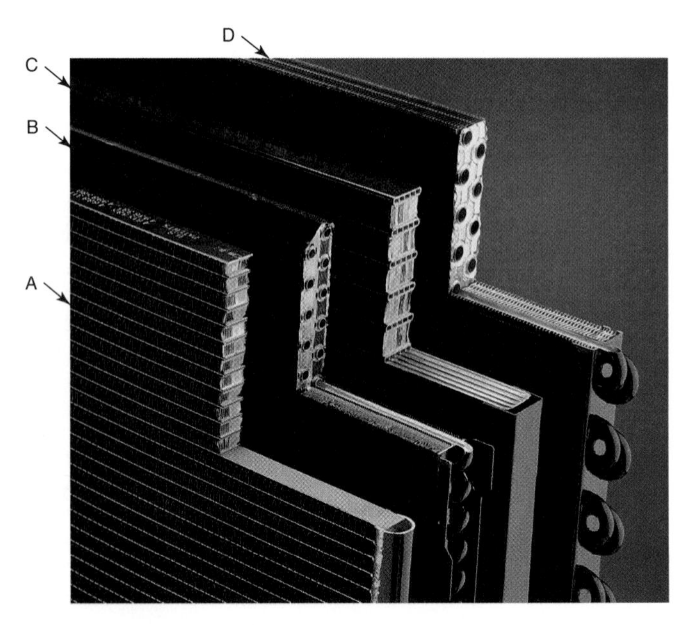

Figure 40-16. Various types of condensers are used in automotive applications. A—Multiflow. B—6mm tube and fin. C—Serpentine. D—Tube and fin. (Modine)

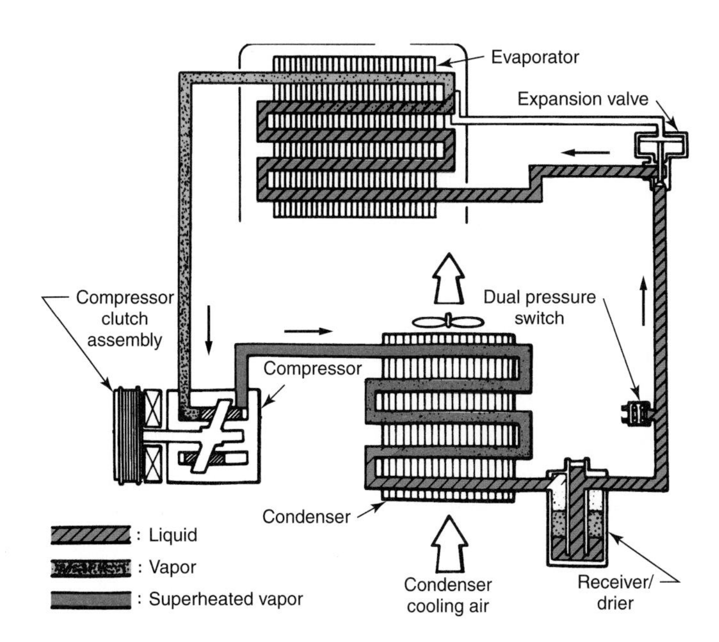

Figure 40-17. The compressor sends superheated high-pressure vapor to the condenser. Incoming cooler air absorbs enough heat from the refrigerant to return it to a liquid state. (General Motors)

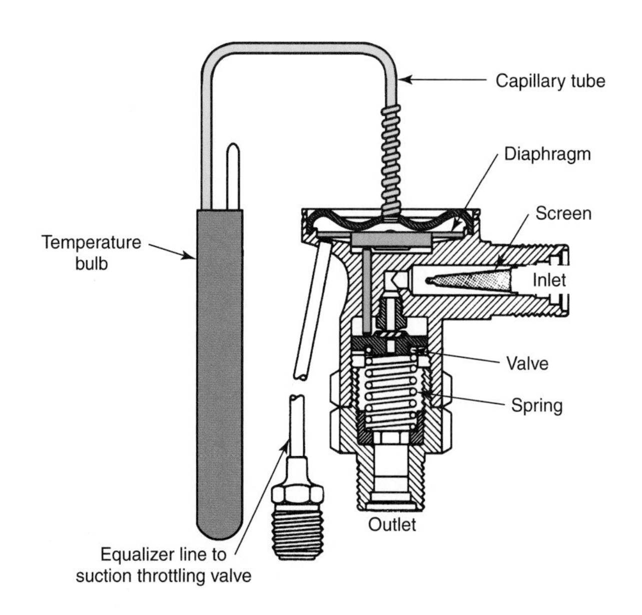

Figure 40-18. A typical thermostatic expansion valve. Changes in temperature at the evaporator outlet control the opening and closing of the valve.(General Motors)

Refrigerant Flow Restrictor

Refrigerant reaches the ***flow restrictor*** as a high-pressure liquid. The restrictor causes the refrigerant to enter the evaporator in small amounts, reducing its pressure. The restrictor can be a thermostatic expansion valve or a fixed orifice tube, depending on the system design.

Thermostatic Expansion Valve

The ***thermostatic expansion valve (TXV)*** admits a metered amount of refrigerant into the evaporator. The refrigerant moves into the evaporator as a relatively low-temperature, low-pressure liquid. A temperature-sensitive bulb is positioned at the evaporator outlet. A capillary tube (length of small-diameter tubing) connects the bulb to the expansion valve. **Figure 40-18** is a cutaway view of a thermostatic expansion valve.

As the evaporator outlet temperature rises, the liquid in the sensing bulb and capillary tube expands, causing the expansion valve to open more. This admits a greater amount of refrigerant to the evaporator. When the outlet temperature drops, the valve begins to close, decreasing the amount of refrigerant flow into the evaporator. The action of the TXV is also affected by spring pressure and by evaporator pressure. Expansion valve servicing is usually limited to replacing the inlet screen and tightening any leaking connections.

Fixed Orifice Tube

Cycling clutch systems use a ***fixed orifice tube,*** **Figure 40-19.** The plastic or metal tube houses a fine screen and a small fixed orifice through which a metered flow of liquid refrigerant can pass. The orifice tube is placed in the line that connects the condenser outlet to the evaporator inlet. The fixed orifice diameter restricts the flow of low-pressure liquid refrigerant into the evaporator.

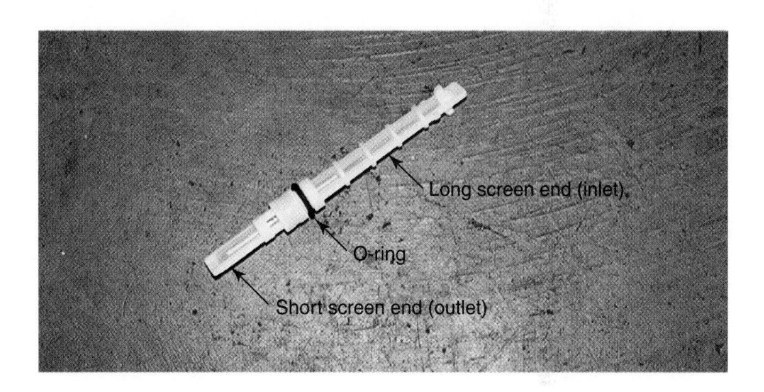

Figure 40-19. A fixed orifice tube meters the flow of refrigerant into the evaporator. The inlet screen traps any small pieces of debris that might plug the orifice.

To control the flow of refrigerant through the evaporator, a cycling clutch compressor is used. The system uses a thermostatic cycling switch to engage or disengage the compressor clutch. The system also has a low-pressure switch to deenergize the compressor clutch if system pressure drops to 25 psi (172 kPa). This protects the compressor if the refrigerant charge is lost. It also prevents system operation when the ambient air temperature is below freezing. Fixed orifice tube service is limited to replacement.

Evaporator

The ***evaporator,*** **Figure 40-20,** transfers heat from the incoming air to the refrigerant. As the low-pressure liquid refrigerant enters the evaporator from the expansion valve, it begins to vaporize. This vaporizing action absorbs heat from the evaporator's tubes and cooling fins. Air passing over the cold tubes and fins gives up heat to the metal surfaces and a stream of cooled air enters the vehicle, **Figure 40-21.**

When the air strikes the cold fins, it is *dehumidified*—some of the moisture in the air condenses to a liquid on the fins and drains off. Particles of dust and pollen in the air tend to stick to the wet fins and drain off with the water. This reduces the amount of pollutants entering the vehicle.

The evaporator contains no moving parts and is usually not subject to internal clogging. However, long use or driving in very dusty areas may cause the evaporator surface to become clogged. A missing intake air screen can allow leaves and other debris to clog the evaporator surface and case. Evaporators can develop refrigerant leaks, usually due to excess moisture in the system freezing in the bottom of the evaporator.

Figure 40-20. In the evaporator, incoming air is cooled by giving up heat as it passes over tubes containing refrigerant vapor. The cooled air then enters the passenger compartment. (Delphi)

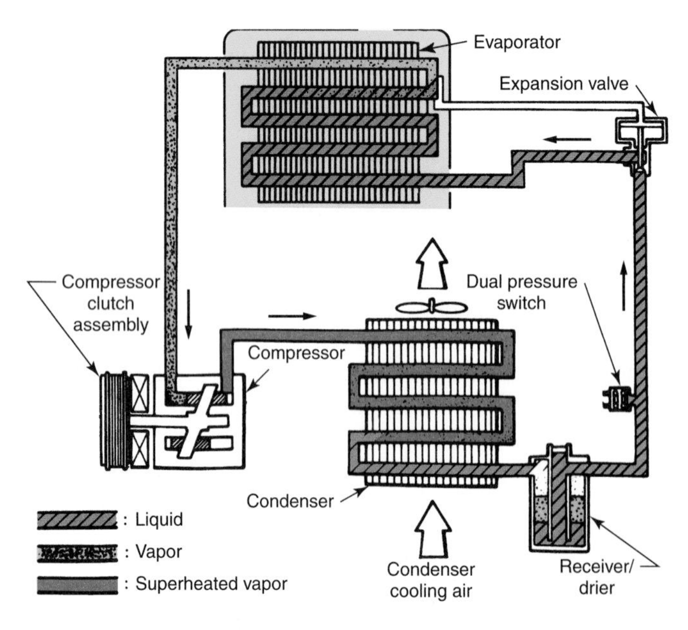

Figure 40-21. Low-pressure liquid refrigerant flows out of the condenser and is metered by a restrictor (in this case, a TXV) into the evaporator. Warm air passing through the evaporator gives up heat, turning the liquid refrigerant back into a vapor. (General Motors)

Interior Air Filter

The air intake side of some blower case assemblies contains an ***interior air filter.*** This filter is usually a conventional filter. In some cases, it may be electrostatically charged. The filter removes dust, pollen, and in some cases, odors from the incoming air. This keeps the inside of the vehicle cleaner and prevents the evaporator from becoming restricted.

Lines and Hoses

To connect the parts of the air conditioning system, ***lines*** and ***hoses*** are used. Connections between parts solidly mounted on the body, such as the condenser and evaporator, are often connected by metal lines made of aluminum or steel tubing. Connections between parts that have relative movement, such as the engine-mounted compressor and the body-mounted condenser, use flexible hoses.

Muffler

A small ***muffler*** is often placed between the compressor and the condenser to reduce pumping noise. The muffler is sometimes part of the hose itself and is usually referred to as a *muffler hose.* A muffler may also be used to reduce line vibrations. Always install mufflers with the outlet side down so refrigerant oil will not be trapped.

Receiver-Drier

The ***receiver-drier*** is used only on air conditioning systems with thermostatic expansion valves and is located between the condenser and the valve. It is a reservoir for refrigerant that has been condensed to liquid form, **Figure 40-22.** The receiver-drier contains a desiccant that filters out dirt and removes moisture from the refrigerant. It is often the

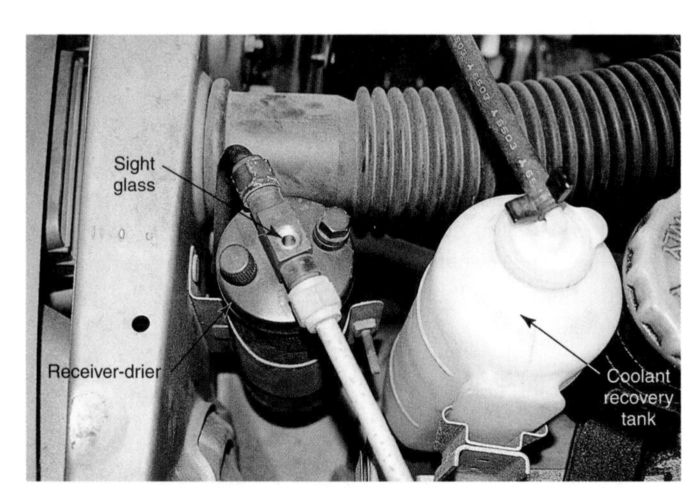

Figure 40-22. The receiver-drier, at left, includes a sight glass for inspecting the refrigerant. The white object at right is the coolant recovery (overflow) tank for the vehicle's cooling system.

location of the sight glass, when one is used. When the receiver-drier becomes inoperative (plugged or moisture-laden), it is usually replaced.

Sight Glass

A ***sight glass*** may be installed in the receiver-drier or in a high-pressure liquid line. The sight glass is used to inspect the liquid refrigerant for the presence of bubbles or foam. One type of sight glass construction is shown in **Figure 40-23.** Many systems, especially those with cycling clutches, do not have a sight glass.

Accumulator

An ***accumulator*** is used only on vehicles with fixed orifice air conditioning systems. The accumulator, **Figure 40-24,** is installed between the evaporator and compressor suction line. The accumulator is designed to prevent liquid refrigerant from entering the compressor, where it would cause damage to the compressor as it cycles on and off. As pressurized refrigerant vapor enters the accumulator, any liquid refrigerant falls to the bottom of the accumulator body. It then will evaporate before it is drawn into the compressor. The accumulator also contains a desiccant for moisture removal.

Some accumulators used on older vehicles may also incorporate a pressure control valve, expansion valve, and a service fitting. When an accumulator becomes clogged or moisture-laden, some models require replacement of the entire unit. On others, the desiccant and filter screen can be replaced.

Service Fittings

To perform various service operations, ***service fittings*** are used. All service fittings connect the refrigeration system to the service hoses and through them to the service equipment. Service fittings are sometimes called service valves or service ports. There are three main types of service fittings.

Push-on Service Fittings

Push-on service fittings are used on R-134a systems. They work like air hose connections. Pushing a service hose connector over the fitting locks the connector in place. Once the service hose connector is installed on the fitting, the knob at the top of the connector is turned clockwise to depress a positive seal inside the fitting. This connects the refrigeration

Figure 40-24. An accumulator prevents liquid refrigerant from damaging the compressor.

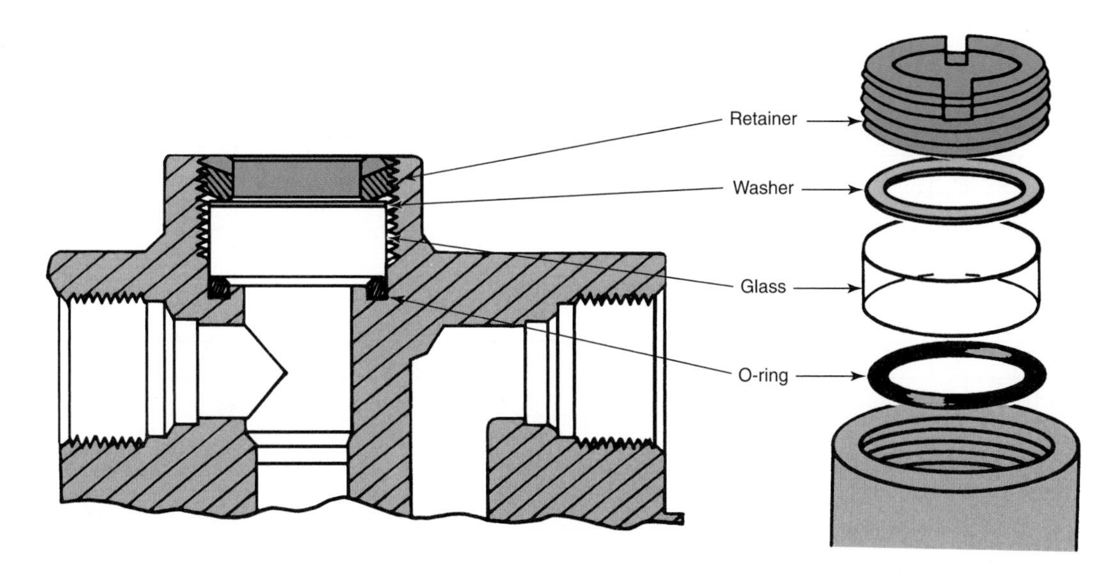

Figure 40-23. An exploded view of sight glass construction. (General Motors)

system to the service hose. See **Figure 40-25.** The service hose connector is removed from the fitting by pulling up on the connector's outer sleeve. The high-side and low-side fittings are different sizes to prevent incorrect connections.

Screw-on Service Fittings

Screw-on service fittings use a Schrader valve and are found only on R-12 systems. A ***Schrader valve*** is a spring-loaded valve similar in operation to a tire valve. The hose connector is threaded onto the valve, causing the pin to be depressed. Removing the hose causes the valve to be closed by spring pressure. A Schrader valve assembly is shown in **Figure 40-26.** Schrader valves on vehicles built after 1986 have different size fittings on the high- and low-pressure sides of the system. The different size threads prevent hose connections from being accidentally reversed.

Manual Service Valves

Manual service valves were found on older vehicles. A manual service valve was opened and closed by turning it with a wrench. These valves were used to isolate the compressor from the rest of the refrigeration system. It allowed the compressor to be removed without discharging the system. Manual valves are not used on modern refrigeration systems.

Note: EPA regulations require air conditioning systems be equipped with fittings specific to the refrigerant installed, Figure 40-27.

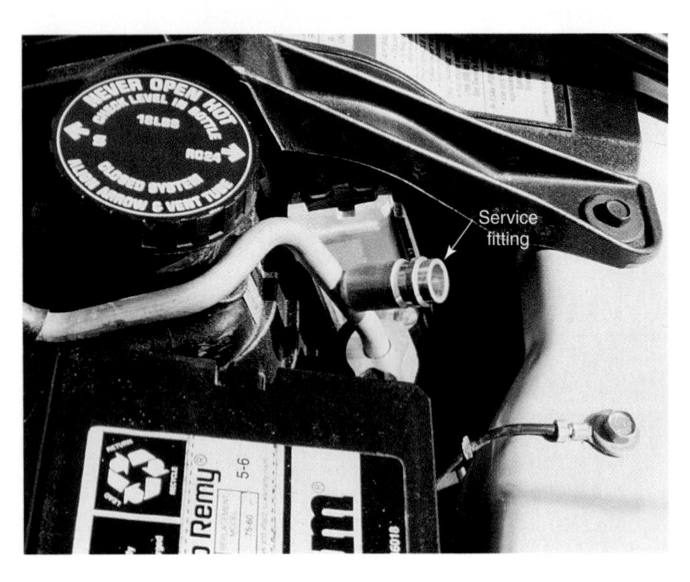

Figure 40-25. A—A push-on service fitting used in an R-134a system. Different-size fittings are used on the high side and low side of the system to prevent incorrect connections. (Nissan)

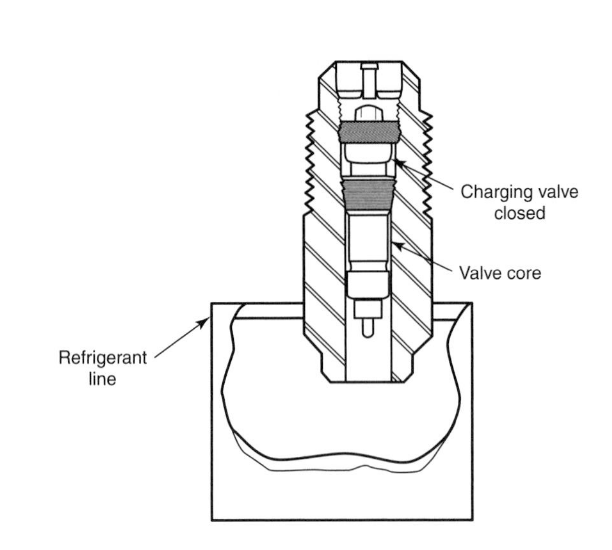

Figure 40-26. A Schrader valve, or screw-on service fitting, is similar in construction to the air valves used on tires.

Refrigerant	High Side Service Port			Low Side Service Port		
	Diameter (inches)	Thread pitch	Thread direction	Diameter (inches)	Thread pitch	Thread direction
HFC-134a	Quick-connect			Quick-connect		
CFC-12 (Pre 1987)	7/16	20	Right	7/16	20	Right
CFC-12 (Post 1987)	6/16	24	Right	7/16	20	Right
Free Zone/RB-276	8/16	13	Right	9/16	18	Right
Hot Shot	10/16	18	Left	10/16	18	Right
GHG-X4/Autofrost	.305	32	Right	.368	26	Right
GHG-X5	1/2	20	Left	9/16	18	Left
R-406A	.305	32	Left	.368	26	Left
Freeze 12	7/16	14	Left	1/2	18	Right
FRIGC FR-12	Quick-connect, different from HFC-134a			Quick-connect, different from HFC-134a		

Figure 40-27. Fittings on air conditioning systems must be specific to the refrigerant used in the system. This chart summarizes fitting types and sizes.

Evaporator Pressure Controls

Many older vehicles using an expansion valve also had an ***evaporator pressure control.*** Evaporator pressure controls are pressure-regulating valves. They operate when the evaporator pressure drops below a set point. The evaporator pressure control was always installed in the suction line between the evaporator and the compressor. The primary purpose of evaporator pressure control was to keep pressure above the point that would allow the temperature to drop below 32°F (0°C). Pressure controls can be thought of as devices that keep condensed moisture from freezing on the evaporator.

Pressure and Temperature Controls

Many vehicles have air conditioning systems that are either monitored or controlled by a computer. Pressure switches and temperature sensors are used to communicate with the computer. The computer uses these inputs to control the operation of the compressor clutch, other parts of the refrigeration system, and airflow.

Pressure Cycling Switch

The ***pressure cycling switch*** is the most common clutch control switch. This switch opens and closes the electrical circuit to the clutch based on low-side pressure. Most pressure switches are similar to the one shown in **Figure 40-28,** although they may have different shapes and connectors. Some pressure switches can be adjusted.

The pressure cycling switch is installed on the low-pressure side of the system, often on the accumulator. When low-side pressure is above a certain value, the switch contacts close and the compressor clutch engages. If pressure falls below the minimum value, the diaphragm contracts, the switch opens, and the clutch is disengaged.

Thermostatic Temperature Cycling Switch

The ***thermostatic temperature cycling switch*** is not as widely used as the pressure cycling switch. However, thermostatic temperature switches are used on a few imported vehicles, many aftermarket air conditioners, and some older domestic cars. The thermostatic temperature switch looks like an expansion valve with an electrical connector. Instead of operating a valve, the diaphragm operates a set of contact points. The sensing bulb is sometimes located on the evaporator outlet, or may be located between the evaporator coils. Some older factory air conditioners used an *adjustable thermostatic temperature cycling switch.*

Figure 40-28. A typical pressure cycling switch used for compressor clutch control.

Pressure Cutoff Switches

Pressure cutoff switches are designed to interrupt, or cut off, power to the compressor clutch electromagnet when certain pressures are present in the refrigeration system. Pressure cutoff switches are sometimes called ***cut-out switches.***

High-Pressure Cutoff Switch

A ***high-pressure cutoff switch,*** **Figure 40-29,** is always installed on the high side of the refrigeration system. A high-pressure cutoff switch is similar to a cycling clutch pressure switch. The switch is wired in series with the compressor clutch electromagnet. The switch contacts remain closed when system pressures are normal. When pressures become too high, the switch contacts will open, cutting off current to the compressor clutch. Normal opening pressure is about 450 psi (3110 kPa). The cutoff switch will usually reclose at about 250 psi (1723 kPa).

Low-Pressure Cutoff Switch

A ***low-pressure cutoff switch*** can be installed on the low or high side of the refrigeration system. No matter where the switch is installed, it performs the same job. If a leak develops and the system is low on refrigerant, system pressures will drop. Compressor damage will result if the system runs with a low charge of refrigerant or oil. To prevent damage,

Figure 40-29. The high-pressure cutoff switch will open if system pressures become too high, cutting off power to the compressor clutch.

the low-pressure cutoff switch will open when refrigerant pressure falls below a certain point. The compressor clutch will disengage and the system will shut down.

When the system is recharged, the cutoff switch will close and the refrigeration system will begin operating normally. In newer vehicles, a low-pressure reading causes the ECM to prevent compressor operation until the system is serviced. Opening and closing pressures vary, depending on which side of the system cutoff switch is installed.

Temperature Sensors

Temperature sensors are used to detect two conditions: refrigerant temperature that is too hot, and outside air that is too cold. Temperature sensors are sometimes called ***thermistors.*** A thermistor is a kind of resistor that changes electrical resistance as its temperature changes. These sensors are sometimes called *temperature cut-out switches.*

The ***high temperature sensor*** has been used on a few vehicles. It is clamped on the condenser outlet tube. If the condenser outlet temperature rises above a certain valve, the switch breaks the circuit to the compressor clutch.

The ***ambient air sensor,*** sometimes called the *ambient air switch,* or *ambient temperature switch,* prevents compressor clutch engagement when the outside air temperature is too low. Ambient air is simply the air surrounding the vehicle. The usual ambient air switch setting is at or below freezing (32°F or 0°C). A few ambient air switches are set at slightly above freezing at roughly 35–40°F (1–4°C).

Pressure Relief Devices

Older air conditioning systems are equipped with a ***pressure relief valve.*** Under unusual conditions, the compressor pressure may exceed safe limits. If this happens, the relief valve will open and reduce pressure to a safe level. If the relief valve has opened, a possible restriction or other condition is present and should be corrected.

Some systems use a ***fusible plug,*** which resembles a pipe plug with a center section made of a material softer and more heat-sensitive than any other part of the refrigeration system. This plug may also be referred to as a *melting bolt.* Both pressure relief valves and fusible plugs are no longer used because of the potential to discharge refrigerant to the atmosphere.

Air Conditioning System Service

For the remainder of this chapter, we will be concerned with common air conditioning service procedures. Most of the procedures will apply to any automotive system, regardless of manufacturer or vehicle model. Refrigerant can be extremely dangerous. Study the following safety rules concerning the air conditioning system. Memorize and observe all of these precautions.

Caution: Always follow all rules regarding proper handling of refrigerant. Do not discharge refrigerant into the atmosphere. This is a violation of federal and state laws. R-12 and R-134a refrigerants, their oils, and most of their service tools are not interchangeable.

Safety Rules

Always wear protective goggles when servicing an air conditioning system. When refrigerant is released into the atmosphere, it will evaporate so fast it will freeze the surface of most objects it contacts. If it strikes the eyes, this rapid freezing action can cause serious eye injury or blindness. If refrigerant gets in your eyes, take the following steps:

1. Do not panic.
2. Splash large amounts of water (90-100°F or 32-38°C) into the eyes to raise temperature. *Do not rub.*
3. Apply several drops of sterile mineral oil to each eye.
4. Consult an eye specialist immediately—even if the pain has passed.

Keep refrigerant away from the skin. If refrigerant contacts your skin, treat it in the same manner recommended for the eyes. When opening fittings or connections, wear a pair of protective gloves and cover the fitting with a loose cloth.

Never discharge refrigerant directly into the atmosphere. Not only is this illegal, it can be deadly. Keep the service area well ventilated. Refrigerant gas is heavier than air—if discharged into a small area without proper ventilation, it can displace the air and cause suffocation. There is also the possibility it may contact an open flame and produce poisonous ***phosgene gas.*** Never breathe the smoke produced when refrigerant contacts a flame. Never subject the air conditioning system to high temperatures by steam cleaning, welding, or baking body finishes on or near the system. To do so can cause a dangerous rise in system refrigerant pressure.

Do not subject refrigerant cylinders or small cans to excessive heat. Dangerous internal pressures can be reached quickly if they are subjected to excessive heat, causing the container to burst. If it is necessary to heat a cylinder or can during system charging, use only warm water or warm wet rags (not over 125°F or 52°C). Never use a torch or stove.

When filling a small cylinder from a larger one, do not completely fill the cylinder. Allow ample space for refrigerant expansion due to heating. A full cylinder is extremely dangerous. Never connect a refrigerant container to the system high-pressure side. Refrigerant can flow back from the system into the container, causing it to explode violently. When transporting refrigerant containers, place them in the car's trunk or pickup bed. Never place refrigerant containers in a vehicle's passenger compartment. If the container is in an open truck bed, protect the container from the sun to prevent overheating.

Proper System Operation

Ask the owner to operate the system. Observe how he or she sets the controls. You will discover occasionally the driver either sets the controls wrong or operates the system with one or more windows open. Advise the driver as to proper operation of the system. Explain that cooling efficiency will vary with ambient air temperature and humidity. On a hot or humid day, the system may fail to cool the vehicle to the point to which it is normally capable. Inspect the compressor drive belt for looseness, excessive wear, or breakage. Adjust or repair as needed.

Compressor

Check the compressor for proper operation. Check the operation of the compressor magnetic clutch to make certain the compressor is being driven. The compressor clutch should not cycle excessively. The compressor should not be excessively noisy. Check the compressor drive belt for proper tension.

Sight Glass

Note: Not every vehicle has a sight glass. A sight glass is not accurate on R-134a systems. The following information applies to R-12 systems only.

Start the system and allow it to operate for at least 5 minutes with controls set for maximum cooling. Examine the refrigerant flow through the sight glass for bubbles. Bubbles are normal when the temperature is below 70°F (21°C). If the ambient air temperature is above 70°F (21°C), however, there should not be any foam or bubbles visible. If bubbles or foam shows in the glass, the system may be low on refrigerant. Exceptionally high temperatures can occasionally cause the appearance of foam or bubbles.

When the system is empty, no foam or bubbles will be visible. The sight glass, however, will have an oily look and will not be as clear as it would be with the system charged. If loss of refrigerant is suspected, check for leaks. **Figure 40-30** is a manufacturer's chart, listing various sight glass conditions, their meanings, and causes.

Lines and Hoses

Examine hoses and lines to make certain they are not kinked or flattened. Restrictions will often cause cold or frosty spots just beyond the point of restriction. Bulges or bubbles are signs that a hose is defective. Leaking hoses will usually be evident by oil at the leak site.

Air Distribution System

Inspect blower motor for proper operation. Check operation of air blending doors (these doors allow mixing of heater air with evaporator air in varying proportions). The air system should be free of obstructions and leaks. The evaporator drain must be open. The air output should be consistent and even from all vents and should be free of odors.

Condenser and Radiator

Check the front and rear surfaces of both the radiator and condenser for signs of clogging from insects, dirt, and other debris. Remove any material that may impede airflow. If the vehicle is equipped with air dams used to direct airflow, make sure they are in place and in good condition. The use of bug screens mounted in front of the radiator and condenser can cause problems. If the vehicle has an electric cooling fan, make sure it is in good condition and operating properly.

Testing System Performance

It is sometimes difficult to determine just how well the system is functioning by merely depending upon the driver's opinion (based on physical reaction to temperature inside the vehicle). As mentioned, temperature and humidity affect system efficiency. To gain a true picture of system efficiency, it is essential it be performance tested. Performance testing generally involves checking system operating pressures (low- and high-side), and the temperature of the air being discharged into the vehicle.

The pressure and temperature readings are then related to the ambient air temperature and relative humidity to determine system efficiency under ideal operating conditions. Test techniques and specifications vary with the different makes and models.

System Diagnostic Charts

Many vehicle manufacturers provide very helpful diagnostic charts covering all aspects of their systems. These

Item to check	Adequate	Insufficient	Almost no refrigerant	Too much refrigerant
State in sight glass	CLEAR Vapor bubbles sometimes appear when engine speed in increased or decreased.	FOAMY or BUBBLY Vapor bubbles always appear.	FROSTY Frost appears.	NO FOAM No vapor bubbles appear.
Temperature of high- and low-pressure lines	High-pressure side is hot while low-pressure side is cold. (A big temperature difference between high- and low-pressure side.)	High-pressure side is warm and low-pressure side is slightly cold. (Not so large a temperature difference between high- and low-pressure side.)	There is almost no temperature difference between high- and low-pressure side.	High-pressure side is hot and low-pressure side is slightly warm. (Slight temperature difference between high- and low-pressure side.
Pressure of system	Both pressures on high- and low-pressure sides are normal.	Both pressures on high- and low-pressure sides are slightly normal.	High-pressure side is abnormally low.	Both pressure on high- and low-pressure sides are abnormally high.

Note: The condition of the bubbles in the sight glass, temperatures, and pressure are affected by ambient temperature and relative humidity.

Figure 40-30. The refrigerant's state and condition can be assessed by observing it through the sight glass. (Nissan)

charts provide tests and other procedures helpful in pinpointing specific problems quickly and accurately, and should be used whenever they are available. A diagnostic chart for air conditioning problems is located at the end of this chapter. Always obtain and use the diagnostic charts in the vehicle's service manual.

Functional and Performance Tests

A ***functional test*** checks for proper system operation at different settings. The ***performance test*** checks the refrigeration and heating system components for proper pressures and temperatures. Some test procedures do not apply to all systems. Always make sure you obtain and use the manufacturer's procedures and specifications for functional and performance tests. A typical functional test procedure is shown in **Figure 40-31.**

The functional test can be performed without gauges or a refrigerant service center. Start the engine and allow it to run for five minutes. Then, perform the functional test steps outlined in the service manual. To make the performance test, shut off the engine, make sure the transmission is in Park or Neutral, and set the parking brake. Attach gauges or a refrigerant service center. Make sure the high- and low-side hoses are attached properly.

Once the gauges are attached, check static pressure. A normally charged system will have 70–125 PSI (482–861 kPa) when it has been inactive for about one hour. If the gauges show low or no pressure in the system, you can be sure there is a leak somewhere. Be sure the hoses do not contact any moving parts.

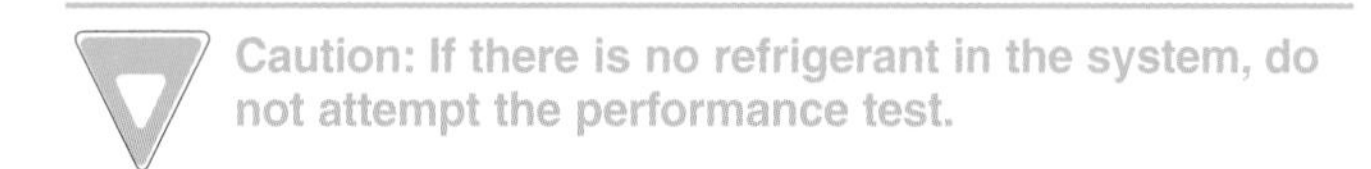

Caution: If there is no refrigerant in the system, do not attempt the performance test.

Install a temperature gauge in the vent nearest the evaporator. Then start the engine and set it to run at approximately 1500 to 2000 rpm (this will vary by manufacturer). Turn the HVAC control panel settings to the maximum cooling position and set the temperature switch to the maximum cold position. Turn the blower speed switch to the high position and open the front windows.

Check the compressor clutch to make sure it is engaged. If the clutch does not engage, shut off the HVAC system and engine and check the clutch, relay, switches, and wiring.

If the compressor clutch engages, allow the refrigeration system to operate for about five minutes to stabilize the gauge readings. Monitor the cooling system gauge or light to make sure the engine does not overheat. Observe the fan clutch or fan motor(s) to be certain they are operating and moving air through the condenser and radiator.

Caution: If the cooling system fans are not operating, or if the high-side pressure exceeds 325 psi (2467 kPa), stop the performance test immediately and determine the cause.

Go through all the steps outlined in the service manual for testing the system. A typical performance test chart is shown in **Figure 40-32.**

System Service Following Collision

When a vehicle has been involved in an accident that could have damaged the air conditioning system, the system should be checked as soon as possible. Examine the compressor and the compressor clutch pulley for damage. Remove the compressor clutch energizing wire before the vehicle is operated, if damage to the compressor or other air conditioning system parts is apparent. Replace any air conditioning parts that were damaged. Do not attempt to repair

	Control Settings			System Response					
Step	Mode	Temp set	Blower switch	Blower speed	Heater outlet	A/C outlet	Def. outlet	Side wind def	A/C comp
1	Defr	Full hot	Off	Off	No air flow	No air flow	No air flow	No air flow	Off
2	Deft	Full hot	Hi	Hi	Some hot air flow	No air flow	Hot air flow	Some hot air flow	On
3	Defog	Full hot	Hi	Hi	Hot air flow	No air flow	Hot air flow	Some hot air flow	Off
4	Heat	Full hot	Hi	Hi	Hot air flow	No air flow	Some hot air flow	Some hot air flow	Off
5	Vent	Full hot	Hi	Hi	No air flow	Hot air flow	No air flow	No air flow	Off
6	Bi-lev	Full cold	Hi	Hi	Air flow	Cold air flow	Some cold air flow	Some cold air flow	On
7	A/C	Full cold	Hi	Hi	No air flow	Cold air flow	No air flow	No air flow	On
8	A/C	Full cold	Lo	Lo	No air flow	Cold air flow	No air flow	No air flow	On
9	Max	Full cold	Hi	Hi	No air flow	Cold air flow	No air flow	No air flow	On
10	Max	Full cold	Lo	Lo	No air flow	Cold air flow	No air flow	No air flow	On

Figure 40-31. A manufacturer's chart outlining steps for a functional test. (General Motors)

Relative Humidity (%)	Ambient Air Temperature °F	°C	Maximum Low Side Pressure PSIG	kPaG	Engine Speed (rpm)	Maximum Right Center Air Outlet Temperature °F	°C	Maximum Highside Pressure PSIG	kPaG
20	70	21	32	221	2000	43	6	175	1207
	80	27	32	221		44	7	225	1551
	90	32	32	221		50	10	275	1896
	100	38	33	228		51	11	275	1896
30	70	21	32	221	2000	45	7	190	1310
	80	27	32	221		47	8	235	1620
	90	32	34	234		54	12	290	2000
	100	38	38	262		57	14	310	2137
40	70	21	32	221	2000	45	8	210	1448
	80	27	32	221		50	10	255	1758
	90	32	37	255		57	14	305	2103
	100	38	44	303		63	17	345	2379
50	70	21	32	221	2000	48	9	225	1551
	80	27	34	234		53	12	270	1862
	90	32	41	283		60	16	325	2241
	100	38	49	338		69	21	380	2620
60	70	21	32	221	2000	50	10	240	1655
	80	27	37	255		56	13	290	2000
	90	32	44	303		63	17	340	2344
	100	38	55	379		75	24	395	2724
70	70	21	32	221	2000	52	11	255	1758
	80	27	43	276		59	15	305	2103
	90	32	48	331		67	19	355	2448
80	70	21	36	248	2000	53	12	270	1862
	80	27	43	296		62	17	320	2206
	90	32	52	356		70	21	370	2551
90	70	21	40	276	2000	55	13	285	1965
	80	27	47	324		65	18	335	2310

Figure 40-32. A manufacturer's chart outlining steps for a performance test. (General Motors)

parts by welding, soldering, or other means. If the system was damaged to the point the refrigerant charge was lost, replace the receiver-drier before recharging.

Refrigerant Identification

When beginning air conditioning work on any vehicle, look for the refrigerant label, **Figure 40-33A.** The label will indicate what refrigerant is in the system. A retrofit label indicates the original refrigerant has been replaced with a substitute. If the retrofit was done properly, the service fittings should be different from the originals.

However, you should always use a refrigerant identifier, **Figure 40-33B,** whether a retrofit label is present or not. The ***refrigerant identifier*** will confirm the refrigerant in the system matches the label, as well as indicating its level of purity. The identifier has a probe that is able to contact the refrigerant at the port. Most probe connectors consist of a hose attached directly to the port. To use a refrigerant identifier, turn it on and allow it to warm up. Make sure the display panel is operating, and, if necessary, calibrate the identifier. Always follow manufacturer's instructions for calibration. Then attach the identifier to one of the service ports. Most refrigerant identifiers will then display the percentages of R-134a, R-12, and unknown refrigerants.

Leak Detection

Modern HVAC shops use several leak testing devices. At one time, the flame-type halide leak detector was widely used. Today, however, it has been largely replaced by electronic and

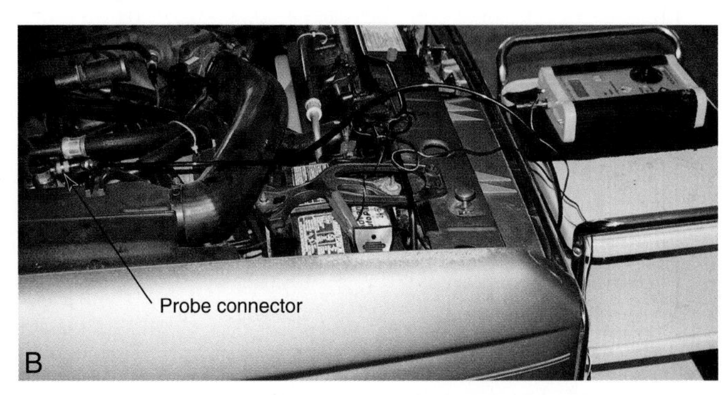

Figure 40-33. Refrigerant identification. A—The refrigerant label identifies the refrigerant used in the system. B—A refrigerant identifier will determine the type of refrigerant actually in the system and whether it matches the label.

dye-detection devices. The following sections explain how to use various types of leak testing devices.

Warning: *Do not* use a halide torch to check for leaks. Some blended refrigerants contain propane or butane, and the leak detector flame may cause a fire or explosion.

Preparing the System for Leak Detection

It is a good idea to leak-test the entire system, even if an obvious leak has already been identified, before any repairs are formed. If the test identifies additional leaks, they can be corrected while the system is open. **Figure 40-34** shows the potential leak points in a typical air conditioning system. Any leak detection device will produce a false leak signal if it contacts refrigerant vapors built up under the hood or in the shop. Before starting the leak checking procedure, run the engine briefly to remove any vapors from the engine compartment. If you suspect refrigerant vapor has built up in the shop, clear the vapor using fans or the shop ventilation system.

If there is no refrigerant in the system, none can leak out to be detected. To make a leak check, there should be a minimum low-side refrigerant charge of 50 psi (345 kPa) with the engine off. Some leaks, especially those on the high side of the system, may require a greater charge. Add a partial charge to the system, if needed, then leak-test. If a leak is so severe the system will not hold any pressure, repair that leak first, then pressurize the system.

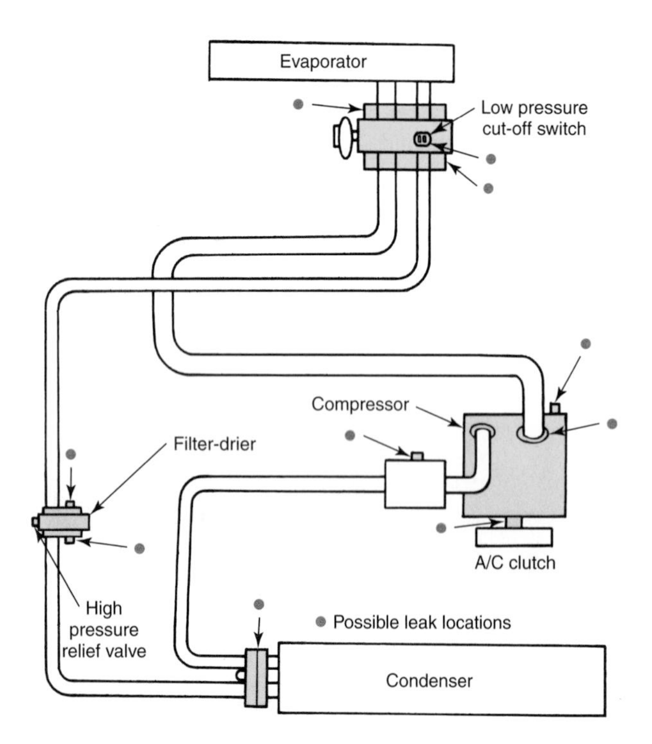

Figure 40-34. There are many potential leak locations in an air conditioning system, especially at points where lines are connected. (DaimlerChrysler)

Note: The compressors on some vehicles are disabled by the engine control computer if the refrigeration system loses its charge. When this occurs, a trouble code is usually set. Use of a scan tool is usually required to clear this code so the compressor will operate.

Some technicians prefer to pressurize completely empty systems with nitrogen. If the system has only recently begun leaking, there may be enough refrigerant left to be detected by a sensitive leak detector. Pressurizing with nitrogen will also allow the technician to find a relatively large leak by using soap solution. Refrigeration systems can be pressurized up to about 150 psi (1033 kPa) without damaging any of the low-side components.

Electronic Leak Detectors

The ***electronic leak detector,*** **Figure 40-35,** is the fastest and best method of leak detection. Modern electronic detectors are extremely sensitive and can locate a leak as small as 1/2 oz. (15 ml) of refrigerant per year.

Begin the leak detection process by turning on the detector and allowing it to warm up for about one minute away from the refrigeration system components. Most leak detectors will make a ticking noise that increases when the probe encounters refrigerant. Many leak detectors have a display that indicates the leak rate.

After setting sensitivity, slowly pass the sensing tip closely around possible leak areas and check for an increase in the ticking noise. Large leaks raise the ticking to a high-pitched squeal. In some cases, the electronic detector's sensitivity may need to be reduced when a large leak is present or if other engine fumes trigger the detector. Also remember to pass the tip under suspected leak areas, since refrigerant is heavier than air and settles downward.

Dyes

Another leak detection method involves using a ***dye.*** The dye is injected into the refrigeration system and allowed to circulate. The material will leak out along with any refrigerant, leaving a color stain at the site of the leak.

Refrigerant dyes may be contained in small cans resembling a one-pound refrigerant can or may be packaged as a

Figure 40-35. An electronic leak detector. (Robinair)

set with an injection tool. Modern dye injectors are designed to insert a fluorescent dye directly into the refrigeration system. The injector is attached to one of the system service ports and the handle is turned to force the dye into the system. The engine and HVAC system are started and the dye allowed to circulate for a few minutes. An ultraviolet light ("black light"), **Figure 40-36,** is then used to make the dye fluoresce (light up or shine) identifying any leaks.

Note: Some manufacturers of R-134a refrigerant add fluorescent dye to their refrigerant. The dye-incorporated materials are marketed under several names.

Soap Solution

The ***soap solution*** method will find only large leaks, and should not be relied upon to locate small leaks or leaks in inaccessible locations. Soap solution is often used with nitrogen pressurizing to check for leaks. It is also an easy way to confirm what appears to be an obvious leak. To make a soap solution test, first make sure the refrigeration system has pressure. Then mix a small amount of soap with water. Dishwashing liquid works best, but almost any kind of soap will do. Spray or pour the soap solution on the area of the suspected leak. Leaking refrigerant will form bubbles.

Figure 40-36. The dye shows as fluorescent bright green under black light illumination, indicating that the system is leaking at that point. (Tracerline)

Refrigerant Recovery/Recycling

The only recommended procedure for ***discharging*** the air conditioning system is to ***recover*** all the refrigerant in the system using a refrigerant service center. *Refrigerant must not be released into the atmosphere.* When recovering the refrigerant, follow the manufacturer's directions. *Do not mix* different types of refrigerant in the recovery equipment.

Refrigerant recycling means to reuse a portion or all the refrigerant taken from a refrigeration system. The refrigerant can be used to recharge the refrigeration system it came from, or in another system. Most recovery and recycling machines clean and remove moisture from refrigerant to allow immediate reuse. The refrigerant may be stored in a standard 30-pound cylinder or in a separate charging tank. Only pure refrigerant should be recycled. Never attempt to recycle contaminated refrigerant.

Always check the system for leaks before recovering the refrigerant. The system charge must be removed from the system before removing and replacing any part containing refrigerant. Always use a refrigerant identifier before performing any recovery operations.

Using Refrigerant Service Center

Select the type of refrigerant recovery and recycling machine (refrigerant service center) and hoses as indicated by the identifier, then turn the master switch to the *on* position. Attach the hose assemblies to the air conditioning system, **Figure 40-37.** Be sure all shutoff valves are in the closed position. Once the connections are made, open both shutoff valves.

Set the machine to recover the refrigerant charge. Most machines have pushbutton controls. A light will indicate the machine is drawing refrigerant from the system. Wait while the machine completely removes the refrigerant. As the machine draws in the refrigerant, it will dry and filter it. This prepares the refrigerant for reuse. When the gauges read zero pressure or a vacuum, the system is empty. Allow the machine to continue operating for ten minutes to be sure all refrigerant has been removed. When you are sure the system is empty, shut off the machine and proceed to perform other refrigeration system repair procedures.

Checking Compressor Oil Level

A leak in the system, careless discharging (discharging too fast), or part replacement can cause oil to be lost. Some compressors have a small oil level valve near the bottom of the compressor body. If the compressor is so equipped, operate the system for fifteen minutes at maximum cooling. Shut off the engine and air conditioning system and wait five minutes. Open the valve cap slightly. If oil drips out, tighten the cap. Wait a few moments and reopen. If oil comes out in a steady stream, sufficient oil is present. If refrigerant vapor hisses out, the oil level is low. Leak test the system and make repairs. A number of compressors have provisions for checking the oil level by using a dipstick, **Figure 40-38.**

When replacing an evaporator, receiver, condenser, or compressor, a specific amount of oil should be added to replenish the system. The expansion valve, fixed orifice tube,

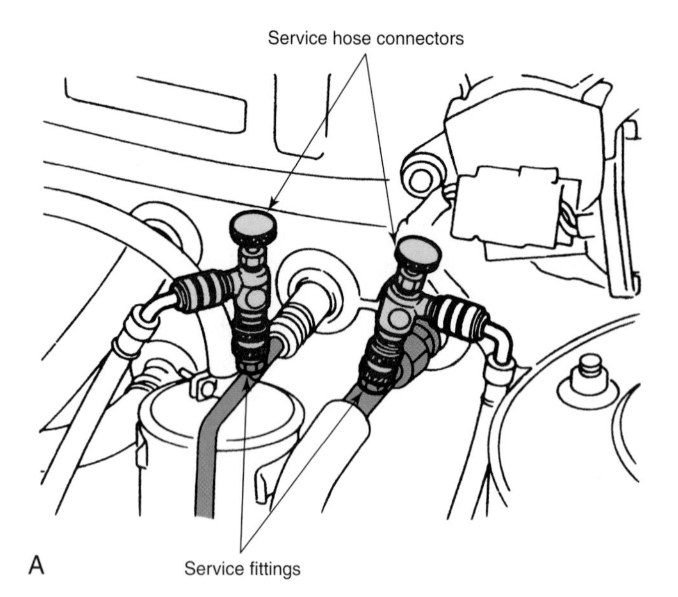

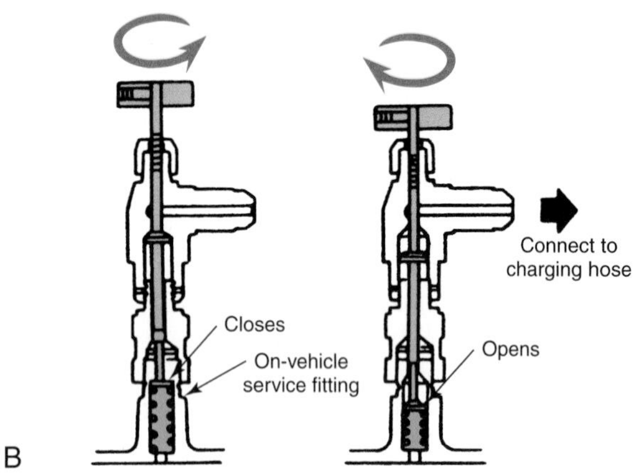

Figure 40-37. Service center connections. A— The hose assemblies are connected to the push-on fittings of an R-134a system. B—The connector depresses the valve pin only after being seated properly on the fitting. (Nissan)

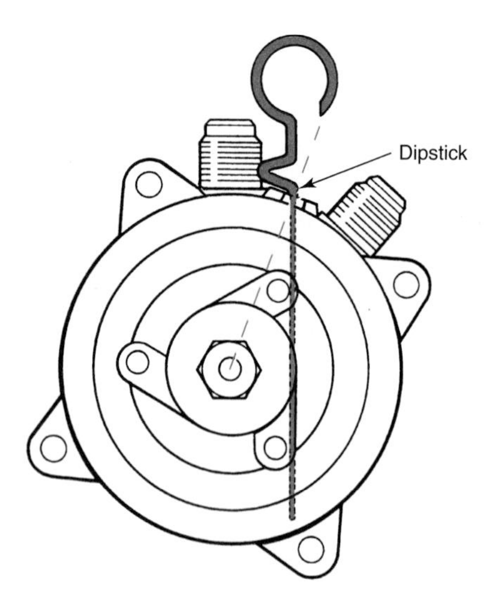

Mounting Angle in Degrees	Acceptable Oil Level in Increments
0	2-4
10	4-5
20	5-6
30	6-7
40	7-9
50	9-10
60	10-12
90	12-13

Figure 40-38. Compressor mounting angles will produce different oil levels on the dipstick. (Jaguar)

and lines may be replaced without adding oil, however. Check and replace the O-ring, if needed, and replace the filler plug. Evacuate the compressor by drawing a vacuum on the high-pressure fitting gauge connection. Shut the vacuum pump line valve and fully open both service fitting hand valves. Remove the vacuum line and replace the gauge connection cap.

Evacuating the System

Evacuating the system means attaching a ***vacuum pump*** to remove any air and moisture remaining after the refrigerant has been recovered. This is absolutely vital before recharging. Any air in the system will cause reduced operating efficiency. Water vapor carried in with the air will cause severe corrosion of internal parts as well as oil sludging. Many refrigerant service units have provisions for automatically evacuating the system before the refrigerant is reinstalled. This eliminates the separate manifold gauge, vacuum pump, refrigerant cylinder, and lines, which must be hooked up each time. Follow the instructions on the refrigerant service center.

If a vacuum pump must be used, begin by attaching the manifold gauge set. Both gauge valves must be off. Attach a vacuum pump to yellow center connection of the manifold gauge set. The refrigerant cylinder valve must be closed. **Figure 40-39** illustrates one type or setup for evacuating a system. The same setup will be used for charging.

Slowly open the high-pressure valve of the manifold gauge to relieve any pressure buildup. Close the high-pressure valve and start the vacuum pump. Slowly open both the high- and low-pressure manifold gauge valves. If there are manual valves in the refrigerant lines, open them. Open the vacuum pump shutoff valve slowly to prevent oil being drawn out of the pump.

Run the vacuum pump until the low-pressure gauge indicates 29″ (711 mm) of vacuum. A vacuum pump, in proper condition, should draw a vacuum of 29″ (711 mm) at sea level. Subtract 1″ (25.4 mm) for every thousand feet (304 m) of elevation. For example: at 5000 feet (1520 m), the pump should draw 24″ (29″ minus 5″) or 584 mm (711 mm minus 127 mm) of vacuum.

Note: Some of the latest air conditioning systems and service equipment may call for measuring vacuum in microns. If a conversion chart is not provided, the system should be evacuated until a reading of 300 microns is reached.

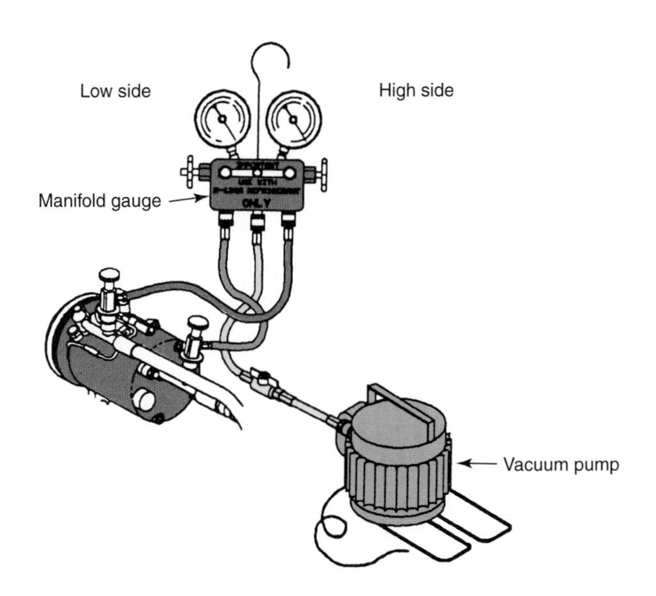

Figure 40-39. Evacuating a system with a manifold gauge set. The yellow center hose is attached to the vacuum pump, and the red and blue hoses to, respectively, the high and low sides of the system.

If the vacuum cannot be drawn to 29″ (711 mm) at sea level, the pump is defective, the desiccant contains excessive moisture, or there is a leak in the system or hookup lines. When the gauge reads 29″ (711 mm) of vacuum, run the pump for an additional 15 minutes. Then shut off both the high- and low-pressure gauges. Shut off the valve in the vacuum line (if equipped) and stop the vacuum pump. The system should hold vacuum with no more than a 2″ (51 mm) drop in a five-minute period.

Refrigerant Charging

After repairs are made, the system must be refilled with the proper amount of refrigerant. Before recharging, make sure the system has the proper amount of oil, has been evacuated, and has been thoroughly checked for leaks. Then determine the proper amount of refrigerant before starting the recharging procedure. Amounts vary from 1.5 lbs. (3.3kg) on a small car to as much as 8 lbs. (17.6kg) on a large vehicle with rear air conditioning. Also make sure you know for certain what type of refrigerant should be used. Once the system has been evacuated, do not release the vacuum before charging. Allow the vacuum to draw in the refrigerant.

Caution: If the system is being recharged with a blended refrigerant, the refrigerant must be charged into the system as a liquid. If a blended refrigerant is charged as a vapor, composition change may occur, resulting in system damage.

Using a Refrigerant Service Center

Turn the master switch to the *on* position, and attach the hose assemblies to the refrigeration system. Be sure the shutoff valves are in the closed position. Once connections are made, fully open both shutoff valves. Set the refrigerant service center controls to the proper amount of refrigerant. Consult the service material to determine the correct amount of refrigerant. Without starting the engine, push the charge button of the refrigerant service center. Allow the service center to charge the system with the proper amount of refrigerant.

Then start the engine and HVAC system and set the blower speed to high. Wait about five minutes for pressures and temperatures to stabilize. If desired, place a temperature gauge in the vent closest to the evaporator to check outlet temperature. Observe the gauges to ensure the high- and low-side pressures are correct. Monitor outlet temperature and other operating conditions to be sure the refrigeration system is operating properly.

Charging with the Manifold Gauge

Make sure the manifold hand valves are closed. Attach the hoses to the appropriate service fittings, and then attach the center manifold gauge hose to the refrigerant container. Purge air from the hoses, if necessary.

Weighing Method, using a Refrigerant Tank

With the engine off, open the high-pressure gauge valve so a charge of liquid refrigerant will enter the high side of the system. The low-pressure valve must be closed. Liquid refrigerant may be obtained by drawing from the bottom of the cylinder or by turning the cylinder upside-down. Allow the liquid refrigerant to flow into the system until the desired amount has entered. Check the scale frequently during the operation. See **Figure 40-40.**

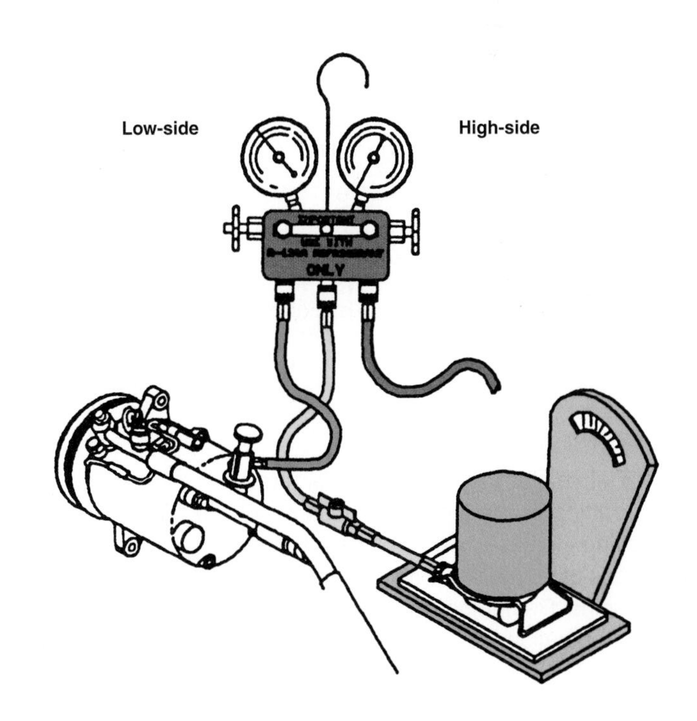

Figure 40-40. Using the weighing method when charging with the manifold gauge.

If sufficient liquid refrigerant will not enter the system, close the high-pressure gauge valve. Turn the cylinder right-side-up so refrigerant vapor will be available. Start the engine, turn on the air conditioner, and set it to maximum cold. Open the low-pressure gauge valve and allow refrigerant vapor to flow into the low side of the system until the required charge has entered. Shut the low-pressure gauge valve. Stop the engine and close the refrigerant cylinder valve. Disconnect the charge setup from system. Restart the engine and air conditioning system and check for proper operation.

Caution: Never allow liquid refrigerant to enter the low-pressure side of the system, unless the procedure is specifically recommended. Charge the high-pressure side with liquid. Charge the low-pressure side with vapor. Remember when the refrigerant container is upright, it will discharge vapor. When container is upside down or on its side, it will discharge liquid refrigerant.

Using One-Pound Cans

One-pound cans are used just as if they were regular refrigerant tanks. Attach the gauge set center hose to the can dispenser fitting. Following proper system evacuation, close the high-side gauge valve and leave it closed. With the air conditioning controls set to off, start the engine and operate it at normal idle speed.

Invert cans to allow liquid to flow. Open one can and allow it to flow into the system via the low-pressure fitting. After the refrigerant from the first can has entered the system, close the dispenser fitting, and turn on the compressor. Repeat the charging procedure using a second can. Continue idling the engine until the recommended charge has entered the system. Shut off the manifold valve and continue engine operation for about 30 seconds. This will clear the lines and manifold gauge. If less than one-pound is called for, open the can valve. When the frost line on the outside of the can reaches the halfway mark, shut off the valve.

Warning: When using one-pound cans, do not open the high-side service valve—the can will explode from system pressure backing up into it. Never remove the charge hose until the manifold gauge hoses are first removed completely from the vehicle's service fitting as this may result in the complete loss of the refrigerant charge.

Before shutting off the engine, remove the low-pressure hose fitting from the accumulator service fitting. When removing the hose fitting, unscrew it quickly to prevent excessive loss of refrigerant. Cap all the service fittings and shut off the engine. Remember to check the system for leaks and for proper operation.

Refrigeration System Purging

Excess air (sometimes called *noncondensible gases*) raises pressures since it cannot condense at normal system pressures. The best indication of excess air in a system is high static pressures (pressures measured with the system not operating) in a system that has not been overcharged. Air enters refrigeration systems often enough that you should know how to remove it. Air removal is called ***purging.***

The simplest way to remove air is by very slightly opening (*cracking*) a valve or hose fitting at the highest point in the system. Watch the connection and close the valve or fitting as soon as refrigerant begins to come out of the connection. This procedure should be done after the system has been idle overnight to allow the air to separate from the refrigerant.

The disadvantage of this method is it releases a small amount of refrigerant to the atmosphere. Another problem is most refrigeration systems do not have a convenient fitting at the highest point in the system. Also, air pockets may develop at two or more locations. Another method uses a refrigerant recovery and recycling machine with an automatic air-purging device.

Adding Oil by Injection

To add oil to a charged refrigeration system, a special oil injector is needed. To add oil by this method, attach the pump to the high-side service port of the system. The engine must be off when adding oil. Use the high-side fitting to reduce the possibility of liquid oil entering the compressor. Pump in the needed amount of oil, remove the pump, and reinstall the service port cap. Then start the engine and turn the HVAC controls to maximum air conditioning to circulate the oil.

A different type of injector is used with a manifold gauge set, **Figure 40-41.** To use this type of injector, fill it with the proper oil and connect it to the gauge set, as shown. Then

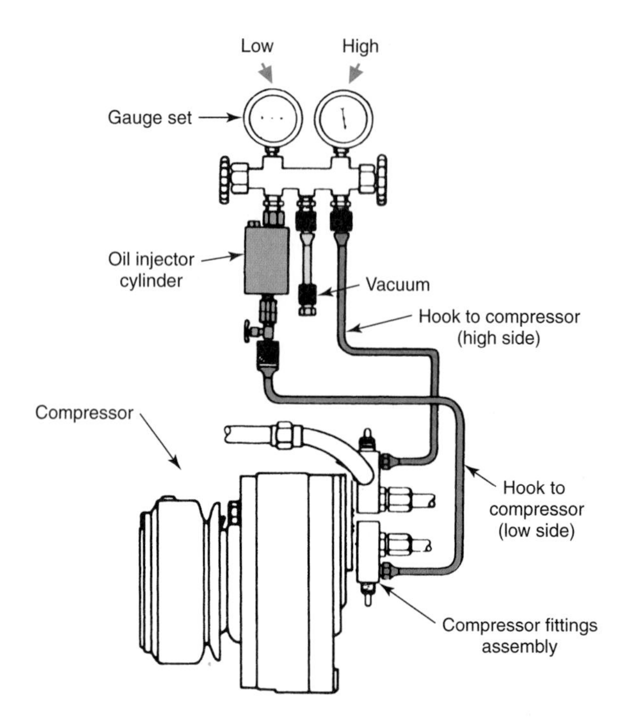

Figure 40-41. Adding oil to a charged system, using an oil injector cylinder setup. (General Motors)

connect the manifold gauge set to the system, start the engine, and turn the HVAC controls to maximum air conditioning. Crack open the low- and high-side hand valves. This allows the high pressure to push the oil into the low side. Do not open the valves more than a small amount. Allow the oil to enter the system; then shut both valves. Allow the refrigeration system to operate for a few more minutes to distribute the oil. Then stop the engine, remove the hoses, and replace the fitting caps.

Air Conditioning Component Service

When a major component of the air conditioning system must be replaced, the system's refrigerant charge must be recovered. Replacement procedures for such units as the condenser, receiver-drier, and compressor are usually straightforward. Remember to use new O-rings when reconnecting fittings. Evaporator replacement, however, can be lengthy and complex. In many cases, the heater core, ductwork, and some dash and passenger compartment components must be removed to replace an evaporator. Always obtain the proper service manual and follow all procedures exactly.

Warning: When working on a vehicle equipped with a supplemental restraint system (air bag), always follow the manufacturer's service, repair, and handling instructions. Failure to do so can result in personal injury and vehicle damage caused by the unintentional activation of the system. Neglecting the manufacturer's precautions can also prevent air bag deployment in an accident, which could result in occupant injury or death.

Service Fitting Precautions

The following are general rules for removing and replacing air conditioning system service fittings. Always recover the system's refrigerant charge before opening a connection. To avoid system contamination with foreign matter, always clean connections before opening them. If any pressure is evident, allow it to seep out slowly before completely separating the connection. Cap or plug lines and parts as soon as they are disconnected. Remove caps from either new or used components just before installation.

Use new O-rings when replacing parts, **Figure 40-42.** Make certain the new O-rings are of the correct size and properly positioned. Coat both the fittings and the O-rings with refrigerant oil before assembly. If a connection is made without applying refrigerant oil, it will probably leak. The connection must be properly aligned, clean, and free of nicks or burrs. Torque all connections to specifications. Remember that aluminum connections require lower torque than steel connections. If one end of the connection is steel and the other is aluminum, torque to aluminum specifications only.

When tightening or loosening connections, always use two wrenches to avoid twisting the tubing. Note that **Figure 40-43** shows one nut being held with an open-end wrench while the other nut is being tightened with an adjustable wrench. A flare-nut wrench would hold better than an open-end type, and would be less likely to deform the nut.

Some systems may employ spring-lock couplings. This coupling type holds the two parts together by using a garter spring that snaps over a flare on the female fitting end. O-rings

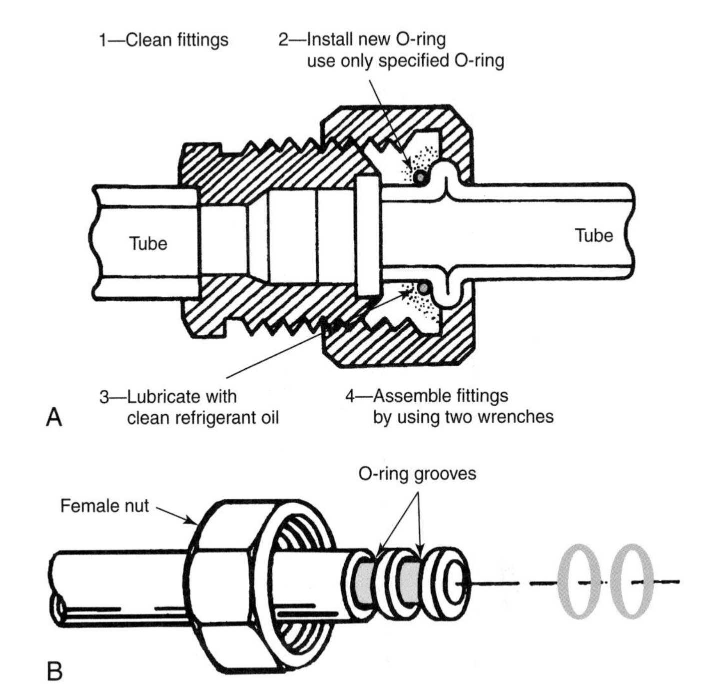

Figure 40-42. When servicing lines, fittings, etc., always use new O-ring seals. A—Single O-ring seal fitting. (Hyundai) B—A joint using a dual O-ring arrangement. (General Motors)

Figure 40-43. Use two wrenches when loosening a connection. Use a torque wrench when tightening. (General Motors)

provide a seal against leakage. The separation of these spring-lock couplings requires the use of a special removal tool. **Figure 40-44** illustrates the special tool and the techniques for connecting and disconnecting spring-lock couplings. When replacing a hose, be sure to use a type of hose specified for air conditioning systems. Reinstall all hose clamps, tubing clamps, and insulators to prevent vibration and other damage.

Preventing Air Conditioning System Contamination

The air conditioning system will not tolerate dirt, air, or moisture. The system must be chemically stable (contain only pure refrigerant and a small quantity of pure refrigerant oil) to function as designed. The presence of air, dirt, or moisture can cause sludging, corrosion, freezing of the expansion valve, and other problems. In order to ensure the chemical stability of the system, always observe the following general service precautions:

- To prevent the entry of moisture, plan the work and lay out parts so the system will be open for the shortest possible time.
- Service tools should be kept dry and spotlessly clean.

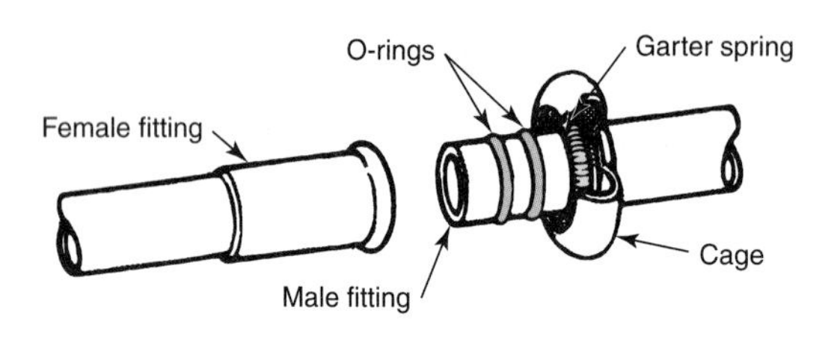

Note parts of spring-lock coupling.

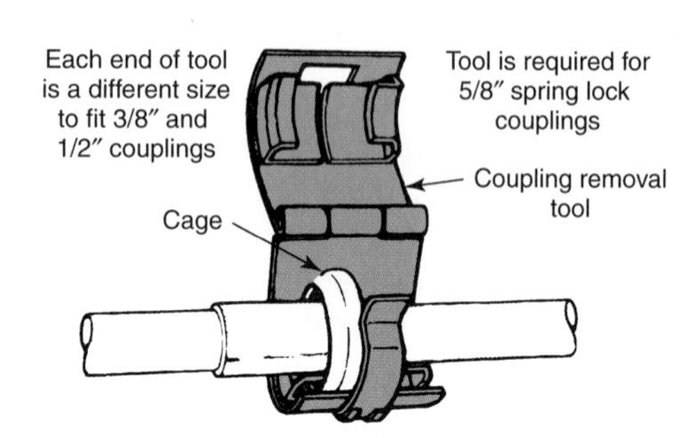

To disconnect coupling, fit tool to coupling so that tool can enter cage to release garter spring.

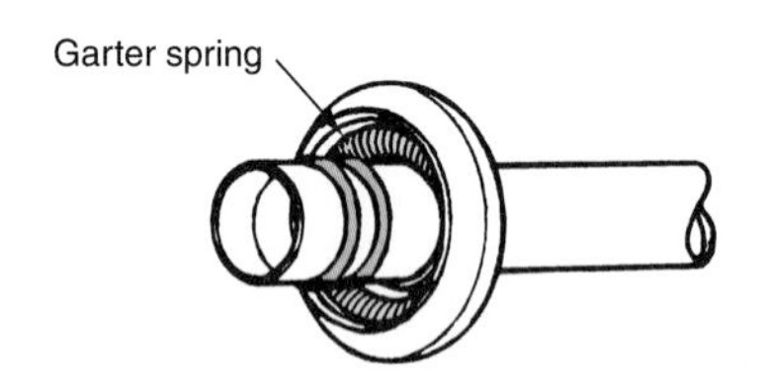

Before connecting, check for missing or damaged garter spring. Remove damaged spring with small hooked wire. Install new spring if damaged or missing.

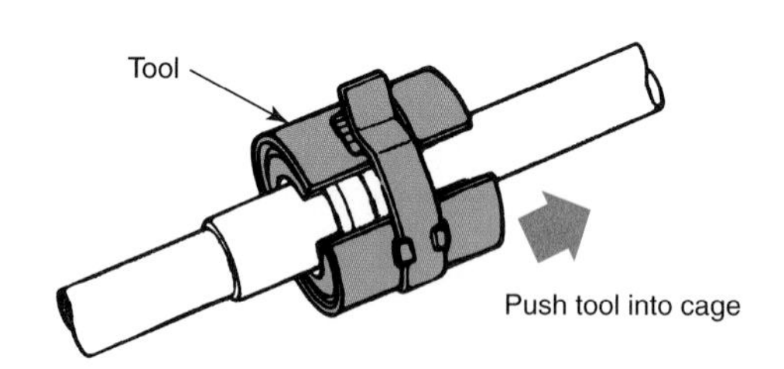

Close tool and push the tool into the cage opening to release the female fitting from the garter spring.

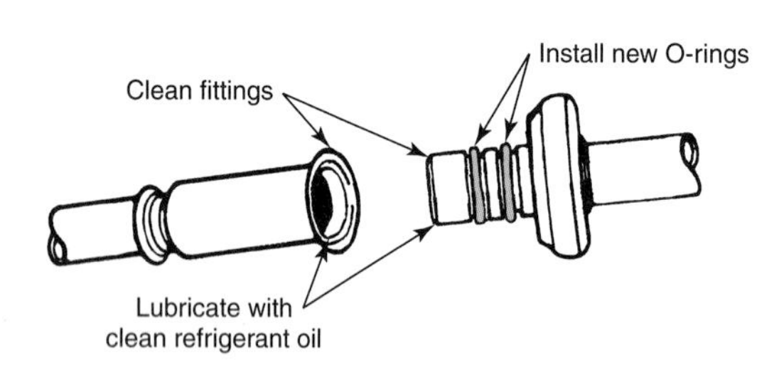

Assemble fitting by pushing with a slight twisting motion.

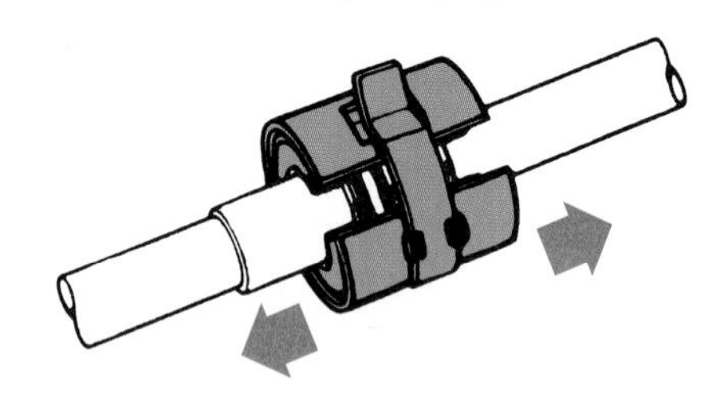

Pull coupling male and female fittings apart.

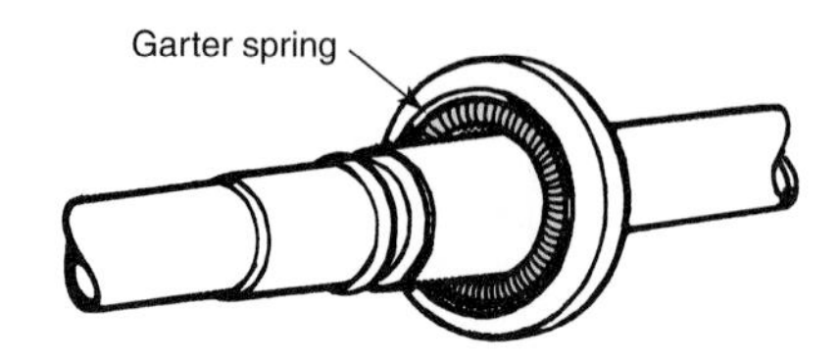

To ensure coupling engagement, check to be sure garter spring is over flared end of female fitting.

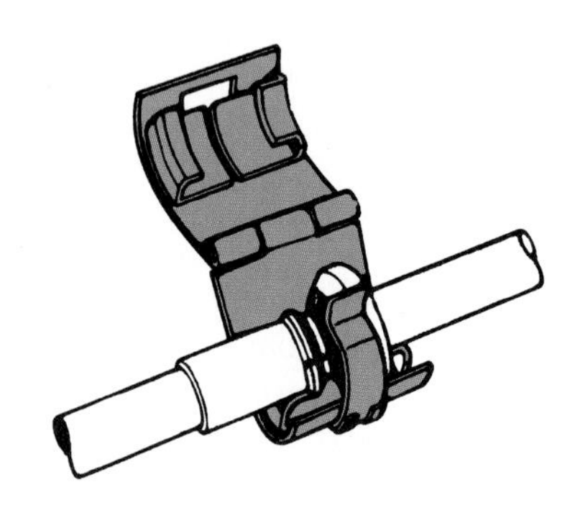

Remove tool from disconnected spring lock coupling.

Figure 40-44. Inspecting, connecting, and disconnecting a spring-lock connector. Note use of special coupling removal tool. (Ford)

- Before disconnecting any fitting, clean around it thoroughly.
- Allow refrigerant lines to become close to room temperature before disconnecting them. This will help prevent condensation from forming inside the line.
- If parts have become contaminated and flushing is indicated, flush with dry nitrogen or other approved agent.
- Replacement lines and parts should be at room temperature before removing sealing caps to prevent moisture contamination.
- Do not remove sealing caps from lines or parts until ready to connect into the system.
- Avoid keeping the system open (part or line disconnected) for longer than five minutes.
- Connect the receiver-drier into the system last. This will ensure maximum moisture protection.
- Keep compressor oil free of moisture. When using a container for compressor oil, the container must be clean and dry. Compressor oil containers should be kept capped. Do not open the oil container until ready to use. Cap immediately after use to prevent entry of dirt or moisture.
- Always evacuate the system to remove air and moisture before charging.

Compressor Replacement

Once the refrigerant has been recovered, remove and cap the compressor inlet and outlet lines. Loosen the belt adjuster and remove the belt. Remove or relocate any parts as needed to access the compressor. If the compressor is mounted low, you may need to raise the vehicle. Remove the clutch electrical connector, and connectors to any switches on the compressor body. Remove the bolts holding the compressor to the engine and remove the compressor from the engine compartment, **Figure 40-45.** Transfer to the new compressor any switches installed on the compressor and the clutch.

To reinstall, place the compressor in position and install the mounting bolts. Install all the bolts before tightening any of them. Reconnect the clutch and any pressure switch electrical connectors. Install the compressor refrigerant lines, using new gaskets or O-rings. If needed, reinstall any other parts that were removed and lower the vehicle. Install the belt on the pulleys. Adjust the belt if needed, and retighten all compressor bolts. Evacuate and recharge the refrigeration system; check compressor operation.

Checking Oil Level after Compressor Removal

When replacing a compressor, you must drain the oil from the old unit before installing the new compressor. Remember that compressor repairs are not complete until the correct amount of oil has been added to the system. In some cases, checking the oil can be done with a dipstick.

Once the compressor is removed, allow the oil to drain into a measuring cup or graduated cylinder, **Figure 40-46.** Carefully measure the oil removed to determine the exact number of fluid ounces. Examine the condition of the oil. If metal chips, water, sludge, or other debris is present, the system should be flushed and a filter added. If the oil is okay, install needed amount of new oil to bring the volume up to specification. Do not add oil to the compressor until you are ready to reinstall the unit. Often, the new compressor may come prefilled with refrigerant oil.

Compressor Clutch Replacement

Some compressors do not come with a clutch installed. If this is the case, you will either need to transfer the clutch from the old compressor or install a new clutch. If the compressor clutch is inoperative or weak, or if the clutch faces are worn, the clutch and magnetic coil can be replaced without discharging the system. Some systems require a

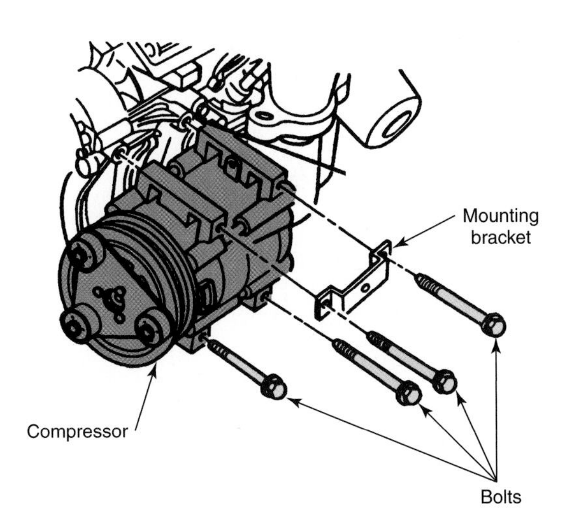

Figure 40-45. Compressor removal is usually a fairly simple procedure. (Ford)

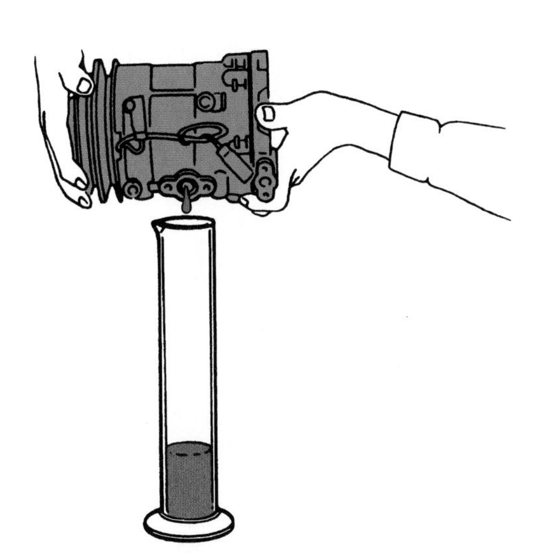

Figure 40-46. Drain compressor oil into a graduated cylinder to measure the volume and compare it to specifications. (Subaru)

special tool to remove the pulley, while other pulleys can be removed by loosening a large bolt on the front of the compressor shaft. Reinstallation is the reverse of removal. Refer to **Figure 40-47.**

Caution: Do not install the clutch assembly without checking the clearance between the drive and driven plates. This specification and the method of checking it are given in the appropriate service manual.

Additional Filtration Devices

Air conditioning systems will not tolerate foreign materials. When a compressor has been badly damaged, a liquid line filter should be installed to remove contamination from the system, **Figure 40-48.** Adding the extra filter will eliminate the need to flush the system following compressor repairs.

Flushing

If a compressor seizes or if an internal compressor part fails, small metal particles are usually distributed throughout the system. To remove these particles, it is necessary to clean the system by ***flushing.*** When flushing to remove contamination, always wear goggles, gloves, and protective clothing.

There are two kinds of flushing. ***Open loop flushing*** is done with the flushing agent blown through the individual component. Most shops now use commercial refrigeration system cleaning solvents such as R-141b or ester-oil-based solvents for open flushing. ***Closed loop flushing*** involves flushing the entire system with a solvent that is recycled. It is

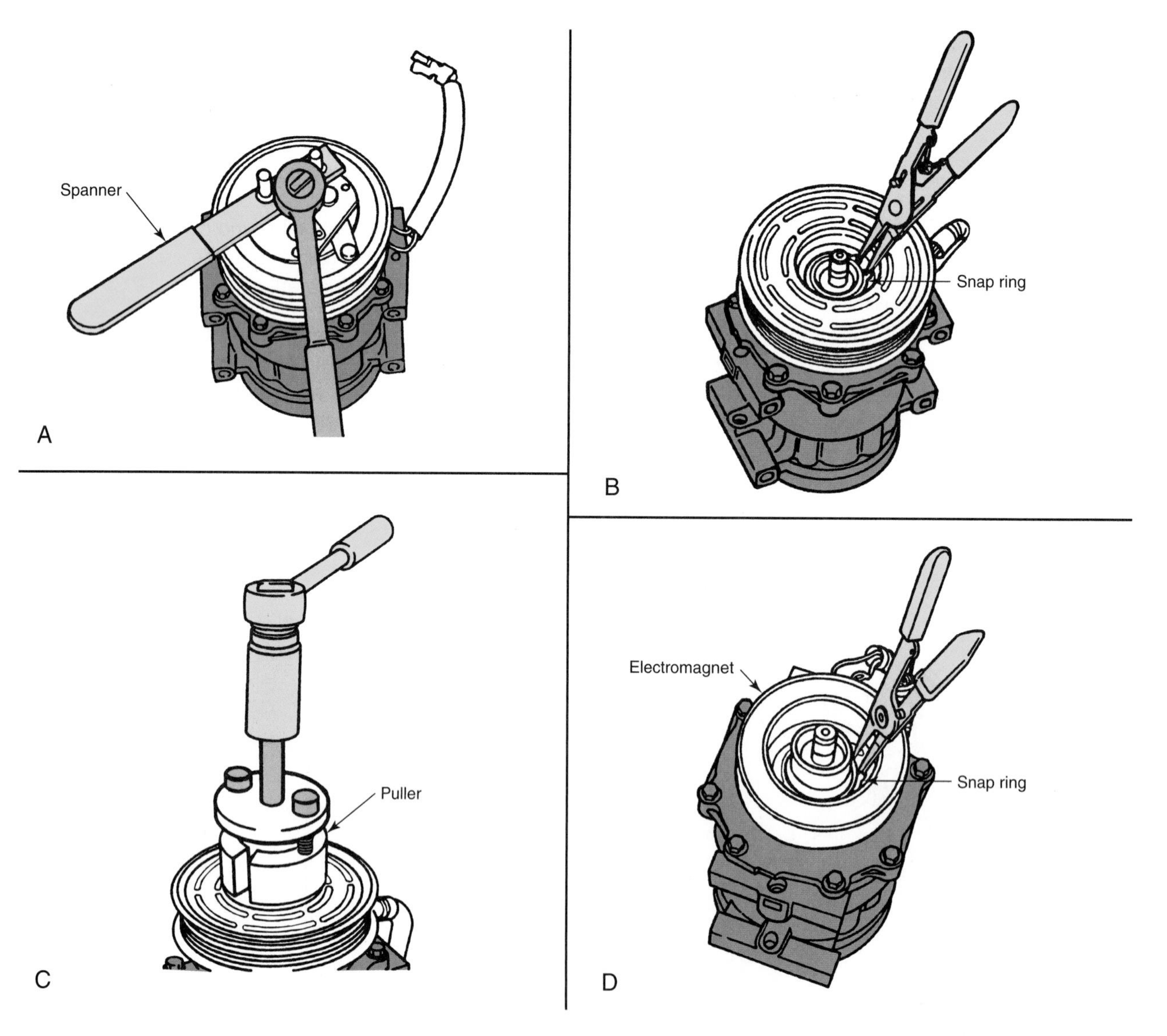

Figure 40-47. Removing the compressor clutch often requires special tools. A—Removing the nut. B— Releasing a snap ring that holds the pulley to the electromagnet. C—A special puller is often used to remove the pulley. D—Removing the electromagnet. Do not pry the clutch loose. (DaimlerChrysler)

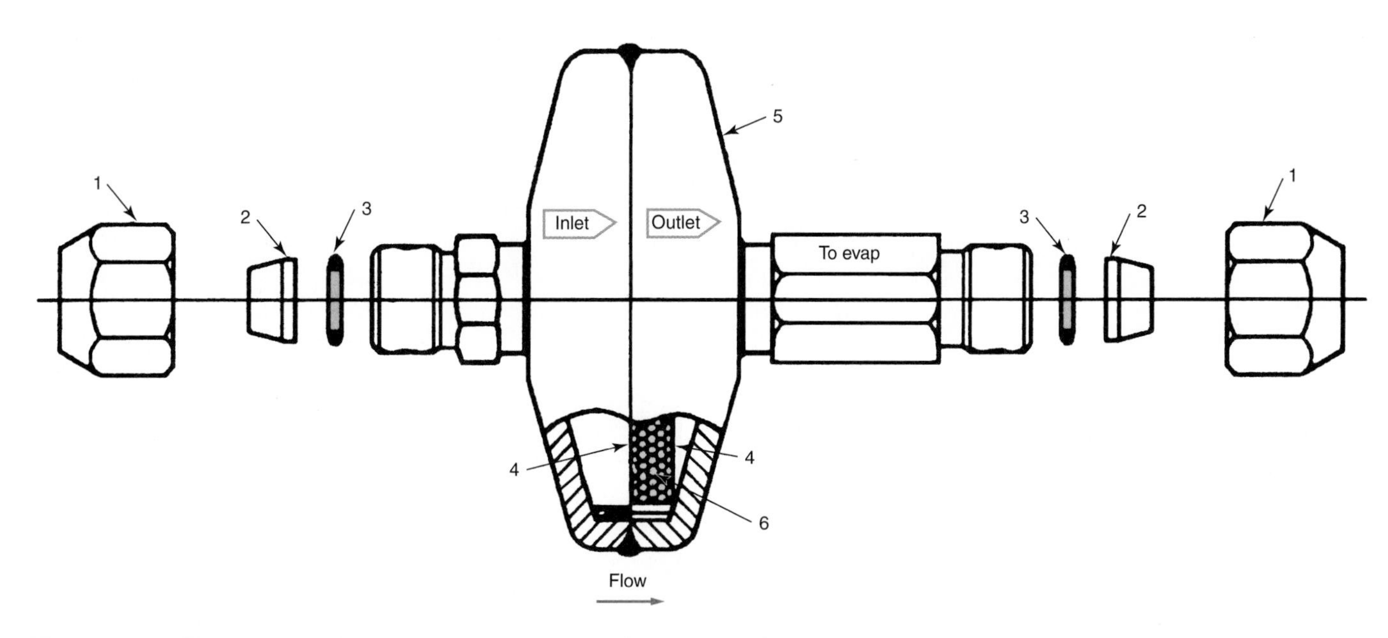

Figure 40-48. This liquid line filter eliminates the need to flush the line following repairs. Follow the manufacturer's installation instructions. 1—Nut. 2—Ferrule. 3—O-ring. 4—Screen. 5—Filter housing. 6—Filtering pad. (General Motors)

usually done with a refrigerant or other liquid, using a dedicated flushing machine. The refrigerant commonly used for flushing is R-134a.

Expansion Valve and Orifice Tube Service

Most expansion valves and orifice tubes have small screens installed ahead of them. These screens often become plugged with metal particles or other debris. The system charge must be recovered to service the restrictor. If system diagnosis indicates the expansion valve or orifice tube is restricted, replacement may be necessary. If the particles are fine and gritty, they can be flushed away. Expansion valves and orifice tubes are not repairable. Each must be replaced as a unit if clogged or defective.

Replacing Hose and Tubing

One of the most common air conditioning service procedures is replacing leaking hoses. When replacing a hose, avoid sharp bends. General practice calls for hose bends with a radius from five to ten times the diameter of the hose, depending upon hose construction. Keep hose at least 3″ (80mm) away from the exhaust system unless special heat shields are used.

Some hoses are assembled with sleeves using either beadlock or barb fittings. To replace a defective fitting on a sleeve arrangement, cut off the hose directly behind the sleeve. Lubricate a new service fitting with refrigerant oil and insert it into the hose. Crimp the sleeve for the new fitting. If a hose must be replaced, be sure to use ***barrier hose.*** Barrier hose has an inner lining that prevents refrigerant molecules from passing through. **Figure 40-49** illustrates the repair procedure involved in hose and fitting replacement.

If metal tubing must be bent, use a suitable tubing bender to avoid kinks. Do not try to rebend a formed line. Always check all connections for leaks after the system has been charged.

Condenser Service

Locate the condenser in front of the engine compartment. To clean the exterior of the condenser, cover any exposed vehicle paint or trim, then spray detergent solution on the fins. Allow the detergent to work for 10–15 minutes, and then clean the fins with a strong stream of water.

Condenser Replacement

To gain access to the condenser on some vehicles, the radiator may need to be relocated or removed. Before removing the condenser, recover the system refrigerant and ensure all residual pressure has been released. Then disconnect the inlet and outlet fittings. Remove the bracket fasteners and lift out the condenser. See **Figure 40-50.**

Add the recommended amount of refrigerant oil to the new condenser. Position the new condenser and reinstall the bracket fasteners. Connect the inlet and outlet lines using new O-rings. If the radiator was relocated or removed, reinstall it. Evacuate and recharge the refrigeration system. Start the HVAC system and check the condenser and fittings for leaks.

Accumulator/Receiver-Drier Service

To remove the accumulator, first recover the system refrigerant and release all residual pressure. Then, use two wrenches to disconnect the inlet and outlet fittings. Remove the single screw holding the bracket around the accumulator and remove the accumulator. Transfer the clutch cycling switch

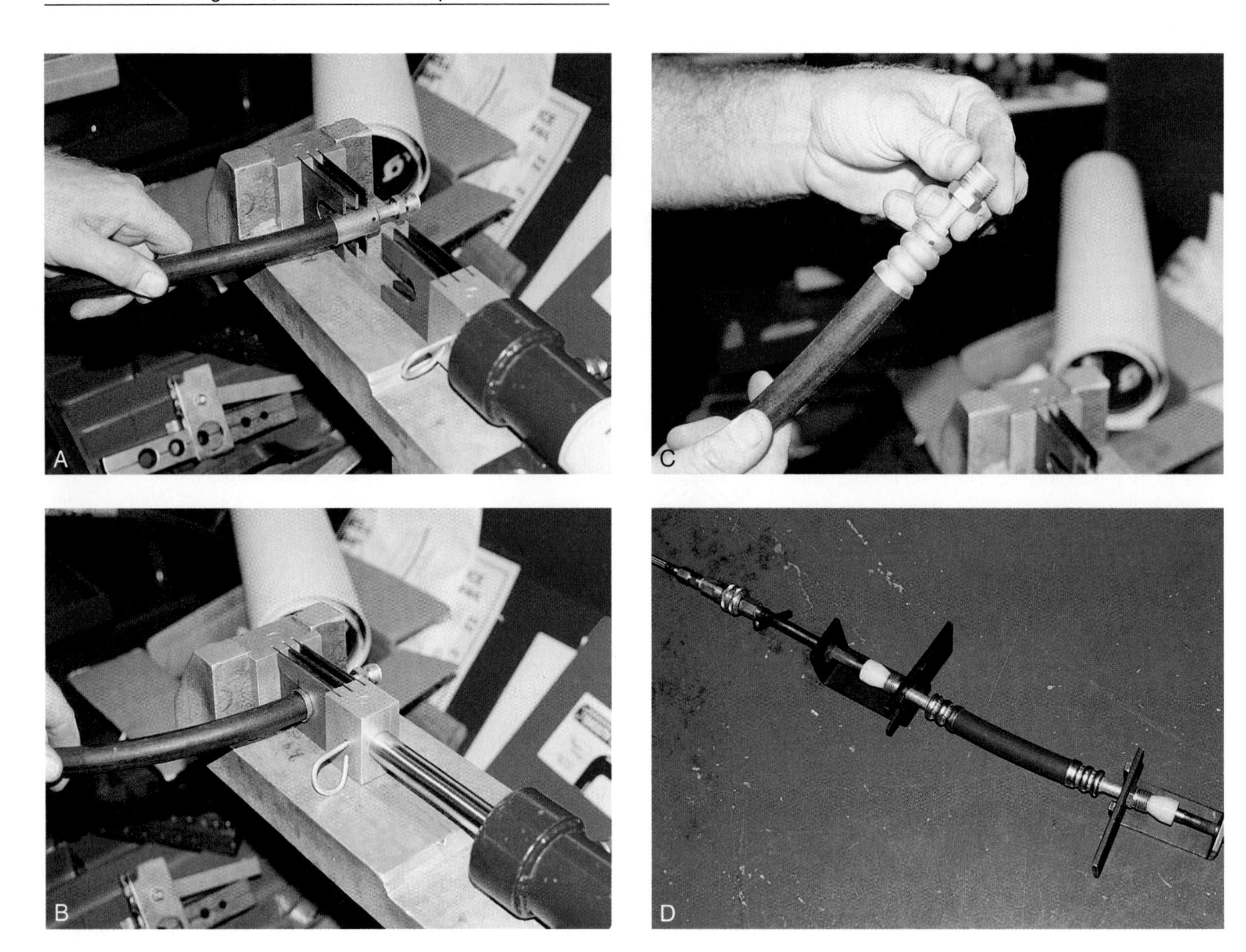

Figure 40-49. Assembling a hose with crimped connections. A—Fully insert the hose end into the fitting. B—Operate the crimping machine. C—Inspect the crimped fitting. The crimped areas must extend completely around the fitting. D—Leak-test the completed hose under pressure.

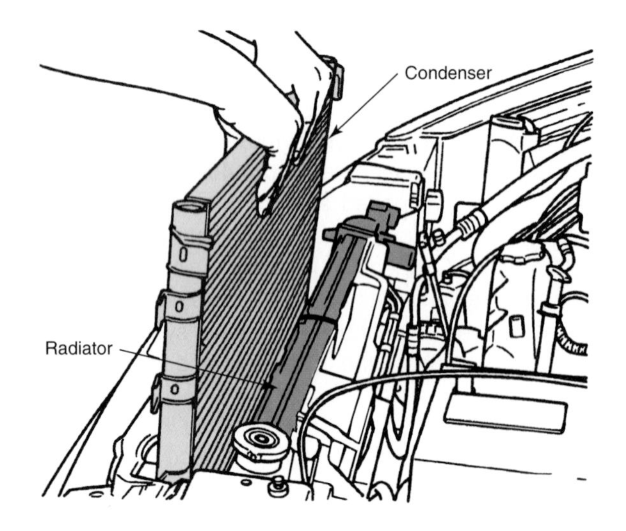

Figure 40-50. On some vehicles, the condenser can be lifted out the top after removing fasteners. On others, the condenser must be removed from below. (DaimlerChrysler)

and any other switches from the old to the new accumulator as necessary. Add the recommended amount of refrigerant oil to the new accumulator. Then, position the new accumulator on the bracket. Loosely connect the inlet and outlet fittings using new O-rings. Reinstall and tighten the bracket fastener. Tighten the inlet and outlet lines, then evacuate and recharge the refrigeration system. Start the HVAC system and check the accumulator and fittings for leaks.

Receiver-drier removal usually involves disconnecting the inlet and outlet lines and removing the attaching bracket. To remove the receiver-drier, first recover the system refrigerant and release all residual pressure. Use two wrenches to disconnect the inlet and outlet fittings. Remove the tensioning screws holding the brackets around the receiver-drier and remove the receiver-drier. Transfer any pressure switches to the replacement receiver-drier, if necessary. Add the recommended amount of refrigerant oil to the new receiver-drier and position it in the brackets. Loosely connect the inlet and outlet fittings using new O-rings. Reinstall and tighten the bracket tensioning screws. Tighten the inlet and outlet lines, then evacuate and recharge the refrigeration system. Start the HVAC system and check for leaks.

Control Head Replacement

Control head service is usually limited to replacement. To begin panel removal, disconnect the battery negative cable. Remove dashboard or center console components as necessary. Remove the control head, disconnecting any electrical and vacuum connectors. Installation is in the reverse order. Be sure to follow static discharge guidelines when handling the new control head.

In some cases, what seems like a major problem with an automatic control head may be caused by a poor connection between the control head and its wiring harness. You may want to try this before condemning an electronic control head. Remove the control head from the dash and carefully clean the electrical contacts. Apply a chemical contact enhancer and reinstall the control head. If this does not solve the problem, replace the unit.

Evaporator Service

Some evaporators must be serviced by removing one side of the evaporator case to expose the evaporator core. In some cases, the only way to access the evaporator core is to remove the entire blower case from the vehicle, **Figure 40-51.** Other evaporators can be accessed by removing a cover in the engine compartment and sliding the evaporator from its case.

Begin by determining what covers and other vehicle components must be removed to gain access to the evaporator. Once you have determined the evaporator core removal method, recover the refrigerant. Disconnect any electrical connectors attached to electrical devices on the evaporator cover and disconnect the evaporator tubing connections. On some vehicles, it may be necessary to disconnect the heater hoses, as well. Remove other components as necessary to gain access to the evaporator.

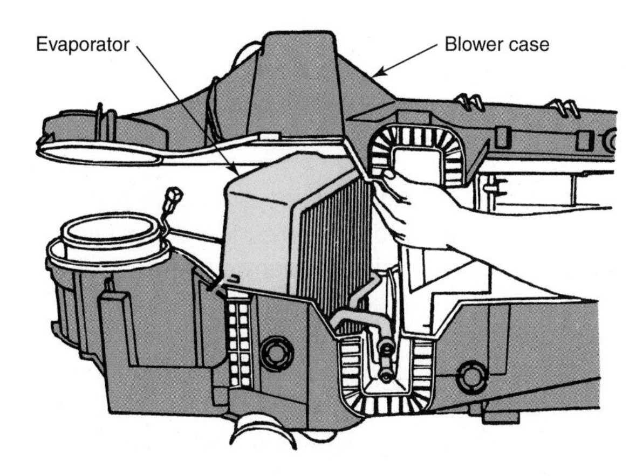

Figure 40-51. To access the evaporator core, it is sometimes necessary to remove the blower case from the vehicle. The case can then be disassembled, allowing the evaporator to be lifted out. (DaimlerChrysler)

After the evaporator has been removed, clean the inside of the case and check it for damage. Replace any internal seals if necessary. Make sure the evaporator drain hole(s) are open.

Compare the old and new evaporators to be sure you have the proper replacement. Place the new evaporator into position. Reconnect the evaporator connections using new O-rings and/or gaskets where needed. Reconnect all electrical wiring and reinstall any other vehicle parts that were removed. Evacuate the system and recharge the refrigerant. Start the HVAC system and check the evaporator and fittings for leaks.

Disinfecting Evaporator Core using a Biocide

A damp evaporator surface is an excellent place for various microorganisms to grow. These microorganisms, usually mold or mildew, can become so numerous they cause a musty smell inside the vehicle. To dispose of the microorganisms, a ***biocide*** (microorganism killer) should be used. Wear protective gloves and safety goggles while performing this procedure. Be sure to work in a well-ventilated area.

Warning: Most biocides can cause eye irritation. If the disinfectant gets into your eyes, flush with water for 15 minutes and consult a physician.

Mix the biocide if necessary. Remove the blower resistor or module to access the evaporator core; do not disconnect any wiring. Shine a light into the evaporator core and check for debris. If the debris is heavy or impacted on the evaporator surface, you will have to remove the core for cleaning and disinfectant (biocide) application. Place a drain pan below the evaporator drain.

Open all vehicle windows and doors and turn on the ignition key, but do not start the engine. Turn the blower motor control to low and the mode to vent. Using a cleaning spray gun, insert the siphon hose into the disinfectant and the nozzle into the evaporator case, **Figure 40-52.** Spray the disinfectant into the evaporator core, making sure the entire core is treated. During this process, you should use the entire bottle of disinfectant. You should also spray a little into the vents to kill any bacteria that may have migrated into the ductwork. Shut off the ignition and allow the disinfectant to stay on the core for at least five minutes. Then, turn the ignition key back to run. Using the spray gun with a quart container of water, thoroughly rinse the core of all disinfectant. Turn off the ignition and reinstall the resistor or power module.

Retrofitting

Automotive shops commonly perform R-12 to R-134a ***retrofitting.*** As R-12 becomes harder to get, almost all R-12 systems will be converted to R-134a or another refrigerant. The only vehicles that should not be retrofitted are those with nonadjustable low-side pressure controls. Very few modern vehicles have such pressure controls.

Remember that although some replacement refrigerants have been approved by the EPA, it does not mean the

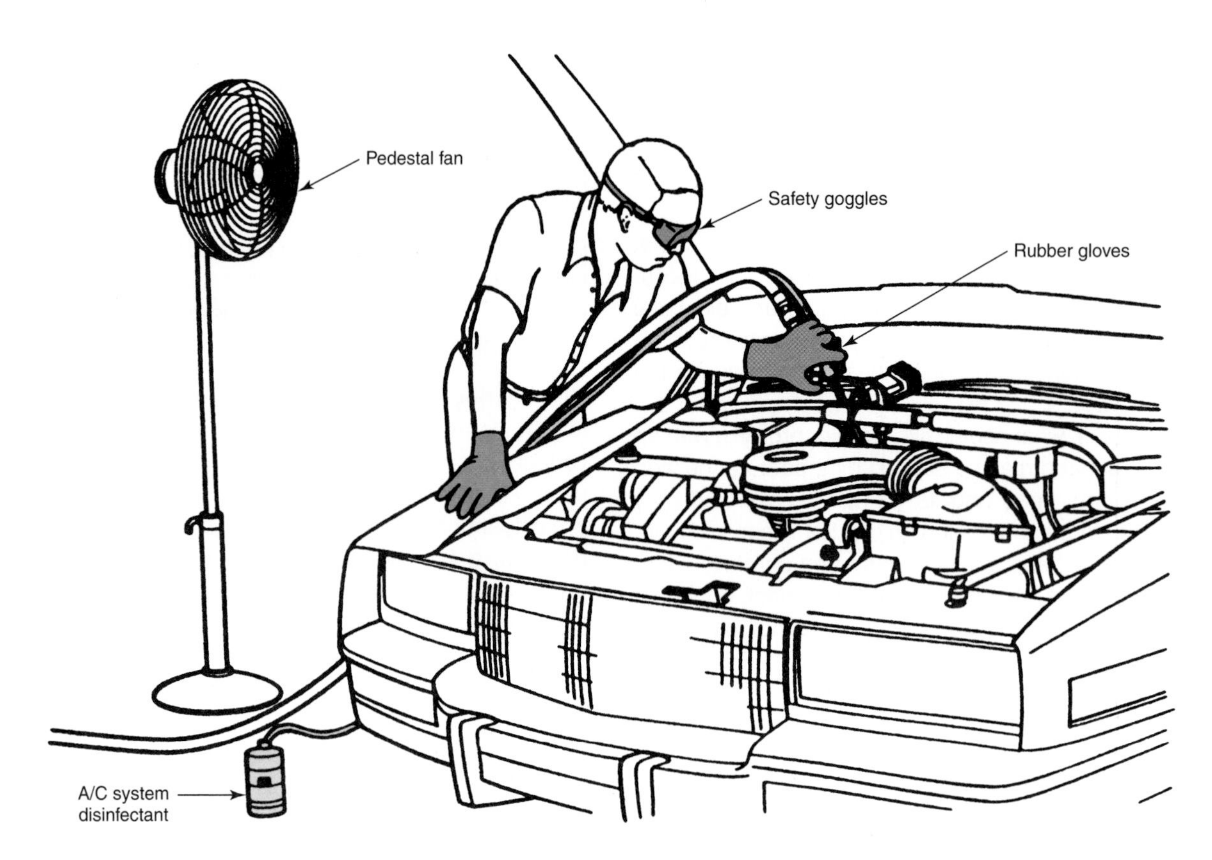

Figure 40-52. A pressure spray gun can be used to spray a disinfectant (biocide) through the evaporator core. Follow all safety procedures. (General Motors)

refrigerant is an efficient substitute for R-12 or R-134a, or that it will not damage refrigeration system parts. EPA approval means only the refrigerant is not ozone-depleting, flammable, or toxic.

If any substitute refrigerant is used, the EPA requires the original service fittings be converted to fittings specific to that refrigerant. This is to keep a future technician from accidentally topping off a system with the wrong refrigerant. Also, if the system is not equipped with a high-pressure cutoff switch, the law requires that one be installed as part of any retrofit.

Simple retrofit packages are sold that claim to contain all that is needed to convert from R-12 to R-134a. Although a retrofit in which only the refrigerant and oil are changed will produce cold air, the system will eventually fail because the original oil, desiccant, and rubber components are not compatible with R-134a, and because high-side pressures may become excessive. Unless the vehicle owner insists on a quick fix, perform a complete retrofit. A complete retrofit means replacing the O-rings, accumulator or receiver-drier (desiccant), and as much of the original oil as possible. Repairing any system leaks is a vital part of a complete retrofit. R-134a molecules are smaller than R-12 molecules, and a slight leak with an R-12 system will become a large leak when R-134a is substituted.

Before beginning a complete retrofit, check the refrigeration system and determine what other repairs may be needed. Complete other refrigeration system repairs as necessary, using R-134a compatible parts. Then replace the original mineral oil with polyalkylene glycol (PAG) or polyol ester (POE) oil.

Many retrofits require the clutch cycling switch be changed. R-134a evaporates most efficiently at a lower pressure than R-12, and the cycling switch should be replaced with a type that will allow the compressor to pull evaporator pressures to a lower value.

Note: When retrofitting, a small amount of the old oil will be left in the system. Usually, this does not affect the efficiency of the new oil.

In addition, R-134a condenses at a higher pressure, and it may be necessary to replace the condenser or add an extra cooling fan to increase airflow. Many OEM and aftermarket manufacturers offer replacement condensers for R-134a retrofitting. The fan can be wired into the air conditioning system so it runs whenever the compressor is turned on, or wired in parallel to operate with the original cooling fan.

Finally, evacuate, leak test, and charge the system. For best operation, most systems should be charged with R-134a to about 75% of the original R-12 charge. For instance, if the original R-12 charge was 2 pounds (0.9 kg), use about 1 1/2 pounds (0.675 kg) of R-134a. Before completing the job, start the system and check pressures and vent temperatures.

Heater System

The heater depends on hot engine coolant flowing through a metal core to provide heat to the passenger compartment. On modern vehicles, the ***heater core*** is integrated into the overall air conditioning system, **Figure 40-53.** On vehicles without air conditioning, the heater system consists of a heater core, blower motor, and associated ductwork.

Checking Heater Operation

When possible, question the owner as to apparent problem, any unusual sounds, or other symptoms. Inspect the entire cooling system and the heater system for leakage. Check hoses for softening, cracking, kinks, or hardening. Inspect the air intake for clogging or bent or loose ducting. Bring the engine to its normal operating temperature and operate the heater controls while checking for proper response. Check heat control valve operation, blower speeds, defroster, and vent operation.

As required, check blower fuse, operation of vacuum motors, and mechanical controls. See Chapter 23 for more on hoses, antifreeze, and other cooling system related information.

Heater Shut-Off Valve

Many heater systems are equipped with a ***shut-off valve.*** This valve stops the flow of engine coolant through the heater core when heating is not needed. If the heater blows cold air, check the operation of the shut-off valve by operating the external linkage. If the heater begins working when the linkage is moved to opposite its original position, the valve was stuck. Check for cooling system debris in the valve, or for a defective vacuum diaphragm. On cable operated shut-off valves, check for misadjustment or a broken cable.

Heater Core Service

If you suspect a heater core leak, conduct a cooling system pressure test. Make certain there are no leaks at the hose-to-core attachment points. If the heater core is clogged, output will be low when the engine is at normal operating temperature. The heater core is usually located under the dashboard, with its connections passing through the vehicle firewall. Replacement procedures are somewhat similar to evaporator core replacement and may involve removing parts of the ductwork and other underdash components. Always obtain the proper service manual and follow all procedures exactly.

Blower Motor Service

If the blower motor is inoperative, first check the fuse. If the fuse is okay, check the blower motor relays, usually located on the engine firewall. The relay is often the cause of an inoperative blower at some speed settings. Many vehicles use resistor assemblies, **Figure 40-54,** to obtain various blower speeds. Always check for burned-out resistors, especially if the blower works on some speeds but not others. Check any electronically controlled blower motor circuits, using procedures outlined in the appropriate service manual. If all other components are good, test the blower motor for voltage and amperage draw. Perform all tests as specified by the manufacturer.

Blower Motor Replacement

Blower motors are typically replaced rather than repaired. To begin removal, first find the blower motor. Many blower motors are located under the hood where they are easily accessible. Others are located under side panels, between the firewall and inner fender, or under the dashboard. In almost all cases, the blower can be removed without removing the HVAC case.

Note: The blower motor may be located in a difficult-to-reach spot on some vehicles. Check the service manual and any technical service bulletins for proper removal procedures.

Once the blower is located, make sure the ignition key is in the *off* position, and then remove any parts necessary to gain access to the blower. Remove the electrical connector, mounting screws, and then the blower motor, **Figure 40-55.**

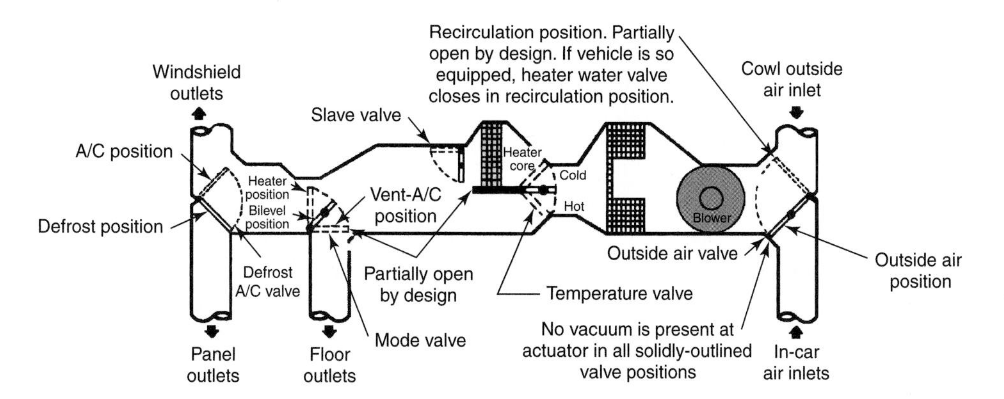

Figure 40-53. One type of heater-air conditioning system assembly, showing location of the heater core in relation to other parts. (General Motors)

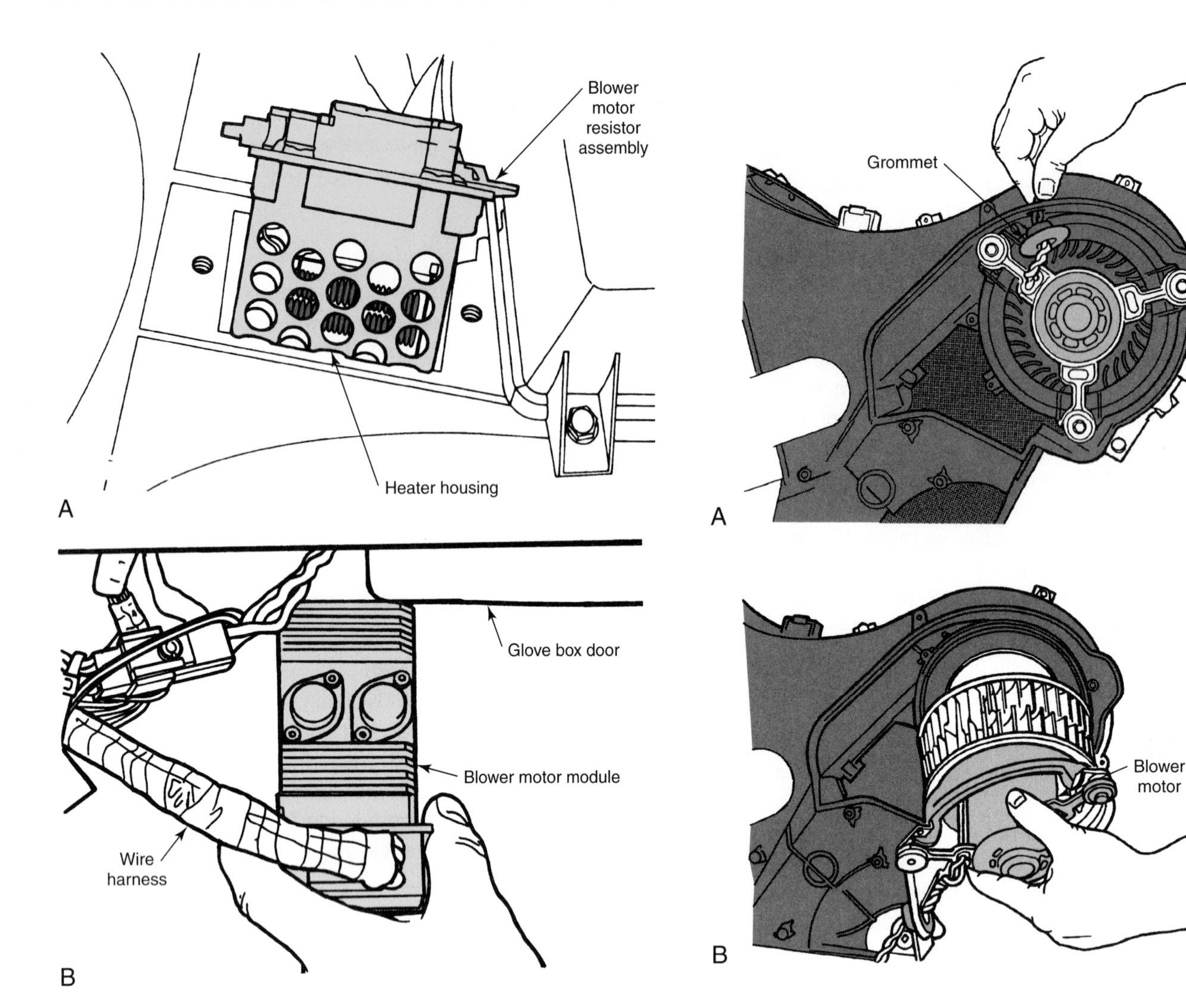

Figure 40-54. Blower motor speed control methods. A—This blower motor resistor assembly is located in the heater housing, and used with a nonautomatic temperature control system. B—A blower motor module, located in the heater housing, that is used with an automatic temperature control system. This module receives control signals from the body control module (BCM). The power module varies the voltage traveling to the blower motor. This unit provides fourteen different speeds. (DaimlerChrysler)

Figure 40-55. Blower motor removal. A—Remove the electrical connector. B—Remove the mounting bolts and pull the motor and squirrel cage free of the housing. (DaimlerChrysler)

After the blower is removed, the blower wheel usually must be dismounted from the old motor and installed on the new one. Some factory and aftermarket blower assemblies are provided with a new wheel already installed. Blower wheels commonly are held to the motor by a threaded nut or push-on clip. Some blower wheel nuts have left-hand threads. If the wheel is hard to remove, hold the wheel and lightly tap the motor shaft with a rubber hammer.

To install the replacement blower, position it properly and install the mounting screws. Make sure any ground wires are connected. Install the power wire and check blower operation. Then, reinstall any other vehicle parts that had been removed.

Air Distribution and Control System Service

Modern heating and air conditioning systems have elaborate control arrangements using electrical switches and relays; electronic sensors and controls; mechanical linkages, and vacuum diaphragms, reservoirs, and hoses. All of these controls work together to operate the heating and air conditioning system and control air distribution. The control system operates the blower motor, as well as doors in the air distribution system, **Figure 40-56.** The doors are built into the ductwork and direct airflow to the vents, floor, or windshield as needed. A separate door, usually called a ***blend door,*** controls the temperature of the incoming air by blending air from the evaporator and the heater core.

Manual control systems are operated by the vehicle occupants through a series of vacuum and electrical switches, cables and/or levers. Modern auto air conditioning systems, especially those that are computer-controlled, have complex wiring and vacuum control systems. In such a system, a central computer (control module) controls the air distribution system based on inputs from thermistors (temperature-sensitive resistors) and pressure sensors, as well as from temperature settings made by the passengers. **Figure 40-57** shows a dual-zone air distribution system. This system allows the driver and the passenger to separately control airflow and temperature.

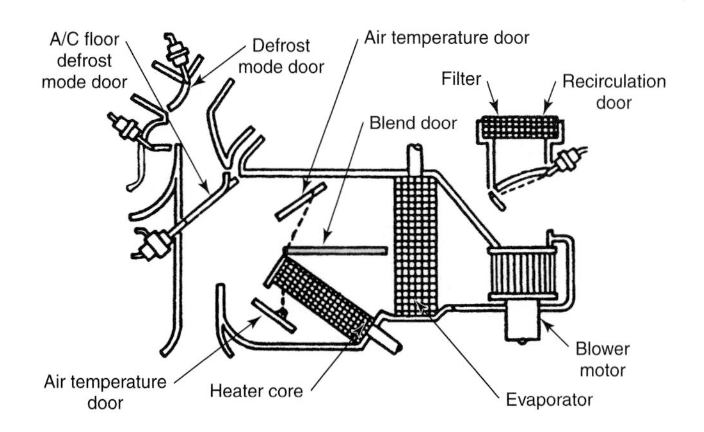

Figure 40-56. Positions of the blower case doors in a manually controlled air conditioning system when controls are in the off position. (Ford)

Air Distribution Wiring and Vacuum Circuits

The air mixing (blending) abilities of an airflow control system permits the discharge of temperature-regulated air into the vehicle. In hot weather, this blending system will permit the evaporator to operate constantly near the freezing point, thus permitting full efficiency. In colder weather, the controls can be adjusted to produce warm air from the same ducts. When the weather is cold and humid, causing the windshield to fog, the system can be adjusted to use the evaporator to remove humidity from the incoming air and the heater core to warm it. This air is then directed to the windshield to reduce fogging.

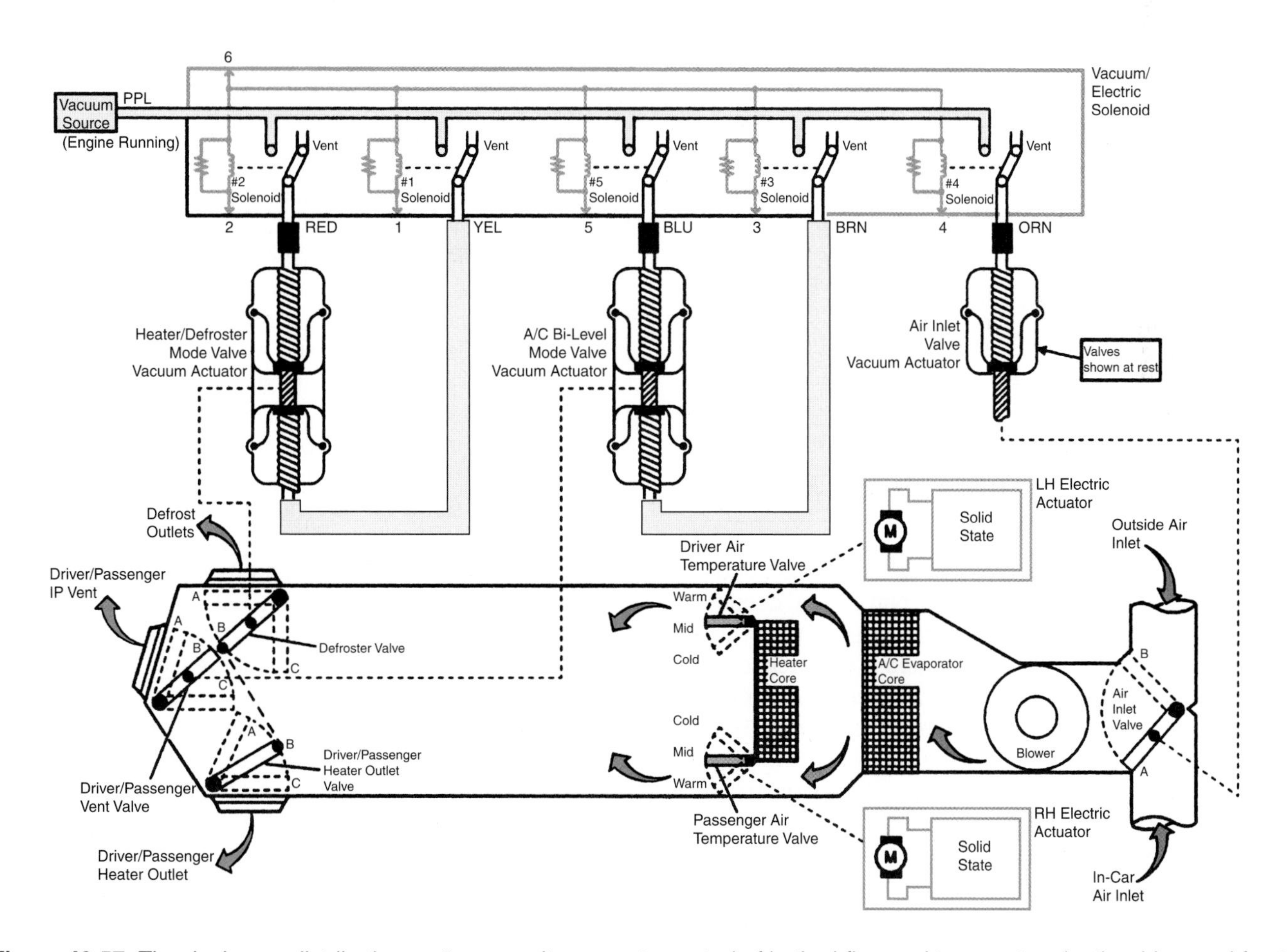

Figure 40-57. The dual-zone distribution system permits separate control of both airflow and temperature by the driver and front-seat passenger. (General Motors)

Basic checks of the control system consist of making sure all vacuum hoses are properly connected and that linkage cables are correctly adjusted. Also check for blown fuses and disconnected wiring. More complex checking will require using the vehicle's service manual.

Air Distribution System Ductwork

The system ductwork must be properly aligned and clear of blockages for the blending abilities of the air distribution system to be fully utilized. Check all ductwork for damage or misalignment. Also check any gaskets and seals for damage. One often-overlooked area is the intake ductwork at the front cowl. This often becomes clogged with dirt and leaves, reducing the ability of the blower to pull in air.

Tech Talk

Improper disposal of refrigerant is a serious concern. Whether or not you believe that refrigerants can damage the ozone layer, there is one overriding reason to recycle refrigerants: if you do not, you are breaking the law. Improper refrigerant disposal is covered by federal and, in some cases, state laws.

A local salvage yard sold an air conditioner compressor from a wrecked vehicle. When the owner of the salvage yard removed the compressor, he simply opened the refrigerant lines, discharging the refrigerant to the atmosphere. Unfortunately, the compressor buyer was an agent of the Environmental Protection Agency, investigating just this sort of activity. In court, the owner was ordered to pay a substantial fine. You do not want this to happen to you.

Also, replacing wasted refrigerants costs money. For instance, R-12 refrigerant costs over seven times as much as it did just a few years ago. R-12 has not been manufactured since January 1, 1996. As supplies of R-12 decrease, the price will become very high. Once R-12 is no longer available, an air conditioner will require a retrofit to accept another type of refrigerant. It makes good economic and environmental sense to recover and reuse refrigerants.

Summary

Air conditioning systems may be either manually or electronically controlled. The automatic system cleans, dehumidifies, and adjusts the temperature of the incoming air. The air is first cooled and then, if needed, heated to the desired level. Air conditioning involves using a refrigerant (R-134a or R-12) that vaporizes and condenses as pressures change. The air conditioning system uses this change in pressures to remove heat from air entering the passenger compartment and transfer it to the outside air.

Modern air conditioners can be classified according to whether they have a variable or fixed restrictor. Variable restrictors are called expansion valves, while fixed restrictors are called orifice tubes. Some components are common to all air conditioning systems, while other components are used only on some systems. The compressor pumps and compresses the incoming refrigerant vapor. The condenser changes the refrigerant vapor into a liquid. The restrictor allows the refrigerant liquid to enter the evaporator at low pressure. Heat from the incoming air is absorbed by the liquid refrigerant in the evaporator. This causes the refrigerant to boil, becoming a vapor. The evaporator also dehumidifies and cleans the incoming air. The entire system is connected by lines and hoses.

The receiver-drier, located between the condenser and evaporator, acts as a storage cylinder for high-pressure liquid refrigerant. The accumulator holds excess refrigerant on the low side of the system. It prevents the entry of liquid refrigerant into the compressor. It also cleans and removes moisture.

On many systems, evaporator pressure is controlled by a valve placed between the evaporator and compressor. On newer vehicles, either a pressure or temperature switch is used to cycle the compressor on and off to control evaporator icing. One or more mufflers may be used to quiet system noises. A sight glass may be used to view the stream of liquid refrigerant.

Always follow safety rules when servicing air conditioning systems. Always wear goggles and gloves and keep the area well ventilated. Do not discharge refrigerant directly into the service area. In addition to being illegal, releasing refrigerant vapor into the air is hazardous: the vapor can form phosgene gas if it contacts a flame. Do not subject the system to high temperatures. Do not overheat or completely fill a cylinder. Never connect refrigerant containers to the system high side. Never transport refrigerant containers in the passenger compartment of a vehicle.

Observe recommended service precautions to prevent system contamination. Remember the enemies of air conditioning systems are dirt, water, and air. Recover the system's refrigerant charge before opening any connection. The manifold gauge set is connected to the system by means of Schrader valves or screw-on service fittings. Manifold gauge valves must be in the closed position before connecting the set. Use only electronic leak detection tools. Detector probe should be used slowly and close to test area. Set up and use as directed by manufacturer. Always use a refrigerant recovery unit. The system must be evacuated for the recommended length of time at 29″ of vacuum. Evacuation removes air and moisture from the system.

Following evacuation and test for leaks, fill the system with a partial charge Then charge the system with the specified amount of refrigerant. Before removing any part for service, follow the series of steps given in the text. Always use new O-rings. Protect hose and tubing from vibration, kinks, sharp bends, and heat. Properly torque all connections.

The amount of compressor oil in the system is critical. In some systems, it may be checked by measuring the level in the compressor. In others, the compressor must be removed and the oil drained and measured. When an evaporator, condenser, receiver, or other part is replaced, a specific amount of oil must be added to the new unit to replace oil that remains in the old part. The oil level need not be checked in a system in which no leaks are present. Some systems permit oil to be injected into the system without discharging.

A number of simple checks can be used to locate the trouble in a malfunctioning system. Performance testing involves comparing system pressure and discharge air temperature with specified pressures and temperatures related to ambient air temperature and humidity. Inspect the system as soon as possible following collision damage.

Review Questions—Chapter 40

Do not write in this book. Write your answers on a separate sheet of paper.

1. Name three things the air conditioning system does to the air entering the vehicle.
2. Name the two types of air conditioning system restrictors.
3. The condenser performs which of the following?
   (A) Changes refrigerant vapor into liquid.
   (B) Changes liquid refrigerant into vapor.
   (C) Controls the evaporator temperature.
   (D) Removes moisture from the air entering the vehicle.
4. The manifold gauge assists in ______.
   (A) checking system pressures
   (B) evacuating the system
   (C) recovering the refrigerant
   (D) All of the above.
5. The sight glass permits ______.
   (A) viewing liquid refrigerant
   (B) viewing refrigerant vapor
   (C) checking refrigerant for icing
   (D) checking oil level in compressor
6. Some R-12 systems have different high- and low-side _____ valve fittings.
7. When torquing a connection that uses aluminum for one end and steel for the other, use the torque values given for _____.
8. List ten safety rules regarding working with, on, or around air conditioning systems.
9. Describe the first aid procedure involved when refrigerant gets in the eyes.
10. Before connecting the manifold gauge set to the system, the gauge valves must be in the _____ position.
11. To evacuate the system properly, the system must be subjected to a vacuum of around _____.
    (A) 8″
    (B) 29″
    (C) 48″
    (D) 58″
12. Performance testing involves checking _____.
    (A) discharge air temperature
    (B) system pressures
    (C) ambient air temperature and humidity
    (D) All of the above.
13. The most common clutch control switch is the _____.
    (A) thermostatic temperature cycling switch
    (B) high-pressure cutoff switch
    (C) pressure cycling switch
    (D) low-pressure cutoff switch
14. Name the door that is used to control air output temperature.
15. List four things that should be checked when inspecting the air distribution system.

ASE-Type Questions

1. Technician A says refrigerant will eventually wear out from endless cycles through the system and should be replaced at specified intervals. Technician B says as the pressure is increased on the refrigerant, the boiling point goes up. Who is right?
   (A) A only.
   (B) B only.
   (C) Both A and B.
   (D) Neither A nor B.
2. R-12 is gradually being replaced for use in air conditioners with _____.
   (A) butane
   (B) nitrogen
   (C) R-134a
   (D) R-22
3. Technician A says some systems require complete discharging in order to check compressor oil level. Technician B says a high-quality polyol ester (POE) oil is satisfactory for use in all air conditioning systems. Who is right?
   (A) A only.
   (B) B only.
   (C) Both A and B.
   (D) Neither A nor B.
4. Refrigerant leaves the compressor as a _____.
   (A) cold, high-pressure vapor
   (B) hot, high-pressure vapor
   (C) cold, low-pressure vapor
   (D) hot, high-pressure liquid

5. Technician A says the evaporator changes refrigerant vapor back into liquid refrigerant. Technician B says a thermostatic expansion valve is controlled by evaporator temperature. Who is right?
   (A) A only.
   (B) B only.
   (C) Both A and B.
   (D) Neither A nor B.
6. On a fixed orifice system, cycling the compressor clutch does all of the following, *except:*
   (A) controls compressor speed.
   (B) controls flow of refrigerant through the evaporator.
   (C) keeps evaporator temperature from becoming too low.
   (D) keeps evaporator pressure from becoming too low.
7. The receiver-drier performs all of the following functions, *except:*
   (A) stores liquid refrigerant.
   (B) removes debris and dirt from the refrigerant.
   (C) causes the refrigerant to condense into a liquid.
   (D) removes moisture from the refrigerant.
8. Phosgene gas is created when refrigerant comes in contact with ______.
   (A) moisture
   (B) refrigerant oil
   (C) flame
   (D) dirt
9. Technician A says moisture can cause corrosion of the internal air conditioning system parts. Technician B says moisture can cause refrigerant oil sludging. Who is right?
   (A) A only.
   (B) B only.
   (C) Both A and B.
   (D) Neither A nor B.
10. When charging a system, never add liquid refrigerant to the ______ unless specifically required.
   (A) system high side
   (B) system low side
   (C) evaporator
   (D) Any of the above.

Suggested Activities

1. Draw a sketch showing the flow of refrigerant through a simple air conditioning system. Indicate major parts and how the system controls refrigerant flow.
2. From the drivers' seat of a vehicle, perform an operational check on the heating and air conditioning system. Use a thermometer to check air conditioner output. Perform two tests, one in the shop and one outside.
3. Attach gauges to an air conditioning system and check system pressures.
4. Using the procedure in the appropriate service manual, check adjustment of the various doors in the blower case.

Air Conditioning System Diagnosis	
Problem: Excessive high-side pressure	
Possible cause	**Correction**
1. Air in system.	1. Leak test. Recover and correct leak. Evacuate and charge system.
2. Overcharge of refrigerant.	2. Recover. Evacuate and charge with correct amount.
3. Engine overheating.	3. Correct cause of overheating.
4. Water pump belt slipping.	4. Adjust or replace belt.
5. Clogged condenser core.	5. Remove bugs, leaves, dirt, and other debris.
6. Excessive oil in system.	6. Remove excess oil.
7. Restriction in lines, condenser, or receiver-dehydrator.	7. Remove part and clean or replace as needed.
8. Expansion valve superheat setting too low.	8. Replace unit.
9. Filters or screens plugged.	9. Remove and clean or replace as needed.
10. Low fan speed.	10. Test and correct.
Problem: Insufficient high-side pressure	
Possible cause	**Correction**
1. Insufficient refrigerant charge.	1. Charge system with recommended amount or refrigerant.
2. Defective compressor valves.	2. Replace valves or compressor.
3. Expansion valve or evaporator pressure valve stuck open.	3. Replace valve.
4. Extremely cold ambient temperature.	4. Retest on warmer day.
Problem: Excessive low-side pressure	
Possible cause	**Correction**
1. Defective expansion valve.	1. Replace expansion valve.
2. Insufficient oil in system.	2. Add oil.
3. Expansion valve thermal bulb not in good contact with evaporator.	3. Clean connection and tighten. Insulate outlet pipe as required.
4. Defective evaporator pressure control valve.	4. Replace control valve.
5. Expansion valve frozen.	5. Replace receiver-dehydrator. Recharge system.
6. Compressor clutch slipping.	6. Repair or replace clutch.
7. Restricted suction line.	7. Clean or replace suction line.
8. Slipping compressor drive belt.	8. Adjust or replace belt.
9. Defective compressor valves.	9. Replace valves or compressor.
10. Moisture in system.	10. Repair leaks. Replace receiver-drier. Evacuate and recharge system.
Problem: Insufficient low-side pressure	
Possible cause	**Correction**
1. Insufficient charge.	1. Charge with recommended amount of refrigerant.
2. Insufficient airflow.	2. Clean evaporator core. Check blower operation.
3. Defective evaporator pressure control valve.	3. Repair. Adjust or replace as needed.
4. Defective expansion valve.	4. Replace expansion valve.
5. Liquid line clogged.	5. Replace line.
6. Restricted suction line, receiver-drier, or expansion valve.	6. Replace line, receiver-dehydrator, or expansion valve.
7. Temperature control thermostat does not cut out.	7. Replace thermostat.
8. Compressor clutch will not disengage.	8. Repair or replace compressor clutch.
Problem: Water discharged with airflow	
Possible cause	**Correction**
1. Clogged evaporator drain.	1. Clean evaporator drain.

(continued)

Air Conditioning System Diagnosis *(continued)*	
Problem: System noisy	
Possible cause	**Correction**
1. Compressor mounting loose. 2. Compressor belt slipping. 3. Refrigeration system lines vibrating. 4. Blower motor defective. 5. Loose air ducts. 6. Excessive oil in system. 7. Blower blades striking housing. 8. Obstructions in airflow system. 9. Defective compressor. 10. Defective expansion valve.	1. Tighten mounting fasteners. 2. Adjust belt tension. 3. Install clamps and insulators. 4. Replace blower motor. 5. Tighten air ducts. 6. Drain and install correct amount of oil. 7. Adjust for clearance. 8. Remove obstructions. 9. Repair or replace as required. 10. Replace expansion valve.
Problem: Airflow contains objectionable odor	
Possible cause	**Correction**
1. Odor-producing material on evaporator core. 2. Outside odors drawn in by airflow system.	1. Clean evaporator core, add disinfectant. 2. Advise driver as to reason.
Problem: Insufficient airflow	
Possible cause	**Correction**
1. Defective blower. 2. Clogged ducts. 3. Evaporator core icing. 4. Loose duct hose connections. 5. Shut-off valves in air discharge outlets closed. 6. Dirty evaporator core. 7. Airflow system control doors malfunctioning. 8. Blower disconnected or circuit fuse blown.	1. Replace blower. 2. Clean ducts. 3. Replace control valve or thermostatic switch. 4. Attach flexible hose securely. 5. Advise driver as to proper operation. 6. Clean core. 7. Check vacuum or electronic control system. 8. Connect or replace fuse.
Problem: Evaporator icing	
Possible cause	**Correction**
1. Defective or improperly adjusted evaporator pressure valve. 2. Defective thermostatic switch. 3. Compressor clutch will not disengage. 4. Thermostat capillary tube not in proper contact with evaporator.	1. Adjust or replace valve. 2. Replace switch. 3. Repair clutch. 4. Place tube in proper contact with core.
Problem: No or insufficient cooling	
Possible cause	**Correction**
1. Defective thermostatic switch. 2. Defective evaporator pressure valve. 3. Broken or slipping compressor belt. 4. Compressor defective. 5. Compressor clutch inoperative. 6. Defective expansion valve. 7. Insufficient refrigerant charge. 8. Excessive oil in system. 9. Expansion valve screen clogged. 10. Bugs, leaves, other debris on condenser. 11. Fan belt slipping or broken. 12. Excessive refrigerant charge. 13. Moisture in system. 14. Air in system.	1. Replace switch. 2. Replace valve. 3. Replace or adjust tension. 4. Replace compressor. 5. Repair clutch. 6. Replace expansion valve. 7. Charge system. 8. Drain system and add correct amount. 9. Clean screen or replace valve. 10. Clean condenser. 11. Adjust tension or replace belt. 12. Charge correctly. 13. Repair leaks. Install new receiver-dehydrator and charge. 14. Repair leaks. Evacuate and charge.

(continued)

Air Conditioning System Diagnosis *(continued)*	
15. Evaporator core dirty.	15. Clean core.
16. Clogged or kinked lines.	16. Clean or replace lines.
17. Clogged receiver-dehydrator.	17. Install new receiver-dehydrator.
18. Engine overheating.	18. Correct cause.
19. Clogged evaporator drain.	19. Clean drain.
20. Evaporator icing.	20. See *Evaporator icing*.
21. Clogged orifice tube.	21. Replace orifice tube.

Air Distribution System Diagnosis	
Problem: Blower does not operate correctly	
Possible cause	**Correction**
1. Blown fuse. 2. Defective motor. 3. Faulty blower resistor. 4. Defective switch. 5. Loose connections. 6. Faulty wiring.	1. Replace fuse. Fix short if needed. 2. Replace motor. 3. Replace resistor. 4. Replace switch. 5. Clean and tighten. 6. Repair wiring.
Problem: Inadequate heating	
Possible cause	**Correction**
1. Faulty blower motor. 2. Heat valve control inoperative. 3. Defective heat control valve. 4. Clogged air inlet. 5. Bent, kinked inlet ducting. 6. Debris on heater core. 7. Control doors inoperative. 8. Faulty controls 9. Faulty heater operation. 10. Faulty engine thermostat. 11. Plugged or kinked heater hoses. 12. Low engine coolant level. 13. Clogged heater core.	1. Replace blower motor. 2. Free control and make certain valve control functions. 3. Replace heat control valve. 4. Clean air inlet, replace screen if missing. 5. Straighten or replace. 6. Clean heater core. 7. Free and make certain they operate properly. 8. Adjust or replace as needed. 9. Advise owner as to proper operation. 10. Replace thermostat. 11. Straighten or replace hoses. 12. Add coolant and check for leaks. 13. Reverse flush or replace heater core.
Problem: Inadequate defrost	
Possible cause	**Correction**
1. Defective blower switch. 2. Defective or inoperative blower motor. 3. Defroster ducts are disconnected. 4. Improper control door operation. 5. Clogged defroster outlets.	1. Replace switch. 2. Repair or replace. 3. Connect ducts. 4. Free, adjust, and check control door. 5. Clean outlets.
Problem: Defrost causes windshield fogging	
Possible cause	**Correction**
1. Leaking heater core. 2. Loose hose to core fitting. 3. Water (from rain or washing) entering system.	1. Replace heater core. 2. Tighten fitting. 3. Check seals for leaks.
Problem: Excessive heat	
Possible cause	**Correction**
1. Faulty operation of controls. 2. Controls loose or stuck.	1. Advise driver in correct use. 2. Connect and check for proper operation.

41

Repair Orders and Cost Estimates

After studying this chapter, you will be able to:

- Explain the purpose of repair orders.
- Fill out a repair order.
- Explain the purpose of cost estimates.
- Complete a cost estimate.

Technical Terms

Repair order	Markup
Customer pay order	Core charge
Warranty order	Disposal fee
Hourly rate	Sales tax
Flat rate manual	Cost estimate

This chapter explains the use of repair orders and cost estimates. Repair orders tell the technician about a vehicle's problem and the needed repair. They also provide a record for billing the customer. The technician must be able to prepare cost estimates so the customer can have a basis for determining whether or not a repair should be performed.

Note: This is an important chapter. It will help you develop skills needed for many positions in the auto repair industry. Study it carefully.

Repair Orders

A ***repair order,*** also called a shop work order, is used to keep a record of the services performed on a particular vehicle. It is used to determine the final bill and may be used to determine the technician's pay. If the customer is going to pay for the vehicle service, the repair order is termed a ***customer pay order.*** If the service is covered by a warranty or guarantee, the repair order is called a ***warranty order.*** Many repair orders contain both customer pay and warranty work. See **Figure 41-1.**

The service manager, service writer, or shop manager initiates a repair order. This person fills in the customer's name, address, and telephone number, and completes the vehicle information, such as the make, model, year, mileage, and number of cylinders.

Note: If the customer is dropping off the vehicle, be sure to get a telephone number where he or she can be reached should additional problems be discovered during repairs.

The vehicle's problem is also listed on the repair order to aid the technician. If the customer complained of a cold start driveability problem, for example, this provides the technician with a starting point for diagnosis. If the vehicle is in for routine maintenance, there will be no problem listed.

Note: It is the job of the person who initiates the order to determine the actual problem area. If, for instance, the customer comes in wanting a "tune-up," the actual vehicle performance or driveability complaint should be determined.

An estimate of the anticipated repair costs (explained below) is shown to the customer. If he or she agrees to the cost, the repair order is then given to a technician to be used for diagnosing, testing, and correcting the problem. If costs will exceed the estimate, the customer should be notified. The customer should give permission for the increased cost before the job is completed.

At the beginning of the repair job, the technician writes or punches in the start time and does the same when finished, allowing time off for lunch, breaks, or other jobs. When defective parts are replaced, they are listed in the space provided on the repair order. The part number, description, and price are also included. In some cases, it is necessary to state whether the part is new, used, or rebuilt.

After the repair is complete, the labor cost is computed. Then, the part prices, the costs of outside repair services, and the prices of other supplies are added to the order. All prices are totaled and sales tax is added. This is the customer's total bill. The calculation of prices will probably be the job of the service manager or the shop cashier. No matter what you are required to fill in on the repair order, make all necessary additions neatly and correctly. This gives a "professional look" to your overall work. In many cases, the repair order will be used to calculate your pay. Accuracy and legibility are critical.

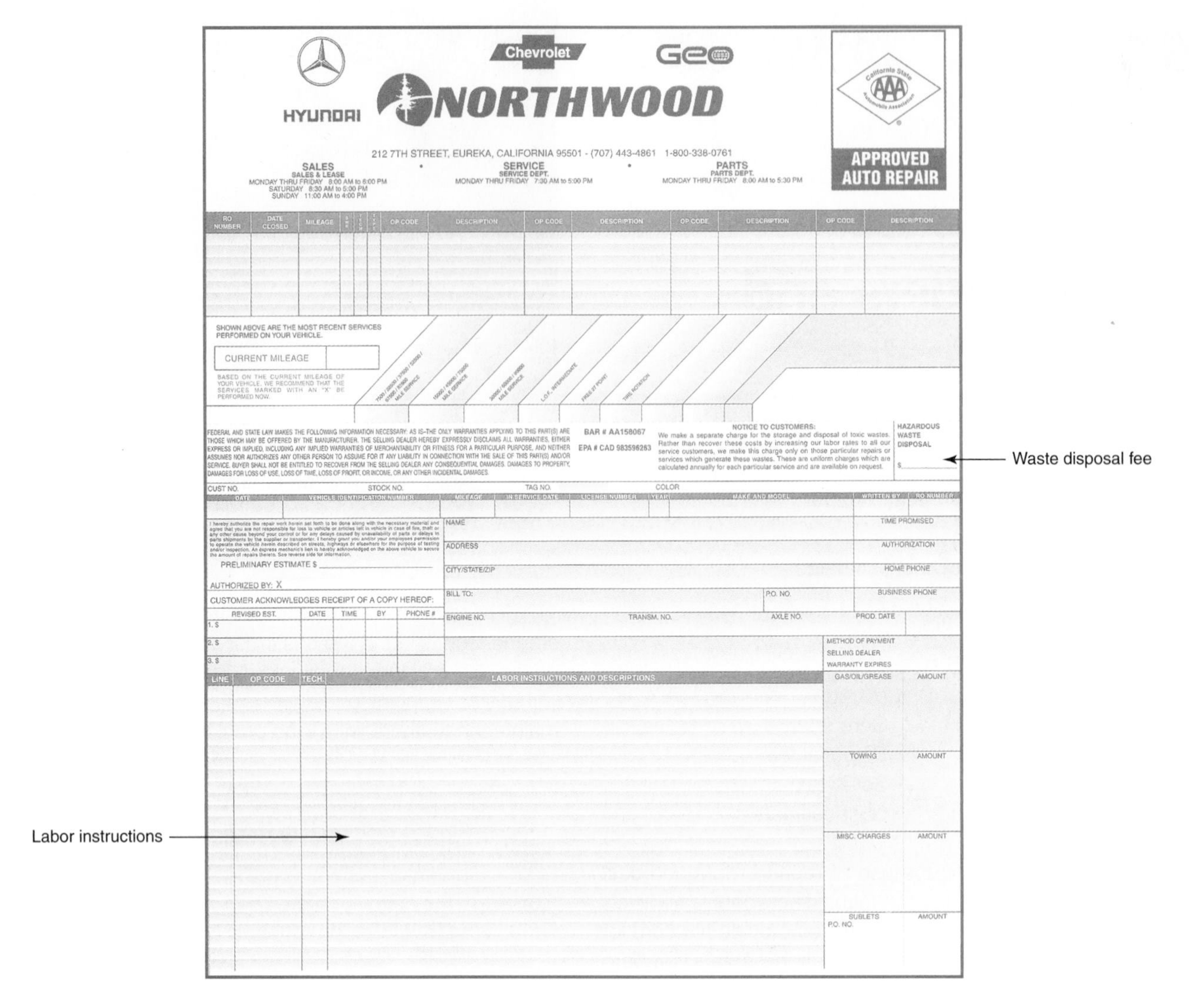

HYUNDAI Chevrolet Geo NORTHWOOD

212 7TH STREET, EUREKA, CALIFORNIA 95501 - (707) 443-4861 1-800-338-0761

SALES
SALES & LEASE
MONDAY THRU FRIDAY 8:00 AM to 6:00 PM
SATURDAY 8:30 AM to 5:00 PM
SUNDAY 11:00 AM to 4:00 PM

SERVICE
SERVICE DEPT.
MONDAY THRU FRIDAY 7:30 AM to 5:00 PM

PARTS
PARTS DEPT.
MONDAY THRU FRIDAY 8:00 AM to 5:30 PM

California State AAA Automobile Association
APPROVED AUTO REPAIR

RO NUMBER	DATE CLOSED	MILEAGE				OP CODE	DESCRIPTION	OP CODE	DESCRIPTION	OP CODE	DESCRIPTION	OP CODE	DESCRIPTION

SHOWN ABOVE ARE THE MOST RECENT SERVICES PERFORMED ON YOUR VEHICLE.

CURRENT MILEAGE

BASED ON THE CURRENT MILEAGE OF YOUR VEHICLE, WE RECOMMEND THAT THE SERVICES MARKED WITH AN "X" BE PERFORMED NOW.

7500 / 22500 / 37500 / 52500 / 67500 / 82500 MILE SERVICE
15000 / 45000 / 75000 MILE SERVICE
30000 / 60000 / 90000 MILE SERVICE
L.O.F., INTERMEDIATE
FREE 27 POINT
TIRE ROTATION

FEDERAL AND STATE LAW MAKES THE FOLLOWING INFORMATION NECESSARY: AS IS–THE ONLY WARRANTIES APPLYING TO THIS PART(S) ARE THOSE WHICH MAY BE OFFERED BY THE MANUFACTURER. THE SELLING DEALER HEREBY EXPRESSLY DISCLAIMS ALL WARRANTIES, EITHER EXPRESS OR IMPLIED, INCLUDING ANY IMPLIED WARRANTIES OF MERCHANTABILITY OR FITNESS FOR A PARTICULAR PURPOSE, AND NEITHER ASSUMES NOR AUTHORIZES ANY OTHER PERSON TO ASSUME FOR IT ANY LIABILITY IN CONNECTION WITH THE SALE OF THIS PART(S) AND/OR SERVICE. BUYER SHALL NOT BE ENTITLED TO RECOVER FROM THE SELLING DEALER ANY CONSEQUENTIAL DAMAGES. DAMAGES TO PROPERTY, DAMAGES FOR LOSS OF USE, LOSS OF TIME, LOSS OF PROFIT, OR INCOME, OR ANY OTHER INCIDENTAL DAMAGES.

BAR # AA158067

EPA # CAD 983596263

NOTICE TO CUSTOMERS:
We make a separate charge for the storage and disposal of toxic wastes. Rather than recover these costs by increasing our labor rates to all our service customers, we make this charge only on those particular repairs or services which generate these wastes. These are uniform charges which are calculated annually for each particular service and are available on request.

HAZARDOUS WASTE DISPOSAL
$

CUST NO. STOCK NO. TAG NO. COLOR

DATE	VEHICLE IDENTIFICATION NUMBER	MILEAGE	IN SERVICE DATE	LICENSE NUMBER	YEAR	MAKE AND MODEL	WRITTEN BY	RO NUMBER

I hereby authorize the repair work herein set forth to be done along with the necessary material and agree that you are not responsible for loss to vehicle or articles left in vehicle in case of fire, theft or any other cause beyond your control or for any delays caused by unavailability of parts or delays in parts shipments by the supplier or transporter. I hereby grant you and/or your employees permission to operate the vehicle herein described on streets, highways or elsewhere for the purpose of testing and/or inspection. An express mechanic's lien is hereby acknowledged on the above vehicle to secure the amount of repairs thereto. See reverse side for information.

PRELIMINARY ESTIMATE $

AUTHORIZED BY: X

CUSTOMER ACKNOWLEDGES RECEIPT OF A COPY HEREOF:

REVISED EST.	DATE	TIME	BY	PHONE #
1. $				
2. $				
3. $				

NAME — TIME PROMISED
ADDRESS — AUTHORIZATION
CITY/STATE/ZIP — HOME PHONE
BILL TO: — P.O. NO. — BUSINESS PHONE
ENGINE NO. — TRANSM. NO. — AXLE NO. — PROD. DATE

METHOD OF PAYMENT
SELLING DEALER
WARRANTY EXPIRES

LINE	OP CODE	TECH.	LABOR INSTRUCTIONS AND DESCRIPTIONS

GAS/OIL/GREASE — AMOUNT
TOWING — AMOUNT
MISC. CHARGES — AMOUNT
SUBLETS — AMOUNT
P.O. NO.

Figure 41-1. One type of shop repair order. Study the various sections carefully. Note the section for labor instructions and the box for the hazardous waste disposal fee. (Northwood Auto Plaza)

One or two copies of the repair order are given to the customer. The shop retains the remaining copies for its records. Warranty repair orders are similar to standard repair orders and usually use the same order form. The customer also receives a copy of this order, but is not billed for repairs covered under the warranty.

Determining Labor Costs

Most shops use one of two methods to determine correct labor costs. In the first method, labor costs are calculated using an ***hourly rate*** (amount charged per hour of technician's time). The time actually spent working on the vehicle multiplied by the labor rate is equal to labor cost. If, for example, the repair took two hours and the hourly rate is \$50.00, the bill would be \$100.00 (\$50.00 × 2 hours).

The other method uses a ***flat rate manual.*** This manual gives an estimate of how much time the technician should need to complete certain repairs. If the technician is to replace a set of spark plugs, he or she simply looks up plug replacement for the engine being serviced. The manual may indicate one hour for the job. Again, the number of hours multiplied by the labor rate will give the cost (1 hour × \$50.00 = \$50.00). If unexpected problems arise and the plug replacement takes two hours, the customer is only required to pay the flat rate amount. This prevents overcharging.

In addition to the regular hourly wage, some shops pay their technicians the difference between flat rate time and actual repair time, if the actual repair time is less than the quoted flat rate. This allows a fast technician to earn more per hour.

Determining Part Costs

When calculating part costs on a repair order, a percentage increase, or ***markup,*** is generally added to the wholesale price of each part. Wholesale prices will vary, depending on suppliers, brand names, and whether the part is new, rebuilt, or used. The percentage is added to cover the shop's cost for handling the part, paperwork, and margin of profit. If, for instance, the markup at your shop is 20% and the wholesale price of the part is $45.00, multiply the part cost by the percentage:

$45.00 × .20 = $9.00

Then, add the markup to the wholesale price of the part:

Wholesale Price:	$45.00
Markup:	+ $ 9.00
Total	$54.00

Some rebuilt parts have a ***core charge.*** The core charge is a deposit that is held by the parts supplier until the old part is returned for rebuilding. When the old part is returned to the parts outlet, the core charge is refunded to the shop. Therefore, the customer should not pay the core charge. Generally, core charges are not refunded for parts that are too damaged to be rebuilt.

Determining Outside Service Costs

Certain repairs require the services of an outside repair facility, such as a machine shop or radiator repair shop. In some cases, outside service personnel come to the shop to perform the repairs directly on the vehicle. Examples are windshield repairs and installing custom radios or other electronic equipment. These costs have to be added to the repair order. The shop may or may not mark up the services of the outside facility.

Other Charges

Some shops charge an extra amount for supplies, such as rags, solvents, or lubricants. This may be a percentage of the total bill or a flat dollar amount. Some shops are required to charge a ***disposal fee*** to cover the costs of disposing of such waste as used motor oil and scrap tires. Again, this may be a flat amount or a percentage of the total bill. Such fees are usually turned over to local or state government agencies to cover the costs of waste disposal, and the shop must remember to get the fee from the customer.

Calculating Sales Tax

Sales tax is levied almost everywhere. The automotive repair shop must pay taxes, even if the shop forgets to charge the customer. Therefore, it is very important to add the sales tax to the bill. In some places, the sales tax applies to parts but not labor. Some of your customers, such as agencies of the federal government, churches, or charitable organizations, may be tax exempt. Check with your local taxing agencies to determine the exact tax application.

To compute the sales tax, multiply the tax rate by the total of the taxable sale. For example, in an area where the total parts and labor cost of $75.00 is taxable and the sales tax rate is 6%, the tax is determined by multiplying $75.00 by .06:

$75.00 × .06 = $4.50

Next, the tax is added to the parts and labor total:

Parts and Labor:	$75.00
Tax	+ $ 4.50
Total:	$79.50

The final total for the job is $79.50. When making these computations, add any other applicable sales or use taxes (city and state) that may apply in your area.

Cost Estimates

A ***cost estimate*** is similar to a repair order, but is written to show the vehicle owner what a repair is expected to cost. See **Figure 41-2.** It is always good practice to give the customer an estimate on all work to be done.

Caution: Do not assume the vehicle owner will always want the work done. Some owners may prefer to trade in the vehicle, shop for a better price, or simply leave the problem uncorrected.

Preparing a cost estimate involves diagnosing the vehicle's problem and determining what needs to be done to correct it. The needed parts and time to make repairs must be determined to the best of the technician's ability. In many cases, the shop manager or service manager determines the cost of the parts and labor. In other situations, the technician must determine the total cost estimate.

Estimating Labor

Begin by determining exactly what service operations need to be performed to complete the job. Do not forget related operations that must be performed to complete the job. For example, you may need to discharge the air conditioner to gain access to the heater core. If you are not using a flat rate manual to calculate labor, the additional work of evacuating and recharging the air conditioner should be added to the total cost.

If you are using a flat rate manual, all preparatory and collateral work is included in the time listed in the manual. For example, the time needed to discharge and recharge the air conditioning system should not be added to the total labor time to replace the compressor. To determine the total labor cost, multiply the estimated time by your shop labor rate

Estimating Parts Prices

Determine what parts will be needed to complete the job. Do not forget related parts and supplies, such as oil and coolant. Also determine the need for outside services as outlined above.

After the needed parts have been determined, refer to a flat rate manual to obtain parts prices or call local parts outlets to get the latest prices. Also, call outside service suppliers to obtain their prices for needed work that cannot be performed in your shop. When listing parts and outside services on the estimate, do not forget to add the normal shop markup.

Adding Other Costs

As mentioned previously, there may be extra charges for supplies, disposal fees, and sales taxes. Do not forget to add them to your estimate.

Finishing the Cost Estimate

After all expenses have been calculated, carefully add the labor costs, parts prices, and any other costs to reach a total cost estimate. Be sure to inform the customer that the

PATTON CHEVROLET OLDSMOBILE PONTIAC

P.O. Box 428 •195 Fortuna Blvd. • Fortuna, CA 95540

ESTIMATE OF REPAIRS

Date 5-16, 19

Name JOHN DOE Address 123 MAIN ST. City FORTUNA Phone

Make CHEV Year Serial No. 163AT89LOGC511 Body Style WAGON Style No. S10

Mileage 11,949.6 License No. 7AMI5CK Paint No. — Trim No. — Insurance Co.

REPAIR	RE-PLACE	ESTIMATE OF REPAIR COSTS	LABOR HOURS		PARTS		SUBLET
	X	SHADED FRONT WINDSHIELD AND TRIM	3	0	399	65	—
		TOTAL					

** Part prices subject to invoice **

REMARKS

$ 50.00 Insurance Deductible

By Bob

This estimate is based on our inspection and does not cover additional parts or labor which may be required after the work has been started. After the work has started, worn or damaged parts which are not evident on first inspection may be discovered, naturally this estimate cannot cover such contingencies. Parts prices subject to change without notice. This estimate is for immediate acceptance.

THIS WORK AUTHORIZED BY John Doe

3.0 HRS. OF LABOR AT $ 40.00 PER HR. $	120	00
PARTS $	399	65
PAINT MATERIALS $	10	00
SUBLET $	—	—
SALES TAX $	24	58
ESTIMATE TOTAL $	554	23
ADVANCE CHARGE $	—	
GRAND TOTAL $	554	23

Figure 41-2. One type of body shop cost estimate.

cost estimate is subject to change and that you will call him or her if additional work is needed.

Get an OK for Additional Costs

Often, the technician will discover that additional parts and labor are necessary to do the job correctly. If you determine that the cost of the repairs will exceed the estimate, immediately notify the customer. Be sure to get the customer's permission before performing any additional work.

Tech Talk

One of the best automotive technicians the authors knew was one of the worst at communicating with vehicle owners. He figured that they were as familiar with the causes of part failures as he was. He would always say that a failed part was "toast." While this is a snappy line in action movies, it fails to explain what is wrong with a part. Witty customers would reply, "Toast? I didn't know it was ever bread." This would make the technician mad, and things would go downhill from there.

Failure to communicate properly is probably the biggest cause of conflict between technicians and customers. When talking to customers, try to describe the problem exactly. Do not use words like "shot" or "wasted." Simply describe the problem, explaining that a part is loose, scored, shorted, grounded, misaligned, or came apart. Even the phrases "burned out" and "broken" are more informative than "toast."

Remember, communication is a two-way process. Listen closely when your customer tells you about the problem. Ask about any sounds, at what speed and in what gear the problem happens, and whether the problem happens when accelerating or decelerating. Also, ask about the weather conditions when it happens (approximate temperature and humidity). Ask whether the problems occur occasionally or all the time. Finally, be sure to ask about any recent work done on the vehicle and changes in the brands or grades of gasoline the owners use.

Be sure that you and the owner agree on what is wrong with the vehicle: engine stalling, hard starting, transmission or clutch slipping, or other definite problem. Also, before starting any repairs, make sure the customer knows what the repairs will be and what they will cost.

Summary

Repair orders are used to keep a record of services performed on vehicles, provide billing information, and to determine the technician's pay. The repair order begins with information about the customer, vehicle, and the problem to be corrected.

After the technician finishes the repairs, he or she completes the repair order, listing the labor and parts used. This information is used to bill the customer. The price of the labor and parts, the cost of outside services and supplies, and sales taxes are added to create the final bill.

Labor costs are determined by multiplying the actual time taken (or the time listed in the flat rate manual) by the shop labor rate. The standard shop markup is added to the to the wholesale price to determine the parts costs. Charges for outside services, supplies, and sales taxes, are added to the total of parts and labor.

A cost estimate is similar to a repair order. It is used to show the owner what a repair is expected to cost. To prepare a cost estimate, the technician must diagnose the vehicle problem and determine what needs to be done to correct it. The needed parts and labor must be determined accurately. Do not assume the vehicle owner will always want the work done.

Always bill customers accurately and honestly. The finest advertisement for any shop—and any technician—is scrupulous honesty at all times.

Review Questions—Chapter 41

Do not write in this book. Write your answers on a separate sheet of paper.

1. If the flat rate manual says a job should take 3 hours and the shop labor charge is $35.00 per hour, the labor cost of the job is _____.
   (A) $35.00
   (B) $70.00
   (C) $105.00
   (D) $350.00
2. If the shop markup on parts is 25% and the wholesale price of a part is $18.00, what is the price to the customer?
   (A) $4.50
   (B) $13.50
   (C) $22.50
   (D) $24.50
3. If the labor cost to install the part in Question 2 is $22.00, a $2.50 supply charge is added, and the local sales tax is 5%, what is the total bill?
   (A) $25.72
   (B) $49.35
   (C) $55.35
   (D) $235.00
4. If an alternator belt cannot be replaced without removing the air conditioner and power steering belts, the estimate for alternator belt replacement would include labor for removing and replacing _____.
   (A) all three belts
   (B) the air conditioner and alternator belts only
   (C) the power steering and alternator belts only
   (D) the alternator belt only
5. Sales taxes are paid to the shop by the customer. What does the shop do with the collected sales taxes?
   (A) Keeps them to pay for disposal fees.
   (B) The shop pays state or local sales tax.
   (C) Uses them to pay its income taxes.
   (D) Returns them to the customer at the end of the year.

ASE-Type Questions

1. All of the following statements about repair orders are true, *except:*
   (A) repair orders are used to keep a record of services performed on a vehicle.
   (B) repair orders may be used to pay the technician for work performed.
   (C) repair orders are used for warranty jobs only.
   (D) repair orders are sometimes called shop work orders.
2. All of the following statements about core charges are true *except:*
   (A) the core charge is refunded when the core is returned to the parts store.
   (B) the core charge is made on parts that can be reconditioned.
   (C) core charges are typically applied to alternators, water pumps, and brake shoes.
   (D) the core charge will be refunded even if the part is too badly damaged to rebuild.
3. Technician A says the customer should pay the core charge on a rebuilt part. Technician B says the core charge is refunded when the old part is turned in. Who is right?
   (A) A only.
   (B) B only.
   (C) Both A and B.
   (D) Neither A nor B.

4. Technician A says the vehicle problem must be diagnosed before preparing a cost estimate. Technician B says the vehicle owner will always want the work done. Who is right?
   (A) A only.
   (B) B only.
   (C) Both A and B.
   (D) Neither A nor B.
5. Technician A says flat rate manuals give the approximate time needed to make a specified repair. Technician B says flat rate manual times must be multiplied by the shop labor rate to obtain the actual labor price. Who is right?
   (A) A only.
   (B) B only.
   (C) Both A and B.
   (D) Neither A nor B.

Suggested Activities

1. List some of the things that should be placed on a repair order or cost estimate. Compare your list with an actual repair order.
2. Fill out a repair order, listing the following information:
   - Owner's name and address.
   - Owner's home and work telephone numbers.
   - Vehicle identification number (VIN).
   - Body style.
   - Number of doors.
   - License plate number.
   - Color.
   - Engine size and manufacturer.
   - Transmission type and manufacturer.
   - Tire size.
   - Whether the vehicle is equipped with air conditioning, power steering, or power brakes.
   - Owner's complaint or work to be done.
3. With other class members, conduct a role-playing session dealing with a customer complaint. Take turns playing the technician and the dissatisfied customer.
4. Write five estimates for the repair of different vehicle components. Be sure to include parts, prices, and appropriate labor time. Total the combined parts and labor as if you were going to sell the work to a customer.

42

ASE Certification

After studying this chapter, you will be able to:

- Explain why technician certification is necessary.
- Explain the process of registering for ASE tests.
- Explain how to take the ASE tests.
- Identify typical ASE test questions.
- Explain what is done with ASE test results.

Technical Terms

National Institute for Automotive Service Excellence
ASE
Certification
Standardized tests
Master technician
ACT
Application form
Registration booklet
Admission ticket
Multiple-choice
One-part questions
Two-part questions
Negative questions
Incomplete sentence questions
Pass/fail letter
Test score report
Certificate in evidence of competence
ASE Preparation Guide
Recertification test

This chapter explains the reasons for the National Institute for Automotive Service Excellence (ASE) certification and the advantages of being ASE certified. This chapter also explains how to apply for and take the ASE tests. When you have finished studying this chapter, you will know the purposes of ASE and the ASE tests. You will also have a good understanding of the testing methods and the test results.

Reasons for ASE Tests

The concept of setting standards of excellence for skilled jobs is not new. Many ancient societies had associations of skilled workers who set standards and enforced rules of conduct. Many modern labor unions are descended from early associations of skilled workers. Certification processes for aircraft mechanics, aerospace workers, and electronics technicians have existed since the beginnings of these industries.

Due to the fragmented, decentralized nature of the automotive repair industry, standards for the industry were difficult to establish and enforce. Anyone, no matter how unqualified, could claim to be an automotive technician. A large segment of the public came to regard technicians as dishonest, unintelligent, or both.

The ***National Institute for Automotive Service Excellence,*** now called ***ASE,*** was established in 1975 to provide a ***certification*** process for automobile technicians. ASE is a nonprofit corporation formed to encourage and promote high standards of automotive service and repair. ASE does this by providing a series of written tests on various subjects in the automotive repair, heavy truck repair, auto body/paint, and engine machinist areas.

These tests are called ***standardized tests,*** which means the same test in a particular subject is given to everyone throughout the United States. Any person passing one of these tests and meeting certain experience requirements, is certified in the subject covered by that test. If a technician can pass all of the tests in the automotive or heavy truck, or engine machinist areas, he or she is certified as a ***master technician*** in that area.

The ASE certification test program identifies and rewards skilled and knowledgeable technicians. Periodic recertification provides an incentive for updating skills and also provides guidelines for keeping up with current technology. The test program allows potential employers and the driving public to identify good technicians and helps the technician advance his or her career. The program is not mandatory, but many employers now hire only ASE-certified technicians. Over 500,000 persons are now ASE certified in one or more areas.

Other ASE programs encourage the development of effective automotive service training programs, conduct research on the best teaching methods, and publicize the advantages of technician certification. Persons from the automotive service industry, vehicle manufacturers, government agencies, educational groups, and consumer associations make up the board that manages ASE.

The ASE certification program has brought many advantages to the automotive industry, including increased respect and trust of automotive technicians, at least of those who are ASE certified. This has resulted in better pay and benefits for technicians and increased standing in the community. Because of ASE, automotive technicians are taking their place next to other skilled artisans.

Applying for the ASE Tests

Anyone may apply for and take any ASE test. However, to become certified the applicant must have two years of work experience as an engine machinist or as an automobile, truck, auto body, or paint technician. In some cases, training programs or courses, an apprenticeship program, or time spent performing similar work may be substituted for all or part of the work experience.

ASE tests are given twice each year, in the spring and fall. Tests are usually held during a two-week period at night during the workweek. The tests are given by a separate organization called ***ACT.*** ACT is a nonprofit organization experienced in administering standardized tests. The tests are given at designated test centers in over 300 places in the United States. If necessary, special test centers can be set up in remote locations. However, there must be enough potential applicants for the establishment of a special test center to be practical. To apply for the ASE tests, begin by obtaining an ***application form*** like the one shown in **Figure 42-1.** To obtain the most current application form, contact ASE at the following address:

National Institute for Automotive Service Excellence
13505 Dulles Technology Drive
Suite 2
Herndon, VA 20171-3421

ASE will send the proper form inside a ***registration booklet.*** The booklet explains how to complete the registration form. When you get the booklet, fill the form out carefully, supplying all the necessary information. You may apply to take as many tests as are being given, fewer tests, or only one test if desired. Proof of work experience or qualified substitutes should also be included as directed. If there is any doubt about what should be placed in a particular space, consult the registration booklet. Determine the closest test center and record its number in the appropriate space. Most test centers are located at local colleges, high schools, or vocational schools. In addition to the application, you must include a check, money order, or credit card number to cover all of the

ASE

Registration Form
ASE Tests

If extra copies are needed, this form may be reproduced.
Important: Read the Booklet first. Then print neatly when you fill out this form.

1. a. Social Security Number: ___-__-____

1. b. Telephone Number: (During the day) Area Code / Number

1. c. Previous Tests: Have you ever registered for any ASE certification tests before? ❑ Yes ❑ No

2. Last Name: / **First** / **Middle Initial**

3. Mailing Address:
Number, Street, and Apt. Number
City
State / ZIP Code

4. Date of Birth: Month / Day / Year

5. Sex: ❍ Male ❍ Female

6. Race or Ethnic Group: Blacken one circle. (For research purposes. It will not be reported to anyone and you need not answer if you prefer.)
1 ❍ American Indian 2 ❍ African American 3 ❍ Caucasian/White 4 ❍ Hispanic/Mexican
5 ❍ Oriental/Asian 6 ❍ Puerto Rican 7 ❍ Other (Specify)

7. Education: Blacken only one circle for the highest grade or year you completed.
Grade School and High School (including Vocational): ❍7 ❍8 ❍9 ❍10 ❍11 ❍12
After High School: Trade or Technical (Vocational) School: ❍1 ❍2 ❍3 ❍4
College: ❍1 ❍2 ❍3 ❍4 ❍More

8. Employer: Which of the below best describes your current employer? (Blacken one circle and fill in subcodes from pp. 19-20 as appropriate.)
1 ❍ New Car/Truck Dealer/Distributor, Domestic—
2 ❍ New Car/Truck Dealer/Distributor, Import— } Enter code for make of car or truck
3 ❍ Service Station—Enter code
4 ❍ Military—Enter Code
5 ❍ Indp't Repair Shop— Enter code for
6 ❍ Fleet Repair Shop— vehicle type
7 ❍ Specialty Shop—Enter code
8 ❍ Government/Civil Service
9 ❍ Student
10 ❍ Educator/Instructor
11 ❍ Manufacturer
12 ❍ Independent Collision Shop
13 ❍ Volume Retailer—Enter code
14 ❍ Tire Dealer—Enter code
16 ❍ Utility
17 ❍ Lift Truck Dealer/Repair Shop
18 ❍ Machinist Facility—Enter code
19 ❍ Leasing & Rental Shop—Enter code
15 ❍ Other

9. Test Center: Center Number / City / State

10. Tests: Blacken the circles (regular) or squares (recertification) for the test(s) you plan to take. (Do not register for a recertification test unless you passed that regular test before.) **Note:** Do not register for more than four regular tests on any single test day.

Regular	Recertification	Regular	Recertification
❍ A1 Auto. Engine Repair	❑ A	❍ A2 Auto: Automatic Trans/Transaxle	❑ E
❍ A8 Auto: Engine Performance	❑ B	❍ A3 Auto: Manual Drive Train and Axles	❑ F
❍ A4 Auto: Suspension & Steering	❑ C	❍ A6 Auto: Electrical/Electronic Systems	❑ G
❍ A5 Auto: Brakes	❑ D	❍ A7 Auto: Heating & Air Conditioning	❑ H
❍ F1 Alt. Fuels: Lt. Vehicle CNG		❍ L1 Adv. Level: Auto Adv. Engine Perf. Spec.	
❍ S4 School Bus: Brakes		❍ T1 Med/Hvy Truck: Gasoline Engines	❑ I
❍ S5 School Bus: Suspension/Steering		❍ T2 Med/Hvy Truck: Diesel Engines	❑ J
		❍ T6 M/Hvy Truck: Elec./Electronic Systems	❑ K
❍ T3 Med/Hvy Truck: Drive Train	❑ L	❍ S6 School Bus: Elec./Electronic Systems	
❍ T4 Med/Hvy Truck: Brakes	❑ M	❍ B2 Body: Painting & Refinish	❑ P
❍ T5 Med/Hvy Truck: Suspension/Steering	❑ N	❍ B3 Body: Nonstructural Analysis	
❍ T8 Med/Hvy Truck: PMI		❍ B4 Body: Structural Analysis	
❍ M1 Machinist: Cylinder Head Specialist	❑ Q	❍ B5 Body: Mechanical & Electrical Components	
❍ M2 Machinist: Cylinder Block Specialist	❑ R		
❍ M3 Machinist: Assembly Specialist	❑ S		

11. Fees: Number of **Regular** tests **(except L1)** marked above_____ x $18* = $_____
If you are taking the L1 Advanced Level Test, add $40 here + $_____
If you marked **any** of the **Recertification** tests, add $20 here + $_____
Registration Fee + $ **20**
TOTAL FEE = $_____

****Important:*** *School Bus Technicians, see page 8 for a money-saving offer! If you qualify, check here ❑ and adjust your fees (Item 11, first line) accordingly.*

❑ MasterCard ❑ VISA
Credit Card # — Expiration Date: Month / Year
Signature of Cardholder

12. Fee Paid By: 1 ❍ Employer 2 ❍ Technician

13. Experience: Blacken one circle. (To substitute training or other appropriate experience, see page 3.)
1 ❍ I certify that I have two years or more of full-time experience (or equivalent) as an automobile, medium/heavy truck, school bus or collision repair/refinish technician or as an engine machinist. (Fill out #14 on opposite side)
2 ❍ I don't yet have the required experience. (Skip 14.)

14. Job History: Required. Provide job history and job details on the opposite side of this form.

15. Authorization: By signing and sending in this form, I accept the terms set forth in the *Registration Booklet* about the tests and the reporting of results.

Signature of Registrant / Date

Do not send cash. Use credit card, or enclose a check or money order for the total fee, made payable to ASE/ACT and mail with this form to:
ASE/ACT, P.O. Box 4007, Iowa City, IA 52243

Figure 42-1. An ASE certification test registration form. Be sure to fill in all required information and include payment for all test fees. (ASE)

necessary fees. A fee is charged to register for the test series, and a separate fee is charged for each test to be taken. See the latest registration booklet for the current fee structure. In some cases, your employer may pay the registration and test fees. Check with your employer before sending in your application. If you need to take the ASE tests in a language other than English, indicate this on the application form. In addition to English, test booklets are available in Spanish and French.

It has recently become possible to register over the Internet for ASE tests. ASE's World Wide Web site (www.asecert.org) contains information on ASE, the certification process, study materials, etc. See **Figure 42-2.**

To be accepted for either the spring or fall ASE tests, your application and payment must arrive at ASE headquarters by the registration deadline, which is at least one month before the first test date. To ensure that you can take the test at the test center of your choice, send in the application as early as possible. After sending the application and fees, you will receive an ***admission ticket*** to the test center. See **Figure 43-3.** This should arrive by mail within two weeks of sending the application. If your admission ticket has not arrived within two weeks of the first test date, contact ASE using the phone number given in the registration booklet. If the desired test center is filled when ASE receives your application, you will be directed to report to the nearest center that has an opening. If you cannot go to the alternate test center that was assigned, contact ACT immediately using the phone number given in the latest ASE registration booklet.

Taking the ASE Tests

Be sure to bring your admission ticket with you when reporting to the test center. When you arrive at the test center, you will be asked to produce the admission ticket and a driver's license or other photographic identification. In addition to these items, bring some extra number 2 pencils. Although pencils will be made available at the test center, some extra pencils may save you time if the original pencil breaks. After you enter the test center and are seated, listen to and follow all instructions given by the test administrators. During the test, read each question carefully before choosing an answer. The ASE tests are designed to measure your knowledge of three things:

- The function of automotive systems and components.
- Diagnosis and testing of systems and components.
- Repairing automotive systems and components.

Each ASE test contains between 40 and 80 test questions, depending on the subject to be tested. All test questions are ***multiple-choice,*** with four possible answers. These types of multiple-choice questions are similar to the multiple-choice questions used in this textbook. Questions can be written in one or two parts. Samples of these types of test questions are given below.

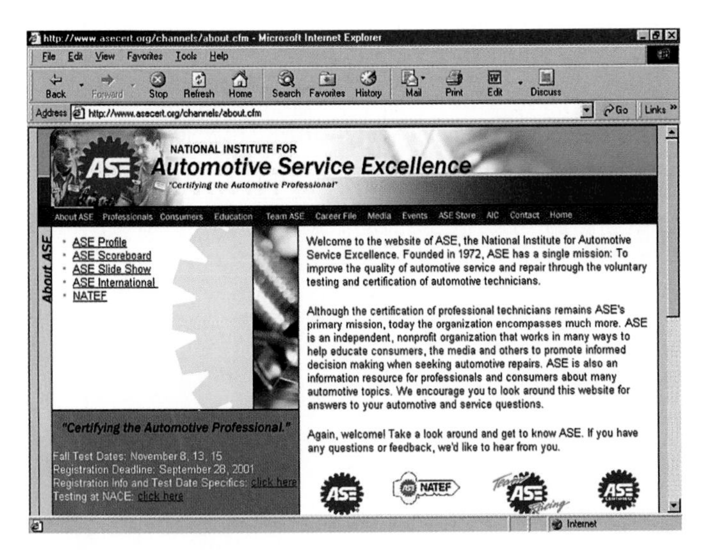

Figure 42-2. You can register for ASE tests or find out more about ASE at ASE's official web site. (ASE)

One-Part Questions

One-part questions require you to answer a single question. For example:

1. The cooling system part that allows the engine to warm up quickly is the _____.
   (A) water pump
   (B) coolant recovery reservoir
   (C) thermostat
   (D) radiator

Notice the question calls for the best answer out of all of the possibilities. The thermostat is the part that remains closed to prevent coolant flow through the radiator until the engine warms up. Therefore, (C) is correct.

Two-Part Questions

Two-part questions used in the ASE tests require you to analyze the statements made by two technicians, Technician A and Technician B. You are asked to determine whether each of the statements is true. For example:

1. Technician A says the ignition coil changes high voltage into low voltage. Technician B says the distributor pickup coil produces a low voltage signal. Who is right?
   (A) A only.
   (B) B only.
   (C) Both A and B.
   (D) Neither A nor B.

Notice that both statements can be true, both can be false, or only one of them can be false. In this case, the statement of Technician A is wrong, since the ignition coil produces high voltage from low voltage. The statement of Technician B is correct, since the pickup coil does produce a low voltage signal. Therefore, the correct answer is (B).

Negative Questions

Some questions are called ***negative questions.*** These questions ask you to identify the wrong answer. They usually have the word "except" in the question.

National Institute for Automotive Service Excellence

ACT, P.O. Box 4007, Iowa City, Iowa 52243, Phone: (319) 337-1433 017910042 T

Admission Ticket

Test Center to which you are assigned:

A

John Smith
123 Main Street
Edens, Il. 60000

REGULAR TESTS (Late arrivals may not be admitted.)		
DATE	REPORTING TIME	TEST(S)
11/14	7:00 PM	A1, A2, A8

RECERTIFICATION TESTS (Late arrivals may not be admitted.)

TEST CODE KEY

A1 Auto: Engine Repair
A2 Auto: Automatic Trans/Transaxle
A3 Auto: Manual Drive Train & Axles
A4 Auto: Suspension & Steering
A5 Auto: Brakes
A6 Auto: Electrical/Electronic Systems
A7 Auto: Heating & Air Conditioning
A8 Auto: Engine Performance
M1 Machinist: Cylinder Head Specialist
M2 Machinist: Cylinder Block Specialist
M3 Machinist: Assembly Specialist
T1 Med/Hvy Truck: Gasoline Engines
T2 Med/Hvy Truck: Diesel Engines
T3 Med/Hvy Truck: Drive Train
T4 Med/Hvy Truck: Brakes
T5 Med/Hvy Truck: Suspension & Steering
T6 Med/Hvy Truck: Elec./Electronic Systems
T8 Med/Hvy Truck: Preventive Main. Inspec.
B2 Coll.: Painting & Refinishing
B3 Coll.: Non-structural Analysis
B4 Coll.: Structural Analysis
B5 Coll.: Mechanical & Elec. Components
B6 Coll.: Damage Analysis & Estimating
P1 Parts: Med/Hvy Truck Parts Specialist
P2 Parts: Automobile Parts Specialist
F1 Alt. Fuels: Lt. Veh. Comprsd. Nat. Gas
L1 Adv. Level: Adv. Engine Perf. Spec.
S1 School Bus: Body Sys. & Spec. Equip.
S4 School Bus: Brakes
S5 School Bus: Suspension & Steering
S6 School Bus: Elec./Electronic Systems

See Notes and Ticketing Rules on reverse side. An asterisk (*) indicates your certification in these areas is expiring.

SPECIAL MESSAGES

-REVIEW ALL INFORMATION ON THIS TICKET. CALL IMMEDIATELY TO REPORT AN ERROR OR IF YOU HAVE QUESTIONS.
-IF YOU MISS ANY EXAMS, FOLLOW THE REFUND INSTRUCTIONS ON THE BACK OF THIS SHEET. THE REFUND DEADLINE IS
-YOU HAVE BEEN ASSIGNED TO AN ALTERNATE TEST CENTER. THE CENTER ORIGINALLY REQUESTED IS FULL.
8010-IL/LOCAL 150 IS LOCATED ON JOLIET AVE, THREE DOORS W. OF LAGRANGE RD ON SOUTH SIDE OF THE ST. ENTER THROUGH BACK DOOR. NO ALCOHOL ON PREMISES.

MATCHING INFORMATION: The information printed in blocks B and C at the right was obtained from your registration form. It will be used to match your registration information and your test information. Therefore, the information at the right must be copied EXACTLY (even if it is in error) onto your answer booklet on the day of the test. If the information is not copied exactly as shown, it may cause a delay in reporting your test results to you.

IF THERE ARE ERRORS: If there are any errors or if any information is missing in block A above or in blocks B and C at the right, you must contact ACT immediately. DO NOT SEND THIS ADMISSION TICKET TO ACT TO MAKE SUCH CORRECTIONS.

Check your tests and test center to be sure they are what you requested. If either is incorrect, call 319/337-1433 immediately. Tests cannot be changed at the test center. **ON THE DAY OF THE TEST**, be sure to bring this admission ticket, positive identification, several sharpened No. 2 pencils, and a watch if you wish to pace yourself.

B FIRST FIVE LETTERS OF LAST NAME

S	M	I	T	H

C SOCIAL SECURITY NUMBER OR ACT IDENTIFICATION NUMBER

1	2	3		4	5		6	7	8	9

SIDE 1

Figure 42-3. You will receive your admission ticket approximately two weeks after sending in your registration form.

1. Low oil pressure could be caused by all of the following problems, *except:*
   (A) a low oil level.
   (B) a worn oil pump.
   (C) a clogged oil screen.
   (D) tight bearing clearances.

Since tight bearing clearances would not cause low oil pressure, while the other defects listed could, the correct answer is (D).

A variation of the negative question uses the word "least," such as the one below.

1. The computer-controlled engine of a late-model car knocks during acceleration. Which of these defects is the *least* likely cause?
   (A) Incorrect timing.
   (B) A defective knock sensor.
   (C) A plugged fuel filter.
   (D) Low octane gasoline.

In this case, the least likely cause of engine knocking was a plugged fuel filter, which is much more likely to cause engine stalling or poor performance instead of knocking. Therefore, the correct answer is (C).

Incomplete Sentence Questions

Some test questions are ***incomplete sentence questions,*** with one of the four possible answers correctly completing the sentence. An example of an incomplete sentence question is given below.

1. The coolant temperature sensor is used to measure _____ temperature.
   (A) exhaust gas
   (B) engine
   (C) incoming air
   (D) ambient (outside) air

Once again, the question calls for the best answer. The coolant temperature sensor measures the temperature in the engine by monitoring coolant temperature, so (B) is correct.

After completing all of the questions in a particular test, recheck all of your answers one time to ensure that you did not miss anything that would change your answer, or that you did not make a careless error on the answer sheet. In most cases, rechecking your answers more than once is unnecessary and may lead you to change correct answers to incorrect ones. The time allowed for each test session is about four hours. Take as long as you need on any one test but work as quickly as you can. You may leave after you complete your last test and hand in all test material.

Test Results

ACT takes about six to eight weeks to process the tests from the various centers and to mail the results. Initially, you will receive a ***pass/fail letter.*** This letter will tell you only whether or not you have passed each test. A typical pass/fail letter is shown in **Figure 20-4.** Two weeks after receiving the pass/fail letter, you will receive a ***test score report.*** The test score report is a confidential report of your performance on the tests. The report will list the number of questions that must be answered correctly to pass the test and the number of questions you have answered correctly. The test questions are also divided into general areas to help you determine which areas require more study. For example, the suspension and steering systems test questions may be divided into the following subsections: steering systems diagnosis and repair; suspension systems diagnosis and repair; wheel alignment diagnosis, adjustment, and repair; and wheel and tire diagnosis and repair. A typical test score report is shown in **Figure 20-5.**

Included with the test score report is a ***certificate in evidence of competence.*** This certificate lists the areas in which you have been certified. In addition, a pocket card and a wallet card are provided. Like the certificate, they list all the areas in which you are certified. Also included is an order form for shoulder patches, wall plates, and other ASE promotional material.

Note: If you did not indicate that you have two years of automotive experience on the test application, you will not receive a certificate in evidence of competence. After you have met the experience requirement, you must provide ASE with the necessary information to receive your certificate.

All ASE test results are confidential and are provided only to the person who took the test. Test results will be mailed to your home address and will not be provided to anyone else. This is done to protect your privacy. The only test information ASE will release is to confirm to an employer that you are certified in a particular area. This is true even if your employer has paid the test fees. If you wish your employer to know exactly how you performed on the tests, you must provide him or her with a copy of your test results.

If you fail a certification test, you can retake it as many times as you like. However, you (or your employer) must pay all applicable registration and test fees again. You should study all available information in the areas in which you did poorly. The ***ASE Preparation Guide*** may help you sharpen your skills in these areas. The ASE Preparation Guide is free and can be obtained by filling out the coupon at the back of the information bulletin.

Other ASE Tests

There are three ASE test areas other than the well-known battery of tests for automotive technicians. ASE now offers an advanced level test in engine performance. This specialty test is designed to comply with amendments to the Clean Air Act for emission inspection and maintenance. Previous certification in regular automobile engine performance is required to take this test. A test in alternative fuels, specializing in light-vehicle compressed natural gas (CNG), is also available. Engine machinists can become certified in three separate skill areas.

National Institute for
AUTOMOTIVE SERVICE EXCELLENCE

December 20, XXXX 032527

John Smith
123 Main Street
Edens, Il 60000

Dear ASE Test Taker:

Listed below are the results of your November XXXX ASE Tests. You will soon be receiving a more detailed report.

If your test result is "Pass", and if you have fulfilled the two-year "hands-on" experience requirement, you will receive a certificate and credential cards for the tests you passed.

If your test result is "More Preparation Needed", you did not attain a passing score. Check your detailed score report when it arrives. This information may help you prepare for your next attempt.

If you do not receive your detailed report within the next three weeks, please call.

Thank you for participating in the ASE program.

A1	ENGINE REPAIR	PASS
A2	AUTOMATIC TRANSMISSION/TRANSAXLE	PASS
A8	ENGINE PERFORMANCE	PASS

123-45-6789

13505 Dulles Technology Drive • Herndon, Virginia 22071-3415 • (703) 713-3800

Figure 42-4. A pass/fail letter will be sent to you shortly after taking the ASE tests. Note that this letter indicates that all three tests taken were passed. (ASE)

Your Score is 38. (Passed)
The total score needed to pass A2 is 34 out of 50.

Test A2 Automatic Transmission/Transaxle Content area	Number of questions answered correctly	Total number of questions
General transmission/transaxle diagnosis	19	25
Transmission/transaxle maintenance and adjustment	4	5
In-vehicle transmission/transaxle repair	7	9
Off-vehicle transmission/ transaxle repair	8	11
Total test	38	50

Figure 42-5. You will receive a test score report approximately two weeks after receiving the pass/fail letter. The test score report shows how you performed on each section of individual tests. (ASE)

Recertification Tests

Once you have passed the certification test in any area, you must take a ***recertification test*** every five years to keep your certification. This assures that you have kept up with current technology. The process of applying to take the recertification tests is similar to that for the original certification tests. Use the same form and enclose the proper recertification test fees. If you allow any of your certifications to lapse, you must take the regular certification test to regain your certification.

Recertification tests use the same test format (multiple-choice questions) as other ASE tests. Recertification questions emphasize new developments in automotive service, such as recently introduced vehicle systems, new repair techniques, and new government regulations.

Tech Talk

School may frustrate you, but do not give up. The fact that you are not great in math or language classes does not mean that you lack ability. It just means that you have aptitudes in different areas. Many automotive technicians are not very good with pencil and paper, even though they may be great at diagnosis and repair.

Modern vehicles, however, require a level of reading ability that was not needed in the past. Factory service manuals and technical service bulletins must be read and understood in order to troubleshoot computer controls and locate driveability problems. The ASE tests are another example. Instead of proving your knowledge by actually working on a vehicle, you must read and comprehend a series of questions to pass the tests. In addition, these materials inevitably include technical terms that are even harder to comprehend if you are still having trouble with the nontechnical words. This is why it is important to increase your reading ability as much as possible.

So what do you do to increase your reading ability? The answer is simple. Read. Practice in anything, especially reading, improves your ability. You do not have to read great works of literature. General reading sources include your local newspaper, news magazines, and a vast array of novels, biographies, histories, and reference books. There are many sources of interesting reading material related to cars, trucks, and service, including a large number of automotive magazines devoted to every possible part of the automotive culture. There are also various automotive technical and marketing publications, service manuals, and training materials.

Summary

The automotive industry was one of the few major industries that did not have testing and certification programs. This lack of professionalism in the automobile industry led to poor or unneeded repairs and decreased the status and pay for automobile technicians. The National Institute for Automotive Service Excellence, or ASE, was started in 1975 to overcome these problems. ASE tests and certifies automotive technicians in major areas of automotive repair. This has increased the skill level of technicians, resulting in better service and increased benefits for technicians.

ASE tests are given in the spring and fall. Anyone can register to take the tests by filling out the proper registration form and paying the registration and test fees. The registrant must also select the test center that he or she would like to go to. To be considered for certification, the registrant must have two years of hands-on experience. About three weeks after applying for the test, the technician will receive an admission ticket, which he or she must bring to the test center.

The test questions challenge your knowledge of general system operation, problem diagnosis, and repair techniques. All of the questions are multiple-choice. The questions must be read carefully. The entire test should be reviewed once to catch any mistakes.

Test results will arrive within six to eight weeks after the test session. Results are confidential and will be sent only to the home address of the person who took the test. If a test was passed and the experience requirement has been met, the technician will be certified in that skill area for five years. Anyone who fails a test can take it again in the next session. Tests can be taken as many times as necessary. Recertification tests can be taken at the end of the five-year certification period.

Review Questions—Chapter 42

Do not write in this book. Write your answers on a separate sheet of paper.

1. Name all of the categories of ASE tests.
2. If a technician can pass all of the tests in the automotive or heavy truck areas, he or she is certified as a _____.
   (A) trainee
   (B) general technician
   (C) master technician
   (D) knowledgeable technician
3. List the optional tests now administered by ASE to automotive technicians.
4. You can take the ASE tests, but ASE will not certify you until you fulfill what requirement?
5. You should begin to register for the ASE test by obtaining a _____.
6. If your admission ticket has not arrived and there is less than two weeks before the first test date, whom should you contact?
7. ASE tests are designed to measure your knowledge of what three things?
8. ASE test questions resemble the ones in _____.
   (A) college level courses
   (B) essay-type tests
   (C) verbal examinations
   (D) this book
9. ASE provides test results to _____.
   (A) the technician who took the test.
   (B) the person that paid for the test.
   (C) the technician's employer.
   (D) Both A & C.
10. An ASE certification is good for _____.

ASE-Type Questions

1. Technician A says ASE encourages high standards of automotive service and repair by providing a series of written tests. Technician B says ASE encourages high standards of automotive service and repair by providing a series of hands-on tests. Who is right?
   (A) A only.
   (B) B only.
   (C) Both A and B.
   (D) Neither A nor B.
2. In some cases, all of the following may be substituted for all or part of the work experience, *except:*
   (A) training programs or courses.
   (B) a promise to work in the automotive field for two years minimum.
   (C) time spent performing similar work.
   (D) an apprenticeship program.
3. The advantages that ASE certification has brought to automotive technicians include all of the following, *except:*
   (A) increased respect.
   (B) better working conditions.
   (C) lower pay scales.
   (D) increased standing in the community.
4. ASE tests are given _____ each year.
   (A) once
   (B) twice
   (C) four times
   (D) twelve times
5. ASE tests are held _____.
   (A) during normal working hours
   (B) at night during the work week
   (C) on national holidays
   (D) anytime
6. A technician who is not certified in any area may take all of the following ASE tests *except:*
   (A) Engine Performance.
   (B) Advanced Engine Performance.
   (C) Light Vehicle, Compressed Natural Gas.
   (D) Engine Repair.
7. A technician can retake any certification test _____.
   (A) two times
   (B) four times
   (C) five times
   (D) any number of times
8. Technician A says the ASE test session lasts about four hours. Technician B says test takers can leave as soon as they are done and turn in all test materials. Who is right?
   (A) A only.
   (B) B only.
   (C) Both A and B.
   (D) Neither A nor B.
9. When taking ASE tests, you should do all of the following *except:*
   (A) arrive at the test center early.
   (B) work quickly, but carefully.
   (C) bring several number 2 pencils.
   (D) check your answers at least three times.
10. Technician A says all ASE test scores are confidential information and can only be released with the technician's permission. Technician B says if the technician's employer pays for the tests, they are entitled to a copy of the test results from ASE. Who is right?
   (A) A only.
   (B) B only.
   (C) Both A and B.
   (D) Neither A nor B.

Suggested Activities

1. Obtain an ASE registration booklet and determine the dates on which each of the eight regular automotive tests is offered. List those tests that you think you could pass.
2. Determine where the nearest ASE test center has been located in the past and whether it is likely to be used as a test center again.
3. Make a copy of the ASE test application form and fill it out, indicating your actual experience as if you were applying for certification.
4. Take the sample test in the ASE registration booklet and use the results to identify the areas in which you are weak. Discuss with your instructor what should be done to raise your skill and knowledge levels in these areas.

This sports car is equipped with a 280-horsepower V6 engine, 6-speed manual transmission, 4-wheel independent suspension, and 18″ alloy wheels. (Nissan)

Appendix A

Carburetor Service

Although they are no longer installed on new vehicles, there are still millions of carburetors in use. This chapter covers the adjustment, disassembly, and repair of carburetors. Carburetor service should be done carefully since carburetors are precision devices, which are easily damaged.

STOP Warning: Carburetor testing, inspecting, and repair involves the possibility of fire or explosion from spilled gasoline. Dispose of spilled gasoline and gasoline-soaked rags properly. Do not allow flames, sparks, or smoking materials anywhere near the fuel system when it is being serviced.

Carburetor Functions

There are many sizes and types of carburetors. However, all carburetors have the same basic function: they atomize gasoline, mix it with air, and deliver the mixture to the engine. The typical carburetor is a metal casting with internal passages for fuel and airflow. Valves, springs, vacuum diaphragms, linkages, gaskets, and seals are typical parts of the carburetor assembly. Sensors and control devices are installed on modern carburetors used with electronic engine control systems.

All carburetors have these basic systems:

- A ***float system*** that controls the level of fuel in the carburetor bowl.
- A ***main metering system*** that uses ***venturi*** vacuum (vacuum developed in the carburetor throat) to pull fuel from the bowl at cruising speeds.
- ***Idle*** and ***off-idle systems*** that use intake manifold vacuum to pull fuel from the bowl at idle and low speeds.
- A ***power system*** that delivers extra fuel when the engine is under heavy load. An ***accelerator pump*** supplies extra fuel into the air horn when the throttle is opened.
- A ***choke system*** that enriches the air-fuel mixture when the engine is cold.

Older carburetor systems are operated by manifold and venturi vacuum, gravity, calibrated springs, and mechanical linkage. Newer carburetors use motors or solenoids, operated by the ECM, to regulate idle speed and mixture. Modern carburetors also contain input sensors.

Carburetor Service

The carburetor is a complex device, but can be adjusted and repaired as long as careful attention is paid. The following sections detail the general procedures for servicing carburetors.

Pre-Service Checks

Before performing carburetor diagnosis and service, make sure that all related systems are functioning properly. What appears to be a carburetor problem quite often turns out to be a problem in another engine system. Inspect all related emission control systems, computer sensors, and electrical connections. Spark plugs must be in good condition and correctly gapped. Check ignition timing and ignition secondary wiring. Engine compression should be up to specifications in all cylinders. The fuel delivery system must be functioning properly. The air cleaner must be clean, and any intake air heat control devices must be operating correctly.

Vacuum leaks are a common source of idle and low-speed problems (especially rough idle, surging, and hesitation). Check for possible vacuum leaks around the intake manifold, carburetor mounting gasket, and vacuum hose connections. When an intake manifold leak is suspected, first torque the manifold fasteners. Next, check for a leak by squirting oil along the gasket edge and other areas where a leak could occur, to act as a temporary seal, **Figure A-1.** If a leak exists, the idle speed and vacuum readings will change. These checks, along with any others related to the specific vehicle, will pay off in time saved. They will often prevent expensive, unnecessary repair attempts that fail to correct the problem. Unless the carburetor is definitely known to be the source of the problem, always check these other areas before starting extensive repairs, adjustments, or replacement.

Caution: Avoid dropping anything into the carburetor while the air cleaner is off.

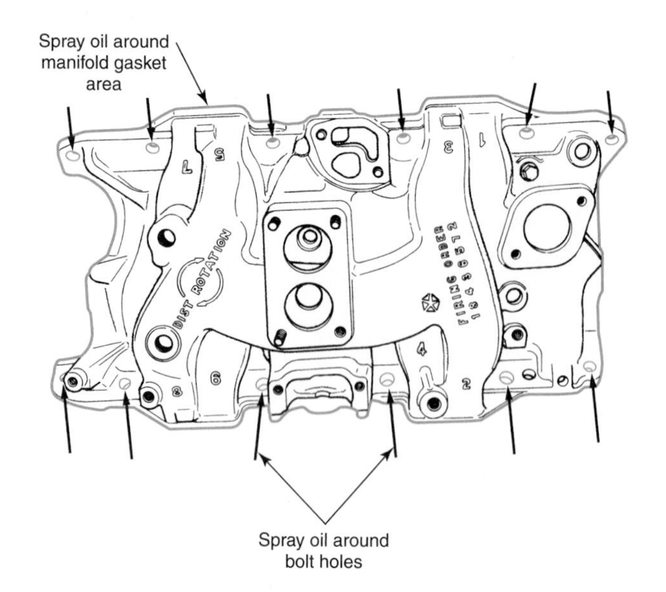

Figure A-1. Spray oil around the intake manifold gasket and bolt areas to check for any vacuum leaks. (DaimlerChrysler)

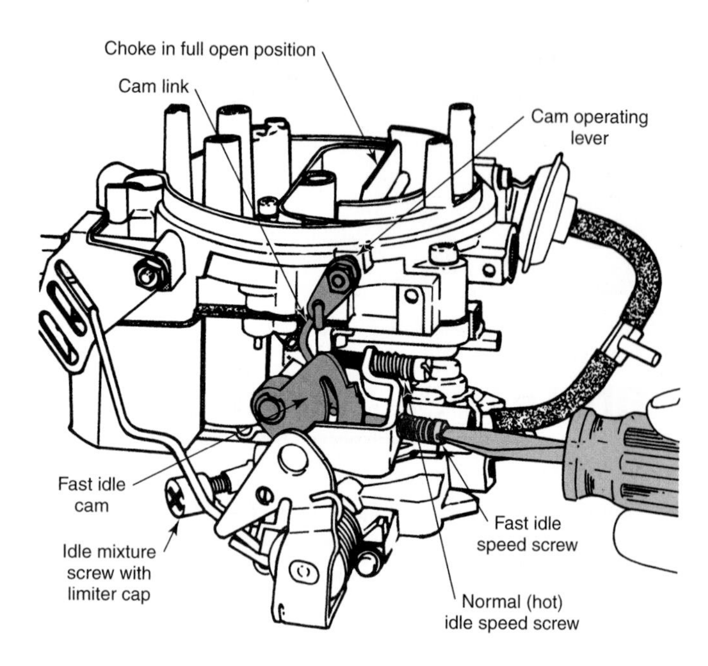

Figure A-2. Note the positions of the fast idle adjusting screw, fast idle cam, and the cam operating lever and link. (DaimlerChrysler)

Carburetor Adjustments

Many adjustments can be made with the carburetor still on the engine. Others require carburetor removal and partial or complete disassembly. Any carburetor adjustment can affect emissions. Newer carburetors do not provide any means for adjusting idle mixture. Even if adjustment were possible, the engine control module in newer model vehicles would compensate for any change in the fuel mixture, canceling out the adjustment attempt. Since there are many different models of carburetors in use, the following procedures will be general and applicable to the adjustment of a typical carburetor. The technician should consult a manual covering the specific carburetor to be adjusted or repaired.

STOP Warning: Use extreme caution when the engine is operating. Do not stand directly in line with the fan. Do not place your hands near the pulleys, belts, or fan. Do not wear loose clothing. If you have long hair, tie it back or wear it under a hat. Keep face and hands away from the carburetor when the engine is running—a backfire through the carburetor could produce a serious burn.

Fast Idle Speed Adjustment

A ***fast idle*** setting is necessary to keep the engine from stalling when it is cold. Bring the engine to normal operating temperature. Shut off the engine and connect a tachometer. Open the throttle and adjust the fast idle cam so that the ***fast idle adjusting screw*** contacts the recommended step or index mark. See **Figure A-2.**

Some specifications call for the automatic transmission to be in drive. If this is the case and if the car is equipped with a vacuum emergency brake release, disconnect the vacuum line. Attach a vacuum gauge to the vacuum line or plug the line with a golf tee. Failure to do this can cause the brake to release, allowing the car to lunge forward. You should also block the wheels. Start the car and turn the fast idle adjusting screw in or out to bring the engine rpm to specifications. Specifications typically are about 750 rpm when the screw is resting in the first (lowest) step of the cam.

Adjusting Fast Idle Cam Linkage

Because the ***fast idle cam*** position is determined by the degree of choke valve opening, it is important that the cam and choke valve relationship be accurate. With the engine stopped, hold the throttle open and close the choke valve.

Note: Some carburetors require the choke to be open a measured distance.

Allow the throttle lever to return to the idle position. Observe the point where the fast idle screw contacts the fast idle cam, **Figure A-3.** Specifications call for the contact to be on a certain step or in line with an index mark. If the contact is off, correct the problem, then bend the choke control rod as needed to produce the specified valve closure, If any adjustment is made, reset the fast idle rpm setting.

Hot Idle Speed Adjustment

The ***hot idle*** adjustment is made while the engine is at normal operating temperature. Check specifications to determine if the automatic transmission or transaxle should be in drive or neutral. If it is to be in drive, set the hand brake and block the wheels. If a vacuum emergency brake release is used, disconnect vacuum line at the diaphragm and insert a vacuum gauge. Attach a tachometer to monitor engine speed.

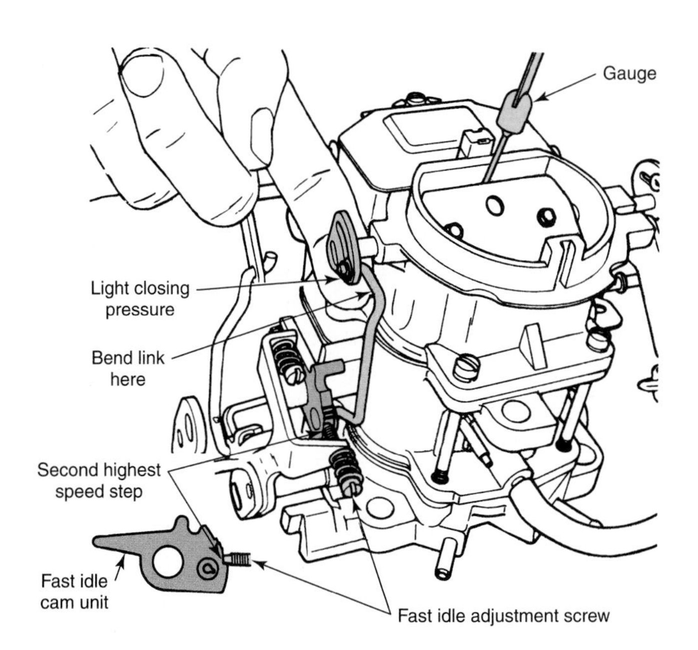

Figure A-3. Checking fast idle cam adjustment. The fast idle adjustment screw should contact the fast idle cam on the specified step or in line with an index mark. (DaimlerChrysler)

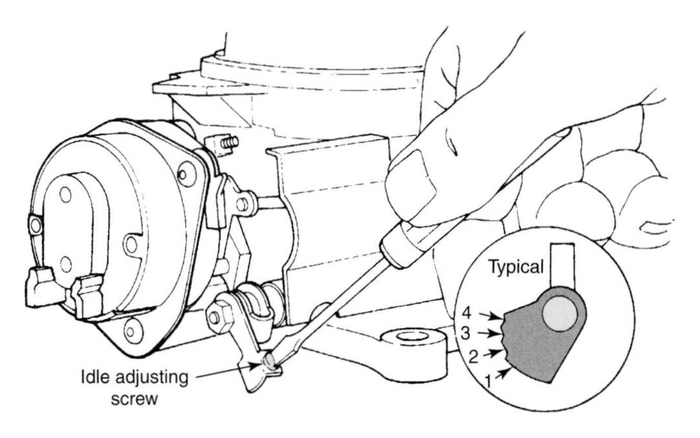

Figure A-4. Idle rpm adjustment. Note the four different cam steps. Each step will change the speed setting. (Ford)

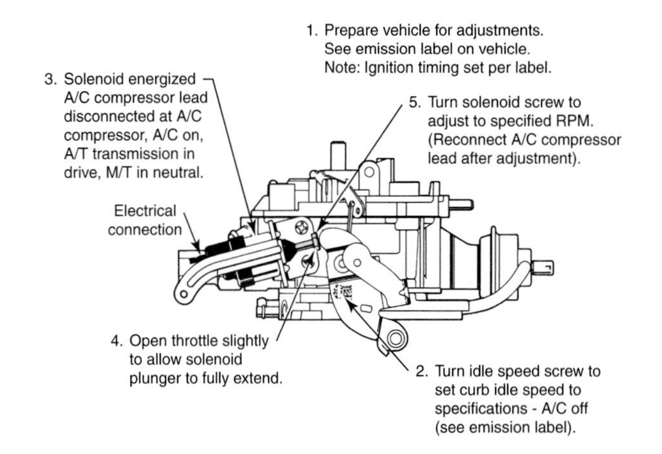

Figure A-5. This figure shows a typical idle speed solenoid. This unit is adjusted by turning the plunger. (General Motors)

Check to see if the air conditioning compressor should be turned on. Some specifications require that the headlights be turned on to load the alternator. Remove the air cleaner and plug any vacuum lines. Check the choke to make sure it is fully open. The secondary throttle valves (when used) must be fully closed at idle. If a hot idle compensator is used, it must be held closed. To prevent permanent damage to the compensator valve, press on the valve end and not on the bimetallic strip. Adjust the idle speed screw to produce the specified rpm.

On older carburetors, readjust the idle mixture after making the hot idle speed adjustment. In some carburetors, a single idle speed adjustment screw is used. If this is the case, adjust for slow (hot) idle with the screw contacting the lowest part of the fast idle cam, **Figure A-4.**The fast idle speed will be automatically adjusted as the choke pulls the cam upward.

Idle Control Devices

Idle speed control solenoids or ***motors*** are used to change the engine idle speed if the load changes while idling. This can occur when the air conditioner is turned on or when the automatic transmission is placed in gear. Other idle control devices are used to hold the throttle open when decelerating to reduce exhaust emissions. Some idle speed control solenoids can be adjusted, **Figure A-5.** Adjustment procedures are similar to those for idle speed screws. Check the solenoid by turning the ignition switch to the on position without starting the engine. The solenoid should extend when the throttle plate is slightly opened. If the solenoid is an air conditioner speed-up solenoid, it should extend when the air conditioner compressor turns on. On many late model engines, no solenoid adjustments are possible.

Antidieseling Solenoids

Antidieseling solenoids are used to set the hot idle on some engines. When the engine is turned off, the solenoid deenergizes, closing the throttle completely. This prevents engine dieseling or *run-on* caused by the hotter combustion chamber temperatures found in emission-controlled vehicles. Operation of the antidieseling solenoid can be checked by turning the engine off while observing the throttle shaft. The throttle should close completely when the ignition is switched off. If the throttle does not close completely, the solenoid is defective or the idle speed has been set with the carburetor idle screw. If the engine idle speed is too low, check to see if the solenoid extends when the ignition switch is turned on. If the solenoid is working, it can be adjusted to obtain the proper idle speed. Adjustment procedures are similar to those for idle speed screws.

The ***anti-stall dashpot*** prevents engine stalling when the throttle must be closed suddenly at low speeds. The dashpot creates a gradually reduced resistance to throttle closing. When the dashpot is actuated, it should move slowly and with resistance. If it fails to do this, repair or replace it as required.

To adjust the dashpot, move the throttle to the fully closed position. Make sure the fast idle cam is not holding the throttle partially open. While holding throttle in the full-closed position, push the dashpot plunger rod away from the throttle contact area as far as possible. Measure the distance between the plunger and the throttle contact. Adjust the dashpot (or dashpot plunger in some cases) to provide specified clearance.

Adjusting Idle Mixture

The ***idle mixture*** can be adjusted to vary the air-fuel ratio at idle. All carburetors produced after 1968 have some provision for limiting the amount that the mixture can be enriched. Many later carburetors cannot be adjusted without removing the carburetor and altering the carburetor throttle body or plate, or by removing special sealing plugs, **Figure A-6.** Other carburetors have a ***mixture control solenoid,*** and have no provision for adjusting the idle mixture.

Note: Adjusting carburetor idle mixture settings must be done according to manufacturer's procedures and specifications. Failure to do so is a violation of federal and state emission control laws.

Idle Mixture Adjustment Procedures

These adjustment procedures apply to vehicles built before 1968. They can be made with a vacuum gauge and tachometer and are used to obtain the highest, smoothest idle. Make sure that the engine is fully warmed up. Install a vacuum gauge and tachometer on the engine.

With the engine off, turn each ***idle mixture screw*** inward until it is lightly seated. Never seat a mixture screw tightly — this will groove the tip and prevent achieving a smooth idle. Next, back each screw out about two turns. Start the engine and adjust each screw in or out to produce the highest rpm and vacuum reading. If the engine begins to miss, this is due to a lean mixture. If the engine begins to run roughly, or *roll,* **the mixture is rich.**

If the idle speed has changed following the mixture adjustment, reset with the idle speed screw. Repeat the process of adjusting the mixture and setting the idle speed until the vacuum gauge shows the highest reading when the idle speed is as specified. Carefully count the turns to ensure that both screws are turned out an equal amount. Screws should be adjusted to be within one-quarter to one-half turn of each other.

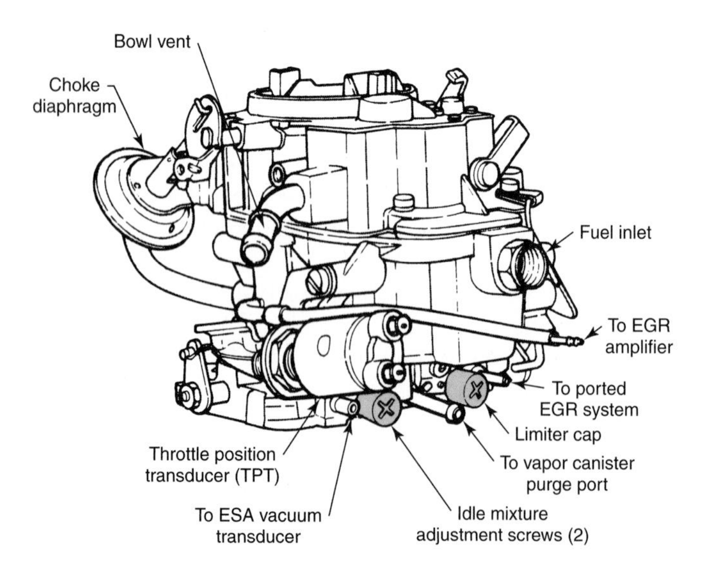

Figure A-6. Idle mixture adjustment screws covered with limiter caps that must be removed to allow for adjustment. They should then be reinstalled. (DaimlerChrysler)

Adjusting Idle Mixture on Emission—Controlled Carburetors

Idle adjustments on vehicles made after the beginning of the 1968 model year are subject to federal law and, in some areas, state law. Always check the emissions label in the engine compartment and follow the procedures and specifications given on this label. The three most common methods used to adjust mixture are the ***idle drop method,*** the ***propane enrichment method,*** and the ***exhaust gas analyzer method.***

These vehicles have some form of ***limiter cap*** or ***seal plug.*** Some idle limiter caps allow a small amount of mixture screw adjustment while others prevent any adjustment. Some idle mixture screws are recessed into the body of the carburetor and covered with a plug. The plug must be removed to access the idle mixture screws. Most plastic limiter caps can be removed with the carburetor installed. Most seal plugs, however, can only be removed after the carburetor is taken off the engine.

On later model vehicles, a control module adjusts the air-fuel ratio through a mixture control solenoid installed in the carburetor. In addition, the catalytic converter will attempt to clean up a rich mixture. Incorrect air-fuel ratios therefore, may not be obvious. For this reason, always check the service literature for the exact procedures and for proper specifications for the year and make of vehicle before beginning carburetor adjustments.

Idle Drop Method

This adjustment procedure applies only to some older engines and can be made using a vacuum gauge and tachometer. Begin by removing the restrictor caps from the idle mixture screws and adjusting the mixture to obtain the highest, smoothest idle at the rpm listed on the emissions label. Then, turn in the mixture screws until the idle speed drops by the amount listed on the emissions label, usually between 50 and 100 rpm. Reinstall the restrictor caps or use color-coded replacement caps when required by law.

Propane Enrichment Method

Idle mixture adjustments on some carburetors require the use of a propane enrichment device.

Warning: Propane can be dangerous. Follow all service procedures carefully.

Begin by attaching the propane enrichment device to the air cleaner or intake manifold as indicated in **Figure A-7.** Also attach a tachometer and remove the restrictor caps from the idle mixture screws. Start the engine and adjust the propane valve opening according to instructions. With the propane valve open, adjust the mixture screws to obtain the highest, smoothest idle at the rpm listed on the emissions label. Readjust both mixture and idle speed screws as needed to obtain the smoothest idle at the specific rpm. Then close the propane valve. The idle speed should drop by the amount listed on the emissions label. If the idle drops properly, reinstall the restrictor caps, or color-coded replacement caps

Exhaust Gas Analyzer Method

The only sure method of determining the air-fuel ratio is by using an exhaust gas analyzer. There are a number of exhaust gas analyzers on the market and the following information is only a general guide. Always follow the equipment and vehicle manufacturer's instructions exactly.

Note: The exhaust gas analyzer can be used to check air-fuel ratios at any engine speed. Consult the analyzer instruction manual for exact testing procedures. This is covered in more detail in Chapter 22.

Begin by turning on and calibrating the exhaust gas analyzer. Start the engine and run it long enough to warm to normal operating temperature. Place the analyzer probe in the vehicle tailpipe, and allow the engine to idle as analyzer readings stabilize. Note the air-fuel ratio and compare it with specifications for the engine. If possible, turn the idle mixture adjustment screws to obtain the proper readings. On newer carburetors, the idle mixture cannot be easily adjusted. The carburetor may require rebuilding if the readings cannot be corrected. After the idle mixture has been properly adjusted, replace the limiter caps or plugs with new ones.

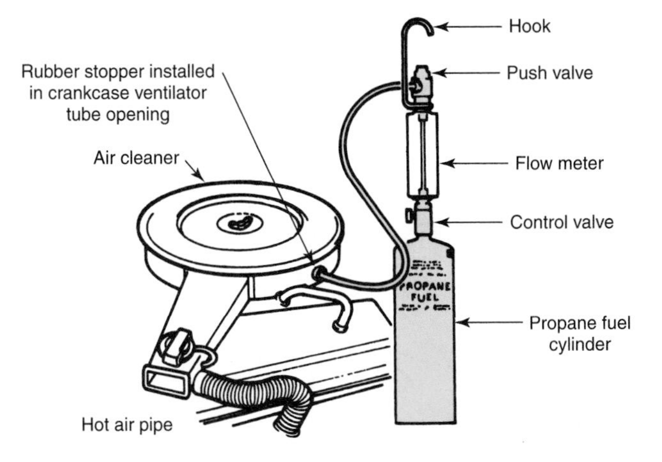

Figure A-7. The propane idle enrichment mixture setup shown here is used to adjust idle mixture on some vehicles. Be extremely careful while using propane gas, since it is highly flammable. (AC Delco)

Choke Service

All modern carburetors are equipped with ***automatic chokes,*** which open and close depending on engine temperature. This is accomplished through the use of a ***thermostatic spring,*** sometimes called a *bimetal spring.* Chokes are usually located on the carburetor, as shown in **Figure A-8.** Newer chokes are heated by an ***electric heating element*** located on the housing. This heating element is energized by the alternator through a special temperature relay or by the control module. The choke system shown in **Figure A-9** places the thermostatic spring directly over the hot exhaust crossover passage instead of on the carburetor itself. This positioning is called a ***divorced choke.*** Some older vehicles use a ***manual choke*** that is operated by the vehicle driver.

Checking Choke Operation

The simplest way to check choke operation is to observe the choke when the engine is cold. When the throttle is moved slightly, the choke should snap closed. When the engine is started, the choke should open slightly. As the engine warms up, the choke should gradually open until it opens completely on a thoroughly warmed-up engine. If the choke does not perform as expected, check for misadjustment, binding or dirty linkage, or a defective thermostatic coil. On an electric choke, check for current to the heating element. Check the element with an ohmmeter to ensure that it has not burned out. Also check the operation and adjustment of the vacuum break.

Adjusting an Automatic Choke

The automatic choke is operated by the tension of a bimetallic spring. The spring coils and uncoils as engine temperature changes. The amount of spring tension is determined by the positioning of the choke housing and cover index marks. Many chokes are not adjustable, or can be adjusted only by drilling out rivets holding the housing and replacing them with screws. If the choke has no provision for adjustment and proves defective, it must be replaced. If the choke *can* be adjusted, align the marks as recommended by the manufacturer. This initial setting, although usually very close, may require a slight adjustment after testing the choke operation. The choke in **Figure A-10** is adjusted one notch lean.

Some covers are marked to indicate the direction in which to turn for a lean or rich setting. Hard starting, stalling, backfiring and sluggish operation during warm-up may indicate the need of a richer setting. Engine rolling and black smoke from the exhaust indicate the need for a leaner setting. Remember the choke valve should be closed with the engine cold and wide open with the engine at normal operating temperature. One divorced choke is adjusted by loosening a locknut and turning with a screwdriver until the index mark is aligned as desired. Other divorced chokes are adjusted by bending the linkage rod that connects it to the choke shaft.

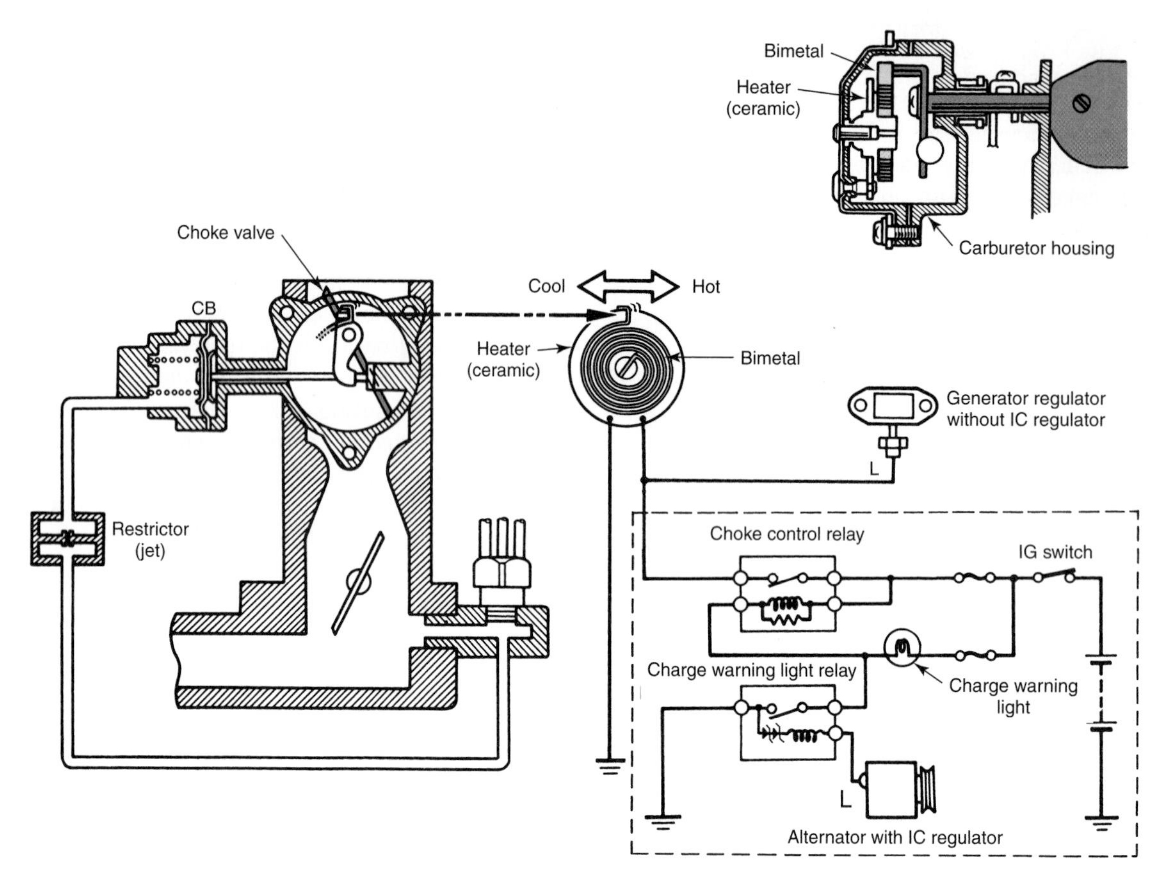

Figure A-8. This figure shows a bimetal (bimetallic) thermostatic spring choke that mounts directly to the carburetor housing. This one is heated electronically. (Toyota)

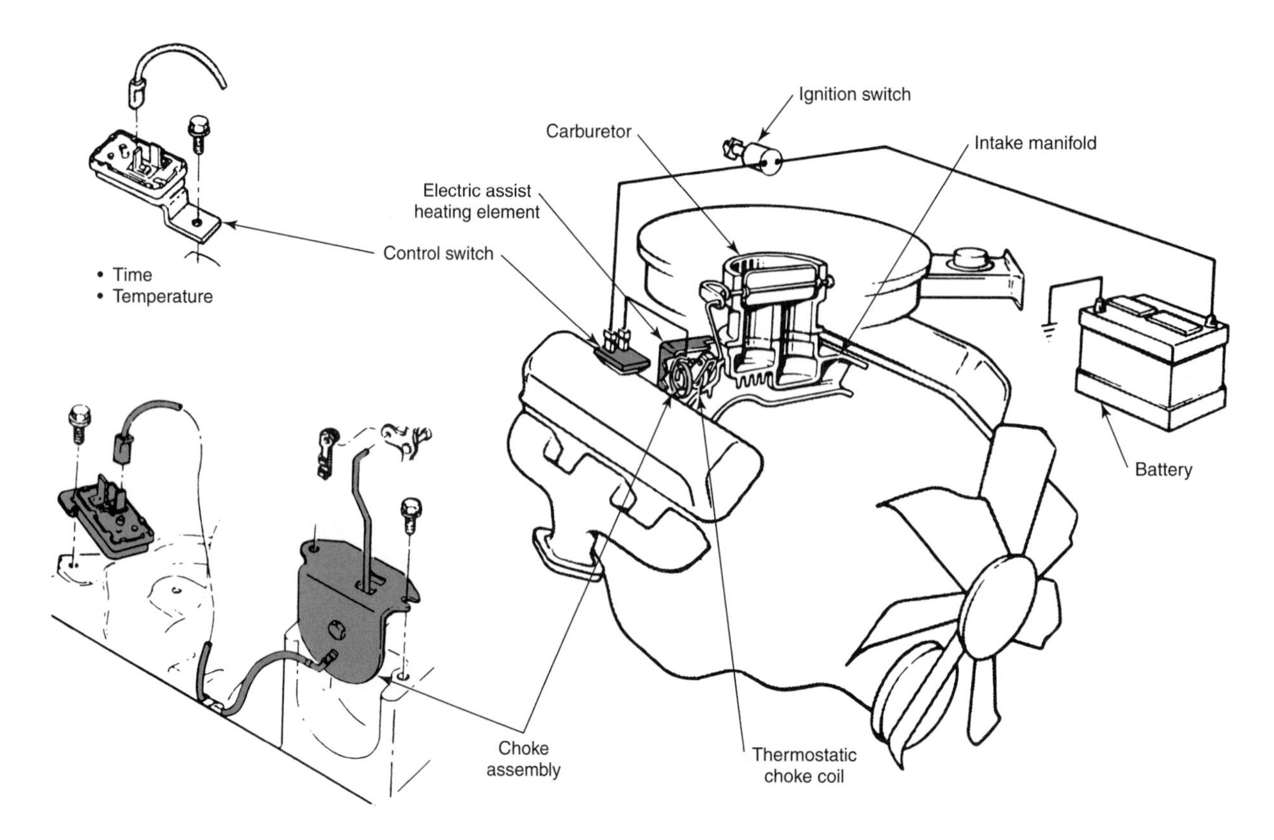

Figure A-9. A divorced choke system. To make use of engine heat, the thermostatic choke coil spring is placed in a well located in the intake manifold. (Colt Industries)

Cleaning an Automatic Choke

Chokes occasionally require cleaning, especially if they are the type heated by air passing through the choke housing. To clean a choke, remove the air cleaner and note the position of the thermostatic spring housing index mark in relation to the choke housing marks. Then remove the choke cover, thermostatic spring, and related parts. If the cover is secured with rivets, **Figure A-11,** they must be drilled out.

Clean all metal parts in recommended cleaner and blow them dry. Do not wash electrical parts. Be careful not to distort the thermostatic spring. Apply a few drops of clean carbon solvent to the choke shaft bearings and the external linkage while moving the choke valve. If the linkage is dirty, use a small brush to remove the dirt. Reassemble all choke parts and use a new cover gasket. Make certain the thermostatic spring is positioned correctly and that it engages the choke shaft lever. Align the housing and cover index marks and tighten cover fasteners securely.

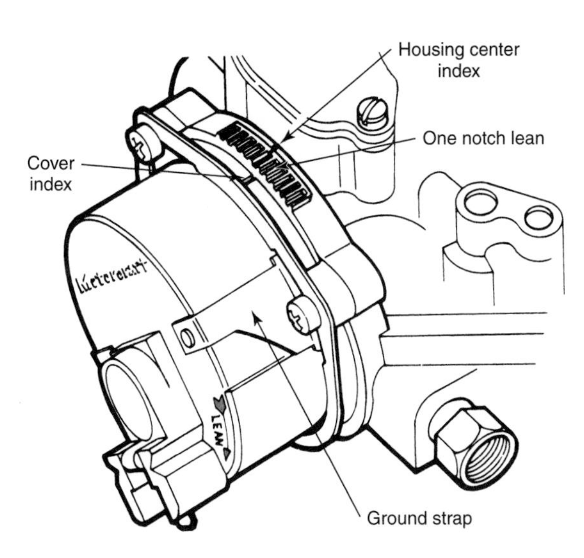

Figure A-10. This particular application specifies an initial setting of the electric choke one notch (index mark) on the lean side. (Ford)

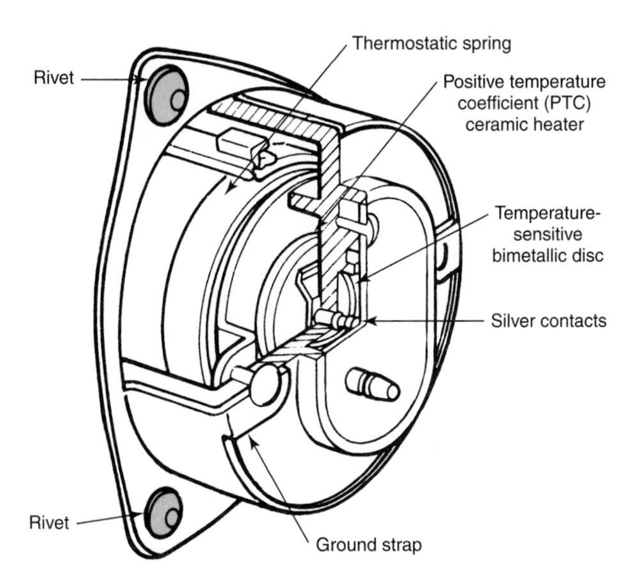

Figure A-11. This choke cover is secured with rivets. Remove the cover by drilling out the rivets. (Ford)

Manual Choke Adjustment

The manual choke is operated by the driver. Some older vehicles equipped with manual chokes are still in use. Aftermarket manual chokes can be used to replace an automatic choke when replacement parts cannot be located. To check the operation of a manual choke, remove the air cleaner. Then pull the choke knob out as far as it will go while stepping on the accelerator pedal. The choke plate should close completely. Next, push the choke knob completely in. The choke valve should be open fully. If necessary, adjust the choke by loosening the swivel mounted on the choke shaft.

Vacuum Break Adjustment

The vacuum break is used to open the choke slightly once the engine starts. If the choke plate is not opened slightly, the engine will run too rich and flood before the thermostatic spring warms up and opens the choke fully. Most vacuum breaks will operate satisfactorily unless the vacuum diaphragm begins to leak. The vacuum diaphragm can be checked with a hand-held vacuum pump. A defective vacuum diaphragm must be replaced. Always consult the appropriate service literature to adjust a vacuum break.

Accelerator Pump Checks and Adjustment

The ***accelerator pump*** supplies extra fuel to the carburetor air horn to prevent flat spots when accelerating. Accelerator pumps can be the plunger type, **Figure A-12,** or the diaphragm-pump type. To check accelerator pump operation, remove the air cleaner and open the throttle quickly. Observe the pump discharge nozzle in the air horn. Depending on the pump nozzle design, one or two streams of gas should be evident. The fine gas stream should be strong

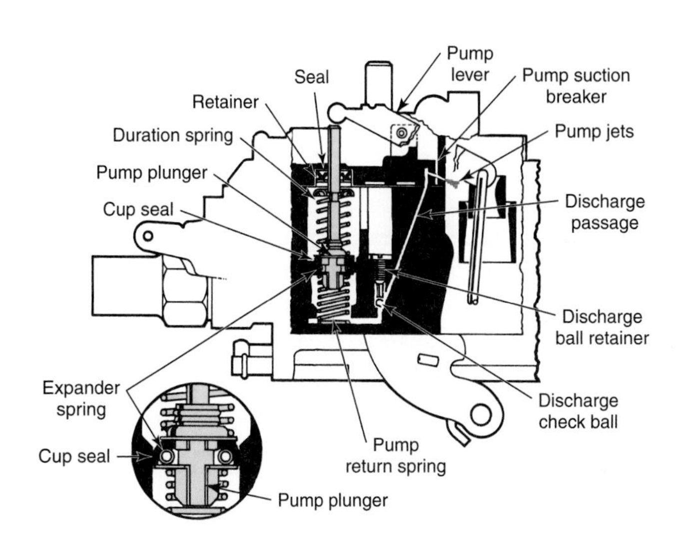

Figure A-12. A cutaway view of a plunger-type accelerator pump. (General Motors)

and should last for a short time even after the throttle reaches full open. If little or no gas output is observed, the accelerator pump piston or diaphragm could be cracked or worn. Check valves might be stuck open by dirt or the discharge nozzle feed system could be clogged. Clean and repair as needed.

Accelerator Pump Adjustment

Most late-model accelerator pumps are not adjustable. To adjust the accelerator pump on an older carburetor, close the throttle and check the position of the accelerator pump linkage in relation to a specified portion of the carburetor. Bend the link or lever as required. By placing the rod in different holes, the pump stroke and amount of gasoline delivered to the air horn can be varied.

Checking and Setting Float Level

The ***float level*** setting is critical since it establishes the level of the fuel in the carburetor bowl, **Figure A-13.** This has a direct effect on the ability of the main metering system to pull gasoline from the bowl, and therefore, on the air-fuel ratio at cruising speeds. A higher-than-specified fuel level will result in poor gas mileage, spark plug fouling, and crankcase dilution. A low fuel level will cause backfiring, surging, and loss of power.

Checking and adjusting the float level with the carburetor disassembled is called ***dry level.*** Dry float level is checked by either measuring from some portion of the float to the cover, or by using a gauge. **Figure A-14** shows a gauge designed for the purpose. Adjust the float level by carefully bending the float arm.

Caution: The carburetor float system is delicate and can be severely damaged by improper handling.

A ***float drop*** setting may be checked with a gauge or by measuring between two specified points. Adjust by careful bending of the float stop tab or lip. When adjusting, make certain the float is not twisted or bent sideways. If it is, it may hang up on the bowl walls. Another method of adjusting the float level on some carburetors is to move an adjustable inlet or float needle seat.

Checking Actual or Wet Fuel Level

The float level and drop settings provide a basic dry setting. On some vehicles, the actual fuel level in the bowl, or ***wet level,*** can be checked. This can be done by running the engine until warm, stopping and measuring the distance from a specified point to the fuel level.

Sometimes a special tool is needed to check wet float level. Another method utilizes a sight plug in the end of the bowl. The plug is removed and the fuel level determined by the relationship between the fuel level height and the bottom of the hole.

Adjusting Wet Fuel Level

The float must be readjusted to provide the exact wet fuel level required. When checking, the car must be level, engine at normal operating temperature, fuel pump pressure normal, and the inlet needle must not be leaking. Where an adjustable needle seat is used, the wet fuel adjustment is easy. Warm up the engine, and then turn it off. Remove the sight plug and check the fuel level. If the fuel level is too low, hold the adjusting nut and loosen the lock screw.

Warning: Do not loosen the lock screw or open the sight plug when the engine is running.

Turn the nut out (counterclockwise) to raise the fuel level. Tighten the lock screw and replace the sight plug. Start the

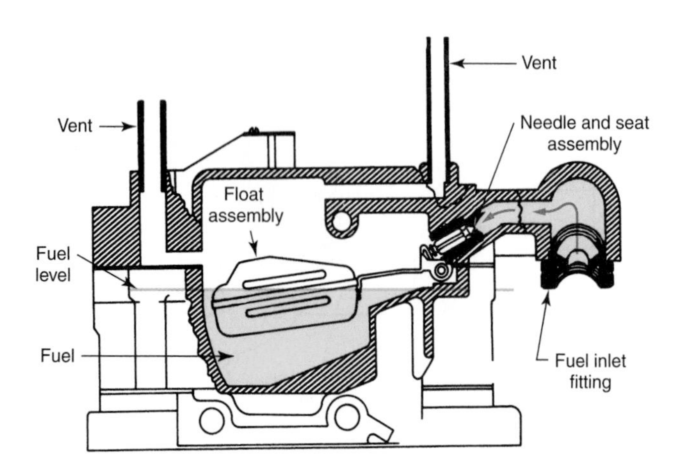

Figure A-13. The float level determines the height of the fuel in the bowl. (DaimlerChrysler)

Figure A-14. Making a dry float level adjustment using a float level gauge. The carburetor has been partly disassembled to provide access to the float and chamber. (Ford)

engine and run it for a minute. Stop the engine, remove the sight plug, and recheck. Repeat until the fuel level is exact. If the fuel level is too high, lower below the specified level and then raise it to the height required.

Checking the Float

The float must be ***airtight;*** that is, no gasoline must enter the float. Occasionally a float will leak, which will cause it to sink and become useless. Shake a hollow metal float to test for gasoline inside. If there is gas in the float, replace it. Most solid plastic floats cannot be checked for leakage and must be replaced if leakage is suspected. One method of checking a solid float is to place it on a flat surface and observe whether the float is pulled off balance by the metal hinge. If it is, the float is probably OK.

Replace Needle and Seat

The fuel bowl inlet ***needle and seat*** (the control valve assembly for fuel entering the bowl) will eventually wear and start to leak. If the needle and seat are leaking, no amount of float adjustment will correct the fuel level. Replace the needle and seat as a matched set. Be sure to readjust the float level and drop after replacing the needle and seat.

Variable Venturi Opening Check

Some carburetors employ a ***variable venturi valve*** that alters the size of the venturi opening to meet various engine speed and load conditions. There are many checks to be made on a variable venturi carburetor, so the factory service manual should be consulted. A common check and adjustment concerns the maximum venturi valve opening at the wide-open throttle position.

Mixture Control Solenoids

A number of carburetors utilize a mixture control solenoid. The mixture control solenoid is operated by the engine control module. Energizing the solenoid causes a valve to open and close. This valve can control the flow of gasoline or air, depending on the type of carburetor and engine control system.

Throttle Position Sensors

Many late model carburetors have a throttle position sensor, or TPS, installed on the throttle shaft or operated by the accelerator pump linkage. This sensor transmits an electrical input signal to the engine control module. The control module reads this input as the amount the throttle is opened. This helps the control module decide which commands to send to various output devices. The operation and service of throttle position sensors was covered in detail in Chapter 15.

Altitude Compensator Adjustment

Carburetors intended for use at high elevations are sometimes equipped with an ***altitude-compensating device.*** High elevations provide less oxygen per unit of air, as well as less atmospheric pressure. By providing additional amounts of air, the correct air-fuel ratio can be maintained. The altitude compensator makes use of an ***aneroid,*** which is a sealed bellows under partial vacuum.

The aneroid expands with a drop in barometric pressure and contracts with a pressure increase. As the altitude increases, the aneroid capsule expands, forcing open an air valve in the carburetor. The air valve admits additional air to the fuel mixture. The altitude compensator must be set as specified by the manufacturer.

Other Carburetor Adjustments

In addition to those covered, other adjustments are needed on some carburetors. These include secondary throttle opening point, secondary throttle choke lockout, power valve opening, and metering rod position. To ensure complete and accurate work, always consult a service manual covering the exact carburetor being serviced.

Note: When making adjustments on computer-controlled carburetors, it is helpful to use a diagnostic scan tool, if available, so that you can monitor the actual amount of adjustment.

Carburetor Overhaul

Reasons to remove and overhaul a carburetor include buildup of gum and varnish, excessive dirt and water in the carburetor, leaky gaskets, and failure of mechanical parts such as the accelerator pump, needle and seat, vacuum diaphragms, and linkage or throttle valve shafts.

Carburetor Removal and Overhaul

Remove the carburetor fuel lines, vacuum lines, electrical connectors, and throttle linkage. Then, remove the bolts holding the carburetor to the intake manifold and lift the carburetor from the manifold. Carefully remove any old gasket material from the intake opening and plug it with a clean cloth. Begin disassembly by attaching *repair legs* to the carburetor to prevent damage to the throttle plates or valves, **Figure A-15.** Use a sharp scribe to mark parts before disassembly. If jets are to be removed, mark the jet and adjacent area.

Disassemble the carburetor and soak all metal parts in fresh carburetor cleaner. Do not soak electrical parts, power valve, accelerator pump plunger, antistall dashpot, diaphragm units, fuel enrichment valve, plastic float, or any parts made of rubber, leather, fabric, or fiber. Parts that cannot be soaked in cleaner should be wiped with a clean cloth. Soak the metal parts no longer than fifteen minutes, since the cleaner will begin to remove needed coatings. Remove the parts, rinse as directed, and blow them dry. Blow out all passageways. Never use a wire or drill bit to probe into jets, air bleeds, or other passages. To prevent damage, use the air blast only. Do not direct the air blast into any diaphragm units. Inspect all parts for cracks. Check all parting surfaces for nicks and burrs. Examine choke and throttle shafts and bearings for excessive looseness or out-of-roundness. Check idle mixture screws and the float needle and seat; replace if grooved or worn. Test the float for leakage. Check float arm-to-needle surface for roughness or grooving. Replace all defective parts, stripped

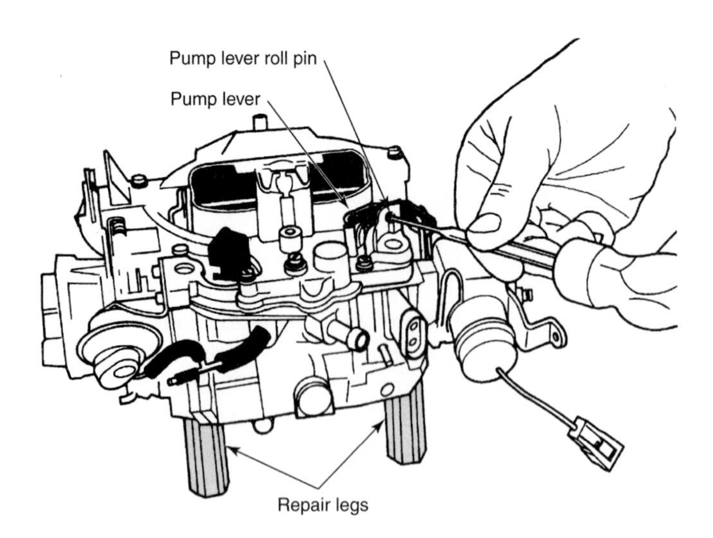

Figure A-15. Mounting a carburetor on repair legs greatly facilitates overhaul and prevents damage to the throttle plates. (General Motors)

fasteners, and distorted springs. Discard all other old parts that will be replaced by those in the overhaul kit.

The carburetor part number is needed to obtain a carburetor overhaul kit and other necessary parts. This number is either stamped on the carburetor housing or on a metal label attached to the carburetor. Most carburetor overhaul kits contain all necessary gaskets, plus a new needle and seat, accelerator pump, and minor hardware parts such as clips and plugs. A simple float-adjusting gauge will also be included. While overhauling a computer-controlled carburetor, you might want to replace the throttle position sensor and mixture control solenoids. These parts are usually located inside the bowl of computer-controlled carburetors, making replacement after an overhaul difficult.

Note: Be sure that the overhaul kit is the correct one for the carburetor that you are servicing. Always use new gaskets when overhauling a carburetor.

Carburetor Assembly

To ensure that the carburetor is properly reassembled, refer to the appropriate service manual, or use the exploded view of the carburetor included in the overhaul kit. **Figure A-16** is an exploded view of a typical two-barrel carburetor employing an electrically operated mixture control solenoid.

Reassemble the carburetor on a clean work surface, using clean tools. Do not use gasket cement unless directed to do so in the manufacturer's instructions. Assemble the unit carefully, avoiding the use of excessive force. As required, use special tools designed for the job. Tighten all fasteners securely. If torque specifications are given, use them. Perform all required adjustment checks, including float level and drop, accelerator pump stroke, and initial idle speed and mixture settings. Fill the fuel bowl and check accelerator pump operation.

Carburetor Installation

Clean the mounting area of the intake manifold and install a new mounting gasket. Reinstall the heat spacers, if used. Torque the carburetor fasteners and attach fuel and vacuum lines, electrical connectors, and all linkages. Start the engine and check for leaks, choke operation, and idle speed. Perform all required adjustments. Road test the vehicle to check performance.

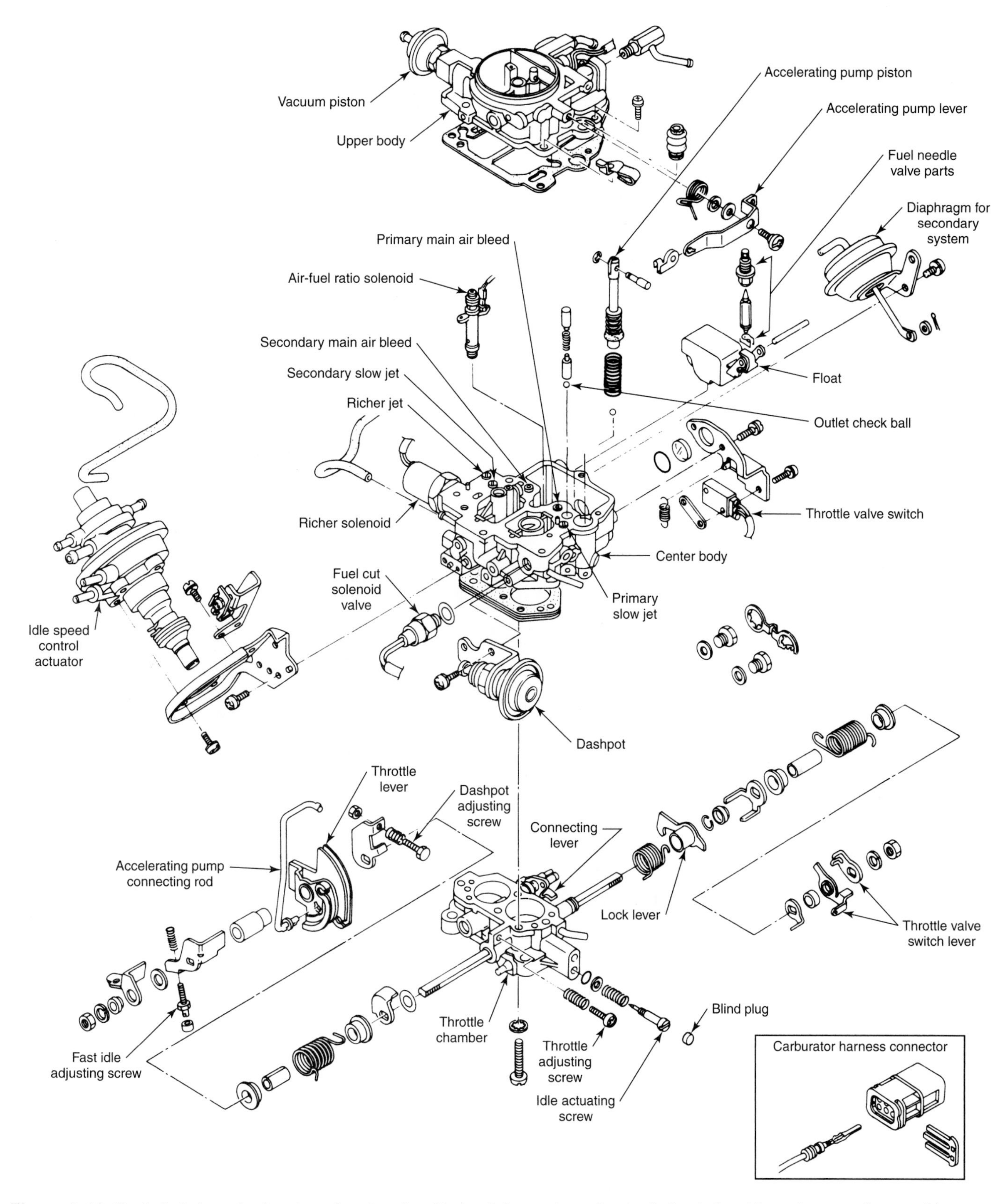

Figure A-16. Exploded view of a two-barrel carburetor. Study all the parts and note their relationship to the overall assembly. (Nissan)

Cutaway of a late-model sedan equipped with a 3.0 liter, V-6 engine and a 5-speed manual transmission. (Ford)

Appendix B

OBD II Trouble Codes

This appendix is provided as a convenient reference for the standardized OBD II codes you may encounter while diagnosing vehicles in the shop. Manufacturer-specific codes (codes that begin with a P1 rather than P0) should be looked up in the service manual.

The list of codes shown here was current at the time of publication. Goodheart-Willcox Publisher cannot be held responsible for any errors or omissions. All codes generated by a vehicle's computer control system should be checked in the appropriate service literature. OBD I ECMs on cars built from 1993–1995 may provide these codes to some scan tools, along with the normal two-digit codes.

P01XX Fuel and Air Metering

Note: For systems with single O_2 sensors, use codes for bank 1 sensor. Bank 1 contains cylinder 1. Sensor 1 is closest to the engine.

P0101 Mass or volume airflow circuit range/performance problem
P0102 Mass or volume airflow circuit low input
P0103 Mass or volume airflow circuit high input
P0104 Mass or volume airflow circuit intermittent
P0105 Manifold absolute pressure/barometric pressure circuit malfunction
P0106 Manifold absolute pressure/barometric pressure circuit range/performance problem
P0107 Manifold absolute pressure/barometric pressure circuit low input
P0108 Manifold absolute pressure/barometric pressure circuit high input
P0109 Manifold absolute pressure/barometric pressure circuit intermittent
P0110 Intake air temperature circuit malfunction
P0111 Intake air temperature circuit range/performance problem
P0112 Intake air temperature circuit low input
P0113 Intake air temperature circuit high input
P0114 Intake air temperature circuit intermittent
P0115 Engine coolant temperature circuit malfunction
P0116 Engine coolant temperature circuit range/performance problem
P0117 Engine coolant temperature circuit low input
P0118 Engine coolant temperature circuit high input
P0119 Engine coolant temperature circuit intermittent
P0120 Throttle/pedal position sensor/switch A circuit malfunction
P0121 Throttle/pedal position sensor/switch A circuit range/performance problem
P0122 Throttle/pedal position sensor/switch A circuit low input
P0123 Throttle/pedal position sensor/switch A circuit high input
P0124 Throttle/pedal position sensor/switch A circuit intermittent
P0125 Insufficient coolant temperature for closed loop fuel control
P0126 Insufficient coolant temperature for stable operation
P0130 Oxygen sensor circuit malfunction (bank 1, sensor 1)
P0131 Oxygen sensor circuit low voltage (bank 1, sensor 1)
P0132 Oxygen sensor circuit high voltage (bank 1, sensor 1)
P0133 Oxygen sensor circuit slow response (bank 1, sensor 1)
P0134 Oxygen sensor circuit no activity detected (bank 1, sensor 1)
P0135 Oxygen sensor heater circuit malfunction (bank 1, sensor 2)
P0136 Oxygen sensor circuit malfunction (bank 1, sensor 2)
P0137 Oxygen sensor circuit low voltage (bank 1, sensor 2)
P0138 Oxygen sensor circuit high voltage (bank 1, sensor 2)

P0139 Oxygen sensor circuit slow response (bank 1, sensor 2)
P0140 Oxygen sensor circuit no activity detected (bank 1, sensor 2)
P0141 Oxygen sensor heater circuit malfunction (bank 1, sensor 2)
P0142 Oxygen sensor circuit malfunction (bank 1, sensor 3)
P0143 Oxygen sensor circuit low voltage (bank 1, sensor 3)
P0144 Oxygen sensor circuit high voltage (bank 1, sensor 3)
P0145 Oxygen sensor circuit slow response (bank 1, sensor 3)
P0146 Oxygen sensor circuit no activity detected (bank 1, sensor 3)
P0147 Oxygen sensor heater circuit malfunction (bank 1, sensor 3)
P0150 Oxygen sensor circuit malfunction (bank 2, sensor 1)
P0151 Oxygen sensor circuit low voltage (bank 2, sensor 1)
P0152 Oxygen sensor circuit high voltage (bank 2, sensor 1)
P0153 Oxygen sensor circuit slow response (bank 2, sensor 1)
P0154 Oxygen sensor circuit no activity detected (bank 2, sensor 1)
P0155 Oxygen sensor heater circuit malfunction
P0156 Oxygen sensor circuit malfunction (bank 2, sensor 1)
P0157 Oxygen sensor circuit low voltage (bank 2, sensor 2)
P0158 Oxygen sensor circuit high voltage (bank 2, sensor 2)
P0159 Oxygen sensor circuit slow response (bank 2, sensor 2)
P0160 Oxygen sensor circuit no activity detected (bank 2, sensor 1)
P0161 Oxygen sensor heater circuit malfunction (bank 2, sensor 2)
P0162 Oxygen sensor circuit malfunction (bank 2, sensor 2)
P0163 Oxygen sensor circuit low voltage (bank 2, sensor 3)
P0164 Oxygen sensor circuit high voltage (bank 2, sensor 3)
P0165 Oxygen sensor circuit slow response (bank 2, sensor 3)
P0166 Oxygen sensor circuit no activity detected (bank 2, sensor 3)
P0167 Oxygen sensor heater circuit malfunction (bank 2, sensor 3)
P0170 Fuel trim malfunction (bank 1)
P0171 System too lean (bank 1)
P0172 System too rich (bank 1)
P0173 Fuel trim malfunction (bank 1)
P0174 System too lean (bank 2)
P0175 System too rich (bank 2)
P0176 Fuel composition sensor circuit malfunction
P0177 Fuel composition sensor circuit range/performance
P0178 Fuel composition sensor circuit low input
P0179 Fuel composition sensor circuit high input
P0180 Fuel temperature sensor A circuit malfunction
P0181 Fuel temperature sensor A circuit range/performance
P0182 Fuel temperature sensor A circuit low input
P0183 Fuel temperature sensor A circuit high input
P0184 Fuel temperature sensor A circuit intermittent
P0185 Fuel temperature sensor B circuit malfunction
P0186 Fuel temperature sensor B circuit range/performance
P0187 Fuel temperature sensor B circuit low input
P0188 Fuel temperature sensor B circuit high input
P0189 Fuel temperature sensor B circuit intermittent
P0190 Fuel rail pressure sensor circuit malfunction
P0191 Fuel rail pressure sensor circuit range/performance
P0192 Fuel rail pressure sensor circuit low input
P0193 Fuel rail pressure sensor circuit high input
P0194 Fuel rail pressure sensor circuit intermittent
P0195 Engine oil temperature sensor malfunction
P0196 Engine oil temperature sensor range/performance
P0197 Engine oil temperature sensor low
P0198 Engine oil temperature sensor high
P0199 Engine oil temperature sensor intermittent

P02XX Fuel and Air Metering

P0200 Injector circuit malfunction
P0201 Injector circuit malfunction—cylinder 1
P0202 Injector circuit malfunction—cylinder 2
P0203 Injector circuit malfunction—cylinder 3
P0204 Injector circuit malfunction—cylinder 4
P0205 Injector circuit malfunction—cylinder 5
P0206 Injector circuit malfunction—cylinder 6
P0207 Injector circuit malfunction—cylinder 7
P0208 Injector circuit malfunction—cylinder 8
P0209 Injector circuit malfunction—cylinder 9
P0210 Injector circuit malfunction—cylinder 10
P0211 Injector circuit malfunction—cylinder 11
P0212 Injector circuit malfunction—cylinder 12
P0213 Cold start injector 1 malfunction
P0214 Cold start injector 2 malfunction
P0215 Engine shutoff solenoid malfunction

P0216 Injection timing control circuit malfunction
P0217 Engine over temperature condition
P0218 Transmission over temperature condition
P0219 Engine overspeed condition
P0220 Throttle/pedal position sensor/switch B circuit malfunction
P0221 Throttle/pedal position sensor/switch B circuit range/performance problem
P0222 Throttle/pedal position sensor/switch B circuit low input
P0223 Throttle/pedal position sensor/switch B circuit high input
P0224 Throttle/pedal position sensor/switch B circuit intermittent
P0225 Throttle/pedal position sensor/switch C circuit malfunction
P0226 Throttle/pedal position sensor/switch C circuit range/performance problem
P0227 Throttle/pedal position sensor/switch C circuit low input
P0228 Throttle/pedal position sensor/switch C circuit high input
P0229 Throttle/pedal position sensor/switch C circuit intermittent
P0230 Fuel pump primary circuit malfunction
P0231 Fuel pump secondary circuit low
P0232 Fuel pump secondary circuit high
P0233 Fuel pump secondary circuit intermittent
P0235 Turbocharger boost sensor A circuit malfunction
P0236 Turbocharger boost sensor A circuit range/performance
P0237 Turbocharger boost sensor A circuit low
P0238 Turbocharger boost sensor A circuit high
P0239 Turbocharger boost sensor B circuit malfunction
P0240 Turbocharger boost sensor B circuit range/performance
P0241 Turbocharger boost sensor B circuit low
P0242 Turbocharger boost sensor B circuit high
P0243 Turbocharger wastegate solenoid A malfunction
P0244 Turbocharger wastegate solenoid A range/performance
P0245 Turbocharger wastegate solenoid A low
P0246 Turbocharger wastegate solenoid A high
P0247 Turbocharger wastegate solenoid B malfunction
P0248 Turbocharger wastegate solenoid B range/performance
P0249 Turbocharger wastegate solenoid B low
P0250 Turbocharger wastegate solenoid B high
P0251 Injection pump A rotor/cam malfunction
P0252 Injection pump A rotor/cam range/performance
P0253 Injection pump A rotor/cam low
P0254 Injection pump A rotor/cam high
P0255 Injection pump A rotor/cam intermittent
P0256 Injection pump B rotor/cam malfunction
P0257 Injection pump B rotor/cam range/performance
P0258 Injection pump B rotor/cam low
P0259 Injection pump B rotor/cam high
P0260 Injection pump B rotor/cam intermittent
P0261 Cylinder 1 injector circuit low
P0262 Cylinder 1 injector circuit high
P0263 Cylinder 1 contribution/balance fault
P0264 Cylinder 2 injector circuit low
P0265 Cylinder 2 injector circuit high
P0266 Cylinder 2 contribution/balance fault
P0267 Cylinder 3 injector circuit low
P0268 Cylinder 3 injector circuit high
P0269 Cylinder 3 contribution/balance fault
P0270 Cylinder 4 injector circuit low
P0271 Cylinder 4 injector circuit high
P0272 Cylinder 4 contribution/balance fault
P0273 Cylinder 5 injector circuit low
P0274 Cylinder 5 injector circuit high
P0275 Cylinder 5 contribution/balance fault
P0276 Cylinder 6 injector circuit low
P0277 Cylinder 6 injector circuit high
P0278 Cylinder 6 contribution/balance fault
P0279 Cylinder 7 injector circuit low
P0280 Cylinder 7 injector circuit high
P0281 Cylinder 7 contribution/balance fault
P0282 Cylinder 8 injector circuit low
P0283 Cylinder 8 injector circuit high
P0284 Cylinder 8 contribution/balance fault
P0285 Cylinder 9 injector circuit low
P0286 Cylinder 9 injector circuit high
P0287 Cylinder 9 contribution/balance fault
P0288 Cylinder 10 injector circuit low
P0289 Cylinder 10 injector circuit high
P0290 Cylinder 10 contribution/balance fault
P0291 Cylinder 11 injector circuit low
P0292 Cylinder 11 injector circuit high
P0293 Cylinder 11 contribution/balance fault
P0294 Cylinder 12 injector circuit low
P0295 Cylinder 12 injector circuit high
P0296 Cylinder 12 contribution/balance fault

P03XX Ignition System or Misfire

Note: Bank 1 contains cylinder #1.

P0300 Random/multiple cylinder misfire detected
P0301 Cylinder 1 misfire detected
P0302 Cylinder 2 misfire detected
P0303 Cylinder 3 misfire detected
P0304 Cylinder 4 misfire detected
P0305 Cylinder 5 misfire detected
P0306 Cylinder 6 misfire detected
P0307 Cylinder 7 misfire detected
P0308 Cylinder 8 misfire detected
P0309 Cylinder 9 misfire detected
P0310 Cylinder 10 misfire detected
P0311 Cylinder 11 misfire detected
P0312 Cylinder 12 misfire detected
P0320 Ignition/distributor engine speed input circuit malfunction
P0321 Ignition/distributor engine speed input circuit range/performance
P0322 Ignition/distributor engine speed input circuit no signal
P0323 Ignition/distributor engine speed input circuit intermittent
P0325 Knock sensor 1 circuit malfunction (bank 1 or single sensor)
P0326 Knock sensor 1 circuit range/performance (bank 1 or single sensor)
P0327 Knock sensor 1 circuit low input (bank 1 or single sensor)
P0328 Knock sensor 1 circuit high input (bank 1 or single sensor)
P0329 Knock sensor 1 circuit input intermittent (bank 1 or single sensor)
P0330 Knock sensor 2 circuit malfunction (bank 2)
P0331 Knock sensor 2 circuit range/performance (bank 2)
P0332 Knock sensor 2 circuit low input (bank 2)
P0333 Knock sensor 2 circuit high input (bank 2)
P0334 Knock sensor 2 circuit input intermittent (bank 2)
P0335 Crankshaft position sensor A circuit malfunction
P0336 Crankshaft position sensor A circuit range/performance
P0337 Crankshaft position sensor A circuit low input
P0338 Crankshaft position sensor A circuit high input
P0339 Crankshaft position sensor A circuit intermittent
P0340 Camshaft position sensor circuit malfunction
P0341 Camshaft position sensor circuit range/performance
P0342 Camshaft position sensor circuit low input
P0343 Camshaft position sensor circuit high input
P0344 Camshaft position sensor circuit intermittent
P0350 Ignition coil primary/secondary circuit malfunction
P0351 Ignition coil A primary/secondary circuit malfunction
P0352 Ignition coil B primary/secondary circuit malfunction
P0353 Ignition coil C primary/secondary circuit malfunction
P0354 Ignition coil D primary/secondary circuit malfunction
P0355 Ignition coil E primary/secondary circuit malfunction
P0356 Ignition coil F primary/secondary circuit malfunction
P0357 Ignition coil G primary/secondary circuit malfunction
P0358 Ignition coil H primary/secondary circuit malfunction
P0359 Ignition coil I primary/secondary circuit malfunction
P0360 Ignition coil J primary/secondary circuit malfunction
P0361 Ignition coil K primary/secondary circuit malfunction
P0362 Ignition coil L primary/secondary circuit malfunction
P0370 Timing reference high resolution signal A malfunction
P0371 Timing reference high resolution signal A too many pulses
P0372 Timing reference high resolution signal A malfunction
P0373 Timing reference high resolution signal intermittent/erratic pulses
P0374 Timing reference high resolution signal A no pulse
P0375 Timing reference high resolution signal B malfunction
P0376 Timing reference high resolution signal B too many pulses
P0377 Timing reference high resolution signal B too few pulses
P0378 Timing reference high resolution signal B intermittent/erratic pulses
P0379 Timing reference high resolution signal B no pulse
P0380 Glow plug/heater circuit malfunction
P0381 Glow plug/heater indicator circuit malfunction
P0385 Crankshaft position sensor B circuit malfunction
P0386 Crankshaft position sensor B circuit range/performance
P0387 Crankshaft position sensor B circuit low input
P0388 Crankshaft position sensor B circuit high input
P0389 Crankshaft position sensor B circuit intermittent

P04XX Auxiliary Emission Controls

Note: Bank 1 contains cylinder #1.

P0400 Exhaust gas recirculation flow malfunction
P0401 Exhaust gas recirculation flow insufficient detected
P0402 Exhaust gas recirculation flow excessive detected
P0403 Exhaust gas recirculation circuit malfunction
P0404 Exhaust gas recirculation circuit range/performance
P0405 Exhaust gas recirculation sensor A circuit low

P0406 Exhaust gas recirculation sensor A circuit high
P0407 Exhaust gas recirculation sensor B circuit low
P0408 Exhaust gas recirculation sensor B circuit high
P0410 Secondary air injection system malfunction
P0411 Secondary air injection system incorrect flow detected
P0412 Secondary air injection system switching valve A circuit malfunction
P0413 Secondary air injection system switching valve A circuit open
P0414 Secondary air injection system switching valve A circuit shorted
P0415 Secondary air injection system switching valve B circuit malfunction
P0416 Secondary air injection system switching valve B circuit open
P0417 Secondary air injection system switching valve B circuit shorted
P0420 Catalyst system efficiency below threshold (bank 1)
P0421 Warm up catalyst efficiency below threshold (bank 1)
P0422 Main catalyst efficiency below threshold (bank 1)
P0423 Heated catalyst efficiency below threshold (bank 1)
P0424 Heated catalyst temperature below threshold (bank 1)
P0430 Catalyst system efficiency below threshold (bank 2)
P0431 Warm up catalyst efficiency below threshold (bank 2)
P0432 Main catalyst efficiency below threshold (bank 2)
P0433 Heated catalyst efficiency below threshold (bank 2)
P0434 Heated catalyst temperature below threshold (bank 2)
P0440 Evaporative emission control system malfunction
P0441 Evaporative emission control system incorrect purge flow
P0442 Evaporative emission control system leak detected (small leak)
P0443 Evaporative emission control system purge control valve circuit malfunction
P0444 Evaporative emission control system purge control valve circuit open
P0445 Evaporative emission control system purge control valve circuit shorted
P0450 Evaporative emission control system pressure sensor malfunction
P0451 Evaporative emission control system pressure sensor range/performance
P0452 Evaporative emission control system pressure sensor low input
P0453 Evaporative emission control system pressure sensor high input
P0454 Evaporative emission control system pressure sensor intermittent
P0455 Evaporative emission control system leak detected (gross leak)
P0460 Fuel level sensor circuit malfunction
P0461 Fuel level sensor circuit range/performance
P0462 Fuel level sensor circuit low input
P0463 Fuel level sensor circuit high input
P0464 Fuel level sensor circuit intermittent
P0465 Purge flow sensor circuit malfunction
P0466 Purge flow sensor circuit range/performance
P0467 Purge flow sensor circuit low input
P0468 Purge flow sensor circuit high input
P0469 Purge flow sensor circuit intermittent
P0470 Exhaust pressure sensor malfunction
P0471 Exhaust pressure sensor range/performance
P0472 Exhaust pressure sensor low
P0473 Exhaust pressure sensor high
P0474 Exhaust pressure sensor intermittent
P0475 Exhaust pressure control valve malfunction
P0476 Exhaust pressure control valve range/performance
P0477 Exhaust pressure control valve low
P0478 Exhaust pressure control valve high
P0479 Exhaust pressure control valve intermittent

P05XX Vehicle Speed, Idle Control, and Auxiliary Inputs

P0500 Vehicle speed sensor malfunction
P0501 Vehicle speed sensor range/performance
P0502 Vehicle speed sensor circuit low input
P0503 Vehicle speed sensor intermittent/erratic/high
P0505 Idle control system malfunction
P0506 Idle control system rpm lower than expected
P0507 Idle control system rpm higher than expected
P0510 Closed throttle position switch malfunction
P0530 A/C refrigerant pressure sensor circuit malfunction
P0531 A/C refrigerant pressure sensor circuit range/performance
P0532 A/C refrigerant pressure sensor circuit low input
P0533 A/C refrigerant pressure sensor circuit high input
P0534 Air conditioner refrigerant charge loss
P0550 Power steering pressure sensor circuit malfunction
P0551 Power steering pressure sensor circuit range/performance
P0552 Power steering pressure sensor circuit low input
P0553 Power steering pressure sensor circuit high input
P0554 Power steering pressure sensor circuit intermittent
P0560 System voltage malfunction
P0561 System voltage unstable

P0562 System voltage low
P0563 System voltage high
P0565 Cruise control on signal malfunction
P0566 Cruise control off signal malfunction
P0567 Cruise control resume signal malfunction
P0568 Cruise control set signal malfunction
P0569 Cruise control coast signal malfunction
P0570 Cruise control acceleration signal malfunction
P0571 Cruise control/brake switch A circuit malfunction
P0572 Cruise control/brake switch A circuit low
P0573 Cruise control/brake switch A circuit high
P0574–P0580 Reserved for cruise control system codes

P06XX Computer and Auxiliary Outputs

P0600 Serial communication link modification
P0601 Internal control module memory check sum error
P0602 Control module programming error
P0603 Internal control module keep alive memory (KAM) error
P0604 Internal control module random access memory (RAM) error
P0605 Internal control module read only memory (ROM) error
P0606 PCM processor fault

P07XX Transmission

P0700 Transmission control system malfunction
P0701 Transmission control system range/performance
P0702 Transmission control system electrical
P0703 Torque converter/brake switch B circuit malfunction
P0704 Clutch switch input circuit malfunction
P0705 Transmission range sensor circuit malfunction (*PRNDL* input)
P0706 Transmission range sensor circuit range/performance
P0707 Transmission range sensor circuit low input
P0708 Transmission range sensor circuit high input
P0709 Transmission range sensor circuit intermittent
P0710 Transmission fluid temperature sensor circuit malfunction
P0711 Transmission fluid temperature sensor circuit range/performance
P0712 Transmission fluid temperature sensor low input
P0713 Transmission fluid temperature sensor circuit high input
P0714 Transmission fluid temperature sensor circuit intermittent
P0715 Input/turbine speed sensor circuit malfunction
P0716 Input/turbine speed sensor circuit range/performance
P0717 Input/turbine speed sensor circuit no signal
P0718 Input/turbine speed sensor circuit intermittent
P0719 Torque converter/brake switch B circuit low
P0720 Output speed sensor circuit malfunction
P0721 Output speed sensor circuit range/performance
P0722 Output speed sensor circuit no signal
P0723 Output speed sensor circuit intermittent
P0724 Torque converter/brake switch B circuit high
P0725 Engine speed input circuit malfunction
P0726 Engine speed input circuit range/performance
P0727 Engine speed input circuit no signal
P0728 Engine speed input circuit intermittent
P0730 Incorrect gear ratio
P0731 Gear 1 incorrect ratio
P0732 Gear 2 incorrect ratio
P0733 Gear 3 incorrect ratio
P0734 Gear 4 incorrect ratio
P0735 Gear 5 incorrect ratio
P0736 Reverse incorrect ratio
P0740 Torque converter clutch circuit malfunction
P0741 Torque converter clutch circuit performance or stuck off
P0742 Torque converter clutch circuit stuck on
P0743 Torque converter clutch circuit electrical
P0744 Torque converter clutch circuit intermittent
P0745 Pressure control solenoid malfunction
P0746 Pressure control solenoid performance or stuck off
P0747 Pressure control solenoid stuck on
P0748 Pressure control solenoid electrical
P0749 Pressure control solenoid intermittent
P0750 Shift solenoid A malfunction
P0751 Shift solenoid A performance or stuck off
P0752 Shift solenoid A stuck on
P0753 Shift solenoid A electrical
P0754 Shift solenoid A intermittent
P0755 Shift solenoid B malfunction
P0756 Shift solenoid B performance or stuck off
P0757 Shift solenoid B stuck on
P0758 Shift solenoid B electrical
P0759 Shift solenoid B intermittent
P0760 Shift solenoid C malfunction
P0761 Shift solenoid C performance or stuck off
P0762 Shift solenoid C stuck on
P0763 Shift solenoid C electrical
P0764 Shift solenoid C intermittent
P0765 Shift solenoid D malfunction
P0766 Shift solenoid D performance or stuck off
P0767 Shift solenoid D stuck on

P0768 Shift solenoid D electrical
P0769 Shift solenoid D intermittent
P0770 Shift solenoid E malfunction
P0771 Shift solenoid E performance or stuck off
P0772 Shift solenoid E stuck on
P0773 Shift solenoid E electrical
P0774 Shift solenoid E intermittent
P0780 Shift malfunction
P0781 1–2 shift malfunction
P0782 2–3 shift malfunction
P0783 3–4 shift malfunction
P0784 4–5 shift malfunction
P0785 Shift/timing solenoid malfunction
P0786 Shift/timing solenoid range/performance
P0787 Shift/timing solenoid low
P0788 Shift/timing solenoid high
P0789 Shift/timing solenoid intermittent
P0790 Normal/performance switch circuit malfunction

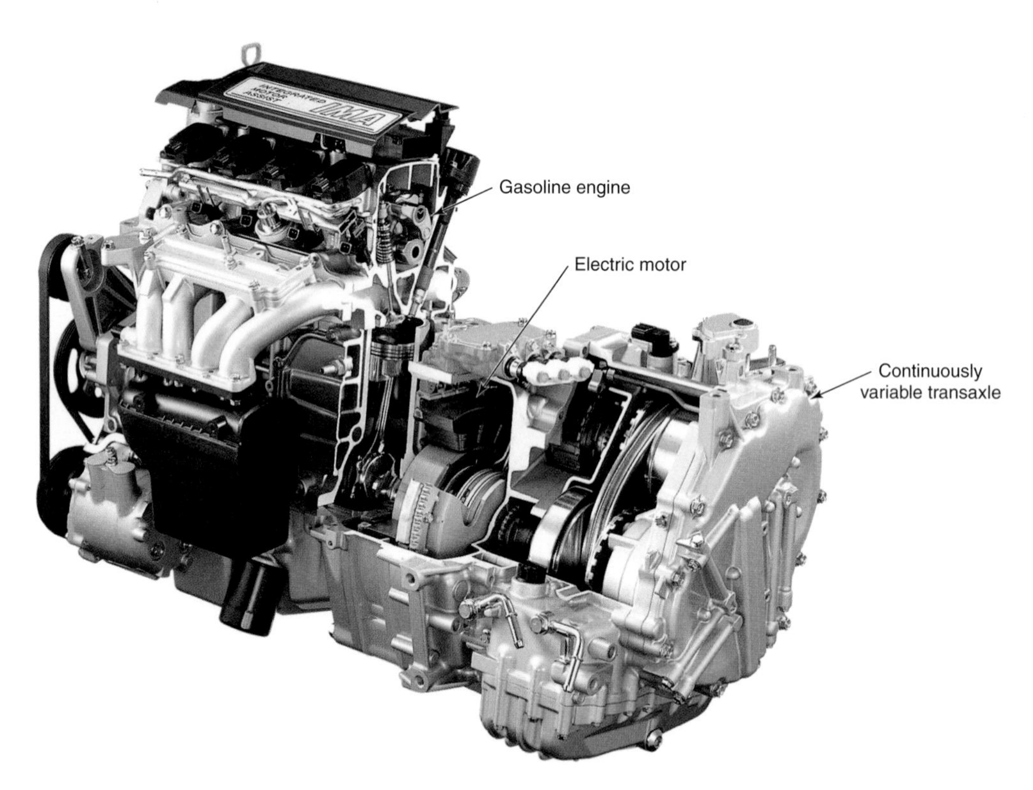

Power train for a hybrid vehicle. In addition to providing power to help drive the vehicle, the electric motor serves as a generator during braking and coasting to recharge the vehicle's batteries. (Honda)

Appendix C

Conversion Charts

Some Common Abbreviations/Symbols			
US Customary		Metric	
Unit	Abbreviation	Unit	Abbreviation
inch	in.	kilometer	km
feet	ft.	hectometer	hm
yard	yd.	dekameter	dkm
mile	mi.	meter	m
grain	gr.	decimeter	dm
ounce	oz.	centimeter	cm
pound	lb.	millimeter	mm
teaspoon	tsp.	cubic centimeter	cm^3
tablespoon	tbsp.	kilogram	kg
fluid ounce	fl. oz.	hectogram	hg
cup	c.	dekagram	dkg
pint	pt.	gram	g
quart	qt.	decigram	dg
gallon	gal.	centigram	cg
cubic inch	$in.^3$	milligram	mg
cubic foot	$ft.^3$	kiloliter	kl
cubic yard	$yd.^3$	hectoliter	hl
square inch	$in.^2$	dekaliter	dal
square foot	$ft.^2$	liter	L
square yard	$yd.^2$	centiliter	cl
square mile	$mi.^2$	milliliter	ml
Fahrenheit	F	square kilometer	km^2
barrel	bbl.	hectare	ha
fluid dram	fl. dr.	are	a
board foot	bd. ft.	centare	ca
rod	rd.	tonne	t
dram	dr.	Celsius	C
bushel	bu.		

Measuring Systems

US Customary | Metric

Length

US Customary	Metric
12 inches = 1 foot	1 kilometer = 1000 meters
36 inches = 1 yard	1 hectometer = 100 meters
3 feet = 1 yard	1 dekameter = 10 meters
5,280 feet = 1 mile	1 meter = 1 meter
16.5 feet = 1 rod	1 decimeter = 0.1 meter
320 rods = 1 mile	1 centimeter = 0.01 meter
6 feet = 1 fathom	1 millimeter = 0.001 meter

Weight

US Customary	Metric
27.34 grains = 1 dram	1 tonne =1,000,000 grams
438 grains = 1 ounce	1 kilogram = 1000 grams
16 drams = 1 ounce	1 hectogram = 100 grams
16 ounces = 1 pound	1 dekagram = 10 grams
2000 pounds = 1 short ton	1 gram = 1 gram
2240 pounds = 1 long ton	1 decigram = 0.1 gram
25 pounds = 1 quarter	1 centigram = 0.01 gram
4 quarters = 1 cwt	1 milligram = 0.001 gram

Volume

US Customary	Metric
8 ounces = 1 cup	1 hectoliter = 00 liters
16 ounces = 1 pint	1 dekaliter = 10 liters
32 ounces = 1 quart	1 liter = 1 liter
2 cups = 1 pint	1 deciliter = 0.1 liter
2 pints = 1 quart	1 centiliter = 0.01 liter
4 quarts = 1 gallon	1 milliliter = 0.001 liter
8 pints = 1 gallon	1000 milliliter = 1 liter

Area

US Customary	Metric
144 sq. inches = 1 sq. foot	100 sq. millimeters = 1 sq. centimeter
9 sq. feet = 1 sq. yard	100 sq. centimeters = 1 sq. decimeter
43,560 sq. ft. = 160 sq. rods	100 sq. decimeters = 1 sq. meter
160 sq. rods = 1 acre	10,000 sq. meters = 1 hectare
640 acres = 1 sq. mile	

Temperature

Fahrenheit		Celsius
32° F	Water freezes	0° C
68° F	Reasonable room temperature	20° C
98.6° F	Normal body temperature	37° C
173° F	Alcohol boils	78.34° C
212° F	Water boils	100° C

Useful Conversions

When You Know:	Multiply By:	To Find:
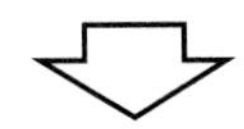		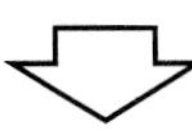
	Torque	
Pound - inch Pound - foot	0.11298 1.3558	newton-meter (N•m) newton-meters
	Light	
Foot candles	1.0764	lumen/meters^2 (lm/m^2)
	Fuel Performance	
Miles/gallon	0.4251	kilometers/liter (km/L)
	Speed	
Miles/hour	1.6093	kilometers/hr (km/h)
	Force	
kilogram ounce pound	9.807 0.278 4.448	newtons (n) newtons newtons
	Power	
Horsepower	0.746	kilowatts (kw)
	Pressure of Stress	
Inches of water pounds/sq. in.	0.2491 6.895	kilopascals (kPa) kilopascals
	Energy or work	
Btu Foot - pound Kilowatt-hour	1055.0 1.3558 3600000.0	joules (J) joules joules

Conversion Table
Metric to US Conventional

When You Know	Multiple by: * = Exact		To Find
	Very Accurate	Approximate	
Length			
millimeters	0.0393701	0.04	inches
centimeters	0.3937008	0.4	inches
meters	3.280840	3.3	feet
meters	1.093613	1.1	yards
kilometers	0.621371	0.6	miles
Weight			
grains	0.00228571	0.0023	ounces
grams	0.03527396	0.035	ounces
kilograms	2.204623	2.2	pounds
tonnes	1.1023113	1.1	short tons
Volume			
milliliters	0.20001	0.2	teaspoons
milliliters	0.06667	0.067	tablespoons
milliliters	0.03381402	0.03	fluid ounces
liters	61.02374	61.024	cubic inches
liters	2.113376	2.1	pints
liters	1.056688	1.06	quarts
liters	0.26417205	0.26	gallons
liters	0.03531467	0.035	cubic feet
cubic meters	61023.74	61023.7	cubic inches
cubic meters	35.31467	35.0	cubic feet
cubic meters	1.3079506	1.3	cubic yards
cubic meters	264.17205	264.0	gallons
Area			
square centimeters	0.1550003	0.16	square inches
square centimeters	0.00107639	0.001	square feet
square meters	10.76391	10.8	square feet
square meters	1.195990	1.2	square yards
square kilometers	0.3861019	0.4	square miles
hectares	2.471054	2.5	acres
Temperature			
Celsius	*9/5 (then add 32)		Fahrenheit

Conversion Table
US Conventional to Metric

When You Know ⬇	Multiply by: * = Exact		To Find ⬇
	Very Accurate	Approximate	
Length			
inches	*25.4		millimeters
inches	*2.54		centimeters
feet	*0.3048		meters
feet	*30.48		centimeters
yards	*0.9144	0.9	meters
miles	*1.609344	1.6	kilometers
Weight			
grains	15.43236	15.4	grams
ounces	*28.349523125	28.0	grams
ounces	*0.028349523125	.028	kilograms
pounds	*0.45359237	0.45	kilograms
short ton	*0.90718474	0.9	tonnes
Volume			
teaspoons	*4.97512	5.0	milliliters
tablespoons	*14.92537	15.0	milliliters
fluid ounces	29.57353	30.0	milliliters
cups	*0.236588240	0.24	liters
pints	*0.473176473	0.47	liters
quarts	*0.946352946	0.95	liters
gallons	*3.785411784	3.8	liters
cubic inches	*0.016387064	0.02	liters
cubic feet	*0.028316846592	0.03	cubic meters
cubic yards	*0.764554857984	0.76	cubic meters
Area			
square inches	*6.4516	6.5	square centimeters
square feet	*0.09290304	0.09	square meters
square yards	*0.83612736	0.8	square meters
square miles	*2.589989	2.6	square kilometers
acres	*0.40468564224	0.4	hectares
Temperature			
Fahrenheit	* 5/9 (after subtracting 32)		Celsius

Metric Tables

Dimensional and Temperature Conversion Chart

Inches			Decimals	Millimeters	Inches to Millimeters		Millimeters to Inches		Fahrenheit & Celsius			
					in	mm	mm	in	°F	°C	°C	°F
		1/64	.015625	.3969	.0001	.00254	0.001	.000039	-20	-28.9	-30	-22
	1/32		.03125	.7937	.0002	.00508	0.002	.000079	-15	-26.1	-28	-18.4
		3/64	.046875	1.1906	.0003	.00762	0.003	.000118	-10	-23.3	-26	-14.8
1/16			.0625	1.5875	.0004	.01016	0.004	.000157	-5	-20.6	-24	-11.2
		5/64	.078125	1.9844	.0005	.01270	0.005	.000197	0	-17.8	-22	-7.6
	3/32		.09375	2.3812	.0006	.01524	0.006	.000236	1	-17.2	-20	-4
		7/64	.109375	2.7781	.0007	.01778	0.007	.000276	2	-16.7	-18	-0.4
1/8			.125	3.1750	.0008	.02032	0.008	.000315	3	-16.1	-16	3.2
		9/64	.140625	3.5719	.0009	.02286	0.009	.000354	4	-15.6	-14	6.8
	5/32		.15625	3.9687	.001	.0254	0.01	.00039	5	-15.0	-12	10.4
		11/64	.171875	4.3656	.002	.0508	0.02	.00079	10	-12.2	-10	14
3/16			.1875	4.7625	.003	.0762	0.03	.00118	15	-9.4	-8	17.6
		13/64	.203125	5.1594	.004	.1016	0.04	.00157	20	-6.7	-6	21.2
	7/32		.21875	5.5562	.005	.1270	0.05	.00197	25	-3.9	-4	24.8
		15/64	.234375	5.9531	.006	.1524	0.06	.00236	30	-1.1	-2	28.4
1/4			.25	6.3500	.007	.1778	0.07	.00276	35	1.7	0	32
		17/64	.265625	6.7469	.008	.2032	0.08	.00315	40	4.4	2	35.6
	9/32		.28125	7.1437	.009	.2286	0.09	.00354	45	7.2	4	39.2
		19/64	.296875	7.5406	.01	.254	0.1	.00394	50	10.0	6	42.8
5/16			.3125	7.9375	.02	.508	0.2	.00787	55	12.8	8	46.4
		21/64	.328125	8.3344	.03	.762	0.3	.01181	60	15.6	10	50
	11/32		.34375	8.7312	.04	1.016	0.4	.01575	65	18.3	12	53.6
		23/64	.359375	9.1281	.05	1.270	0.5	.01969	70	21.1	14	57.2
3/8			.375	9.5250	.06	1.524	0.6	.02362	75	23.9	16	60.8
		25/64	.390625	9.9219	.07	1.778	0.7	.02756	80	26.7	18	64.4
	13/32		.40625	10.3187	.08	2.032	0.8	.03150	85	29.4	20	68
		27/64	3421875	10.7156	.09	2.286	0.9	.03543	90	32.2	22	71.6
7/16			.4375	11.1125	.1	2.54	1	.03937	95	35.0	24	75.2
		29/64	.453125	11.5094	.2	5.08	2	.07874	100	37.8	26	78.8
	15/32		.46875	11.9062	.3	7.62	3	.11811	105	40.6	28	82.4
		31/64	.484375	12.3031	.4	10.16	4	.15748	110	43.3	30	86
1/2			.5	12.7000	.5	12.70	5	.19685	115	46.1	32	89.6
		33/64	.515625	13.0969	.6	15.24	6	.23622	120	48.9	34	93.2
	17/32		.53125	13.4937	.7	17.78	7	.27559	125	51.7	36	96.8
		35/64	.546875	13.8906	.8	20.32	8	.31496	130	54.4	38	100.4
9/16			.5625	14.2875	.9	22.86	9	.35433	135	57.2	40	104
		37/64	.578125	14.6844	1	25.4	10	.39370	140	60.0	42	107.6
	19/32		.59375	15.0812	2	50.8	11	.43307	145	62.8	44	112.2
		39/64	.609375	15.4781	3	76.2	12	.47244	150	65.6	46	114.8
5/8			.625	15.8750	4	101.6	13	.51181	155	68.3	48	118.4
		41/64	.640625	16.2719	5	127.0	14	.55118	160	71.1	50	122
	21/32		.65625	16.6687	6	152.4	15	.59055	165	73.9	52	125.6
		43/64	.671875	17.0656	7	177.8	16	.62992	170	76.7	54	129.2
11/16			.6875	17.4625	8	203.2	17	.66929	175	79.4	56	132.8
		45/64	.703125	17.8594	9	228.6	18	.70866	180	82.2	58	136.4
	23/32		.71875	18.2562	10	254.0	19	.74803	185	85.0	60	140
		47/64	.734375	18.6531	11	279.4	20	.78740	190	87.8	62	143.6
3/4			.75	19.0500	12	304.8	21	.82677	195	90.6	64	147.2
		49/64	.765625	19.4469	13	330.2	22	.86614	200	93.3	66	150.8
	25/32		.78125	19.8437	14	355.6	23	.90551	205	96.1	68	154.4
		51/64	.796875	20.2406	15	381.0	24	.94488	210	98.9	70	158
13/16			.8125	20.6375	16	406.4	25	.98425	215	100.0	75	167
		53/64	.828125	21.0344	17	431.8	26	1.02362	212	101.7	80	176
	27/32		.84375	21.4312	18	457.2	27	1.06299	220	104.4	85	185
		55/64	.859375	21.8281	19	482.6	28	1.10236	225	107.2	90	194
7/8			.875	22.2250	20	508.0	29	1.14173	230	110.0	95	203
		57/64	.890625	22.6219	21	533.4	30	1.18110	235	112.8	100	212
	29/32		.90625	23.0187	22	558.8	31	1.22047	240	115.6	105	221
		59/64	.921875	23.4156	23	584.2	32	1.25984	245	118.3	110	230
15/16			.9375	23.8125	24	609.6	33	1.29921	250	121.1	115	239
		61/64	.953125	24.2094	25	635.0	34	1.33858	255	123.9	120	248
	31/32		.96875	24.6062	26	660.4	35	1.37795	260	126.6	125	257
		63/64	.984375	25.0031	27	690.6	36	1.41732	265	129.4	130	266

Capacity Conversion US Gallons to Liters

Gallons	0 Liters	1 Liters	2 Liters	3 Liters	4 Liters	5 Liters
0	00.0000	3.7853	7.5707	11.3560	15.1413	18.9267
10	37.8533	41.6387	45.4240	49.2098	52.9947	56.7800
20	75.7066	79.4920	83.2773	87.0626	90.8480	94.6333
30	113.5600	117.3453	121.1306	124.9160	128.7013	132.4866
40	151.4133	155.1986	158.9840	162.7693	166.5546	170.3400

Millimeter/Inch Conversion Chart

mm	in	15	= .5905	30	=1.1811	45	=1.7716	60	=2.3622	75	=2.9527	90	=3.5433	105	=4.1338	120	=4.7244
.25	=.0098	15.25	= .6004	30.25	=1.1909	45.25	=1.7815	60.25	=2.3720	75.25	=2.9626	90.25	=3.5531	105.25	=4.1437	120.25	=4.7342
.50	=.0197	15.50	= .6102	30.50	=1.2008	45.50	=1.7913	60.50	=2.3819	75.50	=2.9724	80.50	=3.5630	105.50	=4.1535	120.50	=4.7441
.75	=.0295	15.75	= .6201	30.75	=1.2106	45.75	=1.8012	60.75	=2.3917	75.75	=2.9823	90.75	=3.5728	105.75	=4.1634	120.75	=4.7539
1	=.0394	16	= .6299	31	=1.2205	46	=1.8110	61	=2.4016	76	=2.9921	91	=3.5827	106	=4.1732	121	=4.7638
1.25	=.0492	16.25	= .6398	31.25	=1.2302	46.25	=1.8209	61.25	=2.4114	76.25	=3.0020	91.25	=3.5925	106.25	=4.1831	121.25	=4.7736
1.50	=.0591	16.50	= .6496	31.50	=1.2402	46.50	=1.8307	61.50	=2.4213	76.50	=3.0118	91.50	=3.6024	106.50	=4.1929	121.50	=4.7885
1.75	=.0689	16.75	= .6594	31.75	=1.2500	46.75	=1.8405	61.75	=2.4311	76.75	=3.0216	91.75	=3.6122	106.75	=4.2027	127.75	=4.7933
2	=.0787	17	= .6693	32	=1.2598	47	=1.8504	62	=2.4409	77	=3.0315	92	=3.6220	107	=4.2126	122	=4.8031
2.25	=.0886	17.25	= .6791	32.25	=1.2697	47.25	=1.8602	62.25	=2.4508	77.25	=3.0413	92.25	=3.6319	107.25	=4.2224	122.25	=4.8130
2.50	=.0984	17.50	= .6890	32.50	=1.2795	47.50	=1.8701	62.50	=2.4606	77.50	=3.0512	92.50	=3.6417	107.50	=4.2323	122.50	=4.8228
2.75	=.1083	17.75	= .6988	32.75	=1.2894	47.75	=1.8799	62.75	=2.4705	77.75	=3.0610	92.75	=3.6516	107.75	=4.2421	122.75	=4.8327
3	=.1181	18	= .7087	33	=1.2992	48	=1.8898	63	=2.4803	78	=3.0709	93	=3.6614	108	=4.2520	123	=4.8425
3.25	=.1280	18.25	= .7185	33.25	=1.3091	48.25	=1.8996	63.25	=2.4901	78.25	=3.0807	93.25	=3.6713	108.25	=4.2618	123.25	=4.8524
3.50	=.1378	18.50	= .7283	33.50	=1.3189	48.50	=1.9094	63.50	=2.5000	78.50	=3.0905	93.50	=3.6811	108.50	=4.2716	123.50	=4.8622
3.75	=.1476	18.75	= .7382	33.75	=1.3287	48.75	=1.9193	63.75	=2.5098	78.75	=3.1004	93.75	=3.6909	108.75	=4.2815	123.75	=4.8720
4	=.1575	19	= .7480	34	=1.3386	49	=1.9291	64	=2.5197	79	=3.1102	94	=3.7008	109	=4.2913	124	=4.8819
4.25	=.1673	19.25	= .7579	34.25	=1.3484	49.25	=1.9390	64.25	=2.5295	79.25	=3.1201	94.25	=3.7106	109.25	=4.3012	124.25	=4.8917
4.50	=.1772	19.50	= .7677	34.50	=1.3583	49.50	=1.9488	64.50	=2.5394	79.50	=3.1299	94.50	=3.7205	109.50	=4.3110	124.50	=4.9016
4.75	=.1870	19.75	= .7776	34.75	=1.3681	49.75	=1.9587	64.75	=2.5492	79.75	=3.1398	94.75	=3.7303	109.75	=4.3209	124.75	=4.9114
5	=.1968	20	= .7874	35	=1.3779	50	=1.9685	65	=2.5590	80	=3.1496	95	=3.7401	110	=4.3307	125	=4.9212
5.25	=.2067	20.25	= .7972	35.25	=1.3878	50.25	=1.9783	65.25	=2.5689	80.25	=3.1594	95.25	=3.7500	110.25	=4.3405	125.25	=4.9311
5.50	=.2165	20.50	= .8071	35.50	=1.3976	50.50	=1.9882	65.50	=2.5787	80.50	=3.1693	95.50	=3.7598	110.50	=4.3504	125.50	=4.9409
5.75	=.2264	20.75	= .8169	35.75	=1.4075	50.75	=1.9980	65.75	=2.5886	80.75	=3.1791	95.75	=3.7697	110.75	=4.3602	125.75	=4.9508
6	=.2362	21	= .8268	36	=1.4173	51	=2.0079	66	=2.5984	81	=3.1890	96	=3.7795	111	=4.3701	126	=4.9606
6.25	=.2461	21.25	= .8366	36.25	=1.4272	51.25	=2.0177	66.25	=2.6083	81.25	=3.1988	96.25	=3.7894	111.25	=4.3799	126.25	=4.9705
6.50	=.2559	21.50	= .8465	36.50	=1.4370	51.50	=2.0276	66.50	=2.6181	81.50	=3.2087	96.50	=3.7992	111.50	=4.3898	126.50	=4.9803
6.75	=.2657	21.75	= .8563	36.75	=1.4468	51.75	=2.0374	66.75	=2.6279	81.75	=3.2185	96.75	=3.8090	111.75	=4.3996	126.75	=4.9901
7	=.2756	22	= .8661	37	=1.4567	52	=2.0472	67	=2.6378	82	=3.2283	97	=3.8189	112	=4.4094	127	=5.000
7.25	=.2854	22.25	= .8760	37.25	=1.4665	52.25	=2.0571	67.25	=2.6476	82.25	=3.2382	97.25	=3.8287	112.25	=4.4193		
7.50	=.2953	22.50	= .8858	37.50	=1.4764	52.50	=2.0669	67.50	=2.6575	82.50	=3.2480	97.50	=3.8386	112.50	=4.4291		
7.75	=.3051	22.75	= .8957	37.75	=1.4862	52.75	=2.0768	67.75	=2.6673	82.75	=3.2579	97.75	=3.8484	112.75	=4.4390		
8	=.3150	23	= .9055	38	=1.4961	53	=2.0866	68	=2.6772	83	=3.2677	98	=3.8583	113	=4.4488		
8.25	=.3248	23.25	= .9153	38.25	=1.5059	53.25	=2.0965	68.25	=2.6870	83.25	=3.2776	98.25	=3.8681	113.25	=4.4587		
8.50	=.3346	23.50	= .9252	38.50	=1.5157	53.50	=2.1063	68.50	=2.6968	83.50	=3.2874	98.50	=3.8779	113.50	=4.4685		
8.75	=.3445	23.75	= .9350	38.75	=1.5256	53.75	=2.1161	68.75	=2.7067	83.75	=3.2972	98.75	=3.8878	113.75	=4.4783		
9	=.3543	24	= .9449	39	=1.5354	54	=2.1260	69	=2.7165	84	=3.3071	99	=3.8976	114	=4.4882		
9.25	=.3642	24.25	= .9547	39.25	=1.5453	54.25	=2.1358	69.25	=2.7264	84.25	=3.3169	99.25	=3.9075	114.25	=4.4980		
9.50	=.3740	24.50	= .9646	39.50	=1.5551	54.50	=2.1457	69.50	=2.7362	84.50	=3.3268	99.50	=3.9173	114.50	=4.5079		
9.75	=.3839	24.75	= .9744	39.75	=1.5650	54.75	=2.1555	69.75	=2.7461	84.72	=3.3366	99.75	=3.9272	114.75	=4.5177		
10	=.3937	25	= .9842	40	=1.5748	55	=2.1653	70	=2.7559	85	=3.3464	100	=3.9370	115	=4.5275		
10.25	=.4035	25.25	= .9941	40.25	=1.5846	55.25	=2.1752	70.25	=2.7657	85.25	=3.3563	100.25	=3.9468	115.25	=4.5374		
10.50	=.4134	25.50	=1.0039	40.50	=1.5945	55.50	=2.1850	70.50	=2.7756	85.50	=3.3661	100.50	=3.9567	115.50	=4.5472		
10.75	=.4232	25.75	=1.0138	40.75	=1.6043	55.75	=2.1949	70.75	=2.7854	85.75	=3.3760	100.75	=3.9665	115.75	=4.5571		
11	=.4331	26	=1.0236	41	=1.6142	56	=2.2047	71	=2.7953	86	=3.3858	101	=3.9764	116	=4.5669		
11.25	=.4429	26.25	=1.0335	41.25	=1.6240	56.25	=2.2146	71.25	=2.8051	86.25	=3.3957	101.25	=3.9862	116.25	=4.5768		
11.50	=.4528	26.50	=1.0433	41.50	=1.6339	56.50	=2.2244	71.50	=2.8150	86.50	=3.4055	101.50	=3.9661	116.50	=4.5866		
11.75	=.4626	26.75	=1.0531	41.75	=1.6437	56.75	=2.2342	71.75	=2.8248	86.75	=3.4153	101.75	=4.0059	116.75	=4.5964		
12	=.4724	27	=1.0630	42	=1.6535	57	=2.2441	72	=2.8346	87	=3.4252	102	=4.0157	117	=4.6063		
12.25	=.4823	27.25	=1.0728	42.25	=1.6634	57.25	=2.2539	72.25	=2.8445	87.25	=3.4350	102.25	=4.0256	117.25	=4.6161		
12.50	=.4921	27.50	=1.0827	42.50	=1.6732	57.50	=2.2638	72.50	=2.8543	87.50	=3.4449	102.50	=4.0354	117.50	=4.6260		
12.75	=.5020	27.75	=1.0925	42.75	=1.6831	57.75	=2.2736	72.75	=2.8642	87.75	=3.4547	102.75	=4.0453	117.75	=4.6358		
13	=.5118	28	=1.1024	43	=1.6929	58	=2.2835	73	=2.8740	88	=3.4646	103	=4.0551	118	=4.6457		
13.25	=.5217	28.25	=1.1122	43.25	=1.7028	58.25	=2.2933	73.25	=2.8839	88.25	=3.4744	103.25	=4.0650	118.25	=4.6555		
13.50	=.5315	28.50	=1.1220	43.50	=1.7126	58.50	=2.3031	73.50	=2.8937	88.50	=3.4842	103.50	=4.0748	118.50	=4.6653		
13.75	=.5413	28.75	=1.1319	43.75	=1.7224	58.75	=2.3130	73.75	=2.9035	88.75	=3.4941	103.75	=4.0846	118.75	=4.6752		
14	=.5512	29	=1.1417	44	=1.7323	59	=2.3228	74	=2.9134	89	=3.5039	104	=4.0945	119	=4.6850		
14.25	=.5610	29.25	=1.1516	44.25	=1.7421	59.25	=2.3327	74.25	=2.9232	89.25	=3.5138	104.25	=4.1043	119.25	=4.6949		
14.50	=.5709	29.50	=1.1614	44.50	=1.7520	59.50	=2.3425	74.50	=2.9331	89.50	=3.5236	104.50	=4.1142	119.50	=4.7047		
14.75	=.5807	29.75	=1.1713	44.75	=1.7618	59.75	=2.3524	74.75	=2.9429	89.75	=3.5335	104.75	=4.1240	119.75	=4.7146		

Metric – Inch Equivalents

Inches		
Fractions	Decimals	Millimeters
	.00394	.1
	.00787	.2
	.01181	.3
1/64	.015625	.3969
	.01575	.4
	.01969	.5
	.02362	.6
	.02756	.7
1/32	.03125	.7938
	.0315	.8
	.03543	.9
	.03937	1.00
3/64	.046875	1.1906
1/16	.0625	1.5875
5/64	.078125	1.9844
	.07874	2.00
3/32	.09375	2.3813
7/64	.109375	2.7781
	.11811	3.00
1/8	.125	3.175
9/64	.140625	3.5719
5/32	.15625	3.9688
	.15748	4.00
11/64	.171875	4.3656
3/16	.1875	4.7625
	.19685	5.00
13/64	.203125	5.1594
7/32	.21875	5.5563
15/64	.234375	5.9531
	.23622	6.00
1/4	.2500	6.35
17/64	.265625	6.7469
	.27559	7.00
9/32	.28125	7.1438
19/64	.296875	7.5406
5/16	.3125	7.9375
	.31496	8.00
21/64	.328125	8.3344
11/32	.34375	8.7313
	.35433	9.00
23/64	.359375	9.1281
3/8	.375	9.525
25/64	.390625	9.9219
	.3937	10.00
13/32	.40625	10.3188
27/64	.421875	10.7156
	.43307	11.00
7/16	.4375	11.1125
29/64	.453125	11.5094

Inches		
Fractions	Decimals	Millimeters
15/32	.46875	11.9063
	.47244	12.00
31/64	.484375	12.3031
1/2	.5000	12.70
	.51181	13.00
33/64	.515625	13.0969
17/32	.53125	13.4938
35/64	.546875	13.8907
	.55118	14.00
9/16	.5625	14.2875
37/64	.578125	14.6844
	.59055	15.00
19/32	.59375	15.0813
39/64	.609375	15.4782
5/8	.625	15.875
	.62992	16.00
41/64	.640625	16.2719
21/32	.65625	16.6688
	.66929	17.00
43/64	.671875	17.0657
11/16	.6875	17.4625
45/64	.703125	17.8594
	.70866	18.00
23/32	.71875	18.2563
47/64	.734375	18.6532
	.74803	19.00
3/4	.7500	19.05
49/64	.765625	19.4469
25/32	.78125	19.8438
	.7874	20.00
51/64	.796875	20.2407
13/16	.8125	20.6375
	.82677	21.00
53/64	.828125	21.0344
27/32	.84375	21.4313
55/64	.859375	21.8282
	.86614	22.00
7/8	.875	22.225
57/64	.890625	22.6219
	.90551	23.00
29/32	.90625	23.0188
59/64	.921875	23.4157
15/16	.9375	23.8125
	.94488	24.00
61/64	.953125	24.2094
31/32	.96875	24.6063
	.98425	25.00
63/64	.984375	25.0032
1	1.0000	25.4001

Dictionary of Automotive Terms

A

AAA: American Automobile Association.

ABDC: After Bottom Dead Center.

ABS: Anti-lock Braking System.

ABS warning light: Amber indicator lamp mounted in the instrument cluster. Illuminates when there is a problem with the anti-lock brake system.

Absolute zero: A state in which no heat is present. Believed to be –459.7°F or –273.16°C.

ACT: American College Testing program. Administers the ASE tests.

Accelerator: Floor pedal used to control the throttle valve through linkage, cable, or, on newer vehicles, electrical wiring.

Accelerator pump: Small pump, located in carburetor, that sprays additional gasoline into air stream during acceleration.

Accident: An unplanned event that results in damage and/or injury.

Accumulator (air conditioning): Receiver-dehydrator combination.

Accumulator (anti-lock brakes): Gas-charged cylinder or chamber used to store brake fluid. Normally used in anti-lock brake systems.

Accumulator (automatic transmission): Unit designed to cushion shifts by pressurizing a holding member softly at first, followed by heavy pressure to prevent slippage.

Acetone: A very flammable, volatile solvent. Mixes with water.

Acetylene: Gas commonly used in welding or cutting operations.

Acid core solder: Wire solder with a central core containing an acid flux.

Actuator: Device that converts electrical signals into physical actions. Most actuators are classified as motors, solenoids, or relays.

Active suspension: A ride control system using input sensors, output devices, and a body computer to adjust ride height and feel.

Additive: Chemical added to gasoline, oil, or greases to improve the characteristics of original product.

Add-on transmission cooler: Heat exchanger installed in front of the radiator to provide extra cooling for an automatic transmission or transaxle.

Adjustable pilot: An expandable device used to position and guide either cutter or stone mandrels (sleeve) for valve grinding.

Adjuster sleeves: Devices for adjusting toe on most vehicles.

Advance: When the ignition spark occurs earlier or more degrees before TDC.

Aeration: Process of mixing air into a solution such as engine coolant.

Aftermarket fuel injection system: A multiport fuel injection system that can be installed to replace the carburetor or throttle body fuel injection system.

Aftermarket seminars: Classes held by aftermarket parts distributors.

Agitation: Shaking or vibration applied to a cleaning solution to assist cleaning process.

Air bag system: Restraint system that uses an inflatable bag to protect occupants during a severe frontal or side collision. Also called Supplemental Inflatable Restraint or SIR.

Air bleed: An orifice or small passageway designed to allow a specific amount of air to enter a moving column of liquid (such as fuel).

Air cleaner: Device containing a filter. Used to remove contaminants from the air being drawn into engines, compressors, and power brake diaphragms.

Air conditioning: System used to control the temperature, movement, cleanliness, and humidity of air in a vehicle.

Air cooled: An object cooled by passing a stream of air over its surface.

Air dam: A device placed beneath the front bumper to reduce amount of air turbulence or direct air to the radiator.

Air drill: Drill operated by pneumatic (air) pressure.

Air filter: A device used to remove contaminants from the air being drawn into an engine, compressor, or power brake diaphragm. Newer vehicles have filters to clean the air entering the passenger compartment

Airflow sensor: Device to measure volume and speed of air moving into intake manifold.

Air-fuel ratio: Ratio by weight between air and gasoline in the fuel mixture.

Air gap (magnetic pickup): Distance between pickup coil and reluctor (stator).

Air gap (spark plugs): Distance between center and side electrodes.

Air horn: Warning horn operated by compressed air.

Air injection reaction (AIR): An emission control system used to lower levels of carbon monoxide and hydrocarbon emissions. Accomplished by injecting a stream of air into exhaust stream near exhaust valve.

Air pollution: Contamination of earth's atmosphere by various natural and manufactured pollutants.

Air pressure check: Using compressed air to check automatic transmission clutch and band operation.

Air spring: Container and plunger separated by air under pressure. When container and plunger attempt to squeeze together, air compresses and produces a spring effect. Air spring has been used on some suspension systems.

Aligning bar: A precision tool (bar) used to check engine main bearing alignment.

Aligning punch: Punch with a long, gradual taper that may be passed through holes in two objects, thus aligning the holes for insertion of fasteners.

Alloy: Mixture of two or more materials.

Alternating current (ac): Electric current that first flows one way in circuit and then other. Type normally used in homes.

Alternator: Device that produces AC current. The AC is changed to DC by diodes before reaching car's electrical system. Some manufacturers call the alternator a generator: however, all modern vehicles use alternators.

Altitude compensation: Altering the adjustment of some vacuum or pneumatic unit so that unit operation remains the same when the altitude changes.

Ambient temperature: Temperature of air surrounding an object.

Ammeter: Instrument used to measure rate of current flow in amperes.

Ampere: Unit of measurement used in expressing rate of current flow in a circuit.

Ampere hour capacity: Measurement of storage battery ability to deliver specified current over specified length of time.

Angular contact bearing: An antifriction bearing, such as a ball bearing, designed to carry both radial and thrust loads.

Anneal: To remove hardness from metal. Heat steel to a red color, then allow it to cool slowly. Unlike steel, copper is annealed by heating, and then plunging it immediately into cold water.

Annealed stainless steel: Stainless steel that has been heat treated to remove stress and to soften.

Annular groove: A groove that runs completely around a bearing or shaft.

Annulus gear: A ring shaped gear having internal teeth.

Antibackfire valve: Valve used in air injection reaction (exhaust emission control) system to prevent backfiring during period immediately following sudden deceleration.

Antifreeze: Chemical added to cooling system to prevent coolant from freezing in cold weather.

Antifriction bearing: Bearing containing rollers or balls plus an inner and outer race. Bearing is designed to roll, thus minimizing friction.

Antiknock: Indicates various substances that can be added to gasoline to improve its resistance to knocking (spark knock, preignition, or detonation).

Antilash plate: A plate designed to maintain constant and close contact between two gears. This removes play or backlash.

Anti-lock brakes: Computerized brake system that prevents wheel lock-up by pulsing fluid pressure to affected wheels.

Antiseize compound: A compound designed to prevent galling, sticking, or seizing of fastener threads.

Antistall dashpot: A diaphragm operated unit designed to retard final closing of throttle valve to prevent stalling.

API: American Petroleum Institute.

Apply servo: The hydraulically operated pistons used to apply bands in an automatic transmission or transaxle.

Arbor press: A hand-operated press used to insert or remove bearings, bushings, collars, and other parts.

Arc or electric welding: Welding by using electric current to melt both metal to be welded and welding rod or electrode that is being added.

Arcing: Electricity leaping the gap between two electrodes.

Armature: The starter component that revolves between the pole shoes, made up of wire windings on an iron core.

Artificial lean mixture: The result of allowing excess additional air into the engine as part of a test procedure.

Artificial rich mixture: The result of reducing the amount of air entering the engine (or introducing propane into the fuel system) as part of a test procedure.

Asbestos: Heat resistant and nonburning fibrous mineral once widely used for brake shoes and clutch linings. Do not inhale asbestos fibers or dust as it can cause cancer.

ASE: National Institute for Automotive Service Excellence.

ASME: American Society of Mechanical Engineers.

Aspect ratio: Ratio between height and width of a tire.

ASTM: American Society of Testing Materials.

Asymmetric tire tread: Tire tread with a non-symmetrical (uneven) pattern. Designed to reduce tire vibration and noise.

ATDC: After Top Dead Center.

Atmospheric pressure: Pressure exerted by atmosphere on all things exposed to it. Around 14.7 pounds per square inch or 101 kPa at sea level.

Atomize: To break a liquid into tiny droplets.

Automatic choke: A carburetor device that improves operation when the engine is cold.

Automatic level control: Pressure system that maintains proper body height with changes in load.

Automatic transmission: A transmission that shifts itself. Shift points are determined by road speed and throttle position.

Automatic transmission fluid (ATF): Special lubricating fluid designed for use in automatic transmissions.

Automatic transaxle: A transaxle that shifts itself. Shift points are determined by road speed and throttle position.

AWG: American Wire Gauge. United States standard for measurement of nonferrous wires, sheet material, and rods.

Awl: Sharp pointed steel tool used to probe cuts, and holes in tires.

Axial: Direction parallel to shaft or bearing hole.

Axial compressor: An air conditioning compressor in which a series of pistons are moved in an axial direction by a swash plate.

Axis: The line about which a revolving object rotates.

Axle end gears: Two gears, one per axle, that are splined to the inner ends of drive axles.

Axle end play: Axle shaft movement in an endwise direction.

Axle housing: Enclosure for axles and differential. It is partially filled with differential oil.

Axle ratio: Relationship or ratio between the number of times the propeller shaft must revolve to turn the axle drive shafts one turn.

Axle retainer: Device such as a nut, collar, or C-lock designed to secure axle in place.

Axle side gear: Gear on drive axle inner end. It mates with differential pinion gears. In turn, through the pinion shaft, it connects axles to differential case.

B

Babbitted bearing: A bearing in which bearing material, in this case Babbitt metal, is bonded directly to bearing housing.

Backfire (exhaust system): Passage of unburned fuel mixture into exhaust system where it is ignited and causes an explosion.

Backfire (intake system): Burning of fuel mixture in intake manifold. May be caused by faulty timing, crossed plug wires, a leaky intake valve, a lean mixture, and other engine problems.

Backhand welding: A method of oxyacetylene welding in which torch tip faces into puddle, AWAY from direction of welding.

Backlash: Amount of movement between two parts. In the case of gears, it refers to how much one gear can be moved back and forth without moving gear into which it is meshed.

Back pressure: Resistance to flow of exhaust gases through exhaust system.

Back pressure transducer valve (BPV): A valve, sensitive to engine exhaust system back pressure, that adjusts or modulates action of EGR (Exhaust Gas Recirculation) valve.

Backup lights: White lights attached to the rear of a vehicle. Lights are turned on whenever transmission is placed in reverse. Provides illumination for backing at night.

Baffle: Obstruction used to slow down or divert the flow of gases, liquids, or sounds.

Balance: See *Static Balance* and *Dynamic Balance*.

Balance point: The point or location at which an object may be lifted while retaining proper balance (equal weight on both sides of lift point).

Balance weight: A weight that is added to a rotating shaft or wheel to bring it into balance and prevent vibration.

Ball bearing: Bearing consisting of an inner and outer hardened steel race separated by a series of hardened steel balls.

Ball joint: Flexible joint utilizing ball-and-socket construction. Used in steering linkages and suspensions.

Ball joint rocker arms: Rocker arms that are mounted on ball-shaped devices on the ends of studs instead of being mounted on a shaft.

Ball nut: That portion of a steering gear riding on the worm shaft (on a series of steel balls) and meshed with pitman shaft sector gear.

Ball peen hammer: A steel hammer with round peen head on one end. Useful for general striking and for peening pins and rivets.

Ball race: That portion of a ball bearing (inner and outer) that contains and bears against bearing balls.

Ball screw: A component of one type of anti-lock brake system.

Ball stud: The stud upon which rocker arm pivot ball is affixed.

Ballast resistor: Resistor constructed of special type wire, properties of which tend to increase or decrease voltage in direct proportion to heat of wire.

Band: Holding member in an automatic transmission. Used to stop drum rotation during certain transmission gear ranges.

Band and clutch application chart: Diagnostic chart used to identify transmission member operating conditions in each gear range.

Banjo-type rear axle: Rear axle housing from which the ring and pinion and differential unit may be removed while housing remains in place on car. Housing is solid from side to side.

Barometric pressure: Atmospheric pressure as determined by a barometer.

Barometric pressure sensitive bellows: An evacuated (air removed) bellows that expands or contracts with changes in atmospheric pressure.

Barrier hose: Air conditioning hose with an internal liner to prevent the escape of R-134a refrigerant.

Base brakes: The hydraulic and friction members of the automotive brake system. Does not include boosters or ABS/traction control components.

Base circle: Lowest spot on camshaft. Area of cam directly opposite lobe.

Base metal: Metal underneath a coating or plate. Can refer to metal of parts being welded as opposed to filler metal being added.

Battery: Electrochemical device for producing electricity.

Battery capacity: Rating of current output of battery. Determined by plate size, number of plates, and amount of acid in electrolyte.

Battery charging: Process of renewing battery by passing electric current through battery in reverse direction.

Battery rating: Standardized measurement of battery's ability to deliver electrical energy under specified conditions.

Battery reserve capacity: Measurement (in minutes) of battery's ability to supply electrical system energy in the event generating system is inoperative.

Battery reserve capacity rating: The measure of number of minutes a fully charged battery at 80°F (26.7°C) can be discharged at 25 amperes and still maintain a minimum voltage of 1.75 volts per cell.

Battery shedding: Grids losing active materials that fall off and settle in bottom of battery case.

Battery state of charge: Electrical energy available from a given battery in relation to that which would be normally available if battery were fully charged.

Battery temperature sensor: Sensor installed in some batteries to control overcharging.

Battery voltage: Determined by the number of cells. Each cell has 2.1V. Three cells will produce a 6V battery and six cells a 12V battery.

BBDC: Before Bottom Dead Center.

BCI: Battery Council International.

BDC: Bottom Dead Center.

Bead: Steel wire reinforced portion of tire that engages the wheel rim.

Bead expander: A device used to expand tire beads out against rim flange to permit inflation of tubeless tire.

Bead wire: Steel wire used in tire bead area to prevent bead expansion.

Bearing: Area of unit in which contacting surface of a revolving part rests.

Bearing axis: A line through center of a bearing, parallel to length of shaft upon which bearing operates.

Bearing bore: Center opening in a bearing inner cone or ring. Bore diameter determines shaft size applicable to a given bearing.

Bearing cap: Removable portion of a split (two piece) friction bearing. May contain either an insert or integral bearing surface.

Bearing cap cross bolt: A strengthening bolt passed through engine crankcase and main bearing cap. When secured, adds rigidity to main bearing cap.

Bearing clearance: Space left between shaft and bearing surface to allow lubricating oil to enter.

Bearing cone creep: Slow rotational movement of the antifriction bearing inner cone or ring around shaft.

Bearing contact area: That area contacted by sliding or rolling elements of a bearing.

Bearing crush: In a precision insert bearing, having bearing ends protrude a very small distance above bearing parting surface. When cap is drawn up, bearing halves are crushed into tight contact with bore.

Bearing journal: That portion of a shaft engaging a friction bearing surface.

Bearing oven: A special oven designed to heat and expand antifriction bearing inner cone to aid installation.

Bearing packer: A device used to inject grease into an antifriction bearing.

Bearing preload: Adjusting a bearing to place it under an initial load. This helps prevent bearing and part misalignment when unit is subjected to normal load stresses.

Bearing spread: In a precision insert bearing, having bearing parting surface diameter slightly larger (positive spread) than bore diameter. This allows bearing to snap into place and holds halves in place during assembly.

Belleville clutch: A positive traction differential clutch disc and plate assembly constructed to impart friction producing pressure when assembled.

Bell housing or clutch housing: Metal covering around flywheel and clutch, or torque converter assembly.

Belt slack: A measured amount of belt looseness or deflection.

Belt tension: Amount of contact pressure produced between pulley and belt when belt is tightened. Determined by measuring belt deflection or slack.

Bench bleeding: Removing air from brake master cylinder before installing new or rebuilt unit on the vehicle.

Bending spring: A special spring that is placed over a section of tubing to permit bending without kinking.

Bendix starter drive: A self-engaging starter drive gear. Gear moves into engagement when starter starts spinning and automatically disengages when starter stops.

Bevel gear: Gear in which teeth are cut in a cone shape, as found in axle end gears.

Bezel: Crimped edge of metal that secures glass face to an instrument.

BHP: Brake horsepower. Measurement of actual power produced by engine.

Bias ply: Tire plies crisscross; belts not used under tread area.

Bias tire: A tire in which body cord plies are arranged in a cross-biased pattern at an angle to tire centerline.

Black light: Instrument used in detecting leaks and cracks. Emits invisible (ultraviolet) light that when directed on fluorescent materials, causes them to glow.

Bleeder screw: A screw that, when loosened, opens a small hole to permit escape of air trapped in a wheel cylinder or master cylinder.

Bleeding: Removing air or fluid from a closed system.

Bleeding the brakes: Refers to removal of air from hydraulic system. Bleeder screws are loosened at each wheel cylinder, (one at a time) and brake fluid is forced from master cylinder through lines until all air is expelled.

Bleed valve: A threaded valve located on the top of the engine and connected to the engine coolant passages. When loosened, the valve permits the escape of air trapped in the cooling system. Used on vehicles where the radiator filler opening is lower than the engine coolant passages.

Blend door: Door installed in the air conditioning blower case. Controls temperature of the air going to the passenger compartment.

Blended refrigerant: Refrigerant made with two or more chemical compounds.

Blind hole: A hole drilled in a part without going completely through the part. Since it has only one opening, it is sometimes difficult to cut threads to the bottom of a blind hole.

Block: Part of engine containing cylinders.

Block distortion: Twisting or bending of parts of the engine block. Can be caused from overheating or improper fastener torque. Can cause cylinder bore, crank bearing bore, or camshaft bore distortion, and misalignment.

Block guide pin: Smooth headed pins that are inserted into cylinder block to provide proper block-to-gasket and cylinder head alignment.

Blow-by: The escape of exhaust gases past piston rings.

Blower: See *Supercharger.*

Blueprinting: Dismantling an engine and reassembling it to exact specifications.

BMEP: Brake Mean Effective Pressure.

Body control module (BCM): On-board computer responsible for controlling such functions as interior climate, radio, instrument cluster readings, and on some vehicles, cellular telephones. May also interact with the engine's electronic control unit.

Boiling point: Exact temperature at which a liquid begins to boil.

Bolt length: A measurement taken from end of bolt to underside of bolt head.

Bolt-on eccentric: An eccentric (cam-shaped) device bolted to camshaft or crankshaft. Converts rotary motion into reciprocating motion to operate mechanical fuel pumps or vacuum pumps.

Bonded brake lining: Brake lining that is attached to brake shoe by adhesive.

Booster: Device incorporated into the brake system to increase pressure output or decrease amount of effort required to operate or both.

Booster battery: An extra battery, connected in parallel to regular vehicle battery. Used as an emergency measure to provide additional cranking capacity when regular battery is damaged, discharged, or of insufficient size.

Bore: Any hole drilled, cut, or formed in a component. This term may be used to refer to cylinder itself or to diameter of the cylinder.

Bore alignment: How well the centerline of each one of a series of bores (such as camshaft bearing bores) aligns with a common centerline through them all. A misaligned bore can be either off-center, out-of-round, or tipped.

Bore diameter: Diameter of cylinders.

Bore face: Surface area surrounding open end of a cylinder or bearing bore.

Bore misalignment: When true centerline of two or more bores do not coincide with their common centerline. Can also refer to bore true centerline not being in desired alignment with some other part.

Boring: Renewing cylinders by cutting them out to a specified size. Boring bar is used to make cut.

Boring bar: Machine used to cut engine cylinders to specific size. As used in garages, to cut worn cylinders to a new diameter.

Boss: A rib or enlarged area designed to strengthen a certain portion or area of an object.

Bowed crankcase: A condition in which engine block warpage or distortion has actually bent crankcase, thus throwing crank bearing bores out of alignment.

Bowl vent: A small pressure relief hole near top of a carburetor float bowl.

Braided cover: A woven shielding placed around a wire or hose to provide additional strength and resistance to abrasion.

Brake anchor: Steel stud upon which one end of brake shoes is either attached to or rests against. Anchor is firmly affixed to backing plate.

Brake backing plate: Rigid steel plate upon which brake shoes are attached. Braking force applied to shoes is absorbed by backing plate.

Brake band: Band, faced with brake lining that encircles a brake drum. Used on older parking brake installations.

Brake bleeding: See *Bleeding the Brakes.*

Brake caliper: See *Caliper.*

Brake cones: Tapered cones used to center brake drums and discs for refinishing on a brake grinder or lathe.

Brake cylinder: See *Wheel cylinder.*

Brake disc pad: Friction device used in disc brake system.

Brake drum: Cast iron or aluminum housing, bolted to wheel. Rotates around brake shoes. When shoes are expanded, they rub against machined inner surface of brake drum and exert braking effect upon wheel.

Brake drum lathe: Machine to refinish inside of a brake drum.

Brake fade: Reduction in braking force due to loss of friction between brake shoes and drum. Caused by heat buildup.

Brake feel: Discernible relationship between the amount of brake pedal pressure and the actual braking force being exerted. Special device is used in power brake installations to give driver this feel.

Brake fluid: Special fluid used in hydraulic brake systems. Never use anything else in place of regular fluid.

Brake flushing: Cleaning brake system by flushing with alcohol or brake fluid. Done to remove water, dirt, or any other contaminant. Flushing fluid is placed in master cylinder and forced through lines and wheel cylinders where it exits at cylinder bleed screws.

Brake horsepower (bhp): Measurement of actual usable horsepower delivered at crankshaft. Commonly computed using an engine on a chassis dynamometer.

Brake hose: Flexible hose that connects brake lines to wheel cylinders. It allows suspension movement without damage.

Brake line: Steel tubing that carries brake fluid from master cylinder to wheel cylinders.

Brake lining: Friction material fastened to brake shoes. Brake lining is pressed against rotating brake drum, stopping car.

Brake pedal height: A measurement of distance between top face of brake pedal and floor.

Brake shoe heel: End of brake shoe adjacent to anchor bolt or pin.

Brake shoe toe: Free end of shoe, not attached to or resting against an anchor pin.

Brake shoes: Part of brake system, located at wheels, upon which brake lining is attached. When wheel cylinders are actuated by hydraulic pressure, they force brake shoes apart.

Brake system warning light: Red light mounted in the instrument cluster. Indicates a brake hydraulic system malfunction.

Brass drift punch: Punch made of brass and used to drive shafts and pins in or out of their respective bores without damage.

Brass hammer: Hammer made of brass. Useful for striking jobs in which a steel hammer could damage work.

Braze: To join two pieces of metal together by heating edges to be joined and then melting drops of brass or bronze on area. Unlike welding, this operation is similar to soldering, only a higher melting point material is used.

Brazing flux: A powder or liquid used in brazing, that tends to remove scale and oxidization, allowing brazing metal to both flow and bond readily.

Brazing rod: Bronze, manganese bronze, silver alloy, or other rods used in braze welding. Melts at a lower temperature than base metal.

Breakaway torque: When tightening a fastener, measurement of twisting force (torque) required to start a stuck or already tightened fastener turning. This torque is not a true indication of actual fastener torque as it will be higher than that used in initial tightening and higher than that required to keep it moving.

Break-in: Period of operation between installation of new or rebuilt parts and time in which parts are worn to the correct fit. Driving at reduced and varying speed for a specified mileage to permit parts to wear to the correct fit.

Break-in coating: A special oil or grease used on a new part to prevent scoring, overheating, or seizing during the break-in period.

Breaker: Rubber or fabric (or both) strip placed under a tire's tread to provide additional protection for main tire carcass.

Breaker points: See *Contact points.*

Breather port: A small hole in master cylinder cap or cover that allows air to move in or out of reservoir. This prevents either vacuum or pressure forming above reservoir brake fluid.

Brinelling: A condition in which an antifriction bearing race or ring has a series of dents or grooves worn in bearing surface.

Brinell test: A test to determine relative hardness of a given material, such as steel. A 1 centimeter hardened steel ball is pressed into material under a specific pressure and surface area of resulting dent is determined, divided into load force, and indicated as a Brinell number.

Bristle brush: A brush used for general cleaning. May use steel, brass, plastic, or fiber bristles.

Bronze: An alloy of tin and copper.

Brush: Pieces of carbon or copper that rub against the commutator on starter motors.

BTDC: Before Top Dead Center.

BTU (British thermal unit): Measurement of the amount of heat required to raise temperature of one pound of water, one degree Fahrenheit.

Bulkhead union: A pipe union (connector) designed to connect two pipes on opposite sides of a bulkhead (divider) such as an engine compartment fire wall.

Buna: Synthetic (manufactured) rubber.

Burnish: To bring a surface to a high shine by rubbing with hard, smooth object.

Bus bar: Common electrical connection from which several electrical devices operate. May be a metal bar or another type of common electrical power terminal.

Bushing: Bearing for shaft, spring shackle, piston pin, etc., of one piece construction, which may be removed from part.

Butane: Petroleum gas that is liquid, when under pressure. Often used as fuel in trucks.

Butt connector: An insulated connecting device (either crimp or solder type) used to join ends of two wires together.

Butt weld: Joining two pieces of metal by placing them edge to edge and welding along junction line.

Butterfly valve: Valve in throttle assembly that is so named due to its resemblance to insect of same name.

Bypass: To move around or detour regular route or circuit taken by air, fluid, or electricity.

Bypass valve: Valve that can open and allow fluid to pass through in other than its normal channel.

C

CAFE: Corporate Average Fuel Economy.

Cage: An enclosure, often made of stamped steel, which retains, separates, and spaces bearing balls or rollers.

Calibrate: Applied to test instruments—adjusting dial needle to correct zero or load setting.

Caliper: Disc brake component that forms cylinder and houses piston and brake pads (linings). It produces clamping action on rotating disc to stop car.

Caliper alignment: Brake caliper position in relation to brake disc (rotor). Caliper pistons-brake pads must be at right (90°) angles to disc braking surface and centered over disc.

Calipers (inside and outside): Adjustable measuring tool placed around or within an object and adjusted until it just contacts. It is then withdrawn and distance measured between contacting points.

Calorie: A unit of heat. Amount of heat required to raise the temperature of one gram of water 1 deg. centigrade.

Cam: Offset portion of shaft that will, when shaft turns, impart motion to another part such as valve lifters.

Cam angle: Number of degrees breaker cam rotates from time breaker points close until they open again. Commonly referred to as dwell.

Cam flank: That portion of a cam lobe between cam nose and point at which opening action occurs or closing action stops.

Cam ground: Piston ground slightly egg-shaped. When heated, it becomes round.

Cam lobe lift: A measurement of amount of travel cam will impart to lifter from full closed (lifter on base circle) to full open (lifter on tip of cam nose).

Cam nose: Tip or highest point on a cam lobe.

Camber: Tire centerline as seen from the front of the vehicle. When the tops of the tires are farther apart than the bottoms, camber is positive. When the bottoms of the tires are further apart than the tops, camber is negative.

Camshaft: Shaft with cam lobes (bumps) used to operate valves.

Camshaft end play: A measurement of longitudinal (end to end) movement of camshaft.

Camshaft gear: Gear that is used to drive camshaft.

Camshaft position sensor: A magnetic or Hall-effect switch that sends a signal to the ECM as the camshaft rotates. The ECM uses this signal to determine camshaft position.

Camshaft runout: A measurement of amount of bend existing in a camshaft.

Camshaft sprocket: Toothed device attached to camshaft and meshed with drive chain or belt.

Camshaft thrust plate: A flat plate designed to control thrust force and amount of end movement or play.

Candle power: Measurement of light producing ability of lightbulb.

Capacitance: Property of capacitor that permits it to receive and retain an electrical charge.

Capacitor: An electrical device with the ability to absorb and retain surges of electricity.

Capillary action: Movement of liquid along surface of a solid. Caused by state of surface tension (attraction) existing between molecules of liquid and solid.

Capillary tube: A very fine tube with a tiny hole. Capillary action (liquid movement into tube) can take place.

Cap parting edges: That portion of a bearing cap that presses against other bearing half when cap is installed.

Cap screw: A headed fastener that passes through one part and threads into another. When cap screw is tightened, parts are drawn together. No nut is required.

CARB: California Air Resources Board.

Carbon: Used to describe hard or soft black deposits found in combustion chamber, on plugs, under rings, and elsewhere in an engine.

Carbon dioxide (CO_2): A tasteless, odorless, colorless gas. Used in carbonated drinks.

Carbonize: Building up of carbon on objects. Conversion of another material to carbon.

Carbon monoxide (CO): Deadly, colorless, odorless, and tasteless gas found in engine exhaust. Formed by incomplete burning of hydrocarbons.

Carbon pile: Refers to amperage or voltage regulator utilizing a stack of carbon discs in its construction.

Carbon solvent: A cleaning solvent especially formulated to loosen and remove carbon deposits.

Carbon tracking: Carbon trail along a path of electrical current flashover. Commonly formed by current flow between insulated terminals of a distributor cap.

Carburetor: Device used to mix gasoline and air in correct proportions. Replaced by fuel injection on late-model vehicles.

Carburetor adapter: Adapter used to fit or place one type of carburetor on an intake manifold that may not be originally designed for it.

Carburetor circuits: Series of passageways and units designed to perform a specific function.

Carburizing flame: Welding torch flame with an excess of acetylene.

Carcass ply: Tire main body plies.

Carcinogen: Any substance that can cause cancer.

Cardan joint: Type of universal joint.

Carrier bearings: Bearings upon which differential case is mounted.

Case-hardened: Piece of steel that has had outer surface hardened while inner portion remains relatively soft.

Caster: Angle through centerline of wheel as seen from the side of the vehicle. When the bottom of the centerline is ahead of the top, the wheel has positive caster. When the bottom of the centerline is behind the top, the wheel has negative caster.

Casting: Pouring metal into a mold to form an object.

Castle or castellated nut: Nut having series of slots cut into one end. Cotter pin may be passed through to secure nut.

Catalytic converter: Device used in exhaust system to reduce harmful emissions. Catalyst in converter may be coated with palladium, platinum, and rhodium. Catalyst may be of oxidizing and/or reducing design.

Catalyst monitor: Oxygen sensor placed after the catalytic converter in the exhaust stream to check converter catalyst operation.

Caustic: Compound or solution that will cause burning and corrosion.

Cell: Individual (separate) compartments in battery that contain positive and negative plates suspended in electrolyte. A six-volt battery has three cells, twelve-volt battery has six cells.

Cell connector: Lead strap or connection between battery cell groups.

Celsius: A temperature scale based on the freezing point of water as 0° and boiling point as 100°.

Centering sleeve: Sleeve passed over engine crankshaft front portion to align timing cover before tightening fasteners.

Centering socket (cross-and-roller U-joint): Central socket of a constant speed universal joint. It tends to keep both halves of joint operating in a plane that forms one-half of operating angle.

Centerline: Imaginary line drawn lengthwise through center of an object.

Center link: See *Relay rod.*

Center of gravity: Point in object through which, if an imaginary pivot line were drawn, would leave object in balance. In a vehicle, the closer the weight to the ground, the lower the center of gravity.

Center punch: A sharp nosed punch used to form a small "V" shaped depression so as to properly align a drill bit prior to actual drilling.

Center support bearing: A bearing used for central support of long, two-piece propeller shafts.

Central fuel injection system: Fuel injection system that uses one or two injectors located at the top of the intake manifold.

Centrifugal advance: Unit designed to advance and retard ignition timing through action of centrifugal force.

Centrifugal clutch: Clutch that utilizes centrifugal force to expand a friction device on driving shaft until it is locked to a drum on driven shaft.

Centrifugal force: Force that tends to keep moving objects traveling in straight line. When a moving vehicle is forced to make a turn, centrifugal force attempts to keep it moving in straight line. If the vehicle is turning at too high a speed, centrifugal force will be greater than frictional force between tires and road and the car will slide off the road.

Ceramic filter: Filtering device utilizing porous ceramic as filtering agent.

Ceramic substrate: A ceramic material, covered with aluminum oxide, that forms base upon which palladium and platinum are applied. Used in one type of catalytic converter construction.

Certified technician: Technician who is certified by ASE and generally handles diagnosis and heavy repair tasks.

Cetane number: Measurement of diesel fuel performance characteristics.

CFM: Cubic feet per minute. A measure of airflow.

Chain hoist: A lifting device employing an endless chain running through two or more chain sheaves.

Chain link pin: Pin passing through engine timing chain link sections.

Chain slack: A measurement of amount of looseness in a timing chain. Generally determined by measuring chain deflection.

Chain tensioner: A device used to reduce slap and flutter from a timing chain by removing slack. Device can be either fixed or spring-loaded.

Chamfer: To bevel (or a bevel on) edge of an object.

Change of state: Condition in which substance changes from a solid to a liquid, a liquid to a gas, a liquid to a solid, or a gas to a liquid.

Charcoal canister: Emission control device containing activated charcoal granules. Used to store gasoline vapors from fuel system. When engine is started, stored vapors are drawn into engine and burned.

Charge (Air conditioning): A given or specified (by weight) amount of refrigerant.

Charge: Passing electric current through battery to restore it to active (charged) state.

Charge rate: Electrical rate of flow, in amperes, passing through the battery during charging.

Charging: Inserting the specified charge (amount) of refrigerant into the air conditioning system.

Charging station: A complete portable setup refrigerant tank with gauges, hoses, and other tools needed to service automotive air conditioning systems.

Chase: To repair damaged threads.

Chassis: Generally, chassis refers to frame, engine, front and rear axles, springs, steering system, and gas tank. Everything but body and fenders.

Chassis dynamometer: See *Dynamometer.*

Chatter: A vibrating, bouncing motion. When applied to cutting tools, tool will leave an uneven surface made up of fine ridges.

Check valve: Valve that opens to permit passage of fluid or air in one direction and closes to prevent passage in opposite direction.

Chilled iron: Cast iron possessing hardened outer skin.

Chloride flux: An acid- (hydrochloric) type soldering flux. Do not use on wiring.

Choke: Carburetor valve used to create a rich mixture when the engine is cold.

Choke break: A device that partially opens carburetor choke when the engine starts.

Choke stove: Heating compartment in or on exhaust manifold from which hot air is drawn to automatic choke.

CID: Cubic Inch Displacement.

Circuit: Source of electricity (battery), resistance unit (electrical device) and wires that form path for flow of electricity from source, through unit and back to source.

Circuit breaker: Protective device that will make and break flow of current when current draw becomes excessive. Unlike fuse, it does not blow out but switches on and off.

Class: A description of thread fit between bolt and nut. Of three classes (1, 2, and 3), class 1 is quite loose, class 2 normal fit as used on automotive applications, and class 3 is very close.

Clearance: Given amount of space between two parts such as between a piston and cylinder or bearing and journal.

Clockwise: Rotation to right as that of clock hands.

Closed cooling system: Type of system using an overflow tank.

Closed loop fuel system: A fuel system in which the air-fuel ratio is constantly adjusted in relationship to the amount of oxygen in the exhaust. The oxygen sensor(s) and ECM alter fuel injection system pulse width.

Cluster or counter gear: Cluster of gears that are all cut on one long gear blank. Cluster gears provide a connection between transmission input shaft and output shaft.

Clutch: Device used to connect or disconnect flow of power from one unit to another.

Clutch arbor: A round steel shaft, one end of which is slipped into flywheel clutch pilot bearing or bushing. Diameter of remainder fits inside of clutch disc splines. Arbor aligns disc with flywheel friction surface while pressure plate is installed.

Clutch cover: That portion of clutch that surrounds and secures pressure plate assembly. Cover bolts to flywheel.

Clutch diaphragm spring: Round dish-shaped piece of flat spring steel. Used to force pressure place against clutch disc in some clutches.

Clutch disc: Part of clutch assembly splined to transmission clutch or input shaft. Faced with friction material. When clutch is engaged, disc is squeezed between flywheel and clutch pressure plate.

Clutch disc damper spring: A series of coil springs located between disc hub and disc proper. Springs soften impact to clutch shaft when disc is seized between pressure plate and flywheel.

Clutch disc hub: Central splined portion of clutch disc.

Clutch explosion: When a clutch literally flies apart (explodes) when subjected to high rpm. Scatter shield is used on competition cars to protect driver and spectators from flying parts if the clutch explodes.

Clutch friction surface: Areas of clutch that rub against disc friction facing.

Clutch housing or bell housing: Cast iron or aluminum housing that surrounds flywheel and clutch mechanism.

Clutch linkage: Mechanism which transfers movement from clutch pedal to throw-out fork.

Clutch pedal free travel: Specified distance clutch pedal may be depressed before throw-out bearing actually contacts clutch release fingers.

Clutch pedal height: A measurement from top face of pedal to floor mat.

Clutch pilot bearing: Small bronze bushing, or in some cases ball bearing, placed in end of crankshaft or in center of flywheel depending on car, used to support outboard end of transmission input shaft.

Clutch pressure plate: Part of a clutch assembly that through spring pressure, squeezes clutch disc against flywheel thereby transmitting driving force through the assembly. To disengage clutch, pressure plate is drawn away from flywheel via linkage.

Clutch release fingers: Steel fingers on pressure plate assembly that, when depressed by throw-out bearing, pull pressure plate away from clutch disc.

Clutch semi-centrifugal release fingers: Clutch release fingers that have a weight attached to them so that at high rpm release fingers place additional pressure on clutch pressure plate.

Clutch throw-out fork: Device or fork that straddles throw-out bearing and used to force throw-out bearing against clutch release fingers.

Coast pattern: Contact pattern (area of contact between differential ring and pinion gear) when vehicle is coasting (car driving engine).

Coil: Used to step up battery voltage to point necessary to fire spark plugs.

Coil core: Multilayered mass of iron around which coil windings are wrapped. Some designs place core outside of windings.

Coil polarity: Refers to ignition coil (Neg. or Pos.) primary connection hookup. Battery and coil polarity must be same. When battery negative is grounded, coil negative must be connected to ground.

Coil spring: Section of spring steel rod wound in spiral pattern or shape. Widely used in both front and rear suspension.

Cold: Little or no perceptible heat.

Cold bending: Bending a part without heating it first.

Cold clearance setting: Adjusting valve lash or clearance when engine is at room temperature.

Cold cranking rating: Measurement of cranking amperes that a battery can deliver over a period of 30 seconds at 0°F (–17.8°C) and still maintain a minimum cell voltage of 1.2 volts.

Cold expanding: Expanding an engine piston by mechanical means and without use of heat.

Cold flow: Excessive fastener torque causing gasket to flatten and extrude outward.

Cold patching: Repair method used to seal leaks in plastic fuel tanks.

Cold soak cleaning: Immersing parts in a cold cleaner bath for recommended time period to help loosen and remove carbon, dirt, and grease.

Cold solder: An electrical joint that has had solder applied to it before it reached proper soldering temperature. Characteristics include clumping of solder on the joint.

Cold start injector: Fuel injector, separate from the other engine fuel injectors, that sprays additional fuel for cold engine starting.

Collapsed piston: Piston whose skirt diameter has been reduced due to heat and forces imposed upon it during service.

Collapsible preload spacer: A special spacer as used between bearings on some differential drive pinion shafts. When nut is torqued to specs, spacer will collapse a specified amount and will thus provide an initial (before torque loading) load (preload) on bearings.

Collector ring: See *Slip ring*.

Color coding (wiring): Using various colors to identify specific wiring circuits. Wiring diagrams show circuits and specify color.

Combination valve: A dual purpose brake system valve that can combine proportioning, metering, and pressure differential warning functions.

Combustion: Chemical reaction that takes place during burning.

Combustion chamber: Area above piston with piston on TDC. Head of piston, cylinder, and head form the chamber.

Combustion chamber volume: Volume of combustion chamber (space above piston with piston on TDC) measured in cc (cubic centimeters).

Commutator: Series of copper bars connected to armature windings in starters. Bars are insulated from each other and from armature. Brushes rub against the turning commutator.

Commutator segment or bar: Copper bars connecting armature windings to brushes in a starter.

Compensating port: Small hole in brake master cylinder to permit fluid to return to reservoir.

Compound: Two or more ingredients mixed together.

Compound planetary gearset: Refers to use of combination of two or more planetary gearsets in a single assembly.

Compressibility: Ability of a material to decrease in size when pressure is applied to it. The compressibility of steel is low, while the compressibility of cork is high.

Compression: Applying pressure to a substance, causing it to be reduced in size in direction of compressing force.

Applying pressure to gas, thus causing a reduction in volume.

Compression check: Testing compression in all cylinders at cranking speed. All plugs are removed, compression gauge placed in one plug hole, throttle wide open, and engine cranked until gauge no longer climbs. Compression check is a fine way in which to determine condition of valves, rings, and cylinders.

Compression connection: A flareless tubing fitting using a tapered edge sleeve that, when squeezed between the fitting body and nut, pinches tubing to produce a seal. Another type uses sleeve function as an integral nose on nut.

Compression gauge: Gauge used to test compression in engine cylinders.

Compression loading: A front suspension ball joint arrangement that, when under loading forces, causes ball joint parts to be compressed together.

Compression ratio: Relationship between cylinder volume (clearance volume) when piston is on TDC and cylinder volume when piston is on BDC.

Compression ring groove: Groove cut into piston head to accept a compression ring.

Compression rings: Top piston rings, generally two, designed to seal between piston and cylinder to prevent escape of gas from combustion chamber.

Compression stroke: Portion of piston's movement devoted to compressing the fuel mixture trapped in engine's cylinder.

Compression tester: Gauge used to check engine compression pressure.

Compressor: Device used to raise pressure of refrigerant in an air conditioning system. Also causes refrigerant to flow through system.

Compressor protection switch: A heat and/or pressure sensitive switch used in air conditioning systems to protect the compressor from damage caused by the loss of oil, excessive pressures, and overheating.

Computer controlled ignition: Ignition system using sensors that feed electrical information to the ECM, which then controls ignition timing for maximum efficiency.

Computerized ride control (active suspension): Advanced suspension system that can adapt to specific driving conditions.

Concentric: Two or more circles so placed as to share common center.

Concentric grinding: Grinding brake lining (mounted on shoe) contour so that it matches that of brake drum. Gives full contact between drum and shoe.

Condense: To turn vapor back into liquid.

Condenser: Unit installed between breaker points and coil to prevent arcing at breaker points.

Condenser: Unit in air conditioning system that cools hot compressed refrigerant and turns it from vapor into liquid.

Condensation: Moisture, from air, deposited on a cool surface.

Conduction: Transfer of heat from one object to another by having objects in physical contact.

Conductor: Material forming path for flow of electric current.

Connecting rod: Connecting link between piston and crankshaft.

Connecting rod aligner: Tool designed to check rod for twist and bend.

Connecting rod bearings: Inserts that fit into connecting rod and ride on crankshaft journals.

Connecting rod bend: A condition in which centerlines of rod pin bore and big end bearing bore are not parallel as viewed from edge of an upright rod.

Connecting rod cap: Lower removable part of rod that holds lower bearing insert.

Connecting rod journal: That portion of crankshaft riding in and contacting connecting rod bearing.

Connecting rod side play: Amount of sliding movement determined by measuring distance rod can be moved from full forward to full rearward position.

Connecting rod spit hole: Small hole drilled through upper portion of connecting rod big end. Allows oil to squirt or spurt out to lubricate another engine area or component, such as the camshaft.

Connecting rod straightener: Tool used to remove connecting rod twist, bend, or offset.

Connecting rod twist: A condition in which centerline of rod pin bore is turned either to left or right, as related to centerline of rod big end bore as viewed looking downward on an upright rod.

Constant mesh gears: Gears that are always in mesh with each other.

Constant velocity universal joint: Universal joint designed to allow smooth transfer of torque from driven shaft to driving shaft without any fluctuations in speed of driven shaft.

Contact points (breaker points): Two removable points or areas that when pressed together, complete a circuit. These points are usually made of tungsten, platinum, or silver.

Continuity: Refers to an electrical circuit in which all contacting parts are clean and tight, so as to permit desired current flow.

Continuous fuel injection: A fuel injection system in which injectors are always open and as such, feed fuel constantly. The amount of fuel delivered is adjusted by varying fuel system pressure.

Contraction: Reduction in size of object when cooled.

Control arm: A pivoting arm that maintains alignment of the wheel assembly and the vehicle chassis while permitting movement.

Control arm bushing: Rubber bushing attaching the control arm to the vehicle's frame.

Control module: Small computer used to control specific systems within an automobile. Control modules are generally used to govern the air bag and anti-lock brake systems. Sometimes used as a term for the ECM.

Control rack: Toothed rod inside mechanical injection pump that rotates pump plunger to control quantity of injected fuel. Used on diesel engines.

Control valve: A movable valve in an air or fluid line that is designed to start, stop, or alter movement, speed, volume, or direction.

Convection: Transfer of heat from one object to another when hotter object heats surrounding air and air heats other object.

Conventional steering: See *Parallelogram steering*.

Coolant: Liquid in cooling system.

Coolant temperature override switch (CTO): Coolant controlled vacuum switch that cuts off engine vacuum to EGR valve when engine temperature is below a specified point.

Coolant test strips: Chemically treated strips that react with coolant to change color. The final color of the test strip determines the amount of antifreeze in mixture.

Cooling system: System, air or water, designed to remove excess heat from engine.

Cooling system scale: A coating of rust or minerals deposited in both engine coolant jacket and radiator. If heavy, interferes with coolant flow and causes overheating. In some areas, impurities in the water supply make this problem more common.

Core: When referring to casting—sand unit placed inside mold so that when metal is poured, core will leave a hollow shape.

Core hole plug: A metal plug pressed into holes in engine water jacket casting sand core holes. Sometimes called freeze plug or welch plug.

Corona: Luminous discharge of electricity visible near surface of an electrical conductor under high voltage.

Corrode: Removal of surface material from object by chemical action.

Cotter key (cotter pin): A split pin that is inserted through a hole in a shaft or bolt. Ends are spread apart to prevent key from falling out. Used to secure castellated nuts.

Counterbalance: Weight attached to some moving part so part will be in balance.

Counterbore: Enlarging hole to certain depth.

Counterclockwise: Rotation to the left as opposed to that of clock hands.

Countershaft: Intermediate shaft that receives motion from one shaft and transfers it to another. It may be fixed (gears turn on it) or it may be free to revolve.

Countersink: To make a counterbore so that head of a screw may set flush, or below the surface.

Counterweight: A weight, integral or attached, affixed to a part such as a crankshaft, to bring revolving mass into balance.

Coupling: Connecting device used between two objects so motion of one will be imparted to other.

Coupling point: This refers to point at which both pump and turbine in torque converter are traveling at same speed. The drive is almost direct at this point.

Cowl: Part of car body between engine firewall and front of dash panel. Sometimes called the bulkhead.

Crack a fitting: To slowly open a fitting a small amount.

Crack detection: Using X ray, chemical solution, magnetic field, or another method to locate otherwise unseen fractures or cracks in a metal part.

Crack pinning: Repairing a crack, such as in a cylinder head, by installing a series of overlapping threaded pins. They are cut off and filed flush with surface.

Crankcase: Part of engine that surrounds crankshaft. Not to be confused with the pan which is a thin steel cover that is bolted to crankcase.

Crankcase dilution: Accumulation of unburned gasoline in crankcase. Excessively rich fuel mixture or poor combustion will allow some of gasoline to pass down between pistons and cylinder walls.

Crankcase ventilation: Process of drawing clean air through interior of engine to remove blow-by gases and other fumes.

Crankcase web: A thin, reinforcing wall cast as integral part of the crankcase to provide rigidity for main bearing bores.

Crank grinder: A machine used to grind crankshaft journals to highly accurate dimensions.

Cranking motor: Starter. Device to revolve engine crankshaft to start engine. Works through a gear engaging another gear on flywheel.

Cranking vacuum test: An engine test that determines basic engine (piston, ring, cylinder, valve) condition by measuring amount of vacuum developed in engine cylinders during cranking.

Crankshaft: Shaft running length of engine. Portions of shaft are offset to form throws to which connecting rods are attached. Crankshaft is supported by main bearings.

Crankshaft gear: Gear mounted on front of crankshaft. Used to drive camshaft gear.

Crankshaft micrometer: A micrometer designed to measure crankshaft main bearing journal diameter while the crankshaft is in place (main cap off).

Crankshaft position sensor: A magnetic or Hall-effect switch that sends a signal to the ECM as the crankshaft rotates. The ECM uses this signal to determine crankshaft and piston position, as well as engine RPM.

Crankshaft sprocket: Timing chain sprocket attached to one end of crankshaft.

Crankshaft throw: Offset part of crankshaft where connecting rods fasten.

Crankshaft thrust face: Precision ground portion of crankshaft, at right angles to ends of main bearing journal, that operates against main bearing thrust surface. Controls end play.

Crimped terminal: An electrical connector into which wire is inserted. Connector body is then forced inward at one or more points, thus gripping wire tightly.

Crocus cloth: A very fine abrasive cloth using a coating of iron oxide particles.

Cross and roller: Type of universal joint using a center cross (spider) mounted in needle bearings. Usually used on rear-wheel drive vehicles.

Cross bolt spacer: On engines employing a transverse stiffening bolt through both main bearing cap and crankcase, this spacer is used between two sides of cap and crankcase walls. Prevents crankcase damage when tightening bolt.

Cross firing: A condition in which firing of a given spark plug fires an additional plug. Caused by voltage from one wire being imparted to another by poor insulation or by improper routing of plug wires.

Cross shaft: Shaft in steering gearbox that engages steering shaft worm. Cross shaft is splined to pitman arm.

Crosshatch finish: Pattern left in an engine cylinder bore by honing. Fine scratch lines cross one another at an angle. Angle depends upon hone rpm and number of up and down strokes in a given period of time.

Cross-threading: Mismating of screw threads caused by engaging threads when their respective centerlines are not aligned. Will cause threads to jam and continued turning will ruin one or both threads.

Crude oil: Petroleum in its raw or unrefined state. It forms the basis of all petroleum products used in vehicles.

Cruise control: System that automatically maintains vehicle speed. Useful for long distance driving, but should not be used in heavy traffic.

Cubes: Cubic inches, or cubic inch displacement of an engine.

Cup driver: A tool designed to install bearing cups (outer ring) in bearing housings.

Cup raceway: The portion of bearing cup (ring) designed to operate against bearing balls or rollers.

Curb height: Indicates height of an automobile without passengers but with full fuel load. Measured from roadway surface to a specified location on car undercarriage.

Curb weight: Full operating (oil, coolant installed) weight of a vehicle without passengers but including a full tank of fuel.

Current: Movement of free electrons through conductor.

Current draw: Amount of current flow through a given electrical device when connected to a circuit.

Cushion spring: A series of coil springs located between clutch disc hub and clutch disc. Springs soften shock when clutch is engaged.

Cutout (regulator): Device to connect or disconnect generator from battery circuit. When generator is charging, cutout makes circuit. When generator stops, cutout breaks circuit. Also referred to as cutout relay and circuit breaker.

Cycle: Reoccurring period during which series of events take place in definite order.

Cycling clutch orifice tube: Refrigeration system with a fixed (does not change size) orifice tube. Evaporator pressure is controlled by turning the compressor clutch on and off.

Cylinder: Hole, or holes, in cylinder block that contain pistons.

Cylinder balance test: Process of checking engine rpm and vacuum while operating engine on selected pairs of cylinders at a time.

Cylinder block: See *Block.*

Cylinder bore: See *Bore.*

Cylinder dial gauge: A dial gauge measuring instrument designed to check engine cylinder bore diameters.

Cylinder head: Metal section bolted on top of block. Used to cover tops of cylinders. The cylinder head contains the valves. Also forms part of combustion chamber.

Cylinder hone: Tool that uses an abrasive to smooth out and bring to exact measurement items such as engine cylinders and wheel cylinders.

Cylinder leakage tester: A special test tool used to check engine mechanical condition by pressurizing a given cylinder (piston TDC on compression) to a given point with compressed air. Gauge then checks percentage of leakage.

Cylinder liner: See *Cylinder Sleeve.*

Cylinder sleeve: Replaceable cylinder. It is made of a pipe-like section that is either pressed or pushed into the block.

Cylinder stroke: See *Stroke.*

D

Damper: A unit or device designed to remove or reduce vibration or oscillation of a moving part.

Damper spring: A coil spring, installed inside regular valve spring, that helps to reduce spring vibration.

Dashboard: Part of body containing driving instruments and controls.

Dashpot: Unit utilizing cylinder and piston, or cylinder and diaphragm, with small vent hole, to retard or slow down movement of some part.

Data link connector (DLC): A built-in electrical connector used to attach a diagnostic test instrument.

Datastream values: Computer system operating conditions and readings obtained directly from the ECM as the vehicle operates.

DC (piston position): Dead Center. Piston at extreme top or bottom of its stroke.

Dead axle: Axle that does not rotate but merely forms base upon which to attach wheels.

Dead center: Point at which piston reaches its uppermost or downmost position in cylinder.

Deceleration: Process of reducing speed.

Deceleration valve: It feeds air into intake manifold to prevent backfiring during deceleration.

Decibel: A unit of measurement used to indicate a sound level or to indicate the difference in specific sound levels.

Deep groove ball: A ball bearing in which balls engage a rather deep groove in both inner and outer rings. Bearing will handle heavy radial and moderate thrust loads.

Deglazer: Abrasive tool used to remove glaze from cylinder walls so a new set of rings will seat.

Degree wheel: Wheel-like unit attached to engine crankshaft. Used to time valves to a high degree of accuracy.

Dehumidify: To remove moisture from air.

Dehydrate: To dry out. Remove moisture.

Demagnetize: To remove residual magnetism from an object.

Deploy: Term used to describe the activation of an air bag system.

Depolarize: Removal of residual magnetism thereby destroying or removing the magnetic poles.

Depth-type filter: An engine oil filter utilizing a relatively thick layer of fibrous materials to filter out contaminants. Entire layer acts as a filter with particles lodging at different depths.

Desiccant: Material, such as silica-gel, placed within a container to absorb and retain moisture.

Detent ball and spring: Spring loaded ball that snaps into a groove or notch to hold some sliding object in position.

Detergent: Chemicals added to engine oil to hold contaminants in suspension until they are caught by the oil filter.

Detergent oil: Oil with additives to hold contaminants in suspension. Used in all modern engines.

Detonation: Fuel charge exploding in the cylinder instead of burning smoothly.

Developer solution: Solution used to draw fluorescent crack detection penetrant to surface so that the application of black light will make presence of any cracks readily visible.

Dexron III/Mercon: Automatic transmission fluid commonly used in newer vehicles.

Diagnosis: Process of analyzing symptoms and test results to determine underlying reason for problem.

Diagnostic control module: Electronic computer that controls the air bag system. Also referred to as an Diagnostic Energy Reserve Module (DERM).

Dial gauge or indicator: Often used precision micrometer-type instrument that indicates exact reading via needle moving across dial face.

Diamond tipped dresser: A stone dresser using a small piece of diamond attached to tip. Used to smooth and reface grinding stones.

Diaphragm: Flexible sheet stretched across an area thereby separating two different compartments.

Diaphragm spring: A clutch pressure plate spring device that, instead of using several coil springs, employs a single, cone-shaped piece of sheet spring steel. Depressing cone center lifts outer edge removing pressure. Release center and outer circumference pushes on pressure plate.

Die: Tool for cutting threads.

Die casting: Formation of object by forcing molten metal or plastic into die.

Dielectric: A material, such as glass or rubber that resists flow of electricity. An insulator.

Diesel engine: Internal combustion engine that uses diesel oil for fuel. True diesel does not use an ignition system but injects diesel fuel into cylinders. Piston compresses air so tightly that air is hot enough to ignite diesel fuel without spark.

Diesel injector: A nozzle device for spraying diesel fuel into precombustion chamber.

Diesel fuel grade: Classification of diesel fuel operating characteristics, 1-D is for cold weather and 2-D is for normal temperatures.

Diesel injection pump: Mechanically operated fuel pump that develops high pressure to force fuel out of injectors and into combustion chambers.

Dieseling: Condition in which a gasoline engine continues to run after ignition switch is turned off. Also called running on.

Differential: Unit that will drive both rear axles at same time but will allow them to turn at different speeds in turns.

Differential case (carrier): Steel unit to which the ring gear is attached. Case drives spider gears and forms an inner bearing surface for axle end gears.

Digital EGR valve: EGR valve that can be operated by the vehicle's engine ECU.

Digital storage oscilloscope (DSO): See *Waveform meter.*

Dimmer switch: Foot or hand operated switch for headlight low and high beams.

Diode: Electrical device that allows current to flow in one direction but blocks current flow in the other direction. Used to change alternating current from the alternator into direct current. Widely used in vehicle electronic components. Sometimes called a rectifier.

Dipstick: Metal rod that passes into oil sump. Used to determine quantity of oil in engine.

Direct current (dc): Electric current that flows steadily in one direction only.

Direct drive: Such as high gear when crankshaft and drive shaft revolve at same speed.

Direct fuel injection: Fuel is sprayed into combustion chamber.

Direct ignition system: A type of computer-controlled ignition system similar to the distributorless system, but with the coils directly over the spark plug.

Directional stability: Ability of car to move forward in straight line with minimum of driver control.

Direct programming: Downloading new information into the ECM by attaching a shop recalibration device programming computer directly to the vehicle's diagnostic connector.

Dirt wear: Bearing wear caused by abrasive action of dirt.

Discharge: Drawing electric current from battery.

Discharge pressure: Pressure of refrigerant as it leaves compressor.

Discharge side: The high-pressure section of the air conditioning system extending from the compressor to the expansion valve.

Discharging: Procedure for draining a vehicle's air conditioning system of its refrigerant charge.

Disc brakes: Braking system that uses steel disc with caliper type lining application. When brakes are applied, section of lining on each side of spinning disc is forced against disc thus imparting braking force. This type is very resistant to brake fade.

Disc wheel: Wheel constructed of stamped steel.

Displacement: Volume of air displaced by piston traveling from BDC to TDC.

Display pattern: Oscilloscope waveform showing all cylinders in an engine in a horizontal pattern, with cylinders following each other the ignition system firing order.

Distillation: Heating a liquid and then catching and condensing the vapors given off by heating process.

Distilled water: Water from which all impurities have been removed through an evaporative process.

Distortion: Unintended alteration in the shape or size of an object.

Distribution block: A multiple outlet tubing (pipe) fitting that is connected to a supply line so as to provide outlet connections for two or more branch lines.

Distribution tubes: Tubes used in engine cooling area to guide and direct flow of coolant to vital areas.

Distributor: Unit designed to distribute high-voltage spark to proper spark plug. On some vehicles, the distributor contains the device to make and break the ignition primary circuit

Distributor cap: Insulated cap containing central terminal with series (one per cylinder) of terminals that are evenly spaced in circular pattern around central terminal. Secondary voltage travels to central terminal where it is then channeled to one of outer terminals by the rotor.

Distributorless ignition system: A type of computer-controlled ignition system that eliminates the distributor by using a sensor mounted on the crankshaft to provide timing. Distributorless ignition systems will have more than one coil.

Diverter valve: A device used in the air injection system to divert air away from the injection nozzles during periods of deceleration. Prevents backfiring.

Documentation: Repair orders or other means used to record work performed on a vehicle.

Dog tracking: Condition where the rear wheels of a vehicle are not aligned with the front wheels.

DOHC: An engine with double (two) overhead camshaft.

DOT: Department of Transportation.

DOT 3, 4, 5: Department of Transportation approval numbers for brake fluid.

Double compression fitting: A tubing fitting in which, upon tightening, nose of the nut is forced tightly against tubing by a corresponding angle on fitting body. This provides seal.

Double flare: End of tubing, especially brake tubing, has a flare so made that flare area utilizes two wall thicknesses. This makes a much stronger joint and from safety standpoint, it is a must.

Double post lift: Vehicle lift utilizing two lifting rams. Can be of suspension-type (front and back ram) or frame-type (side to side rams) design.

Dowel pin: Steel pin, passed through or partly through, two parts to provide proper alignment.

Downshift: Shifting to lower gear.

Downshift switch: A switch that actuates an electrical solenoid that causes automatic transmission to downshift when throttle valve reaches a predetermined angle.

Drag link: See *Relay rod.*

Drain cock: A faucet-like fitting used to drain fluid from a container. Usually found on radiators and fuel tanks.

Drain plug: A plug that can be removed to drain oil pan, rear axle, or transmission.

Draw (electrical): Amount of electrical current required to operate electrical device.

Draw (forming): To form (such as wire) by pulling wire stock through series of hardened dies.

Draw (temper): Process of removing hardness from a piece of metal.

Draw-filing: Passing file, at right angles, up and down length of work.

Drier (receiver-drier): Tank inserted in air conditioning system to absorb and retain moisture and hold extra refrigerant. See *Desiccant.*

Drill: Tool used to bore holes.

Drill gauge: A flat metal gauge used to check twist drill lip angle and width. Useful for accurate sharpening.

Drill press: Nonportable machine used for accurate drilling.

Drive axle: A steel shaft used to transmit driving force (torque) from differential to driving wheels. May be of single or multiple piece construction.

Drive-cycle test: Made on OBD II vehicles only. Performed by attaching a scan tool to the vehicle and driving the vehicle through a set of specific acceleration, cruising, and deceleration steps. The drive-cycle test will tell the technician whether the OBD II system is operating and whether vehicle can pass an emissions test.

Drive-fit: Fit between two parts when they must be literally driven together.

Driveability: The overall operation of the vehicle, with emphasis on performance, smoothness, and fuel economy.

Driveability diagnosis: The process of diagnosing, troubleshooting, isolating, and repairing a vehicle drivability problem.

Driveline: Propeller shaft and universal joints connecting transmission output shaft to axle pinion gear shaft on rear wheel drive vehicles.

Drive or propeller shaft safety strap: A metal strap or straps, surrounding drive shaft to prevent shaft from falling to ground in event of a universal joint or shaft failure.

Drive pattern: Contact area between driving and driven gear tooth under driving (engine driving car) conditions—as opposed to coast (car driving engine) conditions.

Drive ratio: Numerical ratio indicated by number of turns a driving gear must revolve to turn the driven gear one full revolution. For example, if differential pinion must turn four times to turn ring gear once, differential ratio is 4 to 1.

Drive shaft: Shaft connecting transmission output shaft to differential pinion shaft.

Drive train: All parts that generate power (engine) and transmit it to road wheels (transmission, clutch, drive shaft, differential, drive axles).

Driving lights: Auxiliary headlights, often very bright, that can be used to increase amount of illumination provided by regular headlights.

Drop center rim: Center section of rim being lower than two outer edges. This allows bead of tire to be pushed into low area on one side while the other side is pulled over and off the flange.

Drop forged: Part that has been formed by heating steel blank red hot and pounding it into shape with a powerful drop hammer.

Drop in refrigerant: Term for a refrigerant that can be mixed with the existing refrigerant in an air conditioner. No refrigerant is approved as a drop in refrigerant and it is illegal to mix refrigerants.

Dry air: Compressed air that does not contain water vapor.

Dry charged battery: Battery with plates charged but lacking electrolyte. To be placed in service, electrolyte is added.

Dry disc clutch: A clutch design in which clutch disc friction surface operates without lubrication of any type. All modern clutches are dry disc clutches.

Dry friction: Resistance to movement between two unlubricated surfaces.

Dry setting: A carburetor float level setting made with no fuel in fuel bowl.

Dry sleeve: Cylinder sleeve application in which sleeve is supported in block metal over its entire length. Coolant does not touch sleeve itself.

Dry sump: Instead of letting oil throw-off drain into a regular oil pan sump, system collects and pumps this oil to a remote (separate) container or sump.

Dry weight: Weight of vehicle without fluid (oil, fuel, water) in various units.

Dual master cylinder: Master cylinder with two pressure pistons and pressure chambers. Each piston and camber is sealed from the other. Provides separate brake system for both front and rear, or one front and one rear wheel of vehicle.

Duals: Two sets of exhaust pipes and mufflers—one for each bank of cylinders.

Dummy shaft: A temporary shaft that is inserted through a cluster gear or other part to hold the bearings in place while regular shaft is inserted.

Dust boot: A flexible cover over CV joints or other moving parts to prevent entry of water or dirt into the unit.

Dwell: See *Cam angle*.

Dye penetrant: A special penetrating dye used in crack detection. Dye flows into even tiny, invisible cracks. When surface is cleaned and developer applied, dye is drawn from cracks and spreads over surface far enough to make crack area very visible.

Dynamic balance: When centerline of weight mass of a revolving object is in same plane as centerline of object. For example, weight mass of the tire must be in the same plane as centerline of wheel.

Dynamometer: Machine used to measure engine horsepower output. Engine dynamometer measures horsepower at crankshaft and chassis dynamometer measures horsepower output at wheels.

Dynamometer break-in: Running a new or rebuilt engine, for a specific length of time, while engine is attached to a dynamometer. This permits controlled loading and careful monitoring of engine performance to ensure proper break-in.

E

Early fuel evaporation (EFE): System that passes hot exhaust gases over a certain portion of intake manifold to provide better fuel atomization during engine warm-up.

Ear protection: Safety device that keeps ears from being damaged by excessive noise.

Eccentric: Two circles, one within the other, neither sharing the same center. A protrusion on a shaft that rubs against or is connected to another part.

Eccentric grinding: Grinding contact surface of brake lining to a different contour (smaller radius) than that of brake drum. When applied, center of shoe contacts before ends. Additional braking pressure springs shoe so that the entire lining surface contacts drum.

ECM (electronic control module): A group of electronic components in a series of complex circuits, used to produce specific electrical outputs in response to external sensor inputs.

ECU (electronic control unit): See *ECM*.

Educated guess: Not a guess, but process of making a diagnosis based on logical diagnostic steps.

EEC: Evaporative Emission Control.

EGR: Exhaust Gas Recirculation valve and/or system.

Elastic limit: A measurement of how far material can be deformed and still return to exact shape when the deforming force is removed.

Elasticity: Property of a material that allows it to stretch, compress, bend, without breaking and still return to its original size and shape. Rubber possesses great elasticity.

Elbow: A curved pipe fitting designed to produce a change in line direction. Common angles are 90° and 45°.

Electric arc: A flow of electricity across an air gap between two conductors. When voltage is high enough, air becomes ionized and then will conduct electrical current.

Electric assist choke: A carburetor choke using an electric heating unit to speed up its opening time.

Electric fuel pump: Fuel pump operated by electric motor. Normally mounted in or near fuel tank. Commonly used on fuel injection systems.

Electrical pitting: Erosion of a bearing surface caused by a flow of electricity through a revolving bearing. Tiny electrical arcs between moving parts burn pinholes in bearing surface.

Electrochemical: Chemical (battery) production of electricity.

Electrochemical degradation (ECD): An electrochemical process occurring in the cooling system. Causes deterioration of rubber components such as hoses.

Electrode (spark plug): Center rod passing through insulator forms one electrode. The rod welded to shell forms another. They are referred to as center and side electrodes.

Electrode (welding): Metal rod used in arc welding.

Electrode holder: In arc welding, handle device that grasps welding electrode.

Electrolyte: Sulfuric acid and water solution in battery.

Electromagnet: Magnet produced by placing coil of wire around steel or iron bar. When current flows through coil, bar becomes magnetized and will remain so as long as current continues to flow.

Electromagnetic: Magnetic (alternator) production of electricity.

Electromagnetic regulator: Voltage regulator using contact points and electromagnetic.

Electromotive force (EMF): Another term for voltage.

Electron: Negatively charged particle that makes up part of the atom.

Electron theory: Accepted theory that electricity is flow of electrons from one area to another.

Electronic: Refers to electrical circuits or units using electronic components such as transistors and diodes.

Electronic control: Using an electronic device, such as a computer, to operate or control action of some part or system.

Electronic fuel injection: Fuel injection system using electronic circuits to control the operation of the fuel injectors based on inputs from input sensors.

Electronic fuel injector: A fuel injector using a solenoid winding. The solenoid is energized to open the injector valve. When the solenoid is de-energized, an internal spring closes the valve.

Electronic ignition: Ignition system using electronic devices to trigger coil firing. First used to define an ignition system without breaker points.

Electronic leak detector: An electrically operated instrument used to test air conditioning system for refrigerant leakage.

Electronic regulator: Voltage regulator using all electronic components.

Elongate: Lengthening or stretching a material by application of heat or pressure.

Emergency drive lockout: A device, as employed on some full-time four-wheel drive vehicles, used to lock up transfer case differential to provide direct mechanical torque transfer to both front and rear drive differentials.

Emissions: Byproducts of automotive engine combustion that are discharged into atmosphere. Major pollutants are oxides of nitrogen, carbon monoxide, hydrocarbons, and various particulates. Term also includes vapor (hydrocarbon) loss from fuel tank and carburetor.

Emissions information label: Label normally located in the engine compartment that gives timing, idle speed, and vacuum hose routing information.

Emissions tune-up: See *Maintenance tune-up*.

End lift: A lifting device designed to elevate one end of vehicle.

End play: Amount of axial (lengthwise) movement between two parts.

Energy: Capacity for doing work.

Engine: Device that converts heat energy into mechanical motion.

Engine adapter: Unit that allows a different engine or transmission to be installed in a vehicle.

Engine block: Main engine casting that contains cylinders, piston and crankshaft assemblies. Head bolts to top. Oil sump or pan bolts to bottom.

Engine centerline: A line, equidistant from top, bottom, and sides, running through engine in a direction parallel to crankshaft.

Engine displacement: Volume of space through which head of piston moves in full length of its stroke—multiplied by number of cylinders in engine. Result is given in liters or cubic inches.

Engine mounts: Pads made of metal and rubber that hold engine to frame.

Engine sequence tests: Laboratory engine tests to determine how well a specific engine oil will prevent scuffing, corrosion, oxidation, wear, and sludge buildup. Also referred to as "Car Manufacturer's Sequence Tests."

Engine stand: A holding fixture for engine so that repair is made easier by having engine at standing height and capable of being turned to different positions.

Enhanced emissions test: An emission in which the vehicle is operated on a dynamometer at different speeds and loads to simulate road conditions. This test is sometimes called an IM test or a transient test.

Entrepreneur: Person with the ability to run his or her own business in the short term while planning for the long term.

EP lubricant: Lubricant compounded to withstand very heavy loads imposed on gear teeth.

EPA: Environmental Protection Agency. A Federal agency concerned with air, water, and ground pollution reduction.

EPROM: Electronically erasable programmable read-only memory. Type of PROM that is updated from an external computer. See *PROM.*

Equalizer line: On air conditioning systems employing an STV, POA, or VIR suction throttling valve, a tube connecting this valve to spot underneath expansion valve control diaphragm. This modulates expansion valve action.

Ermeto fitting: A tubing fitting designed for high-pressure application. A special compression sleeve cuts into tubing and forces it against body seat. Often used on heavy walled tubing that is hard to flare.

ESC: Electronic Spark Control.

Etching: Dulling or corroding of bearing surface through action of water (rusting) or presence of a corrosive substance. Bearings that are inactive for prolonged periods are particularly susceptible.

Ethylene glycol: Chemical solution added to cooling system to protect against freezing.

Evacuating: Process of removing oxygen from an air conditioning system by pumping air out of the system, creating a vacuum.

Evaporation: Process of a liquid turning into a vapor.

Evaporation control system: Emission control system designed to prevent gasoline vapors from escaping into atmosphere from tank and carburetor.

Evaporator: Unit in air conditioning system used to transform refrigerant from a liquid to a gas. It is at this point that cooling takes place.

Excite: To pass an electric current through a unit such as field coils in generator.

Exhaust gas analyzer: Instrument used to check exhaust gases to determine combustion efficiency.

Exhaust gas recirculation (EGR): Admitting a controlled amount of exhaust gas into intake manifold during certain periods of engine operation. This lowers combustion flame temperature, thus reducing level of nitrogen oxide emission.

Exhaust manifold: Connecting pipes between cylinder head exhaust ports and exhaust pipe.

Exhaust oxygen sensor: See *Oxygen sensor.*

Exhaust pipe: Pipe connecting exhaust manifold to catalytic converter. A second exhaust pipe may be used to connect the catalytic converter to the muffler.

Exhaust stroke: Portion of piston's movement devoted to expelling burned gases from cylinder.

Exhaust system: Parts that carry engine exhaust to rear of car—exhaust manifold, pipes, muffler, and catalytic converter.

Exhaust valve: Valve through which burned fuel charge passes on its way from cylinder to exhaust manifold.

Expansion tank: A tank, connected to cooling system, into which water can enter or leave as needed during coolant heating (expansion) or cooling (contraction).

Expansion valve: Air conditioning system component used to reduce pressure and meter flow of refrigerant into evaporator. Sometimes called a TVX (thermostatic expansion valve)

Extension housing: A housing attached to rear of a transmission. Encloses the output shaft and other transmission parts.

Extension jack: A lifting jack with a very long, adjustable stem making it suitable for supporting engines, transmissions/transaxles, and fuel tanks while vehicle is in air on a lift.

Eye bolt: A bolt with threads on one end and a round loop (eye) on other. Useful for engine removal from car.

Eye protection: Safety device to protect the eyes from flying particles.

F

Fabric ply: A layer of fabric making up one of plies (layers) involved in construction of a tire. May refer to either sidewall or to tread plies.

Face: That portion of intake or exhaust valve that contacts valve seat.

Face indented wear: A grooved wear area on valve face. Caused by wear and pounding action of valve face slamming against valve seat.

Face shield: A clear, protective shield that covers entire face for protection during grinding or other repair operations that cause metal or dirt to be thrown with great force.

Fahrenheit: Thermometer on which boiling point of water is 212° and freezing point is 32°.

Fan: A device designed to create a moving stream of air. Generally employed for cooling purposes.

Farad: Unit of capacitance; used to rate capacitors or condensers.

Fast charging: Charging a battery by passing a relatively heavy current through it. This will restore battery charge reasonably well in one to two hours.

Fast idle: Engine idle rpm when throttle is held partially open by a fast idle cam solenoid, or idle air control valve when the engine is cold.

Fault designators: The last two numbers of an OBD II trouble code. Identify the specific component or system causing the trouble code to set.

Feeder wire: A wire supplying current to several circuits—such as the feeder wire to a terminal block to which several other circuit wires connect.

Feeler gauge: Thin strip of brass or hardened steel, ground to an exact thickness, used to check clearances between parts.

FEPROM: Flash electronically erasable programmable read-only memory. Type of PROM that is updated from an external computer.

Fender cover: A protective pad that is laid over car fender area to prevent marring paint while leaning over to work on underhood areas.

Fender skirt: Plate designed to cover portion of rear fender wheel opening.

Ferromagnetic: That property of certain metals, such as steel, iron, and nickel, that cause it to form molecular alignment within magnetic domains when material is magnetized. A material that is magnetic, capable of being magnetized, or that responds to a magnetic field.

Ferrous metal: Metal containing iron or steel.

F-head engine: Engine having one valve in the head and the other in the block.

Fiber gear: A gear constructed of resin impregnated fibers that are compressed and hardened. Generally provide quiet operation.

Fiber optic: A path for electricity or data transmission in which light acts as the carrier.

Fiberglass: Mixture of glass fibers and resin that when cured (hardened) produces a very light and strong material.

Field: Area covered or filled with a magnetic force.

Field coil: Insulated wire wrapped around an iron or steel core. When current flows through wire, strong magnetic force field is built up.

Filament: Fine wire inside lightbulb that heats to incandescence when current passes through it. The filament produces the light.

Filler metal: Molten metal added to base metal during welding or brazing process.

Fillet: Rounding joint between two parts connected at an angle.

Fillet radius: Degree of curve (radius) involved in a curved edge or corner such as used on crankshaft journals where journal end meets crankshaft sides. Fillet prevents corner cracking better than a sharp right angle corner. Indicates size of fillet.

Fillet weld: A weld in which filler metal is placed in corner where two metal pieces join at angles to each other.

Filter: Device designed to remove foreign substances. Air, oil, gasoline, and transmission filters are always installed on vehicles.

Final drive ratio: Overall gear reduction (includes transmission and rear axle ratio) at rear wheels.

Finishing stone: Fine stone used for final finishing during honing.

Fire wall: Metal partition between driver's compartment and engine compartment. May be called a bulkhead.

Firing order: Order in which cylinders must be fired.

Fit: Contact area between two parts.

Fixed caliper: A disc brake design in which caliper body is rigidly attached and thus is unable to move in any direction. It cannot, as is done in a sliding caliper, center itself over disc or rotor.

Flange: A flared, collar-like section formed on a shaft or pipe. Often located on one end to facilitate fastening sections together or to provide a base for attaching some other part.

Flanged end axle: A drive axle with outer end formed into a flange to mount wheel.

Flange height: A distance determined by measuring from bottom of wheel bead to top of flange. Flange retains tire bead.

Flare angle: The angle formed when end of tubing is spread (flared) open for use in a flare fitting.

Flare connection: A tubing connection that both secures and seals by gripping flared end of tubing between two similar angled areas formed in fitting body and nut.

Flaring cone: A tapered cone that is forced into open end of a piece of tubing to produce a flared end suitable for use in a flare fitting.

Flaring tool: Tool used to form flare connections on tubing.

Flashback: A condition in which oxygen has entered acetylene hose or vice versa and due to failure to correctly purge both lines, a combustible mixture exists in hoses. Upon lighting torch, flame, instead of burning just at tip, moves inside and a fire starts burning in hoses.

Flashover: A condition in which electrical current, instead of moving through an intended conductor, jumps across an open space or moves along surface of an insulator, such as a coil tower or distributor cap, either directly to ground or to some other circuit.

Flash point: The point in the temperature range at which a given fuel or lubricant will ignite and flash into flame.

Flash programming: Method of programming the ECM by erasing old information and entering new programming.

Flat head: Engine with all the valves in block.

Flat rate: Payment system in which the employee is paid by the amount of work performed instead of an hourly rate.

Flat spot: Refers to a spot experienced during an acceleration period where the engine seems to "fall on its face" for a second or so and will then begin to pull again.

Flex test: Method of testing wire and connector condition by bending and wiggling wires and connectors.

Flexible drive plate: A thin, bendable drive plate attached both to crankshaft and transmission torque converter. Drives torque converter. Many drive plates contain the ring gear.

Flint: Stone-like material that is used in a spark lighter for welding torches. The flint, when scratched across a rough, hardened metal surface, gives off a heavy shower of sparks.

Float: Unit in carburetor bowl that floats on top of fuel. It controls inlet needle valve to produce proper fuel level in bowl.

Float bowl: The part of the carburetor that acts as a reservoir for gasoline and in which the float is placed.

Float drop: A measurement of distance a carburetor fuel bowl float moves from full up to full down positions with no fuel in bowl.

Floating caliper: A disc brake in which caliper assembly is mounted on pins (rods) to allow some lateral (side) movement. This permits caliper to automatically maintain a centralized position over disc or rotor.

Floating insert: A precision insert friction bearing that has bearing material on both inner and outer surfaces. Bearing slides against both journal and bore surfaces. It is seldom used.

Flooding: Condition where fuel mixture is overly rich or an excessive amount has reached cylinders. Starting will be difficult and sometimes impossible until condition is corrected.

Flow meter: Sensing device that measures flow of air or liquid.

Fluid coupling: Unit that transfers engine torque to transmission input shaft through use of two vaned units (called a torus) operating very close together in a bath of oil. Replaced by torque converter.

Fluid drive fan: A cooling fan that uses a liquid, such as silicone oil, to permit a controlled amount of slippage to limit maximum fan speed. Also controls fan speed, through thermostatic valve action, in relation to engine temperature.

Fluorescent penetrant: A crack detection solution that penetrates into even minute cracks. When surface is wiped clean, a developer will draw penetrant to surface where it will spread out wider than crack. When subjected to black light, penetrant glows, disclosing crack.

Flute: Groove in cutting tool that forms a passageway for exit of chips removed during the cutting process.

Flux (magnetic): Lines of magnetic force moving through magnetic field.

Flux (soldering, brazing): Ingredient placed on metal being soldered or brazed, to remove and prevent formation of surface oxidation.

Flux core wire solder: Wire solder with a hollow center that is filled with either rosin or acid flux.

Flywheel: Relatively large wheel that is attached to crankshaft to smooth out firing impulses. It provides inertia to keep crankshaft turning smoothly during periods when no power is being applied. It also forms a base for starter ring gear and in many instances, for clutch assembly. The lightweight flex plates that connect the crankshaft to the torque converter are usually called flywheels.

Flywheel ring gear: Gear on outer circumference of flywheel. Starter drive gear engages ring gear and cranks engine.

Flywheel runout: Measured amount of either side to side (lateral) movement or radial (up and down) movement present in a revolving flywheel.

Fog lights: Amber or clear lamps specially designed to provide better visibility in fog. Are usually mounted as close to the road as is feasible.

Follower ball joint: The ball joint that does not transfer the vehicle weight to the wheel. Its function is to help keep the wheel assembly in place.

Foot-pound: Measurement of work involved in lifting one pound one foot. Also, one pound of pull one foot from center of an object.

Force: Pull or push acting upon body that tends to change state of motion, or rest, of the body.

Force-fit: See *Drive-fit.*

Forcing an actuator: Method of diagnosing an actuator by bypassing the ECM and directly operating the actuator.

Forehand welding: An oxyacetylene welding technique in which torch tip is aimed slightly away from molten puddle and towards direction of welding.

Forge: To force piece of hot metal into desired shape by hammering.

Four bolt main: A crankshaft main bearing employing two fasteners on each side.

Four-stroke cycle engine: Engine requiring two complete revolutions of crankshaft to fire each piston once.

Four-wheel alignment: Process in which all the wheels on a vehicle are aligned with each other.

Four-wheel drive: Vehicle in which front and rear wheels are driven.

Four-wheel steering: System used to provide limited steering for the rear wheels. Operates in relation to the front wheels.

Fractional drill: A twist drill, size of which is indicated in fractions of an inch.

Frame: Portion of automobile upon which body rests and to which engine and springs are attached. Generally constructed of steel channels.

Frame: Strong, steel members run from front to rear of body.

Frame rails: Structural sections of the car frame. Often specifically used to refer to two outside longitudinal sections.

Free electrons: Electrons in outer orbits around nucleus of atom. They can be moved out of orbit fairly easily.

Free play: Amount of unimpeded movement between two parts.

Freewheel: Usually refers to action of car on downgrade when an overrunning clutch in an automatic transmission or transaxle is unlocked, or the manual clutch is disengaged. No engine braking occurs during freewheeling.

Freeze plug: See *Core hole plug.*

Freezing: When two parts that are rubbing together heat up and force lubricant out of area, they will gall and finally freeze or stick together.

Frequency: Rate of change in direction, oscillation, or cycles, in given time span.

Fretting: Corrosive-like surface damage to bearing face or to outer ring, outer surface or inner ring, inner surface. These are portions of bearing touching bore walls and shaft surface. There is little, or in some cases, no movement between these surfaces.

Friction: Resistance to movement between any two objects when placed in contact with each other. Friction is not constant but depends on type of surface, contact pressure, lubrication, and temperature.

Friction bearing: A bearing with no moving parts. The shaft that turns in the bearing rubs or slides against the bearing.

Frictional horsepower: Amount of horsepower lost to engine friction.

FTC: Federal Trade Commission.

Fuel: Combustible substance that is burned in an engine to impart motion to the engine pistons.

Fuel accumulator: Spring loaded diaphragm device that dampens fuel pressure pulsations, muffles noise, and helps maintain residual pressure with engine off.

Fuel bowl: Storage area in carburetor for extra fuel.

Fuel distributor: Device in a mechanical fuel injection system that meters fuel to injectors at correct rate of flow for engine conditions.

Fuel filter: A device that removes dirt, rust particles, and in some cases, water from fuel before it moves into carburetor or fuel injection system.

Fuel gauge: A device to indicate the approximate amount of fuel in tank.

Fuel injection: Fuel system that sprays fuel either directly into cylinders or into intake manifold just ahead of the intake valve. Central fuel injection systems spray the fuel into the air horn ahead of the throttle valve.

Fuel injector: Fuel injection system part that squirts (injects) measured amount of gasoline into intake manifold in vicinity of intake valve. In diesel engines, the fuel is injected directly into the combustion chamber.

Fuel level: Height of fuel in carburetor fuel bowl. Determined by float setting.

Fuel line: Tubing and hose that carry fuel from tank to carburetor or injection system.

Fuel mixture: Mixture of gasoline and air. The ideal mixture, by weight, would contain 14.7 parts of air to one part of gasoline.

Fuel pressure regulator: Fuel injection system device that controls pressure at the injectors. May be operated by spring pressure alone or a combination of fuel pressure and intake manifold vacuum.

Fuel pulsation: Fuel pressure variations due to fuel pump action.

Fuel pump: Vacuum device, operated either mechanically or electrically, used to draw gasoline from tank and force it into carburetor or fuel injectors.

Fuel pump pressure: Pressure, in pounds per square inch (PSI) or kilopascals (kPA), developed by a fuel pump.

Fuel pump vacuum: Amount of vacuum, in inches of mercury, developed by a fuel pump.

Fuel pump volume: Amount of fuel a pump will deliver in a specified period of time. Engine driven fuel pumps are tested at a specified rpm.

Fuel rail: In an electronic gasoline injection system, hollow pipe that is connected to and supplies fuel for injectors. Can also refer to a common feeder pipe for multi-carburetor installation.

Fuel tank: A large tank of steel or plastic, used to store a supply of fuel aboard vehicle.

Fuel trim: The action of the ECM in maintaining the proper or air-fuel ratio of the fuel injection system. Short-term fuel trim adjusts for brief changes in engine operation. Long-term fuel trim compensates for ongoing changing in operating conditions.

Fulcrum: Support on which a lever pivots in raising an object.

Full-floating axle: Rear drive axle that does not hold wheel on nor does it hold wheel in line or support any weight. It merely drives wheel. Used primarily on trucks.

Full-floating piston pin: A piston pin that is free to turn in both piston and rod. Snap rings at each end secure the pin.

Full-flow lubrication system: An engine lubrication system in which all engine oil from oil pump must first pass through a filter before reaching bearings. If filter clogs, a bypass permits oil to reach bearings.

Full-flow oil filter: Oil filter that filters all oil passing through engine.

Full pressure lubrication system: Type of oiling or lubrication system using an oil pump to draw oil out of a sump and force it through passages in engine.

Full rebound: In a suspension system, position of a spring or shock at precise time that frame (or body) is separated from axles (or control arms) the maximum distance.

Full-time four-wheel drive: Setup in which all four wheels are driven all the time off road or on. Addition of a third differential, located at transfer case, permits front and rear wheels to operate at different speeds.

Full-time transfer case: Four-wheel drive transfer case that drives all four wheels all the time. Two-wheel drive is not possible. Such systems permit four-wheel drive on dry, hard surfaced roads by incorporating a differential in transfer case unit.

Fuse: Protective device that will break flow of current by melting when current draw exceeds capacity of fuse.

Fuse block: A central block or area for various circuit fuses. May also contain relays and circuit breakers.

Fusible link: A special wire inserted into a circuit to provide protection in event of overloading or short circuits. Overloads will melt wire and break circuit. Unlike a regular fuse, fusible link will permit overloading for a short time before melting.

Fusion: Two metals reaching the melting point and flowing or welding themselves together.

G

Galled: Bearing surface damage caused by overheating, lack of lubrication, or improper lubrication. Metal surface in numerous high spots literally melts from heat and attempts to weld together, thus pulling out and smearing small chunks of metal. Advanced galling will cause bearing seizure.

Galvanometer: Instrument used to measure pressure, amount of, and direction of an electric current.

Gap bridging: A buildup of deposits on spark plug electrodes. Buildup can advance to completely fill (bridge) the gap and plug becomes inoperative.

Gas: A nonsolid material such as air. It can be compressed. When heated, it will expand and when cooled, it will contract.

Gaseous shield: Shield of inert gases produced when arc welding with coated electrodes, T.I.G., or M.I.G. welding setup. Gases shield molten metal being welded from air, thus preventing oxidization of weld metal.

Gasket: Material placed between two parts to ensure proper sealing.

Gasket shellac: A shellac sealing compound placed on certain gasket installations to assist in sealing and often, as an aid in positioning gasket during part assembly.

Gasohol: Automotive engine fuel made up of around 90% gasoline and 10% ethanol alcohol.

Gasoline: Hydrocarbon fuel used in the internal combustion engine.

Gassing: Small hydrogen bubbles rising to top of battery electrolyte during battery charging.

Gear: Circular object, usually flat edged or cone-shaped, upon which a series of teeth have been cut. These are meshed with teeth of another gear and when one turns, it also drives the other.

Gear backlash: Amount of measured movement between mating gear teeth when one gear is held and other is moved to limit of travel, first one way, then the other.

Gear pump: A pump, utilizing two meshed gears, that draw in and force out oil. Capable of building high pressure.

Gear ratio: Relationship between number of turns made by driving gear to complete one full turn of driven gear. If driving gear turns four times to turn driven gear once, gear ratio would be 4 to 1.

Gear runout: A measured amount of either gear lateral movement (side to side wobble) or radial (up and down) movement that exists when a gear is revolved.

Generator: Electromagnetic device for producing electricity.

G-force sensor: Component of anti-lock brake system that senses rapid acceleration or deceleration.

Glass: Term used for the material Fiberglass.

Glass pack muffler: Straight through (no baffles) muffler utilizing fiberglass packing around perforated pipe to deaden exhaust sound.

Glaze: Highly smooth, glassy finish on cylinder walls. This glaze must be removed by honing to ensure proper seating of new piston rings.

Glaze breaker or deglazer: Abrasive tool used to remove glaze from cylinder walls before installation of new piston rings.

Glow plug: A heating device placed in a diesel engine precombustion chamber to facilitate cold engine starting. When engine is cold, an electric current is passed through plug causing it to glow red hot. This helps ignite compressed fuel. Glow plugs remain on until the engine is warm enough to run on its own.

GMC: General Motors Corporation.

Governor: Device designed to automatically control speed or position of some part. Also used as a pressure control in automatic transmissions and transaxles.

GPM: Gallons Per Minute.

Gradient: Angle of hill. A 20% gradient would be a hill that would rise two feet for every ten forward feet of travel. This is determined by rise as a percentage of forward travel.

Grid: Lead screen or plate to which battery plate active material is affixed.

Grind: Remove metal from object by means of revolving abrasive wheel, disc, or belt.

Grommet: A rubber or plastic doughnut shaped object, with a slot cut around outer edge. Used to snap into holes in sheet metal to provide protection for wires or tubing passing through hole.

Gross axle weight rating (GAWR): Load carrying capacity of a given front or rear axle. Weight rating can be expressed as rating at springs (total load on springs) or at ground (total load measured where tire meets ground). Weight at ground rating includes weight of tires, wheels, axle, and springs.

Gross Horsepower: Maximum horsepower developed by engine without the drag from the fan, alternator, coolant pump, or exhaust system. Usually the air cleaner is removed and the timing is adjusted manually.

Ground: Terminal of battery connected to metal framework of car. Most modern vehicles ground the negative terminal.

Ground clamp: Clamp on end of ground wire. Must be attached to work to complete a circuit when arc welding.

Growler: Electrical device for testing starter armatures.

Gum: Oxidized portions of fuel that form deposits in fuel system or engine parts.

H

Hacksaw: Hand operated metal cutting saw. Can use various blade lengths and tooth configurations.

Half-hard copper tubing: Copper tubing that has not been fully annealed (softened).

Half-moon key: Driving key serving same purpose as regular key but it is shaped like a half circle.

Halide torch: A gas torch used to detect refrigerant leaks in air conditioning systems. Has been replaced by electronic leak detectors.

Hall effect: The effect on current flowing through a coil when a metal object is placed between the north and south poles.

Hall-effect pickup coil: A pickup coil that creates a magnetic field whenever current passes through it. A rotating shutter passing between north and south poles of the coil influences the magnetic field. This causes a voltage fluctuation which the ignition module or ECM uses to trigger the coil.

Halogen bulb: Lightbulb in which tungsten filament is surrounded by a halogen gas. Bulb glass is quartz to withstand intense heat.

Hand brake: Hand operated brake that prevents vehicle movement while parked by applying rear wheel brakes or transmission brake.

Hand drill: A drill, either manual or electric, that is held in hands to operate.

Hard codes: Diagnostic trouble codes that indicate an ongoing defect. After being cleared will reset within a few minutes of engine operation.

Hardening sealant: A sealant material that upon full curing, becomes hard.

Harmonic balancer: See *Vibration damper.*

HC: Symbol for hydrocarbon, consisting of chemical combinations of hydrogen and carbon.

Headlights: Main driving lights used on front of vehicle.

Head marking: Marks placed on a bolt or capscrew to indicate fastener grade (material strength).

Head mating surface: That portion of engine cylinder block that mates (contacts) with cylinder head.

Head pressure: See *Discharge pressure.*

Heat baffle: A shield placed between a component such as the starter motor, a wiring harness, or a fuel line, to protect it from a heat source such as an exhaust manifold.

Heat checking: Small surface cracks in brake drums or brake discs (rotors) caused from heat generated during braking.

Heat control valve: A valve in exhaust manifold that directs a portion of hot exhaust gases to a selected area of intake manifold to provide heat for improved cold engine operation. Controlled by vacuum motor or thermostatic spring. Also called heat riser or early fuel evaporation (EFE) valve.

Heat crossover: Passage from one exhaust manifold up, over, and under carburetor and on to other manifold. Crossover provides heat to fuel-air intake system during engine warmup.

Heat discoloration: Metal surface coloring that is caused by application of heat, either from friction or from an outside heat source.

Heat engine: Engine operated by heat energy released from burning fuel.

Heater core: Finned device for heating vehicle interiors. Resembles a small radiator. Hot coolant passes through the heater core. Air is forced through the fins of the heater core and picks up heat from the coolant.

Heat exchanger: Device, such as radiator, either used to cool or heat by transferring heat from one object to another.

Heat range: Refers to operating temperature of a spark plug. Plugs are made to operate at different temperatures depending upon thickness and length of porcelain insulator as measured from sealing ring down to tip.

Heat riser: Area, surrounding portion of the intake manifold, through which exhaust gases can pass to heat fuel mixture during warmup.

Heat-shrink tubing: Plastic tube used to insulate electrical solder joints. Heating the plastic causes it to shrink around the connection.

Heat sink: Device used to prevent overheating of electrical device by absorbing heat and transferring it to atmosphere.

Heat stove: Sheet metal housing around a portion of exhaust manifold. An intake pipe from housing provides hot air to air cleaner when needed. Can also mean a small shrouded depression in exhaust manifold from which hot air may be drawn to automatic choke housing.

Heat transfer: Movement of heat from one object to another by convection, conduction, or radiation. Can also refer to movement of heat through a given object by conduction.

Heat treating: Application of controlled heat to metal object to alter its hardness.

Heavy repair: Generally refers to shop work involving complete dismantling, rebuilding, or replacement of major items such as the engine or transmission.

Heel (brake): End of brake shoe that rests against anchor pin.

Heel (gear tooth): Wide end of tapered gear tooth such as found in differential gears.

Height control sensor: Part of an automatic level control system that senses the height of the vehicle by measuring the distance between the frame and rear axle.

Height control valve: Pneumatic valve that controls the amount of compressed air in the air shock absorbers of the automatic level control system.

Helical: Spiraling shape such as that made by a coil spring.

Helical gear: Gear that has teeth cut at an angle to centerline of gear.

Heli-coil: A coil spring insert used to create new threads with the same diameter and thread as original threads in a hole with stripped threads.

Helper: Technician who performs basic service and maintenance.

Hemi: Engine using hemispherical-shaped (half of globe) combustion chambers.

Hemispherical combustion chamber: A round, dome-shaped combustion chamber that is considered by many to be one of the finest shapes ever developed. Hemispherical-shape lends itself to use of large valves for improved breathing and suffers less heat loss than other shapes.

Hg: Abbreviation for the word mercury. Vacuum is measured in inches of mercury.

High compression heads: Cylinder head with smaller combustion chamber area thereby raising compression. Head can be custom built or can be a stock head milled (cut) down.

High flash point: A liquid such as cleaning solution that must be heated to fairly high temperatures before vapors will spontaneously ignite.

High-pressure cutoff switch: Air conditioning system switch that cuts off the refrigeration system pressure if the pressure becomes so high that the compressor could be damaged or refrigerant lost.

High-pressure lubricant: A lubricant containing special additives that provide great resistance to film rupture under heavy contact pressure, such as found in some traction type differentials.

High-pressure spray cleaning: Cleaning process employing a solution under heavy pressure, discharged at high velocity from a spray nozzle.

High reach jack: A lifting jack in which the lift arm can move over a very long range making it capable of raising objects to a considerable height.

High-rise manifold: Intake manifold designed to mount carburetor or carburetors, considerably higher above engine than is done in standard manifold. Done to improve angle at which fuel is delivered.

High side: The portion of an air conditioning system between compressor and expansion valve that operates under high pressure.

High tension: High voltage from ignition coil. May also indicate secondary wire from the coil to distributor and wires from distributor to plugs.

Hoist: See *Chain hoist.*

Hold-down clip: A metal or plastic device (clip) used to secure wires or tubing to prevent movement.

Holding jig: A device for holding an object while performing various operations such as grinding, bending, or machining.

Holding member: Devices that hold certain parts of the planetary gearsets of automatic transmissions to obtain various gear ratios. Can be hydraulic or mechanical.

Hone: To remove metal with fine grit abrasive stone to precise tolerances.

Honed to size: Bringing a bore to final finished size by honing (removal of a small amount of metal with an abrasive material such as a stone).

Honing machine: Machine used to perform honing.

Honing oil: Special oil used in honing process. Flushes stones to assist in cutting and reduces heat buildup.

Hook rule: Small, short flat steel rule with a hook attached to scale starting end.

Hooke's law: Law stating that amount of distortion (bending, twisting, lengthening, compressing) caused in a solid, as long as it is kept within elastic limits of material, will be directly proportional to applied force.

Horizontal-opposed engine: Engine possessing two banks of cylinders that are placed flat or 180 deg. apart. Sometimes called a pancake engine.

Horsepower: Measurement of engine's ability to perform work. One horsepower is defined as ability to lift 33,000 pounds one foot in one minute. To find horsepower, total rate of work in foot pounds

accomplished is divided by 33,000. If a machine was lifting 100 pounds 660 feet per minute, its total rate of work would be 66,000 foot pounds.
Dividing this by 33,000 foot pounds (1 horsepower) you find the machine is rated as 2 horsepower (hp).

Horsepower—weight factor: Relationship between total weight of car and horsepower available. By dividing weight by horsepower, number of pounds to be moved by one horsepower is determined. This factor has a great effect on acceleration, gas mileage, and all around performance.

Hose clamps: Devices used to secure hoses to their fittings.

Hoses: Flexible rubber tubes for carrying water, oil, air, and other fluids.

Hot idle: Engine idle speed during normal operating temperature range. It is slower than the idle speed when the engine is cold.

Hot patch: Application of a sealing patch to a tire or tube through use of heat and pressure.

Hot spot: Localized area in which temperature is considerably higher than surrounding area.

Hot tank: Tank filled with a hot, caustic solution (often alkaline based). Used for heavy duty cleaning of radiators and engine parts.

Hot wire: A current-carrying wire, usually a positive wire.

Hotchkiss drive: Method of connecting transmission output shaft to differential pinion by using open drive shafts. Driving force of rear wheels is transmitted to frame through rear springs or through link arms connecting rear axle housing to frame.

Housing spreader: A tool used to spread differential housing a small amount to permit removal or installation of differential case and bearing assembly.

Hub: Unit to which wheel is bolted.

Hunting gearset: A ring and pinion gearset in which the same teeth on pinion and ring mesh every so many revolutions. Caused by even gear ratios such as 3 to 1 or 2.5 to 1. If the ratio is uneven, for example 3.11 to 1 or 2.53 to 1, this would not occur and the gearset would then become "nonhunting."

Hydraulic: Refers to fluids in motion. Hydraulics is science of fluid in motion.

Hydraulic actuator: Any device that converts hydraulic pressure to another form of energy.

Hydraulic brakes: Brakes operated by hydraulic pressure. Master cylinder provides operating pressure transmitted via steel tubing to wheel cylinders or pistons that apply brake shoes to brake drums and/or discs.

Hydraulic floor jack: A low hydraulic jack, equipped with wheels for easy moving, used for lifting front, side, or rear of car.

Hydraulic hand jack: A compact, powerful, portable hydraulic jack consisting of base, reservoir, and short ram.

Hydraulic lifter: Valve lifter that utilizes hydraulic pressure from engine's oiling system to keep it in constant contact with both camshaft and valve stem. They automatically adjust to any variation in valve stem length.

Hydraulic press: A press with a large, suspended hydraulic cylinder and ram. Ram is actuated by pressure generated in a small cylinder and transmitted through a connecting pipe to large cylinder. Used for applying force to remove or install press fit parts.

Hydraulic pump: Device for transferring and pressurizing liquids. The average vehicle has many hydraulic pumps.

Hydraulic puller: A pulling device utilizing a hydraulic cylinder to generate force required.

Hydraulic ram: A round steel shaft, one end attached to a piston operating in a sealed cylinder. Other end passes out of cylinder and is placed against workpiece. When hydraulic pressure is applied to piston, ram exerts pressure on workpiece. Used in hydraulic jacks, presses, and lifts.

Hydraulics: The science of liquid in motion.

Hydro-boost: Name for one type of hydraulically operated (steering pump pressure) brake booster.

Hydrocarbons (HC): Combination of hydrogen and carbon atoms. All petroleum based fuels (gasoline, diesel fuel, kerosene) consist of hydrocarbons.

Hydrometer: Float device for determining specific gravity of electrolyte in a battery. This will determine the state of charge.

Hygroscopic: Having the ability to absorb moisture from air.

Hypoid gearing: System of gearing wherein pinion gear meshes with ring gear below centerline of ring gear. This allows a lower drive line, reducing hump in the floor of car. Hypoid gearing is used in the differential of most vehicles.

I

ICEI: Internal Combustion Engine Institute.

Icing: Formation of ice on throttle plate or valve. Vacuum at the throttle plate causes freezing point to lower and ice forms around the throttle plate. Common when weather conditions are cold and humid.

ID: Inside diameter.

Idle: Indicates engine operating at its normal slow speed with throttle closed.

Idle air control valve: Device that controls idle speed by bypassing air around the throttle plate. Operated by the ECM.

Idle mixture: Air-fuel mixture delivered to engine during idle.

Idler arm: Steering system part that supports one end of center link on parallelogram-type steering systems.

Idler gear: A gear, between two other gears, that is driven by one and drives other. This permits both driving and driven gear to rotate in same direction.

Idler sprocket: An additional sprocket, generally used on long runs of timing chain, that takes up excess chain slack to prevent harmful chain flap or distortion.

Idle screw: Adjusting screw used to control amount of and/or ratio of fuel and air delivered to engine during idle.

Idle speed screw: A screw used to adjust throttle plate position to secure desired idle rpm.

Idle valve or idle needle: See *Idle screw.*

Ignition: Lighting or igniting fuel charge by means of a spark (gas engine) or my heat of compression (diesel engine).

Ignition switch: Key operated switch in driver compartment (usually on steering column) for connecting and disconnecting power to ignition and electrical system.

Ignition system: Portion of car electrical system, designed to produce a spark within cylinders to ignite fuel charge.

Ignition timing: Refers to relationship between exact time a plug is fired and position of piston in degrees of crankshaft rotation.

I-head engine: Engine having both valves in the head.

IHP: Indicated Horsepower.

IMEP: Indicated Mean Effective Pressure.

IMI: Ignition Manufacturer's Institute.

Impact sensors: An open switch used in air bag systems. The switch is designed to close when vehicle is involved in a severe frontal impact.

Impact wrench: An air or electrical driven wrench that tightens or loosens fasteners with a series of sharp, rapid blows.

Impeller: Wheel-like device upon which fins are attached. It is turned to pump coolant in the cooling system, and to move and slightly compress air.

Inch-pounds: English unit of measure indicating amount of twisting or turning force being applied to a fastener or shaft. Equivalent to 12 inch-ounces.

Included angle: Angle formed by centerlines drawn through steering axis and camber angle as viewed from front of car. Combines both steering axis and camber angles.

Independent suspension: A suspension system that allows each wheel to move up and down without undue influence on other wheels.

Indirect programming: Method of downloading information to the ECM by first downloading programming information from a separate programming computer to a scan tool. Once the programming information has been downloaded, the scan tool is disconnected from the programming computer, taken to the vehicle, and connected to the data link connector. Programming information is then downloaded from the scan tool to the ECM through the DLC.

Indicated horsepower (IHP): Measure of power developed by burning fuel within cylinders.

Induction: Imparting of electricity from one object to another by magnetic fields.

Inertia: Force that tends to keep stationary object from being moved, and tends to keep moving objects in motion.

Inertia switch: An electrical switch designed to be operated by a sudden movement, such as that caused by a collision.

Inflator module: Air bag system component that houses the inflatable bag, initiator, and gas generating material.

Inhibited cleaning solution: A caustic cleaning solution that has been weakened to prevent surface erosion of aluminum parts while soaking.

Inhibitor: Substance added to oil, water, gas, or another liquid to prevent action such as foaming, or rusting.

Injector balance test: Test made to locate an injector that is allowing too much or too little fuel to enter the cylinder. Performed by cutting off the electrical power to each fuel injector in turn.

Injector pump: A mechanical pump that forces diesel fuel, under high pressure, to and through injector nozzles. Pump provides proper injection timing.

Injector timing: Relationship between instant of fuel injection in any one cylinder, to position of piston in degrees of crankshaft rotation.

Inline engine: Engine in which all cylinders are arranged in straight row.

Inline fuse: A fuse placed directly in an individual circuit wire. Installed as close to electrical source as possible.

Inline injector pump: An injector pump utilizing a number of plungers operating in cylinders. Plunger-cylinder assemblies (one per engine cylinder) are positioned in a straight line and are operated by a single camshaft.

Inner flame cone: Small cone of flame visible at the tip opening of an oxyacetylene torch.

Input shaft: Shaft delivering power into mechanism. Shaft from clutch into transmission is transmission input shaft.

Insert bearing: Removable, precision made bearing, which ensures specified clearance between bearing and shaft.

Insert recess cutter: A cutting tool used to counterbore head to install a valve seat insert. Removes integral seat and leaves a precise recess to accept insert.

Insert removal plug: A tool used to grasp underside of a valve seat insert so that it may be pulled out.

Inside caliper: A precision measuring tool designed to measure inside diameter of cylinders and other bores.

Installed height: The height of the valve and spring when reassembled.

Installer: Technician who removes and installs parts.

Insulation: Material used to reduce transfer of noise (sound insulation), heat (heat insulation), electricity (electrical insulation).

Insulator: Material that will not (readily) conduct electricity.

Intake manifold: Connecting tubes between throttle valve and port openings to intake valves.

Intake stroke: Portion of piston's movement devoted to drawing fuel mixture into engine cylinder.

Intake valve: Valve through which fuel mixture is admitted to cylinder.

Integral: Part of. (The cam lobe is an integral part of camshaft.)

Integral carrier: A drive axle differential assembly in which carrier (supports differential case) is constructed as an integral part of housing and as such, cannot be removed as a separate unit.

Integral guide: A valve guide machined right in head metal itself.

Integral seat: Valve seat cut in head metal itself with no insert.

Interference angle: A valve face-to-valve seat contact in which either valve face or valve seat is ground to a slightly different angle (around 1°) than other. Aids in quick seating.

Intermediate bearing: A bearing on a multiple bearing shaft, such as a camshaft or crankshaft, located between two end bearings.

Intermediate gear: Any gear in auto transmission between 1st and high.

Intermittent: Not constant but occurring at intervals.

Intermittent codes: Computer diagnostic code that does not return immediately after it has been cleared.

Internal combustion engine: Engine that burns fuel within itself as means of developing power.

Internal gear: A gear with teeth cut on an inward facing surface. Example: outer gear is a planetary gearset. Teeth face inward towards center.

Inverted flare: A tubing 45° angle flare fitting in which nut (male threads) threads into body (female).

Ion: Electrically charged atom or molecule produced by electrical field.

Ionize: To convert wholly or partly into ions. This causes air to become a conductor of electricity.

I.S.O. flare: A tubing flare that looks somewhat like a regular double-lap flare following initial belling operation. Belled end section is compressed between nut and body producing a secure seal.

J

Jab saw: A special hacksaw used for cutting in restricted areas. Consists of a handle and rigid section to secure a short section of hacksaw blade.

Jackshaft: A shaft used between two other shafts.

Jack stand: A fixed or adjustable height, metal stand placed under a raised vehicle. Prevents car from falling in event of jack failure or slippage.

Jam nut: Two nuts securely tightened against each other on a stud. A wrench can then be placed on lower nut to remove stud or top nut to install. Can also be used to prevent a nut from loosening.

Jet: Small hole or orifice used to control flow of gasoline or air in a carburetor.

Joule: The equivalent of 0.737324 foot-pounds.

Journal: Part of shaft prepared to accept a bearing. (Connecting rod, main bearing.)

Journal polishing: Removing minute sharp edges following grinding of a bearing journal. Leaves extremely smooth finish (around 7 micro inches).

Journal taper: Condition in which one end of a bearing journal is smaller in diameter than other. Difference in two readings indicates amount of taper.

Jump starting: Starting a vehicle with a discharged battery by using jumper battery or vehicle with a charged battery. Batteries are connected with jumper cables. Connect in parallel. Follow safety rules!

Jumper cables: Two large diameter (8 gage or heavier) insulated cables with clamps. Used to temporarily connect a charged battery to a discharged one to provide extra cranking capacity. Use caution!

Jumper wire: A length of automotive wire, used for bypassing a defective electrical unit to test the condition of that unit.

Junction block: A connection point (block) for a number of wires. Can also contain receptacles for fuses. Generally incorporates a nonconducting base for mounting.

K

Keensert: A hollow plug with internal and external threads. Used to bring a threaded, stripped hole back to a usable condition.

Kerf: Name commonly given to the cut (slot) made with an oxyacetylene cutting torch.

Kerosene: Flammable hydrocarbon oil produced from distilling petroleum.

Key: Parallel-sided piece inserted into groove cut part way into each of two parts, which prevents slippage between two parts.

Keyway: Slot cut in shaft or hub. Square key is placed in slot and engages a similar keyway in mating piece. Key prevents slippage between two parts.

Kickdown switch: electrical switch that will cause transmission, or overdrive unit, to shift down to lower gear. Often used to secure fast acceleration.

Kill switch: Special switch designed to shut off ignition in case of emergency.

Kilometer: Metric measurement equivalent to 5/8 of mile.

Kilopascals (kPa): Metric equivalent to English psi (pounds per square inch).

Kingpin: Hardened steel pin that is passed through the steering knuckle and axle end. The steering knuckle pivots about the kingpin.

Kingpin inclination: See *Steering axis inclination.*

Knife file: A file that has a blade that tapers to a sharp edge on one side. End view resembles an elongated triangle shape.

Knocking (bearing): Noise created by part movement in a loose or worn bearing.

Knocking (fuel): Condition, accompanied by audible noise, occurring when gasoline in cylinders burns too quickly. Also referred to as detonation.

Knuckle: A part utilizing a hinge pin (kingpin, swivel pin, ball joint) that allows one part to swivel around another part. An example is a steering knuckle.

Knurl: To roughen surface of piece of metal by pressing series of cross-hatched lines into the surface and thereby raising area between these lines.

L

Lacquer: Fast drying automotive body paint.

Laminated: Something made up of many layers.

Land: Metal separating a series of grooves.

Lap or lapping: To fit two surfaces together by coating them with abrasive and then rubbing them together.

Latent heat: Amount of heat beyond boiling or melting point, required to change liquid to a gas, or a solid to a liquid. Latent heat of evaporation is the amount of heat that must be added to change a liquid to a vapor without elevating vapor temperature above that of the liquid. Latent heat of condensation is the amount of heat that must be removed to change a vapor to a liquid without reducing liquid temperature above that of the vapor. Latent heat of melting is the amount of heat that must be added to change a solid to a liquid without elevating liquid temperature above that of the solid. Latent heat of freezing (or solidification) is the amount of heat that must be removed to change a liquid to a solid without reducing solid temperature above that of the liquid.

Lateral accelerometer: Component of traction control system that senses rapid sideways movement.

Lateral movement: Side to side movement (wobble or runout) such as may be exhibited by side surface of a revolving object. Movement is in a direction nearly parallel to centerline of the turning part that the object is attached to.

Lead hammer: A striking tool using a head made of lead to prevent marring object being struck.

Leaded gasoline: Gasoline containing tetraethyl lead, an antiknock additive. It cannot be used in vehicles with catalytic converters, and has been almost completely replaced by unleaded gasoline.

Leaf spring: Suspension spring made up of several pieces of flat spring steel. Varying numbers of leaves (individual pieces) are used depending on intended use. One car uses single leaf in each rear spring.

Lean mixture: Mixture with an excessive amount of air in relation to fuel.

Leakdown rate: Rate at which a gas or fluid, under pressure, escapes from container. In an engine hydraulic valve lifter, how long it takes to collapse lifter a specified distance under a specified pressure.

Left-hand thread: A thread that is cut so that fastener must be turned counterclockwise to tighten, clockwise to loosen.

Lever: A rigid bar or shaft pivoting about a fixed fulcrum. It is used to increase force or to transmit or change motion.

Lever bender: A mechanical, hand-operated tubing bender using a lever arm to form a smooth bend in tubing.

Leverage: Increasing force by utilizing one or more levers.

L-head engine: Engine having both valves in block and on same side of cylinder.

Lift saddle: That portion of a lift designed to engage vehicle frame, axle, or body projection for lifting.

Lift strap: A strap specifically designed for engine removal. Has an adjustable, non-slip pull point attachment.

Light repair: Automotive repair work in which parts are relatively small and easily portable.

Lightened valves: Valves in which all possible metal has been ground away to reduce weight. This will allow higher rpm without valve float.

Limited-slip differential: Differential unit designed to provide superior traction by transferring driving torque, when one wheel is spinning, to wheel that is not slipping.

Line boring: Boring a series of holes that all are accurately aligned with a common centerline.

Linkage: Movable bars or links connecting one unit to another.

Liquid charging: Charging air conditioning system with liquid refrigerant instead of refrigerant vapor.

Liquid line: High-pressure liquid refrigerant line between receiver-dehydrator and expansion valve.

Liquid traction: Special liquid applied to tires of drag racers to provide superior traction.

Liquid-vapor separator: Tank with internal baffles to prevent liquid fuel from entering vapor line.

Liquid withdrawal: Drawing LPG from bottom of tank to ensure delivery of liquid LPG. Withdrawal from top of tank will deliver LPG in the gaseous state.

Liter: Metric measurement of capacity—equivalent to 2.11 pints. Five liters equals 1.32 gallon.

Live axle: Axle upon which wheels are firmly affixed. Axle drives the wheels.

Live wire: See *Hot wire.*

Load-carrying ball joint: The ball joint that transfers the vehicle weight from the vehicle frame to wheel spindle.

Lobe: Projection (hump-like portion) of a cam, such as on engine camshaft, that extends beyond base circle.

Locating lug: A small projecting tab on a precision insert bearing that holds it in place to prevent either insert rotation or end movement.

Locking hub: Device installed on the front wheels of a four-wheel drive vehicle. Turning the locking hubs places the vehicle in two- or four-wheel drive.

Locking plate: A metal plate, one section of which is affixed to part and other to a nut, bolt or capscrew, to prevent fastener loosening.

Lock up torque converter: A torque converter equipped with a hydraulically operated clutch that, when applied, locks impeller (pump) and turbine together, thus eliminating fluid slippage.

Lockwasher: A washer, either split or with locking tabs, placed between a fastener and part. When fastener is tightened, washer cuts into both fastener and part and prevents loosening.

Long and short arm suspension: Suspension system utilizing upper and lower control arm. Upper arm is shorter than lower. This is done so as to allow wheel to deflect in a vertical direction with a minimum change in camber.

Longitudinal leaf spring: Leaf spring mounted parallel to length of car.

Long block: Engine assembly with heads and oil pan installed.

Long-life antifreeze: Antifreeze that can be left in the vehicle cooling system for as long as 50,000 miles (80,000 km). Identified by its orange color.

Louver: Ventilation slots such as sometimes found in hood of automobile.

Low brake pedal: Condition where brake pedal approaches too close to floorboard before actuating the brakes.

Low flash point: That property of a liquid that will allow vapors given off during heating to ignite spontaneously at a relatively low liquid temperature.

Low lead fuel: Gasoline containing not much more than 0.5 grams of tetraethyl lead per gallon.

Low pivot swing axle: Rear axle setup that attaches differential housing to frame via a pivot mount. Conventional type of housing and axle extend from differential to one wheel. The other side of differential is connected to other driving wheel by a housing and axle that is pivoted at a point in line with differential to frame pivot point.

Low-pressure adapter: An adapter placed on a high-pressure lube gun that reduces pressure to prevent universal joint seal and bearing retainer damage. Used on other low-pressure applications also.

Low-pressure cutoff switch: Switch that cuts off the refrigeration system pressure if the pressure becomes so low that the compressor could be damaged.

Low-pressure line: Refrigerant line between evaporator outlet and compressor, normally under low pressure when the system is operating.

Low-pressure spray cleaning: Part cleaning through use of an air stream that, passing through nozzle, draws in cleaning solution. Solution then leaves nozzle in form of a spray.

Low side: That portion of an air conditioning system that is subjected to low-pressure refrigerant. Extends from expansion valve to suction side of compressor.

LPG: Liquefied petroleum gas.

Lubricant: Any material, usually a petroleum product that is placed between two moving parts in an effort to reduce friction. Examples are engine oil, gear oil, transmission fluid, and chassis grease.

Lubrication: Reducing friction between parts by coating them with a lubricant.

Lubrication passage: A hole, channel, or groove designed to permit flow of lubricant.

Lug: To cause engine to labor by failing to shift to a lower gear when necessary.

Lumen: A measurement of light intensity.

M

Machine screw: A nontapered screw, using a standard screw thread. May have a round or hex head. The head may be formed to accept an box end, Allen, Phillips, Fluted, or Clutch-type wrench or driver.

MacPherson strut: Front end suspension system in which wheel assembly is attached to a long, telescopic strut that replaces the shock absorber. The front spring is usually installed around the strut. Strut permits wheels to pivot and move up and down.

MacPherson strut damper: McPherson strut component that absorbs spring oscillations in the same manner as a shock absorber on a conventional suspension.

MacPherson strut suspension: Type of suspension consisting of a lower control arm and a shock absorbing strut installed between the control arm and the vehicle body. The coil is usually installed around the strut.

Magnaflux: Special chemical process used to check parts for cracks.

Magnet: Piece of magnetized steel that will attract all ferrous material. Permanent magnet does not need electricity to function and will retain its magnetism over a period of years.

Magnetic clutch: Electromagnetic clutch that engages or disengages air conditioning compressor pulley.

Magnetic field: Area encompassed by magnetic lines of force surrounding either a permanent magnet or electromagnet.

Magnetic pickup coil: A coil that produces a small electrical signal when a metal object passes nearby.

Magnetic pulse distributor: An electronic distributor that makes and breaks coil primary circuit through means of a magnetic pickup coil that is triggered by an armature (reluctor, trigger wheel) tooth passing close by. Induced current (pulse) triggers a transistor thus breaking the circuit.

Maintenance-free battery: A battery so designed as to not need any additional water during its normal service life.

Maintenance tune-up: Tune-up that includes replacing the spark plugs and air, fuel, and emissions filters.

Maintenance indicator light (MIL): Instrument panel light that illuminates to indicate a malfunction in the computer control system. Used on both OBD I and OBD II systems.

Maintenance schedule: List of necessary maintenance services for a particular vehicle and the time intervals between each service.

Main bearing bore: Cylindrical hole in which precision insert crankshaft main bearings are placed.

Main bearing cap: Removable section (cap) that contains lower half of crankshaft main bearing insert.

Main bearings: Bearings supporting crankshaft in cylinder block.

Main bearing supports: Steel plate installed over main bearing caps to increase their strength for racing purposes.

Main discharge tube: Carburetor fuel passage from bowl to air horn.

Main journal gauge: A gauge designed to check crankshaft main bearing journals for size, taper, and out-of-roundness, while crank is still mounted in block.

Main leaf spring: Long leaf on which ends are turned to form an eye to receive shackle.

Major diameter: Diameter of a fastener threaded area, as determined by distance from top of one thread, at right angles through center of fastener, to top of a corresponding thread.

Mandrel: Round shaft used to mount a rotary cutting or grinding device.

Manifold: Pipe or number of pipes connecting series of holes or outlets to common opening. See *Exhaust manifold* and *Intake manifold*.

Manifold absolute pressure sensor (MAP): A sensing device that determines intake manifold pressure as a factor of engine speed and load changes. Sensor sends appropriate electrical signals to an electronic control module (ECM). ECM adjusts engine settings accordingly.

Manifold air temperature sensors (MAT): Computer sensor used to measure the temperature of the air coming into the intake manifold.

Manifold gauge set: A manifold assembly containing two pressure gauges (one high, one low), two gauge valves, and three outlet connections. Used to service (discharge, charge, evacuate) air conditioning system.

Manifold heat control valve: Valve placed in exhaust manifold, or in exhaust pipe, that deflects certain amount of hot gas around intake manifold to aid in warmup.

Manual choke: Carburetor choke operated by hand.

Manual control valve: Hand (linkage) operated valve that controls oil flow and transmission gear selection.

Mass airflow sensor (MAF): Sensor used to measure the amount of air entering the intake manifold. Also called an airflow meter. Sensor sends appropriate electrical signals to an electronic control module (ECM). ECM adjusts engine settings according to MAF input.

Master cylinder: Part of hydraulic brake system in which pressure is generated when driver presses on the brake pedal. Pedal movement is transferred to master cylinder pistons by a brake rod.

Material Safety Data Sheet (MSDS): Information on a chemical or material that must be provided by the material's manufacturer. Lists potential health risks and proper handling procedures.

Mating surface: A surface upon an object that is placed in contact with a surface on another object. Can also refer to a specific area being used as a contact surface.

Matter: Substance making up physical things occupying space, having weight, and perceptible to the senses.

Maximum wear: Generally refers to greatest amount of wear of a moving part that can be tolerated before part must either be rebuilt or replaced. Can also refer to a section of a bore or shaft that has greatest amount of wear.

Mean Effective Pressure (MEP): Pressure of burning fuel (average) on power stroke subtracted by average pressure on other three strokes. Pressure is in pounds per square inch or kilopascals.

Mechanical efficiency: Engine's rating as to how much potential horsepower is wasted through friction within moving parts of engine.

Mechanical fuel injection: A mechanically driven pump forces fuel into engine through injector nozzles.

Mechanical fuel pump: Engine mounted pump operated by eccentric on camshaft.

Mechanical lifter: A solid, nonhydraulic valve lifter. Does not self-adjust and requires periodic clearance adjustment.

Mechanical puller: A nonhydraulic pulling device utilizing either a screw thread, lever, or hammer blow as a means of developing pulling pressure.

Megohm: 1,000,000 ohms.

Melting point: Temperature point at which a material changes from a solid into a liquid.

MEMA: Motor and Equipment Manufacturers' Association.

Mesh: To engage teeth of one gear with those of another.

Metal fatigue: Crystallizing of metal due to vibration, twisting, or bending. Unit will eventually break. Bending a piece of wire back and forth to break it is a good example of metal fatigue.

Metal inert gas (MIG) welding: A procedure in which inert gas is used to shield weld metal from atmosphere.

Metal shot: Very small, metallic balls that are projected, by means of an air blast, against an object so as to remove rust or paint. May also be used for light surface peening to relieve surface tension.

Metering rod: Movable rod used to vary opening area through carburetor jet.

Metering valve: Valve in brake line that limits fluid pressure to disc brakes. This assures that the rear (drum) brakes apply at about the same time as front (disc) brakes.

Metric size: Units made to metric system measurement.

Metric system: A decimal system of measurement based on meter (length, area, volume), liter (capacity), and gram (weight and mass).

Mica: Insulation material used between commutator segments in starter motors.

Micro: When the word "micro" precedes measurement units, such as watt and ampere, it means one-millionth of that unit.

Micro inch: A linear measurement unit equivalent to one millionth of an inch. Often used in describing relative smoothness of a metal surface finish as measured in micro inches.

Microfarad: 1/1,000,000 farad.

Micrometer: Precision measuring tool that will give readings accurate to within fraction of one thousandth of an inch.

Micrometer depth gauge: A micrometer specially designed to measure hole depth.

Mike: Either refers to micrometer or to using micrometer to measure an object.

MIL: See *Maintenance indicator light.*

Mill: To remove metal through use of rotating toothed cutter.

Mill file: A general purpose single-cut, flat file available in different lengths and cuts.

Millimeter: Metric measurement equivalent to .039370 of an inch.

Milling machine: Machine that uses variety of rotating cutter wheels to cut splines, gears, keyways, and other slots in metal.

Mineral oil: Petroleum based oil used in R-12 systems. Should never be used in R-134a systems.

Miniature fuse: A very compact, plug-in–type fuse. Uses two prongs connected by a wire-like element designed to melt when circuit loading exceeds a specified amperage.

Minimum wear: Can be used to indicate a small amount of wear that leaves part in serviceable condition. Can also be used to indicate, for example, that portion of a shaft or bore showing least amount of dimensional change from wear.

Minor diameter: Diameter as determined by measuring from bottom of one thread "V", at right angles through the center of fastener, to bottom of "V" of the corresponding thread.

Misfire: Fuel charge in one or more engine cylinders fails to fire or ignite at proper time.

Miscible: Oil that can be mixed and be carried with the refrigerant to lubricate moving parts.

Misfire monitoring: OBD II system whereby the ECM detects and stores the number of misfires on each engine cylinder. This information is helpful when diagnosing a drivability problem.

Mixing handle: Handle of an oxyacetylene welding torch, which contains both an acetylene and an oxygen control valve. Valves adjust flow and mixing occurs in handle area.

Modulator: Vacuum operated device that controls pressure in the hydraulic system of automatic transmission or transaxle.

Mold: Hollow unit into which molten metal is poured to form a casting.

Molecule: Smallest portion that matter may be divided into and still retain all properties of original matter.

Monolithic substrate: A ceramic honeycomb mass, surface areas of which can be treated with platinum, palladium, and rhodium. Used in one type of catalytic converter.

Motor: Power unit driven by an outside power source, such as electricity or air pressure. Term is often incorrectly applied to internal combustion engine.

Mounting bullet: A thin, smooth, round nose cap placed over nose of a shaft so that it may be passed through a seal without damage to seal lip.

Movable pole shoe starter: A starter design with a movable pole shoe attached, through linkage, to starter drive pinion assembly. When starter is energized, magnetic field forces movable shoe to actuate linkage, thus engaging starter drive pinion.

MPH: Miles per hour.

Muffler (air conditioning): Device installed on the compressor outlet line. Reduces pumping noise and vibration in system.

Muffler (exhaust): Unit through which exhaust gases are passed to quiet sounds of running engine.

Multiport injection: Fuel injection system in which there is one injector per cylinder.

Multiple disc clutch: Clutch assembly using several clutch discs.

Multiple link chain: Drive chain employing, in addition to outside links, additional inner links. Permits use of a wide chain for great strength and good wear characteristics.

Multiple-ply hose: A hose using two or more layers of reinforcing fabric.

Multiple row bearing: A bearing containing two or more rows of balls, rollers, or needles.

Multi-port fuel injection: Injection system having one injector per cylinder, located just ahead of each intake valve. Also called multi-point fuel injection.

Multi-viscosity oils: Oils meeting SAE requirements for both low temperature requirements of light oil and high temperature requirements of heavy oil. For example, SAE 10W-30 means that the viscosity is 10 cold (W for winter) and 30 at normal operating temperatures.

Mutual induction: Creating voltage in one coil by altering current in another nearby coil.

N

NC threads: National Coarse thread sizes.

Needle bearing: Roller-type bearing in which rollers have very narrow diameter in relation to their length.

Needle nose fitting: A grease fitting, such as used in some universal joints, that requires use of a needle nose (thin, pointed) tip on grease gun.

Needle valve: Valve with long, thin, tapered point that operates in small hole or jet. Hole size is changed by moving needle valve in or out.

NF threads: National Fine thread sizes.

Negative spread: A two-piece insert bearing with diameter across outside of parting edges slightly smaller than bearing bore diameter. When inserted in the cap, insert will touch bottom but sides at parting edges will have some clearance.

Negative terminal: Terminal (such as on battery) from which current flows on its path to positive terminal.

Neoprene: Synthetic rubber. More resistant to oils, grease, ozone, and sunlight than natural rubber.

Net horsepower: Maximum horsepower developed by engine equipped with fan, air conditioning, air cleaner, exhaust system, and all other systems and items normally present when engine is installed in car.

Neutral flame: An oxyacetylene flame in which proportion of acetylene and oxygen are such that flame is neither reducing (carburizing—excess acetylene) nor oxidizing (excess oxygen).

Neutralize: To stop or destroy some force or chemical action, such as by applying baking soda solution to stop action of battery acid.

Neutral safety switch: A switch that opens (disconnects) starter circuit when transmission is in gear.

Neutron: Neutral charge particle forming part of an atom.

Newton-meters: A unit of force, as used in metric system, to indicate torque.

Newton's law: For every action there is an equal and opposite reaction.

NHRA: National Hot Rod Association.

NHTSA: National Highway Traffic Safety Administration.

Nitrogen oxides (NO_x): In combustion process, nitrogen from air combines with oxygen to form nitrogen oxides.

NLGI: National Lubricating Grease Institute.

Noid light: Light used to determine whether voltage is available at the fuel injector wiring harness and connectors.

Non-detergent oil: Engine oil without the ability to keep contaminants in suspension. Not recommended for any modern engine.

Nonferrous metals: All metals containing no iron, except in very minute quantities.

Nonflammable: A substance that will not burn.

Nonhardening sealant: A sealant, that when fully cured, remains in a soft, pliable state.

Nonhunting gearset : A ring and pinion gearset in which same teeth on ring and pinion do not mesh on a regular basis every so many revolutions. Brought about by using uneven ratios such as 3.21 to 1 and 4.11 to 1. The nonhunting feature improves wear characteristics.

Nonreinforced hose: Hose that uses no wires or fabric to strengthen it.

Nonseparable bearing: An antifriction bearing that cannot be taken apart.

Nonserviceable: A part or device whose design and construction does not permit rebuilding. Can also be used to indicate a part of device that is no longer fit for use.

Nonservo brake: A brake design in which each shoe (primary and secondary) works independently and as such, neither one assists in application of other.

North pole: Magnetic pole from which lines of force emanate; travel is from north to south pole.

Nylon: A synthetic thermoplastic made up of long-chain polymeric amide molecules.

O

OBD I: On-board diagnostics-generation one. Designation given to diagnostic system found on vehicles produced before 1996.

OBD II: On-board diagnostics-generation two. Protocol adapted by vehicle manufacturers for standardization of diagnostic trouble codes and automotive terminology.

OBD II self-monitoring: Capability of an OBD II system to periodically check the operation of the EGR valve, air injection system, catalytic converter and evaporative emissions system. If any of the self-checks indicate a problem, the ECM sets a trouble code and turns on the MIL.

Octane rating: Rating that indicates a specific gasoline's ability to resist detonation.

OD: Outside diameter.

Odometer: Device used to measure and register number of miles traveled by vehicle.

OEM: Original Equipment Manufacturer.

Offset dowel pin: A locating steel dowel pin in which around one-half of cylindrical body is offset somewhat from other. Following insertion in, for example, the rear face of block, they can be turned in either direction to correct clutch housing bore to crankshaft misalignment. Fasteners are then tightened.

Ohm: Unit of measurement used to indicate amount of resistance to flow of electricity in a given circuit.

Ohmmeter: Instrument used to measure amount of resistance in given unit or circuit. (In ohms.)

Ohm's law: Formula for calculating electrical values in a circuit.

Oil baffle: A thin plate or diaphragm (often steel) arranged to restrict and direct oil movement from one area to another. Often used in oil pan sump to prevent oil slosh and pump starvation during engine steep angle operation.

Oil bath air cleaner: Air cleaner that utilizes a pool of oil to ensure removal of impurities from air entering carburetor. Formerly common, now only used on large trucks.

Oil burner: Engine that consumes an excessive quantity of oil. Slang term for diesel engines.

Oil clearance: Amount of operating clearance (between bearing and journal) needed to provide proper lubricating oil circulation.

Oil control ring: A piston ring designed to prevent excessive oil consumption by scraping excess oil from cylinder. Usually lower (furthest down in cylinder) ring or rings.

Oil cooler: Device used to remove excess heat from engine and/or transmission oil. Can be air or water cooled design.

Oil dipstick: See *Dipstick.*

Oil filter: Device used to strain oil in engine thus removing abrasive particles.

Oil gallery: Pipe or drilled passageway in engine used to carry engine oil from one area to another.

Oil gauge: A dash mounted device that indicates engine oil pressure in pounds per square inch (psi) or kilopascals (kPa). Most modern oil pressure gauges are electrical.

Oil nozzle: A short section of pipe or nozzle used to direct a stream of oil against a moving part inside of the engine.

Oil pan: See *Pan.*

Oil pickup: Connects to oil pump and extends into bottom of oil pan. Oil is drawn through pickup into pump.

Oilproof paper: A special paper used to wrap greased antifriction bearings for long-term storage. Prevents leaching of lube from bearings.

Oil pump: Device used to force oil, under pressure, to various parts of the engine. Usually driven by gear on camshaft, although some oil pumps are turned by the crankshaft.

Oil pumping: Condition wherein an excessive quantity of oil passes piston rings and is consumed in combustion chamber. See *Oil burning.*

Oil control ring: Piston ring designed to scrape oil from cylinder wall. Ring is of such design as to allow oil to pass through ring and then through holes or slots in groove. In this way, oil is returned to pan. There are many shapes and special designs used on oil control rings.

Oil seal: Device used to prevent oil leakage past certain area.

Oil slinger: Device attached to revolving shaft so any oil passing that point will be thrown outward where it will return to point of origin.

Oil strainer: A fine wire mesh screen through which oil entering oil pump is drawn. It will remove larger particles of dirt or other abrasives.

Oil sump: That portion of oil pan that holds supply of engine oil.

On-vehicle disc brake lathe: Lathe that will refinish a disc brake rotor without the need to remove the rotor from the vehicle.

On-vehicle injector cleaner: Injector cleaning device that attaches to the fuel injection system. Cleans the injectors as the engine runs.

One-way clutch: See *Overrunning clutch.*

Open circuit: Circuit in which a wire is broken or disconnected.

Open circuit voltage: Cell voltage when battery has no completed circuit across posts and is not receiving or delivering energy.

Open windings: A winding circuit that has been broken, thus stopping current flow.

Orifice: A small hole or restricted opening used to control flow of gasoline, air, engine oil, transmission fluid, or power steering fluid.

Orifice tube expansion valve: A form of air conditioning system expansion valve that utilizes a fixed orifice (hole) to control flow rate of refrigerant to evaporator. Compressor clutch is cycled on and off to start and stop refrigerant flow.

O-ring: A round ring, often of neoprene, that is used as a seal.

O-ring connector: A tubing fitting that uses an O-ring to produce a seal between tubing and fitting.

Oscillating action: Swinging action such as that in pendulum of a clock.

Oscillating piston pin: A piston pin secured to rod and thus forced to oscillate (rotate back and forth) in piston.

Oscilloscope: Testing unit which projects visual reproduction of ignition and other electrical and electronic systems onto the screen of cathode-ray tube.

Otto cycle: Four-stroke cycle consisting of intake, compression, firing, and exhaust strokes.

Out-of-round: A piston, shaft, journal or other part in which diameter is greater in one direction than another, thus forming an egg shape.

Output shaft: Shaft delivering power from within mechanism. Shaft leaving transmission, attached to propeller shaft, is transmission output shaft.

Outside caliper: An instrument used to check outside diameter of an object. Caliper leg tips are carefully adjusted to workpiece then removed. Distance between tips is measured with a steel rule.

Outside diameter: Diameter of a cylinder or shaft, as measured in a straight line from outside surface on one side, through center, to outside surface on other side.

Overcenter adjustment: Adjustment between pitman shaft gear sector teeth and ball nut teeth in straight ahead (center) position. Adjustment made by testing torque required to turn wormshaft from one side of straight-ahead position, through straight-ahead position (overcenter) on other side.

Overdrive transmission/transaxle: A transmission or transaxle in which highest gear utilizes a ratio that drives output shaft faster than input, thus allowing reduced engine rpm.

Overhead camshaft: Camshaft mounted above the head, driven by long timing chain or belt.

Overhead valves: Valves located in head.

Overheating: A condition in which any vehicle part or system has attained a temperature well above normal operating temperatures.

Over plate : A very thin layer, such as a tin-lead mix, applied over another layer of different material on a precision insert bearing.

Overrunning clutch: Clutch mechanism that will drive in one direction only. If torque direction is reversed, clutch does not transmit power. Also called one-way clutch, sprag clutch.

Overrunning clutch starter drive: Starter drive that is mechanically engaged. When engine starts, overrunning clutch operates until drive is mechanically disengaged.

Oversize bore: A cylinder that has been bored out to a size that is larger than normal. Requires appropriate oversize piston.

Oversize pin: A piston pin that is larger in diameter than one of standard size. Used to compensate for pin boss, rod, or bushing wear.

Oversize valve stem: An engine valve with a stem that is larger in diameter than one of standard size. Used to compensate for valve guide wear.

Oversquare engine: Engine in which bore diameter is larger than length of stroke.

Oversteer: Tendency for car, when negotiating a corner, to turn more sharply than driver intends.

Oxidation: Surface of a material, such as steel, combining with oxygen in air, thus forming a very thin layer of oxide.

Oxides of nitrogen (NO_x): Undesirable exhaust emission, especially prevalent when combustion chamber flame temperatures are high.

Oxidize: Action where surface of object is combined with oxygen in air to produce corrosion.

Oxidizing flame: Welding torch flame with an excess of oxygen. Free or unburned oxygen tends to burn molten metal.

Oxyacetylene welding: Welding in which required heat is produced by a torch burning a mixture of oxygen and acetylene.

Oxygen: Gas, used in welding, made up of colorless, tasteless, odorless, gaseous element oxygen found in atmosphere.

Oxygen sensor: A device installed in the exhaust system to measure oxygen content. Sometimes called O^2 sensor. Works with other sensors and the ECM to keep air fuel ratio as near to ideal as possible. Most newer vehicles have two or more oxygen sensors.

P

Pads: Another term for disc brake lining assembly.

Palladium: Rare metallic element (Pd) often used as a catalyst coating in catalytic converters.

Pan: Thin stamped cover bolted to bottom of crankcase, forms a sump for engine oil and keeps dirt from entering engine.

Pancake engine: See *Horizontal-opposed engine.*

Paper air cleaner: Air cleaner that uses special filter paper. Air passes through the paper before entering the engine. Almost all modern engines use paper filters.

Parabolic reflector: A light reflector (concave mirror) that emits parallel light rays. Bulb filament must be located at focal point of parabola.

Parallel circuit: Electrical circuit with two or more resistance units so wired as to permit current to flow through both units at same time. Unlike series circuit, current in parallel circuit does not have to pass through one unit to reach the other.

Parallelogram steering: Steering system utilizing two short tie rods connected to steering arms and to a long center link. The link is supported on one end on an idler arm and the other end is attached directly to pitman arm. Arrangement forms a parallelogram shape.

Parasitic load: Normal electrical load from the ECM, radio, and other electrical components placed on a vehicle's battery when the engine is not operating.

Parking brake: Hand operated brake that prevents vehicle movement while parked by locking rear wheels, or transmission output shaft.

Parking lights: Small lights on sides, front, and rear of vehicle. Usually red or amber color in rear and amber in front. Used so that vehicle will be more visible during dark hours. Lights are turned on whenever headlight switch is operated.

Parking pawl: A plunger or toothed segment that engages a notched gear to lock up (prevent rotation) transmission output shaft when placed in "park." Prevents car from rolling when automatic transmission is in park.

Particulates: Tiny particles of lead found in engine exhaust emissions when leaded fuel is used. This term is also applied to other types of exhaust particles.

Parting edge: That edge or surface that forms a juncture (connecting line) between upper and lower halves of a friction bearing.

Parting surface: Surface forming a juncture (connecting point) between two parts. When two parts are disassembled, they will separate at parting surfaces.

Parts washer: A device used to clean parts by a combination of spraying soaking and agitation.

Part-time four-wheel drive: A four-wheel drive system that can be operated in either a two-wheel or four-wheel drive mode. This setup cannot be operated in four-wheel drive on dry, hard surfaced roads.

Part-time transfer case: Four-wheel drive transfer case that permits either four-wheel or two-wheel drive.

Pascal's law: "When pressure is exerted on confined liquid, it is transmitted undiminished."

Pass-key system: An anti-theft device using a calibrated resistor in the ignition key.

Paste state: That state in which a material, such as solder, is neither solid nor liquid but is soft and plastic.

Pattern failure: A defect that occurs in a large number of a particular vehicle or subsystem. Once a pattern failure has been identified on a particular vehicle, the technician can then look for the same problem on all similar vehicles.

Pawl: Stud or pin that can be moved or pivoted into engagement with teeth cut on another part, such as parking pawl on automatic transmission that can be slid into contact with teeth on another part to lock rear wheels.

Payload: Amount of weight that may be carried by vehicle. Computed by subtracting vehicle curb weight from GVW.

Pedal free travel: Distance that a fully released brake or clutch pedal can be readily depressed before linkage actuates either master cylinder or clutch release fingers. Free travel ensures that brake or clutch is fully released.

Pedal travel switch: Component of some antilock brake and traction control systems.

Pedestal grinder: A grinder mounted atop a column to place grinder at a convenient working height.

Peen: To flatten out end of a rivet, by pounding with round end of a hammer.

Penetrating oil: Special oil used to free rusted parts so they can be removed.

Personal protective equipment: Any safety device that lessens the chance of injury or damage to health.

Permanent magnet: Magnet capable of retaining its magnetic properties over very long periods of time.

Petroleum: Raw material from which gasoline, kerosene, and lubricating oils and greases are made. Consists of hydrogen and carbon.

Ph level: Scale that indicates whether a liquid is acid or alkaline (sometimes called base). On the Ph scale, 7 is neutral, neither acid nor alkaline. Lower numbers indicate an acid condition. Higher numbers indicate an alkali condition.

Phillips head screw: Screw having a cross slot instead of single slot as used in conventional screws.

Phosgene gas: A very toxic (poisonous) gas. Can be formed when refrigerant is exposed to an open flame.

Phosphate coating: A thin coating of phosphate applied to surface of metal. Sometimes applied to camshafts to aid in proper break-in. Often applied to other metallic surfaces to prevent corrosion.

Phosphor-bronze: Bearing material composed of tin, lead, and copper.

Photochemical: Relates to branch of chemistry where sunlight produces various chemical changes.

Photochemical smog: Fog-like condition produced by sunlight acting upon hydrocarbon and carbon monoxide exhaust emissions in atmosphere.

Pickup coil: Device in electronic distributor that senses engine speed from distributor rotation and sends electrical pulses to control unit.

Piezoelectric ignition: System of ignition that employs use of small section of ceramic-like material. When this material is compressed, even a very tiny amount, it emits a high voltage that will fire plugs.

Pilot bearing: A small bearing (either antifriction or bushing type) located in center of flywheel or crankshaft end. Bearing aligns and supports flywheel end of clutch shaft (transmission input shaft).

Pilot operated absolute (POA): A modified form of a suction throttling valve. Used in air conditioning system to control evaporator pressure to allow low evaporator temperatures without freezing (icing).

Pilot shaft: Dummy shaft placed in a mechanism as a means of aligning parts. It is then removed and regular shaft installed.

Pinging: Metallic rattling sound produced by the engine during heavy acceleration when ignition timing is too far advanced for grade of fuel being burned.

Pinion: Small gear either driven by or driving a larger gear.

Pinion carrier: Part of rear axle assembly that supports and contains pinion gear shaft.

Pinion flange: Unit (flange) splined to outer end of differential pinion gear shaft. Used as an attaching point for propeller shaft universal joint.

Pinion shaft : Shaft that supports, aligns, and drives differential pinion gear.

Pin boss: That section of engine piston that aligns and supports piston (wrist) pin.

Pin fitting: Process involved in honing piston or rod pin bores until a correct pin fit is attained.

Pin knock: Sound (sharp, double-knock) produced by excessively worn piston pins and/or bearings.

Pin punch: A punch with a long, narrow, nontapered nose. Used to drive pins from their holes.

Pintle: A round pin about which some part pivots.

Pintle valve: Diaphragm controlled valve used to control flow of exhaust gas through exhaust gas recirculation valve.

Pipes: Exhaust system pipes.

Piston: Round plug, open at one end, that slides up and down in cylinder. It is attached to connecting rod and when fuel charge is fired, will transfer force of explosion to connecting rod then to crankshaft.

Piston boss: Built-up area around piston pin hole.

Piston collapse: Reduction in diameter of piston skirt caused by heat and constant impact stresses.

Piston displacement: Amount (volume) of air displaced by piston when moved through full length of its stroke.

Piston expansion: Increase in diameter of piston due to normal piston heating.

Piston head: Portion of piston above top ring.

Piston lands: Portion of piston between ring grooves.

Piston pin: Steel pin that is passed through piston. Used as base upon which to fasten upper end of connecting rod. It is round and is usually hollow.

Piston ring: Split ring installed in a groove in piston. Ring contacts sides of ring groove and also rubs against cylinder wall thus sealing space between piston and wall.

Piston ring end gap: Distance left between ends of the ring when installed in cylinder.

Piston ring expander: See *Ring expander.*

Piston ring groove: Slots or grooves cut in piston head to receive piston rings.

Piston ring side clearance: Space between sides of ring and ring lands.

Piston skirt: Portion of piston below rings. (Some older engines have an oil ring in skirt area.)

Piston skirt expander: Spring device placed inside piston skirt to produce an outward pressure that increases diameter of skirt.

Piston skirt expanding: Enlarging diameter of piston skirt by inserting an expander, by knurling outer skirt surface, or by peening inside of piston.

Piston slap: Slamming side to side (tipping) movement of piston when piston-to-cylinder clearance is excessive. Can produce a very audible hollow, clattering sound.

Piston thrust pressure: Pressure applied to cylinder wall by one side of piston. During compression stroke, piston is thrust against one side of cylinder and during firing stroke, against other.

Pitch: Distance from center of one screw thread to center of adjacent (next to) thread. Pitch determines number of threads per inch.

Pitman arm: Short lever arm splined to steering gear cross shaft. Pitman arm transmits steering force from cross shaft to steering linkage system.

Pitman shaft: Steering gearbox shaft upon which pitman arm is attached.

Pivot: Pin or shaft about which a part moves.

Planet carrier: Part of a planetary gearset upon which planet gears are affixed. Planet gears are free to turn on hardened pins set into carrier.

Planet gears: Gears in planetary gearset that are in mesh with both ring and sun gear. Referred to as planet gears in that they orbit or move around central or sun gear.

Planetary gearset: Gearing unit consisting of ring gear with internal teeth, sun or central pinion gear with external teeth, and series of planet gears that are meshed with both the ring and the sun gear.

Plastic range: Temperature range over which a material, such as solder, remains neither liquid nor solid.

Plastigage: A soft plastic, in wire form, used to measure bearing clearance.

Plates (battery): Thin sections of lead peroxide or porous lead. There are two kinds of plates—positive and negative. The plates are arranged in groups, in an alternate fashion, called elements. They are completely submerged in the electrolyte.

Platinum: Precious metal sometimes used in the construction of breaker points. It conducts well and is highly resistant to burning.

Play: Movement between two parts.

Plies: Layers of rubber impregnated fabric that make up carcass or body of tire.

Plug gapping: Adjusting side electrode on spark plug to provide proper air gap between it and the center electrode.

Plug-in connector: Connector that is pushed together to connect. Pulled apart to disconnect after a locking tab is disengaged.

Plug weld: A welding technique in which one steel plate, containing one or more holes, is placed in contact with another. They then are joined by applying weld metal in holes, thus fusing both parts together.

Ply rating: Indication of tire strength (load carrying capacity). Does not necessarily indicate actual number of plies. Two-ply four-ply rating tire would have load capacity of a four-ply tire of same size but would have only two actual plies.

P-metric: Tire size designation based on international standards. Example: P 155/80R13. P = passenger car use. 155 = section width in millimeters. 80 = height to width ratio. R = radial construction. 13 = wheel rim diameter in inches.

Pneumatic: Pertaining to air. Operated by air pressure.

Pocket tape: A flexible, retracting measuring tape, available in different lengths, designed to be carried in pocket.

Polarity (battery terminals): Indicates if the battery terminal (either one) is positive or negative (plus or minus) (+ or –).

Polarity (magnet): Indicates if end of a magnet is north or south pole (N or S).

Pole: One end, either north or south, of a magnet.

Pole shoes: Metal pieces about which field coil windings are placed. When current passes through windings, pole shoes become powerful magnets. Used in starter motors.

Pollution: Any substance that causes environmental damage.

Polyalkylene glycol (PAG): Oil used in air conditioning systems that have R-134a as a refrigerant.

Polyethylene: Thermoplastic resin. Used for its flexible, tough, oil resistant and insulation properties.

Polyol ester (POE): Oil that can be used in R-134a or R-12 air conditioning systems. Some manufacturers do not allow POE to be used in their AC systems.

Poppet valve: Valve used to open and close valve port to engine cylinders.

Porcelain: Material used to insulate center electrode of spark plug. It is hard and resistant to damage by heat.

Porosity: Small air or gas pockets, or voids, in metal.

Porous bronze: A material made by compressing tiny particles of bronze, under extremely high pressure, until they form a rigid but porous (full of holes) mass. Used in filters and oil retaining bushings.

Port: Openings in engine cylinder blocks for exhaust and intake valves and water connections. To smooth out, align, and enlarge intake passageway to the valves.

Portable crane: A hydraulic, electric, or manual lifting device equipped with wheels to facilitate moving about.

Portable steam cleaner: A relatively small steam cleaner equipped with wheels for easy portability.

Ported vacuum: A vacuum source connected to a hole (or port) located above the throttle valve when in closed position. As throttle is opened, port will be uncovered and vacuum applied to a vacuum device.

Positive Crankcase Ventilation (PCV): System that prevents crankcase vapors from being discharged directly into atmosphere.

Positive terminal: Terminal (such as on battery), to which current flows.

Post: Round, tapered lead posts protruding above top of battery to which battery cables are attached.

Potential: An indication of amount of available energy.

Potentiometer: Variable resistor with three connections. One connection (called wiper) slides along resistive unit. Can be used as a voltage divider.

Pour point: Lowest temperature at which fluid will flow.

Power: Time rate at which energy is converted into work.

Power booster: Engine vacuum or power steering fluid–operated device on fire wall which increases brake pedal force on master cylinder during stops.

Power brakes: A brake system in which a vacuum or hydraulic booster is used to greatly multiply foot pressure to master cylinder.

Power drill: An electric or pneumatic tool designed to hold and rotate twist drills. Can also be used for wire brushing, sanding, and polishing with correct accessories.

Power plant: The vehicle engine.

Power steering: Steering system utilizing hydraulic pressure to increase the driver's turning effort. Pressure is utilized either in gearbox itself or in hydraulic cylinder attached to steering linkage.

Power steering pump: Belt driven pump that produces pressure for power steering system.

Power or firing stroke: Portion of piston's movement that transmits power of burning fuel mixture to crankshaft.

Powertrain control module (PCM): Computer that controls the operation of the engine and transmission/transaxle shift points. Often combined with the engine control module (ECM) into a single unit.

PPM (Parts-Per-Million): Term used in determining extent of pollution in given sample of air.

Practical efficiency: Amount of horsepower delivered to drive wheels.

Precision insert bearing: Very accurately made replaceable bearing. It consists of an upper and lower shell. The shells are made of steel to which a friction-type bearing material has been bonded. Connecting rod and main bearings are generally of precision insert type.

Preheat flame: Ring of small flames emitted at tip of an oxyacetylene cutting torch. They are used to heat metal to a red heat so as to facilitate cutting with a central jet of oxygen.

Preheating: Application of some heat prior to later application of more heat. Cast iron is preheated to avoid cracking when welding process is started. A coil (ignition) is preheated prior to testing.

Preignition: Fuel charge being ignited before proper time.

Preloading: Adjusting antifriction bearing so it is under mild pressure. This prevents bearing looseness under a driving stress.

Prelubricator: A pressurized, oil-filled tank used to force lubricating oil through engine lubrication system prior to actually starting engine. Lubes all bearings and prevents possible delay in pressure buildup following first starting of a rebuilt engine.

Press-fit: Condition of fit (contact) between two parts that requires pressure to force parts together. Also referred to as drive or force fit.

Pressure bleeder: Device that forces brake fluid, under pressure, into master cylinder so that by opening bleeder screws at wheel cylinders, all air will be removed from brake system.

Pressure cap: Special cap for radiator. It holds a predetermined amount of pressure in cooling system. This enables coolant to run hotter without boiling.

Pressure differential switch: Hydraulic switch in brake system that operates brake warning light in dashboard.

Pressure plate: Circular plate, driven by flywheel, which forces clutch disc friction surface tightly against flywheel. When clutch is disengaged, pressure plate releases clutch disc.

Pressure regulator valve: A valve used in gasoline injection systems to maintain a predetermined pressure at the fuel injectors.

Pressure relief valve: Valve designed to open at specific pressure. This will prevent pressures in system from exceeding certain limits.

Pressure tester: A hand-operated air pump used to pressurize cooling system to check for leaks. Can also be used to check pressure cap for proper opening pressure and sealing leaks.

Prestart pressurization: See *Prelubricator*.

Primary circuit: 12-volt part of ignition system.

Primary, forward, or leading brake shoe: Brake shoe installed facing front of vehicle. It will be a self-energizing shoe.

Primary winding: Low-voltage winding in ignition coil. The primary winding is heavy wire; secondary winding uses fine wire.

Primary wires: Wiring that serves low-voltage part of ignition system.

Printed circuit: Electrical circuit made by connecting units with electrically conductive lines printed on a panel. This eliminates actual wire and task of connecting it.

Probe: A thin, long necked tool that is often expandable. Used to reach into relatively inaccessible areas. May utilize a magnet or movable fingers for part insertion or removal or a mirror for viewing blind areas.

Programmable read-only memory (PROM): An ECM memory chip that can be updated. The PROM can be reprogrammed or replaced.

Prony brake: Device utilizing friction brake to measure horsepower output of engine.

Propane (LPG): Petroleum product, similar to and often mixed with butane for use as engine fuel. May be referred to as LP-Gas.

Propelyene glycol: Non-harmful antifreeze compound. Used propelyene glycol has picked up heavy metals from the engine and is no longer safe.

Propeller shaft: Shaft connecting transmission output shaft to differential pinion shaft.

Proportioning valve: Valve in brake line that keeps rear wheels from locking up during rapid stops.

Proton: Positive charge particle, part of atom.

Prussian blue: A deep blue pigment (dye) mixed with a grease-like carrier. By spreading a thin film on one part and then placing other part firmly in position, then removing, it is possible to check contact surfaces for high and low spots.

PSI: Pounds per square inch.

PTO: Power take off. A spot or place on transmission or transfer case from which an operating shaft from another unit (such as a winch) can be driven. Removing a plate on the side of the transmission exposes a drive gear.

Puddle: Area of base metal that is brought to molten state by electric arc or by oxyacetylene flame. Where required, filler rod is melted into puddle.

Puller bar: A rigid steel bar to which puller arms or jaws are attached. Bar allows different positioning of jaws to provide wide range of adjustment.

Puller strap: A length of cable or chain with attachment device at each end, used to pull (raise) engines.

Pull point: When lifting engines, that point, in relation to static balance, at which lifting hook is attached to puller strap or bar. Pull point must be secure and located to keep engine properly balanced and at correct angle while lifting.

Pulsation: Rhythmic cycling of a force or sensation. Pulsation of one kind or another occurs in many vehicle systems. May be harmful to the vehicle or simply an annoyance to the driver.

Pulsation damper: Device to smooth out fuel pulsations or surges from pump to carburetor.

Pulse air injection: Emission control system that feeds air into exhaust gases using pressure pulsations of exhaust system.

Pulse width: The length of time a fuel injector is held open. Pulse being electric current applied to the injector winding and width being length of time current is allowed to flow. The wider the pulse, the more fuel delivered.

Pulsed fuel injection: Fuel system in which injectors are only open for a short period and remain closed the rest of the time. The amount of fuel delivered is controlled by how long the injector is open.

Pump: A device designed to cause movement of liquids or air from one area to another.

Purge: Removing impurities from system. See *Bleeding*.

Push rod: Rod that connects valve lifter to rocker arm. Used on valve-in-head installations.

Push rod runout: A measurement of amount of bend present in push rod. By placing each end in a V-block and then rotating with a dial indicator contacting the center area, exact amount of runout is determined.

Push rod seat: That portion of both valve lifter and rocker arm engaged by their respective ends of push rod.

Pylon: Marker for controlling traffic.

Q

Quadrant: A gear position indicator using a shift lever actuated pointer.

Quenched: Flame front in combustion chamber being extinguished as it contacts colder cylinder walls. This sharply elevates hydrocarbon emissions.

Quenching: Dipping heated object into water, oil, or other substance, to quickly reduce temperature.

Quick take-up master cylinder: A brake master cylinder that supplies a large amount of fluid (under low pressure) during first part of brake application. This supplies extra fluid needed to quickly return no-drag–type brake caliper pistons so pads engage disc (rotor).

R

R-12 (CFC-12): Refrigerant used in older air conditioning systems. Gradually being replaced by R-134a in newer vehicles.

R-134a (HFC-134a): Refrigerant used in the air conditioning systems of most vehicles manufactured after 1992. Replaced R-12 due to environmental concerns.

Race: Inner or outer ring that provides a contact surface for balls or rollers in bearing.

Rack and pinion gearbox: Steering gear utilizing pinion gear on end of steering shaft. Pinion engages long rack (bar with teeth along one edge). Rack is connected to steering arms through rods.

Rack piston: In a recirculating ball–type power steering gear, large piston that rides on worm. One side of piston has a series of teeth (called rack) that engage pitman shaft sector gear teeth. When worm is turned, rack piston slides (does not rotate), thus rotating pitman shaft.

Radial: Line of force at right angles perpendicular to the centerline of a rotating part.

Radial compressor: A small air conditioning compressor using reciprocating pistons working at right angles to shaft and spaced around shaft in radial fashion.

Radial loading: A load (pressure) placed on a bearing in a direction that would be at right angles to shaft centerline.

Radial movement: Up and down movement (runout) of the outer surface of a revolving unit. Movement is at right angles to axis of rotation.

Radial pressure: Pressure on an object applied at a right angle to its centerline.

Radial runout: Up and down movement of a rotating part, as measured at a right angle to centerline of the part.

Radial tire: Plies parallel and at right angle to tread, belts under tread area.

Radiation: Transfer of heat from one object to another when hotter object sends out invisible rays or waves that, upon striking colder object, cause it to heat.

Radiator: A device used to remove heat from engine coolant. It consists of a series of finned passageways. As coolant moves through passages, heat is conducted to fins where it transfers to a stream of air forced through fins.

Radiator cap: Pressure cap that fits on radiator neck. It keeps coolant from boiling.

Radius: Distance (in a straight line) from center of a circle or circular motion, to a point on edge (circumference).

Radius rods: Rods attached to axle and pivoted on frame. Used to keep axle at right angles to frame and yet permit an up and down motion.

Ram air: Air scooped into the engine compartment by an opening in the front of the vehicle during vehicle forward motion.

Ram induction: Using forward momentum of car to scoop air and force it into engine via a suitable passageway.

Rasp cut file: A file utilizing rows of relatively large sharp, tooth-like projections. Used for woodwork, filing soft metals, such as solder, brass and aluminum.

Raster pattern: On an oscilloscope, a display of all engine cylinders, one above other, with number one cylinder at bottom of screen.

Ratchet handle: Socket wrench handle with a ratcheting head to allow short back and forth handle movement without disengaging socket from fastener. Can be set to drive in either direction.

Rated horsepower: Indication of horsepower load that may safely be placed upon engine for prolonged periods of time. This would be somewhat less than the engine maximum horsepower.

Ratio: Fixed relationship between things in number, quantity, or degree. For example, if fuel mixture contains one part of gas for fifteen parts of air, ratio would be 15 to 1.

Rawhide hammer: A striking tool with a head formed of rawhide. Used for lighter striking on objects that would be damaged by a regular steel hammer.

Rayon braid: A woven rayon reinforcing layer, either in hose or applied to outside.

Reactor: See *Stator.*

Ream: To enlarge or smooth hole by using round cutting tool with fluted edges.

Reamed finish: A finished surface in a bore, produced by a reamer.

Reamer: A cylindrical cutting tool with a series of longitudinal fluted cutting edges (fixed or adjustable). Used to enlarge, bring to size, or finish a bore.

Rebore: To bore out engine cylinders to a larger size. New pistons will be required.

Receiver-drier: See *Drier.*

Reciprocating action: Back-and-forth movement such as action of pistons.

Recirculating ball worm and nut: Very popular steering gear. It utilizes series of ball bearings that feed through and around and back through grooves in worm and nut.

Rectifier: See *Diode.*

Red line: Top recommended engine rpm. If a tachometer is used, it will have a mark (Red line) indicating maximum rpm.

Reducing flame: Welding flame in which there is an excess of acetylene.

Reduction gear: A gear that increases torque by reducing rpm of a driven shaft in relation to that of driving shaft.

Refrigerant: Liquid used in refrigeration systems to remove heat from evaporator coils and carry it to condenser.

Refrigerant oil: Special oil that lubricates air conditioning compressor.

Refrigerant recovery unit: Electronic station that combines a refrigerant storage tank, vacuum pump, gauges, and service valves. Used to recover, recycle, and recharge refrigerant in automotive air conditioning systems.

Refrigerant recycling: The capture and reuse of refrigerant by a machine dedicated to that purpose.

Refrigeration cycle: Series of events or actions that take place in air conditioning system as refrigerant moves through system.

Reground: Refers to a crankshaft or camshaft which bearing journals have been resurfaced by grinding. Unless built up before grinding, reground journals will have a smaller diameter.

Regulator (electrical): Device used to control alternator voltage and current output.

Regulator (gas or liquid): Device used to reduce and control pressure.

Relative humidity: Actual amount of moisture in a given sample of air compared to total amount that sample could hold (at same temperature).

Relay: Magnetically operated switch used to make and break flow of current in circuit.

Relay rod: Steering component connecting the pitman arm, idler arm, and tie rods on a parallelogram steering system. Sometimes called a center link or drag link.

Relief valve: A spring-loaded valve designed to open and relieve pressure when pressure exceeds certain limits.

Relieve: Removing, by grinding, small lip of metal between valve seat area and cylinder—and removing any other metal deemed necessary to improve flow of fuel mixture into cylinder. Porting is generally done at same time.

Reluctor: A component in some electronic ignition system distributors. It is affixed to the distributor shaft and triggers magnetic pickup. The signal from the pickup triggers the electronic control device to break the coil primary circuit causing coil to fire.

Remote choke: A bimetallic choke device, usually set in a choke stove well, located some distance from and connected to the choke plate by linkage.

Remote keyless entry: Electronic system that is added to some vehicles to enable the vehicle's owner to lock and unlock the doors and open the trunk using a key fob transmitter.

Remote programming: Method of reprogramming the ECM by removing it from the vehicle. The ECM is then connected to the programming device, usually a computer in the shop. When programming is complete, the ECM is reinstalled in the vehicle.

Replaceable valve guide: An engine valve guide that can be removed and replaced with a new guide.

Residual magnetism: Magnetism remaining in an object after removal of a magnetic field.

Residual tension: A pulling force or strain that remains constantly applied to some object even though other major forces have been removed.

Resistance: Measure of conductor's ability to retard flow of electricity.

Resistance wire: A wire designed to provide a carefully calculated resistance to electrical current flow. Can be used to limit current flow and to reduce voltage.

Resistor: Device placed in circuit to lower voltage. It will also decrease flow of current.

Resistor spark plug: Spark plug containing resistor to suppress radio interference and lengthen electrode life.

Resonator: Small muffler-like device that is placed into exhaust system near end of tail pipe. Used to provide additional silencing of exhaust.

Respiratory protection: A device to filter air before it enters the lungs.

Resurfacing: Process of grinding or planing the engine block to cylinder head contact surfaces. Done to eliminate warping or physical damage, and to increase compression.

Retard (ignition timing): To set the ignition timing so that spark occurs later or less degrees before TDC.

Retorque: Following initial torquing, a second torquing of fasteners to compensate for gasket crush between parts, or loosening caused by vibration or heating cycles.

Retrofitting: Process of converting an air conditioning system that uses R-12 to handle R-134a refrigerant.

Returnless fuel system: A fuel injection system that does not use a pressure regulator. The ECM regulates fuel pressure by varying electric fuel pump output.

Reverse flush: Cleaning cooling system by pumping a powerful cleaning agent through system in a direction opposite to that of normal flow.

Reverse idler gear: Gear used in transmission to produce a reverse rotation of transmission output shaft.

Reverse polarity: In arc welding, when electrode is connected to machine positive terminal, ground wire to negative terminal.

Rheostat: A variable resistor used to control current flow.

Rhodium: A metallic element (Rh), sometimes employed as a catalyst in catalytic converters.

Ribbed belt: A flexible, fabric reinforced belt that uses a series of longitudinal "V" shaped ribs that mesh with corresponding grooves in pulleys.

Rich mixture: Mixture with an excessive amount of fuel in relation to air.

Riding the clutch: Riding the clutch refers to driver resting a foot on clutch pedal while car is being driven.

Rigid hone: A honing tool in which stone holders, although capable of in and out adjustment, cannot spring inward, tip, or flex. Stones thus remain parallel at all times.

Rim: The outer portion of a wheel upon which tire is mounted.

Ring break-in: Wearing-in process involved following installation of new piston rings. Tiny ridges on both rings and cylinder wall quickly wear off producing a smooth and proper fit.

Ring compressor: A tool designed to force piston rings inward in their grooves so that piston and rings can be inserted into engine cylinder bore.

Ring expander: Spring device placed under rings to hold them snugly against cylinder wall.

Ring float: That situation in which piston rings are not fully pressed against cylinder wall. Caused by excessive cylinder bore taper. Ring is squeezed in at bore bottom and piston travels to top and before ring can fully expand, returns to bottom.

Ring gap: Distance between ends of piston ring when installed in cylinder.

Ring gap spacing: Locating ring gaps equidistant around piston. By spacing them so they do not line up, gas leakage through gaps is reduced and vertical wear pattern in cylinder is avoided. Also applies to gaps in a single, multiple piece ring.

Ring gear: Large gear with internal teeth, attached to differential carrier or to outer gear in planetary gear setup.

Ring gear centerline: A line that would, if drawn across face of gear, pass through exact center of gear.

Ring gear runout: Measured amount of wobble (side-to-side movement) produced when a ring gear is rotated through one full turn.

Ring groove cleaner: A scraping tool designed to remove carbon from piston ring grooves.

Ring groove gauge: A measuring device (gauge) used to check piston ring groove depth to insure depth is compatible with replacement rings.

Ring groove reconditioning tool: A tool, hand or power driven, used to cut a piston ring groove to a wider size, thus removing tapered, battered sides. A thin spacer is then inserted into top of recut groove returning it to proper size.

Ring grooves: Grooves cut into piston to accept rings.

Ring groove spacer: A thin spacer ring that is inserted in a groove cut on top side of a piston ring groove. Ring takes place of battered material machined from damaged groove and produces a groove of proper shape and size.

Ring job: Reconditioning cylinders and installing new rings.

Ring lands: Piston metal between ring grooves.

Ring ridge: Portion of cylinder above top limit of ring travel. In a worn cylinder, this area is of smaller diameter than remainder of cylinder and will leave ledge or ridge that must be removed.

Ring ridge reamer: A tool used to cut away narrow band of unworn cylinder wall (ring ridge) found between top of cylinder and highest point of ring travel.

Ring side clearance: That space existing between the side of piston ring and groove in which it operates.

Ring tipping: A condition in which a piston ring operating in a badly tapered groove, tips from one side to other during engine operation.

Ring travel: The total distance a ring moves in cylinder from BDC to TDC. Ring travel and length of piston stroke are same.

Rivet: Metal pin used to hold two objects together. One end of the pin has head and other end must be set or peened over.

RMA: Rubber Manufacturer's Association.

Road crown: Roads are raised in center so that water will run off readily. This angled surface is termed road camber. A vehicle will tend to drift to the right when driving on a crowned road, and alignment must be set to compensate for this tendency.

Road feel: Feeling imparted to steering wheel by wheels of car in motion. This feeling can be very important in sensing and predetermining vehicle steering response.

Road test: Process of driving vehicle upon road for purpose of diagnosing problems and checking efficiency of repairs.

Rocker arm: Arm used to direct upward motion of push rod into a downward or opening motion of valve stem. Used in overhead valve installations.

Rocker arm cover: Aluminum, stamped steel, or plastic covering or shroud fastened to engine cylinder head so as to cover valve operating mechanisms.

Rocker arm shaft: Shaft upon which rocker arms are mounted.

Rocker panel: Section of car body between front and rear fenders and beneath doors.

Rockwell hardness: Measurement of the degree of hardness of given substance.

Rod: See *Connecting rod.*

Rod cap: Lower removable half of connecting rod big end.

Rod heater: A device used to heat connecting rod small end to facilitate installation of piston pin.

Roll bar: Heavy steel bar that goes from one side of frame, up and around in back of the driver, and back down to the other side of frame. It is used to protect driver in the event that the car rolls over.

Roll cabinet: A portable tool cabinet on small wheels or casters to facilitate moving about. Tool chest of drawers often is placed atop cabinet.

Roller bearing: Bearing utilizing a series of straight, cupped, or tapered rollers engaging an inner and outer ring or race.

Roller clutch: Clutch, utilizing series of rollers placed in ramps, which will provide drive power in one direction but will slip or freewheel in the other direction.

Roller lifter: A valve lifter incorporating a small roller on bottom end. Roller engages camshaft and allows lifter operation with a minimum of friction between lifter and cam lobe.

Roller tappets or lifters: Valve lifters that have roller placed on end contacting camshaft. This is done to reduce friction between lobe and lifter.

Roller-vane pump: A vane pump, such as some fuel injection pumps, utilizing round rollers as operating vanes.

Rolling contact: Contact between moving parts, such as a roller against bearing race, where one part, instead of sliding, is rolling against other.

Rolling friction: Resistance to movement (friction) created by two moving objects in contact with each other—when one of the units is rolling across other. Will be less than sliding friction.

Rolling radius: Distance from road surface to center of wheel with vehicle moving under normal load. Rolling radius is dependent on tire size.

Rollover valve: Valve in fuel delivery line to prevent escape of raw fuel during an accident in which car is upside down.

Room temperature: An enclosed space air temperature of around 68–72°F (20.1–22.2°C).

Room temperature vulcanizing (RTV) silicone: A silicone sealing compound used as a gasket. Will set up or harden (vulcanize) at normal room temperature.

Rosin core solder: A wire solder with a hollow center filled with rosin flux. Used for soldering electrical connections.

Rotary engine: Internal combustion engine that does not use pistons. Central rotor turns in one direction only and produces required intake, compression, firing, and exhaust strokes.

Rotary injection pump: A diesel injection pump that uses a single pumping unit (two radial opposed pistons) to generate pressure for fuel distribution to all injectors. Fuel from pump passes through a distributor unit for proper injector delivery.

Rotary motion: Continual motion in circular direction such as performed by crankshaft.

Rotary valve: A power steering gear valve, that instead of moving back and forth, is rotated by steering gear shaft to open appropriate ports for hydraulic assist control.

Rotor: Cap-like unit placed on end of distributor shaft. It is in constant contact with distributor cap central terminal and as it turns, it will conduct secondary voltage to one of the other terminals.

Rotor lateral runout: The amount of wobble or side-to-side (lateral) movement present when brake disc (rotor) is turned through one full revolution.

Rotor parallelism: Relates to how well one side of brake disc or rotor aligns (parallel) with other. Perfect parallelism is present when rotor thickness is constant at all points around rotor.

Rotor pump: A liquid pump employing a multiple lobe roller operating inside of a rotating ring with corresponding rotor recesses. Liquid is trapped, carried around and squeezed out by enlargement and then reduction of spaces between inner and outer rotor.

Roughing stone: Coarse stone used for quick removal of material during honing.

RPM: Revolutions per minute.

Rubbing block: Insulated block attached to movable distributor point arm. Arm rubs against distributor cam and opens and closes points.

Runaway diesel engine: Diesel engine that will not stop. Usually caused by failure of the fuel cutoff system.

Run-down torque: That torque required to turn fastener before it exerts closing pressure to parts. Caused by friction between fastener threads.

Running-fit: Fit in which sufficient clearance has been provided to enable parts to turn freely and to receive lubrication.

Running on: See *Dieseling.*

Runout: Rotating object, surface of which is not revolving in a true circle or plane. Runout can be measured in a radial (at right angles to centerline of object) direction or in a lateral (lengthwise to centerline) direction.

S

SAE: Society of Automotive Engineers.

SAE or rated horsepower: A simple formula of long standing is used to determine what is commonly referred to as the SAE or Rated Horsepower. The **formula is:**

$$\frac{\text{Bore Diameter}^2 \times \text{Number of Cylinders}}{2.5}$$

This formula is used primarily for licensing purposes and is not an accurate means of determining actual brake horsepower.

Safe edge file: A file in which one or more surface areas are smooth (no cutting edges at all). This allows filing one particular surface without danger of accidentally cutting an adjacent area.

Safety factor: Providing strength beyond that needed, as an extra margin of insurance against part failure.

Safety hubs: Device installed on the rear axle to prevent wheels from leaving car in event of a broken axle.

Safety rim: Rim having two safety ridges, one on each lip, to prevent tire beads from entering drop center area in event of a blowout. This feature keeps tire on rim.

Safety shoes: Shoes with a steel protective insert over the toes. Protects the toes from crushing hazards.

Safety valve: Valve designed to open and relieve pressure within a container when container pressure exceeds predetermined level.

Safety wire: Soft wire used to secure fasteners to prevent unscrewing from vibration.

Sand blast: Cleaning by the use of sand propelled at high speeds in an air blast.

Saybolt viscometer: Instrument used to determine fluidity or viscosity (resistance to flow) of an oil.

Scale: Accumulation of rust and minerals within cooling system.

Scan tool: A diagnostic tool used to extract trouble codes from an automobile's computer. Most scan tools perform other diagnostic functions.

Scatter shield: Steel or nylon guard placed around bell or clutch housing to protect driver and spectator from flying parts in event of part failure at high rpm. Such a shield is often placed around transmissions and differential units.

Scavenging: Referring to the vacuum pulses occurring in the exhaust system.

Schematic: Electrical circuit diagram.

Schrader valve: Valve, similar to spring loaded valve used in tire stem, used in car air conditioning and fuel injection system service valves.

Score: Scratch or groove on finished surface.

Scored piston: A piston with vertical grooves and deep scratches in piston to cylinder contact surface. Caused by lack of lubrication or abrasive material.

Scoring: Producing deep scratches and grooves in two moving parts. Caused by insufficient lubrication or abrasive particles.

Screw clamp: A hose clamp that is tightened by means of a threaded screw.

Screw extractor: Device used to remove broken fasteners from holes.

Screw-pitch gauge: See *Thread-pitch gauge.*

Scuffing: Very similar to scoring in many respects. Caused by lack of lubrication, improper clearance, or abrasive particles. Metal particles are torn from one contact surface such as piston, and deposited on other, such as cylinder. Contact surfaces often discolored by heat produced.

Seal: Device that prevents oil leakage around moving part.

Seal driver: A tool used to install seals.

Sealed beam headlight: Headlight lamp in which lens, reflector, and filament are fused together to form single unit.

Sealed bearing: Bearing that has been lubricated at factory and then sealed. It cannot be lubricated during service.

Sea level: Elevation as measured at ocean surface.

Seat: Surface upon which another part rests or seats. Example: Valve seat is matched surface upon which valve face rests.

Secondary circuit: High-voltage part of ignition system.

Secondary piston: In a brake system dual master cylinder, that piston actuated by pressure generated by primary piston. Primary is actuated by either brake pedal linkage or power booster rod.

Secondary, reverse, or trailing brake shoe: Brake shoe that is installed facing rear of vehicle.

Secondary wires: High-voltage wire from coil to distributor central tower and from outer towers to spark plugs.

Section height: Distance, at right angles, from center of inside surface of tread area to a line intersecting (crossing through) two bead bottom edges. Calculated with tire mounted and at normal pressure.

Section width: Measurement of a tires' inside width at widest point between two sidewalls, when mounted and at normal pressure.

Sector gear: A partial gear, such as used on a steering gear pitman shaft, that is actually only a portion (pie shaped slice) of normal 360° gear. Gear can only be operated through a limited arc.

Sector shaft: Another term for pitman shaft. Has a sector gear (partial gear) on one end and pitman arm attached to other.

Sediment: Accumulation of matter in a liquid, which settles to bottom of the container.

Seize: See *Freezing.*

Seized: A term describing condition of a moving part that has overheated. Through scuffing and galling, it is stuck fast and is thus unable to turn or slide.

Self-adjusting: A mechanism in which a required clearance is automatically maintained without need for periodic manual adjustment.

Self-aligning bearing: A bearing that is secured in such a manner that it is free to pivot or swivel, thus permitting a certain amount of shaft lateral movement without binding bearing.

Self-centering brakes: Wheel brake design in which both brake shoe assemblies float and are free to move about so that when applied, center themselves in relation to brake drum.

Self-diagnostics: The ability of a computer to continuously monitor the operation of a specific system and send warning signals when an abnormal condition is detected.

Self-energizing: Brake shoe (sometimes both shoes) that when applied develops wedging action that actually assists or boost braking force applied by wheel cylinder.

Self-induction: Creation of voltage in a circuit by varying current in circuit.

Self-locking: A fastener utilizing built-in thread friction produced by tight threads or plastic inserts, so that when tightened it will not loosen under normal use.

Self-starting rod: An electric arc welding rod that utilizes a special flux coating that makes it very easy to get arc started and tends to automatically maintain proper rod to work distance as rod is consumed.

Self-tapping: Fastener designed to cut its own thread as it is turned into hole.

SEMA: Specialty Equipment Manufacturer's Association.

Semi-elliptical spring: Spring, such as commonly used on truck rear axles, consisting of one main leaf and number of progressively shorter leaf springs.

Semifloating drive axle: A drive axle in which drive shaft secures, aligns, and drives wheel. Supports weight of vehicle.

Sensor: A device used to detect a change in temperature, pressure, exhaust gases, rpm or vehicle speed, and to send a suitable signal to a control unit.

Separable bearing: A bearing capable of being disassembled.

Separators: Nonconductive sheets inserted between positive and negative battery plates to prevent contact.

Sequence of assembly: Recommended order of operations involved in assembly of a complex part.

Series circuit: Circuit with two or more resistance units so wired that current must pass through one unit before reaching other.

Series-parallel circuit: Circuit of three or more resistance units in which a series and a parallel circuit are combined.

Serviceable: A part or unit whose design and construction permit disassembly for purposes of rebuilding. Can also be used to indicate a part or unit whose condition is such that it can still be used.

Service classification: A listing indicating ability of a specific oil to provide proper lubrication under specific engine operating conditions.

Service manager: Person in shop who greets customers, determines their needs, and assigns mechanics as required. Prepares cost estimates, handles service complaints, and in general, coordinates shop activity.

Service literature: Diagnosis and repair manuals, CDs, or Internet resources. Contain specifications and diagnosis and repair instructions.

Service valve: A valve (hand, Schrader or coupling type) used to open sealed air conditioning system to perform service operations.

Service writer: Person who greets customer, determines their needs, and writes up necessary information. Service writer is primarily concerned with paperwork involved. Beyond this point, other persons take over.

Servo: Oil pressure–operated device used to push or pull another part, such as tightening the transmission brake bands.

Servo action: Brakes so constructed as to have one end of primary shoe bearing against end of secondary shoe. When brakes are applied, primary shoe attempts to move in the direction of the rotating drum and in so doing applies force to the secondary shoe. Servo action makes less brake pedal pressure necessary and is widely used in drum brake construction.

Setscrew: A screw threaded through one part until it jams solidly against another, thus locking them together—such as a pulley to a shaft.

Shackle: Device used to attach ends of a leaf spring to frame.

Shaft drive angle: Angle formed between center lines of propeller shaft and differential pinion shaft.

Shaft runout: Amount of radial (up and down) movement that takes place during one complete revolution of shaft.

Shave: Removal of some chrome or decorative part.

Shave: Removal of metal from contact surface of cylinder head or block.

Shift forks: Devices that straddle slots cut in sliding gears. Fork is used to move gear back and forth on shaft.

Shift mechanism: Device for changing transmission gear range.

Shift point: Point, either in engine rpm or road speed, at which transmission should be shifted to next gear.

Shift rails: Sliding rods upon which shift forks are attached. Used for shifting the transmission (manual).

Shim: A thin piece of brass or steel inserted between two parts so as to adjust distance between them. Sometimes used to adjust bearing clearance.

Shim pack: A bearing shim made up of numerous thin (.001—.002 in.) laminations (layers). Bearing can be adjusted by removing a layer at a time until specified clearance is obtained.

Shock absorber: Oil filled device used to control spring oscillation in suspension system by forcing fluid through small openings.

Shop supervisor: Person responsible for overall shop operation. Checks progress and quality of work, observance of safety rules, utilization and condition of shop, coordination between various departments, customer relations, and documentation of all jobs and other financial transactions.

Short (short circuit): Refers to some "hot" portion of the electrical system that has become grounded. (Wire touching a ground and providing a completed circuit to the battery.)

Short block: Engine block complete with crankshaft and piston assemblies, but without some other components, such as the cylinder heads, oil pump, and oil pan.

Shorting harness: An arrangement of wires that are used to short out (ground) spark plugs so as to perform engine cylinder balance test.

Shot peening: A process utilizing a blast of small metal or plastic shot that is directed against surface of some part for cleaning.

Shrink-fit: Fit between two parts that is so tight, outer or encircling piece must be expanded by heating so it will fit over inner piece. In cooling, outer part shrinks and grasps inner part securely.

Shroud: Metal enclosure around fan, engine, or engine compartment to guide and facilitate airflow.

Shunt: An alternate or bypass portion of an electrical circuit.

Shunt winding: Wire coil forming an alternate or by-pass circuit through which current may flow.

Side air bag: Air bag located in the side door or header panel above the door. Provided protection for a side impact.

Sidewall: Part of tire between tread and bead, has size and rating information.

Sight glass: Clear glass window in R12 air conditioning line, which lets mechanic check refrigerant for bubbles indicating low refrigerant.

Silver soldering: Similar to brazing except that special silver solder metal is used.

Single-barrel, double-barrel, and four-barrel carburetors: Number of throttle openings from the carburetor to the intake manifold.

Single cut file: A file utilizing a single row of angled cuts that are all parallel.

Single lap flare: A tubing flare in which a single layer of tubing is secured between fitting body and nut seal angles.

Single plate clutch: A clutch utilizing a single clutch disc secured between flywheel and pressure plate.

Single post frame lift: A vehicle lift utilizing one centrally located ram.

Sintered bronze: Tiny particles of bronze pressed tightly together so that they form a solid piece. The piece is highly porous and is often used for filtering purposes.

Sintering: Process involved in forming a part, such as a bushing, by heating (no melting involved) a compressed mass of powdered metal.

Sipe: Small slits in tire tread designed to increase traction. Also called kerfs.

Siphon: Process of removing liquid from container by use of atmospheric pressure and gravity acting on a column of liquid in a hose or pipe. One end of hose or pipe is submerged in liquid and other (discharge end) is held at a point below level of liquid in container.

SIR: See *Air bag system.*

Skid plate: Metal plate or plates attached to underside of vehicle to protect oil pan, transmission, or fuel tank from damage caused by "grounding out" on rocks, curbs, and road surface.

Skirt collapse: Condition in which diameter across piston skirt area has been reduced through action of heat and shock.

Skirt thrust surface: That portion of piston skirt at right angles to piston pin when viewing piston from top.

Skiving: Cutting away a portion of tire tread to correct out-of-round problem.

Skiving hose: Removing one or more outer layers of hose rubber so that skived section can be pushed into a hose fitting. Usually involves removal of rubber down to first fabric layer.

Skiving knife: A special knife used to skive hose prior to insertion in fitting.

Slag: Accumulation of hardened and burned flux left on weld area following brazing or arc welding in which flux was used.

Slag inclusions: A weld in which weld metal contains particles of slag.

Slave cylinder: In a hydraulic system, a cylinder containing a piston (such as a brake wheel cylinder) that is forced to move by pressure generated in another (connected) cylinder (such as brake master cylinder).

Sleeve: See *Cylinder sleeve.*

Sleeve puller: Tool used to withdraw and install engine cylinder sleeves.

Slide hammer: A pulling device that utilizes a sliding weight that can be hammered against a portion of handle to generate pulling force.

Sliding fit: See *Running-fit.*

Sliding friction: Friction (resistance to movement) between two parts in which surface of one is sliding over surface of other.

Sliding gear: Transmission gear splined to the shaft. It may be moved back and forth for shifting purposes.

Slip angle: Difference in actual path taken by a car making a turn and path it would have taken if it had followed exactly as wheels were pointed.

Slip joint: Joint that will transfer driving torque from one shaft to another while allowing longitudinal movement between two shafts.

Slip rings: Rings mounted on alternator rotor. Brushes rub against slip rings to deliver field current to rotor windings.

Slipper pump: A pump using spring-loaded, slipper-like vanes riding in grooves in a rotor. Rotor operates inside a cam-shaped ring.

Slow charging: Recharging a battery by passing a low current (5 to 7 amps) through it for a relatively long period (14 to 24 hours).

Slow idle: Normal hot engine idle speed.

Sludge: Black, mushy deposits throughout interior of the engine. Caused from mixture of dust, oil, and water being whipped together by moving parts.

Small hole gauge: A device used to measure inside diameter of small holes. Gauge is adjusted to hole and gauge is then measured with an outside micrometer.

Smog: Fog made darker and heavier by chemical fumes and smoke.

SNAP program: Significant New Alternatives Program of the Environmental

Protection Agency. A program to identify refrigerants that can be used to replace R-12. Refrigerants found acceptable for their impact on global warming might not perform well in an air conditioning system.

Snap ring: Split ring snapped into a groove in a shaft or in a groove in a hole. It is used to hold many types of automotive parts in place.

Snapshot: A method of recording engine operating conditions at the exact moment that a malfunction occurs. Done with a scan tool.

Soaking tank: A cleaning tank in which parts are submerged for a recommended period in some type of cleaner.

Sodium valve: Valve in which stem has been partially filled with metallic sodium to speed up transfer of heat from valve head, to stem and then to guide and block.

Soft water: Water containing a minimum amount of minerals such as calcium and magnesium.

SOHC: Engine with single overhead camshaft.

Soldering: Joining two pieces of metal together with lead-tin mixture. Both pieces of metal must be heated to ensure proper adhesion of melted solder.

Soldering flux: A cleaning agent (rosin or acid type) that is placed on parts to be soldered. Prevents oxidization of surface and greatly assists in solder flow and adhesion.

Soldering gun: Electrically heated soldering iron shaped somewhat like a pistol. Trigger switch sends current through copper tip, raising it to soldering heat. Usually used for light soldering such as done on wiring.

Solenoid: Electrically operated magnetic device used to move a plunger.

Solid axle: Single beam runs between both wheels. May be used on either front or rear of car.

Solid state: An electrical device with no moving parts. Electronic components perform all electrical functions.

Solvent: Liquid used to dissolve or thin other material. Examples: Alcohol thins shellac; paint thinner dissolves grease.

Spade terminal: A male electrical connector with a single, slotted, flat blade.

Spalling: A condition in which tiny areas of bearing surface have flaked off. As flaking accelerates, large craters can be formed.

Spark: Bridging or jumping of a gap between two electrodes by current of electricity.

Spark advance: Causing spark plug to fire earlier.

Spark arrestor: Device used to prevent sparks (burning particles of carbon) from being discharged from exhaust pipe. Usually used on off-road equipment to prevent forest fires.

Spark gap: Space between center and side electrode tips on a spark plug.

Spark knock: See *Preignition.*

Spark lighter: A flint and steel device used to produce sparks with which to light an oxyacetylene torch.

Spark plug: Device containing two electrodes across which electricity jumps to produce a spark to fire air-fuel mixture.

Spark plug heat range: An indication of how fast plug transfers heat from insulator tip, through insulator to shell and into head and atmosphere. Longer or narrower insulator produces hotter tip.

Spark plug oil fouling: Excessive amount of oil entering combustion chamber and building up on plug until it misfires.

Spark plug reach: Distance from plug-to-head contact surface (gasket or taper) to end of threaded nose section.

Spark plug size: Relates to outside diameter (major thread diameter) of threaded nose section.

Spark plug type: Refers to plug construction, such as resistor, nonresistor, single electrode, thread length and diameter, heat range, and sealing method.

Specialization: Mechanic concentrated in one particular area of auto repair—brakes, drivability, transmission, engines.

Specific gravity: Relative weight of a given volume of a liquid, compared to weight of an equal volume of water.

Specific gravity test: Method of testing a battery by measuring the specific gravity of the electrolyte. See *Hydrometer.*

Specifications: The published service tolerances and settings for a vehicle.

Speed: Time rate of motion without regard to direction. Forward speed (mph or km/h) of a vehicle, rotational speed (rpm) of an engine.

Speed density: Method of determining the amount of air going into the intake manifold by monitoring sensor input and calculating the amount of airflow based on the sensor readings.

Speed handle: A socket hand tool with an offset (crank) area that permits rapid and full rotation of socket.

Speedometer: Instrument used to determine forward speed of an auto in miles per hour or kilometers per hour.

Spherical roller: A bearing roller in which center outside diameter exceeds that of either end giving roller a convex (barrel shape).

Spider gears: Small gears mounted on shaft pinned to differential case. They mesh with, and drive, the axle end gears.

Spindle: Machined shaft upon which inside races of front wheel bearings rest. Spindle is an integral part of steering knuckle.

Spin-on filter: An engine oil filter that is a complete unit (case and filter) in itself. It is attached by threading it onto the engine.

Spiral bevel gear: Ring and pinion setup widely used in automobile differentials. Teeth of both ring and pinion are tapered and are cut on a spiral so that they are at an angle to centerline of pinion shaft.

Spirit level: A leveling device using a floating bubble visible inside a glass tube.

Splash lubrication system: Engine oiling system that depends on connecting rods to dip into oil troughs and splash oil to all moving parts.

Spline: Metal (land) remaining between two grooves. Used to connect parts.

Splined joint: Joint between two parts in which each part has splines cut along contact area. Splines on each part slide into grooves between splines on other part.

Split keepers: Valve spring retainer lock made up of a two piece tapered cone that wedges between retainer and valve stem grooves.

Split manifold: Exhaust manifold that has a baffle placed near its center. An exhaust pipe leads out of each half.

Spongy pedal: When there is air in brake lines, or shoes that are not properly centered in drums, brake pedal will have a springy or spongy feeling when brakes are applied. Pedal normally will feel hard when applied.

Spool balance valve: Hydraulic valve that balances incoming oil pressure against spring control pressure to produce a steady pressure to some control unit.

Spool valve: Hydraulic control valve shaped somewhat like spool upon which thread is wound.

Spot weld: Fastening parts together by fusing, at various spots. Heavy surge of electricity is passed through the parts held in firm contact by electrodes.

Sprag clutch: A type of overrunning clutch. Uses small metal pieces called sprags to allow rotation in one direction only.

Spring booster: Device used to compensate for sagging springs or to increase the load capacity of standard springs.

Spring capacity at ground: Total vehicle weight (sprung and unsprung) that will be carried by spring bent or deflected to its maximum normal loaded position.

Spring capacity at pad: Total vehicle sprung weight that will be carried by spring bent or deflected to its normal fully loaded position.

Spring free length: Overall length of a coil spring when all compression forces are removed.

Spring loaded: Device held in place, or under pressure from a spring or springs.

Spring oscillation: Rapid compression and rebound movement that exists in a coil spring following a sudden change in loading pressure.

Spring rebound: Reverse direction spring movement following sudden compression (loading). Spring attempts to return to its original position.

Spring scale: A weighing device using a coil spring that is stretched as pressure is applied. Stretching is indicated by a needle moving along a scale.

Spring skirt expander: A spring steel device inserted inside a piston to apply outward (expansion) pressure to skirt thrust surfaces.

Spring squareness: Refers to angle of spring ends as related to spring centerline. Should be perfectly square (90°).

Spring steel: Heat treated steel having the ability to stand a great amount of deflection and yet return to its original shape or position.

Spring windup: Curved shape assumed by rear leaf springs during acceleration or braking.

Sprocket: Toothed wheel used to drive chain or cogged belt.

Sprung weight: Weight of all parts of car that are supported by suspension system.

Spur gear: Gear on which teeth are cut parallel to shaft.

Spurt or squirt hole: Small hole in connecting rod big end that indexes (aligns) with oil hole in crank journal. When holes index, oil spurts out.

Square engine: Engine in which bore diameter and stroke are of equal dimensions.

Stabilizer bar: Transverse mounted spring steel bar that controls and minimizes body lean or tipping on corners. Each end of the bar is attached to the right and left lower control arms.

Stabilizer bar bushing: Flexible bushing that attaches the stabilizer bar to the frame and suspension while cushioning movement.

Staked insert: A valve seat insert that is held in recess by extruding some of head metal over seat edge by hammering metal with a special punch.

Staked nut: Nut held in position by staking.

Staking: Holding a fastener in place by hammering a portion of part into a groove or cutout area in mating part. This prevents loosening due to vibration. A staking punch, with an appropriate nose, is needed.

Stall: To stop rotation or operation.

Stall test: Testing torque converter-transmission action by applying full engine load (wide open throttle) with transmission in gear and brakes firmly applied. Must be very brief test (5 seconds). Engine rpm is noted at full stall.

Stall torque: Full torque developed by starter motor when pinion is prevented from turning.

Stamping: Sheet metal part formed by pressing between metal dies.

Star wheel: A small brake shoe adjusting wheel with a series of radial prongs that can be engaged by tip of a screwdriver so as to turn wheel.

Starter: Electric motor that cranks (turns) the engine for starting.

Starter drive: Starter pinion gear and related engagement and disengagement mechanisms.

Starter pinion gear: Small gear on end of starter shaft that engages and turns large flywheel ring gear.

Starter solenoid: Large electric relay that makes and breaks the electrical connection between the battery and starting motor.

Starting punch: A punch, with a tapered nose, used to start pins out of their holes.

Starting system: Parts (starter motor, switch, solenoid, battery, and connecting wires) used to crank car for starting.

State of charge: Refers to amount of potential electrical energy present in battery at time of testing. Indicated by specific gravity of electrolyte.

Static balance: When a tire, flywheel, crankshaft, or other rotating part has an absolutely even distribution of weight mass around axis of rotation, it will be in static balance. For example, if front wheel is jacked up and tire, regardless of where it is placed, always slowly turns and stops with the same spot down, it would not be in static balance. If, however, wheel remains in any position in which it is placed, it would be in static balance. (Bearings must be free and there must be no brake drag)

Static electricity: Electricity generated by friction between two objects. It will remain in one object until discharged.

Static emissions test: Test of exhaust emissions as engine idles. Sometimes called basic test or idle only test.

Static pressure: Pressure that always exists in brake lines—even with brake pedal released. Static pressure is maintained by a check valve.

Static radius: Distance from road surface to center of wheel with vehicle normally loaded, at rest.

Static suppression: Removal or minimizing of unwanted electromagnetic waves that cause radio static.

Stator: Small hub, upon which series of vanes are affixed in radial position, that is so placed that oil leaving torque converter turbine strikes stator vanes and is redirected into pump at an angle conducive to high efficiency. Stator makes torque multiplication possible. Torque multiplication is highest at stall when the engine speed is at its highest and the turbine is standing still.

Steam cleaning: Cleaning by using a jet of steam directed against surface to be cleaned.

Steel drift: A steel bar or heavy punch that is placed against an object so that it can be moved by hammering against drift.

Steel pack muffler: Straight-through (no baffles) muffler utilizing metal shavings surrounding a perforated pipe. Quiets exhaust sound.

Steering arms: Arms, either bolted to, or forged as an integral part of steering knuckles. They transmit steering force from tie rods to knuckles, thus causing wheels to pivot.

Steering axis inclination: An imaginary line passing through the upper and lower ball joint or top of the strut mount and lower ball joint. Setting the upper ball joint or top of the strut mount towards the center of the vehicle places the line of steering axis nearer centerline of tire-road contact area.

Steering gear: Gears, mounted on lower end of steering column, used to multiply turning force.

Steering geometry: Term used to describe various angles assumed by components making up front wheel turning arrangement, primarily camber, caster, and toe-in. Also used to describe related angles assumed by front wheels when car is negotiating a curve.

Steering knuckle: Inner portion of spindle affixed to and pivoting on either a kingpin or on upper and lower ball joints.

Steering knuckle angle: Angle formed between steering axis and centerline of spindle. This angle is sometimes referred to as Included Angle.

Steering linkage: Various components connecting steering gear to front wheels.

Steering system: All parts used in transferring motion of steering wheel to front wheels.

Stethoscope: Device (as used by doctors) to detect and locate abnormal engine noises. Handy tool for troubleshooting.

Stick shift: Transmission that is shifted manually through use of various forms of linkage. Often refers to upright gearshift stick that protrudes through floor. Either floor or steering column mounted manual shift device for transmission.

Stoichiometric fuel mixture: A fuel mixture in which proportions of air and fuel are such as to permit complete burning. The ideal mixture for any given engine and set of conditions, usually 14.7 to 1.

Stone dresser: A device used to cut a new surface on a grinding wheel or stone. May be diamond tipped or mechanical star wheel type.

Stone sleeve: Cylindrical unit upon which engine valve seat grinding stone is attached. Sleeve slips over and is guided by pilot. Also called a mandrel.

Stone sleeve pilot: Rigid metal rod that is twisted into valve guide with upper protruding portion acting as a guide or pilot for grinding stone sleeve.

Stone truing: Cutting a fresh, correct surface on a grinding stone by passing a diamond tipped dresser across stone at correct angle.

Stoplight: Warning lights, red in color, attached to rear of vehicle. Stoplights come on whenever brake pedal is depressed.

Storage battery: Another term for vehicle battery. See *Battery*.

Straightedge: Long, flat, relatively thin, steel strip with perfectly straight edges. Used for drawing straight lines, checking surfaces for warping.

Straight polarity: In DC arc welding, having current flow from machine negative terminal (rod holder cable attached to negative terminal) through electrode to work and then on through ground cable to machine positive terminal.

Straight roller bearing: A bearing roller in which roller outside diameter is constant from one end to other.

Stress: To apply force to an object. Force or pressure an object is subjected to.

Striking an arc: Scratching arc welding electrode along workpiece to establish an arc producing current flow.

Striping tool: Tool used to apply paint in long narrow lines.

Stroboscope: See *Timing light*.

Stroke: Distance piston moves when traveling from TDC to BDC.

Stroked crankshaft: Crankshaft, either special new one or stock crank reworked, that has connecting rod throws offset so that length of stroke is increased.

Strut rod: Metal bar that connects the lower control arm to the frame. Used to prevent back and forth motion of the lower control arm.

Stud: Metal rod with threads on both ends.

Stud puller: Tool used to install or remove studs.

Suction: See *Vacuum*.

Suction line: See *Low-pressure line*.

Suction throttling valve (STV): Valve placed between air conditioning evaporator and compressor on some older vehicles. Controls evaporator pressure to provide maximum cooling without icing evaporator core.

Sulfation: Formation of lead sulfate on battery plates.

Sulfuric acid: Acid (36%) that is mixed with water (64%) to form battery electrolyte. A powerful acid; wear goggles and rubber gloves.

Sump: Part of oil pan that contains oil.

Sun gear: Center gear around which planet gears revolve.

Supercharger: Unit designed to force air, under pressure, into cylinders. Can be mounted between throttle valve and cylinders or between throttle valve and atmosphere.

Superheat switch: See *Compressor protection switch*.

Supplemental inflatable restraint: See *Air bag system*.

Sure-grip differential: High traction differential which causes both axles to rotate under power.

Surface charge: Condition occurring immediately after charging when the battery cells have a higher than normal charge. Surface charge must be removed before testing the battery.

Surface-type filter: A filter in which filtering medium traps dirt on surface. A paper filter without additional depth filtering material would be a surface type filter.

Surge bleeding: Rapid stroke of brake pedal (with pressure bleeder attached to master cylinder reservoir and wheel bleed screw open) to cause a violent surge of fluid into wheel cylinder in order to remove trapped air.

Surge tank: Fuel injection component used to damp out fuel pump pulsations.

Suspension height: A specified distance from one or more suspension components (such as suspension lower arm) to either floor or to some spot on body. This will establish correct suspension heights and/or angles in relation to vehicle frame or body.

Suspension lift: A lift that raises vehicle by engaging only front and rear suspension systems (such as front, lower suspension arms and rear axle housing).

Swaged insert: An engine valve seat insert that is held in its recess by swaging (rolling pressure that forces metal to flow or extrude), thus locking seat by forcing head metal over a chamfer cut on seat outside diameter.

Sway bar: See *Stabilizer bar*.

Sweating: Joining two pieces of metal together by placing solder between them and then clamping them tightly together while heat, sufficient to melt the solder, is applied.

Swing axle: Independent rear suspension system in which each driving wheel can move up or down independently of other. Differential unit is bolted to frame and various forms of linkage are used upon which to mount wheels. Drive axles, utilizing one or more universal joints, connect differential to drive wheels.

Switch: A device to make (complete) or break (interrupt) flow of current through a circuit.

Synchronize: To cause two or more events to occur simultaneously. Examples are plug firing when the piston is in correct position, speed of two shafts being the same, valve opening when piston is in correct position.

T

Tachometer: Device used to indicate speed of engine in rpm.

Taillight: Lights, usually red or amber, attached to rear of vehicle. Lights operate in conjunction with headlights.

Tail pipe: Exhaust piping running from muffler to rear of vehicle.

Tank gauge unit: Variable resistor device in fuel tank. It operates fuel gauge in dashboard.

Tap: To cut threads in a hole, or can be used to indicate fluted tool used to cut threads.

Tap and die set: Set of taps and dies for internal and external threading. Usually covers a range of the most popular sizes.

Tap water: Water as drawn from a typical household faucet.

Tapered end axle: Axle using a tapered end upon which wheel hub is secured by means of a key and attaching nut.

Tapered roller bearing: Bearing utilizing series of tapered, hardened steel rollers operating between an outer and inner hardened steel race.

Tappet: Screw used to adjust clearance between valve stem and lifter or rocker arm.

Tappet clearance: On non-hydraulic lifter setups, operating clearance that exists in valve train to prevent valve from being held open when parts heat up and elongate.

Tappet noise: Noise caused by too much lash or clearance between valve stem and rocker arm or between valve stem and valve lifter.

TDC: Top Dead Center.

Technical service bulletins (TSB): Information published by vehicle manufacturers in response to vehicle problems that may not be diagnosed by normal methods.

Teflon: Plastic with excellent self-lubricating (slippery) bearing properties.

Telescoping gauge: Measuring tool consisting of a spring loaded collapsible rod affixed at right angles to a locking handle. Used to measure inside diameter of holes.

Temper: To effect a change in physical structure of piece of steel through use of heat and cold.

Temperature gauge: Dash mounted instrument to indicate temperature of engine coolant.

Tensile strength: An indication of how much stretching force a material can withstand before it breaks.

Tension: Pulling or stretching stress applied to an object.

Tension loaded ball joint: Suspension design that places forces on the ball joint that attempt to pull ball out of socket.

Tension loading: Placing a load, in such a manner, as to apply a pulling or stretching force to an object. A torqued fastener is under tension loading.

Terminal: Connecting point in electric circuit. When referring to battery, it would indicate two battery posts.

Terminal block: A central distribution block that receives current from one wire and through means of a bus bar, conducts it to a number of others.

Test lamp: A lightbulb wired in series with a battery and two test wire leads. Used to check electrical circuits.

T-fitting: A T-shaped fitting that may be inserted into a vacuum or oil line for purpose of attaching an additional line at same point.

T-handle: Socket wrench extension with a round rod handle passed through it at right angles at one end. Other end has square socket drive connection tip.

T-head engine: Engine having intake valve on one side of cylinder and exhaust on other.

Thermal bulb: A temperature sensitive device consisting of a hollow bulb filled with liquid, connected by means of a capillary tube to a control unit.

Thermal efficiency: Percentage of heat developed in burning fuel charge that is actually used to develop power determines thermal efficiency. Efficiency will vary according to engine design. If an engine utilizes a great deal of heat to produce power, its thermal efficiency would be high.

Thermal sensor: A temperature sensitive sensor.

Thermal time switch: A time controlled switch that is also temperature sensitive. Used with some fuel injection system cold start valves.

Thermistor: Resistor that changes its resistance in relation to temperature.

Thermostat: Temperature sensitive device used in cooling system to control flow of coolant in relation to temperature.

Thermostatically controlled air cleaner (TAC): An emission control device used to control temperature of air entering air cleaner. Cleaner receives heated air during engine warm-up.

Thermostatic spring: A heat sensitive, bimetallic (two kinds of metal) spring that winds or unwinds as temperature changes take place.

Thermostatic switch: A switch that is actuated by temperature changes.

Thread length: A measurement of length of threaded area on a bolt, capscrew, or stud.

Thread-pitch gauge: A gauge used to determine number of threads per inch on a fastener. Consists of many leaves, each with a specific thread cut on one edge.

Three-quarter floating axle: A drive axle that secures, aligns, and drives wheel. Does not support vehicle weight.

Three-way catalytic converter: Converter, sometimes called dual converter, that combines both an oxidizing and reducing catalyst. Controls NO_X, CO, and HC emissions.

Throttle body: Part of the air intake system that contains the throttle plate(s). Also contains sensors and vacuum connectors.

Throttle body injection: See *Central fuel injection.*

Throttle return dashpot: Carburetor device that slows throttle closing and prevents stalling.

Throttle valve: Valve in throttle body. It is used to control the amount of air that reaches cylinders.

Throttle valve: Valve in automatic transmission that controls oil pressure to downshift side of shift valves.

Throttle valve sensor and throttle position sensor: Sensor that measures amount of throttle valve opening and provides information for fuel injection computer.

Throw: Offset portion of crankshaft designed to accept connecting rod.

Throw off oil: Lubricating oil that is thrown outward from rotating parts.

Throw-out bearing: Bearing that is forced against clutch throw-out fingers to release clutch.

Throw-out lever: Pivoted lever that actuates clutch throw-out bearing.

Throw-out sleeve: Sliding sleeve upon which clutch throw-out bearing is mounted. Actuated by throw-out lever.

Thrust: A pushing or shoving force exerted against one body by another.

Thrust angle: Imaginary lines of force that cross lengthwise through vehicle's tires.

Thrust bearing: Bearing designed so as to resist side pressure.

Thrust flange: A shoulder on a shaft that rides against a thrust bearing to control end play and thrust forces.

Thrust loading: Load force applied parallel to the centerline of a rotating part.

Thrust surface: That portion (surface) of a part that either receives or transmits a force from or to another part.

Thrust washer: Bronze or hardened steel washer placed between two moving parts. The washer prevents longitudinal movement and provides a bearing surface for thrust surfaces of parts.

Tie rod: Rod, or rods, connecting steering arms together. When tie rod is moved, wheels pivot.

TIG: Gas tungsten arc welding (tungsten inert gas).

Tightening sequence: Order in which fasteners should be tightened or torqued to prevent part distortion.

Time fuel injection: Fuel injection is timed to occur when intake valve opens.

Timing: The act of coordinating two or more separate events or actions in relation to each other. Example: Timing the firing of plug to piston position on compression stroke.

Timing belt: A flexible, toothed belt used to rotate camshaft.

Timing chain: Drive chain that operates camshaft by engaging sprockets on camshaft and crankshaft.

Timing cover: Cover over front of engine. Houses timing chain or gear mechanism and front crankshaft oil seal.

Timing gears: Both the gear attached to the camshaft and the gear on the crankshaft. They provide a means of driving the camshaft.

Timing light: Stroboscopic unit that is connected to secondary circuit to produce flashes of light in unison with firing of specific spark plug. By directing these flashes of light on whirling timing marks, marks appear to stand still. By adjusting distributor, timing marks may be properly aligned, thus setting timing.

Timing marks (ignition): Marks, usually located on vibration damper, used to synchronize ignition system so plugs will fire at precise time.

Timing marks (valves): One tooth on either the camshaft or crankshaft gear will be marked with an indentation or some other mark. Another mark will be found on other gear between two of teeth. Two gears must be meshed so that marked tooth meshes with marked spot on other gear.

Timing sprockets: Chain- or belt-type sprockets on crankshaft and camshaft.

Tinning: Coating piece of metal with a very thin layer of solder.

Tin-plated: Covered with a very thin coating of tin.

Tire balance: In that tires turn at relatively high speeds, they must be carefully balanced both for static balance and for dynamic balance.

Tire bead: Portion of tire that bears against rim flange. Bead has a number of turns of steel wire in it to provide great strength.

Tire casing: Main body of tire exclusive of tread.

Tire inflation pressure: Recommended maximum tire air pressure as given on tire sidewall.

Tire load range: Letter system (A, B, C) used to indicate specific tire load and inflation limit.
Tire plies: Layers of fabric or steel used to form tire casing.
Tire rotation: Moving front tires to rear and rear to front to equalize any wear irregularities.
Tire sidewall: Portion of tire between tread and bead.
Tire size: Given on tire sidewall as coded letter-number designation of size, section height, and diameter across bead.
Tire tread: Part of tire that contacts road.
Titania-type oxygen sensor: Sensor that produces a resistance reading to indicate exhaust gas condition.
Toe-in: Having front of wheels closer together than back (front wheels). Difference in measurement across front of wheels and the back will give amount of toe-in.
Toe-out: Having front of wheels further apart than the back.
Toe-out on turns: When car negotiates a curve, inner wheel turns more sharply and while wheels remain in this position, a condition of toe-out exists.
Toggle switch: Switch actuated by flipping a small lever either up and down or from side to side.
Tolerance: Amount of variation permitted from an exact size or measurement. Actual amount from smallest acceptable dimension to largest acceptable dimension.
Tooth contact pattern: Refers to shape, size, and location of actual contact area between two mating gear teeth.
Tooth heel: Wider outside end of tooth.
Tooth toe: Narrower inside end of tooth.
Top off: Fill a container to full capacity.
Torque: Turning or twisting force such as force imparted on drive line by engine.
Torque arm: An arm or lever attached to some part to control amount and direction of movement caused by torque (twisting) forces.
Torque bar or rod: An articulated bar between frame and drive axle housing designed to relieve leaf springs of axle torque (twisting) strain. Prevents axle windup and/or hop during heavy acceleration or braking. Also called traction bar.
Torque converter: Unit, similar to fluid coupling, that transfers engine torque to transmission input shaft. Unlike fluid coupling, torque converter can multiply engine torque. This is done by installing one or more stators between torus members. Driving torus is referred to as pump or impeller and driven torus as turbine.
Torque multiplication: Increasing engine torque through the use of a torque converter.
Torque-plus-angle: Fastener installation method that calls for tightening a fastener to a specified torque, then tightening an additional part of a turn, such as 90 degrees (1/4 turn).
Torque stick: Calibrated tool used with an impact wrench to remove wheel lug nuts.
Torque-to-yield-fastener: Fastener that, when tightened to the specified torque, stretches (yields) to maintain the proper tension on the parts being fastened. Torque-to-yield-fasteners cannot be reused.
Torque tube drive: Method of connecting transmission output shaft to differential pinion shaft by using an enclosed drive shaft. Driving force of rear wheels is transferred to frame through torque tube. Used on only a few vehicles.
Torque wrench: Wrench used to tighten fasteners to specified tension by measuring torque (turning force) being applied.
Torquing sequence: See *Tightening sequence.*
Torsional vibration: Twisting and untwisting action developed in shaft. It is caused either by intermittent applications of power or load.
Torsion bar: Long spring steel rod attached in such a way that one end is anchored while other is free to twist. If an arm is attached at right angles to free end, any movement of arm will cause rod or bar to twist. Bar's resistance to twisting provides a spring action. Torsion bar replaces both coil and leaf springs in some suspension systems.
Torsion bar suspension: Suspension system that makes use of torsion bars in place of leaf or coil spring.
Torus: Fluid coupling rotating member. There are two—driving and driven torus.
Tote board: Portable board or rack containing a series of tools such as pullers and accessories.
Toxic: Poisonous.
TPC: Tire Performance Criteria.
Track: Distance between front wheels or distance between rear wheels. They are not always the same.
Traction: Frictional force generated between tire and road. Necessary for acceleration, braking, and steering.
Traction bar: Articulated bar or link attached to frame and rear axle housing to prevent spring windup during heavy acceleration or braking.
Traction control system: Computerized system that governs engine throttle and brake system operation to increase vehicle traction on low-friction surfaces.
Traction differential: See *Limited-slip differential.*
Trammel: Steel bar or rod with adjustable, sliding indicator arms. Used to check measurements between parts. Often used for checking wheel and tire toe-in.
Transaxle: Drive setup in which transmission and differential are combined into a single unit.
Transducer: Vacuum regulator actuated or controlled electrically. A device that converts an input signal (electrical) into an output signal (diaphragm movement) of a different form.
Transfer case: Gearbox, driven by transmission, that will provide driving force to both front and rear propeller shafts on four-wheel drive vehicle. Some transfer cases are manually operated. Most modern transfer cases are automatic.
Transformer: Electrical device used to increase or decrease voltage. Car ignition coil transforms voltage from 12 volts to as much as 100,000 volts.
Transistor: Electrical device made of semi-conducting material and using at least three electrical connections. Often used as a switching device.
Transistorized: Electronic device employing transistors. Can also refer to electronic, as opposed to mechanical, control or operation of some function, such as making and breaking ignition primary circuit.
Transmission: Device that uses gearing or torque conversion to cause a change in ratio between engine rpm and driving wheel rpm. When engine rpm goes up in relation to wheel rpm, more torque but less speed is produced. Reduction in engine rpm in relation to wheel rpm produces a higher road speed but delivers less torque.
Transmission adapter: A unit that allows a different make or year transmission to be bolted up to original engine.
Transverse leaf spring: Leaf spring mounted at right angles to length of car.
Transverse mounted: Mounted crosswise or at right angles to vehicle centerline.
Traps: Area over which car is raced for timing purposes.
Tread: Distance between two front or two rear wheels.
Tread: Portion of tire which contacts roadway.
Tread ply: Reinforcing plies lying beneath tread area. These are in addition to body plies that also pass under tread.
Tread separation: Loosening of the bond between the tire tread and the tire casing. In extreme cases, the tread may come completely off the casing.
Tread width: Distance between outside edges of tread as measured across tread surface.
Trickle charging: Maintaining a charge in a stored battery by passing a constant, but very low current, through the battery.
Trip odometer: Auxiliary odometer that may be reset to zero at option of driver. Used for keeping track of mileage on trips up to one thousand miles.

Trouble codes: Numerical value that represents an abnormal operating condition as detected by a self-diagnostic system. These codes are usually extracted from the computer's memory during service.

Troubleshooting: Diagnosing vehicle problems by various tests and observations.

Troubleshooting chart: Diagnostic flow chart that provides step-by-step procedures to test automotive systems.

Tube cutter: Tool used to cut tubing by passing a sharp wheel around tube.

Tune-up: Process of replacing parts and making adjustments to obtain maximum performance from engine.

Turbine: Wheel upon which series of angled vanes are affixed so moving column of air or liquid will impart a turning motion to wheel.

Turbocharger: Exhaust powered supercharger.

Turbulence: Violent, broken movement or agitation of a fluid or gas.

Turning radius: Diameter of circle transcribed by outer front wheel when making a full turn.

TV: Throttle valve rod or cable that extends from foot throttle linkage to throttle valve in automatic transmission.

TVS: Thermostatic Vacuum Switch.

Twist drill: Metal cutting drill with spiral flutes (grooves) to permit exit of chips while cutting.

Two-stroke cycle engine: Engine requiring one complete revolution of crankshaft to fire each piston once. Not used in vehicles.

U

UAW: United Auto Workers.

U-bolt: A horseshoe-shaped bolt, with threads on each end, that passes around axle housing to secure leaf spring.

Umbrella seal: A cover placed over valve stem end and extending out over valve spring. Directs dripping oil away from stem and guide area to minimize oil leakage past valve guides.

UNC: Unified National Coarse thread sizes.

Unburned hydrocarbons: Hydrocarbons that were not burned during the normal engine combustion process. Unburned hydrocarbons make up about 0.1 percent of engine exhaust emission.

Undercoating: Soft deadening material sprayed on underside of vehicle.

Under-square engine: Engine in which bore diameter is smaller than length of stroke.

Understeer: Tendency for car, when negotiating a corner, to turn less sharply than driver intends.

UNF: Unified National Fine thread size.

Unit body: Car body in which the body itself acts as frame.

Unit loading: Applying a loading force to very specific area or part, such as is done by adding a crimp or wire to a gasket to increase pressure around an opening.

Universal joint: Flexible joint that will permit changes in driving angle between driving and driven shaft.

Unleaded gasoline: Gasoline not containing tetraethyl lead. Must be used with vehicles equipped with a catalytic converter.

Unsprung weight: All parts of car not supported by suspension system; wheels, wheel bearings, tires, spindles, brake parts.

Upshift: Shifting to a higher gear.

V

Vacuum: Enclosed area in which air pressure is below that of surrounding atmospheric pressure.

Vacuum advance: Unit designed to advance and retard ignition timing through action of engine vacuum working on a diaphragm.

Vacuum amplifier: A device used to increase force or action produced by a vacuum.

Vacuum booster: Small diaphragm vacuum pump, generally in combination with fuel pump, used to bolster engine vacuum during acceleration so vacuum operated devices continue to operate.

Vacuum gauge: Gauge used to determine amount of vacuum in a chamber.

Vacuum modulator: Device that uses engine vacuum to control throttle valve in automatic transmission.

Vacuum motor: A device, utilizing a vacuum operated diaphragm, which causes movement of some other unit.

Vacuum pump: Diaphragm pump used to produce vacuum.

Vacuum runout point: Point reached when vacuum brake power piston has built up all the braking force it is capable of with vacuum available.

Vacuum switch: An electrical switch that is operated by vacuum.

Vacuum tank: Tank in which vacuum exists. Generally used to provide vacuum to power brake installation in event engine vacuum cannot be obtained. Tank will supply several brake applications before vacuum is exhausted.

Valve: Device used to either open or close an opening. There are many different types.

Valve body: Unit containing various shift valves, springs, solenoids, and passageways. Used to control transmission shift operation.

Valve clearance: Space between end of valve stem and actuating mechanism (rocker arm, lifter, lash adjuster.)

Valve duration: Length of time, measured in degrees of engine crankshaft rotation that valve remains open.

Valve face: Outer lower edge of valve head. The face contacts the valve seat when the valve is closed.

Valve face angle: See *Face angle.*

Valve float: Condition where valves in engine are forced back open before they have had a chance to seat. Brought about (usually) by extremely high rpm.

Valve grinding: Renewing valve face area by grinding on special grinding machine.

Valve guide: Hole through which stem of poppet valve passes. It is designed to keep valve in proper alignment. Some guides are pressed into place and others are merely drilled in block or in head metal.

Valve head: Portion of valve above stem.

Valve-in-head engine: Engine in which both intake and exhaust valve are mounted in the cylinder head and are driven by push rods or by an overhead camshaft.

Valve keeper or **valve key** or **valve retainer:** Small unit that snaps into a groove in end of valve stem. Designed to secure valve spring, valve spring retaining washer, and valve stem together. Some are of split design and some are horseshoe shaped.

Valve lash: Valve tappet clearance or total clearance in the valve operating train with cam follower on camshaft base circle.

Valve lift: Distance a valve moves from full closed to full open position.

Valve lifter or **cam follower:** Unit that contacts end of valve stem and camshaft. Follower rides on camshaft. Cam lobes move it upward to open valve.

Valve margin: Width of edge of valve head between top of valve and edge of face. Too narrow a margin results in preignition and valve damage through overheating.

Valve oil seal: Neoprene rubber ring placed in groove in valve stem to prevent excess oil entering area between stem and guide. There are other types of these seals.

Valve overlap: Certain period in which both intake and exhaust valve are partially open. (Intake is starting to open while exhaust is not yet closed.)

Valve port: Opening, through head or block, from intake or exhaust manifold to valve seat.

Valve rotator: Unit that is placed on end of valve stem so that when valve is opened and closed, the valve will rotate a small amount with each opening and closing. This gives longer valve life.

Valve seat: Area onto which face of valve seats when closed. Two common angles for this seat are forty-five and thirty degrees.

Valve seat grinding: Renewing valve seat area by grinding with a stone mounted upon a special mandrel.

Valve seat insert: Hardened steel valve seat may be removed and replaced.

Valve seat runout: A measurement of valve seat out-of-roundness (concentricity). Also provides an indication of seat squareness with valve guide centerline.

Valve spring: Coil spring used to keep valves closed.

Valve spring compressor: Tool used to collapse valve spring so that retainer may be removed.

Valve stem: Portion of valve below head. The stem rides in the guide.

Valve stem clearance: Clearance between stem and guide.

Valve tappet: Adjusting screw to obtain specified clearance at end of valve stem (tappet clearance). Screw may be in top of lifter, in rocker arm, or in the case of ball joint rocker arm, nut on mounting stud acts in place of a tappet screw.

Valve timing: Adjusting position of camshaft to crankshaft so that valves will open and close at the proper time.

Valve train: Various parts making up valve and its operating mechanism.

Valve umbrella: Washer-like unit placed over end of the valve stem to prevent entry of excess oil between stem and guide. Used in valve-in-head installations.

Vane: Thin plate affixed to rotating unit to either throw off air or liquid, or to receive thrust imparted by moving air or liquid striking the vane. In the first case, it would be acting as a pump and in the second case as a turbine.

Vane pump: Pump using two or more spring loaded vanes (flat, rectangular strips) operating in slots in a rotor. As rotor spins, vane ends rotate and rub against pump walls. Rotor is offset in relation to pump walls.

Vapor: Gaseous state of a substance usually a liquid or solid. Example: Steam.

Vapor charging: Charging air conditioning system with refrigerant vapor.

Vaporization: Breaking gasoline into fine particles and mixing it with air.

Vapor-liquid separator: A device used to separate liquid from a vapor—such as can be used in an engine PCV system to prevent engine oil being drawn out of engine along with crankcase vapors.

Vapor lock: Boiling or vaporizing of the fuel in the lines from excess heat. Boiling will interfere with movement of the fuel and will in some cases, completely stop the flow.

Vapor separator: A device used on cars equipped with air conditioning to prevent vapor lock by feeding vapors back to the gas tank via a separate line.

Variable pitch stator: Stator that has vanes that may be adjusted to various angles depending on load conditions. Vane adjustment will increase or decrease efficiency of stator.

Variable valve timing: ECM controlled device that can advance and retard engine valve timing in response to engine and vehicle conditions. Usually consists of a hydraulic advance mechanism controlled by solenoids.

Variable venturi: A carburetor venturi whose opening size can be varied to meet changing engine speed and load.

Varnish: Deposit on interior of engine caused by engine oil breaking down under prolonged heat and use. Certain portions of oil deposit themselves in hard coatings of varnish.

V-belt: V shaped belt commonly used to spin alternator, water pump, power steering pump, and air conditioning compressor.

V-block: A block of steel with a deep "V" groove cut in one or more spots. Can be used in pairs to support a shaft while it is turned to check for runout.

Velocity: Time rate of motion. Speed with which an object moves. Usually measured in feet per second or miles per hour.

Venturi: That part of a tube, channel, or pipe so tapered as to form a smaller or constricted area. Liquid, or a gas, moving through this area will speed up and as it passes narrowest point, a partial vacuum will be formed. Taper facing flow of air is much steeper than taper facing away from flow of air. Venturi principle is used in carburetor.

Vibration damper: Round weighted device attached to front of crankshaft to minimize torsional vibration.

Viscosimeter: Device used to determine viscosity of a given sample of oil. Oil is heated to specific temperature and then allowed to flow through set orifice. Length of time required for certain amount of flow determines oil's viscosity.

Viscosity: Measure of oil's ability to pour.

Viscosity index: Measure of oil's ability to resist changes in viscosity when heated.

Viscosity rating: A rating of how "thick" or "thin" a sample of oil may be. This refers to how fast a given amount will flow through a specific size hole at a given temperature.

Viscous coupling or viscous clutch: A nondirect coupling (driving unit) between two drive components that utilizes a fluid (such as silicone) to produce transfer of power. Allows for some slippage.

Volatile: A substance that evaporates (turns to vapor) easily. Example: Gasoline.

Volatility: Property of liquids to evaporate quickly and at relatively low temperatures.

Volt: Unit of electrical pressure or force that will move a current of one ampere through a resistance of one ohm.

Voltage: Difference in electrical potential between one end of a circuit and the other. Also called EMF (electromotive force). Voltage causes current to flow.

Voltage drop: Lowering of voltage due to excess length of wire, undersize wire, or another source of resistance.

Voltage drop test: Test method for determining the resistance in a connection. A voltmeter is connected across the connector and the circuit is energized. If the voltmeter reads above a certain value (usually about .2 volt) the connection has high resistance.

Voltage regulator: See *Regulator.*

Voltmeter: Instrument used to measure voltage in given circuit (in volts).

Volume: Measurement, in cubic inches, cubic feet, cubic centimeters, or cubic liters of amount of space within a certain object or area.

Volumetric efficiency: Comparison between actual volume of fuel mixture drawn in on intake stroke and what would be drawn in if cylinder were to be completely filled.

Vortex: Mass of whirling liquid or gas.

V-type engine: An engine with cylinders arranged in two separate banks (rows) and set at an angle (V-shape) to each other.

Vulcanization: Process of heating compounded rubber to alter its characteristics—making it tough and resilient.

W

Wandering: Condition in which front wheels tend to steer one way and then another.

Wankel engine: Rotary combustion engine having one or more three-sided rotors mounted on drive shaft operating in specially shaped chambers. Rotor turns constantly in one direction yet produces an intake, compression, firing, and exhaust stroke.

Warpage: Bending or twisting of an object caused by heat, improper fastener torque, or other source of stress.

Water detector: Sensor in diesel fuel system that warns driver of water contamination of fuel.

Water jacket: Area around cylinders and valves that is left hollow so that water may be admitted for cooling.

Water pump: The pump, usually a centrifugal type, used to circulate coolant throughout cooling system.

Water sensor: A device in a diesel fuel system to detect presence of water in fuel.

Watt: A unit of electrical power amounting to one joule per second.

Waveform: Single line pattern produced on an oscilloscope or waveform meter screen. Comparing the waveform with a known good waveform allows the technician to locate electrical problems.

Waveform meter: Device for displaying electrical activity in a circuit as a single line pattern, or waveform. The vertical direction indicates voltage; the horizontal direction indicates time.

Wedge combustion chamber: Combustion chamber having a wedge shape.

Weight distribution: Percentage of total vehicle weight as carried by each axle (front and rear).

Welch plug: See *Core hole plug.*

Weld: To join two pieces of metal together by raising area to be joined to point hot enough for two sections to melt and flow together. Additional metal is usually added by melting small drops from end of metal rod while welding is in progress.

Weld bead: Layer of metal deposited during a welding operation.

Wet friction: Resistance to movement between two lubricated surfaces.

Wet sleeve: Cylinder sleeve application in which water in cooling system contacts a major portion of sleeve itself.

Wet steam: Steam containing heavier water droplets that have not been fully vaporized.

Wheel aligner: Device used to check camber, caster, toe-in, and other alignment angles.

Wheel alignment: Refers to checking or adjusting various angles involved in proper placement or alignment of both front and rear wheels.

Wheel balancer: Machine used to check wheel and tire assembly for static and dynamic balance.

Wheelbase: Distance between center of front wheels and center of rear wheels.

Wheel bearing: Ball or roller bearings on which wheel hub rotates.

Wheel brake: A brake operating at wheel, either drum or disc design.

Wheel cylinder: Part of hydraulic brake system that receives pressure from master cylinder and in turn applies brake shoes to drums.

Wheel dolly: A wheeled device used to elevate and install heavy truck wheel and tire assemblies.

Wheel hop: Hopping action of rear wheels during heavy acceleration.

Wheel hub: That unit upon which wheel is fastened.

Wheel lug or **lug bolt:** Bolts used to fasten wheel to hub.

Wheel rim locks: Locking lug nuts or bolts used to deter theft of custom wheels.

Wheel sensor: Magnetic speed sensor used in an anti-lock brake system to measure wheel speed.

Wheel shimmy: Lateral (side-to-side) vibration of a tire and wheel assembly.

Wheel tracking: Ability of the rear wheels to follow directly behind the front wheels.

Wheel tramp: Hopping (up and down) vibration of a tire and wheel assembly.

Winch: An electrically or mechanically driven drum that will wind in a length of cable. Used to remove vehicles from mud, ascend very steep slopes, pull logs.

Wire gauge: A steel plate with a number of various size openings. Used to determine wire diameter and thickness of sheet metal.

Wire loom: A woven covering through which one or more wires can be passed. Secures and protects wires.

Wiring diagram: Drawing showing various electrical units and wiring arrangement necessary for proper operation.

Wiring harness: A series of wires bound together to form a handy, compact, and protected unit. Often sheathed in a protective covering.

Witness marks: Punch marks used to position or locate some part properly.

Wobble plate: A round, flat plate with a shaft passing through its center. Plate is affixed to shaft at an angle to shaft centerline. When shaft turns, plate rotates and also wobbles from side to side. Used on axial air conditioning compressor to operate pistons.

Woodruff key: A half round key inserted in a circular slot in a shaft. Protrudes above shaft into a part keyway to prevent turning.

Work hardening: Hardening of a material by bending, pounding, or other mechanical means.

Worm and roller: Steering gear utilizing a worm gear on steering shaft. A roller on one end of cross shaft engages worm.

Worm and sector: Type of steering gear utilizing worm gear engaging sector (a portion of a gear) on cross shaft.

Worm and taper pin: Steering gear utilizing worm gear on steering shaft. End of cross shaft engages worm via taper pin.

Worm gear: A long, cylindrical gear containing a continuous spiral tooth (thread-like). Gear centerline is at right angles to centerline of worm wheel.

Worm shaft: Steering gear shaft upon which a worm gear is formed. Turns pitman shaft.

Windup: Internal stresses between the front and rear wheel drive trains on a four-wheel drive vehicle. Occurs on dry pavement.

Wrist pin: See *Piston pin.*

Y

Yield strength: Maximum force (in pounds per square inch) that can be sustained by given member and have that member return to its original shape when force or pressure is removed.

Yoke: Slotted or split end of an object that straddles and is fastened to another, such as a universal joint yoke straddles cross or spider.

Z

Zener diode: A silicone diode that allows electrical current to flow in one direction only until voltage being applied to diode increases to a certain amount. After this level is reached, the diode will allow current to flow in both directions.

Zero defects: In manufacturing, taking extreme care to make something so well that there will be no (zero) failures caused by faulty workmanship or parts.

Zinc: A metal used in plating steel to help prevent rusting and corrosion.

Zirconia-type oxygen sensor: Sensor that produces voltage to indicate exhaust gas condition.

Acknowledgments

The production of a book of this nature would not be possible without the cooperation of the automotive industry. In preparing the manuscript for **Auto Diagnosis, Service, and Repair,** the industry has been most cooperative. The authors acknowledge the cooperation of these companies with great appreciation:

Aaladin Industries; A.L.C. Co.; Accurate Products, Inc.; AC-Delco; ACF Industries, Inc., Carter Fuel Pump Div.; Aeroquip Corp.; AFBMA; AIRCO; Aimco; Air Lift Co.; Air Reduction; Al-Beck Forbes, Inc.; Albertson and Co.; Alfa Romeo Cars; All-Lock Co., Inc.; Alondra, Inc.; Aluminum Co. of America; Amco Mfg. Corp.; American Association of Battery Manufacturers (AABM); American Bosch Arma Corp.; American Brake Shoe Co.; American Hammered Corp., Automotive Replacement Div.; American Iron and Steel Institute; American Manufacturers Assn.; American Optical Co.; American Petroleum Institute; American Standards Assn., Inc.; American-Standard; Amerimac, Inc.; Ammco Tools, Inc.; Anti-Friction Bearing Manufacturers Assn., Inc.; AP Parts Corp.; Appleton Electric Co.; Armstrong Patents Co., Ltd.; Armstrong Tire & Rubber; Armstrong Tool Co.; Arnolt Corp.; Aro Corp.; Audi; Ausco Co.; Automotive Electric Assn.; Automotive Products, Inc.; Automotive Service Industry Assn.; B & B Electronics; B & M Automotive Products; Baldwin, J. A., Mfg. Co.; Band-it Co. ; Barbee Co., Inc.; Battery Council International; Beach Precision Parts Co.; Bear Mfg. Co.; Beckman Instruments, Inc.; Belden Mfg. Co.; Bendix Corp., Automotive Service Div.; Benwil Industries; Bethlehem Steel Co.; BF Goodrich; Big Four Industries, Inc.; Bilstein Corp. of America; Binks Mfg. Co.; Black and Decker Mfg. Co.; Blackhawk Mfg. Co.; BMW of America; Bonney Forge and Tool Works; Borg & Beck; Borg-Warner Corp., Marvell-Schebler Products Div.; Borg-Warner Corp., Rockford Clutch Div.; Bosch, Robert, Corp.; Bowes Mfg., Inc.; BP, Gulf Oil Div.; Branich Mfg. Co., Inc.; Breed Technologies, Inc.; Breeze Corp., Inc.; Bremen Bearing Co.; Bridgestone/Firestone Tire and Rubber Co.; British Leyland; British Motor Corp.-Hambro, Inc.; Brown and Sharpe, Indus. Prod. Div.; Bundy Tubing; Burke Co.; Cedar Rapids Eng. Co.; Central Tool Co.; Champion Pneumatic Machinery Co.; Champion Spark Plug Co.; Chicago Rawhide Mfg. Co.; Chief Industries, Inc.; Citroen Cars Corp.; Clayton Associates, Inc.; Cleveland Graphite; Clevite Corp., Bronze Div.; Clevite Corp., Clevite Service Div.; Cole-Hersee Co.; Colt Industries, Holley Carburetor Div.; Continental Air Tools; Continental Motors Corp.; Cook, Brian; Cooper Tire and Rubber Co.; Corbin Co.; Cornell, William Co.; Corning, Cox Instrument; CPI Engineering Services, Inc.; CR Industries.; CRC Chemicals; Cummins Engine Co., Inc.; Daihatsu; DaimlerChrysler, Chrysler Div.; DaimlerChrysler, Dodge Div.; DaimlerChrysler, Jeep Div.; DaimlerChrysler, Mercedes-Benz Div.; Dana Corp.; Deere & Co.; Delphi Automotive, Harrison Radiator; DeVilbiss Co.; Dole Valve Co.; Doug Nash Equipment; Dover Corp.; Dow Corning Corp.; Dreyer, H. E., Inc.; Driveshaft Technology, Inc.; Dual Drive, Inc.; Duetz Corp.; Duff-Norton; Dunlop Tire Co. ; duPont de Nemours, E.I., and Co.; Dura Corp., Weaver Mfg. Div.; Dura-Bond Engine Parts Co.; Duralcan USA; Durke - Atwood Co.; Eagle Manufacturing; Easco Tools; Eaton Corp.; Echlin Mfg. Co.; Echlin Mfg. Co., United Parts Div. ; Edelmann, E., and Co.; EIS Automotive Corp.; Electrodyne; Environmental Systems Products, Inc.; ESAB Welding and Cutting Products; ESB Brands, Inc.; Ethyl Corp.; Eutectic Welding Alloys Corp.; Everco Industries, Inc.; Exxon Company USA; Exxon-Mobil Corp.; Fafnir Bearing Co.; FAG Bearing, LTD.; Federal-Mogul Corp.; Fel-Pro, Inc.; Ferrari Cars; Ferrett Instruments; Fiat; Fiske Brothers Refining Co.; Fletch/Air Inc.; Fluke Corp.; FMC Corp; Ford Motor Co., Ford Div. ; Ford Motor Co., Jaguar Div.; Ford Motor Co., LandRover Div.; Ford Motor Co., Lincoln Div.; Ford Motor Co., Motorcraft Div.; Fox Valley Instrument; Fram Corp.; G.H. Meiser & Co.; Gates Rubber Co.; Gatke Corp.; General Electric; General Instrument Corp.; General Motors Corp.; General Motors, Detroit Diesel Allison Div.; General Motors, Guide Lamp Div.; General Motors, Rochester Div.; General Motors, Saab Div.; General Motors, Saginaw Div.; General Motors, Service Technology Group; General Tire & Rubber Co.; Girling Ltd.; GKN Automotive, Inc.; Glassinger & Co. ; Globe Hoist Co.; Goodall Mfg. Co.; Goodrich Co.; Goodyear Tire and Rubber Co.; Gould Inc.; Graco, Inc.; Gray Co., Inc.; Graymills Corp.; Guaranteed Parts Co.; Halibrand Eng. Corp.; Hamilton Test Systems; Harper Motors; Hastings Mfg. Co.; Hayden, Inc.; Hein-Werner Corp.; Heli-Coil Products; Hickok Automotive Group; Homestead Industries, Inc. ; Honda Motor Co., Acura Div.; Honda Motor Co., Honda Div.; Hub City Iron Co.; Huck Mfg. Co.; Hunter Eng. Co.; Hyundai Motor America; Ideal Corp.; Ignition Manufacturer's Inst.; Imperial Eastman Corp.; Ingersoll-Rand; Inland Mfg. Co.; International Automotive Technicians Network (iATN); International/Navistar; International Mfg. Co.; Iskenderian Racing Cams; Isuzu of America, Inc.; ITT Automotive; Johns Manville Corp.; Johnson Bronze Co.; Justrite Manufacturing Co.; Kal-Equip. Co.; K-D Mfg. Co.; Kelly-Springfield Tire Co.; Kelsey-Hayes Co., Gunite Foundries Div.; Kelsey-Hayes Co., Utica-Herbrand Div.; Ken Tools; Kern Manufacturing, Inc.; Kester Solder Co.; KIA Motors; Klasey, Jack; Kleer-Flo Co.; Koni America, Inc.; Kwik-Way Mfg. Co.; Lear Siegler, Inc.; Lee, K. O., Co.; Lenroc Co.; Libby-Owens-Ford Co.; Lincoln Automotive; Lincoln Electric Co.; Lincoln Eng. Co.; Lisle Corporation; Littlefuse, Inc.; Loctite Corporation; Lucas, Joseph, Ltd.; Lufkin Rule Co.; Mac Tools; Mack Trucks, Inc.; MacMillan Petroleum Corp.; Magnaflux Corp.; Mansfield Tire & Rubber Co.; Maremont Corp.; Marquette Corp.; Martin Senour Paints; Maserati; Master Pneumatic-Detroit, Inc.; Mazda; McCartney Manufacturing Co., Inc.; McCord Corp.; McCreary Tire & Rubber Corp.; Meco, Inc.; Merit Industries, Inc.; Meyer Hydraulics; Micro Test; Midland-Ross Corp.; Miller Electric Mfg. Co.; Mitsubishi; Modine Manufacturing; Monitor Manufacturing; Monroe Auto Equipment Co.; Moog Industries, Inc.; Morton-Norwich Products, Inc.; Motor Wheel Corp.; Motorola Automotive Products, Inc.; Murray Corp.; Muskegon Piston Ring Co.; Mustang Dynamometers; Napa-Belden; National Board of Fire Underwriters; National Engines Co.; National Institute for Automotive Service Excellence (ASE); Nice Ball Bearing Co.; Nicholson File Co.; Niehoff; Nissan Motor Co., Nissan Div. ; Nissan Motor Co., Infiniti Div.; Nugier, F.A., Co.; Oakite Products, Inc.; Omega Mfg. Co.; P and G Mfg. Co.; Packard Electric; Parker Fluid Connectors; Paxton Products; Pennsylvania Refining Co.; Perfect Circle Corp.; Perfect Equipment Corp.; Permatex Co., Inc.; Peugeot, Inc.; Phillips Temco, Inc.; Porsche; Porter, H.K., Inc.; Prestolite Co.; Pro-Cut International; Proto Tool Co.; Pruit, Larry, Pruit Cylinder Head Service; Purolator Products, Inc.; Pyroil Co.; Quaker State Corp.; Questor; Racetech/SDS; Radiator Specialty Co., Gunk Chemical Div.; Raybestos-Manhattan, Inc., Grey-Rock Div.; Realmarket Associates; Rexnord; Rinck-McIlwaine, Inc.; Robertshaw Controls Co.; Rockwell International; Rodac Corp.; Rootes Motors, Inc.; Rottler Boring Bar Co.; RT Brake Lining Co., Inc.; RTI Technologies, Inc.; Rubber Manufacturers Assn.; Ruger Equipment Co.; Safety-Kleen; Saginaw Steering Gear; Salisbury Corp.; Scovill Mfg. Co., Inc., Schrader Div.; Sealed Power Corp.; Semperit of America, Inc.; Shell Oil Co.; Sherwin-Williams Co.; Shim-A-Line Inc.; Simpson Electric Co.; Sioux Tools Inc.; SKF Industries, Inc.; Skil; Slep Electronics; Snap-on Tools Corp., Snap-on Tools Div.; Snap-on Tools Corp., Sun Electric Div.; Society of Automotive Eng., Inc.; Solex Ltd.; Spicer; SPX, Corp, Allen Test Products Div.; SPX Corp., OTC Division; SPX Corp., Robinair Division; SPX, Kent Moore Div.; Standard Motor Products; Standard-Thomson Corp.; Stant Mfg. Co., Inc.; Star Machine and Tool Co.; Starrett, L.S., Co.; Stemco Mfg. Co.; Sterling Motors; Stewart-Warner, Alemite Div.; Storm-Vulcan, Inc.; Straza Industries; Sturtevant, P.A., Co.; Subaru Motor Co.; Sunnen Products Co.; Suzuki Motor Co.; Takata Total Safety Systems; Tescom Corp, Smith Equipment Div.; Testing Systems, Inc.; Texaco, Inc.; The Aluminum Association; Thexton Mfg. Co., Inc.; Thor Power Tool Co.; 3M Co. ; TIF Instruments; Timken Roller Bearing Co.; Tomco Coupler; Torrington; Toyota Motor Co., Lexus Div.; Toyota Motor Co., Toyota Div.; Tracerline; Traction Master Co.; Triad; Trucut (Frank Wood and Co.); TRW, Inc., Thompson Products Replacement Div.; U.S. Chemicals; U.S. Cleaner Corp.; Ultra-Violet Products, Inc.; Union Carbide Corp.; Uniroyal, Inc.; United Tool Processes Corp.; UOP, Inc.; Vaco Products Co.; Valvoline Oil Co.; Van Norman Machine Co.; Vellumoid Co.; Vetronix Vehicle Electronics; Victor Mfg. and Gasket Co.; Volkswagen of America, Inc.; Volvo of America Corp.; Voss Inc.; Wagner Electric Corp.; Walker Mfg. Co., Walbro Div.; Warner Gear-Warner Motive; Weatherhead Co.; Werther International; Wessels Co.; Westberg Mfg. Co.; Wheelabrator-Frye Inc.; Whitaker Cable Corp.; White Engine Co.; Williams, J.H., and Co.; Wilton Corp.; Wix Corp.; Woodhill Permatex; Wright-Austin Co.; Wudel Mfg. Co.; Young Radiator Co.; ZF Transmission Group

Index

B

C

F

N

O

P

Q

R

S

U

V

W

Y

Z